FORMULAS FROM GEOMETRY

Triangle

$h = a \sin \theta$

$\text{Area} = \dfrac{1}{2}bh$

(Law of Cosines)
$c^2 = a^2 + b^2 - 2ab \cos \theta$

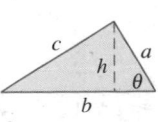

Right Triangle

(Pythagorean Theorem)
$c^2 = a^2 + b^2$

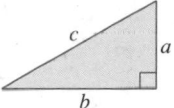

Equilateral Triangle

$h = \dfrac{\sqrt{3}s}{2}$

$\text{Area} = \dfrac{\sqrt{3}s^2}{4}$

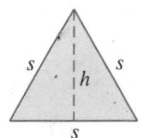

Parallelogram

$\text{Area} = bh$

Trapezoid

$\text{Area} = \dfrac{h}{2}(a + b)$

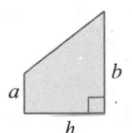

Circle

$\text{Area} = \pi r^2$
$\text{Circumference} = 2\pi r$

Sector of Circle

(θ in radians)

$\text{Area} = \dfrac{\theta r^2}{2}$

$s = r\theta$

Circular Ring

(p = average radius,
w = width of ring)
$\text{Area} = \pi(R^2 - r^2)$
$\quad\quad = 2\pi pw$

Sector of Circular Ring

(p = average radius,
w = width of ring,
θ in radians)
$\text{Area} = \theta pw$

Ellipse

$\text{Area} = \pi ab$

$\text{Circumference} \approx 2\pi \sqrt{\dfrac{a^2 + b^2}{2}}$

Cone

(A = area of base)

$\text{Volume} = \dfrac{Ah}{3}$

Right Circular Cone

$\text{Volume} = \dfrac{\pi r^2 h}{3}$

$\text{Lateral Surface Area} = \pi r \sqrt{r^2 + h^2}$

Frustum of Right Circular Cone

$\text{Volume} = \dfrac{\pi(r^2 + rR + R^2)h}{3}$

$\text{Lateral Surface Area} = \pi s(R + r)$

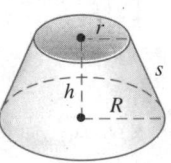

Right Circular Cylinder

$\text{Volume} = \pi r^2 h$
$\text{Lateral Surface Area} = 2\pi rh$

Sphere

$\text{Volume} = \dfrac{4}{3}\pi r^3$

$\text{Surface Area} = 4\pi r^2$

Wedge

(A = area of upper face,
B = area of base)

$A = B \sec \theta$

use Leibniz's technique
to find ~~$\frac{d[uv]}{dx}$~~ $\frac{d}{dx}\left[\frac{u}{v}\right]$

from $d[uv] = u\,dv + v\,du$

$$\frac{v\,du - u\,dv}{v^2}$$

CALCULUS
WITH ANALYTIC GEOMETRY

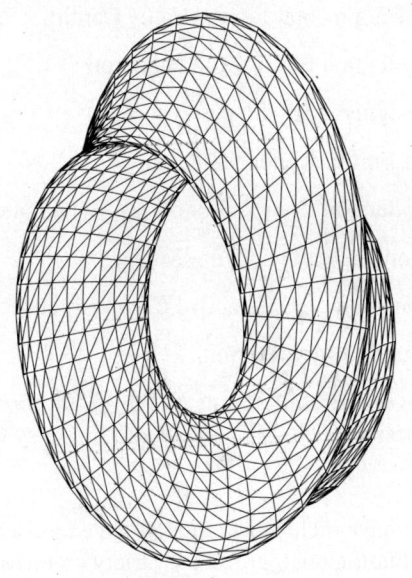

FIFTH EDITION

ROLAND E. LARSON
ROBERT P. HOSTETLER
The Pennsylvania State University
The Behrend College

BRUCE H. EDWARDS
University of Florida

with the assistance of DAVID E. HEYD
The Pennsylvania State University
The Behrend College

D. C. Heath and Company
Lexington, Massachusetts Toronto

Address editorial correspondence to:
D. C. Heath and Company
125 Spring Street
Lexington, MA 02173

Acquisitions Editors: Ann Marie Jones and Charles Hartford

Developmental Editor: Cathy Cantin

Production Editor: Ron Hampton

Designer: Sally Steele

Art Editor: Elinor Stapleton

Production Coordinators: Lisa Merrill and Richard Tonachel

Photo Researcher: Toni Michaels

Composition: Interactive Composition Corporation

Technical Art: Folium

Cover: "Umbilic Torus NC" by Sculptor/Mathematician Helaman Ferguson
(Bronze 27" x 27" x 9"). Photo by Terry Clough.

We have included examples and exercises that use real-life data as well as
technology output from a variety of software. This would not have been possible
without the help of many people and organizations. Our wholehearted thanks goes to
all for their time and effort.

International Standard Book Number: 0-669-35335-3

Library of Congress Catalog Card Number: 93-71417

10 9 8 7 6 5 4 3 2

Preface

Calculus with Analytic Geometry, Fifth Edition, is designed for use in a calculus course for students in engineering, mathematics, the physical and life sciences, and economics.

Calculus Reform

During the past several years, much has been written about the need for reform in mathematics education. Most suggestions for reform have fallen into four categories: (1) problem solving, (2) technology, (3) communicating mathematics, and (4) real-life applications. Every portion of the Fifth Edition was carefully examined and revised in the spirit of reform.

PROBLEM SOLVING Helping students learn to be efficient and creative problem solvers is the primary goal of any mathematics course. Much of the time spent on this revision was devoted to enhancing the exercise sets—with many new exercises, the text now contains nearly 10,000 exercises. The range of exercises is much broader than in previous editions. While the text still contains an abundance of exercises that are designed to develop skills, the text now has many other types of exercises as well: some have several parts that lead students through problem-solving strategies (Exercises 21–24, Section 1.3), some ask for interpretations (Exercises 65, 75–78, Section 2.2), some have many correct answers (Exercises 57–62, Section 3.6), and there are *many* other new types. Often, students are asked to consider solving a problem *numerically*, *graphically*, and *analytically* (see Page 63).

TECHNOLOGY Much of calculus reform has centered around the appropriate integration of technology in calculus instruction. *Calculus*, Fifth Edition, addresses the use of technology *much* more aggressively than previous editions. Because graphing utilities and symbolic algebra systems are now readily available, we have incorporated their use throughout the prose and exercises of the text. Sometimes technology is used as a means of discovery (Exercises 51 and 52, Section 2.4), sometimes technology is used to reinforce concepts (Exercises 21–28, Section 3.3), and sometimes it is used as an efficient problem-solving tool (Exercises 43 and 44, Section 3.6). In addition to describing the benefits of using technology, the text also pays special attention to its possible misuse. For example, look at the discussions on pages 174 and 206.

Calculus, Fifth Edition, is also available in an interactive edition. This complete learning package for calculus is a Mathcad-based, multimedia resource offered in a variety of formats including CD-ROM and floppy disk.

COMMUNICATING MATHEMATICS   The third important aspect of calculus reform is a renewed emphasis on development of communication skills. To address this, we have included many exercises that ask for interpretations, descriptions, discussions, justifications, and conjectures. For instance, look at Exercises 33 and 34 in Section 1.2 and Exercise 63 in Section 4.2.

REAL-LIFE APPLICATIONS   The fourth important aspect of calculus reform is a renewed dedication to the applied nature of calculus. Almost all of the early development of calculus was written in the context of real-life problem solving. In writing the Fifth Edition, we have included many new and interesting real-life applications. To our knowledge, many of these have never before appeared in a calculus text. For instance, see Example 6 on page 22, Example 5 on page 88, and Exercise 66 on page 188.

Improved Coverage of Calculus

Improved Coverage of Calculus

ORDER OF TOPICS   The new Fifth Edition is very flexible with respect to the order of coverage of the core calculus topics. Some users of the Fourth Edition preferred to cover applications of integration (Chapter 6) before covering transcendental functions (Chapter 5). The Prerequisites chapter can be used as an individual review. The ε–δ discussion of limits in Chapter 1 can be given minimal coverage. L'Hôpital's Rule can be covered after Section 5.4. Sections 6.6 and 6.7 can be covered later in the course. Chapter 8 can be covered any time after Chapter 7. The coverage of Section 11.7 can be delayed until just before Section 14.8.

PREREQUISITES   The chapters in the text have been renumbered to fit the way most instructors use the text. Chapter 1 now begins with an introduction of the tangent line problem (see page 61). The old Chapter 1 was revised and relabeled as Prerequisites. To see the type of revision in this chapter, compare Section 1.3 in the Fourth Edition with the Prerequisites Section 3 in the Fifth Edition—you will notice greater attention to graphing utilities and a much stronger discussion of mathematical modeling.

EARLY TRANSCENDENTAL FUNCTIONS   Coverage of transcendental functions occurs earlier than in previous editions. Trigonometric functions are reviewed in the Prerequisites chapter, and then used from the beginnning of Chapter 1. Moreover, coverage of logarithmic and exponential functions occurs in Chapter 5—one chapter earlier than in the Fourth Edition.

INTRODUCTION TO DIFFERENTIATION   Many users of the Fourth Edition have told us that the old Section 3.2 on velocity and acceleration was difficult to teach because it was sandwiched between the introduction of derivatives and the discussion of differentiation rules. In response to this, we have dispersed the material on velocity and acceleration throughout the chapter. For example, see the discussion of velocity on pages 121 and 122.

EARLY DIFFERENTIAL EQUATIONS   Many users have indicated that they would welcome an early introduction to the language of differential equations. To accommodate this, we have incorporated the language of differential equations throughout the material on integration. For instance, see Exercises 45–48 in Section 4.1 and Exercises 27–30 in Section 5.2. An introduction to separation of variables was added to Section 5.6, which is now entitled Differential Equations: Growth and Decay.

POLAR COORDINATES   The Fourth Edition devoted two sections to graphs of equations in polar coordinates. Because of the availability of technology to sketch polar graphs, we have condensed the material to one section (Section 10.3), which is now entitled Polar Coordinates and Polar Graphs.

PARAMETRIC SURFACES   At the request of several users, the Fifth Edition contains a new section (Section 15.5) on parametric surfaces. This topic fits well with the text's enhanced attention to technology. The use of parametric surfaces allows students to make use of programs such as Derive, Maple, Axiom, and Mathematica to sketch surfaces that are not of the form $z = f(x, y)$. For instance, Example 7 in Section 15.5 shows how to sketch a torus as a parametric surface.

THIRD-SEMESTER ENHANCEMENTS   Many of our users have commented that the number of examples and exercises in the third portion of the text (Chapters 11–16) did not match that of the first two thirds. In this revision, we paid special attention to this portion of the text, adding 42 examples, 600 exercises, 145 graphs, and 121 applications.

PARAMETRIC EQUATIONS   Thanks to one of our reviewers, Jack Ceder of the University of California—Santa Barbara, we now have an improved presentation of parametric equations. The Fifth Edition is now careful to distinguish between a graph (the set of points) and a curve (the points with their defining parametric equations). This distinction helps students understand that concepts such as smoothness cannot be determined by considering a graph only. For instance, a line may or may not be smooth. As a curve given by $x = t$ and $y = t$, it is smooth, but as a curve given by $x = t^3$ and $y = t^3$, it is not smooth. (For other examples, see Exercises 29–32, Section 10.1.)

RIGOR, DEFINITIONS, AND THEOREMS   One feature of other calculus reform projects that this edition has not embraced is the informality of definitions and theorems. For instance, one of the current reform projects defines a graph to be concave up if it "curves upward." We do not find this type of casualness to be helpful to students. In fact, we have made a special effort to write the definitions and theorems in the Fifth Edition with greater care than in previous editions. For example, the new statement of the Product and Quotient Rules given in Section 2.3 now makes it clear that the product (or quotient) of two differentiable functions is differentiable. (This improvement was suggested by Leonard Gillman of the University of Texas.)

PROOFS   The text includes proofs of all theorems that are within the grasp of a beginning calculus student. Some of the proofs are presented in outline form, with an emphasis on the essence of the argument. (See the proof of the Product Rule in Section 2.3.) Some of the more complicated proofs are placed in Appendix A. (See the proof of the Chain Rule in Section 2.4.)

ACCURACY   Throughout the entire revision process for the Fifth Edition, special care has been given to creating a text that is accurate. We thank the many users who have helped us to improve the accuracy of the text.

WHAT IS CALCULUS?   Pages xxxv–xxxvii of the front matter contain a student introduction to the text. We encourage instructors to ask their students to read this introduction—it was designed to show students how precalculus mathematics is transformed by the limit process to form the new mathematics of calculus. This theme is carried throughout the text. For instance, look at the discussion on page 407.

END PAPERS   The end papers of the text are used for summaries of common formulas from calculus, geometry, algebra, and trigonometry. (A more comprehensive list of integration formulas is given in Appendix C.)

Fifth Edition Features

The Fifth Edition has many new and improved design features that enhance the relevance of the text and make it an even more effective teaching tool. These are shown on the following pages.

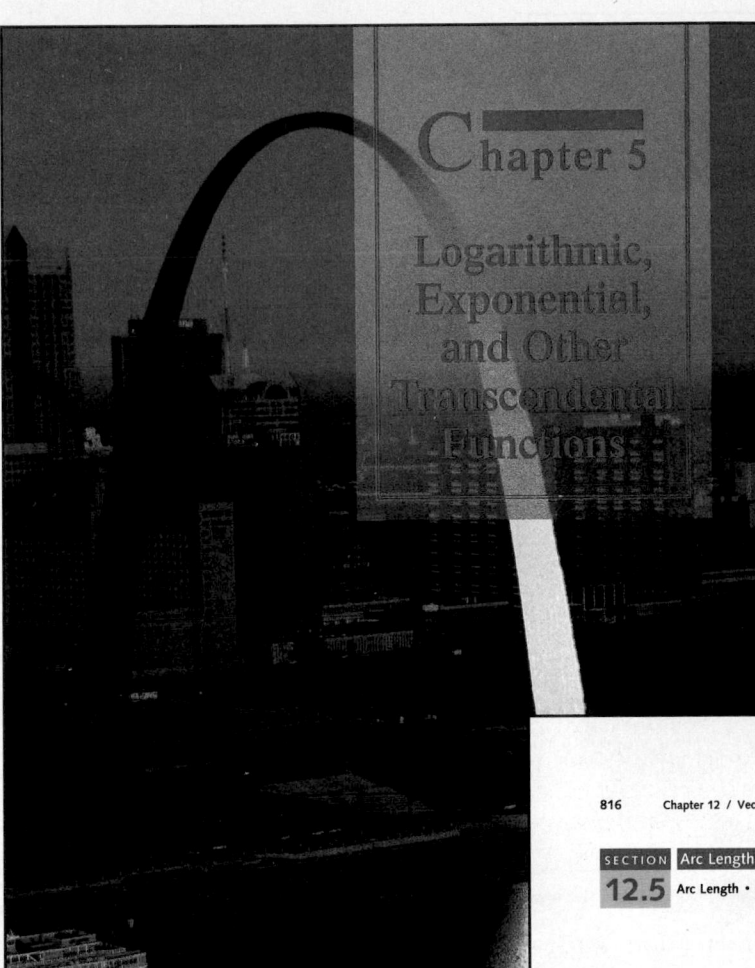

Chapter Openers

Each chapter opens with a photograph of a mathematical application that is keyed to an exercise in the chapter.

Section Topics

Each section begins with a list of sub-section topics. This outline will help instructors with class planning and students with studying and synthesizing the material in the section.

Examples

To increase the usefulness of the text as a study tool, many new examples were added. The text now contains over 1000 examples, each titled for easy reference. Many of the examples include side comments that clarify the steps of the solution.

Definitions and Theorems

All definitions and theorems are shown in color for emphasis and easy reference.

---

816   Chapter 12 / Vector-Valued Functions

SECTION 12.5   Arc Length and Curvature

Arc Length • Arc Length Parameter • Curvature • Applications

Arc Length

In Section 10.2, you saw that the arc length of a smooth *plane* curve C given by the parametric equations $x = x(t)$ and $y = y(t)$, $a \le t \le b$, is

$$s = \int_a^b \sqrt{[x'(t)]^2 + [y'(t)]^2}\, dt.$$

In vector form, where C is given by $\mathbf{r}(t) = x(t)\mathbf{i} + y(t)\mathbf{j}$, you can rewrite this equation for arc length as

$$s = \int_a^b \|\mathbf{r}'(t)\|\, dt.$$

EXAMPLE 1   Finding the Arc Length of a Plane Curve

Find the arc length of the curve given by $\mathbf{r}(t) = \cos t\, \mathbf{i} + \sin t\, \mathbf{j}$ from $t = 0$ to $t = 2\pi$.

Solution   Using $x(t) = \cos t$ and $y(t) = \sin t$, you obtain $x'(t) = -\sin t$ and $y'(t) = \cos t$. Thus, the arc length from $t = 0$ to $t = 2\pi$ is

$$s = \int_0^{2\pi} \sqrt{[x'(t)]^2 + [y'(t)]^2}\, dt = \int_0^{2\pi} \sqrt{\sin^2 t + \cos^2 t}\, dt$$

$$= \int_0^{2\pi} dt$$

$$= t \Big]_0^{2\pi} = 2\pi.$$

In Figure 12.26, notice that the curve is a circle of radius 1. Thus, the arc length formula confirms that the circumference of the circle is 2π.

The formula for the arc length of a plane curve has a natural extension to a smooth curve in *space*, as stated in the following theorem.

THEOREM 12.6   Arc Length of Space Curve

If C is a smooth curve given by $\mathbf{r}(t) = x(t)\mathbf{i} + y(t)\mathbf{j} + z(t)\mathbf{k}$, on an interval $[a, b]$, then the arc length of C on the interval is

$$s = \int_a^b \sqrt{[x'(t)]^2 + [y'(t)]^2 + [z'(t)]^2}\, dt = \int_a^b \|\mathbf{r}'(t)\|\, dt.$$

$\mathbf{r}(t) = \cos t\, \mathbf{i} + \sin t\, \mathbf{j}$

FIGURE 12.26
The arc length of the circle is 2π.

Area of a Surface of Revolution

In Sections 6.2 and 6.3, integration was used to calculate the volume of a solid of revolution. We now look at a procedure for finding the area of a surface of revolution.

Definition of a Surface of Revolution

If the graph of a continuous function is revolved about a line, the resulting surface is a **surface of revolution.**

The area of a surface of revolution is derived from the formula for the lateral surface area of the frustum of a right circular cone. Consider the line segment in Figure 6.45, where L is the length of the line segment, r_1 is the radius at the left end of the line segment, and r_2 is the radius at the right end of the line segment. When the line segment is revolved about its axis of revolution, it forms a frustum of a right circular cone, with

$$S = 2\pi r L \qquad \text{Lateral surface area of frustum}$$
$$r = \frac{1}{2}(r_1 + r_2). \qquad \text{Average radius of frustum}$$

FIGURE 6.45

(In Exercise 46 you are asked to verify the formula for S.)

Suppose the graph of a function f, having a continuous derivative on the interval $[a, b]$, is revolved about the x-axis to form a surface of revolution, as shown in Figure 6.46. Let Δ be a partition of $[a, b]$, with subintervals of width Δx_i. Then the line segment of length

$$\Delta L_i = \sqrt{\Delta x_i^2 + \Delta y_i^2}$$

generates a frustum of a cone. Let r_i be the average radius of this frustum. By the Intermediate Value Theorem, a point d_i exists (in the ith subinterval) such that $r_i = f(d_i)$. The lateral surface area, ΔS_i, of the frustum is

$$\Delta S_i = 2\pi r_i \Delta L_i = 2\pi f(d_i)\sqrt{\Delta x_i^2 + \Delta y_i^2}$$
$$= 2\pi f(d_i)\sqrt{1 + \left(\frac{\Delta y_i}{\Delta x_i}\right)^2}\,\Delta x_i.$$

FIGURE 6.46

Graphics

The Fifth Edition has over 2800 figures. Of these, over 1200 are in the examples and exposition, over 850 are in the exercise sets, and over 750 are in the answers to odd-numbered exercises. Computer-generated for accuracy and designed with full color, this art program will help students better visualize mathematical concepts. This is particularly true in the presentation of complex, three-dimensional material, for which the art was drawn in true perspective.

Derivatives of Sine and Cosine Functions

In Section 1.3, you studied the following limits.

$$\lim_{\Delta x \to 0} \frac{\sin \Delta x}{\Delta x} = 1 \quad \text{and} \quad \lim_{\Delta x \to 0} \frac{1 - \cos \Delta x}{\Delta x} = 0$$

These two limits can be used to prove differentiation rules for the sine and cosine functions. (The derivatives of the other four trigonometric functions will be discussed in the next section.)

THEOREM 2.6   Derivatives of Sine and Cosine Functions

$$\frac{d}{dx}[\sin x] = \cos x \qquad \frac{d}{dx}[\cos x] = -\sin x$$

Proof

$$\frac{d}{dx}[\sin x] = \lim_{\Delta x \to 0} \frac{\sin(x + \Delta x) - \sin x}{\Delta x}$$
$$= \lim_{\Delta x \to 0} \frac{\sin x \cos \Delta x + \cos x \sin \Delta x - \sin x}{\Delta x}$$
$$= \lim_{\Delta x \to 0} \frac{\cos x \sin \Delta x - (\sin x)(1 - \cos \Delta x)}{\Delta x}$$
$$= \lim_{\Delta x \to 0} \left[(\cos x)\left(\frac{\sin \Delta x}{\Delta x}\right) - (\sin x)\left(\frac{1 - \cos \Delta x}{\Delta x}\right)\right]$$
$$= \cos x \left[\lim_{\Delta x \to 0} \frac{\sin \Delta x}{\Delta x}\right] - \sin x \left[\lim_{\Delta x \to 0} \frac{1 - \cos \Delta x}{\Delta x}\right]$$
$$= (\cos x)(1) - (\sin x)(0)$$
$$= \cos x$$

This differentiation rule is shown graphically in Figure 2.17. Note that for each x the *slope* of the sine curve is equal to the *value* of the cosine. The proof of the second rule is left as an exercise (see Exercise 95).

FIGURE 2.17
The derivative of the sine function is the cosine function.

Technology

Students are encouraged to use a graphing utility or symbolic algebra system as a tool for exploration and discovery as well as a problem-solving tool to execute complicated computations, to visualize theoretical concepts, to offer alternative approaches, and to verify the results of other solution methods. In addition to describing the benefits of using technology, the text also pays special attention to its possible misuse.

EXAMPLE 8   Derivatives Involving Sines and Cosines

Function	Derivative
a. $y = 2\sin x$	$y' = 2\cos x$
b. $y = \dfrac{\sin x}{2} = \dfrac{1}{2}\sin x$	$y' = \dfrac{1}{2}\cos x = \dfrac{\cos x}{2}$
c. $y = x + \cos x$	$y' = 1 - \sin x$

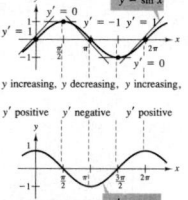
FIGURE 2.18
$$\frac{d}{dx}[a \sin x] = a \cos x$$

TECHNOLOGY A graphing utility can provide insight into the interpretation of a derivative. For instance, Figure 2.18 shows the graphs of $y = a \sin x$ for $a = \frac{1}{2}$, 1, $\frac{3}{2}$, and 2. Estimate the slope of each graph at the point $(0, 0)$. Then verify your estimate analytically by evaluating the derivative of each function when $x = 0$.

EXAMPLE 5   Finding a Power Series by Integration

Find a power series for $g(x) = \arctan x$, centered at 0.

Solution   Because $D_x[\arctan x] = 1/(1 + x^2)$, you can use the series

$$f(x) = \frac{1}{1 + x} = \sum_{n=0}^{\infty} (-1)^n x^n.$$

Substituting x^2 for x produces

$$f(x^2) = \frac{1}{1 + x^2} = \sum_{n=0}^{\infty} (-1)^n x^{2n}.$$

Finally, by integrating, you obtain

$$\arctan x = \int \frac{1}{1 + x^2}\, dx + C$$

$$= C + \sum_{n=0}^{\infty} (-1)^n \frac{x^{2n+1}}{2n + 1}$$

$$= \sum_{n=0}^{\infty} (-1)^n \frac{x^{2n+1}}{2n + 1}$$

$$= x - \frac{x^3}{3} + \frac{x^5}{5} - \frac{x^7}{7} + \cdots.$$

It can be shown that the power series developed for $\arctan x$ in Example 5 also converges (to $\arctan x$) for $x = \pm 1$. For instance, when $x = 1$, you can write

$$\arctan 1 = 1 - \frac{1}{3} + \frac{1}{5} - \frac{1}{7} + \cdots = \frac{\pi}{4}.$$

However, this series (developed by James Gregory in 1671) does not give us a practical way of approximating π because it converges so slowly that hundreds of terms would have to be used to obtain reasonable accuracy. Example 6 shows how to use *two* different arctangent series to obtain a very good approximation of π using only a few terms. This approximation was developed by John Machin in 1706.

EXAMPLE 6   Approximating π with a Se

Use the trigonometric identity

$$4 \arctan \frac{1}{5} - \arctan \frac{1}{239} = \frac{\pi}{4}$$

(see Exercise 38b) to approximate the numbe

Solution   By using only five terms from arctan(1/239), you obtain

$$4\left(4 \arctan \frac{1}{5} - \arctan \frac{1}{239}\right) \approx 3.141593$$

which agrees with the decimal representation

SRINIVASA RAMANUJAN

Series that can be used to approximate π have interested mathematicians for the past 300 years. An amazing series for approximating $1/\pi$ was discovered by the Indian mathematician Srinivasa Ramanujan (1887–1920) in 1914. Each successive term of Ramanujan's series adds roughly eight more correct digits to the value of $1/\pi$. (For more information about Ramanujan's work, see the article "Ramanujan and Pi" by Jonathan M. Borwein and Peter B. Borwein in the February, 1988 issue of *Scientific American*.)

Historical Notes
To help students understand that calculus has a past, historical notes are integrated throughout the text.

Remarks
The text contains many special instructional notes to students in the form of "Remarks." These notes appear after definitions, theorems, or examples and are designed to give additional insight, help avoid common errors, or describe generalizations.

Summaries
Many sections have summaries that identify core ideas and procedures. In some instances, an entire section summarizes the preceding topics.

FIGURE 15.68

EXAMPLE 3   Motion of a Swirling Liquid

A liquid is swirling around in a cylindrical container of radius 2 such that its motion is described by the velocity field

$$\mathbf{F}(x, y, z) = -y\sqrt{x^2 + y^2}\,\mathbf{i} + x\sqrt{x^2 + y^2}\,\mathbf{j}$$

as shown in Figure 15.68. Find

$$\iint_S (\operatorname{curl} \mathbf{F}) \cdot \mathbf{N}\, dS$$

where S is the upper surface of the cylindrical tank.

Solution   The curl of $\mathbf{F}$ is given by

$$\operatorname{curl} \mathbf{F} = \begin{vmatrix} \mathbf{i} & \mathbf{j} & \mathbf{k} \\ \dfrac{\partial}{\partial x} & \dfrac{\partial}{\partial y} & \dfrac{\partial}{\partial z} \\ -y\sqrt{x^2 + y^2} & x\sqrt{x^2 + y^2} & 0 \end{vmatrix}$$

$$= 3\sqrt{x^2 + y^2}\,\mathbf{k}.$$

Letting $\mathbf{N} = \mathbf{k}$, you have

$$\iint_S (\operatorname{curl} \mathbf{F}) \cdot \mathbf{N}\, dS = \iint_R 3\sqrt{x^2 + y^2}\, dA$$

$$= \int_0^{2\pi} \int_0^2 (3r) r\, dr\, d\theta$$

$$= \int_0^{2\pi} r^3 \Big]_0^2 d\theta$$

$$= \int_0^{2\pi} 8\, d\theta$$

$$= 16\pi.$$

REMARK   If $\operatorname{curl} \mathbf{F} = \mathbf{0}$ throughout region Q, then the rotation of $\mathbf{F}$ about each unit normal $\mathbf{N}$ is zero. That is, $\mathbf{F}$ is irrotational. From earlier work, you know that this is a characteristic of conservative vector fields.

Summary of Integration Formulas

Fundamental Theorem of Calculus

$$\int_a^b F'(x)\, dx = F(b) - F(a)$$

Fundamental Theorem of Line Integrals

$$\int_C \mathbf{F} \cdot d\mathbf{r} = \int_C \nabla f \cdot d\mathbf{r} = f(x(b), y(b)) - f(x(a), y(a))$$

Green's Theorem

$$\int_C M\, dx + N\, dy = \iint_R \left(\frac{\partial N}{\partial x} - \frac{\partial M}{\partial y}\right) dA = \int_C \mathbf{F} \cdot \mathbf{T}\, ds = \int_C \mathbf{F} \cdot d\mathbf{r} = \iint_R (\operatorname{curl} \mathbf{F}) \cdot \mathbf{k}\, dA$$

$$\int_C \mathbf{F} \cdot \mathbf{N}\, ds = \iint_R \operatorname{div} \mathbf{F}\, dA$$

Divergence Theorem

$$\iint_S \mathbf{F} \cdot \mathbf{N}\, dS = \iiint_Q \operatorname{div} \mathbf{F}\, dV$$

Stokes's Theorem

$$\int_C \mathbf{F} \cdot d\mathbf{r} = \iint_S (\operatorname{curl} \mathbf{F}) \cdot \mathbf{N}\, dS$$

Exercises

The text now contains nearly 10,000 exercises. The exercises are graded, progressing from skill-development problems to more challenging problems involving applications and proofs. Many exercise sets begin with a group of exercises that provide the graphs of functions involved. Review exercises are included at the end of each chapter. Answers to all odd-numbered exercises are included in the back of the text.

Graphing Utilities

Many exercises in the text can be solved using technology; however, the symbol C identifies all exercises in which students are specifically instructed to use a graphing utility or a symbolic algebra system.

Technology Labs

A *Technology Laboratory Guide*, referenced at the beginning of most exercise sets, offers additional technology assignments, explorations, and projects. Appropriate for any graphing technology, each lab opens with a list of the graphing capabilities required.

---

82   Chapter 1 / Limits and Their Properties

EXERCISES for Section 1.3

TECHNOLOGY
Laboratory Guide
Lab 1.3

In Exercises 1–4, use the graph to determine each limit visually, if it exists. When possible, identify two functions that agree in all but one point.

1. $g(x) = \dfrac{-2x^2 + x}{x}$

a. $\lim\limits_{x \to 0} g(x)$
b. $\lim\limits_{x \to -1} g(x)$

2. $h(x) = \dfrac{x^2 - 3x}{x}$

a. $\lim\limits_{x \to 0} h(x)$
b. $\lim\limits_{x \to -1} h(x)$

3. $g(x) = \dfrac{x^3 - x}{x - 1}$

a. $\lim\limits_{x \to 1} g(x)$
b. $\lim\limits_{x \to -1} g(x)$

4. $f(x) = \dfrac{1}{x - 1}$

a. $\lim\limits_{x \to 0} f(x)$
b. $\lim\limits_{x \to -1} f(x)$

In Exercises 5–8, find the limit (if it exists) of the function. Identify two functions that agree in all but one point and sketch the graph of the given function.

5. $\lim\limits_{x \to -1} \dfrac{x^2 - 1}{x + 1}$

6. $\lim\limits_{x \to -1} \dfrac{2x^2 - x - 3}{x + 1}$

7. $\lim\limits_{x \to -2} \dfrac{x^3 + 8}{x + 2}$

8. $\lim\limits_{x \to -1} \dfrac{x^3 + 1}{x + 1}$

In Exercises 9–20, find the limit (if it exists).

9. $\lim\limits_{x \to 5} \dfrac{x - 5}{x^2 - 25}$

10. $\lim\limits_{x \to 2} \dfrac{2 - x}{x^2 - 4}$

11. $\lim\limits_{x \to 1} \dfrac{x^2 + x - 2}{x^2 - 1}$

12. $\lim\limits_{x \to 0} \dfrac{\sqrt{2 + x} - \sqrt{2}}{x}$

13. $\lim\limits_{x \to 0} \dfrac{\sqrt{3 + x} - \sqrt{3}}{x}$

14. $\lim\limits_{x \to 0} \dfrac{[1/(x + 4)] - (1/4)}{x}$

15. $\lim\limits_{x \to 0} \dfrac{[1/(2 + x)] - (1/2)}{x}$

16. $\lim\limits_{x \to 3} \dfrac{\sqrt{x + 1} - 2}{x - 3}$

17. $\lim\limits_{\Delta x \to 0} \dfrac{2(x + \Delta x) - 2x}{\Delta x}$

18. $\lim\limits_{\Delta x \to 0} \dfrac{(x + \Delta x)^2 - x^2}{\Delta x}$

19. $\lim\limits_{\Delta x \to 0} \dfrac{(x + \Delta x)^2 - 2(x + \Delta x) + 1 - (x^2 - 2x + 1)}{\Delta x}$

20. $\lim\limits_{\Delta x \to 0} \dfrac{(x + \Delta x)^3 - 1}{\Delta x}$

C In Exercises 21–24, use a graphing utility to graph the function and use the graph to estimate the limit. Use a table to reinforce your conclusion. Then, find the limit by analytic methods.

21. $\lim\limits_{x \to 0} \dfrac{\sqrt{x + 2} - \sqrt{2}}{x}$

22. $\lim\limits_{x \to 16} \dfrac{4 - \sqrt{x}}{x - 16}$

23. $\lim\limits_{x \to 2} \dfrac{x^5 - 32}{x - 2}$

24. $\lim\limits_{x \to 0} \dfrac{[1/(2 + x)] - (1/2)}{x}$

In Exercises 25–40, determine the limit of the trigonometric function (if it exists).

25. $\lim\limits_{x \to 0} \dfrac{\sin x}{5x}$

26. $\lim\limits_{x \to 0} \dfrac{3(1 - \cos x)}{x}$

27. $\lim\limits_{\theta \to 0} \dfrac{\sec \theta - 1}{\theta \sec \theta}$

28. $\lim\limits_{\theta \to 0} \dfrac{\cos \theta \tan \theta}{\theta}$

29. $\lim\limits_{x \to 0} \dfrac{\sin^2 x}{x}$

30. $\lim\limits_{x \to 0} \dfrac{\tan^2 x}{x}$

31. $\lim\limits_{h \to 0} \dfrac{(1 - \cos h)^2}{h}$

32. $\lim\limits_{\phi \to \pi} \phi \sec \phi$

33. $\lim\limits_{x \to \pi/2} \dfrac{\cos x}{\cot x}$

34. $\lim\limits_{x \to \pi/4} \dfrac{1 - \tan x}{\sin x - \cos x}$

35. $\lim\limits_{t \to 0} \dfrac{\sin^2 t}{t^2}$

$\left[\textit{Hint: Find } \lim\limits_{t \to 0} \left(\dfrac{\sin t}{t} \right)^2 . \right]$

36. $\lim\limits_{x \to 0} \dfrac{\sin 2x}{\sin 3x}$

$\left[\textit{Hint: Find } \lim\limits_{x \to 0} \left(\dfrac{2 \sin 2x}{2x} \right) \left(\dfrac{3x}{3 \sin 3x} \right) . \right]$

37. $\lim\limits_{t \to 0} \dfrac{\sin 3t}{t}$

39. $\lim\limits_{x \to 0} \dfrac{\sin x^2}{x}$

In Exercises 41
$\lim\limits_{x \to c} f(x)$.

41. $c = 0$
    $4 - x^2 \le f($

42. $c = a$
    $b - |x - a|$

---

398   Chapter 5 / Logarithmic, Exponential, and Other Transcendental Functions

80. *St. Louis Arch* The Gateway Arch in St. Louis, Missouri was constructed in using the hyperbolic cosine function. The equation used to construct the arch was

$$y = 693.8597 - 68.7672 \cosh 0.0100333x,$$
$$-299.2239 \le x \le 299.2239$$

where x and y are measured in feet. Cross sections of the arch are equilateral triangles and (x, y) traces the path of the centers of mass of the cross-sectional triangles. For each value of x, the area of the cross-sectional triangle is $A = 125.1406 \cosh 0.0100333x$.

a. How high above ground is the center of the highest triangle? (At ground level, $y = 0$.)
b. What is the height of the arch? (*Hint:* For an equilateral triangle $A = \sqrt{3} c^2$, where c is one-half the base of the triangle, and the center of mass of the triangle is located at two-thirds the height of the triangle.)
c. How wide is the arch at ground level?
(*Source: Owner's Manual for the Gateway Arch, Saint Louis, MO, by William Thayer.*)

In Exercises 81–85, verify the differentiation formula.

81. $\dfrac{d}{dx}[\cosh x] = \sinh x$

82. $\dfrac{d}{dx}[\operatorname{sech} x] = -\operatorname{sech} x \tanh x$

83. $\dfrac{d}{dx}[\cosh^{-1} x] = \dfrac{1}{\sqrt{x^2 - 1}}$

84. $\dfrac{d}{dx}[\sinh^{-1} x] = \dfrac{1}{\sqrt{x^2 + 1}}$

85. $\dfrac{d}{dx}[\operatorname{sech}^{-1} x] = \dfrac{-1}{x\sqrt{1 - x^2}}$

86. Show that $\arctan(\sinh x) = \arcsin(\tanh x)$.

REVIEW EXERCISES for Chapter 5

In Exercises 1 and 2, sketch the graph of the function without using a graphing utility. Identify any asymptotes to the graph.

1. $f(x) = \ln x + 3$

2. $f(x) = \ln(x - 3)$

In Exercises 3 and 4, use the properties of logarithms to write the expression as a sum, difference, and/or multiple of logarithms.

3. $\ln \sqrt[5]{\dfrac{4x^2 - 1}{4x^2 + 1}}$

4. $\ln[(x^2 + 1)(x - 1)]$

In Exercises 5 and 6, write the expression as the logarithm of a single quantity.

5. $\ln 3 + \frac{1}{3}\ln(4 - x^2) - \ln x$

6. $3[\ln x - 2\ln(x^2 + 1)] + 2\ln 5$

True or False In Exercises 7 and 8, determine whether the statement or equation is true or false.

7. The domain of the function $f(x) = \ln x$ is the set of all real numbers.

8. $\ln(x + y) = \ln x + \ln y$

In Exercises 9 and 10, solve the equation for x.

9. $\ln \sqrt{x + 1} = 2$

10. $\ln x + \ln(x - 3) = 0$

In Exercises 11–18, find the derivative of the function.

11. $g(x) = \ln \sqrt{x}$

12. $h(x) = \ln \dfrac{x(x - 1)}{x - 2}$

13. $f(x) = x\sqrt{\ln x}$

14. $f(x) = \ln[x(x^2 - 2)^{2/3}]$

15. $y = \dfrac{1}{b^2}\left[\ln(a + bx) + \dfrac{a}{a + bx} \right]$

16. $y = \dfrac{1}{b^2}[a + bx - a\ln(a + bx)]$

17. $y = -\dfrac{1}{a}\ln \dfrac{a + bx}{x}$

18. $y = -\dfrac{1}{ax} + \dfrac{b}{a^2}\ln \dfrac{a + bx}{x}$

In Exercises 19–26, evaluate the integral.

19. $\displaystyle\int \dfrac{1}{7x - 2}\,dx$

20. $\displaystyle\int \dfrac{x}{x^2 - 1}\,dx$

42. *Tidal Energy* Tidal power plants use "tidal energy" to produce electrical energy. To construct a tidal power plant, a dam is built to separate a bay from the sea. Electrical energy is produced as the water flows back and forth between the bay and the sea. The amount of "natural energy" produced depends on the volume of the basin and the tidal range—the vertical distance between high and low tides. (Throughout the world, several natural bays have tidal ranges in excess of 15 feet—the Bay of Fundy in Nova Scotia has a tidal range of 47.5 feet!)

FIGURE FOR 42

a. Consider a bay with a rectangular base as shown in the figure. The bay has a tidal range of 25 feet, with low tide corresponding to $y = 0$. How much water does the bay hold at high tide?

b. The amount of energy produced between filling (or emptying) of the bay is proportional to the amount of work required to fill or empty the bay. How much work is required to fill the bay with sea water? (Use a sea-water density of 64 pounds per cubic foot.)

The natural energy from tides can be increased by pumping the water at certain times during the tidal cycle. (For more information on tidal power, see the article "LaRance: Six Years of Operating a Tidal Power Plant in France" by J. Cotillon in the October, 1984 issue of *Water Power Magazine.*

Challenging Problems
New to the Fifth Edition are several exercises that can be assigned as special projects to individual students or groups of students. All of these exercises are real-life applications. Color photographs accompanied by descriptive captions emphasize the relevance of the mathematical concepts involved.

Journal References
References to articles in readily available journals help students understand that calculus is a current, dynamic field.

Computer Graphics
The usefulness of technology in representing graphs in two and three dimensions is illustrated by a variety of computer-generated graphs. Some are graphing calculator screens, and some were generated from commercial software such as Derive, Maple, Mathematica, and Axiom.

54. Find the volume of the solid of intersection (the solid common to both) of the two right circular cylinders of radius r whose axes meet at right angles (see figure).

FOR FURTHER INFORMATION For more information on this problem, see the article "Estimating the Volumes of Solid Figures with Curved Surfaces" by Donald Cohen in the May, 1991 issue of *Mathematics Teacher.*

Two intersecting cylinders   Solid of intersection

FIGURE FOR 54

55. Two planes are cut to the axis of a right circular cylinder to form a wedge. One plane is perpendicular to the axis of the cylinder and the second makes an angle of θ degrees with the first (see figure).

a. Find the volume of the wedge if $\theta = 45°$.

b. Find the volume of the wedge for an arbitrary but fixed angle $\theta = \theta_0$. Assuming the cylinder has sufficient length, how does the volume of the wedge change as θ_0 increases in the interval $0 < \theta_0 < 90°$?

FIGURE FOR 55

56. *Volume of Oil* A barrel of diameter d lays on a stand so that its axis makes an angle of 20° with its horizontal (see figure). The amount of motor oil in the barrel just covers the bottom of the barrel. Approximate the volume of oil.

FIGURE FOR 56

57. *Cavalieri's Theorem* Prove that if two solids have equal altitudes and all plane sections parallel to their bases and at equal distances from their bases have equal areas, then the solids have the same volume (see figure).

Area of R_1 = area of R_2

FIGURE FOR 57

58. A manufacturer drills a hole through the center of a metal sphere of radius R. The hole has a radius r. Find the volume of the resulting ring.

59. For the metal sphere in Exercise 58, let $R = 5$. What value of r will produce a ring whose volume is exactly half the volume of the sphere?

60. The solid shown in the figure has cross sections bounded by the graph of

$$|x|^a + |y|^a = 1$$

where $1 \le a \le 2$.

a. Describe the cross section when $a = 1$ and $a = 2$.

b. Describe a procedure for approximating the volume of the solid.

Generated by Mathematica

FIGURE FOR 60

124   Chapter 2 / Differentiation

$g(t) = t^2 - \dfrac{4}{t}$ **34.** $f(x) = x + \dfrac{1}{x^2}$

$f(x) = \dfrac{x^3 - 3x^2 + 4}{x^2}$

36. $h(x) = \dfrac{2x^2 - 3x + 1}{x}$

37. $y = x(x^2 + 1)$ **38.** $f(x) = \sqrt[3]{x} + \sqrt[5]{x}$

39. $h(s) = s^{4/5}$ **40.** $f(t) = t^{1/3} - 1$

$f(x) = 4\sqrt{x} + 3 \cos x$

42. $f(x) = 2 \sin x + 3 \cos x$

In Exercises 43–46, find an equation of the tangent line to the graph of the function at the indicated point.

Function	Point
$y = x^4 - 3x^2 + 2$	$(1, 0)$
44. $y = x^3 + x$	$(-1, -2)$
45. $f(x) = \dfrac{1}{\sqrt[3]{x^2}}$	$\left(8, \dfrac{1}{4}\right)$
46. $y = (x^2 + 2x)(x + 1)$	$(1, 6)$

In Exercises 47–52, determine the point(s) (if any) at which the function has a horizontal tangent line.

$y = x^4 - 3x^2 + 2$ **48.** $y = x^3 + x$

49. $y = \dfrac{1}{x^2}$ **50.** $y = x^2 + 1$

51. $y = x + \sin x$, $0 \le x < 2\pi$

52. $y = \sqrt{3}x + 2 \cos x$, $0 \le x < 2\pi$

Essay   In Exercises 53 and 54, the graphs of a function f and its derivative f' are given on the same coordinate axes. Label the graphs as f or f' and write a short paragraph stating the criteria used in making the selection.

53.   **54.**

Sketch the graphs of the two equations $y = x^2$ and $y = -x^2 + 6x - 5$, and sketch the two lines that are tangent to both graphs. Find the equations of these lines.

56. Show that the graphs of the two equations $y = x$ and $y = 1/x$ have tangent lines that are perpendicular to each other at their point of intersection.

In Exercises 57 and 58, find an equation of the tangent line to the graph of the function f through the point (x_0, y_0) not on the graph. To find the point of tangency (x, y) on the graph of f, solve the equation

$$f'(x) = \frac{y_0 - y}{x_0 - x}.$$

57. $f(x) = \sqrt{x}$ **58.** $f(x) = \dfrac{2}{x}$

$(x_0, y_0) = (-4, 0)$ $(x_0, y_0) = (5, 0)$

59. *Linear Approximation*   Consider the function $f(x) = x^{3/2}$ with the solution point $(4, 8)$.
  a. Use a graphing utility to graph f. Use the "zoom in" feature to obtain successive magnifications of the graph in the neighborhood of the point $(4, 8)$. After zooming in a few times, the graph should appear almost linear.
  b. Find the equation of the line

$$T(x) = f'(4)(x - 4) + f(4)$$

  tangent to the graph of f passing through the given point.
  c. Use a graphing utility to obtain the graphs of f and T on the same set of coordinate axes. Note that T is a "good" approximation of f when x is "close to" 4. What happens to the accuracy of the approximation as you move farther away from the point of tangency?
  d. Demonstrate the conclusion of part c by completing the table.

Δx	-3	-2	-1	-0.5	-0.1	0
$f(4 + \Delta x)$						
$T(4 + \Delta x)$						

Δx	0.1	0.5	1	2	3
$f(4 + \Delta x)$					
$T(4 + \Delta x)$					

60. *Linear Approximation*   Repeat Exercise 59 for the function $f(x) = x^3$ where $T(x)$ is the line tangent to the graph at the point $(1, 1)$. Explain why the accuracy of the linear approximation decreases more rapidly than it did in Exercise 59.

True or False   In Exercises 61–64, determine whether the statement is true or false. If it is false, explain why or give an example that shows it is false.

61. If $f'(x) = g'(x)$, then $f(x) = g(x)$.

62. If $f(x) = g(x) + c$, then $f'(x) = g'(x)$.

63. If $y = \pi^2$, then $dy/dx = 2\pi$.

64. If $y = x/\pi$, then $dy/dx = 1/\pi$.

71. *Cornu Spiral*   The cornu spiral is given by

$$x = \int_0^s \cos \frac{\pi u^2}{2}\, du, \qquad y = \int_0^s \sin \frac{\pi u^2}{2}\, du.$$

The spiral shown in the figure was plotted over the interval $-\pi \le s \le \pi$.
  a. Find the arc length of this curve from $s = 0$ to $s = a$.
  b. Find the curvature of the graph when $s = a$.
  c. The cornu spiral was discovered by James Bernoulli. He found that the spiral has an amazing relationship between curvature and arc length. What is this relationship?

FIGURE FOR 71

72. A sphere of radius 4 is dropped into the paraboloid given by

Variety of Problems

Many new types of problems have been added to the exercise sets, offering opportunities for writing, for individual and group projects, for solving multi-part applications, for working quantitative as well as qualitative problems, and for working with real data. To help instructors make homework assignments, many of the exercises in the text are labeled to indicate the area of application (as in Cornu Spiral) or the type of exercise (as in Essay or Linear Approximation).

True or False Questions

To help students understand the logical structure of calculus, a set of "true or false" questions are included toward the end of most exercise sets. These questions help students focus on concepts, common errors, and the correct statements of definitions and theorems.

Career Interviews

Career interviews with engineers, scientists, mathematicians, and other professionals who use calculus in their jobs encourage students to understand that calculus is a modern, problem-solving language.

CAREER INTERVIEW

Donna F. Stroup, Ph.D., M.Sc.

Director, Surveillance and Epidemiology

Centers for Disease Control and Prevention

Atlanta, GA

At the CDC, I direct a group of scientists who are trained in mathematics, statistics, and epidemiology, and whose job it is to reduce the incidence of preventable diseases, disabilities, and injuries. By analyzing information about the numbers of people infected by a particular disease, for instance, we can assess the need for a specific vaccination program. Furthermore, our data analysis on injuries associated with automobile crashes has led to recommendations on laws governing required safety features for automobiles, such as seat belts or content of gasoline.

Because much of our work at CDC is in statistics, my understanding of the calculus that is the basis of many statistical tests is an enormous asset. I assure policymakers that the application of a method is valid and that incorrect conclusions or inappropriate recommendations are avoided. The careful and systematic approach to problems that underlies calculus is essential to prevention and intervention efforts in public health, and to the acceptance of public health policy.

Supplements

Calculus with Analytic Geometry, Fifth Edition, by Larson, Hostetler, and Edwards is accompanied by a comprehensive supplements package with ancillaries for students, for instructors, and for classroom resources. Most items are keyed directly to the book.

PRINTED RESOURCES

For the instructor

Instructor's Guide by Ann R. Kraus, Pennsylvania State University, The Behrend College
- Notes to the new teacher
- Chapter summaries
- Ready-made chapter tests
- Teaching strategies

Complete Solutions Guide, Volumes I, II, and III by Dianna L. Zook, Indiana University—Purdue University at Fort Wayne
- Detailed solutions to all text exercises

Test Item File by Ann R. Kraus, Pennsylvania State University, The Behrend College
- Printed test bank
- Approximately 3000 test items
- Multiple-choice and open-ended questions coded by level of difficulty
- Technology required to solve some test questions
- Also available as test-generating software

Graphing Calculator Demonstration Problems by August J. Zarcone and Russell Lundstrom, College of DuPage
- Classroom demonstration problems using TI-81 and TI-82 graphing calculators

Transparency package
- 120 color transparencies of figures from the text

For the student

Study and Solutions Guide by David E. Heyd, The Pennsylvania State University, The Behrend College
- Detailed solutions to all odd-numbered text exercises
- Algebra review
- Study strategies

Technology Laboratory Guide by David E. Heyd, The Pennsylvania State University, The Behrend College
- Supplementary exercises using graphics calculators and graphing utility software
- Worksheet format

Graphing Technology Guide for Calculus by Benjamin N. Levy
- Keystroke-level commands and instructions
- Step-by-step examples with numerous screen displays
- Technology tips

Computer Projects in Calculus by Bruce H. Edwards, University of Florida
- Guided exploration labs
- Brief introductions to Mathematica, Maple, and Derive

Insights into Calculus with the Graphics Calculator by Herbert A. Hollister, Bowling Green State University
- Examples with step-by-step solutions and exercises with answers to odds
- Graphics screen displays
- Graphing calculator programs

Calculus Applications in Engineering and Science by Stuart Goldenberg and Harvey Greenwald, California Polytechnic State University—San Luis Obispo
- Extended applications

The Algebra of Calculus by Eric J. Braude
- Review of algebra, trigonometry, and analytic geometry required for calculus
- Over 200 examples with solutions
- Pretests and exercises

Technology Resources

Software

PC-81 Emulation Software
Simulates the functions of the TI-81 graphics calculator

Computerized Testing
- Test-generating software
- Approximately 3000 test items
- Also available as a printed test bank

Derive
- Symbolic mathematics package
- Available to adopters at a discount

BestGrapher by George Best
- Function grapher
- Simultaneous display of equation, graph, and table of values
- Zoom and print features

Workout! by Joe Mazur
- Self-study tutorial software
- On-line hints and guidance
- Generates problems at four levels of difficulty

Mathlab Interactive Calculus
- Graphing program
- Lab manual with 80 assignments

Videotapes

Graphing Calculator Instructional Videotape by Dana Mosely

Calculus Videotapes by Video Tutorial Services
- Interactive videotapes covering the major first- and second-semester topics
- Computer-generated graphics

For the convenience of software users, a technical support telephone number is available with all D. C. Heath software products: (617) 860-1218

Acknowledgements

We would like to thank the many people who have helped us at various stages of this project during the past seventeen years. Their encouragement, criticisms, and suggestions have been invaluable to us.

FIFTH EDITION REVIEWERS   James Angelos, Central Michigan University; Norman Biernes, University of Regina; Brian Blank, Washington University; Jack Ceder, University of California—Santa Barbara; Kathy Davis, University of Texas; Russell Euler, Northwest Missouri State University; Michael Frantz, University of La Verne; Elgin Johnston, Iowa State University; Toni Kasper, Borough of Manhattan Community College; Ronnie Khuri, University of Florida; Murray Lieb, New Jersey Institute of Technology; Mauricio Marroquin, Los Angeles Valley College; Robert McMaster, John Abbott College; Philip Montgomery, University of Kansas; Charlotte Newsom, Tidewater Community College; Luis Ortiz-Franco, Chapman University; Donald Poulson, Mesa Community College; Enid Steinbart, University of New Orleans; Marjorie Valentine, North Side ISD; Carroll G. Wells, Western Kentucky University; Paul D. Zahn, Borough of Manhattan Community College

FOURTH EDITION FOCUS GROUP   Homer F. Bechtell, University of New Hampshire; K. Elayn Gay, University of New Orleans; Hideaki Kaneko, Old Dominion University; Judith A. Palagallo, University of Akron; John Tweed, Old Dominion University

FOURTH EDITION REVIEWERS   Keith Bergeron, United States Air Force Academy; Jorge Cossio, Miami-Dade Community College; Rosario Diprizio, Oakton Community College; Ali Hajjafar, University of Akron; Ransom Van B. Lynch, Phillips Exeter Academy; Bennet Manvel, Colorado State University; Duff A. Muir, United States Air Force Academy; Charlotte J. Newsom, Tidewater Community College; Terry J. Newton, United States Air Force Academy; Wayne J. Peeples, University of Texas; Jorge A. Perez, LaGuardia Community College; Barry J. Sarnacki, United States Air Force Academy; George W. Schultz, St. Petersburg Junior College; Frank Soler, De Anza College; Michael Steuer, Nassau Community College; John Tweed, Old Dominion University; Jay Wiestling, Palomar College; August J. Zarcone, College of DuPage; Li Fong, Johnson County Community College

THIRD EDITION REVIEWERS   Dennis Albér, Palm Beach Junior College; Garret J. Etgen, University of Houston; William R. Fuller, Purdue University; Timothy J. Kearns, Boston College; Norbert Lerner, State University of New York at Cortland; Robert L. Maynard, Tidewater Community College; Barbara L. Osofsky, Rutgers University; Jean E. Rubin, Purdue University; Lawrence A. Trivieri, Mohawk Valley Community College; J. Philip Smith, Southern Connecticut State University

SECOND EDITION REVIEWERS   Harry L. Baldwin, Jr., San Diego City College; Phillip A. Ferguson, Fresno City College; Thomas M. Green, Contra Costa College; Arnold J. Insel, Illinois State University; William J. Keane, Boston College; David C. Lantz, Colgate University; Richard E. Shermoen,

Washburn University; Thomas W. Shilgalis, Illinois State University; Florence A. Warfel, University of Pittsburgh

FIRST EDITION REVIEWERS   Paul W. Davis, Worcester Polytechnic Institute; Eric R. Immel, Georgia Institute of Technology; Frank T. Kocher, Jr., Pennsylvania State University; Joseph F. Krebs, Boston College; Maurice L. Monahan, South Dakota State University; Robert A. Nowlan, Southern Connecticut State University; N. James Schoonmaker, University of Vermont; Bert K. Waits, Ohio State University

We would also like to thank users of the Third Edition who answered a questionnaire concerning changes they wanted in the new edition. They are George Anderson, Rhode Island College; Frank P. Battles, Massachusetts Maritime Academy; Derek I. Bloomfield, Orange County Community College; Karen J. Edwards, Paul Smith's College; Theodore Hanley, State University of New York at Utica/Rome; Peter Herron, Suffolk County Community College; Ann M. Joyce, Chestnut Hill College; Arthur Kaufman, College of Staten Island; Alan Levine, Franklin and Marshall College; James Magliano, Union County College; Frank Morgan, Castleton State College; Raymond Pluta, Castleton State College; M. Susan Richman, Pennsylvania State University; Carmen Vlad, Pace University; Christopher White, Castleton State College.

A special thanks to all the people at D. C. Heath and Company who worked with us in the development of the Fifth Edition, especially Ann Marie Jones, Mathematics Acquisitions Editor; Cathy Cantin, Developmental Editor; Ron Hampton, Production Editor; Sally Steele, Designer; Elinor Stapleton, Art Editor; Wing-Harn Chen, Developmental Assistant; Carolyn Johnson, Editorial Associate; Mike O'Dea, Production Manager; and Toni Michaels, Photo Researcher.

We would also like to thank the staff at Larson Texts, Inc. who assisted with proofreading the manuscript; preparing and proofreading the art package; and checking and typesetting the supplements.

A special note of thanks goes to the over 1,000,000 students who have used earlier editions of the text.

On a personal level, we are grateful to our wives, Deanna Gilbert Larson, Eloise Hostetler, and Consuelo Edwards, for their love, patience, and support. Also, a special thanks goes to R. Scott O'Neil.

If you have suggestions for improving this text, please feel free to write to us. Over the past two decades we have received many useful comments from both instructors and students, and we value these very much.

Roland E. Larson, Robert P. Hostetler, Bruce H. Edwards

Calculus

To accommodate the different methods of teaching calculus, D. C. Heath also offers the three programs described below. The following is a brief discussion of how each program differs from *Calculus with Analytic Geometry,* Fifth Edition.

Calculus with Analytic Geometry, Alternate Fifth Edition with Late Trigonometry
Larson/Hostetler/Edwards

This text, with late trigonometry, is also designed for a three-semester course. Additionally, the text offers a different treatment of the following topics: limits, applications of integration, exponential and logarithmic functions, and vectors. This text has its own supplement package.

Calculus of a Single Variable, Fifth Edition
Larson/Hostetler/Edwards

This single-variable text is designed for a two-semester course. All six trigonometic functions are reviewed in the Prerequisites chapter and then used throughout the text.

Interactive Calculus for Windows
Larson/Hostetler/Edwards

Interactive Calculus is a multi-media environment that unites the text of *Calculus with Analytic Geometry*, Fifth Edition, with a symbolic algebra system, graphics, video, and tools for problem solving. In a guided exploration of the traditional calculus course, students can manipulate equations and see the results immediately as the computer solves equations and redraws graphs.

Contents

Chapter 1   Limits and Their Properties   60

Chapter 2   Differentiation   104

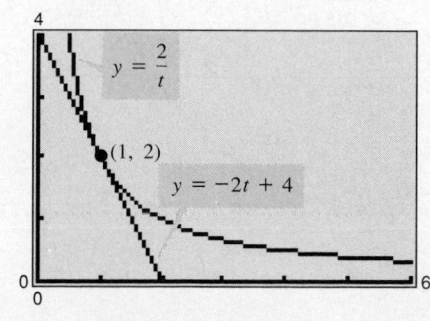

Chapter 3   Applications of Differentiation   164

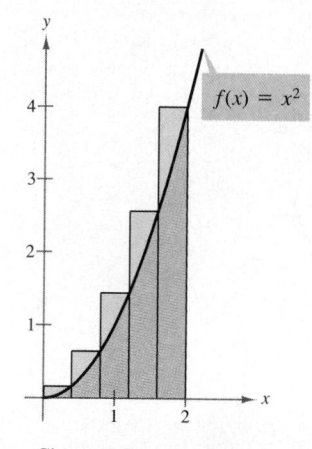

$f(x) = x^4 - 12x^3 + 48x^2 - 64x$

(0, 0)

(4, 0)
Point of inflection

(2, −16)
Point of inflection

(1, −27)
Minimum

Chapter 4   Integration   248

$f(x) = x^2$

Circumscribed rectangles

Chapter 5   Logarithmic, Exponential, and Other Transcendental Functions   316

Number of fruit flies

(4, 300)

$y = Ce^{kt}$

(2, 100)

(0, 33)

Time (in days)

Chapter 6   Applications of Integration   402

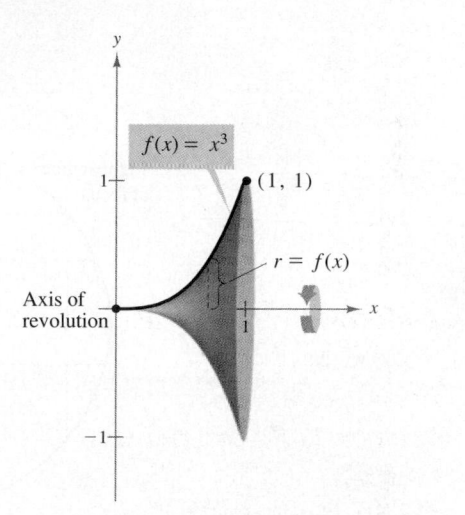

Chapter 7   Integration Techniques, L'Hôpital's Rule, and Improper Integrals 470

Chapter 12   Vector-Valued Functions   782

$a_N = b$

b

Chapter 13   Functions of Several Variables   830

Surface:
$f(x, y) = \dfrac{5x^2y}{x^2 + y^2}$

Chapter 14   Multiple Integration   924

Surface:
$f(x, y) = 1 - \frac{1}{2}x^2 - \frac{1}{2}y^2$

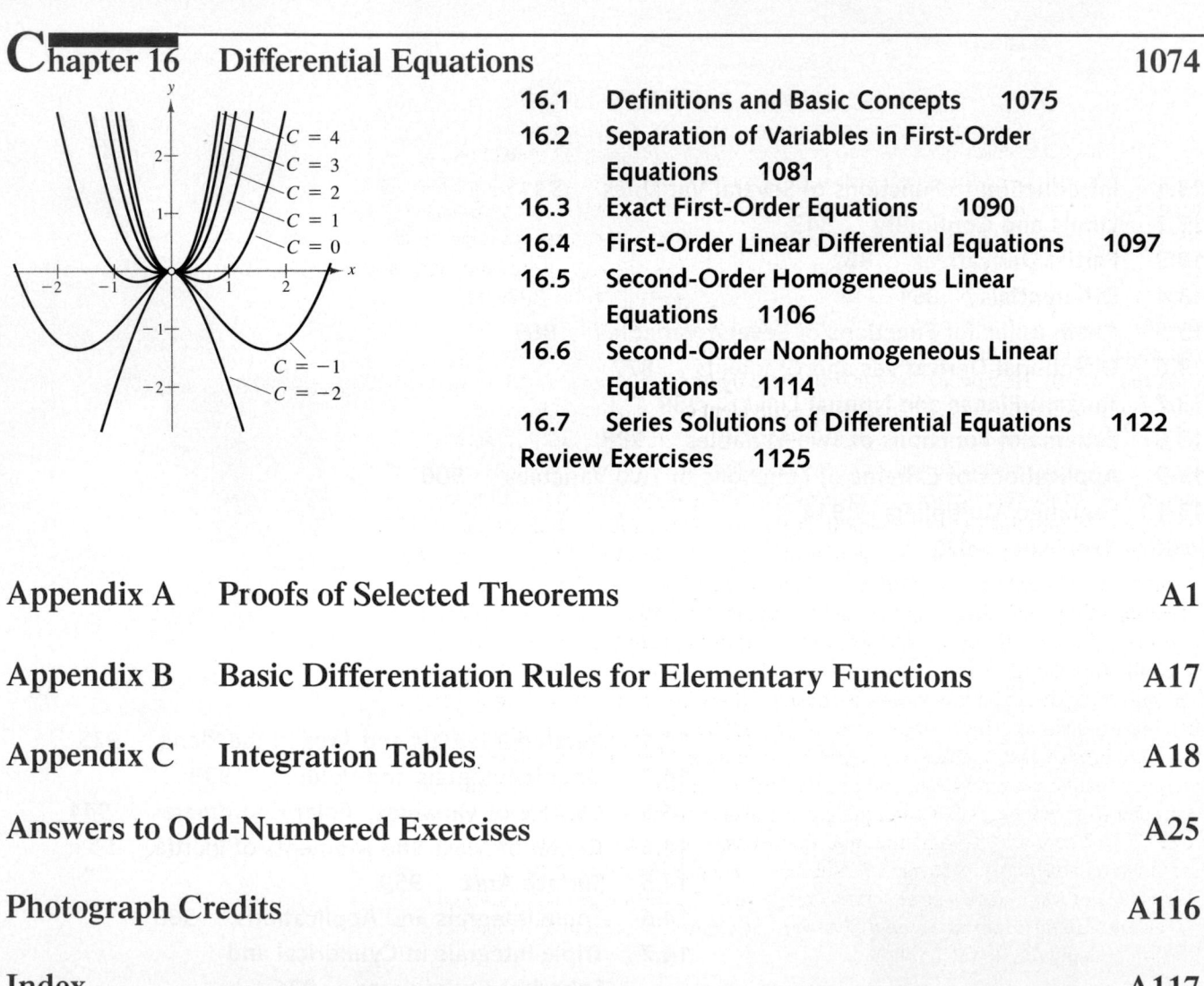

ABOUT THE COVER

The "Umbilic Torus NC" is a sculptural expression of timeless mathematical theorems in new silicon bronze created by Helaman Ferguson in 1987–88. Ferguson thought about the binary quadratic forms $ax^2 + bxy + cy^2$ where $b^2 - 4ac$ tells when the coefficients (a, b, c) represent the quadratic part of conic sections for hyperbolas ($b^2 - 4ac > 0$), ellipses ($b^2 - 4ac < 0$), or parabolas ($b^2 - 4ac = 0$). Then he thought about the binary cubic forms $ax^3 + bx^2y + cxy^2 + dy^3$ where the discriminant $\Delta = b^2c^2 - 4ac^3 - 4b^3d + 18abcd - 27a^2d^2$ tells when the coefficients (a, b, c, d) represent hyperbolic umbilics ($\Delta < 0$), elliptic umbilics ($\Delta > 0$), and parabolic umbilics ($\Delta = 0$). He saw an umbilic torus here by looking at the real cubic form with four real coefficients a, b, c, d as an equivalent complex cubic form with two complex coefficients u, v. He substituted z by $e^{i\theta}z$ and changed (u, v) into $(uc^{3i\theta}, vc^{i\theta})$, so u rotates thrice while v rotates once. He set $u = 1$, differentiated, and solved for the torus cross-section curve $v = 2e^{i\emptyset} - e^{-2i\emptyset}$, a hypocycloid with three cusps. From this he got the umbilic torus in two complex dimensions as $(e^{3i\theta}, 2e^{i(\theta + \emptyset)} - e^{i(\theta - 2\emptyset)})$. He felt how the space curve of cusps of this hypocycloid wound the umbilic torus three times the long way and once the short way. He developed three real dimensional equations to get the umbilic torus in a sculpture-like space, computed normal vectors, tool offset vectors, and an NC (numerically controlled) tool path. For surface texture he chose the computer-driven cutting ball end mill tool path to be a stage of the surface filling Hilbert curve or 2-adic space filling Peano curve. This positive form so carved, he put through the traditional lost-wax bronze casting process: a negative silicone rubber mold, positive wax image, negative ceramic flask, and positive silicon bronze which was chased and corroded to give a permanent antique green finish. For more about this sculpture read "Two Theorems, Two Sculptures, Two Posters" appearing in the *American Mathematical Monthly*, Volume 97, Number 7, August–September 1990, pp. 589–610. For more about the sculptor's work, read the full-color book *Helaman Ferguson: Mathematics in Stone and Bronze* by Claire Ferguson, published by Meridian Creative Group, 1994.

Applications

THE LIFE SCIENCES

GENERAL

What Is Calculus?

We begin to answer this question by saying that calculus is the reformulation of elementary mathematics through the use of a limit process. If limit processes are unfamiliar to you, then this answer is, at least for now, somewhat less than illuminating. From an elementary point of view, we may think of calculus as a "limit machine" that generates new formulas from old. Actually, the study of calculus involves three distinct stages of mathematics: precalculus mathematics (the length of a line segment, the area of a rectangle, and so forth), the limit process, and new calculus formulations (derivatives, integrals, and so forth).

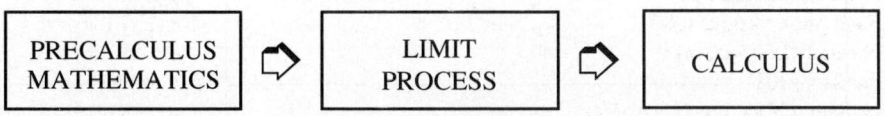

Some students try to learn calculus as if it were simply a collection of new formulas. This is unfortunate. When students reduce calculus to the memorization of differentiation and integration formulas, they miss a great deal of understanding, self-confidence, and satisfaction.

On the following two pages we have listed some familiar precalculus concepts coupled with their more powerful calculus versions. Throughout this text, our goal is to show you how precalculus formulas and techniques are used as building blocks to produce the more general calculus formulas and techniques. Don't worry if you are unfamiliar with some of the "old formulas" listed on the following two pages—we will be reviewing all of them.

As you proceed through this text, we suggest that you come back to this discussion repeatedly. Try to keep track of where you are relative to the three stages involved in the study of calculus. For example, the first three chapters break down as follows: precalculus (Prerequisites), the limit process (Chapter 1), and new calculus formulas (Chapter 2). This cycle is repeated many times on a smaller scale throughout the text. We wish you well in your venture into calculus.

WITHOUT CALCULUS	WITH DIFFERENTIAL CALCULUS
value of $f(x)$ when $x = c$	limit of $f(x)$ as x approaches c
slope of a line	slope of a curve
secant line to a curve	tangent line to a curve
average rate of change between $t = a$ and $t = b$	instantaneous rate of change at $t = c$
curvature of a circle	curvature of a curve
height of a curve when $x = c$	maximum height of a curve on an interval
tangent plane to a sphere	tangent plane to a surface
direction of motion along a straight line	direction of motion along a curved line

WITHOUT CALCULUS	WITH INTEGRAL CALCULUS
area of a rectangle	area under a curve
work done by a constant force	work done by a variable force
center of a rectangle	centroid of a region
length of a line segment	length of an arc
surface area of a cylinder	surface area of a solid of revolution
mass of a solid of constant density	mass of a solid of variable density
volume of a rectangular solid	volume of a region under a surface
sum of a finite number of terms $a_1 + a_2 + \cdots + a_n = S$	sum of an infinite number of terms $a_1 + a_2 + a_3 \cdots = S$

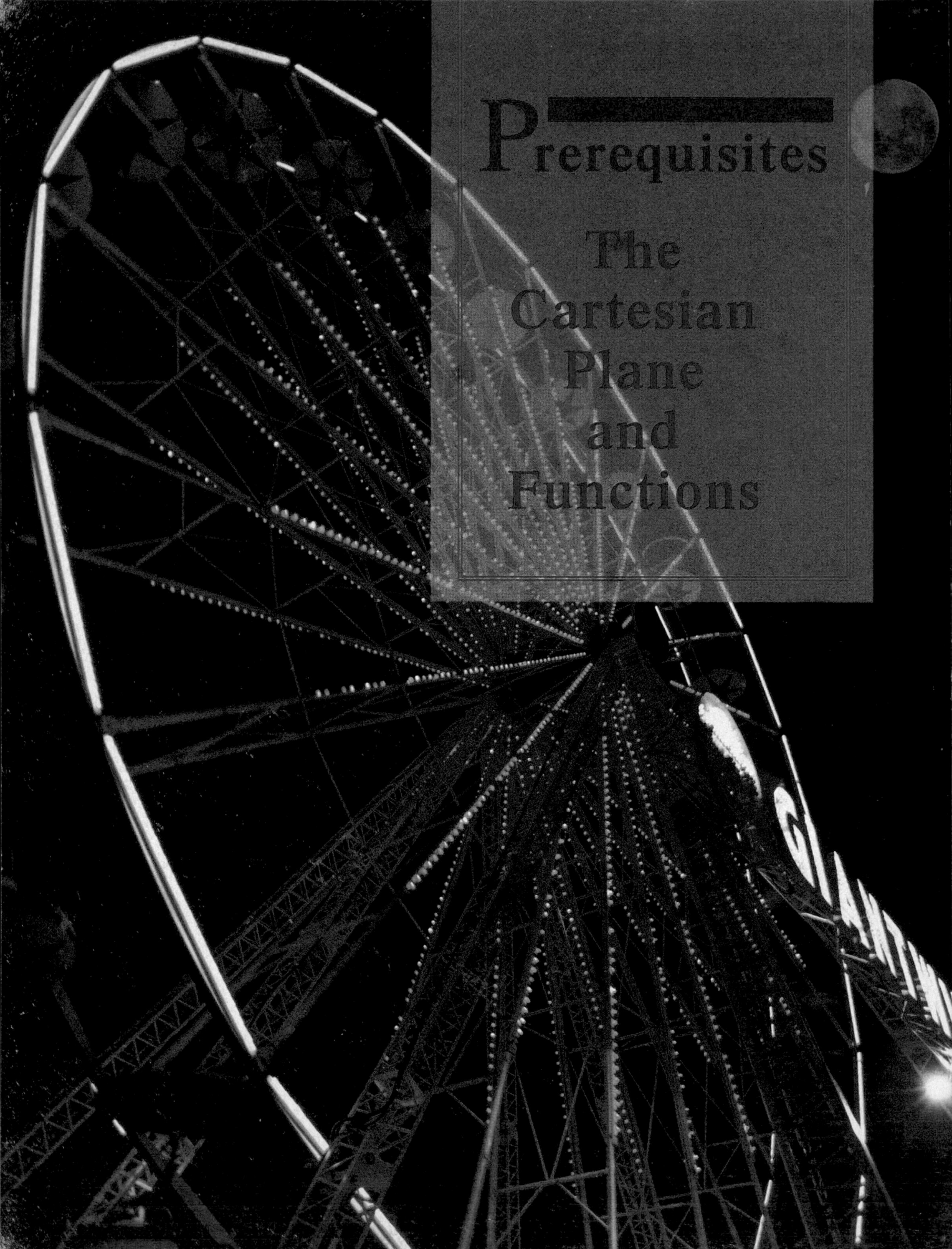

Prerequisites

The Cartesian Plane and Functions

Real Numbers and the Real Line

Real numbers can be represented by a coordinate system called the **real line** or x-axis (see Figure 1). The real number corresponding to a point on the real line is the **coordinate** of the point. As Figure 1 shows, it is customary to identify those points whose coordinates are integers.

FIGURE 1
The real line

The point on the real line corresponding to zero is the **origin** and is denoted by 0. The **positive direction** (to the right) is denoted by an arrowhead and is the direction of increasing values of x. Numbers to the right of the origin are **positive**. Numbers to the left of the origin are **negative**. The term **nonnegative** describes a number that is positive or zero. The term **nonpositive** describes a number that is negative or zero.

Each point on the real line corresponds to one and only one real number, and each real number corresponds to one and only one point on the real line. This type of relationship is called a **one-to-one correspondence**.

FIGURE 2
Rational numbers

FIGURE 3
Irrational numbers

Each of the four points in Figure 2 corresponds to a **rational number**—one that can be expressed as the ratio of two integers. (Note that $4.5 = \frac{9}{2}$ and $-2.6 = \frac{-13}{5}$.) Rational numbers can be represented either by *terminating decimals* such as $\frac{2}{5} = 0.4$, or by *repeating decimals* such as $\frac{1}{3} = 0.333 \ldots = 0.\overline{3}$.

Real numbers that are not rational are **irrational.** Irrational numbers cannot be represented as terminating or repeating decimals. In computations, irrational numbers are represented by decimal approximations. Here are three familiar examples.

$$\sqrt{2} \approx 1.414213562$$
$$\pi \approx 3.141592654$$
$$e \approx 2.718281828$$

(See Figure 3.)

◀ *The height of a Ferris wheel is modeled in Section 6, Exercise 78.*

Order and Inequalities

One important property of real numbers is that they can be **ordered.** If a and b are real numbers, then a is **less than** b if $b - a$ is positive. This order is denoted by the **inequality**

$$a < b.$$

The statement b is **greater than** a is equivalent to saying a is less than b. When three real numbers a, b, and c are ordered so that $a < b$ and $b < c$, we say that b is **between** a and c and $a < b < c$.

Geometrically, $a < b$ if and only if a lies to the *left* of b on the real line (see Figure 4). For example, $1 < 2$ because 1 lies to the left of 2 on the real line.

The following properties are used in working with inequalities. Similar properties are obtained if $<$ is replaced by $\leq$ and $>$ is replaced by $\geq$. (The symbols $\leq$ and $\geq$ mean **less than or equal to** and **greater than or equal to,** respectively.)

FIGURE 4
$a < b$ if and only if a lies to the left of b.

Properties of Inequalities

Let a, b, c, d, and k be real numbers.
1. If $a < b$ and $b < c$, then $a < c$.   Transitive Property
2. If $a < b$ and $c < d$, then $a + c < b + d$.   Add inequalities
3. If $a < b$, then $a + k < b + k$.   Add a constant
4. If $a < b$ and $k > 0$, then $ak < bk$.   Multiply by a positive constant
5. If $a < b$ and $k < 0$, then $ak > bk$.   Multiply by a negative constant

REMARK   Note that you *reverse the inequality* when you multiply by a negative number. For example, if $x < 3$, then $-4x > -12$. This also applies to division by a negative number. Thus, if $-2x > 4$, then $x < -2$.

A **set** is a collection of elements. Two common sets are the set of real numbers and the set of points on the real line. Many problems in calculus involve **subsets** of one of these two sets. In such cases it is convenient to use **set notation** of the form $\{x:$ condition on $x\}$, which is read as follows.

The set of all x such that a certain condition is true.

$$\{ \quad x \quad : \quad \text{condition on } x\}$$

For example, you can describe the set of positive real numbers as

$$\{x: 0 < x\}. \quad \text{Set of positive real numbers}$$

Similarly, you can describe the set of nonnegative real numbers as

$$\{x: 0 \leq x\}. \quad \text{Set of nonnegative real numbers}$$

The **union** of two sets A and B, denoted by $A \cup B$, is the set of elements that are members of A *or B or both*. The **intersection** of two sets A and B, denoted by $A \cap B$, is the set of elements that are members of A *and B*. Two sets are **disjoint** if they have no elements in common.

The most commonly used subsets are **intervals** on the real line. For example, the **open** interval

$$(a, b) = \{x: a < x < b\} \qquad \text{Open interval}$$

is the set of all real numbers greater than a and less than b, where a and b are the **endpoints** of the interval. Note that the endpoints are not included in an open interval. Intervals that include their endpoints are **closed** and are denoted by

$$[a, b] = \{x: a \leq x \leq b\}. \qquad \text{Closed interval}$$

The nine basic types of intervals on the real line are shown in Table 1. The first four are **bounded intervals** and the remaining five are **unbounded intervals.** Unbounded intervals are also classified as open or closed. The intervals $(-\infty, b)$ and (a, ∞) are open, the intervals $(-\infty, b]$ and $[a, \infty)$ are closed, and the interval $(-\infty, \infty)$ is considered to be both open *and* closed.

TABLE 1

Intervals on the Real Line

	Interval notation	*Set notation*	*Graph*
Bounded open interval	(a, b)	$\{x: a < x < b\}$	
Bounded closed interval	$[a, b]$	$\{x: a \leq x \leq b\}$	
Bounded half-open intervals	$[a, b)$	$\{x: a \leq x < b\}$	
	$(a, b]$	$\{x: a < x \leq b\}$	
Unbounded open intervals	$(-\infty, b)$	$\{x: x < b\}$	
	(a, ∞)	$\{x: a < x\}$	
Unbounded closed intervals	$(-\infty, b]$	$\{x: x \leq l\}$	
	$[a, \infty)$	$\{x: a \leq x\}$	
Entire real line	$(-\infty, \infty)$	$\{x: x \text{ is a real number}\}$	

REMARK　The symbols ∞ and $-\infty$ refer to positive and negative infinity. These symbols do not denote real numbers. They simply enable you to describe unbounded conditions more concisely. For instance, the interval $[a, \infty)$ is unbounded to the right because it includes *all* real numbers that are greater than or equal to a.

EXAMPLE 1   Liquid and Gaseous States of Water

Describe the intervals on the real line that correspond to the temperature x (in degrees Celsius) for water in

a. A liquid state   **b.** A gaseous state.

Solution

a. Water is in a liquid state at temperatures greater than 0° and less than 100°, as shown in Figure 5(a).

$$(0, 100) = \{x: 0 < x < 100\}$$

b. Water is in a gaseous state (steam) at temperatures greater than or equal to 100°, as shown in Figure 5(b).

$$[100, \infty) = \{x: 100 \le x\}$$

(a) Temperature range of water (in degrees Celsius)   (b) Temperature range of steam (in degrees Celsius)

FIGURE 5

A real number a is a **solution** of an inequality if the inequality is **satisfied** (is true) when a is substituted for x. The set of all solutions is the **solution set** of the inequality.

EXAMPLE 2   Solving an Inequality

Solve $2x - 5 < 7$.

Solution

$$2x - 5 < 7 \qquad \text{Original inequality}$$
$$2x - 5 + 5 < 7 + 5 \qquad \text{Add 5 to both sides}$$
$$2x < 12 \qquad \text{Simplify}$$
$$\frac{1}{2}(2x) < \frac{1}{2}(12) \qquad \text{Multiply both sides by } \tfrac{1}{2}$$
$$x < 6 \qquad \text{Simplify}$$

The solution set is $(-\infty, 6)$.

REMARK   In Example 2, all five inequalities listed as steps in the solution are called **equivalent** because they have the same solution set.

Once you have solved an inequality, check some x-values in your solution set to verify that they satisfy the original inequality. You should also check some values outside your solution set to verify that they *do not* satisfy the inequality. For example, Figure 6 shows that when $x = 0$ or $x = 5$ the inequality $2x - 5 < 7$ is satisfied, but when $x = 7$ the inequality $2x - 5 < 7$ is not satisfied.

If $x = 0$, then $2(0) - 5 = -5 < 7$.

If $x = 5$, then $2(5) - 5 = 5 < 7$.

If $x = 7$, then $2(7) - 5 = 9 > 7$.

FIGURE 6

Checking solutions of $2x - 5 < 7$.

EXAMPLE 3 Solving a Double Inequality

Solve $-3 \leq 2 - 5x \leq 12$.

Solution

$-3 \leq$	$2 - 5x$	≤ 12		Original inequality
$-3 - 2 \leq$	$2 - 5x - 2$	$\leq 12 - 2$		Subtract 2
$-5 \leq$	$-5x$	≤ 10		Simplify
$\dfrac{-5}{-5} \geq$	$\dfrac{-5x}{-5}$	$\geq \dfrac{10}{-5}$		Divide by -5 and reverse both inequalities
$1 \geq$	x	≥ -2		Simplify

FIGURE 7

Solution set of $-3 \leq 2 - 5x \leq 12$.

The solution set is $[-2, 1]$, as shown in Figure 7.

The inequalities in Examples 2 and 3 are **linear inequalities**—that is, they involve first-degree polynomials. To solve inequalities involving polynomials of higher degree, use the fact that a polynomial can change signs *only* at its real **zeros** (the numbers that make the polynomial zero). Between two consecutive real zeros a polynomial must be either entirely positive or entirely negative. This means that when the real zeros of a polynomial are put in order, they divide the real line into **test intervals** in which the polynomial has no sign changes. Thus, if a polynomial has the factored form

$$(x - r_1)(x - r_2) \cdots (x - r_n), \qquad r_1 < r_2 < r_3 < \cdots < r_n$$

then the test intervals are

$$(-\infty, r_1), \quad (r_1, r_2), \quad \ldots, \quad (r_{n-1}, r_n), \quad \text{and} \quad (r_n, \infty).$$

To determine the sign of the polynomial in each test interval, you need to test only *one value* from the interval.

EXAMPLE 4 Solving a Quadratic Inequality

Solve $x^2 < x + 6$.

Solution

	$x^2 < x + 6$	Original inequality
$x^2 - x - 6 < 0$		Write in standard form
$(x - 3)(x + 2) < 0$		Factor

The polynomial $x^2 - x - 6$ has $x = -2$ and $x = 3$ as its zeros. Thus, you can solve the inequality by testing the sign of $x^2 - x - 6$ in each of the test intervals $(-\infty, -2)$, $(-2, 3)$, and $(3, \infty)$. To test an interval, choose any number in the interval and compute the sign of $x^2 - x - 6$. After doing this, you will find that the polynomial is positive for all real numbers in the first and third intervals and negative for all real numbers in the second interval. The solution of the original inequality is therefore $(-2, 3)$, as shown in Figure 8.

Choose $x = -3$.
$(x - 3)(x + 2) > 0$

Choose $x = 4$.
$(x - 3)(x + 2) > 0$

Choose $x = 0$.
$(x - 3)(x + 2) < 0$

FIGURE 8

Testing an interval.

Absolute Value and Distance

If a is a real number, then the **absolute value** of a is

$$|a| = \begin{cases} a, & \text{if } a \geq 0 \\ -a, & \text{if } a < 0. \end{cases}$$

The absolute value of a number cannot be negative. For example, let $a = -4$. Then, because $-4 < 0$, you have

$$|a| = |-4| = -(-4) = 4.$$

Remember that the symbol $-a$ does not necessarily mean that $-a$ is negative.

Operations with Absolute Value

Let a and b be real numbers and let n be a positive integer.

1. $|ab| = |a|\,|b|$ **2.** $\left|\dfrac{a}{b}\right| = \dfrac{|a|}{|b|}, \quad b \neq 0$

3. $|a| = \sqrt{a^2}$ **4.** $|a^n| = |a|^n$

REMARK You are asked to prove these properties in Exercises 65, 67, 68, and 69.

Properties of Inequalities and Absolute Value

Let a and b be real numbers, and let k be a positive real number.

1. $-|a| \leq a \leq |a|$
2. $|a| \leq k$ if and only if $-k \leq a \leq k$.
3. $k \leq |a|$ if and only if $k \leq a$ or $a \leq -k$.
4. *Triangle Inequality:* $|a + b| \leq |a| + |b|$

Properties 2 and 3 are also true if $\leq$ is replaced by $<$.

EXAMPLE 5   Solving an Absolute Value Inequality

Solve $|x - 3| \leq 2$.

Solution Using the second property of inequalities and absolute value, you can rewrite the original inequality as a double inequality.

$$
\begin{array}{lll}
-2 \leq & x - 3 & \leq 2 & \quad \text{Write as double inequality} \\
-2 + 3 \leq & x - 3 + 3 & \leq 2 + 3 & \quad \text{Add 3} \\
1 \leq & x & \leq 5 & \quad \text{Simplify}
\end{array}
$$

The solution set is $[1, 5]$, as shown in Figure 9.

2 units   2 units

FIGURE 9
Solution set of $|x - 3| \leq 2$.

FIGURE 10
Solution set of $3 < |x + 2|$.

Solution set of $|x - a| \leq d$

Solution set of $|x - a| \geq d$

FIGURE 11

EXAMPLE 6 **A Two-Interval Solution Set**

Solve $3 < |x + 2|$.

Solution   Using the third property of inequalities and absolute value, you can rewrite the original inequality as two linear inequalities.

$$3 < x + 2 \quad \text{or} \quad x + 2 < -3$$
$$1 < x \qquad \text{or} \qquad x < -5$$

The solution set is the union of the disjoint intervals $(-\infty, -5)$ and $(1, \infty)$, as shown in Figure 10.   ■

Examples 5 and 6 illustrate the general results shown in Figure 11. Note that if $d > 0$, the solution set for the inequality $|x - a| \leq d$ is a *single* interval, whereas the solution set for the inequality $|x - a| \geq d$ is the union of *two* disjoint intervals.

The **distance between two points** a and b on the real line is given by

$$\boxed{d = |a - b| = |b - a|.}$$

The **directed distance from a to b** is $b - a$ and the **directed distance from b to a** is $a - b$, as shown in Figure 12.

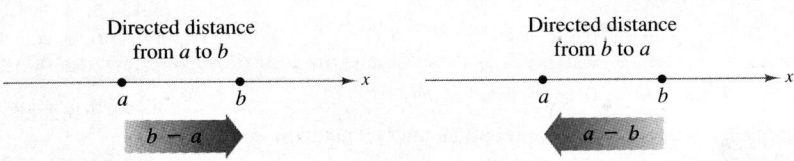

FIGURE 12

EXAMPLE 7 **Distance on the Real Line**

a. The distance between -3 and 4 is

$$|4-(-3)| = |7| = 7 \quad \text{or} \quad |-3 - 4| = |-7| = 7.$$

(See Figure 13.)

b. The directed distance from -3 to 4 is $4 - (-3) = 7$.

c. The directed distance from 4 to -3 is $-3 - 4 = -7$.   ■

FIGURE 13

The **midpoint** of an interval with endpoints a and b is the average value of a and b. That is,

$$\boxed{\text{Midpoint of interval } (a, b) = \frac{a + b}{2}.}$$

To show that this is the midpoint, you need only show that $(a + b)/2$ is equidistant from a and b.

EXERCISES for Section 1

In Exercises 1–10, determine whether the real number is rational or irrational.

1. 0.7

2. -3678

3. $\dfrac{3\pi}{2}$

4. $3\sqrt{2} - 1$

5. $4.3451\overline{451}$

6. $\frac{22}{7}$

7. $\sqrt[3]{64}$

8. 0.81778177

9. $4\frac{5}{8}$

10. $(\sqrt{2})^3$

In Exercises 11–14, express the repeating decimal as a ratio of integers using the following procedure. If $x = 0.6363\ldots$, then $100x = 63.6363\ldots$. Subtracting the first equation from the second produces $99x = 63$ or $x = \frac{63}{99} = \frac{7}{11}$.

11. $0.36\overline{36}$

12. $0.318\overline{18}$

13. $0.297\overline{297}$

14. 0.99009900

15. Given $a < b$, determine which of the following are true.

a. $a + 2 < b + 2$ **b.** $5b < 5a$

c. $5 - a > 5 - b$ **d.** $\dfrac{1}{a} < \dfrac{1}{b}$

e. $(a - b)(b - a) > 0$ **f.** $a^2 < b^2$

16. Complete the table with the appropriate interval notation, set notation, and graph on the real line.

Interval notation	*Set notation*	*Graph*
		![graph from -2]
$(-\infty, -4]$		
	$\{x : 3 \le x \le \frac{11}{2}\}$	
$(-1, 7)$		

In Exercises 17–40, solve the inequality and sketch the graph of the solution on the real line.

17. $4x + 1 < 2x$

18. $2x + 7 < 3$

19. $2x - 1 \ge 0$

20. $3x + 1 \ge 2x + 2$

21. $-4 < 2x - 3 < 4$

22. $0 \le x + 3 < 5$

23. $\dfrac{3}{4}x > x + 1$

24. $-1 < -\dfrac{x}{3} < 1$

25. $\dfrac{x}{2} + \dfrac{x}{3} > 5$

26. $x > \dfrac{1}{x}$

27. $|x| < 1$

28. $\dfrac{x}{2} - \dfrac{x}{3} > 5$

29. $\left|\dfrac{x - 3}{2}\right| \ge 5$

30. $\left|\dfrac{x}{2}\right| > 3$

31. $|x - a| < b, b > 0$

32. $|x + 2| < 5$

33. $|2x + 1| < 5$

34. $|3x + 1| \ge 4$

35. $|1 - \frac{2}{3}x| < 1$

36. $|9 - 2x| < 1$

37. $x^2 \le 3 - 2x$

38. $x^4 - x \le 0$

39. $x^2 + x - 1 \le 5$

40. $2x^2 + 1 < 9x - 3$

In Exercises 41–44, find the directed distance from a to b, the directed distance from b to a, and the distance between a and b.

41.

42.

43. a. $a = 126, b = 75$
    b. $a = -126, b = -75$

44. a. $a = 9.34, b = -5.65$
    b. $a = \frac{16}{5}, b = \frac{112}{75}$

In Exercises 45–48, find the midpoint of the interval.

45.

46.

47. a. $[7, 21]$ **48. a.** $[-6.85, 9.35]$
    b. $[8.6, 11.4]$ **b.** $[-4.6, -1.3]$

In Exercises 49–54, use absolute values to define the interval (or pair of intervals) on the real line.

49.

50.

51.

52.

53. **a.** All numbers that are at most 10 units from 12.
    b. All numbers that are at least 10 units from 12.

54. **a.** y is at most two units from a.
    b. y is less that δ units from c.

55. *Simple Interest*   The balance in an account after t years is

$$A = P + Prt$$

where P dollars is the initial investment and r is the simple interest rate (in decimal form). For an investment of $1000 to attain a balance greater than $1250 in 2 years, the interest rate must be greater than what value?

56. *Profit*   The revenue for selling x units of a product is

$$R = 115.95x$$

and the cost of producing x units is

$$C = 95x + 750.$$

To make a (positive) profit, R must be greater than C. For what values of x will the product return a profit?

57. *Fleet Costs*   A utility company has a fleet of vans. The annual operating cost of each van is estimated to be

$$C = 0.32m + 2300$$

where C is measured in dollars and m is measured in miles. The company wants the annual operating cost of each van to be less than $10,000. To do this, m must be less than what value?

58. *Height*   Two-thirds of the members of a population have heights that satisfy the inequality

$$\left| \frac{h - 68.5}{2.7} \right| \leq 1$$

where h is measured in inches. Determine the interval on the real line in which these heights lie.

59. *Fair Coin*   To determine whether a coin is fair (has an equal probability of landing tails up or heads up), an experimenter tosses it 100 times and records the number of heads x. The coin is declared unfair if

$$\left| \frac{x - 50}{5} \right| \geq 1.645.$$

For what values of x will the coin be declared unfair?

60. *Daily Production*   The estimated daily production p at a refinery is

$$|p - 2,250,000| < 125,000$$

where p is measured in barrels of oil. Determine the high and low production levels.

In Exercises 61 and 62, determine which of the two real numbers is greater.

61. **a.** π or $\frac{355}{113}$ **62.** **a.** $\frac{224}{151}$ or $\frac{144}{97}$
    b. π or $\frac{22}{7}$ **b.** $\frac{73}{81}$ or $\frac{6427}{7132}$

63. *Estimation—Powers of 10*   The speed of light is 2.998×10^8 meters per second. Which best estimates the distance in meters that light travels in a year?
    a. 9.5×10^5 **b.** 9.5×10^{15}
    c. 9.5×10^{12} **d.** 9.5×10^{16}

64. *Essay*   The accuracy of an approximation to a number is related to how many significant digits there are in the approximation. Write a definition for significant digits and illustrate the concept with examples.

In Exercises 65–72, prove the property.

65. $|ab| = |a| \, |b|$

66. $|a - b| = |b - a|$
    (*Hint:* $(a - b) = (-1)(b - a)$)

67. $\left| \dfrac{a}{b} \right| = \dfrac{|a|}{|b|}, \, b \neq 0$

68. $|a| = \sqrt{a^2}$

69. $|a^n| = |a|^n, \, n = 1, 2, 3, \ldots$

70. $-|a| \leq a \leq |a|$

71. $|a| \leq k$ if and only if $-k \leq a \leq k, \, k > 0$.

72. $k \leq |a|$ if and only if $k \leq a$ or $a \leq -k, \, k > 0$.

True or False   **In Exercises 73–78, determine whether the statement is true or false. If it is false, explain why or give an example that shows it is false.**

73. The reciprocal of a nonzero integer is an integer.

74. The reciprocal of a nonzero rational number is a rational number.

75. Each real number is either rational or irrational.

76. The absolute value of each real number is positive.

77. If $x < 0$, then $\sqrt{x^2} = -x$.

78. If a and b are any two distinct real numbers, then $a < b$ or $a > b$.

79. Find an example for which $|a - b| > |a| - |b|$, and an example for which $|a - b| = |a| - |b|$. Then prove that $|a - b| \geq |a| - |b|$ for all a, b.

80. Show that the maximum of two numbers a and b is given by the formula

$$\max(a, b) = \tfrac{1}{2}(a + b + |a - b|).$$

Derive a similar formula for $\min(a, b)$.

RENÉ DESCARTES

Descartes made many contributions to philosophy, science, and mathematics. The idea of representing points in the plane by pairs of real numbers and representing curves in the plane by equations was described by Descartes in his book titled *La Géométrie*, published in 1637.

The Cartesian Plane

An **ordered pair** (x, y) of real numbers has x as its *first* member and y as its *second* member. The model for representing ordered pairs is called the **rectangular coordinate system,** or the **Cartesian plane,** after the French mathematician René Descartes (1596–1650). It is developed by considering two real lines intersecting at right angles (see Figure 14).

The horizontal real line is usually called the **x-axis,** and the vertical real line is usually called the **y-axis.** Their point of intersection is the **origin.** The two axes divide the plane into four **quadrants.**

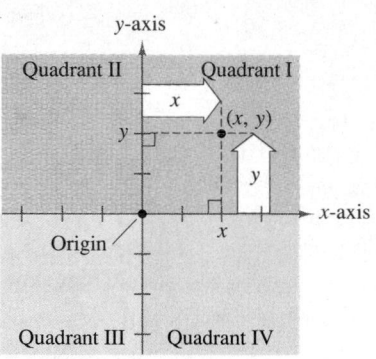

FIGURE 14
The Cartesian plane

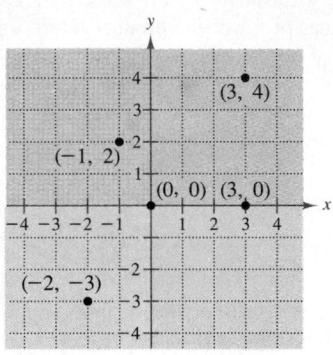

FIGURE 15
Points represented by ordered pairs.

Each point in the plane is identified by an ordered pair (x, y) of real numbers x and y, called **coordinates** of the point. The number x represents the directed distance from the y-axis to the point, and the number y represents the directed distance from the x-axis to the point (see Figure 14). For the point (x, y), the first coordinate is the x-coordinate or **abscissa,** and the second coordinate is the y-coordinate or **ordinate.** For example, Figure 15 shows the locations of the points $(-1, 2)$, $(3, 4)$, $(0, 0)$, $(3, 0)$, and $(-2, -3)$ in the Cartesian plane.

REMARK   The signs of the coordinates of a point determine the quadrant in which the point lies. For instance, if $x > 0$ and $y < 0$, then (x, y) lies in Quadrant IV.

Note that an ordered pair (a, b) is used to denote either a point in the plane *or* an open interval on the real line. This, however, should not be confusing—the nature of the problem should clarify whether a point in the plane or an open interval is being discussed.

TECHNOLOGY   Note that Figure 15 shows only a small portion of the Cartesian plane—the portion for which $-4 \le x \le 4$ and $-4 \le y \le 4$. Such a portion is called a **viewing rectangle.** If you are using a graphing utility in this course, you need to become familiar with the steps required to change the viewing rectangle on your graphing utility.

The Distance and Midpoint Formulas

Recall from the Pythagorean Theorem that, in a right triangle, the hypotenuse c and sides a and b are related by $a^2 + b^2 = c^2$. Conversely, if $a^2 + b^2 = c^2$, then the triangle is a right triangle (see Figure 16).

FIGURE 16
The Pythagorean Theorem:
$a^2 + b^2 = c^2$.

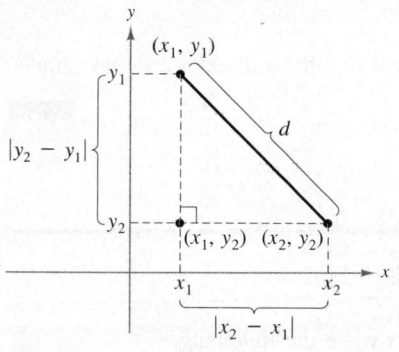

FIGURE 17
The distance between two points.

Suppose you want to determine the distance d between the two points (x_1, y_1) and (x_2, y_2) in the plane. If the points lie on a horizontal line, then $y_1 = y_2$ and the distance between the points is $|x_2 - x_1|$. If the points lie on a vertical line, then $x_1 = x_2$ and the distance between the points is $|y_2 - y_1|$. If the two points do not lie on a horizontal or vertical line, then they can be used to form a right triangle, as shown in Figure 17. The length of the vertical side of the triangle is $|y_2 - y_1|$, and the length of the horizontal side is $|x_2 - x_1|$. By the Pythagorean Theorem, it follows that

$$d^2 = |x_2 - x_1|^2 + |y_2 - y_1|^2$$
$$d = \sqrt{|x_2 - x_1|^2 + |y_2 - y_1|^2}.$$

Replacing $|x_2 - x_1|^2$ and $|y_2 - y_1|^2$ by the equivalent expressions $(x_2 - x_1)^2$ and $(y_2 - y_1)^2$ produces the following result.

Distance Formula

The distance d between the points (x_1, y_1) and (x_2, y_2) in the plane is given by
$$d = \sqrt{(x_2 - x_1)^2 + (y_2 - y_1)^2}.$$

EXAMPLE 1   Finding the Distance Between Two Points

Find the distance between the points $(-2, 1)$ and $(3, 4)$.

Solution

$$d = \sqrt{[3 - (-2)]^2 + (4 - 1)^2} \qquad \text{Distance Formula}$$
$$= \sqrt{(5)^2 + (3)^2}$$
$$= \sqrt{25 + 9}$$
$$= \sqrt{34}$$
$$\approx 5.83$$

EXAMPLE 2   Verifying a Right Triangle

Verify that the points $(2, 1)$, $(4, 0)$, and $(5, 7)$ form the vertices of a right triangle.

Solution   Figure 18 shows the triangle formed by the three points. The lengths of the three sides are as follows.

$$d_1 = \sqrt{(5 - 2)^2 + (7 - 1)^2} = \sqrt{9 + 36} = \sqrt{45}$$
$$d_2 = \sqrt{(4 - 2)^2 + (0 - 1)^2} = \sqrt{4 + 1} = \sqrt{5}$$
$$d_3 = \sqrt{(5 - 4)^2 + (7 - 0)^2} = \sqrt{1 + 49} = \sqrt{50}$$

Because

$$d_1^2 + d_2^2 = 45 + 5 = 50 \qquad \text{Sum of squares of sides}$$

and

$$d_3^2 = 50 \qquad \text{Square of hypotenuse}$$

you can apply the Pythagorean Theorem to conclude that the triangle is a right triangle.

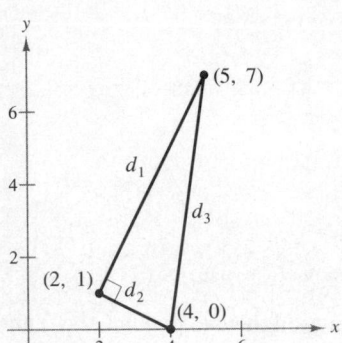

FIGURE 18
Verifying a right triangle.

EXAMPLE 3   Using the Distance Formula

Find x so that the distance between $(x, 3)$ and $(2, -1)$ is 5.

Solution   Using the Distance Formula, you can write the following.

$$5 = \sqrt{(x - 2)^2 + [3 - (-1)]^2} \qquad \text{Distance Formula}$$
$$25 = (x^2 - 4x + 4) + 16 \qquad \text{Square both sides}$$
$$0 = x^2 - 4x - 5 \qquad \text{Write in standard form}$$
$$0 = (x - 5)(x + 1) \qquad \text{Factor}$$

Therefore, $x = 5$ or $x = -1$, and you can conclude that there are two solutions. That is, each of the points $(5, 3)$ and $(-1, 3)$ lies 5 units from the point $(2, -1)$, as shown in Figure 19.

FIGURE 19
Given a distance, find a point.

The coordinates of the **midpoint** of the line segment joining two points can be found by "averaging" the x-coordinates of the two points and "averaging" the y-coordinates of the two points. That is, the midpoint of the line segment joining the points (x_1, y_1) and (x_2, y_2) in the plane is

$$\left(\frac{x_1 + x_2}{2}, \frac{y_1 + y_2}{2} \right). \qquad \text{Midpoint Formula}$$

For instance, the midpoint of the line segment joining the points $(-5, -3)$ and $(9, 3)$ is

$$\left(\frac{-5 + 9}{2}, \frac{-3 + 3}{2} \right) = (2, 0)$$

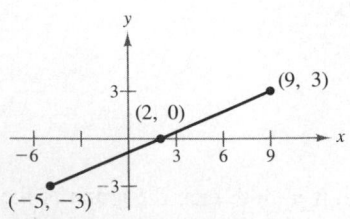

FIGURE 20
Midpoint of a line segment.

as shown in Figure 20.

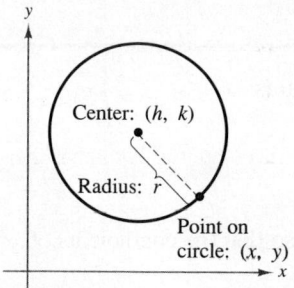

FIGURE 21
Definition of a circle

FOR FURTHER INFORMATION
Can you recognize the graph of the equation

$$x^4 + 2x^2y^2 + y^4 - 5x^2 - 5y^2 = -4?$$

For the solution, see the article "Single Equations Can Draw Pictures" by Keith M. Kendig in the March, 1991 issue of *The College Mathematics Journal*.

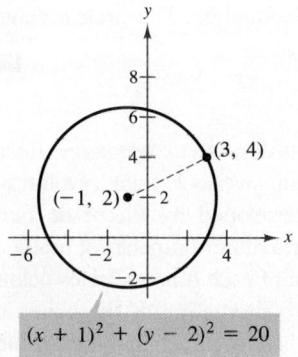

$(x + 1)^2 + (y - 2)^2 = 20$

FIGURE 22
Standard form of the equation of a circle.

Equations of Circles

A **circle** can be defined as the set of all points in a plane that are equidistant from a fixed point. The fixed point is the **center** of the circle, and the distance between the center and a point on the circle is the **radius** (see Figure 21).

You can use the Distance Formula to write an equation for the circle with center (h, k) and radius r. Let (x, y) be any point on the circle. Then the distance between (x, y) and the center (h, k) is given by

$$\sqrt{(x - h)^2 + (y - k)^2} = r.$$

By squaring both sides of this equation, you obtain the **standard form of the equation of a circle.**

Standard Form of the Equation of a Circle

The point (x, y) lies on the circle of radius r and center (h, k) if and only if

$$(x - h)^2 + (y - k)^2 = r^2.$$

The standard form of the equation of a circle with center at the origin, $(h, k) = (0, 0)$, is

$$x^2 + y^2 = r^2.$$

If $r = 1$, then the circle is called the **unit circle.**

EXAMPLE 4   Finding the Equation of a Circle

The point $(3, 4)$ lies on a circle whose center is at $(-1, 2)$, as shown in Figure 22. Find an equation for the circle.

Solution   The radius of the circle is the distance between $(-1, 2)$ and $(3, 4)$.

$$r = \sqrt{[3 - (-1)]^2 + (4 - 2)^2} = \sqrt{16 + 4} = \sqrt{20}$$

You can write the standard form of the equation of this circle as

$$[x - (-1)]^2 + (y - 2)^2 = (\sqrt{20})^2$$
$$(x + 1)^2 + (y - 2)^2 = 20. \qquad \text{Standard form}$$

By squaring and simplifying, the equation $(x - h)^2 + (y - k)^2 = r^2$ can be written in the following **general form of the equation of a circle.**

$$Ax^2 + Ay^2 + Cx + Dy + F = 0, \qquad A \neq 0$$

To convert such an equation to the standard form

$$(x - h)^2 + (y - k)^2 = p$$

you can use a process called **completing the square.** If $p > 0$, then the graph of the equation is a circle. If $p = 0$, then the graph is the single point (h, k). If $p < 0$, then the equation has no graph.

EXAMPLE 5   Completing the Square

Sketch the graph of the circle whose general equation is

$$4x^2 + 4y^2 + 20x - 16y + 37 = 0.$$

Solution   To complete the square, first divide by 4 so that the coefficients of x^2 and y^2 are both 1.

$$4x^2 + 4y^2 + 20x - 16y + 37 = 0 \qquad \text{General form}$$

$$x^2 + y^2 + 5x - 4y + \frac{37}{4} = 0 \qquad \text{Divide by 4}$$

$$(x^2 + 5x + \quad) + (y^2 - 4y + \quad) = -\frac{37}{4} \qquad \text{Group terms}$$

$$\left(x^2 + 5x + \frac{25}{4}\right) + (y^2 - 4y + 4) = -\frac{37}{4} + \frac{25}{4} + 4 \qquad \begin{array}{l}\text{Complete the square} \\ \text{by adding } \frac{25}{4} \text{ and} \\ \text{4 to both sides}\end{array}$$

$$\underbrace{\qquad}_{(\text{half})^2} \qquad \underbrace{\qquad}_{(\text{half})^2}$$

$$\left(x + \frac{5}{2}\right)^2 + (y - 2)^2 = 1 \qquad \text{Standard form}$$

Note that you complete the square by adding the square of half the coefficient of x *and* the square of half the coefficient of y to both sides of the equation. The circle is centered at $(-\frac{5}{2}, 2)$ and its radius is 1, as shown in Figure 23.

$$\left(x + \tfrac{5}{2}\right)^2 + (y - 2)^2 = 1$$

FIGURE 23
A circle with a radius of 1 and center at $(-\frac{5}{2}, 2)$.

We have now reviewed some fundamental concepts of *analytic geometry*. Because these concepts are in common use today, it is easy to overlook their revolutionary nature. At the time analytic geometry was being developed by Pierre de Fermat (1601–1665) and René Descartes (1596–1650), the two major branches of mathematics—geometry and algebra—were largely independent of each other. Circles belonged to geometry and equations belonged to algebra. The coordination of the points on a circle and the solutions of an equation belongs to what is now called analytic geometry.

It is important to become skilled in analytic geometry so that you can move easily between geometry and algebra. For instance, in Example 4, you were given a geometric description of a circle and were asked to find an algebraic equation for the circle. Thus, you were moving from geometry to algebra. Similarly, in Example 5 you were given an algebraic equation and asked to sketch a geometric picture. In this case, you were moving from algebra to geometry. These two examples illustrate the two most common problems in analytic geometry.

1. Given a graph, find its equation.

$$\boxed{\text{Geometry}} \ \rightarrow \ \boxed{\text{Algebra}}$$

2. Given an equation, find its graph.

$$\boxed{\text{Algebra}} \ \rightarrow \ \boxed{\text{Geometry}}$$

PIERRE DE FERMAT

Fermat was trained as a lawyer—mathematics was more of a hobby than a profession. Nevertheless, Fermat made many contributions to number theory, calculus, and optics.

In the next two sections, you will review other examples of these two types of problems.

EXERCISES for Section 2

In Exercises 1–6, (a) plot the points, (b) find the distance between the points, and (c) find the midpoint of the line segment joining the points.

1. $(2, 1), (4, 5)$ **2.** $(-3, 2), (3, -2)$

3. $(\frac{1}{2}, 1), (-\frac{3}{2}, -5)$ **4.** $(\frac{2}{3}, -\frac{1}{3}), (\frac{5}{6}, 1)$

5. $(1, \sqrt{3}), (-1, 1)$ **6.** $(-2, 0), (0, \sqrt{2})$

In Exercises 7–10, show that the points are the vertices of the polygon. (A rhombus is a quadrilateral whose sides are all of the same length.)

Vertices	Figure
7. $(4, 0), (2, 1), (-1, -5)$	Right triangle
8. $(1, -3), (3, 2), (-2, 4)$	Isosceles triangle
9. $(0, 0), (1, 2), (2, 1), (3, 3)$	Rhombus
10. $(0, 1), (3, 7), (4, 4), (1, -2)$	Parallelogram

11. *Federal Debt*   The table lists the per capita federal debt for the United States from 1950 to 1990. (*Source:* U. S. Treasury Department)

Year	1950	1960	1970
Per capita debt	$1688	$1572	$1807

Year	1980	1985	1990
Per capita debt	$3981	$7614	$12,848

Select reasonable scales on the coordinate axes and plot the points (x, y) where y is the per capita debt and x is the time in years, with $x = 0$ corresponding to 1950.

12. *Life Expectancy*   The table gives the life expectancy of a child (at birth) from 1920 to 1990. (*Source:* Department of Health and Human Services)

Year	1920	1930	1940	1950
Life expectancy	54.1	59.7	62.9	68.2

Year	1960	1970	1980	1990
Life expectancy	69.7	70.8	73.7	75.2

Select reasonable scales on the coordinate axes and plot the points (x, y) where y is the life expectancy and x is the time in years, with $x = 0$ corresponding to 1950.

In Exercises 13–16, use the Distance Formula to determine whether the points lie on the same line.

13. $(0, -4), (2, 0), (3, 2)$

14. $(0, 4), (7, -6), (-5, 11)$

15. $(-2, 1), (-1, 0), (2, -2)$

16. $(-1, 1), (3, 3), (5, 5)$

In Exercises 17 and 18, find x so that the distance between the points is 5.

17. $(0, 0), (x, -4)$ **18.** $(2, -1), (x, 2)$

In Exercises 19 and 20, find y so that the distance between the points is 8.

19. $(0, 0), (3, y)$ **20.** $(5, 1), (5, y)$

In Exercises 21 and 22, find the relationship between x and y so that (x, y) is equidistant from the two given points.

21. $(4, -1), (-2, 3)$ **22.** $(3, \frac{5}{2}), (-7, -1)$

23. Use the Midpoint Formula to find the three points that divide the line segment joining (x_1, y_1) and (x_2, y_2) into four equal parts.

24. Use the result of Exercise 23 to find the points that divide the line segment joining the given points into four equal parts.
   a. $(1, -2), (4, -1)$ **b.** $(-2, -3), (0, 0)$

In Exercises 25 and 26, complete the square for each expression.

25. a. $x^2 + 5x$ **26. a.** $4x^2 - 4x - 39$
    b. $x^2 + 8x + 7$ **b.** $5x^2 + x$

In Exercises 27–30, match the equation with its graph. [Graphs are labeled (a), (b), (c), and (d).]

27. $x^2 + y^2 = 1$

28. $(x - 1)^2 + (y - 3)^2 = 4$

29. $(x - 1)^2 + y^2 = 0$

30. $(x + \frac{1}{2})^2 + (y - \frac{3}{4})^2 = \frac{1}{4}$

(a) **(b)**

(c) **(d)**

In Exercises 31–40, write the equation of the circle in general form.

31. Center: $(0, 0)$
Radius: 3

32. Center: $(0, 0)$
Radius: 5

33. Center: $(2, -1)$
Radius: 4

34. Center: $(-4, 3)$
Radius: $\frac{5}{8}$

35. Center: $(-1, 2)$
Point on circle: $(0, 0)$

36. Center: $(3, -2)$
Point on circle: $(-1, 1)$

37. Endpoints of diameter: $(2, 5)$, $(4, -1)$

38. Endpoints of diameter: $(1, 1)$, $(-1, -1)$

39. Points on circle: $(0, 0)$, $(0, 8)$, $(6, 0)$

40. Points on circle: $(1, -1)$, $(2, -2)$, $(0, -2)$

41. *Satellite Communication*   Write an equation for the path of a communications satellite in a circular orbit 22,000 miles above the earth. (Assume that the radius of the earth is 4000 miles.)

42. *Building Design*   A circular air duct of diameter D is fit firmly into the right angle corner where a basement wall meets the floor (see figure). Find the diameter of the largest water pipe that can be run in the right angle corner behind the air duct.

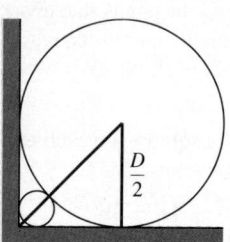

In Exercises 43–50, write the equation of the circle in standard form and sketch its graph.

43. $x^2 + y^2 - 2x + 6y + 6 = 0$

44. $x^2 + y^2 - 2x + 6y - 15 = 0$

45. $x^2 + y^2 - 2x + 6y + 10 = 0$

46. $3x^2 + 3y^2 - 6y - 1 = 0$

47. $2x^2 + 2y^2 - 2x - 2y - 3 = 0$

48. $4x^2 + 4y^2 - 4x + 2y - 1 = 0$

49. $16x^2 + 16y^2 + 16x + 40y - 7 = 0$

50. $x^2 + y^2 - 4x + 2y + 3 = 0$

C   **In Exercises 51 and 52, use a computer algebra system to (a) solve the equation for y, and (b) graph the resulting two equations on the same set of coordinate axes.**

51. $4x^2 + 4y^2 - 4x + 24y - 63 = 0$

52. $x^2 + y^2 - 8x - 6y - 11 = 0$

In Exercises 53 and 54, sketch the set of all points satisfying the inequality.

53. $x^2 + y^2 - 4x + 2y + 1 \leq 0$

54. $(x - 1)^2 + (y - \frac{1}{2})^2 > 1$

55. Prove that

$$\left(\frac{2x_1 + x_2}{3}, \frac{2y_1 + y_2}{3}\right)$$

is one of the points of trisection of the line segment joining (x_1, y_1) and (x_2, y_2). Find the midpoint of the line segment joining

$$\left(\frac{2x_1 + x_2}{3}, \frac{2y_1 + y_2}{3}\right)$$

and (x_2, y_2) to find the second point of trisection.

56. Use the results of Exercise 55 to find the points of trisection of the line segment joining the following points.
a. $(1, -2)$, $(4, 1)$ **b.** $(-2, -3)$, $(0, 0)$

In Exercises 57–60, prove the statement.

57. The line segments joining the midpoints of the opposite sides of a quadrilateral bisect each other.

58. The perpendicular bisector of a chord of a circle passes through the center of the circle.

59. An angle inscribed in a semicircle is a right angle.

60. The midpoint of the line segment joining the points (x_1, y_1) and (x_2, y_2) is

$$\left(\frac{x_1 + x_2}{2}, \frac{y_1 + y_2}{2}\right).$$

True or False   **In Exercises 61–64, determine whether the statement is true or false. If it is false, explain why or give an example that shows it is false.**

61. If $ab < 0$, then the point (a, b) lies in either the second quadrant or the fourth quadrant.

62. The distance between the points $(a + b, a)$ and $(a - b, a)$ is $2b$.

63. If the distance between two points is zero, then the two points must coincide.

64. If $ab = 0$, then the point (a, b) lies on the x-axis or on the y-axis.

65. Let a and b be two nonnegative real numbers. Show that their arithmetic mean, $\frac{1}{2}(a + b)$, is greater than or equal to their geometric mean, $\sqrt{ab}$.

66. Find the point that is equidistant from the three points $(0, 0)$, $(15, 3)$, and $(4, 6)$.

*A **C** indicates an exercise in which you are instructed to use graphing technology or a symbolic computer algebra system. The solutions of other exercises may also be facilitated by appropriate technology.

3

The Graph of an Equation • Intercepts of a Graph • Symmetry of a Graph • Points of Intersection • Mathematical Models

The Graph of an Equation

Consider the equation $3x + y = 7$. The point $(2, 1)$ is a **solution point** of the equation because the equation is satisfied (is true) when 2 is substituted for x and 1 is substituted for y. This equation has many other solutions, such as $(1, 4)$ and $(0, 7)$. To systematically find other solutions, solve the original equation for y.

$$y = 7 - 3x \qquad \text{Solve original equation for } y$$

Then construct a **table of values** by substituting several values of x, as shown in Table 2. From the table, you can see that $(0, 7)$, $(1, 4)$, $(2, 1)$, $(3, -2)$, and $(4, -5)$ are solutions of the original equation

$$3x + y = 7. \qquad \text{Original equation}$$

Like many equations, this equation has an infinite number of solutions. The set of all solution points is the **graph** of the equation, as shown in Figure 24.

REMARK   Even though we refer to the sketch shown in Figure 24 as the graph of $3x + y = 7$, it really represents only a *portion* of the graph. The entire graph would extend beyond the page.

In this course, you will study many sketching aids. The simplest is point plotting—that is, plotting enough points so that the basic shape of the graph becomes apparent.

TABLE 2

x	0	1	2	3	4
y	7	4	1	-2	-5

FIGURE 24
Solution points of $3x + y = 7$.

EXAMPLE 1   Sketching a Graph by Point Plotting

Sketch the graph of $y = x^2 - 2$.

Solution   First, construct a table of values, as shown in Table 3.

TABLE 3

x	-2	-1	0	1	2	3
y	2	-1	-2	-1	2	7

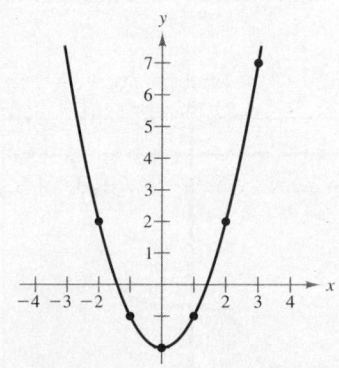

FIGURE 25
The parabola: $y = x^2 - 2$

Next, plot the points given by the table. Finally, connect the points by a *smooth curve*, as shown in Figure 25. This particular graph is a **parabola.** It is one of the conic sections you will study in Chapter 9.

One disadvantage of point plotting is that to get a good idea about the shape of a graph, you may need to plot many points. With only a few points, you could badly misrepresent the graph. For instance, suppose that to sketch the graph of

$$y = \frac{1}{30}x(39 - 10x^2 + x^4)$$

you plotted only five points: $(-3, -3)$, $(-1, -1)$, $(0, 0)$, $(1, 1)$, and $(3, 3)$ as shown in Figure 26(a). From these five points, you might conclude that the graph is a line. This, however, is not correct. By plotting several more points, you can see that the graph is more complicated, as shown in Figure 26(b).

(a)   (b)

FIGURE 26
Plotting only a few points can misrepresent a graph.

TECHNOLOGY Modern technology has made sketching graphs easier. Even with technology, however, it is possible to badly misrepresent a graph. For instance, each of the graphing utility screens shown in Figure 27 shows a portion of the graph of

$$y = x^3 - x^2 - 25.$$

From the screen on the left, you might assume that the graph is a line. From the screen on the right, however, you can see that the graph is not a line. Thus, whether you are sketching a graph by hand or using a graphing utility, you must realize that different "viewing rectangles" can produce very different views of a graph. In choosing a viewing rectangle, your goal is to show a view of the graph that fits well in the context of the problem.

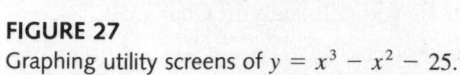

FIGURE 27
Graphing utility screens of $y = x^3 - x^2 - 25$.

Intercepts of a Graph

Two types of solution points that are especially useful are those having zero as their x-coordinate or y-coordinate. Such points are called **intercepts** because they are the points at which the graph intersects the x-axis or y-axis. The point $(a, 0)$ is an **x-intercept** of the graph of an equation if it is a solution point of the equation. To find the x-intercepts of a graph, let y be zero and solve the equation for x. The point $(0, b)$ is a **y-intercept** of the graph of an equation if it is a solution point of the equation. To find the y-intercepts of a graph, let x be zero and solve the equation for y.

REMARK   Some texts denote the x-intercept as the x-coordinate of the point $(a, 0)$ rather than the point itself. Unless it is necessary to make a distinction, we will use the term *intercept* to mean either the point or the coordinate.

It is possible for a graph to have no intercepts, or it might have several. For instance, consider the four graphs shown in Figure 28.

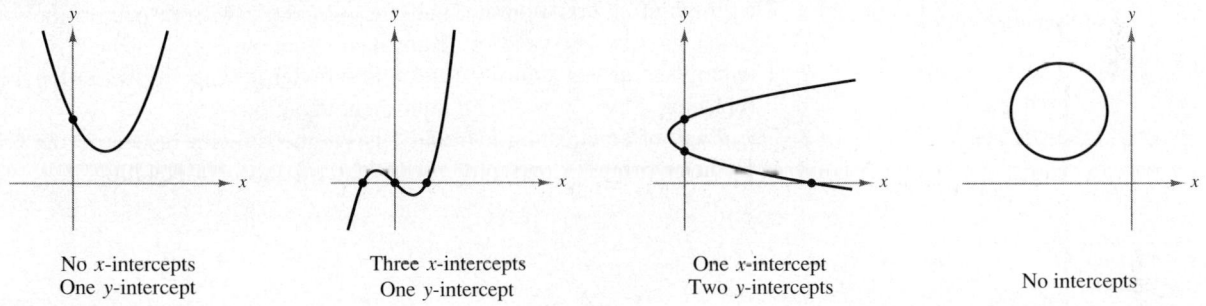

No x-intercepts
One y-intercept

Three x-intercepts
One y-intercept

One x-intercept
Two y-intercepts

No intercepts

FIGURE 28

EXAMPLE 2   Finding *x*- and *y*-Intercepts

Find the x- and y-intercepts of the graph of $y = x^3 - 4x$.

Solution   To find the x-intercepts, let y be zero and solve for x.

$$x^3 - 4x = 0 \qquad \text{Let } y \text{ be zero}$$
$$x(x - 2)(x + 2) = 0 \qquad \text{Factor}$$
$$x = 0, 2, \text{ or } -2 \qquad \text{Solve for } x$$

Because this equation has three solutions, you can conclude that the graph has three x-intercepts:

$$(0, 0), (2, 0), \text{ and } (-2, 0). \qquad \text{x-intercepts}$$

To find the y-intercepts, let x be zero. Doing this produces $y = 0$. Thus, the y-intercept is

$$(0, 0). \qquad \text{y-intercept}$$

(See Figure 29.)

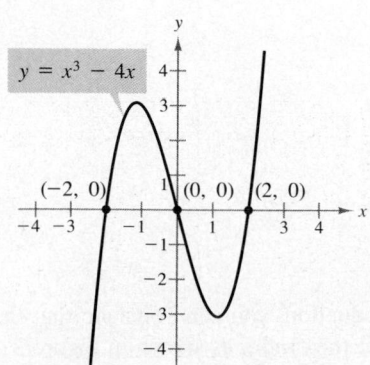

$y = x^3 - 4x$

$(-2, 0)$ $(0, 0)$ $(2, 0)$

FIGURE 29
Intercepts

y-Axis Symmetry

x-Axis Symmetry

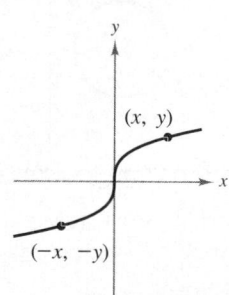

Origin Symmetry

FIGURE 30

Symmetry of a Graph

The following three types of symmetry can be used to help sketch the graph of an equation (see Figure 30).

1. A graph is **symmetric with respect to the y-axis** if, whenever (x, y) is a point on the graph, $(-x, y)$ is also a point on the graph. This means that the portion of the graph to the left of the y-axis is a mirror image of the portion to the right of the y-axis.
2. A graph is **symmetric with respect to the x-axis** if, whenever (x, y) is a point on the graph, $(x, -y)$ is also a point on the graph. This means that the portion of the graph above the x-axis is a mirror image of the portion below the x-axis.
3. A graph is **symmetric with respect to the origin** if, whenever (x, y) is a point on the graph, $(-x, -y)$ is also a point on the graph. This means that the graph is unchanged by a rotation of 180° about the origin.

Knowing that a graph has symmetry *before* attempting to sketch it is helpful because it reduces the number of solution points you need to find.

Tests for Symmetry

1. The graph of an equation in x and y is symmetric with respect to the y-axis if replacing x by $-x$ yields an equivalent equation.
2. The graph of an equation in x and y is symmetric with respect to the x-axis if replacing y by $-y$ yields an equivalent equation.
3. The graph of an equation in x and y is symmetric with respect to the origin if replacing x by $-x$ and y by $-y$ yields an equivalent equation.

EXAMPLE 3   Testing for Origin Symmetry

Show that the graph of

$$y = 2x^3 - x$$

is symmetric with respect to the origin.

Solution

$$y = 2x^3 - x \qquad \text{Original equation}$$

$$-y = 2(-x)^3 - (-x) \qquad \text{Replace } x \text{ by } -x \text{ and } y \text{ by } -y$$

$$-y = -2x^3 + x \qquad \text{Simplify}$$

$$y = 2x^3 - x \qquad \text{Equivalent equation}$$

Because the replacement produces an equivalent equation, you can conclude that the graph of $y = 2x^3 - x$ is symmetric with respect to the origin, as shown in Figure 31.

FIGURE 31
Origin symmetry

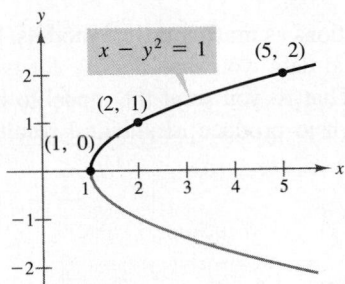

FIGURE 32
First, plot the points above the x-axis, then use symmetry to complete the graph.

EXAMPLE 4   Using Symmetry to Sketch a Graph

Sketch the graph of $x - y^2 = 1$.

Solution   The graph is symmetric with respect to the x-axis because replacing y by $-y$ yields an equivalent equation.

$$x - y^2 = 1 \qquad \text{Original equation}$$
$$x - (-y)^2 = 1 \qquad \text{Replace } y \text{ by } -y$$
$$x - y^2 = 1 \qquad \text{Equivalent equation}$$

This means that the portion of the graph below the x-axis is a mirror image of the portion above the x-axis. To sketch the graph, first sketch the portion above the x-axis. Then reflect in the x-axis to obtain the entire graph, as shown in Figure 32.

Points of Intersection

A **point of intersection** of the graphs of two equations is a point that satisfies both equations. You can find the points of intersection of two graphs by solving their equations simultaneously.

EXAMPLE 5   Finding Points of Intersection

Find all points of intersection of the graphs of

$$x^2 - y = 3 \quad \text{and} \quad x - y = 1.$$

Solution   It is helpful to begin by sketching the graphs of both equations on the *same* coordinate plane, as shown in Figure 33. When this is done, it appears that the two graphs have two points of intersection. To find these two points, you can use the following steps.

$$y = x^2 - 3 \qquad \text{Solve first equation for } y$$
$$y = x - 1 \qquad \text{Solve second equation for } y$$
$$x^2 - 3 = x - 1 \qquad \text{Equate } y\text{-values}$$
$$x^2 - x - 2 = 0 \qquad \text{Write in standard form}$$
$$(x - 2)(x + 1) = 0 \qquad \text{Factor}$$
$$x = 2 \text{ or } -1 \qquad \text{Solve for } x$$

The corresponding values of y are obtained by substituting $x = 2$ and $x = -1$ into either of the original equations. Doing this produces two points of intersection:

$$(2, 1) \quad \text{and} \quad (-1, -2). \qquad \text{Points of intersection}$$

You can check these points by substituting into *both* of the original equations.

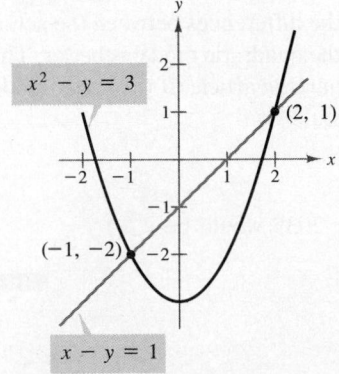

FIGURE 33
Two points of intersection.

The increasing concentration of carbon dioxide in the earth's atmosphere is contributing to global warming.

Mathematical Models

Real-life applications of mathematics often use equations as **mathematical models.** In developing a mathematical model to represent actual data, you should strive for two (often conflicting) goals: accuracy and simplicity. That is, you want the model to be simple enough to be workable, yet accurate enough to produce meaningful results.

EXAMPLE 6   The Rise in Atmospheric Carbon Dioxide

Between 1960 and 1990, the Mauna Loa Observatory in Hawaii recorded the carbon dioxide concentration y (in parts per million) in the earth's atmosphere. The January readings for each year are shown in Figures 34 and 35. In Figure 34, a linear model,

$$y = 313.6 + 1.24t, \qquad \text{Linear model}$$

where $t = 0$ represents 1960, has been fit to the data. In Figure 35, a quadratic model,

$$y = 316.2 + 0.70t + 0.018t^2, \qquad \text{Quadratic model}$$

has been fit to the data. Which model better represents the data? In the July, 1990 issue of *Scientific American,* these data were used to predict the carbon dioxide level in the earth's atmosphere in the year 2035. The prediction was 470 parts per million. Which, if either, of the two models could have been used to make this prediction?

FIGURE 34

FIGURE 35

Solution   To answer the first question, you need to define what "better" means. When a statistical definition involving the squares of the differences between the actual y-values and the y-values given by the model is used, the quadratic model is better. This conclusion is shared by the author of the *Scientific American* article. If the linear model with $t = 75$ (for 2035) were used, the prediction would be

$$y = 313.6 + 1.24(75) = 406.6. \qquad \text{Linear model}$$

If the quadratic model were used, the prediction for 2035 would be

$$y = 316.2 + 0.70(75) + 0.018(75)^2 = 469.95. \qquad \text{Quadratic model}$$

REMARK   These models were developed using a procedure called least squares regression (see Section 13.9). The linear model has a correlation given by $r^2 = 0.984$. The quadratic model has a correlation given by $r^2 = 0.997$.

EXERCISES for Section 3

In Exercises 1–6, match the equation with its graph. [Graphs are labeled (a), (b), (c), (d), (e), and (f).]

1. $y = x - 2$ **2.** $y = -\frac{1}{2}x + 2$

3. $y = x^2 + 2x$ **4.** $y = \sqrt{9 - x^2}$

5. $y = 4 - x^2$ **6.** $y = x^3 - x$

(a)   **(b)**

(c)   **(d)**

(e)   **(f)**

In Exercises 7–16, find the intercepts.

7. $y = 2x - 3$ **8.** $y = (x - 1)(x - 3)$

9. $y = x^2 + x - 2$ **10.** $y^2 = x^3 - 4x$

11. $y = x^2\sqrt{9 - x^2}$ **12.** $xy = 4$

13. $y = \dfrac{x - 1}{x - 2}$ **14.** $y = \dfrac{x^2 + 3x}{(3x + 1)^2}$

15. $x^2y - x^2 + 4y = 0$ **16.** $y = 2x - \sqrt{x^2 + 1}$

In Exercises 17–26, check for symmetry with respect to both axes and to the origin.

17. $y = x^2 - 2$ **18.** $y = x^4 - x^2 + 3$

19. $x^2y - x^2 + 4y = 0$ **20.** $xy - \sqrt{4 - x^2} = 0$

21. $y^2 = x^3 - 4x$ **22.** $xy^2 = -10$

23. $y = x^3 + x$ **24.** $xy = 1$

25. $y = \dfrac{x}{x^2 + 1}$ **26.** $y = x^3 + x - 3$

In Exercises 27–30, determine whether the points lie on the graph of the equation.

27. Equation: $2x - y - 3 = 0$
Points: $(1, 2)$, $(1, -1)$, $(4, 5)$

28. Equation: $x^2 + y^2 = 4$
Points: $\left(1, -\sqrt{3}\right)$, $\left(\frac{1}{2}, -1\right)$, $\left(\frac{3}{2}, \frac{7}{2}\right)$

29. Equation: $x^2y - x^2 + 4y = 0$
Points: $\left(1, \frac{1}{5}\right)$, $\left(2, \frac{1}{2}\right)$, $(-1, -2)$

30. Equation: $x^2 - xy + 4y = 3$
Points: $(0, 2)$, $\left(-2, -\frac{1}{6}\right)$, $(3, -6)$

In Exercises 31–46, sketch the graph of the equation. Identify any intercepts and test for symmetry.

31. $y = x$ **32.** $y = x - 2$

33. $y = x + 3$ **34.** $y = 2x - 3$

35. $y = -3x + 2$ **36.** $y = -\dfrac{1}{2}x + 2$

37. $y = \dfrac{1}{2}x - 4$ **38.** $y = x^2 + 3$

39. $y = 1 - x^2$ **40.** $y = 2x^2 + x$

41. $y = x^3 + 2$ **42.** $y = \sqrt{9 - x^2}$

43. $y = (x + 2)^2$ **44.** $x = y^2 - 4$

45. $y = \dfrac{1}{x}$ **46.** $y = 2x^4$

C **In Exercises 47–52, use a graphing utility to graph the equation. (It may be necessary to solve for y and plot two equations.) Identify any intercepts and test for symmetry.**

47. $y = -2x^2 + x + 1$ **48.** $y = x^3 - 1$

49. $y = x\sqrt{25 - x^2}$ **50.** $y = \dfrac{5}{x^2 + 1} - 1$

51. $x^2 + 4y^2 = 4$ **52.** $9x^2 + y^2 = 9$

In Exercises 53–56, create an equation whose graph has the indicated property. (There is more than one correct answer.)

53. The graph has intercepts at $x = -2$, $x = 4$, and $x = 6$.

54. The graph has intercepts at $x = -\frac{5}{2}$, $x = 2$, and $x = \frac{3}{2}$.

55. The graph is symmetric with respect to the origin.

56. The graph is symmetric with respect to the x-axis.

In Exercises 57–64, find the points of intersection of the graphs of the equations and check your results.

57. $x + y = 2$ **58.** $2x - 3y = 13$
  $2x - y = 1$ $5x + 3y = 1$

59. $x + y = 7$ **60.** $x^2 + y^2 = 25$
  $3x - 2y = 11$ $2x + y = 10$

61. $x^2 + y^2 = 5$ **62.** $x^2 + y = 4$
  $x - y = 1$ $2x - y = 1$

63. $y = x^3$ **64.** $x = 3 - y^2$
  $y = x$ $y = x - 1$

C In Exercises 65 and 66, use a graphing utility to graph the equations. Then find the point(s) of intersection of the graphs.

65. $y = x^3 - 2x^2 + x - 1$
    $y = -x^2 + 3x - 1$

66. $y = x^4 - 2x^2 + 1$
    $y = 1 - x^2$

Break-Even Point In Exercises 67 and 68, find the sales necessary to break even ($R = C$) for the cost C of x units and the revenue R obtained by selling x units.

67. $C = 8650x + 250,000$
    $R = 9950x$

68. $C = 5.5\sqrt{x} + 10,000$
    $R = 3.29x$

69. For what values of k does the graph of $y = kx^3$ pass through the point?
    a. $(1, 4)$ b. $(-2, 1)$
    c. $(0, 0)$ d. $(-1, -1)$

70. Each of the tables lists solution points of one of the equations $y = kx + 5$, $y = x^2 + k$, $y = kx^{3/2}$, and $xy = k$. Match the equation to the correct table and find k. [Tables are labeled (a), (b), (c), and (d).]

(a)
x	1	4	9
y	3	24	81

(b)
x	1	4	9
y	7	13	23

(c)
x	1	4	9
y	36	9	4

(d)
x	1	4	9
y	-9	6	71

C 71. *The Cost of Living* The table lists the consumer price index (1982–84 = 100) for selected years. (*Source:* Bureau of Labor Statistics)

Year	1970	1975	1980	1985	1990
CPI	38.8	53.8	82.4	107.6	130.7

A mathematical model for the CPI during this time period is

$y = 0.05t^2 + 3.63t + 37.68, \qquad 0 \le t \le 20$

where y represents the consumer price index and t represents the year, with $t = 0$ corresponding to 1970.
a. Use a graphing utility to graph the model and compare the data with the model.
b. During this 20-year period, the cost of living has increased by approximately what factor?
c. Use the model to predict the CPI for the year 2000.

C 72. *Agriculture* The table shows the average number of acres per farm in the United States for selected years. (*Source:* Department of Agriculture)

Year	1950	1960	1970	1980	1990
No. of acres	213	297	374	427	461

A mathematical model for this data is

$y = -0.09t^2 + 9.80t + 211.49, \qquad 0 \le t \le 40$

where y represents the average acreage and t represents the year, with $t = 0$ corresponding to 1950.

a. Use a graphing utitlity to graph the model and compare the data with the model.
b. Use the model to predict the average number of acres per farm in the United States in the year 2000.

C 73. *Saturated Steam* The temperature y (°F) of saturated steam increases as pressure increases. This relationship is approximated by the model

$y = 75.82 - 2.11x + 43.51\sqrt{x}, \qquad 5 \le x \le 40$

where x is the absolute presssure in pounds per square inch.
a. Use a graphing utility to graph the model.
b. The temperature of steam at sea level ($x \doteq 14.696$) is 212°. Evaluate the model at this pressure.
c. Use the model to approximate the pressure if the temperature of the steam is 240°.

C 74. *Copper Wire* The resistance y (in ohms) of 1000 feet of solid copper wire at 77° Fahrenheit can be approximated by the mathematical model

$y = \dfrac{10,770}{x^2} - 0.37, \qquad 5 \le x \le 100$

where x is the diameter of the wire in mils (0.001 in.). Use a graphing utility to graph the model. If the diameter of the wire is doubled, the resistance is changed by approximately what factor?

75. Prove that if a graph is symmetric with respect to the x-axis and to the y-axis, then it is symmetric with respect to the origin. Give an example to show that the converse is not true.

76. Prove that if a graph is symmetric with respect to one axis and the origin, then it is symmetric with respect to the other axis also.

True or False In Exercises 77–80, determine whether the statement is true or false. If it is false, explain why or give an example that shows it is false.

77. If $(1, -2)$ is a point on a graph that is symmetric with respect to the x-axis, then $(-1, -2)$ is also a point on the graph.

78. If $(1, -2)$ is a point on a graph that is symmetric with respect to the y-axis, then $(-1, -2)$ is also a point on the graph.

79. If $b^2 - 4ac > 0$ and $a \ne 0$, then the graph of $y = ax^2 + bx + c$ has two x-intercepts.

80. If $b^2 - 4ac = 0$ and $a \ne 0$, then the graph of $y = ax^2 + bx + c$ has only one x-intercept.

81. Find an equation the graph of which consists of all points (x, y) whose distance to the origin is K times ($K \ne 1$) the distance to $(2, 0)$.

C 82. Let $x_0 = 1$ and consider the sequence x_n given by the formula

$x_n = \dfrac{1}{2}x_{n-1} + \dfrac{1}{x_{n-1}}, \qquad n = 1, 2, \ldots.$

For instance

$x_1 = \dfrac{1}{2}x_0 + \dfrac{1}{x_0} + \dfrac{1}{2}(1) + \dfrac{1}{1} = 1.5.$

Use a computer or calculator to compute x_2, x_3, and x_4. What is x_n getting close to as n gets larger?

4 The Slope of a Line • Equations of Lines • Sketching the Graph of a Line • Parallel and Perpendicular Lines

The Slope of a Line

The **slope** of a nonvertical line is a measure of the number of units a line rises (or falls) vertically for each unit of horizontal change from left to right. Consider the two points (x_1, y_1) and (x_2, y_2) on the line in Figure 36. As you move from left to right along this line, a vertical change of

$$\Delta y = y_2 - y_1 \qquad \text{Change in } y$$

units corresponds to a horizontal change of

$$\Delta x = x_2 - x_1 \qquad \text{Change in } x$$

units. (Δ is the Greek uppercase letter *delta*, and the symbols Δy and Δx are read "delta y" and "delta x.")

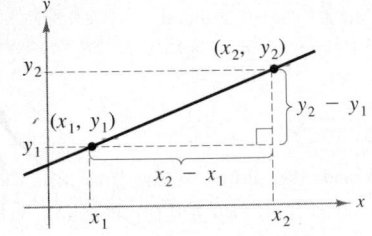

FIGURE 36
$\Delta y = y_2 - y_1 =$ change in y
$\Delta x = x_2 - x_1 =$ change in x

Definition of the Slope of a Line

The **slope** m of a nonvertical line passing through the points (x_1, y_1) and (x_2, y_2) is

$$m = \frac{\Delta y}{\Delta x} = \frac{y_2 - y_1}{x_2 - x_1}, \qquad x_1 \neq x_2.$$

REMARK When using the formula for slope, note that

$$\frac{y_2 - y_1}{x_2 - x_1} = \frac{-(y_1 - y_2)}{-(x_1 - x_2)} = \frac{y_1 - y_2}{x_1 - x_2}.$$

It does not matter in which order you substract *as long as* you are consistent and both "subtracted coordinates" come from the same point.

Figure 37 shows four lines: one has a positive slope, one has a slope of zero, one has a negative slope, and one has an "undefined slope."

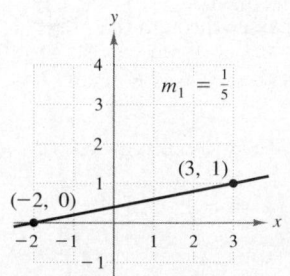

If m is positive, then the line rises.

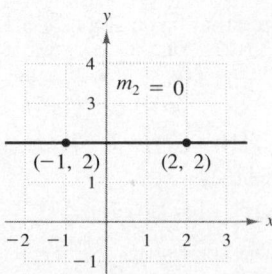

If m is zero, then the line is horizontal.

If m is negative, then the line falls.

If the line is vertical, then the slope is undefined.

FIGURE 37

Equations of Lines

Any two points on a nonvertical line can be used to calculate its slope. This can be verified from the similar triangles shown in Figure 38. (Recall that the ratios of corresponding sides of similar triangles are equal.)

$$m = \frac{y_2^* - y_1^*}{x_2^* - x_1^*} = \frac{y_2 - y_1}{x_2 - x_1}$$

FIGURE 38
Any two points on a line can be used
to determine its slope.

You can write an equation of a line if you know the slope of the line and the coordinates of one point on the line. Suppose the slope is m and the point is (x_1, y_1). If (x, y) is any other point on the line, then

$$\frac{y - y_1}{x - x_1} = m.$$

This equation, involving the two variables x and y, can be rewritten in the form $y - y_1 = m(x - x_1)$, which is called the **point-slope equation of a line.**

Point-Slope Equation of a Line

An equation of the line with slope m passing through the point (x_1, y_1) is given by

$$y - y_1 = m(x - x_1).$$

FIGURE 39
Point-slope equation of a
line.

EXAMPLE 1   Finding an Equation of a Line

Find an equation of the line that has a slope of 3 and passes through the point $(1, -2)$.

Solution

$$
\begin{aligned}
y - y_1 &= m(x - x_1) && \text{Point-slope form} \\
y - (-2) &= 3(x - 1) && \text{Substitute } -2 \text{ for } y_1, 1 \text{ for } x_1, \text{ and } 3 \text{ for } m \\
y + 2 &= 3x - 3 && \text{Simplify} \\
y &= 3x - 5 && \text{Solve for } y
\end{aligned}
$$

(See Figure 39.)

The slope of a line can be interpreted as either a *ratio* or a *rate*. If the *x*-axis and *y*-axis have the same unit of measure, then the slope has no units and is a **ratio.** If the *x*-axis and *y*-axis have different units of measure, then the slope is a rate or **rate of change.** As you study this text, you will encounter applications involving both interpretations of slope.

EXAMPLE 2   Population Growth and Engineering Design

a. The population of Arizona was 1,775,000 in 1970 and 2,718,000 in 1980. Over this 10-year period, the average rate of change of the population was

$$\text{Rate of change} = \frac{\text{Change in population}}{\text{Change in years}}$$

$$= \frac{2{,}718{,}000 - 1{,}775{,}000}{1980 - 1970}$$

$$= 94{,}300 \text{ people per year.}$$

If Arizona's population had continued to increase at this same rate for the next 10 years, it would have had a 1990 population of 3,661,000. In the 1990 census, however, Arizona's population was determined to be 3,665,000, so the population's rate of change from 1980 to 1990 was a little greater than in the previous decade (see Figure 40).

b. In tournament water-ski jumping, the ramp rises to a height of 6 feet on a raft that is 21 feet long, as shown in Figure 41. The slope of the ski ramp is the ratio of its height (the rise) to the length of its base (the run).

$$\text{Slope of ramp} = \frac{\text{Rise}}{\text{Run}} = \frac{6 \text{ feet}}{21 \text{ feet}} = \frac{2}{7}$$

In this case, note that the slope is a ratio and has no units.

FIGURE 40

FIGURE 41
Dimensions of a water-ski ramp.

REMARK   The rate of change found in Example 2a is an **average rate of change.** An average rate of change is always calculated over an interval. In this case, the interval is [1970, 1980]. Later in the text you will study another type of rate of change called an *instantaneous rate of change.*

Sketching the Graph of a Line

In Section 2, we mentioned that many problems in analytic geometry can be classified in two basic categories: (1) Given a graph, what is its equation? and (2) Given an equation, what is its graph? The point-slope equation of a line can be used to solve problems in the first category. However, this form is not especially useful for solving problems in the second category. The form that is better suited to sketching the graph of a line is the **slope-intercept** form for the equation of a line.

> ### The Slope-Intercept Equation of a Line
>
> The graph of the linear equation
>
> $$y = mx + b$$
>
> is a line having a *slope* of m and a *y-intercept* at $(0, b)$.

EXAMPLE 3   Sketching Lines in the Plane

Sketch the graphs of the equations.

a. $y = 2x + 1$ **b.** $y = 2$ **c.** $3y + x - 6 = 0$

Solution

a. Because $b = 1$, the y-intercept is $(0, 1)$. Because the slope is $m = 2$, you know that the line rises 2 units for each unit it moves to the right, as shown in Figure 42(a).

b. Because $b = 2$, the y-intercept is $(0, 2)$. Because the slope is $m = 0$, you know that the line is horizontal, as shown in Figure 42(b).

c. Begin by writing the equation in slope-intercept form.

$$3y + x - 6 = 0 \qquad \text{Original equation}$$

$$3y = -x + 6 \qquad \text{Isolate } y\text{-term on the left}$$

$$y = -\frac{1}{3}x + 2 \qquad \text{Slope-intercept form}$$

In this form, you can see that the y-intercept is $(0, 2)$ and the slope is $m = -\frac{1}{3}$. This means that the line falls 1 unit for every 3 units it moves to the right, as shown in Figure 42(c).

(a) $m = 2$; line rises.

(b) $m = 0$; line is horizontal.

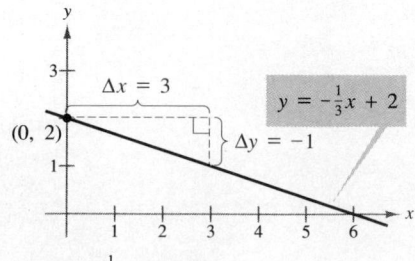

(c) $m = -\frac{1}{3}$; line falls.

FIGURE 42

Because the slope of a vertical line is not defined, its equation cannot be written in the slope-intercept form. However, the equation of *any* line can written in the **general form**

$$Ax + By + C = 0$$

where A and B are not *both* zero. For instance, the vertical line given by $x = a$ can be represented by the general form $x - a = 0$.

Summary of Equations of Lines

1. General form: $\qquad\qquad Ax + By + C = 0$
2. Vertical line: $\qquad\qquad x = a$
3. Horizontal line: $\qquad\quad\; y = b$
4. Point-slope form: $\qquad\; y - y_1 = m(x - x_1)$
5. Slope-intercept form: $\quad y = mx + b$

REMARK   Three points are **collinear** if they lie on the same line.

Parallel and Perpendicular Lines

The slope of a line is a convenient tool for determining whether two lines are parallel or perpendicular, as shown in Figure 43.

Parallel and Perpendicular Lines

1. Two distinct nonvertical lines are **parallel** if and only if their slopes are equal.
2. Two nonvertical lines are **perpendicular** if and only if their slopes are negative reciprocals of each other. That is, if and only if

$$m_1 = -\frac{1}{m_2}.$$

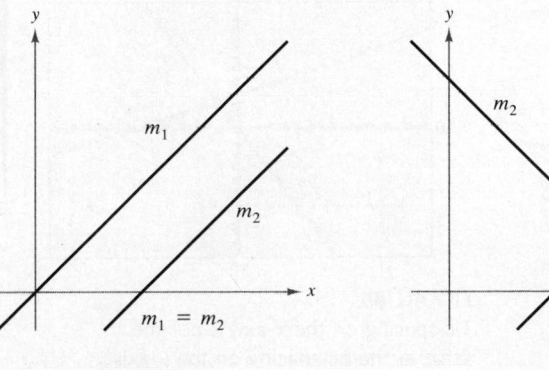

Parallel Lines   Perpendicular Lines

FIGURE 43

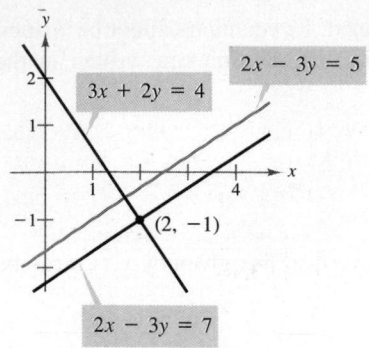

FIGURE 44
Lines parallel and perpendicular to
$2x - 3y = 5$.

EXAMPLE 4   Finding Parallel and Perpendicular Lines

Find the general forms of the equations of the lines that pass through the point $(2, -1)$ and are

a. parallel to the line $2x - 3y = 5$
b. perpendicular to the line $2x - 3y = 5$

as shown in Figure 44.

Solution   By writing the linear equation $2x - 3y = 5$ in slope-intercept form $y = \frac{2}{3}x - \frac{5}{3}$, you can see that the given line has a slope of $m = \frac{2}{3}$.

a. The line through $(2, -1)$ that is parallel to the given line also has a slope of $\frac{2}{3}$.

$$y - (-1) = \frac{2}{3}(x - 2) \qquad \text{Point-slope form}$$

$$3(y + 1) = 2(x - 2) \qquad \text{Simplify}$$

$$2x - 3y - 7 = 0 \qquad \text{General form}$$

Note the similarity to the original equation.

b. Using the negative reciprocal of the slope of the given line, you can determine that the slope of a line perpendicular to the given line is $-\frac{3}{2}$. Therefore, the line through the point $(2, -1)$ that is perpendicular to the given line has the following equation.

$$y - (-1) = -\frac{3}{2}(x - 2) \qquad \text{Point-slope form}$$

$$2(y + 1) = -3(x - 2) \qquad \text{Simplify}$$

$$3x + 2y - 4 = 0 \qquad \text{General form}$$

TECHNOLOGY   The apparent slope of a line will be distorted if you use different tic-spacing on the x- and y-axes. For instance, the graphing calculator screens in Figures 45 and 46 both show the lines given by $y = 2x$ and $y = -\frac{1}{2}x + 3$. Because these lines have slopes that are negative reciprocals, they must be perpendicular. In Figure 45, however, the lines don't appear to be perpendicular because the tic-spacing on the x-axis is not the same as the tic-spacing on the y-axis.

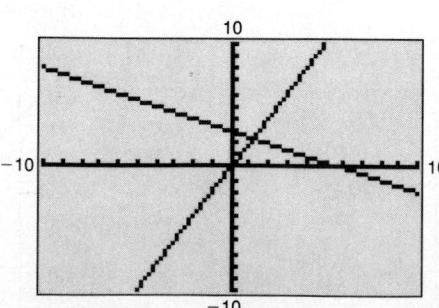

FIGURE 45
Tic-spacing on the x-axis is not the same as the tic-spacing on the y-axis.

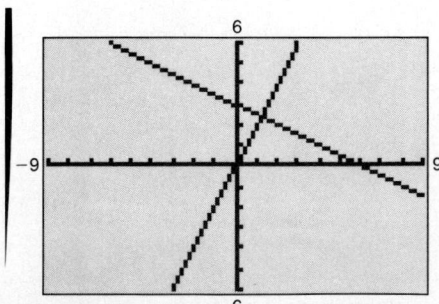

FIGURE 46
Tic-spacing on the x-axis is the same as the tic-spacing on the y-axis.

EXERCISES for Section 4

In Exercises 1–6, estimate the slope of the line from its graph.

1.

2.

3.

4.

5.

6.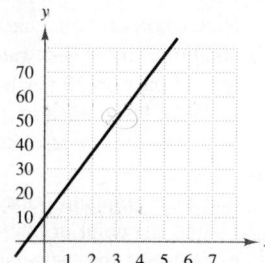

In Exercises 7 and 8, sketch the graphs of the lines through the point with the indicated slopes. Make the sketches for each exercise on the same set of coordinate axes.

Point	Slopes	
7. $(2, 3)$	**a.** 0	**b.** 1
	c. -2	**d.** Undefined
8. $(-4, 1)$	**a.** 3	**b.** -3
	c. $\frac{1}{3}$	**d.** 0

In Exercises 9–12, plot the pair of points and find the slope of the line passing through them.

9. $(3, -4), (5, 2)$ **10.** $(2, 1), (2, 5)$

11. $(1, 2), (-2, 4)$ **12.** $\left(\frac{7}{8}, \frac{3}{4}\right), \left(\frac{5}{4}, -\frac{1}{4}\right)$

In Exercises 13–16, use the point on the line and the slope of the line to find three additional points that the line passes through. (There is more than one correct answer.)

Point	Slope	Point	Slope
13. $(2, 1)$	$m = 0$	**14.** $(-3, 4)$	m undefined
15. $(1, 7)$	$m = -3$	**16.** $(-2, -2)$	$m = 2$

17. *Investment Growth* In 1989 an investor bought shares of common stock in Automatic Data Processing, Inc. The table lists the earnings per share y for three consecutive years. (*Source:* ADP Report to Shareholders, March 31, 1992)

t	1989	1990	1991
y	\$1.26	\$1.44	\$1.63

  a. Plot the points to obtain a graphical representation of the data.

  b. The data are approximately linear. Approximate the slope of a line passing through the points and use the result to estimate the earnings per share in 1992.

18. *Conveyor Design* A moving conveyor is built so that it rises 1 meter for each 3 meters it moves horizontally.

  a. Find the slope of the conveyor.

  b. Suppose the conveyor runs between two floors in a factory. The vertical distance between floors is 10 feet. Find the length of the conveyor.

In Exercises 19–22, find the slope and the y-intercept (if possible) of the line.

19. $x + 5y = 20$ **20.** $6x - 5y = 15$

21. $x = 4$ **22.** $y = -1$

In Exercises 23–28, find an equation for the line that passes through the points, and sketch the line.

23. $(2, 1), (0, -3)$ **24.** $(-3, -4), (1, 4)$

25. $(0, 0), (-1, 3)$ **26.** $(-3, 6), (1, 2)$

27. $(1, -2), (3, -2)$ **28.** $\left(\frac{7}{8}, \frac{3}{4}\right), \left(\frac{5}{4}, -\frac{1}{4}\right)$

In Exercises 29–34, find an equation of the line that passes through the point and has the indicated slope. Sketch the line.

Point	Slope	Point	Slope
29. $(0, 3)$	$m = \frac{3}{4}$	**30.** $(-1, 2)$	m undefined
31. $(0, 0)$	$m = \frac{2}{3}$	**32.** $(-2, 4)$	$m = -\frac{3}{5}$
33. $(0, 2)$	$m = 4$	**34.** $(0, 4)$	$m = 0$

35. Find an equation of the vertical line with x-intercept at 3.

36. Show that the line with intercepts $(a, 0)$ and $(0, b)$ has the following equation.

$$\frac{x}{a} + \frac{y}{b} = 1, \qquad a \neq 0, b \neq 0$$

In Exercises 37–40, use the result of Exercise 36 to write an equation of the indicated line.

37. x-intercept: $(2, 0)$
   y-intercept: $(0, 3)$

38. x-intercept $\left(-\frac{2}{3}, 0\right)$
   y-intercept: $(0, -2)$

39. Point on line: $(1, 2)$
   x-intercept: $(a, 0)$
   y-intercept: $(0, a)$
   $(a \neq 0)$

40. Point on line: $(-3, 4)$
   x-intercept: $(a, 0)$
   y-intercept: $(0, a)$
   $(a \neq 0)$

In Exercises 41–46, write equations of the lines through the point (a) parallel to the given line and (b) perpendicular to the given line.

	Point	Given Line
41.	$(2, 1)$	$4x - 2y = 3$
42.	$(-3, 2)$	$x + y = 7$
43.	$\left(\frac{7}{8}, \frac{3}{4}\right)$	$5x + 3y = 0$
44.	$(-6, 4)$	$3x + 4y = 7$
45.	$(2, 5)$	$x = 4$
46.	$(-1, 0)$	$y = -3$

In Exercises 47–52, sketch the graph of the equation.

47. $y = -3$

48. $x = 4$

49. $2x - y - 3 = 0$

50. $x + 2y + 6 = 0$

51. $y = -2x + 1$

52. $y - 1 = 3(x + 4)$

In Exercises 53 and 54, find an equation of the line determined by the points of intersection of the parabolas.

53. $y = x^2$
   $y = 4x - x^2$

54. $y = x^2 - 4x + 3$
   $y = -x^2 + 2x + 3$

In Exercises 55 and 56, determine whether the three points are collinear.

55. $(-2, 1), (-1, 0), (2, -2)$

56. $(0, 4), (7, -6), (-5, 11)$

In Exercises 57–60, refer to the triangle in the figure.

57. Find the coordinates of the point of intersection of the perpendicular bisectors of the sides.

58. Find the coordinates of the point of intersection of the medians.

59. Find the coordinates of the point of intersection of the altitudes.

60. Show that the points of intersection in Exercises 57, 58, and 59, are collinear.

FIGURE FOR 57–60

61. *Temperature Conversion* Write a linear equation that expresses the relationship between the temperature in degrees Celsius C and degrees Fahrenheit F. Use the fact that water freezes at $0°$ Celsius ($32°$ Fahrenheit) and boils at $100°$ Celsius ($212°$ Fahrenheit).

62. *Temperature Conversion* Use the result of Exercise 61 to complete the table.

C		$-10°$	$10°$			$177°$
F	$0°$			$68°$	$90°$	

63. *Reimbursed Expenses* A company reimburses its sales representatives $125 per day for lodging and meals, plus 27¢ per mile driven. Write a linear equation for the daily cost C in terms of x, the number of miles driven.

C 64. *Career Choice* An employee has two options for positions in a large corporation. One position pays $9.50 per hour *plus* an additional piecework rate of $0.75 per unit produced. The second pays $6.20 per hour *plus* a piecework rate of $1.30.
   a. Write linear equations for the hourly wages W in terms of x, the number of units produced per hour, for each of the options.
   b. Use a graphing utility to graph the linear equations and find the point of intersection.
   c. Interpret the meaning of the point of intersection of the graphs in part (b). How would you use this information to select the correct option if the goal is to obtain the highest hourly wage?

65. *Annual Salary* Suppose your salary was $28,500 in 1992 and $32,900 in 1994. Assume your salary follows a linear growth pattern.
   a. Write a linear equation giving your salary S in terms of the year t where $t = 0$ corresponds to the year 1992.
   b. Use the linear equation to predict your salary in 1997.

C 66. *Straight-Line Depreciation* A small business purchases a piece of equipment for $875. After 5 years the equipment will be outdated, having no value.
   a. Write a linear equation giving the value of the equipment y in terms of the time x, $0 \leq x \leq 5$.
   b. Use a graphing utility to graph the equation.
   c. Move the cursor along the graph and estimate (to two decimal place accuracy) the value of the equipment when $x = 2$.
   d. Move the cursor along the graph and estimate (to two decimal place accuracy) the time when the value of the equipment is $200.

67. *Apartment Rental*   A real estate office handles an apartment complex with 50 units. When the rent is $380 per month, all 50 units are occupied. However, when the rent is $425, the average number of occupied units drops to 47. Assume that the relationship between the monthly rent p and the demand x is linear. (Note: Here we use the term *demand* to refer to the number of occupied units.)

   a. Write a linear equation expressing the demand x in terms of the rent p.

   b. *Linear extrapolation*   Use the equation determined in part (a) to predict the number of units occupied if the rent is raised to $455.

   c. *Linear interpolation*   Predict the number of units occupied if the rent is lowered to $395.

68. *Cable Television*   The numbers of subscribers to cable TV in the years 1986 and 1990 were 37.5 million and 54.0 million, respectively. Assume that the relationship between the year t and the number of subscribers y is linear.

   a. Write the equation that gives the number of subscribers y in terms of t. (Let $t = 0$ represent 1986.)

   b. *Linear extrapolation*   Use this equation to estimate the number of subscribers in 1995.

   c. *Linear interpolation*   Estimate the number of subscribers in 1988.

   d. What information does the slope of the line in part (a) convey?

In Exercises 69–74, find the distance between the point and line, or between the two lines, using the following formula for the distance between the point (x_1, y_1) and the line $Ax + By + C = 0$.

$$\text{Distance} = \frac{|Ax_1 + By_1 + C|}{\sqrt{A^2 + B^2}}$$

69. Point: $(0, 0)$
   Line: $4x + 3y = 10$

70. Point: $(2, 3)$
   Line: $4x + 3y = 10$

71. Point: $(-2, 1)$
   Line: $x - y - 2 = 0$

72. Point: $(6, 2)$
   Line: $x = -1$

73. Line: $x + y = 1$
   Line: $x + y = 5$

74. Line: $3x - 4y = 1$
   Line: $3x - 4y = 10$

75. Prove that the diagonals of a rhombus intersect at right angles.

76. Prove that the figure formed by connecting consecutive midpoints of the sides of any quadrilateral is a parallelogram.

77. Prove that if the points (x_1, y_1) and (x_2, y_2) lie on the same line as (x_1^*, y_1^*) and (x_2^*, y_2^*), then

$$\frac{y_2^* - y_1^*}{x_2^* - x_1^*} = \frac{y_2 - y_1}{x_2 - x_1}.$$

Assume $x_1 \neq x_2$ and $x_1^* \neq x_2^*$.

78. Prove that if the slopes of two nonvertical lines are negative reciprocals of each other, then the lines are perpendicular.

True or False   In Exercises 79 and 80, determine whether the statement is true or false. If it is false, explain why or give an example that shows it is false.

79. If $a \neq 0$ and $b \neq 0$, then the lines represented by $ax + by = c_1$ and $bx - ay = c_2$ are perpendicular.

80. It is possible for two lines with positive slopes to be perpendicular to each other.

81. Find the equations of the two lines passing through the point $(4, 0)$ and tangent to the circle $x^2 - 2x + y^2 = 0$.

82. Show that the distance from (x_0, y_0) to the line $ax + by + c = 0$ is

$$\text{Distance} = \frac{|ax_0 + by_0 + c|}{\sqrt{a^2 + b^2}}.$$

CAREER INTERVIEW

Jina Chung

Wellman Laboratories

of Photomedicine

Mass. General Hospital

Boston, MA 02114

As a medical researcher, I can attest that technology in the medical field today is advancing at a breathtaking pace. I am currently involved in two projects. The first is to develop a diagnostic photon detector that measures the light emitted by white blood cells engulfing foreign particles. The second is to study the effects of various laser parameters that will safely remove dead skin for large-area skin burns with minimal blood loss and allow quick healing.

Particularly in the second project, I rely on the calculus I studied as a physics major. Calculating the diameter of a laser beam requires the double integration of intensity over a specific area. Usually these calculations can be simplified into algebraic equations, but, without a thorough knowledge of calculus, the concepts of laser engineering would be impossible to grasp.

Calculus has proved to be an essential tool in my physics and medical research experiences (almost like a second nature).

5 Functions and Function Notation • The Domain and Range of a Function • The Graph of a Function • Transformations of Functions • Classifications and Combinations of Functions

Functions and Function Notation

A **relation** between two sets X and Y is a set of ordered pairs, each of the form (x, y) where x is a member of X and y is a member of Y. A **function** from X to Y is a relation between X and Y that has the property that any two ordered pairs with the same x-value also have the same y-value. The variable x is the **independent variable,** and the variable y is the **dependent variable.**

Many real-life situations can be modeled by functions. For instance, the area A of a circle is a function of the circle's radius r. In this case r is the independent variable and A is the dependent variable.

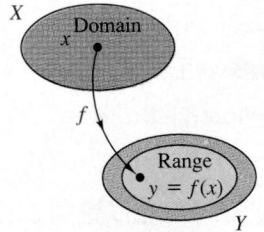

FIGURE 47
A real-valued function f of a real variable.

Definition of a Real-Valued Function of a Real Variable

Let X and Y be sets of real numbers. A **real-valued function f of a real variable** x from X to Y is a correspondence that assigns to each number x in X exactly one number y in Y.

The **domain** of f is the set X. The number y is the **image** of x under f and is denoted by $f(x)$. The **range** of f is a subset of Y and consists of all images of numbers in X (see Figure 47).

Functions can be specified in a variety of ways. In this text, however, we will concentrate primarily on functions that are given by equations involving the dependent and independent variables. For instance, the equation $x + 2y = 1$ defines y, the dependent variable, as a function of x, the independent variable. To **evaluate** this function (that is, to find the y-value that corresponds to a given x-value), it is convenient to isolate y on the left side of the equation.

$$y = \frac{1}{2}(1 - x)$$

Using f as the name of the function, you can write this equation as

$$f(x) = \frac{1}{2}(1 - x). \quad \text{Function notation}$$

Function notation has the advantage of clearly identifying the dependent variable as $f(x)$ while at the same time telling you that x is the independent variable and that the function itself is "f." The notation $f(x)$ is read "f of x." Function notation allows you to be less wordy. Instead of asking "What is the value of y that corresponds to $x = 3$?" you can ask "What is $f(3)$?"

In an equation that defines a function, the role of the variable x is simply that of a placeholder. For instance, the function given by

$$f(x) = 2x^2 - 4x + 1$$

can be described by the form

$$f(\ ) = 2(\ )^2 - 4(\ ) + 1$$

where parentheses are used instead of x. To evaluate $f(-2)$, simply place -2 in each set of parentheses.

$$f(-2) = 2(-2)^2 - 4(-2) + 1$$
$$= 2(4) + 8 + 1$$
$$= 17$$

REMARK   Although f is often used as a convenient function name and x as the independent variable, you can use other symbols. For instance, the following equations all define the same function.

$$f(x) = x^2 - 4x + 7, \qquad f(t) = t^2 - 4t + 7, \qquad g(s) = s^2 - 4s + 7$$

EXAMPLE 1   Evaluating a Function

For the function f defined by $f(x) = x^2 + 7$, evaluate the following.

a. $f(3a)$ **b.** $f(b-1)$ **c.** $\dfrac{f(x + \Delta x) - f(x)}{\Delta x}, \qquad \Delta x \neq 0$

Solution

a. $f(3a) = (3a)^2 + 7$ Replace x with $3a$
   $\quad\quad\ = 9a^2 + 7$ Simplify

b. $f(b-1) = (b-1)^2 + 7$ Replace x with $b-1$
   $\quad\quad\quad = b^2 - 2b + 1 + 7$ Expand binomial
   $\quad\quad\quad = b^2 - 2b + 8$ Simplify

c. $\dfrac{f(x + \Delta x) - f(x)}{\Delta x} = \dfrac{[(x + \Delta x)^2 + 7] - (x^2 + 7)}{\Delta x}$

$\quad\quad\quad = \dfrac{x^2 + 2x\Delta x + (\Delta x)^2 + 7 - x^2 - 7}{\Delta x}$

$\quad\quad\quad = \dfrac{2x\Delta x + (\Delta x)^2}{\Delta x}$

$\quad\quad\quad = \dfrac{\Delta x\,(2x + \Delta x)}{\Delta x}$

$\quad\quad\quad = 2x + \Delta x, \qquad \Delta x \neq 0$

REMARK   The expression in Example 1c is called a difference quotient and has special significance in calculus. We will say more about this in Chapter 2.

The Domain and Range of a Function

The domain of a function can be described explicitly, or it may be described *implicitly* by an equation used to define the function. The implied domain is the set of all real numbers for which the equation is defined. For example, the function given by

$$f(x) = \frac{1}{x^2 - 4}, \qquad 4 \le x \le 5$$

has an explicitly defined domain given by $\{x: 4 \le x \le 5\}$. On the other hand, the function given by

$$g(x) = \frac{1}{x^2 - 4}$$

has an implied domain which is the set $\{x: x \ne \pm 2\}$.

EXAMPLE 2   Finding the Domain and Range of a Function

a. The domain of the function

$$f(x) = \sqrt{x - 1}$$

is the set of all x-values for which $x - 1 \ge 0$, which is the interval $[1, \infty)$. To find the range, observe that $f(x) = \sqrt{x - 1}$ is never negative. Moreover, as x takes on the various values in the domain, $f(x)$ takes on all nonnegative values. Thus, the range is the interval $[0, \infty)$, as indicated in Figure 48.

b. The domain of the function

$$f(x) = \sqrt{x^2 - x - 6}$$

is the set of all x-values such that $x^2 - x - 6 \ge 0$. Using the techniques shown in Example 4 in Section 1, you can conclude that the domain is

$$(-\infty, -2] \cup [3, \infty).$$

FIGURE 48
The domain of $f(x)$ is $[1, \infty)$ and ' .e range is $[0, \infty)$.

EXAMPLE 3   A Function Defined by More than One Equation

Determine the domain and range of the following function.

$$f(x) = \begin{cases} 1 - x, & \text{if } x < 1 \\ \sqrt{x - 1}, & \text{if } x \ge 1 \end{cases}$$

Solution   Because f is defined for $x < 1$ and $x \ge 1$, the domain is the entire set of real numbers. On the portion of the domain for which $x \ge 1$, the function behaves as in Example 2. For $x < 1$, the values of $1 - x$ are positive. Therefore, the range of the function is the interval

$$[0, \infty). \quad \text{Range}$$

(See Figure 49.)

FIGURE 49
The domain of $f(x)$ is $(-\infty, \infty)$ and the range is $[0, \infty)$.

A function from X to Y is **one-to-one** if to each y-value in the range there corresponds exactly one x-value in the domain. For instance, the function given in Example 2a is one-to-one, whereas the function given in Example 3 is not one-to-one. If the range consists of all of Y, then the function is called **onto.**

FIGURE 50
The graph of a function.

The Graph of a Function

The graph of the function $y = f(x)$ consists of all points $(x, f(x))$, where x is in the domain of f. In Figure 50, note that

x = the directed distance from the y-axis

$f(x)$ = the directed distance from the x-axis.

A vertical line can intersect the graph of a function of x at most *once*. This observation provides a convenient visual test (called the **vertical line test**) for functions of x. For example, in Figure 51(a), you can see that the graph does not define y as a function of x because a vertical line intersects the graph twice.

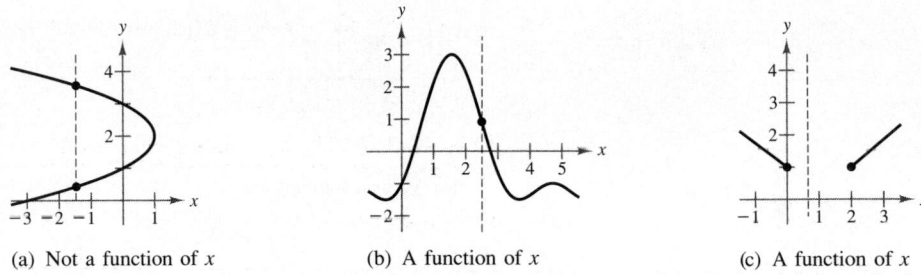

(a) Not a function of x (b) A function of x (c) A function of x

FIGURE 51
The vertical line test for functions.

Figure 52 shows the graphs of six basic functions. You should be able to recognize these graphs.

(a) Identity function   (b) Squaring function   (c) Cubing function

(d) Square root function   (e) Absolute value function   (f) Rational function

FIGURE 52
The graphs of six basic functions.

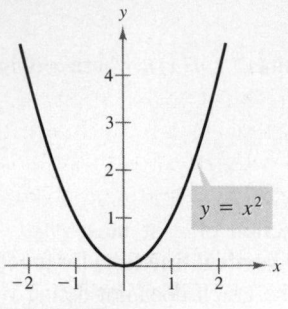

FIGURE 53
The original graph.

Transformations of Functions

Some families of graphs have the same basic shape. For example, consider the graph of $y = x^2$, as shown in Figure 53. Now compare this graph with those shown in Figure 54.

(a) Vertical shift upward

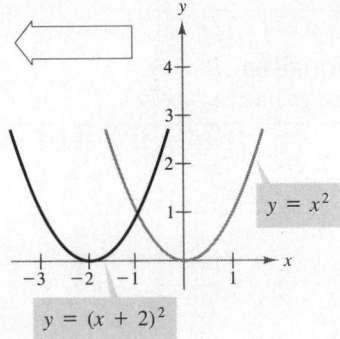

(b) Horizontal shift to the left

(c) Reflection

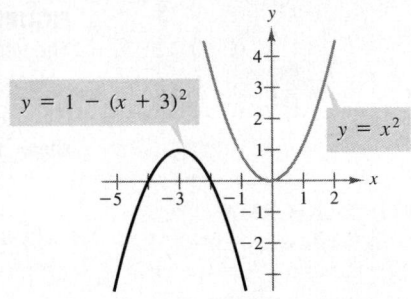

(d) Shift to the left, reflect, and then shift upward

FIGURE 54
Transformations of the graph of $y = x^2$.

Each of the graphs in Figure 54 is a **transformation** of the graph of $y = x^2$. The three basic types of transformations illustrated by these graphs are vertical shifts, horizontal shifts, and reflections. Function notation lends itself well to describing transformations of graphs in the plane.

Basic Types of Transformations ($c > 0$)

Original graph:	$y = f(x)$
Horizontal shift c units to the **right:**	$y = f(x - c)$
Horizontal shift c units to the **left:**	$y = f(x + c)$
Vertical shift c units **downward:**	$y = f(x) - c$
Vertical shift c units **upward:**	$y = f(x) + c$
Reflection (about the x-axis):	$y = -f(x)$
Reflection (about the y-axis):	$y = f(-x)$

LEONHARD EULER

In addition to making major contributions to almost every branch of mathematics, Euler was one of the first to apply calculus to real-life problems in physics. His extensive published writings include such topics as shipbuilding, acoustics, optics, astronomy, mechanics, and magnetism.

FOR FURTHER INFORMATION For more on the history of the concept of a function, see the article "Evolution of the Function Concept: A Brief Survey" by Israel Kleiner in the September, 1989 issue of *The College Mathematics Journal.*

Classifications and Combinations of Functions

The modern notion of a function was derived from the efforts of many seventeenth- and eighteenth-century mathematicians. Of particular note was Leonhard Euler (1707–1783), to whom we are indebted for the function notation $y = f(x)$. By the end of the eighteenth century, mathematicians and scientists had concluded that most real-world phenomena could be represented by mathematical models taken from a basic collection of functions called **elementary functions.**

Elementary functions are divided into three categories: (1) algebraic, (2) trigonometric, and (3) logarithmic and exponential. We will review the trigonometric functions in Section 6 of this preliminary chapter and introduce the remaining elementary functions in Chapter 5.

The most common type of algebraic function is a **polynomial function**

$$f(x) = a_n x^n + a_{n-1} x^{n-1} + \cdots + a_2 x^2 + a_1 x + a_0, \qquad a_n \neq 0$$

where the positive integer n is the **degree** of the polynomial function. The numbers a_i are **coefficients,** with a_n the **leading coefficient** and a_0 the **constant term** of the polynomial function. It is common practice to use subscript notation for coefficients of general polynomial functions, but for polynomial functions of low degree, the following simpler forms are often used.

Zeroth degree:	$f(x) = a$	Constant function
First degree:	$f(x) = ax + b$	Linear function
Second degree:	$f(x) = ax^2 + bx + c$	Quadratic function
Third degree:	$f(x) = ax^3 + bx^2 + cx + d$	Cubic function

Although the graph of a polynomial function can have several turns, eventually the graph will rise or fall without bound as x moves to the right or left. Whether the graph of

$$f(x) = a_n x^n + a_{n-1} x^{n-1} + \cdots + a_2 x^2 + a_1 x + a_0$$

eventually rises or falls can be determined by the function's degree (odd or even) and by the leading coefficient a_n, as indicated in Figure 55. Note that the dashed portions of the graphs indicate that the **leading coefficient test** determines *only* the right and left behavior of the graph.

Graphs of polynomial functions of even degree   Graphs of polynomial functions of odd degree

FIGURE 55
The leading coefficient test for polynomial functions.

Just as a rational number can be written as the quotient of two integers, a **rational function** can be written as the quotient of two polynomials. Specifically, a function f is rational if it has the form

$$f(x) = \frac{p(x)}{q(x)}, \qquad q(x) \neq 0$$

where $p(x)$ and $q(x)$ are polynomials.

Polynomial functions and rational functions are examples of **algebraic functions.** An algebraic function of x is one that can be expressed as a finite number of sums, differences, multiples, quotients, and radicals involving x^n. For example, $f(x) = \sqrt{x + 1}$ is algebraic. Functions that are not algebraic are **transcendental.** For instance, the trigonometric functions discussed in Section 6 are transcendental.

Two functions can be combined in various ways to create new functions. For example, given $f(x) = 2x - 3$ and $g(x) = x^2 + 1$, you can form the following functions.

$$f(x) + g(x) = (2x - 3) + (x^2 + 1) = x^2 + 2x - 2 \qquad \text{Sum}$$
$$f(x) - g(x) = (2x - 3) - (x^2 + 1) = -x^2 + 2x - 4 \qquad \text{Difference}$$
$$f(x)g(x) = (2x - 3)(x^2 + 1) = 2x^3 - 3x^2 + 2x - 3 \qquad \text{Product}$$
$$\frac{f(x)}{g(x)} = \frac{2x - 3}{x^2 + 1} \qquad \text{Quotient}$$

You can combine two functions in yet another way to form a **composite function.**

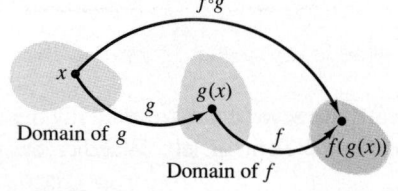

FIGURE 56
The domain of the composite function $f \circ g$.

Definition of Composite Function

Let f and g be functions. The function given by $(f \circ g)(x) = f(g(x))$ is called the **composite** of f with g. The domain of $f \circ g$ is the set of all x in the domain of g such that $g(x)$ is in the domain of f (see Figure 56).

The composite of f with g may not be equal to the composite of g with f.

EXAMPLE 4   Finding Composites of Functions

Given $f(x) = 2x - 3$ and $g(x) = x^2 + 1$, find

a. $f \circ g$ **b.** $g \circ f$.

Solution

a. $(f \circ g)(x) = f(g(x)) = 2(g(x)) - 3 = 2(x^2 + 1) - 3$
$$= 2x^2 - 1$$

b. $(g \circ f)(x) = g(f(x)) = (f(x))^2 + 1 = (2x - 3)^2 + 1$
$$= 4x^2 - 12x + 10$$

Note that $(f \circ g)(x) \neq (g \circ f)(x)$.

In Section 3, we defined an x-intercept of a graph to be a point $(a, 0)$ at which the graph crosses the x-axis. If the graph represents a function f, then the number a is a **zero** of f. In other words, the zeros of a function f are the solutions of the equation $f(x) = 0$. For example, the function $f(x) = x - 4$ has a zero at $x = 4$ because $f(4) = 0$.

In Section 3 we also discussed different types of symmetry. In the terminology of functions, a function is **even** if its graph is symmetric with respect to the y-axis, and is **odd** if its graph is symmetric with respect to the origin. The symmetry tests in Section 3 yield the following test for even and odd functions.

Test for Even and Odd Functions

The function $y = f(x)$ is **even** if $f(-x) = f(x)$.
The function $y = f(x)$ is **odd** if $f(-x) = -f(x)$.

REMARK   Except for the constant function $f(x) = 0$, the graph of a function cannot have symmetry with respect to the x-axis because it then would fail the vertical line test for the graph of a function.

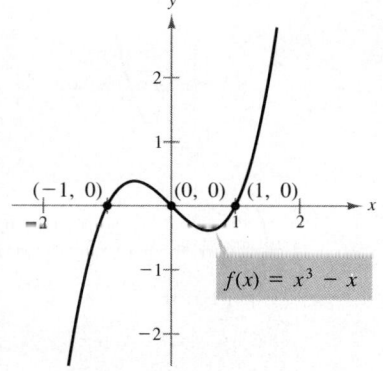

(a) Odd function

(b) Even function

FIGURE 57

EXAMPLE 5   Even and Odd Functions and Zeros of Functions

Determine whether the following functions are even, odd, or neither. Then find the zeros of the functions.

a. $f(x) = x^3 - x$ **b.** $g(x) = x^2 + 1$

Solution

a. This function is odd because

$$f(-x) = (-x)^3 - (-x)$$
$$= -x^3 + x$$
$$= -(x^3 - x)$$
$$= -f(x).$$

The zeros of f are found as follows.

$$x^3 - x = 0 \qquad \text{Let } f(x) = 0$$
$$x(x^2 - 1) = x(x - 1)(x + 1) = 0 \qquad \text{Factor}$$
$$x = 0, 1, -1 \qquad \text{Zeros of } f$$

See Figure 57(a).

b. This function is even because

$$g(-x) = (-x)^2 + 1 = x^2 + 1 = g(x).$$

It has no zeros because $x^2 + 1$ is positive for all x. See Figure 57(b).

REMARK   Each of the functions in Example 5 is either even or odd. However, some functions, such as $f(x) = x^2 + x + 1$, are neither even nor odd.

EXERCISES for Section 5

TECHNOLOGY
Laboratory Guide
Lab 0.5

In Exercises 1–12, evaluate (if possible) the function at the specified value(s) of the independent variable and simplify the results.

1. $f(x) = 2x - 3$
  a. $f(0)$
  b. $f(-3)$
  c. $f(b)$
  d. $f(x - 1)$

2. $f(x) = x^2 - 2x + 2$
  a. $f(\frac{1}{2})$
  b. $f(-1)$
  c. $f(c)$
  d. $f(x + \Delta x)$

3. $f(x) = \sqrt{x + 3}$
  a. $f(-2)$
  b. $f(6)$
  c. $f(c)$
  d. $f(x + \Delta x)$

4. $f(x) = |x| + 4$
  a. $f(2)$
  b. $f(-2)$
  c. $f(x^2)$
  d. $f(x + \Delta x) - f(x)$

5. $f(x) = \begin{cases} 2x + 1, & x < 0 \\ 2x + 2, & x \geq 0 \end{cases}$
  a. $f(-1)$ **b.** $f(0)$ **c.** $f(2)$ **d.** $f(t^2 + 1)$

6. $f(x) = \begin{cases} x^2 + 2, & x \leq 1 \\ 2x^2 + 2, & x > 1 \end{cases}$
  a. $f(-2)$ **b.** $f(0)$ **c.** $f(1)$ **d.** $f(s^2 + 2)$

7. $f(x) = x^2 - x + 1$
$$\frac{f(2 + \Delta x) - f(2)}{\Delta x}$$

8. $f(x) = \frac{1}{x}$
$$\frac{f(1 + \Delta x) - f(1)}{\Delta x}$$

9. $f(x) = x^3$
$$\frac{f(x + \Delta x) - f(x)}{\Delta x}$$

10. $f(x) = 3x - 1$
$$\frac{f(x) - f(1)}{x - 1}$$

11. $f(x) = \frac{1}{\sqrt{x - 1}}$
$$\frac{f(x) - f(2)}{x - 2}$$

12. $f(x) = x^3 - x$
$$\frac{f(x) - f(1)}{x - 1}$$

In Exercises 13–20, find the domain and range of the function, and sketch its graph.

13. $f(x) = 4 - x$

14. $g(x) = \frac{4}{x}$

15. $h(x) = \sqrt{x - 1}$

16. $f(x) = \frac{1}{2}x^3 + 2$

17. $f(x) = \sqrt{9 - x^2}$

18. $h(x) = \sqrt{25 - x^2}$

19. $f(x) = |x - 2|$

20. $f(x) = \frac{|x|}{x}$

[C] In Exercises 21 and 22, use a graphing utility to graph the function. Estimate the domain and range of the function from the graph. Guess the coordinates of any intercepts of the graph and then find them analytically.

21. $f(x) = x + \sqrt{4 - x^2}$

22. $f(x) = \frac{4}{\sqrt{x + 3} + 1}$

[C] **23.** *Essay*   Use a graphing utility to graph the polynomial functions $p_1(x) = x^3 - x + 1$ and $p_2(x) = x^3 - x$. How many zeros does each have? Can you think of a cubic polynomial having no zeros? Give a written explanation of your answer.

[C] **24.** *Essay*   Use a graphing utility to graph the functions. You will notice that as x moves without bound to the left or to the right, the graph of the polynomial function eventually rises or falls. Write a short paragraph stating how the right and left behavior of a polynomial function is determined by the degree of the polynomial and the sign of the leading coefficient.
  a. $f(x) = 3x^4 + 4x^3$
  b. $f(x) = 2x^5 - 3x^2 + 1$
  c. $f(x) = -2x^3 + 3x - 4$
  d. $f(x) = -10x^6 - 3x^5 + 2x$

In Exercises 25–28, use the vertical line test to determine whether y is a function of x.

25. $x - y^2 = 0$

26. $y = x^3 - 1$

27. $\sqrt{x^2 - 4} - y = 0$

28. $x^2 + y^2 = 9$

    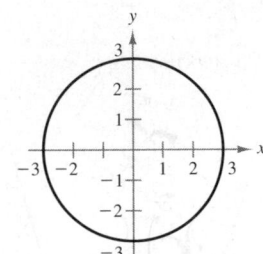

In Exercises 29–36, determine whether y is a function of x.

29. $x^2 + y^2 = 4$

30. $x = y^2$

31. $x^2 + y = 4$

32. $x + y^2 = 4$

33. $2x + 3y = 4$

34. $x^2 + y^2 - 4y = 0$

35. $y^2 = x^2 - 1$

36. $x^2y - x^2 + 4y = 0$

37. *Beam Loads*   Uniformly distributed loads are placed on beams 1 inch wide and 6 inches high. The table lists experimental data for the maximum safe load L (in pounds) of a beam of length d (in feet).

d	5	10	15	20
L	800	400	270	200

  a. Plot the points. What information (if any) can you determine from the plot?
  b. Is L a function of d? If so, what are its domain and range?

38. *Distance*   A student who commutes 27 miles to attend college remembers, after driving a few minutes, that a term paper that is due has been left at home. Driving at a higher speed than usual, the student goes back home, picks up the paper, and once again starts toward school. Sketch a graph of the student's distance from home as a function of time.

39. Use the graph of $f(x) = \sqrt{x}$ to sketch the graph of each function. In each case describe the transformation.
a. $y = \sqrt{x} + 2$ b. $y = -\sqrt{x}$
c. $y = \sqrt{x} - 2$ d. $y = \sqrt{-x}$

40. Use the graph of $f(x) = x^2$ to write an equation for the function.

(a)

(b)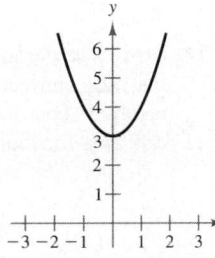

In Exercises 41 and 42, specify a sequence of transformations that will yield the graph of each function from the graph of the function $f(x) = x^3$.

41. a. $g(x) = 4 - x^3$
    b. $g(x) = (x - 4)^3 + 2$

42. a. $h(x) = (x + 2)^3 + 1$
    b. $h(x) = 5 - (x - 1)^3$

43. Given $f(x) = \sqrt{x}$ and $g(x) = x^2 - 1$, find the composite functions.
a. $f(g(1))$ b. $g(f(1))$
c. $g(f(0))$ d. $f(g(-4))$
e. $f(g(x))$ f. $g(f(x))$

44. Given $f(x) = 1/x$ and $g(x) = x^2 - 1$, find the composite functions.
a. $f(g(2))$ b. $g(f(2))$
c. $f\left(g\left(\dfrac{1}{\sqrt{2}}\right)\right)$ d. $g\left(f\left(\dfrac{1}{\sqrt{2}}\right)\right)$
e. $g(f(x))$ f. $f(g(x))$

In Exercises 45–48, find the composite functions $(f \circ g)$ and $(g \circ f)$. What is the domain of each function? Are the two composite functions equal?

45. $f(x) = x^2$ **46.** $f(x) = x^3$
    $g(x) = \sqrt{x}$ $g(x) = \sqrt[3]{x}$

47. $f(x) = x + 1$ **48.** $f(x) = x^2 - 1$
    $g(x) = \dfrac{1}{x}$ $g(x) = x$

49. *Ripples*   A pebble is dropped into a calm pond, causing ripples in the form of concentric circles. The radius (in feet) of the outer ripple is $r(t) = 0.6t$, where t is the time in seconds after the pebble strikes the water. The area of the circle is $A(r) = \pi r^2$. Find and interpret $(A \circ r)(t)$.

C 50. *Automobile Aerodynamics*   The number of horsepower H required to overcome wind drag on an automobile is approximated by

$$H(x) = 0.002x^2 + 0.005x - 0.029, \qquad 10 \le x \le 100$$

where x is the speed of the car in miles per hour.
a. Use a graphing utility to graph H.
b. Rewrite the function so that x represents the speed in kilometers per hour. [Find $H(1.6x)$.]

In Exercises 51 and 52, find the (real) zeros of the function.

51. $f(x) = \dfrac{3}{x - 1} + \dfrac{4}{x - 2}$ **52.** $f(x) = x^3 - x$

In Exercises 53–56, determine whether the function is even, odd, or neither.

53. $f(x) = 4 - x^2$ **54.** $f(x) = \sqrt[3]{x}$

55. $f(x) = x(4 - x^2)$ **56.** $f(x) = 4x - x^2$

57. Show that the function is odd.

$$f(x) = a_{2n+1}x^{2n+1} + \cdots + a_3x^3 + a_1x$$

58. Show that the function is even.

$$f(x) = a_{2n}x^{2n} + a_{2n-2}x^{2n-2} + \cdots + a_2x^2 + a_0$$

59. Show that the product of two even (or two odd) functions is even.

60. Show that the product of an odd function and an even function is odd.

C 61. Use a graphing utility to graphically demonstrate the results of Exercises 59 and 60. Use functions of your choice.

62. What can be said about the sum or difference of (a) two even functions, (b) two odd functions, and (c) an odd function and an even function? Demonstrate your conclusions graphically.

In Exercises 63–66, express the indicated values as functions of x.

63. R and r **64.** R and r

65. *h* and *p*

66. *h* and *p*

67. *Area* A rectangle has a perimeter of 100 feet (see figure). Express the area *A* of the rectangle as a function of *x*.

68. *Area* You have 200 feet of fencing to enclose two adjacent rectangular fields (see figure). Express the area *A* of the enclosures as a function of *x*.

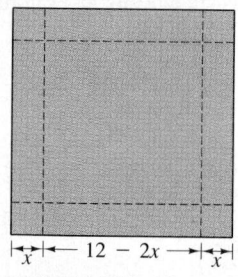

FIGURE FOR 67   **FIGURE FOR 68**

69. *Volume* An open box is to be made from a square piece of material 12 inches on a side by cutting equal squares from each corner and turning up the sides (see figure). Express the volume *V* as a function of *x*.

70. *Area* A rectangle is bounded by the *x*-axis and the semicircle $y = \sqrt{25 - x^2}$ (see figure). Write the area *A* of the rectangle as a function of *x*.

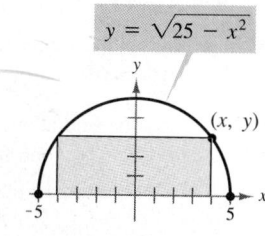

FIGURE FOR 69   **FIGURE FOR 70**

71. *Volume* A rectangular package with square cross sections has a combined length and girth (perimeter of a cross section) of 108 inches. Express the volume *V* as a function of *x* (see figure).

72. *Volume* A closed box with a square base of side *x* has a surface area of 100 square feet (see figure). Express the volume *V* of the box as a function of *x*.

FIGURE FOR 71   **FIGURE FOR 72**

73. *Total Time* You are in a boat 2 miles from the nearest point on the coast. You are to go to a point *Q*, 3 miles down the coast and 1 mile inland (see figure). You can row at 2 mph and walk at 4 mph. Express the total time *T* of the trip as a function of *x*.

74. *Area* The portion of the vertical line through the point $(x, 0)$ that lies between the *x*-axis and the graph of $y = \sqrt{x}$ is revolved about the *x*-axis. Express the area *A* of the resulting disk as a function of *x* (see figure).

          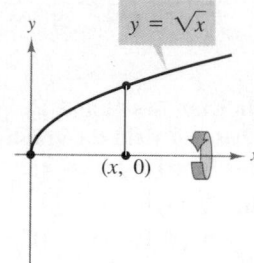

FIGURE FOR 73   **FIGURE FOR 74**

True or False **In Exercises 75–78, determine whether the statement is true or false. If it is false, explain why or give an example that shows it is false.**

75. If $f(a) = f(b)$, then $a = b$.

76. A vertical line can intersect the graph of a function at most once.

77. If $f(x) = f(-x)$ for all *x* in the domain of *f*, then the graph of *f* is symmetric with respect to the *y*-axis.

78. If *f* is a function, then $f(ax) = af(x)$.

79. Show that any function *f* defined for all real numbers can be written as the sum of an odd function and an even function.

80. Let $g(x) = 1/(1 - x)$. What are the domain and range of *g*? Find the domain and range of the composition $g \circ g \circ g$.

81. Express the following functions without using absolute value signs.
  a. $f(x) = |x - 3| - 2$
  b. $g(x) = |x| + |x - 2|$

Review of Trigonometric Functions

6

Angles and Degree Measure • Radian Measure • The Trigonometric Functions • Evaluating Trigonometric Functions • Solving Trigonometric Equations • Graphs of Trigonometric Functions

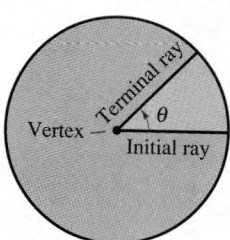

FIGURE 58
Standard position of an angle.

Angles and Degree Measure

An **angle** has three parts: an **initial ray,** a **terminal ray,** and a **vertex** (the point of intersection of the two rays), as shown in Figure 58. An angle is in **standard position** if its initial ray coincides with the positive x-axis and its vertex is at the origin. We assume that you are familiar with the degree measure of an angle.* It is common practice to use θ (the Greek lowercase letter *theta*) to represent both an angle and its measure. Angles between 0° and 90° are **acute** and angles between 90° and 180° are **obtuse.**

Positive angles are measured *counterclockwise,* and negative angles are measured *clockwise.* For instance, Figure 59 shows an angle whose measure is −45°. You cannot assign a measure to an angle by simply knowing where its initial and terminal rays are located. To measure an angle, you must also know how the terminal ray was revolved. For example, Figure 59 shows that the angle measuring −45° has the same terminal ray as the angle measuring 315°. Such angles are **coterminal.** In general, if θ is any angle, then

$$\theta + n(360), \qquad n \text{ is a nonzero integer}$$

is coterminal with θ.

An angle that is larger then 360° is one whose terminal ray has been revolved more than one full revolution counterclockwise, as shown in Figure 60. You can form an angle whose measure is less than −360° by revolving a terminal ray more than one full revolution clockwise.

FIGURE 59
Coterminal angles

FIGURE 60
Coterminal angles

REMARK   It is common to use the symbol θ to refer to both an *angle* and its *measure.* For instance, in Figure 60, you can write the measure of the smaller angle as $\theta = 45°$.

---

*For a more complete review of trigonometry, see *Precalculus,* 3rd edition, by Larson and Hostetler (Lexington, Mass., D. C. Heath and Company, 1993).

Radian Measure

To assign a radian measure to an angle θ, consider θ to be a central angle of a circle of radius 1, as shown in Figure 61. The **radian measure** of θ is then defined to be the length of the arc of the sector. Because the circumference of a circle is $2\pi r$, the circumference of a **unit circle** (of radius 1) is 2π. This implies that the radian measure of an angle measuring $360°$ is 2π. In other words, $360° = 2\pi$ radians.

Using radian measure for θ, the length s of a circular arc of radius r is $s = r\theta$, as shown in Figure 62.

FIGURE 61
Unit circle

FIGURE 62
Circle of radius r

You should know the conversions of the common angles shown in Figure 63. For other angles, use the fact that $180°$ is equal to π radians.

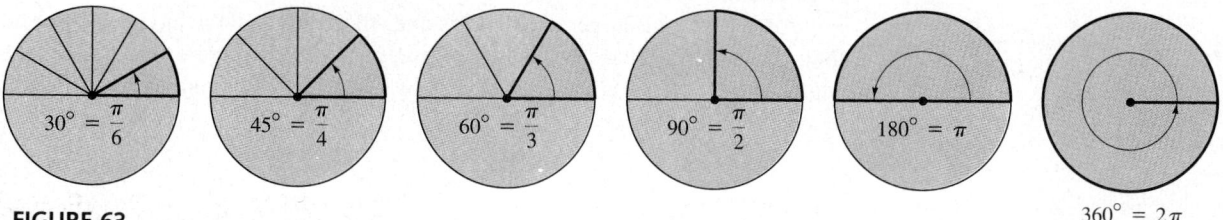

FIGURE 63
Radian and degree measure for several common angles.

EXAMPLE 1   Conversions Between Degrees and Radians

a. $40° = (40 \text{ deg})\left(\dfrac{\pi \text{ rad}}{180 \text{ deg}}\right) = \dfrac{2\pi}{9} \text{ radians}$

b. $-270° = (-270 \text{ deg})\left(\dfrac{\pi \text{ rad}}{180 \text{ deg}}\right) = -\dfrac{3\pi}{2} \text{ radians}$

c. $-\dfrac{\pi}{2} \text{ radians} = \left(-\dfrac{\pi}{2} \text{ rad}\right)\left(\dfrac{180 \text{ deg}}{\pi \text{ rad}}\right) = -90°$

d. $\dfrac{9\pi}{2} \text{ radians} = \left(\dfrac{9\pi}{2} \text{ rad}\right)\left(\dfrac{180 \text{ deg}}{\pi \text{ rad}}\right) = 810°$

FIGURE 64
Sides of a right triangle.

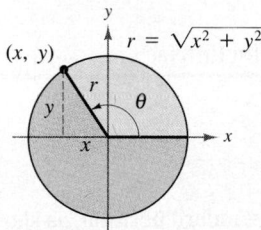

FIGURE 65
An angle in standard position.

The Trigonometric Functions

There are two common approaches to the study of trigonometry. In one, the trigonometric functions are defined as ratios of two sides of a right triangle. In the other, these functions are defined in terms of a point on the terminal side of an angle in standard position. We define the six trigonometric functions, **sine, cosine, tangent, cotangent, secant,** and **cosecant** (abbreviated as sin, cos, etc.), from both viewpoints.

Definition of the Six Trigonometric Functions

Right triangle definitions, where $0 < \theta < \dfrac{\pi}{2}$ *(see Figure 64).*

$$\sin\theta = \frac{\text{opp.}}{\text{hyp.}} \qquad \cos\theta = \frac{\text{adj.}}{\text{hyp.}} \qquad \tan\theta = \frac{\text{opp.}}{\text{adj.}}$$

$$\csc\theta = \frac{\text{hyp.}}{\text{opp.}} \qquad \sec\theta = \frac{\text{hyp.}}{\text{adj.}} \qquad \cot\theta = \frac{\text{adj.}}{\text{opp.}}$$

Circular function definitions, where θ *is any angle (see Figure 65).*

$$\sin\theta = \frac{y}{r} \qquad \cos\theta = \frac{x}{r} \qquad \tan\theta = \frac{y}{x}$$

$$\csc\theta = \frac{r}{y} \qquad \sec\theta = \frac{r}{x} \qquad \cot\theta = \frac{x}{y}$$

The following trigonometric identities are direct consequences of the definitions. (ϕ is the Greek letter *phi*.)

Trigonometric Identities [*Note that* $\sin^2\theta$ *is used to represent* $(\sin\theta)^2$.]

Pythagorean identities:

$\sin^2\theta + \cos^2\theta = 1$

$\tan^2\theta + 1 = \sec^2\theta$

$\cot^2\theta + 1 = \csc^2\theta$

Sum or difference of two angles:

$\sin(\theta \pm \phi) = \sin\theta\cos\phi \pm \cos\theta\sin\phi$

$\cos(\theta \pm \phi) = \cos\theta\cos\phi \mp \sin\theta\sin\phi$

$\tan(\theta \pm \phi) = \dfrac{\tan\theta \pm \tan\phi}{1 \mp \tan\theta\tan\phi}$

Law of Cosines:

$a^\cdot = b^2 + c^2 - 2bc\cos A$

Law of Cosines

Reduction formulas:

$\sin(-\theta) = -\sin\theta$

$\cos(-\theta) = \cos\theta$

$\tan(-\theta) = -\tan\theta$

$\sin\theta = -\sin(\theta - \pi)$

$\cos\theta = -\cos(\theta - \pi)$

$\tan\theta = \tan(\theta - \pi)$

Half-angle formulas:

$\sin^2\theta = \dfrac{1}{2}(1 - \cos 2\theta)$

$\cos^2\theta = \dfrac{1}{2}(1 + \cos 2\theta)$

Double angle formulas:

$\sin 2\theta = 2\sin\theta\cos\theta$

$\cos 2\theta = 2\cos^2\theta - 1$

$\qquad = 1 - 2\sin^2\theta$

$\qquad = \cos^2\theta - \sin^2\theta$

Reciprocal identities:

$\csc\theta = \dfrac{1}{\sin\theta}$

$\sec\theta = \dfrac{1}{\cos\theta}$

$\cot\theta = \dfrac{1}{\tan\theta}$

Quotient identities:

$\tan\theta = \dfrac{\sin\theta}{\cos\theta}$

$\cot\theta = \dfrac{\cos\theta}{\sin\theta}$

Evaluating Trigonometric Functions

There are two ways to evaluate trigonometric functions: (1) decimal approximations with a calculator (or a table of trigonometric values) and (2) exact evaluations using trigonometric identities and formulas from geometry. When using a calculator to evaluate a trigonometric function, remember to set the calculator to the appropriate mode—degree mode or radian mode.

EXAMPLE 2   Exact Evaluation of Trigonometric Functions

Evaluate the sine, cosine, and tangent of $\dfrac{\pi}{3}$.

Solution   Begin by drawing the angle $\theta = \pi/3$ in the standard position, as shown in Figure 66. Then, because $60° = \pi/3$ radians, you can draw an equilateral triangle with sides of length 1 and θ as one of its angles. Because the altitude of this triangle bisects its base, you know that $x = \frac{1}{2}$. Using the Pythagorean Theorem, you obtain

$$y = \sqrt{r^2 - x^2} = \sqrt{1 - \left(\frac{1}{2}\right)^2} = \sqrt{\frac{3}{4}} = \frac{\sqrt{3}}{2}.$$

Now, knowing the values of x, y, and r, you can write the following.

$$\sin\frac{\pi}{3} = \frac{y}{r} = \frac{\sqrt{3}/2}{1} = \frac{\sqrt{3}}{2}$$

$$\cos\frac{\pi}{3} = \frac{x}{r} = \frac{1/2}{1} = \frac{1}{2}$$

$$\tan\frac{\pi}{3} = \frac{y}{x} = \frac{\sqrt{3}/2}{1/2} = \sqrt{3}$$

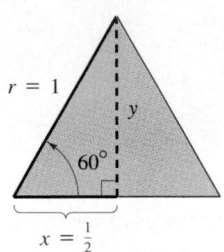

FIGURE 66
The angle $\pi/3$
in standard position.

REMARK   All angles in the remainder of this text are measured in radians unless stated otherwise. For example, when we write $\sin 3$, we mean the sine of three radians, and when we write $\sin 3°$, we mean the sine of three degrees.

The degree and radian measures of several common angles are given in Table 4, along with the corresponding values of the sine, cosine, and tangent (see Figure 67).

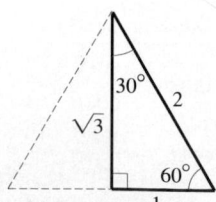

FIGURE 67
Common angles.

TABLE 4
Common First Quadrant Angles

Degrees	0	30°	45°	60°	90°
Radians	0	$\dfrac{\pi}{6}$	$\dfrac{\pi}{4}$	$\dfrac{\pi}{3}$	$\dfrac{\pi}{2}$
$\sin\theta$	0	$\dfrac{1}{2}$	$\dfrac{\sqrt{2}}{2}$	$\dfrac{\sqrt{3}}{2}$	1
$\cos\theta$	1	$\dfrac{\sqrt{3}}{2}$	$\dfrac{\sqrt{2}}{2}$	$\dfrac{1}{2}$	0
$\tan\theta$	0	$\dfrac{\sqrt{3}}{3}$	1	$\sqrt{3}$	Undefined

Quadrant II sin θ: + cos θ: − tan θ: −	Quadrant I sin θ: + cos θ: + tan θ: +
Quadrant III sin θ: − cos θ: − tan θ: +	Quadrant IV sin θ: − cos θ: + tan θ: −

FIGURE 68

Quadrant signs for trigonometric functions.

The quadrant signs of the sine, cosine, and tangent functions are shown in Figure 68. To extend the use of Table 4 to angles in quadrants other than the first quadrant, you can use the concept of a **reference angle** (see Figure 69), with the appropriate quadrant sign. For instance, the reference angle for $3\pi/4$ is $\pi/4$, and because the sine is positive in the second quadrant, you can write

$$\sin\frac{3\pi}{4} = +\sin\frac{\pi}{4} = \frac{\sqrt{2}}{2}.$$

Similarly, because the reference angle for 330° is 30°, and the tangent is negative in the fourth quadrant, you can write

$$\tan 330° = -\tan 30° = -\frac{\sqrt{3}}{3}.$$

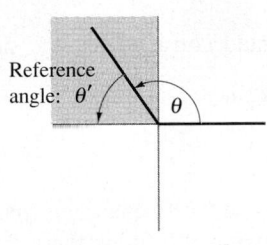

Quadrant II
$\theta' = \pi - \theta$ (radians)
$\theta' = 180° - \theta$ (degrees)

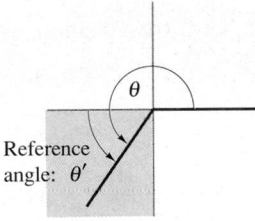

Quadrant III
$\theta' = \theta - \pi$ (radians)
$\theta' = \theta - 180°$ (degrees)

Quadrant IV
$\theta' = 2\pi - \theta$ (radians)
$\theta' = 360° - \theta$ (degrees)

FIGURE 69

EXAMPLE 3   Trigonometric Identities and Calculators

Evaluate the trigonometric expression.

a. $\sin\left(-\dfrac{\pi}{3}\right)$ **b.** $\sec 60°$ **c.** $\cos(1.2)$

Solution

a. Using the reduction formula $\sin(-\theta) = -\sin\theta$, you can write

$$\sin\left(-\frac{\pi}{3}\right) = -\sin\frac{\pi}{3} = -\frac{\sqrt{3}}{2}.$$

b. Using the reciprocal identity $\sec\theta = 1/\cos\theta$, you can write

$$\sec 60° = \frac{1}{\cos 60°} = \frac{1}{1/2} = 2.$$

c. Using a calculator, you can obtain

$$\cos(1.2) \approx 0.3624.$$

Remember that 1.2 is given in *radian* measure. Consequently, your calculator must be set in radian mode.

Solving Trigonometric Equations

How would you solve the equation $\sin \theta = 0$? You know that $\theta = 0$ is one solution, but this is not the only solution. Any one of the following values of θ is also a solution.

$$\ldots, -3\pi, -2\pi, -\pi, 0, \pi, 2\pi, 3\pi, \ldots$$

You can write this infinite solution set as $\{n\pi : n \text{ is an integer}\}$.

EXAMPLE 4   Solving a Trigonometric Equation

Solve the equation

$$\sin \theta = -\frac{\sqrt{3}}{2}.$$

Solution   To solve the equation, you should consider that the sine is negative in Quadrants III and IV and that

$$\sin \frac{\pi}{3} = \frac{\sqrt{3}}{2}.$$

Thus, you are seeking values of θ in the third and fourth quadrants that have a reference angle of $\pi/3$. In the interval $[0, 2\pi]$, the two angles fitting these criteria are

$$\theta = \pi + \frac{\pi}{3} = \frac{4\pi}{3} \quad \text{and} \quad \theta = 2\pi - \frac{\pi}{3} = \frac{5\pi}{3}.$$

By adding integer multiples of 2π to each of these solutions, you obtain the following general solution.

$$\theta = \frac{4\pi}{3} + 2n\pi \quad \text{or} \quad \theta = \frac{5\pi}{3} + 2n\pi, \qquad \text{where } n \text{ is an integer}$$

(See Figure 70.)

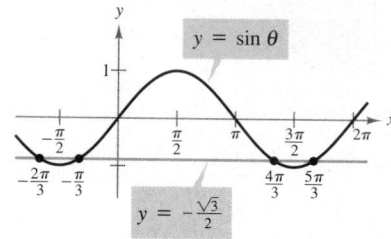

FIGURE 70

Solution points of $\sin \theta = \dfrac{-\sqrt{3}}{2}$.

EXAMPLE 5   Solving a Trigonometric Equation

Solve $\cos 2\theta = 2 - 3 \sin \theta$, where $0 \le \theta \le 2\pi$.

Solution   Using the double angle identity $\cos 2\theta = 1 - 2 \sin^2 \theta$, you can rewrite the equation as follows.

$$\cos 2\theta = 2 - 3 \sin \theta \qquad \text{Given equation}$$
$$1 - 2 \sin^2 \theta = 2 - 3 \sin \theta \qquad \text{Trigonometric identity}$$
$$0 = 2 \sin^2 \theta - 3 \sin \theta + 1 \qquad \text{Quadratic form}$$
$$0 = (2 \sin \theta - 1)(\sin \theta - 1) \qquad \text{Factor}$$

If $2 \sin \theta - 1 = 0$, then $\sin \theta = 1/2$ and $\theta = \pi/6$ or $\theta = 5\pi/6$. If $\sin \theta - 1 = 0$, then $\sin \theta = 1$ and $\theta = \pi/2$. Thus, for $0 \le \theta \le 2\pi$, there are three solutions.

$$\theta = \frac{\pi}{6}, \quad \frac{5\pi}{6}, \quad \text{or} \quad \frac{\pi}{2}$$

Graphs of Trigonometric Functions

A function f is **periodic** if there exists a nonzero number p such that $f(x + p) = f(x)$ for all x in the domain of f. The smallest such positive value of p (if it exists) is the **period** of f. The sine, cosine, secant, and cosecant functions each have a period of 2π, and the other two trigonometric functions have a period of π, as shown in Figure 71.

FIGURE 71
The graphs of the six trigonometric functions.

Note in Figure 71 that the maximum value of $\sin x$ and $\cos x$ is 1 and the minimum value is -1. The graphs of the functions $y = a \sin bx$ and $y = a \cos bx$ oscillate between $-a$ and a, and hence have an **amplitude** of $|a|$. Furthermore, because $bx = 0$ when $x = 0$ and $bx = 2\pi$ when $x = 2\pi/b$, it follows that the functions $y = a \sin bx$ and $y = a \cos bx$ each have a period of $2\pi/|b|$. Table 5 summarizes the amplitudes and periods for some types of trigonometric functions.

TABLE 5

Function	*Period*	*Amplitude*
$y = a \sin bx$ or $y = a \cos bx$	$\dfrac{2\pi}{|b|}$	$|a|$
$y = a \tan bx$ or $y = a \cot bx$	$\dfrac{\pi}{|b|}$	Not applicable
$y = a \sec bx$ or $y = a \csc bx$	$\dfrac{2\pi}{|b|}$	Not applicable

FIGURE 72

EXAMPLE 6   Sketching the Graph of a Trigonometric Function

Sketch the graph of $f(x) = 3\cos 2x$.

Solution   The graph of $f(x) = 3\cos 2x$ has an amplitude of 3 and a period of $2\pi/2 = \pi$. Using the basic shape of the graph of the cosine function, sketch one period of the function on the interval $[0, \pi]$, using the following pattern.

Maximum: $(0, 3)$ Minimum: $\left(\dfrac{\pi}{2}, -3\right)$ Maximum: $[\pi, 3]$

By continuing this pattern, you can sketch several cycles of the graph, as shown in Figure 72.

The discussion of horizontal shifts, vertical shifts, and reflections given in Section 5 can be applied to the graphs of trigonometric functions, as illustrated in Example 7.

EXAMPLE 7   Shifts of Graphs of Trigonometric Functions

Sketch the graph of the following functions.

a. $f(x) = \sin\left(x + \dfrac{\pi}{2}\right)$ **b.** $f(x) = 2 + \sin x$ **c.** $f(x) = 2 + \sin\left(x - \dfrac{\pi}{4}\right)$

Solution

a. To sketch the graph of $f(x) = \sin(x + \pi/2)$, shift the graph of $y = \sin x$ to the left $\pi/2$ units, as shown in Figure 73(a).
b. To sketch the graph of $f(x) = 2 + \sin x$, shift the graph of $y = \sin x$ up 2 units, as shown in Figure 73(b).
c. To sketch the graph of $f(x) = 2 + \sin(x - \pi/4)$, shift the graph of $y = \sin x$ up 2 units and to the right $\pi/4$ units, as shown in Figure 73(c).

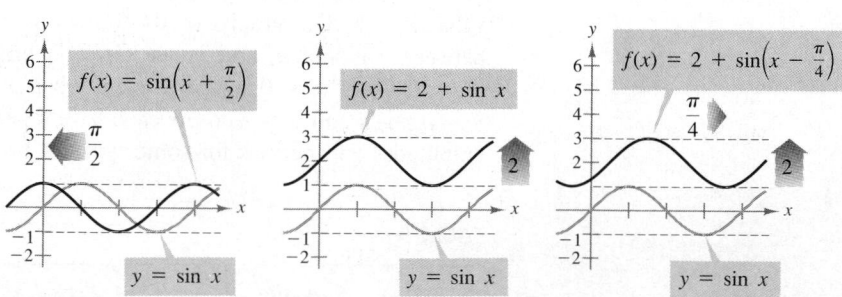

(a) Horizontal shift to the left   (b) Vertical shift upward   (c) Horizontal and vertical shift

FIGURE 73
Transformations of the graph of $y = \sin x$.

EXERCISES for Section 6

In Exercises 1 and 2, determine two coterminal angles (one positive and one negative) for each given angle. Express your answers in degrees.

1. a.

$\theta = 36°$

b.

$\theta = -120°$

2. a.

$\theta = 300°$

b.

$\theta = -420°$

In Exercises 3 and 4, determine two coterminal angles (one positive and one negative) for each given angle. Express your answers in radians.

3. a.

$\theta = \dfrac{\pi}{9}$

b.
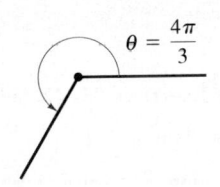
$\theta = \dfrac{4\pi}{3}$

4. a.

$\theta = -\dfrac{9\pi}{4}$

b.

$\theta = \dfrac{8\pi}{9}$

In Exercises 5 and 6, express the angles in radian measure as multiples of π and as decimals accurate to three decimal places.

5. a. 30°   **b.** 150°   **c.** 315°   **d.** 120°

6. a. −20°   **b.** −240°   **c.** −270°   **d.** 144°

In Exercises 7 and 8, express the angles in degree measure.

7. a. $\dfrac{3\pi}{2}$ **b.** $\dfrac{7\pi}{6}$ **c.** $-\dfrac{7\pi}{12}$ **d.** −2.367

8. a. $\dfrac{7\pi}{3}$ **b.** $-\dfrac{11\pi}{30}$ **c.** $\dfrac{11\pi}{6}$ **d.** 0.438

9. Let r represent the radius of a circle, θ the central angle (measured in radians), and s the length of the arc subtended by the angle. Use the relationship $s = r\theta$ to complete the table.

r	8 ft	15 in.	85 cm		
s	12 ft			96 in.	8642 mi
θ		1.6	$\dfrac{3\pi}{4}$	4	$\dfrac{2\pi}{3}$

10. *Instrumentation* The pointer on a voltmeter is 2 inches in length (see figure). Find the angle through which the pointer rotates when it moves $\frac{1}{2}$ inch on the scale.

11. *Electric Hoist* An electric hoist is being used to lift a piece of equipment (see figure). The diameter of the drum on the hoist is 8 inches and the equipment must be raised 1 foot. Find the number of degrees through which the drum must rotate.

FIGURE FOR 10   **FIGURE FOR 11**

12. *Angular Speed* A car is moving at the rate of 50 miles per hour, and the diameter of its wheels is 2.5 feet.
   a. Find the number of revolutions per minute that the wheels are rotating.
   b. Find the angular speed of the wheels in radians per minute.

In Exercises 13 and 14, determine all six trigonometric functions for the angle θ.

13. a.

b.

14. a.

b.
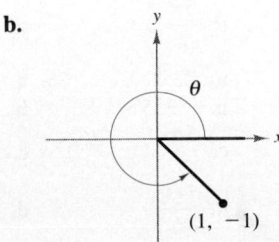

In Exercises 15 and 16, determine the quadrant in which θ lies.

15. a. $\sin \theta < 0$ and $\cos \theta < 0$
    b. $\sec \theta > 0$ and $\cot \theta < 0$

16. a. $\sin \theta > 0$ and $\cos \theta < 0$
    b. $\csc \theta < 0$ and $\tan \theta > 0$

In Exercises 17–22, evaluate the trigonometric function.

17. $\sin \theta = \frac{1}{2}$
    $\cos \theta = ?$

18. $\sin \theta = \frac{1}{3}$
    $\tan \theta = ?$

19. $\cos \theta = \frac{4}{5}$
    $\cot \theta = ?$

20. $\sec \theta = \frac{13}{5}$
    $\csc \theta = ?$

21. $\cot \theta = \frac{15}{8}$
    $\sec \theta = ?$

22. $\tan \theta = \frac{1}{2}$
    $\sin \theta = ?$

In Exercises 23–26, evaluate the sine, cosine, and tangent of each angle *without* using a calculator.

23. a. $60°$
    b. $120°$
    c. $\dfrac{\pi}{4}$
    d. $\dfrac{5\pi}{4}$

24. a. $-30°$
    b. $150°$
    c. $-\dfrac{\pi}{6}$
    d. $\dfrac{\pi}{2}$

25. a. $225°$
    b. $-225°$
    c. $\dfrac{5\pi}{3}$
    d. $\dfrac{11\pi}{6}$

26. a. $750°$
    b. $510°$
    c. $\dfrac{10\pi}{3}$
    d. $\dfrac{17\pi}{3}$

In Exercises 27–30, use a calculator to evaluate the trigonometric functions to four significant digits.

27. a. $\sin 10°$
    b. $\csc 10°$

28. a. $\sec 225°$
    b. $\sec 135°$

29. a. $\tan \dfrac{\pi}{9}$
    b. $\tan \dfrac{10\pi}{9}$

30. a. $\cot(1.35)$
    b. $\tan(1.35)$

In Exercises 31–34, find two solutions of each equation. Express the results in radians ($0 \le \theta < 2\pi$). Do not use a calculator.

31. a. $\cos \theta = \dfrac{\sqrt{2}}{2}$
    b. $\cos \theta = -\dfrac{\sqrt{2}}{2}$

32. a. $\sec \theta = 2$
    b. $\sec \theta = -2$

33. a. $\tan \theta = 1$
    b. $\cot \theta = -\sqrt{3}$

34. a. $\sin \theta = \dfrac{\sqrt{3}}{2}$
    b. $\sin \theta = -\dfrac{\sqrt{3}}{2}$

In Exercises 35–42, solve the equation for θ ($0 \le \theta < 2\pi$).

35. $2 \sin^2 \theta = 1$

36. $\tan^2 \theta = 3$

37. $\tan^2 \theta - \tan \theta = 0$

38. $2 \cos^2 \theta - \cos \theta = 1$

39. $\sec \theta \csc \theta = 2 \csc \theta$

40. $\sin \theta = \cos \theta$

41. $\cos^2 \theta + \sin \theta = 1$

42. $\cos \dfrac{\theta}{2} - \cos \theta = 1$

In Exercises 43–46, solve for x, y, or r as indicated.

43. Solve for y.

44. Solve for x.

45. Solve for x.

46. Solve for r.

47. *Airplane Ascent* An airplane leaves the runway climbing at 18° with a speed of 275 feet per second (see figure). Find the altitude of the plane after 1 minute.

FIGURE FOR 47

48. *River Width* A biologist wants to know the width w of a river in order to set instruments properly to study the pollutants in the water. From point A, the biologist walks downstream 100 feet and sights point C to determine that $\theta = 50°$ (see figure). How wide is the river?

49. *Machine Shop Calculations* A tapered shaft has a diameter of 2 inches at the small end and is 6 inches long (see figure). The taper is 3°. Find the diameter d of the large end of the shaft.

FIGURE FOR 48   **FIGURE FOR 49**

50. *Height of a Mountain* In traveling across flat land, you notice a mountain directly in front of you. Its angle of elevation (to the peak) is 3.5°. After you drive 13 miles closer to the mountain, the angle of elevation is 9°. Approximate the height of the mountain.

FIGURE FOR 50

In Exercises 51–54, determine the period and amplitude of each function.

51. a. $y = 2 \sin 2x$ **b.** $y = \frac{1}{2} \sin \pi x$

52. a. $y = \frac{3}{2} \cos \frac{x}{2}$ **b.** $y = -2 \sin \frac{x}{3}$

53. $y = 3 \sin 4\pi x$ **54.** $y = \frac{2}{3} \cos \frac{\pi x}{10}$

In Exercises 55–58, find the period of the function.

55. $y = 5 \tan 2x$ **56.** $y = 7 \tan 2\pi x$

57. $y = \sec 5x$ **58.** $y = \csc 4x$

C *Essay* **In Exercises 59 and 60, use a graphing utility to graph each function f on the same set of coordinate axes when $c = -2$, $c = -1$, $c = 1$, and $c = 2$. Give a written description of the change in the graph caused by changing c.**

59. a. $f(x) = c \sin x$ **60. a.** $f(x) = \sin x + c$
   b. $f(x) = \cos(cx)$ **b.** $f(x) = -\sin(2\pi x - c)$
   c. $f(x) = \cos(\pi x - c)$ **c.** $f(x) = c \cos x$

In Exercises 61–72, sketch the graph of the function.

61. $y = \sin \frac{x}{2}$ **62.** $y = 2 \cos 2x$

63. $y = -\sin \frac{2\pi x}{3}$ **64.** $y = 2 \tan x$

65. $y = \csc \frac{x}{2}$ **66.** $y = \tan 2x$

67. $y = 2 \sec 2x$ **68.** $y = \csc 2\pi x$

69. $y = \sin(x + \pi)$ **70.** $y = \cos\left(x - \frac{\pi}{3}\right)$

71. $y = 1 + \cos\left(x - \frac{\pi}{2}\right)$

72. $y = 1 + \sin\left(x + \frac{\pi}{2}\right)$

C **73.** Use a graphing utility to graph the functions $f(x) = 2 \cos 2x + 3 \sin 3x$ and $g(x) = 2 \cos 2x + 3 \sin 4x$.
   a. Use the graphs to find the period of each function.
   b. If α and β are positive integers, is the function $h(x) = A \cos \alpha x + B \sin \beta x$ periodic? Explain your answer.

74. Two trigonometric functions f and g have periods of 2, and their graphs intersect when $x = 5.35$.
   a. Give one smaller and one larger positive value of x where the functions have the same value.
   b. Determine one negative value of x where the graphs intersect.
   c. Is it true that $f(13.35) = g(-4.65)$? Give a reason for your answer.

In Exercises 75 and 76, find $a, b,$ and c so that the graph of the function matches the graph in the figure.

75. $y = a\cos(bx - c)$

76. $y = a\sin(bx - c)$

C **77.** *Temperature* The temperature T (°F) in Orlando, Florida for a 24-hour period is listed in the table (time t in hours, with $t = 0$ corresponding to midnight).

t	0	1	2	3	4	5
T	57	56	55	55	57	59

t	6	7	8	9	10	11
T	63	66	70	74	77	81

t	12	13	14	15	16	17
T	83	84	85	84	83	81

t	18	19	20	21	22	23
T	77	74	70	66	63	59

a. Select an appropriate scale and plot the data on a rectangular coordinate system.

b. Estimate the constants a, b, ω, and δ so that the function

$$T(t) = a + b\sin(\omega t - \delta)$$

is a reasonable model for the data.

c. Use a graphing utility to plot the data and the model found in part b.

78. *Ferris Wheel* The model for the height h of a Ferris wheel car is

$$h = 50 + 50\sin 8\pi t$$

where t is measured in minutes. (The Ferris wheel has a radius of 50 feet.) This model yields a height of 50 feet when $t = 0$. Alter the model so that the height of the car is 0 feet when $t = 0$.

79. *Piano Tuning* When tuning a piano, a technician strikes a tuning fork for the A above middle C. This creates a sound (a type of wave motion) that can be approximated by

$$y = 0.001\sin 880\pi t$$

where y is measured in inches and t is the time in seconds.

a. What is the period p of this function?

b. What is the frequency f of this note ($f = 1/p$)?

c. Sketch the graph of this function.

80. *Blood Pressure* The function

$$P = 100 - 20\cos\frac{5\pi t}{3}$$

approximates the blood pressure P in millimeters of mercury at time t in seconds for a person at rest.

a. Find the period of the function.

b. Find the number of heartbeats per minute.

c. Sketch the graph of the pressure function.

C In Exercises 81 and 82, use a graphing utility to compare the graph of f with the given graph. In each case, the graph of f is a Fourier approximation of the given graph. Try to improve the approximation by adding a term to $f(x)$. Use a graphing utility to verify that your new approximation is better than the original. Can you find other terms to add to make the approximation even better? What is the pattern? (In Exercise 81, sine terms can be used to improve the approximation and in Exercise 82, cosine terms can be used.)

81. $f(x) = \dfrac{4}{\pi}\left(\sin\pi x + \dfrac{1}{3}\sin 3\pi x\right)$

FIGURE FOR 81

82. $f(x) = \dfrac{1}{2} - \dfrac{4}{\pi^2}\left(\cos\pi x + \dfrac{1}{9}\cos 3\pi x\right)$

FIGURE FOR 82

C **83.** *Normal Temperatures* The normal monthly high temperature for Erie, Pennsylvania is approximated by

$$H(t) = 54.33 - 20.38\cos\frac{\pi t}{6} - 15.69\sin\frac{\pi t}{6}$$

and the normal monthly low temperature is approximated by

$$L(t) = 39.36 - 15.70\cos\frac{\pi t}{6} - 14.16\sin\frac{\pi t}{6}$$

where t is the time in months with $t = 1$ corresponding to January. (*Source:* NOAA) Use a graphing utility to graph the functions over a period of one year, and use the graphs to answer the following questions.

a. During what part of the year is the difference between the normal high and low temperatures greatest? When is it smallest?

b. The sun is the farthest north in the sky around June 21, but the graph shows the warmest temperatures at a later date. Approximate the lag time of the temperatures relative to the position of the sun.

C **84.** *Sales* Sales S, in thousands of units, of a seasonal product are modeled by

$$S = 58.3 + 32.5 \cos \frac{\pi t}{6}$$

where t is the time in months (with $t = 1$ corresponding to January and $t = 12$ corresponding to December). Use a graphing utility to graph the model for S and determine the months when sales exceed 75,000 units.

85. *Essay* Write a paragraph that relates the slope of a line to the angle the line makes with the x-axis.

86. Let L_1 and L_2 be two nonvertical lines. Use the formula for the tangent of a difference of two angles to derive the following formula for the angle α between L_1 and L_2

$$\tan \alpha = \left| \frac{m_2 - m_1}{1 + m_2 m_1} \right|$$

where m_1 and m_2 are the slopes of L_1 and L_2, respectively. Assume that $m_1 m_2 \neq -1$.

87. Sketch the graphs of $f(x) = \sin x$, $g(x) = |\sin x|$, and $h(x) = \sin(|x|)$. In general, how are the graphs of $|f(x)|$ and $f(|x|)$ related to the graph of f?

C **88.** Use a graphing utility to graph $y = x$ and $y = \cos x$ to convince yourself that there exists an x between 0 and 1 for which $x = \cos x$. Starting with $x_0 = 1$, generate a sequence x_1, x_2, x_3 using the formula $x_n = \cos(x_{n-1})$. Explain why the sequence is converging to the solution of $x = \cos x$.

C **89.** *Musical Sound* A sound wave is modeled by

$$p(t) = \frac{1}{4\pi} \left(p_1(t) + 30 p_2(t) + p_3(t) + p_5(t) + 30 p_6(t) \right)$$

where $p_n(t) = (\sin 524 \, n\pi t)/n$. The following graphs represent the function p and each of its sine components.

a. What is the value of $p(t)$ when $t = 0.0025$?
b. Use a graphing utility to verify that the graph of p is as represented in the right column.
c. Is the function p periodic? If so, what is its period?
d. The frequency of a sound wave is the reciprocal of the period. What is the frequency of this sound?
e. The sound whose wave is given by p sounds much like "high C," the frequency of which is 524. Why?

REVIEW EXERCISES for Prerequisites Chapter

In Exercises 1–4, sketch the interval(s) defined by the inequality.

1. $|x - 2| \leq 3$

2. $|3x - 2| \leq 0$

3. $4 < (x + 3)^2$

4. $\dfrac{1}{|x|} < 1$

In Exercises 5 and 6, find the midpoint of the interval.

5. $\left[\frac{7}{8}, \frac{10}{4} \right]$

6. $\left[-1, \frac{3}{2} \right]$

7. Find the midpoints of the sides of the triangle whose vertices are $(1, 4)$, $(-3, 2)$, and $(5, 0)$.

8. Find the vertices of the triangle whose sides have midpoints $(0, 2)$, $(1, -1)$, and $(2, 1)$.

In Exercises 9–12, determine the radius and center of the circle and sketch its graph.

9. $x^2 + y^2 + 6x - 2y + 1 = 0$

10. $4x^2 + 4y^2 - 4x + 8y = 11$

11. $x^2 + y^2 + 6x - 2y + 10 = 0$

12. $x^2 - 6x + y^2 + 8y = 0$

13. Determine the value of c so that the circle has a radius of 2.

$$x^2 - 6x + y^2 + 8y = c$$

14. Find an equation in x and y such that the distance between (x, y) and $(-2, 0)$ is twice the distance between (x, y) and $(3, 1)$.

15. Find an equation for the circle whose center is $(1, 2)$ and whose radius is 3. Then determine whether the points are inside, outside, or on the circle.
    a. $(1, 5)$ b. $(0, 0)$
    c. $(-2, 1)$ d. $(0, 4)$

16. Find an equation for the circle whose center is $(2, 1)$ and whose radius is 2. Then determine whether the points are inside, outside, or on the circle.
    a. $(1, 1)$ b. $(4, 2)$
    c. $(0, 1)$ d. $(3, 1)$

In Exercises 17 and 18, use the concept of slope to find t so that the three points are collinear.

17. $(-2, 5), (0, t), (1, 1)$

18. $(-3, 3), (t, -1) (8, 6)$

In Exercises 19–26, sketch the graph of the equation.

19. $y = \dfrac{-x + 3}{2}$ 20. $4x - 2y = 6$

21. $-\frac{1}{3}x + \frac{5}{6}y = 1$ 22. $0.02x + 0.15y = 0.25$

23. $y = 7 - 6x - x^2$ 24. $y = 6x - x^2$

25. $y = \sqrt{5 - x}$ 26. $y = |x - 4| - 4$

27. Find equations of the lines passing through $(-2, 4)$ and having the indicated characteristics.
    a. Slope of $\frac{7}{16}$
    b. Parallel to the line $5x - 3y = 3$
    c. Passing through the origin
    d. Parallel to the y-axis

28. Find equations of the lines passing through $(1, 3)$ and having the indicated characteristics.
    a. Slope of $-\frac{2}{3}$
    b. Perpendicular to the line $x + y = 0$
    c. Passing through the point $(2, 4)$
    d. Parallel to the x-axis

29. The midpoint of a line segment is $(-1, 4)$. If one end of the line segment is $(2, 3)$, find the other end.

30. Find the point that is equidistant from $(0, 0)$, $(2, 3)$, and $(3, -2)$.

In Exercises 31 and 32, find the point(s) of intersection of the graphs of the equations.

31. $3x - 4y = 8$ 32. $x - y + 1 = 0$
    $x + y = 5$ $y - x^2 = 7$

33. *Air Travel* Consider a plane flying at a constant rate on a direct route between two cities. The distance s (in miles) traveled in t hours is $s = 560t$. What information does the slope of this line convey?

34. *Break-Even Analysis* A contractor purchases a piece of equipment for \$36,500 that costs an average of \$9.25 per hour for fuel and maintenance. The equipment operator is paid \$13.50 per hour, and customers are charged \$30 per hour.
    a. Write an equation for the cost C of operating this equipment for t hours.
    b. Write an equation for the revenue R derived from t hours of use.
    c. Find the break-even point for this equipment by finding the time at which $R = C$.

In Exercises 35–38, find a formula for the function and find its domain.

35. *Dimensions of a Rectangle* A wire 24 inches long is to be cut into four pieces to form a rectangle whose shortest side has a length of x. The area A of the rectangle is a function of x.

36. *Area* The area A of an equilateral triangle is a function of the length of one of its sides x.

37. *Distance* The distance d traveled by a car at a speed of 45 miles per hour is a function of the time traveled t.

38. *Real Estate* The value v of a farm at \$1250 per acre, with buildings, livestock, and equipment worth \$500,000, is a function of the number of acres a.

In Exercises 39–42, sketch the graph of the equation and use the vertical line test to determine whether the equation expresses y as a function of x.

39. $x - y^2 = 0$ 40. $x^2 - y = 0$

41. $y = x^2 - 2x$ 42. $x = 9 - y^2$

43. For each function f, sketch (on the same set of coordinate axes) a graph of f for $c = -2, 0$, and 2.
    a. $f(x) = x^3 + c$ b. $f(x) = (x - c)^3$
    c. $f(x) = (x - 2)^3 + c$ d. $f(x) = cx^3$

C 44. *Essay* Use a graphing utility to graph the functions f, g, and h in the same viewing rectangle. Write a description of any similarities and differences you observe among the graphs.
    a. $f(x) = x^2, g(x) = x^4, h(x) = x^6$
    b. $f(x) = x, g(x) = x^3, h(x) = x^5$

45. Use the results of Exercise 44 to sketch the graphs of f and g *without* the aid of a graphing utility.
    a. $f(x) = x^7, g(x) = x^8$
    b. $f(x) = (x - 3)^3, g(x) = (x + 1)^3$

C **46.** Use the results of Exercises 44 and 45 to guess the shape of the graphs of the functions f, g, and h. Then use a graphing utility to graph each function and compare the result with your guess.
 a. $f(x) = x^2(x - 6)^2$ **b.** $f(x) = x^3(x - 6)^2$
 c. $f(x) = x^3(x - 6)^3$

47. Given $f(x) = 1 - x^2$ and $g(x) = 2x + 1$, find
 a. $f(x) - g(x)$ **b.** $f(x)g(x)$ **c.** $g(f(x))$

48. Given $f(x) = 2x - 3$ and $g(x) = \sqrt{x + 1}$, find
 a. $f(x) + g(x)$ **b.** $f(x)/g(x)$ **c.** $f(g(x))$

49. What is the minimum degree of a polynomial function whose graph approximates the given graph? What sign must the leading coefficient have?

(a)

(b)
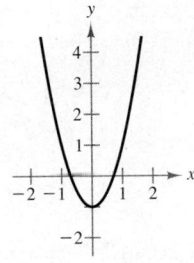

50. *Essay* The following graphs give the profit P for a small company over a period of 2 years. Create a story to describe the behavior of each profit function for some hypothetical product the company produces.

(a)

(b)

51. *Distance Between Towns* A passenger in an airplane flying at 35,000 feet sees two towns directly to the left of the airplane. The angles of depression to the towns are 32° and 76° (see figure). How far apart are the towns?

FIGURE FOR 51

52. *Machine Shop Calculations* A steel plate has the form of one-fourth of a circle with radius 24 inches. Two $\frac{3}{8}$-inch holes are to be drilled in the plate positioned as shown in the figure. Find the coordinates of the center of each hole.

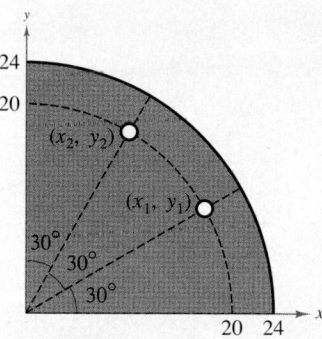

FIGURE FOR 52

In Exercises 53–60, sketch a graph showing two periods of the function.

53. $f(x) = 2 \sin \dfrac{2x}{3}$ **54.** $f(x) = \dfrac{1}{2} \cos \dfrac{x}{3}$

55. $f(x) = \cos\left(2x - \dfrac{\pi}{3}\right)$ **56.** $f(x) = -\sin\left(2x + \dfrac{\pi}{2}\right)$

57. $f(x) = \tan \dfrac{x}{2}$ **58.** $f(x) = \csc 2x$

59. $f(x) = \sec\left(x - \dfrac{\pi}{4}\right)$ **60.** $f(x) = \cot 3x$

61. *Seasonal Sales* The monthly sales S (in thousands of units) of a seasonal product are approximated by

$$S = 74.50 + 43.75 \sin \dfrac{\pi t}{6}$$

where t is the time in months with $t = 1$ corresponding to January. Sketch the graph of this sales function over 1 year.

62. *Truss Members* Find the length of each unknown member of the truss shown in the figure.

FIGURE FOR 62

Chapter 1

Limits and Their Properties

1.1

The Tangent Line Problem • An Introduction to Limits • Limits that Fail to Exist • A Formal Definition of a Limit

The Tangent Line Problem

The notion of a limit is *fundamental* to the study of calculus. To give you some idea of the way limits are used, we begin this section with a brief introduction to a classic problem in calculus—*the tangent line problem*. In this problem, you are given a function f and a point P on its graph, and are asked to find an equation of the tangent to the graph at point P, as shown in Figure 1.1.

Except for cases involving a vertical tangent line, the problem of finding the **tangent line** at a point P is equivalent to finding the *slope* of the tangent line at P. You can approximate this slope by using a line through the point of tangency and a second point on the curve, as shown in Figure 1.2(a). Such a line is called a **secant line.** If $P = (c, f(c))$ is the point of tangency and

$$Q = (c + \Delta x, f(c + \Delta x))$$

is a second point on the graph of f, then the slope of the secant line through these two points is given by

$$m_{\text{sec}} = \frac{f(c + \Delta x) - f(c)}{c + \Delta x - c} = \frac{f(c + \Delta x) - f(c)}{\Delta x}.$$

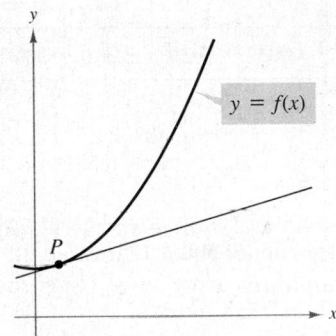

FIGURE 1.1
The tangent line to the graph of f at P.

(a) The secant line through
$(c, f(c))$ and $(c + \Delta x, f(c + \Delta x))$

(b) As Q approaches P, the secant lines approach the tangent line.

FIGURE 1.2

As point Q approaches point P, the slope of the secant line will approach the slope of the tangent line (see Figure 1.2(b)). When such a "limiting position" exists, the slope of the tangent line is said to be the **limit** of the slope of the secant line. (Much more will be said about this important problem in Chapter 2.)

◀ *The velocity of an object dropped by a construction worker is discussed in Section 1.3, Exercises 51 and 52.*

An Introduction to Limits

Suppose you are asked to sketch the graph of the function f given by

$$f(x) = \frac{x^3 - 1}{x - 1}, \qquad x \neq 1.$$

For all values other than $x = 1$, you can use standard curve-sketching techniques. However, at $x = 1$, it is not clear what to expect. To get an idea of the behavior of the graph of f near $x = 1$, you can use two sets of x-values—one set that approaches 1 from the left and one set that approaches 1 from the right, as shown in Table 1.1.

TABLE 1.1

	x approaches 1 from the left.					x approaches 1 from the right.			
x	0.75	0.9	0.99	0.999	1	1.001	1.01	1.1	1.25
$f(x)$	2.313	2.710	2.970	2.997	?	3.003	3.030	3.310	3.813

	$f(x)$ approaches 3.			$f(x)$ approaches 3.		

When you plot these points, it appears that the graph of f is a parabola that has a gap at point $(1, 3)$, as shown in Figure 1.3. Although x cannot equal 1, you can move arbitrarily close to 1, and as a result $f(x)$ moves arbitrarily close to 3. Using limit notation, you can write

$$\lim_{x \to 1} f(x) = 3,$$

read as "*the limit of $f(x)$ as x approaches 1 is 3.*"

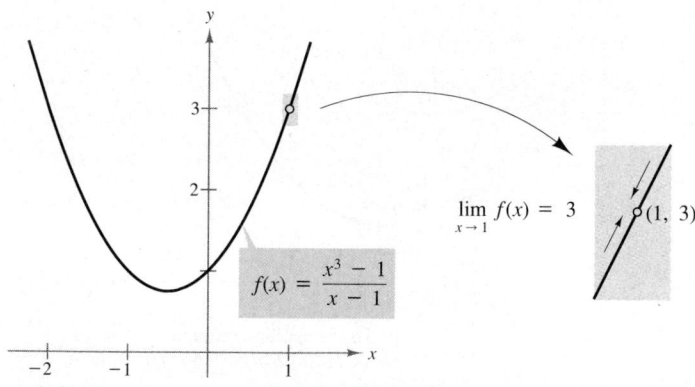

FIGURE 1.3
The limit of $f(x)$ as x approaches 1 is 3.

REMARK For many functions, you can estimate the value of the limit L by using a calculator and evaluating the function at several points near c, as was done in Table 1.1.

This discussion leads to an informal description of a limit. If $f(x)$ becomes arbitrarily close to a single number L as x approaches c from either side, then the **limit** of $f(x)$, as x approaches c, is L. This limit is written as

$$\lim_{x \to c} f(x) = L.$$

EXAMPLE 1　Using a Calculator to Estimate a Limit

Evaluate the function

$$f(x) = \frac{x}{\sqrt{x+1} - 1}$$

at several points near $x = 0$ and use the result to estimate the limit

$$\lim_{x \to 0} \frac{x}{\sqrt{x+1} - 1}.$$

Solution　Table 1.2 lists the values of $f(x)$ for several x-values near 0.

TABLE 1.2

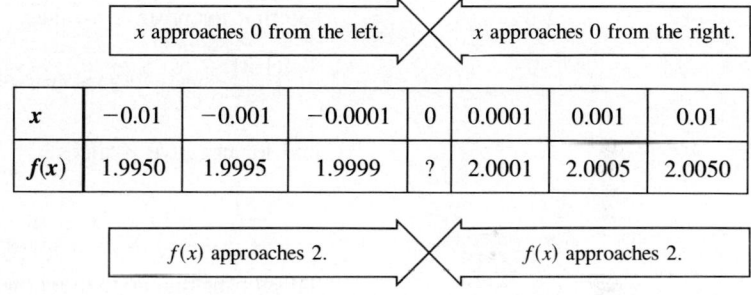

	x approaches 0 from the left.				x approaches 0 from the right.		
x	-0.01	-0.001	-0.0001	0	0.0001	0.001	0.01
$f(x)$	1.9950	1.9995	1.9999	?	2.0001	2.0005	2.0050

$f(x)$ approaches 2.		$f(x)$ approaches 2.

From the results shown in this table, you can estimate the limit to be 2. This limit is reinforced by the graph of f (see Figure 1.4).

When studying calculus (or any branch of mathematics), remember that there are several ways to solve most problems. Often, a problem can be solved *numerically, graphically,* or *analytically.* For instance, the limit in Example 1 was found numerically (by constructing a table) and graphically. In Section 1.3, you will study an analytic technique for evaluating limits. If you have a graphing utility, try using it to estimate the following limit. Then construct a table to reinforce your conclusion.

$$\lim_{x \to 0} \frac{\sqrt{2x+1} - 1}{x}$$

In Example 1, note that the function is undefined at $x = 0$ and yet $f(x)$ appears to be approaching a limit as x approaches 0. This often happens, and it is important to realize that *the existence or nonexistence of $f(x)$ at $x = c$ has no bearing on the existence of the limit of $f(x)$ as x approaches c.* This is further demonstrated in the next example.

EXAMPLE 2　Finding a Limit

Find the limit of $f(x)$ as x approaches 2 where f is defined by

$$f(x) = \begin{cases} 1, & x \neq 2 \\ 0, & x = 2. \end{cases}$$

Solution　Because $f(x) = 1$ for all x other than $x = 2$, you can conclude that the limit is 1, as shown in Figure 1.5. The fact that $f(2) = 0$ has no bearing on the existence or value of the limit as x approaches 2.

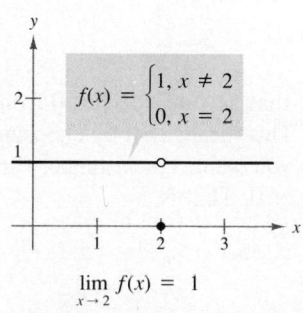

$$\lim_{x \to 0} \frac{x}{\sqrt{x+1} - 1} = 2$$

FIGURE 1.4
The limit of $f(x)$ as x approaches 0 is 2.

$$\lim_{x \to 2} f(x) = 1$$

FIGURE 1.5
The limit of $f(x)$ as x approaches 2 is 1.

Limits that Fail to Exist

In the next three examples, you will examine some functions for which limits do not exist.

EXAMPLE 3   Behavior that Differs from the Right and the Left

Show that the following limit does not exist.

$$\lim_{x \to 0} \frac{|x|}{x}$$

Solution   Consider the graph of the function $f(x) = |x|/x$. In Figure 1.6, you can see that for positive x-values

$$\frac{|x|}{x} = 1, \qquad x > 0$$

and for negative x-values

$$\frac{|x|}{x} = -1, \qquad x < 0.$$

This means that no matter how close x gets to 0, there will be both positive and negative x-values that yield $f(x) = 1$ and $f(x) = -1$. Specifically, if δ (the lowercase Greek letter *delta*) is a positive number, then for x-values satisfying the inequality $0 < |x| < \delta$, you can classify the values of $|x|/x$ as follows.

$(-\delta, 0)$ $(0, \delta)$

| Negative x-values yield $|x|/x = -1$. | Positive x-values yield $|x|/x = 1$. |

This implies that the limit does not exist. ■

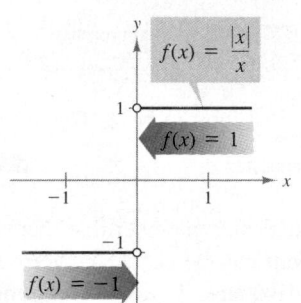

FIGURE 1.6
$\lim_{x \to 0} f(x)$ does not exist.

EXAMPLE 4   Unbounded Behavior

Discuss the existence of the limit

$$\lim_{x \to 0} \frac{1}{x^2}.$$

Solution   Let $f(x) = 1/x^2$. In Figure 1.7, you can see that as x approaches 0 from either the right or the left, $f(x)$ increases without bound. This means that by choosing x close enough to 0, you can force $f(x)$ to be as large as you want. For instance, $f(x)$ will be larger than 100 if you choose x that is within $\frac{1}{10}$ of 0. That is,

$$0 < |x| < \frac{1}{10} \quad \rightarrow \quad f(x) = \frac{1}{x^2} > 100.$$

Similarly, you can force $f(x)$ to be larger than 1,000,000, as follows.

$$0 < |x| < \frac{1}{1000} \quad \rightarrow \quad f(x) = \frac{1}{x^2} > 1{,}000{,}000$$

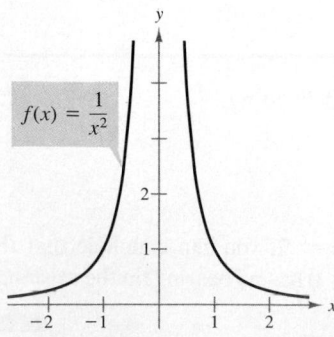

FIGURE 1.7
$\lim_{x \to 0} f(x)$ does not exist.

Because $f(x)$ is not approaching a real number L as x approaches 0, you can conclude that the limit does not exist. ■

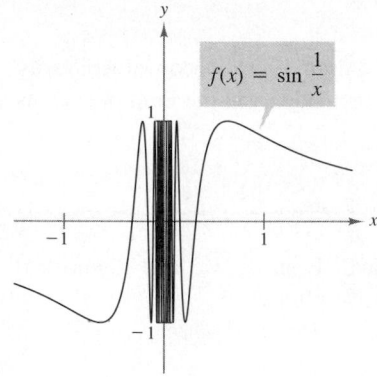

FIGURE 1.8
$\lim\limits_{x \to 0} f(x)$ does not exist.

EXAMPLE 5   Oscillating Behavior

Discuss the existence of the limit $\lim\limits_{x \to 0} \sin\left(\dfrac{1}{x}\right)$.

Solution   Let $f(x) = \sin(1/x)$. In Figure 1.8, you can see that as x approaches 0, $f(x)$ oscillates between -1 and 1. Therefore, the limit does not exist, because no matter how small you choose δ, it is possible to choose x_1 and x_2 within δ units of 0 such that $\sin(1/x_1) = 1$ and $\sin(1/x_2) = -1$ as indicated in Table 1.3.

TABLE 1.3

x	$\dfrac{2}{\pi}$	$\dfrac{2}{3\pi}$	$\dfrac{2}{5\pi}$	$\dfrac{2}{7\pi}$	$\dfrac{2}{9\pi}$	$\dfrac{2}{11\pi}$	$x \to 0$
$\sin\left(\dfrac{1}{x}\right)$	1	-1	1	-1	1	-1	Limit does not exist.

Examples 3, 4, and 5 show three of the most common types of behavior associated with the nonexistence of a limit.

1. $f(x)$ approaches a different number from the right side of c than it approaches from the left side.
2. $f(x)$ increases or decreases without bound as x approaches c.
3. $f(x)$ oscillates between two fixed values as x approaches c.

There are many other interesting functions that have unusual limit behavior. An often cited one is the *Dirichlet function*

$$f(x) = \begin{cases} 0, & \text{if } x \text{ is rational} \\ 1, & \text{if } x \text{ is irrational.} \end{cases}$$

This function has *no limit* at any real number c.

TECHNOLOGY   You can use a graphing utility to investigate the behavior of a function near the x-value at which you are trying to evaluate a limit. When you do this, however, you should realize that you can't always trust the pictures that graphing utilities draw. For instance, if you use a graphing utility to sketch the graph of the function in Example 5 over an interval containing 0, you will most likely obtain an incorrect graph—such as that shown in Figure 1.9. The reason that a graphing utility can't show the correct graph is that the graph has infinitely many oscillations over any interval that contains 0.

PETER GUSTAV DIRICHLET

In the early development of calculus, the definition of a function was much more restricted than it is today, and "functions" such as the Dirichlet function would not have been considered. The modern definition of a function was given by the German mathematician Peter Gustav Dirichlet (1805–1859).

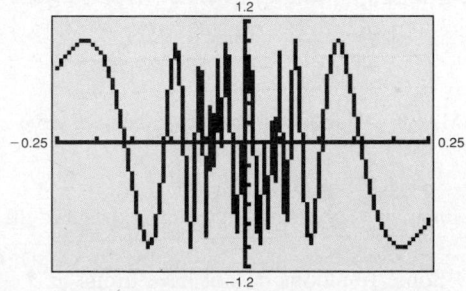

FIGURE 1.9
Incorrect graph of $f(x) = \sin(1/x)$.

AUGUSTIN-LOUIS CAUCHY

Cauchy began his mathematical work at age 22. During his brilliant career he wrote almost 800 papers on subjects ranging from mechanics to astronomy.

FOR FURTHER INFORMATION
For more on the introduction of rigor to calculus, see "Who Gave You the Epsilon? Cauchy and the Origins of Rigorous Calculus," by Judith V. Grabiner, in the March, 1983 Issue of the *American Mathematical Monthly*.

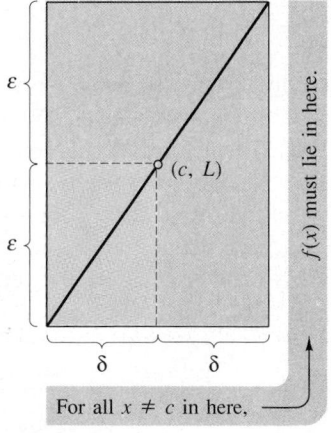

FIGURE 1.10
The ϵ–δ definition of a limit: $\lim\limits_{x \to c} f(x) = L$.

A Formal Definition of a Limit

Let's take another look at the informal description of a limit. If $f(x)$ becomes arbitrarily close to a single number L as x approaches c from either side, then the limit of $f(x)$, as x approaches c, is L, written as

$$\lim_{x \to c} f(x) = L.$$

At first glance, this description looks fairly technical. Even so, we call it informal because we have yet to give exact meanings to the two phrases

"$f(x)$ becomes arbitrarily close to L"

and

"x approaches c."

The first person to assign mathematically rigorous meanings to these two phrases was Augustin-Louis Cauchy (1789–1857). His **ε-δ definition of a limit** is the standard used today.

In Figure 1.10, let ε represent a (small) positive number. Then the phrase "$f(x)$ becomes arbitrarily close to L" means that $f(x)$ lies in the interval $(L - \varepsilon, L + \varepsilon)$. Using absolute value, you can write this as

$$| f(x) - L | < \varepsilon.$$

Similarly, the phrase "x approaches c" means that there exists a positive number δ such that x lies in either the interval $(c - \delta, c)$ or the interval $(c, c + \delta)$. This fact can be concisely expressed by the double inequality

$$0 < | x - c | < \delta.$$

The first inequality, $0 < | x - c |$, expresses the fact that $x \neq c$, whereas the second, $| x - c | < \delta$, says that x is within a distance δ of c.

Definition of a Limit

Let f be a function defined on an open interval containing c (except possibly at c) and let L be a real number. The statement

$$\lim_{x \to c} f(x) = L$$

means that for each $\varepsilon > 0$ there exists a $\delta > 0$ such that if

$$0 < | x - c | < \delta, \quad \text{then} \quad | f(x) - L | < \varepsilon.$$

REMARK   Throughout this text, when we write

$$\lim_{x \to c} f(x) = L$$

we imply two statements—the limit **exists** *and* the limit is L.

Some functions do not have limits as $x \to c$, but those that do cannot have two different limits as $x \to c$. That is, *if the limit of a function exists, then it is unique* (see Exercise 38).

FIGURE 1.11
The limit of $f(x)$ as x approaches 3 is 1.

The next two examples should help you develop a better understanding of the ε-δ definition of a limit.

EXAMPLE 6   Finding a δ for a Given ε

Given the limit

$$\lim_{x \to 3}(2x - 5) = 1$$

find δ such that $|(2x - 5) - 1| < 0.01$ whenever $0 < |x - 3| < \delta$.

Solution   In this problem, you are working with a given value of ε, namely $\varepsilon = 0.01$. To find an appropriate δ, notice that

$$|(2x - 5) - 1| = |2x - 6| = 2|x - 3|.$$

Because the inequality $|(2x - 5) - 1| < 0.01$ is equivalent to $2|x - 3| < 0.01$, you can choose $\delta = \frac{1}{2}(0.01) = 0.005$. This choice works because

$$0 < |x - 3| < 0.005$$

implies that

$$|(2x - 5) - 1| = 2|x - 3| < 2(0.005) = 0.01$$

as shown in Figure 1.11.

REMARK   Note in Example 6 that 0.005 is the *largest* value of δ that will guarantee $|(2x - 5) - 1| < 0.01$ whenever $0 < |x - 3| < \delta$. Any *smaller* positive value of δ would, of course, also work.

In Example 6, you found a δ-value for a *given* ε. This does not prove the existence of the limit. To do that, you must prove that you can find a δ for *any* ε, as demonstrated in the next example.

EXAMPLE 7   Using the ε-δ Definition of a Limit

Use the ε-δ definition of a limit to prove that

$$\lim_{x \to 2}(3x - 2) = 4.$$

Solution   You must show that for each $\varepsilon > 0$, there exists a $\delta > 0$ such that $|(3x - 2) - 4| < \varepsilon$ whenever $0 < |x - 2| < \delta$. Because your choice for δ depends on ε, you need to establish a connection between the absolute values $|(3x - 2) - 4|$ and $|x - 2|$.

$$|(3x - 2) - 4| = |3x - 6| = 3|x - 2|$$

Thus, you can choose $\delta = \varepsilon/3$. This choice works because

$$0 < |x - 2| < \delta = \frac{\varepsilon}{3}$$

implies that

$$|(3x - 2) - 4| = 3|x - 2| < 3\left(\frac{\varepsilon}{3}\right) = \varepsilon$$

FIGURE 1.12
The limit of $f(x)$ as x approaches 2 is 4.

as shown in Figure 1.12.

FIGURE 1.13
The limit of $f(x)$ as x approaches 2 is 4.

EXAMPLE 8   Using the ε-δ Definition of a Limit

Use the ε-δ definition of a limit to prove that

$$\lim_{x \to 2} x^2 = 4.$$

Solution   You must show that for each $\varepsilon > 0$, there exists a $\delta > 0$ such that

$$|x^2 - 4| < \varepsilon \quad \text{when} \quad 0 < |x - 2| < \delta.$$

To find an appropriate δ, begin by writing $|x^2 - 4| = |x - 2||x + 2|$. For all x in the interval $(1, 3)$, you know that $|x + 2| < 5$. Thus, letting δ be the minimum of $\varepsilon/5$ and 1, it follows that whenever $0 < |x - 2| < \delta$, you have

$$|x^2 - 4| = |x - 2||x + 2| < \left(\frac{\varepsilon}{5}\right)(5) = \varepsilon$$

as shown in Figure 1.13.

Throughout this chapter you will use the ε-δ definition of a limit primarily to prove theorems about limits and to establish the existence or nonexistence of particular types of limits. For *finding* limits, you will learn techniques that are easier to use than the ε-δ definition of a limit.

EXERCISES for Section 1.1

TECHNOLOGY
Laboratory Guide
Lab 1.1

In Exercises 1–8, complete the table and use the result to estimate the limit.

1. $\lim\limits_{x \to 2} \dfrac{x - 2}{x^2 - x - 2}$

x	1.9	1.99	1.999	2.001	2.01	2.1
$f(x)$						

2. $\lim\limits_{x \to 2} \dfrac{x - 2}{x^2 - 4}$

x	1.9	1.99	1.999	2.001	2.01	2.1
$f(x)$						

3. $\lim\limits_{x \to 0} \dfrac{\sqrt{x + 3} - \sqrt{3}}{x}$

x	-0.1	-0.01	-0.001	0.001	0.01	0.1
$f(x)$						

4. $\lim\limits_{x \to -3} \dfrac{\sqrt{1 - x} - 2}{x + 3}$

x	-3.1	-3.01	-3.001	-2.999	-2.99	-2.9
$f(x)$						

5. $\lim\limits_{x \to 3} \dfrac{[1/(x + 1)] - (1/4)}{x - 3}$

x	2.9	2.99	2.999	3.001	3.01	3.1
$f(x)$						

6. $\lim\limits_{x \to 4} \dfrac{[x/(x + 1)] - (4/5)}{x - 4}$

x	3.9	3.99	3.999	4.001	4.01	4.1
$f(x)$						

7. $\lim\limits_{x \to 0} \dfrac{\sin x}{x}$

x	-0.1	-0.01	-0.001	0.001	0.01	0.1
$f(x)$						

8. $\lim\limits_{x \to 0} \dfrac{\cos x - 1}{x}$

x	-0.1	-0.01	-0.001	0.001	0.01	0.1
$f(x)$						

In Exercises 9–18, use the graph to find the limit (if it exists).

9. $\lim_{x \to 3} (4 - x)$

10. $\lim_{x \to 1} (x^2 + 2)$

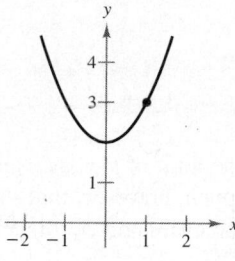

11. $\lim_{x \to 2} f(x)$

$$f(x) = \begin{cases} 4 - x, & x \neq 2 \\ 0, & x = 2 \end{cases}$$

12. $\lim_{x \to 1} f(x)$

$$f(x) = \begin{cases} x^2 + 2, & x \neq 1 \\ 1, & x = 1 \end{cases}$$

13. $\lim_{x \to 5} \dfrac{|x - 5|}{x - 5}$

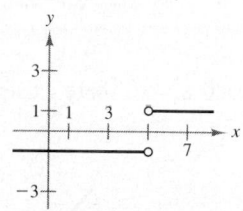

14. $\lim_{x \to 3} \dfrac{1}{x - 3}$

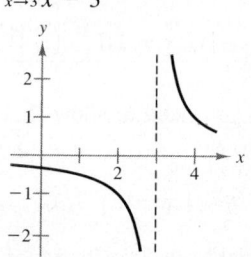

15. $\lim_{x \to \pi/2} \tan x$

16. $\lim_{x \to 0} \sec x$

17. $\lim_{x \to 0} \cos\left(\dfrac{1}{x}\right)$

18. $\lim_{x \to 1} \sin \pi x$

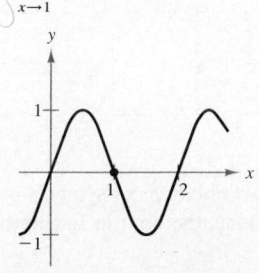

In Exercises 19–22, find the limit L. Then find $\delta > 0$ such that $|f(x) - L| < 0.01$ whenever $0 < |x - c| < \delta$.

19. $\lim_{x \to 2} (3x + 2)$

20. $\lim_{x \to 4} \left(4 - \dfrac{x}{2}\right)$

21. $\lim_{x \to 2} (x^2 - 3)$

22. $\lim_{x \to 5} (x^2 + 4)$

In Exercises 23–30, find the limit L. Then use the definition of a limit to prove that the limit is L.

23. $\lim_{x \to 2} (x + 3)$

24. $\lim_{x \to -3} (2x + 5)$

25. $\lim_{x \to 6} 3$

26. $\lim_{x \to 2} (-1)$

27. $\lim_{x \to 0} \sqrt[3]{x}$

28. $\lim_{x \to 3} |x - 3|$

29. $\lim_{x \to 1} (x^2 + 1)$

30. $\lim_{x \to -3} (x^2 + 3x)$

True or False   **In Exercises 31–34, determine whether the statement is true or false. If it is false, explain why or give an example that shows it is false.**

31. If f is undefined at $x = c$, then the limit of $f(x)$ as $x \to c$ does not exist.

32. If $\lim_{x \to c} f(x) = 0$, then there must exist a number d such that $f(d) < 0.001$.

33. If $f(c) = L$, then $\lim_{x \to c} f(x) = L$.

34. If $\lim_{x \to c} f(x) = L$, then $f(c) = L$.

C In Exercises 35 and 36, use a graphing utility to graph the function f and find $\lim_{x \to 4} f(x)$. Is the graph displayed correct at $x = 4$? Explain why or why not.

35. $f(x) = \dfrac{\sqrt{x + 5} - 3}{x - 4}$

36. $f(x) = \dfrac{x - 4}{x^2 - 5x + 4}$

C **37.** Write a computer program or use a spreadsheet to approximate $\lim_{x \to 4} f(x)$ where

$$f(x) = \dfrac{x^2 - x - 12}{x - 4}.$$

[*Hint*: Generate two columns whose entries form the ordered pairs $(4 \pm [0.1]^n, f(4 \pm [0.1]^n))$ for $n = 1, 2, 3,$ and 4.]

38. Prove that if the limit of $f(x)$ as $x \to c$ exists, then the limit must be unique. [*Hint:* Let $\lim_{x \to c} f(x) = L_1$ and $\lim_{x \to c} f(x) = L_2$, and prove that $L_1 = L_2$.]

39. Let $f(x) = mx + b$, $m \neq 0$, be a line. Use the ε-δ definition of the limit to prove that $\lim_{x \to c} f(x) = mc + b$.

40. Prove that $\lim_{x \to c} f(x) = L$ is equivalent to $\lim_{x \to c} [f(x) - L] = 0$.

41. Consider the function $f(x) = (1 + x)^{1/x}$. Estimate the limit $\lim_{x \to 0} (1 + x)^{1/x}$ by evaluating f at x-values near 0. Sketch the graph of f.

* **C** indicates an exercise in which you are instructed to use graphing technology or a symbolic computer algebra system. The solutions of other exercises may also be facilitated by appropriate technology.

SECTION **Properties of Limits**

1.2   **Limits of Algebraic Functions • Limits of Trigonometric Functions**

Limits of Algebraic Functions

In Section 1.1, you learned that the limit of $f(x)$ as x approaches c does not depend on the value of f at $x = c$. It may happen, however, that the limit is precisely $f(c)$. In such cases, we say that the limit can be evaluated by **direct substitution.** That is,

$$\lim_{x \to c} f(x) = f(c).$$ Substitute c for x

Such *well-behaved* functions are **continuous at c** —you will examine this concept more closely in Section 1.4. We devote the remainder of this section to the study of limits that can be evaluated by direct substitution. In the next section, we will discuss techniques for evaluating limits for which direct substitution fails.

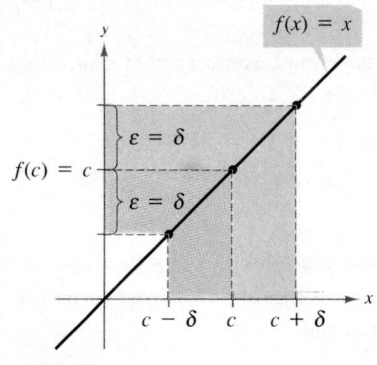

FIGURE 1.14
$\lim_{x \to c} x = c$

THEOREM 1.1   Some Basic Limits

Let b and c be real numbers and let n be a positive integer.

1. $\lim_{x \to c} b = b$ **2.** $\lim_{x \to c} x = c$ **3.** $\lim_{x \to c} x^n = c^n$

Proof   To prove Property 2, you need to show that for each $\varepsilon > 0$ there exists a $\delta > 0$ such that

$$|x - c| < \varepsilon \quad \text{whenever} \quad 0 < |x - c| < \delta.$$

Because the inequality on the right is a stricter version of the one on the left, you can simply choose $\delta = \varepsilon$, as shown in Figure 1.14. The proofs of the other two properties are left as exercises. (See Exercises 37 and 38.)

EXAMPLE 1   Evaluating Basic Limits

a. $\lim_{x \to 2} 3 = 3$

b. $\lim_{x \to -4} x = -4$

c. $\lim_{x \to 2} x^2 = 2^2 = 4$

REMARK   When you encounter new notations or symbols in mathematics, be sure you know how the notations are read. For instance, the limit in Example 1c is read as "the limit of x^2 as x approaches 2, is 4."

By combining Theorem 1.1 with the following theorem, you can find limits for a wide variety of algebraic functions.

THEOREM 1.2   Properties of Limits

Let b and c be real numbers, let n be a positive integer, and let f and g be functions with the following limits.

$$\lim_{x \to c} f(x) = L \quad \text{and} \quad \lim_{x \to c} g(x) = K$$

1. Scalar multiple: $\displaystyle\lim_{x \to c}[b\, f(x)] = bL$

2. Sum or difference: $\displaystyle\lim_{x \to c}[f(x) \pm g(x)] = L \pm K$

3. Product: $\displaystyle\lim_{x \to c}[f(x)g(x)] = LK$

4. Quotient: $\displaystyle\lim_{x \to c}\frac{f(x)}{g(x)} = \frac{L}{K}, \qquad$ provided $K \neq 0$

5. Power: $\displaystyle\lim_{x \to c}[f(x)]^n = L^n$

Proof   To prove Property 2, choose $\varepsilon > 0$. Because $\varepsilon/2 > 0$, you know that there exists $\delta_1 > 0$ such that $0 < |x - c| < \delta_1$ implies $|f(x) - L| < \varepsilon/2$. You also know that there exists $\delta_2 > 0$ such that $0 < |x - c| < \delta_2$ implies $|g(x) - K| < \varepsilon/2$. Let δ be the smaller of δ_1 and δ_2, then $0 < |x - c| < \delta$ implies that

$$|f(x) - L| < \frac{\varepsilon}{2} \quad \text{and} \quad |g(x) - K| < \frac{\varepsilon}{2}.$$

Thus, you can apply the Triangle Inequality to conclude that

$$|[f(x) + g(x)] - (L + K)| \leq |f(x) - L| + |g(x) - K| < \frac{\varepsilon}{2} + \frac{\varepsilon}{2} = \varepsilon$$

which implies that

$$\lim_{x \to c}[f(x) + g(x)] = L + K = \lim_{x \to c} f(x) + \lim_{x \to c} g(x).$$

The proof that

$$\lim_{x \to c}[f(x) - g(x)] = L - K$$

is similar. The proof of Property 1 is left as an exercise (see Exercise 39). For proofs of Properties 3, 4, and 5, see Appendix A.

EXAMPLE 2   The Limit of a Polynomial

$$
\begin{aligned}
\lim_{x \to 2}(4x^2 + 3) &= \lim_{x \to 2} 4x^2 + \lim_{x \to 2} 3 && \text{Property 2}\\
&= 4\left[\lim_{x \to 2} x^2\right] + \lim_{x \to 2} 3 && \text{Property 1}\\
&= 4(2^2) + 3 && \text{Example 1}\\
&= 19 && \text{Simplify}
\end{aligned}
$$

In Example 2, note that the limit (as $x \to 2$) of the *polynomial function* $p(x) = 4x^2 + 3$ is simply the value of p at $x = 2$.

$$\lim_{x \to 2} p(x) = p(2) = 4(2^2) + 3 = 19$$

This *direct substitution* property is valid for all polynomial and rational functions.

THEOREM 1.3   Limits of Polynomial and Rational Functions

1. If p is a polynomial function and c is a real number, then

$$\lim_{x \to c} p(x) = p(c).$$

2. If r is a rational function given by $r(x) = p(x)/q(x)$ and c is a real number such that $q(c) \neq 0$, then

$$\lim_{x \to c} r(x) = r(c) = \frac{p(c)}{q(c)}.$$

Proof   Let the polynomial function p be given by

$$p(x) = a_n x^n + \cdots + a_1 x + a_0.$$

Repeated applications of the sum and scalar multiple properties produce

$$\lim_{x \to c} p(x) = a_n \left[\lim_{x \to c} x^n \right] + \cdots + a_1 \left[\lim_{x \to c} x \right] + \lim_{x \to c} a_0.$$

Properties 1, 2, and 3, of Theorem 1.1 produce

$$\lim_{x \to c} p(x) = a_n c^n + \cdots + a_1 c + a_0 = p(c)$$

which proves Property 1 of Theorem 1.3. For the polynomial functions p and q, you can write

$$\lim_{x \to c} p(x) = p(c) \quad \text{and} \quad \lim_{x \to c} q(x) = q(c).$$

Because $q(c) \neq 0$, you can apply Property 4 of Theorem 1.2 to conclude that

$$\lim_{x \to c} r(x) = \lim_{x \to c} \frac{p(x)}{q(x)} = \frac{\lim_{x \to c} p(x)}{\lim_{x \to c} q(x)} = \frac{p(c)}{q(c)} = r(c).$$

EXAMPLE 3   The Limit of a Rational Function

Find the limit: $\displaystyle \lim_{x \to 1} \frac{x^2 + x + 2}{x + 1}$

Solution   Because the denominator is not 0 when $x = 1$, you can apply Theorem 1.3 to obtain

$$\lim_{x \to 1} \frac{x^2 + x + 2}{x + 1} = \frac{1^2 + 1 + 2}{1 + 1} = \frac{4}{2} = 2.$$

Polynomial functions and rational functions are two of the three basic types of algebraic functions. The following theorem deals with the limit of the third type of algebraic function—one that involves radicals. A proof of this theorem is given in Appendix A.

THEOREM 1.4   The Limit of a Function Involving a Radical

Let n be a positive integer. The following limit is valid for all c if n is odd, and is valid for $c > 0$ if n is even.

$$\lim_{x \to c} \sqrt[n]{x} = \sqrt[n]{c}$$

The next theorem involves the limit of a composite function. Its proof is also given in Appendix A.

THEOREM 1.5   The Limit of a Composite Function

If f and g are functions such that $\lim_{x \to c} g(x) = L$ and $\lim_{x \to L} f(x) = f(L)$, then

$$\lim_{x \to c} f(g(x)) = f(L).$$

The next two examples show how to use Theorem 1.5 to find the limit of a composite function.

EXAMPLE 4   The Limit of a Composite Function

Find the limit: $\lim_{x \to 0} \sqrt{x^2 + 4}$

Solution
Let $g(x) = x^2 + 4$ and let $f(x) = \sqrt{x}$. Because

$$\lim_{x \to 0} g(x) = 4 \quad \text{and} \quad \lim_{x \to 4} f(x) = 2$$

it follows that

$$\lim_{x \to 0} \sqrt{x^2 + 4} = \lim_{x \to 0} f(g(x)) = f(4) = \sqrt{4} = 2.$$

EXAMPLE 5   The Limit of a Composite Function

Find the limit: $\lim_{x \to 3} \sqrt[3]{2x^2 - 10}$

Solution   Because

$$\lim_{x \to 3}(2x^2 - 10) = 2(3^2) - 10 = 8 \quad \text{and} \quad \lim_{x \to 8} \sqrt[3]{x} = 2$$

you can write

$$\lim_{x \to 3} \sqrt[3]{2x^2 - 10} = 2.$$

Limits of Trigonometric Functions

You have seen that the limits of many algebraic functions can be evaluated by direct substitution. Each of the six basic trigonometric functions also possesses this desirable quality, as shown in the next theorem (presented without proof).

THEOREM 1.6   Limits of Trigonometric Functions

Let c be a real number in the domain of the given trigonometric function.

1. $\lim\limits_{x \to c} \sin x = \sin c$ 2. $\lim\limits_{x \to c} \cos x = \cos c$ 3. $\lim\limits_{x \to c} \tan x = \tan c$

4. $\lim\limits_{x \to c} \cot x = \cot c$ 5. $\lim\limits_{x \to c} \sec x = \sec c$ 6. $\lim\limits_{x \to c} \csc x = \csc c$

EXAMPLE 6   Limits of Trigonometric Functions

a. $\lim\limits_{x \to 0} \sin x = \sin(0) = 0$

b. $\lim\limits_{x \to \pi} (x \cos x) = \left[\lim\limits_{x \to \pi} x\right]\left[\lim\limits_{x \to \pi} \cos x\right]$

$$= \pi \cos(\pi)$$

$$= -\pi$$

c. $\lim\limits_{x \to 0} \dfrac{\tan x}{x^2 + 1} = \dfrac{\lim\limits_{x \to 0} \tan x}{\lim\limits_{x \to 0}(x^2 + 1)}$

$$= \frac{0}{0^2 + 1}$$

$$= 0$$

d. $\lim\limits_{x \to 0} \sin^2 x = \lim\limits_{x \to 0} (\sin x)^2$

$$= 0^2$$

$$= 0$$

EXERCISES for Section 1.2

In Exercises 1–4, use the graph to visually determine the limits.

1. $f(x) = -2x + 1$ **2.** $f(x) = x^3 - 3x^2$ **3.** $g(x) = \sin x$ **4.** $h(x) = \tan x$

1.
  a. $\lim\limits_{x \to 0} f(x)$
  b. $\lim\limits_{x \to -1} f(x)$

2.
  a. $\lim\limits_{x \to 1} f(x)$
  b. $\lim\limits_{x \to 3} f(x)$

3.
  a. $\lim\limits_{x \to \pi/3} g(x)$
  b. $\lim\limits_{x \to \pi/2} g(x)$

4.
  a. $\lim\limits_{x \to 0} h(x)$
  b. $\lim\limits_{x \to 3\pi/4} h(x)$

In Exercises 5–28, find the limit.

5. $\lim\limits_{x \to 4} x^2$

6. $\lim\limits_{x \to -3} (3x + 2)$

7. $\lim\limits_{x \to 0} (2x - 1)$

8. $\lim\limits_{x \to 1} (-x^2 + 1)$

9. $\lim\limits_{x \to 2} (-x^2 + x - 2)$

10. $\lim\limits_{x \to 1} (3x^3 - 2x^2 + 4)$

11. $\lim\limits_{x \to 3} \sqrt{x + 1}$

12. $\lim\limits_{x \to 4} \sqrt[3]{x + 4}$

13. $\lim\limits_{x \to -4} (x + 3)^2$

14. $\lim\limits_{x \to 0} (2x - 1)^3$

15. $\lim\limits_{x \to 2} \dfrac{1}{x}$

16. $\lim\limits_{x \to -3} \dfrac{2}{x + 2}$

17. $\lim\limits_{x \to -1} \dfrac{x^2 + 1}{x}$

18. $\lim\limits_{x \to 3} \dfrac{\sqrt{x + 1}}{x - 4}$

19. $\lim\limits_{x \to \pi/2} \sin x$

20. $\lim\limits_{x \to \pi} \tan x$

21. $\lim\limits_{x \to 1} \cos \pi x$

22. $\lim\limits_{x \to 1} \sin \dfrac{\pi x}{2}$

23. $\lim\limits_{x \to 0} \sec 2x$

24. $\lim\limits_{x \to \pi} \cos 3x$

25. $\lim\limits_{x \to 5\pi/6} \sin x$

26. $\lim\limits_{x \to 5\pi/3} \cos x$

27. $\lim\limits_{x \to 3} \tan\left(\dfrac{\pi x}{4}\right)$

28. $\lim\limits_{x \to 7} \sec\left(\dfrac{\pi x}{6}\right)$

In Exercises 29–32, use the given information to evaluate the limits.

29. $\lim\limits_{x \to c} f(x) = 2$
$\lim\limits_{x \to c} g(x) = 3$

a. $\lim\limits_{x \to c} [5g(x)]$

b. $\lim\limits_{x \to c} [f(x) + g(x)]$

c. $\lim\limits_{x \to c} [f(x)g(x)]$

d. $\lim\limits_{x \to c} \dfrac{f(x)}{g(x)}$

30. $\lim\limits_{x \to c} f(x) = \frac{3}{2}$
$\lim\limits_{x \to c} g(x) = \frac{1}{2}$

a. $\lim\limits_{x \to c} [4f(x)]$

b. $\lim\limits_{x \to c} [f(x) + g(x)]$

c. $\lim\limits_{x \to c} [f(x)g(x)]$

d. $\lim\limits_{x \to c} \dfrac{f(x)}{g(x)}$

31. $\lim\limits_{x \to c} f(x) = 4$

a. $\lim\limits_{x \to c} [f(x)]^3$

b. $\lim\limits_{x \to c} \sqrt{f(x)}$

c. $\lim\limits_{x \to c} [3f(x)]$

d. $\lim\limits_{x \to c} [f(x)]^{3/2}$

32. $\lim\limits_{x \to c} f(x) = 27$

a. $\lim\limits_{x \to c} \sqrt[3]{f(x)}$

b. $\lim\limits_{x \to c} \dfrac{f(x)}{18}$

c. $\lim\limits_{x \to c} [f(x)]^2$

d. $\lim\limits_{x \to c} [f(x)]^{2/3}$

C *Essay* **In Exercises 33 and 34, use a graphing utility to graph the function and estimate the limit (if it exists). What is the domain of the function? Can you detect a possible danger in determining the domain of a function solely by analyzing the graph generated by a graphing utility? Write a short paragraph about the importance of examining a function analytically as well as graphically.**

33. $f(x) = \dfrac{x - 9}{\sqrt{x} - 3}$

$\lim\limits_{x \to 9} f(x)$

34. $f(x) = \dfrac{x - 3}{x^2 - 9}$

$\lim\limits_{x \to 3} f(x)$

35. Find two functions f and g such that

$\lim\limits_{x \to 0} f(x)$ and $\lim\limits_{x \to 0} g(x)$ do not exist, but

$\lim\limits_{x \to 0} [f(x) + g(x)]$ does exist.

36. Prove that if

$$\lim\limits_{x \to c} f(x)$$

exists and

$$\lim\limits_{x \to c} [f(x) + g(x)]$$

does not exist, then $\lim\limits_{x \to c} g(x)$ does not exist.

37. Prove Property 1 of Theorem 1.1.

38. Prove Property 3 of Theorem 1.1. (You may use Property 3 of Theorem 1.2.)

39. Prove Property 1 of Theorem 1.2.

40. If $f(x) < g(x)$ for all $x \neq a$, is it true that

$$\lim\limits_{x \to a} f(x) < \lim\limits_{x \to a} g(x)?$$

If it is false, explain why or give an example that shows it is false.

CAREER INTERVIEW

Wesley L. Harris

Associate Administrator

for Aeronautics

NASA

Washington, DC

I taught and performed research for more than 25 years before I became an administrator at NASA. My professional academic career included 13 years as a faculty member at MIT, five years at the University of Connecticut, and four years at the University of Virginia. My undergraduate and graduate students worked with me to solve a series of problems in reducing the noise generated by helicopters in flight. We also solved problems which have lead to improved aerodynamic performance of fixed-wing aircraft in flight at very high speeds.

In solving these complex technical problems, we found it was critical to first formulate the problems correctly. Second, we applied a variety of techniques to obtain a solution. These techniques included experimental modeling, numerical simulation, and applied mathematics. The applied mathematics techniques consisted of calculus, differential equations, integral equations, and special functions and transforms.

My students and I gained a deep respect for the strength of applied mathematics techniques. These strengths taught us a better appreciation for logic, accuracy, precision, and discipline. For us, mathematics is much more than finding an answer to a complex problem.

A Strategy for Finding Limits

In Section 1.2, you studied several types of functions whose limits can be evaluated by direct substitution. In this section, you will use that knowledge with the following theorem to develop a strategy for finding limits.

THEOREM 1.7   Functions that Agree at All But One Point

Let c be a real number and let $f(x) = g(x)$ for all $x \neq c$ in an open interval containing c. If the limit of $g(x)$ as x approaches c exists, then the limit of $f(x)$ also exists and

$$\lim_{x \to c} f(x) = \lim_{x \to c} g(x).$$

Proof   Let L be the limit of $g(x)$ as x approaches c. Then, for each $\varepsilon > 0$ there exists a $\delta > 0$ such that $f(x) = g(x)$ in the open intervals $(c - \delta, c)$ and $(c, c + \delta)$, and

$$|g(x) - L| < \varepsilon \quad \text{whenever} \quad 0 < |x - c| < \delta.$$

Because $f(x) = g(x)$ for all x in the open interval other than $x = c$, it follows that

$$|f(x) - L| < \varepsilon \quad \text{whenever} \quad 0 < |x - c| < \delta.$$

Thus, the limit of $f(x)$ as x approaches c is also L.   ∎

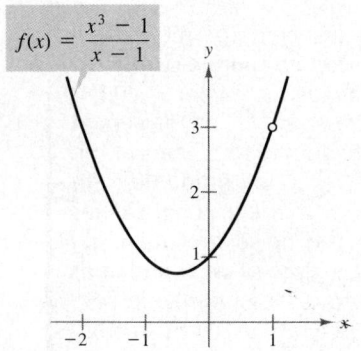

$f(x) = \dfrac{x^3 - 1}{x - 1}$

EXAMPLE 1   Two Functions that Agree at All But One Point

Show that the functions $f(x) = (x^3 - 1)/(x - 1)$ and $g(x) = x^2 + x + 1$ have the same values for all x other than $x = 1$.

Solution   Factoring the numerator of f produces

$$f(x) = \frac{x^3 - 1}{x - 1} = \frac{(x - 1)(x^2 + x + 1)}{x - 1}.$$

If $x \neq 1$, you can cancel like factors to obtain

$$f(x) = \frac{\cancel{(x - 1)}(x^2 + x + 1)}{\cancel{x - 1}} = x^2 + x + 1 = g(x), \qquad x \neq 1.$$

Thus, for all points other than $x = 1$, the graphs of f and g are identical, as shown in Figure 1.15.   ∎

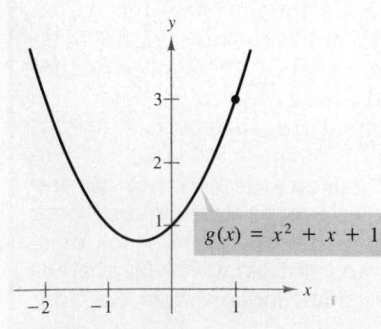

$g(x) = x^2 + x + 1$

FIGURE 1.15
f and g agree at all but one point.

"Incorrect" graph of f.

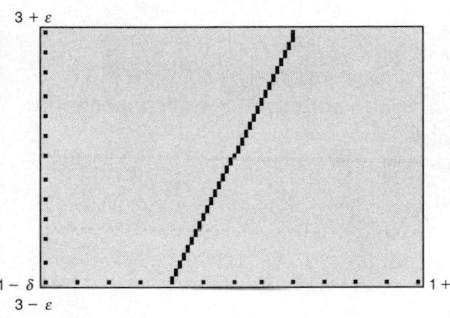

"Correct" graph of g.

FIGURE 1.16

: TECHNOLOGY   In Figure 1.15, note that the only difference between the graphs of
: f and g is that the graph of f has a gap at the point (1, 3). This gap consists of a single
: point—hence you should not be able to see the difference in the two graphs. However,
: because of the pixel configuration and round-off error of a graphing utility, you should
: be able to find screen settings that distinguish between the graphs of f and g, as shown
: in Figure 1.16.

EXAMPLE 2   Finding the Limit of a Function

Find the limit: $\displaystyle\lim_{x \to 1} \frac{x^3 - 1}{x - 1}$

Solution   From Example 1, you know that the functions

$$f(x) = \frac{x^3 - 1}{x - 1} \quad \text{and} \quad g(x) = x^2 + x + 1$$

agree at all but one point (when $x = 1$). By applying Theorem 1.7, you can conclude
that each function has the same limit for all values of x (including $x = 1$). You can
therefore write the following.

$$\lim_{x \to 1} \frac{x^3 - 1}{x - 1} = \lim_{x \to 1} \frac{(x - 1)(x^2 + x + 1)}{x - 1} \qquad \text{Factor}$$

$$= \lim_{x \to 1} \frac{\cancel{(x - 1)}(x^2 + x + 1)}{\cancel{x - 1}} \qquad \text{Cancel like factors}$$

$$= \lim_{x \to 1} (x^2 + x + 1) \qquad \text{Apply Theorem 1.7}$$

$$= (1^2 + 1 + 1) \qquad \text{Use direct substitution}$$

$$= 3 \qquad \text{Simplify}$$

A Strategy for Finding Limits

1. Learn to recognize which limits can be evaluated by direct substitution.
   (You studied these in Section 1.2.)
2. If the limit of $f(x)$ as x approaches c *cannot* be evaluated by direct substitution, try to find a function g that agrees with f for all x other than $x = c$.
   [Choose g so that the limit of $g(x)$ *can* be evaluated by direct substitution.]
3. Apply Theorem 1.7 to conclude that
   $$\lim_{x \to c} f(x) = \lim_{x \to c} g(x) = g(c).$$
4. Use a graph or a table to reinforce your conclusion.

REMARK   When applying this strategy for finding limits, remember that some functions do not
have limits (as x approaches c). For instance, the following limit does not exist.

$$\lim_{x \to 1} \frac{x^3 + 1}{x - 1}$$

Cancellation and Rationalization Techniques

Two techniques for evaluating a limit are shown in Examples 3 and 4. The first technique involves canceling common factors, and the second technique involves rationalizing the numerator of a fractional expression.

EXAMPLE 3   Cancellation Technique

Find the limit: $\displaystyle\lim_{x \to -3} \frac{x^2 + x - 6}{x + 3}$

Solution   Although you are taking the limit of a rational function, you *cannot* apply Theorem 1.3 because the limit of the denominator is 0.

$$\lim_{x \to -3} \frac{x^2 + x - 6}{x + 3} \quad \nearrow \lim_{x \to -3}(x^2 + x - 6) = 0$$
$$\searrow \lim_{x \to -3}(x + 3) = 0$$

Direct substitution fails

Because the limit of the numerator is also 0, the numerator and denominator have a *common factor* of $(x + 3)$. Thus, for all $x \neq -3$, you can cancel this factor to obtain

$$\frac{x^2 + x - 6}{x + 3} = \frac{\cancel{(x + 3)}(x - 2)}{\cancel{x + 3}} = x - 2, \quad x \neq -3.$$

Using Theorem 1.7, it follows that

$$\lim_{x \to -3} \frac{x^2 + x - 6}{x + 3} = \lim_{x \to -3}(x - 2) = -5.$$

This result is shown graphically in Figure 1.17. Note that the graph of the function f coincides with the graph of the function $g(x) = x - 2$, except that the graph of f has a gap at the point $(-3, -5)$.

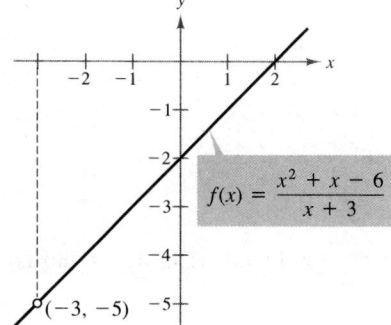

$$f(x) = \frac{x^2 + x - 6}{x + 3}$$

$(-3, -5)$

FIGURE 1.17
f is undefined when $x = -3$.

REMARK   In the solution of Example 3, be sure you see the usefulness of the Factor Theorem of Algebra. This theorem states that if c is a zero of a polynomial function, then $(x - c)$ is a factor of the polynomial. Thus, if you apply direct substitution to a rational function and obtain

$$r(c) = \frac{p(c)}{q(c)} = \frac{0}{0}$$

then you can conclude that the factor $(x - c)$ must be common to both $p(x)$ and $q(x)$.

In Example 3, direct substitution produced the meaningless fractional form 0/0. Such an expression is called an **indeterminate form** because you cannot (from the form alone) determine the limit. When you try to evaluate a limit and encounter this form, remember that you must rewrite the fraction so that the new denominator does not have 0 as its limit. One way to do this is to *cancel like factors,* as shown in Example 3. A second way is to *rationalize the numerator,* as shown in Example 4.

EXAMPLE 4   Rationalization Technique

Find the limit: $\displaystyle\lim_{x \to 0} \frac{\sqrt{x + 1} - 1}{x}$

Solution   By direct substitution, you obtain the indeterminate form 0/0.

$$\lim_{x \to 0} \frac{\sqrt{x + 1} - 1}{x} \qquad \begin{array}{l} \nearrow \lim_{x \to 0}\left(\sqrt{x + 1} - 1\right) = 0 \\[1em] \searrow \lim_{x \to 0} x = 0 \end{array}$$

Direct substitution fails

In this case, you can rewrite the fraction by rationalizing the numerator.

$$\frac{\sqrt{x + 1} - 1}{x} = \left(\frac{\sqrt{x + 1} - 1}{x}\right)\left(\frac{\sqrt{x + 1} + 1}{\sqrt{x + 1} + 1}\right)$$

$$= \frac{(x + 1) - 1}{x\left(\sqrt{x + 1} + 1\right)}$$

$$= \frac{\cancel{x}}{\cancel{x}\left(\sqrt{x + 1} + 1\right)}$$

$$= \frac{1}{\sqrt{x + 1} + 1}, \qquad x \neq 0$$

Now, using Theorem 1.7, you can evaluate the limit as follows.

$$\lim_{x \to 0} \frac{\sqrt{x + 1} - 1}{x} = \lim_{x \to 0} \frac{1}{\sqrt{x + 1} + 1}$$

$$= \frac{1}{1 + 1}$$

$$= \frac{1}{2}$$

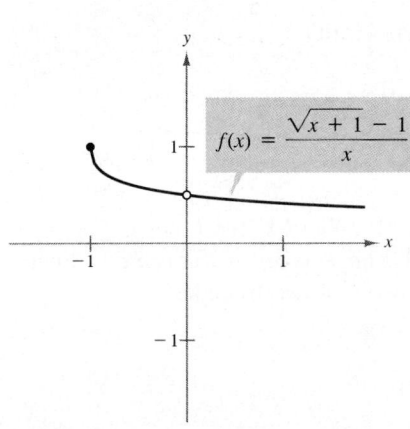

$$f(x) = \frac{\sqrt{x + 1} - 1}{x}$$

FIGURE 1.18
The limit of $f(x)$ as x approaches 0 is $\frac{1}{2}$.

A table such as Table 1.4, or a graph such as that shown in Figure 1.18, will reinforce your conclusion that the limit is $\frac{1}{2}$.

TABLE 1.4

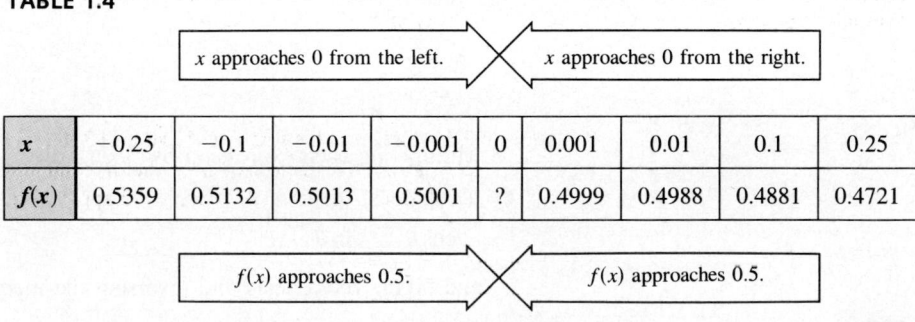

x approaches 0 from the left.					x approaches 0 from the right.			
x	−0.25	−0.1	−0.01	−0.001	0	0.001	0.01	0.1	0.25
f(x)	0.5359	0.5132	0.5013	0.5001	?	0.4999	0.4988	0.4881	0.4721

| f(x) approaches 0.5. | | | | | f(x) approaches 0.5. | | | |

REMARK   The rationalization technique for evaluating limits is based on multiplication by a convenient form of 1. In Example 4, the convenient form is

$$1 = \frac{\sqrt{x + 1} + 1}{\sqrt{x + 1} + 1}.$$

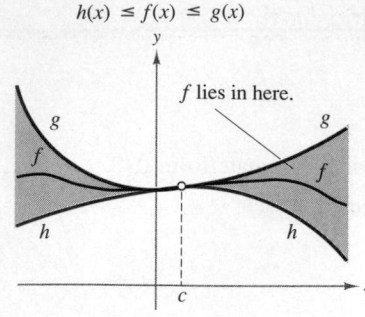

$h(x) \le f(x) \le g(x)$

f lies in here.

FIGURE 1.19
The Squeeze Theorem

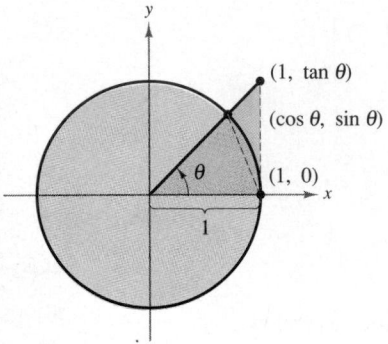

FIGURE 1.20

The Squeeze Theorem

The next theorem concerns the limit of a function that is squeezed between two other functions, each of which has the same limit at a given *x*-value, as shown in Figure 1.19. (The proof of this theorem is given in Appendix A.)

THEOREM 1.8   The Squeeze Theorem

If $h(x) \le f(x) \le g(x)$ for all *x* in an open interval containing *c*, except possibly at *c* itself, and if

$$\lim_{x \to c} h(x) = L = \lim_{x \to c} g(x)$$

then $\lim_{x \to c} f(x)$ exists and is equal to *L*.

You can see the usefulness of the Squeeze Theorem in the proof of Theorem 1.9.

THEOREM 1.9   Two Special Trigonometric Limits

1. $\lim\limits_{x \to 0} \dfrac{\sin x}{x} = 1$ **2.** $\lim\limits_{x \to 0} \dfrac{1 - \cos x}{x} = 0$

Proof   To avoid the confusion of two different uses of *x*, the proof is presented using the variable θ, where θ is an acute positive angle *measured in radians*. Figure 1.20 shows a circular sector that is squeezed between two triangles.

FOR FURTHER INFORMATION
For more information on the function $f(x) = (\sin x)/x$ see the article "The Function $(\sin x)/x$" by William B. Gearhart and Harris S. Shultz in the March, 1990 issue of *The College Mathematics Journal*.

Area of triangle	$\ge$	Area of sector	$\ge$	Area of triangle
$\dfrac{\tan \theta}{2}$	$\ge$	$\dfrac{\theta}{2}$	$\ge$	$\dfrac{\sin \theta}{2}$

Multiplying each expression by $2/\sin \theta$ produces

$$\frac{1}{\cos \theta} \ge \frac{\theta}{\sin \theta} \ge 1$$

and taking reciprocals and reversing the inequalities yields

$$\cos \theta \le \frac{\sin \theta}{\theta} \le 1.$$

Because $\cos \theta = \cos(-\theta)$ and $(\sin \theta)/\theta = [\sin(-\theta)]/(-\theta)$, we can conclude that this inequality is valid for *all* nonzero θ in the open interval $(-\pi/2, \pi/2)$. Finally, because $\lim\limits_{\theta \to 0} \cos \theta = 1$ and $\lim\limits_{\theta \to 0} 1 = 1$, you can apply the Squeeze Theorem to conclude that $\lim\limits_{\theta \to 0} (\sin \theta)/\theta = 1$. The proof of the second limit is left as an exercise (see Exercise 58).

EXAMPLE 5   A Limit Involving a Trigonometric Function

Find the limit: $\lim\limits_{x \to 0} \dfrac{\tan x}{x}$

Solution   Direct substitution yields the indeterminate form 0/0. To solve this problem, you can write $\tan x$ as $(\sin x)/(\cos x)$ and obtain

$$\lim_{x \to 0} \frac{\tan x}{x} = \lim_{x \to 0} \left(\frac{\sin x}{x}\right)\left(\frac{1}{\cos x}\right).$$

Now, because

$$\lim_{x \to 0} \frac{\sin x}{x} = 1 \quad \text{and} \quad \lim_{x \to 0} \frac{1}{\cos x} = 1$$

you can obtain

$$\lim_{x \to 0} \frac{\tan x}{x} = \left[\lim_{x \to 0} \frac{\sin x}{x}\right]\left[\lim_{x \to 0} \frac{1}{\cos x}\right]$$

$$= (1)(1)$$

$$= (1).$$

(See Figure 1.21.)

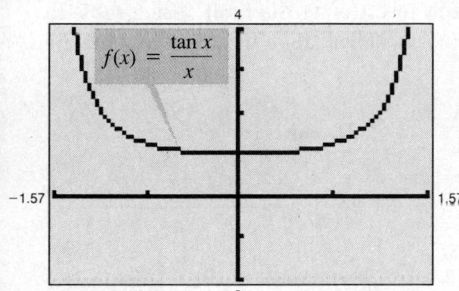

FIGURE 1.21
The limit of $f(x)$ as x approaches 0 is 1.

EXAMPLE 6   A Limit Involving a Trigonometric Function

Find the limit: $\lim\limits_{x \to 0} \dfrac{\sin 4x}{x}$

Solution   Direct substitution yields the indeterminate form 0/0. To solve this problem, you can rewrite the limit as

$$\lim_{x \to 0} \frac{\sin 4x}{x} = 4\left[\lim_{x \to 0} \frac{\sin 4x}{4x}\right].$$

Now, by letting $y = 4x$ and observing that $x \to 0$ if and only if $y \to 0$, you can write

$$\lim_{x \to 0} \frac{\sin 4x}{x} = 4\left[\lim_{x \to 0} \frac{\sin 4x}{4x}\right]$$

$$= 4\left[\lim_{y \to 0} \frac{\sin y}{y}\right]$$

$$= 4(1)$$

$$= 4.$$

(See Figure 1.22.)

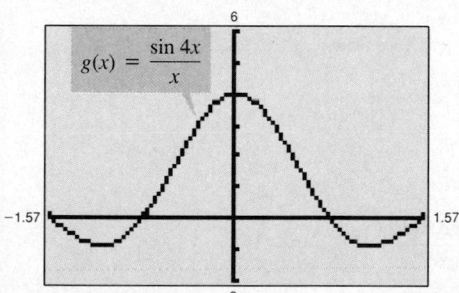

FIGURE 1.22
The limit of $g(x)$ as x approaches 0 is 4.

TECHNOLOGY   If you have access to a graphing utility, try using it to confirm the limits in the examples and exercise set. For instance, Figures 1.21 and 1.22 show the graphs of

$$f(x) = \frac{\tan x}{x} \quad \text{and} \quad g(x) = \frac{\sin 4x}{x}.$$

Note that the first graph appears to contain the point (0, 1) and the second graph appears to contain the point (0, 4), which lends support to the conclusions obtained in Examples 5 and 6.

EXERCISES for Section 1.3

In Exercises 1–4, use the graph to determine each limit visually, if it exists. When possible, identify two functions that agree in all but one point.

1. $g(x) = \dfrac{-2x^2 + x}{x}$

2. $h(x) = \dfrac{x^2 - 3x}{x}$

a. $\lim\limits_{x \to 0} g(x)$

b. $\lim\limits_{x \to -1} g(x)$

a. $\lim\limits_{x \to -2} h(x)$

b. $\lim\limits_{x \to 0} h(x)$

3. $g(x) = \dfrac{x^3 - x}{x - 1}$

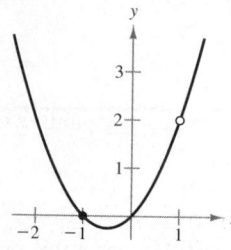

a. $\lim\limits_{x \to 1} g(x)$

b. $\lim\limits_{x \to -1} g(x)$

4. $f(x) = \dfrac{1}{x - 1}$

a. $\lim\limits_{x \to 1} f(x)$

b. $\lim\limits_{x \to 2} f(x)$

In Exercises 5–8, find the limit (if it exists) of the function. Identify two functions that agree in all but one point and sketch the graph of the given function.

5. $\lim\limits_{x \to -1} \dfrac{x^2 - 1}{x + 1}$

6. $\lim\limits_{x \to -1} \dfrac{2x^2 - x - 3}{x + 1}$

7. $\lim\limits_{x \to -2} \dfrac{x^3 + 8}{x + 2}$

8. $\lim\limits_{x \to -1} \dfrac{x^3 + 1}{x + 1}$

In Exercises 9–20, find the limit (if it exists).

9. $\lim\limits_{x \to 5} \dfrac{x - 5}{x^2 - 25}$

10. $\lim\limits_{x \to 2} \dfrac{2 - x}{x^2 - 4}$

11. $\lim\limits_{x \to 1} \dfrac{x^2 + x - 2}{x^2 - 1}$

12. $\lim\limits_{x \to 0} \dfrac{\sqrt{2 + x} - \sqrt{2}}{x}$

13. $\lim\limits_{x \to 0} \dfrac{\sqrt{3 + x} - \sqrt{3}}{x}$

14. $\lim\limits_{x \to 0} \dfrac{[1/(x + 4)] - (1/4)}{x}$

15. $\lim\limits_{x \to 0} \dfrac{[1/(2 + x)] - (1/2)}{x}$

16. $\lim\limits_{x \to 3} \dfrac{\sqrt{x + 1} - 2}{x - 3}$

17. $\lim\limits_{\Delta x \to 0} \dfrac{2(x + \Delta x) - 2x}{\Delta x}$

18. $\lim\limits_{\Delta x \to 0} \dfrac{(x + \Delta x)^2 - x^2}{\Delta x}$

19. $\lim\limits_{\Delta x \to 0} \dfrac{(x + \Delta x)^2 - 2(x + \Delta x) + 1 - (x^2 - 2x + 1)}{\Delta x}$

20. $\lim\limits_{\Delta x \to 0} \dfrac{(x + \Delta x)^3 - 1}{\Delta x}$

C In Exercises 21–24, use a graphing utility to graph the function and use the graph to estimate the limit. Use a table to reinforce your conclusion. Then, find the limit by analytic methods.

21. $\lim\limits_{x \to 0} \dfrac{\sqrt{x + 2} - \sqrt{2}}{x}$

22. $\lim\limits_{x \to 1} \dfrac{4 - \sqrt{x}}{x - 16}$

23. $\lim\limits_{x \to 2} \dfrac{x^5 - 32}{x - 2}$

24. $\lim\limits_{x \to 0} \dfrac{[1/(2 + x)] - (1/2)}{x}$

In Exercises 25–40, determine the limit of the trigonometric function (if it exists).

25. $\lim\limits_{x \to 0} \dfrac{\sin x}{5x}$

26. $\lim\limits_{x \to 0} \dfrac{3(1 - \cos x)}{x}$

27. $\lim\limits_{\theta \to 0} \dfrac{\sec \theta - 1}{\theta \sec \theta}$

28. $\lim\limits_{\theta \to 0} \dfrac{\cos \theta \tan \theta}{\theta}$

29. $\lim\limits_{x \to 0} \dfrac{\sin^2 x}{x}$

30. $\lim\limits_{x \to 0} \dfrac{\tan^2 x}{x}$

31. $\lim\limits_{h \to 0} \dfrac{(1 - \cos h)^2}{h}$

32. $\lim\limits_{\phi \to \pi} \phi \sec \phi$

33. $\lim\limits_{x \to \pi/2} \dfrac{\cos x}{\cot x}$

34. $\lim\limits_{x \to \pi/4} \dfrac{1 - \tan x}{\sin x - \cos x}$

35. $\lim\limits_{t \to 0} \dfrac{\sin^2 t}{t^2}$

$\left[\text{\textit{Hint}: Find } \lim\limits_{t \to 0} \left(\dfrac{\sin t}{t} \right)^2. \right]$

36. $\lim\limits_{x \to 0} \dfrac{\sin 2x}{\sin 3x}$

$\left[\text{\textit{Hint}: Find } \lim\limits_{x \to 0} \left(\dfrac{2 \sin 2x}{2x} \right) \left(\dfrac{3x}{3 \sin 3x} \right). \right]$

37. $\lim\limits_{t \to 0} \dfrac{\sin 3t}{t}$

38. $\lim\limits_{h \to 0} (1 + \cos 2h)$

39. $\lim\limits_{x \to 0} \dfrac{\sin x^2}{x}$

40. $\lim\limits_{x \to 0} \dfrac{\sin x}{\sqrt[3]{x}}$

In Exercises 41 and 42, use the Squeeze Theorem to find $\lim\limits_{x \to c} f(x)$.

41. $c = 0$
$4 - x^2 \le f(x) \le 4 + x^2$

42. $c = a$
$b - |x - a| \le f(x) \le b + |x - a|$

C In Exercises 43–48, use a graphing utility to graph the given function and the equations $y = x$ and $y = -x$ in the same viewing rectangle. Using the graphs to visually observe the Squeeze Theorem, find $\lim\limits_{x \to 0} f(x)$.

43. $f(x) = x \cos x$

44. $f(x) = |x \sin x|$

45. $f(x) = |x| \sin x$

46. $f(x) = |x| \cos x$

47. $f(x) = x \sin \dfrac{1}{x}$

48. $h(x) = x \cos \dfrac{1}{x}$

C **49.** *Essay*   Use a graphing utility to graph

$$f(x) = x, \quad g(x) = \sin x, \quad \text{and} \quad h(x) = \frac{\sin x}{x}$$

in the same viewing rectangle. Compare the magnitudes of $f(x)$ and $g(x)$ when x is "close to" 0. Use the comparison to write a short paragraph explaining why $\lim\limits_{x \to 0} h(x) = 1$.

C **50.** *Essay*   Use a graphing utility to graph

$$f(x) = x, \quad g(x) = \sin^2 x, \quad \text{and} \quad h(x) = \frac{\sin^2 x}{x}$$

in the same viewing rectangle. Compare the magnitudes of $f(x)$ and $g(x)$ when x is "close to" 0. Use the comparison to write a short paragraph explaining why $\lim\limits_{x \to 0} h(x) = 0$.

Free-Falling Object   In Exercises 51 and 52, use the position function $s(t) = -16t^2 + 1000$, which gives the height (in feet) of an object that has fallen for t seconds from a height of 1000 feet. The velocity at time $t = a$ seconds is

$$\lim_{t \to a} \frac{s(a) - s(t)}{a - t}.$$

51. If a construction worker drops a wrench from a height of 1000 feet, how fast will the wrench be falling after 5 seconds?

52. If a construction worker drops a wrench from a height of 1000 feet, when will the wrench hit the ground? With what velocity will the wrench impact the ground?

53. Prove that if $\lim\limits_{x \to c} f(x) = 0$, then

$$\lim_{x \to c} |f(x)| = 0.$$

54. Prove that if $\lim\limits_{x \to c} f(x) = 0$ and $|g(x)| \le M$ for a fixed number M and all $x \neq c$, then

$$\lim_{x \to c} f(x) g(x) = 0.$$

55. Prove that if $\lim\limits_{x \to c} |f(x)| = 0$, then

$$\lim_{x \to c} f(x) = 0.$$

(*Note:* This is the converse of Exercise 53.)

56. Prove that if $\lim\limits_{x \to c} f(x) = L$, then

$$\lim_{x \to c} |f(x)| = |L|.$$

[*Hint:* Use $\big| |f(x)| - |L| \big| \le |f(x) - L|$.]

57. Find a function f to show that the converse of Exercise 56 is not true.

[*Hint:* Find a function f such that $\lim\limits_{x \to c} |f(x)| = |L|$ but $\lim\limits_{x \to c} f(x)$ does not exist.]

58. Complete the proof of Theorem 1.9 by proving that

$$\lim_{x \to 0} \frac{1 - \cos x}{x} = 0.$$

59. Let $f(x) = \begin{cases} 0, & \text{if } x \text{ is rational} \\ 1, & \text{if } x \text{ is irrational} \end{cases}$

and

$$g(x) = \begin{cases} 0, & \text{if } x \text{ is rational} \\ x, & \text{if } x \text{ is irrational} \end{cases}$$

Find $\lim\limits_{x \to 0} f(x)$ and $\lim\limits_{x \to 0} g(x)$.

60. Evaluate the limit $\lim\limits_{x \to 0} \dfrac{\sec x - 1}{x^2}$, and confirm your answer with a calculator.

61. Find $\lim\limits_{x \to 0} \dfrac{1 - \cos x}{x^2}$. Use this limit to derive the approximation $\cos x \approx 1 - \frac{1}{2}x^2$ for x near 0. Use this estimate of $\cos x$ to approximate $\cos(0.1)$.

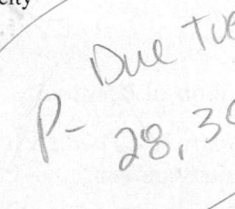

1.4

Continuity at a Point and on an Open Interval • One-Sided Limits and Continuity on a Closed Interval • Properties of Continuity • The Intermediate Value Theorem

Continuity at a Point and on an Open Interval

In mathematics the term *continuous* has much the same meaning as it has in everyday usage. To say that a function is continuous at $x = c$ means that there is no interruption in the graph of f at c. That is, its graph is unbroken at c and there are no holes, jumps, or gaps. Figure 1.23 identifies three values of x at which the graph of f is *not* continuous. At all other points in the interval (a, b), the graph of f is uninterrupted and **continuous.**

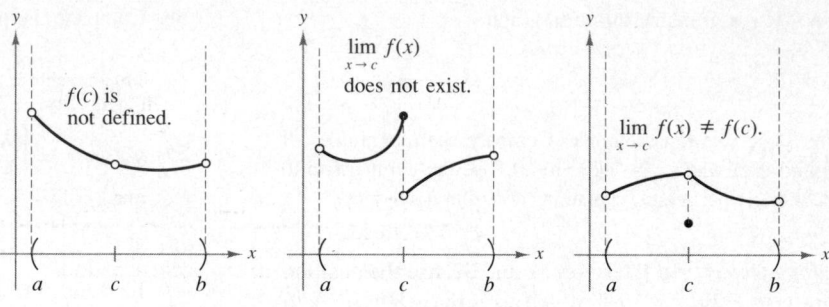

FIGURE 1.23

In Figure 1.23, it appears that continuity at $x = c$ can be destroyed by any one of the following conditions.

1. The function is not defined at $x = c$.

2. The limit of $f(x)$ does not exist at $x = c$.

3. The limit of $f(x)$ exists at $x = c$, but it is not equal to $f(c)$.

Definition of Continuity

Continuity at a Point: A function f is **continuous at c** if the following three conditions are met.

1. $f(c)$ is defined.

2. $\lim\limits_{x \to c} f(x)$ exists.

3. $\lim\limits_{x \to c} f(x) = f(c)$.

Continuity on an Open Interval: A function is **continuous on an open interval (a, b)** if it is continuous at each point in the interval. A function that is continuous on the entire real line $(-\infty, \infty)$ is **everywhere continuous.**

Consider an open interval I that contains a real number c. If a function f is defined on I (except possibly at c), and f is not continuous at c, then f is said to have a **discontinuity** at c. Discontinuities fall into two categories: **removable** and **nonremovable**. A discontinuity at c is called removable if f can be made continuous by appropriately defining (or redefining) $f(c)$. For instance, the functions shown in Figure 1.24(a) and (c) have removable discontinuities at c and the function shown in Figure 1.24(b) has a nonremovable discontinuity at c.

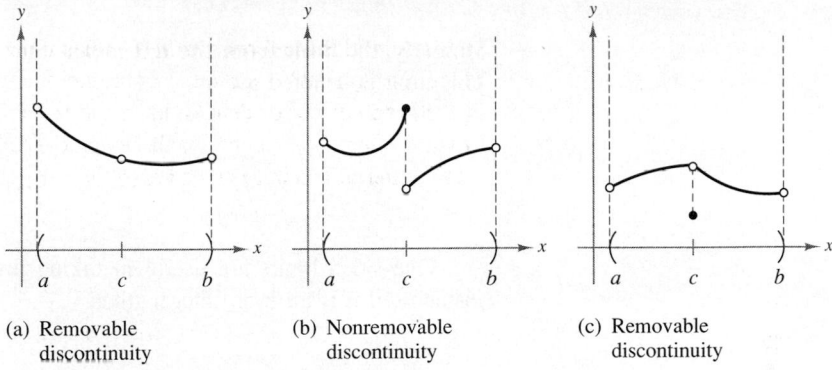

(a) Removable discontinuity

(b) Nonremovable discontinuity

(c) Removable discontinuity

FIGURE 1.24

EXAMPLE 1   Continuity of a Function

Discuss the continuity of each function.

a. $f(x) = \dfrac{1}{x}$ **b.** $g(x) = \dfrac{x^2 - 1}{x - 1}$ **c.** $h(x) = \sin x$

Solution

a. The domain of f is all nonzero real numbers. From Theorem 1.3, you can conclude that f is continuous at every x-value in its domain. At $x = 0$, the function has a nonremovable discontinuity (see Figure 1.25a). In other words, there is no way to define $f(0)$ so as to make the function continuous at $x = 0$.

b. The domain of g is all real numbers except $x = 1$. From Theorem 1.3, you can conclude that f is continuous at every x-value in its domain. At $x = 1$, the function has a removable discontinuity (see Figure 1.25b). If $g(1)$ is defined as 2, the "newly defined" function is continuous for all real numbers.

c. The domain of h is all real numbers. From Theorem 1.6, you can conclude that h is continuous on its entire domain (see Figure 1.25c).

FIGURE 1.25

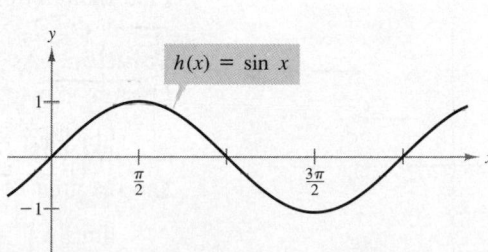

(a) Nonremovable discontinuity at $x = 0$

(b) Removable discontinuity at $x = 1$

(c) Continuous on the entire real line

One-Sided Limits and Continuity on a Closed Interval

To understand continuity on a closed interval, you first need to look at a different type of limit called a **one-sided limit.** For example, the **limit from the right** means that x approaches c from values greater than c. This limit is denoted as

$$\lim_{x \to c^+} f(x) = L. \qquad \text{Limit from the right}$$

Similarly, the **limit from the left** means that x approaches c from values less than c. This limit is denoted as

$$\lim_{x \to c^-} f(x) = L. \qquad \text{Limit from the left}$$

One-sided limits are useful in taking limits of functions involving radicals. For instance, if n is an even integer, then

$$\lim_{x \to 0^+} \sqrt[n]{x} = 0.$$

EXAMPLE 2   A One-Sided Limit

Find the limit of $f(x) = \sqrt{x - 1}$ as x approaches 1 from the right.

Solution   As indicated in Figure 1.26, the limit as x approaches 1 from the right is

$$\lim_{x \to 1^+} \sqrt{x - 1} = 0.$$

Another use of one-sided limits is in the investigation of the behavior of **step functions.** One common type of step function is the **greatest integer function $[\![x]\!]$,** defined by

$$[\![x]\!] = \text{greatest integer } n \text{ such that } n \le x.$$

For instance, $[\![2.5]\!] = 2$ and $[\![-2.5]\!] = -3$.

EXAMPLE 3   The Greatest Integer Function

Find the limit of $f(x) = [\![x]\!]$ as x approaches 0 from the left and from the right.

Solution   As shown in Figure 1.27, the limit as x approaches 0 *from the left* is given by

$$\lim_{x \to 0^-} [\![x]\!] = -1$$

and the limit as x approaches 0 *from the right* is given by

$$\lim_{x \to 0^+} [\![x]\!] = 0.$$

The greatest integer function is not continuous at 0 because the left and right limits at zero are different. By similar reasoning, you can see that the greatest integer function is not continuous at any integer n.

$f(x) = \sqrt{x - 1}$

(1, 0)

FIGURE 1.26
The limit of $f(x)$ as x approaches 1 from the right is 0.

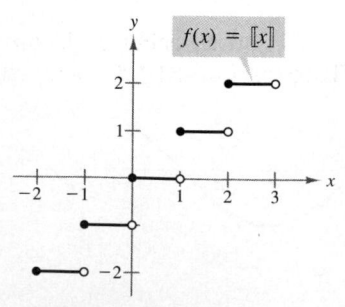

$f(x) = [\![x]\!]$

FIGURE 1.27
Greatest integer function

When the limit from the left is not equal to the limit from the right, the (two-sided) limit *does not exist*. The next theorem makes this more explicit. The proof of this theorem follows directly from the definition of a limit and the definition of a one-sided limit.

THEOREM 1.10   The Existence of a Limit

Let f be a function and let c and L be real numbers. The limit of $f(x)$ as x approaches c is L if and only if

$$\lim_{x \to c^-} f(x) = L \quad \text{and} \quad \lim_{x \to c^+} f(x) = L.$$

The concept of a one-sided limit allows you to extend the definition of continuity to closed intervals. Basically, a function is continuous on a closed interval if it is continuous in the interior of the interval and possesses one-sided continuity at the endpoints. We state this formally as follows.

Definition of Continuity on a Closed Interval

A function f is **continuous on the closed interval** $[a, b]$ if it is continuous on the open interval (a, b) and

$$\lim_{x \to a^+} f(x) = f(a) \quad \text{and} \quad \lim_{x \to b^-} f(x) = f(b).$$

The function f is **continuous from the right** at a and **continuous from the left** at b (see Figure 1.28).

FIGURE 1.28
Continuous function on a closed interval.

Similar definitions can be made to cover continuity on half-open intervals of the form $(a, b]$ and $[a, b)$, or on infinite intervals. For example, the function $f(x) = \sqrt{x}$ is continuous on the infinite interval $[0, \infty)$.

EXAMPLE 4   Continuity on a Closed Interval

Discuss the continuity of $f(x) = \sqrt{1 - x^2}$.

Solution   The domain of f is the closed interval $[-1, 1]$. At all points in the open interval $(-1, 1)$ the continuity of f follows from Theorems 1.4 and 1.5. Moreover, because

$$\lim_{x \to -1^+} \sqrt{1 - x^2} = 0 = f(-1)$$

and

$$\lim_{x \to 1^-} \sqrt{1 - x^2} = 0 = f(1)$$

you can conclude that f is continuous on the closed interval $[-1, 1]$, as shown in Figure 1.29.

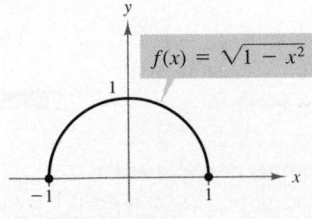

FIGURE 1.29
Continuous function on $[-1, 1]$.

The space shuttle uses liquid hydrogen and liquid oxygen as fuel. The liquid hydrogen must be stored at a temperature between $-262°$ C and $-253°$ C. At lower temperatures, the hydrogen would be a solid and at higher temperatures, it would be a gas.

The next example shows how a one-sided limit can be used to determine the value of absolute zero on the Kelvin scale.

EXAMPLE 5   Charles's Law and Absolute Zero

On the Kelvin scale, *absolute zero* is the temperature 0 K. Although temperatures of approximately 0.0001 K have been produced in laboratories, absolute zero has never been attained. In fact, evidence suggests that absolute zero *cannot* be attained. How did scientists determine that 0 K is the "lower limit" of the temperature of matter?

Solution   The determination of absolute zero stems from the work of the French physicist, Jacques Charles (1746–1823). Charles discovered that the volume of a gas at a constant pressure increases linearly with the temperature of the gas. Table 1.5 illustrates this relationship between volume and temperature. In the table, one mole of hydrogen is held at a constant pressure of one atmosphere. The volume V is measured in liters and the temperature T is measured in degrees Celsius.

TABLE 1.5

T	-40	-20	0	20	40	60	80
V	19.1482	20.7908	22.4334	24.0760	25.7186	27.3612	29.0038

The points represented by the table are shown in Figure 1.30. Moreover, by using the points in the table, you can determine that T and V are related by the linear equation

$$V = 0.08213T + 22.4334$$

or

$$T = \frac{V - 22.4334}{0.08213}.$$

By reasoning that the volume of the gas can approach 0 (but never go below 0), you can determine that the "least possible temperature" is given by

$$\lim_{V \to 0^+} T = \lim_{V \to 0^+} \frac{V - 22.4334}{0.08213}$$

$$= \frac{0 - 22.4334}{0.08213} \qquad \text{Use direct substitution}$$

$$\approx -273.15.$$

Thus, absolute zero on the Kelvin scale is $-273.15°$ on the Celsius scale.

FIGURE 1.30
The volume of hydrogen gas depends on its temperature.

$V = 0.08213T + 22.4334$

$(-273.15, 0)$

REMARK   Charles's Law for gases (assuming constant pressure) can be stated as

$$V = RT \qquad \text{Charles's Law}$$

where V is volume, R is a constant, and T is temperature. In the statement of this law, what property must the temperature scale have?

Properties of Continuity

In Section 1.2, you studied several properties of limits. Each of those properties yields a corresponding property pertaining to the continuity of a function. For instance, Theorem 1.11 follows directly from Theorem 1.2.

THEOREM 1.11   Properties of Continuity

If b is a real number and f and g are continuous at $x = c$, then the following functions are also continuous at c.

1. Scalar multiple: $b\,f$ **2.** Sum and difference: $f \pm g$

3. Product: $f\,g$ **4.** Quotient: $\dfrac{f}{g}$,   if $g(c) \neq 0$

The following types of functions are continuous at every point in their domains.

1. Polynomial functions: $p(x) = a_n x^n + a_{n-1} x^{n-1} + \cdots + a_1 x + a_0$

2. Rational functions: $r(x) = \dfrac{p(x)}{q(x)}$, $q(x) \neq 0$

3. Radical functions: $f(x) = \sqrt[n]{x}$

4. Trigonometric functions: $\sin x,\ \cos x,\ \tan x,\ \cot x,\ \sec x,\ \csc x$

By combining Theorem 1.11 with this summary, you can conclude that a wide variety of elementary functions are continuous at every point in their domains.

EXAMPLE 6   Applying Properties of Continuity

By Theorem 1.11, it follows that each of the following functions is continuous at every point in its domain.

$$f(x) = x + \sin x, \qquad f(x) = 3\tan x, \qquad f(x) = \frac{x^2 + 1}{\cos x}$$

For instance, the first function is continuous at every real number because the functions $y = x$ and $y = \sin x$ are continuous at every real number and the sum of continuous functions is continuous.

The next theorem, which is a consequence of Theorem 1.5, allows you to determine the continuity of *composite* functions such as

$$f(x) = \sin 3x, \qquad f(x) = \sqrt{x^2 + 1}, \qquad f(x) = \tan \frac{1}{x}.$$

THEOREM 1.12   Continuity of a Composite Function

If g is continuous at c and f is continuous at $g(c)$, then the composite function given by $(f \circ g)(x) = f(g(x))$ is continuous at c.

REMARK   One consequence of Theorem 1.12 is that if f and g satisfy the given conditions, you can determine the limit of $f(g(x))$ as x approaches c to be

$$\lim_{x \to c} f(g(x)) = f(g(c)).$$

EXAMPLE 7   Testing for Continuity

Describe the intervals on which each function is continuous.

a. $f(x) = \tan x$ **b.** $g(x) = \begin{cases} \sin\dfrac{1}{x}, & x \neq 0 \\ 0, & x = 0 \end{cases}$ **c.** $h(x) = \begin{cases} x\sin\dfrac{1}{x}, & x \neq 0 \\ 0, & x = 0 \end{cases}$

Solution

a. The tangent function, $f(x) = \tan x$, is undefined at

$$x = \frac{\pi}{2} + n\pi, \qquad n \text{ is an integer.}$$

At all other points it is continuous. Thus, $f(x) = \tan x$ is continuous on the open intervals

$$\ldots, \left(-\frac{3\pi}{2}, -\frac{\pi}{2}\right), \left(-\frac{\pi}{2}, \frac{\pi}{2}\right), \left(\frac{\pi}{2}, \frac{3\pi}{2}\right), \ldots$$

as shown in Figure 1.31(a).

b. Because $y = 1/x$ is continuous except at $x = 0$ and the sine function is continuous for all real values of x, it follows that

$$y = \sin\frac{1}{x}$$

is continuous at all real values except $x = 0$. At $x = 0$, the limit of $g(x)$ does not exist (see Example 5, Section 1.1). Therefore, g is continuous on the intervals $(-\infty, 0)$ and $(0, \infty)$, as indicated in Figure 1.31(b).

c. This function is similar to that in part b except that the oscillations are damped by the factor x. Using the Squeeze Theorem, you obtain

$$-|x| \leq x\sin\frac{1}{x} \leq |x|, \qquad x \neq 0$$

and you can conclude that

$$\lim_{x \to 0} h(x) = 0.$$

Thus, h is continuous on the entire real line, as indicated in Figure 1.31(c).

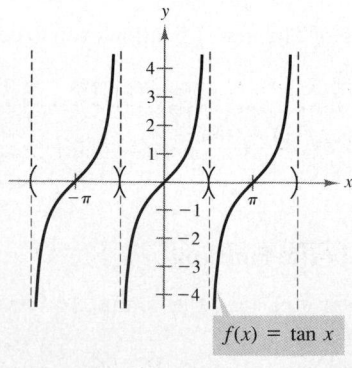

$f(x) = \tan x$

(a)

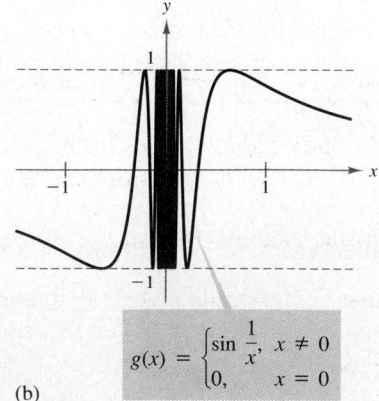

$g(x) = \begin{cases} \sin\dfrac{1}{x}, & x \neq 0 \\ 0, & x = 0 \end{cases}$

(b)

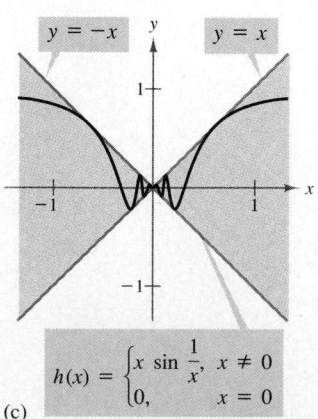

$y = -x \qquad y = x$

$h(x) = \begin{cases} x\sin\dfrac{1}{x}, & x \neq 0 \\ 0, & x = 0 \end{cases}$

(c)

FIGURE 1.31

FIGURE 1.32

f is continuous on $[a, b]$. (For k, there exist 3 c's.)

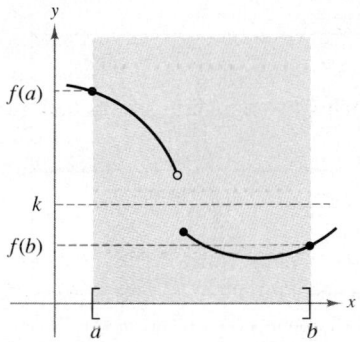

FIGURE 1.33

f is not continuous on $[a, b]$. (For k, there are no c's.)

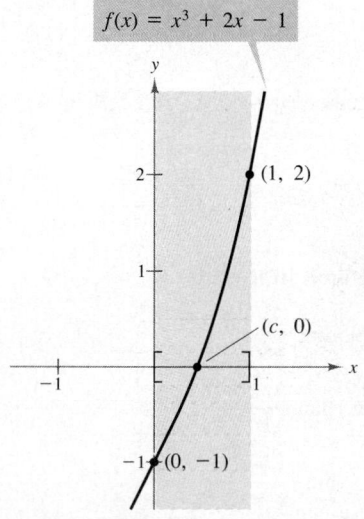

FIGURE 1.34

f is continuous on $[0, 1]$ with $f(0) < 0$ and $f(1) > 0$.

The Intermediate Value Theorem

We conclude this section with an important theorem concerning the behavior of functions that are continuous on a closed interval.

THEOREM 1.13   Intermediate Value Theorem

If f is continuous on the closed interval $[a, b]$ and k is any number between $f(a)$ and $f(b)$, then there is at least one number c in $[a, b]$ such that $f(c) = k$.

REMARK   Note that the Intermediate Value Theorem tells you that at least one c exists, but it does not give a method for finding c. Such theorems are called **existence theorems.**

By referring to a text on advanced calculus, you will find that a proof of this theorem is based on a property of real numbers called *completeness*. The Intermediate Value Theorem states that for a continuous function f, if x takes on all values between a and b, then $f(x)$ must take on all values between $f(a)$ and $f(b)$.

As a simple example of this theorem, consider a person's height. Suppose that a girl is 5 feet tall on her thirteenth birthday and 5 feet 7 inches tall on her fourteenth birthday. Then, for any height h between 5 feet and 5 feet 7 inches, there must have been a time t when her height was exactly h. This seems reasonable because human growth is continuous and a person's height does not abruptly change from one value to another.

The Intermediate Value Theorem guarantees the existence of *at least one* number c in the closed interval $[a, b]$. There may, of course, be more than one number c such that $f(c) = k$, as shown in Figure 1.32. A function that is not continuous does not necessarily possess the intermediate value property. For example, the graph of the function shown in Figure 1.33 jumps over the horizontal line given by $y = k$ and for this function there is no value of c in $[a, b]$ such that $f(c) = k$.

The Intermediate Value Theorem often can be used to locate the zeros of a function that is continuous on a closed interval. Specifically, if f is continuous on $[a, b]$ and $f(a)$ and $f(b)$ differ in sign, then the Intermediate Value Theorem guarantees the existence of at least one zero of f in the closed interval $[a, b]$.

EXAMPLE 8   An Application of the Intermediate Value Theorem

Use the Intermediate Value Theorem to show that the polynomial function $f(x) = x^3 + 2x - 1$ has a zero in the interval $[0, 1]$.

Solution   Note that f is continuous on the closed interval $[0, 1]$. Because

$$f(0) = 0^3 + 2(0) - 1 = -1 \quad \text{and} \quad f(1) = 1^3 + 2(1) - 1 = 2$$

it follows that $f(0) < 0$ and $f(1) > 0$. You can therefore apply the Intermediate Value Theorem to conclude that there must be some c in $[0, 1]$ such that $f(c) = 0$, as shown in Figure 1.34.

The **bisection method** for approximating the real zeros of a continuous function is similar to the method used in Example 8. If you know that a zero exists in the closed interval $[a, b]$, then the zero must lie in the interval

$$\left[a, \frac{a + b}{2} \right] \quad \text{or} \quad \left[\frac{a + b}{2}, b \right].$$

From the sign of $f([a + b]/2)$, you can determine which interval contains the zero. By repeatedly bisecting the interval, you can "close in" on the zero of the function.

: TECHNOLOGY   You can also use the zoom feature of a graphing utility to approximate the real zeros of a continuous function. By repeatedly zooming in on the point where the graph crosses the x-axis, and adjusting the x-axis scale, you can approximate the zero of the function to any desired accuracy, as shown in Figure 1.35.

FIGURE 1.35
Zooming in on the zero of f.

EXERCISES for Section 1.4

TECHNOLOGY
Laboratory Guide
Lab 1.4

In Exercises 1–6, use the graph to determine the limit.
a. $\lim\limits_{x \to c^+} f(x)$ **b.** $\lim\limits_{x \to c^-} f(x)$ **c.** $\lim\limits_{x \to c} f(x)$

1.

2.

3.

4.

5.

6.

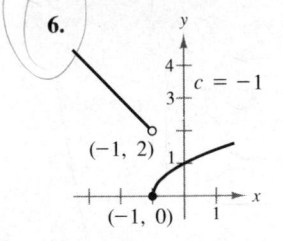

In Exercises 7–22, find the limit (if it exists).

7. $\lim\limits_{x \to 5^+} \dfrac{x - 5}{x^2 - 25}$

8. $\lim\limits_{x \to 2^+} \dfrac{2 - x}{x^2 - 4}$

9. $\lim\limits_{x \to 2^+} \dfrac{x}{\sqrt{x^2 - 4}}$

10. $\lim\limits_{x \to 4^-} \dfrac{\sqrt{x} - 2}{x - 4}$

11. $\lim\limits_{x \to 0} \dfrac{|x|}{x}$

12. $\lim\limits_{x \to 2} \dfrac{|x - 2|}{x - 2}$

13. $\lim\limits_{\Delta x \to 0^-} \dfrac{\dfrac{1}{x + \Delta x} - \dfrac{1}{x}}{\Delta x}$

14. $\lim\limits_{\Delta x \to 0^+} \dfrac{(x + \Delta x)^2 + x + \Delta x - (x^2 + x)}{\Delta x}$

15. $\lim\limits_{x \to 3} f(x)$, where $f(x) = \begin{cases} \dfrac{x+2}{2}, & x \le 3 \\[2mm] \dfrac{12 - 2x}{3}, & x > 3 \end{cases}$

16. $\lim\limits_{x \to 2} f(x)$, where $f(x) = \begin{cases} x^2 - 4x + 6, & x < 2 \\ -x^2 + 4x - 2, & x \ge 2 \end{cases}$

17. $\lim\limits_{x \to 1} f(x)$, where $f(x) = \begin{cases} x^3 + 1, & x < 1 \\ x + 1, & x \ge 1 \end{cases}$

18. $\lim\limits_{x \to 1} f(x)$, where $f(x) = \begin{cases} x, & x \le 1 \\ 1 - x, & x > 1 \end{cases}$

19. $\lim\limits_{x \to \pi} \cot x$ **20.** $\lim\limits_{x \to \pi/2} \sec x$

21. $\lim\limits_{x \to 3^-} (2[\![x]\!] - 1)$ **22.** $\lim\limits_{x \to 2^+} (2x - [\![x]\!])$

In Exercises 23–26 find the x-values (if any) at which f is not continuous.

23. $f(x) = \dfrac{1}{x^2 - 4}$ **24.** $f(x) = \dfrac{x^2 - 1}{x + 1}$

25. $f(x) = \dfrac{[\![x]\!]}{2} + x$ **26.** $f(x) = \begin{cases} x, & x < 1 \\ 2, & x = 1 \\ 2x - 1, & x > 1 \end{cases}$

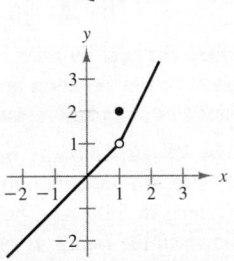

In Exercises 27–48, find the x-values (if any) at which f is not continuous. Which of the discontinuities are removable?

27. $f(x) = x^2 - 2x + 1$ **28.** $f(x) = \dfrac{1}{x^2 + 1}$

29. $f(x) = x + \sin x$ **30.** $f(x) = \cos\dfrac{\pi x}{2}$

31. $f(x) = \dfrac{1}{x - 1}$ **32.** $f(x) = \dfrac{x}{x^2 - 1}$

33. $f(x) = \dfrac{x}{x^2 + 1}$ **34.** $f(x) = \dfrac{x - 3}{x^2 - 9}$

35. $f(x) = \dfrac{x - 1}{x^2 + x - 2}$ **36.** $f(x) = \dfrac{x + 2}{x^2 - 3x - 10}$

37. $f(x) = \dfrac{|x + 2|}{x + 2}$ **38.** $f(x) = \dfrac{|x - 3|}{x - 3}$

39. $f(x) = \begin{cases} x, & x \le 1 \\ x^2, & x > 1 \end{cases}$

40. $f(x) = \begin{cases} -2x + 3, & x < 1 \\ x^2, & x \ge 1 \end{cases}$

41. $f(x) = \begin{cases} \dfrac{x}{2} + 1, & x \le 2 \\[2mm] 3 - x, & x > 2 \end{cases}$

42. $f(x) = \begin{cases} -2x, & x \le 2 \\ x^2 - 4x + 1, & x > 2 \end{cases}$

43. $f(x) = \begin{cases} \csc\dfrac{\pi x}{6}, & |x - 3| \le 2 \\[2mm] 2, & |x - 3| > 2 \end{cases}$

44. $f(x) = \begin{cases} \tan\dfrac{\pi x}{4}, & |x| < 1 \\[2mm] x, & |x| \ge 1 \end{cases}$

45. $f(x) = \csc 2x$ **46.** $f(x) = \tan\dfrac{\pi x}{2}$

47. $f(x) = [\![x - 1]\!]$ **48.** $f(x) = x - [\![x]\!]$

In Exercises 49–52, find the constants a and b so that the function is continuous on the entire real line.

49. $f(x) = \begin{cases} x^3, & x \le 2 \\ ax^2, & x > 2 \end{cases}$

50. $f(x) = \begin{cases} 2, & x \le -1 \\ ax + b, & -1 < x < 3 \\ -2, & x \ge 3 \end{cases}$

51. $g(x) = \begin{cases} \dfrac{4 \sin x}{x}, & x < 0 \\[2mm] a - 2x, & x \ge 0 \end{cases}$

52. $g(x) = \begin{cases} \dfrac{x^2 - a^2}{x - a}, & x \ne a \\[2mm] 8, & x = a \end{cases}$

In Exercises 53–56, discuss the continuity of the composite function $h(x) = f(g(x))$.

53. $f(x) = x^2$
$g(x) = x - 1$

54. $f(x) = \dfrac{1}{\sqrt{x}}$
$g(x) = x - 1$

55. $f(x) = \dfrac{1}{(x - 6)}$
$g(x) = x^2 + 5$

56. $f(x) = \sin x$
$g(x) = x^2$

In Exercises 57–60, find the interval(s) for which the function is continuous.

57. $f(x) = \dfrac{x}{x^2 + 1}$

58. $f(x) = x\sqrt{x + 3}$

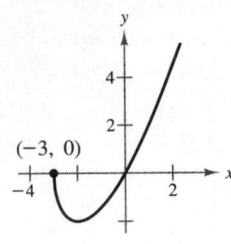

59. $f(x) = \csc \dfrac{x}{2}$

60. $f(x) = \dfrac{x + 1}{\sqrt{x}}$

C **In Exercises 61–64, use a graphing utility to graph the function. Then use the graph to determine any x-values at which the function is not continuous.**

61. $f(x) = [\![x]\!] - x$

62. $h(x) = \dfrac{1}{x^2 - x - 2}$

63. $g(x) = \begin{cases} 2x - 4, & x \le 3 \\ x^2 - 2x, & x > 3 \end{cases}$

64. $f(x) = \begin{cases} \dfrac{\cos x - 1}{x}, & x < 0 \\ 5x, & x \ge 0 \end{cases}$

C **Essay   In Exercises 65 and 66, use a graphing utility to graph the function on the interval $[-4, 4]$. Does the graph of the function appear continuous on this interval? Is the function continuous on $[-4, 4]$? Write a short paragraph about the importance of examining a function analytically as well as graphically.**

65. $f(x) = \dfrac{\sin x}{x}$

66. $f(x) = \dfrac{x^3 - 8}{x - 2}$

Essay   In Exercises 67 and 68, give a written explanation of why the function has a zero in the indicated interval.

67. $f(x) = x^2 - 4x + 3$, $[2, 4]$

68. $f(x) = x^3 + 3x - 2$, $[0, 1]$

C **In Exercises 69 and 70, use the Intermediate Value Theorem and a graphing utility to approximate the zero of the function in the interval $[0, 1]$. Repeatedly "zoom in" on the graph of the function to approximate the zero accurate to two decimal places.**

69. $f(x) = x^3 + x - 1$

70. $f(x) = x^3 + 3x - 2$

In Exercises 71–74, verify that the Intermediate Value Theorem applies in the indicated interval and find the value of c guaranteed by the theorem.

71. $f(x) = x^2 + x - 1$, $[0, 5]$, $f(c) = 11$

72. $f(x) = x^2 - 6x + 8$, $[0, 3]$, $f(c) = 0$

73. $f(x) = x^3 - x^2 + x - 2$, $[0, 3]$, $f(c) = 4$

74. $f(x) = \dfrac{x^2 + x}{x - 1}$, $\left[\dfrac{5}{2}, 4\right]$, $f(c) = 6$

75. *Salary Contract*   A union contract guarantees a 9 percent increase for 5 years. For an initial annual salary of $28,500, the salary S is

$$S = 28{,}500(1.09)^{[\![t]\!]}$$

where $t = 0$ corresponds to 1990. Sketch a graph of this function and discuss its continuity.

76. *Telephone Rates*   A dial-direct long distance call between two cities costs $1.04 for the first 2 minutes and $0.36 for each additional minute or fraction thereof. Use the greatest integer function to write the cost C of a call in terms of the time t (in minutes). Sketch a graph of this function and discuss its continuity.

77. *Inventory Management*   The number of units in inventory in a small company is

$$N(t) = 25\left(2\left[\!\left[\dfrac{t + 2}{2}\right]\!\right] - t\right), \qquad 0 \le t \le 12$$

where the real number t is the time in months. Sketch the graph of this function and discuss its continuity. How often must this company replenish its inventory?

78. *Déjà Vu*   At 8:00 A.M. on Saturday a man begins running up the side of a mountain to his weekend campsite. On Sunday morning at 8:00 A.M. he runs back down the mountain, as shown in the figure. It takes him 20 minutes to run up, but only 10 minutes to run down. At some point on the way down, he realizes that he passed the same place at exactly the same time on Saturday. Prove that he is correct. [*Hint:* Let $s(t)$ and $r(t)$ be the position functions for the run up and down, respectively, and apply the Intermediate Value Theorem to the function $f(t) = s(t) - r(t)$.]

Saturday 8:00 A.M.

Sunday 8:00 A.M.

FIGURE FOR 78

79. Prove that if f is continuous and has no zeros in $[a, b]$, then either

$f(x) > 0$ for all x in $[a, b]$

or

$f(x) < 0$ for all x in $[a, b]$.

80. Show that the Dirichlet function

$$f(x) = \begin{cases} 0, & x \text{ rational} \\ 1, & x \text{ irrational} \end{cases}$$

is not continuous at any real number.

True or False **In Exercises 81–84, determine whether the statement is true or false. If it is false, explain why or give an example that shows it is false.**

81. If $\lim\limits_{x \to c} f(x) = L$ and $f(c) = L$, then f is continuous at c.

82. If $f(x) = g(x)$ for $x \neq c$ and $f(c) \neq g(c)$, then either f or g is not continuous at c.

83. A rational function can have infinitely many x-values at which it is not continuous.

84. The function $f(x) = |x - 1|/(x - 1)$ is continuous.

85. Prove that if $\lim\limits_{\Delta x \to 0} f(x_0 + \Delta x) = f(x_0)$, then f is continuous at x_0.

86. The **signum function** is defined by

$$\operatorname{sgn}(x) = \begin{cases} -1, & \text{if } x < 0 \\ 0, & \text{if } x = 0 \\ 1, & \text{if } x > 0. \end{cases}$$

Sketch the graph of $\operatorname{sgn}(x)$ and find the following limits.

a. $\lim\limits_{x \to 0^-} \operatorname{sgn}(x)$

b. $\lim\limits_{x \to 0^+} \operatorname{sgn}(x)$

c. $\lim\limits_{x \to 0} \operatorname{sgn}(x)$

d. Is $f(x) = |\operatorname{sgn}(x)|$ continuous on $(-20, 20)$?

87. Prove that for any real number y, there exists x in $(-\pi/2, \pi/2)$ such that $\tan x = y$.

88. Let $f(x) = (\sqrt{x + c^2} - c)/x$, $c > 0$. What is the domain of f? How can you define f at $x = 0$ in order for f to be continuous there?

89. Discuss the continuity of the function $h(x) = x[\![x]\!]$.

90. Let $f_1(x)$ and $f_2(x)$ be continuous on the closed interval $[a, b]$. If $f_1(a) < f_2(a)$ and $f_1(b) > f_2(b)$, prove that there exists c between a and b such that $f_1(c) = f_2(c)$.

Infinite Limits

Let f be the function given by $f(x) = 3/(x - 2)$. From Figure 1.36 and Table 1.6, you can see that $f(x)$ *decreases without bound* as x approaches 2 from the left, and $f(x)$ *increases without bound* as x approaches 2 from the right. This behavior is denoted as

$$\lim_{x \to 2^-} \frac{3}{x - 2} = -\infty \quad \text{and} \quad \lim_{x \to 2^+} \frac{3}{x - 2} = \infty.$$

TABLE 1.6

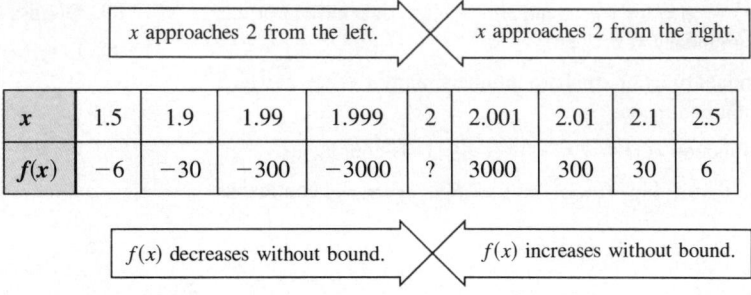

	x approaches 2 from the left.					x approaches 2 from the right.			
x	1.5	1.9	1.99	1.999	2	2.001	2.01	2.1	2.5
$f(x)$	−6	−30	−300	−3000	?	3000	300	30	6

	$f(x)$ decreases without bound.				$f(x)$ increases without bound.	

A limit in which $f(x)$ increases or decreases without bound as x approaches c is called an **infinite limit**.

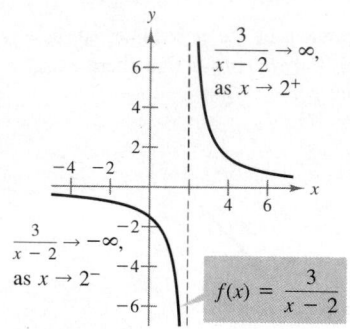

FIGURE 1.36

$f(x)$ increases and decreases without bound as x approaches 2.

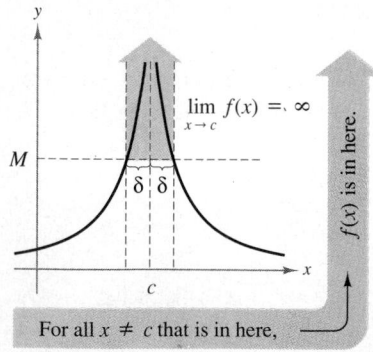

FIGURE 1.37

Infinite limits

Definition of Infinite Limits

Let f be a function that is defined at every real number in some open interval containing c (except possibly at c itself). The statement

$$\lim_{x \to c} f(x) = \infty$$

means that for each $M > 0$ there exists a $\delta > 0$ such that $f(x) > M$ whenever $0 < |x - c| < \delta$ (see Figure 1.37). Similarly, the statement

$$\lim_{x \to c} f(x) = -\infty$$

means that for each $N < 0$ there exists a $\delta > 0$ such that $f(x) < N$ whenever $0 < |x - c| < \delta$. To define the **infinite limit from the left,** replace $0 < |x - c| < \delta$ by $c - \delta < x < c$. To define the **infinite limit from the right,** replace $0 < |x - c| < \delta$ by $c < x < c + \delta$.

Be sure you see that the equal sign in the statement $\lim f(x) = \infty$ does not mean that the limit exists! On the contrary, it tells you how the limit *fails to exist* by denoting the unbounded behavior of $f(x)$ as x approaches c.

EXAMPLE 1   Determining Infinite Limits from a Graph

Use Figure 1.38 to determine the limit of each function as $x \to 1$ from the left and from the right.

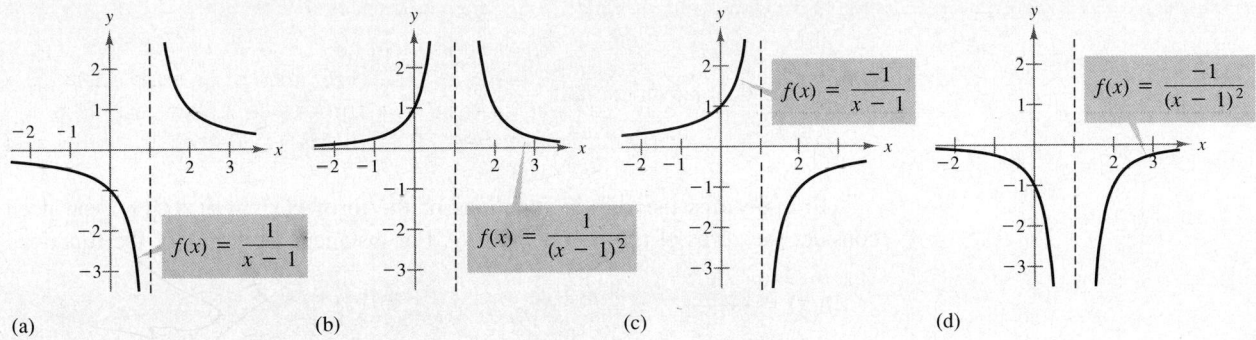

(a)   (b)   (c)   (d)

FIGURE 1.38
Each graph has an asymptote at $x = 1$.

Solution

a. $\displaystyle \lim_{x \to 1^-} \frac{1}{x-1} = -\infty$ and $\displaystyle \lim_{x \to 1^+} \frac{1}{x-1} = \infty$

b. $\displaystyle \lim_{x \to 1} \frac{1}{(x-1)^2} = \infty$

c. $\displaystyle \lim_{x \to 1^-} \frac{-1}{x-1} = \infty$ and $\displaystyle \lim_{x \to 1^+} \frac{-1}{x-1} = -\infty$

d. $\displaystyle \lim_{x \to 1} \frac{-1}{(x-1)^2} = -\infty$

Vertical Asymptotes

If it were possible to extend the graphs in Figure 1.38 toward infinity, you would see that each graph becomes arbitrarily close to the vertical line $x = 1$. This line is a **vertical asymptote** of the graph of f.

> ### Definition of Vertical Asymptote
>
> If $f(x)$ approaches infinity (or negative infinity) as x approaches c from the right or the left, then the line $x = c$ is a **vertical asymptote** of the graph of f.

REMARK   If a function f has a vertical asymptote at $x = c$, then f is *not continuous* at c.

In Example 1, note that each of the functions is a *quotient* and that the vertical asymptote occurs at a number where the denominator is 0 (and the numerator is not 0). The next theorem generalizes this observation. (A proof of the theorem is given in Appendix A.)

THEOREM 1.14   Vertical Asymptotes

Let f and g be continuous on an open interval containing c. If $f(c) \neq 0$, $g(c) = 0$, and there exists an open interval containing c such that $g(x) \neq 0$ for all $x \neq c$ in the interval, then the graph of the function given by

$$h(x) = \frac{f(x)}{g(x)}$$

has a vertical asymptote at $x = c$.

To apply this theorem to functions of the form $h(x) = f(x)/g(x)$, you need to consider the zeros of the denominator, g. For instance, the graph of the function

$$h(x) = \frac{2}{x - 3}$$

has a vertical asymptote at $x = 3$ because the denominator is zero (and the numerator is not zero) when $x = 3$.

EXAMPLE 2   Finding Vertical Asymptotes

Determine all vertical asymptotes of the graphs of the functions.

a. $f(x) = \dfrac{1}{2(x + 1)}$ **b.** $f(x) = \dfrac{x^2 + 1}{x^2 - 1}$ **c.** $f(x) = \cot x$

Solution

a. When $x = -1$, the denominator is 0 and the numerator is not 0. Hence, by Theorem 1.14, you can conclude that $x = -1$ is a vertical asymptote, as shown in Figure 1.39(a).

b. By factoring the denominator as

$$(x^2 - 1) = (x - 1)(x + 1)$$

you can see that the denominator is zero at $x = -1$ and $x = 1$. Moreover, because the numerator is not 0 at these two points, you can apply Theorem 1.14 to conclude that the graph of f has two vertical asymptotes, as shown in Figure 1.39(b).

c. By writing the cotangent function in the form

$$f(x) = \cot x = \frac{\cos x}{\sin x}$$

you can apply Theorem 1.14 to conclude that vertical asymptotes occur at all values of x such that $\cos x \neq 0$ and $\sin x = 0$, as shown in Figure 1.39(c). Hence, the graph of this function has infinitely many vertical asymptotes. These occur when $x = n\pi$, where n is an integer.

Theorem 1.14 requires that the value of the numerator at $x = c$ be nonzero. If both the numerator and the denominator are 0 at $x = c$, then you obtain the *indeterminate form* 0/0, and you cannot determine the limit behavior at $x = c$ without further investigation.

(a)

(b)

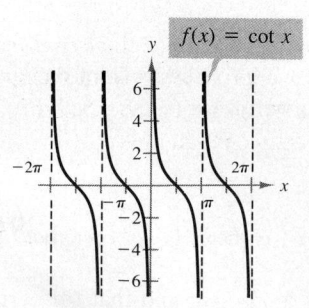

(c)

FIGURE 1.39

Functions with vertical asymptotes

EXAMPLE 3   A Rational Function with Common Factors

Determine all vertical asymptotes of the graph of

$$f(x) = \frac{x^2 + 2x - 8}{x^2 - 4}.$$

Solution   Begin by simplifying the expression, as follows.

$$f(x) = \frac{x^2 + 2x - 8}{x^2 - 4}$$

$$= \frac{(x + 4)(x - 2)}{(x + 2)(x - 2)}$$

$$= \frac{x + 4}{x + 2}, \qquad x \neq 2$$

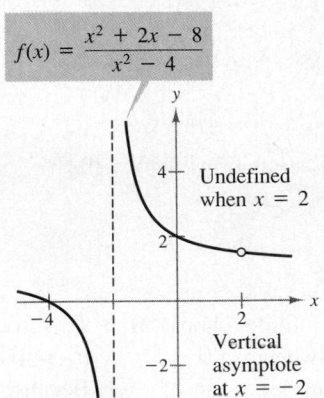

$$f(x) = \frac{x^2 + 2x - 8}{x^2 - 4}$$

Undefined when $x = 2$

Vertical asymptote at $x = -2$

FIGURE 1.40
$f(x)$ increases and decreases without bound as x approaches -2.

At all x-values other than $x = 2$, the graph of f coincides with the graph of $g(x) = (x + 4)/(x + 2)$. Thus, you can apply Theorem 1.14 to g to conclude that there is a vertical asymptote at $x = -2$, as shown in Figure 1.40. From the graph, you can see that

$$\lim_{x \to -2^-} \frac{x^2 + 2x - 8}{x^2 - 4} = -\infty \quad \text{and} \quad \lim_{x \to -2^+} \frac{x^2 + 2x - 8}{x^2 - 4} = \infty.$$

Note that $x = 2$ is *not* a vertical asymptote.

EXAMPLE 4   Determining Infinite Limits

Find the limits.

$$\lim_{x \to 1^-} \frac{x^2 - 3x}{x - 1} \quad \text{and} \quad \lim_{x \to 1^+} \frac{x^2 - 3x}{x - 1}$$

$$f(x) = \frac{x^2 - 3x}{x - 1}$$

FIGURE 1.41
f has a vertical asymptote at $x = 1$.

Solution   Because the denominator is 0 when $x = 1$ (and the numerator is not zero), you know that the graph of

$$f(x) = \frac{x^2 - 3x}{x - 1}$$

has a vertical asymptote at $x = 1$. This means that each of the given limits is either ∞ or $-\infty$. A graphing utility can help determine the result. From the graph of f shown in Figure 1.41, you can see that the graph approaches ∞ from the left of $x = 1$ and approaches $-\infty$ from the right of $x = 1$. Thus, you can conclude that

$$\lim_{x \to 1^-} \frac{x^2 - 3x}{x - 1} = \infty \quad \text{and} \quad \lim_{x \to 1^+} \frac{x^2 - 3x}{x - 1} = -\infty.$$

REMARK   When using a graphing calculator or graphing software, be careful to correctly interpret the graph of a function with a vertical asymptote—graphing utilities often have difficulty drawing this type of graph.

THEOREM 1.15   Properties of Infinite Limits

Let c and L be real numbers and let f and g be functions such that

$$\lim_{x \to c} f(x) = \infty \quad \text{and} \quad \lim_{x \to c} g(x) = L.$$

1. Sum or difference: $\displaystyle \lim_{x \to c}[f(x) \pm g(x)] = \infty$

2. Product: $\displaystyle \lim_{x \to c}[f(x)g(x)] = \infty, \qquad L > 0$

   $\displaystyle \lim_{x \to c}[f(x)g(x)] = -\infty, \qquad L < 0$

3. Quotient: $\displaystyle \lim_{x \to c} \frac{g(x)}{f(x)} = 0$

Similar properties hold for one-sided limits and for functions for which the limit of $f(x)$ as x approaches c is $-\infty$.

Proof   To show that the limit of $f(x) + g(x)$ is infinite, choose $M > 0$. You then need to find $\delta > 0$ such that $[f(x) + g(x)] > M$ whenever $0 < |x - c| < \delta$. For simplicity's sake, you can assume L is positive and let $M_1 = M + 1$. Because the limit of $f(x)$ is infinite, there exists δ_1 such that $f(x) > M_1$ whenever $0 < |x - c| < \delta_1$. Also, because the limit of $g(x)$ is L, there exists δ_2 such that $|g(x) - L| < 1$ whenever $0 < |x - c| < \delta_2$. By letting δ be the smaller of δ_1 and δ_2, you can conclude that $0 < |x - c| < \delta$ implies $f(x) > M + 1$ and $|g(x) - L| < 1$. The second of these two inequalities implies that $g(x) > L - 1$, and, adding this to the first inequality, you can write $f(x) + g(x) > M + L > M$. Thus, you can conclude that

$$\lim_{x \to c}[f(x) + g(x)] = \infty.$$

We leave the proofs of the remaining properties as exercises (see Exercise 59).

EXAMPLE 5   Determining Limits

a. Because $\displaystyle \lim_{x \to 0} 1 = 1$ and $\displaystyle \lim_{x \to 0} \frac{1}{x^2} = \infty$, you can write

$$\lim_{x \to 0}\left(1 + \frac{1}{x^2}\right) = \infty. \qquad \text{Property 1, Theorem 1.15}$$

b. Because $\displaystyle \lim_{x \to 1^-}(x^2 + 1) = 2$ and $\displaystyle \lim_{x \to 1^-} \frac{1}{x - 1} = -\infty$, you can write

$$\lim_{x \to 1^-} \frac{x^2 + 1}{1/(x - 1)} = 0. \qquad \text{Property 3, Theorem 1.15}$$

c. Because $\displaystyle \lim_{x \to 0^+} 3 = 3$ and $\displaystyle \lim_{x \to 0^+} \cot x = \infty$, you can write

$$\lim_{x \to 0^+} 3 \cot x = \infty. \qquad \text{Property 2, Theorem 1.15}$$

EXERCISES for Section 1.5

In Exercises 1–4, determine whether $f(x)$ approaches ∞ or $-\infty$ as x approaches -2 from the left and from the right.

1. $f(x) = \dfrac{1}{(x + 2)^2}$

2. $f(x) = \dfrac{1}{x + 2}$

3. $f(x) = \tan \dfrac{\pi x}{4}$

4. $f(x) = \sec \dfrac{\pi x}{4}$

In Exercises 5–8, determine whether $f(x)$ approaches ∞ or $-\infty$ as x approaches -3 from the left and from the right.

5. $f(x) = \dfrac{1}{x^2 - 9}$

6. $f(x) = \dfrac{x}{x^2 - 9}$

7. $f(x) = \dfrac{x^2}{x^2 - 9}$

8. $f(x) = \sec \dfrac{\pi x}{6}$

In Exercises 9–24, find the vertical asymptotes (if any) of the function.

9. $f(x) = \dfrac{1}{x^2}$

10. $f(x) = \dfrac{4}{(x - 2)^3}$

11. $f(x) = \dfrac{x^2 - 2}{x^2 - x - 2}$

12. $f(x) = \dfrac{2 + x}{1 - x}$

13. $f(x) = \dfrac{x^3}{x^2 - 1}$

14. $f(x) = \dfrac{-4x}{x^2 + 4}$

15. $f(x) = 1 - \dfrac{4}{x^2}$

16. $f(x) = \dfrac{-2}{(x - 2)^2}$

17. $f(x) = \dfrac{x}{x^2 + x - 2}$

18. $f(x) = \dfrac{1}{(x + 3)^4}$

19. $f(x) = \tan 2x$

20. $f(x) = \sec \pi x$

21. $f(x) = \dfrac{x^3 + 1}{x + 1}$

22. $f(x) = \dfrac{x^2 - 4}{x^3 + 2x^2 + x + 2}$

23. $f(x) = \dfrac{x}{\sin x}$

24. $f(x) = \dfrac{\tan x}{x}$

In Exercises 25–28, determine whether the function has a vertical asymptote or a removable discontinuity at $x = -1$.

25. $f(x) = \dfrac{x^2 - 1}{x + 1}$

26. $f(x) = \dfrac{x^2 - 6x - 7}{x + 1}$

27. $f(x) = \dfrac{x^2 + 1}{x + 1}$

28. $f(x) = \dfrac{\sin(x + 1)}{x + 1}$

In Exercises 29–42, find the limit.

29. $\displaystyle\lim_{x \to 2^+} \dfrac{x - 3}{x - 2}$

30. $\displaystyle\lim_{x \to 1^+} \dfrac{2 + x}{1 - x}$

31. $\displaystyle\lim_{x \to 4^+} \dfrac{x^2}{x^2 - 16}$

32. $\displaystyle\lim_{x \to 4} \dfrac{x^2}{x^2 + 16}$

33. $\displaystyle\lim_{x \to -3^-} \dfrac{x^2 + 2x - 3}{x^2 + x - 6}$

34. $\displaystyle\lim_{x \to (-1/2)^+} \dfrac{6x^2 + x - 1}{4x^2 - 4x - 3}$

35. $\displaystyle\lim_{x \to 0^-} \left(1 + \dfrac{1}{x}\right)$

36. $\displaystyle\lim_{x \to 0^-} \left(x^2 - \dfrac{1}{x}\right)$

37. $\displaystyle\lim_{x \to 0^+} \dfrac{2}{\sin x}$

38. $\displaystyle\lim_{x \to (\pi/2)^+} \dfrac{-2}{\cos x}$

39. $\displaystyle\lim_{x \to 1} \dfrac{x^2 - x}{(x^2 + 1)(x - 1)}$

40. $\displaystyle\lim_{x \to 3} \dfrac{x - 2}{x^2}$

41. $\displaystyle\lim_{x \to 1^+} \dfrac{x^2 + x + 1}{x^3 - 1}$

42. $\displaystyle\lim_{x \to 1} \dfrac{x^3 - 1}{x^2 + x + 1}$

C In Exercises 43 and 44, use a graphing utility to graph the function and determine the one-sided limit.

43. $f(x) = \dfrac{1}{x^2 - 25}$

$\displaystyle\lim_{x \to 5^-} f(x)$

44. $f(x) = \sec \dfrac{\pi x}{6}$

$\displaystyle\lim_{x \to 3^+} f(x)$

45. Suppose that Q varies inversely as the square root of $t - 4$. Find the limit of Q as $t \to 4^+$.

46. *Boyle's Law*   For a quantity of gas at a constant temperature, the pressure P is inversely proportional to its volume V. Find the limit of P as $V \to 0^+$.

47. *Moving Ladder*   A 25-foot ladder is leaning against a house (see figure). If the base of the ladder is pulled away from the house at a rate of 2 feet per second, the top will move down the wall at a rate of

$$r = \dfrac{2x}{\sqrt{625 - x^2}} \text{ ft/sec.}$$

Let x be the distance the base of the ladder is from the house.

a. Find the rate when x is 7 feet.

b. Find the rate when x is 15 feet.

c. Find the limit of r as $x \to 25^-$.

FIGURE FOR 47

25 ft

2 ft/sec

48. *Linear vs. Angular Speed* A patrol car is parked 50 feet from a long warehouse (see figure). The revolving light on top of the car turns at a rate of $\frac{1}{2}$ revolution per second. The rate at which the light beam is moving along the wall is

$$r = 50\pi \sec^2\theta \text{ ft/sec}.$$

a. Find the rate when θ is $\pi/6$.
b. Find the rate when θ is $\pi/3$.
c. Find the limit of r as $\theta \to \pi/2^-$.

50 ft

|← x →|

FIGURE FOR 48

49. *Illegal Drugs* The cost in millions of dollars for the federal government to seize x percent of a certain illegal drug as it enters the country is

$$C = \frac{528x}{100 - x}, \quad 0 \le x < 100.$$

a. Find the cost of seizing 25 percent.
b. Find the cost of seizing 50 percent.
c. Find the cost of seizing 75 percent.
d. Find the limit of C as $x \to 100^-$.

50. *Average Speed* On a trip of d miles to another city, a truck driver's average speed was x mi/hr. On the return trip the average speed was y mi/hr and the average speed for the round trip was 50 mi/hr.
a. Verify that

$$y = \frac{25x}{x - 25}.$$

What is the domain of this function?
b. Complete the table.

x	30	40	50	60
y				

Are the values of y different from what you expected? Explain.
c. Find the limit of y as $x \to 25^+$.

51. Use the graph of the function f (see figure) to sketch a graph of $g(x) = 1/f(x)$ on the interval $[-2, 3]$.

FIGURE FOR 51

52. Find functions f and g such that

$$\lim_{x \to c} f(x) = \infty \quad \text{and} \quad \lim_{x \to c} g(x) = \infty$$

but

$$\lim_{x \to c}[f(x) - g(x)] \ne 0.$$

True or False In Exercises 53–58, determine whether the statement is true or false. If it is false, explain why or give an example that shows it is false.

53. If $p(x)$ is a polynomial, then the function $f(x) = p(x)/(x - 1)$ has a vertical asymptote at $x = 1$.

54. If $\lim_{x \to c} f(x) = \infty$ and $\lim_{x \to c} g(x) = \infty$, then $\lim_{x \to c}[f(x) - g(x)] = 0$.

55. A rational function has at least one vertical asymptote.

56. Polynomial functions have no vertical asymptotes.

57. If f has a vertical asymptote at $x = 0$, then f is undefined at $x = 0$.

58. If f is continuous on $(-\infty, \infty)$, then it has no vertical asymptotes.

59. Prove the remaining parts of Theorem 1.15.

60. Prove that if $\lim_{x \to c} f(x) = \infty$, then

$$\lim_{x \to c} \frac{1}{f(x)} = 0.$$

61. Prove that if $\lim_{x \to c} \dfrac{1}{f(x)} = 0$, then $\lim_{x \to c} f(x)$ does not exist.

REVIEW EXERCISES for Chapter 1

In Exercises 1–16, find the limit (if it exists).

1. $\lim\limits_{x \to 2}(5x - 3)$

2. $\lim\limits_{x \to 2}(3x + 5)$

3. $\lim\limits_{x \to 2}(5x - 3)(3x + 5)$

4. $\lim\limits_{x \to 2}\dfrac{3x + 5}{5x - 3}$

5. $\lim\limits_{t \to 3}\dfrac{t^2 + 1}{t}$

6. $\lim\limits_{t \to 3}\dfrac{t^2 - 9}{t - 3}$

7. $\lim\limits_{t \to -2}\dfrac{t + 2}{t^2 - 4}$

8. $\lim\limits_{x \to 0}\dfrac{\sqrt{4 + x} - 2}{x}$

9. $\lim\limits_{x \to 0}\dfrac{[1/(x + 1)] - 1}{x}$

10. $\lim\limits_{s \to 0}\dfrac{(1/\sqrt{1 + s}) - 1}{s}$

11. $\lim\limits_{x \to -5}\dfrac{x^3 + 125}{x + 5}$

12. $\lim\limits_{x \to -2}\dfrac{x^2 - 4}{x^3 + 8}$

13. $\lim\limits_{x \to 0^+}\left(x - \dfrac{1}{x^3}\right)$

14. $\lim\limits_{x \to 2}\dfrac{1}{\sqrt[3]{x^2 - 4}}$

15. $\lim\limits_{\Delta x \to 0}\dfrac{\sin[(\pi/6) + \Delta x] - (1/2)}{\Delta x}$

[*Hint:* $\sin(\theta + \phi) = \sin\theta\cos\phi + \cos\theta\sin\phi$]

16. $\lim\limits_{\Delta x \to 0}\dfrac{\cos(\pi + \Delta x) + 1}{\Delta x}$

[*Hint:* $\cos(\theta + \phi) = \cos\theta\cos\phi - \sin\theta\sin\phi$]

In Exercises 17–26, find the one-sided limit.

17. $\lim\limits_{x \to -2^-} \dfrac{2x^2 + x + 1}{x + 2}$ 18. $\lim\limits_{x \to (1/2)^+} \dfrac{x}{2x - 1}$

19. $\lim\limits_{x \to -1^+} \dfrac{x + 1}{x^3 + 1}$ 20. $\lim\limits_{x \to -1^-} \dfrac{x + 1}{x^4 - 1}$

21. $\lim\limits_{x \to 1^-} \dfrac{x^2 + 2x + 1}{x - 1}$ 22. $\lim\limits_{x \to -1^+} \dfrac{x^2 - 2x + 1}{x + 1}$

23. $\lim\limits_{x \to 0^+} \dfrac{\sin 4x}{5x}$ 24. $\lim\limits_{x \to 0^+} \dfrac{\sec x}{x}$

25. $\lim\limits_{x \to 0^+} \dfrac{\csc 2x}{x}$ 26. $\lim\limits_{x \to 0^-} \dfrac{\cos^2 x}{x}$

In Exercises 27 and 28, (a) complete the table to approximate the limit and (b) rationalize the numerator to find the exact value of the limit.

x	1.1	1.01	1.001	1.0001
$f(x)$				

27. $\lim\limits_{x \to 1^+} \dfrac{\sqrt{2x + 1} - \sqrt{3}}{x - 1}$

28. $\lim\limits_{x \to 1^+} \dfrac{1 - \sqrt[3]{x}}{x - 1}$

[*Hint:* $a^3 - b^3 = (a - b)(a^2 + ab + b^2)$]

In Exercises 29–38, determine the intervals on which the function is continuous.

29. $f(x) = [\![x + 3]\!]$ 30. $f(x) = \dfrac{3x^2 - x - 2}{x - 1}$

31. $f(x) = \begin{cases} \dfrac{3x^2 - x - 2}{x - 1}, & x \neq 1 \\ 0, & x = 1 \end{cases}$

32. $f(x) = \begin{cases} 5 - x, & x \leq 2 \\ 2x - 3, & x > 2 \end{cases}$

33. $f(x) = \dfrac{1}{(x - 2)^2}$ 34. $f(x) = \sqrt{\dfrac{x + 1}{x}}$

35. $f(x) = \dfrac{3}{x + 1}$ 36. $f(x) = \dfrac{x + 1}{2x + 2}$

37. $f(x) = \csc \dfrac{\pi x}{2}$ 38. $f(x) = \tan 2x$

39. Determine the value of c so that the function is continuous on the entire real line.

$$f(x) = \begin{cases} x + 3, & x \leq 2 \\ cx + 6, & x > 2 \end{cases}$$

40. Determine the values of b and c so that the function is continuous on the entire real line.

$$f(x) = \begin{cases} x + 1, & 1 < x < 3 \\ x^2 + bx + c, & |x - 2| \geq 1 \end{cases}$$

41. *Compound Interest*   A sum of $5000 is deposited in a savings account that pays 12 percent compounded semiannually. After t years the amount of money saved is $A = 5000(1.06)^{[\![2t]\!]}$. Sketch a graph of this function and discuss its continuity.

42. *Cost of Overnight Delivery*   Suppose the cost of sending an overnight package from New York to Atlanta is $9.80 for the first pound and $2.50 for each additional pound. Use the greatest integer function to create a model for the cost C of overnight delivery of a package weighing x pounds. Sketch the graph of the function and discuss its continuity.

In Exercises 43–46, find the vertical asymptotes (if any) of the function.

43. $g(x) = 1 + \dfrac{2}{x}$ 44. $h(x) = \dfrac{4x}{4 - x^2}$

45. $f(x) = \dfrac{8}{(x - 10)^2}$ 46. $f(x) = \csc \pi x$

47. *Cost of Clear Air*   A utility company burns coal to generate electricity. The cost C in dollars of removing p percent of the air pollutants in the stack emissions is

$$C = \dfrac{80{,}000 p}{100 - p}, \qquad 0 \leq p < 100.$$

Find the cost of removing

a. 15 percent.   **b.** 50 percent.   **c.** 90 percent.
d. Find the limit of C as $p \to 100^-$.

48. The function f is defined as follows.

$$f(x) = \dfrac{\tan 2x}{x}, \qquad x \neq 0$$

a. Find $\lim\limits_{x \to 0} \dfrac{\tan 2x}{x}$ (if it exists).

b. Can the function f be defined at 0 so that it is continuous at $x = 0$?

True or False   **In Exercises 49–55, determine whether the statement is true or false. If it is false, explain why or give an example that shows it is false.**

49. $\lim\limits_{x \to 0} \dfrac{|x|}{x} = 1$ 50. $\lim\limits_{x \to 0} x^3 = 0$

51. If $f(x) = g(x)$ for all real numbers other than $x = 0$, and $\lim\limits_{x \to 0} f(x) = L$, then $\lim\limits_{x \to 0} g(x) = L$.

52. If $\lim\limits_{x \to c} f(x) = L$, then $f(c) = L$.

53. For polynomial functions, the limits from the right and the left must exist and must be equal.

54. $\lim\limits_{x \to 2^-} f(x) = 3$, where $f(x) = \begin{cases} 3, & x \leq 2 \\ 0, & x > 2 \end{cases}$

55. $\lim\limits_{x \to 3} f(x) = 1$, where $f(x) = \begin{cases} x - 2, & x \leq 3 \\ -x^2 + 8x - 14, & x > 3 \end{cases}$

56. Let $f(x) = \dfrac{x^2 - 4}{|x - 2|}$. Find $\lim\limits_{x \to 2^-} f(x)$, $\lim\limits_{x \to 2^+} f(x)$, and $\lim\limits_{x \to 2} f(x)$.

57. Let $f(x) = \sqrt{(x - 1)x}$. Find the domain of f. Find $\lim\limits_{x \to 0^-} f(x)$ and $\lim\limits_{x \to 1^+} f(x)$.

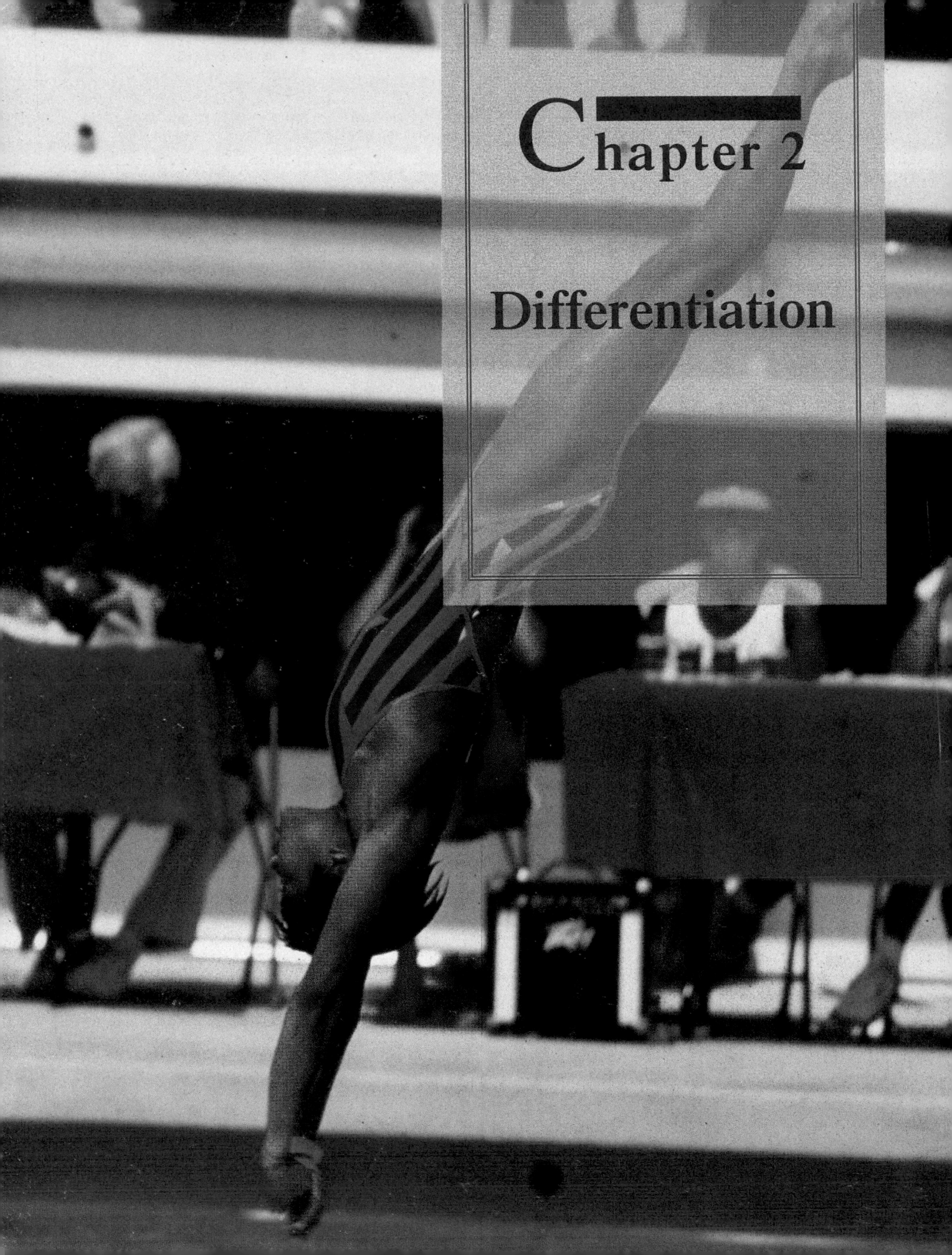

Chapter 2

Differentiation

ISAAC NEWTON

In addition to his work in calculus, Newton made revolutionary contributions to physics—these include the Universal Law of Gravitation and the three laws of motion.

The Tangent Line Problem

Calculus grew out of four major problems that European mathematicians were working on during the seventeenth century.

1. The tangent line problem (Sections 1.1 and 2.1)
2. The velocity and acceleration problem (Sections 2.2 and 2.3)
3. The minimum and maximum problem (Section 3.1)
4. The area problem (Section 4.2)

Each problem involves the notion of a limit, and we could introduce calculus with any of the four problems.

We gave a brief introduction to the tangent line problem in Section 1.1. Although partial solutions to this problem were given by Pierre de Fermat (1601–1655), René Descartes (1596–1650), Christian Huygens (1629–1695), and Isaac Barrow (1630–1677), credit for the first general solution is usually given to Isaac Newton (1642–1727) and Gottfried Leibniz (1646–1716). (For more information on the crediting of mathematical discoveries to the first "discoverer," see the article "Mathematical Firsts—Who Done It?" by Richard H. Williams and Roy D. Mazzagatti in the May, 1986 issue of *Mathematics Teacher*.)

What does it mean to say that a line is tangent to a curve at a point? For a circle, the tangent line at point P is the line that is perpendicular to the radial line at point P, as shown in Figure 2.1. For a general curve, however, the problem is more difficult. For example, how would you define the tangent lines shown in Figure 2.2?

You might say that a line is tangent to a curve at point P if it touches, but does not cross, the curve at point P. This definition would work for the first curve shown in Figure 2.2, but not for the second. *Or* you might say that a line is tangent to a curve if the line touches or intersects the curve at exactly one point—this definition would work for a circle but not for more general curves, as the third curve in Figure 2.2 shows.

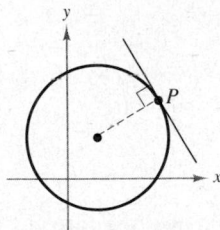

FIGURE 2.1
Tangent line to a circle

FIGURE 2.2
Tangent line to a curve at a point

◄ The velocity of a diver is discussed in Section 2.2, Example 10.

Essentially, the problem of finding the tangent line at a point P boils down to the problem of finding the *slope* of the tangent line at P. You can approximate this slope using a **secant line*** through the point of tangency and a second point on the curve, as shown in Figure 2.3. If $(c, f(c))$ is the point of tangency and $(c + \Delta x, f(c + \Delta x))$ is a second point on the graph of f, then the slope of the secant line through the two points is given by

$$
\begin{aligned}
m_{\text{sec}} &= \frac{f(c + \Delta x) - f(c)}{(c + \Delta x) - c} \qquad \frac{\text{Change in } y}{\text{Change in } x} \\[2mm]
&= \frac{f(c + \Delta x) - f(c)}{\Delta x}. \qquad \text{Slope of secant line}
\end{aligned}
$$

The right-hand side of this equation is a **difference quotient.** The denominator Δx is the **change in x,** and the numerator $\Delta y = f(c + \Delta x) - f(c)$ is the **change in y.**

The beauty of this procedure is that you can obtain more and more accurate approximations to the slope of the tangent line by choosing points closer and closer to the point of tangency, as shown in Figure 2.4.

FIGURE 2.3

The secant line through $(c, f(c))$ and $(c + \Delta x, f(c + \Delta x))$

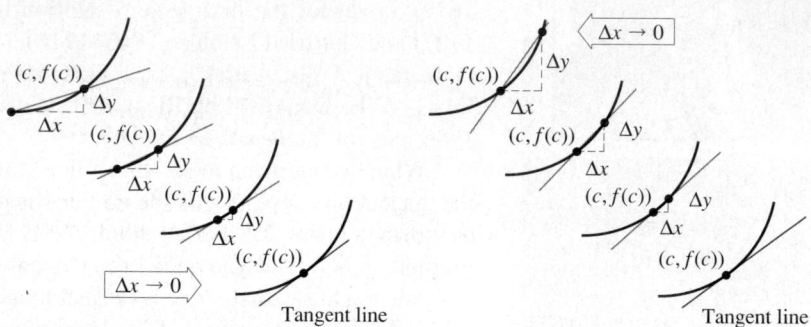

FIGURE 2.4

Definition of Tangent Line with Slope m

If f is defined on an open interval containing c, and if the limit

$$
\lim_{\Delta x \to 0} \frac{\Delta y}{\Delta x} = \lim_{\Delta x \to 0} \frac{f(c + \Delta x) - f(c)}{\Delta x} = m
$$

exists, then the line passing through $(c, f(c))$ with slope m is the **tangent line** to the graph of f at the point $(c, f(c))$.

The slope of the tangent line to the graph of f at the point $(c, f(c))$ is also called the **slope of the graph of f at $x = c$.**

---

* This use of the word *secant* comes from the Latin *secare,* meaning to cut, and is not a reference to the trigonometric function of the same name.

EXAMPLE 1   The Slope of the Graph of a Linear Function

Find the slope of the graph of $f(x) = 2x - 3$ at the point $(2, 1)$.

Solution   To find the slope of the graph of f when $c = 2$, you can apply the definition of the slope of a tangent line, as follows.

$$\lim_{\Delta x \to 0} \frac{f(2 + \Delta x) - f(2)}{\Delta x} = \lim_{\Delta x \to 0} \frac{[2(2 + \Delta x) - 3] - [2(2) - 3]}{\Delta x}$$

$$= \lim_{\Delta x \to 0} \frac{4 + 2\Delta x - 3 - 4 + 3}{\Delta x}$$

$$= \lim_{\Delta x \to 0} \frac{2\Delta x}{\Delta x}$$

$$= \lim_{\Delta x \to 0} 2$$

$$= 2$$

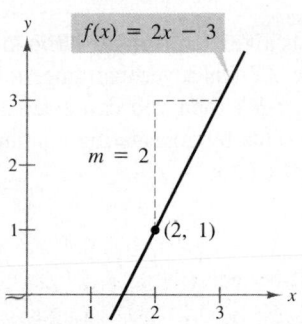

FIGURE 2.5
The slope of f at $(2, 1)$ is $m = 2$.

The slope of f at $(2, 1)$ is $m = 2$, as shown in Figure 2.5.

REMARK   In Example 1, the limit definition of the slope of f agrees with the definition of the slope of a line as discussed in Section 4 of the Prerequisites chapter.

The graph of a linear function has the same slope at any point. This is not true of nonlinear functions, as can be seen in the following example.

EXAMPLE 2   Tangent Lines to the Graph of a Nonlinear Function

Find the slope of the tangent line to the graph of

$$f(x) = x^2 + 1$$

at the points $(0, 1)$ and $(-1, 2)$, as shown in Figure 2.6.

Solution   Let $(x, f(x))$ represent an arbitrary point on the graph of f. Then the slope of the tangent line at $(x, f(x))$ is given by

$$\lim_{\Delta x \to 0} \frac{f(x + \Delta x) - f(x)}{\Delta x} = \lim_{\Delta x \to 0} \frac{(x + \Delta x)^2 + 1 - (x^2 + 1)}{\Delta x}$$

$$= \lim_{\Delta x \to 0} \frac{x^2 + 2x(\Delta x) + (\Delta x)^2 + 1 - x^2 - 1}{\Delta x}$$

$$= \lim_{\Delta x \to 0} \frac{2x(\Delta x) + (\Delta x)^2}{\Delta x}$$

$$= \lim_{\Delta x \to 0} (2x + \Delta x)$$

$$= 2x.$$

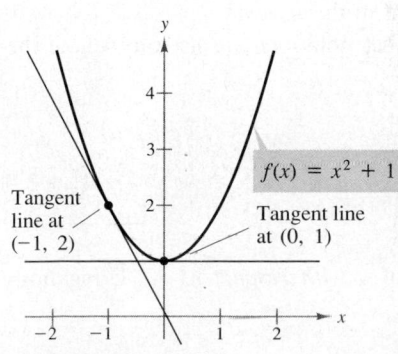

FIGURE 2.6
The slope of f at any point is $m = 2x$.

Therefore, the slope at *any* point $(x, f(x))$ on the graph of f is $m = 2x$. At the point $(0, 1)$, the slope is $m = 2(0) = 0$, and at the point $(-1, 2)$, the slope is $m = 2(-1) = -2$.

REMARK   In Example 2, note that x is held constant in the limit process (as $\Delta x \to 0$).

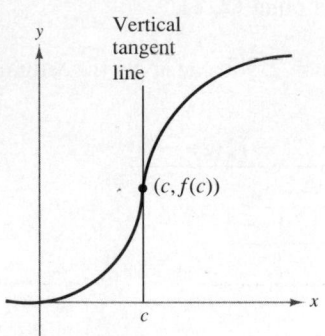

FIGURE 2.7
The graph of f has a vertical tangent line at $(c, f(c))$.

The definition of a tangent line to a curve does not cover the possibility of a vertical tangent line. For vertical tangent lines, you can use the following definition. If f is continuous at c and

$$\lim_{\Delta x \to 0} \left| \frac{f(c + \Delta x) - f(c)}{\Delta x} \right| = \infty$$

then the vertical line, $x = c$, passing through $(c, f(c))$ is a **vertical tangent line** to the graph of f. For example, the function shown in Figure 2.7 has a vertical tangent line at $(c, f(c))$. If the domain of f is the closed interval $[a, b]$, then you can extend the definition of a vertical tangent line to include the endpoints by considering continuity and limits from the right (for $x = a$) and from the left (for $x = b$).

$$\lim_{\Delta x \to 0^+} \left| \frac{f(a + \Delta x) - f(a)}{\Delta x} \right| = \infty \qquad \text{Limit from the right}$$

$$\lim_{\Delta x \to 0^-} \left| \frac{f(b + \Delta x) - f(b)}{\Delta x} \right| = \infty \qquad \text{Limit from the left}$$

The Derivative of a Function

You have now arrived at a crucial point in the study of calculus. The limit used to define the slope of a tangent line is also used to define one of the two fundamental operations of calculus—*differentiation.*

Definition of the Derivative of a Function

The **derivative** of f at x is given by

$$f'(x) = \lim_{\Delta x \to 0} \frac{f(x + \Delta x) - f(x)}{\Delta x}$$

provided the limit exists.

The process of finding the derivative of a function is called **differentiation**. A function is **differentiable** at x if its derivative exists at x and **differentiable on an open interval** (a, b) if it is differentiable at every point in the interval.

In addition to $f'(x)$, read "f prime of x," other notations are used to denote the derivative of $y = f(x)$. The most common are

$$f'(x), \qquad \frac{dy}{dx}, \qquad y', \qquad \frac{d}{dx}[f(x)], \qquad D_x[y]. \qquad \text{Notation for derivatives}$$

The notation dy/dx is read as "the derivative of y *with respect to x*." Using limit notation, you can write

$$\frac{dy}{dx} = \lim_{\Delta x \to 0} \frac{\Delta y}{\Delta x}$$

$$= \lim_{\Delta x \to 0} \frac{f(x + \Delta x) - f(x)}{\Delta x}$$

$$= f'(x).$$

EXAMPLE 3   Finding the Derivative by the Limit Process

Find the derivative of $f(x) = x^3 + 2x$.

Solution

$$f'(x) = \lim_{\Delta x \to 0} \frac{f(x + \Delta x) - f(x)}{\Delta x}$$

$$= \lim_{\Delta x \to 0} \frac{(x + \Delta x)^3 + 2(x + \Delta x) - (x^3 + 2x)}{\Delta x}$$

$$= \lim_{\Delta x \to 0} \frac{x^3 + 3x^2 \Delta x + 3x(\Delta x)^2 + (\Delta x)^3 + 2x + 2\Delta x - x^3 - 2x}{\Delta x}$$

$$= \lim_{\Delta x \to 0} \frac{3x^2 \Delta x + 3x(\Delta x)^2 + (\Delta x)^3 + 2\Delta x}{\Delta x}$$

$$= \lim_{\Delta x \to 0} \frac{\Delta x[3x^2 + 3x\Delta x + (\Delta x)^2 + 2]}{\Delta x}$$

$$= \lim_{\Delta x \to 0} [3x^2 + 3x\Delta x + (\Delta x)^2 + 2]$$

$$= 3x^2 + 2$$

Remember that the derivative of a function f is itself a function, which can be used to find the slope of the tangent line at the point $(x, f(x))$ on the graph of f.

EXAMPLE 4   Using the Derivative to Find the Slope at a Point

Find $f'(x)$ for $f(x) = \sqrt{x}$. Then find the slope of the graph of f at the points $(1, 1)$ and $(4, 2)$. Discuss the behavior of f at $(0, 0)$.

Solution   Use the procedure for rationalizing numerators, as discussed in Section 1.3.

$$f'(x) = \lim_{\Delta x \to 0} \frac{f(x + \Delta x) - f(x)}{\Delta x}$$

$$= \lim_{\Delta x \to 0} \frac{\sqrt{x + \Delta x} - \sqrt{x}}{\Delta x}$$

$$= \lim_{\Delta x \to 0} \left(\frac{\sqrt{x + \Delta x} - \sqrt{x}}{\Delta x} \right)\left(\frac{\sqrt{x + \Delta x} + \sqrt{x}}{\sqrt{x + \Delta x} + \sqrt{x}} \right)$$

$$= \lim_{\Delta x \to 0} \frac{(x + \Delta x) - x}{\Delta x(\sqrt{x + \Delta x} + \sqrt{x})}$$

$$= \lim_{\Delta x \to 0} \frac{\Delta x}{\Delta x(\sqrt{x + \Delta x} + \sqrt{x})}$$

$$= \lim_{\Delta x \to 0} \frac{1}{\sqrt{x + \Delta x} + \sqrt{x}}$$

$$= \frac{1}{2\sqrt{x}}$$

At the point $(1, 1)$, the slope is $f'(1) = \frac{1}{2}$. At the point $(4, 2)$, the slope is $f'(4) = \frac{1}{4}$. (See Figure 2.8.) At the point $(0, 0)$ the slope is undefined. Moreover, because the limit of $f'(x)$ as $x \to 0$ from the right is infinite, the graph of f has a vertical tangent line at $(0, 0)$.

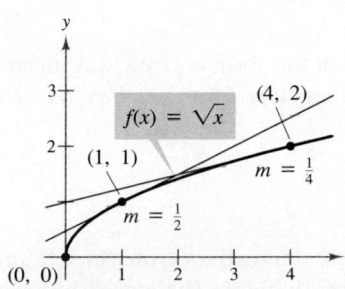

FIGURE 2.8
The slope of f at $(x, f(x))$, $x > 0$, is $m = 1/(2\sqrt{x})$.

In many applications, it is convenient to use a variable other than x as the independent variable, as shown in Example 5.

EXAMPLE 5   Finding the Derivative of a Function

Find the derivative with respect to t for the function $y = \dfrac{2}{t}$.

Solution   Considering $y = f(t)$, you obtain the following.

$$\dfrac{dy}{dt} = \lim_{\Delta t \to 0} \dfrac{f(t + \Delta t) - f(t)}{\Delta t} \qquad \text{Definition of derivative}$$

$$= \lim_{\Delta t \to 0} \dfrac{\dfrac{2}{t + \Delta t} - \dfrac{2}{t}}{\Delta t} \qquad f(t + \Delta t) = 2/(t + \Delta t) \text{ and } f(t) = 2/t$$

$$= \lim_{\Delta t \to 0} \dfrac{\dfrac{2t - 2(t + \Delta t)}{t(t + \Delta t)}}{\Delta t} \qquad \text{Combine fractions in numerator}$$

$$= \lim_{\Delta t \to 0} \dfrac{-2\Delta t}{\Delta t(t)(t + \Delta t)} \qquad \text{Simplify}$$

$$= \lim_{\Delta t \to 0} \dfrac{-2}{t(t + \Delta t)} \qquad \text{Cancel common factor of } \Delta t$$

$$= -\dfrac{2}{t^2} \qquad \text{Evaluate limit as } \Delta t \to 0$$

FIGURE 2.9
At the point (1, 2) the line
$y = -2t + 4$ is tangent to the graph
of $y = 2/t$.

TECHNOLOGY   A graphing utility can be used to reinforce the result given in Example 5. For instance, using the formula $dy/dt = -2/t^2$, you know that the slope of the graph of $y = 2/t$ at the point (1, 2) is $m = -2$. This implies that an equation of the tangent line to the graph at (1, 2) is $y - 2 = -2(t - 1)$ or $y = -2t + 4$, as shown in Figure 2.9.

Differentiability and Continuity

The following alternative limit form of the derivative is useful in investigating this relationship. The derivative of f at c is given by

$$f'(c) = \lim_{x \to c} \dfrac{f(x) - f(c)}{x - c} \qquad \text{Alternative form of derivative}$$

provided this limit exists. (A proof of the equivalence of this form is given in Appendix A.) Note that the existence of the limit in this alternative form requires that the one-sided limits

$$\lim_{x \to c^-} \dfrac{f(x) - f(c)}{x - c} \quad \text{and} \quad \lim_{x \to c^+} \dfrac{f(x) - f(c)}{x - c}$$

exist and are equal. These one-sided limits are called the **derivatives from the left and from the right,** respectively. We say that f is **differentiable on the closed interval** $[a, b]$ if it is differentiable on (a, b) and if the derivative from the right at a and the derivative from the left at b both exist.

If a function is not continuous at $x = c$, then it is also not differentiable at $x = c$. For instance, the greatest integer function

$$f(x) = [\![x]\!]$$

is not continuous at $x = 0$—hence, is not differentiable at $x = 0$ (see Figure 2.10). You can verify this by observing that

$$\lim_{x \to 0^-} \frac{f(x) - f(0)}{x - 0} = \lim_{x \to 0^-} \frac{[\![x]\!] - 0}{x} = -\infty \qquad \text{Derivative from the left}$$

and

$$\lim_{x \to 0^+} \frac{f(x) - f(0)}{x - 0} = \lim_{x \to 0^+} \frac{[\![x]\!] - 0}{x} = 0. \qquad \text{Derivative from the right}$$

Although it is true that differentiability implies continuity (as we will show in Theorem 2.1), the converse is not true. That is, it is possible for a function to be continuous at $x = c$ and *not* differentiable at $x = c$. Examples 6 and 7 illustrate this possibility.

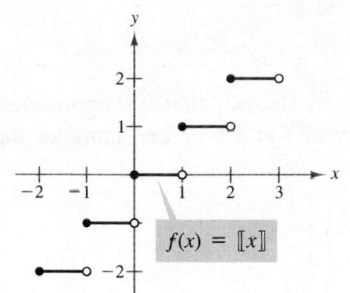

FIGURE 2.10
Not differentiable at $x = 0$, because f is not continuous at $x = 0$.

EXAMPLE 6   A Graph with a Sharp Turn

The function $f(x) = |x - 2|$ shown in Figure 2.11 is continuous at $x = 2$. However, the one-sided limits

$$\lim_{x \to 2^-} \frac{f(x) - f(2)}{x - 2} = \lim_{x \to 2^-} \frac{|x - 2| - 0}{x - 2} = -1 \qquad \text{Derivative from the left}$$

and

$$\lim_{x \to 2^+} \frac{f(x) - f(2)}{x - 2} = \lim_{x \to 2^+} \frac{|x - 2| - 0}{x - 2} = 1 \qquad \text{Derivative from the right}$$

are not equal. Therefore, f is not differentiable at $x = 2$ and the graph of f does not have a tangent line at the point $(2, 0)$.   ∎

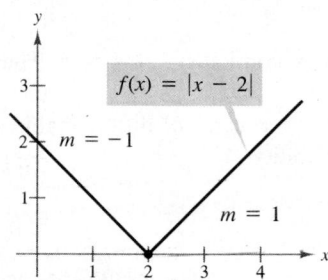

FIGURE 2.11
Not differentiable at $x = 2$, because the one-sided derivatives are not equal.

EXAMPLE 7   A Graph with a Vertical Tangent Line

The function $f(x) = x^{1/3}$ is continuous at $x = 0$, as shown in Figure 2.12. However, because the following limit is infinite

$$\lim_{x \to 0} \frac{f(x) - f(0)}{x - 0} = \lim_{x \to 0} \frac{x^{1/3} - 0}{x}$$

$$= \lim_{x \to 0} \frac{1}{x^{2/3}}$$

$$= \infty$$

you can conclude that the tangent line is vertical at $x = 0$. Therefore, f is not differentiable at $x = 0$.   ∎

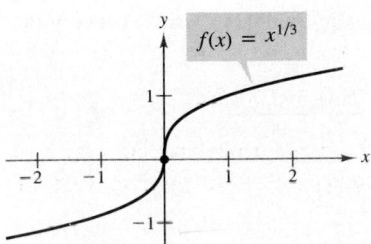

FIGURE 2.12
Not differentiable at $x = 0$, because f has a vertical tangent at $x = 0$.

From Examples 6 and 7, you can see that a function is not differentiable at a point at which its graph has a sharp turn *or* a vertical tangent.

THEOREM 2.1   Differentiability Implies Continuity

If f is differentiable at $x = c$, then f is continuous at $x = c$.

Proof   You can prove that f is continuous at $x = c$ by showing that $f(x)$ approaches $f(c)$ as $x \to c$. To do this, use the differentiability of f at $x = c$ and consider the following limit.

$$\lim_{x \to c}[f(x) - f(c)] = \lim_{x \to c}\left[(x - c)\left(\frac{f(x) - f(c)}{x - c}\right)\right]$$

$$= \left[\lim_{x \to c}(x - c)\right]\left[\lim_{x \to c}\frac{f(x) - f(c)}{x - c}\right]$$

$$= (0)[f'(c)]$$

$$= 0$$

Because the difference $[f(x) - f(c)]$ approaches zero as $x \to c$, you can conclude that $\lim_{x \to c} f(x) = f(c)$. Therefore, f is continuous at $x = c$.   ▬

You can summarize the relationship between continuity and differentiability as follows.

1. If a function is differentiable at $x = c$, then it is continuous at $x = c$. Thus, differentiability implies continuity.
2. It is possible for a function to be continuous at $x = c$ and *not* differentiable at $x = c$. Thus, continuity does not imply differentiability.

EXERCISES for Section 2.1

TECHNOLOGY
Laboratory Guide
Lab 2.1

In Exercises 1 and 2, estimate the slope of the curve at the point (x, y).

1. a.   b.

2. a.   b.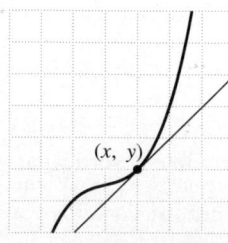

In Exercises 3 and 4, use the graph shown in the figure.

3. Identify or sketch each of the quantities on the figure.

   a. $f(1)$ and $f(4)$ b. $f(4) - f(1)$

   c. $y = \dfrac{f(4) - f(1)}{4 - 1}(x - 1) + f(1)$

4. Insert the proper inequality symbol ($<$ or $>$) between the expressions.

   a. $\dfrac{f(4) - f(1)}{4 - 1}$ ▨ $\dfrac{f(4) - f(3)}{4 - 3}$

   b. $\dfrac{f(4) - f(1)}{4 - 1}$ ▨ $f'(1)$

FIGURE FOR 3 AND 4

In Exercises 5–16, use the definition of the derivative to find $f'(x)$.

5. $f(x) = 3$

6. $f(x) = 3x + 2$

7. $f(x) = -5x$

8. $f(x) = 9 - \frac{1}{2}x$

9. $f(x) = 2x^2 + x - 1$

10. $f(x) = 1 - x^2$

11. $f(x) = x^3 - 12x$

12. $f(x) = x^3 + x^2$

13. $f(x) = \dfrac{1}{x - 1}$

14. $f(x) = \dfrac{1}{x^2}$

15. $f(x) = \sqrt{x - 4}$

16. $f(x) = \dfrac{1}{\sqrt{x}}$

In Exercises 17–22, find an equation of the tangent line to the graph of f at the indicated point. Then verify your answer by sketching both the graph of f and the tangent line.

17. $f(x) = x^2 + 1,\ (2, 5)$

18. $f(x) = x^2 + 2x + 1,\ (-3, 4)$

19. $f(x) = x^3,\ (2, 8)$

20. $f(x) = \sqrt{x},\ (1, 1)$

21. $f(x) = x + \dfrac{1}{x},\ (1, 2)$

22. $f(x) = \dfrac{1}{x + 1},\ (0, 1)$

In Exercises 23 and 24, find an equation of the line that is tangent to the graph of f and parallel to the given line.

Function	Line
23. $f(x) = x^3$	$3x - y + 1 = 0$
24. $f(x) = \dfrac{1}{\sqrt{x}}$	$x + 2y - 6 = 0$

In Exercises 25 and 26, find equations of the two tangent lines to the graph of f that pass through the indicated point.

25. $f(x) = 4x - x^2$

26. $f(x) = x^2$

In Exercises 27–32, use your knowledge of the graph of the function and the geometric interpretation of the derivative to match the function with the graph of its *derivative*. It is not necessary to find the derivative of the function. [The graphs are labeled (a), (b), (c), (d), (e), and (f).]

27. $f(x) = x$

28. $f(x) = x^2$

29. $f(x) = \sqrt{x}$

30. $f(x) = \dfrac{1}{x}$

31. $f(x) = |x|$

32. $f(x) = \sin x$

(a)

(b)

(c)

(d)

(e)

(f)

* C In Exercises 33 and 34, use a graphing utility to graph f over the interval $[-2, 2]$. Then complete the table by *graphically* estimating the slope of the graph at the indicated points. Finally, evaluate the slopes *analytically* and compare your results with those obtained graphically.

x	-2	-1.5	-1	-0.5	0	0.5	1	1.5	2
$f(x)$									
$f'(x)$									

33. $f(x) = \frac{1}{4}x^3$ **34.** $f(x) = \frac{1}{2}x^2$

C *Essay* In Exercises 35 and 36, consider the functions $f(x)$ and $S_{\Delta x}(x)$ where

$$S_{\Delta x}(x) = \frac{f(2 + \Delta x) - f(2)}{\Delta x}(x - 2) + f(2).$$

(a) Use a graphing utility to graph f and $S_{\Delta x}$ on the same coordinate axes for $\Delta x = 1$, 0.5, and 0.1.

(b) Give a written description of the graphs of S for the different values of Δx.

35. $f(x) = 4 - (x - 3)^2$ **36.** $f(x) = x + \dfrac{1}{x}$

In Exercises 37–42, use the alternative form of the derivative to find the derivative at $x = c$ (if it exists).

37. $f(x) = x^2 - 1,\quad c = 2$

38. $f(x) = x^3 + 2x,\quad c = 1$

*A C indicates an exercise in which you are instructed to use graphing technology or a symbolic computer algebra system. The solutions of other exercises may also be facilitated by appropriate technology.

39. $f(x) = x^3 + 2x^2 + 1$, $c = -2$

40. $f(x) = \dfrac{1}{x}$, $c = 3$

41. $f(x) = (x - 1)^{2/3}$, $c = 1$

42. $f(x) = |x - 2|$, $c = 2$

In Exercises 43–52, find every x-value at which the function is differentiable.

43. $f(x) = |x + 3|$ **44.** $f(x) = |x^2 - 9|$

45. $f(x) = \dfrac{1}{x + 1}$ **46.** $f(x) = \dfrac{2x}{x - 1}$

47. $f(x) = (x - 3)^{2/3}$ **48.** $f(x) = x^{2/5}$

49. $f(x) = \sqrt{x - 1}$ **50.** $f(x) = \dfrac{x^2}{x^2 - 4}$

51. $f(x) = \begin{cases} 4 - x^2, & 0 < x \\ x^2 - 4, & x \le 0 \end{cases}$

52. $f(x) = \begin{cases} x^2 - 2x, & x > 1 \\ x^3 - 3x^2 + 3x, & x \le 1 \end{cases}$

FIGURE FOR 51   **FIGURE FOR 52**

In Exercises 53–56, find the derivatives from the left and from the right at $x = 1$ (if they exist). Is the function differentiable at $x = 1$?

53. $f(x) = |x - 1|$ **54.** $f(x) = \sqrt{1 - x^2}$

55. $f(x) = \begin{cases} (x - 1)^3, & x \le 1 \\ (x - 1)^2, & x > 1 \end{cases}$

56. $f(x) = \begin{cases} x, & x \le 1 \\ x^2, & x > 1 \end{cases}$

In Exercises 57 and 58, determine whether the function is differentiable at $x = 2$.

57. $f(x) = \begin{cases} x^2 + 1, & x \le 2 \\ 4x - 3, & x > 2 \end{cases}$

58. $f(x) = \begin{cases} \frac{1}{2}x + 1, & x < 2 \\ \sqrt{2x}, & x \ge 2 \end{cases}$

59. Assume $f'(c) = 3$. Find $f'(-c)$ given the following.
   a. f is an odd function.   **b.** f is an even function.

60. Sketch the graphs of f and f' on the same set of axes.
   a. $f(x) = x^2$ **b.** $f(x) = x^3$

True or False In Exercises 61–64, determine whether the statement is true or false. If it is false, explain why or give an example that shows it is false.

61. The slope of the graph of $y = x^2$ is different at every point on the curve.

62. If a function is continuous at a point, then it is differentiable at that point.

63. If a function is differentiable at a point, then it is continuous at that point.

64. If a function has derivatives from both the right and the left of a point, then it is differentiable at that point.

65. Let
$$f(x) = \begin{cases} x \sin \dfrac{1}{x}, & x \ne 0 \\ 0, & x = 0 \end{cases}$$
and
$$g(x) = \begin{cases} x^2 \sin \dfrac{1}{x}, & x \ne 0 \\ 0, & x = 0. \end{cases}$$

Show that f is continuous, but not differentiable, at $x = 0$. Show that g is differentiable at 0, and find $g'(0)$.

Basic Differentiation Rules and Rates of Change

The Constant Rule • The Power Rule • The Constant Multiple Rule • The Sum and Difference Rules • Derivatives of Sine and Cosine Functions • Rates of Change

The Constant Rule

In Section 2.1 you used the limit definition to find derivatives. In this and the next two sections you will be introduced to several "differentiation rules" that allow you to find derivatives without the *direct* use of the limit definition.

THEOREM 2.2   The Constant Rule

The derivative of a constant function is 0. That is, if c is a real number, then

$$\frac{d}{dx}[c] = 0.$$

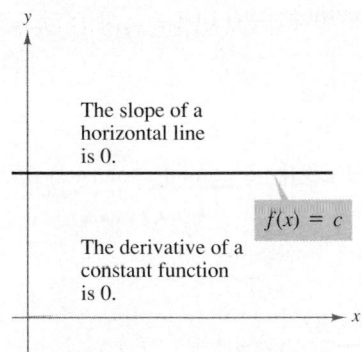

The slope of a horizontal line is 0.

$f(x) = c$

The derivative of a constant function is 0.

FIGURE 2.13
The Constant Rule

Proof   Let $f(x) = c$. Then, by the limit definition of the derivative,

$$\frac{d}{dx}[c] = f'(x)$$

$$= \lim_{\Delta x \to 0} \frac{f(x + \Delta x) - f(x)}{\Delta x}$$

$$= \lim_{\Delta x \to 0} \frac{c - c}{\Delta x}$$

$$= 0.$$

REMARK   Note in Figure 2.13 that the Constant Rule is equivalent to saying that the slope of a horizontal line is 0.

EXAMPLE 1   Using the Constant Rule

Function	Derivative
a. $y = 7$	$\dfrac{dy}{dx} = 0$
b. $f(x) = 0$	$f'(x) = 0$
c. $s(t) = -3$	$s'(t) = 0$
d. $y = k\pi^2$, k is constant	$y' = 0$

The Power Rule

Before proving the next rule, we review the procedure for expanding a binomial.

$$(x + \Delta x)^2 = x^2 + 2x\Delta x + (\Delta x)^2$$
$$(x + \Delta x)^3 = x^3 + 3x^2\Delta x + 3x(\Delta x)^2 + (\Delta x)^3$$

The general binomial expansion for a positive integer n is

$$(x + \Delta x)^n = x^n + nx^{n-1}\Delta x + \underbrace{\frac{n(n-1)x^{n-2}}{2}(\Delta x)^2 + \cdots + (\Delta x)^n}.$$

$(\Delta x)^2$ is a factor of these terms

This binomial expansion is used in proving a special case of the Power Rule.

THEOREM 2.3   The Power Rule

If n is a rational number, then the function $f(x) = x^n$ is differentiable and

$$\frac{d}{dx}[x^n] = nx^{n-1}.$$

For f to be differentiable at $x = 0$, n must be a number such that x^{n-1} is defined on an interval containing 0.

Proof   If n is a positive integer greater than 1, then the binomial expansion produces the following.

$$\frac{d}{dx}[x^n] = \lim_{\Delta x \to 0} \frac{(x + \Delta x)^n - x^n}{\Delta x}$$

$$= \lim_{\Delta x \to 0} \frac{x^n + nx^{n-1}(\Delta x) + \dfrac{n(n-1)x^{n-2}}{2}(\Delta x)^2 + \cdots + (\Delta x)^n - x^n}{\Delta x}$$

$$= \lim_{\Delta x \to 0} \left[nx^{n-1} + \frac{n(n-1)x^{n-2}}{2}(\Delta x) + \cdots + (\Delta x)^{n-1} \right]$$

$$= nx^{n-1} + 0 + \cdots + 0$$

$$= nx^{n-1}$$

This proves the case for which n is a positive integer greater than 1. We leave it to you to prove the case for $n = 1$. Example 7 in Section 2.3 proves the case for which n is a negative integer. Exercise 55 in Section 2.5 proves the case for which n is rational. (In Chapter 5, the Power Rule will be extended to cover irrational values of n.)

In the Power Rule, the case for which $n = 1$ is best thought of as a separate differentiation rule. That is

$$\frac{d}{dx}[x] = 1.$$ Power Rule when $n = 1$

FIGURE 2.14
The slope of the line $y = x$ is 1.

This rule is consistent with the fact that the slope of the line $y = x$ is 1, as shown in Figure 2.14.

EXAMPLE 2 Using the Power Rule

Function	Derivative
a. $f(x) = x^3$	$f'(x) = 3x^2$
b. $g(x) = \sqrt[3]{x}$	$g'(x) = \dfrac{d}{dx}[x^{1/3}] = \dfrac{1}{3}x^{-2/3} = \dfrac{1}{3x^{2/3}}$
c. $y = \dfrac{1}{x^2}$	$\dfrac{dy}{dx} = \dfrac{d}{dx}[x^{-2}] = (-2)x^{-3} = -\dfrac{2}{x^3}$

In Example 2c, note that *before* differentiating, $1/x^2$ was rewritten as x^{-2}. Rewriting is the first step in *many* differentiation problems.

$$\boxed{\text{Given: } y = \frac{1}{x^2}} \rightarrow \boxed{\text{Rewrite: } y = x^{-2}} \rightarrow \boxed{\text{Differentiate: } \frac{dy}{dx} = (-2)x^{-3}} \rightarrow \boxed{\text{Simplify: } \frac{dy}{dx} = -\frac{2}{x^3}}$$

EXAMPLE 3 Finding the Slope of a Graph

Find the slope of the graph of $f(x) = x^4$ when

a. $x = -1$ b. $x = 0$ c. $x = 1$.

Solution The derivative of f is $f'(x) = 4x^3$.

a. When $x = -1$, the slope is $f'(-1) = 4(-1)^3 = -4$.
b. When $x = 0$, the slope is $f'(0) = 4(0)^3 = 0$.
c. When $x = 1$, the slope is $f'(1) = 4(1)^3 = 4$.

In Figure 2.15, note that the slope of the graph is negative at the point $(-1, 1)$, the slope is zero at the point $(0, 0)$, and the slope is positive at the point $(1, 1)$.

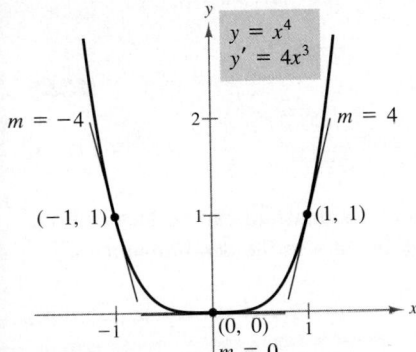

FIGURE 2.15
The slope of a graph at a point is the value of the derivative at that point.

EXAMPLE 4 Finding an Equation of a Tangent Line

Find an equation of the tangent line to the graph of $f(x) = x^2$ when $x = -2$.

Solution To find the *point* on the graph of f, evaluate the original function at $x = -2$.

$$(-2, f(-2)) = (-2, 4) \qquad \text{Point on graph}$$

To find the *slope* of the graph when $x = -2$, evaluate the derivative, $f'(x) = 2x$, at $x = -2$.

$$m = f'(-2) = -4 \qquad \text{Slope of graph at } (-2, 4)$$

Now, using the point-slope form of the equation of a line, you can write

$$y - y_1 = m(x - x_1) \qquad \text{Point-slope form}$$
$$y - 4 = -4[x - (-2)] \qquad \text{Substitute for } y_1, m, \text{ and } x_1$$
$$y = -4x - 4. \qquad \text{Simplify}$$

(See Figure 2.16.)

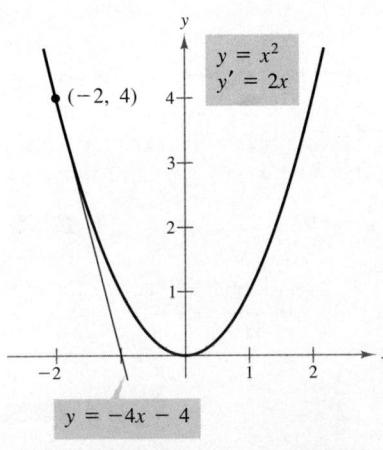

FIGURE 2.16
The line $y = -4x - 4$ is tangent to the graph of $y = x^2$ at the point $(-2, 4)$.

The Constant Multiple Rule

THEOREM 2.4   The Constant Multiple Rule

If f is a differentiable function and c is a real number, then cf is also differentiable and

$$\frac{d}{dx}[cf(x)] = cf'(x).$$

Proof

$$\frac{d}{dx}[cf(x)] = \lim_{\Delta x \to 0} \frac{cf(x + \Delta x) - cf(x)}{\Delta x}$$

$$= \lim_{\Delta x \to 0} c\left[\frac{f(x + \Delta x) - f(x)}{\Delta x}\right]$$

$$= c\left[\lim_{\Delta x \to 0} \frac{f(x + \Delta x) - f(x)}{\Delta x}\right]$$

$$= cf'(x)$$

Informally, the Constant Multiple Rule states that a constant can be factored out of the differentiation process, even if the constant appears in the denominator.

$$\frac{d}{dx}[cf(x)] = c\frac{d}{dx}[\bigcirc f(x)] = cf'(x)$$

$$\frac{d}{dx}\left[\frac{f(x)}{c}\right] = \frac{d}{dx}\left[\left(\frac{1}{c}\right)f(x)\right] = \left(\frac{1}{c}\right)\frac{d}{dx}[\bigcirc f(x)] = \left(\frac{1}{c}\right)f'(x)$$

EXAMPLE 5   Using the Constant Multiple Rule

Function	Derivative
a. $y = \dfrac{2}{x}$	$\dfrac{dy}{dx} = \dfrac{d}{dx}[2x^{-1}] = 2\dfrac{d}{dx}[x^{-1}] = 2(-1)x^{-2} = -\dfrac{2}{x^2}$
b. $f(t) = \dfrac{4t^2}{5}$	$f'(t) = \dfrac{d}{dt}\left[\dfrac{4}{5}t^2\right] = \dfrac{4}{5}\dfrac{d}{dt}[t^2] = \dfrac{4}{5}(2t) = \dfrac{8}{5}t$
c. $y = 2\sqrt{x}$	$\dfrac{dy}{dx} = \dfrac{d}{dx}[2x^{1/2}] = 2\left(\dfrac{1}{2}x^{-1/2}\right) = x^{-1/2} = \dfrac{1}{\sqrt{x}}$
d. $y = \dfrac{1}{2\sqrt[3]{x^2}}$	$\dfrac{dy}{dx} = \dfrac{d}{dx}\left[\dfrac{1}{2}x^{-2/3}\right] = \dfrac{1}{2}\left(-\dfrac{2}{3}\right)x^{-5/3} = -\dfrac{1}{3x^{5/3}}$
e. $y = -\dfrac{3x}{2}$	$y' = \dfrac{d}{dx}\left[-\dfrac{3}{2}x\right] = -\dfrac{3}{2}(1) = -\dfrac{3}{2}$

REMARK The Constant Multiple and Power Rules can be combined into one rule. The combination rule is $D_x[cx^n] = cnx^{n-1}$.

EXAMPLE 6   Using Parentheses when Differentiating

Given Function	Rewrite	Differentiate	Simplify
a. $y = \dfrac{5}{2x^3}$	$y = \dfrac{5}{2}(x^{-3})$	$y' = \dfrac{5}{2}(-3x^{-4})$	$y' = -\dfrac{15}{2x^4}$
b. $y = \dfrac{5}{(2x)^3}$	$y = \dfrac{5}{8}(x^{-3})$	$y' = \dfrac{5}{8}(-3x^{-4})$	$y' = -\dfrac{15}{8x^4}$
c. $y = \dfrac{7}{3x^{-2}}$	$y = \dfrac{7}{3}(x^2)$	$y' = \dfrac{7}{3}(2x)$	$y' = \dfrac{14x}{3}$
d. $y = \dfrac{7}{(3x)^{-2}}$	$y = 63(x^2)$	$y' = 63(2x)$	$y' = 126x$

The Sum and Difference Rules

THEOREM 2.5   The Sum and Difference Rules

The sum (or difference) of two differentiable functions is differentiable and is the sum (or difference) of their derivatives.

$$\frac{d}{dx}[f(x) + g(x)] = f'(x) + g'(x) \qquad \text{Sum Rule}$$

$$\frac{d}{dx}[f(x) - g(x)] = f'(x) - g'(x) \qquad \text{Difference Rule}$$

Proof   A proof of the Sum Rule follows from Theorem 1.2. (The Difference Rule can be proved in a similar way.)

$$\frac{d}{dx}[f(x) + g(x)] = \lim_{\Delta x \to 0} \frac{[f(x + \Delta x) + g(x + \Delta x)] - [f(x) + g(x)]}{\Delta x}$$

$$= \lim_{\Delta x \to 0} \frac{f(x + \Delta x) + g(x + \Delta x) - f(x) - g(x)}{\Delta x}$$

$$= \lim_{\Delta x \to 0} \left[\frac{f(x + \Delta x) - f(x)}{\Delta x} + \frac{g(x + \Delta x) - g(x)}{\Delta x} \right]$$

$$= \lim_{\Delta x \to 0} \frac{f(x + \Delta x) - f(x)}{\Delta x} + \lim_{\Delta x \to 0} \frac{g(x + \Delta x) - g(x)}{\Delta x}$$

$$= f'(x) + g'(x)$$

The Sum and Difference rules can be extended to any finite number of functions. For instance, if $F(x) = f(x) + g(x) - h(x) - k(x)$, then $F'(x) = f'(x) + g'(x) - h'(x) - k'(x)$.

EXAMPLE 7   Using the Sum and Difference Rules

Function	Derivative
a. $f(x) = x^3 - 4x + 5$	$f'(x) = 3x^2 - 4$
b. $g(x) = -\dfrac{x^4}{2} + 3x^3 - 2x$	$g'(x) = -2x^3 + 9x^2 - 2$

Derivatives of Sine and Cosine Functions

In Section 1.3, you studied the following limits.

$$\lim_{\Delta x \to 0} \frac{\sin \Delta x}{\Delta x} = 1 \quad \text{and} \quad \lim_{\Delta x \to 0} \frac{1 - \cos \Delta x}{\Delta x} = 0$$

These two limits can be used to prove differentiation rules for the sine and cosine functions. (The derivatives of the other four trigonometric functions will be discussed in the next section.)

THEOREM 2.6 Derivatives of Sine and Cosine Functions

$$\frac{d}{dx}[\sin x] = \cos x \qquad \frac{d}{dx}[\cos x] = -\sin x$$

Proof

$$\frac{d}{dx}[\sin x] = \lim_{\Delta x \to 0} \frac{\sin(x + \Delta x) - \sin x}{\Delta x}$$

$$= \lim_{\Delta x \to 0} \frac{\sin x \cos \Delta x + \cos x \sin \Delta x - \sin x}{\Delta x}$$

$$= \lim_{\Delta x \to 0} \frac{\cos x \sin \Delta x - (\sin x)(1 - \cos \Delta x)}{\Delta x}$$

$$= \lim_{\Delta x \to 0} \left[(\cos x)\left(\frac{\sin \Delta x}{\Delta x}\right) - (\sin x)\left(\frac{1 - \cos \Delta x}{\Delta x}\right) \right]$$

$$= \cos x \left[\lim_{\Delta x \to 0} \frac{\sin \Delta x}{\Delta x} \right] - \sin x \left[\lim_{\Delta x \to 0} \frac{1 - \cos \Delta x}{\Delta x} \right]$$

$$= (\cos x)(1) - (\sin x)(0)$$

$$= \cos x$$

This differentiation rule is shown graphically in Figure 2.17. Note that for each x the *slope* of the sine curve is equal to the *value* of the cosine. The proof of the second rule is left as an exercise (see Exercise 95).

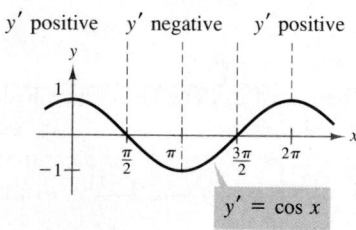

FIGURE 2.17
The derivative of the sine function is the cosine function.

EXAMPLE 8 Derivatives Involving Sines and Cosines

Function	Derivative
a. $y = 2\sin x$	$y' = 2\cos x$
b. $y = \dfrac{\sin x}{2} = \dfrac{1}{2}\sin x$	$y' = \dfrac{1}{2}\cos x = \dfrac{\cos x}{2}$
c. $y = x + \cos x$	$y' = 1 - \sin x$

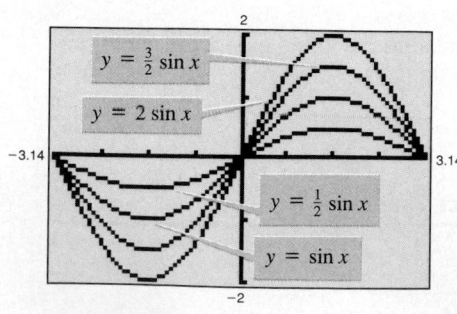

FIGURE 2.18
$\dfrac{d}{dx}[a\sin x] = a\cos x$

TECHNOLOGY A graphing utility can provide insight into the interpretation of a derivative. For instance, Figure 2.18 shows the graphs of $y = a\sin x$ for $a = \frac{1}{2}, 1, \frac{3}{2},$ and 2. Estimate the slope of each graph at the point $(0, 0)$. Then verify your estimate analytically by evaluating the derivative of each function when $x = 0$.

Rates of Change

You have seen how the derivative is used to determine slope. The derivative can also be used to determine the rate of change of one variable with respect to another. Applications involving rates of change occur in a wide variety of fields. A few examples are population growth rates, production rates, the rate of water flow, velocity, and acceleration.

A common use of rate of change is to describe the motion of an object moving in a straight line. In such problems, it is customary to use either a horizontal or vertical line with a designated origin to represent the line of motion. On such lines, movement to the right (or upward) is considered to be in the positive direction, and movement to the left (or down) is considered to be in the negative direction.

The function s that gives the position (relative to the origin) of an object as a function of time t is called a **position function.** If, over a period of time Δt, the object changes its position by the amount $\Delta s = s(t + \Delta t) - s(t)$, then, by the familiar formula

$$\text{Rate} = \frac{\text{Distance}}{\text{Time}}$$

the **average velocity** is

$$\frac{\text{Change in distance}}{\text{Change in time}} = \frac{\Delta s}{\Delta t}. \qquad \text{Average velocity}$$

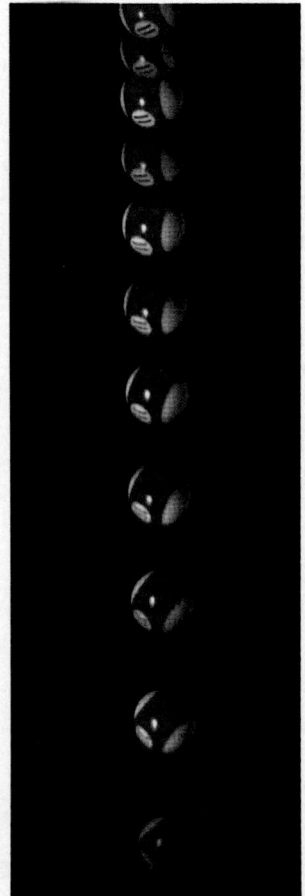

FIGURE 2.19

Time-lapse photo of a free-falling billiard ball.

EXAMPLE 9   Finding Average Velocities for a Falling Object

If a billiard ball is dropped from a height of 100 feet, its height s at time t is given by the position function $s = -16t^2 + 100$, where s is measured in feet and t is measured in seconds (see Figure 2.19). Find the average velocity over the following time intervals.

a. $[1, 2]$ **b.** $[1, 1.5]$ **c.** $[1, 1.1]$

Solution

a. For the interval $[1, 2]$, the object falls from a height of 84 feet to a height of 36 feet. The average velocity is

$$\frac{\Delta s}{\Delta t} = \frac{36 - 84}{2 - 1} = \frac{-48}{1} = -48 \text{ ft/sec.}$$

b. For the interval $[1, 1.5]$, the object falls from a height of 84 feet to a height of 64 feet. The average velocity is

$$\frac{\Delta s}{\Delta t} = \frac{64 - 84}{1.5 - 1} = \frac{-20}{0.5} = -40 \text{ ft/sec.}$$

c. For the interval $[1, 1.1]$, the object falls from a height of 84 feet to a height of 80.64 feet. The average velocity is

$$\frac{\Delta s}{\Delta t} = \frac{80.64 - 84}{1.1 - 1} = \frac{-3.36}{0.1} = -33.6 \text{ ft/sec.}$$

Note that the average velocities are *negative,* indicating that the object is moving downward.

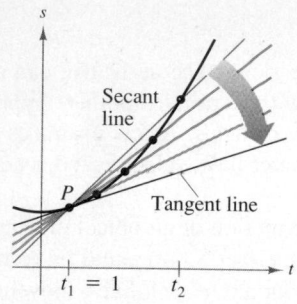

FIGURE 2.20
The average velocity between t_1 and t_2 is the slope of the secant line, and the instantaneous velocity at t_1 is the slope of the tangent line.

Suppose that in Example 9, you wanted to find the *instantaneous* velocity (or simply the velocity) of the object when $t = 1$. Just as you can approximate the slope of the tangent line by calculating the slope of the secant line, you can approximate the velocity at $t = 1$ by calculating the average velocity over a small interval $[1, 1 + \Delta t]$, as indicated in Figure 2.20. By taking the limit as Δt approaches zero, you obtain the velocity when $t = 1$. Try doing this—you will find that the velocity when $t = 1$ is -32 feet per second.

In general, if $s = s(t)$ is the position function for an object moving along a straight line, then the **velocity** of the object at time t is

$$v(t) = \lim_{\Delta t \to 0} \frac{s(t + \Delta t) - s(t)}{\Delta t} = s'(t). \qquad \text{Velocity}$$

In other words, the velocity function is the derivative of the position function. (Velocity can be negative, zero, or positive. The **speed** of an object is the absolute value of its velocity. Speed cannot be negative.)

The position of a free-falling object (neglecting air resistance) under the influence of gravity can be represented by the equation

$$s(t) = \frac{1}{2}gt^2 + v_0 t + s_0 \qquad \text{Position function}$$

where s_0 is the initial height of the object, v_0 is the initial velocity of the object, and g is the acceleration due to gravity. On earth, the value of g is approximately -32 feet per second per second or -9.8 meters per second per second.

EXAMPLE 10   Using the Derivative to Find Velocity

At time $t = 0$, a diver jumps from a diving board that is 32 feet above the water (see Figure 2.21). The position of the diver is given by

$$s(t) = -16t^2 + 16t + 32$$

where s is measured in feet and t is measured in seconds.

a. When does the diver hit the water?
b. What is the diver's velocity at impact?

Solution

a. To find when the diver hits the water, let $s = 0$ and solve for t.

$$-16t^2 + 16t + 32 = 0$$
$$-16(t + 1)(t - 2) = 0$$
$$t = -1 \text{ or } 2$$

Because $t \geq 0$, choose the positive value to conclude that the diver hits the water at $t = 2$ seconds.

b. The velocity at time t is given by the derivative $s'(t) = -32t + 16$. Therefore, the velocity at time $t = 2$ is

$$s'(2) = -32(2) + 16 = -48 \text{ ft/sec.}$$

FIGURE 2.21

REMARK   In Figure 2.21, note that the diver moves upward for the first half-second because the velocity is positive for $0 < t < \frac{1}{2}$. When the velocity is 0, the diver has reached his maximum height.

EXERCISES for Section 2.2

In Exercises 1 and 2, find the slope of the tangent line to $y = x^n$ at point $(1, 1)$.

1. a. $y = x^{1/2}$ **b.** $y = x^{3/2}$

c. $y = x^2$ **d.** $y = x^3$

2. a. $y = x^{-1/2}$ **b.** $y = x^{-1}$

c. $y = x^{-3/2}$ **d.** $y = x^{-2}$

In Exercises 3–16, find the derivative of the function.

3. $y = 3$ **4.** $f(x) = -2$

5. $f(x) = x + 1$ **6.** $g(x) = 3x - 1$

7. $g(x) = x^2 + 4$ **8.** $y = t^2 + 2t - 3$

9. $f(t) = -2t^2 + 3t - 6$ **10.** $y = x^3 - 9$

11. $s(t) = t^3 - 2t + 4$ **12.** $f(x) = 2x^3 - x^2 + 3x$

13. $y = x^2 - \frac{1}{2}\cos x$ **14.** $y = 5 + \sin x$

15. $y = \frac{1}{x} - 3\sin x$ **16.** $g(t) = \pi \cos t$

In Exercises 17–22, complete the table, using Example 6 as a model.

Function	Rewrite	Differentiate	Simplify
17. $y = \dfrac{1}{3x^3}$			
18. $y = \dfrac{2}{3x^2}$			
19. $y = \dfrac{1}{(3x)^3}$			
20. $y = \dfrac{\pi}{(3x)^2}$			
21. $y = \dfrac{\sqrt{x}}{x}$			
22. $y = \dfrac{4}{x^{-3}}$			

In Exercises 23–30, find the value of the derivative of the function at the indicated point.

Function	Point
23. $f(x) = \dfrac{1}{x}$	$(1, 1)$
24. $f(t) = 3 - \dfrac{3}{5t}$	$\left(\frac{3}{5}, 2\right)$
25. $f(x) = -\frac{1}{2} + \frac{7}{5}x^3$	$\left(0, -\frac{1}{2}\right)$
26. $y = 3x\left(x^2 - \dfrac{2}{x}\right)$	$(2, 18)$
27. $y = (2x + 1)^2$	$(0, 1)$
28. $f(x) = 3(5 - x)^2$	$(5, 0)$
29. $f(\theta) = 4\sin\theta - \theta$	$(0, 0)$
30. $g(t) = 2 + 3\cos t$	$(\pi, -1)$

In Exercises 31–42, find the derivative of the function.

31. $f(x) = x^3 - 3x - 2x^{-4}$

32. $f(x) = x^2 - 3x - 3x^{-2}$

33. $g(t) = t^2 - \dfrac{4}{t}$

34. $f(x) = x + \dfrac{1}{x^2}$

35. $f(x) = \dfrac{x^3 - 3x^2 + 4}{x^2}$

36. $h(x) = \dfrac{2x^2 - 3x + 1}{x}$

37. $y = x(x^2 + 1)$

38. $f(x) = \sqrt[3]{x} + \sqrt[5]{x}$

39. $h(s) = s^{4/5}$

40. $f(t) = t^{1/3} - 1$

41. $f(x) = 4\sqrt{x} + 3\cos x$

42. $f(x) = 2\sin x + 3\cos x$

In Exercises 43–46, find an equation of the tangent line to the graph of the function at the indicated point.

Function	Point
43. $y = x^4 - 3x^2 + 2$	$(1, 0)$
44. $y = x^3 + x$	$(-1, -2)$
45. $f(x) = \dfrac{1}{\sqrt[3]{x^2}}$	$\left(8, \dfrac{1}{4}\right)$
46. $y = (x^2 + 2x)(x + 1)$	$(1, 6)$

In Exercises 47–52, determine the point(s) (if any) at which the function has a horizontal tangent line.

47. $y = x^4 - 3x^2 + 2$

48. $y = x^3 + x$

49. $y = \dfrac{1}{x^2}$

50. $y = x^2 + 1$

51. $y = x + \sin x, \quad 0 \le x < 2\pi$

52. $y = \sqrt{3}x + 2\cos x, \quad 0 \le x < 2\pi$

Essay **In Exercises 53 and 54, the graphs of a function f and its derivative f' are given on the same coordinate axes. Label the graphs as f or f' and write a short paragraph stating the criteria used in making the selection.**

53.

54.

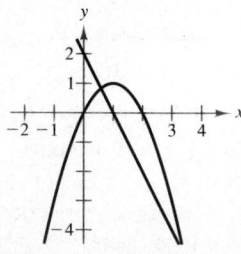

55. Sketch the graphs of the two equations $y = x^2$ and $y = -x^2 + 6x - 5$, and sketch the two lines that are tangent to both graphs. Find the equations of these lines.

56. Show that the graphs of the two equations $y = x$ and $y = 1/x$ have tangent lines that are perpendicular to each other at their point of intersection.

In Exercises 57 and 58, find an equation of the tangent line to the graph of the function f through the point (x_0, y_0) not on the graph. To find the point of tangency (x, y) on the graph of f, solve the equation

$$f'(x) = \frac{y_0 - y}{x_0 - x}.$$

57. $f(x) = \sqrt{x}$

$(x_0, y_0) = (-4, 0)$

58. $f(x) = \dfrac{2}{x}$

$(x_0, y_0) = (5, 0)$

C **59.** *Linear Approximation* Consider the function $f(x) = x^{3/2}$ with the solution point $(4, 8)$.

  a. Use a graphing utility to graph f. Use the "zoom in" feature to obtain successive magnifications of the graph in the neighborhood of the point $(4, 8)$. After zooming in a few times, the graph should appear almost linear.

  b. Find the equation of the line

$$T(x) = f'(4)(x - 4) + f(4)$$

  tangent to the graph of f passing through the given point.

  c. Use a graphing utility to obtain the graphs of f and T on the same set of coordinate axes. Note that T is a "good" approximation of f when x is "close to" 4. What happens to the accuracy of the approximation as you move farther away from the point of tangency?

  d. Demonstrate the conclusion of part c by completing the table.

Δx	-3	-2	-1	-0.5	-0.1	0
$f(4 + \Delta x)$						
$T(4 + \Delta x)$						

Δx	0.1	0.5	1	2	3
$f(4 + \Delta x)$					
$T(4 + \Delta x)$					

60. *Linear Approximation* Repeat Exercise 59 for the function $f(x) = x^3$ where $T(x)$ is the line tangent to the graph at the point $(1, 1)$. Explain why the accuracy of the linear approximation decreases more rapidly than it did in Exercise 59.

True or False **In Exercises 61–64, determine whether the statement is true or false. If it is false, explain why or give an example that shows it is false.**

61. If $f'(x) = g'(x)$, then $f(x) = g(x)$.

62. If $f(x) = g(x) + c$, then $f'(x) = g'(x)$.

63. If $y = \pi^2$, then $dy/dx = 2\pi$.

64. If $y = x/\pi$, then $dy/dx = 1/\pi$.

65. Use the graph of *f* to answer each question.

a. Between which two consecutive points is the average rate of change of the function greatest?

b. Is the average rate of change of the function between *A* and *B* greater than or less than the instantaneous rate of change at *B*?

c. Sketch a tangent line to the graph between the points *C* and *D* so that the slope of the tangent line is the same as the average rate of change of the function between *C* and *D*.

d. Give any sets of consecutive points for which the average rates of change of the function are approximately equal.

66. Sketch the graph of a function *f* such that $f' > 0$ for all *x* and the rate of change of the function is decreasing.

In Exercises 67–70, find the average rate of change of the function over the indicated interval. Compare this average rate of change to the instantaneous rates of change at the endpoints of the interval.

Function	Interval
67. $f(t) = 2t + 7$	$[1, 2]$
68. $f(t) = t^2 - 3$	$[2, 2.1]$
69. $f(x) = \dfrac{-1}{x}$	$[1, 2]$
70. $f(x) = \sin x$	$\left[0, \dfrac{\pi}{6}\right]$

Vertical Motion **In Exercises 71–74, use the position function $s(t) = -16t^2 + v_0 t + s_0$ for free-falling objects.**

71. A silver dollar is dropped from the World Trade Center, which is 1362 feet high.
  a. Determine the position and velocity functions for the coin.
  b. Determine the average velocity on the interval $[1, 2]$.
  c. Find the instantaneous velocity when $t = 1$ and $t = 2$.
  d. Find the time required for the coin to reach ground level.
  e. Find the velocity of the dollar just prior to hitting the ground.

72. A ball is thrown straight down from the top of a 220-foot building with an initial velocity of -22 feet per second. What is its velocity after 3 seconds? What is its velocity after falling 108 feet?

73. A projectile is shot upward from earth's surface with an initial velocity of 384 feet per second. What is its velocity after 5 seconds? After 10 seconds?

74. To estimate the height of a building, a stone is dropped from the top of the building into a pool of water at ground level. How high is the building if the splash is seen 6.8 seconds after the stone is dropped?

In Exercises 75 and 76, the graphs of position functions are given. They represent the distance in miles that a person drives during a 10-minute trip to work. Make sketches of the corresponding velocity functions.

75.

76.

In Exercises 77 and 78, the graphs of velocity functions are given. They represent the velocity in miles per hour during a 10-minute drive to work. Make sketches of the corresponding position functions.

77.

78.

C **79.** *Stopping Distance* The stopping distance of an automobile traveling at a certain speed is the distance the car travels during the reaction time of the driver, plus the distance the car travels after the brakes are applied. In an experiment, these distances are measured (in feet) when the automobile is traveling at a speed of *x* miles per hour. The distance traveled during the reaction time is

$$R = 1.1x$$

and the braking distance is

$$B = 0.14x^2 - 4.43x + 58.40.$$

a. Determine the polynomial giving the total stopping distance *T*.

b. Use a graphing utility to graph the functions *R*, *B*, and *T* on the same set of coordinate axes.

c. Find the rate of change of the total stopping distance when $x = 30$, $x = 40$, and $x = 55$.

80. Velocity Verify that the average velocity over the time interval $[t_0 - \Delta t, \, t_0 + \Delta t]$ is the same as the instantaneous velocity at $t = t_0$ for the position function

$$s(t) = -\tfrac{1}{2} at^2 + c.$$

81. Area The area of a square with sides of length s is $A = s^2$. Find the rate of change of the area with respect to s when $s = 4$ m.

82. Volume The volume of a cube with sides of length s is $V = s^3$. Find the rate of change of the volume with respect to s when $s = 4$ cm.

83. Drug Effectiveness Suppose the effectiveness E of a pain-killing drug t hours after entering the bloodstream is

$$E = \tfrac{1}{27}(9t + 3t^2 - t^3), \qquad 0 \le t \le 4.5.$$

Find the rate of change of E with respect to t when
a. $t = 1$ **b.** $t = 2$
c. $t = 3$ **d.** $t = 4$.

84. Profit Suppose the profit P obtained in selling x units of an item each week is

$$P = 50\sqrt{x} - 0.5x - 500, \qquad 0 \le x \le 8000.$$

Find the rate of change of P with respect to x when
a. $x = 900$ **b.** $x = 1600$
c. $x = 2500$ **d.** $x = 3600$.
(Note that the eventual decline in profit occurs because the only way to sell larger quantities is to decrease the price per unit.)

85. Inventory Management The annual inventory cost for a manufacturer is

$$C = \frac{1,008,000}{Q} + 6.3Q$$

where Q is the order size when the inventory is replenished. Find the change in annual cost when Q is increased from 350 to 351, and compare this with the instantaneous rate of change when $Q = 350$.

86. Fuel Cost A car is driven 15,000 miles a year and gets x miles per gallon. The average fuel cost is \$1.10 per gallon. Find the annual cost of fuel C as a function of x and use this function to complete the table.

x	10	15	20	25	30	35	40
C							
$\dfrac{dC}{dx}$							

Who would benefit more from a 1 mile per gallon increase in fuel efficiency—the driver of a car that gets 15 miles per gallon *or* the driver of a car that gets 35 miles per gallon? Explain.

87. Newton's Law of Gravitation This law states that the gravitational force between two particles of masses m_1 and m_2 is

$$F = K\frac{m_1 m_2}{r^2}, \qquad r > 0$$

where r is the distance between the particles. Find the rate of change of force between the particles with respect to r, and explain why the rate of change is negative.

88. Newton's Law of Cooling This law states that the rate of change of the temperature of an object is proportional to the difference between the object's temperature T and the temperature T_a of the surrounding medium. Write an equation for this law.

89. Find the equation of the parabola $y = ax^2 + bx + c$ that passes through $(0, 1)$ and is tangent to the line $x = y + 1$ at $(1, 0)$.

90. Let (a, b) be an arbitrary point on the graph of $y = 1/x$, $x > 0$. Prove that the area of the triangle formed by the tangent line through (a, b) and the coordinate axes is 2.

91. Find the tangent line(s) to the curve $y = x^3 - 9x$ through the point $(1, -9)$.

92. Find the equations of the tangent line(s) to the parabola $y = x^2$ through the given point.
a. $(0, a)$ **b.** $(a, 0)$
Are there any restrictions on the constant a?

93. Find a and b such that

$$f(x) = \begin{cases} ax^3, & x \le 2 \\ x^2 + b, & x > 2 \end{cases}$$

is differentiable everywhere.

94. Where are the functions $f_1(x) = |\sin x|$ and $f_2(x) = \sin|x|$ differentiable?

95. Prove that $\dfrac{d}{dx}[\cos x] = -\sin x$.

The Product and Quotient Rules and Higher-Order Derivatives

2.3

The Product Rule • The Quotient Rule • Derivatives of Trigonometric Functions • Higher-Order Derivatives

The Product Rule

In Section 2.2 you learned that the derivative of a sum of two functions is simply the sum of their derivatives. The rules for the derivatives of a product and quotient of two functions are not as simple.

> **THEOREM 2.7 The Product Rule**
>
> The product of two differentiable functions, f and g, is itself differentiable. Moreover, the derivative of fg is the first function times the derivative of the second, plus the second function times the derivative of the first.
>
> $$\frac{d}{dx}[f(x)g(x)] = f(x)g'(x) + g(x)f'(x)$$

Proof Some mathematical proofs, such as the proof of the Sum Rule, are straightforward. Others involve clever steps that may appear unmotivated to a reader. This proof involves such a step—adding and subtracting the same quantity.

$$\frac{d}{dx}[f(x)g(x)] = \lim_{\Delta x \to 0} \frac{f(x + \Delta x)(g + \Delta x) - f(x)g(x)}{\Delta x}$$

$$= \lim_{\Delta x \to 0} \frac{f(x + \Delta x)g(x + \Delta x) - f(x + \Delta x)g(x) + f(x + \Delta x)g(x) - f(x)g(x)}{\Delta x}$$

$$= \lim_{\Delta x \to 0} \left[f(x + \Delta x)\frac{g(x + \Delta x) - g(x)}{\Delta x} + g(x)\frac{f(x + \Delta x) - f(x)}{\Delta x} \right]$$

$$= \lim_{\Delta x \to 0} \left[f(x + \Delta x)\frac{g(x + \Delta x) - g(x)}{\Delta x} \right] + \lim_{\Delta x \to 0} \left[g(x)\frac{f(x + \Delta x) - f(x)}{\Delta x} \right]$$

$$= \lim_{\Delta x \to 0} f(x + \Delta x) \cdot \lim_{\Delta x \to 0} \frac{g(x + \Delta x) - g(x)}{\Delta x} + \lim_{\Delta x \to 0} g(x) \cdot \lim_{\Delta x \to 0} \frac{f(x + \Delta x) - f(x)}{\Delta x}$$

$$= f(x)g'(x) + g(x)f'(x)$$

The Product Rule can be extended to cover products involving more than two terms. For example, if f, g, and h are differentiable functions of x, then

$$\frac{d}{dx}[f(x)g(x)h(x)] = f'(x)g(x)h(x) + f(x)g'(x)h(x) + f(x)g(x)h'(x).$$

The derivative of a product of two functions is not (in general) given by the product of the derivatives of the two functions. To see this, try comparing the product of the derivatives of $f(x) = 3x - 2x^2$ and $g(x) = 5 + 4x$ with the derivative in Example 1.

EXAMPLE 1   Using the Product Rule

Find the derivative of $h(x) = (3x - 2x^2)(5 + 4x)$.

Solution

$$
h'(x) = \overbrace{(3x - 2x^2)}^{\text{(First)}} \overbrace{\frac{d}{dx}[5 + 4x]}^{\substack{\text{(Derivative} \\ \text{of second)}}} + \overbrace{(5 + 4x)}^{\text{(Second)}} \overbrace{\frac{d}{dx}[3x - 2x^2]}^{\substack{\text{(Derivative} \\ \text{of first)}}}
$$

$$
\begin{aligned}
&= (3x - 2x^2)(4) + (5 + 4x)(3 - 4x) \\
&= (12x - 8x^2) + (15 - 8x - 16x^2) \\
&= -24x^2 + 4x + 15
\end{aligned}
$$

In Example 1, you have the option of finding the derivative with or without the Product Rule. To find the derivative without the Product Rule, you can write

$$
\begin{aligned}
D_x[(3x - 2x^2)(5 + 4x)] &= D_x[-8x^3 + 2x^2 + 15x] \\
&= -24x^2 + 4x + 15.
\end{aligned}
$$

In the next example, you must use the Product Rule.

EXAMPLE 2   Using the Product Rule

$$
\begin{aligned}
\frac{d}{dx}[x \sin x] &= x \frac{d}{dx}[\sin x] + \sin x \frac{d}{dx}[x] \\
&= x \cos x + (\sin x)(1) \\
&= x \cos x + \sin x
\end{aligned}
$$

EXAMPLE 3   Using the Product Rule

Find the derivative of $y = 2x \cos x - 2 \sin x$.

Solution

$$
\frac{dy}{dx} = \overbrace{(2x)\left(\frac{d}{dx}[\cos x]\right) + (\cos x)\left(\frac{d}{dx}[2x]\right)}^{\text{Product Rule}} - \overbrace{2\frac{d}{dx}[\sin x]}^{\text{Constant Multiple Rule}}
$$

$$
\begin{aligned}
&= (2x)(-\sin x) + (\cos x)(2) - 2(\cos x) \\
&= -2x \sin x
\end{aligned}
$$

REMARK   In Example 3, notice that you use the Product Rule when both factors of the product are variable, and you use the Constant Multiple Rule when one of the factors is a constant.

The Quotient Rule

THEOREM 2.8   The Quotient Rule

The quotient f/g, of two differentiable functions, f and g, is itself differentiable at all values of x for which $g(x) \neq 0$. Moreover, the derivative of f/g is given by the denominator times the derivative of the numerator minus the numerator times the derivative of the denominator divided by the square of the denominator.

$$\frac{d}{dx}\left[\frac{f(x)}{g(x)}\right] = \frac{g(x)f'(x) - f(x)g'(x)}{[g(x)]^2}, \qquad g(x) \neq 0$$

Proof

$$\frac{d}{dx}\left[\frac{f(x)}{g(x)}\right] = \lim_{\Delta x \to 0} \frac{\dfrac{f(x + \Delta x)}{g(x + \Delta x)} - \dfrac{f(x)}{g(x)}}{\Delta x}$$

$$= \lim_{\Delta x \to 0} \frac{g(x)f(x + \Delta x) - f(x)g(x + \Delta x)}{\Delta x g(x)g(x + \Delta x)}$$

$$= \lim_{\Delta x \to 0} \frac{g(x)f(x + \Delta x) - f(x)g(x) + f(x)g(x) - f(x)g(x + \Delta x)}{\Delta x g(x)g(x + \Delta x)}$$

$$= \frac{\displaystyle\lim_{\Delta x \to 0} \frac{g(x)[f(x + \Delta x) - f(x)]}{\Delta x} - \lim_{\Delta x \to 0} \frac{f(x)[g(x + \Delta x) - g(x)]}{\Delta x}}{\displaystyle\lim_{\Delta x \to 0} [g(x)g(x + \Delta x)]}$$

$$= \frac{g(x)\left[\displaystyle\lim_{\Delta x \to 0} \frac{f(x + \Delta x) - f(x)}{\Delta x}\right] - f(x)\left[\displaystyle\lim_{\Delta x \to 0} \frac{g(x + \Delta x) - g(x)}{\Delta x}\right]}{\displaystyle\lim_{\Delta x \to 0} [g(x)g(x + \Delta x)]}$$

$$= \frac{g(x)f'(x) - f(x)g'(x)}{[g(x)]^2}$$

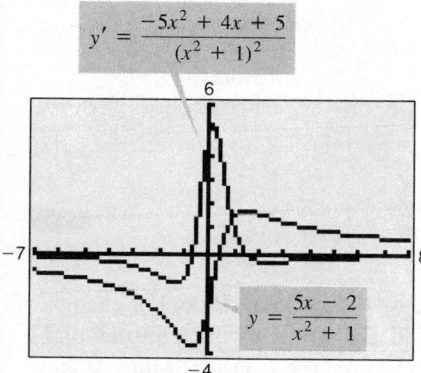

$$y' = \frac{-5x^2 + 4x + 5}{(x^2 + 1)^2}$$

$$y = \frac{5x - 2}{x^2 + 1}$$

FIGURE 2.22

TECHNOLOGY Graphing utilities can be used to compare the graph of a function with the graph of its derivative. For instance, in Figure 2.22, the graph of the function in Example 4 appears to have two points that have horizontal tangent lines. What is the value of y' at these two points?

EXAMPLE 4   Using the Quotient Rule

$$\frac{d}{dx}\left[\frac{5x - 2}{x^2 + 1}\right] = \frac{(x^2 + 1)\dfrac{d}{dx}[5x - 2] - (5x - 2)\dfrac{d}{dx}[x^2 + 1]}{(x^2 + 1)^2}$$

$$= \frac{(x^2 + 1)(5) - (5x - 2)(2x)}{(x^2 + 1)^2}$$

$$= \frac{(5x^2 + 5) - (10x^2 - 4x)}{(x^2 + 1)^2}$$

$$= \frac{-5x^2 + 4x + 5}{(x^2 + 1)^2}$$

Note the use of parentheses in Example 4. A liberal use of parentheses is recommended for *all* types of differentiation problems. For instance, with the Quotient Rule, it is a good idea to enclose all factors and derivatives in parentheses, and to pay special attention to the subtraction required in the numerator.

When we introduced differentiation rules in the previous section, we emphasized the need for rewriting *before* differentiating. The next example illustrates this point with the Quotient Rule.

EXAMPLE 5   Rewriting Before Differentiating

$$y = \frac{3 - (1/x)}{x + 5} \qquad \text{Given function}$$

$$= \frac{(3x - 1)/x}{x + 5}$$

$$= \frac{3x - 1}{x(x + 5)}$$

$$= \frac{3x - 1}{x^2 + 5x} \qquad \text{Rewrite}$$

$$\frac{dy}{dx} = \frac{(x^2 + 5x)(3) - (3x - 1)(2x + 5)}{(x^2 + 5x)^2} \qquad \text{Quotient Rule}$$

$$= \frac{(3x^2 + 15x) - (6x^2 + 13x - 5)}{(x^2 + 5x)^2}$$

$$= \frac{-3x^2 + 2x + 5}{(x^2 + 5x)^2} \qquad \text{Simplify}$$

Not every quotient needs to be differentiated by the Quotient Rule. For example, each quotient in the next example can be considered as the product of a constant times a function of x. In such cases it is more convenient to use the Constant Multiple Rule than the Quotient Rule.

EXAMPLE 6   Using the Constant Multiple Rule

	Given Function	*Rewrite*	*Differentiate*	*Simplify*
a.	$y = \dfrac{x^2 + 3x}{6}$	$y = \dfrac{1}{6}(x^2 + 3x)$	$y' = \dfrac{1}{6}(2x + 3)$	$y' = \dfrac{2x + 3}{6}$
b.	$y = \dfrac{5x^4}{8}$	$y = \dfrac{5}{8}x^4$	$y' = \dfrac{5}{8}(4x^3)$	$y' = \dfrac{5}{2}x^3$
c.	$y = \dfrac{-3(3x - 2x^2)}{7x}$	$y = -\dfrac{3}{7}(3 - 2x)$	$y' = -\dfrac{3}{7}(-2)$	$y' = \dfrac{6}{7}$
d.	$y = \dfrac{9}{5x^2}$	$y = \dfrac{9}{5}(x^{-2})$	$y' = \dfrac{9}{5}(-2x^{-3})$	$y' = -\dfrac{18}{5x^3}$

REMARK   To see the benefit of using the Constant Multiple Rule for some quotients, try using the Quotient Rule to differentiate the functions in Example 6—you should obtain the same result, but with more work.

In Section 2.2, we proved the Power Rule only for the case where the exponent n is a positive integer greater than 1. The next example extends the proof to include negative integer exponents.

EXAMPLE 7   Proof of the Power Rule (Negative Integer Exponents)

If n is a negative integer, then there exists a positive integer k such that $n = -k$. Thus, by the Quotient Rule, you can write

$$\frac{d}{dx}[x^n] = \frac{d}{dx}\left[\frac{1}{x^k}\right]$$

$$= \frac{x^k(0) - (1)(kx^{k-1})}{(x^k)^2} \qquad \text{Quotient Rule}$$

$$= \frac{0 - kx^{k-1}}{x^{2k}}$$

$$= -kx^{-k-1}$$

$$= nx^{n-1}. \qquad\qquad n = -k$$

Thus, the Power Rule, $D_x[x^n] = nx^{n-1}$, is valid for any integer. (You will be asked to prove that the Power Rule is valid for any rational number in Exercise 55 in Section 2.5.)

Derivatives of Trigonometric Functions

Knowing the derivatives of the sine and cosine functions, you can use the Quotient Rule to find the derivatives of the four remaining trigonometric functions.

THEOREM 2.9   Derivatives of Trigonometric Functions

$$\frac{d}{dx}[\tan x] = \sec^2 x \qquad\qquad \frac{d}{dx}[\cot x] = -\csc^2 x$$

$$\frac{d}{dx}[\sec x] = \sec x \tan x \qquad\qquad \frac{d}{dx}[\csc x] = -\csc x \cot x$$

Proof   Considering $\tan x = (\sin x)/(\cos x)$ and applying the Quotient Rule, you obtain

$$\frac{d}{dx}[\tan x] = \frac{(\cos x)(\cos x) - (\sin x)(-\sin x)}{\cos^2 x}$$

$$= \frac{\cos^2 x + \sin^2 x}{\cos^2 x}$$

$$= \frac{1}{\cos^2 x}$$

$$= \sec^2 x.$$

You are asked to prove the other three parts of the theorem in Exercise 69.

EXAMPLE 8   Differentiating Trigonometric Functions

Function	*Derivative*
a. $y = x - \tan x$	$\dfrac{dy}{dx} = 1 - \sec^2 x$
b. $y = x \sec x$	$y' = x(\sec x \tan x) + (\sec x)(1)$
	$\quad = (\sec x)(1 + x \tan x)$

Because of trigonometric identities, the derivative of a trigonometric function can take many forms. This presents a challenge when you are trying to match your answers to those given in the back of the text.

EXAMPLE 9   Different Forms of a Derivative

Differentiate both forms of $y = \dfrac{1 - \cos x}{\sin x} = \csc x - \cot x$.

Solution

First form $y = \dfrac{1 - \cos x}{\sin x}$

$$y' = \frac{(\sin x)(\sin x) - (1 - \cos x)(\cos x)}{\sin^2 x}$$

$$= \frac{\sin^2 x + \cos^2 x - \cos x}{\sin^2 x} = \frac{1 - \cos x}{\sin^2 x}$$

Second form $y = \csc x - \cot x$

$$y' = -\csc x \cot x + \csc^2 x$$

To show that the two derivatives are equal, you can write

$$\frac{1 - \cos x}{\sin^2 x} = \frac{1}{\sin^2 x} - \left(\frac{1}{\sin x}\right)\left(\frac{\cos x}{\sin x}\right) = \csc^2 x - \csc x \cot x.$$

The summary in Table 2.1 shows that much of the work in obtaining a simplified form of a derivative occurs *after* differentiating. Note that two characteristics of a simplified form are the absence of negative exponents and the combining of like terms.

TABLE 2.1

	$f'(x)$ *after differentiating*	$f'(x)$ *after simplifying*
Example 1	$(3x - 2x^2)(4) + (5 + 4x)(3 - 4x)$	$-24x^2 + 4x + 15$
Example 3	$(2x)(-\sin x) + (\cos x)(2) - 2(\cos x)$	$-2x \sin x$
Example 4	$\dfrac{(x^2 + 1)(5) - (5x - 2)(2x)}{(x^2 + 1)^2}$	$\dfrac{-5x^2 + 4x + 5}{(x^2 + 1)^2}$
Example 5	$\dfrac{(x^2 + 5x)(3) - (3x - 1)(2x + 5)}{(x^2 + 5x)^2}$	$\dfrac{-3x^2 + 2x + 5}{(x^2 + 5x)^2}$
Example 9	$\dfrac{(\sin x)(\sin x) - (1 - \cos x)(\cos x)}{\sin^2 x}$	$\dfrac{1 - \cos x}{\sin^2 x}$

Higher-Order Derivatives

Just as you can obtain a velocity function by differentiating a position function, you can obtain an **acceleration** function by differentiating a velocity function. Another way of looking at this is that you can obtain an acceleration function by differentiating a position function *twice*.

$$s(t) \qquad \text{Position function}$$
$$v(t) = s'(t) \qquad \text{Velocity function}$$
$$a(t) = v'(t) = s''(t) \qquad \text{Acceleration function}$$

The function given by $a(t)$ is the **second derivative** of $s(t)$ and is denoted by $s''(t)$.

The second derivative is an example of a **higher-order derivative.** You can define derivatives of any positive integer order. For instance, the **third derivative** is the derivative of the second derivative. Higher-order derivatives are denoted as follows.

First derivative	y',	$f'(x)$,	$\dfrac{dy}{dx}$,	$\dfrac{d}{dx}[f(x)]$,	$D_x[y]$
Second derivative	y'',	$f''(x)$,	$\dfrac{d^2y}{dx^2}$,	$\dfrac{d^2}{dx^2}[f(x)]$,	$D_x^2[y]$
Third derivative	y''',	$f'''(x)$,	$\dfrac{d^3y}{dx^3}$,	$\dfrac{d^3}{dx^3}[f(x)]$,	$D_x^3[y]$
Fourth derivative	$y^{(4)}$,	$f^{(4)}(x)$,	$\dfrac{d^4y}{dx^4}$,	$\dfrac{d^4}{dx^4}[f(x)]$,	$D_x^4[y]$
$\vdots$					
nth derivative	$y^{(n)}$,	$f^{(n)}(x)$,	$\dfrac{d^ny}{dx^n}$,	$\dfrac{d^n}{dx^n}[f(x)]$,	$D_x^n[y]$

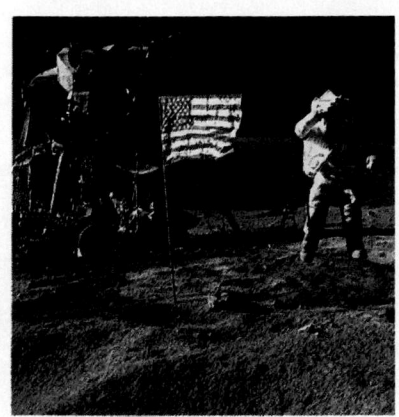

EXAMPLE 10   Finding the Acceleration Due to Gravity

Because the moon has no atmosphere, a falling object on the moon encounters no air resistance. In 1971, astronaut David Scott demonstrated that a feather and a hammer fall at the same rate on the moon. The position function for each of these falling objects was given by $s(t) = -0.81t^2 + 2$, where $s(t)$ is the height in meters and t is the time in seconds. What is the ratio of earth's gravitational force to the moon's?

The moon's mass is 7.354×10^{22} kilograms, and earth's mass is 5.979×10^{24} kilograms. The moon's radius is 1738 kilometers, and earth's radius is 6371 kilometers. Because the gravitational force on the surface of a planet is directly proportional to its mass and inversely proportional to the square of its radius, the ratio of the gravitational force on earth to the gravitational force on the moon is

$$\frac{(5.979 \times 10^{24})/6371^2}{(7.354 \times 10^{22})/1738^2} \approx 6.05.$$

Solution   To find the acceleration, differentiate the position function twice.

$$s(t) = -0.81t^2 + 2 \qquad \text{Position function}$$
$$s'(t) = -1.62t \qquad \text{Velocity function}$$
$$s''(t) = -1.62 \qquad \text{Acceleration function}$$

Thus, the acceleration due to gravity on the moon is -1.62 meters per second per second. Because the acceleration due to gravity on earth is -9.8 meters per second per second, the ratio of earth's gravitational force to the moon's is

$$\frac{\text{Earth's gravitational force}}{\text{Moon's gravitational force}} = \frac{-9.8}{-1.62} \approx 6.05.$$

EXERCISES for Section 2.3

In Exercises 1–6, find $f'(x)$ and $f'(c)$.

Function	Value of c
1. $f(x) = \dfrac{1}{3}(2x^3 - 4)$	$c = 0$
2. $f(x) = (x^2 - 2x + 1)(x^3 - 1)$	$c = 1$
3. $f(x) = (x^3 - 3x)(2x^2 + 3x + 5)$	$c = 0$
4. $f(x) = \dfrac{x + 1}{x - 1}$	$c = 2$
5. $f(x) = x \cos x$	$c = \dfrac{\pi}{4}$
6. $f(x) = \dfrac{\sin x}{x}$	$c = \dfrac{\pi}{6}$

In Exercises 7–12, complete the table without using the Quotient Rule (see Example 6).

Function	Rewrite	Differentiate	Simplify
7. $y = \dfrac{x^2 + 2x}{x}$			
8. $y = \dfrac{4x^{3/2}}{x}$			
9. $y = \dfrac{7}{3x^3}$			
10. $y = \dfrac{4}{5x^2}$			
11. $y = \dfrac{3x^2 - 5}{7}$			
12. $y = \dfrac{x^2 - 4}{x + 2}$			

In Exercises 13–28, differentiate the algebraic function.

13. $f(x) = \dfrac{3x - 2}{2x - 3}$

14. $f(x) = \dfrac{x^3 + 3x + 2}{x^2 - 1}$

15. $f(x) = \dfrac{3 - 2x - x^2}{x^2 - 1}$

16. $f(x) = x^4\left(1 - \dfrac{2}{x + 1}\right)$

17. $f(x) = \dfrac{x + 1}{\sqrt{x}}$

18. $f(x) = \sqrt[3]{x}(\sqrt{x} + 3)$

19. $h(s) = (s^3 - 2)^2$

20. $h(x) = (x^2 - 1)^2$

21. $h(t) = \dfrac{t + 1}{t^2 + 2t + 2}$

22. $f(x) = \dfrac{x(x^2 - 1)}{x + 3}$

23. $f(x) = (3x^3 + 4x)(x - 5)(x + 1)$

24. $f(x) = (x^2 - x)(x^2 + 1)(x^2 + x + 1)$

25. $g(x) = \left(\dfrac{x + 1}{x + 2}\right)(2x - 5)$

26. $f(x) = \left(\dfrac{x^2 - x - 3}{x^2 + 1}\right)(x^2 + x + 1)$

27. $f(x) = \dfrac{x^2 + c^2}{x^2 - c^2},$ c is a constant

28. $f(x) = \dfrac{c^2 - x^2}{c^2 + x^2},$ c is a constant

In Exercises 29–46, find the derivative of the trigonometric function.

29. $f(t) = t^2 \sin t$

30. $f(\theta) = (\theta + 1)\cos\theta$

31. $f(t) = \dfrac{\cos t}{t}$

32. $f(x) = \dfrac{\sin x}{x}$

33. $f(x) = -x + \tan x$

34. $y = x + \cot x$

35. $g(t) = \sqrt{t} + 4\sec t$

36. $h(s) = \dfrac{1}{s} - 10\csc s$

37. $y = 5x \csc x$

38. $y = \dfrac{\sec x}{x}$

39. $y = -\csc x - \sin x$

40. $y = x\sin x + \cos x$

41. $y = x^2 \sin x + 2x\cos x$

42. $f(x) = \sin x \cos x$

43. $f(x) = x^2 \tan x$

44. $h(\theta) = 5\sec\theta + \tan\theta$

45. $g(\theta) = \dfrac{\theta}{1 - \sin\theta}$

46. $f(\theta) = \dfrac{\sin\theta}{1 - \cos\theta}$

In Exercises 47–50, evaluate (if possible) the derivative of the function at the indicated point.

Function	Point
47. $y = \dfrac{1 + \csc x}{1 - \csc x}$	$\left(\dfrac{\pi}{6}, -3\right)$
48. $f(x) = \tan x \cot x$	$(1, 1)$
49. $h(t) = \dfrac{\sec t}{t}$	$\left(\pi, -\dfrac{1}{\pi}\right)$
50. $f(x) = \sin x(\sin x + \cos x)$	$\left(\dfrac{\pi}{4}, 1\right)$

In Exercises 51–56, find an equation of the tangent line to the graph of the function at the indicated point.

Function	Point
51. $f(x) = \dfrac{x}{x - 1}$	$(2, 2)$
52. $f(x) = (x - 1)(x^2 - 2)$	$(0, 2)$
53. $f(x) = (x^3 - 3x + 1)(x + 2)$	$(1, -3)$
54. $f(x) = \dfrac{(x - 1)}{(x + 1)}$	$\left(2, \dfrac{1}{3}\right)$
55. $f(x) = \tan x$	$\left(\dfrac{\pi}{4}, 1\right)$
56. $f(x) = \sec x$	$\left(\dfrac{\pi}{3}, 2\right)$

In Exercises 57 and 58, determine the point(s) at which the graph of the function has a horizontal tangent.

57. $f(x) = \dfrac{x^2}{x - 1}$

58. $f(x) = \dfrac{x^2}{x^2 + 1}$

In Exercises 59–62, find $f'(2)$ given that $g(2) = 3$, $g'(2) = -2$, $h(2) = -1$, and $h'(2) = 4$.

59. $f(x) = 2g(x) + h(x)$

60. $f(x) = 4 - h(x)$

61. $f(x) = \dfrac{g(x)}{h(x)}$

62. $f(x) = g(x)h(x)$

In Exercises 63 and 64, find the derivative of the function f for $n = 1, 2, 3,$ and 4. Use the result to write a general rule for $f'(x)$ in terms of n.

63. $f(x) = x^n \sin x$

64. $f(x) = \dfrac{\cos x}{x^n}$

65. *Inventory Replenishment* The ordering and transportation cost C of the components used in manufacturing a product is

$$C = 100\left(\dfrac{200}{x^2} + \dfrac{x}{x + 30}\right), \quad 1 \leq x$$

where C is measured in thousands of dollars and x is the order size in hundreds. Find the rate of change of C with respect to x when (a) $x = 10$, (b) $x = 15$, and (c) $x = 20$. What do these rates of change imply about increasing order size?

66. *Boyle's Law* This law states that if the temperature of a gas remains constant, its pressure is inversely proportional to its volume. Use the derivative to show that the rate of change of the pressure is inversely proportional to the square of the volume.

67. *Population Growth* A population of 500 bacteria is introduced into a culture and grows in number according to the equation

$$P(t) = 500\left(1 + \dfrac{4t}{50 + t^2}\right)$$

where t is measured in hours. Find the rate at which the population is growing when $t = 2$.

68. Determine whether there exist any values of x in the interval $[0, 2\pi)$ such that the rate of change of $f(x) = \sec x$ and the rate of change of $g(x) = \csc x$ are equal.

69. Prove the following differentiation rules.

a. $\dfrac{d}{dx}[\sec x] = \sec x \tan x$

b. $\dfrac{d}{dx}[\csc x] = -\csc x \cot x$

c. $\dfrac{d}{dx}[\cot x] = -\csc^2 x$

70. Sketch a graph of a differentiable function f such that $f > 0$ and $f' < 0$ for all real numbers x.

In Exercises 71–76, find the second derivative of the function.

71. $f(x) = 4x^{3/2}$

72. $f(x) = \dfrac{x^2 + 2x - 1}{x}$

73. $f(x) = \dfrac{x}{x - 1}$

74. $f(x) = x + \dfrac{32}{x^2}$

75. $f(x) = 3 \sin x$

76. $f(x) = \sec x$

In Exercises 77–80, find the higher-order derivative.

Given	Find
77. $f'(x) = x^2$	$f''(x)$
78. $f''(x) = 2 - \dfrac{2}{x}$	$f'''(x)$
79. $f''' = 2\sqrt{x}$	$f^{(4)}(x)$
80. $f^{(4)}(x) = 2x + 1$	$f^{(6)}(x)$

81. Consider the function $f(x) = g(x)h(x)$.
   a. Use the Product Rule to generate rules for finding $f''(x)$, $f'''(x)$, and $f^{(4)}(x)$.
   b. Use the results of part a to write a general rule for $f^{(n)}(x)$.

82. Develop a general rule for $f^{(n)}(x)$ if
   a. $f(x) = x^n$ **b.** $f(x) = \dfrac{1}{x}$.

83. *Acceleration* The velocity of an object in meters per second is

$$v(t) = 36 - t^2, \quad 0 \leq t \leq 6.$$

Find the velocity and acceleration of the object when $t = 3$. What can be said about the speed of the object when the velocity and acceleration have opposite signs?

84. *Acceleration* An automobile's velocity starting from rest is

$$v(t) = \dfrac{100t}{2t + 15}$$

where v is measured in feet per second. Find the acceleration at the following times.
   a. 5 seconds **b.** 10 seconds **c.** 20 seconds

85. *Stopping Distance* A car is traveling at a rate of 66 feet per second (45 miles per hour) when the brakes are applied. The position function for the car is $s(t) = -8.25t^2 + 66t$, where s is measured in feet and t is measured in seconds. Use this function to complete the table, and find the average velocity during each time interval.

t	0	1	2	3	4
$s(t)$					
$v(t)$					
$a(t)$					

86. *Acceleration on the Moon*   An astronaut standing on the moon throws a rock into the air. The height of the rock is

$$s = -\frac{27}{10}t^2 + 27t + 6$$

where *s* is measured in feet and *t* is measured in seconds.

a. Find expressions for the velocity and acceleration of the rock.

b. Find the time when the rock is at its highest point by finding the time when the velocity is 0. What is its height at this time?

c. How does the acceleration of the rock compare with the acceleration due to gravity on earth?

True or False   **In Exercises 87–92, determine whether the statement is true or false. If it is false, explain why or give an example that shows it is false.**

87. If $y = f(x)g(x)$, then $dy/dx = f'(x)g'(x)$.

88. If $y = (x + 1)(x + 2)(x + 3)(x + 4)$, then $d^5y/dx^5 = 0$.

89. If $f'(c)$ and $g'(c)$ are zero and $h(x) = f(x)g(x)$, then $h'(c) = 0$.

90. If $f(x)$ is an *n*th-degree polynomial, then $f^{(n+1)}(x) = 0$.

91. The second derivative represents the rate of change of the first derivative.

92. If the velocity of an object is constant, then its acceleration is 0.

93. Find the derivative of the function $f(x) = x\,|\,x\,|$. Does $f''(0)$ exist?

94. Let *f* and *g* be functions whose first and second derivatives exist on an interval *I*. Which of the following formulas is (are) true?

a. $fg'' - f''g = (fg' - f'g)'$

b. $fg'' + f''g = (fg)''$

In Exercises 95 and 96, the graphs of f, f', and f'' are shown together. Which is which?

95.

96.

The Chain Rule • The General Power Rule • Simplifying Derivatives • Trigonometric Functions and the Chain Rule

The Chain Rule

We have yet to discuss one of the most powerful differentiation rules—the **Chain Rule.** This rule deals with composite functions and adds a surprising versatility to the rules discussed in the two previous sections. For example, compare the following functions. Those on the left can be differentiated without the Chain Rule, and those on the right are best done with the Chain Rule.

Without the Chain Rule	*With the Chain Rule*
$y = x^2 + 1$	$y = \sqrt{x^2 + 1}$
$y = \sin x$	$y = \sin 6x$
$y = 3x + 2$	$y = (3x + 2)^5$
$y = x + \tan x$	$y = x + \tan x^2$

Basically, the Chain Rule states that if y changes dy/du times as fast as u, and u changes du/dx times as fast as x, then y changes $(dy/du)(du/dx)$ times as fast as x.

EXAMPLE 1   The Derivative of a Composite Function

A set of gears is constructed as shown in Figure 2.23 so that the second and third gears are on the same axle. As the first axle revolves, it drives the second axle, which in turn drives the third axle. Let y, u, and x represent the numbers of revolutions per minute of the first, second, and third axles, respectively. Find dy/du, du/dx, and dy/dx, and show that

$$\frac{dy}{dx} = \frac{dy}{du} \cdot \frac{du}{dx}.$$

Solution   Because the circumference of the second gear is three times that of the first, the first axle must make three revolutions to turn the second axle once. Similarly, the second gear must make two revolutions to turn the third axle once, and you can write

$$\frac{dy}{du} = 3 \quad \text{and} \quad \frac{du}{dx} = 2.$$

Combining these two results, you know that the first axle must make six revolutions to turn the third axle once. Thus, you can write

$$\frac{dy}{dx} = \frac{dy}{du} \cdot \frac{du}{dx} = 3 \cdot 2 = 6.$$

In other words, the rate of change of y with respect to x is the product of the rate of change of y with respect to u and the rate of change of u with respect to x.

FIGURE 2.23
Axle 1: y revolutions per minute
Axle 2: u revolutions per minute
Axle 3: x revolutions per minute

Example 1 illustrates a simple case of the Chain Rule. The general rule is stated below.

THEOREM 2.10   The Chain Rule

If $y = f(u)$ is a differentiable function of u and $u = g(x)$ is a differentiable function of x, then $y = f(g(x))$ is a differentiable function of x and

$$\frac{dy}{dx} = \frac{dy}{du} \cdot \frac{du}{dx}$$

or, equivalently,

$$\frac{d}{dx}[f(g(x))] = f'(g(x))g'(x).$$

Proof   Let $h(x) = f(g(x))$. Then, using the alternative form of the derivative, you need to show that for $x = c$

$$h'(c) = f'(g(c))g'(c).$$

An important consideration in this proof is the behavior of g as x approaches c. A problem occurs if there are values of x, other than c, such that $g(x) = g(c)$. In Appendix A we show how to use the differentiability of f and g to overcome this problem. For now, assume that $g(x) \neq g(c)$ for values of x other than c. In the proof of the Product and Quotient Rules, we added and subtracted the same quantity to obtain the desired form. This proof uses a similar technique—multiplying and dividing by the same (nonzero) quantity. Note that because g is differentiable, it is also continuous, and it follows that $g(x) \rightarrow g(c)$ as $x \rightarrow c$.

$$h'(c) = \lim_{x \to c} \frac{f(g(x)) - f(g(c))}{x - c}$$

$$= \lim_{x \to c} \left[\frac{f(g(x)) - f(g(c))}{g(x) - g(c)} \cdot \frac{g(x) - g(c)}{x - c} \right], \qquad g(x) \neq g(c)$$

$$= \left[\lim_{x \to c} \frac{f(g(x)) - f(g(c))}{g(x) - g(c)} \right]\left[\lim_{x \to c} \frac{g(x) - g(c)}{x - c} \right]$$

$$= f'(g(c))g'(c)$$

When applying the Chain Rule, it is helpful to think of the composite function $f \circ g$ as having two parts—an inner part and an outer part.

$$y = \underset{\text{Inner function}}{\overset{\text{Outer function}}{f(g(x))}} = f(u)$$

The derivative of $y = f(u)$ is the derivative of the outer function (at the inner function u) *times* the derivative of the inner function.

$$y' = f'(u) \cdot u'$$

EXAMPLE 2   Decomposition of a Composite Function

$y = f(g(x))$	$u = g(x)$	$y = f(u)$
a. $y = \dfrac{1}{x + 1}$	$u = x + 1$	$y = \dfrac{1}{u}$
b. $y = \sin 2x$	$u = 2x$	$y = \sin u$
c. $y = \sqrt{3x^2 - x + 1}$	$u = 3x^2 - x + 1$	$y = \sqrt{u}$
d. $y = \tan^2 x$	$u = \tan x$	$y = u^2$

EXAMPLE 3   Using the Chain Rule

Find dy/dx for $y = (x^2 + 1)^3$.

Solution   For this function, you can consider the inside function to be $u = x^2 + 1$. By the Chain Rule, you obtain

$$\frac{dy}{dx} = \underbrace{3(x^2 + 1)^2}_{\frac{dy}{du}} \underbrace{(2x)}_{\frac{du}{dx}} = 6x(x^2 + 1)^2.$$

The General Power Rule

The function in Example 3 is an example of one of the most common types of composite functions, $y = [u(x)]^n$. The rule for differentiating such functions is called the **General Power Rule,** and it is a special case of the Chain Rule.

THEOREM 2.11   The General Power Rule

If $y = [u(x)]^n$, where u is a differentiable function of x and n is a rational number, then

$$\frac{dy}{dx} = n[u(x)]^{n-1}\frac{du}{dx}$$

or, equivalently,

$$\frac{d}{dx}[u^n] = nu^{n-1}u'.$$

Proof   Because $y = u^n$, you apply the Chain Rule to obtain

$$\frac{dy}{dx} = \left(\frac{dy}{du}\right)\left(\frac{du}{dx}\right) = \frac{d}{du}[u^n]\frac{du}{dx}.$$

By the (simple) Power Rule in Section 2.2, you have $D_u[u^n] = nu^{n-1}$, and it follows that $dy/dx = nu^{n-1}(du/dx)$.

EXAMPLE 4 Applying the General Power Rule

Find the derivative of $f(x) = (3x - 2x^2)^3$.

Solution Let $u = 3x - 2x^2$. Then $f(x) = (3x - 2x^2)^3 = u^3$, and by the General Power Rule the derivative is

$$f'(x) = \overbrace{3}^{n}\overbrace{(3x - 2x^2)^2}^{u^{n-1}}\overbrace{\frac{d}{dx}[3x - 2x^2]}^{u'} \qquad \text{Apply General Power Rule}$$

$$= 3(3x - 2x^2)^2(3 - 4x). \qquad \text{Differentiate } 3x - 2x^2$$

EXAMPLE 5 Differentiating Functions Involving Radicals

Find all points on the graph of $f(x) = \sqrt[3]{(x^2 - 1)^2}$ for which $f'(x) = 0$ and $f'(x)$ does not exist.

Solution Begin by rewriting the function as

$$y = (x^2 - 1)^{2/3}.$$

Then, applying the General Power Rule (with $u = x^2 - 1$) produces

$$f'(x) = \overbrace{\frac{2}{3}}^{n}\overbrace{(x^2 - 1)^{-1/3}}^{u^{n-1}}\overbrace{(2x)}^{u'} \qquad \text{Apply General Power Rule}$$

$$= \frac{4x}{3\sqrt[3]{x^2 - 1}} \qquad \text{Write in radical form}$$

Thus, $f'(x) = 0$ when $x = 0$ and $f'(x)$ does not exist when $x = \pm 1$, as indicated in Figure 2.24.

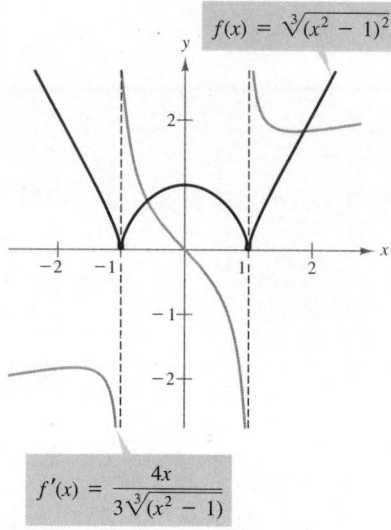

$$f(x) = \sqrt[3]{(x^2 - 1)^2}$$

$$f'(x) = \frac{4x}{3\sqrt[3]{(x^2 - 1)}}$$

FIGURE 2.24
The derivative of f is 0 at $x = 0$ and undefined at $x = \pm 1$.

EXAMPLE 6 Differentiating Quotients with Constant Numerators

Differentiate $g(t) = \dfrac{-7}{(2t - 3)^2}$.

Solution Begin by rewriting the function as

$$g(t) = -7(2t - 3)^{-2}.$$

Then, applying the General Power Rule produces

$$g'(t) = \underbrace{(-7)}_{\substack{\text{Constant} \\ \text{Multiple Rule}}} \overbrace{(-2)}^{n}\overbrace{(2t - 3)^{-3}}^{u^{n-1}}\overbrace{(2)}^{u'} \qquad \text{Apply General Power Rule}$$

$$= 28(2t - 3)^{-3} \qquad \text{Simplify}$$

$$= \frac{28}{(2t - 3)^3}. \qquad \text{Write with positive exponent}$$

REMARK Try differentiating the function in Example 6 with the Quotient Rule. You should obtain the same result, but using the Quotient Rule is less efficient than using the General Power Rule.

Simplifying Derivatives

The next three examples illustrate some techniques for simplifying the "raw derivatives" of functions involving products, quotients, and composites.

EXAMPLE 7   Simplifying by Factoring Out the Least Powers

$$f(x) = x^2\sqrt{1 - x^2} = x^2(1 - x^2)^{1/2}$$ Rewrite

$$f'(x) = x^2\frac{d}{dx}[(1 - x^2)^{1/2}] + (1 - x^2)^{1/2}\frac{d}{dx}[x^2]$$ Product Rule

$$= x^2\left[\frac{1}{2}(1 - x^2)^{-1/2}(-2x)\right] + (1 - x^2)^{1/2}(2x)$$ General Power Rule

$$= -x^3(1 - x^2)^{-1/2} + 2x(1 - x^2)^{1/2}$$ Simplify

$$= x(1 - x^2)^{-1/2}[-x^2(1) + 2(1 - x^2)]$$ Factor

$$= \frac{x(2 - 3x^2)}{\sqrt{1 - x^2}}$$ Simplify

EXAMPLE 8   Simplifying the Derivative of a Quotient

TECHNOLOGY Symbolic differentiation utilities are capable of differentiating very complicated functions. Often, however, the result is given in unsimplified form. If you have access to such a utility, use it to find the derivatives of the functions given in Examples 7, 8, and 9. Then compare the results with those given on this page.

$$f(x) = \frac{x}{\sqrt[3]{x^2 + 4}} = \frac{x}{(x^2 + 4)^{1/3}}$$ Rewrite

$$f'(x) = \frac{(x^2 + 4)^{1/3}(1) - x(1/3)(x^2 + 4)^{-2/3}(2x)}{(x^2 + 4)^{2/3}}$$ Quotient Rule

$$= \frac{1}{3}(x^2 + 4)^{-2/3}\left[\frac{3(x^2 + 4) - (2x^2)(1)}{(x^2 + 4)^{2/3}}\right]$$ Factor

$$= \frac{x^2 + 12}{3(x^2 + 4)^{4/3}}$$ Simplify

EXAMPLE 9   Simplifying the Derivative of a Power

$$y = \left(\frac{3x - 1}{x^2 + 3}\right)^2$$ Original function

$$y' = \overbrace{2}^{n}\overbrace{\left(\frac{3x - 1}{x^2 + 3}\right)}^{u^{n-1}}\overbrace{\frac{d}{dx}\left[\frac{3x - 1}{x^2 + 3}\right]}^{u'}$$ General Power Rule

$$= \left[\frac{2(3x - 1)}{x^2 + 3}\right]\left[\frac{(x^2 + 3)(3) - (3x - 1)(2x)}{(x^2 + 3)^2}\right]$$ Quotient Rule

$$= \frac{2(3x - 1)(3x^2 + 9 - 6x^2 + 2x)}{(x^2 + 3)^3}$$ Multiply

$$= \frac{2(3x - 1)(-3x^2 + 2x + 9)}{(x^2 + 3)^3}$$ Simplify

Trigonometric Functions and the Chain Rule

The "Chain Rule versions" of the derivatives of the six trigonometric functions are as follows.

$$\frac{d}{dx}[\sin u] = (\cos u)u' \qquad\qquad \frac{d}{dx}[\cos u] = -(\sin u)u'$$

$$\frac{d}{dx}[\tan u] = (\sec^2 u)u' \qquad\qquad \frac{d}{dx}[\cot u] = -(\csc^2 u)u'$$

$$\frac{d}{dx}[\sec u] = (\sec u \tan u)u' \qquad \frac{d}{dx}[\csc u] = -(\csc u \cot u)u'$$

EXAMPLE 10   Applying the Chain Rule to Trigonometric Functions

$$\overbrace{u}$$

$$\overbrace{\cos u}\ \overbrace{u'}$$

a. $y = \sin 2x$ $\qquad\qquad y' = \cos 2x \dfrac{d}{dx}[2x] = (\cos 2x)(2) = 2\cos 2x$

b. $y = \cos(x - 1)$ $\qquad y' = -\sin(x - 1)$

c. $y = \tan 3x$ $\qquad\qquad y' = 3\sec^2 3x$

Be sure that you understand the mathematical conventions regarding parentheses and trigonometric functions. For instance, in Example 10a, $\sin 2x$ is written to mean $\sin(2x)$.

EXAMPLE 11   Parentheses and Trigonometric Functions

a. $y = \cos 3x^2 = \cos(3x^2)$ $\qquad y' = (-\sin 3x^2)(6x) = -6x\sin 3x^2$

b. $y = (\cos 3)x^2$ $\qquad\qquad\quad y' = (\cos 3)(2x) = 2x\cos 3$

c. $y = \cos(3x)^2 = \cos(9x^2)$ $\qquad y' = (-\sin 9x^2)(18x) = -18x\sin 9x^2$

d. $y = \cos^2 3x = (\cos 3x)^2$ $\qquad y' = 2(\cos 3x)\,D_x[\cos 3x]$

$$= 2(\cos 3x)(-\sin 3x)(3)$$

$$= -6\cos 3x\sin 3x$$

To find the derivative of a function of the form $k(x) = f(g(h(x)))$, you need to apply the Chain Rule twice, as shown in Example 12.

EXAMPLE 12   Repeated Application of the Chain Rule

$$f(t) = \sin^3 4t = (\sin 4t)^3 \qquad \text{Rewrite}$$

$$f'(t) = 3(\sin 4t)^2 \frac{d}{dt}[\sin 4t] \qquad \text{Apply Chain Rule once}$$

$$= 3(\sin 4t)^2(\cos 4t)\frac{d}{dx}[4t] \qquad \text{Apply Chain Rule second time}$$

$$= 3(\sin 4t)^2(\cos 4t)(4)$$

$$= 12\sin^2 4t\cos 4t \qquad \text{Simplify}$$

We conclude this section with a summary of the differentiation rules studied so far. To become skilled at differentiation, you should memorize each rule.

Summary of Differentiation Rules

General Differentiation Rules   Let u and v be differentiable functions of x.

Constant Multiple Rule:

$$\frac{d}{dx}[cu] = cu'$$

Sum or Difference Rule:

$$\frac{d}{dx}[u \pm v] = u' \pm v'$$

Product Rule:

$$\frac{d}{dx}[uv] = uv' + vu'$$

Quotient Rule:

$$\frac{d}{dx}\left[\frac{u}{v}\right] = \frac{vu' - uv'}{v^2}$$

Derivatives of Algebraic Functions

Constant Rule:

$$\frac{d}{dx}[c] = 0$$

Simple Power Rule:

$$\frac{d}{dx}[x^n] = nx^{n-1}, \qquad \frac{d}{dx}[x] = 1$$

Derivatives of Trigonometric Functions

$$\frac{d}{dx}[\sin x] = \cos x$$

$$\frac{d}{dx}[\tan x] = \sec^2 x$$

$$\frac{d}{dx}[\sec x] = \sec x \tan x$$

$$\frac{d}{dx}[\cos x] = -\sin x$$

$$\frac{d}{dx}[\cot x] = -\csc^2 x$$

$$\frac{d}{dx}[\csc x] = -\csc x \cot x$$

Chain Rule

Chain Rule:

$$\frac{d}{dx}[f(u)] = f'(u)u'$$

General Power Rule:

$$\frac{d}{dx}[u^n] = nu^{n-1}u'$$

REMARK   As an aid to memorization, note that the cofunctions (cosine, cotangent, and cosecant) require a negative sign as part of their derivatives.

EXERCISES for Section 2.4

TECHNOLOGY
Laboratory Guide
Lab 2.4

In Exercises 1–6, complete the table using Example 2 as a model.

$y = f(g(x))$	$u = g(x)$	$y = f(u)$
1. $y = (6x - 5)^4$		
2. $y = \dfrac{1}{\sqrt{x+1}}$		
3. $y = \sqrt{x^2 - 1}$		
4. $y = \tan(\pi x + 1)$		
5. $y = \csc^3 x$		
6. $y = \cos\dfrac{3x}{2}$		

In Exercises 7–30, find the first derivative of the algebraic function.

7. $y = (2x - 7)^3$

8. $y = (3x^2 + 1)^4$

9. $g(x) = 3(4 - 9x)^4$

10. $f(x) = 2(1 - x^2)^3$

11. $f(x) = (9 - x^2)^{2/3}$

12. $f(t) = (9t + 2)^{2/3}$

13. $f(t) = \sqrt{1 - t}$

14. $g(x) = \sqrt{3 - 2x}$

15. $y = \sqrt[3]{9x^2 + 4}$

16. $g(x) = \sqrt{x^2 - 2x + 1}$

17. $y = 2\sqrt{4 - x^2}$

18. $f(x) = -3\sqrt[4]{2 - 9x}$

19. $y = \dfrac{1}{x - 2}$

20. $s(t) = \dfrac{1}{t^2 + 3t - 1}$

21. $f(t) = \left(\dfrac{1}{t-3}\right)^2$

22. $y = -\dfrac{4}{(t+2)^2}$

23. $y = \dfrac{1}{\sqrt{x+2}}$

24. $g(t) = \sqrt{\dfrac{1}{t^2-2}}$

25. $f(x) = x^2(x-2)^4$

26. $f(x) = x(3x-9)^3$

27. $y = x\sqrt{1-x^2}$

28. $y = x^2\sqrt{9-x^2}$

29. $y = \dfrac{x}{\sqrt{x^2+1}}$

30. $y = \sqrt{x+\sqrt{x}}$

C In Exercises 31–38, use a symbolic differentiation utility to find the first derivative of the function. Then use the utility to graph the function and its derivative on the same set of coordinate axes. Describe the behavior of the function that corresponds to any zeros of the graph of the derivative.

31. $y = \dfrac{\sqrt{x}+1}{x^2+1}$

32. $y = \sqrt{\dfrac{2x}{x+1}}$

33. $g(t) = \dfrac{3t^2}{\sqrt{t^2+2t-1}}$

34. $f(x) = \sqrt{x}(2-x)^2$

35. $y = \sqrt{\dfrac{x+1}{x}}$

36. $y = (t^2-9)\sqrt{t+2}$

37. $s(t) = \dfrac{-2(2-t)\sqrt{1+t}}{3}$

38. $g(x) = \sqrt{x-1}+\sqrt{x+1}$

In Exercises 39 and 40, find the slope of the tangent line to the sine curve at the origin. Compare this value to the number of complete cycles in the interval $[0, 2\pi]$.

39. a. $y = \sin x$

b. $y = \sin 2x$

40. a. $y = \sin 3x$

b. $y = \sin\dfrac{x}{2}$

In Exercises 41–50, find the derivative of the function.

41. $y = \cos 3x$

42. $y = \sin \pi x$

43. $g(x) = 3\tan 4x$

44. $h(x) = \sec(x^2)$

45. $f(\theta) = \dfrac{1}{4}\sin^2 2\theta$

46. $g(t) = 5\cos^2 \pi t$

47. $y = \sqrt{x} + \dfrac{1}{4}\sin(2x)^2$

48. $y = 3x - 5\cos(\pi x)^2$

49. $y = \sin(\cos x)$

50. $y = \sin\sqrt{x} + \sqrt{\sin x}$

C In Exercises 51 and 52, use a symbolic differentiation utility to find the derivative of the function. Then use the utility to graph the function and its derivative on the same set of coordinate axes. Describe the behavior of the function that corresponds to any zeros of the graph of the derivative.

51. $y = \dfrac{\cos \pi x + 1}{x}$

52. $y = x^2 \tan\dfrac{1}{x}$

In Exercises 53–60, evaluate the derivative of the function at the indicated point.

Function	Point
53. $s(t) = \sqrt{t^2+2t+8}$	$(2, 4)$
54. $y = \sqrt[5]{3x^3+4x}$	$(2, 2)$
55. $f(x) = \dfrac{3}{x^3-4}$	$\left(-1, -\dfrac{3}{5}\right)$
56. $f(x) = \dfrac{1}{(x^2-3x)^2}$	$\left(4, \dfrac{1}{16}\right)$
57. $f(t) = \dfrac{3t+2}{t-1}$	$(0, -2)$
58. $f(x) = \dfrac{x+1}{2x-3}$	$(2, 3)$
59. $y = 37 - \sec^3(2x)$	$(0, 36)$
60. $y = \dfrac{1}{x} + \sqrt{\cos x}$	$\left(\dfrac{\pi}{2}, \dfrac{2}{\pi}\right)$

In Exercises 61–64, find an equation of the tangent line to the graph of f at the indicated point.

Function	Point
61. $f(x) = \sqrt{3x^2-2}$	$(3, 5)$
62. $f(x) = x\sqrt{x^2+5}$	$(2, 6)$
63. $f(x) = \sin 2x$	$(\pi, 0)$
64. $f(x) = \tan^2 x$	$\left(\dfrac{\pi}{4}, 1\right)$

Essay In Exercises 65–68, the graphs of a function f and its derivative f' are given. Label the graphs as f or f' and write a short paragraph stating the criteria used in making the selection.

65.

66.

67.

68.

In Exercises 69–72, find the second derivative of the function.

69. $f(x) = 2(x^2 - 1)^3$

70. $f(x)\dfrac{1}{x-2}$

71. $f(x) = \sin x^2$

72. $f(x) = \sec^2 \pi x$

73. Given that $g(5) = -3$, $g'(5) = 6$, $h(5) = 3$, and $h'(5) = -2$, find $f'(5)$ (if possible) for each of the following. If it is not possible, state what additional information is required.

a. $f(x) = g(x)h(x)$

b. $f(x) = g(h(x))$

c. $f(x) = \dfrac{g(x)}{h(x)}$

d. $f(x) = [g(x)]^3$

74. Let u be a differentiable function of x. Use the fact that $|u| = \sqrt{u^2}$ to prove that

$$\frac{d}{dx}[\,|u|\,] = u'\frac{u}{|u|}, \qquad u \neq 0.$$

In Exercises 75–78, use the result of Exercise 74 to find the derivative of the function.

75. $g(x) = |\,2x - 3\,|$

76. $f(x) = |\,x^2 - 4\,|$

77. $h(x) = |\,x\,|\cos x$

78. $f(x) = |\sin x|$

79. *Doppler Effect* The frequency f of a fire truck siren heard by a stationary observer is

$$f = \frac{132{,}400}{331 \pm v}$$

where $\pm v$ represents the velocity of the accelerating fire truck (see figure). Find the rate of change of f with respect to v when

a. the fire truck is approaching at a velocity of 30 m/s (use $-v$).

b. the fire truck is moving away at a velocity of 30 m/s (use $+v$).

$$f = \frac{132{,}400}{331 + v} \qquad\qquad f = \frac{132{,}400}{331 - v}$$

FIGURE FOR 79

80. *Harmonic Motion* The displacement from equilibrium of an object in harmonic motion on the end of a spring is

$$y = \tfrac{1}{3}\cos 12t - \tfrac{1}{4}\sin 12t$$

where y is measured in feet and t is time in seconds. Determine the position and velocity of the object when $t = \pi/8$.

81. *Pendulum* A 6-inch pendulum moves according to the equation

$$\theta = 0.2\cos 8t$$

where θ is the angular displacement from the vertical in radians and t is time in seconds. Determine the maximum angular displacement and the rate of change of θ when $t = 3$ seconds.

82. *Wave Motion* A buoy oscillates in simple harmonic motion ($y = A\cos \omega t$) as waves move past it. The buoy moves a total of 3.5 feet (vertically) from its low point to its high point. It returns to the high point every 10 seconds.

a. Write an equation describing the motion of the buoy if it is at its high point at $t = 0$.

b. Determine the velocity of the buoy as a function of t.

83. *Circulatory System* The speed S of blood that is r centimeters from the center of an artery is $S = C(R^2 - r^2)$, where C is a constant, R is the radius of the artery, and S is measured in centimeters per second. Suppose a drug is administered and the artery begins dilating at the rate of dR/dt. At a constant distance r, find the rate at which S changes with respect to t for $C = 1.76 \times 10^5$, $R = 1.2 \times 10^{-2}$, and $dR/dt = 10^{-5}$.

84. For $f(x) = \sec^2 x$ and $g(x) = \tan^2 x$, show that $f'(x) = g'(x)$.

85. Consider the function $f(x) = \sin \beta x$ where β is a constant.

a. Find the first, second, third, and fourth derivatives of the function.

b. Verify that the second derivative satisfies the equation $f''(x) + \beta^2 f(x) = 0$.

c. Use the results of part a to write general rules for the even- and odd-order derivatives

$$f^{(2k)}(x) \quad \text{and} \quad f^{(2k-1)}(x).$$

[*Hint:* $(-1)^k$ is positive if k is even and negative if k is odd.]

86. Find the derivative of the function $g(x) = \sin^2 x + \cos^2 x$.

True or False **In Exercises 87–89, determine whether the statement is true or false. If it is false, explain why or give an example that shows it is false.**

87. If $y = (1 - x)^{1/2}$, then $y' = \tfrac{1}{2}(1 - x)^{-1/2}$.

88. If $f(x) = \sin^2(2x)$, then $f'(x) = 2(\sin 2x)(\cos 2x)$.

89. If y is a differentiable function of u, u is a differentiable function of v, and v is a differentiable function of x, then

$$\frac{dy}{dx} = \frac{dy}{du}\frac{du}{dv}\frac{dv}{dx}.$$

90. Show that the derivative of an odd function is even. That is, if $f(-x) = -f(x)$, then $f'(-x) = f'(x)$.

91. The geometric mean of x and $x + n$ is $g = \sqrt{x(x+n)}$, and the arithmetic mean is $a = [x + (x + n)]/2$. Show that

$$\frac{dg}{dx} = \frac{a}{g}.$$

SECTION Implicit Differentiation

2.5 Implicit and Explicit Functions • Implicit Differentiation

Implicit and Explicit Functions

Up to this point in the text, most functions have been expressed in **explicit form.** For example, in the equation $y = 3x^2 - 5$, the variable y is explicitly written as a function of x. Some functions, however, are only implied by an equation. For instance, the function $y = 1/x$ is defined **implicitly** by the equation $xy = 1$. Suppose you were asked to find dy/dx in this equation. For this equation, you could begin by writing y explicitly as a function of x and then differentiating.

Implicit form	Explicit form	Derivative
$xy = 1$	$y = \dfrac{1}{x} = x^{-1}$	$\dfrac{dy}{dx} = -x^{-2} = -\dfrac{1}{x^2}$

This strategy works well whenever you can solve for the function explicitly. You cannot, however, use this procedure when you are unable to solve for y as a function of x. For instance, how would you find dy/dx in the equation $x^2 - 2y^3 + 4y = 2$, where it is very difficult to express y as a function of x explicitly? To do this, you can use **implicit differentiation.**

To understand how to find dy/dx implicitly, you must realize that the differentiation is taking place *with respect to x.* This means that when you differentiate terms involving x alone, you can differentiate as usual. *But* when you differentiate terms involving y, you must apply the Chain Rule because you are assuming that y is defined implicitly as a function of x.

EXAMPLE 1   Differentiating with Respect to x

a. $\dfrac{d}{dx}[x^3] = 3x^2$ Variables agree—use Simple Power Rule

Variables agree

b. $\overbrace{\dfrac{d}{dx}[y^3]}^{u^n} = \overbrace{3y^2}^{nu^{n-1}}\,\overbrace{\dfrac{dy}{dx}}^{u'} = 3y^2\dfrac{dy}{dx}$ Variables disagree—use Chain Rule

Variables disagree

c. $\dfrac{d}{dx}[x + 3y] = 1 + 3\dfrac{dy}{dx}$ Chain Rule: $\dfrac{d}{dx}[3y] = 3y'$

d. $\dfrac{d}{dx}[xy^2] = x\dfrac{d}{dx}[y^2] = y^2\dfrac{d}{dx}[x]$ Product Rule

$\qquad\quad = x\left(2y\dfrac{dy}{dx}\right) + y^2(1)$ Chain Rule

$\qquad\quad = 2xy\dfrac{dy}{dx} + y^2$ Simplify

Implicit Differentiation

Given an equation involving x and y, and assuming y is a differentiable function of x, you can find dy/dx as follows.

Guidelines for Implicit Differentiation	1. Differentiate both sides of the equation *with respect to x*.
	2. Collect all terms involving dy/dx on the left side of the equation and move all other terms to the right side of the equation.
	3. Factor dy/dx out of the left side of the equation.
	4. Solve for dy/dx by dividing both sides of the equation by the left-hand factor that does not contain dy/dx.

EXAMPLE 2   Implicit Differentiation

Find dy/dx given that $y^3 + y^2 - 5y - x^2 = -4$.

Solution

1. Differentiate both sides of the equation with respect to x.

$$\frac{d}{dx}[y^3 + y^2 - 5y - x^2] = \frac{d}{dx}[-4]$$

$$\frac{d}{dx}[y^3] + \frac{d}{dx}[y^2] - \frac{d}{dx}[5y] - \frac{d}{dx}[x^2] = \frac{d}{dx}[-4]$$

$$3y^2\frac{dy}{dx} + 2y\frac{dy}{dx} - 5\frac{dy}{dx} - 2x = 0$$

2. Collect the dy/dx terms on the left side of the equation.

$$3y^2\frac{dy}{dx} + 2y\frac{dy}{dx} - 5\frac{dy}{dx} = 2x$$

3. Factor dy/dx out of the left side of the equation.

$$\frac{dy}{dx}(3y^2 + 2y - 5) = 2x$$

4. Solve for dy/dx by dividing by $(3y^2 + 2y - 5)$.

$$\frac{dy}{dx} = \frac{2x}{3y^2 + 2y - 5}$$

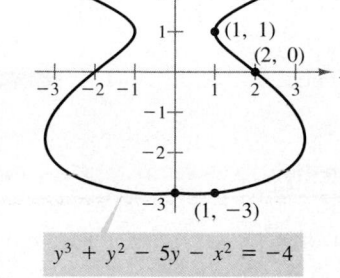

$y^3 + y^2 - 5y - x^2 = -4$

FIGURE 2.25

The derivative is $\dfrac{dy}{dx} = \dfrac{2x}{3y^2 + 2y - 5}$.

Point on Graph	Slope of Graph
$(2, 0)$	$-\frac{4}{5}$
$(1, -3)$	$\frac{1}{8}$
$x = 0$	0
$(1, 1)$	Undefined

To see how you can use an *implicit derivative,* consider the graph shown in Figure 2.25. From the graph, you can see that y is not a function of x. Even so, the derivative found in Example 2 gives a formula for the slope of the tangent line at a point on this graph. The slopes at several points on the graph are shown at the left.

TECHNOLOGY With most graphing utilities, it is easy to sketch the graph of an equation that explicitly represents y as a function of x. Sketching graphs of other equations, however, can require some ingenuity. For instance, to sketch the graph of the equation given in Example 2, try using a graphing utility, set in *parametric mode,* to sketch the graphs given by $x = \sqrt{t^3 + t^2 - 5t + 4}$, $y = t$, and $x = -\sqrt{t^3 + t^2 - 5t + 4}$, $y = t$. How does the result compare with the graph shown in Figure 2.25?

(a)

(b)

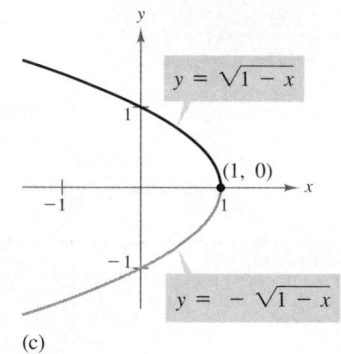

(c)

FIGURE 2.26
Some graph segments can be represented by differentiable functions.

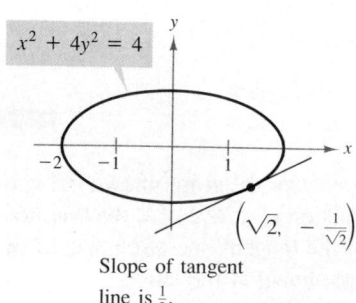

Slope of tangent
line is $\frac{1}{2}$.

FIGURE 2.27
The derivative is $\dfrac{dy}{dx} = \dfrac{-x}{4y}$.

It is meaningless to solve for dy/dx in an equation that has no solution points. (For example, $x^2 + y^2 = -4$ has no solution points.) If, however, a segment of a graph can be represented by a differentiable function, then dy/dx will have meaning as the slope at each point on the segment. Recall that a function is not differentiable at (1) points with vertical tangents, and (2) points at which the function is not continuous.

EXAMPLE 3   Representing a Graph by Differentiable Functions

If possible, represent the graph of the equation by differentiable functions (see Figure 2.26).

a. $x^2 + y^2 = 0$ **b.** $x^2 + y^2 = 1$ **c.** $x + y^2 = 1$

Solution

a. The graph of this equation is a single point. Therefore, it does not define y as a differentiable function of x.

b. The graph of this equation is the unit circle, centered at $(0, 0)$. The upper semicircle is given by the differentiable function

$$y = \sqrt{1 - x^2}, \qquad -1 < x < 1$$

and the lower semicircle is given by the differentiable function

$$y = -\sqrt{1 - x^2}, \qquad -1 < x < 1.$$

At the points $(-1, 0)$ and $(1, 0)$, the slope of the graph is undefined.

c. The upper half of this parabola is given by the differentiable function

$$y = \sqrt{1 - x}, \qquad x < 1$$

and the lower half is given by the differentiable function

$$y = -\sqrt{1 - x}, \qquad x < 1.$$

At the point $(1, 0)$, the slope of the graph is undefined.

EXAMPLE 4   Finding the Slope of a Graph Implicitly

Determine the slope of the tangent line to the graph of $x^2 + 4y^2 = 4$ at the point $\left(\sqrt{2}, -1/\sqrt{2}\right)$ (see Figure 2.27).

Solution   Implicit differentiation of the equation $x^2 + 4y^2 = 4$ with respect to x yields

$$2x + 8y \frac{dy}{dx} = 0$$

$$\frac{dy}{dx} = \frac{-2x}{8y} = \frac{-x}{4y}.$$

Therefore, at $\left(\sqrt{2}, -1/\sqrt{2}\right)$, the slope is

$$\frac{dy}{dx} = \frac{-\sqrt{2}}{-4/\sqrt{2}} = \frac{1}{2}.$$

REMARK   To see the benefit of implicit differentiation, try doing Example 4 using the explicit function $y = -\frac{1}{2}\sqrt{4 - x^2}$.

EXAMPLE 5   Finding the Slope of a Graph Implicitly

Determine the slope of the graph of $3(x^2 + y^2)^2 = 100xy$ at the point $(3, 1)$.

Solution

$$\frac{d}{dx}[3(x^2 + y^2)^2] = \frac{d}{dx}[100xy]$$

$$3(2)(x^2 + y^2)\left(2x + 2y\frac{dy}{dx}\right) = 100\left[x\frac{dy}{dx} + y(1)\right]$$

$$12y(x^2 + y^2)\frac{dy}{dx} - 100x\frac{dy}{dx} = 100y - 12x(x^2 + y^2)$$

$$[12y(x^2 + y^2) - 100x]\frac{dy}{dx} = 100y - 12x(x^2 + y^2)$$

$$\frac{dy}{dx} = \frac{100y - 12x(x^2 + y^2)}{-100x + 12y(x^2 + y^2)}$$

$$= \frac{25y - 3x(x^2 + y^2)}{-25x + 3y(x^2 + y^2)}$$

At the point $(3, 1)$ the slope of the graph is

$$\frac{dy}{dx} = \frac{25(1) - 3(3)(3^2 + 1^2)}{-25(3) + 3(1)(3^2 + 1^2)} = \frac{25 - 90}{-75 + 30} = \frac{-65}{-45} = \frac{13}{9}$$

as shown in Figure 2.28. This graph is called a **lemniscate.**

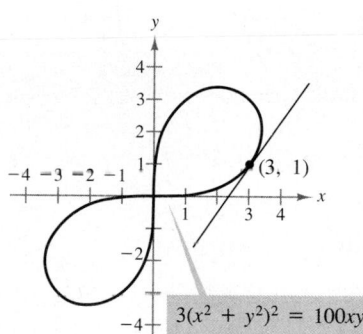

FIGURE 2.28
Lemniscate

$3(x^2 + y^2)^2 = 100xy$

EXAMPLE 6   Determining a Differentiable Function

Find dy/dx implicitly for the equation $\sin y = x$. Then find the largest interval of the form $-a < y < a$ such that y is a differentiable function of x (see Figure 2.29).

Solution

$$\frac{d}{dx}[\sin y] = \frac{d}{dx}[x]$$

$$\cos y\frac{dy}{dx} = 1$$

$$\frac{dy}{dx} = \frac{1}{\cos y}$$

The largest interval about the origin for which y is a differentiable function of x is $-\pi/2 < y < \pi/2$. To see this, note that $\cos y$ is positive for all y in this interval and 0 at the endpoints. If you restrict y to the interval $-\pi/2 < y < \pi/2$, you should be able to write dy/dx explicitly as a function of x. To do this, you can use

$$\cos y = \sqrt{1 - \sin^2 y}$$

$$= \sqrt{1 - x^2}, \qquad -\frac{\pi}{2} < y < \frac{\pi}{2}$$

and conclude that

$$\frac{dy}{dx} = \frac{1}{\sqrt{1 - x^2}}.$$

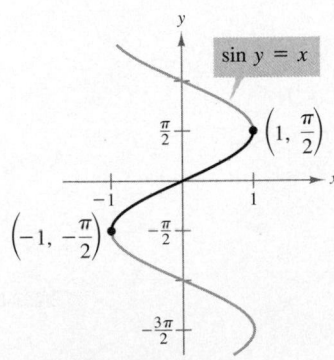

FIGURE 2.29
The derivative is $\dfrac{dy}{dx} = \dfrac{1}{\sqrt{1 - x^2}}$.

$\sin y = x$

$\left(1, \dfrac{\pi}{2}\right)$

$\left(-1, -\dfrac{\pi}{2}\right)$

With implicit differentiation, the form of the derivative often can be simplified or changed to a single variable (as in Example 6) by an appropriate use of the *original equation*. A similar technique can be used to find and simplify higher-order derivatives obtained implicitly.

EXAMPLE 7   Finding the Second Derivative Implicitly

Given $x^2 + y^2 = 25$, find y''.

Solution   Differentiating each term with respect to x produces

$$2x + 2yy' = 0$$
$$2yy' = -2x$$
$$y' = \frac{-2x}{2y}$$
$$= -\frac{x}{y}.$$

Differentiating a second time with respect to x yields

$$y'' = -\frac{(y)(1) - (x)(y')}{y^2} \qquad \text{Quotient Rule}$$
$$= -\frac{y - (x)(-x/y)}{y^2} \qquad \text{Substitute } -x/y \text{ for } y'$$
$$= -\frac{y^2 + x^2}{y^3}$$
$$= -\frac{25}{y^3}. \qquad \text{Substitute 25 for } x^2 + y^2$$

EXAMPLE 8   Finding a Tangent Line to a Graph

Find the tangent line to the graph given by $x^2(x^2 + y^2) = y^2$ at the point $(\sqrt{2}/2, \sqrt{2}/2)$, as shown in Figure 2.30.

Solution   By rewriting and differentiating implicitly, you obtain

$$x^4 + x^2y^2 - y^2 = 0$$
$$4x^3 + x^2(2yy') + 2xy^2 - 2yy' = 0$$
$$2y(x^2 - 1)y' = -2x(2x^2 + y^2)$$
$$y' = \frac{x(2x^2 + y^2)}{y(1 - x^2)}.$$

At the point $(\sqrt{2}/2, \sqrt{2}/2)$, the slope is

$$m = \frac{(\sqrt{2}/2)[2(1/2) + (1/2)]}{(\sqrt{2}/2)[1 - (1/2)]} = \frac{3/2}{1/2} = 3$$

and the equation of the tangent line at this point is

$$y - \frac{\sqrt{2}}{2} = 3\left(x - \frac{\sqrt{2}}{2}\right)$$
$$y = 3x - \sqrt{2}.$$

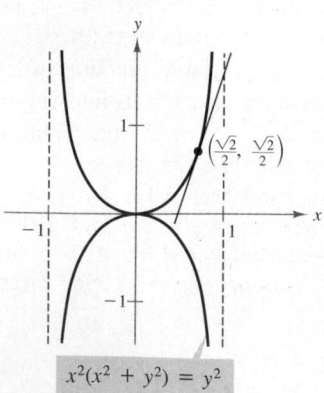

$$x^2(x^2 + y^2) = y^2$$

FIGURE 2.30
The kappa curve

EXERCISES for Section 2.5

TECHNOLOGY
Laboratory Guide
Lab 2.5

In Exercises 1–16, find dy/dx by implicit differentiation.

1. $x^2 + y^2 = 16$

2. $x^2 - y^2 = 16$

3. $x^{1/2} + y^{1/2} = 9$

4. $x^3 + y^3 = 8$

5. $x^3 - xy + y^2 = 4$

6. $x^2y + y^2x = -2$

7. $x^3y^3 - y = x$

8. $\sqrt{xy} = x - 2y$

9. $x^3 - 2x^2y + 3xy^2 = 38$

10. $2\sin x \cos y = 1$

11. $\sin x + 2\cos 2y = 1$

12. $(\sin \pi x + \cos \pi y)^2 = 2$

13. $\sin x = x(1 + \tan y)$

14. $\cot y = x - y$

15. $y = \sin(xy)$

16. $x = \sec \dfrac{1}{y}$

In Exercises 17–24, find dy/dx by implicit differentiation and evaluate the derivative at the indicated point.

Equation	Point
17. $xy = 4$	$(-4, -1)$
18. $x^2 - y^3 = 0$	$(1, 1)$
19. $y^2 = \dfrac{x^2 - 9}{x^2 + 9}$	$(3, 0)$
20. $(x + y)^3 = x^3 + y^3$	$(-1, 1)$
21. $x^{2/3} + y^{2/3} = 5$	$(8, 1)$
22. $x^3 + y^3 = 2xy$	$(1, 1)$
23. $\tan(x + y) = x$	$(0, 0)$
24. $x \cos y = 1$	$\left(2, \dfrac{\pi}{3}\right)$

In Exercises 25–28, find the slope of the tangent line to the graph at the indicated point.

25. Witch of Agnesi:
$(x^2 + 4)y = 8$
Point: $(2, 1)$

26. Cissoid:
$(4 - x)y^2 = x^3$
Point: $(2, 2)$

27. Bifolium:
$(x^2 + y^2)^2 = 4x^2y$
Point: $(1, 1)$

28. Folium of Descartes:
$x^3 + y^3 - 6xy = 0$
Point: $\left(\frac{4}{3}, \frac{8}{3}\right)$

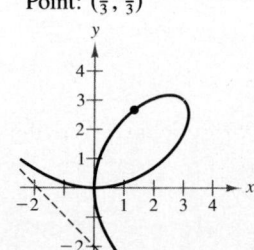

C **In Exercises 29 and 30, use a graphing utility to graph the equation. Find the equation of the tangent line to the graph at the indicated point and sketch its graph.**

29. $\sqrt{x} + \sqrt{y} = 3$

$(4, 1)$

30. $y^2 = \dfrac{x - 1}{x^2 + 1}$

$\left(2, \dfrac{\sqrt{5}}{5}\right)$

In Exercises 31–34, (a) find two explicit functions by solving the equation for y in terms of x, (b) sketch the graph of the equation and label the parts given by the explicit functions, (c) differentiate the explicit functions, and (d) find dy/dx implicitly and show that the result is equivalent to that of part (c).

31. $x^2 + y^2 = 16$

32. $x^2 + y^2 - 4x + 6y + 9 = 0$

33. $9x^2 + 16y^2 = 144$

34. $4y^2 - x^2 = 4$

In Exercises 35–40, find d^2y/dx^2 in terms of x and y.

35. $x^2 + xy = 5$

36. $x^2y^2 - 2x = 3$

37. $x^2 - y^2 = 16$

38. $1 - xy = x - y$

39. $y^2 = x^3$

40. $y^2 = 4x$

In Exercises 41 and 42, find equations for the tangent line and normal line to the circle at the indicated points. (The *normal line* at a point is perpendicular to the tangent line at the point.)

41. $x^2 + y^2 = 25$
$(4, 3), (-3, 4)$

42. $x^2 + y^2 = 9$
$(0, 3), \left(2, \sqrt{5}\right)$

43. Show that the normal line at any point on the circle $x^2 + y^2 = r^2$ passes through the origin.

44. Two circles of radius 4 are tangent to the graph of $y^2 = 4x$ at the point $(1, 2)$. Find the equations for these two circles.

In Exercises 45 and 46, find the points at which the graph of the equation has a vertical or horizontal tangent line.

45. $25x^2 + 16y^2 + 200x - 160y + 400 = 0$

46. $4x^2 + y^2 - 8x + 4y + 4 = 0$

Orthogonal Trajectories **In Exercises 47–50, sketch the intersecting graphs of the equations and show that they are *orthogonal*. [Two graphs are orthogonal if at their point(s) of intersection, their tangent lines are perpendicular to each other.]**

47. $2x^2 + y^2 = 6$
$y^2 = 4x$

48. $y^2 = x^3$
$2x^2 + 3y^2 = 5$

49. $x + y = 0$
$x = \sin y$

50. $x^3 = 3(y - 1)$
$x(3y - 29) = 3$

C *Orthogonal Trajectories* In Exercises 51 and 52, verify that the two families of curves are orthogonal where C and K are positive real numbers. Use a graphing utility to graph the two families for two values of C and two values of K.

51. $xy = C$
$x^2 - y^2 = K$

52. $x^2 + y^2 = C^2$
$y = Kx$

53. Use Example 6 as a model to find dy/dx implicitly for the equation $\tan y = x$ and find the largest interval of the form $-a < y < a$ such that y is a differentiable function of x. Then express dy/dx as a function of x.

54. Show that

$$\frac{dy}{dx} = \frac{-(x - h)}{y - k}$$

for the circle

$$(x - h)^2 + (y - k)^2 = r^2.$$

55. Prove (Theorem 2.3) that

$$\frac{d}{dx}[x^n] = nx^{n-1}$$

for the case that n is a rational number. (*Hint:* Write $y = x^{p/q}$ in the form $y^q = x^p$ and differentiate implicitly. Assume p and q are integers where $q > 0$.)

56. Let L be any tangent line to the curve $\sqrt{x} + \sqrt{y} = \sqrt{c}$. Show that the sum of the x- and y-intercepts of L is c.

Optical Illusions In Exercises 57–60, an optical illusion is created by having lines intersect a family of curves. In each case, the lines appear to be curved rather than straight. Find the value of dy/dx for the indicated values of x and y. [For more information on the mathematics of optical illusions, see the article "Descriptive Models for Perception of Optical Illusions" by David A. Smith in the 1985 (Volume 2) issue of the *UMAP Journal.*]

57. Circles: $x^2 + y^2 = C^2$
$x = 3$, $y = 4$, $C = 5$

58. Hyperbolas: $xy = C$
$x = 1$, $y = 4$, $C = 4$

59. Lines: $ax = by$
$x = \sqrt{3}$, $y = 3$,
$a = \sqrt{3}$, $b = 1$

60. Cosine curves: $y = C \cos x$
$x = \dfrac{\pi}{3}$, $y = \dfrac{1}{3}$, $C = \dfrac{2}{3}$

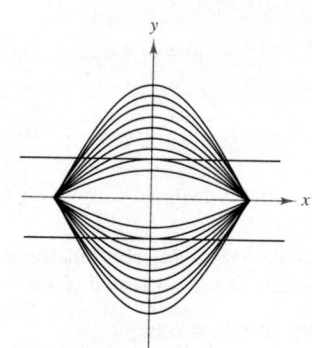

SECTION 2.6

Related Rates

Finding Related Rates • Problem Solving with Related Rates

FIGURE 2.31
Volume is related to radius and height.

Finding Related Rates

You have seen how the Chain Rule can be used to find dy/dx implicitly. Another important use of the Chain Rule is to find the rates of change of two or more related variables that are changing with respect to *time*.

For example, when water is drained out of a conical tank (see Figure 2.31), the volume V, the radius r, and the height h of the water level are all functions of time t. Knowing that these variables are related by the equation

$$V = \frac{\pi}{3} r^2 h$$

you can differentiate implicitly with respect to t to obtain the **related rate** equation

$$\frac{dV}{dt} = \frac{\pi}{3} \left[r^2 \frac{dh}{dt} + h \left(2r \frac{dr}{dt} \right) \right]$$

$$= \frac{\pi}{3} \left[r^2 \frac{dh}{dt} + 2rh \frac{dr}{dt} \right].$$

From this equation you can see that the rate of change of V is related to the rates of change of both h and r.

EXAMPLE 1   Two Rates that Are Related

Suppose x and y are both differentiable functions of t and are related by the equation

$$y = x^2 + 3.$$

Find dy/dt when $x = 1$, given that $dx/dt = 2$ when $x = 1$.

Solution   Using the Chain Rule, you can differentiate both sides of the equation *with respect to t.*

$$y = x^2 + 3 \qquad \text{Given equation}$$

$$\frac{d}{dt}[y] = \frac{d}{dt}[x^2 + 3] \qquad \text{Differentiate with respect to } t$$

$$\frac{dy}{dt} = 2x \frac{dx}{dt} \qquad \text{Chain Rule}$$

When $x = 1$ and $dx/dt = 2$, you have

$$\frac{dy}{dt} = 2(1)(2) = 4.$$

Problem Solving with Related Rates

In Example 1, you were *given* an equation that relates the variables x and y, and asked to find the rate of change of y when $x = 1$.

Equation $y = x^2 + 3$

Given rate $\dfrac{dx}{dt} = 2$ when $x = 1$

Find $\dfrac{dy}{dt}$ when $x = 1$

In the remaining examples in this section, you must *create* an equation from a verbal description.

EXAMPLE 2   Ripples in a Pond

A pebble is dropped into a calm pond, causing ripples in the form of concentric circles, as shown in Figure 2.32. The radius r of the outer ripple is increasing at a constant rate of 1 foot per second. When the radius is 4 feet, at what rate is the total area A of the disturbed water changing?

Solution   The variables r and A are related by $A = \pi r^2$. The rate of change of the radius r is $dr/dt = 1$.

Given equation $A = \pi r^2$

Given rate $\dfrac{dr}{dt} = 1$

Find $\dfrac{dA}{dt}$ when $r = 4$

With this information, you can proceed as in Example 1.

$$\frac{d}{dt}[A] = \frac{d}{dt}[\pi r^2]$$ Differentiate with respect to t

$$\frac{dA}{dt} = 2\pi r \frac{dr}{dt}$$ Chain Rule

$$\frac{dA}{dt} = 2\pi(4)(1) = 8\pi$$ Substitute 1 for dr/dt and 4 for r

When $r = 4$, the area is changing at a rate of 8π ft^2 per second.

FIGURE 2.32
Total area increases as outer radius increases.

Guidelines for Solving Related-Rate Problems

1. Identify all *given* quantities and quantities *to be determined.* Make a sketch and label the quantities.
2. Write an equation involving the variables whose rates of change either are given or are to be determined.
3. Using the Chain Rule, implicitly differentiate both sides of the equation *with respect to time t.*
4. *After* completing Step 3, substitute into the resulting equation all known values for the variables and their rates of change. Then solve for the required rate of change.

REMARK   In these guidelines, be sure you perform Step 3 *before* Step 4. Substituting the known values of the variables before differentiating will produce an inappropriate derivative.

Table 2.2 shows the mathematical models for some rates of change.

TABLE 2.2

Verbal statement	Mathematical model
The velocity of a car after traveling 1 hour is 50 miles per hour.	x = Distance traveled $\dfrac{dx}{dt} = 50$ when $t = 1$
Water is being pumped into a swimming pool at the rate of 10 cubic feet per minute.	V = Volume of water in pool $\dfrac{dV}{dt} = 10$ ft³/min
A gear is revolving at the rate of 25 revolutions per minute (1 rev = 2π rad).	θ = Angle of revolution $\dfrac{d\theta}{dt} = 25(2\pi)$ rad/min

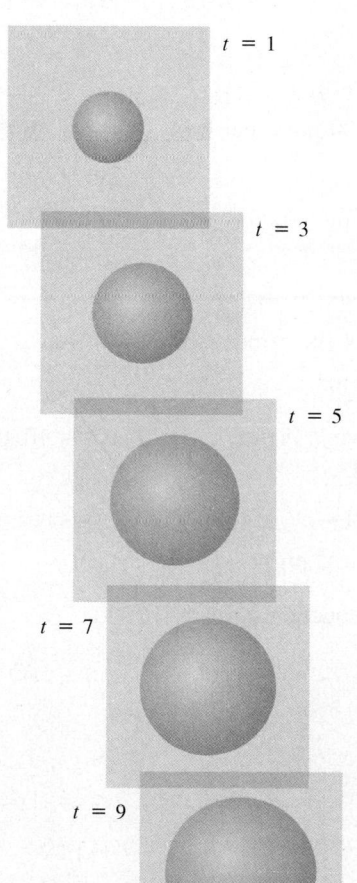

$t = 1$

$t = 3$

$t = 5$

$t = 7$

$t = 9$

FIGURE 2.33
Expanding balloon

EXAMPLE 3   An Inflating Balloon

Air is being pumped into a spherical balloon at the rate of 4.5 cubic inches per minute, as shown in Figure 2.33. Find the rate of change of the radius when the radius is 2 inches.

Solution   Let V be the volume of the balloon and let r be its radius. Because the volume is increasing at the rate of 4.5 cubic inches per minute, you know that at time t the rate of change of the volume is $dV/dt = \frac{9}{2}$. Thus, the problem can be stated as follows.

Given rate $\dfrac{dV}{dt} = \dfrac{9}{2}$ (constant rate)

Find $\dfrac{dr}{dt}$ when $r = 2$

To find the rate of change of the radius, you must find an equation that relates the radius r to the volume V.

$$V = \frac{4}{3}\pi r^3 \qquad \text{Volume of a sphere}$$

Implicit differentiation *with respect to t* produces

$$\frac{dV}{dt} = 4\pi r^2 \frac{dr}{dt} \qquad \text{Differentiate with respect to } t$$

$$\frac{dr}{dt} = \frac{1}{4\pi r^2}\left(\frac{dV}{dt}\right). \qquad \text{Solve for } dr/dt$$

Finally, when $r = 2$, the rate of change of the radius is

$$\frac{dr}{dt} = \frac{1}{16\pi}\left(\frac{9}{2}\right) \approx 0.09 \text{ in./min.}$$

In Example 3, note that the volume is increasing at a *constant* rate but the radius is increasing at a *variable* rate. Just because two rates are related does not mean that they are proportional. In this particular case, the radius is growing more and more slowly as t increases.

EXAMPLE 4   The Speed of an Airplane Tracked by Radar

An airplane is flying on a flight path that will take it directly over a radar tracking station, as shown in Figure 2.34. If s is decreasing at a rate of 400 miles per hour when s is 10 miles, what is the speed of the plane?

Solution   Label the distance x as in Figure 2.34. Notice that when $s = 10$, $x = \sqrt{10^2 - 36} = 8$.

> *Given rate* $ds/dt = -400$ when $s = 10$
>
> *Find* dx/dt when $s = 10$ and $x = 8$

You can find the velocity of the plane as follows.

$$x^2 + 6^2 = s^2 \qquad\qquad \text{Pythagorean Theorem}$$

$$2x\frac{dx}{dt} = 2s\frac{ds}{dt} \qquad\qquad \text{Differentiate with respect to } t$$

$$\frac{dx}{dt} = \frac{s}{x}\left(\frac{ds}{dt}\right) \qquad\qquad \text{Solve for } dx/dt$$

$$\frac{dx}{dt} = \frac{10}{8}(-400) = -500 \text{ mph} \qquad \text{Substitute for } x, s, \text{ and } ds/dt$$

Because the velocity is -500 mph, the *speed* is 500 miles per hour.

FIGURE 2.34
An airplane is flying at an altitude of 6 miles.

EXAMPLE 5   A Changing Angle of Elevation

Find the rate of change in the angle of elevation of the camera shown in Figure 2.35 at 10 seconds after lift-off.

Solution   Let θ be the angle of elevation, as shown in Figure 2.35. When $t = 10$, the height s of the rocket is $s = 50(10^2) = 5000$ feet.

> *Given rate* $ds/st = 100t =$ Velocity of rocket
>
> *Find* $d\theta/dt$ when $t = 10$ and $s = 5000$

Using Figure 2.35, you can relate s and θ by the equation $\tan\theta = s/2000$.

$$\tan\theta = \frac{s}{2000} \qquad\qquad \text{See Figure 2.35}$$

$$(\sec^2\theta)\frac{d\theta}{dt} = \frac{1}{2000}\left(\frac{ds}{dt}\right) \qquad\qquad \text{Differentiate with respect to } t$$

$$\frac{d\theta}{dt} = \cos^2\theta \frac{100t}{2000} \qquad\qquad \text{Substitute } 100t \text{ for } ds/dt$$

$$= \left(\frac{2000}{\sqrt{s^2 + 2000^2}}\right)^2 \frac{100t}{2000} \qquad \cos\theta = 2000/\sqrt{s^2 + 2000^2}$$

$$\frac{d\theta}{dt} = \frac{2000(100)(10)}{5000^2 + 2000^2} \qquad\qquad \text{Substitute for } s \text{ and } t$$

$$= \frac{2}{29} \text{ rad/sec.} \qquad\qquad \text{Simplify}$$

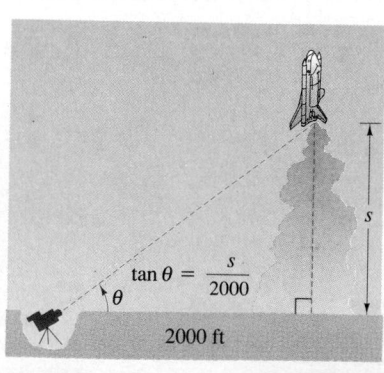

FIGURE 2.35
A television camera at ground level is filming the lift-off of a space shuttle that is rising vertically according to the position equation $s = 50t^2$, where s is measured in feet and t is measured in seconds. The camera is 2000 feet from the launchpad.

When $t = 10$, θ is changing at the rate of $\frac{2}{29}$ radians per second.

EXAMPLE 6   The Velocity of a Piston

For the engine shown in Figure 2.36, a 7-inch connecting rod is fastened to a crank of radius 3 inches. The crankshaft rotates counterclockwise at a constant rate of 200 revolutions per minute. Find the velocity of the piston when $\theta = \pi/3$.

Solution   Label the distances as shown in Figure 2.36. Because a complete revolution corresponds to 2π radians, it follows that $d\theta/dt = 200(2\pi) = 400\pi$ radians per minute.

FIGURE 2.36
The velocity of a piston is related to the angle of the crankshaft.

Given rate $\dfrac{d\theta}{dt} = 400\pi$ (constant rate)

Find $\dfrac{dx}{dt}$ when $\theta = \dfrac{\pi}{3}$

You can use the Law of Cosines (Figure 2.37) to find an equation that relates x and θ.

$$7^2 = 3^2 + x^2 - 2(3)(x)\cos\theta$$

$$0 = 2x\frac{dx}{dt} - 6\left[-x\sin\theta\frac{d\theta}{dt} + \cos\theta\frac{dx}{dt}\right]$$

$$(6\cos\theta - 2x)\frac{dx}{dt} = 6x\sin\theta\frac{d\theta}{dt}$$

$$\frac{dx}{dt} = \frac{6x\sin\theta}{6\cos\theta - 2x}\left(\frac{d\theta}{dt}\right)$$

FIGURE 2.37
Law of Cosines:
$b^2 = a^2 + c^2 - 2ac\cos\theta$.

When $\theta = \pi/3$, you can solve for x as follows.

$$7^2 = 3^2 + x^2 - 2(3)(x)\cos\frac{\pi}{3}$$

$$49 = 9 + x^2 - 6x\left(\frac{1}{2}\right)$$

$$0 = x^2 - 3x - 40$$

$$0 = (x - 8)(x + 5)$$

$$x = 8 \qquad \text{Choose positive solution}$$

Thus, when $x = 8$ and $\theta = \pi/3$, the velocity of the piston is

$$\frac{dx}{dt} = \frac{6(8)\left(\sqrt{3}/2\right)}{6(1/2) - 16}(400\pi)$$

$$= \frac{9600\pi\sqrt{3}}{-13}$$

$$\approx -4018 \text{ in./min.}$$

REMARK   Note that the velocity in Example 6 is negative, because x represents a distance that is decreasing.

1-11 odd 17-23 odd 31, 41

EXERCISES for Section 2.6

TECHNOLOGY
Laboratory Guide
Lab 2.6

In Exercises 1–4, assume that x and y are both differentiable functions of t and find the indicated values of dy/dt and dx/dt.

Equation	Find		Given
1. $y = \sqrt{x}$	a. $\dfrac{dy}{dt}$ when $x = 4$		$\dfrac{dx}{dt} = 3$
	b. $\dfrac{dx}{dt}$ when $x = 25$		$\dfrac{dy}{dt} = 2$
2. $y = x^2 - 3x$	a. $\dfrac{dy}{dt}$ when $x = 3$		$\dfrac{dx}{dt} = 2$
	b. $\dfrac{dx}{dt}$ when $x = 1$		$\dfrac{dy}{dt} = 5$
3. $xy = 4$	a. $\dfrac{dy}{dt}$ when $x = 8$		$\dfrac{dx}{dt} = 10$
	b. $\dfrac{dx}{dt}$ when $x = 1$		$\dfrac{dy}{dt} = -6$
4. $x^2 + y^2 = 25$	a. $\dfrac{dy}{dt}$ when $x = 3,$ $y = 4$		$\dfrac{dx}{dt} = 8$
	b. $\dfrac{dx}{dt}$ when $x = 4,$ $y = 3$		$\dfrac{dy}{dt} = -2$

In Exercises 5–8, a point is moving along the graph of the function so that dx/dt is 2 centimeters per second. Find dy/dt for the indicated values of x.

Function	Values of x	
5. $y = x^2 + 1$	a. $x = -1$	b. $x = 0$
	c. $x = 1$	d. $x = 3$
6. $y = \dfrac{1}{1 + x^2}$	a. $x = -2$	b. $x = 0$
	c. $x = 2$	d. $x = 10$
7. $y = \tan x$	a. $x = -\dfrac{\pi}{3}$	b. $x = -\dfrac{\pi}{4}$
	c. $x = 0$	d. $x = 1$
8. $y = \sin x$	a. $x = \dfrac{\pi}{6}$	b. $x = \dfrac{\pi}{4}$
	c. $x = \dfrac{\pi}{3}$	d. $x = \dfrac{\pi}{2}$

9. Use the conditions of Exercise 5 to find the rate of change of the distance between the origin and the moving point on the graph of the function.

10. Use the conditions of Exercise 8 to find the rate of change of the distance between the origin and the moving point on the graph of the function.

11. *Area* The radius r of a circle is increasing at a rate of 2 inches per minute. Find the rate of change of the area when (a) $r = 6$ inches and (b) $r = 24$ inches.

12. *Area* Let A be the area of a circle of radius r that is changing with respect to time. If dr/dt is constant, is dA/dt constant? Explain why or why not.

13. *Area* The included angle of the two sides of equal length s in an isosceles triangle is θ.
    a. Show that the area of the triangle is
    $$A = \frac{s^2}{2}\sin\theta.$$
    b. If θ is increasing at the rate of $\frac{1}{2}$ radian per minute, find the rate of change of the area when $\theta = \pi/6$ and $\theta = \pi/3$.
    c. Explain why the rate of change of the area of the triangle is not constant even though $d\theta/dt$ is constant.

14. *Volume* The radius r of a sphere is increasing at a rate of 2 inches per minute.
    a. Find the rate of change of the volume when $r = 6$ inches and $r = 24$ inches.
    b. Explain why the rate of change of the volume of the sphere is not constant even though dr/dt is constant.

15. *Volume* A spherical balloon is inflated with gas at the rate of 20 cubic feet per minute. How fast is the radius of the balloon increasing at the instant the radius is (a) 1 foot and (b) 2 feet?

16. *Volume* All edges of a cube are expanding at the rate of 3 centimeters per second. How fast is the volume changing when each edge is (a) 1 centimeter and (b) 10 centimeters?

17. *Surface Area* The conditions are the same as in Exercise 16. Now measure how fast the *surface* area is changing when each edge is (a) 1 centimeter and (b) 10 centimeters.

18. *Volume* The formula for the volume of a cone is
    $$V = \tfrac{1}{3}\pi r^2 h.$$
    Find the rate of change of the volume if dr/dt is 2 inches per minute and $h = 3r$ when (a) $r = 6$ inches and (b) $r = 24$ inches.

19. *Volume* At a sand and gravel plant, sand is falling off a conveyor and onto a conical pile at the rate of 10 cubic feet per minute. The diameter of the base of the cone is approximately three times the altitude. At what rate is the height of the pile changing when it is 15 feet high?

20. *Depth* A conical tank (with vertex down) is 10 feet across the top and 12 feet deep. If water is flowing into the tank at the rate of 10 cubic feet per minute, find the rate of change of the depth of the water the instant it is 8 feet deep.

21. *Depth* A swimming pool is 40 feet long, 20 feet wide, 4 feet deep at the shallow end, and 9 feet deep at the deep end (see figure). Water is being pumped into the pool at 10 cubic feet per minute and there is 4 feet of water at the deep end.

a. What percentage of the pool is filled?
b. At what rate is the water level rising?

22. *Depth*   A trough is 12 feet long and 3 feet across the top (see figure). Its ends are isosceles triangles with an altitude of 3 feet. Water is being pumped into the trough at 2 cubic feet per minute. How fast is the water level rising when it is 1 foot deep?

FIGURE FOR 21   **FIGURE FOR 22**

23. *Moving Ladder*   A ladder 25 feet long is leaning against the wall of a house (see figure). The base of the ladder is pulled away from the wall at a rate of 2 feet per second.
  a. How fast is the top moving down the wall when the base of the ladder is 7 feet, 15 feet, and 24 feet from the wall?
  b. Consider the triangle formed by the side of the house, the ladder, and the ground. Find the rate at which the area of the triangle is changing when the base of the ladder is 7 feet from the wall.
  c. Find the rate at which the angle between the top of the ladder and the wall of the house is changing when the base of the ladder is 7 feet from the wall.

24. *Construction*   A construction worker pulls a 16-foot plank up the side of a building under construction by means of a rope tied to the end of the plank (see figure). Assume the opposite end of the plank follows a path perpendicular to the wall of the building and the worker pulls the rope at the rate of 0.5 feet per second. How fast is the end of the plank sliding along the ground when it is 8 feet from the wall of the building?

FIGURE FOR 23   **FIGURE FOR 24**

25. *Construction*   A winch at the top of a 40-foot building pulls a pipe of the same length to a vertical position using the method shown in the accompanying figure. The winch pulls in rope at the rate of −0.5 feet per second. Find the rate of vertical change and the rate of horizontal change at the end of the pipe when $y = 20$.

26. *Boating*   A boat is pulled in by means of a winch on the dock 12 feet above the deck of the boat (see figure). The winch pulls in rope at the rate of 4 feet per second. Determine the speed of the boat when there is 13 feet of rope out. What happens to the speed of the boat as it gets closer to the dock?

FIGURE FOR 25   **FIGURE FOR 26**

27. *Air Traffic Control*   An air traffic controller spots two planes at the same altitude converging on a point as they fly at right angles to each other (see figure). One plane is 150 miles from the point and is moving at 450 miles per hour. The other plane is 200 miles from the point and has a speed of 600 miles per hour.
  a. At what rate is the distance s between the planes decreasing?
  b. How much time does the traffic controller have to get one of the planes on a different flight path?

28. *Air Traffic Control*   An airplane is flying at an altitude of 6 miles and passes directly over a radar antenna (see figure). When the plane is 10 miles away ($s = 10$), the radar detects that the distance s is changing at a rate of 240 miles per hour. What is the speed of the plane?

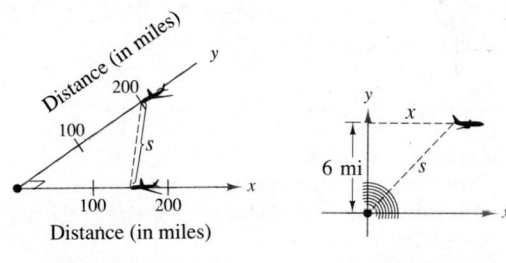

FIGURE FOR 27   **FIGURE FOR 28**

29. *Baseball*   A baseball diamond has the shape of a square with sides 90 feet long (see figure). A player 30 feet from third base is running at a speed of 28 feet per second. At what rate is the player's distance s from home plate changing?

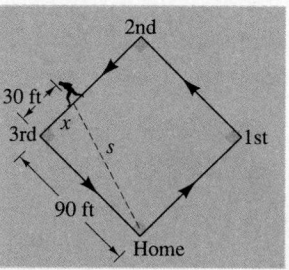

FIGURE FOR 29

30. *Baseball* For the baseball diamond in Exercise 29, suppose the player is running from first to second at a speed of 28 feet per second. Find the rate at which the distance from home plate is changing when the player is 30 feet from second base.

31. *Shadow Length* A man 6 feet tall walks at a rate of 5 feet per second away from a light that is 15 feet above the ground (see figure). When he is 10 feet from the base of the light,
a. at what rate is the tip of his shadow changing?
b. at what rate is the length of his shadow changing?

32. *Mechanical Design* One end of a mechanical arm of variable length pivots at the point $(2, 0)$ while the other end moves on a vertical slide (see figure). The position of the end on the vertical slide is given by $(0, y)$ where

$$y = 2 \sin \frac{\pi t}{2}$$

and t is time in seconds. Find the rate of change of the length $\dfrac{ds}{dt}$ of the mechanical arm when $t = \dfrac{3}{2}$ and $t = 3$.

FIGURE FOR 31   **FIGURE FOR 32**

33. *Mechanical Design* The endpoints of a movable rod of length 3 feet have coordinates $(x, 0)$ and $(0, y)$ (see figure). The position of the end on the x-axis is

$$x(t) = 2 \sin \frac{\pi t}{6}$$

where t is time in seconds.
a. Find the time of one complete cycle of the rod.
b. What is the lowest point reached by the end of the rod on the y-axis?
c. Find the speed of the y-axis endpoint when the x-axis endpoint is $(1, 0)$.

34. *Mechanical Design* Repeat Exercise 33 if the position function is

$$x(t) = 2 \sin \pi t.$$

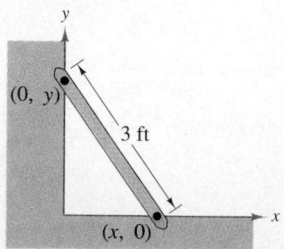

FIGURE FOR 33 AND 34

35. *Evaporation* As a spherical raindrop falls, it reaches a layer of dry air and begins to evaporate at a rate that is proportional to its surface area $(S = 4\pi r^2)$. Show that the radius decreases at a constant rate.

36. *Electricity* The combined electrical resistance R of R_1 and R_2, connected in parallel, is

$$\frac{1}{R} = \frac{1}{R_1} + \frac{1}{R_2}$$

where R, R_1, and R_2 are measured in ohms. R_1 and R_2 are increasing at rates of 1 and 1.5 ohms per second, respectively. At what rate is R changing when $R_1 = 50$ ohms and $R_2 = 75$ ohms?

37. *Adiabatic Expansion* When a certain polyatomic gas undergoes adiabatic expansion, its pressure p and volume v satisfy the equation

$$pv^{1.3} = k$$

where k is a constant. Find the relationship between the related rates dp/dt and dv/dt.

38. *Roadway Design* Cars on a certain roadway travel on a circular arc of radius r. In order not to rely on friction alone to overcome the centrifugal force, the road is banked at an angle of magnitude θ from the horizontal. The banking angle must satisfy the equation

$$rg \tan \theta = v^2$$

where v is the velcity of the cars and $g = 32$ ft/sec^2 is the acceleration due to gravity. Find the relationship between the related rates dv/dt and $d\theta/dt$.

39. *Angle of Elevation* A balloon rises at the rate of 10 feet per second from a point on the ground 100 feet from an observer. Find the rate of change of the angle of elevation of the balloon from the observer when the balloon is 100 feet above the ground.

40. *Angle of Elevation* A fishing line is reeled in at a rate of 1 foot per second from a bridge 15 feet above the water (see figure). At what rate is the angle between the line and the water changing when 25 feet of line is out?

41. *Angle of Elevation* An airplane flies at an altitude of 5 miles toward a point directly over an observer (see figure). The speed of the plane is 600 miles per hour. Find the rate at which the angle of elevation θ is changing when the angle is (a) $\theta = 30°$, (b) $\theta = 60°$, and (c) $\theta = 75°$.

FIGURE FOR 40   **FIGURE FOR 41**

42. *Linear vs. Angular Speed*   A patrol car is parked 50 feet from a long warehouse (see figure). The revolving light on top of the car turns at a rate of 30 revolutions per minute. How fast is the light beam moving along the wall when the beam makes the following angles with the line, perpendicular to the wall, from the light to the wall?
a. $\theta = 30°$ **b.** $\theta = 60°$ **c.** $\theta = 70°$

43. *Linear vs. Angular Speed*   A wheel of radius 1 foot revolves at a rate of 10 revolutions per second. A dot is painted at a point P on the rim of the wheel (see figure). Find the rate of horizontal movement of the dot for the following angles.
a. $\theta = 0°$ **b.** $\theta = 30°$ **c.** $\theta = 60°$

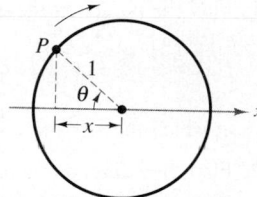

FIGURE FOR 42   **FIGURE FOR 43**

44. *Flight Control*   An airplane is flying in still air with an airspeed of 240 miles per hour. If it is climbing at an angle of 22°, find the rate at which it is gaining altitude.

Acceleration   **In Exercises 45 and 46, find the acceleration of the specified object. (*Hint:* Remember that if a variable is changing at a constant rate, then its acceleration is 0.)**

45. Find the acceleration of the top of the ladder described in Exercise 23 when $x = 7$.

46. Find the acceleration of the boat in Exercise 26 when there is 13 feet of rope out.

REVIEW EXERCISES for Chapter 2

In Exercises 1 and 2, find the derivative of the function by using the definition of the derivative.

1. $f(x) = x^2 - 2x + 3$ **2.** $f(x) = \dfrac{x + 1}{x - 1}$

3. *Essay*   Use the function $f(x) = x\sqrt{4 - x}$ for each of the following.
a. Use a graphing utility to graph the function.
b. Find an equation of the tangent line to the graph of f at $(0, 0)$.
c. Use a computer algebra system or a spreadsheet to complete the table for $x = 0$.

Δx	$f(x + \Delta x)$	$f(x)$	$\dfrac{f(x + \Delta x) - f(x)}{\Delta x}$
2			
1			
0.5			
0.1			

d. Write a paragraph giving the geometric interpretation of the last column of the table. How does it relate to the result of part b?

4. Each of the figures shows the graphs of a function and its derivative. Identify the graphs of f and f'.

(a)

(b)

In Exercises 5–20, find the derivative of the algebraic function.

5. $f(x) = x^3 - 3x^2$ **6.** $f(x) = x^{1/2} - x^{-1/2}$

7. $f(x) = \dfrac{2x^3 - 1}{x^2}$ **8.** $f(x) = \dfrac{x + 1}{x - 1}$

9. $g(t) = \dfrac{2}{3t^2}$ **10.** $h(x) = \dfrac{2}{(3x)^2}$

11. $f(x) = \sqrt{1 - x^3}$ **12.** $f(x) = \sqrt[3]{x^2 - 1}$

13. $f(x) = (3x^2 + 7)(x^2 - 2x + 3)$

14. $f(x) = \left(x^2 + \dfrac{1}{x}\right)^5$

15. $f(s) = (s^2 - 1)^{5/2}(s^3 + 5)$

16. $h(\theta) = \dfrac{\theta}{(1 - \theta)^3}$

17. $f(x) = \dfrac{x^2 + x - 1}{x^2 - 1}$ **18.** $f(x) = \dfrac{6x - 5}{x^2 + 1}$

19. $f(x) = \dfrac{1}{4 - 3x^2}$ **20.** $f(x) = \dfrac{9}{3x^2 - 2x}$

In Exercises 21–32, find the derivative of the trigonometric function.

21. $y = 3\cos(3x + 1)$ **22.** $y = 1 - \cos 2x + 2\cos^2 x$

23. $y = \frac{1}{2}\csc 2x$ **24.** $y = \csc 3x + \cot 3x$

25. $y = \dfrac{x}{2} - \dfrac{\sin 2x}{4}$ **26.** $y = \dfrac{1 + \sin x}{1 - \sin x}$

27. $y = \frac{2}{3}\sin^{3/2} x - \frac{2}{7}\sin^{7/2} x$

28. $y = \dfrac{\sec^7 x}{7} - \dfrac{\sec^5 x}{5}$

29. $y = -x\tan x$ **30.** $y = x\cos x - \sin x$

31. $y = \dfrac{\sin x}{x^2}$ **32.** $y = \dfrac{\cos(x - 1)}{x - 1}$

C **In Exercises 33–40, use a symbolic differentiation utility to find the derivative of the function. Then use the utility to graph the function and its derivative on the same set of coordinate axes. Describe the behavior of the function that corresponds to any zeros of the graph of the derivative.**

33. $f(t) = t^2(t - 1)^5$

34. $f(x) = [(x - 2)(x + 4)]^2$

35. $g(x) = \dfrac{2x}{\sqrt{x + 1}}$ **36.** $g(x) = x\sqrt{x^2 + 1}$

37. $f(t) = \sqrt{t + 1}\sqrt[3]{t + 1}$ **38.** $y = \sqrt{3x}(x + 2)^3$

39. $y = \tan\sqrt{1 - x}$ **40.** $y = 2\csc^3\left(\sqrt{x}\right)$

In Exercises 41–44, find the second derivative of the function.

41. $y = 2x^2 + \sin 2x$ **42.** $y = \dfrac{1}{x} + \tan x$

43. $f(x) = \cot x$ **44.** $y = \sin^2 x$

C **In Exercises 45–48, use a symbolic differentiation utility to find the second derivative of the function.**

45. $f(t) = \dfrac{t}{(1 - t)^2}$ **46.** $g(x) = \dfrac{6x - 5}{x^2 + 1}$

47. $g(x) = x\tan x$ **48.** $h(x) = x\sqrt{x^2 - 1}$

In Exercises 49–54, use implicit differentiation to find dy/dx.

49. $x^2 + 3xy + y^3 = 10$

50. $x^2 + 9y^2 - 4x + 3y - 7 = 0$

51. $y\sqrt{x} - x\sqrt{y} = 16$ **52.** $y^2 = (x - y)(x^2 + y)$

53. $x\sin y = y\cos x$ **54.** $\cos(x + y) = x$

In Exercises 55–60, find an equation of the tangent line and the normal line to the graph of the equation at the indicated point.

Function	Point
55. $y = (x + 3)^3$	$(-2, 1)$
56. $y = (x - 2)^2$	$(2, 0)$
57. $x^2 + y^2 = 20$	$(2, 4)$
58. $x^2 - y^2 = 16$	$(5, 3)$
59. $y = \sqrt[3]{(x - 2)^2}$	$(3, 1)$
60. $y = \dfrac{2x}{1 - x^2}$	$(0, 0)$

61. Find the points on the graph of

$$f(x) = \frac{1}{3}x^3 + x^2 - x - 1$$

at which the slope is (a) -1, (b) 2, and (c) 0.

62. Find the points on the graph of $f(x) = x^2 + 1$ at which the slope is (a) -1, (b) 0, and (c) 1.

63. Sketch the graph of $f(x) = 4 - |x - 2|$.
   a. Is f continuous at $x = 2$?
   b. Is f differentiable at $x = 2$? Why or why not?

64. Sketch the graph of

$$f(x) = \begin{cases} x^2 + 4x + 2, & x < -2 \\ 1 - 4x - x^2, & x \ge -2. \end{cases}$$

   a. Is f continuous at $x = -2$?
   b. Is f differentiable at $x = -2$? Why or why not?

In Exercises 65 and 66, show that the function satisfies the given equation.

Function	Equation
65. $y = 2\sin x + 3\cos x$	$y'' + y = 0$
66. $y = \dfrac{10 - \cos x}{x}$	$xy' + y = \sin x$

67. *Refrigeration* Suppose that the temperature T of food placed in a freezer is

$$T = \frac{700}{t^2 + 4t + 10}$$

where t is the time in hours. Find the rate of change of T with respect to t at the following times.
   a. $t = 1$ **b.** $t = 3$ **c.** $t = 5$ **d.** $t = 10$

68. *Fluid Flow* The emergent velocity v of a liquid flowing from a hole in the bottom of a tank is $v = \sqrt{2gh}$, where g is the acceleration due to gravity (32 ft/sec^2) and h is the depth of the liquid in the tank. Find the rate of change of v with respect to h when (a) $h = 9$ and (b) $h = 4$.

69. *Vibrating String* When a guitar string is plucked, it vibrates with a frequency of $F = 200\sqrt{T}$, where F is measured in vibrations per second and the tension T is measured in pounds. Find the rate of change of the frequency when (a) $T = 4$ and (b) $T = 9$.

70. *Vertical Motion*   A ball is dropped from a height of 100 feet. One second later another ball is dropped from a height of 75 feet. Which ball hits the ground first?

71. *Vertical Motion*   What is the smallest initial velocity that is required to throw a stone 49 feet up to the top of a silo?

72. *Vertical Motion*   A bomb is dropped from an airplane at an altitude of 14,400 feet. How long will it take to reach the ground? (Even though it will not be a vertical fall because of the motion of the plane, the time will be the same as that for a vertical fall.) The plane is moving at 600 miles per hour. How far will the bomb move horizontally after it is released from the plane?

73. *Projectile Motion*   A ball is thrown and follows a path described by $y = x - 0.02x^2$.
  a. Sketch a graph of the path.
  b. Find the total horizontal distance the ball was thrown.
  c. For what x-value does the ball reach its maximum height? (Use the symmetry of the path.)
  d. Find the equation that gives the instantaneous rate of change in the height of the ball with respect to the horizontal change, and evaluate this equation at $x = 0, 10, 25, 30,$ and 50.
  e. What is the instantaneous rate of change of the height when the ball reaches its maximum height?

74. *Projectile Motion*   The path of a projectile thrown at an angle of $45°$ with level ground is

$$y = x - \frac{32}{v_0^2}(x^2)$$

where the initial velocity is v_0 feet per second.
  a. Sketch the path followed by the projectile.
  b. Find the x-coordinate of the point where the projectile strikes the ground. Use the symmetry of the path of the projectile to locate the x-coordinate of the point where the projectile reaches its maximum height.
  c. What is the instantaneous rate of change of the height when the projectile is at its maximum height?

75. *Projectile Motion*   The path of a projectile thrown at an angle of $45°$ with the ground is given by

$$y = x - \frac{32}{v_0^2}(x^2)$$

where the initial velocity is v_0 feet per second. Show that doubling the initial velocity of the projectile multiplies both the maximum height and the range by a factor of 4.

76. *Projectile Motion*   Use the equation given in Exercise 75 to find the maximum height and range of a projectile thrown with an initial velocity of 70 feet per second.

77. A point moves along the curve $y = \sqrt{x}$ in such a way that the y-value is increasing at the rate of two units per second. At what rate is x changing for the following values?
  a. $x = \frac{1}{2}$ **b.** $x = 1$ **c.** $x = 4$

78. The same conditions exist as in Exercise 77. Find the rate of change of the distance between (x, y) and the origin for the following x-values.
  a. $x = \frac{1}{2}$ **b.** $x = 1$ **c.** $x = 4$

79. *Changing Depth*   The cross section of a 5-foot trough is an isosceles trapezoid with a 2-foot lower base, a 3-foot upper base, and an altitude of 2 feet. Water is running into the trough at the rate of 1 cubic foot per minute. How fast is the water level rising when the water is 1 foot deep?

80. *Linear and Angular Velocity*   A rotating beacon is located 1 mile off a straight shoreline. If the beacon rotates at the rate of 3 revolutions per minute, how fast (in miles per hour) does the beam of light appear to be moving to a viewer who is $\frac{1}{2}$ mile down the shoreline?

81. *Moving Shadow*   A sandbag is dropped from a balloon at a height of 200 feet when the angle of elevation to the sun is $30°$ (see figure). Find the rate at which the shadow of the sandbag is traveling along the ground when the sandbag is at a height of 125 feet. (*Hint:* The position of the sandbag is given by $s(t) = 200 - 16t^2$.)

FIGURE FOR 81

Chapter 3

Applications
of
Differentiation

Extrema on an Interval

Extrema of a Function • Relative Extrema and Critical Numbers • Finding Extrema on a Closed Interval

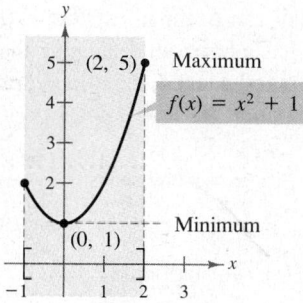

(a) f is continuous.
  $[-1, 2]$ is closed.

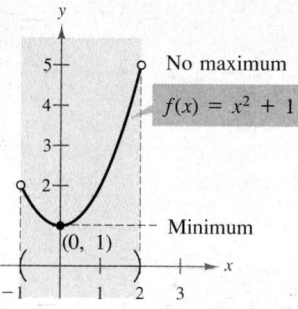

(b) f is continuous.
  $(-1, 2)$ is open.

(c) g is discontinuous.
  $[-1, 2]$ is closed.

FIGURE 3.1

Extrema can occur at interior points or endpoints of an interval. Extrema that occur at the endpoints are called **endpoint extrema.**

Extrema of a Function

Many questions in calculus concern the behavior of a function f on an interval I. Does f have a maximum value on I? Does it have a minimum value? Where is the function increasing? Where is it decreasing? In this chapter you will learn how derivatives can be used to answer these questions. You will also see why these questions are important in real-life applications.

Definition of Extrema

Let f be defined on an interval I containing c.

1. $f(c)$ is the **minimum of f on I** if $f(c) \leq f(x)$ for all x in I.
2. $f(c)$ is the **maximum of f on I** if $f(c) \geq f(x)$ for all x in I.

The minimum and maximum of a function on an interval are the **extreme values,** or **extrema,** of the function on the interval. The minimum and maximum of a function on an interval are also called the **absolute minimum** and **absolute maximum** on the interval, respectively.

A function need not have a minimum or a maximum on an interval. For instance, in Figure 3.1(a) and (b), you can see that the function $f(x) = x^2 + 1$ has both a minimum and a maximum on the closed interval $[-1, 2]$, but does not have a maximum on the open interval $(-1, 2)$. Moreover, in Figure 3.1(c), you can see that continuity (or the lack of it) can affect the existence of an extremum on the interval. This suggests the following theorem, which identifies conditions that guarantee the existence of both a minimum and a maximum of a function on a closed interval. (A proof of this theorem is beyond the scope of this text.)

THEOREM 3.1   The Extreme Value Theorem

If f is continuous on a closed interval $[a, b]$, then f has both a minimum and a maximum on the interval.

REMARK   Note that the Extreme Value Theorem (like the Intermediate Value Theorem) is an *existence theorem* because it tells of the existence of minimum and maximum values, but does not show how to find these values.

◀ *The velocity of a ball bearing rolling down an inclined plane is discussed in Section 3.3, Exercise 47.*

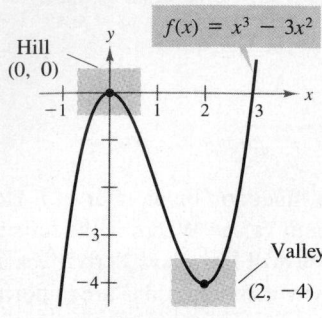

FIGURE 3.2

f has a relative maximum at $(0, 0)$ and a relative minimum at $(2, -4)$.

(a)

(b)

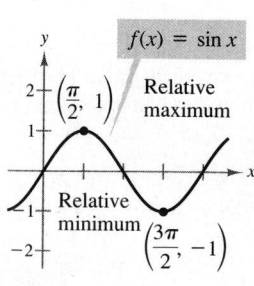

(c)

FIGURE 3.3

At the relative extrema the derivative is zero or is undefined.

Relative Extrema and Critical Numbers

In Figure 3.2, the graph of $f(x) = x^3 - 3x^2$ has a **relative maximum** at the point $(0, 0)$ and a **relative minimum** at the point $(2, -4)$. Informally, you can think of a relative maximum as occurring on a "hill" on the graph, and a relative minimum as occurring in a "valley" on the graph. Such a hill and valley can occur in two ways. If the hill (or valley) is smooth and rounded, then the graph has a horizontal tangent line at the high point (or low point). If the hill (or valley) is sharp and peaked, then the graph represents a function that is not differentiable at the high point (or low point).

> **Definition of Relative Extrema**
>
> 1. If there is an open interval containing c on which $f(c)$ is a maximum, then $f(c)$ is called a **relative maximum** of f.
> 2. If there is an open interval containing c on which $f(c)$ is a minimum, then $f(c)$ is called a **relative minimum** of f.
>
> The plural of relative maximum is relative maxima, and the plural of relative minimum is relative minima.

Example 1 examines the derivative of a function at some *given* relative extrema. (Much more is said about *finding* the relative extrema of a function in Section 3.3.)

EXAMPLE 1   The Value of the Derivative at Relative Extrema

Find the value of the derivative at each of the relative extrema shown in Figure 3.3.

Solution

a. The derivative of

$$f(x) = \frac{9(x^2 - 3)}{x^3}$$

is

$$f'(x) = \frac{x^3(18x) - (9)(x^2 - 3)(3x^2)}{(x^3)^2} = \frac{9(9 - x^2)}{x^4}.$$

At the point $(3, 2)$, the value of the derivative is $f'(3) = 0$.

b. At $x = 0$, the derivative of $f(x) = |x|$ *does not exist* because the following one-sided limits differ.

$$\lim_{x \to 0^-} \frac{f(x) - f(0)}{x - 0} = \lim_{x \to 0^-} \frac{|x|}{x} = -1 \qquad \text{Limit from the left}$$

$$\lim_{x \to 0^+} \frac{f(x) - f(0)}{x - 0} = \lim_{x \to 0^+} \frac{|x|}{x} = +1 \qquad \text{Limit from the right}$$

c. The derivative of $f(x) = \sin x$ is $f'(x) = \cos x$. At the point $(\pi/2, 1)$, the value of the derivative is $f'(\pi/2) = \cos(\pi/2) = 0$. At the point $(3\pi/2, -1)$, the value of the derivative is $f'(3\pi/2) = \cos(3\pi/2) = 0$.

Note in Example 1 that at a relative extremum the derivative is either 0 or undefined. The *x*-values at these special points are called **critical numbers.** Figure 3.4 illustrates the two types of critical numbers.

Definition of Critical Number

Let *f* be defined at *c*. If $f'(c) = 0$ or if f' is undefined at *c*, then *c* is a **critical number** of *f*.

FIGURE 3.4
c is a critical number of *f*.

THEOREM 3.2 Relative Extrema Occur Only at Critical Numbers

If *f* has a relative minimum or relative maximum at $x = c$, then *c* is a critical number of *f*.

REMARK   Theorem 3.2 is sometimes called Fermat's Theorem, after the French mathematician Pierre de Fermat (1601–1665).

Proof   *Case 1:* If *f* is *not* differentiable at $x = c$, then by definition *c* is a critical number of *f* and the theorem is valid.

Case 2: If *f* is differentiable at $x = c$, then $f'(c)$ must be either positive, negative, or 0. Suppose $f'(c)$ is positive. Then

$$f'(c) = \lim_{x \to c} \frac{f(x) - f(c)}{x - c} > 0$$

which implies that there exists an interval (a, b) containing *c* such that

$$\frac{f(x) - f(c)}{x - c} > 0, \qquad \text{for all } x \neq c \text{ in } (a, b).$$

Because this quotient is positive, the signs of the denominator and numerator must agree. This produces the following inequalities for *x*-values in the interval (a, b).

Left of c $x < c$ and $f(x) < f(c) \Rightarrow f(c)$ is not a relative minimum

Right of c $x > c$ and $f(x) > f(c) \Rightarrow f(c)$ is not a relative maximum

Thus, the assumption that $f'(c) > 0$ contradicts the hypothesis that $f(c)$ is a relative extremum. Assuming that $f'(c) < 0$ produces a similar contradiction, you are left with only one possibility, namely $f'(c) = 0$.

Finding Extrema on a Closed Interval

Theorem 3.2 states that the relative extrema of a function can occur *only* at the critical numbers of the function. Knowing this, you can use the following guidelines to find extrema on a closed interval.

Guidelines for Finding Extrema on a Closed Interval

To find the extrema of a continuous function f on a closed interval $[a, b]$, use the following steps.

1. Find the critical numbers of f in (a, b).
2. Evaluate f at each critical number in (a, b).
3. Evaluate f at each endpoint of $[a, b]$.
4. The least of these values is the minimum. The greatest is the maximum.

EXAMPLE 2   Finding Extrema on a Closed Interval

Find the extrema of $f(x) = 3x^4 - 4x^3$ on the interval $[-1, 2]$.

Solution   Begin by differentiating the given function.

$$f(x) = 3x^4 - 4x^3 \qquad \text{Given function}$$
$$f'(x) = 12x^3 - 12x^2 \qquad \text{Derivative}$$

To find the critical numbers of f, you must find all x-values for which $f'(x) = 0$ *and* all x-values for which $f'(x)$ is undefined.

$$f'(x) = 12x^3 - 12x^2 = 0 \qquad \text{Set } f'(x) = 0$$
$$12x^2(x - 1) = 0 \qquad \text{Factor}$$
$$x = 0, 1 \qquad \text{Critical numbers}$$

Because f' is defined for all x, you can conclude that these are the only critical numbers of f. By evaluating f at these two critical numbers and at the endpoints of $[-1, 2]$, you can determine that the maximum is $f(2) = 16$ and the minimum is $f(1) = -1$, as indicated in Table 3.1. The graph of f is shown in Figure 3.5.

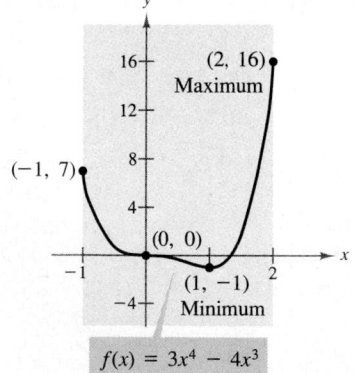

FIGURE 3.5
On the closed interval $[-1, 2]$, f has a minimum at $(1, -1)$ and a maximum at $(2, 16)$.

TABLE 3.1

Left endpoint	Critical number	Critical number	Right endpoint
$f(-1) = 7$	$f(0) = 0$	$f(1) = -1$ Minimum	$f(2) = 16$ Maximum

REMARK   Note in Figure 3.5 that the critical number $x = 0$ does not yield a relative minimum or a relative maximum. This tells you that the converse of Theorem 3.2 is not true. In other words, *the critical numbers of a function need not produce relative extrema.*

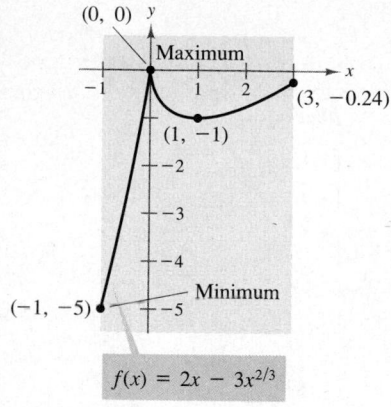

FIGURE 3.6
On the closed interval $[-1, 3]$, f has a minimum at $(-1, -5)$ and a maximum at $(0, 0)$.

EXAMPLE 3   Finding Extrema on a Closed Interval

Find the extrema of $f(x) = 2x - 3x^{2/3}$ on the interval $[-1, 3]$.

Solution   Differentiating produces

$$f'(x) = 2 - \frac{2}{x^{1/3}} = 2\left(\frac{x^{1/3} - 1}{x^{1/3}}\right)$$

From this derivative, you can see that the function has two critical numbers in the interval $[-1, 3]$. The number 1 is a critical number because $f'(1) = 0$, and the number 0 is a critical number because $f'(0)$ is undefined. By evaluating f at these two numbers and at the endpoints of the interval, you can conclude that the minimum is $f(-1) = -5$ and the maximum is $f(0) = 0$, as indicated in Table 3.2. The graph of f is shown in Figure 3.6.

TABLE 3.2

Left endpoint	Critical number	Critical number	Right endpoint
$f(-1) = -5$ Minimum	$f(0) = 0$ Maximum	$f(1) = -1$	$f(3) = 6 - 3\sqrt[3]{9} \approx -0.24$

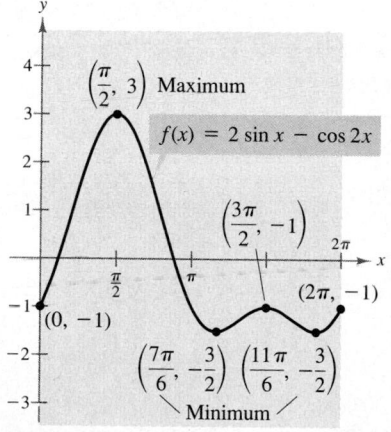

FIGURE 3.7
On the closed interval $[0, 2\pi]$, f has two minima at $(7\pi/6, -3/2)$ and $(11\pi/6, -3/2)$ and a maximum at $(\pi/2, 3)$.

EXAMPLE 4   Finding Extrema on a Closed Interval

Find the extrema of $f(x) = 2\sin x - \cos 2x$ on the interval $[0, 2\pi]$.

Solution   This function is differentiable for all real x, so you can find all critical numbers by setting $f'(x)$ equal to 0, as follows.

$$f'(x) = 2\cos x + 2\sin 2x = 0$$
$$2\cos x + 4\cos x \sin x = 0 \qquad \text{\small $\sin 2x = 2\cos x \sin x$}$$
$$2(\cos x)(1 + 2\sin x) = 0 \qquad \text{\small Factor}$$

In the interval $[0, 2\pi]$, the factor $\cos x$ is 0 when $x = \pi/2$ and when $x = 3\pi/2$. The factor $(1 + 2\sin x)$ is 0 when $x = 7\pi/6$ and when $x = 11\pi/6$. By evaluating f at these four critical numbers and at the endpoints of the interval, you can conclude that the maximum is $f(\pi/2) = 3$ and the minimum occurs at *two* points, $f(7\pi/6) = -3/2$ and $f(11\pi/6) = -3/2$, as indicated in Table 3.3. The graph is shown in Figure 3.7.

TABLE 3.3

Left endpoint	Critical number	Critical number	Critical number	Critical number	Right endpoint
$f(0) = -1$	$f\left(\frac{\pi}{2}\right) = 3$ Maximum	$f\left(\frac{7\pi}{6}\right) = -\frac{3}{2}$ Minimum	$f\left(\frac{3\pi}{2}\right) = -1$	$f\left(\frac{11\pi}{6}\right) = -\frac{3}{2}$ Minimum	$f(2\pi) = -1$

EXERCISES for Section 3.1

In Exercises 1–6, find the value of the derivative (if it exists) at the indicated extrema.

1. $f(x) = \dfrac{x^2}{x^2 + 4}$

2. $f(x) = \cos \dfrac{\pi x}{2}$

3. $f(x) = x + \dfrac{32}{x^2}$

4. $f(x) = -3x\sqrt{x + 1}$

5. $f(x) = (x + 2)^{2/3}$

6. $f(x) = 4 - |x|$

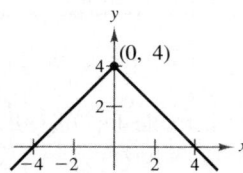

In Exercises 7–12, find the critical numbers of the function.

7. $f(x) = x^2(x - 3)$

8. $g(x) = x^2(x^2 - 4)$

9. $g(t) = t\sqrt{4 - t}$

10. $f(x) = \dfrac{4x}{x^2 + 1}$

11. $h(x) = \sin^2 x + \cos x$
$0 \le x < 2\pi$

12. $f(\theta) = 2 \sec \theta + \tan \theta$
$0 \le \theta < 2\pi$

In Exercises 13–26, locate the absolute extrema of the function on the closed interval.

Function	Interval
13. $f(x) = 2(3 - x)$	$[-1, 2]$
14. $f(x) = \dfrac{2x + 5}{3}$	$[0, 5]$
15. $f(x) = -x^2 + 3x$	$[0, 3]$
16. $f(x) = x^2 + 2x - 4$	$[-1, 1]$
17. $f(x) = x^3 - 3x^2$	$[-1, 3]$
18. $f(x) = x^3 - 12x$	$[0, 4]$
19. $f(x) = 3x^{2/3} - 2x$	$[-1, 1]$

Function	Interval
20. $g(x) = \sqrt[3]{x}$	$[-1, 1]$
21. $h(t) = 4 - |t - 4|$	$[1, 6]$
22. $g(t) = \dfrac{t^2}{t^2 + 3}$	$[-1, 1]$
23. $h(s) = \dfrac{1}{s - 2}$	$[0, 1]$
24. $h(t) = \dfrac{t}{t - 2}$	$[3, 5]$
25. $f(x) = \cos \pi x$	$\left[0, \dfrac{1}{6}\right]$
26. $g(x) = \csc x$	$\left[\dfrac{\pi}{6}, \dfrac{\pi}{3}\right]$

27. *Essay* Explain why the function $f(x) = \tan x$ has a maximum on $[0, \pi/4]$, but not on $[0, \pi]$.

28. *Essay* Write a short paragraph explaining why a continuous function on an open interval may not have a maximum or minimum. Illustrate your explanation with a sketch of the graph of a function.

In Exercises 29–32, determine from the graph whether f possesses a minimum on the open interval (a, b).

29. a.

b.

30. a.

b.

31. a.

b.

32. a.

b.

In Exercises 33 and 34, locate the absolute extrema of the function (if any exist) over the indicated interval.

33. $f(x) = 2x - 3$
   a. $[0, 2]$
   b. $[0, 2)$
   c. $(0, 2]$
   d. $(0, 2)$

34. $f(x) = 5 - x$
   a. $[1, 4]$
   b. $[1, 4)$
   c. $(1, 4]$
   d. $(1, 4)$

In Exercises 35–38, locate the absolute extrema of the function on the indicated interval.

Function	Interval
35. $f(x) = \begin{cases} 2x + 2, & 0 \le x \le 1 \\ 4x^2, & 1 < x \le 3 \end{cases}$	$[0, 3]$
36. $f(x) = \begin{cases} 2 - x^2, & 1 \le x < 3 \\ 2 - 3x, & 3 \le x \le 5 \end{cases}$	$[1, 5]$
37. $f(x) = \dfrac{3}{x - 1}$	$(1, 4]$
38. $f(x) = \dfrac{2}{2 - x}$	$[0, 2)$

C In Exercises 39 and 40, use a symbolic differentiation utility to find the maximum value of $|f''(x)|$ on the indicated interval. (This value is used in the error estimate for the Trapezoidal Rule, as discussed in Section 4.6.)

Function	Interval
39. $f(x) = \sqrt{1 + x^3}$	$[0, 2]$
40. $f(x) = \dfrac{1}{x^2 + 1}$	$\left[\dfrac{1}{2}, 3\right]$

C In Exercises 41 and 42, use a symbolic differentiation utility to find the maximum value of $|f^{(4)}|$ on the indicated interval. (This value is used in the error estimate for Simpson's Rule, as discussed in Section 4.6.)

Function	Interval
41. $f(x) = (x + 1)^{2/3}$	$[0, 2]$
42. $f(x) = \dfrac{1}{x^2 + 1}$	$[-1, 1]$

43. *Power* The formula for the power output P of a battery is

$$P = VI - RI^2$$

where V is the electromotive force in volts, R is the resistance in ohms, and I is the current in amperes. Find the current that corresponds to a maximum value of P in a battery for which $V = 12$ volts and $R = 0.5$ ohm. (Assume that a 15-amp fuse bounds the output in the interval $0 \le I \le 15$.)

44. *Inventory Cost* A retailer has determined that the cost C for ordering and storing x units of a product is

$$C = 2x + \frac{300,000}{x}, \qquad 1 \le x \le 300.$$

The delivery truck can bring at most 300 units per order. Find the order size that will minimize cost.

45. *Lawn Sprinkler* A lawn sprinkler is constructed in such a way that $d\theta/dt$ is constant, where θ ranges between $45°$ and $135°$ (see figure). The distance the water travels horizontally is

$$x = \frac{v^2 \sin 2\theta}{32}, \qquad \frac{\pi}{4} \le \theta \le \frac{3\pi}{4}$$

where v is the speed of the water. Find dx/dt and explain why this model of lawn sprinkler does not water evenly. What part of the lawn receives the most water?

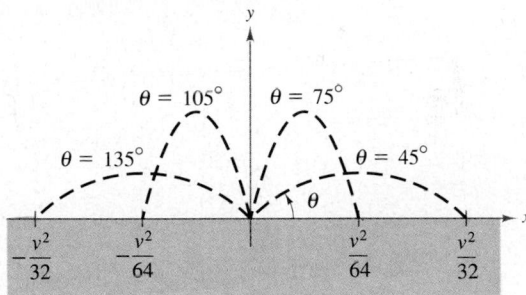

Water Sprinkler: $45° \le \theta \le 135°$

FIGURE FOR 45

FOR FURTHER INFORMATION For more information on the "calculus of lawn sprinklers," see the article "Design of an Oscillating Sprinkler" by Bart Braden in the January, 1985 issue of *Mathematics Magazine.*

46. *Honeycomb* The surface area of a cell in a honeycomb is

$$S = 6hs + \frac{3s^2}{2}\left(\frac{\sqrt{3} - \cos\theta}{\sin\theta}\right)$$

where h and s are positive constants and θ is the angle at which the upper faces meet the altitude of the cell. Find the angle $\theta(\pi/6 \le \theta \le \pi/2)$ that minimizes the surface area S.

*A **C** indicates an exercise in which you are instructed to use graphing technology or a symbolic computer algebra system. The solutions of other exercises may also be facilitated by appropriate technology.

C In Exercises 47 and 48, use a symbolic differentiation utility to analyze the function f over the specified closed interval. Then, (a) graph the function and approximate the absolute extrema from the graph and (b) find any critical numbers. Use them to locate any absolute extrema not located at the endpoints of the interval. Compare the results with those of part (a).

Function	Interval
47. $f(x) = 3.2x^5 + 5x^3 - 3.5x$	$[0, 1]$
48. $f(x) = \frac{4}{3}x\sqrt{3 - x}$	$[0, 3]$

True or False In Exercises 49–52, determine whether the statement is true or false. If it is false, explain why or give an example that shows it is false.

49. The maximum of a function that is continuous on a closed interval can occur at two different values in the interval.

50. If a function is continuous on a closed interval, then it must have a minimum on the interval.

51. If $x = c$ is a critical number of the function f, then it is also a critical number of the function $g(x) = f(x) + k$, where k is a constant.

52. If $x = c$ is a critical number of the function f, then it is also a critical number of the function $g(x) = f(x - k)$, where k is a constant.

53. Find all critical numbers of the greatest integer function, $f(x) = [\![x]\!]$.

Rolle's Theorem and the Mean Value Theorem

Rolle's Theorem • The Mean Value Theorem

Rolle's Theorem

The Extreme Value Theorem (Section 3.1) states that a continuous function on a closed interval $[a, b]$ must have both a minimum and a maximum on the interval. Both of these values, however, can occur at the endpoints. **Rolle's Theorem,** named after the French mathematician Michel Rolle (1652–1719), gives conditions that guarantee the existence of an extreme value in the interior of a closed interval.

THEOREM 3.3   Rolle's Theorem

Let f be continuous on the closed interval $[a, b]$ and differentiable on the open interval (a, b). If

$$f(a) = f(b)$$

then there is at least one number c in (a, b) such that $f'(c) = 0$.

Proof   Let $f(a) = d = f(b)$.

Case 1:   If $f(x) = d$ for all x in $[a, b]$, then f is constant on the interval, and by Theorem 2.2, $f'(x) = 0$ for all x in (a, b).

Case 2:   Suppose $f(x) > d$ for some x in (a, b). By the Extreme Value Theorem, you know that f has a maximum at some c in the interval. Moreover, because $f(c) > d$, this maximum does not occur at either endpoint. Therefore, f has a maximum in the *open* interval (a, b). This implies that $f(c)$ is a *relative* maximum, and by Theorem 3.2, c is a critical number of f. Finally, because f is differentiable at c, you can conclude that $f'(c) = 0$.

Case 3:   If $f(x) < d$ for some x in (a, b), then you can use an argument similar to that in Case 2, but involving the minimum instead of the maximum.   ■

From Rolle's Theorem, you can see that if a function f is continuous on $[a, b]$ and differentiable on (a, b), and if $f(a) = f(b)$, then there must be at least one x-value between a and b at which the graph of f has a horizontal tangent, as shown in Figure 3.8(a).

If the differentiability requirement is dropped from Rolle's Theorem, then f will still have a critical number in (a, b), but it may not yield a horizontal tangent. Such a case is shown in Figure 3.8(b).

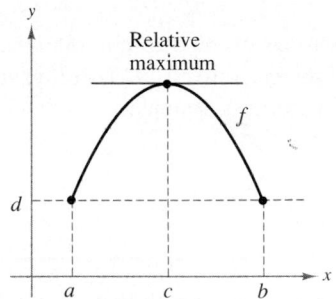

(a) f is continuous on $[a, b]$ and differentiable on (a, b).

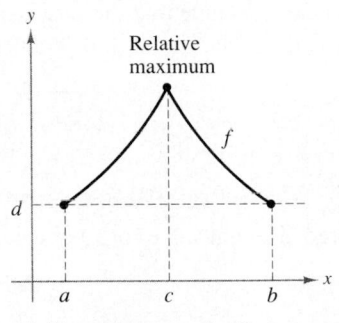

(b) f is continuous on $[a, b]$.

FIGURE 3.8

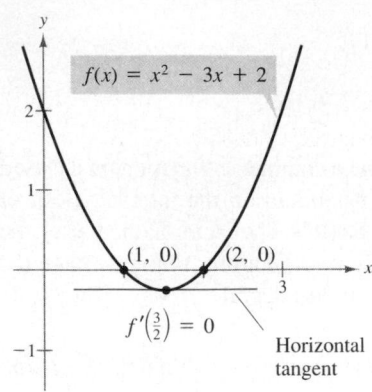

FIGURE 3.9
The x-value for which $f'(x) = 0$ is between the two x-intercepts.

EXAMPLE 1   Illustrating Rolle's Theorem

Find the two x-intercepts of $f(x) = x^2 - 3x + 2$ and show that $f'(x) = 0$ at some point between the two intercepts.

Solution   Note that f is differentiable on the entire real line. Setting $f(x)$ equal to 0 produces

$$x^2 - 3x + 2 = 0 \qquad \text{Set } f(x) \text{ equal to } 0$$
$$(x - 1)(x - 2) = 0. \qquad \text{Factor}$$

Thus, $f(1) = f(2) = 0$, and from Rolle's Theorem you know that there *exists* at least one c in the interval $(1, 2)$ such that $f'(c) = 0$. To *find* such a c, you can solve the equation

$$f'(x) = 2x - 3 = 0 \qquad \text{Set } f'(x) \text{ equal to } 0$$

and determine that $f'(x) = 0$ when $x = \frac{3}{2}$. Note that the x-value lies in the open interval $(1, 2)$, as shown in Figure 3.9.

Rolle's Theorem states that if f satisfies the conditions of the theorem, then there must be *at least* one point between a and b at which the derivative is 0. There may of course be more than one such point, as illustrated in the next example.

EXAMPLE 2   Illustrating Rolle's Theorem

Let $f(x) = x^4 - 2x^2$. Find all values of c in the interval $(-2, 2)$ such that $f'(c) = 0$.

Solution   To begin, note that the function satisfies the conditions of Rolle's Theorem. That is, f is continuous on the interval $[-2, 2]$ and differentiable on the interval $(-2, 2)$. Moreover, because $f(-2) = 8 = f(2)$, you can conclude that there exists at least one c in $(-2, 2)$ such that $f'(c) = 0$. Setting the derivative equal to 0 produces

$$f'(x) = 4x^3 - 4x = 0$$
$$4x(x^2 - 1) = 0$$
$$x = 0, 1, -1.$$

Thus, in the interval $(-2, 2)$, the derivative is 0 at three different values of x, as shown in Figure 3.10.

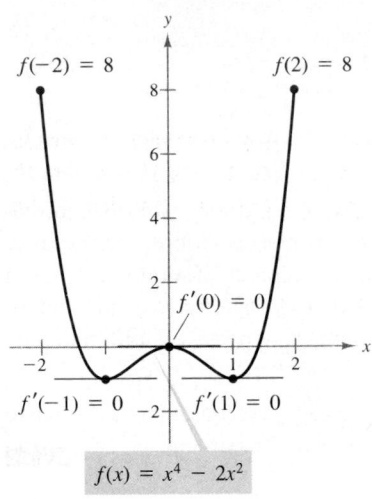

FIGURE 3.10
$f'(x) = 0$ for more than one x-value in the interval $(-2, 2)$.

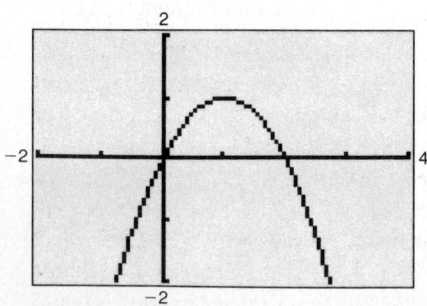

FIGURE 3.11

TECHNOLOGY   A graphing utility can be used to indicate whether the points on the graphs in Examples 1 and 2 are relative minima or relative maxima of the functions. When using a graphing utility, however, you should keep in mind that it can give misleading pictures of graphs. For example, try using a graphing utility to graph

$$f(x) = 1 - (x - 1)^2 - \frac{1}{1000(x - 1)^{1/7} + 1}.$$

With most viewing rectangles, it appears that the function has a maximum of 1 when $x = 1$ (see Figure 3.11). By evaluating the function at $x = 1$, however, you can see that $f(1) = 0$. To examine the behavior of this function near $x = 1$, you need to use calculus.

The Mean Value Theorem

Rolle's Theorem can be used to prove another theorem—the **Mean Value Theorem.**

THEOREM 3.4   The Mean Value Theorem

If f is continuous on the closed interval $[a, b]$ and differentiable on the open interval (a, b), then there exists a number c in (a, b) such that

$$f'(c) = \frac{f(b) - f(a)}{b - a}.$$

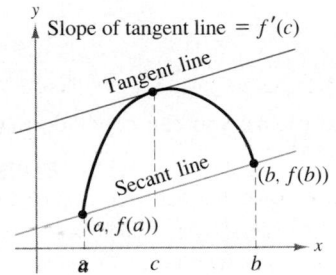

Slope of tangent line $= f'(c)$

Tangent line

Secant line

$(b, f(b))$

$(a, f(a))$

FIGURE 3.12

Proof   Refer to Figure 3.12. The equation of the secant line containing the points $(a, f(a))$ and $(b, f(b))$ is

$$y = \left[\frac{f(b) - f(a)}{b - a}\right](x - a) + f(a).$$

Let $g(x)$ be the difference between $f(x)$ and y. Then

$$g(x) = f(x) - y = f(x) - \left[\frac{f(b) - f(a)}{b - a}\right](x - a) - f(a).$$

By evaluating g at a and b, you can see that $g(a) = 0 = g(b)$. Furthermore, because f is differentiable, g is also differentiable, and you can apply Rolle's Theorem to the function g. Thus, there exists a point c in (a, b) such that $g'(c) = 0$, which implies that

$$0 = g'(c) = f'(c) - \frac{f(b) - f(a)}{b - a}.$$

Therefore, there exists a point c in (a, b) such that

$$f'(c) = \frac{f(b) - f(a)}{b - a}.$$

JOSEPH-LOUIS LAGRANGE

The Mean Value Theorem was first proved by the famous mathematician Joseph-Louis Lagrange (1736–1813).

REMARK   The "mean" in the Mean Value Theorem refers to the mean (or average) rate of change of f in the interval $[a, b]$.

Although the Mean Value Theorem can be used directly in problem solving, it is used more often to prove other theorems. In fact, some people consider this to be the most important theorem in calculus—it is closely related to the Fundamental Theorem of Calculus discussed in Chapter 4. For now, you can get an idea of the versatility of this theorem by looking at the results stated in Exercises 49–54 in this section.

The Mean Value Theorem has implications for both basic interpretations of the derivative. Geometrically, the theorem guarantees the existence of a tangent line that is parallel to the secant line through the points $(a, f(a))$ and $(b, f(b))$, as shown in Figure 3.12. Example 3 illustrates this geometrical interpretation of the Mean Value Theorem. In terms of rates of change, the Mean Value Theorem implies that there must be a point in the open interval (a, b) at which the instantaneous rate of change is equal to the average rate of change over the interval $[a, b]$. This is illustrated in Example 4.

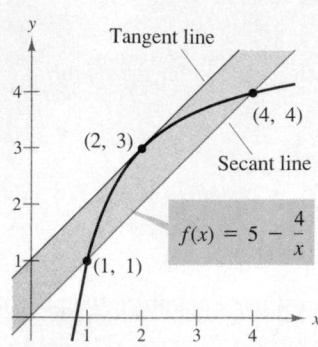

FIGURE 3.13
The tangent line at (2, 3) is parallel to the secant line through (1, 1) and (4, 4).

EXAMPLE 3   Finding a Tangent Line

Given $f(x) = 5 - (4/x)$, find all values of c in the open interval (1, 4) such that

$$f'(c) = \frac{f(4) - f(1)}{4 - 1}.$$

Solution   The slope of the secant line through $(1, f(1))$ and $(4, f(4))$ is

$$\frac{f(4) - f(1)}{4 - 1} = \frac{4 - 1}{4 - 1} = 1.$$

Because f satisfies the conditions of the Mean Value Theorem, there exists at least one number c in (1, 4) such that $f'(c) = 1$. Solving the equation $f'(x) = 1$ yields

$$f'(x) = \frac{4}{x^2} = 1$$

which implies that $x = \pm 2$. Therefore, in the interval (1, 4), you can conclude that $c = 2$, as shown in Figure 3.13.

EXAMPLE 4   Finding an Instantaneous Rate of Change

Two stationary patrol cars equipped with radar are 5 miles apart on a highway, as shown in Figure 3.14. As a truck passes the first patrol car, its speed is clocked at 55 miles per hour. Four minutes later, when the truck passes the second patrol car, its speed is clocked at 50 miles per hour. Prove that the truck must have exceeded the speed limit (of 55 miles per hour) at some time during the 4 minutes.

Solution   Let $t = 0$ be the time (in hours) when the truck passes the first patrol car. The time when the truck passes the second patrol car is

$$t = \frac{4}{60} = \frac{1}{15} \text{ hr.}$$

By letting $s(t)$ represent the distance (in miles) traveled by the truck, you have $s(0) = 0$ and $s(\frac{1}{15}) = 5$. Therefore, the average velocity of the truck over the 5-mile stretch of highway is

$$\text{Average velocity} = \frac{s(1/15) - s(0)}{(1/15) - 0} = \frac{5}{1/15} = 75 \text{ mph.}$$

Assuming that the position function is differentiable, we can apply the Mean Value Theorem to conclude that the truck must have been traveling at a rate of 75 miles per hour sometime during the 4 minutes.

FIGURE 3.14
At some time t the instantaneous velocity is equal to the average velocity over 4 minutes.

REMARK   A useful alternative form of the Mean Value Theorem is as follows: If f is continuous on $[a, b]$ and differentiable on (a, b), then there exists a number c in (a, b) such that $f(b) = f(a) + (b - a)f'(c)$.

When working the exercises for this section, keep in mind that polynomial functions, rational functions, and trigonometric functions are differentiable at all points in their domains.

EXERCISES for Section 3.2

In Exercises 1 and 2, state why Rolle's Theorem does not apply to the function even though there exist a and b such that $f(a) = f(b)$.

1. $f(x) = 1 - |x - 1|$ **2.** $f(x) = \cot \dfrac{x}{2}$

In Exercises 3–20, determine whether Rolle's Theorem can be applied to f on the indicated interval. If Rolle's Theorem can be applied, find all values of c in the interval such that $f'(c) = 0$.

Function	Interval
3. $f(x) = x^2 - 2x$	$[0, 2]$
4. $f(x) = x^2 - 3x + 2$	$[1, 2]$
5. $f(x) = (x - 1)(x - 2)(x - 3)$	$[1, 3]$
6. $f(x) = (x - 3)(x + 1)^2$	$[-1, 3]$
7. $f(x) = |x| - 1$	$[-1, 1]$
8. $f(x) = 3 - |x - 3|$	$[0, 6]$
9. $f(x) = x^{2/3} - 1$	$[-8, 8]$
10. $f(x) = x - x^{1/3}$	$[0, 1]$
11. $f(x) = \dfrac{x^2 - 2x - 3}{x + 2}$	$[-1, 3]$
12. $f(x) = \dfrac{x^2 - 1}{x}$	$[-1, 1]$
13. $f(x) = \sin x$	$[0, 2\pi]$
14. $f(x) = \cos x$	$[0, 2\pi]$
15. $f(x) = \sin 2x$	$\left[\dfrac{\pi}{6}, \dfrac{\pi}{3}\right]$
16. $f(x) = 4x - \tan \pi x$	$[-\tfrac{1}{4}, \tfrac{1}{4}]$
17. $f(x) = \dfrac{x}{2} - \sin \dfrac{\pi x}{6}$	$[-1, 0]$
18. $f(x) = \dfrac{6x}{\pi} - 4\sin^2 x$	$\left[0, \dfrac{\pi}{6}\right]$
19. $f(x) = \tan x$	$[0, \pi]$
20. $f(x) = \sec x$	$\left[-\dfrac{\pi}{4}, \dfrac{\pi}{4}\right]$

21. *Vertical Motion* The height of a ball t seconds after it is thrown is

$$f(t) = -16t^2 + 48t + 32.$$

a. Verify that $f(1) = f(2)$.
b. According to Rolle's Theorem, what must be the velocity at some time in the interval $[1, 2]$?

22. *Reorder Costs* The ordering and transportation cost C of components used in a manufacturing process is approximated by

$$C(x) = 10\left(\dfrac{1}{x} + \dfrac{x}{x + 3}\right)$$

where C is measured in thousands of dollars and x is the order size in hundreds.
a. Verify that $C(3) = C(6)$.
b. According to Rolle's Theorem, the rate of change of cost must be 0 for some order size in the interval $[3, 6]$. Find that order size.

23. Let f be continuous on $[a, b]$ and differentiable on (a, b). If there exists c in (a, b) such that $f'(c) = 0$, does it follow that $f(a) = f(b)$?

24. Let f be continuous on the closed interval $[a, b]$ and differentiable on the open interval (a, b). Also suppose $f(a) = f(b)$ and c is a real number in the interval such that $f'(c) = 0$. Use this information to find an interval for the function g over which Rolle's Theorem can be applied, and the corresponding critical number of g. (k is a constant.)
a. $g(x) = f(x) + k$ **b.** $g(x) = f(x - k)$
c. $g(x) = f(kx)$

In Exercises 25–34, apply the Mean Value Theorem to f on the indicated interval, $[a, b]$. In each case, find all values of c in the interval (a, b) such that

$$f'(c) = \dfrac{f(b) - f(a)}{b - a}.$$

Function	Interval
25. $f(x) = x^2$	$[-2, 1]$
26. $f(x) = x(x^2 - x - 2)$	$[-1, 1]$
27. $f(x) = x^{2/3}$	$[0, 1]$
28. $f(x) = \dfrac{x + 1}{x}$	$\left[\dfrac{1}{2}, 2\right]$
29. $f(x) = \dfrac{x}{x + 1}$	$\left[-\dfrac{1}{2}, 2\right]$
30. $f(x) = \sqrt{x - 2}$	$[2, 6]$
31. $f(x) = x^3$	$[0, 1]$
32. $f(x) = \sin x$	$[0, \pi]$
33. $f(x) = x - 2\sin x$	$[-\pi, \pi]$
34. $f(x) = 2\sin x + \sin 2x$	$[0, \pi]$

Essay **In Exercises 35 and 36, explain why the Mean Value Theorem does not apply to the function on the interval $[0, 6]$.**

35. $f(x) = \dfrac{1}{x - 3}$ **36.** $f(x) = |x - 3|$

In Exercises 37 and 38, sketch the graph of a function f that satisfies the given condition but *does not* satisfy the conditions of the Mean Value Theorem on the interval $[-5, 5]$.

37. f is continuous on $[-5, 5]$

38. f is not continuous on $[-5, 5]$

39. *Vertical Motion* The height of an object t seconds after it is dropped from a height of 500 feet is

$$s(t) = -16t^2 + 500.$$

  a. Find the average velocity of the object during the first 3 seconds.

  b. Use the Mean Value Theorem to verify that at some time during the first 3 seconds of fall the instantaneous velocity equals the average velocity. Find that time.

40. *Sales* A company introduces a new product for which the number of units sold S is

$$S(t) = 200\left(5 - \frac{9}{2 + t}\right)$$

where t is the time in months.

  a. Find the average rate of change of $S(t)$ during the first year.

  b. During what month does $S'(t)$ equal the average value during the first year?

C **In Exercises 41 and 42, use a graphing utility to graph the function f, the secant line, and any tangent lines to the graph of f that are parallel to the secant line on the specified interval.**

Function	Interval
41. $f(x) = \sqrt{x}$	$[1, 9]$
42. $f(x) = -x^4 + 4x^3 + 8x^2 + 5$	$[0, 5]$

True or False **In Exercises 43–48, determine whether the statement is true or false. If it is false, explain why or give an example that shows it is false.**

43. The Mean Value Theorem can be applied to $f(x) = 1/x$ on the interval $[-1, 1]$.

44. If the graph of a function has three x-intercepts, then it must have at least two points at which its tangent line is horizontal.

45. If the graph of a polynomial function has three x-intercepts, then it must have at least two points at which its tangent line is horizontal.

46. If $f'(x) = 0$ for all x in the domain of f, then f is a constant function.

47. The average rate of change of a linear function over a closed interval is equal to the instantaneous rate of change at any point in the interval.

48. In driving between Erie, Pennsylvania and Cleveland, Ohio, there exists at least one time when the speed of the car is the same as the average speed over the entire trip.

49. Prove that if $a > 0$ and n is any positive integer, then the polynomial function

$$p(x) = x^{2n+1} + ax + b$$

cannot have two real roots.

50. Let p be a *nonconstant* polynomial function.

  a. Prove that between any two consecutive zeros of p', there is at most one zero of p.

  b. If p has three distinct zeros in the interval $[a, b]$, prove that $p''(c) = 0$ for some real number c in (a, b).

51. Prove that if $f'(x) = 0$ for all x in an interval (a, b), then f is constant on $[a, b]$.

52. Let $p(x) = Ax^2 + Bx + C$. Prove that for any interval $[a, b]$, the value c guaranteed by the Mean Value Theorem is the midpoint of the interval.

53. Prove that if two functions f and g have the same derivatives on an interval (a, b), then they must differ only by a constant on the interval. (*Hint:* Let $h(x) = f(x) - g(x)$ and use the result of Exercise 51.)

54. Prove that if f is differentiable on $(-\infty, \infty)$ and $f'(x) < 1$ for all real numbers, then f has at most one fixed point. A **fixed point** of a function f is a real number c such that $f(c) = c$.

55. Use the result of Exercise 54 to show that $f(x) = \frac{1}{2}\cos x$ has at most one fixed point.

56. Prove that $|\cos x - \cos y| \leq |x - y|$ for all x and y.

57. Suppose that f and g satisfy $f(x) = g'(x)$ and $g(x) = -f'(x)$. Prove that $f(x)^2 + g(x)^2$ is a constant. Find examples of f and g that satisfy these conditions.

SECTION   Increasing and Decreasing Functions and the First Derivative Test

3.3   Increasing and Decreasing Functions • The First Derivative Test

Increasing and Decreasing Functions

In this section you will learn how derivatives can be used to *classify* relative extrema as either relative minima or relative maxima. We begin by defining increasing and decreasing functions.

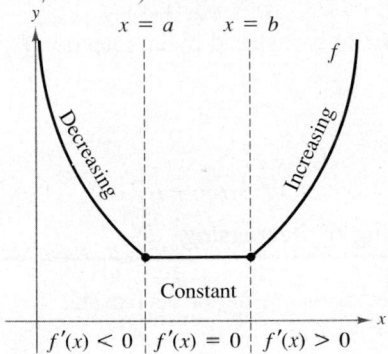

FIGURE 3.15

> **Definition of Increasing and Decreasing Functions**
>
> A function f is **increasing** on an interval if for any two numbers x_1 and x_2 in the interval, $x_1 < x_2$ implies $f(x_1) < f(x_2)$.
>
> A function f is **decreasing** on an interval if for any two numbers x_1 and x_2 in the interval, $x_1 < x_2$ implies $f(x_1) > f(x_2)$.

A function is increasing if, *as x moves to the right,* its graph moves up, and is decreasing if its graph moves down. For example, the function in Figure 3.15 is decreasing on the interval $(-\infty, a)$, is constant on the interval (a, b), and is increasing on the interval (b, ∞). As we will show below, a positive derivative implies that the function is increasing; a negative derivative implies that the function is decreasing; and a zero derivative on an entire interval implies that the function is constant on the interval.

> **THEOREM 3.5   Test for Increasing or Decreasing Functions**
>
> Let f be a function that is continuous on the closed interval $[a, b]$ and differentiable on the open interval (a, b).
>
> 1. If $f'(x) > 0$ for all x in (a, b), then f is increasing on $[a, b]$.
> 2. If $f'(x) < 0$ for all x in (a, b), then f is decreasing on $[a, b]$.
> 3. If $f'(x) = 0$ for all x in (a, b), then f is constant on $[a, b]$.

REMARK The conclusions in the first two cases of Theorem 3.5 are valid even if $f'(x) = 0$ at a finite number of x-values in (a, b).

Proof To prove the first case, assume that $f'(x) > 0$ for all x in the interval (a, b) and let $x_1 < x_2$ be any two points in the interval. By the Mean Value Theorem, you know there exists a number c such that $x_1 < c < x_2$, and $f'(c) = [f(x_2) - f(x_1)]/(x_2 - x_1)$. Because $f'(c) > 0$ and $x_2 - x_1 > 0$, you know that $f(x_2) - f(x_1) > 0$, which implies that $f(x_1) < f(x_2)$. Thus, f is increasing on the interval. The second case has a similar proof (see Exercise 63), and the third case was proved in Exercise 51 in Section 3.2.

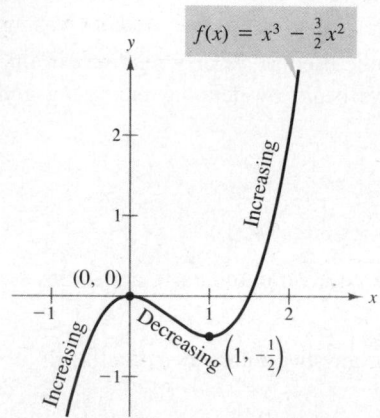

$$f(x) = x^3 - \tfrac{3}{2}x^2$$

(0, 0)

Decreasing $\left(1, -\tfrac{1}{2}\right)$

Increasing

Increasing

Decreasing

FIGURE 3.16

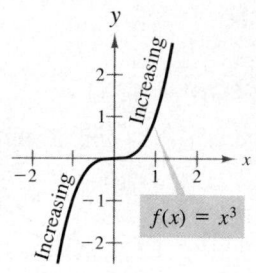

$$f(x) = x^3$$

Increasing

Increasing

(a) Strictly monotonic function

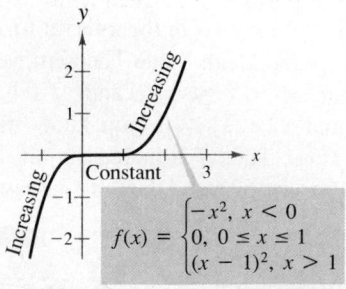

Increasing

Increasing

Constant

Increasing

$$f(x) = \begin{cases} -x^2, & x < 0 \\ 0, & 0 \le x \le 1 \\ (x-1)^2, & x > 1 \end{cases}$$

(b) Not a strictly monotonic function

FIGURE 3.17

Guidelines for Finding Intervals on which a Function is Increasing or Decreasing

Let f be continuous on the interval (a, b). To find the open intervals on which f is increasing or decreasing, use the following steps.

1. Locate the critical numbers of f in (a, b), and use these numbers to determine test intervals.

2. Determine the sign of $f'(x)$ at one value in each of the test intervals.

3. Use Theorem 3.5 to decide whether f is increasing or decreasing on each interval.

These guidelines are also valid if the interval (a, b) is replaced by an interval of the form $(-\infty, b)$, (a, ∞), or $(-\infty, \infty)$.

EXAMPLE 1   Intervals on which *f* is Increasing or Decreasing

Find the open intervals on which $f(x) = x^3 - \tfrac{3}{2}x^2$ is increasing or decreasing.

Solution   Note that f is continuous on the entire real line. To determine the critical numbers of f, set $f'(x)$ equal to 0.

$$f'(x) = 3x^2 - 3x = 0 \qquad \text{Let } f'(x) = 0$$

$$3(x)(x - 1) = 0 \qquad \text{Factor}$$

$$x = 0, 1 \qquad \text{Critical numbers}$$

Because there are no points for which f' is undefined, you can conclude that $x = 0$ and $x = 1$ are the only critical numbers. Table 3.4 summarizes the testing of the three intervals determined by these two critical numbers. The graph of f is shown in Figure 3.16.

TABLE 3.4

Interval	$-\infty < x < 0$	$0 < x < 1$	$1 < x < \infty$
Test value	$x = -1$	$x = \tfrac{1}{2}$	$x = 2$
Sign of $f'(x)$	$f'(-1) = 6 > 0$	$f'(\tfrac{1}{2}) = -\tfrac{3}{4} < 0$	$f'(2) = 6 > 0$
Conclusion	Increasing	Decreasing	Increasing

A function is **strictly monotonic** on an interval if it is either increasing on the entire interval or decreasing on the entire interval. For instance, the function $f(x) = x^3$ is strictly monotonic on the entire real line because it is increasing on the entire real line, as shown in Figure 3.17(a). The function shown in Figure 3.17(b) is not strictly monotonic on the entire real line because it is constant on the interval $[0, 1]$.

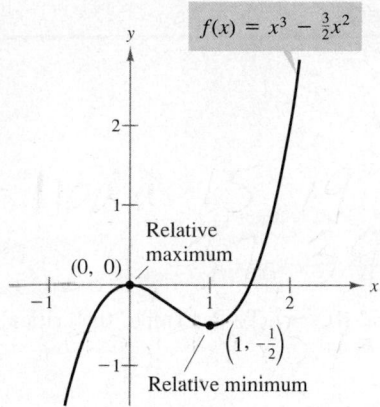

FIGURE 3.18
Relative extrema of f

The First Derivative Test

After you have determined the intervals on which a function is increasing or decreasing, it is not difficult to locate the relative extrema of the function. For instance, in Figure 3.18 (from Example 1), the function $f(x) = x^3 - \frac{3}{2}x^2$ has a relative maximum at the point $(0, 0)$ because f is increasing immediately to the left of $x = 0$ and decreasing immediately to the right of $x = 0$. Similarly, f has a relative minimum at the point $(1, -\frac{1}{2})$ because f is decreasing immediately to the left of $x = 1$ and increasing immediately to the right of $x = 1$. The following theorem, called the First Derivative Test, makes this more explicit.

THEOREM 3.6 The First Derivative Test

Let c be a critical number of a function f that is continuous on an open interval I containing c. If f is differentiable on the interval, except possibly at c, then $f(c)$ can be classified as follows.

1. If $f'(x)$ changes from negative to positive at c, then $f(c)$ is a **relative minimum** of f.
2. If $f'(x)$ changes from positive to negative at c, then $f(c)$ is a **relative maximum** of f.

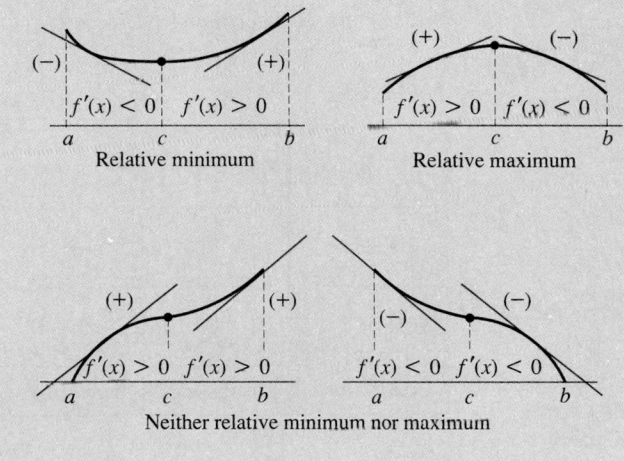

Proof Assume that $f'(x)$ changes from negative to positive at c. Then there exist a and b in I such that

$$f'(x) < 0 \qquad \text{for all } x \text{ in } (a, c)$$

and

$$f'(x) > 0 \qquad \text{for all } x \text{ in } (c, b).$$

By Theorem 3.5, f is decreasing on (a, c) and increasing on (c, b). Therefore, $f(c)$ is a minimum of f on the open interval (a, b) and, consequently, a relative minimum of f. This proves the first case of the theorem. The second case can be proved in a similar way (see Exercise 64).

EXAMPLE 2   Applying the First Derivative Test

Find the relative extrema of the function

$$f(x) = \frac{1}{2}x - \sin x$$

in the interval $(0, 2\pi)$.

Solution   Note that f is continuous on the interval $(0, 2\pi)$. To determine the critical numbers of f in this interval, set $f'(x)$ equal to 0.

$$f'(x) = \frac{1}{2} - \cos x = 0 \qquad \text{Let } f'(x) = 0$$

$$\cos x = \frac{1}{2}$$

$$x = \frac{\pi}{3}, \frac{5\pi}{3} \qquad \text{Critical numbers}$$

Because there are no points for which f' is undefined, you can conclude that $x = \pi/3$ and $x = 5\pi/3$ are the only critical numbers. Table 3.5 summarizes the testing of the three intervals determined by these two critical numbers.

TABLE 3.5

Interval	$0 < x < \dfrac{\pi}{3}$	$\dfrac{\pi}{3} < x < \dfrac{5\pi}{3}$	$\dfrac{5\pi}{3} < x < 2\pi$
Test value	$x = \dfrac{\pi}{4}$	$x = \pi$	$x = \dfrac{7\pi}{4}$
Sign of $f'(x)$	$f'\left(\dfrac{\pi}{4}\right) < 0$	$f'(\pi) > 0$	$f'\left(\dfrac{7\pi}{4}\right) < 0$
Conclusion	Decreasing	Increasing	Decreasing

By applying the First Derivative Test, you can conclude that f has a relative minimum at $x = \pi/3$ and a relative maximum at $x = 5\pi/3$, as shown in Figure 3.19.

FIGURE 3.19
A relative minimum occurs where f changes from decreasing to increasing, and a relative maximum occurs where f changes from increasing to decreasing.

⋮ TECHNOLOGY   In Section 3.2, we pointed out that, *by itself,* a graphing utility can give misleading information about the relative extrema of a graph. *Used in conjunction with calculus,* however, a graphing utility can provide a good way to reinforce your conclusions. Try using a graphing utility to graph the function in Example 2. Then use the zoom and trace features to estimate the relative extrema. How close are your graphical approximations?

Note that in Examples 1 and 2 the given functions are differentiable on the entire real line. For such functions, the only critical numbers are those for which $f'(x) = 0$. The next example concerns a function that has two types of critical numbers—those for which $f'(x) = 0$ and those for which f' is undefined.

EXAMPLE 3   Applying the First Derivative Test

Find the relative extrema of

$$f(x) = (x^2 - 4)^{2/3}.$$

Solution   Begin by noting that f is continuous on the entire real line. The derivative of f

$$f'(x) = \frac{2}{3}(x^2 - 4)^{-1/3}(2x) \qquad \text{General Power Rule}$$

$$= \frac{4x}{3(x^2 - 4)^{1/3}} \qquad \text{Simplify}$$

is 0 when $x = 0$ and undefined when $x = \pm 2$. Thus, the critical numbers are $x = -2$, $x = 0$, and $x = 2$. Table 3.6 summarizes the testing of the four intervals determined by these three critical numbers.

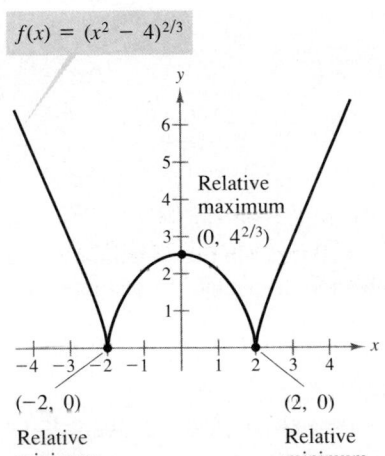

$f(x) = (x^2 - 4)^{2/3}$

Relative
maximum
$(0, 4^{2/3})$

$(-2, 0)$
Relative
minimum

$(2, 0)$
Relative
minimum

FIGURE 3.20

TABLE 3.6

Interval	$-\infty < x < -2$	$-2 < x < 0$	$0 < x < 2$	$2 < x < \infty$
Test value	$x = -3$	$x = -1$	$x = 1$	$x = 3$
Sign of $f'(x)$	$f'(-3) < 0$	$f'(-1) > 0$	$f'(1) < 0$	$f'(3) > 0$
Conclusion	Decreasing	Increasing	Decreasing	Increasing

By applying the First Derivative Test, you can conclude that f has a relative minimum at the point $(-2, 0)$, a relative maximum at the point $(0, 4^{2/3})$, and a relative minimum at the point $(2, 0)$, as shown in Figure 3.20.

TECHNOLOGY   When using a graphing utility to graph a function involving radicals or rational exponents, be sure you understand the way the utility evaluates radical expressions. For instance, even though

$$f(x) = (x^2 - 4)^{2/3} \quad \text{and} \quad g(x) = [(x^2 - 4)^2]^{1/3}$$

are the same algebraically, many graphing utilities distinguish between the two functions. Which of the graphs shown in Figure 3.21 is incorrect? Why did the graphing utility produce an incorrect graph?

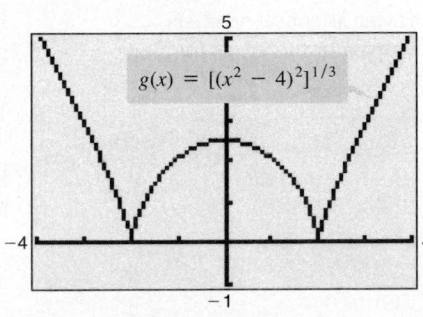

FIGURE 3.21
Which graph is incorrect?

When using the First Derivative Test, be sure to consider the domain of the function. For instance, in the next example the function $f(x) = (x^4 + 1)/x^2$ is not defined when $x = 0$. This x-value must be used with the critical numbers to determine the test intervals.

EXAMPLE 4   Applying the First Derivative Test

Find the relative extrema of $f(x) = \dfrac{x^4 + 1}{x^2}$.

Solution

$$
\begin{aligned}
f(x) &= x^2 + x^{-2} & &\text{Rewrite} \\
f'(x) &= 2x - 2x^{-3} & &\text{Differentiate} \\
&= 2x - \frac{2}{x^3} \\
&= \frac{2(x^4 - 1)}{x^3} & &\text{Simplify} \\
&= \frac{2(x^2 + 1)(x - 1)(x + 1)}{x^3}. & &\text{Factor}
\end{aligned}
$$

Thus, $f'(x)$ is zero at $x = \pm 1$. Moreover, because $x = 0$ is not in the domain of f, you should use this x-value along with the critical numbers to determine the test intervals.

$\quad x = \pm 1 \qquad$ Critical numbers, $f'(\pm 1) = 0$

$\quad x = 0 \qquad\;\;$ 0 is not in the domain of f

Table 3.7 summarizes the testing of the four intervals determined by these three x-values.

TABLE 3.7

Interval	$-\infty < x < -1$	$-1 < x < 0$	$0 < x < 1$	$1 < x < \infty$
Test value	$x = -2$	$x = -\frac{1}{2}$	$x = \frac{1}{2}$	$x = 2$
Sign of $f'(x)$	$f'(-2) < 0$	$f'(-\frac{1}{2}) > 0$	$f'(\frac{1}{2}) < 0$	$f'(2) > 0$
Conclusion	Decreasing	Increasing	Decreasing	Increasing

By applying the First Derivative Test, you can conclude that f has one relative minimum at the point $(-1, 2)$ and another at the point $(1, 2)$, as shown in Figure 3.22.

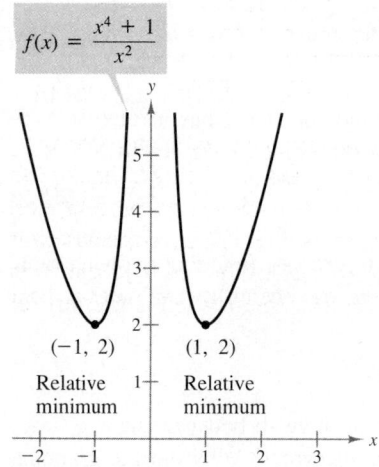

$f(x) = \dfrac{x^4 + 1}{x^2}$

$(-1, 2)$ Relative minimum

$(1, 2)$ Relative minimum

FIGURE 3.22

x-values that are not in the domain of f, as well as critical numbers, determine test intervals for f'.

TECHNOLOGY The most difficult step in applying the First Derivative Test is finding the values for which the derivative is equal to 0. For instance, the values of x for which the derivative of

$$ f(x) = \frac{x^4 + 1}{x^2 + 1} $$

is equal to 0 are 0 and $\pm\sqrt{\sqrt{2} - 1}$. If you have access to technology that can perform symbolic differentiation and solve equations, try using it to apply the First Derivative Test to this function.

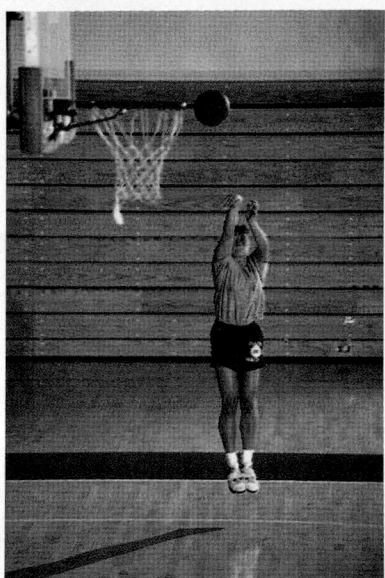

FIGURE 3.23
If a projectile is propelled from ground level and air resistance is neglected, then the object will travel farthest with an initial angle of 45°. If, however, the projectile is propelled from a point above ground level, then the angle that yields a maximum horizontal distance is not 45° (see Example 5).

EXAMPLE 5   The Path of a Projectile

Neglecting air resistance, the path of a projectile that is propelled at an angle θ is

$$y = -\frac{g \sec^2 \theta}{2v_0^2} x^2 + (\tan \theta)x + h, \qquad 0 \le \theta \le \frac{\pi}{2}$$

where y is the height, x is the horizontal distance, g is the acceleration due to gravity, v_0 is the initial velocity, and h is the initial height. (This equation is derived in Section 12.3.) Let $g = 32$ feet per second per second, $v_0 = 24$ feet per second, and $h = 9$ feet. What value of θ will produce a maximum horizontal distance?

Solution   To find the distance the projectile travels, let $y = 0$, and use the quadratic formula to solve for x.

$$-\frac{g \sec^2 \theta}{2v_0^2} x^2 + (\tan \theta)x + h = 0$$

$$-\frac{32 \sec^2 \theta}{2(24^2)} x^2 + (\tan \theta)x + 9 = 0$$

$$-\frac{\sec^2 \theta}{36} x^2 + (\tan \theta)x + 9 = 0$$

Using the Quadratic Formula, we solve for x as follows.

$$x = \frac{-\tan \theta \pm \sqrt{\tan^2 \theta + \sec^2 \theta}}{-\sec^2 \theta / 18}$$

$$x = 18 \cos \theta (\sin \theta + \sqrt{\sin^2 \theta + 1}), \qquad (x \ge 0)$$

At this point, you need to find the value of θ that produces a maximum value of x. Applying the First Derivative Test by hand would be very tedious. Using technology to solve the equation $dx/d\theta = 0$, however, eliminates most of the messy computations. The result is that the maximum value of x occurs when $\theta \approx 35.3°$. This conclusion is reinforced by sketching the path of the projectile for different values of θ, as shown in Figure 3.24. Of the three paths shown, note that the distance traveled is greatest for $\theta = 35°$.

FIGURE 3.24
The path of a projectile with initial angle θ.

EXERCISES for Section 3.3

In Exercises 1–6, identify the open intervals on which the function is increasing or decreasing.

1. $f(x) = x^2 - 6x + 8$

2. $y = -(x + 1)^2$

3. $y = \dfrac{x^3}{4} - 3x$

4. $f(x) = x^4 - 2x^2$

5. $f(x) = \dfrac{1}{x^2}$

6. $y = \dfrac{x^2}{x + 1}$

In Exercises 7–20, find the critical numbers of f (if any), find the open intervals on which the algebraic function is increasing or decreasing, and locate all relative extrema.

7. $f(x) = -2x^2 + 4x + 3$

8. $f(x) = x^2 + 8x + 10$

9. $f(x) = x^2 - 6x$

10. $f(x) = (x - 1)^2(x + 2)$

11. $f(x) = 2x^3 + 3x^2 - 12x$

12. $f(x) = (x - 3)^3$

13. $f(x) = \dfrac{x^5 - 5x}{5}$

14. $f(x) = x^4 - 32x + 4$

15. $f(x) = x^{1/3} + 1$

16. $f(x) = x^{2/3}(x - 5)$

17. $f(x) = 5 - |x - 5|$

18. $f(x) = |x + 3| - 1$

19. $f(x) = \dfrac{x^2}{x^2 - 9}$

20. $f(x) = \dfrac{x + 3}{x^2}$

C **In Exercises 21–28, find the critical numbers of f (if any), find the open intervals on which the algebraic function is increasing or decreasing, and locate all relative extrema. Use a graphing utility to confirm your results.**

21. $f(x) = x^3 - 6x^2 + 15$

22. $f(x) = x^4 - 2x^3$

23. $f(x) = (x - 1)^{2/3}$

24. $f(x) = (x - 1)^{1/3}$

25. $f(x) = x + \dfrac{1}{x}$

26. $f(x) = \dfrac{x}{x + 1}$

27. $f(x) = \dfrac{x^2 - 2x + 1}{x + 1}$

28. $f(x) = \dfrac{x^2 - 3x - 4}{x - 2}$

C **In Exercises 29–32, consider the trigonometric function on the interval $(0, 2\pi)$. Find the critical numbers, find the open intervals on which the function is increasing or decreasing, and locate all relative extrema. Use a graphing utility to confirm your results.**

29. $f(x) = \dfrac{x}{2} + \cos x$

30. $f(x) = \sin x \cos x$

31. $f(x) = \sin^2 x + \sin x$

32. $f(x) = \dfrac{\cos x}{1 + \sin^2 x}$

C **In Exercises 33–36, use a symbolic differentiation utility to (a) differentiate the function, (b) sketch the graphs of f and f' on the same coordinate axes over the specified interval, (c) find the critical numbers of f in the open interval, and (d) find the interval(s) on which f' is positive and the interval(s) on which it is negative. State the relationship between the behavior of f and the sign of f'.**

	Function	*Interval*
33.	$f(x) = 2x\sqrt{9 - x^2}$	$[-3, 3]$
34.	$f(x) = 10(5 - \sqrt{x^2 - 3x + 16})$	$[0, 5]$
35.	$f(t) = t^2 \sin t$	$[0, 2\pi]$
36.	$f(x) = \dfrac{x}{2} + \cos\dfrac{x}{2}$	$[0, 4\pi]$

In Exercises 37–42, assume that f is differentiable for all x. The sign of f' is as follows.

$$f'(x) > 0 \text{ on } (-\infty, -4)$$
$$f'(x) < 0 \text{ on } (-4, 6)$$
$$f'(x) > 0 \text{ on } (6, \infty)$$

Supply the appropriate inequality for the indicated value of c.

	Function	*Sign of $g'(c)$*
37.	$g(x) = f(x) + 5$	$g'(0) \underline{\quad} 0$
38.	$g(x) = 3f(x) - 3$	$g'(-5) \underline{\quad} 0$
39.	$g(x) = -f(x)$	$g'(-6) \underline{\quad} 0$
40.	$g(x) = -f(x)$	$g'(0) \underline{\quad} 0$
41.	$g(x) = f(x - 10)$	$g'(0) \underline{\quad} 0$
42.	$g(x) = f(x - 10)$	$g'(8) \underline{\quad} 0$

43. Sketch the graph of a function f such that

$$f'(x) \begin{cases} > 0, & x < 4 \\ \text{Undefined}, & x = 4 \\ < 0, & x > 4 \end{cases}$$

44. A differentiable function f has only one critical number: $x = 5$. Identify the relative extrema of f at the critical number if $f'(4) = -2.5$ and $f'(6) = 3$.

45. The function f is differentiable on the interval $[-1, 1]$. The table gives the values of f' for selected values of x. Sketch the graph of f, approximate the critical numbers, and identify the relative extrema.

x	-1	-0.75	-0.50	-0.25
$f'(x)$	-10	-3.2	-0.5	0.8

x	0	0.25	0.50	0.75	1
$f'(x)$	5.6	3.6	-0.2	-6.7	-20.1

46. Consider the functions $f(x) = x$ and $g(x) = \sin x$ on the interval $(0, \pi)$.
 a. Prove that $f(x) > g(x)$. (*Hint:* Show that $h'(x) > 0$ where $h = f - g$.)
 b. Sketch the graphs of f and g on the same set of axes.

47. *Rolling Ball Bearing* A ball bearing is placed on an inclined plane and begins to roll. The angle of elevation of the plane is θ. The acceleration (in feet per second per second) of the ball bearing is $a(t) = -16(\sin \theta)t^2$. You are measuring the velocity of the ball bearing after 2 seconds. Can you change the velocity by changing θ? If so, is there a value of θ that will produce a maximum velocity?

48. *Drug Concentration* The concentration C of a certain chemical in the bloodstream t hours after injection into muscle tissue is given by

$$C = \frac{3t}{27 + t^3}.$$

When is the concentration greatest?

49. *Trachea Contraction* Coughing forces the trachea (windpipe) to contract, which affects the velocity v of the air passing through the trachea. Suppose the velocity of the air during coughing is

$$v = k(R - r)r^2, \qquad 0 \le r \le R$$

where k is a constant, R is the normal radius of the trachea, and r is the radius during coughing. What radius will produce the maximum air velocity?

50. *Profit* The profit P (in dollars) made by a fast-food restaurant selling x hamburgers is

$$P = 2.44x - \frac{x^2}{20,000} - 5000, \qquad 0 \le x \le 35,000.$$

Find the open intervals on which P is increasing or decreasing.

51. *Newborn Weight Change* After birth, an infant normally will lose weight for a few days and then start gaining. A model for the average weight W (in pounds) of infants over the first 2 weeks following birth is

$$W = 0.033t^2 - 0.3974t + 7.3032, \qquad 0 \le t \le 14$$

where t is measured in days. Find the open intervals on which W is increasing or decreasing.

52. *Power* The electric power P (in watts) in a direct current circuit with two resistors R_1 and R_2 connected in series is

$$P = \frac{vR_1R_2}{(R_1 + R_2)^2}$$

where v is the voltage. If v and R_1 are held constant, what resistance R_2 produces maximum power?

C 53. *Electrical Resistance* The resistance R of a resistor is

$$R = \sqrt{0.001T^4 - 4T + 100}$$

where R is measured in ohms and the temperature T is measured in degrees Celsius.
 a. Use a symbolic differentiation utility to find dR/dT and the critical number of the function. Then determine the minimum resistance for this type of resistor.
 b. Use a graphing utility to graph the function R.

C 54. *Meteorology* The normal average daily temperature (in degrees Fahrenheit) for a city is

$$T = 45 - 23 \cos \left[\frac{2\pi(t - 32)}{365} \right]$$

where t is the time in days, with $t = 1$ corresponding to January 1.
 a. Use a symbolic differentiation utility to find dT/dt and the critical numbers of the function. Then determine the expected date of the warmest temperature and the expected date of the coldest temperature.
 b. Use a graphing utility to graph the function T.

55. Find a, b, c, and d so that the function

$$f(x) = ax^3 + bx^2 + cx + d$$

has a relative minimum at $(0, 0)$ and a relative maximum at $(2, 2)$.

56. Find a, b, c, and d so that the function

$$f(x) = ax^3 + bx^2 + cx + d$$

has a relative minimum at $(0, 0)$ and a relative maximum at $(4, 1000)$.

True or False In Exercises 57–62, determine whether the statement is true or false. If it is false, explain why or give an example that shows it is false.

57. The sum of two increasing functions is increasing.

58. The product of two increasing functions is increasing.

59. If $f(x) = ax^3 + b$ and f is increasing on $(-1, 1)$, then $a > 0$.

60. Every nth-degree polynomial has $(n - 1)$ critical numbers.

61. An nth-degree polynomial has at most $(n - 1)$ critical numbers.

62. If $a < c < b$ and c is a critical number of f, then f has a relative extremum in (a, b).

63. Prove the second case of Theorem 3.5.

64. Prove the second case of Theorem 3.6.

65. Let $x > 0$ and $n > 1$ be real numbers. Prove that $(1 + x)^n > 1 + nx$.

C **66.** Rainbows are formed when light strikes raindrops and is reflected and refracted, as shown in the figure. (This figure shows a cross section of a spherical raindrop.) The Law of Refraction states that

$$\frac{\sin \alpha}{\sin \beta} = k$$

where $k \approx 1.33$ (for water). The angle of deflection is given by

$$D = \pi + 2\alpha - 4\beta.$$

a. Sketch the graph of D for $0 \le \alpha \le \pi/2$. Use a graphing calculator with

$$D = \pi + 2\alpha - 4\sin^{-1}\left(\frac{1}{k}\sin \alpha\right).$$

b. Prove that the minimum angle of deflection occurs when

$$\cos \alpha = \sqrt{\frac{k^2 - 1}{3}}.$$

For water, what is the minimum angle of deflection, $D_{\min}$? (The angle $\pi - D_{\min}$ is called the **rainbow angle.**) What value of α produces this minimum angle? (A ray of sunlight that strikes a raindrop at this angle, α, is called a **rainbow ray.)**

FOR FURTHER INFORMATION   For more information about the mathematics of rainbows, see the article "Somewhere Within the Rainbow" by Steven Janke in the *UMAP Journal,* Volume 13, Number 2, 1992.

Rainbows are formed as rays of light hit raindrops and are reflected and refracted.

Concavity and the Second Derivative Test

Concavity • Points of Inflection • The Second Derivative Test

Concavity

You have already seen that locating the intervals in which a function f increases or decreases helps to describe its graph. In this section, you will see how locating the intervals in which f' increases or decreases can be used to determine where the graph of f is *curving upward* or *curving downward*.

Definition of Concavity

Let f be differentiable on an open interval, I. The graph of f is **concave upward** on I if f' is increasing on the interval and **concave downward** on I if f' is decreasing on the interval.

The following graphical interpretation of concavity is useful. (See Appendix A for a proof of these results.)

1. Let f be differentiable at c. If the graph of f is concave upward at $(c, f(c))$, then the graph of f lies *above* the tangent line at $(c, f(c))$ on some open interval containing c (see Figure 3.25a).
2. Let f be differentiable at c. If the graph of f is concave downward at $(c, f(c))$, then the graph of f lies *below* the tangent line at $(c, f(c))$ on some open interval containing c (see Figure 3.25b).

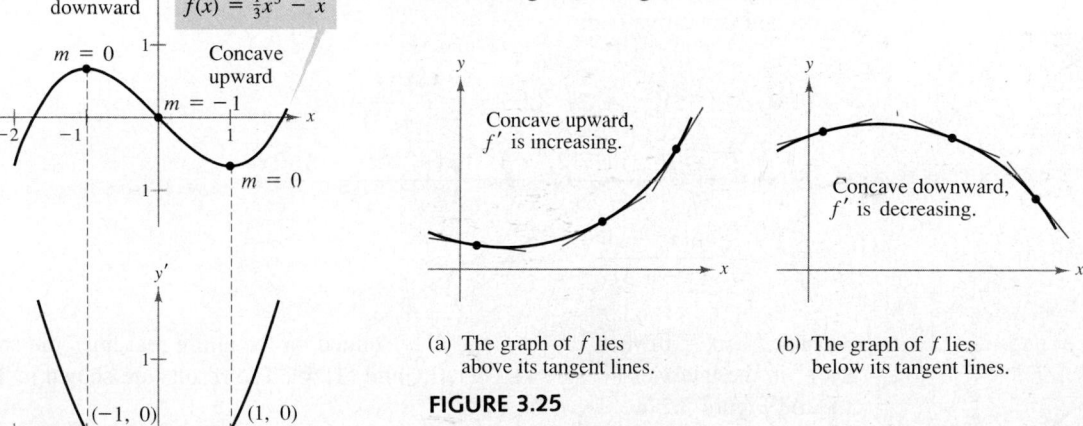

Concave downward

$m = 0$

$f(x) = \frac{1}{3}x^3 - x$

Concave upward

$m = -1$

$m = 0$

$(-1, 0)$ $(1, 0)$

$f'(x) = x^2 - 1$ $(0, -1)$

f' is decreasing. f' is increasing.

FIGURE 3.26

(a) Concave upward, f' is increasing. The graph of f lies above its tangent lines.

(b) Concave downward, f' is decreasing. The graph of f lies below its tangent lines.

FIGURE 3.25

To find the open intervals on which the graph of a function f is concave upward or downward, you need to find the intervals on which f' is increasing or decreasing. For instance, the graph of $f(x) = \frac{1}{3}x^3 - x$ is concave downward on the open interval $(-\infty, 0)$ because $f' = x^2 - 1$ is decreasing there (see Figure 3.26). Similarly, the graph of f is concave upward on the interval $(0, \infty)$ because f' is increasing on $(0, \infty)$.

The following theorem shows how to use the *second* derivative of a function f to determine intervals on which the graph of f is concave upward or downward. A proof of this theorem follows directly from Theorem 3.5 and the definition of concavity.

THEOREM 3.7   Test for Concavity

Let f be a function whose second derivative exists on an open interval I.

1. If $f''(x) > 0$ for all x in I, then the graph of f is concave upward.
2. If $f''(x) < 0$ for all x in I, then the graph of f is concave downward.

REMARK   A third case of Theorem 3.7 could be that if $f''(x) = 0$ for all x in I, then f is linear. Note, however, that concavity is not defined for a line. In other words, a straight line is neither concave upward nor concave downward.

To apply Theorem 3.7, locate the x-values at which $f'' = 0$ or f'' is undefined. Second, use these x-values to determine test intervals. Finally, test the sign of $f''(x)$ in each of the test intervals.

EXAMPLE 1   Determining Concavity

Determine the open intervals on which the graph of $f(x) = 6(x^2 + 3)^{-1}$ is concave upward or downward.

Solution   Begin by observing that f is continuous on the entire real line. Next, find the second derivative of f.

$$f'(x) = (-6)(x^2 + 3)^{-2}(2x) = \frac{-12x}{(x^2 + 3)^2}$$

$$f''(x) = \frac{(x^2 + 3)^2(-12) - (-12x)(2)(x^2 + 3)(2x)}{(x^2 + 3)^4}$$

$$= \frac{36(x^2 - 1)}{(x^2 + 3)^3}$$

Because $f''(x) = 0$ when $x = \pm 1$ and f'' is defined on the entire real line, you should test f'' in the intervals $(-\infty, -1)$, $(-1, 1)$, and $(1, \infty)$. The results are shown in Table 3.8 and Figure 3.27.

TABLE 3.8

Interval	$-\infty < x < -1$	$-1 < x < 1$	$1 < x < \infty$
Test value	$x = -2$	$x = 0$	$x = 2$
Sign of $f''(x)$	$f''(-2) > 0$	$f''(0) < 0$	$f''(2) > 0$
Conclusion	Concave upward	Concave downward	Concave upward

FIGURE 3.27
From the sign of f'' you can determine the concavity of the graph of f.

The function given in Example 1 is continuous on the entire real line. If there are x-values at which the function is not continuous, then these values should be used along with the points at which $f''(x) = 0$ or is undefined to form the test intervals.

EXAMPLE 2 Determining Concavity

Determine the open intervals in which the graph of

$$f(x) = \frac{x^2 + 1}{x^2 - 4}$$

is concave upward or downward.

Solution

$$f'(x) = \frac{(x^2 - 4)(2x) - (x^2 + 1)(2x)}{(x^2 - 4)^2} = \frac{-10x}{(x^2 - 4)^2}$$

$$f''(x) = \frac{(x^2 - 4)^2(-10) - (-10x)(2)(x^2 - 4)(2x)}{(x^2 - 4)^4} = \frac{10(3x^2 + 4)}{(x^2 - 4)^3}$$

There are no points at which $f'' = 0$, but at $x = \pm 2$ the function f is not continuous, so we test for concavity in the intervals $(-\infty, -2)$, $(-2, 2)$, and $(2, \infty)$, as shown in Table 3.9. The graph of f is shown in Figure 3.28.

TABLE 3.9

Interval	$-\infty < x < -2$	$-2 < x < 2$	$2 < x < \infty$
Test value	$x = -3$	$x = 0$	$x = 3$
Sign of $f''(x)$	$f''(-3) > 0$	$f''(0) < 0$	$f''(3) > 0$
Conclusion	Concave upward	Concave downward	Concave upward

Concave upward · Concave upward

Concave downward

$$f(x) = \frac{x^2 + 1}{x^2 - 4}$$

FIGURE 3.28

Points of Inflection

The graph in Figure 3.27 has two points at which the concavity changes. If the tangent line to the graph exists at such a point, then the point is a **point of inflection**. Three types of points of inflection are shown in Figure 3.29. Note that a graph crosses its tangent line at a point of inflection.

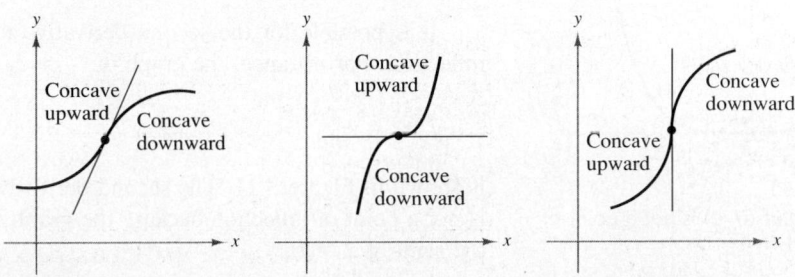

FIGURE 3.29
The concavity of f changes at a point of inflection.

To locate possible points of inflection, you need only determine the values of x for which $f''(x) = 0$ or for which f'' is undefined. This is similar to the procedure for locating relative extrema of f.

THEOREM 3.8   Points of Inflection

If $(c, f(c))$ is a point of inflection of the graph of f, then either $f''(c) = 0$ or f'' is undefined at $x = c$.

EXAMPLE 3   Finding Points of Inflection

Determine the points of inflection and discuss the concavity of the graph of

$$f(x) = x^4 - 4x^3.$$

Solution   Differentiating twice produces

$$f'(x) = 4x^3 - 12x^2$$

$$f''(x) = 12x^2 - 24x$$

$$= 12x(x - 2).$$

Possible points of inflection occur at $x = 0$ and $x = 2$. By testing the intervals determined by these x-values, you can conclude that they both yield points of inflection. A summary of this testing is shown in Table 3.10, and the graph of f is shown in Figure 3.30.

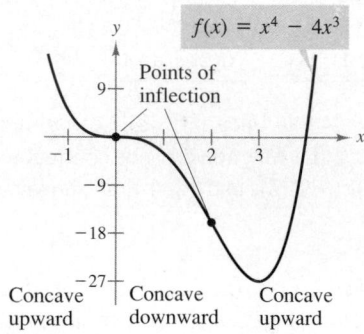

FIGURE 3.30
Points of inflection can occur where $f''(x) = 0$ or is undefined.

TABLE 3.10

Interval	$-\infty < x < 0$	$0 < x < 2$	$2 < x < \infty$
Test value	$x = -1$	$x = 1$	$x = 3$
Sign of $f''(x)$	$f''(-1) > 0$	$f''(1) < 0$	$f''(3) > 0$
Conclusion	Concave upward	Concave downward	Concave upward

It is possible for the second derivative to be 0 at a point that is *not* a point of inflection. For instance, the graph of

$$f(x) = x^4$$

is shown in Figure 3.31. The second derivative is 0 when $x = 0$, but the point $(0, 0)$ is not a point of inflection because the graph of f is concave upward to the left and to the right of $x = 0$.

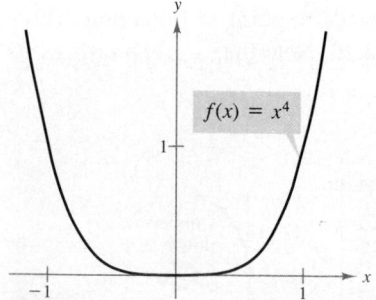

FIGURE 3.31
$f''(0) = 0$, but $(0, 0)$ is *not* a point of inflection.

The Second Derivative Test

In addition to testing for concavity, the second derivative can be used to perform a simple test for relative maxima and minima. The test is based on the fact that if the graph of a function f is concave upward on an open interval containing c, and $f'(c) = 0$, then $f(c)$ must be a relative minimum of f. Similarly, if the graph of a function f is concave downward on an open interval containing c, and $f'(c) = 0$, then $f(c)$ must be a relative maximum of f (see Figure 3.32).

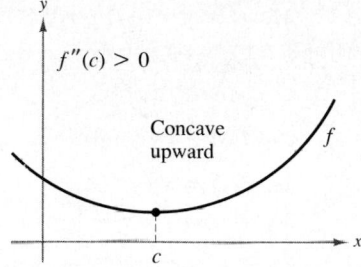

If $f'(c) = 0$ and $f''(c) > 0$ then $f(c)$ is a relative minimum.

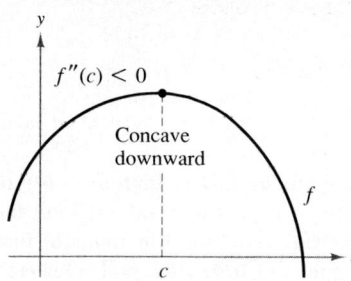

If $f'(c) = 0$ and $f''(c) < 0$ then $f(c)$ is a relative maximum.

FIGURE 3.32

THEOREM 3.9   Second Derivative Test

Let f be a function such that $f'(c) = 0$ and the second derivative of f exists on an open interval containing c.

1. If $f''(c) > 0$, then $f(c)$ is a relative minimum.
2. If $f''(c) < 0$, then $f(c)$ is a relative maximum.

If $f''(c) = 0$, then the test fails. In such cases, you can use the First Derivative Test.

Proof   If $f'(c) = 0$ and $f''(c) > 0$, then there exists an open interval I containing c for which

$$\frac{f'(x) - f'(c)}{x - c} = \frac{f'(x)}{x - c} > 0$$

for all $x \neq c$ in I. If $x < c$, then $x - c < 0$ and $f'(x) < 0$. Also, if $x > c$, then $x - c > 0$ and $f'(x) > 0$. Thus, $f'(x)$ changes from negative to positive at c, and the First Derivative Test implies that $f(c)$ is a relative minimum. A proof of the second case is left to you.

EXAMPLE 4   Using the Second Derivative Test

Find the relative extrema for $f(x) = -3x^5 + 5x^3$.

Solution   Begin by finding the critical numbers of f.

$$f'(x) = -15x^4 + 15x^2 = 15x^2(1 - x^2) = 0 \qquad \text{Set } f'(x) = 0$$
$$x = -1, 0, 1 \qquad \text{Critical numbers}$$

Using $f''(x) = -60x^3 + 30x = 30(-2x^3 + x)$, you can apply the Second Derivative Test as follows.

Point	Sign of f''		Conclusion
$(-1, -2)$	$f''(-1) = 30 > 0$	$\longrightarrow$	Relative minimum
$(1, 2)$	$f''(1) = -30 < 0$	$\longrightarrow$	Relative maximum
$(0, 0)$	$f''(0) = 0$	$\longrightarrow$	Test fails

Because the Second Derivative Test fails at $(0, 0)$, you can use the First Derivative Test and observe that f increases to the left and right of $x = 0$. Thus, $(0, 0)$ is neither a relative minimum nor a relative maximum (even though the graph has a horizontal tangent line at this point). The graph of f is shown in Figure 3.33.

FIGURE 3.33

$(0, 0)$ is neither a relative minimum nor a relative maximum.

EXERCISES for Section 3.4

In Exercises 1–6, find the open intervals on which the graph of the function is concave upward and those on which it is concave downward.

1. $y = x^2 - x - 2$

2. $f(x) = \dfrac{24}{x^2 + 12}$

Generated by Derive

Generated by Derive

3. $y = -x^3 + 3x^2 - 2$

4. $f(x) = \dfrac{x^2 - 1}{2x + 1}$

Generated by Derive

Generated by Derive

5. $f(x) = \dfrac{x^2 + 1}{x^2 - 1}$

6. $y = \dfrac{-3x^5 + 40x^3 + 135x}{270}$

Generated by Derive

Generated by Derive

In Exercises 7–20, find all relative extrema. Use the Second Derivative Test where applicable.

7. $f(x) = 6x - x^2$

8. $f(x) = x^2 + 3x - 8$

9. $f(x) = (x - 5)^2$

10. $f(x) = -(x - 5)^2$

11. $f(x) = x^3 - 3x^2 + 3$

12. $f(x) = 5 + 3x^2 - x^3$

13. $f(x) = x^4 - 4x^3 + 2$

14. $f(x) \doteq x^3 - 9x^2 + 27x$

15. $f(x) = x^{2/3} - 3$

16. $f(x) = \sqrt{x^2 + 1}$

17. $f(x) = x + \dfrac{4}{x}$

18. $f(x) = \dfrac{x}{x - 1}$

19. $f(x) = \cos x - x$
$0 \le x \le 4\pi$

20. $f(x) = 2 \sin x + \cos 2x$
$0 \le x \le 2\pi$

In Exercises 21–36, find all relative extrema and points of inflection and sketch the graph of the function.

21. $f(x) = x^3 - 12x$

22. $f(x) = x^3 + 1$

23. $f(x) = x^3 - 6x^2 + 12x - 8$

24. $f(x) = 2x^3 - 3x^2 - 12x + 8$

25. $f(x) = \dfrac{1}{4}x^4 - 2x^2$

26. $f(x) = 2x^4 - 8x + 3$

27. $f(x) = x(x - 4)^3$

28. $f(x) = x^3(x - 4)$

29. $f(x) = x\sqrt{x + 3}$

30. $f(x) = x\sqrt{x + 1}$

31. $f(x) = \sin\dfrac{x}{2}$
$0 < x < 4\pi$

32. $f(x) = 2\csc\dfrac{3x}{2}$
$0 < x < 2\pi$

33. $f(x) = \sec\left(x - \dfrac{\pi}{2}\right)$
$0 < x < 4\pi$

34. $f(x) = \sin x + \cos x$
$0 \le x \le 2\pi$

35. $f(x) = 2\sin x + \sin 2x$
$0 \le x \le 2\pi$

36. $f(x) = x - \sin x$
$0 \le x \le 4\pi$

C In Exercises 37–40, use a symbolic differentiation utility to analyze the function over the specified interval. (a) Find the first- and second-order derivatives of the function, (b) find any relative extrema and points of inflection, and (c) sketch the graphs of f, f', and f'' on the same coordinate axes and state the relationship between the behavior of f and the signs of f' and f''.

Function	Interval
37. $f(x) = 0.2x^2(x - 3)^3$	$[-1, 4]$
38. $f(x) = x^2\sqrt{6 - x^2}$	$[-\sqrt{6}, \sqrt{6}]$
39. $f(x) = \sin x - \dfrac{1}{3}\sin 3x + \dfrac{1}{5}\sin 5x$	$[0, \pi]$
40. $f(x) = \sqrt{2x}\sin x$	$[0, 2\pi]$

41. Sketch the graph of a function f such that f' is an increasing function.

42. Sketch the graph of a function f such that f' is negative and decreasing.

In Exercises 43–46, sketch the graph of a function f having the indicated characteristics.

43. $f(2) = f(4) = 0$
$f'(x) < 0$ if $x < 3$
$f'(3)$ is undefined
$f'(x) > 0$ if $x > 3$
$f''(x) < 0, x \ne 3$

44. $f(0) = f(2) = 0$
$f'(x) > 0$ if $x < 1$
$f'(1) = 0$
$f'(x) < 0$ if $x > 1$
$f''(x) < 0$

45. $f(2) = f(4) = 0$
$f'(x) > 0$ if $x < 3$
$f'(3)$ is undefined
$f'(x) < 0$ if $x > 3$
$f''(x) > 0, x \ne 3$

46. $f(0) = f(2) = 0$
$f'(x) < 0$ if $x < 1$
$f'(1) = 0$
$f'(x) > 0$ if $x > 1$
$f''(x) > 0$

In Exercises 47–50, trace the graph of *f*. On the same set of axes, sketch the graphs of *f′* and *f″*.

47.

48.

49.

50.

51. The figure is the graph of the second derivative of a function *f*. Sketch a graph of *f*. (The answer is not unique.)

52. Water is running into the vase shown in the figure at a constant rate.
   a. Sketch a graph of the depth *d* of water in the vase as a function of time.
   b. Does the function have any extrema? Explain your answer.
   c. Give an interpretation of the inflection points of the graph of *d*.

FIGURE FOR 51

FIGURE FOR 52

53. Find the inflection points (if any) and sketch the graph of $f(x) = (x - c)^n$ for $n = 1, 2, 3,$ and 4. What conclusion can be made about the value of *n* and the existence of inflection points on the graph of *f*?

54. a. Sketch the graph of $f(x) = \sqrt[3]{x}$ and identify the inflection point.
   b. Does $f''(x)$ exist at the inflection point?

In Exercises 55 and 56, find *a*, *b*, *c*, and *d* such that the cubic polynomial function $f(x) = ax^3 + bx^2 + cx + d$ satisfies the indicated conditions.

55. Relative maximum: (3, 3)
    Relative minimum: (5, 1)
    Inflection point: (4, 2)

56. Relative maximum: (2, 4)
    Relative minimum: (4, 2)
    Inflection point: (3, 3)

57. *Aircraft Glide Path*   A small aircraft starts its descent from an altitude of 1 mile, 4 miles west of the runway (see figure).
   a. Find the cubic polynomial function $f(x) = ax^3 + bx^2 + cx + d$ on the interval $[-4, 0]$, which describes a smooth glide path for the landing.
   b. If the glide path of the plane is described by the function of part a, when would the plane be descending at the greatest rate?

FIGURE FOR 57

FOR FURTHER INFORMATION   For more information on this type of modeling, see the article "How Not to Land at Lake Tahoe!" by Richard Barshinger in the May, 1992 issue of *The American Mathematical Monthly*.

58. Show that the point of inflection of
$$f(x) = x(x - 6)^2$$
lies midway between the relative extrema of *f*.

59. Prove that a cubic function with three distinct real zeros has a point of inflection whose *x*-coordinate is the average of the three zeros.

60. *Specific Gravity*   A model for the specific gravity of water *S* is
$$S = \frac{5.755T^3}{10^8} - \frac{8.521T^2}{10^6} + \frac{0.654T}{10^4} + 0.99987,$$
$$0 < T < 25$$
where *T* is the water temperature in degrees Celsius.
   a. Use a symbolic differentiation utility to find the coordinates of the maximum value of the function.
   b. Sketch a graph of the function over the specified domain. (Use a setting in which $0.996 \le S \le 1.001$.) Estimate the specific gravity of water when $T = 20°$.

61. *Beam Deflection*   The deflection *D* of a particular beam of length *L* is
$$D = 2x^4 - 5Lx^3 + 3L^2x^2$$
where *x* is the distance from one end of the beam. Find the value of *x* that yields the maximum deflection.

62. *Electric Field Intensity*   The equation
$$E = \frac{kqx}{(x^2 + a^2)^{3/2}}$$
gives the electric field intensity on the axis of a uniformly charged ring, where *q* is the total charge on the ring, *k* is a constant, and *a* is the radius of the ring. At what value of *x* is *E* maximum?

63. *Average Cost* A manufacturer has determined that the total cost C of operating a factory is

$$C = 0.5x^2 + 15x + 5000$$

where x is the number of units produced. At what level of production will the average cost per unit be minimum? (The average cost per unit is given by C/x.)

64. *Inventory Cost* The total cost C for ordering and storing x units is

$$C = 2x + \frac{300,000}{x}.$$

What order size will produce a minimum cost?

65. *Engine Design* For the engine shown in the figure, a connecting rod 7 inches long is fastened to a crank of radius 3 inches at point P. The crankshaft rotates counterclockwise at a constant rate of 200 revolutions per minute. The horizontal velocity (in./min) of point P is

$$v = -1200\pi \sin\theta$$

where θ is the central angle of the crankshaft. What values of θ produce a maximum horizontal velocity?

66. The graph of $y = x\sin(1/x)$ is shown in the figure. Show that the graph is concave downward to the right of $x = 1/\pi$.

FIGURE FOR 65   **FIGURE FOR 66**

C | *Linear and Quadratic Approximation* **In Exercises 67–70, use a graphing utility to graph the function. Then graph**

$$P_1(x) = f(a) + f'(a)(x - a)$$

and

$$P_2(x) = f(a) + f'(a)(x - a) + \frac{1}{2}f''(a)(x - a)^2$$

on the same viewing rectangle. Compare the values of f, P_1, and P_2 and their first derivative at $x = a$.

Function	Value of a
67. $f(x) = 2(\sin x + \cos x)$	$a = 0$
68. $f(x) = \sqrt{1 - x}$	$a = 0$
69. $f(x) = 2(\sin x + \cos x)$	$a = \dfrac{\pi}{4}$
70. $f(x) = \dfrac{\sqrt{x}}{x - 1}$	$a = 2$

71. *Sales* S represents the weekly sales of a product. What can be said of S' and S'' for each of the following?
   a. The rate of change of sales is increasing.
   b. The rate of change of sales is decreasing.
   c. The rate of change of sales is constant.
   d. Sales are steady.
   e. Sales are declining, but at a lower rate.
   f. Sales have bottomed out and have started to rise.

72. Sketch the graph of an arbitrary function that does *not* have a point of inflection at $(c, f(c))$ even though $f''(c) = 0$.

True or False **In Exercises 73–76, determine whether the statement is true or false. If it is false, explain why or give an example that shows it is false.**

73. The graph of every cubic polynomial has precisely one point of inflection.

74. The graph of $f(x) = 1/x$ is concave downward for $x < 0$ and concave upward for $x > 0$, and thus it has a point of inflection at $x = 0$.

75. The maximum value of $y = 3\sin x + 2\cos x$ is 5.

76. The maximum slope of the graph of $y = \sin(bx)$ is b.

77. *Essay* Discuss the advantages and disadvantages of the First and Second Derivative Tests. Illustrate your discussion with examples.

78. Prove Darboux's Theorem: Let $f(x)$ be differentiable on $[a, b]$, $f'(a) = y_1$, and $f'(b) = y_2$. If d lies between y_1 and y_2, then there exists c in (a, b) such that $f'(c) = d$.

79. Show that the cubic polynomial $p(x) = ax^3 + bx^2 + cx + d$ has exactly one point of inflection (x_0, y_0), where

$$x_0 = \frac{-b}{3a} \quad \text{and} \quad y_0 = \frac{2b^3}{27a^2} - \frac{bc}{3a} + d.$$

Use this formula to find the point of inflection of $p(x) = x^3 - 3x^2 + 2$.

80. The logistics curve $y = f(t)$ satisfies the equation

$$\frac{dy}{dt} = Ky(L - y)$$

where K and L are positive constants. Find the y-coordinate of the point of inflection of y in terms of L.

Limits at Infinity

3.5 Limits at Infinity • Horizontal Asymptotes

Limits at Infinity

FIGURE 3.34
The limit of $f(x)$ as x approaches $-\infty$ or ∞ is 3.

This section discusses the "end behavior" of a function on an *infinite* interval. Consider the graph of

$$f(x) = \frac{3x^2}{x^2 + 1}$$

as shown in Figure 3.34. Table 3.11 suggests that the value of $f(x)$ approaches 3 as x increases without bound ($x \to \infty$). Similarly, $f(x)$ approaches 3 as $x \to -\infty$. These **limits at infinity** are denoted by

$$\lim_{x \to -\infty} f(x) = 3 \quad \text{and} \quad \lim_{x \to \infty} f(x) = 3.$$

TABLE 3.11

	x decreases without bound					x increases without bound			

x	$-\infty \leftarrow$	-100	-10	-1	0	1	10	100	$\to \infty$
$f(x)$	$3 \leftarrow$	2.9997	2.97	1.5	0	1.5	2.97	2.9997	$\to 3$

	$f(x)$ approaches 3					$f(x)$ approaches 3			

To say that a statement is true as x increases *without bound* means that for any (large) real number M, the statement is true for *all* x in the interval $\{x: x > M\}$. The following definition uses this concept.

Definition of Limits at Infinity

(Let L be a real number.)

1. The statement $\lim\limits_{x \to \infty} f(x) = L$ means that for each $\varepsilon > 0$ there exists an $M > 0$ such that $|f(x) - L| < \varepsilon$ whenever $x > M$.
2. The statement $\lim\limits_{x \to -\infty} f(x) = L$ means that for each $\varepsilon > 0$ there exists an $N < 0$ such that $|f(x) - L| < \varepsilon$ whenever $x < N$.

FIGURE 3.35

REMARK   By writing $\lim\limits_{x \to -\infty} f(x) = L$ or $\lim\limits_{x \to \infty} f(x) = L$, we mean that the limit exists *and* the limit is equal to the number L.

The definition of a limit at infinity is illustrated in Figure 3.35. In this figure, note that for a given positive number ε there exists a positive number M such that, *to the right of* $x = M$, the graph of f will lie between the horizontal lines given by $y = L + \varepsilon$ and $y = L - \varepsilon$.

Horizontal Asymptotes

In Figure 3.34, the graph of f approaches the line $y = L$ as x increases without bound. We call the line $y = L$ a **horizontal asymptote** of the graph of f.

Definition of Horizontal Asymptote

The line $y = L$ is a **horizontal asymptote** of the graph of f if

$$\lim_{x \to -\infty} f(x) = L \quad \text{or} \quad \lim_{x \to \infty} f(x) = L.$$

REMARK   From this definition, it follows that the graph of a *function* of x can have at most two horizontal asymptotes—one to the right and one to the left.

Limits at infinity have many of the same properties of limits discussed in Section 1.2. For example, if $\lim_{x \to \infty} f(x)$ and $\lim_{x \to \infty} g(x)$ both exist, then

$$\lim_{x \to \infty}[f(x) + g(x)] = \lim_{x \to \infty} f(x) + \lim_{x \to \infty} g(x)$$

and

$$\lim_{x \to \infty}[f(x)g(x)] = \left[\lim_{x \to \infty} f(x)\right]\left[\lim_{x \to \infty} g(x)\right].$$

Similar properties hold for limits at $-\infty$.

When evaluating limits at infinity, the following theorem is helpful. (A proof of this theorem is given in Appendix A.)

THEOREM 3.10   Limits at Infinity

If r is a positive rational number and c is any real number, then

$$\lim_{x \to \infty} \frac{c}{x^r} = 0.$$

Furthermore, if x^r is defined when $x < 0$, then $\lim_{x \to -\infty} \frac{c}{x^r} = 0.$

EXAMPLE 1   Evaluating a Limit at Infinity

Find the limit: $\lim_{x \to \infty}\left(5 - \dfrac{2}{x^2}\right).$

Solution   Using Theorem 3.10, you can write the following.

$$\lim_{x \to \infty}\left(5 - \frac{2}{x^2}\right) = \lim_{x \to \infty} 5 - \lim_{x \to \infty} \frac{2}{x^2}$$
$$= 5 - 0$$
$$= 5$$

EXAMPLE 2   Evaluating a Limit at Infinity

Find the limit: $\lim\limits_{x \to \infty} \dfrac{2x - 1}{x + 1}$.

Solution   Note that both the numerator and the denominator approach infinity as x approaches infinity.

$$\lim_{x \to \infty} \frac{2x - 1}{x + 1}$$

$$\lim_{x \to \infty}(2x - 1) \to \infty$$

$$\lim_{x \to \infty}(x + 1) \to \infty$$

To resolve this problem, you can divide both the numerator and the denominator by x. After dividing, the limit may be evaluated as follows.

$$\lim_{x \to \infty} \frac{2x - 1}{x + 1} = \lim_{x \to \infty} \frac{\dfrac{2x - 1}{x}}{\dfrac{x + 1}{x}} \qquad \text{Divide numerator and denominator by } x$$

$$= \lim_{x \to \infty} \frac{2 - (1/x)}{1 + (1/x)}$$

$$= \frac{\lim\limits_{x \to \infty} 2 - \lim\limits_{x \to \infty} \dfrac{1}{x}}{\lim\limits_{x \to \infty} 1 + \lim\limits_{x \to \infty} \dfrac{1}{x}} \qquad \text{Take limit of numerator and denominator}$$

$$= \frac{2 - 0}{1 + 0} \qquad \text{Apply Theorem 3.10}$$

$$= 2$$

Thus, the line $y = 2$ is a horizontal asymptote to the right. By taking the limit as $x \to -\infty$, you can see that $y = 2$ is also a horizontal asymptote to the left. The graph of the function is shown in Figure 3.36.

FIGURE 3.36
$y = 2$ is a horizontal asymptote.

FIGURE 3.37
As x increases, the graph of f moves closer and closer to the line $y = 2$.

TECHNOLOGY   You can test the reasonableness of the limit found in Example 2 by evaluating $f(x)$ for a few large positive values of x. For instance,

$$f(100) \approx 1.9703, \quad f(1000) \approx 1.9970, \quad \text{and} \quad f(10,000) \approx 1.9997.$$

Another way to test the reasonableness of the limit is to use a graphing utility. For instance, in Figure 3.37, the graph of

$$f(x) = \frac{2x - 1}{x + 1}$$

is shown with the horizontal line $y = 2$. Note that as x increases, the graph of f moves closer and closer to its horizontal asymptote.

In Example 2, the first attempt to evaluate the limit resulted in the **indeterminate form** ∞/∞. We were able to resolve the difficulty by rewriting the given expression in an equivalent form by dividing both the numerator and the denominator by x. In general, we suggest dividing by the highest power of x in the *denominator*.

MARIA AGNESI

Agnesi was one of a handful of women to receive credit for significant contributions to mathematics before the 20th century. In her early twenties she wrote the first text that included both differential and integral calculus. By age 30, she was an honorary member of the faculty at the University of Bologna.

$$\lim_{x \to -\infty} f(x) = 0 \qquad \lim_{x \to \infty} f(x) = 0$$

FIGURE 3.38

f has a horizontal asymptote at $y = 0$.

FOR FURTHER INFORMATION
For more information on the contributions of women to mathematics, see the article "Why Women Succeed in Mathematics" by Mona Fabricant, Sylvia Svitak, and Patricia Clark Kenschaft in the February, 1990 issue of *Mathematics Teacher*.

EXAMPLE 3   A Comparison of Three Rational Functions

Find the limits.

a. $\displaystyle \lim_{x \to \infty} \frac{2x + 5}{3x^2 + 1}$ **b.** $\displaystyle \lim_{x \to \infty} \frac{2x^2 + 5}{3x^2 + 1}$ **c.** $\displaystyle \lim_{x \to \infty} \frac{2x^3 + 5}{3x^2 + 1}$

Solution

a. Divide both the numerator and the denominator by x^2.

$$\lim_{x \to \infty} \frac{2x + 5}{3x^2 + 1} = \lim_{x \to \infty} \frac{(2/x) + (5/x^2)}{3 + (1/x^2)} = \frac{0 + 0}{3 + 0} = \frac{0}{3} = 0$$

b. Divide both the numerator and the denominator by x^2.

$$\lim_{x \to \infty} \frac{2x^2 + 5}{3x^2 + 1} = \lim_{x \to \infty} \frac{2 + (5/x^2)}{3 + (1/x^2)} = \frac{2 + 0}{3 + 0} = \frac{2}{3}$$

c. Divide both the numerator and the denominator by x^2.

$$\lim_{x \to \infty} \frac{2x^3 + 5}{3x^2 + 1} = \lim_{x \to \infty} \frac{2x + (5/x^2)}{3 + (1/x^2)}$$

$$\lim_{x \to \infty} \left(2x + \frac{5}{x^2} \right) \to \infty$$

$$\lim_{x \to \infty} \left(3 + \frac{1}{x^2} \right) \to 3$$

You can conclude that the limit *does not exist* because the numerator increases without bound while the denominator approaches 3.

Compare the three rational functions in Example 3. In part a, the degree of the numerator is *less* than the degree of the denominator and the limit of the rational function is zero. In part b, the degrees of the numerator and the denominator are *equal* and the limit is the ratio of the two leading coefficients 2 and 3. In part c, the degree of the numerator is *greater* than that of the denominator and the limit does not exist.

These limits seem reasonable when you consider that for large values of x, the highest-power term of the rational function is the most "influential" in determining the limit. For instance, the limit as x approaches infinity of the function given by

$$f(x) = \frac{1}{x^2 + 1}$$

is 0 because the denominator overpowers the numerator as x increases or decreases without bound, as shown in Figure 3.38.

The function shown in Figure 3.38 is a special case of a type of curve studied by the Italian mathematician Maria Gaetana Agnesi (1718–1799). The general form of this function is $f(x) = a^3/(x^2 + a^2)$ and, through a mistranslation of the Italian word *vertéré*, the curve has come to be known as the witch of Agnesi. Agnesi's work with this curve first appeared in a comprehensive text on calculus that was published in 1748.

In Figure 3.38, you can see that the function $f(x) = 1/(x^2 + 1)$ approaches the same horizontal asymptote to the right and to the left. This is always true of rational functions. Functions that are not rational, however, may approach different horizontal asymptotes to the right and to the left.

EXAMPLE 4   A Function with Two Horizontal Asymptotes

Determine the limits.

a. $\displaystyle\lim_{x \to \infty} \frac{3x - 2}{\sqrt{2x^2 + 1}}$ **b.** $\displaystyle\lim_{x \to -\infty} \frac{3x - 2}{\sqrt{2x^2 + 1}}$

Solution

a. For $x > 0$, you can write $x = \sqrt{x^2}$. Thus, dividing both the numerator and the denominator by x produces

$$\frac{3x - 2}{\sqrt{2x^2 + 1}} = \frac{\dfrac{3x - 2}{x}}{\dfrac{\sqrt{2x^2 + 1}}{\sqrt{x^2}}} = \frac{3 - \dfrac{2}{x}}{\sqrt{\dfrac{2x^2 + 1}{x^2}}} = \frac{3 - \dfrac{2}{x}}{\sqrt{2 + \dfrac{1}{x^2}}}$$

and you can take the limit as follows.

$$\lim_{x \to \infty} \frac{3x - 2}{\sqrt{2x^2 + 1}} = \lim_{x \to \infty} \frac{3 - \dfrac{2}{x}}{\sqrt{2 + \dfrac{1}{x^2}}} = \frac{3 - 0}{\sqrt{2 + 0}} = \frac{3}{\sqrt{2}}$$

b. For $x < 0$, you can write $x = -\sqrt{x^2}$. Thus, dividing both the numerator and the denominator by x produces

$$\frac{3x - 2}{\sqrt{2x^2 + 1}} = \frac{\dfrac{3x - 2}{x}}{\dfrac{\sqrt{2x^2 + 1}}{-\sqrt{x^2}}} = \frac{3 - \dfrac{2}{x}}{-\sqrt{\dfrac{2x^2 + 1}{x^2}}} = \frac{3 - \dfrac{2}{x}}{-\sqrt{2 + \dfrac{1}{x^2}}}$$

and you can take the limit as follows.

$$\lim_{x \to -\infty} \frac{3x - 2}{\sqrt{2x^2 + 1}} = \lim_{x \to -\infty} \frac{3 - \dfrac{2}{x}}{-\sqrt{2 + \dfrac{1}{x^2}}} = \frac{3 - 0}{-\sqrt{2 + 0}} = -\frac{3}{\sqrt{2}}$$

The graph of $f(x) = (3x - 2)/\sqrt{2x^2 + 1}$ is shown in Figure 3.39.

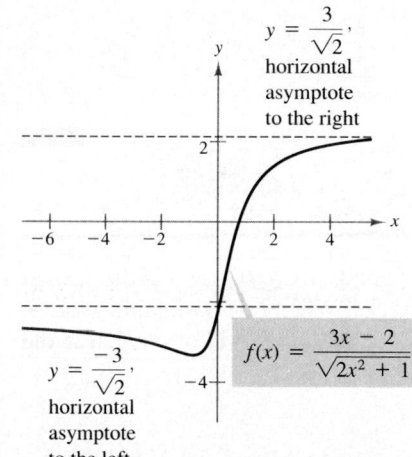

$y = \dfrac{3}{\sqrt{2}}$, horizontal asymptote to the right

$y = \dfrac{-3}{\sqrt{2}}$, horizontal asymptote to the left

$f(x) = \dfrac{3x - 2}{\sqrt{2x^2 + 1}}$

FIGURE 3.39
Functions that are not rational may have different right and left horizontal asymptotes.

TECHNOLOGY   If you use a graphing utility to help estimate a limit, be sure that you also confirm the estimate analytically—the pictures shown by a graphing utility can be misleading. For instance, Figure 3.40 shows one view of the graph of

$$y = \frac{2x^3 + 1000x^2 + x}{x^3 + 1000x^2 + x + 1000}.$$

From this view, one could be convinced that the graph has $y = 1$ as a horizontal asymptote. Using an analytic approach, however, you can see that the horizontal asymptote is actually $y = 2$.

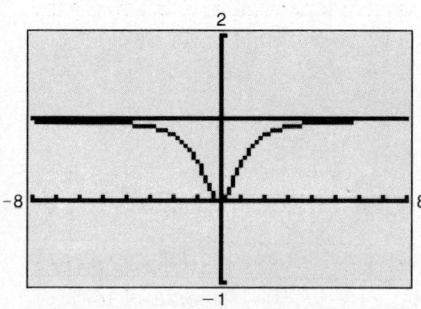

FIGURE 3.40
The horizontal asymptote appears to be the line $y = 1$, but is actually $y = 2$.

In Section 1.4 (Example 6), you saw how the Squeeze Theorem can be used to evaluate limits involving trigonometric functions. This theorem is also valid for limits at infinity.

EXAMPLE 5   Limits Involving Trigonometric Functions

Determine the limits.

a. $\lim\limits_{x \to \infty} \sin x$ **b.** $\lim\limits_{x \to \infty} \dfrac{\sin x}{x}$

Solution

a. As x approaches infinity, the sine function oscillates between 1 and -1. Thus, this limit does not exist.

b. Because $-1 \le \sin x \le 1$, it follows that

$$-\frac{1}{x} \le \frac{\sin x}{x} \le \frac{1}{x}$$

where $\lim\limits_{x \to \infty}(-1/x) = 0$ and $\lim\limits_{x \to \infty}(1/x) = 0$. Therefore, by the Squeeze Theorem, you can obtain

$$\lim_{x \to \infty} \frac{\sin x}{x} = 0$$

as indicated in Figure 3.41.   ▬

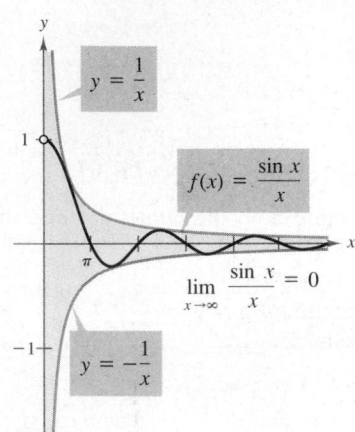

FIGURE 3.41

As x increases without bound, $f(x)$ approaches 0.

EXAMPLE 6   Oxygen Level in a Pond

Suppose that $f(t)$ measures the level of oxygen in a pond, where $f(t) = 1$ is the normal (unpolluted) level and the time t is measured in weeks. When $t = 0$, organic waste is dumped into the pond, and as the waste material oxidizes, the level of oxygen in the pond is given by

$$f(t) = \frac{t^2 - t + 1}{t^2 + 1}.$$

What percentage of the normal level of oxygen exists in the pond after 1 week? After 2 weeks? After 10 weeks? What is the limit as t approaches infinity?

Solution   When $t = 1$, 2, and 10, the levels of oxygen are as follows.

$$f(1) = \frac{1^2 - 1 + 1}{1^2 + 1} = \frac{1}{2} = 50\% \qquad \text{1 week}$$

$$f(2) = \frac{2^2 - 2 + 1}{2^2 + 1} = \frac{3}{5} = 60\% \qquad \text{2 weeks}$$

$$f(10) = \frac{10^2 - 10 + 1}{10^2 + 1} = \frac{91}{101} \approx 90.1\% \qquad \text{10 weeks}$$

To take the limit as t approaches infinity, divide the numerator and the denominator by t^2 to obtain

$$\lim_{t \to \infty} \frac{t^2 - t + 1}{t^2 + 1} = \lim_{t \to \infty} \frac{1 - (1/t) + (1/t^2)}{1 + (1/t^2)} = \frac{1 - 0 + 0}{1 + 0} = 1 = 100\%.$$

(See Figure 3.42.)   ▬

FIGURE 3.42

The level of oxygen in a pond approaches the normal level as t approaches ∞.

TECHNOLOGY
Laboratory Guide
Lab 3.5

EXERCISES for Section 3.5

In Exercises 1–8, match the function to one of the graphs (a), (b), (c), (d), (e), (f), (g), or (h), using horizontal asymptotes as an aid.

1. $f(x) = \dfrac{3x^2}{x^2 + 2}$ **2.** $f(x) = \dfrac{2x}{\sqrt{x^2 + 2}}$

3. $f(x) = \dfrac{x}{x^2 + 2}$ **4.** $f(x) = 2 + \dfrac{x^2}{x^4 + 1}$

5. $f(x) = \dfrac{-6x}{\sqrt{4x^2 + 5}}$ **6.** $f(x) = 5 - \dfrac{1}{x^2 + 1}$

7. $f(x) = \dfrac{4 \sin x}{x^2 + 1}$ **8.** $f(x) = \dfrac{2x^2 - 3x + 5}{x^2 + 1}$

(a)

(b)

(c)

(d)

(e)

(f)

(g)

(h)

In Exercises 9–28, find the limit.

9. $\displaystyle \lim_{x \to \infty} \dfrac{2x - 1}{3x + 2}$ **10.** $\displaystyle \lim_{x \to \infty} \dfrac{5x^3 + 1}{10x^3 - 3x^2 + 7}$

11. $\displaystyle \lim_{x \to \infty} \dfrac{x}{x^2 - 1}$ **12.** $\displaystyle \lim_{x \to \infty} \dfrac{2x^{10} - 1}{10x^{11} - 3}$

13. $\displaystyle \lim_{x \to -\infty} \dfrac{5x^2}{x + 3}$ **14.** $\displaystyle \lim_{x \to \infty} \dfrac{x^3 - 2x^2 + 3x + 1}{x^2 - 3x + 2}$

15. $\displaystyle \lim_{x \to \infty} \left(2x - \dfrac{1}{x^2} \right)$ **16.** $\displaystyle \lim_{x \to \infty} (x + 3)^{-2}$

17. $\displaystyle \lim_{x \to -\infty} \left(\dfrac{2x}{x - 1} + \dfrac{3x}{x + 1} \right)$ **18.** $\displaystyle \lim_{x \to \infty} \left(\dfrac{2x^2}{x - 1} + \dfrac{3x}{x + 1} \right)$

19. $\displaystyle \lim_{x \to -\infty} \dfrac{x}{\sqrt{x^2 - x}}$ **20.** $\displaystyle \lim_{x \to \infty} \dfrac{x}{\sqrt{x^2 + 1}}$

21. $\displaystyle \lim_{x \to \infty} \dfrac{2x + 1}{\sqrt{x^2 - x}}$ **22.** $\displaystyle \lim_{x \to -\infty} \dfrac{-3x + 1}{\sqrt{x^2 + x}}$

23. $\displaystyle \lim_{x \to \infty} \dfrac{x^2 - x}{\sqrt{x^4 + x}}$ **24.** $\displaystyle \lim_{x \to \infty} \dfrac{2x}{\sqrt{4x^2 + 1}}$

25. $\displaystyle \lim_{x \to \infty} \dfrac{\sin 2x}{x}$ **26.** $\displaystyle \lim_{x \to \infty} \dfrac{x - \cos x}{x}$

27. $\displaystyle \lim_{x \to \infty} \dfrac{1}{2x + \sin x}$ **28.** $\displaystyle \lim_{x \to \infty} \sin \dfrac{1}{x}$

In Exercises 29 and 30, find the limit. (*Hint:* Let $x = 1/t$ and find the limit as $t \to 0^+$.)

29. $\displaystyle \lim_{x \to \infty} x \sin \dfrac{1}{x}$ **30.** $\displaystyle \lim_{x \to \infty} x \tan \dfrac{1}{x}$

In Exercises 31–34, find the limit. (*Hint:* Treat the expression as a fraction whose denominator is 1, and rationalize the numerator.)

31. $\displaystyle \lim_{x \to -\infty} \left(x + \sqrt{x^2 + 3} \right)$ **32.** $\displaystyle \lim_{x \to \infty} \left(2x - \sqrt{4x^2 + 1} \right)$

33. $\displaystyle \lim_{x \to \infty} \left(x - \sqrt{x^2 + x} \right)$ **34.** $\displaystyle \lim_{x \to -\infty} \left(3x + \sqrt{9x^2 - x} \right)$

In Exercises 35–46, sketch the graph of the equation. Look for extrema, intercepts, symmetry, and asymptotes as necessary.

35. $y = \dfrac{2 + x}{1 - x}$ **36.** $y = \dfrac{x - 3}{x - 2}$

37. $y = \dfrac{x^2}{x^2 + 9}$ **38.** $y = \dfrac{x^2}{x^2 - 9}$

39. $xy^2 = 4$ **40.** $x^2 y = 4$

41. $y = \dfrac{2x}{1 - x}$ **42.** $y = \dfrac{2x}{1 - x^2}$

43. $y = 2 - \dfrac{3}{x^2}$ **44.** $y = 1 + \dfrac{1}{x}$

45. $y = \dfrac{x^3}{\sqrt{x^2 - 4}}$ **46.** $y = \dfrac{x}{\sqrt{x^2 - 4}}$

C **In Exercises 47–54, use a symbolic differentiation utility to analyze the graph of the function. Label any extrema and/or asymptotes that exist.**

47. $f(x) = 5 - \dfrac{1}{x^2}$ **48.** $f(x) = \dfrac{x^2}{x^2 - 1}$

49. $f(x) = \dfrac{x}{x^2 - 4}$

50. $f(x) = \dfrac{1}{x^2 - x - 2}$

51. $f(x) = \dfrac{x - 2}{x^2 - 4x + 3}$

52. $f(x) = \dfrac{x + 1}{x^2 + x + 1}$

53. $f(x) = \dfrac{3x}{\sqrt{4x^2 + 1}}$

54. $f(x) = \dfrac{2 \sin 2x}{x}$

C **In Exercises 55 and 56 (a) use a graphing utility to graph f and g on the same coordinate axes, (b) verify algebraically that f and g represent the same function, (c) zoom out sufficiently far so that the graph appears as a line. What is this line? (Note that the points of discontinuity are not readily seen when you zoom out.)**

55. $f(x) = \dfrac{x^3 - 3x^2 + 2}{x(x - 3)}$

56. $f(x) = -\dfrac{x^3 - 2x^2 + 2}{2x^2}$

$g(x) = x + \dfrac{2}{x(x - 3)}$

$g(x) = -\dfrac{1}{2}x + 1 - \dfrac{1}{x^2}$

In Exercises 57–60, complete the table

x	10^0	10^1	10^2	10^3	10^4	10^5	10^6
$f(x)$							

and estimate the limit of $f(x)$ as x approaches infinity. Then find the limit analytically and compare your results.

57. $f(x) = x - \sqrt{x(x - 1)}$

58. $f(x) = x^2 - x\sqrt{x(x - 1)}$

59. $f(x) = x \sin \dfrac{1}{2x}$

60. $f(x) = \dfrac{x + 1}{x\sqrt{x}}$

61. *Average Cost*   A business has a cost of $C = 0.5x + 500$ for producing x units. The average cost per unit is $\overline{C} = C/x$. Find the limit of $\overline{C}$ as x approaches infinity.

62. *Relativity*   According to the theory of relativity, the mass m of a particle depends on its velocity v. That is,

$$m = \dfrac{m_0}{\sqrt{1 - (v^2/c^2)}}$$

where m_0 is the mass when the particle is at rest and c is the speed of light. Find the limit of the mass as v approaches c.

63. *Engine Efficiency*   The efficiency of an internal combustion engine is

$$\text{Efficiency (\%)} = 100\left[1 - \dfrac{1}{(v_1/v_2)^c}\right]$$

where v_1/v_2 is the ratio of the uncompressed gas to the compressed gas and c is a positive constant dependent upon the engine design. Find the limit of the efficiency as the compression ratio approaches infinity.

64. *Demographics*   The figure shows a scatter plot of the number of high school graduates for representative years from 1900 to 1989. (*Source:* Department of Education) A model for approximating this data is also shown on the graph. The equation of the model is

$$N = \dfrac{109.25 + 4.15t}{1000 - 23.88t + 0.164t^2}, \qquad 0 \le t \le 89$$

where N is the number of graduates in millions and t is the time in years with $t = 0$ corresponding to 1900.

a. Approximate the year when the number of graduates was greatest.

b. Approximate the time when the rate of increase in the number of graduates was greatest.

c. Why should this model *not* be used to predict the number of graduates in future years?

FIGURE FOR 64

True or False   **In Exercises 65 and 66, determine whether the statement is true or false. If it is false, explain why or give an example that shows it is false.**

65. If $f'(x) > 0$ for all real numbers x, then f increases without bound.

66. If $f''(x) < 0$ for all real numbers x, then f decreases without bound.

67. Verify that each of the following functions has two horizontal asymptotes.

a. $f(x) = \dfrac{|x|}{x + 1}$ **b.** $f(x) = \dfrac{2x}{\sqrt{x^2 + 1}}$

68. Given $f(x) = 5x^3 - 3x^2 + 10$, find $\lim\limits_{x \to \infty} h(x)$ (if possible).

a. $h(x) = \dfrac{f(x)}{x^2}$ **b.** $h(x) = \dfrac{f(x)}{x^3}$

c. $h(x) = \dfrac{f(x)}{x^4}$

69. Sketch the graph of a differentiable function f that satisfies the following conditions and has $x = 2$ as its only critical number.

$f'(x) < 0$ for $x < 2$

$f'(x) > 0$ for $x > 2$

$\lim\limits_{x \to -\infty} f(x) = \lim\limits_{x \to \infty} f(x) = 6$

70. Is it possible to sketch the graph of a function that satisfies the conditions of Exercise 69 and has *no* points of inflection? Explain your answer.

71. Prove that if $p(x) = a_n x^n + \cdots + a_1 x + a_0$ and $q(x) = b_m x^m + \cdots + b_1 x + b_0$ ($a_n \ne 0,\ b_m \ne 0$), then

$$\lim_{x \to \infty} \dfrac{p(x)}{q(x)} = \begin{cases} 0, & n < m \\ \dfrac{a_n}{b_m}, & n = m \\ \pm\infty, & n > m. \end{cases}$$

A Summary of Curve Sketching

3.6   Summary of Curve-Sketching Techniques

33   51-63, 64

Summary of Curve-Sketching Techniques

It would be difficult to overstate the importance of using graphs in mathematics. Descartes' introduction of analytic geometry contributed significantly to the rapid advances in calculus that began during the mid-seventeenth century. In the words of Lagrange, "As long as algebra and geometry traveled separate paths their advance was slow and their applications limited. But when these two sciences joined company, they drew from each other fresh vitality and thenceforth marched on at a rapid pace toward perfection."

So far, you have studied several concepts that are useful in analyzing the graph of a function.

- *x*-intercepts and *y*-intercepts   (Section 3 of the Prerequisites chapter)
- Symmetry   (Section 3 of the Prerequisites chapter)
- Domain and Range   (Section 5 of the Prerequisites chapter)
- Continuity   (Section 1.4)
- Vertical asymptotes   (Section 1.5)
- Differentiability   (Section 2.1)
- Relative extrema   (Section 3.3)
- Concavity   (Section 3.4)
- Points of inflection   (Section 3.4)
- Horizontal asymptotes   (Section 3.5)

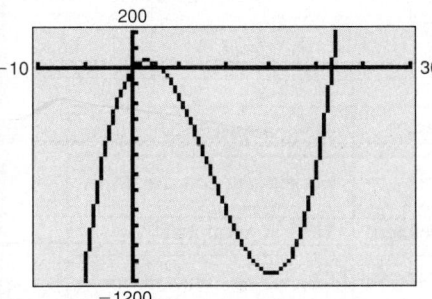

FIGURE 3.43
Different viewing rectangles for the same graph.

When you are sketching the graph of a function, either by hand or with a graphing utility, remember that you cannot normally show the *entire* graph. The decision as to which part of the graph you choose to show is often crucial. For instance, which of the viewing rectangles in Figure 3.43 better represents the graph of $f(x) = x^3 - 25x^2 + 74x - 20$? By seeing both views, it is clear that the second viewing rectangle gives a more complete representation of the graph. But would a third viewing rectangle reveal other interesting portions of the graph? To answer this, you need to use calculus to interpret the first and second derivatives. Here are some guidelines for determining a good viewing rectangle for the graph of a function.

Guidelines for Analyzing the Graph of a Function

1. Determine the domain and range of the function.
2. Determine the intercepts and asymptotes of the graph.
3. Locate the *x*-values where $f'(x)$ and $f''(x)$ are either zero or undefined. Use the results to determine relative extrema and points of inflection.

REMARK   Note in these guidelines the importance of *algebra* (as well as calculus) for solving the equations $f(x) = 0$, $f'(x) = 0$, and $f''(x) = 0$.

$$f(x) = \frac{2(x^2 - 9)}{x^2 - 4}$$

FIGURE 3.44
Using calculus you can be certain that you have determined all characteristics of the graph of f.

$$g(x) = \frac{2(x^2 - 9)(x - 20)}{(x^2 - 4)(x - 21)}$$

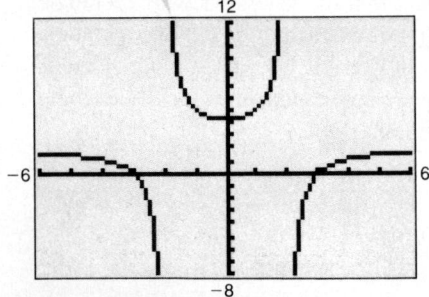

FIGURE 3.45
By not using calculus you may overlook important characteristics of the graph of g.

FOR FURTHER INFORMATION
For more information on the use of technology to graph rational functions, see the article "Graphs of Rational Functions for Computer Assisted Calculus" by Stan Byrd and Terry Walters in the September, 1991 issue of the *College Mathematics Journal.*

EXAMPLE 1   Analyzing the Graph of a Rational Function

Analyze the graph of $f(x) = \dfrac{2(x^2 - 9)}{x^2 - 4}$.

Solution

First derivative	$f'(x) = \dfrac{20x}{(x^2 - 4)^2}$
Second derivative	$f'' = \dfrac{-20(3x^2 + 4)}{(x^2 - 4)^3}$
x-intercepts	$(-3, 0), (3, 0)$
y-intercept	$\left(0, \dfrac{9}{2}\right)$
Vertical asymptotes	$x = -2, x = 2$
Horizontal asymptote	$y = 2$
Critical number	$x = 0$
Possible points of inflection	None
Domain	All real numbers except $x = \pm 2$
Symmetry	With respect to y-axis
Test intervals	$(-\infty, -2), (-2, 0), (0, 2), (2, \infty)$

Table 3.12 shows how the test intervals are used to determine several characteristics of the graph. The graph of f is shown in Figure 3.44.

TABLE 3.12

	$f(x)$	$f'(x)$	$f''(x)$	*Characteristic of graph*
$-\infty < x < -2$		$-$	$-$	Decreasing, concave down
$x = -2$	Undefined	Undefined	Undefined	Vertical asymptote
$-2 < x < 0$		$-$	$+$	Decreasing, concave up
$x = 0$	$\frac{9}{2}$	0	$+$	Relative minimum
$0 < x < 2$		$+$	$+$	Increasing, concave up
$x = 2$	Undefined	Undefined	Undefined	Vertical asymptote
$2 < x < \infty$		$+$	$-$	Increasing, concave down

Be sure you understand all of the implications of creating a table such as that shown in Example 1. Because of the use of calculus, you can *be sure* that the graph has no relative extrema or points of inflection other than those indicated in Figure 3.44.

Without using the type of analysis outlined in Example 1, it is easy to obtain an incomplete view of a graph's basic characteristics. For instance, Figure 3.45 shows a view of the graph of

$$g(x) = \frac{2(x^2 - 9)(x - 20)}{(x^2 - 4)(x - 21)}.$$

From this view, it appears that the graph of g is about the same as the graph of f shown in Figure 3.44. The graphs of these two functions, however, differ significantly. Try enlarging the viewing rectangle to see the differences.

EXAMPLE 2   Analyzing the Graph of a Rational Function

Analyze the graph of $f(x) = \dfrac{x^2 - 2x + 4}{x - 2}$.

Solution

$$\textit{First derivative} \quad f'(x) = \frac{x(x - 4)}{(x - 2)^2}$$

$$\textit{Second derivative} \quad f''(x) = \frac{8}{(x - 2)^3}$$

x-intercepts	None
y-intercept	$(0, -2)$
Vertical asymptote	$x = 2$
Horizontal asymptotes	None
Critical numbers	$x = 0, x = 4$
Possible points of inflection	None
Domain	All real numbers except $x = 2$
Test intervals	$(-\infty, 0), (0, 2), (2, 4), (4, \infty)$

The analysis of the graph of f is shown in Table 3.13, and the graph is shown in Figure 3.46.

$f(x) = \dfrac{x^2 - 2x + 4}{x - 2}$

FIGURE 3.46

TABLE 3.13

	$f(x)$	$f'(x)$	$f''(x)$	*Characteristic of graph*
$-\infty < x < 0$		$+$	$-$	Increasing, concave down
$x = 0$	-2	0	$-$	Relative maximum
$0 < x < 2$		$-$	$-$	Decreasing, concave down
$x = 2$	Undefined	Undefined	Undefined	Vertical asymptote
$2 < x < 4$		$-$	$+$	Decreasing, concave up
$x = 4$	6	0	$+$	Relative minimum
$4 < x < \infty$		$+$	$+$	Increasing, concave up

Although the graph of the function in Example 2 has no horizontal asymptote, it does have a slant asymptote. The graph of a rational function (having no common factors) has a **slant asymptote** if the degree of the numerator exceeds the degree of the denominator by one. To find the slant asymptote, use division to rewrite the rational function as the sum of a first-degree polynomial and another rational function.

$$f(x) = \frac{x^2 - 2x + 4}{x - 2} = x + \frac{4}{x - 2} \qquad y = x \text{ is a slant asymptote}$$

In Figure 3.47, note that the graph of f approaches the slant asymptote $y = x$ as x approaches $-\infty$ or ∞.

$f(x) = \dfrac{x^2 - 2x + 4}{x - 2}$

FIGURE 3.47

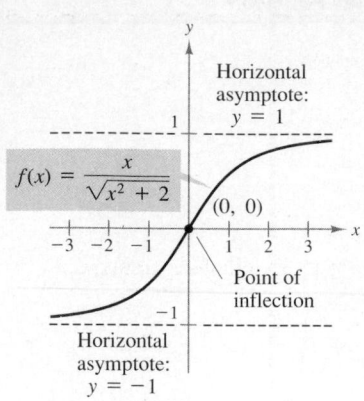

FIGURE 3.48

EXAMPLE 3   Analyzing the Graph of a Radical Function

Analyze the graph of $f(x) = \dfrac{x}{\sqrt{x^2 + 2}}$.

Solution

$$f'(x) = \frac{2}{(x^2 + 2)^{3/2}} \qquad f''(x) = -\frac{6x}{(x^2 + 2)^{5/2}}$$

The graph has only one intercept, $(0, 0)$. It has no vertical asymptotes, but it has two horizontal asymptotes: $y = 1$ (to the right) and $y = -1$ (to the left). The function has no critical numbers and one possible point of inflection (when $x = 0$). The domain of the function is all real numbers, and the graph is symmetric with respect to the origin. The analysis of the graph of f is shown in Table 3.14, and the graph is shown in Figure 3.48.

TABLE 3.14

	$f(x)$	$f'(x)$	$f''(x)$	*Characteristic of graph*
$-\infty < x < 0$		+	+	Increasing, concave up
$x = 0$	0	$\dfrac{1}{\sqrt{2}}$	0	Point of inflection
$0 < x < \infty$		+	−	Increasing, concave down

EXAMPLE 4   Analyzing the Graph of a Radical Function

Analyze the graph of $f(x) = 2x^{5/3} - 5x^{4/3}$.

Solution

$$f'(x) = \frac{10}{3}x^{1/3}(x^{1/3} - 2) \qquad f''(x) = \frac{20(x^{1/3} - 1)}{9x^{2/3}}$$

The function has two intercepts: $(0, 0)$ and $(\frac{125}{8}, 0)$. There are no horizontal or vertical asymptotes. The function has two critical numbers ($x = 0$ and $x = 8$) and two possible points of inflection ($x = 0$ and $x = 1$). The domain is all real numbers. The analysis of the graph of f is shown in Table 3.15, and the graph is shown in Figure 3.49.

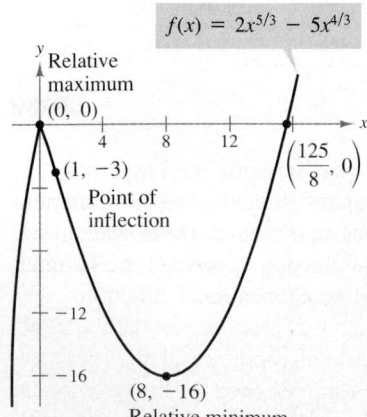

FIGURE 3.49

TABLE 3.15

	$f(x)$	$f'(x)$	$f''(x)$	*Characteristic of graph*
$-\infty < x < 0$		+	−	Increasing, concave down
$x = 0$	0	0	Undefined	Relative maximum
$0 < x < 1$		−	−	Decreasing, concave down
$x = 1$	−3	−	0	Point of inflection
$1 < x < 8$		−	+	Decreasing, concave up
$x = 8$	−16	0	+	Relative minimum
$8 < x < \infty$		+	+	Increasing, concave up

(a)

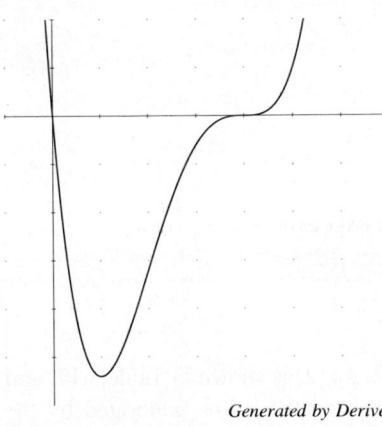

Generated by Derive

(b)

FIGURE 3.50

A polynomial function of even degree must have at least one relative extremum.

EXAMPLE 5   Analyzing the Graph of a Polynomial Function

Analyze the graph of $f(x) = x^4 - 12x^3 + 48x^2 - 64x$.

Solution   Begin by factoring to obtain

$$f(x) = x^4 - 12x^3 + 48x^2 - 64x$$
$$= x(x - 4)^3.$$

Then, using the factored form of $f(x)$, you can perform the following analysis.

First derivative	$f'(x) = 4(x - 1)(x - 4)^2$
Second derivative	$f''(x) = 12(x - 4)(x - 2)$
x-intercepts	$(0, 0), (4, 0)$
y-intercept	$(0, 0)$
Vertical asymptotes	None
Horizontal asymptotes	None
Critical numbers	$x = 1, x = 4$
Possible points of inflection	$x = 2, x = 4$
Domain	All real numbers
Test intervals	$(-\infty, 1), (1, 2), (2, 4), (4, \infty)$

The analysis of the graph of f is shown in Table 3.16, and the graph is shown in Figure 3.50(a). Using a computer algebra system like Derive (see Figure 3.50(b)) can help you verify your analysis.

TABLE 3.16

	$f(x)$	$f'(x)$	$f''(x)$	*Characteristic of graph*
$-\infty < x < 1$		$-$	$+$	Decreasing, concave up
$x = 1$	-27	0	$+$	Relative minimum
$1 < x < 2$		$+$	$+$	Increasing, concave up
$x = 2$	-16	$+$	0	Point of inflection
$2 < x < 4$		$+$	$-$	Increasing, concave down
$x = 4$	0	0	0	Point of inflection
$4 < x < \infty$		$+$	$+$	Increasing, concave up

The fourth-degree polynomial function in Example 5 has one relative minimum and no relative maxima. In general, a polynomial function of degree n can have *at most* $n - 1$ relative extrema, and *at most* $n - 2$ points of inflection. Moreover, polynomial functions of even degree must have *at least* one relative extremum.

Remember from the leading coefficient test described in Section 5 of the Prerequisites chapter that the "end behavior" of the graph of a polynomial function is determined by its leading coefficient and its degree. For instance, because the polynomial in Example 5 has a positive leading coefficient, the graph moves up to the right. Moreover, because the degree is even, the graph also moves up to the left.

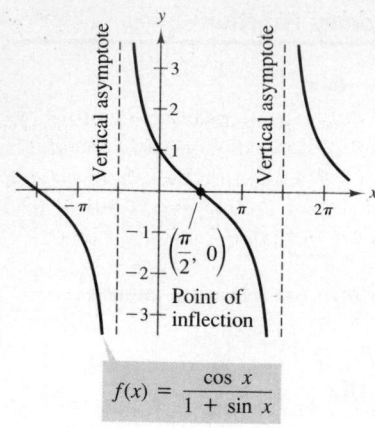

$$f(x) = \frac{\cos x}{1 + \sin x}$$

(a)

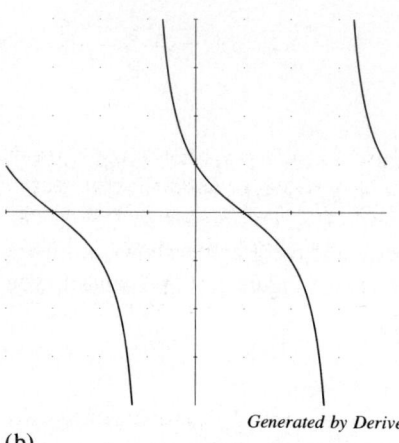

Generated by Derive

(b)

FIGURE 3.51

EXAMPLE 6   Analyzing the Graph of a Trigonometric Function

Analyze the graph of $f(x) = \dfrac{\cos x}{1 + \sin x}$.

Solution Because the function has a period of 2π, you can restrict the analysis of the graph to the interval $(-\pi/2, 3\pi/2)$.

First derivative	$f'(x) = -\dfrac{1}{1 + \sin x}$
Second derivative	$f''(x) = \dfrac{\cos x}{(1 + \sin x)^2}$
Period	2π
x-intercept	$\left(\dfrac{\pi}{2}, 0\right)$
y-intercept	$(0, 1)$
Vertical asymptotes	$x = -\dfrac{\pi}{2}, x = \dfrac{3\pi}{2}$
Horizontal asymptotes	None
Critical numbers	None
Possible points of inflection	$x = \dfrac{\pi}{2}$
Domain	All real numbers except $x = \dfrac{3 + 4n}{2}\pi$
Test intervals	$\left(-\dfrac{\pi}{2}, \dfrac{\pi}{2}\right), \left(\dfrac{\pi}{2}, \dfrac{3\pi}{2}\right)$

The analysis of the graph of f on the interval $(-\pi/2, 3\pi/2)$ is shown in Table 3.17, and the graph is shown in Figure 3.51(a). Compare this to the graph generated by the computer algebra system Derive in Figure 3.51(b).

REMARK The work involved in sketching the graph of a trigonometric function can be lessened sometimes by using trigonometric identities. For instance, the function in Example 6 can be rewritten as

$$f(x) = \frac{\cos x}{1 + \sin x} = \cot\left(\frac{x}{2} + \frac{\pi}{4}\right).$$

In this form, you can recognize the familiar cotangent graph shown in Figure 3.51.

TABLE 3.17

	$f(x)$	$f'(x)$	$f''(x)$	*Characteristic of graph*
$x = -\dfrac{\pi}{2}$	Undefined	Undefined	Undefined	Vertical asymptote
$-\dfrac{\pi}{2} < x < \dfrac{\pi}{2}$		$-$	$+$	Decreasing, concave up
$x = \dfrac{\pi}{2}$	0	$-\dfrac{1}{2}$	0	Point of inflection
$\dfrac{\pi}{2} < x < \dfrac{3\pi}{2}$		$-$	$-$	Decreasing, concave down
$x = \dfrac{3\pi}{2}$	Undefined	Undefined	Undefined	Vertical asymptote

TECHNOLOGY
Laboratory Guide
Lab 3.6

EXERCISES for Section 3.6

In Exercises 1–22, sketch the graph of the function. Choose a scale that allows all relative extrema and points of inflection to be identified on the sketch.

1. $y = x^3 - 3x^2 + 3$

2. $y = -\frac{1}{3}(x^3 - 3x + 2)$

3. $y = 2 - x - x^3$

4. $f(x) = \frac{1}{3}(x - 1)^3 + 2$

5. $f(x) = 3x^3 - 9x + 1$

6. $f(x) = (x + 1)(x - 2)(x - 5)$

7. $y = 3x^4 + 4x^3$

8. $y = 3x^4 - 6x^2$

9. $f(x) = x^4 - 4x^3 + 16x$

10. $f(x) = x^4 - 8x^3 + 18x^2 - 16x + 5$

11. $y = x^5 - 5x$

12. $y = (x - 1)^5$

13. $y = |2x - 3|$

14. $y = |x^2 - 6x + 5|$

15. $y = \dfrac{x^2}{x^2 + 3}$

16. $y = \dfrac{x}{x^2 + 1}$

17. $y = x\sqrt{4 - x}$

18. $y = x\sqrt{4 - x^2}$

19. $y = 3x^{2/3} - 2x$

20. $y = 3x^{2/3} - x^2$

21. $f(x) = \dfrac{x}{\sqrt{x^2 + 7}}$

22. $f(x) = \dfrac{4x}{\sqrt{x^2 + 15}}$

In Exercises 23–26, sketch a graph of the function over the specified interval.

Function	Interval
23. $y = \sin x - \frac{1}{18}\sin 3x$	$0 \le x \le 2\pi$
24. $y = \cos x - \frac{1}{2}\cos 2x$	$0 \le x \le 2\pi$
25. $y = 2x - \tan x$	$-\dfrac{\pi}{2} < x < \dfrac{\pi}{2}$
26. $y = 2x + \cot x$	$0 < x < \pi$

In Exercises 27–36, sketch the graph of the function. In each case label any intercepts, relative extrema, points of inflection, or asymptotes.

27. $y = \dfrac{1}{x - 2} - 3$

28. $y = \dfrac{x^2 + 1}{x^2 - 2}$

29. $y = \dfrac{2x}{x^2 - 1}$

30. $f(x) = \dfrac{x + 2}{x}$

31. $g(x) = x + \dfrac{4}{x^2 + 1}$

32. $f(x) = x + \dfrac{32}{x^2}$

33. $f(x) = \dfrac{x^2 + 1}{x}$

34. $f(x) = \dfrac{x^3}{x^2 - 1}$

35. $y = \dfrac{x^2 - 6x + 12}{x - 4}$

36. $y = \dfrac{2x^2 - 5x + 5}{x - 2}$

C In Exercises 37 and 38, use a graphing utility to graph the function.

37. $f(x) = \dfrac{20x}{x^2 + 1} - \dfrac{1}{x}$

38. $f(x) = 5\left(\dfrac{1}{x - 4} - \dfrac{1}{x + 2}\right)$

C *Essay* In Exercises 39 and 40, use a graphing utility to graph the function. Use the graph to determine whether or not it is possible for the graph of a function to cross its horizontal asymptote. Do you think it is possible for the graph of a function to cross its vertical asymptote? Why or why not?

39. $f(x) = \dfrac{4(x - 1)^2}{x^2 - 4x + 5}$

40. $g(x) = \dfrac{3x^4 - 5x + 3}{x^4 + 1}$

C *Essay* In Exercises 41 and 42, use a graphing utility to graph the function. Explain why there is no vertical asymptote when a superficial examination of the function may indicate that there should be one.

41. $h(x) = \dfrac{6 - 2x}{3 - x}$

42. $g(x) = \dfrac{x^2 + x - 2}{x - 1}$

C *Essay* In Exercises 43 and 44, use a graphing utility to graph the function and determine the slant asymptote of the graph. Zoom out repeatedly and describe how the graph on the display appears to change. Why does this occur?

43. $f(x) = -\dfrac{x^2 - 3x - 1}{x - 2}$

44. $g(x) = \dfrac{2x^2 - 8x - 15}{x - 5}$

In Exercises 45–48, create a function whose graph has the indicated characteristics. (The answer is not unique.)

45. Vertical asymptote: $x = 5$
  Horizontal asymptote: $y = 0$

46. Vertical asymptote: $x = -3$
  Horizontal asymptote: None

47. Vertical asymptote: $x = 5$
  Slant asymptote: $y = 3x + 2$

48. Vertical asymptote: $x = 0$
  Slant asymptote: $y = -x$

49. Consider the function

$$f(x) = \dfrac{ax}{x - b}.$$

  a. Determine the effect on the graph of f if $b \ne 0$ and a is varied. Consider cases where a is positive and a is negative.

  b. Determine the effect on the graph of f if $a \ne 0$ and b is varied.

50. Consider the function

$$f(x) = \tfrac{1}{2}(ax)^2 - (ax), \qquad a \ne 0.$$

  a. Determine the changes (if any) in the intercepts, extrema, and concavity of the graph of f when a is varied.

  b. On one set of coordinate axes, sketch the function for four different values of a.

In Exercises 51–56, determine conditions for the coefficients of

$$f(x) = ax^3 + bx^2 + cx + d$$

such that the graph of f resembles the given graph.

51.

52.

53.

54.

55.

56.

In Exercises 57–62, use the graph (of f' or f'') to sketch a graph of the function f. (*Hint:* There is more than one correct answer.)

57.

58.

59.

60.

61.

62.

63. The graphs of f, f', and f'' are shown together. Which is which?

(a)

(b)

Generated by Maple

64. Match the graph of f in the left column with that of its derivative in the right column.

f f'

a.

A.

b.

B.

c.

C.

d.

D.

3.7 Applied Minimum and Maximum Problems

Applied Minimum and Maximum Problems

One of the most common applications of calculus involves the determination of minimum and maximum values. Consider how frequently you hear or read terms such as greatest profit, least cost, least time, greatest voltage, optimum size, least size, greatest strength, and greatest distance. Before outlining a general problem-solving strategy for such problems, let's look at an example.

FIGURE 3.52
Open box with square base:
$S = x^2 + 4xh = 108$.

EXAMPLE 1   Finding the Maximum Volume

A manufacturer wants to design an open box having a square base and a surface area of 108 square inches, as shown in Figure 3.52. What dimensions will produce a box with maximum volume?

Solution   Because the box has a square base, its volume is

$$V = x^2 h. \qquad \text{Primary equation}$$

(This equation is called the **primary equation** because it gives a formula for the quantity to be optimized.) The surface area of the box is

$$S = (\text{Area of base}) + (\text{Area of four sides})$$
$$S = x^2 + 4xh = 108. \qquad \text{Secondary equation}$$

Because V is to be maximized, you want to express V as a function of just one variable. To do this, you can solve the equation $x^2 + 4xh = 108$ for h in terms of x to obtain $h = (108 - x^2)/(4x)$. Substituting in the primary equation produces

$$V = x^2 h = x^2 \left(\frac{108 - x^2}{4x} \right) = 27x - \frac{x^3}{4}. \qquad \text{Function of one variable}$$

Before finding which x-value will yield a maximum value of V, you should determine the *feasible domain*. That is, what values of x make sense in this problem? You know that x must be nonnegative and that the area of the base $(A = x^2)$ is at most 108. Thus, the feasible domain is

$$0 \le x \le \sqrt{108}. \qquad \text{Feasible domain}$$

To maximize V, you can find the critical numbers of the volume function.

$$\frac{dV}{dx} = 27 - \frac{3x^2}{4} = 0$$
$$3x^2 = 108 \quad \longrightarrow \quad x = \pm 6 \qquad \text{Critical numbers}$$

Evaluating V at the critical number $(x = 6)$ in the domain and at the endpoints of the domain produces $V(0) = 0$, $V(6) = 108$, and $V(\sqrt{108}) = 0$. Thus, V is maximum when $x = 6$ and the dimensions of the box are $6 \times 6 \times 3$ inches.

In Example 1, you should realize that there are infinitely many open boxes having 108 square inches of surface area. To begin solving the problem, you might ask yourself which basic shape would seem to yield a maximum volume. Should the box be tall, squat, or more cubical?

You might even try calculating a few volumes, as shown in Figure 3.53, to see if you can get a better feeling for what the optimum dimensions should be. Remember that you are not ready to begin solving a problem until you have clearly identified what the problem is.

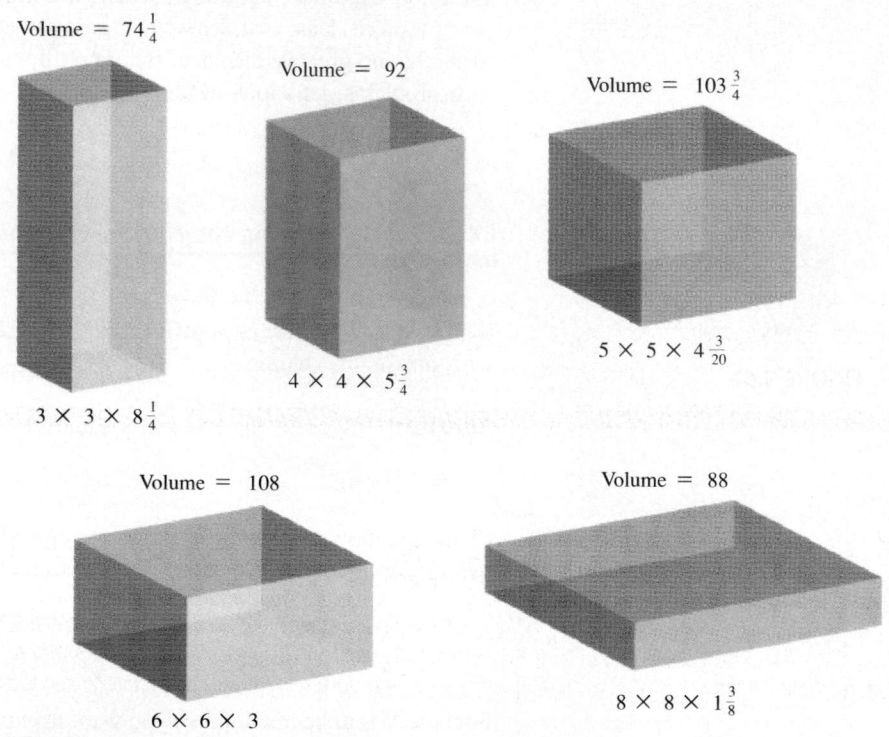

Volume $= 74\frac{1}{4}$

Volume $= 92$

Volume $= 103\frac{3}{4}$

$3 \times 3 \times 8\frac{1}{4}$

$4 \times 4 \times 5\frac{3}{4}$

$5 \times 5 \times 4\frac{3}{20}$

Volume $= 108$

Volume $= 88$

$6 \times 6 \times 3$

$8 \times 8 \times 1\frac{3}{8}$

FIGURE 3.53
Which box has the greatest volume?

Example 1 illustrates the following strategy for solving applied minimum and maximum problems.

Problem-Solving Strategy for Applied Minimum and Maximum Problems

1. Assign symbols to all *given* quantities and quantities *to be determined.* When feasible, make a sketch.
2. Write a **primary equation** for the quantity that is to be maximized (or minimized). (A review of several useful formulas from geometry is given inside the front cover of the text.)
3. Reduce the primary equation to one having a single independent variable. This may involve the use of **secondary equations** relating the independent variables of the primary equation.
4. Determine the domain of the primary equation. That is, determine the values for which the stated problem makes sense.
5. Determine the desired maximum or minimum value by the calculus techniques discussed in Sections 3.1 through 3.4.

REMARK When performing Step 5, recall that to determine the maximum or minimum value of a continuous function f on a closed interval, you should compare the values of f at its critical numbers to the values of f at the endpoints of the interval.

EXAMPLE 2   Finding the Minimum Distance

Which points on the graph of $y = 4 - x^2$ are closest to the point $(0, 2)$?

Solution Figure 3.54 indicates that there are two points at a minimum distance from the point $(0, 2)$. The distance between the point $(0, 2)$ and a point (x, y) on the graph of $y = 4 - x^2$ is given by

$$d = \sqrt{(x - 0)^2 + (y - 2)^2}.$$ Primary equation

Using the secondary equation $y = 4 - x^2$, you can rewrite the primary equation as

$$d = \sqrt{x^2 + (4 - x^2 - 2)^2} = \sqrt{x^4 - 3x^2 + 4}.$$

Because d is smallest when the expression inside the radical is smallest, you need only find the critical numbers of $f(x) = x^4 - 3x^2 + 4$. Note that the domain of f is the entire real line. Moreover, differentiation yields

$$f'(x) = 4x^3 - 6x = 2x(2x^2 - 3) = 0$$

$$x = 0, \ \sqrt{\frac{3}{2}}, \ -\sqrt{\frac{3}{2}}.$$

The First Derivative Test verifies that $x = 0$ yields a relative maximum, whereas both $x = \sqrt{3/2}$ and $x = -\sqrt{3/2}$ yield a minimum distance. Hence, the closest points are $\left(\sqrt{3/2}, 5/2\right)$ and $\left(-\sqrt{3/2}, 5/2\right)$.

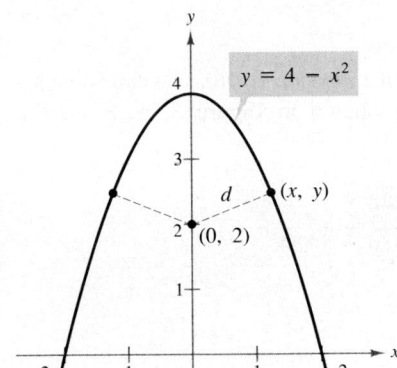

FIGURE 3.54
The quantity to be minimized is distance: $d = \sqrt{(x - 0)^2 + (y - 2)^2}$.

EXAMPLE 3   Finding the Minimum Area

A rectangular page is to contain 24 square inches of print. The margins at the top and bottom of the page are each $1\frac{1}{2}$ inches. The margins on each side are 1 inch (see Figure 3.55). What should the dimensions of the page be so that the least amount of paper is used?

Solution Let A be the area to be minimized.

$$A = (x + 3)(y + 2)$$ Primary equation

The printed area inside the margins is given by

$$24 = xy.$$ Secondary equation

Solving this equation for y produces $y = 24/x$. Substitution into the primary equation produces

$$A = (x + 3)\left(\frac{24}{x} + 2\right) = 30 + 2x + \frac{72}{x}.$$ Function of one variable

Because x must be positive, you are interested only in values of A when $x > 0$. To find the critical numbers, differentiate with respect to x.

$$\frac{dA}{dx} = 2 - \frac{72}{x^2} = 0 \ \rightarrow \ x^2 = 36.$$

Thus, the critical numbers are $x = \pm 6$. You do not have to consider -6 because it is outside the domain. The First Derivative Test confirms that A is a minimum when $x = 6$. Therefore, $y = \frac{24}{6} = 4$ and the dimensions of the page should be $x + 3 = 9$ inches by $y + 2 = 6$ inches.

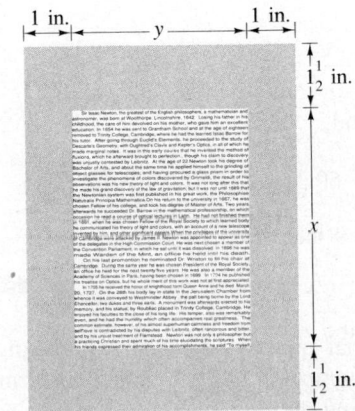

FIGURE 3.55
The quantity to be minimized is area: $A = (x + 3)(y + 2)$.

EXAMPLE 4   Finding the Minimum Length

Two posts, one 12 feet high and the other 28 feet high, stand 30 feet apart. They are to be stayed by two wires, attached to a single stake, running from ground level to the top of each post. Where should the stake be placed to use the least wire?

Solution   Let W be the wire length to be minimized. Using Figure 3.56, you can write

$$W = y + z. \quad \text{Primary equation}$$

In this problem, rather than solving for y in terms of z (or vice versa), you can solve for both y and z in terms of a third variable x, as shown in Figure 3.56. From the Pythagorean Theorem, you obtain

$$x^2 + 12^2 = y^2 \quad \rightarrow \quad y = \sqrt{x^2 + 144}$$
$$(30 - x)^2 + 28^2 = z^2 \quad \rightarrow \quad z = \sqrt{x^2 - 60x + 1684}.$$

Thus, W is given by

$$W = y + z$$
$$= \sqrt{x^2 + 144} + \sqrt{x^2 - 60x + 1684}, \quad 0 \le x \le 30.$$

Differentiating W with respect to x yields

$$\frac{dW}{dx} = \frac{x}{\sqrt{x^2 + 144}} + \frac{x - 30}{\sqrt{x^2 - 60x + 1684}}.$$

By letting $dW/dx = 0$, you obtain

$$\frac{x}{\sqrt{x^2 + 144}} + \frac{x - 30}{\sqrt{x^2 - 60x + 1684}} = 0$$
$$x\sqrt{x^2 - 60x + 1684} = (30 - x)\sqrt{x^2 + 144}$$
$$x^2(x^2 - 60x + 1684) = (30 - x)^2(x^2 + 144)$$
$$x^4 - 60x^3 + 1684x^2 = x^4 - 60x^3 + 1044x^2 - 8640x + 129{,}600$$
$$640x^2 + 8640x - 129{,}600 = 0$$
$$320(x - 9)(2x + 45) = 0$$
$$x = 9, -22.5.$$

Because $x = -22.5$ is not in the interval $[0, 30]$ and

$$W(0) \approx 53.04, \quad W(9) = 50, \quad \text{and} \quad W(30) \approx 60.31$$

you can conclude that the wire should be staked at 9 feet from the 12-foot pole.

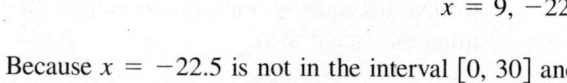

TECHNOLOGY   From Example 4, you can see that applied optimization problems can involve a lot of algebra. If you have access to a graphing utility, you can confirm that $x = 9$ yields a minimum value of W by sketching the graph of

$$W = \sqrt{x^2 + 144} + \sqrt{x^2 - 60x + 1684}$$

as shown in Figure 3.57.

FIGURE 3.56
The quantity to be minimized is length. From the diagram, you can see that x varies between 0 and 30.

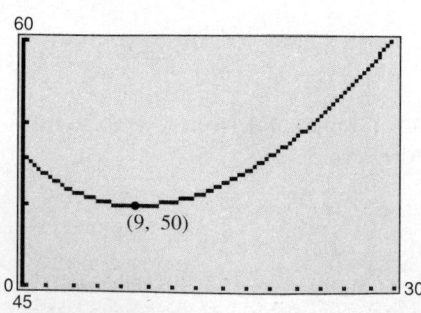

FIGURE 3.57
You can confirm the minimum value of W with a graphing utility.

In each of the first four examples, the extreme value occurs at a critical number. Although this happens often, remember that an extreme value can occur at an endpoint of an interval.

EXAMPLE 5   An Endpoint Maximum

Four feet of wire is to be used to form a square and a circle. How much of the wire should be used for the square and how much should be used for the circle to enclose the maximum total area?

Solution   From Figure 3.58, the total area is given by

$$A = (\text{Area of square}) + (\text{Area of circle})$$
$$A = x^2 + \pi r^2. \qquad \text{Primary equation}$$

Because the total amount of wire is 4 feet, you obtain

$$4 = (\text{Perimeter of square}) + (\text{Circumference of circle})$$
$$4 = 4x + 2\pi r.$$

Thus, $r = 2(1 - x)/\pi$, and by substituting in the primary equation you have

$$A = x^2 + \pi \left[\frac{2(1 - x)}{\pi} \right]^2$$

$$= x^2 + \frac{4(1 - x)^2}{\pi}$$

$$= \frac{1}{\pi}[(\pi + 4)x^2 - 8x + 4].$$

The feasible domain is $0 \le x \le 1$. Because $dA/dx = [2(\pi + 4)x - 8]/\pi$, the only critical number is $x = 4/(\pi + 4) \approx 0.56$. Therefore, using

$$A(0) \approx 1.273, \quad A(0.56) \approx 0.56, \quad \text{and} \quad A(1) = 1$$

you can conclude that the maximum area occurs when $x = 0$. That is, *all* the wire is used for the circle.

x

x Area: x^2

Perimeter: $4x$

?

4 feet

r

Area: πr^2

Circumference: $2\pi r$

$$4x + 2\pi r = 4$$

FIGURE 3.58
The quantity to be maximized is area: $A = x^2 + \pi r^2$.

Let's review the primary equations developed in the first five examples. As applications go, these five examples are fairly simple, and yet the resulting primary equations are quite complicated.

$$V = 27x - \frac{x^3}{4} \qquad\qquad W = \sqrt{x^2 + 144} + \sqrt{x^2 - 60x + 1684}$$

$$d = \sqrt{x^4 - 3x^2 + 4} \qquad A = \frac{1}{\pi}[(\pi + 4)x^2 - 8x + 4]$$

$$A = 30 + 2x + \frac{72}{x}$$

You must expect that real-life applications often involve equations that are at least as complicated as these five. Remember that one of the main goals of this course is to teach you to use calculus to analyze equations that initially seem formidable.

For our last example, we look at a problem in which the primary equation is familiar enough that we do not have to use calculus to discover the extrema.

EXAMPLE 6   Finding the Minimum and Maximum Rate of Growth

Because their growth is affected by sunlight, plants do not grow at constant rates during a normal 24-hour period. Suppose the growth of a plant in a controlled environment is given by the model $h = 0.2t + 0.03 \sin 2\pi t$, where h is the height of the plant in inches and t is the time in days with $t = 0$ corresponding to midnight.

a. During what time of day is the *rate of growth* a maximum?
b. During what time of day is the *rate of growth* a minimum?

Solution   Figure 3.59 shows the growth pattern for this plant. Because you are interested in the *rate* of growth, the primary equation is given by dh/dt.

$$h = 0.2t + 0.03 \sin 2\pi t$$

$$\frac{dh}{dt} = 0.2 + (2\pi)(0.03)\cos 2\pi t \qquad \text{Primary equation}$$

In this case the primary equation is a familiar one, and without taking its derivative we know that $\cos 2\pi t$ has a maximum of 1 and a minimum of -1.

a. If $t = 0$, the rate is a maximum because $\cos[2\pi(0)] = 1$.
b. If $t = \frac{1}{2}$, the rate is a minimum because $\cos[2\pi(\frac{1}{2})] = -1$.

Thus, the maximum rate of growth occurs at midnight and the minimum rate at noon.

Plant height (in inches) vs **Time (in days)**

$h = 0.2t + 0.03 \sin 2\pi t$

$t = \dfrac{1}{2}$

FIGURE 3.59

 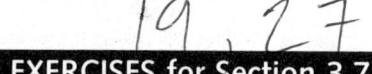

EXERCISES for Section 3.7

TECHNOLOGY
Laboratory Guide
Lab 3.7

C 1. Find two positive numbers whose sum is 110 and whose product is a maximum.
   a. Use paper and pencil to complete six rows of a table like the one below. (The first two rows are shown.) Use the result to guess the two numbers required.

First number, x	Second number	Product, P
10	$110 - 10$	$10(110 - 10) = 1000$
20	$110 - 20$	$20(110 - 20) = 1800$

   b. Write the product P as a function of x.
   c. Use calculus to find the critical number of the function of part b. Then, find the two numbers.
   d. Use a computer to generate additional rows of the table. Use the table to estimate the solution.
   e. Use a graphing utility to graph the function in part b and estimate the solution from the graph.

In Exercises 2–6, find two positive numbers that satisfy the given requirements.

2. The sum is S and the product is maximum.

3. The product is 192 and the sum is minimum.

4. The product is 192 and the sum of the first plus three times the second is minimum.

5. The second number is the reciprocal of the first and their sum is minimum.

6. The sum of the first and twice the second is 100 and the product is maximum.

In Exercises 7 and 8, find the length and width of a rectangle that has the given perimeter and has a maximum area.

7. Perimeter: 100 feet   8. Perimeter: P units

In Exercises 9 and 10, find the length and width of a rectangle that has the given area and has a minimum perimeter.

9. Area: 64 square feet   10. Area: A square feet

In Exercises 11 and 12, find the point on the graph of the function closest to the given point.

11. $f(x) = \sqrt{x}$, $(4, 0)$

12. $f(x) = x^2$, $(2, \frac{1}{2})$

13. *Chemical Reaction*   In an autocatalytic chemical reaction, the product formed is a catalyst for the reaction. If Q_0 is the amount of the original substance and x is the amount of catalyst formed, then the rate of chemical reaction is

$$\frac{dQ}{dx} = kx(Q_0 - x).$$

For what value of x is the rate of chemical reaction greatest?

14. *Traffic Control* The flow rate F (cars/hour) on a congested roadway is $F = v/(22 + 0.02v^2)$, where v is the speed of the traffic in miles per hour. What speed will maximize the flow rate on the road?

15. *Area* A dairy farmer plans to fence in a rectangular pasture adjacent to a river. The pasture must contain 180,000 square meters in order to provide enough grass for the herd. What dimensions would require the least amount of fencing if no fencing is needed along the river?

16. *Area* A rancher has 200 feet of fencing with which to enclose two adjacent rectangular corrals (see figure). What dimensions should be used so that the enclosed area will be a maximum?

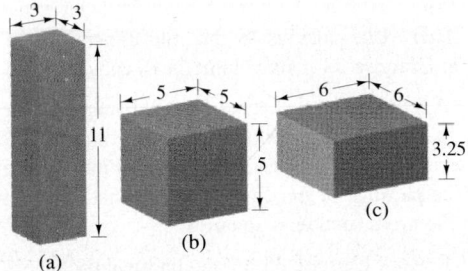

FIGURE FOR 16

17. *Volume* Verify that each of the rectangular solids shown in the figure has a surface area of 150 square inches. Find the volume of each.

FIGURE FOR 17

18. *Volume* Verify that each of the right circular cylinders shown in the figure has a surface area of 24π square inches. Find the volume of each.

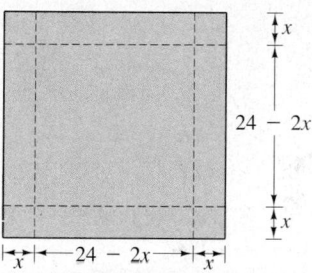

FIGURE FOR 18

C **19.** *Volume* An open box of maximum volume is to be made from a square piece of material, 24 inches on a side, by cutting equal squares from each corner and turning up the sides (see figure).

a. Use paper and pencil to complete six rows of a table like the one below. (The first two rows are shown.) Use the result to guess the maximum volume.

Height, x	Length and width	Volume
1	$24 - 2(1)$	$1[24 - 2(1)]^2 = 484$
2	$24 - 2(2)$	$2[24 - 2(2)]^2 = 800$

b. Write the volume V as a function of x.

c. Use calculus to find the critical number of the function in part b and find the maximum volume.

d. Use a graphing utility to graph the function of part b and estimate the maximum volume from the graph.

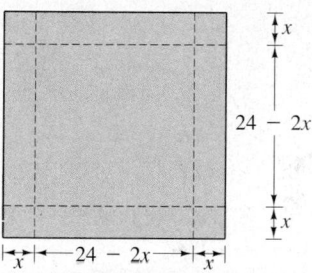

FIGURE FOR 19

20. a. Solve Exercise 19 given that the square piece of material is s inches on a side.
 b. If the dimensions of the square piece of material are doubled, how does the volume change?

21. *Volume* An open box is to be made from a rectangular piece of material (2 ft by 3 ft) by cutting equal squares from each corner and turning up the sides. Find the dimensions of the box of maximum volume.

22. *Surface Area* A net enclosure for practicing golf is open at one end (see figure). Find the dimensions that require the least amount of netting if the volume of the enclosure is to be $83\frac{1}{3}$ cubic meters.

23. *Area* A Norman window is constructed by adjoining a semicircle to the top of an ordinary rectangular window (see figure). Find the dimensions of a Norman window of maximum area if the total perimeter is 16 feet.

FIGURE FOR 22   **FIGURE FOR 23**

24. *Area* A physical fitness room consists of a rectangular region with a semicircle on each end. If the perimeter of the room is to be a 200-meter running track, find the dimensions that will make the area of the rectangular region as large as possible.

25. *Area*   A rectangle is bounded by the *x*- and *y*-axes and the graph of $y = (6 - x)/2$ (see figure). What length and width should the rectangle have so that its area is a maximum?

26. *Length*   A right triangle is formed in the first quadrant by the *x*- and *y*-axes and a line through the point $(1, 2)$ (see figure).
  a. Write the length L of the hypotenuse as a function of *x*.
  b. Use a graphing utility to graph the function in part a and use the graph to approximate *x* so that the length of the hypotenuse is minimum.

FIGURE FOR 25   **FIGURE FOR 26**

27. *Area*   A right triangle is formed in the first quadrant by the *x*- and *y*-axes and a line through the point $(2, 3)$. Find the vertices of the triangle so that its area is minimum.

28. *Area*   Find the dimensions of the largest isosceles triangle that can be inscribed in a circle of radius 4 (see figure).

29. *Area*   A rectangle is bounded by the *x*-axis and the semicircle $y = \sqrt{25 - x^2}$ (see figure). What length and width should the rectangle have so that its area is a maximum?

FIGURE FOR 28   **FIGURE FOR 29**

30. *Area*   Find the dimensions of the largest rectangle that can be inscribed in a semicircle of radius *r* (see Exercise 29).

31. *Area*   Find the dimensions of the trapezoid of greatest area that can be inscribed in a semicircle of radius *r*.

32. *Area*   A page is to contain 30 square inches of print. The margins on each side are 1 inch. Find the dimensions of the page so that the least paper is used.

C **33.** *Surface Area*   A right circular cylinder is to be designed to hold 22 cubic inches of a soft drink (approximately 12 fluid ounces) and to use a minimum of material in its construction (see figure).
  a. Use paper and pencil to complete four rows of the table in the next column. (The first two rows are shown.) Use the result to guess the minimum amount of material.

Radius, *r*	Height	Surface area	
0.2	$\dfrac{22}{\pi(0.2)^2}$	$2\pi(0.2)\left[0.2 + \dfrac{22}{\pi(0.2)^2}\right]$	≈ 220.3
0.4	$\dfrac{22}{\pi(0.4)^2}$	$2\pi(0.4)\left[0.4 + \dfrac{22}{\pi(0.4)^2}\right]$	≈ 111.1

  b. Write the surface area S as a function of *r*.
  c. Use calculus to find the critical number of the function in part b and find the dimensions that will yield the minimum surface area.
  d. Use a computer to generate additional rows of the table in part a and estimate the minimum surface area from the table.
  e. Use a graphing utility to obtain the graph of the function in part b and estimate the required dimensions from the graph.

FIGURE FOR 33

34. *Surface Area*   Use calculus to find the dimensions of the cylinder in Exercise 33 if its volume is V_0 cubic units.

35. *Volume*   A rectangular package to be sent by a postal service can have maximum combined length and girth (perimeter of a cross section) of 108 inches (see figure). Find the dimensions of the package of maximum volume that can be sent. (Assume the cross section is square.)

36. *Volume*   Rework Exercise 35 for a cylindrical package. (The cross sections are circular.)

37. *Volume*   Find the volume of the largest right circular cone that can be inscribed in a sphere of radius *r* (see figure).

FIGURE FOR 35   **FIGURE FOR 37**

38. *Volume*   Find the volume of the largest right circular cylinder that can be inscribed in a sphere of radius *r*.

39. *Surface Area*   A solid is formed by adjoining a hemisphere to each end of a right circular cylinder. The total volume of the figure is 12 cubic inches. Find the radius of the cylinder that produces the minimum surface area.

40. *Cost* An industrial tank of the shape described in Exercise 39 must have a volume of 3000 cubic feet. If the construction cost of the hemispherical ends is twice as much per square foot of surface area as the sides, find the dimensions that will minimize cost.

41. *Area* The combined perimeter of an equilateral triangle and a square is 10. Find the dimensions of the triangle and square that produce a minimum total area.

42. *Area* The combined perimeter of a circle and a square is 16. Find the dimensions of the circle and square that produce a minimum total area.

43. *Area* Ten feet of wire is to be used to form an isosceles right triangle and a circle. Determine how much of the wire should be used for the circle if the total area enclosed is to be
a. Minimum.
b. Maximum.

44. *Area* Twenty feet of wire is to be used to form two figures. In each of the following cases, how much should be used for each figure so that the total enclosed area is a maximum?
a. Equilateral triangle and square.
b. Square and regular pentagon.
c. Regular pentagon and hexagon.
d. Regular hexagon and circle.
What can you conclude from this pattern? (*Hint:* The area of a regular polygon with n sides of length x is $A = (n/4)(\cot \pi/n)x^2$.)

45. *Beam Strength* A wooden beam has a rectangular cross section of height h and width w (see figure). The strength S of the beam is directly proportional to the width and the square of the height. What are the dimensions of the strongest beam that can be cut from a round log of diameter 24 inches? (*Hint:* $S = kh^2w$, where k is the proportionality constant.)

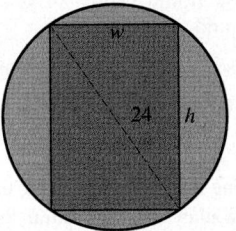

FIGURE FOR 45

46. *Minimum Length* Two factories are located at the coordinates $(-x, 0)$ and $(x, 0)$ with their power supply located at the point $(0, h)$ (see figure). Find y so that the total amount of power line from the power supply to the factories is minimum.

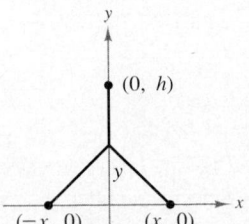

FIGURE FOR 46

47. *Illumination* A light source is located over the center of a circular table of diameter 4 feet (see figure). Find the height h of the light source so that the illumination I at the circumference of the table is maximum if $I = k \sin \alpha/s^2$, where s is the slant height, α is the angle at which the light strikes the table, and k is a constant.

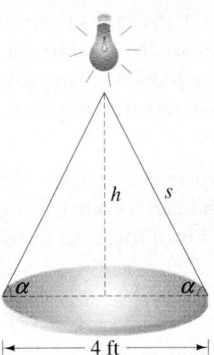

FIGURE FOR 47

48. *Illumination* The illumination from a light source is directly proportional to the strength of the source and inversely proportional to the square of the distance from the source. Two light sources of intensities I_1 and I_2 are d units apart. At what point on the line segment joining the two sources is the illumination least?

49. *Minimum Time* A man is in a boat 2 miles from the nearest point on the coast. He is to go to a point Q, 3 miles down the coast and 1 mile inland (see figure). If he can row at 2 miles per hour and walk at 4 miles per hour, toward what point on the coast should he row in order to reach point Q in the least time?

50. *Minimum Time* The conditions are the same as in Exercise 49 except that the man can row at 4 miles per hour. How does this change the solution?

51. *Minimum Time* The conditions are the same as in Exercise 49 except that the man can row at v_1 miles per hour and walk at v_2 miles per hour. If θ_1 and θ_2 are the magnitudes of the angles shown in the figure, show that the man will reach point Q in the least time when
$$\frac{\sin \theta_1}{v_1} = \frac{\sin \theta_2}{v_2}.$$

FIGURE FOR 49–51

52. *Minimum Time*   When light waves, traveling in a transparent medium, strike the surface of a second transparent medium, they begin to travel in a different direction. This change of direction is called refraction and is defined by **Snell's Law of Refraction,**

$$\frac{\sin\theta_1}{v_1} = \frac{\sin\theta_2}{v_2}$$

where θ_1 and θ_2 are the magnitudes of the angles shown in the accompanying figure and v_1 and v_2 are the velocities of light in the two mediums. Show that this problem is equivalent to that of Exercise 51, and that light waves traveling from P to Q follow the path of minimum time.

53. *Minimum Force*   A component is designed to slide a block of steel of weight W across a table and into a chute (see figure). The motion of the block is resisted by a frictional force proportional to its apparent weight. (Let k be the constant of proportionality.) Find the minimum force F needed to slide the block, and find the corresponding value of θ. (*Hint:* $F\cos\theta$ is the force in the direction of motion, and $F\sin\theta$ is the amount of force tending to lift the block. Therefore, the apparent weight of the block is $W - F\sin\theta$.)

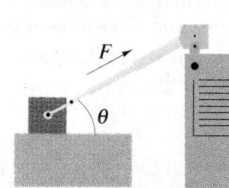

FIGURE FOR 52   **FIGURE FOR 53**

54. *Projectile Range*   The range R of a projectile fired with an initial velocity v_0 at an angle θ (with the horizontal) is $R = v_0^2 \sin 2\theta/g$, where g is the acceleration due to gravity. Find the angle θ so that the range is maximum.

55. *Volume*   A sector with central angle θ is cut from a circle of radius 12 inches, and the resulting edges are brought together to form a cone (see figure in right column). Find the magnitude of θ so that the volume of the cone is maximum.

C **56.** *Area*   The cross sections of an irrigation canal are isosceles trapezoids of which three of the sides are 8 feet long (see figure). Determine the angle of elevation θ of the sides so that the area of the cross section is maximum by completing the following.

a. Use paper and pencil to complete six rows of a table like the one below. (The first two rows are shown.) Use the result to guess the maximum cross-sectional area.

Base 1	Base 2	Altitude	Area
8	$8 + 16\cos 10°$	$8\sin 10°$	≈ 22.1
8	$8 + 16\cos 20°$	$8\sin 20°$	≈ 42.5

b. Use the table to write the area A as a function of θ.
c. Use calculus to find the critical number of the function in part b and find the angle that will yield the maximum cross-sectional area.
d. Use a computer to generate additional rows of the table in part a and estimate the maximum cross-sectional area.
e. Use a graphing utility to graph the function in part b and estimate the maximum cross-sectional area.

FIGURE FOR 55   **FIGURE FOR 56**

C In Exercises 57 and 58, use a graphing utility to graph the primary equation and its first derivative in the applied extrema problem. Use the graphs to answer the questions.

57. *Area*   Approximate the dimensions of the rectangle of maximum area that can be inscribed in a semicircle of radius 10.

58. Approximate the length of the longest pipe that can be carried level around a right-angle corner if the widths of the two intersecting corridors are 5 feet and 8 feet.

59. *Friction*   The efficiency E of a screw with square threads is

$$E = \frac{\tan\phi(1 - \mu\tan\phi)}{\mu + \tan\phi}$$

where μ is the coefficient of sliding friction and ϕ is the angle of inclination of the threads to a plane perpendicular to the axis of the screw. Find the angle ϕ that yields maximum efficiency when $\mu = 0.1$.

Newton's Method

Newton's Method • Algebraic Solutions of Polynomial Equations

Newton's Method

In the first seven sections of this chapter, you frequently needed to find the zeros of a function. Up to now we chose most of the examples and exercises so that they involved functions whose zeros could be found by algebraic techniques. For instance, the zeros of $f(x) = x^3 - 2x^2 - x + 2$ can be found easily by factoring. In applied problems, however, you frequently encounter functions whose zeros are more difficult to find. For instance, the zeros of a function as simple as

$$f(x) = x^3 - x + 1$$

cannot be found by elementary algebraic methods. (More is said about this at the end of this section.)

To introduce Newton's Method for approximating the real zeros of a function, consider a function f that is continuous on the interval $[a, b]$ and differentiable on the interval (a, b). If $f(a)$ and $f(b)$ differ in sign, then by the Intermediate Value Theorem f must possess at least one zero in the interval (a, b). Suppose you estimate this zero to occur at $x = x_1$, as shown in Figure 3.60. Newton's Method is based on the assumption that the graph of f and the tangent line at $(x_1, f(x_1))$ both cross the x-axis at *about* the same point. Because you can easily calculate the x-intercept for this tangent line, you can use it as a second (and, usually, better) estimate for the zero of f. The tangent line passes through the point $(x_1, f(x_1))$ with a slope of $f'(x_1)$. In point-slope form, the equation of the tangent line is therefore

$$y - f(x_1) = f'(x_1)(x - x_1)$$
$$y = f'(x_1)(x - x_1) + f(x_1).$$

Letting $y = 0$ and solving for x produces

$$x = x_1 - \frac{f(x_1)}{f'(x_1)}.$$

Thus, from the initial estimate, x_1, you obtain a new estimate

$$x_2 = x_1 - \frac{f(x_1)}{f'(x_1)}.$$

You can improve upon x_2 and calculate yet a third estimate

$$x_3 = x_2 - \frac{f(x_2)}{f'(x_2)}.$$

Repeated application of this process is called **Newton's Method.**

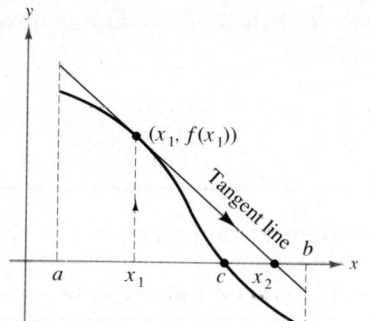

FIGURE 3.60

The x-intercept of the tangent line approximates the zero of f.

Newton's Method for Approximating the Zeros of a Function

Let $f(c) = 0$, where f is differentiable on an open interval containing c. Then, to approximate c, use the following steps.

1. Make an initial estimate x_1 that is "close" to c. (A graph is helpful.)
2. Determine a new approximation

$$x_{n+1} = x_n - \frac{f(x_n)}{f'(x_n)}.$$

3. If $|x_n - x_{n+1}|$ is less than the desired accuracy, let x_{n+1} serve as the final approximation. Otherwise, return to Step 2 and calculate a new approximation.

Each successive application of this procedure is called an **iteration.**

For many functions, just a few iterations of Newton's Method will produce approximations having very small errors, as shown in Example 1.

EXAMPLE 1   Using Newton's Method

Calculate three iterations of Newton's Method to approximate a zero of $f(x) = x^2 - 2$. Use $x_1 = 1$ as the initial guess.

Solution   Because $f(x) = x^2 - 2$, you have $f'(x) = 2x$ and the iterative process is given by the formula

$$x_{n+1} = x_n - \frac{f(x_n)}{f'(x_n)}$$

$$= x_n - \frac{x_n^2 - 2}{2x_n}.$$

The calculations for three iterations are shown in Table 3.18.

TABLE 3.18

n	x_n	$f(x_n)$	$f'(x_n)$	$\dfrac{f(x_n)}{f'(x_n)}$	$x_n - \dfrac{f(x_n)}{f'(x_n)}$
1	1.000000	-1.000000	2.000000	-0.500000	1.500000
2	1.500000	0.250000	3.000000	0.083333	1.416667
3	1.416667	0.006945	2.833334	0.002451	1.414216
4	1.414216				

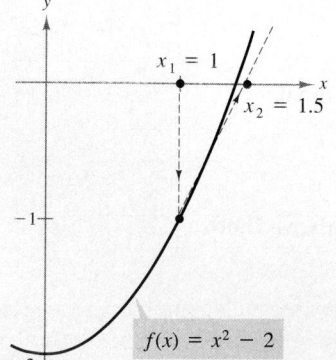

FIGURE 3.61

The first iteration of Newton's Method.

Of course, in this case you know that the two zeros of the function are $\pm\sqrt{2}$. To six decimal places, $\sqrt{2} = 1.414214$. Thus, after only three iterations of Newton's Method, you obtained an approximation that is within 0.000002 of an actual root. The first iteration of this process is shown in Figure 3.61.

EXAMPLE 2   Using Newton's Method

Use Newton's Method to approximate the zeros of $f(x) = 2x^3 + x^2 - x + 1$. Continue the iterations until two successive approximations differ by less than 0.0001.

Solution   Begin by sketching a graph of f, as shown in Figure 3.62. From the graph, you can observe that the function has only one real zero, which occurs near $x = -1.2$. Next, differentiate f and form the iterative formula

$$x_{n+1} = x_n - \frac{f(x_n)}{f'(x_n)} = x_n - \frac{2x_n^3 + x_n^2 - x_n + 1}{6x_n^2 + 2x_n - 1}.$$

The calculations are shown in Table 3.19.

TABLE 3.19

n	x_n	$f(x_n)$	$f'(x_n)$	$\dfrac{f(x_n)}{f'(x_n)}$	$x_n - \dfrac{f(x_n)}{f'(x_n)}$
1	-1.20000	0.18400	5.24000	0.03511	-1.23511
2	-1.23511	-0.00771	5.68276	-0.00136	-1.23375
3	-1.23375	0.00001	5.66533	0.00000	-1.23375
4	-1.23375				

Because two successive approximations differ by less than the required 0.0001, you can estimate the zero of f to be -1.23375.

$f(x) = 2x^3 + x^2 - x + 1$

FIGURE 3.62
After 3 iterations of Newton's Method the zero of f is approximated to the desired accuracy.

TECHNOLOGY   Newton's Method is especially useful in conjunction with a computer or programmable calculator. The following program is for a *Texas Instruments TI-81* graphing calculator. Programs for other computers and calculators would involve similar steps. (To run the program, enter the function as Y_1 and enter the initial guess as X.)

```
Prgm:NEWTON
:(Xmax-Xmin)/100→D           :I+1→I
:1→I                         :Goto 1
:Lbl 1                       :Lbl 2
:X-Y₁/NDeriv(Y₁,D)→R         :Disp "ROOT="
:If abs(X-R)≤abs(X/1E10)     :Disp R
:Goto 2                      :Disp "ITER="
:R→X                         :Disp I
```

When, as in Examples 1 and 2, the approximations approach a limit, the sequence $x_1, x_2, x_3, \ldots, x_n, \ldots$ is said to **converge.** Moreover, if the limit is c, then it can be shown that c must be a zero of f.

Newton's Method does not always yield a convergent sequence. One way this can happen is shown in Figure 3.63. Because Newton's Method involves division by $f'(x_n)$, it is clear that the method will fail if the derivative is zero for any x_n in the sequence. When you encounter this problem, you can usually overcome it by choosing a different value for x_1.

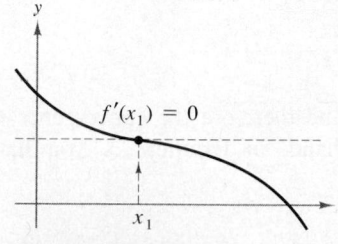

$f'(x_1) = 0$

x_1

FIGURE 3.63
Newton's Method fails to converge if $f'(x_n) = 0$.

Another way Newton's Method can fail is illustrated in the next example.

EXAMPLE 3   An Example in Which Newton's Method Fails

Using $x_1 = 0.1$, show that Newton's Method fails to converge for the function $f(x) = x^{1/3}$.

Solution   Because $f'(x) = \frac{1}{3}x^{-2/3}$, the iterative formula is

$$x_{n+1} = x_n - \frac{f(x_n)}{f'(x_n)} = x_n - \frac{x_n^{1/3}}{\frac{1}{3}x_n^{-2/3}} = x_n - 3x_n = -2x_n.$$

The calculations are given in Table 3.20. This table and Figure 3.64 indicate that x_n increases in magnitude as $n \to \infty$, and thus the limit of the sequences does not exist.

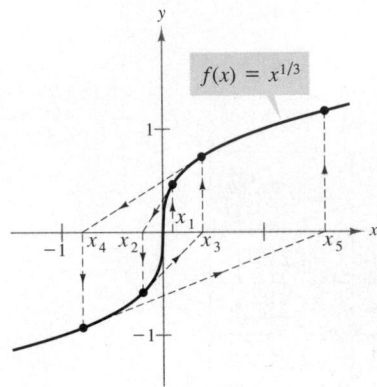

FIGURE 3.64
Newton's Method fails to converge for every x-value other than the actual zero of f.

TABLE 3.20

n	x_n	$f(x_n)$	$f'(x_n)$	$\dfrac{f(x_n)}{f'(x_n)}$	$x_n - \dfrac{f(x_n)}{f'(x_n)}$
1	0.10000	0.46416	1.54720	0.30000	−0.20000
2	−0.20000	−0.58480	0.97467	−0.60000	0.40000
3	0.40000	0.73681	0.61401	1.20000	−0.80000
4	−0.80000	−0.92832	0.38680	−2.40000	1.60000

REMARK   In Example 3, the initial estimate $x_1 = 0.1$ fails to produce a convergent sequence. Try showing that Newton's Method also fails for every other choice of x_1 (other than the actual zero).

It can be shown that a condition sufficient to produce convergence of Newton's Method to a zero of f is that

$$\left| \frac{f(x)f''(x)}{[f'(x)]^2} \right| < 1$$

on an open interval containing the zero. For instance, in Example 1 this test would yield $f(x) = x^2 - 2, f'(x) = 2x, f''(x) = 2$, and

$$\left| \frac{f(x)f''(x)}{[f'(x)]^2} \right| = \left| \frac{(x^2 - 2)(2)}{4x^2} \right| = \left| \frac{1}{2} - \frac{1}{x^2} \right|. \qquad \text{Example 1}$$

On the interval $(1, 3)$ this quantity is less than 1 and therefore the convergence of Newton's Method is guaranteed. On the other hand, in Example 3, you have $f(x) = x^{1/3}, f'(x) = \frac{1}{3}x^{-2/3}, f''(x) = -\frac{2}{9}x^{-5/3}$, and

$$\left| \frac{f(x)f''(x)}{[f'(x)]^2} \right| = \left| \frac{x^{1/3}(-2/9)(x^{-5/3})}{(1/9)(x^{-4/3})} \right| = 2 \qquad \text{Example 3}$$

which is not less than 1 for any value of x.

FIGURE 3.65

NIELS HENRIK ABEL

EVARISTE GALOIS

Although the lives of both Abel and Galois were brief, their work in the fields of analysis and abstract algebra was far-reaching.

EXAMPLE 4   Approximating a Point of Intersection

Approximate the point of intersection of the graphs of $y = \tan x$ and $y = 2x$ between 0 and $\pi/2$, as shown in Figure 3.65. Use Newton's Method and continue the iterations until two successive approximations differ by less than 0.0001.

Solution   Let $\tan x = 2x$. You can approximate the solutions of this equation by approximating the zeros of the function

$$f(x) = 2x - \tan x.$$

The iterative formula for Newton's Method takes the form

$$x_{n+1} = x_n - \frac{f(x_n)}{f'(x_n)} = x_n - \frac{2x_n - \tan x_n}{2 - \sec^2 x_n}.$$

From the calculations shown in Table 3.21, you can conclude that the point of intersection occurs when $x \approx 1.16556$.

TABLE 3.21

n	x_n	$f(x_n)$	$f'(x_n)$	$\dfrac{f(x_n)}{f'(x_n)}$	$x_n - \dfrac{f(x_n)}{f'(x_n)}$
1	1.25000	−0.50957	−8.05751	0.06324	1.18676
2	1.18676	−0.10110	−5.12374	0.01973	1.16703
3	1.16703	−0.00653	−4.47831	0.00146	1.16557
4	1.16557	−0.00003	−4.43436	0.00001	1.16556

Algebraic Solutions of Polynomial Equations

At the beginning of this section we mentioned that the zeros of the cubic function $f(x) = x^3 - x + 1$ cannot be found by *elementary* algebraic methods. This particular function has only one real zero, and by using more advanced algebraic techniques you can determine the zero to be

$$x = -\sqrt[3]{\frac{3 - \sqrt{23/3}}{6}} - \sqrt[3]{\frac{3 + \sqrt{23/3}}{6}}.$$

Because the *exact* solution is written in terms of square roots and cube roots, it is called a **solution by radicals.**

The determination of radical solutions of a polynomial equation is one of the fundamental problems of algebra. The earliest such result is the Quadratic Formula, which dates back at least to Babylonian times. The general formula for the zeros of a cubic function was developed much later. In the sixteenth century an Italian mathematician, Jerome Cardan, published a method for finding radical solutions to cubic and quartic equations. Then, for 300 years the problem of finding a general quintic formula remained open. Finally, in the nineteenth century, the problem was answered independently by two young mathematicians. Niels Henrik Abel (1802–1829), a Norwegian mathematician, and Evariste Galois (1811–1832), a French mathematician, proved that it is not possible to solve a *general* fifth (or higher) degree polynomial equation by radicals. Of course, you can solve particular fifth-degree equations such as $x^5 - 1 = 0$, but Abel and Galois were able to show that no general *radical* solution exists.

1 (by hand) 5-17, 21-24, 25, 27, 33, 34

EXERCISES for Section 3.8

In Exercises 1–4, complete two iterations of Newton's Method for the given function using the indicated initial guess.

1. $f(x) = x^2 - 3$
$x_1 = 1.7$

2. $f(x) = 3x^2 - 2$
$x_1 = 1$

3. $f(x) = \sin x$
$x_1 = 3$

4. $f(x) = \tan x$
$x_1 = 0.1$

In Exercises 5–12, approximate the zero(s) of the function. Use Newton's Method and continue the process until two successive approximations differ by less than 0.001.

5. $f(x) = x^3 + x - 1$

6. $f(x) = x^5 + x - 1$

7. $f(x) = 3\sqrt{x - 1} - x$

8. $f(x) = x^3 - 3.9x^2 + 4.79x - 1.881$

9. $f(x) = x^4 - 10x^2 - 11$

10. $f(x) = x^3 + 3$

11. $f(x) = x + \sin(x + 1)$

12. $f(x) = x^3 - \cos x$

In Exercises 13–16, apply Newton's Method to approximate the x-value of the indicated point(s) of intersection of the two graphs. Continue the process until two successive approximations differ by less than 0.001.

13. $f(x) = 2x + 1;$
$g(x) = \sqrt{x + 4}$

14. $f(x) = 3 - x;$
$g(x) = \dfrac{1}{x^2 + 1}$

15. $f(x) = x;$
$g(x) = \tan x$

16. $f(x) = x^2;$
$g(x) = \cos x$

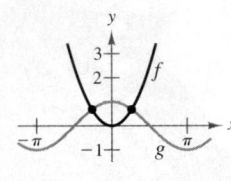

Fixed Point **In Exercises 17 and 18, approximate to two decimal places the *fixed point* of the function. [A fixed point x_0 of a function f is a value of x such that $f(x_0) = x_0$.]**

17. $f(x) = \cos x$

18. $f(x) = \cot x$
$0 < x < \pi$

19. *Essay* Consider the function $f(x) = x^3 - 3x^2 + 3$.
   a. Use a graphing utility to graph f.
   b. Use Newton's Method with $x_1 = 1$ as an initial guess.
   c. Repeat part b using $x_1 = \frac{1}{4}$ as an initial guess and observe that the result is different.
   d. To understand why the results of parts b and c are different, sketch the tangent lines to the graph of f at the points $(1, f(1))$ and $(\frac{1}{4}, f(\frac{1}{4}))$. Find the x-intercept of each tangent line and compare the intercepts with the first iteration of Newton's Method with the respective initial guesses.
   e. Write a short paragraph summarizing how Newton's Method works. Use the results of this exercise to describe why it is important to select the initial guess carefully.

20. Repeat the steps of Exercise 19 for the function $f(x) = \sin x$ with initial guesses of $x_1 = 1.8$ and $x_1 = 3$.

In Exercises 21–24, apply Newton's Method, using the indicated initial guess, and explain why the method fails.

21. $y = 2x^3 - 6x^2 + 6x - 1$
$x_1 = 1$

22. $y = 4x^3 - 12x^2 + 12x - 3$
$x_1 = \dfrac{3}{2}$

FIGURE FOR 21

FIGURE FOR 22

23. $y = -x^3 + 3x^2 - x + 1$
$x_1 = 1$

24. $f(x) = 2\sin x + \cos 2x$
$x_1 = \dfrac{3\pi}{2}$

FIGURE FOR 23

FIGURE FOR 24

In Exercises 25 and 26, use Newton's Method to obtain a general rule for approximating the required radical.

25. $\sqrt{a}$ (*Hint:* Consider $f(x) = x^2 - a$.)

26. $\sqrt[n]{a}$ (*Hint:* Consider $f(x) = x^n - a$.)

In Exercises 27–30, use the results of Exercises 25 and 26 to approximate the indicated radical to three decimal places.

27. $\sqrt{7}$ **28.** $\sqrt{5}$

29. $\sqrt[4]{6}$ **30.** $\sqrt[3]{15}$

31. Use Newton's Method to show that the equation

$$x_{n+1} = x_n(2 - ax_n)$$

can be used to approximate $1/a$ if x_1 is an initial guess of the reciprocal of a. Note that this method of approximating reciprocals uses only the operations of multiplication and subtraction. (*Hint:* Consider $f(x) = (1/x) - a$.)

32. Use the result of Exercise 31 to approximate the indicated reciprocal to three decimal places.

a. $\dfrac{1}{3}$ **b.** $\dfrac{1}{11}$

In Exercises 33 and 34, approximate π to three decimal places using Newton's Method and the given function.

33. $f(x) = 1 + \cos x$ **34.** $f(x) = \tan x$

In Exercises 35 and 36, approximate the critical number of f on the indicated interval accurate to three decimal places. Sketch the graph of f, labeling any extrema.

35. $f(x) = x\cos x$
$0 \le x \le \pi$

36. $f(x) = x\sin x$
$0 \le x \le \pi$

In Exercises 37–40, we review some typical problems from the previous sections of this chapter. In each case, use Newton's Method to approximate the solution.

37. *Minimum Distance* Find the point on the graph of $f(x) = 4 - x^2$ that is closest to the point $(1, 0)$.

38. *Minimum Distance* Find the point on the graph of $f(x) = x^2$ that is closest to the point $(4, -3)$.

39. *Minimum Time* A woman is in a boat 2 miles from the nearest point on the coast (see figure). She is to go to a point Q, which is 3 miles down the coast and 1 mile inland. She can row at 3 miles per hour and walk at 4 miles per hour. Toward what point on the coast should she row in order to reach Q in the least time?

FIGURE FOR 39

40. *Medicine* The concentration C of a certain chemical in the bloodstream t hours after injection into muscle tissue is given by

$$C = \frac{3t^2 + t}{50 + t^3}.$$

When is the concentration greatest?

41. *Advertising Costs* A company that produces portable cassette players estimates that the profit for selling a particular model is given by

$$P = -76x^3 + 4830x^2 - 320{,}000, \qquad 0 \le x \le 60$$

where P is the profit in dollars and x is the advertising expense in 10,000s of dollars (see figure at top of next page). According to this model, find the smaller of two advertising amounts that yield a profit of $2,500,000.

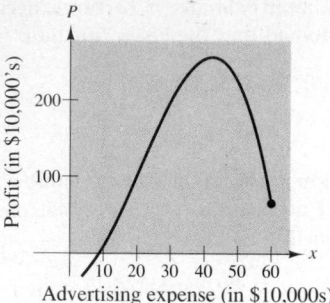

FIGURE FOR 41

42. Engine Power The torque produced by a compact automobile engine is approximated by the model

$$T = 0.808x^3 - 17.974x^2 + 71.248x + 110.843,$$
$$1 \le x \le 5$$

where T is the torque in foot-pounds and x is the engine speed in thousands of revolutions per minute (see figure). Approximate the two engine speeds that yield a torque of 170 foot-pounds.

FIGURE FOR 42

In Exercises 43 and 44, write a computer program or use a spreadsheet that can be used to find the zeros of a function by using Newton's Method. The output should be a table with the following headings.

$$n \quad x_n \quad f(x_n) \quad f'(x_n) \quad \frac{f(x_n)}{f'(x_n)} \quad x_n - \frac{f(x_n)}{f'(x_n)}$$

Use the computer program or spreadsheet to approximate the zeros of the function accurate to three decimal places. (You may need to sketch a graph of the function to obtain the initial guess.)

43. $f(x) = \frac{1}{4}x^3 - 3x^2 + \frac{3}{4}x - 2$

44. $f(x) = \sqrt{4 - x^2}\sin(x - 2)$

True or False In Exercises 45–48, determine whether the statement is true or false. If it is false, explain why or give an example that shows it is false.

45. The zeros of $f(x) = p(x)/q(x)$ coincide with the zeros of $p(x)$.

46. If the coefficients of a polynomial function are all positive, then the polynomial has no positive zeros.

47. If $f(x)$ is a cubic polynomial such that $f'(x)$ is never 0, then any initial guess will force Newton's Method to converge to the zero of f.

48. The solutions of $\sqrt{f(x)} = 0$ coincide with the solutions of $f(x) = 0$.

49. Use Newton's Method to approximate the second-largest positive solution of the equation $x = \tan x$. Show that there is no solution of this equation in the interval $(0, \pi/2)$.

3.9

Linear Approximations • Differentials • Error Propagation • Calculating Differentials

Linear Approximations

Newton's Method (Section 3.8) is an example of using a tangent line to a graph to approximate the graph. In this section, you will study other situations in which the graph of a function can be approximated by a straight line.

To begin, consider a function f that is differentiable at c. The equation for the tangent line at the point $(c, f(c))$ is given by

$$y - f(c) = f'(c)(x - c)$$
$$y = f(c) + f'(c)(x - c).$$

Because c is a constant, y is a linear function of x. Moreover, by restricting the values of x to be sufficiently close to c, the values of y can be used as approximations (to any desired accuracy) of the values of the function f. In other words, as $x \to c$, the limit of y is $f(c)$.

EXAMPLE 1   Using a Tangent Line Approximation

Find the tangent line approximation of $f(x) = 1 + \sin x$ at the point $(0, 1)$. Then use a table to compare the y-values of the linear function with those of $f(x)$ in an open interval containing $x = 0$.

Solution   The derivative of f is $f'(x) = \cos x$. Thus, the equation of the tangent line of the graph of f at the point $(0, 1)$ is

$$y - f(0) = f'(0)(x - 0)$$
$$y - 1 = (1)(x - 0)$$
$$y = 1 + x. \qquad \text{Tangent line approximation}$$

Table 3.22 compares the values of y given by this linear approximation with the values of $f(x)$ near $x = 0$. Notice that the closer x is to 0, the better the approximation is. This conclusion is reinforced by the graph shown in Figure 3.66.

TABLE 3.22

x	-0.5	-0.1	-0.01	0	0.01	0.1	0.5
$f(x) = 1 + \sin x$	0.521	0.9002	0.9900002	1	1.0099998	1.0998	1.479
$y = 1 + x$	0.5	0.9	0.99	1	1.01	1.1	1.5

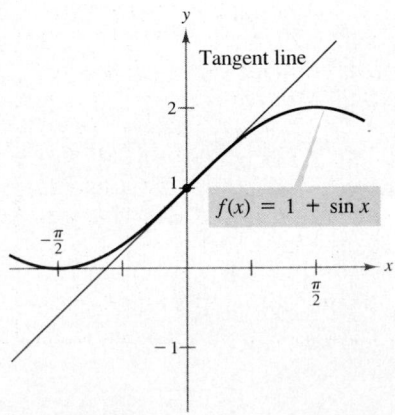

FIGURE 3.66
The tangent line approximation of f at the point $(0, 1)$.

Be sure you see that this linear approximation of $f(x) = 1 + \sin x$ depends on the point of tangency. At a different point on the graph of f, you would obtain a different tangent line approximation.

Differentials

When the tangent line to the graph of f at the point $(c, f(c))$

$$y = f(c) + f'(c)(x - c)$$

is used as an approximation to the graph of f, the quantity $x - c$ is called the *change in x*, and is denoted by Δx, as shown in Figure 3.67. When Δx is small, the change in y (denoted by Δy) can be approximated as follows

$$\Delta y = f(c + \Delta x) - f(c) \approx f'(c)\Delta x.$$

For such an approximation, the quantity Δx is traditionally denoted by dx, and is called the **differential of x.** The expression $f'(x)dx$ is denoted by dy, and is called the **differential of y.**

FIGURE 3.67
When Δx is small, $f(c + \Delta x) - f(c)$ is approximated by $f'(c)\Delta x$.

> ### Definition of Differentials
>
> Let $y = f(x)$ represent a function that is differentiable in an open interval containing x. The **differential of x** (denoted by dx) is any nonzero real number. The **differential of y** (denoted by dy) is
>
> $$dy = f'(x)dx.$$

In many types of applications, the differential of y can be used as an approximation of the change in y. That is,

$$\Delta y \approx dy \quad \text{or} \quad \Delta y \approx f'(x)dx.$$

EXAMPLE 2   Comparing Δy and dy

Let $y = x^2$. Find dy when $x = 1$ and $dx = 0.01$. Compare this value to Δy when $x = 1$ and $\Delta x = 0.01$.

Solution   Because $y = f(x) = x^2$, you have $f'(x) = 2x$, and the differential dy is given by

$$
\begin{aligned}
dy &= f'(x)dx && \text{Differential of } y\\
&= f'(1)(0.01)\\
&= 2(0.01)\\
&= 0.02.
\end{aligned}
$$

Now, using $\Delta x = 0.01$, the change in y is

$$
\begin{aligned}
\Delta y &= f(x + \Delta x) - f(x) && \text{Change in } y\\
&= f(1.01) - f(1)\\
&= (1.01)^2 - 1^2\\
&= 0.0201.
\end{aligned}
$$

Figure 3.68 shows the geometric comparison of dy and Δy. Try comparing other values of dy and Δy. You will see that the values become closer to each other as dx (or Δx) approaches 0.

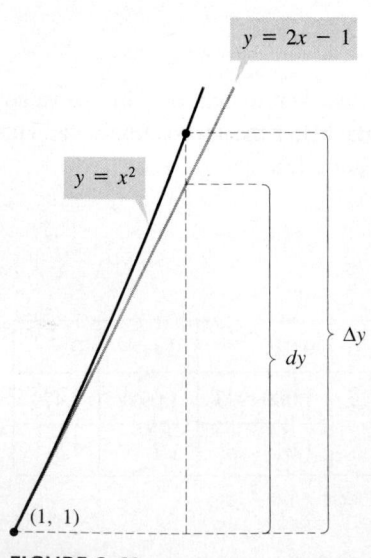

FIGURE 3.68
The change in y, Δy, is approximated by the differential of y, dy.

Error Propagation

Physicists and engineers tend to make liberal use of the approximation of Δy by dy. One way this occurs in practice is in the estimation of errors propagated by physical measuring devices. For example, if you let x represent the measured value of a variable and let $x + \Delta x$ represent the exact value, then Δx is the *error in measurement*. Finally, if the measured value x is used to compute another value $f(x)$, then the difference between $f(x + \Delta x)$ and $f(x)$ is the **propagated error.**

$$\underbrace{f(\overbrace{x + \Delta x}^{\substack{\text{Measurement} \\ \text{error}}})}_{\substack{\text{Exact} \\ \text{value}}} - \underbrace{f(x)}_{\substack{\text{Measured} \\ \text{value}}} = \underbrace{\Delta y}_{\substack{\text{Propagated} \\ \text{error}}}$$

EXAMPLE 3   Estimation of Error

The radius of a ball bearing is measured to be 0.7 inch, as shown in Figure 3.69. If the measurement is correct to within 0.01 inch, estimate the propagated error in the volume V of the ball bearing.

FIGURE 3.69
Ball bearing with measured radius
that is correct to within 0.01 inch.

Solution   The formula for the volume of a sphere is $V = \frac{4}{3}\pi r^3$, where r is the radius of the sphere. Thus, you can write

$$r = 0.7 \qquad \text{Measured radius}$$

and

$$-0.01 \leq \Delta r \leq 0.01. \qquad \text{Possible error}$$

To approximate the propagated error in the volume, differentiate V to obtain $dV/dr = 4\pi r^2$ and write

$$\begin{aligned} \Delta V &\approx dV \\ &= 4\pi r^2\, dr \\ &= 4\pi (0.7)^2 (\pm 0.01) \\ &\approx \pm 0.06158 \text{ in}^3. \end{aligned}$$

Would you say that the propagated error in Example 3 is large or small? The answer is best given in *relative* terms by comparing dV to V. The ratio

$$\frac{dV}{V} = \frac{4\pi r^2\, dr}{\frac{4}{3}\pi r^3} = \frac{3\, dr}{r} \approx \frac{3}{0.7}(\pm 0.01) \approx \pm 0.0429$$

is called the **relative error.** The corresponding **percentage error** is

$$\frac{dV}{V}(100) \approx 4.29\%.$$

Calculating Differentials

Each of the differentiation rules that you studied in Chapter 2 can be written in **differential form.** For example, suppose u and v are differentiable functions of x. By the definition of differentials, you have $du = u' \, dx$ and $dv = v' \, dx$. Therefore, you can write the differential form of the Product Rule as follows.

$$
\begin{aligned}
d[uv] &= \frac{d}{dx}[uv] \, dx && \text{Differential of } uv \\
&= [uv' + vu'] \, dx && \text{Product Rule} \\
&= uv' \, dx + vu' \, dx \\
&= u \, dv + v \, du
\end{aligned}
$$

Differential Formulas

Let u and v be differentiable functions ot x.

Constant multiple	$d[cu] = c \, du$
Sum or difference	$d[u \pm v] = du \pm dv$
Product	$d[uv] = u \, dv + v \, du$
Quotient	$d\left[\dfrac{u}{v}\right] = \dfrac{v \, du - u \, dv}{v^2}$

EXAMPLE 4   Finding Differentials

Function	Derivative	Differential
a. $y = x^2$	$\dfrac{dy}{dx} = 2x$	$dy = 2x \, dx$
b. $y = 2 \sin x$	$\dfrac{dy}{dx} = 2 \cos x$	$dy = 2 \cos x \, dx$
c. $y = x \cos x$	$\dfrac{dy}{dx} = -x \sin x + \cos x$	$dy = (-x \sin x + \cos x) \, dx$
d. $y = \dfrac{1}{x}$	$\dfrac{dy}{dx} = -\dfrac{1}{x^2}$	$dy = -\dfrac{dx}{x^2}$

GOTTFRIED WILHELM LEIBNIZ

Both Leibniz and Newton are credited with creating calculus. It was Leibniz, however, who tried to broaden calculus by developing rules and formal notation. He often spent days choosing an appropriate notation for a new concept.

The notation in Example 4 is called the **Leibniz notation** for derivatives and differentials, named after the German mathematician Gottfried Wilhelm Leibniz (1646–1716). The beauty of this notation is that it provides an easy way to remember several important calculus formulas by making it seem as though the formulas were derived from algebraic manipulations of differentials. For instance, in Leibniz notation, the *Chain Rule*

$$
\frac{dy}{dx} = \frac{dy}{du} \frac{du}{dx}
$$

would appear to be true because the du's cancel. Even though this reasoning is *incorrect,* the notation does help one remember the Chain Rule.

EXAMPLE 5   Finding the Differential of a Composite Function

$$y = f(x) = \sin 3x \qquad \text{Given function}$$
$$f'(x) = 3\cos 3x \qquad \text{Apply Chain Rule}$$
$$dy = f'(x)dx = 3\cos 3x\, dx \qquad \text{Differential form}$$

EXAMPLE 6   Finding the Differential of a Composite Function

$$y = f(x) = (x^2 + 1)^{1/2} \qquad \text{Given function}$$
$$f'(x) = \frac{1}{2}(x^2 + 1)^{-1/2}(2x) = \frac{x}{\sqrt{x^2 + 1}} \qquad \text{Apply Chain Rule}$$
$$dy = f'(x)dx = \frac{x}{\sqrt{x^2 + 1}}\, dx \qquad \text{Differential form}$$

Differentials can be used to approximate function values. To do this for the function given by $y = f(x)$, you use the formula

$$f(x + \Delta x) \approx f(x) + dy = f(x) + f'(x)dx$$

which is derived from the approximation $\Delta y = f(x + \Delta x) - f(x) \approx dy$. The key to using this formula is to choose a value for x that makes the calculations easier, as shown in Example 7.

EXAMPLE 7   Approximating Function Values

Use differentials to approximate $\sqrt{16.5}$.

Solution   Using $f(x) = \sqrt{x}$, you can write

$$f(x + \Delta x) \approx f(x) + f'(x)dx = \sqrt{x} + \frac{1}{2\sqrt{x}}\, dx.$$

Now, choosing $x = 16$ and $dx = 0.5$, you obtain the following approximation.

$$f(x + \Delta x) = \sqrt{16.5}$$
$$\approx \sqrt{16} + \frac{1}{2\sqrt{16}}(0.5)$$
$$= 4 + \left(\frac{1}{8}\right)\left(\frac{1}{2}\right)$$
$$= 4 + \frac{1}{16}$$
$$= 4.0625.$$

REMARK   The use of differentials to approximate function values has diminished with the availability of calculators and computers. Using a calculator, however, you can determine that

$$\sqrt{16.5} \approx 4.0620$$

which indicates the accuracy of the differential method in Example 7.

3, 7, 9, ~~~~23, 25, 27, 31
37, 53, 47-50

EXERCISES for Section 3.9

In Exercises 1–6, find the equation of the tangent line T to the function f at the specified point. Use this linear approximation to complete the table.

x	1.9	1.99	2	2.01	2.1
$f(x)$					
$T(x)$					

	Function	*Point*
1.	$f(x) = x^2$	$(2, 4)$
2.	$f(x) = \dfrac{1}{x^2}$	$(2, 1/4)$
3.	$f(x) = x^5$	$(2, 32)$
4.	$f(x) = \sqrt{x}$	$(2, \sqrt{2})$
5.	$f(x) = \sin x$	$(2, \sin 2)$
6.	$f(x) = \csc x$	$(2, \csc 2)$

In Exercises 7–10, use the information to evaluate and compare Δy and dy.

7. $y = x^3$ $x = 1$ $\Delta x = dx = 0.1$

8. $y = 1 - 2x^2$ $x = 0$ $\Delta x = dx = -0.1$

9. $y = x^4 + 1$ $x = -1$ $\Delta x = dx = 0.01$

10. $y = 2x + 1$ $x = 2$ $\Delta x = dx = 0.01$

In Exercises 11–20, find the differential dy of the function.

11. $y = 3x^2 - 4$

12. $y = 2x^{3/2}$

13. $y = \dfrac{x + 1}{2x - 1}$

14. $y = \sqrt{x^2 - 4}$

15. $y = x\sqrt{1 - x^2}$

16. $y = \sqrt{x} + \dfrac{1}{\sqrt{x}}$

17. $y = \dfrac{\sec^2 x}{x^2 + 1}$

18. $y = x \sin x$

19. $y = \dfrac{1}{3} \cos\left(\dfrac{6\pi x - 1}{2}\right)$

20. $y = x - \tan^2 x$

21. *Area* The measurement of the side of a square is found to be 12 inches, with a possible error of $\frac{1}{64}$ inch. Use differentials to approximate the possible error in computing the area of the square.

22. *Area* The measurements of the base and altitude of a triangle are 36 and 50 centimeters, respectively. The possible error in each measurement is 0.25 centimeter. Use differentials to approximate the possible error in computing the area of the triangle.

23. *Area* The measurement of the radius of the end of a log is found to be 14 inches, with a possible error of $\frac{1}{4}$ inch. Use differentials to approximate the possible error in computing the area of the end of the log.

24. *Volume and Surface Area* The measurement of the edge of a cube is found to be 12 inches, with a possible error of 0.03 inch. Use differentials to approximate the maximum possible error in computing
a. the volume of the cube.
b. the surface area of the cube.

25. *Area* The measurement of a side of a square is found to be 15 centimeters.
a. The possible error in measuring the side is 0.05 centimeter. Approximate the percentage error in computing the area of the square.
b. Estimate the maximum allowable percentage error in measuring the side if the error in computing the area cannot exceed 2.5%.

26. *Circumference* The measurement of the circumference of a circle is found to be 56 centimeters.
a. Approximate the percentage error in computing the area of the circle if the possible error in measuring the circumference is 1.2 centimeters.
b. Estimate the maximum allowable percentage error in measuring the circumference if the error in computing the area cannot exceed 3%.

27. *Volume and Surface Area* The radius of a sphere is claimed to be 6 inches, with a possible error of 0.02 inch.
a. Use differentials to approximate the maximum possible error in calculating the volume of the sphere.
b. Use differentials to approximate the maximum possible error in calculating the surface area of the sphere.
c. What is the relative error in parts a and b?

28. *Profit* The profit P for a company is

$$P = (500x - x^2) - \left(\frac{1}{2}x^2 - 77x + 3000\right).$$

Approximate the change and percentage change in profit as production changes from $x = 115$ to $x = 120$ units.

In Exercises 29 and 30, the thickness of the shell is 0.2 centimeter. Use differentials to approximate the volume of the shell.

29. A cylindrical shell with height 40 centimeters and radius 5 centimeters.

30. A spherical shell of radius 100 centimeters.

0.2 cm

40 cm

5 cm

0.2 cm

|← 100 cm →|

FIGURE FOR 29   FIGURE FOR 30

31. *Pendulum*   The period of a pendulum is

$$T = 2\pi\sqrt{\frac{L}{g}}$$

where L is the length of the pendulum in feet, g is the acceleration due to gravity, and T is the time in seconds. Suppose that the pendulum has been subjected to an increase in temperature so that the length increases by $\frac{1}{2}$ percent.
   a. Find the approximate percentage change in the period.
   b. Using the result of part a, find the approximate error in this pendulum clock in one day.

32. *Pendulum*   Suppose a pendulum of length L is to be used to find the acceleration of gravity at a point on the earth's surface. Use the formula

$$T = 2\pi\sqrt{\frac{L}{g}}$$

to approximate the percentage error in the value of g. Assume that you can determine the period T to within 0.1 percent of its true value.

33. *Ohm's Law*   A current of I amps passes through a resistor of R ohms. **Ohm's Law** states that the voltage E applied to the resistor is

$P = IV$

$E = IR$.

If the voltage is constant, show that the magnitude of the relative error in R caused by a change in I is equal in magnitude to the relative error in I.

34. *Environment*   The cost in dollars of removing $p\%$ of the air pollutants in the stack emission of a utility company that burns coal to generate electricity is

$$C = \frac{80,000p}{100 - p}, \quad 0 \le p < 100.$$

Use differentials to approximate the increase in cost given that the government requires the utility company to remove 2% more of the pollutants and p is currently
   a. 40%   **b.** 75%.

35. *Triangle Measurements*   The measurement of one side of a right triangle is found to be 9.5 inches, and the angle opposite that side is $26°45'$ with a possible error of $15'$.
   a. Approximate the percentage error in computing the length of the hypotenuse.
   b. Estimate the maximum allowable percentage error in measuring the angle if the error in computing the length of the hypotenuse cannot exceed 2%.

36. *Area*   The measurement of one side of a right triangle is found to be 9.5 inches, and the angle opposite that side is $26°45'$ with a possible error of $15'$. Approximate the percentage error in computing its area.

37. *Projectile Motion*   The range of a projectile is

$$R = \frac{v_0^2}{32}(\sin 2\theta)$$

where v_0 is the initial velocity in feet per second and θ is the angle of elevation. If $v_0 = 2200$ feet per second and θ is changed from $10°$ to $11°$, use differentials to approximate the change in the range.

38. *Surveying*   A surveyor is standing 50 feet from the base of a large tree. He measures the angle of elevation to the top of the tree as $71.5°$. How accurately must the angle be measured given that the percentage error in estimating the height of the tree is to be less than 6 percent?

In Exercises 39–42, use differentials to approximate the value of the function. Compare your answer to that of a calculator.

39. $\sqrt{99.4}$ **40.** $\sqrt[3]{28}$

41. $\sqrt[4]{624}$ **42.** $(2.99)^3$

Essay   **In Exercises 43–46, write a short note explaining why the approximation is valid.**

43. $0.99^4 \approx 1 - 4(0.01)$ **44.** $\sqrt{4.02} \approx 2 + \dfrac{1}{4}(0.02)$

45. $\sec 0.03 \approx 1 + 0(0.03)$ **46.** $\tan 0.05 \approx 0 + 1(0.05)$

True or False   **In Exercises 47–50, determine whether the statement is true or false. If it is false, explain why or give an example that shows it is false.**

47. If $y = x + c$, then $dy = dx$.

48. If $y = ax + b$, then $\Delta y/\Delta x = dy/dx$.

49. If y is differentiable, then $\lim\limits_{\Delta x \to 0} (\Delta y - dy) = 0$.

50. If $y = f(x)$, f is increasing and differentiable, and $\Delta x > 0$, then $\Delta y \ge dy$.

51. The area of a square of side x is $A(x) = x^2$.
   a. Compute dA and ΔA in terms of x and Δx.
   b. Use the figure to identify the region whose area is dA.
   c. Use the figure to identify the region whose area is $\Delta A - dA$.

FIGURE FOR 51

52. Show that if $y = f(x)$ is a differentiable function, then

$$\Delta y - dy = \varepsilon \Delta x$$

where $\varepsilon \to 0$ as $\Delta x \to 0$.

53. Use differentials to show that $f(x) = 1/(1 - x)$ is approximately equal to $g(x) = 1 + x$ for x near 0. Graph f and g on the same set of coordinate axes.

Business and Economics Applications

In Section 2.6 you learned that one of the most common ways to measure change is with respect to time. In this section you will study some important rates of change in economics that are not measured with respect to time. For example, economists refer to **marginal profit, marginal revenue,** and **marginal cost** as the rates of change of the profit, revenue, and cost with respect to the number of units produced or sold.

We begin with a summary of some basic terms and formulas.

Summary of Business Terms and Formulas

Basic Terms   *Basic Formulas*

x is the number of units produced (or sold)
p is the price per unit
R is the total revenue from selling x units $R = xp$
C is the total cost of producing x units

$\overline{C}$ is the average cost per unit $\overline{C} = \dfrac{C}{x}$

P is the total profit from selling x units $P = R - C$

The **break-even point** is the number of units for which $R = C$.

Marginals

$\dfrac{dR}{dx}$ = (Marginal revenue) $\approx$ (*Extra* revenue for selling one additional unit)

$\dfrac{dC}{dx}$ = (Marginal cost) $\approx$ (*Extra* cost of producing one additional unit)

$\dfrac{dP}{dx}$ = (Marginal profit) $\approx$ (*Extra* profit for selling one additional unit)

In this summary, note that marginals can be used to approximate the *extra* revenue, cost, or profit associated with selling or producing one additional unit. This is illustrated graphically for marginal revenue in Figure 3.70.

FIGURE 3.70
Revenue function

EXAMPLE 1 Using Marginals as Approximations

A manufacturer determines that the profit derived from selling x units of a certain item is given by $P = 0.0002x^3 + 10x$.

a. Find the marginal profit for a production level of 50 units.
b. Compare this to the actual gain in profit obtained by increasing the production from 50 to 51 units (see Figure 3.71).

Solution

a. The marginal profit is given by

$$\frac{dP}{dx} = 0.0006x^2 + 10.$$

When $x = 50$, the marginal profit is

$$\frac{dP}{dx} = (0.0006)(50)^2 + 10 = \$11.50. \qquad \text{Marginal profit}$$

b. For $x = 50$ and 51, the actual profit is

$$P = (0.0002)(50)^3 + 10(50) = 25 + 500 = \$525.00$$
$$P = (0.0002)(51)^3 + 10(51) = 26.53 + 510 = \$536.53.$$

Thus, the additional profit obtained by increasing the production level from 50 to 51 units is

$$536.53 - 525.00 = \$11.53. \qquad \text{Extra profit for one unit}$$

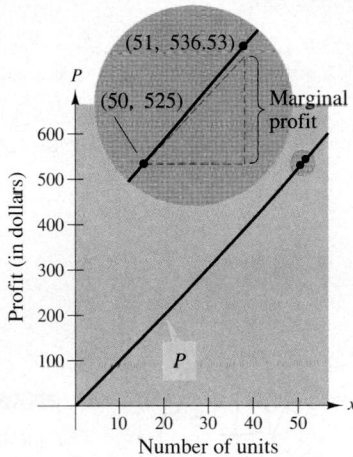

FIGURE 3.71
Marginal profit is the extra profit from selling one additional unit.

The profit function in Example 1 is unusual in that the profit continues to increase as long as the number of units sold increases. In practice, it is more common to encounter situations in which sales can be increased only by lowering the price per item. Such reductions in price ultimately cause the profit to decline. The number of units x that consumers are willing to purchase at a given price p per unit is defined as the **demand function**

$$p = f(x). \qquad \text{Demand function}$$

EXAMPLE 2 Finding the Demand Function

A business sells 2000 items per month at a price of $10 each. It is predicted that monthly sales will increase by 250 items for each $0.25 reduction in price. Find the demand function corresponding to this prediction.

Solution From the given prediction, x increases 250 units each time p drops $0.25 from the original cost of $10. This is described by the equation

$$x = 2000 + 250\left(\frac{10 - p}{0.25}\right) = 12,000 - 1000p$$

or

$$p = 12 - \frac{x}{1000}, \qquad x \geq 2000. \qquad \text{Demand function}$$

FIGURE 3.72

The graph of the demand function is shown in Figure 3.72.

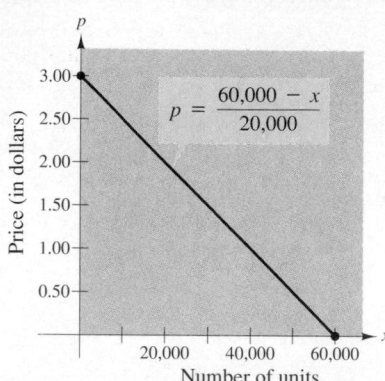

FIGURE 3.73
As the price decreases, more hamburgers are sold.

EXAMPLE 3   Finding the Marginal Revenue

A fast-food restaurant has determined that the monthly demand for its hamburgers is given by

$$p = \frac{60,000 - x}{20,000}.$$

Find the increase in revenue per hamburger for monthly sales of 20,000 hamburgers. In other words, find the marginal revenue when $x = 20,000$.

Solution   Because the total revenue is given by $R = xp$, you have

$$R = xp = x\left(\frac{60,000 - x}{20,000}\right) = \frac{1}{20,000}(60,000x - x^2)$$

and the marginal revenue is

$$\frac{dR}{dx} = \frac{1}{20,000}(60,000 - 2x).$$

When $x = 20,000$, the marginal revenue is

$$\frac{dR}{dx} = \frac{1}{20,000}[60,000 - 2(20,000)] = \frac{20,000}{20,000} = \$1/\text{unit}.$$

REMARK   The demand function in Example 3 is typical in that a high demand corresponds to a low price, as shown in Figure 3.73.

EXAMPLE 4   Finding the Marginal Profit

Suppose that in Example 3 the cost of producing x hamburgers is

$$C = 5000 + 0.56x.$$

Find the total profit and the marginal profit for 20,000, for 24,400, and for 30,000 units.

Solution   Because $P = R - C$, you can use the revenue function in Example 3 to obtain

$$P = \frac{1}{20,000}(60,000x - x^2) - 5000 - 0.56x$$

$$= 2.44x - \frac{x^2}{20,000} - 5000.$$

Thus, the marginal profit is

$$\frac{dP}{dx} = 2.44 - \frac{x}{10,000}.$$

Table 3.23 shows the total profit and the marginal profit for the three indicated demands. Figure 3.74 shows the graph of the profit function.

FIGURE 3.74
The maximum profit corresponds to the point where the marginal profit is 0. When more than 24,400 hamburgers are sold, the marginal profit is negative—increasing production beyond this point will *reduce* rather than increase profit.

TABLE 3.23

Demand	20,000	24,400	30,000
Profit	\$23,800	\$24,768	\$23,200
Marginal profit	\$0.44	\$0.00	-\$0.56

EXAMPLE 5   Finding the Maximum Profit

In marketing a certain item, a business has discovered that the demand for the item is represented by

$$p = \frac{50}{\sqrt{x}}.$$ Demand function

The cost of producing x items is given by $C = 0.5x + 500$. Find the price per unit that yields a maximum profit.

Solution   From the given cost function, you obtain

$$P = R - C = xp - (0.5x + 500).$$ Primary equation

Substituting for p (from the demand function) produces

$$P = x\left(\frac{50}{\sqrt{x}}\right) - (0.5x + 500) = 50\sqrt{x} - 0.5x - 500.$$

Setting the marginal profit equal to 0

$$\frac{dP}{dx} = \frac{25}{\sqrt{x}} - 0.5 = 0$$

yields $x = 2500$. From this, you can conclude that the maximum profit occurs when the price is

$$p = \frac{50}{\sqrt{2500}} = \frac{50}{50} = \$1.00.$$ ■

REMARK   To find the maximum profit in Example 5, we differentiated the profit function, $P = R - C$, and set dP/dx equal to 0. From the equation

$$\frac{dP}{dx} = \frac{dR}{dx} - \frac{dC}{dx} = 0$$

it follows that the maximum profit occurs when the marginal revenue is equal to the marginal cost, as shown in Figure 3.75.

FIGURE 3.75

EXAMPLE 6   Minimizing the Average Cost

A company estimates that the cost (in dollars) of producing x units of a certain product is given by $C = 800 + 0.04x + 0.0002x^2$. Find the production level that minimizes the average cost per unit.

Solution   Substituting from the given equation for C produces

$$\overline{C} = \frac{C}{x} = \frac{800 + 0.04x + 0.0002x^2}{x} = \frac{800}{x} + 0.04 + 0.0002x.$$

Setting the derivative $d\overline{C}/dx$ equal to 0 yields

$$\frac{d\overline{C}}{dx} = -\frac{800}{x^2} + 0.0002 = 0$$

$$x^2 = \frac{800}{0.0002} = 4{,}000{,}000 \quad \Rightarrow \quad x = 2000 \text{ units.}$$

(See Figure 3.76.)   ■

FIGURE 3.76
Minimum average cost occurs when
$\dfrac{d\overline{C}}{dx} = 0.$

EXERCISES for Section 3.10

1. The figure shows the cost C of producing x units of a product.

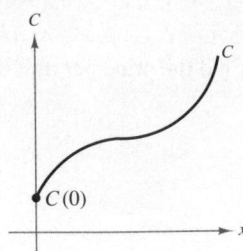

   a. What is $C(0)$ called?
   b. Sketch a graph of the marginal cost function.
   c. Does the marginal cost function have an extremum? If so, describe what it means in economic terms.

2. The figure shows the cost C and revenue R for producing and selling x units of a product.

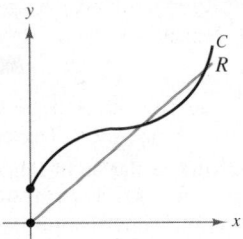

   a. Sketch a graph of the marginal revenue function.
   b. Sketch a graph of the profit function. Approximate the position of the value of x for which profit is maximum.

In Exercises 3–6, find the number of units x that produce a maximum revenue R.

3. $R = 900x - 0.1x^2$

4. $R = 600x^2 - 0.02x^3$

5. $R = \dfrac{1,000,000x}{0.02x^2 + 1800}$

6. $R = 30x^{2/3} - 2x$

In Exercises 7–10, find the number of units x that produce the minimum average cost per unit $\overline{C}$.

7. $C = 0.125x^2 + 20x + 5000$

8. $C = 0.001x^3 - 5x + 250$

9. $C = 3000x - x^2\sqrt{300 - x}$

10. $C = \dfrac{2x^3 - x^2 + 5000x}{x^2 + 2500}$

In Exercises 11–14, find the price per unit p that produces the maximum profit P.

Cost function	Demand function
11. $C = 100 + 30x$	$p = 90 - x$
12. $C = 2400x + 5200$	$p = 6000 - 0.4x^2$
13. $C = 4000 - 40x + 0.02x^2$	$p = 50 - \dfrac{x}{100}$
14. $C = 35x + 2\sqrt{x - 1}$	$p = 40 - \sqrt{x - 1}$

Average Cost   In Exercises 15 and 16, use the cost function to find the value of x for which the average cost is minimum. For that value of x, show that the marginal cost and average cost are equal.

15. $C = 2x^2 + 5x + 18$ 16. $C = x^3 - 6x^2 + 13x$

17. Prove that the average cost is minimum at the value of x where the average cost equals the marginal cost.

18. *Maximum Profit*   The profit for a certain company is

$$P = 230 + 20s - \frac{1}{2}s^2$$

where s is the amount (in hundreds of dollars) spent on advertising. What amount of advertising gives the maximum profit?

C 19. *Maximum Profit*   The cost per unit for the production of a certain radio model is $60. The manufacturer charges $90 per unit for orders of 100 or less. To encourage large orders, the manufacturer reduces the charge by $0.15 per radio for each unit ordered in excess of 100 (for example, there would be a charge of $87 per radio for an order size of 120).
   a. Use paper and pencil to complete four rows of the table. (The first two rows are shown.)

Order size, x	Price	Profit
102	$90 - 2(0.15)$	$102[90 - 2(0.15)] - 102(60) = 3029.40$
104	$90 - 4(0.15)$	$104[90 - 4(0.15)] - 104(60) = 3057.60$

   b. Write the profit as a function of x.
   c. Use calculus to find the critical number of the function in part b and find the required order size.
   d. Use a computer to generate additional rows of the table in part a and estimate the maximum profit from the table.
   e. Use a graphing utility to graph the function in part b and estimate the maximum profit from the graph.

20. *Maximum Profit*   A real estate office handles 50 apartment units. When the rent is $540 per month, all units are occupied. However, on the average, for each $30 increase in rent, one unit becomes vacant. Each occupied unit requires an average of $36 per month for service and repairs. What rent should be charged to realize the most profit?

Minimum Cost In Exercises 21 and 22, find the speed v in miles per hour that will minimize costs on a 110–mile delivery trip. The cost per hour for fuel for the van is C dollars and the driver is paid W dollars per hour. (Assume there are no costs other than wages and fuel.)

21. Fuel cost: $C = \dfrac{v^2}{600}$

Driver: $W = \$5$

22. Fuel cost: $C = \dfrac{v^2}{500}$

Driver: $W = \$7.50$

23. *Minimum Cost* A power station is on one side of a river that is $\frac{1}{2}$ mile wide, and a factory is 6 miles downstream on the other side. It costs \$6 per foot to run power lines overland and \$8 per foot to run them underwater. Find the most economical path for the transmission line from the power station to the factory.

24. *Minimum Cost* An offshore oil well is 1 mile off the coast. The refinery is 2 miles down the coast. If laying pipe in the ocean is twice as expensive as on land, what path should the pipe follow in order to minimize the cost?

Minimum Cost In Exercises 25–28, consider a fuel distribution center that is located at the origin of the rectangular coordinate system (units in miles; see figures). The center supplies three factories with coordinates (4, 1), (5, 6), and (10, 3). A trunk line will run from the distribution center along the line $y = mx$ and feeder lines will run to the three factories. The objective is to find m so that the lengths of the feeder lines are minimized.

25. Minimize the sum of the squares of the lengths of the vertical feeder lines

$$S_1 = (4m - 1)^2 + (5m - 6)^2 + (10m - 3)^2.$$

Find the equation for the trunk line by this method and then determine the sum of the lengths of the feeder lines.

C **26.** Minimize the sum of the absolute values of the lengths of the vertical feeder lines

$$S_2 = |4m - 1| + |5m - 6| + |10m - 3|.$$

Find the equation for the trunk line by this method and then determine the sum of the lengths of the feeder lines. (*Hint:* Use a graphing utility to obtain the graph of S_2 and approximate the required critical number.)

FIGURE FOR 25 AND 26

C **27.** Minimize the sum of the perpendicular distances (see Exercises in Section 4 in the Prerequisites chapter) from the trunk line to the factories given by

$$S_3 = \frac{|4m - 1|}{\sqrt{m^2 + 1}} + \frac{|5m - 6|}{\sqrt{m^2 + 1}} + \frac{|10m - 3|}{\sqrt{m^2 + 1}}.$$

Find the equation for the trunk line by this method and then determine the sum of the lengths of the feeder lines. (*Hint:* Use a graphing utility to obtain the graph of S_3 and approximate the required critical number.)

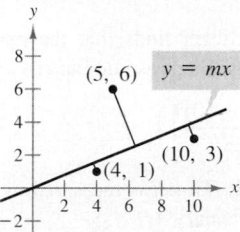

FIGURE FOR 27

28. *Essay* Write a short paragraph comparing the methods of Exercises 25–27. Which method is easier to use? Which method best meets the stated objective of minimizing the amount of feeder line required? Explain.

29. *Maximum Profit* Assume that the amount of money deposited in a bank is proportional to the square of the interest rate the bank pays on this money. Furthermore, the bank can reinvest this money at 12 percent. Find the interest rate the bank should pay to maximize profit. (Use the simple interest formula.)

30. *Maximum Revenue* When a wholesaler sold a certain product at \$25 per unit, sales were 800 units per week. After a price increase of \$5, the average number of units sold dropped to 775 per week. Assume that the demand function is linear and find the price that will maximize the total revenue.

31. *Minimum Cost* The ordering and transportation cost C of the components used in manufacturing a certain product is

$$C = 100\left(\frac{200}{x^2} + \frac{x}{x + 30}\right), \qquad 1 \le x$$

where C is measured in thousands of dollars and x is the order size in hundreds. Find the order size that minimizes the cost. (*Hint:* Use Newton's Method.)

32. *Average Cost* A company estimates that the cost in dollars of producing x units of a certain product is

$$C = 800 + 0.4x + 0.02x^2 + 0.0001x^3.$$

Find the production level that minimizes the average cost per unit. (*Hint:* Use Newton's Method.)

33. *Revenue* The revenue R for a company selling x units is

$$R = 900x - 0.1x^2.$$

Use differentials to approximate the change in revenue if sales increase from $x = 3000$ to $x = 3100$ units.

34. *Profit* The profit P for a company selling x units is

$$P = (500x - x^2) - \left(\frac{1}{2}x^2 - 77x + 3000\right).$$

Use differentials to approximate the change and percentage change in profit as production changes from $x = 175$ to $x = 180$ units.

35. *Sales* A manufacturer of fertilizer finds that the national sales of fertilizer roughly follow the seasonal pattern

$$F = 100{,}000\left(1 + \sin\left[\frac{2\pi(t - 60)}{365}\right]\right)$$

where F is measured in pounds. Time t is measured in days, with $t = 1$ corresponding to January 1.
a. Use calculus to determine the day of the year when the maximum amount of fertilizer is sold.
b. Use a graphing utility to obtain the graph of the function and approximate the day of the year when sales are minimum.

C **36.** *Natural Gas Sales* The figure shows the domestic consumption (United States) of natural gas G in billions of cubic feet for the years 1989 and 1990. (*Source:* Department of Energy.) A model for this data is

$$G = 1600 + 418\sin\left(\frac{\pi t}{6} + 7\right)$$

where t is the time in months, with $t = 1$ corresponding to January 1989.
a. Use the model to approximate the month when natural gas sales are least.
b. What factor causes the seasonal variation in sales of natural gas? How could this factor make the model very inaccurate for predicting how much gas is needed in a year?
c. Suppose that the Department of Energy uses the model

$$G = 1600 + 10t + 418\sin\left(\frac{\pi t}{6} + 7\right)$$

in forecasting natural gas consumption through the year 2000. What does the inclusion of the term $10t$ mean? Use this model to estimate the maximum monthly consumption in the year 2000.

FIGURE FOR 36

The relative responsiveness of consumers to a change in the price of an item is called the *price elasticity of demand.* If $p = f(x)$ is a differentiable demand function, then the price elasticity of demand is

$$\eta = \frac{p/x}{dp/dx}.$$

For a given price, if $|\eta| < 1$, the demand is *inelastic,* and if $|\eta| > 1$, the demand is *elastic.*

Elasticity In Exercises 37–40, find η for the demand function at the indicated x-value. Is the demand elastic, inelastic, or neither at the indicated x-value?

37. $p = 400 - 3x$ **38.** $p = 5 - 0.03x$
       $x = 20$ $x = 100$

39. $p = 400 - 0.5x^2$ **40.** $p = \dfrac{500}{x + 2}$
       $x = 20$ $x = 23$

In Exercises 41 and 42, find the amount of advertising s (in thousands of dollars) that yields the maximum profit P (in thousands of dollars) and find the point of diminishing returns. (The *point of diminishing returns* is the point at which the rate of growth of the profit function begins to decline.)

41. $P = -2s^3 + 35s^2 - 100s + 200$

42. $P = -\dfrac{1}{10}s^3 + 6s^2 + 400$

REVIEW EXERCISES for Chapter 3

In Exercises 1–4, find the critical numbers (if any) and the open intervals on which the function is increasing or decreasing.

1. $f(x) = (x - 1)^2(x - 3)$ **2.** $g(x) = (x + 1)^3$
3. $h(x) = \sqrt{x}(x - 3)$ **4.** $f(x) = \sin x + \cos x$
                                          $0 \le x \le 2\pi$

In Exercises 5 and 6, locate the absolute extrema of the function on the closed interval.

5. $g(x) = 2x + 5\cos x$ **6.** $f(x) = \dfrac{x}{\sqrt{x^2 + 1}}$
       $[0, 2\pi]$ $[0, 2]$

In Exercises 7 and 8, use the First Derivative Test to find any relative extrema of the function.

7. $h(t) = \dfrac{1}{4}t^4 - 8t$

8. $g(\theta) = \dfrac{3}{2}\sin\left(\dfrac{\pi\theta}{2} - 1\right)$
$[0, 4]$

In Exercises 9 and 10, find any points of inflection of the function.

9. $f(x) = x + \cos x$
$0 \le x \le 2\pi$

10. $f(x) = (x + 2)^2(x - 4)$

In Exercises 11–14, find any vertical and horizontal asymptotes of the graph of the function.

11. $h(x) = \dfrac{2x + 3}{x - 4}$

12. $g(x) = \dfrac{5x^2}{x^2 + 2}$

13. $f(x) = \dfrac{3}{x} - 2$

14. $f(x) = \dfrac{3x}{\sqrt{x^2 + 2}}$

In Exercises 15–32, make use of domain, range, symmetry, asymptotes, intercepts, relative extrema, and/or points of inflection to obtain an accurate graph of the function.

15. $f(x) = 4x - x^2$

16. $f(x) = 4x^3 - x^4$

17. $f(x) = x\sqrt{16 - x^2}$

18. $f(x) = (x^2 - 4)^2$

19. $f(x) = (x - 1)^3(x - 3)^2$

20. $f(x) = (x - 3)(x + 2)^3$

21. $f(x) = x^{1/3}(x + 3)^{2/3}$

22. $f(x) = (x - 2)^{1/3}(x + 1)^{2/3}$

23. $f(x) = \dfrac{x + 1}{x - 1}$

24. $f(x) = \dfrac{2x}{1 + x^2}$

25. $f(x) = \dfrac{4}{1 + x^2}$

26. $f(x) = \dfrac{x^2}{1 + x^4}$

27. $f(x) = x^3 + x + \dfrac{4}{x}$

28. $f(x) = x^2 + \dfrac{1}{x}$

29. $f(x) = |x^2 - 9|$

30. $f(x) = |x - 1| + |x - 3|$

31. $f(x) = x + \cos x$
$0 \le x \le 2\pi$

32. $f(x) = \dfrac{1}{\pi}(2\sin \pi x - \sin 2\pi x)$
$-1 \le x \le 1$

In Exercises 33–36, use a graphing utility to graph the function. Use the graphs to approximate any relative extrema.

33. $f(x) = x^3 + \dfrac{243}{x}$

34. $f(x) = |x^3 - 3x^2 + 2x|$

35. $f(x) = \dfrac{x - 1}{1 + 3x^2}$

36. $g(x) = \dfrac{\pi^2}{3} - 4\cos x + \cos 2x$

In Exercises 37–40, find the point(s) guaranteed by the Mean Value Theorem for the indicated interval.

37. $f(x) = x^{2/3}$
$1 \le x \le 8$

38. $f(x) = \dfrac{1}{x}$
$1 \le x \le 4$

39. $f(x) = x - \cos x$
$-\dfrac{\pi}{2} \le x \le \dfrac{\pi}{2}$

40. $f(x) = \sqrt{x} - 2x$
$0 \le x \le 4$

41. Can the Mean Value Theorem be aplied to the function $f(x) = 1/x^2$ on the interval $[-2, 1]$? Explain your answer.

42. If $f(x) = 3 - |x - 4|$, verify that $f(1) = f(7)$ and yet $f'(x)$ is not equal to 0 for any x in $[1, 7]$. Explain why this does not contradict Rolle's Theorem.

43. For the function

$$f(x) = Ax^2 + Bx + C$$

determine the value of c guaranteed by the Mean Value Theorem on the interval $[x_1, x_2]$.

44. Demonstrate the result of Exercise 43 for

$$f(x) = 2x^2 - 3x + 1$$

on the interval $[0, 4]$.

45. *Minimum Distance* At noon, ship A was 100 miles due east of ship B. Ship A is sailing west at 12 miles per hour, and ship B is sailing south at 10 miles per hour. At what time will the ships be nearest to each other, and what will this distance be?

46. *Maximum Area* Find the dimensions of the rectangle of maximum area, with sides parallel to the coordinate axes, that can be inscribed in the ellipse given by

$$\dfrac{x^2}{144} + \dfrac{y^2}{16} = 1.$$

47. *Minimum Length* A right triangle in the first quadrant has the coordinate axes as sides, and the hypotenuse passes through the point $(1, 8)$. Find the vertices of the triangle so that the length of the hypotenuse is minimum.

48. *Minimum Length* The wall of a building is to be braced by a beam that must pass over a parallel fence 5 feet high and 4 feet from the building. Find the length of the shortest beam that can be used.

49. *Maximum Area* Three sides of a trapezoid have the same length s. Of all such possible trapezoids, show that the one of maximum area has a fourth side of length $2s$.

50. *Maximum Area* Show that the greatest area of any rectangle inscribed in a triangle is one-half that of the triangle.

51. *Minimum Distance* Find the length of the longest pipe that can be carried level around a right-angle corner if the two intersecting corridors are of width 4 feet and 6 feet. (Do not use trigonometry.)

52. *Minimum Distance* Rework Exercise 51, given corridors of widths a feet and b feet. (Do not use trigonometry.)

53. *Minimum Distance* A hallway of width 6 feet meets a hallway of width 9 feet at right angles. Find the length of the longest pipe that can be carried level around this corner. [*Hint:* If L is the length of the pipe, show that

$$L = 6\csc\theta + 9\csc\left(\frac{\pi}{2} - \theta\right)$$

where θ is the angle between the pipe and the wall of the narrower hallway.]

54. *Minimum Distance* Rework Exercise 53, given that one of the hallways is of width a feet and one is of width b feet. Show that the result is the same as in Exercise 52.

55. *Harmonic Motion* The height of an object attached to a spring is given by the harmonic equation

$$y = \frac{1}{3}\cos 12t - \frac{1}{4}\sin 12t$$

where y is measured in inches and t is measured in seconds.
a. Calculate the height and velocity of the object when $t = \pi/8$ second.
b. Show that the maximum displacement of the object is $\frac{5}{12}$ inch.
c. Find the period P of y. Find the frequency f (number of oscillations per second) if $f = 1/P$.

56. *Essay* The general equation giving the displacement of an oscillating object attached to a spring is

$$y = A\sin\sqrt{\frac{k}{m}}t + B\cos\sqrt{\frac{k}{m}}t$$

where k is the spring constant and m is the mass of the object.
a. Show that the maximum displacement of the object is $\sqrt{A^2 + B^2}$.
b. Show that the object oscillates with a frequency of $(1/2\pi)\sqrt{k/m}$.
c. How is the frequency changed if the stiffness k of the spring is increased?
d. How is the frequency changed if the mass m of the object is increased?

In Exercises 57 and 58, approximate any real zeros of the function accurate to three decimal places.

57. $f(x) = x^3 - 3x - 1$ **58.** $f(x) = x^3 + 2x + 1 = 0$

In Exercises 59 and 60, approximate to three decimal places the x-value of the points of intersection of the equations.

59. $y = x^4$ **60.** $y = \sin \pi x$
$y = x + 3$ $y = 1 - x$

61. *Surface Area and Volume* The diameter of a sphere is measured to be 18 inches with a maximum possible error of 0.05 inch. Use the differential to approximate the possible error and percentage error in calculating the surface area and volume of the sphere.

62. A company finds that the demand for its commodity is

$$p = 75 - \frac{1}{4}x.$$

If x changes from 7 to 8, find the corresponding change in p. Compare the value of Δp and dp.

63. *Profit* Find the maximum profit if the demand equation is $p = 36 - 4x$ and the total cost is $C = 2x^2 + 6$.

64. *Profit* The cost C of producing x units per day is

$$C = \frac{1}{4}x^2 + 62x + 125$$

and the price p per unit is

$$p = 75 - \frac{1}{3}x.$$

a. What daily output produces maximum profit?
b. What daily output produces minimum average cost?
c. Find the elasticity of demand.

65. *Revenue* For groups of 80 or more, a charter bus company determines the rate per person according to the following formula.

Rate $= \$8.00 - \$0.05(n - 80),\qquad n \geq 80$

What number of passengers will give the bus company maximum revenue?

66. *Cost* The cost of fuel for running a locomotive is proportional to the $\frac{3}{2}$ power of the speed $(s^{3/2})$ and is \$50 per hour for a speed of 25 miles per hour. Other fixed costs amount to an average of \$100 per hour. Find the speed that will minimize the cost per mile.

67. *Inventory Cost* The cost of inventory depends on ordering cost and storage cost, according to the following inventory model.

$$C = \left(\frac{Q}{x}\right)s + \left(\frac{x}{2}\right)r$$

Determine the order size that will minimize the cost, assuming that sales occur at a constant rate, Q is the number of units sold per year, r is the cost of storing one unit for one year, s is the cost of placing an order, and x is the number of units per order.

68. *Taxes* The demand and cost equations for a certain product are $p = 600 - 3x$ and $C = 0.3x^2 + 6x + 600$, where p is the price per unit, x is the number of units, and C is the cost of producing x units. If t is the excise tax per unit, the profit for producing x units is $P = xp - C - xt$. Find the maximum profit for

a. $t = 5$ **b.** $t = 10$ **c.** $t = 20$.

69. Find the maximum and minimum points on the graph of

$$x^2 + 4y^2 - 2x - 16y + 13 = 0$$

a. Without using calculus.
b. Using calculus.

70. Consider the function $f(x) = x^n$ for positive integer values of n.

a. For what values of n does the function have a relative minimum at the origin?

b. For what values of n does the function have a point of inflection at the origin?

True or False **In Exercises 71 and 72, determine whether the statement is true or false. If it is false, explain why or give an example that shows it is false.**

71. If c is a critical number of f, then f has a relative extremum at $x = c$.

72. The function $y = f(x)$ can have at most one horizontal asymptote.

73. *Essay* A newspaper headline states that "the rate of growth of the national deficit is increasing." What does this mean? What does it say about the graph of the deficit as a function of time?

Chapter 4

Integration

Antiderivatives and Indefinite Integration

Antiderivatives • Notation for Antiderivatives • Basic Integration Rules • Initial Conditions and Particular Solutions

Antiderivatives

Suppose you were asked to find a function F whose derivative is

$$f(x) = 3x^2.$$

From your knowledge of derivatives, you would probably say that

$$F(x) = x^3 \quad \text{because} \quad \frac{d}{dx}[x^3] = 3x^2.$$

The function F is an *antiderivative* of f. In general, a function F is an **antiderivative** of f on an interval I if $F'(x) = f(x)$ for all x in I.

Note that F is called *an* antiderivative of f, rather than *the* antiderivative of f. To see why, consider that $F_1(x) = x^3$, $F_2(x) = x^3 - 5$, and $F_3(x) = x^3 + 97$ are all antiderivatives of $f(x) = 3x^2$. In fact, for any constant C, the function given by $F(x) = x^3 + C$ is an antiderivative of f.

THEOREM 4.1   Representation of Antiderivatives

If F is an antiderivative of f on an interval I, then G is an antiderivative of f on the interval I if and only if G is of the form

$$G(x) = F(x) + C, \quad \text{for all } x \text{ in } I$$

where C is a constant.

Proof   The proof of one direction is straightforward. That is, if $G(x) = F(x) + C$, $F'(x) = f(x)$ and c is a constant, then

$$G'(x) = \frac{d}{dx}[F(x) + C] = F'(x) + 0 = f(x).$$

To prove the other direction, you can define a function H such that $H(x) = G(x) - F(x)$. If H is not constant on the interval I, then there must exist a and b ($a < b$) in the interval such that $H(a) \neq H(b)$. Moreover, because H is differentiable on (a, b), you can apply the Mean Value Theorem to conclude that there exists some c in (a, b) such that $H'(c) = (H(b) - H(a))/(b - a)$. Because $H(b) \neq H(a)$, it follows that $H'(c) \neq 0$. However, because $G'(c) = F'(c)$, you know that $H'(c) = G'(c) - F'(c) = 0$, which contradicts the fact that $H'(c) \neq 0$. Consequently, you can conclude that $H(x)$ is a constant, C. Therefore, $G(x) - F(x) = C$ and it follows that $G(x) = F(x) + C$.

◄ *In 1992 a crew of the space shuttle* Endeavor *attached a booster rocket to a stranded communication satellite to propel it to its correct orbit 22,300 miles above earth. The velocity required to place a satellite in orbit is less than the velocity required to escape earth's gravitational pull. The escape velocity is discussed in Section 4.1, Exercise 60.*

Using Theorem 4.1, you can represent the entire family of antiderivatives of a function by adding a constant to a *known* antiderivative. For example, knowing that $D_x[x^2] = 2x$, you can represent the family of *all* antiderivatives of $f(x) = 2x$ by

$$G(x) = x^2 + C$$

where C is a constant. The constant C is called the **constant of integration.** The family of functions represented by G is the **general antiderivative** of f, and $G(x) = x^2 + C$ is the **general solution** of the *differential equation* $G'(x) = 2x$.

A **differential equation** in x and y is an equation that involves x, y, and derivatives of y. For instance, $y' = 3x$ and $y' = x^2 + 1$ are examples of simple differential equations.

FIGURE 4.1
Functions of the form $y = 2x + C$.

EXAMPLE 1   Solving a Differential Equation

Find the general solution of $y' = 2$.

Solution   To begin, you need to find a function whose derivative is 2. One such function is

$$y = 2x.$$

Now, you can use Theorem 4.1 to conclude that the general solution of the differential equation is

$$y = 2x + C.$$

The graph of several functions of the form $y = 2x + C$ is shown in Figure 4.1.

Notation for Antiderivatives

When solving a differential equation of the form

$$\frac{dy}{dx} = f(x)$$

it is convenient to write it in the equivalent differential form

$$dy = f(x)\,dx.$$

The operation of finding all solutions of this equation is called **antidifferentiation** (or **indefinite integration**) and is denoted by an integral sign $\int$. The general solution is denoted by

$$y = \int f(x)\,dx = F(x) + C.$$

The expression $\int f(x)\,dx$ is read as the *antiderivative of f with respect to x.* Thus, the differential dx serves to identify x as the variable of integration. The term **indefinite integral** is a synonym for antiderivative.

In this text, whenever we write $\int f(x)\,dx = F(x) + C$, we will mean that F is an antiderivative of f *on an interval.*

Basic Integration Rules

The inverse nature of integration and differentiation can be verified by substituting $F'(x)$ for $f(x)$ in this definition to obtain

$$\int F'(x)\,dx = F(x) + C.$$ Integration is the "inverse" of differentiation

Moreover, if $\int f(x)\,dx = F(x) + C$, then

$$\frac{d}{dx}\left[\int f(x)\,dx\right] = f(x).$$ Differentiation is the "inverse" of integration

These two equations allow you to obtain integration formulas directly from differentiation formulas, as shown in the following summary.

Basic Integration Rules

Differentiation formula	Integration formula
$\dfrac{d}{dx}[C] = 0$	$\displaystyle\int 0\,dx = C$
$\dfrac{d}{dx}[kx] = k$	$\displaystyle\int k\,dx = kx + C$
$\dfrac{d}{dx}[kf(x)] = kf'(x)$	$\displaystyle\int kf(x)\,dx = k\int f(x)\,dx$
$\dfrac{d}{dx}[f(x) \pm g(x)] = f'(x) \pm g'(x)$	$\displaystyle\int [f(x) \pm g(x)]\,dx =$ $\displaystyle\int f(x)\,dx \pm \int g(x)\,dx$
$\dfrac{d}{dx}[x^n] = nx^{n-1}$	$\displaystyle\int x^n\,dx = \frac{x^{n+1}}{n+1} + C,$ $n \neq -1$ Power Rule
$\dfrac{d}{dx}[\sin x] = \cos x$	$\displaystyle\int \cos x\,dx = \sin x + C$
$\dfrac{d}{dx}[\cos x] = -\sin x$	$\displaystyle\int \sin x\,dx = -\cos x + C$
$\dfrac{d}{dx}[\tan x] = \sec^2 x$	$\displaystyle\int \sec^2 x\,dx = \tan x + C$
$\dfrac{d}{dx}[\sec x] = \sec x \tan x$	$\displaystyle\int \sec x \tan x\,dx = \sec x + C$
$\dfrac{d}{dx}[\cot x] = -\csc^2 x$	$\displaystyle\int \csc^2 x\,dx = -\cot x + C$
$\dfrac{d}{dx}[\csc x] = -\csc x \cot x$	$\displaystyle\int \csc x \cot x\,dx = -\csc x + C$

REMARK   Note that the Power Rule for integration has the restriction that $n \neq -1$. The evaluation of $\int 1/x\,dx$ must wait until the introduction of the natural logarithm function in Chapter 5.

EXAMPLE 2   Applying the Basic Integration Rules

$$\int 3x\,dx = 3\int x\,dx \qquad \text{Constant Multiple Rule}$$

$$= 3\int x^1\,dx \qquad \text{Rewrite } (x = x^1)$$

$$= 3\left(\frac{x^2}{2}\right) + C \qquad \text{Power Rule } (n = 1)$$

$$= \frac{3}{2}x^2 + C \qquad \text{Simplify}$$

When indefinite integrals are evaluated, a strict application of the basic integration rules tends to produce complicated constants of integration. For instance, in Example 2, we could have written

$$\int 3x\,dx = 3\int x\,dx$$

$$= 3\left(\frac{x^2}{2} + C\right)$$

$$= \frac{3}{2}x^2 + 3C.$$

However, because C represents *any* constant, it is both cumbersome and unnecessary to write $3C$ as the constant of integration, and we choose the simpler form, $\frac{3}{2}x^2 + C$.

In Example 2, note that the general pattern of integration is similar to that of differentiation.

$$\boxed{\text{Given integral}} \rightarrow \boxed{\text{Rewrite}} \rightarrow \boxed{\text{Integrate}} \rightarrow \boxed{\text{Simplify}}$$

EXAMPLE 3   Rewriting Before Integrating

Given integral	Rewrite	Integrate	Simplify
a. $\int \dfrac{1}{x^3}\,dx$	$\int x^{-3}\,dx$	$\dfrac{x^{-2}}{-2} + C$	$-\dfrac{1}{2x^2} + C$
b. $\int \sqrt{x}\,dx$	$\int x^{1/2}\,dx$	$\dfrac{x^{3/2}}{3/2} + C$	$\dfrac{2}{3}x^{3/2} + C$
c. $\int 2\sin x\,dx$	$2\int \sin x\,dx$	$2(-\cos x) + C$	$-2\cos x + C$

Remember that you can check your answer to an antidifferentiation problem by differentiating. For instance, in Example 3b, you can check that $\frac{2}{3}x^{3/2}$ is the correct antiderivative by differentiating the answer to obtain

$$D_x\left[\frac{2}{3}x^{3/2}\right] = \left(\frac{2}{3}\right)\left(\frac{3}{2}\right)x^{1/2} = \sqrt{x}.$$

The basic integration rules listed earlier in this section allow you to integrate *any* polynomial function, as demonstrated in Example 4.

EXAMPLE 4 Integrating Polynomial Functions

a. $\displaystyle\int dx = \int 1\,dx = x + C$

b. $\displaystyle\int (x + 2)\,dx = \int x\,dx + \int 2\,dx$

$$= \frac{x^2}{2} + C_1 + 2x + C_2$$

$$= \frac{x^2}{2} + 2x + C \qquad\qquad C = C_1 + C_2$$

The second line in the solution is usually omitted.

c. $\displaystyle\int (3x^4 - 5x^2 + x)\,dx = 3\left(\frac{x^5}{5}\right) - 5\left(\frac{x^3}{3}\right) + \frac{x^2}{2} + C$

$$= \frac{3}{5}x^5 - \frac{5}{3}x^3 + \frac{1}{2}x^2 + C$$

EXAMPLE 5 Rewriting Before Integrating

$$\int \frac{x + 1}{\sqrt{x}}\,dx = \int \left(\frac{x}{\sqrt{x}} + \frac{1}{\sqrt{x}}\right)dx \qquad \text{Rewrite as two fractions}$$

$$= \int \left(x^{1/2} + x^{-1/2}\right)dx \qquad \text{Rewrite with fractional exponents}$$

$$= \frac{x^{3/2}}{3/2} + \frac{x^{1/2}}{1/2} + C \qquad \text{Integrate}$$

$$= \frac{2}{3}x^{3/2} + 2x^{1/2} + C \qquad \text{Simplify}$$

REMARK When integrating quotients, do not integrate the numerator and denominator separately. This is no more valid in integration than it is in differentiation. For instance, in Example 5, be sure you understand that

$$\int \frac{x + 1}{\sqrt{x}}\,dx \neq \frac{\int (x + 1)\,dx}{\int \sqrt{x}\,dx}.$$

EXAMPLE 6 Rewriting Before Integrating

$$\int \frac{\sin x}{\cos^2 x}\,dx = \int \left(\frac{1}{\cos x}\right)\left(\frac{\sin x}{\cos x}\right)dx \qquad \text{Rewrite as product}$$

$$= \int \sec x \tan x\,dx \qquad \text{Rewrite using trigonometric identity}$$

$$= \sec x + C \qquad \text{Integrate}$$

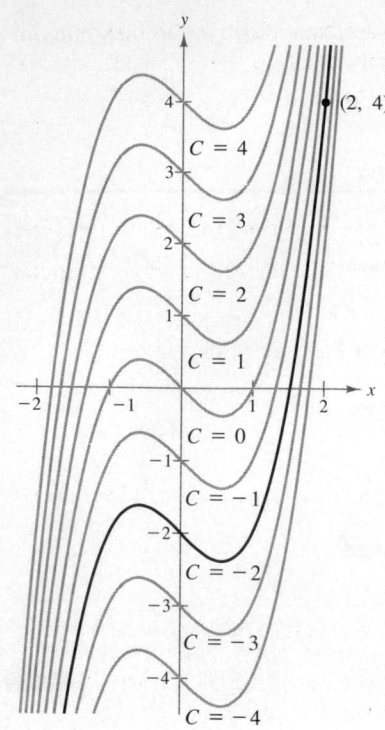

$$F(x) = x^3 - x + C$$

FIGURE 4.2
The particular solution that satisfies
the initial condition $F(2) = 4$ is
$F(x) = x^3 - x - 2$.

Initial Conditions and Particular Solutions

You have already seen that the equation $y = \int f(x)\,dx$ has many solutions (each differing from the others by a constant). This means that the graphs of any two antiderivatives of f are vertical translations of each other. For example, Figure 4.2 shows the graphs of several antiderivatives of the form

$$y = \int (3x^2 - 1)\,dx = x^3 - x + C \qquad \text{General solution}$$

for various integer values of C. Each of these antiderivatives is a solution of the differential equation

$$\frac{dy}{dx} = 3x^2 - 1.$$

In many applications of integration, you are given enough information to determine a **particular solution.** To do this you need only know the value of $F(x)$ for one value of x. (This information is called an **initial condition.**) For example, in Figure 4.2, only one curve passes through the point $(2, 4)$. To find this curve, you can use the following information.

$$F(x) = x^3 - x + C \qquad \text{General solution}$$
$$F(2) = 4 \qquad \text{Initial condition}$$

By using the initial condition in the general solution, you can determine that $F(2) = 8 - 2 + C = 4$, which implies that $C = -2$. Thus, you obtain

$$F(x) = x^3 - x - 2. \qquad \text{Particular solution}$$

EXAMPLE 7   Finding a Particular Solution

Find the general solution of

$$F'(x) = \frac{1}{x^2}, \quad x > 0$$

and find the particular solution that satisfies the initial condition $F(1) = 0$.

Solution   To find the general solution, integrate to obtain

$$F(x) = \int \frac{1}{x^2}\,dx = \int x^{-2}\,dx$$
$$= \frac{x^{-1}}{-1} + C$$
$$= -\frac{1}{x} + C. \qquad \text{General solution}$$

Using the initial condition $F(1) = 0$, you can solve for C as follows.

$$F(1) = -\frac{1}{1} + C = 0 \quad \rightarrow \quad C = 1$$

Thus, the particular solution, as shown in Figure 4.3, is

$$F(x) = -\frac{1}{x} + 1, \qquad x > 0. \qquad \text{Particular solution}$$

$$F(x) = -\frac{1}{x} + C$$

FIGURE 4.3
The particular solution
that satisfies the initial
condition $F(1) = 0$ is
$F(x) = -\dfrac{1}{x} + 1, x > 0$.

So far in this section we have been using x as the variable of integration. In applications, it is often convenient to use a different variable. For instance, in the following example involving *time,* the variable of integration is t.

EXAMPLE 8   Solving a Vertical Motion Problem

A ball is thrown upward with an initial velocity of 64 feet per second from an initial height of 80 feet, as shown in Figure 4.4.

a. Find the position function giving the height s as a function of the time t.
b. When does the ball hit the ground?

Solution

a. Let $t = 0$ represent the initial time. The two given initial conditions can be written as follows.

$$s(0) = 80 \qquad \text{Initial height is 80 ft}$$
$$s'(0) = 64 \qquad \text{Initial velocity is 64 ft/sec}$$

Using -32 feet per second per second as the acceleration due to gravity, you can write

$$s''(t) = -32$$
$$s'(t) = \int s''(t)\, dt = \int -32\, dt = -32t + C_1.$$

Using the initial velocity, you obtain $s'(0) = 64 = -32(0) + C_1$, which implies that $C_1 = 64$. Next, by integrating $s'(t)$, you obtain

$$s(t) = \int s'(t)\, dt = \int (-32t + 64)\, dt = -16t^2 + 64t + C_2$$

and using the initial height, you obtain

$$s(0) = 80 = -16(0^2) + 64(0) + C_2$$

which implies that $C_2 = 80$. Therefore, the position function is

$$s(t) = -16t^2 + 64t + 80.$$

b. Using the position function found in part a, you can find the time that the ball hits the ground by solving the equation $s(t) = 0$.

$$s(t) = -16t^2 + 64t + 80 = 0$$
$$-16(t + 1)(t - 5) = 0$$
$$t = -1, 5$$

Because t must be positive, you conclude that the ball hits the ground 5 seconds after it was thrown.

FIGURE 4.4
Position of a ball at time t.

REMARK   In Example 8, note that the position function has the form $s(t) = \frac{1}{2}gt^2 + v_0 t + s_0$, where $g = -32$, v_0 is the initial velocity, and s_0 is the initial height, as presented in Section 2.2.

Before you begin the exercise set for this section, be sure you realize that one of the most important steps in integration is *rewriting the integrand* in a form that fits the basic integration rules. To further illustrate this point, we list several additional examples in Table 4.1.

TABLE 4.1

Given	Rewrite	Integrate	Simplify
$\displaystyle\int \frac{2}{\sqrt{x}}\,dx$	$\displaystyle 2\int x^{-1/2}\,dx$	$\displaystyle 2\left(\frac{x^{1/2}}{1/2}\right) + C$	$4x^{1/2} + C$
$\displaystyle\int (t^2 + 1)^2\,dt$	$\displaystyle\int (t^4 + 2t^2 + 1)\,dt$	$\displaystyle \frac{t^5}{5} + 2\left(\frac{t^3}{3}\right) + t + C$	$\displaystyle \frac{1}{5}t^5 + \frac{2}{3}t^3 + t + C$
$\displaystyle\int \frac{x^3 + 3}{x^2}\,dx$	$\displaystyle\int (x + 3x^{-2})\,dx$	$\displaystyle \frac{x^2}{2} + 3\left(\frac{x^{-1}}{-1}\right) + C$	$\displaystyle \frac{1}{2}x^2 - \frac{3}{x} + C$
$\displaystyle\int \sqrt[3]{x}\,(x - 4)\,dx$	$\displaystyle\int (x^{4/3} - 4x^{1/3})\,dx$	$\displaystyle \frac{x^{7/3}}{7/3} - 4\left(\frac{x^{4/3}}{4/3}\right) + C$	$\displaystyle \frac{3}{7}x^{4/3}(x - 7) + C$

1-33 odd

EXERCISES for Section 4.1

TECHNOLOGY
Laboratory Guide
Lab 4.1

In Exercises 1–4, find the general solution of the differential equation and check the result by differentiation.

1. $\dfrac{dy}{dt} = 3t^2$ 2. $\dfrac{dr}{d\theta} = \pi$

3. $\dfrac{dy}{dx} = x^{3/2}$ 4. $\dfrac{dy}{dx} = 3x^{-4}$

In Exercises 5–10, complete the table using Table 4.1 as a model.

Given	Rewrite	Integrate	Simplify
5. $\displaystyle\int \sqrt[3]{x}\,dx$			
6. $\displaystyle\int \frac{1}{x^2}\,dx$			
7. $\displaystyle\int \frac{1}{x\sqrt{x}}\,dx$			
8. $\displaystyle\int x(x^2 + 3)\,dx$			
9. $\displaystyle\int \frac{1}{2x^3}\,dx$			
10. $\displaystyle\int \frac{1}{(2x)^3}\,dx$			

In Exercises 11–26, evaluate the indefinite integral and check the result by differentiation.

11. $\displaystyle\int (x^3 + 2)\,dx$ 12. $\displaystyle\int (x^2 - 2x + 3)\,dx$

13. $\displaystyle\int (x^{3/2} + 2x + 1)\,dx$ 14. $\displaystyle\int \left(\sqrt{x} + \frac{1}{2\sqrt{x}}\right)\,dx$

15. $\displaystyle\int \sqrt[3]{x^2}\,dx$ 16. $\displaystyle\int \left(\sqrt[4]{x^3} + 1\right)\,dx$

17. $\displaystyle\int \frac{1}{x^3}\,dx$ 18. $\displaystyle\int \frac{1}{x^4}\,dx$

19. $\displaystyle\int \frac{x^2 + x + 1}{\sqrt{x}}\,dx$ 20. $\displaystyle\int \frac{x^2 + 1}{x^2}\,dx$

21. $\displaystyle\int (x + 1)(3x - 2)\,dx$ 22. $\displaystyle\int (2t^2 - 1)^2\,dt$

23. $\displaystyle\int y^2 \sqrt{y}\,dy$ 24. $\displaystyle\int (1 + 3t)t^2\,dt$

25. $\displaystyle\int dx$ 26. $\displaystyle\int 3\,dt$

In Exercises 27–34, evaluate the trigonometric integral and check the result by differentiation.

27. $\displaystyle\int (2\sin x + 3\cos x)\,dx$ 28. $\displaystyle\int (t^2 - \sin t)\,dt$

29. $\int (1 - \csc t \cot t)\, dt$

30. $\int (\theta^2 + \sec^2 \theta)\, d\theta$

31. $\int (\sec^2 \theta - \sin \theta)\, d\theta$

32. $\int \sec y(\tan y - \sec y)\, dy$

33. $\int (\tan^2 y + 1)\, dy$

34. $\int \dfrac{\sin x}{1 - \sin^2 x}\, dx$

In Exercises 35 and 36, sketch the graphs of the function $g(x) = f(x) + C$ **on the same coordinate axes for** $C = -2$, $C = 0$, **and** $C = 3$.

35. $f(x) = \cos x$

36. $f(x) = \sqrt{x}$

In Exercises 37–40, the graph of the derivative of a function is given. Sketch the graphs of *two* **functions that have the given derivative. (There is more than one correct answer.)**

37.

38.

39.

40.
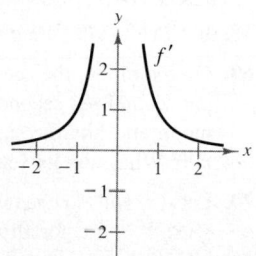

In Exercises 41–44, find the equation for *y*, **given the derivative and the indicated point on the curve.**

41. $\dfrac{dy}{dx} = 2x - 1$

42. $\dfrac{dy}{dx} = 2(x - 1)$

43. $\dfrac{dy}{dx} = \cos x$

44. $\dfrac{dy}{dx} = -\dfrac{1}{x^2}, \quad x > 0$

In Exercises 45–48, solve the differential equation.

45. $f''(x) = 2, \ f'(2) = 5, \ f(2) = 10$

46. $f''(x) = x^2, \ f'(0) = 6, \ f(0) = 3$

47. $f''(x) = x^{-3/2}, \ f'(4) = 2, \ f(0) = 0$

48. $f''(x) = \sin x, \ f'(0) = 1, \ f(0) = 6$

49. *Tree Growth* An evergreen nursery usually sells a certain shrub after 6 years of growth and shaping. The growth rate during those 6 years is approximated by

$$\frac{dh}{dt} = 0.5t + 2$$

where t is the time in years and h is the height in inches. The seedlings are 5 inches tall when planted ($t = 0$).
a. Find the height after t years.
b. How tall are the shrubs when they are sold?

50. *Population Growth* The rate of growth dP/dt of a population of bacteria is proportional to the square root of t where P is the population size and t is the time in days ($0 \le t \le 10$). The initial size of the population is 500. After 1 day, the population has grown to 600. Estimate the population after 7 days.

Vertical Motion **In Exercises 51–54, use** $a(t) = -32$ **ft/sec²** **as the acceleration due to gravity. (Neglect air resistance.)**

51. Show that the height above the ground of an object thrown upward from a point s_0 feet above the ground with an initial velocity of v_0 feet per second is given by the function

$$f(t) = -16t^2 + v_0 t + s_0.$$

52. A ball is thrown vertically upward from the ground with an initial velocity of 60 feet per second. How high will the ball go?

53. With what initial velocity must an object be thrown upward (from ground level) to reach a maximum height of 550 feet (approximate height of the Washington Monument)?

54. A balloon, rising vertically with a velocity of 16 feet per second, releases a sandbag at the instant when the balloon is 64 feet above the ground.
a. How many seconds after its release will the bag strike the ground?
b. With what velocity will it reach the ground?

55. *Acceleration*   The maker of a certain automobile advertises that the automobile takes 13 seconds to accelerate from 15 to 50 miles per hour. Assuming constant acceleration, compute

a. The acceleration in feet per second per second.

b. The distance the car travels during the 13 seconds.

56. *Deceleration*   A car that has been traveling at 45 miles per hour is brought to a stop, at constant deceleration, 132 feet from where the brakes have been applied.

a. How far has the car moved when its speed has been reduced to 30 miles per hour?

b. How far has the car moved when its speed has been reduced to 15 miles per hour?

c. Draw the real number line from 0 to 132, and plot the points found in parts a and b. What conclusions can you draw?

57. *Acceleration*   At the instant the traffic light turns green, an automobile that has been waiting at an intersection starts moving forward with a constant acceleration of 6 feet per second per second. At the same instant, a truck traveling with a constant velocity of 30 feet per second overtakes and passes the car.

a. How far beyond its starting point will the automobile overtake the truck?

b. How fast will the automobile be traveling?

58. *Acceleration*   Assume that a fully loaded airplane starting from rest has a constant acceleration while moving down the runway. The plane requires 0.7 mile of runway and a speed of 160 miles per hour before lifting off. What is the plane's acceleration?

59. *Acceleration*   Galileo Galilei (1564–1642) stated the following proposition concerning falling objects: The time in which any space is traversed by a uniformly accelerating body is equal to the time in which that same space would be traversed by the same body moving at a uniform speed whose value is the mean of the highest speed (of the accelerating body) and the speed just before acceleration began. Use the techniques of this section to verify the statement.

60. *Escape Velocity*   The minimum velocity required for an object to escape earth's gravitational pull is obtained from the solution of the equation

$$\int v\,dv = -GM \int \frac{1}{y^2}\,dy$$

where v is the velocity of an object projected from earth, y is the distance from the center of earth, G is the gravitational constant, and M is the mass of earth. Show that v and y are related by the equation

$$v^2 = v_0^2 + 2GM\left(\frac{1}{y} - \frac{1}{R}\right)$$

where v_0 is the initial velocity of the object and R is the radius of earth.

Marginal Cost   In Exercises 61 and 62, find the cost function and average cost for the given marginal cost and fixed cost (cost when $x = 0$).

	Marginal Cost	Fixed Cost
61.	$\dfrac{dC}{dx} = 2x - 12$	\$50
62.	$\dfrac{dC}{dx} = \dfrac{\sqrt[4]{x}}{10} + 10$	\$2300

In Exercises 63 and 64, find the revenue and demand functions for the given marginal revenue.

63. $\dfrac{dR}{dx} = 100 - 5x$ **64.** $\dfrac{dR}{dx} = 100 - 6x - 2x^2$

True or False   In Exercises 65–68, determine whether the statement is true or false. If it is false, explain why or give an example that shows it is false.

65. Each antiderivative of an nth-degree polynomial function is an $(n + 1)$st-degree polynomial function.

66. If $p(x)$ is a polynomial function, then p has exactly one antiderivative whose graph contains the origin.

67. If $F(x)$ and $G(x)$ are antiderivatives of f on an interval I, then $F(x) = G(x) + C$.

68. If $f'(x) = g(x)$, then $\int g(x)\,dx = f(x) + C$.

69. On the moon, the acceleration due to gravity is -1.6 meters per second per second. A stone is dropped off a cliff on the moon and hits the surface 20 seconds later. How far did it fall? What was its velocity at impact?

70. Let $s(x)$ and $c(x)$ be two functions satisfying $s'(x) = c(x)$ and $c'(x) = -s(x)$ for all x. If $s(0) = 0$ and $c(0) = 1$, prove that $[s(x)]^2 + [c(x)]^2 = 1$.

71. The graphs of f and f' each pass through the origin. Use the graph of f'' to sketch the graphs of f and f'.

FIGURE FOR 71

SECTION **Area**

4.2   Sigma Notation • Area • The Area of a Plane Region • Upper and Lower Sums

Sigma Notation

In the previous section, you studied antidifferentiation. In this section, you will be introduced to a new problem—that of finding the area of a region in the plane. At first glance, these two ideas seem unrelated, but you will discover in Section 4.4 that they are closely related by an important theorem called the Fundamental Theorem of Calculus.

We begin this section by introducing a concise notation for sums. This notation is called **sigma notation** because it uses the uppercase Greek letter sigma, written as Σ.

Sigma Notation

The sum of n terms $a_1, a_2, a_3, \ldots, a_n$ is written as

$$\sum_{i=1}^{n} a_i = a_1 + a_2 + a_3 + \cdots + a_n$$

where i is the **index of summation,** a_i is the **ith term** of the sum, and the **upper and lower bounds of summation** are n and 1, respectively.

REMARK   The upper and lower bounds must be constant *with respect to the index of summation.* However, the lower bound doesn't have to be 1. Any integer less than or equal to the upper bound is legitimate.

EXAMPLE 1   Examples of Sigma Notation

a. $\displaystyle\sum_{i=1}^{6} i = 1 + 2 + 3 + 4 + 5 + 6$

b. $\displaystyle\sum_{j=3}^{7} j^2 = 3^2 + 4^2 + 5^2 + 6^2 + 7^2$

c. $\displaystyle\sum_{k=1}^{n} \frac{1}{n}(k^2 + 1) = \frac{1}{n}(1^2 + 1) + \frac{1}{n}(2^2 + 1) + \cdots + \frac{1}{n}(n^2 + 1)$

d. $\displaystyle\sum_{i=1}^{n} f(x_i)\Delta x = f(x_1)\Delta x + f(x_2)\Delta x + \cdots + f(x_n)\Delta x$ ■

Although any variable can be used as the index of summation, $i, j,$ and k are often used. Note in Example 1 that the index of summation does not appear in the terms of the expanded sum.

The following properties of summation can be derived using the associative and commutative properties of addition and the distributive property of addition over multiplication. (In the first property, k is a constant.)

$$1. \ \sum_{i=1}^{n} ka_i = k\sum_{i=1}^{n} a_i \qquad\qquad 2. \ \sum_{i=1}^{n} (a_i \pm b_i) = \sum_{i=1}^{n} a_i \pm \sum_{i=1}^{n} b_i$$

The next theorem lists some useful formulas for sums of powers. A proof of this theorem is given in Appendix A.

REMARK   Theorem 4.2 allows you to easily evaluate some sums. For example,

$$\sum_{i=1}^{100} i = 1 + 2 + \cdots + 100$$

$$= \frac{100(101)}{2}$$

$$= 5050.$$

THEOREM 4.2   Summation Formulas

$$1. \ \sum_{i=1}^{n} c = cn \qquad\qquad 2. \ \sum_{i=1}^{n} i = \frac{n(n+1)}{2}$$

$$3. \ \sum_{i=1}^{n} i^2 = \frac{n(n+1)(2n+1)}{6} \qquad 4. \ \sum_{i=1}^{n} i^3 = \frac{n^2(n+1)^2}{4}$$

EXAMPLE 2   Evaluating a Sum

Evaluate $\sum_{i=1}^{n} \dfrac{i+1}{n^2}$ for $n = 10, 100, 1000,$ and $10,000.$

Solution   Applying Theorem 4.2, you can write

$$\sum_{i=1}^{n} \frac{i+1}{n^2} = \frac{1}{n^2}\sum_{i=1}^{n} (i+1) \qquad\quad \text{Factor constant } 1/n^2 \text{ out of sum}$$

$$= \frac{1}{n^2}\left(\sum_{i=1}^{n} i + \sum_{i=1}^{n} 1\right) \qquad \text{Write as two sums}$$

$$= \frac{1}{n^2}\left[\frac{n(n+1)}{2} + n\right] \qquad \text{Apply Theorem 4.2}$$

$$= \frac{1}{n^2}\left(\frac{n^2 + 3n}{2}\right)$$

$$= \frac{n+3}{2n}. \qquad\qquad \text{Simplify}$$

Now you can evaluate the sum by substituting the appropriate values of n as shown in Table 4.2.

TABLE 4.2

n	$\sum\limits_{i=1}^{n}\dfrac{i+1}{n^2} = \dfrac{n+3}{2n}$
10	0.65000
100	0.51500
1000	0.50150
10,000	0.50015

In Table 4.2, note that the sum appears to approach a limit as n increases. Although the discussion of limits at infinity in Section 3.5 applies to a variable x, where x can be any real number, many of the same results hold true for limits involving the variable n, where n is restricted to positive integer values. Thus, to find the limit of $(n+3)/2n$ as n approaches infinity, you can write

$$\lim_{n\to\infty} \frac{n+3}{2n} = \frac{1}{2}.$$

FIGURE 4.5
Rectangle: $A = bh$

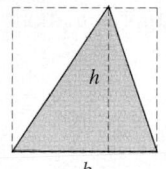

FIGURE 4.6
Triangle: $A = \frac{1}{2}bh$

Area

In Euclidean geometry, the simplest type of plane region is a rectangle. Although people often say that the *formula* for the area of a rectangle is $A = bh$, as shown in Figure 4.5, it is actually more proper to say that this is the *definition* of the **area of a rectangle.**

From this definition, you can develop formulas for the areas of many other plane regions. For example, to determine the area of a triangle, you can form a rectangle whose area is twice that of the triangle, as shown in Figure 4.6. Once you know how to find the area of a triangle, you can determine the area of any polygon by subdividing the polygon into triangular regions, as shown in Figure 4.7.

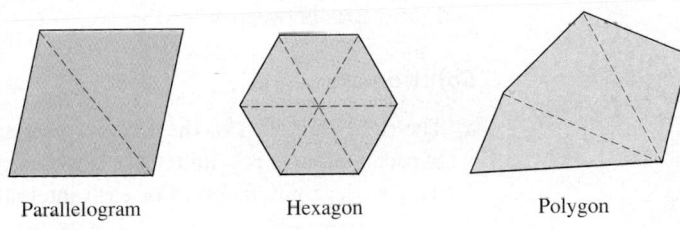

Parallelogram   Hexagon   Polygon

FIGURE 4.7

ARCHIMEDES

Archimedes used the method of exhaustion to derive formulas for the areas of ellipses, parabolic segments, and sectors of a spiral. He is considered to be the greatest applied mathematician of antiquity.

FOR FURTHER INFORMATION
For an alternative development of the formula for the area of a circle, see the article "Proof Without Words" by Russell Jay Hendel in the June, 1990 issue of *Mathematics Magazine.*

Finding the areas of regions other than polygons is more difficult. The ancient Greeks were able to determine formulas for the areas of some general regions (principally those bounded by conics) by the *exhaustion* method. The clearest description of this method was given by Archimedes (287–212 B.C.). Essentially, the method is a limiting process in which the area is squeezed between two polygons—one inscribed in the region and one circumscribed about the region.

For instance, in Figure 4.8 the area of a circular region is approximated by an *n*-sided inscribed polygon and an *n*-sided circumscribed polygon. For each value of *n* the area of the inscribed polygon is less than the area of the circle, and the area of the circumscribed polygon is greater than the area of the circle. Moreover, as *n* increases, the areas of both polygons become better and better approximations of the area of the circle.

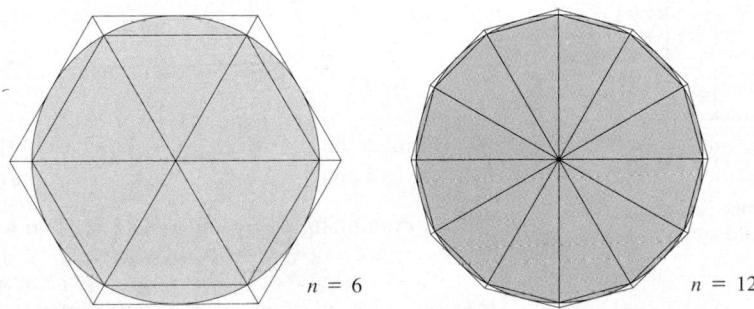

$n = 6$ $n = 12$

FIGURE 4.8
The exhaustion method for finding the area of a circular region.

In the remaining examples in this section, we use a process that is similar to that used by Archimedes to determine the area of a plane region.

The Area of a Plane Region

We introduce the general problem of finding the area of a region in the plane with an example.

EXAMPLE 3   Approximating the Area of a Plane Region

Use the five rectangles in Figures 4.9(a) and (b) to find *two* approximations of the area of the region lying between the graph of

$$f(x) = -x^2 + 5$$

and the x-axis between $x = 0$ and $x = 2$.

Solution

a. The right endpoints of the five intervals are $\frac{2}{5}i$, where $i = 1, 2, 3, 4, 5$. The width of each rectangle is $\frac{2}{5}$, and the height of each rectangle can be obtained by evaluating the f at the right endpoint of each interval.

$$\left[0, \frac{2}{5}\right], \left[\frac{2}{5}, \frac{4}{5}\right], \left[\frac{4}{5}, \frac{6}{5}\right], \left[\frac{6}{5}, \frac{8}{5}\right], \left[\frac{8}{5}, \frac{10}{5}\right]$$

Evaluate $f(x)$ at the right endpoints of these intervals.

The sum of the areas of the five rectangles is

$$\sum_{i=1}^{5} \overbrace{f\left(\frac{2i}{5}\right)}^{\text{Height}} \overbrace{\left(\frac{2}{5}\right)}^{\text{Width}} = \sum_{i=1}^{5} \left[-\left(\frac{2i}{5}\right)^2 + 5\right]\left(\frac{2}{5}\right) = \frac{162}{25} = 6.48.$$

Because each of the five rectangles lies *inside* the parabolic region, you can conclude that the area of the parabolic region is *greater* than 6.48.

b. The left endpoints of the five intervals are $\frac{2}{5}(i - 1)$, where $i = 1, 2, 3, 4, 5$. The width of each rectangle is $\frac{2}{5}$, and the height of each rectangle can be obtained by evaluating the f at the left endpoint of each interval.

$$\sum_{i=1}^{5} \overbrace{f\left(\frac{2i - 2}{5}\right)}^{\text{Height}} \overbrace{\left(\frac{2}{5}\right)}^{\text{Width}} = \sum_{i=1}^{5} \left[-\left(\frac{2i - 2}{5}\right)^2 + 5\right]\left(\frac{2}{5}\right) = \frac{202}{25} = 8.08.$$

Because the parabolic region lies within the union of the five rectangular regions, you can conclude that the area of the parabolic region is *less* than 8.08.

By combining the results of parts a and b, you conclude that

$$6.48 < (\text{Area of region}) < 8.08.$$

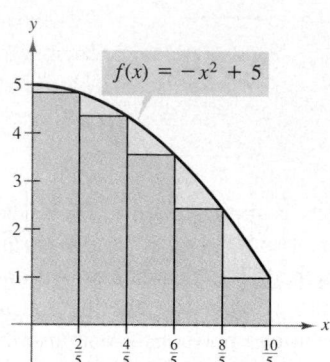

(a) The area of the parabolic region is greater than the area of the rectangles.

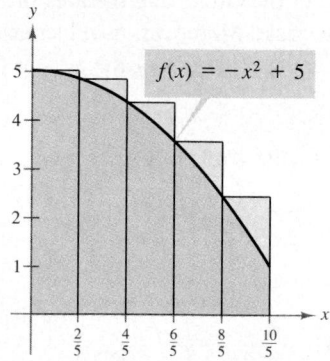

(b) The area of the parabolic region is less than the area of the rectangles.

FIGURE 4.9

REMARK   By increasing the number of rectangles used in Example 3, you can obtain closer and closer approximations of the area of the region. For instance, using 25 rectangles of width $\frac{2}{25}$ each, you can conclude that

$$7.17 < (\text{Area of region}) < 7.49.$$

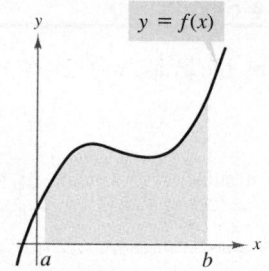

FIGURE 4.10
The region under a curve.

FIGURE 4.11
The interval $[a, b]$ is divided into n subintervals of length $\Delta x = \dfrac{b - a}{n}$

Upper and Lower Sums

The procedure used in Example 3 can be generalized as follows. Consider a plane region bounded above by the graph of a nonnegative, continuous function $y = f(x)$, as shown in Figure 4.10. The region is bounded below by the x-axis, and the left and right boundaries of the region are the vertical lines $x = a$ and $x = b$.

To approximate the area of the region, begin by subdividing the interval $[a, b]$ into n subintervals, each of length $\Delta x = (b - a)/n$, as shown in Figure 4.11. The endpoints of the intervals are as follows.

$$\overbrace{a + 0(\Delta x)}^{a\,=\,x_0} < \overbrace{a + 1(\Delta x)}^{x_1} < \overbrace{a + 2(\Delta x)}^{x_2} < \cdots < \overbrace{a + n(\Delta x)}^{x_n\,=\,b}$$

Because f is continuous, the Extreme Value Theorem guarantees the existence of a minimum and a maximum value of $f(x)$ in *each* subinterval.

$f(m_i) = $ Minimum value of $f(x)$ in ith subinterval

$f(M_i) = $ Maximum value of $f(x)$ in ith subinterval

Next, define an **inscribed rectangle** lying *inside* the ith subregion and **a circumscribed rectangle** extending *outside* the ith subregion. The height of the ith inscribed rectangle is $f(m_i)$ and the height of the ith circumscribed rectangle is $f(M_i)$. For each i, the area of the inscribed rectangle is less than or equal to the area of the circumscribed rectangle.

$$\begin{pmatrix} \text{Area of inscribed} \\ \text{rectangle} \end{pmatrix} = f(m_i)\,\Delta x \leq f(M_i)\,\Delta x = \begin{pmatrix} \text{Area of circumscribed} \\ \text{rectangle} \end{pmatrix}$$

The sum of the areas of the inscribed rectangles is called a **lower sum** and the sum of the areas of the circumscribed rectangles is called an **upper sum.**

$$\text{Lower sum} = s(n) = \sum_{i=1}^{n} f(m_i)\,\Delta x \qquad \text{Area of inscribed rectangles}$$

$$\text{Upper sum} = S(n) = \sum_{i=1}^{n} f(M_i)\,\Delta x \qquad \text{Area of circumscribed rectangles}$$

From Figure 4.12, you can see that the lower sum $s(n)$ is less than or equal to the upper sum $S(n)$. Moreover, the actual area of the region lies between these two sums.

$$s(n) \leq (\text{Area of region}) \leq S(n)$$

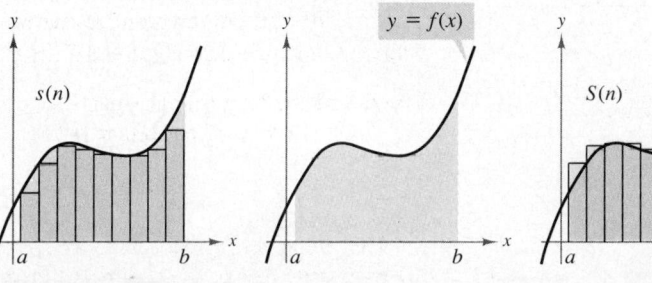

Area of inscribed rectangles
is less than area of region.

Area of circumscribed rectangles
is greater than area of region.

FIGURE 4.12

Inscribed rectangles

Circumscribed rectangles

FIGURE 4.13

EXAMPLE 4 Finding Upper and Lower Sums for a Region

Find the upper and lower sums for the region bounded by the graph of $f(x) = x^2$ and the x-axis between $x = 0$ and $x = 2$.

Solution To begin, partition the interval $[0, 2]$ into n subintervals, each of length

$$\Delta x = \frac{b - a}{n} = \frac{2 - 0}{n} = \frac{2}{n}.$$

Figure 4.13 shows the endpoints of the subintervals and several inscribed and circumscribed rectangles. Because f is increasing on the interval $[0, 2]$, the minimum value on each subinterval occurs at the left endpoint, and the maximum value occurs at the right endpoint.

Left endpoints   *Right endpoints*

$$m_i = 0 + (i - 1)\left(\frac{2}{n}\right) = \frac{2(i - 1)}{n} \qquad M_i = 0 + i\left(\frac{2}{n}\right) = \frac{2i}{n}$$

Using the left endpoints, the lower sum is

$$\begin{aligned}
s(n) &= \sum_{i=1}^{n} f(m_i)\,\Delta x = \sum_{i=1}^{n} f\left[\frac{2(i-1)}{n}\right]\left(\frac{2}{n}\right) \\
&= \sum_{i=1}^{n} \left[\frac{2(i-1)}{n}\right]^2 \left(\frac{2}{n}\right) \\
&= \sum_{i=1}^{n} \left(\frac{8}{n^3}\right)(i^2 - 2i + 1) \\
&= \frac{8}{n^3}\left(\sum_{i=1}^{n} i^2 - 2\sum_{i=1}^{n} i + \sum_{i=1}^{n} 1\right) \\
&= \frac{8}{n^3}\left\{\frac{n(n + 1)(2n + 1)}{6} - 2\left[\frac{n(n + 1)}{2}\right] + n\right\} \\
&= \frac{4}{3n^3}(2n^3 - 3n^2 + n) \\
&= \frac{8}{3} - \frac{4}{n} + \frac{4}{3n^2}.
\end{aligned}$$

Using the right endpoints, the upper sum is

$$\begin{aligned}
S(n) &= \sum_{i=1}^{n} f(M_i)\,\Delta x = \sum_{i=1}^{n} f\left(\frac{2i}{n}\right)\left(\frac{2}{n}\right) \\
&= \sum_{i=1}^{n} \left(\frac{2i}{n}\right)^2 \left(\frac{2}{n}\right) \\
&= \sum_{i=1}^{n} \left(\frac{8}{n^3}\right)i^2 \\
&= \frac{8}{n^3}\left[\frac{n(n + 1)(2n + 1)}{6}\right] \\
&= \frac{4}{3n^3}(2n^3 + 3n^2 + n) \\
&= \frac{8}{3} + \frac{4}{n} + \frac{4}{3n^2}.
\end{aligned}$$

Example 4 illustrates some important things about lower and upper sums. First, notice that for any value of n, the lower sum is less than (or equal to) the upper sum.

$$s(n) = \frac{8}{3} - \frac{4}{n} + \frac{4}{3n^2} < \frac{8}{3} + \frac{4}{n} + \frac{4}{3n^2} = S(n)$$

Second, the difference between these two sums lessens as n increases. In fact, if you take the limit as $n \to \infty$, both the upper and the lower sums approach $\frac{8}{3}$.

$$\lim_{n\to\infty} s(n) = \lim_{n\to\infty}\left(\frac{8}{3} - \frac{4}{n} + \frac{4}{3n^2}\right) = \frac{8}{3} \qquad \text{Lower sum limit}$$

$$\lim_{n\to\infty} S(n) = \lim_{n\to\infty}\left(\frac{8}{3} + \frac{4}{n} + \frac{4}{3n^2}\right) = \frac{8}{3} \qquad \text{Upper sum limit}$$

The next theorem shows that the equivalence of the limit (as $n \to \infty$) of the upper and lower sums is not mere coincidence. It is true for all functions that are continuous and nonnegative on the closed interval $[a, b]$. The proof of this theorem is best left to a course in advanced calculus.

THEOREM 4.3   Limit of the Lower and Upper Sums

Let f be continuous and nonnegative on the interval $[a, b]$. The limits as $n \to \infty$ of both the lower and upper sums exist and are equal to each other. That is,

$$\lim_{n\to\infty} s(n) = \lim_{n\to\infty}\sum_{i=1}^{n} f(m_i)\,\Delta x = \lim_{n\to\infty}\sum_{i=1}^{n} f(M_i)\,\Delta x = \lim_{n\to\infty} S(n)$$

where $\Delta x = (b - a)/n$ and $f(m_i)$ and $f(M_i)$ are the minimum and maximum values of f on the ith subinterval.

Because the same limit is attained for both the minimum value $f(m_i)$ and the maximum value $f(M_i)$, it follows from the Squeeze Theorem (Theorem 1.8) that the choice of x in the ith subinterval does not affect the limit. This means that you are free to choose an *arbitrary* x-value in the ith subinterval, as in the following *definition of the area of a region in the plane.*

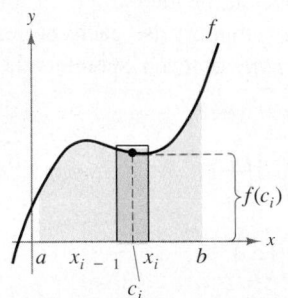

FIGURE 4.14
The length of the ith subinterval is
$\Delta x = x_i - x_{i-1}$.

Definition of the Area of a Region in the Plane

Let f be continuous and nonnegative on the interval $[a, b]$. The **area** of the region bounded by the graph of f, the x-axis, and the vertical lines $x = a$ and $x = b$ is

$$\text{Area} = \lim_{n\to\infty}\sum_{i=1}^{n} f(c_i)\,\Delta x, \qquad x_{i-1} \le c_i \le x_i$$

where $\Delta x = (b - a)/n$ (see Figure 4.14).

EXAMPLE 5   Finding Area by the Limit Definition

Find the area of the region bounded by the graph of $f(x) = x^3$, the x-axis, and the vertical lines $x = 0$ and $x = 1$, as shown in Figure 4.15.

Solution   Begin by noting that f is continuous and nonnegative on the interval $[0, 1]$. Next, partition the interval $[0, 1]$ into n equal subintervals, each of length $\Delta x = 1/n$. According to the definition of area, you can choose any x-value in the ith subinterval. For this example, the right endpoints $(c_i = i/n)$ are convenient.

$$\text{Area} = \lim_{n \to \infty} \sum_{i=1}^{n} f(c_i)\,\Delta x = \lim_{n \to \infty} \sum_{i=1}^{n} \left(\frac{i}{n}\right)^3 \left(\frac{1}{n}\right)$$

$$= \lim_{n \to \infty} \frac{1}{n^4} \sum_{i=1}^{n} i^3$$

$$= \lim_{n \to \infty} \frac{1}{n^4} \left[\frac{n^2(n+1)^2}{4} \right]$$

$$= \lim_{n \to \infty} \left(\frac{1}{4} + \frac{1}{2n} + \frac{1}{4n^2} \right)$$

$$= \frac{1}{4}$$

Thus, the area of the region is $\frac{1}{4}$.

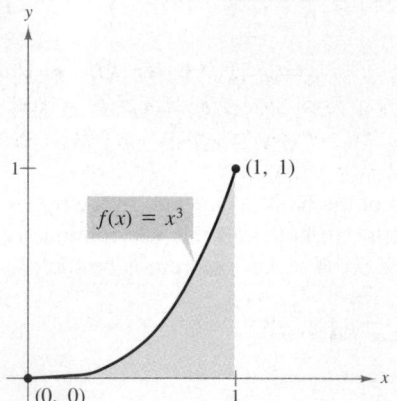

FIGURE 4.15
The area of the region bounded by the graph of f, the x-axis, $x = 0$ and $x = 1$ is $\frac{1}{4}$.

EXAMPLE 6   Finding Area by the Limit Definition

Find the area of the region bounded by the graph of $f(x) = 4 - x^2$, the x-axis, and the vertical lines $x = 1$ and $x = 2$, as shown in Figure 4.16.

Solution   The function f is continuous and nonnegative on the interval $[1, 2]$, and so we begin by partitioning the interval into n equal subintervals, each of length $\Delta x = 1/n$. Choosing the right endpoint, $c_i = 1 + (i/n)$, of each subinterval, you obtain the following.

$$\text{Area} = \lim_{n \to \infty} \sum_{i=1}^{n} f(c_i)\,\Delta x = \lim_{n \to \infty} \sum_{i=1}^{n} \left[4 - \left(1 + \frac{i}{n}\right)^2 \right]\left(\frac{1}{n}\right)$$

$$= \lim_{n \to \infty} \sum_{i=1}^{n} \left(3 - \frac{2i}{n} - \frac{i^2}{n^2} \right)\left(\frac{1}{n}\right)$$

$$= \lim_{n \to \infty} \left(\frac{1}{n}\sum_{i=1}^{n} 3 - \frac{2}{n^2}\sum_{i=1}^{n} i - \frac{1}{n^3}\sum_{i=1}^{n} i^2 \right)$$

$$= \lim_{n \to \infty} \left[3 - \left(1 + \frac{1}{n}\right) - \left(\frac{1}{3} + \frac{1}{2n} + \frac{1}{6n^2}\right) \right]$$

$$= 3 - 1 - \frac{1}{3}$$

$$= \frac{5}{3}$$

Thus, the area of the region is $\frac{5}{3}$.

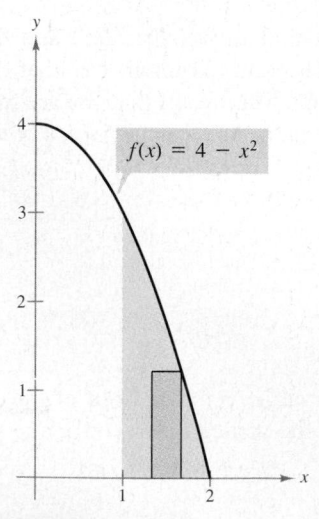

FIGURE 4.16
The area of the region bounded by the graph of f, the x-axis, $x = 1$ and $x = 2$ is $\frac{5}{3}$.

The last example in this section looks at a region that is bounded by the y-axis (rather than the x-axis).

EXAMPLE 7   A Region Bounded by the y-Axis

Find the area of the region bounded by the graph of $f(y) = y^2$ and the y-axis for $0 \le y \le 1$, as shown in Figure 4.17.

Solution   When f is a continuous, nonnegative function of y, you still can use the same basic procedure illustrated in Examples 5 and 6. Begin by partitioning the interval $[0, 1]$ into n equal subintervals, each of length $\Delta y = 1/n$. Then, using the upper endpoints $c_i = i/n$, you obtain the following.

$$\text{Area} = \lim_{n \to \infty} \sum_{i=1}^{n} f(c_i)\, \Delta y = \lim_{n \to \infty} \sum_{i=1}^{n} \left(\frac{i}{n}\right)^2 \left(\frac{1}{n}\right)$$

$$= \lim_{n \to \infty} \frac{1}{n^3} \sum_{i=1}^{n} i^2$$

$$= \lim_{n \to \infty} \frac{n(n+1)(2n+1)}{6n^3}$$

$$= \lim_{n \to \infty} \left(\frac{1}{3} + \frac{1}{2n} + \frac{1}{6n^2}\right)$$

$$= \frac{1}{3}$$

Thus, the area of the region is $\frac{1}{3}$.

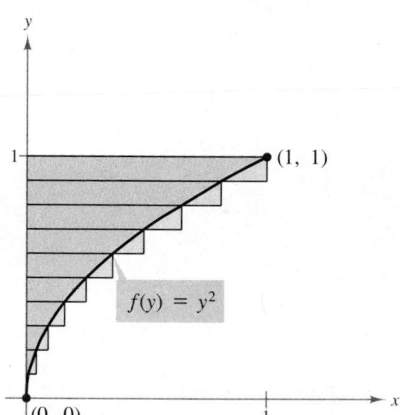

FIGURE 4.17
The area of the region bounded by the graph of f and the y-axis for $0 \le y \le 1$ is $\frac{1}{3}$.

EXERCISES for Section 4.2

In Exercises 1–6, find the sum.

1. $\displaystyle\sum_{i=1}^{5} (2i + 1)$ 2. $\displaystyle\sum_{k-2}^{5} (k + 1)(k - 3)$

3. $\displaystyle\sum_{k=0}^{4} \frac{1}{k^2 + 1}$ 4. $\displaystyle\sum_{j=3}^{5} \frac{1}{j}$

5. $\displaystyle\sum_{k=1}^{4} c$ 6. $\displaystyle\sum_{i=1}^{4} [(i-1)^2 + (i+1)^3]$

In Exercises 7–14, use sigma notation to write the sum.

7. $\dfrac{1}{3(1)} + \dfrac{1}{3(2)} + \dfrac{1}{3(3)} + \cdots + \dfrac{1}{3(9)}$

8. $\dfrac{5}{1+1} + \dfrac{5}{1+2} + \dfrac{5}{1+3} + \cdots + \dfrac{5}{1+15}$

9. $\left[2\left(\dfrac{1}{8}\right) + 3\right] + \left[2\left(\dfrac{2}{8}\right) + 3\right] + \cdots + \left[2\left(\dfrac{8}{8}\right) + 3\right]$

10. $\left[1 - \left(\dfrac{1}{4}\right)^2\right] + \left[1 - \left(\dfrac{2}{4}\right)^2\right] + \cdots + \left[1 - \left(\dfrac{4}{4}\right)^2\right]$

11. $\left[\left(\dfrac{2}{n}\right)^3 - \dfrac{2}{n}\right]\left(\dfrac{2}{n}\right) + \cdots + \left[\left(\dfrac{2n}{n}\right)^3 - \dfrac{2n}{n}\right]\left(\dfrac{2}{n}\right)$

12. $\left[1 - \left(\dfrac{2}{n} - 1\right)^2\right]\left(\dfrac{2}{n}\right) + \cdots + \left[1 - \left(\dfrac{2n}{n} - 1\right)^2\right]\left(\dfrac{2}{n}\right)$

13. $\left[2\left(1 + \dfrac{3}{n}\right)^2\right]\left(\dfrac{3}{n}\right) + \cdots + \left[2\left(1 + \dfrac{3n}{n}\right)^2\right]\left(\dfrac{3}{n}\right)$

14. $\left(\dfrac{1}{n}\right)\sqrt{1 + \left(\dfrac{0}{n}\right)^2} + \cdots + \left(\dfrac{1}{n}\right)\sqrt{1 - \left(\dfrac{n-1}{n}\right)^2}$

In Exercises 15–20, use the properties of sigma notation and Theorem 4.2 to evaluate the sum.

15. $\displaystyle\sum_{i=1}^{20} 2i$ 16. $\displaystyle\sum_{i=1}^{15} (2i - 3)$

17. $\displaystyle\sum_{i=1}^{20} (i - 1)^2$ 18. $\displaystyle\sum_{i=1}^{10} (i^2 - 1)$

19. $\displaystyle\sum_{i=1}^{15} \frac{1}{n^3}(i - 1)^2$ 20. $\displaystyle\sum_{i=1}^{10} i(i^2 + 1)$

In Exercises 21–26, find the limit of $s(n)$ as $n \to \infty$.

21. $s(n) = \left(\dfrac{4}{3n^3}\right)(2n^3 + 3n^2 + n)$

22. $s(n) = \left(\dfrac{8}{3} + \dfrac{4}{n} + \dfrac{4}{3n^2}\right)$

23. $s(n) = \dfrac{81}{n^4}\left[\dfrac{n^2(n+1)^2}{4}\right]$

24. $s(n) = \dfrac{64}{n^3}\left[\dfrac{n(n+1)(2n+1)}{6}\right]$

25. $s(n) = \dfrac{18}{n^2}\left[\dfrac{n(n+1)}{2}\right]$

26. $s(n) = \dfrac{1}{n^2}\left[\dfrac{n(n+1)}{2}\right]$

In Exercises 27–32, find a formula for the sum of n terms. Then use the formula to find the limit as $n \to \infty$.

27. $\displaystyle\lim_{n\to\infty}\sum_{i=1}^{n}\dfrac{16i}{n^2}$

28. $\displaystyle\lim_{n\to\infty}\sum_{i=1}^{n}\left(\dfrac{2i}{n}\right)\left(\dfrac{2}{n}\right)$

29. $\displaystyle\lim_{n\to\infty}\sum_{i=1}^{n}\dfrac{1}{n^3}(i-1)^2$

30. $\displaystyle\lim_{n\to\infty}\sum_{i=1}^{n}\left(1+\dfrac{2i}{n}\right)^2\left(\dfrac{2}{n}\right)$

31. $\displaystyle\lim_{n\to\infty}\sum_{i=1}^{n}\left(1+\dfrac{2i}{n}\right)^3\left(\dfrac{2}{n}\right)$

32. $\displaystyle\lim_{n\to\infty}\sum_{i=1}^{n}\left(1+\dfrac{i}{n}\right)\left(\dfrac{2}{n}\right)$

In Exercises 33–38, use the upper and lower sums to approximate the area of the region using the indicated number of subintervals (of equal length).

33. $y = \sqrt{x}$

34. $y = \sqrt{x} + 1$

35. $y = \dfrac{1}{x}$

36. $y = \dfrac{1}{x-2}$

37. $y = \sqrt{1-x^2}$

38. $y = x^2 - 1$

39. Consider the triangle of area 2 bounded by the graphs of $y = x$, $y = 0$, and $x = 2$.

a. Sketch the graph of the region.

b. Divide the interval $[0, 2]$ into n equal subintervals and show that the endpoints are

$$0 < 1\left(\dfrac{2}{n}\right) < \cdots < (n-1)\left(\dfrac{2}{n}\right) < n\left(\dfrac{2}{n}\right).$$

c. Show that $s(n) = \displaystyle\sum_{i=1}^{n}\left[(i-1)\left(\dfrac{2}{n}\right)\right]\left(\dfrac{2}{n}\right).$

d. Show that $S(n) = \displaystyle\sum_{i=1}^{n}\left[i\left(\dfrac{2}{n}\right)\right]\left(\dfrac{2}{n}\right).$

e. Complete the table.

n	5	10	50	100
$s(n)$				
$S(n)$				

f. Show that $\displaystyle\lim_{n\to\infty} s(n) = \lim_{n\to\infty} S(n) = 2.$

40. Consider the trapezoid of area 4 bounded by the graphs of $y = x$, $y = 0$, $x = 1$, and $x = 3$.

a. Sketch the graph of the region.

b. Divide the interval $[1, 3]$ into n equal subintervals and show that the endpoints are

$$1 < 1 + 1\left(\dfrac{2}{n}\right) < \cdots < 1 + (n-1)\left(\dfrac{2}{n}\right) < 1 + n\left(\dfrac{2}{n}\right).$$

c. Show that $s(n) = \displaystyle\sum_{i=1}^{n}\left[1+(i-1)\left(\dfrac{2}{n}\right)\right]\left(\dfrac{2}{n}\right).$

d. Show that $S(n) = \displaystyle\sum_{i=1}^{n}\left[1+i\left(\dfrac{2}{n}\right)\right]\left(\dfrac{2}{n}\right).$

e. Complete the table.

n	5	10	50	100
$s(n)$				
$S(n)$				

f. Show that $\displaystyle\lim_{n\to\infty} s(n) = \lim_{n\to\infty} S(n) = 4.$

In Exercises 41–48, use the limit process to find the area of the region between the graph of the function and the x-axis over the indicated interval. Sketch the region.

	Function	Interval
41.	$y = -2x + 3$	$[0, 1]$
42.	$y = 3x - 4$	$[2, 5]$
43.	$y = x^2 + 2$	$[0, 1]$
44.	$y = 1 - x^2$	$[-1, 1]$
45	$y = 27 - x^3$	$[1, 3]$
46.	$y = 2x - x^3$	$[0, 1]$
47.	$y = x^2 - x^3$	$[-1, 1]$
48.	$y = x^2 - x^3$	$[-1, 0]$

In Exercises 49 and 50, use the limit process to find the area of the region between the graph of the function and the y-axis over the indicated y-interval. Sketch the region.

Function	Interval
49. $f(y) = 3y$	$[0, 2]$
50. $f(y) = y^2$	$[0, 3]$

In Exercises 51–54, use the *Midpoint Rule*

$$\text{Area} \approx \sum_{i=1}^{n} f\left(\frac{x_i + x_{i-1}}{2}\right) \Delta x$$

with $n = 4$ to approximate the area of the region bounded by the graph of the function and the x-axis over the indicated interval.

Function	Interval
51. $f(x) = x^2 + 3$	$[0, 2]$
52. $f(x) = x^2 + 4x$	$[0, 4]$
53. $f(x) = \tan x$	$\left[0, \dfrac{\pi}{4}\right]$
54. $f(x) = \sin x$	$\left[0, \dfrac{\pi}{2}\right]$

C **In Exercises 55–58, write a computer program or use a spreadsheet to approximate areas by using the Midpoint Rule. Use the program or spreadsheet to approximate the area between the graph of the function and the x-axis over the specified interval and complete the table.**

n	4	8	12	16	20
Approximate area					

Function	Interval
55. $f(x) = \sqrt{x}$	$[0, 4]$
56. $f(x) = \dfrac{8}{x^2 + 1}$	$[2, 6]$
57. $f(x) = \tan\left(\dfrac{\pi x}{8}\right)$	$[1, 3]$
58. $f(x) = \cos \sqrt{x}$	$[0, 2]$

Approximation **In Exercises 59 and 60, determine which given value best approximates the area of the region between the x-axis and the graph of the function over the indicated interval. (Make your selection on the basis of a sketch of the region and *not* by performing any calculations.)**

59. $f(x) = 4 - x^2$ $[0, 2]$
   a. -2 **b.** 6   **c.** 10   **d.** 3   **e.** 8

60. $f(x) = \sin \dfrac{\pi x}{4}$ $[0, 4]$
   a. 3   **b.** 1   **c.** -2 **d.** 8   **e.** 6

*A **C** indicates an exercise in which you are instructed to use graphing technology or a symbolic computer algebra system. The solutions of other exercises may also be facilitated by appropriate technology.

True or False   **In Exercises 61 and 62, determine whether the statement is true or false. If it is false, explain why or give an example that shows it is false.**

61. The sum of the first n positive integers is $n(n + 1)/2$.

62. If f is continuous and nonnegative on $[a, b]$, then the limits as $n \to \infty$ of its lower sum $s(n)$ and upper sum $S(n)$ both exist and are equal to each other.

C **63.** *Monte Carlo Method*   The following program was adapted from the article "Approximation of Area under a Curve" by James M. Sconyers in the February, 1984 issue of *Mathematics Teacher*. The program approximates the area of the region lying under the graph of a monotonic function and above the x-axis between $x = a$ and $x = b$. Run the program for $a = 0$ and $b = \pi/2$ for several values of N2. Explain why the Monte Carlo Method works.

```
10 DEF FNF(X)=SIN(X)
20 A=0
30 B=1.570796
40 PRINT "Input Number of Random Points"
50 INPUT N2
60 N1=0
70 IF FNF(A)>FNF(B) THEN YMAX=FNF(A) ELSE YMAX=FNF(B)
80 FOR I=1 TO N2
90 X=A+(B-A)*RND(1)
100 Y=YMAX*RND(1)
110 IF Y>=FNF(X) THEN GOTO 130
120 N1=N1+1
130 NEXT I
140 AREA=(N1/N2)*(B-A)*YMAX
150 PRINT "Approximate Area:"; AREA
160 END
```

64. Consider an n-sided regular polygon inscribed in a circle of radius r. Join the vertices of the polygon to the center of the circle, forming n congruent triangles (see figure).
   a. Determine the central angle θ (in terms of n).
   b. Show that the area of each triangle is $\frac{1}{2}(r^2 \sin \theta)$.
   c. Let A_n be the sum of the areas of the n triangles. Find $\lim_{n \to \infty} A_n$.

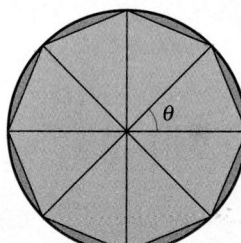

FIGURE FOR 64

65. Verify the formula $\displaystyle\sum_{i=1}^{n} i^2 = \frac{n(n + 1)(2n + 1)}{6}$ by showing the following.
   a. $(i + 1)^3 - i^3 = 3i^2 + 3i + 1$
   b. $-1 + (n + 1)^3 = \displaystyle\sum_{i=1}^{n} (3i^2 + 3i + 1)$
   c. $\displaystyle\sum_{i=1}^{n} i^2 = \frac{n(n + 1)(2n + 1)}{6}$

Riemann Sums

In the definition of area given in Section 4.2, the partitions have subintervals of *equal* length. This was done only for computational convenience. We begin this section with an example that shows that it is not necessary to have subintervals of equal length.

EXAMPLE 1   A Partition with Subintervals of Unequal Lengths

Consider the region bounded by the graph of $f(x) = \sqrt{x}$ and the x-axis for $0 \le x \le 1$, as shown in Figure 4.18. Evaluate the limit

$$\lim_{n \to \infty} \sum_{i=1}^{n} f(c_i) \, \Delta x_i$$

where c_i is the right endpoint of the partition given by $x_i = i^2/n^2$ and Δx_i is the width of the ith interval.

Solution   The length of the ith interval is given by

$$\Delta x_i = \frac{i^2}{n^2} - \frac{(i-1)^2}{n^2} = \frac{i^2 - i^2 + 2i - 1}{n^2} = \frac{2i - 1}{n^2}.$$

Thus, the limit is

$$\lim_{n \to \infty} \sum_{i=1}^{n} f(c_i) \, \Delta x_i = \lim_{n \to \infty} \sum_{i=1}^{n} \sqrt{\frac{i^2}{n^2}} \left(\frac{2i - 1}{n^2} \right)$$

$$= \lim_{n \to \infty} \frac{1}{n^3} \sum_{i=1}^{n} (2i^2 - i)$$

$$= \lim_{n \to \infty} \frac{1}{n^3} \left[2 \frac{n(n+1)(2n+1)}{6} - \frac{n(n+1)}{2} \right]$$

$$= \lim_{n \to \infty} \frac{4n^3 + 3n^2 - n}{6n^3}$$

$$= \frac{2}{3}.$$

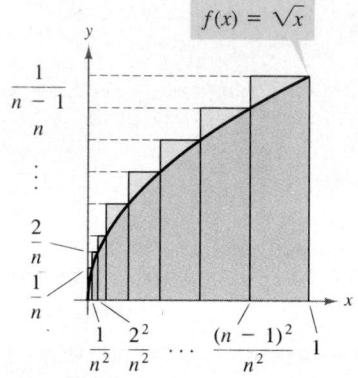

FIGURE 4.18
The subintervals do not have equal length.

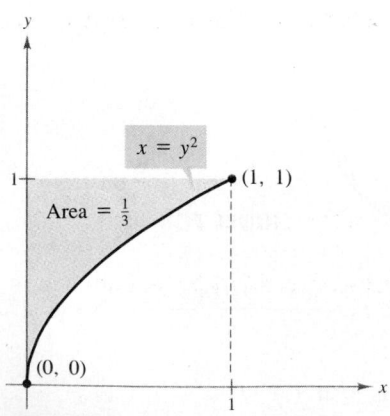

FIGURE 4.19
The area of the region bounded by the graph of $x = y^2$ and the y-axis for $0 \le y \le 1$ is $\frac{1}{3}$.

From Example 7 in Section 4.2, you know that the region shown in Figure 4.19 has an area of $\frac{1}{3}$. Because the square bounded by $0 \le x \le 1$ and $0 \le y \le 1$ has an area of 1, you can conclude that the area of the region shown in Figure 4.18 has an area of $\frac{2}{3}$. This agrees with the limit found in Example 1, even though that example used a partition having subintervals of unequal lengths. The reason this particular partition gave the proper area is that as n increases, the *length of the largest subinterval approaches zero*. This is a key feature of the development of definite integrals.

In the previous section, the limit of a sum was used to define the area of a region in the plane. Finding area by this means is only one of *many* applications involving the limit of a sum. A similar approach can be used to determine quantities as diverse as arc length, average value, centroids, volumes, work, and surface areas. The following development is named after Georg Friedrich Bernhard Riemann (1826–1866). Although the definite integral had been defined and used long before the time of Riemann, he generalized the concept to cover a broader category of functions.

In the following definition of a Riemann sum, note that the function f has no restrictions other than being defined on the interval $[a, b]$. (In the previous section, the function f was assumed to be continuous and nonnegative because we were dealing with the area under a curve.)

GEORG FRIEDRICH BERNHARD RIEMANN

Riemann, a German mathematician and professor at Göttingen, did his most famous work in the areas of non-Euclidean geometry, differential equations, and number theory.

Definition of a Riemann Sum

Let f be defined on the closed interval $[a, b]$, and let Δ be a partition of $[a, b]$ given by $a = x_0 < x_1 < x_2 < \cdots < x_{n-1} < x_n = b$, where Δx_i is the length of the ith subinterval. If c_i is *any* point in the ith subinterval, then the sum

$$\sum_{i=1}^{n} f(c_i) \Delta x_i, \qquad x_{i-1} \leq c_i \leq x_i$$

is called a **Riemann sum** of f for the partition Δ.

REMARK   The sums in Section 4.2 are examples of Riemann sums, but there are more general Riemann sums than those covered there.

The length of the largest subinterval of a partition Δ is the **norm** of the partition, and is denoted by $\|\Delta\|$. If every subinterval is of equal length, then the partition is **regular** and the norm is denoted by

$$\|\Delta\| = \Delta x = \frac{b - a}{n}. \qquad \text{Regular partition}$$

For a general partition, the norm is related to the number of subintervals of $[a, b]$ in the following way.

$$\frac{b - a}{\|\Delta\|} \leq n \qquad \text{General partition}$$

Thus, the number of subintervals in a partition approaches infinity as the norm of the partition approaches 0. That is, $\|\Delta\| \to 0$ implies that $n \to \infty$.

The converse of this statement is not true. For example, let Δ_n be the partition of the interval $[0, 1]$ given by

$$0 < \frac{1}{2^n} < \frac{1}{2^{n-1}} < \cdots < \frac{1}{8} < \frac{1}{4} < \frac{1}{2} < 1.$$

For any positive value of n, the norm of the partition Δ_n is $\frac{1}{2}$. Thus, letting n approach infinity does not force $\|\Delta\|$ to approach 0. In a regular partition, the statements $\|\Delta\| \to 0$ and $n \to \infty$ are equivalent.

Definite Integrals

To define the definite integral, consider the following limit.

$$\lim_{\|\Delta\|\to 0} \sum_{i=1}^{n} f(c_i)\,\Delta x_i = L$$

To say that this limit exists means that for $\varepsilon > 0$ there exists a $\delta > 0$ such that for every partition with $\|\Delta\| < \delta$ it follows that

$$\left| L - \sum_{i=1}^{n} f(c_i)\,\Delta x_i \right| < \varepsilon.$$

(This must be true for any choice of c_i in the ith subinterval of Δ.)

Definition of a Definite Integral

If f is defined on the closed interval $[a, b]$ and the limit

$$\lim_{\|\Delta\|\to 0} \sum_{i=1}^{n} f(c_i)\,\Delta x_i$$

exists, then f is **integrable** on $[a, b]$ and the limit is denoted by

$$\lim_{\|\Delta\|\to 0} \sum_{i=1}^{n} f(c_i)\,\Delta x_i = \int_{a}^{b} f(x)\,dx.$$

The limit is called the **definite integral** of f from a to b. The number a is the **lower limit** of integration, and the number b is the **upper limit** of integration.

It is not a coincidence that the notation for definite integrals is similar to that used for indefinite integrals. You will see why in the next section when we discuss the Fundamental Theorem of Calculus. For now it is important to see that definite integrals and indefinite integrals are different entities. A definite integral is a *number,* whereas an indefinite integral is a *family of functions.*

A sufficient condition for a function f to be integrable on $[a, b]$ is that it is continuous on $[a, b]$. A proof of this theorem is beyond the scope of this text.

THEOREM 4.4   Continuity Implies Integrability

If a function f is continuous on the closed interval $[a, b]$, then f is integrable on $[a, b]$.

REMARK   The converse of Theorem 4.4 is not true. That is, a function can be integrable on a closed interval without being continuous on the interval. For instance, the greatest integer function is integrable on the interval $[0, 2]$ (see Exercise 55).

EXAMPLE 2   Evaluating a Definite Integral as a Limit

Evaluate the definite integral $\displaystyle\int_{-2}^{1} 2x\, dx$.

Solution   The function $f(x) = 2x$ is integrable on the interval $[-2, 1]$, because it is continuous on $[-2, 1]$. Moreover, the definition of integrability implies that any partition whose norm approaches 0 can be used to determine the limit. For computational convenience, define Δ by subdividing $[-2, 1]$ into n subintervals of equal length

$$\Delta x_i = \Delta x = \frac{b - a}{n} = \frac{3}{n}.$$

Choosing c_i as the right endpoint of each subinterval produces

$$c_i = a + i(\Delta x) = -2 + \frac{3i}{n}.$$

Thus, the definite integral is given by

$$\int_{-2}^{1} 2x\, dx = \lim_{\|\Delta\|\to 0} \sum_{i=1}^{n} f(c_i)\, \Delta x_i = \lim_{n\to\infty} \sum_{i=1}^{n} f(c_i)\Delta x$$

$$= \lim_{n\to\infty} \sum_{i=1}^{n} 2\left(-2 + \frac{3i}{n}\right)\left(\frac{3}{n}\right)$$

$$= \lim_{n\to\infty} \frac{6}{n}\sum_{i=1}^{n}\left(-2 + \frac{3i}{n}\right)$$

$$= \lim_{n\to\infty} \frac{6}{n}\left\{-2n + \frac{3}{n}\left[\frac{n(n + 1)}{2}\right]\right\}$$

$$= \lim_{n\to\infty}\left(-12 + 9 + \frac{9}{n}\right)$$

$$= -3.$$

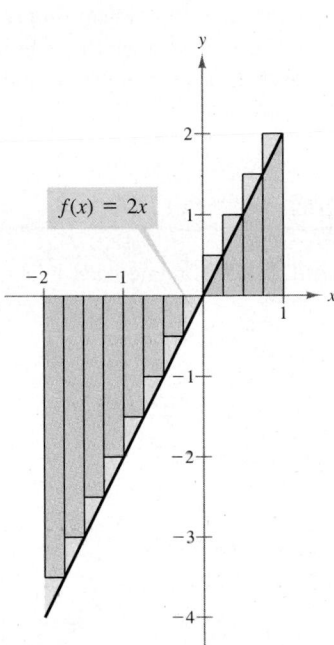

FIGURE 4.20
Because the definite integral is negative, it does not represent the area of the region.

Because the definite integral in Example 2 is negative, it *does not* represent the area of the region shown in Figure 4.20. Definite integrals can be positive, negative, or 0. For a definite integral to be interpreted as an area (as defined in Section 4.2), the function f must be continuous and nonnegative on $[a, b]$, as stated in the following theorem. (Proof of this theorem is straightforward—you simply use the definition of area given in Section 4.2.)

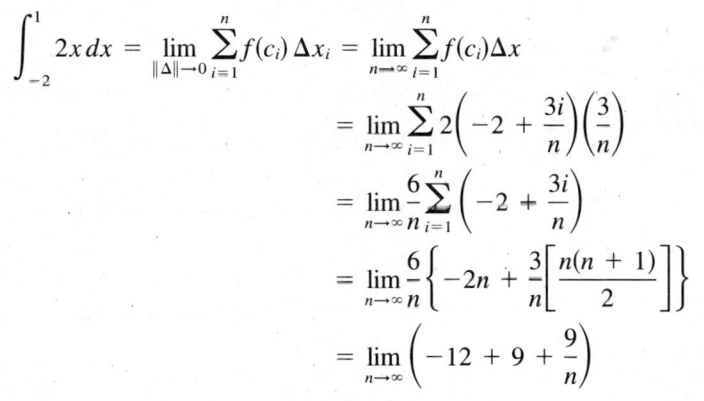

THEOREM 4.5   The Definite Integral as the Area of a Region

If f is continuous and nonnegative on the closed interval $[a, b]$, then the area of the region bounded by the graph of f, the x-axis, and the vertical lines $x = a$ and $x = b$ is given by

$$\text{Area} = \int_{a}^{b} f(x)\, dx.$$

(See Figure 4.21.)

FIGURE 4.21
The region bounded by the graph of f, the x-axis, $x = a$, and $x = b$.

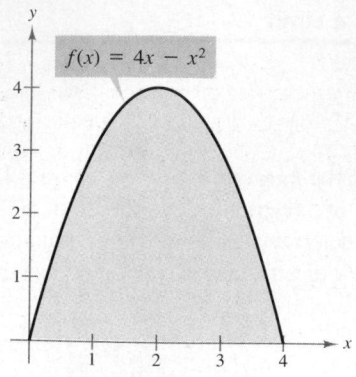

FIGURE 4.22

$$\text{Area} = \int_0^4 (4x - x^2)\, dx$$

REMARK   The variable of integration in a definite integral is sometimes called a *dummy variable* because it can be replaced by any other variable without changing the value of the integral. For instance, the definite integrals

$$\int_0^3 (x + 2)\, dx$$

and

$$\int_0^3 (t + 2)\, dt$$

have the same value.

As an example of Theorem 4.5, consider the region bounded by the graph of

$$f(x) = 4x - x^2$$

and the *x*-axis, as shown in Figure 4.22. Because *f* is continuous and nonnegative on the closed interval $[0, 4]$, the area of the region is

$$\text{Area} = \int_0^4 (4x - x^2)\, dx.$$

A straightforward technique for evaluating a definite integral such as this will be discussed in Section 4.4. For now, however, you can evaluate a definite integral in two ways—you can use the limit definition *or* you can check to see whether the definite integral represents the area of a common geometric region such as a rectangle, triangle, or semicircle.

EXAMPLE 3   Areas of Common Geometric Figures

Sketch the region corresponding to each definite integral. Then evaluate each integral using a geometric formula.

a. $\displaystyle\int_1^3 4\, dx$ **b.** $\displaystyle\int_0^3 (x + 2)\, dx$ **c.** $\displaystyle\int_{-2}^2 \sqrt{4 - x^2}\, dx$

Solution   A sketch of each region is shown in Figure 4.23.

a. This region is a rectangle of height 4 and width 2.

$$\int_1^3 4\, dx = (\text{Area of rectangle}) = 4(2) = 8$$

b. This region is a trapezoid with an altitude of 3 and parallel bases of lengths 2 and 5. The formula for the area of a trapezoid is $\frac{1}{2}h(b_1 + b_2)$.

$$\int_0^3 (x + 2)\, dx = (\text{Area of trapezoid}) = \frac{1}{2}(3)(2 + 5) = \frac{21}{2}$$

c. This region is a semicircle of radius 2. The formula for the area of a semicircle is $\frac{1}{2}\pi r^2$.

$$\int_{-2}^2 \sqrt{4 - x^2}\, dx = (\text{Area of semicircle}) = \frac{1}{2}\pi(2^2) = 2\pi$$

FIGURE 4.23

Properties of Definite Integrals

The definition of the definite integral of f on the interval $[a, b]$ specified that $a < b$. Now, however, it is convenient to extend the definition to cover cases in which $a = b$ or $a > b$. Geometrically, the following two special definitions seem reasonable. For instance, it makes sense to define the area of a region of zero width and finite height to be 0.

Definition of Two Special Definite Integrals

1. If f is defined at $x = a$, then $\displaystyle\int_a^a f(x)\, dx = 0$.

2. If f is integrable on $[a, b]$, then $\displaystyle\int_b^a f(x)\, dx = -\int_a^b f(x)\, dx$.

EXAMPLE 4   Evaluating Definite Integrals

Evaluate the definite integrals.

a. $\displaystyle\int_\pi^\pi \sin x\, dx$ **b.** $\displaystyle\int_3^0 (x + 2)\, dx$

Solution

a. Because the sine function is defined at $x = \pi$, and the upper and lower limits of integration are equal, you can write

$$\int_\pi^\pi \sin x\, dx = 0.$$

b. This integral is the same as that given in Example 3b except that the upper and lower limits are interchanged. Because the integral in Example 3b has a value of $\frac{21}{2}$, you can write

$$\int_3^0 (x + 2)\, dx = -\int_0^3 (x + 2)\, dx = -\frac{21}{2}.$$

THEOREM 4.6   Additive Interval Property

If f is integrable on the three closed intervals determined by a, b, and c, then

$$\int_a^b f(x)\, dx = \int_a^c f(x)\, dx + \int_c^b f(x)\, dx.$$

This theorem is valid for any integrable function and for any three numbers a, b, and c. Rather than give a formal proof for the general case, however, it seems more instructive to give a geometric argument for the case in which $a < c < b$ and f is continuous and nonnegative, as indicated in Figure 4.24.

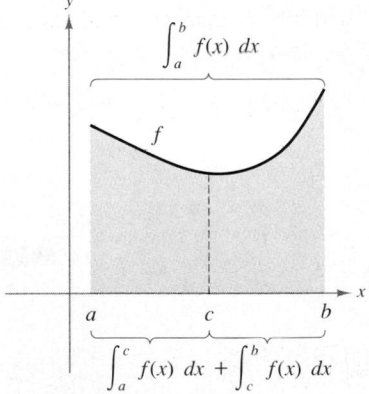

FIGURE 4.24
The larger region can be divided at $x = c$ into two subregions whose intersection is a line segment. Because the line segment has zero area, it follows that the area of the larger region is equal to the sum of the areas of the two smaller regions.

Because the definite integral is defined as the limit of a sum, it inherits the properties of summation given before in Theorem 4.2.

THEOREM 4.7   Properties of Definite Integrals

If f and g are integrable on $[a, b]$ and k is a constant, then the functions kf and $f \pm g$ are integrable on $[a, b]$.

1. $\displaystyle\int_a^b k f(x)\, dx = k \int_a^b f(x)\, dx$

2. $\displaystyle\int_a^b [f(x) \pm g(x)]\, dx = \int_a^b f(x)\, dx \pm \int_a^b g(x)\, dx$

REMARK   Property 2 of Theorem 4.7 can be extended to cover any finite number of functions. For example,

$$\int_a^b [f(x) + g(x) + h(x)]\, dx = \int_a^b f(x)\, dx + \int_a^b g(x)\, dx + \int_a^b h(x)\, dx.$$

EXAMPLE 5   Evaluation of a Definite Integral

Evaluate $\displaystyle\int_1^3 (-x^2 + 4x - 3)\, dx$ using the following values.

$$\int_1^3 x^2\, dx = \frac{26}{3}, \quad \int_1^3 x\, dx = 4, \quad \text{and} \quad \int_1^3 dx = 2$$

Solution

$$\int_1^3 (-x^2 + 4x - 3)\, dx = \int_1^3 (-x^2)\, dx + \int_1^3 4x\, dx + \int_1^3 (-3)\, dx$$

$$= -\int_1^3 x^2\, dx + 4\int_1^3 x\, dx - 3\int_1^3 dx$$

$$= -\left(\frac{26}{3}\right) + 4(4) - 3(2)$$

$$= \frac{4}{3}$$

If f and g are continuous on the closed interval $[a, b]$ and

$$0 \le f(x) \le g(x)$$

for $a \le x \le b$, then the following properties are true. First, the area of the region bounded by the graph of f and the x-axis (between a and b) must be nonnegative. Second, this area must be less than or equal to the area of the region bounded by the graph of g and the x-axis (between a and b), as shown in Figure 4.25. These two results are generalized in the following theorem. (A proof of this theorem is given in Appendix A.)

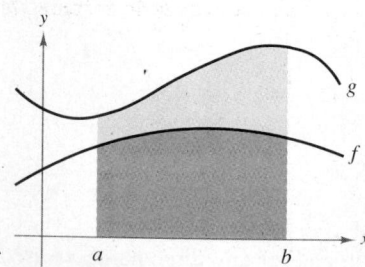

FIGURE 4.25
$\int_a^b f(x)\, dx \le \int_a^b g(x)\, dx$

THEOREM 4.8   Preservation of Inequality

1. If f is integrable and nonnegative on the closed interval $[a,b]$, then

$$0 \le \int_a^b f(x)\,dx.$$

2. If f and g are integrable on the closed interval $[a,b]$ and $f(x) \le g(x)$ for every x in $[a,b]$, then

$$\int_a^b f(x)\,dx \le \int_a^b g(x)\,dx.$$

EXERCISES for Section 4.3

TECHNOLOGY
Laboratory Guide
Lab 4.3

In Exercises 1–10, set up a definite integral that yields the area of the region. (Do not evaluate the integral.)

1. $f(x) = 3$

2. $f(x) = 4 - 2x$

3. $f(x) = 4 - |x|$

4. $f(x) = x^2$

5. $f(x) = 4 - x^2$

6. $f(x) = \dfrac{1}{x^2 + 1}$

7. $f(x) = \sin x$

8. $f(x) = \tan x$

9. $g(y) = y^3$

10. $f(y) = (y - 2)^2$

In Exercises 11–20, sketch the region whose area is indicated by the definite integral. Then use a geometric formula to evaluate the integral ($a > 0, r > 0$).

11. $\displaystyle\int_0^3 4\,dx$

12. $\displaystyle\int_{-a}^a 4\,dx$

13. $\displaystyle\int_0^4 x\,dx$

14. $\displaystyle\int_0^4 \frac{x}{2}\,dx$

15. $\displaystyle\int_0^2 (2x + 5)\,dx$

16. $\displaystyle\int_0^5 (5 - x)\,dx$

17. $\displaystyle\int_{-1}^1 (1 - |x|)\,dx$

18. $\displaystyle\int_{-a}^a (a - |x|)\,dx$

19. $\int_{-3}^{3} \sqrt{9 - x^2}\,dx$ **20.** $\int_{-r}^{r} \sqrt{r^2 - x^2}\,dx$

21. Given $\int_0^5 f(x)\,dx = 10$ and $\int_5^7 f(x)\,dx = 3$, find

a. $\int_0^7 f(x)\,dx$ **b.** $\int_5^0 f(x)\,dx$

c. $\int_5^5 f(x)\,dx$ **d.** $\int_0^5 3f(x)\,dx$

22. Given $\int_0^3 f(x)\,dx = 4$ and $\int_3^6 f(x)\,dx = -1$, find

a. $\int_0^6 f(x)\,dx$ **b.** $\int_6^3 f(x)\,dx$

c. $\int_4^4 f(x)\,dx$ **d.** $\int_3^6 -5f(x)\,dx$

23. Given $\int_2^6 f(x)\,dx = 10$ and $\int_2^6 g(x)\,dx = -2$, find

a. $\int_2^6 [f(x) + g(x)]\,dx$ **b.** $\int_2^6 [g(x) - f(x)]\,dx$

c. $\int_2^6 2g(x)\,dx$ **d.** $\int_2^6 3f(x)\,dx$

24. Given $\int_{-1}^1 f(x)\,dx = 0$ and $\int_0^1 f(x)\,dx = 5$, find

a. $\int_{-1}^0 f(x)\,dx$ **b.** $\int_0^1 f(x)\,dx - \int_{-1}^0 f(x)\,dx$

c. $\int_{-1}^1 3f(x)\,dx$ **d.** $\int_0^1 3f(x)\,dx$

In Exercises 25–30, evaluate the definite integral by the limit definition.

25. $\int_4^{10} 6\,dx$ **26.** $\int_{-2}^3 x\,dx$

27. $\int_{-1}^1 x^3\,dx$ **28.** $\int_0^1 x^3\,dx$

29. $\int_1^2 (x^2 + 1)\,dx$ **30.** $\int_1^2 4x^2\,dx$

In Exercises 31–34, express the limit as a definite integral on the interval $[a, b]$ where c_i is any point in the ith subinterval.

Limit	Interval
31. $\displaystyle\lim_{\|\Delta\| \to 0} \sum_{i=1}^n (3c_i + 10)\,\Delta x_i$	$[-1, 5]$
32. $\displaystyle\lim_{\|\Delta\| \to 0} \sum_{i=1}^n 6c_i(4 - c_i)^2\,\Delta x_i$	$[0, 4]$
33. $\displaystyle\lim_{\|\Delta\| \to 0} \sum_{i=1}^n \sqrt{c_i^2 + 4}\,\Delta x_i$	$[0, 3]$
34. $\displaystyle\lim_{\|\Delta\| \to 0} \sum_{i=1}^n \left(\frac{3}{c_i^2}\right) \Delta x_i$	$[1, 3]$

C **Write a computer program or use a spreadsheet to approximate a definite integral by using the Riemann sum**

$$\sum_{i=1}^n f(c_i)\,\Delta x_i$$

where the subintervals are of equal width. The output should give three approximations of the integral where c_i is the left-

hand endpoint, midpoint, and right-hand endpoint of each subinterval. Denote these by $L(n)$, $M(n)$, and $R(n)$, respectively. In Exercises 35–38, use the program or spreadsheet to approximate the definite integral and complete the table.

n	4	8	12	16	20
$L(n)$					
$M(n)$					
$R(n)$					

35. $\int_0^3 x\sqrt{3 - x}\,dx$ **36.** $\int_0^3 \frac{5}{x^2 + 1}\,dx$

37. $\int_0^{\pi/2} \sin^2 x\,dx$ **38.** $\int_0^3 x\sin x\,dx$

In Exercises 39 and 40, determine which given value best approximates the definite integral. (Make your selection on the basis of a sketch.)

39. $\int_0^4 \sqrt{x}\,dx$

a. 5   **b.** -3 **c.** 10   **d.** 2   **e.** 8

40. $\int_0^{1/2} 4\cos \pi x\,dx$

a. 4   **b.** $\frac{4}{3}$ **c.** 16   **d.** 2π **e.** -6

41. Find the Riemann sum for $f(x) = x^2 + 3x$ over the interval $[0, 8]$ where $x_0 = 0$, $x_1 = 1$, $x_2 = 3$, $x_3 = 7$, $x_4 = 8$, and $c_1 = 1$, $c_2 = 2$, $c_3 = 5$, and $c_4 = 8$.

42. Find the Riemann sum for $f(x) = \sin x$ over the interval $[0, 2\pi]$ where $x_0 = 0$, $x_1 = \pi/4$, $x_2 = \pi/3$, $x_3 = \pi$, $x_4 = 2\pi$, and $c_1 = \pi/6$, $c_2 = \pi/3$, $c_3 = 2\pi/3$, and $c_4 = 3\pi/2$.

In Exercises 43–48, determine whether the statement is true or false. If it is false, explain why or give an example that shows it is false.

43. $\int_a^b [f(x) + g(x)]\,dx = \int_a^b f(x)\,dx + \int_a^b g(x)\,dx$

44. $\int_a^b f(x)g(x)\,dx = \left[\int_a^b f(x)\,dx\right]\left[\int_a^b g(x)\,dx\right]$

45. If the norm of a partition approaches 0, then the number of subintervals approaches infinity.

46. If f is increasing on $[a, b]$, then the minimum value of $f(x)$ on $[a, b]$ is $f(a)$.

47. The value of $\int_a^b f(x)\,dx$ must be positive.

48. If $\int_a^b f(x)\,dx > 0$, then f is nonnegative for all x in $[a, b]$.

In Exercises 49 and 50, use Example 1 as a model to evaluate the limit

$$\lim_{n \to \infty} \sum_{i=1}^n f(c_i)\,\Delta x_i$$

over the region bounded by the graphs of the equations.

49. $f(x) = \sqrt{x}$, $y = 0$, $x = 0$, $x = 2$

(*Hint*: Let $c_i = 2i^2/n^2$.)

50. $f(x) = \sqrt[3]{x}$, $y = 0$, $x = 0$, $x = 1$

(*Hint*: Let $c_i = i^3/n^3$.)

51. Prove that if f is a continuous function on a closed interval $[a, b]$, then

$$\left| \int_a^b f(x) \, dx \right| \leq \int_a^b |f(x)| \, dx.$$

52. Determine whether the function

$$f(x) = \frac{1}{x - 4}$$

is integrable on the interval $[3, 5]$, and give a reason for your answer.

53. Determine whether the function

$$f(x) = \begin{cases} 1, & x \text{ is rational} \\ 0, & x \text{ is irrational} \end{cases}$$

is integrable on the interval $[0, 1]$, and give a reason for your answer.

54. Give an example of a function that is integrable on the interval $[-1, 1]$ but not continuous on $[-1, 1]$.

55. Evaluate, if possible, the integral

$$\int_0^2 [\![x]\!] \, dx.$$

56. Suppose f is integrable on $[a, b]$ and $m \leq f(x) \leq M$ for all x in $[a, b]$, and $m > 0$, $M > 0$. Prove that

$$m(b - a) \leq \int_a^b f(x) \, dx \leq M(b - a).$$

Use this result to estimate $\int_0^1 \sqrt{1 + x^4} \, dx$.

57. Determine $\lim\limits_{n \to \infty} \dfrac{1}{n^3} [1^2 + 2^2 + 3^2 + \cdots + n^2]$ by using an appropriate Riemann sum.

The Fundamental Theorem of Calculus

The Fundamental Theorem of Calculus • The Mean Value Theorem for Integrals • Average Value of a Function • The Second Fundamental Theorem of Calculus

The Fundamental Theorem of Calculus

You have now been introduced to the two major branches of calculus: differential calculus (introduced with the tangent line problem) and integral calculus (introduced with the area problem). At this point, these two poblems might seem unrelated—but there is a very close connection. The connection was discovered independently by Isaac Newton and Gottfried Leibniz. The connection is stated in a theorem that is appropriately called the **Fundamental Theorem of Calculus.**

Roughly, the theorem states that differentiation and (definite) integration are inverse operations, in the same sense that division and multiplication are inverse operations. To see how Newton and Leibniz might have anticipated this relationship, consider the approximations shown in Figure 4.26. When we defined the slope of the tangent line, we used the *quotient* $\Delta y/\Delta x$ (the slope of the secant line). Similarly, when we defined the area of a region under a curve, we used the *product* $\Delta y \Delta x$ (the area of a rectangle). Thus, at least in the primitive approximation stage, the operations of differentiation and definite integration appear to have an inverse relationship in the same sense that division and multiplication are inverse operations. The Fundamental Theorem of Calculus states that the limit processes (used to define the derivative and the definite integral) preserve this inverse relationship.

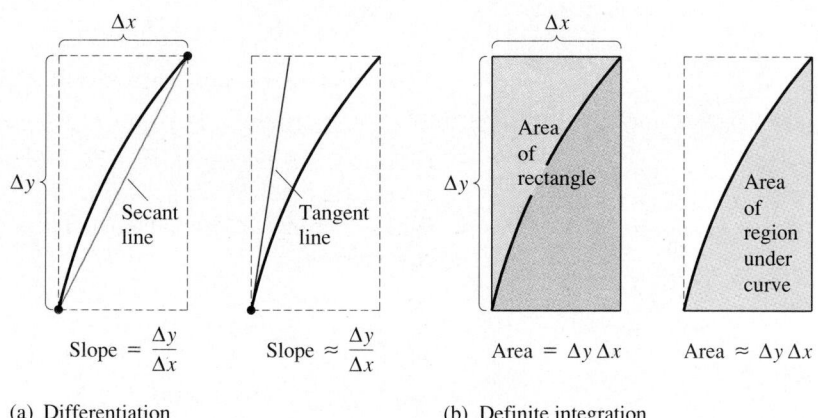

(a) Differentiation   (b) Definite integration

FIGURE 4.26

THEOREM 4.9   The Fundamental Theorem of Calculus

If a function f is continuous on the closed interval $[a, b]$ and F is an antiderivative of f on the interval $[a, b]$, then

$$\int_a^b f(x)\,dx = F(b) - F(a).$$

Proof   The key to the proof is in writing the difference $F(b) - F(a)$ in a convenient form. Let Δ be the following partition of $[a, b]$.

$$a = x_0 < x_1 < x_2 < \cdots < x_{n-1} < x_n = b$$

By pairwise subtraction and addition of like terms, you can write

$$F(b) - F(a) = F(x_n) - F(x_{n-1}) + F(x_{n-1}) - \cdots - F(x_1) + F(x_1) - F(x_0)$$

$$= \sum_{i=1}^{n} [F(x_i) - F(x_{i-1})].$$

By the Mean Value Theorem, you know that there exists a number c_i in the ith subinterval such that

$$F'(c_i) = \frac{F(x_i) - F(x_{i-1})}{x_i - x_{i-1}}$$

Because $F'(c_i) = f(c_i)$, you can let $\Delta x_i = x_i - x_{i-1}$ and obtain

$$F(b) - F(a) = \sum_{i=1}^{n} f(c_i) \, \Delta x_i.$$

This important equation tells you that by applying the Mean Value Theorem you can always find a collection of c_i's such that the *constant* $F(b) - F(a)$ is a Riemann sum of f on $[a, b]$. Taking the limit (as $\| \Delta \| \to 0$) produces

$$F(b) - F(a) = \int_a^b f(x) \, dx. \qquad \blacksquare$$

Here are three observations regarding the use of the Fundamental Theorem of Calculus.

1. *Provided you can find* an antiderivative of f, you now have a way to evaluate a definite integral without having to use the limit of a sum.

2. When applying the Fundamental Theorem of Calculus, the following notation is convenient.

$$\int_a^b f(x) \, dx = \left[F(x) \right]_a^b$$

$$= F(b) - F(a)$$

For instance, to evaluate $\int_1^3 x^3 \, dx$, you can write

$$\int_1^3 x^3 \, dx = \left[\frac{x^4}{4} \right]_1^3 = \frac{3^4}{4} - \frac{1^4}{4} = \frac{81}{4} - \frac{1}{4} = 20.$$

3. It is not necessary to include a constant of integration, C, in the antiderivative because

$$\int_a^b f(x) \, dx = \left[F(x) + C \right]_a^b$$

$$= [F(b) + C] - [F(a) + C]$$

$$= F(b) - F(a).$$

EXAMPLE 1 **Evaluating a Definite Integral**

a. $\displaystyle\int_{1}^{2}(x^2-3)\,dx = \left[\frac{x^3}{3}-3x\right]_{1}^{2} = \left(\frac{8}{3}-6\right)-\left(\frac{1}{3}-3\right) = -\frac{2}{3}$

b. $\displaystyle\int_{1}^{4}3\sqrt{x}\,dx = 3\int_{1}^{4}x^{1/2}\,dx = 3\left[\frac{x^{3/2}}{3/2}\right]_{1}^{4} = 2(4)^{3/2}-2(1)^{3/2} = 14$

c. $\displaystyle\int_{0}^{\pi/4}\sec^2 x\,dx = \tan x \Big]_{0}^{\pi/4} = 1-0 = 1$

EXAMPLE 2 **A Definite Integral Involving Absolute Value**

Evaluate $\displaystyle\int_{0}^{2}|2x-1|\,dx$.

Solution Using Figure 4.27 and the definition of absolute value, you can rewrite the integrand as follows.

$$|2x-1| = \begin{cases} -(2x-1), & x < \dfrac{1}{2} \\[2mm] 2x-1, & x \geq \dfrac{1}{2} \end{cases}$$

From this, you can rewrite the integral in two parts.

$$\int_{0}^{2}|2x-1|\,dx = \int_{0}^{1/2}-(2x-1)\,dx + \int_{1/2}^{2}(2x-1)\,dx$$

$$= \left[-x^2+x\right]_{0}^{1/2} + \left[x^2-x\right]_{1/2}^{2}$$

$$= \left(-\frac{1}{4}+\frac{1}{2}\right)-(0+0)+(4-2)-\left(\frac{1}{4}-\frac{1}{2}\right)$$

$$= \frac{5}{2}$$

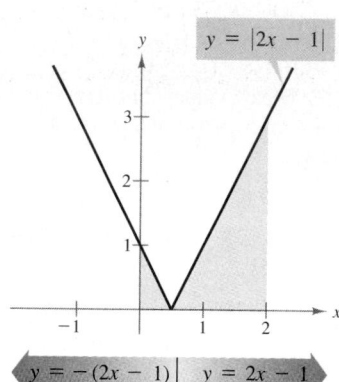

FIGURE 4.27
The definite integral of y on $[0, 2]$ is $\frac{5}{2}$.

EXAMPLE 3 **Using the Fundamental Theorem to Find Area**

Find the area of the region bounded by the graph of $y = 2x^2 - 3x + 2$, the x-axis, and the vertical lines $x = 0$ and $x = 2$, as shown in Figure 4.28.

Solution Note that $y > 0$ on the interval $[0, 2]$.

$$\text{Area} = \int_{0}^{2}(2x^2-3x+2)\,dx \qquad \text{Integrate between } x = 0 \text{ and } x = 2$$

$$= \left[\frac{2x^3}{3}-\frac{3x^2}{2}+2x\right]_{0}^{2} \qquad \text{Find antiderivative, } F(x)$$

$$= \left(\frac{16}{3}-6+4\right)-(0-0+0) \qquad \text{Evaluate } F(2)-F(0)$$

$$= \frac{10}{3} \qquad \text{Simplify}$$

FIGURE 4.28
The area of the region bounded by the graph of y, the x-axis, $x = 0$ and $x = 2$ is $\frac{10}{3}$.

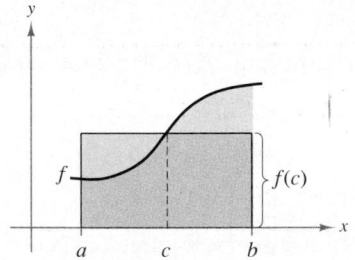

FIGURE 4.29
Mean value rectangle:
$f(c)(b - a) = \int_a^b f(x)\,dx.$

The Mean Value Theorem for Integrals

In Section 4.2, you saw that the area of a region under a curve is greater than the area of an inscribed rectangle and less than the area of a circumscribed rectangle. The Mean Value Theorem for Integrals states that somewhere "between" the inscribed and circumscribed rectangles there is a rectangle whose area is precisely equal to the area of the region under the curve, as shown in Figure 4.29.

THEOREM 4.10   Mean Value Theorem for Integrals

If f is continuous on the closed interval $[a, b]$, then there exists a number c in the closed interval $[a, b]$ such that

$$\int_a^b f(x)\,dx = f(c)(b - a).$$

Proof   *Case 1:*   If f is constant on the interval $[a, b]$, the theorem is clearly valid because c can be any point in $[a, b]$.

Case 2:   If f is not constant on $[a, b]$, then, by the Extreme Value Theorem, you can choose $f(m)$ and $f(M)$ to be the minimum and maximum values of f on $[a, b]$. Because $f(m) \le f(x) \le f(M)$ for all x in $[a, b]$, you can apply Theorem 4.8 to write the following.

$$\int_a^b f(m)\,dx \le \int_a^b f(x)\,dx \le \int_a^b f(M)\,dx \qquad \text{See Figure 4.30}$$

$$f(m)(b - a) \le \int_a^b f(x)\,dx \le f(M)(b - a)$$

$$f(m) \le \frac{1}{b - a}\int_a^b f(x)\,dx \le f(M)$$

From the third inequality, you can apply the Intermediate Value Theorem to conclude that there exists some c in $[a, b]$ such that

$$f(c) = \frac{1}{b - a}\int_a^b f(x)\,dx \quad \text{or} \quad f(c)(b - a) = \int_a^b f(x)\,dx.$$

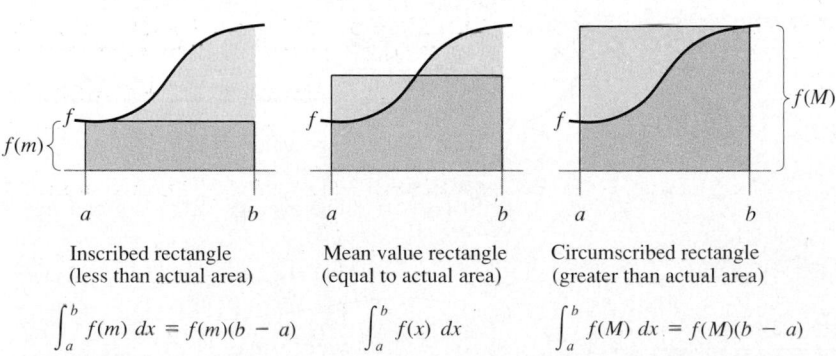

Inscribed rectangle (less than actual area)	Mean value rectangle (equal to actual area)	Circumscribed rectangle (greater than actual area)
$\int_a^b f(m)\,dx = f(m)(b - a)$	$\int_a^b f(x)\,dx$	$\int_a^b f(M)\,dx = f(M)(b - a)$

FIGURE 4.30

REMARK   Note that Theorem 4.10 does not specify how to determine c. It merely guarantees the existence of at least one number c in the interval.

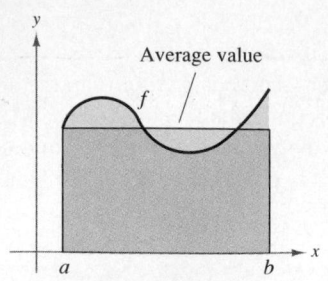

FIGURE 4.31

Average value $= \dfrac{1}{b-a} \displaystyle\int_a^b f(x)\,dx.$

Average Value of a Function

The value of $f(c)$, given in the Mean Value Theorem for Integrals, is called the **average value** of f on the interval $[a, b]$.

Definition of the Average Value of a Function on an Interval

If f is integrable on the closed interval $[a, b]$, then the **average value** of f on the interval is

$$\frac{1}{b-a}\int_a^b f(x)\,dx.$$

REMARK Note in Figure 4.31 that the area of the region under the graph of f is equal to the area of the rectangle whose height is the average value.

To see why the average value of f is defined this way, suppose that you partition $[a, b]$ into n subintervals of equal width $\Delta x = (b - a)/n$. If c_i is any point in the ith subinterval, then the arithmetic average (or mean) of the function values at the c_i's is given by

$$a_n = \frac{1}{n}[\, f(c_1) + f(c_2) + \cdots + f(c_n)].\qquad \text{Average of } f(c_1), \ldots, f(c_n)$$

By multiplying and dividing by $(b - a)$, you can write the average as

$$a_n = \frac{1}{n}\sum_{i=1}^{n} f(c_i)\left(\frac{b-a}{b-a}\right) = \frac{1}{b-a}\sum_{i=1}^{n} f(c_i)\left(\frac{b-a}{n}\right)$$

$$= \frac{1}{b-a}\sum_{i=1}^{n} f(c_i)\,\Delta x.$$

Finally, taking the limit as $n \to \infty$ produces the average value of f on the interval $[a, b]$, as given in the above definition.

This development of the average value of a function on an interval is only one of many practical uses of definite integrals to represent summation processes. In Chapter 6 you will study other applications, such as volume, arc length, centers of mass, and work.

EXAMPLE 4 Finding the Average Value of a Function

Find the average value of $f(x) = 3x^2 - 2x$ on the interval $[1, 4]$.

Solution The average value is given by

$$\frac{1}{b-a}\int_a^b f(x)\,dx = \frac{1}{3}\int_1^4 (3x^2 - 2x)\,dx$$

$$= \frac{1}{3}\Big[x^3 - x^2\Big]_1^4$$

$$= \frac{1}{3}[64 - 16 - (1 - 1)] = \frac{48}{3} = 16.$$

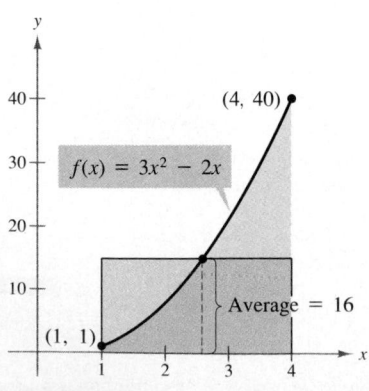

FIGURE 4.32

(See Figure 4.32.)

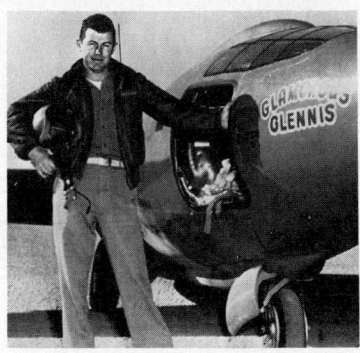

The first person to fly at a speed greater than the speed of sound was Charles Yeager. On October 14, 1947, flying in an X-1 rocket plane at an altitude of 12.8 kilometers, Yeager was clocked at 299.5 meters per second. If Yeager had been flying at an altitude under 10.375 kilometers, his speed of 299.5 meters per second would not have "broken the sound barrier."

EXAMPLE 5   The Speed of Sound

At different altitudes in earth's atmosphere, sound travels at different speeds. The speed of sound $s(x)$ (in meters per second) can be modeled by

$$s(x) = \begin{cases} -4x + 341, & 0 \le x < 11.5 \\ 295, & 11.5 \le x < 22 \\ \frac{3}{4}x + 278.5, & 22 \le x < 32 \\ \frac{3}{2}x + 254.5, & 32 \le x < 50 \\ -\frac{3}{2}x + 404.5, & 50 \le x < 80 \end{cases}$$

where x is the altitude in kilometers (see Figure 4.33). What is the average speed of sound over the interval $[0, 80]$?

Solution   Begin by integrating $s(x)$ over the interval $[0, 80]$. To do this, you can break the integral into five parts.

$$\int_0^{11.5} s(x)\, dx = \int_0^{11.5} (-4x + 341)\, dx = 3657$$

$$\int_{11.5}^{22} s(x)\, dx = \int_{11.5}^{22} (295)\, dx = 3097.5$$

$$\int_{22}^{32} s(x)\, dx = \int_{22}^{32} (\tfrac{3}{4}x + 278.5)\, dx = 2987.5$$

$$\int_{32}^{50} s(x)\, dx = \int_{32}^{50} (\tfrac{3}{2}x + 254.5)\, dx = 5688$$

$$\int_{50}^{80} s(x)\, dx = \int_{50}^{80} (-\tfrac{3}{2}x + 404.5)\, dx = 9210$$

By adding the values of the five integrals, you have

$$\int_0^{80} s(x)\, dx = 24,640.$$

Therefore, the average speed of sound from an altitude of 0 kilometers to 80 kilometers is

$$\text{Average speed } \frac{1}{80}\int_0^{80} s(x)\, dx = \frac{24,640}{80} = 308 \text{ meters per second.}$$

FIGURE 4.33

The Second Fundamental Theorem of Calculus

When we defined the definite integral of f on the interval $[a, b]$, we used the constant b as the upper limit of integration and x as the variable of integration. We now look at a slightly different situation in which the variable x is used as the upper limit of integration. To avoid the confusion of using x in two different ways, we temporarily switch to using t as the variable of integration. (Remember that the definite integral is *not* a function of its variable of integration.)

EXAMPLE 6　The Definite Integral as a Function

Evaluate the function

$$F(x) = \int_0^x \cos t \, dt$$

at $x = 0$, $\pi/6$, $\pi/4$, $\pi/3$, and $\pi/2$.

Solution　You could evaluate five different definite integrals, one for each of the given upper limits. However, it is much simpler to fix x (as a constant) temporarily and apply the Fundamental Theorem once, to obtain

$$\int_0^x \cos t \, dt = \left[\sin t \right]_0^x$$
$$= \sin x - \sin 0$$
$$= \sin x.$$

Now using $F(x) = \sin x$, you can obtain the results shown in Figure 4.34.

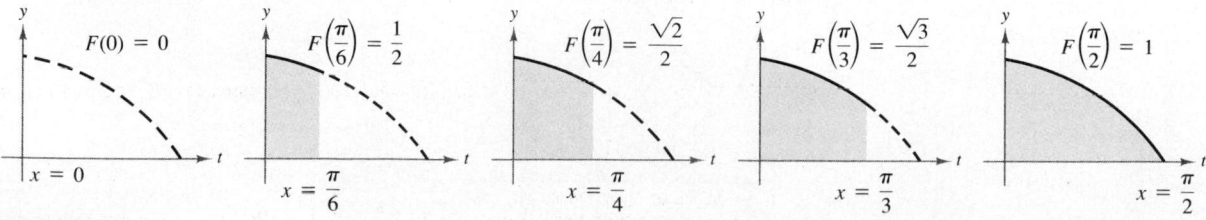

FIGURE 4.34
$F(x) = \int_0^x \cos t \, dt$

In Example 6, note that the derivative of F is the original integrand (with only the variable changed). That is,

$$\frac{d}{dx}[F(x)] = \frac{d}{dx}[\sin x] = \frac{d}{dx}\left[\int_0^x \cos t\,dt\right] = \cos x.$$

We generalize this result in the following theorem, called the **Second Fundamental Theorem of Calculus.**

THEOREM 4.11　The Second Fundamental Theorem of Calculus

If f is continuous on an open interval I containing a, then for every x in the interval,

$$\frac{d}{dx}\left[\int_a^x f(t)\,dt\right] = f(x).$$

Proof　Begin by defining F as

$$F(x) = \int_a^x f(t)\,dt.$$

Then, by the definition of the derivative, you can write

$$F'(x) = \lim_{\Delta x \to 0} \frac{F(x + \Delta x) - F(x)}{\Delta x}$$

$$= \lim_{\Delta x \to 0} \frac{1}{\Delta x}\left[\int_a^{x+\Delta x} f(t)\,dt - \int_a^x f(t)\,dt\right]$$

$$= \lim_{\Delta x \to 0} \frac{1}{\Delta x}\left[\int_a^{x+\Delta x} f(t)\,dt + \int_x^a f(t)\,dt\right]$$

$$= \lim_{\Delta x \to 0} \frac{1}{\Delta x}\left[\int_x^{x+\Delta x} f(t)\,dt\right].$$

From the Mean Value Theorem for Integrals (assuming $\Delta x > 0$), you know there exists a number c in the interval $[x,\, x + \Delta x]$ such that the integral in the above expression is equal to $f(c)\,\Delta x$. Moreover, because $x \le c \le x + \Delta x$, it follows that $c \to x$ as $\Delta x \to 0$. Thus, you obtain

$$F'(x) = \lim_{\Delta x \to 0}\left[\frac{1}{\Delta x}f(c)\,\Delta x\right] = \lim_{\Delta x \to 0} f(c) = f(x).$$

A similar argument can be made for $\Delta x < 0$.　　████

REMARK　Using the area model for definite integrals, you can view the approximation

$$f(x)\,\Delta x \approx \int_x^{x+\Delta x} f(t)\,dt$$

as saying that the area of the rectangle of height $f(x)$ and width Δx is approximately equal to the area of the region lying between the graph of f and the x-axis on the interval $[x,\, x + \Delta x]$, as shown in Figure 4.35.

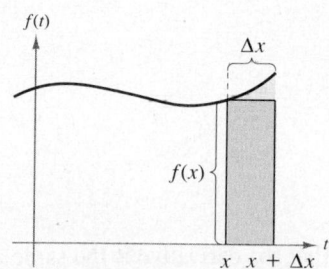

FIGURE 4.35
$f(x)\,\Delta x \approx \int_x^{x+\Delta x} f(t)\,dt$

Note that the Second Fundamental Theorem of Calculus tells you that if a function is continuous, then you can be sure that it has an antiderivative. This antiderivative need not, however, be an elementary function. (Recall the discussion of elementary functions in Section 5 of the Prerequisites chapter.)

EXAMPLE 7   Using the Second Fundamental Theorem of Calculus

Evaluate $\dfrac{d}{dx}\left[\displaystyle\int_0^x \sqrt{t^2 + 1}\, dt\right]$.

Solution   Note that $f(t) = \sqrt{t^2 + 1}$ is continuous on the entire real line. Thus, using the Second Fundamental Theorem of Calculus, you can write

$$\frac{d}{dx}\left[\int_0^x \sqrt{t^2 + 1}\, dt\right] = \sqrt{x^2 + 1}. \qquad \blacksquare$$

The differentiation shown in Example 7 is a straightforward application of the Second Fundamental Theorem of Calculus. The next example shows how this theorem can be combined with the Chain Rule to find the derivative of a function.

EXAMPLE 8   Using the Second Fundamental Theorem of Calculus

Find the derivative of $F(x) = \displaystyle\int_{\pi/2}^{x^3} \cos t\, dt$.

Solution   Using $u = x^3$, you can apply the Second Fundamental Theorem of Calculus with the Chain Rule as follows.

$$
\begin{aligned}
F'(x) &= \frac{dF}{du}\frac{du}{dx} \\[4pt]
&= \frac{d}{du}\left[\int_{\pi/2}^{u} \cos t\, dt\right]\frac{du}{dx} \\[4pt]
&= (\cos u)(3x^2) \\[4pt]
&= (\cos x^3)(3x^2) \qquad \blacksquare
\end{aligned}
$$

Because the integrand in Example 8 is easily integrated, you can verify the derivative as follows.

$$
\begin{aligned}
F(x) &= \int_{\pi/2}^{x^3} \cos t\, dt \\[4pt]
&= \sin t \;\Big]_{\pi/2}^{x^3} \\[4pt]
&= \sin x^3 - \sin\frac{\pi}{2} \\[4pt]
&= (\sin x^3) - 1
\end{aligned}
$$

In this form, you can apply the Chain Rule to verify that the derivative is the same as that obtained in Example 7,

$$F'(x) = (\cos x^3)(3x^2).$$

EXERCISES for Section 4.4

In Exercises 1–20, evaluate the definite integral of the algebraic function.

1. $\displaystyle\int_0^1 2x\,dx$

2. $\displaystyle\int_2^7 3\,dv$

3. $\displaystyle\int_{-1}^0 (x - 2)\,dx$

4. $\displaystyle\int_2^5 (-3v + 4)\,dv$

5. $\displaystyle\int_{-1}^1 (t^2 - 2)\,dt$

6. $\displaystyle\int_0^3 (3x^2 + x - 2)\,dx$

7. $\displaystyle\int_0^1 (2t - 1)^2\,dt$

8. $\displaystyle\int_{-1}^1 (t^3 - 9t)\,dt$

9. $\displaystyle\int_1^2 \left(\frac{3}{x^2} - 1\right)dx$

10. $\displaystyle\int_{-2}^{-1} \left(u - \frac{1}{u^2}\right)du$

11. $\displaystyle\int_1^4 \frac{u - 2}{\sqrt{u}}\,du$

12. $\displaystyle\int_{-3}^3 v^{1/3}\,dv$

13. $\displaystyle\int_{-1}^1 \left(\sqrt[3]{t} - 2\right)dt$

14. $\displaystyle\int_1^8 \sqrt{\frac{2}{x}}\,dx$

15. $\displaystyle\int_0^1 \frac{x - \sqrt{x}}{3}\,dx$

16. $\displaystyle\int_0^2 (2 - t)\sqrt{t}\,dt$

17. $\displaystyle\int_{-1}^0 (t^{1/3} - t^{2/3})\,dt$

18. $\displaystyle\int_{-8}^{-1} \frac{x - x^2}{2\sqrt[3]{x}}\,dx$

19. $\displaystyle\int_0^3 |2x - 3|\,dx$

20. $\displaystyle\int_0^4 |x^2 - 4x + 3|\,dx$

In Exercises 21–26, evaluate the definite integral of the trigonometric function.

21. $\displaystyle\int_0^\pi (1 + \sin x)\,dx$

22. $\displaystyle\int_0^{\pi/4} \frac{1 - \sin^2\theta}{\cos^2\theta}\,d\theta$

23. $\displaystyle\int_{-\pi/6}^{\pi/6} \sec^2 x\,dx$

24. $\displaystyle\int_{\pi/4}^{\pi/2} (2 - \csc^2 x)\,dx$

25. $\displaystyle\int_{-\pi/3}^{\pi/3} 4\sec\theta\tan\theta\,d\theta$

26. $\displaystyle\int_{-\pi/2}^{\pi/2} (2t + \cos t)\,dt$

27. *Depreciation* A company purchases a new machine for which the rate of depreciation is

$$\frac{dV}{dt} = 10{,}000(t - 6), \qquad 0 \le t \le 5$$

where V is the value of the machine after t years. Set up and evaluate the definite integral that yields the total loss of value of the machine over the first 3 years.

28. *Buffon's Needle Experiment* A horizontal plane is ruled with parallel lines 2 inches apart. If a 2-inch needle is tossed randomly onto the plane, it can be shown that the probability that the needle will touch a line is

$$P = \frac{2}{\pi}\int_0^{\pi/2} \sin\theta\,d\theta$$

where θ is the acute angle between the needle and any one of the parallel lines. Find this probability.

In Exercises 29–34, determine the area of the indicated region.

29. $y = x - x^2$

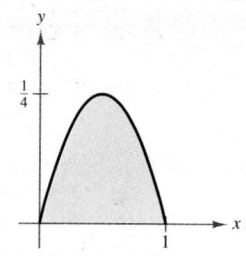

30. $y = 1 - x^4$

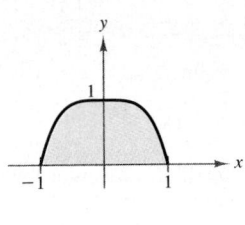

31. $y = (3 - x)\sqrt{x}$

32. $y = \dfrac{1}{x^2}$

33. $y = \cos x$

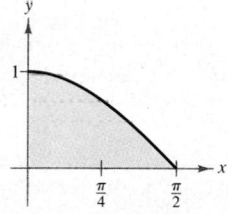

34. $y = x + \sin x$

In Exercises 35–38, find the area of the region bounded by the graphs of the equations.

35. $y = 3x^2 + 1$, $\quad x = 0$, $\quad x = 2$, $\quad y = 0$

36. $y = 1 + \sqrt{x}$, $\quad x = 0$, $\quad x = 4$, $\quad y = 0$

37. $y = x^3 + x$, $\quad x = 2$, $\quad y = 0$

38. $y = -x^2 + 3x$, $\quad y = 0$

In Exercises 39–42, find the value of c guaranteed by the *Mean Value Theorem for Integrals* for the function over the specified interval.

Function	Interval
39. $f(x) = x - 2\sqrt{x}$	$[0, 2]$
40. $f(x) = \dfrac{9}{x^3}$	$[1, 3]$
41. $f(x) = 2\sec^2 x$	$\left[-\dfrac{\pi}{4}, \dfrac{\pi}{4}\right]$
42. $f(x) = \cos x$	$\left[-\dfrac{\pi}{3}, \dfrac{\pi}{3}\right]$

In Exercises 43–46, sketch the graph of the function over the specified interval. Find the average value of the function over the interval and all values of x in the interval for which the function equals its average value.

Function	*Interval*
43. $f(x) = 4 - x^2$	$[-2, 2]$
44. $f(x) = \dfrac{x^2 + 1}{x^2}$	$\left[\dfrac{1}{2}, 2\right]$
45. $f(x) = \sin x$	$[0, \pi]$
46. $f(x) = \cos x$	$\left[0, \dfrac{\pi}{2}\right]$

47. The graph of a function f is shown in the figure.

a. Evaluate $\displaystyle\int_1^7 f(x)\,dx$.

b. Determine the average value of the function on the interval $[1, 7]$.

c. Determine the answers to parts a and b if the graph is translated two units upward.

48. *Average Profit* A company introduces a new product, and the profit in thousands of dollars over the first 6 months is approximated by the model

$$P = 5(\sqrt{t} + 30), \qquad t = 1, 2, 3, 4, 5, 6.$$

a. Use the model to complete the table and use the entries to calculate (arithmetically) the average profit over the first 6 months.

t	1	2	3	4	5	6
P						

b. Find the average value of the profit function by integration and compare the result with that of part a. (Integrate over the interval $[0.5, 6.5]$.)

c. What, if any, is the advantage of using the approximation of the average given by the definite integral? (Note that the integral approximation uses all real values of t in the interval rather than just integers.)

49. *Death Rate* A life insurance company needs a model to approximate the death rate of citizens during the years they are in the work force. The table gives the death rate R per 1000 for individuals of age x. (*Source:* Department of Health and Human Services)

x	20	30	40	50	60
R	1.1	1.3	2.2	5.0	12.6

A model for this data is

$$R = -90.767 - 6.434x + 0.037x^2 + 46.004\sqrt{x}$$

$20 \le x \le 60$.

a. Use a graphing utility to graph the model and compare it with the data.

b. Find the rate of increase of the death rate when $x = 40$ and $x = 50$.

c. Find the average death rate for people between the ages of 30 and 40 and for people between the ages of 50 and 60.

50. *Blood Flow* The velocity v of the flow of blood at a distance r from the central axis of an artery of radius R is

$$v = k(R^2 - r^2)$$

where k is the constant of proportionality. Find the average rate of flow of blood along a radius of the artery. (Use 0 and R as the limits of integration.)

51. *Force* The force F (in pounds) of a hydraulic cylinder in a press is proportional to the square of $\sec x$ where x is the distance (in feet) the cylinder is extended in its cycle. The domain of F is $[0, \pi/3]$ and $F(0) = 100$.

a. Find F as a function of x.

b. Find the average force exerted by the press over the interval $[0, \pi/3]$.

52. *Respiratory Cycle* The volume V in liters of air in the lungs during a 5-second respiratory cycle is approximated by the model

$$V = 0.1729t + 0.1522t^2 - 0.0374t^3$$

where t is the time in seconds. Approximate the average volume of air in the lungs during one cycle.

In Exercises 53–58, (a) integrate to find F as a function of x and (b) demonstrate the Second Fundamental Theorem of Calculus by differentiating the result of part (a).

53. $F(x) = \displaystyle\int_0^x (t + 2)\,dt$

54. $F(x) = \displaystyle\int_0^x t(t^2 + 1)\,dt$

55. $F(x) = \displaystyle\int_8^x \sqrt[3]{t}\,dt$

56. $F(x) = \displaystyle\int_4^x \sqrt{t}\,dt$

57. $F(x) = \displaystyle\int_{\pi/4}^x \sec^2 t\,dt$

58. $F(x) = \displaystyle\int_{\pi/3}^x \sec t \tan t\,dt$

In Exercises 59–64, use the Second Fundamental Theorem of Calculus to find $F'(x)$.

59. $F(x) = \displaystyle\int_{-2}^x (t^2 - 2t + 5)\,dt$

60. $F(x) = \displaystyle\int_1^x \sqrt[4]{t}\,dt$

61. $F(x) = \displaystyle\int_{-1}^x \sqrt{t^4 + 1}\,dt$

62. $F(x) = \displaystyle\int_0^x \tan^4 t\,dt$

63. $F(x) = \displaystyle\int_0^x t \cos t\,dt$

64. $F(x) = \displaystyle\int_1^x \dfrac{t^2}{t^2 + 1}\,dt$

In Exercises 65–70, find $F'(x)$.

65. $F(x) = \displaystyle\int_{x}^{x+2} (4t + 1)\, dt$

66. $F(x) = \displaystyle\int_{-x}^{x} t^3\, dt$

67. $F(x) = \displaystyle\int_{0}^{\sin x} \sqrt{t}\, dt$

68. $F(x) = \displaystyle\int_{2}^{x^2} \frac{1}{t^2}\, dt$

69. $F(x) = \displaystyle\int_{0}^{x^3} \sin t^2\, dt$

70. $F(x) = \displaystyle\int_{0}^{3x} \sqrt{1 + t^3}\, dt$

In Exercises 71 and 72, sketch a graph of the function

$$F(x) = \int_{0}^{x} f(t)\, dt.$$

State any relationship that may exist between the extrema and inflection points on the graphs of f and F.

71.

72.

73. *Cost*   The total cost of purchasing and maintaining a piece of equipment for x years is

$$C(x) = 5000\left(25 + 3 \int_{0}^{x} t^{1/4}\, dt \right).$$

a. Perform the integration to write C as a function of x.

b. Find $C(1)$, $C(5)$, and $C(10)$.

74. *Area*   The area A between the graph of the function $g(t) = 4 - 4/t^2$ and the t-axis over the interval $[1, x]$ is

$$A(x) = \int_{1}^{x} \left(4 - \frac{4}{t^2} \right) dt.$$

a. Find the horizontal asymptote of the graph of g.

b. Integrate to find A as a function of x. Does the graph of A have a horizontal asymptote? Why or why not?

In Exercises 75–77, determine whether the statement is true or false. If it is false, explain why or give an example that shows it is false.

75. If $F'(x) = G'(x)$ on the interval $[a, b]$, then $F(b) - F(a) = G(b) - G(a)$.

76. If f is continuous on $[a, b]$, then f is integrable on $[a, b]$.

77. $\displaystyle\int_{-1}^{1} x^{-2}\, dx = [-x^{-1}]_{-1}^{1} = (-1) - 1 = -2$

78. Prove: $\displaystyle\frac{d}{dx}\left[\int_{u(x)}^{v(x)} f(t)\, dt \right] = f(v(x))v'(x) - f(u(x))u'(x).$

79. Show that the function

$$f(x) = \int_{0}^{1/x} \frac{1}{t^2 + 1}\, dt + \int_{0}^{x} \frac{1}{t^2 + 1}\, dt$$

is constant for $x > 0$.

80. Let $G(x) = \displaystyle\int_{0}^{x} \left[s \int_{0}^{s} f(t)\, dt \right] ds$, where f is continuous for all real t. Find

a. $G(0)$ **b.** $G'(x)$ **c.** $G'(0)$ **d.** $G''(x)$ **e.** $G''(0)$.

SECTION	Integration by Substitution
4.5	Pattern Recognition • Change of Variables • The General Power Rule for Integration • Change of Variables for Definite Integrals • Integration of Even and Odd Functions

Pattern Recognition

In this section you will study techniques for integrating composite functions. The discussion is split into two parts—*pattern recognition* and *change of variables*. Both techniques involve a *u-substitution.* With pattern recognition you perform the substitution mentally, and with change of variables you write the substitution steps.

The role of substitution in integration is comparable to the role of the Chain Rule in differentiation. Recall that for differentiable functions given by $y = F(u)$ and $u = g(x)$, the Chain Rule states that

$$\frac{d}{dx}[F(g(x))] = F'(g(x))g'(x).$$

From the definition of an antiderivative, it follows that

$$\int F'(g(x))g'(x)\,dx = F(g(x)) + C$$

$$= F(u) + C.$$

We state these results in the following theorem.

THEOREM 4.12   Antidifferentiation of a Composite Function

Let g be a function whose range is an interval I, and let f be a function that is continuous on I. If g is differentiable on its domain and F is an antiderivative of f on I, then

$$\int f(g(x))g'(x)\,dx = F(g(x)) + C.$$

If $u = g(x)$, then

$$du = g'x\,dx$$

and

$$\int f(u)\,du = F(u) + C.$$

There are several techniques for applying substitution, each differing slightly from the others. However, you should remember that the goal is the same with every technique—*you are trying to find an antiderivative of the integrand.*

Note that the statement of Theorem 4.12 doesn't tell how to distinguish between $f(g(x))$ and $g'(x)$ in the integrand. As you become more experienced at integration, your skill in doing this will increase. Of course, part of the key is familiarity with derivatives.

Examples 1 and 2 show how to apply the theorem *directly,* by recognizing the presence of $f(g(x))$ and $g'(x)$. Note that the composite function in the integrand has an *outside function f* and an *inside function g.* Moreover, the derivative $g'(x)$ is present as a factor of the integrand.

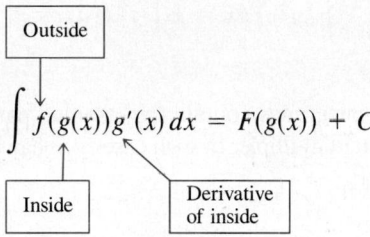

EXAMPLE 1   Recognizing the $f(g(x))g'(x)$ Pattern

Evaluate $\displaystyle\int (x^2 + 1)^2(2x)\,dx.$

Solution   Letting $g(x) = x^2 + 1$, you obtain

$$g'(x) = 2x \quad \text{and} \quad f(g(x)) = [g(x)]^2.$$

From this, you can recognize that the integrand follows the $f(g(x))g'(x)$ pattern. Thus, using the Power Rule for integration and Theorem 4.12, you can write

$$\int \overbrace{(x^2 + 1)^2}^{[g(x)]^2}\overbrace{(2x)}^{g'(x)}\,dx = \frac{1}{3}(x^2 + 1)^3 + C.$$

Try using the Chain Rule to check that the derivative of $\frac{1}{3}(x^2 + 1)^3 + C$ is the integrand of the original integral.   ■

EXAMPLE 2   Recognizing the $f(g(x))g'(x)$ Pattern

Evaluate $\displaystyle\int 5\cos 5x\,dx.$

Solution   Letting $g(x) = 5x$, you obtain

$$g'(x) = 5 \quad \text{and} \quad f(g(x)) = \cos[g(x)].$$

From this you can recognize that the integrand follows the $f(g(x))g'(x)$ pattern. Using the Sine Rule for integration and Theorem 4.12, you can write

$$\int \overbrace{(\cos 5x)}^{\cos[g(x)]}\overbrace{(5)}^{g'(x)}\,dx = \sin 5x + C.$$

You can check this by differentiating $\sin 5x + C$ to obtain the original integrand.   ■

Both of the integrands in Examples 1 and 2 fit the $f(g(x))g'(x)$ pattern exactly—you only had to recognize the pattern. You can extend this technique considerably with the Constant Multiple Rule

$$\int k f(x)\, dx = k \int f(x)\, dx.$$

Many integrands contain the essential part (the variable part) of $g'(x)$, but are missing a constant multiple. In such cases you can multiply and divide by the necessary constant multiple.

EXAMPLE 3   Multiplying and Dividing by a Constant

Evaluate $\int x(x^2 + 1)^2\, dx$.

Solution   This is similar to the integral given in Example 1, except that the integrand is missing a factor of 2. Recognizing that $2x$ is the derivative of $x^2 + 1$, you can let $g(x) = x^2 + 1$ and supply the $2x$ as follows.

$$\int x(x^2 + 1)^2\, dx = \int (x^2 + 1)^2 \left(\frac{1}{2}\right)(2x)\, dx \qquad \text{Multiply and divide by 2}$$

$$= \frac{1}{2} \int (x^2 + 1)^2 (2x)\, dx \qquad \text{Constant Multiple Rule}$$

$$= \frac{1}{2} \int [g(x)]^2 g'(x)\, dx$$

$$= \frac{1}{2} \frac{[g(x)]^3}{3} + C \qquad \text{Integrate}$$

$$= \frac{1}{6}(x^2 + 1)^3 + C$$

In practice, most people would not write as many steps as are shown in Example 3. For instance, you could evaluate the integral by simply writing

$$\int x(x^2 + 1)^2\, dx = \frac{1}{2} \int (x^2 + 1)^2 2x\, dx$$

$$= \frac{1}{2} \frac{(x^2 + 1)^3}{3} + C$$

$$= \frac{1}{6}(x^2 + 1)^3 + C.$$

REMARK   Be sure you see that the *Constant* Multiple Rule applies only to constants. You cannot multiply and divide by a variable and then move the variable outside the integral sign. For instance,

$$\int (x^2 + 1)^2\, dx \neq \frac{1}{2x} \int (x^2 + 1)^2 (2x)\, dx.$$

After all, if it were legitimate to move variable quantities outside the integral sign, you could move the entire integrand out and simplify the whole process! But the result would be incorrect.

Change of Variables

With a formal **change of variables,** you completely rewrite the integral in terms of u and du (or any other convenient variable). Although this procedure can involve more written steps than the pattern recognition illustrated in Examples 1–3, it is useful for complicated integrands. The change of variable technique uses the Leibniz notation for the differential. That is, if $u = g(x)$, then $du = g'(x)\,dx$, and the integral in Theorem 4.12 takes the form

$$\int f(g(x))g'(x)\,dx = \int f(u)\,du = F(u) + C.$$

EXAMPLE 4   Change of Variable

Evaluate $\displaystyle\int \sqrt{2x-1}\,dx$.

Solution   First, let u be the inner function, $u = 2x - 1$. Then calculate the differential du to be $du = 2\,dx$. Now, using $\sqrt{2x-1} = \sqrt{u}$ and $dx = du/2$, substitute to obtain the following.

$$\int \sqrt{2x-1}\,dx = \int \sqrt{u}\left(\frac{du}{2}\right) \qquad \text{Integral in terms of } u$$

$$= \frac{1}{2}\int u^{1/2}\,du$$

$$= \frac{1}{2}\left(\frac{u^{3/2}}{3/2}\right) + C \qquad \text{Antiderivative in terms of } u$$

$$= \frac{1}{3}u^{3/2} + C$$

$$= \frac{1}{3}(2x - 1)^{3/2} + C \qquad \text{Antiderivative in terms of } x$$

EXAMPLE 5   Change of Variable

Evaluate $\displaystyle\int x\sqrt{2x-1}\,dx$.

Solution   As in the previous example, let $u = 2x - 1$ and obtain $dx = du/2$. Because the integrand contains a factor of x, you must also solve for x in terms of u as follows.

$$u = 2x - 1 \;\rightarrow\; x = \frac{u+1}{2} \qquad \text{Solve for } x \text{ in terms of } u$$

Now, using substitution, you obtain the following.

$$\int x\sqrt{2x-1}\,dx = \int \left(\frac{u+1}{2}\right)u^{1/2}\left(\frac{du}{2}\right)$$

$$= \frac{1}{4}\int (u^{3/2} + u^{1/2})\,du$$

$$= \frac{1}{4}\left(\frac{u^{5/2}}{5/2} + \frac{u^{3/2}}{3/2}\right) + C$$

$$= \frac{1}{10}(2x - 1)^{5/2} + \frac{1}{6}(2x - 1)^{3/2} + C$$

To complete the change of variable in Example 5, we solved for x in terms of u. Sometimes this is very difficult. Fortunately it is not always necessary, as shown in the next example.

EXAMPLE 6   Change of Variable

Evaluate $\displaystyle\int \sin^2 3x \cos 3x \, dx$.

Solution   Because $\sin^2 3x = (\sin 3x)^2$, you can let $u = \sin 3x$. Then

$$du = (\cos 3x)(3) \, dx.$$

Now, because $\cos 3x \, dx$ is part of the given integral, you can write

$$\frac{du}{3} = \cos 3x \, dx.$$

Substituting u and $du/3$ in the given integral yields the following.

$$\int \sin^2 3x \cos 3x \, dx = \int u^2 \frac{du}{3}$$

$$= \frac{1}{3} \int u^2 \, du$$

$$= \frac{1}{3}\left(\frac{u^3}{3}\right) + C$$

$$= \frac{1}{9} \sin^3 3x + C$$

You can check this by differentiating.

$$\frac{d}{dx}\left[\frac{1}{9}\sin^3 3x\right] = \left(\frac{1}{9}\right)(3)(\sin 3x)^2(\cos 3x)(3)$$

$$= \sin^2 3x \cos 3x$$

Because differentiation produces the original integrand, you know that you obtained the correct antiderivative. ▬▬▬

We summarize the steps used for integration by substitution in the following guidelines.

Guidelines for Making a Change of Variables

1. Choose a substitution $u = g(x)$. Usually, it is best to choose the *inner* part of a composite function, such as a quantity raised to a power.
2. Compute $du = g'(x) \, dx$.
3. Rewrite the integral in terms of the variable u.
4. Evaluate the resulting integral in terms of u.
5. Replace u by $g(x)$ to obtain an antiderivative in terms of x.

The General Power Rule for Integration

One of the most common u-substitutions involves quantities in the integrand that are raised to a power. Because of the importance of this type of substitution, it is given a special name—the **General Power Rule.** A proof of this rule follows directly from the (simple) Power Rule for integration, together with Theorem 4.12.

THEOREM 4.13   The General Power Rule for Integration

If g is a differentiable function of x, then

$$\int [g(x)]^n g'(x)\, dx = \frac{[g(x)]^{n+1}}{n+1} + C, \qquad n \neq -1.$$

Equivalently, if $u = g(x)$, then

$$\int u^n\, du = \frac{u^{n+1}}{n+1} + C, \qquad n \neq -1.$$

EXAMPLE 7   Substitution and the General Power Rule

a. $\displaystyle \int 3(3x-1)^4\, dx = \int \overbrace{(3x-1)^4}^{u^4}\overbrace{(3)\, dx}^{du} = \overbrace{\frac{(3x-1)^5}{5}}^{u^5/5} + C$

b. $\displaystyle \int (2x+1)(x^2+x)\, dx = \int \overbrace{(x^2+x)^1}^{u^1}\overbrace{(2x+1)\, dx}^{du} = \overbrace{\frac{(x^2+x)^2}{2}}^{u^2/2} + C$

c. $\displaystyle \int 3x^2\sqrt{x^3-2}\, dx = \int \overbrace{(x^3-2)^{1/2}}^{u^{1/2}}\overbrace{(3x^2)\, dx}^{du} = \overbrace{\frac{(x^3-2)^{3/2}}{3/2}}^{u^{3/2}/(3/2)} + C$

d. $\displaystyle \int \frac{-4x}{(1-2x^2)^2}\, dx = \int \overbrace{(1-2x^2)^{-2}}^{u^{-2}}\overbrace{(-4x)\, dx}^{du} = \overbrace{\frac{(1-2x^2)^{-1}}{-1}}^{u^{-1}/(-1)} + C$

e. $\displaystyle \int \cos^2 x \sin x\, dx = -\int \overbrace{(\cos x)^2}^{u^2}\overbrace{(-\sin x)\, dx}^{du} = -\overbrace{\frac{(\cos x)^3}{3}}^{u^3/3} + C$ ■

Some integrals whose integrands involve a quantity raised to a power cannot be evaluated by the General Power Rule. Consider the two integrals

$$\int x(x^2+1)^2\, dx \quad \text{and} \quad \int (x^2+1)^2\, dx.$$

The substitution $u = x^2 + 1$ works in the first integral but not in the second. (In the second, the substitution fails because the integrand lacks the factor x needed for du.) Fortunately, *for this particular integral,* you can expand the integrand as $(x^2+1)^2 = x^4 + 2x^2 + 1$ and use the Power Rule to integrate each term.

Change of Variables for Definite Integrals

When using u-substitution with a definite integral, it is often convenient to determine the limits of integration for the variable u rather than to convert the antiderivative back to the variable x and evaluate at the original limits. This change of variables is stated explicitly in the next theorem. The proof follows from Theorem 4.12 combined with the Fundamental Theorem of Calculus.

THEOREM 4.14   Change of Variables for Definite Integrals

If the function $u = g(x)$ has a continuous derivative on the closed interval $[a, b]$ and f is continuous on the range of g, then

$$\int_a^b f(g(x))g'(x)\,dx = \int_{g(a)}^{g(b)} f(u)\,du.$$

EXAMPLE 8   Change of Variables

Evaluate $\displaystyle\int_0^1 x(x^2 + 1)^3\,dx$.

Solution   To evaluate this integral, let $u = x^2 + 1$. Then, you obtain

$$u = x^2 + 1 \;\rightarrow\; du = 2x\,dx.$$

Before substituting, determine the new upper and lower limits of integration.

Lower limit	*Upper limit*
When $x = 0$, $u = 0^2 + 1 = 1$.	When $x = 1$, $u = 1^2 + 1 = 2$.

Now, you can substitute to obtain

$$\int_0^1 x(x^2 + 1)^3\,dx = \frac{1}{2}\int_0^1 (x^2 + 1)^3(2x)\,dx \qquad \text{Integration limits for } x$$

$$= \frac{1}{2}\int_1^2 u^3\,du \qquad \text{Integration limits for } u$$

$$= \frac{1}{2}\left[\frac{u^4}{4}\right]_1^2$$

$$= \frac{1}{2}\left(4 - \frac{1}{4}\right)$$

$$= \frac{15}{8}.$$

Try converting the antiderivative $\frac{1}{2}(u^4/4)$ back to the variable x and evaluate the definite integral at the original limits of integration. You should obtain the same result.

EXAMPLE 9   Change of Variables

Evaluate $A = \displaystyle\int_1^5 \frac{x}{\sqrt{2x-1}}\,dx$.

Solution   To evaluate this integral, let $u = \sqrt{2x-1}$. Then, you obtain

$$u^2 = 2x - 1 \to x = \frac{u^2+1}{2} \to dx = u\,du.$$

Before substituting, determine the new upper and lower limits of integration.

Lower limit

When $x = 1$, $u = \sqrt{2-1} = 1$.

Upper limit

When $x = 5$, $u = \sqrt{10-1} = 3$.

Now, substitute to obtain

$$\int_1^5 \frac{x}{\sqrt{2x-1}}\,dx = \int_1^3 \frac{1}{2}\left(\frac{u^2+1}{u}\right)u\,du$$

$$= \frac{1}{2}\int_1^3 (u^2+1)\,du$$

$$= \frac{1}{2}\left[\frac{u^3}{3} + u\right]_1^3$$

$$= \frac{1}{2}\left(9 + 3 - \frac{1}{3} - 1\right)$$

$$= \frac{16}{3}.$$

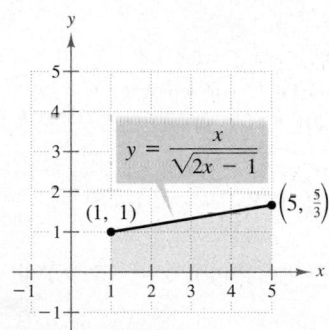

FIGURE 4.36
Region before substitution.

Geometrically, you can interpret the equation

$$\int_1^5 \frac{x}{\sqrt{2x-1}}\,dx = \int_1^3 \frac{u^2+1}{2}\,du$$

to mean that the two *different* regions shown in Figures 4.36 and 4.37 have the *same* area.

When evaluating definite integrals by substitution, it is possible for the upper limit of integration of the u-variable form to be smaller than the lower limit. If this happens, don't rearrange the limits. Simply evaluate as usual. For example, after substituting $u = \sqrt{1-x}$ in the integral

$$\int_0^1 x^2(1-x)^{1/2}\,dx$$

you obtain $u = \sqrt{1-1} = 0$ when $x = 1$, and $u = \sqrt{1-0} = 1$ when $x = 0$. Thus, the correct u-variable form of this integral is

$$-2\int_1^0 (1-u^2)^2 u^2\,du.$$

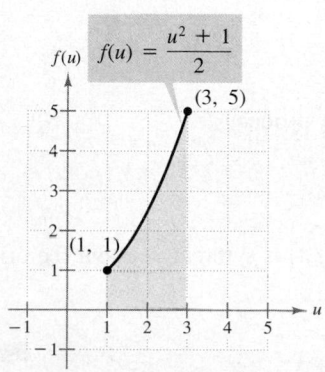

FIGURE 4.37
Region after substitution.

Even Function

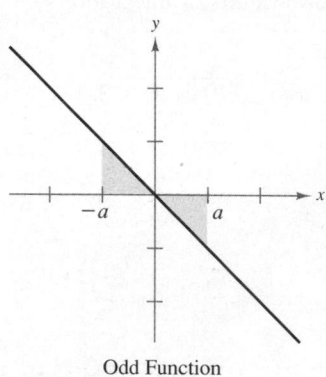

Odd Function

FIGURE 4.38

Integration of Even and Odd Functions

Even by making a change of variables, integration can be difficult. Occasionally, you will be able to simplify the evaluation of a definite integral (over an interval that is symmetric about the origin) by recognizing the integrand to be an even or odd function (see Figure 4.38).

THEOREM 4.15   Integration of Even and Odd Functions

Let f be integrable on the closed interval $[-a, a]$.

1. If f is an *even* function, then $\displaystyle\int_{-a}^{a} f(x)\,dx = 2\int_{0}^{a} f(x)\,dx$.

2. If f is an *odd* function, then $\displaystyle\int_{-a}^{a} f(x)\,dx = 0$.

Proof   Because f is even, you know that $f(x) = f(-x)$. Using Theorem 4.12 with the substitution $u = -x$ produces

$$\int_{-a}^{0} f(x)\,dx = \int_{a}^{0} f(-u)(-du)$$

$$= -\int_{a}^{0} f(u)\,du = \int_{0}^{a} f(u)\,du = \int_{0}^{a} f(x)\,dx.$$

Finally, using Theorem 4.6, you obtain

$$\int_{-a}^{a} f(x)\,dx = \int_{-a}^{0} f(x)\,dx + \int_{0}^{a} f(x)\,dx$$

$$= \int_{0}^{a} f(x)\,dx + \int_{0}^{a} f(x)\,dx = 2\int_{0}^{a} f(x)\,dx.$$

This proves the first property. The proof of the second property is left to you. ∎

EXAMPLE 10   Integration of an Odd Function

Evaluate $\displaystyle\int_{-\pi/2}^{\pi/2} (\sin^3 x \cos x + \sin x \cos x)\,dx$.

Solution   Letting $f(x) = \sin^3 x \cos x + \sin x \cos x$ produces

$$f(-x) = \sin^3(-x)\cos(-x) + \sin(-x)\cos(-x)$$

$$= -\sin^3 x \cos x - \sin x \cos x = -f(x).$$

Thus, f is an odd function, and because $[-\pi/2,\ \pi/2]$ is symmetric about the origin, you can apply Theorem 4.15 to conclude that

$$\int_{-\pi/2}^{\pi/2} (\sin^3 x \cos x + \sin x \cos x)\,dx = 0.$$

$f(x) = \sin^3 x \cos x + \sin x \cos x$

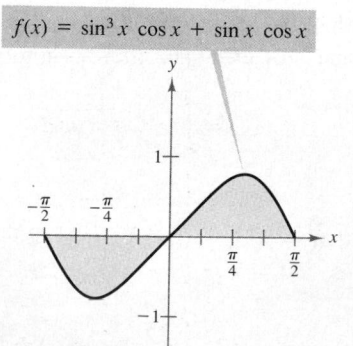

FIGURE 4.39
Because f is an odd function, $\int_{-\pi/2}^{\pi/2} f(x)\,dx = 0$.

REMARK   From Figure 4.39, you see that the two regions on either side of the y-axis have the same area. However, because one lies below the x-axis and one lies above, integration produces a cancellation effect. (We will say more about this in Section 6.1.)

TECHNOLOGY
Laboratory Guide
Lab 4.5

EXERCISES for Section 4.5

In Exercises 1–6, complete the table by identifying u and du for the integral.

$$\int f(g(x))g'(x)\,dx \qquad u = g(x) \qquad du = g'(x)\,dx$$

1. $\int (5x^2 + 1)^2(10x)\,dx$

2. $\int x^2\sqrt{x^3 + 1}\,dx$

3. $\int \dfrac{x}{\sqrt{x^2 + 1}}\,dx$

4. $\int \sec 2x \tan 2x\,dx$

5. $\int \tan^2 x \sec^2 x\,dx$

6. $\int \dfrac{\cos x}{\sin^2 x}\,dx$

In Exercises 7–26, evaluate the indefinite integral and check the result by differentiation.

7. $\int (1 + 2x)^4(2)\,dx$

8. $\int (x^2 - 1)^3(2x)\,dx$

9. $\int \sqrt{9 - x^2}(-2x)\,dx$

10. $\int (1 - 2x^2)^3(-4x)\,dx$

11. $\int x^2(x^3 - 1)^4\,dx$

12. $\int x(4x^2 + 3)^3\,dx$

13. $\int 5x\sqrt[3]{1 - x^2}\,dx$

14. $\int u^3\sqrt{u^4 + 2}\,du$

15. $\int \dfrac{x^2}{(1 + x^3)^2}\,dx$

16. $\int \dfrac{x^2}{(16 - x^3)^2}\,dx$

17. $\int \left(1 + \dfrac{1}{t}\right)^3\left(\dfrac{1}{t^2}\right)dt$

18. $\int \left[x^2 + \dfrac{1}{(3x)^2}\right]dx$

19. $\int \dfrac{1}{\sqrt{2x}}\,dx$

20. $\int \dfrac{1}{2\sqrt{x}}\,dx$

21. $\int \dfrac{x^2 + 3x + 7}{\sqrt{x}}\,dx$

22. $\int \dfrac{t + 2t^2}{\sqrt{t}}\,dt$

23. $\int t^2\left(t - \dfrac{2}{t}\right)dt$

24. $\int \left(\dfrac{t^3}{3} + \dfrac{1}{4t^2}\right)dt$

25. $\int (9 - y)\sqrt{y}\,dy$

26. $\int 2\pi y(8 - y^{3/2})\,dy$

In Exercises 27–30, solve the differential equation.

27. $\dfrac{dy}{dx} = 4x + \dfrac{4x}{\sqrt{16 - x^2}}$

28. $\dfrac{dy}{dx} = \dfrac{10x^2}{\sqrt{1 + x^3}}$

29. $\dfrac{dy}{dx} = \dfrac{x + 1}{(x^2 + 2x - 3)^2}$

30. $\dfrac{dy}{dx} = \dfrac{x - 4}{\sqrt{x^2 - 8x + 1}}$

In Exercises 31–42, evaluate the indefinite integral.

31. $\int \sin 2x\,dx$

32. $\int x \sin x^2\,dx$

33. $\int \dfrac{1}{\theta^2}\cos\dfrac{1}{\theta}\,d\theta$

34. $\int \cos 6x\,dx$

35. $\int \sin 2x \cos 2x\,dx$

36. $\int \sec(1 - x)\tan(1 - x)\,dx$

37. $\int \tan^4 x \sec^2 x\,dx$

38. $\int \sqrt{\cot x}\,\csc^2 x\,dx$

39. $\int \dfrac{\csc^2 x}{\cot^3 x}\,dx$

40. $\int \dfrac{\sin x}{\cos^2 x}\,dx$

41. $\int \cot^2 x\,dx$

42. $\int \csc^2\left(\dfrac{x}{2}\right)dx$

In Exercises 43–50, evaluate the indefinite integral by the method shown in Example 5.

43. $\int x\sqrt{x + 2}\,dx$

44. $\int x\sqrt{2x + 1}\,dx$

45. $\int x^2\sqrt{1 - x}\,dx$

46. $\int (x + 1)\sqrt{2 - x}\,dx$

47. $\int \dfrac{x^2 - 1}{\sqrt{2x - 1}}\,dx$

48. $\int \dfrac{2x - 1}{\sqrt{x + 3}}\,dx$

49. $\int \dfrac{-x}{(x + 1) - \sqrt{x + 1}}\,dx$

50. $\int t\sqrt[3]{t - 4}\,dt$

In Exercises 51–64, evaluate the definite integral.

51. $\displaystyle\int_{-1}^{1} x(x^2 + 1)^3\,dx$

52. $\displaystyle\int_{0}^{1} x\sqrt{1 - x^2}\,dx$

53. $\displaystyle\int_{0}^{4} \dfrac{1}{\sqrt{2x + 1}}\,dx$

54. $\displaystyle\int_{0}^{2} \dfrac{x}{\sqrt{1 + 2x^2}}\,dx$

55. $\displaystyle\int_{1}^{9} \dfrac{1}{\sqrt{x}(1 + \sqrt{x})^2}\,dx$

56. $\displaystyle\int_{0}^{2} x\sqrt[3]{4 + x^2}\,dx$

57. $\displaystyle\int_{1}^{2} (x - 1)\sqrt{2 - x}\,dx$

58. $\displaystyle\int_{0}^{4} \dfrac{x}{\sqrt{2x + 1}}\,dx$

59. $\displaystyle\int_{0}^{7} x\sqrt[3]{x + 1}\,dx$

60. $\displaystyle\int_{-2}^{6} x^2\sqrt[3]{x + 2}\,dx$

61. $\displaystyle\int_{0}^{\pi/2} \cos\left(\dfrac{2x}{3}\right)dx$

62. $\displaystyle\int_{\pi/3}^{\pi/2} (x + \cos x)\,dx$

63. $\displaystyle\int_{\pi/2}^{2\pi/3} \sec^2\left(\dfrac{x}{2}\right)dx$

64. $\displaystyle\int_{\pi/12}^{\pi/4} \csc 2x \cot 2x\,dx$

C In Exercises 65–70, use a symbolic integration utility to evaluate the integral.

65. $\displaystyle\int \frac{x}{\sqrt{2x+1}}\,dx$

66. $\displaystyle\int x^3\sqrt{x+2}\,dx$

67. $\displaystyle\int_3^7 x\sqrt{x-3}\,dx$

68. $\displaystyle\int_0^1 \frac{1}{\sqrt{x}+\sqrt{x+1}}\,dx$

69. $\displaystyle\int_1^5 x^2\sqrt{x-1}\,dx$

70. $\displaystyle\int_0^{\pi/2} \sin 2x\,dx$

Essay In Exercises 71 and 72, perform the integration in two ways. Explain any difference in the forms of the answers.

71. $\displaystyle\int (2x-1)^2\,dx$

72. $\displaystyle\int \sin x\cos x\,dx$

In Exercises 73 and 74, find the area of the region.

73. $y = 2\sin x + \sin 2x$

74. $y = \sin x + \cos 2x$

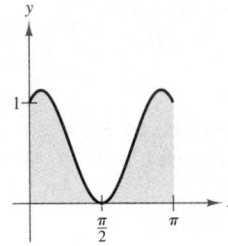

75. Use the fact that

$$\int_0^2 x^2\,dx = \frac{8}{3}$$

to evaluate the definite integrals without using the Fundamental Theorem of Calculus.

a. $\displaystyle\int_{-2}^0 x^2\,dx$

b. $\displaystyle\int_{-2}^2 x^2\,dx$

c. $\displaystyle\int_0^2 -x^2\,dx$

d. $\displaystyle\int_{-2}^0 3x^2\,dx$

76. Use the symmetry of the graphs of the sine and cosine functions as an aid in evaluating the integrals.

a. $\displaystyle\int_{-\pi/4}^{\pi/4} \sin x\,dx$

b. $\displaystyle\int_{-\pi/4}^{\pi/4} \cos x\,dx$

c. $\displaystyle\int_{-\pi/2}^{\pi/2} \cos x\,dx$

d. $\displaystyle\int_{-\pi/2}^{\pi/2} \sin x\cos x\,dx$

In Exercises 77 and 78, write the integral as the sum of an integral of an odd function and the integral of an even function. Use this simplification to evaluate the integral.

77. $\displaystyle\int_{-4}^4 (x^3 + 6x^2 - 2x - 3)\,dx$

78. $\displaystyle\int_{-\pi}^{\pi} (\sin 3x + \cos 3x)\,dx$

In Exercises 79 and 80, find an equation for the function f that has the indicated derivative and whose graph passes through the given point.

Derivative	Point
79. $f'(x) = x\sqrt{1-x^2}$	$(0, \frac{7}{3})$
80. $f'(x) = \pi \sec \pi x \tan \pi x$	$(\frac{1}{3}, 1)$

81. *Depreciation* The rate of depreciation dV/dt of a machine is inversely proportional to the square of $t + 1$, where V is the value of the machine t years after it was purchased. If the initial value of the machine was \$500,000 and its value dropped \$100,000 in the first year, estimate its value after 4 years.

82. *Cash Flow* The rate of disbursement dQ/dt of a 2 million dollar federal grant is proportional to the square of $100 - t$. Time t is measured in days $(0 \le t \le 100)$, and Q is the amount that remains to be disbursed. Find the amount that remains to be disbursed after 50 days. Assume all the money will be disbursed in 100 days.

83. *Marginal Cost* The marginal cost for a certain commodity has been determined to be

$$\frac{dC}{dx} = \frac{12}{\sqrt[3]{12x+1}}.$$

a. Find the cost function if $C = 100$ when $x = 13$.
b. Graph the marginal cost function and the cost function on the same set of axes.

84. *Gasoline Reserves* The minimum stockpile level of gasoline in the United States can be approximated by the model

$$Q = 217 + 13\cos\frac{\pi(t-3)}{6}$$

where Q is measured in millions of barrels of gasoline and t is the time in months, with $t = 1$ corresponding to January. Find the average minimum level given by this model during the following periods.
a. The first quarter $(0 \le t \le 3)$
b. The second quarter $(3 \le t \le 6)$
c. The entire year $(0 \le t \le 12)$

85. *Sales* The sales of a seasonal product are given by the model

$$S = 74.50 + 43.75\sin\frac{\pi t}{6}$$

where S is measured in thousands of units and t is the time in months, with $t = 1$ corresponding to January. Find the average sales for the following periods.
a. The first quarter $(0 \le t \le 3)$
b. The second quarter $(3 \le t \le 6)$
c. The entire year $(0 \le t \le 12)$

86. *Electricity* The oscillating current in an electrical circuit is

$$I = 2\sin(60\pi t) + \cos(120\pi t)$$

where I is measured in amperes and t is measured in seconds. Find the average current for the following time intervals.

a. $0 \leq t \leq \frac{1}{60}$ **b.** $0 \leq t \leq \frac{1}{240}$

c. $0 \leq t \leq \frac{1}{30}$

True or False **In Exercises 87–92, determine whether the statement is true or false. If it is false, explain why or give an example that shows it is false.**

87. $\displaystyle\int (2x + 1)^2\, dx = \frac{1}{3}(2x + 1)^3 + C$

88. $\displaystyle\int x(x^2 + 1)\, dx = (\frac{1}{2}x^2)(\frac{1}{3}x^3 + x) + C$

89. $\displaystyle\int_{-10}^{10} (ax^3 + bx^2 + cx + d)\, dx = 2\int_0^{10} (bx^2 + d)\, dx$

90. $\displaystyle\int_a^b \sin x\, dx = \int_a^{b+2\pi} \sin x\, dx$

91. $\displaystyle 4\int \sin x \cos x\, dx = -\cos 2x + C$

92. $\displaystyle\int \sin^2 2x \cos 2x\, dx = \frac{1}{3}\sin^3 2x + C$

93. Show that if f is continuous on the entire real line, then
$$\int_a^b f(x + h)\, dx = \int_{a+h}^{b+h} f(x)\, dx.$$

94. Let f be continuous on the interval $[0, b]$. Show that
$$\int_0^b \frac{f(x)}{f(x) + f(b - x)}\, dx = \frac{b}{2}.$$

Use this result to evaluate
$$\int_0^1 \frac{\sin(x)}{\sin(1 - x) + \sin(x)}\, dx.$$

SECTION

4.6 Numerical Integration

The Trapezoidal Rule • Simpson's Rule • Error Analysis

The Trapezoidal Rule

Some elementary functions simply do not possess antiderivatives that are elementary functions. For example, there is no elementary function that has any of the following functions as its derivative.

$$\sqrt[3]{x}\sqrt{1-x} \qquad \sqrt{x}\cos x \qquad \frac{\cos x}{x}$$

$$\sqrt{1-x^3} \qquad \sqrt{1+\sin^2 x} \qquad \sin x^2$$

If you need to evaluate a definite integral involving a function whose antiderivative cannot be found, then the Fundamental Theorem of Calculus cannot be applied, and you must resort to an approximation technique. Two such techniques are described in this section.

One way to approximate a definite integral is to use n trapezoids, as shown in Figure 4.40. In the development of this method, assume that f is continuous and positive on the interval $[a, b]$. Thus, the definite integral

$$\int_a^b f(x)\,dx$$

represents the area of the region bounded by the graph of f and the x-axis, from $x = a$ to $x = b$.

First, partition the interval $[a, b]$ into n equal subintervals, each of width $\Delta x = (b - a)/n$, such that

$$a = x_0 < x_1 < x_2 < \cdots < x_n = b.$$

Then, form a trapezoid for each subinterval, as shown in Figure 4.41. The area of the ith trapezoid is

$$\text{Area of } i\text{th trapezoid} = \left[\frac{f(x_{i-1}) + f(x_i)}{2}\right]\left(\frac{b - a}{n}\right).$$

This implies that the sum of the areas of the n trapezoids is

$$\text{Area} = \left(\frac{b - a}{n}\right)\left[\frac{f(x_0) + f(x_1)}{2} + \cdots + \frac{f(x_{n-1}) + f(x_n)}{2}\right]$$

$$= \left(\frac{b - a}{2n}\right)[f(x_0) + f(x_1) + f(x_1) + f(x_2) + \cdots + f(x_{n-1}) + f(x_n)]$$

$$= \left(\frac{b - a}{2n}\right)[f(x_0) + 2f(x_1) + 2f(x_2) + \cdots + 2f(x_{n-1}) + f(x_n)].$$

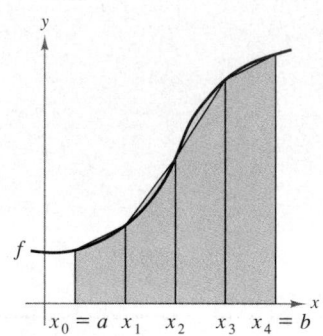

FIGURE 4.40
The area of the region can be approximated using 4 trapezoids.

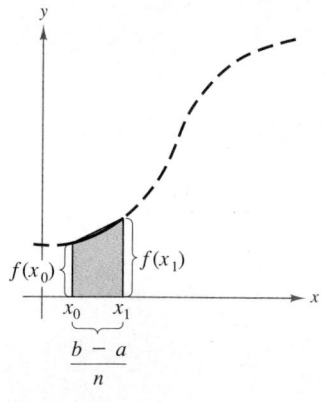

FIGURE 4.41
The area of the first trapezoid is
$$\left[\frac{f(x_0) + f(x_1)}{2}\right]\left(\frac{b - a}{n}\right).$$

Letting $\Delta x = (b - a)/n$, you can take the limit as $n \to \infty$, to obtain

$$\lim_{n \to \infty} \left(\frac{b - a}{2n} \right) [f(x_0) + 2f(x_1) + \cdots + 2f(x_{n-1}) + f(x_n)]$$

$$= \lim_{n \to \infty} \left[\frac{[f(a) - f(b)] \Delta x}{2} + \sum_{i=1}^{n} f(x_i) \Delta x \right]$$

$$= \lim_{n \to \infty} \frac{[f(a) - f(b)](b - a)}{2n} + \lim_{n \to \infty} \sum_{i=1}^{n} f(x_i) \Delta x$$

$$= 0 + \int_a^b f(x) \, dx.$$

This result is summarized in the following theorem.

> **THEOREM 4.16 The Trapezoidal Rule**
>
> Let f be continuous on $[a, b]$. The Trapezoidal Rule for approximating $\int_a^b f(x) \, dx$ is given by
>
> $$\int_a^b f(x) \, dx \approx \frac{b - a}{2n} [f(x_0) + 2f(x_1) + 2f(x_2) + \cdots + 2f(x_{n-1}) + f(x_n)].$$
>
> Moreover, as $n \to \infty$, the right-hand side approaches $\int_a^b f(x) \, dx$.

EXAMPLE 1   Approximation with the Trapezoidal Rule

Use the Trapezoidal Rule to approximate

$$\int_0^{\pi} \sin x \, dx.$$

Compare the results for $n = 4$ and $n = 8$, as shown in Figure 4.42.

Solution   When $n = 4$, $\Delta x = \pi/4$, and you obtain

$$\int_0^{\pi} \sin x \, dx \approx \frac{\pi}{8} \left(\sin 0 + 2 \sin \frac{\pi}{4} + 2 \sin \frac{\pi}{2} + 2 \sin \frac{3\pi}{4} + \sin \pi \right)$$

$$= \frac{\pi}{8} (0 + \sqrt{2} + 2 + \sqrt{2} + 0) = \frac{\pi(1 + \sqrt{2})}{4} \approx 1.896.$$

When $n = 8$, $\Delta x = \pi/8$, and you obtain

$$\int_0^{\pi} \sin x \, dx \approx \frac{\pi}{16} \left(\sin 0 + 2 \sin \frac{\pi}{8} + 2 \sin \frac{\pi}{4} + 2 \sin \frac{3\pi}{8} + 2 \sin \frac{\pi}{2} \right.$$

$$\left. + 2 \sin \frac{5\pi}{8} + 2 \sin \frac{3\pi}{4} + 2 \sin \frac{7\pi}{8} + \sin \pi \right)$$

$$= \frac{\pi}{16} \left(2 + 2\sqrt{2} + 4 \sin \frac{\pi}{8} + 4 \sin \frac{3\pi}{8} \right) \approx 1.974.$$

For this particular integral, you could have found an antiderivative and determined that the exact area of the region is 2.

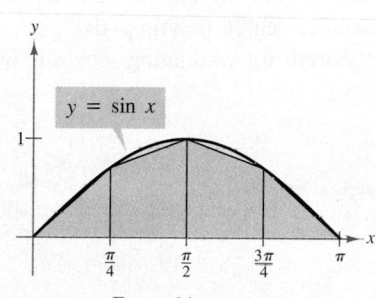

$y = \sin x$

Four subintervals

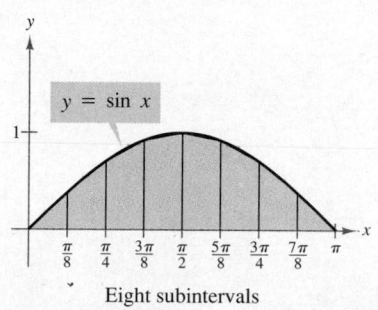

$y = \sin x$

Eight subintervals

FIGURE 4.42

It is interesting to compare the Trapezoidal Rule to the Midpoint Rule given in Section 4.2 (Exercises 51–54). For the Trapezoidal Rule, you average the function values at the endpoints of the subintervals, but for the Midpoint Rule you take the function value of the subinterval midpoints.

Midpoint Rule

$$\int_a^b f(x)\,dx \approx \sum_{i=1}^n f\left(\frac{x_i + x_{i-1}}{2}\right)\Delta x$$

Trapezoidal Rule

$$\int_a^b f(x)\,dx \approx \sum_{i=1}^n \left(\frac{f(x_i) + f(x_{i-1})}{2}\right)\Delta x$$

There are two important points that should be made concerning the Trapezoidal Rule (or the Midpoint Rule). First, the approximation tends to become more accurate as n increases. For instance, in Example 1, if $n = 16$, the Trapezoidal Rule yields an approximation of 1.994. Second, although you could have used the Fundamental Theorem to evaluate the integral in Example 1, this theorem cannot be used to evaluate an integral as simple as

$$\int_0^\pi \sin x^2\,dx$$

because $\sin x^2$ has no elementary antiderivative. Yet, the Trapezoidal Rule can be applied easily to this integral.

Simpson's Rule

One way to view the trapezoidal approximation of a definite integral is to say that on each subinterval you approximate f by a *first*-degree polynomial. In Simpson's Rule, named after the English mathematician Thomas Simpson (1710–1761), we take this procedure one step further and approximate f by *second*-degree polynomials.

Before presenting Simpson's Rule, we list a theorem for evaluating integrals of polynomials of degree 2 (or less).

THEOREM 4.17   Integral of $p(x) = Ax^2 + Bx + C$

If $p(x) = Ax^2 + Bx + C$, then

$$\int_a^b p(x)\,dx = \left(\frac{b - a}{6}\right)\left[p(a) + 4p\left(\frac{a + b}{2}\right) + p(b)\right].$$

Proof

$$\int_a^b p(x)\,dx = \int_a^b (Ax^2 + Bx + C)\,dx$$

$$= \left[\frac{Ax^3}{3} + \frac{Bx^2}{2} + Cx\right]_a^b$$

$$= \frac{A(b^3 - a^3)}{3} + \frac{B(b^2 - a^2)}{2} + C(b - a)$$

$$= \left(\frac{b - a}{6}\right)[2A(a^2 + ab + b^2) + 3B(b + a) + 6C]$$

By expansion and collection of terms, the expression inside the brackets becomes

$$\underbrace{(Aa^2 + Ba + C)}_{p(a)} + \underbrace{4\left[A\left(\frac{b+a}{2}\right)^2 + B\left(\frac{b+a}{2}\right) + C\right]}_{4p\left(\frac{a+b}{2}\right)} + \underbrace{(Ab^2 + Bb + C)}_{p(b)}$$

and you can write

$$\int_a^b p(x)\,dx = \left(\frac{b-a}{6}\right)\left[p(a) + 4p\left(\frac{a+b}{2}\right) + p(b)\right].$$

To develop Simpson's Rule for approximating a definite integral, you again partition the interval $[a, b]$ into n equal subintervals, each of width $\Delta x = (b - a)/n$. This time, however, n is required to be even, and the subintervals are grouped in pairs such that

$$a = \underbrace{x_0 < x_1 < }_{[x_0, x_2]}\underbrace{x_2 < x_3 < x_4}_{[x_2, x_4]} < \cdots < \underbrace{x_{n-2} < x_{n-1} < x_n}_{[x_{n-2}, x_n]} = b.$$

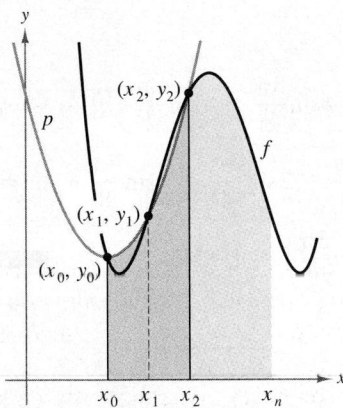

FIGURE 4.43
$\int_{x_0}^{x_2} p(x)\,dx \approx \int_{x_0}^{x_2} f(x)\,dx$

On each (double) subinterval $[x_{i-2}, x_i]$, you can approximate f by a polynomial p of degree less than or equal to 2. For example, on the subinterval $[x_0, x_2]$, choose the polynomial of least degree passing through the points (x_0, y_0), (x_1, y_1), and (x_2, y_2), as shown in Figure 4.43. Now, using p as an approximation for f on this subinterval, you have

$$\int_{x_0}^{x_2} f(x)\,dx \approx \int_{x_0}^{x_2} p(x)\,dx = \frac{x_2 - x_0}{6}\left[p(x_0) + 4p\left(\frac{x_0 + x_2}{2}\right) + p(x_2)\right]$$

$$= \frac{2[(b-a)/n]}{6}\left[p(x_0) + 4p(x_1) + p(x_2)\right]$$

$$= \frac{b-a}{3n}\left[f(x_0) + 4f(x_1) + f(x_2)\right].$$

Repeating this procedure on the entire interval $[a, b]$ produces the following theorem.

THEOREM 4.18   Simpson's Rule (*n* is even)

Let f be continuous on $[a, b]$. Simpson's Rule for approximating $\int_a^b f(x)\,dx$ is

$$\int_a^b f(x)\,dx \approx \frac{b-a}{3n}\left[f(x_0) + 4f(x_1) + 2f(x_2) + 4f(x_3) + \cdots + 4f(x_{n-1}) + f(x_n)\right].$$

Moreover, as $n \rightarrow \infty$, the right-hand side approaches $\int_a^b f(x)\,dx$.

REMARK   Note that the coefficients in Simpson's Rule have the following pattern.

1   4   2   4   2   4 $\cdots$ 4   2   4   1

In Example 1, the Trapezoidal Rule was used to estimate $\int_0^\pi \sin x\, dx$. In the next example, Simpson's Rule is applied to the same integral.

EXAMPLE 2   Approximation with Simpson's Rule

Use Simpson's Rule to approximate

$$\int_0^\pi \sin x\, dx.$$

Compare the results for $n = 4$ and $n = 8$.

Solution   When $n = 4$, you have

$$\int_0^\pi \sin x\, dx \approx \frac{\pi}{12}\left(\sin 0 + 4\sin\frac{\pi}{4} + 2\sin\frac{\pi}{2} + 4\sin\frac{3\pi}{4} + \sin\pi\right)$$

$$= \frac{\pi}{12}\left(4\sqrt{2} + 2\right) \approx 2.005.$$

When $n = 8$, you have

$$\int_0^\pi \sin x\, dx \approx \frac{\pi}{24}\left(\sin 0 + 4\sin\frac{\pi}{8} + 2\sin\frac{\pi}{4} + 4\sin\frac{3\pi}{8} + 2\sin\frac{\pi}{2}\right.$$

$$\left. + 4\sin\frac{5\pi}{8} + 2\sin\frac{3\pi}{4} + 4\sin\frac{7\pi}{8} + \sin\pi\right)$$

$$= \frac{\pi}{24}\left(2 + 2\sqrt{2} + 8\sin\frac{\pi}{8} + 8\sin\frac{3\pi}{8}\right) \approx 2.0003.\qquad\blacksquare$$

Error Analysis

In Examples 1 and 2, you could have calculated the exact values of the integrals and compared those values to the approximations to see how close they were. In practice, of course, you would not bother with an approximation if it were possible to evaluate the integral exactly. However, if you must use an approximation technique, it is important to know how accurate you can expect the approximation to be. The following theorem, which we list without proof, gives the formulas for estimating the errors involved in the use of Simpson's Rule and the Trapezoidal Rule.

THEOREM 4.19   Errors in the Trapezoidal Rule and Simpson's Rule

If f has a continuous second derivative on $[a, b]$, then the error E in approximating $\int_a^b f(x)\, dx$ by the Trapezoidal Rule is

$$E \leq \frac{(b - a)^3}{12n^2}\left[\max |f''(x)|\right], \qquad a \leq x \leq b. \qquad \text{Trapezoidal Rule}$$

Moreover, if f has a continuous fourth derivative on $[a, b]$, then the error E in approximating $\int_a^b f(x)\, dx$ by Simpson's Rule is

$$E \leq \frac{(b - a)^5}{180n^4}\left[\max |f^{(4)}(x)|\right], \qquad a \leq x \leq b. \qquad \text{Simpson's Rule}$$

Theorem 4.19 states that the errors generated by the Trapezoidal Rule and Simpson's Rule have upper bounds dependent on the extreme values of $f''(x)$ and $f^{(4)}(x)$, respectively, in the interval $[a, b]$. Furthermore, it is evident from this theorem that these errors can be made arbitrarily small by *increasing n*, provided that f'' and $f^{(4)}$ are continuous and therefore bounded in $[a, b]$. The next two examples show how to determine a value of n that will bound the error within a predetermined **tolerance.**

EXAMPLE 3   The Approximate Error in the Trapezoidal Rule

Determine a value of n so that the Trapezoidal Rule will approximate the value of the following integral with an error that is less than 0.01.

$$\int_0^1 \sqrt{1 + x^2}\, dx$$

Solution   Begin by letting $f(x) = \sqrt{1 + x^2}$ and finding the second derivative of f.

$$f' = x(1 + x^2)^{-1/2}$$

$$f'' = x\left(-\frac{1}{2}\right)(2x)(1 + x^2)^{-3/2} + (1 + x^2)^{-1/2}$$

$$= (1 + x^2)^{-3/2}$$

The maximum value of $|f''(x)|$ on the interval $[0, 1]$ is $|f''(0)| = 1$. Thus, by Theorem 4.19, you can write

$$E \le \frac{(b - a)^3}{12n^2}\, |f''(0)| \le \frac{1}{12n^2}(1) = \frac{1}{12n^2},$$

To obtain an error E that is less than 0.01, you must choose n so that $1/(12n^2) \le 1/100$. Thus,

$$100 \le 12n^2 \rightarrow 2.89 \approx \sqrt{\frac{100}{12}} \le n.$$

Therefore, you can choose $n = 3$ (because n must be greater than or equal to 2.89) and apply the Trapezoidal Rule, as shown in Figure 4.44, to obtain

$$\int_0^1 \sqrt{1 + x^2}\, dx \approx \frac{1}{6}\left[\sqrt{1 + 0^2} + 2\sqrt{1 + (\tfrac{1}{3})^2} + 2\sqrt{1 + (\tfrac{2}{3})^2} + \sqrt{1 + 1^2}\right]$$

$$\approx 1.154.$$

Thus, with an error no larger than 0.01, you know that

$$1.144 \le \int_0^1 \sqrt{1 + x^2}\, dx \le 1.164.$$

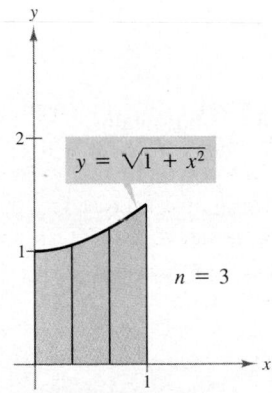

y

$y = \sqrt{1 + x^2}$

$n = 3$

x

FIGURE 4.44

$1.144 \le \int_0^1 \sqrt{1 + x^2}\, dx \le 1.164$

TECHNOLOGY   If you have access to a symbolic integrating utility, try using it to evaluate the definite integral in Example 3. You should obtain a value of

$$\int_0^1 \sqrt{1 + x^2}\, dx = \tfrac{1}{2}\left(\sqrt{2} + \ln\left(1 + \sqrt{2}\right)\right) \approx 1.14779.$$

("ln" represents the natural logarithmic function, which you will study in Section 5.1.)

EXAMPLE 4   The Approximate Error in Simpson's Rule

Determine a value of n so that Simpson's Rule will approximate the value of the following integral with an error that is less than 0.001.

$$\int_0^1 \cos x^2 \, dx$$

Solution   Begin by finding the fourth derivative of the integrand.

$$f(x) = \cos x^2$$
$$f'(x) = -2x \sin x^2$$
$$f''(x) = -2(2x^2 \cos x^2 + \sin x^2)$$
$$f'''(x) = 4(2x^3 \sin x^2 - 3x \cos x^2)$$
$$f^{(4)}(x) = 4(4x^4 \cos x^2 + 12x^2 \sin x^2 - 3 \cos x^2)$$
$$= 4[(4x^4 - 3)(\cos x^2) + 12x^2 \sin x^2]$$

In the interval $[0, 1]$, $|f^{(4)}(x)|$ is greatest when $x = 1$. Therefore, you know that

$$|f^{(4)}(x)| \le |f^{(4)}(1)| = 4[\cos 1 + 12 \sin 1] \approx 42.6 < 45.$$

This implies that

$$E \le \frac{(b-a)^5}{180n^4} |f^{(4)}(1)| = \frac{1}{180n^4} |f^{(4)}(1)| < \frac{45}{180n^4} = \frac{1}{4n^4}.$$

Now, choosing n so that $(1/4n^4) < 1/1000$, you obtain

$$\frac{1000}{4} < n^4 \rightarrow 3.98 < n.$$

Therefore, you can choose $n = 4$ as shown in Figure 4.45 and obtain

$$\int_0^1 \cos x^2 \, dx \approx \frac{1}{12} \left(\cos 0 + 4 \cos \frac{1}{16} + 2 \cos \frac{1}{4} + 4 \cos \frac{9}{16} + \cos 1 \right)$$
$$\approx 0.9045$$

and conclude that

$$0.9035 \le \int_0^1 \cos x^2 \, dx \le 0.9055.$$

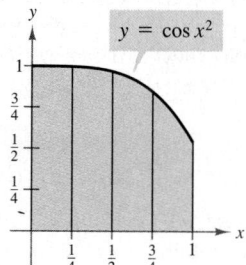

FIGURE 4.45
Choosing $n = 4$, the approximation of the definite integral using Simpson's Rule has an error of less than 0.001.

TECHNOLOGY   As you might imagine, technology can be very helpful in performing the many calculations required by the Trapezoidal Rule or by Simpson's Rule. For instance, using a computer program, it is a simple matter to approximate the definite integral in Example 4 to be

$$\int_0^1 \cos x^2 \, dx \approx 0.9045242375$$

which has an error of less than 0.0000000001!

EXERCISES for Section 4.6

In Exercises 1–10, use the Trapezoidal Rule and Simpson's Rule to approximate the value of the definite integral for the indicated value of *n*. Round your answer to four decimal places and compare the results with the exact value of the definite integral.

1. $\int_0^2 x^2\,dx$ **2.** $\int_0^1 \left(\frac{x^2}{2} + 1\right)dx$

3. $\int_0^2 x^3\,dx$ **4.** $\int_1^2 \frac{1}{x^2}\,dx$

     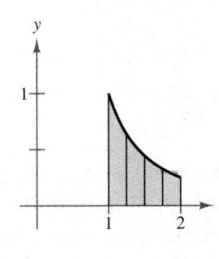

5. $\int_0^2 x^3\,dx,\qquad n = 8$ **6.** $\int_0^8 \sqrt[3]{x}\,dx,\qquad n = 8$

7. $\int_4^9 \sqrt{x}\,dx,\qquad n = 8$ **8.** $\int_1^3 (4 - x^2)\,dx,\qquad n = 4$

9. $\int_1^2 \frac{1}{(x + 1)^2}\,dx,\qquad n = 4$

10. $\int_0^2 x\sqrt{x^2 + 1}\,dx,\qquad n = 4$

In Exercises 11–20, approximate the definite integral with *n* = 4, using (a) the Trapezoidal Rule and (b) Simpson's Rule.

11. $\int_0^2 \sqrt{1 + x^3}\,dx$ **12.** $\int_0^2 \frac{1}{\sqrt{1 + x^3}}$

13. $\int_0^1 \sqrt{x}\sqrt{1 - x}\,dx$ **14.** $\int_{\pi/2}^{\pi} \sqrt{x}\sin x\,dx$

15. $\int_0^{\sqrt{\pi/2}} \cos x^2\,dx$ **16.** $\int_0^{\sqrt{\pi/4}} \tan x^2\,dx$

17. $\int_1^{1.1} \sin x^2\,dx$ **18.** $\int_0^{\pi/2} \sqrt{1 + \cos^2 x}\,dx$

19. $\int_0^{\pi/4} x\tan x\,dx$

20. $\int_0^{\pi} f(x)\,dx,\qquad f(x) = \begin{cases} \dfrac{\sin x}{x}, & x > 0 \\ 1, & x = 0 \end{cases}$

In Exercises 21 and 22, use the error formulas to find the maximum possible error in approximating the integral, with *n* = 4, by (a) The Trapezoidal Rule and (b) Simpson's Rule.

21. $\int_0^2 x^3\,dx$ **22.** $\int_0^1 \frac{1}{x + 1}\,dx$

In Exercises 23 and 24, use the error formulas to find *n* so that the error in the approximation of the definite integral is less than 0.00001 using (a) The Trapezoidal Rule and (b) Simpson's Rule.

23. $\int_1^3 \frac{1}{x}\,dx$ **24.** $\int_0^1 \frac{1}{1 + x}\,dx$

C In Exercises 25–28, use a symbolic differentiation utility and the error formulas to find *n* so that the error in the approximation of the definite integral is less than 0.00001 using (a) The Trapezoidal Rule and (b) Simpson's Rule.

25. $\int_0^2 \sqrt{1 + x}\,dx$ **26.** $\int_0^2 (x + 1)^{2/3}\,dx$

27. $\int_0^1 \tan x^2\,dx$ **28.** $\int_0^1 \sin x^2\,dx$

29. Prove that Simpson's Rule is exact when approximating the integral of a cubic polynomial function, and demonstrate the result for

$$\int_0^1 x^3\,dx,\qquad n = 2.$$

C 30. Write a computer program to approximate a definite integral using the Trapezoidal Rule and Simpson's Rule. Start with the program written in Section 4.3 exercises and note that the Trapezoidal Rule can be written as

$$T(n) = \tfrac{1}{2}[L(n) + R(n)]$$

and Simpson's Rule can be written as

$$S(n) = \tfrac{1}{3}[T(n/2) + 2M(n/2)].$$

[Recall that $L(n)$, $M(n)$, and $R(n)$ represent the Riemann sums using the left-hand endpoint, midpoint, and right-hand endpoint of subintervals of equal width.]

In Exercises 31–34, use the computer program from Exercise 30 to approximate the definite integral and complete the table.

n	L(n)	M(n)	R(n)	T(n)	S(n)
4					
8					
10					
12					
16					
20					

31. $\int_0^4 \sqrt{2 + 3x^2}\, dx$ **32.** $\int_0^1 \sqrt{1 - x^2}\, dx$

33. $\int_0^4 \sin \sqrt{x}\, dx$ **34.** $\int_1^2 \frac{\sin x}{x}\, dx$

35. *Area*   Use Simpson's Rule with $n = 14$ to approximate the area of the region bounded by the graphs of $y = \sqrt{x}\cos x$, $y = 0$, $x = 0$, and $x = \pi/2$.

36. *Circumference*   The **elliptic integral**

$$8\sqrt{3} \int_0^{\pi/2} \sqrt{1 - \tfrac{2}{3}\sin^2 \theta}\, d\theta$$

gives the circumference of an ellipse whose major axis has length $4\sqrt{3}$ and whose minor axis has length 4. Use Simpson's Rule with $n = 8$ to approximate the circumference of the ellipse.

37. *Approximation of Pi*   Use Simpson's Rule with $n = 6$ to approximate π using the equation

$$\pi = \int_0^1 \frac{4}{1 + x^2}\, dx.$$

(In Section 5.8, you will be able to evaluate this integral using the inverse tangent function.)

38. *Work*   To determine the size of motor required to operate a press, a company must know the amount of work done when the press moves an object linearly 5 feet. The variable force to move the object is given by

$$F(x) = 100x\sqrt{125 - x^3}$$

where F is given in pounds and x gives the position of the unit in feet. Use Simpson's Rule with $n = 12$ to approximate the work W (in foot-pounds) done through one cycle if

$$W = \int_0^5 F(x)\, dx.$$

Area   In Exercises 39 and 40, use the Trapezoidal Rule to estimate the number of square feet of land in a lot where x and y are measured in feet, as shown in the figures. In each case, the land is bounded by a stream and two straight roads that meet at right angles.

39.

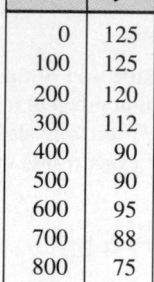

x	y
0	125
100	125
200	120
300	112
400	90
500	90
600	95
700	88
800	75
900	35
1000	0

40.

x	y
0	75
10	81
20	84
30	76
40	67
50	68
60	69
70	72
80	68
90	56
100	42
110	23
120	0

41. *Area*   To estimate the surface area of a pond, a surveyor takes several measurements, as shown in the figure. Estimate the surface area of this pond using **(a)** The Trapezoidal Rule and **(b)** Simpson's Rule.

FIGURE FOR 41

42. The table lists several physical measurements gathered in an experiment to approximate a continuous function $y = f(x)$. Approximate the integral $\int_0^2 f(x)\, dx$ using **(a)** The Trapezoidal Rule and **(b)** Simpson's Rule.

x	0.00	0.25	0.50	0.75	1.00
y	4.32	4.36	4.58	5.79	6.14

x	1.25	1.50	1.75	2.00
y	7.25	7.64	8.08	8.14

C **43.** Use Simpson's Rule (with $n = 10$) and a computer to approximate t in the integral equation

$$\int_0^t \sin \sqrt{x}\, dx = 2.$$

44. Determine whether the Trapezoidal Rule overestimates or underestimates a definite integral if the graph of the integrand is

a. Concave up.

b. Concave down.

45. Let $L(x) = \int_1^x \dfrac{1}{t}\, dt$ for all $x > 0$.

a. Find $L(1)$.

b. Find $L'(x)$ and $L'(1)$.

c. Use the Trapezoidal Rule to approximate the value of x (to three decimal places) for which $L(x) = 1$.

d. Show that L is an increasing function on $(0, \infty)$.

e. Prove that $L(x_1 x_2) = L(x_1) + L(x_2)$, for $x_1 > 0$ and $x_2 > 0$.

REVIEW EXERCISES for Chapter 4

In Exercises 1 and 2, use the graph of f' to sketch a graph of f.

In Exercises 3–12, find the indefinite integral.

3. $\displaystyle\int (2x^2 + x - 1)\, dx$

4. $\displaystyle\int \frac{2}{\sqrt[3]{3x}}\, dx$

5. $\dfrac{dy}{dx} = (x^2 + 1)^3$

6. $\dfrac{dy}{dx} = \left(x + \dfrac{1}{x}\right)^2$

7. $\displaystyle\int \frac{x^3 + 1}{x^2}\, dx$

8. $\displaystyle\int \frac{x^3 - 2x^2 + 1}{x^2}\, dx$

9. $\displaystyle\int \frac{x^2}{\sqrt{x^3 + 3}}\, dx$

10. $\displaystyle\int x^2 \sqrt{x^3 + 3}\, dx$

11. $\displaystyle\int x(1 - 3x^2)^4\, dx$

12. $\displaystyle\int \frac{x + 3}{(x^2 + 6x - 5)^2}\, dx$

In Exercises 13–20, find the indefinite integral of the trigonometric function.

13. $\dfrac{dy}{dx} = \sin^3 x \cos x$

14. $\dfrac{dy}{dx} = x \sin 3x^2$

15. $\displaystyle\int \frac{\sin \theta}{\sqrt{1 - \cos \theta}}\, d\theta$

16. $\displaystyle\int \frac{\cos x}{\sqrt{\sin x}}\, dx$

17. $\displaystyle\int \tan^n x \sec^2 x\, dx, \quad n \neq -1$

18. $\displaystyle\int \sec 2x \tan 2x\, dx$

19. $\displaystyle\int (1 + \sec \pi x)^2 \sec \pi x \tan \pi x\, dx$

20. $\displaystyle\int \cot^4 \alpha \csc^2 \alpha\, d\alpha$

21. Find the particular solution of the differential equation $f'(x) = -2x$ whose graph passes through the point $(-1, 1)$.

22. Find the particular solution of the differential equation $f''(x) = 6(x - 1)$. The graph of the solution passes through the point $(2, 1)$ and at that point is tangent to the line $3x - y - 5 = 0$.

23. *Velocity and Acceleration*   An airplane taking off from a runway travels 3600 feet before lifting off. If it starts from rest, moves with constant acceleration, and makes the run in 30 seconds, with what speed does it lift off?

24. *Velocity and Deceleration*   The speed of a car traveling in a straight line is reduced from 45 to 30 miles per hour in a distance of 264 feet. Find the distance in which the car can be brought to rest from 30 miles per hour, assuming the same constant deceleration.

25. *Velocity and Acceleration*   A ball is thrown vertically upward from ground level with an initial velocity of 96 feet per second.

a. How long will it take the ball to rise to its maximum height?

b. What is the maximum height?

c. When is the velocity of the ball one-half the initial velocity?

d. What is the height of the ball when its velocity is one-half the initial velocity?

26. *Velocity and Acceleration*   Repeat Exercise 25 for an initial velocity of 128 feet per second.

27. Write in sigma notation.

a. The sum of the first ten positive odd integers

b. The sum of the cubes of the first n positive integers

c. $6 + 10 + 14 + 18 + \cdots + 42$

28. Evaluate the sums for $x_1 = 2$, $x_2 = -1$, $x_3 = 5$, $x_4 = 3$, and $x_5 = 7$.

a. $\dfrac{1}{5} \displaystyle\sum_{i=1}^{5} x_i$

b. $\displaystyle\sum_{i=1}^{5} \frac{1}{x_i}$

c. $\displaystyle\sum_{i=1}^{5} (2x_i - x_i^2)$

d. $\displaystyle\sum_{i=2}^{5} (x_i - x_{i-1})$

29. Consider the region bounded by $y = mx$, $y = 0$, $x = 0$, and $x = b$, $m > 0$.

    a. Find the upper and lower sums to approximate the area of the region when $\Delta x = b/4$.

    b. Find the upper and lower sums to approximate the area of the region when $\Delta x = b/n$.

    c. Find the area of the region by letting n approach infinity in both sums of part b. Show that in each case you obtain the formula for the area of a triangle.

    d. Find the area of the region by using the Fundamental Theorem of Calculus.

30. a. Find the area of the region bounded by the graphs of $y = x^3$, $y = 0$, $x = 1$, and $x = 3$ by the limit definition.

    b. Find the area of the region by using the Fundamental Theorem of Calculus.

In Exercises 31 and 32, use the given values to find the values of the indicated definite integrals.

31. If $\displaystyle\int_2^6 f(x)\,dx = 10$ and $\displaystyle\int_2^6 g(x)\,dx = 3$, find

    a. $\displaystyle\int_2^6 [f(x) + g(x)]\,dx$ **b.** $\displaystyle\int_2^6 [f(x) - g(x)]\,dx$

    c. $\displaystyle\int_2^6 [2f(x) - 3g(x)]\,dx$ **d.** $\displaystyle\int_2^6 5f(x)\,dx$

32. If $\displaystyle\int_0^3 f(x)\,dx = 4$ and $\displaystyle\int_3^6 f(x)\,dx = -1$, find

    a. $\displaystyle\int_0^6 f(x)\,dx$ **b.** $\displaystyle\int_6^3 f(x)\,dx$

    c. $\displaystyle\int_4^4 f(x)\,dx$ **d.** $\displaystyle\int_3^6 -10f(x)\,dx$

In Exercises 33–44, use the Fundamental Theorem of Calculus to evaluate the definite integral.

33. $\displaystyle\int_0^4 (2 + x)\,dx$ **34.** $\displaystyle\int_{-1}^1 (t^2 + 2)\,dt$

35. $\displaystyle\int_{-1}^1 (4t^3 - 2t)\,dt$ **36.** $\displaystyle\int_3^6 \frac{x}{3\sqrt{x^2 - 8}}\,dx$

37. $\displaystyle\int_0^3 \frac{1}{\sqrt{1 + x}}\,dx$ **38.** $\displaystyle\int_0^1 x^2(x^3 + 1)^3\,dx$

39. $\displaystyle\int_4^9 x\sqrt{x}\,dx$ **40.** $\displaystyle\int_1^2 \left(\frac{1}{x^2} - \frac{1}{x^3}\right)dx$

41. $\displaystyle 2\pi \int_0^1 (y + 1)\sqrt{1 - y}\,dy$

42. $\displaystyle 2\pi \int_{-1}^0 x^2\sqrt{x + 1}\,dx$

43. $\displaystyle\int_0^\pi \cos\frac{x}{2}\,dx$ **44.** $\displaystyle\int_{-\pi/4}^{\pi/4} \sin 2x\,dx$

In Exercises 45–50, sketch the graph of the region whose area is given by the integral and find the area.

45. $\displaystyle\int_1^3 (2x - 1)\,dx$ **46.** $\displaystyle\int_0^2 (x + 4)\,dx$

47. $\displaystyle\int_3^4 (x^2 - 9)\,dx$ **48.** $\displaystyle\int_{-1}^2 (-x^2 + x + 2)\,dx$

49. $\displaystyle\int_0^1 (x - x^3)\,dx$ **50.** $\displaystyle\int_0^1 \sqrt{x}(1 - x)\,dx$

In Exercises 51–54, sketch the region bounded by the graphs of the equations and determine its area.

51. $y = \dfrac{4}{\sqrt{x + 1}}$, $y = 0$, $x = 0$, $x = 8$

52. $y = x$, $y = x^5$

53. $y = \sec^2 x$, $y = 0$, $x = 0$, $x = \dfrac{\pi}{3}$

54. $y = \cos x$, $y = 0$, $x = -\dfrac{\pi}{4}, \dfrac{\pi}{4}$

In Exercises 55–58, find the average value of the function over the interval. Find the values of x where the function assumes its average value and sketch the graph of the function.

Function	Interval
55. $f(x) = \dfrac{1}{\sqrt{x - 1}}$	$[5, 10]$
56. $f(x) = x^3$	$[0, 2]$
57. $f(x) = x$	$[0, 4]$
58. $f(x) = x^2 - \dfrac{1}{x^2}$	$[1, 2]$

In Exercises 59 and 60, use Simpson's Rule with $n = 4$ to approximate the definite integral.

59. $\displaystyle\int_1^2 \frac{1}{1 + x^3}\,dx$ **60.** $\displaystyle\int_0^1 \frac{x^{3/2}}{3 - x^2}\,dx$

61. *Respiratory Cycle*   For a person at rest, the rate of air intake v, in liters per second, during a respiratory cycle is

$$v = 0.85 \sin\frac{\pi t}{3}$$

where t is the time in seconds. Find the volume in liters of air inhaled during one cycle by integrating this function over the interval $[0, 3]$.

62. *Respiratory Cycle*   After exercising a few minutes, a person has a respiratory cycle for which the rate of air intake is

$$v = 1.75 \sin\frac{\pi t}{2}.$$

How much does the person's lung capacity increase as a result of exercising? (Compare your answer to that found in Exercise 61.)

63. *Air Conditioning Costs*   The temperature in degrees Fahrenheit is

$$T = 72 + 12 \sin \frac{\pi(t - 8)}{12}$$

where t is the time in hours, with $t = 0$ representing midnight. Suppose the hourly cost of cooling a house is \$0.10 per degree.

a. Find the cost C of cooling this house if its thermostat is set at 72° by evaluating the integral

$$C = 0.1 \int_8^{20} \left[72 + 12 \sin \frac{\pi(t - 8)}{12} - 72 \right] dt.$$

(See figure.)

FIGURE FOR 63a

b. Find the savings realized by resetting the thermostat to 78° by evaluating the integral

$$C = 0.1 \int_{10}^{18} \left[72 + 12 \sin \frac{\pi(t - 8)}{12} - 78 \right] dt.$$

(See figure.)

FIGURE FOR 63b

64. *Production Scheduling*   A manufacturer of fertilizer finds that national sales of fertilizer roughly follow the seasonal pattern

$$F = 100,000 \left[1 + \sin \frac{2\pi(t - 60)}{365} \right]$$

where F is measured in pounds and t is the time in days, with $t = 1$ representing January 1. The manufacturer wants to set up a schedule to produce a uniform amount each day. What should this amount be?

65. Suppose that gasoline is increasing in price according to the equation

$$p = 1.10 + 0.04t$$

where p is the dollar price per gallon and $t = 0$ represents the year 1990. If an automobile is driven 15,000 miles a year and gets M miles per gallon, then the annual fuel cost is

$$C = \frac{15,000}{M} \int_t^{t+1} p \, ds.$$

Find the annual fuel cost for the years
a. 1995   **b.** 2000

Probability   **In Exercises 66 and 67, the function**

$$f(x) = kx^n(1 - x)^m, \qquad 0 \le x \le 1$$

where $n > 0$, $m > 0$, and k is a constant, can be used to represent various probability distributions. If k is chosen so that

$$\int_0^1 f(x) \, dx = 1$$

the probability that x will fall between a and b is

$$P_{a,b} = \int_a^b f(x) \, dx.$$

66. The probability of recall in a certain experiment is found to be

$$P_{a,b} = \int_a^b \frac{15}{4} x \sqrt{1 - x} \, dx$$

where x represents the percentage of recall (see figure).
a. For a randomly chosen individual, what is the probability that he or she will recall between 50% and 75% of the material?
b. What is the median percentage recall? That is, for what value of b is it true that the probability from 0 to b is 0.5?

67. The probability of finding between a and b percentage of iron in ore samples taken from a certain region is

$$P_{a,b} = \int_a^b \frac{1155}{32} x^3(1 - x)^{3/2} \, dx.$$

(See figure.) What is the probability that a sample will contain between
a. 0% and 25%   **b.** 50% and 100%?

  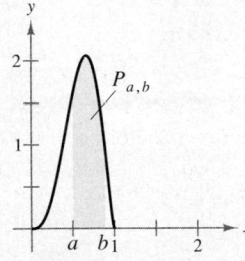

FIGURE FOR 66   **FIGURE FOR 67**

True or False   **In Exercises 68–72, determine whether the statement is true or false. If it is false, explain why or give an example that shows it is false.**

68. $\displaystyle\int x f(x) \, dx = x \int f(x) \, dx$

69. $\displaystyle\int \left(\frac{1}{x}\right) dx = -\left(\frac{1}{x^2}\right) + C$

70. If $f(x) = -f(-x)$ on the interval $[-a, a]$, then $\int_{-a}^a f(x) \, dx = 0$.

71. The average value of the sine function over an interval of length 2π is 0.

72. $\displaystyle\int \tan x \, dx = \sec^2 x + C.$

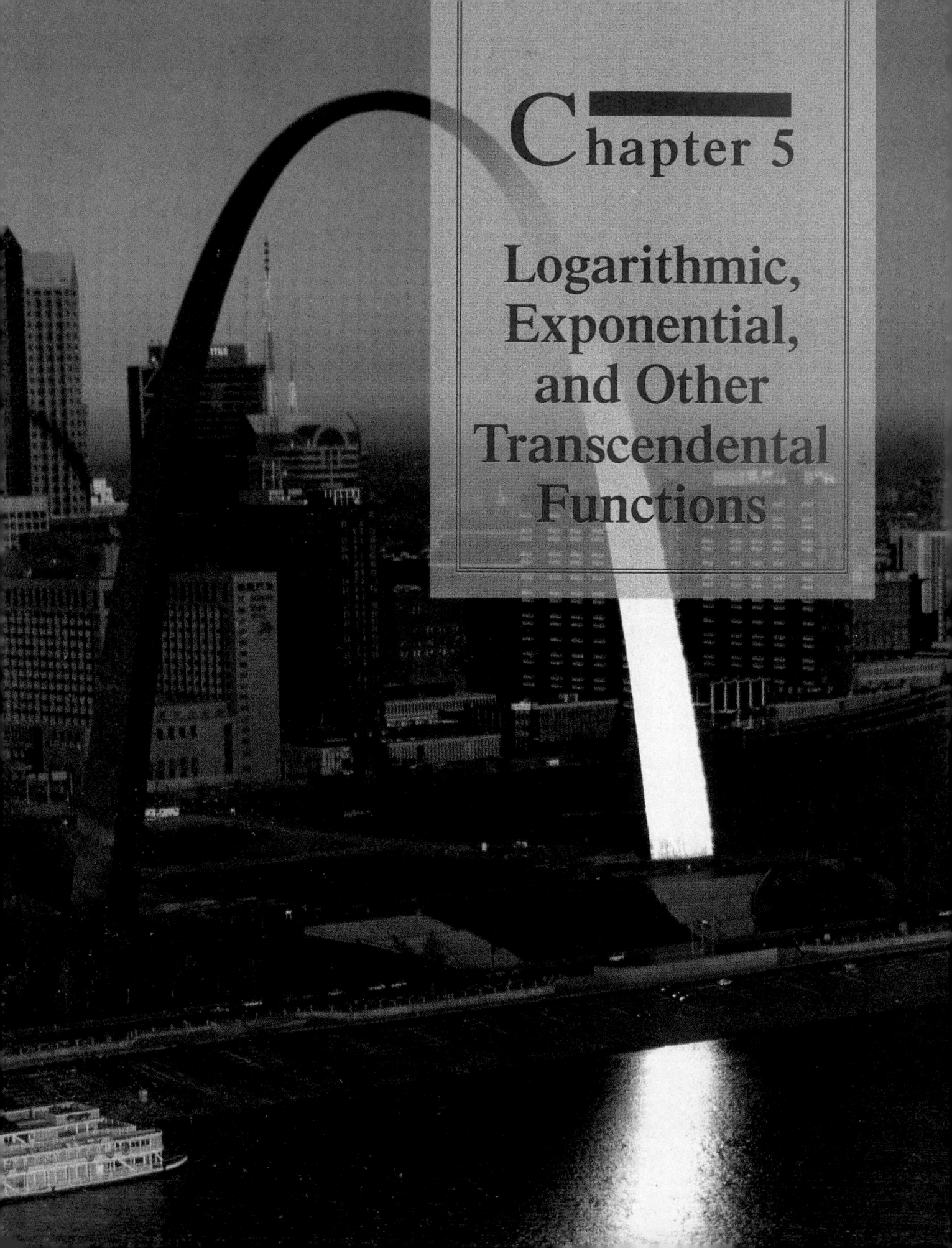

Chapter 5

Logarithmic, Exponential, and Other Transcendental Functions

JOHN NAPIER

Logarithms were invented by the Scottish mathematician John Napier (1550–1617). Although he did not introduce the *natural* logarithmic function, it is sometimes called the *Napierian* logarithm.

◀ *The mathematical model used to construct the St. Louis Arch is discussed in Section 5.9, Exercise 80.*

The Natural Logarithmic Function

Recall that the Power Rule

$$\int x^n \, dx = \frac{x^{n+1}}{n+1} + C, \qquad n \neq -1$$

has an important disclaimer—it doesn't apply when $n = -1$. Consequently, we have not yet found an antiderivative for the function $f(x) = 1/x$. In this section, we will use the Second Fundamental Theorem of Calculus to *define* such a function. This antiderivative is a function that you have not encountered previously in the text. It is neither algebraic nor trigonometric, but falls into a new class of functions called *logarithmic functions.* This particular function is the **natural logarithmic function.**

Definition of the Natural Logarithmic Function

The **natural logarithmic function** is defined by

$$\ln x = \int_1^x \frac{1}{t} \, dt, \qquad 0 < x.$$

The domain of the natural logarithmic function is the set of all positive real numbers.

From the definition, you can see that $\ln x$ is positive for $x > 1$ and negative for $0 < x < 1$, as shown in Figure 5.1. Moreover, $\ln(1) = 0$, because the upper and lower limits of integration are equal when $x = 1$.

If $x > 1$, then $\ln x > 0$. If $0 < x < 1$, then $\ln x < 0$.

FIGURE 5.1

FIGURE 5.2
Each small line segment has a slope of $\dfrac{1}{x}$.

To sketch the graph of $y = \ln x$, you can think of the natural logarithmic function as an *antiderivative* given by the differential equation

$$\frac{dy}{dx} = \frac{1}{x}.$$

Figure 5.2 is a computer-generated graph showing small line segments of slope $1/x$. The graph of $y = \ln x$ is the solution that passes through the point $(1, 0)$.

The following theorem lists some basic properties of the natural logarithmic function.

THEOREM 5.1 Properties of the Natural Logarithmic Function

The natural logarithmic function has the following properties.

1. The domain is $(0, \infty)$ and the range is $(-\infty, \infty)$.
2. The function is continuous, increasing, and one-to-one.
3. The graph is concave downward.

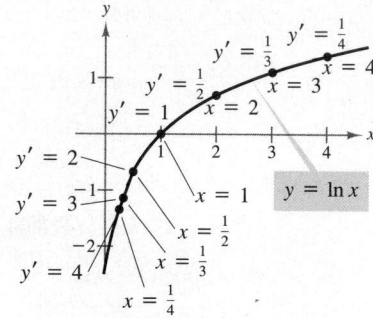

FIGURE 5.3
The natural logarithmic function is increasing, and its graph is concave downward.

Proof The domain of $f(x) = \ln x$ is $(0, \infty)$ by definition. Moreover, the function is continuous because it is differentiable. It is increasing because its derivative $f'(x) = 1/x$ is positive for $x > 0$, as shown in Figure 5.3. It is concave downward because

$$f''(x) = -\frac{1}{x^2}$$

is negative. We leave the proof that f is one-to-one as an exercise (see Exercise 97). The following limits imply that its range is the entire real line.

$$\lim_{x \to 0+} \ln x = -\infty \quad \text{and} \quad \lim_{x \to \infty} \ln x = \infty$$

Verification of these two limits is given in Appendix A. ■

Using the definition of the natural logarithmic function, you can prove several important properties involving operations with natural logarithms. If you are already familiar with logarithms, you will recognize that these properties are characteristic of all logarithms.

THEOREM 5.2 Logarithmic Properties

If a and b are positive numbers and n is rational, then the following properties are true.

1. $\ln(1) = 0$
2. $\ln(ab) = \ln a + \ln b$
3. $\ln(a^n) = n \ln a$
4. $\ln\left(\dfrac{a}{b}\right) = \ln a - \ln b$

Proof   We have already discussed the first property. The proof of the second property follows from the fact that two antiderivatives of the same function differ at most by a constant. From the Second Fundamental Theorem of Calculus and the definition of the natural logarithmic function, you know that

$$\frac{d}{dx}[\ln x] = \frac{1}{x}.$$

Thus, consider the two derivatives

$$\frac{d}{dx}[\ln(ax)] = \frac{a}{ax} = \frac{1}{x} \quad \text{and} \quad \frac{d}{dx}[\ln a + \ln x] = 0 + \frac{1}{x} = \frac{1}{x}.$$

Because $\ln(ax)$ and $(\ln a + \ln x)$ are both antiderivatives of $1/x$, they must differ at most by a constant

$$\ln(ax) = \ln a + \ln x + C.$$

By letting $x = 1$, you can see that $C = 0$. The third property can be proved similarly by comparing the derivatives of $\ln(x^n)$ and $n \ln x$. Finally, using the second and third properties, you can prove the fourth property.

$$\ln\left(\frac{a}{b}\right) = \ln[a(b^{-1})]$$
$$= \ln a + \ln(b^{-1})$$
$$= \ln a - \ln b$$

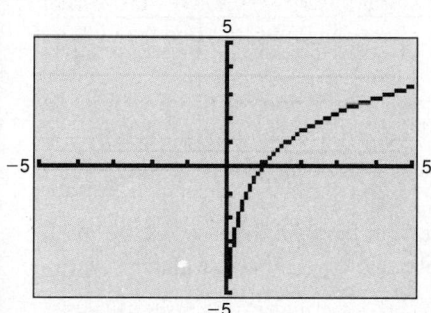

FIGURE 5.4

Try using a graphing utility to compare the graphs of

$$f(x) = \ln x^2$$

and

$$g(x) = 2 \ln x.$$

Which of the above is the graph of f? Which is the graph of g?

EXAMPLE 1   Expanding Logarithmic Expressions

a. $\ln\dfrac{10}{9} = \ln 10 - \ln 9$ Property 4

b. $\ln \sqrt{3x + 2} = \ln(3x + 2)^{1/2}$ Rewrite with rational exponent
$$= \tfrac{1}{2}\ln(3x + 2) \quad \text{Property 3}$$

c. $\ln\dfrac{6x}{5} = \ln(6x) - \ln 5$ Property 4
$$= \ln 6 + \ln x - \ln 5 \quad \text{Property 2}$$

d. $\ln\dfrac{(x^2 + 3)^2}{x\sqrt[3]{x^2 + 1}} = \ln(x^2 + 3)^2 - \ln(x\sqrt[3]{x^2 + 1})$
$$= 2\ln(x^2 + 3) - [\ln x + \ln(x^2 + 1)^{1/3}]$$
$$= 2\ln(x^2 + 3) - \ln x - \ln(x^2 + 1)^{1/3}$$
$$= 2\ln(x^2 + 3) - \ln x - \tfrac{1}{3}\ln(x^2 + 1)$$

REMARK   When using the properties of logarithms to rewrite logarithmic functions, you must check to see whether the domain of the rewritten function is the same as the domain of the original. For instance, the domain of $f(x) = \ln x^2$ is all real numbers except $x = 0$, and the domain of $g(x) = 2 \ln x$ is all positive real numbers. (See Figure 5.4.)

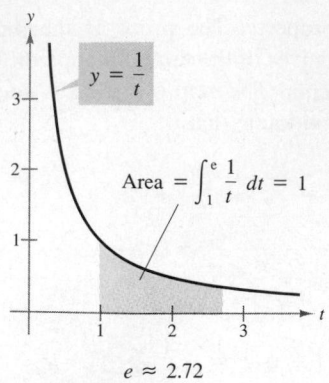

e ≈ 2.72

FIGURE 5.5
e is the base for the natural logarithm because $\ln e = 1$.

FOR FURTHER INFORMATION
For more information about the number e, see the article "Unexpected Occurrences of the Number e" by Harris S. Shultz and Bill Leonard in the October, 1989 issue of *Mathematics Magazine*.

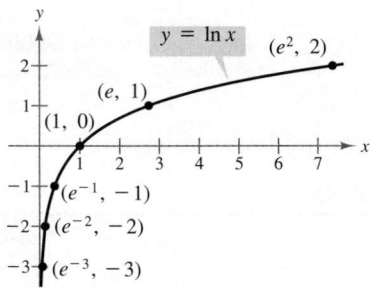

FIGURE 5.6
If $x = e^n$, then $\ln x = n$.

The Number e

It is likely that you have previously studied logarithms in an algebra course. There, without the benefit of calculus, logarithms would have been defined in terms of a **base** number. For example, common logarithms have a base of 10 because $\log_{10} 10 = 1$. (We will say more about this in Section 5.5.)

To define the **base for the natural logarithm,** we use the fact that the natural logarithmic function is continuous, one-to-one, and has a range of $(-\infty, \infty)$. Hence, there must be a unique real number x such that $\ln x = 1$, as shown in Figure 5.5. This number is denoted by the letter e. It can be shown that e is irrational and has the following decimal approximation.

$$e \approx 2.71828182846$$

Definition of e

The letter e denotes the positive real number such that

$$\ln e = \int_1^e \frac{1}{t}\, dt = 1.$$

Once you know that $\ln e = 1$, you can use logarithmic properties to evaluate the natural logarithm of several other numbers. For example, by using the property

$$\ln(e^n) = n \ln e = n(1) = n$$

you can evaluate $\ln(e^n)$ for various powers of n, as shown in Table 5.1 and Figure 5.6.

TABLE 5.1

x	$\frac{1}{e^3} \approx 0.050$	$\frac{1}{e^2} \approx 0.135$	$\frac{1}{e} \approx 0.368$	$e^0 = 1$	$e \approx 2.718$	$e^2 \approx 7.389$
$\ln x$	-3	-2	-1	0	1	2

The logarithms given in Table 5.1 are convenient because the x-values are integer powers of e. Most logarithmic expressions are, however, best evaluated with a calculator.

EXAMPLE 2 Evaluating Natural Logarithmic Expressions

a. $\ln 2 \approx 0.693$
b. $\ln 32 \approx 3.466$
c. $\ln 0.1 \approx -2.303$

The Derivative of the Natural Logarithmic Function

The derivative of the natural logarithmic function is given in Theorem 5.3. The first part of the theorem follows from the definition of the natural logarithmic function as an antiderivative. The second part of the theorem is simply the Chain Rule version of the first part.

THEOREM 5.3   Derivative of the Natural Logarithmic Function

Let u be a differentiable function of x.

1. $\dfrac{d}{dx}[\ln x] = \dfrac{1}{x}, \qquad x > 0$ **2.** $\dfrac{d}{dx}[\ln u] = \dfrac{1}{u}\dfrac{du}{dx} = \dfrac{u'}{u}, \qquad u > 0$

EXAMPLE 3   Differentiation of Logarithmic Functions

a. $\dfrac{d}{dx}[\ln(2x)] = \dfrac{u'}{u} = \dfrac{2}{2x} = \dfrac{1}{x}$ $u = 2x$

b. $\dfrac{d}{dx}[\ln(x^2 + 1)] = \dfrac{u'}{u} = \dfrac{2x}{x^2 + 1}$ $u = x^2 + 1$

c. $\dfrac{d}{dx}[x \ln x] = x\left(\dfrac{d}{dx}[\ln x]\right) + (\ln x)\left(\dfrac{d}{dx}[x]\right)$ Product Rule

$\qquad\qquad = x\left(\dfrac{1}{x}\right) + (\ln x)(1)$

$\qquad\qquad = 1 + \ln x$

d. $\dfrac{d}{dx}[(\ln x)^3] = 3(\ln x)^2 \dfrac{d}{dx}[\ln x]$ Chain Rule

$\qquad\qquad = 3(\ln x)^2 \dfrac{1}{x}$

Napier used logarithmic properties to simplify *calculations* involving products, quotients, and powers. Of course, given the availability of calculators, there is now little need for this particular application of logarithms. However, there is great value in using logarithmic properties to simplify *differentiation* involving products, quotients, and powers.

EXAMPLE 4   Logarithmic Properties as Aids to Differentiation

Differentiate $f(x) = \ln \sqrt{x + 1}$.

Solution   Because

$$f(x) = \ln \sqrt{x + 1} = \ln(x + 1)^{1/2} = \frac{1}{2}\ln(x + 1)$$

you can write the following.

$$f'(x) = \frac{1}{2}\left(\frac{1}{x + 1}\right) = \frac{1}{2(x + 1)}$$

EXAMPLE 5   Logarithmic Properties as Aids to Differentiation

Differentiate $f(x) = \ln \dfrac{x(x^2 + 1)^2}{\sqrt{2x^3 - 1}}$.

Solution

$$f(x) = \ln \frac{x(x^2 + 1)^2}{\sqrt{2x^3 - 1}} \qquad \text{Original function}$$

$$= \ln x + 2\ln(x^2 + 1) - \frac{1}{2}\ln(2x^3 - 1) \qquad \text{Rewrite before differentiating}$$

$$f'(x) = \frac{1}{x} + 2\left(\frac{2x}{x^2 + 1}\right) - \frac{1}{2}\left(\frac{6x^2}{2x^3 - 1}\right) \qquad \text{Differentiate}$$

$$= \frac{1}{x} + \frac{4x}{x^2 + 1} - \frac{3x^2}{2x^3 - 1} \qquad \text{Simplify}$$

REMARK   In Examples 4 and 5, be sure to see the benefit of applying logarithmic properties *before* differentiating. Consider, for instance, the difficulty of direct differentiation of the function given in Example 5.

On occasion, it is convenient to use logarithms as an aid in differentiating *non* logarithmic functions. This procedure is called **logarithmic differentiation.**

EXAMPLE 6   Logarithmic Differentiation

Find the derivative of $y = \dfrac{(x - 2)^2}{\sqrt{x^2 + 1}}$.

Solution   Note that $y > 0$ and hence, $\ln y$ is defined. Begin by taking the natural logarithm of both sides of the equation. Then apply logarithmic properties and differentiate implicitly. Finally, solve for y'.

$$\ln y = \ln \frac{(x - 2)^2}{\sqrt{x^2 + 1}} \qquad \text{Take ln of both sides}$$

$$\ln y = 2\ln(x - 2) - \frac{1}{2}\ln(x^2 + 1) \qquad \text{Logarithmic properties}$$

$$\frac{y'}{y} = 2\left(\frac{1}{x - 2}\right) - \frac{1}{2}\left(\frac{2x}{x^2 + 1}\right) \qquad \text{Differentiate}$$

$$= \frac{2}{x - 2} - \frac{x}{x^2 + 1}$$

$$y' = y\left(\frac{2}{x - 2} - \frac{x}{x^2 + 1}\right) \qquad \text{Solve for } y'$$

$$= \frac{(x - 2)^2}{\sqrt{x^2 + 1}}\left[\frac{x^2 + 2x + 2}{(x - 2)(x^2 + 1)}\right] \qquad \text{Substitute for } y$$

$$= \frac{(x - 2)(x^2 + 2x + 2)}{(x^2 + 1)^{3/2}} \qquad \text{Simplify}$$

Because the natural logarithm is undefined for negative numbers, you will often encounter expressions of the form $\ln |u|$. The following theorem states that you can differentiate functions of the form $y = \ln |u|$ as if the absolute value sign were not present.

THEOREM 5.4   Derivative Involving Absolute Value

If u is a differentiable function of x such that $u \neq 0$, then

$$\frac{d}{dx}[\ln |u|] = \frac{u'}{u}.$$

Proof   If $u > 0$, then $|u| = u$, and the result follows from Theorem 5.3. If $u < 0$, then $|u| = -u$, and you have

$$\frac{d}{dx}[\ln |u|] = \frac{d}{dx}[\ln(-u)]$$

$$= \frac{-u'}{-u}$$

$$= \frac{u'}{u}.$$

EXAMPLE 7   Derivative Involving Absolute Value

Find the derivative of $f(x) = \ln |\cos x|$.

Solution   Using Theorem 5.4, let $u = \cos x$ and write

$$\frac{d}{dx}[\ln |\cos x|] = \frac{u'}{u}$$

$$= \frac{-\sin x}{\cos x}$$

$$= -\tan x.$$

EXAMPLE 8   Finding Relative Extrema

Locate the relative extrema of $y = \ln(x^2 + 2x + 3)$.

Solution   Differentiating y, you obtain

$$\frac{dy}{dx} = \frac{2x + 2}{x^2 + 2x + 3}.$$

Because $dy/dx = 0$ when $x = -1$, you can apply the First Derivative Test and conclude that the point $(-1, \ln 2)$ is a relative minimum. Because there are no other critical points, it follows that this is the only relative extremum (see Figure 5.7).

$y = \ln(x^2 + 2x + 3)$

$(-1, \ln 2)$

Relative minimum

FIGURE 5.7
The derivative of y changes from negative to positive at $x = -1$.

EXERCISES for Section 5.1

TECHNOLOGY
Laboratory Guide
Lab 5.1

C **1.** Use a computer or calculator and Simpson's Rule with $n = 10$ to approximate the integral

$$\int_1^x \frac{1}{t}\, dt$$

to complete the table.

x	0.5	1.5	2	2.5	3	3.5	4
$\int_1^x \frac{1}{t}\, dt$							

2. Plot the points generated in Exercise 1 and connect them with a smooth curve. Compare the result with the graph of $y = \ln x$.

In Exercises 3–6, use the graph of $y = \ln x$ to match the function with its graph. (The graphs are labeled (a), (b), (c), and (d).)

3. $f(x) = \ln x + 2$ **4.** $f(x) = -\ln x$

5. $f(x) = \ln(x - 1)$ **6.** $f(x) = -\ln(-x)$

(a)   **(b)**

(c)   **(d)**

In Exercises 7–12, sketch the graph of the function and state its domain.

7. $f(x) = 3 \ln x$ **8.** $f(x) = -2 \ln x$

9. $f(x) = \ln 2x$ **10.** $f(x) = \ln |x|$

11. $f(x) = \ln(x - 1)$ **12.** $g(x) = 2 + \ln x$

In Exercises 13 and 14, use the properties of logarithms to approximate the indicated logarithm given that $\ln 2 \approx 0.6931$ and $\ln 3 \approx 1.0986$.

13. a. $\ln 6$ **b.** $\ln \frac{2}{3}$ **c.** $\ln 81$ **d.** $\ln \sqrt{3}$

14. a. $\ln 0.25$ **b.** $\ln 24$ **c.** $\ln \sqrt[3]{12}$ **d.** $\ln \frac{1}{72}$

In Exercises 15–24, use the properties of logarithms to write each expression as a sum, difference, and/or multiple of logarithms.

15. $\ln \frac{2}{3}$ **16.** $\ln \frac{1}{5}$

17. $\ln \frac{xy}{z}$ **18.** $\ln(xyz)$

19. $\ln \sqrt{2^3}$ **20.** $\ln \sqrt{a - 1}$

21. $\ln \left(\frac{x^2 - 1}{x^3} \right)^3$ **22.** $\ln 3e^2$

23. $\ln z(z - 1)^2$ **24.** $\ln \frac{1}{e}$

In Exercises 25–30, write the expression as a logarithm of a single quantity.

25. $\ln(x - 2) - \ln(x + 2)$

26. $3 \ln x + 2 \ln y - 4 \ln z$

27. $\frac{1}{3}[2 \ln(x + 3) + \ln x - \ln(x^2 - 1)]$

28. $2[\ln x - \ln(x + 1) - \ln(x - 1)]$

29. $2 \ln 3 - \frac{1}{2} \ln(x^2 + 1)$

30. $\frac{3}{2}[\ln(x^2 + 1) - \ln(x + 1) - \ln(x - 1)]$

In Exercises 31–34, find the slope of the tangent line to the graph of the logarithmic function at the point $(1, 0)$.

31. $y = \ln x^3$ **32.** $y = \ln x^{3/2}$

33. $y = \ln x^2$ **34.** $y = \ln x^{1/2}$

In Exercises 35–60, find the derivative of the function.

35. $g(x) = \ln x^2$ **36.** $h(x) = \ln(x^2 + 3)$

37. $y = (\ln x)^4$ **38.** $y = x \ln x$

39. $y = \ln(x\sqrt{x^2 - 1})$ **40.** $y = \ln \sqrt{x^2 - 4}$

41. $f(x) = \ln\left(\dfrac{x}{x^2 + 1}\right)$

42. $f(x) = \ln\left(\dfrac{x}{x + 1}\right)$

43. $g(t) = \dfrac{\ln t}{t^2}$

44. $h(t) = \dfrac{\ln t}{t}$

45. $y = \ln(\ln x^2)$

46. $y = \ln(\ln x)$

47. $y = \ln\sqrt{\dfrac{x + 1}{x - 1}}$

48. $y = \ln\sqrt{\dfrac{x - 1}{x + 1}}$

49. $f(x) = \ln\left(\dfrac{\sqrt{4 + x^2}}{x}\right)$

50. $f(x) = \ln\left(x + \sqrt{4 + x^2}\right)$

51. $y = \dfrac{-\sqrt{x^2 + 1}}{x} + \ln\left(x + \sqrt{x^2 + 1}\right)$

52. $y = \dfrac{-\sqrt{x^2 + 4}}{2x^2} - \dfrac{1}{4}\ln\left(\dfrac{2 + \sqrt{x^2 + 4}}{x}\right)$

53. $y = \ln|\sin x|$

54. $y = \ln|\sec x|$

55. $y = \ln\left|\dfrac{\cos x}{\cos x - 1}\right|$

56. $y = \ln|\sec x + \tan x|$

57. $y = \ln\left|\dfrac{-1 + \sin x}{2 + \sin x}\right|$

58. $y = \ln\sqrt{1 + \sin^2 x}$

59. $f(x) = \sin 2x \ln x^2$

60. $g(x) = \displaystyle\int_1^{\ln x} (t^2 + 3)\,dt$

In Exercises 61 and 62, use implicit differentiation to find dy/dx.

61. $x^2 - 3\ln y + y^2 = 10$

62. $\ln xy + 5x = 30$

In Exercises 63 and 64, find an equation of the tangent line to the graph of the equation at the indicated point.

Function	Point
63. $y = 3x^2 - \ln x$	$(1, 3)$
64. $x^2 + \ln(x + 1) + y^2 = 4$	$(0, 2)$

In Exercises 65 and 66, show that the function is a solution of the differential equation.

Function	Differential Equation
65. $y = 2\ln x + 3$	$xy'' + y' = 0$
66. $y = x\ln x - 4x$	$x + y - xy' = 0$

In Exercises 67–72, find any relative extrema and inflection points, and sketch the graph of the function.

67. $y = \dfrac{x^2}{2} - \ln x$

68. $y = x - \ln x$

69. $y = x\ln x$

70. $y = \dfrac{\ln x}{x}$

71. $y = \dfrac{x}{\ln x}$

72. $y = x^2\ln x$

Linear and Quadratic Approximation   **In Exercises 73 and 74, use a graphing utility to graph the function. Then graph**

$$P_1(x) = f(1) + f'(1)(x - 1)$$

and

$$P_2(x) = f(1) + f'(1)(x - 1) + \tfrac{1}{2}f''(1)(x - 1)^2$$

on the same viewing rectangle. Compare the values of f, P_1 and P_2 and their first derivatives at $x = 1$.

73. $f(x) = \ln x$

74. $f(x) = x\ln x$

In Exercises 75 and 76, use Newton's Method to approximate, to three decimal places, the x-coordinate of the point of intersection of the graphs of the two equations.

75. $y = \ln x$
    $y = -x$

76. $y = \ln x$
    $y = 3 - x$

In Exercises 77–82, find dy/dx using logarithmic differentiation.

77. $y = x\sqrt{x^2 - 1}$

78. $y = \sqrt{(x - 1)(x - 2)(x - 3)}$

79. $y = \dfrac{x^2\sqrt{3x - 2}}{(x - 1)^2}$

80. $y = \sqrt[3]{\dfrac{x^2 + 1}{x^2 - 1}}$

81. $y = \dfrac{x(x - 1)^{3/2}}{\sqrt{x + 1}}$

82. $y = \dfrac{(x + 1)(x + 2)}{(x - 1)(x - 2)}$

83. *Sound Intensity*   The relationship between the number of decibels β and the intensity of a sound I in watts per meter squared is

$$\beta = \dfrac{10}{\ln 10}\ln\left(\dfrac{I}{10^{-16}}\right).$$

Use properties of logarithms to write the formula in simpler form, and determine the number of decibels of a sound with an intensity of 10^{-10} watts per meter squared.

84. *Home Mortgage*   The term t (in years) of a $120,000 home mortgage (at 10% interest) can be approximated by

$$t = \dfrac{5.315}{-6.7968 + \ln x}, \qquad 1000 < x$$

where x is the monthly payment in dollars.

a. Use a graphing utility to graph the model.
b. Use the model to approximate the term of a home mortgage for which monthly payment is $1167.41. What is the total amount paid?
c. Use the model to approximate the term of a home mortgage for which monthly payment is $1068.45. What is the total amount paid?
d. Find the instantaneous rate of change of t with respect to x when $x = 1167.41$ and $x = 1068.45$.
e. Write a short paragraph describing the benefit of the higher monthly payment.

85. *Boiling Temperature*   The table gives the temperature T (°C) at which water boils at selected pressures p (pounds per square inch). (*Source: Standard Handbook of Mechanical Engineers*)

p	5	10	14.696 (1 atm)	20
T	162.24°	193.21°	212.00°	227.96°

p	30	40	60	80	100
T	250.33°	267.25°	292.71°	312.03°	327.81°

A model that approximates these data is

$$T = 87.97 + 34.96 \ln p + 7.91\sqrt{p}.$$

a. Plot the data and sketch the graph of the model on the same set of coordinate axes.

b. Find the rate of change of T with respect to p when $p = 10$ and $p = 70$.

86. Use Simpson's Rule with $n = 4$ to show that

$$\int_1^{2.7} \frac{1}{t}\, dt < 1 < \int_1^{2.8} \frac{1}{t}\, dt.$$

Then use this inequality to show that

$$\ln 2.7 < \ln e < \ln 2.8$$

and therefore $2.7 < e < 2.8$.

87. Let f be a function that is positive and differentiable on the entire real line. Let $g(x) = \ln f(x)$.

a. If the graph of g is increasing, must the graph of f be increasing? Explain.

b. If the graph of f is increasing, must the graph of g be increasing? Explain.

C 88. *Tractrix*   A person walking along a dock drags a boat by a 10-foot rope. The boat travels along a path called a *tractrix* (see figure). The equation of this path is

$$y = 10 \ln\left(\frac{10 + \sqrt{100 - x^2}}{x}\right) - \sqrt{100 - x^2}.$$

a. Use a graphing utility to graph the function.

b. What is the slope of this path when $x = 5$ and $x = 9$?

c. What does the slope of the path approach as x approaches 10?

FIGURE FOR 88

C 89. *Essay*   Use a graphing utility to graph f and g in the same viewing rectangle and determine which is increasing at the higher rate for "large" values of x. What can you conclude about the rate of growth of the natural logarithmic function?

a. $f(x) = \ln x$, $g(x) = \sqrt{x}$

b. $f(x) = \ln x$, $g(x) = \sqrt[4]{x}$

90. There are 25 prime numbers less than 100. The **Prime Number Theorem** states that if $p(x)$ is the number of primes less than x, then the ratio of $p(x)$ to $x/(\ln x)$ approaches 1 as x approaches infinity. Compute $x/(\ln x)$ for $x = 1000, x = 10^6$, and $x = 10^9$. Then compute the ratio of $p(x)$ to $x/(\ln x)$ given that $p(1000) = 168$, $p(10^6) = 78{,}498$, and $p(10^9) = 50{,}847{,}478$.

C In Exercises 91 and 92, illustrate that $f = g$ by using a graphing utility to graph f and g on the same set of coordinate axes. (Assume $x > 0$.)

91. $f(x) = \ln \dfrac{x^2}{4}$

$g(x) = 2 \ln x - \ln 4$

92. $f(x) = \ln \sqrt{x(x^2 + 1)}$

$g(x) = \frac{1}{2}[\ln x + \ln(x^2 + 1)]$

True or False   **In Exercises 93–96, determine whether the statement is true or false. If it is false, explain why or give an example that shows it is false.**

93. $\ln(x + 25) = \ln x + \ln 25$

94. The derivative of $f(x) = \ln x$ is positive for all x in the domain of f.

95. If $f(x) = \ln(ax)$ where $a > 0$, and $g(x) = \ln(bx)$ where $b > 0$, then $f'(x) = g'(x)$.

96. If $y = \ln \pi$, then $y' = 1/\pi$.

97. Prove that the natural logarithmic function is one-to-one.

98. Prove that $\ln\left(x + \sqrt{x^2 - 1}\right) + \ln\left(x - \sqrt{x^2 - 1}\right) = 0$.

SECTION
5.2
The Natural Logarithmic Function and Integration

Log Rule for Integration • Integrals of Trigonometric Functions

Log Rule for Integration

The differentiation rules

$$\frac{d}{dx}[\ln |x|] = \frac{1}{x} \quad \text{and} \quad \frac{d}{dx}[\ln |u|] = \frac{u'}{u}$$

that you studied in the previous section produce the following integration rules.

THEOREM 5.5   Log Rule for Integration

Let u be a differentiable function of x.

1. $\displaystyle\int \frac{1}{x} dx = \ln |x| + C$ 2. $\displaystyle\int \frac{1}{u} du = \ln |u| + C$

Because $du = u' dx$, the second formula can also be written as

$$\int \frac{u'}{u} dx = \ln |u| + C. \qquad \text{Alternative form of Log Rule}$$

EXAMPLE 1   Using the Log Rule for Integration

$$\int \frac{2}{x} dx = 2 \int \frac{1}{x} dx = 2 \ln |x| + C = \ln(x^2) + C$$

Because x^2 cannot be negative, the absolute value is unnecessary in the final form of the antiderivative.

EXAMPLE 2   Using the Log Rule with a Change of Variable

Evaluate $\displaystyle\int \frac{1}{4x - 1} dx$.

Solution   Let $u = 4x - 1$, then $du = 4 dx$.

$$\int \frac{1}{4x - 1} dx = \frac{1}{4} \int \left(\frac{1}{4x - 1} \right) 4 dx \qquad \text{Multiply and divide by 4}$$

$$= \frac{1}{4} \int \frac{1}{u} du \qquad \text{Substitute: } u = 4x - 1$$

$$= \frac{1}{4} \ln |u| + C \qquad \text{Apply Log Rule}$$

$$= \frac{1}{4} \ln |4x - 1| + C \qquad \text{Back-substitute}$$

Example 3 uses the alternative form of the Log Rule. To apply this rule, look for quotients in which the numerator is the derivative of the denominator.

EXAMPLE 3   Finding Area with the Log Rule

Find the area of the region bounded by the graph of $y = x/(x^2 + 1)$, the x-axis, and the line $x = 3$.

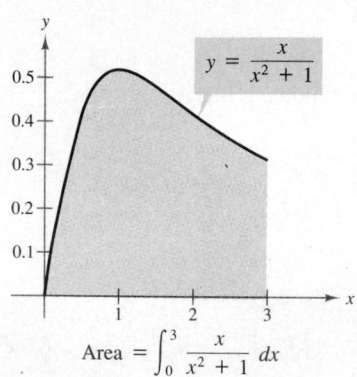

$$\text{Area} = \int_0^3 \frac{x}{x^2 + 1}\, dx$$

FIGURE 5.8

The area of the region bounded by the graph of y, the x-axis, and $x = 3$ is $\frac{1}{2}\ln 10$.

Solution   From Figure 5.8, you can see that the area of the region is given by the definite integral

$$\int_0^3 \frac{x}{x^2 + 1}\, dx.$$

Let $u = x^2 + 1$, then $u' = 2x$. To apply the Log Rule, multiply and divide by 2 as follows.

$$\int_0^3 \frac{x}{x^2 + 1}\, dx = \frac{1}{2}\int_0^3 \frac{2x}{x^2 + 1}\, dx \qquad \int \frac{u'}{u}\, dx = \ln|u| + C$$

$$= \frac{1}{2}\left[\ln(x^2 + 1)\right]_0^3$$

$$= \frac{1}{2}(\ln 10 - \ln 1)$$

$$= \frac{1}{2}\ln 10$$

$$\approx 1.151$$

EXAMPLE 4   Recognizing Quotient Forms of the Log Rule

a. $\displaystyle\int \frac{3x^2 + 1}{x^3 + x}\, dx = \ln|x^3 + x| + C \qquad u = x^3 + x$

b. $\displaystyle\int \frac{\sec^2 x}{\tan x}\, dx = \ln|\tan x| + C \qquad u = \tan x$

c. $\displaystyle\int \frac{x + 1}{x^2 + 2x}\, dx = \frac{1}{2}\int \frac{2x + 2}{x^2 + 2x}\, dx \qquad u = x^2 + 2x$

$$= \frac{1}{2}\ln|x^2 + 2x| + C$$

d. $\displaystyle\int \frac{1}{3x + 2}\, dx = \frac{1}{3}\int \frac{3}{3x + 2}\, dx \qquad u = 3x + 2$

$$= \frac{1}{3}\ln|3x + 2| + C$$

Integrals to which the Log Rule can be applied often appear in disguised form. For instance, if a rational function has a *numerator of degree greater than or equal to that of the denominator*, division may reveal a form to which we can apply the Log Rule. This is illustrated in Example 5.

EXAMPLE 5   Using Long Division Before Integrating

Evaluate $\int \dfrac{x^2 + x + 1}{x^2 + 1}\, dx$.

Solution   Begin by using long division to rewrite the integrand.

$$\frac{x^2 + x + 1}{x^2 + 1} = 1 + \frac{x}{x^2 + 1}$$

Now, you can integrate to obtain

$$\int \frac{x^2 + x + 1}{x^2 + 1}\, dx = \int \left(1 + \frac{x}{x^2 + 1}\right) dx$$

$$= \int dx + \frac{1}{2}\int \frac{2x}{x^2 + 1}\, dx$$

$$= x + \frac{1}{2}\ln(x^2 + 1) + C.$$

Try checking this result by differentiating to obtain the original integrand.   ▬▬▬

The next example gives another instance in which the use of the Log Rule is disguised. In this case, a change of variables helps you recognize the Log Rule.

EXAMPLE 6   Change of Variables with the Log Rule

Evaluate $\int \dfrac{2x}{(x + 1)^2}\, dx$.

Solution   Let $u = x + 1$, then $du = dx$ and $x = u - 1$.

$$\int \frac{2x}{(x + 1)^2}\, dx = \int \frac{2(u - 1)}{u^2}\, du \qquad \text{Substitute: } u = x + 1$$

$$= 2\int \left[\frac{u}{u^2} - \frac{1}{u^2}\right] du \qquad \text{Rewrite as two fractions}$$

$$= 2\int \frac{du}{u} - 2\int u^{-2}\, du \qquad \text{Rewrite as two integrals}$$

$$= 2\ln|u| - 2\left(\frac{u^{-1}}{-1}\right) + C \qquad \text{Integrate}$$

$$= 2\ln|u| + \frac{2}{u} + C \qquad \text{Simplify}$$

$$= 2\ln|x + 1| + \frac{2}{x + 1} + C \qquad \text{Back-substitute} \quad ▬▬▬$$

TECHNOLOGY   If you have access to a symbolic integration utility, try using it to evaluate the indefinite integrals in Examples 5 and 6. How does the form of the antiderivative compare to that given in Examples 5 and 6?

As you study the methods shown in Examples 5 and 6, be aware that both methods involve rewriting a disguised integrand so that it fits one or more of the basic integration formulas. In the remaining sections of Chapter 5 and in Chapter 7, we will devote much time to integration techniques. To master these techniques, you must recognize the "form-fitting" nature of integration. In this sense, integration is not nearly as straightforward as differentiation. Differentiation takes the form

"Here is the question; what is the answer?"

Integration is more like

"Here is the answer; what is the question?"

We suggest the following guidelines for integration.

Guidelines for Integration

1. Memorize a basic list of integration formulas. (Including those given in this section, we now have 12 formulas: the Power Rule, the Log Rule, and ten trigonometric rules. By the end of Section 5.8 we will have expanded this list to 20 basic formulas.)

2. Find an integration formula that resembles all or part of the integrand, and by trial and error find a choice of u that will make the integrand conform to the formula.

3. If you cannot find a u-substitution that works, try altering the integrand. You might try a trigonometric identity, multiplication and division by the same quantity, or addition and subtraction of the same quantity. Be creative!

4. If you have access to computer software that will find antiderivatives symbolically, use it!

EXAMPLE 7 u-Substitution and the Log Rule

Evaluate $\displaystyle\int \frac{1}{x \ln x} dx$.

Solution   Because the integrand is a quotient whose denominator is raised to the first power, you should try the Log Rule. There are three basic choices for u. The choices $u = x$ and $u = x \ln x$ fail to fit the u'/u form of the Log Rule. However, the third choice does fit. Letting $u = \ln x$ produces $u' = 1/x$, and you obtain the following.

$$\int \frac{1}{x \ln x} dx = \int \frac{1/x}{\ln x} dx \qquad \text{Divide numerator and denominator by } x$$

$$= \int \frac{u'}{u} dx \qquad \text{Substitute: } u = \ln x$$

$$= \ln |u| + C \qquad \text{Apply Log Rule}$$

$$= \ln |\ln x| + C \qquad \text{Back-substitute}$$

Integrals of Trigonometric Functions

In Section 4.1 we looked at six trigonometric integration rules—the six that correspond directly to differentiation rules. With the Log Rule, we can now complete the set of basic trigonometric integration formulas.

EXAMPLE 8   Using a Trigonometric Identity

Evaluate $\int \tan x \, dx$.

Solution   This integral does not seem to fit any formulas on our basic list. However, by using a trigonometric identity, you obtain the following.

$$\int \tan x \, dx = \int \frac{\sin x}{\cos x} \, dx$$

Knowing that $D_x[\cos x] = -\sin x$, let $u = \cos x$ and write

$$\int \tan x \, dx = -\int \frac{-\sin x}{\cos x} \, dx \qquad \text{Trigonometric identity}$$

$$= -\int \frac{u'}{u} \, dx \qquad \text{Substitute: } u = \cos x$$

$$= -\ln |u| + C \qquad \text{Apply Log Rule}$$

$$= -\ln |\cos x| + C. \qquad \text{Back-substitute}$$

Example 8 used a trigonometric identity to derive an integration rule for the tangent function. In the next example we take a rather unusual step (multiplying and dividing by the same quantity) to derive an integration rule for the secant function.

EXAMPLE 9   Derivation of the Secant Formula

Evaluate $\int \sec x \, dx$.

Solution   Consider the following procedure.

$$\int \sec x \, dx = \int \sec x \left(\frac{\sec x + \tan x}{\sec x + \tan x} \right) dx$$

$$= \int \frac{\sec^2 x + \sec x \tan x}{\sec x + \tan x} \, dx$$

Letting u be the denominator of this quotient produces

$$u = \sec x + \tan x \longrightarrow u' = \sec x \tan x + \sec^2 x.$$

Therefore, you can conclude that

$$\int \sec x \, dx = \int \frac{\sec^2 x + \sec x \tan x}{\sec x + \tan x} \, dx \qquad \text{Rewrite integrand}$$

$$= \int \frac{u'}{u} \, dx \qquad \text{Substitute: } u = \sec x + \tan x$$

$$= \ln |u| + C \qquad \text{Apply Log Rule}$$

$$= \ln |\sec x + \tan x| + C. \qquad \text{Back-substitute}$$

With the results of Examples 8 and 9, you now have integration formulas for $\sin x$, $\cos x$, $\tan x$, and $\sec x$. All six trigonometric rules are summarized in the following list.

Integrals of the Six Basic Trigonometric Functions

$$\int \sin u \, du = -\cos u + C \qquad \int \cos u \, du = \sin u + C$$

$$\int \tan u \, du = -\ln |\cos u| + C \qquad \int \cot u \, du = \ln |\sin u| + C$$

$$\int \sec u \, du = \ln |\sec u + \tan u| + C \qquad \int \csc u \, du = -\ln |\csc u + \cot u| + C$$

REMARK Using trigonometric identities and properties of logarithms, you could rewrite these six integration rules in other forms. For instance, you could write

$$\int \csc u \, du = \ln |\csc u - \cot u| + C.$$

(See Exercises 45–48.)

EXAMPLE 10   Integrating Trigonometric Functions

Evaluate $\displaystyle\int_0^{\pi/4} \sqrt{1 + \tan^2 x} \, dx$.

Solution   Using $1 + \tan^2 x = \sec^2 x$, you can write

$$\int_0^{\pi/4} \sqrt{1 + \tan^2 x} \, dx = \int_0^{\pi/4} \sqrt{\sec^2 x} \, dx = \int_0^{\pi/4} \sec x \, dx$$

$$= \Big[\ln |\sec x + \tan x| \Big]_0^{\pi/4}$$

$$= \ln(\sqrt{2} + 1) - \ln 1 \approx 0.8814. \qquad \blacksquare$$

EXAMPLE 11   Electromotive Force

The electromotive force E of a particular electrical circuit is given by $E = 3 \sin 2t$, where E is measured in volts and t is measured in seconds. Find the average value of E as t ranges from 0 to 0.5 second.

Solution   The average value of E is given by

$$\text{Average value} = \frac{1}{0.5 - 0} \int_0^{0.5} 3 \sin 2t \, dt$$

$$= 6 \int_0^{0.5} \sin 2t \, dt$$

$$= 6 \left(\frac{1}{2} \right) \int_0^{0.5} (\sin 2t)(2) \, dt \qquad u = 2t, \, du = 2 \, dt$$

$$= \Big[3(-\cos 2t) \Big]_0^{0.5}$$

$$\approx 1.379 \text{ volts.} \qquad \blacksquare$$

EXERCISES for Section 5.2

In Exercises 1–18, find the indefinite integral.

1. $\int \dfrac{1}{x+1}\,dx$

2. $\int \dfrac{1}{x-5}\,dx$

3. $\int \dfrac{1}{3-2x}\,dx$

4. $\int \dfrac{1}{6x+1}\,dx$

5. $\int \dfrac{x}{x^2+1}\,dx$

6. $\int \dfrac{x^2}{3-x^3}\,dx$

7. $\int \dfrac{x^2-4}{x}\,dx$

8. $\int \dfrac{x}{\sqrt{9-x^2}}\,dx$

9. $\int \dfrac{x^2+2x+3}{x^3+3x^2+9x}\,dx$

10. $\int \dfrac{x+3}{x^2+6x+7}\,dx$

11. $\int \dfrac{(\ln x)^2}{x}\,dx$

12. $\int \dfrac{1}{x\ln(x^2)}\,dx$

13. $\int \dfrac{1}{\sqrt{x}+1}\,dx$

14. $\int \dfrac{1}{x^{2/3}(1+x^{1/3})}\,dx$

15. $\int \dfrac{\sqrt{x}}{\sqrt{x}-3}\,dx$

16. $\int \dfrac{1}{1+\sqrt{2x}}\,dx$

17. $\int \dfrac{2x}{(x-1)^2}\,dx$

18. $\int \dfrac{x(x-2)}{(x-1)^3}\,dx$

In Exercises 19–26, find the indefinite integral of the trigonometric function.

19. $\int \dfrac{\cos\theta}{\sin\theta}\,d\theta$

20. $\int \tan 5\theta\,d\theta$

21. $\int \csc 2x\,dx$

22. $\int \sec\dfrac{x}{2}\,dx$

23. $\int \dfrac{\cos t}{1+\sin t}\,dt$

24. $\int \dfrac{\sin x}{1+\cos x}\,dx$

25. $\int \dfrac{\sec x\tan x}{\sec x-1}\,dx$

26. $\int (\sec t+\tan t)\,dt$

In Exercises 27–30, solve the differential equation.

27. $\dfrac{dy}{dx}=\dfrac{3}{2-x}$

28. $\dfrac{dy}{dx}=\dfrac{2x}{x^2-9}$

29. $\dfrac{ds}{d\theta}=\tan 2\theta$

30. $\dfrac{dr}{dt}=\dfrac{\sec^2 t}{\tan t+1}$

In Exercises 31–38, evaluate the definite integral.

31. $\displaystyle\int_0^4 \dfrac{5}{3x+1}\,dx$

32. $\displaystyle\int_{-1}^1 \dfrac{1}{5-2x}\,dx$

33. $\displaystyle\int_1^e \dfrac{(1+\ln x)^2}{x}\,dx$

34. $\displaystyle\int_e^{e^2} \dfrac{1}{x\ln x}\,dx$

35. $\displaystyle\int_0^2 \dfrac{x^2-2}{x+1}\,dx$

36. $\displaystyle\int_0^1 \dfrac{x-1}{x+1}\,dx$

37. $\displaystyle\int_1^2 \dfrac{1-\cos\theta}{\theta-\sin\theta}\,d\theta$

38. $\displaystyle\int_{0.1}^{0.2} (\csc 2\theta-\cot 2\theta)^2\,d\theta$

C In Exercises 39–44, use a symbolic integration utility to evaluate the integral.

39. $\int \dfrac{1}{1+\sqrt{x}}\,dx$

40. $\int \dfrac{1-\sqrt{x}}{1+\sqrt{x}}\,dx$

41. $\int \cos(1-x)\,dx$

42. $\int \dfrac{\tan^2 2x}{\sec 2x}\,dx$

43. $\displaystyle\int_{\pi/4}^{\pi/2} (\csc x-\sin x)\,dx$

44. $\displaystyle\int_{-\pi/4}^{\pi/4} \dfrac{\sin^2 x-\cos^2 x}{\cos x}\,dx$

In Exercises 45–48, show that the two formulas are equivalent.

45. $\int \tan x\,dx=-\ln|\cos x|+C$

$\int \tan x\,dx=\ln|\sec x|+C$

46. $\int \cot x\,dx=\ln|\sin x|+C$

$\int \cot x\,dx=-\ln|\csc x|+C$

47. $\int \sec x\,dx=\ln|\sec x+\tan x|+C$

$\int \sec x\,dx=-\ln|\sec x-\tan x|+C$

48. $\int \csc x\,dx=-\ln|\csc x+\cot x|+C$

$\int \csc x\,dx=\ln|\csc x-\cot x|+C$

In Exercises 49–52, find $F'(x)$.

49. $F(x)=\displaystyle\int_1^x \dfrac{1}{t}\,dt$

50. $F(x)=\displaystyle\int_0^x \tan t\,dt$

51. $F(x)=\displaystyle\int_x^{3x} \dfrac{1}{t}\,dt$

52. $F(x)=\displaystyle\int_1^{x^2} \dfrac{1}{t}\,dt$

Approximation **In Exercises 53 and 54, determine which best approximates the area of the region between the x-axis and the function over the given interval. (Make your selection on the basis of a sketch of the region and *not* by performing any calculations.)**

53. $f(x)=\sec x,\quad [0,1]$

   a. 6   **b.** -6 **c.** $\tfrac{1}{2}$ **d.** 1.25   **e.** 3

54. $f(x)=\dfrac{2x}{x^2+1},\quad [0,4]$

   a. 3   **b.** 7   **c.** -2 **d.** 5   **e.** 1

Area   In Exercises 55 and 56, find the area of the indicated region.

55. $y = \dfrac{x^2 + 4}{x}$

56. $y = \dfrac{x + 5}{x}$

57. *Population Growth*   A population of bacteria is changing at the rate of

$$\frac{dP}{dt} = \frac{3000}{1 + 0.25t}$$

where t is the time in days. The initial population (when $t = 0$) is 1000. Write an equation that gives the population at any time t, and then find the population when $t = 3$ days.

58. *Heat Transfer*   Find the time required for an object to cool from $300°$ to $250°$ by evaluating

$$t = \frac{10}{\ln 2} \int_{250}^{300} \frac{1}{T - 100} \, dT.$$

59. *Average Price*   The demand equation for a product is given by

$$p = \frac{90,000}{400 + 3x}.$$

Find the *average* price p on the interval $40 \le x \le 50$.

60. *Sales*   The rate of change in sales S is inversely proportional to time t ($t > 1$) measured in weeks. Find S as a function of t if sales after 2 and 4 weeks were 200 and 300 units, respectively.

True or False   In Exercises 61–64, determine whether the statement is true or false. If it is false, explain why or give an example that shows it is false.

61. $(\ln x)^{1/2} = \frac{1}{2}(\ln x)$

62. $\displaystyle \int \ln x \, dx = \left(\frac{1}{x}\right) + C$

63. $\displaystyle \int \frac{1}{x} \, dx = \ln |cx|, \qquad$ for $c \ne 0$

64. $\displaystyle \int_{-1}^{e} \frac{1}{x} = \ln |x| \, \Big]_{-1}^{e} = \ln e - \ln 1 = 1$

65. Graph the function

$$f_k(x) = \frac{x^k - 1}{k}$$

for $k = 1, 0.5$, and 0.1 on $[0, 10]$. Find $\lim\limits_{k \to 0} f_k(x)$.

CAREER INTERVIEW

Donna F. Stroup, Ph.D., M.Sc.

Director, Surveillance and Epidemiology

Centers for Disease Control and Prevention

Atlanta, GA

At the CDC, I direct a group of scientists who are trained in mathematics, statistics, and epidemiology, and whose job it is to reduce the incidence of preventable diseases, disabilities, and injuries. By analyzing information about the numbers of people infected by a particular disease, for instance, we can assess the need for a specific vaccination program. Furthermore, our data analysis on injuries associated with automobile crashes has led to recommendations on laws governing required safety features for automobiles, such as seat belts or content of gasoline.

Because much of our work at CDC is in statistics, my understanding of the calculus that is the basis of many statistical tests is an enormous asset. I assure policymakers that the application of a method is valid and that incorrect conclusions or inappropriate recommendations are avoided. The careful and systematic approach to problems that underlies calculus is essential to prevention and intervention efforts in public health, and to the acceptance of public health policy.

Inverse Functions

5.3 Inverse Functions • Existence of an Inverse Function • Derivative of an Inverse Function

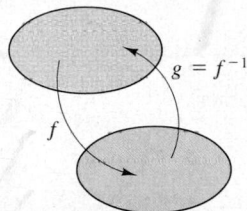

FIGURE 5.9
Domain of f = Range of g
Domain of g = Range of f

Inverse Functions

Recall from Section 5 of the Prerequisites chapter that a function can be represented by a set of ordered pairs. For instance, the function $f(x) = x + 3$ from $A = \{1, 2, 3, 4\}$ to $B = \{4, 5, 6, 7\}$ can be written as

$$f: \{(1, 4), (2, 5), (3, 6), (4, 7)\}.$$

By interchanging the first and second coordinates of each ordered pair, you can form the **inverse function** of f. This function is denoted by f^{-1}. It is a function from B to A, and can be written as

$$f^{-1}: \{(4, 1), (5, 2), (6, 3), (7, 4)\}.$$

Note that the domain of f is equal to the range of f^{-1}, and vice versa, as shown in Figure 5.9. The functions f and f^{-1} have the effect of "undoing" each other. That is, when you form the composition of f with f^{-1} or the composition of f^{-1} with f, you obtain the identity function.

$$f(f^{-1}(x)) = x \quad \text{and} \quad f^{-1}(f(x)) = x$$

Definition of Inverse Function

A function g is the **inverse** of the function f if

$$f(g(x)) = x \quad \text{for each } x \text{ in the domain of } g$$

and

$$g(f(x)) = x \quad \text{for each } x \text{ in the domain of } f.$$

The function g is denoted by f^{-1} (read "f inverse").

REMARK Although the notation used to denote an inverse function resembles *exponential notation*, it is a different use of -1 as a superscript. That is, in general, $f^{-1}(x) \neq 1/f(x)$.

Here are some important observations about inverse functions.

1. If g is the inverse of f, then f is the inverse of g.
2. The domain of f^{-1} is equal to the range of f, and the range of f^{-1} is equal to the domain of f.
3. A function need not possess an inverse, but if it does, the inverse is unique (see Exercise 79).

As mentioned on the preceding page, you can think of f^{-1} as undoing what has been done by f. For example, subtraction can be used to undo addition, and division can be used to undo multiplication. Use the definition of an inverse function to check the following inverses.

$f(x) = x + c$ and $f^{-1}(x) = x - c$ are inverses of each other.

$f(x) = cx$ and $f^{-1}(x) = \dfrac{x}{c}, \quad c \neq 0,$ are inverses of each other.

EXAMPLE 1   Verifying Inverse Functions

Show that the following functions are inverses of each other.

$$f(x) = 2x^3 - 1 \quad \text{and} \quad g(x) = \sqrt[3]{\frac{x + 1}{2}}$$

Solution   Because the domains and ranges of both f and g consist of all real numbers, you can conclude that both composite functions exist. The composite of f with g is given by

$$f(g(x)) = 2\left(\sqrt[3]{\frac{x + 1}{2}}\right)^3 - 1$$

$$= 2\left(\frac{x + 1}{2}\right) - 1$$

$$= x + 1 - 1$$

$$= x.$$

The composite of g with f is given by

$$g(f(x)) = \sqrt[3]{\frac{(2x^3 - 1) + 1}{2}} = \sqrt[3]{\frac{2x^3}{2}} = \sqrt[3]{x^3} = x.$$

Because $f(g(x)) = g(f(x)) = x$, you can conclude that f and g are inverses of each other.

FIGURE 5.10
f and g are inverses of each other.

In Figure 5.10, the graphs of f and $g = f^{-1}$ appear to be mirror images of each other with respect to the line $y = x$. The graph of f^{-1} is a **reflection** of the graph of f in the line $y = x$. This idea is generalized in the following theorem.

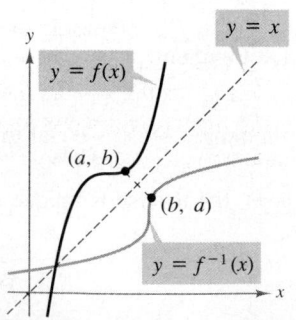

FIGURE 5.11
The graph of f^{-1} is a reflection of the graph of f in the line $y = x$.

THEOREM 5.6   Reflective Property of Inverse Functions

The graph of f contains the point (a, b) if and only if the graph of f^{-1} contains the point (b, a).

Proof   If (a, b) is on the graph of f, then $f(a) = b$ and you can write

$$f^{-1}(b) = f^{-1}(f(a)) = a.$$

Thus, (b, a) is on the graph of f^{-1}, as shown in Figure 5.11. A similar argument will prove the theorem in the other direction.

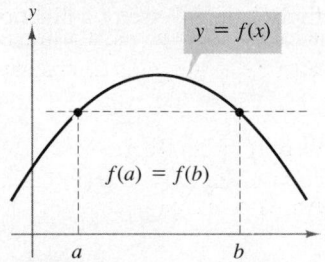

FIGURE 5.12
If a horizontal line intersects the graph of f twice, then f is not one-to-one.

Existence of an Inverse Function

Not every function has an inverse, and Theorem 5.6 suggests a graphical test for those that do—the **horizontal line test** for an inverse function. The test states that a function f has an inverse if and only if every horizontal line intersects the graph of f at most once (see Figure 5.12). The following theorem formally states why the horizontal line test is valid. (Recall from Section 3.3 that a function is *strictly monotonic* if it is either increasing on its entire domain or decreasing on its entire domain.)

THEOREM 5.7   The Existence of an Inverse Function

1. A function possesses an inverse if and only if it is one-to-one.
2. If f is strictly monotonic on its entire domain, then it is one-to-one and, hence, possesses an inverse.

Proof   To prove the second part of the theorem, recall from Section 5 of the Prerequisites chapter that f is one-to-one if for x_1 and x_2 in its domain

$$f(x_1) = f(x_2) \longrightarrow x_1 = x_2.$$

The *contrapositive* of this implication is logically equivalent and it states that

$$x_1 \neq x_2 \longrightarrow f(x_1) \neq f(x_2).$$

Now, choose x_1 and x_2 in the domain of f. If $x_1 \neq x_2$, then, because f is strictly monotonic, it follows that either $f(x_1) < f(x_2)$ or $f(x_1) > f(x_2)$. In either case, $f(x_1) \neq f(x_2)$. Thus, f is one-to-one on the interval. The proof of the first part is left as an exercise (see Exercise 80). ∎

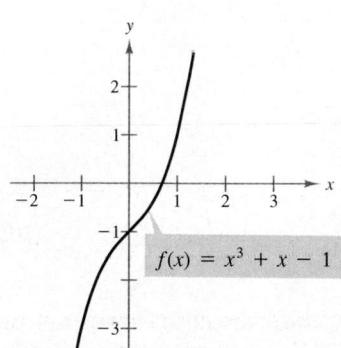

(a) Because f is increasing over its domain, it has an inverse.

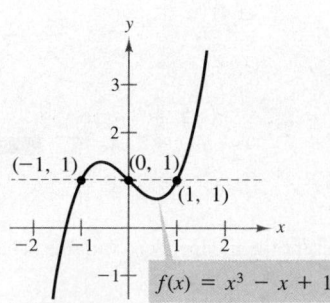

(b) Because f is not one-to-one, it does not have an inverse.

FIGURE 5.13

EXAMPLE 2   The Existence of an Inverse

Which of the following functions has an inverse?

a. $f(x) = x^3 + x - 1$ b. $f(x) = x^3 - x + 1$

Solution

a. From the graph of f given in Figure 5.13(a), it appears that f is increasing over its entire domain. To verify this, note that the derivative, $f'(x) = 3x^2 + 1$, is positive for all real values of x. Therefore, f is strictly monotonic and it must have an inverse.
b. From the graph in Figure 5.13(b), you can see that the function does not pass the horizontal line test. In other words, it is not one-to-one. For instance, f has the same value when $x = -1$, 0, and 1.

$$f(-1) = f(1) = f(0) = 1 \qquad \text{Not one-to-one}$$

Therefore, by Theorem 5.7, f does not have an inverse.   ∎

REMARK   Often it is easier to prove that a function has an inverse than to find the inverse. For instance, it would be difficult algebraically to determine the inverse of the function in Example 2a.

The following guidelines suggest a procedure for finding the inverse of a function.

Guidelines for Finding the Inverse of a Function

1. Use Theorem 5.7 to determine whether the function given by $y = f(x)$ has an inverse.

2. Solve for x as a function of y.

3. Interchange x and y. The resulting equation is $y = f^{-1}(x)$.

4. Define the domain of f^{-1} to be the range of f.

5. Verify that $f(f^{-1}(x)) = x$ and $f^{-1}(f(x)) = x$.

EXAMPLE 3 Finding the Inverse of a Function

Find the inverse of $f(x) = \sqrt{2x - 3}$.

Solution The function has an inverse because it is increasing on its entire domain (see Figure 5.14). To find an equation for the inverse, let $y = f(x)$ and solve for x in terms of y.

$$\sqrt{2x - 3} = y \qquad \text{Let } y = f(x)$$

$$2x - 3 = y^2 \qquad \text{Square both sides}$$

$$x = \frac{y^2 + 3}{2} \qquad \text{Solve for } x$$

$$y = \frac{x^2 + 3}{2} \qquad \text{Interchange } x \text{ and } y$$

$$f^{-1}(x) = \frac{x^2 + 3}{2} \qquad \text{Replace } y \text{ by } f^{-1}(x)$$

The domain of f^{-1} is the range of f, which is $[0, \infty)$. You can verify this result as follows.

$$f(f^{-1}(x)) = \sqrt{2\left(\frac{x^2 + 3}{2}\right) - 3} = \sqrt{x^2} = x, \qquad x \geq 0$$

$$f^{-1}(f(x)) = \frac{(\sqrt{2x - 3})^2 + 3}{2} = \frac{2x - 3 + 3}{2} = x, \qquad x \geq \frac{3}{2}$$

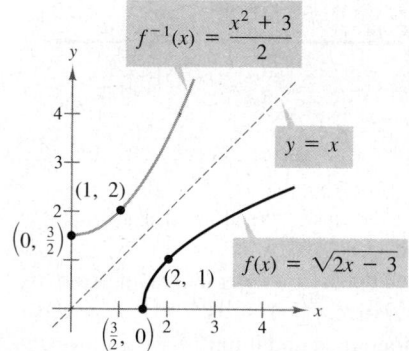

FIGURE 5.14
The domain of f^{-1}, $[0, \infty)$, is the range of f.

REMARK Remember that any letter can be used to represent the independent variable. Thus,

$$f^{-1}(y) = \frac{y^2 + 3}{2}, \quad f^{-1}(x) = \frac{x^2 + 3}{2}, \quad \text{and} \quad f^{-1}(s) = \frac{s^2 + 3}{2}$$

all represent the same function.

Theorem 5.7 is useful in the following type of problem. Suppose you are given a function that is *not* one-to-one on its domain. By restricting the domain to an interval on which the function is strictly monotonic, you can conclude that the new function *is* one-to-one on the restricted domain.

EXAMPLE 4 Testing Whether a Function is One-to-One

Show that the sine function $f(x) = \sin x$ is not one-to-one on the entire real line. Then show that $[-\pi/2, \pi/2]$ is the largest interval, centered at the origin, for which f is strictly monotonic.

Solution It is clear that f is not one-to-one, because many different x-values yield the same y-value. For instance, $\sin(0) = 0 = \sin(\pi)$. Moreover, f is increasing on the open interval $(-\pi/2, \pi/2)$, because its derivative

$$f'(x) = \cos x$$

is positive there. Finally, because the left and right endpoints correspond to relative extrema of the sine function, you can conclude that f is increasing on the closed interval $[-\pi/2, \pi/2]$ *and* that in any larger interval the function would not be strictly monotonic (see Figure 5.15).

FIGURE 5.15
f is one-to-one on the interval $[-\pi/2, \pi/2]$.

Derivative of an Inverse Function

The next two theorems discuss the derivative of an inverse function. The reasonableness of Theorem 5.8 follows from the reflective property of inverse functions as shown in Figure 5.11. Proofs of the two theorems are given in Appendix A.

THEOREM 5.8 Continuity and Differentiability of Inverse Functions

Let f be a function whose domain is an interval I. If f possesses an inverse, then the following statements are true.

1. If f is continuous on its domain, then f^{-1} is continuous on its domain.
2. If f is increasing on its domain, then f^{-1} is increasing on its domain.
3. If f is decreasing on its domain, then f^{-1} is decreasing on its domain.
4. If f is differentiable at c and $f'(c) \neq 0$, then f^{-1} is differentiable at $f(c)$.

THEOREM 5.9 The Derivative of an Inverse Function

Let f be a function that is differentiable on an interval I. If f possesses an inverse function g, then g is differentiable at any x for which $f'(g(x)) \neq 0$. Moreover,

$$g'(x) = \frac{1}{f'(g(x))}, \qquad f'(g(x)) \neq 0.$$

EXAMPLE 5 Evaluating the Derivative of an Inverse Function

Let $f(x) = \frac{1}{4}x^3 + x - 1$.

a. What is the value of $f^{-1}(x)$ when $x = 3$?
b. What is the value of $(f^{-1})'(x)$ when $x = 3$?

Solution Notice that f is one-to-one and hence has an inverse.

a. Because $f(2) = 3$, you know that $f^{-1}(3) = 2$.
b. Because the function f is differentiable and has an inverse, you can apply Theorem 5.9 to write

$$(f^{-1})'(3) = \frac{1}{f'(f^{-1}(3))} = \frac{1}{f'(2)}.$$

Moreover, using $f'(x) = \frac{3}{4}x^2 + 1$, you can conclude that

$$(f^{-1})'(3) = \frac{1}{f'(2)} = \frac{1}{\frac{3}{4}(2^2) + 1} = \frac{1}{4}.$$

FIGURE 5.16
The graphs of the inverse functions f and f^{-1} have reciprocal slopes at points (a, b) and (b, a).

In Example 5, note that at the point $(2, 3)$ the slope of the graph of f is 4 and at the point $(3, 2)$ the slope of the graph of f^{-1} is $\frac{1}{4}$ (see Figure 5.16). This reciprocal relationship (which follows from Theorem 5.9) is sometimes written as

$$\frac{dy}{dx} = \frac{1}{dx/dy}.$$

EXAMPLE 6 Graphs of Inverse Functions Have Reciprocal Slopes

Let $f(x) = x^2$ (for $x \geq 0$) and let $f^{-1}(x) = \sqrt{x}$. Show that the slopes of the graphs of f and f^{-1} are reciprocals at the following points.

a. $(2, 4)$ and $(4, 2)$ **b.** $(3, 9)$ and $(9, 3)$

Solution The derivatives of f and f^{-1} are given by

$$f'(x) = 2x \quad \text{and} \quad (f^{-1})'(x) = \frac{1}{2\sqrt{x}}.$$

a. At $(2, 4)$, the slope of the graph of f is $f'(2) = 2(2) = 4$. At $(4, 2)$, the slope of the graph of f^{-1} is

$$(f^{-1})'(4) = \frac{1}{2\sqrt{4}} = \frac{1}{2(2)} = \frac{1}{4}.$$

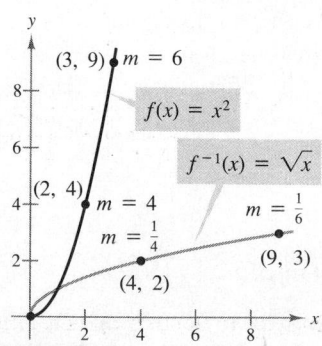

FIGURE 5.17
At $(0, 0)$ the derivative of f is 0, and the derivative of f^{-1} does not exist.

b. At $(3, 9)$, the slope of the graph of f is $f'(3) = 2(3) = 6$. At $(9, 3)$, the slope of the graph of f^{-1} is

$$(f^{-1})'(9) = \frac{1}{2\sqrt{9}} = \frac{1}{2(3)} = \frac{1}{6}.$$

Thus, in both cases, the slopes are reciprocals, as shown in Figure 5.17.

REMARK In Example 6, note that f^{-1} is not differentiable at $(0, 0)$. This is consistent with Theorem 5.9 because the derivative of f at $(0, 0)$ is 0.

EXERCISES for Section 5.3

C In Exercises 1–8, show that f and g are inverses of each other by showing that $f(g(x)) = x$ and $g(f(x)) = x$. Then use a graphing utility to graph f and g in the same viewing rectangle. Observe that the graph of g is a reflection of the graph of f in the line $y = x$.

1. $f(x) = 5x + 1$, $g(x) = \dfrac{x - 1}{5}$

2. $f(x) = 3 - 4x$, $g(x) = \dfrac{3 - x}{4}$

3. $f(x) = x^3$, $g(x) = \sqrt[3]{x}$

4. $f(x) = 1 - x^3$, $g(x) = \sqrt[3]{1 - x}$

5. $f(x) = \sqrt{x - 4}$, $g(x) = x^2 + 4$, $x \geq 0$

6. $f(x) = 9 - x^2$, $x \geq 0$, $g(x) = \sqrt{9 - x}$

7. $f(x) = \dfrac{1}{x}$, $g(x) = \dfrac{1}{x}$

8. $f(x) = \dfrac{1}{1 + x}$, $x \geq 0$, $g(x) = \dfrac{1 - x}{x}$, $0 < x \leq 1$

In Exercises 9–22, find the inverse of f. Then graph both f and f^{-1}.

9. $f(x) = 2x - 3$ **10.** $f(x) = 3x$

11. $f(x) = x^5$ **12.** $f(x) = x^3 + 1$

13. $f(x) = \sqrt{x}$ **14.** $f(x) = x^2$, $x \geq 0$

15. $f(x) = \sqrt{4 - x^2}$, $0 \leq x$

16. $f(x) = \sqrt{x^2 - 4}$, $x \geq 2$

17. $f(x) = \sqrt[3]{x - 1}$ **18.** $f(x) = 3\sqrt[5]{2x - 1}$

19. $f(x) = x^{2/3}$, $x \geq 0$ **20.** $f(x) = x^{3/5}$

21. $f(x) = \dfrac{x}{\sqrt{x^2 + 7}}$ **22.** $f(x) = \dfrac{x + 2}{x}$

C In Exercises 23 and 24, find the inverse function of f over the specified interval. Use a graphing utility to graph f and f^{-1} in the same viewing rectangle. Observe that the graph of f^{-1} is a reflection of the graph of f in the line $y = x$.

Function	Interval
23. $f(x) = \dfrac{x}{x^2 - 4}$	$-2 < x < 2$
24. $f(x) = 2 - \dfrac{3}{x^2}$	$0 < x < 10$

In Exercises 25 and 26, use the graph of the function f to complete the table and to sketch the graph of f^{-1}.

25.

x	1	2	3	4
$f^{-1}(x)$				

26.

x	0	2	4
$f^{-1}(x)$			

In Exercises 27–30, use the horizontal line test to determine whether the function is one-to-one on its entire domain and therefore has an inverse (on its entire domain).

27. $f(x) = \frac{3}{4}x + 6$ **28.** $f(x) = 5x - 3$

29. $f(\theta) = \sin \theta$ **30.** $f(x) = \dfrac{x^2}{x^2 + 4}$

C In Exercises 31–36, use a graphing utility to graph the function. From the graph, determine whether the function is one-to-one on its entire domain.

31. $h(s) = \dfrac{1}{s - 2} - 3$ **32.** $g(t) = \dfrac{1}{\sqrt{t^2 + 1}}$

33. $f(x) = \ln x$ **34.** $f(x) = 3x\sqrt{x + 1}$

35. $h(x) = |x + 4| - |x - 4|$ **36.** $g(x) = (x + 5)^3$

In Exercises 37–42, use the derivative to determine whether the function is strictly monotonic on its entire domain and therefore has an inverse.

37. $f(x) = (x + a)^3 + b$

38. $f(x) = \cos \dfrac{3x}{2}$

39. $f(x) = \dfrac{x^4}{4} - 2x^2$

40. $f(x) = x^3 - 6x^2 + 12x$

41. $f(x) = 2 - x - x^3$

42. $f(x) = \ln(x - 3)$

In Exercises 43–48, show that f is strictly monotonic on the indicated interval and therefore has an inverse on that interval.

Function	Interval
43. $f(x) = (x - 4)^2$	$[4, \infty)$
44. $f(x) = \|x + 2\|$	$[-2, \infty)$
45. $f(x) = \dfrac{4}{x^2}$	$(0, \infty)$
46. $f(x) = \tan x$	$\left(-\dfrac{\pi}{2}, \dfrac{\pi}{2}\right)$
47. $f(x) = \cos x$	$[0, \pi]$
48. $f(x) = \sec x$	$\left[0, \dfrac{\pi}{2}\right)$

In Exercises 49 and 50, the derivative of the function has the same sign for all x in its domain, but the function is not one-to-one. Explain why.

49. $f(x) = \tan x$

50. $f(x) = \dfrac{x}{x^2 - 4}$

In Exercises 51–54, determine whether the function is one-to-one. If it is, find its inverse.

51. $f(x) = \sqrt{x - 2}$

52. $f(x) = -3$

53. $f(x) = |x - 2|, \quad x \leq 2$

54. $f(x) = ax + b, \quad a \neq 0$

In Exercises 55–58, restrict the domain of f to an infinite interval on which f is one-to-one. Find the inverse of the restricted function and give the domain of the inverse. (*Note:* There is more than one correct answer.)

55. $f(x) = (x - 3)^2$

56. $f(x) = 16 - x^4$

57. $f(x) = |x + 3|$

58. $f(x) = |x - 3|$

In Exercises 59–64, find $(f^{-1})'(a)$ for the function f and real number a.

Function	Real number
59. $f(x) = x^3 + 2x - 1$	$a = 2$
60. $f(x) = 2x^5 + x^3 + 1$	$a = -2$
61. $f(x) = \sin x, \quad -\dfrac{\pi}{2} \leq x \leq \dfrac{\pi}{2}$	$a = \dfrac{1}{2}$
62. $f(x) = \cos 2x, \quad 0 \leq x \leq \dfrac{\pi}{2}$	$a = 1$
63. $f(x) = x^3 - \dfrac{4}{x}$	$a = 6$
64. $f(x) = \sqrt{x - 4}$	$a = 2$

In Exercises 65–68, show that the slopes of the graphs of f and f^{-1} are reciprocals at the indicated points.

Functions	Points
65. $f(x) = x^3$	$\left(\dfrac{1}{2}, \dfrac{1}{8}\right)$
$\quad f^{-1}(x) = \sqrt[3]{x}$	$\left(\dfrac{1}{8}, \dfrac{1}{2}\right)$
66. $f(x) = 3 - 4x$	$(1, -1)$
$\quad f^{-1}(x) = \dfrac{3 - x}{4}$	$(-1, 1)$
67. $f(x) = \sqrt{x - 4}$	$(5, 1)$
$\quad f^{-1}(x) = x^2 + 4$	$(1, 5)$
68. $f(x) = \dfrac{1}{1 + x^2}$	$\left(1, \dfrac{1}{2}\right)$
$\quad f^{-1}(x) = \sqrt{\dfrac{1 - x}{x}}$	$\left(\dfrac{1}{2}, 1\right)$

In Exercises 69–72, use the functions

$$f(x) = \tfrac{1}{8}x - 3 \quad \text{and} \quad g(x) = x^3$$

to find the indicated value.

69. $(f^{-1} \circ g^{-1})(1)$

70. $(g^{-1} \circ f^{-1})(-3)$

71. $(f^{-1} \circ f^{-1})(6)$

72. $(g^{-1} \circ g^{-1})(-4)$

In Exercises 73–76, use the functions

$$f(x) = x + 4 \quad \text{and} \quad g(x) = 2x - 5$$

to find the specified function.

73. $g^{-1} \circ f^{-1}$

74. $f^{-1} \circ g^{-1}$

75. $(f \circ g)^{-1}$

76. $(g \circ f)^{-1}$

77. Prove that if f and g are one-to-one functions, then $(f \circ g)^{-1}(x) = (g^{-1} \circ f^{-1})(x)$.

78. Prove that if f has an inverse, then $(f^{-1})^{-1} = f$.

79. Prove that if a function has an inverse, then the inverse is unique.

80. Prove that a function has an inverse if and only if it is one-to-one.

True or False In Exercises 81–84, determine whether the statement is true or false. If it is false, explain why or give an example that shows it is false.

81. If f is an even function on the interval $[-1, 1]$, then f^{-1} exists on the interval $[-1, 1]$.

82. If the inverse of f exists, then the y-intercept of f is an x-intercept of f^{-1}.

83. If $f(x) = x^n$ where n is odd, then f^{-1} exists.

84. There exists no function f such that $f = f^{-1}$.

85. Is the converse of part 2 of Theorem 5.7 true? That is, if a function is one-to-one (and hence possesses an inverse) then must the function be strictly monotonic? If so, prove it. If not, give a counter example.

86. Let f be twice-differentiable and one-to-one on an open interval I. Show that its inverse g satisfies

$$g''(x) = -\frac{f''(g(x))}{[f'(g(x))]^3}.$$

If f is increasing and concave downward, what is the concavity of $f^{-1} = g$?

87. If $f(x) = \int_2^x \frac{dt}{\sqrt{1 + t^4}}$, find $(f^{-1})'(0)$.

FIGURE 5.18
The inverse of the natural logarithmic function is the natural exponential function.

The Natural Exponential Function

The function $f(x) = \ln x$ is increasing on its entire domain—hence, it has an inverse f^{-1}. The domain of f^{-1} is the set of all reals, and the range is the set of positive reals, as shown in Figure 5.18. Thus, for any real number x,

$$f(f^{-1}(x)) = \ln(f^{-1}(x)) = x. \qquad \text{x is any real number}$$

If x happens to be rational, then

$$\ln(e^x) = x \ln e = x(1) = x. \qquad \text{x is a rational number}$$

Because the natural logarithmic function is one-to-one, you can conclude that $f^{-1}(x)$ and e^x agree for *rational* values of x. The following definition extends the meaning of e^x to include *all* real values of x.

Definition of the Natural Exponential Function

The inverse of the natural logarithmic function $f(x) = \ln x$ is called the **natural exponential function** and is denoted by

$$f^{-1}(x) = e^x.$$

That is,

$$y = e^x \quad \text{if and only if} \quad x = \ln y.$$

The inverse relationship between the natural logarithmic function and the natural exponential function can be summarized as follows.

$$\ln(e^x) = x \quad \text{and} \quad e^{\ln x} = x \qquad \text{Inverse relationships}$$

EXAMPLE 1   Solving Exponential Equations

Solve $7 = e^{x+1}$.

Solution   You can convert from exponential form to logarithmic form by *taking the natural log of both sides* of the equation.

$$
\begin{aligned}
7 &= e^{x+1} & \text{Given equation}\\
\ln 7 &= \ln(e^{x+1}) & \text{Take natural log of both sides}\\
\ln 7 &= x + 1 & \text{Apply inverse property}\\
-1 + \ln 7 &= x & \text{Solve for x}\\
0.946 &\approx x & \text{Use a calculator}
\end{aligned}
$$

EXAMPLE 2   Solving a Logarithmic Equation

Solve $\ln(2x - 3) = 5$.

Solution   To convert from logarithmic form to exponential form, you can *exponentiate both sides* of the logarithmic equation.

$$\ln(2x - 3) = 5 \qquad \text{Given equation}$$

$$e^{\ln(2x-3)} = e^5 \qquad \text{Exponentiate both sides}$$

$$2x - 3 = e^5 \qquad \text{Apply inverse property}$$

$$x = \frac{1}{2}(e^5 + 3) \qquad \text{Solve for } x$$

$$x \approx 75.707 \qquad \text{Use a calculator}$$

The familiar rules for operating with rational exponents can be extended to the natural exponential function, as indicated in the following theorem.

THEOREM 5.10   Operations with Exponential Functions

Let a and b be any real numbers.

1. $e^a e^b = e^{a+b}$ $\qquad$ **2.** $\dfrac{e^a}{e^b} = e^{a-b}$

Proof   To prove Property 1, you can write

$$\ln(e^a e^b) = \ln(e^a) + \ln(e^b) = a + b = \ln(e^{a+b}).$$

Because the natural log function is one-to-one, you can conclude that

$$e^a e^b = e^{a+b}.$$

The proof of the second property is left to you (see Exercise 97).

In Section 5.3, you learned that an inverse function f^{-1} shares many properties with f. Thus, the natural exponential function inherits the following properties from the natural logarithmic function (see Figure 5.19).

FIGURE 5.19
The natural exponential function is increasing and its graph is concave upward.

Properties of the Natural Exponential Function

1. The domain of $f(x) = e^x$ is $(-\infty, \infty)$, and the range is $(0, \infty)$.
2. The function $f(x) = e^x$ is continuous, increasing, and one-to-one on its entire domain.
3. The graph of $f(x) = e^x$ is concave upward on its entire domain.
4. $\lim\limits_{x \to -\infty} e^x = 0$ and $\lim\limits_{x \to \infty} e^x = \infty$

Derivatives of Exponential Functions

One of the most intriguing (and useful) characteristics of the natural exponential function is that *it is its own derivative*. In other words, it is a solution to the differential equation $y' = y$. This result is stated in the next theorem.

THEOREM 5.11   The Derivative of the Natural Exponential Function

Let u be a differential function of x.

1. $\dfrac{d}{dx}[e^x] = e^x$ 　　　　 **2.** $\dfrac{d}{dx}[e^u] = e^u \dfrac{du}{dx}$

Proof　Let $f(x) = \ln x$ and $g(x) = e^x$. Then $f'(x) = 1/x$, and, by Theorem 5.9, you can write

$$\frac{d}{dx}[e^x] = \frac{d}{dx}[g(x)] = \frac{1}{f'(g(x))} = \frac{1}{f'(e^x)} = \frac{1}{(1/e^x)} = e^x.$$

The derivative of e^u follows from the Chain Rule. ■

REMARK　You can interpret this theorem geometrically by saying that the slope of the graph of $f(x) = e^x$ at any point (x, e^x) is equal to the y-coordinate of the point.

EXAMPLE 3　Differentiating Exponential Functions

a. $\dfrac{d}{dx}[e^{2x-1}] = e^u \dfrac{du}{dx} = 2e^{2x-1}$ 　　　　 $u = 2x - 1$

b. $\dfrac{d}{dx}[e^{-3/x}] = e^u \dfrac{du}{dx} = \left(\dfrac{3}{x^2}\right)e^{-3/x} = \dfrac{3e^{-3/x}}{x^2}$ 　　 $u = -\dfrac{3}{x}$

EXAMPLE 4　Locating Relative Extrema

Find the relative extrema of $f(x) = xe^x$.

Solution　The derivative of f is given by

$$f'(x) = x(e^x) + e^x(1) \qquad \text{Product Rule}$$
$$= e^x(x + 1).$$

Because e^x is never 0, the derivative is 0 only when $x = -1$. Moreover, by the First Derivative Test, you can determine that this corresponds to a relative minimum, as shown in Figure 5.20. Because the derivative $f'(x) = e^x(x + 1)$ is defined for all x, there are no other critical points. ■

FIGURE 5.20
The derivative of f changes from negative to positive at $x = -1$.

EXAMPLE 5 The Normal Probability Density Function

Show that the *normal probability density function*

$$f(x) = \frac{1}{\sqrt{2\pi}} e^{-x^2/2}$$

has points of inflection when $x = \pm 1$.

Solution To locate possible points of inflection, find the *x*-values for which the second derivative is 0.

$$f(x) = \frac{1}{\sqrt{2\pi}} e^{-x^2/2} \qquad \text{Original function}$$

$$f'(x) = \frac{1}{\sqrt{2\pi}}(-x)e^{-x^2/2} \qquad \text{First derivative}$$

$$f''(x) = \frac{1}{\sqrt{2\pi}}[(-x)(-x)e^{-x^2/2} + (-1)e^{-x^2/2}] \qquad \text{Product Rule}$$

$$= \frac{1}{\sqrt{2\pi}}(e^{-x^2/2})(x^2 - 1) \qquad \text{Second derivative}$$

Therefore, $f''(x) = 0$ when $x = \pm 1$, and you can apply the techniques of Chapter 3 to conclude that these values yield the two points of inflection shown in Figure 5.21.

Two points of inflection

$$f(x) = \frac{1}{\sqrt{2\pi}} e^{-x^2/2}$$

FIGURE 5.21
The bell-shaped curve given by a normal probability density function.

REMARK The general form of a normal probability density function (whose mean is 0) is given by

$$f(x) = \frac{1}{\sigma\sqrt{2\pi}} e^{-x^2/2\sigma^2}$$

where σ is the standard deviation (σ is the lowercase Greek letter sigma). This "bell-shaped curve" has points of inflection when $x = \pm \sigma$.

EXAMPLE 6 M.D.'s in the United States

For 1970 through 1990, the number *y* of medical doctors in the United States can be modeled by

$$y = 348,000e^{0.033666t}$$

where $t = 0$ represents 1970. At what rate was the number of M.D.'s changing in 1988? (*Source:* U.S. Bureau of Labor Statistics)

Solution The derivative of the given model is

$$y' = (0.033666)(348,000)e^{0.033666t}$$
$$\approx 11,716e^{0.033666t}.$$

By evaluating the derivative when $t = 18$, you can conclude that the rate of change in 1988 was about

21,476 doctors per year.

The graph of this model is shown in Figure 5.22.

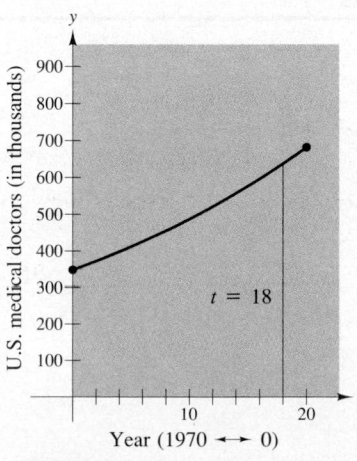

FIGURE 5.22

Integrals of Exponential Functions

Each differentiation formula in Theorem 5.11 has a corresponding integration formula.

THEOREM 5.12   Integration Rules for Exponential Functions

Let u be a differentiable function of x.

1. $\displaystyle \int e^x \, dx = e^x + C$ **2.** $\displaystyle \int e^u \, du = e^u + C$

EXAMPLE 7   Integrating Exponential Functions

Evaluate $\int e^{3x+1} \, dx$.

Solution   Let $u = 3x + 1$, then $du = 3 \, dx$.

$$\int e^{3x+1} \, dx = \frac{1}{3} \int e^{3x+1}(3) \, dx \qquad \text{Multiply and divide by 3}$$

$$= \frac{1}{3} \int e^u \, du \qquad \text{Substitute}$$

$$= \frac{1}{3} e^u + C \qquad \text{Apply Exponential Rule}$$

$$= \frac{e^{3x+1}}{3} + C \qquad \text{Back-substitute}$$

REMARK   In Example 7, the missing *constant* factor 3 was introduced to create $du = 3 \, dx$. However, remember that you cannot introduce a missing *variable* factor in the integrand. For instance,

$$\int e^{-x^2} \, dx \neq \frac{1}{x} \int e^{-x^2}(x \, dx).$$

EXAMPLE 8   Integrating Exponential Functions

Evaluate $\int 5xe^{-x^2} \, dx$.

Solution   Let $u = -x^2$, then $du = -2x \, dx$ or $x \, dx = -du/2$.

$$\int 5xe^{-x^2} \, dx = \int 5e^{-x^2}(x \, dx) \qquad \text{Regroup integrand}$$

$$= \int 5e^u \left(-\frac{du}{2} \right) \qquad \text{Substitute}$$

$$= -\frac{5}{2} \int e^u \, du \qquad \text{Factor } -\tfrac{5}{2} \text{ out of integral}$$

$$= -\frac{5}{2} e^u + C \qquad \text{Apply Exponential Rule}$$

$$= -\frac{5}{2} e^{-x^2} + C \qquad \text{Back-substitute}$$

EXAMPLE 9   Integrating Exponential Functions

a. $\displaystyle\int \frac{e^{1/x}}{x^2}\,dx = -\int \overbrace{e^{1/x}}^{e^u}\overbrace{\left(-\frac{1}{x^2}\right)dx}^{du}$ $u = \dfrac{1}{x}$

$\qquad\qquad = -e^{1/x} + C$

b. $\displaystyle\int \sin x\, e^{\cos x}\,dx = -\int \overbrace{e^{\cos x}}^{e^u}\overbrace{(-\sin x\,dx)}^{du}$ $u = \cos x$

$\qquad\qquad = -e^{\cos x} + C$

EXAMPLE 10   Finding Areas Bounded by Exponential Functions

a. $\displaystyle\int_0^1 e^{-x}\,dx = \left[-e^{-x}\right]_0^1$ See Figure 5.23(a)

$\qquad\qquad = -e^{-1} - (-1)$

$\qquad\qquad = 1 - \dfrac{1}{e}$

$\qquad\qquad \approx 0.632$

b. $\displaystyle\int_0^1 \frac{e^x}{1 + e^x}\,dx = \left[\ln(1 + e^x)\right]_0^1$ See Figure 5.23(b)

$\qquad\qquad = \ln(1 + e) - \ln 2$

$\qquad\qquad \approx 0.620$

c. $\displaystyle\int_{-1}^0 \left[e^x \cos(e^x)\right]dx = \left[\sin(e^x)\right]_{-1}^0$ See Figure 5.23(c)

$\qquad\qquad = \sin 1 - \sin(e^{-1})$

$\qquad\qquad \approx 0.482$

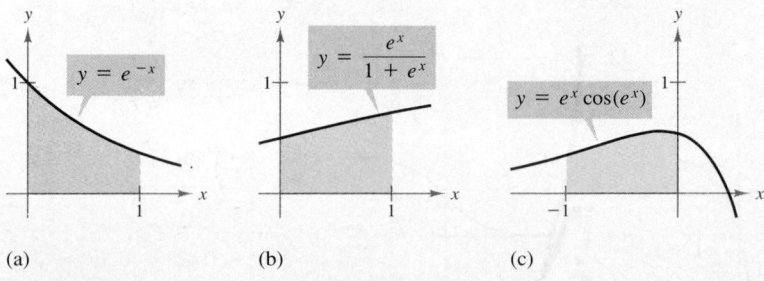

FIGURE 5.23
Areas bounded by exponential functions.

EXERCISES for Section 5.4

TECHNOLOGY
Laboratory Guide
Lab 5.4

In Exercises 1–4, write the exponential equation as a logarithmic equation or vice versa.

1. a. $e^0 = 1$
   b. $e^2 = 7.389 \ldots$
2. a. $e^{-2} = 0.1353 \ldots$
   b. $e^{-1} = 0.3679 \ldots$

3. a. $\ln 2 = 0.6931 \ldots$
   b. $\ln 8.4 = 2.128 \ldots$
4. a. $\ln 0.5 = -0.6931 \ldots$
   b. $\ln 1 = 0$

In Exercises 5–8, solve for x.

5. a. $e^{\ln x} = 4$
   b. $\ln e^{2x} = 3$
6. a. $e^{\ln 2x} = 12$
   b. $\ln e^{-x} = 0$

7. a. $\ln x = 2$
   b. $e^x = 4$
8. a. $\ln x^2 = 10$
   b. $e^{-4x} = 5$

In Exercises 9–12, sketch the graph of the function.

9. $y = e^{-x}$
10. $y = \frac{1}{2}e^x$
11. $y = e^{-x^2}$
12. $y = e^{-x/2}$

C 13. Use a graphing utility to graph $f(x) = e^x$ and the given function in the same viewing rectangle. How are the two graphs related?
   a. $g(x) = f(x - 2) = e^{x-2}$
   b. $h(x) = -\frac{1}{2}f(x) = -\frac{1}{2}e^x$
   c. $q(x) = f(-x) + 3 = e^{-x} + 3$

C 14. Use a graphing utility to graph the function. Use the graph to determine any asymptotes of the function.
   a. $f(x) = \dfrac{8}{1 + e^{-0.5x}}$
   b. $g(x) = \dfrac{8}{1 + e^{-0.5/x}}$

C In Exercises 15–18, illustrate that the functions are inverses of each other by graphing both functions in the same viewing rectangle.

15. $f(x) = e^{2x}$
    $g(x) = \ln \sqrt{x}$
16. $f(x) = e^{x/3}$
    $g(x) = \ln x^3$

17. $f(x) = e^x - 1$
    $g(x) = \ln(x + 1)$
18. $f(x) = e^{x-1}$
    $g(x) = 1 + \ln x$

In Exercises 19–22, match the equation with the correct graph. Assume that a and C are arbitrary real numbers such that $a > 0$. (The graphs are labeled (a), (b), (c), and (d).)

19. $y = Ce^{ax}$
20. $y = Ce^{-ax}$
21. $y = C(1 - e^{-ax})$
22. $y = \dfrac{C}{1 + e^{-ax}}$

(a)

(b)

(c)

(d)

In Exercises 23 and 24, compare the given number to the number e. Is the number less than or greater than e?

23. $\left(1 + \frac{1}{1,000,000}\right)^{1,000,000}$

24. $1 + 1 + \frac{1}{2} + \frac{1}{6} + \frac{1}{24} + \frac{1}{120} + \frac{1}{720} + \frac{1}{5040}$

In Exercises 25 and 26, find the slope of the tangent line to the function at the point $(0, 1)$.

25. a. $y = e^{3x}$

    b. $y = e^{-3x}$

26. a. $y = e^{2x}$

    b. $y = e^{-2x}$

In Exercises 27–46, find the derivative of the function.

27. $f(x) = e^{2x}$
28. $f(x) = e^{1-x}$
29. $y = e^{-2x+x^2}$
30. $y = e^{-x^2}$
31. $y = e^{\sqrt{x}}$
32. $y = x^2 e^{-x}$
33. $g(t) = (e^{-t} + e^t)^3$
34. $g(t) = e^{-1/t^2}$
35. $y = \ln(e^{x^2})$
36. $y = \ln\left(\dfrac{1 + e^x}{1 - e^x}\right)$
37. $y = \ln(1 + e^{2x})$
38. $y = \ln\dfrac{e^x + e^{-x}}{2}$
39. $y = \dfrac{2}{e^x + e^{-x}}$
40. $y = \dfrac{e^x - e^{-x}}{2}$
41. $y = x^2 e^x - 2xe^x + 2e^x$
42. $y = xe^x - e^x$
43. $f(x) = e^{-x} \ln x$
44. $f(x) = e^3 \ln x$
45. $y = e^x(\sin x + \cos x)$
46. $y = \ln e^x$

In Exercises 47 and 48, use implicit differentiation to find dy/dx.

47. $xe^y - 10x + 3y = 0$　　　**48.** $e^{xy} + x^2 - y^2 = 10$

In Exercises 49 and 50, find the second derivative of the function.

49. $f(x) = (3 + 2x)e^{-3x}$　　**50.** $g(x) = \sqrt{x} + e^x \ln x$

In Exercises 51 and 52, show that the function $y = f(x)$ is a solution of the differential equation.

51. $y = e^x(\cos \sqrt{2}x + \sin \sqrt{2}x)$
$y'' - 2y' + 3y = 0$

52. $y = e^x(3\cos 2x - 4\sin 2x)$
$y'' - 2y' + 5y = 0$

In Exercises 53–58, find the extrema and the points of inflection (if any exist), and sketch the graph of the function.

53. $f(x) = \dfrac{1}{\sqrt{2\pi}} e^{-(x^2/2)}$　　**54.** $f(x) = \dfrac{e^x - e^{-x}}{2}$

55. $f(x) = \dfrac{e^x + e^{-x}}{2}$　　**56.** $f(x) = xe^{-x}$

57. $f(x) = x^2 e^{-x}$

58. $f(x) = -2 + e^{3x}(4 - 2x)$

[C] *Linear and Quadratic Approximation*　In Exercises 59 and 60, use a graphing utility to graph the function. Then graph

$$P_1(x) = f(0) + f'(0)(x - 0)$$

and

$$P_2(x) = f(0) + f'(0)(x - 0) + \tfrac{1}{2}f''(0)(x - 0)^2$$

in the same viewing rectangle. Compare the values of f, P_1, and P_2 and their first derivatives at $x = 0$.

59. $f(x) = e^{x/2}$　　　**60.** $f(x) = e^{-x^2/2}$

61. *Area*　Find the area of the largest rectangle that can be inscribed under the curve $y = e^{-x^2}$ in the first and second quadrants.

62. Find the point on the graph of $y = e^{-x}$ where the normal line to the curve will pass through the origin. (Use Newton's Method.)

63. Find, to three decimal places, the value of x such that $e^{-x} = x$. (Use Newton's Method.)

[C] 64. *Depreciation*　The value V of an item t years after it is purchased is

$$V = 15{,}000e^{-0.6286t}, \qquad 0 \le t \le 10.$$

　a. Use a graphing utility to graph the function.
　b. Find the rate of change of V with respect to t when $t = 1$ and $t = 5$.
　c. Sketch the tangent line to the graph of the function when $t = 1$ and $t = 5$.

[C] 65. *Essay*　Consider the function

$$f(x) = \frac{2}{1 + e^{1/x}}.$$

　a. Use a graphing utility to graph f.
　b. Write a short paragraph explaining why the graph has a horizontal asymptote $y = 1$ and why the function has a nonremovable discontinuity at $x = 0$.

[C] 66. *Harmonic Motion*　The displacement from equilibrium of a mass oscillating on the end of a spring suspended from a ceiling is

$$y = 1.56e^{-0.22t}\cos 4.9t$$

where y is the displacement in feet and t is the time in seconds. Use a graphing utility to graph the displacement function on the interval $[0, 10]$. Find a value of t past which the displacement is less than 3 inches from equilibrium.

In Exercises 67–84, evaluate the integral.

67. $\displaystyle\int e^{5x}(5)\,dx$　　**68.** $\displaystyle\int e^{-x^4}(-4x^3)\,dx$

69. $\displaystyle\int_0^1 e^{-2x}\,dx$　　**70.** $\displaystyle\int_1^2 e^{1-x}\,dx$

71. $\displaystyle\int \frac{e^{-x}}{1 + e^{-x}}\,dx$　　**72.** $\displaystyle\int \frac{e^{2x}}{1 + e^{2x}}\,dx$

73. $\displaystyle\int_1^3 \frac{e^{3/x}}{x^2}\,dx$　　**74.** $\displaystyle\int_0^{\sqrt{2}} xe^{-(x^2/2)}\,dx$

75. $\displaystyle\int e^x\sqrt{1 - e^x}\,dx$　　**76.** $\displaystyle\int \frac{e^x - e^{-x}}{e^x + e^{-x}}\,dx$

77. $\displaystyle\int \frac{e^x + e^{-x}}{e^x - e^{-x}}\,dx$　　**78.** $\displaystyle\int \frac{2e^x - 2e^{-x}}{(e^x + e^{-x})^2}\,dx$

79. $\displaystyle\int \frac{5 - e^x}{e^{2x}}\,dx$　　**80.** $\displaystyle\int \frac{e^{2x} + 2e^x + 1}{e^x}\,dx$

81. $\displaystyle\int e^{\sin \pi x}\cos \pi x\,dx$　　**82.** $\displaystyle\int e^{\tan 2x}\sec^2 2x\,dx$

83. $\displaystyle\int e^{-x}\tan(e^{-x})\,dx$　　**84.** $\displaystyle\int \ln(e^{2x-1})\,dx$

In Exercises 85 and 86, solve the differential equation.

85. $y' = xe^{ax^2}$　　　**86.** $y' = (e^x - e^{-x})^2$

In Exercises 87 and 88, find the particular solution of the differential equation that satisfies the initial conditions.

87. $f''(x) = \tfrac{1}{2}(e^x + e^{-x})$,　**88.** $f''(x) = \sin x + e^{2x}$,
$f(0) = 1$, $f'(0) = 0$　　$f(0) = \tfrac{1}{4}$, $f'(0) = \tfrac{1}{2}$

Area　In Exercises 89–92, find the area of the region bounded by the graphs of the equations.

89. $y = e^x$, $y = 0$, $x = 0$, $x = 5$

90. $y = e^{-x}$, $y = 0$, $x = a$, $x = b$

91. $y = xe^{-(x^2/2)}$, $y = 0$, $x = 0$, $x = \sqrt{2}$

92. $y = e^{-2x} + 2$, $y = 0$, $x = 0$, $x = 2$

C **93.** Use a computer program to approximate each integral using the Midpoint Rule, the Trapezoidal Rule, and Simpson's Rule with $n = 12$.

a. $\displaystyle\int_0^4 \sqrt{x} e^x \, dx$ **b.** $\displaystyle\int_0^2 2xe^{-x} \, dx$

C **94.** *Probability* A car battery has an average lifetime of 48 months with a standard deviation of 6 months. The battery lives are normally distributed. The probability that a given battery lasts between 48 months and 60 months is

$$0.0665 \int_{48}^{60} e^{-0.0139(t-48)^2} \, dt.$$

Use Simpson's Rule with $n = 24$ to find this probability.

95. Given $e^x \geq 1$ for $x \geq 0$, it follows that

$$\int_0^x e^t \, dt \geq \int_0^x 1 \, dt.$$

Perform this integration to derive the inequality $e^x \geq 1 + x$ for $x \geq 0$.

C **96.** *Approximation* Integrate each term of the inequality to obtain a new inequality. Then use a graphing utility to graph $y = e^x$ and the right-hand member of each inequality. Write a short paragraph about the accuracy of the polynomials as approximations of the exponential function.

a. $e^x \geq 1 + x$

b. $e^x \geq 1 + x + \dfrac{x^2}{2}$

c. $e^x \geq 1 + x + \dfrac{x^2}{2} + \dfrac{x^3}{6}$

d. $e^x \geq 1 + x + \dfrac{x^2}{2} + \dfrac{x^3}{6} + \dfrac{x^4}{24}$

97. Prove that $\dfrac{e^a}{e^b} = e^{a-b}$.

98. Let $f(x) = \dfrac{\ln x}{x}$.

a. Graph f on $(0, \infty)$ and show that f is strictly decreasing on $[e, \infty)$.

b. Show that if $e \leq A < B$, then $A^B > B^A$.

c. Show that $e^\pi > \pi^e$.

5.5
Bases Other than *e* • Differentiation and Integration • Applications of Exponential Functions

Bases Other than *e*

The **base** of the natural exponential function is *e*. We now assign a meaning to a general base *a*.

Definition of Exponential Function to Base *a*

If *a* is a positive real number ($a \neq 1$) and *x* is any real number, then the **exponential function to the base *a*** is denoted by a^x and is defined as

$$a^x = e^{(\ln a)x}.$$

If $a = 1$, then $y = 1^x = 1$ is a constant function.

These functions obey the usual laws of exponents. For instance, here are some familiar properties.

1. $a^0 = 1$ **2.** $a^x a^y = a^{x+y}$

3. $\dfrac{a^x}{a^y} = a^{x-y}$ **4.** $(a^x)^y = a^{xy}$

EXAMPLE 1 Radioactive Half-Life Model

The half-life of carbon-14 is about 5730 years. If 1 gram of carbon-14 is present in a sample, how much will be in the sample in 10,000 years?

Solution Let $t = 0$ represent the present time and let *y* represent the amount (in grams) of carbon-14 in the sample. Using a base of $\frac{1}{2}$, you can model *y* by the equation

$$y = \left(\frac{1}{2}\right)^{t/5730}$$

Notice that when $t = 5730$, the amount is reduced to half of the original amount. When $t = 11,460$, the amount is reduced to a quarter of the original amount, and so on. To find the amount of carbon-14 after 10,000 years, substitute 10,000 for *t*.

$$y = \left(\frac{1}{2}\right)^{10,000/5730}$$

$$\approx 0.30 \text{ grams}$$

The graph of *y* is shown in Figure 5.24.

FIGURE 5.24
The half-life of carbon-14 is about 5730 years.

Logarithmic functions to bases other than e can be defined in much the same way that exponential functions to other bases were defined.

> ### Definition of Logarithmic Function to Base a
>
> If a is a positive real number $(a \neq 1)$ and x is any positive real number, then the **logarithmic function to the base a** is denoted by $\log_a x$ and is defined as
>
> $$\log_a x = \frac{1}{\ln a} \ln x.$$

Logarithmic functions to the base a have properties similar to those of the natural logarithmic function given in Theorem 5.2.

1. $\log_a 1 = 0$ Log of 1
2. $\log_a xy = \log_a x + \log_a y$ Log of a product
3. $\log_a \dfrac{x}{y} = \log_a x - \log_a y$ Log of a quotient
4. $\log_a x^n = n \log_a x$ Log of a power

From the definition of the exponential and logarithmic functions to the base a, it follows that $f(x) = a^x$ and $g(x) = \log_a x$ are inverse functions of each other.

> $y = a^x$ if and only if $x = \log_a y$.
>
> $a^{\log_a x} = x,$ for $x > 0$
>
> $\log_a a^x = x,$ for all x

The logarithmic function to the base 10 is called the **common logarithmic function.** Thus, for common logarithms, $y = 10^x$ if and only if $x = \log_{10} y$.

EXAMPLE 2   Bases Other than e

Solve for x in the following equations.

a. $3^x = \dfrac{1}{27}$ **b.** $\log_2 x = -4$

Solution

a. To solve this equation, you can apply the logarithmic function to the base 3 to both sides of the equation.

$$3^x = \frac{1}{27}$$

$$\log_3 3^x = \log_3 \frac{1}{27}$$

$$x = \log_3 3^{-3}$$

$$x = -3$$

b. To solve this equation, you can apply the exponential function to the base 2 to both sides of the equation.

$$\log_2 x = -4$$

$$2^{\log_2 x} = 2^{-4}$$

$$x = \frac{1}{2^4}$$

$$x = \frac{1}{16}$$

Differentiation and Integration

To differentiate exponential and logarithmic functions to other bases, you have three options: (1) use the definitions of a^x and $\log_a x$ and differentiate using the rules for the natural exponential and logarithmic functions, (2) use logarithmic differentiation, or (3) use the following differentiation rules for bases other than e.

THEOREM 5.13   Derivatives for Bases Other than e

Let a be a positive real number ($a \neq 1$) and let u be a differentiable function of x.

1. $\dfrac{d}{dx}[a^x] = (\ln a)a^x$ **2.** $\dfrac{d}{dx}[a^u] = (\ln a)a^u \dfrac{du}{dx}$

3. $\dfrac{d}{dx}[\log_a x] = \dfrac{1}{(\ln a)x}$ **4.** $\dfrac{d}{dx}[\log_a u] = \dfrac{1}{(\ln a)u} \dfrac{du}{dx}$

Proof   By definition, $a^x = e^{(\ln a)x}$. Therefore, you can prove the first rule by letting $u = (\ln a)x$ and differentiating with base e to obtain

$$\frac{d}{dx}[a^x] = \frac{d}{dx}[e^{(\ln a)x}] = e^u \frac{du}{dx} = e^{(\ln a)x}(\ln a) = (\ln a)a^x.$$

To prove the third rule, you can write

$$\frac{d}{dx}[\log_a x] = \frac{d}{dx}\left[\frac{1}{\ln a}\ln x\right] = \frac{1}{\ln a}\left(\frac{1}{x}\right) = \frac{1}{(\ln a)x}.$$

The second and fourth rules are simply the Chain Rule versions of the first and third rules.   ▬

REMARK   Note that these differentiation rules are similar to those for the natural exponential function and natural logarithmic function. In fact, they differ only by the constant factors $\ln a$ and $1/\ln a$, respectively. This points out one reason why, for calculus, e is the most convenient base.

EXAMPLE 3   Differentiating Functions to Other Bases

Find the derivatives of the following.

a. $y = 2^x$ **b.** $y = 2^{3x}$ **c.** $y = \log_{10} \cos x$

Solution

a. $y' = \dfrac{d}{dx}[2^x] = (\ln 2)2^x$

b. $y' = \dfrac{d}{dx}[2^{3x}] = (\ln 2)2^{3x}(3) = (3 \ln 2)2^{3x}$

Try writing 2^{3x} as 8^x and differentiating to see that you obtain the same result.

c. $y' = \dfrac{d}{dx}[\log_{10} \cos x] = \dfrac{-\sin x}{(\ln 10)\cos x} = -\dfrac{1}{\ln 10}\tan x$ ▬

Occasionally, an integrand involves an exponential function to a base other than e. When this occurs, there are two options: (1) convert to base e using the formula $a^x = e^{(\ln a)x}$ and then integrate, or (2) integrate directly, using the integration formula (which follows from Theorem 5.13).

$$\int a^x \, dx = \left(\frac{1}{\ln a}\right) a^x + C$$

EXAMPLE 4   Integrating an Exponential Function to Another Base

Evaluate $\int 2^x \, dx$.

Solution

$$\int 2^x \, dx = \frac{1}{\ln 2} 2^x + C$$

When we introduced the Power Rule, $D_x[x^n] = nx^{n-1}$, in Chapter 2, we required the exponent n to be a rational number. We now extend the rule to cover any real value of n. Try to prove this theorem using logarithmic differentiation.

THEOREM 5.14   The Power Rule for Real Exponents

Let n be any real number and let u be a differentiable function of x.

1. $\dfrac{d}{dx}[x^n] = nx^{n-1}$ **2.** $\dfrac{d}{dx}[u^n] = nu^{n-1}\dfrac{du}{dx}$

The next example compares the derivatives of four types of functions. Each function uses a different differentiation formula, depending on whether the base and exponent are constants or variables.

EXAMPLE 5   Comparing Variables and Constants

a. $\dfrac{d}{dx}[e^e] = 0$ Constant Rule

b. $\dfrac{d}{dx}[e^x] = e^x$ Exponential Rule

c. $\dfrac{d}{dx}[x^e] = ex^{e-1}$ Power Rule

d. $y = x^x$ Logarithmic differentiation

$\ln y = x \ln x$

$\dfrac{y'}{y} = x\left(\dfrac{1}{x}\right) + (\ln x)(1) = 1 + \ln x$

$y' = y(1 + \ln x) = x^x(1 + \ln x)$

Applications of Exponential Functions

TABLE 5.2

n	A
1	$1080.00
2	$1081.60
4	$1082.43
12	$1083.00
365	$1083.28

Suppose P dollars is deposited at an annual interest rate r (in decimal form). If interest accumulates in the account, what is the balance in the account at the end of 1 year? The answer depends on the number of times, n, the interest is compounded according to the formula

$$A = P\left(1 + \frac{r}{n}\right)^n.$$

For instance, the result for a deposit of $1000 at 8 percent interest compounded n times a year is shown in Table 5.2.

As n increases, the balance A approaches a limit. To develop this limit, we use the following theorem. To test the reasonableness of this theorem, try evaluating $[(x + 1)/x]^x$ for several values of x, as shown in Table 5.3. (A proof of the theorem is given in Appendix A.)

THEOREM 5.15   A Limit Involving *e*

$$\lim_{x \to \infty}\left(1 + \frac{1}{x}\right)^x = \lim_{x \to \infty}\left(\frac{x + 1}{x}\right)^x = e$$

TABLE 5.3

x	$\left(\dfrac{x+1}{x}\right)^x$
10	2.59374
100	2.70481
1,000	2.71692
10,000	2.71815
100,000	2.71827
1,000,000	2.71828

Now, let's take another look at the formula for the balance, A, in an account in which the interest is compounded n times per year. By taking the limit as n approaches infinity, you obtain the following.

$$A = \lim_{x \to \infty} P\left(1 + \frac{r}{n}\right)^n \qquad \text{Take limit as } n \to \infty$$

$$= P \lim_{n \to \infty}\left[\left(1 + \frac{1}{n/r}\right)^{n/r}\right]^r \qquad \text{Rewrite}$$

$$= P\left[\lim_{x \to \infty}\left(1 + \frac{1}{x}\right)^x\right]^r \qquad \text{Let } x = n/r. \text{ Then } x \to \infty \text{ as } n \to \infty.$$

$$= Pe^r \qquad \text{Apply Theorem 5.15}$$

This limit produces the balance after 1 year of **continuous compounding.** Thus, for a deposit of $1000 at 8 percent interest compounded continuously, the balance at the end of 1 year would be

$$A = 1000e^{0.08} \approx \$1083.29.$$

These results are summarized as follows.

Summary of Compound Interest Formulas

Let P = amount of deposit, t = number of years, A = balance after t years, r = annual interest rate (decimal form), and n = the number of compoundings per year.

1. Compounded n times per year: $A = P\left(1 + \dfrac{r}{n}\right)^{nt}$
2. Compounded continuously: $A = Pe^{rt}$

EXAMPLE 6   Comparing Continuous and Quarterly Compounding

A deposit of $2500 is made in an account that pays an annual interest rate of 10 percent. Find the balance in the account at the end of 5 years if the interest is compounded (a) quarterly, (b) monthly, and (c) continuously.

Solution

a. $A = P\left(1 + \dfrac{r}{n}\right)^{nt} = 2500\left(1 + \dfrac{0.1}{4}\right)^{4(5)}$ Quarterly compounding

$\qquad\qquad\qquad = 2500(1.025)^{20}$

$\qquad\qquad\qquad = \$4096.54$

b. $A = P\left(1 + \dfrac{r}{n}\right)^{nt} = 2500\left(1 + \dfrac{0.1}{12}\right)^{12(5)}$ Monthly compounding

$\qquad\qquad\qquad = 2500(1.00833)^{60}$

$\qquad\qquad\qquad = \$4113.27$

c. $A = Pe^{rt} = 2500[e^{0.1(5)}]$ Continuous compounding

$\qquad\quad = 2500e^{0.5}$

$\qquad\quad = \$4121.80$

Figure 5.25 shows how the balance increases over the 5-year period. Notice that the scale used in the figure does not graphically distinguish among the three types of exponential growth.

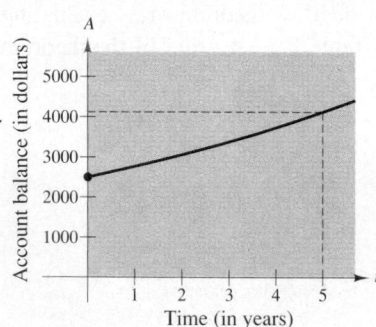

FIGURE 5.25
The balance in a savings account grows exponentially.

EXAMPLE 7   Bacterial Culture Growth

A bacterial culture is growing according to the *logistics growth function*

$$y = \frac{1.25}{1 + 0.25e^{-0.4t}}, \qquad 0 \le t$$

where y is the weight of the culture in grams and t is the time in hours. Find the weight of the culture after (a) 0 hours, (b) 1 hour, and (c) 10 hours. (d) What is the limit as t approaches infinity?

Solution

a. When $t = 0$, $y = \dfrac{1.25}{1 + 0.25e^{0}} = \dfrac{1.25}{1.25} = 1$ g.

b. When $t = 1$, $y = \dfrac{1.25}{1 + 0.25e^{-0.4}} \approx 1.071$ g.

c. When $t = 10$, $y = \dfrac{1.25}{1 + 0.25e^{-4}} \approx 1.244$ g.

d. Finally, taking the limit as t approaches infinity, you obtain

$$\lim_{t \to \infty} \frac{1.25}{1 + 0.25e^{-0.4t}} = \frac{1.25}{1 + 0} = 1.25 \text{ g}.$$

The graph of the function is shown in Figure 5.26.

FIGURE 5.26
The limit of the weight of the culture as $t \to \infty$ is 1.25 grams.

EXERCISES for Section 5.5

TECHNOLOGY
Laboratory Guide
Lab 5.5

In Exercises 1–4, evaluate the expression without using a calculator.

1. $\log_2(\frac{1}{8})$ **2.** $\log_{27} 9$

3. $\log_7 1$ **4.** $\log_a \frac{1}{a}$

In Exercises 5–8, write the exponential equation as a logarithmic equation or vice versa.

5. a. $2^3 = 8$ **6. a.** $27^{2/3} = 9$
   b. $3^{-1} = \frac{1}{3}$ **b.** $16^{3/4} = 8$

7. a. $\log_{10} 0.01 = -2$ **8. a.** $\log_3 \frac{1}{9} = -2$
   b. $\log_{0.5} 8 = -3$ **b.** $49^{1/2} = 7$

In Exercises 9–14, solve for x (or b).

9. a. $\log_{10} 1000 = x$ **10. a.** $\log_4 \frac{1}{64} = x$
   b. $\log_{10} 0.1 = x$ **b.** $\log_5 25 = x$

11. a. $\log_3 x = -1$ **12. a.** $\log_b 27 = 3$
    b. $\log_2 x = -4$ **b.** $\log_b 125 = 3$

13. a. $x^2 - x = \log_5 25$
    b. $3x + 5 = \log_2 64$

14. a. $\log_3 x + \log_3(x - 2) = 1$
    b. $\log_{10}(x + 3) - \log_{10} x = 1$

In Exercises 15–20, sketch the graph of the function.

15. $y = 3^x$ **16.** $y = 3^{x-1}$

17. $y = (\frac{1}{3})^x$ **18.** $y = 2^{x^2}$

19. $h(x) = 5^{x-2}$ **20.** $y = 3^{-|x|}$

In Exercises 21 and 22, illustrate that the functions are inverses of each other by sketching their graphs on the same coordinate system.

21. $f(x) = 4^x$ **22.** $f(x) = 3^x$
    $g(x) = \log_4 x$ $g(x) = \log_3 x$

23. The table of values was obtained from evaluating a function. Determine which of the statements may be true and which must be false.
   a. y is an exponential function of x.
   b. y is a logarithmic function of x.
   c. x is an exponential function of y.
   d. y is a linear function of x.

x	1	2	8
y	0	1	3

24. Consider the function $f(x) = \log_{10} x$.
   a. What is the domain of f?
   b. Find f^{-1}.
   c. If x is a real number between 1000 and 10,000, then determine the interval in which $f(x)$ will be found.

   d. Determine the interval in which x will be found if $f(x)$ is negative.
   e. If $f(x)$ is increased by one unit, then x must have been increased by what factor?
   f. If $f(x_1) = 3n$ and $f(x_2) = n$, then find the ratio of x_1 to x_2.

In Exercises 25–36, find the derivative of the function.

25. $f(x) = 4^x$ **26.** $g(x) = 2^{-x}$

27. $y = 5^{x-2}$ **28.** $y = x(7^{-3x})$

29. $g(t) = t^2 2^t$ **30.** $f(t) = \frac{3^{2t}}{t}$

31. $y = \log_3 x$ **32.** $y = \log_{10} 2x$

33. $f(x) = \log_2 \frac{x^2}{x - 1}$ **34.** $h(x) = \log_3 \frac{x\sqrt{x - 1}}{2}$

35. $y = \log_5 \sqrt{x^2 - 1}$ **36.** $y = \log_{10} \frac{x^2 - 1}{x}$

In Exercises 37–40, use logarithmic differentiation to find dy/dx.

37. $y = x^{2/x}$ **38.** $y = x^{x-1}$

39. $y = (x - 2)^{x+1}$ **40.** $y = (1 + x)^{1/x}$

41. Order the functions $f(x) = \log_2 x$, $g(x) = x^x$, $h(x) = x^2$, and $k(x) = 2^x$ from the one with the greatest rate of growth to the one with the smallest rate of growth for "large" values of x.

42. Given the exponential function $f(x) = a^x$, show that
    a. $f(u + v) = f(u) \cdot f(v)$ **b.** $f(2x) = [f(x)]^2$.

43. *Inflation*   If the annual rate of inflation averages 5% over the next 10 years, then the approximate cost C of goods or services during any year in that decade is

$$C(t) = P(1.05)^t$$

where t is the time in years and P is the present cost.
   a. If the price of an oil change for your car is presently $24.95, estimate the price 10 years from now.
   b. Find the rate of change of C with respect to t when $t = 1$ and $t = 8$.
   c. Verify that the rate of change of C is proportional to C. What is the constant of proportionality?

44. *Depreciation*   After t years, the value of a car purchased for $20,000 is

$$V(t) = 20,000\left(\frac{3}{4}\right)^t.$$

   a. Sketch a graph of the function and determine the value of the car 2 years after it was purchased.
   b. Find the rate of change of V with respect to t when $t = 1$ and $t = 4$.
   c. Sketch the graph of $V'(t)$ and determine its horizontal asymptote. Interpret its meaning relative to this depreciation problem.

In Exercises 45–48, complete the table to determine the balance A for P dollars invested at rate r for t years and compounded n times per year.

n	1	2	4	12	365	Continuous compounding
A						

45. $P = \$1000$
$r = 7\frac{1}{2}\%$
$t = 10$ years

46. $P = \$2500$
$r = 12\%$
$t = 20$ years

47. $P = \$1000$
$r = 10\%$
$t = 30$ years

48. $P = \$2500$
$r = 10\%$
$t = 40$ years

In Exercises 49–52, complete the table to determine the amount of money P (present value) that should be invested at rate r to produce a final balance of $\$100,000$ in t years.

t	1	10	20	30	40	50
P						

49. $r = 9\%$
Compounded continuously

50. $r = 12\%$
Compounded continuously

51. $r = 10\%$
Compounded monthly

52. $r = 7\%$
Compounded daily

53. *Compound Interest*   Assume that you can earn 9% on an investment, compounded daily. Which of the following options would yield the greatest balance in 8 years?
a. $20,000 now
b. $30,000 in 8 years
c. $8000 now and $20,000 in 4 years
d. $9000 now, $9000 in 4 years, and $9000 in 8 years

54. *Trust Fund*   On the day of your grandchild's birth, you deposited $25,000 in a trust fund that pays 8.75% interest, compounded continuously. Determine the balance in this account on your grandchild's 25th birthday.

55. *Compound Interest*   Consider a deposit of $100 placed in an account for 20 years at $r\%$ interest compounded continuously. Use a graphing utility to compare the balances in the account for the following interest rates.
a. $r = 6\%$ **b.** $r = 9\%$ **c.** $r = 12\%$

56. *Timber Yield*   The yield V (in millions of cubic feet per acre) for a stand of timber at age t is

$$V = 6.7e^{(-48.1)/t}$$

where t is measured in years.
a. Find the limiting volume of wood per acre as t approaches infinity.
b. Find the rate at which the yield is changing when $t = 20$ and $t = 60$ years.

57. *Population Growth*   A lake is stocked with 500 fish, and their population increases according to the logistics curve

$$p(t) = \frac{10,000}{1 + 19e^{-t/5}}$$

where t is measured in months. At what rate is the fish population changing at the end of 1 month and at the end of 10 months? After how many months is the population increasing most rapidly? (See figure.)

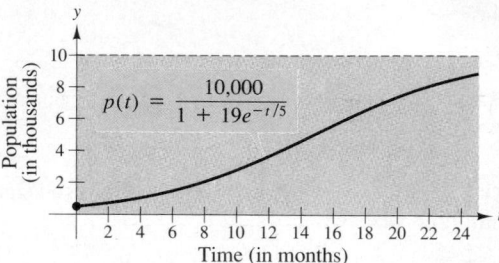

FIGURE FOR 57

58. *Learning Theory*   In a group project in learning theory, a mathematical model for the proportion P of correct responses after n trials was found to be

$$P = \frac{0.83}{1 + e^{-0.2n}}.$$

a. Find the limiting proportion of correct responses as n approaches infinity.
b. Find the rate at which P is changing after $n = 3$ and $n = 10$ trials.

C 59. *Forest Defoliation*   To estimate the amount of defoliation caused by the gypsy moth during a year, a forester counts the number of egg masses on $\frac{1}{40}$ of an acre the preceding fall. The percentage of defoliation y is approximated by

$$y = \frac{300}{3 + 17e^{-1.57x}}$$

where x is the number of egg masses in thousands. (*Source:* National Forest Service)
a. Use a graphing utility to graph the function.
b. Estimate the percentage of defoliation if 2000 egg masses are counted.
c. Estimate the number of egg masses that existed if you observe that approximately $\frac{2}{3}$ of a forest is defoliated.
d. Use calculus to estimate the value of x for which y is increasing most rapidly.

In Exercises 60–65, evaluate the integral.

60. $\displaystyle\int 3^x \, dx$

61. $\displaystyle\int 4^{-x} \, dx$

62. $\displaystyle\int_{-1}^{2} 2^x \, dx$

63. $\displaystyle\int_{-2}^{0} (3^3 - 5^2) \, dx$

64. $\displaystyle\int x(5^{-x^2}) \, dx$

65. $\displaystyle\int (3 - x)7^{(3-x)^2} \, dx$

66. *Area*   Find the area of the region bounded by the graphs of $y = 3^x$, $y = 0$, $x = 0$, and $x = 3$.

67. *Average Balance*   A deposit of $2500 is made in a savings account earning 12% interest compounded continuously. Find the average balance in the account during the first 2 years.

68. *Continuous Cash Flow* The present value P of a continuous cash flow of \$2000 per year earning 10% interest compounded continuously over 10 years is

$$\int_0^{10} 2000e^{-0.10t}\, dt.$$

Find P.

69. Complete the table to demonstrate that e can also be defined as

$$\lim_{x \to 0^+} (1 + x)^{1/x}.$$

x	1	10^{-1}	10^{-2}	10^{-4}	10^{-6}
$(1 + x)^{1/x}$					

In Exercises 70 and 71, find an exponential function that fits the experimental data collected over time t. (*Hint:* Begin by dividing the measurement for each time period by the measurement for the previous time period.)

70.

t	0	1	2	3	4
y	1200.00	720.00	432.00	259.20	155.52

71.

t	0	1	2	3	4
y	600.00	630.00	661.50	694.58	729.30

True or False In Exercises 72–77, determine whether the statement is true or false. If it is false, explain why or give an example that shows it is false.

72. $e = \frac{271{,}801}{99{,}990}$

73. If $f(x) = \ln x$, then $f(e^{n+1}) - f(e^n) = 1$ for any value of n.

74. The functions $f(x) = 2 + e^x$ and $g(x) = \ln(x - 2)$ are inverses of each other.

75. The exponential function $y = Ce^x$ is a solution of the differential equation $d^n y/dx^n = y$, $n = 1, 2, 3, \ldots$.

76. The graphs of $f(x) = e^x$ and $g(x) = e^{-x}$ meet at right angles.

77. If $f(x) = g(x)e^x$, then the only zeros of f are the zeros of g.

78. Solve the logistics differential equation

$$\frac{dy}{dt} = \frac{8}{25} y \left(\frac{5}{4} - y \right), \qquad y(0) = 1$$

and obtain the logistics growth function of Example 7, Section 5.5. $\left(\textit{Hint: } \dfrac{1}{y\left(\frac{5}{4} - y\right)} = \dfrac{4}{5}\left[\dfrac{1}{y} + \dfrac{1}{\frac{5}{4} - y} \right] \right)$

5.6 Differential Equations • Growth and Decay Models

Differential Equations

Up to now in the text, you have learned to solve only two types of differential equations—those of the forms

$$y' = f(x) \quad \text{and} \quad y'' = f(x).$$

In this section, you will learn how to solve a more general type. The strategy is to rewrite the equation so that each variable occurs on only one side of the equation. This strategy is called **separation of variables.**

EXAMPLE 1 Using Separation of Variables

Solve the differential equation

$$y' = \frac{2x}{y}.$$

Solution

$$y' = \frac{2x}{y} \qquad \text{Given equation}$$

$$yy' = 2x \qquad \text{Multiply both sides by } y$$

$$\int yy'\,dx = \int 2x\,dx \qquad \text{Integrate with respect to } x$$

$$\int y\,dy = \int 2x\,dx \qquad dy = y'\,dx$$

$$\frac{1}{2}y^2 = x^2 + C_1 \qquad \text{Apply Power Rule}$$

$$y^2 - 2x^2 = C \qquad \text{Rewrite, letting } C = 2C_1$$

Thus, the general solution is given by $y^2 - 2x^2 = C$. ∎

In practice, most people prefer to use Leibniz notation and differentials when applying separation of variables. With this notation, the solution of Example 1 is as follows.

$$\frac{dy}{dx} = \frac{2x}{y}$$

$$y\,dy = 2x\,dx$$

$$\int y\,dy = \int 2x\,dx$$

$$\frac{1}{2}y^2 = x^2 + C_1$$

$$y^2 - 2x^2 = C$$

Growth and Decay Models

In many applications, the rate of change of a variable y is proportional to the value of y. If y is a function of time t, then the proportionality can be written as follows.

$$\frac{dy}{dt} = ky$$

The general solution of this differential equation is given in the following theorem.

THEOREM 5.16   Exponential Growth and Decay Model

If y is a differentiable function of t such that $y > 0$ and $y' = ky$, for some constant k, then

$$y = Ce^{kt}.$$

C is the **initial value** of y, and k is the **proportionality constant.** Exponential growth occurs when $k > 0$, and **exponential decay** when $k < 0$.

Proof

$$y' = ky \qquad \text{Given equation}$$

$$\frac{y'}{y} = k \qquad \text{Divide by } y \text{ to separate variables}$$

$$\int \frac{y'}{y}\, dt = \int k\, dt \qquad \text{Integrate with respect to } t$$

$$\int \frac{1}{y}\, dy = \int k\, dt \qquad dy = y'\, dt$$

$$\ln y = kt + C_1 \qquad \text{Find antiderivative of each side}$$

$$y = e^{C_1}e^{kt} \qquad \text{Solve for } y$$

$$y = Ce^{kt} \qquad \text{Let } C = e^{C_1}$$

EXAMPLE 2   Using an Exponential Growth Model

The rate of change of y is proportional to y. When $t = 0$, $y = 2$. When $t = 2$, $y = 4$. What is the value of y when $t = 3$?

Solution   Because $y' = ky$, you know that y and t are related by the equation $y = Ce^{kt}$. You can find the values of the constants C and k by applying the initial conditions.

$$2 = Ce^0 \to C = 2 \qquad \text{When } t = 0,\ y = 2$$

$$4 = 2e^{2k} \to k = \frac{1}{2}\ln 2 \approx 0.3466 \qquad \text{When } t = 2,\ y = 4$$

Therefore, the model is $y \approx 2e^{0.3466t}$, and when $t = 3$, the value of y is $2e^{0.3466(3)} \approx 5.657$ (see Figure 5.27).

FIGURE 5.27
If the rate of change of y is proportional to y, then y follows an exponential model.

Radioactive decay is measured in terms of *half-life*—the number of years required for half of the atoms in a sample of radioactive material to decay. The half-lives of some common radioactive isotopes are as follows.

Uranium (U^{238})	4,510,000,000 years
Plutonium (Pu^{239})	24,360 years
Carbon (C^{14})	5730 years
Radium (Ra^{226})	1620 years
Einsteinium (Es^{254})	276 days
Nobelium (No^{257})	23 seconds

EXAMPLE 3   Radioactive Decay

Suppose that 10 grams of the plutonium isotope Pu-239 were released in the Chernobyl nuclear accident. How long will it take for the 10 grams to decay to 1 gram?

Solution   Let y represent the mass (in grams) of the plutonium. Because the rate of decay is proportional to y, you know that

$$y = Ce^{kt}$$

where t is the time in years. To find the values of the constants C and k, apply the initial conditions. Using the fact that $y = 10$ when $t = 0$, you can write

$$10 = Ce^0$$

which implies that $C = 10$. Next, using the fact that $y = 5$ when $t = 24,360$, you can write

$$5 = 10e^{24,360k}$$

$$\frac{1}{2} = e^{24,360k}$$

$$\frac{1}{24,360}\ln\frac{1}{2} = k$$

Therefore, the model is

$$y \approx 10e^{-0.000028454t}.$$

To find the time it would take for the 10 grams to decay to 1 gram, you can solve the equation

$$1 = 10e^{-0.000028454t}.$$

The solution is approximately 80,922 years!   ■

REMARK   The exponential decay model in Example 3 could also be written as $y = 10(\frac{1}{2})^{t/24,360}$. This model is much easier to derive, but for some applications, it is not as convenient to use.

From Example 3, notice that in an exponential growth or decay problem, it is easy to solve for C when you are given the value of y at $t = 0$. The next example demonstrates a procedure for solving for C and k when you do not know the value of y at $t = 0$.

EXAMPLE 4   Population Growth

Suppose an experimental population of fruit flies increases according to the law of exponential growth. There were 100 flies after the second day of the experiment and 300 flies after the fourth day. Approximately how many flies were in the original population?

Solution   Let $y = Ce^{kt}$ be the number of flies at time t, where t is measured in days. Because $y = 100$ when $t = 2$ and $y = 300$ when $t = 4$, you can write

$$100 = Ce^{2k} \quad \text{and} \quad 300 = Ce^{4k}.$$

From the first equation, you know that $C = 100e^{-2k}$. Substituting this value into the second equation produces the following.

$$300 = 100e^{-2k}e^{4k}$$
$$300 = 100e^{2k}$$
$$\ln 3 = 2k$$
$$\frac{1}{2}\ln 3 = k$$
$$0.5493 \approx k$$

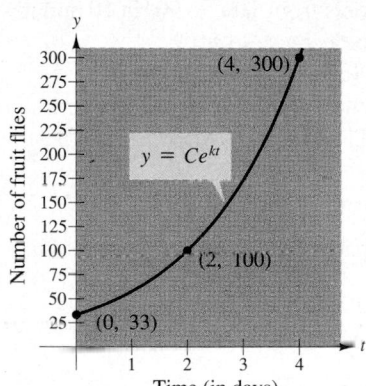

Therefore, the exponential growth model is $y = Ce^{0.5493t}$. To solve for C, reapply the condition $y = 100$ when $t = 2$ and obtain

$$100 = Ce^{0.5493(2)} \rightarrow C = 100e^{-1.0986} \approx 33,$$

Thus, the original population (when $t = 0$) consisted of approximately $y = C = 33$ flies, as indicated in Figure 5.28.

FIGURE 5.28

EXAMPLE 5   Declining Sales

Four months after it stopped advertising, a manufacturing company noticed that its sales had dropped from 100,000 units per month to 80,000 units per month. If the sales follow an exponential pattern of decline, what will they be after another 2 months?

Solution   Use the exponential decay model $y = Ce^{kt}$, where t is measured in months, as shown in Figure 5.29. From the initial condition ($t = 0$), you know that $C = 100,000$. Moreover, because $y = 80,000$ when $t = 4$, you have

$$80,000 = 100,000e^{4k}$$
$$0.8 = e^{4k}$$
$$\ln(0.8) = 4k$$
$$-0.0558 \approx k.$$

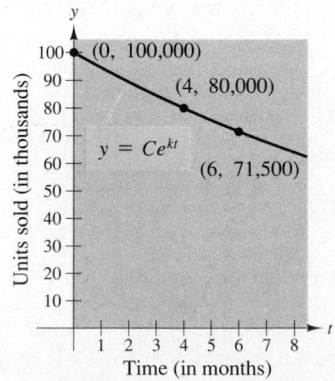

FIGURE 5.29

Thus, after 2 more months ($t = 6$), you can expect the monthly sales rate to be

$$y \approx 100,000e^{-0.0558(6)} \approx 71,500 \text{ units.}$$

In Examples 2 to 5, we did not actually have to solve the differential equation

$$y' = ky.$$

(We did this once in the proof of Theorem 5.16.) In the next example, we demonstrate a problem whose solution involves the separation of variables technique. The example concerns **Newton's Law of Cooling,** which states that the rate of change in the temperature of an object is proportional to the difference between the object's temperature and the temperature of the surrounding medium.

EXAMPLE 6 Newton's Law of Cooling

Let y represent the temperature (in degrees Fahrenheit) of an object in a room whose temperature is kept at a constant 60°. If the object cools from 100° to 90° in 10 minutes, how much longer will it take for its temperature to decrease to 80°?

Solution   From Newton's Law of Cooling, you know that the rate of change in y is proportional to the difference between y and 60. This can be written as

$$y' = k(y - 60), \qquad 80 \le y \le 100.$$

To solve this differential equation, you can use separation of variables, as follows.

$$\frac{dy}{dt} = k(y - 60) \qquad \text{Differential equation}$$

$$\left(\frac{1}{y - 60}\right) dy = k\, dt \qquad \text{Separate variables}$$

$$\int \frac{1}{y - 60}\, dy = \int k\, dt \qquad \text{Integrate both sides}$$

$$\ln|y - 60| = kt + C \qquad \text{Find antiderivative of each side}$$

Because $y > 60$, $|y - 60| = y - 60$, and you can omit the absolute value signs. Using $y = 100$ when $t = 0$ yields $C = \ln 40$, which implies that

$$kt = \ln(y - 60) - \ln 40$$

$$= \ln\left(\frac{y - 60}{40}\right).$$

Because $y = 90$ when $t = 10$, you know that $k = \frac{1}{10}\ln\frac{3}{4}$, and thus

$$t = \left(\frac{10}{\ln(3/4)}\right)\ln\left(\frac{y - 60}{40}\right).$$

Finally, when $y = 80$

$$t = \frac{10\ln(1/2)}{\ln(3/4)} \approx 24.09 \text{ min.}$$

Therefore, it will require about 14.09 *more* minutes for the object to cool to a temperature of 80° (see Figure 5.30).

FIGURE 5.30

TECHNOLOGY
Laboratory Guide
Lab 5.6

EXERCISES for Section 5.6

In Exercises 1–6, solve the differential equation.

1. $y' = \dfrac{5x}{y}$

2. $y' = \dfrac{\sqrt{x}}{2y}$

3. $y' = \sqrt{xy}$

4. $y' = x(1 + y)$

5. $(1 + x^2)y' - 2xy = 0$

6. $xy + y' = 100x$

In Exercises 7–10, write and solve a differential equation that models the verbal statement.

7. The rate of change of Q with respect to t is inversely proportional to the square of t.

8. The rate of change of P with respect to t is proportional to $10 - t$.

9. The rate of change of N with respect to s is proportional to $250 - s$.

10. The rate of change of y with respect to x varies jointly as x and $L - y$.

In Exercises 11–14, find the function $y = f(t)$ passing through the point $(0, 10)$ that has the given first derivative.

11. $\dfrac{dy}{dt} = \dfrac{1}{2}t$

12. $\dfrac{dy}{dt} = -\dfrac{3}{4}\sqrt{t}$

13. $\dfrac{dy}{dt} = -\dfrac{1}{2}y$

14. $\dfrac{dy}{dt} = \dfrac{3}{4}y$

In Exercises 15–18, find the exponential function $y = Ce^{kt}$ that passes through the two given points.

15.

16.

17.

18.

Radioactive Decay In Exercises 19–24, complete the table for the radioactive isotope.

Isotope	Half-life (in yrs)	Initial quantity	Amount after 1000 yrs	Amount after 10,000 yrs
19. Ra^{226}	1,620	10 g		
20. Ra^{226}	1,620		1.5 g	
21. C^{14}	5,730			2.0 g
22. C^{14}	5,730	3.0 g		
23. Pu^{239}	24,360		2.1 g	
24. Pu^{239}	24,360			0.4 g

25. *Radioactive Decay* Radioactive radium has a half-life of approximately 1620 years. What percentage of a given amount remains after 100 years?

26. *Carbon Dating* C^{14} dating assumes that the carbon dioxide on earth today has the same radioactive content as it did centuries ago. If this is true, then the amount of C^{14} absorbed by a tree that grew several centuries ago should be the same as the amount of C^{14} absorbed by a tree growing today. A piece of ancient charcoal contains only 15% as much of the radioactive carbon as a piece of modern charcoal. How long ago was the tree burned to make the ancient charcoal? (The half-life of C^{14} is 5730 years.)

Compound Interest In Exercises 27–32, complete the table for a savings account in which interest is compounded continuously.

	Initial investment	Annual rate	Time to double	Amount after 10 yrs
27.	$1,000	12%		
28.	$20,000	$10\frac{1}{2}\%$		
29.	$750		$7\frac{3}{4}$ years	
30.	$10,000		5 years	
31.	$500			$1292.85
32.	$2,000			$5436.56

Compound Interest In Exercises 33 and 34, determine the principal P that must be invested at rate r, compounded monthly, so that $500,000 will be available for retirement in t years.

33. $r = 7\frac{1}{2}\%$, $t = 20$

34. $r = 12\%$, $t = 40$

Compound Interest In Exercises 35 and 36, determine the time necessary for $1000 to double if it is invested at interest rate r compounded (a) annually, (b) monthly, (c) daily, and (d) continuously.

35. $r = 11\%$

36. $r = 10\frac{1}{2}\%$

37. *Population Growth*   Dhaka, Bangladesh had a population of 4.22 million in 1990, and its projected population for the year 2000 is 6.49 million. (*Source:* U.S. Bureau of the Census) Find the exponential growth model $y = Ce^{kt}$ for the population growth of Dhaka by letting $t = 0$ correspond to 1990. Use the model to predict the population of the city in 2010.

38. *Population Growth*   Houston, Texas had a population of 2.30 million in 1990, and its projected population for the year 2000 is 2.65 million. (*Source:* U.S. Bureau of the Census) Find the exponential growth model $y = Ce^{kt}$ for the population growth of Houston by letting $t = 0$ correspond to 1990. Use the model to predict the population of the city in 2010.

39. *Essay*   From Exercises 37 and 38, you can see that the populations of Dhaka and Houston are growing at different rates. What constant in the equation $y = Ce^{kt}$ is affected by the difference in growth rates? Discuss the relationship between the different growth rates and the magnitude of the constant.

40. *Population Growth*   The number of a certain type of bacteria increases continuously at a rate proportional to the number present. If there are 100 present at a certain time and 300 present 5 hours later, how many will there be 10 hours after the initial time? How long does it take the number of bacteria to double?

41. *Atmospheric Pressure*   Atmospheric pressure P (measured in millimeters of mercury) decreases exponentially with increasing altitude x (measured in meters). The pressure is 760 millimeters of mercury at sea level ($x = 0$) and 672.71 millimeters of mercury at an altitude of 1000 meters. Find the pressure at an altitude of 3000 meters.

42. *Revenue*   Because of a slump in the economy, a company finds that its annual revenues have dropped from \$742,000 in 1992 to \$632,000 in 1994. If the revenue follows an exponential pattern of decline, what is the expected revenue for 1995? (Let $t = 0$ represent 1992.)

43. *Learning Curve*   The management at a certain factory has found that a worker can produce at most 30 units in a day. The learning curve for the number of units N produced per day after a new employee has worked t days is

$$N = 30(1 - e^{kt}).$$

After 20 days on the job, a particular worker produced 19 units.
  a. Find the learning curve for this worker.
  b. How many days should pass before this worker is producing 25 units per day?

44. *Learning Curve*   If in Exercise 43, the management requires a new employee to produce at least 20 units per day after 30 days on the job, find
  a. the learning curve that describes this minimum requirement.
  b. the number of days before a minimal achiever is producing 25 units per day.

45. *Sales*   The sales S (in thousands of units) of a new product after it has been on the market for t years is

$$S = Ce^{k/t}.$$

  a. Find S as a function of t if 5000 units have been sold after 1 year and the saturation point for the market is 30,000 (that is, $\lim_{t \to \infty} S = 30$).
  b. How many units have been sold after 5 years?
  c. Use a graphing utility to graph this sales function.

46. *Sales*   The sales S (in thousands of units) of a new product after it has been on the market for t years is

$$S = 30(1 - e^{kt}).$$

  a. Find S as a function of t if 5000 units have been sold after 1 year.
  b. How many units will saturate this market?
  c. How many units will have been sold after 5 years?
  d. Use a graphing utility to graph this sales function.

47. *Forestry*   The value of a tract of timber is

$$V(t) = 100,000e^{0.8\sqrt{t}}$$

where t is the time in years with $t = 0$ corresponding to 1994. If money earns interest continuously at 10%, then the present value of the timber at any time t is

$$A(t) = V(t)e^{-0.10t}.$$

Find the year in which the timber should be harvested to maximize the present value function.

48. *Federal Debt*   The table gives the net receipts and the amounts required to service the national debt of the United States for the years 1986 to 1990. The monetary amounts are listed in millions of dollars. (*Source:* Department of the Treasury)

Year	1986	1987	1988	1989	1990
Receipts	769,091	854,143	908,953	990,691	1,031,46
Interest	135,284	138,519	151,711	169,100	183,79

Mathematical models for the receipts $R(t)$ and the amount required to service the debt $I(t)$ are

$$R(t) = 502,826e^{0.07t} \quad \text{and} \quad I(t) = 80,983e^{0.08t}$$

where t is the time in years with $t = 0$ corresponding to 1980.
  a. Plot the points corresponding to the data and use a graphing utility to obtain the graphs of the two models.
  b. Based on the 5 years of data, what are the continuous rates of growth of the two models? Write a short paragraph giving the implications for the future if these rates are not changed.
  c. Find a function $P(t)$ that approximates the percentage of the receipts that is required to service the national debt. Use a graphing utility to graph this function.

49. *Sound Intensity*   The level of sound β, in decibels, with an intensity of I is

$$\beta(I) = 10 \log_{10} \frac{I}{I_0}$$

where I_0 is an intensity of 10^{-16} watts per square centimeter, corresponding roughly to the faintest sound that can be heard. Determine $\beta(I)$ for the following.

a. $I = 10^{-14}$ watts per square centimeter (whisper)
b. $I = 10^{-9}$ watts per square centimeter (busy street corner)
c. $I = 10^{-6.5}$ watts per square centimeter (air hammer)
d. $I = 10^{-4}$ watts per square centimeter (threshold of pain)

50. *Noise Level*   Due to the installation of noise suppression materials, the noise level in an auditorium was reduced from 93 to 80 decibels. Use the function in Exercise 49 to find the percentage decrease in the intensity level of the noise because of the installation of these materials.

51. *Earthquake Intensity*   On the Richter scale, the magnitude R of an earthquake of intensity I is

$$R = \frac{\ln I - \ln I_0}{\ln 10}$$

where I_0 is the minimum intensity used for comparison. Assume $I_0 = 1$.

a. Find the intensity of the 1906 San Francisco earthquake when $R = 8.3$.
b. Find the factor by which the intensity is increased if the Richter scale measurement is doubled.
c. Find dR/dI.

52. *Newton's Law of Cooling*   When an object is removed from a furnace and placed in an environment with a constant temperature of $90°$ F, its core temperature is $1500°$ F. One hour after it is removed, the core temperature is $1120°$ F. Find the core temperature 5 hours after the unit is removed from the furnace.

53. *Newton's Law of Cooling*   A thermometer is taken from a room at $72°$ F to the outdoors where the temperature is $20°$ F. Determine what the reading on the thermometer will be after 5 minutes, if the reading drops to $48°$ F after 1 minute.

54. *Newton's Law of Cooling*   A thermometer is taken outdoors from a room where the temperature is $20°$ C. After $\frac{1}{2}$ minute and 1 minute, the thermometer reads $12°$ C and $5.5°$ C, respectively. Find the outdoor temperature.

SECTION | **Inverse Trigonometric Functions and Differentiation**

5.7 Inverse Trigonometric Functions • Derivatives of Inverse Trigonometric Functions • Review of Basic Differentiation Rules

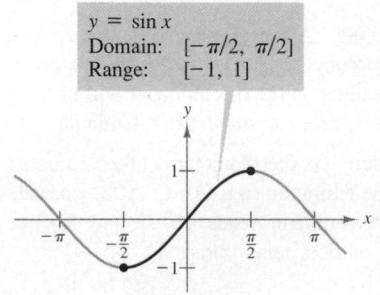

$y = \sin x$
Domain: $[-\pi/2, \pi/2]$
Range: $[-1, 1]$

FIGURE 5.31
The sine function is one-to-one on $[-\pi/2, \pi/2]$.

Inverse Trigonometric Functions

This section begins with a rather startling statement: *None of the six basic trigonometric functions has an inverse.* This statement is true because all six trigonometric functions are periodic, and hence not one-to-one. In this section you will examine these six functions to see whether their domains can be redefined in such a way that they will have inverses on the *restricted domains*.

In Example 4 of Section 5.3, you saw that the sine function is increasing (and therefore is one-to-one) on the interval $[-\pi/2, \pi/2]$ (see Figure 5.31). On this interval you can define the inverse of the *restricted* sine function to be

$$y = \arcsin x \quad \text{if and only if} \quad \sin y = x$$

where $-1 \le x \le 1$ and $-\pi/2 \le \arcsin x \le \pi/2$.

Under suitable restrictions, each of the six trigonometric functions is one-to-one and so possesses an inverse, as indicated in the following definition. (The term "iff" is used to represent the phrase "if and only if.")

Definition of Inverse Trigonometric Functions

Function	Domain	Range
$y = \mathbf{arcsin}\,x$ iff $\sin y = x$	$-1 \le x \le 1$	$-\dfrac{\pi}{2} \le y \le \dfrac{\pi}{2}$
$y = \mathbf{arccos}\,x$ iff $\cos y = x$	$-1 \le x \le 1$	$0 \le y \le \pi$
$y = \mathbf{arctan}\,x$ iff $\tan y = x$	$-\infty < x < \infty$	$-\dfrac{\pi}{2} < y < \dfrac{\pi}{2}$
$y = \mathbf{arccot}\,x$ iff $\cot y = x$	$-\infty < x < \infty$	$0 < y < \pi$
$y = \mathbf{arcsec}\,x$ iff $\sec y = x$	$\lvert x \rvert \ge 1$	$0 \le y \le \pi, y \ne \dfrac{\pi}{2}$
$y = \mathbf{arccsc}\,x$ iff $\csc y = x$	$\lvert x \rvert \ge 1$	$-\dfrac{\pi}{2} \le y \le \dfrac{\pi}{2}, y \ne 0$

REMARK The term $\arcsin x$ is read as the "arcsine of x" or sometimes the "angle whose sine is x." An alternative notation for the inverse sine function is $\sin^{-1} x$.

The graphs of the six inverse functions are shown in Figure 5.32.

Domain: $[-1, 1]$
Range: $[-\pi/2, \pi/2]$

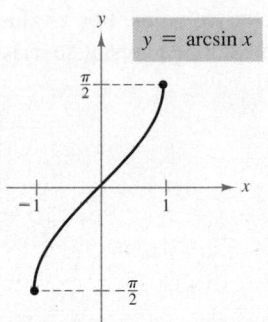

Domain: $(-\infty, -1] \cup [1, \infty)$
Range: $[-\pi/2, 0) \cup (0, \pi/2]$

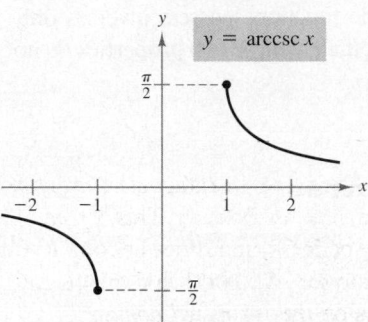

Domain: $(-\infty, \infty)$
Range: $(-\pi/2, \pi/2)$

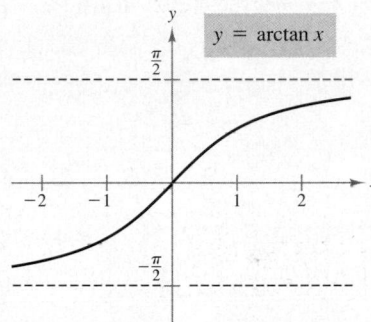

Domain: $[-1, 1]$
Range: $[0, \pi]$

Domain: $(-\infty, -1] \cup [1, \infty)$
Range: $[0, \pi/2) \cup (\pi/2, \pi]$

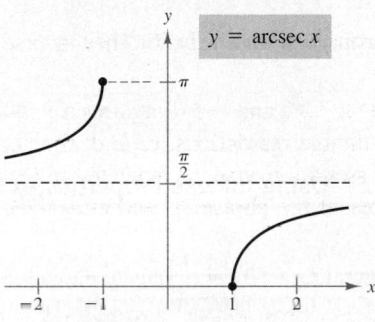

Domain: $(-\infty, \infty)$
Range: $(0, \pi)$

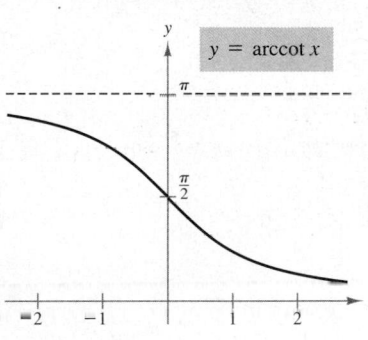

FIGURE 5.32

REMARK   When evaluating inverse trigonometric functions, remember that they denote *angles in radian measure*.

EXAMPLE 1   Evaluating Inverse Trigonometric Functions

Evaluate the following.

a. $\arcsin(-\frac{1}{2})$ **b.** $\arccos 0$ **c.** $\arctan \sqrt{3}$ **d.** $\arcsin(0.3)$

Solution

a. By definition, $y = \arcsin(-\frac{1}{2})$ implies that $\sin y = -\frac{1}{2}$. In the interval $[-\pi/2, \pi/2]$, the correct value of y is $-\pi/6$.

$$\arcsin\left(-\frac{1}{2}\right) = -\frac{\pi}{6}$$

b. By definition, $y = \arccos 0$ implies that $\cos y = 0$. In the interval $[0, \pi]$, you have $y = \pi/2$.

$$\arccos 0 = \frac{\pi}{2}$$

c. By definition, $y = \arctan \sqrt{3}$ implies that $\tan y = \sqrt{3}$. In the interval $(-\pi/2, \pi/2)$, you have $y = \pi/3$.

$$\arctan \sqrt{3} = \frac{\pi}{3}$$

d. Using a calculator set in *radian mode* produces $\arcsin(0.3) \approx 0.3047$.

Inverse functions possess the properties

$$f(f^{-1}(x)) = x \quad \text{and} \quad f^{-1}(f(x)) = x.$$

When applying these properties to inverse trigonometric functions, remember that the trigonometric functions possess inverses only in restricted domains. For x-values outside these domains, these two properties do not hold. For example, $\arcsin(\sin \pi)$ is equal to 0, not π.

Inverse Properties

If $-1 \le x \le 1$ and $-\pi/2 \le y \le \pi/2$, then

$$\sin(\arcsin x) = x \quad \text{and} \quad \arcsin(\sin y) = y.$$

If $-\pi/2 < y < \pi/2$, then

$$\tan(\arctan x) = x \quad \text{and} \quad \arctan(\tan y) = y.$$

If $|x| \ge 1$ and $0 \le y < \pi/2$ or $\pi/2 < y \le \pi$, then

$$\sec(\arcsec x) = x \quad \text{and} \quad \arcsec(\sec y) = y.$$

Similar properties hold for the other inverse trigonometric functions.

EXAMPLE 2   Using Inverse Properties to Solve an Equation

$$\arctan(2x - 3) = \frac{\pi}{4} \qquad \text{Given equation}$$

$$\tan[\arctan(2x - 3)] = \tan\frac{\pi}{4} \qquad \text{Take tangent of both sides}$$

$$2x - 3 = 1 \qquad \text{Apply inverse property}$$

$$x = 2 \qquad \text{Solve for } x$$

Some problems in calculus require you to evaluate expressions such as $\sec(\arctan x)$, as illustrated in Example 3.

EXAMPLE 3   Using Right Triangles

a. Given $y = \arcsin x$, where $0 < y < \pi/2$, find $\cos y$.
b. Given $y = \arcsec\left(\sqrt{5}/2\right)$, find $\tan y$.

Solution

a. Because $y = \arcsin x$, you know that $\sin y = x$. This relationship between x and y can be represented by a right triangle, as shown in Figure 5.33.

$$\cos y = \cos(\arcsin x) = \frac{\text{adj.}}{\text{hyp.}} = \sqrt{1 - x^2}$$

(This result is also valid for $-\pi/2 < y < 0$.)
b. Use the right triangle shown in Figure 5.34.

$$\tan y = \tan\left[\arcsec\left(\frac{\sqrt{5}}{2}\right)\right] = \frac{\text{opp.}}{\text{adj.}} = \frac{1}{2}$$

FIGURE 5.33
$y = \arcsin x$

FIGURE 5.34
$y = \arcsec \dfrac{\sqrt{5}}{2}$

Derivatives of Inverse Trigonometric Functions

In Section 5.1 you saw that the derivative of the *transcendental* function $f(x) = \ln x$ is the *algebraic* function $f'(x) = 1/x$. You will now see that the derivatives of the inverse trigonometric functions also are algebraic (even though the inverse trigonometric functions are themselves transcendental).

The following theorem lists the derivatives of the six inverse trigonometric functions. Note that the derivatives of $\arccos u$, $\text{arccot}\, u$, and $\text{arccsc}\, u$ are the *negatives* of the derivatives of $\arcsin u$, $\arctan u$, and $\text{arcsec}\, u$, respectively.

THEOREM 5.17   Derivatives of Inverse Trigonometric Functions

Let u be a differentiable function of x.

$$\frac{d}{dx}[\arcsin u] = \frac{u'}{\sqrt{1 - u^2}} \qquad \frac{d}{dx}[\arccos u] = \frac{-u'}{\sqrt{1 - u^2}}$$

$$\frac{d}{dx}[\arctan u] = \frac{u'}{1 + u^2} \qquad \frac{d}{dx}[\text{arccot}\, u] = \frac{-u'}{1 + u^2}$$

$$\frac{d}{dx}[\text{arcsec}\, u] = \frac{u'}{|u|\sqrt{u^2 - 1}} \qquad \frac{d}{dx}[\text{arccsc}\, u] = \frac{-u'}{|u|\sqrt{u^2 - 1}}$$

Proof Let $y = \arcsin x$, $-\pi/2 \le y \le \pi/2$. Hence, $\sin y = x$, and use implicit differentiation as follows.

$$\sin y = x$$

$$(\cos y)\left(\frac{dy}{dx}\right) = 1$$

$$\frac{dy}{dx} = \frac{1}{\cos y} = \frac{1}{\sqrt{1 - \sin^2 y}} = \frac{1}{\sqrt{1 - x^2}}$$

Because u is a differentiable function of x, you can use the Chain Rule to write

$$\frac{d}{dx}[\arcsin u] = \frac{u'}{\sqrt{1 - u^2}}, \qquad \text{where } u' = \frac{du}{dx}.$$

Proofs of the other differentiation rules are left as exercises (see Exercise 75).

There is no common agreement on the definition of $\text{arcsec}\, x$ (or $\text{arccsc}\, x$) for negative values of x. When we defined the range of the arcsecant, we chose to preserve the reciprocal identity $\text{arcsec}\, x = \arccos(1/x)$. For example, to evaluate $\text{arcsec}(-2)$, you can write

$$\text{arcsec}(-2) = \arccos(-0.5) \approx 2.09.$$

One of the consequences of the definition of the inverse secant function given in this text is that its graph has a positive slope at every x-value in its domain. (See Figure 5.32.) This accounts for the absolute value sign in the formula for the derivative of $\text{arcsec}\, x$.

TECHNOLOGY If your graphing utility does not have the arcsecant function, you can obtain its graph using

$$f(x) = \text{arcsec}\, x = \arccos\frac{1}{x}.$$

EXAMPLE 4   Differentiating Inverse Trigonometric Functions

a. $\dfrac{d}{dx}[\arcsin(2x)] = \dfrac{2}{\sqrt{1 - (2x)^2}}$ $u = 2x$

$\qquad\qquad\qquad\;\; = \dfrac{2}{\sqrt{1 - 4x^2}}$

b. $\dfrac{d}{dx}[\arctan(3x)] = \dfrac{3}{1 + (3x)^2}$ $u = 3x$

$\qquad\qquad\qquad\;\; = \dfrac{3}{1 + 9x^2}$

c. $\dfrac{d}{dx}[\arcsin\sqrt{x}] = \dfrac{(1/2)x^{-1/2}}{\sqrt{1 - x}}$ $u = \sqrt{x}$

$\qquad\qquad\qquad\;\; = \dfrac{1}{2\sqrt{x}\sqrt{1 - x}}$

$\qquad\qquad\qquad\;\; = \dfrac{1}{2\sqrt{x - x^2}}$

d. $\dfrac{d}{dx}[\operatorname{arcsec} e^{2x}] = \dfrac{2e^{2x}}{e^{2x}\sqrt{(e^{2x})^2 - 1}}$ $u = e^{2x}$

$\qquad\qquad\qquad\;\; = \dfrac{2e^{2x}}{e^{2x}\sqrt{e^{4x} - 1}}$

$\qquad\qquad\qquad\;\; = \dfrac{2}{\sqrt{e^{4x} - 1}}$

The absolute value sign is not necessary because $e^{2x} > 0$.   ▬▬

EXAMPLE 5   A Derivative that Can Be Simplified

Differentiate $y = \arcsin x + x\sqrt{1 - x^2}$.

Solution

$$y' = \frac{1}{\sqrt{1 - x^2}} + x\left(\frac{1}{2}\right)(-2x)(1 - x^2)^{-1/2} + \sqrt{1 - x^2}$$

$$= \frac{1}{\sqrt{1 - x^2}} - \frac{x^2}{\sqrt{1 - x^2}} + \sqrt{1 - x^2}$$

$$= \sqrt{1 - x^2} + \sqrt{1 - x^2}$$

$$= 2\sqrt{1 - x^2}$$ ▬▬

REMARK   From Example 5, you can see one of the benefits of inverse trigonometric functions—they can be used to integrate common algebraic functions. For instance, from the result shown in the example, it follows that

$$\int \sqrt{1 - x^2}\, dx = \frac{1}{2}\left(\arcsin x + x\sqrt{1 - x^2}\right).$$

(This and several other integration formulas are summarized in a table of integrals in Appendix C.)

EXAMPLE 6   Sketching the Graph of an Inverse Trigonometric Function

Sketch the graph of $y = (\arctan x)^2$.

Solution   From the derivative

$$y' = 2(\arctan x)\left(\frac{1}{1 + x^2}\right) = \frac{2 \arctan x}{1 + x^2}$$

you can see that the only critical number is $x = 0$. By the First Derivative Test, this value corresponds to a relative minimum. From the second derivative

$$y'' = \frac{(1 + x^2)\left(\dfrac{2}{1 + x^2}\right) - (2 \arctan x)(2x)}{(1 + x^2)^2}$$

$$= \frac{2(1 - 2x \arctan x)}{(1 + x^2)^2}$$

it follows that points of inflection occur when $2x \arctan x = 1$. Using Newton's Method, these points occur when $x \approx \pm 0.765$. Finally, because

$$\lim_{x \to \pm\infty} (\arctan x)^2 = \frac{\pi^2}{4}$$

it follows that the graph has a horizontal asymptote at $y = \pi^2/4$. The graph is shown in Figure 5.35.

FIGURE 5.35
The graph of $y = (\arctan x)^2$ has a horizontal asymptote at $y = \pi^2/4$.

EXAMPLE 7   Maximizing an Angle

A photographer is taking a picture of a 4-foot painting hanging in an art gallery. The camera lens is 1 foot below the lower edge of the painting, as shown in Figure 5.36. How far should the camera be from the painting to maximize the angle subtended by the camera lens?

Solution   In Figure 5.36, let β be the angle to be maximized.

$$\beta = \theta - \alpha = \operatorname{arccot}\frac{x}{5} - \operatorname{arccot} x$$

Differentiating produces

$$\frac{d\beta}{dx} = \frac{-1/5}{1 + (x^2/25)} - \frac{-1}{1 + x^2}$$

$$= \frac{-5}{25 + x^2} + \frac{1}{1 + x^2}$$

$$= \frac{4(5 - x^2)}{(25 + x^2)(1 + x^2)}.$$

Because $d\beta/dx = 0$ when $x = \sqrt{5}$, you can conclude from the First Derivative Test that this distance yields a maximum value of β. Thus, the distance is $x \approx 2.236$ feet and the angle is $\beta \approx 41.81°$.

FIGURE 5.36
To maximize the angle β the camera should be 2.236 ft from the painting.

Review of Basic Differentiation Rules

In the 1600s, Europe was ushered into the scientific age by such great thinkers as Descartes, Galileo, Huygens, Newton, and Kepler. These men believed that nature is governed by basic laws—laws that can, for the most part, be written in terms of mathematical equations. One of the most influential publications during this period was written by Galileo Galilei (1564–1642). It is called *Dialogue on the Great World Systems,* and it has become a classic description of modern scientific thought.

As mathematics developed during the past few hundred years, a list of elementary functions has proven sufficient for modeling most phenomena in physics, chemistry, biology, engineering, economics, and a variety of other fields.* An **elementary function** is one that can be formed as the sum, product, quotient, or composition of functions from the following list.

Algebraic functions	*Transcendental functions*
Polynomial functions	Logarithmic functions
Rational functions	Exponential functions
Functions involving radicals	Trigonometric functions
	Inverse trigonometric functions

With the differentiation rules introduced so far in the text, you can differentiate *any* elementary function. For convenience, we summarize these differentiation rules here.

*Some important functions used in engineering and science (such as Bessel functions or gamma functions) are not elementary functions.

GALILEO GALILEI

Galileo's approach to science departed from the accepted Aristotelian view that nature had describable *qualities,* such as "fluidity" and "potentiality." He chose to describe the physical world in terms of measurable *quantities,* such as time, distance, force, and mass.

Basic Differentiation Rules for Elementary Functions

1. $\dfrac{d}{dx}[cu] = cu'$

2. $\dfrac{d}{dx}[u \pm v] = u' \pm v'$

3. $\dfrac{d}{dx}[uv] = uv' + vu'$

4. $\dfrac{d}{dx}\left[\dfrac{u}{v}\right] = \dfrac{vu' - uv'}{v^2}$

5. $\dfrac{d}{dx}[c] = 0$

6. $\dfrac{d}{dx}[u^n] = nu^{n-1}u'$

7. $\dfrac{d}{dx}[x] = 1$

8. $\dfrac{d}{dx}\left[|u|\right] = \dfrac{u}{|u|}(u'), \quad u \neq 0$

9. $\dfrac{d}{dx}[\ln u] = \dfrac{u'}{u}$

10. $\dfrac{d}{dx}[e^u] = e^u u'$

11. $\dfrac{d}{dx}[\log_a u] = \dfrac{u'}{(\ln a)u}$

12. $\dfrac{d}{dx}[a^u] = (\ln a)a^u u'$

13. $\dfrac{d}{dx}[\sin u] = (\cos u)u'$

14. $\dfrac{d}{dx}[\cos u] = -(\sin u)u'$

15. $\dfrac{d}{dx}[\tan u] = (\sec^2 u)u'$

16. $\dfrac{d}{dx}[\cot u] = -(\csc^2 u)u'$

17. $\dfrac{d}{dx}[\sec u] = (\sec u \tan u)u'$

18. $\dfrac{d}{dx}[\csc u] = -(\csc u \cot u)u'$

19. $\dfrac{d}{dx}[\arcsin u] = \dfrac{u'}{\sqrt{1 - u^2}}$

20. $\dfrac{d}{dx}[\arccos u] = \dfrac{-u'}{\sqrt{1 - u^2}}$

21. $\dfrac{d}{dx}[\arctan u] = \dfrac{u'}{1 + u^2}$

22. $\dfrac{d}{dx}[\text{arccot } u] = \dfrac{-u'}{1 + u^2}$

23. $\dfrac{d}{dx}[\text{arcsec } u] = \dfrac{u'}{|u|\sqrt{u^2 - 1}}$

24. $\dfrac{d}{dx}[\text{arccsc } u] = \dfrac{-u'}{|u|\sqrt{u^2 - 1}}$

EXERCISES for Section 5.7

In Exercises 1–8, evaluate the expression without using a calculator.

1. $\arcsin \frac{1}{2}$

2. $\arcsin 0$

3. $\arccos \frac{1}{2}$

4. $\arccos 0$

5. $\arctan \frac{\sqrt{3}}{3}$

6. $\operatorname{arccot}(-1)$

7. $\operatorname{arccsc}(-\sqrt{2})$

8. $\arccos\left(-\frac{\sqrt{3}}{2}\right)$

C In Exercises 9–12, use a calculator to approximate the inverse trigonometric function. Round your answers to two decimal places.

9. $\arccos(-0.8)$

10. $\arcsin(-0.39)$

11. $\operatorname{arcsec} 1.269$

12. $\arctan(-3)$

13. *Essay* Explain why $\tan \pi = 0$ does not imply that $\arctan 0 = \pi$.

14. *Writing* Write a sentence describing the meaning of the equation $y = \arcsin x$.

C In Exercises 15 and 16, use a graphing utility to confirm that f and g are inverse functions. (Remember to properly restrict the domain of f.)

15. $f(x) = \tan x$
    $g(x) = \arctan x$

16. $f(x) = \sin x$
    $g(x) = \arcsin x$

In Exercises 17 and 18, use the properties of inverse trigonometric functions to evaluate the expression.

17. $\cos[\arccos(-0.1)]$

18. $\arcsin(\sin 3\pi)$

In Exercises 19–24, evaluate the expression without using a calculator. [*Hint:* Make a sketch of a right triangle, as illustrated in Example 3.]

19. a. $\sin\left(\arcsin \frac{1}{2}\right)$

    b. $\cos\left(\arcsin \frac{1}{2}\right)$

20. a. $\tan\left(\arccos \frac{\sqrt{2}}{2}\right)$

    b. $\cos\left(\arcsin \frac{5}{13}\right)$

21. a. $\sin\left(\arctan \frac{3}{4}\right)$

    b. $\sec\left(\arcsin \frac{4}{5}\right)$

22. a. $\tan(\operatorname{arccot} 2)$

    b. $\cos(\operatorname{arcsec} \sqrt{5})$

23. a. $\cot\left[\arcsin\left(-\frac{1}{2}\right)\right]$

    b. $\csc\left[\arctan\left(-\frac{5}{12}\right)\right]$

24. a. $\sec\left[\arctan\left(-\frac{3}{5}\right)\right]$

    b. $\tan\left[\arcsin\left(-\frac{5}{6}\right)\right]$

In Exercises 25–34, write the expression in algebraic form.

25. $\tan(\arctan x)$

26. $\sin(\arccos x)$

27. $\cos(\arcsin 2x)$

28. $\sec(\arctan 3x)$

29. $\sin(\operatorname{arcsec} x)$

30. $\cos(\operatorname{arccot} x)$

31. $\tan\left(\operatorname{arcsec} \frac{x}{3}\right)$

32. $\sec[\arcsin(x - 1)]$

33. $\csc\left(\arctan \frac{x}{\sqrt{2}}\right)$

34. $\cos\left(\arcsin \frac{x - h}{r}\right)$

In Exercises 35 and 36, fill in the blank.

35. $\arctan \dfrac{9}{x} = \arcsin(\blacksquare), \qquad x > 0$

36. $\arcsin \dfrac{\sqrt{36 - x^2}}{6} = \arccos(\blacksquare)$

In Exercises 37 and 38, verify each identity.

37. a. $\operatorname{arccsc} x = \arcsin \dfrac{1}{x}, \qquad |x| \ge 1$

    b. $\arctan x + \arctan \dfrac{1}{x} = \dfrac{\pi}{2}, \qquad x > 0$

38. a. $\arcsin(-x) = -\arcsin x, \qquad |x| \le 1$

    b. $\arccos(-x) = \pi - \arccos x, \qquad |x| \le 1$

In Exercises 39–42, sketch the graph of the function.

39. $f(x) = \arcsin(x - 1)$

40. $f(x) = \arctan x + \dfrac{\pi}{2}$

41. $f(x) = \operatorname{arcsec} 2x$

42. $f(x) = \arccos \dfrac{x}{4}$

In Exercises 43–46, solve the equation for x.

43. $\arcsin(3x - \pi) = \frac{1}{2}$

44. $\arctan 2x = -1$

45. $\arcsin \sqrt{2x} = \arccos \sqrt{x}$

46. $\arccos x = \operatorname{arcsec} x$

In Exercises 47–64, find the derivative of the function.

47. $f(x) = 2\arcsin(x - 1)$

48. $f(t) = \arcsin t^2$

49. $g(x) = 3\arccos \dfrac{x}{2}$

50. $g(x) = \arccos \sqrt{x}$

51. $f(x) = \arctan \dfrac{x}{a}$

52. $f(x) = \arctan \sqrt{x}$

53. $g(x) = \dfrac{\arcsin 3x}{x}$

54. $h(x) = x\arctan x$

55. $h(x) = \operatorname{arccot} 6x$

56. $f(x) = \operatorname{arcsec} 2x$

57. $h(t) = \sin(\arccos t)$

58. $g(t) = \tan(\arcsin t)$

59. $f(x) = \arcsin x + \arccos x$

60. $f(x) = \operatorname{arcsec} x + \operatorname{arccsc} x$

61. $y = \dfrac{1}{2}\left(\dfrac{1}{2}\ln \dfrac{x + 1}{x - 1} + \arctan x\right)$

62. $y = \frac{1}{2}(x\sqrt{1 - x^2} + \arcsin x)$

63. $y = x \arcsin x + \sqrt{1 - x^2}$

64. $y = x \arctan 2x - \frac{1}{4}\ln(1 + 4x^2)$

65. Find equations of all tangent lines to the graph of $f(x) = \arccos x$ that have slope -2.

66. Use differentials to approximate the change in the function $g(x) = \arctan x$ when x is increased from $x = 2$ to $x = 2.1$.

C *Linear and Quadratic Approximation* In Exercises 67 and 68, use a symbolic differentiation utility to find the linear approximation

$$P_1(x) = f(a) + f'(a)(x - a)$$

and quadratic approximation

$$P_2(x) = f(a) + f'(a)(x - a) + \frac{1}{2}f''(a)(x - a)^2$$

to the function f at $x = a$. Sketch the graph of the function and its linear and quadratic approximations.

67. $f(x) = \arcsin x$
$a = \frac{1}{2}$

68. $f(x) = \arctan x$
$a = 1$

In Exercises 69 and 70, find any relative extrema of the function.

69. $f(x) = \text{arcsec } x - x$

70. $f(x) = \arcsin x - 2x$

In Exercises 71 and 72, find the point of intersection of the graphs of the functions.

71. $y = \arccos x$
$y = \arctan x$

72. $y = \arcsin x$
$y = \arccos x$

73. *Falling Object* In a free-fall experiment, an object is dropped from a height of 256 feet. A camera on the ground 500 feet from the point of impact records the fall of the object.
a. Find the position function giving the height of the object at time t assuming the object is released at time $t = 0$. At what time will the object reach ground level?
b. Find the rate of change of the angle of elevation of the camera when $t = 1$ and $t = 2$.

74. *Rising Balloon* An observer is standing 300 feet from the point at which a balloon is released. The balloon rises at a rate of 5 feet per second. How fast is the angle of elevation of the observer's line of sight increasing when the balloon is 100 feet high?

75. Verify the differentiation formulas.

a. $\dfrac{d}{dx}[\arctan u] = \dfrac{u'}{1 + u^2}$

b. $\dfrac{d}{dx}[\text{arcsec } u] = \dfrac{u'}{|u|\sqrt{u^2 - 1}}$

c. $\dfrac{d}{dx}[\arccos u] = \dfrac{-u'}{\sqrt{1 - u^2}}$

d. $\dfrac{d}{dx}[\text{arccot } u] = \dfrac{-u'}{1 + u^2}$

e. $\dfrac{d}{dx}[\text{arccsc } u] = \dfrac{-u'}{|u|\sqrt{u^2 - 1}}$

76. Determine the values of k such that the function $f(x) = kx + \sin x$ has an inverse.

C **77.** Use a graphing utility to graph $f(x) = \sin x$ and $g(x) = \arcsin(\sin x)$.
a. Why isn't the graph of g the line $y = x$?
b. Determine the extrema of g.

78. Repeat Exercise 77 for the functions $f(x) = \cos x$ and $g(x) = \arccos(\cos x)$.

True or False In Exercises 79–82, determine whether the statement is true or false. If it is false, explain why or give an example that shows it is false.

79. The slope of the graph of the inverse tangent is positive for all x.

80. The range of $y = \arcsin x$ is $[0, \pi]$.

81. $\dfrac{d}{dx}[\arctan(\tan x)] = 1$ for all x.

82. $\arcsin^2 x + \arccos^2 x = 1$

83. Show that the function

$$f(x) = \arcsin\frac{x - 2}{2} - 2\arcsin\left(\frac{\sqrt{x}}{2}\right)$$

is constant for $0 \le x \le 4$.

84. Compare the graphs of $y_1 = \sin(\arcsin x)$ and $y_2 = \arcsin(\sin x)$. What are the domains and ranges of y_1 and y_2?

85. Prove that

$$\arctan x + \arctan y = \arctan\frac{x + y}{1 - xy}, \qquad xy \ne 1.$$

Use this formula to show that

$$\arctan\frac{1}{2} + \arctan\frac{1}{3} = \frac{\pi}{4}.$$

Inverse Trigonometric Functions: Integration and Completing the Square

Integrals Involving Inverse Trigonometric Functions • Completing the Square • Review of Basic Integration Rules

Integrals Involving Inverse Trigonometric Functions

The derivatives of the six inverse trigonometric functions occur in three pairs. In each pair the derivative of one function is the negative of the other. For example,

$$\frac{d}{dx}[\arcsin x] = \frac{1}{\sqrt{1 - x^2}} \quad \text{and} \quad \frac{d}{dx}[\arccos x] = -\frac{1}{\sqrt{1 - x^2}}.$$

When listing the *antiderivative* that corresponds to each of the inverse trigonometric functions, you need use only one member from each pair. It is conventional to use $\arcsin x$ as the antiderivative of $1/\sqrt{1 - x^2}$, rather than $-\arccos x$. The next theorem gives one antiderivative formula for each of the three pairs.

THEOREM 5.18   Integrals Involving Inverse Trigonometric Functions

Let u be a differentiable function of x, and let $a > 0$.

1. $\displaystyle\int \frac{du}{\sqrt{a^2 - u^2}} = \arcsin\frac{u}{a} + C$

2. $\displaystyle\int \frac{du}{a^2 + u^2} = \frac{1}{a}\arctan\frac{u}{a} + C$

3. $\displaystyle\int \frac{du}{u\sqrt{u^2 - a^2}} = \frac{1}{a}\operatorname{arcsec}\frac{|u|}{a} + C$

Proof Let $y = \arcsin(u/a)$. Then

$$y' = \frac{1}{\sqrt{1 - (u/a)^2}}\left(\frac{u'}{a}\right)$$

$$= \frac{u'}{a\sqrt{(a^2 - u^2)/a^2}}$$

$$= \frac{u'}{\sqrt{a^2 - u^2}}.$$

Consequently, by the definition of an antiderivative, you can write

$$\int \frac{du}{\sqrt{a^2 - u^2}} = \int \frac{u'}{\sqrt{a^2 - u^2}}\,dx = \arcsin\frac{u}{a} + C.$$

In Exercise 58 you are asked to verify the other two integration rules.

FOR FURTHER INFORMATION For an alternative proof of part 2 of Theorem 5.18, see the article "A Direct Proof of the Integral Formula for Arctangent" by Arnold J. Insel in the May, 1989 issue of the *College Mathematics Journal*.

EXAMPLE 1   Integration with Inverse Trigonometric Functions

a. $\displaystyle\int \frac{dx}{\sqrt{4-x^2}} = \arcsin\frac{x}{2} + C$

b. $\displaystyle\int \frac{dx}{2+9x^2} = \frac{1}{3}\int \frac{3dx}{(\sqrt{2})^2 + (3x)^2}$ $u = 3x,\, a = \sqrt{2}$

$\displaystyle\qquad\qquad\quad = \frac{1}{3\sqrt{2}}\arctan\frac{3x}{\sqrt{2}} + C$

The integrals in Example 1 are fairly straightforward applications of integration formulas. Unfortunately, this is not typical. The inverse trigonometric integration formulas can be disguised in many ways.

EXAMPLE 2   Integration by Substitution

Evaluate $\displaystyle\int \frac{dx}{\sqrt{e^{2x}-1}}$.

Solution   As it stands, this integral doesn't fit any of the three inverse trigonometric formulas. Using the substitution $u = e^x$, however, produces the following.

$$u = e^x \rightarrow du = e^x\, dx \rightarrow dx = \frac{du}{e^x} = \frac{du}{u}$$

With this substitution, you can integrate as follows.

$\displaystyle\int \frac{dx}{\sqrt{e^{2x}-1}} = \int \frac{dx}{\sqrt{(e^x)^2-1}}$ Write e^{2x} as $(e^x)^2$

$\displaystyle\qquad\qquad\quad = \int \frac{du/u}{\sqrt{u^2-1}}$ Substitute

$\displaystyle\qquad\qquad\quad = \int \frac{du}{u\sqrt{u^2-1}}$ Rewrite to fit Arcsecant Rule

$\displaystyle\qquad\qquad\quad = \operatorname{arcsec}\frac{|u|}{1} + C$ Apply Arcsecant Rule

$\displaystyle\qquad\qquad\quad = \operatorname{arcsec} e^x + C$ Back-substitute

TECHNOLOGY   Computer software that can perform symbolic integration is useful for integrating functions such as the one in Example 2. When using such software, however, you must remember that it can fail to find an antiderivative for two reasons. First, some elementary functions simply do not have antiderivatives that are elementary functions. Second, every symbolic integrator has limitations—you might have entered a function that the software was not programmed to handle. You should also remember that antiderivatives involving trigonometric functions or logarithmic functions can be written in many different forms. For instance, when we used a symbolic integrator to evaluate the integral in Example 2, we obtained

$$\int \frac{dx}{\sqrt{e^{2x}-1}} = \arctan\sqrt{e^{2x}-1} + C.$$

Try showing that this antiderivative is equivalent to that obtained in Example 2.

A useful integration technique is to rewrite the integrand as the sum of two quotients.

EXAMPLE 3   Rewriting as the Sum of Two Quotients

Evaluate $\int \dfrac{x + 2}{\sqrt{4 - x^2}}\, dx.$

Solution   This integral does not appear to fit any of the basic integration formulas. By splitting the integrand into two parts, however, you can see that the first part can be evaluated with the Power Rule and the second part yields an inverse sine function.

$$
\int \frac{x + 2}{\sqrt{4 - x^2}}\, dx = \int \frac{x}{\sqrt{4 - x^2}}\, dx + \int \frac{2}{\sqrt{4 - x^2}}\, dx
$$

$$
= -\frac{1}{2} \int (4 - x^2)^{-1/2}(-2x)\, dx + 2 \int \frac{1}{\sqrt{4 - x^2}}\, dx
$$

$$
= -\frac{1}{2}\left[\frac{(4 - x^2)^{1/2}}{1/2} \right] + 2\arcsin\frac{x}{2} + C
$$

$$
= -\sqrt{4 - x^2} + 2\arcsin\frac{x}{2} + C
$$

Another disguised use of integration formulas occurs when the integrand is an improper rational function. (That is, the integrand is a rational function in which the degree of the numerator is greater than or equal to the degree of the denominator.) For such functions, long division can be used to write the integrand as the sum of a polynomial and a proper rational function.

EXAMPLE 4   Integrating an Improper Rational Function

Evaluate $\int \dfrac{3x^3 - 2}{x^2 + 4}\, dx.$

Solution   The integrand is an improper rational function because the degree of the numerator is greater than the degree of the denominator. You can use long division to rewrite the integrand.

$$
\begin{array}{r}
3x \\
x^2 + 4\,\overline{)\,3x^3 - 2\,} \\
\underline{3x^3 + 12x } \\
-12x - 2
\end{array}
$$

Then, you can integrate as follows.

$$
\int \frac{3x^3 - 2}{x^2 + 4}\, dx = \int \left(3x - \frac{12x + 2}{x^2 + 4} \right) dx
$$

$$
= \int 3x\, dx - 6 \int \frac{2x}{x^2 + 4}\, dx - 2 \int \frac{dx}{x^2 + 4}
$$

$$
= \frac{3x^2}{2} - 6\ln(x^2 + 4) - \arctan\frac{x}{2} + C
$$

Completing the Square

Completing the square helps when quadratic functions are involved in the integrand. For example, the quadratic $x^2 + bx + c$ can be written as the difference of two squares by adding and subtracting $(b/2)^2$.

$$x^2 + bx + c = x^2 + bx + \left(\frac{b}{2}\right)^2 - \left(\frac{b}{2}\right)^2 + c$$

$$= \left(x + \frac{b}{2}\right)^2 - \left(\frac{b}{2}\right)^2 + c$$

EXAMPLE 5   Completing the Square

Evaluate $\displaystyle\int \frac{dx}{x^2 - 4x + 7}$.

Solution You can write the denominator as the sum of two squares as follows.

$$x^2 - 4x + 7 = (x^2 - 4x + 4) - 4 + 7 = (x - 2)^2 + 3 = u^2 + a^2$$

Now, in this completed square form, let $u = x - 2$ and $a = \sqrt{3}$.

$$\int \frac{dx}{x^2 - 4x + 7} = \int \frac{dx}{(x - 2)^2 + 3}$$

$$= \frac{1}{\sqrt{3}} \arctan \frac{x - 2}{\sqrt{3}} + C$$

If the leading coefficient is not 1, it helps to factor before completing the square. For instance, you can complete the square of $2x^2 - 8x + 10$ as follows.

$$2x^2 - 8x + 10 = 2(x^2 - 4x + 5)$$

$$= 2(x^2 - 4x + 4 - 4 + 5)$$

$$= 2[(x - 2)^2 + 1]$$

EXAMPLE 6   Completing the Square (Leading Coefficient Not 1)

Evaluate $\displaystyle\int \frac{dx}{2x^2 - 8x + 10}$.

Solution Using the result given above, you can integrate as follows.

$$\int \frac{dx}{2x^2 - 8x + 10} = \frac{1}{2} \int \frac{dx}{(x - 2)^2 + 1}$$

$$= \frac{1}{2} \int \frac{du}{u^2 + 1} \qquad u = x - 2, \quad a = 1$$

$$= \frac{1}{2} \arctan u + C$$

$$= \frac{1}{2} \arctan(x - 2) + C$$

To complete the square when the coefficient of x^2 is negative, use the same "factoring process" illustrated in Example 6. For instance, you can complete the square for $3x - x^2$ as follows.

$$3x - x^2 = -(x^2 - 3x)$$

$$= -\left[x^2 - 3x + \left(\frac{3}{2}\right)^2 - \left(\frac{3}{2}\right)^2 \right]$$

$$= \left(\frac{3}{2}\right)^2 - \left(x - \frac{3}{2}\right)^2$$

EXAMPLE 7   Completing the Square (Negative Leading Coefficient)

Find the area of the region bounded by the graph of

$$f(x) = \frac{1}{\sqrt{3x - x^2}},$$

the x-axis, and the lines $x = \frac{3}{2}$ and $x = \frac{9}{4}$.

Solution   From Figure 5.37, you can see that the area is given by

$$\text{Area} = \int_{3/2}^{9/4} \frac{1}{\sqrt{3x - x^2}} \, dx.$$

Using the completed square form derived above, you can integrate as follows.

$$\int_{3/2}^{9/4} \frac{dx}{\sqrt{3x - x^2}} = \int_{3/2}^{9/4} \frac{dx}{\sqrt{(3/2)^2 - [x - (3/2)]^2}}$$

$$= \left[\arcsin \frac{x - (3/2)}{3/2} \right]_{3/2}^{9/4}$$

$$= \arcsin \frac{1}{2} - \arcsin 0$$

$$= \frac{\pi}{6} - 0$$

$$= \frac{\pi}{6} \approx 0.524$$

You can use the graph shown in Figure 5.37 to check the reasonableness of this value. For instance, in the figure, the shaded area has a width of $\frac{3}{4}$ and a height of about 0.7, which implies that the area is about $\frac{3}{4}(0.7) = 0.525$.

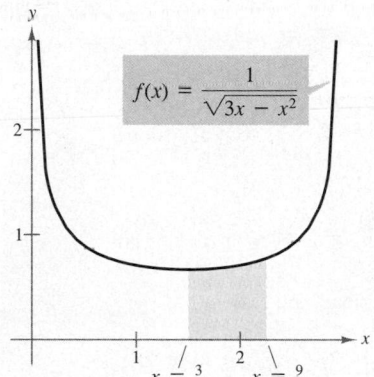

FIGURE 5.37
The area of the region bounded by the graph of f, the x-axis, $x = \frac{3}{2}$ and $x = \frac{9}{4}$ is $\pi/6$.

TECHNOLOGY   With definite integrals such as the one given in Example 7, remember that you can resort to a numerical solution. For instance, applying Simpson's Rule (with $n = 6$) to the integral in the example, you obtain

$$\int_{3/2}^{9/4} \frac{1}{\sqrt{3x - x^2}} \, dx \approx 0.523599.$$

This differs from the exact value of the integral ($\pi/6 \approx 0.5235987$) by less than one millionth!

Review of Basic Integration Rules

We have now completed the introduction of the **basic integration rules.** To be efficient at applying the rules, you should have practiced enough so that each rule is committed to memory.

Basic Integration Rules ($a > 0$)

1. $\displaystyle \int k f(u)\, du = k \int f(u)\, du$

2. $\displaystyle \int [f(u) \pm g(u)]\, du = \int f(u)\, du \pm \int g(u)\, du$

3. $\displaystyle \int du = u + C$

4. $\displaystyle \int u^n\, du = \frac{u^{n+1}}{n+1} + C, \quad n \neq -1$

5. $\displaystyle \int \frac{du}{u} = \ln |u| + C$

6. $\displaystyle \int e^u\, du = e^u + C$

7. $\displaystyle \int a^u\, du = \left(\frac{1}{\ln a}\right) a^u + C$

8. $\displaystyle \int \sin u\, du = -\cos u + C$

9. $\displaystyle \int \cos u\, du = \sin u + C$

10. $\displaystyle \int \tan u\, du = -\ln |\cos u| + C$

11. $\displaystyle \int \cot u\, du = \ln |\sin u| + C$

12. $\displaystyle \int \sec u\, du = \ln |\sec u + \tan u| + C$

13. $\displaystyle \int \csc u\, du = -\ln |\csc u + \cot u| + C$

14. $\displaystyle \int \sec^2 u\, du = \tan u + C$

15. $\displaystyle \int \csc^2 u\, du = -\cot u + C$

16. $\displaystyle \int \sec u \tan u\, du = \sec u + C$

17. $\displaystyle \int \csc u \cot u\, du = -\csc u + C$

18. $\displaystyle \int \frac{du}{\sqrt{a^2 - u^2}} = \arcsin \frac{u}{a} + C$

19. $\displaystyle \int \frac{du}{a^2 + u^2} = \frac{1}{a} \arctan \frac{u}{a} + C$

20. $\displaystyle \int \frac{du}{u\sqrt{u^2 - a^2}} = \frac{1}{a} \operatorname{arcsec} \frac{|u|}{a} + C$

You can learn quite a bit about the nature of integration by comparing this list to the summary of differentiation rules given in the previous section. For differentiation, you now have rules that allow you to differentiate *any* elementary function. For integration, this is far from true.

The integration rules listed above are primarily those that we happened upon when developing differentiation rules. We do not find an integration rule for the antiderivative of a general product or quotient, the natural logarithmic function, or the inverse trigonometric functions. More importantly, you cannot apply any of the rules in this list unless you can create the proper *du* corresponding to the *u* in the formula. The point is that we need to work more on integration techniques, which we will do in Chapter 7. The next two examples should give you a better feeling for the integration problems that you *can* and *cannot* do with the techniques and rules you now know.

EXAMPLE 8 Comparing Integration Problems

Evaluate as many of the following integrals as you can using the formulas and techniques you have studied so far in the text.

a. $\displaystyle\int \frac{dx}{x\sqrt{x^2 - 1}}$ **b.** $\displaystyle\int \frac{x\,dx}{\sqrt{x^2 - 1}}$ **c.** $\displaystyle\int \frac{dx}{\sqrt{x^2 - 1}}$

Solution

a. You *can* evaluate this integral (it fits the Arcsecant Rule).

$$\int \frac{dx}{x\sqrt{x^2 - 1}} = \text{arcsec}\,|x| + C$$

b. You *can* also evaluate this integral (it fits the Power Rule).

$$\int \frac{x\,dx}{\sqrt{x^2 - 1}} = \frac{1}{2}\int (x^2 - 1)^{-1/2}(2x)\,dx$$

$$= \frac{1}{2}\left(\frac{(x^2 - 1)^{1/2}}{1/2}\right) + C$$

$$= \sqrt{x^2 - 1} + C$$

c. You *cannot* evaluate this integral using present techniques. (You should scan the list of basic integration rules to verify this conclusion.)

EXAMPLE 9 Comparing Integration Problems

Evaluate as many of the following integrals as you can using the formulas and techniques you have studied so far in the text.

a. $\displaystyle\int \frac{dx}{x \ln x}$ **b.** $\displaystyle\int \frac{\ln x\,dx}{x}$ **c.** $\displaystyle\int \ln x\,dx$

Solution

a. You *can* evaluate this integral (it fits the Log Rule).

$$\int \frac{dx}{x \ln x} = \int \frac{1/x}{\ln x}\,dx = \ln|\ln x| + C$$

b. You *can* also evaluate this integral (it fits the Power Rule).

$$\int \frac{\ln x}{x}\,dx = \int \left(\frac{1}{x}\right)(\ln x)^1\,dx$$

$$= \frac{(\ln x)^2}{2} + C$$

c. You *cannot* evaluate this integral using present techniques.

REMARK Note in Examples 8 and 9 that the *simplest* functions are the ones that we cannot yet integrate.

EXERCISES for Section 5.8

In Exercises 1–26, evaluate the integral.

1. $\displaystyle\int_0^{1/6} \frac{1}{\sqrt{1 - 9x^2}}\, dx$

2. $\displaystyle\int_0^1 \frac{dx}{\sqrt{4 - x^2}}$

3. $\displaystyle\int_0^{\sqrt{3}/2} \frac{1}{1 + 4x^2}\, dx$

4. $\displaystyle\int_{\sqrt{3}}^3 \frac{1}{9 + x^2}\, dx$

5. $\displaystyle\int \frac{1}{x\sqrt{4x^2 - 1}}\, dx$

6. $\displaystyle\int \frac{1}{4 + (x - 1)^2}\, dx$

7. $\displaystyle\int \frac{x^3}{x^2 + 1}\, dx$

8. $\displaystyle\int \frac{x^4 - 1}{x^2 + 1}\, dx$

9. $\displaystyle\int \frac{1}{\sqrt{1 - (x + 1)^2}}\, dx$

10. $\displaystyle\int \frac{t}{t^4 + 16}\, dt$

11. $\displaystyle\int \frac{t}{\sqrt{1 - t^4}}\, dt$

12. $\displaystyle\int \frac{1}{x\sqrt{x^4 - 4}}\, dx$

13. $\displaystyle\int \frac{\arctan x}{1 + x^2}\, dx$

14. $\displaystyle\int \frac{1}{(x - 1)\sqrt{(x - 1)^2 - 4}}\, dx$

15. $\displaystyle\int_0^{1/\sqrt{2}} \frac{\arcsin x}{\sqrt{1 - x^2}}\, dx$

16. $\displaystyle\int_0^{1/\sqrt{2}} \frac{\arccos x}{\sqrt{1 - x^2}}\, dx$

17. $\displaystyle\int_{-1/2}^0 \frac{x}{\sqrt{1 - x^2}}\, dx$

18. $\displaystyle\int_{-\sqrt{3}}^0 \frac{x}{1 + x^2}\, dx$

19. $\displaystyle\int \frac{e^x}{\sqrt{1 - e^{2x}}}\, dx$

20. $\displaystyle\int \frac{\cos x}{\sqrt{4 - \sin^2 x}}\, dx$

21. $\displaystyle\int \frac{1}{9 + (x - 3)^2}\, dx$

22. $\displaystyle\int \frac{x + 1}{x^2 + 1}\, dx$

23. $\displaystyle\int \frac{e^{2x}}{4 + e^{4x}}\, dx$

24. $\displaystyle\int_1^2 \frac{1}{3 + (x - 2)^2}\, dx$

25. $\displaystyle\int_{\pi/2}^{\pi} \frac{\sin x}{1 + \cos^2 x}\, dx$

26. $\displaystyle\int_0^{\pi} \frac{\cos x}{1 + \sin^2 x}\, dx$

In Exercises 27–40, evaluate the integral. (Complete the square, if necessary.)

27. $\displaystyle\int_0^2 \frac{dx}{x^2 - 2x + 2}$

28. $\displaystyle\int_{-3}^{-1} \frac{dx}{x^2 + 6x + 13}$

29. $\displaystyle\int \frac{2x}{x^2 + 6x + 13}\, dx$

30. $\displaystyle\int \frac{2x - 5}{x^2 + 2x + 2}\, dx$

31. $\displaystyle\int \frac{1}{\sqrt{-x^2 - 4x}}\, dx$

32. $\displaystyle\int \frac{1}{\sqrt{-x^2 + 2x}}\, dx$

33. $\displaystyle\int \frac{x + 2}{\sqrt{-x^2 - 4x}}\, dx$

34. $\displaystyle\int \frac{x - 1}{\sqrt{x^2 - 2x}}\, dx$

35. $\displaystyle\int_2^3 \frac{2x - 3}{\sqrt{4x - x^2}}\, dx$

36. $\displaystyle\int \frac{1}{(x - 1)\sqrt{x^2 - 2x}}\, dx$

37. $\displaystyle\int \frac{x}{x^4 + 2x^2 + 2}\, dx$

38. $\displaystyle\int \frac{x}{\sqrt{9 + 8x^2 - x^4}}\, dx$

39. $\displaystyle\int \frac{1}{\sqrt{-16x^2 + 16x - 3}}\, dx$

40. $\displaystyle\int \frac{1}{(x - 1)\sqrt{9x^2 - 18x + 5}}\, dx$

In Exercises 41–44, use substitution to evaluate the integral.

41. $\displaystyle\int \frac{\sqrt{x - 1}}{x}\, dx$

42. $\displaystyle\int \frac{\sqrt{x - 2}}{x + 1}\, dx$

43. $\displaystyle\int \sqrt{e^t - 3}\, dt$

44. $\displaystyle\int \frac{1}{t\sqrt{t} + \sqrt{t}}\, dt$

In Exercises 45–48, solve the differential equation.

45. $(1 + x^6)y' = 3x^2$

46. $\sqrt{1 - 9x^2}\, y' = 1$

47. $\dfrac{dy}{dx} = \dfrac{xy}{1 + x^2}$

48. $\dfrac{dy}{dx} = x\sec y$

In Exercises 49–52, find the area of the region bounded by the graphs of the given equations.

49. $y = \dfrac{1}{1 + x^2}$, $y = 0$, $x = 0$, $x = 1$

50. $y = \dfrac{1}{\sqrt{4 - x^2}}$, $y = 0$, $x = 0$, $x = 1$

51. $y = \dfrac{1}{x^2 - 2x + 5}$, $y = 0$, $x = 1$, $x = 3$

52. $y = \dfrac{1}{\sqrt{3 + 2x - x^2}}$, $y = 0$, $x = 0$, $x = 2$

53. Consider the function

$$F(x) = \frac{1}{2}\int_x^{x+2} \frac{2}{t^2 + 1}\, dt.$$

a. Write a short paragraph giving a geometric interpretation of the function F relative to the function

$$f(x) = \frac{2}{x^2 + 1}.$$

Use what you have written to guess the value of x that will make F maximum.

b. Perform the specified integration to find an alternative form of $F(x)$. Use calculus to locate the value of x that will make F maximum and compare the result with your guess in part a.

54. *Approximation* Determine which value best approximates the area of the region between the x-axis and the function $f(x) = 1/\sqrt{1 - x^2}$ over the given interval $[-0.5,\ 0.5]$. (Make your selection on the basis of a sketch of the region and *not* by performing any calculations.)

a. 4   **b.** −3   **c.** 1   **d.** 2   **e.** 3

55. *Vertical Motion* An object is projected upward from ground level with an initial velocity of 500 feet per second. In this exercise, the goal is to analyze the motion of the object during its upward flight.

  a. If air resistance is neglected, express the velocity of the object as a function of time. Sketch the graph of this function.

  b. Use the result of part a to find the position function and determine the maximum height attained by the object.

  c. If the air resistance is proportional to the square of the velocity, you obtain the equation

$$\frac{dv}{dt} = -(32 + kv^2)$$

  where 32 feet per second per second is the acceleration due to gravity and k is a constant. Find the velocity as a function of time by solving the equation

$$\int \frac{dv}{32 + kv^2} = -\int dt.$$

  d. Use a graphing utility to graph the velocity function v from part c if $k = 0.001$. Use the graph to approximate the time t_0 at which the object reaches its maximum height.

  e. Use Simpson's Rule with $n = 10$ to approximate the integral

$$\int_0^{t_0} v(t)\, dt$$

  where $v(t)$ and t_0 are those found in part d. This is the approximation of the maximum height of the object.

  f. Explain the difference between the results of part b and part e.

FOR FURTHER INFORMATION For more information on this topic, see "What Goes Up Must Come Down; Will Air Resistance Make It Return Sooner, or Later?" by John Lekner in the January, 1982 issue of *Mathematics Magazine*.

56. *Harmonic Motion* A weight of mass m is attached to a spring and oscillates with simple harmonic motion (see figure). By Hooke's Law we can determine that

$$\int \frac{dy}{\sqrt{A^2 - y^2}} = \int \sqrt{\frac{k}{m}}\, dt$$

where A is the maximum displacement, t is the time, and k is a constant. Find y as a function of t, given that $y = 0$ when $t = 0$.

FIGURE FOR 56

57. a. Show that

$$\int_0^1 \frac{4}{1 + x^2}\, dx = \pi.$$

  b. Approximate the number π using Simpson's Rule (with $n = 6$) and the integral in part a.

58. Verify the following rules by differentiating ($a > 0$).

  a. $\displaystyle\int \frac{du}{a^2 + u^2} = \frac{1}{a}\arctan\frac{u}{a} + C$

  b. $\displaystyle\int \frac{du}{u\sqrt{u^2 - a^2}} = \frac{1}{a}\operatorname{arcsec}\frac{|u|}{a} + C$

In Exercises 59–62, determine which of the given integrals can be evaluated using the basic integration rules you have studied so far.

59. a. $\displaystyle\int \frac{1}{\sqrt{1 - x^2}}\, dx$ **60. a.** $\displaystyle\int e^{x^2}\, dx$

  b. $\displaystyle\int \frac{x}{\sqrt{1 - x^2}}\, dx$ **b.** $\displaystyle\int x e^{x^2}\, dx$

  c. $\displaystyle\int \frac{1}{x\sqrt{1 - x^2}}\, dx$ **c.** $\displaystyle\int \frac{1}{x^2} e^{1/x}\, dx$

61. a. $\displaystyle\int \sqrt{x - 1}\, dx$ **62. a.** $\displaystyle\int \frac{1}{1 + x^4}\, dx$

  b. $\displaystyle\int x\sqrt{x - 1}\, dx$ **b.** $\displaystyle\int \frac{x}{1 + x^4}\, dx$

  c. $\displaystyle\int \frac{x}{\sqrt{x - 1}}\, dx$ **c.** $\displaystyle\int \frac{x^3}{1 + x^4}\, dx$

63. Sketch the region whose area is represented by the integral

$$\int_0^1 \arcsin(x)\, dx$$

and approximate its value.

64. Graph $y_1 = \dfrac{x}{1 + x^2}$, $y_2 = \arctan x$, and $y_3 = x$ on $[0, 10]$. Prove that

$$\frac{x}{1 + x^2} < \arctan x < x \quad \text{for} \quad x > 0.$$

Hyperbolic Functions • Differentiation and Integration of Hyperbolic Functions • Inverse Hyperbolic Functions • Differentiation and Integration Involving Inverse Hyperbolic Functions

Hyperbolic Functions

In this section we look briefly at a special class of exponential functions called the **hyperbolic functions.** The name *hyperbolic function* arose from comparison of the area of a semicircular region, as shown in Figure 5.38, to the area of a region under a hyperbola, as shown in Figure 5.39. The integral for the semicircular region involves an *inverse trigonometric (circular) function:*

$$\int_{-1}^{1} \sqrt{1 - x^2}\, dx = \frac{1}{2}\left[x\sqrt{1 - x^2} + \arcsin x \right]_{-1}^{1} = \frac{\pi}{2} \approx 1.571.$$

The integral for the hyperbolic region involves an inverse hyperbolic function:

$$\int_{-1}^{1} \sqrt{1 + x^2}\, dx = \frac{1}{2}\left[x\sqrt{1 + x^2} + \sinh^{-1} x \right]_{-1}^{1} \approx 2.296.$$

This is only one of many ways in which the hyperbolic functions are similar to the trigonometric functions.

FIGURE 5.38
Circle: $x^2 + y^2 = 1$

FIGURE 5.39
Hyperbola: $-x^2 + y^2 = 1$

JOHANN HEINRICH LAMBERT

The first person to publish a comprehensive study on hyperbolic functions was Johann Heinrich Lambert (1728–1777), a Swiss-German mathematician and colleague of Euler.

FOR FURTHER INFORMATION
For more information on the development of hyperbolic functions, see the article "An Introduction to Hyperbolic Functions in Elementary Calculus" by Jerome Rosenthal in the April, 1986 issue of *Mathematics Teacher*.

Definition of the Hyperbolic Functions

$$\sinh x = \frac{e^x - e^{-x}}{2} \qquad \cosh x = \frac{e^x + e^{-x}}{2} \qquad \tanh x = \frac{\sinh x}{\cosh x}$$

$$\operatorname{csch} x = \frac{1}{\sinh x}, \quad x \neq 0 \qquad \operatorname{sech} x = \frac{1}{\cosh x} \qquad \coth x = \frac{1}{\tanh x}, \quad x \neq 0$$

REMARK $\sinh x$ is read as "the hyperbolic sine of x," $\cosh x$ as "the hyperbolic cosine of x," and so on.

The graphs of the six hyperbolic functions and their domains and ranges are shown in Figure 5.40. Note that the graphs of $\sinh x$ and $\cosh x$ can be obtained by *addition of ordinates* using the exponential functions $f(x) = \frac{1}{2}e^x$ and $g(x) = \frac{1}{2}e^{-x}$.

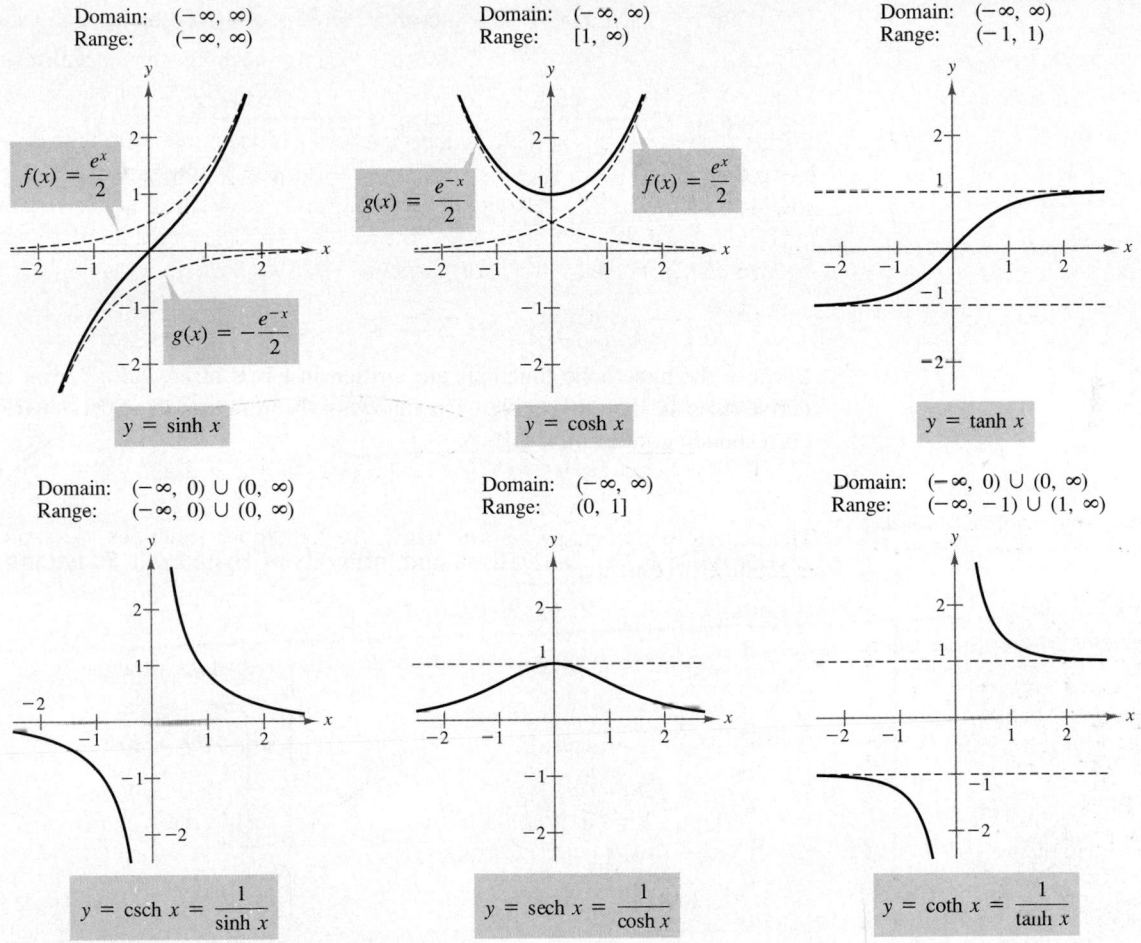

FIGURE 5.40
Graphs of the six hyperbolic functions.

Many of the trigonometric identities have corresponding *hyperbolic identities*. For instance

$$\cosh^2 x - \sinh^2 x = \left(\frac{e^x + e^{-x}}{2}\right)^2 - \left(\frac{e^x - e^{-x}}{2}\right)^2$$

$$= \frac{e^{2x} + 2 + e^{-2x}}{4} - \frac{e^{2x} - 2 + e^{-2x}}{4} = \frac{4}{4} = 1$$

and

$$2\sinh x \cosh x = 2\left(\frac{e^x - e^{-x}}{2}\right)\left(\frac{e^x + e^{-x}}{2}\right) = \frac{e^{2x} - e^{-2x}}{2} = \sinh 2x.$$

Hyperbolic Identities

$$\cosh^2 x - \sinh^2 x = 1 \qquad \sinh(x + y) = \sinh x \cosh y + \cosh x \sinh y$$

$$\tanh^2 x + \text{sech}^2 x = 1 \qquad \sinh(x - y) = \sinh x \cosh y - \cosh x \sinh y$$

$$\coth^2 x - \text{csch}^2 x = 1 \qquad \cosh(x + y) = \cosh x \cosh y + \sinh x \sinh y$$

$$\cosh(x - y) = \cosh x \cosh y - \sinh x \sinh y$$

$$\sinh^2 x = \frac{-1 + \cosh 2x}{2} \qquad \cosh^2 x = \frac{1 + \cosh 2x}{2}$$

$$\sinh 2x = 2 \sinh x \cosh x \qquad \cosh 2x = \cosh^2 x + \sinh^2 x$$

Differentiation and Integration of Hyperbolic Functions

Because the hyperbolic functions are written in terms of e^x and e^{-x}, you can easily derive rules for their derivatives. The following theorem lists these derivatives with the corresponding integration rules.

THEOREM 5.19   Derivatives and Integrals of Hyperbolic Functions

Let u be a differentiable function of x.

$$\frac{d}{dx}[\sinh u] = (\cosh u)u' \qquad \int \cosh u \, du = \sinh u + C$$

$$\frac{d}{dx}[\cosh u] = (\sinh u)u' \qquad \int \sinh u \, du = \cosh u + C$$

$$\frac{d}{dx}[\tanh u] = (\text{sech}^2 u)u' \qquad \int \text{sech}^2 u \, du = \tanh u + C$$

$$\frac{d}{dx}[\coth u] = -(\text{csch}^2 u)u' \qquad \int \text{csch}^2 u \, du = -\coth u + C$$

$$\frac{d}{dx}[\text{sech } u] = -(\text{sech } u \tanh u)u' \qquad \int \text{sech } u \tanh u \, du = -\text{sech } u + C$$

$$\frac{d}{dx}[\text{csch } u] = -(\text{csch } u \coth u)u' \qquad \int \text{csch } u \coth u \, du = -\text{csch } u + C$$

Proof

$$\frac{d}{dx}[\sinh x] = \frac{d}{dx}\left[\frac{e^x - e^{-x}}{2}\right] = \frac{e^x + e^{-x}}{2} = \cosh x$$

$$\frac{d}{dx}[\tanh x] = \frac{d}{dx}\left[\frac{\sinh x}{\cosh x}\right] = \frac{\cosh x(\cosh x) - \sinh x(\sinh x)}{\cosh^2 x}$$

$$= \frac{1}{\cosh^2 x} = \text{sech}^2 x$$

In Exercises 81 and 82, you are asked to prove the other rules.

EXAMPLE 1   Differentiation of Hyperbolic Functions

a. $\dfrac{d}{dx}[\sinh(x^2 - 3)] = 2x\cosh(x^2 - 3)$

b. $\dfrac{d}{dx}[\ln(\cosh x)] = \dfrac{\sinh x}{\cosh x}$

$\qquad\qquad\qquad = \tanh x$

c. $\dfrac{d}{dx}[x\sinh x - \cosh x] = x\cosh x + \sinh x - \sinh x$

$\qquad\qquad\qquad\qquad\qquad = x\cosh x$

EXAMPLE 2   Finding Relative Extrema

Find the relative extrema of $f(x) = (x - 1)\cosh x - \sinh x$.

Solution   Begin by setting the first derivative of f equal to 0.

$$f'(x) = (x - 1)\sinh x + \cosh x - \cosh x = 0$$
$$(x - 1)\sinh x = 0$$

Thus, the critical numbers are $x = 1$ and $x = 0$. Using the Second Derivative Test, you can verify that the point $(0, -1)$ yields a relative maximum and the point $(1, -\sinh 1)$ yields a relative minimum, as shown in Figure 5.41.

When a uniform flexible cable, such as a telephone wire, is suspended from two points, it takes the shape of a **catenary,** as discussed in Example 3.

EXAMPLE 3   Hanging Power Cables

Power cables are suspended between two towers, forming the catenary in Figure 5.42. The equation for this catenary is

$$y = a\cosh\frac{x}{a}.$$

The distance between the two towers is $2b$. Find the slope of the catenary at the point where the cable meets the right-hand tower.

Solution   Differentiating produces

$$y' = a\left(\frac{1}{a}\right)\sinh\frac{x}{a}$$
$$= \sinh\frac{x}{a}.$$

At the point $(b, a\cosh(x/a))$, the slope (from the left) is given by

$$m = \sinh\frac{b}{a}.$$

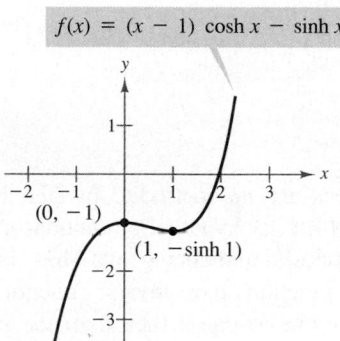

$f(x) = (x - 1)\cosh x - \sinh x$

FIGURE 5.41
$f''(0) < 0$, so $(0, -1)$ is a relative maximum. $f''(1) > 0$, so $(1, -\sinh 1)$ is a relative minimum.

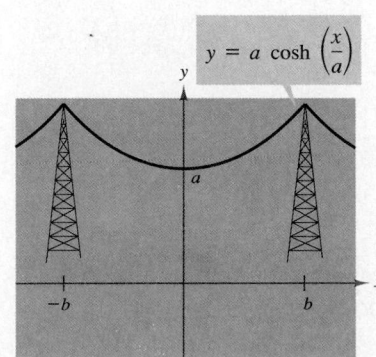

$y = a\cosh\left(\dfrac{x}{a}\right)$

FIGURE 5.42
Catenary

EXAMPLE 4   **Integrating a Hyperbolic Function**

$$\int \cosh 2x \sinh^2 2x \, dx = \frac{1}{2} \int (\sinh 2x)^2 (2 \cosh 2x) \, dx \qquad u = \sinh 2x$$

$$= \frac{1}{2} \left[\frac{(\sinh 2x)^3}{3} \right] + C$$

$$= \frac{\sinh^3 2x}{6} + C$$

Inverse Hyperbolic Functions

Unlike trigonometric functions, hyperbolic functions are *not* periodic. In fact, by looking back at Figure 5.40, you can see that four of the six hyperbolic functions are actually one-to-one (the hyperbolic sine, tangent, cosecant, and cotangent). Thus, you can apply Theorem 5.7 to conclude that these four functions have inverse functions. The other two (the hyperbolic cosine and secant) are one-to-one if their domains are restricted to the positive real numbers, and for this restricted domain they also have inverse functions. Because the hyperbolic functions are defined in terms of exponential functions, it is not surprising to find that the inverse hyperbolic functions can be written in terms of logarithmic functions, as shown in Theorem 5.20.

THEOREM 5.20   Inverse Hyperbolic Functions

Function	*Domain*
$\sinh^{-1} x = \ln(x + \sqrt{x^2 + 1})$	$(-\infty, \infty)$
$\cosh^{-1} x = \ln(x + \sqrt{x^2 - 1})$	$[1, \infty)$
$\tanh^{-1} x = \dfrac{1}{2} \ln \dfrac{1 + x}{1 - x}$	$(-1, 1)$
$\coth^{-1} x = \dfrac{1}{2} \ln \dfrac{x + 1}{x - 1}$	$(-\infty, -1) \cup (1, \infty)$
$\text{sech}^{-1} x = \ln \dfrac{1 + \sqrt{1 - x^2}}{x}$	$(0, 1]$
$\text{csch}^{-1} x = \ln\left(\dfrac{1}{x} + \dfrac{\sqrt{1 + x^2}}{|x|}\right)$	$(-\infty, 0) \cup (0, \infty)$

Proof   The proof of this theorem is a straightforward application of the properties of the exponential and logarithmic functions. For example, if

$$f(x) = \sinh x = \frac{e^x - e^{-x}}{2} \quad \text{and} \quad g(x) = \ln(x + \sqrt{x^2 + 1})$$

then you can show that $f(g(x)) = x = g(f(x))$, which implies that g is the inverse of f.

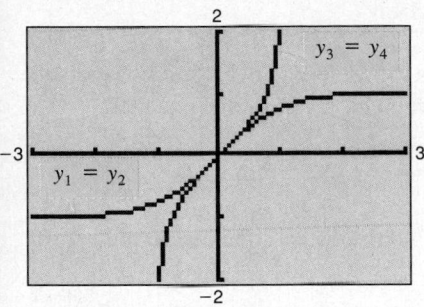

FIGURE 5.43
Graphs of hyperbolic tangent function and inverse hyperbolic tangent function.

TECHNOLOGY   You can use a graphing utility to graphically confirm the results of Theorem 5.20. For instance, try sketching the graphs of the following functions.

$y_1 = \tanh x$ Hyperbolic tangent

$y_2 = \dfrac{e^x - e^{-x}}{e^x + e^{-x}}$ Definition of hyperbolic tangent

$y_3 = \tanh^{-1} x$ Inverse hyperbolic tangent

$y_4 = \dfrac{1}{2}\ln\dfrac{1 + x}{1 - x}$ Definition of inverse hyperbolic tangent

The resulting display is shown in Figure 5.43. As you watch the graphs being traced out, notice that $y_1 = y_2$ and $y_3 = y_4$. Also notice that the graph of y_1 is the reflection of the graph of y_3 in the line $y = x$.

The graphs of the inverse hyperbolic functions are shown in Figure 5.44.

Domain: $[1, \infty)$
Range: $[0, \infty)$

$y = \cosh^{-1} x$

Domain: $(-1, 1)$
Range: $(-\infty, \infty)$

$y = \tanh^{-1} x$

Domain: $(-\infty, \infty)$
Range: $(-\infty, \infty)$

$y = \sinh^{-1} x$

Domain: $(0, 1]$
Range: $[0, \infty)$

$y = \operatorname{sech}^{-1} x$

Domain: $(-\infty, -1) \cup (1, \infty)$
Range: $(-\infty, 0) \cup (0, \infty)$

$y = \coth^{-1} x$

Domain: $(-\infty, 0) \cup (0, \infty)$
Range: $(-\infty, 0) \cup (0, \infty)$

$y = \operatorname{csch}^{-1} x$

FIGURE 5.44
Graphs of the six inverse hyperbolic functions.

The inverse hyperbolic secant can be used to define a curve called a *tractrix* or *pursuit curve,* as discussed in Example 5.

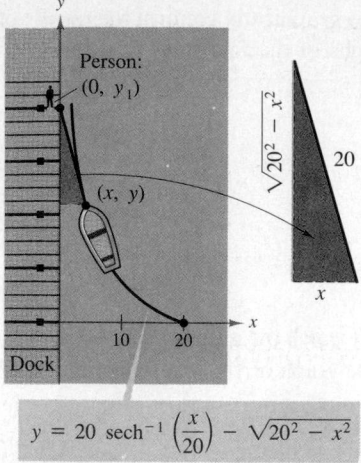

$$y = 20 \operatorname{sech}^{-1}\left(\frac{x}{20}\right) - \sqrt{20^2 - x^2}$$

FIGURE 5.45
A person must walk 41.27 ft. to bring the boat 5 feet from the dock.

EXAMPLE 5   A Tractrix

A person is holding a rope that is tied to a boat, as shown in Figure 5.45. As the person walks along the dock, the boat travels along a **tractrix,** given by the equation

$$y = a \operatorname{sech}^{-1}\frac{x}{a} - \sqrt{a^2 - x^2}$$

where a is the length of the rope. If $a = 20$ feet, find the distance the person must walk to bring the boat 5 feet from the dock.

Solution   In Figure 5.45, notice that the distance the person has walked is given by

$$y_1 = y + \sqrt{20^2 - x^2} = \left(20 \operatorname{sech}^{-1}\frac{x}{20} - \sqrt{20^2 - x^2}\right) + \sqrt{20^2 - x^2}$$

$$= 20 \operatorname{sech}^{-1}\frac{x}{20}.$$

When $x = 5$, this distance is

$$y_1 = 20 \operatorname{sech}^{-1}\frac{5}{20} = 20\ln\frac{1 + \sqrt{1 - (1/4)^2}}{1/4}$$

$$= 20\ln\left(4 + \sqrt{15}\right) \approx 41.27 \text{ ft.} \quad \blacksquare$$

Differentiation and Integration Involving Inverse Hyperbolic Functions

The derivatives of the inverse hyperbolic functions resemble the derivatives of the inverse trigonometric functions. These are listed in Theorem 5.21 with the corresponding integration formulas (in logarithmic form). You can verify each of these formulas by applying the logarithmic definitions of the inverse hyperbolic functions.

THEOREM 5.21   Differentiation and Integration Involving Inverse Hyperbolic Functions

Let u be a differentiable function of x.

$$\frac{d}{dx}[\sinh^{-1}u] = \frac{u'}{\sqrt{u^2 + 1}} \qquad \frac{d}{dx}[\cosh^{-1}u] = \frac{u'}{\sqrt{u^2 - 1}}$$

$$\frac{d}{dx}[\tanh^{-1}u] = \frac{u'}{1 - u^2} \qquad \frac{d}{dx}[\coth^{-1}u] = \frac{u'}{1 - u^2}$$

$$\frac{d}{dx}[\operatorname{sech}^{-1}u] = \frac{-u'}{u\sqrt{1 - u^2}} \qquad \frac{d}{dx}[\operatorname{csch}^{-1}u] = \frac{-u'}{|u|\sqrt{1 + u^2}}$$

$$\int \frac{du}{\sqrt{u^2 \pm a^2}} = \ln(u + \sqrt{u^2 \pm a^2}) + C$$

$$\int \frac{du}{a^2 - u^2} = \frac{1}{2a}\ln\left|\frac{a + u}{a - u}\right| + C$$

$$\int \frac{du}{u\sqrt{a^2 \pm u^2}} = -\frac{1}{a}\ln\frac{a + \sqrt{a^2 \pm u^2}}{|u|} + C$$

EXAMPLE 6   More About a Tractrix

For the tractrix given in Example 5, show that the boat is always pointing toward the person.

Solution   For a point (x, y) on the tractrix, the slope of the graph gives the direction of the boat, as shown in Figure 5.45.

$$y' = \frac{d}{dx}\left[20\,\text{sech}^{-1}\frac{x}{20} - \sqrt{20^2 - x^2} \right]$$

$$= -20\left(\frac{1}{20}\right)\left(\frac{1}{(x/20)\sqrt{1 - (x/20)^2}}\right) - \left(\frac{1}{2}\right)\left(\frac{-2x}{\sqrt{20^2 - x^2}}\right)$$

$$= \frac{-20^2}{x\sqrt{20^2 - x^2}} + \frac{x}{\sqrt{20^2 - x^2}}$$

$$= -\frac{\sqrt{20^2 - x^2}}{x}.$$

However, from Figure 5.45, you can see that the slope of the line segment connecting the point $(0, y_1)$ with the point (x, y) is also

$$m = -\frac{\sqrt{20^2 - x^2}}{x}.$$

Thus, the boat is always pointing toward the person. (It is because of this property that a tractrix is called a *pursuit curve*.)

EXAMPLE 7   Integration Using Inverse Hyperbolic Functions

Evaluate $\displaystyle\int \frac{dx}{x\sqrt{4 - 9x^2}}$.

Solution   Let $a = 2$ and $u = 3x$.

$$\int \frac{dx}{x\sqrt{4 - 9x^2}} = \int \frac{3\,dx}{(3x)\sqrt{4 - 9x^2}} \qquad \int \frac{du}{u\sqrt{a^2 - u^2}}$$

$$= -\frac{1}{2}\ln\frac{2 + \sqrt{4 - 9x^2}}{|3x|} + C \qquad -\frac{1}{a}\ln\frac{a + \sqrt{a^2 - u^2}}{|u|} + C$$

EXAMPLE 8   Integration Using Inverse Hyperbolic Functions

Evaluate $\displaystyle\int \frac{dx}{5 - 4x^2}$.

Solution   Let $a = \sqrt{5}$ and $u = 2x$.

$$\int \frac{dx}{5 - 4x^2} = \frac{1}{2}\int \frac{2\,dx}{(\sqrt{5})^2 - (2x)^2} \qquad \int \frac{du}{a^2 - u^2}$$

$$= \frac{1}{2}\frac{1}{2\sqrt{5}}\ln\left|\frac{\sqrt{5} + 2x}{\sqrt{5} - 2x}\right| + C \qquad \frac{1}{2a}\ln\left|\frac{a + u}{z - u}\right| + C$$

$$= \frac{1}{4\sqrt{5}}\ln\left|\frac{\sqrt{5} + 2x}{\sqrt{5} - 2x}\right| + C$$

EXERCISES for Section 5.9

C In Exercises 1–6, evaluate the function. If the value is not a rational number, list the answer to three-decimal-place accuracy.

1. a. $\sinh 3$
  b. $\tanh(-2)$

2. a. $\cosh 0$
  b. $\operatorname{sech} 1$

3. a. $\operatorname{csch}(\ln 2)$
  b. $\coth(\ln 5)$

4. a. $\sinh^{-1} 0$
  b. $\tanh^{-1} 0$

5. a. $\cosh^{-1} 2$
  b. $\operatorname{sech}^{-1} \frac{2}{3}$

6. a. $\operatorname{csch}^{-1} 2$
  b. $\coth^{-1} 3$

In Exercises 7–12, verify the identity.

7. $\tanh^2 x + \operatorname{sech}^2 x = 1$

8. $\cosh^2 x = \dfrac{1 + \cosh 2x}{2}$

9. $\sinh(x + y) = \sinh x \cosh y + \cosh x \sinh y$

10. $\sinh 2x = 2 \sinh x \cosh x$

11. $\sinh 3x = 3 \sinh x + 4 \sinh^3 x$

12. $\cosh x + \cosh y = 2 \cosh \dfrac{x + y}{2} \cosh \dfrac{x - y}{2}$

In Exercises 13 and 14, use the value of the given hyperbolic function to find the other hyperbolic functions.

13. $\sinh x = \frac{3}{2}$
  $\cosh x = \underline{\hspace{1cm}}$
  $\tanh x = \underline{\hspace{1cm}}$
  $\operatorname{csch} x = \underline{\hspace{1cm}}$
  $\operatorname{sech} x = \underline{\hspace{1cm}}$
  $\coth x = \underline{\hspace{1cm}}$

14. $\sinh x = \underline{\hspace{1cm}}$
  $\cosh x = \underline{\hspace{1cm}}$
  $\tanh x = \frac{1}{2}$
  $\operatorname{csch} x = \underline{\hspace{1cm}}$
  $\operatorname{sech} x = \underline{\hspace{1cm}}$
  $\coth x = \underline{\hspace{1cm}}$

In Exercises 15–28, find the derivative of the function.

15. $y = \sinh(1 - x^2)$

16. $y = \coth 3x$

17. $f(x) = \ln(\sinh x)$

18. $g(x) = \ln(\cosh x)$

19. $y = \ln\left(\tanh \dfrac{x}{2}\right)$

20. $y = x \sinh x - \cosh x$

21. $h(x) = \dfrac{1}{4} \sinh 2x - \dfrac{x}{2}$

22. $h(t) = t - \coth t$

23. $f(t) = \arctan(\sinh t)$

24. $f(x) = e^{\sinh x}$

25. $g(x) = x^{\cosh x}$

26. $g(x) = \operatorname{sech}^2 3x$

27. $y = (\cosh x - \sinh x)^2$

28. $y = \operatorname{sech}(x + 1)$

In Exercises 29 and 30, find and graph any relative extrema of the function.

29. $f(x) = \sin x \sinh x - \cos x \cosh x, \quad -4 \le x \le 4$

30. $f(x) = x \cosh(x - 1) - \sinh(x - 1)$

C In Exercises 31 and 32, use a graphing utility to graph the function and approximate any relative extrema of the function.

31. $g(x) = x \operatorname{sech} x$

32. $h(x) = 2 \tanh x - x$

In Exercises 33 and 34, show that the function satisfies the indicated differential equation.

Function	Differential equation
33. $y = a \sinh x$	$y''' - y' = 0$
34. $y = a \cosh x$	$y'' - y = 0$

In Exercises 35–50, evaluate the integral.

35. $\displaystyle \int \sinh(1 - 2x)\, dx$

36. $\displaystyle \int \dfrac{\cosh \sqrt{x}}{\sqrt{x}}\, dx$

37. $\displaystyle \int \cosh^2(x - 1)\sinh(x - 1)\, dx$

38. $\displaystyle \int \dfrac{\sinh x}{1 + \sinh^2 x}\, dx$

39. $\displaystyle \int \dfrac{\cosh x}{\sinh x}\, dx$

40. $\displaystyle \int \operatorname{sech}^2(2x - 1)\, dx$

41. $\displaystyle \int x \operatorname{csch}^2 \dfrac{x^2}{2}\, dx$

42. $\displaystyle \int \operatorname{sech}^3 x \tanh x\, dx$

43. $\displaystyle \int \dfrac{\operatorname{csch}(1/x)\coth(1/x)}{x^2}\, dx$

44. $\displaystyle \int \sinh^2 x\, dx$

45. $\displaystyle \int_0^4 \dfrac{1}{25 - x^2}\, dx$

46. $\displaystyle \int_0^4 \dfrac{1}{\sqrt{25 - x^2}}\, dx$

47. $\displaystyle \int_0^{\sqrt{2}/4} \dfrac{2}{\sqrt{1 - 4x^2}}\, dx$

48. $\displaystyle \int \dfrac{2}{x\sqrt{1 + 4x^2}}\, dx$

49. $\displaystyle \int \dfrac{x}{x^4 + 1}\, dx$

50. $\displaystyle \int \dfrac{\cosh x}{\sqrt{9 - \sinh^2 x}}\, dx$

In Exercises 51–58, find the first derivative of the function.

51. $y = \cosh^{-1}(3x)$

52. $y = \tanh^{-1} \dfrac{x}{2}$

53. $y = \sinh^{-1}(\tan x)$

54. $y = \operatorname{sech}^{-1}(\cos 2x)$, $0 < x < \dfrac{\pi}{4}$

55. $y = \coth^{-1}(\sin 2x)$

56. $y = (\operatorname{csch}^{-1} x)^2$

57. $y = 2x \sinh^{-1}(2x) - \sqrt{1 + 4x^2}$

58. $y = x \tanh^{-1} x + \ln \sqrt{1 - x^2}$

In Exercises 59 and 60, use the equation of the tractrix

$$y = a \operatorname{sech}^{-1} \frac{x}{a} - \sqrt{a^2 - x^2}, \qquad a > 0.$$

59. Find dy/dx.

60. Let L be the tangent line at the point P to the tractrix. If L intersects the y-axis at the point Q, show that the distance between P and Q is a.

In Exercises 61–68, find the indefinite integral using the formulas of Theorem 5.21.

61. $\displaystyle\int \frac{1}{\sqrt{1 + e^{2x}}}\,dx$

62. $\displaystyle\int \frac{e^x}{1 - e^{2x}}\,dx$

63. $\displaystyle\int \frac{1}{\sqrt{x}\sqrt{1 + x}}\,dx$

64. $\displaystyle\int \frac{\sqrt{x}}{\sqrt{1 + x^3}}\,dx$

65. $\displaystyle\int \frac{1}{(x - 1)\sqrt{x^2 - 2x + 2}}\,dx$

66. $\displaystyle\int \frac{-1}{4x - x^2}\,dx$

67. $\displaystyle\int \frac{1}{1 - 4x - 2x^2}\,dx$

68. $\displaystyle\int \frac{1}{(x + 1)\sqrt{2x^2 + 4x + 8}}\,dx$

In Exercises 69–72, solve the differential equation.

69. $\dfrac{dy}{dx} = \dfrac{1}{\sqrt{80 + 8x - 16x^2}}$

70. $\dfrac{dy}{dx} = \dfrac{1}{(x - 1)\sqrt{-4x^2 + 8x - 1}}$

71. $\dfrac{dy}{dx} = \dfrac{x^3 - 21x}{5 + 4x - x^2}$

72. $\dfrac{dy}{dx} = \dfrac{1 - 2x}{4x - x^2}$

In Exercises 73–76, find the area of the region bounded by the graphs of the equations.

73. $y = \operatorname{sech}\dfrac{x}{2}$, $y = 0$, $x = -4$, $x = 4$

74. $y = \tanh 2x$, $y = 0$, $x = 2$

75. $y = \dfrac{5x}{\sqrt{x^4 + 1}}$, $y = 0$, $x = 2$

76. $y = \dfrac{6}{\sqrt{x^2 - 4}}$, $y = 0$, $x = 3$, $x = 5$

77. *Chemical Reactions* Suppose that two chemicals, A and B, combine in a 3-to-1 ratio to form a compound. The amount of compound x being produced at any time t is proportional to the unchanged amounts of A and B remaining in the solution. Thus, if 3 kilograms of A is mixed with 2 kilograms of B, we have

$$\frac{dx}{dt} = k\left(3 - \frac{3x}{4}\right)\left(2 - \frac{x}{4}\right) = \frac{3k}{16}(x^2 - 12x + 32).$$

If 1 kilogram of the compound is formed after 10 minutes, find the amount formed after 20 minutes by solving the integral

$$\int \frac{3k}{16}\,dt = \int \frac{dx}{x^2 - 12x + 32}.$$

78. *Vertical Motion* An object is dropped from a height of 400 feet.

 a. Express the velocity as a function of time—neglect air resistance on the object.

 b. Use the result of part a to find the position function.

 c. Consider that air resistance on the object is proportional to the square of the velocity of the object. This is expressed by the differential equation

$$\frac{dv}{dt} = -32 + kv^2$$

 where 32 feet per second is the acceleration due to gravity and k is a constant. Show that the velocity v as a function of time is

$$v(t) = -\frac{\sqrt{32}}{\sqrt{k}}\tanh\left(\sqrt{32k}\,t\right)$$

 by performing the following integration and simplifying the result.

$$\int \frac{dv}{32 - kv^2} = -\int dt$$

 d. Use the result of part c to find $\lim\limits_{t \to \infty} v(t)$ and give its interpretation.

 e. Integrate the velocity function in part c and find the position s of the object as a function of t. Use a graphing utility to graph the position function when $k = 0.01$ and the position function in part b on the same display. Estimate the additional time required for the object to reach ground level when air resistance is not neglected.

79. *Essay* Give a written description of what you believe would happen if k were increased in Exercise 78. Then test your assertion with a particular value of k.

80. *St. Louis Arch*   The Gateway Arch in St. Louis, Missouri was constructed using the hyperbolic cosine function. The equation used to construct the arch was

$$y = 693.8597 - 68.7672 \cosh 0.0100333x,$$
$$-299.2239 \leq x \leq 299.2239$$

where x and y are measured in feet. Cross sections of the arch are equilateral triangles and (x, y) traces the path of the centers of mass of the cross-sectional triangles. For each value of x, the area of the cross-sectional triangle is $A = 125.1406 \cosh 0.0100333x$.

a. How high above ground is the center of the highest triangle? (At ground level, $y = 0$.)

b. What is the height of the arch? (*Hint:* For an equilateral triangle $A = \sqrt{3} \, c^2$, where c is one-half the base of the triangle, and the center of mass of the triangle is located at two-thirds the height of the triangle.)

c. How wide is the arch at ground level?
(*Source: Owner's Manual for the Gateway Arch, Saint Louis, MO*, by William Thayer.)

In Exercises 81–85, verify the differentiation formula.

81. $\dfrac{d}{dx}[\cosh x] = \sinh x$

82. $\dfrac{d}{dx}[\operatorname{sech} x] = -\operatorname{sech} x \tanh x$

83. $\dfrac{d}{dx}[\cosh^{-1} x] = \dfrac{1}{\sqrt{x^2 - 1}}$

84. $\dfrac{d}{dx}[\sinh^{-1} x] = \dfrac{1}{\sqrt{x^2 + 1}}$

85. $\dfrac{d}{dx}[\operatorname{sech}^{-1} x] = \dfrac{-1}{x\sqrt{1 - x^2}}$

86. Show that $\arctan(\sinh x) = \arcsin(\tanh x)$.

REVIEW EXERCISES for Chapter 5

In Exercises 1 and 2, sketch the graph of the function without using a graphing utility. Identify any asymptotes to the graph.

1. $f(x) = \ln x + 3$ **2.** $f(x) = \ln(x - 3)$

In Exercises 3 and 4, use the properties of logarithms to write the expression as a sum, difference, and/or multiple of logarithms.

3. $\ln \sqrt[5]{\dfrac{4x^2 - 1}{4x^2 + 1}}$ **4.** $\ln[(x^2 + 1)(x - 1)]$

In Exercises 5 and 6, write the expression as the logarithm of a single quantity.

5. $\ln 3 + \frac{1}{3}\ln(4 - x^2) - \ln x$

6. $3[\ln x - 2\ln(x^2 + 1)] + 2\ln 5$

True or False   **In Exercises 7 and 8, determine whether the statement or equation is true or false.**

7. The domain of the function $f(x) = \ln x$ is the set of all real numbers.

8. $\ln(x + y) = \ln x + \ln y$

In Exercises 9 and 10, solve the equation for x.

9. $\ln \sqrt{x + 1} = 2$ **10.** $\ln x + \ln(x - 3) = 0$

In Exercises 11–18, find the derivative of the function.

11. $g(x) = \ln \sqrt{x}$ **12.** $h(x) = \ln \dfrac{x(x - 1)}{x - 2}$

13. $f(x) = x\sqrt{\ln x}$ **14.** $f(x) = \ln[x(x^2 - 2)^{2/3}]$

15. $y = \dfrac{1}{b^2}\left[\ln(a + bx) + \dfrac{a}{a + bx}\right]$

16. $y = \dfrac{1}{b^2}[a + bx - a\ln(a + bx)]$

17. $y = -\dfrac{1}{a}\ln \dfrac{a + bx}{x}$

18. $y = -\dfrac{1}{ax} + \dfrac{b}{a^2}\ln \dfrac{a + bx}{x}$

In Exercises 19–26, evaluate the integral.

19. $\displaystyle\int \dfrac{1}{7x - 2}\,dx$ **20.** $\displaystyle\int \dfrac{x}{x^2 - 1}\,dx$

21. $\displaystyle\int \frac{\sin x}{1 + \cos x} dx$ **22.** $\displaystyle\int \frac{\ln \sqrt{x}}{x} dx$

23. $\displaystyle\int_1^4 \frac{x + 1}{x} dx$ **24.** $\displaystyle\int_1^e \frac{\ln x}{x} dx$

25. $\displaystyle\int_0^{\pi/3} \sec \theta \, d\theta$ **26.** $\displaystyle\int_0^{\pi/4} \tan\left(\frac{\pi}{4} - x\right) dx$

In Exercises 27 and 28, solve the differential equation.

27. $\dfrac{dy}{dx} = \dfrac{x^2 + 3}{x}$ **28.** $\dfrac{dy}{dx} = \dfrac{x + 2}{2x + 3}$

In Exercises 29–36, (a) find the inverse of the function, (b) sketch the graphs of $f(x)$ and $f^{-1}(x)$ on the same axes, and (c) verify that $f^{-1}[f(x)] = f[f^{-1}(f(x))] = x$.

29. $f(x) = \frac{1}{2}x - 3$ **30.** $f(x) = 5x - 7$

31. $f(x) = \sqrt{x + 1}$ **32.** $f(x) = x^3 + 2$

33. $f(x) = x^2 - 5, \quad x \geq 0$ **34.** $f(x) = \sqrt[3]{x + 1}$

35. $f(x) = \ln \sqrt{x}$ **36.** $f(x) = e^{1-x}$

In Exercises 37 and 38, the function does not have an inverse. Give a restriction on the domain so that the restricted function has an inverse, and then find the inverse.

37. $f(x) = 2(x - 4)^2$ **38.** $f(x) = |x - 2|$

In Exercises 39–56, find the first derivative of the function.

39. $f(x) = \ln(e^{-x^2})$ **40.** $g(x) = \ln \dfrac{e^x}{1 + e^x}$

41. $g(t) = t^2 e^t$ **42.** $h(z) = e^{-z^2/2}$

43. $y = \sqrt{e^{2x} + e^{-2x}}$ **44.** $y = x^{2x+1}$

45. $f(x) = 3^{x-1}$ **46.** $f(x) = (4e)^x$

47. $g(x) = \dfrac{x^2}{e^x}$ **48.** $f(\theta) - \dfrac{1}{2} e^{\sin 2\theta}$

49. $y = \tan(\arcsin x)$ **50.** $y = \arctan(x^2 - 1)$

51. $y = x \operatorname{arcsec} x$ **52.** $y = \frac{1}{2} \arctan e^{2x}$

53. $y = x(\arcsin x)^2 - 2x + 2\sqrt{1 - x^2} \arcsin x$

54. $y = \sqrt{x^2 - 4} - 2 \operatorname{arcsec} \dfrac{x}{2}, \quad 2 < x < 4$

55. $y = 2x - \cosh \sqrt{x}$ **56.** $y = x \tanh^{-1} 2x$

In Exercises 57–60, use implicit differentiation to find dy/dx.

57. $y \ln x + y^2 = 0$ **58.** $\ln(x + y) = x$

59. $\cos x^2 = xe^y$ **60.** $ye^x + xe^y = xy$

61. Find the derivative of each function, given that a is constant.
 a. $y = x^a$ **b.** $y = a^x$
 c. $y = x^x$ **d.** $y = a^a$

62. Show that $y = e^x(a \cos 3x + b \sin 3x)$ satisfies the differential equation $y'' - 2y' + 10y = 0$.

In Exercises 63–78, evaluate the integral.

63. $\displaystyle\int xe^{-3x^2} dx$ **64.** $\displaystyle\int \frac{e^{1/x}}{x^2} dx$

65. $\displaystyle\int \frac{e^{4x} - e^{2x} + 1}{e^x} dx$ **66.** $\displaystyle\int \frac{e^{2x} - e^{-2x}}{e^{2x} + e^{-2x}} dx$

67. $\displaystyle\int \frac{e^x}{e^x - 1} dx$ **68.** $\displaystyle\int x^2 e^{x^3+1} dx$

69. $\displaystyle\int \frac{1}{e^{2x} + e^{-2x}} dx$ **70.** $\displaystyle\int \frac{1}{3 + 25x^2} dx$

71. $\displaystyle\int \frac{x}{\sqrt{1 - x^4}} dx$ **72.** $\displaystyle\int \frac{1}{16 + x^2} dx$

73. $\displaystyle\int \frac{x}{16 + x^2} dx$ **74.** $\displaystyle\int \frac{4 - x}{\sqrt{4 - x^2}} dx$

75. $\displaystyle\int \frac{\arctan(x/2)}{4 + x^2} dx$ **76.** $\displaystyle\int \frac{\arcsin x}{\sqrt{1 - x^2}} dx$

77. $\displaystyle\int \frac{x}{\sqrt{x^4 - 1}} dx$ **78.** $\displaystyle\int x^2 \operatorname{sech}^2 x^3 dx$

In Exercises 79–82, solve the differential equation.

79. $\dfrac{dy}{dt} = te^{-t^2/2}$ **80.** $\dfrac{dy}{dx} = \dfrac{x - 1}{3x^2 - 6x - 1}$

81. $\dfrac{dy}{dx} = \dfrac{e^{-2x}}{1 + e^{-2x}}$ **82.** $\dfrac{dr}{dt} = \dfrac{\tan(1/t)}{t^2}$

In Exercises 83 and 84, sketch the graph of the region whose area is given by the integral, and find the area.

83. $\displaystyle\int_0^{\pi/3} \tan x \, dx$ **84.** $\displaystyle\int_0^1 \frac{1}{x + 1} dx$

In Exercises 85 and 86, find the area of the region bounded by the graphs of the equations.

85. $y = xe^{-x^2}, y = 0, x = 0, x = 4$

86. $y = 3e^{-x/2}, y = 0, x = 0, x = 4$

C **In Exercises 87 and 88, use Simpson's Rule with $n = 4$ to approximate the definite integral.**

87. $\displaystyle\int_0^1 e^{x^3} dx$ **88.** $\displaystyle\int_0^1 e^{-x^2} dx$

89. *Compound Interest* A deposit of $500 earns interest at the rate of 5 percent compounded continuously. Find its value after
 a. 1 year.
 b. 10 years.
 c. 100 years.

90. *Compound Interest*   A deposit earns interest at the rate of r percent compounded continuously and doubles in value in 10 years. Find r.

91. *Compound Interest*   How large a deposit, at 7 percent interest compounded continuously, must be made to obtain a balance of $10,000 in 15 years?

92. *Compound Interest*   A deposit of $2500 is made in a savings account at an annual interest rate of 12 percent compounded continuously. Find the average balance in this account during the first 5 years.

93. *Drug Decomposition*   A solution of a certain drug contains 500 units per milliliter when it is prepared. After 40 days it contains 300 units per milliliter. Assuming that the rate of decomposition is proportional to the amount present, find an equation giving the amount A after t days.

94. *Air Pressure*   Under ideal conditions, air pressure decreases continuously with height above sea level at a rate proportional to the pressure at that height. If the barometer reads 30 inches at sea level and 15 inches at 18,000 feet, find the barometric pressure at 35,000 feet.

Human Memory Model   In Exercises 95 and 96, use the human memory model

$$p(t) = 80e^{-0.5t} + 20$$

where $p(t)$ is the percentage retained after t weeks (see figure).

95. At what rate is information being retained
    a. after 1 week?   b. after 2 weeks?

96. Find the average percentage retained during
    a. the first 2 weeks.   b. the second 2 weeks.

FIGURE FOR 95 AND 96

97. *Probability*   The **exponential density function** is $P = (1/\mu)e^{-t/\mu}$, where t is the time in minutes and μ is the average time between successive events. The definite integral

$$\int_{t_1}^{t_2} \frac{1}{\mu} e^{-t/\mu} \, dt$$

gives the probability that the elapsed time before the next occurrence lies between t_1 and t_2 units of time. Trucks arrive at a terminal at an average rate of 3 per hour ($\mu = 20$

minutes is the average time between arrivals). If a truck has just arrived, find the probability that the next arrival will occur
a. within 10 minutes.
b. within 30 minutes.
c. between 15 and 30 minutes.
d. within 60 minutes.

98. *Probability*   Two numbers between 0 and 10 are chosen at random. The probability that their product is less than $n(0 < n < 100)$ is

$$P = \frac{1}{100}\left(n + \int_{n/10}^{10} \frac{n}{x}\,dx\right).$$

a. What is the probability that the product is less than 25?
b. What is the probability that the product is less than 50?

99. *Vertical Motion*   An object falling from rest encounters air resistance that is proportional to its velocity. If the acceleration due to gravity is considered to be -32 feet per second per second, the net change in velocity is

$$\frac{dv}{dt} = kv - 32.$$

a. Find the velocity of the object as a function of time if the initial velocity is v_0.
b. Use the result of part a to find the limit of the velocity as t approaches infinity.
c. Integrate the velocity function found in part a to find the position function s.

100. *Fuel Economy*   A certain automobile gets 28 miles per gallon of gasoline for speeds up to 50 miles per hour. Over 50 miles per hour, the number of miles per gallon drops at the rate of 12 percent for each 10 miles per hour.
a. If s is the speed and y is the miles per gallon, find y as a function of s by solving the differential equation

$$\frac{dy}{ds} = -0.012y, \quad s > 50.$$

b. Use the function in part a to complete the table.

Speed	50	55	60	65	70
Miles per gallon					

101. *Typing Speed*   The average typing speed (in number of words per minute) after t weeks of lessons is

$$N = \frac{157}{1 + 5.4e^{-0.12t}}.$$

a. Find the limiting number of words per minute as t approaches infinity.
b. Find the rates at which typing speed is changing when $t = 5$ and $t = 14$ weeks.

C 102. *Climb Rate* The time t, in minutes, for a small plane to climb to an altitude of h feet is

$$t = 50 \log_{10} \frac{18{,}000}{18{,}000 - h}$$

where 18,000 feet is the plane's absolute ceiling.

a. Determine the domain of the function appropriate for the context of the problem.

b. Use a graphing utility to graph the time function and identify any asymptotes.

c. As the plane approaches its absolute ceiling, what can be concluded about the time required to further increase its altitude?

d. Find the time at which the altitude is increasing at the greatest rate.

Chapter 6

Applications
of
Integration

6.1

Area of a Region Between Two Curves • Area of a Region Between Intersecting Curves

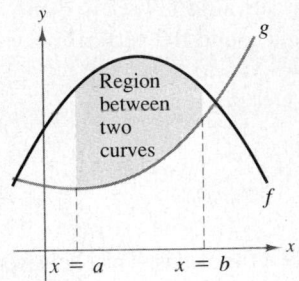

FIGURE 6.1

Area of a Region Between Two Curves

With a few modifications you can extend the application of definite integrals from the area of a region *under* a curve to the area of a region *between* two curves. Consider two functions f and g that are continuous on the interval $[a, b]$. If, as in Figure 6.1, the graphs of both f and g lie above the x-axis, and the graph of g lies below the graph of f, you can geometrically interpret the area of the region between the graphs as the area of the region under the graph of g subtracted from the area of the region under the graph of f, as shown in Figure 6.2.

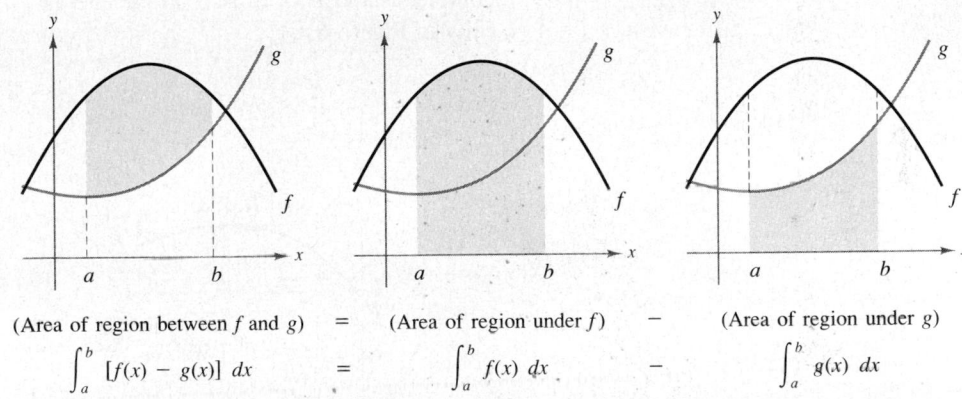

$$(\text{Area of region between } f \text{ and } g) \quad = \quad (\text{Area of region under } f) \quad - \quad (\text{Area of region under } g)$$

$$\int_a^b [f(x) - g(x)] \, dx \quad = \quad \int_a^b f(x) \, dx \quad - \quad \int_a^b g(x) \, dx$$

FIGURE 6.2

To verify the reasonableness of the result shown in Figure 6.2, you can partition the interval $[a, b]$ into n subintervals, each of width Δx. Then, as shown in Figure 6.3, sketch a **representative rectangle** of width Δx and height $f(x_i) - g(x_i)$, where x_i is in the ith interval. The area of the representative rectangle is

$$\Delta A_i = (\text{Height})(\text{Width}) = [f(x_i) - g(x_i)] \Delta x.$$

By adding the areas of the n rectangles and taking the limit as $\| \Delta \| \to 0 \; (n \to \infty)$, you obtain

$$\lim_{n \to \infty} \sum_{i=1}^{n} [f(x_i) - g(x_i)] \Delta x.$$

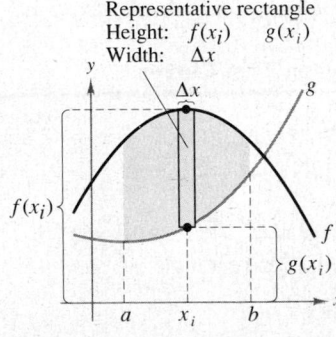

FIGURE 6.3

Because f and g are continuous on $[a, b]$, $f - g$ is also continuous on $[a, b]$ and the limit exists. Therefore, the area of the given region is

$$\text{Area} = \lim_{n \to \infty} \sum_{i=1}^{n} [f(x_i) - g(x_i)] \Delta x = \int_a^b [f(x) - g(x)] \, dx.$$

◀ *The volume of a glass lab container is discussed in Section 6.2, Exercise 44.*

Area of a Region Between Two Curves

If f and g are continuous on $[a, b]$ and $g(x) \leq f(x)$ for all x in $[a, b]$, then the area of the region bounded by the graphs of f and g and the vertical lines $x = a$ and $x = b$ is

$$A = \int_a^b [f(x) - g(x)]\, dx.$$

In Figure 6.1, the graphs of f and g are shown above the x-axis. This, however, is not necessary. The same integrand $[f(x) - g(x)]$ can be used as long as f and g are continuous and $g(x) \leq f(x)$ on the interval $[a, b]$. This result is summarized graphically in Figure 6.4.

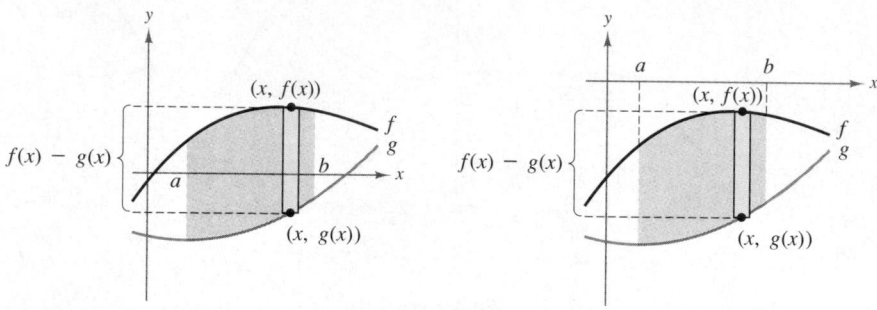

FIGURE 6.4
Height of representative rectangle is $f(x) - g(x)$ regardless of the relative position of the x-axis.

Representative rectangles are used throughout this chapter in various applications of integration. A vertical rectangle (of width Δx) implies integration with respect to x, whereas a horizontal rectangle (of width Δy) implies integration with respect to y.

EXAMPLE 1 . Finding the Area of a Region Between Two Curves

Find the area of the region bounded by the graphs of $y = x^2 + 2$, $y = -x$, $x = 0$, and $x = 1$.

Solution   Let $g(x) = -x$ and $f(x) = x^2 + 2$. Then $g(x) \leq f(x)$ for all x in $[0, 1]$, as shown in Figure 6.5. Thus, the area of the representative rectangle is

$$\Delta A = [f(x) - g(x)]\Delta x = [(x^2 + 2) - (-x)]\Delta x$$

and the area of the region is

$$A = \int_a^b [f(x) - g(x)]\, dx = \int_0^1 [(x^2 + 2) - (-x)]\, dx$$

$$= \left[\frac{x^3}{3} + \frac{x^2}{2} + 2x \right]_0^1$$

$$= \frac{1}{3} + \frac{1}{2} + 2$$

$$= \frac{17}{6}.$$

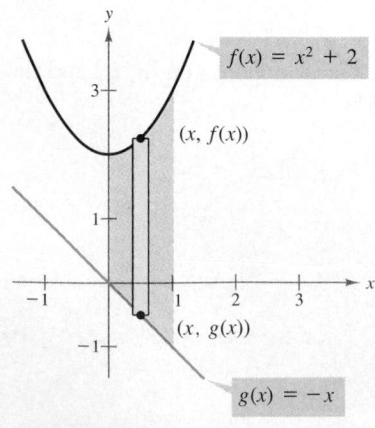

FIGURE 6.5
Region bounded by the graph of f, the graph of g, $x = 0$, and $x = 1$.

Area of a Region Between Intersecting Curves

In Example 1, the graphs of $f(x) = x^2 + 2$ and $g(x) = -x$ do not intersect, and the values of a and b are given explicitly. A more common problem involves the area of a region bounded by two *intersecting* graphs, where the values of a and b must be calculated.

EXAMPLE 2   A Region Lying Between Two Intersecting Graphs

Find the area of the region bounded by the graphs of $f(x) = 2 - x^2$ and $g(x) = x$.

Solution   In Figure 6.6, notice that the graphs of f and g have two points of intersection. To find the x-coordinates of these points, set $f(x)$ and $g(x)$ equal to each other and solve for x.

$$2 - x^2 = x \qquad \text{Set } f(x) \text{ equal to } g(x)$$

$$-x^2 - x + 2 = 0 \qquad \text{Write in standard form}$$

$$-(x + 2)(x - 1) = 0 \qquad \text{Factor}$$

$$x = -2 \text{ or } 1 \qquad \text{Solve for } x$$

Thus, $a = -2$ and $b = 1$. Because $g(x) \leq f(x)$ on the interval $[-2, 1]$, the representative rectangle has an area of

$$\Delta A = [f(x) - g(x)]\Delta x = [(2 - x^2) - x]\Delta x$$

and the area of the region is

$$A = \int_{-2}^{1} [(2 - x^2) - x]\,dx = \left[-\frac{x^3}{3} - \frac{x^2}{2} + 2x \right]_{-2}^{1} = \frac{9}{2}.$$

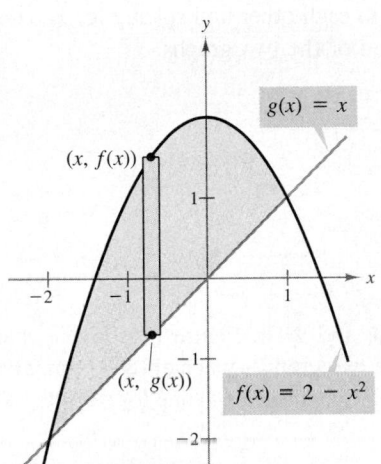

FIGURE 6.6
Region bounded by the graph of f and the graph of g.

EXAMPLE 3   A Region Lying Between Two Intersecting Graphs

The sine and cosine curves intersect infinitely many times bounding regions of equal areas, as shown in Figure 6.7. Find the area of one of these regions.

Solution

$$\sin x = \cos x \qquad \text{Set } f(x) \text{ equal to } g(x)$$

$$\frac{\sin x}{\cos x} = 1 \qquad \text{Divide both sides by } \cos x$$

$$\tan x = 1 \qquad \text{Trigonometric identity}$$

$$x = \frac{\pi}{4} \text{ or } \frac{5\pi}{4}, \qquad 0 \leq x \leq 2\pi \qquad \text{Solve for } x$$

Thus, $a = \pi/4$ and $b = 5\pi/4$. Because $\sin x \geq \cos x$ on the interval $[\pi/4, 5\pi/4]$, the area of the region is

$$A = \int_{\pi/4}^{5\pi/4} [\sin x - \cos x]\,dx = \left[-\cos x - \sin x \right]_{\pi/4}^{5\pi/4}$$

$$= 2\sqrt{2}.$$

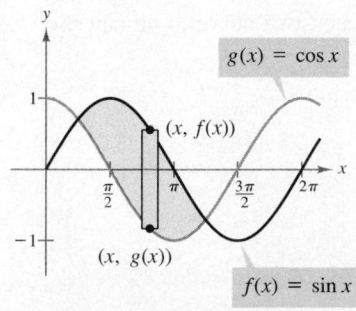

FIGURE 6.7
Region bounded by the graphs of the sine and cosine functions.

If two curves intersect at *more* than two points, then to find the area of the region between the curves, you must find all points of intersection and check to see which curve is above the other in each interval determined by these points.

EXAMPLE 4   Curves that Intersect at More than Two Points

Find the area of the region between the graphs of $f(x) = 3x^3 - x^2 - 10x$ and $g(x) = -x^2 + 2x$.

Solution   Begin by setting $f(x)$ and $g(x)$ equal to each other and solving for x. This will yield the x-values of each point of intersection of the two graphs.

$$f(x) = g(x) \qquad \text{Set } f(x) \text{ equal to } g(x)$$
$$3x^3 - x^2 - 10x = -x^2 + 2x$$
$$3x^3 - 12x = 0 \qquad \text{Write in standard form}$$
$$3x(x^2 - 4) = 0 \qquad \text{Factor}$$
$$x = -2, 0, 2 \qquad \text{Solve}$$

Thus, the two graphs intersect when $x = -2$, 0, and 2. In Figure 6.8, notice that $g(x) \le f(x)$ on the interval $[-2, 0]$. However, the two graphs switch at the origin, and $f(x) \le g(x)$ on the interval $[0, 2]$. Hence, you need two integrals—one for the interval $[-2, 0]$ and one for $[0, 2]$.

$$A = \int_{-2}^{0} [f(x) - g(x)]\,dx + \int_{0}^{2} [g(x) - f(x)]\,dx$$

$$= \int_{-2}^{0} (3x^3 - 12x)\,dx + \int_{0}^{2} (-3x^3 + 12x)\,dx$$

$$= \left[\frac{3x^4}{4} - 6x^2\right]_{-2}^{0} + \left[\frac{-3x^4}{4} + 6x^2\right]_{0}^{2}$$

$$= -(12 - 24) + (-12 + 24)$$

$$= 24$$

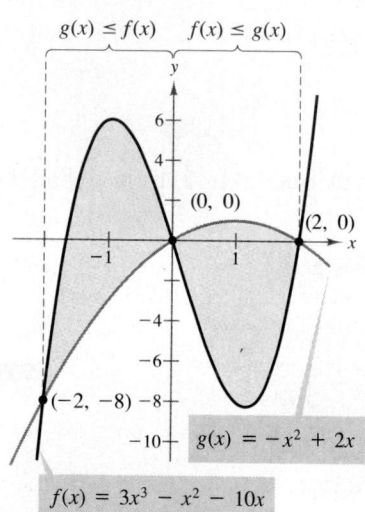

FIGURE 6.8
In $[-2, 0]$ $g(x) \le f(x)$ and in $[0, 2]$ $f(x) \le g(x)$.

If the graph of a function of y is a boundary of a region, it is often convenient to use representative rectangles that are *horizontal* and find the area by integrating with respect to y. In general, to determine the area between two curves, you can use

$$A = \int_{x_1}^{x_2} \underbrace{[(\text{Top curve}) - (\text{Bottom curve})]}_{\text{in variable } x}\,dx \qquad \text{Vertical rectangles}$$

$$A = \int_{y_1}^{y_2} \underbrace{[(\text{Right curve}) - (\text{Left curve})]}_{\text{in variable } y}\,dy \qquad \text{Horizontal rectangles}$$

where (x_1, y_1) and (x_2, y_2) are either adjacent points of intersection of the two curves involved or points on the specified boundary lines.

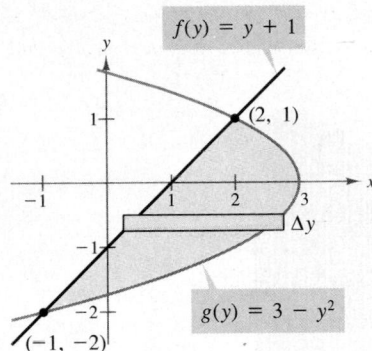

FIGURE 6.9
Horizontal rectangles (Integration with respect to y)

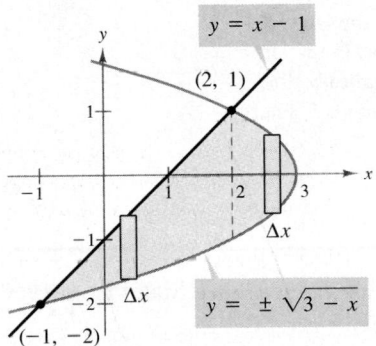

FIGURE 6.10
Vertical rectangles (Integration with respect to x)

EXAMPLE 5   Horizontal Representative Rectangles

Find the area of the region bounded by the graphs of $x = 3 - y^2$ and $y = x - 1$.

Solution   Consider $g(y) = 3 - y^2$ and $f(y) = y + 1$. These two curves intersect when $y = -2$ and $y = 1$, as shown in Figure 6.9. Because $f(y) \le g(y)$ on this interval, you have

$$\Delta A = [g(y) - f(y)]\Delta y = [(3 - y^2) - (y + 1)]\Delta y.$$

Hence, the area is

$$
\begin{aligned}
A &= \int_{-2}^{1} [(3 - y^2) - (y + 1)]\,dy \\
&= \int_{-2}^{1} (-y^2 - y + 2)\,dy \\
&= \left[\frac{-y^3}{3} - \frac{y^2}{2} + 2y \right]_{-2}^{1} \\
&= \left(-\frac{1}{3} - \frac{1}{2} + 2 \right) - \left(\frac{8}{3} - 2 - 4 \right) \\
&= \frac{9}{2}.
\end{aligned}
$$

REMARK   Note in Example 5 that by integrating with respect to y we need only one integral. If we had integrated with respect to x, we would have needed two integrals,

$$A = \int_{-1}^{2} \left[(x - 1) + \sqrt{3 - x} \right] dx + \int_{2}^{3} \left[\sqrt{3 - x} + \sqrt{3 - x} \right] dx$$

as shown in Figure 6.10.

In this section, we developed the integration formula for the area between two curves by using a rectangle as the *representative element*. For each new application in the remaining sections of this chapter, we will construct an appropriate representative element using precalculus formulas you already know. Each integration formula then will be obtained by summing these representative elements.

Known precalculus formula	→	Representative element	→	New integration formula

For example, in this section we developed the area formula as follows.

$$A = (\text{Height})(\text{Width})$$
$$\downarrow$$
$$\Delta A = [f(x) - g(x)]\Delta x$$
$$\downarrow$$
$$A = \int_{a}^{b} [f(x) - g(x)]\,dx$$

FIGURE 6.11

From 1960 to 1979, the consumption of petroleum in the United States followed a pattern that was approximately linear. In the late 1970s, however, when crude oil prices increased dramatically, the consumption pattern changed, as shown above. (*Source:* U.S. Energy Information Administration)

EXAMPLE 6   Consumption of Petroleum

The consumption (in billions of barrels) of petroleum in the United States from 1960 to 1979 can be modeled by

$$f(t) = 0.19t + 5.36, \qquad -10 \le t \le 9$$

where $t = 0$ corresponds to January 1, 1970. From 1979 through 1990, the consumption can be modeled by

$$g(t) = -0.0064t^3 + 0.309t^2 - 4.76t + 29.4, \qquad 9 \le t \le 20$$

as shown in Figure 6.12. Find the total amount of fuel saved from 1979 through 1990 as a result of fuel being consumed at the post-1979 rate rather than at the pre-1979 rate.

Solution   Because the graph of the pre-1979 model lies above the post-1979 graph on the interval [9, 20], the amount of petroleum saved is given by

$$\text{Fuel saved} = \int_{9}^{20} [f(t) - g(t)]\, dt.$$

Using the given models for f and g, you can find this amount to be

$$\text{Fuel saved} = \int_{9}^{20} (0.0064t^3 - 0.309t^2 + 4.95t - 24.04)\, dt$$

$$= \left[0.0016t^4 - 0.103t^3 + 2.475t^2 - 24.04t \right]_{9}^{20}$$

$$\approx 21.7 \text{ billion barrels.}$$

Therefore, approximately 21.7 billion barrels of petroleum was saved.

f: Pre-1979 consumption rate
g: Post-1979 consumption rate

FIGURE 6.12

EXERCISES for Section 6.1

In Exercises 1–6, find the area of the region.

1. $f(x) = x^2 - 6x$
$g(x) = 0$

2. $f(x) = x^2 + 2x + 1$
$g(x) = 2x + 5$

3. $f(x) = x^2 - 4x + 3$
$g(x) = -x^2 + 2x + 3$

4. $f(x) = x^2$
$g(x) = x^3$

5. $f(x) = 3(x^3 - x)$
$g(x) = 0$

6. $f(x) = (x - 1)^3$
$g(x) = x - 1$

In Exercises 7–10, the integrand of the definite integral is a difference of two functions. Sketch the graph of each function and shade the region whose area is represented by the integral.

7. $\int_0^4 \left[(x + 1) - \dfrac{x}{2} \right] dx$

8. $\int_{-1}^1 [(1 - x^2) - (x^2 - 1)] dx$

9. $\int_0^6 \left[4(2^{-x/3}) - \dfrac{x}{6} \right] dx$

10. $\int_{-\pi/3}^{\pi/3} [2 - \sec x] dx$

Approximation **In Exercises 11 and 12, determine which value best approximates the area of the region bounded by the graphs of f and g. (Make your selection on the basis of a sketch of the region and *not* by performing any calculations.)**

11. $f(x) = x + 1, g(x) = (x - 1)^2$
  a. -2 **b.** 2 **c.** 10 **d.** 4 **e.** 8

12. $f(x) = 2 - \frac{1}{2}x, g(x) = 2 - \sqrt{x}$
  a. 1 **b.** 6 **c.** -3 **d.** 3 **e.** 4

In Exercises 13–32, sketch the region bounded by the graphs of the algebraic functions and find the area of the region.

13. $f(x) = x^2 - 4x, g(x) = 0$

14. $f(x) = 3 - 2x - x^2, g(x) = 0$

15. $f(x) = x^2 + 2x + 1, g(x) = 3x + 3$

16. $f(x) = -x^2 + 4x + 2, g(x) = x + 2$

17. $y = x, y = 2 - x, y = 0$

18. $y = \dfrac{1}{x^2}, y = 0, x = 1, x = 5$

19. $f(x) = x(x^2 - 3x + 3), g(x) = x^2$

20. $f(x) = x^3 - 2x + 1, g(x) = -2x, x = 1$

21. $y = x^2 - 4x + 3, y = 3 + 4x - x^2$

22. $y = x^4 - 2x^2, y = 2x^2$

23. $f(x) = \sqrt{3x} + 1, g(x) = x + 1$

24. $f(x) = \sqrt[3]{x}, g(x) = x$

25. $f(x) = \dfrac{1}{1 + x^2}, g(x) = \dfrac{x^2}{2}$

26. $f(x) = \dfrac{6x}{x^2 + 1}, y = 0, 0 \le x \le 3$

27. $f(y) = y^2, g(y) = y + 2$

28. $f(y) = y(2 - y), g(y) = -y$

29. $f(y) = y^2 + 1, g(y) = 0, y = -1, y = 2$

30. $f(y) = \dfrac{y}{\sqrt{16 - y^2}}, g(y) = 0, y = 3$

31. $f(x) = \dfrac{4}{x}, x = 0, y = 1, y = 4$

32. $g(x) = \dfrac{4}{2 - x}, y = 4, x = 0$

* C In Exercises 33 and 34, use a symbolic integration utility to graph the region bounded by the graphs of the functions and find the area of the region.

33. $f(x) = x^4, g(x) = 3x + 4$

34. $f(x) = x^6, g(x) = x + 2$

*A **C** indicates an exercise in which you are instructed to use graphing technology or a symbolic computer algebra system. The solutions of other exercises may also be facilitated by appropriate technology.

In Exercises 35–42, sketch the region bounded by the graphs of the transcendental functions and find the area of the region.

35. $f(x) = 2 \sin x$, $g(x) = \tan x$, $-\dfrac{\pi}{3} \le x \le \dfrac{\pi}{3}$

36. $f(x) = \sin 2x$, $g(x) = \cos x$, $\dfrac{\pi}{6} \le x \le \dfrac{5\pi}{6}$

37. $f(x) = 2 \sin x + \sin 2x$, $y = 0$, $0 \le x \le \pi$

38. $f(x) = 2 \sin x + \cos 2x$, $y = 0$, $0 \le x \le \pi$

39. $f(x) = xe^{-x^2}$, $y = 0$, $0 \le x \le 1$

40. $f(x) = \dfrac{1}{x^2} e^{1/x}$, $y = 0$, $1 \le x \le 3$

41. $g(x) = \dfrac{4 \ln x}{x}$, $y = 0$, $x = 5$

42. $f(x) = 3^x$, $g(x) = 2x + 1$

In Exercises 43 and 44, use integration to find the area of the triangle having the given vertices.

43. $(0, 0)$, $(a, 0)$, (b, c) **44.** $(2, -3)$, $(4, 6)$, $(6, 1)$

In Exercises 45 and 46, find b so that the line $y = b$ divides the region bounded by the graphs of the two equations into two regions of equal area.

45. $y = 9 - x^2$, $y = 0$ **46.** $y = 9 - |x|$, $y = 0$

47. *Essay* The graphs of $y = x^4 - 2x^2 + 1$ and $y = 1 - x^2$ intersect at three points. However, the area between the curves *can* be found by a single integral. Explain why this is so, and write an integral for this area.

48. *Essay* The area of the region bounded by the graphs of $y = x^3$ and $y = x$ *cannot* be found by the single integral

$$\int_{-1}^{1} (x^3 - x)\, dx.$$

Explain why this is so. Use symmetry to write a single integral that does represent the area.

In Exercises 49 and 50, find the area of the region bounded by the graph of f and the tangent line to the graph of f at the specified point.

Function	Point
49. $f(x) = x^3$	$(1, 1)$
50. $f(x) = \dfrac{1}{x^2 + 1}$	$\left(1, \dfrac{1}{2}\right)$

C In Exercises 51–54, use Simpson's Rule with $n = 10$ to approximate the area of the region bounded by the graphs of the equations.

51. $y = \sqrt{1 + x^3}$, $y = \tfrac{1}{2}x + 2$, $x = 0$

52. $y = \sqrt{xe^x}$, $y = 0$, $x = 0$, $x = 1$

53. $y = \sqrt{\dfrac{x^3}{4 - x}}$, $y = 0$, $x = 3$

54. $y = x\sqrt{\dfrac{4 - x}{4 + x}}$, $y = 0$, $x = 4$

In Exercises 55 and 56, evaluate the limit and sketch the graph of the region whose area is represented by the limit.

55. $\displaystyle \lim_{\|\Delta\| \to 0} \sum_{i=1}^{n} (x_i - x_i^2)\, \Delta x$

where $x_i = i/n$ and $\Delta x = 1/n$

56. $\displaystyle \lim_{\|\Delta\| \to 0} \sum_{i=1}^{n} (4 - x_i^2)\, \Delta x$

where $x_i = -2 + (4i/n)$ and $\Delta x = 4/n$

Revenue In Exercises 57 and 58, two models R_1 and R_2 are given for revenue (in billions of dollars) for a large corporation. Both models project the annual revenue from 1995 through 2000, with $t = 0$ representing 1995. The second model, however, assumes a decrease in growth of corporate sales over the period. Approximate the total reduction in revenue if corporate sales are actually closer to the model R_2.

57. $R_1 = 7.21 + 0.58t$ **58.** $R_1 = 7.21 + 0.26t + 0.02t^2$
     $R_2 = 7.21 + 0.45t$ $R_2 = 7.21 + 0.1t + 0.01t^2$

59. *Beef Consumption* For the years 1980 through 1990, the consumption of beef (in billions of pounds) in the United States can be modeled by

$$f(t) = \begin{cases} 23.613 + 0.270t + 0.019t^2, & 0 \le t < 5 \\ 23.333 + 0.784t - 0.073t^2, & 5 \le t \le 10. \end{cases}$$

a. Use a graphing utility to graph this model.
b. Suppose the beef consumption from 1985 through 1990 had continued to follow the model given for the years 1980 through 1985. How much additional beef would have been consumed from 1985 through 1990?

60. *Fuel Cost* For the years 1995 to 2005, the projected fuel cost C (in millions of dollars) for a corporation is

$$C_1 = 568.50 + 7.15t$$

where t is time in years with $t = 0$ corresponding to 1995. Because of the installation of fuel-saving equipment, a more accurate model for fuel costs for the period is

$$C_2 = 525.60 + 6.43t.$$

Approximate the savings for the 10-year period due to the installation of the new equipment.

61. *Building Design* Precast concrete sections for a new educational building have the dimensions (in meters) and shape shown in the figure.
   a. Find the area of the face of the section superimposed on the rectangular coordinate system.
   b. Find the volume of concrete in one of the sections by multiplying the area of part a by 2 meters.
   c. One cubic meter of concrete weighs 5000 pounds. Find the weight of a section.

FIGURE FOR 61

62. *Building Design* To decrease weight and to aid in the hardening process, sections such as those in Exercise 61 often are not solid. Rework Exercise 61 so as to allow for cylindrical openings such as those shown in the accompanying figure.

FIGURE FOR 62

In Exercises 63–66, find the consumer surplus and producer surplus for the supply and demand curves. The consumer surplus and producer surplus are represented by the areas shown in the figure.

Demand Function	*Supply Function*
63. $p_1(x) = 50 - 0.5x$	$p_2(x) = 0.125x$
64. $p_1(x) = 1000 - 0.4x^2$	$p_2(x) = 42x$
65. $p_1(x) = \dfrac{10,000}{\sqrt{x + 100}}$	$p_2(x) = 100\sqrt{0.05x + 10}$
66. $p_1(x) = \sqrt{25 - 0.1x}$	$p_2(x) = \sqrt{9 + 0.1x} - 2$

FIGURE FOR 63–66

True or False **In Exercises 67 and 68, determine whether the statement is true or false. If it is false, explain why or give an example that shows it is false.**

67. If the area of the region bounded by the graphs of f and g is 1, then the area of the region bounded by the graphs of $h(x) = f(x) + C$ and $k(x) = g(x) + C$ is also 1.

68. If $\displaystyle\int_a^b [f(x) - g(x)]\,dx = A,$

then $\displaystyle\int_a^b [g(x) - f(x)]\,dx = -A.$

69. *Essay* The **distance** between two continuous functions f and g over the closed interval $[a, b]$ is defined as

$$D(f, g) = \sqrt{\frac{1}{b - a}\int_a^b [f(x) - g(x)]^2\,dx}.$$

What is the distance between the constant functions $f(x) = c$ and $g(x) = d$ over the closed interval $[a, b]$? Determine the distances between other functions and discuss your results.

FOR FURTHER INFORMATION For more information on this definition, see the article "Integrals, With Applications" by Clark Kimberling in the April, 1987 issue of *Mathematics Teacher*.

70. Evaluate the integral

$$\int_{-2}^2 \left| x^3 + x^2 - 2x \right| dx.$$

Give a geometric interpretation of the integral.

SECTION

6.2 Volume: The Disc Method

The Disc Method • The Washer Method • Solids with Known Cross Sections

The Disc Method

In Chapter 4, we mentioned that area is only *one* of the many applications of the definite integral. Another important application is its use in finding the volume of a three-dimensional solid. In this section you will study a particular type of three-dimensional solid—one whose cross sections are similar. We begin with solids of revolution. Such solids are used commonly in engineering and manufacturing. Some examples are axles, funnels, pills, bottles, and pistons, as indicated in Figure 6.13.

FIGURE 6.13   Solids of Revolution
Solids of Revolution

FIGURE 6.14
Volume of a disc: $\pi R^2 w$

If a region in the plane is revolved about a line, the resulting solid is a **solid of revolution,** and the line is called the **axis of revolution.** The simplest such solid is a right circular cylinder or **disc,** which is formed by revolving a rectangle about an axis adjacent to one side of the rectangle, as shown in Figure 6.14. The volume of such a disc is

Volume of disc = (Area of disc)(Width of disc)

$$= \pi R^2 w$$

where R is the radius of the disc and w is the width.

FIGURE 6.15
The Disc Method

To see how to use the volume of a disc to find the volume of a general solid of revolution, consider a solid of revolution formed by revolving the plane region in Figure 6.15 about the indicated axis. To determine the volume of this solid, consider a representative rectangle in the plane region. When this rectangle is revolved about the axis of revolution, it generates a representative disc whose volume is

$$\Delta V = \pi R^2 \, \Delta x.$$

Approximating the volume of the solid by n such discs of width Δx and radius $R(x_i)$ produces

$$\text{Volume of solid } \approx \sum_{i=1}^{n} \pi \, [R(x_i)]^2 \, \Delta x = \pi \sum_{i=1}^{n} [R(x_i)]^2 \, \Delta x.$$

This approximation appears to become better and better as $\|\Delta\| \to 0$ $(n \to \infty)$. Therefore, we define the volume of the solid as

$$\text{Volume of solid } = \lim_{\|\Delta\| \to 0} \pi \sum_{i=1}^{n} [R(x_i)]^2 \, \Delta x$$

$$= \pi \int_{a}^{b} [R(x)]^2 \, dx.$$

Schematically, the disc method looks like this.

Known precalculus formula		*Representative element*		*New integration formula*
Volume of disc $V = \pi R^2 w$	$\rightarrow$	$\Delta V = \pi [R(x_i)]^2 \, \Delta x$	$\rightarrow$	Solid of revolution $V = \pi \int_{a}^{b} [R(x)]^2 \, dx$

A similar formula can be derived if the axis of revolution is vertical.

$$V = \pi \int_{a}^{b} [R(x)]^2 \, dx$$

$R(x)$

a b

Horizontal Axis of Revolution

$$V = \pi \int_{c}^{d} [R(y)]^2 \, dy$$

d

c

$R(y)$

Vertical Axis of Revolution

FIGURE 6.16

The Disc Method

To find the volume of a solid of revolution with the **disc method,** use one of the following as indicated in Figure 6.16.

Horizontal axis of revolution	*Vertical axis of revolution*
Volume $= V = \pi \displaystyle\int_{a}^{b} [R(x)]^2 \, dx$	Volume $= V = \pi \displaystyle\int_{c}^{d} [R(y)]^2 \, dy$

REMARK In Figure 6.16, note that you can determine the variable of integration by placing a representative rectangle in the *plane* region "perpendicular" to the axis of revolution. If the width of the rectangle is Δx, integrate with respect to x, and if the width of the rectangle is Δy, integrate with respect to y.

The simplest application of the disc method involves a plane region bounded by the graph of f and the x-axis. If the axis of revolution is the x-axis, then the radius $R(x)$ is simply $f(x)$.

EXAMPLE 1   Using the Disc Method

Find the volume of the solid formed by revolving the region bounded by the graph of

$$f(x) = \sqrt{\sin x}$$

and the x-axis $(0 \le x \le \pi)$ about the x-axis.

Solution   From the representative rectangle in Figure 6.17, you can see that the radius of this solid is

$$R(x) = f(x) = \sqrt{\sin x}.$$

Thus, the volume of the solid of revolution is

$$
\begin{aligned}
V = \pi \int_a^b [R(x)]^2 \, dx &= \pi \int_0^\pi \left(\sqrt{\sin x} \right)^2 dx \\
&= \pi \int_0^\pi \sin x \, dx \\
&= -\pi \cos x \Big]_0^\pi \\
&= \pi(1 + 1) \\
&= 2\pi.
\end{aligned}
$$

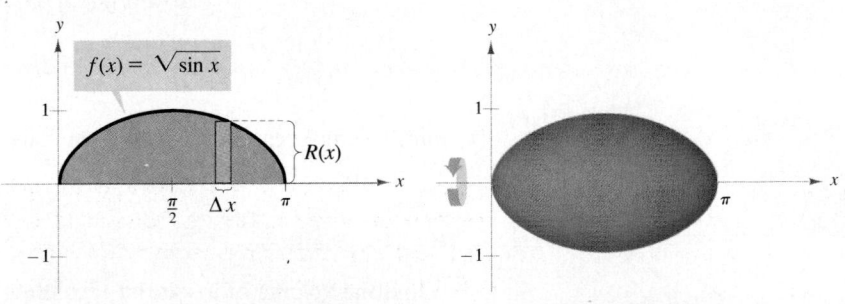

Plane Region   Solid of Revolution

FIGURE 6.17

REMARK   Note in Example 1 that the problem was solved *without* referring to the three-dimensional portion of the sketch in Figure 6.17. In general, to set up an integral to find the volume of a solid of revolution, a sketch of the plane region is more useful than a sketch of the solid, because the radius is more easily visualized in the plane region.

TECHNOLOGY   Some graphing utilities will allow you to view a solid of revolution. If you have access to such software, try using it to sketch the solid of revolution described in Example 1. Figure 6.18 shows an example of the type of display you might see.

Generated by Mathematica

FIGURE 6.18

EXAMPLE 2   Revolving About a Line that is Not a Coordinate Axis

Find the volume of the solid formed by revolving the region bounded by $f(x) = 2 - x^2$ and $g(x) = 1$ about the line $y = 1$, as shown in Figure 6.19.

Solution   Begin by finding the points of intersection of f and g.

$$2 - x^2 = 1 \qquad \text{Set } f(x) \text{ equal to } g(x)$$
$$x^2 = 1$$
$$x = \pm 1 \qquad \text{Solve for } x$$

To find the radius, subtract $g(x)$ from $f(x)$.

$$\begin{aligned} R(x) &= f(x) - g(x) \\ &= (2 - x^2) - 1 \\ &= 1 - x^2 \end{aligned}$$

Finally, integrate between -1 and 1 to find the volume.

$$\begin{aligned} V &= \pi \int_a^b [R(x)]^2\, dx = \pi \int_{-1}^{1} (1 - x^2)^2\, dx \\ &= \pi \int_{-1}^{1} (1 - 2x^2 + x^4)\, dx \\ &= \left[\pi \left(x - \frac{2x^3}{3} + \frac{x^5}{5} \right) \right]_{-1}^{1} \\ &= \frac{16\pi}{15} \end{aligned}$$

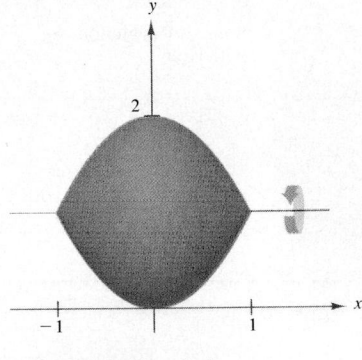

Plane Region

Solid of Revolution

FIGURE 6.19

The Washer Method

The disc method can be extended to cover solids of revolution with a hole by replacing the representative disc with a representative **washer.** The washer is formed by revolving a rectangle about an axis, shown in Figure 6.20. If r and R are the inner and outer radii of the washer and w is the width of the washer, then the volume is given by

$$\text{Volume of washer} = \pi(R^2 - r^2)w.$$

Axis of Revolution

Solid of Revolution

FIGURE 6.20

To see how this concept can be used to find the volume of a solid of revolution, consider a region bounded by an **outer radius** $R(x)$ and an **inner radius** $r(x)$, as shown in Figure 6.21. If the region is revolved about its axis of revolution, then the volume of the resulting solid is given by

$$V = \pi \int_a^b [R(x)]^2 \, dx - \pi \int_a^b [r(x)]^2 \, dx$$

$$V = \pi \int_a^b \left([R(x)]^2 - [r(x)]^2 \right) dx. \qquad \text{Washer method}$$

Note that the integral involving the inner radius represents the volume of the hole and is *subtracted* from the integral involving the outer radius.

FIGURE 6.21

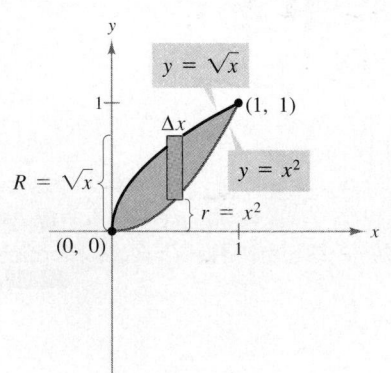

Plane Region

EXAMPLE 3   Using the Washer Method

Find the volume of the solid formed by revolving the region bounded by the graphs of $y = \sqrt{x}$ and $y = x^2$ about the x-axis, as shown in Figure 6.22.

Solution   In Figure 6.22, you can see that the outer and inner radii are as follows.

$$R(x) = \sqrt{x} \qquad\qquad \text{Outer radius}$$
$$r(x) = x^2 \qquad\qquad \text{Inner radius}$$

Integrating between 0 and 1 produces

$$V = \pi \int_a^b \left([R(x)]^2 - [r(x)]^2 \right) dx$$

$$= \pi \int_0^1 (x - x^4) \, dx$$

$$= \pi \left[\frac{x^2}{2} - \frac{x^5}{5} \right]_0^1$$

$$= \frac{3\pi}{10}.$$

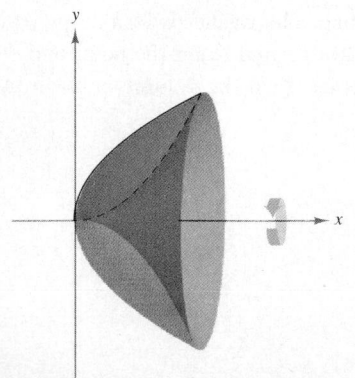

Solid of Revolution

FIGURE 6.22

In each example so far, the axis of revolution has been *horizontal* and we have integrated with respect to *x*. In the next example, the axis of revolution is *vertical* and (using the disc method) you must integrate with respect to *y*. In this example, you need two separate integrals to compute the volume.

EXAMPLE 4   Integrating with Respect to *y*, Two-Integral Case

Find the volume of the solid formed by revolving the region bounded by the graphs of $y = x^2 + 1$, $y = 0$, $x = 0$, and $x = 1$ about the *y*-axis, as shown in Figure 6.23.

For $1 \le y \le 2$:
$R = 1$
$r = \sqrt{y - 1}$

For $0 \le y \le 1$:
$R = 1$
$r = 0$

Plane Region   Solid of Revolution

FIGURE 6.23

Solution   For the region shown in Figure 6.23, the outer radius is simply $R = 1$. There is, however, no convenient formula to represent the inner radius. When $0 \le y \le 1$, then $r = 0$, but when $1 \le y \le 2$, r is determined by the equation $y = x^2 + 1$, which implies that $r = \sqrt{y - 1}$.

$$r(y) = \begin{cases} 0, & 0 \le y \le 1 \\ \sqrt{y - 1}, & 1 \le y \le 2 \end{cases}$$

Using this definition of the inner radius, you can use two integrals to find the volume.

$$V = \pi \int_0^1 (1^2 - 0^2) \, dy + \pi \int_1^2 \left[1^2 - \left(\sqrt{y - 1} \right)^2 \right] dy$$

$$= \pi \int_0^1 1 \, dy + \pi \int_1^2 (2 - y) \, dy$$

$$= \pi y \Big]_0^1 + \pi \left[2y - \frac{y^2}{2} \right]_1^2$$

$$= \pi \left(1 + 4 - 2 - 2 + \frac{1}{2} \right)$$

$$= \frac{3\pi}{2}$$

Note that the first integral $\pi \int_0^1 1 \, dy$ represents the volume of a right circular cylinder of radius 1 and height 1. This portion of the volume could have been determined without using calculus.

EXAMPLE 5   Manufacturing

A manufacturer drills a hole through the center of a metal sphere of radius 5 inches, as shown in Figure 6.24(a). The hole has a radius of 3 inches. What is the volume of the resulting metal ring?

Solution   You can imagine the ring to be generated by a segment of the circle whose equation is $x^2 + y^2 = 25$, as shown in Figure 6.24(b). Because the radius of the hole is 3 inches, you can let $y = 3$ and solve the equation $x^2 + y^2 = 25$ to determine that the limits of integration are $x = \pm 4$. Thus, the inner and outer radii are

$$r(x) = 3 \quad \text{and} \quad R(x) = \sqrt{25 - x^2}$$

and the volume is given by

$$
\begin{aligned}
V &= \pi \int_a^b ([R(x)]^2 - [r(x)]^2)\, dx \\
&= \pi \int_{-4}^4 \left[\left(\sqrt{25 - x^2}\right)^2 - (3)^2 \right] dx \\
&= \pi \int_{-4}^4 (16 - x^2)\, dx \\
&= \pi \left[16x - \frac{x^3}{3} \right]_{-4}^4 \\
&= \frac{256\pi}{3} \, \text{in}^3.
\end{aligned}
$$

(a) Solid of revolution

(b) Plane region

FIGURE 6.24

Solids with Known Cross Sections

With the disc method, you can find the volume of a solid having a circular cross section whose area is

$$A = \pi R^2.$$

This method can be generalized to solids of any shape, as long as you know a formula for the area of an arbitrary cross section. Some common cross sections are squares, rectangles, triangles, semicircles, and trapezoids.

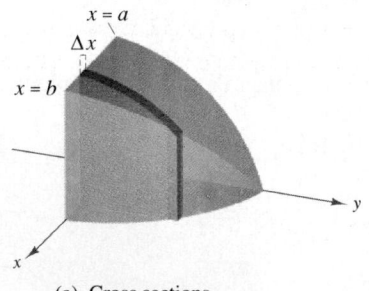

(a) Cross sections perpendicular to x-axis

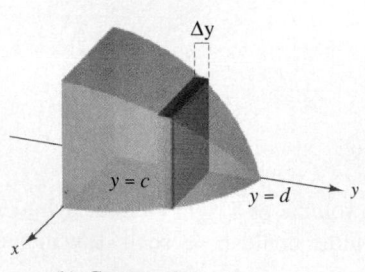

(b) Cross sections perpendicular to y-axis

FIGURE 6.25

Volumes of Solids with Known Cross Sections

1. For cross sections of area $A(x)$ taken perpendicular to the x-axis,

$$\text{Volume} = \int_a^b A(x)\, dx$$

as shown in Figure 6.25(a).

2. For cross sections of area $A(y)$ taken perpendicular to the y-axis,

$$\text{Volume} = \int_c^d A(y)\, dy$$

as shown in Figure 6.25(b).

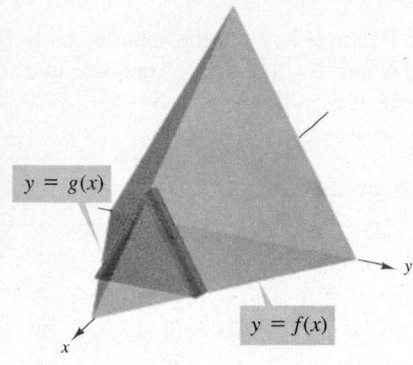

$y = g(x)$

$y = f(x)$

x

y

Cross sections are equilateral triangles.

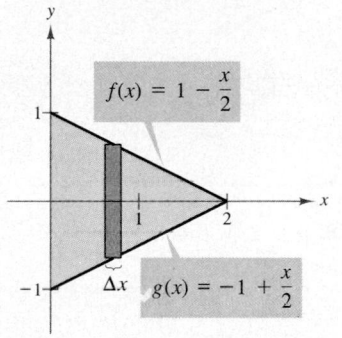

$f(x) = 1 - \dfrac{x}{2}$

$g(x) = -1 + \dfrac{x}{2}$

Δx

Triangular base in xy-plane

FIGURE 6.26

EXAMPLE 6 Triangular Cross Sections

Find the volume of the solid shown in Figure 6.26. The base of the solid is the region bounded by the lines

$$f(x) = 1 - \frac{x}{2}, \quad g(x) = -1 + \frac{x}{2}, \quad \text{and} \quad x = 0.$$

The cross sections perpendicular to the x-axis are equilateral triangles.

Solution The base and area of each triangular cross section are

$$\text{Base} = \left(1 - \frac{x}{2}\right) - \left(-1 + \frac{x}{2}\right) = 2 - x \qquad \text{Length of base}$$

$$\text{Area} = \frac{\sqrt{3}}{4}(\text{Base})^2 \qquad \text{Area of equilateral triangle}$$

$$A(x) = \frac{\sqrt{3}}{4}(2 - x)^2. \qquad \text{Area of cross section}$$

Because x ranges from 0 to 2, the volume of the solid is

$$V = \int_a^b A(x)\, dx = \int_0^2 \frac{\sqrt{3}}{4}(2 - x)^2 \, dx$$

$$= -\frac{\sqrt{3}}{4}\left[\frac{(2 - x)^3}{3}\right]_0^2$$

$$= \frac{2\sqrt{3}}{3}.$$

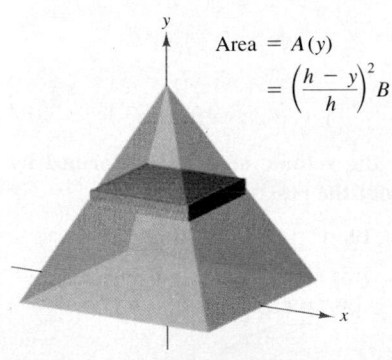

Area = $A(y)$

$= \left(\dfrac{h - y}{h}\right)^2 B$

x

y

Area of base $= B = b^2$

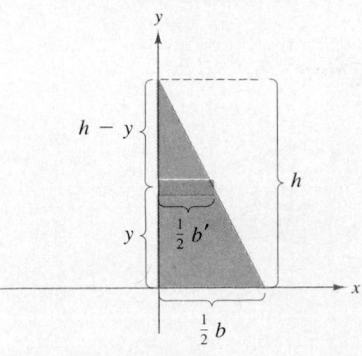

$h - y$

y

$\frac{1}{2} b'$

h

$\frac{1}{2} b$

FIGURE 6.27

EXAMPLE 7 An Application to Geometry

Prove that the volume of a pyramid with a square base is $V = \frac{1}{3}hB$, where h is the height of the pyramid and B is the area of the base.

Solution As shown in Figure 6.27, you can intersect the pyramid with a plane parallel to the base at height y to form a square cross section whose sides are of length b'. Using similar triangles, you can show that

$$\frac{b'}{b} = \frac{h - y}{h} \quad \text{or} \quad b' = \frac{b}{h}(h - y)$$

where b is the length of the sides of the base of the pyramid. Thus,

$$A(y) = (b')^2 = \frac{b^2}{h^2}(h - y)^2.$$

Integrating between 0 and h produces

$$V = \int_0^h \frac{b^2}{h^2}(h - y)^2 \, dy = \frac{b^2}{h^2}\int_0^h (h - y)^2 \, dy$$

$$= -\left(\frac{b^2}{h^2}\right)\frac{(h - y)^3}{3}\bigg]_0^h$$

$$= \frac{b^2}{h^2}\left(\frac{h^3}{3}\right)$$

$$= \frac{1}{3}hB. \qquad B = b^2$$

An interesting generalization of the formula in Example 7 is that the volume of *any* cone can be determined by the formula $V = \frac{1}{3}hB$, where B is the area of the base and h is the height of the cone, as shown in Figure 6.28.

Square base of area B Circular base of area B Irregular base of area B

EXERCISES for Section 6.2

TECHNOLOGY
Laboratory Guide
Labs 6.2–6.3

In Exercises 1–8, find the volume of the solid formed by revolving the region about the x-axis.

1. $y = -x + 1$

2. $y = 4 - x^2$

7. $y = x^2,\ y = x^3$

8. $y = 2,\ y = 4 - \dfrac{x^2}{2}$

3. $y = \sqrt{4 - x^2}$

4. $y = x^2$

In Exercises 9–12, find the volume of the solid formed by revolving the region about the y-axis.

9. $y = x^2$

10. $y = \sqrt{16 - x^2}$

5. $y = \sqrt{x}$

6. $y = \sqrt{4 - x^2}$

11. $y = x^{2/3}$

12. $x = -y^2 + 4y$

In Exercises 13–16, find the volume of the solid generated by revolving the region bounded by the graphs of the equations about the indicated lines.

13. $y = \sqrt{x}$, $y = 0$, $x = 4$
  a. the x-axis   **b.** the y-axis
  c. the line $x = 4$ **d.** the line $x = 6$

14. $y = 2x^2$, $y = 0$, $x = 2$
  a. the y-axis   **b.** the x-axis
  c. the line $y = 8$ **d.** the line $x = 2$

15. $y = x^2$, $y = 4x - x^2$
  a. the x-axis   **b.** the line $y = 6$

16. $y = 6 - 2x - x^2$, $y = x + 6$
  a. the x-axis   **b.** the line $y = 3$

In Exercises 17–20, find the volume of the solid generated by revolving the region bounded by the graphs of the equations about the line $y = 4$.

17. $y = x$, $y = 3$, $x = 0$ **18.** $y = x^2$, $y = 4$

19. $y = \dfrac{1}{x}$, $y = 0$, $x = 1$, $x = 4$

20. $y = \sec x$, $y = 0$, $0 \le x \le \dfrac{\pi}{3}$

In Exercises 21–24, find the volume of the solid generated by revolving the region bounded by the graphs of the equations about the line $x = 6$.

21. $y = x$, $y = 0$, $y = 4$, $x = 6$

22. $y = 6 - x$, $y = 0$, $y = 4$, $x = 0$

23. $x = y^2$, $x = 4$ **24.** $xy = 6$, $y = 2$, $y = 6$, $x = 6$

In Exercises 25–32, find the volume of the solid generated by revolving the region bounded by the graphs of the equations about the x-axis.

25. $y = \dfrac{1}{\sqrt{x + 1}}$, $y = 0$, $x = 0$, $x = 3$

26. $y = x\sqrt{4 - x^2}$, $y = 0$

27. $y = \dfrac{1}{x}$, $y = 0$, $x = 1$, $x = 4$

28. $y = \dfrac{3}{x + 1}$, $y = 0$, $x = 0$, $x = 8$

29. $y = e^{-x}$, $y = 0$, $x = 0$, $x = 1$

30. $y = e^{x/2}$, $y = 0$, $x = 0$, $x = 4$

31. $y = \sqrt{\sin x}$, $y = 0$, $x = 0$, $x = \dfrac{\pi}{2}$

32. $y = \sqrt{\cos x}$, $y = 0$, $x = 0$, $x = \dfrac{\pi}{2}$

Approximation **In Exercises 33 and 34, determine which value best approximates the volume of the solid generated by revolving the region bounded by the graphs of the equations**

about the x-axis. (Make your selection on the basis of a sketch of the solid and *not* by performing any calculations.)

33. $y = e^{-x^2/2}$, $y = 0$, $x = 0$, $x = 2$
  a. 3   **b.** −5   **c.** 10   **d.** 7   **e.** 20

34. $y = \arctan x$, $y = 0$, $x = 0$, $x = 1$
  a. 10   **b.** $\frac{3}{4}$ **c.** 5   **d.** −6   **e.** 15

35. a. The region bounded by the parabola $y = 4x - x^2$ and the x-axis is revolved about the x-axis. Find the volume of the resulting solid.
  b. If the equation of the parabola in part a were changed to $y = 4 - x^2$, would the volume of the solid generated be different? Why or why not?

36. Find the volume of the solid generated if the upper half of the ellipse $9x^2 + 25y^2 = 225$ is revolved about
  a. the x-axis to form a prolate spheroid (shaped like a football).
  b. the y-axis to form an oblate spheroid (shaped like an M&M candy).

37. If the portion of the line $y = \frac{1}{2}x$ lying in the first quadrant is revolved about the x-axis, a cone is generated. Find the volume of the cone extending from $x = 0$ to $x = 6$.

38. Use the disc method to verify that the volume of a right circular cone is $\frac{1}{3}\pi r^2 h$, where r is the radius of the base and h is the height.

39. Use the disc method to verify that the volume of a sphere of radius r is $\frac{4}{3}\pi r^3$.

40. A sphere of radius r is cut by a plane h ($h < r$) units above the equator. Find the volume of the solid (spherical segment) above the plane.

41. A cone with a base of radius r and height H is cut by a plane parallel to and h units above the base. Find the volume of the solid (frustum of a cone) below the plane.

42. The region bounded by $y = \sqrt{x}$, $y = 0$, $x = 0$, and $x = 4$ is revolved about the x-axis.
  a. Find the value of x on the interval $[0, 4]$ that divides the solid into two parts of equal volume.
  b. Find the values of x on the interval $[0, 4]$ that divide the solid into three parts of equal volume.

43. *Volume of a Fuel Tank*   A tank on the wing of a jet is formed by revolving the region bounded by the graph of $y = \frac{1}{8}x^2\sqrt{2 - x}$ and the x-axis about the x-axis (see figure), where x and y are measured in meters. Find the volume of the tank.

$y = \frac{1}{8}x^2\sqrt{2 - x}$

FIGURE FOR 43

44. *Volume of Lab Glass*   A glass container can be modeled by revolving the graph of

$$y = \begin{cases} \sqrt{0.1x^3 - 2.2x^2 + 10.9x + 22.2} & 0 \le x \le 11.5 \\ 2.95 & 11.5 < x \le 15 \end{cases}$$

about the *x*-axis, where *x* and *y* are measured in centimeters. Sketch this container and find its volume.

45. *Volume of a Water Tank*   A tank on a water tower is a sphere of radius 50 feet. Determine the depth of the water when the tank is filled to one-fourth and three-fourths of its total capacity. (*Note:* Use Newton's Method after evaluating the definite integral.)

46. *Mechanical Design*   A draftsman is asked to determine the amount of material required to produce a machine part (see figure). The diameters *d* of the part at equally spaced points *x* are listed in the table. Use these data with Simpson's Rule to approximate the volume of the part. The measurements are listed in centimeters.

x	0	1	2	3	4	5
d	4	4.1	4.2	4.7	5.2	5.7

x	6	7	8	9	10
d	6.0	5.9	4.9	4.4	4.6

FIGURE FOR 46

C   In Exercises 47–49, use Simpson's Rule with *n* = 10 to approximate the volume of the solid generated by revolving the region bounded by the graphs of the equations about the *x*-axis.

47. $y = e^{-x^2}$, $y = 0$, $x = 0$, $x = 2$

48. $y = \ln x$, $y = 0$, $x = 1$, $x = 3$

49. $y = e^{x/2} + e^{-x/2}$, $y = 0$, $x = -1$, $x = 2$

50. Match each of the integrals with the solid whose volume it represents, and give the dimensions of each solid.
   a. Right circular cylinder   **b.** Ellipsoid
   c. Sphere   **d.** Right circular cone
   e. Torus

_____ $\pi \int_0^h \left(\dfrac{rx}{h}\right)^2 dx$

_____ $\pi \int_0^h r^2 \, dx$

_____ $\pi \int_{-r}^{r} \left(\sqrt{r^2 - x^2}\right)^2 dx$

_____ $\pi \int_{-b}^{b} \left(a\sqrt{1 - \dfrac{x^2}{b^2}}\right)^2 dx$

_____ $\pi \int_{-r}^{r} \left[(R + \sqrt{r^2 - x^2})^2 - (R - \sqrt{r^2 - x^2})^2\right] dx$

51. Find the volume of the solid whose base is bounded by the circle $x^2 + y^2 = 4$, with the indicated cross sections taken perpendicular to the *x*-axis.
   a. Squares   **b.** Equilateral triangles

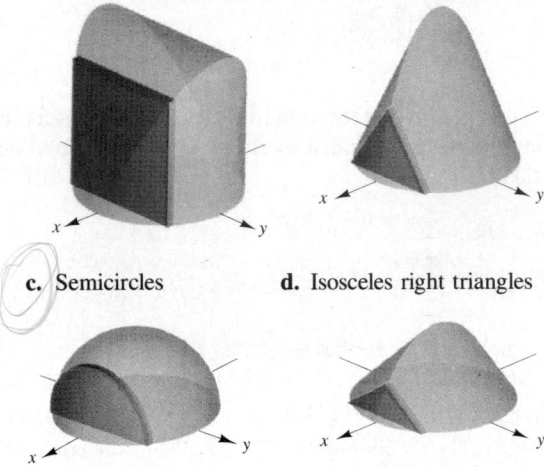

   c. Semicircles   **d.** Isosceles right triangles

52. Find the volume of the solid whose base is bounded by the graphs of $y = x + 1$ and $y = x^2 - 1$, with the indicated cross sections taken perpendicular to the *x*-axis.
   a. Squares   **b.** Rectangles of height 1

   c. Semiellipses of height   **d.** Equilateral triangles
       2 ($A = \pi ab$)

53. The base of a solid is bounded by $y = x^3$, $y = 0$, and $x = 1$. Find the volume of the solid for the following cross sections (taken perpendicular to the *y*-axis).
   a. Squares
   b. Semicircles
   c. Equilateral triangles
   d. Trapezoids for which $h = b_1 = \frac{1}{2}b_2$, where b_1 and b_2 are the lengths of the upper and lower bases
   e. Semiellipses whose heights are twice the lengths of their bases

54. Find the volume of the solid of intersection (the solid common to both) of the two right circular cylinders of radius r whose axes meet at right angles (see figure).

FOR FURTHER INFORMATION For more information on this problem, see the article "Estimating the Volumes of Solid Figures with Curved Surfaces" by Donald Cohen in the May, 1991 issue of *Mathematics Teacher*.

Two intersecting cylinders   Solid of intersection

FIGURE FOR 54

55. Two planes are cut to the axis of a right circular cylinder to form a wedge. One plane is perpendicular to the axis of the cylinder and the second makes an angle of θ degrees with the first (see figure).
  a. Find the volume of the wedge if $\theta = 45°$.
  b. Find the volume of the wedge for an arbitrary but fixed angle $\theta = \theta_0$. Assuming the cylinder has sufficient length, how does the volume of the wedge change as θ_0 increases in the interval $0 < \theta_0 < 90°$?

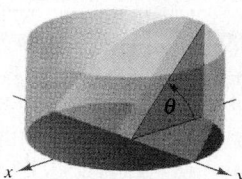

FIGURE FOR 55

56. *Volume of Oil*   A barrel of diameter d lays on a stand so that its axis makes an angle of $20°$ with its horizontal (see figure). The amount of motor oil in the barrel just covers the bottom of the barrel. Approximate the volume of oil.

FIGURE FOR 56

57. *Cavalieri's Theorem*   Prove that if two solids have equal altitudes and all plane sections parallel to their bases and at equal distances from their bases have equal areas, then the solids have the same volume (see figure).

Area of R_1 = area of R_2

FIGURE FOR 57

58. A manufacturer drills a hole through the center of a metal sphere of radius R. The hole has a radius r. Find the volume of the resulting ring.

59. For the metal sphere in Exercise 58, let $R = 5$. What value of r will produce a ring whose volume is exactly half the volume of the sphere?

60. The solid shown in the figure has cross sections bounded by the graph of

$$|x|^a + |y|^a = 1$$

where $1 \le a \le 2$.
  a. Describe the cross section when $a = 1$ and $a = 2$.
  b. Describe a procedure for approximating the volume of the solid.

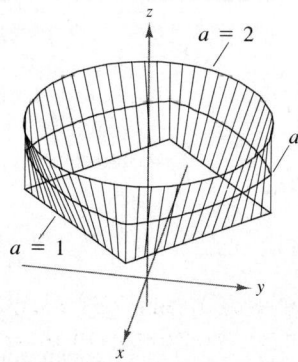

Generated by Mathematica

FIGURE FOR 60

The Shell Method

In this section, you will study an alternative method for finding the volume of a solid of revolution. This method is called the **shell method** because it uses cylindrical shells. We will compare the advantages of the disc and shell methods later in this section.

To begin, consider a representative rectangle as shown in Figure 6.29, where w is the width of the rectangle, h is the height of the rectangle, and p is the distance between the axis of revolution and the *center* of the rectangle. When this rectangle is revolved about its axis of revolution, it forms a cylindrical shell (or tube) of thickness w. To find the volume of this shell, consider two cylinders. The radius of the larger cylinder corresponds to the outer radius of the shell, and the radius of the smaller cylinder corresponds to the inner radius of the shell. Because p is the average radius of the shell, you know the outer radius is $p + (w/2)$ and the inner radius is $p - (w/2)$. Thus, the volume of the shell is

$$\text{Volume of shell} = (\text{Volume of cylinder}) - (\text{Volume of hole})$$
$$= \pi\left(p + \frac{w}{2}\right)^2 h - \pi\left(p - \frac{w}{2}\right)^2 h$$
$$= 2\pi phw$$
$$= 2\pi(\text{Average radius})(\text{Height})(\text{Thickness}).$$

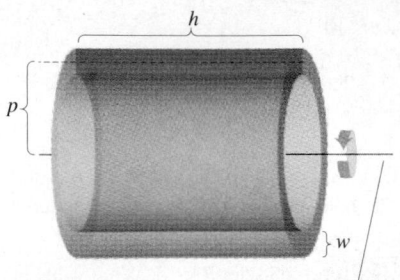

FIGURE 6.29

You can use this formula to find the volume of a solid of revolution as follows. Assume that the plane region in Figure 6.30 is revolved about a line to form the indicated solid. If you consider a horizontal rectangle of width Δy, then as the plane region is revolved about a line parallel to the x-axis, the rectangle generates a representative shell whose volume is

$$\Delta V = 2\pi[p(y)h(y)]\Delta y.$$

Plane Region

You can approximate the volume of the solid by n such shells of thickness Δy, height $h(y_i)$, and average radius $p(y_i)$.

$$\text{Volume of solid} \approx \sum_{i=1}^{n} 2\pi[p(y_i)h(y_i)]\Delta y = 2\pi\sum_{i=1}^{n}[p(y_i)h(y_i)]\Delta y$$

This approximation appears to become better and better as $\|\Delta\| \to 0$ $(n \to \infty)$. Therefore, we define the volume of the solid to be

$$\text{Volume of solid} = \lim_{\|\Delta\| \to 0} 2\pi\sum_{i=1}^{n}[p(y_i)h(y_i)]\Delta y$$
$$= 2\pi\int_{c}^{d}[p(y)h(y)]\,dy.$$

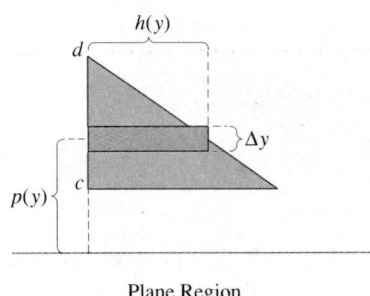

Solid of Revolution

FIGURE 6.30

The Shell Method

To find the volume of a solid of revolution with the **shell method,** use one of the following, as shown in Figure 6.31.

Horizontal axis of revolution

$$\text{Volume} = V = 2\pi \int_{c}^{d} p(y)h(y)\, dy$$

Vertical axis of revolution

$$\text{Volume} = V = 2\pi \int_{a}^{b} p(x)h(x)\, dx$$

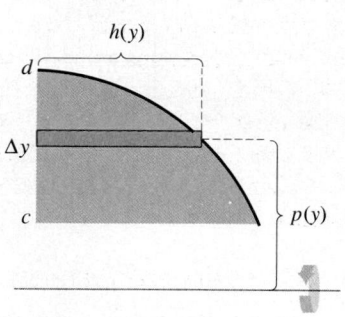

Horizontal Axis of Revolution   Vertical Axis of Revolution

FIGURE 6.31

EXAMPLE 1   Using the Shell Method to Find Volume

Find the volume of the solid of revolution formed by revolving the region bounded by

$$y = x - x^3$$

and the x-axis ($0 \le x \le 1$) about the y-axis.

Solution   Because the axis of revolution is vertical, use a vertical representative rectangle, as shown in Figure 6.32. The width Δx indicates that x is the variable of integration. The distance from the center of the rectangle to the axis of revolution is $p(x) = x$, and the height of the rectangle is $h(x) = x - x^3$. Because x ranges from 0 to 1, the volume of the solid is

$$V = 2\pi \int_{a}^{b} p(x)h(x)\, dx = 2\pi \int_{0}^{1} x(x - x^3)\, dx$$

$$= 2\pi \int_{0}^{1} (-x^4 + x^2)\, dx$$

$$= 2\pi \left[-\frac{x^5}{5} + \frac{x^3}{3} \right]_{0}^{1}$$

$$= 2\pi \left(-\frac{1}{5} + \frac{1}{3} + 0 - 0 \right)$$

$$= \frac{4\pi}{15}.$$

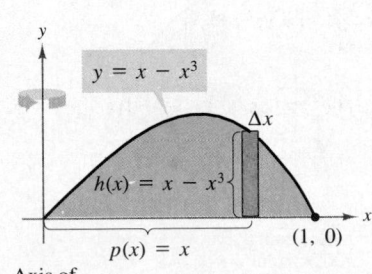

Axis of
revolution

FIGURE 6.32

EXAMPLE 2   Using the Shell Method to Find Volume

Find the volume of the solid of revolution formed by revolving the region bounded by the graph of $x = e^{-y^2}$ and the y-axis ($0 \le y \le 1$) about the x-axis.

Solution   Because the axis of revolution is horizontal, use a horizontal representative rectangle, as shown in Figure 6.33. The width Δy indicates that y is the variable of integration. The distance from the center of the rectangle to the axis of revolution is $p(y) = y$, and the height of the rectangle is $h(y) = e^{-y^2}$. Because y ranges from 0 to 1, the volume of the solid is

$$
V = 2\pi \int_c^d p(y)h(y)\, dy = 2\pi \int_0^1 ye^{-y^2}\, dy
$$

$$
= \left[-\pi e^{-y^2} \right]_0^1
$$

$$
= \pi\left(1 - \frac{1}{e} \right)
$$

$$
\approx 1.986.
$$

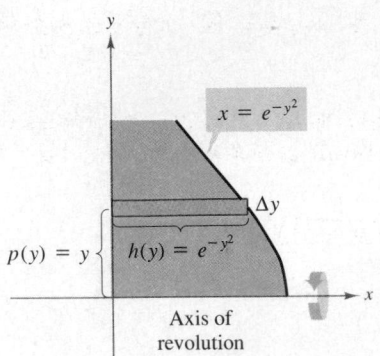

FIGURE 6.33

REMARK   To see the advantage of using the shell method in Example 2, solve the equation $x = e^{-y^2}$ for y.

$$
y = \begin{cases} 1, & 0 \le x \le 1/e \\ \sqrt{-\ln x}, & 1/e < x \le 1 \end{cases}
$$

Then use this equation to find the volume using the disc method.

Comparison of Disc and Shell Methods

The disc and shell methods can be distinguished as follows. For the disc method, the representative rectangle is always *perpendicular* to the axis of revolution, whereas for the shell method, the representative rectangle is always *parallel* to the axis of revolution, as shown in Figure 6.34.

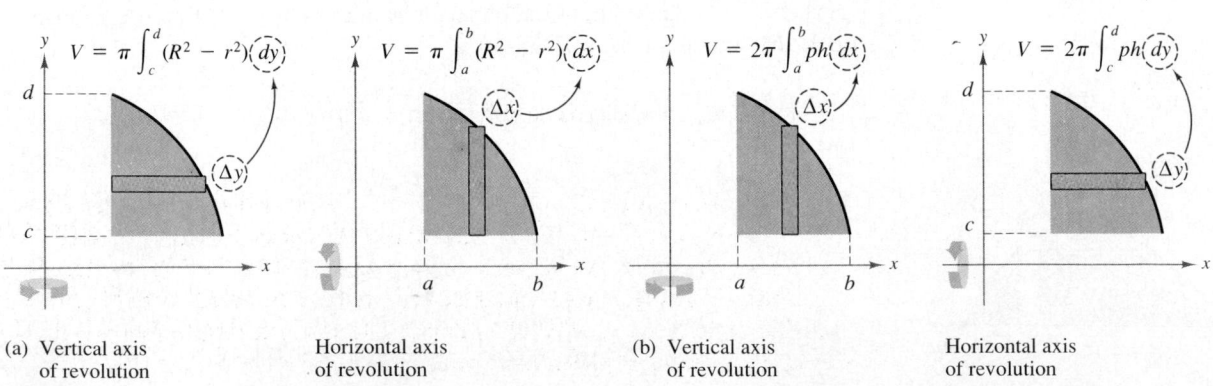

(a) Vertical axis of revolution   Horizontal axis of revolution

Disc Method: Representative rectangle is perpendicular to the axis of revolution.

(b) Vertical axis of revolution   Horizontal axis of revolution

Shell Method: Representative rectangle is parallel to the axis of revolution.

FIGURE 6.34

Often, one method is more convenient to use than the other. The following example illustrates a case in which the shell method is preferable.

EXAMPLE 3   Shell Method Preferable

Find the volume of the solid formed by revolving the region bounded by the graphs of $y = x^2 + 1$, $y = 0$, $x = 0$, and $x = 1$ about the y-axis.

Solution   In Example 4 in the previous section, you saw that the disc method requires two integrals to determine the volume of this solid. See Figure 6.35(a).

$$V = \pi \int_0^1 (1^2 - 0^2)\, dy + \pi \int_1^2 \left[1^2 - \left(\sqrt{y-1}\right)^2 \right] dy \qquad \text{Disc method}$$

$$= \pi \int_0^1 1\, dy + \pi \int_1^2 (2 - y)\, dy$$

$$= \pi y \Big]_0^1 + \pi \left[2y - \frac{y^2}{2} \right]_1^2$$

$$= \pi \left(1 + 4 - 2 - 2 + \frac{1}{2} \right)$$

$$= \frac{3\pi}{2}$$

From Figure 6.35(b), you can see that the shell method requires only one integral to find the volume.

$$V = 2\pi \int_a^b p(x)h(x)\, dx \qquad \text{Shell method}$$

$$= 2\pi \int_0^1 x(x^2 + 1)\, dx$$

$$= 2\pi \left[\frac{x^4}{4} + \frac{x^2}{2} \right]_0^1$$

$$= 2\pi \left(\frac{3}{4} \right)$$

$$= \frac{3\pi}{2}$$

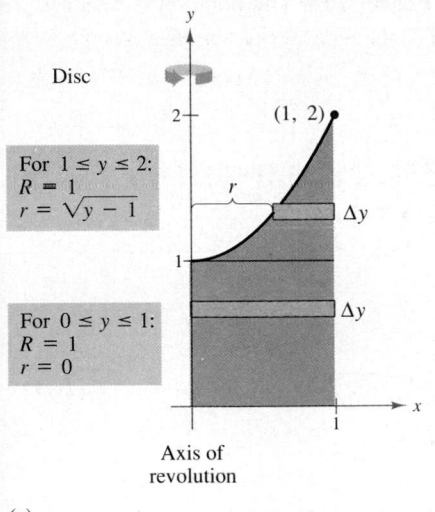

Disc

For $1 \le y \le 2$:
$R = 1$
$r = \sqrt{y-1}$

For $0 \le y \le 1$:
$R = 1$
$r = 0$

(a)

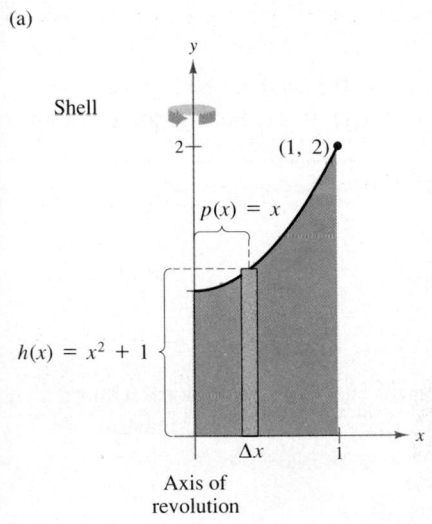

Shell

$p(x) = x$

$h(x) = x^2 + 1$

(b)

FIGURE 6.35

Suppose the region in Example 3 were revolved about the vertical line $x = 1$. Would the resulting solid of revolution have a greater volume than the solid in Example 3 or a smaller volume? Without integrating, you should be able to reason that the resulting solid would have a smaller volume because "more" of the revolved region would be closer to the axis of revolution. To confirm this, try solving the following integral, which gives the volume of the solid.

$$V = 2\pi \int_0^1 (1 - x)(x^2 + 1)\, dx \qquad p(x) = 1 - x$$

In Example 3, the shell method is preferable. In the next example, the preferred method is the disc method.

EXAMPLE 4   Volume of a Pontoon

Pontoon

FIGURE 6.36

A pontoon is to be made in the shape shown in Figure 6.36. The pontoon is designed by rotating the graph of

$$y = 1 - \frac{x^2}{16}, \qquad -4 \le x \le 4$$

about the x-axis, where x and y are measured in feet. Find the volume of the pontoon.

Solution

FIGURE 6.37

Figure 6.37 compares the shell and disc methods. Using the shell method, you can see that Δy is the width of the rectangle, $p(y) = y$, and $h(y) = 2x$. Solving for x in the equation $y = 1 - (x^2/16)$ produces

$$x = 4\sqrt{1 - y} \quad \text{or} \quad h(y) = 2x = 8\sqrt{1 - y}.$$

Hence, the shell method results in the integral

$$V = 2\pi \int_0^1 8y\sqrt{1 - y}\, dy. \qquad \text{Shell method}$$

Although this is not a particularly difficult integral, its evaluation does require a u-substitution. By contrast, the disc method yields the relatively simple integral

$$V = \pi \int_{-4}^4 \left(1 - \frac{x^2}{16}\right)^2 dx$$

$$= \pi \int_{-4}^4 \left(1 - \frac{x^2}{8} + \frac{x^4}{256}\right) dx \qquad \text{Disc method}$$

$$= \pi \left[x - \frac{x^3}{24} + \frac{x^5}{1280}\right]_{-4}^4$$

$$= \frac{64\pi}{15} \approx 13.4 \text{ ft}^3.$$

For the shell method in Example 4 you had to solve for x in terms of y in the equation $y = 1 - (x^2/16)$. Sometimes solving for x is very difficult (or even impossible). In such cases you must use a vertical rectangle (of width Δx), thus making x the variable of integration. The position (horizontal or vertical) of the axis of revolution then determines the method to be used. This is illustrated in Example 5.

EXAMPLE 5   Shell Method Necessary

Find the volume of the solid formed by revolving the region bounded by the graphs of $y = x^3 + x + 1$, $y = 1$, and $x = 1$ about the line $x = 2$, as shown in Figure 6.38.

Solution   In the equation $y = x^3 + x + 1$, you cannot easily solve for x in terms of y. (See the discussion at the end of Section 3.8.) Therefore, the variable of integration must be x, and you should choose a vertical representative rectangle. Because the rectangle is parallel to the axis of revolution, use the shell method and obtain

$$V = 2\pi \int_a^b p(x)h(x)\,dx = 2\pi \int_0^1 (2 - x)(x^3 + x + 1 - 1)\,dx$$

$$= 2\pi \int_0^1 (-x^4 + 2x^3 - x^2 + 2x)\,dx$$

$$= 2\pi\left[-\frac{x^5}{5} + \frac{x^4}{2} - \frac{x^3}{3} + x^2 \right]_0^1$$

$$= 2\pi\left(-\frac{1}{5} + \frac{1}{2} - \frac{1}{3} + 1 \right)$$

$$= \frac{29\pi}{15}.$$

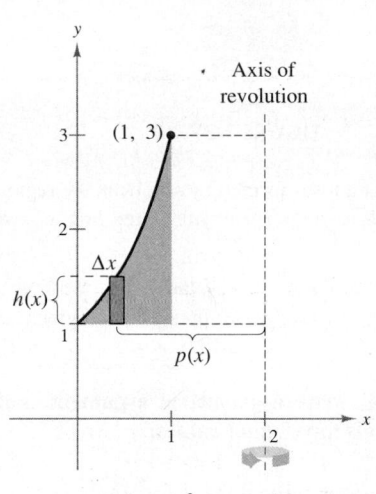

$h(x) = (x^3 + x + 1) - 1$
$p(x) = 2 - x$

FIGURE 6.38

EXERCISES for Section 6.3

TECHNOLOGY
Laboratory Guide
Labs 6.2–6.3

In Exercises 1–12, use the shell method to find the volume of the solid generated by revolving the region about the y-axis.

1. $y = x$

2. $y = 1 - x$

3. $y = \sqrt{x}$

4. $y = x^2 + 4$

5. $y = x^2$
   $y = 0$
   $x = 2$

6. $y = x^2$
   $y = 0$
   $x = 4$

7. $y = x^2$
   $y = 4x - x^2$

8. $y = 4 - x^2$
   $y = 0$

9. $y = 4x - x^2$
   $x = 0$
   $y = 4$

10. $y = 2x$
   $y = 4$
   $x = 0$

11. $y = \dfrac{1}{\sqrt{2\pi}}e^{-x^2/2}$
   $y = 0$
   $x = 0$
   $x = 1$

12. $y = \begin{cases} \dfrac{\sin x}{x}, & x > 0 \\ 1, & x = 0 \end{cases}$
   $y = 0$
   $x = 0$
   $x = \pi$

In Exercises 13–16, use the shell method to find the volume of the solid generated by revolving the region about the x-axis.

13. $y = x$

14. $y = 2 - x$

15. $y = \dfrac{1}{x}$
   $x = 1$
   $x = 2$
   $y = 0$

16. $x + y^2 = 9$
   $x = 0$

In Exercises 17–20, use the shell method to find the volume of the solid generated by revolving the region about the specified line.

17. $y = x^2$, $y = 4x - x^2$, about the line $x = 4$

18. $y = x^2$, $y = 4x - x^2$, about the line $x = 2$

19. $y = 4x - x^2$, $y = 0$, about the line $x = 5$

20. $y = \sqrt{x}$, $y = 0$, $x = 4$, about the line $x = 6$

In Exercises 21–24, use the disc or shell method to find the volume of the solid generated by revolving the region bounded by the graphs of the equations about the specified line.

21. $y = x^3$, $y = 0$, $x = 2$
   a. The x-axis   b. The y-axis
   c. The line $x = 4$ d. The line $y = 8$

22. $y = \dfrac{1}{x^2}$, $y = 0$, $x = 1$, $x = 4$
   a. The x-axis   b. The y-axis
   c. The line $x = 4$ d. The line $y = 1$

23. $x^{1/2} + y^{1/2} = a^{1/2}$, $x = 0$, $y = 0$
   a. The x-axis   b. The y-axis
   c. The line $x = a$ d. The line $y = a$

24. $x^{2/3} + y^{2/3} = a^{2/3}$, $a > 0$ (hypocycloid)
   a. The x-axis   b. The y-axis

Approximation **In Exercises 25 and 26, determine which value best approximates the volume of the solid generated by revolving the region bounded by the graphs of the equations about the y-axis. (Make your selection on the basis of a sketch of the solid and *not* by performing any calculations.)**

25. $y = 2e^{-x}$, $y = 0$, $x = 0$, $x = 2$
   a. $\frac{3}{2}$ b. -2 c. 4 d. 7.5 e. 15

26. $y = \tan x$, $y = 0$, $x = 0$, $x = \dfrac{\pi}{4}$
   a. 3.5 b. $-\frac{9}{4}$ c. 8 d. 10 e. 1

27. *Machine Part* A solid is generated by revolving the region bounded by $y = \frac{1}{2}x^2$ and $y = 2$ about the y-axis. A hole, centered along the axis of revolution, is drilled through this solid so that one-quarter of the volume is removed. Find the diameter of the hole.

28. *Machine Part* A solid is generated by revolving the region bounded by $y = \sqrt{9 - x^2}$ and $y = 0$ about the y-axis. A hole, centered along the axis of revolution, is drilled through this solid so that one-third of the volume is removed. Find the diameter of the hole.

29. A hole is cut through the center of a sphere of radius r. The height of the remaining spherical ring is h, as shown in the figure. Show that the volume of the ring is $V = \pi h^3/6$. (*Note:* The volume is independent of r.)

30. A **torus** is formed by revolving the region bounded by the circle $x^2 + y^2 = 1$ about the line $x = 2$, as shown in the figure. Find the volume of this "doughnut-shaped" solid. (*Hint:* The integral $\int_{-1}^{1} \sqrt{1 - x^2}\, dx$ represents the area of a semicircle.)

FIGURE FOR 29

FIGURE FOR 30

31. Repeat Exercise 30 for a torus formed by revolving the region bounded by the circle $x^2 + y^2 = r^2$ about the line $x = R$ where $r < R$.

32. Let a sphere of radius r be cut by a plane, thus forming a segment of height h. Show that the volume of this segment is $\frac{1}{3}\pi h^2(3r - h)$.

In Exercises 33 and 34, write a geometric argument that explains why the integrals have equal values.

33. $\pi\displaystyle\int_{1}^{5}(x - 1)\,dx$, $2\pi\displaystyle\int_{0}^{2}y[5 - (y^2 + 1)]\,dy$

34. $\pi\displaystyle\int_{0}^{2}[16 - (2y)^2]\,dy$, $2\pi\displaystyle\int_{0}^{4}x\left(\dfrac{x}{2}\right)dx$

C **In Exercises 35–38, use Simpson's Rule with $n = 10$ to approximate the volume of the solid. The solid is generated by revolving the region bounded by the graphs of the equations about the y-axis.**

35. $x^{4/3} + y^{4/3} = 1$, $y = 0$, $x = 0$

36. $y = \sqrt{1 - x^3}$, $y = 0$, $x = 0$

37. $y = \sqrt[3]{(x - 2)^2(x - 6)^2}$, $y = 0$, $x = 2$, $x = 6$

38. $y = \dfrac{2}{1 + e^{1/x}}$, $y = 0$, $x = 1$, $x = 3$

39. *Volume of a Storage Shed* A storage shed has a circular base of diameter 80 feet (see figure). Starting at the center, the interior height is measured every 10 feet and recorded in the table. Use Simpson's Rule to approximate the volume of the building.

x	0	10	20	30	40
Height	50	46	40	20	0

40. *Volume of a Pond* A pond is approximately circular with a diameter of 400 feet (see figure). Starting at the center, the depth of the water is measured every 25 feet and recorded in the table. Use Simpson's Rule to approximate the volume of water in the pond.

x	0	25	50	75	100
Depth	20	19	19	17	15

x	125	150	175	200
Depth	14	10	6	0

FIGURE FOR 39   **FIGURE FOR 40**

41. Match each of the integrals with the solid whose volume it represents and give the dimensions of each solid.

    a. Right circular cone   **b.** Torus
    c. Sphere   **d.** Right circular cylinder
    e. Ellipsoid

  _____ $2\pi \displaystyle\int_0^r hx\,dx$

  _____ $2\pi \displaystyle\int_0^r hx\left(1 - \frac{x}{r}\right)dx$

  _____ $2\pi \displaystyle\int_0^r 2x\sqrt{r^2 - x^2}\,dx$

  _____ $2\pi \displaystyle\int_0^b 2ax\sqrt{1 - \frac{x^2}{b^2}}\,dx$

  _____ $2\pi \displaystyle\int_{-r}^r (R - x)\left(2\sqrt{r^2 - x^2}\right)dx$

42. *The Oblateness of Saturn*   Saturn is the most oblate of the nine planets in our solar system. Its equatorial radius is 60,268 kilometers and its polar radius is 54,364 kilometers. Find the ratio of the volumes of the oblate ellipsoid and the sphere shown below.

 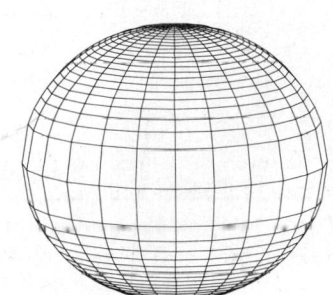

Generated by Mathematica

Computer model of "spherical Saturn" whose equatorial radius is equal to its polar radius. The equation of the cross section passing through the pole is

$$x^2 + y^2 = 60{,}268^2.$$

Computer model of "oblate Saturn" whose equatorial radius is greater than its polar radius. The equation of the cross section passing through the pole is

$$\frac{x^2}{60{,}268^2} + \frac{y^2}{54{,}364^2} = 1.$$

This color-enhanced photograph of Saturn was taken by *Voyager 1*. In the photograph, the oblateness of Saturn is clearly visible.

CHRISTIAN HUYGENS

The Dutch mathematician Christian Huygens (1629–1695), who invented the pendulum clock, and James Gregory (1638–1675), a Scottish mathematician, both made early contributions to the problem of finding the length of a rectifiable curve.

Arc Length

In this section, definite integrals are used to find the arc length of a plane curve and the area of a surface of revolution. In both cases, we approximate an arc (a segment of a curve) by straight line segments whose lengths are given by the familiar distance formula

$$d = \sqrt{(x_2 - x_1)^2 + (y_2 - y_1)^2}.$$

A **rectifiable** curve is one that has a finite arc length. You will see that a sufficient condition for the graph of a function f to be rectifiable between $(a, f(a))$ and $(b, f(b))$ is that f' be continuous on $[a, b]$. Such a function is **continuously differentiable** on $[a, b]$, and its graph on the interval $[a, b]$ is a **smooth curve.**

Consider a function, $y = f(x)$, that is continuously differentiable on the interval $[a, b]$. You can approximate the graph of f by n line segments whose endpoints are determined by the partition

$$a = x_0 < x_1 < x_2 < \cdots < x_n = b$$

as shown in Figure 6.39. By letting $\Delta x_i = x_i - x_{i-1}$ and $\Delta y_i = y_i - y_{i-1}$, you can approximate the length of the graph by

$$s \approx \sum_{i=1}^{n} \sqrt{(\Delta x_i)^2 + (\Delta y_i)^2} = \sum_{i=1}^{n} \sqrt{1 + \left(\frac{\Delta y_i}{\Delta x_i}\right)^2}\,(\Delta x_i).$$

This approximation appears to become better and better as $\|\Delta\| \to 0$ $(n \to \infty)$. Therefore, we define the length of the graph to be

$$s = \lim_{\|\Delta\| \to 0} \sum_{i=1}^{n} \sqrt{1 + \left(\frac{\Delta y_i}{\Delta x_i}\right)^2}\,(\Delta x_i).$$

Because $f'(x)$ exists for each x in (x_{i-1}, x_i), the Mean Value Theorem guarantees the existence of c_i in (x_{i-1}, x_i) such that

$$f(x_i) - f(x_{i-1}) = f'(c_i)(x_i - x_{i-1}) \longrightarrow \frac{\Delta y_i}{\Delta x_i} = f'(c_i).$$

Because f' is continuous on $[a, b]$, it follows that $\sqrt{1 + [f'(x)]^2}$ is also continuous (and hence integrable) on $[a, b]$, which implies that

$$s = \lim_{\|\Delta\| \to 0} \sum_{i=1}^{n} \sqrt{1 + [f'(c_i)]^2}\,(\Delta x_i)$$

$$= \int_a^b \sqrt{1 + [f'(x)]^2}\,dx.$$

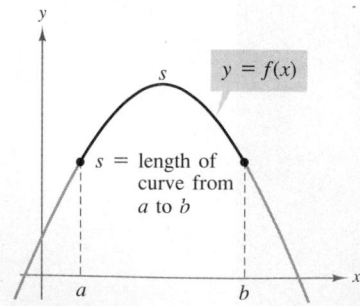

FIGURE 6.39

We call s the **arc length** of f between a and b.

Definition of Arc Length

Let the function given by $y = f(x)$ represent a smooth curve on the interval $[a, b]$. The **arc length** of f between a and b is

$$s = \int_a^b \sqrt{1 + [f'(x)]^2}\, dx.$$

Similarly, for a smooth curve given by $x = g(y)$, the **arc length** of g between c and d is

$$s = \int_c^d \sqrt{1 + [g'(y)]^2}\, dy.$$

Definite integrals representing arc length often are very difficult to evaluate. In this section we present a few examples. In the next chapter, with more advanced integration techniques, you will be able to tackle more difficult arc length problems.

Because the definition of arc length can be applied to a linear function, we can check to see that this new definition agrees with the standard distance formula for the length of a line segment. We do this in Example 1.

EXAMPLE 1   The Length of a Line Segment

Find the arc length from (x_1, y_1) to (x_2, y_2) on the graph of

$$f(x) = mx + b$$

as shown in Figure 6.40.

Solution   Because

$$m = f'(x) = \frac{y_2 - y_1}{x_2 - x_1}$$

it follows that

$$
\begin{aligned}
s &= \int_{x_1}^{x_2} \sqrt{1 + [f'(x)]^2}\, dx \\
&= \int_{x_1}^{x_2} \sqrt{1 + \left(\frac{y_2 - y_1}{x_2 - x_1}\right)^2}\, dx \\
&= \left[\sqrt{\frac{(x_2 - x_1)^2 + (y_2 - y_1)^2}{(x_2 - x_1)^2}}\, (x) \right]_{x_1}^{x_2} \\
&= \sqrt{\frac{(x_2 - x_1)^2 + (y_2 - y_1)^2}{(x_2 - x_1)^2}}\, (x_2 - x_1) \\
&= \sqrt{(x_2 - x_1)^2 + (y_2 - y_1)^2}
\end{aligned}
$$

which is the formula for the distance between two points in the plane.

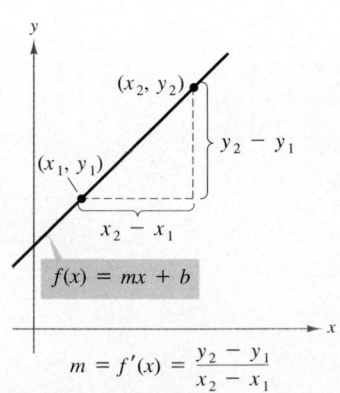

FIGURE 6.40

EXAMPLE 2   Finding Arc Length

Find the arc length of the graph of

$$y = \frac{x^3}{6} + \frac{1}{2x}$$

on the interval $[\frac{1}{2}, 2]$, as shown in Figure 6.41.

Solution   Using

$$\frac{dy}{dx} = \frac{3x^2}{6} - \frac{1}{2x^2} = \frac{1}{2}\left(x^2 - \frac{1}{x^2}\right)$$

yields an arc length of

$$
\begin{aligned}
s &= \int_a^b \sqrt{1 + \left(\frac{dy}{dx}\right)^2}\, dx = \int_{1/2}^2 \sqrt{1 + \left[\frac{1}{2}\left(x^2 - \frac{1}{x^2}\right)\right]^2}\, dx \\
&= \int_{1/2}^2 \sqrt{\frac{1}{4}\left(x^4 + 2 + \frac{1}{x^4}\right)}\, dx \\
&= \int_{1/2}^2 \frac{1}{2}\left(x^2 + \frac{1}{x^2}\right) dx \\
&= \left[\frac{1}{2}\left(\frac{x^3}{3} - \frac{1}{x}\right)\right]_{1/2}^2 \\
&= \frac{1}{2}\left(\frac{13}{6} + \frac{47}{24}\right) = \frac{33}{16}.
\end{aligned}
$$

FIGURE 6.41
The arc length of the graph of y on $[\frac{1}{2}, 2]$ is $\frac{33}{16}$.

EXAMPLE 3   Finding Arc Length

Find the arc length of the graph of $(y - 1)^3 = x^2$ on the interval $[0, 8]$, as shown in Figure 6.42.

Solution   Begin by solving for x in terms of y: $x = \pm(y - 1)^{3/2}$. Choosing the positive value for x produces

$$\frac{dx}{dy} = \frac{3}{2}(y - 1)^{1/2}.$$

Therefore, the arc length is

$$
\begin{aligned}
s &= \int_c^d \sqrt{1 + \left(\frac{dx}{dy}\right)^2}\, dy = \int_1^5 \sqrt{1 + \left[\frac{3}{2}(y - 1)^{1/2}\right]^2}\, dy \\
&= \int_1^5 \sqrt{\frac{9}{4}y - \frac{5}{4}}\, dy \\
&= \frac{1}{2}\int_1^5 \sqrt{9y - 5}\, dy \\
&= \frac{1}{18}\left[\frac{(9y - 5)^{3/2}}{3/2}\right]_1^5 \\
&= \frac{1}{27}(40^{3/2} - 4^{3/2}) \\
&= \frac{8}{27}(10^{3/2} - 1) \approx 9.0734.
\end{aligned}
$$

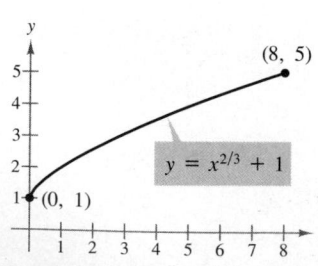

FIGURE 6.42
The arc length of the graph of y on $[0, 8]$ is approximately 9.0734.

EXAMPLE 4   Finding Arc Length

Find the arc length of the graph of $y = \ln(\cos x)$ from $x = 0$ to $x = \pi/4$, as shown in Figure 6.43.

Solution   Using

$$\frac{dy}{dx} = -\frac{\sin x}{\cos x} = -\tan x$$

yields an arc length of

$$
\begin{aligned}
s = \int_a^b \sqrt{1 + \left(\frac{dy}{dx}\right)^2}\, dx &= \int_0^{\pi/4} \sqrt{1 + \tan^2 x}\, dx \\
&= \int_0^{\pi/4} \sqrt{\sec^2 x}\, dx \\
&= \int_0^{\pi/4} \sec x\, dx \\
&= \Big[\, \ln|\sec x + \tan x| \,\Big]_0^{\pi/4} \\
&= \ln(\sqrt{2} + 1) - \ln 1 \\
&\approx 0.8814.
\end{aligned}
$$

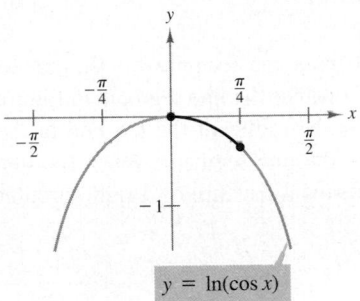

FIGURE 6.43
The arc length of the graph of y on $[0, \frac{\pi}{4}]$ is approximately 0.8814.

EXAMPLE 5   Length of a Cable

An electric cable is hung between two towers that are 200 feet apart, as shown in Figure 6.44. The cable takes the shape of a **catenary** whose equation is

$$y = 75(e^{x/150} + e^{-x/150}) = 150 \cosh\frac{x}{150}.$$

Find the arc length of the cable between the two towers.

Solution   Because $y' = \frac{1}{2}(e^{x/150} - e^{-x/150})$, you can write

$$(y')^2 = \frac{1}{4}(e^{x/75} - 2 + e^{-x/75})$$

and

$$1 + (y')^2 = \frac{1}{4}(e^{x/75} + 2 + e^{-x/75}) = \left[\frac{1}{2}(e^{x/150} + e^{-x/150})\right]^2.$$

Therefore, the arc length of the cable is

$$
\begin{aligned}
s = \int_a^b \sqrt{1 + (y')^2}\, dx &= \frac{1}{2}\int_{-100}^{100} (e^{x/150} + e^{-x/150})\, dx \\
&= \Big[\, 75(e^{x/150} - e^{-x/150}) \,\Big]_{-100}^{100} \\
&= 150(e^{2/3} - e^{-2/3}) \\
&\approx 215 \text{ ft.}
\end{aligned}
$$

Catenary:
$y = 150 \cosh \dfrac{x}{150}$

FIGURE 6.44
The arc length of the cable is approximately 215 feet.

Area of a Surface of Revolution

In Sections 6.2 and 6.3, integration was used to calculate the volume of a solid of revolution. We now look at a procedure for finding the area of a surface of revolution.

Definition of a Surface of Revolution

If the graph of a continuous function is revolved about a line, the resulting surface is a **surface of revolution.**

The area of a surface of revolution is derived from the formula for the lateral surface area of the frustum of a right circular cone. Consider the line segment in Figure 6.45, where L is the length of the line segment, r_1 is the radius at the left end of the line segment, and r_2 is the radius at the right end of the line segment. When the line segment is revolved about its axis of revolution, it forms a frustum of a right circular cone, with

$$S = 2\pi r L \qquad \text{Lateral surface area of frustum}$$

$$r = \frac{1}{2}(r_1 + r_2). \qquad \text{Average radius of frustum}$$

(In Exercise 46 you are asked to verify the formula for S.)

Suppose the graph of a function f, having a continuous derivative on the interval $[a, b]$, is revolved about the x-axis to form a surface of revolution, as shown in Figure 6.46. Let Δ be a partition of $[a, b]$, with subintervals of width Δx_i. Then the line segment of length

$$\Delta L_i = \sqrt{\Delta x_i^2 + \Delta y_i^2}$$

generates a frustum of a cone. Let r_i be the average radius of this frustum. By the Intermediate Value Theorem, a point d_i exists (in the ith subinterval) such that $r_i = f(d_i)$. The lateral surface area, ΔS_i, of the frustum is

$$\Delta S_i = 2\pi r_i \Delta L_i = 2\pi f(d_i)\sqrt{\Delta x_i^2 + \Delta y_i^2}$$

$$= 2\pi f(d_i)\sqrt{1 + \left(\frac{\Delta y_i}{\Delta x_i}\right)^2}\, \Delta x_i.$$

FIGURE 6.45

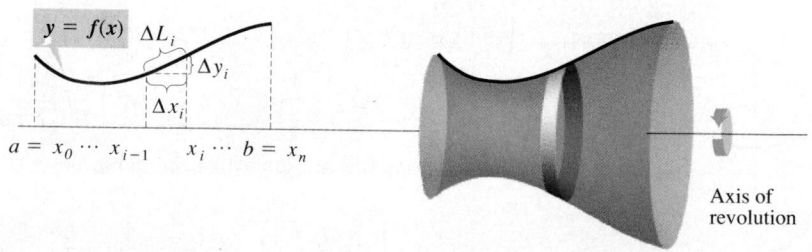

FIGURE 6.46

By the Mean Value Theorem, a point c_i exists in (x_{i-1}, x_i) such that

$$f'(c_i) = \frac{f(x_i) - f(x_{i-1})}{x_i - x_{i-1}} = \frac{\Delta y_i}{\Delta x_i}.$$

Therefore, $\Delta S_i = 2\pi f(d_i)\sqrt{1 + [f'(c_i)]^2}\,\Delta x_i$, and the total surface area can be approximated by

$$S \approx 2\pi \sum_{i=1}^{n} f(d_i)\sqrt{1 + [f'(c_i)]^2}\,\Delta x_i.$$

It can be shown that the limit of the right side as $\|\Delta\| \to 0$ $(n \to \infty)$ is

$$S = 2\pi \int_a^b f(x)\sqrt{1 + [f'(x)]^2}\,dx.$$

In a similar manner, it follows that if the graph of f is revolved about the y-axis then S is

$$S = 2\pi \int_a^b x\sqrt{1 + [f'(x)]^2}\,dx.$$

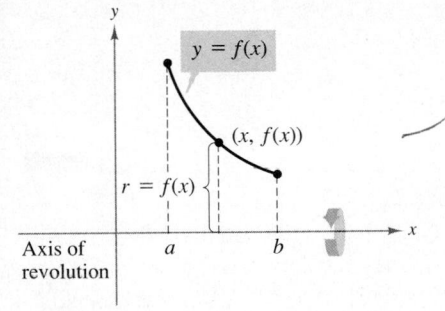

FIGURE 6.47

In both formulas for S, you can regard the products $2\pi f(x)$ and $2\pi x$ as the circumference of the circle traced by a point (x, y) on the graph of f as it is revolved about the x- or y-axis (Figure 6.47). In one case the radius is $r = f(x)$, and in the other case the radius is $r = x$. Moreover, by appropriately adjusting r, you can generalize the formula for surface area to cover *any* horizontal or vertical axis of revolution, as indicated in the following definition.

Definition of the Area of a Surface of Revolution

Let $y = f(x)$ have a continuous derivative on the interval $[a, b]$. The area S of the surface of revolution formed by revolving the graph of f about a horizontal or vertical axis is

$$S = 2\pi \int_a^b r(x)\sqrt{1 + [f'(x)]^2}\,dx$$

where $r(x)$ is the distance between the graph of f and the axis of revolution. If $x = g(y)$ on the interval $[c, d]$, then the surface area is

$$S = 2\pi \int_c^d r(y)\sqrt{1 + [g'(y)]^2}\,dy$$

where $r(y)$ is the distance between the graph of g and the axis of revolution.

REMARK   The formulas in this definition are sometimes written as $S = 2\pi \int_a^b r(x)\,ds$ and $S = 2\pi \int_c^d r(y)\,ds$, where $ds = \sqrt{1 + [f'(x)]^2}\,dx$ and $ds = \sqrt{1 + [g'(y)]^2}\,dy$, respectively.

EXAMPLE 6   The Area of a Surface of Revolution

Find the area of the surface formed by revolving the graph of

$$f(x) = x^3$$

on the interval $[0, 1]$ about the x-axis, as shown in Figure 6.48.

Solution   The distance between the x-axis and the graph of f is $r(x) = f(x)$, and because $f'(x) = 3x^2$, the surface area is

$$S = 2\pi \int_a^b r(x)\sqrt{1 + [f'(x)]^2}\,dx$$

$$= 2\pi \int_0^1 x^3\sqrt{1 + (3x^2)^2}\,dx$$

$$= \frac{2\pi}{36} \int_0^1 (36x^3)(1 + 9x^4)^{1/2}\,dx$$

$$= \frac{\pi}{18}\left[\frac{(1 + 9x^4)^{3/2}}{3/2} \right]_0^1$$

$$= \frac{\pi}{27}(10^{3/2} - 1)$$

$$\approx 3.563.$$

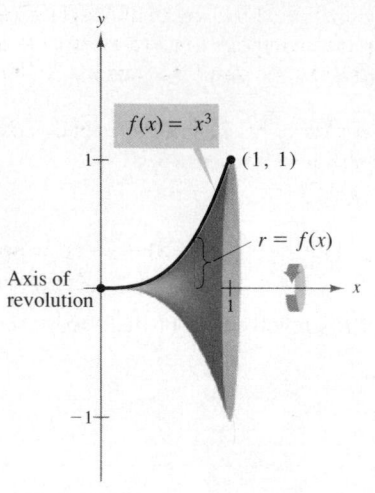

$f(x) = x^3$

(1, 1)

$r = f(x)$

Axis of revolution

FIGURE 6.48

EXAMPLE 7   The Area of a Surface of Revolution

Find the area of the surface formed by revolving the graph of

$$f(x) = x^2$$

on the interval $[0, \sqrt{2}]$ about the y-axis, as shown in Figure 6.49.

Solution   In this case, the distance between the graph of f and the y-axis is $r(x) = x$. Using $f'(x) = 2x$, you can determine that the surface area is

$$S = 2\pi \int_a^b r(x)\sqrt{1 + [f'(x)]^2}\,dx$$

$$= 2\pi \int_0^{\sqrt{2}} x\sqrt{1 + (2x)^2}\,dx$$

$$= \frac{2\pi}{8} \int_0^{\sqrt{2}} (1 + 4x^2)^{1/2}(8x)\,dx$$

$$= \frac{2\pi}{8}\left[\frac{(1 + 4x^2)^{3/2}}{3/2} \right]_0^{\sqrt{2}}$$

$$= \frac{\pi}{6}[(1 + 8)^{3/2} - (1)^{3/2}]$$

$$= \frac{13\pi}{3}.$$

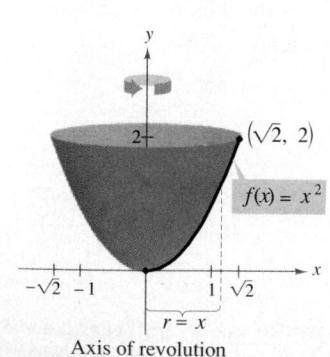

$(\sqrt{2}, 2)$

$f(x) = x^2$

$r = x$

Axis of revolution

FIGURE 6.49

EXERCISES for Section 6.4

In Exercises 1 and 2, find the distance between the points by (a) Using the distance formula and (b) determining the equation of the line through the points and using the formula for arc length.

1. $(0, 0)$, $(5, 12)$ **2.** $(1, 2)$, $(7, 10)$

In Exercises 3–8, find the arc length of the graph of the function over the indicated interval.

Function	Interval
3. $y = \frac{2}{3}x^{3/2} + 1$	$[0, 1]$
4. $y = x^{3/2} - 1$	$[0, 4]$
5. $y = \frac{3}{2}x^{2/3}$	$[1, 8]$
6. $y = \frac{x^5}{10} + \frac{1}{6x^3}$	$[1, 2]$
7. $y = \frac{x^4}{8} + \frac{1}{4x^2}$	$[1, 2]$
8. $y = \frac{1}{2}(e^x + e^{-x})$	$[0, 2]$

In Exercises 9–14, find a definite integral that represents the arc length of the curve over the indicated interval. (Do not evaluate the integral.)

Function	Interval
9. $y = x^2 + x - 2$	$-2 \le x \le 1$
10. $y = \frac{1}{x + 1}$	$0 \le x \le 1$
11. $y = 4 - x^2$	$0 \le x \le 2$
12. $y = \cos x$	$-\frac{\pi}{2} \le x \le \frac{\pi}{2}$
13. $x = e^{-y}$	$0 \le y \le 2$
14. $x = \sqrt{a^2 - y^2}$	$0 \le y \le \frac{a}{2}$

Approximation In Exercises 15 and 16, determine which value best approximates the length of the arc represented by the integral. (Make your selection on the basis of a sketch of the arc and *not* by performing any calculations.)

15. $\int_0^2 \sqrt{1 + \left[\frac{d}{dx} \left(\frac{5}{x^2 + 1} \right) \right]^2} \, dx$

   a. 25   **b.** 5   **c.** 2   **d.** -4 **e.** 3

16. $\int_0^{\pi/4} \sqrt{1 + \left[\frac{d}{dx} (\tan x) \right]^2} \, dx$

   a. 3   **b.** -2 **c.** 4   **d.** $\frac{4\pi}{3}$ **e.** 1

In Exercises 17 and 18, approximate the arc length of the graph of the function over the interval $[0, 4]$ in three ways.
a. Use the distance formula to find the distance between the endpoints of the arc.
b. Use the distance formula to find the length of the four line segments connecting the points on the arc when $x = 0$, $x = 1$, $x = 2$, $x = 3$, and $x = 4$. Find the sum of the four lengths.
c. Use Simpson's Rule with $n = 10$ to approximate the integral yielding the specified arc length.

17. $f(x) = x^3$ **18.** $f(x) = (x^2 - 4)^2$

C In Exercises 19–22, use Simpson's Rule with $n = 10$ to approximate the arc length of the function over the indicated interval.

Function	Interval
19. $y = \frac{1}{x}$	$[1, 3]$
20. $y = x^2$	$[0, 1]$
21. $y = \sin x$	$[0, \pi]$
22. $y = \ln x$	$[1, 5]$

23. *Length of Pursuit* A fleeing object leaves the origin and moves up the y-axis (see figure). At the same time, a pursuer leaves the point $(1, 0)$ and moves always toward the fleeing object. If the pursuer's speed is twice that of the fleeing object, the equation of the path is

$$y = \frac{1}{3}(x^{3/2} - 3x^{1/2} + 2).$$

How far has the fleeing object traveled when it is caught? Show that the pursuer has traveled twice as far.

24. *Roof Area* A barn is 100 feet long and 40 feet wide (see figure). A cross section of the roof is the inverted catenary

$$y = 31 - 10(e^{x/20} + e^{-x/20}).$$

Find the number of square feet of roofing on the barn. (*Hint:* Find the arc length of the catenary and multiply by the length of the barn.)

$y = \frac{1}{3}(x^{3/2} - 3x^{1/2} + 2)$

FIGURE FOR 23

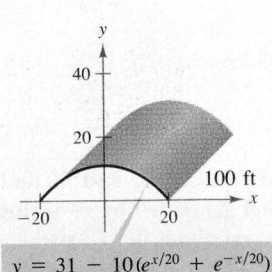

$y = 31 - 10(e^{x/20} + e^{-x/20})$

FIGURE FOR 24

25. *Length of a Catenary* Electric wires suspended between two towers form a catenary (see figure) modeled by the equation

$$y = 60\cosh\frac{x}{60}, \qquad -60 \le x \le 60$$

where x and y are measured in feet. Find the length of the suspended cable if the towers are 120 feet apart.

26. *Length of Gateway Arch* The Gateway Arch in St. Louis, Missouri is modeled by

$$y = 693.8597 - 68.7672\cosh 0.0100333x,$$
$$-299.2239 \le x \le 299.2239.$$

(See Exercise 80 in Section 5.9.) Find the length of this curve.

FIGURE FOR 25

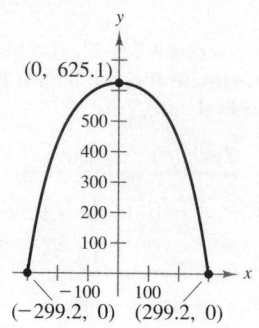

FIGURE FOR 26

27. Find the arc length from $(0, 3)$ clockwise to $(2, \sqrt{5})$ along the circle $x^2 + y^2 = 9$.

28. Find the arc length from $(-3, 4)$ clockwise to $(4, 3)$ along the circle $x^2 + y^2 = 25$. Show that the result is one-fourth of the circumference of the circle.

In Exercises 29–32, find the area of the surface of revolution generated by revolving the curve about the x-axis.

Function	Interval
29. $y = \dfrac{x^3}{3}$	$[0, 3]$
30. $y = \sqrt{x}$	$[1, 4]$
31. $y = \dfrac{x^3}{6} + \dfrac{1}{2x}$	$[1, 2]$
32. $y = \dfrac{x}{2}$	$[0, 6]$

In Exercises 33 and 34, find the area of the surface of revolution generated by revolving the curve over the indicated interval about the y-axis.

Function	Interval
33. $y = \sqrt[3]{x} + 2$	$[1, 8]$
34. $y = 4 - x^2$	$[0, 2]$

In Exercises 35 and 36, use Simpson's Rule with $n = 10$ to approximate the definite integral that yields the surface area of the solid of revolution.

Function	Interval
35. $y = \sin x$ revolved about the x-axis	$[0, \pi]$
36. $y = \ln x$ revolved about the y-axis	$[1, e]$

37. A right circular cone is generated by revolving the region bounded by $y = hx/r$, $y = h$, and $x = 0$ about the y-axis. Verify that the lateral surface area of the cone is

$$S = \pi r\sqrt{r^2 + h^2}.$$

38. A sphere of radius r is generated by revolving the graph of $y = \sqrt{r^2 - x^2}$ about the x-axis. Verify that the surface area of the sphere is $4\pi r^2$.

39. Find the area of the zone of a sphere formed by revolving the graph of $y = \sqrt{9 - x^2}$, $0 \le x \le 2$, about the y-axis.

40. Find the area of the zone of a sphere formed by revolving the graph of $y = \sqrt{r^2 - x^2}$, $0 \le x \le a$, about the y-axis. Assume that $a < r$.

41. *Bulb Design* An ornamental light bulb is designed by revolving the graph of

$$y = \frac{1}{3}x^{1/2} - x^{3/2}, \qquad 0 \le x \le \frac{1}{3}$$

about the x-axis, where x and y are measured in feet (see figure). Find the surface area of the bulb and use the result to approximate the amount of glass needed to make the bulb. (Assume the glass is 0.015 inch thick.)

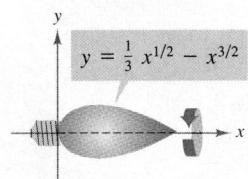

FIGURE FOR 41

42. *Analyzing a Vase* The circumference C (inches) of a vase is measured at 3-inch intervals starting at its base. The measurements are listed in the table, where y is the vertical distance in inches from the base.

y	0	3	6	9	12	15	18
C	50	65.5	67	68	65	51	48

a. Use the data to approximate the volume of the vase.
b. Use the data to approximate the outside surface area (excluding the base) of the vase.

43. *Individual Project* Select a solid of revolution from everyday life. Measure the radius of the solid at a minimum of seven points along its axis. Use the data to approximate the volume of the solid and the surface area of the lateral sides of the solid.

44. Given a circular sector with radius L and central angle θ (see figure), show that its area is

$$S = \frac{1}{2}L^2\theta.$$

45. By joining the straight line edges of the sector in Exercise 44, a right circular cone is formed (see figure) and the lateral surface area of the cone is the same as the area of the sector. Show that the area is

$$S = \pi r L$$

where r is the radius of the base of the cone. (*Hint:* The arc length of the sector equals the circumference of the base of the cone.)

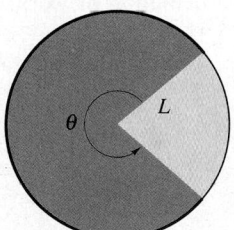

FIGURE FOR 44   **FIGURE FOR 45**

46. Use the result of Exercise 45 to verify that the formula for the lateral surface area of the frustum of a cone with slant height L and radii r_1 and r_2 (see figure) is

$$S = \pi(r_1 + r_2)L.$$

(*Note:* This formula was used to develop the integral for finding the surface area of a surface of revolution.)

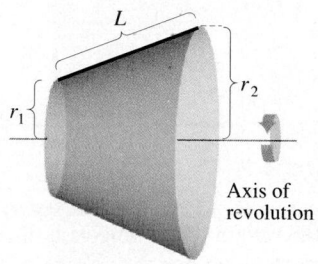

FIGURE FOR 46

47. *Essay* Read the article "Arc Length, Area, and the Arcsine Function" by Andrew M. Rockett in the March, 1983 issue of *Mathematics Magazine*. Then write a paragraph explaining how the arcsine function can be defined in terms of an arc length.

48. Let R be the region bounded by $y = 1/x$, the x-axis, $x = 1$, and $x = b$, where $b > 1$. Let D be the solid formed when R is revolved about the x-axis.
   a. Find the volume V of D.
   b. Express the surface area S as an integral.
   c. Show that V approaches a finite limit as $b \to \infty$.
   d. Show that $S \to \infty$ as $b \to \infty$.

49. Let f be rectifiable on the closed interval $[a, b]$, and let

$$s(x) = \int_a^x \sqrt{1 + [f'(t)]^2}\, dt.$$

   a. Find $\dfrac{ds}{dx}$.
   b. Find ds and $(ds)^2$.
   c. If $f(t) = t^{3/2}$, find $s(x)$ on $[1, 3]$. What is $s(2)$?

50. Show that the length of one arch of the sine curve is equal to the length of the upper portion of the ellipse $x^2 + \frac{1}{2}y^2 = 1$.

SECTION Work

6.5   Work Done by a Constant Force • Work Done by a Variable Force

Work Done by a Constant Force

The concept of work is important to scientists and engineers for determining the energy needed to perform various jobs. For instance, it is useful to know the amount of work done when a crane lifts a steel girder, when a spring is compressed, when a rocket is propelled into the air, or when a truck pulls a load along a highway.

In general, we say that **work** is done by a force when it moves an object. If the force applied to the object is *constant,* we have the following definition of work.

Definition of Work Done by a Constant Force

If an object is moved a distance D in the direction of an applied constant force F, then the **work** W done by the force is defined as

$$W = FD.$$

There are many types of forces—centrifugal, electromotive, and gravitational, to name a few. A **force** can be thought of as a *push* or a *pull;* a force changes the state of rest or state of motion of a body. For gravitational forces on the earth, it is common to use units of measure corresponding to the weight of an object.

EXAMPLE 1   Lifting an Object

Determine the work done in lifting a 150-pound object 4 feet.

Solution   The magnitude of the required force F is the weight of the object, as shown in Figure 6.50. Thus, the work done in lifting the object 4 feet is

$$W = FD \qquad \text{(Work) = (Force)(Distance)}$$
$$= 150(4) \qquad \text{Force = 150 lb, Distance = 4 ft}$$
$$= 600 \text{ ft} \cdot \text{lb}.$$

Hence, the work done in lifting a 150-pound object 4 feet is 600 ft·lb.   ■

REMARK   In the U.S. system of measurement, work is typically expressed in foot-pounds (ft·lb), inch-pounds, or foot-tons. In the centimeter-gram-second (C-G-S) system, the basic unit of force is the **dyne**—the force required to produce an acceleration of 1 centimeter per second per second on a mass of 1 gram. In this system, work is typically expressed in dyne-centimeters (ergs) or newton-meters (joules), where 1 joule $= 10^7$ ergs.

FIGURE 6.50
The work done in lifting a 150-pound object 4 feet is 600 ft · lb.

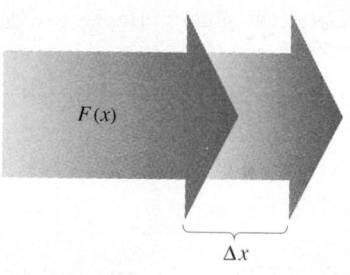

FIGURE 6.51
The amount of force changes as an object changes position (Δx).

EMILIE DE BRETEUIL

Another major work by de Breteuil was the translation of Newton's "Philosophiae Naturalis Principia Mathematica" into French. Her translations and commentaries greatly contributed to the acceptance of Newtonian science in Europe.

Work Done by a Variable Force

In Example 1, the force involved is *constant*. If a *variable* force is applied to an object, then calculus is needed to determine the work done, because the amount of force changes as the object changes positions. For instance, the force required to compress a spring increases as the spring is compressed.

Suppose that an object is moved along a straight line from $x = a$ to $x = b$ by a continuously varying force $F(x)$. Let Δ be a partition that divides the interval $[a, b]$ into n subintervals determined by

$$a = x_0 < x_1 < x_2 < \cdots < x_n = b$$

and let $\Delta x_i = x_i - x_{i-1}$. For each i, choose c_i such that $x_{i-1} \le c_i \le x_i$. Then at c_i the force is given by $F(c_i)$. Because F is continuous, you can approximate the work done in moving the object through the ith subinterval by the increment

$$\Delta W_i = F(c_i) \, \Delta x_i$$

as shown in Figure 6.51. Thus, the total work done as the object moves from a to b is approximated by

$$W \approx \sum_{i=1}^{n} \Delta W_i = \sum_{i=1}^{n} F(c_i) \, \Delta x_i.$$

This approximation appears to become better and better as $\|\Delta\| \to 0$ ($n \to \infty$). Therefore, we define the work to be

$$W = \lim_{\|\Delta\| \to 0} \sum_{i=1}^{n} F(c_i) \, \Delta x_i$$

$$= \int_a^b F(x) \, dx.$$

Definition of Work Done by a Variable Force

If an object is moved along a straight line by a continuously varying force $F(x)$, then the **work** W done by the force as the object is moved from $x = a$ to $x = b$ is

$$W = \lim_{\|\Delta\| \to 0} \sum_{i=1}^{n} \Delta W_i = \int_a^b F(x) \, dx.$$

The remaining examples in this section use some well-known physical laws. The discovery of many of these laws occurred during the same period in which calculus was being developed. In fact, during the seventeenth and eighteenth centuries, there was little difference between physicists and mathematicians. One such physicist-mathematician was Emilie de Breteuil (1706–1749). Breteuil was instrumental in synthesizing the work of many other scientists, including Newton, Leibniz, Huygens, Kepler, and Descartes. Her physics text *Institutions* was widely used for many years.

The following three laws of physics were developed by Robert Hooke (1635–1703), Isaac Newton (1642–1727), and Charles Coulomb (1736–1806).

1. **Hooke's Law:**   The force F required to compress or stretch a spring (within its elastic limits) is proportional to the distance d that the spring is compressed or stretched from its original length. That is,

$$F = kd$$

where the constant of proportionality k (the spring constant) depends on the specific nature of the spring.

2. **Law of Universal Gravitation:**   The force F of attraction between two particles of masses m_1 and m_2 is proportional to the product of the masses and inversely proportional to the square of the distance d between the two particles. That is,

$$F = k\frac{m_1 m_2}{d^2}.$$

If m_1 and m_2 are given in grams and d in centimeters, then F will be in dynes for a value of $k = 6.670 \times 10^{-8}$ cm^3 per gram sec^2.

3. **Coulomb's Law:**   The force between two charges q_1 and q_2 in a vacuum is proportional to the product of the charges and inversely proportional to the square of the distance d between the two charges. That is,

$$F = k\frac{q_1 q_2}{d^2}.$$

If q_1 and q_2 are given in electrostatic units and d in centimeters, then F will be in dynes for a value of $k = 1$.

EXAMPLE 2   Compressing a Spring

A force of 750 pounds compresses a spring 3 inches from its natural length of 15 inches. Find the work done in compressing the spring an additional 3 inches.

Solution   By Hooke's Law, the force $F(x)$ required to compress the spring x units (from its natural length) is $F(x) = kx$. Using the given data, it follows that $F(3) = 750 = (k)(3)$ and thus $k = 250$ and $F(x) = 250x$, as shown in Figure 6.52. To find the increment of work, assume that the force required to compress the spring over a small increment Δx is nearly constant. Thus, the increment of work is

$$\Delta W = (\text{Force})(\text{Distance increment}) = (250x)\,\Delta x.$$

Because the spring is compressed from $x = 3$ to $x = 6$ inches less than its natural length, the work required is

$$W = \int_3^6 250x\,dx = 125x^2 \bigg]_3^6 = 4500 - 1125 = 3375 \text{ in}\cdot\text{lb}.$$

Note that you do not integrate from $x = 0$ to $x = 6$ because you were asked to determine the work done in compressing the spring an *additional* 3 inches (not including the first 3 inches).

Natural length $(F = 0)$

Compressed 3 inches $(F = 750)$

Compressed x inches $(F = 250x)$

FIGURE 6.52

EXAMPLE 3   Moving a Space Module into Orbit

A space module weighs 15 tons on the surface of the earth. How much work is done in propelling the module to a height of 800 miles above the earth, as shown in Figure 6.53? (Do not consider the effect of air resistance or the weight of the propellant.)

Solution   Because the weight of a body varies inversely as the square of its distance from the center of the earth, the force $F(x)$ exerted by gravity is

$$F(x) = \frac{C}{x^2}.$$

Because the module weighs 15 tons on the surface of the earth and the radius of the earth is approximately 4000 miles, you have

$$15 = \frac{C}{(4000)^2} \quad \longrightarrow \quad C = 240,000,000.$$

Thus, the increment of work is

$$\Delta W = \text{(Force) (Distance increment)}$$
$$= \frac{240,000,000}{x^2}\Delta x.$$

Finally, because the module is propelled from $x = 4000$ to $x = 4800$ miles, the total work done is

$$W = \int_{4000}^{4800} \frac{240,000,000}{x^2}\,dx$$
$$= \frac{-240,000,000}{x}\Bigg]_{4000}^{4800}$$
$$= -50,000 + 60,000$$
$$= 10,000 \text{ mile-tons}$$
$$\approx 1.056 \times 10^{11} \text{ ft} \cdot \text{lb}.$$

In the C-G-S system, using a conversion of 1 ft·lb ≈ 1.35582 joules, the work done is

$$W = 1.432 \times 10^{11} \text{ joules.} \qquad \blacksquare$$

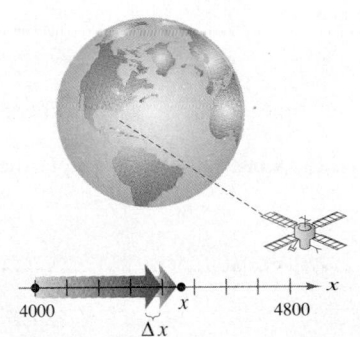

FIGURE 6.53
A space module is moved 800 miles above the earth.

The solutions to Examples 2 and 3 conform to our development of work as the summation of increments in the form

$$\Delta W = \text{(Force)(Distance increment)} = (F)\,(\Delta x).$$

Another way to formulate the increment of work is

$$\Delta W = \text{(Force increment)(Distance)} = (\Delta F)(x).$$

This second interpretation of ΔW is useful in problems involving the movement of nonrigid substances such as fluids or chains.

EXAMPLE 4   Emptying a Tank of Oil

A spherical tank of radius 8 feet is half full of oil that weighs 50 pounds per cubic foot. Find the work required to pump the oil out through a hole in the top of the tank.

Solution   Consider the oil to be subdivided into discs of thickness Δy and radius x, as shown in Figure 6.54. Because the increment of force for each disc is given by its weight, you have

$$\Delta F = \text{Weight} = \left(\frac{50 \text{ lb}}{\text{ft}^3}\right)(\text{Volume}) = 50(\pi x^2 \, \Delta y).$$

For a circle of radius 8 and center at $(0, 8)$, you have

$$x^2 + (y - 8)^2 = 8^2$$
$$x^2 = 16y - y^2$$

and you can write the force increment as

$$\Delta F = 50(\pi x^2 \, \Delta y)$$
$$= 50\pi(16y - y^2) \, \Delta y.$$

In Figure 6.54, note that a disc y feet from the bottom of the tank must be moved a distance of $(16 - y)$ feet. Therefore, the increment of work is

$$\Delta W = \Delta F(16 - y)$$
$$= 50\pi(16y - y^2) \, \Delta y(16 - y)$$
$$= 50\pi(256y - 32y^2 + y^3) \, \Delta y.$$

Because the tank is half full, y ranges from 0 to 8, and the work required to empty the tank is

$$W = \int_0^8 50\pi(256y - 32y^2 + y^3) \, dy$$
$$= 50\pi\left[128y^2 - \frac{32}{3}y^3 + \frac{y^4}{4}\right]_0^8$$
$$= 50\pi\left(\frac{11{,}264}{3}\right)$$
$$\approx 589{,}782 \text{ ft} \cdot \text{lb}.$$

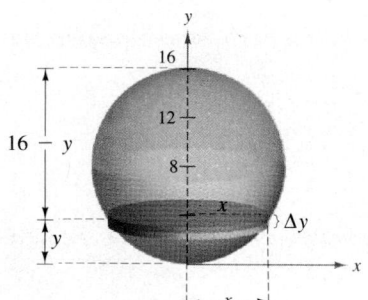

FIGURE 6.54
The work required to pump oil out through a hole in the top of the tank is approximately 589,782 ft · lb.

To estimate the reasonableness of the result in Example 4, consider that the weight of the oil in the tank is

$$\left(\frac{1}{2}\right)(\text{Volume})(\text{Density}) = \frac{1}{2}\left(\frac{4}{3}\pi 8^3\right)(50) \approx 53{,}616.5 \text{ pounds.}$$

Lifting the entire half-tank of oil 8 feet would involve a work of $8(53{,}616.5) \approx 428{,}932$ ft · lb. Because the oil is actually lifted between 8 and 16 feet, it seems reasonable that the work done is 589,782 ft · lb.

FIGURE 6.55
The work required to raise one end of the chain 20 feet is 1000 ft · lb.

FIGURE 6.56
Work done by expanding gas.

EXAMPLE 5   Lifting a Chain

A 20-foot chain, weighing 5 pounds per foot, is lying coiled on the ground. How much work is required to raise one end of the chain to a height of 20 feet so that it is fully extended, as shown in Figure 6.55?

Solution   Imagine that the chain is divided into small sections, each of length Δy. Then the weight of each section is the increment of force

$$\Delta F = (\text{Weight}) = \left(\frac{5 \text{ lb}}{\text{ft}}\right)(\text{Length}) = 5 \, \Delta y.$$

Because a typical section (initially on the ground) is raised to a height of y, the increment of work is

$$\Delta W = (\text{Force increment})(\text{Distance}) = (5 \, \Delta y)y = 5y\Delta y.$$

Because y ranges from 0 to 20, the total work is

$$W = \int_0^{20} 5y \, dy = \frac{5y^2}{2}\bigg]_0^{20} = \frac{5(400)}{2} = 1000 \text{ ft} \cdot \text{lb.} \qquad \blacksquare$$

In the next example we consider a piston of radius r in a cylindrical casing, as shown in Figure 6.56. As the gas in the cylinder expands, the piston moves and work is done. If p represents the pressure of the gas (in pounds per square foot) against the piston head and V represents the volume of the gas (in cubic feet), then the work increment involved in moving the piston Δx feet is

$$\Delta W = (\text{Force})(\text{Distance increment}) = F(\Delta x) = p(\pi r^2) \, \Delta x = p\Delta V.$$

Thus, as the volume of the gas expands from V_0 to V_1, the work done in moving the piston is

$$W = \int_{V_0}^{V_1} p \, dV.$$

Assuming the pressure of the gas to be inversely proportional to its volume, we have $p = k/V$ and the integral for work becomes

$$W = \int_{V_0}^{V_1} \frac{k}{V} \, dV.$$

EXAMPLE 6   Work Done by an Expanding Gas

A quantity of gas with an initial volume of 1 cubic foot and pressure of 500 pounds per square foot expands to a volume of 2 cubic feet. Find the work done by the gas. (Assume the pressure is inversely proportional to the volume.)

Solution   Because $p = k/V$ and $p = 500$ and $V = 1$, we have $k = 500$. Thus, the work is

$$W = \int_{V_0}^{V_1} \frac{k}{V} \, dV = \int_1^2 \frac{500}{V} \, dV = 500 \ln |V| \bigg]_1^2 \approx 346.6 \text{ ft} \cdot \text{lb.} \qquad \blacksquare$$

EXERCISES for Section 6.5

Constant Force   **In Exercises 1–4, determine the work done by the constant force.**

1. A 100-pound bag of sugar is lifted 10 feet.

2. An electric hoist lifts a 2400-pound car 6 feet.

3. A force of 25 pounds is required to slide a cement block 12 feet in a construction project.

4. The locomotive of a freight train pulls its cars with a constant force of 9 tons a distance of one-half mile.

Hooke's Law   **In Exercises 5–10, use Hooke's Law to determine the variable force in the spring problem.**

5. A force of 5 pounds compresses a 15-inch spring a total of 4 inches. How much work is done in compressing the spring 7 inches?

6. How much work is done in compressing the spring in Exercise 5 from a length of 10 inches to a length of 6 inches?

7. A force of 60 pounds stretches a spring 1 foot. How much work is done in stretching the spring from 9 inches to 15 inches?

8. A force of 200 pounds stretches a spring 2 feet on a mechanical device for driving fence posts. Find the work done in stretching the spring the required 2 feet.

9. A force of 15 pounds stretches a spring 6 inches in an exercise machine. Find the work done in stretching the spring 1 foot from its natural position.

10. An overhead garage door has two springs, one on each side of the door. A force of 15 pounds is required to stretch each spring 1 foot. Because of the pulley system, the springs only stretch one-half the distance the door travels. Find the work done by the pair of springs if the door moves a total of 8 feet and the springs are at their natural length when the door is open.

11. *Propulsion*   Neglecting air resistance and the weight of the propellant, determine the work done in propelling a 4-ton satellite to a height of
a. 200 miles above the earth.
b. 400 miles above the earth.

12. *Propulsion*   Use the information from Exercise 11 to write the work W of the propulsion system as a function of the height h of the satellite above the earth. Find the limit (if it exists) of W as h approaches infinity.

13. *Propulsion*   Neglecting air resistance, determine the work done in propelling a 10-ton satellite to a height of
a. 11,000 miles above the earth.
b. 22,000 miles above the earth.

14. *Propulsion*   If a lunar module weighs 12 tons on the surface of the earth, how much work is done in propelling the module from the surface of the moon to a height of 50 miles? Consider the radius of the moon to be 1100 miles and its force of gravity to be one-sixth that of the earth's.

15. *Pumping Water*   A rectangular tank with a base 4 feet by 5 feet and a height of 4 feet is full of water (see figure). (The water weighs 62.4 pounds per cubic foot.) How much work is done in pumping water out over the top edge in order to empty
a. half of the tank?
b. all of the tank?

4 ft

4 ft

5 ft

FIGURE FOR 15

16. *Essay*   Explain why the answer to part b of Exercise 11 is not twice that of part a.

17. *Pumping Water*   A cylindrical water tank 12 feet high with a radius of 8 feet is buried so that the top of the tank is 3 feet below ground level (see figure). How much work is done in pumping a full tank of water up to ground level?

Ground level

12

15 − y

Δy

−8   8

FIGURE FOR 17

18. *Pumping Water*   Suppose the tank in Exercise 17 is located on a tower so that the bottom of the tank is 20 feet above the level of a stream (see figure). How much work is done in filling the tank half full of water through a hole in the bottom, using water from the stream?

Δy

20 ft

y

FIGURE FOR 18

19. *Pumping Water*   A hemispherical tank of radius 6 feet is positioned so that its base is circular. How much work is required to fill the tank with water through a hole in the base if the water source is at the base?

20. *Pumping Diesel Fuel*   The fuel tank on a large truck has trapezoidal cross sections with dimensions (in feet) shown in the accompanying figure. Assume that the engine is approximately 2 feet above the top of the fuel tank and that diesel fuel weighs approximately 55.6 pounds per cubic foot. Find the work done by the fuel pump in raising a full tank of fuel to the level of the engine.

21. *Pumping Water*   An open tank has the shape of a right circular cone (see figure). The tank is 8 feet across the top and 6 feet high. How much work is done in emptying the tank by pumping the water over the top edge?

FIGURE FOR 29

30. Repeat Exercise 29 raising the bottom of the chain to the 12-foot level.

Demolition Crane   **In Exercises 31 and 32, consider a demolition crane with a 500-pound ball suspended from a 40-foot cable that weighs 1 pound per foot.**

31. Find the work required to wind up 15 feet of the apparatus.

32. Find the work required to wind up all 40 feet of the apparatus.

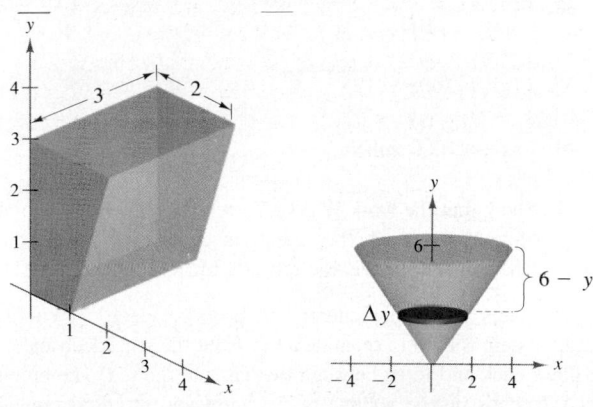

FIGURE FOR 20   FIGURE FOR 21

22. *Pumping Water*   If water is pumped in through the bottom of the tank in Exercise 21, how much work is done in filling the tank
a. to a depth of 2 feet?
b. from a depth of 4 feet to a depth of 6 feet?

Pumping Gasoline   **In Exercises 23 and 24, find the work done in pumping gasoline that weighs 42 pounds per cubic foot.** (*Hint:* **Evaluate one integral by a geometric formula and the other by observing that the integrand is an odd function.**)

23. A cylindrical gasoline tank 3 feet in diameter and 4 feet long is carried on the back of a truck and is used to fuel tractors in the field. The axis of the tank is horizontal. Find the work done in pumping the entire contents of the full tank into a tractor if the opening on the tractor tank is 5 feet above the top of the tank in the truck.

24. The top of a cylindrical storage tank for gasoline at a service station is 4 feet below ground level. The axis of the tank is horizontal and its diameter and length are 5 feet and 12 feet, respectively. Find the work done in pumping the entire contents of the full tank to a height of 3 feet above ground level.

Lifting a Chain   **In Exercises 25–28, consider a 15-foot chain hanging from a winch 15 feet above ground level. Find the work done by the winch in winding up the required amount of chain, if the chain weighs 3 pounds per foot.**

25. Wind up the entire chain.

26. Wind up one-third of the chain.

27. Run the winch until the bottom of the chain is at the 10-foot level.

28. Wind up the entire chain with a 100-pound load attached.

Lifting a Chain   **In Exercises 29 and 30, consider a 15-foot hanging chain that weighs 3 pounds per foot. Find the work done in lifting the chain vertically to the required position.**

29. Take the bottom of the chain and raise it to the 15-foot level, leaving the chain doubled and still hanging vertically (see figure).

Boyle's Law   **In Exercises 33 and 34, find the work done by the gas for the given volumes and pressures. Assume the pressure is inversely proportional to the volume. (Use Example 6 as a model.)**

33. A quantity of gas with an initial volume of 2 cubic feet and pressure of 1000 pounds per square foot expands to a volume of 3 cubic feet.

34. A quantity of gas with an initial volume of 1 cubic foot and pressure of 2000 pounds per square foot expands to a volume of 4 cubic feet.

35. *Electric Force*   Two electrons repel each other with a force that varies inversely as the square of the distance between them. If one electron is fixed at the point $(2, 4)$, find the work done in moving a second electron from $(-2, 4)$ to $(1, 4)$.

36. *Wood Splitting*   The hydraulic cylinder on a wood splitter has a 4-inch bore (diameter) and a stroke of 2 feet. The hydraulic pump creates a maximum pressure of 2000 pounds per square inch. Therefore, the maximum force created by the cylinder is $2000(\pi 2^2) = 8000\pi$ pounds.
a. Find the work done through one extension of the cylinder given that the maximum force is required.
b. In splitting a piece of wood, the force is variable. Measurements of the force were obtained when a piece of wood was split and are given in the table. The variable x measures the extension of the cylinder in feet, and F is the force in pounds.

x	0	$\frac{1}{3}$	$\frac{2}{3}$	1
$F(x)$	0	20,000	22,000	15,000

x	$\frac{4}{3}$	$\frac{5}{3}$	2
$F(x)$	10,000	5,000	0

Use Simpson's Rule to approximate the work done in splitting the piece of wood.

c. Using a curve fitting program and the data from part b, a model for the force is given by

$$F(x) = 258 - 41,957x + 58,188\sqrt{x}, \qquad 0 \leq x \leq 2.$$

Plot the data and sketch a graph of the model on the same set of coordinate axes. Use the model to approximate the extension of the cylinder when the force is maximum.

d. Use this model of part b to approximate the work done in splitting the piece of wood.

Hydraulic Press **In Exercises 37–40, use Simpson's Rule with $n = 10$ to approximate the work done by a press in a manufacturing process. A model is given for the variable force F (pounds) and the distance x (feet) the press moves.**

Force	Interval
37. $F(x) = 1000[1.8 - \ln(x + 1)]$	$0 \leq x \leq 5$
38. $F(x) = \dfrac{e^{x^2} - 1}{100}$	$0 \leq x \leq 4$
39. $F(x) = 100x\sqrt{125 - x^3}$	$0 \leq x \leq 5$
40. $F(x) = 1000 \sinh x$	$0 \leq x \leq 2$

41. Show that the work W done by a force F acting on an object of mass m equals the change in the body's kinetic energy, where $K = \frac{1}{2}mv^2$ is the formula for kinetic energy.

42. *Tidal Energy* Tidal power plants use "tidal energy" to produce electrical energy. To construct a tidal power plant, a dam is built to separate a bay from the sea. Electrical energy is produced as the water flows back and forth between the bay and the sea. The amount of "natural energy" produced depends on the volume of the basin and the tidal range—the vertical distance between high and low tides. (Throughout the world, several natural bays have tidal ranges in excess of 15 feet—the Bay of Fundy in Nova Scotia has a tidal range of 47.5 feet!)

$$y = \frac{1}{40,000}x^2$$

FIGURE FOR 42

a. Consider a bay with a rectangular base as shown in the figure. The bay has a tidal range of 25 feet, with low tide corresponding to $y = 0$. How much water does the bay hold at high tide?

b. The amount of energy produced between filling (or emptying) of the bay is proportional to the amount of work required to fill or empty the bay. How much work is required to fill the bay with sea water? (Use a sea-water density of 64 pounds per cubic foot.)

The natural energy from tides can be increased by pumping the water at certain times during the tidal cycle. (For more information on tidal power, see the article "LaRance: Six Years of Operating a Tidal Power Plant in France" by J. Cotillon in the October, 1984 issue of *Water Power Magazine*.

SECTION | Fluid Pressure and Fluid Force

6.6   Fluid Pressure and Fluid Force

Fluid Pressure and Fluid Force

Swimmers know that the deeper an object is submerged in a fluid, the greater the pressure on the object. **Pressure** is defined as the force per unit of area over the surface of a body. For example, because a column of water that is 10 feet in height and 1 inch square weighs 4.3 pounds, the *fluid* pressure at a depth of 10 feet of water is 4.3 pounds per square inch.* At 20 feet, this would increase to 8.6 pounds per square inch, and in general the pressure is proportional to the depth of the object in the fluid.

BLAISE PASCAL

Pascal is well known for his work in many areas of mathematics and physics and also for his influence on Leibniz. Although much of Pascal's work in calculus was intuitive and lacked the rigor of modern mathematics, he nevertheless anticipated many important results.

> ### Definition of Fluid Pressure
>
> The **pressure** on an object at depth h in a liquid is
>
> Pressure $= P = wh$
>
> where w is the weight-density of the liquid per unit of volume.

Listed below are some commonly used weight-densities of fluids in pounds per cubic foot.

Ethyl alcohol	49.4
Gasoline	41.0–43.0
Glycerin	78.6
Kerosene	51.2
Mercury	849.0
Sea water	64.0
Water	62.4

When calculating fluid pressure, we use an important (and rather surprising) physical law called **Pascal's Principle,** named after the French mathematician Blaise Pascal (1623–1662). Pascal's Principle states that the pressure exerted by a fluid at a depth h is transmitted equally *in all directions*. For example, in Figure 6.57, the pressure at the indicated depth is the same for all three objects. Because fluid pressure is given in terms of force per unit area ($P = F/A$), the fluid force on a *submerged horizontal* surface of area A is

Fluid force $= F = PA =$ (Pressure) (Area).

FIGURE 6.57
The pressure at h is the same for all three objects.

* The total pressure on an object in 10 feet of water would also include the pressure due to the earth's atmosphere. At sea level, atmospheric pressure is approximately 14.7 pounds per square inch.

FIGURE 6.58
The fluid force on a
horizontal metal sheet is
equal to the fluid pressure
times the area.

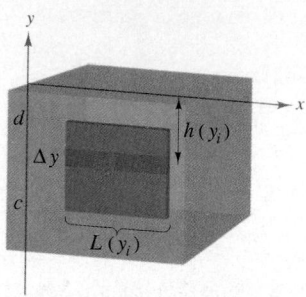

FIGURE 6.59
Calculus methods must be
used to find the fluid force on
a vertical metal plate.

EXAMPLE 1   Fluid Force on a Submerged Sheet

Find the fluid force on a rectangular metal sheet measuring 3 feet by 4 feet that is submerged in 6 feet of water, as shown in Figure 6.58.

Solution   Because the weight-density of water is 62.4 pounds per cubic foot and the sheet is submerged in 6 feet of water, the fluid pressure is

$$P = (62.4)(6) = 374.4 \text{ lb/ft}^2.$$

Because the total area of the sheet is $A = (3)(4) = 12$ square feet, the fluid force is

$$F = PA = \left(374.4 \ \frac{\text{lb}}{\text{ft}^2} \right)(12 \ \text{ft}^2)$$

$$= 4492.8 \text{ lb}.$$

This result is independent of the size of the body of water. The fluid force would be the same in a swimming pool or a lake.

In Example 1, the fact that the sheet is rectangular and horizontal means that you do not need the methods of calculus to solve the problem. We now look at a surface that is submerged vertically in a fluid. This problem is more difficult because the pressure is not constant over the surface.

Suppose a vertical plate is submerged in a fluid of weight-density w (per unit of volume), as shown in Figure 6.59. To determine the total force against *one side* of the region from depth c to depth d, you can subdivide the interval $[c, d]$ into n subintervals, each of width Δy. Next, consider the representative rectangle of width Δy and length $L(y_i)$, where y_i is in the ith subinterval. The force against this representative rectangle is

$$\Delta F_i = w\,(\text{Depth})(\text{Area}) = wh(y_i)L(y_i)\,\Delta y.$$

The force against n such rectangles is

$$\sum_{i=1}^{n} \Delta F_i = w\sum_{i=1}^{n} h(y_i)L(y_i)\,\Delta y.$$

Note that w is considered to be constant and is factored out of the summation. Therefore, taking the limit as $\|\Delta\| \to 0$ ($n \to \infty$) suggests the following definition.

Definition of Force Exerted by a Fluid

The **force F exerted by a fluid** of constant weight-density w (per unit of volume) against a submerged vertical plane region from $y = c$ to $y = d$ is

$$F = w \lim_{\|\Delta\| \to 0} \sum_{i=1}^{n} h(y_i)L(y_i)\,\Delta y = w \int_{c}^{d} h(y)L(y)\,dy$$

where $h(y)$ is the depth of the fluid at y and $L(y)$ is the horizontal length of the region at y.

EXAMPLE 2   Fluid Force on a Vertical Surface

A vertical gate in a dam has the shape of an isosceles trapezoid 8 feet across the top and 6 feet across the bottom, with a height of 5 feet, as shown in Figure 6.60(a). What is the fluid force on the gate if the top of the gate is 4 feet below the surface of the water?

Solution   In setting up a mathematical model for this problem, you are at liberty to locate the x- and y-axes in several different ways. A convenient approach is to let the y-axis bisect the gate and place the x-axis at the surface of the water, as shown in Figure 6.60(b). Thus, the depth of the water at y is

$$\text{Depth} = h(y) = -y.$$

To find the length $L(y)$ of the region at y, we find the equation of the line forming the right side of the gate. Because this line passes through the two points $(3, -9)$ and $(4, -4)$, its equation is

$$y - (-9) = \frac{-4 - (-9)}{4 - 3}(x - 3)$$

$$y + 9 = 5(x - 3)$$

$$y = 5x - 24$$

$$x = \frac{y + 24}{5}.$$

From Figure 6.60(b), you can see that the length of the region at y is

$$\text{Length} = 2x = \frac{2}{5}(y + 24) = L(y).$$

Finally, by integrating from $y = -9$ to $y = -4$, you can calculate the fluid force to be

$$F = w \int_{c}^{d} h(y)L(y)\,dy = 62.4 \int_{-9}^{-4} (-y)\left(\frac{2}{5}\right)(y + 24)\,dy$$

$$= -62.4\left(\frac{2}{5}\right)\int_{-9}^{-4} (y^2 + 24y)\,dy$$

$$= -62.4\left(\frac{2}{5}\right)\left[\frac{y^3}{3} + 12y^2\right]_{-9}^{-4}$$

$$= -62.4\left(\frac{2}{5}\right)\left(\frac{-1675}{3}\right)$$

$$= 13,936 \text{ lb.}$$

(a) Water gate in a dam

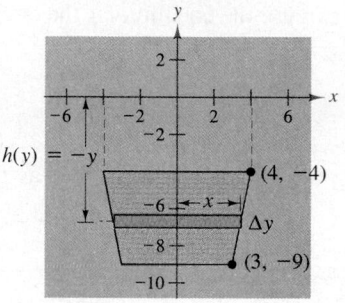

(b) The fluid force against the gate is 13,936 lb.

FIGURE 6.60

REMARK   In Example 2, we let the x-axis coincide with the surface of the water. This was convenient, but arbitrary. In choosing a coordinate system to represent a physical situation, you should consider various possibilities. Often, you can simplify the calculations in a problem by locating the coordinate system so as to take advantage of special characteristics of the problem, such as symmetry.

(a)

(b)

The fluid force on the window is 1608.5 lb.

FIGURE 6.61

EXAMPLE 3   Fluid Force on a Vertical Surface

A circular observation window on a marine science ship has a radius of 1 foot, and the center of the window is 8 feet below water level, as shown in Figure 6.61(a) and (b). What is the fluid force on the window?

Solution   To take advantage of symmetry, locate a coordinate system so that the origin coincides with the center of the window, as shown in Figure 6.61(b). The depth at y is then

$$\text{Depth} = h(y) = 8 - y.$$

The horizontal length of the window is $2x$, and you can use the equation for the circle, $x^2 + y^2 = 1$, to solve for x as follows.

$$\text{Length} = 2x = 2\sqrt{1 - y^2} = L(y)$$

Finally, because y ranges from -1 to 1, and using 64 pounds per cubic foot as the weight-density of sea water, you have

$$F = w \int_c^d h(y)L(y)\, dy = 64 \int_{-1}^1 (8 - y)(2)\sqrt{1 - y^2}\, dy.$$

Initially it looks as if this integral would be difficult to solve. However, if you break the integral into two parts and apply symmetry, the solution is simple.

$$F = 64(16) \int_{-1}^1 \sqrt{1 - y^2}\, dy - 64(2) \int_{-1}^1 y\sqrt{1 - y^2}\, dy$$

The second integral is 0 (because the integrand is odd and the limits of integration are symmetric to the origin). Moreover, by recognizing that the first integral represents the area of a semicircle of radius 1, you obtain

$$F = 64(16)\left(\frac{\pi}{2}\right) - 64(2)(0)$$

$$= 512\pi$$

$$\approx 1608.5 \text{ lb.}$$

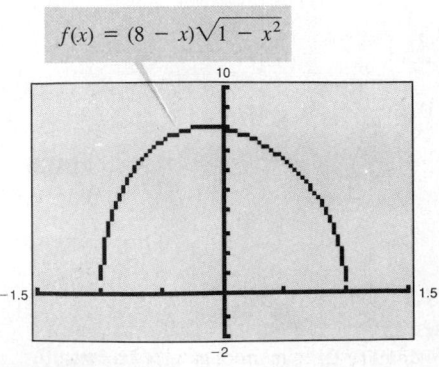

FIGURE 6.62
f is not differentiable at $x = \pm 1$.

TECHNOLOGY   To confirm the result obtained in Example 3, you might consider using Simpson's Rule to approximate the value of

$$128 \int_{-1}^1 (8 - x)\sqrt{1 - x^2}\, dx.$$

From the graph of

$$f(x) = (8 - x)\sqrt{1 - x^2}$$

however, you can see that the f is not differentiable when $x = \pm 1$ (see Figure 6.62). This means that you cannot apply Theorem 4.19 in Section 4.6 to determine the potential error in Simpson's Rule. Without knowing the potential error, the approximation is of little value.

EXAMPLE 4   Fluid Force on a Vertical Surface

A swimming pool is 2 feet deep at one end and 10 feet deep at the other, as shown in Figure 6.63(a). The pool is 40 feet long and 30 feet wide with vertical sides. Find the fluid force against one of the 40-foot sides.

(a)   (b)

The fluid force against one of the 40 foot
sides of the swimming pool is 51,584 lb.

FIGURE 6.63

Solution   By placing the x- and y-axes as shown in Figure 6.63(b), you can see that the depth at y is

$$\text{Depth} = h(y) = 10 - y.$$

The equation representing the base of the side is

$$y = mx + b$$

$$y = \frac{1}{5}x$$

$$x = 5y.$$

Note that the equation $x = 5y$ is only valid for $0 \le y \le 8$. When y is between 8 and 10, x is a constant 40. Thus, the length of the side of the pool is

$$\text{Length} = L(y) = \begin{cases} 5y, & 0 \le y \le 8 \\ 40, & 8 \le y \le 10. \end{cases}$$

Therefore, the fluid force on the side of the pool is given by the following two integrals.

$$F = \int_0^8 62.4(10 - y)(5y)\,dy + \int_8^{10} 62.4(10 - y)(40)\,dy$$

$$= 312 \int_0^8 (10y - y^2)\,dy + 2496 \int_8^{10} (10 - y)\,dy$$

$$= 312 \left[5y^2 - \frac{y^3}{3} \right]_0^8 + 2496 \left[10y - \frac{y^2}{2} \right]_8^{10}$$

$$= 312 \left(\frac{448}{3} \right) + 2496(2)$$

$$= 51{,}584 \text{ lb}$$

REMARK   In Example 4, note that the 30-foot width of the swimming pool is a "red herring." That is, the width of the pool is unnecessary information, and the fluid force against the 40-foot sides can be determined without knowing the width of the pool.

EXERCISES for Section 6.6

Force on Submerged Sheet Metal In Exercises 1 and 2, the area of the top side of a piece of sheet metal is given. The sheet metal is submerged horizontally in 5 feet of water. Find the fluid force on the top side.

1. 3 square feet   **2.** 18 square feet

Buoyant Force In Exercises 3 and 4, find the buoyant force of a rectangular solid of the given dimensions submerged in water so that the top side is parallel to the surface of the water. The buoyant force is the difference between the fluid forces on the top and bottom sides of the solid.

Fluid Force on a Tank Wall In Exercises 5–10, find the fluid force on the indicated vertical side of a tank. Assume that the tank is full of water.

5. Rectangle   **6.** Triangle

7. Trapezoid   **8.** Semicircle

9. Parabola   **10.** Semiellipse

          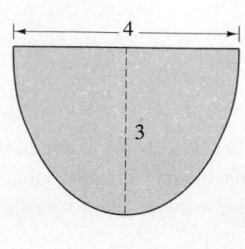

Fluid Force of Water In Exercises 11–14, find the fluid force on the vertical plates submerged in water.

11. Square   **12.** Square

13. Triangle   **14.** Rectangle

Force on a Concrete Form In Exercises 15–18, the figure illustrates the vertical side of a form for poured concrete that weighs 140.7 pounds per cubic foot. Determine the force on this part of the form.

15. Rectangle   **16.** Rectangle

17. Semiellipse   **18.** Triangle

          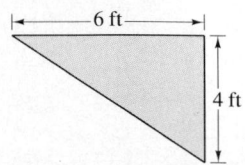

19. *Fluid Force of Gasoline* A cylindrical gasoline tank is placed so that the axis of the cylinder is horizontal. Find the fluid force on a circular end of the tank if the tank is half full, assuming that the diameter of the tank is 3 feet and the gasoline weighs 42 pounds per cubic foot.

20. *Fluid Force of Gasoline* Repeat Exercise 19 for a full tank. (Evaluate one integral by a geometric formula and the other by observing that the integrand is an odd function.)

21. *Fluid Force on a Circular Plate* A circular plate of radius r feet is submerged vertically in a tank of fluid that weighs w pounds per cubic foot. The center of the circle is k feet below the surface of the fluid. Show that the fluid force on the surface of the circle is

$$F = wk(\pi r^2).$$

(Evaluate one integral by a geometric formula and the other by observing that the integrand is an odd function.)

22. *Fluid Force on a Rectangular Plate* A rectangular plate of height h feet and base b feet is submerged vertically in a tank of fluid that weighs w pounds per cubic foot. The center is k feet below the surface of the fluid, where $h \leq k/2$. Show that the fluid force on the surface of the rectangle is

$$F = wkhb.$$

23. *Submarine Porthole* A porthole on a vertical side of a submarine (submerged in sea water) is 1 foot square. Find the fluid force on the porthole, assuming that the center of the square is 15 feet below the surface.

24. *Submarine Porthole* Repeat Exercise 23 for a circular porthole that has a diameter of 1 foot. The center is 15 feet below the surface.

25. *Swimming Pool Design* A swimming pool is 20 feet wide, 40 feet long, 4 feet deep at one end, and 8 feet deep at the other. The bottom is an inclined plane. Find the fluid force on each of the vertical walls.

26. *Boat Design* The figure shows the vertical stern of a boat superimposed on a coordinate system. The table gives the width w of the stern at specified values of y. Find the fluid force on the stern if both measurements are given in feet.

y	0	$\frac{1}{2}$	1	$\frac{3}{2}$	2	$\frac{5}{2}$	3	$\frac{7}{2}$	4
w	0	3	5	8	9	10	10.25	10.5	10.5

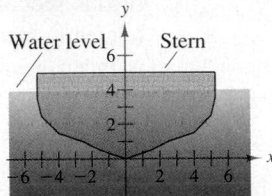

FIGURE FOR 26

C In Exercises 27 and 28, use Simpson's Rule with $n = 10$ to approximate the fluid force on the vertical plate submerged in water bounded by the x-axis and the top half of the graph of the equation. Assume that the base of the plate is 12 feet beneath the surface of the water.

27. $x^{2/3} + y^{2/3} = 4^{2/3}$ **28.** $\dfrac{x^2}{28} + \dfrac{y^2}{16} = 1$

29. The vertical cross section of an irrigation canal is modeled by

$$f(x) = \frac{5x^2}{x^2 + 4}$$

where x is measured in feet and $x = 0$ corresponds to the center of the canal. Use Simpson's Rule with $n = 6$ to approximate the fluid force on a vertical gate used to stop the flow of water if the water is 3 feet deep.

SECTION 6.7   Moments, Centers of Mass, and Centroids

Mass • Center of Mass in a One-Dimensional System • Center of Mass in a Two-Dimensional System • Center of Mass of a Planar Lamina • Theorem of Pappus

Mass

In this section you will study several important applications of integration that are related to **mass.** Mass is a measure of a body's resistance to changes in motion, and is independent of the particular gravitational system in which the body is located. However, because so many applications involving mass occur on the earth's surface, we tend to equate an object's mass with its *weight.* This is not technically correct. Weight is a type of force and as such is dependent on gravity. Force and mass are related by the equation

$$\text{Force} = (\text{Mass})(\text{Acceleration}).$$

Table 6.1 lists some commonly used measures of mass and force, together with their conversion factors.

TABLE 6.1

System of measurement	Measure of mass	Measure of force
U.S.	slug	pound = $(\text{slug})(\text{ft/sec}^2)$
International	kilogram	newton = $(\text{kilogram})(\text{m/sec}^2)$
C-G-S	gram	dyne = $(\text{gram})(\text{cm/sec}^2)$

Conversions:
  1 pound = 4.448 newtons   1 slug = 14.59 kilograms
  1 newton = 0.2248 pound   1 kilogram = 0.06854 slug
    1 dyne = 0.02248 pound   1 gram = 0.00006854 slug

EXAMPLE 1   Mass on the Surface of the Earth

Find the mass (in slugs) of an object whose weight at sea level is 1 pound.

Solution   Using 32 feet per second per second as the acceleration due to gravity produces

$$\text{Mass} = \frac{\text{Force}}{\text{Acceleration}}$$

$$= \frac{1 \text{ lb}}{32 \text{ ft/sec}^2}$$

$$= 0.03125 \, \frac{\text{lb}}{\text{ft/sec}^2}$$

$$= 0.03125 \text{ slug}.$$

Because many applications occur on the earth's surface, this amount of mass is called a **pound mass.**

Center of Mass in a One-Dimensional System

We will consider two types of moments of a mass—the **moment about a point** and the **moment about a line.** To define these two moments, consider an idealized situation in which a mass m is concentrated at a point. If x is the distance between this point-mass and another point P, then the **moment of m about the point P** is

$$\text{Moment} = mx$$

and x is the **length of the moment arm.**

The concept of moment can be demonstrated simply by a seesaw, as illustrated in Figure 6.64. Suppose a child of mass 20 kilograms sits 2 meters to the left of fulcrum P, and an older child of mass 30 kilograms sits 2 meters to the right of P. From experience, you know that the seesaw would begin to rotate clockwise, moving the larger child down. This rotation occurs because the moment produced by the child on the left is less than the moment produced by the child on the right.

Left moment = $(20)(2) = 40$ kg·m

Right moment = $(30)(2) = 60$ kg·m

To balance the seesaw, the two moments must be equal. For example, if the larger child moved to a position $\frac{4}{3}$ meters from the fulcrum, then the seesaw would balance, because both children would produce a moment of 40 kilogram-meters.

To generalize this situation, you can introduce a coordinate line on which the origin corresponds to the fulcrum, as shown in Figure 6.65. Suppose several point masses are located on the x-axis. The measure of the tendency of this system to rotate about the origin is the **moment about the origin,** and it is defined as the sum of the n products $m_i x_i$.

$$M_0 = m_1 x_1 + m_2 x_2 + \cdots + m_n x_n$$

20 kg   **30 kg**

P

|←—2 m—→|←—2 m—→|

FIGURE 6.64

The seesaw will balance when the left and right moments are equal.

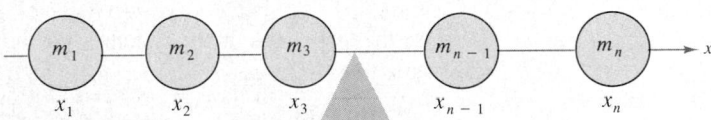

FIGURE 6.65

If $x_1 m_1 + x_2 m_2 + \cdots + x_n m_n = 0$, then the system is in equilibrium.

If M_0 is 0, then the system is said to be in **equilibrium.**

For a system that is not in equilibrium, the **center of mass** is defined as the point $\overline{x}$ at which the fulcrum could be relocated to attain equilibrium. If the system were translated $\overline{x}$ units, each coordinate x_i would become $(x_i - \overline{x})$, and, because the moment of the translated system is 0, you have

$$\sum_{i=1}^{n} m_i(x_i - \overline{x}) = \sum_{i=1}^{n} m_i x_i - \sum_{i=1}^{n} m_i \overline{x} = 0.$$

Solving for $\overline{x}$ produces

$$\overline{x} = \frac{\displaystyle\sum_{i=1}^{n} m_i x_i}{\displaystyle\sum_{i=1}^{n} m_i} = \frac{\text{Moment of system about origin}}{\text{Total mass of system}}.$$

If $x_1 m_1 + x_2 m_2 + \cdots + x_n m_n = 0$, then the system is in equilibrium.

Moments and Center of Mass: One-Dimensional System

Let the point masses $m_1, m_2, \ldots, m_n$ be located at $x_1, x_2, \ldots, x_n$, respectively.
1. The **moment about the origin** is $M_0 = m_1x_1 + m_2x_2 + \cdots + m_nx_n$.
2. The **center of mass** is $\bar{x} = \dfrac{M_0}{m}$ where $m = m_1 + m_2 + \cdots + m_n$ is the **total mass** of the system.

EXAMPLE 2   The Center of Mass of a Linear System

Find the center of mass of the linear system shown in Figure 6.66.

FIGURE 6.66

Solution   The moment about the origin is

$$\begin{aligned}
M_0 &= m_1x_1 + m_2x_2 + m_3x_3 + m_4x_4 \\
&= 10(-5) + 15(0) + 5(4) + 10(7) \\
&= -50 + 0 + 20 + 70 \\
&= 40.
\end{aligned}$$

Because the total mass of the system is $m = 10 + 15 + 5 + 10 = 40$, the center of mass is

$$\bar{x} = \frac{M_0}{m} = \frac{40}{40} = 1.$$

Rather than define the moment of a *mass,* you could define the moment of a *force.* In this context, the center of mass is called the **center of gravity.** Suppose that a system of point masses $m_1, m_2, \ldots, m_n$ is located at $x_1, x_2, \ldots, x_n$. Then, because Force = (Mass)(Acceleration), the total force of the system is

$$F = m_1a + m_2a + \cdots + m_na = ma.$$

The **torque** (moment) about the origin is

$$T_0 = (m_1a)x_1 + (m_2a)x_2 + \cdots + (m_na)x_n = M_0a$$

and the **center of gravity** is

$$\frac{T_0}{F} = \frac{M_0a}{ma} = \frac{M_0}{m} = \bar{x}.$$

Therefore, the center of gravity and the center of mass have the same location.

Center of Mass in a Two-Dimensional System

You can extend the concept of moment to two dimensions by considering a system of masses located in the xy-plane at the points (x_1, y_1), (x_2, y_2), . . . , (x_n, y_n) as shown in Figure 6.67. Rather than defining a single moment (with respect to the origin), we define two moments—one with respect to the x-axis and one with respect to the y-axis.

FIGURE 6.67
In a two-dimensional system, there is a moment about the y-axis, M_y, and a moment about the x-axis, M_x.

Moments and Center of Mass: Two-Dimensional System

Let the point masses $m_1, m_2, \ldots, m_n$ be located at (x_1, y_1), (x_2, y_2), . . . , and (x_n, y_n), respectively.
1. The **moment about the y-axis** is $M_y = m_1x_1 + m_2x_2 + \cdots + m_nx_n$.
2. The **moment about the x-axis** is $M_x = m_1y_1 + m_2y_2 + \cdots + m_ny_n$.
3. The **center of mass** $(\bar{x}, \bar{y})$ (or **center of gravity**) is
$$\bar{x} = \frac{M_y}{m} \quad \text{and} \quad \bar{y} = \frac{M_x}{m}$$
where $m = m_1 + m_2 + \cdots + m_n$ is the **total mass** of the system.

The moment of a system of masses in the plane can be taken about any horizontal or vertical line. In general, the moment about a line is the sum of the product of the masses and the *directed distances* from the points to the line.

$$\text{Moment} = m_1(y_1 - b) + m_2(y_2 - b) + \cdots + m_n(y_n - b) \qquad \text{Horizontal line } y = b$$

$$\text{Moment} = m_1(x_1 - a) + m_2(x_2 - a) + \cdots + m_n(x_n - a) \qquad \text{Vertical line } x = a$$

EXAMPLE 3 The Center of Mass of a Two-Dimensional System

Find the center of mass of a system of point masses $m_1 = 6$, $m_2 = 3$, $m_3 = 2$, and $m_4 = 9$, located at

$$(3, -2), \quad (0, 0), \quad (-5, 3), \quad \text{and} \quad (4, 2)$$

as shown in Figure 6.68.

FIGURE 6.68
The center of mass of the system is $(\frac{11}{5}, \frac{3}{5})$.

Solution

$$
\begin{aligned}
m &= 6 \quad + \quad 3 \quad + 2 \quad + 9 \quad = 20 && \text{Mass} \\
M_y &= 6(3) \quad + \quad 3(0) + 2(-5) + 9(4) = 44 && \text{Moment about } y\text{-axis} \\
M_x &= 6(-2) + \quad 3(0) + 2(3) \quad + 9(2) = 12 && \text{Moment about } x\text{-axis}
\end{aligned}
$$

Therefore,

$$\bar{x} = \frac{M_y}{m} = \frac{44}{20} = \frac{11}{5} \quad \text{and} \quad \bar{y} = \frac{M_x}{m} = \frac{12}{20} = \frac{3}{5}$$

and the center of mass is $(\frac{11}{5}, \frac{3}{5})$.

FIGURE 6.69

You can think of the center of mass $(\bar{x}, \bar{y})$ of a lamina as its balancing point. For a circular lamina, the center of mass is the center of the circle. For a rectangular lamina, the center of mass is the center of the rectangle.

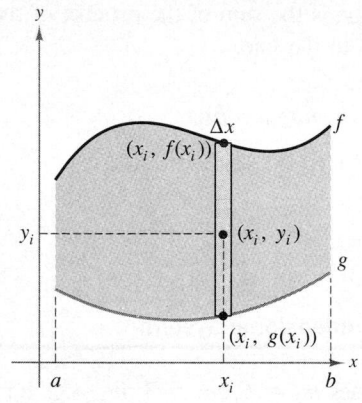

FIGURE 6.70

Planar lamina of uniform density ρ.

Center of Mass of a Planar Lamina

So far in this section, we have assumed the total mass of a system to be distributed at discrete points in a plane or on a line. We now consider a thin flat plate of material of uniform density called a **planar lamina** (see Figure 6.69). **Density** is a measure of mass per unit of volume such as grams per cubic centimeter. For planar laminas, however, density is considered to be a measure of mass per unit of area. Density is denoted by ρ, the lowercase Greek letter rho.

Consider an irregularly shaped planar lamina of uniform density ρ, bounded by the graphs of $y = f(x)$, $y = g(x)$, and $a \le x \le b$, as shown in Figure 6.70. The mass of this region is given by

$$m = (\text{Density})(\text{Area}) = \rho \int_a^b [f(x) - g(x)]\,dx = \rho A$$

where A is the area of the region. To find the center of mass of this lamina, partition the interval $[a, b]$ into n equal subintervals of width Δx. Let x_i be the center of the ith subinterval. You can approximate the portion of the lamina lying in the ith subinterval by a rectangle whose height is $h = f(x_i) - g(x_i)$. Because the density of the rectangle is ρ, its mass is

$$m_i = (\text{Density})(\text{Area}) = \underbrace{\rho}_{\text{Density}} \underbrace{[f(x_i) - g(x_i)]}_{\text{Height}} \underbrace{\Delta x}_{\text{Width}}.$$

Now, considering this mass to be located at the center (x_i, y_i) of the rectangle, the directed distance from the x-axis to (x_i, y_i) is $y_i = [f(x_i) + g(x_i)]/2$. Thus, the moment of m_i about the x-axis is

$$\text{Moment} = (\text{Mass})(\text{Distance})$$
$$= m_i y_i = \rho[f(x_i) - g(x_i)]\Delta x \left[\frac{f(x_i) + g(x_i)}{2}\right].$$

Summing the moments and taking the limit as $n \to \infty$ suggests the following definitions.

Moments and Center of Mass of a Planar Lamina

Let f and g be continuous functions such that $f(x) \ge g(x)$ on $[a, b]$ and consider the planar lamina of uniform density ρ bounded by the graphs of $y = f(x)$, $y = g(x)$, and $a \le x \le b$.

1. The **moments about the x- and y-axes** are

$$M_x = \rho \int_a^b \left[\frac{f(x) + g(x)}{2}\right][f(x) - g(x)]\,dx$$

$$M_y = \rho \int_a^b x[f(x) - g(x)]\,dx.$$

2. The **center of mass** $(\bar{x}, \bar{y})$ is given by $\bar{x} = \dfrac{M_y}{m}$ and $\bar{y} = \dfrac{M_x}{m}$ where $m = \rho \int_a^b [f(x) - g(x)]\,dx$ is the mass of the lamina.

EXAMPLE 4   The Center of Mass of a Planar Lamina

Find the center of mass of the lamina of uniform density ρ bounded by the graph of $f(x) = 4 - x^2$ and the x-axis.

Solution   Because the center of mass lies on the axis of symmetry, you know that $\bar{x} = 0$. Moreover, the mass of the lamina is

$$m = \rho \int_{-2}^{2} (4 - x^2)\,dx = \rho \left[4x - \frac{x^3}{3} \right]_{-2}^{2} = \frac{32\rho}{3}.$$

To find the moment about the x-axis, place a representative rectangle in the region as shown in Figure 6.71. The distance from the x-axis to the center of this rectangle is

$$y_i = \frac{f(x)}{2} = \frac{4 - x^2}{2}.$$

Because the mass of the representative rectangle is

$$\rho f(x)\,\Delta x = \rho(4 - x^2)\,\Delta x$$

you have

$$M_x = \rho \int_{-2}^{2} \frac{4 - x^2}{2}(4 - x^2)\,dx$$

$$= \frac{\rho}{2} \int_{-2}^{2} (16 - 8x^2 + x^4)\,dx$$

$$= \frac{\rho}{2} \left[16x - \frac{8x^3}{3} + \frac{x^5}{5} \right]_{-2}^{2}$$

$$= \frac{256\rho}{15}$$

and $\bar{y}$ is given by

$$\bar{y} = \frac{M_x}{m}$$

$$= \frac{256\rho/15}{32\rho/3}$$

$$= \frac{8}{5}.$$

Thus, the center of mass (the balancing point) of the lamina is $(0, \frac{8}{5})$, as shown in Figure 6.72.

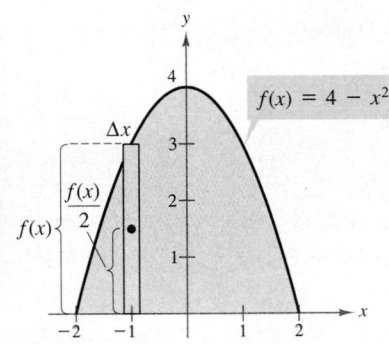

$f(x) = 4 - x^2$

FIGURE 6.71

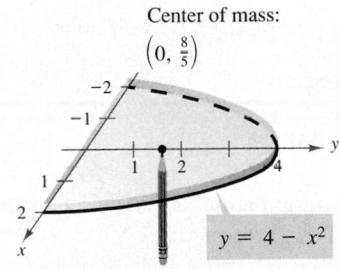

Center of mass: $\left(0, \frac{8}{5}\right)$

$y = 4 - x^2$

FIGURE 6.72
Center of mass is the balancing point.

The density ρ in Example 4 is a common factor of both the moments and the mass, and as such cancels out of the quotients representing the coordinates of the center of mass. Thus, the center of mass of a lamina of *uniform* density depends only on the shape of the lamina and not on its density. For this reason, the point $(\bar{x}, \bar{y})$ is sometimes called the center of mass of a *region* in the plane, or the **centroid** of the region. In other words, to find the centroid of a region in the plane, you simply assume that the region has a constant density of $\rho = 1$, and compute the corresponding center of mass.

EXAMPLE 5   The Centroid of a Plane Region

Find the centroid of the region bounded by the graphs of $f(x) = 4 - x^2$ and $g(x) = x + 2$.

Solution   The two graphs intersect at the points $(-2, 0)$ and $(1, 3)$, as shown in Figure 6.73. Thus, the area of the region is

$$A = \int_{-2}^{1} [f(x) - g(x)]\, dx = \int_{-2}^{1} (2 - x - x^2)\, dx = \frac{9}{2}.$$

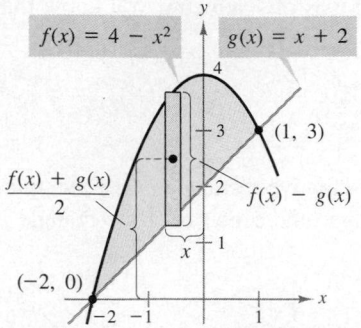

$f(x) = 4 - x^2$ $g(x) = x + 2$

$\dfrac{f(x) + g(x)}{2}$ $f(x) - g(x)$

$(1, 3)$

$(-2, 0)$

FIGURE 6.73
The centroid of the region is $(-\frac{1}{2}, \frac{12}{5})$.

The centroid $(\bar{x}, \bar{y})$ of the region has the following coordinates.

$$\bar{x} = \frac{1}{A}\int_{-2}^{1} x[(4 - x^2) - (x + 2)]\, dx = \frac{2}{9}\int_{-2}^{1}(-x^3 - x^2 + 2x)\, dx$$

$$= \frac{2}{9}\left[-\frac{x^4}{4} - \frac{x^3}{3} + x^2\right]_{-2}^{1} = -\frac{1}{2}$$

$$\bar{y} = \frac{1}{A}\int_{-2}^{1}\left[\frac{(4 - x^2) + (x + 2)}{2}\right][(4 - x^2) - (x + 2)]\, dx$$

$$= \frac{2}{9}\left(\frac{1}{2}\right)\int_{-2}^{1}(-x^2 + x + 6)(-x^2 - x + 2)\, dx$$

$$= \frac{1}{9}\int_{-2}^{1}(x^4 - 9x^2 - 4x + 12)\, dx$$

$$= \frac{1}{9}\left[\frac{x^5}{5} - 3x^3 - 2x^2 + 12x\right]_{-2}^{1} = \frac{12}{5}.$$

Thus, the centroid of the region is $(\bar{x}, \bar{y}) = (-\frac{1}{2}, \frac{12}{5})$.

For simple plane regions, you may be able to find the centroid without resorting to integration.

(a)

EXAMPLE 6   The Centroid of a Simple Plane Region

Find the centroid of the region shown in Figure 6.74(a).

Solution   By superimposing a coordinate system on the region, as shown in Figure 6.74(b). you can locate the centroids of the three rectangles at

$$\left(\frac{1}{2}, \frac{3}{2}\right), \quad \left(\frac{5}{2}, \frac{1}{2}\right), \quad \text{and} \quad (5, 1).$$

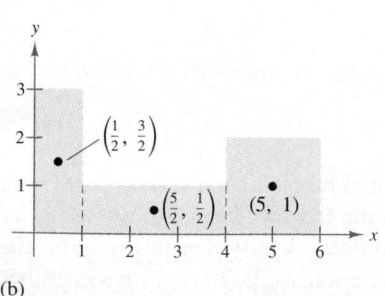

$\left(\frac{1}{2}, \frac{3}{2}\right)$

$\left(\frac{5}{2}, \frac{1}{2}\right)$ $(5, 1)$

(b)

The centroid of the region is $(2.9, \ 1)$

FIGURE 6.74

Using these three points, you can find the centroid of the region.

$$A = \text{Area of region} = 3 + 3 + 4 = 10$$

$$\bar{x} = \frac{(1/2)(3) + (5/2)(3) + (5)(4)}{10} = \frac{29}{10} = 2.9$$

$$\bar{y} = \frac{(3/2)(3) + (1/2)(3) + (1)(4)}{10} = \frac{10}{10} = 1$$

Thus, the centroid of the region is $(2.9, 1)$.

Theorem of Pappus

The final topic in this section is a useful theorem credited to Pappus of Alexandria (ca. 300 A.D.), a Greek mathematician whose eight-volume *Mathematical Collection* is a record of much of classical Greek mathematics. We delay the proof of this theorem until Section 14.4 (Exercise 45).

Centroid of R

R   **r**   **L**

FIGURE 6.75
Volume: $2\pi r$(area of region R)

THEOREM 6.1   The Theorem of Pappus

Let R be a region in a plane and let L be a line in the same plane such that L does not intersect the interior of R, as shown in Figure 6.75. If r is the distance between the centroid of R and the line, then the volume V of the solid of revolution formed by revolving R about the line is

$$V = 2\pi r A$$

where A is the area of R. (Note that $2\pi r$ is the distance traveled by the centroid as the region is revolved about the line.)

The Theorem of Pappus can be used to find the volume of a torus, as shown in the following example. Recall that a **torus** is a doughnut-shaped solid formed by revolving a circular region about a line that lies in the same plane as the circle (but does not intersect the circle).

EXAMPLE 7   Finding Volume by the Theorem of Pappus

Find the volume of the torus formed by revolving the circular region bounded by $(x - 2)^2 + y^2 = 1$ about the y-axis, as shown in Figure 6.76(a).

Torus

(a)   (b)

FIGURE 6.76

Solution From Figure 6.76(b), you see that the centroid of the circular region is $(2, 0)$. Thus, the distance between the centroid and the axis of revolution is $r = 2$. Because the area of the circular region is $A = \pi$, the volume of the torus is

$$V = 2\pi r A = 2\pi(2)(\pi) = 4\pi^2 \approx 39.5.$$

EXERCISES for Section 6.7

In Exercises 1–4, find the center of mass of the point masses lying on the *x*-axis.

1. $m_1 = 6, m_2 = 3, m_3 = 5$
   $x_1 = -5, x_2 = 1, x_3 = 3$

2. $m_1 = 7, m_2 = 4, m_3 = 3, m_4 = 8$
   $x_1 = -3, x_2 = -2, x_3 = 5, x_4 = 6$

3. $m_1 = 1, m_2 = 1, m_3 = 1, m_4 = 1, m_5 = 1$
   $x_1 = 7, x_2 = 8, x_3 = 12, x_4 = 15, x_5 = 18$

4. $m_1 = 12, m_2 = 1, m_3 = 6, m_4 = 3, m_5 = 11$
   $x_1 = -3, x_2 = -2, x_3 = -1, x_4 = 0, x_5 = 4$

5. In Exercise 3, $\bar{x}$ is the arithmetic mean of the *x*-coordinates. Translate each point mass five units to the right and compare the resulting center of mass with that obtained in Exercise 3.

6. Translate each point mass in Exercise 4 to the left three units and compare the resulting center of mass with that obtained in Exercise 4.

In Exercises 7–10, find the center of mass of the system of point masses.

7.

m_i	5	1	3
(x_i, y_i)	(2, 2)	(−3, 1)	(1, −4)

8.

m_i	10	2	5
(x_i, y_i)	(1, −1)	(5, 5)	(−4, 0)

9.

m_i	3	4
(x_i, y_i)	(−2, −3)	(−1, 0)

m_i	2	1	6
(x_i, y_i)	(7, 1)	(0, 0)	(−3, 0)

10.

m_i	4	2	5/2	5
(x_i, y_i)	(2, 3)	(−1, 5)	(6, 8)	(2, −2)

In Exercises 11–22, find M_x, M_y, and $(\bar{x}, \bar{y})$ for the laminas of uniform density ρ bounded by the graphs of the equations.

11. $y = \sqrt{x}, y = 0, x = 4$

12. $y = x^2, y = 0, x = 4$

13. $y = \sqrt{x}, y = x$

14. $y = x^2, y = x^3$

15. $y = -x^2 + 4x + 2, y = x + 2$

16. $y = \sqrt{3x} + 1, y = x + 1$

17. $y = x^{2/3}, y = 0, x = 8$

18. $y = x^{2/3}, y = 4$

19. $x = 4 - y^2, x = 0$

20. $x = 2y - y^2, x = 0$

21. $x = -y, x = 2y - y^2$

22. $x = y + 2, x = y^2$

In Exercises 23–26, find the centroid of the region bounded by the graphs of the equations.

23. $y = x^2, y = x$

24. $y = \dfrac{1}{x}, y = 0, 1 \le x \le 4$

25. $y = 2x + 4, y = 0, 0 \le x \le 3$

26. $y = x^2 - 4, y = 0$

In Exercises 27–32, find and/or verify the centroid of common regions used in engineering.

27. *Triangle* Show that the centroid of the triangle with vertices $(-a, 0)$, $(a, 0)$, and (b, c), is the point of intersection of the medians (see figure).

28. *Parallelogram* Show that the centroid of the parallelogram with vertices $(0, 0)$, $(a, 0)$, (b, c), and $(a + b, c)$ is the point of intersection of the diagonals (see figure).

FIGURE FOR 27

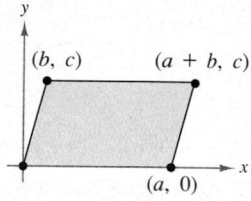

FIGURE FOR 28

29. *Trapezoid* Find the centroid of the trapezoid with vertices $(0, 0)$, $(0, a)$, (c, b), and $(c, 0)$. Show that it is the intersection of the line connecting the midpoints of the parallel sides and the line connecting the extended parallel sides as shown in the figure.

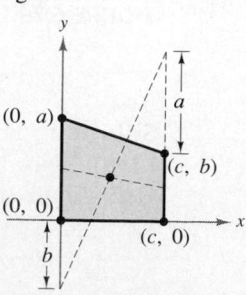

FIGURE FOR 29

30. *Semicircle* Find the centroid of the semicircular region bounded by the graphs of $y = \sqrt{r^2 - x^2}$ and $y = 0$ (see figure).

FIGURE FOR 30

31. *Semiellipse* Find the centroid of the semielliptical region bounded by the graphs of $y = \frac{b}{a}\sqrt{a^2 - x^2}$ and $y = 0$ (see figure).

32. *Parabolic Spandrel* Find the centroid of the **parabolic spandrel** shown in the figure.

FIGURE FOR 31

FIGURE FOR 32

C In Exercises 33–36, use a computer or calculator and Simpson's Rule with $n = 10$ to approximate the centroid of the region bounded by the graphs of the equations.

33. $y = 10x\sqrt{125 - x^3}$, $y = 0$, $x = 0$

34. $y = xe^{-x/2}$, $y = 0$, $x = 0$, $x = 4$

35. *Prefabricated End Section of a Building*

$y = 5\sqrt[3]{400 - x^2}$, $y = 0$

36. *Witch of Agnesi*

$y = 8/(x^2 + 4)$, $y = 0$, $x = -2$, $x = 2$.

37. *Conversion Van Window* The manufacturer of glass for a window in a conversion van needs to approximate the window's center of mass. A coordinate system is superimposed on a prototype of the glass (see figure). The measurements (in centimeters) for the right half of the symmetric piece of glass are listed in the table.

x	0	10	20	30	40
y	30	29	26	20	0

Approximate the center of mass of the glass.

FIGURE FOR 37

38. *Boat Construction* The manufacturer of a boat needs to approximate the center of mass of a section of the hull. A coordinate system is superimposed on a prototype (see figure). The measurements (in feet) for the right half of the symmetric prototype are listed in the table.

x	0	0.5	1.0	1.5	2.0
l	1.50	1.45	1.30	0.99	0
d	0.50	0.48	0.43	0.33	0

Approximate the center of mass of the section of the hull.

  FIGURE FOR 38

In Exercises 39–42, introduce an appropriate coordinate system and find the coordinates of the center of mass of the planar lamina. (The answer depends on the position of the coordinate system.)

39.

40.

41.

42.

43. Find the center of mass of the lamina in Exercise 39 if the circular lamina has twice the density of the square lamina.

44. Find the center of mass of the lamina in Exercise 39 if the square lamina has twice the density of the circular lamina.

In Exercises 45–48, use the Theorem of Pappus to find the volume of the solid of revolution.

45. Torus formed by revolving the circle $(x - 5)^2 + y^2 = 16$ about the y-axis.

46. Torus formed by revolving the circle $(x - a)^2 + y^2 = b^2$, where $0 < b < a$, about the y-axis.

47. Solid formed by revolving the region bounded by the graphs of $y = x$, $y = 4$, and $x = 0$ about the x-axis.

48. Solid formed by revolving the region bounded by the graphs of $y = \sqrt{x - 1}$, $y = 0$, and $x = 5$ about the y-axis.

In Exercises 49 and 50, use the *Second Theorem of Pappus*, which is stated as follows. If a segment of a plane curve C is revolved about an axis that does not intersect the curve (except possibly at its endpoints), then the area S of the resulting surface of revolution is given by the product of the length of C times the distance d traveled by the centroid of C.

49. A sphere is formed by revolving the graph of $y = \sqrt{r^2 - x^2}$ about the x-axis. Use the formula for surface area, $S = 4\pi r^2$, to find the centroid of the semicircle $y = \sqrt{r^2 - x^2}$.

50. A torus is formed by revolving the graph of $(x - 1)^2 + y^2 = 1$ about the y-axis. Find the surface area of the torus.

51. Let $n \geq 1$ be constant, and consider the region bounded by $f(x) = x^n$, the x-axis, and $x = 1$. Find the centroid of this region. As $n \to \infty$, what does the region look like, and where is its centroid?

REVIEW EXERCISES for Chapter 6

Area In Exercises 1–14, sketch the region bounded by the graphs of the equations and determine the area of the region.

1. $y = \dfrac{1}{x^2}, y = 0, x = 1, x = 5$

2. $y = \dfrac{1}{x^2}, y = 4, x = 5$

3. $y = \dfrac{1}{x^2 + 1}, y = 0, x = -1, x = 1$

4. $x = y^2 - 2y, x = -1, y = 0$

5. $y = x, y = x^3$

6. $x = y^2 + 1, x = y + 3$

7. $y = x^2 - 8x + 3, y = 3 + 8x - x^2$

8. $y = x^2 - 4x + 3, y = x^3, x = 0$
   (*Hint:* Use Newton's Method to approximate the point of intersection.)

9. $\sqrt{x} + \sqrt{y} = 1, y = 0, x = 0$

10. $y = x^4 - 2x^2, y = 2x^2$

11. $y = e^x, y = e^2, x = 0$

12. $y = \csc x, y = 2$ (one region)

13. $y = \sin x, y = \cos x, \dfrac{\pi}{4} \leq x \leq \dfrac{5\pi}{4}$

14. $x = \cos y, x = \dfrac{1}{2}, \dfrac{\pi}{3} \leq y \leq \dfrac{7\pi}{3}$

Area In Exercises 15–18, set up integrals for finding the area of the region bounded by the graphs of the equations using vertical and horizontal representative rectangles. Find the area of the region by evaluating the easier of the two integrals.

15. $x = y^2 - 2y, x = 0$

16. $y = 1 - \dfrac{x}{2}, y = x - 2, y = 1$

17. $y = \sqrt{x - 1}, y = 2, y = 0, x = 0$

18. $y = \sqrt{x - 1}, y = \dfrac{x - 1}{2}$

Volume In Exercises 19–26, find the volume of the solid generated by revolving the plane region bounded by the equations about the indicated line.

19. $y = x, y = 0, x = 4$
   a. The x-axis   **b.** The y-axis
   c. The line $x = 4$ **d.** The line $x = 6$

20. $y = \sqrt{x}, y = 2, x = 0$
   a. The x-axis   **b.** The line $y = 2$
   c. The y-axis   **d.** The line $x = -1$

21. $\dfrac{x^2}{16} + \dfrac{y^2}{9} = 1$
   a. The y-axis (oblate spheroid)
   b. The x-axis (prolate spheroid)

22. $\dfrac{x^2}{a^2} + \dfrac{y^2}{b^2} = 1$
   a. The y-axis (oblate spheroid)
   b. The x-axis (prolate spheroid)

23. $y = \dfrac{1}{x^4 + 1}, y = 0, x = 0, x = 1$
   revolved about the y-axis

24. $y = \dfrac{1}{\sqrt{1 + x^2}}, y = 0, x = -1, x = 1$
   revolved about the x-axis

25. $y = 1/(1 + \sqrt{x - 2}), y = 0, x = 2, x = 6$
   revolved about the y-axis.

26. $y = e^{-x}, y = 0, x = 0, x = 1$
   revolved about the x-axis

In Exercises 27 and 28, consider the region bounded by the graphs of the equations $y = x\sqrt{x + 1}$ and $y = 0$.

27. *Area* Find the area of the region.

28. *Volume* Find the volume of the solid generated by revolving the region about
   a. The x-axis.
   b. The y-axis.

29. *Volume of a Gasoline Tank* A gasoline tank is an oblate spheroid generated by revolving the region bounded by the graph of

$$\frac{x^2}{16} + \frac{y^2}{9} = 1$$

about the *y*-axis, where *x* and *y* are measured in feet. Find the depth of the gasoline in the tank when it is filled to one-fourth its capacity. (*Note:* Use Newton's Method after evaluating the definite integral.)

30. *Magnitude of the Base* The base of a solid is a circle of radius *a* and its vertical cross sections are equilateral triangles. Find the radius of the circle if the volume of the solid is 10 cubic inches.

Arc Length **In Exercises 31 and 32, find the arc length of the graph of the function over the specified interval.**

Function	Interval
31. $f(x) = \frac{4}{5}x^{5/4}$	$[0, 4]$
32. $y = \frac{1}{6}x^3 + \frac{1}{2x}$	$[1, 3]$

33. *Surface Area* Use integration to find the lateral surface area of a right circular cone of height 4 and radius 3.

34. *Surface Area* The region bounded by $y = 2\sqrt{x}, y = 0$, and $x = 3$ is revolved about the *x*-axis. Find the surface area of the solid generated.

35. *Work* Find the work done in stretching a spring from its natural length of 10 inches to a length of 15 inches, if a force of 4 pounds is needed to stretch it 1 inch from its natural position.

36. *Work* Find the work done in stretching a spring from its natural length of 9 inches to double that length, if a force of 50 pounds is required to hold the spring at double its natural length.

37. *Work* A water well has an 8-inch casing (diameter) and is 175 feet deep. If the water is 25 feet from the top of the well, determine the amount of work done in pumping it dry, assuming that no water enters the well while it is being pumped.

38. *Work* Repeat Exercise 37, assuming that water enters the well at the rate of 4 gallons per minute and the pump works at the rate of 12 gallons per minute. How many gallons are pumped in this case?

39. *Work* A chain 10 feet long weighs 5 pounds per foot and is suspended from a platform 20 feet above the ground. How much work is required to raise the entire chain to the 20-foot level?

40. *Work* A windlass, 200 feet above ground level on the top of a building, uses a cable weighing 4 pounds per foot. Find the work done in winding up the cable if
a. one end is at ground level.
b. there is a 300-pound load attached to the end of the cable.

41. *Fluid Force* A swimming pool is 5 feet deep at one end and 10 feet deep at the other, and the bottom is an inclined plane. The length and width of the pool are 40 feet and 20 feet, respectively. If the pool is full of water, what is the fluid force on each of the vertical walls?

42. *Fluid Force* Show that the fluid force against any vertical region in a liquid is the product of the weight per cubic volume of the liquid, the area of the region, and the depth of the centroid of the region.

43. *Fluid Force* Using the result of Exercise 42, find the fluid force on one side of a vertical circular plate of radius 4 feet that is submerged in water so that its center is 5 feet below the surface.

44. *Fluid Force* How far must the water level be raised to double the fluid force on one side of the plate in Exercise 43?

Centroids **In Exercises 45–48, find the centroid of the region bounded by the graphs of the equations.**

45. $\sqrt{x} + \sqrt{y} = \sqrt{a}, x = 0, y = 0$

46. $y = x^2, y = 2x + 3$

47. $y = a^2 - x^2, y = 0$

48. $y = x^{2/3}, y = \frac{1}{2}x$

49. *Centroid* A blade on an industrial fan has the configuration of a semicircle attached to a trapezoid (see figure). Find the centroid of the blade.

FIGURE FOR 49

50. *Centroid* Cut an object of irregular shape from a piece of cardboard.
a. Hold a pencil vertically and move the object on the pencil point until the centroid is located.
b. Divide the object into representative elements. Make the necessary measurements and numerically approximate the centroid. Compare this result with that of part a.

51. *Length of a Catenary* A cable of a suspension bridge forms a catenary modeled by the equation

$$y = 300 \cosh\left(\frac{x}{2000}\right) - 280, \qquad -2000 \le x \le 2000$$

where *x* and *y* are measured in feet. Approximate the length of the cable.

Chapter 7

Integration Techniques, L'Hôpital's Rule, and Improper Integrals

Basic Integration Rules

In this chapter you will study several integration techniques that greatly expand the set of integrals to which the basic integration rules can be applied.

Basic Integration Rules ($a > 0$)

1. $\displaystyle\int k f(u)\,du = k\int f(u)\,du$

2. $\displaystyle\int [f(u) \pm g(u)]\,du = \int f(u)\,du \pm \int g(u)\,du$

3. $\displaystyle\int du = u + C$

4. $\displaystyle\int u^n\,du = \frac{u^{n+1}}{n+1} + C, \qquad n \neq -1$

5. $\displaystyle\int \frac{du}{u} = \ln|u| + C$

6. $\displaystyle\int e^u\,du = e^u + C$

7. $\displaystyle\int a^u\,du = \left(\frac{1}{\ln a}\right)a^u + C$

8. $\displaystyle\int \sin u\,du = -\cos u + C$

9. $\displaystyle\int \cos u\,du = \sin u + C$

10. $\displaystyle\int \tan u\,du = -\ln|\cos u| + C$

11. $\displaystyle\int \cot u\,du = \ln|\sin u| + C$

12. $\displaystyle\int \sec u\,du = \ln|\sec u + \tan u| + C$

13. $\displaystyle\int \csc u\,du = -\ln|\csc u + \cot u| + C$

14. $\displaystyle\int \sec^2 u\,du = \tan u + C$

15. $\displaystyle\int \csc^2 u\,du = -\cot u + C$

16. $\displaystyle\int \sec u \tan u\,du = \sec u + C$

17. $\displaystyle\int \csc u \cot u\,du = -\csc u + C$

18. $\displaystyle\int \frac{du}{\sqrt{a^2 - u^2}} = \arcsin\frac{u}{a} + C$

19. $\displaystyle\int \frac{du}{a^2 + u^2} = \frac{1}{a}\arctan\frac{u}{a} + C$

20. $\displaystyle\int \frac{du}{u\sqrt{u^2 - a^2}} = \frac{1}{a}\operatorname{arcsec}\frac{|u|}{a} + C$

A major step in solving any integration problem is recognizing the proper basic integration rule to be used. To do this, you must memorize the basic rules and practice using them. In this first section, we review the use of basic integration rules.

One of the challenging things about integration is the fact that slight differences in the integrand can lead to very different solution techniques.

◀ *The velocity of the rocket that propels the space shuttle into orbit is discussed in Section 7.2, Exercise 83.*

Fitting Integrands to Basic Rules

EXAMPLE 1   A Comparison of Three Similar Integrals

Evaluate the following integrals.

a. $\displaystyle\int \frac{4}{x^2 + 9}\,dx$ **b.** $\displaystyle\int \frac{4x}{x^2 + 9}\,dx$ **c.** $\displaystyle\int \frac{4x^2}{x^2 + 9}\,dx$

Solution

a. Use the Arctangent Rule and let $u = x$ and $a = 3$.

$$\int \frac{4}{x^2 + 9}\,dx = 4 \int \frac{1}{x^2 + 3^2}\,dx$$

$$= 4\left(\frac{1}{3}\arctan\frac{x}{3}\right) + C$$

$$= \frac{4}{3}\arctan\frac{x}{3} + C$$

b. Here the Arctangent Rule does not apply because the numerator contains a factor of x. Consider the Log Rule and let $u = x^2 + 9$. Then, $du = 2x\,dx$, and you have

$$\int \frac{4x}{x^2 + 9}\,dx = 2 \int \frac{2x\,dx}{x^2 + 9}$$

$$= 2 \int \frac{du}{u}$$

$$= 2 \ln |u| + C$$

$$= 2 \ln(x^2 + 9) + C.$$

c. Because the degree of the numerator is equal to the degree of the denominator, you should first divide to obtain

$$\frac{4x^2}{x^2 + 9} = 4 - \frac{36}{x^2 + 9}.$$

Now, you can integrate as follows.

$$\int \frac{4x^2}{x^2 + 9}\,dx = \int \left(4 - \frac{36}{x^2 + 9}\right) dx$$

$$= \int 4\,dx - 36 \int \frac{1}{x^2 + 9}\,dx$$

$$= 4x - 36\left(\frac{1}{3}\arctan\frac{x}{3}\right) + C$$

$$= 4x - 12\arctan\frac{x}{3} + C$$

Notice in Example 1c that some preliminary algebra was required before applying the rules for integration, and that subsequently more than one rule was needed to evaluate the resulting integral.

EXAMPLE 2 **Using Two Basic Rules to Solve a Single Integral**

Evaluate $\displaystyle\int_0^1 \frac{x+3}{\sqrt{4-x^2}}\,dx$.

Solution Begin by writing the integral as the sum of two integrals. Then apply the Power Rule and the Arcsine Rule as follows.

$$\int_0^1 \frac{x+3}{\sqrt{4-x^2}}\,dx = \int_0^1 \frac{x}{\sqrt{4-x^2}}\,dx + \int_0^1 \frac{3}{\sqrt{4-x^2}}\,dx$$

$$= -\frac{1}{2}\int_0^1 (4-x^2)^{-1/2}(-2x)\,dx + 3\int_0^1 \frac{1}{\sqrt{2^2-x^2}}\,dx$$

$$= \left[-(4-x^2)^{1/2} + 3\arcsin\frac{x}{2} \right]_0^1$$

$$= \left(-\sqrt{3} + \frac{1}{2}\pi \right) - (-2 + 0)$$

$$\approx 1.839 \quad \text{(See Figure 7.1.)}$$

FIGURE 7.1
The area of the region is approximately 1.839.

TECHNOLOGY Simpson's Rule can be used to give a good approximation of the value of the integral in Example 2 (for $n = 10$, the approximation is 1.839). When using numerical integration, however, you should be aware that Simpson's Rule does not give good approximations when one or both of the limits of integration are near a vertical asymptote. Using the Fundamental Theorem of Calculus, you can obtain

$$\int_0^{1.99} \frac{x+3}{\sqrt{4-x^2}}\,dx \approx 6.213.$$

Applying Simpson's Rule (with $n = 10$) produces an approximation of 6.889.

EXAMPLE 3 **Breaking a Quotient into Two Parts**

Evaluate $\displaystyle\int \frac{1+\cos(e^{-2x})}{e^{2x}}\,dx$.

Solution As it stands, the integrand does not appear to fit any of the basic integration rules. However, because the numerator is a sum, you can break the integrand into two fractions and integrate as follows.

$$\int \frac{1+\cos e^{-2x}}{e^{2x}}\,dx = \int \frac{dx}{e^{2x}} + \int \frac{\cos e^{-2x}}{e^{2x}}\,dx$$

$$= \int \underbrace{e^{-2x}}_{u\,=\,-2x}\,dx + \int (\cos \underbrace{e^{-2x}}_{u\,=\,e^{-2x}})\,(e^{-2x})\,dx$$

$$= -\frac{1}{2}\int e^{-2x}(-2)\,dx - \frac{1}{2}\int (\cos e^{-2x})(-2e^{-2x})\,dx$$

$$= -\frac{1}{2}\int e^u\,du - \frac{1}{2}\int \cos u\,du$$

$$= -\frac{1}{2}e^u - \frac{1}{2}\sin u + C$$

$$= -\frac{1}{2}e^{-2x} - \frac{1}{2}\sin e^{-2x} + C$$

Note that two *different* u-substitutions were used in the second step.

Rules 18, 19, and 20 all have expressions involving the sum or difference of two squares.

$$a^2 - [f(x)]^2 \qquad a^2 + [f(x)]^2 \qquad [f(x)]^2 - a^2$$

With such an expression, consider the substitution $u = f(x)$.

EXAMPLE 4   A Substitution Involving $a^2 - u^2$

Evaluate $\displaystyle\int \frac{x^2}{\sqrt{16 - x^6}}\,dx$.

Solution Because the radical in the denominator can be written in the form $\sqrt{a^2 - u^2} = \sqrt{4^2 - (x^3)^2}$, you can try the substitution $u = x^3$. Then $du = 3x^2\,dx$, and you have

$$
\begin{aligned}
\int \frac{x^2}{\sqrt{16 - x^6}}\,dx &= \frac{1}{3}\int \frac{3x^2\,dx}{\sqrt{16 - (x^3)^2}} \\
&= \frac{1}{3}\int \frac{du}{\sqrt{4^2 - u^2}} \\
&= \frac{1}{3}\arcsin\frac{u}{4} + C \\
&= \frac{1}{3}\arcsin\frac{x^3}{4} + C.
\end{aligned}
$$

Also, when applying Rules 18, 19, and 20, recall from Chapter 5 that completing the square can be helpful. For instance,

$$\int \frac{dx}{x^2 + 2x + 5} = \int \frac{dx}{(x + 1)^2 + 2^2} = \frac{1}{2}\arctan\frac{x + 1}{2} + C.$$

Surprisingly, two of the most commonly overlooked integration rules are the Log Rule and the Power Rule. Notice in the next two examples how these two integration rules can be disguised.

EXAMPLE 5   A Disguised Form of the Log Rule

Evaluate $\displaystyle\int \frac{1}{1 + e^x}\,dx$.

Solution The integral does not appear to fit any of the basic rules. However, the quotient form suggests the Log Rule. If you let $u = 1 + e^x$, then $du = e^x\,dx$. You can obtain the required du by adding and subtracting e^x in the numerator as follows.

$$
\begin{aligned}
\int \frac{1}{1 + e^x}\,dx &= \int \frac{1 + e^x - e^x}{1 + e^x}\,dx && \text{Add and subtract } e^x \text{ in numerator} \\
&= \int \left(\frac{1 + e^x}{1 + e^x} - \frac{e^x}{1 + e^x}\right) dx \\
&= \int dx - \int \frac{e^x\,dx}{1 + e^x} \\
&= x - \ln(1 + e^x) + C
\end{aligned}
$$

REMARK There is usually more than one way to solve an integration problem. For instance, in Example 5, try integrating by multiplying the numerator and denominator by e^{-x} to obtain an integral of the form $-\int (u'/u)\,dx$. See if you can get the same answer by this procedure. (Be careful; the answer will appear in a different form.)

EXAMPLE 6   A Disguised Form of the Power Rule

Evaluate $\int (\cot x)[\ln(\sin x)]\,dx$.

Solution   Again, the integral does not appear to fit any of the basic rules. However, considering the two primary choices for u ($u = \cot x$ or $u = \ln \sin x$), you can see that the second choice is the appropriate one because

$$u = \ln \sin x \rightarrow du = \frac{\cos x}{\sin x}\,dx = \cot x\,dx.$$

Thus, you have

$$\int (\cot x)[\ln(\sin x)]\,dx = \int u\,du = \frac{u^2}{2} + C$$

$$= \frac{1}{2}[\ln(\sin x)]^2 + C.$$

REMARK   In Example 6, try *checking* that the derivative of

$$\frac{1}{2}[\ln(\sin x)]^2 + C$$

is the integrand of the original integral.

Trigonometric identities can often be used to fit integrals to one of the basic integration rules.

EXAMPLE 7   Using Trigonometric Identities

Evaluate $\int \tan^2 2x\,dx$.

Solution   Note that $\tan^2 u$ is not in the list of basic integration rules. However, $\sec^2 u$ is in the list. This suggests the trigonometric identity $\tan^2 u = \sec^2 u - 1$. If you let $u = 2x$, then $du = 2\,dx$ and

$$\int \tan^2 2x\,dx = \frac{1}{2}\int \tan^2 u\,du$$

$$= \frac{1}{2}\int (\sec^2 u - 1)\,du$$

$$= \frac{1}{2}\int \sec^2 u\,du - \frac{1}{2}\int du$$

$$= \frac{1}{2}\tan u - \frac{u}{2} + C$$

$$= \frac{1}{2}\tan 2x - x + C.$$

TECHNOLOGY   If you have access to symbolic integrating software, try using it to evaluate the integrals in this section. Compare the *form* of the antiderivative given by the software to the form obtained by hand. Sometimes the forms will be the same, but often they will differ. For instance, why is the antiderivative $\ln 2x + C$ equivalent to the antiderivative $\ln x + C$?

We conclude with a summary of some common procedures for fitting integrands to the basic integration rules.

Procedures for Fitting Integrands to Basic Rules

Technique	*Example*
Expand (numerator)	$(1 + e^x)^2 = 1 + 2e^x + e^{2x}$
Separate numerator	$\dfrac{1 + x}{x^2 + 1} = \dfrac{1}{x^2 + 1} + \dfrac{x}{x^2 + 1}$
Complete the square	$\dfrac{1}{\sqrt{2x - x^2}} = \dfrac{1}{\sqrt{1 - (x - 1)^2}}$
Divide improper rational function	$\dfrac{x^2}{x^2 + 1} = 1 - \dfrac{1}{x^2 + 1}$
Add and subtract terms in numerator	$\dfrac{2x}{x^2 + 2x + 1} = \dfrac{2x + 2 - 2}{x^2 + 2x + 1} = \dfrac{2x + 2}{x^2 + 2x + 1} - \dfrac{2}{(x + 1)^2}$
Use trigonometric identities	$\cot^2 x = \csc^2 x - 1$
Multiply and divide by Pythagorean conjugate	$\dfrac{1}{1 + \sin x} = \left(\dfrac{1}{1 + \sin x}\right)\left(\dfrac{1 - \sin x}{1 - \sin x}\right) = \dfrac{1 - \sin x}{1 - \sin^2 x}$
	$= \dfrac{1 - \sin x}{\cos^2 x} = \sec^2 x - \dfrac{\sin x}{\cos^2 x}$

REMARK   Watch out for this common error when fitting integrands to basic rules.

$$\frac{1}{x^2 + 1} \neq \frac{1}{x^2} + \frac{1}{1} \qquad \text{Do not separate denominators}$$

EXERCISES for Section 7.1

In Exercises 1–4, select the correct antiderivative.

1. $\displaystyle \int \frac{x}{\sqrt{x^2 + 1}}\,dx$

   a. $2\sqrt{x^2 + 1} + C$ **b.** $\sqrt{x^2 + 1} + C$

   c. $\frac{1}{2}\sqrt{x^2 + 1} + C$ **d.** $\ln(x^2 + 1) + C$

2. $\displaystyle \int \frac{x}{x^2 + 1}\,dx$

   a. $\ln\sqrt{x^2 + 1} + C$ **b.** $\dfrac{2x}{(x^2 + 1)^2} + C$

   c. $\arctan x + C$ **d.** $\ln(x^2 + 1) + C$

3. $\displaystyle \int \frac{1}{x^2 + 1}\,dx$

   a. $\ln\sqrt{x^2 + 1} + C$ **b.** $\dfrac{2x}{(x^2 + 1)^2} + C$

   c. $\arctan x + C$ **d.** $\ln(x^2 + 1) + C$

4. $\displaystyle \int x\cos(x^2 + 1)\,dx$

   a. $2x\sin(x^2 + 1) + C$ **b.** $-\frac{1}{2}\sin(x^2 + 1) + C$

   c. $\frac{1}{2}\sin(x^2 + 1) + C$ **d.** $-2x\sin(x^2 + 1) + C$

In Exercises 5–14, select the basic integration formula you would use to evaluate the integral, and identify u and a when appropriate.

5. $\displaystyle \int (3x - 2)^4\,dx$

6. $\displaystyle \int \frac{2t - 1}{t^2 - t + 2}\,dt$

7. $\displaystyle \int \frac{1}{\sqrt{x}(1 - 2\sqrt{x})}\,dx$

8. $\displaystyle \int \frac{2}{(2t - 1)^2 + 4}\,dt$

9. $\displaystyle \int \frac{3}{\sqrt{1 - t^2}}\,dt$

10. $\displaystyle \int \frac{-2x}{\sqrt{x^2 - 4}}\,dx$

11. $\displaystyle \int t\sin t^2\,dt$

12. $\displaystyle \int \sec 3x \tan 3x\,dx$

13. $\displaystyle \int \cos x\, e^{\sin x}\,dx$

14. $\displaystyle \int \frac{1}{x\sqrt{x^2 - 4}}\,dx$

In Exercises 15–46, evaluate the indefinite integral.

15. $\displaystyle \int (-2x + 5)^{3/2}\,dx$

16. $\displaystyle \int \frac{2}{(t - 9)^2}\,dt$

17. $\displaystyle \int \left[v + \frac{1}{(3v - 1)^3}\right]\,dv$

18. $\displaystyle \int x\sqrt{4 - 2x^2}\,dx$

19. $\displaystyle\int \frac{t^2 - 3}{-t^3 + 9t + 1} dt$

20. $\displaystyle\int \frac{2x}{x - 4} dx$

21. $\displaystyle\int \frac{x^2}{x - 1} dx$

22. $\displaystyle\int \frac{x + 1}{\sqrt{x^2 + 2x - 4}} dx$

23. $\displaystyle\int \left(\frac{1}{3x - 1} - \frac{1}{3x + 1} \right) dx$

24. $\displaystyle\int \frac{e^x}{1 + e^x} dx$

25. $\displaystyle\int (1 + 2x^2)^2 dx$

26. $\displaystyle\int x \left(1 + \frac{1}{x} \right)^3 dx$

27. $\displaystyle\int x(\cos 2\pi x^2) dx$

28. $\displaystyle\int \sec 4u \, du$

29. $\displaystyle\int \csc \pi x \cot \pi x \, dx$

30. $\displaystyle\int \frac{\sin x}{\sqrt{\cos x}} dx$

31. $\displaystyle\int e^{5x} dx$

32. $\displaystyle\int \csc^2 x e^{\cot x} dx$

33. $\displaystyle\int \frac{2}{e^{-x} + 1} dx$

34. $\displaystyle\int \frac{1}{2e^x - 3} dx$

35. $\displaystyle\int \frac{1 + \sin x}{\cos x} dx$

36. $\displaystyle\int \frac{1}{\sec x - 1} dx$

37. $\displaystyle\int \frac{2t - 1}{t^2 + 4} dt$

38. $\displaystyle\int \frac{3}{t^2 + 1} dt$

39. $\displaystyle\int \frac{-1}{\sqrt{1 - (2t - 1)^2}} dt$

40. $\displaystyle\int \frac{1}{4 + 3x^2} dx$

41. $\displaystyle\int \frac{\tan(2/t)}{t^2} dt$

42. $\displaystyle\int \frac{e^{1/t}}{t^2} dt$

43. $\displaystyle\int \frac{3}{\sqrt{6x - x^2}} dx$

44. $\displaystyle\int \frac{1}{(x - 1)\sqrt{4x^2 - 8x + 3}} dx$

45. $\displaystyle\int \frac{4}{4x^2 + 4x + 65} dx$

46. $\displaystyle\int \frac{1}{\sqrt{2 - 2x - x^2}} dx$

In Exercises 47–52, solve the differential equation.

47. $\displaystyle\frac{dy}{dx} = (1 + e^x)^2$

48. $\displaystyle\frac{dr}{dt} = \frac{(1 + e^t)^2}{e^t}$

49. $\displaystyle\frac{ds}{dt} = \frac{t}{\sqrt{1 - t^4}}$

50. $\displaystyle\frac{dy}{dx} = \frac{1}{x\sqrt{4x^2 - 1}}$

51. $(4 + \tan^2 x)y' = \sec^2 x$

52. $y' = \tan^2 2x$

In Exercises 53–60, evaluate the definite integral.

53. $\displaystyle\int_0^{\pi/4} \cos 2x \, dx$

54. $\displaystyle\int_0^{\pi} \sin^2 t \cos t \, dt$

55. $\displaystyle\int_0^1 xe^{-x^2} dx$

56. $\displaystyle\int_1^e \frac{1 - \ln x}{x} dx$

57. $\displaystyle\int_0^4 \frac{2x}{\sqrt{x^2 + 9}} dx$

58. $\displaystyle\int_1^2 \frac{x - 2}{x} dx$

59. $\displaystyle\int_0^{2/\sqrt{3}} \frac{1}{4 + 9x^2} dx$

60. $\displaystyle\int_0^4 \frac{1}{\sqrt{25 - x^2}} dx$

* C **In Exercises 61–64, use a symbolic integration utility to evaluate the integral. Then verify the result "by hand."**

61. $\displaystyle\int \frac{1}{x^2 + 4x + 13} dx$

62. $\displaystyle\int \frac{x - 2}{x^2 + 4x + 13} dx$

63. $\displaystyle\int_0^{\pi/4} \frac{1}{1 + \sin \theta} d\theta$

64. $\displaystyle\int_0^2 \left(\frac{e^x + e^{-x}}{2} \right)^3 dx$

Estimating Area **In Exercises 65 and 66, determine which value best approximates the area of the region between the x-axis and the function over the given interval. (Make your selection on the basis of a sketch of the region and *not* by performing any calculations.)**

65. $f(x) = \dfrac{4x}{x^2 + 1}$ $[0, 2]$

  a. 3   **b.** 1   **c.** −8   **d.** 8   **e.** 10

66. $f(x) = \dfrac{4}{x^2 + 1}$ $[0, 2]$

  a. 3   **b.** 1   **c.** −4   **d.** 4   **e.** 10

Area **In Exercises 67 and 68, find the area of the region bounded by the graph(s) of the equation(s).**

67. $y^2 = x^2(1 - x^2)$

68. $y = \sin 2x,\ y = 0,\ x = 0,\ x = \dfrac{\pi}{2}$

69. *Area* The graphs of $f(x) = x$ and $g(x) = ax^2$ intersect at the points $(0, 0)$ and $(1/a, 1/a)$. Find a so that the area of the region bounded by the graphs of these two functions is $\frac{2}{3}$.

70. *Interpreting an Integral* You are given the integral

$\int_0^2 2\pi x^2 \, dx,$

but are not told what it represents. (There is more than one correct answer for each part.)
  a. Sketch a region whose area is given by the integral.
  b. Sketch a solid whose volume is given by the integral if the Disc Method is used.
  c. Sketch a solid whose volume is given by the integral if the Shell Method is used.

71. *Volume* The region bounded by $y = e^{-x^2}$, $y = 0$, $x = 0$, and $x = b$ is revolved around the y-axis.
  a. Find the volume of the solid generated if $b = 1$.
  b. Find b so that the volume of the generated solid is $\frac{4}{3}$ cubic units.

72. *Average Value* Compute the average value of each function over the indicated interval.
  a. $f(x) = \sin nx,\qquad 0 \le x \le \pi/n,\ n$ is a positive integer
  b. $f(x) = \dfrac{1}{1 + x^2},\qquad -3 \le x \le 3$

---

*A **C** indicates an exercise in which you are instructed to use graphing technology or a symbolic computer algebra system. The solutions of other exercises may also be facilitated by appropriate technology.

73. *Centroid* Find the *x*-coordinate of the centroid of the region bounded by the graphs of

$$y = \frac{5}{\sqrt{25 - x^2}}, \qquad y = 0, \qquad x = 0, \qquad x = 4.$$

74. *Surface Area* Find the area of the surface formed by revolving the graph of $y = 2\sqrt{x}$ on the interval $[0, 9]$ about the *x*-axis.

C **In Exercises 75 and 76, use Simpson's Rule with $n = 12$ to approximate the arc length of the curve over the indicated interval.**

Function	Interval
75. $y = \tan \pi x$	$[0, \frac{1}{4}]$
76. $y = x^{2/3}$	$[1, 8]$

True or False **In Exercises 77–80, determine whether the statement is true or false. If it is false, explain why or give an example that shows it is false.**

77. $\displaystyle \int \frac{x^3}{x^4 - 1} dx = \frac{1}{4} \int \frac{du}{u}, \qquad u = x^4 - 1$

78. $\displaystyle \int \frac{dx}{1 + \sin^2 x} = \int \frac{du}{a^2 + u^2}, \qquad u = \sin x, a = 1$

79. $\displaystyle \int \frac{dx}{\sqrt{x - 1}} = \int \frac{du}{u}, \qquad u = \sqrt{x - 1}$

80. If $u = \sqrt{x - 1}$, then

$$\int_1^5 x\sqrt{x - 1} \, dx = 2\int_0^2 u^2(u^2 + 1) \, du.$$

81. Determine constants a and b such that

$$\sin x + \cos x = a \sin(x + b).$$

Use this result to integrate

$$\int \frac{dx}{\sin x + \cos x}.$$

Integration by Parts

In this section you will study an important integration technique called **integration by parts.** This technique applies to a wide variety of functions and is particularly useful for integrands involving a *product* of algebraic and transcendental functions. For instance, integration by parts works well with integrals such as

$$\int x \ln x \, dx, \quad \int x^2 e^x \, dx, \quad \text{and} \quad \int e^x \sin x \, dx.$$

Integration by parts is based on the formula for the derivative of a product

$$\frac{d}{dx}[uv] = u\frac{dv}{dx} + v\frac{du}{dx} = uv' + vu'$$

where both u and v are differentiable functions of x. If u' and v' are continuous, then you can integrate both sides of this equation to obtain

$$uv = \int uv' \, dx + \int vu' \, dx$$

$$= \int u \, dv + \int v \, du.$$

By rewriting this equation, you obtain the following theorem.

FOR FURTHER INFORMATION
For a different approach to proving the formula for integration by parts, see the article "Proof Without Words" by Roger B. Nelson in the April, 1991 issue of *Mathematics Magazine.*

THEOREM 7.1 Integration by Parts

If u and v are functions of x and have continuous derivatives, then

$$\int u \, dv = uv - \int v \, du.$$

This formula expresses the original integral in terms of another integral. Depending on the choices for u and dv, it may be easier to evaluate the second integral than the original one. Because the choices of u and dv are critical in the integration by parts process, we provide the following guidelines.

Guidelines for Integration by Parts

1. Try letting dv be the most complicated portion of the integrand that fits a basic integration rule. Then u will be the remaining factor(s) of the integrand.
2. Try letting u be the portion of the integrand whose derivative is a simpler function than u. Then dv will be the remaining factor(s) of the integrand.

EXAMPLE 1 Integration by Parts

Evaluate $\int xe^x \, dx$.

Solution To apply integration by parts, you need to write the integral in the form $\int u \, dv$. There are several ways to do this.

$$\int \underset{u}{(x)}\underset{dv}{\underbrace{(e^x \, dx)}}, \qquad \int \underset{u}{\underbrace{(e^x)}}\underset{dv}{\underbrace{(x \, dx)}}, \qquad \int \underset{u}{\underbrace{(1)}}\underset{dv}{\underbrace{(xe^x \, dx)}}, \qquad \int \underset{u}{\underbrace{(xe^x)}}\underset{dv}{\underbrace{(dx)}}$$

The guidelines suggest choosing the first option because the derivative of $u = x$ is simpler than x and $dv = e^x \, dx$ is the most complicated portion of the integrand that fits a basic integration formula.

$$dv = e^x \, dx \quad \longrightarrow \quad v = \int dv = \int e^x \, dx = e^x$$

$$u = x \qquad \longrightarrow \quad du = dx$$

Now, integration by parts produces the following.

$$\int u \, dv = uv - \int v \, du$$

$$\int xe^x \, dx = xe^x - \int e^x \, dx$$

$$= xe^x - e^x + C$$

To check this result, try differentiating $xe^x - e^x + C$ to see that you obtain the original integrand.

EXAMPLE 2 Integration by Parts

Evaluate $\int x^2 \ln x \, dx$.

Solution In this case x^2 is more easily integrated than $\ln x$. Furthermore, the derivative of $\ln x$ is simpler than $\ln x$. Therefore, you should let $dv = x^2 \, dx$.

$$dv = x^2 \, dx \quad \longrightarrow \quad v = \int x^2 \, dx = \frac{x^3}{3}$$

$$u = \ln x \qquad \longrightarrow \quad du = \frac{1}{x} \, dx$$

Integration by parts produces the following.

$$\int x^2 \ln x \, dx = \frac{x^3}{3} \ln x - \int \left(\frac{x^3}{3}\right)\left(\frac{1}{x}\right) dx$$

$$= \frac{x^3}{3} \ln x - \frac{1}{3} \int x^2 \, dx$$

$$= \frac{x^3}{3} \ln x - \frac{x^3}{9} + C$$

You can check this result by differentiating.

$$\frac{d}{dx}\left[\frac{x^3}{3} \ln x - \frac{x^3}{9}\right] = \frac{x^3}{3}\left(\frac{1}{x}\right) + (\ln x)(x^2) - \frac{x^2}{3} = x^2 \ln x$$

One surprising application of integration by parts involves integrands consisting of a single factor, such as $\int \ln x \, dx$ or $\int \arcsin x \, dx$. In such cases, you should let $dv = dx$, as illustrated in the next example.

EXAMPLE 3   An Integrand with a Single Term

Evaluate $\displaystyle\int_0^1 \arcsin x \, dx$.

Solution   Let $dv = dx$.

$$dv = dx \qquad \longrightarrow \qquad v = \int dx = x$$

$$u = \arcsin x \qquad \longrightarrow \qquad du = \frac{1}{\sqrt{1 - x^2}} dx$$

Integration by parts now produces the following.

$$\int \arcsin x \, dx = x \arcsin x - \int \frac{x}{\sqrt{1 - x^2}} dx$$

$$= x \arcsin x + \frac{1}{2} \int (1 - x^2)^{-1/2}(-2x) \, dx$$

$$= x \arcsin x + \sqrt{1 - x^2} + C$$

Using this antiderivative, you can evaluate the definite integral as follows.

$$\int_0^1 \arcsin x \, dx = \left[x \arcsin x + \sqrt{1 - x^2} \right]_0^1$$

$$= \frac{\pi}{2} - 1$$

$$\approx 0.571$$

The area represented by this definite integral is shown in Figure 7.2.

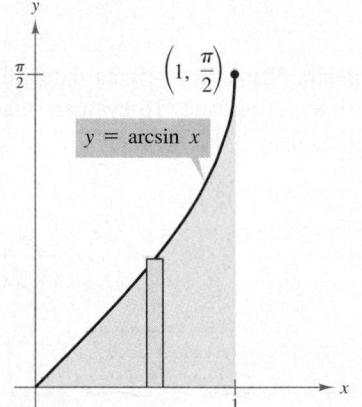

FIGURE 7.2
The area of the region is approximately 0.571.

TECHNOLOGY   Remember that there are two ways to use technology to evaluate a definite integral: (1) you can use a numerical approximation such as the Trapezoidal Rule or Simpson's Rule, or (2) you can use a symbolic integrator to find the antiderivative and then apply the Fundamental Theorem of Calculus. *Both* methods have shortcomings. To find the possible error in a numerical method, the integrand must possess a second derivative (Trapezoidal Rule) or fourth derivative (Simpson's Rule) in the interval of integration—the integrand in Example 3 fails to meet this requirement. To apply the Fundamental Theorem of Calculus, the symbolic integrator must be able to find the antiderivative.

Which method would you use to evaluate

$$\int_0^1 \arctan x \, dx\,?$$

Which method would you use to evaluate

$$\int_0^1 \arctan x^2 \, dx\,?$$

Some integrals require repeated use of the integration by parts formula. This is demonstrated in Examples 4 and 5.

EXAMPLE 4   Repeated Use of Integration by Parts

Evaluate $\int x^2 \sin x\,dx$.

Solution   The factors x^2 and $\sin x$ are equally easy to integrate. However, the derivative of x^2 becomes simpler, whereas the derivative of $\sin x$ does not. Therefore, you should let $u = x^2$.

$$dv = \sin x\,dx \quad \rightarrow \quad v = \int \sin x\,dx = -\cos x$$

$$u = x^2 \qquad\quad \rightarrow \quad du = 2x\,dx$$

Now, integration by parts produces the following.

$$\int x^2 \sin x\,dx = -x^2 \cos x + \int 2x \cos x\,dx \qquad \text{First use of parts}$$

This first use of integration by parts has succeeded in simplifying the original integral, but the integral on the right still doesn't fit a basic integration rule. To evaluate that integral, you can apply integration by parts again. This time, let $u = 2x$.

$$dv = \cos x\,dx \quad \rightarrow \quad v = \int \cos x\,dx = \sin x$$

$$u = 2x \qquad\quad \rightarrow \quad du = 2\,dx$$

Now, integration by parts produces

$$\int 2x \cos x\,dx = 2x \sin x - \int 2 \sin x\,dx \qquad \text{Second use of parts}$$

$$= 2x \sin x + 2 \cos x + C.$$

Combining these two results, you can write

$$\int x^2 \sin x\,dx = -x^2 \cos x + 2x \sin x + 2 \cos x + C. \qquad \blacksquare$$

When making repeated applications of integration by parts, you need to be careful not to interchange the substitutions in successive applications. For instance, in Example 4 the first substitution was $u = x^2$ and $dv = \sin x\,dx$. If, in the second application, you had switched the substitution to $u = \cos x$ and $dv = 2x$, you would have obtained

$$\int x^2 \sin x\,dx = -x^2 \cos x + \int 2x \cos x\,dx$$

$$= -x^2 \cos x + x^2 \cos x + \int x^2 \sin x\,dx$$

$$= \int x^2 \sin x\,dx$$

thus undoing the previous integration and returning to the *original* integral.

When making repeated applications of integration by parts, you should also watch for the appearance of a *constant multiple* of the original integral.

EXAMPLE 5   Repeated Use of Integration by Parts

Evaluate $\int e^x \cos 2x \, dx$.

Solution   The guidelines fail to help with a choice of u and dv, so you can arbitrarily choose $dv = e^x \, dx$ and $u = \cos 2x$. (Try verifying that the choice of $dv = \cos 2x \, dx$ and $u = e^x$ works equally well.)

$$dv = e^x \, dx \quad \longrightarrow \quad v = \int e^x \, dx = e^x$$

$$u = \cos 2x \quad \longrightarrow \quad du = -2 \sin 2x \, dx$$

Integration by parts produces the following.

$$\int e^x \cos 2x \, dx = e^x \cos 2x + 2 \int e^x \sin 2x \, dx \qquad \text{First use of parts}$$

Make the same type of substitutions for the next application of integration by parts.

$$dv = e^x \, dx \quad \longrightarrow \quad v = \int e^x \, dx = e^x$$

$$u = \sin 2x \quad \longrightarrow \quad du = 2 \cos 2x \, dx$$

Integration by parts now produces

$$\int e^x \sin 2x \, dx = e^x \sin 2x - 2 \int e^x \cos 2x \, dx. \qquad \text{Second use of parts}$$

Therefore, you have

$$\int e^x \cos 2x \, dx = e^x \cos 2x + 2e^x \sin 2x - 4 \int e^x \cos 2x \, dx.$$

Because the right-hand integral is a constant multiple of the original integral, you can add it to both sides of the equation to obtain

$$5 \int e^x \cos 2x \, dx = e^x \cos 2x + 2e^x \sin 2x$$

$$\int e^x \cos 2x \, dx = \frac{1}{5} e^x \cos 2x + \frac{2}{5} e^x \sin 2x + C. \qquad \text{Divide by 5}$$

The technique used in Example 5 can be used to compile general integration rules. For instance, try using integration by parts to obtain the following.

$$\int e^{ax} \sin bx \, dx = \frac{e^{ax}}{a^2 + b^2} (a \sin bx - b \cos bx) + C$$

$$\int e^{ax} \cos bx \, dx = \frac{e^{ax}}{a^2 + b^2} (a \cos bx + b \sin bx) + C$$

The integral in the next example is an important one. In Section 7.4 (Example 7), you will see that it is used to find the arc length of a parabolic segment.

EXAMPLE 6 Integration by Parts

Evaluate $\int \sec^3 x \, dx$.

Solution The most complicated portion of the integrand that can be easily integrated is $\sec^2 x$, so you should let $dv = \sec^2 x \, dx$ and $u = \sec x$.

$$dv = \sec^2 x \, dx \quad \longrightarrow \quad v = \int \sec^2 x \, dx = \tan x$$

$$u = \sec x \quad \longrightarrow \quad du = \sec x \tan x \, dx$$

Integration by parts produces the following.

$$\int \sec^3 x \, dx = \sec x \tan x - \int \sec x \tan^2 x \, dx$$

$$= \sec x \tan x - \int \sec x (\sec^2 x - 1) \, dx$$

$$= \sec x \tan x - \int \sec^3 x \, dx + \int \sec x \, dx$$

$$2 \int \sec^3 x \, dx = \sec x \tan x + \int \sec x \, dx \qquad \text{Collect like integrals}$$

$$\int \sec^3 x \, dx = \frac{1}{2} \sec x \tan x + \frac{1}{2} \ln |\sec x + \tan x| + C \qquad \blacksquare$$

So far, we have used integration by parts to find antiderivatives of products. It can also be used to find antiderivatives of quotients.

EXAMPLE 7 Integration by Parts Applied to a Quotient

Evaluate $\int \dfrac{xe^x}{(x+1)^2} \, dx$.

Solution Let $u = xe^x$ and $dv = dx/(x+1)^2$. After solving for du and v, you can apply integration by parts to obtain

$$\int \frac{xe^x}{(x+1)^2} \, dx = xe^x \left(\frac{-1}{x+1} \right) - \int (x+1)e^x \left(\frac{-1}{x+1} \right) dx$$

$$= -\frac{xe^x}{x+1} + \int e^x \, dx$$

$$= -\frac{xe^x}{x+1} + e^x + C$$

$$= \frac{e^x}{x+1} + C.$$

Check this result by differentiating. $\blacksquare$

EXAMPLE 8   Finding a Centroid

A machine part is modeled by the region bounded by the graph of $y = \sin x$ and the x-axis, $0 \le x \le \pi/2$, as shown in Figure 7.3. Find the centroid of this region.

Solution   Begin by finding the area of the region.

$$A = \int_0^{\pi/2} \sin x \, dx = -\cos x \Big]_0^{\pi/2} = 1$$

Now, you can find the coordinates of the centroid as follows.

$$\bar{y} = \frac{1}{A} \int_0^{\pi/2} \frac{\sin x}{2} (\sin x) \, dx$$

$$= \frac{1}{4} \int_0^{\pi/2} (1 - \cos 2x) \, dx \qquad \text{Half-angle formula}$$

$$= \frac{1}{4} \left[x - \frac{\sin 2x}{2} \right]_0^{\pi/2} = \frac{\pi}{8}$$

You can evaluate the integral for $\bar{x}$, $(1/A) \int_0^{\pi/2} x \sin x \, dx$, with integration by parts. To do this, let $dv = \sin x \, dx$ and $u = x$. This produces $v = -\cos x$ and $du = dx$ and you can write

$$\int x \sin x \, dx = -x \cos x + \int \cos x \, dx = -x \cos x + \sin x + C.$$

Finally, you can determine $\bar{x}$ to be

$$\bar{x} = \frac{1}{A} \int_0^{\pi/2} x \sin x \, dx = \left[-x \cos x + \sin x \right]_0^{\pi/2} = 1.$$

Therefore, the centroid of the region is $(1, \pi/8)$.   ▄▄▄

As you gain experience in using integration by parts, your skill in determining u and dv will increase. The following summary lists several common integrals with suggestions for the choice of u and dv.

FIGURE 7.3
The centroid of the region, $(\bar{x}, \bar{y})$, is its balancing point.

Summary of Common Integrals Using Integration by Parts

1. (Examples 1 and 4) For integrals of the form

$$\int x^n e^{ax} \, dx, \quad \int x^n \sin ax \, dx, \quad \text{or} \quad \int x^n \cos ax \, dx$$

let $u = x^n$ and $dv = e^{ax} \, dx$, $\sin ax \, dx$, or $\cos ax \, dx$.

2. (Examples 2 and 3) For integrals of the form

$$\int x^n \ln x \, dx, \quad \int x^n \arcsin ax \, dx, \quad \text{or} \quad \int x^n \arctan ax \, dx$$

let $u = \ln x$, $\arcsin ax$, or $\arctan ax$ and $dv = x^n \, dx$.

3. (Example 5) For integrals of the form

$$\int e^{ax} \sin bx \, dx \quad \text{or} \quad \int e^{ax} \cos bx \, dx$$

let $u = \sin bx$ or $\cos bx$ and $dv = e^{ax} \, dx$.

Tabular Method

In problems involving repeated applications of integration by parts, a tabular method, illustrated in Example 9, can help to organize work. The method works well for integrals of the form

$$\int x^n \sin ax \, dx, \quad \int x^n \cos ax \, dx, \quad \text{or} \quad \int x^n e^{ax} \, dx.$$

EXAMPLE 9 Using the Tabular Method

FOR FURTHER INFORMATION
For more information on the tabular method, see the article "More on Tabular Integration by Parts" by Leonard Gillman in the November, 1991 issue of *The College Mathematics Journal*.

Evaluate $\int x^2 \sin 4x \, dx$.

Solution Begin as usual by letting $u = x^2$ and $dv = v' \, dx = \sin 4x \, dx$. Next, create a table consisting of three columns, as follows.

Alternate signs	u and its derivatives	v' and its antiderivatives
+	x^2	$\sin 4x$
−	$2x$	$-\frac{1}{4}\cos 4x$
+	2	$-\frac{1}{16}\sin 4x$
−	0	$\frac{1}{64}\cos 4x$

Differentiate until you obtain 0 as a derivative.

Finally, the solution is given by adding the *signed* products of the diagonal entries to obtain

$$\int x^2 \sin 4x \, dx = -\frac{1}{4}x^2 \cos 4x + \frac{1}{8}x \sin 4x + \frac{1}{32}\cos 4x + C.$$

TECHNOLOGY
Laboratory Guide
Lab 7.2

EXERCISES for Section 7.2

1. Find the derivative of each function.
   a. $y = \sin x - x \cos x$
   b. $y = x^2 \sin x + 2x \cos x - 2 \sin x$
   c. $y = x^2 e^x - 2x e^x + 2e^x$
   d. $y = -x + x \ln x$

2. Use the results of Exercise 1 to evaluate each integral.
   a. $\int \ln x \, dx$ b. $\int x \sin x \, dx$
   c. $\int x^2 e^x \, dx$ d. $\int x^2 \cos x \, dx$

In Exercises 3–8, identify u and dv for evaluating the integral by integration by parts. (Do not evaluate the integral.)

3. $\int x e^{2x} \, dx$ 4. $\int x^2 e^{2x} \, dx$

5. $\int (\ln x)^2 \, dx$ 6. $\int \ln 3x \, dx$

7. $\int x \sec^2 x \, dx$ 8. $\int x^2 \cos x \, dx$

In Exercises 9–30, evaluate the given integral. (*Note:* Solve by the simplest method—not all require integration by parts.)

9. $\int x e^{-2x} \, dx$ 10. $\int \frac{x}{e^x} \, dx$

11. $\int x^3 e^x \, dx$ 12. $\int \frac{e^{1/t}}{t^2} \, dt$

13. $\int x^2 e^{x^3} \, dx$ 14. $\int x^3 \ln x \, dx$

15. $\int t \ln(t + 1) \, dt$ 16. $\int \frac{1}{x(\ln x)^3} \, dx$

17. $\int \frac{(\ln x)^2}{x} \, dx$ 18. $\int \frac{\ln x}{x^2} \, dx$

19. $\int \frac{x e^{2x}}{(2x + 1)^2} \, dx$ 20. $\int \frac{x^3 e^{x^2}}{(x^2 + 1)^2} \, dx$

21. $\int (x^2 - 1)e^x \, dx$ 22. $\int \frac{\ln 2x}{x^2} \, dx$

23. $\displaystyle\int x\sqrt{x-1}\,dx$

24. $\displaystyle\int \frac{x}{\sqrt{2+3x}}\,dx$

25. $\displaystyle\int x\cos x\,dx$

26. $\displaystyle\int \theta \sec\theta\tan\theta\,d\theta$

27. $\displaystyle\int \arctan x\,dx$

28. $\displaystyle\int \arccos x\,dx$

29. $\displaystyle\int e^{2x}\sin x\,dx$

30. $\displaystyle\int e^{x}\cos 2x\,dx$

In Exercises 31–36, solve the differential equation.

31. $y' = xe^{x^2}$

32. $y' = \ln x$

33. $\dfrac{dy}{dt} = \dfrac{t^2}{\sqrt{2+3t}}$

34. $\dfrac{dy}{dx} = x^2\sqrt{x-1}$

35. $(\cos y)y' = 2x$

36. $y' = \arctan\dfrac{x}{2}$

In Exercises 37–42, evaluate the definite integral.

37. $\displaystyle\int_0^{\pi} x\sin 2x\,dx$

38. $\displaystyle\int_0^1 x\arcsin x^2\,dx$

39. $\displaystyle\int_0^1 e^{x}\sin x\,dx$

40. $\displaystyle\int_0^1 x^2 e^{x}\,dx$

41. $\displaystyle\int_0^{\pi/2} x\cos x\,dx$

42. $\displaystyle\int_0^1 \ln(1+x^2)\,dx$

In Exercises 43–48, use the tabular method to evaluate the integral.

43. $\displaystyle\int x^2 e^{2x}\,dx$

44. $\displaystyle\int x^4 e^{-x}\,dx$

45. $\displaystyle\int x^3 \sin x\,dx$

46. $\displaystyle\int x^3 \cos 2x\,dx$

47. $\displaystyle\int x\sec^2 x\,dx$

48. $\displaystyle\int x^2(x-2)^{3/2}\,dx$

C **In Exercises 49–52, use a symbolic integration utility to evaluate the integral.**

49. $\displaystyle\int t^3 e^{-4t}\,dt$

50. $\displaystyle\int \alpha^4 \sin \pi\alpha\,d\alpha$

51. $\displaystyle\int_0^{\pi/2} e^{-2x}\sin 3x\,dx$

52. $\displaystyle\int_0^5 x^4(25-x^2)^{3/2}\,dx$

53. Integrate $\int 2x\sqrt{2x-3}\,dx$
   a. by parts, letting $dv = \sqrt{2x-3}\,dx$.
   b. by substitution, letting $u = 2x-3$.

54. Integrate $\int x\sqrt{4+x}\,dx$
   a. by parts, letting $dv = \sqrt{4+x}\,dx$.
   b. by substitution, letting $u = 4+x$.

55. Integrate $\displaystyle\int \frac{x^3}{\sqrt{4+x^2}}\,dx$
   a. by parts, letting $dv = \dfrac{x}{\sqrt{4+x^2}}\,dx$.
   b. by substitution, letting $u = 4+x^2$.

56. Integrate $\int x\sqrt{4-x}\,dx$
   a. by parts, letting $dv = \sqrt{4-x}\,dx$.
   b. by substitution, letting $u = 4-x$.

C **In Exercises 57 and 58, use a symbolic integration utility to evaluate the integral for $n = 0$, 1, 2, and 3. Use the result to obtain a general rule for the integral for any positive integer n and test your result for $n = 4$.**

57. $\displaystyle\int x^n \ln x\,dx$

58. $\displaystyle\int x^n e^{x}\,dx$

In Exercises 59–64, use integration by parts to verify the formula. (For Exercises 59–62, assume that n is a positive integer.)

59. $\displaystyle\int x^n \sin x\,dx = -x^n\cos x + n\int x^{n-1}\cos x\,dx$

60. $\displaystyle\int x^n \cos x\,dx = x^n\sin x - n\int x^{n-1}\sin x\,dx$

61. $\displaystyle\int x^n \ln x\,dx = \frac{x^{n+1}}{(n+1)^2}[-1+(n+1)\ln x] + C$

62. $\displaystyle\int x^n e^{ax}\,dx = \frac{x^n e^{ax}}{a} - \frac{n}{a}\int x^{n-1}e^{ax}\,dx$

63. $\displaystyle\int e^{ax}\sin bx\,dx = \frac{e^{ax}(a\sin bx - b\cos bx)}{a^2+b^2} + C$

64. $\displaystyle\int e^{ax}\cos bx\,dx = \frac{e^{ax}(a\cos bx + b\sin bx)}{a^2+b^2} + C$

In Exercises 65–68, evaluate the integral using the appropriate formula from Exercises 59–64.

65. $\displaystyle\int x^3 \ln x\,dx$

66. $\displaystyle\int x^2 \cos x\,dx$

67. $\displaystyle\int e^{2x}\cos 3x\,dx$

68. $\displaystyle\int x^3 e^{2x}\,dx$

Area **In Exercises 69–72, find the area of the region bounded by the graphs of the equations.**

69. $y = xe^{-x}$, $y = 0$, $x = 4$

70. $y = \frac{1}{9}xe^{-x/3}$, $y = 0$, $x = 0$, $x = 3$

71. $y = e^{-x}\sin\pi x$, $y = 0$, $x = 0$, $x = 1$

72. $y = x\sin x$, $y = 0$, $x = 0$, $x = \pi$

73. *Area, Volume, and Centroid* Given the region bounded by the graphs of $y = \ln x$, $y = 0$, and $x = e$, find
   a. the area of the region.
   b. the volume of the solid generated by revolving the region about the x-axis.
   c. the volume of the solid generated by revolving the region about the y-axis.
   d. the centroid of the region.

74. *Average Displacement* A damping force affects the vibration of a spring so that the displacement of the spring is

$$y = e^{-4t}(\cos 2t + 5\sin 2t).$$

Find the average value of y on the interval from $t = 0$ to $t = \pi$.

75. *Memory Model* A model for the ability M of a child to memorize, measured on a scale from 0 to 10, is

$$M = 1 + 1.6t \ln t, \qquad 0 < t \leq 4$$

where t is the child's age in years. Find the average value of this function
a. between the child's first and second birthdays.
b. between the child's third and fourth birthdays.

Present Value   In Exercises 76 and 77, find the present value P of a continuous income flow of $c(t)$ dollars per year if

$$P = \int_0^{t_1} c(t)e^{-rt}\,dt$$

where t_1 is time in years and r is the annual interest rate compounded continuously.

76. $c(t) = 100{,}000 + 4000t$, $r = 9\%$, $t_1 = 10$

77. $c(t) = 30{,}000 + 500t$, $r = 7\%$, $t_1 = 5$

Integrals Used to Find Fourier Coefficients   In Exercises 78 and 79, verify the value of the definite integral for positive values of n.

78. $\displaystyle\int_{-\pi}^{\pi} x \sin nx\,dx = \begin{cases} -\dfrac{2\pi}{n}, & n \text{ is even} \\[2mm] \dfrac{2\pi}{n}, & n \text{ is odd} \end{cases}$

79. $\displaystyle\int_{-\pi}^{\pi} x^2 \cos nx\,dx = \dfrac{(-1)^n 4\pi}{n^2}$

80. *Vibrating String* A string stretched between two points $(0, 0)$ and $(0, 2)$ is plucked by displacing the string h units at its midpoint. The motion of the string is modeled by a **Fourier Sine Series** whose coefficients are given by

$$b_n = h \int_0^1 x \sin\frac{n\pi x}{2}\,dx + h \int_1^2 (-x + 2)\sin\frac{n\pi x}{2}\,dx.$$

Find b_n.

81. Find the fallacy in the following argument that $0 = 1$.

$$dv = dx \;\rightarrow\; v = x$$

$$u = \frac{1}{x} \;\rightarrow\; du = -\frac{1}{x^2}dx$$

$$0 + \int \frac{dx}{x} = \left(\frac{1}{x}\right)(x) - \int \left(-\frac{1}{x^2}\right)(x)\,dx$$

$$= 1 + \int \frac{dx}{x}$$

Hence $0 = 1$.

82. Let $y = f(x)$ be positive and strictly increasing on the interval $0 < a \leq x \leq b$. Consider the region R bounded by the graphs of $y = f(x)$, the x-axis, $x = a$, and $x = b$. If R is revolved about the y-axis, show that the Disc Method and Shell Method yield the same volume.

83. *Rocket Velocity* The velocity (in ft/sec) of a rocket whose initial mass is m (including fuel) is

$$v = gt + u \ln\frac{m}{m - rt}, \qquad t < \frac{m}{r}$$

where u is the expulsion speed of the fuel, r is the rate at which the fuel is consumed, and $g = -32$ feet per second per second is the acceleration due to gravity. Find the position equation for a rocket for which $m = 50{,}000$ pounds, $u = 12{,}000$ feet per second, and $r = 400$ pounds per second. What is the height of the rocket when $t = 100$ seconds? (Assume it was fired from ground level and is moving straight up.)

The space shuttle *Columbia*, after four test flights, made its first operational flight on November 16, 1982. It stayed in orbit 5 days. The second space shuttle, the *Challenger*, made its maiden flight in April, 1983. On September 29, 1988, the *Discovery* was launched. It returned 4 days later to Edwards Air Force Base in California.

Trigonometric Integrals

Integrals Involving Powers of Sine and Cosine • Integrals Involving Powers of Secant and Tangent • Integrals Involving Sine-Cosine Products with Different Angles

Integrals Involving Powers of Sine and Cosine

In this section you will study techniques for evaluating integrals of the form

$$\int \sin^m x \cos^n x \, dx \quad \text{and} \quad \int \sec^m x \tan^n x \, dx$$

where either m or n is a positive integer. To find antiderivatives for these forms, try to break them into combinations of trigonometric integrals to which you can apply the Power Rule.

For instance, you can evaluate $\int \sin^5 x \cos x \, dx$ with the Power Rule by letting $u = \sin x$. Then, $du = \cos x \, dx$ and you have

$$\int \sin^5 x \cos x \, dx = \int u^5 \, du$$

$$= \frac{u^6}{6} + C$$

$$= \frac{\sin^6 x}{6} + C.$$

Similarly, to evaluate $\int \sec^4 x \tan x \, dx$ by the Power Rule, you can let $u = \sec x$. Then, $du = \sec x \tan x \, dx$ and you have

$$\int \sec^4 x \tan x \, dx = \int \sec^3 x (\sec x \tan x) \, dx$$

$$= \int u^3 \, du$$

$$= \frac{u^4}{4} + C$$

$$= \frac{\sec^4 x}{4} + C.$$

To break up $\int \sin^m x \cos^n x \, dx$ into forms to which you can apply the Power Rule, use the following identities.

$\sin^2 x + \cos^2 x = 1$	Pythagorean identity
$\sin^2 x = \dfrac{1 - \cos 2x}{2}$	Half-angle identity for $\sin^2 x$
$\cos^2 x = \dfrac{1 + \cos 2x}{2}$	Half-angle identity for $\cos^2 x$

Integrals Involving Sine and Cosine

1. If the power of the sine is odd and positive, save one sine factor and convert the remaining factors to cosine. Then, expand and integrate.

$$\int \overbrace{\sin^{2k+1} x}^{\text{Odd}} \cos^n x \, dx = \int \overbrace{(\sin^2 x)^k}^{\text{Convert to cosine}} \cos^n x \overbrace{\sin x \, dx}^{\text{Save for } du} = \int (1 - \cos^2 x)^k \cos^n x \sin x \, dx$$

2. If the power of the cosine is odd and positive, save one cosine factor and convert the remaining factors to sine. Then, expand and integrate.

$$\int \sin^m x \overbrace{\cos^{2k+1} x}^{\text{Odd}} \, dx = \int \sin^m x \overbrace{(\cos^2 x)^k}^{\text{Convert to sine}} \overbrace{\cos x \, dx}^{\text{Save for } du} = \int \sin^m x (1 - \sin^2 x)^k \cos x \, dx$$

3. If the powers of *both* the sine and cosine are even and nonnegative, make repeated use of the identities

$$\sin^2 x = \frac{1 - \cos 2x}{2} \quad \text{and} \quad \cos^2 x = \frac{1 + \cos 2x}{2}$$

to convert the integrand to odd powers of the cosine. Then, proceed as in case 2.

EXAMPLE 1   Power of Sine is Odd and Positive

Evaluate $\displaystyle\int \sin^3 x \cos^4 x \, dx$.

Solution   Because you expect to use the Power Rule with $u = \cos x$, *save one sine factor* to form *du* and convert the remaining sine factors to cosines.

$$\int \sin^3 x \cos^4 x \, dx = \int \sin^2 x \cos^4 x (\sin x) \, dx$$

$$= \int (1 - \cos^2 x)\cos^4 x \sin x \, dx$$

$$= \int (\cos^4 x - \cos^6 x)\sin x \, dx$$

$$= \int \cos^4 x \sin x \, dx - \int \cos^6 x \sin x \, dx$$

$$= -\int \cos^4 x(-\sin x) \, dx + \int \cos^6 x(-\sin x) \, dx$$

$$= -\frac{\cos^5 x}{5} + \frac{\cos^7 x}{7} + C$$

TECHNOLOGY   Try using a symbolic integration utility to evaluate the integral in Example 1. When we did this, we obtained

$$\int \sin^3 x \cos^4 x \, dx = -\cos^5 x\left(\frac{1}{7}\sin^2 x + \frac{2}{35}\right) + C.$$

Is this equivalent to the result obtained in Example 1?

EXAMPLE 2   **Power of Cosine is Odd and Positive**

Evaluate $\displaystyle\int \sin^2 x \cos^5 x \, dx$.

Solution

$$
\begin{aligned}
\int \sin^2 x \cos^5 dx &= \int \sin^2 x \cos^4 x \cos x \, dx \\
&= \int \sin^2 x (\cos^2 x)^2 \cos x \, dx \\
&= \int \sin^2 x (1 - \sin^2 x)^2 \cos x \, dx \\
&= \int \sin^2 x (1 - 2 \sin^2 x + \sin^4 x) \cos x \, dx \\
&= \int [\sin^2 x - 2 \sin^4 x + \sin^6 x] \cos x \, dx \\
&= \frac{\sin^3 x}{3} - \frac{2 \sin^5 x}{5} + \frac{\sin^7 x}{7} + C
\end{aligned}
$$

In Examples 1 and 2, *both* of the powers m and n happened to be positive integers. However, the same strategy will work as long as either m or n is odd and positive. For instance, in the next example the power of the sine is 3, but the power of the cosine is $-\frac{1}{2}$.

EXAMPLE 3   **Power of Sine is Odd and Positive**

Evaluate $\displaystyle\int_0^{\pi/3} \frac{\sin^3 x}{\sqrt{\cos x}} \, dx$.

Solution

$$
\begin{aligned}
\int_0^{\pi/3} \frac{\sin^3 x}{\sqrt{\cos x}} dx &= \int_0^{\pi/3} \frac{(1 - \cos^2 x)(\sin x)}{\sqrt{\cos x}} dx \\
&= \int_0^{\pi/3} [(\cos x)^{-1/2} \sin x - (\cos x)^{3/2} \sin x] \, dx \\
&= \int_0^{\pi/3} [-(\cos x)^{-1/2}(-\sin x) + (\cos x)^{3/2}(-\sin x)] \, dx \\
&= \left[-\frac{(\cos x)^{1/2}}{1/2} + \frac{(\cos x)^{5/2}}{5/2} \right]_0^{\pi/3} \\
&= \left(-\frac{2}{\sqrt{2}} + \frac{2}{5(2^{5/2})} + 2 - \frac{2}{5} \right) \\
&= \frac{32 - 19\sqrt{2}}{20} \\
&\approx 0.256.
\end{aligned}
$$

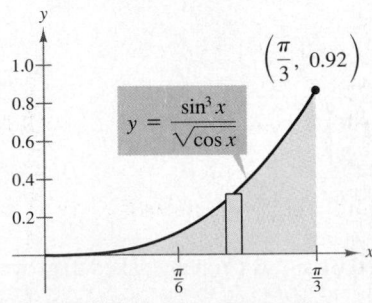

FIGURE 7.4
The area of the region is approximately 0.256.

Figure 7.4 shows the region whose area is represented by this integral.

EXAMPLE 4   Power of Cosine is Even and Nonnegative

Evaluate $\displaystyle\int \cos^4 x\, dx$.

Solution   Because m and n are both even and nonnegative ($m = 0$), you can replace $\cos^4 x$ by $[(1 + \cos 2x)/2]^2$.

$$\int \cos^4 x\, dx = \int \left(\frac{1 + \cos 2x}{2}\right)^2 dx$$

$$= \int \left(\frac{1}{4} + \frac{\cos 2x}{2} + \frac{\cos^2 2x}{4}\right) dx$$

$$= \int \left[\frac{1}{4} + \frac{\cos 2x}{2} + \frac{1}{4}\left(\frac{1 + \cos 4x}{2}\right)\right] dx$$

$$= \frac{3}{8}\int dx + \frac{1}{4}\int 2\cos 2x\, dx + \frac{1}{32}\int 4\cos 4x\, dx$$

$$= \frac{3x}{8} + \frac{\sin 2x}{4} + \frac{\sin 4x}{32} + C$$

In Example 4, suppose that you were to evaluate the *definite* integral from 0 to $\pi/2$. You would obtain

$$\int_0^{\pi/2} \cos^4 x\, dx = \left[\frac{3x}{8} + \frac{\sin 2x}{4} + \frac{\sin 4x}{32}\right]_0^{\pi/2}$$

$$= \left(\frac{3\pi}{16} + 0 + 0\right) - (0 + 0 + 0)$$

$$= \frac{3\pi}{16}.$$

JOHN WALLIS

Wallis did much of his work prior to Newton and Leibniz, and he influenced the thinking of both of these men.

Note that the only term that contributes to the solution is $3x/8$. This observation is generalized in the following formulas developed by John Wallis (1616–1703).

Wallis's Formulas

1. If n is odd ($n \geq 3$), then

$$\int_0^{\pi/2} \cos^n x\, dx = \left(\frac{2}{3}\right)\left(\frac{4}{5}\right)\left(\frac{6}{7}\right)\cdots\left(\frac{n-1}{n}\right).$$

2. If n is even ($n \geq 2$), then

$$\int_0^{\pi/2} \cos^n x\, dx = \left(\frac{1}{2}\right)\left(\frac{3}{4}\right)\left(\frac{5}{6}\right)\cdots\left(\frac{n-1}{n}\right)\left(\frac{\pi}{2}\right)$$

These formulas are also valid if $\cos^n x$ is replaced by $\sin^n x$. (You are asked to prove both formulas in Exercise 81.)

Integrals Involving Powers of Secant and Tangent

The following guidelines can help you evaluate integrals of the form $\int \sec^m x \tan^n x \, dx$.

Intergrals Involving Secants and Tangents

1. If the power of the secant is even and positive, save a secant-squared factor and convert the remaining factors to tangents. Then, expand and integrate.

$$\int \overbrace{\sec^{2k} x}^{\text{Even}} \tan^n x \, dx = \int \overbrace{(\sec^2 x)^{k-1}}^{\text{Convert to tangents}} \tan^n x \overbrace{\sec^2 x}^{\text{Save for } du} \, dx = \int (1 + \tan^2 x)^{k-1} \tan^n x \sec^2 x \, dx$$

2. If the power of the tangent is odd and positive, save a secant-tangent factor and convert the remaining factors to secants. Then, expand and integrate.

$$\int \sec^m x \overbrace{\tan^{2k+1} x}^{\text{Odd}} \, dx = \int \sec^{m-1} x \overbrace{(\tan^2 x)^k}^{\text{Convert to secants}} \overbrace{\sec x \tan x}^{\text{Save for } du} \, dx = \int \sec^{m-1} x (\sec^2 x - 1)^k \sec x \tan x \, dx$$

3. If there are no secant factors and the power of the tangent is even and positive, convert a tangent-squared factor to a secant-squared factor; then expand and repeat if necessary.

$$\int \tan^n x \, dx = \int \tan^{n-2} x \overbrace{(\tan^2 x)}^{\text{Convert to secants}} \, dx = \int \tan^{n-2} x (\sec^2 x - 1) \, dx$$

4. If the integral is of the form $\int \sec^m x \, dx$, where m is odd and positive, use integration by parts, as illustrated in Example 6 in the previous section.

5. If none of the first four cases applies, try converting to sines and cosines.

EXAMPLE 5   Power of Tangent is Odd and Positive

Evaluate $\displaystyle\int \frac{\tan^3 x}{\sqrt{\sec x}} \, dx$.

Solution

$$\int \frac{\tan^3 x}{\sqrt{\sec x}} \, dx = \int (\sec x)^{-1/2} \tan^3 x \, dx$$

$$= \int (\sec x)^{-3/2} (\tan^2 x)(\sec x \tan x) \, dx$$

$$= \int (\sec x)^{-3/2} (\sec^2 x - 1)(\sec x \tan x) \, dx$$

$$= \int [(\sec x)^{1/2} - (\sec x)^{-3/2}](\sec x \tan x) \, dx$$

$$= \frac{2}{3} (\sec x)^{3/2} + 2(\sec x)^{-1/2} + C$$

EXAMPLE 6   Power of Secant is Even and Positive

Evaluate $\displaystyle\int \sec^4 3x \tan^3 3x\,dx$.

Solution   Let $u = \tan 3x$, then

$$du = 3\sec^2 3x\,dx$$

and you can write

$$\int \sec^4 3x \tan^3 3x\,dx = \int \sec^2 3x \tan^3 3x(\sec^2 3x)\,dx$$

$$= \int (1 + \tan^2 3x)\tan^3 3x(\sec^2 3x)\,dx$$

$$= \frac{1}{3}\int (\tan^3 3x + \tan^5 3x)(3\sec^2 3x)\,dx$$

$$= \frac{1}{3}\left(\frac{\tan^4 3x}{4} + \frac{\tan^6 3x}{6}\right) + C$$

$$= \frac{\tan^4 3x}{12} + \frac{\tan^6 3x}{18} + C.$$

REMARK   In Example 6, the power of the tangent is odd and positive. Thus, you could also evaluate the integral with the procedure described in case 2 of the boxed feature on page 493. In Exercise 67, you are asked to show that the results obtained by these two procedures differ only by a constant.

EXAMPLE 7   Power of Tangent is Even

Evaluate $\displaystyle\int_0^{\pi/4} \tan^4 x\,dx$.

Solution   Because there are no secant factors, you can begin by converting a tangent-squared factor to a secant-squared factor.

$$\int \tan^4 x\,dx = \int \tan^2 x(\tan^2 x)\,dx$$

$$= \int \tan^2 x(\sec^2 x - 1)\,dx$$

$$= \int \tan^2 x \sec^2 x\,dx - \int \tan^2 x\,dx$$

$$= \int \tan^2 x \sec^2 x\,dx - \int (\sec^2 x - 1)\,dx$$

$$= \frac{\tan^3 x}{3} - \tan x + x + C$$

You can evaluate the definite integral as follows.

$$\int_0^{\pi/4} \tan^4 x\,dx = \left[\frac{\tan^3 x}{3} - \tan x + x\right]_0^{\pi/4}$$

$$= \frac{\pi}{4} - \frac{2}{3}$$

$$\approx 0.119$$

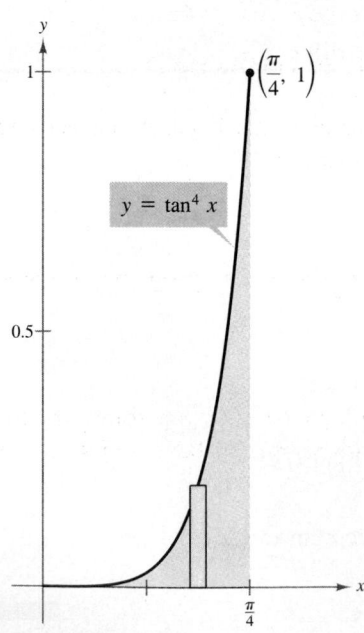

FIGURE 7.5
The area of the region is approximately 0.119.

The area represented by the definite integral is shown in Figure 7.5. Try using Simpson's Rule to approximate this integral. With $n = 10$, you should obtain an approximation that is within 0.00001 of the actual value.

For integrals involving powers of cotangents and cosecants, you can follow a strategy similar to that used for powers of tangents and secants, as illustrated in the next example.

EXAMPLE 8   Powers of Cotangents and Cosecants

Evaluate $\int \csc^4 x \cot^4 x \, dx$.

Solution   Because the power of the cosecant is even, you can write the following.

$$\int \csc^4 x \cot^4 x \, dx = \int \csc^2 x \cot^4 x (\csc^2 x) \, dx$$

$$= \int (1 + \cot^2 x) \cot^4 x (\csc^2 x) \, dx$$

$$= -\int (\cot^4 x + \cot^6 x)(-\csc^2 x) \, dx$$

$$= -\left(\frac{\cot^5 x}{5} + \frac{\cot^7 x}{7} \right) + C$$

$$= -\frac{\cot^5 x}{5} - \frac{\cot^7 x}{7} + C$$

TECHNOLOGY   When you use a symbolic integration utility to evaluate antiderivatives, remember that trigonometric antiderivatives can differ greatly in form. For instance, when we used a symbolic integrator to evaluate the integral in Example 8, we obtained

$$\int \csc^4 x \cot^4 x \, dx$$

$$= \left(-\frac{2}{35 \sin x} - \frac{1}{35 \sin^3 x} - \frac{3}{140 \sin^5 x} + \frac{3}{28 \sin^7 x} \right) \cos x - \frac{\cos^3 x}{4 \sin^7 x} + C.$$

Try showing that this form is equivalent to that obtained in Example 8.

When integrating trigonometric functions, remember that it sometimes helps to convert the entire integrand to powers of sines and cosines.

EXAMPLE 9   Converting to Sines and Cosines

Evaluate $\int \frac{\sec x}{\tan^2 x} \, dx$.

Solution   Because the first four cases on page 493 do not apply, try converting the integrand to sines and cosines. In this case, you are able to integrate the resulting powers of sine and cosine as follows.

$$\int \frac{\sec x}{\tan^2 x} \, dx = \int \left(\frac{1}{\cos x} \right) \left(\frac{\cos x}{\sin x} \right)^2 dx$$

$$= \int (\sin x)^{-2} (\cos x) \, dx$$

$$= -(\sin x)^{-1} + C$$

$$= -\csc x + C$$

Integrals Involving Sine-Cosine Products with Different Angles

Integrals involving the products of sines and cosines of two *different* angles occur in many applications. In such instances you can use the following product-to-sum identities.

$$\sin mx \sin nx = \frac{1}{2}(\cos[(m - n)x] - \cos[(m + n)x])$$

$$\sin mx \cos nx = \frac{1}{2}(\sin[(m - n)x] + \sin[(m + n)x])$$

$$\cos mx \cos nx = \frac{1}{2}(\cos[(m - n)x] + \cos[(m + n)x])$$

EXAMPLE 10   Using Product-to-Sum Identities

Evaluate $\int \sin 5x \cos 4x\, dx$.

Solution   Considering the second product-to-sum identity above, you can write the following.

$$\int \sin 5x \cos 4x\, dx = \frac{1}{2}\int (\sin 9x + \sin x)\, dx$$

$$= \frac{1}{2}\left(-\frac{\cos 9x}{9} - \cos x\right) + C$$

$$= -\frac{\cos 9x}{18} - \frac{\cos x}{2} + C$$

EXERCISES for Section 7.3

TECHNOLOGY
Laboratory Guide
Lab 7.3

In Exercises 1–10, evaluate the integral involving sine and cosine.

1. $\int \cos^3 x \sin x\, dx$

2. $\int \cos^3 x \sin^2 x\, dx$

3. $\int \sin^5 2x \cos 2x\, dx$

4. $\int \sin^3 x\, dx$

5. $\int \sin^5 x \cos^2 x\, dx$

6. $\int \cos^3 \frac{x}{3}\, dx$

7. $\int \cos^2 3x\, dx$

8. $\int \sin^2 2x\, dx$

9. $\int x \sin^2 x\, dx$

10. $\int x^2 \sin^2 x\, dx$

In Exercises 11–14, verify Wallis's Formulas by evaluating the integral.

11. $\int_0^{\pi/2} \cos^3 x\, dx = \frac{2}{3}$

12. $\int_0^{\pi/2} \cos^5 x\, dx = \frac{8}{15}$

13. $\int_0^{\pi/2} \cos^7 x\, dx = \frac{16}{35}$

14. $\int_0^{\pi/2} \sin^2 x\, dx = \frac{\pi}{4}$

In Exercises 15–30, evaluate the integral involving secant and tangent.

15. $\int \sec 3x\, dx$

16. $\int \sec^2(2x - 1)\, dx$

17. $\int \sec^4 5x\, dx$

18. $\int \sec^6 \frac{x}{2}\, dx$

19. $\int \sec^3 \pi x\, dx$

20. $\int \tan^2 x\, dx$

21. $\int \tan^5 \frac{x}{4}\, dx$

22. $\int \tan^3 \frac{\pi x}{2} \sec^2 \frac{\pi x}{2}\, dx$

23. $\int \sec^2 x \tan x\, dx$

24. $\int \tan^3 t \sec^3 t\, dt$

25. $\int \tan^2 x \sec^2 x\, dx$

26. $\int \tan^5 2x \sec^2 2x\, dx$

27. $\int \sec^6 4x \tan 4x\, dx$

28. $\int \sec^2 \frac{x}{2} \tan \frac{x}{2}\, dx$

29. $\int \sec^3 x \tan x\, dx$

30. $\int \tan^3 3x\, dx$

In Exercises 31–34, solve the differential equation.

31. $\dfrac{dr}{d\theta} = \sin^4 \pi\theta$

32. $\dfrac{ds}{d\alpha} = \sin^2 \dfrac{\alpha}{2} \cos^2 \dfrac{\alpha}{2}$

33. $y' = \tan^3 3x \sec 3x$

34. $y' = \sqrt{\tan x} \sec^4 x$

In Exercises 35–38, evaluate the integral involving sine-cosine products with different angles.

35. $\displaystyle\int \sin 3x \cos 2x \, dx$

36. $\displaystyle\int \cos 3\theta \cos(-2\theta) \, d\theta$

37. $\displaystyle\int \sin \theta \sin 3\theta \, d\theta$

38. $\displaystyle\int \sin(-4x) \cos 3x \, dx$

In Exercises 39–48, evaluate the integral.

39. $\displaystyle\int \cot^3 2x \, dx$

40. $\displaystyle\int \tan^4 \dfrac{x}{2} \sec^4 \dfrac{x}{2} \, dx$

41. $\displaystyle\int \csc^4 \theta \, d\theta$

42. $\displaystyle\int \csc^2 3x \cot 3x \, dx$

43. $\displaystyle\int \dfrac{\cot^2 t}{\csc t} \, dt$

44. $\displaystyle\int \dfrac{\cot^3 t}{\csc t} \, dt$

45. $\displaystyle\int \dfrac{1}{\sec x \tan x} \, dx$

46. $\displaystyle\int \dfrac{\sin^2 x - \cos^2 x}{\cos x} \, dx$

47. $\displaystyle\int (\tan^4 t - \sec^4 t) \, dt$

48. $\displaystyle\int \dfrac{1 - \sec t}{\cos t - 1} \, dt$

In Exercises 49–56, evaluate the definite integral.

49. $\displaystyle\int_{-\pi}^{\pi} \sin^2 x \, dx$

50. $\displaystyle\int_{0}^{\pi/4} \tan^2 x \, dx$

51. $\displaystyle\int_{0}^{\pi/4} \tan^3 x \, dx$

52. $\displaystyle\int_{0}^{\pi/4} \sec^2 t \sqrt{\tan t} \, dt$

53. $\displaystyle\int_{0}^{\pi/2} \dfrac{\cos t}{1 + \sin t} \, dt$

54. $\displaystyle\int_{-\pi}^{\pi} \sin 3\theta \cos \theta \, d\theta$

55. $\displaystyle\int_{-\pi/2}^{\pi/2} \cos^3 x \, dx$

56. $\displaystyle\int_{-\pi/2}^{\pi/2} (\sin^2 x + 1) \, dx$

C **In Exercises 57–66, use a symbolic integration utility to evaluate the integral.**

57. $\displaystyle\int \cos^4 \dfrac{x}{2} \, dx$

58. $\displaystyle\int \sin^2 x \cos^2 x \, dx$

59. $\displaystyle\int \sec^5 \pi x \, dx$

60. $\displaystyle\int \tan^3(1 - x) \, dx$

61. $\displaystyle\int \sec^5 \pi x \tan \pi x \, dx$

62. $\displaystyle\int \sec^4(1 - x) \tan(1 - x) \, dx$

63. $\displaystyle\int_{0}^{\pi/4} \sin 2\theta \sin 3\theta \, d\theta$

64. $\displaystyle\int_{0}^{\pi/2} (1 - \cos \theta)^2 \, d\theta$

65. $\displaystyle\int_{0}^{\pi/2} \sin^4 x \, dx$

66. $\displaystyle\int_{0}^{\pi/2} \sin^6 x \, dx$

In Exercises 67 and 68, find the indefinite integral in two different ways and show that the results differ only by a constant.

67. $\displaystyle\int \sec^4 3x \tan^3 3x \, dx$

68. $\displaystyle\int \sec^2 x \tan x \, dx$

69. *Area* Find the area of the region bounded by the graphs of the equations $y = \sin^2 \pi x$, $y = 0$, $x = 0$, and $x = 1$.

70. *Volume* Find the volume of the solid generated by revolving the region bounded by the graphs of the equations $y = \tan x$, $y = 0$, $x = -\pi/4$, and $x = \pi/4$ about the x-axis.

Volume and Centroid In Exercises 71 and 72, for the region bounded by the graphs of the equations, find (a) the volume of the solid formed by revolving the region about the x-axis and (b) the centroid of the region.

71. $y = \sin x$, $y = 0$, $x = 0$, $x = \pi$

72. $y = \cos x$, $y = 0$, $x = 0$, $x = \dfrac{\pi}{2}$

In Exercises 73–76, use integration by parts to verify the reduction formula.

73. $\displaystyle\int \sin^n x \, dx = -\dfrac{\sin^{n-1} x \cos x}{n} + \dfrac{n-1}{n} \int \sin^{n-2} x \, dx$

74. $\displaystyle\int \cos^n x \, dx = \dfrac{\cos^{n-1} x \sin x}{n} + \dfrac{n-1}{n} \int \cos^{n-2} x \, dx$

75. $\displaystyle\int \cos^m x \sin^n x \, dx = -\dfrac{\cos^{m+1} x \sin^{n-1} x}{m+n}$
$\qquad\qquad + \dfrac{n-1}{m+n} \int \cos^m x \sin^{n-2} x \, dx$

76. $\displaystyle\int \sec^n x \, dx = \dfrac{1}{n-1} \sec^{n-2} x \tan x + \dfrac{n-2}{n-1} \int \sec^{n-2} x \, dx$

In Exercises 77–80, use the results of Exercises 73–76 to evaluate the integral.

77. $\displaystyle\int \sin^5 x \, dx$

78. $\displaystyle\int \cos^4 x \, dx$

79. $\displaystyle\int \sec^4 \dfrac{2\pi x}{5} \, dx$

80. $\displaystyle\int \sin^4 x \cos^2 x \, dx$

81. *Wallis's Formula* Use the result of Exercise 74 to prove the following versions of Wallis's Formula.

a. If n is odd $(n \geq 3)$, then

$$\int_{0}^{\pi/2} \cos^n x \, dx = \left(\dfrac{2}{3}\right)\left(\dfrac{4}{5}\right)\left(\dfrac{6}{7}\right) \cdots \left(\dfrac{n-1}{n}\right).$$

b. If n is even $(n \geq 2)$, then

$$\int_{0}^{\pi/2} \cos^n x \, dx = \left(\dfrac{1}{2}\right)\left(\dfrac{3}{4}\right)\left(\dfrac{5}{6}\right) \cdots \left(\dfrac{n-1}{n}\right)\left(\dfrac{\pi}{2}\right).$$

82. The **inner product** of two functions f and g on $[a, b]$ is given by $\langle f, g \rangle = \int_a^b f(x)g(x) \, dx$. Two distinct functions f and g are said to be **orthogonal** if $\langle f, g \rangle = 0$. Show that the following functions are orthogonal on the specified interval.

Functions		Interval
a. $f_n(x) = \cos nx$,	$n = 0, 1, 2, \ldots$	$[0, \pi]$
b. $f_n(x) = \sin nx$,	$n = 1, 2, 3, \ldots$	$[-\pi, \pi]$
$\quad g_n(x) = \cos nx$,	$n = 0, 1, 2, \ldots$	

C 83. *Temperature Modeling* The following table gives the normal high and low temperatures for Erie, Pennsylvania for each month of the year.

Month	Jan	Feb	Mar	Apr	May	Jun
Max	30.9	32.2	41.1	53.7	64.6	74.0
Min	18.0	17.7	25.8	36.1	45.4	55.2

Month	Jul	Aug	Sep	Oct	Nov	Dec
Max	78.2	77.0	71.0	60.1	47.1	35.7
Min	59.9	59.4	53.1	43.2	34.3	24.2

The maximum and minimum temperatures can be modeled by

$$f(t) = a_0 + a_1 \cos \frac{\pi t}{6} + b_1 \sin \frac{\pi t}{6}$$

where a_0, a_1, and b_1 are as follows.

$$a_0 = \frac{1}{12} \int_0^{12} f(t)\, dt$$

$$a_1 = \frac{1}{6} \int_0^{12} f(t) \cos \frac{\pi t}{6}\, dt$$

$$b_1 = \frac{1}{6} \int_0^{12} f(t) \sin \frac{\pi t}{6}\, dt$$

a. Approximate the model for the maximum temperatures. Let $t = 0$ correspond to January. [*Hint:* Use Simpson's Rule to approximate the integrals and use the January data twice.]

b. Repeat part a for the minimum temperature data.

c. Use a graphing utility to compare each model with the actual data. During what part of the year is the difference between the maximum and minimum temperatures greatest?

84. Use a trigonometric substitution and Wallis's Formulas to show that

$$\int_{-1}^{1} (1 - x^2)^n\, dx = \frac{2^{2n+1}(n!)^2}{(2n+1)!}$$

where n is a positive integer.

85. *Power Lines* Power lines are constructed by stringing wire between supports and adjusting the tension on each span. The wire hangs between supports in the shape of a catenary, as shown in the figure.

FIGURE FOR 85

Let T be the tension (in pounds) on a span of wire, let u be the density (in pounds per foot) of the wire, let $g \approx 32.2$ be the acceleration due to gravity (in feet per second per second), and let L be the distance (in feet) between the supports. Then the equation of the catenary is

$$y = \frac{T}{ug}\left(\cosh \frac{ugx}{T} - 1\right)$$

where x and y are measured in feet.

a. Find the length of the wire between two spans.

b. To measure the tension in a span, power line workers use the *return wave method*. The wire is struck at one support, creating a wave in the line, and the time t (in seconds) it takes for the wave to make a round trip is measured. The velocity v (in feet per second) is given by $v = \sqrt{T/u}$. How long does it take the wave to make a round trip between supports?

c. The sag s (in inches) can be obtained by evaluating y when $x = L/2$ in the equation for the catenary (and multiplying by 12). In practice, however, power line workers use the "lineman's equation" given by $s \approx 12.075t^2$. Use the fact that $[\cosh(ugL/2T) + 1] \approx 2$ to derive this equation.

FOR FURTHER INFORMATION For more information on the mathematics of power lines, see the article "Constructing Power Lines" by Thomas O'Neil in the *UMAP Journal*, Volume 4, Number 3, 1983.

Trigonometric Substitution

Now that you can evaluate integrals involving powers of trigonometric functions, you can use **trigonometric substitution** to evaluate integrals involving the radicals

$$\sqrt{a^2 - u^2}, \quad \sqrt{a^2 + u^2}, \quad \text{and} \quad \sqrt{u^2 - a^2}.$$

The objective with trigonometric substitution is to eliminate the radical in the integrand. You do this with the Pythagorean identities

$$\cos^2 \theta = 1 - \sin^2 \theta$$
$$\sec^2 \theta = 1 + \tan^2 \theta$$
$$\tan^2 \theta = \sec^2 \theta - 1.$$

For example, if $a > 0$, let $u = a \sin \theta$, where $-\pi/2 \le \theta \le \pi/2$. Then

$$\sqrt{a^2 - u^2} = \sqrt{a^2 - a^2 \sin^2 \theta} = \sqrt{a^2(1 - \sin^2 \theta)}$$
$$= \sqrt{a^2 \cos^2 \theta} = a \cos \theta.$$

Note that $\cos \theta \ge 0$, because $-\pi/2 \le \theta \le \pi/2$.

Trigonometric Substitution ($a > 0$)

1. For integrals involving $\sqrt{a^2 - u^2}$, let

   $u = a \sin \theta.$

   Then $\sqrt{a^2 - u^2} = a \cos \theta$ where $-\pi/2 \le \theta \le \pi/2$.

2. For integrals involving $\sqrt{a^2 + u^2}$, let

   $u = a \tan \theta.$

   Then $\sqrt{a^2 + u^2} = a \sec \theta$ where $-\pi/2 < \theta < \pi/2$.

3. For integrals involving $\sqrt{u^2 - a^2}$, let

   $u = a \sec \theta.$

   Then $\sqrt{u^2 - a^2} = \pm a \tan \theta$ where $0 \le \theta < \pi/2$ or $\pi/2 < \theta \le \pi$. Use the positive value if $u > a$ and the negative value if $u < -a$.

REMARK The restrictions on θ ensure that the function that defines the substitution is one-to-one. In fact, these are the same intervals over which the arcsine, arctangent, and arcsecant are defined.

EXAMPLE 1   Trigonometric Substitution: $u = a \sin \theta$

Evaluate $\displaystyle \int \frac{dx}{x^2 \sqrt{9 - x^2}}$.

Solution   First, note that none of the basic integration rules in Section 7.1 applies. To use trigonometric substitution, you should observe that $\sqrt{9 - x^2}$ is of the form $\sqrt{a^2 - u^2}$. Hence, you can use the substitution

$$x = a \sin \theta = 3 \sin \theta.$$

Using differentiation and the triangle shown in Figure 7.6, you obtain

$$dx = 3 \cos \theta \, d\theta, \quad \sqrt{9 - x^2} = 3 \cos \theta, \quad \text{and} \quad x^2 = 9 \sin^2 \theta.$$

Therefore, trigonometric substitution yields the following.

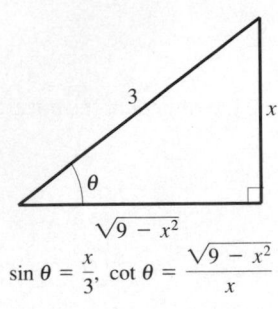

$$\sin \theta = \frac{x}{3}, \quad \cot \theta = \frac{\sqrt{9 - x^2}}{x}$$

FIGURE 7.6

$$\int \frac{dx}{x^2 \sqrt{9 - x^2}} = \int \frac{3 \cos \theta \, d\theta}{(9 \sin^2 \theta)(3 \cos \theta)} \qquad \text{Substitute}$$

$$= \frac{1}{9} \int \frac{d\theta}{\sin^2 \theta} \qquad \text{Simplify}$$

$$= \frac{1}{9} \int \csc^2 \theta \, d\theta \qquad \text{Trigonometric identity}$$

$$= -\frac{1}{9} \cot \theta + C \qquad \text{Apply Cosecant Rule}$$

$$= -\frac{1}{9} \left(\frac{\sqrt{9 - x^2}}{x} \right) + C \qquad \text{Substitute for } \cot \theta$$

$$= -\frac{\sqrt{9 - x^2}}{9x} + C$$

Note that the triangle in Figure 7.6 can be used to convert the θ's back to x's as follows.

$$\cot \theta = \frac{\text{adj.}}{\text{opp.}} = \frac{\sqrt{9 - x^2}}{x} \qquad \blacksquare$$

TECHNOLOGY   Try using a symbolic integration utility to integrate each of the following.

$$\int \frac{dx}{\sqrt{9 - x^2}} \qquad \int \frac{dx}{x \sqrt{9 - x^2}} \qquad \int \frac{dx}{x^2 \sqrt{9 - x^2}} \qquad \int \frac{dx}{x^3 \sqrt{9 - x^2}}$$

Then try using trigonometric substitution to duplicate the results obtained with the symbolic integration utility.

In Chapter 5, you saw how the inverse hyperbolic functions can be used to evaluate the integrals

$$\int \frac{du}{\sqrt{u^2 \pm a^2}}, \quad \int \frac{du}{a^2 - u^2}, \quad \text{and} \quad \int \frac{du}{u \sqrt{a^2 \pm u^2}}.$$

You can also evaluate these integrals using trigonometric substitution. This is illustrated in the next example.

EXAMPLE 2   Trigonometric Substitution: $u = a\tan\theta$

Evaluate $\displaystyle\int \frac{dx}{\sqrt{4x^2 + 1}}$.

Solution   Let $u = 2x$, $a = 1$, and $2x = \tan\theta$, as shown in Figure 7.7. Then,

$$dx = \frac{1}{2}\sec^2\theta\, d\theta \quad \text{and} \quad \sqrt{4x^2 + 1} = \sec\theta.$$

Trigonometric substitution produces the following.

$$\int \frac{1}{\sqrt{4x^2 + 1}}\, dx = \frac{1}{2}\int \frac{\sec^2\theta\, d\theta}{\sec\theta} \qquad \text{Substitute}$$

$$= \frac{1}{2}\int \sec\theta\, d\theta \qquad \text{Simplify}$$

$$= \frac{1}{2}\ln\left|\sec\theta + \tan\theta\right| + C \qquad \text{Apply Secant Rule}$$

$$= \frac{1}{2}\ln\left|\sqrt{4x^2 + 1} + 2x\right| + C \qquad \text{Back-substitute}$$

You can extend the use of trigonometric substitution to cover integrals involving expressions such as $(a^2 - u^2)^{n/2}$ by writing the expression as

$$(a^2 - u^2)^{n/2} = \left(\sqrt{a^2 - u^2}\right)^n.$$

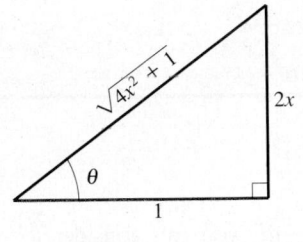

$\tan\theta = 2x$, $\sec\theta = \sqrt{4x^2 + 1}$

FIGURE 7.7

EXAMPLE 3   Trigonometric Substitution: Rational Powers

Evaluate $\displaystyle\int \frac{dx}{(x^2 + 1)^{3/2}}$.

Solution   Begin by writing $(x^2 + 1)^{3/2}$ as $\left(\sqrt{x^2 + 1}\right)^3$. Then, let $a = 1$ and $u = x = \tan\theta$, as shown in Figure 7.8. Using

$$dx = \sec^2\theta\, d\theta \quad \text{and} \quad \sqrt{x^2 + 1} = \sec\theta$$

you can apply trigonometric substitution as follows.

$$\int \frac{dx}{(x^2 + 1)^{3/2}} = \int \frac{dx}{\left(\sqrt{x^2 + 1}\right)^3} \qquad \text{Rewrite denominator}$$

$$= \int \frac{\sec^2\theta\, d\theta}{\sec^3\theta} \qquad \text{Substitute}$$

$$= \int \frac{d\theta}{\sec\theta} \qquad \text{Simplify}$$

$$= \int \cos\theta\, d\theta \qquad \text{Trigonometric identity}$$

$$= \sin\theta + C \qquad \text{Apply Cosine Rule}$$

$$= \frac{x}{\sqrt{x^2 + 1}} + C \qquad \text{Back-substitute}$$

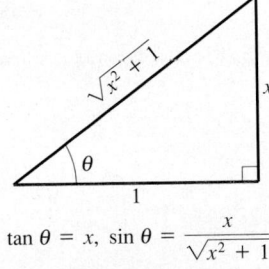

$\tan\theta = x$, $\sin\theta = \dfrac{x}{\sqrt{x^2 + 1}}$

FIGURE 7.8

For definite integrals, it is often convenient to determine the integration limits for θ. This avoids converting back to x. You might want to review this procedure in Section 4.5, Examples 8 and 9.

EXAMPLE 4   Converting the Limits of Integration

Evaluate $\displaystyle\int_{\sqrt{3}}^{2} \frac{\sqrt{x^2 - 3}}{x}\,dx.$

Solution   Because $\sqrt{x^2 - 3}$ has the form $\sqrt{u^2 - a^2}$, you can consider $u = x$, $a = \sqrt{3}$, and let $x = \sqrt{3}\sec\theta$, as shown in Figure 7.9. Then,

$$dx = \sqrt{3}\sec\theta\tan\theta\,d\theta \quad \text{and} \quad \sqrt{x^2 - 3} = \sqrt{3}\tan\theta.$$

To determine the upper and lower limits of integration, use the substitution $x = \sqrt{3}\sec\theta$ as follows.

Lower limit	*Upper limit*
When $x = \sqrt{3}$, $\sec\theta = 1$ and $\theta = 0$.	When $x = 2$, $\sec\theta = \dfrac{2}{\sqrt{3}}$ and $\theta = \dfrac{\pi}{6}$.

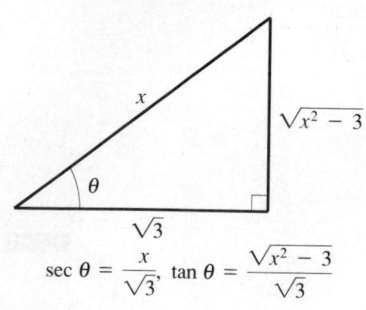

$\sec\theta = \dfrac{x}{\sqrt{3}},\ \tan\theta = \dfrac{\sqrt{x^2 - 3}}{\sqrt{3}}$

FIGURE 7.9

Therefore, we have

$$\underset{\substack{\uparrow\\ \text{Integration}\\ \text{limits for } x}}{\int_{\sqrt{3}}^{2} \frac{\sqrt{x^2 - 3}}{x}\,dx} = \underset{\substack{\uparrow\\ \text{Integration}\\ \text{limits for } \theta}}{\int_{0}^{\pi/6} \frac{(\sqrt{3}\tan\theta)(\sqrt{3}\sec\theta\tan\theta)\,d\theta}{\sqrt{3}\sec\theta}}$$

$$= \int_{0}^{\pi/6} \sqrt{3}\tan^2\theta\,d\theta$$

$$= \sqrt{3}\int_{0}^{\pi/6} (\sec^2\theta - 1)\,d\theta$$

$$= \sqrt{3}\Big[\tan\theta - \theta\Big]_{0}^{\pi/6}$$

$$= \sqrt{3}\left(\frac{1}{\sqrt{3}} - \frac{\pi}{6}\right)$$

$$= 1 - \frac{\sqrt{3}\,\pi}{6}$$

$$\approx 0.0931.$$

REMARK   In Example 4, try converting back to variable x and evaluating the antiderivative at the original limits of integration. You should obtain

$$\int_{\sqrt{3}}^{2} \frac{\sqrt{x^2 - 3}}{x}\,dx = \sqrt{3}\left[\frac{\sqrt{x^2 - 3}}{\sqrt{3}} - \text{arcsec}\frac{x}{\sqrt{3}}\right]_{\sqrt{3}}^{2}.$$

When using trigonometric substitution to evaluate definite integrals, you must be careful to check that the values of θ lie in the intervals discussed at the beginning of this section. For instance, if in Example 4 you had been asked to evaluate the definite integral

$$\int_{-2}^{-\sqrt{3}} \frac{\sqrt{x^2-3}}{x} dx$$

then using $u = x$ and $a = \sqrt{3}$ in the interval $\left[-2, -\sqrt{3}\right]$ would imply that $u < -a$. Thus, when determining the upper and lower limits of integration, you would have to choose θ such that $\pi/2 < \theta \leq \pi$. In this case the integral would be evaluated as follows.

$$\int_{-2}^{-\sqrt{3}} \frac{\sqrt{x^2-3}}{x} dx = \int_{5\pi/6}^{\pi} \frac{\left(-\sqrt{3}\tan\theta\right)\left(\sqrt{3}\sec\theta\tan\theta\right)d\theta}{\sqrt{3}\sec\theta}$$

$$= \int_{5\pi/6}^{\pi} -\sqrt{3}\tan^2\theta\, d\theta$$

$$= -\sqrt{3}\int_{5\pi/6}^{\pi} (\sec^2\theta - 1)\, d\theta$$

$$= -\sqrt{3}\left[\tan\theta - \theta\right]_{5\pi/6}^{\pi}$$

$$= -\sqrt{3}\left[(0-\pi) - \left(-\frac{1}{\sqrt{3}} - \frac{5\pi}{6}\right)\right]$$

$$= -1 + \frac{\sqrt{3}\pi}{6} \approx -0.0931$$

EXAMPLE 5 Completing the Square

Evaluate $\displaystyle\int \frac{dx}{(x^2 - 4x)^{3/2}}, \qquad x > 4.$

Solution Completing the square produces

$$x^2 - 4x = x^2 - 4x + 4 - 4 = (x-2)^2 - 2^2 = u^2 - a^2.$$

Let $a = 2$ and $u = x - 2 = 2\sec\theta$, as shown in Figure 7.10. (Note that $x > 4$ implies that $u > a$.) Then

$$dx = 2\sec\theta\tan\theta\, d\theta \quad \text{and} \quad \sqrt{(x-2)^2 - 2^2} = 2\tan\theta.$$

Trigonometric substitution produces

$$\int \frac{dx}{(x^2 - 4x)^{3/2}} = \int \frac{2\sec\theta\tan\theta\, d\theta}{(2\tan\theta)^3}$$

$$= \frac{2}{8}\int \frac{\sec\theta}{\tan^2\theta} d\theta$$

$$= \frac{1}{4}\int (\sin\theta)^{-2}(\cos\theta)\, d\theta$$

$$= -\frac{1}{4}\csc\theta + C$$

$$= -\frac{x-2}{4\sqrt{x^2-4x}} + C.$$

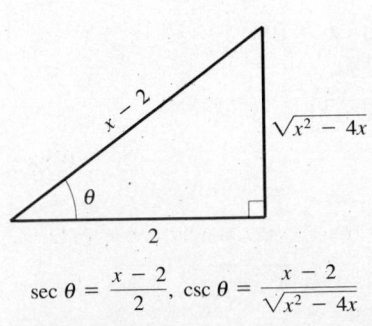

$$\sec\theta = \frac{x-2}{2}, \quad \csc\theta = \frac{x-2}{\sqrt{x^2-4x}}$$

FIGURE 7.10

In the next section, you will study techniques for evaluating rational functions. Some such integrals can be evaluated with trigonometric substitution, as shown in the following example.

EXAMPLE 6 Trigonometric Substitution for Rational Functions

Evaluate $\displaystyle\int \frac{dx}{(x^2 + 1)^2}$.

Solution Let $a = 1$ and $u = x = \tan\theta$, as shown in Figure 7.11. Then

$$dx = \sec^2\theta\, d\theta \quad \text{and} \quad x^2 + 1 = \sec^2\theta$$

and trigonometric substitution produces the following.

$$\int \frac{dx}{(x^2 + 1)^2} = \int \frac{\sec^2\theta\, d\theta}{\sec^4\theta}$$

$$= \int \cos^2\theta\, d\theta$$

$$= \frac{1}{2}\int (1 + \cos 2\theta)\, d\theta$$

$$= \frac{\theta}{2} + \frac{\sin 2\theta}{4} + C$$

$$= \frac{\theta}{2} + \frac{1}{2}\sin\theta\cos\theta + C$$

$$= \frac{\arctan x}{2} + \frac{1}{2}\left(\frac{x}{\sqrt{x^2 + 1}}\right)\left(\frac{1}{\sqrt{x^2 + 1}}\right) + C$$

$$= \frac{1}{2}\left(\arctan x + \frac{x}{x^2 + 1}\right) + C$$

Note that substitutions for $\sin\theta$ and $\cos\theta$ are obtained from the triangle shown in Figure 7.11.

Trigonometric substitution can be used to evaluate the three integrals listed in the following theorem. We will encounter these integrals several times in the remainder of the text, and when we do, we will simply refer to this theorem. (In Exercise 66, you are asked to verify the formulas given in the theorem.)

$\tan\theta = x, \ \sin\theta = \dfrac{x}{\sqrt{x^2 + 1}},$

$\cos\theta = \dfrac{1}{\sqrt{x^2 + 1}}$

FIGURE 7.11

THEOREM 7.2   Special Integration Formulas ($a > 0$)

1. $\displaystyle\int \sqrt{a^2 - u^2}\, du = \frac{1}{2}\left(a^2 \arcsin\frac{u}{a} + u\sqrt{a^2 - u^2}\right) + C$

2. $\displaystyle\int \sqrt{u^2 - a^2}\, du = \frac{1}{2}\left(u\sqrt{u^2 - a^2} - a^2 \ln\left|u + \sqrt{u^2 - a^2}\right|\right) + C, \ u > a$

3. $\displaystyle\int \sqrt{u^2 + a^2}\, du = \frac{1}{2}\left(u\sqrt{u^2 + a^2} + a^2 \ln\left|u + \sqrt{u^2 + a^2}\right|\right) + C$

Applications

EXAMPLE 7　Finding Arc Length

Find the arc length of the graph of $f(x) = \frac{1}{2}x^2$ from $x = 0$ to $x = 1$, as shown in Figure 7.12.

Solution　(Refer to the arc length formula in Section 6.4.)

$$s = \int_0^1 \sqrt{1 + [f'(x)]^2}\, dx \qquad \text{Formula for arc length}$$

$$= \int_0^1 \sqrt{1 + x^2}\, dx \qquad f'(x) = x$$

$$= \int_0^{\pi/4} \sec^3\theta\, d\theta \qquad \text{Let } a = 1 \text{ and } x = \tan\theta$$

$$= \frac{1}{2}\Big[\sec\theta\tan\theta + \ln\left|\sec\theta + \tan\theta\right| \Big]_0^{\pi/4} \qquad \text{Example 6, Section 7.2}$$

$$= \frac{1}{2}\Big[\sqrt{2} + \ln(\sqrt{2} + 1)\Big] \approx 1.148$$

FIGURE 7.12
The arc length of the curve from (0, 0) to $\left(1, \frac{1}{2}\right)$ is approximately 1.148.

EXAMPLE 8　Comparing Two Fluid Forces

A sealed barrel of oil (weighing 48 pounds per cubic foot) is floating in sea water (weighing 64 pounds per cubic foot), as shown in Figures 7.13 and 7.14. (Note that the barrel is not completely full of oil—on its side, the top 0.2 feet of the barrel is empty.) Compare the fluid forces against one end of the barrel from the inside and from the outside.

Solution　In Figure 7.14, locate the coordinate system with the origin at the center of the circle given by $x^2 + y^2 = 1$. To find the fluid force against an end of the barrel *from the inside*, integrate between -1 and 0.8 (using a weight of $w = 48$).

$$F_{\text{inside}} = 48\int_{-1}^{0.8} (0.8 - y)(2)\sqrt{1 - y^2}\, dy$$

$$= 76.8\int_{-1}^{0.8} \sqrt{1 - y^2}\, dy - 96\int_{-1}^{0.8} y\sqrt{1 - y^2}\, dy$$

To find the fluid force *from the outside*, integrate between -1 and 0.4 (using a weight of $w = 64$).

$$F_{\text{outside}} = 64\int_{-1}^{0.4} (0.4 - y)(2)\sqrt{1 - y^2}\, dy$$

$$= 51.2\int_{-1}^{0.4} \sqrt{1 - y^2}\, dy - 128\int_{-1}^{0.4} y\sqrt{1 - y^2}\, dy$$

We leave the details of integration for you to complete in Exercise 61. Intuitively, would you say that the force from the oil (the inside) or the force from the sea water (the outside) is greater? By evaluating these two integrals, you can determine that

$$F_{\text{inside}} \approx 121.3 \text{ lb} \quad \text{and} \quad F_{\text{outside}} \approx 93.0 \text{ lb}.$$

FIGURE 7.13
The barrel is not quite full of oil—the top 0.2 feet of the barrel is empty.

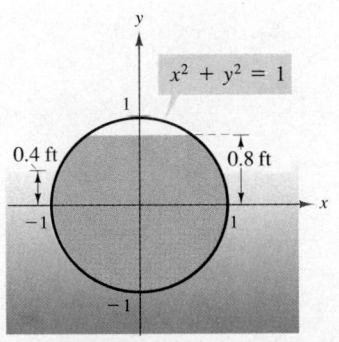

FIGURE 7.14

TECHNOLOGY
Laboratory Guide
Lab 7.4

EXERCISES for Section 7.4

In Exercises 1–4, write the indefinite integral that corresponds to the antiderivative.

1. $4 \ln \left| \dfrac{\sqrt{x^2 + 16} - 4}{x} \right| + \sqrt{x^2 + 16} + C$

2. $8 \ln \left| \sqrt{x^2 - 16} + x \right| + \dfrac{x\sqrt{x^2 - 16}}{2} + C$

3. $x \arcsin \dfrac{x}{4} + \sqrt{16 - x^2} + C$

4. $8 \arcsin \dfrac{x - 3}{4} + \dfrac{(x - 3)\sqrt{7 + 6x - x^2}}{2} + C$

In Exercises 5–8, evaluate the integral using the substitution $x = 5 \sin \theta$.

5. $\displaystyle\int \dfrac{1}{(25 - x^2)^{3/2}} dx$

6. $\displaystyle\int \dfrac{1}{x^2 \sqrt{25 - x^2}} dx$

7. $\displaystyle\int \dfrac{\sqrt{25 - x^2}}{x} dx$

8. $\displaystyle\int \dfrac{x^2}{\sqrt{25 - x^2}} dx$

In Exercises 9–12, evaluate the integral using the substitution $x = 2 \sec \theta$.

9. $\displaystyle\int \dfrac{1}{\sqrt{x^2 - 4}} dx$

10. $\displaystyle\int \dfrac{\sqrt{x^2 - 4}}{x} dx$

11. $\displaystyle\int x^3 \sqrt{x^2 - 4} \, dx$

12. $\displaystyle\int \dfrac{x^3}{\sqrt{x^2 - 4}} dx$

In Exercises 13–16, evaluate the integral using the substitution $x = \tan \theta$.

13. $\displaystyle\int x\sqrt{1 + x^2} \, dx$

14. $\displaystyle\int \dfrac{x^3}{\sqrt{1 + x^2}} dx$

15. $\displaystyle\int \dfrac{1}{(1 + x^2)^2} dx$

16. $\displaystyle\int \dfrac{x^2}{(1 + x^2)^2} dx$

In Exercises 17 and 18, use Theorem 7.2 to evaluate the integral.

17. $\displaystyle\int \sqrt{4 + 9x^2} \, dx$

18. $\displaystyle\int \sqrt{1 + x^2} \, dx$

In Exercises 19–38, evaluate the integral.

19. $\displaystyle\int \dfrac{x}{\sqrt{x^2 + 9}} dx$

20. $\displaystyle\int \dfrac{1}{\sqrt{25 - x^2}} dx$

21. $\displaystyle\int_0^2 \sqrt{16 - 4x^2} \, dx$

22. $\displaystyle\int_0^2 x\sqrt{16 - 4x^2} \, dx$

23. $\displaystyle\int \dfrac{1}{\sqrt{x^2 - 9}} dx$

24. $\displaystyle\int \dfrac{t}{(1 - t^2)^{3/2}} dt$

25. $\displaystyle\int \dfrac{\sqrt{1 - x^2}}{x^4} dx$

26. $\displaystyle\int \dfrac{\sqrt{4x^2 + 9}}{x^4} dx$

27. $\displaystyle\int \dfrac{1}{x\sqrt{4x^2 + 9}} dx$

28. $\displaystyle\int \dfrac{1}{(x^2 + 3)^{3/2}} dx$

29. $\displaystyle\int \dfrac{x}{(x^2 + 3)^{3/2}} dx$

30. $\displaystyle\int \dfrac{1}{x\sqrt{4x^2 + 16}} dx$

31. $\displaystyle\int e^{2x} \sqrt{1 + e^{2x}} \, dx$

32. $\displaystyle\int (x + 1)\sqrt{x^2 + 2x + 2} \, dx$

33. $\displaystyle\int e^x \sqrt{1 - e^{2x}} \, dx$

34. $\displaystyle\int \dfrac{\sqrt{1 - x}}{\sqrt{x}} dx$

35. $\displaystyle\int \dfrac{1}{4 + 4x^2 + x^4} dx$

36. $\displaystyle\int \dfrac{x^3 + x + 1}{x^4 + 2x^2 + 1} dx$

37. $\displaystyle\int \operatorname{arcsec} 2x \, dx$

38. $\displaystyle\int x \arcsin x \, dx$

In Exercises 39–42, complete the square and evaluate the integral.

39. $\displaystyle\int \dfrac{1}{\sqrt{4x - x^2}} dx$

40. $\displaystyle\int \dfrac{x^2}{\sqrt{2x - x^2}} dx$

41. $\displaystyle\int \dfrac{x}{\sqrt{x^2 + 4x + 8}} dx$

42. $\displaystyle\int \dfrac{x}{\sqrt{x^2 - 6x + 5}} dx$

In Exercises 43–46, evaluate the given integral using (a) the given integration limits and (b) the limits obtained by trigonometric substitution.

43. $\displaystyle\int_0^{\sqrt{3}/2} \dfrac{t^2}{(1 - t^2)^{3/2}} dt$

44. $\displaystyle\int_0^{\sqrt{3}/1} \dfrac{1}{(1 - t^2)^{5/2}} dt$

45. $\displaystyle\int_0^3 \dfrac{x^3}{\sqrt{x^2 + 9}} dx$

46. $\displaystyle\int_0^{5/3} \sqrt{25 - 9x^2} \, dx$

C In Exercises 47–50, use a symbolic integration utility to evaluate the integral.

47. $\displaystyle\int \dfrac{x^2}{\sqrt{x^2 + 10x + 9}} dx$

48. $\displaystyle\int (x^2 + 2x + 11)^{3/2} dx$

49. $\displaystyle\int \dfrac{x^2}{\sqrt{x^2 - 1}} dx$

50. $\displaystyle\int x^2 \sqrt{x^2 - 4} \, dx$

Area In Exercises 51 and 52, find the area enclosed by the graph of the equation.

51. $x^2 + y^2 = r^2$

52. $\dfrac{x^2}{a^2} + \dfrac{y^2}{b^2} = 1$

FIGURE FOR 51

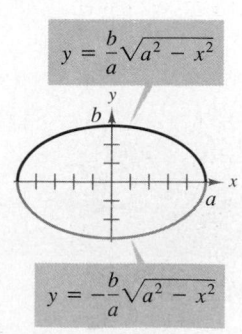

FIGURE FOR 52

Volume of a Torus In Exercises 53 and 54, find the volume of the torus generated by revolving the region bounded by the graph of the circle about the *y*-axis.

53. $(x - 3)^2 + y^2 = 1$ (see figure)

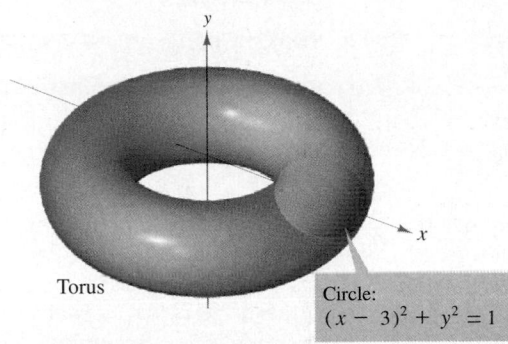

Torus

Circle:
$(x - 3)^2 + y^2 = 1$

FIGURE FOR 53

54. $(x - h)^2 + y^2 = r^2$, $h > r$

Arc Length In Exercises 55 and 56, find the arc length of the plane curve over the indicated interval.

Function	Interval
55. $y = \ln x$	$[1, 5]$
56. $y = x^2$	$[0, 3]$

Projectile Motion In Exercises 57 and 58, find the distance that a projectile travels when it follows the path given by the graph of the equation over the specified interval.

Function	Interval
57. $y = x - 0.005x^2$	$[0, 200]$
58. $y = x - \dfrac{x^2}{72}$	$[0, 72]$

Fluid Force In Exercises 59 and 60, find the fluid force on the circular observation window of radius 1 foot in a vertical wall of a large water-filled tank at a fish hatchery. Use trigonometric substitution to evaluate the one integral. (In a similar problem in Section 6.6, you evaluated the one integral by a geometric formula and the other by observing that the integrand was odd.)

59. The center of the window is 3 feet below the surface of the water (see figure).

$x^2 + y^2 = 1$

FIGURE FOR 59

60. The center of the window is *d* feet below the surface of the water ($d > 1$).

61. *Fluid Force* Evaluate the integrals—they yield the fluid forces given in Example 8.

a. $F_{\text{inside}} = 48 \displaystyle\int_{-1}^{0.8} (0.8 - y)(2)\sqrt{1 - y^2}\, dy$

b. $F_{\text{outside}} = 64 \displaystyle\int_{-1}^{0.4} (0.4 - y)(2)\sqrt{1 - y^2}\, dy$

62. *Centroid* Find the centroid of the region bounded by the graphs of $y = 3/\sqrt{x^2 + 9}$, $y = 0$, $x = -4$, and $x = 4$.

63. *Surface Area* Find the surface area of the solid generated by revolving the region bounded by the graphs of $y = x^2$, $y = 0$, $x = 0$, and $x = \sqrt{2}$ about the *x*-axis.

64. *Average Field Strength* The field strength *H* of a magnet of length 2*L* on a particle *r* units from the center of the magnet is given by

$$H = \frac{2mL}{(r^2 + L^2)^{3/2}}$$

where $\pm m$ are the poles of the magnet (see figure). Find the average field strength as the particle moves from 0 to *R* units from the center by evaluating the integral

$$\frac{1}{R}\int_0^R \frac{2mL}{(r^2 + L^2)^{3/2}}\, dr.$$

C **65.** *Tractrix* A person moves from the origin along the positive *y*-axis pulling a weight at the end of a 12-foot rope (see figure). Initially the weight is located at the point $(12, 0)$.

a. Show that the slope of the tangent line of the path of the weight is

$$\frac{dy}{dx} = -\frac{\sqrt{144 - x^2}}{x}.$$

b. Use the result of part a to find the equation of the path of the weight. Use a graphing utility to graph the path and compare it with the figure.

c. Find any vertical asymptotes of the graph in part b.

d. When the person has reached the point $(0, 12)$, how far has the weight moved?

FIGURE FOR 64   **FIGURE FOR 65**

66. Use trigonometric substitution to verify the integration formulas given in Theorem 7.2.

True or False In Exercises 67–70, determine whether the statement is true or false. If it is false, explain why or give an example that shows it is false.

67. If $x = \sin\theta$, then $\displaystyle\int \frac{dx}{\sqrt{1 - x^2}} = \int d\theta$.

68. If $x = \sec\theta$, then $\displaystyle\int \frac{\sqrt{x^2 - 1}}{x}\,dx = \int \sec\theta\tan\theta\,d\theta$.

69. If $x = \tan\theta$, then $\displaystyle\int_0^{\sqrt{3}} \frac{dx}{(1 + x^2)^{3/2}} = \int_0^{4\pi/3} \cos\theta\,d\theta$.

70. If $x = \sin\theta$, then $\displaystyle\int_{-1}^{1} x^2\sqrt{1 - x^2}\,dx =$

$$2\int_0^{\pi/2} \sin^2\theta\cos^2\theta\,d\theta.$$

CAREER INTERVIEW

Robert Chidlaw

Chief Scientist

Young Chang

Research & Development Inst.

Waltham, MA

I apply my background in electrical engineering to the design of music synthesizers. A synthesizer is a specialized computer, with microprocessors and custom-designed integrated circuits. A great deal of engineering effort goes into designing the product, as well as considerations of cost, sound quality, and facility of use. My instruments can produce the sounds of acoustic instruments in addition to all sorts of strange electronic noises. They are sold in the home market, and also to professional musicians.

My specialty, digital signal processing, involves a fair amount of mathematics. I have recently used calculus to find the frequency at which a digital filter has its maximum response. I have also used it to find functions which can simulate guitar amplifier distortion.

When designing a new synthesizer, I first build a "simulation" of it on my computer, allowing me to evaluate the output with Fourier transforms. Only when we are sure that everything is working right does my team commit to the design. When a new design is completed, you know it's successful if you can enjoy playing it, and forget that the sound is really being made up out of hundreds of millions of arithmetic operations per second.

7.5

Partial Fractions • Linear Factors • Quadratic Factors

JOHN BERNOULLI

The method of partial fractions was introduced by John Bernoulli (1667–1748), a Swiss mathematician who was instrumental in the early development of calculus. John Bernoulli was a professor at the University of Basel and taught many outstanding students, the most famous of whom was Leonhard Euler.

Partial Fractions

This section examines a procedure for decomposing a rational function into simpler rational functions to which you can apply the basic integration formulas. This procedure is called the **method of partial fractions.**

To see the benefit of the method of partial fractions, consider the integral of $1/(x^2 - 5x + 6)$. To evaluate this integral *without* partial fractions, you can complete the square and use trigonometric substitution (see Figure 7.15) to obtain the following.

$$\int \frac{1}{x^2 - 5x + 6}\,dx = \int \frac{dx}{(x - 5/2)^2 - (1/2)^2} \qquad a = \tfrac{1}{2},\, x - \tfrac{5}{2} = \tfrac{1}{2}\sec\theta$$

$$= \int \frac{(1/2)\sec\theta\tan\theta\,d\theta}{(1/4)\tan^2\theta} \qquad dx = \tfrac{1}{2}\sec\theta\tan\theta\,d\theta$$

$$= 2\int \csc\theta\,d\theta$$

$$= 2\ln\left|\csc\theta - \cot\theta\right| + C$$

$$= 2\ln\left|\frac{2x - 5}{2\sqrt{x^2 - 5x + 6}} - \frac{1}{2\sqrt{x^2 - 5x + 6}}\right| + C$$

$$= 2\ln\left|\frac{x - 3}{\sqrt{x^2 - 5x + 6}}\right| + C$$

$$= 2\ln\left|\frac{\sqrt{x - 3}}{\sqrt{x - 2}}\right| + C$$

$$= \ln\left|\frac{x - 3}{x - 2}\right| + C$$

$$= \ln\left|x - 3\right| - \ln\left|x - 2\right| + C$$

Now, suppose you had observed that

$$\frac{1}{x^2 - 5x + 6} = \frac{1}{x - 3} - \frac{1}{x - 2}. \qquad \text{Partial fraction decomposition}$$

Then you could evaluate the integral easily as follows.

$$\int \frac{1}{x^2 - 5x + 6}\,dx = \int\left(\frac{1}{x - 3} - \frac{1}{x - 2}\right)dx$$

$$= \ln\left|x - 3\right| - \ln\left|x - 2\right| + C$$

This method is clearly preferable to trigonometric substitution. However, its use depends on the ability to factor the denominator, $x^2 - 5x + 6$, and to find the **partial fractions** $1/(x - 3)$ and $-1/(x - 2)$.

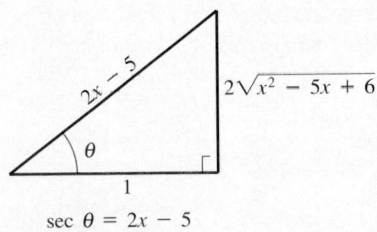

$\sec\theta = 2x - 5$

FIGURE 7.15

Recall from algebra that every polynomial with real coefficients can be factored into linear and irreducible quadratic factors.* For instance, the polynomial $x^5 + x^4 - x - 1$ can be written as

$$
\begin{aligned}
x^5 + x^4 - x - 1 &= x^4(x + 1) - (x + 1) \\
&= (x^4 - 1)(x + 1) \\
&= (x^2 + 1)(x^2 - 1)(x + 1) \\
&= (x^2 + 1)(x + 1)(x - 1)(x + 1) \\
&= (x - 1)(x + 1)^2(x^2 + 1)
\end{aligned}
$$

where $(x - 1)$ is a linear factor, $(x + 1)^2$ is a repeated linear factor, and $(x^2 + 1)$ is an irreducible quadratic factor. Using this factorization, you can write the partial fraction decomposition of the rational expression

$$
\frac{N(x)}{x^5 + x^4 - x - 1}
$$

where $N(x)$ is a polynomial of degree less than 5, as follows.

$$
\frac{N(x)}{(x - 1)(x + 1)^2(x^2 + 1)} = \frac{A}{x - 1} + \frac{B}{x + 1} + \frac{C}{(x + 1)^2} + \frac{Dx + E}{x^2 + 1}
$$

Decomposition of $N(x)/D(x)$ into Partial Fractions

1. **Divide if improper:** If $N(x)/D(x)$ is an improper fraction (that is, if the degree of the numerator is greater than or equal to the degree of the denominator), then divide the denominator into the numerator to obtain

   $$
   \frac{N(x)}{D(x)} = (\text{a polynomial}) + \frac{N_1(x)}{D(x)}
   $$

   where the degree of $N_1(x)$ is less than the degree of $D(x)$. Then apply steps 2, 3, and 4 to the proper rational expression $N_1(x)/D(x)$.
2. **Factor denominator:** Completely factor the denominator into factors of the form

   $$
   (px + q)^m \quad \text{and} \quad (ax^2 + bx + c)^n
   $$

   where $ax^2 + bx + c$ is irreducible.
3. **Linear factors:** For *each* factor of the form $(px + q)^m$, the partial fraction decomposition must include the following sum of m fractions.

   $$
   \frac{A_1}{(px + q)} + \frac{A_2}{(px + q)^2} + \cdots + \frac{A_m}{(px + q)^m}
   $$
4. **Quadratic factors:** For *each* factor of the form $(ax^2 + bx + c)^n$, the partial fraction decomposition must include the following sum of n fractions.

   $$
   \frac{B_1x + C_1}{ax^2 + bx + c} + \frac{B_2x + C_2}{(ax^2 + bx + c)^2} + \cdots + \frac{B_nx + C_n}{(ax^2 + bx + c)^n}
   $$

*For a review of factorization techniques, see *Precalculus,* 3rd edition, by Larson and Hostetler (Lexington, Mass.: D. C. Heath and Company, 1993).

Linear Factors

Algebraic techniques for determining the constants in the numerators of a partial fraction decomposition with linear or repeated linear factors are demonstrated in Examples 1 and 2.

EXAMPLE 1   Distinct Linear Factors

Write the partial fraction decomposition for $\dfrac{1}{x^2 - 5x + 6}$.

Solution Because

$$x^2 - 5x + 6 = (x - 3)(x - 2)$$

you should include one partial fraction for each factor and write

$$\frac{1}{x^2 - 5x + 6} = \frac{A}{x - 3} + \frac{B}{x - 2}$$

where A and B are to be determined. Multiplying this equation by the lowest common denominator, $(x - 3)(x - 2)$, leads to the **basic equation**

$$1 = A(x - 2) + B(x - 3). \qquad \text{Basic equation}$$

Because this equation is to be true for all x, you can substitute any *convenient* values for x to obtain equations in A and B. The most convenient values are the ones that make particular factors equal to 0.

To solve for A, let $x = 3$ and obtain

$$1 = A(3 - 2) + B(3 - 3) \qquad \text{Let } x = 3 \text{ in basic equation}$$
$$1 = A(1) + B(0)$$
$$A = 1.$$

To solve for B, let $x = 2$ and obtain

$$1 = A(2 - 2) + B(2 - 3) \qquad \text{Let } x = 2 \text{ in basic equation}$$
$$1 = A(0) + B(-1)$$
$$B = -1.$$

Therefore, the decomposition is

$$\frac{1}{x^2 - 5x + 6} = \frac{1}{x - 3} - \frac{1}{x - 2}$$

as indicated at the beginning of this section.

REMARK Note that the substitutions for x in Example 1 are chosen for their convenience in determining values for A and B: $x = 2$ is chosen so as to eliminate the term $A(x - 2)$, and $x = 3$ is chosen to eliminate the term $B(x - 3)$. The goal is to make *convenient* substitutions whenever possible.

Be sure you see that the method of partial fractions is practical only for integrals of rational functions whose denominators factor "nicely." For instance, if the denominator in Example 1 were changed to $x^2 - 5x + 5$, then its factorization as $x^2 - 5x + 5 = \left[x - (5 + \sqrt{5})/2\right]\left[x - (5 - \sqrt{5})/2\right]$ would be too cumbersome to use with partial fractions. In such cases, you should use completing the square or a symbolic integration utility to perform the integration. If you do this, you should obtain

$$\int \frac{1}{x^2 - 5x + 5}\, dx = \frac{\sqrt{5}}{5}\ln\left(2x - \sqrt{5} - 5\right) - \frac{\sqrt{5}}{5}\ln\left(2x + \sqrt{5} - 5\right) + C.$$

EXAMPLE 2 Repeated Linear Factors

Evaluate $\int \dfrac{5x^2 + 20x + 6}{x^3 + 2x^2 + x}\, dx.$

Solution

$$
\begin{aligned}
x^3 + 2x^2 + x &= x(x^2 + 2x + 1) \\
&= x(x + 1)^2
\end{aligned}
$$

FOR FURTHER INFORMATION
For an alternative approach to using partial fractions, see the article "A Shortcut in Partial Fractions" by Xun-Cheng Huang in the November, 1991 issue of *The College Mathematics Journal.*

you should include one fraction for *each power* of x and $(x + 1)$ and write

$$
\frac{5x^2 + 20x + 6}{x(x + 1)^2} = \frac{A}{x} + \frac{B}{x + 1} + \frac{C}{(x + 1)^2}.
$$

Multiplying by the least common denominator, $x(x + 1)^2$, leads to the *basic equation*

$$
5x^2 + 20x + 6 = A(x + 1)^2 + Bx(x + 1) + Cx. \qquad \text{Basic equation}
$$

To solve for A, let $x = 0$. This eliminates the B and C terms and yields

$$
6 = A(1) + 0 + 0
$$
$$
A = 6.
$$

To solve for C, let $x = -1$. This eliminates the A and B terms and yields

$$
5 - 20 + 6 = 0 + 0 - C
$$
$$
C = 9.
$$

The most convenient choices for x have been used, so to find the value of B, you can use *any other value* for x along with the calculated values of A and C. Using $x = 1$, $A = 6$, and $C = 9$ produces

$$
\begin{aligned}
5 + 20 + 6 &= A(4) + B(2) + C \\
31 &= 6(4) + 2B + 9 \\
-2 &= 2B \\
B &= -1.
\end{aligned}
$$

Therefore, if follows that

$$
\begin{aligned}
\int \frac{5x^2 + 20x + 6}{x(x + 1)^2}\, dx &= \int \left(\frac{6}{x} - \frac{1}{x + 1} + \frac{9}{(x + 1)^2} \right) dx \\
&= 6 \ln |x| - \ln |x + 1| + 9 \frac{(x + 1)^{-1}}{-1} + C \\
&= \ln \left| \frac{x^6}{x + 1} \right| - \frac{9}{x + 1} + C.
\end{aligned}
$$

REMARK Note that it is necessary to make as many substitutions for x as there are unknowns (A, B, C, . . .) to be determined. For instance, in Example 2 we made three substitutions ($x = -1$, $x = 0$, and $x = 1$) to solve for C, A, and B, respectively.

Quadratic Factors

When using the method of partial fractions with *linear* factors, a convenient choice of x immediately yields a value for one of the coefficients. With *quadratic* factors, a system of linear equations usually has to be solved, regardless of the choice of x.

EXAMPLE 3   Distinct Linear and Quadratic Factors

Evaluate $\displaystyle\int \frac{2x^3 - 4x - 8}{(x^2 - x)(x^2 + 4)}\, dx$.

Solution   Because $(x^2 - x)(x^2 + 4) = x(x - 1)(x^2 + 4)$, you should include one partial fraction for each factor and write

$$\frac{2x^3 - 4x - 8}{x(x - 1)(x^2 + 4)} = \frac{A}{x} + \frac{B}{x - 1} + \frac{Cx + D}{x^2 + 4}.$$

Multiplying by the least common denominator, $x(x - 1)(x^2 + 4)$, yields the *basic equation*

$$2x^3 - 4x - 8 = A(x - 1)(x^2 + 4) + Bx(x^2 + 4) + (Cx + D)(x)(x - 1).$$

To solve for A, let $x = 0$ and obtain

$$-8 = A(-1)(4) + 0 + 0 \rightarrow A = 2.$$

To solve for B, let $x = 1$ and obtain

$$-10 = 0 + B(5) + 0 \rightarrow B = -2.$$

At this point, C and D are yet to be determined. You can find these remaining constants by choosing two other values for x and solving the resulting system of linear equations. If $x = -1$, then, using $A = 2$ and $B = -2$, you can write

$$-6 = (2)(-2)(5) + (-2)(-1)(5) + (-C + D)(-1)(-2)$$
$$2 = -C + D.$$

If $x = 2$, then you have

$$0 = (2)(1)(8) + (-2)(2)(8) + (2C + D)(2)(1)$$
$$8 = 2C + D.$$

Solving the linear system

$$-C + D = 2$$
$$2C + D = 8$$

yields $C = 2$. Consequently, $D = 4$, and it follows that

$$\int \frac{2x^3 - 4x - 8}{x(x - 1)(x^2 + 4)}\, dx$$

$$= \int \left(\frac{2}{x} - \frac{2}{x - 1} + \frac{2x}{x^2 + 4} + \frac{4}{x^2 + 4} \right) dx$$

$$= 2 \ln |x| - 2 \ln |x - 1| + \ln(x^2 + 4) + 2 \arctan \frac{x}{2} + C.$$

In Examples 1, 2, and 3, we began the solution of the basic equation by substituting values of x that made the linear factors equal to 0. This method works well when the partial fraction decomposition involves linear factors. However, if the decomposition involves only quadratic factors, then an alternative procedure is often more convenient.

EXAMPLE 4   Repeated Quadratic Factors

Evaluate $\displaystyle\int \frac{8x^3 + 13x}{(x^2 + 2)^2}\, dx.$

Solution   Include one partial fraction for each power of $(x^2 + 2)$ and write

$$\frac{8x^3 + 13x}{(x^2 + 2)^2} = \frac{Ax + B}{x^2 + 2} + \frac{Cx + D}{(x^2 + 2)^2}.$$

Multiplying by the least common denominator, $(x^2 + 2)^2$, yields the *basic equation*

$$8x^3 + 13x = (Ax + B)(x^2 + 2) + Cx + D.$$

Expanding the basic equation and collecting like terms produces

$$8x^3 + 13x = Ax^3 + 2Ax + Bx^2 + 2B + Cx + D$$
$$8x^3 + 13x = Ax^3 + Bx^2 + (2A + C)x + (2B + D).$$

Now, you can equate the coefficients of like terms on opposite sides of the equation.

$$8 = A \qquad 13 = 2A + C$$
$$8x^3 + 0x^2 + 13x + 0 = Ax^3 + Bx^2 + (2A + C)x + (2B + D)$$
$$0 = B \qquad 0 = 2B + D$$

Using the known values $A = 8$ and $B = 0$, you can write the following.

$$13 = 2A + C = 2(8) + C \rightarrow C = -3$$
$$0 = 2B + D = 2(0) + D \rightarrow D = 0$$

Finally, you can conclude that

$$\int \frac{8x^3 + 13x}{(x^2 + 2)^2}\, dx = \int \left(\frac{8x}{x^2 + 2} + \frac{-3x}{(x^2 + 2)^2} \right) dx$$

$$= 4\ln(x^2 + 2) + \frac{3}{2(x^2 + 2)} + C.$$ ▬

TECHNOLOGY   Try using a symbolic integration utility to evaluate the integral in Example 4—you might find that the form of the antiderivative is different. For instance, when we used a symbolic integrator to work Example 4, we obtained

$$\int \frac{8x^3 + 13x}{(x^2 + 2)^2}\, dx = \ln(x^8 + 8x^6 + 24x^4 + 32x^2 + 16) + \frac{3}{2(x^2 + 2)} + C.$$

Is this result equivalent to that obtained in Example 4?

EXAMPLE 5   Repeated Quadratic Factors

Evaluate $\displaystyle\int \frac{x^2}{(x^2 + 1)^2}\,dx$.

Solution   Begin by writing

$$\frac{x^2}{(x^2 + 1)^2} = \frac{Ax + B}{x^2 + 1} + \frac{Cx + D}{(x^2 + 1)^2}.$$

Then multiply by $(x^2 + 1)^2$ to obtain

$$x^2 = (Ax + B)(x^2 + 1) + Cx + D$$
$$= Ax^3 + Bx^2 + (A + C)x + (B + D).$$

Now, you can equate coefficients to obtain $A = 0, B = 1, C = 0$, and $D = -1$. Thus, you can write the following.

$$\int \frac{x^2}{(x^2 + 1)^2}\,dx = \int \left[\frac{1}{x^2 + 1} - \frac{1}{(x^2 + 1)^2} \right] dx$$

$$= \int \frac{dx}{x^2 + 1} - \int \frac{dx}{(x^2 + 1)^2}$$

$$= \arctan x - \frac{1}{2}\left(\arctan x + \frac{x}{x^2 + 1} \right) + C$$

$$= \frac{1}{2}\left(\arctan x - \frac{x}{x^2 + 1} \right) + C$$

The first integral was evaluated by the Arctangent Rule, and the second integral was solved in Section 7.4 (see Example 6).   ▬

EXAMPLE 6   Linear and Quadratic Factors

Evaluate $\displaystyle\int \frac{3x + 4}{x^3 - 2x - 4}\,dx$.

Solution   Because $x^3 - 2x - 4 = (x - 2)(x^2 + 2x + 2)$, you can write

$$\frac{3x + 4}{x^3 - 2x - 4} = \frac{A}{x - 2} + \frac{Bx + C}{x^2 + 2x + 2}$$

$$3x + 4 = A(x^2 + 2x + 2) + (Bx + C)(x - 2) \qquad \text{Basic equation}$$

$$3x + 4 = (A + B)x^2 + (2A - 2B + C)x + (2A - 2C).$$

If $x = 2$, then in the basic equation you have $10 = 10A$ and $A = 1$. From the expanded equation, you obtain $A + B = 0$, which implies that $B = -1$, and $2A - 2C = 4$ which implies that $C = -1$. Therefore,

$$\int \frac{3x + 4}{x^3 - 2x - 4}\,dx = \int \left(\frac{A}{x - 2} + \frac{Bx + C}{x^2 + 2x + 2} \right) dx$$

$$= \int \left(\frac{1}{x - 2} + \frac{-x - 1}{x^2 + 2x + 2} \right) dx$$

$$= \ln|x - 2| - \frac{1}{2}\ln(x^2 + 2x + 2) + C. \qquad \blacksquare$$

When integrating rational expressions, keep in mind that for *improper* rational expressions such as

$$\frac{N(x)}{D(x)} = \frac{2x^3 + x^2 - 7x + 7}{x^2 + x - 2}$$

you must first divide to obtain

$$\frac{N(x)}{D(x)} = 2x - 1 + \frac{-2x + 5}{x^2 + x - 2}.$$

The proper rational expression is then decomposed into its partial fractions by the usual methods. Here are some guidelines for solving the basic equation that is obtained in a partial fraction decomposition.

Guidelines for Solving the Basic Equation

Linear Factors

1. Substitute the *roots* of the distinct linear factors into the basic equation.
2. For repeated linear factors, use the coefficients determined in part 1 to rewrite the basic equation. Then substitute *other* convenient values of x and solve for the remaining coefficients.

Quadratic Factors

1. Expand the basic equation.
2. Collect terms according to powers of x.
3. Equate the coefficients of like powers to obtain a system of linear equations involving A, B, C, and so on.
4. Solve the system of linear equations.

Before concluding this section, here are a few things you should remember. First, it is not necessary to use the partial fractions technique on all rational functions. For instance, the following integral is evaluated more easily by the Log Rule.

$$\int \frac{x^2 + 1}{x^3 + 3x - 4} dx = \frac{1}{3} \int \frac{3x^2 + 3}{x^3 + 3x - 4} dx = \frac{1}{3} \ln |x^3 + 3x - 4| + C$$

Second, if the integrand is not in reduced form, reducing it may eliminate the need for partial fractions, as shown in the following integral.

$$\int \frac{x^2 - x - 2}{x^3 - 2x - 4} dx = \int \frac{(x + 1)(x - 2)}{(x - 2)(x^2 + 2x + 2)} dx$$

$$= \int \frac{x + 1}{x^2 + 2x + 2} dx$$

$$= \frac{1}{2} \ln |x^2 + 2x + 2| + C$$

Finally, partial fractions can be used with some quotients involving transcendental functions. For instance, the substitution $u = \sin x$ allows you to write

$$\int \frac{\cos x}{\sin x(\sin x - 1)} dx = \int \frac{du}{u(u - 1)} = \int \left(-\frac{1}{u} + \frac{1}{u - 1} \right) du$$

$$= -\ln |u| + \ln |u - 1| + C$$

$$= \ln \left| \frac{\sin x - 1}{\sin x} \right| + C.$$

EXERCISES for Section 7.5

TECHNOLOGY
Laboratory Guide
Lab 7.5

In Exercises 1–6, write the form of the partial fraction decomposition of the rational expression. Do not solve for the constants.

1. $\dfrac{5}{x^2 - 10x}$

2. $\dfrac{4x^2 + 3}{(x - 5)^3}$

3. $\dfrac{2x - 3}{x^3 + 10x}$

4. $\dfrac{x - 2}{x^2 + 4x + 3}$

5. $\dfrac{16x}{x^3 - 10x^2}$

6. $\dfrac{2x - 1}{x(x^2 + 1)^2}$

In Exercises 7–18, use partial fractions (linear factors) to evaluate the integral.

7. $\displaystyle\int \dfrac{1}{x^2 - 1}\,dx$

8. $\displaystyle\int \dfrac{1}{4x^2 - 9}\,dx$

9. $\displaystyle\int \dfrac{3}{x^2 + x - 2}\,dx$

10. $\displaystyle\int \dfrac{x + 1}{x^2 + 4x + 3}\,dx$

11. $\displaystyle\int \dfrac{5 - x}{2x^2 + x - 1}\,dx$

12. $\displaystyle\int \dfrac{3x^2 - 7x - 2}{x^3 - x}\,dx$

13. $\displaystyle\int \dfrac{x^2 + 12x + 12}{x^3 - 4x}\,dx$

14. $\displaystyle\int \dfrac{x^3 - x + 3}{x^2 + x - 2}\,dx$

15. $\displaystyle\int \dfrac{2x^3 - 4x^2 - 15x + 5}{x^2 - 2x - 8}\,dx$

16. $\displaystyle\int \dfrac{x + 2}{x^2 - 4x}\,dx$

17. $\displaystyle\int \dfrac{4x^2 + 2x - 1}{x^3 + x^2}\,dx$

18. $\displaystyle\int \dfrac{2x - 3}{(x - 1)^2}\,dx$

In Exercises 19–26, use partial fractions (linear and quadratic factors) to evaluate the integral.

19. $\displaystyle\int \dfrac{x^2 - 1}{x^3 + x}\,dx$

20. $\displaystyle\int \dfrac{x}{x^3 - 1}\,dx$

21. $\displaystyle\int \dfrac{x^2}{x^4 - 2x^2 - 8}\,dx$

22. $\displaystyle\int \dfrac{2x^2 + x + 8}{(x^2 + 4)^2}\,dx$

23. $\displaystyle\int \dfrac{x}{16x^4 - 1}\,dx$

24. $\displaystyle\int \dfrac{x^2 - 4x + 7}{x^3 - x^2 + x + 3}\,dx$

25. $\displaystyle\int \dfrac{x^2 + 5}{x^3 - x^2 + x + 3}\,dx$

26. $\displaystyle\int \dfrac{x^2 + x + 3}{x^4 + 6x^2 + 9}\,dx$

In Exercises 27–30, solve the differential equation.

27. $\dfrac{dy}{dx} = \dfrac{x^4}{(x - 1)^3}$

28. $\dfrac{dy}{dx} = \dfrac{4x^2 - 1}{(2x)(x^2 + 2x + 1)}$

29. $\dfrac{dy}{dx} = \dfrac{6x^2 - 3x + 14}{x^3 - 2x^2 + 4x - 8}$

30. $\dfrac{dy}{dx} = \dfrac{x^3 - 6x^2 + x - 4}{1 - x^4}$

In Exercises 31–34, evaluate the definite integral.

31. $\displaystyle\int_0^1 \dfrac{3}{2x^2 + 5x + 2}\,dx$

32. $\displaystyle\int_1^5 \dfrac{x - 1}{x^2(x + 1)}\,dx$

33. $\displaystyle\int_1^2 \dfrac{x + 1}{x(x^2 + 1)}\,dx$

34. $\displaystyle\int_0^1 \dfrac{x^2 - x}{x^2 + x + 1}\,dx$

C In Exercises 35–42, use partial fractions to evaluate the integral. Then use a symbolic integration utility to verify your answer.

35. $\displaystyle\int \dfrac{3x}{x^2 - 6x + 9}\,dx$

36. $\displaystyle\int \dfrac{6x^2 + 1}{x^2(x - 1)^3}\,dx$

37. $\displaystyle\int \dfrac{x^2 + x + 2}{(x^2 + 2)^2}\,dx$

38. $\displaystyle\int \dfrac{x^3}{(x^2 - 4)^2}\,dx$

39. $\displaystyle\int \dfrac{2x^2 - 2x + 3}{x^3 - x^2 - x - 2}\,dx$

40. $\displaystyle\int \dfrac{x(2x - 9)}{x^3 - 6x^2 + 12x - 8}\,dx$

41. $\displaystyle\int_2^3 \dfrac{x^2 - x + 2}{x^3 - x^2 + x - 1}\,dx$

42. $\displaystyle\int_3^4 \dfrac{1}{x^2 - 4}\,dx$

In Exercises 43–48, evaluate the integral by using an appropriate substitution.

43. $\displaystyle\int \dfrac{\sin x}{\cos x(\cos x - 1)}\,dx$

44. $\displaystyle\int \dfrac{\sin x}{\cos x + \cos^2 x}\,dx$

45. $\displaystyle\int \dfrac{3\cos x}{\sin^2 x + \sin x - 2}\,dx$

46. $\displaystyle\int \dfrac{\sec^2 x}{\tan x(\tan x + 1)}\,dx$

47. $\displaystyle\int \dfrac{e^x}{(e^x - 1)(e^x + 4)}\,dx$

48. $\displaystyle\int \dfrac{e^x}{(e^{2x} + 1)(e^x - 1)}\,dx$

In Exercises 49–52, use the method of partial fractions to verify the integration formula.

49. $\displaystyle\int \dfrac{1}{x(a + bx)}\,dx = \dfrac{1}{a}\ln\left|\dfrac{x}{a + bx}\right| + C$

50. $\displaystyle\int \dfrac{1}{a^2 - x^2}\,dx = \dfrac{1}{2a}\ln\left|\dfrac{a + x}{a - x}\right| + C$

51. $\displaystyle\int \dfrac{x}{(a + bx)^2}\,dx = \dfrac{1}{b^2}\left(\dfrac{a}{a + bx} + \ln|a + bx|\right) + C$

52. $\displaystyle\int \dfrac{1}{x^2(a + bx)}\,dx = -\dfrac{1}{ax} - \dfrac{b}{a^2}\ln\left|\dfrac{x}{a + bx}\right| + C$

53. *Approximation* Determine which value best approximates the area of the region between the x-axis and the graph of the function $10/[x(x^2 + 1)]$ over the interval $[1, 3]$. (Make your selection on the basis of a sketch of the region and *not* by performing any calculations.)
a. -6 **b.** 6 **c.** 3 **d.** 5 **e.** 8

54. *Area* Find the area of the region bounded by the graphs of $y = 7/(16 - x^2)$ and $y = 1$.

55. *Volume and Centroid* Consider the region bounded by the graphs of $y = 2x/(x^2 + 1)$, $y = 0$, $x = 0$, and $x = 3$.

a. Find the volume of the solid generated by revolving the region about the x-axis.

b. Find the centroid of the region.

56. *Limited Growth* In Chapter 5 the exponential growth equation was derived from the assumption that the rate of growth was proportional to the existing quantity. In practice, there often exists some upper limit L past which growth cannot occur. In such cases, we assume the rate of growth to be proportional not only to the existing quantity but also to the difference between the existing quantity y and the upper limit L. That is,

$$\frac{dy}{dt} = ky(L - y).$$

In integral form we can express this relationship as

$$\int \frac{dy}{y(L - y)} = \int k \, dt.$$

a. Evaluate the two integrals and solve for y as a function of t, where y_0 is the initial quantity.

b. The graph of the function y is called a **logistics curve** (see figure). Show that the rate of growth is a maximum at the point of inflection, and that this occurs when $y = L/2$.

Logistics curve

FIGURE FOR 56

57. *Epidemic Model* A single infected individual enters a community of n susceptible individuals. Let x be the number of newly infected individuals at time t. The common epidemic model assumes that the disease spreads at a rate proportional to the product of the total number infected and the number not yet infected. Thus

$$\frac{dx}{dt} = k(x + 1)(n - x)$$

and we obtain

$$\int \frac{1}{(x + 1)(n - x)} \, dx = \int k \, dt.$$

Solve for x as a function of t

58. *Chemical Reactions* In a chemical reaction, one unit of compound Y and one unit of compound Z are converted into a single unit of compound X. If x is the amount of compound X formed, and the rate of formation of X is proportional to the product of the amounts of unconverted compounds Y and Z, then

$$\frac{dx}{dt} = k(y_0 - x)(z_0 - x)$$

where y_0 and z_0 are the initial amounts of substances Y and Z, respectively. From the above equation we obtain

$$\int \frac{1}{(y_0 - x)(z_0 - x)} \, dx = \int k \, dt.$$

a. Do the two integrations and solve for x in terms of t.

b. Use the result of part a to find x as $t \to \infty$ if (1) $y_0 < z_0$ and (2) $y_0 > z_0$. If $y_0 = z_0$, what is the limit of x as $t \to \infty$?

59. Use partial fractions to evaluate

$$\int_0^1 \frac{x}{1 + x^4} \, dx.$$

60. Suppose the denominator of a rational function can be factored into distinct linear factors

$$D(x) = (x - c_1)(x - c_2) \ldots (x - c_n),$$

for a positive integer n, and distinct real numbers $c_1, c_2, \ldots, c_n$. If $N(x)$ is a polynomial of degree less than n, show that

$$\frac{N(x)}{D(x)} = \frac{P_1}{x - c_1} + \frac{P_2}{x - c_2} + \cdots + \frac{P_n}{x - c_n},$$

where $P_k = N(c_k)/D'(c_k)$ for $k = 1, 2, \ldots, n$. Note that this is the partial fraction decomposition of $\dfrac{N(x)}{D(x)}$.

61. Use the results of Exercise 60 to find the partial fraction decomposition of

$$\frac{x^3 - 3x^2 + 1}{x^4 - 13x^2 + 12x}.$$

Integration by Tables

So far in this chapter you have studied several integration techniques to use with the basic integration rules. But merely knowing *how* to use the various techniques is not enough. You also need to know *when* to use them. Integration is first and foremost a problem of recognition. That is, you must recognize which rule or technique to apply to obtain an antiderivative. Frequently, a slight alteration of an integrand will require a different integration technique (or produce a function whose antiderivative is not an elementary function), as shown below.

$$\int x \ln x \, dx = \frac{x^2}{2} \ln x - \frac{x^2}{4} + C \qquad \text{Integration by parts}$$

$$\int \frac{\ln x}{x} \, dx = \frac{(\ln x)^2}{2} + C \qquad \text{Power Rule}$$

$$\int \frac{1}{x \ln x} \, dx = \ln |\ln x| + C \qquad \text{Log Rule}$$

$$\int \frac{x}{\ln x} \, dx = ? \qquad \text{Not an elementary function}$$

Many people find a table of integrals to be a valuable supplement to the integration techniques discussed in this chapter. A table of common integrals can be found in Appendix C of this text. **Integration by tables** is not a "cure-all" for all of the difficulties that can accompany integration—using a table of integrals requires considerable thought and insight and often involves substitution.

Each integration formula in the tables in this text can be developed using one or more of the techniques in this chapter. You should try to verify several of the formulas. For instance, Formula 4

$$\int \frac{u}{(a + bu)^2} \, du = \frac{1}{b^2} \left(\frac{a}{a + bu} + \ln |a + bu| \right) + C \qquad \text{Formula 4}$$

can be verified using the method of partial fractions, and Formula 19

$$\int \frac{\sqrt{a + bu}}{u} \, du = 2\sqrt{a + bu} + a \int \frac{du}{u \sqrt{a + bu}} \qquad \text{Formula 19}$$

can be verified using integration by parts.

Note that the integrals in the tables in this text are classified according to forms involving the following: u^n, $(a + bu)$, $(a + bu + cu^2)$, $\sqrt{a + bu}$, $(a^2 \pm u^2)$, $\sqrt{u^2 \pm a^2}$, $\sqrt{a^2 - u^2}$, trigonometric functions, inverse trigonometric functions, exponential functions, and logarithmic functions.

TECHNOLOGY A symbolic integrator consists, in part, of a database of integration tables. The primary difference between using a symbolic integrator and using a table of integrals is that with a symbolic integrator the computer searches through the database to find a fit. With a table of integrals, *you* must do the searching.

REMARK Try using the table of integrals in Appendix C and the substitution $u = \sqrt{x - 1}$ to evaluate the integral in Example 1. If you do this, you should obtain

$$\int \frac{dx}{x\sqrt{x - 1}} = \int \frac{2\,du}{u^2 - 1}.$$

Does this produce the same result that was obtained in Example 1?

EXAMPLE 1 Integration by Tables

Evaluate $\displaystyle\int \frac{dx}{x\sqrt{x - 1}}$.

Solution Because the expression inside the radical is linear, you should consider forms involving $\sqrt{a + bu}$.

$$\int \frac{du}{u\sqrt{a + bu}} = \frac{2}{\sqrt{-a}}\arctan\sqrt{\frac{a + bu}{-a}} + C \qquad \text{Formula 17 } (a < 0)$$

Let $a = -1$, $b = 1$, and $u = x$. Then $du = dx$, and you can write

$$\int \frac{dx}{x\sqrt{x - 1}} = 2\arctan\sqrt{x - 1} + C.$$

EXAMPLE 2 Integration by Tables

Evaluate $\int x\sqrt{x^4 - 9}\,dx$.

Solution Because the radical has the form $\sqrt{u^2 - a^2}$, you should consider Formula 26.

$$\int \sqrt{u^2 - a^2}\,du = \frac{1}{2}\left(u\sqrt{u^2 - a^2} - a^2\ln\left|u + \sqrt{u^2 - a^2}\right|\right) + C$$

Let $u = x^2$ and $a = 3$. Then $du = 2x\,dx$, and you have

$$\int x\sqrt{x^4 - 9}\,dx = \frac{1}{2}\int \sqrt{(x^2)^2 - 3^2}\,(2x)\,dx$$

$$= \frac{1}{4}\left(x^2\sqrt{x^4 - 9} - 9\ln\left|x^2 + \sqrt{x^4 - 9}\right|\right) + C.$$

EXAMPLE 3 Integration by Tables

Evaluate $\displaystyle\int \frac{x}{1 + e^{-x^2}}\,dx$.

Solution Of the forms involving e^u, consider the following.

$$\int \frac{du}{1 + e^u} = u - \ln(1 + e^u) + C \qquad \text{Formula 84}$$

Let $u = -x^2$. Then, $du = -2x\,dx$, and you have

$$\int \frac{x}{1 + e^{-x^2}}\,dx = -\frac{1}{2}\int \frac{-2x\,dx}{1 + e^{-x^2}}$$

$$= -\frac{1}{2}\left[-x^2 - \ln\left(1 + e^{-x^2}\right)\right] + C$$

$$= \frac{1}{2}\left[x^2 + \ln\left(1 + e^{-x^2}\right)\right] + C.$$

TECHNOLOGY Example 3 shows the importance of having several solution techniques at your disposal. This integral is not difficult to solve with a table, but when we entered it into a well-known symbolic integrator, the integrator was unable to find the antiderivative.

Reduction Formulas

Several of the integrals in the integration tables have the form $\int f(x)\,dx = g(x) + \int h(x)\,dx$. Such integration formulas are called **reduction formulas** because they reduce a given integral to the sum of a function and a simpler integral.

EXAMPLE 4   Using a Reduction Formula

Evaluate $\displaystyle\int x^3 \sin x\,dx$.

Solution   Consider the following three formulas.

$$\int u \sin u\,du = \sin u - u\cos u + C \qquad\qquad \text{Formula 52}$$

$$\int u^n \sin u\,du = -u^n\cos u + n\int u^{n-1}\cos u\,du \qquad\qquad \text{Formula 54}$$

$$\int u^n \cos u\,du = u^n\sin u - n\int u^{n-1}\sin u\,du \qquad\qquad \text{Formula 55}$$

Using Formula 54, Formula 55, and then Formula 52 produces

$$\int x^3 \sin x\,dx = -x^3\cos x + 3\int x^2\cos x\,dx$$

$$= -x^3\cos x + 3\left(x^2\sin x - 2\int x\sin x\,dx\right)$$

$$= -x^3\cos x + 3x^2\sin x + 6x\cos x - 6\sin x + C. \qquad \blacksquare$$

EXAMPLE 5   Using a Reduction Formula

Evaluate $\displaystyle\int \frac{\sqrt{3 - 5x}}{2x}\,dx$.

Solution   Consider the following two formulas.

$$\int \frac{du}{u\sqrt{a + bu}} = \frac{1}{\sqrt{a}}\ln\left|\frac{\sqrt{a + bu} - \sqrt{a}}{\sqrt{a + bu} + \sqrt{a}}\right| + C \qquad \text{Formula 17 } (0 < a)$$

$$\int \frac{\sqrt{a + bu}}{u}\,du = 2\sqrt{a + bu} + a\int \frac{du}{u\sqrt{a + bu}} \qquad \text{Formula 19}$$

Using Formula 19, with $a = 3$, $b = -5$, and $u = x$, produces

$$\frac{1}{2}\int \frac{\sqrt{3 - 5x}}{x}\,dx = \frac{1}{2}\left(2\sqrt{3 - 5x} + 3\int \frac{dx}{x\sqrt{3 - 5x}}\right)$$

$$= \sqrt{3 - 5x} + \frac{3}{2}\int \frac{dx}{x\sqrt{3 - 5x}}.$$

Using Formula 17, with $a = 3$, $b = -5$, and $u = x$, produces

$$\int \frac{\sqrt{3 - 5x}}{2x}\,dx = \sqrt{3 - 5x} + \frac{3}{2}\left(\frac{1}{\sqrt{3}}\ln\left|\frac{\sqrt{3 - 5x} - \sqrt{3}}{\sqrt{3 - 5x} + \sqrt{3}}\right|\right) + C$$

$$= \sqrt{3 - 5x} + \frac{\sqrt{3}}{2}\ln\left|\frac{\sqrt{3 - 5x} - \sqrt{3}}{\sqrt{3 - 5x} + \sqrt{3}}\right| + C. \qquad \blacksquare$$

Rational Functions of Sine and Cosine

EXAMPLE 6   Integration by Tables

Evaluate $\displaystyle\int \frac{\sin 2x}{2 + \cos x}\, dx.$

Solution   Substituting $2 \sin x \cos x$ for $\sin 2x$ produces

$$\int \frac{\sin 2x}{2 + \cos x}\, dx = 2\int \frac{\sin x \cos x}{2 + \cos x}\, dx.$$

A check of the forms involving $\sin u$ or $\cos u$ shows that none of those listed applies. Therefore, you can consider forms involving $a + bu$. For example,

$$\int \frac{u\, du}{a + bu} = \frac{1}{b^2}(bu - a \ln |a + bu|) + C. \qquad \text{Formula 3}$$

Let $a = 2, b = 1$, and $u = \cos x$. Then, $du = -\sin x\, dx$, and you have

$$
\begin{aligned}
2\int \frac{\sin x \cos x}{2 + \cos x}\, dx &= -2\int \frac{\cos x(-\sin x\, dx)}{2 + \cos x}\\
&= -2(\cos x - 2 \ln |2 + \cos x|) + C\\
&= -2 \cos x + 4 \ln |2 + \cos x| + C.
\end{aligned}
$$

Example 6 involves a rational expression of $\sin x$ and $\cos x$. If you are unable to find an integral of this form in the integration tables, try using the following special substitution to convert the trigonometric expression to a standard rational expression.

Substitution for Rational Functions of Sine and Cosine

For integrals involving rational functions of sine and cosine, the substitution

$$u = \frac{\sin x}{1 + \cos x} = \tan \frac{x}{2}$$

yields

$$\cos x = \frac{1 - u^2}{1 + u^2}, \quad \sin x = \frac{2u}{1 + u^2}, \quad \text{and} \quad dx = \frac{2\, du}{1 + u^2}.$$

Proof   From the substitution for u, it follows that

$$u^2 = \frac{\sin^2 x}{(1 + \cos x)^2} = \frac{1 - \cos^2 x}{(1 + \cos x)^2} = \frac{1 - \cos x}{1 + \cos x}.$$

Solving for $\cos x$ produces $\cos x = (1 - u^2)/(1 + u^2)$. To find $\sin x$, write $u = \sin x/(1 + \cos x)$ as

$$\sin x = u(1 + \cos x) = u\left(1 + \frac{1 - u^2}{1 + u^2}\right) = \frac{2u}{1 + u^2}.$$

Finally, to find dx, consider $u = \tan (x/2)$. Then you have $\arctan u = x/2$ and $dx = (2\, du)/(1 + u^2)$.

EXAMPLE 7   Rational Function of Sine and Cosine

Evaluate $\displaystyle\int \frac{dx}{1 + \sin x - \cos x}$.

Solution Let $u = \sin x/(1 + \cos x)$.

$$\int \frac{dx}{1 + \sin x - \cos x} = \int \frac{2\,du/(1 + u^2)}{1 + [2u/(1 + u^2)] - [(1 - u^2)/(1 + u^2)]}$$

$$= \int \frac{2\,du}{(1 + u^2) + 2u - (1 - u^2)}$$

$$= \int \frac{2\,du}{2u + 2u^2}$$

$$= \int \frac{du}{u(1 + u)}$$

$$= \int \frac{1}{u}\,du - \int \frac{1}{1 + u}\,du \qquad \text{Partial fractions}$$

$$= \ln |u| - \ln |1 + u| + C$$

$$= \ln \left| \frac{u}{1 + u} \right| + C$$

$$= \ln \left| \frac{\sin x/(1 + \cos x)}{1 + [\sin x/(1 + \cos x)]} \right| + C$$

$$= \ln \left| \frac{\sin x}{1 + \sin x + \cos x} \right| + C$$

EXERCISES for Section 7.6

In Exercises 1 and 2, use a table of integrals with forms involving $a + bu$ to evaluate the integral.

1. $\displaystyle\int \frac{x^2}{1 + x}\,dx$ **2.** $\displaystyle\int \frac{1}{2x^2(2x - 1)^2}\,dx$

In Exercises 3 and 4, use a table of integrals with forms involving $\sqrt{u^2 \pm a^2}$ to evaluate the integral.

3. $\displaystyle\int e^x\sqrt{1 + e^{2x}}\,dx$ **4.** $\displaystyle\int \frac{\sqrt{x^2 - 4}}{x}\,dx$

In Exercises 5 and 6, use a table of integrals with forms involving $\sqrt{a^2 - u^2}$ to evaluate the integral.

5. $\displaystyle\int \frac{1}{x^2\sqrt{1 - x^2}}\,dx$ **6.** $\displaystyle\int \frac{x}{\sqrt{9 - x^4}}\,dx$

In Exercises 7–10, use a table of integrals with forms involving the trigonometric functions to evaluate the integral.

7. $\displaystyle\int \sin^4 2x\,dx$ **8.** $\displaystyle\int \frac{\cos^3 \sqrt{x}}{\sqrt{x}}\,dx$

9. $\displaystyle\int \frac{1}{\sqrt{x}(1 - \cos \sqrt{x})}\,dx$ **10.** $\displaystyle\int \frac{1}{1 - \tan 5x}\,dx$

In Exercises 11 and 12, use a table of integrals with forms involving e^u to evaluate the integral.

11. $\displaystyle\int \frac{1}{1 + e^{2x}}\,dx$ **12.** $\displaystyle\int e^{-2x}\cos 3x\,dx$

In Exercises 13 and 14, use a table of integrals with forms involving $\ln u$ to evaluate the integral.

13. $\displaystyle\int x^3 \ln x\,dx$ **14.** $\displaystyle\int (\ln x)^3\,dx$

In Exercises 15–46, use a table of integrals to evaluate the integral.

15. $\displaystyle\int xe^{x^2}\,dx$ **16.** $\displaystyle\int \frac{x}{\sqrt{1 + x}}\,dx$

17. $\displaystyle\int x\,\text{arcsec}(x^2 + 1)\,dx$ **18.** $\displaystyle\int \text{arcsec}\,2x\,dx$

19. $\displaystyle\int x^2 \ln x\,dx$ **20.** $\displaystyle\int x\sin x\,dx$

21. $\displaystyle\int \frac{1}{x^2\sqrt{x^2 - 4}}\,dx$ **22.** $\displaystyle\int \frac{x^2}{(3x - 5)^2}\,dx$

23. $\displaystyle\int \frac{2x}{(1 - 3x)^2}\, dx$

24. $\displaystyle\int \frac{1}{x^2 + 2x + 2}\, dx$

25. $\displaystyle\int e^x \arccos e^x\, dx$

26. $\displaystyle\int \frac{\theta^2}{1 - \sin \theta^3}\, d\theta$

27. $\displaystyle\int \frac{x}{1 - \sec x^2}\, dx$

28. $\displaystyle\int \frac{e^x}{1 - \tan e^x}\, dx$

29. $\displaystyle\int \frac{\cos x}{1 + \sin^2 x}\, dx$

30. $\displaystyle\int \frac{1}{t[1 + (\ln t)^2]}\, dt$

31. $\displaystyle\int \frac{\cos \theta}{3 + 2\sin \theta + \sin^2 \theta}\, d\theta$

32. $\displaystyle\int \sqrt{3 + x^2}\, dx$

33. $\displaystyle\int \frac{1}{x^2\sqrt{2 + 9x^2}}\, dx$

34. $\displaystyle\int x^2\sqrt{2 + 9x^2}\, dx$

35. $\displaystyle\int t^4 \cos t\, dt$

36. $\displaystyle\int \sqrt{x} \arctan x^{3/2}\, dx$

37. $\displaystyle\int \frac{\ln x}{x(3 + 2\ln x)}\, dx$

38. $\displaystyle\int \frac{e^x}{(1 - e^{2x})^{3/2}}\, dx$

39. $\displaystyle\int \frac{x}{(x^2 - 6x + 10)^2}\, dx$

40. $\displaystyle\int (2x - 3)^2\sqrt{(2x - 3)^2 + 4}\, dx$

41. $\displaystyle\int \frac{x}{\sqrt{x^4 - 6x^2 + 5}}\, dx$

42. $\displaystyle\int \frac{\cos x}{\sqrt{\sin^2 x + 1}}\, dx$

43. $\displaystyle\int \frac{x^3}{\sqrt{4 - x^2}}\, dx$

44. $\displaystyle\int \sqrt{\frac{3 - x}{3 + x}}\, dx$

45. $\displaystyle\int \frac{e^{3x}}{(1 + e^x)^3}\, dx$

46. $\displaystyle\int \sec^5 \theta\, d\theta$

In Exercises 47–52, verify the integration formula.

47. $\displaystyle\int \frac{u^2}{(a + bu)^2}\, du = \frac{1}{b^3}\left(bu - \frac{a^2}{a + bu} - 2a \ln |a + bu| \right) + C$

48. $\displaystyle\int \frac{u^n}{\sqrt{a + bu}}\, du =$
$$\frac{2}{(2n + 1)b}\left(u^n\sqrt{a + bu} - na \int \frac{u^{n-1}}{\sqrt{a + bu}}\, du \right)$$

49. $\displaystyle\int \frac{1}{(u^2 \pm a^2)^{3/2}}\, du = \frac{\pm u}{a^2\sqrt{u^2 \pm a^2}} + C$

50. $\displaystyle\int u^n \cos u\, du = u^n \sin u - n \int u^{n-1} \sin u\, du$

51. $\displaystyle\int \arctan u\, du = u \arctan u - \ln \sqrt{1 + u^2} + C$

52. $\displaystyle\int (\ln u)^n\, du = u(\ln u)^n - n \int (\ln u)^{n-1}\, du$

In Exercises 53–60, evaluate the integral.

53. $\displaystyle\int \frac{1}{2 - 3\sin \theta}\, d\theta$

54. $\displaystyle\int \frac{\sin \theta}{1 + \cos^2 \theta}\, d\theta$

55. $\displaystyle\int_0^{\pi/2} \frac{1}{1 + \sin \theta + \cos \theta}\, d\theta$

56. $\displaystyle\int_0^{\pi/2} \frac{1}{3 - 2\cos \theta}\, d\theta$

57. $\displaystyle\int \frac{\sin \theta}{3 - 2\cos \theta}\, d\theta$

58. $\displaystyle\int \frac{\sin \theta}{1 + \sin \theta}\, d\theta$

59. $\displaystyle\int \frac{\cos \sqrt{\theta}}{\sqrt{\theta}}\, d\theta$

60. $\displaystyle\int \frac{1}{\sec \theta - \tan \theta}\, d\theta$

C **In Exercises 61–66, use a symbolic integration utility to evaluate the integral.**

61. $\displaystyle\int \frac{1}{x^{3/2}\sqrt{1 - x}}\, dx$

62. $\displaystyle\int x\sqrt{x^2 + 2x}\, dx$

63. $\displaystyle\int \frac{\sqrt{2 - 2x - x^2}}{x + 1}\, dx$

64. $\displaystyle\int \frac{1}{(x^2 - 6x + 10)^2}\, dx$

65. $\displaystyle\int \frac{1}{\sin \theta \tan \theta}\, d\theta$

66. $\displaystyle\int \frac{\sin \theta}{(\cos \theta)(1 + \sin \theta)}\, d\theta$

Area **In Exercises 67 and 68, find the area of the region bounded by the graphs of the given equations.**

67. $y = \dfrac{x}{\sqrt{x + 1}},\ y = 0,\ x = 8$

68. $y = \dfrac{x}{1 + e^{x^2}},\ y = 0,\ x = 2$

69. *Work*   A hydraulic cylinder on an industrial machine pushes a steel block a distance of x feet $(0 \le x \le 5)$, where the variable force required is

$$F(x) = 2000xe^{-x} \text{ lb.}$$

Find the work done in pushing the block the full 5 feet through the machine.

70. *Work*   Repeat Exercise 69, using a force of

$$F(x) = \frac{500x}{\sqrt{26 - x^2}} \text{ lb.}$$

71. *Building Design*   The cross section of a precast concrete beam for a building is bounded by the graphs of the equations

$$x = \frac{2}{\sqrt{1 + y^2}}, \quad x = \frac{-2}{\sqrt{1 + y^2}}, \quad y = 0, \quad y = 3$$

where x and y are measured in feet. The length of the beam is 20 feet (see figure).

a. Find the volume and weight of the concrete in the beam. Assume the concrete weighs 148 pounds per cubic foot.

b. Find the centroid of a cross section of the beam.

FIGURE FOR 71

72. *Average Population Size*   A population is growing according to the logistics model

$$N = \frac{5000}{1 + e^{4.8 - 1.9t}}$$

where t is time in days. Find the average population over the interval $[0, 2]$.

Indeterminate Forms and L'Hôpital's Rule

Indeterminate Forms • L'Hôpital's Rule

Indeterminate Forms

Recall from Chapter 1 that the forms $0/0$ and ∞/∞ are called *indeterminate* because they do not guarantee that a limit exists, nor do they indicate what the limit is, if one does exist. When we encountered one of these indeterminate forms earlier in the text, we attempted to rewrite the expression by using various algebraic techniques.

Indeterminate form	*Limit*	*Algebraic technique*
$\dfrac{0}{0}$	$\displaystyle\lim_{x \to -1} \frac{2x^2 - 2}{x + 1} = \lim_{x \to -1} 2(x - 1) = -4$	Divide numerator and denominator by $(x + 1)$.
$\dfrac{\infty}{\infty}$	$\displaystyle\lim_{x \to \infty} \frac{3x^2 - 1}{2x^2 + 1} = \lim_{x \to \infty} \frac{3 - (1/x^2)}{2 + (1/x^2)} = \frac{3}{2}$	Divide numerator and denominator by x^2.

Occasionally, you can extend these algebraic techniques to find limits of transcendental functions. For instance, the limit

$$\lim_{x \to 0} \frac{e^{2x} - 1}{e^x - 1}$$

produces the indeterminate form $0/0$. Factoring and then dividing produces

$$\lim_{x \to 0} \frac{e^{2x} - 1}{e^x - 1} = \lim_{x \to 0} \frac{(e^x + 1)(e^x - 1)}{e^x - 1} = \lim_{x \to 0}(e^x + 1) = 2.$$

However, not all indeterminate forms can be evaluated by algebraic manipulation. This is particularly true when *both* algebraic and transcendental functions are involved. For instance, the limit

$$\lim_{x \to 0} \frac{e^{2x} - 1}{x}$$

produces the indeterminate form $0/0$. Dividing the numerator and denominator by x to obtain

$$\lim_{x \to 0}\left(\frac{e^{2x}}{x} - \frac{1}{x}\right)$$

merely produces another indeterminate form, $\infty - \infty$. Of course, you could use technology to estimate the limit, as shown in Table 7.1. From the table, the limit appears to be 2. (This limit will be verified in Example 1.)

TABLE 7.1

x	$\dfrac{e^{2x} - 1}{x}$
-1.0	0.865
-0.1	1.813
-0.01	1.980
-0.001	1.998
0	$?$
0.001	2.002
0.01	2.020
0.1	2.214
1.0	6.389

GUILLAUME L'HÔPITAL

L'Hôpital's Rule is named after the French mathematician Guillaume Francois Antoine De L'Hôpital (1661–1704). L'Hôpital is credited with writing the first text on differential calculus (in 1696).

L'Hôpital's Rule

To find the limit illustrated in Table 7.1, you can use a theorem called **L'Hôpital's Rule.** This theorem states that under certain conditions the limit of the quotient $f(x)/g(x)$ is determined by the limit of the quotient $f'(x)/g'(x)$. To prove this theorem, you can use a more general result called the **Extended Mean Value Theorem.**

THEOREM 7.3 The Extended Mean Value Theorem

If f and g are differentiable on an open interval (a, b) and continuous on $[a, b]$ such that $g'(x) \neq 0$ for any x in (a, b), then there exists a point c in (a, b) such that

$$\frac{f'(c)}{g'(c)} = \frac{f(b) - f(a)}{g(b) - g(a)}.$$

REMARK To see why this is called the Extended Mean Value Theorem, consider the special case in which $g(x) = x$. For this case, you obtain the "standard" Mean Value Theorem as presented in Section 3.2.

The Extended Mean Value Theorem and L'Hôpital's Rule are both proved in Appendix A.

THEOREM 7.4 L'Hôpital's Rule

Let f and g be functions that are differentiable on an open interval (a, b) containing c, except possibly at c itself. Assume that $g'(x) \neq 0$ for all x in (a, b), except possibly at c itself. If the limit of $f(x)/g(x)$ as x approaches c produces the indeterminate form $0/0$, then

$$\lim_{x \to c} \frac{f(x)}{g(x)} = \lim_{x \to c} \frac{f'(x)}{g'(x)}$$

provided the limit on the right exists (or is infinite). This result also applies if the limit of $f(x)/g(x)$ as x approaches c produces any one of the indeterminate forms ∞/∞, $(-\infty)/\infty$, $\infty/(-\infty)$, or $(-\infty)/(-\infty)$.

FOR FURTHER INFORMATION
For more information on the necessity of the restriction that $g'(x)$ be nonzero for all x in (a, b), except possibly at c, see the article "Counterexamples to L'Hôpital's Rule" by R. P. Boas in the October, 1986 issue of *The American Mathematical Monthly.*

REMARK People occasionally use L'Hôpital's Rule incorrectly by applying the Quotient Rule to $f(x)/g(x)$. Be sure you see that the rule involves $f'(x)/g'(x)$, not the derivative of $f(x)/g(x)$.

L'Hôpital's Rule can also be applied to one-sided limits. For instance, if the limit of $f(x)/g(x)$ as x approaches c *from the right* produces the indeterminate form $0/0$, then

$$\lim_{x \to c^+} \frac{f(x)}{g(x)} = \lim_{x \to c^+} \frac{f'(x)}{g'(x)}$$

provided the limit exists (or is infinite).

EXAMPLE 1   Indeterminate Form 0/0

Evaluate $\displaystyle\lim_{x\to 0}\frac{e^{2x}-1}{x}$.

Solution   Because direct substitution results in the indeterminate form 0/0

$$\lim_{x\to 0}\frac{e^{2x}-1}{x}\qquad\begin{array}{l}\lim_{x\to 0}(e^{2x}-1)=0\\[2ex]\lim_{x\to 0}x=0\end{array}$$

you can apply L'Hôpital's Rule as follows.

$$\lim_{x\to 0}\frac{e^{2x}-1}{x}=\lim_{x\to 0}\frac{\dfrac{d}{dx}[e^{2x}-1]}{\dfrac{d}{dx}[x]}\qquad\text{Apply L'Hôpital's Rule}$$

$$=\lim_{x\to 0}\frac{2e^{2x}}{1}\qquad\text{Differentiate numerator and denominator}$$

$$=2\qquad\text{Evaluate the limit}$$

REMARK   In writing the string of equations in Example 1, you actually do not know that the first limit is equal to the second until you have shown that the second limit exists. In other words, if the second limit had not existed, it would not have been permissible to apply L'Hôpital's Rule.

Another form of L'Hôpital's Rule states that if the limit of $f(x)/g(x)$ as x approaches ∞ (or $-\infty$) produces the indeterminate form 0/0 or ∞/∞, then

$$\lim_{x\to\infty}\frac{f(x)}{g(x)}=\lim_{x\to\infty}\frac{f'(x)}{g'(x)}$$

provided the limit on the right exists.

EXAMPLE 2   Indeterminate Form ∞/∞

Evaluate $\displaystyle\lim_{x\to\infty}\frac{\ln x}{x}$.

Solution   Because direct substitution results in the indeterminate form ∞/∞, you can apply L'Hôpital's Rule to obtain

$$\lim_{x\to\infty}\frac{\ln x}{x}=\lim_{x\to\infty}\frac{\dfrac{d}{dx}[\ln x]}{\dfrac{d}{dx}[x]}\qquad\text{Apply L'Hôpital's Rule}$$

$$=\lim_{x\to\infty}\frac{1}{x}\qquad\text{Differentiate numerator and denominator}$$

$$=0\qquad\text{Evaluate the limit}$$

Occasionally it is necessary to apply L'Hôpital's Rule more than once to remove an indeterminate form, as illustrated in Example 3.

EXAMPLE 3   Applying L'Hôpital's Rule More Than Once

Evaluate $\displaystyle\lim_{x\to-\infty}\frac{x^2}{e^{-x}}$.

Solution   Because direct substitution results in the indeterminate form ∞/∞, you can apply L'Hôpital's Rule.

$$\lim_{x\to-\infty}\frac{x^2}{e^{-x}}=\lim_{x\to-\infty}\frac{\dfrac{d}{dx}[x^2]}{\dfrac{d}{dx}[e^{-x}]}=\lim_{x\to-\infty}\frac{2x}{-e^{-x}}$$

This limit yields the indeterminate form $(-\infty)/(-\infty)$, so you can apply L'Hôpital's Rule again to obtain

$$\lim_{x\to-\infty}\frac{2x}{-e^{-x}}=\lim_{x\to-\infty}\frac{\dfrac{d}{dx}[2x]}{\dfrac{d}{dx}[-e^{-x}]}=\lim_{x\to-\infty}\frac{2}{e^{-x}}=0.$$

In addition to the forms $0/0$ and ∞/∞, there are other indeterminate forms such as $0\cdot\infty$, 1^{∞}, ∞^0, 0^0, and $\infty-\infty$. For example, consider the following four limits that lead to the indeterminate form $0\cdot\infty$.

$$\underbrace{\lim_{x\to0}(x)\left(\frac{1}{x}\right)}_{\text{Limit is }1},\qquad \underbrace{\lim_{x\to0}(x)\left(\frac{2}{x}\right)}_{\text{Limit is }2},\qquad \underbrace{\lim_{x\to\infty}(x)\left(\frac{1}{e^x}\right)}_{\text{Limit is }0},\qquad \underbrace{\lim_{x\to\infty}(e^x)\left(\frac{1}{x}\right)}_{\text{Limit is }\infty}$$

Because each limit is different, it is clear that the form $0\cdot\infty$ is indeterminate in the sense that it does not determine the value (or even the existence) of the limit. The following examples indicate methods for evaluating these forms. Basically, you attempt to convert each of these forms to $0/0$ or ∞/∞ so that L'Hôpital's Rule can be applied.

EXAMPLE 4   Indeterminate Form $0\cdot\infty$

Evaluate $\displaystyle\lim_{x\to\infty}e^{-x}\sqrt{x}$.

Solution   Because direct substitution produces the indeterminate form $0\cdot\infty$, you should try to rewrite the limit to fit the form $0/0$ or ∞/∞. In this case, you can rewrite the limit to fit the second form.

$$\lim_{x\to\infty}e^{-x}\sqrt{x}=\lim_{x\to\infty}\frac{\sqrt{x}}{e^x}$$

Now, by L'Hôpital's Rule, you have

$$\lim_{x\to\infty}\frac{\sqrt{x}}{e^x}=\lim_{x\to\infty}\frac{1/(2\sqrt{x})}{e^x}=\lim_{x\to\infty}\frac{1}{2\sqrt{x}\,e^x}=0.$$

If rewriting a limit in one of the forms $0/0$ or ∞/∞ does not seem to work, try the other form. For instance, in Example 4 you can write the limit as

$$\lim_{x \to \infty} e^{-x} \sqrt{x} = \lim_{x \to \infty} \frac{e^{-x}}{x^{-1/2}}$$

which yields the indeterminate form $0/0$. As it happens, applying L'Hôpital's Rule to this limit produces

$$\lim_{x \to \infty} \frac{e^{-x}}{x^{-1/2}} = \lim_{x \to \infty} \frac{-e^{-x}}{-1/(2x^{3/2})}$$

which also yields the indeterminate form $0/0$.

The indeterminate forms 1^∞, ∞^0, and 0^0 arise from limits of functions that have a variable base and a variable exponent. When we encountered this type of function in Chapter 5 we used logarithmic differentiation to find the derivative. You can use a similar procedure when taking limits, as indicated in the next example. (Note that this example serves as an alternative proof of the theorem "A Limit Involving e.")

EXAMPLE 5   Indeterminate Form 1^∞

Evaluate $\displaystyle \lim_{x \to \infty} \left(1 + \frac{1}{x} \right)^x$.

Solution Because direct substitution yields the indeterminate form 1^∞, you can proceed as follows. To begin, assume the limit exists and is equal to y.

$$y = \lim_{x \to \infty} \left(1 + \frac{1}{x} \right)^x.$$

Taking the natural logarithm of both sides produces

$$\ln y = \ln \left[\lim_{x \to \infty} \left(1 + \frac{1}{x} \right)^x \right].$$

Because the natural logarithmic function is continuous, you can write the following.

$$\ln y = \lim_{x \to \infty} \left[x \ln \left(1 + \frac{1}{x} \right) \right] \qquad \text{Indeterminate form } \infty \cdot 0$$

$$= \lim_{x \to \infty} \left\{ \frac{\ln[1 + (1/x)]}{1/x} \right\} \qquad \text{Indeterminate form } 0/0$$

$$= \lim_{x \to \infty} \left[\frac{(-1/x^2)\{1/[1 + (1/x)]\}}{-1/x^2} \right] \qquad \text{L'Hôpital's Rule}$$

$$= \lim_{x \to \infty} \frac{1}{1 + (1/x)} = 1.$$

Now, because you have shown that $\ln y = 1$, you can conclude that $y = e$ and obtain

$$\lim_{x \to \infty} \left(1 + \frac{1}{x} \right)^x = e.$$

L'Hôpital's Rule can also be applied to one-sided limits, as demonstrated in Examples 6 and 7.

EXAMPLE 6   Indeterminate Form 0^0

Evaluate $\displaystyle\lim_{x \to 0^+} (\sin x)^x$.

Solution   Because direct substitution produces the indeterminate form 0^0, you can proceed as follows. To begin, assume the limit exists and is equal to y.

$$y = \lim_{x \to 0^+} (\sin x)^x \qquad \text{Indeterminate form } 0^0$$

$$\ln y = \ln\left[\lim_{x \to 0^+} (\sin x)^x\right] \qquad \text{Take log of both sides}$$

$$= \lim_{x \to 0^+}\left[\ln(\sin x)^x\right] \qquad \text{Continuity}$$

$$= \lim_{x \to 0^+}\left[x \ln(\sin x)\right] \qquad \text{Indeterminate form } 0 \cdot (-\infty)$$

$$= \lim_{x \to 0^+} \frac{\ln(\sin x)}{1/x} \qquad \text{Indeterminate form } -\infty/\infty$$

$$= \lim_{x \to 0^+} \frac{\cot x}{-1/x^2} \qquad \text{L'Hôpital's Rule}$$

$$= \lim_{x \to 0^+} \frac{-x^2}{\tan x} \qquad \text{Indeterminate form } 0/0$$

$$= \lim_{x \to 0^+} \frac{-2x}{\sec^2 x}$$

$$= 0 \qquad \text{L'Hôpital's Rule}$$

Now, because $\ln y = 0$, you can conclude that $y = e^0 = 1$, and it follows that

$$\lim_{x \to 0^+} (\sin x)^x = 1.$$

FIGURE 7.16
The limit of $(\sin x)^x$ is 1 as x approaches 0 from the right.

TECHNOLOGY   When evaluating complicated limits such as the one in Example 6, it is helpful to check the reasonableness of the solution with a calculator or with a graphing utility. For instance, the calculations shown in Table 7.2 and the graph shown in Figure 7.16 are consistent with the conclusion that $(\sin x)^x$ approaches 1 as x approaches 0 from the right.

TABLE 7.2

x	1.0	0.1	0.01	0.001	0.0001	0.00001
$(\sin x)^x$	0.8415	0.7942	0.9550	0.9931	0.9991	0.9999

Try using a computer or graphing utility to estimate the following limits.

$$\lim_{x \to 0}(1 - \cos x)^x \quad \text{and} \quad \lim_{x \to 0^+} (\tan x)^x$$

Then see if you can verify your estimate analytically.

EXAMPLE 7   Indeterminate Form ∞ − ∞

Evaluate $\lim\limits_{x\to 1^+}\left(\dfrac{1}{\ln x}-\dfrac{1}{x-1}\right)$.

Solution   Because direct substitution yields the indeterminate form $\infty - \infty$, you should try to rewrite the expression to produce a form to which you can apply L'Hôpital's Rule. In this case, you can combine the two fractions to obtain

$$\lim_{x\to 1^+}\left(\frac{1}{\ln x}-\frac{1}{x-1}\right)=\lim_{x\to 1^+}\left[\frac{x-1-\ln x}{(x-1)\ln x}\right].$$

Now, because direct substitution produces the indeterminate form $0/0$, you can apply L'Hôpital's Rule to obtain

$$\lim_{x\to 1^+}\left(\frac{1}{\ln x}-\frac{1}{x-1}\right)=\lim_{x\to 1^+}\frac{\dfrac{d}{dx}[x-1-\ln x]}{\dfrac{d}{dx}[(x-1)\ln x]}$$

$$=\lim_{x\to 1^+}\left[\frac{1-(1/x)}{(x-1)(1/x)+\ln x}\right]$$

$$=\lim_{x\to 1^+}\left(\frac{x-1}{x-1+x\ln x}\right).$$

This limit also yields the indeterminate form $0/0$, so you can apply L'Hôpital's Rule again to obtain

$$\lim_{x\to 1^+}\left(\frac{1}{\ln x}-\frac{1}{x-1}\right)=\lim_{x\to 1^+}\left[\frac{1}{1+x(1/x)+\ln x}\right]$$

$$=\frac{1}{2}.$$

In each of the examples presented so far in this section, we have used L'Hôpital's Rule to find a limit that exists. L'Hôpital's Rule can also be used to conclude that a limit is infinite, and this is demonstrated in the last example.

EXAMPLE 8   An Infinite Limit

Evaluate $\lim\limits_{x\to\infty}\dfrac{e^x}{x}$.

Solution   Because direct substitution produces the indeterminate form ∞/∞, you can apply L'Hôpital's Rule.

$$\lim_{x\to\infty}\frac{e^x}{x}=\lim_{x\to\infty}\frac{\dfrac{d}{dx}[e^x]}{\dfrac{d}{dx}[x]}$$

$$=\lim_{x\to\infty}\frac{e^x}{1}$$

$$=\lim_{x\to\infty}e^x=\infty$$

Because $e^x\to\infty$ as $x\to\infty$, you can conclude that the limit of e^x/x as $x\to\infty$ is also infinite.

We have identified the forms $0/0$, ∞/∞, $\infty - \infty$, $0 \cdot \infty$, 0^0, 1^∞, and ∞^0 as *indeterminate*. There are similar forms that you should recognize as "determinate."

$$\infty + \infty \to \infty \qquad -\infty - \infty \to -\infty \qquad 0^\infty \to 0 \qquad 0^{-\infty} \to \infty$$

(You are asked to verify two of these in Exercises 51 and 52.)

As a final comment, we remind you that L'Hôpital's Rule can be applied only to quotients leading to the indeterminate forms $0/0$ and ∞/∞. For instance, the following application of L'Hôpital's Rule is *incorrect*.

$$\lim_{x\to 0}\frac{e^x}{x} \overset{(=)}{=} \lim_{x\to 0}\frac{e^x}{1} = 1 \qquad \text{Incorrect use of L'Hôpital's Rule}$$

The reason this application is incorrect is that, even though the limit of the denominator is 0, the limit of the numerator is 1, which means that the hypotheses of L'Hôpital's Rule have not been satisfied.

EXERCISES for Section 7.7

In Exercises 1–6, evaluate the limit (a) using techniques from Chapters 1 and 3 and (b) using L'Hôpital's Rule.

1. $\lim\limits_{x\to 3}\dfrac{2(x-3)}{x^2-9}$

2. $\lim\limits_{x\to -1}\dfrac{2x^2-x-3}{x+1}$

3. $\lim\limits_{x\to 3}\dfrac{\sqrt{x+1}-2}{x-3}$

4. $\lim\limits_{x\to 0}\dfrac{\sin 4x}{2x}$

5. $\lim\limits_{x\to \infty}\dfrac{5x^2-3x+1}{3x^2-5}$

6. $\lim\limits_{x\to \infty}\dfrac{2x+1}{4x^2+x}$

In Exercises 7–26, evaluate the limit, using L'Hôpital's Rule if necessary. (In Exercise 13, n is a positive integer.)

7. $\lim\limits_{x\to 2}\dfrac{x^2-x-2}{x-2}$

8. $\lim\limits_{x\to -1}\dfrac{x^2-x-2}{x+1}$

9. $\lim\limits_{x\to 0}\dfrac{\sqrt{4-x^2}-2}{x}$

10. $\lim\limits_{x\to 2^-}\dfrac{\sqrt{4-x^2}}{x-2}$

11. $\lim\limits_{x\to 0}\dfrac{e^x-(1-x)}{x}$

12. $\lim\limits_{x\to 0^+}\dfrac{e^x-(1+x)}{x^3}$

13. $\lim\limits_{x\to 0^+}\dfrac{e^x-(1+x)}{x^n}$

14. $\lim\limits_{x\to 1}\dfrac{\ln x}{x^2-1}$

15. $\lim\limits_{x\to 0}\dfrac{\sin 2x}{\sin 3x}$

16. $\lim\limits_{x\to 0}\dfrac{\sin ax}{\sin bx}$

17. $\lim\limits_{x\to 0}\dfrac{\arcsin x}{x}$

18. $\lim\limits_{x\to 1}\dfrac{\arctan x - (\pi/4)}{x-1}$

19. $\lim\limits_{x\to \infty}\dfrac{3x^2-2x+1}{2x^2+3}$

20. $\lim\limits_{x\to \infty}\dfrac{x-1}{x^2+2x+3}$

21. $\lim\limits_{x\to \infty}\dfrac{x^2+2x+3}{x-1}$

22. $\lim\limits_{x\to \infty}\dfrac{x^2}{e^x}$

23. $\lim\limits_{x\to \infty}\dfrac{x}{\sqrt{x^2+1}}$

24. $\lim\limits_{x\to \infty}\dfrac{\sin x}{x-\pi}$

25. $\lim\limits_{x\to \infty}\dfrac{\ln x}{x}$

26. $\lim\limits_{x\to \infty}\dfrac{e^x}{x}$

In Exercises 27–40, describe the type of indeterminate form (if any) that is obtained with direct substitution. Then evaluate the limit, using L'Hôpital's Rule when necessary. (For a geometric approach to Exercise 27, see the article by John H. Mathews in the May, 1992 issue of *The College Mathematics Journal*.)

27. $\lim\limits_{x\to 0^+}(-x\ln x)$

28. $\lim\limits_{x\to 0^+}x^2\cot x$

29. $\lim\limits_{x\to 0}\left(x\sin\dfrac{1}{x}\right)$

30. $\lim\limits_{x\to \infty}x\tan\dfrac{1}{x}$

31. $\lim\limits_{x\to 0^+}x^{1/x}$

32. $\lim\limits_{x\to 0^+}(e^x+x)^{1/x}$

33. $\lim\limits_{x\to \infty}x^{1/x}$

34. $\lim\limits_{x\to \infty}\left(1+\dfrac{1}{x}\right)^x$

35. $\lim\limits_{x\to 0^+}(1+x)^{1/x}$

36. $\lim\limits_{x\to \infty}(1+x)^{1/x}$

37. $\lim\limits_{x\to 2^+}\left(\dfrac{8}{x^2-4}-\dfrac{x}{x-2}\right)$

38. $\lim\limits_{x\to 2^+}\left(\dfrac{1}{x^2-4}-\dfrac{\sqrt{x-1}}{x^2-4}\right)$

39. $\lim\limits_{x\to 1^+}\left(\dfrac{3}{\ln x}-\dfrac{2}{x-1}\right)$

40. $\lim\limits_{x\to 0^+}\left(\dfrac{1}{x}-\dfrac{1}{x^2}\right)$

41. Find differentiable functions f and g that satisfy the specified condition such that $\lim\limits_{x\to 5}f(x)=0$ and $\lim\limits_{x\to 5}g(x)=0$.

a. $\lim\limits_{x\to 5}\dfrac{f(x)}{g(x)}=10$ **b.** $\lim\limits_{x\to 5}\dfrac{f(x)}{g(x)}=0$

c. $\lim\limits_{x\to 5}\dfrac{f(x)}{g(x)}=\infty$

(*Note*: There are many correct answers.)

42. Find differentiable functions f and g such that

$$\lim\limits_{x\to \infty}f(x)=\lim\limits_{x\to \infty}g(x)=\infty \text{ and } \lim\limits_{x\to \infty}[f(x)-g(x)]=25.$$

(*Note*: There are many correct answers.)

In Exercises 43–48, use L'Hôpital's Rule to determine the comparative rates of increase of the functions

$$f(x) = x^m, \quad g(x) = e^{nx}, \quad \text{and} \quad h(x) = (\ln x)^n$$

where $0 < n, 0 < m$, and $x \to \infty$. The limits obtained in these exercises suggest that $(\ln x)^n$ approaches infinity more slowly than x^m, which, in turn, approaches infinity more slowly than e^{nx}.

43. $\lim\limits_{x \to \infty} \dfrac{x^2}{e^{5x}}$

44. $\lim\limits_{x \to \infty} \dfrac{x^3}{e^{2x}}$

45. $\lim\limits_{x \to \infty} \dfrac{(\ln x)^3}{x}$

46. $\lim\limits_{x \to \infty} \dfrac{(\ln x)^2}{x^3}$

47. $\lim\limits_{x \to \infty} \dfrac{(\ln x)^n}{x^m}$

48. $\lim\limits_{x \to \infty} \dfrac{x^m}{e^{nx}}$

49. Complete the table to show that x eventually "overpowers" $(\ln x)^4$.

x	10	10^2	10^4	10^6	10^8	10^{10}
$\dfrac{(\ln x)^4}{x}$						

50. Complete the table to show that e^x eventually "overpowers" x^5.

x	1	5	10	20	30	40	50	100
$\dfrac{e^x}{x^5}$								

51. Prove that if $f(x) \geq 0$, $\lim\limits_{x \to a} f(x) = 0$, and $\lim\limits_{x \to a} g(x) = \infty$, then $\lim\limits_{x \to a} f(x)^{g(x)} = 0$.

52. Prove that if $f(x) \geq 0$, $\lim\limits_{x \to a} f(x) = 0$, and $\lim\limits_{x \to a} g(x) = -\infty$, then

$$\lim\limits_{x \to a} f(x)^{g(x)} = \infty.$$

In Exercises 53–56, find any asymptotes and relative extrema that may exist and sketch the graph of the function. (*Hint:* Some of the limits required in finding asymptotes have been found in preceding exercises.)

53. $y = x^{1/x}, \quad x > 0$

54. $y = x^x, \quad x > 0$

55. $y = 2xe^{-x}$

56. $y = \dfrac{\ln x}{x}$

In Exercises 57–60, L'Hôpital's Rule is used *incorrectly*. Describe the error.

57. $\lim\limits_{x \to 0} \dfrac{e^{2x} - 1}{e^x} = \lim\limits_{x \to 0} \dfrac{2e^{2x}}{e^x} = \lim\limits_{x \to 0} 2e^x = 2$

58. $\lim\limits_{x \to 0} \dfrac{\sin \pi x - 1}{x} = \lim\limits_{x \to 0} \dfrac{\pi \cos \pi x}{1} = \pi$

59. $\lim\limits_{x \to \infty} x \cos \dfrac{1}{x} = \lim\limits_{x \to \infty} \dfrac{\cos(1/x)}{1/x}$

$$= \lim\limits_{x \to \infty} \dfrac{[-\sin(1/x)](1/x^2)}{-1/x^2} = 0$$

60. $\lim\limits_{x \to \infty} \dfrac{e^{-x}}{1 + e^{-x}} = \lim\limits_{x \to \infty} \dfrac{-e^{-x}}{-e^{-x}} = \lim\limits_{x \to \infty} 1 = 1$

61. Show that L'Hôpital's Rule fails for the limit

$$\lim\limits_{x \to \infty} \dfrac{x}{\sqrt{x^2 + 1}}.$$

What is this limit?

62. *Compound Interest* The formula for the amount A in a savings account compounded n times a year for t years at an interest rate r and an initial deposit of P is

$$A = P\left(1 + \dfrac{r}{n}\right)^{nt}.$$

Use L'Hôpital's Rule to show that the limiting formula as the number of compoundings per year becomes infinite is

$$A = Pe^{rt}.$$

63. *Velocity in a Resisting Medium* The velocity of an object falling through a resisting medium such as air or water is given by

$$v = \dfrac{32}{k}\left(1 - e^{-kt} + \dfrac{v_0 k e^{-kt}}{32}\right)$$

where v_0 is the initial velocity, t is time in seconds, and k is the resistance constant of the medium. Use L'Hôpital's Rule to find the formula for the velocity of a falling body in a vacuum by fixing v_0 and t and letting k approach 0. (Assume the downward direction is positive.)

64. *The Gamma Function* The Gamma Function $\Gamma(n)$ is defined in terms of the integral of the function

$$f(x) = x^{n-1}e^{-x}, \quad n > 0.$$

Show that for any fixed value of n, the limit of $f(x)$ as x approaches infinity is 0.

65. *Area* Find the limit as x approaches 0 of the ratio of the area of the triangle to the total shaded area in the accompanying figure.

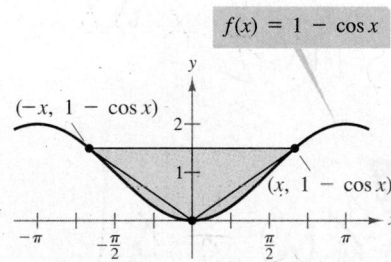

$f(x) = 1 - \cos x$

$(-x, 1 - \cos x)$

$(x, 1 - \cos x)$

FIGURE FOR 65

66. Use a graphing utility to graph

$$f(x) = \dfrac{x^k - 1}{k}$$

for $k = 1, 0.1$, and 0.01. Then evaluate the limit

$$\lim\limits_{k \to 0^+} \dfrac{x^k - 1}{k}.$$

C In Exercises 67–70, use a graphing utility to (a) graph the function, and (b) find the required limit (if it exists).

67. $\lim\limits_{x \to 3} \dfrac{x - 3}{\ln(2x - 5)}$

68. $\lim\limits_{x \to 0^+} (\sin x)^x$

69. $\lim\limits_{x \to \infty} \left(\sqrt{x^2 + 5x + 2} - x\right)$

70. $\lim\limits_{x \to \infty} \dfrac{x^3}{e^{2x}}$

In Exercises 71–74, apply the Extended Mean Value Theorem to the functions f and g on the specified interval. Find all values c in the interval (a, b) such that

$$\frac{f'(c)}{g'(c)} = \frac{f(b) - f(a)}{g(b) - g(a)}.$$

Functions	Interval
71. $f(x) = x^3$ $\quad$ $g(x) = x^2 + 1$	$[0, 1]$
72. $f(x) = \dfrac{1}{x}$ $\quad$ $g(x) = x^2 - 4$	$[1, 2]$
73. $f(x) = \sin x$ $\quad$ $g(x) = \cos x$	$\left[0, \dfrac{\pi}{2}\right]$
74. $f(x) = \ln x$ $\quad$ $g(x) = x^3$	$[1, 4]$

True or False In Exercises 75–78, determine whether the statement is true or false. If it is false, explain why or give an example that shows it is false.

75. $\lim\limits_{x \to 0} \left[\dfrac{x^2 + x + 1}{x}\right] = \lim\limits_{x \to 0} \left[\dfrac{2x + 1}{1}\right] = 1$

76. If $y = e^x/x^2$, then $y' = e^x/2x$.

77. If $p(x)$ is a polynomial, then $\lim\limits_{x \to \infty} [p(x)/e^x] = 0$.

78. If $\lim\limits_{x \to \infty} \dfrac{f(x)}{g(x)} = 1$, then $\lim\limits_{x \to \infty} [f(x) - g(x)] = 0$.

79. In Chapter 1 we used a geometrical argument (see figure) to prove that

$$\lim\limits_{\theta \to 0} \frac{\sin \theta}{\theta} = 1.$$

a. Express the area of triangle $\triangle ABD$ in terms of θ.
b. Express the area of the shaded region in terms of θ.
c. Find the ratio R of the area of $\triangle ABD$ to that of the shaded region.
d. What is $\lim\limits_{\theta \to 0} R$?

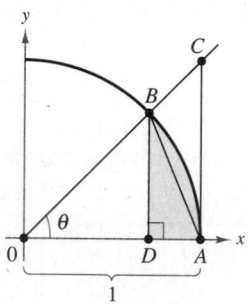

FIGURE FOR 79

80. Sketch the graph of

$$g(x) = \begin{cases} e^{-1/x^2}, & x \neq 0 \\ 0, & x = 0 \end{cases}$$

and then determine $g'(0)$.

81. Prove the following generalization of the Mean Value Theorem. If f is twice differentiable on the closed interval $[a, b]$, then

$$f(b) - f(a) = f'(a)(b - a) - \int_a^b f''(t)(t - b)\, dt.$$

82. Show that the indeterminate form 0^0 is not always equal to 1 by evaluating the following limit.

$$\lim\limits_{x \to 0^+} x^{\ln 2/(1 + \ln x)}$$

Improper Integrals

The definition of a definite integral

$$\int_a^b f(x)\, dx$$

requires that the interval $[a, b]$ be finite. Furthermore, the Fundamental Theorem of Calculus, by which we have been evaluating definite integrals, requires that f be continuous on $[a, b]$. In this section you will study a limit procedure for evaluating integrals that do not satisfy these requirements—usually because either one or both of the limits of integration are infinite, or f has a finite number of infinite discontinuities in the interval $[a, b]$. Integrals that possess either property are **improper integrals.** Note that a function f is said to have an **infinite discontinuity** at c if, *from the right or left,*

$$\lim_{x \to c} f(x) = \infty$$

or

$$\lim_{x \to c} f(x) = -\infty.$$

EXAMPLE 1   Examples of Improper Integrals

a. The integrals

$$\int_1^\infty \frac{dx}{x} \quad \text{and} \quad \int_{-\infty}^\infty \frac{dx}{x^2 + 1}$$

are improper because one or both of the limits of integration are infinite.

b. The integrals

$$\int_1^5 \frac{dx}{\sqrt{x - 1}} \quad \text{and} \quad \int_{-2}^2 \frac{dx}{(x + 1)^2}$$

are improper because the integrands have infinite discontinuities somewhere in the interval of integration. The function

$$f(x) = \frac{1}{\sqrt{x - 1}}$$

has an infinite discontinuity at $x = 1$, and the function

$$f(x) = \frac{1}{(x + 1)^2}$$

has an infinite discontinuity at $x = -1$.

FIGURE 7.17
The unbounded region has an area of 1.

FIGURE 7.18

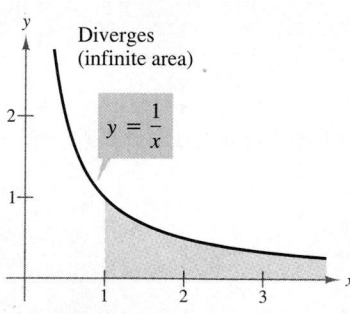

FIGURE 7.19

REMARK   Notice that although the areas indicated in Figures 7.18 and 7.19 appear to be similar, the area indicated in Figure 7.18 is finite (the improper integral converges) whereas the area indicated in Figure 7.19 is infinite (the improper integral diverges).

Infinite Limits of Integration

To get an idea of how to evaluate an improper integral, consider the integral

$$\int_1^b \frac{dx}{x^2} = \left[-\frac{1}{x}\right]_1^b = -\frac{1}{b} + 1 = 1 - \frac{1}{b}$$

which can be interpreted as the area of the shaded region shown in Figure 7.17. Taking the limit as $b \to \infty$ produces

$$\int_1^\infty \frac{dx}{x^2} = \lim_{b\to\infty}\left(\int_1^b \frac{dx}{x^2}\right) = \lim_{b\to\infty}\left(1 - \frac{1}{b}\right) = 1.$$

This improper integral can be interpreted as the area of the *unbounded* region between the graph of $f(x) = 1/x^2$ and the x-axis (to the right of $x = 1$).

Definition of Improper Integrals with Infinite Integration Limits

1. If f is continuous on the interval $[a, \infty)$, then

$$\int_a^\infty f(x)\,dx = \lim_{b\to\infty}\int_a^b f(x)\,dx.$$

2. If f is continuous on the interval $(-\infty, b]$, then

$$\int_{-\infty}^b f(x)\,dx = \lim_{a\to-\infty}\int_a^b f(x)\,dx.$$

3. If f is continuous on the interval $(-\infty, \infty)$, then

$$\int_{-\infty}^\infty f(x)\,dx = \int_{-\infty}^c f(x)\,dx + \int_c^\infty f(x)\,dx$$

where c is any real number.

In the first two cases, the improper integral **converges** if the limit exists—otherwise, the improper integral **diverges.** In the third case, the improper integral on the left diverges if either of the improper integrals on the right diverges.

EXAMPLE 2   An Improper Integral that Diverges

Evaluate $\displaystyle\int_1^\infty \frac{dx}{x}$.

Solution

$$\int_1^\infty \frac{dx}{x} = \lim_{b\to\infty}\int_1^b \frac{dx}{x} \qquad \text{Take limit as } b \to \infty$$

$$= \lim_{b\to\infty}\left[\ln x\right]_1^b \qquad \text{Apply Log Rule}$$

$$= \lim_{b\to\infty}(\ln b - 0) \qquad \text{Apply Fundamental Theorem of Calculus}$$

$$= \infty \qquad \text{Evaluate limit}$$

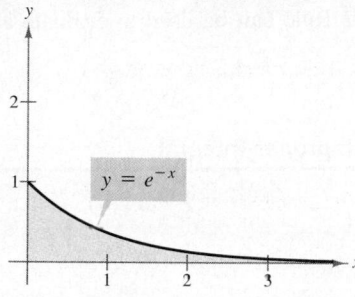

FIGURE 7.20
The area of the unbounded region is 1.

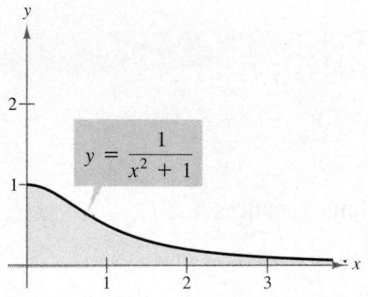

FIGURE 7.21
The area of the unbounded region is $\pi/2$.

EXAMPLE 3   Improper Integrals that Converge

Evaluate the improper integrals.

a. $\displaystyle\int_0^\infty e^{-x}\,dx$ **b.** $\displaystyle\int_0^\infty \frac{1}{x^2+1}\,dx$

Solution

a. $\displaystyle\int_0^\infty e^{-x}\,dx = \lim_{b\to\infty}\int_0^b e^{-x}\,dx$

$$= \lim_{b\to\infty}\left[-e^{-x}\right]_0^b$$

$$= \lim_{b\to\infty}(-e^{-b}+1)$$

$$= 1$$

See Figure 7.20.

b. $\displaystyle\int_0^\infty \frac{1}{x^2+1}\,dx = \lim_{b\to\infty}\int_0^b \frac{1}{x^2+1}\,dx$

$$= \lim_{b\to\infty}\left[\arctan x\right]_0^b$$

$$= \lim_{b\to\infty}\arctan b$$

$$= \frac{\pi}{2}$$

See Figure 7.21.

EXAMPLE 4   An Improper Integral that Diverges

Evaluate $\displaystyle\int_0^\infty \sin x\,dx$.

Solution

$$\int_0^\infty \sin x\,dx = \lim_{b\to\infty}\int_0^b \sin x\,dx$$

$$= \lim_{b\to\infty}\left[-\cos x\right]_0^b$$

$$= \lim_{b\to\infty}(1-\cos b)$$

Because $\cos b$ does not approach a limit as b approaches infinity, the improper integral diverges.

REMARK   Note that the improper integrals in Examples 2 and 4 diverge for different reasons. In Example 2, the improper integral $\int_1^\infty (1/x)\,dx$ diverges because the region (to the right of 1) between $y = 1/x$ and the x-axis has an infinite area. In Example 4, the improper integral $\int_0^\infty \sin x\,dx$ diverges because of the oscillation of the sine function. This type of divergence is sometimes called *divergence by oscillation*.

In the following example, note how L'Hôpital's Rule can be used to evaluate an improper integral.

EXAMPLE 5   Using L'Hôpital's Rule with an Improper Integral

Evaluate $\displaystyle\int_{1}^{\infty} (1 - x)e^{-x}\, dx.$

Solution   Use integration by parts, with $dv = e^{-x}\, dx$ and $u = (1 - x).$

$$\int (1 - x)e^{-x}\, dx = -e^{-x}(1 - x) - \int e^{-x}\, dx$$

$$= -e^{-x} + xe^{-x} + e^{-x} + C$$

$$= xe^{-x} + C$$

Now, apply the definition of an improper integral.

$$\int_{1}^{\infty} (1 - x)e^{-x}\, dx = \lim_{b \to \infty}\left[xe^{-x} \right]_{1}^{b} = \left(\lim_{b \to \infty} \frac{b}{e^{b}} \right) - \frac{1}{e}.$$

Finally, using L'Hôpital's Rule on the right-hand limit produces

$$\lim_{b \to \infty} \frac{b}{e^{b}} = \lim_{b \to \infty} \frac{1}{e^{b}} = 0$$

from which you can conclude that

$$\int_{1}^{\infty} (1 - x)e^{-x}\, dx = -\frac{1}{e}.$$

(See Figure 7.22.)

FIGURE 7.22
The area of the unbounded region is $|-1/e|.$

EXAMPLE 6   Infinite Upper and Lower Limits of Integration

Evaluate $\displaystyle\int_{-\infty}^{\infty} \frac{e^{x}}{1 + e^{2x}}\, dx.$

Solution   Note that the integrand is continuous on $(-\infty, \infty).$ To evaluate the integral, you can break it into two parts, choosing $c = 0$ as a convenient value.

$$\int_{-\infty}^{\infty} \frac{e^{x}}{1 + e^{2x}}\, dx = \int_{-\infty}^{0} \frac{e^{x}}{1 + e^{2x}}\, dx + \int_{0}^{\infty} \frac{e^{x}}{1 + e^{2x}}\, dx$$

$$= \lim_{b \to -\infty}\left[\arctan e^{x} \right]_{b}^{0} + \lim_{b \to \infty}\left[\arctan e^{x} \right]_{0}^{b}$$

$$= \lim_{b \to -\infty}\left(\frac{\pi}{4} - \arctan e^{b} \right) + \lim_{b \to \infty}\left(\arctan e^{b} - \frac{\pi}{4} \right)$$

$$= \frac{\pi}{4} - 0 + \frac{\pi}{2} - \frac{\pi}{4}$$

$$= \frac{\pi}{2}$$

(See Figure 7.23.)

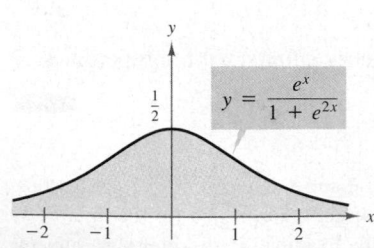

FIGURE 7.23
The area of the unbounded region is $\pi/2.$

EXAMPLE 7   Sending a Space Module into Orbit

In Example 3 of Section 6.5, we found that it would require 10,000 mile-tons of work to propel a 15-ton space module to a height of 800 miles above the earth. How much work is required to propel the module an unlimited distance away from the earth's surface? (See Figure 7.24.)

Solution   At first you might think that an infinite amount of work would be required. But if this were the case, then it would be impossible to send rockets into outer space. Because this has been done, the work required must be finite. You can determine the work in the following manner. Using the integral of Example 3, Section 6.5, replace the upper bound of 4800 miles by ∞ and write

$$W = \int_{4000}^{\infty} \frac{240{,}000{,}000}{x^2}\,dx$$

$$= \lim_{b\to\infty}\left[-\frac{240{,}000{,}000}{x}\right]_{4000}^{b}$$

$$= \lim_{b\to\infty}\left(-\frac{240{,}000{,}000}{b} + \frac{240{,}000{,}000}{4000}\right)$$

$$= 60{,}000 \text{ mile-tons}$$

$$\approx 6.336 \times 10^{11} \text{ ft}\cdot\text{lb.}$$

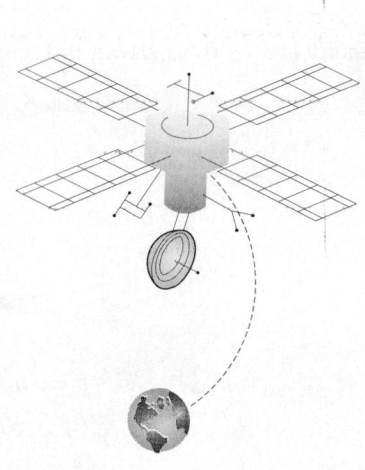

FIGURE 7.24
The space module is propelled an unlimited distance away from the earth.

Infinite Discontinuities

The second basic type of improper integral is one that has an infinite discontinuity *at or between* the limits of integration.

Definition of Improper Integrals with Infinite Discontinuities

1. If f is continuous on the interval $[a, b)$ and has an infinite discontinuity at b, then

$$\int_{a}^{b} f(x)\,dx = \lim_{c\to b^-} \int_{a}^{c} f(x)\,dx.$$

2. If f is continuous on the interval $(a, b]$ and has an infinite discontinuity at a, then

$$\int_{a}^{b} f(x)\,dx = \lim_{c\to a^+} \int_{c}^{b} f(x)\,dx.$$

3. If f is continuous on the interval $[a, b]$, except for some c in (a, b) at which f has an infinite discontinuity, then

$$\int_{a}^{b} f(x)\,dx = \int_{a}^{c} f(x)\,dx + \int_{c}^{b} f(x)\,dx.$$

In the first two cases, the improper integral **converges** if the limit exists—otherwise, the improper integral **diverges.** In the third case, the improper integral on the left diverges if either of the improper integrals on the right diverges.

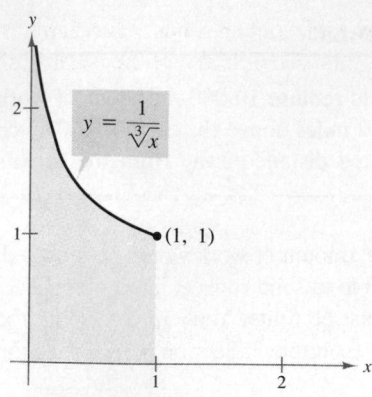

FIGURE 7.25
The area of the unbounded region is 3/2.

EXAMPLE 8 An Improper Integral with an Infinite Discontinuity

Evaluate $\displaystyle\int_0^1 \frac{dx}{\sqrt[3]{x}}$.

Solution The integrand has an infinite discontinuity at $x = 0$, as shown in Figure 7.25. You can evaluate this integral as follows.

$$\int_0^1 x^{-1/3}\, dx = \lim_{b \to 0^+}\left[\frac{x^{2/3}}{2/3}\right]_b^1$$

$$= \lim_{b \to 0^+}\frac{3}{2}(1 - b^{2/3})$$

$$= \frac{3}{2}$$

EXAMPLE 9 An Improper Integral that Diverges

Evaluate $\displaystyle\int_0^2 \frac{dx}{x^3}$.

Solution Because the integrand has an infinite discontinuity at $x = 0$, you can write the following.

$$\int_0^2 \frac{dx}{x^3} = \lim_{b \to 0^+}\left[-\frac{1}{2x^2}\right]_b^2 = \lim_{b \to 0^+}\left(-\frac{1}{8} + \frac{1}{2b^2}\right) = \infty$$

Thus, you can conclude that the improper integral diverges.

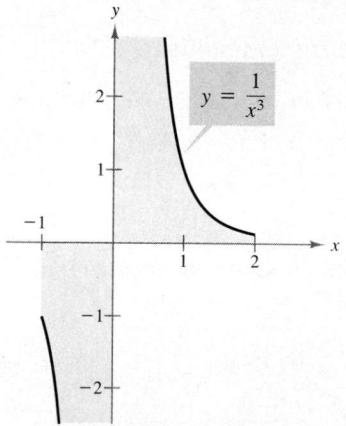

FIGURE 7.26
The improper integral $\int_{-1}^2 1/x^3\, dx$ diverges.

EXAMPLE 10 An Improper Integral with an Interior Discontinuity

Evaluate $\displaystyle\int_{-1}^2 \frac{dx}{x^3}$.

Solution This integral is improper because the integrand has an infinite discontinuity at the interior point $x = 0$, as shown in Figure 7.26. Thus, you can write the following.

$$\int_{-1}^2 \frac{dx}{x^3} = \int_{-1}^0 \frac{dx}{x^3} + \int_0^2 \frac{dx}{x^3}$$

From Example 9 you know that the second integral diverges. Therefore, the original improper integral also diverges.

REMARK Remember to check for infinite discontinuities at interior points as well as endpoints when determining whether an integral is improper. For instance, if you had not recognized that the integral in Example 10 was improper, you would have obtained the *incorrect* result

$$\int_{-1}^2 \frac{dx}{x^3} \stackrel{?}{=} \left[\frac{-1}{2x^2}\right]_{-1}^2 = -\frac{1}{8} + \frac{1}{2} = \frac{3}{8} \quad \text{Incorrect evaluation}$$

The integral in the next example is improper for *two* reasons. One limit of integration is infinite, and the integrand has an infinite discontinuity at the other limit of integration, as shown in Figure 7.27.

EXAMPLE 11   A Doubly Improper Integral

Evaluate $\displaystyle \int_0^\infty \frac{dx}{\sqrt{x}(x + 1)}$.

Solution   To evaluate this integral, split it at a convenient point (say, $x = 1$) and write

$$\int_0^\infty \frac{dx}{\sqrt{x}(x + 1)} = \int_0^1 \frac{dx}{\sqrt{x}(x + 1)} + \int_1^\infty \frac{dx}{\sqrt{x}(x + 1)}$$

$$= \lim_{b \to 0^+} \left[2\arctan\sqrt{x} \right]_b^1 + \lim_{c \to \infty} \left[2\arctan\sqrt{x} \right]_1^c$$

$$= 2\left(\frac{\pi}{4}\right) - 0 + 2\left(\frac{\pi}{2}\right) - 2\left(\frac{\pi}{4}\right)$$

$$= \pi.$$

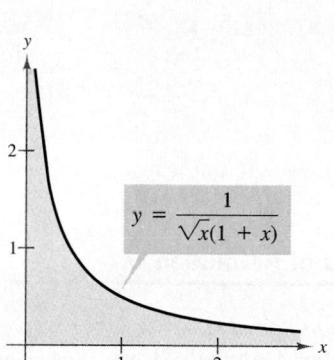

FIGURE 7.27
The area of the unbounded region is π.

EXAMPLE 12   An Application Involving Arc Length

Use the formula for arc length to show that the circumference of the circle $x^2 + y^2 = 1$ is 2π.

Solution   To simplify the work, consider the quarter circle given by $y = \sqrt{1 - x^2}$, where $0 \le x \le 1$. The function y is differentiable for any x in this interval except $x = 1$. Therefore, the arc length of the quarter circle is given by

$$s = \lim_{b \to 1^-} \int_0^b \sqrt{1 + (y')^2}\, dx$$

$$= \int_0^1 \sqrt{1 + (y')^2}\, dx$$

$$= \int_0^1 \sqrt{1 + \left(\frac{-x}{\sqrt{1 - x^2}}\right)^2}\, dx$$

$$= \int_0^1 \frac{dx}{\sqrt{1 - x^2}}.$$

This integral is improper because it has an infinite discontinuity at $x = 1$. Thus, you can write

$$s = \int_0^1 \frac{dx}{\sqrt{1 - x^2}}$$

$$= \lim_{b \to 1^-} \left[\arcsin x \right]_0^b$$

$$= \frac{\pi}{2} - 0 = \frac{\pi}{2}.$$

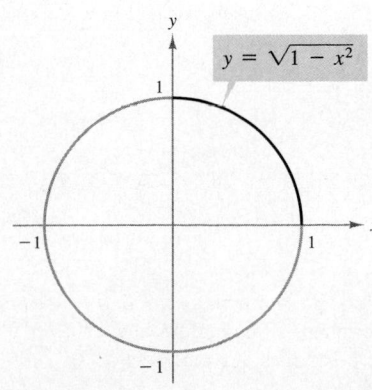

FIGURE 7.28
The circumference of the circle is 2π.

Finally, multiplying by 4, you can conclude that the circumference of the circle is $4s = 2\pi$, as shown in Figure 7.28.

We conclude this section with a useful theorem describing the convergence or divergence of a common type of improper integral. The proof of this theorem is left as an exercise (see Exercise 33).

THEOREM 7.5   A Special Type of Improper Integral

$$\int_{1}^{\infty} \frac{dx}{x^p} = \begin{cases} \dfrac{1}{p-1}, & \text{if } p > 1 \\ \text{diverges}, & \text{if } p \le 1 \end{cases}$$

EXAMPLE 13   An Application Involving a Solid of Revolution

FOR FURTHER INFORMATION
For more information on solids that have a finite volume and an infinite surface area, see the article "Supersolids: Solids Having Finite Volume and Infinite Surfaces" by William P. Love in the January, 1989 issue of *Mathematics Teacher*.

The solid formed by revolving (about the x-axis) the *unbounded* region lying between the graph of $f(x) = 1/x$ and the x-axis ($x \ge 1$) is called **Gabriel's Horn.** (See Figure 7.29.) Show that this solid has a finite volume and an infinite surface area.

Solution   Using the disc method and Theorem 7.5, you can determine the volume to be

$$V = \pi \int_{1}^{\infty} \left(\frac{1}{x}\right)^2 dx \qquad \text{Theorem 7.5, } p = 2 > 1$$

$$= \pi\left(\frac{1}{2-1}\right)$$

$$= \pi.$$

The surface area is given by

$$S = 2\pi \int_{1}^{\infty} f(x)\sqrt{1 + [f'(x)]^2}\, dx$$

$$= 2\pi \int_{1}^{\infty} \frac{1}{x}\sqrt{1 + \frac{1}{x^4}}\, dx.$$

Because

$$\sqrt{1 + \frac{1}{x^4}} > 1$$

on the interval $[1, \infty)$, and the improper integral

$$\int_{1}^{\infty} \frac{1}{x}\, dx$$

diverges, you can conclude that the improper integral

$$\int_{1}^{\infty} \frac{1}{x}\sqrt{1 + \frac{1}{x^4}}\, dx$$

also diverges. (See Exercise 36.) Thus, the surface area is infinite.

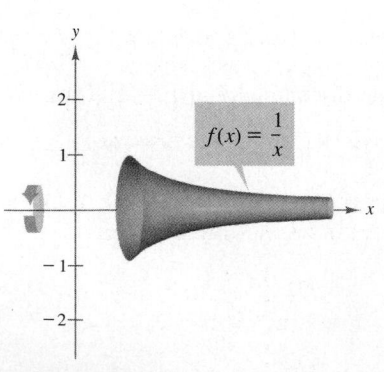

FIGURE 7.29
Gabriel's Horn has a finite volume and an infinite surface area.

1-33 every 3rd, 55, 56

EXERCISES for Section 7.8

In Exercises 1–6, state why the integral is improper and determine its divergence or convergence. Evaluate the integral if it converges.

1. $\int_0^4 \frac{1}{\sqrt{x}}dx$

2. $\int_3^4 \frac{1}{(x-3)^{3/2}}dx$

3. $\int_0^2 \frac{1}{(x-1)^2}dx$

4. $\int_0^2 \frac{1}{(x-1)^{2/3}}dx$

5. $\int_0^\infty e^{-x}dx$

6. $\int_{-\infty}^0 e^{2x}dx$

In Exercises 7–20, determine the divergence or convergence of the improper integral. Evaluate the integral if it converges.

7. $\int_{-\infty}^0 xe^{-2x}dx$

8. $\int_0^\infty xe^{-x}dx$

9. $\int_0^\infty x^2 e^{-x}dx$

10. $\int_0^\infty (x-1)e^{-x}dx$

11. $\int_1^\infty \frac{1}{x^2}dx$

12. $\int_1^\infty \frac{1}{\sqrt{x}}dx$

13. $\int_0^\infty e^{-x}\cos x\,dx$

14. $\int_0^\infty e^{-ax}\sin bx\,dx, \quad a>0$

15. $\int_{-\infty}^\infty \frac{1}{1+x^2}dx$

16. $\int_0^\infty \frac{x^3}{(x^2+1)^2}dx$

17. $\int_0^\infty \frac{1}{e^x+e^{-x}}dx$

18. $\int_0^\infty \frac{e^x}{1+e^x}dx$

19. $\int_0^\infty \cos \pi x\,dx$

20. $\int_0^\infty \sin\frac{x}{2}dx$

In Exercises 21–32, determine the divergence or convergence of the improper integral. Evaluate the integral if it converges.

21. $\int_0^1 \frac{1}{x^2}dx$

22. $\int_0^1 \frac{1}{x}dx$

23. $\int_0^8 \frac{1}{\sqrt[3]{8-x}}dx$

24. $\int_0^e \ln x\,dx$

25. $\int_0^1 x\ln x\,dx$

26. $\int_0^{\pi/2} \sec\theta\,d\theta$

27. $\int_0^{\pi/2} \tan\theta\,d\theta$

28. $\int_0^2 \frac{1}{\sqrt{4-x^2}}dx$

29. $\int_2^4 \frac{1}{\sqrt{x^2-4}}dx$

30. $\int_0^2 \frac{1}{4-x^2}dx$

31. $\int_0^2 \frac{1}{\sqrt[3]{x-1}}dx$

32. $\int_0^2 \frac{1}{(x-1)^{4/3}}dx$

33. Determine all values of p for which the improper integral converges.

$$\int_1^\infty \frac{1}{x^p}dx$$

34. Determine all values of p for which the improper integral converges.

$$\int_0^1 \frac{1}{x^p}dx$$

35. Use mathematical induction to verify that the improper integral converges for any positive integer n.

$$\int_0^\infty x^n e^{-x}dx$$

36. Given continuous functions f and g such that $0 \le f(x) \le g(x)$ on the interval $[a, \infty)$, prove the following.
   a. If $\int_a^\infty g(x)\,dx$ converges, then $\int_a^\infty f(x)\,dx$ converges.
   b. If $\int_a^\infty f(x)\,dx$ diverges, then $\int_a^\infty g(x)\,dx$ diverges.

In Exercises 37–46, use the results of Exercises 33–36 to determine whether the improper integral converges.

37. $\int_0^1 \frac{1}{x^3}dx$

38. $\int_0^1 \frac{1}{\sqrt[3]{x}}dx$

39. $\int_1^\infty \frac{1}{x^3}dx$

40. $\int_0^\infty x^4 e^{-x}dx$

41. $\int_1^\infty \frac{1}{x^2+5}dx$

42. $\int_2^\infty \frac{1}{\sqrt{x-1}}dx$

43. $\displaystyle\int_2^\infty \frac{1}{\sqrt[3]{x(x-1)}}\,dx$ **44.** $\displaystyle\int_1^\infty \frac{1}{\sqrt{x}(x+1)}\,dx$

45. $\displaystyle\int_0^\infty e^{-x^2}\,dx$ **46.** $\displaystyle\int_2^\infty \frac{1}{\sqrt{x}\ln x}\,dx$

Laplace Transforms   Let $f(t)$ be a function defined for all positive values of t. The Laplace Transform of $f(t)$ is defined by

$$F(s) = \int_0^\infty e^{-st}f(t)\,dt$$

if the improper integral exists. In Exercises 47–54, find the Laplace Transform of the function.

47. $f(t) = 1$ **48.** $f(t) = t$

49. $f(t) = t^2$ **50.** $f(t) = e^{at}$

51. $f(t) = \cos at$ **52.** $f(t) = \sin at$

53. $f(t) = \cosh at$ **54.** $f(t) = \sinh at$

55. *Area and Volume*   Given the region bounded by the graphs of $y = 1/x^2$ and $y = 0\,(x \geq 1)$, find
   a. the area of the region.
   b. the volume of the solid generated by revolving the region about the x-axis.
   c. the volume of the solid generated by revolving the region about the y-axis.

56. *Area and Volume*   Given the region bounded by the graphs of $y = e^{-x}$ and $y = 0\,(x \geq 0)$, find
   a. the area of the region.
   b. the volume of the solid generated by revolving the region about the y-axis.

57. *Arc Length*   Sketch the graph of the hypocycloid of four cusps

$$x^{2/3} + y^{2/3} = 1$$

and find its perimeter.

58. *Surface Area*   The region bounded by

$$(x - 2)^2 + y^2 = 1$$

is revolved about the y-axis to form a torus. Find the surface area of this torus.

59. *The Gamma Function*   The Gamma Function $\Gamma(n)$ is defined by

$$\Gamma(n) = \int_0^\infty x^{n-1}e^{-x}\,dx, \qquad n > 0.$$

   a. Find $\Gamma(1)$, $\Gamma(2)$, and $\Gamma(3)$.
   b. Use integration by parts to show that $\Gamma(n + 1) = n\Gamma(n)$.
   c. Express $\Gamma(n)$ in terms of factorial notation where n is a positive integer.

60. *Work*   A 5-ton rocket is fired from the surface of the earth into outer space.
   a. How much work is required to overcome the earth's gravitational force?
   b. How far has the rocket traveled when half the total work has occurred?

Probability   A nonnegative function f is called a *probability density function* if

$$\int_{-\infty}^\infty f(t)\,dt = 1.$$

The *probability* that x lies between a and b is given by

$$P(a \leq x \leq b) = \int_a^b f(t)\,dt.$$

The *expected value* of x is given by

$$E(x) = \int_{-\infty}^\infty t\,f(t)\,dt.$$

In Exercises 61 and 62, (a) show that the nonnegative function is a probability density function, (b) find $P(0 \leq x \leq 4)$, and (c) find $E(x)$.

61. $f(t) = \begin{cases} \frac{1}{7}e^{-t/7}, & t \geq 0 \\ 0, & t < 0 \end{cases}$

62. $f(t) = \begin{cases} \frac{2}{5}e^{-2t/5}, & t \geq 0 \\ 0, & t < 0 \end{cases}$

Capitalized Cost   In Exercises 63 and 64, find the capitalized cost C of an asset (a) for $n = 5$ years, (b) for $n = 10$ years, and (c) forever. The capitalized cost is given by

$$C = C_0 + \int_0^n c(t)e^{-rt}\,dt$$

where C_0 is the original investment, t is the time in years, r is the annual rate compounded continuously, and $c(t)$ is the annual cost of maintenance.

63. $C_0 = \$650,000$ **64.** $C_0 = \$650,000$
   $c(t) = \$25,000$ $c(t) = \$25,000(1 + 0.08t)$
   $r = 0.12$ $r = 0.12$

65. *Electromagnetic Theory*   Find the value of the following integral used in electromagnetic theory.

$$P = k\int_1^\infty \frac{1}{(a^2 + x^2)^{3/2}}\,dx$$

66. *Essay*
   a. The improper integrals

$$\int_1^\infty \frac{1}{x}\,dx \quad\text{and}\quad \int_1^\infty \frac{1}{x^2}\,dx$$

   diverge and converge, respectively. Describe the essential differences between the integrands that cause one integral to converge and the other to diverge.
   b. Sketch a graph of the function $y = \sin x/x$ over the interval $(1, \infty)$. Use your knowledge of the definite integral to make an inference as to whether or not the integral

$$\int_1^\infty \frac{\sin x}{x}\,dx$$

   converges. Give reasons for your answer.
   c. Use one iteration of integration by parts on the integral in part b to determine its divergence or convergence.

True or False **In Exercises 67–70, determine whether the statement is true or false. If it is false, explain why or give an example that shows it is false.**

67. If f is continuous on $[0, \infty)$ and $\lim_{x \to \infty} f(x) = 0$, then $\int_0^\infty f(x)\,dx$ converges.

68. If f is continuous on $[0, \infty)$ and $\int_0^\infty f(x)\,dx$ converges, then $\lim_{x \to \infty} f(x) = 0$.

69. If f' is continuous on $[0, \infty)$ and $\lim_{x \to \infty} f(x) = 0$, then $\int_0^\infty f'(x)\,dx = -f(0)$.

70. If the graph of f is symmetric with respect to the origin or the y-axis, then $\int_0^\infty f(x)\,dx$ converges if and only if $\int_{-\infty}^\infty f(x)\,dx$ converges.

71. Let R be the region that lies between the x-axis and the graph of

$$f(x) = \frac{1}{\sqrt{x}}, \qquad 0 < x \le 1.$$

a. Sketch the region R and determine its area.

b. Find the volume of the solid generated by revolving R about the x-axis.

72. Let

$$I_n = \int_0^\infty \frac{x^{2n-1}}{(x^2 + 1)^{n+3}}\,dx, \qquad n \ge 1.$$

Prove that

$$I_n = \left(\frac{n-1}{n+2}\right) I_{n-1}$$

and then evaluate the following.

a. $\displaystyle\int_0^\infty \frac{x}{(x^2 + 1)^4}\,dx$

b. $\displaystyle\int_0^\infty \frac{x^3}{(x^2 + 1)^5}\,dx$

c. $\displaystyle\int_0^\infty \frac{x^5}{(x^2 + 1)^6}\,dx$

REVIEW EXERCISES for Chapter 7

In Exercises 1 and 2, use integration by parts to evaluate the integral.

1. $\displaystyle\int e^{2x} \sin 3x\,dx$

2. $\displaystyle\int (x^2 - 1)e^x\,dx$

In Exercises 3–6, evaluate the trigonometric integral.

3. $\displaystyle\int \cos^3(\pi x - 1)\,dx$

4. $\displaystyle\int \sin^2 \frac{\pi x}{2}\,dx$

5. $\displaystyle\int \sec^4 \frac{x}{2}\,dx$

6. $\displaystyle\int \tan \theta \sec^4 \theta\,d\theta$

In Exercises 7 and 8, use trigonometric substitution to evaluate the integral.

7. $\displaystyle\int \frac{-12}{x^2\sqrt{4 - x^2}}\,dx$

8. $\displaystyle\int \frac{\sqrt{x^2 - 9}}{x}\,dx$

In Exercises 9 and 10, use partial fractions to evaluate the integral.

9. $\displaystyle\int \frac{x^2 + 2x}{x^3 - x^2 + x - 1}\,dx$

10. $\displaystyle\int \frac{4x - 2}{3(x - 1)^2}\,dx$

In Exercises 11–36, evaluate the integral.

11. $\displaystyle\int \frac{x^2}{x^2 + 2x - 15}\,dx$

12. $\displaystyle\int \frac{3}{2x\sqrt{9x^2 - 1}}\,dx$

13. $\displaystyle\int \frac{1}{1 - \sin \theta}\,d\theta$

14. $\displaystyle\int x^2 \sin 2x\,dx$

15. $\displaystyle\int \frac{\ln(2x)}{x^2}\,dx$

16. $\displaystyle\int 2x\sqrt{2x - 3}\,dx$

17. $\displaystyle\int \sqrt{4 - x^2}\,dx$

18. $\displaystyle\int \frac{\sec^2 \theta}{\tan \theta(\tan \theta - 1)}\,d\theta$

19. $\displaystyle\int \frac{3x^3 + 4x}{(x^2 + 1)^2}\,dx$

20. $\displaystyle\int \sqrt{\frac{x - 2}{x + 2}}\,dx$

21. $\displaystyle\int \frac{16}{\sqrt{16 - x^2}}\,dx$

22. $\displaystyle\int \frac{\sin \theta}{1 + 2\cos^2 \theta}\,d\theta$

23. $\displaystyle\int \frac{x}{x^2 + 4x + 8}\,dx$

24. $\displaystyle\int \frac{x}{x^2 - 4x + 8}\,dx$

25. $\displaystyle\int \theta \sin \theta \cos \theta\,d\theta$

26. $\displaystyle\int \frac{\csc \sqrt{2x}}{\sqrt{x}}\,dx$

27. $\displaystyle\int (\sin \theta + \cos \theta)^2\,d\theta$

28. $\displaystyle\int \cos 2\theta(\sin \theta + \cos \theta)^2\,d\theta$

29. $\displaystyle\int \sqrt{1 + \cos x}\,dx$

30. $\displaystyle\int \ln\sqrt{x^2 - 1}\,dx$

31. $\displaystyle\int \cos x \ln(\sin x)\,dx$

32. $\displaystyle\int \frac{x^4 + 2x^2 + x + 1}{(x^2 + 1)^2}\,dx$

33. $\displaystyle\int x \arcsin 2x\,dx$

34. $\displaystyle\int e^x \arctan e^x\,dx$

35. $\displaystyle\int \frac{x^{1/4}}{1 + x^{1/2}}\,dx$

36. $\displaystyle\int \sqrt{1 + \sqrt{x}}\,dx$

In Exercises 37–40, solve the differential equation.

37. $\dfrac{dy}{dx} = \dfrac{9}{x^2 - 9}$

38. $\dfrac{dy}{dx} = \dfrac{\sqrt{4 - x^2}}{2x}$

39. $y' = \ln(x^2 + x)$

40. $y' = \sqrt{1 - \cos \theta}$

In Exercises 41–44, evaluate the given integral using the specified methods.

41. $\displaystyle\int \frac{1}{x^2\sqrt{4+x^2}}\,dx$

   a. Trigonometric substitution

   b. Substitution: $x = 2/u$

42. $\displaystyle\int \frac{1}{x\sqrt{4+x^2}}\,dx$

   a. Trigonometric substitution

   b. Substitution: $u^2 = 4 + x^2$

43. $\displaystyle\int \frac{x^3}{\sqrt{4+x^2}}\,dx$

   a. Trigonometric substitution

   b. Substitution: $u^2 = 4 + x^2$

   c. Integration by parts: $dv = \left(x/\sqrt{4+x^2}\right)dx$

44. $\displaystyle\int x\sqrt{4+x}\,dx$

   a. Trigonometric substitution

   b. Substitution: $u^2 = 4 + x$

   c. Substitution: $u = 4 + x$

   d. Integration by parts: $dv = \sqrt{4+x}\,dx$

In Exercises 45–50, evaluate the definite integral.

45. $\displaystyle\int_2^{\sqrt{5}} x(x^2-4)^{3/2}\,dx$ **46.** $\displaystyle\int_0^1 \frac{x}{(x-2)(x-4)}\,dx$

47. $\displaystyle\int_1^4 \frac{\ln x}{x}\,dx$ **48.** $\displaystyle\int_0^2 xe^{3x}\,dx$

49. $\displaystyle\int_0^{\pi} x\sin x\,dx$ **50.** $\displaystyle\int_0^3 \frac{x}{\sqrt{1+x}}\,dx$

Area **In Exercises 51 and 52, find the area of the region enclosed by the graphs of the equations.**

51. $y = x\sqrt{4-x}, \ y = 0$

52. $y = \dfrac{1}{25 - x^2}, \ y = 0, \ x = 0, \ x = 4$

In Exercises 53 and 54, find the centroid of the region bounded by the graphs of the equations.

53. $y = \sqrt{1 - x^2}, \ y = 0$

54. $(x-1)^2 + y^2 = 1, \ (x-4)^2 + y^2 = 4$

In Exercises 55 and 56, evaluate each definite integral to two-decimal-place accuracy.

55. a. $\displaystyle\int_0^1 e^x\,dx$ **b.** $\displaystyle\int_0^1 xe^x\,dx$

   c. $\displaystyle\int_0^1 xe^{x^2}\,dx$ **d.** $\displaystyle\int_0^1 e^{x^2}\,dx$

56. a. $\displaystyle\int_0^{\pi/2} \cos x\,dx$ **b.** $\displaystyle\int_0^{\pi/2} \cos^2 x\,dx$

   c. $\displaystyle\int_0^{\pi/2} \cos x^2\,dx$ **d.** $\displaystyle\int_0^{\pi/2} \cos\sqrt{x}\,dx$

In Exercises 57 and 58, approximate to two decimal places the arc length of the curve over the given interval.

Function	Interval
57. $y = \sin x$	$[0, \pi]$
58. $y = \sin^2 x$	$[0, \pi]$

In Exercises 59–66, use L'Hôpital's Rule to evaluate the limit.

59. $\displaystyle\lim_{x\to1} \frac{(\ln x)^2}{x-1}$ **60.** $\displaystyle\lim_{x\to0} \frac{\sin \pi x}{\sin 2\pi x}$

61. $\displaystyle\lim_{x\to\infty} \frac{e^{2x}}{x^2}$ **62.** $\displaystyle\lim_{x\to\infty} xe^{-x^2}$

63. $\displaystyle\lim_{x\to\infty} (\ln x)^{2/x}$ **64.** $\displaystyle\lim_{x\to1^+} (x-1)^{\ln x}$

65. $\displaystyle\lim_{n\to\infty} 1000\left(1 + \frac{0.09}{n}\right)^n$ **66.** $\displaystyle\lim_{x\to1^+} \left(\frac{2}{\ln x} - \frac{2}{x-1}\right)$

In Exercises 67–70, determine the convergence or divergence of the improper integral. Evaluate the integral if it converges.

67. $\displaystyle\int_0^{16} \frac{1}{\sqrt[4]{x}}\,dx$ **68.** $\displaystyle\int_0^1 \frac{6}{x-1}\,dx$

69. $\displaystyle\int_1^{\infty} x^2 \ln x\,dx$ **70.** $\displaystyle\int_0^{\infty} \frac{e^{-1/x}}{x^2}\,dx$

71. *Present Value* The board of directors of a corporation is calculating the price to pay for a business that is forecast to yield a continuous flow of profit of $500,000 per year. If money will earn a nominal rate of 9% per year compounded continuously, what is the present value of the business

   a. for 20 years?

   b. forever (in perpetuity)?

   (*Note:* The present value for t_0 years is $\int_0^{t_0} 500{,}000e^{-0.09t}\,dt$.)

72. *Volume* Find the volume of the solid generated by revolving the region bounded by the graphs of $y = xe^{-x}$, $y = 0$, and $x = 0$ about the x-axis.

73. Approximate

$$\int_2^{\infty} \frac{1}{x^5 - 1}\,dx$$

using the inequality

$$\frac{1}{x^5} + \frac{1}{x^{10}} + \frac{1}{x^{15}} < \frac{1}{x^5 - 1} < \frac{1}{x^5} + \frac{1}{x^{10}} + \frac{2}{x^{15}}$$

for $x \geq 2$.

74. Verify the reduction formula

$$\int (\ln x)^n\,dx = x(\ln x)^n - n\int (\ln x)^{n-1}\,dx.$$

75. Verify the reduction formula

$$\int \tan^n x \, dx = \frac{1}{n-1} \tan^{n-1} x - \int \tan^{n-2} x \, dx.$$

76. Show that

$$\int_x^1 \frac{1}{1+t^2} \, dt = \int_1^{1/x} \frac{1}{1+t^2} \, dt$$

by evaluating each integral and then using the identity

$$\arctan x + \arctan \frac{1}{x} = \frac{\pi}{2}, \qquad x > 0.$$

True or False In Exercises 77–80, determine whether the statement is true or false. If it is false, explain why or give an example that shows it is false.

77. $\displaystyle \int \frac{\ln x^2}{x} \, dx = \int u \, du, \qquad u = \ln x^2$

78. If $u = \sqrt[3]{3x + 1}$, then $dx = u^2 \, du$.

79. $\displaystyle \int_{-1}^1 \sqrt{x^2 - x^3} \, dx = \int_{-1}^1 x\sqrt{1 - x} \, dx$

80. $\displaystyle \int_1^\infty \frac{1}{x^p} \, dx$ converges if $p > 1$.

Chapter 8

Infinite Series

Sequences

In mathematics, the word "sequence" is used in much the same way as in ordinary English. To say that a collection of objects or events is *in sequence* usually means that the collection is ordered so that it has an identified first member, second member, third member, and so on.

Mathematically, a **sequence** is defined as a function whose domain is the set of positive integers. Although a sequence is a function, it is common to represent sequences by subscript notation, rather than the standard function notation. For instance, in the sequence

$$
\begin{array}{ccccccc}
1, & 2, & 3, & 4, & \ldots, & n, & \ldots \\
\downarrow & \downarrow & \downarrow & \downarrow & & \downarrow & \\
a_1, & a_2, & a_3, & a_4, & \ldots, & a_n, & \ldots
\end{array}
$$

Sequence

REMARK Occasionally, it is convenient to begin a sequence with a_0, so that the terms of the sequence become

$$a_0, a_1, a_2, a_3, \ldots, a_n, \ldots.$$

1 is mapped onto a_1, 2 is mapped onto a_2, and so on. The numbers $a_1, a_2, a_3, \ldots, a_n, \ldots$ are the **terms** of the sequence. The number a_n is the **nth term** of the sequence and the entire sequence is denoted by $\{a_n\}$.

EXAMPLE 1 Listing the Terms of a Sequence

a. The terms of the sequence $\{a_n\} = \{3 + (-1)^n\}$ are

$$3 + (-1)^1, 3 + (-1)^2, 3 + (-1)^3, 3 + (-1)^4, \ldots$$
$$2, 4, 2, 4, \ldots$$

b. The terms of the sequence $\{b_n\} = \left\{\dfrac{2n}{1 + n}\right\}$ are

$$\frac{2 \cdot 1}{1 + 1}, \frac{2 \cdot 2}{1 + 2}, \frac{2 \cdot 3}{1 + 3}, \frac{2 \cdot 4}{1 + 4}, \ldots$$
$$\frac{2}{2}, \frac{4}{3}, \frac{6}{4}, \frac{8}{5}, \ldots$$

c. The terms of the sequence $\{c_n\} = \left\{\dfrac{n^2}{2^n - 1}\right\}$ are

$$\frac{1^2}{2^1 - 1}, \frac{2^2}{2^2 - 1}, \frac{3^2}{2^3 - 1}, \frac{4^2}{2^4 - 1}, \ldots$$
$$\frac{1}{1}, \frac{4}{3}, \frac{9}{7}, \frac{16}{15}, \ldots$$

◄ *In Section 8.2, Exercise 70 you are asked to show that this sphereflake fractal has an infinite surface area.*

Limit of a Sequence

The primary focus of this chapter concerns sequences whose terms approach limiting values. Such sequences are said to **converge.** For instance, the sequence $\{1/2^n\}$

$$\frac{1}{2}, \frac{1}{4}, \frac{1}{8}, \frac{1}{16}, \frac{1}{32}, \cdots$$

converges to 0, as indicated in the following definition.

FIGURE 8.1
For $n > M$ the terms of the sequence all lie within ε units of L.

Definition of the Limit of a Sequence

Let L be a real number. The **limit** of a sequence $\{a_n\}$ is L, written as

$$\lim_{n \to \infty} a_n = L$$

if for each $\varepsilon > 0$, there exists $M > 0$ such that $|a_n - L| < \varepsilon$ whenever $n > M$. Sequences that have limits **converge,** whereas sequences that do not have limits **diverge.**

Graphically, this definition says that eventually (for $n > M$) the terms of a sequence that converges to L will lie within the band between the lines $y = L + \varepsilon$ and $y = L - \varepsilon$, as illustrated in Figure 8.1.

If a sequence $\{a_n\}$ agrees with a function f at every positive integer, and if $f(x)$ approaches a limit L as $x \to \infty$, then the sequence must converge to the same limit L.

THEOREM 8.1   Limit of a Sequence

Let f be a function of a real variable such that

$$\lim_{x \to \infty} f(x) = L.$$

If $\{a_n\}$ is a sequence such that $f(n) = a_n$ for every positive integer n, then

$$\lim_{n \to \infty} a_n = L.$$

EXAMPLE 2   Finding the Limit of a Sequence

Find the limit of the sequence whose nth term is $a_n = \left(1 + \dfrac{1}{n}\right)^n$.

Solution   You know from Chapter 5 that

$$\lim_{x \to \infty} \left(1 + \frac{1}{x}\right)^x = e.$$

Therefore, you can apply Theorem 8.1 to conclude that

$$\lim_{n \to \infty} a_n = \lim_{n \to \infty} \left(1 + \frac{1}{n}\right)^n = e.$$

The following properties of limits of sequences parallel those given for limits of functions of a real variable in Section 3.5.

THEOREM 8.2   Properties of Limits of Sequences

Let $\lim_{n \to \infty} a_n = L$ and $\lim_{n \to \infty} b_n = K$.

1. $\lim_{n \to \infty} (a_n \pm b_n) = L \pm K$

2. $\lim_{n \to \infty} ca_n = cL,$ c is any real number

3. $\lim_{n \to \infty} (a_n b_n) = LK$

4. $\lim_{n \to \infty} \dfrac{a_n}{b_n} = \dfrac{L}{K},$ $b_n \neq 0$ and $K \neq 0$

EXAMPLE 3   Determining Convergence or Divergence

a. Because the sequence $\{a_n\} = \{3 + (-1)^n\}$ has terms

$$2, 4, 2, 4, \ldots$$

that alternate between 2 and 4, the limit

$$\lim_{n \to \infty} a_n$$

does not exist. Thus the sequence diverges.

b. For $\{b_n\} = \{n/(1 - 2n)\}$, you can divide the numerator and denominator by n to obtain

$$\lim_{n \to \infty} \frac{n}{1 - 2n} = \lim_{n \to \infty} \left[\frac{1}{(1/n) - 2} \right] = -\frac{1}{2}$$

which implies that the sequence converges to $-\frac{1}{2}$.

■

Theorem 8.1 opens up the possibility of using L'Hôpital's Rule to determine the limit of a sequence.

TECHNOLOGY Use a graphing utility to graph the function in Example 4. Notice that as x approaches infinity, the value of the function gets closer and closer to 0. If you have access to a graphing utility that can generate terms of a sequence, try using it to calculate the first 20 terms of the sequence in Example 4. Then view the terms to observe numerically that the sequence converges to 0.

EXAMPLE 4   Using L'Hôpital's Rule to Determine Convergence

Show that the sequence whose nth term is $a_n = \dfrac{n^2}{2^n - 1}$ converges.

Solution   Consider the function of a real variable

$$f(x) = \frac{x^2}{2^x - 1}.$$

Applying L'Hôpital's Rule twice produces

$$\lim_{x \to \infty} \frac{x^2}{2^x - 1} = \lim_{x \to \infty} \frac{2x}{(\ln 2)2^x} = \lim_{x \to \infty} \frac{2}{(\ln 2)^2 2^x} = 0.$$

Because $f(n) = a_n$ for every positive integer, you can apply Theorem 8.1 to conclude that

$$\lim_{n \to \infty} \frac{n^2}{2^n - 1} = 0.$$

Thus, the sequence converges to 0.

■

To simplify some of the formulas developed in this chapter, we use the symbol $n!$ (read "n factorial"). Let n be a positive integer; then **n factorial** is given by

$$n! = 1 \cdot 2 \cdot 3 \cdot 4 \cdots (n - 1) \cdot n.$$

Zero factorial is given by $0! = 1$. From this definition, you can see that $1! = 1$, $2! = 1 \cdot 2 = 2$, $3! = 1 \cdot 2 \cdot 3 = 6$, and so on. Factorials follow the same conventions for order of operation as exponents. That is, just as $2x^3$ and $(2x)^3$ imply different orders of operations, $2n!$ and $(2n)!$ imply the following orders.

$$2n! = 2(n!) = 2(1 \cdot 2 \cdot 3 \cdot 4 \cdots n)$$

and

$$(2n)! = 1 \cdot 2 \cdot 3 \cdot 4 \cdots n \cdot (n + 1) \cdots 2n$$

Another useful limit theorem that can be rewritten for sequences is the Squeeze Theorem from Chapter 1.

THEOREM 8.3   Squeeze Theorem for Sequences

If

$$\lim_{n \to \infty} a_n = L = \lim_{n \to \infty} b_n$$

and there exists an integer N such that $a_n \le c_n \le b_n$ for all $n > N$, then $\lim_{n \to \infty} c_n = L$.

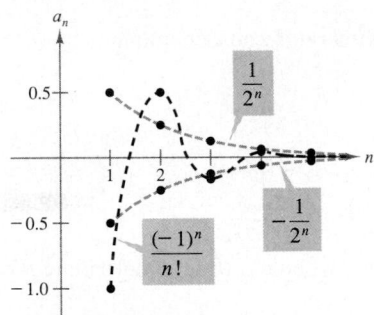

FIGURE 8.2
For $n \ge 4$, $(-1)^n/n!$ is squeezed between $-1/2^n$ and $1/2^n$.

REMARK   Example 5 suggests something about the rate at which $n!$ increases as $n \to \infty$. From Figure 8.2 you can see that both $1/2^n$ and $1/n!$ approach 0 as $n \to \infty$. Yet $1/n!$ approaches 0 so much faster than $1/2^n$ does that

$$\lim_{n \to \infty} \frac{1/n!}{1/2^n} = \lim_{n \to \infty} \frac{2^n}{n!} = 0.$$

In fact, it can be shown that for any fixed number k,

$$\lim_{n \to \infty} \frac{k^n}{n!} = 0.$$

This means that *the factorial function grows faster than any exponential function.*

EXAMPLE 5   Using the Squeeze Theorem

Show that the sequence $\{c_n\} = \left\{ (-1)^n \dfrac{1}{n!} \right\}$ converges, and find its limit.

Solution   To apply the Squeeze Theorem, you must find two convergent sequences that can be related to the given sequence. Two possibilities are $a_n = -1/2^n$ and $b_n = 1/2^n$, both of which converge to 0. By comparing the term $n!$ with 2^n, you see that

$$n! = 1 \cdot 2 \cdot 3 \cdot 4 \cdot 5 \cdot 6 \cdots n = 24 \cdot \underbrace{5 \cdot 6 \cdots n}_{n - 4 \text{ factors}} \qquad (n \ge 4)$$

and

$$2^n = 2 \cdot 2 \cdot 2 \cdot 2 \cdot 2 \cdot 2 \cdots 2 = 16 \cdot \underbrace{2 \cdot 2 \cdots 2.}_{n - 4 \text{ factors}} \qquad (n \ge 4)$$

This implies that for $n \ge 4$, $2^n < n!$, and you have

$$\frac{-1}{2^n} \le (-1)^n \frac{1}{n!} \le \frac{1}{2^n}, \qquad n \ge 4$$

as illustrated in Figure 8.2. Therefore, by the Squeeze Theorem it follows that

$$\lim_{n \to \infty} (-1)^n \frac{1}{n!} = 0.$$

In Example 5, the sequence $\{c_n\}$ has both positive and negative terms. For this sequence, it happens that the sequence of absolute values, $\{|c_n|\}$, also converges to 0. You can show this by the Squeeze Theorem using the inequality

$$0 \le \frac{1}{n!} \le \frac{1}{2^n}, \qquad n \ge 4.$$

In such cases, it is often convenient to consider the sequence of absolute values—then apply Theorem 8.4, which states that if the absolute-value sequence converges to 0, the original signed sequence also converges to 0.

THEOREM 8.4   Absolute Value Theorem

For the sequence $\{a_n\}$, if

$$\lim_{n \to \infty} |a_n| = 0 \quad \text{then} \quad \lim_{n \to \infty} a_n = 0.$$

Proof   Consider the two sequences $\{|a_n|\}$ and $\{-|a_n|\}$. Because both of these sequences converge to 0 and

$$-|a_n| \le a_n \le |a_n|$$

you can use the Squeeze Theorem to conclude that $\{a_n\}$ converges to 0. ∎

Pattern Recognition for Sequences

Sometimes the terms of a sequence are generated by some rule that does not explicitly identify the nth term of the sequence. In such cases, you may be required to discover a *pattern* in the sequence and to describe the nth term. Once the nth term is specified, you can investigate the convergence or divergence of the sequence.

EXAMPLE 6   Finding the nth Term of a Sequence

Find a sequence $\{a_n\}$ whose first five terms are

$$\frac{2}{1}, \frac{4}{3}, \frac{8}{5}, \frac{16}{7}, \frac{32}{9}, \ldots$$

and then determine whether the particular sequence you have chosen converges or diverges.

Solution   First, note that the numerators are successive powers of 2, and the denominators form the sequence of positive odd integers. By comparing a_n with n, you have the following pattern.

$$\frac{2^1}{1}, \frac{2^2}{3}, \frac{2^3}{5}, \frac{2^4}{7}, \frac{2^5}{9}, \ldots, \frac{2^n}{2n-1}$$

Using L'Hôpital's Rule to evaluate the limit of $f(x) = 2^x/(2x - 1)$, you obtain

$$\lim_{x \to \infty} \frac{2^x}{2x - 1} = \lim_{x \to \infty} \frac{2^x(\ln 2)}{2} = \infty \quad \longrightarrow \quad \lim_{n \to \infty} \frac{2^n}{2n - 1} = \infty.$$

Hence, the sequence *diverges*.

Without a specific rule for generating the terms of a sequence or some knowledge of the context in which the terms of the sequence are obtained, it is not possible to determine the convergence or divergence of the sequence merely from its first several terms. For instance, although the first three terms of the four sequences given next are identical, the first two sequences converge to 0, the third sequence converges to $\frac{1}{9}$, and the fourth sequence diverges.

$$\{a_n\}: \quad \frac{1}{2}, \frac{1}{4}, \frac{1}{8}, \frac{1}{16}, \ldots, \frac{1}{2^n}, \ldots$$

$$\{b_n\}: \quad \frac{1}{2}, \frac{1}{4}, \frac{1}{8}, \frac{1}{15}, \ldots, \frac{6}{(n+1)(n^2-n+6)}, \ldots$$

$$\{c_n\}: \quad \frac{1}{2}, \frac{1}{4}, \frac{1}{8}, \frac{7}{62}, \ldots, \frac{n^2-3n+3}{9n^2-25n+18}, \ldots$$

$$\{d_n\}: \quad \frac{1}{2}, \frac{1}{4}, \frac{1}{8}, 0, \ldots, \frac{-n(n+1)(n-4)}{6(n^2+3n-2)}, \ldots$$

The process of determining an nth term from the pattern observed in the first several terms of a sequence is an example of *inductive reasoning*.

EXAMPLE 7   Finding the nth Term of a Sequence

Determine an nth term for a sequence whose first five terms are

$$-\frac{2}{1}, \frac{8}{2}, -\frac{26}{6}, \frac{80}{24}, -\frac{242}{120}, \ldots$$

and then decide if your sequence converges or diverges.

Solution   Note that the numerators are 1 less than 3^n. Hence, you can reason that the numerators are given by the rule $3^n - 1$. Factoring the denominators produces

$$1 = 1$$
$$2 = 1 \cdot 2$$
$$6 = 1 \cdot 2 \cdot 3$$
$$24 = 1 \cdot 2 \cdot 3 \cdot 4$$
$$120 = 1 \cdot 2 \cdot 3 \cdot 4 \cdot 5 \cdots.$$

This suggests that the denominators are represented by $n!$. Finally, because the signs alternate, you can write the nth term as

$$a_n = (-1)^n \left(\frac{3^n - 1}{n!} \right).$$

From the discussion about the growth of $n!$, it follows that

$$\lim_{n \to \infty} |a_n| = \lim_{n \to \infty} \frac{3^n - 1}{n!} = 0.$$

Applying Theorem 8.4, you can conclude that

$$\lim_{n \to \infty} a_n = 0.$$

Thus, the sequence $\{a_n\}$ converges to 0.

Monotonic Sequences and Bounded Sequences

So far we have determined the convergence of a sequence by finding its limit. Even if you cannot determine the limit of a particular sequence, it still may be useful to know whether the sequence converges. Theorem 8.5 identifies a test for convergence of sequences without determining the limit. First, we give some preliminary definitions.

Definition of a Monotonic Sequence

A sequence $\{a_n\}$ is **monotonic** if its terms are nondecreasing

$$a_1 \leq a_2 \leq a_3 \leq \cdots \leq a_n \leq \cdots$$

or if its terms are nonincreasing

$$a_1 \geq a_2 \geq a_3 \geq \cdots \geq a_n \geq \cdots .$$

(a) Not monotonic

(b) Monotonic

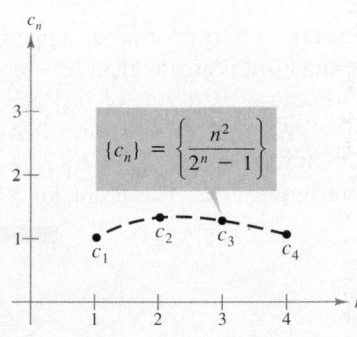

(c) Not monotonic

FIGURE 8.3

EXAMPLE 8   Determining Whether a Sequence is Monotonic

Determine whether the sequence having the given nth term is monotonic.

a. $a_n = 3 + (-1)^n$ **b.** $b_n = \dfrac{2n}{1 + n}$ **c.** $c_n = \dfrac{n^2}{2^n - 1}$

Solution

a. This sequence alternates between 2 and 4. Therefore, it is not monotonic.
b. This sequence is monotonic because each successive term is larger than its predecessor. To see this, compare the terms b_n and b_{n+1}. [Note that, because n is positive, you can multiply both sides of the inequality by $(1 + n)$ and $(2 + n)$ without reversing the inequality sign.]

$$b_n = \frac{2n}{1 + n} \overset{?}{<} \frac{2(n + 1)}{1 + (n + 1)} = b_{n+1}$$

$$2n(2 + n) \overset{?}{<} (1 + n)(2n + 2)$$

$$4n + 2n^2 \overset{?}{<} 2 + 4n + 2n^2$$

$$0 < 2$$

Starting with the final inequality, which is valid, you can reverse the steps to conclude that the original inequality is also valid.
c. This sequence is not monotonic because the second term is larger than the first term, and larger than the third. [Note that if we drop the first term, the remaining sequence $c_2, c_3, c_4, \ldots$ is monotonic.]

Figure 8.3 graphically illustrates these three sequences.   ◼

REMARK   In Example 8b, another way to see that the sequence is monotonic is to argue that the derivative of the corresponding differentiable function $f(x) = 2x/(1 + x)$ is positive for all x. This implies that f is increasing, which in turn implies that $\{a_n\}$ is increasing.

REMARK All three sequences shown in Figure 8.3 are bounded. To see this, consider the following.

$$2 \le a_n \le 4$$
$$1 \le b_n \le 2$$
$$0 \le c_n \le \frac{4}{3}$$

Definition of a Bounded Sequence

1. A sequence $\{a_n\}$ is **bounded above** if there is a real number M such that $a_n \le M$ for all n. The number M is called an **upper bound** of the sequence.
2. A sequence $\{a_n\}$ is **bounded below** if there is a real number N such that $N \le a_n$ for all n. The number N is called a **lower bound** of the sequence.
3. A sequence $\{a_n\}$ is **bounded** if it is bounded above and bounded below.

One important property of the real numbers is that they are **complete.** Informally, this means that there are no holes or gaps on the real number line. (The set of rational numbers does not have the completeness property.) The completeness axiom for real numbers can be used to conclude that if a sequence has an upper bound, then it must have a **least upper bound** (an upper bound that is smaller than all other upper bounds for the sequence). For example, the least upper bound of the sequence $\{a_n\} = \{n/(n + 1)\}$,

$$\frac{1}{2}, \frac{2}{3}, \frac{3}{4}, \frac{4}{5}, \ldots, \frac{n}{n + 1}, \ldots$$

is 1. We use the completeness axiom in the proof of Theorem 8.5.

THEOREM 8.5 Bounded Monotonic Sequences

If a sequence $\{a_n\}$ is bounded and monotonic, then it converges.

Proof Assume that the sequence is nondecreasing, as shown in Figure 8.4. For the sake of simplicity, also assume that each term in the sequence is positive. Because the sequence is bounded, there must exist an upper bound M such that

$$a_1 \le a_2 \le a_3 \le \cdots \le a_n \le \cdots \le M.$$

From the completeness axiom, it follows that there is a least upper bound L such that

$$a_1 \le a_2 \le a_3 \le \cdots \le a_n \le \cdots \le L.$$

For $\varepsilon > 0$, it follows that $L - \varepsilon < L$, and therefore $L - \varepsilon$ cannot be an upper bound for the sequence. Consequently, at least one term of $\{a_n\}$ is greater than $L - \varepsilon$. That is, $L - \varepsilon < a_N$ for some positive integer N. Because the terms of $\{a_n\}$ are nondecreasing, it follows that $a_N \le a_n$ for $n > N$. You now know that $L - \varepsilon < a_N \le a_n \le L < L + \varepsilon$, for every $n > N$. It follows that $|a_n - L| < \varepsilon$ for $n > N$, which by definition means that $\{a_n\}$ converges to L. The proof for a nonincreasing sequence is similar.

FIGURE 8.4
Every bounded nondecreasing sequence converges.

EXAMPLE 9   **Bounded and Monotonic Sequences**

a. The sequence

$$\{a_n\} = \left\{\frac{1}{n}\right\}$$

is both bounded and monotonic and thus, by Theorem 8.5, must converge.

b. The divergent sequence

$$\{b_n\} = \left\{\frac{n^2}{n+1}\right\}$$

is monotonic, but not bounded. (It *is* bounded below.)

c. The divergent sequence

$$\{c_n\} = \{(-1)^n\}$$

is bounded, but not monotonic.

1-68 every 3rd   73   85-88

EXERCISES for Section 8.1

In Exercises 1–10, write out the first five terms of the sequence

1. $a_n = 2^n$

2. $a_n = \dfrac{n}{n+1}$

3. $a_n = \left(-\dfrac{1}{2}\right)^n$

4. $a_n = \sin\dfrac{n\pi}{2}$

5. $a_n = \dfrac{(-1)^{n(n+1)/2}}{n^2}$

6. $a_n = 5 - \dfrac{1}{n} + \dfrac{1}{n^2}$

7. $a_n = \dfrac{3^n}{n!}$

8. $a_n = \dfrac{3n!}{(n-1)!}$

9. $a_1 = 3$ and $a_{n+1} = 2(a_n - 1)$

10. $a_1 = 4$ and $a_{n+1} = \left(\dfrac{n+1}{2}\right)a_n$

In Exercises 11–14, write the next two *apparent* terms of the sequence. Describe the pattern you used to find the next two terms.

11. $2, 5, 8, 11, \ldots$

12. $\frac{7}{2}, 4, \frac{9}{2}, 5, \ldots$

13. $3, -\frac{3}{2}, \frac{3}{4}, -\frac{3}{8}, \ldots$

14. $5, 10, 20, 40, \ldots$

In Exercises 15–20, simplify the ratio of factorials.

15. $\dfrac{10!}{8!}$

16. $\dfrac{25!}{23!}$

17. $\dfrac{(n+1)!}{n!}$

18. $\dfrac{(n+2)!}{n!}$

19. $\dfrac{(2n-1)!}{(2n+1)!}$

20. $\dfrac{(2n+2)!}{(2n)!}$

In Exercises 21–34, write an expression for the *n*th term of the sequence. (There is more than one correct answer.)

21. $1, 4, 7, 10, \ldots$

22. $3, 7, 11, 15, \ldots$

23. $-1, 2, 7, 14, 23, \ldots$

24. $1, -\frac{1}{4}, \frac{1}{9}, -\frac{1}{16}, \ldots$

25. $\frac{2}{3}, \frac{3}{4}, \frac{4}{5}, \frac{5}{6}, \ldots$

26. $2, \frac{3}{3}, \frac{4}{5}, \frac{5}{7}, \frac{6}{9}, \ldots$

27. $2, -1, \frac{1}{2}, -\frac{1}{4}, \frac{1}{8}, \ldots$

28. $-\frac{1}{2}, \frac{1}{3}, -\frac{2}{9}, \frac{4}{27}, -\frac{8}{81}, \ldots$

29. $2, 1 + \frac{1}{2}, 1 + \frac{1}{3}, 1 + \frac{1}{4}, 1 + \frac{1}{5}, \ldots$

30. $1 + \frac{1}{2}, 1 + \frac{3}{4}, 1 + \frac{7}{8}, 1 + \frac{15}{16}, 1 + \frac{31}{32}, \ldots$

31. $\dfrac{1}{2 \cdot 3}, \dfrac{2}{3 \cdot 4}, \dfrac{3}{4 \cdot 5}, \dfrac{4}{5 \cdot 6}, \ldots$

32. $1, \frac{1}{2}, \frac{1}{6}, \frac{1}{24}, \frac{1}{120}, \ldots$

33. $1, -\dfrac{1}{1 \cdot 3}, \dfrac{1}{1 \cdot 3 \cdot 5}, -\dfrac{1}{1 \cdot 3 \cdot 5 \cdot 7}, \ldots$

34. $1, x, \dfrac{x^2}{2}, \dfrac{x^3}{6}, \dfrac{x^4}{24}, \dfrac{x^5}{120}, \ldots$

In Exercises 35–54, determine the convergence or divergence of the sequence with the given *n*th term. If the sequence converges, find its limit.

35. $a_n = \dfrac{n+1}{n}$

36. $a_n = \dfrac{1}{n^{3/2}}$

37. $a_n = (-1)^n\left(\dfrac{n}{n+1}\right)$

38. $a_n = 3 - \dfrac{1}{2^n}$

39. $a_n = \dfrac{3n^2 - n + 4}{2n^2 + 1}$

40. $a_n = \dfrac{\sqrt{n}}{\sqrt{n}+1}$

41. $a_n = \cos\dfrac{n\pi}{2}$

42. $a_n = 1 + (-1)^n$

43. $a_n = \dfrac{1 + (-1)^n}{n}$

44. $a_n = \dfrac{\ln(n^2)}{n}$

45. $a_n = \dfrac{3^n}{4^n}$

46. $a_n = (0.5)^n$

47. $a_n = \dfrac{(n+1)!}{n!}$

48. $a_n = \dfrac{(n-2)!}{n!}$

49. $a_n = \dfrac{n-1}{n} - \dfrac{n}{n-1}$

50. $a_n = \dfrac{n^2}{2n+1} - \dfrac{n^2}{2n-1}$

51. $a_n = \dfrac{n^p}{e^n} \; (p > 0, \, n \geq 2)$

52. $a_n = n \sin \dfrac{1}{n}$

53. $a_n = \left(1 + \dfrac{k}{n}\right)^n$

54. $a_n = 2^{1/n}$

In Exercises 55–64, determine whether the sequence with the given nth term is monotonic. Discuss the boundedness of the sequence.

55. $a_n = 4 - \dfrac{1}{n}$

56. $a_n = \dfrac{4n}{n+1}$

57. $a_n = \dfrac{\cos n}{n}$

58. $a_n = ne^{-n/2}$

59. $a_n = (-1)^n \left(\dfrac{1}{n}\right)$

60. $a_n = \left(-\dfrac{2}{3}\right)^n$

61. $a_n = (\tfrac{2}{3})^n$

62. $a_n = (\tfrac{3}{2})^n$

63. $a_n = \sin \dfrac{n\pi}{6}$

64. $a_n = \dfrac{n}{2^{n+2}}$

In Exercises 65–68, use Theorem 8.5 to show that the sequence with the given nth term converges, and find the limit.

65. $a_n = 5 + \dfrac{1}{n}$

66. $a_n = 3 - \dfrac{4}{n}$

67. $a_n = \dfrac{1}{3}\left(1 - \dfrac{1}{3^n}\right)$

68. $a_n = 4 + \dfrac{1}{2^n}$

69. Give an example of a sequence satisfying the given condition, or explain why no such sequence exists. (There is more than one correct answer.)
  a. A monotonic sequence that converges to 10.
  b. A monotonic, bounded sequence that does not converge.
  c. A sequence that converges to $\tfrac{3}{4}$.
  d. An unbounded sequence that converges to 100.

70. Find a sequence that converges to $\tfrac{3}{8}$. (There are many such sequences.) What is the first term in the sequence that is within 0.001 of the limit?

71. *Compound Interest*   Consider the sequence $\{A_n\}$, whose nth term is given by

$$A_n = P\left(1 + \dfrac{r}{12}\right)^n$$

where P is the principal, A_n is the amount at compound interest after n months, and r is the annual percentage rate.
  a. Is $\{A_n\}$ a convergent sequence?
  b. Find the first 10 terms of the sequence if $P = \$9000$ and $r = 0.115$.

72. *Investment*   A deposit of $100 is made at the beginning of each month in an account at an annual interest rate of 12 percent compounded monthly. The balance in the account after n months is given by

$$A_n = 100(101)[(1.01)^n - 1].$$

  a. Compute the first six terms of the sequence $\{A_n\}$.
  b. Find the balance after 5 years by computing the 60th term of the sequence.
  c. Find the balance after 20 years by computing the 240th term of the sequence.

73. *Government Expenditures*   A government program that currently costs taxpayers $2.5 billion per year is to be cut back by 20 percent per year.
  a. Write an expression for the amount budgeted for this program after n years.
  b. Compute the budgets for the first 4 years.
  c. Determine the convergence or divergence of the sequence of reduced budgets. If the sequence converges, find its limit.

74. *Inflation*   If the rate of inflation is $4\tfrac{1}{2}$ percent per year and the average price of a car is currently $16,000, then the average price after n years is

$$P_n = \$16{,}000(1.045)^n.$$

Compute the average price for the next 5 years.

75. *Hospital Costs*   The average cost per day for a semiprivate hospital room from 1980 to 1990 is approximated by the model

$$a_n = 127.42 + 17.37n, \qquad n = 0, 1, 2, \ldots, 10$$

where a_n is the average cost in dollars and n is the year with $n = 0$ corresponding to 1980. (*Source:* American Hospital Association) Find the terms of this finite sequence and construct a bar graph that represents the sequence. Use the model to predict the cost in 1995.

76. *Federal Debt*   It took more than 200 years for the United States to accumulate a $1 trillion debt. Then it took just 8 years to get to $3 trillion. (*Source:* Treasury Department) The federal debt during the decade of the 1980s is approximated by the model

$$a_n = 0.917e^{0.087n}, \qquad n = 0, 1, 2, \ldots, 10$$

where a_n is the debt in trillions and n is the year with $n = 0$ corresponding to 1980. Find the terms of this finite sequence and construct a bar graph that represents the sequence.

77. *Population Growth*   Consider an idealized population with the characteristic that each member of the population produces one offspring at the end of every time period. If each member has a life span of 3 time periods and the population begins with 10 newborn members, then the following table gives the population during the first 5 time periods.

Age bracket	Time period				
	1	2	3	4	5
0–1	10	10	20	40	70
1–2		10	10	20	40
2–3			10	10	20
Total	10	20	40	70	130

The sequence for the total population has the property that

$$S_n = S_{n-1} + S_{n-2} + S_{n-3}, \qquad n > 3.$$

Find the total population during the next 5 time periods.

78. Consider the sequence $\{x_n\}$ defined by

$$x_n = x_{n-1} - \frac{f(x_{n-1})}{f'(x_{n-1})}.$$

a. What does this sequence represent and what is the method called?

b. If $f(x) = x^2 + x - 1$ and $x_1 = 0.5$, find the next three terms of the sequence.

c. Find the limit of the sequence.

79. Consider the sequence $a_n = \dfrac{10^n}{n!}$.

a. Find two consecutive terms that are equal in magnitude.

b. Are the terms following those found in part a increasing or decreasing?

c. In Section 7.7, Exercises 43–48, it was shown that for "large" values of the independent variable an exponential function increased more rapidly than a polynomial. From the result of part b, what inference can you make about the rate of growth of an exponential function versus a factorial function for "large" integer values on n?

80. Compute the first 6 terms of the sequence

$$\{a_n\} = \left\{ \left(1 + \frac{1}{n}\right)^n \right\}.$$

If the sequence converges, find its limit.

81. Compute the first six terms of the sequence $\{a_n\} = \left\{ \sqrt[n]{n} \right\}$. If the sequence converges, find its limit.

82. Prove that if $\{s_n\}$ converges to L and $L > 0$, then there exists a number N such that $s_n > 0$ for $n > N$.

83. *Fibonacci Sequence*   In the study of the progeny of rabbits, Fibonacci (ca. 1175–1250) encountered the now famous sequence bearing his name. It is defined recursively by

$$a_{n+2} = a_n + a_{n+1}, \qquad \text{where } a_1 = 1 \text{ and } a_2 = 1.$$

a. Write out the first 12 terms of the sequence.

b. Write out the first 10 terms of the sequence defined by

$$b_n = \frac{a_{n+1}}{a_n}, \qquad \text{for } n \geq 1.$$

c. Using the definition of part b, show that

$$b_n = 1 + \frac{1}{b_{n-1}}.$$

e. The **golden ratio** ρ can be defined to be $\lim\limits_{n \to \infty} b_n = \rho$. Show that

$$\rho = 1 + \frac{1}{\rho}$$

and solve this equation for ρ.

84. Complete the proof of Theorem 8.5.

True or False   **In Exercises 85–88, determine whether the statement is true or false. If it is false, explain why or give an example that shows it is false.**

85. If $\{a_n\}$ converges to 3 and $\{b_n\}$ converges to 2, then $\{a_n + b_n\}$ converges to 5.

86. If $\{a_n\}$ converges, then $\lim\limits_{n \to \infty} (a_n - a_{n+1}) = 0$.

87. If $n > 1$, then $n! = n(n-1)!$.

88. If $\{a_n\}$ converges, then $\{a_n/n\}$ converges to 0.

89. Consider the sequence

$$\sqrt{2}, \ \sqrt{2 + \sqrt{2}}, \ \sqrt{2 + \sqrt{2 + \sqrt{2}}}, \ \ldots$$

where $a_n = \sqrt{2 + a_{n-1}}$ for $n \geq 2$. Compute the first five terms of this sequence. Find the limit

$$\lim_{n \to \infty} a_n.$$

Infinite Series

One important application of infinite sequences is in representing "infinite summations." Informally, if $\{a_n\}$ is an infinite sequence, then

$$\sum_{n=1}^{\infty} a_n = a_1 + a_2 + a_3 + \cdots + a_n + \cdots$$ Infinite series

is called an **infinite series** (or simply a **series**). The numbers $a_1, a_2, a_3, \ldots$ are called the **terms** of the series. For some series it is convenient to begin the index at $n = 0$ (or some other integer). As a typesetting convention, it is common to represent an infinite series as simply $\Sigma\, a_n$. In such cases the starting value for the index must be taken from the context of the statement.

To find the sum of an infinite series, consider the following **sequence of partial sums.**

$$S_1 = a_1$$
$$S_2 = a_1 + a_2$$
$$S_3 = a_1 + a_2 + a_3$$
$$\vdots$$
$$S_n = a_1 + a_2 + a_3 + \cdots + a_n$$

If this sequence of partial sums converges, then the series is said to converge and has the sum indicated in the following definition.

Definition of Convergent and Divergent Series

For the infinite series $\Sigma\, a_n$, the **nth partial sum** is given by
$$S_n = a_1 + a_2 + \cdots + a_n.$$
If the sequence of partial sums $\{S_n\}$ converges to S, then the series $\Sigma\, a_n$ **converges.** The limit S is called the **sum of the series.**
$$S = a_1 + a_2 + \cdots + a_n + \cdots$$
If $\{S_n\}$ diverges, then the series **diverges.**

As you study this chapter, you will see that there are two basic questions involving infinite series. Does a series converge or does it diverge? If a series converges, what is its sum? These questions are not always easy to answer, especially the second one. We begin with some simple examples.

EXAMPLE 1   Convergent and Divergent Series

a. The series

$$\sum_{n=1}^{\infty} \frac{1}{2^n} = \frac{1}{2} + \frac{1}{4} + \frac{1}{8} + \frac{1}{16} + \cdots$$

has the following partial sums.

$$S_1 = \frac{1}{2}$$

$$S_2 = \frac{1}{2} + \frac{1}{4} = \frac{3}{4}$$

$$S_3 = \frac{1}{2} + \frac{1}{4} + \frac{1}{8} = \frac{7}{8}$$

$$\vdots$$

$$S_n = \frac{1}{2} + \frac{1}{4} + \frac{1}{8} + \cdots + \frac{1}{2^n} = \frac{2^n - 1}{2^n}$$

Because

$$\lim_{n \to \infty} \frac{2^n - 1}{2^n} = 1$$

it follows that the series converges and its sum is 1.

b. The nth partial sum of the series

$$\sum_{n=1}^{\infty} \left(\frac{1}{n} - \frac{1}{n+1} \right) = \left(1 - \frac{1}{2} \right) + \left(\frac{1}{2} - \frac{1}{3} \right) + \left(\frac{1}{3} - \frac{1}{4} \right) + \cdots$$

is given by

$$S_n = 1 - \frac{1}{n+1}.$$

Because the limit of S_n is 1, the series converges and its sum is 1.

c. The series

$$\sum_{n=1}^{\infty} 1 = 1 + 1 + 1 + 1 + \cdots$$

diverges because $S_n = n$ and the sequence of partial sums diverges.

The series in Example 1b is a **telescoping series.** That is, it is of the form

$$(b_1 - b_2) + (b_2 - b_3) + (b_3 - b_4) + (b_4 - b_5) + \cdots . \qquad \text{Telescoping series}$$

Note that b_2 is canceled by the second term, b_3 is canceled by the third term, and so on. Because the nth partial sum of this series is $S_n = b_1 - b_{n+1}$, it follows that a telescoping series will converge if and only if b_n approaches a finite number as $n \to \infty$. Moreover, if the series converges, then its sum is

$$S = b_1 - \lim_{n \to \infty} b_{n+1}.$$

if b_n approaches finite #, does it always converge?

EXAMPLE 2   Writing a Series in Telescoping Form

Find the sum of the series $\displaystyle\sum_{n=1}^{\infty} \frac{2}{4n^2 - 1}$.

Solution   Using partial fractions, you can write

$$a_n = \frac{2}{4n^2 - 1} = \frac{2}{(2n - 1)(2n + 1)} = \frac{1}{2n - 1} - \frac{1}{2n + 1}.$$

From this telescoping form, you can see that the nth partial sum is

$$S_n = \left(\frac{1}{1} - \frac{1}{3}\right) + \left(\frac{1}{3} - \frac{1}{5}\right) + \cdots + \left(\frac{1}{2n - 1} - \frac{1}{2n + 1}\right) = 1 - \frac{1}{2n + 1}.$$

Thus, the series converges and its sum is 1. That is,

$$\sum_{n=1}^{\infty} \frac{2}{4n^2 - 1} = \lim_{n\to\infty} S_n = \lim_{n\to\infty}\left(1 - \frac{1}{2n + 1}\right) = 1. \qquad \blacksquare$$

Geometric Series

The series given in Example 1a is a **geometric series.** In general, the series given by

$$\sum_{n=0}^{\infty} ar^n = a + ar + ar^2 + \cdots + ar^n + \cdots, \qquad a \neq 0 \qquad \text{Geometric series}$$

is called a **geometric series** with ratio r.

THEOREM 8.6   Convergence of a Geometric Series

A geometric series with ratio r diverges if $|r| \geq 1$. If $0 < |r| < 1$, then the series converges to the sum

$$\sum_{n=0}^{\infty} ar^n = \frac{a}{1 - r}, \qquad 0 < |r| < 1.$$

FOR FURTHER INFORMATION
For a different approach to proving this theorem, see the article "Proof Without Words" by Benjamin G. Klein and Irl C. Bivens, in the October, 1988 issue of *Mathematics Magazine.*

Proof   It is easy to see that the series diverges if $r = \pm 1$. If $r \neq \pm 1$, then $S_n = a + ar + ar^2 + \cdots + ar^{n-1}$. Multiplication by r yields

$$rS_n = ar + ar^2 + ar^3 + \cdots + ar^n.$$

Subtracting the second equation from the first produces $S_n - rS_n = a - ar^n$. Therefore, $S_n(1 - r) = a(1 - r^n)$, and the nth partial sum is

$$S_n = \frac{a}{1 - r}(1 - r^n).$$

If $0 < |r| < 1$, it follows that $r^n \to 0$ as $n \to \infty$, and you obtain

$$\lim_{n\to\infty} S_n = \lim_{n\to\infty}\left[\frac{a}{1 - r}(1 - r^n)\right] = \frac{a}{1 - r}\left[\lim_{n\to\infty}(1 - r^n)\right] = \frac{a}{1 - r}$$

which means the series *converges* and its sum is $a/(1 - r)$. We leave it to you to show that the series diverges if $|r| > 1$. $\qquad \blacksquare$

EXAMPLE 3   A Convergent Geometric Series

The geometric series

$$\sum_{n=0}^{\infty} \frac{3}{2^n} = \sum_{n=0}^{\infty} 3\left(\frac{1}{2}\right)^n = 3(1) + 3\left(\frac{1}{2}\right) + 3\left(\frac{1}{2}\right)^2 + \cdots$$

has a ratio of $r = \frac{1}{2}$ with $a = 3$. Because $0 < |r| < 1$, the series converges and its sum is

$$S = \frac{a}{1 - r} = \frac{3}{1 - (1/2)} = 6.$$

TECHNOLOGY Try writing a computer program that computes the sum of the first 20 terms of the sequence in Example 3. You should obtain a sum of about 5.999997.

EXAMPLE 4   A Divergent Geometric Series

The geometric series

$$\sum_{n=0}^{\infty} \left(\frac{3}{2}\right)^n = 1 + \frac{3}{2} + \frac{9}{4} + \frac{27}{8} + \cdots$$

has a ratio of $r = \frac{3}{2}$. Because $|r| \geq 1$, the series diverges.

The formula for the sum of a geometric series can be used to write a repeating decimal as the ratio of two integers, as demonstrated in the next example.

EXAMPLE 5   A Geometric Series for a Repeating Decimal

Use a geometric series to express $0.080\overline{808}$ as the ratio of two integers.

Solution   For the repeating decimal $0.080\overline{808}$, you can write

$$0.080808 \ldots = \frac{8}{10^2} + \frac{8}{10^4} + \frac{8}{10^6} + \frac{8}{10^8} + \cdots = \sum_{n=0}^{\infty} \left(\frac{8}{10^2}\right)\left(\frac{1}{10^2}\right)^n.$$

For this series, you have $a = 8/10^2$ and $r = 1/10^2$. Thus,

$$0.080808 \ldots = \frac{a}{1 - r} = \frac{8/10^2}{1 - (1/10^2)} = \frac{8}{99}.$$

Try dividing 8 by 99 on a calculator to see that it produces $0.080\overline{808}$.

The convergence of a series is not affected by removal of a finite number of terms from the beginning of the series. For instance, the geometric series

$$\sum_{n=4}^{\infty} \left(\frac{1}{2}\right)^n \quad \text{and} \quad \sum_{n=0}^{\infty} \left(\frac{1}{2}\right)^n$$

both converge. Furthermore, because the sum of the second series is $a/(1 - r) = 2$, you can conclude that the sum of the first series is

$$S = 2 - \left[\left(\frac{1}{2}\right)^0 + \left(\frac{1}{2}\right)^1 + \left(\frac{1}{2}\right)^2 + \left(\frac{1}{2}\right)^3 \right] = 2 - \frac{15}{8} = \frac{1}{8}.$$

The following properties are direct consequences of the corresponding properties of limits of sequences.

THEOREM 8.7   Properties of Infinite Series

If $\Sigma\, a_n = A$, $\Sigma\, b_n = B$, and c is a real number, then the following series converge to the indicated sums.

1. $\displaystyle\sum_{n=1}^{\infty} ca_n = cA$ **2.** $\displaystyle\sum_{n=1}^{\infty} (a_n + b_n) = A + B$ **3.** $\displaystyle\sum_{n=1}^{\infty} (a_n - b_n) = A - B$

---

nth-Term Test for Divergence

The following theorem states that if a series converges, then the limit of its nth term must be 0.

THEOREM 8.8   Limit of nth Term of a Convergent Series

If the series $\Sigma\, a_n$ converges, then the sequence $\{a_n\}$ converges to 0.

Proof   Assume that

$$\sum_{n=1}^{\infty} a_n = \lim_{n\to\infty} S_n = L.$$

Then, because $S_n = S_{n-1} + a_n$ and

$$\lim_{n\to\infty} S_n = \lim_{n\to\infty} S_{n-1} = L$$

it follows that

$$L = \lim_{n\to\infty} S_n = \lim_{n\to\infty} (S_{n-1} + a_n) = \lim_{n\to\infty} S_{n-1} + \lim_{n\to\infty} a_n$$

$$= L + \lim_{n\to\infty} a_n$$

which implies that $\{a_n\}$ converges to 0.   ■

The contrapositive of Theorem 8.8 provides a useful test for *divergence*. This **nth-Term Test for Divergence** states that if the limit of the nth term of a series does *not* converge to 0, then the series must diverge.

THEOREM 8.9 nth-Term Test for Divergence

If the sequence $\{a_n\}$ does not converge to 0, then the series $\Sigma\, a_n$ diverges.

REMARK   Be sure you see that this theorem does *not* state that the series $\Sigma\, a_n$ converges if $\{a_n\}$ converges to 0.

EXAMPLE 6   Using the nth-Term Test for Divergence

a. For the series $\displaystyle\sum_{n=0}^{\infty} 2^n$, you have

$$\lim_{n\to\infty} 2^n = \infty.$$

Thus, the limit of the nth term is not 0, and you can conclude that the series *diverges*.

b. For the series $\displaystyle\sum_{n=1}^{\infty} \frac{n!}{2n! + 1}$, you have

$$\lim_{n\to\infty} \frac{n!}{2n! + 1} = \frac{1}{2}.$$

Thus, the limit of the nth term is not 0, and you can conclude that the series *diverges*.

c. For the series $\displaystyle\sum_{n=1}^{\infty} \frac{1}{n}$, you have

$$\lim_{n\to\infty} \frac{1}{n} = 0.$$

Because the limit of the nth term is 0, the nth-Term Test for Divergence does *not* apply and you can draw no conclusions about convergence or divergence. (In the next section, you will see that this particular series diverges.)

EXAMPLE 7   Bouncing Ball Problem

A ball is dropped from a height of 6 feet and begins bouncing, as shown in Figure 8.5. The height of each bounce is three-fourths the height of the previous bounce. Find the total vertical distance traveled by the ball.

Solution   When the ball hits the ground for the first time, it has traveled a distance of $D_1 = 6$. For subsequent bounces, let D_n be the distance traveled up *and* down. For example, D_2 and D_3 are as follows.

$$D_2 = \underbrace{6\left(\frac{3}{4}\right)}_{\text{Up}} + \underbrace{6\left(\frac{3}{4}\right)}_{\text{Down}} = 12\left(\frac{3}{4}\right)$$

$$D_3 = \underbrace{6\left(\frac{3}{4}\right)\left(\frac{3}{4}\right)}_{\text{Up}} + \underbrace{6\left(\frac{3}{4}\right)\left(\frac{3}{4}\right)}_{\text{Down}} = 12\left(\frac{3}{4}\right)^2$$

By continuing this process, it can be determined that the total vertical distance is

$$\begin{aligned}
D &= 6 + 12\left(\frac{3}{4}\right) + 12\left(\frac{3}{4}\right)^2 + 12\left(\frac{3}{4}\right)^3 + \cdots \\
&= 6 + 12 \sum_{n=0}^{\infty} \left(\frac{3}{4}\right)^{n+1} \\
&= 6 + 12\left(\frac{3}{4}\right) \sum_{n=0}^{\infty} \left(\frac{3}{4}\right)^n \\
&= 6 + 9\left(\frac{1}{1 - \dfrac{3}{4}}\right) = 6 + 9(4) = 42 \text{ feet.}
\end{aligned}$$

FIGURE 8.5
The height of each bounce is three-fourths the height of the previous bounce.

EXERCISES for Section 8.2

In Exercises 1–6, find the first five terms of the sequence of partial sums.

1. $1 + \frac{1}{4} + \frac{1}{9} + \frac{1}{16} + \frac{1}{25} + \cdots$

2. $\frac{1}{2 \cdot 3} + \frac{2}{3 \cdot 4} + \frac{3}{4 \cdot 5} + \frac{4}{5 \cdot 6} + \frac{5}{6 \cdot 7} + \cdots$

3. $3 - \frac{9}{2} + \frac{27}{4} - \frac{81}{8} + \frac{243}{16} - \cdots$

4. $\frac{1}{1} + \frac{1}{3} + \frac{1}{5} + \frac{1}{7} + \frac{1}{9} + \frac{1}{11} + \cdots$

5. $\sum_{n=1}^{\infty} \frac{3}{2^{n-1}}$ **6.** $\sum_{n=1}^{\infty} \frac{(-1)^{n+1}}{n!}$

In Exercises 7–16, verify that the infinite series diverges.

7. $\sum_{n=1}^{\infty} \frac{n}{n + 1} = \frac{1}{2} + \frac{2}{3} + \frac{3}{4} + \frac{4}{5} + \cdots$

8. $\sum_{n=1}^{\infty} \frac{n}{2n + 3} = \frac{1}{5} + \frac{2}{7} + \frac{3}{9} + \frac{4}{11} + \cdots$

9. $\sum_{n=1}^{\infty} \frac{n^2}{n^2 + 1}$ **10.** $\sum_{n=1}^{\infty} \frac{n}{\sqrt{n^2 + 1}}$

11. $\sum_{n=0}^{\infty} 3\left(\frac{3}{2}\right)^n$ **12.** $\sum_{n=0}^{\infty} \left(\frac{4}{3}\right)^n$

13. $\sum_{n=0}^{\infty} 1000(1.055)^n$ **14.** $\sum_{n=0}^{\infty} 2(-1.03)^n$

15. $\sum_{n=1}^{\infty} \frac{2^n + 1}{2^{n+1}}$ **16.** $\sum_{n=1}^{\infty} \frac{n!}{2^n}$

In Exercises 17–22, verify that the infinite series converges.

17. $\sum_{n=0}^{\infty} 2\left(\frac{3}{4}\right)^n = 2 + \frac{3}{2} + \frac{9}{8} + \frac{27}{32} + \frac{81}{128} + \cdots$

18. $\sum_{n=1}^{\infty} 2\left(-\frac{1}{2}\right)^n = 2 - 1 + \frac{1}{2} - \frac{1}{4} + \frac{1}{8} - \cdots$

19. $\sum_{n=0}^{\infty} (0.9)^n = 1 + 0.9 + 0.81 + 0.729 + \cdots$

20. $\sum_{n=0}^{\infty} (-0.6)^n = 1 - 0.6 + 0.36 - 0.216 + \cdots$

21. $\sum_{n=1}^{\infty} \frac{1}{n(n + 1)}$ (Use partial fractions.)

22. $\sum_{n=1}^{\infty} \frac{1}{n(n + 2)}$ (Use partial fractions.)

In Exercises 23–36, find the sum of the convergent series.

23. $\sum_{n=0}^{\infty} \left(\frac{1}{2}\right)^n$ **24.** $\sum_{n=0}^{\infty} 2\left(\frac{2}{3}\right)^n$

25. $\sum_{n=0}^{\infty} \left(\frac{-1}{2}\right)^n$ **26.** $\sum_{n=0}^{\infty} 2\left(\frac{-2}{3}\right)^n$

27. $1 + 0.1 + 0.01 + 0.001 + \cdots$

28. $8 + 6 + \frac{9}{2} + \frac{27}{8} + \cdots$

29. $3 - 1 + \frac{1}{3} - \frac{1}{9} + \cdots$

30. $4 - 2 + 1 - \frac{1}{2} + \cdots$

31. $\sum_{n=2}^{\infty} \frac{1}{n^2 - 1}$ **32.** $\sum_{n=1}^{\infty} \frac{1}{n(n + 1)}$

33. $\sum_{n=1}^{\infty} \frac{4}{n(n + 2)}$ **34.** $\sum_{n=1}^{\infty} \frac{1}{(2n + 1)(2n + 3)}$

35. $\sum_{n=0}^{\infty} \left(\frac{1}{2^n} - \frac{1}{3^n}\right)$ **36.** $\sum_{n=1}^{\infty} [(0.7)^n + (0.9)^n]$

In Exercises 37–40, express the repeating decimal as a geometric series, and write its sum as the ratio of two integers.

37. $0.\overline{4}$ **38.** $0.23\overline{23}$

39. $0.075\overline{75}$ **40.** $0.215\overline{15}$

In Exercises 41–52, determine the convergence or divergence of the series.

41. $\sum_{n=1}^{\infty} \frac{n + 10}{10n + 1}$ **42.** $\sum_{n=1}^{\infty} \frac{n + 1}{2n - 1}$

43. $\sum_{n=1}^{\infty} \left(\frac{1}{n} - \frac{1}{n + 2}\right)$ **44.** $\sum_{n=1}^{\infty} \frac{1}{n(n + 3)}$

45. $\sum_{n=1}^{\infty} \frac{3n - 1}{2n + 1}$ **46.** $\sum_{n=1}^{\infty} \frac{2^n}{n^2}$

47. $\sum_{n=0}^{\infty} \frac{4}{2^n}$ **48.** $\sum_{n=0}^{\infty} \frac{1}{4^n}$

49. $\sum_{n=0}^{\infty} (1.075)^n$ **50.** $\sum_{n=1}^{\infty} \frac{2^n}{100}$

51. $\sum_{n=2}^{\infty} \frac{n}{\ln n}$ **52.** $\sum_{n=1}^{\infty} \left(1 + \frac{k}{n}\right)^n$

In Exercises 53 and 54, write a formula for the sum of the geometric series. For what values of x is your formula valid?

53. $1 + x + x^2 + x^3 + \cdots$

54. $1 - \frac{x}{2} + \frac{x^2}{4} - \frac{x^3}{8} + \cdots$

* C **In Exercises 55 and 56, use a graphing utility to graph the function. Identify the horizontal asymptote of the graph and determine its relationship to the sum of the series.**

Function	Series
55. $f(x) = 3\left[\dfrac{1 - (0.5)^x}{1 - 0.5}\right]$	$\sum_{n=0}^{\infty} 3\left(\dfrac{1}{2}\right)^n$
56. $f(x) = 2\left[\dfrac{1 - (0.8)^x}{1 - 0.8}\right]$	$\sum_{n=0}^{\infty} 2\left(\dfrac{4}{5}\right)^n$

*A **C** indicates an exercise in which you are instructed to use graphing technology or a symbolic computer algebra system. The solutions of other exercises may also be facilitated by appropriate technology.

In Exercises 57 and 58, determine the first term that is less than 0.0001 in each of the convergent series. Note that the answers are very different. Explain how this will affect the rate at which the series converges.

57. $\displaystyle\sum_{n=1}^{\infty} \frac{1}{n(n+1)}, \qquad \sum_{n=1}^{\infty} \left(\frac{1}{8}\right)^n$

58. $\displaystyle\sum_{n=1}^{\infty} \frac{1}{2^n}, \qquad \sum_{n=1}^{\infty} (0.01)^n$

C *Essay* In Exercises 59–62, find the sum of the series. Then use a computer or graphing utility to find the specified partial sum S_n and complete the table. Explain the relationship between the magnitude of the terms of the series and the rate at which the sequence of partial sums approaches the sum of the series.

n	5	10	20	50	100
S_n					

59. $\displaystyle\sum_{n=1}^{\infty} 2(0.9)^{n-1}$

60. $\displaystyle\sum_{n=1}^{\infty} \frac{4}{n(n+4)}$

61. $\displaystyle\sum_{n=1}^{\infty} 10(0.25)^{n-1}$

62. $\displaystyle\sum_{n=1}^{\infty} 5\left(-\frac{1}{3}\right)^{n-1}$

63. *Marketing* A company producing a new product estimates the annual sales to be 8000 units. Each year 10% of the units that have been sold become inoperative. Thus, after 1 year 8000 are in use, after 2 years $[8000 + 0.9(8000)]$ units are in use, and so on. How many units will be in use after n years?

64. *Multiplier Effect* The annual spending by tourists in a resort city is 100 million dollars. Approximately 75% of that revenue is again spent in the resort city, and of that amount approximately 75% is again spent in the same city, and so on. Write the geometric series that gives the total amount of spending generated by the 100 million dollars and find the sum of the series.

65. *Distance* A ball is dropped from a height of 16 feet. Each time it drops h feet, it rebounds $0.81h$ feet. Find the total distance traveled by the ball.

66. *Time* The ball in Exercises 65 takes the following times for each fall.

$s_1 = -16t^2 + 16,$ $s_1 = 0$ if $t = 1$

$s_2 = -16t^2 + 16(0.81),$ $s_2 = 0$ if $t = 0.9$

$s_3 = -16t^2 + 16(0.81)^2,$ $s_3 = 0$ if $t = (0.9)^2$

$s_4 = -16t^2 + 16(0.81)^3,$ $s_4 = 0$ if $t = (0.9)^3$

$\vdots$ $\vdots$

$s_n = -16t^2 + 16(0.81)^{n-1},$ $s_n = 0$ if $t = (0.9)^{n-1}$

Beginning with s_2, the ball takes the same amount of time to bounce up as it does to fall, and thus the total time elapsed before it comes to rest is

$$t = 1 + 2\sum_{n=1}^{\infty} (0.9)^n.$$

Find this total.

Probability In Exercises 67 and 68, the random variable n represents the number of units of a certain product sold per day in a store. The probability distribution of n is given by $P(n)$. Find the probability that two units are sold in a given day $[P(2)]$ and show that

$$\sum_{n=0}^{\infty} P(n) = 1.$$

67. $P(n) = \frac{1}{2}\left(\frac{1}{2}\right)^n$ **68.** $P(n) = \frac{1}{3}\left(\frac{2}{3}\right)^n$

69. *Area* The sides of a square are 16 inches in length. A new square is formed by connecting the midpoints of the sides of the original square, and two of the triangles outside the second square are shaded (see figure). Determine the area of the shaded region (a) if this process is continued five more times, and (b) if this pattern of shading is continued indefinitely.

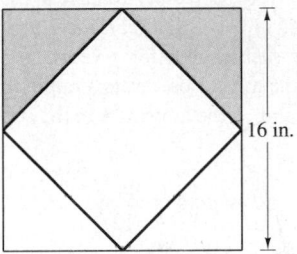

FIGURE FOR 69

16 in.

In Exercises 70–74, use the formula for the nth partial sum of a geometric series.

$$\sum_{i=0}^{n-1} ar^i = \frac{a(1-r^n)}{1-r}.$$

70. *Sphereflake* The sphereflake shown below is a computer-generated fractal that was created by Eric Haines, 3D/Eye Inc. The radius of the large sphere is 1. To the large sphere, nine spheres of radius $\frac{1}{3}$ are attached. To each of these nine spheres, nine spheres of radius $\frac{1}{9}$ are attached. This process is continued infinitely. Prove that the sphereflake has an infinite surface area. *prove that the vol. is finite & find it.*

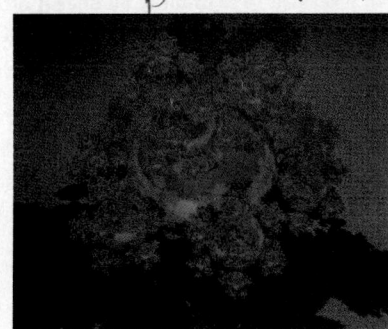

This sphereflake was created by Eric Haines, using a recursive program.

71. *Salary* You accept a job with a salary of $30,000 for the first year. Suppose, during the next 39 years, you receive a 5% raise each year. What would be your total compensation over the 40-year period?

72. *Would You Take This Job?* Suppose you went to work at a company that pays $0.01 for the first day, $0.02 for the second day, $0.04 for the third day, and so on. If the daily wage keeps doubling, what would your total income be for working
  a. 29 days?
  b. 30 days?
  c. 31 days?

73. *Profit* The annual profit for the H. J. Heinz Company from 1980 through 1989 can be approximated by the model

$$a_n = 167.5e^{0.12n}, \qquad n = 0, 1, 2, \ldots, 9$$

where a_n is the annual profit in millions of dollars and n represents the year with $n = 0$ corresponding to 1980. Use the formula for the sum of a geometric sequence to approximate the total profit earned during this 10-year period.

74. *Annuities* When an employee receives a paycheck at the end of each month, P dollars is invested in a retirement account. These deposits are made each month for t years, and the account earns interest at the annual percentage rate r. If the interest is compounded monthly, the amount A in the account at the end of t years is

$$A = P + P\left(1 + \frac{r}{12}\right) + \cdots + P\left(1 + \frac{r}{12}\right)^{12t-1}$$

$$= P\left(\frac{12}{r}\right)\left[\left(1 + \frac{r}{12}\right)^{12t} - 1\right]$$

If the interest is compounded continuously, the amount A in the account after t years is

$$A = P + Pe^{r/12} + Pe^{2r/12} + \cdots + Pe^{(12t-1)r/12}$$

$$= \frac{P(e^{rt} - 1)}{e^{r/12} - 1}.$$

Verify that each of the sums given above equals the right-hand member of the respective equation.

Annuities   In Exercises 75–78, consider making monthly deposits of P dollars into a savings account at an annual interest rate r. Use the results of Exercise 74 to find the balance A after t years if the interest is compounded (a) monthly and (b) continuously.

75. $P = \$50$ $r = 7\%$ $t = 20$ years

76. $P = \$75$ $r = 9\%$ $t = 25$ years

77. $P = \$100$ $r = 10\%$ $t = 40$ years

78. $P = \$20$ $r = 6\%$ $t = 50$ years

79. Prove that $0.75 = 0.749999 \ldots$

80. Prove that every decimal with a repeating pattern of digits is a rational number.

81. Show that the series

$$\sum_{n=1}^{\infty} a_n$$

can be written in the telescoping form

$$\sum_{n=1}^{\infty} [(c - S_{n-1}) - (c - S_n)]$$

where $S_0 = 0$ and S_n is the nth partial sum.

82. Let Σa_n be a convergent series, and let

$$R_N = a_{N+1} + a_{N+2} + \cdots$$

be the remainder of the series after the first N terms. Prove that

$$\lim_{N \to \infty} R_N = 0.$$

83. Find two divergent series Σa_n and Σb_n such that $\Sigma (a_n + b_n)$ converges.

84. Given two infinite series Σa_n and Σb_n such that Σa_n converges and Σb_n diverges, prove that $\Sigma (a_n + b_n)$ diverges.

True or False   **In Exercises 85–88, determine whether the statement is true or false. If it is false, explain why or give an example that shows it is false.**

85. If $\lim\limits_{n \to \infty} a_n \neq 0$, then $\sum\limits_{n=1}^{\infty} a_n$ diverges.

86. If $\sum\limits_{n=1}^{\infty} a_n = L$, then $\sum\limits_{n=0}^{\infty} a_n = L + a_0$.

87. If $|r| < 1$, then $\sum\limits_{n=1}^{\infty} ar^n = a/(1 - r)$.

88. The series $\sum\limits_{n=1}^{\infty} \dfrac{n}{1000(n + 1)}$ diverges.

89. *Essay* Read the article "The Exponential-Decay Law Applied to Medical Dosages" by Gerald M. Armstrong and Calvin P. Midgley in the February, 1987 issue of *Mathematics Teacher*. Then write a paragraph on how a geometric sequence can be used to find the total amount of a drug that remains in a patient's system after n equal dosages are administered (at equal time intervals).

90. Prove that

$$\frac{1}{r} + \frac{1}{r^2} + \frac{1}{r^3} + \cdots = \frac{1}{r - 1}$$

for $|r| > 1$.

91. *Cantor's Disappearing Table*   The following procedure shows how to make a table disappear by removing only half of the table!

a. Original table has a length of L.

b. Remove $\frac{1}{4}$ of the table centered at the midpoint. Each remaining piece has a length that is less than $\frac{1}{2}L$.

c. Remove $\frac{1}{8}$ of the table by taking sections of length $\frac{1}{16}L$ from the centers of the two remaining pieces. Now, you have removed $\frac{1}{4} + \frac{1}{8}$ of the table. Each remaining piece has a length that is less than $\frac{1}{4}L$.

d. Remove $\frac{1}{16}$ of the table by taking sections of length $\frac{1}{64}L$ from the centers of each of the four remaining pieces. Now, you have removed $\frac{1}{4} + \frac{1}{8} + \frac{1}{16}$ of the table. Each remaining piece has a length that is less than $\frac{1}{8}L$.

FIGURE FOR 91

Will continuing this process cause the table to disappear, even though you have only removed half of the table? Why? (For more information on this problem, see the article "Cantor's Disappearing Table" by Larry E. Knop in the November, 1985 issue of the *College Mathematics Journal.*)

SECTION
8.3

The Integral Test and *p*-Series

The Integral Test • *p*-Series and Harmonic Series

The Integral Test

In this and the following section, you will study several convergence tests that apply to series with *positive* terms.

THEOREM 8.10   The Integral Test

If f is positive, continuous, and decreasing for $x \geq 1$ and $a_n = f(n)$, then

$$\sum_{n=1}^{\infty} a_n \quad \text{and} \quad \int_{1}^{\infty} f(x)\, dx$$

either both converge or both diverge.

Proof   Begin by partitioning the interval $[1, n]$ into $n - 1$ unit intervals, as shown in Figure 8.6. The total areas of the inscribed rectangles and the circumscribed rectangles are as follows.

$$\sum_{i=2}^{n} f(i) = f(2) + f(3) + \cdots + f(n) \qquad \text{Inscribed area}$$

$$\sum_{i=1}^{n-1} f(i) = f(1) + f(2) + \cdots + f(n - 1) \qquad \text{Circumscribed area}$$

The exact area under the graph of f from $x = 1$ to $x = n$ lies between the inscribed and circumscribed areas.

$$\sum_{i=2}^{n} f(i) \leq \int_{1}^{n} f(x)\, dx \leq \sum_{i=1}^{n-1} f(i)$$

Using the nth partial sum, $S_n = f(1) + f(2) + \cdots + f(n)$, you can write this inequality as

$$S_n - f(1) \leq \int_{1}^{n} f(x)\, dx \leq S_{n-1}.$$

Now, assuming that $\int_{1}^{\infty} f(x)\, dx$ converges to L, it follows that for $n \geq 1$

$$S_n - f(1) \leq L \qquad \longrightarrow \qquad S_n \leq L + f(1).$$

Consequently, $\{S_n\}$ is bounded and monotonic, and by Theorem 8.5 it converges. Thus, $\Sigma\, a_n$ converges. For the other direction of the proof, assume that the improper integral diverges. Then $\int_{1}^{n} f(x)\, dx$ approaches infinity as $n \to \infty$, and the inequality $S_{n-1} \geq \int_{1}^{n} f(x)\, dx$ implies that $\{S_n\}$ diverges. Thus, $\Sigma\, a_n$ diverges.

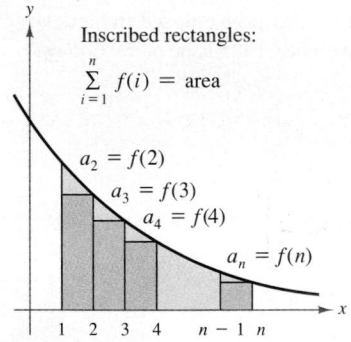

Inscribed rectangles:

$$\sum_{i=1}^{n} f(i) = \text{area}$$

$a_2 = f(2)$
$a_3 = f(3)$
$a_4 = f(4)$
$a_n = f(n)$

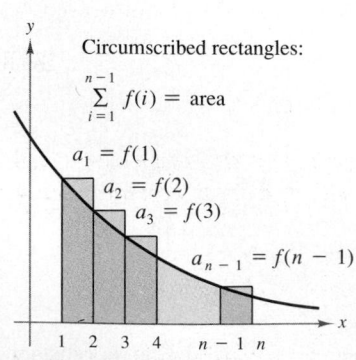

Circumscribed rectangles:

$$\sum_{i=1}^{n-1} f(i) = \text{area}$$

$a_1 = f(1)$
$a_2 = f(2)$
$a_3 = f(3)$
$a_{n-1} = f(n - 1)$

FIGURE 8.6

We noted earlier that the convergence or divergence of $\Sigma\, a_n$ is not affected by deleting the first N terms. Similarly, if the conditions for the Integral Test are satisfied for all $x \geq N > 1$, you can simply use the integral $\int_N^\infty f(x)\, dx$ to test for convergence or divergence. (This is illustrated in Example 4.)

EXAMPLE 1   Using the Integral Test

Apply the Integral Test to the series $\displaystyle\sum_{n=1}^{\infty} \frac{n}{n^2 + 1}$.

Solution   Because $f(x) = x/(x^2 + 1)$ satisfies the conditions for the Integral Test (check this), you can integrate to obtain

$$\int_1^\infty \frac{x}{x^2 + 1}\, dx = \frac{1}{2} \int_1^\infty \frac{2x}{x^2 + 1}\, dx = \frac{1}{2} \lim_{b \to \infty} \int_1^b \frac{2x}{x^2 + 1}\, dx$$

$$= \frac{1}{2} \lim_{b \to \infty} \left[\ln(x^2 + 1) \right]_1^b$$

$$= \frac{1}{2} \lim_{b \to \infty} \left[\ln(b^2 + 1) - \ln 2 \right]$$

$$= \infty.$$

Thus, the series *diverges*.

EXAMPLE 2   Using the Integral Test

Apply the Integral Test to the series $\displaystyle\sum_{n=1}^{\infty} \frac{1}{n^2 + 1}$.

Solution   Because $f(x) = 1/(x^2 + 1)$ satisfies the conditions for the Integral Test, you can integrate to obtain

$$\int_1^\infty \frac{1}{x^2 + 1}\, dx = \lim_{b \to \infty} \int_1^b \frac{1}{x^2 + 1}\, dx = \lim_{b \to \infty} \left[\arctan x \right]_1^b$$

$$= \lim_{b \to \infty} (\arctan b - \arctan 1)$$

$$= \frac{\pi}{2} - \frac{\pi}{4} = \frac{\pi}{4}.$$

Thus, the series *converges* (see Figure 8.7).

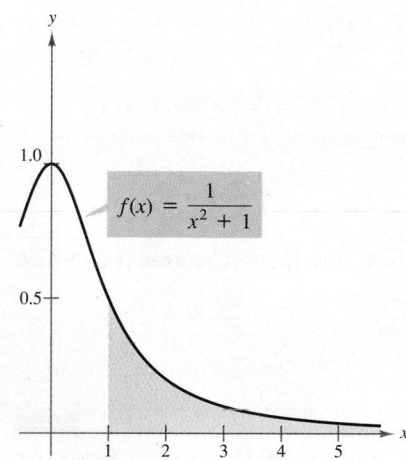

FIGURE 8.7
Because the improper integral converges, the infinite series also converges.

$f(x) = \dfrac{1}{x^2 + 1}$

: **TECHNOLOGY**   In Example 2, note that the improper integral converges to $\pi/4$— this does not imply that the infinite series converges to $\pi/4$. To approximate the sum of the series, you can use the inequality

$$\sum_{n=1}^{N} \frac{1}{n^2 + 1} \leq \sum_{n=1}^{\infty} \frac{1}{n^2 + 1} \leq \sum_{n=1}^{N} \frac{1}{n^2 + 1} + \int_N^\infty \frac{1}{x^2 + 1}\, dx.$$

(See Exercise 38.) The larger the value of N, the better the approximation. For instance, using $N = 200$ produces $1.071 \leq \Sigma\, 1/(n^2 + 1) \leq 1.077$.

p-Series and Harmonic Series

In the rest of this section, we investigate a second type of series that has a simple arithmetic test for convergence or divergence. A series of the form

$$\sum_{n=1}^{\infty} \frac{1}{n^p} = \frac{1}{1^p} + \frac{1}{2^p} + \frac{1}{3^p} + \cdots \qquad \text{\textit{p}-Series}$$

is a **p-series,** where *p* is a positive constant. For *p* = 1, the series

$$\sum_{n=1}^{\infty} \frac{1}{n} = 1 + \frac{1}{2} + \frac{1}{3} + \cdots \qquad \text{Harmonic series}$$

is the **harmonic series.** A **general harmonic series** is of the form $\Sigma\, 1/(an + b)$. In music, strings of the same material, diameter, and tension, whose lengths form a harmonic series, produce harmonic tones.

The Integral Test is convenient for establishing the convergence or divergence of *p*-series. This is shown in the proof of Theorem 8.11.

THEOREM 8.11   Convergence of *p*-Series

The *p*-series

$$\sum_{n=1}^{\infty} \frac{1}{n^p} = \frac{1}{1^p} + \frac{1}{2^p} + \frac{1}{3^p} + \frac{1}{4^p} + \cdots$$

1. converges if $p > 1$,   and   **2.** diverges if $0 < p \leq 1$.

Proof   The proof follows from the Integral Test and from Theorem 7.5, which states that

$$\int_{1}^{\infty} \frac{1}{x^p}\,dx$$

converges if $p > 1$, and diverges if $0 < p \leq 1$.   ∎

REMARK   The sum of the series in Example 3b can be shown to be $\pi^2/6$. (This was shown by Leonhard Euler, but the proof is too difficult to present here.) Be sure you see that the Integral Test does not tell you that the sum of the series is equal to the value of the integral. For instance, the sum of the series in Example 3b is

$$\sum_{n=1}^{\infty} \frac{1}{n^2} = \frac{\pi^2}{6} \approx 1.645$$

but the value of the corresponding improper integral is

$$\int_{1}^{\infty} \frac{1}{x^2}\,dx = 1.$$

EXAMPLE 3   Convergent and Divergent *p*-Series

a. From Theorem 8.11, it follows that the *harmonic series*

$$\sum_{n=1}^{\infty} \frac{1}{n} = \frac{1}{1} + \frac{1}{2} + \frac{1}{3} + \cdots \qquad p = 1$$

diverges.

b. From Theorem 8.11, it follows that the *p*-series

$$\sum_{n=1}^{\infty} \frac{1}{n^2} = \frac{1}{1^2} + \frac{1}{2^2} + \frac{1}{3^2} + \cdots \qquad p = 2$$

converges.

EXAMPLE 4   **Testing a Series for Convergence**

Determine whether the following series converges or diverges.

$$\sum_{n=2}^{\infty} \frac{1}{n \ln n}$$

Solution   This series is similar to the divergent harmonic series. If its terms were larger than those of the harmonic series, you would expect it to diverge. However, because its terms are smaller, you are not sure what to expect. Using the Integral Test, with $f(x) = 1/(x \ln x)$, you can see that the series diverges.

$$\int_{2}^{\infty} \frac{1}{x \ln x} dx = \int_{2}^{\infty} \frac{1/x}{\ln x} dx$$

$$= \lim_{b \to \infty} \left[\ln(\ln x) \right]_{2}^{b}$$

$$= \lim_{b \to \infty} \left[\ln(\ln b) - \ln(\ln 2) \right]$$

$$= \infty$$

EXERCISES for Section 8.3

TECHNOLOGY
Laboratory Guide
Lab 8.3

In Exercises 1–10, use the Integral Test to determine the convergence or divergence of the series.

1. $\sum_{n=1}^{\infty} \dfrac{1}{n+1}$ **2.** $\sum_{n=1}^{\infty} n e^{-n}$

3. $\sum_{n=1}^{\infty} e^{-n}$ **4.** $\sum_{n=1}^{\infty} \dfrac{1}{4n+1}$

5. $\dfrac{1}{2} + \dfrac{1}{5} + \dfrac{1}{10} + \dfrac{1}{17} + \dfrac{1}{26} + \cdots$

6. $\dfrac{1}{3} + \dfrac{1}{5} + \dfrac{1}{7} + \dfrac{1}{9} + \dfrac{1}{11} + \cdots$

7. $\dfrac{\ln 2}{2} + \dfrac{\ln 3}{3} + \dfrac{\ln 4}{4} + \dfrac{\ln 5}{5} + \dfrac{\ln 6}{6} + \cdots$

8. $\dfrac{1}{4} + \dfrac{2}{7} + \dfrac{3}{12} + \cdots + \dfrac{n}{n^2 + 3} + \cdots$

9. $\sum_{n=1}^{\infty} \dfrac{n^{k-1}}{n^k + c}$, k is a positive integer

10. $\sum_{n=1}^{\infty} n^k e^{-n}$, k is a positive integer

In Exercises 11 and 12, use the Integral Test to determine the convergence or divergence of the *p*-series.

11. $\sum_{n=1}^{\infty} \dfrac{1}{n^3}$ **12.** $\sum_{n=1}^{\infty} \dfrac{1}{n^{1/3}}$

In Exercises 13–20, use Theorem 8.11 to determine the convergence or divergence of the *p*-series.

13. $\sum_{n=1}^{\infty} \dfrac{1}{\sqrt[5]{n}}$ **14.** $\sum_{n=1}^{\infty} \dfrac{1}{n^{4/3}}$

15. $1 + \dfrac{1}{\sqrt{2}} + \dfrac{1}{\sqrt{3}} + \dfrac{1}{\sqrt{4}} + \cdots$

16. $1 + \dfrac{1}{4} + \dfrac{1}{9} + \dfrac{1}{16} + \dfrac{1}{25} + \cdots$

17. $1 + \dfrac{1}{2\sqrt{2}} + \dfrac{1}{3\sqrt{3}} + \dfrac{1}{4\sqrt{4}} + \dfrac{1}{5\sqrt{5}} + \cdots$

18. $1 + \dfrac{1}{\sqrt[3]{4}} + \dfrac{1}{\sqrt[3]{9}} + \dfrac{1}{\sqrt[3]{16}} + \dfrac{1}{\sqrt[3]{25}} + \cdots$

19. $\sum_{n=1}^{\infty} \dfrac{1}{n^{1.04}}$ **20.** $\sum_{n=1}^{\infty} \dfrac{1}{n^{\pi}}$

In Exercises 21–32, determine the convergence or divergence of the series.

21. $\sum_{n=1}^{\infty} \dfrac{1}{2n-1}$ **22.** $\sum_{n=2}^{\infty} \dfrac{1}{n\sqrt{n^2-1}}$

23. $\sum_{n=1}^{\infty} \dfrac{1}{n\sqrt[4]{n}}$ **24.** $3\sum_{n=1}^{\infty} \dfrac{1}{n^{0.95}}$

25. $\sum_{n=0}^{\infty} \left(\dfrac{2}{3}\right)^n$ **26.** $\sum_{n=0}^{\infty} (1.075)^n$

27. $\sum_{n=1}^{\infty} \dfrac{n}{\sqrt{n^2+1}}$ **28.** $\sum_{n=1}^{\infty} \left(\dfrac{1}{n^2} - \dfrac{1}{n^3}\right)$

29. $\sum_{n=1}^{\infty} \left(1 + \dfrac{1}{n}\right)^n$ **30.** $\sum_{n=2}^{\infty} \ln n$

31. $\sum_{n=2}^{\infty} \dfrac{1}{n(\ln n)^3}$ **32.** $\sum_{n=2}^{\infty} \dfrac{\ln n}{n^3}$

In Exercises 33 and 34, find the positive values of *p* for which the series converges.

33. $\sum_{n=2}^{\infty} \dfrac{1}{n(\ln n)^p}$ **34.** $\sum_{n=2}^{\infty} \dfrac{\ln n}{n^p}$

C **35.** Because the harmonic series diverges, it follows that for any positive real number M there exists a positive integer N such that the partial sum

$$\sum_{n=1}^{N} \frac{1}{n} > M.$$

a. Use a computer to complete the table.

M	2	4	6	8
N				

b. As the real number M increases in equal increments, does the number N increase in equal increments? Why or why not?

36. The **Riemann zeta function** for real numbers is defined for all x for which the series

$$\zeta(x) = \sum_{n=1}^{\infty} n^{-x}$$

converges. Find the domain of the function.

37. Let f be a positive, continuous, and decreasing function for $x \geq 1$, such that $a_n = f(n)$. Prove that if the series

$$\sum_{n=1}^{\infty} a_n$$

converges to S, then the remainder $R_N = S - S_N$ is bounded by

$$0 \leq R_N \leq \int_{N}^{\infty} f(x)\, dx.$$

38. Show that the result of Exercise 37 can be written as

$$\sum_{n=1}^{N} a_n \leq \sum_{n=1}^{\infty} a_n \leq \sum_{n=1}^{N} a_n + \int_{N}^{\infty} f(x)\, dx.$$

C **In Exercises 39–44, use the result of Exercise 37 to approximate the sum of the convergent series using the indicated number of terms. Include an estimate of the maximum error for your approximation.**

39. $\displaystyle\sum_{n=1}^{\infty} \frac{1}{n^4},$ six terms   **40.** $\displaystyle\sum_{n=1}^{\infty} \frac{1}{n^5},$ four terms

41. $\displaystyle\sum_{n=1}^{\infty} \frac{1}{n^2 + 1},$ ten terms

42. $\displaystyle\sum_{n=1}^{\infty} \frac{1}{(n + 1)[\ln(n + 1)]^3},$ ten terms

43. $\displaystyle\sum_{n=1}^{\infty} ne^{-n^2},$ four terms   **44.** $\displaystyle\sum_{n=1}^{\infty} e^{-n},$ four terms

In Exercises 45–48, use the result of Exercise 37 to find N so that $R_N \leq 0.001$ for the (convergent) series.

45. $\displaystyle\sum_{n=1}^{\infty} \frac{1}{n^4}$ **46.** $\displaystyle\sum_{n=1}^{\infty} \frac{1}{n^{3/2}}$

47. $\displaystyle\sum_{n=1}^{\infty} e^{-5n}$ **48.** $\displaystyle\sum_{n=1}^{\infty} \frac{1}{n^2 + 1}$

C **49.** *Essay* Use a computer to find the specified partial sum S_n and complete the table. Compare the rate at which the sequence of partial sums approaches the sum of the series.

n	5	10	20	50	100
S_n					

a. $\displaystyle\sum_{n=1}^{\infty} 3\left(\frac{1}{5}\right)^{n-1} = \frac{15}{4}$

b. $\displaystyle\sum_{n=1}^{\infty} \frac{1}{n^2} = \frac{\pi^2}{6}$

C **50.** Ten terms are used to approximate a convergent p-series. Therefore, the remainder is a function of p and is given by

$$0 \leq R_{10}(p) \leq \int_{10}^{\infty} \frac{1}{x^p}\, dx, \qquad p > 1.$$

a. Perform the integration in the above inequality.
b. Use a graphing utility to graphically represent the inequality.
c. Identify any asymptotes of the error function and interpret their meaning.

51. a. Show that

$$\sum_{n=2}^{\infty} \frac{1}{n^{1.1}}$$

converges and

$$\sum_{n=2}^{\infty} \frac{1}{n \ln n}$$

diverges.

b. Compare the first five terms of each series.
c. Find $n > 3$ such that

$$\frac{1}{n^{1.1}} < \frac{1}{n \ln n}.$$

52. *Euler's Constant* Let

$$S_n = \sum_{k=1}^{n} \frac{1}{k} = 1 + \frac{1}{2} + \cdots + \frac{1}{n}.$$

a. Show that $\ln(n + 1) \leq S_n \leq 1 + \ln n$.
b. Show that the sequence $\{a_n\} = \{S_n - \ln n\}$ is bounded.
c. Show that the sequence $\{a_n\}$ is decreasing.
d. Show that a_n converges to a limit γ (called Euler's constant).
e. Approximate γ using a_{100}.

53. Find the sum of the series.

$$\sum_{n=2}^{\infty} \ln\left(1 - \frac{1}{n^2}\right)$$

SECTION
8.4

Comparisons of Series

Direct Comparison Test • Limit Comparison Test

Direct Comparison Test

For the convergence tests developed so far, the terms of the series had to be fairly simple and the series *had to possess* special characteristics in order for the convergence tests to be applied. A slight deviation from these special characteristics can make a test nonapplicable. For example, in the following pairs, the second series cannot be tested by the same convergence test as the first series even though it is similar to the first.

1. $\sum_{n=0}^{\infty} \dfrac{1}{2^n}$ is geometric, but $\sum_{n=0}^{\infty} \dfrac{n}{2^n}$ is not.

2. $\sum_{n=1}^{\infty} \dfrac{1}{n^3}$ is a *p*-series, but $\sum_{n=1}^{\infty} \dfrac{1}{n^3 + 1}$ is not.

3. $a_n = \dfrac{n}{(n^2 + 3)^2}$ is easily integrated, but $b_n = \dfrac{n^2}{(n^2 + 3)^2}$ is not.

In this section you will study two additional tests for positive-term series. These two tests greatly expand the variety of series you are able to test for convergence or divergence by allowing you to *compare* a series having complicated terms with a simpler series whose convergence or divergence is known.

THEOREM 8.12   Direct Comparison Test

Let $0 \le a_n \le b_n$ for all *n*.

1. If $\displaystyle\sum_{n=1}^{\infty} b_n$ converges, then $\displaystyle\sum_{n=1}^{\infty} a_n$ converges.

2. If $\displaystyle\sum_{n=1}^{\infty} a_n$ diverges, then $\displaystyle\sum_{n=1}^{\infty} b_n$ diverges.

REMARK   As stated, the Direct Comparison Test requires that $0 \le a_n \le b_n$ for all *n*. Because the convergence of a series is not dependent on its first several terms, you could modify the test to require only that $0 \le a_n \le b_n$ for all *n* greater than some integer *N*.

Proof   To prove the first property, let $L = \displaystyle\sum_{n=1}^{\infty} b_n$ and let $S_n = a_1 + a_2 + \cdots + a_n$. Because $0 \le a_n \le b_n$, the sequence $S_1, S_2, S_3, \ldots$ is nondecreasing and bounded above by *L*; hence it must converge. Because

$$\lim_{n \to \infty} S_n = \sum_{n=1}^{\infty} a_n$$

it follows that $\Sigma\, a_n$ converges. The second property is logically equivalent to the first.

EXAMPLE 1 Using the Direct Comparison Test

Determine the convergence or divergence of $\displaystyle\sum_{n=1}^{\infty} \frac{1}{2 + 3^n}$.

Solution This series resembles

$$\sum_{n=1}^{\infty} \frac{1}{3^n}. \qquad \text{Convergent geometric series}$$

Term-by-term comparison yields

$$a_n = \frac{1}{2 + 3^n} < \frac{1}{3^n} = b_n, \qquad n \geq 1.$$

Thus, by the Direct Comparison Test, the series converges.

EXAMPLE 2 Using the Direct Comparison Test

Determine the convergence or divergence of $\displaystyle\sum_{n=1}^{\infty} \frac{1}{2 + \sqrt{n}}$.

Solution This series resembles

$$\sum_{n=1}^{\infty} \frac{1}{n^{1/2}}. \qquad \text{Divergent } p\text{-series}$$

Term-by-term comparison yields

$$\frac{1}{2 + \sqrt{n}} \leq \frac{1}{\sqrt{n}}, \qquad n \geq 1$$

which *does not* meet the requirements for divergence. (Remember that if term-by-term comparison reveals a series that is *smaller* than a divergent series, then the Direct Comparison Test tells you nothing.) Still expecting the series to diverge, you can compare the given series with

$$\sum_{n=1}^{\infty} \frac{1}{n}. \qquad \text{Divergent harmonic series}$$

In this case, term-by-term comparison yields

$$a_n = \frac{1}{n} \leq \frac{1}{2 + \sqrt{n}} = b_n, \qquad n \geq 4$$

REMARK To verify the last inequality in Example 2, try showing that $2 + \sqrt{n} \leq n$ whenever $n \geq 4$.

and, by the Direct Comparison Test, the given series diverges.

Remember that both parts of the Direct Comparison Test require that $0 \leq a_n \leq b_n$. Informally, the test says the following about two series with nonnegative terms.

1. If the "larger" series $\sum b_n$ converges, then the "smaller" series $\sum a_n$ must also converge.
2. If the "smaller" series $\sum a_n$ diverges, then the "larger" series $\sum b_n$ must also diverge.

Limit Comparison Test

Often a given series closely resembles a p-series or a geometric series, yet you cannot establish the term-by-term comparison necessary to apply the Direct Comparison Test. Under these circumstances you may be able to apply a second comparison test, called the **Limit Comparison Test.**

THEOREM 8.13   Limit Comparison Test

Suppose that $a_n > 0$, $b_n > 0$, and

$$\lim_{n \to \infty} \left(\frac{a_n}{b_n} \right) = L$$

where L is *finite and positive*. Then the two series $\Sigma \, a_n$ and $\Sigma \, b_n$ either both converge or both diverge.

Proof   Because $a_n > 0$, $b_n > 0$, and $(a_n/b_n) \to L$ as $n \to \infty$, there exists $N > 0$ such that

$$0 < (a_n/b_n) < (L + 1), \qquad \text{for } n \geq N.$$

This implies that

$$0 < a_n < (L + 1)b_n.$$

Hence, by the Direct Comparison Test, the convergence of $\Sigma \, b_n$ implies the convergence of $\Sigma \, a_n$. Similarly, the fact that

$$\lim_{n \to \infty} \left(\frac{b_n}{a_n} \right) = \frac{1}{L}$$

can be used to show that the convergence of $\Sigma \, a_n$ implies the convergence of $\Sigma \, b_n$. ▄

REMARK   As with the Direct Comparison Test, the Limit Comparison Test could be modified to require only that a_n and b_n be positive for all n greater than some integer N.

The versatility of the Limit Comparison Test is demonstrated in the next example, which shows that a general harmonic series diverges.

EXAMPLE 3   Using the Limit Comparison Test

Show that the following general harmonic series diverges.

$$\sum_{n=1}^{\infty} \frac{1}{an + b}, \qquad a > 0, b > 0$$

Solution   By comparison with

$$\sum_{n=1}^{\infty} \frac{1}{n} \qquad \text{Divergent harmonic series}$$

we have

$$\lim_{n \to \infty} \frac{1/(an + b)}{1/n} = \lim_{n \to \infty} \frac{n}{an + b} = \frac{1}{a}.$$

Because this limit is greater than 0, you can conclude from the Limit Comparison Test that the given series diverges. ▄

The Limit Comparison Test works well for comparing a "messy" algebraic series with a p-series. In choosing an appropriate p-series, you must choose one with an nth term of the same magnitude as the nth term of the given series.

Given series	Comparison series	Conclusion
$\displaystyle\sum_{n=1}^{\infty} \frac{1}{3n^2 - 4n + 5}$	$\displaystyle\sum_{n=1}^{\infty} \frac{1}{n^2}$	Both series converge.
$\displaystyle\sum_{n=1}^{\infty} \frac{1}{\sqrt{3n - 2}}$	$\displaystyle\sum_{n=1}^{\infty} \frac{1}{\sqrt{n}}$	Both series diverge.
$\displaystyle\sum_{n=1}^{\infty} \frac{n^2 - 10}{4n^5 + n^3}$	$\displaystyle\sum_{n=1}^{\infty} \frac{n^2}{n^5} = \sum_{n=1}^{\infty} \frac{1}{n^3}$	Both series converge.

In other words, when choosing a series for comparison, you can disregard all but the *highest powers of n* in both the numerator and the denominator.

EXAMPLE 4   Using the Limit Comparison Test

Determine the convergence or divergence of $\displaystyle\sum_{n=1}^{\infty} \frac{\sqrt{n}}{n^2 + 1}$.

Solution   Disregarding all but the highest powers of n in the numerator and the denominator, you can compare the series with

$$\sum_{n=1}^{\infty} \frac{\sqrt{n}}{n^2} = \sum_{n=1}^{\infty} \frac{1}{n^{3/2}}. \qquad \text{Convergent } p\text{-series}$$

Because

$$\lim_{n \to \infty} \frac{a_n}{b_n} = \lim_{n \to \infty} \left(\frac{\sqrt{n}}{n^2 + 1} \right) \left(\frac{n^{3/2}}{1} \right) = \lim_{n \to \infty} \frac{n^2}{n^2 + 1} = 1$$

you conclude by the Limit Comparison Test that the given series converges.

EXAMPLE 5   Using the Limit Comparison Test

Determine the convergence or divergence of $\displaystyle\sum_{n=1}^{\infty} \frac{n 2^n}{4n^3 + 1}$.

Solution   A reasonable comparison would be with the series

$$\sum_{n=1}^{\infty} \frac{2^n}{n^2}.$$

Note that this series diverges by the nth-Term Test. From the limit

$$\lim_{n \to \infty} \frac{a_n}{b_n} = \lim_{n \to \infty} \left(\frac{n 2^n}{4n^3 + 1} \right) \left(\frac{n^2}{2^n} \right) = \lim_{n \to \infty} \frac{1}{4 + (1/n^3)} = \frac{1}{4}$$

you can conclude that the given series diverges.

1-25 odd, 27-34, 47-50

EXERCISES for Section 8.4

In Exercises 1–12, use the Direct Comparison Test to determine the convergence or divergence of the series.

1. $\displaystyle\sum_{n=1}^{\infty} \frac{1}{n^2 + 1}$

2. $\displaystyle\sum_{n=1}^{\infty} \frac{1}{3n^2 + 2}$

3. $\displaystyle\sum_{n=2}^{\infty} \frac{1}{n - 1}$

4. $\displaystyle\sum_{n=2}^{\infty} \frac{1}{\sqrt{n - 1}}$

5. $\displaystyle\sum_{n=0}^{\infty} \frac{1}{3^n + 1}$

6. $\displaystyle\sum_{n=0}^{\infty} \frac{2^n}{3^n + 5}$

7. $\displaystyle\sum_{n=2}^{\infty} \frac{\ln n}{n + 1}$

8. $\displaystyle\sum_{n=1}^{\infty} \frac{1}{\sqrt{n^3 + 1}}$

9. $\displaystyle\sum_{n=0}^{\infty} \frac{1}{n!}$

10. $\displaystyle\sum_{n=1}^{\infty} \frac{1}{3\sqrt[4]{n} - 1}$

11. $\displaystyle\sum_{n=0}^{\infty} e^{-n^2}$

12. $\displaystyle\sum_{n=1}^{\infty} \frac{4^n}{3^n - 1}$

In Exercises 13–26, use the Limit Comparison Test to determine the convergence or divergence of the series.

13. $\displaystyle\sum_{n=1}^{\infty} \frac{n}{n^2 + 1}$

14. $\displaystyle\sum_{n=2}^{\infty} \frac{1}{\sqrt{n^2 - 1}}$

15. $\displaystyle\sum_{n=0}^{\infty} \frac{1}{\sqrt{n^2 + 1}}$

16. $\displaystyle\sum_{n=1}^{\infty} \frac{1}{2^n - 5}$

17. $\displaystyle\sum_{n=1}^{\infty} \frac{2n^2 - 1}{3n^5 + 2n + 1}$

18. $\displaystyle\sum_{n=1}^{\infty} \frac{5n - 3}{n^2 - 2n + 5}$

19. $\displaystyle\sum_{n=1}^{\infty} \frac{n + 3}{n(n + 2)}$

20. $\displaystyle\sum_{n=1}^{\infty} \frac{1}{n(n^2 + 1)}$

21. $\displaystyle\sum_{n=1}^{\infty} \frac{1}{n\sqrt{n^2 + 1}}$

22. $\displaystyle\sum_{n=1}^{\infty} \frac{n}{(n + 1)2^{n-1}}$

23. $\displaystyle\sum_{n=1}^{\infty} \frac{n^{k-1}}{n^k + 1}, \quad k > 2$

24. $\displaystyle\sum_{n=1}^{\infty} \frac{1}{n + \sqrt{n^2 + 1}}$

25. $\displaystyle\sum_{n=1}^{\infty} \sin\frac{1}{n}$

26. $\displaystyle\sum_{n=1}^{\infty} \tan\frac{1}{n}$

In Exercises 27–34, test for convergence or divergence, using each of the seven tests at least once.

a. nth-Term Test
b. Geometric Series Test
c. p-Series Test
d. Telescoping Series Test
e. Integral Test
f. Direct Comparison Test
g. Limit Comparison Test

27. $\displaystyle\sum_{n=1}^{\infty} \frac{\sqrt{n}}{n}$

28. $\displaystyle\sum_{n=0}^{\infty} 5\left(-\frac{1}{5}\right)^n$

29. $\displaystyle\sum_{n=1}^{\infty} \frac{1}{3^n + 2}$

30. $\displaystyle\sum_{n=4}^{\infty} \frac{1}{3n^2 - 2n - 15}$

31. $\displaystyle\sum_{n=1}^{\infty} \frac{n}{2n + 3}$

32. $\displaystyle\sum_{n=1}^{\infty} \left(\frac{1}{n + 1} - \frac{1}{n + 2}\right)$

33. $\displaystyle\sum_{n=1}^{\infty} \frac{n}{(n^2 + 1)^2}$

34. $\displaystyle\sum_{n=1}^{\infty} \frac{3}{n(n + 3)}$

35. Use the Limit Comparison Test with the harmonic series to show that the series $\sum a_n$ diverges if

$$\lim_{n \to \infty} na_n = L \neq 0.$$

36. Prove that if $P(n)$ and $Q(n)$ are polynomials of degree j and k, respectively, then the series

$$\sum_{n=1}^{\infty} \frac{P(n)}{Q(n)}$$

converges if $j < k - 1$, and diverges if $j \geq k - 1$.

In Exercises 37–40, use the polynomial test as given in Exercise 36 to determine whether the series converges or diverges.

37. $\frac{1}{2} + \frac{2}{5} + \frac{3}{10} + \frac{4}{17} + \frac{5}{26} + \cdots$

38. $\frac{1}{3} + \frac{1}{8} + \frac{1}{15} + \frac{1}{24} + \frac{1}{35} + \cdots$

39. $\displaystyle\sum_{n=1}^{\infty} \frac{1}{n^3 + 1}$

40. $\displaystyle\sum_{n=1}^{\infty} \frac{n^2}{n^3 + 1}$

In Exercises 41 and 42, use the divergence test given in Exercise 35 to show that each series diverges.

41. $\displaystyle\sum_{n=1}^{\infty} \frac{n^3}{5n^4 + 3}$

42. $\displaystyle\sum_{n=2}^{\infty} \frac{1}{\ln n}$

C **43.** Consider the following series and its sum.

$$\sum_{n=1}^{\infty} \frac{1}{(2n - 1)^2} = \frac{\pi^2}{8}$$

a. Verify that the series converges.

b. Use a computer or calculator to find the partial sum S_n and complete the table.

n	5	10	20	50	100
S_n					

c. Find the sum of the series

$$\sum_{n=3}^{\infty} \frac{1}{(2n - 1)^2}.$$

d. Use a computer or calculator to find the sum of the series

$$\sum_{n=10}^{\infty} \frac{1}{(2n - 1)^2}.$$

e. Does excluding the first finite number of terms of an infinite series change the divergence or convergence of the series? Why or why not?

44. It appears that the terms of the series

$$\tfrac{1}{1000} + \tfrac{1}{1001} + \tfrac{1}{1002} + \tfrac{1}{1003} + \cdots$$

are less than the corresponding terms of the convergent series

$$1 + \tfrac{1}{4} + \tfrac{1}{9} + \tfrac{1}{16} + \cdots.$$

If the statement above is correct, then the first series converges. Is this correct? Why or why not? As in Exercise 43, make a statement about how the divergence or convergence of a series is affected by inclusion or exclusion of the first finite number of terms.

45. Prove that if the nonnegative series

$$\sum_{n=1}^{\infty} a_n \quad \text{and} \quad \sum_{n=1}^{\infty} b_n$$

converge, then so does the series

$$\sum_{n=1}^{\infty} a_n b_n.$$

46. Use the result of Exercise 45 to prove that if the nonnegative series

$$\sum_{n=1}^{\infty} a_n$$

converges, then so does the series

$$\sum_{n=1}^{\infty} a_n^2.$$

True or False **In Exercises 47–50, determine whether the statement is true or false. If it is false, explain why or give an example that shows it is false.**

47. If $0 \le a_n \le b_n$ and $\displaystyle\sum_{n=1}^{\infty} a_n$ converges, then $\displaystyle\sum_{n=1}^{\infty} b_n$ diverges.

48. If $0 \le a_{n+10} \le b_n$ and $\displaystyle\sum_{n=1}^{\infty} b_n$ converges, then $\displaystyle\sum_{n=1}^{\infty} a_n$ converges.

49. If $a_n + b_n \le c_n$ and $\displaystyle\sum_{n=1}^{\infty} c_n$ converges, then the series $\displaystyle\sum_{n=1}^{\infty} a_n$ and $\displaystyle\sum_{n=1}^{\infty} b_n$ both converge. (Assume that the terms of all three series are positive.)

50. If $a_n \le b_n + c_n$ and $\displaystyle\sum_{n=1}^{\infty} a_n$ diverges, then the series $\displaystyle\sum_{n=1}^{\infty} b_n$ and $\displaystyle\sum_{n=1}^{\infty} c_n$ both diverge. (Assume that the terms of all three series are positive.)

51. Suppose that Σa_n and Σb_n are series with positive terms. Prove the following.

a. If $\displaystyle\lim_{n \to \infty} \frac{a_n}{b_n} = 0$ and Σb_n converges, then Σa_n also converges.

b. If $\displaystyle\lim_{n \to \infty} \frac{a_n}{b_n} = \infty$ and Σb_n diverges, then Σa_n also diverges.

52. *Solera Method* Most wines are produced entirely from grapes grown in a single year. Sherry, however, is a complex mixture of older wines with new wines. This is done with a sequence of barrels (called a solera) stacked on top of each other, as shown in the figure.

The oldest wine is in the bottom tier of barrels, and the newest is in the top tier. Each year, half of each barrel in the bottom tier is bottled as sherry. The bottom barrels are then refilled with wine from the barrels above. This process is repeated throughout the solera, with new wine being added to the top barrels. A mathematical model for the amount of n-year-old wine that is removed from a solera (with k tiers) each year is

$$f(n, k) = \binom{n-1}{k-1}\left(\frac{1}{2}\right)^{n+1}, \qquad k \leq n.$$

a. Consider a solera that has five tiers, numbered $k = 1, 2, 3, 4,$ and 5. In 1970 ($n = 0$), half of each barrel in the top tier (tier 1) was refilled with new wine. How much of this wine was removed from the solera in 1971? In 1972? In 1973? . . . In 1985? During which year(s) was the greatest amount of the 1970 wine removed from the solera?

b. In part a, let a_n be the amount of 1970 wine that is removed from the solera in year n. Evaluate

$$\sum_{n=0}^{\infty} a_n.$$

(For more information on this problem, see the article "Finding Vintage Concentrations in a Sherry Solera" by Rhodes Peele and John T. MacQueen in the 1990 *UMAP Modules*.)

SECTION **Alternating Series**

8.5   **Alternating Series • Alternating Series Remainder • Absolute and Conditional Convergence • Rearrangement of Series**

Alternating Series

So far, most series we have dealt with have had positive terms. In this section and the following section, you will study series that contain both positive and negative terms. The simplest such series is an **alternating series,** whose terms alternate in sign. For example, the geometric series

$$\sum_{n=0}^{\infty} \left(-\frac{1}{2} \right)^n = \sum_{n=0}^{\infty} (-1)^n \frac{1}{2^n} = 1 - \frac{1}{2} + \frac{1}{4} - \frac{1}{8} + \frac{1}{16} - \cdots$$

is an *alternating geometric series* with $r = -\frac{1}{2}$. Alternating series occur in two ways: either the odd terms are negative or the even terms are negative.

THEOREM 8.14   Alternating Series Test

Let $a_n > 0$. The alternating series

$$\sum_{n=1}^{\infty} (-1)^n a_n \quad \text{and} \quad \sum_{n=1}^{\infty} (-1)^{n+1} a_n$$

converge, if the following two conditions are met.

1. $a_{n+1} \leq a_n$, for all n **2.** $\lim_{n \to \infty} a_n = 0$

Proof   Consider the alternating series $\sum (-1)^{n+1} a_n$. For this series, the partial sum (where $2n$ is even)

$$S_{2n} = (a_1 - a_2) + (a_3 - a_4) + (a_5 - a_6) + \cdots + (a_{2n-1} - a_{2n})$$

has all nonnegative terms, and therefore $\{S_{2n}\}$ is a nondecreasing sequence. But you can also write

$$S_{2n} = a_1 - (a_2 - a_3) - (a_4 - a_5) - \cdots - (a_{2n-2} - a_{2n-1}) - a_{2n}$$

which implies that $S_{2n} \leq a_1$ for every integer n. Thus $\{S_{2n}\}$ is a bounded, nondecreasing sequence that converges to some value L. Because $S_{2n-1} - a_{2n} = S_{2n}$ and $a_{2n} \to 0$, you have

REMARK   The first condition in the Alternating Series Test can be modified to require only that $0 < a_{n+1} \leq a_n$ for all n greater than some integer N.

$$\lim_{n \to \infty} S_{2n-1} = \lim_{n \to \infty} S_{2n} + \lim_{n \to \infty} a_{2n} = L + \lim_{n \to \infty} a_{2n} = L.$$

Because both S_{2n} and S_{2n-1} converge to the same limit L, it follows that $\{S_n\}$ also converges to L. Consequently, the given alternating series converges.

EXAMPLE 1   Using the Alternating Series Test

Determine the convergence or divergence of $\sum_{n=1}^{\infty} (-1)^{n+1} \dfrac{1}{n}$.

Solution   Because $1/(n + 1) \le 1/n$ for all n and the limit (as $n \to \infty$) of $1/n$ is 0, you can apply the Alternating Series Test to conclude that the series converges. (This series is called the *alternating harmonic series*—more is said about this series in Example 6.)

EXAMPLE 2   Using the Alternating Series Test

Determine the convergence or divergence of $\sum_{n=1}^{\infty} \dfrac{n}{(-2)^{n-1}}$.

Solution   To apply the Alternating Series Test, note that, for $n \ge 1$,

$$\frac{1}{2} \le \frac{n}{n + 1}$$

$$\frac{2^{n-1}}{2^n} \le \frac{n}{n + 1}$$

$$(n + 1)2^{n-1} \le n2^n$$

$$\frac{n + 1}{2^n} \le \frac{n}{2^{n-1}}.$$

Hence, $a_{n+1} = (n + 1)/2^n \le n/2^{n-1} = a_n$ for all n. Furthermore, by L'Hôpital's Rule,

$$\lim_{x \to \infty} \frac{x}{2^{x-1}} = \lim_{x \to \infty} \frac{1}{2^{x-1}(\ln 2)} = 0 \quad \rightarrow \quad \lim_{n \to \infty} \frac{n}{2^{n-1}} = 0.$$

Therefore, by the Alternating Series Test, the given series converges.

EXAMPLE 3   Cases for which the Alternating Series Test Fails

a. The alternating series

$$\sum_{n=1}^{\infty} \frac{(-1)^{n+1}(n + 1)}{n} = \frac{2}{1} - \frac{3}{2} + \frac{4}{3} - \frac{5}{4} + \frac{6}{5} - \cdots$$

passes the first condition in the Alternating Series Test because $a_{n+1} \le a_n$ for all n. You cannot apply the Alternating Series Test, however, because the series does not pass the second condition. In fact, the series diverges by the nth-Term Test for Divergence.

b. The alternating series

$$\frac{2}{1} - \frac{1}{1} + \frac{2}{2} - \frac{1}{2} + \frac{2}{3} - \frac{1}{3} + \frac{2}{4} - \frac{1}{4} + \cdots$$

passes the second condition because a_n approaches 0 as $n \to \infty$. You cannot apply the Alternating Series Test, however, because the series does not pass the first condition. To conclude that the series diverges, you can argue that S_{2N} equals the Nth partial sum of the divergent harmonic series. This implies that the sequence of partial sums diverges and, hence, the series diverges.

REMARK   In Example 3a, remember that whenever a series does not pass the second condition of the Alternating Series Test, you can use the nth-Term Test for Divergence to conclude that the series diverges.

Alternating Series Remainder

For a convergent alternating series, the partial sum S_N can be a useful approximation for the sum S of the series. Just how close S_N is to S is stated in the following theorem.

THEOREM 8.15   Alternating Series Remainder

If a convergent alternating series satisfies the condition $a_{n+1} \leq a_n$, then the absolute value of the remainder R_N involved in approximating the sum S by S_n is less than (or equal to) the first neglected term. That is,

$$|S - S_N| = |R_N| \leq a_{N+1}.$$

Proof   The series obtained by deleting the first N terms of the given series satisfies the conditions of the Alternating Series Test and has a sum of R_N.

$$R_N = S - S_N = \sum_{n=1}^{\infty} (-1)^{n+1} a_n - \sum_{n=1}^{N} (-1)^{n+1} a_n$$

$$= (-1)^N a_{N+1} + (-1)^{N+1} a_{N+2} + (-1)^{N+2} a_{N+3} + \cdots$$

$$= (-1)^N (a_{N+1} - a_{N+2} + a_{N+3} - \cdots)$$

$$|R_N| = a_{N+1} - a_{N+2} + a_{N+3} - a_{N+4} + a_{N+5} - \cdots$$

$$= a_{N+1} - (a_{N+2} - a_{N+3}) - (a_{N+4} - a_{N+5}) - \cdots \leq a_{N+1}$$

Consequently, $|S - S_N| = |R_N| \leq a_{N+1}$, which establishes the theorem. ∎

EXAMPLE 4   Approximating the Sum of an Alternating Series

Approximate the sum of the following series by its first six terms.

$$\sum_{n=1}^{\infty} (-1)^{n+1} \left(\frac{1}{n!} \right) = \frac{1}{1!} - \frac{1}{2!} + \frac{1}{3!} - \frac{1}{4!} + \frac{1}{5!} - \frac{1}{6!} + \cdots$$

Solution   The series converges by the Alternating Series Test because

$$\frac{1}{(n+1)!} \leq \frac{1}{n!} \quad \text{and} \quad \lim_{n \to \infty} \frac{1}{n!} = 0.$$

The sum of the first six terms is

$$S_6 = 1 - \frac{1}{2} + \frac{1}{6} - \frac{1}{24} + \frac{1}{120} - \frac{1}{720} \approx 0.63194$$

and, by the Alternating Series Remainder, you have

$$|S - S_6| = |R_6| \leq a_7 = \frac{1}{5040} \approx 0.0002.$$

Therefore, the sum S lies between $0.63194 - 0.0002$ and $0.63194 + 0.0002$, and you have

$$0.63174 \leq S \leq 0.63214.$$

∎

TECHNOLOGY   Later, in Section 8.10 you will be able to show that the series in Example 4 converges to

$$\frac{e-1}{e} \approx 0.63212.$$

For now, try using a computer to obtain an approximation of the sum of the series. How many terms do you need to obtain an approximation that is within 0.00001 units of the actual sum?

Absolute and Conditional Convergence

Occasionally, a series may have both positive and negative terms and not be an alternating series. For instance, the series

$$\sum_{n=1}^{\infty} \frac{\sin n}{n^2} = \frac{\sin 1}{1} + \frac{\sin 2}{4} + \frac{\sin 3}{9} + \dots$$

has both positive and negative terms, yet it is not an alternating series. One way to obtain some information about the convergence of this series is to investigate the convergence of the series

$$\sum_{n=1}^{\infty} \left| \frac{\sin n}{n^2} \right|.$$

By direct comparison, you have $|\sin n| \le 1$ for all n, so

$$\left| \frac{\sin n}{n^2} \right| \le \frac{1}{n^2}, \qquad n \ge 1.$$

Thus, by the Direct Comparison Test, the series $\Sigma \left| (\sin n)/n^2 \right|$ converges. But the question still is "Does the original series converge?" The next theorem tells us that the answer is yes.

THEOREM 8.16   Absolute Convergence

If the series $\Sigma \left| a_n \right|$ converges, then the series Σa_n also converges.

Proof   Because $0 \le a_n + |a_n| \le 2|a_n|$ for all n, the series

$$\sum_{n=1}^{\infty} \left(a_n + |a_n| \right)$$

converges by comparison with the convergent series

$$\sum_{n=1}^{\infty} 2 |a_n|.$$

Furthermore, because $a_n = (a_n + |a_n|) - |a_n|$, you can write

$$\sum_{n=1}^{\infty} a_n = \sum_{n=1}^{\infty} \left(a_n + |a_n| \right) - \sum_{n=1}^{\infty} |a_n|$$

where both series on the right converge. Hence it follows that Σa_n converges.

The converse of Theorem 8.16 is not true. For instance, the **alternating harmonic series**

$$\sum_{n=1}^{\infty} \frac{(-1)^{n+1}}{n} = \frac{1}{1} - \frac{1}{2} + \frac{1}{3} - \frac{1}{4} + \dots$$

converges by the Alternating Series Test. Yet the harmonic series diverges. This type of convergence is called **conditional.**

Definition of Absolute and Conditional Convergence

1. $\Sigma\, a_n$ is **absolutely convergent** if $\Sigma\, |a_n|$ converges.
2. $\Sigma\, a_n$ is **conditionally convergent** if $\Sigma\, a_n$ converges but $\Sigma\, |a_n|$ diverges.

EXAMPLE 5   Absolute and Conditional Convergence

Determine whether the following series are convergent or divergent. Classify any convergent series as absolutely or conditionally convergent.

a. $\displaystyle\sum_{n=1}^{\infty} \frac{(-1)^{n(n+1)/2}}{3^n} = -\frac{1}{3} - \frac{1}{9} + \frac{1}{27} + \frac{1}{81} - \cdots$

b. $\displaystyle\sum_{n=1}^{\infty} \frac{(-1)^n}{\ln(n+1)} = -\frac{1}{\ln 2} + \frac{1}{\ln 3} - \frac{1}{\ln 4} + \frac{1}{\ln 5} - \cdots$

c. $\displaystyle\sum_{n=0}^{\infty} \frac{(-1)^n n!}{2^n} = \frac{0!}{2^0} - \frac{1!}{2^1} + \frac{2!}{2^2} - \frac{3!}{2^3} + \cdots$

d. $\displaystyle\sum_{n=1}^{\infty} \frac{(-1)^n}{\sqrt{n}} = -\frac{1}{\sqrt{1}} + \frac{1}{\sqrt{2}} - \frac{1}{\sqrt{3}} + \frac{1}{\sqrt{4}} - \cdots$

Solution

a. This is *not* an alternating series. However, because

$$\sum_{n=1}^{\infty} \left| \frac{(-1)^{n(n+1)/2}}{3^n} \right| = \sum_{n=1}^{\infty} \frac{1}{3^n}$$

is a convergent geometric series, you can apply Theorem 8.16 to conclude that the given series is *absolutely* convergent (and hence convergent).

b. In this case, the Alternating Series Test indicates that the given series converges. However, the series

$$\sum_{n=1}^{\infty} \left| \frac{(-1)^n}{\ln(n+1)} \right| = \frac{1}{\ln 2} + \frac{1}{\ln 3} + \frac{1}{\ln 4} + \cdots$$

diverges by direct comparison with the terms of the harmonic series. Therefore, the given series is *conditionally* convergent.

c. By the *n*th-Term Test for Divergence, you can conclude that this series diverges.

d. The given series converges by the Alternating Series Test. Moreover, because the *p*-series

$$\sum_{n=1}^{\infty} \left| \frac{(-1)^n}{\sqrt{n}} \right| = \frac{1}{\sqrt{1}} + \frac{1}{\sqrt{2}} + \frac{1}{\sqrt{3}} + \frac{1}{\sqrt{4}} + \cdots$$

diverges, the given series is *conditionally* convergent.

Rearrangement of Series

A finite sum such as $(1 + 3 - 2 + 5 - 4)$ can be rearranged without changing the value of the sum. This is not necessarily true of an infinite series—it depends on whether the series is absolutely convergent or conditionally convergent.

1. If a series is *absolutely convergent,* then its terms can be rearranged in any order without changing the sum of the series.
2. If a series is *conditionally convergent,* then the series can be rearranged to give a different sum.

The second case is illustrated in Example 6.

EXAMPLE 6   Rearrangement of a Series

The alternating harmonic series converges to $\ln 2$. That is,

$$\sum_{n=1}^{\infty} (-1)^{n+1} \frac{1}{n} = \frac{1}{1} - \frac{1}{2} + \frac{1}{3} - \frac{1}{4} + \cdots = \ln 2.$$

(See Exercise 41, Section 8.10.) Find a rearrangement of the series that produces a different sum.

Solution   Consider the following rearrangement.

$$1 - \frac{1}{2} - \frac{1}{4} + \frac{1}{3} - \frac{1}{6} - \frac{1}{8} + \frac{1}{5} - \frac{1}{10} - \frac{1}{12} + \frac{1}{7} - \frac{1}{14} - \cdots$$

$$= \left(1 - \frac{1}{2}\right) - \frac{1}{4} + \left(\frac{1}{3} - \frac{1}{6}\right) - \frac{1}{8} + \left(\frac{1}{5} - \frac{1}{10}\right) - \frac{1}{12} + \left(\frac{1}{7} - \frac{1}{14}\right) - \cdots$$

$$= \frac{1}{2} - \frac{1}{4} + \frac{1}{6} - \frac{1}{8} + \frac{1}{10} - \frac{1}{12} + \frac{1}{14} - \cdots$$

$$= \frac{1}{2}\left(1 - \frac{1}{2} + \frac{1}{3} - \frac{1}{4} + \frac{1}{5} - \frac{1}{6} + \frac{1}{7} - \cdots\right)$$

$$= \frac{1}{2}(\ln 2)$$

FOR FURTHER INFORMATION Georg Friedrich Riemann (1826–1866) proved that if $\Sigma\, a_n$ is conditionally convergent and S is any real number, then the terms of the series can be rearranged to converge to S. For more on this topic, see the article "Riemann's Rearrangement Theorem" by Stewart Galanor in the November, 1987 issue of *Mathematics Teacher.*

Thus, by rearranging the terms, you can obtain a sum that is half the sum of the original series.

EXERCISES for Section 8.5

TECHNOLOGY
Laboratory Guide
Labs 8.4–8.5

In Exercises 1–20, use the Alternating Series Test to determine the convergence or divergence of the series.

1. $\displaystyle\sum_{n=1}^{\infty} \frac{(-1)^{n+1}}{n}$

2. $\displaystyle\sum_{n=1}^{\infty} \frac{(-1)^{n+1} n}{2n - 1}$

3. $\displaystyle\sum_{n=1}^{\infty} \frac{(-1)^{n+1}}{2n - 1}$

4. $\displaystyle\sum_{n=2}^{\infty} \frac{(-1)^{n}}{\ln n}$

5. $\displaystyle\sum_{n=1}^{\infty} \frac{(-1)^{n} n^2}{n^2 + 1}$

6. $\displaystyle\sum_{n=1}^{\infty} \frac{(-1)^{n+1} n}{n^2 + 1}$

7. $\displaystyle\sum_{n=1}^{\infty} \frac{(-1)^{n}}{\sqrt{n}}$

8. $\displaystyle\sum_{n=1}^{\infty} \frac{(-1)^{n+1} n^3}{n^3 + 6}$

9. $\displaystyle\sum_{n=1}^{\infty} \frac{(-1)^{n+1}(n + 1)}{\ln(n + 1)}$

10. $\displaystyle\sum_{n=1}^{\infty} \frac{(-1)^{n+1} \ln(n + 1)}{n + 1}$

11. $\displaystyle\sum_{n=1}^{\infty} \sin \frac{(2n - 1)\pi}{2}$

12. $\displaystyle\sum_{n=1}^{\infty} \cos n\pi$

13. $\displaystyle\sum_{n=1}^{\infty} \frac{1}{n} \sin \frac{(2n - 1)\pi}{2}$

14. $\displaystyle\sum_{n=1}^{\infty} \frac{1}{n} \cos n\pi$

15. $\displaystyle\sum_{n=0}^{\infty} \frac{(-1)^n}{n!}$

16. $\displaystyle\sum_{n=0}^{\infty} \frac{(-1)^n}{(2n)!}$

17. $\displaystyle\sum_{n=1}^{\infty} \frac{(-1)^{n+1}\sqrt{n}}{n+2}$

18. $\displaystyle\sum_{n=1}^{\infty} \frac{(-1)^{n+1}\sqrt{n}}{\sqrt[3]{n}}$

19. $\displaystyle\sum_{n=1}^{\infty} \frac{2(-1)^{n+1}}{e^n - e^{-n}} = \sum_{n=1}^{\infty}(-1)^{n+1}\operatorname{csch} n$

20. $\displaystyle\sum_{n=1}^{\infty} \frac{2(-1)^{n+1}}{e^n + e^{-n}} = \sum_{n=1}^{\infty}(-1)^{n+1}\operatorname{sech} n$

In Exercises 21–36, determine whether the series converges conditionally or absolutely, or diverges.

21. $\displaystyle\sum_{n=1}^{\infty} \frac{(-1)^{n+1}}{(n+1)^2}$

22. $\displaystyle\sum_{n=1}^{\infty} \frac{(-1)^{n+1}}{n+1}$

23. $\displaystyle\sum_{n=1}^{\infty} \frac{(-1)^{n+1}}{\sqrt{n}}$

24. $\displaystyle\sum_{n=1}^{\infty} \frac{(-1)^{n+1}}{n\sqrt{n}}$

25. $\displaystyle\sum_{n=1}^{\infty} \frac{(-1)^{n+1}n^2}{(n+1)^2}$

26. $\displaystyle\sum_{n=1}^{\infty} \frac{(-1)^{n+1}(2n+3)}{n+10}$

27. $\displaystyle\sum_{n=2}^{\infty} \frac{(-1)^n}{\ln n}$

28. $\displaystyle\sum_{n=0}^{\infty} (-1)^n e^{-n^2}$

29. $\displaystyle\sum_{n=2}^{\infty} \frac{(-1)^n n}{n^3 - 1}$

30. $\displaystyle\sum_{n=1}^{\infty} \frac{(-1)^{n+1}}{n^{1.5}}$

31. $\displaystyle\sum_{n=0}^{\infty} \frac{(-1)^n}{(2n+1)!}$

32. $\displaystyle\sum_{n=0}^{\infty} \frac{(-1)^n}{\sqrt[3]{n+1}}$

33. $\displaystyle\sum_{n=0}^{\infty} \frac{\cos n\pi}{n+1}$

34. $\displaystyle\sum_{n=1}^{\infty} (-1)^{n+1}\arctan n$

35. $\displaystyle\sum_{n=1}^{\infty} \frac{\cos n\pi}{n^2}$

36. $\displaystyle\sum_{n=1}^{\infty} \frac{\sin[(2n-1)\pi/2]}{n}$

In Exercises 37–42, approximate the sum of the series with an error of less than 0.001. (Use Theorem 8.15.)

37. $\displaystyle\sum_{n=0}^{\infty} \frac{(-1)^n}{n!} \left(\text{the sum is } \frac{1}{e}\right)$

38. $\displaystyle\sum_{n=0}^{\infty} \frac{(-1)^n}{2^n n!} \left(\text{the sum is } \frac{1}{\sqrt{e}}\right)$

39. $\displaystyle\sum_{n=0}^{\infty} \frac{(-1)^n}{(2n+1)!}$ (the sum is $\sin 1$)

40. $\displaystyle\sum_{n=0}^{\infty} \frac{(-1)^n}{(2n)!}$ (the sum is $\cos 1$)

41. $\displaystyle\sum_{n=1}^{\infty} \frac{(-1)^{n+1}}{n}$ (the sum is $\ln 2$)

42. $\displaystyle\sum_{n=1}^{\infty} \frac{(-1)^{n+1}}{n4^n} \left(\text{the sum is } \ln\frac{5}{4}\right)$

In Exercises 43 and 44, find the number of terms necessary to approximate the sum of the series with an error less than 0.001. (Use Theorem 8.15.)

43. $\displaystyle\sum_{n=1}^{\infty} \frac{(-1)^{n+1}}{2n^3 - 1}$

44. $\displaystyle\sum_{n=1}^{\infty} \frac{(-1)^{n+1}}{n^4}$

C **In Exercises 45 and 46, use the convergent series and its sum to perform the following.**

a. Construct a rectangular coordinate system in which the vertical axis represents the magnitude of the nth partial sum S_n of the series and the horizontal axis represents n. Sketch a horizontal line at the height equal to the sum of the series.

b. Use a computer or calculator to complete the table of partial sums.

n	1	2	3	4	5	6	7	8
S_n								

c. Plot the points (n, S_n) from the table in part b on the coordinate system in part a.

d. What pattern exists between the plot of the successive points and the horizontal line representing the sum of the series? Do the distances between the successive points and the horizontal line increase or decrease? How do the answers to the two preceding questions relate to the alternating series remainder as given in Theorem 8.15?

45. $\displaystyle\sum_{n=1}^{\infty} \frac{(-1)^{n-1}}{2n-1} = \frac{\pi}{4}$

46. $\displaystyle\sum_{n=1}^{\infty} \frac{(-1)^{n-1}}{n^2} = \frac{\pi^2}{12}$

47. Prove that the alternating p-series

$$\sum_{n=1}^{\infty}(-1)^n\left(\frac{1}{n^p}\right)$$

converges if $p > 0$.

48. Prove that if $\Sigma |a_n|$ converges, then Σa_n^2 converges. Is the converse true? If not, give an example that shows it is false.

49. Find all values of x for which the series

$$\sum_{n=1}^{\infty} \frac{x^n}{n}$$

a. converges absolutely.

b. converges conditionally.

50. Determine the error in the following argument that $0 = 1$.

$0 = 0 + 0 + 0 + \cdots$

$= (1 - 1) + (1 - 1) + (1 - 1) + \cdots$

$= 1 + (-1 + 1) + (-1 + 1) + \cdots$

$= 1 + 0 + 0 + \cdots$

$= 1$

True or False **In Exercises 51–54, determine whether the statement is true or false. If it is false, explain why or give an example that shows it is false.**

51. If $0 < a_{n+1} < a_n$ for $n > 100$ and $\lim_{n\to\infty} a_n = 0$, then $\displaystyle\sum_{n=1}^{\infty}(-1)^n a_n$ converges.

52. If $\displaystyle\sum_{n=1}^{\infty}(-1)^n a_n$ converges, then $\displaystyle\sum_{n=1}^{\infty}(-1)^{n-1} a_n$ converges.

53. If both $\displaystyle\sum_{n=1}^{\infty} a_n$ and $\displaystyle\sum_{n=1}^{\infty}(-a_n)$ converge, then $\displaystyle\sum_{n=1}^{\infty}|a_n|$ converges.

54. If $\displaystyle\sum_{n=1}^{\infty} a_n$ does not converge, then $\displaystyle\sum_{n=1}^{\infty}|a_n|$ does not converge.

The Ratio and Root Tests

8.6 The Ratio Test • The Root Test • Strategies for Testing Series

The Ratio Test

This section begins with a test for absolute convergence—the **Ratio Test.**

THEOREM 8.17   Ratio Test

1. $\Sigma\, a_n$ converges absolutely if $\lim\limits_{n\to\infty}\left|\dfrac{a_{n+1}}{a_n}\right| < 1$.

2. $\Sigma\, a_n$ diverges if $\lim\limits_{n\to\infty}\left|\dfrac{a_{n+1}}{a_n}\right| > 1$ or $\lim\limits_{n\to\infty}\left|\dfrac{a_{n+1}}{a_n}\right| = \infty$.

3. The Ratio Test is inconclusive if $\lim\limits_{n\to\infty}\left|\dfrac{a_{n+1}}{a_n}\right| = 1$.

Proof   To prove Property 1, assume that

$$\lim_{n\to\infty}\left|\frac{a_{n+1}}{a_n}\right| = r < 1$$

and choose R such that $0 \le r < R < 1$. By the definition of the limit of a sequence, there exists some $N > 0$ such that $|a_{n+1}/a_n| < R$ for all $n > N$. Therefore, you can write the following inequalities.

$$|a_{N+1}| < |a_N|\,R$$
$$|a_{N+2}| < |a_{N+1}|\,R < |a_N|\,R^2$$
$$|a_{N+3}| < |a_{N+2}|\,R < |a_{N+1}|\,R^2 < |a_N|\,R^3$$
$$\vdots$$

REMARK   The fact that the Ratio Test is inconclusive when $|a_{n+1}/a_n| \to 1$ can be seen by comparing the two series

$$\sum_{n=1}^{\infty}\frac{1}{n} \quad\text{and}\quad \sum_{n=1}^{\infty}\frac{1}{n^2}.$$

The first series diverges and the second one converges, but in both cases

$$\lim_{n\to\infty}\left|\frac{a_{n+1}}{a_n}\right| = 1.$$

The geometric series $\Sigma\,|a_N|\,R^n = |a_N|\,R + |a_N|\,R^2 + \cdots + |a_N|\,R^n + \cdots$ converges, and so, by the Direct Comparison Test, the series

$$\sum_{n=1}^{\infty}|a_{N+n}| = |a_{N+1}| + |a_{N+2}| + \cdots + |a_{N+n}| + \cdots$$

also converges. This in turn implies that the series $\Sigma\,|a_n|$ converges, because discarding a finite number of terms ($n = N - 1$) does not affect convergence. Consequently, by Theorem 8.16, the series $\Sigma\,a_n$ converges absolutely. The proof of Property 2 is similar and is left as an exercise (see Exercise 60).

Although the Ratio Test is not a cure for all ills related to tests for convergence, it is particularly useful for series that *converge rapidly*. Series involving factorials or exponentials are frequently of this type.

EXAMPLE 1   Using the Ratio Test

Determine the convergence or divergence of $\displaystyle\sum_{n=0}^{\infty} \frac{2^n}{n!}$.

Solution   Because $a_n = 2^n/n!$, you can write the following.

$$\lim_{n \to \infty} \left| \frac{a_{n+1}}{a_n} \right| = \lim_{n \to \infty} \left[\frac{2^{n+1}}{(n + 1)!} \div \frac{2^n}{n!} \right]$$

$$= \lim_{n \to \infty} \left[\frac{2^{n+1}}{(n + 1)!} \cdot \frac{n!}{2^n} \right]$$

$$= \lim_{n \to \infty} \frac{2}{n + 1}$$

$$= 0$$

Therefore, the series converges.

EXAMPLE 2   Using the Ratio Test

Determine whether the following series converge or diverge.

a. $\displaystyle\sum_{n=0}^{\infty} \frac{n^2 2^{n+1}}{3^n}$ **b.** $\displaystyle\sum_{n=1}^{\infty} \frac{n^n}{n!}$

Solution

a. This series converges because the limit of $|a_{n+1}/a_n|$ is less than 1.

$$\lim_{n \to \infty} \left| \frac{a_{n+1}}{a_n} \right| = \lim_{n \to \infty} \left[(n + 1)^2 \left(\frac{2^{n+2}}{3^{n+1}} \right) \left(\frac{3^n}{n^2 2^{n+1}} \right) \right]$$

$$= \lim_{n \to \infty} \frac{2(n + 1)^2}{3n^2}$$

$$= \frac{2}{3} < 1$$

b. This series diverges because the limit of $|a_{n+1}/a_n|$ is greater than 1.

$$\lim_{n \to \infty} \left| \frac{a_{n+1}}{a_n} \right| = \lim_{n \to \infty} \left[\frac{(n + 1)^{n+1}}{(n + 1)!} \left(\frac{n!}{n^n} \right) \right]$$

$$= \lim_{n \to \infty} \left[\frac{(n + 1)^{n+1}}{(n + 1)} \left(\frac{1}{n^n} \right) \right]$$

$$= \lim_{n \to \infty} \frac{(n + 1)^n}{n^n}$$

$$= \lim_{n \to \infty} \left(1 + \frac{1}{n} \right)^n$$

$$= e > 1$$

EXAMPLE 3   A Failure of the Ratio Test

Determine the convergence or divergence of $\displaystyle\sum_{n=1}^{\infty} (-1)^n \frac{\sqrt{n}}{n+1}$.

Solution   The limit of $|a_{n+1}/a_n|$ is equal to 1.

$$\lim_{n\to\infty}\left|\frac{a_{n+1}}{a_n}\right| = \lim_{n\to\infty}\left[\left(\frac{\sqrt{n+1}}{n+2}\right)\left(\frac{n+1}{\sqrt{n}}\right)\right]$$

$$= \lim_{n\to\infty}\left[\sqrt{\frac{n+1}{n}}\left(\frac{n+1}{n+2}\right)\right]$$

$$= \sqrt{1}(1)$$

$$= 1$$

Thus, the Ratio Test is inconclusive. To determine whether the series converges, you need to try a different test. In this case, you can apply the Alternating Series Test. To show that $a_{n+1} \le a_n$, let

$$f(x) = \frac{\sqrt{x}}{x+1}.$$

Then the derivative is

$$f'(x) = \frac{-x+1}{2\sqrt{x}(x+1)^2}.$$

Because the derivative is negative for $x > 1$, you know that f is a decreasing function. Also, by L'Hôpital's Rule,

$$\lim_{x\to\infty}\frac{\sqrt{x}}{x+1} = \lim_{x\to\infty}\frac{1/(2\sqrt{x})}{1}$$

$$= \lim_{x\to\infty}\frac{1}{2\sqrt{x}}$$

$$= 0.$$

Therefore, by the Alternating Series Test, the series converges. ∎

The series in Example 3 is *conditionally convergent*. This follows from the fact that the series

$$\sum_{n=1}^{\infty} |a_n|$$

diverges $\left(\text{by the Limit Comparison Test with } \Sigma\, 1/\sqrt{n}\right)$, but the series

$$\sum_{n=1}^{\infty} a_n$$

converges.

TECHNOLOGY   A computer or programmable calculator can reinforce the conclusion that the series in Example 3 converges *conditionally*. By adding the first 100 terms of the series, you obtain a sum of about -0.2. (The sum of the first 100 terms of the series $\Sigma\, |a_n|$ is about 17.)

The Root Test

The next test for convergence or divergence of series works especially well for series involving nth powers. The proof of this theorem is similar to that given for the Ratio Test, and we leave it as an exercise (see Exercise 61).

THEOREM 8.18   Root Test

Let Σa_n be a series with nonzero terms.

1. Σa_n converges absolutely if $\lim\limits_{n \to \infty} \sqrt[n]{|a_n|} < 1$.

2. Σa_n diverges if $\lim\limits_{n \to \infty} \sqrt[n]{|a_n|} > 1$ or $\lim\limits_{n \to \infty} \sqrt[n]{|a_n|} = \infty$.

3. The Root Test is inconclusive if $\lim\limits_{n \to \infty} \sqrt[n]{|a_n|} = 1$.

EXAMPLE 4   Using the Root Test

Determine the convergence or divergence of $\sum\limits_{n=1}^{\infty} \dfrac{e^{2n}}{n^n}$.

Solution   You can apply the Root Test as follows.

$$
\begin{aligned}
\lim_{n \to \infty} \sqrt[n]{|a_n|} &= \lim_{n \to \infty} \sqrt[n]{\frac{e^{2n}}{n^n}} \\
&= \lim_{n \to \infty} \frac{e^{2n/n}}{n^{n/n}} \\
&= \lim_{n \to \infty} \frac{e^2}{n} \\
&= 0 < 1
\end{aligned}
$$

Because this limit is less than 1, you can conclude that the series converges absolutely (and hence converges).

To see the usefulness of the Root Test for the series in Example 4, try applying the Ratio Test to the series. When you do this, you obtain the following.

$$
\begin{aligned}
\lim_{n \to \infty} \left| \frac{a_{n+1}}{a_n} \right| &= \lim_{n \to \infty} \left[\frac{e^{2(n+1)}}{(n+1)^{n+1}} \div \frac{e^{2n}}{n^n} \right] \\
&= \lim_{n \to \infty} e^2 \frac{n^n}{(n+1)^{n+1}} \\
&= \lim_{n \to \infty} e^2 \left(\frac{n}{n+1} \right)^n \left(\frac{1}{n+1} \right) \\
&= 0
\end{aligned}
$$

Note that this limit is not as easily evaluated as the limit obtained by the Root Test in Example 4.

FOR FURTHER INFORMATION
For more information on the usefulness of the Root Test, see the article "*N!* and the Root Test" by Charles C. Mumma II in the August-September, 1986 issue of the *American Mathematical Monthly*.

Strategies for Testing Series

You have now studied 10 tests for determining the convergence or divergence of an infinite series. (See the summary in Table 8.1 on page 594.) Skill in choosing and applying the various tests will come only with practice. Below is a useful checklist for choosing an appropriate test.

Guidelines for Testing a Series for Convergence or Divergence

1. Does the nth term approach 0? If not, the series diverges.
2. Is the series one of the special types—geometric, p-series, telescoping, or alternating?
3. Can the Integral, Root, or Ratio Test be applied?
4. Can the series be compared favorably to one of the special types?

In some instances, more than one test is applicable. However, your objective should be to learn to choose the most efficient test.

EXAMPLE 5 Applying the Strategies for Testing Series

Determine the convergence or divergence of the series.

a. $\displaystyle\sum_{n=1}^{\infty} \frac{n+1}{3n+1}$ b. $\displaystyle\sum_{n=1}^{\infty} \left(\frac{\pi}{6}\right)^n$ c. $\displaystyle\sum_{n=1}^{\infty} ne^{-n^2}$ d. $\displaystyle\sum_{n=1}^{\infty} \frac{1}{3n+1}$

e. $\displaystyle\sum_{n=1}^{\infty} (-1)^n \frac{3}{4n+1}$ f. $\displaystyle\sum_{n=1}^{\infty} \frac{n!}{10^n}$ g. $\displaystyle\sum_{n=1}^{\infty} \left(\frac{n+1}{2n+1}\right)^n$

Solution

a. For this series, the limit of the nth term is not 0 ($a_n \to \frac{1}{3}$ as $n \to \infty$). Thus, by the nth-Term Test, the series diverges.

b. This series is geometric. Moreover, because the common ratio of the terms is less than 1 in absolute value ($r = \pi/6$), you can conclude that the series converges.

c. Because the function $f(x) = xe^{-x^2}$ is easily integrated, you can use the Integral Test to conclude that the series converges.

d. The nth term of this series can be compared to the nth term of the harmonic series. After using the Limit Comparison Test, you can conclude that the series diverges.

e. This is an alternating series whose nth term approaches 0. Because $a_{n+1} \le a_n$, you can use the Alternating Series Test to conclude that the series converges.

f. The nth term of this series involves a factorial, which indicates that the Ratio Test may work well. After applying the Ratio Test, you can conclude that the series diverges.

g. The nth term of this series involves a variable that is raised to the nth power, which indicates that the Root Test may work well. After applying the Root Test, you can conclude that the series converges.

TABLE 8.1
Summary of Tests for Series

Test	Series	Converges	Diverges	Comment
nth-Term	$\displaystyle\sum_{n=1}^{\infty} a_n$		$\displaystyle\lim_{n\to\infty} a_n \neq 0$	This test cannot be used to show convergence.
Geometric Series	$\displaystyle\sum_{n=0}^{\infty} ar^n$	$\lvert r \rvert < 1$	$\lvert r \rvert \geq 1$	Sum: $S = \dfrac{a}{1-r}$
Telescoping Series	$\displaystyle\sum_{n=1}^{\infty} (b_n - b_{n+1})$	$\displaystyle\lim_{n\to\infty} b_n = L$		Sum: $S = b_1 - L$
p-Series	$\displaystyle\sum_{n=1}^{\infty} \dfrac{1}{n^p}$	$p > 1$	$p \leq 1$	
Alternating Series	$\displaystyle\sum_{n=1}^{\infty} (-1)^{n-1} a_n$	$0 < a_{n+1} \leq a_n$ and $\displaystyle\lim_{n\to\infty} a_n = 0$		Remainder: $\lvert R_N \rvert \leq a_{N+1}$
Integral (f is continuous, positive, and decreasing)	$\displaystyle\sum_{n=1}^{\infty} a_n,$ $a_n = f(n) \geq 0$	$\displaystyle\int_1^{\infty} f(x)\,dx$ converges	$\displaystyle\int_1^{\infty} f(x)\,dx$ diverges	Remainder: $0 < R_N < \displaystyle\int_N^{\infty} f(x)\,dx$
Root	$\displaystyle\sum_{n=1}^{\infty} a_n$	$\displaystyle\lim_{n\to\infty} \sqrt[n]{\lvert a_n \rvert} < 1$	$\displaystyle\lim_{n\to\infty} \sqrt[n]{\lvert a_n \rvert} > 1$	Test is inconclusive if $\displaystyle\lim_{n\to\infty} \sqrt[n]{\lvert a_n \rvert} = 1$.
Ratio	$\displaystyle\sum_{n=1}^{\infty} a_n$	$\displaystyle\lim_{n\to\infty} \left\lvert \dfrac{a_{n+1}}{a_n} \right\rvert < 1$	$\displaystyle\lim_{n\to\infty} \left\lvert \dfrac{a_{n+1}}{a_n} \right\rvert > 1$	Test is inconclusive if $\displaystyle\lim_{n\to\infty} \left\lvert \dfrac{a_{n+1}}{a_n} \right\rvert = 1$.
Direct Comparison ($a_n, b_n > 0$)	$\displaystyle\sum_{n=1}^{\infty} a_n$	$0 \leq a_n \leq b_n$ and $\displaystyle\sum_{n=1}^{\infty} b_n$ converges	$0 \leq b_n \leq a_n$ and $\displaystyle\sum_{n=1}^{\infty} b_n$ diverges	
Limit Comparison ($a_n, b_n > 0$)	$\displaystyle\sum_{n=1}^{\infty} a_n$	$\displaystyle\lim_{n\to\infty} \dfrac{a_n}{b_n} = L > 0$ and $\displaystyle\sum_{n=1}^{\infty} b_n$ converges	$\displaystyle\lim_{n\to\infty} \dfrac{a_n}{b_n} = L > 0$ and $\displaystyle\sum_{n=1}^{\infty} b_n$ diverges	

$1-52$ odd

EXERCISES for Section 8.6

TECHNOLOGY
Laboratory Guide
Lab 8.6

In Exercises 1–4, verify the formula.

1. $\dfrac{(n+1)!}{(n-2)!} = (n+1)(n)(n-1)$

2. $\dfrac{(2k-2)!}{(2k)!} = \dfrac{1}{(2k)(2k-1)}$

3. $1 \cdot 3 \cdot 5 \cdots (2k-1) = \dfrac{(2k)!}{2^k k!}$

4. $\dfrac{1}{1 \cdot 3 \cdot 5 \cdots (2k-5)} = \dfrac{2^k k!(2k-3)(2k-1)}{(2k)!}, \quad k \geq 3$

In Exercises 5–24, use the Ratio Test to test for convergence or divergence of the series.

5. $\displaystyle\sum_{n=0}^{\infty} \dfrac{n!}{3^n}$ **6.** $\displaystyle\sum_{n=1}^{\infty} n\left(\dfrac{2}{3}\right)^n$

7. $\displaystyle\sum_{n=0}^{\infty} \dfrac{3^n}{n!}$ **8.** $\displaystyle\sum_{n=1}^{\infty} n\left(\dfrac{3}{2}\right)^n$

9. $\displaystyle\sum_{n=1}^{\infty} \dfrac{n}{2^n}$ **10.** $\displaystyle\sum_{n=1}^{\infty} \dfrac{n^2}{2^n}$

11. $\displaystyle\sum_{n=1}^{\infty} \frac{2^n}{n^2}$

12. $\displaystyle\sum_{n=1}^{\infty} \frac{(-1)^{n+1}(n+2)}{n(n+1)}$

13. $\displaystyle\sum_{n=0}^{\infty} \frac{(-1)^n 2^n}{n!}$

14. $\displaystyle\sum_{n=1}^{\infty} \frac{(-1)^{n-1}(3/2)^n}{n^2}$

15. $\displaystyle\sum_{n=1}^{\infty} \frac{n!}{n 3^n}$

16. $\displaystyle\sum_{n=1}^{\infty} \frac{(2n)!}{n^5}$

17. $\displaystyle\sum_{n=0}^{\infty} \frac{4^n}{n!}$

18. $\displaystyle\sum_{n=1}^{\infty} \frac{n^n}{n!}$

19. $\displaystyle\sum_{n=0}^{\infty} \frac{3n}{(n+1)^n}$

20. $\displaystyle\sum_{n=0}^{\infty} \frac{(n!)^2}{(3n)!}$

21. $\displaystyle\sum_{n=0}^{\infty} \frac{4^n}{3^n+1}$

22. $\displaystyle\sum_{n=0}^{\infty} \frac{(-1)^n 2^{4n}}{(2n+1)!}$

23. $\displaystyle\sum_{n=0}^{\infty} \frac{(-1)^{n+1} n!}{1 \cdot 3 \cdot 5 \cdots (2n+1)}$

24. $\displaystyle\sum_{n=1}^{\infty} \frac{(-1)^n 2 \cdot 4 \cdot 6 \cdots (2n)}{2 \cdot 5 \cdot 8 \cdots (3n-1)}$

In Exercises 25 and 26, verify that the Ratio Test is inconclusive for the p-series.

25. $\displaystyle\sum_{n=1}^{\infty} \frac{1}{n^{3/2}}$

$\displaystyle\sum_{n=1}^{\infty} \frac{1}{n^{1/2}}$

26. $\displaystyle\sum_{n=1}^{\infty} \frac{1}{n^3}$

$\displaystyle\sum_{n=1}^{\infty} \frac{1}{n^p}$

In Exercises 27–34, use the Root Test to test for convergence or divergence of the series.

27. $\displaystyle\sum_{n=1}^{\infty} \left(\frac{n}{2n+1}\right)^n$

28. $\displaystyle\sum_{n=1}^{\infty} \left(\frac{2n}{n+1}\right)^n$

29. $\displaystyle\sum_{n=2}^{\infty} \frac{(-1)^n}{(\ln n)^n}$

30. $\displaystyle\sum_{n=1}^{\infty} \left(\frac{-2n}{3n+1}\right)^{3n}$

31. $\displaystyle\sum_{n-1}^{\infty} \left(2\sqrt[n]{n}+1\right)^n$

32. $\displaystyle\sum_{n=0}^{\infty} e^{-n}$

33. $\dfrac{1}{(\ln 3)^3} + \dfrac{1}{(\ln 4)^4} + \dfrac{1}{(\ln 5)^5} + \dfrac{1}{(\ln 6)^6} + \cdots$

34. $1 + \dfrac{2}{3} + \dfrac{3}{3^2} + \dfrac{4}{3^3} + \dfrac{5}{3^4} + \dfrac{6}{3^5} + \cdots$

In Exercises 35–52, test for convergence or divergence using any appropriate test from this chapter. Identify the test used.

35. $\displaystyle\sum_{n=1}^{\infty} \frac{(-1)^{n+1} 5}{n}$

36. $\displaystyle\sum_{n=1}^{\infty} \frac{5}{n}$

37. $\displaystyle\sum_{n=1}^{\infty} \frac{3}{n\sqrt{n}}$

38. $\displaystyle\sum_{n=1}^{\infty} \left(\frac{\pi}{4}\right)^n$

39. $\displaystyle\sum_{n=1}^{\infty} \frac{2n}{n+1}$

40. $\displaystyle\sum_{n=1}^{\infty} \frac{n}{2n^2+1}$

41. $\displaystyle\sum_{n=1}^{\infty} \frac{(-1)^n 3^{n-2}}{2^n}$

42. $\displaystyle\sum_{n=1}^{\infty} \frac{10}{3\sqrt{n^3}}$

43. $\displaystyle\sum_{n=1}^{\infty} \frac{10n+3}{n 2^n}$

44. $\displaystyle\sum_{n=1}^{\infty} \frac{2n}{4n^2-1}$

45. $\displaystyle\sum_{n=1}^{\infty} \frac{\cos n}{2^n}$

46. $\displaystyle\sum_{n=2}^{\infty} \frac{(-1)^n}{n \ln n}$

47. $\displaystyle\sum_{n=1}^{\infty} \frac{n 7^n}{n!}$

48. $\displaystyle\sum_{n=1}^{\infty} \frac{\ln n}{n^2}$

49. $\displaystyle\sum_{n=1}^{\infty} \frac{(-1)^n 3^{n-1}}{n!}$

50. $\displaystyle\sum_{n=1}^{\infty} \frac{(-1)^n 3^n}{n 2^n}$

51. $\displaystyle\sum_{n=1}^{\infty} \frac{(-3)^n}{3 \cdot 5 \cdot 7 \cdots (2n+1)}$

52. $\displaystyle\sum_{n=1}^{\infty} \frac{3 \cdot 5 \cdot 7 \cdots (2n+1)}{18^n(2n-1)n!}$

In Exercises 53 and 54, identify the two series that are the same.

53. a. $\displaystyle\sum_{n=1}^{\infty} \frac{n 5^n}{n!}$

b. $\displaystyle\sum_{n=0}^{\infty} \frac{n 5^n}{n!}$

c. $\displaystyle\sum_{n=0}^{\infty} \frac{(n+1) 5^{n+1}}{(n+1)!}$

54. a. $\displaystyle\sum_{n=4}^{\infty} n\left(\frac{3}{4}\right)^n$

b. $\displaystyle\sum_{n=0}^{\infty} (n+1)\left(\frac{3}{4}\right)^n$

c. $\displaystyle\sum_{n=1}^{\infty} n\left(\frac{3}{4}\right)^{n-1}$

In Exercises 55 and 56, write an equivalent series with the index of summation beginning at $n = 0$.

55. $\displaystyle\sum_{n=1}^{\infty} \frac{n}{4^n}$

56. $\displaystyle\sum_{n=2}^{\infty} \frac{2n}{(n-2)!}$

C **In Exercises 57 and 58, use a computer or calculator to approximate the sum of the series with an error less than 0.0001.**

57. $\displaystyle\sum_{k=1}^{\infty} \frac{(-3)^k}{2^k k!}$

58. $\displaystyle\sum_{k=1}^{\infty} \frac{(-3)^k}{1 \cdot 3 \cdot 5 \cdots (2k+1)}$

59. *Essay* You are told that the terms of a positive series appear to approach 0 rapidly as n approaches infinity. In fact, $a_7 \le 0.0001$. Given no other information, does this imply that the series converges? Support your conclusion with examples.

60. Prove Property 2 of Theorem 8.17.

61. Prove Theorem 8.18. (*Hint for Property 1:* If the limit equals $r < 1$, choose a real number R such that $r < R < 1$. By the definition of the limit, there exists some $N > 0$ such that $\sqrt[n]{|a_n|} < R$ for $n > N$.)

62. *Essay* Read the article "A Differentiation Test for Absolute Convergence" by Yaser S. Abu-Mostafa in the September, 1984 issue of *Mathematics Magazine*. Then write a paragraph that describes the test. Support your answer with examples of series that converge and examples of series that diverge.

Taylor Polynomials and Approximations

Polynomial Approximations of Elementary Functions • Taylor and Maclaurin
Polynomials • Remainder of a Taylor Polynomial

Polynomial Approximations of Elementary Functions

The goal of this section is to show how polynomial functions can be used as approximations for other elementary functions. To find a polynomial function P that approximates another function f, begin by choosing a number c in the domain of f at which f and P have the same value. That is,

$$P(c) = f(c). \qquad \text{Graphs of } f \text{ and } P \text{ pass through } (c, f(c))$$

The approximating polynomial is said to be **expanded about** c or **centered at** c. Geometrically, the requirement that $P(c) = f(c)$ means that the graph of P passes through the point $(c, f(c))$. Of course, there are many polynomials whose graphs pass through the point $(c, f(c))$. Your task is to find a polynomial whose graph resembles the graph of f near this point. One way to do this is to impose the additional requirement that the slope of the polynomial function be the same as the slope of the graph of f at the point $(c, f(c))$.

$$P'(c) = f'(c) \qquad \text{Graphs of } f \text{ and } P \text{ have the same slope at } (c, f(c))$$

With these two requirements, you can obtain a simple linear approximation of f, as shown in Figure 8.8.

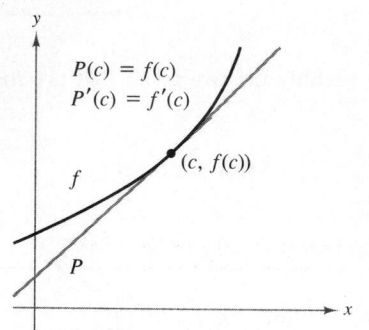

FIGURE 8.8
Near $(c, f(c))$, the graph of P can be used to approximate the graph of f.

EXAMPLE 1   First-Degree Polynomial Approximation of $f(x) = e^x$

For the function $f(x) = e^x$, find a first-degree polynomial function

$$P_1(x) = a_1 x + a_0$$

whose value and slope agree with the value and slope of f at $x = 0$.

Solution   Because $f(x) = e^x$ and $f'(x) = e^x$, the value and slope of f at $x = 0$ are given by

$$f(0) = e^0 = 1 \quad \text{and} \quad f'(0) = e^0 = 1.$$

Now, because $P_1(x) = a_1 x + a_0$, you can impose the condition that $P_1(0) = f(0)$ to conclude that $a_0 = 1$. Moreover, because $P_1'(x) = a_1$, you can use the condition that $P_1'(0) = f'(0)$ to conclude that $a_1 = 1$. Therefore,

$$P_1(x) = x + 1.$$

Figure 8.9 shows the graphs of $P_1(x) = x + 1$ and $f(x) = e^x$.

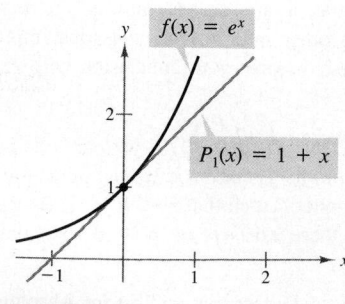

FIGURE 8.9
P_1 is the first-degree polynomial approximation of $f(x) = e^x$.

From Figure 8.9 you can see that, at points near $(0, 1)$, the graph of

$$P_1(x) = x + 1 \qquad \text{First-degree approximation}$$

is reasonably close to the graph of $f(x) = e^x$. However, as you move away from $(0, 1)$, the graphs move farther from each other and the accuracy of the approximation decreases. To improve the approximation, you can impose yet another requirement—that the values of the second derivatives of P and f agree when $x = 0$. The polynomial P_2 of least degree that satisfies all three requirements, $P_2(0) = f(0)$, $P_2'(0) = f'(0)$, and $P_2''(0) = f''(0)$, can be shown to be

$$P_2(x) = 1 + x + \frac{1}{2}x^2. \qquad \text{Second-degree approximation}$$

y
$f(x) = e^x$

$P_2(x) = 1 + x + \frac{1}{2}x^2$

FIGURE 8.10
P_2 is the second-degree polynomial approximation of $f(x) = e^x$.

Moreover, from Figure 8.10, you can see that P_2 is a better approximation of f than P. If you continue this pattern, requiring that the values of $P_n(x)$ and its first n derivatives match those of $f(x) = e^x$ at $x = 0$, you obtain the following.

$$P_n(x) = 1 + x + \frac{1}{2}x^2 + \frac{1}{3!}x^3 + \cdots + \frac{1}{n!}x^n \qquad n\text{th-degree approximation}$$

$$\approx e^x$$

EXAMPLE 2   Third-Degree Polynomial Approximation of $f(x) = e^x$

Construct a table comparing the values of the polynomial

$$P_3(x) = 1 + x + \frac{1}{2}x^2 + \frac{1}{3!}x^3$$

with those of $f(x) = e^x$ for several values of x near 0.

Solution   Using a calculator or a computer, you can obtain the results shown in Table 8.2. Note that for $x = 0$, the two functions have the same value, but that as x moves farther away from 0, the accuracy of the approximating polynomial $P_3(x)$ decreases.

TABLE 8.2

x	-1.0	-0.2	-0.1	0.0	0.1	0.2	1.0
e^x	0.3679	0.81873	0.904837	1	1.105171	1.2214	2.7183
$P_3(x)$	0.3333	0.81867	0.904833	1	1.105167	1.2213	2.6667

9

f

P_3

f

-3 3

-1

P_3

FIGURE 8.11
P_3 is the third-degree polynomial approximation of $f(x) = e^x$.

TECHNOLOGY   A graphing utility can be used to compare the graphs of the approximating polynomial with the graph of the function f. For instance, in Figure 8.11, the graph of

$$P_3(x) = 1 + x + \frac{1}{2}x^2 + \frac{1}{6}x^3$$

is compared with the graph of $f(x) = e^x$. If you have access to a graphing utility, try comparing the graphs of P_4, P_5, and P_6 with the graph of f. What do you notice?

Taylor and Maclaurin Polynomials

The polynomial approximation of $f(x) = e^x$ given in Example 2 is expanded about $c = 0$. For expansions about an arbitrary value of c, it is convenient to write the polynomial in the form

$$P_n(x) = a_0 + a_1(x - c) + a_2(x - c)^2 + a_3(x - c)^3 + \cdots + a_n(x - c)^n.$$

In this form, repeated differentiation produces

$$P_n{}'(x) = a_1 + 2a_2(x - c) + 3a_3(x - c)^2 + \cdots + na_n(x - c)^{n-1}$$
$$P_n{}''(x) = 2a_2 + 2(3a_3)(x - c) + \cdots + n(n - 1)a_n(x - c)^{n-2}$$
$$P_n{}'''(x) = 2(3a_3) + \cdots + n(n - 1)(n - 2)a_n(x - c)^{n-3}$$
$$\vdots$$
$$P_n^{(n)}(x) = n(n - 1)(n - 2) \cdots (2)(1)a_n.$$

Letting $x = c$, you then obtain

$$P_n(c) = a_0, \qquad P_n{}'(c) = a_1, \qquad P_n{}''(c) = 2a_2, \qquad \ldots, \qquad P_n^{(n)}(c) = n!a_n$$

and because the value of f and its first n derivatives must agree with the value of P_n and its first n derivatives at $x = c$, it follows that

$$f(c) = a_0, \qquad f'(c) = a_1, \qquad \frac{f''(c)}{2!} = a_2, \qquad \ldots, \qquad \frac{f^{(n)}(c)}{n!} = a_n.$$

With these coefficients, you can obtain the following definition of **Taylor polynomials,** named after the English mathematician Brook Taylor (1685–1731).

BROOK TAYLOR

Although Taylor was not the first to seek polynomial approximations of transcendental functions, his published account in 1715 was one of the first comprehensive works on the subject.

Definition of nth Taylor Polynomial and Maclaurin Polynomial

If f has n derivatives at c, then the polynomial

$$P_n(x) = f(c) + f'(c)(x - c) + \frac{f''(c)}{2!}(x - c)^2 + \cdots + \frac{f^{(n)}(c)}{n!}(x - c)^n$$

is called the **nth Taylor polynomial for f at c.** If $c = 0$, then

$$P_n(x) = f(0) + f'(0)x + \frac{f''(0)}{2!}x^2 + \frac{f'''(0)}{3!}x^3 + \cdots + \frac{f^{(n)}(0)}{n!}x^n$$

is called the **nth Maclaurin polynomial for f.**

REMARK   The nth Maclaurin polynomial for f is named after the English mathematician, Colin Maclaurin (1698–1746).

EXAMPLE 3   A Maclaurin Polynomial for $f(x) = e^x$

From the discussion earlier in this section, the nth Maclaurin polynomial for $f(x) = e^x$ is given by

$$P_n(x) = 1 + x + \frac{1}{2}x^2 + \frac{1}{3!}x^3 + \cdots + \frac{1}{n!}x^n.$$

EXAMPLE 4   Finding Taylor Polynomials for ln x

Find the Taylor polynomials P_0, P_1, P_2, P_3, and P_4 for $f(x) = \ln x$ centered at $c = 1$.

Solution   Expanding about $c = 1$ yields the following.

$$f(x) = \ln x \qquad\qquad f(1) = 0$$
$$f'(x) = \frac{1}{x} \qquad\qquad f'(1) = 1$$
$$f''(x) = -\frac{1}{x^2} \qquad\qquad f''(1) = -1$$
$$f'''(x) = \frac{2!}{x^3} \qquad\qquad f'''(1) = 2$$
$$f^{(4)}(x) = -\frac{3!}{x^4} \qquad\qquad f^{(4)}(1) = -6$$

Therefore, the Taylor polynomials are as follows.

$$P_0(x) = f(1) = 0$$
$$P_1(x) = f(1) + f'(1)(x - 1) = (x - 1)$$
$$P_2(x) = f(1) + f'(1)(x - 1) + \frac{f''(1)}{2!}(x - 1)^2$$
$$ = (x - 1) - \frac{1}{2}(x - 1)^2$$
$$P_3(x) = f(1) + f'(1)(x - 1) + \frac{f''(1)}{2!}(x - 1)^2 + \frac{f'''(1)}{3!}(x - 1)^3$$
$$ = (x - 1) - \frac{1}{2}(x - 1)^2 + \frac{1}{3}(x - 1)^3$$
$$P_4(x) = f(1) + f'(1)(x - 1) + \frac{f''(1)}{2!}(x - 1)^2 + \frac{f'''(1)}{3!}(x - 1)^3$$
$$ + \frac{f^{(4)}(1)}{4!}(x - 1)^4$$
$$ = (x - 1) - \frac{1}{2}(x - 1)^2 + \frac{1}{3}(x - 1)^3 - \frac{1}{4}(x - 1)^4$$

Figure 8.12 compares the graphs of P_1, P_2, P_3, and P_4 with the graph of $f(x) = \ln x$. Note that near $x = 1$ the graphs are nearly indistinguishable. For instance, $P_4(0.9) \approx -0.105358$ and $\ln(0.9) \approx -0.105361$.

FIGURE 8.12
As n increases, the graph of P_n becomes a better and better approximation of the graph of $f(x) = \ln x$.

EXAMPLE 5   Finding Maclaurin Polynomials for cos x

Find the Maclaurin polynomials P_0, P_2, P_4, and P_6 for $f(x) = \cos x$. Use $P_6(x)$ to approximate the value of $\cos(0.1)$.

Solution   Expanding about $c = 0$ yields the following.

$$
\begin{array}{ll}
f(x) = \cos x & f(0) = 1 \\
f'(x) = -\sin x & f'(0) = 0 \\
f''(x) = -\cos x & f''(0) = -1 \\
f'''(x) = \sin x & f'''(0) = 0
\end{array}
$$

Through repeated differentiation, you can see that the pattern $1, 0, -1, 0$ continues, and you obtain the following Maclaurin polynomials.

$$P_0(x) = 1$$

$$P_2(x) = 1 - \frac{1}{2!}x^2$$

$$P_4(x) = 1 - \frac{1}{2!}x^2 + \frac{1}{4!}x^4$$

$$P_6(x) = 1 - \frac{1}{2!}x^2 + \frac{1}{4!}x^4 - \frac{1}{6!}x^6$$

Using $P_6(x)$, you obtain the approximation $\cos(0.1) \approx 0.995004165$, which coincides with the calculator value to nine decimal places. Figure 8.13 compares the graphs of $f(x) = \cos x$ and P_6. ■

Note in Example 5 that the Maclaurin polynomials for $\cos x$ have only even powers of x. Similarly, the Maclaurin polynomials for $\sin x$ have only odd powers of x (see Exercise 9). This is not generally true of the Taylor polynomials for $\sin x$ and $\cos x$ expanded about $c \neq 0$, as you can see in the next example.

EXAMPLE 6   Finding a Taylor Polynomial for sin x

Find the third Taylor polynomial for $f(x) = \sin x$, expanded about $c = \pi/6$.

Solution   Expanding about $c = \pi/6$ yields the following.

$$
\begin{array}{ll}
f(x) = \sin x & f\left(\dfrac{\pi}{6}\right) = \dfrac{1}{2} \\[2mm]
f'(x) = \cos x & f'\left(\dfrac{\pi}{6}\right) = \dfrac{\sqrt{3}}{2} \\[2mm]
f''(x) = -\sin x & f''\left(\dfrac{\pi}{6}\right) = -\dfrac{1}{2} \\[2mm]
f'''(x) = -\cos x & f'''\left(\dfrac{\pi}{6}\right) = -\dfrac{\sqrt{3}}{2}
\end{array}
$$

Thus, the third Taylor polynomial for $f(x) = \sin x$, expanded about $c = \pi/6$, is as follows.

$$P_3(x) = f\left(\frac{\pi}{6}\right) + f'\left(\frac{\pi}{6}\right)\left(x - \frac{\pi}{6}\right) + \frac{f''\left(\frac{\pi}{6}\right)}{2!}\left(x - \frac{\pi}{6}\right)^2 + \frac{f'''\left(\frac{\pi}{6}\right)}{3!}\left(x - \frac{\pi}{6}\right)^3$$

$$= \frac{1}{2} + \frac{\sqrt{3}}{2}\left(x - \frac{\pi}{6}\right) - \frac{1}{2(2!)}\left(x - \frac{\pi}{6}\right)^2 - \frac{\sqrt{3}}{2(3!)}\left(x - \frac{\pi}{6}\right)^3$$

Figure 8.14 compares the graphs of $f(x) = \sin x$ and the graph of P_3. ■

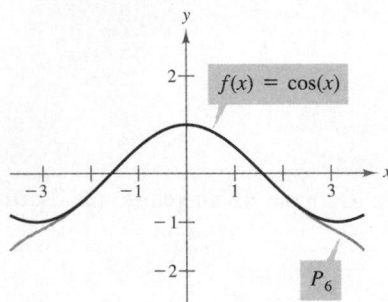

FIGURE 8.13

Near $(0, 0)$, the graph of P_6 can be used to approximate the graph of $f(x) = \cos x$.

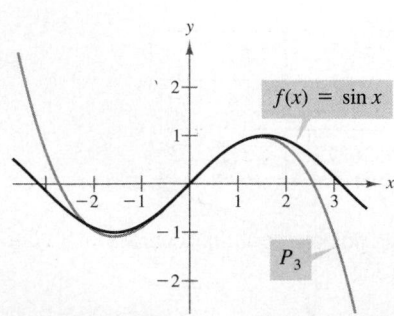

FIGURE 8.14

Near $(\pi/6, 1/2)$, the graph of P_3 can be used to approximate the graph of $f(x) = \sin x$.

Taylor polynomials and Maclaurin polynomials can be used to approximate the value of a function at a specific point. For instance, to approximate the value of $\ln(1.1)$, you can use Taylor polynomials for $f(x) = \ln x$ expanded about $c = 1$, as shown in Example 4. Or you can use Maclaurin polynomials, as shown in Example 7.

EXAMPLE 7   Approximation Using Maclaurin Polynomials

Use a fourth Maclaurin polynomial to approximate the value of $\ln(1.1)$.

Solution   Because 1.1 is closer to 1 than to 0, you should consider Maclaurin polynomials for the function $g(x) = \ln(1 + x)$.

$$
\begin{array}{ll}
g(x) = \ln(1 + x) & g(0) = 0 \\
g'(x) = (1 + x)^{-1} & g'(0) = 1 \\
g''(x) = -(1 + x)^{-2} & g''(0) = -1 \\
g'''(x) = 2(1 + x)^{-3} & g'''(0) = 2 \\
g^{(4)}(x) = -6(1 + x)^{-4} & g^{(4)}(0) = -6
\end{array}
$$

Note that you obtain the same coefficients as in Example 4. Therefore, the fourth Maclaurin polynomial for $g(x) = \ln(1 + x)$ is

$$
P_4(x) = g(0) + g'(0)x + \frac{g''(0)}{2!}x^2 + \frac{g'''(0)}{3!}x^3 + \frac{g^{(4)}(0)}{4!}x^4
$$

$$
= x - \frac{1}{2}x^2 + \frac{1}{3}x^3 - \frac{1}{4}x^4.
$$

Consequently,

$$
\ln(1.1) = \ln(1 + 0.1) \approx P_4(0.1) \approx 0.0953083.
$$

Check to see that the fourth Taylor polynomial (from Example 4), evaluated at $x = 1.1$, yields the same result. ◼

TABLE 8.3

n	$P_n(1.1)$
1	0.1000000
2	0.0950000
3	0.0953333
4	0.0953083

Table 8.3 illustrates the accuracy of the Taylor polynomial approximation from Example 4 to the calculator value of $\ln(1.1)$. You can see that as n becomes larger, $P_n(1.1)$ approaches the calculator value of 0.0953102.

On the other hand, Table 8.4 illustrates that as you move away from the expansion point $c = 1$, the accuracy of the approximation decreases.

TABLE 8.4
Fourth Taylor polynomial approximation of ln x

x	1.0	1.1	1.5	1.75	2.0
ln x	0.0000000	0.0953102	0.4054651	0.5596158	0.6931472
$P_4(x)$	0.0000000	0.0953083	0.4010417	0.5302734	0.5833333

Tables 8.3 and 8.4 illustrate two very important points about the accuracy of Taylor (or Maclaurin) polynomials for use in approximations.

1. The approximation is usually better at x-values close to c than at x-values far from c.
2. The approximation is usually better for higher-degree Taylor (or Maclaurin) polynomials than for those of lower degree.

Remainder of a Taylor Polynomial

An approximation technique is of little value without some idea of its accuracy. To measure the accuracy of approximating a functional value $f(x)$ by the Taylor polynomial $P_n(x)$, you can use the concept of a **remainder,** $R_n(x)$, defined as follows.

$$f(x) = P_n(x) + R_n(x)$$

Thus, $R_n(x) = f(x) - P_n(x)$. The absolute value of $R_n(x)$ is called the **error** associated with the approximation. That is,

$$\text{Error} = |R_n(x)| = |f(x) - P_n(x)|.$$

The next theorem gives a general procedure for estimating the remainder associated with a Taylor polynomial. This important theorem is called **Taylor's Theorem** and the remainder given in the theorem is called the **Lagrange form of the remainder.** (The proof of the theorem is lengthy, and is given in Appendix A.)

THEOREM 8.19   Taylor's Theorem

If a function f is differentiable through order $n + 1$ in an interval I containing c, then for each x in I there exists z between x and c such that

$$f(x) = f(c) + f'(c)(x - c) + \frac{f''(c)}{2!}(x - c)^2 + \cdots +$$

$$\frac{f^{(n)}(c)}{n!}(x - c)^n + R_n(x)$$

where

$$R_n(x) = \frac{f^{(n+1)}(z)}{(n + 1)!}(x - c)^{n+1}.$$

REMARK   For $n = 0$, Taylor's Theorem states that if f is differentiable in an interval I containing c, then for each x in I there exists z between x and c such that

$$f(x) = f(c) + f'(z)(x - c)$$

or

$$f'(z) = \frac{f(x) - f(c)}{x - c}.$$

Do you recognize this special case of Taylor's Theorem? (It is the Mean Value Theorem!)

When applying Taylor's Theorem, you should not expect to be able to find the exact value of z. (If you could do that, an approximation would not be necessary.) Rather, you try to find bounds for $f^{(n+1)}(z)$ from which you are able to tell how large the remainder $R_n(x)$ is.

$$y = e^{-2x}$$

$$\frac{(-1)^n 2^n x^n}{n!}$$

EXAMPLE 8   Determining the Accuracy of an Approximation

The third Maclaurin polynomial for $\sin x$ is given by

$$P_3(x) = x - \frac{x^3}{3!}.$$

Use Taylor's Theorem to approximate $\sin(0.1)$ by $P_3(0.1)$ and determine the accuracy of the approximation.

Solution   Using Taylor's Theorem, you have

$$\sin x = x - \frac{x^3}{3!} + R_3(x) = x - \frac{x^3}{3!} + \frac{f^{(4)}(z)}{4!} x^4$$

where $0 < z < 0.1$. Therefore,

$$\sin(0.1) \approx 0.1 - \frac{(0.1)^3}{3!} \approx 0.1 - 0.000167 = 0.099833.$$

Because $f^{(4)}(z) = \sin z$, it follows that the error $|R_3(0.1)|$ can be bounded as follows.

$$0 < R_3(0.1) = \frac{\sin z}{4!}(0.1)^4 < \frac{0.0001}{4!} \approx 0.000004$$

This implies that

$$0.099833 < \sin(0.1) = 0.099833 + R_3(x) < 0.099833 + 0.000004$$
$$0.099833 < \sin(0.1) < 0.099837.$$

EXAMPLE 9   Approximating a Value to a Desired Accuracy

Determine the degree of the Taylor polynomial $P_n(x)$ expanded about $c = 1$ that should be used to approximate $\ln(1.2)$ so that the error is less than 0.001.

REMARK   Try using a calculator to verify the results obtained in Examples 8 and 9. For Example 8, you obtain

$\sin(0.1) \approx 0.0998334.$

For Example 9, you obtain

$P_3(1.2) \approx 0.1827$

and

$\ln(1.2) \approx 0.1823.$

Solution   Following the pattern of Example 4, you can see that the $(n + 1)$st derivative of $f(x) = \ln x$ is given by

$$f^{(n+1)}(x) = (-1)^n \frac{n!}{x^{n+1}}.$$

Using Taylor's Theorem, you know that the error $|R_n(1.2)|$ is given by

$$|R_n(1.2)| = \left| \frac{f^{(n+1)}(z)}{(n + 1)!}(1.2 - 1)^{n+1} \right| = \frac{n!}{z^{n+1}} \left(\frac{1}{(n + 1)!} \right)(0.2)^{n+1}$$
$$= \frac{(0.2)^{n+1}}{z^{n+1}(n + 1)}.$$

where $1 < z < 1.2$. In this interval, $|R_n(1.2)|$ is largest when $z = 1$. Thus, you are seeking a value of n such that

$$\frac{(0.2)^{n+1}}{(1)^{n+1}(n + 1)} < 0.001 \quad \longrightarrow \quad 1000 < (n + 1)5^{n+1}.$$

By trial and error, you can determine that the smallest value of n satisfying this inequality is $n = 3$. Thus, you would need the third Taylor polynomial to achieve the desired accuracy in approximating $\ln(1.2)$.

1-20 odd   39-42 (handwritten)

TECHNOLOGY
Laboratory Guide
Lab 8.7

In Exercises 1–4, match the Taylor polynomial approximation of the function $f(x) = e^{-x^2/2}$ with its graph. [The graphs are labeled (a), (b), (c), and (d).]

d **1.** $y = -\frac{1}{2}x^2 + 1$

2. $y = \frac{1}{8}x^4 - \frac{1}{2}x^2 + 1$ *c*

a **3.** $y = e^{-1/2}[(x + 1) + 1]$

b **4.** $y = e^{-1/2}[\frac{1}{3}(x - 1)^3 - (x - 1) + 1]$

(a)

(b)

(c)

(d)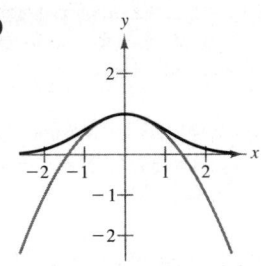

In Exercises 5–16, find the Maclaurin polynomial of degree n for the function.

5. $f(x) = e^{-x}$, $n = 3$

6. $f(x) = e^{-x}$, $n = 5$

7. $f(x) = e^{2x}$, $n = 4$

8. $f(x) = e^{3x}$, $n = 4$

9. $f(x) = \sin x$, $n = 5$

10. $f(x) = \sin \pi x$, $n = 3$

11. $f(x) = xe^x$, $n = 4$

12. $f(x) = x^2 e^{-x}$, $n = 4$

13. $f(x) = \dfrac{1}{x + 1}$, $n = 4$

14. $f(x) = \sec x$, $n = 2$

15. $f(x) = 2 - 3x^3 + x^4$, $n = 4$

16. $f(x) = 3 - 2x^2 + x^3$, $n = 3$

In Exercises 17–20, find the Taylor polynomial of degree n centered at c.

17. $f(x) = \dfrac{1}{x}$, $n = 4, c = 1$

18. $f(x) = \sqrt{x}$, $n = 4, c = 4$

19. $f(x) = \ln x$, $n = 4, c = 1$

20. $f(x) = x^2 \cos x$, $n = 2, c = \pi$

[C] In Exercises 21 and 22, use a symbolic differentiation utility to find the indicated Taylor polynomials for the function f. Graph the function and the Taylor polynomials.

21. $f(x) = \tan x$

a. $n = 3, c = 0$
b. $n = 5, c = 0$
c. $n = 3, c = \pi/4$

22. $f(x) = \dfrac{1}{x^2 + 1}$

a. $n = 2, c = 0$
b. $n = 4, c = 0$
c. $n = 4, c = 1$

[C] **23. a.** Use the Maclaurin polynomials $P_1(x)$, $P_3(x)$, $P_5(x)$, and $P_7(x)$ for $f(x) = \sin x$ to complete the table.

x	0	0.25	0.50	0.75	1.00
$\sin x$	0	0.2474	0.4794	0.6816	0.8415
$P_1(x)$					
$P_3(x)$					
$P_5(x)$					
$P_7(x)$					

b. Use a graphing utility to graph $f(x) = \sin x$ and the Maclaurin polynomials from part a.

[C] **24. a.** Use the Taylor polynomials $P_1(x)$ and $P_4(x)$ for $f(x) = \ln x$ centered at $c = 1$ to complete the table.

x	1.00	1.25	1.50	1.75	2.00
$\ln x$	0	0.2231	0.4055	0.5596	0.6931
$P_1(x)$					
$P_4(x)$					

b. Use a graphing utility to graph $f(x) = \ln x$ and the Taylor polynomials from part a.

In Exercises 25 and 26, (a) find the Maclaurin polynomial $P_3(x)$ for $f(x)$, (b) complete the table for $f(x)$ and $P_3(x)$, and (c) sketch the graphs of $f(x)$ and $P_3(x)$ on the same axes.

25. $f(x) = \arcsin x$ **26.** $f(x) = \arctan x$

x	-1	-0.75	-0.50	-0.25
$f(x)$				
$P_3(x)$				

x	0	0.25	0.50	0.75	1
$f(x)$					
$P_3(x)$					

In Exercises 27–30, approximate the function at the given value of x, using the polynomial found in the indicated exercise.

27. $f(x) = e^{-x}$, $f(\frac{1}{2})$, Exercise 5

28. $f(x) = x^2 e^{-x}$, $f(\frac{1}{4})$, Exercise 12

29. $f(x) = \ln x$, $f(1.2)$, Exercise 19

30. $f(x) = x^2 \cos x$, $f\left(\dfrac{7\pi}{8}\right)$, Exercise 20

In Exercises 31–34, use Taylor's Theorem to determine the accuracy of the approximation.

31. $\cos(0.3) \approx 1 - \dfrac{(0.3)^2}{2!} + \dfrac{(0.3)^4}{4!}$

32. $e \approx 1 + 1 + \dfrac{1^2}{2!} + \dfrac{1^3}{3!} + \dfrac{1^4}{4!} + \dfrac{1^5}{5!}$

33. $\arcsin(0.4) \approx 0.4 + \dfrac{(0.4)^3}{2 \cdot 3}$

34. $\arctan(0.5) \approx 0.5 - \dfrac{(0.5)^3}{3}$

In Exercises 35 and 36, determine the degree of the Maclaurin polynomial required so that the error in the function at the specified value of x is less than 0.001.

35. $g(x) = \sin x$,   Approximate $g(0.3)$.

36. $f(x) = e^x$,   Approximate $f(0.75)$.

[C] In Exercises 37 and 38, determine the degree of the Maclaurin polynomial required so that the error in the function at the specified value of x will be less than 0.0001. Use a symbolic differentiation utility to obtain and evaluate the required derivatives.

37. $f(x) = \ln(x + 1)$,   Approximate $f(1.5)$.

38. $f(x) = \cos(\pi x^2)$,   Approximate $f(0.6)$.

In Exercises 39 and 40, determine the values of x for which the given function can be replaced by the Taylor polynomial if the error cannot exceed 0.001.

39. $f(x) = e^x \approx 1 + x + \dfrac{x^2}{2!} + \dfrac{x^3}{3!}$ $x < 0$

40. $f(x) = \sin x \approx x - \dfrac{x^3}{3!}$

41. Compare the Maclaurin polynomials of degree 4 for the functions

$$f(x) = e^x \quad \text{and} \quad g(x) = xe^x.$$

What is the relationship between them?

42. Differentiate the Maclaurin polynomial of degree 5 for $f(x) = \sin x$ and compare the result with the Maclaurin polynomial of degree 4 for $g(x) = \cos x$.

43. Prove that if f is an odd function, then the nth Maclaurin polynomial for the function will contain only terms with odd powers of x.

44. Prove that if f is an even function, then the nth Maclaurin polynomial for the function will contain only terms with even powers of x.

45. Let $P_n(x)$ be the nth Taylor polynomial for f at c. Prove that $P_n(c) = f(c)$ and $P^{(k)}(c) = f^{(k)}(c)$ for $1 \le k \le n$.

46. *Essay* The proof in Exercise 45 guarantees that the Taylor polynomial and its derivatives agree with the function and its derivatives at $x = c$. Use the graphs and tables in Exercises 21–26 to discuss what happens to the accuracy of the Taylor polynomial as you move away from $x = c$.

[C] In Exercises 47 and 48, the graph of $y = f(x)$ is shown with seven of its Maclaurin polynomials. Identify the Maclaurin polynomials and use a graphing utility to confirm your results.

47. $f(x) = \cos x$

$y = \cos x$

48. $f(x) = \arctan x$

$y = \arctan x$

8.8

Power Series • Radius and Interval of Convergence • Endpoint Convergence • Differentiation and Integration of Power Series

Power Series

In Section 8.7, we introduced the concept of approximating functions by Taylor polynomials. For instance, the function

$$f(x) = e^x$$

can be *approximated* by its Maclaurin polynomials as follows.

$e^x \approx 1 + x$ 1st-degree polynomial

$e^x \approx 1 + x + \dfrac{x^2}{2}$ 2nd-degree polynomial

$e^x \approx 1 + x + \dfrac{x^2}{2} + \dfrac{x^3}{3!}$ 3rd-degree polynomial

$e^x \approx 1 + x + \dfrac{x^2}{2} + \dfrac{x^3}{3!} + \dfrac{x^4}{4!}$ 4th-degree polynomial

In Section 8.7, you saw that the higher the degree of the approximating polynomial, the better the approximation becomes.

In this section and the next two sections of this chapter, you will see that several important types of functions, including $f(x) = e^x$, can be represented *exactly* by an infinite series called a **power series.** For example, the power series representation for e^x is

$$e^x = 1 + x + \frac{x^2}{2} + \frac{x^3}{3!} + \cdots + \frac{x^n}{n!} + \cdots.$$

For each real number x, it can be shown that the infinite series on the right converges to the number e^x. Before doing this, however, we will discuss some preliminary results dealing with power series—beginning with the following definition.

Definition of Power Series

If x is a variable, then an infinite series of the form

$$\sum_{n=0}^{\infty} a_n x^n = a_0 + a_1 x + a_2 x^2 + a_3 x^3 + \cdots + a_n x^n + \cdots$$

is called a **power series.** More generally, a series of the form

$$\sum_{n=0}^{\infty} a_n (x - c)^n = a_0 + a_1(x - c) + a_2(x - c)^2 + \cdots + a_n(x - c)^n + \cdots$$

is called a **power series centered at c,** where c is a constant.

REMARK   To simplify the notation for power series, we agree that $(x - c)^0 = 1$, even if $x = c$.

EXAMPLE 1 Power Series

a. The following power series is centered at 0.

$$\sum_{n=0}^{\infty} \frac{x^n}{n!} = 1 + x + \frac{x^2}{2} + \frac{x^3}{3!} + \cdots$$

b. The following power series is centered at -1.

$$\sum_{n=0}^{\infty} (-1)^n (x + 1)^n = 1 - (x + 1) + (x + 1)^2 - (x + 1)^3 + \cdots$$

c. The following power series is centered at 1.

$$\sum_{n=1}^{\infty} \frac{1}{n}(x - 1)^n = (x - 1) + \frac{1}{2}(x - 1)^2 + \frac{1}{3}(x - 1)^3 + \cdots$$

A Single Point

An Interval

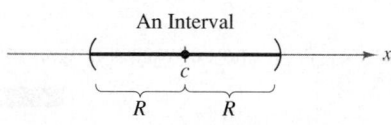

The Real Line

FIGURE 8.15
The domain of a power series has only three basic forms: a single point, an interval centered at c, or the entire real line.

Radius and Interval of Convergence

A power series in x can be viewed as a function of x

$$f(x) = \sum_{n=0}^{\infty} a_n (x - c)^n$$

where the *domain of f* is the set of all x for which the power series converges. Determination of the domain of a power series is the primary concern in this section. Of course, every power series converges at its center c because

$$\begin{aligned} f(c) &= \sum_{n=0}^{\infty} a_n (c - c)^n \\ &= a_0(1) + 0 + 0 + \cdots + 0 + \cdots \\ &= a_0. \end{aligned}$$

Thus, c always lies in the domain of f. The following important theorem (which we present without proof) states that the domain of a power series can take three basic forms: a single point, an interval centered at c, or the entire real line, as shown in Figure 8.15.

THEOREM 8.20 Convergence of a Power Series

For a power series centered at c, precisely one of the following is true.

1. The series converges only at c.

2. There exists a real number $R > 0$ such that the series converges absolutely for $|x - c| < R$, and diverges for $|x - c| > R$.

3. The series converges absolutely for all x.

The number R is the **radius of convergence** of the power series. If the series converges only at c, then the radius of convergence is $R = 0$, and if the series converges for all x, then the radius of convergence is $R = \infty$. The set of all values of x for which the power series converges is the **interval of convergence** of the power series.

To determine the radius of convergence of a power series, we use the Ratio Test, as shown in Examples 2, 3, and 4.

EXAMPLE 2　Finding the Radius of Convergence

Find the radius of convergence of $\displaystyle\sum_{n=0}^{\infty} n!x^n$.

Solution　For $x = 0$, you obtain

$$f(0) = \sum_{n=0}^{\infty} n!0^n = 1 + 0 + 0 + \cdots = 1.$$

For any fixed value of x such that $|x| > 0$, let $u_n = n!x^n$. Then

$$\lim_{n\to\infty} \left| \frac{u_{n+1}}{u_n} \right| = \lim_{n\to\infty} \left| \frac{(n+1)!x^{n+1}}{n!x^n} \right|$$
$$= |x| \lim_{n\to\infty}(n+1)$$
$$= \infty.$$

Therefore, by the Ratio Test, the series diverges for $|x| > 0$, and converges only at its center, 0. Hence, the radius of convergence is $R = 0$. ▮

EXAMPLE 3　Finding the Radius of Convergence

Find the radius of convergence of $\displaystyle\sum_{n=0}^{\infty} 3(x - 2)^n$.

Solution　For $x \neq 2$, let $u_n = 3(x - 2)^n$. Then,

$$\lim_{n\to\infty} \left| \frac{u_{n+1}}{u_n} \right| = \lim_{n\to\infty} \left| \frac{3(x-2)^{n+1}}{3(x-2)^n} \right|$$
$$= \lim_{n\to\infty} |x - 2|$$
$$= |x - 2|.$$

By the Ratio Test, the series converges if $|x - 2| < 1$ and diverges if $|x - 2| > 1$. Therefore, the radius of convergence of the series is $R = 1$. ▮

EXAMPLE 4　Finding the Radius of Convergence

Find the radius of convergence of $\displaystyle\sum_{n=0}^{\infty} \frac{(-1)^n x^{2n+1}}{(2n + 1)!}$.

Solution　Let $u_n = (-1)^n x^{2n+1}/(2n + 1)!$. Then

$$\lim_{n\to\infty} \left| \frac{u_{n+1}}{u_n} \right| = \lim_{n\to\infty} \left| \frac{[(-1)^{n+1}x^{2n+3}]/(2n + 3)!}{[(-1)^n x^{2n+1}]/(2n + 1)!} \right|$$
$$= \lim_{n\to\infty} \frac{x^2}{(2n + 3)(2n + 2)}.$$

For any *fixed* value of x, this limit is 0. Thus, by the Ratio Test, the series converges for all x. Therefore, the radius of convergence is $R = \infty$. ▮

Endpoint Convergence

Note that for a power series whose radius of convergence is a finite number R, Theorem 8.20 says nothing about the convergence at the *endpoints* of the interval of convergence. Each endpoint must be tested separately for convergence or divergence. As a result, the interval of convergence of a power series can take any one of the six forms shown in Figure 8.16.

FIGURE 8.16
Intervals of convergence.

EXAMPLE 5   Finding the Interval of Convergence

Find the interval of convergence of $\displaystyle\sum_{n=1}^{\infty} \frac{x^n}{n}$.

Solution   Letting $u_n = x^n/n$ produces

$$\lim_{n \to \infty} \left| \frac{u_{n+1}}{u_n} \right| = \lim_{n \to \infty} \left| \frac{x^{n+1}/(n+1)}{x^n/n} \right|$$

$$= \lim_{n \to \infty} \left| \frac{nx}{n+1} \right|$$

$$= |x|.$$

Therefore, by the Ratio Test, the radius of convergence is $R = 1$. Moreover, because the series is centered at 0, it converges in the interval $(-1, 1)$. This interval, however, is not necessarily the *interval of convergence*. To determine this, you must test for convergence at each endpoint. When $x = 1$, you obtain the *divergent* harmonic series

$$\sum_{n=1}^{\infty} \frac{1}{n} = \frac{1}{1} + \frac{1}{2} + \frac{1}{3} + \cdots. \qquad \text{Diverges when } x = 1$$

When $x = -1$, you obtain the *convergent* alternating harmonic series

$$\sum_{n=1}^{\infty} \frac{(-1)^n}{n} = -1 + \frac{1}{2} - \frac{1}{3} + \frac{1}{4} - \cdots. \qquad \text{Converges when } x = -1$$

Therefore, the interval of convergence for the series is $[-1, 1)$, as shown in Figure 8.17.

Interval: $[-1, 1)$
Radius: $R = 1$

FIGURE 8.17

EXAMPLE 6 Finding the Interval of Convergence

Find the interval of convergence of $\displaystyle\sum_{n=0}^{\infty} \frac{(-1)^n(x+1)^n}{2^n}$.

Solution Letting $u_n = (-1)^n(x+1)^n/2^n$ produces

$$\lim_{n\to\infty}\left|\frac{u_{n+1}}{u_n}\right| = \lim_{n\to\infty}\left|\frac{(-1)^{n+1}(x+1)^{n+1}/2^{n+1}}{(-1)^n(x+1)^n/2^n}\right|$$

$$= \lim_{n\to\infty}\left|\frac{2^n(x+1)}{2^{n+1}}\right|$$

$$= \left|\frac{x+1}{2}\right|.$$

By the Ratio Test, the series converges if $|(x+1)/2| < 1$ or $|x+1| < 2$. Hence the radius of convergence is $R = 2$. Because the series is centered at $x = -1$, it will converge in the interval $(-3, 1)$. Furthermore, at the endpoints you have

$$\sum_{n=0}^{\infty}\frac{(-1)^n(-2)^n}{2^n} = \sum_{n=0}^{\infty}\frac{2^n}{2^n} = \sum_{n=0}^{\infty}1 \qquad \text{Diverges when } x = -3$$

and

$$\sum_{n=0}^{\infty}\frac{(-1)^n(2)^n}{2^n} = \sum_{n=0}^{\infty}(-1)^n \qquad \text{Diverges when } x = 1$$

both of which diverge. Thus, the interval of convergence is $(-3, 1)$, as shown in Figure 8.18.

Interval: $(-3, 1)$
Radius: $R = 2$

FIGURE 8.18

EXAMPLE 7 Finding the Interval of Convergence

Find the interval of convergence of $\displaystyle\sum_{n=1}^{\infty}\frac{x^n}{n^2}$.

Solution Letting $u_n = x^n/n^2$ produces

$$\lim_{n\to\infty}\left|\frac{u_{n+1}}{u_n}\right| = \lim_{n\to\infty}\left|\frac{x^{n+1}/(n+1)^2}{x^n/n^2}\right| = \lim_{n\to\infty}\left|\frac{n^2 x}{(n+1)^2}\right| = |x|.$$

Thus, the radius of convergence is $R = 1$. Because the series is centered at $x = 0$, it converges in the interval $(-1, 1)$. When $x = 1$, you obtain the *convergent p*-series

$$\sum_{n=1}^{\infty}\frac{1}{n^2} = \frac{1}{1^2} + \frac{1}{2^2} + \frac{1}{3^2} + \frac{1}{4^2} + \cdots. \qquad \text{Converges when } x = 1$$

When $x = -1$, you obtain the *convergent* alternating series

$$\sum_{n=1}^{\infty}\frac{(-1)^n}{n^2} = -\frac{1}{1^2} + \frac{1}{2^2} - \frac{1}{3^2} + \frac{1}{4^2} - \cdots. \qquad \text{Converges when } x = -1$$

Therefore, the interval of convergence for the given series is $[-1, 1]$.

Differentiation and Integration of Power Series

Power series representation of functions has played an important role in the development of calculus. In fact, much of Newton's work with differentiation and integration was done in the context of power series—especially his work with complicated algebraic functions and transcendental functions. Euler, Lagrange, Leibniz, and the Bernoullis all used power series extensively in calculus.

Once you have defined a function with a power series, it is natural to wonder how you can determine the characteristics of the function. Is it continuous? Differentiable? Integrable? Theorem 8.21, which we state without proof, answers these questions.

JAMES GREGORY

One of the earliest mathematicians to work with power series was a Scotsman, James Gregory (1638–1675). He developed a power series method for interpolating table values—a method that was later used by Brook Taylor in the development of Taylor polynomials and Taylor series.

THEOREM 8.21   Properties of a Function Defined by a Power Series

If the function given by

$$f(x) = \sum_{n=0}^{\infty} a_n(x - c)^n = a_0 + a_1(x - c) + a_2(x - c)^2 + a_3(x - c)^3 + \cdots$$

has a radius of convergence of $R > 0$, then, on the interval $(c - R, c + R)$, f is continuous, differentiable, and integrable. Moreover, the derivative and antiderivative of f are as follows.

1. $f'(x) = \displaystyle\sum_{n=1}^{\infty} na_n(x - c)^{n-1} = a_1 + 2a_2(x - c) + 3a_3(x - c)^2 + \cdots$

2. $\displaystyle\int f(x)\, dx = C + \sum_{n=0}^{\infty} a_n \frac{(x - c)^{n+1}}{n + 1}$

$$= C + a_0(x - c) + a_1\frac{(x - c)^2}{2} + a_2\frac{(x - c)^3}{3} + \cdots$$

The *radius of convergence* for the series obtained by differentiating or integrating a power series is the same as for the original power series. The *interval of convergence*, however, may differ as a result of the behavior at the endpoints.

Theorem 8.21 states that, in many ways, a function defined by a power series behaves like a polynomial. It is continuous in its interval of convergence, and both its derivative and antiderivative can be determined by differentiating and integrating each term of the given power series. For instance, the derivative of the power series

$$f(x) = \sum_{n=0}^{\infty} \frac{x^n}{n!} = 1 + x + \frac{x^2}{2} + \frac{x^3}{3!} + \frac{x^4}{4!} + \cdots$$

is

$$f'(x) = 1 + (2)\frac{x}{2} + (3)\frac{x^2}{3!} + (4)\frac{x^3}{4!} + \cdots$$

$$= 1 + x + \frac{x^2}{2} + \frac{x^3}{3!} + \frac{x^4}{4!} + \cdots$$

$$= f(x).$$

EXAMPLE 8   Intervals of Convergence for $f(x)$, $f'(x)$, and $\int f(x)\,dx$

Find the intervals of convergence of

a. $\int f(x)\,dx$ **b.** $f(x)$ **c.** $f'(x)$

where $f(x) = \displaystyle\sum_{n=1}^{\infty} \frac{x^n}{n} = x + \frac{x^2}{2} + \frac{x^3}{3} + \cdots$.

Solution   By Theorem 8.21, you have

$$f'(x) = \sum_{n=1}^{\infty} x^{n-1} = 1 + x + x^2 + x^3 + \cdots$$

and

$$\int f(x)\,dx = C + \sum_{n=1}^{\infty} \frac{x^{n+1}}{n(n+1)}$$

$$= C + \frac{x^2}{1\cdot 2} + \frac{x^3}{2\cdot 3} + \frac{x^4}{3\cdot 4} + \cdots.$$

By the Ratio Test, you can show that each series has a radius of convergence of $R = 1$. Considering the interval $(-1, 1)$, you have the following.

a. For $\int f(x)\,dx$, the series

$$\sum_{n=1}^{\infty} \frac{x^{n+1}}{n(n+1)} \qquad \text{Interval of convergence } [-1, 1]$$

converges for $x = \pm 1$, and its interval of convergence is $[-1, 1]$.

b. For $f(x)$, the series

$$\sum_{n=1}^{\infty} \frac{x^n}{n} \qquad \text{Interval of convergence } [-1, 1)$$

converges for $x = -1$ and diverges for $x = 1$. Hence, its interval of convergence is $[-1, 1)$.

c. For $f'(x)$, the series

$$\sum_{n=1}^{\infty} x^{n-1} \qquad \text{Interval of convergence } (-1, 1)$$

diverges for $x = \pm 1$, and its interval of convergence is $(-1, 1)$.   ■

REMARK   From Example 8, it appears that of the three series, the one for the derivative $f'(x)$ is the least likely to converge at the endpoints. In fact, it can be shown that if the series for $f'(x)$ converges at the endpoints $x = c \pm R$, then the series for $f(x)$ will also converge there.

EXERCISES for Section 8.8

In Exercises 1–6, find the radius of convergence of the power series.

1. $\displaystyle\sum_{n=0}^{\infty} (-1)^n \frac{x^n}{n+1}$

2. $\displaystyle\sum_{n=0}^{\infty} (4x)^n$

3. $\displaystyle\sum_{n=1}^{\infty} \frac{(2x)^n}{n^2}$

4. $\displaystyle\sum_{n=0}^{\infty} \frac{(-1)^n x^n}{2^n}$

5. $\displaystyle\sum_{n=0}^{\infty} \frac{(2x)^n}{n!}$

6. $\displaystyle\sum_{n=0}^{\infty} \frac{(2n)! x^n}{n!}$

In Exercises 7–30, find the interval of convergence of the power series. (Be sure to include a check for convergence at the endpoints of the interval.)

7. $\displaystyle\sum_{n=0}^{\infty} \left(\frac{x}{2}\right)^n$

8. $\displaystyle\sum_{n=0}^{\infty} \left(\frac{x}{k}\right)^n$

9. $\displaystyle\sum_{n=1}^{\infty} \frac{(-1)^n x^n}{n}$

10. $\displaystyle\sum_{n=0}^{\infty} (-1)^{n+1} n x^n$

11. $\displaystyle\sum_{n=0}^{\infty} \frac{x^n}{n!}$

12. $\displaystyle\sum_{n=0}^{\infty} \frac{(3x)^n}{(2n)!}$

13. $\displaystyle\sum_{n=0}^{\infty} (2n)! \left(\frac{x}{2}\right)^n$

14. $\displaystyle\sum_{n=0}^{\infty} \frac{(-1)^n x^n}{(n+1)(n+2)}$

15. $\displaystyle\sum_{n=1}^{\infty} \frac{(-1)^{n+1} x^n}{4^n}$

16. $\displaystyle\sum_{n=0}^{\infty} \frac{(-1)^n n! (x-4)^n}{3^n}$

17. $\displaystyle\sum_{n=1}^{\infty} \frac{(-1)^{n+1}(x-5)^n}{n 5^n}$

18. $\displaystyle\sum_{n=0}^{\infty} \frac{(x-2)^{n+1}}{(n+1)3^{n+1}}$

19. $\displaystyle\sum_{n=0}^{\infty} \frac{(-1)^{n+1}(x-1)^{n+1}}{n+1}$

20. $\displaystyle\sum_{n=1}^{\infty} \frac{(-1)^{n+1}(x-c)^n}{n c^n}$

21. $\displaystyle\sum_{n=1}^{\infty} \frac{(x-c)^{n-1}}{c^{n-1}}, \quad 0 < c$

22. $\displaystyle\sum_{n=1}^{\infty} \frac{(-1)^{n+1} x^{2n-1}}{2n-1}$

23. $\displaystyle\sum_{n=1}^{\infty} \frac{n}{n+1} (-2x)^{n-1}$

24. $\displaystyle\sum_{n=0}^{\infty} \frac{(-1)^n x^{2n}}{n!}$

25. $\displaystyle\sum_{n=0}^{\infty} \frac{x^{2n+1}}{(2n+1)!}$

26. $\displaystyle\sum_{n=1}^{\infty} \frac{n! x^n}{(2n)!}$

27. $\displaystyle\sum_{n=1}^{\infty} \frac{k(k+1)(k+2)\cdots(k+n-1)x^n}{n!}, \quad k \geq 1$

28. $\displaystyle\sum_{n=1}^{\infty} \left(\frac{2\cdot 4\cdot 6\cdots 2n}{3\cdot 5\cdot 7\cdots(2n+1)}\right) x^{2n+1}$

29. $\displaystyle\sum_{n=1}^{\infty} \frac{(-1)^{n+1} 3\cdot 7\cdot 11\cdots(4n-1)(x-3)^n}{4^n}$

30. $\displaystyle\sum_{n=1}^{\infty} \frac{n!(x-c)^n}{1\cdot 3\cdot 5\cdots(2n-1)}$

In Exercises 31–34, find the interval of convergence of (a) $f(x)$, (b) $f'(x)$, (c) $f''(x)$, and (d) $\int f(x)\,dx$. Include a check for convergence at the endpoints.

31. $f(x) = \displaystyle\sum_{n=0}^{\infty} \left(\frac{x}{2}\right)^n$

32. $f(x) = \displaystyle\sum_{n=1}^{\infty} \frac{(-1)^{n+1}(x-5)^n}{n 5^n}$

33. $f(x) = \displaystyle\sum_{n=0}^{\infty} \frac{(-1)^{n+1}(x-1)^{n+1}}{n+1}$

34. $f(x) = \displaystyle\sum_{n=1}^{\infty} \frac{(-1)^{n+1}(x-1)^n}{n}$

35. Let

$$f(x) = \sum_{n=0}^{\infty} \frac{(-1)^n x^{2n+1}}{(2n+1)!}$$

and

$$g(x) = \sum_{n=0}^{\infty} \frac{(-1)^n x^{2n}}{(2n)!}$$

a. Find the interval of convergence of f and g.
b. Show that $f'(x) = g(x)$.
c. Show that $g'(x) = -f(x)$.
d. Identify the functions f and g.

36. Let

$$f(x) = \sum_{n=0}^{\infty} \frac{x^n}{n!}.$$

a. Find the interval of convergence of f.
b. Show that $f'(x) = f(x)$.
c. Show that $f(0) = 1$.
d. Identify the function f.

In Exercises 37 and 38, show that the function represented by the power series is a solution of the differential equation.

37. $y = \displaystyle\sum_{n=0}^{\infty} \frac{x^{2n}}{2^n n!}$

$y'' - xy' - y = 0$

38. $y = 1 + \displaystyle\sum_{n=1}^{\infty} \frac{(-1)^n x^{4n}}{2^{2n} n! \cdot 3\cdot 7\cdot 11\cdots(4n-1)}$

$y'' + x^2 y = 0$

C **39.** *Bessel Function* The Bessel function of order 0 is

$$J_0(x) = \sum_{k=0}^{\infty} \frac{(-1)^k x^{2k}}{2^{2k}(k!)^2}.$$

a. Show that the series converges for all x.
b. Show that the series is a solution of the differential equation

$$x^2 J_0'' + x J_0' + x^2 J_0 = 0.$$

c. Use a graphing utility to graph the polynomial composed of the first four terms of J_0.
d. Approximate $\int_0^1 J_0\,dx$ accurate to two decimal places.

C 40. *Bessel Function* The Bessel function of order 1 is

$$J_1(x) = x \sum_{k=0}^{\infty} \frac{(-1)^k x^{2k}}{2^{2k+1} k! (k+1)!}.$$

  a. Show that the series converges for all x.

  b. Show that the series is a solution of the differential equation

$$x^2 J_1'' + x J_1' + (x^2 - 1) J_1 = 0.$$

  c. Use a graphing utility to graph the polynomial composed of the first four terms of J_1.

  d. Show that $J_0'(x) = -J_1(x)$.

C 41. In Exercise 7, you found that the interval of convergence of the geometric series

$$\sum_{n=0}^{\infty} \left(\frac{x}{2} \right)^n \quad \text{is } (-2, 2).$$

  a. Find the sum of the series when $x = \frac{3}{4}$. Create a rectangular coordinate system with the vertical axis representing the partial sums of the series and the horizontal axis representing n. Sketch a horizontal line at the level of the sum when $x = \frac{3}{4}$. On the same coordinate system, plot the points representing the partial sums S_1, S_2, S_3, S_4, S_5, and S_6.

  b. Repeat part a for $x = -\frac{3}{4}$.

  c. Write a short paragraph comparing the rate of convergence of the partial sums to the sum of the series in parts a and b. How do the plots of the partial sums differ as they converge toward the sum of the series?

  d. Given any positive real number M, there exists a positive integer N such that

$$\sum_{n=0}^{N} \left(\frac{3}{2} \right)^n > M.$$

  Use a computer to complete the table.

M	10	100	1000	10,000
N				

42. Write a series equivalent to

$$\sum_{n=0}^{\infty} \frac{x^{2n+1}}{(2n+1)!}$$

where the index of summation has been adjusted to begin at $n = 1$.

C In Exercises 43–46, use a graphing utility to graph the power series. Each of the four power series represents a well-known function. Identify the function from the graph of its power series.

43. $f(x) = \sum_{n=0}^{\infty} (-1)^n x^n, \qquad -1 < x < 1$

44. $f(x) = \sum_{n=0}^{\infty} (-1)^n \frac{x^{2n+1}}{(2n+1)!}$ 45. $f(x) = \sum_{n=0}^{\infty} (-1)^n \frac{x^{2n}}{(2n)!}$

46. $f(x) = \sum_{n=0}^{\infty} (-1)^n \frac{x^{2n+1}}{2n+1}, \qquad -1 \le x \le 1$

True or False **In Exercises 47–50, determine whether the statement is true or false. If it is false, explain why or give an example that shows it is false.**

47. If the power series $\sum_{n=0}^{\infty} a_n x^n$ converges for $x = 2$, then it also converges for $x = -2$.

48. If the power series $\sum_{n=0}^{\infty} a_n x^n$ converges for $x = 2$, then it also converges for $x = -1$.

49. If the interval of convergence for $\sum_{n=0}^{\infty} a_n x^n$ is $(-1, 1)$, then the interval of convergence for $\sum_{n=0}^{\infty} a_n (x-1)^2$ is $(0, 2)$.

50. If $f(x) = \sum_{n=0}^{\infty} a_n x^n$ converges for $|x| < 2$, then $\int_0^1 f(x)\,dx = \sum_{n=0}^{\infty} \frac{a_n}{n+1}$.

C A R E E R   I N T E R V I E W

Ravi Chanmugam

Actuary

Equitable Life

New York, NY

As a student with a keen interest in business and a knack for math, I became involved in the actuarial profession—a branch of finance that combines the two. For those unfamiliar with the work of actuaries, it is a type of quantitative analysis that evaluates the financial implications of insurable events such as fire, retirement, and death. Simply put, actuaries place price tags on future risks.

My responsibility is the investment of our policyholder's money so that the money is available at the proper time and a good return on the investment is ensured. This area of actuarial work, known as asset/liability matching, is a rapidly growing one that draws directly on calculus. I solve first and second derivatives to measure the sensitivity of the price of bonds to changes in interest rates. Using these measures, I can then create a portfolio of securities that best satisfies the financial promises we make to our customers.

Calculations are no longer done by hand; we have computer software to do them for us now. But understanding the calculus-based financial theory is, without a doubt, key to analysis of this complexity. The unpredictability of today's financial markets demands increasingly sophisticated mathematical knowledge.

SECTION

8.9

Representation of Functions by Power Series

Geometric Power Series • Operations with Power Series

JOSEPH FOURIER

Some of the early work in representing functions by series was done by the French mathematician Joseph Fourier (1768–1830). Fourier's work is important in the history of calculus, partly because it forced eighteenth century mathematicians to question the then-prevailing narrow concept of a function. Both Cauchy and Dirichlet were motivated by Fourier's work with series, and in 1837 Dirichlet published the general definition of a function that is used today.

Geometric Power Series

In this section and the next, you will study several techniques for finding a power series to represent a given function.

Consider the function given by $f(x) = 1/(1 - x)$. The form of f closely resembles the sum of a geometric series

$$\sum_{n=0}^{\infty} ar^n = \frac{a}{1 - r}, \qquad |r| < 1.$$

In other words, if you let $a = 1$ and $r = x$, then a power series representation for $1/(1 - x)$, centered at 0, is

$$\frac{1}{1 - x} = \sum_{n=0}^{\infty} x^n = 1 + x + x^2 + x^3 + \cdots, \qquad |x| < 1.$$

Of course, this series represents $f(x) = 1/(1 - x)$ only on the interval $(-1, 1)$, whereas f is defined for all $x \neq 1$, as shown in Figure 8.19. To represent f on another interval, you must develop a different series. For instance, to obtain the power series centered at -1, you could write

$$\frac{1}{1 - x} = \frac{1}{2 - (x + 1)} = \frac{1/2}{1 - [(x + 1)/2]} = \frac{a}{1 - r}$$

which implies that $a = \frac{1}{2}$ and $r = (x + 1)/2$. Thus, for $|x + 1| < 2$, you have

$$\frac{1}{1 - x} = \sum_{n=0}^{\infty} \left(\frac{1}{2}\right)\left(\frac{x + 1}{2}\right)^n$$

$$= \frac{1}{2}\left[1 + \frac{(x + 1)}{2} + \frac{(x + 1)^2}{4} + \frac{(x + 1)^3}{8} + \cdots\right], \quad |x + 1| < 2$$

which converges on the interval $(-3, 1)$.

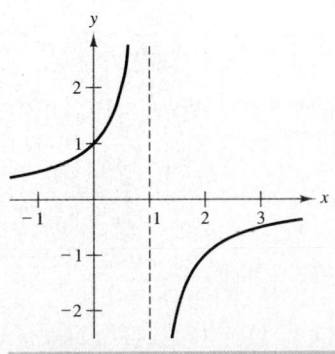

$$f(x) = \frac{1}{1 - x}, \quad \text{Domain: all } x \neq 1$$

$$f(x) = \sum_{n=0}^{\infty} x^n, \quad \text{Domain: } -1 < x < 1$$

FIGURE 8.19

EXAMPLE 1   Finding a Geometric Power Series Centered at 0

Find a power series for $f(x) = \dfrac{4}{x + 2}$, centered at 0.

Solution   Writing $f(x)$ in the form $a/(1 - r)$ produces

$$\frac{4}{2 + x} = \frac{2}{1 - (-x/2)} = \frac{a}{1 - r}$$

which implies that $a = 2$ and $r = -x/2$. Therefore, the power series for $f(x)$ is given by

$$\frac{4}{x + 2} = \sum_{n=0}^{\infty} ar^n$$

$$= \sum_{n=0}^{\infty} 2\left(-\frac{x}{2}\right)^n$$

$$= 2\left(1 - \frac{x}{2} + \frac{x^2}{4} - \frac{x^3}{8} + \cdots\right).$$

This power series converges when

$$\left|-\frac{x}{2}\right| < 1$$

which implies that the interval of convergence is $(-2, 2)$.

Another way to determine a power series for a rational function such as the one in Example 1 is to use long division. For instance, by dividing $2 + x$ into 4, you obtain the result shown in the margin at the left.

Long Division

$$
\begin{array}{r}
2 - x + \frac{1}{2}x^2 - \frac{1}{4}x^3 + \cdots \\
2 + x \overline{)4} \\
\underline{4 + 2x} \\
-2x \\
\underline{-2x - x^2} \\
x^2 \\
\underline{x^2 + \frac{1}{2}x^3} \\
-\frac{1}{2}x^3 \\
\underline{-\frac{1}{2}x^3 - \frac{1}{4}x^4}
\end{array}
$$

EXAMPLE 2   Finding a Geometric Power Series Centered at 1

Find a power series for $f(x) = \dfrac{1}{x}$, centered at 1.

Solution   Writing $f(x)$ in the form $a/(1 - r)$ produces

$$\frac{1}{x} = \frac{1}{1 - (-x + 1)} = \frac{a}{1 - r}$$

which implies that $a = 1$ and $r = 1 - x = -(x - 1)$. Therefore, the power series for $f(x)$ is given by

$$\frac{1}{x} = \sum_{n=0}^{\infty} ar^n$$

$$= \sum_{n=0}^{\infty} [-(x - 1)]^n$$

$$= 1 - (x - 1) + (x - 1)^2 - (x - 1)^3 + \cdots.$$

This power series converges when

$$|x - 1| < 1$$

which implies that the interval of convergence is $(0, 2)$.

Operations with Power Series

The versatility of geometric power series will be shown later in this section, following a discussion of power series operations. These operations, used with differentiation and integration, provide a means of developing power series for a variety of elementary functions. (For simplicity, the following properties are stated for a series centered at 0.)

THEOREM 8.22   Operations with Power Series

Let $f(x) = \Sigma a_n x^n$ and $g(x) = \Sigma b_n x^n$.

1. $f(kx) = \displaystyle\sum_{n=0}^{\infty} a_n k^n x^n$ **2.** $f(x^N) = \displaystyle\sum_{n=0}^{\infty} a_n x^{nN}$

3. $f(x) \pm g(x) = \displaystyle\sum_{n=0}^{\infty} (a_n \pm b_n) x^n$

The operations described in Theorem 8.22 can change the interval of convergence for the resulting series. For example, in the following addition, the interval of convergence for the sum is the *intersection* of the intervals of convergence of the two original series.

$$\underbrace{\sum_{n=0}^{\infty} x^n}_{(-1,\,1)} + \underbrace{\sum_{n=0}^{\infty} \left(\frac{x}{2}\right)^n}_{(-2,\,2)} = \underbrace{\sum_{n=0}^{\infty} \left(1 + \frac{1}{2^n}\right) x^n}_{(-1,\,1)}$$

$$(-1, 1) \ \cap \ (-2, 2) \ = \ (-1, 1)$$

EXAMPLE 3   Adding Two Power Series

Find a power series, centered at 0, for $f(x) = \dfrac{3x - 1}{x^2 - 1}$.

Solution   Using partial fractions, you can write $f(x)$ as

$$\frac{3x - 1}{x^2 - 1} = \frac{2}{x + 1} + \frac{1}{x - 1}.$$

By adding the two geometric power series

$$\frac{2}{x + 1} = \frac{2}{1 - (-x)} = \sum_{n=0}^{\infty} 2(-1)^n x^n, \qquad |x| < 1$$

and

$$\frac{1}{x - 1} = \frac{-1}{1 - x} = -\sum_{n=0}^{\infty} x^n, \qquad |x| < 1$$

you obtain the following power series.

$$\frac{3x - 1}{x^2 - 1} = \sum_{n=0}^{\infty} [2(-1)^n - 1] x^n = 1 - 3x + x^2 - 3x^3 + x^4 - \cdots$$

The interval of convergence for this power series is $(-1, 1)$.

EXAMPLE 4   Finding a Power Series by Integration

Find a power series for $f(x) = \ln x$, centered at 1.

Solution   From Example 2, you know that

$$\frac{1}{x} = \sum_{n=0}^{\infty} (-1)^n (x - 1)^n. \qquad \text{Interval of convergence (0, 2)}$$

Integrating this series produces

$$\ln x = \int \frac{1}{x} dx + C$$

$$= C + \sum_{n=0}^{\infty} (-1)^n \frac{(x - 1)^{n+1}}{n + 1}.$$

By letting $x = 1$, you can conclude that $C = 0$. Therefore,

$$\ln x = \sum_{n=0}^{\infty} (-1)^n \frac{(x - 1)^{n+1}}{n + 1}$$

$$= \frac{(x - 1)}{1} - \frac{(x - 1)^2}{2} + \frac{(x - 1)^3}{3} - \frac{(x - 1)^4}{4} + \cdots. \qquad (0, 2]$$

Note that the series converges at $x = 2$. This is consistent with the observation in the previous section that integration of a power series may alter the convergence at the endpoints of the interval of convergence.

TECHNOLOGY   In Section 8.7, we used the fourth-degree Taylor polynomial for the natural logarithmic function

$$\ln x \approx (x - 1) - \frac{(x - 1)^2}{2} + \frac{(x - 1)^3}{3} - \frac{(x - 1)^4}{4}$$

to approximate $\ln(1.1)$.

$$\ln(1.1) \approx (0.1) - \frac{1}{2}(0.1)^2 + \frac{1}{3}(0.1)^3 - \frac{1}{4}(0.1)^4$$

$$\approx 0.0953083$$

You now know from Example 4 that this polynomial represents the first four terms of the power series for $\ln x$. Moreover, using the Alternating Series Remainder, you can determine that the error in this approximation is less than

$$|R_4| \le |a_5|$$

$$= \frac{1}{5}(0.1)^5$$

$$= 0.000002.$$

During the seventeenth and eighteenth centuries, mathematical tables for logarithms and values of other transcendental functions were computed in this manner. The use for such numerical techniques is far from outdated, because it is precisely by such means that modern calculating devices are programmed to evaluate transcendental functions.

EXAMPLE 5   Finding a Power Series by Integration

Find a power series for $g(x) = \arctan x$, centered at 0.

Solution   Because $D_x[\arctan x] = 1/(1 + x^2)$, you can use the series

$$f(x) = \frac{1}{1 + x} = \sum_{n=0}^{\infty} (-1)^n x^n. \qquad (-1, 1)$$

Substituting x^2 for x produces

$$f(x^2) = \frac{1}{1 + x^2} = \sum_{n=0}^{\infty} (-1)^n x^{2n}.$$

Finally, by integrating, you obtain

$$\arctan x = \int \frac{1}{1 + x^2}\, dx + C$$

$$= C + \sum_{n=0}^{\infty} (-1)^n \frac{x^{2n+1}}{2n + 1}$$

$$= \sum_{n=0}^{\infty} (-1)^n \frac{x^{2n+1}}{2n + 1} \qquad \text{Let } x = 0, \text{ then } C = 0$$

$$= x - \frac{x^3}{3} + \frac{x^5}{5} - \frac{x^7}{7} + \cdots. \qquad (-1, 1)$$

It can be shown that the power series developed for $\arctan x$ in Example 5 also converges (to $\arctan x$) for $x = \pm 1$. For instance, when $x = 1$, you can write

$$\arctan 1 = 1 - \frac{1}{3} + \frac{1}{5} - \frac{1}{7} + \cdots = \frac{\pi}{4}.$$

However, this series (developed by James Gregory in 1671) does not give us a practical way of approximating π because it converges so slowly that hundreds of terms would have to be used to obtain reasonable accuracy. Example 6 shows how to use *two* different arctangent series to obtain a very good approximation of π using only a few terms. This approximation was developed by John Machin in 1706.

EXAMPLE 6   Approximating π with a Series

Use the trigonometric identity

$$4 \arctan \frac{1}{5} - \arctan \frac{1}{239} = \frac{\pi}{4}$$

(see Exercise 38b) to approximate the number π.

Solution   By using only five terms from each of the series for $\arctan(1/5)$ and $\arctan(1/239)$, you obtain

$$4 \left(4 \arctan \frac{1}{5} - \arctan \frac{1}{239} \right) \approx 3.141593$$

which agrees with the decimal representation of π with an error of less than 0.000001.

SRINIVASA RAMANUJAN

Series that can be used to approximate π have interested mathematicians for the past 300 years. An amazing series for approximating $1/\pi$ was discovered by the Indian mathematician Srinivasa Ramanujan (1887–1920) in 1914. Each successive term of Ramanujan's series adds roughly eight more correct digits to the value of $1/\pi$. (For more information about Ramanujan's work, see the article "Ramanujan and Pi" by Jonathan M. Borwein and Peter B. Borwein in the February, 1988 issue of *Scientific American*.)

EXERCISES for Section 8.9

In Exercises 1–4, find a geometric power series for the function centered at 0 (a) by the technique shown in Examples 1 and 2, and (b) using long division.

1. $f(x) = \dfrac{1}{2 - x}$ **2.** $f(x) = \dfrac{3}{4 - x}$

3. $f(x) = \dfrac{1}{2 + x}$ **4.** $f(x) = \dfrac{1}{1 + x}$

In Exercises 5–16, find a power series for the function, centered at c, and determine the interval of convergence.

5. $f(x) = \dfrac{1}{2 - x}$, $c = 5$

6. $f(x) = \dfrac{3}{4 - x}$, $c = -2$

7. $f(x) = \dfrac{3}{2x - 1}$, $c = 0$

8. $f(x) = \dfrac{3}{2x - 1}$, $c = 2$

9. $g(x) = \dfrac{1}{2x - 5}$, $c = -3$

10. $h(x) = \dfrac{1}{2x - 5}$, $c = 0$

11. $f(x) = \dfrac{3}{x + 2}$, $c = 0$

12. $f(x) = \dfrac{4}{3x + 2}$, $c = 2$

13. $g(x) = \dfrac{3x}{x^2 + x - 2}$, $c = 0$

14. $g(x) = \dfrac{4x - 7}{2x^2 + 3x - 2}$, $c = 0$

15. $f(x) = \dfrac{2}{1 - x^2}$, $c = 0$

16. $f(x) = \dfrac{4}{4 + x^2}$, $c = 0$

In Exercises 17–26, use the power series

$$\frac{1}{1 + x} = \sum_{n=0}^{\infty} (-1)^n x^n$$

to determine a power series, centered at 0, for the function. Identify the interval of convergence.

17. $h(x) = \dfrac{-2}{x^2 - 1} = \dfrac{1}{1 + x} + \dfrac{1}{1 - x}$

18. $h(x) = \dfrac{2x}{x^2 - 1} = \dfrac{1}{1 + x} - \dfrac{1}{1 - x}$

19. $f(x) = -\dfrac{1}{(x + 1)^2} = \dfrac{d}{dx}\left[\dfrac{1}{x + 1}\right]$

20. $f(x) = \dfrac{2}{(x + 1)^3} = \dfrac{d^2}{dx^2}\left[\dfrac{1}{x + 1}\right]$

21. $f(x) = \ln(x + 1) = \displaystyle\int \dfrac{1}{x + 1}\,dx$

22. $f(x) = \ln(1 - x^2) = \displaystyle\int \dfrac{1}{1 + x}\,dx - \int \dfrac{1}{1 - x}\,dx$

23. $g(x) = \dfrac{1}{x^2 + 1}$ **24.** $f(x) = \ln(x^2 + 1)$

25. $h(x) = \dfrac{1}{4x^2 + 1}$ **26.** $f(x) = \arctan 2x$

C In Exercises 27 and 28, let

$$S_n = x - \frac{x^2}{2} + \frac{x^3}{3} - \frac{x^4}{4} + \cdots + \frac{x^n}{n}.$$

Use a graphing utility to graphically confirm the inequality. Then complete the table to numerically confirm the inequality.

x	0.0	0.2	0.4	0.6	0.8	1.0
S_n						
$\ln(x + 1)$						
S_{n+1}						

27. $S_2 \le \ln(x + 1) \le S_3$
28. $S_4 \le \ln(x + 1) \le S_5$

In Exercises 29–32, use the series for $f(x) = \arctan x$ to approximate the given value, using $R_N \le 0.001$.

29. $\arctan \frac{1}{4}$ **30.** $\displaystyle\int_0^{3/4} \arctan x^2\,dx$

31. $\displaystyle\int_0^{1/2} \dfrac{\arctan x^2}{x}\,dx$ **32.** $\displaystyle\int_0^{1/2} x^2 \arctan x\,dx$

In Exercises 33–36, use the power series

$$\frac{1}{1 - x} = \sum_{n=0}^{\infty} x^n.$$

33. Find the series representation of $1/(1 - x)^2$ and determine its interval of convergence.

34. Adjust the index of summation for the series given to begin with $n = 1$.

35. *Probability*   If a fair coin is tossed repeatedly, the probability that the first head occurs on the nth toss is

$$P(n) = \left(\frac{1}{2}\right)^n.$$

When this game is repeated many times, the average number of tosses required until the first head occurs is

$$E(n) = \sum_{n=1}^{\infty} nP(n).$$

(This value is called the *expected value* of n.) Use the results of Exercises 33 and 34 to find $E(n)$. Is the answer what you expected? Why or why not?

36. Use the results of Exercises 33 and 34 to find the sums of the following series.

a. $\dfrac{1}{3} \displaystyle\sum_{n=1}^{\infty} n\left(\frac{2}{3}\right)^n$ **b.** $\dfrac{1}{10} \displaystyle\sum_{n=1}^{\infty} n\left(\frac{9}{10}\right)^n$

37. Prove that

$$\arctan x + \arctan y = \arctan \frac{x + y}{1 - xy} \quad \text{for } xy \neq 1$$

provided the value of the left side of the equation is between $-\pi/2$ and $\pi/2$.

38. Use the result of Exercise 37 to verify the identity.

a. $\arctan \frac{120}{119} - \arctan \frac{1}{239} = \pi/4$

b. $4 \arctan \frac{1}{5} - \arctan \frac{1}{239} = \pi/4$

[*Hint*: Use Exercise 37 twice to find $4 \arctan \frac{1}{5}$. Then use part a.]

In Exercises 39 and 40 (a) verify the given equation and (b) use the equation and the series for the arctangent to approximate π to two-decimal-place accuracy.

39. $2 \arctan \dfrac{1}{2} - \arctan \dfrac{1}{7} = \dfrac{\pi}{4}$

40. $2 \arctan \dfrac{2}{3} - \arctan \dfrac{7}{17} = \dfrac{\pi}{4}$

In Exercises 41–46, find the sum of the convergent series. Explain how you obtained the sum.

41. $\displaystyle\sum_{n=1}^{\infty} (-1)^{n+1} \frac{1}{2^n n}$

42. $\displaystyle\sum_{n=1}^{\infty} (-1)^{n+1} \frac{1}{3^n n}$

43. $\displaystyle\sum_{n=1}^{\infty} (-1)^{n+1} \frac{2^n}{5^n n}$

44. $\displaystyle\sum_{n=0}^{\infty} (-1)^n \frac{1}{2n + 1}$

45. $\displaystyle\sum_{n=0}^{\infty} (-1)^n \frac{1}{2^{2n+1}(2n + 1)}$

46. $\displaystyle\sum_{n=1}^{\infty} (-1)^{n+1} \frac{1}{3^{2n-1}(2n - 1)}$

C **47.** *Essay*   One of the series in Exercises 41–46 converges to its sum at a much lower rate than the other five series. Which is it? Explain why this series converges so slowly. Use a computer to illustrate the rate of convergence.

48. Prove that

$$\sum_{n=0}^{\infty} \frac{(-1)^n}{3^n(2n + 1)} = \frac{\pi}{2\sqrt{3}}.$$

COLIN MACLAURIN

The development of power series to represent functions is credited to the combined work of many seventeenth and eighteenth century mathematicians. Gregory, Newton, John and James Bernoulli, Leibniz, Euler, Lagrange, Wallis, and Fourier all contributed to this work. However, the two names that are most commonly associated with power series are Brook Taylor (1685–1731) and Colin Maclaurin (1698–1746).

Taylor Series and Maclaurin Series

In Section 8.9, you derived power series for several functions using geometric series with term-by-term differentiation or integration. In this section you will study a *general* procedure for deriving the power series for a function that has derivatives of all orders. The following theorem gives the form that *every* (convergent) power series must take.

THEOREM 8.23   The Form of a Convergent Power Series

If f is represented by a power series $f(x) = \Sigma a_n(x - c)^n$ for all x in an open interval I containing c, then $a_n = f^{(n)}(c)/n!$ and

$$f(x) = f(c) + f'(c)(x - c) + \frac{f''(c)}{2!}(x - c)^2 + \cdots$$

$$+ \frac{f^{(n)}(c)}{n!}(x - c)^n + \cdots .$$

Proof   Suppose the power series $\Sigma a_n(x - c)^n$ has a radius of convergence R. Then, by Theorem 8.21, you know that the nth derivative of f exists for $|x - c| < R$, and by successive differentiation you obtain the following.

$$f^{(0)}(x) = a_0 + a_1(x - c) + a_2(x - c)^2 + a_3(x - c)^3 + a_4(x - c)^4 + \cdots$$

$$f^{(1)}(x) = a_1 + 2a_2(x - c) + 3a_3(x - c)^2 + 4a_4(x - c)^3 + \cdots$$

$$f^{(2)}(x) = 2a_2 + 3!a_3(x - c) + 4 \cdot 3a_4(x - c)^2 + \cdots$$

$$f^{(3)}(x) = 3!a_3 + 4!a_4(x - c) + \cdots$$

$$\vdots$$

$$f^{(n)}(x) = n!a_n + (n + 1)!a_{n+1}(x - c) + \cdots$$

Evaluating each of these derivatives at $x = c$ yields

$$f^{(0)}(c) = 0!a_0, \qquad f^{(1)}(c) = 1!a_1, \qquad f^{(2)}(c) = 2!a_2, \qquad f^{(3)}(c) = 3!a_3$$

and, in general, $f^{(n)}(c) = n!a_n$. By solving for a_n, you can find that the coefficients of the power series representation of $f(x)$ are

$$a_n = \frac{f^{(n)}(c)}{n!} .$$ ■

Notice that the coefficients of the power series in Theorem 8.23 are precisely the coefficients of the Taylor polynomials for $f(x)$ at c, as defined in Section 8.7. For this reason, the series is called the **Taylor series** for $f(x)$ at c.

Definition of Taylor and Maclaurin Series

If a function f has derivatives of all orders at $x = c$, then the series

$$\sum_{n=0}^{\infty} \frac{f^{(n)}(c)}{n!}(x - c)^n = f(c) + f'(c)(x - c) + \cdots + \frac{f^{(n)}(c)}{n!}(x - c)^n + \cdots$$

is called the **Taylor series for $f(x)$ at c.** Moreover, if $c = 0$, then the series is the **Maclaurin series for f.**

If you know the pattern for the coefficients of the Taylor polynomials for a function, then you can extend the pattern easily to form the corresponding Taylor series. For instance, in Example 4 of Section 8.7, we found the fourth Taylor polynomial for $\ln x$, centered at 1, to be

$$P_4(x) = (x - 1) - \frac{1}{2}(x - 1)^2 + \frac{1}{3}(x - 1)^3 - \frac{1}{4}(x - 1)^4.$$

From this pattern, you can obtain the Taylor series for $\ln x$ centered at $c = 1$,

$$(x - 1) - \frac{1}{2}(x - 1)^2 + \cdots + \frac{(-1)^{n+1}}{n}(x - 1)^n + \cdots.$$

EXAMPLE 1   Forming a Power Series

Use the function $f(x) = \sin x$ to form the Maclaurin series

$$\sum_{n=0}^{\infty} \frac{f^{(n)}(0)}{n!}x^n = f(0) + f'(0)x + \frac{f''(0)}{2!}x^2 + \frac{f^{(3)}(0)}{3!}x^3 + \frac{f^{(4)}(0)}{4!}x^4 + \cdots$$

and determine the interval of convergence.

Solution   Successive differentiation of $f(x)$ yields

$$f(x) = \sin x \qquad f(0) = 0$$
$$f'(x) = \cos x \qquad f'(0) = 1$$
$$f''(x) = -\sin x \qquad f''(0) = 0$$
$$f^{(3)}(x) = -\cos x \qquad f^{(3)}(0) = -1$$
$$f^{(4)}(x) = \sin x \qquad f^{(4)}(0) = 0$$
$$f^{(5)}(x) = \cos x \qquad f^{(5)}(0) = 1$$

and so on. The pattern repeats after the third derivative. Hence, the power series is as follows.

$$\sum_{n=0}^{\infty} \frac{f^{(n)}(0)}{n!}x^n = f(0) + f'(0)x + \frac{f''(0)}{2!}x^2 + \frac{f^{(3)}(0)}{3!}x^3 + \frac{f^{(4)}(0)}{4!}x^4 + \cdots$$

$$\sum_{n=0}^{\infty} \frac{(-1)^n x^{2n+1}}{(2n + 1)!} = x - \frac{x^3}{3!} + \frac{x^5}{5!} - \frac{x^7}{7!} + \cdots$$

By the Ratio Test, you can conclude that this series converges for all x.

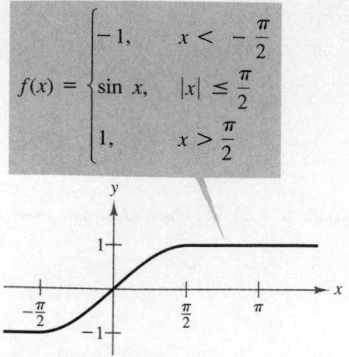

$$f(x) = \begin{cases} -1, & x < -\dfrac{\pi}{2} \\ \sin x, & |x| \leq \dfrac{\pi}{2} \\ 1, & x > \dfrac{\pi}{2} \end{cases}$$

FIGURE 8.20

Notice that we did not conclude that the power series in Example 1 converges to $\sin x$ for all x. We simply concluded that the power series converges to some function, but we are not sure what function it is. This is a subtle, but important, point in dealing with Taylor or Maclaurin series. To persuade yourself that the series

$$f(c) + f'(c)(x - c) + \frac{f''(c)}{2!}(x - c)^2 + \cdots + \frac{f^{(n)}(c)}{n!}(x - c)^n + \cdots$$

might converge to a function other than f, remember that the derivatives are being evaluated at a single point. It can easily happen that another function will agree with the values of $f^{(n)}(x)$ when $x = c$ and disagree at other x-values. For instance, if you formed the power series (centered at 0) for the function shown in Figure 8.20, you would obtain the same series obtained in Example 1. You know that the series converges for all x, and yet it obviously cannot converge to both $f(x)$ and $\sin x$ for all x.

REMARK   Don't confuse this observation with the result given in Theorem 8.23. That theorem says that *if a power series converges to $f(x)$*, then the series must be a Taylor series. The theorem does not say that every series formed with the Taylor coefficients $a_n = f^{(n)}(c)/n!$ will converge to $f(x)$.

THEOREM 8.24   Convergence of Taylor Series

If a function f has derivatives of all orders in an open interval I centered at c, then the equality

$$f(x) = \sum_{n=0}^{\infty} \frac{f^{(n)}(c)}{n!}(x - c)^n$$

holds if and only if

$$\lim_{n \to \infty} R_n(x) = \lim_{n \to \infty} \frac{f^{(n+1)}(z)}{(n + 1)!}(x - c)^{n+1} = 0$$

for every x in I.

Proof   For a Taylor series, the nth partial sum coincides with the nth Taylor polynomial. That is, $S_n(x) = P_n(x)$. Moreover, because

$$P_n(x) = f(x) - R_n(x)$$

it follows that

$$\begin{aligned} \lim_{n \to \infty} S_n(x) &= \lim_{n \to \infty} P_n(x) \\ &= \lim_{n \to \infty}[f(x) - R_n(x)] \\ &= f(x) - \lim_{n \to \infty} R_n(x). \end{aligned}$$

Hence, for a given x, the Taylor series (the sequence of partial sums) converges to $f(x)$ if and only if $R_n(x) \to 0$ as $n \to \infty$.

REMARK   Stated another way, this theorem says that a power series formed with the Taylor coefficients $a_n = f^{(n)}(c)/n!$ converges to the function from which it was derived at precisely those values for which the remainder approaches 0 as $n \to \infty$.

In Example 1, we derived the power series from the sine function and we also concluded that the series converges to some function on the entire real line. In Example 2, you will see that the series actually converges to $\sin x$.

EXAMPLE 2   A Convergent Maclaurin Series

Show that the Maclaurin series for $f(x) = \sin x$ converges to $\sin x$ for all x.

Solution   Using the result in Example 1, you need to show that

$$\sin x = x - \frac{x^3}{3!} + \frac{x^5}{5!} - \frac{x^7}{7!} + \cdots + \frac{(-1)^n x^{2n+1}}{(2n+1)!} + \cdots$$

is true for all x. Because

$$f^{(n+1)}(x) = \pm\sin x \quad \text{or} \quad f^{(n+1)}(x) = \pm\cos x$$

you know that $\left| f^{(n+1)}(z) \right| \le 1$ for every real number z. Therefore, for any fixed x, you can apply Taylor's Theorem (Theorem 8.19) to conclude that

$$0 \le \left| R_n(x) \right| = \left| \frac{f^{(n+1)}(z)}{(n+1)!} x^{n+1} \right| \le \frac{|x|^{n+1}}{(n+1)!}.$$

From the discussion in Section 8.1 regarding the relative rate of convergence of exponential and factorial sequences, it follows that for a fixed x

$$\lim_{n \to \infty} \frac{|x|^{n+1}}{(n+1)!} = 0.$$

Finally, by the Squeeze Theorem, it follows that for all x, $R_n(x) \to 0$ as $n \to \infty$. Hence, by Theorem 8.24, the Maclaurin series for $\sin x$ converges to $\sin x$ for all x. ■

TECHNOLOGY   Figure 8.21 visually illustrates the convergence of the Maclaurin series for $\sin x$ by comparing the graphs of the Maclaurin polynomials $P_1(x)$, $P_3(x)$, $P_5(x)$, and $P_7(x)$ with the graph of the sine function. Notice that as the degree of the polynomial increases, its graph more closely resembles that of the sine function.

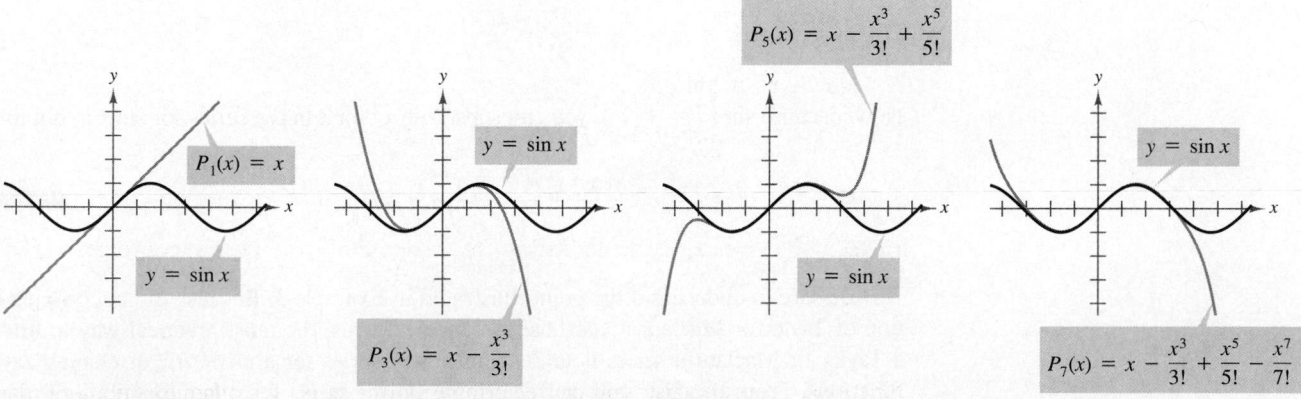

$$P_5(x) = x - \frac{x^3}{3!} + \frac{x^5}{5!}$$

$$P_1(x) = x$$

$$y = \sin x$$

$$P_3(x) = x - \frac{x^3}{3!}$$

$$y = \sin x$$

$$y = \sin x$$

$$P_7(x) = x - \frac{x^3}{3!} + \frac{x^5}{5!} - \frac{x^7}{7!}$$

FIGURE 8.21
As n increases, the graph of P_n more closely resembles the sine function.

The steps for finding a Taylor series for $f(x)$ at c are summarized in the following list.

Guidelines for Finding a Taylor Series

1. Differentiate $f(x)$ several times and evaluate each derivative at c.

$$f(c), f'(c), f''(c), f'''(c), \ldots, f^{(n)}(c), \ldots$$

Try to recognize a pattern for these numbers.

2. Use the sequence developed in the first step to form the Taylor coefficients $a_n = f^{(n)}(c)/n!$, and determine the interval of convergence for the resulting power series

$$f(c) + f'(c)(x - c) + \frac{f''(c)}{2!}(x - c)^2 + \cdots + \frac{f^{(n)}(c)}{n!}(x - c)^n + \cdots.$$

3. Within this interval of convergence, determine whether the series converges to $f(x)$.

The direct determination of Taylor or Maclaurin coefficients using successive differentiation can be difficult, and the next example illustrates a shortcut for finding the coefficients indirectly—using the coefficients of a known Taylor or Maclaurin series.

EXAMPLE 3   Maclaurin Series for a Composite Function

Find the Maclaurin series for $f(x) = \sin x^2$.

Solution   To find the coefficients for this Maclaurin series *directly* you must calculate successive derivatives of $f(x) = \sin x^2$. By calculating just the first two,

$$f'(x) = 2x \cos x^2 \quad \text{and} \quad f''(x) = -4x^2 \sin x^2 + 2 \cos x^2$$

you can see that the task would be quite cumbersome. Fortunately, there is an alternative. Suppose you first consider the Maclaurin series for $\sin x$ found in Example 1.

$$g(x) = \sin x = x - \frac{x^3}{3!} + \frac{x^5}{5!} - \frac{x^7}{7!} + \cdots$$

Now, because $\sin x^2 = g(x^2)$, you can substitute x^2 for x in the series for $\sin x$ to obtain

$$\sin x^2 = g(x^2) = x^2 - \frac{x^6}{3!} + \frac{x^{10}}{5!} - \frac{x^{14}}{7!} + \cdots. \qquad \blacksquare$$

Be sure to understand the point illustrated in Example 3. Because direct computation of Taylor or Maclaurin coefficients can be tedious, the most practical way to find a Taylor or Maclaurin series is to develop power series for a *basic list* of elementary functions. From this list, you can determine power series for other functions by the operations of addition, subtraction, multiplication, division, differentiation, integration, or composition with known power series.

Binomial Series

Before presenting the basic list for elementary functions, we develop one additional formula—for a function of the form $f(x) = (1 + x)^k$. This produces the **binomial series.**

EXAMPLE 4   Binomial Series

Find the Maclaurin series for $f(x) = (1 + x)^k$ and determine its radius of convergence.

Solution   By successive differentiation, you have

$$f(x) = (1 + x)^k \qquad\qquad f(0) = 1$$
$$f'(x) = k(1 + x)^{k-1} \qquad\qquad f'(0) = k$$
$$f''(x) = k(k - 1)(1 + x)^{k-2} \qquad\qquad f''(0) = k(k - 1)$$
$$f'''(x) = k(k - 1)(k - 2)(1 + x)^{k-3} \qquad f'''(0) = k(k - 1)(k - 2)$$
$$\vdots \qquad\qquad\qquad\qquad \vdots$$
$$f^{(n)}(x) = k \cdots (k - n + 1)(1 + x)^{k-n} \qquad f^{(n)}(0) = k(k - 1) \cdots (k - n + 1)$$

which produces the series

$$1 + kx + \frac{k(k - 1)x^2}{2} + \cdots + \frac{k(k - 1) \cdots (k - n + 1)x^n}{n!} + \cdots.$$

Because $a_{n+1}/a_n \to 1$, you can apply the Ratio Test to conclude that the radius of convergence is $R = 1$. Thus, the series converges to *some* function in the interval $(-1, 1)$.

Note that in Example 4 we showed that the Taylor series for $(1 + x)^k$ converges to *some* function in the interval $(-1, 1)$. However, we did not show that the series actually converges to $(1 + x)^k$. To do this, you could show that the remainder $R_n(x)$ converges to 0, as in Example 2.

EXAMPLE 5   Finding a Binomial Series

Find the power series for $f(x) = \sqrt[3]{1 + x}$.

Solution   Using the binomial series

$$(1 + x)^k = 1 + kx + \frac{k(k - 1)x^2}{2!} + \frac{k(k - 1)(k - 2)x^3}{3!} + \cdots$$

let $k = \frac{1}{3}$ and write

$$(1 + x)^{1/3} = 1 + \frac{x}{3} - \frac{2x^2}{3^2 2!} + \frac{2 \cdot 5x^3}{3^3 3!} - \frac{2 \cdot 5 \cdot 8x^4}{3^4 4!} + \cdots$$

which converges for $-1 \le x \le 1$.

TECHNOLOGY   Try using a graphing utility to confirm the result of Example 5. If you graph the functions

$$f(x) = (1 + x)^{1/3} \quad \text{and} \quad P_4(x) = 1 + \frac{x}{3} - \frac{x^2}{9} + \frac{5x^3}{81} - \frac{10x^4}{243}$$

on the same screen, you should obtain the result shown in Figure 8.22.

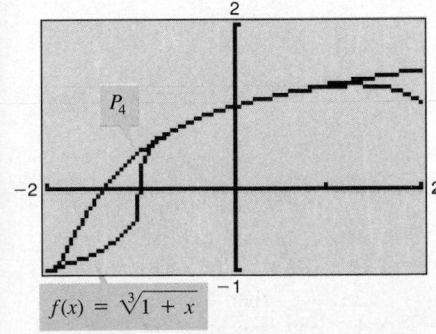

$f(x) = \sqrt[3]{1 + x}$

FIGURE 8.22

Deriving Taylor Series from a Basic List

In the following list, we provide the power series for several elementary functions with the corresponding intervals of convergence.

Power Series for Elementary Functions

Function	Interval of convergence
$\dfrac{1}{x} = 1 - (x - 1) + (x - 1)^2 - (x - 1)^3 + (x - 1)^4 - \cdots + (-1)^n(x - 1)^n + \cdots,$	$0 < x < 2$
$\dfrac{1}{1 + x} = 1 - x + x^2 - x^3 + x^4 - x^5 + \cdots + (-1)^n x^n + \cdots,$	$-1 < x < 1$
$\ln x = (x - 1) - \dfrac{(x - 1)^2}{2} + \dfrac{(x - 1)^3}{3} - \dfrac{(x - 1)^4}{4} + \cdots + \dfrac{(-1)^{n-1}(x - 1)^n}{n} + \cdots,$	$0 < x \leq 2$
$e^x = 1 + x + \dfrac{x^2}{2!} + \dfrac{x^3}{3!} + \dfrac{x^4}{4!} + \dfrac{x^5}{5!} + \cdots + \dfrac{x^n}{n!} + \cdots,$	$-\infty < x < \infty$
$\sin x = x - \dfrac{x^3}{3!} + \dfrac{x^5}{5!} - \dfrac{x^7}{7!} + \dfrac{x^9}{9!} - \cdots + \dfrac{(-1)^n x^{2n+1}}{(2n + 1)!} + \cdots,$	$-\infty < x < \infty$
$\cos x = 1 - \dfrac{x^2}{2!} + \dfrac{x^4}{4!} - \dfrac{x^6}{6!} + \dfrac{x^8}{8!} - \cdots + \dfrac{(-1)^n x^{2n}}{(2n)!} + \cdots,$	$-\infty < x < \infty$
$\arctan x = x - \dfrac{x^3}{3} + \dfrac{x^5}{5} - \dfrac{x^7}{7} + \dfrac{x^9}{9} - \cdots + \dfrac{(-1)^n x^{2n+1}}{2n + 1} + \cdots,$	$-1 \leq x \leq 1$
$\arcsin x = x + \dfrac{x^3}{2 \cdot 3} + \dfrac{1 \cdot 3 x^5}{2 \cdot 4 \cdot 5} + \dfrac{1 \cdot 3 \cdot 5 x^7}{2 \cdot 4 \cdot 6 \cdot 7} + \cdots + \dfrac{(2n)! x^{2n+1}}{(2^n n!)^2 (2n + 1)} + \cdots,$	$-1 \leq x \leq 1$
$(1 + x)^k = 1 + kx + \dfrac{k(k - 1)x^2}{2!} + \dfrac{k(k - 1)(k - 2)x^3}{3!} + \dfrac{k(k - 1)(k - 2)(k - 3)x^4}{4!} + \cdots,$	$-1 < x < 1*$

*The convergence at $x = \pm 1$ depends on the value of k.

REMARK The binomial series is valid for noninteger values of k. Moreover, if k happens to be a positive integer, then the binomial series reduces to a simple binomial expansion.

EXAMPLE 6   Deriving a Power Series from a Basic List

Find the power series for $f(x) = \cos \sqrt{x}$.

Solution   Using the power series

$$\cos x = 1 - \frac{x^2}{2!} + \frac{x^4}{4!} - \frac{x^6}{6!} + \frac{x^8}{8!} - \cdots$$

you can replace x by $\sqrt{x}$ to obtain the series

$$\cos \sqrt{x} = 1 - \frac{x}{2!} + \frac{x^2}{4!} - \frac{x^3}{6!} + \frac{x^4}{8!} - \cdots.$$

This series converges for all x in the domain of $\cos \sqrt{x}$—that is, for $x \geq 0$.   ■

EXAMPLE 7   A Power Series for $\sin^2 x$

Find the power series for $f(x) = \sin^2 x$.

Solution   Consider rewriting $\sin^2 x$ as follows.

$$\sin^2 x = \frac{1 - \cos 2x}{2} = \frac{1}{2} - \frac{\cos 2x}{2}$$

Now, use the series for $\cos x$.

$$\cos x = 1 - \frac{x^2}{2!} + \frac{x^4}{4!} - \frac{x^6}{6!} + \frac{x^8}{8!} - \cdots$$

$$\cos 2x = 1 - \frac{2^2}{2!}x^2 + \frac{2^4}{4!}x^4 - \frac{2^6}{6!}x^6 + \frac{2^8}{8!}x^8 - \cdots$$

$$-\frac{1}{2}\cos 2x = -\frac{1}{2} + \frac{2}{2!}x^2 - \frac{2^3}{4!}x^4 + \frac{2^5}{6!}x^6 - \frac{2^7}{8!}x^8 + \cdots$$

$$\sin^2 x = \frac{1}{2} - \frac{1}{2}\cos 2x = \frac{1}{2} - \frac{1}{2} + \frac{2}{2!}x^2 - \frac{2^3}{4!}x^4 + \frac{2^5}{6!}x^6 - \frac{2^7}{8!}x^8 + \cdots$$

$$= \frac{2}{2!}x^2 - \frac{2^3}{4!}x^4 + \frac{2^5}{6!}x^6 - \frac{2^7}{8!}x^8 + \cdots$$

This series converges for $-\infty < x < \infty$.   ▬

As mentioned in the previous section, power series can be used to obtain tables of values of transcendental functions. They are also useful for estimating the values of definite integrals for which antiderivatives cannot be found. The next example demonstrates this use.

EXAMPLE 8   Power Series Approximation of a Definite Integral

Use a power series to approximate

$$\int_0^1 e^{-x^2}\, dx$$

with an error of less than 0.01.

Solution   Replacing x with $-x^2$ in the series for e^x produces the following.

$$e^{-x^2} = 1 - x^2 + \frac{x^4}{2!} - \frac{x^6}{3!} + \frac{x^8}{4!} - \cdots$$

$$\int_0^1 e^{-x^2}\, dx = \left[x - \frac{x^3}{3} + \frac{x^5}{5\cdot 2!} - \frac{x^7}{7\cdot 3!} + \frac{x^9}{9\cdot 4!} - \cdots \right]_0^1$$

$$= 1 - \frac{1}{3} + \frac{1}{10} - \frac{1}{42} + \frac{1}{216} - \cdots$$

Summing the first *four* terms, you have

$$\int_0^1 e^{-x^2}\, dx \approx 0.74$$

which, by the Alternating Series Test, has an error of less than $\frac{1}{216} \approx 0.005$.   ▬

1-21 every 3rd

EXERCISES for Section 8.10

TECHNOLOGY
Laboratory Guide
Labs 8.8–8.10

In Exercises 1–10, use the definition to find the Taylor series (centered at c) for the function.

1. $f(x) = e^{2x}$, $c = 0$ **2.** $f(x) = e^{-2x}$, $c = 0$

3. $f(x) = \cos x$, $c = \dfrac{\pi}{4}$ **4.** $f(x) = \sin x$, $c = \dfrac{\pi}{4}$

5. $f(x) = \ln x$, $c = 1$ **6.** $f(x) = e^x$, $c = 1$

7. $f(x) = \sin 2x$, $c = 0$

8. $f(x) = \ln(x^2 + 1)$, $c = 0$

9. $f(x) = \sec x$, $c = 0$ (first three nonzero terms)

10. $f(x) = \tan x$, $c = 0$ (first three nonzero terms)

In Exercises 11–16, use the binomial series to find the Maclaurin series of the function.

11. $f(x) = \dfrac{1}{(1 + x)^2}$ **12.** $f(x) = \dfrac{1}{\sqrt{1 - x}}$

13. $f(x) = \dfrac{1}{\sqrt{4 + x^2}}$ **14.** $f(x) = \sqrt{1 + x}$

15. $f(x) = \sqrt{1 + x^2}$ **16.** $f(x) = \sqrt{1 + x^3}$

In Exercises 17–28, find the Maclaurin series for the function. (Use the table of power series for elementary functions.)

17. $f(x) = e^{x^2/2}$ **18.** $g(x) = e^{-3x}$

19. $g(x) = \sin 2x$ **20.** $h(x) = x \cos x$

21. $f(x) = \cos \sqrt{x}$ **22.** $f(x) = \sin x^2$

23. $g(x) = \dfrac{\sin x}{x}$ **24.** $f(x) = \dfrac{\arcsin x}{x}$

25. $f(x) = \frac{1}{2}(e^x - e^{-x}) = \sinh x$

26. $f(x) = \frac{1}{2}(e^x - e^{-x}) = \cosh x$

27. $f(x) = \cos^2 x$
  [*Hint*: $\cos^2 x = \frac{1}{2}(1 + \cos 2x)$]

28. $f(x) = \sinh^{-1} x = \ln\!\left(x + \sqrt{x^2 + 1}\right)$
  $\left(\textit{Hint}: \text{Integrate the series for } \dfrac{1}{\sqrt{x^2 + 1}}.\right)$

In Exercises 29 and 30, use a power series with the fact that $i^2 = -1$ to verify the formula.

29. $g(x) = \dfrac{1}{2i}(e^{ix} - e^{-ix}) = \sin x$

30. $g(x) = \frac{1}{2}(e^{ix} + e^{-ix}) = \cos x$

C In Exercises 31–36, find the first four nonzero terms of the Maclaurin series of the function by multiplying or dividing the appropriate power series in the table of power series for elementary functions. Use a graphing utility to obtain a graph of the function and its corresponding polynomial approximation.

31. $f(x) = e^x \sin x$ **32.** $g(x) = e^x \cos x$

33. $h(x) = \cos x \ln(1 + x)$ **34.** $f(x) = e^x \ln(1 + x)$

35. $g(x) = \dfrac{\sin x}{1 + x}$ **36.** $f(x) = \dfrac{e^x}{1 + x}$

In Exercises 37 and 38, find a Maclaurin series for $f(x)$ defined by the integral.

37. $\displaystyle\int_0^x (e^{-t^2} - 1)\, dt$ **38.** $\displaystyle\int_0^x \sqrt{1 + t^3}\, dt$

C In Exercises 39–42, verify the given sum of the infinite series. Then use a computer to approximate the sum with an error of less than 0.0001.

39. $\displaystyle\sum_{n=0}^{\infty} (-1)^n \left[\dfrac{1}{(2n + 1)!}\right] = \sin 1$

40. $\displaystyle\sum_{n=1}^{\infty} (-1)^{n-1}\left(\dfrac{1}{n!}\right) = \dfrac{e - 1}{e}$

41. $\displaystyle\sum_{n=1}^{\infty} (-1)^{n+1} \dfrac{1}{n} = \ln 2$ **42.** $\displaystyle\sum_{n=0}^{\infty} \dfrac{2^n}{n!} = e^2$

In Exercises 43 and 44, use the series representation of the function f to find (if it exists) $\lim\limits_{x \to 0} f(x)$.

43. $f(x) = \dfrac{1 - \cos x}{x}$ **44.** $f(x) = \dfrac{\sin x}{x}$

In Exercises 45–50, use power series to approximate the definite integral with an error of less than 0.0001. (In Exercises 45, 46, and 50, assume that the integrand is defined as 1 when $x = 0$.)

45. $\displaystyle\int_0^1 \dfrac{\sin x}{x}\, dx$ **46.** $\displaystyle\int_0^{1/2} \dfrac{\arctan x}{x}\, dx$

47. $\displaystyle\int_0^1 \sqrt{x}\cos x\, dx$ **48.** $\displaystyle\int_{0.5}^1 \cos \sqrt{x}\, dx$

49. $\displaystyle\int_{0.1}^{0.3} \sqrt{1 + x^3}\, dx$ **50.** $\displaystyle\int_0^{1/2} \dfrac{\ln(x + 1)}{x}\, dx$

Probability   In Exercises 51 and 52, approximate the normal probability with an error of less than 0.0001 where the probability is given by

$$P(a < x < b) = \dfrac{1}{\sqrt{2\pi}} \int_a^b e^{-x^2/2}\, dx.$$

51. $P(0 < x < 1)$ **52.** $P(1 < x < 2)$

C In Exercises 53–56, use a symbolic differentiation utility to find the fifth-degree Taylor polynomial (centered at c) of the function. Graph the function and the polynomial. Use the graph to obtain the largest interval on which the polynomial is a reasonable approximation of the function.

53. $f(x) = x\cos 2x, \qquad c = 0$

54. $f(x) = \sin\dfrac{x}{2}\ln(1 + x), \qquad c = 0$

55. $g(x) = \sqrt{x}\ln x, \qquad c = 1$

56. $h(x) = \sqrt[3]{x}\arctan x, \qquad c = 1$

57. *Projectile Motion*   A projectile fired from the ground follows the trajectory given by

$$y = \left(\tan\theta - \frac{g}{kv_0\cos\theta}\right)x - \frac{g}{k^2}\ln\left(1 - \frac{kx}{v_0\cos\theta}\right)$$

where v_0 is the initial speed, θ is the angle of projection, g is the acceleration due to gravity, and k is the drag factor caused by air resistance. Using the power series representation

$$\ln(1 + x) = x - \frac{x^2}{2} + \frac{x^3}{3} - \frac{x^4}{4} + \cdots, \qquad -1 < x < 1$$

verify that the trajectory can be rewritten as

$$y = (\tan\theta)x + \frac{gx^2}{2v_0^2\cos^2\theta} + \frac{kgx^3}{3v_0^3\cos^3\theta}$$
$$+ \frac{k^2gx^4}{4v_0^4\cos^4\theta} + \cdots.$$

58. *Projectile Motion*   Use the result of Exercise 57 to determine the series for the path of a projectile projected from ground level at an angle of $\theta = 60°$ with an initial speed of $v_0 = 64$ feet per second and a drag factor of $k = \frac{1}{16}$.

59. Consider the function f defined by

$$f(x) = \begin{cases} e^{-1/x^2}, & x \neq 0 \\ 0, & x = 0. \end{cases}$$

a. Sketch a graph of the function.
b. Use the alternative form of the definition of the derivative (Section 2.1) and L'Hôpital's Rule to show that $f'(0) = 0$. [By continuing this process, it can be shown that $f^{(n)}(0) = 0$ for $n > 1$.]
c. Using the result of part b, find the Maclaurin series for f. Does the series converge to f?

60. Sketch the graph of

$$f(x) = \begin{cases} 0, & x < -\pi/2 \\ \cos x, & -\pi/2 \leq x \leq \pi/2. \\ 0, & x > \pi/2 \end{cases}$$

Does the Maclaurin series for this function converge to the function? Explain.

61. Find the Maclaurin series for $f(x) = xe^x$. Integrate this series term-by-term over the closed interval $[0, 1]$, and show that

$$1 = \sum_{n=0}^{\infty} \frac{1}{(n + 2)n!}.$$

62. Prove that $\lim\limits_{n\to\infty} \dfrac{x^n}{n!} = 0$ for any real x.

63. Prove that e is irrational. (*Hint:* Assume that $e = p/q$ is rational (p, q integers) and consider

$$e = 1 + 1 + \frac{1}{2!} + \cdots + \frac{1}{n!} + \cdots.)$$

REVIEW EXERCISES for Chapter 8

In Exercises 1 and 2, find the general term of the sequence.

1. $1, \frac{1}{2}, \frac{1}{6}, \frac{1}{24}, \frac{1}{120}, \ldots$

2. $\frac{1}{2}, \frac{2}{5}, \frac{3}{10}, \frac{4}{17}, \ldots$

In Exercises 3–10, determine the convergence or divergence of the sequence with the given general terms. (b and c are positive real numbers.)

3. $a_n = \dfrac{n + 1}{n^2}$

4. $a_n = \dfrac{1}{\sqrt{n}}$

5. $a_n = \dfrac{n^3}{n^2 + 1}$

6. $a_n = \dfrac{n}{\ln n}$

7. $a_n = \sqrt{n + 1} - \sqrt{n}$

8. $a_n = \left(1 + \dfrac{1}{2n}\right)^n$

9. $a_n = \dfrac{\sin\sqrt{n}}{\sqrt{n}}$

10. $a_n = (b^n + c^n)^{1/n}$

11. *Compound Interest*   A deposit of \$5000 is made in an account that earns 8% interest compounded quarterly. The balance in the account after n quarters is

$$A_n = 5000\left(1 + \frac{0.08}{4}\right)^n, \qquad n = 1, 2, 3, \ldots.$$

a. Compute the first eight terms of this sequence.
b. Find the balance in this account after 10 years by computing the 40th term of the sequence.

12. *Depreciation*   A company buys a machine for \$120,000. During the next 5 years it will depreciate at the rate of 30% per year. (That is, at the end of each year the depreciated value is 70% of what it was at the beginning of the year.)
a. Find the formula for the nth term of a sequence that gives the value of the machine t full years after it was purchased.
b. Find the depreciated value of the machine at the end of 5 full years.

In Exercises 13–16, find the first five terms of the sequence of partial sums for the series.

13. $\sum_{n=0}^{\infty} \left(\frac{3}{2}\right)^n$

14. $\sum_{n=1}^{\infty} \frac{(-1)^{n+1}}{2n}$

15. $\sum_{n=1}^{\infty} \frac{(-1)^{n+1}}{(2n)!}$

16. $\sum_{n=1}^{\infty} \frac{1}{n(n+1)}$

In Exercises 17–20, find the sum of the series.

17. $\sum_{n=1}^{\infty} \left(\frac{2}{3}\right)^n$

18. $\sum_{n=0}^{\infty} \frac{2^{n+2}}{3^n}$

19. $\sum_{n=0}^{\infty} \left(\frac{1}{2^n} - \frac{1}{3^n}\right)$

20. $\sum_{n=0}^{\infty} \left[\left(\frac{2}{3}\right)^n - \frac{1}{(n+1)(n+2)}\right]$

In Exercises 21 and 22, express the repeating decimal as the ratio of two integers.

21. $0.\overline{09}$

22. $0.\overline{923076}$

23. *Bouncing Ball*   A ball is dropped from a height of 8 feet. Each time it drops h feet, it rebounds $0.7h$ feet. Find the total distance traveled by the ball.

24. *Total Compensation*   Suppose you accept a job that pays a salary of \$32,000 the first year. During the next 39 years you receive a 5.5% raise each year. What would your total salary be over the 40-year period?

25. *Compound Interest*   A deposit of \$200 is made at the end of each month for 2 years in an account that pays 6%, compounded continuously. What is the balance in the account at the end of the 2 years?

26. *Compound Interest*   A deposit of \$100 is made at the end of each month for 10 years in an account that pays 6.5%, compounded monthly. What is the balance in the account at the end of the 10 years?

In Exercises 27–38, determine the convergence or divergence of the series.

27. $\sum_{n=1}^{\infty} \frac{2^n}{n^3}$

28. $\sum_{n=1}^{\infty} \frac{1}{\sqrt[4]{n^3}}$

29. $\sum_{n=1}^{\infty} \frac{1}{\sqrt{n^3 + 2n}}$

30. $\sum_{n=1}^{\infty} \frac{n+1}{n(n+2)}$

31. $\sum_{n=1}^{\infty} \frac{n}{e^{n^2}}$

32. $\sum_{n=1}^{\infty} \frac{n!}{e^n}$

33. $\sum_{n=1}^{\infty} \frac{(-1)^n n}{\ln n}$

34. $\sum_{n=1}^{\infty} \frac{(-1)^n \sqrt{n}}{n+1}$

35. $\sum_{n=1}^{\infty} \left(\frac{1}{n^2} - \frac{1}{n}\right)$

36. $\sum_{n=1}^{\infty} \left(\frac{1}{n^2} - \frac{1}{2^n}\right)$

37. $\sum_{n=1}^{\infty} \frac{(-1)^n 1 \cdot 3 \cdot 5 \cdots (2n-1)}{2 \cdot 4 \cdot 6 \cdots (2n)(2n+1)}$

38. $\sum_{n=1}^{\infty} \frac{1 \cdot 3 \cdot 5 \cdots (2n-1)}{2 \cdot 5 \cdot 8 \cdots (3n-1)}$

C 39. *Essay*   Use a computer to complete the table for
  a. $p = 2$ **b.** $p = 5$
  Write a short paragraph describing and comparing the entries in the tables generated.

N	5	10	20	30	40
$\sum_{n=1}^{N} \dfrac{1}{n^p}$					
$\int_{N}^{\infty} \dfrac{1}{x^p}\,dx$					

40. *Essay*   You are told that the terms of a positive series appear to approach 0 very slowly as $n \to \infty$. (In fact, $a_{75} = 0.7$.) If you are given no other information, can you conclude that the series diverges? Support your answer with examples.

In Exercises 41–46, find the interval of convergence of the power series.

41. $\sum_{n=0}^{\infty} \left(\frac{x}{10}\right)^n$

42. $\sum_{n=0}^{\infty} (2x)^n$

43. $\sum_{n=0}^{\infty} \frac{(-1)^n (x-2)^n}{(n+1)^2}$

44. $\sum_{n=1}^{\infty} \frac{3^n (x-2)^n}{n}$

45. $\sum_{n=0}^{\infty} n!(x-2)^n$

46. $\sum_{n=0}^{\infty} \frac{(x-2)^n}{2^n}$

In Exercises 47–54, find the power series for the function centered at c.

47. $f(x) = \sin x, \quad c = \frac{3\pi}{4}$

48. $f(x) = \cos x, \quad c = -\frac{\pi}{4}$

49. $f(x) = 3^x, \quad c = 0$

50. $f(x) = \csc x, \quad c = \frac{\pi}{2}$
  (first three terms)

51. $f(x) = \frac{1}{x}, \quad c = -1$

52. $f(x) = \sqrt{x}, \quad c = 4$

53. $g(x) = \frac{2}{3-x}, \quad c = 0$

54. $h(x) = \frac{1}{(1+x)^3}, \quad c = 0$

55. Determine the first four terms of the Maclaurin series for e^{2x}
  a. by using the definition of the Maclaurin series and the formula for the coefficient of the nth term $a_n = f^{(n)}(0)/n!$.
  b. by replacing x by $2x$ in the series for e^x.
  c. by multiplying the series for e^x by itself because $e^{2x} = e^x \cdot e^x$.

56. Follow the pattern of Exercise 55 to find the first four terms of the series for $\sin 2x$. [*Hint:* $\sin 2x = 2 \sin x \cos x$.]

In Exercise 57 and 58, find a function represented by the series and give the domain of the function.

57. $1 + \dfrac{2}{3}x + \dfrac{4}{9}x^2 + \dfrac{8}{27}x^3 + \cdots$

58. $8 - 2(x - 3) + \dfrac{1}{2}(x - 3)^2 - \dfrac{1}{8}(x - 3)^3 + \cdots$

In Exercises 59–62, find the series representation of the function defined by the given integral.

59. $\displaystyle\int_0^x \dfrac{\sin t}{t}\,dt$

60. $\displaystyle\int_0^x \cos\dfrac{\sqrt{t}}{2}\,dt$

61. $\displaystyle\int_0^x \dfrac{\ln(t + 1)}{t}\,dt$

62. $\displaystyle\int_0^x \dfrac{e^t - 1}{t}\,dt$

In Exercises 63 and 64, use power series to find (if it exists) the limit. Verify the result by using L'Hôpital's Rule.

63. $\displaystyle\lim_{x \to 0} \dfrac{\arctan x}{\sqrt{x}}$

64. $\displaystyle\lim_{x \to 0} \dfrac{\arcsin x}{x}$

In Exercises 65–68, use a Taylor polynomial to approximate the function with an error of less than 0.001.

65. $\sin 95°$

66. $\cos(0.75)$

67. $\ln(1.75)$

68. $e^{-0.25}$

69. A Taylor polynomial centered at 0 will be used to approximate the cosine function. Find the degree of the polynomial required to obtain the desired accuracy over the specified interval.

Maximum error	Interval
a. 0.001	$[-0.5, 0.5]$
b. 0.001	$[-1, 1]$
c. 0.0001	$[-0.5, 0.5]$
d. 0.0001	$[-2, 2]$

C **70.** Use a graphing utility to graph the cosine function and the Taylor polynomials in Exercise 69.

71. Show that the function defined by

$$y = \sum_{n=0}^{\infty} \dfrac{(-3)^n x^{2n}}{2^n n!}$$

is a solution of the differential equation

$$y'' + 3xy' + 3y = 0.$$

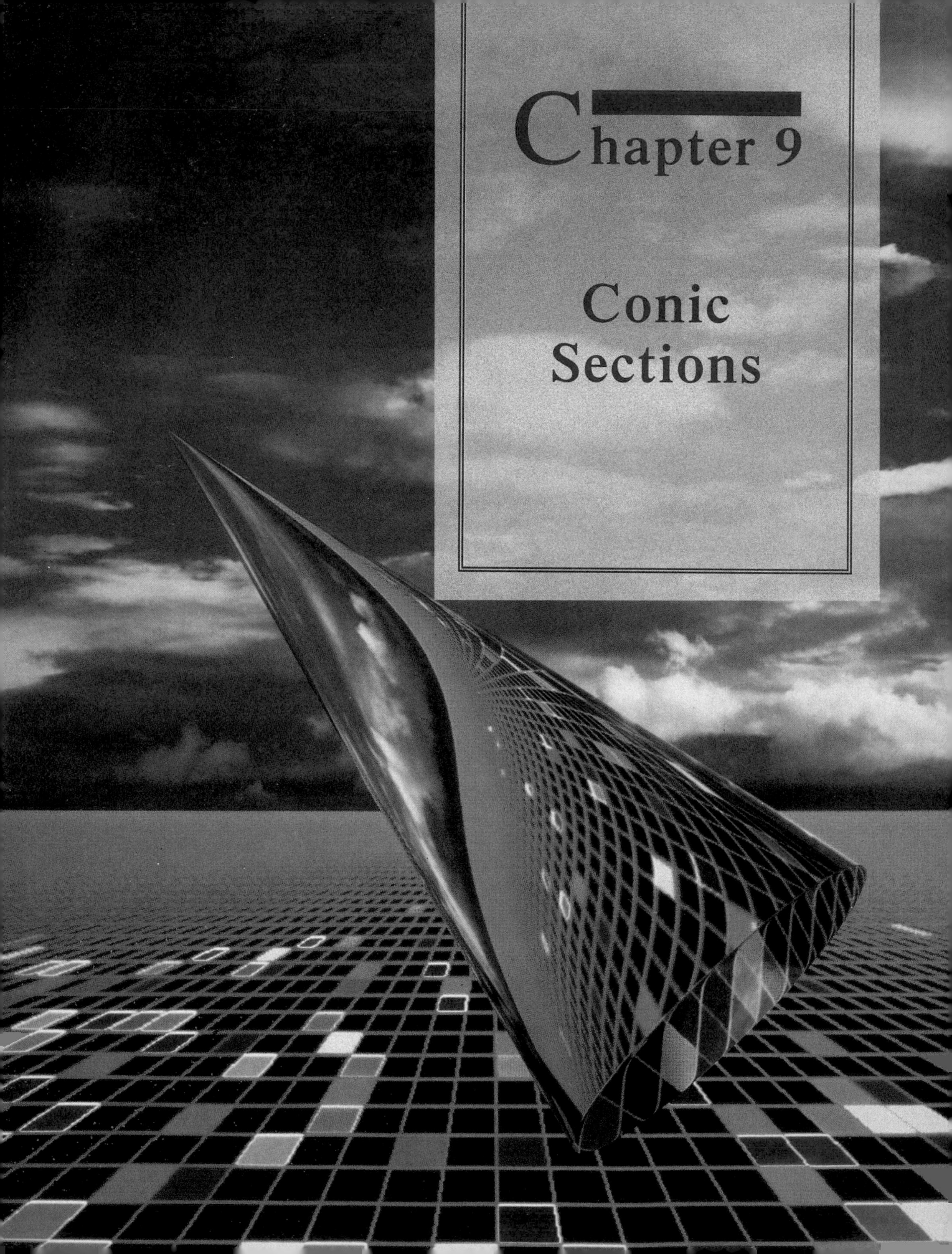

Chapter 9

Conic
Sections

HYPATIA

The Greeks discovered conic sections sometime between 600 and 300 B.C. By the beginning of the Alexandrian period, enough was known about conics for Apollonius (262–190 B.C.) to produce an eight-volume work on the subject. Later, toward the end of the Alexandrian period, Hypatia (370–415 A.D.) wrote a textbook entitled *On the Conics of Apollonius*. Her death marked the end of major mathematical discoveries in Europe for several hundred years.

The early Greeks were largely concerned with the geometric properties of conics. It was not until 1900 years later, in the early seventeenth century, that the broader applicability of conic became apparent, and conics then played a prominent role in the development of calculus.

◀ *This computer-generated photo shows a single-napped cone. Each of the conic sections in this chapter can be formed by intersecting a plane with a double-napped cone.*

Conic Sections

Each **conic section** (or simply **conic**) can be described as the intersection of a plane and a double-napped cone. Notice in Figure 9.1 that for the four basic conics, the intersecting plane does not pass through the vertex of the cone. When the plane does pass through the vertex, the resulting figure is a **degenerate conic,** as shown in Figure 9.2.

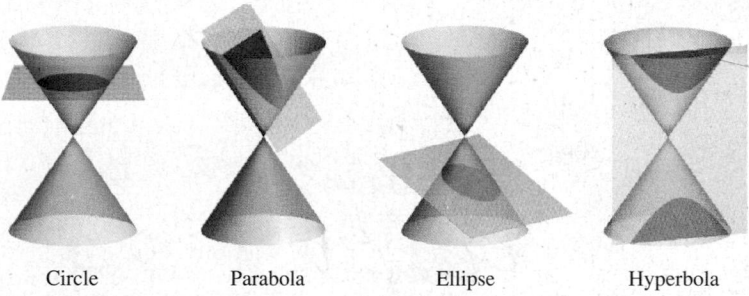

| Circle | Parabola | Ellipse | Hyperbola |

FIGURE 9.1
Conic sections

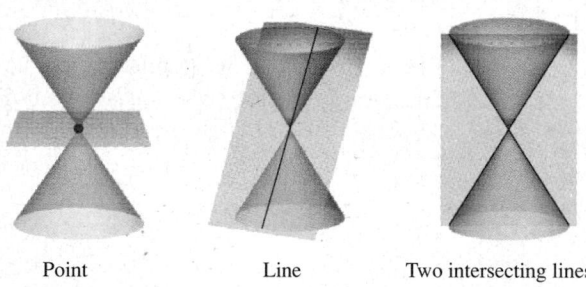

| Point | Line | Two intersecting lines |

FIGURE 9.2
Degenerate conics

There are several ways to study conics. You could begin as the Greeks did by defining the conics in terms of the intersections of planes and cones, or, as discussed in Section 9.4, you could define them algebraically in terms of the general second-degree equation

$$Ax^2 + Bxy + Cy^2 + Dx + Ey + F = 0.$$

General second-degree equation

However, a third approach, in which each of the conics is defined as a **locus** (collection) of points satisfying a certain geometric property, suits our needs best. For example, in the prerequisites chapter, a circle was defined as the collection of all points (x, y) that are equidistant from a fixed point (h, k). This locus definition easily produced the standard equation of a circle,

$$(x - h)^2 + (y - k)^2 = r^2.$$

Standard equation of a circle

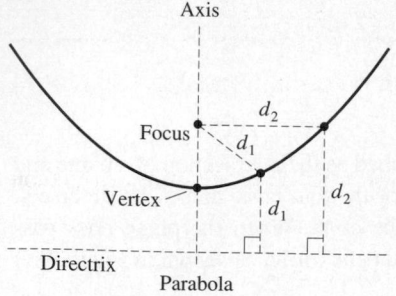

Directrix

Parabola

FIGURE 9.3

Parabolas

In this and the following two sections, you will study properties of parabolas, ellipses, and hyperbolas. You will also identify practical geometric properties used in the construction of objects such as bridges, searchlights, telescopes, and radar detectors.

A **parabola** is the set of all points (x, y) that are equidistant from a fixed line (**directrix**) and a fixed point (**focus**) not on the line. The midpoint between the focus and the directrix is the **vertex,** and the line passing through the focus and the vertex is the **axis** of the parabola. Note in Figure 9.3 that a parabola is symmetric with respect to its axis.

THEOREM 9.1 Standard Equation of a Parabola

The **standard form** of the equation of a parabola with vertex (h, k) and directrix $y = k - p$ is

$$(x - h)^2 = 4p(y - k). \qquad \text{Vertical axis}$$

For directrix $x = h - p$, the equation is

$$(y - k)^2 = 4p(x - h). \qquad \text{Horizontal axis}$$

The focus lies on the axis p units (*directed distance*) from the vertex.

Proof We prove the case for which the directrix is parallel to the x-axis and the focus lies above the vertex, as shown in Figure 9.4(a). If (x, y) is any point on the parabola, then (x, y) is equidistant from the focus $(h, k + p)$ and the directrix $y = k - p$.

$$\sqrt{(x - h)^2 + [y - (k + p)]^2} = |y - (k - p)|$$
$$(x - h)^2 + [y - (k + p)]^2 = [y - (k - p)]^2$$
$$(x - h)^2 + y^2 - 2y(k + p) + (k + p)^2 = y^2 - 2y(k - p) + (k - p)^2$$
$$(x - h)^2 - 2py + 2pk = 2py - 2pk$$
$$(x - h)^2 = 4p(y - k)$$

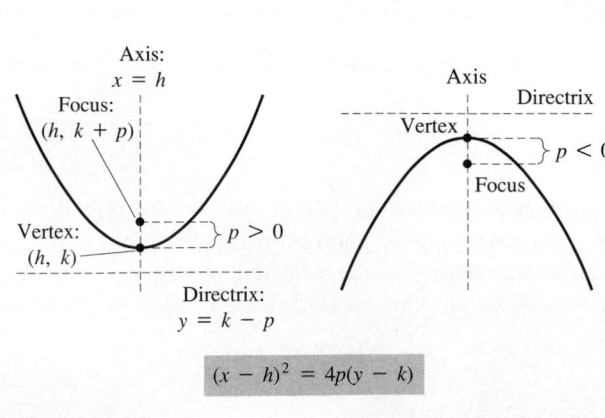

(a) Vertical axis: $p > 0$ (b) Vertical axis: $p < 0$

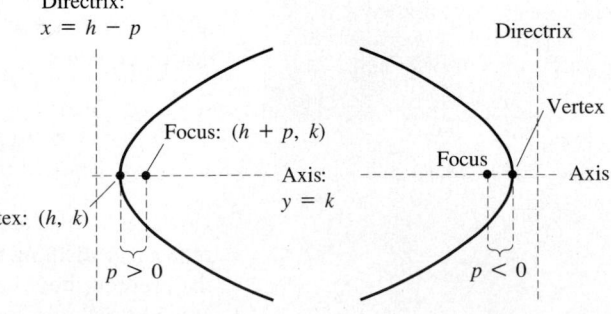

(c) Horizontal axis: $p > 0$ (d) Horizontal axis: $p < 0$

FIGURE 9.4
Parabolic orientations

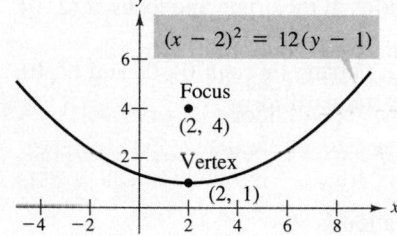

FIGURE 9.5
Parabola with a vertical axis, $p > 0$.

EXAMPLE 1   Finding the Standard Equation of a Parabola

Find the standard form of the equation of the parabola with vertex (2, 1) and focus (2, 4).

Solution   Because the axis of the parabola is vertical, you should use the equation

$$(x - h)^2 = 4p(y - k)$$

where $h = 2$, $k = 1$, and $p = 3$. Thus, the standard form is

$$(x - 2)^2 = 12(y - 1).$$ Standard form

The graph of this parabola is shown in Figure 9.5.

REMARK   By expanding the standard equation in Example 1, you can obtain the more common quadratic form $y = \frac{1}{12}(x^2 - 4x + 16)$.

EXAMPLE 2   Finding the Focus of a Parabola

Find the focus of the parabola given by $y = -\frac{1}{2}x^2 - x + \frac{1}{2}$.

Solution   To find the focus, convert to standard form by completing the square.

$$y = \frac{1}{2} - x - \frac{1}{2}x^2$$ Original equation

$$y = \frac{1}{2}(1 - 2x - x^2)$$ Factor out $\frac{1}{2}$

$$2y = 1 - (x^2 + 2x)$$ Group terms

$$2y = 2 - (x^2 + 2x + 1)$$ Add and subtract 1 on right side

$$x^2 + 2x + 1 = -2y + 2$$

$$(x + 1)^2 = -2(y - 1)$$ Standard form

Comparing this equation with $(x - h)^2 = 4p(y - k)$, you can conclude that

$$h = -1, \quad k = 1, \quad \text{and} \quad p = -\frac{1}{2}.$$

Because p is negative, the parabola opens downward, as shown in Figure 9.6. Therefore, the focus of the parabola is

$$(h, k + p) = \left(-1, \frac{1}{2}\right).$$ Focus

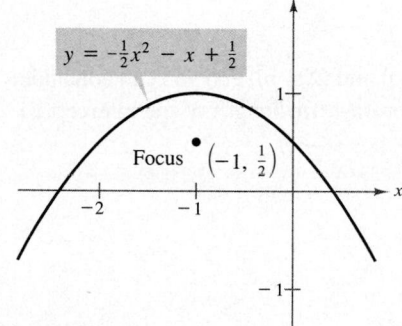

FIGURE 9.6
Parabola with a vertical axis, $p < 0$.

If the vertex of a parabola is at the origin, then the standard form

$$(x - h)^2 = 4p(y - k)$$ Vertex at (h, k)

simplifies to

$$x^2 = 4py$$ Vertex at origin

as demonstrated in the next example.

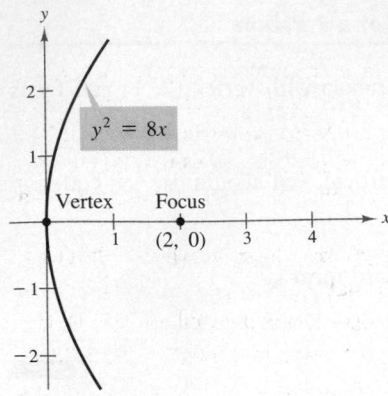

FIGURE 9.7

Parabola with a horizontal axis, $p > 0$.

EXAMPLE 3   Vertex at the Origin

Find the standard equation of the parabola with vertex at the origin and focus at $(2, 0)$.

Solution   The axis of the parabola is horizontal, passing through $(0, 0)$ and $(2, 0)$, as shown in Figure 9.7. Thus, you should use the standard form

$$y^2 = 4px$$

where $h = k = 0$ and $p = 2$. Therefore, the equation is

$$y^2 = 8x.$$

A line segment that passes through the focus of a parabola and has endpoints on the parabola is called a **focal chord.** The specific focal chord perpendicular to the axis of the parabola is the **latus rectum.** In the next example, we determine the length of the latus rectum and the length of the corresponding intercepted arc.

EXAMPLE 4   Focal Chord Length and Arc Length

Find the length of the latus rectum of the parabola given by $x^2 = 4py$. Then find the length of the parabolic arc intercepted by the latus rectum.

Solution   Because the latus rectum passes through the focus $(0, p)$ and is perpendicular to the y-axis, the coordinates of its endpoints are $(-x, p)$ and (x, p). Substituting p for y in the equation of the parabola produces

$$x^2 = 4p(p)$$
$$x^2 = 4p^2$$
$$x = \pm 2p.$$

Thus, the endpoints of the latus rectum are $(-2p, p)$ and $(2p, p)$, and you can conclude that its length is $4p$, as shown in Figure 9.8. In contrast, the length of the intercepted arc is given by the following.

$$
\begin{aligned}
s &= \int_{-2p}^{2p} \sqrt{1 + (y')^2}\, dx \\
&= 2\int_0^{2p} \sqrt{1 + \left(\frac{x}{2p}\right)^2}\, dx \\
&= \frac{1}{p}\int_0^{2p} \sqrt{4p^2 + x^2}\, dx \\
&= \frac{1}{2p}\left[x\sqrt{4p^2 + x^2} + 4p^2 \ln\left| x + \sqrt{4p^2 + x^2} \right| \right]_0^{2p} \quad \text{Theorem 7.2}\\
&= \frac{1}{2p}\left[2p\sqrt{8p^2} + 4p^2 \ln\left(2p + \sqrt{8p^2}\right) - 4p^2 \ln(2p) \right] \\
&= 2p\left[\sqrt{2} + \ln\left(1 + \sqrt{2}\right) \right] \\
&\approx 4.59p
\end{aligned}
$$

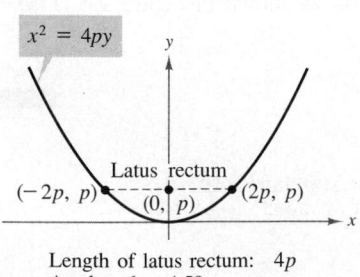

Length of latus rectum: $4p$
Arc length: $4.59p$

FIGURE 9.8

The latus rectum and an intercepted arc.

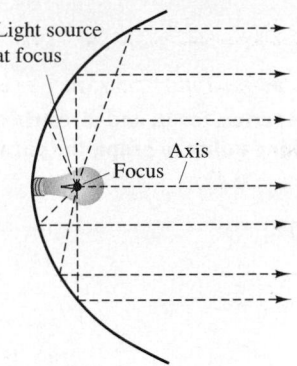

FIGURE 9.9

Parabolic reflector: light is reflected in parallel rays.

Reflective Property of Parabolas

One widely used property of a parabola is its reflective property. In physics, a surface is called **reflective** if the tangent line at any point on the surface makes equal angles with an incoming ray and the resulting outgoing ray. The angle corresponding to the incoming ray is called the **angle of incidence,** and the angle corresponding to the outgoing ray is called the **angle of reflection.** One example of a reflective surface is a flat mirror.

Another type of reflective surface is that formed by revolving a parabola about its axis. A special property of parabolic reflectors is that they allow us to direct all incoming rays parallel to the axis through the focus of the parabola—this is the principle behind the design of the parabolic mirrors used in reflecting telescopes. Conversely, all light rays emanating from the focus of a parabolic reflector used in a flashlight are parallel, as shown in Figure 9.9.

THEOREM 9.2   Reflective Property of a Parabola

Let P be a point on a parabola. The tangent line to the parabola at point P makes equal angles with the following two lines.

1. The line passing through P and the focus.
2. The line passing through P parallel to the axis of the parabola.

Proof   Consider the point $P = (x_0, y_0)$ on the parabola given by $x^2 = 4py$, as shown in Figure 9.10. Using differentiation, you can determine the tangent line at P to have a slope of $x_0/2p$, which implies that the equation of the tangent line is

$$y - y_0 = \frac{x_0}{2p}(x - x_0). \qquad \text{Tangent line at } P$$

If P lies at the vertex of the parabola, then the other two lines referred to in the theorem coincide, and the theorem is trivially valid. Thus, assume that P is a point other than the vertex. This implies that the tangent line intersects the y-axis at a point $Q = (0, q)$, different from P. From the equation of the tangent line at $x = 0$, you know that

$$q = y_0 - \frac{x_0^2}{2p} = \frac{x_0^2}{4p} - \frac{x_0^2}{2p} = -\frac{x_0^2}{4p}$$

which implies that the length of $\overline{FQ}$ is

$$p - q = p + \frac{x_0^2}{4p}. \qquad \text{Length of } \overline{FQ}$$

From the Distance Formula, you can find the length of $\overline{FP}$ to be

$$\sqrt{x_0^2 + (y_0 - p)^2} = \frac{x_0^2}{4p} + p. \qquad \text{Length of } \overline{FP}$$

Therefore, ΔFQP is isosceles, which implies that $\alpha = \beta$.   ■

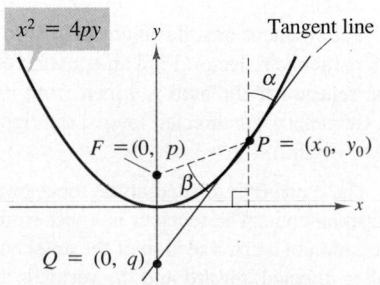

FIGURE 9.10

The tangent line to the parabola $x^2 = 4py$ at (x_0, y_0) has a slope of $x_0/2p$.

EXERCISES for Section 9.1

In Exercises 1–6, match the equation with its graph. [The graphs are labeled (a), (b), (c), (d), (e), and (f).]

1. $y^2 = 4x$ **2.** $x^2 = -2y$

3. $x^2 = 8y$ **4.** $y^2 = -12x$

5. $(y - 1)^2 = 4(x - 2)$ **6.** $(x + 3)^2 = -2(y - 2)$

(a)

(b)

(c)

(d)

(e)

(f)
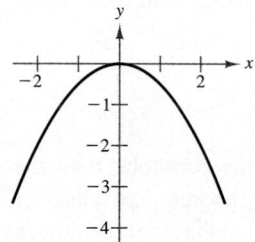

In Exercises 7–22, find the vertex, focus, and directrix of the parabola. Then sketch the parabola.

7. $y = 4x^2$ **8.** $y = 2x^2$

9. $y^2 = -6x$ **10.** $y^2 = 3x$

11. $x^2 + 8y = 0$ **12.** $x + y^2 = 0$

13. $(x - 1)^2 + 8(y + 2) = 0$

14. $(x + 3) + (y - 2)^2 = 0$

15. $(y + \frac{1}{2})^2 = 2(x - 5)$

16. $(x + \frac{1}{2})^2 = 4(y - 3)$

17. $y = \frac{1}{4}(x^2 - 2x + 5)$

18. $4x - y^2 - 2y - 33 = 0$

19. $y^2 - 4y - 4x = 0$

20. $y^2 + 6y + 8x + 25 = 0$

21. $y^2 + 4y + 8x - 12 = 0$

22. $x^2 + 4x + 4y - 4 = 0$

C In Exercises 23–26, find the vertex, focus, and directrix of the parabola. Then use a graphing utility to graph the parabola.

23. $y = -\frac{1}{6}(x^2 + 4x - 2)$

24. $x^2 - 2x + 8y + 9 = 0$

25. $y^2 + x + y = 0$

26. $y^2 - 4x - 4 = 0$

In Exercises 27–40, find an equation of the parabola.

27. Vertex: $(0, 0)$
Focus: $(0, -\frac{3}{2})$

28. Vertex: $(0, 0)$
Focus: $(2, 0)$

29. Vertex: $(3, 2)$
Focus: $(1, 2)$

30. Vertex: $(-1, 2)$
Focus: $(-1, 0)$

31. Vertex: $(0, -4)$
Directrix: $y = 2$

32. Vertex: $(-2, 1)$
Directrix: $x = 1$

33. Focus: $(2, 2)$
Directrix: $x = -2$

34. Focus: $(0, 0)$
Directrix: $y = 4$

35.

36.
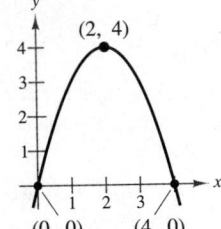

37. Axis is parallel to y-axis; graph passes through $(0, 3)$, $(3, 4)$, and $(4, 11)$.

38. Axis is parallel to x-axis; graph passes through $(4, -2)$, $(0, 0)$, and $(3, -3)$.

39. Directrix: $y = 1$; length of latus rectum is 8; opens downward.

40. Directrix: $y = -2$; endpoints of latus rectum are $(0, 2)$ and $(8, 2)$.

41. *Parabolic Reflector* The filament of a flashlight bulb is located at the focus of a parabolic reflector. Find an equation of a cross section of the reflector if the bulb is $\frac{3}{8}$ inch from its vertex. (Assume that the reflector is directed toward the right and the vertex is at the origin.)

42. *Parabolic Receiver* The receiver is located at the focus of a parabolic television dish antenna. The receiver is 3 feet from the vertex. Find an equation of a cross section of the antenna. (Assume that the dish is directed upward and the vertex is at the origin.)

43. *Solar Collector* A solar collector for heating water is constructed with a sheet of stainless steel that is formed into the shape of a parabola (see figure). The water will flow through a pipe that is located at the focus of the parabola. At what distance is the pipe from the vertex?

44. *Beam Deflection* A simply supported beam that is 100 feet long has a load concentrated at its center (see figure). The deflection of the beam at its center is 3 inches. If the shape of the deflected beam is parabolic, find an equation of the parabola. (Assume that the origin is at the center of the unloaded beam.)

FIGURE FOR 43   **FIGURE FOR 44**

In Exercises 45 and 46, find an equation of the tangent line to the parabola at the specified point.

Parabola	Point
45. $x^2 = 2y$	$(4, 8)$
46. $y^2 = -8x$	$(-2, 4)$

47. Find an equation of the tangent line to the parabola $y = ax^2$ at $x = x_0$. Prove that the x-intercept of this tangent line is $(x_0/2, 0)$.

48. a. Prove that any two distinct tangent lines to a parabola intersect.

  b. Demonstrate the result of part a by finding the point of intersection of the tangent lines to the parabola $x^2 - 4x - 4y = 0$ at the points $(0, 0)$ and $(6, 3)$.

49. a. Prove that if any two tangent lines to a parabola intersect at right angles, then their point of intersection must lie on the directrix.

  b. Demonstrate the result of part a by showing that the tangent lines to the parabola $x^2 - 4x - 4y + 8 = 0$ at the points $(-2, 5)$ and $(3, \frac{5}{4})$ intersect at right angles and that the point of intersection lies on the directrix.

50. Prove that two tangent lines to a parabola intersect at right angles if and only if the focus of the parabola lies on the line segment connecting the two points of tangency.

51. Find the equations of the tangent lines to the graph of the parabola $y^2 = 2x$ that pass through the point $(-4, 1)$.

52. Find the point on the graph of $x^2 = 8y$ that is closest to the focus of the parabola.

53. *Radio and Television Reception* In mountainous areas, reception of radio and television signals is sometimes poor. Consider an idealized case in which a hill is represented by the graph of the parabola $y = x - x^2$, a transmitter is located at the point $(-1, 1)$, and a receiver is located on the other side of the hill at the point $(x_0, 0)$. What is the closest the receiver can be to the hill so that the reception is unobstructed?

C **54.** *Beef Consumption* Beef consumption in the United States in billions of pounds for selected years is listed in the table. (*Source:* U.S. Department of Agriculture)

Year	1980	1985	1987
Consumption	23.561	25.523	25.257

Year	1988	1989	1990
Consumption	25.252	24.330	24.031

A model for this data is given by

$$C = 23.433 + 0.784t - 0.073t^2$$

where C is measured in billions of pounds and t is the time in years with $t = 0$ corresponding to 1980.

  a. Use a graphing utility to graph and plot the data.

  b. Use the model to estimate the year when consumption was maximum.

  c. Assuming that the present trend continues, forecast the year when consumption will be less than that of 1980.

55. *Architecture* A church window is bounded on top by a parabola and below by the arc of a circle (see figure). Find the surface area of the window.

56. *Centroid* Find the centroid of the parabolic region shown in the accompanying figure.

FIGURE FOR 55   **FIGURE FOR 56**

In Exercises 57–60, find the arc length of the parabola over the given interval.

Parabola	Interval
57. $x^2 + 8y = 0$	$0 \le x \le 4$
58. $x + y^2 = 0$	$0 \le y \le 2$
59. $4x - y^2 = 0$	$0 \le y \le 4$
60. $x^2 - 2y = 0$	$0 \le x \le 1$

61. *Bridge Design* A cable of a suspension bridge is suspended (in the shape of a parabola) between two towers that are 400 feet apart and 50 feet above the roadway (see figure). The cables touch the roadway midway between the towers.

  a. Find an equation for the parabolic shape of each cable.

  b. Find the length of the parabolic supporting cable.

FIGURE FOR 61

62. *Satellite Trajectory*   A satellite in a 100-mile-high circular orbit around the earth has a velocity of approximately 17,500 miles per hour. If this velocity is increased by a factor of $\sqrt{2}$, then the satellite will have the minimum velocity necessary to escape the earth's gravity and it will begin to follow a parabolic path with the center of the earth as the focus (see figure).

a. Find the escape velocity of the satellite.

b. Find the equation of its path (assume that the radius of the earth is 4000 miles).

c. Approximate the distance the satellite travels from the point $(0, 4100)$ to the point where it intersects the x-axis in the figure. How much greater is this distance than if the satellite had remained in a circular orbit?

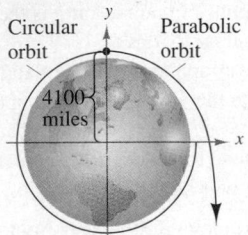

FIGURE FOR 62

63. *Surface Area*   A satellite-signal receiving dish is formed by revolving the parabola given by the graph of $x^2 = 20y$ about the y-axis. If the radius of the dish is r feet, verify that the surface area of the dish is

$$2\pi \int_0^r x\sqrt{1 + \left(\frac{x}{10}\right)^2}\, dx = \frac{\pi}{15}[(100 + r^2)^{3/2} - 1000].$$

64. *Surface Area*   Find the surface area of the satellite dish in Exercise 63 if its radius is

a. $r = 6$ feet.

b. $r = 10$ feet.

65. Prove that the area enclosed by a parabola and a line parallel to its directrix is two-thirds the area of the circumscribed rectangle.

True or False   In Exercises 66 and 67, determine whether the statement is true or false. If it is false, explain why or give an example that shows it is false.

66. It is possible for a parabola to intersect its directrix.

67. The point on a parabola closest to its focus is its vertex.

68. Let A and B be the endpoints of a focal chord on the parabola $x^2 = 4py$. Prove that the tangent lines at A and B are perpendicular.

Ellipses

Ellipses • Standard Equation of an Ellipse • Eccentricity

NICHOLAS COPERNICUS

Copernicus began to study planetary motion when he was asked to revise the calendar. At that time, the exact length of a year could not be accurately predicted using the theory that the earth was the center of the universe.

Ellipses

More than a thousand years after the close of the Alexandrian period of Greek mathematics, Western civilization finally began a Renaissance of mathematical and scientific discovery. One of the principal figures in this rebirth was the Polish astronomer Nicholas Copernicus (1473–1543). In his work *On the Revolutions of the Heavenly Spheres,* Copernicus claimed that all of the planets, including the earth, revolved about the sun in circular orbits. Although some of Copernicus's claims were invalid, the controversy set off by his heliocentric theory motivated astronomers to search for a mathematical model to explain the observed movements of the sun and planets. The first to find the correct model was the German astronomer Johannes Kepler (1571–1630). Kepler discovered that the planets move about the sun in elliptical orbits, with the sun not as the center but as a focal point of the orbit.

The use of ellipses to explain the movement of the planets is only one of their many practical and aesthetic uses. As in the previous section, we begin our study of this second type of conic by defining it as a locus of points. Now, however, we use *two* focal points rather than one.

An **ellipse** is the set of all points (x, y) for which the sum of the distances from two distinct fixed points (**foci**) is constant. (See Figure 9.11.) The line through the foci intersects the ellipse at two points, called the **vertices.** The chord joining the vertices is the **major axis,** and its midpoint is the **center** of the ellipse. The chord perpendicular to the major axis at the center is the **minor axis** of the ellipse.

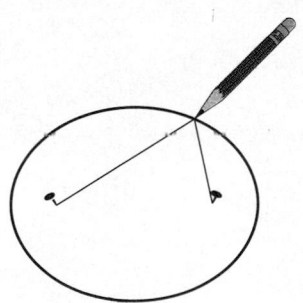

FIGURE 9.12
Sketching an ellipse with a string, thumbtacks, and a pencil.

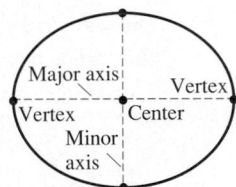

FIGURE 9.11

REMARK   You can visualize the definition of an ellipse by imagining two thumbtacks placed at the foci, as shown in Figure 9.12. If the ends of a fixed length of string are fastened to the thumbtacks and the string is drawn taut with a pencil, the path traced by the pencil will be an ellipse.

Standard Equation of an Ellipse

To derive the standard form of the equation of an ellipse, consider the ellipse in Figure 9.13, which has the following points.

Center: (h, k) Vertices: $(h \pm a, k)$ Foci: $(h \pm c, k)$

The sum of the distances from any point on the ellipse to the two foci is constant. At a vertex, this constant sum is

$$(a + c) + (a - c) = 2a \qquad \text{Length of major axis}$$

or the length of the major axis. Now, for *any* point (x, y) on the ellipse, the sum of the distances between (x, y) and the two foci must also be $2a$:

$$\sqrt{[x - (h - c)]^2 + (y - k)^2} + \sqrt{[x - (h + c)]^2 + (y - k)^2} = 2a$$

which, after expanding and regrouping, reduces to

$$(a^2 - c^2)(x - h)^2 + a^2(y - k)^2 = a^2(a^2 - c^2).$$

Finally, from Figure 9.13, you can see that $b^2 = a^2 - c^2$, which implies that the equation of the ellipse is

$$b^2(x - h)^2 + a^2(y - k)^2 = a^2 b^2$$
$$\frac{(x - h)^2}{a^2} + \frac{(y - k)^2}{b^2} = 1.$$

Choosing a vertical major axis would have produced a similar equation.

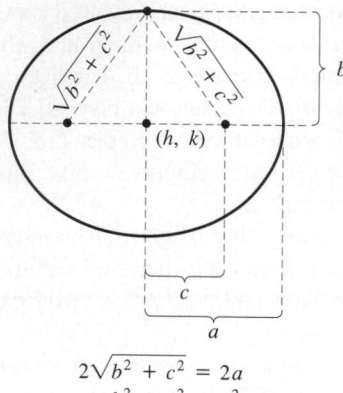

$$2\sqrt{b^2 + c^2} = 2a$$
$$b^2 + c^2 = a^2$$

FIGURE 9.13

REMARK Figure 9.14 shows two orientations for an ellipse—one in which the major axis is horizontal and one in which the major axis is vertical.

THEOREM 9.3   Standard Equation of an Ellipse

The standard form of the equation of an ellipse, with center (h, k) and major and minor axes of lengths $2a$ and $2b$, where $a > b$, is

$$\frac{(x - h)^2}{a^2} + \frac{(y - k)^2}{b^2} = 1 \qquad \text{Major axis is horizontal}$$

$$\frac{(x - h)^2}{b^2} + \frac{(y - k)^2}{a^2} = 1. \qquad \text{Major axis is vertical}$$

The foci lie on the major axis, c units from the center, with $c^2 = a^2 - b^2$.

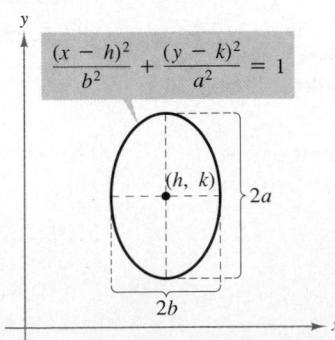

FIGURE 9.14
Standard equations for ellipses.

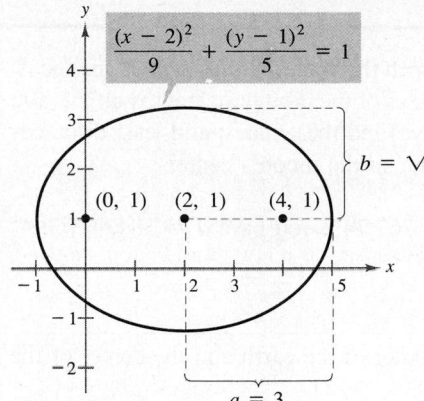

FIGURE 9.15
Ellipse with a horizontal major axis.

EXAMPLE 1 Finding the Standard Equation of an Ellipse

Find the standard form of the equation of an ellipse having foci at $(0, 1)$ and $(4, 1)$ with a major axis of length 6, as shown in Figure 9.15.

Solution   Because the foci occur at $(0, 1)$ and $(4, 1)$, the center of the ellipse is $(2, 1)$. This implies that the distance from the center to one of the foci is $c = 2$, and because $2a = 6$ you know that $a = 3$. Now, using $c^2 = a^2 - b^2$, you have

$$b = \sqrt{a^2 - c^2} = \sqrt{9 - 4} = \sqrt{5}.$$

Because the major axis is parallel to the x-axis, the standard equation is

$$\frac{(x - 2)^2}{9} + \frac{(y - 1)^2}{5} = 1.$$

EXAMPLE 2 Completing the Square

Find the center, vertices, and foci of the ellipse given by

$$4x^2 + y^2 - 8x + 4y - 8 = 0.$$

Solution   By completing the square, you can write the given equation in standard form.

$$
\begin{aligned}
4x^2 + y^2 - 8x + 4y - 8 &= 0 &&\text{Original equation}\\
4x^2 - 8x + y^2 + 4y &= 8\\
4(x^2 - 2x + 1) + (y^2 + 4y + 4) &= 8 + 4 + 4\\
4(x - 1)^2 + (y + 2)^2 &= 16\\
\frac{(x - 1)^2}{4} + \frac{(y + 2)^2}{16} &= 1 &&\text{Standard form}
\end{aligned}
$$

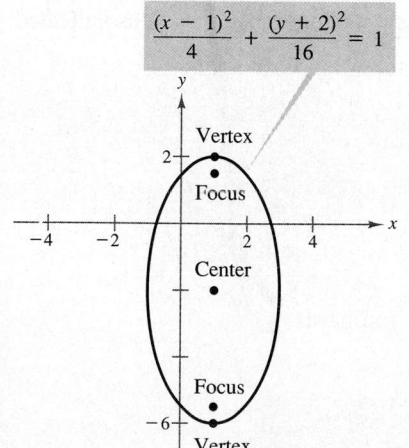

FIGURE 9.16
Ellipse with a vertical major axis.

Thus, the major axis is parallel to the y-axis, where $h = 1, k = -2, a = 4, b = 2$, and $c = \sqrt{16 - 4} = 2\sqrt{3}$. Therefore, you obtain the following.

Center	$(1, -2)$
Vertices	$(1, -6)$ and $(1, 2)$
Foci	$\left(1, -2 - 2\sqrt{3}\right)$ and $\left(1, -2 + 2\sqrt{3}\right)$

The graph of the ellipse is shown in Figure 9.16.

REMARK   If the constant term $F = -8$ in the equation in Example 2 had been greater than or equal to 8, we would have obtained one of the following degenerate cases.

1. Single point, $(1, -2)$: $\dfrac{(x - 1)^2}{4} + \dfrac{(y + 2)^2}{16} = 0$ $F = 8$

2. No solution points: $\dfrac{(x - 1)^2}{4} + \dfrac{(y + 2)^2}{16} < 0$ $F > 8$

EXAMPLE 3   The Orbit of the Moon

The moon orbits the earth in an elliptical path with the center of the earth at one focus, as shown in Figure 9.17. The major and minor axes of the orbit have lengths of 768,806 kilometers and 767,746 kilometers, respectively. Find the greatest and least distances (the apogee and perigee) from the earth's center to the moon's center.

Solution   Because $2a = 768,806$ and $2b = 767,746$, you have $a = 384,403$, $b = 383,873$, and

$$c = \sqrt{a^2 - b^2} \approx 20,179.$$

Therefore, the greatest distance between the center of the earth and the center of the moon is

$$a + c \approx 404,582 \text{ km}$$

and the least distance is

$$a - c \approx 364,224 \text{ km.}$$

Perigee   Apogee

FIGURE 9.17

EXAMPLE 4   Finding the Area of an Ellipse

Find the area of an ellipse whose major and minor axes have lengths of $2a$ and $2b$, respectively.

Solution   For simplicity, choose an ellipse centered at the origin

$$\frac{x^2}{a^2} + \frac{y^2}{b^2} = 1.$$

Then, using symmetry, you can find the area of the entire region lying within the ellipse by finding the area of the region in the first quadrant and multiplying by 4, as indicated in Figure 9.18. In the first quadrant, you have

$$y = \frac{b}{a}\sqrt{a^2 - x^2}$$

which implies that the entire area is

$$A = 4\int_0^a \frac{b}{a}\sqrt{a^2 - x^2}\,dx.$$

Using the trigonometric substitution $x = a\sin\theta$, you have

$$A = \frac{4b}{a}\int_0^{\pi/2} a^2\cos^2\theta\,d\theta$$

$$= 4ab\int_0^{\pi/2}\frac{1 + \cos 2\theta}{2}\,d\theta$$

$$= 2ab\left[\theta + \frac{\sin 2\theta}{2}\right]_0^{\pi/2}$$

$$= 2ab\left(\frac{\pi}{2}\right) = \pi ab.$$

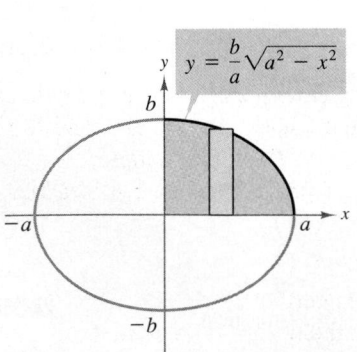

FIGURE 9.18
Area of an ellipse is πab.

REMARK   Note that if $a = b$, then the formula for the area of an ellipse reduces to the area of a circle.

In the previous section we discussed a reflective property of parabolas. Ellipses have a similar reflective property. You are asked to prove the following theorem in Exercise 70.

THEOREM 9.4   Reflective Property of an Ellipse

Let P be a point on an ellipse. The tangent line to the ellipse at point P makes equal angles with the lines through P and the foci.

FOR FURTHER INFORMATION   For more information on some uses of the reflective properties of conics, see the article "Parabolic Mirrors, Elliptic and Hyperbolic Lenses" by Mohsen Maesumi in the June–July, 1992 issue of *The American Mathematical Monthly*. Also see the article "The Geometry of Microwave Antennas" by William R. Parzynski in the April, 1984 issue of *Mathematics Teacher*.

Eccentricity

One of the reasons that astronomers had difficulty in detecting that the orbits of the planets are ellipses is that the foci of the planetary orbits are relatively close to the center of the sun, making the orbits nearly circular. To measure the ovalness of an ellipse, we use the concept of **eccentricity.**

Definition of Eccentricity of an Ellipse

The **eccentricity** e of an ellipse is given by the ratio

$$e = \frac{c}{a}.$$

(a) $\dfrac{c}{a}$ is small.

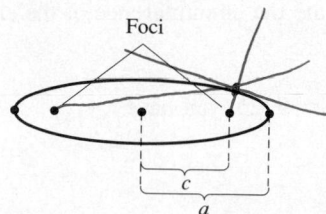

(b) $\dfrac{c}{a}$ is close to 1.

FIGURE 9.19

Eccentricity is the ratio $\dfrac{c}{a}$.

To see how this ratio is used to describe the shape of an ellipse, note that because the foci of an ellipse are located along the major axis between the vertices and the center, it follows that

$$0 < c < a.$$

For an ellipse that is nearly circular, the foci are close to the center and the ratio c/a is small, as shown in Figure 9.19. For an elongated ellipse, the foci are close to the vertices and the ratio is close to 1. Note that $0 < e < 1$ for every ellipse.

The orbit of the moon has an eccentricity of $e = 0.0549$, and the eccentricities of the nine planetary orbits are as follows.

Mercury:	$e = 0.2056$	Saturn:	$e = 0.0543$
Venus:	$e = 0.0068$	Uranus:	$e = 0.0460$
Earth:	$e = 0.0167$	Neptune:	$e = 0.0082$
Mars:	$e = 0.0934$	Pluto:	$e = 0.2481$
Jupiter:	$e = 0.0484$		

In Example 4, we found that the formula for the *area* of an ellipse is $A = \pi ab$. However, it is not so simple to find the *circumference* of an ellipse. The next example shows how to use eccentricity to set up an "elliptic integral" for the circumference of an ellipse.

EXAMPLE 5   Finding the Circumference of an Ellipse

Show that the circumference of the ellipse $(x^2/a^2) + (y^2/b^2) = 1$ is

$$4a \int_0^{\pi/2} \sqrt{1 - e^2 \sin^2 \theta} \, d\theta. \qquad\qquad e = \frac{c}{a}$$

Solution   Because the given ellipse is symmetric with respect to both the x-axis and the y-axis, you know that its circumference C is four times the arc length of $y = (b/a)\sqrt{a^2 - x^2}$ in the first quadrant. The function y is differentiable for all x in the interval $[0, a]$ except at $x = a$. Thus, the circumference is given by

$$C = \lim_{d \to a} 4 \int_0^d \sqrt{1 + (y')^2} \, dx = 4 \int_0^a \sqrt{1 + (y')^2} \, dx$$

$$= 4 \int_0^a \sqrt{1 + \frac{b^2 x^2}{a^2(a^2 - x^2)}} \, dx. \qquad \text{Improper integral}$$

Using the trigonometric substitution $x = a \sin \theta$, you obtain

$$C = 4 \int_0^{\pi/2} \sqrt{1 + \frac{b^2 \sin^2 \theta}{a^2 \cos^2 \theta}} \, (a \cos \theta) \, d\theta$$

$$= 4 \int_0^{\pi/2} \sqrt{a^2 \cos^2 \theta + b^2 \sin^2 \theta} \, d\theta$$

$$= 4 \int_0^{\pi/2} \sqrt{a^2(1 - \sin^2 \theta) + b^2 \sin^2 \theta} \, d\theta$$

$$= 4 \int_0^{\pi/2} \sqrt{a^2 - (a^2 - b^2)\sin^2 \theta} \, d\theta.$$

Because $e^2 = c^2/a^2 = (a^2 - b^2)/a^2$, you can rewrite this integral as

$$C = 4a \int_0^{\pi/2} \sqrt{1 - e^2 \sin^2 \theta} \, d\theta.$$

A great deal of time has been devoted to the study of elliptic integrals. Such integrals generally do not have elementary antiderivatives. To find the circumference of an ellipse, we usually resort to an approximation technique, as illustrated in the following example.

EXAMPLE 6   Approximating the Value of an Elliptic Integral

Use the elliptic integral in Example 5 to approximate the circumference of the ellipse $(x^2/25) + (y^2/16) = 1$.

Solution   Because $e^2 = c^2/a^2 = (a^2 - b^2)/a^2 = 9/25$, you have

$$C = (4)(5) \int_0^{\pi/2} \sqrt{1 - \frac{9 \sin^2 \theta}{25}} \, d\theta.$$

Applying Simpson's Rule with $n = 4$ produces

$$C \approx 20\left(\frac{\pi}{6}\right)\left(\frac{1}{4}\right)[1 + 4(0.9733) + 2(0.9055) + 4(0.8323) + 0.8]$$

$$\approx 28.36.$$

EXERCISES for Section 9.2

TECHNOLOGY
Laboratory Guide
Lab 9.2

In Exercises 1–6, match the equation with its graph. [The graphs are labeled (a), (b), (c), (d), (e), and (f).]

1. $\dfrac{x^2}{1} + \dfrac{y^2}{9} = 1$

2. $\dfrac{x^2}{9} + \dfrac{y^2}{1} = 1$

3. $\dfrac{x^2}{9} + \dfrac{y^2}{4} = 1$

4. $\dfrac{x^2}{9} + \dfrac{y^2}{9} = 1$

5. $\dfrac{(x-2)^2}{16} + \dfrac{(y+1)^2}{4} = 1$

6. $\dfrac{(x+2)^2}{4} + \dfrac{(y+2)^2}{25} = 1$

(a)

(b)

(c)

(d)

(e)

(f)
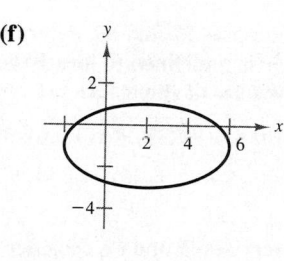

In Exercises 7–22, find the center, foci, vertices, and eccentricity of the ellipse, and sketch its graph.

7. $\dfrac{x^2}{25} + \dfrac{y^2}{16} = 1$

8. $\dfrac{x^2}{144} + \dfrac{y^2}{169} = 1$

9. $\dfrac{x^2}{16} + \dfrac{y^2}{25} = 1$

10. $\dfrac{x^2}{169} + \dfrac{y^2}{144} = 1$

11. $\dfrac{x^2}{9} + \dfrac{y^2}{5} = 1$

12. $\dfrac{x^2}{28} + \dfrac{y^2}{64} = 1$

13. $x^2 + 4y^2 = 4$

14. $5x^2 + 3y^2 = 15$

15. $3x^2 + 2y^2 = 6$

16. $5x^2 + 7y^2 = 70$

17. $\dfrac{(x-1)^2}{9} + \dfrac{(y-5)^2}{25} = 1$

18. $(x+2)^2 + \dfrac{(y+4)^2}{1/4} = 1$

19. $9x^2 + 4y^2 + 36x - 24y + 36 = 0$

20. $9x^2 + 4y^2 - 36x + 8y + 31 = 0$

21. $16x^2 + 25y^2 - 32x + 50y + 31 = 0$

22. $9x^2 + 25y^2 - 36x - 50y + 61 = 0$

C **In Exercises 23–28, find the center, foci, and vertices of the ellipse. Use a graphing utility to graph the ellipse. (Explain how you used the utility to obtain the graph.)**

23. $4x^2 + y^2 = 1$

24. $16x^2 + 25y^2 = 1$

25. $12x^2 + 20y^2 - 12x + 40y - 37 = 0$

26. $36x^2 + 9y^2 + 48x - 36y + 43 = 0$

27. $x^2 + 2y^2 - 3x + 4y + 0.25 = 0$

28. $2x^2 + y^2 + 4.8x - 6.4y + 3.12 = 0$

In Exercises 29–38, find an equation of the ellipse.

29. Center: (0, 0)
Focus: (2, 0)
Vertex: (3, 0)

30. Center: (0, 0)
Vertex: (2, 0)
Minor axis length: 3

31. Vertices: (±5, 0)
Eccentricity: $\frac{3}{5}$

32. Vertices: (0, 2), (4, 2)
Eccentricity: $\frac{1}{2}$

33. Vertices: (3, 1), (3, 9)
Minor axis length: 6

34. Vertices: (0, 2), (4, 2)
Minor axis length: 2

35. Foci: (0, ±5)
Major axis length: 14

36. Foci: (±2, 0)
Major axis length: 8

37. Center: (0, 0)
Major axis: horizontal
Points on ellipse:
(3, 1), (4, 0)

38. Center: (1, 2)
Major axis: vertical
Points on ellipse:
(1, 6), (3, 2)

39. Sketch a graph of the ellipse that consists of all points (x, y) such that the sum of the distances between (x, y) and two fixed points is 16 units and the foci are located at the centers of the two sets of concentric circles in the figure.

40. A line segment through a focus with endpoints on the ellipse and perpendicular to the major axis is called a **latus rectum** of the ellipse (see figure). Show that the length of each latus rectum is $2b^2/a$.

FIGURE FOR 39

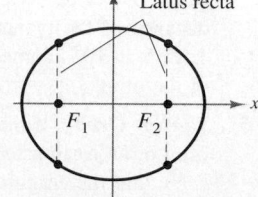

FIGURE FOR 40

In Exercises 41–44, sketch the ellipse. Then sketch the latus recta of the ellipse (see Exercise 40).

41. $\dfrac{x^2}{4} + \dfrac{y^2}{1} = 1$ 42. $\dfrac{x^2}{9} + \dfrac{y^2}{16} = 1$

43. $9x^2 + 4y^2 = 36$ 44. $5x^2 + 3y^2 = 15$

45. *Essay*   At the beginning of this section it was noted that an ellipse can be drawn using two thumbtacks, a string of fixed length (greater than the distance between the tacks), and a pencil (see figure). If the ends of the string are fastened at the tacks and the string is drawn taut with a pencil, the path traced by the pencil will be an ellipse.
   a. What is the length of the string in terms of a?
   b. Explain why the path is an ellipse.

46. *Construction of a Semielliptical Arch*   A fireplace arch is to be constructed in the shape of a semiellipse. The opening is to have a height of 2 feet at the center and a width of 5 feet along the base (see figure). The contractor draws the outline of the ellipse by the method shown in Exercise 45. Where should the tacks be placed and what should be the length of the piece of string?

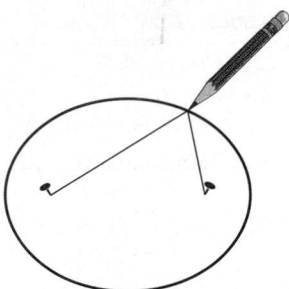

FIGURE FOR 45   **FIGURE FOR 46**

47. *Mountain Tunnel*   A semielliptical arch over a tunnel for a road through a mountain has a major axis of 100 feet and a height at the center of 30 feet. Determine the height of the arch 5 feet from the edge of the tunnel.

48. *Orbit of Earth*   The earth moves in an elliptical orbit with the sun at one of the foci. The length of half of the major axis is 14,957,000 kilometers, and the eccentricity is 0.0167. Find the smallest distance (*perihelion*) and the greatest distance (*aphelion*) of the earth from the sun.

49. *Orbit of Pluto*   Pluto moves in an elliptical orbit with the sun at one of the foci. The length of half of the major axis is 3.666×10^9 miles and the eccentricity is 0.248. Find the smallest distance and the greatest distance of Pluto from the sun.

50. *Orbit of Saturn*   Saturn moves in an elliptical orbit with the sun at one of the foci. The smallest distance and the greatest distance of the planet from the sun are 1.3495×10^9 and 1.5045×10^9 kilometers, respectively. Find the eccentricity of the orbit.

51. *Satellite Orbit*   If the apogee and the perigee of an elliptical orbit of an earth satellite are given by A and P, respectively, show that the eccentricity of the orbit is

$$e = \frac{A - P}{A + P}.$$

52. *Sputnik I*   The first artificial satellite to orbit the earth was Sputnik I (launched by the Soviet Union in 1957). Its highest point above the earth's surface was 583 miles, and its lowest point was 132 miles. Assume that the center of the earth is the focus of the elliptical orbit and the radius of the earth is 4000 miles. Find the eccentricity of the orbit.

53. *Explorer 18*   On November 26, 1963, the United States launched Explorer 18. Its low and high points over the surface of the earth were 119 miles and 122,000 miles, respectively. Find the eccentricity of its elliptical orbit.

54. *Halley's Comet*   Probably the most famous of all comets, Halley's Comet, has an elliptical orbit with the sun at one focus. Its maximum distance from the sun is approximately 35.34 au (astronomical unit $\approx 92.956 \times 10^6$ miles), and its minimum distance (perihelion) is approximately 0.59 au. Find the eccentricity of the orbit.

55. Show that the eccentricity of the ellipse

$$\frac{x^2}{a^2} + \frac{y^2}{b^2} = 1$$

is identical to the eccentricity of

$$\frac{(tx)^2}{a^2} + \frac{(ty)^2}{b^2} = 1$$

for any real t. Give a geometrical explanation of this result.

56. The equation of an ellipse with its center at the origin can be written as

$$\frac{x^2}{a^2} + \frac{y^2}{a^2(1 - e^2)} = 1.$$

Show that as $e \to 0$, with a remaining fixed, the ellipse approaches a circle.

In Exercises 57 and 58, determine the points at which dy/dx is 0 or is undefined to locate the endpoints of the major and minor axes of the ellipse.

57. $16x^2 + 9y^2 + 96x + 36y + 36 = 0$

58. $9x^2 + 4y^2 + 36x - 24y + 36 = 0$

In Exercises 59 and 60, consider a particle traveling clockwise on the given elliptical path. The particle leaves the orbit at the indicated point and travels in a straight line tangent to the ellipse. At what point will the particle cross the y-axis?

Ellipse	Point
59. $\dfrac{x^2}{100} + \dfrac{y^2}{25} = 1$	$(-8, 3)$
60. $\dfrac{x^2}{16} + \dfrac{y^2}{25} = 1$	$\left(3, \dfrac{5\sqrt{7}}{4}\right)$

61. *Area*   Find the dimensions of the rectangle of maximum area that can be inscribed in the ellipse

$$\frac{x^2}{a^2} + \frac{y^2}{b^2} = 1.$$

(Assume that the sides of the rectangle are parallel to the coordinate axes.)

62. *Volume* The water tank on a fire truck is 16 feet long and its cross sections are ellipses. Find the volume of water in the partially filled tank as shown in the accompanying figure.

FIGURE FOR 62

63. *Volume* A solid has an elliptical base given by

$$\frac{x^2}{25} + \frac{y^2}{16} = 1.$$

Cross sections perpendicular to the major axis are isosceles triangles of height 6. Find the volume of the solid.

64. *Volume* Repeat Exercise 63 semicircular cross-sections.

In Exercises 65 and 66, find (a) the area of the region bounded by the given ellipse, (b) the volume and surface area of the solid generated by revolving the region about its major axis (prolate spheroid), and (c) the volume and surface area of the solid generated by revolving the region about its minor axis (oblate spheroid).

65. $\dfrac{x^2}{4} + \dfrac{y^2}{1} = 1$ **66.** $\dfrac{x^2}{16} + \dfrac{y^2}{9} = 1$

C In Exercises 67 and 68, use a computer or calculator and Simpson's Rule with $n = 12$ to approximate the elliptic integral representing the circumference of the given ellipse.

67. $\dfrac{x^2}{9} + \dfrac{y^2}{16} = 1$

68. $\dfrac{(x-3)^2}{4} + \dfrac{(y+1)^2}{3} = 1$

69. Show that the tangent line to the ellipse

$$\frac{x^2}{a^2} + \frac{y^2}{b^2} = 1$$

at the point (x_0, y_0) is given by

$$\frac{x_0}{a^2}x + \frac{y_0}{b^2}y = 1.$$

70. Prove that the tangent line to an ellipse at a point P makes equal angles with lines through P and the foci (see figure). [*Hint:* (1) Find the slope of the tangent line at P, (2) find the slopes of the lines through P and each focus, and (3) use the formula for the tangent of the angle between two lines.]

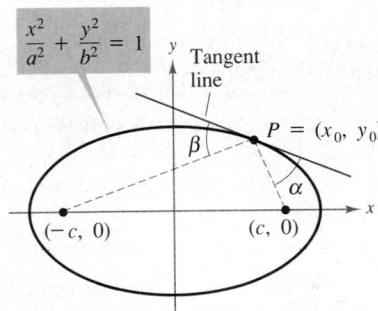

FIGURE FOR 70

71. A line segment 9 inches long moves so that one endpoint is always on the y-axis and the other is always on the x-axis. Find the equation of the curve traced by a point on the line segment 6 inches from the endpoint that is on the y-axis.

True or False In Exercises 72 and 73, determine whether the statement is true or false. If it is false, explain why or give an example that shows it is false.

72. If C is the circumference of the ellipse $(x^2/a^2) + (y^2/b^2) = 1$, then $2\pi b \le C \le 2\pi a$.

73. The graph of $(x^2/4) + y^4 = 1$ is an ellipse.

74. Prove that if a line (other than the axes) normal to the ellipse $x^2/a^2 + y^2/b^2 = 1$ passes through the origin, then the ellipse is a circle.

Hyperbolas • Asymptotes • Applications

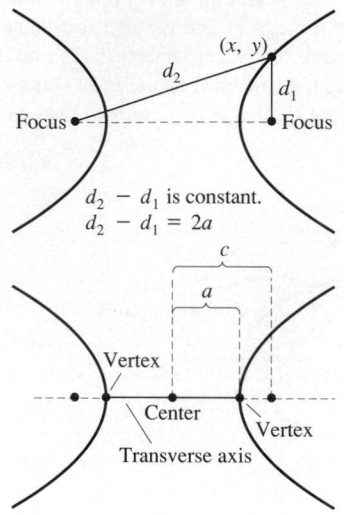

FIGURE 9.20

Hyperbolas

The definition of a hyperbola is similar to that of an ellipse. For an ellipse, the *sum* of the distances between the foci and a point on the ellipse is fixed, whereas for a hyperbola, the *difference* between these distances is fixed.

A **hyperbola** is the set of all points (x, y) for which the difference between the distances from two distinct fixed points (foci) is constant. (See Figure 9.20.) The line through the two foci intersects a hyperbola at two points called the **vertices.** The line segment connecting the vertices is the **transverse axis,** and the midpoint of the transverse axis is the **center** of the hyperbola. One distinguishing feature of a hyperbola is that its graph has two separate *branches.*

THEOREM 9.5   Standard Equation of a Hyperbola

The standard form of the equation of a hyperbola with center at (h, k) is

$$\frac{(x - h)^2}{a^2} - \frac{(y - k)^2}{b^2} = 1 \qquad \text{Transverse axis is horizontal}$$

or

$$\frac{(y - k)^2}{a^2} - \frac{(x - h)^2}{b^2} = 1. \qquad \text{Transverse axis is vertical}$$

The vertices are a units from the center, and the foci are c units from the center. Moreover, $b^2 = c^2 - a^2$.

Figure 9.21 shows both the horizontal and vertical orientations for a hyperbola.

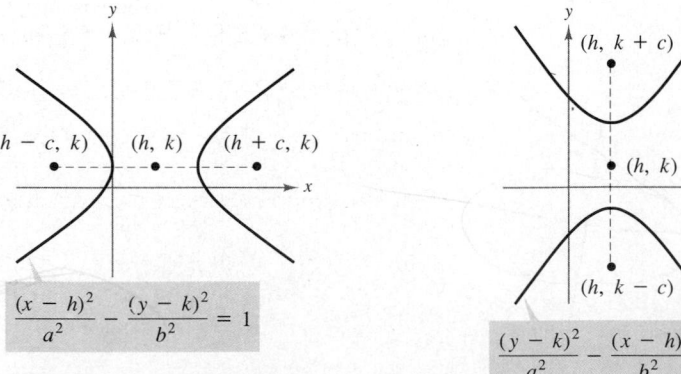

FIGURE 9.21
The standard equations for hyperbolas.

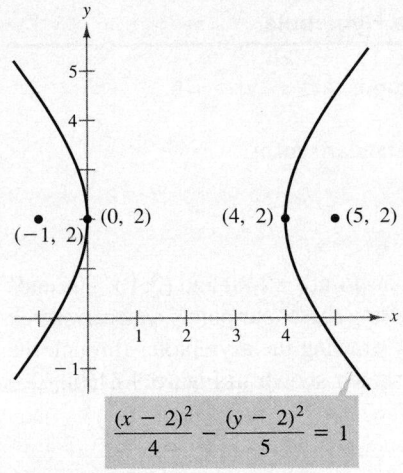

FIGURE 9.22
Hyperbola with a horizontal transverse axis.

EXAMPLE 1   Finding the Standard Equation of a Hyperbola

Find the standard form of the equation of the hyperbola with foci at $(-1, 2)$ and $(5, 2)$ and vertices at $(0, 2)$ and $(4, 2)$.

Solution   By the Midpoint Formula, the center of the hyperbola occurs at the point $(2, 2)$. Furthermore, $c = 3$ and $a = 2$, and it follows that

$$b^2 = 3^2 - 2^2 = 9 - 4 = 5.$$

Thus, the equation of the hyperbola is

$$\frac{(x - 2)^2}{4} - \frac{(y - 2)^2}{5} = 1.$$

Figure 9.22 shows the graph of the hyperbola.

Asymptotes

An important aid in sketching the graph of a hyperbola is the determination of its **asymptotes,** as shown in Figure 9.23. Each hyperbola has two asymptotes that intersect at the center of the hyperbola. The asymptotes pass through the vertices of a rectangle of dimensions $2a$ and $2b$, with its center at (h, k). The line segment of length $2b$ joining $(h, k + b)$ and $(h, k - b)$ is referred to as the **conjugate axis** of the hyperbola.

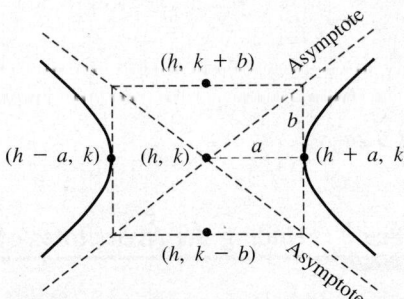

FIGURE 9.23

THEOREM 9.6   Asymptotes of a Hyperbola

For a *horizontal* transverse axis, the equations of the asymptotes are

$$y = k + \frac{b}{a}(x - h) \quad \text{and} \quad y = k - \frac{b}{a}(x - h).$$

For a *vertical* transverse axis, the equations of the asymptotes are

$$y = k + \frac{a}{b}(x - h) \quad \text{and} \quad y = k - \frac{a}{b}(x - h).$$

In Figure 9.23, you can see that the asymptotes coincide with the diagonals of the rectangle with dimensions $2a$ and $2b$, centered at (h, k). This provides you with a quick means of sketching the asymptotes, which in turn aids in sketching the hyperbola.

EXAMPLE 2   Using Asymptotes to Sketch a Hyperbola

Sketch the graph of the hyperbola whose equation is $4x^2 - y^2 = 16$.

Solution   Begin by rewriting the equation in standard form.

$$\frac{x^2}{2^2} - \frac{y^2}{4^2} = 1$$

The transverse axis is horizontal, and the vertices occur at $(-2, 0)$ and $(2, 0)$. The ends of the conjugate axis occur at $(0, -4)$ and $(0, 4)$. Using these four points, you can sketch the rectangle shown in Figure 9.24(a). Finally, by drawing the asymptotes through the corners of this rectangle, you can complete the sketch shown in Figure 9.24(b).

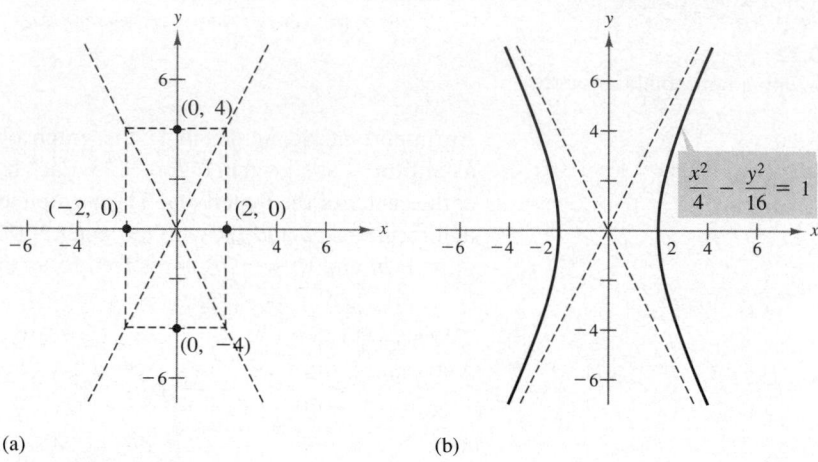

(a)   (b)

FIGURE 9.24

EXAMPLE 3   Finding the Asymptotes of a Hyperbola

Sketch the hyperbola given by $4x^2 - 3y^2 + 8x + 16 = 0$ and find the equations of its asymptotes.

Solution   Begin by rewriting the equation in standard form.

$$4x^2 - 3y^2 + 8x + 16 = 0 \qquad \text{Original equation}$$
$$4(x^2 + 2x) - 3y^2 = -16$$
$$-4(x^2 + 2x + 1) + 3y^2 = 16 - 4$$
$$-4(x + 1)^2 + 3y^2 = 12$$
$$\frac{y^2}{4} - \frac{(x + 1)^2}{3} = 1 \qquad \text{Standard form}$$

The hyperbola is centered at $(-1, 0)$ and has vertices at $(-1, 2)$ and $(-1, -2)$. The ends of the conjugate axis occur at $\left(-1 - \sqrt{3}, 0\right)$ and $\left(-1 + \sqrt{3}, 0\right)$. To sketch the graph of the hyperbola, draw a rectangle through these four points. The asymptotes are the lines passing through the corners of the rectangle, as shown in Figure 9.25. Finally, using $a = 2$ and $b = \sqrt{3}$, you can determine that the equations of the asymptotes are

$$y = \frac{2}{\sqrt{3}}(x + 1) \quad \text{and} \quad y = -\frac{2}{\sqrt{3}}(x + 1).$$

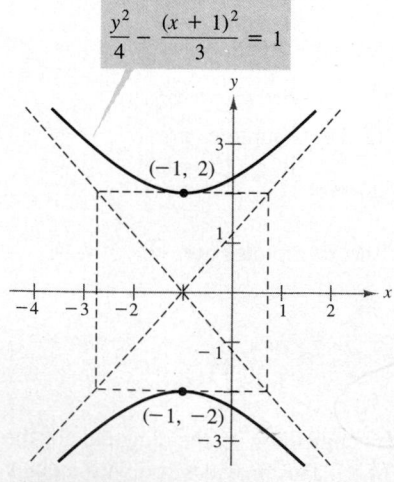

FIGURE 9.25
Hyperbola with a vertical transverse axis.

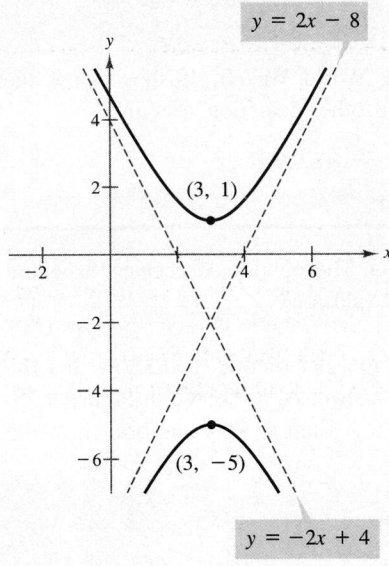

FIGURE 9.26
The asymptotes have equations whose slopes are $\dfrac{a}{b}$ and $-\dfrac{a}{b}$.

EXAMPLE 4 Using Asymptotes to Find the Standard Equation

Find the standard form of the equation of the hyperbola having vertices at $(3, -5)$ and $(3, 1)$ and asymptotes

$$y = 2x - 8 \quad \text{and} \quad y = -2x + 4$$

as shown in Figure 9.26.

Solution By the Midpoint Formula, the center of the hyperbola is at $(3, -2)$. Furthermore, the hyperbola has a vertical transverse axis with $a = 3$. By Theorem 9.6, you can determine that the asymptotes have equations whose slopes are

$$m_1 = \frac{a}{b} \quad \text{and} \quad m_2 = -\frac{a}{b}.$$

From the given equations of the asymptotes, you know that

$$\frac{a}{b} = 2 \quad \text{and} \quad -\frac{a}{b} = -2$$

and because $a = 3$, you can conclude that $b = \frac{3}{2}$. Therefore, the standard equation is

$$\frac{(y + 2)^2}{9} - \frac{(x - 3)^2}{9/4} = 1.$$

As with an ellipse, the **eccentricity** of a hyperbola is $e = c/a$. Because $c > a$, it follows that $e > 1$. If the eccentricity is large, then the branches of the hyperbola are nearly flat. If the eccentricity is close to 1, then the branches of the hyperbola are more pointed, as shown in Figure 9.27.

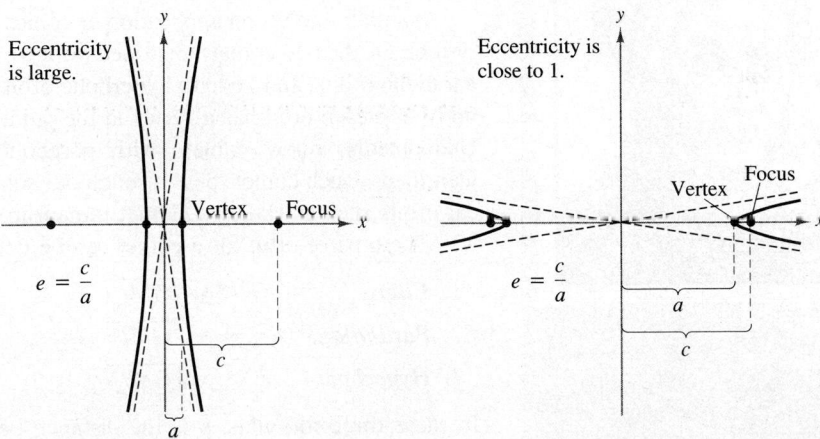

FIGURE 9.27

Definition of Eccentricity of a Hyperbola

The eccentricity e of a hyperbola is given by the ratio

$$e = \frac{c}{a}.$$

Applications

The following application was developed during World War II. It shows how the properties of hyperbolas can be used in radar and other detection systems.

EXAMPLE 5   A Hyperbolic Detection System

Two microphones, 1 mile apart, record an explosion. Microphone A receives the sound 2 seconds before microphone B. Where was the explosion?

Solution   Assuming that sound travels at 1100 feet per second, you know that the explosion took place 2200 feet farther from B than from A, as shown in Figure 9.28. The locus of all points that are 2200 feet closer to A than to B is one branch of the hyperbola $(x^2/a^2) - (y^2/b^2) = 1$, where

$$c = \frac{5280}{2} = 2640 \quad \text{and} \quad a = \frac{2200}{2} = 1100.$$

Thus, $b^2 = c^2 - a^2 = 5,759,600$, and you can conclude that the explosion occurred somewhere on the right branch of the hyperbola given by

$$\frac{x^2}{1,210,000} - \frac{y^2}{5,759,600} = 1.$$

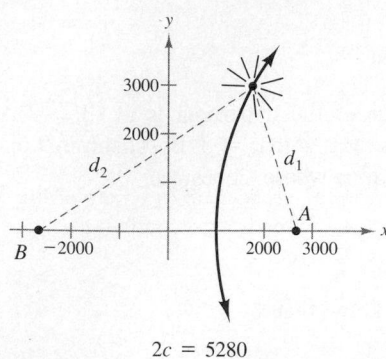

$$2c = 5280$$
$$d_2 - d_1 = 2a = 2200$$

FIGURE 9.28
The locus of all points that are 2200 feet closer to A than to B is one branch of the hyperbola with $c = 2640$ and $a = 1100$.

In Example 5, you were able to determine only the hyperbola on which the explosion occurred, but not the exact location of the explosion. If, however, you had received the sound at a third position, C, then two other hyperbolas could have been determined. The exact location of the explosion would have been the point at which these three hyperbolas intersected.

Another interesting application of conics involves the orbits of comets in our solar system. Of the 610 comets identified prior to 1970, 245 have elliptical orbits, 295 have parabolic orbits, and 70 have hyperbolic orbits. The center of the sun is a focus of each orbit, and each orbit has a vertex at the point at which the comet is closest to the sun. Undoubtedly, many comets with parabolic or hyperbolic orbits have not been identified—such comets pass through our solar system once. Only comets with elliptical orbits such as Halley's Comet remain in our solar system.

The type of orbit for a comet can be determined as follows.

1. *Ellipse* $v < \sqrt{2GM/p}$
2. *Parabola* $v = \sqrt{2GM/p}$
3. *Hyperbola* $v > \sqrt{2GM/p}$

In these three formulas, p is the distance between one vertex and one focus of the comet's orbit (in meters), v is the velocity of the comet at the vertex (in meters per second), $M \approx 1.991 \times 10^{30}$ kilograms is the mass of the sun, and $G \approx 6.67 \times 10^{-11}$ cubic meters per kilogram-second squared is the gravitational constant.

CAROLINE HERSCHEL

The first woman to be credited with detecting a new comet was the English astronomer Caroline Herschel (1750–1848). During her long life, Caroline Herschel discovered a total of eight new comets.

TECHNOLOGY
Laboratory Guide
Lab 6.9

EXERCISES for Section 9.3

In Exercises 1–6, match the equation with its graph. [The graphs are labeled (a), (b), (c), (d), (e), and (f).]

1. $\dfrac{x^2}{9} - \dfrac{y^2}{4} = 1$

2. $\dfrac{y^2}{9} - \dfrac{x^2}{4} = 1$

3. $\dfrac{y^2}{1} - \dfrac{x^2}{16} = 1$

4. $\dfrac{y^2}{16} - \dfrac{x^2}{1} = 1$

5. $\dfrac{(x - 2)^2}{9} - \dfrac{y^2}{4} = 1$

6. $\dfrac{(x + 1)^2}{16} - \dfrac{(y - 3)^2}{9} = 1$

(a)

(b)

(c)

(d)

(e)

(f)
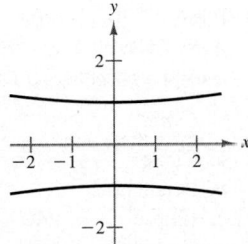

In Exercises 7–22, find the center, foci, and vertices of the hyperbola and sketch its graph using asymptotes as an aid.

7. $x^2 - y^2 = 1$

8. $\dfrac{x^2}{9} - \dfrac{y^2}{16} = 1$

9. $y^2 - \dfrac{x^2}{4} = 1$

10. $\dfrac{y^2}{9} - \dfrac{x^2}{1} = 1$

11. $\dfrac{y^2}{25} - \dfrac{x^2}{144} = 1$

12. $\dfrac{x^2}{36} - \dfrac{y^2}{4} = 1$

13. $5y^2 = 4x^2 + 20$

14. $7x^2 - 3y^2 = 21$

15. $\dfrac{(x - 1)^2}{4} - \dfrac{(y + 2)^2}{1} = 1$

16. $\dfrac{(x + 1)^2}{144} - \dfrac{(y - 4)^2}{25} = 1$

17. $(y + 6)^2 - (x - 2)^2 = 1$

18. $\dfrac{(y - 1)^2}{1/4} - \dfrac{(x + 3)^2}{1/9} = 1$

19. $9x^2 - y^2 - 36x - 6y + 18 = 0$

20. $x^2 - 9y^2 + 36y - 72 = 0$

21. $x^2 - 9y^2 + 2x - 54y - 80 = 0$

22. $16y^2 - x^2 + 2x + 64y + 63 = 0$

 In Exercises 23–28, find the center, foci, and vertices of the hyperbola and sketch a graph of the hyperbola and its asymptotes with the aid of a graphing utility.

23. $2x^2 - 3y^2 = 6$ **24.** $3y^2 = 5x^2 + 15$

25. $9y^2 - x^2 + 2x + 54y + 62 = 0$

26. $9x^2 - y^2 + 54x + 10y + 55 = 0$

27. $3x^2 - 2y^2 - 6x - 12y - 27 = 0$

28. $3y^2 - x^2 + 6x - 12y = 0$

In Exercises 29–38, find an equation of the specified hyperbola.

29. Center: $(0, 0)$
Vertex: $(0, 2)$
Focus: $(0, 4)$

30. Center: $(0, 0)$
Vertex: $(3, 0)$
Focus: $(5, 0)$

31. Vertices $(\pm 1, 0)$
Asymptotes: $y = \pm 3x$

32. Vertices: $(0, \pm 3)$
Asymptotes: $y = \pm 3x$

33. Vertices: $(0, 2), (6, 2)$
Asymptotes: $y = \frac{2}{3}x$
$y = 4 - \frac{2}{3}x$

34. Focus: $(10, 0)$
Asymptotes: $y = \pm \frac{3}{4}x$

35. Vertices: $(2, \pm 3)$
Point on graph: $(0, 5)$

36. Vertices: $(2, \pm 3)$
Foci: $(2, \pm 5)$

37. Find an equation of the hyperbola such that for any point on the hyperbola, the difference between its distances from the points $(2, 2)$ and $(10, 2)$ is 6.

38. Find an equation of the hyperbola such that for any point on the hyperbola, the difference between its distances from the points $(-3, 0)$ and $(-3, 3)$ is 2.

39. Consider a hyperbola centered at the origin with a horizontal transverse axis. Use the definition of a hyperbola to derive its standard form.

40. Sketch a graph of the hyperbola that consists of all points (x, y) such that the difference between the distances from (x, y) to two fixed points is 10 units and the foci are located at the centers of the two sets of concentric circles in the accompanying figure.

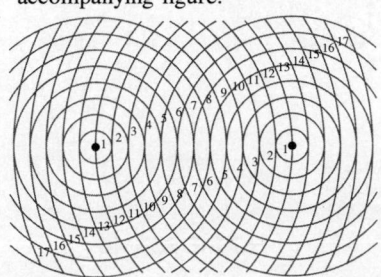

FIGURE FOR 40

41. *Navigation*   LORAN (long distance radio navigation) for aircraft and ships uses synchronized pulses transmitted by widely separated transmitting stations. These pulses travel at the speed of light (186,000 miles per second). The difference in the times of arrival of these pulses at an aircraft or ship is constant on a hyperbola having the transmitting stations as foci. Assume that two stations, 300 miles apart, are positioned on the rectangular coordinate system at points with coordinates $(-150, 0)$ and $(150, 0)$ and that a ship is traveling on a path with coordinates $(x, 75)$ (see figure). Find the x-coordinate of the position of the ship if the time difference between the pulses from the transmitting stations is 1000 microseconds (0.001 second).

FIGURE FOR 41

42. *Sound Location*   A rifle positioned at point $(-c, 0)$ is fired at a target positioned at point $(c, 0)$. A person hears the sound of the rifle and the sound of the bullet hitting the target at the same time. Show that the person is positioned on one branch of the hyperbola given by

$$\frac{x^2}{c^2 v_s^2 / v_m^2} - \frac{y^2}{c^2(v_m^2 - v_s^2)/v_m^2} = 1$$

where v_m is the muzzle velocity of the rifle and v_s is the speed of sound = 1100 feet per second.

43. *Thunder*   You and a friend live 4 miles apart and are talking on the phone. You hear a clap of thunder from lightning in a storm, and 18 seconds later your friend hears the thunder.

a. Find an equation that gives the possible places where the lightning could have occurred using the coordinate system shown in the figure.

b. Could the lightning have hit a baseball stadium that is 1 mile directly north of your home?

44. *Hyperbolic Mirror*   A hyperbolic mirror (used in some telescopes) has the property that a light ray directed at the focus will be reflected to the other focus. The mirror in the accompanying figure has the equation

$$\frac{x^2}{36} - \frac{y^2}{64} = 1.$$

At which point on the mirror will light from the point $(0, 10)$ be reflected to the other focus?

FIGURE FOR 43

FIGURE FOR 44

In Exercises 45 and 46, find equations for (a) the tangent lines and (b) the normal lines to the hyperbola for the given values of x.

45. $\dfrac{x^2}{9} - y^2 = 1,$ **46.** $\dfrac{y^2}{4} - \dfrac{x^2}{2} = 1,$

$\qquad x = 6$ $\qquad\qquad\qquad x = 4$

Volume and Surface Area   **In Exercises 47 and 48, the region bounded by the graphs of the equations revolves about the x-axis. Find the volume and surface area of the resulting solid.**

47. $x^2 - y^2 = 1,$ **48.** $\dfrac{x^2}{16} - \dfrac{y^2}{9} = 1,$

$\qquad y = 0, x = 2$ $\qquad\quad y = 0, x = 5$

49. Show that the equation of the tangent line to $x^2/a^2 - y^2/b^2 = 1$ at the point (x_0, y_0) is $x_0 x/a^2 - y_0 y/b^2 = 1$.

50. Show that the ellipse $x^2/a^2 + 2y^2/b^2 = 1$ and the hyperbola $x^2/(a^2 - b^2) - 2y^2/b^2 = 1$ intersect at right angles.

True or False   **In Exercises 51–54, determine whether the statement is true or false. If it is false, explain why or give an example that shows it is false.**

51. If $D \neq 0$ or $E \neq 0$, then the graph of $y^2 - x^2 + Dx + Ey = 0$ is a hyperbola.

52. If the asymptotes of the hyperbola $(x^2/a^2) - (y^2/b^2) = 1$ intersect at right angles, then $a = b$.

53. The graph of $y^2 = (x^2 + 1)^2$ is a hyperbola.

54. Every tangent line to a hyperbola intersects the hyperbola only at the point of tangency.

55. Prove that the graph of the equation $Ax^2 + Cy^2 + Dx + Ey + F = 0$ is one of the following (except in degenerate cases).

Conic	Condition
a. Circle	$A = C$
b. Parabola	$A = 0$ or $C = 0$ (but not both)
c. Ellipse	$AC > 0$
d. Hyperbola	$AC < 0$

In Exercises 56–64, classify the graph of the equation as a circle, a parabola, an ellipse, or a hyperbola.

56. $4x^2 - y^2 - 4x - 3 = 0$

57. $y^2 - 4y - 4x = 0$

58. $25x^2 - 10x - 200y - 119 = 0$

59. $4x^2 + 4y^2 - 16y + 15 = 0$

60. $y^2 - 4y = x + 5$

61. $9x^2 + 9y^2 - 36x + 6y + 34 = 0$

62. $2x(x - y) = y(3 - y - 2x)$

63. $3(x - 1)^2 = 6 + 2(y + 1)^2$

64. $9(x + 3)^2 = 36 - 4(y - 2)^2$

Rotation of Axes

In previous sections, you learned that equations of conics with axes parallel to one of the coordinate axes can be written in the general form

$$Ax^2 + Cy^2 + Dx + Ey + F = 0.$$ Horizontal or vertical axes

In this section, you will study the equations of conics whose axes are rotated so that they are *not* parallel to the *x*-axis or the *y*-axis. The general equation for such conics contains an *xy term.*

$$Ax^2 + Bxy + Cy^2 + Dx + Ey + F = 0$$ Equation in *xy*-plane

To eliminate this *xy* term, you can use a procedure called **rotation of axes.** You want to rotate the *x*- and *y*-axes until they are parallel to the axes of the conic. (The rotated axes are denoted as the *x'*-axis and the *y'*-axis, as shown in Figure 9.29.) After the rotation has been accomplished, the equation of the conic in the new *x'y'*-plane will have the form

$$A'(x')^2 + C'(y')^2 + D'x' + E'y' + F' = 0.$$ Equation in *x'y'*-plane

Because this equation has no *x'y'* term, you can obtain a standard form by completing the square.

The following theorem identifies how much to rotate the axes to eliminate an *xy* term, and also the equations for determining the new coefficients *A'*, *C'*, *D'*, *E'*, and *F'*. (A proof of Theorem 9.7 is given in Appendix A.)

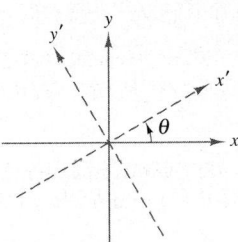

FIGURE 9.29
After rotating the axes counterclockwise through an angle θ, the rotated axes are denoted as the *x'*-axis and *y'*-axis.

THEOREM 9.7   Rotation of Axes

The general equation of the conic

$$Ax^2 + Bxy + Cy^2 + Dx + Ey + F = 0,$$

where $B \neq 0$, can be rewritten as

$$A'(x')^2 + C'(y')^2 + D'x' + E'y' + F' = 0$$

by rotating the coordinate axes through an angle θ, where

$$\cot 2\theta = \frac{A - C}{B}.$$

The coefficients of the new equation are obtained by making the substitutions

$$x = x'\cos\theta - y'\sin\theta$$
$$y = x'\sin\theta + y'\cos\theta.$$

EXAMPLE 1   Rotation of a Hyperbola

Write the equation $xy - 1 = 0$ in standard form.

Solution   Because $A = 0$, $B = 1$, and $C = 0$, you have (for $0 < \theta < \pi/2$)

$$\cot 2\theta = \frac{A - C}{B} = 0 \quad \rightarrow \quad 2\theta = \frac{\pi}{2} \quad \rightarrow \quad \theta = \frac{\pi}{4}.$$

The equation in the $x'y'$-system is obtained by making the following substitutions.

$$x = x' \cos\frac{\pi}{4} - y' \sin\frac{\pi}{4} = x'\left(\frac{\sqrt{2}}{2}\right) - y'\left(\frac{\sqrt{2}}{2}\right) = \frac{x' - y'}{\sqrt{2}}$$

$$y = x' \sin\frac{\pi}{4} + y' \cos\frac{\pi}{4} = x'\left(\frac{\sqrt{2}}{2}\right) + y'\left(\frac{\sqrt{2}}{2}\right) = \frac{x' + y'}{\sqrt{2}}$$

Substituting these expressions into the equation $xy - 1 = 0$ produces

$$\left(\frac{x' - y'}{\sqrt{2}}\right)\left(\frac{x' + y'}{\sqrt{2}}\right) - 1 = 0$$

$$\frac{(x')^2 - (y')^2}{2} - 1 = 0$$

$$\frac{(x')^2}{(\sqrt{2})^2} - \frac{(y')^2}{(\sqrt{2})^2} = 1. \qquad \text{Standard form}$$

This is the equation of a hyperbola centered at the origin with vertices at $\left(\pm\sqrt{2}, 0\right)$ in the $x'y'$-system, as shown in Figure 9.30.

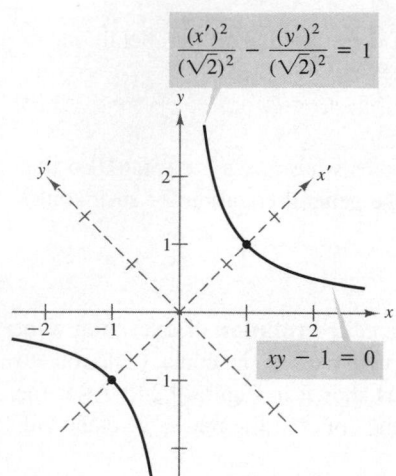

$$\frac{(x')^2}{(\sqrt{2})^2} - \frac{(y')^2}{(\sqrt{2})^2} = 1$$

$xy - 1 = 0$

FIGURE 9.30
Vertices:
$\left(\sqrt{2}, 0\right), \left(-\sqrt{2}, 0\right)$ in $x'y'$-system
$(1, 1), (-1, -1)$ in xy-system

EXAMPLE 2   Rotation of an Ellipse

Sketch the graph of $7x^2 - 6\sqrt{3}xy + 13y^2 - 16 = 0$.

Solution   Because $A = 7$, $B = -6\sqrt{3}$, and $C = 13$, you have (for $0 < \theta < \pi/2$)

$$\cot 2\theta = \frac{A - C}{B} = \frac{7 - 13}{-6\sqrt{3}} = \frac{1}{\sqrt{3}} \quad \rightarrow \quad \theta = \frac{\pi}{6}.$$

Therefore, the equation in the $x'y'$-system is derived by making the following substitutions.

$$x = x' \cos\frac{\pi}{6} - y' \sin\frac{\pi}{6} = x'\left(\frac{\sqrt{3}}{2}\right) - y'\left(\frac{1}{2}\right) = \frac{\sqrt{3}x' - y'}{2}$$

$$y = x' \sin\frac{\pi}{6} + y' \cos\frac{\pi}{6} = x'\left(\frac{1}{2}\right) + y'\left(\frac{\sqrt{3}}{2}\right) = \frac{x' + \sqrt{3}y'}{2}$$

Substituting these expressions into the original equation eventually simplifies (after considerable algebra) to

$$4(x')^2 + 16(y')^2 = 16$$

$$\frac{(x')^2}{4} + \frac{(y')^2}{1} = 1. \qquad \text{Standard form}$$

This is the equation of an ellipse centered at the origin with vertices at $(\pm 2, 0)$ and $(0, \pm 1)$ in the $x'y'$-system, as shown in Figure 9.31.

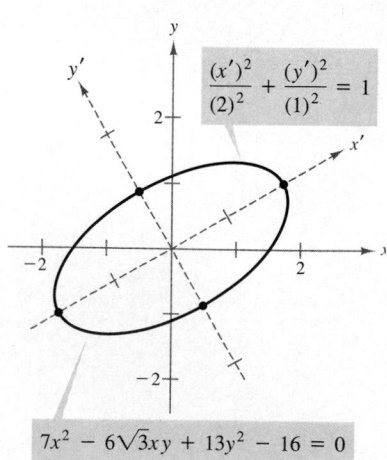

$$\frac{(x')^2}{(2)^2} + \frac{(y')^2}{(1)^2} = 1$$

$7x^2 - 6\sqrt{3}xy + 13y^2 - 16 = 0$

FIGURE 9.31
Vertices:
$(\pm 2, 0), (0, \pm 1)$ in $x'y'$-system
$\left(\pm\sqrt{3}, \pm 1\right), \left(\pm\frac{1}{2}, \pm\frac{\sqrt{3}}{2}\right)$ in xy-system

In writing Examples 1 and 2, we chose the equations so that θ would be one of the common angles 30°, 45°, and so forth. Of course, many second-degree equations do not yield such common solutions to the equation

$$\cot 2\theta = \frac{A - C}{B}.$$

Example 3 illustrates such a case.

EXAMPLE 3   Rotation of a Parabola

Sketch the graph of $x^2 - 4xy + 4y^2 + 5\sqrt{5}y + 1 = 0$.

Solution   Because $A = 1$, $B = -4$, and $C = 4$, you have

$$\cot 2\theta = \frac{A - C}{B} = \frac{1 - 4}{-4} = \frac{3}{4}.$$

The trigonometric identity $\cot 2\theta = (\cot^2 \theta - 1)/(2 \cot \theta)$ produces

$$\cot 2\theta = \frac{3}{4} = \frac{\cot^2 \theta - 1}{2 \cot \theta}$$

from which you can obtain the equation

$$6 \cot \theta = 4 \cot^2 \theta - 4 \quad \rightarrow \quad 4 \cot^2 \theta - 6 \cot \theta - 4 = 0$$
$$(2 \cot \theta - 4)(2 \cot \theta + 1) = 0.$$

Considering $0 < \theta < \pi/2$, it follows that $2 \cot \theta = 4$. Thus,

$$\cot \theta = 2 \quad \rightarrow \quad \theta \approx 26.6°.$$

FIGURE 9.32

From the triangle in Figure 9.32, you can obtain $\sin \theta = 1/\sqrt{5}$ and $\cos \theta = 2/\sqrt{5}$. Consequently, you can write the following.

$$x = x' \cos \theta - y' \sin \theta = x'\left(\frac{2}{\sqrt{5}}\right) - y'\left(\frac{1}{\sqrt{5}}\right) = \frac{2x' - y'}{\sqrt{5}}$$

$$y = x' \sin \theta + y' \cos \theta = x'\left(\frac{1}{\sqrt{5}}\right) + y'\left(\frac{2}{\sqrt{5}}\right) = \frac{x' + 2y'}{\sqrt{5}}$$

Substituting these expressions into the original equation produces

$$\left(\frac{2x' - y'}{\sqrt{5}}\right)^2 - 4\left(\frac{2x' - y'}{\sqrt{5}}\right)\left(\frac{x' + 2y'}{\sqrt{5}}\right) + 4\left(\frac{x' + 2y'}{\sqrt{5}}\right)^2$$
$$+ 5\sqrt{5}\left(\frac{x' + 2y'}{\sqrt{5}}\right) + 1 = 0$$

which simplifies to

$$5(y')^2 + 5x' + 10y' + 1 = 0.$$

By completing the square, you can obtain the standard form

$$5(y' + 1)^2 = -5x' + 4$$
$$(y' + 1)^2 = (-1)\left(x' - \frac{4}{5}\right). \qquad \text{Standard form}$$

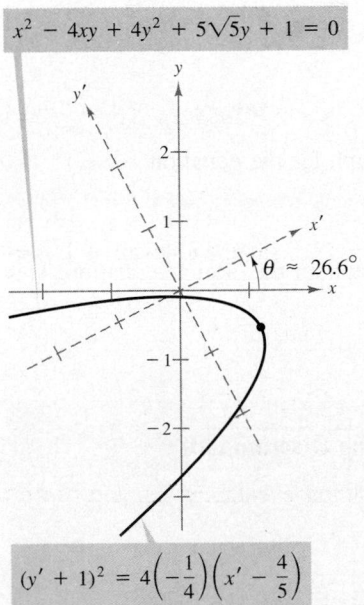

$$x^2 - 4xy + 4y^2 + 5\sqrt{5}y + 1 = 0$$

$$(y' + 1)^2 = 4\left(-\frac{1}{4}\right)\left(x' - \frac{4}{5}\right)$$

FIGURE 9.33

Vertex: $\left(\dfrac{4}{5}, -1\right)$ in $x'y'$-system

$\left(\dfrac{13}{5\sqrt{5}}, -\dfrac{6}{5\sqrt{5}}\right)$ in xy-system

The graph of the equation is a parabola with its vertex at $\left(\frac{4}{5}, -1\right)$ and its axis parallel to the x'-axis in the $x'y'$-system, as shown in Figure 9.33.

Invariants Under Rotation

In Theorem 9.7, note that the constant term $F' = F$ is the same in both equations. Because of this, F is said to be **invariant under rotation.** Theorem 9.8 lists some other rotation invariants. The proof of this theorem is left as an exercise (see Exercise 36).

THEOREM 9.8   Rotation Invariants

The rotation of coordinate axes through an angle θ that transforms the equation $Ax^2 + Bxy + Cy^2 + Dx + Ey + F = 0$ into the form

$$A'(x')^2 + C'(y')^2 + D'x' + E'y' + F' = 0$$

has the following rotation invariants.

1. $F = F'$
2. $A + C = A' + C'$
3. $B^2 - 4AC = (B')^2 - 4A'C'$

You can use this theorem to classify the graph of a second-degree equation *with* an xy term in much the same way you do for a second-degree equation *without* an xy term (see Exercise 55 in Section 9.3). Note that because $B' = 0$, the invariant $B^2 - 4AC$ reduces to

$$B^2 - 4AC = -4A'C' \qquad \text{Discriminant}$$

which is called the **discriminant** of the equation

$$Ax^2 + Bxy + Cy^2 + Dx + Ey + F = 0.$$

Because the sign of $A'C'$ determines the type of graph for the equation

$$A'(x')^2 + C'(y')^2 + D'x' + E'y' + F' = 0$$

the sign of $B^2 - 4AC$ must determine the type of graph for the original equation. This result is stated in Theorem 9.9.

THEOREM 9.9   Classification of Conics by the Discriminant

The graph of the equation

$$Ax^2 + Bxy + Cy^2 + Dx + Ey + F = 0$$

is, except in degenerate cases, determined by its discriminant as follows.

1. *Ellipse or circle* $B^2 - 4AC < 0$

2. *Parabola* $B^2 - 4AC = 0$

3. *Hyperbola* $B^2 - 4AC > 0$

EXAMPLE 4   Using the Discriminant

Classify the graph of each of the following equations.

a. $4xy - 9 = 0$

b. $2x^2 - 3xy + 2y^2 - 2x = 0$

c. $x^2 - 6xy + 9y^2 - 2y + 1 = 0$

d. $3x^2 + 8xy + 4y^2 - 7 = 0$

Solution

a. The graph is a hyperbola because

$$B^2 - 4AC = 16 - 0 > 0.$$

b. The graph is a circle or an ellipse because

$$B^2 - 4AC = 9 - 16 < 0.$$

c. The graph is a parabola because

$$B^2 - 4AC = 36 - 36 = 0.$$

d. The graph is a hyperbola because

$$B^2 - 4AC = 64 - 48 > 0.$$

1– 11 odd 19–33 odd

EXERCISES for Section 9.4

TECHNOLOGY
Laboratory Guide
Lab 9.4

In Exercises 1–12, rotate the axes to eliminate the xy-term. Sketch the graph of the resulting equation, showing both sets of axes.

1. $xy + 1 = 0$

2. $xy - 4 = 0$

3. $x^2 - 10xy + y^2 + 1 = 0$

4. $xy + x - 2y + 3 = 0$

5. $xy - 2y - 4x = 0$

6. $13x^2 + 6\sqrt{3}xy + 7y^2 - 16 = 0$

7. $5x^2 - 2xy + 5y^2 - 12 = 0$

8. $2x^2 - 3xy - 2y^2 + 10 = 0$

9. $3x^2 - 2\sqrt{3}xy + y^2 + 2x + 2\sqrt{3}y = 0$

10. $16x^2 - 24xy + 9y^2 - 60x - 80y + 100 = 0$

11. $9x^2 + 24xy + 16y^2 + 90x - 130y = 0$

12. $9x^2 + 24xy + 16y^2 + 80x - 60y = 0$

C In Exercises 13–18, use a graphing utility to graph the conic. Determine the angle θ through which the axes are rotated. Explain how you used the utility to obtain the graph.

13. $x^2 + xy + y^2 = 10$

14. $x^2 - 4xy + 2y^2 = 6$

15. $17x^2 + 32xy - 7y^2 = 75$

16. $40x^2 + 36xy + 25y^2 = 52$

17. $32x^2 + 50xy + 7y^2 = 52$

18. $4x^2 - 12xy + 9y^2 + \left(4\sqrt{13} - 12\right)x$
$$- \left(6\sqrt{13} + 8\right)y = 91$$

In Exercises 19 and 20, evaluate the derivative of the implicit function at any y-intercepts of its graph.

19. $5x^2 - 2xy + 5y^2 - 12 = 0$

20. $16x^2 - 24xy + 9y^2 - 60x - 80y + 100 = 0$

In Exercises 21–28, use the discriminant to determine whether the graph of the equation is a parabola, an ellipse, or a hyperbola.

21. $16x^2 - 24xy + 9y^2 - 30x - 40y = 0$

22. $x^2 - 4xy - 2y^2 - 6 = 0$

23. $13x^2 - 8xy + 7y^2 - 45 = 0$

24. $2x^2 + 4xy + 5y^2 + 3x - 4y - 20 = 0$

25. $x^2 - 6xy - 5y^2 + 4x - 22 = 0$

26. $36x^2 - 60xy + 25y^2 + 9y = 0$

27. $x^2 + 4xy + 4y^2 - 5x - y - 3 = 0$

28. $x^2 + xy + 4y^2 + x + y - 4 = 0$

In Exercises 29–34, sketch the graph of the degenerate conic.

29. $y^2 - 4x^2 = 0$

30. $x^2 + y^2 - 2x + 6y + 10 = 0$

31. $x^2 + 2xy + y^2 - 1 = 0$

32. $x^2 - 10xy + y^2 = 0$

33. $(x - 2y + 1)(x + 2y - 3) = 0$

34. $(2x + y - 3)^2 = 0$

35. Show that the equation $x^2 + y^2 = r^2$ is invariant under rotation of axes.

36. Prove Theorem 9.8.

37. Consider $Ax^2 + Bxy + Cy^2 = 1$, where $B^2 - 4AC < 0$. Show that the area of this ellipse is

$$\pi ab = \frac{2\pi}{\sqrt{4AC - B^2}}$$

where $2a$ and $2b$ are the lengths of the major and minor axes, respectively.

38. Graph and identify the curve given by

$$\arcsin(x) + \arcsin(y) = \frac{\pi}{2}.$$

What is the graph of $\arcsin(x) + \arcsin(y) = \dfrac{\pi}{3}$?

39. Graph the equation $\sqrt{x} + \sqrt{y} = 1$ and show that it is a portion of a parabola.

REVIEW EXERCISES for Chapter 9

In Exercises 1–10, match the equation with the correct graph. [The graphs are labeled (a), (b), (c), (d), (e), (f), (g), (h), (i), and (j).]

1. $4x^2 + y^2 = 4$

2. $x^2 = 4y$

3. $4x^2 - y^2 = 4$

4. $y^2 = -4x$

5. $x^2 + 4y^2 = 4$

6. $y^2 - 4x^2 = 4$

7. $x^2 = -6y$

8. $x^2 + 5y^2 = 10$

9. $x^2 - 5y^2 = -5$

10. $y^2 - 8x = 0$

(a)

(b)

(c)

(d)

(e)

(f)

(g)

(h)

(i)

(j)

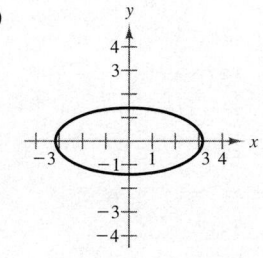

In Exercises 11–22, analyze each equation and sketch its graph.

11. $16x^2 + 16y^2 - 16x + 24y - 3 = 0$

12. $y^2 - 12y - 8x + 20 = 0$

13. $3x^2 - 2y^2 + 24x + 12y + 24 = 0$

14. $4x^2 + y^2 - 16x + 15 = 0$

15. $3x^2 + 2y^2 - 12x + 12y + 29 = 0$

16. $4x^2 - 4y^2 - 4x + 8y - 11 = 0$

17. $x^2 - 6x + 2y + 9 = 0$

18. $x^2 + y^2 - 2x - 4y + 5 = 0$

19. $4x^2 + 9y^2 - 8x + 9y + 4 = 0$

20. $9x^2 + 4y^2 - 36x + 8y + 31 = 0$

21. $x^2 + y^2 + 2xy + 2\sqrt{2}x - 2\sqrt{2}y + 2 = 0$

22. $9x^2 + 6y^2 + 4xy - 20 = 0$

In Exercises 23–26, find an equation of the parabola.

23. Vertex: (4, 2)
Focus: (4, 0)

24. Vertex: (2, 0)
Focus: (0, 0)

25. Vertex: (0, 2)
Directrix: $x = -3$

26. Vertex: (0, 2)
Solution point: $(-1, 0)$

27. Find an equation of the line tangent to the parabola $y = x^2 - 2x + 2$ and perpendicular to the line $y = x - 2$.

28. *Satellite Antenna* A cross section of a large parabolic antenna is modeled by the graph of

$$y = \frac{x^2}{200}, \qquad 0 \le x \le 100.$$

The receiving and transmitting equipment is positioned at the focus.
a. Find the coordinates of the focus.
b. Find the surface area of the antenna.

In Exercises 29–34, find an equation of the ellipse.

29. Vertices: $(-3, 0)$, $(7, 0)$; Foci: $(0, 0)$, $(4, 0)$

30. Vertices: $(2, 0)$, $(2, 4)$; Foci: $(2, 1)$, $(2, 3)$

31. Vertices: $(0, \pm 6)$; Solution point: $(2, 2)$

32. Vertices: $(0, 1)$, $(4, 1)$; Minor axis endpoints: $(2, 0)$, $(2, 2)$

33. Center: $(0, 0)$; Solution points: $(1, 2)$, $(2, 0)$

34. Foci: $(0, 0)$, $(4, 0)$; Sum of the distances from foci to a point on the ellipse: 10

35. *Semielliptical Archway* A semielliptical archway is to be formed over the entrance to an estate. The arch is to be set on pillars that are 10 feet apart and is to have a height (atop the pillars) of 4 feet. Where should the foci be placed in order to sketch a model for the arch?

36. Find the equations of the lines tangent to the ellipse $x^2 + 4y^2 - 4x - 8y - 24 = 0$ and parallel to the line $x - 2y - 12 = 0$.

In Exercises 37–39, consider the region bounded by the ellipse

$$\frac{x^2}{a^2} + \frac{y^2}{b^2} = 1, \qquad \text{eccentricity } e = c/a.$$

37. Show that the area of the region is πab.

38. Show that the solid (oblate spheroid) generated by revolving the region about the minor axis of the ellipse has a volume of $V = 4\pi a^2 b/3$ and a surface area of

$$S = 2\pi a^2 + \pi \left(\frac{b^2}{e}\right) \ln\left(\frac{1 + e}{1 - e}\right).$$

39. Show that the solid (prolate spheroid) generated by revolving the region about the major axis of the ellipse has a volume of $V = 4\pi ab^2/3$ and a surface area of

$$S = 2\pi b^2 + 2\pi \left(\frac{ab}{e}\right) \arcsin e.$$

40. Use the results of Exercises 38 and 39 to find the volume and surface areas of the prolate and oblate spheroids generated by revolving the region bounded by the graph of

$$\frac{x^2}{9} + \frac{y^2}{4} = 1$$

about the appropriate axis.

Water Tank **In Exercises 41–44, consider a fire truck with a water tank 16 feet long whose vertical cross sections are ellipses described by the equation**

$$\frac{x^2}{16} + \frac{y^2}{9} = 1.$$

41. Find the volume of the tank.

42. Find the force on the end of the tank when the tank is full of water.

43. Find the depth of the water in the tank if the tank is $\frac{3}{4}$ full (by volume) and the truck is on level ground.

44. Approximate, to two decimal places, the surface area of the tank.

C **In Exercises 45 and 46, use a graphing utility or computer algebra system to approximate the perimeter of the ellipse.**

45. $\dfrac{x^2}{9} + \dfrac{y^2}{4} = 1$

46. $\dfrac{x^2}{4} + \dfrac{y^2}{25} = 1$

In Exercises 47–52, find an equation of the hyperbola.

47. Vertices: $(0, \pm 1)$
Foci: $(0, \pm 3)$

48. Vertices: $(\pm 2, 2)$
Foci: $(\pm 4, 2)$

49. Foci: $(\pm 4, 0)$; Absolute value of the difference of the distances from a point on the hyperbola to the foci: 4

50. Foci: $(\pm 4, 0)$; Asymptotes: $y = \pm 2x$

51. Foci: $(0, 0)$, $(8, 0)$; Asymptotes: $y = \pm 2(x - 4)$

52. Foci: $(3, \pm 2)$; Asymptotes: $y = \pm 2(x - 3)$

53. Find a so that the hyperbola given by

$$\frac{x^2}{a^2} - \frac{y^2}{4} = 1$$

is tangent to the line $2x - y - 4 = 0$.

54. *Locating an Explosion* Two of your friends live 4 miles apart and on the same "east-west" street, and you live halfway between them. You are talking on a three-way phone call when you hear an explosion. Six seconds later your friend to the east hears the explosion, and your friend to the west hears it 8 seconds after you do. Find the equations of two hyperbolas that would locate the explosion. (Remember that sound travels at a rate of 1100 ft/sec.)

Chapter 10

Plane Curves,
Parametric
Equations,
and Polar
Coordinates

10.1

Plane Curves and Parametric Equations • Eliminating the Parameter • Finding Parametric Equations • The Tautochrone and Brachistochrone Problems

Rectangular equation:
$$y = -\frac{x^2}{72} + x$$

Parametric equations:
$x = 24\sqrt{2}t$
$y = -16t^2 + 24\sqrt{2}t$

FIGURE 10.1
Curvilinear motion: two variables for position, one variable for time.

REMARK At times it is important to distinguish between a graph (the set of points) and a curve (the points together with their defining parametric equations). When it is important, we will make the distinction explicit. When it is not important, we will use C to represent the graph or the curve.

Plane Curves and Parametric Equations

Until now, we have been representing a graph by a single equation involving *two* variables. In this section, you will study situations in which *three* variables are used to represent a curve in the plane.

Consider the path followed by an object that is propelled into the air at an angle of 45°. If the initial velocity of the object is 48 feet per second, the object travels the parabolic path given by

$$y = -\frac{x^2}{72} + x \qquad \text{Rectangular equation}$$

as shown in Figure 10.1. However, this equation does not tell the whole story. Although it does tell you *where* the object has been, it doesn't tell you *when* the object was at a given point (x, y). To determine this time, you can introduce a third variable t, called a **parameter.** By writing both x and y as functions of t, you obtain the **parametric equations**

$$x = 24\sqrt{2}t \quad \text{and} \quad y = -16t^2 + 24\sqrt{2}t. \qquad \text{Parametric equations}$$

From this set of equations, you can determine that at time $t = 0$, the object is at the point $(0, 0)$. Similarly, at time $t = 1$, the object is at the point $(24\sqrt{2}, 24\sqrt{2} - 16)$, and so on. (We will discuss a method for determining this particular set of parametric equations—the equations of motion—later, in Section 12.3.)

For this particular motion problem, x and y are continuous functions of t, and the resulting path is called a **plane curve.**

Definition of a Plane Curve

If f and g are continuous functions of t on an interval I, then the equations

$$x = f(t) \quad \text{and} \quad y = g(t)$$

are called **parametric equations** and t is called a **parameter.** The set of points (x, y) obtained as t varies over the interval I is called the **graph** of the parametric equations. Taken together, the parametric equations and the graph are called a **plane curve,** denoted by C.

◄ The path of a satellite in orbit around earth is discussed in Section 10.5, Exercise 54.

When sketching (by hand) a curve represented by a pair of parametric equations, you can plot points in the xy-plane. Each set of coordinates (x, y) is determined from a value chosen for the parameter t. By plotting the resulting points in the order of *increasing* values of t, the curve is traced out in a specific direction. This is called the **orientation** of the curve.

EXAMPLE 1   Sketching a Curve

Sketch the curve described by the parametric equations

$$x = t^2 - 4 \quad \text{and} \quad y = \frac{t}{2}, \qquad -2 \le t \le 3.$$

Solution   For values of t on the given interval, the parametric equations yield the points (x, y) shown in Table 10.1.

Parametric equations:
$x = t^2 - 4$ and $y = \dfrac{t}{2}$

FIGURE 10.2

TABLE 10.1

t	-2	-1	0	1	2	3
x	0	-3	-4	-3	0	5
y	-1	$-\dfrac{1}{2}$	0	$\dfrac{1}{2}$	1	$\dfrac{3}{2}$

By plotting these points in the order of increasing t and using the continuity of f and g, you obtain the curve C shown in Figure 10.2. Note that the arrows on the curve indicate its orientation as t increases from -2 to 3.

REMARK   From the vertical line test, you can see that the graph shown in Figure 10.2 does not define y as a function of x. This points out one benefit of parametric equations—they can be used to represent graphs that are more general than graphs of functions.

It often happens that two different sets of parametric equations have the same graph. For example, the set of parametric equations

$$x = 4t^2 - 4 \quad \text{and} \quad y = t, \qquad -1 \le t \le \frac{3}{2}$$

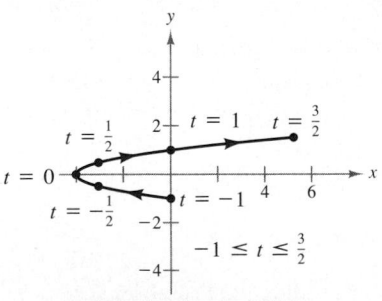

Parametric equations:
$x = 4t^2 - 4$ and $y = t$

FIGURE 10.3

has the same graph as the set given in Example 1. However, comparing the values of t in Figures 10.2 and 10.3, you can see that this second graph is traced out more *rapidly* (considering t as time) than the first graph. Thus, in applications, different parametric representations can be used to represent various *speeds* at which objects travel along a given path.

TECHNOLOGY   Most graphing utilities have a parametric graphing mode. If you have access to such a utility (either a graphing calculator or graphing software), try using it to confirm the graphs shown in Figures 10.2 and 10.3. Does the curve given by $x = 4t^2 - 8t$ and $y = 1 - t$, for $-\frac{1}{2} \le t \le 2$, represent the same graph as that shown in Figures 10.2 and 10.3? What do you notice about the *direction* in which this curve is traced?

Eliminating the Parameter

Finding a rectangular equation that represents the graph of a set of parametric equations is called **eliminating the parameter.** For instance, you can eliminate the parameter from the set of parametric equations in Example 1 as follows.

Parametric equations	→	Solve for t in one equation	→	Substitute into second equation	→	Rectangular equation
$x = t^2 - 4$		$t = 2y$		$x = (2y)^2 - 4$		$x = 4y^2 - 4$
$y = t/2$						

Once you have eliminated the parameter, you can recognize that the equation $x = 4y^2 - 4$ represents a parabola with a horizontal axis and vertex at $(-4, 0)$.

The range of x and y implied by the parametric equations may be altered by the change to rectangular form. In such instances the domain of the rectangular equation must be adjusted so that its graph matches the graph of the parametric equations. Such a situation is demonstrated in the next example.

EXAMPLE 2   Adjusting the Domain after Eliminating the Parameter

Sketch the curve represented by the equations

$$x = \frac{1}{\sqrt{t + 1}} \quad \text{and} \quad y = \frac{t}{t + 1}, \quad -1 < t$$

by eliminating the parameter and adjusting the domain of the resulting rectangular equation.

Solution   Begin by solving one of the parametric equations for t. For instance, you can solve the first equation for t as follows.

$$x = \frac{1}{\sqrt{t + 1}} \qquad \text{Parametric equation for } x$$

$$x^2 = \frac{1}{t + 1} \qquad \text{Square both sides}$$

$$t + 1 = \frac{1}{x^2}$$

$$t = \frac{1}{x^2} - 1 = \frac{1 - x^2}{x^2} \qquad \text{Solve for } t$$

Now, substituting into the parametric equation for y produces the following.

$$y = \frac{t}{t + 1} \qquad \text{Parametric equation for } y$$

$$y = \frac{(1 - x^2)/x^2}{[(1 - x^2)/x^2] + 1} \qquad \text{Substitute } (1 - x^2)/x^2 \text{ for } t$$

$$y = 1 - x^2 \qquad \text{Simplify}$$

The rectangular equation, $y = 1 - x^2$, is defined for all values of x, but from the parametric equation for x you can see that the curve is defined only when $-1 < t$. This implies that you should restrict the domain of x to positive values, as shown in Figure 10.4.

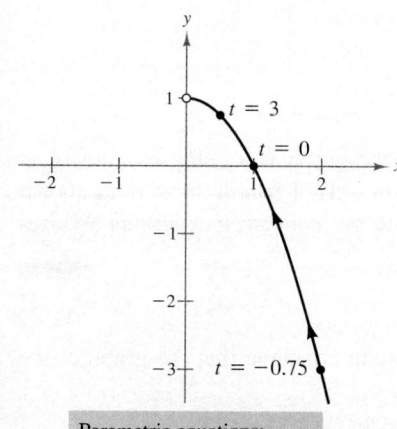

Parametric equations:

$$x = \frac{1}{\sqrt{t + 1}}, \quad y = \frac{t}{t + 1}$$

Rectangular equation:

$$y = 1 - x^2, \quad x > 0$$

FIGURE 10.4

It is not necessary for the parameter in a set of parametric equations to represent time. The next example uses an *angle* as the parameter.

EXAMPLE 3   Using Trigonometry to Eliminate a Parameter

Sketch the curve represented by

$$x = 3\cos\theta \quad \text{and} \quad y = 4\sin\theta, \qquad 0 \le \theta \le 2\pi$$

by eliminating the parameter and finding the corresponding rectangular equation.

Solution   Begin by solving for $\cos\theta$ and $\sin\theta$ in the given equations.

$$\cos\theta = \frac{x}{3} \quad \text{and} \quad \sin\theta = \frac{y}{4} \qquad \text{Solve for } \sin\theta \text{ and } \cos\theta$$

Next, make use of the identity $\sin^2\theta + \cos^2\theta = 1$ to form an equation involving only x and y.

$$\cos^2\theta + \sin^2\theta = 1 \qquad \text{Trigonometric identity}$$

$$\left(\frac{x}{3}\right)^2 + \left(\frac{y}{4}\right)^2 = 1 \qquad \text{Substitute}$$

$$\frac{x^2}{9} + \frac{y^2}{16} = 1 \qquad \text{Rectangular equation}$$

From this rectangular equation you can see that the graph is an ellipse centered at $(0, 0)$, with vertices at $(0, 4)$ and $(0, -4)$ and minor axis of length $2b = 6$, as shown in Figure 10.5. Note that the elliptic curve is traced out *counterclockwise* as θ varies from 0 to 2π.

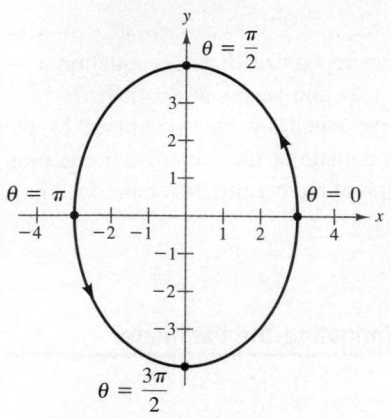

Parametric equations:
$x = 3\cos\theta \quad y = 4\sin\theta$
Rectangular equation:
$$\frac{x^2}{9} + \frac{y^2}{16} = 1$$

FIGURE 10.5

Using the technique shown in Example 3, you can conclude that the graph of the parametric equations

$$x = h + a\cos\theta \quad \text{and} \quad y = k + b\sin\theta, \qquad 0 \le \theta \le 2\pi$$

is the ellipse (traced counterclockwise) given by

$$\frac{(x - h)^2}{a^2} + \frac{(y - k)^2}{b^2} = 1.$$

The graph of the parametric equations

$$x = h + a\sin\theta \quad \text{and} \quad y = k + b\cos\theta, \qquad 0 \le \theta \le 2\pi$$

is also the ellipse (traced clockwise) given by

$$\frac{(x - h)^2}{a^2} + \frac{(y - k)^2}{b^2} = 1.$$

If you have access to a graphing utility, try using it in parametric mode to sketch several ellipses.

In Examples 2 and 3 it is important to realize that eliminating the parameter is primarily an *aid to curve sketching*. If the parametric equations represent the path of a moving object, the graph alone is not sufficient to describe the object's motion. You still need the parametric equations to tell you the *position, direction,* and *speed* at a given time.

Finding Parametric Equations

The first three examples in this section illustrated techniques for sketching the graph represented by a set of parametric equations. We now look at the reverse problem. How can you determine a set of parametric equations for a given graph or a given physical description? From the discussion following Example 1, you know that such a representation is not unique. This is demonstrated further in the following example, which finds two different parametric representations for a given graph.

EXAMPLE 4   Finding Parametric Equations for a Given Graph

Find a set of parametric equations to represent the graph of $y = 1 - x^2$, using the following parameters.

a. $t = x$ **b.** the slope $m = \dfrac{dy}{dx}$ at the point (x, y)

Solution

a. Letting $x = t$ produces the parametric equations

$$x = t \quad \text{and} \quad y = 1 - x^2 = 1 - t^2.$$

b. To express x and y in terms of the parameter m, you can proceed as follows.

$$m = \frac{dy}{dx} = -2x \qquad \text{Differentiate } y = 1 - x^2$$

$$x = -\frac{m}{2} \qquad \text{Solve for } x$$

This produces a parametric equation for x. To obtain a parametric equation for y, substitute $-m/2$ for x in the original equation.

$$y = 1 - x^2 \qquad \text{Original "rectangular" equation}$$

$$y = 1 - \left(-\frac{m}{2}\right)^2 \qquad \text{Substitute } -m/2 \text{ for } x$$

$$y = 1 - \frac{m^2}{4} \qquad \text{Simplify}$$

Thus, the parametric equations are

$$x = -\frac{m}{2} \quad \text{and} \quad y = 1 - \frac{m^2}{4}.$$

In Figure 10.6, note that the resulting curve has a right-to-left orientation as determined by the direction of increasing values of slope m. For part **a** the curve would have the opposite orientation.

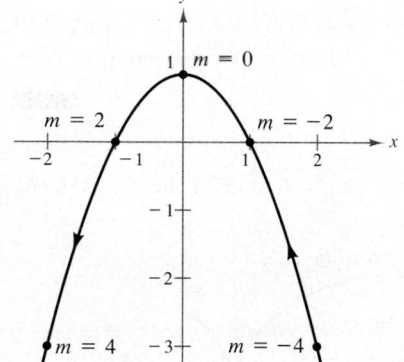

Rectangular equation: $y = 1 - x^2$
Parametric equations:
$x = -\dfrac{m}{2} \quad y = 1 - \dfrac{m^2}{4}$

FIGURE 10.6

TECHNOLOGY   To be efficient at using a graphing utility, it is important that you develop skill in representing a graph by a set of parametric equations. The reason for this is that many graphing utilities have only three graphing modes—(1) functions, (2) parametric equations, and (3) polar equations. Most graphing utilities are not programmed to sketch the graph of a general equation. For instance, suppose you want to sketch the graph of the hyperbola $x^2 - y^2 = 1$. To sketch the graph in function mode, you need two equations: $y = \sqrt{x^2 - 1}$ and $y = -\sqrt{x^2 - 1}$. In parametric mode, you can represent the graph by $x = \sec t$ and $y = \tan t$.

FOR FURTHER INFORMATION
For more information on cycloids, see
the article "The Geometry of Rolling
Curves" by John Bloom and Lee Whitt
in the June–July, 1981 issue of the
American Mathematical Monthly.

EXAMPLE 5   Parametric Equations for a Cycloid

Determine the curve traced by a point P on the circumference of a circle of radius a rolling along a straight line in a plane. Such a curve is called a **cycloid.**

Solution   Let the parameter θ be the measure of the circle's rotation, and let the point $P = (x, y)$ begin at the origin. When $\theta = 0$, P is at the origin. When $\theta = \pi$, P is at a maximum point $(\pi a, 2a)$. When $\theta = 2\pi$, P is back on the x-axis at $(2\pi a, 0)$. From Figure 10.7, you can see that $\angle APC = 180° - \theta$. Hence,

$$\sin\theta = \sin(180° - \theta) = \sin(\angle APC) = \frac{AC}{a} = \frac{BD}{a}$$

$$\cos\theta = -\cos(180° - \theta) = -\cos(\angle APC) = \frac{AP}{-a}$$

which implies that

$$AP = -a\cos\theta \quad \text{and} \quad BD = a\sin\theta.$$

Because the circle rolls along the x-axis, you know that

$$OD = \overset{\frown}{PD}$$
$$= a\theta.$$

Furthermore, because $BA = DC = a$, you have

$$x = OD - BD = a\theta - a\sin\theta$$
$$y = BA + AP = a - a\cos\theta.$$

Therefore, the parametric equations are

$$x = a(\theta - \sin\theta) \quad \text{and}$$
$$y = a(1 - \cos\theta).$$

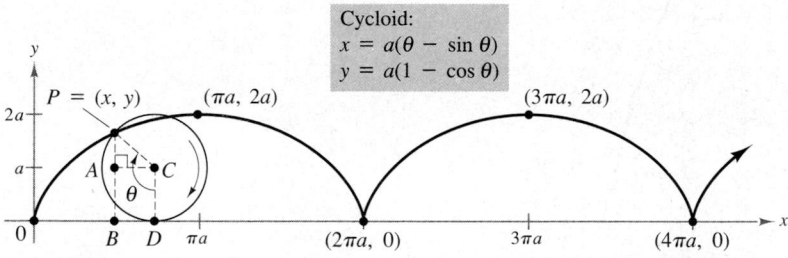

Cycloid:
$x = a(\theta - \sin\theta)$
$y = a(1 - \cos\theta)$

FIGURE 10.7

TECHNOLOGY   Some computer
programs allow you to simulate the
motion of an object that is moving in
the plane or in space. If you have access to such a program, try using it
to trace out the path of the cycloid
shown in Figure 10.7.

The cycloid in Figure 10.7 has sharp corners at the values $x = 2n\pi a$. Between these points, the cycloid is called **smooth.**

Definition of a Smooth Curve

A curve C represented by $x = f(t)$ and $y = g(t)$ on an interval I is called **smooth** if f' and g' are continuous on I and not simultaneously 0, except pos-sibly at the endpoints of I. The curve C is called **piecewise smooth** if it is smooth on each subinterval of some partition of I.

The Tautochrone and Brachistochrone Problems

The type of curve described in Example 5 is related to one of the most famous pairs of problems in the history of calculus. The first problem (called the **tautochrone problem**) began with Galileo's discovery that the time required to complete a full swing of a given pendulum is *approximately* the same whether it makes a large movement at high speeds or a small movement at lower speeds (see Figure 10.8). Late in his life, Galileo (1564–1642) realized that he could use this principle to construct a clock. However, he was not able to conquer the mechanics of actual construction. Christian Huygens (1629–1695) was the first to design and construct a working model. In his work with pendulums, Huygens realized that a pendulum does not take *exactly* the same time to complete swings of varying lengths. (This doesn't affect a pendulum clock, because the length of the circular arc is kept constant by giving the pendulum a slight boost each time it passes its lowest point.) But, in studying the problem, Huygens discovered that a ball rolling back and forth on an inverted cycloid does complete each cycle in exactly the same time.

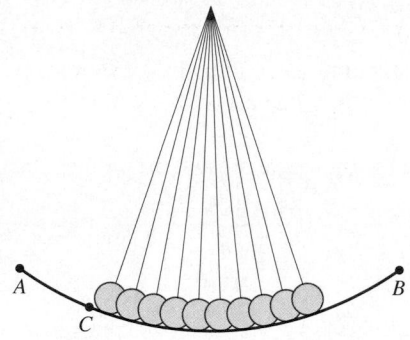

FIGURE 10.8

The time required to complete a full swing of the pendulum when starting from point C is only approximately the same as when starting from point A.

JAMES BERNOULLI

James Bernoulli (1654–1705), also called Jacques, was the older brother of John. He was one of several accomplished mathematicians of the Swiss Bernoulli family. James's mathematical accomplishments have given him a prominent place in the early development of calculus.

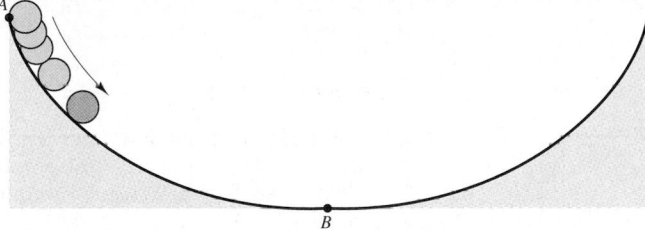

FIGURE 10.9

An inverted cycloid is the path down which a ball will roll in the shortest time.

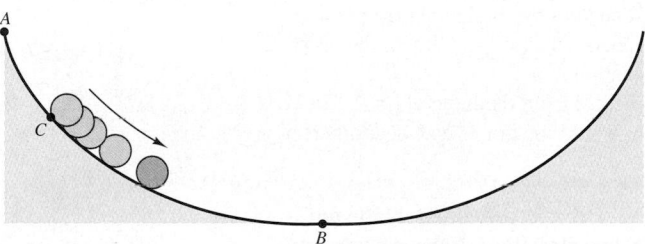

FIGURE 10.10

A ball starting at point C takes the same time to reach point B as one that starts at point A.

The second problem, posed by John Bernoulli in 1696, is called the **brachistochrone problem**—in Greek *brachys* means short and *chronos* means time. The problem was to determine the path down which a particle will slide from point A to point B in the *shortest time*. Several mathematicians took up the challenge, and the following year the problem was solved by Newton, Leibniz, L'Hôpital, John Bernoulli, and James Bernoulli. As it turns out, the solution is not a straight line from A to B, but an inverted cycloid passing through the points A and B, as shown in Figure 10.9. The amazing part of the solution is that a particle starting at rest at *any* other point C of the cycloid between A and B will take exactly the same time to reach B, as indicated in Figure 10.10.

EXERCISES for Section 10.1

TECHNOLOGY
Laboratory Guide
Lab 10.1

1. Consider the parametric equations $x = \sqrt{t}$ and $y = 1 - t$.
   a. Complete the table.

t	0	1	2	3	4
x					
y					

   b. Plot the points (x, y) generated in the table, and sketch a graph of the parametric equations. Indicate the orientation of the graph.
   c. Find the rectangular equation by eliminating the parameter. Compare the graph sketched in part b with the graph of the rectangular equation.

2. Consider the parametric equations $x = 4\cos^2\theta$ and $y = 2\sin\theta$.
   a. Complete the table.

θ	$-\dfrac{\pi}{2}$	$-\dfrac{\pi}{4}$	0	$\dfrac{\pi}{4}$	$\dfrac{\pi}{2}$
x					
y					

   b. Plot the points (x, y) generated in the table, and sketch a graph of the parametric equations. Indicate the orientation of the graph.
   c. Find the rectangular equation by eliminating the parameter. Compare the graph sketched in part b with the graph of the rectangular equation.
   d. If values of θ were selected from the interval $[\pi/2, 3\pi/2]$ for the table in part a, would the graph sketched in part b be different? Why or why not?

In Exercises 3–28, sketch the curve represented by the parametric equations (indicate the direction of the curve), and write the corresponding rectangular equation by eliminating the parameter.

3. $x = 3t - 1$
   $y = 2t + 1$

4. $x = 3 - 2t$
   $y = 2 + 3t$

5. $x = t + 1$
   $y = t^2$

6. $x = \sqrt[3]{t}$
   $y = 1 - t$

7. $x = t^3$
   $y = \dfrac{t^2}{2}$

8. $x = t^2 + t$
   $y = t^2 - t$

9. $x = t - 1$
   $y = \dfrac{t}{t - 1}$

10. $x = 1 + \dfrac{1}{t}$
   $y = t - 1$

11. $x = 2t$
   $y = |t - 2|$

12. $x = |t - 1|$
   $y = t + 2$

13. $x = \sec\theta$
   $y = \cos\theta$

14. $x = \tan^2\theta$
   $y = \sec^2\theta$

15. $x = 3\cos\theta$
   $y = 3\sin\theta$

16. $x = \cos\theta$
   $y = 3\sin\theta$

17. $x = 4\sin 2\theta$
   $y = 2\cos 2\theta$

18. $x = \cos\theta$
   $y = 2\sin 2\theta$

19. $x = 4 + 2\cos\theta$
   $y = -1 + \sin\theta$

20. $x = 4 + 2\cos\theta$
   $y = -1 + 2\sin\theta$

21. $x = 4 + 2\cos\theta$
   $y = -1 + 4\sin\theta$

22. $x = \sec\theta$
   $y = \tan\theta$

23. $x = 4\sec\theta$
   $y = 3\tan\theta$

24. $x = \cos^3\theta$
   $y = \sin^3\theta$

25. $x = t^3$
   $y = 3\ln t$

26. $x = \ln 2t$
   $y = t^2$

27. $x = e^{-t}$
   $y = e^{3t}$

28. $y = e^{2t}$
   $y = e^t$

In Exercises 29–32, describe any differences between the curves of the parametric equations. Are the graphs the same? Are the orientations the same? Are the curves smooth?

29. a. $x = t$
   $y = 2t + 1$
   b. $x = \cos\theta$
   $y = 2\cos\theta + 1$
   c. $x = e^{-t}$
   $y = 2e^{-t} + 1$
   d. $x = e^t$
   $y = 2e^t + 1$

30. a. $x = 2\cos\theta$
   $y = 2\sin\theta$
   b. $x = \dfrac{\sqrt{4t^2 - 1}}{|t|}$
   $y = \dfrac{1}{t}$
   c. $x = \sqrt{t}$
   $y = \sqrt{4 - t}$
   d. $x = -\sqrt{4 - e^{2t}}$
   $y = e^t$

31. a. $x = \cos\theta$
   $y = \sin\theta$
   b. $x = \sin\theta$
   $y = \cos\theta$
   c. $x = \sin\theta^2$
   $y = \cos\theta^2$
   d. $x = -\cos\theta$
   $y = \sin\theta$

32. a. $x = t$
   $y = t$
   b. $x = t^2$
   $y = t^2$
   c. $x = -t$
   $y = -t$
   d. $x = t^3$
   $y = t^3$

In Exercises 33–38, eliminate the parameter and obtain the standard form of the rectangular equation.

33. Line through (x_1, y_1) and (x_2, y_2):
   $x = x_1 + t(x_2 - x_1)$
   $y = y_1 + t(y_2 - y_1)$

34. Circle: $x = r\cos\theta$
   $\quad\quad\quad\, y = r\sin\theta$

35. Circle: $x = h + r\cos\theta$
   $\quad\quad\quad\, y = k + r\sin\theta$

36. Ellipse: $x = h + a\cos\theta$
   $\quad\quad\quad\quad y = k + b\sin\theta$

37. Hyperbola: $x = h + a \sec \theta$
$y = k + b \tan \theta$

38. Hyperbola: $x = h + a\sqrt{t + 1}$
$y = k + b\sqrt{t}$

In Exercises 39–46, use the results of Exercises 33–38 to find a set of parametric equations for the line or conic.

39. Line: Passes through $(0, 0)$ and $(5, -2)$

40. Line: Passes through $(1, 4)$ and $(5, -2)$

41. Circle: Center: $(2, 1)$
Radius: 4

42. Circle: Center: $(-3, 1)$
Radius: 3

43. Ellipse: Vertices: $(\pm 5, 0)$
Foci: $(\pm 4, 0)$

44. Ellipse: Vertices: $(4, 7), (4, -3)$
Foci: $(4, 5), (4, -1)$

45. Hyperbola: Vertices: $(\pm 4, 0)$
Foci: $(\pm 5, 0)$

46. Hyperbola: Vertices: $(0, \pm 1)$
Foci: $(0, \pm 2)$

In Exercises 47–50, find two different sets of parametric equations for the given rectangular equation.

47. $y = 3x - 2$
48. $y = \dfrac{1}{x}$

49. $y = x^3$
50. $y = x^2$

C **In Exercises 51–58, use a graphing utility to graph the curve represented by the parametric equations. At which points are the curves not smooth?**

51. Cycloid: $x = 2(\theta - \sin \theta)$
$y = 2(1 - \cos \theta)$

52. Cycloid: $x = \theta + \sin \theta$
$y = 1 - \cos \theta$

53. Prolate cycloid: $x = \theta - \frac{3}{2} \sin \theta$
$y = 1 - \frac{3}{2} \cos \theta$

54. Prolate cycloid: $x = 2\theta - 4 \sin \theta$
$y = 2 - 4 \cos \theta$

55. Hypocycloid: $x = 3 \cos^3 \theta$
$y = 3 \sin^3 \theta$

56. Curtate cycloid: $x = 2\theta - \sin \theta$
$y = 2 - \cos \theta$

57. Witch of Agnesi: $x = 2 \cot \theta$
$y = 2 \sin^2 \theta$

58. Folium of Descartes: $x = \dfrac{3t}{1 + t^3}$

$y = \dfrac{3t^2}{1 + t^3}$

In Exercises 59–62, match the set of parametric equations with the correct graph. [The graphs are labeled (a), (b), (c), and (d).]

59. Lissajous Curve: $x = 4 \cos \theta$
$y = 2 \sin 2\theta$

60. Evolute of Ellipse: $x = \cos^3 \theta$
$y = 2 \sin^3 \theta$

61. Involute of Circle: $x = \cos \theta + \theta \sin \theta$
$y = \sin \theta - \theta \cos \theta$

62. Serpentine Curve: $x = \cot \theta$
$y = 4 \sin \theta \cos \theta$

(a)

(b)

(c)

(d)

Projectile Motion The path of a projectile launched at an angle θ with the horizontal and with initial velocity v_0 feet per second is modeled by the parametric equations

$$x = (v_0 \cos \theta)t \quad \text{and} \quad y = (v_0 \sin \theta)t - 16t^2.$$

C **In Exercises 63 and 64, use a graphing utility to graph the paths of the projectile for the specified values of θ and v_0. For each case, use the graph to approximate the maximum height and the range of the projectile.**

63. a. $\theta = 20°, v_0 = 88$ ft/sec
b. $\theta = 20°, v_0 = 132$ ft/sec
c. $\theta = 45°, v_0 = 88$ ft/sec
d. $\theta = 45°, v_0 = 132$ ft/sec
e. $\theta = 60°, v_0 = 88$ ft/sec
f. $\theta = 60°, v_0 = 132$ ft/sec

64. a. $\theta = 15°, v_0 = 60$ ft/sec
b. $\theta = 15°, v_0 = 100$ ft/sec
c. $\theta = 30°, v_0 = 60$ ft/sec
d. $\theta = 30°, v_0 = 100$ ft/sec
e. $\theta = 45°, v_0 = 60$ ft/sec
f. $\theta = 45°, v_0 = 100$ ft/sec

*A **C** indicates an exercise in which you are instructed to use graphing technology or a symbolic computer algebra system. The solutions of other exercises may also be facilitated by appropriate technology.

65. *Curtate Cycloid*   A wheel of radius a rolls along a straight line without slipping. The curve traced by a point P that is b units from the center ($b < a$) is called a **curtate cycloid** (see figure). Use the angle θ shown in the figure to find a set of parametric equations for the curve.

66. *Epicycloid*   A circle of radius 1 rolls around the outside of a circle of radius 2 without slipping. The curve traced by a point on the circumference of the smaller circle is called an **epicycloid** (see figure). Use the angle θ shown in the figure to find a set of parametric equations for the curve.

FIGURE FOR 65   **FIGURE FOR 66**

True or False   **In Exercises 67–70, determine whether the statement is true or false. If it is false, explain why or give an example that shows it is false.**

67. The two sets of parametric equations $x = t$, $y = t^2 + 1$ and $x = 3t$, $y = 9t^2 + 1$ correspond to the same rectangular equation.

68. The graph of the parametric equations $x = t^2$ and $y = t^2$ is the line $y = x$.

69. If y is a function of t and x is a function of t, then y is a function of x.

70. If $f(t_1) = 0$ and $g(t_2) = 0$, then the curve represented by the parametric equations $x = f(t)$ and $y = g(t)$ passes through the origin.

71. Consider the curve given by the parametric equations

$$x(t) = \frac{1 - t^2}{1 + t^2} \quad \text{and} \quad y(t) = \frac{2t}{1 + t^2}.$$

Graph this curve in the xy-plane. Use the graph to show that $(t^2 - 1)^2 + (2t)^2 = (t^2 + 1)^2$. Illustrate the formula for $t = 2$ and 3.

In Exercises 72–77, match the hypocycloid or epicycloid with its graph. [The graphs are labeled (a), (b), (c), (d), (e), and (f).] (These exercises are based on the article "Mathematical Discovery via Computer Graphs: Hypocycloids and Epicycloids" by Florence S. Gordon and Sheldon P. Gordon in the November, 1990 issue of the *College Mathematics Journal*.)

Hypocycloid, H(A, B)

Path traced by a fixed point on a circle of radius B as it rolls around the *inside* of a circle of radius A.

$$x = (A - B)\cos t + B\cos\left(\frac{A - B}{B}\right)t$$

$$y = (A - B)\sin t - B\sin\left(\frac{A - B}{B}\right)t$$

72. $H(8, 3)$

74. $H(8, 7)$

76. $H(24, 7)$

Epicycloid, E(A, B)

Path traced by a fixed point on a circle of radius B as it rolls around the *outside* of a circle of radius A.

$$x = (A + B)\cos t - B\cos\left(\frac{A + B}{B}\right)t$$

$$y = (A + B)\sin t - B\sin\left(\frac{A + B}{B}\right)t$$

73. $E(8, 3)$

75. $E(24, 3)$

77. $E(24, 7)$

(a)

(b)

(c)

(d)

(e)

(f)

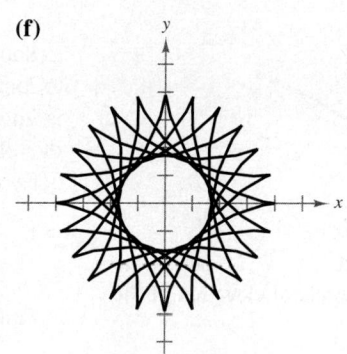

SECTION **Parametric Equations and Calculus**

10.2

Slope and Tangent Lines • Arc Length • Area of a Surface of Revolution

FIGURE 10.11

At time t the angle of elevation of the projectile is θ.

Slope and Tangent Lines

Now that you can represent a graph in the plane by a set of parameteric equations, it is natural to ask how to use calculus to study plane curves. To begin, let's take another look at the projectile represented by the parametric equations.

$$x = 24\sqrt{2}t \quad \text{and} \quad y = -16t^2 + 24\sqrt{2}t$$

as shown in Figure 10.11. From Section 10.1 you know that these equations enable you to locate the position of the projectile at a given time. You also know that the object is initially projected at an angle of 45°. But how can you find the angle θ representing the object's direction at some other time t? The following theorem answers this question by giving a formula for the slope of the tangent line as a function of t.

THEOREM 10.1   Parametric Form of the Derivative

If a smooth curve C is given by the equations $x = f(t)$ and $y = g(t)$, then the slope of C at (x, y) is

$$\frac{dy}{dx} = \frac{dy/dt}{dx/dt}, \qquad \frac{dx}{dt} \neq 0.$$

Proof   In Figure 10.12, consider $\Delta t > 0$ and let

$$\Delta y = g(t + \Delta t) - g(t) \quad \text{and} \quad \Delta x = f(t + \Delta t) - f(t).$$

Because $\Delta x \to 0$ as $\Delta t \to 0$, you can write

$$\frac{dy}{dx} = \lim_{\Delta x \to 0} \frac{\Delta y}{\Delta x} = \lim_{\Delta t \to 0} \frac{g(t + \Delta t) - g(t)}{f(t + \Delta t) - f(t)}.$$

Dividing both the numerator and denominator by Δt, you can use the differentiability of f and g to conclude that

$$\frac{dy}{dx} = \lim_{\Delta t \to 0} \frac{[g(t + \Delta t) - g(t)]/\Delta t}{[f(t + \Delta t) - f(t)]/\Delta t}$$

$$= \frac{\displaystyle\lim_{\Delta t \to 0} \frac{g(t + \Delta t) - g(t)}{\Delta t}}{\displaystyle\lim_{\Delta t \to 0} \frac{f(t + \Delta t) - f(t)}{\Delta t}}$$

$$= \frac{g'(t)}{f'(t)} = \frac{dy/dt}{dx/dt}.$$

FIGURE 10.12

The slope of the secant line through $(f(t), g(t))$ and $(f(t + \Delta t), g(t + \Delta t))$ is $\Delta y / \Delta x$.

EXAMPLE 1   Differentiation and Parametric Form

Find dy/dx for the curve given by $x = \sin t$ and $y = \cos t$.

Solution

$$\frac{dy}{dx} = \frac{dy/dt}{dx/dt} = \frac{-\sin t}{\cos t} = -\tan t$$

Because dy/dx is a function of t, you can use Theorem 10.1 repeatedly to find *higher-order* derivatives. For instance,

$$\frac{d^2y}{dx^2} = \frac{d}{dx}\left[\frac{dy}{dx}\right] = \frac{\dfrac{d}{dt}\left[\dfrac{dy}{dx}\right]}{dx/dt} \qquad \text{Second derivative}$$

$$\frac{d^3y}{dx^3} = \frac{d}{dx}\left[\frac{d^2y}{dx^2}\right] = \frac{\dfrac{d}{dt}\left[\dfrac{d^2y}{dx^2}\right]}{dx/dt}. \qquad \text{Third derivative}$$

EXAMPLE 2   Finding Slope and Concavity

For the curve given by

$$x = \sqrt{t} \quad \text{and} \quad y = \frac{1}{4}(t^2 - 4), \qquad 0 \le t$$

find the slope and concavity at the point $(2, 3)$.

Solution   Because

$$\frac{dy}{dx} = \frac{dy/dt}{dx/dt} = \frac{(1/2)t}{(1/2)t^{-1/2}} = t^{3/2}$$

you can find the second derivative to be

$$\frac{d^2y}{dx^2} = \frac{\dfrac{d}{dt}\left[\dfrac{dy}{dx}\right]}{dx/dt} = \frac{\dfrac{d}{dt}[t^{3/2}]}{dx/dt} = \frac{(3/2)t^{1/2}}{(1/2)t^{-1/2}} = 3t.$$

At $(x, y) = (2, 3)$, it follows that $t = 4$, and the slope is

$$\frac{dy}{dx} = (4)^{3/2} = 8.$$

Moreover, when $t = 4$, the second derivative is

$$\frac{d^2y}{dx^2} = 3(4) = 12 > 0$$

and you can conclude that the graph is concave upward at $(2, 3)$, as shown in Figure 10.13.

FIGURE 10.13
The graph is concave upward at $(2, 3)$.

Because the parametric equations $x = f(t)$ and $y = g(t)$ need not define y as a function of x, it follows that a plane curve can loop and cross itself. At such points the curve may have more than one tangent line, as shown in the next example.

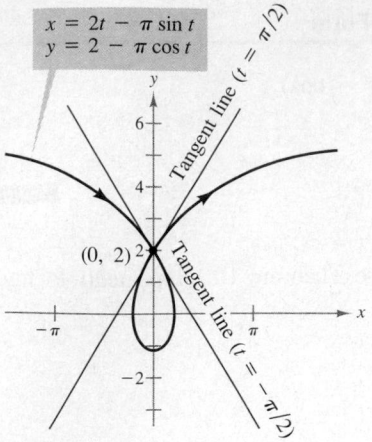

$x = 2t - \pi \sin t$
$y = 2 - \pi \cos t$

Tangent line ($t = \pi/2$)

Tangent line ($t = -\pi/2$)

(0, 2)

FIGURE 10.14
This prolate cycloid has two tangent lines at the point (0, 2).

EXAMPLE 3   A Curve with Two Tangent Lines at a Point

The **prolate cycloid** given by

$$x = 2t - \pi \sin t \quad \text{and} \quad y = 2 - \pi \cos t$$

crosses itself at the point (0, 2), as shown in Figure 10.14. Find the equations of both tangent lines at this point.

Solution   Because $x = 0$ and $y = 2$ when $t = \pm\pi/2$, and

$$\frac{dy}{dx} = \frac{dy/dt}{dx/dt} = \frac{\pi \sin t}{2 - \pi \cos t}$$

you have $dy/dx = -\pi/2$ when $t = -\pi/2$ and $dy/dx = \pi/2$ when $t = \pi/2$. Therefore, the two tangent lines at (0, 2) are

$$y - 2 = -\left(\frac{\pi}{2}\right)x \qquad \text{Tangent line when } t = -\frac{\pi}{2}$$

$$y - 2 = \left(\frac{\pi}{2}\right)x. \qquad \text{Tangent line when } t = \frac{\pi}{2}$$

If $dy/dt = 0$ and $dx/dt \neq 0$ when $t = t_0$, then the curve represented by $x = f(t)$ and $y = g(t)$ has a *horizontal* tangent at $(f(t_0), g(t_0))$. For instance, in Example 3, the given curve has a horizontal tangent at the point $(0, 2 - \pi)$ (when $t = 0$). Similarly, if $dx/dt = 0$ and $dy/dt \neq 0$ when $t = t_0$, then the curve represented by $x = f(t)$ and $y = g(t)$ has a *vertical* tangent at $(f(t_0), g(t_0))$.

Arc Length

You have seen how parametric equations can be used to describe the path of a particle moving in the plane. We now develop a formula for determining the *distance* traveled by the particle along its path.

Recall from Section 6.4 that the formula for the arc length of a curve C given by $y = h(x)$ over the interval $[x_0, x_1]$ is

$$s = \int_{x_0}^{x_1} \sqrt{1 + [h'(x)]^2}\, dx = \int_{x_0}^{x_1} \sqrt{1 + \left(\frac{dy}{dx}\right)^2}\, dx.$$

If C is represented by the parametric equations $x = f(t)$ and $y = g(t)$, $a \leq t \leq b$, and if $dx/dt = f'(t) > 0$, then you can write

$$s = \int_{x_0}^{x_1} \sqrt{1 + \left(\frac{dy}{dx}\right)^2}\, dx = \int_{x_0}^{x_1} \sqrt{1 + \left(\frac{dy/dt}{dx/dt}\right)^2}\, dx$$

$$= \int_a^b \sqrt{\frac{(dx/dt)^2 + (dy/dt)^2}{(dx/dt)^2}} \frac{dx}{dt}\, dt$$

$$= \int_a^b \sqrt{\left(\frac{dx}{dt}\right)^2 + \left(\frac{dy}{dt}\right)^2}\, dt$$

$$= \int_a^b \sqrt{[f'(t)]^2 + [g'(t)]^2}\, dt.$$

The formula for arc length can be shown to be valid for *any* smooth curve C that does not intersect itself.

THEOREM 10.2   Arc Length in Parametric Form

If a smooth curve C is given by $x = f(t)$ and $y = g(t)$ such that C does not intersect itself on the interval $a \le t \le b$ (except possibly at the endpoints), then the arc length of C over the interval is given by

$$s = \int_a^b \sqrt{\left(\frac{dx}{dt}\right)^2 + \left(\frac{dy}{dt}\right)^2}\, dt = \int_a^b \sqrt{[f'(t)]^2 + [g'(t)^2]}\, dt.$$

REMARK When applying the arc length formula to a curve, be sure that the curve is traced out only once on the interval of integration. For instance, the circle given by $x = \cos t$ and $y = \sin t$ is traced out once on the interval $0 \le t \le 2\pi$, but is traced out twice on the interval $0 \le t \le 4\pi$.

In the previous section you saw that if a circle rolls along a line, then a point on its circumference will trace a path called a cycloid. If the circle rolls around the circumference of another circle, then the path of the point is an **epicycloid.** The next example shows how to find the arc length of an epicycloid.

EXAMPLE 4   Finding Arc Length

A circle of radius 1 rolls around the circumference of a larger circle of radius 4, as shown in Figure 10.15. The epicycloid traced by a point on the circumference of the smaller circle is given by

$$x = 5\cos t - \cos 5t \quad \text{and} \quad y = 5\sin t - \sin 5t.$$

Find the distance traveled by the point in one complete trip about the larger circle.

Solution Before applying Theorem 10.2, note in Figure 10.15 that the curve has sharp points when $t = 0$ and $t = \pi/2$. Between these two points, dx/dt and dy/dt are not simultaneously 0. Thus, the portion of the curve generated from $t = 0$ to $t = \pi/2$ is smooth. To find the total distance traveled by the point, you can find the arc length of that portion lying in the first quadrant and multiply by 4.

$$
\begin{aligned}
s &= 4\int_0^{\pi/2} \sqrt{\left(\frac{dx}{dt}\right)^2 + \left(\frac{dy}{dt}\right)^2}\, dt \\
&= 4\int_0^{\pi/2} \sqrt{(-5\sin t + 5\sin 5t)^2 + (5\cos t - 5\cos 5t)^2}\, dt \\
&= 20\int_0^{\pi/2} \sqrt{2 - 2\sin t \sin 5t - 2\cos t \cos 5t}\, dt \\
&= 20\int_0^{\pi/2} \sqrt{2 - 2\cos 4t}\, dt = 20\int_0^{\pi/2} \sqrt{4\sin^2 2t}\, dt \\
&= 40\int_0^{\pi/2} \sin 2t\, dt = -20\Big[\cos 2t\Big]_0^{\pi/2} = 40
\end{aligned}
$$

$$x = 5\cos t - \cos 5t$$
$$y = 5\sin t - \sin 5t$$

FIGURE 10.15

An epicycloid is traced by a point on the smaller circle as it rolls around the larger circle.

For the epicycloid shown in Figure 10.15, an arc length of 40 seems about right because the circumference of a circle of radius 6 is $2\pi r = 12\pi \approx 37.7$.

0.5 in.

0.001 in.

2 in.

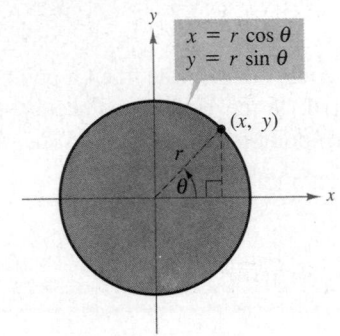

$x = r \cos \theta$
$y = r \sin \theta$

(x, y)

r

θ

FIGURE 10.16
It takes approximately 982 feet of tape to fill the reel.

REMARK   The graph of $r = a\theta$ is called the **Spiral of Archimedes.** The graph of $r = (1/2000\pi)\theta$ (in Example 5) is of this form.

EXAMPLE 5   Length of a Recording Tape

A recording tape 0.001 inch thick is wound around a reel with an inner radius of 0.5 inch and an outer radius of 2 inches, as shown in Figure 10.16. How much tape is required to fill the reel?

Solution   To create a model for this problem, assume that as the tape is wound around the reel its distance r from the center increases linearly at the rate of 0.001 inch per revolution, or

$$r = (0.001)\frac{\theta}{2\pi} = \frac{\theta}{2000\pi}, \qquad 1000\pi \le \theta \le 4000\pi$$

where θ is measured in radians. You can determine the coordinates of the point (x, y) corresponding to a given radius to be

$$x = r \cos \theta$$

and

$$y = r \sin \theta.$$

Substituting for r, you obtain the parametric equations

$$x = \left(\frac{\theta}{2000\pi}\right)\cos \theta \quad \text{and} \quad y = \left(\frac{\theta}{2000\pi}\right)\sin \theta.$$

You can use the arc length formula to determine the total length of the tape to be

$$
\begin{aligned}
s &= \int_{1000\pi}^{4000\pi} \sqrt{\left(\frac{dx}{d\theta}\right)^2 + \left(\frac{dy}{d\theta}\right)^2}\, d\theta \\
&= \frac{1}{2000\pi} \int_{1000\pi}^{4000\pi} \sqrt{(-\theta \sin \theta + \cos \theta)^2 + (\theta \cos \theta + \sin \theta)^2}\, d\theta \\
&= \frac{1}{2000\pi} \int_{1000\pi}^{4000\pi} \sqrt{\theta^2 + 1}\, d\theta \qquad \text{Integration tables, Formula 26} \\
&= \frac{1}{2000\pi}\left(\frac{1}{2}\right)\left[\theta \sqrt{\theta^2 + 1} + \ln \left|\theta + \sqrt{\theta^2 + 1}\right|\right]_{1000\pi}^{4000\pi} \\
&\approx 11,781 \text{ in.} \approx 982 \text{ ft.}
\end{aligned}
$$

For more information on the mathematics of recording tape, see "Tape Counters" by Richard L. Roth in *The American Mathematical Monthly,* Volume 99, Number 7.

The length of the tape in Example 5 can be approximated by adding the circumferences of circular pieces of tape. The smallest circle has a radius of 0.501, and the largest has a radius of 2.

$$
\begin{aligned}
s &\approx 2\pi(0.501) + 2\pi(0.502) + 2\pi(0.503) + \cdots + 2\pi(2.000) \\
&= \sum_{i=1}^{1500} 2\pi(0.5 + 0.001i) \\
&= 2\pi[1500(0.5) + 0.001(1500)(1501)/2] \\
&\approx 11,786 \text{ in.}
\end{aligned}
$$

Area of a Surface of Revolution

You can use the formula for the area of a surface of revolution in rectangular form to develop a formula for surface area in parametric form.

THEOREM 10.3   Area of Surface of Revolution

If a smooth curve C given by $x = f(t)$ and $y = g(t)$ does not cross itself on an interval $a \leq t \leq b$, then the area S of the surface of revolution formed by revolving C about the coordinate axes is given by the following.

1. $\displaystyle S = 2\pi \int_{a}^{b} g(t) \sqrt{\left(\frac{dx}{dt}\right)^{2} + \left(\frac{dy}{dt}\right)^{2}}\, dt$ Revolution about x-axis $g(t) \geq 0$

2. $\displaystyle S = 2\pi \int_{a}^{b} f(t) \sqrt{\left(\frac{dx}{dt}\right)^{2} + \left(\frac{dy}{dt}\right)^{2}}\, dt$ Revolution about y-axis $f(t) \geq 0$

EXAMPLE 6   Finding the Area of a Surface of Revolution

Let C be the arc of the circle

$$x^2 + y^2 = 9$$

from $(3, 0)$ to $\left(3/2, 3\sqrt{3}/2\right)$, as shown in Figure 10.17. Find the area of the surface formed by revolving C about the x-axis.

Solution   You can represent C parametrically by the equations

$$x = 3\cos t \quad \text{and} \quad y = 3\sin t, \qquad 0 \leq t \leq \pi/3.$$

(Note that you can determine the interval for t by observing that $t = 0$ when $x = 3$ and $t = \pi/3$ when $x = 3/2$.) On this interval, C is smooth and y is nonnegative, and you can apply Theorem 10.3 to obtain a surface area of

$$S = 2\pi \int_{0}^{\pi/3} (3\sin t)\sqrt{(-3\sin t)^2 + (3\cos t)^2}\, dt$$

$$= 6\pi \int_{0}^{\pi/3} \sin t \sqrt{9(\sin^2 t + \cos^2 t)}\, dt$$

$$= 6\pi \int_{0}^{\pi/3} 3\sin t\, dt$$

$$= -18\pi \left[\cos t\right]_{0}^{\pi/3}$$

$$= -18\pi \left(\frac{1}{2} - 1\right)$$

$$= 9\pi.$$

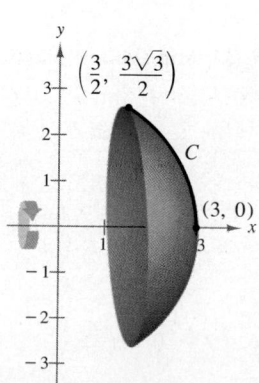

FIGURE 10.17

This surface of revolution has a surface area of 9π.

1-36 every 3rd, 43-48, 53

EXERCISES for Section 10.2

In Exercises 1–10, find dy/dx and d^2y/dx^2, and evaluate each at the specified value of the parameter.

Parametric equations	Point
1. $x = 2t$, $y = 3t - 1$	$t = 3$
2. $x = \sqrt{t}$, $y = 3t - 1$	$t = 1$
3. $x = t + 1$, $y = t^2 + 3t$	$t = -1$
4. $x = t^2 + 3t$, $y = t + 1$	$t = 0$
5. $x = 2\cos\theta$, $y = 2\sin\theta$	$\theta = \dfrac{\pi}{4}$
6. $x = \cos\theta$, $y = 3\sin\theta$	$\theta = 0$
7. $x = 2 + \sec\theta$, $y = 1 + 2\tan\theta$	$\theta = \dfrac{\pi}{6}$
8. $x = \sqrt{t}$, $y = \sqrt{t - 1}$	$t = 2$
9. $x = \cos^3\theta$, $y = \sin^3\theta$	$\theta = \dfrac{\pi}{4}$
10. $x = \theta - \sin\theta$, $y = 1 - \cos\theta$	$\theta = \pi$

In Exercises 11 and 12, find an equation of the tangent line at the specified points on the curve.

11. $x = 2\cot\theta$
$y = 2\sin^2\theta$

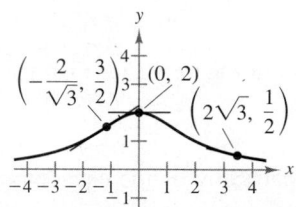

12. $x = 2 - 3\cos\theta$
$y = 3 + 2\sin\theta$

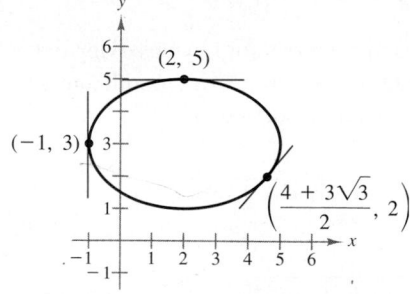

In Exercises 13–16, find an equation of the tangent line to the curve at the specified value of the parameter.

Parametric equations	Point
13. $x = 2t$, $y = t^2 - 1$	$t = 2$
14. $x = t - 1$, $y = \dfrac{1}{t} + 1$	$t = 1$
15. $x = t^2 - t + 2$, $y = t^3 - 3t$	$t = -1$
16. $x = 4\cos\theta$, $y = 3\sin\theta$	$\theta = \dfrac{3\pi}{4}$

In Exercises 17 and 18, find all points (if any) of horizontal and vertical tangency to the portion of the curve shown.

17. Involute of a circle:
$x = \cos\theta + \theta\sin\theta$
$y = \sin\theta - \theta\cos\theta$

18. $x = 2\theta$
$y = 2(1 - \cos\theta)$

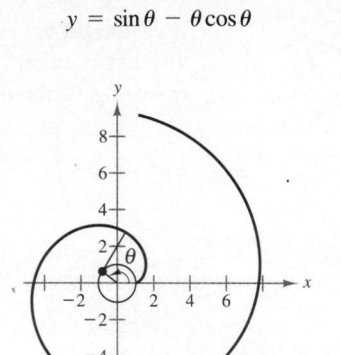

In Exercises 19–28, find all points (if any) of horizontal and vertical tangency to the curve.

19. $x = 1 - t$
$y = t^2$

20. $x = t + 1$
$y = t^2 + 3t$

21. $x = 1 - t$
$y = t^3 - 3t$

22. $x = t^2 - t + 2$
$y = t^3 - 3t$

23. $x = 3\cos\theta$
$y = 3\sin\theta$

24. $x = \cos\theta$
$y = 2\sin 2\theta$

25. $x = 4 + 2\cos\theta$
$y = -1 + \sin\theta$

26. $x = 4\cos^2\theta$
$y = 2\sin\theta$

27. $x = \sec\theta$
$y = \tan\theta$

28. $x = \cos^2\theta$
$y = \cos\theta$

Arc Length In Exercises 29–34, find the arc length of the curve on the specified interval.

Parametric equations	Interval
29. $x = e^{-t}\cos t$, $y = e^{-t}\sin t$	$0 \le t \le \dfrac{\pi}{2}$
30. $x = t^2$, $y = 4t^3 - 1$	$-1 \le t \le 1$
31. $x = t^2$, $y = 2t$	$0 \le t \le 2$
32. $x = \arcsin t$, $y = \ln\sqrt{1 - t^2}$	$0 \le t \le \dfrac{1}{2}$
33. $x = \sqrt{t}$, $y = 3t - 1$	$0 \le t \le 1$
34. $x = t$, $y = \dfrac{t^5}{10} + \dfrac{1}{6t^3}$	$1 \le t \le 2$

$$S = \int_a^b \sqrt{\left(\frac{dx}{dt}\right)^2 + \left(\frac{dy}{dt}\right)^2}$$

horiz.

$$A = 2\pi \int_a^b g(t)\sqrt{\left(\frac{dx}{dt}\right)^2 + \left(\frac{dy}{dt}\right)^2}$$

Arc Length In Exercises 35–38, find the arc length of the given curve on the interval $[0, 2\pi]$.

35. Perimeter of a hypocycloid: $x = a\cos^3 \theta$
$$y = a\sin^3 \theta$$

36. Circumference of a circle: $x = a\cos \theta$
$$y = a\sin \theta$$

37. One arch of a cycloid: $x = a(\theta - \sin \theta)$
$$y = a(1 - \cos \theta)$$

38. Involute of a circle: $x = \cos \theta + \theta \sin \theta$
$$y = \sin \theta - \theta \cos \theta$$

C **39.** *Path of a Projectile* The path of a projectile is modeled by the parametric equations

$$x = (90 \cos 30°)t \quad \text{and} \quad y = (90 \sin 30°)t - 16t^2$$

where x and y are measured in feet. Use a graphing utility to perform the following.
   a. Obtain a graph of the path of the projectile.
   b. Approximate the range of the projectile.
   c. Use Simpson's Rule with $n = 12$ to approximate the arc length of the path (the total distance traveled by the projectile).

C **40.** *Circumference of an Ellipse* Use a graphing utility and Simpson's Rule with $n = 10$ to approximate the circumference of the ellipse given by the parametric equations

$$x = 3\cos \theta \quad \text{and} \quad y = 4\sin \theta.$$

C **41.** *Folium of Descartes* Given the parametric equations

$$x = \frac{4t}{1 + t^3} \quad \text{and} \quad y = \frac{4t^2}{1 + t^3}$$

use a graphing utility to perform the following.
   a. Sketch the curve described by the parametric equations.
   b. Find the points of horizontal tangency to the curve.
   c. Use Simpson's Rule with $n = 10$ to approximate the arc length of the closed loop. (*Hint:* Because of symmetry, we need only integrate over the interval $0 \leq t \leq 1$.)

C **42.** *Essay*
   a. Use a graphing utility to obtain the graphs of the following two sets of parametric equations.

$$x = t - \sin t \qquad x = 2t - \sin(2t)$$
$$y = 1 - \cos t \qquad y = 1 - \cos(2t)$$
$$0 \leq t \leq 2\pi \qquad 0 \leq t \leq \pi$$

   b. Compare the graphs of the two sets of parametric equations in part a. If the curve represents the motion of a particle and t is time, what can you infer about the average speed of the particle on the paths represented by the two sets of parametric equations?
   c. Without graphing the curve, determine the time required for a particle to traverse the same path as in parts a and b if the path is modeled by

$$x = \tfrac{1}{2}t - \sin(\tfrac{1}{2}t) \quad \text{and} \quad y = 1 - \cos(\tfrac{1}{2}t).$$

Surface Area In Exercises 43–48, find the area of the surface generated by revolving the curve about the given axis.

43. $x = t, y = 2t, \quad 0 \leq t \leq 4$
   a. x-axis   **b.** y-axis

44. $x = t, y = 4 - 2t, \quad 0 \leq t \leq 2$
   a. x-axis   **b.** y-axis

45. $x = 4\cos \theta, y = 4\sin \theta, \quad 0 \leq \theta \leq \dfrac{\pi}{2}$
   y-axis

46. $x = t^3, y = t + 2, \quad 1 \leq t \leq 2$
   y-axis

47. $x = a\cos^3 \theta, y = a\sin^3 \theta, \quad 0 \leq \theta \leq \pi$
   x-axis

48. $x = a\cos \theta, y = b\sin \theta, \quad 0 \leq \theta \leq 2\pi$
   a. x-axis   **b.** y-axis

49. *Surface Area* A portion of a sphere of radius r is removed by cutting out a circular cone with its vertex at the center of the sphere. Find the surface area removed from the sphere if the vertex of the cone forms an angle of 2θ.

50. Use integration by substitution to show that if y is a continuous function of x on the interval $a \leq x \leq b$, where $x = f(t)$ and $y = g(t)$, then

$$\int_a^b y\, dx = \int_{t_1}^{t_2} g(t)\, |f'(t)|\, dt$$

where $f(t_1) = a, f(t_2) = b$, and both g and f' are continuous on $[t_1, t_2]$.

Area In Exercises 51–54, find the area of the indicated region. (Use the result of Exercise 50.)

51.

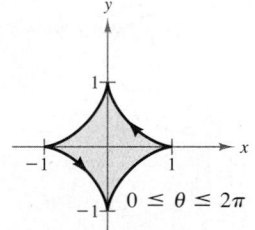

$$0 \leq \theta \leq 2\pi$$

52.

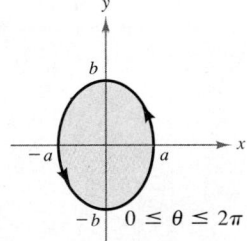

$$0 \leq \theta \leq 2\pi$$

53.

$$0 \leq \theta < \frac{\pi}{2}$$

54.

$$0 < \theta < \pi$$

Centroid **In Exercises 55 and 56, find the centroid of the region bounded by the graph of the parametric equations and the coordinate axes. (Use the result of Exercise 50.)**

55. $x = \sqrt{t}$

$y = 4 - t$

56. $x = \sqrt{4 - t}$

$y = \sqrt{t}$

Volume **In Exercises 57 and 58, find the volume of the solid formed by revolving the region bounded by the graphs of the given equations about the *x*-axis. (Use the result of Exercise 50.)**

57. $x = 3\cos\theta$

$y = 3\sin\theta$

58. $x = \cos\theta$

$y = 3\sin\theta$

C 59. *Tractrix* A person moves from the origin along the positive *y*-axis pulling a weight at the end of a 12-foot rope. Initially the weight is located at the point (12, 0).

a. In Exercise 65 of Section 7.4, it was shown that the path of the weight is modeled by the rectangular equation

$$y = -12\ln\left(\frac{12 - \sqrt{144 - x^2}}{x}\right) - \sqrt{144 - x^2}$$

where $0 < x \le 12$. Use a graphing utility to graph the path.

b. Use a graphing utility to graph the curve given by

$$x = 12\,\text{sech}\frac{t}{12} \quad \text{and} \quad y = t - 12\tanh\frac{t}{12}$$

where $0 \le t$. How does this graph compare with the graph generated in part a? Which graph (if any) do you think is a better representation of the path?

c. Use the parametric equations for the tractrix to verify that the distance from the *y*-intercept of the tangent line to the point of tangency is independent of the location of the point of tangency.

C 60. Use a graphing utility to graph the curve given by

$$x = \frac{1 - t^2}{1 + t^2}, \; y = \frac{2t}{1 + t^2}$$

for $-20 \le t \le 20$.

a. Describe the graph and confirm your result analytically.

b. Discuss the speed at which the curve is traced as *t* increases from -20 to 20.

True or False **In Exercises 61–64, determine whether the statement is true or false. If it is false, explain why or given an example that shows it is false.**

61. If $x = f(t)$ and $y = g(t)$, then $d^2y/dx^2 = g''(t)/f''(t)$.

62. The curve given by $x = t^3$, $y = t^2$ has a horizontal tangent at the origin because $dy/dt = 0$ when $t = 0$.

63. A curve may have more than one tangent line at a given point.

64. If $f'(t)$ and $g'(t)$ are positive for all real values of *t*, then the curve given by $x = f(t)$, $y = g(t)$ does not cross itself.

65. Let $x = g(t)$ and $y = f(t)$ be the parametric equations for a planar curve *C*. Interpret the Extended Mean Value Theorem in this case.

66. For the function $r = f(\theta)$, the *relative extrema* of *r* yield points on the curve whose distances from the pole are relative minima or maxima. Find the relative extrema of $r = f(\theta) = 1 + 2\cos\theta$.

67. Consider the tractrix with parametric equations

$$x(t) = \text{sech}(t) \quad \text{and} \quad y(t) = t - \tanh(t)$$

for $t \ge 0$. Graph this curve. Let $(0, y_1) = B$ be the point where the tangent line to the curve at *A* intersects the *y*-axis. Show that the length of *AB* is independent of *A*.

SECTION Polar Coordinates and Polar Graphs

10.3 Polar Coordinates • Coordinate Conversion • Polar Graphs • Slope and Tangent Lines • Special Polar Graphs

Polar Coordinates

So far, we have been representing graphs as collections of points (x, y) in the rectangular coordinate system. The corresponding equations for these graphs have been in either rectangular or parametric form. In this section we introduce a coordinate system called the **polar coordinate system.**

To form the polar coordinate system in the plane, we fix a point O, called the **pole** (or **origin**), and construct from O an initial ray called the **polar axis,** as shown in Figure 10.18. Then each point P in the plane can be assigned **polar coordinates** (r, θ), as follows.

$r = directed\ distance$ from O to P

$\theta = directed\ angle$, counterclockwise from polar axis to segment $\overline{OP}$

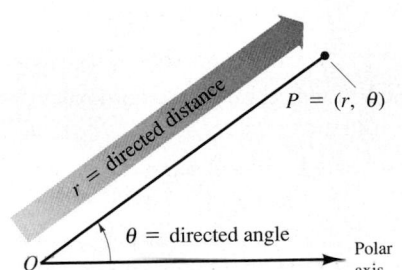

FIGURE 10.18
Polar coordinates

Figure 10.19 shows three points in the polar coordinate system. Notice that in this system, it is convenient to locate points with respect to a grid of concentric circles intersected by **radial lines** through the pole.

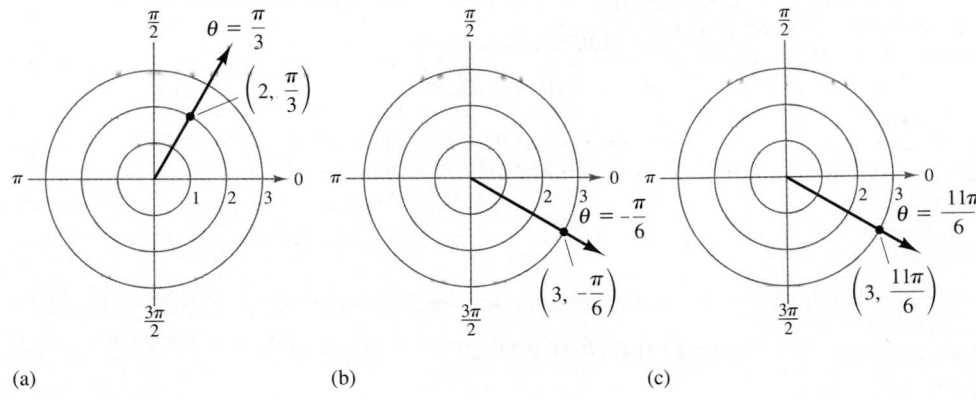

(a)   (b)   (c)

FIGURE 10.19

With rectangular coordinates, each point (x, y) has a unique representation. This is not true with polar coordinates. For instance, the coordinates (r, θ) and $(r, 2\pi + \theta)$ represent the same point (see Figure 10.19b and c). Also, because r is a *directed distance*, the coordinates (r, θ) and $(-r, \theta + \pi)$ represent the same point. In general, the point (r, θ) can be written as

$$(r, \theta) = (r, \theta + 2n\pi) \quad \text{or} \quad (r, \theta) = (-r, \theta + (2n + 1)\pi)$$

where n is any integer. Moreover, the pole is represented by $(0, \theta)$, where θ is any angle.

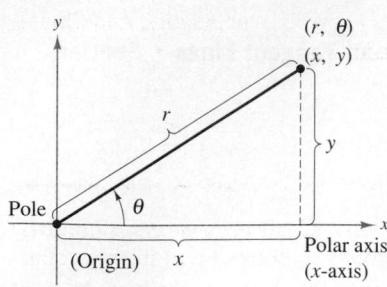

FIGURE 10.20
Relating polar and rectangular coordinates.

Coordinate Conversion

To establish the relationship between polar and rectangular coordinates, let the polar axis coincide with the positive x-axis and the pole with the origin, as shown in Figure 10.20. Because (x, y) lies on a circle of radius r, it follows that $r^2 = x^2 + y^2$. Moreover, for $r > 0$, the definition of the trigonometric functions implies that

$$\tan\theta = \frac{y}{x}, \quad \cos\theta = \frac{x}{r}, \quad \text{and} \quad \sin\theta = \frac{y}{r}.$$

If $r < 0$, you can show that the same relationships hold.

THEOREM 10.4   Coordinate Conversion

The polar coordinates (r, θ) of a point are related to the rectangular coordinates (x, y) of the point as follows.

1. $x = r\cos\theta$ **2.** $\tan\theta = \dfrac{y}{x}$

  $y = r\sin\theta$ $r^2 = x^2 + y^2$

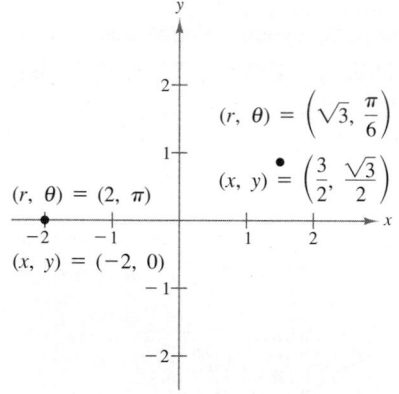

FIGURE 10.21
To convert from polar to rectangular coordinates, let $x = r\cos\theta$ and $y = r\sin\theta$.

EXAMPLE 1   Polar-to-Rectangular Conversion

a. For the point $(r, \theta) = (2, \pi)$,

$$x = r\cos\theta = 2\cos\pi = -2 \quad \text{and} \quad y = r\sin\theta = 2\sin\pi = 0.$$

Thus, the rectangular coordinates are $(x, y) = (-2, 0)$.

b. For the point $(r, \theta) = (\sqrt{3}, \pi/6)$,

$$x = \sqrt{3}\cos\frac{\pi}{6} = \frac{3}{2} \quad \text{and} \quad y = \sqrt{3}\sin\frac{\pi}{6} = \frac{\sqrt{3}}{2}.$$

Thus, the rectangular coordinates are $(x, y) = \left(3/2, \sqrt{3}/2\right)$.

(See Figure 10.21.)

EXAMPLE 2   Rectangular-to-Polar Conversion

a. For the second quadrant point $(x, y) = (-1, 1)$,

$$\tan\theta = \frac{y}{x} = -1 \quad \rightarrow \quad \theta = \frac{3\pi}{4}.$$

Because θ was chosen to be in the same quadrant as (x, y), you should use a positive value of r.

$$r = \sqrt{x^2 + y^2} = \sqrt{(-1)^2 + (1)^2} = \sqrt{2}$$

This implies that *one* set of polar coordinates is $(r, \theta) = \left(\sqrt{2}, 3\pi/4\right)$.

b. Because the point $(x, y) = (0, 2)$ lies on the positive y-axis, we choose $\theta = \pi/2$ and $r = 2$, and one set of polar coordinates is $(r, \theta) = (2, \pi/2)$.

(See Figure 10.22.)

FIGURE 10.22
To convert from rectangular to polar coordinates, let $\tan\theta = y/x$ and $r = \sqrt{x^2 + y^2}$.

Polar Graphs

One way to sketch the graph of a polar equation is to convert to rectangular coordinates and then sketch the graph of the rectangular equation.

EXAMPLE 3 Graphing Polar Equations

Describe the graphs of the following polar equations. Confirm each description by converting to a rectangular equation.

a. $r = 2$ **b.** $\theta = \dfrac{\pi}{3}$ **c.** $r = \sec\theta$

Solution (See Figure 10.23.)

a. The graph of the polar equation $r = 2$ consists of all points that are two units from the pole. In other words, this graph is a circle centered at the origin with a radius of 2. You can confirm this by using the relationship $r^2 = x^2 + y^2$ to obtain the rectangular equation

$$x^2 + y^2 = 2^2. \qquad \text{Rectangular equation}$$

b. The graph of the polar equation $\theta = \pi/3$ consists of all points on the line that makes an angle of $\pi/3$ with the positive x-axis. You can confirm this by using the relationship $\tan\theta = y/x$ to obtain the rectangular equation

$$y = \sqrt{3}x. \qquad \text{Rectangular equation}$$

c. The graph of the polar equation $r = \sec\theta$ is not evident by simple inspection, so you can begin by converting to rectangular form using the relationship $r\cos\theta = x$.

$$r = \sec\theta \qquad \text{Polar equation}$$
$$r\cos\theta = 1$$
$$x = 1 \qquad \text{Rectangular equation}$$

From the rectangular equation, you can see that the graph is a vertical line.

(a) Circle: $r = 2$

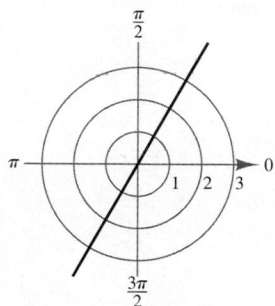

(b) Radial line: $\theta = \dfrac{\pi}{3}$

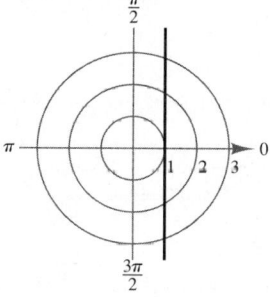

(c) Vertical line: $r = \sec\theta$

FIGURE 10.23

TECHNOLOGY Sketching the graphs of complicated polar equations *by hand* can be tedious. With technology, however, the task is not difficult. If you have access to a graphing utility that has a polar mode, try using it to sketch the graphs in the exercise set. If your graphing utility doesn't have a polar mode, but *does have* a parametric mode, you can sketch the graph of $r = f(\theta)$ by writing the equation as

$$x = f(\theta)\cos\theta \quad \text{and} \quad y = f(\theta)\sin\theta.$$

For instance, the graph of $r = \frac{1}{2}\theta$ shown in Figure 10.24 was produced with a graphing calculator in parametric mode. To sketch the graph, we entered the parametric equations

$$x = \frac{1}{2}\theta\cos\theta \quad \text{and} \quad y = \frac{1}{2}\theta\sin\theta$$

and let the values of θ vary from -4π to 4π. This curve is of the form $r = a\theta$ and is called a **spiral of Archimedes.**

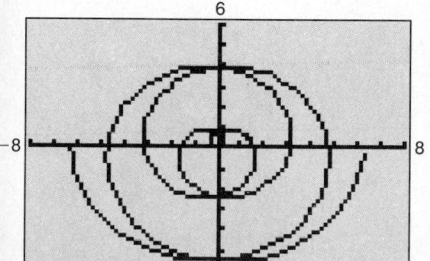

FIGURE 10.24
Spiral of Archimedes

EXAMPLE 4   Sketching a Polar Graph

Use a graphing utility to sketch the graph of $r = 2 \cos 3\theta$.

Solution   Begin by writing the polar equation in parametric form.

$$x = 2 \cos 3\theta \cos \theta \quad \text{and} \quad y = 2 \cos 3\theta \sin \theta$$

After some experimentation, you will find that the entire curve, which is called a **rose curve,** can be sketched by letting θ vary from 0 to π, as shown in Figure 10.25. If you try duplicating this graph with a graphing utility, you will find that by letting θ vary from 0 to 2π, you will actually trace the entire curve *twice.*

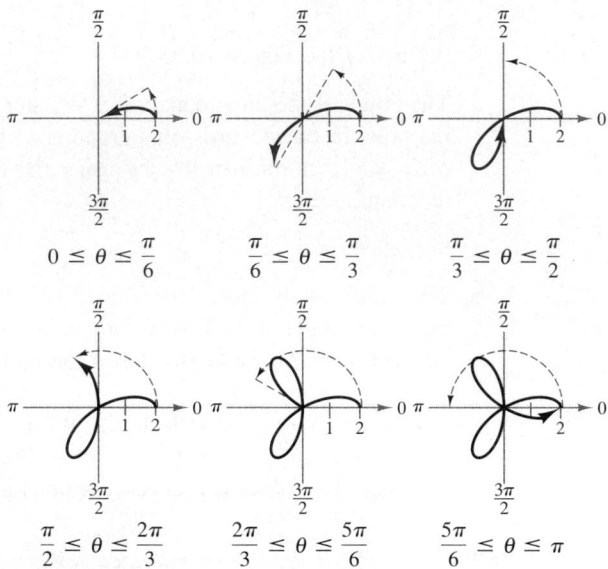

$$0 \le \theta \le \frac{\pi}{6} \qquad \frac{\pi}{6} \le \theta \le \frac{\pi}{3} \qquad \frac{\pi}{3} \le \theta \le \frac{\pi}{2}$$

$$\frac{\pi}{2} \le \theta \le \frac{2\pi}{3} \qquad \frac{2\pi}{3} \le \theta \le \frac{5\pi}{6} \qquad \frac{5\pi}{6} \le \theta \le \pi$$

FIGURE 10.25

If you have access to a graphing utility, try using it to experiment with other rose curves (they are of the form $r = a \cos n\theta$ or $r = a \sin n\theta$). For instance, Figure 10.26 shows the graphs of two other rose curves.

(a)

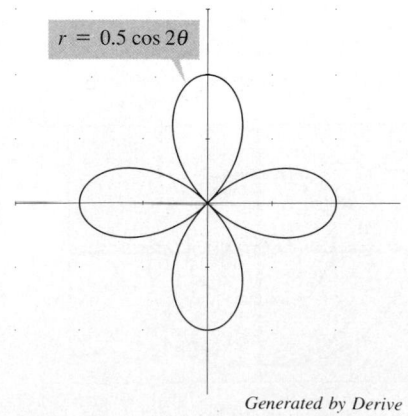

$r = 0.5 \cos 2\theta$

Generated by Derive

(b) Each tic corresponds to 0.25 units.

FIGURE 10.26
Rose curves

Slope and Tangent Lines

To find the slope of a tangent line to a polar graph, consider a differentiable function given by $r = f(\theta)$. To convert to polar form, use the parametric equations

$$x = r\cos\theta = f(\theta)\cos\theta \quad \text{and} \quad y = r\sin\theta = f(\theta)\sin\theta.$$

Using the parametric form of dy/dx given in Theorem 10.1, you have

$$\frac{dy}{dx} = \frac{dy/d\theta}{dx/d\theta} = \frac{f(\theta)\cos\theta + f'(\theta)\sin\theta}{-f(\theta)\sin\theta + f'(\theta)\cos\theta}$$

which establishes the following theorem.

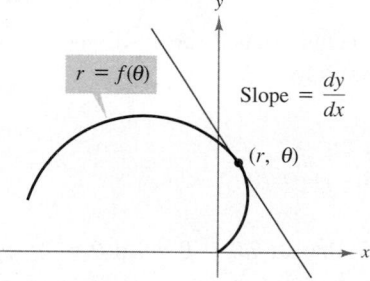

FIGURE 10.27
Tangent line to polar curve.

THEOREM 10.5   Slope in Polar Form

If f is a differentiable function of θ, then the *slope* of the tangent line to the graph of $r = f(\theta)$ at the point (r, θ) is

$$\frac{dy}{dx} = \frac{dy/d\theta}{dx/d\theta} = \frac{f(\theta)\cos\theta + f'(\theta)\sin\theta}{-f(\theta)\sin\theta + f'(\theta)\cos\theta}$$

provided that $dx/d\theta \neq 0$ at (r, θ). (See Figure 10.27.)

From Theorem 10.5, you can make the following observations.

1. Solutions to $\dfrac{dy}{d\theta} = 0$ yield horizontal tangents, provided that $\dfrac{dx}{d\theta} \neq 0$.

2. Solutions to $\dfrac{dx}{d\theta} = 0$ yield vertical tangents, provided that $\dfrac{dy}{d\theta} \neq 0$.

If $dy/d\theta$ and $dx/d\theta$ are *simultaneously* 0, then no conclusion can be drawn about tangent lines.

EXAMPLE 5   Finding Horizontal and Vertical Tangent Lines

Find the horizontal and vertical tangent lines of $r = \sin\theta$, $0 \le \theta \le \pi$.

Solution   Begin by writing the equation in parametric form.

$$x = r\cos\theta = \sin\theta\cos\theta$$

and

$$y = r\sin\theta = \sin\theta\sin\theta = \sin^2\theta$$

Next, differentiate x and y with respect to θ and set each derivative equal to 0.

$$\frac{dx}{d\theta} = \cos^2\theta - \sin^2\theta = \cos 2\theta = 0 \quad \rightarrow \quad \theta = \frac{\pi}{4}, \frac{3\pi}{4}$$

$$\frac{dy}{d\theta} = 2\sin\theta\cos\theta = \sin 2\theta = 0 \quad \rightarrow \quad \theta = 0, \frac{\pi}{2}$$

Thus, the graph has vertical tangent lines at $\left(\sqrt{2}/2, \ \pi/4\right)$ and $\left(\sqrt{2}/2, \ 3\pi/4\right)$, and it has horizontal tangent lines at $(0, 0)$ and $(1, \pi/2)$, as shown in Figure 10.28.

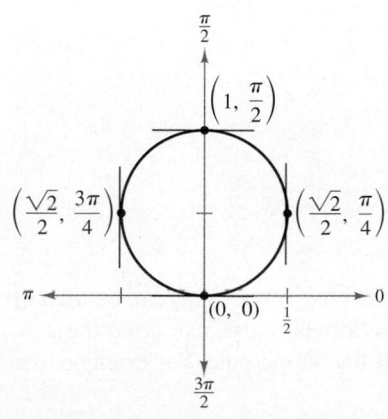

FIGURE 10.28
Horizontal and vertical tangent lines of $r = \sin\theta$.

EXAMPLE 6   Finding Horizontal and Vertical Tangent Lines

Find the horizontal and vertical tangent lines of

$$r = \sin\theta, \ 0 \le \theta \le \pi.$$

Solution   Using $y = r\sin\theta$, differentiate and set $dy/d\theta$ equal to 0.

$$y = r\sin\theta = 2(1 - \cos\theta)\sin\theta$$

$$\frac{dy}{d\theta} = 2[(1 - \cos\theta)(\cos\theta) + \sin\theta(\sin\theta)]$$

$$= -2(2\cos\theta + 1)(\cos\theta - 1) = 0$$

Thus, $\cos\theta = -\frac{1}{2}$ and $\cos\theta = 1$, and you can conclude that $dy/d\theta = 0$ when $\theta = 2\pi/3, 4\pi/3$, and 0. Similarly, using $x = r\cos\theta$, you have

$$x = r\cos\theta = 2\cos\theta - 2\cos^2\theta$$

$$\frac{dx}{d\theta} = -2\sin\theta + 4\cos\theta\sin\theta = 2\sin\theta(2\cos\theta - 1) = 0.$$

Thus, $\sin\theta = 0$ or $\cos\theta = \frac{1}{2}$, and you can conclude that $dx/d\theta = 0$ when $\theta = 0, \pi, \pi/3$, and $5\pi/3$. From these results, and from the graph shown in Figure 10.29, you can conclude that the graph has horizontal tangents at $(3, 2\pi/3)$ and $(3, 4\pi/3)$, and has vertical tangents at $(1, \pi/3), (1, 5\pi/3)$, and $(4, \pi)$. This graph is called a **cardioid**. Note that both derivatives ($dy/d\theta$ and $dx/d\theta$) are 0 when $\theta = 0$. Using this information alone, you don't know whether the graph has a horizontal or vertical tangent line at the pole. From Figure 10.29, however, you can see that the graph has a cusp at the pole.

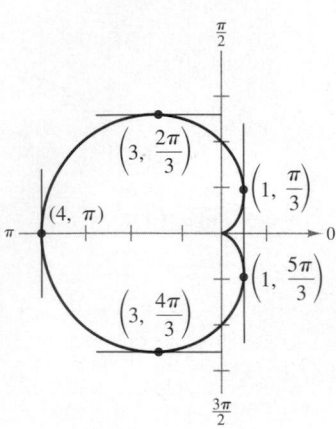

FIGURE 10.29
Horizontal and vertical tangent lines of $r = 2(1 - \cos\theta)$.

Theorem 10.5 has an important consequence. Suppose the graph of $r = f(\theta)$ passes through the pole when $\theta = \alpha$ and $f'(\alpha) \ne 0$. Then the formula for dy/dx simplifies as follows.

$$\frac{dy}{dx} = \frac{f'(\alpha)\sin\alpha + f(\alpha)\cos\alpha}{f'(\alpha)\cos\alpha - f(\alpha)\sin\alpha} = \frac{f'(\alpha)\sin\alpha + 0}{f'(\alpha)\cos\alpha - 0} = \frac{\sin\alpha}{\cos\alpha} = \tan\alpha$$

Thus, the line $\theta = \alpha$ is tangent to the graph at the pole, $(0, \alpha)$.

THEOREM 10.6   Tangent Lines at the Pole

If $f(\alpha) = 0$ and $f'(\alpha) \ne 0$, then the line $\theta = \alpha$ is tangent at the pole to the graph of $r = f(\theta)$.

Theorem 10.6 is useful because it states that the zeros of $r = f(\theta)$ can be used to find the tangent lines at the pole. Note that because a polar curve can cross the pole more than once, it can have more than one tangent line at the pole. For example, the rose curve

$$f(\theta) = 2\cos 3\theta$$

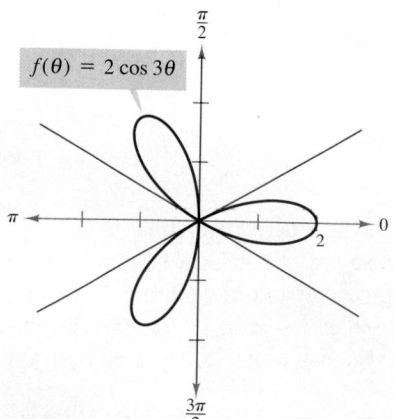

FIGURE 10.30
This rose curve has three tangent lines at the pole.

has three tangent lines at the pole, as shown in Figure 10.30. For this curve, $f(\theta) = 2\cos 3\theta$ is 0 when θ is $\pi/6, \pi/2$, and $5\pi/6$. Moreover, the derivative $f'(\theta) = -6\sin 3\theta$ is not 0 for these values of θ.

Special Polar Graphs

Several important types of graphs have equations that are simpler in polar form than in rectangular form. For example, the polar equation of a circle having a radius of a and centered at the origin is simply $r = a$. Later in the text you will come to appreciate this benefit. For now, we summarize some other types of graphs that have simpler equations in polar form. (Conics are considered in Section 10.5).

Limaçons

$r = a \pm b \cos \theta$

$r = a \pm b \sin \theta$

$(0 < a, 0 < b)$

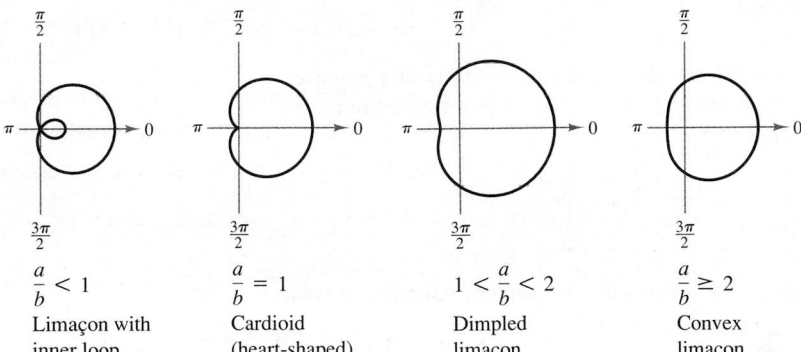

$\dfrac{a}{b} < 1$	$\dfrac{a}{b} = 1$	$1 < \dfrac{a}{b} < 2$	$\dfrac{a}{b} \geq 2$
Limaçon with inner loop	Cardioid (heart-shaped)	Dimpled limaçon	Convex limaçon

Rose Curves

n petals if n is odd

$2n$ petals if n is even

$(n \geq 2)$

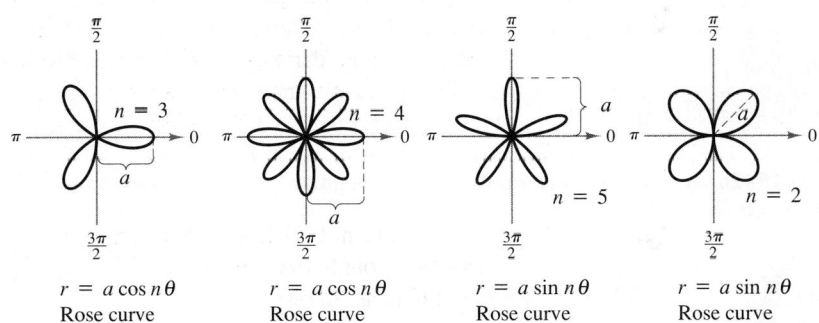

$r = a \cos n\theta$	$r = a \cos n\theta$	$r = a \sin n\theta$	$r = a \sin n\theta$
Rose curve	Rose curve	Rose curve	Rose curve

Circles and Lemniscates

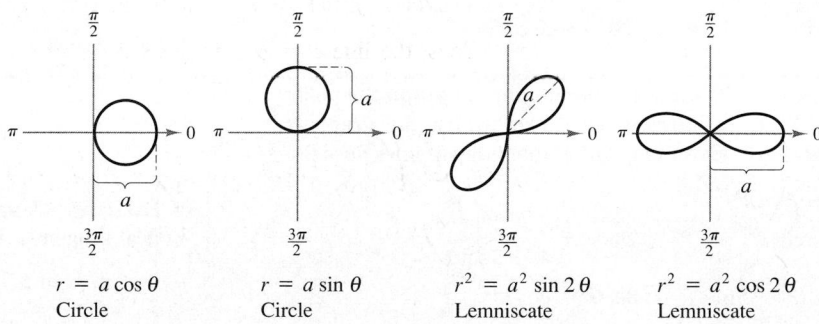

$r = a \cos \theta$	$r = a \sin \theta$	$r^2 = a^2 \sin 2\theta$	$r^2 = a^2 \cos 2\theta$
Circle	Circle	Lemniscate	Lemniscate

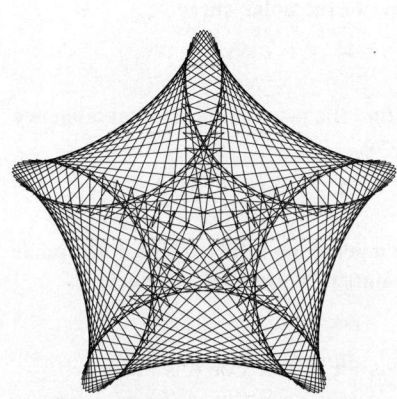

"The Rose"

TECHNOLOGY The rose curves described above are of the form $r = a \cos n\theta$ or $r = a \sin n\theta$, where n is a positive integer that is greater than or equal to 2. Try using a graphing utility to sketch the graph of $r = a \cos n\theta$ or $r = a \sin n\theta$ for some noninteger values of n. Are these graphs also rose curves? For example, try sketching the graph of $r = \cos \frac{2}{3}\theta$, $0 \leq \theta \leq 6\pi$.

For more information on rose curves and related curves, see the article "A Rose is a Rose . . ." by Peter M. Maurer in the August–September, 1987 issue of *The American Mathematical Monthly*. (The computer-generated graph at the left is the result of an algorithm that Maurer calls "The Rose.")

mult. of 3

EXERCISES for Section 10.3

TECHNOLOGY
Laboratory Guide
Lab 10.3

In Exercises 1–6, plot the point in polar coordinates and find the corresponding rectangular coordinates for the point.

1. $(4, \pi/2)$

2. $(-1, 5\pi/4)$

3. $(-4, -\pi/3)$

4. $(0, -7\pi/6)$

5. $(\sqrt{2}, 2.36)$

6. $(-3, -1.57)$

In Exercises 7–10, the rectangular coordinates of a point are given. Plot the point and find *two* sets of polar coordinates for the point for $0 \le \theta < 2\pi$.

7. $(1, 1)$

8. $(0, -5)$

9. $(-3, 4)$

10. $(3, -1)$

In Exercises 11–18 convert the rectangular equation to polar form.

11. $x^2 + y^2 = a^2$

12. $x^2 + y^2 - 2ax = 0$

13. $y = 4$

14. $x = 10$

15. $3x - y + 2 = 0$

16. $xy = 4$

17. $y^2 = 9x$

18. $(x^2 + y^2)^2 - 9(x^2 - y^2) = 0$

In Exercises 19–26, convert the polar equation to rectangular form and sketch its graph.

19. $r = 3$

20. $r = -2$

21. $r = \sin\theta$

22. $r = 3\cos\theta$

23. $r = \theta$

24. $\theta = \dfrac{5\pi}{6}$

25. $r = 3\sec\theta$

26. $r = 2\csc\theta$

C **In Exercises 27–30, use a graphing utility to graph the polar equation. Find an interval for θ over which the graph is traced *only once*. How many horizontal tangent lines does the graph have?**

27. $r = 6\cos\theta$

28. $r = \dfrac{2}{4 - 3\sin\theta}$

29. $r = 3(2 - \sin\theta)$

30. $r = \cos 2\theta$

31. Convert the equation

$$r = 2(h\cos\theta + k\sin\theta)$$

to a rectangular form and verify that it is the equation of a circle. Find the radius and the rectangular coordinates of the center of the circle.

32. Verify that the distance between the two points (r_1, θ_1) and (r_2, θ_2) in polar coordinates is

$$d = \sqrt{r_1^2 + r_2^2 - 2r_1 r_2 \cos(\theta_1 - \theta_2)}.$$

In Exercises 33–36, use the result of Exercise 32 to approximate the distance between the two points in polar coordinates.

33. $\left(4, \dfrac{2\pi}{3}\right)$, $\left(2, \dfrac{\pi}{6}\right)$

34. $\left(10, \dfrac{7\pi}{6}\right)$, $(3, \pi)$

35. $(2, 0.5)$, $(7, 1.2)$

36. $(4, 2.5)$, $(12, 1)$

In Exercises 37 and 38, find dy/dx and the slope of the tangent lines shown on the graph of the polar equation.

37. $r = 2 + 3\sin\theta$

38. $r = 2(1 - \sin\theta)$

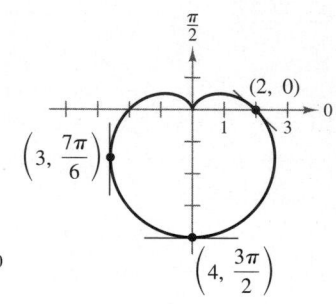

In Exercises 39–42, find dy/dx and the slope of the graph of the polar curve at the given value of θ.

Polar equation	Value of θ
39. $r = 3(1 - \cos\theta)$	$\theta = \dfrac{\pi}{2}$
40. $r = 3 - 2\cos\theta$	$\theta = 0$
41. $r = 3\sin\theta$	$\theta = \dfrac{\pi}{3}$
42. $r = 4$	$\theta = \dfrac{\pi}{4}$

In Exercises 43 and 44, find the points of horizontal and vertical tangency (if any) to the polar curve.

43. $r = 1 + \sin\theta$

44. $r = a\sin\theta$

In Exercises 45 and 46, find the points of horizontal tangency (if any) to the polar curve.

45. $r = 2\csc\theta + 3$

46. $r = a\sin\theta\cos^2\theta$

C **In Exercises 47–50, use a graphing utility to graph the polar equation and find all points of horizontal tangency.**

47. $r = 4\sin\theta\cos^2\theta$

48. $r = 3\cos 2\theta\sec\theta$

49. $r = 2\csc\theta + 5$

50. $r = 2\cos(3\theta - 2)$

In Exercises 51–56, sketch the graph of the polar equation and find the tangents at the pole.

51. $r = 3 \sin \theta$ **52.** $r = 3(1 - \cos \theta)$

53. $r = 2 \cos 3\theta$ **54.** $r = -\sin 5\theta$

55. $r = 3 \sin 2\theta$ **56.** $r = 3 \cos 2\theta$

In Exercises 57–64, sketch the graph of the polar equation.

57. $r^2 = 4 \cos 2\theta$ **58.** $r^2 = 4 \sin \theta$

59. $r = 3 - 2 \cos \theta$ **60.** $r = 5 - 4 \sin \theta$

61. $r = 3 \csc \theta$ **62.** $r = \dfrac{6}{2 \sin \theta - 3 \cos \theta}$

63. $r = 2\theta$ **64.** $r = \dfrac{1}{\theta}$

C **In Exercises 65–72, use a graphing utility to graph the polar equation. Find an interval for θ over which the graph is traced *only once*.**

65. $r = 3 - 4 \cos \theta$ **66.** $r = 2(1 - 2 \sin \theta)$

67. $r = 2 + \sin \theta$ **68.** $r = 4 + 3 \cos \theta$

69. $r = 2 \cos\left(\dfrac{3\theta}{2}\right)$ **70.** $r = 3 \sin\left(\dfrac{5\theta}{2}\right)$

71. $r^2 = 4 \sin 2\theta$ **72.** $r^2 = 3 \cos 4\theta$

C **In Exercises 73–76, use a graphing utility to graph the equation and show that the indicated line is an asymptote of the graph.**

	Name of graph	Polar equation	Asymptote
73.	Conchoid	$r = 2 - \sec \theta$	$x = -1$
74.	Conchoid	$r = 2 + \csc \theta$	$y = 1$
75.	Hyperbolic spiral	$r = \dfrac{2}{\theta}$	$y = 2$
76.	Strophoid	$r = 2 \cos 2\theta \sec \theta$	$x = -2$

77. Sketch the graph of $r = 4 \sin \theta$ over each of the following intervals.

a. $0 \le \theta \le \dfrac{\pi}{2}$ **b.** $\dfrac{\pi}{2} \le \theta \le \pi$

c. $-\dfrac{\pi}{2} \le \theta \le \dfrac{\pi}{2}$ **d.** $\dfrac{\pi}{4} \le \theta \le \dfrac{3\pi}{4}$

C **78.** Use a graphing utility to graph the polar equation

$$r = 6[1 + \cos(\theta - \phi)]$$

for
a. $\phi = 0$. **b.** $\phi = \pi/4$. **c.** $\phi = \pi/2$. Use these graphs to describe the effect of the angle ϕ. Write the equation as a function of $\sin \theta$ for part c.

79. Verify that if the curve whose polar equation is $r = f(\theta)$ is rotated about the pole through an angle ϕ, then an equation for the rotated curve is $r = f(\theta - \phi)$.

80. If the polar form of an equation for a curve is $r = f(\sin \theta)$, show that the form becomes

a. $r = f(-\cos \theta)$ if the curve is rotated counterclockwise $\pi/2$ radians about the pole.

b. $r = f(-\sin \theta)$ if the curve is rotated counterclockwise π radians about the pole.

c. $r = f(\cos \theta)$ if the curve is rotated counterclockwise $3\pi/2$ radians about the pole.

In Exercises 81–84, use the results of Exercises 79 and 80.

C **81.** Write an equation for the limaçon $r = 2 - \sin \theta$ after it has been rotated by the given amount. Verify the result by using a graphing utility to graph the rotated limaçon.

a. $\pi/4$ **b.** $\pi/2$

c. π **d.** $3\pi/2$

C **82.** Write an equation for the rose curve $r = 2 \sin 2\theta$ after it has been rotated by the given amount. Verify the result by using a graphing utility to graph the rotated rose curve.

a. $\pi/6$ **b.** $\pi/2$

c. $2\pi/3$ **d.** π

83. Sketch the graph of each equation.

a. $r = 1 - \sin \theta$ **b.** $r = 1 - \sin\left(\theta - \dfrac{\pi}{4}\right)$

84. Sketch the graph of each equation.

a. $r = 3 \sec \theta$

b. $r = 3 \sec\left(\theta - \dfrac{\pi}{4}\right)$

c. $r = 3 \sec\left(\theta + \dfrac{\pi}{3}\right)$

True or False **In Exercises 85–90, determine whether the statement is true or false. If it is false, explain why or give an example that shows it is false.**

85. If (r_1, θ_1) and (r_2, θ_2) represent the same point in the polar coordinate system, then $|r_1| = |r_2|$.

86. If (r, θ_1) and (r, θ_2) represent the same point in the polar coordinate system, then $\theta_1 = \theta_2 + 2\pi n$, for some integer n.

87. If $x \ge 0$, then the point (x, y) in the rectangular coordinate system can be represented by (r, θ) in the polar coordinate system, where $r = \sqrt{x^2 + y^2}$ and $\theta = \arctan(y/x)$.

88. The polar equations $r = \sin 2\theta$ and $r = -\sin 2\theta$ have the same graph.

89. If the graph of $r = f(\theta)$ has a vertical tangent at (r_1, θ_1), then $f'(\theta_1)$ is undefined.

90. The graph of $r = \sin \theta$ is a circle of radius 1.

C **91.** *The Butterfly Curve*   Use a computer graphing utility to produce the curve shown to the right. The curve is given by

$$r = e^{\cos\theta} - 2\cos 4\theta + \sin^5\frac{\theta}{12}.$$

FOR FURTHER INFORMATION   For more information on this curve, see the article "The Butterfly Curve" by Temple H. Fay in the May, 1989 issue of the *American Mathematical Monthly*. Over what interval must θ vary to produce this curve?

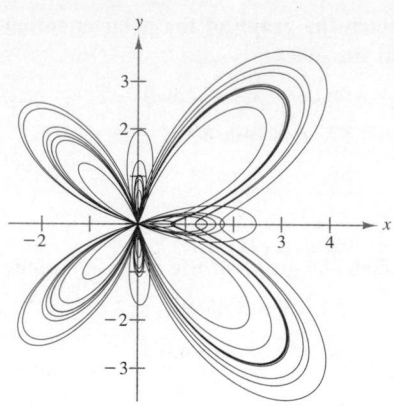

Anamorphic Art   In Exercises 92–95, use the anamorphic transformations

$$r = y + 16 \quad\text{and}\quad \theta = -\frac{\pi}{8}x, \qquad -\frac{3\pi}{4} \le \theta \le \frac{3\pi}{4}$$

to sketch the transformed polar image of the rectangular graph. When the reflection (in a cylindrical mirror centered at the pole) of each polar image is viewed from the polar axis, the viewer will see the original rectangular image.

92. $y = 3$ **93.** $x = 2$

94. $y = x + 5$ **95.** $x^2 + (y - 5)^2 = 5^2$

This example of anamorphic art is from the Manchester Museum of Science and Technology. When the reflection of transformed "polar painting" is viewed in the cylindrical mirror, the viewer sees a recognizable "rectangular painting."

FOR FURTHER INFORMATION   For more information on anamorphic art, see the article "Anamorphisms" by Philip Hickin in the July, 1992 issue of the *Mathematical Gazette*.

10.4

Area of a Polar Region • Points of Intersection of Polar Graphs • Arc Length in Polar Form • Area of a Surface of Revolution

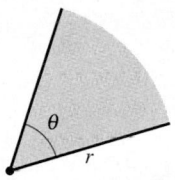

FIGURE 10.31

Area of a Polar Region

The development of a formula for the area of a polar region parallels that for the area of a region in the rectangular coordinate system, but uses *sectors* of a circle instead of rectangles as the basic element of area. In Figure 10.31, note that the area of a circular sector of radius r is given by $\frac{1}{2}\theta r^2$, provided that θ is measured in radians.

Consider the function given by $r = f(\theta)$, where f is continuous and nonnegative in the interval given by $\alpha \le \theta \le \beta$. The region bounded by the graph of f and the radial lines $\theta = \alpha$ and $\theta = \beta$ is shown in Figure 10.32. To find the area of this region, partition the interval $[\alpha, \beta]$ into n equal subintervals,

$$\alpha = \theta_0 < \theta_1 < \theta_2 < \cdots < \theta_{n-1} < \theta_n = \beta.$$

Then, approximate the area of the region by the sum of the areas of the n sectors.

Radius of ith sector $= f(\theta_i)$

Central angle of ith sector $= \dfrac{\beta - \alpha}{n} = \Delta\theta$

$$A \approx \sum_{i=1}^{n}\left(\frac{1}{2}\right)\Delta\theta f(\theta_i)^2$$

Taking the limit as $n \to \infty$ produces

$$A = \lim_{n\to\infty}\frac{1}{2}\sum_{i=1}^{n} f(\theta_i)^2 \Delta\theta = \frac{1}{2}\int_{\alpha}^{\beta}[f(\theta)^2]\,d\theta$$

which leads to the following theorem.

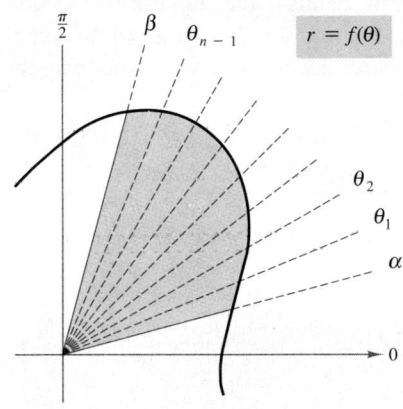

THEOREM 10.7   Area in Polar Coordinates

If f is continuous and nonnegative on the interval $[\alpha, \beta]$, then the area of the region bounded by the graph of $r = f(\theta)$ between the radial lines $\theta = \alpha$ and $\theta = \beta$ is given by

$$A = \frac{1}{2}\int_{\alpha}^{\beta}[f(\theta)]^2\,d\theta = \frac{1}{2}\int_{\alpha}^{\beta} r^2\,d\theta.$$

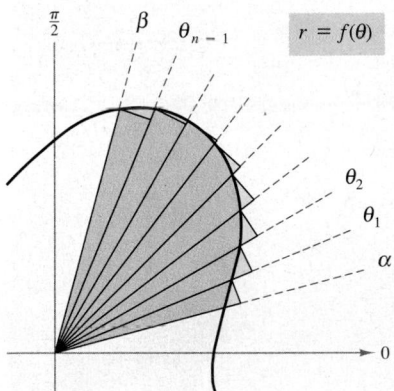

FIGURE 10.32

REMARK   You can use the same formula to find the area of a region bounded by the graph of a continuous *nonpositive* function. However, the formula is not necessarily valid if f takes on both positive *and* negative values in the interval $[\alpha, \beta]$.

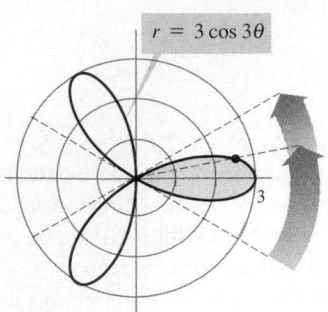

FIGURE 10.33
One petal of the rose curve lies between the radial lines $\theta = -\pi/6$ and $\theta = \pi/6$.

EXAMPLE 1   Finding the Area of a Polar Region

Find the area of *one petal* of the rose curve given by $r = 3\cos 3\theta$.

Solution   In Figure 10.33, you can see that the right petal is traced as θ increases from $-\pi/6$ to $\pi/6$. Thus, the area is

$$A = \frac{1}{2}\int_{\alpha}^{\beta} r^2\,d\theta = \frac{1}{2}\int_{-\pi/6}^{\pi/6} (3\cos 3\theta)^2\,d\theta$$

$$= \frac{9}{2}\int_{-\pi/6}^{\pi/6} \frac{1 + \cos 6\theta}{2}\,d\theta$$

$$= \frac{9}{4}\left[\theta + \frac{\sin 6\theta}{6}\right]_{-\pi/6}^{\pi/6} = \frac{9}{4}\left(\frac{\pi}{6} + \frac{\pi}{6}\right) = \frac{3\pi}{4}.$$

To find the area of the region lying inside all three petals of the rose curve in Example 1, you could not simply integrate between 0 and 2π. In doing this you would obtain $9\pi/2$, which is twice the area of the three petals—the duplication occurs because the rose curve is traced *twice* as θ increases from 0 to 2π. To avoid this type of error, be sure that the graph of $r = f(\theta)$ does not retrace itself between your chosen limits of integration.

EXAMPLE 2   Finding the Area Bounded by a Single Curve

Find the area of the region lying between the inner and outer loops of the limaçon $r = 1 - 2\sin\theta$.

Solution   In Figure 10.34, note that the inner loop is traced as θ increases from $\pi/6$ to $5\pi/6$. Thus, the area inside the *inner loop* is

$$A_1 = \frac{1}{2}\int_{\pi/6}^{5\pi/6} (1 - 2\sin\theta)^2\,d\theta$$

$$= \frac{1}{2}\int_{\pi/6}^{5\pi/6} (1 - 4\sin\theta + 4\sin^2\theta)\,d\theta$$

$$= \frac{1}{2}\int_{\pi/6}^{5\pi/6} \left[1 - 4\sin\theta + 4\left(\frac{1 - \cos 2\theta}{2}\right)\right]d\theta$$

$$= \frac{1}{2}\int_{\pi/6}^{5\pi/6} (3 - 4\sin\theta - 2\cos 2\theta)\,d\theta$$

$$= \frac{1}{2}\left[3\theta + 4\cos\theta - \sin 2\theta\right]_{\pi/6}^{5\pi/6}$$

$$= \frac{1}{2}\left(2\pi - 3\sqrt{3}\right) = \pi - \frac{3\sqrt{3}}{2}.$$

In a similar way, you can integrate from $5\pi/6$ to $13\pi/6$ to find that the area of the region lying inside the *outer loop* is $A_2 = 2\pi + (3\sqrt{3}/2)$. The area of the region lying between the two loops is the difference between A_2 and A_1.

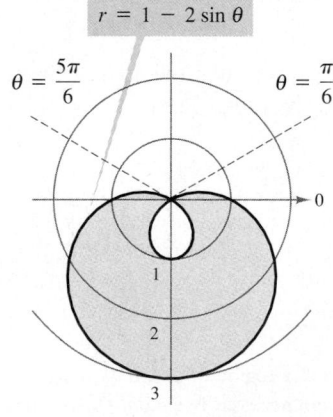

FIGURE 10.34
The area between the inner and outer loops is approximately 8.34.

$$A = A_2 - A_1 = \left(2\pi + \frac{3\sqrt{3}}{2}\right) - \left(\pi - \frac{3\sqrt{3}}{2}\right) = \pi + 3\sqrt{3} \approx 8.34$$

Points of Intersection of Polar Graphs

Because a point may be represented in different ways in polar coordinates, care must be taken in determining the points of intersection of two polar graphs. For example, consider the points of intersection of the graphs of

$$r = 1 - 2\cos\theta$$

$$r = 1,$$

as shown in Figure 10.35. If, as with rectangular equations, you attempted to find the points of intersection by solving the two equations simultaneously, you would obtain the following.

$r = 1 - 2\cos\theta$	First equation
$1 = 1 - 2\cos\theta$	Substitute $r = 1$ from 2nd equation into 1st equation
$\cos\theta = 0$	Simplify
$\theta = \dfrac{\pi}{2}, \dfrac{3\pi}{2}$	Solve for θ

FOR FURTHER INFORMATION
For more information on using technology to find points of intersection, see the article "Finding Points of Intersection of Polar-Coordinate Graphs" by Warren W. Esty in the September, 1991 issue of *Mathematics Teacher*.

The corresponding points of intersection are $(1, \pi/2)$ and $(1, 3\pi/2)$. However, in Figure 10.35 you can see that there is a *third* point of intersection that did not show up when the two polar equations were solved simultaneously. (This is one reason we stress sketching a graph when finding the area of a polar region.) The reason the third point was not found is that it does not occur with the same coordinates in the two graphs. On the graph of $r = 1$, the point occurs with coordinates $(1, \pi)$, but on the graph of $r = 1 - 2\cos\theta$, the point occurs with coordinates $(-1, 0)$.

You can compare the problem of finding points of intersection of two polar graphs with that of finding collision points of two satellites in intersecting orbits about the earth, as shown in Figure 10.36. The satellites will not collide as long as they reach the points of intersection at different times (θ-values). A collision will occur only at the points of intersection that are "simultaneous points"—those reached at the same time (θ-value).

REMARK   Because the pole can be represented by $(0, \theta)$, where θ is *any* angle, you should check separately for the pole when hunting for points of intersection.

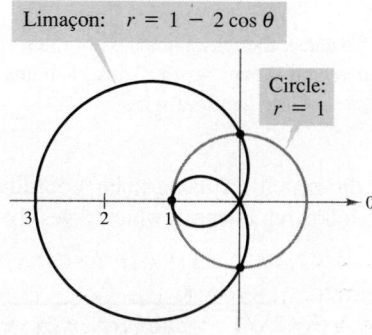

FIGURE 10.35
Three points of intersection: $(1, \pi/2)$, $(-1, 0)$, $(1, 3\pi/2)$

FIGURE 10.36
The paths of satellites can cross without causing a collision.

EXAMPLE 3   Finding the Area of a Region Between Two Curves

Find the area of the region common to the two regions bounded by the following curves.

$$r = -6\cos\theta \qquad \text{Circle}$$
$$r = 2 - 2\cos\theta \qquad \text{Cardioid}$$

Solution   Because both curves are symmetric with respect to the *x*-axis, you can work with the upper half-plane, as shown in Figure 10.37. The gray shaded region lies between the circle and the radial line $\theta = 2\pi/3$. Because the circle has coordinates $(0, \pi/2)$ at the pole, you can integrate between $\pi/2$ and $2\pi/3$ to obtain the area of this region. The region that is shaded red lies between the radial lines $\theta = 2\pi/3$ and $\theta = \pi$ and the cardioid. Thus, you can find the area of this second region by integrating between $2\pi/3$ and π. The sum of these two integrals gives the area of the common region lying *above* the polar axis.

$$\frac{A}{2} = \overbrace{\frac{1}{2}\int_{\pi/2}^{2\pi/3} (-6\cos\theta)^2\, d\theta}^{\substack{\text{Region between circle}\\\text{and radial line } \theta = 2\pi/3}} + \overbrace{\frac{1}{2}\int_{2\pi/3}^{\pi} (2 - 2\cos\theta)^2\, d\theta}^{\substack{\text{Region between cardioid and}\\\text{radial line } \theta = 2\pi/3 \text{ and } \theta = \pi}}$$

$$= 18\int_{\pi/2}^{2\pi/3} \cos^2\theta\, d\theta + \frac{1}{2}\int_{2\pi/3}^{\pi} (4 - 8\cos\theta + 4\cos^2\theta)\, d\theta$$

$$= 9\int_{\pi/2}^{2\pi/3} (1 + \cos 2\theta)\, d\theta + \int_{2\pi/3}^{\pi} (3 - 4\cos\theta + \cos 2\theta)\, d\theta$$

$$= 9\left[\theta + \frac{\sin 2\theta}{2}\right]_{\pi/2}^{2\pi/3} + \left[3\theta - 4\sin\theta + \frac{\sin 2\theta}{2}\right]_{2\pi/3}^{\pi}$$

$$= 9\left(\frac{2\pi}{3} - \frac{\sqrt{3}}{4} - \frac{\pi}{2}\right) + \left(3\pi - 2\pi + 2\sqrt{3} + \frac{\sqrt{3}}{4}\right)$$

$$= \frac{5\pi}{2}$$

Finally, multiplying by 2, you conclude that the total area is 5π. ■

REMARK   To check the reasonableness of the result obtained in Example 3, note that the area of the circular region is $\pi r^2 = 9\pi$. Thus, it seems reasonable that the area of the region lying inside the circle and the cardioid is 5π.

To see the benefit of using polar coordinates for finding the area in Example 3, consider the following integral, which gives the comparable area in rectangular coordinates.

$$\frac{A}{2} = \int_{-4}^{-3/2} \sqrt{2\sqrt{1 - 2x} - x^2 - 2x + 2}\, dx + \int_{-3/2}^{0} \sqrt{-x^2 - 6x}\, dx$$

Try using a computer and numerical integration to show that you obtain the same area as found in Example 3.

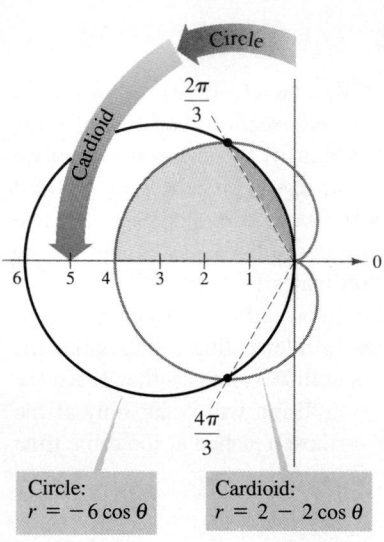

Circle
$\dfrac{2\pi}{3}$

Cardioid

$\dfrac{4\pi}{3}$

Circle:
$r = -6\cos\theta$

Cardioid:
$r = 2 - 2\cos\theta$

FIGURE 10.37

Arc Length in Polar Form

The formula for the length of a polar arc can be obtained from the arc length formula for a curve described by parametric equations. (See Exercise 61.)

THEOREM 10.8 Arc Length of a Polar Curve

Let f be a function whose derivative is continuous on an interval $a \leq \theta \leq \beta$. The length of the graph of $r = f(\theta)$ from $\theta = \alpha$ to $\theta = \beta$ is

$$s = \int_\alpha^\beta \sqrt{[f(\theta)]^2 + [f'(\theta)]^2} \, d\theta = \int_\alpha^\beta \sqrt{r^2 + \left(\frac{dr}{d\theta}\right)^2} \, d\theta.$$

REMARK When applying the arc length formula to a polar curve, be sure that the curve is traced out only once on the interval of integration. For instance, the rose given by $r = \cos 3\theta$ is traced out once on the interval $0 \leq \theta \leq \pi$, but is traced out twice on the interval $0 \leq \theta \leq 2\pi$.

EXAMPLE 4 Finding the Length of a Polar Curve

Find the length of the arc from $\theta = 0$ to $\theta = 2\pi$ for the cardioid

$$r = f(\theta) = 2 - 2\cos\theta$$

as shown in Figure 10.38.

Solution Because $f'(\theta) = 2\sin\theta$, you can find the arc length as follows.

$$
\begin{aligned}
s &= \int_\alpha^\beta \sqrt{[f(\theta)]^2 + [f'(\theta)]^2} \, d\theta &&\text{Formula for arc length} \\
&= \int_0^{2\pi} \sqrt{(2 - 2\cos\theta)^2 + (2\sin\theta)^2} \, d\theta \\
&= 2\sqrt{2} \int_0^{2\pi} \sqrt{1 - \cos\theta} \, d\theta \\
&= 2\sqrt{2} \int_0^{2\pi} \sqrt{2\sin^2\frac{\theta}{2}} \, d\theta \\
&= 4 \int_0^{2\pi} \sin\frac{\theta}{2} \, d\theta &&\sin\frac{\theta}{2} \geq 0 \text{ for } 0 \leq \theta \leq 2\pi \\
&= -8\cos\frac{\theta}{2} \Big]_0^{2\pi} \\
&= 8 + 8 \\
&= 16
\end{aligned}
$$

In the fifth step of the solution, it is legitimate to write

$$\sqrt{2\sin^2(\theta/2)} = \sqrt{2}\sin(\theta/2)$$

rather than $\sqrt{2\sin^2(\theta/2)} = \sqrt{2}\,|\sin(\theta/2)|$ because $\sin(\theta/2) \geq 0$ for $0 \leq \theta \leq 2\pi$.

$r = 2 - 2\cos\theta$

FIGURE 10.38
The arc length of this cardioid is 16.

REMARK Using Figure 10.38, you can determine the reasonableness of this answer by comparing it with the circumference of a circle. For example, a circle of radius $\frac{5}{2}$ has a circumference of $5\pi \approx 15.7$.

Area of a Surface of Revolution

The polar coordinate version of the formulas for the area of a surface of revolution can be obtained from the parametric versions given in Theorem 10.3, using the equations $x = r \cos \theta$ and $y = r \sin \theta$.

REMARK   When using Theorem 10.9, check to see that the graph of $r = f(\theta)$ is traced only once on the interval $\alpha \leq \theta \leq \beta$. For example, the circle given by $r = \cos \theta$ is traced once on the interval $0 \leq \theta \leq \pi$.

THEOREM 10.9   Area of a Surface of Revolution

Let f be a function whose derivative is continuous on an interval $\alpha \leq \theta \leq \beta$. The area of the surface formed by revolving the graph of $r = f(\theta)$ from $\theta = \alpha$ to $\alpha = \beta$ about the indicated line is as follows.

1. $S = 2\pi \displaystyle\int_{\alpha}^{\beta} f(\theta) \sin \theta \sqrt{[f(\theta)]^2 + [f'(\theta)]^2} \, d\theta$ About polar axis

2. $S = 2\pi \displaystyle\int_{\alpha}^{\beta} f(\theta) \cos \theta \sqrt{[f(\theta)]^2 + [f'(\theta)]^2} \, d\theta$ About line $\theta = \dfrac{\pi}{2}$

EXAMPLE 5   Finding the Area of a Surface of Revolution

Find the area of the surface formed by revolving the circle given by

$$r = f(\theta) = \cos \theta$$

about the line $\theta = \pi/2$, as shown in Figure 10.39.

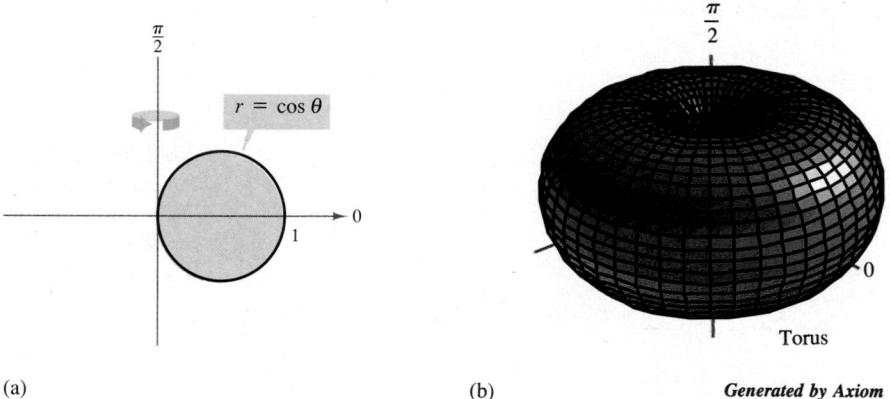

(a)   (b)   *Generated by Axiom*

FIGURE 10.39
The surface area of this torus is π^2.

Solution   You can use the second formula given in Theorem 10.9 with $f'(\theta) = -\sin \theta$. Because the circle is traced once as θ increases from 0 to π, we have the following integral.

$$S = 2\pi \int_{\alpha}^{\beta} f(\theta) \cos \theta \sqrt{[f(\theta)]^2 + [f'(\theta)]^2} \, d\theta$$

$$= 2\pi \int_{0}^{\pi} \cos \theta (\cos \theta) \sqrt{\cos^2 \theta + \sin^2 \theta} \, d\theta$$

$$= 2\pi \int_{0}^{\pi} \cos^2 \theta \, d\theta = \pi \int_{0}^{\pi} (1 + \cos 2\theta) \, d\theta$$

$$= \pi \left[\theta + \frac{\sin 2\theta}{2} \right]_{0}^{\pi} = \pi^2$$

TECHNOLOGY
Laboratory Guide
Lab 10.4

EXERCISES for Section 10.4

In Exercises 1 and 2, find the area of the region bounded by the graph of the polar equation (a) by a geometric formula and (b) by integration.

1. $r = 8 \sin \theta$ 2. $r = 3 \cos \theta$

In Exercises 3–12, find the area of the region.

3. One petal of $r = 2 \cos 3\theta$

4. One petal of $r = 4 \sin 2\theta$

5. One petal of $r = \cos 2\theta$

6. One petal of $r = \cos 5\theta$

7. Interior of $r = 1 - \sin \theta$

8. Interior of $r = 1 - \sin \theta$ (above the polar axis)

9. Inner loop of $r = 1 + 2 \cos \theta$

10. Inner loop of $r = 3 + 4 \sin \theta$

11. Between the loops of $r = 1 + 2 \cos \theta$

12. Between the loops of $r = 2(1 + 2 \sin \theta)$

In Exercises 13–22, find the points of intersection of the graphs of the equations.

13. $r = 1 + \cos \theta$
    $r = 1 - \cos \theta$

14. $r = 3(1 + \sin \theta)$
    $r = 3(1 - \sin \theta)$

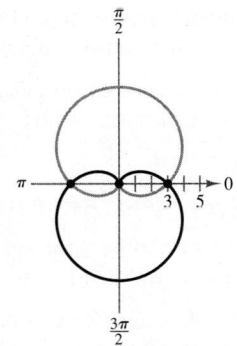

15. $r = 1 + \cos \theta$
    $r = 1 - \sin \theta$

16. $r = 2 - 3 \cos \theta$
    $r = \cos \theta$

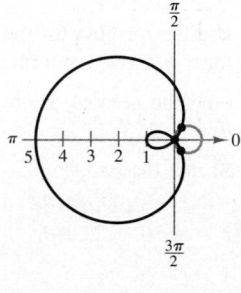

17. $r = 4 - 5 \sin \theta$
    $r = 3 \sin \theta$

18. $r = 1 + \cos \theta$
    $r = 3 \cos \theta$

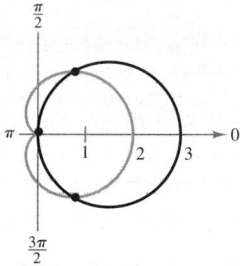

19. $r = \dfrac{\theta}{2}$
    $r = 2$

20. $\theta = \dfrac{\pi}{4}$
    $r = 2$

21. $r = 4 \sin 2\theta$
    $r = 2$

22. $r = 3 + \sin \theta$
    $r = 2 \csc \theta$

C **In Exercises 23 and 24, use a graphing utility to estimate the points of intersection of the graphs of the polar equations. Confirm your results analytically.**

23. $r = 2 + 3 \cos \theta$
    $r = \dfrac{\sec \theta}{2}$

24. $r = 3(1 - \cos \theta)$
    $r = \dfrac{6}{1 - \cos \theta}$

C *Essay* **In Exercises 25 and 26, use a graphing utility to find the point of intersection of the graphs of the polar equations. Watch the graphs as they are traced on the display screen. Explain why the pole is not a point of intersection obtained by solving the equations simultaneously.**

25. $r = \cos \theta$
    $r = 2 - 3 \sin \theta$

26. $r = 4 \sin \theta$
    $r = 2(1 + \sin \theta)$

In Exercises 27–36, find the area of the region.

27. Common interior of $r = 4 \sin 2\theta$ and $r = 2$

28. Common interior of $r = 3(1 + \sin \theta)$ and $r = 3(1 - \sin \theta)$

29. Common interior of $r = 3 - 2 \sin \theta$ and $r = -3 + 2 \sin \theta$

30. Common interior of $r = 3 - 2 \sin \theta$ and $r = 3 - 2 \cos \theta$

31. Common interior of $r = 4 \sin \theta$ and $r = 2$

32. Inside $r = 3 \sin \theta$ and outside $r = 2 - \sin \theta$

33. Inside $r = a(1 + \cos \theta)$ and outside $r = a \cos \theta$

34. Inside $r = 2a \cos \theta$ and outside $r = a$

35. Common interior of $r = a(1 + \cos \theta)$ and $r = a \sin \theta$

36. Region bounded by the graphs of

$$r = \frac{ab}{a \sin \theta + b \cos \theta}, \quad \theta = 0, \quad \text{and} \quad \theta = \frac{\pi}{2}$$

C **37.** *Antenna Radiation* The radiation from a transmitting antenna is not uniform in all directions. The intensity from a particular antenna is modeled by

$$r = a\cos^2\theta.$$

   a. Convert the polar equation to rectangular form.

   b. Use a graphing utility to graph the model for $a = 4$ and $a = 6$.

   c. Find the area of the geographic region between the two curves in part b.

38. *Area* The area inside one or more of the three interlocking circles

$$r = 2a\cos\theta, \quad r = 2a\sin\theta, \quad \text{and} \quad r = a$$

is divided into seven regions. Find the area of each region.

In Exercises 39–46, find the length of the given graph over the indicated interval.

Polar equation	Interval
39. $r = a$	$0 \le \theta \le 2\pi$
40. $r = 2a\cos\theta$	$-\dfrac{\pi}{2} \le \theta \le \dfrac{\pi}{2}$
41. $r = 1 + \sin\theta$	$0 \le \theta \le 2\pi$
42. $r = 5(1 + \cos\theta)$	$0 \le \theta \le 2\pi$
43. $r = 2\theta$	$0 \le \theta \le \dfrac{\pi}{2}$
44. $r = \sec\theta$	$0 \le \theta \le \dfrac{\pi}{3}$
45. $r = \dfrac{1}{\theta}$	$\pi \le \theta \le 2\pi$
46. $r = e^{\theta}$	$0 \le \theta \le \pi$

C **In Exercises 47 and 48, use a graphing utility to graph the equation. Then, use Simpson's Rule with $n = 12$ to approximate the arc length of the curve and the area of the region enclosed.**

47. $r = \sin(3\cos\theta)$ **48.** $r = 2\sin(2\cos\theta)$

In Exercises 49–52, find the area of the surface formed by revolving the curve about the given line.

Polar equation	Interval	Axis of revolution
49. $r = 2\cos\theta$	$0 \le \theta \le \dfrac{\pi}{2}$	Polar axis
50. $r = a\cos\theta$	$0 \le \theta \le \dfrac{\pi}{2}$	$\theta = \dfrac{\pi}{2}$
51. $r = e^{a\theta}$	$0 \le \theta \le \dfrac{\pi}{2}$	$\theta = \dfrac{\pi}{2}$
52. $r = a(1 + \cos\theta)$	$0 \le \theta \le \pi$	Polar axis

C **In Exercises 53 and 54, use a computer or calculator and Simpson's Rule with $n = 10$ to approximate the area of the surface formed by revolving the curve about the polar axis.**

Polar equation	Interval
53. $r = 4\cos 2\theta$	$0 \le \theta \le \dfrac{\pi}{4}$
54. $r = \theta$	$0 \le \theta \le \pi$

55. *Surface Area of a Torus* Find the surface area of the torus generated by revolving the circle given by $r = a$ about the line $r = b\sec\theta$ where $b > a > 0$. [*Hint:* Derive the integral that yields the area of the surface formed by revolving the graph of $r = f(\theta)$ about the line $r = b\sec\theta$.]

56. *Archimedes' Spiral* Let A_n be the area between turns $n - 1$ and $n (n \ge 2)$ of Archimedes' spiral $r = a\theta$ (see figure).

   a. Complete the following table.

n	2	3	4	5	6	7
A_n						

   [*Hint:* Note that

$$A_n = \frac{1}{2}\int_{(n-1)(2\pi)}^{n(2\pi)} (a\theta)^2\, d\theta - \frac{1}{2}\int_{(n-2)(2\pi)}^{(n-1)(2\pi)} (a\theta)^2\, d\theta].$$

   b. Verify that $A_n = (n - 1)A_2$ for $n \ge 3$.

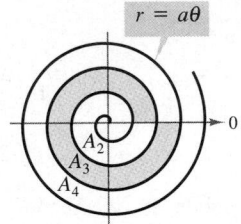

FIGURE FOR 56

True or False **In Exercises 57–60, determine whether the statement is true or false. If it is false, explain why or give an example that shows it is false.**

57. The area of the region enclosed by the circle $r = \sin\theta$ is given by the definite integral

$$\frac{1}{2}\int_{0}^{2\pi} \sin^2\theta\, d\theta.$$

58. If $f(\theta) > 0$ for all θ and $g(\theta) < 0$ for all θ, then the graphs of $r = f(\theta)$ and $r = g(\theta)$ do not intersect.

59. If $f(\theta) = g(\theta)$ for $\theta = 0$, $\pi/2$, π, and $3\pi/2$, then the graphs of $r = f(\theta)$ and $r = g(\theta)$ have at least four points of intersection.

60. If n is an even integer, then the area of the region enclosed by $r = \sin(n\theta)$ is twice the area of the region enclosed by $r = \sin[(n + 1)\theta]$.

61. Use the formula for the arc length of a curve in parametric form to derive the formula for the arc length of a polar curve.

62. Find the area of the region enclosed by $r = a\cos(n\theta)$ for $n = 1, 2, 3, \ldots$.

63. Sketch the strophoid

$$r = \sec\theta - 2\cos\theta, \quad -\frac{\pi}{2} < \theta < \frac{\pi}{2}.$$

Convert this equation to Cartesian coordinates. Find the area enclosed by the loop.

Polar Equations of Conics

In Chapter 9, you saw that the rectangular equations of ellipses and hyperbolas take simple forms when the origin lies at their *centers*. As it happens, there are many important applications of conics in which it is more convenient to use one of the *foci* as the reference point (the origin) for the coordinate system. For example, the sun lies at a focus of the earth's orbit. Similarly, the light source of a parabolic reflector lies at its focus. In this section you will see that polar equations of conics take simple forms if one of the foci lies at the pole.

The following theorem uses the concept of *eccentricity,* as defined in Chapter 9, to classify the three basic types of conics. A proof of this theorem is given in Appendix A.

THEOREM 10.10   Classification of Conics by Eccentricity

The locus of a point in the plane whose distance from a fixed point (*focus*) has a constant ratio to its distance from a fixed line (*directrix*) is a conic. The constant ratio e is the *eccentricity* of the conic.

1. The conic is an ellipse if $0 < e < 1$.
2. The conic is a parabola if $e = 1$.
3. The conic is a hyperbola if $e > 1$.

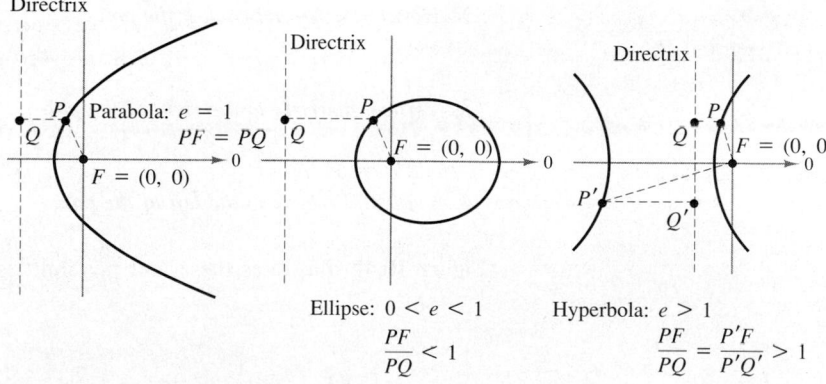

FIGURE 10.40

In Figure 10.40, note that for each type of conic the pole corresponds to the fixed point (focus) given in the definition. The benefit of this location can be seen in the proof of the following theorem.

> **THEOREM 10.11 Polar Equations for Conics**
>
> The graph of a polar equation of the form
>
> $$r = \frac{ed}{1 \pm e\cos\theta} \quad \text{or} \quad r = \frac{ed}{1 \pm e\sin\theta}$$
>
> is a conic, where $e > 0$ is the eccentricity and $|d|$ is the distance between the focus at the pole and its corresponding directrix.

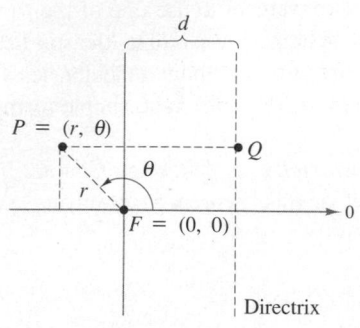

FIGURE 10.41

Proof We give a proof for $r = ed/(1 + e\cos\theta)$ with $d > 0$. In Figure 10.41, consider a vertical directrix d units to the right of the focus $F = (0, 0)$. If $P = (r, \theta)$ is a point on the graph of $r = ed/(1 + e\cos\theta)$, then the distance between P and the directrix can be shown to be

$$PQ = |d - x| = |d - r\cos\theta| = \left| \frac{r(1 + e\cos\theta)}{e} - r\cos\theta \right| = \left| \frac{r}{e} \right|.$$

Because the distance between P and the pole is simply $PF = |r|$, the ratio of PF to PQ is $PF/PQ = |r|/|r/e| = |e| = e$ and, by Theorem 10.10, the graph of the equation must be a conic. The proofs of the other cases are similar. ∎

The four types of equations indicated in Theorem 10.11 can be classified as follows, where $d > 0$.

1. *Horizontal directrix above the pole* $\qquad r = \dfrac{ed}{1 + e\sin\theta}$

2. *Horizontal directrix below the pole* $\qquad r = \dfrac{ed}{1 - e\sin\theta}$

3. *Vertical directrix to the right of the pole* $\qquad r = \dfrac{ed}{1 + e\cos\theta}$

4. *Vertical directrix to the left of the pole* $\qquad r = \dfrac{ed}{1 - e\cos\theta}$

Figure 10.42 illustrates these four possibilities for a parabola.

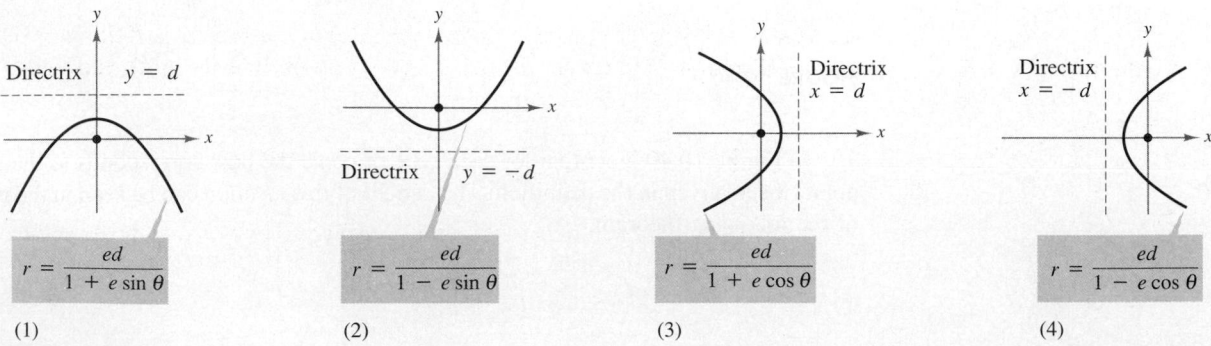

FIGURE 10.42
The four types of polar equations for a parabola.

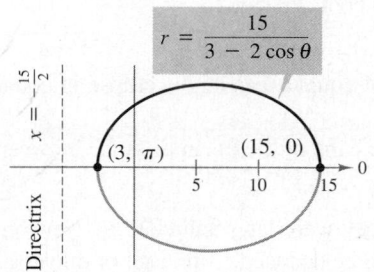

FIGURE 10.43

The graph of the conic is an ellipse with $e = \frac{2}{3}$.

EXAMPLE 1 Determining a Conic from its Equation

Sketch the graph of the conic given by $r = \dfrac{15}{3 - 2\cos\theta}$.

Solution   To determine the type of conic, rewrite the equation as

$$r = \frac{15}{3 - 2\cos\theta}$$

$$= \frac{5}{1 - (2/3)\cos\theta}.$$

Thus the graph is an ellipse with $e = \frac{2}{3}$. You can sketch the upper half of the ellipse by plotting points from $\theta = 0$ to $\theta = \pi$, as shown in Figure 10.43. Then, using symmetry with respect to the polar axis, you can sketch the lower half.

For the ellipse in Figure 10.43, the major axis is horizontal and the vertices lie at $(15, 0)$ and $(3, \pi)$. Thus, the length of the *major* axis is $2a = 18$. To find the length of the *minor* axis, you can use the equations $e = c/a$ and $b^2 = a^2 - c^2$ and conclude that

$$b^2 = a^2 - c^2 = a^2 - (ea)^2 = a^2(1 - e^2). \qquad \text{Ellipse}$$

Because $e = \frac{2}{3}$, you have

$$b^2 = 9^2\left[1 - \left(\frac{2}{3}\right)^2\right] = 45$$

which implies that $b = \sqrt{45} = 3\sqrt{5}$. Thus, the length of the minor axis is $2b = 6\sqrt{5}$. A similar analysis for hyperbolas yields

$$b^2 = c^2 - a^2 = (ea)^2 - a^2 = a^2(e^2 - 1). \qquad \text{Hyperbola}$$

EXAMPLE 2 Sketching a Conic from its Polar Equation

Sketch the graph of the polar equation $r = \dfrac{32}{3 + 5\sin\theta}$.

Solution   Dividing the numerator and denominator by 3 produces

$$r = \frac{32/3}{1 + (5/3)\sin\theta}.$$

Because $e = \frac{5}{3} > 1$, the graph is a hyperbola. Because $d = \frac{32}{5}$, the directrix is the line $y = \frac{32}{5}$. The transverse axis of the hyperbola lies on the line $\theta = \pi/2$, and the vertices occur at

$$\left(4, \frac{\pi}{2}\right) \quad \text{and} \quad \left(-16, \frac{3\pi}{2}\right).$$

Because the length of the transverse axis is 12, you can see that $a = 6$. To find b, write

$$b^2 = a^2(e^2 - 1) = 6^2\left[\left(\frac{5}{3}\right)^2 - 1\right] = 64.$$

Therefore, $b = 8$. Finally, you can use a and b to determine the asymptotes of the hyperbola and obtain the sketch shown in Figure 10.44.

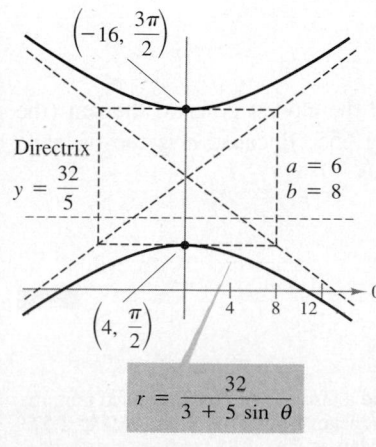

FIGURE 10.44

The graph of the conic is a hyperbola with $e = \frac{5}{3}$.

JOHANNES KEPLER

Kepler formulated his three laws from the extensive data recorded by Danish astronomer Tycho Brahe, and from direct observation of the orbit of Mars.

Kepler's Laws

Kepler's Laws, named after the German astronomer Johannes Kepler (1571–1630), can be used to describe the orbits of the planets about the sun.

1. Each planet moves in an elliptical orbit with the sun as a focus.
2. The ray from the sun to the planet sweeps out equal areas of the ellipse in equal times.
3. The square of the period is proportional to the cube of the mean distance between the planet and the sun.*

Although Kepler derived these laws empirically, they were later validated by Newton. In fact, Newton was able to show that each law can be deduced from a set of universal laws of motion and gravitation that govern the movement of all heavenly bodies, including comets and satellites. This is illustrated in the next example, involving the comet named after the English mathematician and physicist Edmund Halley (1656–1742).

EXAMPLE 3   Halley's Comet

Halley's comet has an elliptical orbit with an eccentricity of $e \approx 0.97$. The length of the major axis of the orbit is approximately 36.18 astronomical units. (An astronomical unit is defined as the mean distance between the earth and the sun, 93 million miles.) Find a polar equation for the orbit. How close does Halley's comet come to the sun?

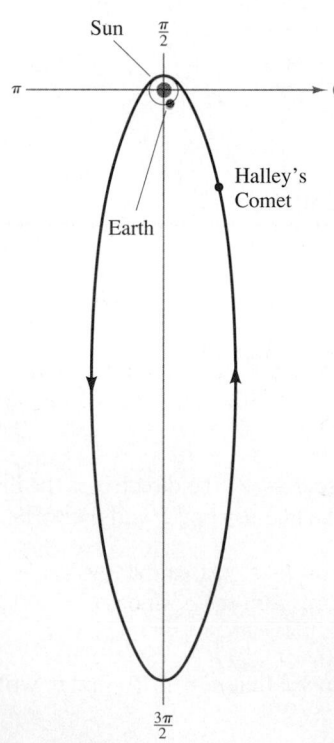

FIGURE 10.45

Solution   Using a vertical axis, you can choose an equation of the form $r = ed/(1 + e \sin\theta)$. Because the vertices of the ellipse occur when $\theta = \pi/2$ and $\theta = 3\pi/2$, you can determine the length of the major axis to be the sum of the r-values of the vertices, as shown in Figure 10.45. That is,

$$2a = \frac{0.97d}{1 + 0.97} + \frac{0.97d}{1 - 0.97} \approx 32.83d \approx 36.18.$$

Thus, $d \approx 1.102$ and $ed \approx (0.97)(1.102) \approx 1.069$. Using this value in the equation produces

$$r = \frac{1.069}{1 + 0.97 \sin\theta}$$

where r is measured in astronomical units. To find the closest point to the sun (the focus), you can write $c = ea \approx (0.97)(18.09) \approx 17.55$. Because c is the distance between the focus and the center, the closest point is

$$a - c \approx 18.09 - 17.55$$
$$\approx 0.54 \text{ AU}$$
$$\approx 50,000,000 \text{ mi.}$$

---

* If the earth is used as a reference with a period of 1 year and a distance of 1 astronomical unit, the proportionality constant is 1. For example, because Mars has a mean distance to the sun of $D = 1.523$ AU, its period P is given by $D^3 = P^2$. Thus, the period for Mars is $P = 1.88$ years.

Kepler's Second Law states that as a planet moves about the sun, a ray from the sun to the planet sweeps out equal areas in equal times. This law can also be applied to comets or asteroids with elliptical orbits. For example, Figure 10.46 shows the orbit of the asteroid Apollo about the sun. Applying Kepler's Second Law to this asteroid, we know that the closer it is to the sun, the greater its velocity, because a short ray must be moving quickly to sweep out as much area as a long ray.

FIGURE 10.46
A ray from the sun to the asteroid sweeps out equal areas in equal times.

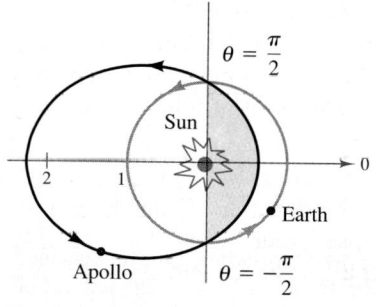

FIGURE 10.47

EXAMPLE 4   The Asteroid Apollo

The asteroid Apollo has a period of 478 earth days, and its orbit is approximated by the ellipse

$$r = \frac{1}{1 + (5/9)\cos\theta} = \frac{9}{9 + 5\cos\theta}$$

where r is measured in astronomical units. How long does it take Apollo to move from the position given by $\theta = -\pi/2$ to $\theta = \pi/2$, as shown in Figure 10.47?

Solution   Begin by finding the area swept out as θ increases from $-\pi/2$ to $\pi/2$.

$$A = \frac{1}{2}\int_{\alpha}^{\beta} r^2\, d\theta$$

$$= \frac{1}{2}\int_{-\pi/2}^{\pi/2} \left(\frac{9}{9 + 5\cos\theta}\right)^2 d\theta$$

Using the substitution $u = \tan(\theta/2)$, as discussed in Section 7.6, you obtain

$$A = \frac{81}{112}\left[\frac{-5\sin\theta}{9 + 5\cos\theta} + \frac{18}{\sqrt{56}}\arctan\frac{\sqrt{56}\tan(\theta/2)}{14}\right]_{-\pi/2}^{\pi/2} \approx 0.90429.$$

Because the major axis of the ellipse has length $2a = 81/28$ and the eccentricity is $e = 5/9$, you can determine that $b = a\sqrt{1 - e^2} = 9/\sqrt{56}$. Thus, the area of the ellipse is

$$\text{Area of ellipse} = \pi ab = \pi\left(\frac{81}{56}\right)\left(\frac{9}{\sqrt{56}}\right) \approx 5.46507.$$

Because the time required to complete the orbit is 478 days, you can apply Kepler's Second Law to conclude that the time t required to move from the position $\theta = -\pi/2$ to $\theta = \pi/2$ is given by

$$\frac{t}{478} = \frac{\text{area of elliptical segment}}{\text{area of ellipse}} \approx \frac{0.90429}{5.46507}$$

which implies that $t \approx 79$ days.

1-36 every 3rd look at 46, 47-51 odd

EXERCISES for Section 10.5

TECHNOLOGY
Laboratory Guide
Lab 10.5

C In Exercises 1–4, use a graphing utility to graph the polar equations on the same viewing rectangle—use (a) $e = 1$, (b) $e = 0.5$, and (c) $e = 1.5$.

1. $r = \dfrac{2e}{1 + e\cos\theta}$

2. $r = \dfrac{2e}{1 - e\cos\theta}$

3. $r = \dfrac{2e}{1 - e\sin\theta}$

4. $r = \dfrac{2e}{1 + e\sin\theta}$

In Exercises 5–10, match the polar equation with the correct graph. [The graphs are labeled (a), (b), (c), (d), (e), and (f).]

5. $r = \dfrac{6}{1 - \cos\theta}$

6. $r = \dfrac{2}{2 - \cos\theta}$

7. $r = \dfrac{3}{1 - 2\sin\theta}$

8. $r = \dfrac{2}{1 + \sin\theta}$

9. $r = \dfrac{6}{2 - \sin\theta}$

10. $r = \dfrac{2}{2 + 3\cos\theta}$

(a)

(b)

(c)

(d)

(e)

(f)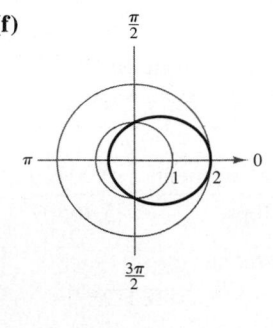

In Exercises 11–20, sketch and identify the graph.

11. $r = \dfrac{-1}{1 - \sin\theta}$

12. $r = \dfrac{6}{1 + \cos\theta}$

13. $r = \dfrac{6}{2 + \cos\theta}$

14. $r = \dfrac{3}{3 + 2\sin\theta}$

15. $r(2 + \sin\theta) = 4$

16. $r(3 - 2\cos\theta) = 6$

17. $r = \dfrac{5}{-1 + 2\cos\theta}$

18. $r = \dfrac{-3}{2 + 4\sin\theta}$

19. $r = \dfrac{3}{2 + 6\sin\theta}$

20. $r = \dfrac{4}{1 + 2\cos\theta}$

C In Exercises 21–24, use a graphing utility to graph the rotated conic. Use the graph to determine the angle of rotation.

21. $r = \dfrac{-1}{1 - \sin(\theta - \pi/4)}$ (See Exercise 11)

22. $r = \dfrac{6}{1 + \cos(\theta - \pi/3)}$ (See Exercise 12)

23. $r = \dfrac{6}{2 + \cos(\theta + \pi/6)}$ (See Exercise 13)

24. $r = \dfrac{-3}{2 + 4\sin(\theta + 2\pi/3)}$ (See Exercise 18)

In Exercises 25–36, find a polar equation for the specified conic. In each case, consider the focus to be at the pole. (For convenience, the equation for the directrix is given in rectangular form.)

Conic	Eccentricity	Directrix
25. Parabola	$e = 1$	$x = -1$
26. Parabola	$e = 1$	$y = 1$
27. Ellipse	$e = \frac{1}{2}$	$y = 1$
28. Ellipse	$e = \frac{3}{4}$	$y = -2$
29. Hyperbola	$e = 2$	$x = 1$
30. Hyperbola	$e = \frac{3}{2}$	$x = -1$

Conic	Vertex or vertices
31. Parabola	$\left(1, -\dfrac{\pi}{2}\right)$
32. Parabola	$(5, \pi)$
33. Ellipse	$(2, 0), (8, \pi)$
34. Ellipse	$\left(2, \dfrac{\pi}{2}\right), \left(4, \dfrac{3\pi}{2}\right)$
35. Hyperbola	$\left(1, \dfrac{3\pi}{2}\right), \left(9, \dfrac{3\pi}{2}\right)$
36. Hyperbola	$(2, 0), (10, 0)$

37. Show that the polar equation for the ellipse

$$\frac{x^2}{a^2} + \frac{y^2}{b^2} = 1$$

is

$$r^2 = \frac{b^2}{1 - e^2 \cos^2 \theta}.$$

38. Show that the polar equation for the hyperbola

$$\frac{x^2}{a^2} - \frac{y^2}{b^2} = 1$$

is

$$r^2 = \frac{-b^2}{1 - e^2 \cos^2 \theta}.$$

In Exercises 39–42, use the results of Exercises 37 and 38 to write the polar form of the equation of the conic.

39. Ellipse: focus at (4, 0), vertices at (5, 0), (5, π)

40. Hyperbola: focus at (5, 0), vertices at (4, 0) (4, π)

41. $\dfrac{x^2}{9} - \dfrac{y^2}{16} = 1$ **42.** $\dfrac{x^2}{4} + y^2 = 1$

[C] **In Exercises 43 and 44, use a computer or calculator and Simpson's Rule with $n = 10$ to approximate the area of the region bounded by the graph of the polar equation.**

43. $r = \dfrac{3}{2 - \cos \theta}$ **44.** $r = \dfrac{2}{3 - 2 \sin \theta}$

45. Show that the graphs of the equations

$$r = \frac{ed}{1 + \sin \theta} \quad \text{and} \quad r = \frac{ed}{1 - \sin \theta}$$

intersect at right angles.

46. Prove that the tangent of the angle ψ ($0 \le \psi \le \pi/2$) between the radial line and the tangent line at the point (r, θ) on the graph of $r = f(\theta)$ (see figure) is given by

$$\tan \psi = \frac{r}{dr/d\theta}.$$

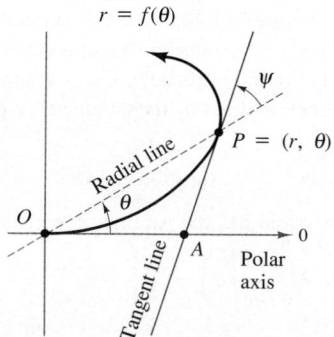

Angle Between Radial
and Tangent Line

FIGURE FOR 46

In Exercises 47–52, use the result of Exercise 46 to find the angle ψ between the radial and tangent lines to the graph at the given point.

Polar equation	Value of θ
47. $r = 2(1 - \cos \theta)$	$\theta = \pi$
48. $r = 3(1 - \cos \theta)$	$\theta = \dfrac{3\pi}{4}$
49. $r = 2 \cos 3\theta$	$\theta = \dfrac{\pi}{6}$
50. $r = 4 \sin 2\theta$	$\theta = \dfrac{\pi}{6}$
51. $r = \dfrac{6}{1 - \cos \theta}$	$\theta = \dfrac{2\pi}{3}$
52. $r = 5$	$\theta = \dfrac{\pi}{6}$

53. *Satellite Tracking* A satellite in a 100-mile-high circular orbit around the earth has a velocity of approximately 17,500 miles per hour. If this velocity is increased by a factor of $\sqrt{2}$, then the satellite will have the minimum velocity necessary to escape the earth's gravity and it will begin to follow a parabolic path with the center of the earth as the focus (see figure). Find a polar equation of the parabolic path of the satellite (assume the radius of the earth is 4000 miles). Find the distance between the surface of the earth and the satellite when $\theta = 30°$.

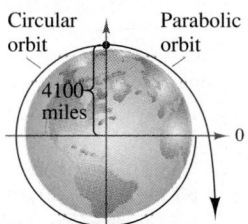

FIGURE FOR 53

54. *Explorer 18* On November 26, 1963, the United States launched Explorer 18. Its low and high points over the surface of the earth were 119 miles and 122,000 miles, respectively (see figure). The center of the earth is the focus of the orbit. Find the polar equation for the orbit and find the distance between the surface of the earth and the satellite when $\theta = 60°$.

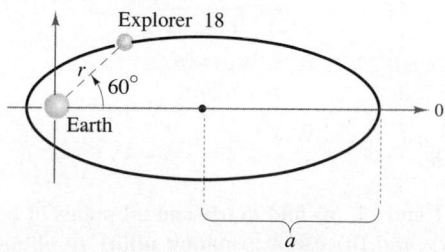

FIGURE FOR 54

55. *Planetary Motion* The planets travel in elliptical orbits with the sun as a focus. Assume that the focus is at the pole, the major axis lies on the polar axis, and the length of the major axis is $2a$ (see figure). Show that the polar equation of the orbit is given by

$$r = \frac{(1 - e^2)a}{1 - e\cos\theta}$$

where e is the eccentricity.

56. *Planetary Motion* Use the result of Exercise 55 to show that the minimum distance (*perihelion distance*) from the sun to the planet is $r = a(1 - e)$ and the maximum distance (*aphelion distance*) is $r = a(1 + e)$.

In Exercises 57–60, use the results of Exercises 55 and 56 to find the polar equation of the planet and the perihelion and aphelion distances.

57. Earth $a = 92.957 \times 10^6$ miles
             $e = 0.0167$

58. Saturn $a = 1.427 \times 10^9$ kilometers
             $e = 0.0543$

59. Pluto $a = 5.900 \times 10^9$ kilometers
             $e = 0.2481$

60. Mercury $a = 36.0 \times 10^6$ miles
             $e = 0.206$

C **61.** *Planetary Motion* In Exercise 59, the polar equation for the elliptical orbit of Pluto was found. Use the equation and a computer or calculator to perform each of the following.

a. Approximate the area swept out by a ray from the sun to the planet as θ increases from 0 to $\pi/9$. Use this result to determine the number of years required for the planet to move through this arc if the period of one revolution around the sun is 248 years.

b. By trial and error, approximate the angle α so that the area swept out by a ray from the sun to the planet as θ increases from π to α equals the area found in part a (see figure). Does the ray sweep through a larger or smaller angle than in part a to generate the same area? Why is this the case?

c. Approximate the distances the planet traveled in parts a and b. Use these distances to approximate the average numbers of kilometers per year the planet traveled in the two cases.

62. What conic section does the polar equation

$$r = a\sin\theta + b\cos\theta$$

represent?

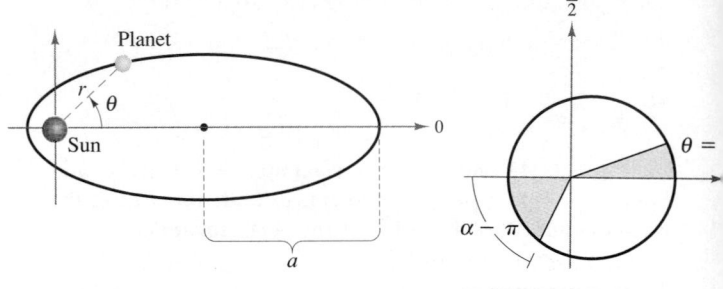

FIGURE FOR 55   **FIGURE FOR 61**

REVIEW EXERCISES for Chapter 10

In Exercises 1–10, (a) find dy/dx and all points of horizontal tangency, (b) eliminate the parameter where possible, and (c) sketch the curve represented by the parametric equations.

1. $x = 1 + 4t$
   $y = 2 - 3t$

2. $x = t + 4$
   $y = t^2$

3. $x = \dfrac{1}{t}$
   $y = 2t + 3$

4. $x = \dfrac{1}{t}$
   $y = t^2$

5. $x = \dfrac{1}{2t + 1}$
   $y = \dfrac{1}{t^2 - 2t}$

6. $x = 2t - 1$
   $y = \dfrac{1}{t^2 - 2t}$

7. $x = 3 + 2\cos\theta$
   $y = 2 + 5\sin\theta$

8. $x = 6\cos\theta$
   $y = 6\sin\theta$

9. $x = \cos^3\theta$
   $y = 4\sin^3\theta$

10. $x = e^t$
    $y = e^{-t}$

C **In Exercises 11 and 12, (a) find dy/dx and all points of horizontal tangency and (b) use a graphing utility to obtain a graph of the curve represented by the parametric equations.**

11. $x = \cot\theta$
    $y = \sin 2\theta$

12. $x = 2\theta - \sin\theta$
    $y = 2 - \cos\theta$

In Exercises 13 and 14, find a parametric representation of the conic.

13. Ellipse: Center at $(-3, 4)$, horizontal major axis of length 8, and minor axis of length 6

14. Hyperbola: Vertices at $(0, \pm 4)$ and foci at $(0, \pm 5)$

C **15.** *Rotary Engine* The rotary engine was developed by Felix Wankel in the 1950s. It features a rotor, which is basically a modified equilateral triangle. The rotor moves in a chamber which, in two dimensions, is an epitrochoid. Use a graphing utility to graph the chamber modeled by the parametric equations

$$x = \cos 3\theta + 5\cos\theta \quad \text{and} \quad y = \sin 3\theta + 5\sin\theta.$$

C **16.** *Hypocycloids* A hypocycloid has the parametric equations

$$x = (a - b)\cos t + b\cos\left(\frac{a - b}{b}t\right)$$

and

$$y = (a - b)\sin t - b\sin\left(\frac{a - b}{b}t\right).$$

Use a graphing utility to obtain graphs of the hypocycloids for the following values of a and b.

a. $a = 2, b = 1$ **b.** $a = 3, b = 1$
c. $a = 4, b = 1$ **d.** $a = 10, b = 1$
e. $a = 3, b = 2$ **f.** $a = 4, b = 3$

17. *Cycloids* Eliminate the parameter from

$$x = a(\theta - \sin\theta)$$

and

$$y = a(1 - \cos\theta)$$

to show that the rectangular equation of a cycloid is

$$x = a\arccos\left(\frac{a - y}{a}\right) \pm \sqrt{2ay - y^2}.$$

18. *Involute of a Circle* The involute of a circle is described by the endpoint P of a string that is held taut as it is unwound from a spool that does not turn (see figure). Show that a parametric representation of the involute is given by

$$x = r(\cos\theta + \theta\sin\theta)$$

and

$$y = r(\sin\theta - \theta\cos\theta).$$

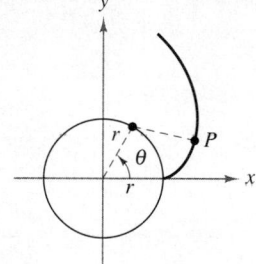

FIGURE FOR 18

In Exercises 19 and 20, find the length of the curve represented by the parametric equations over the given interval.

19. $x = r(\cos\theta + \theta\sin\theta)$ **20.** $x = 6\cos\theta$
$y = r(\sin\theta - \theta\cos\theta)$ $y = 6\sin\theta$
$0 \le \theta \le \pi$ $0 \le \theta \le \pi$

In Exercises 21–34, sketch a graph of the polar equation.

21. $r = 4$ **22.** $\theta = \dfrac{\pi}{12}$

23. $r = -\sec\theta$ **24.** $r = 3\csc\theta$

25. $r = -2(1 + \cos\theta)$ **26.** $r = 3 - 4\cos\theta$

27. $r = 4 - 3\cos\theta$ **28.** $r = 2\theta$

29. $r = -3\cos 2\theta$ **30.** $r = \cos 5\theta$

31. $r^2 = 4\sin^2 2\theta$ **32.** $r^2 = \cos 2\theta$

33. $r = \dfrac{2}{1 - \sin\theta}$ **34.** $r = \dfrac{4}{5 - 3\cos\theta}$

C In Exercises 35–38, use a graphing utility to graph the polar equation.

35. $r = \dfrac{3}{\cos(\theta - \pi/4)}$ **36.** $r = 2\sin\theta\cos^2\theta$

37. $r = 4\cos 2\theta\sec\theta$ **38.** $r = 4(\sec\theta - \cos\theta)$

In Exercises 39–46, convert the polar equation to rectangular form.

39. $r = 3\cos\theta$ **40.** $r = 10$

41. $r = -2(1 + \cos\theta)$ **42.** $r = \dfrac{1}{2 - \cos\theta}$

43. $r^2 = \cos 2\theta$ **44.** $r = 4\sec\left(\theta - \dfrac{\pi}{3}\right)$

45. $r = 4\cos 2\theta\sec\theta$ **46.** $\theta = \dfrac{3\pi}{4}$

In Exercises 47–50, convert the rectangular equation to polar form.

47. $(x^2 + y^2)^2 = ax^2 y$

48. $x^2 + y^2 - 4x = 0$

49. $x^2 + y^2 = a^2\left(\arctan\dfrac{y}{x}\right)^2$

50. $(x^2 + y^2)\left(\arctan\dfrac{y}{x}\right)^2 = a^2$

In Exercises 51–56, find a polar equation for the line or conic.

51. Circle   Center: $(5, \pi/2)$
                 Solution point: $(0, 0)$

52. Line   Solution point: $(0, 0)$
                Slope: $\sqrt{3}$

53. Parabola   Vertex: $(2, \pi)$
                    Focus: $(0, 0)$

54. Parabola   Vertex: $(2, \pi/2)$
                    Focus: $(0, 0)$

55. Ellipse   Vertices: $(5, 0), (1, \pi)$
                   One focus: $(0, 0)$

56. Hyperbola   Vertices: $(1, 0), (7, 0)$
                     One focus: $(0, 0)$

In Exercises 57 and 58, (a) find the tangents at the pole, (b) Find all points of horizontal and vertical tangency, And (c) sketch the graph of the equation.

57. $r = 1 - 2\cos\theta$ **58.** $r^2 = 4\sin 2\theta$

In Exercises 59 and 60, show that the graphs of the given polar equations are orthogonal at the points of intersection.

59. $r = 1 + \cos\theta$ **60.** $r = a\sin\theta$
$r = 1 - \cos\theta$ $r = a\cos\theta$

In Exercises 61–68, find the area of the region.

61. Interior of $r = 2 + \cos\theta$

62. Interior of $r = 5(1 - \sin\theta)$

63. Interior of $r = \sin\theta\cos^2\theta$

64. Interior of $r = 4\sin 3\theta$

65. Interior of $r^2 = a^2\sin 2\theta$

66. Common interior of $r = a$ and $r^2 = 2a^2\sin 2\theta$

67. Common interior of $r = 4\cos\theta$ and $r = 2$

68. Region bounded by the polar axis and $r = e^\theta$ for $0 \le \theta \le \pi$

In Exercises 69 and 70, find the perimeter of the curve represented by the polar equation.

69. $r = a(1 - \cos\theta)$ **70.** $r = a\cos 2\theta$

71. Find the angle between the circle $r = 3\sin\theta$ and the limaçon $r = 4 - 5\sin\theta$ at the point of intersection $(3/2, \pi/6)$.

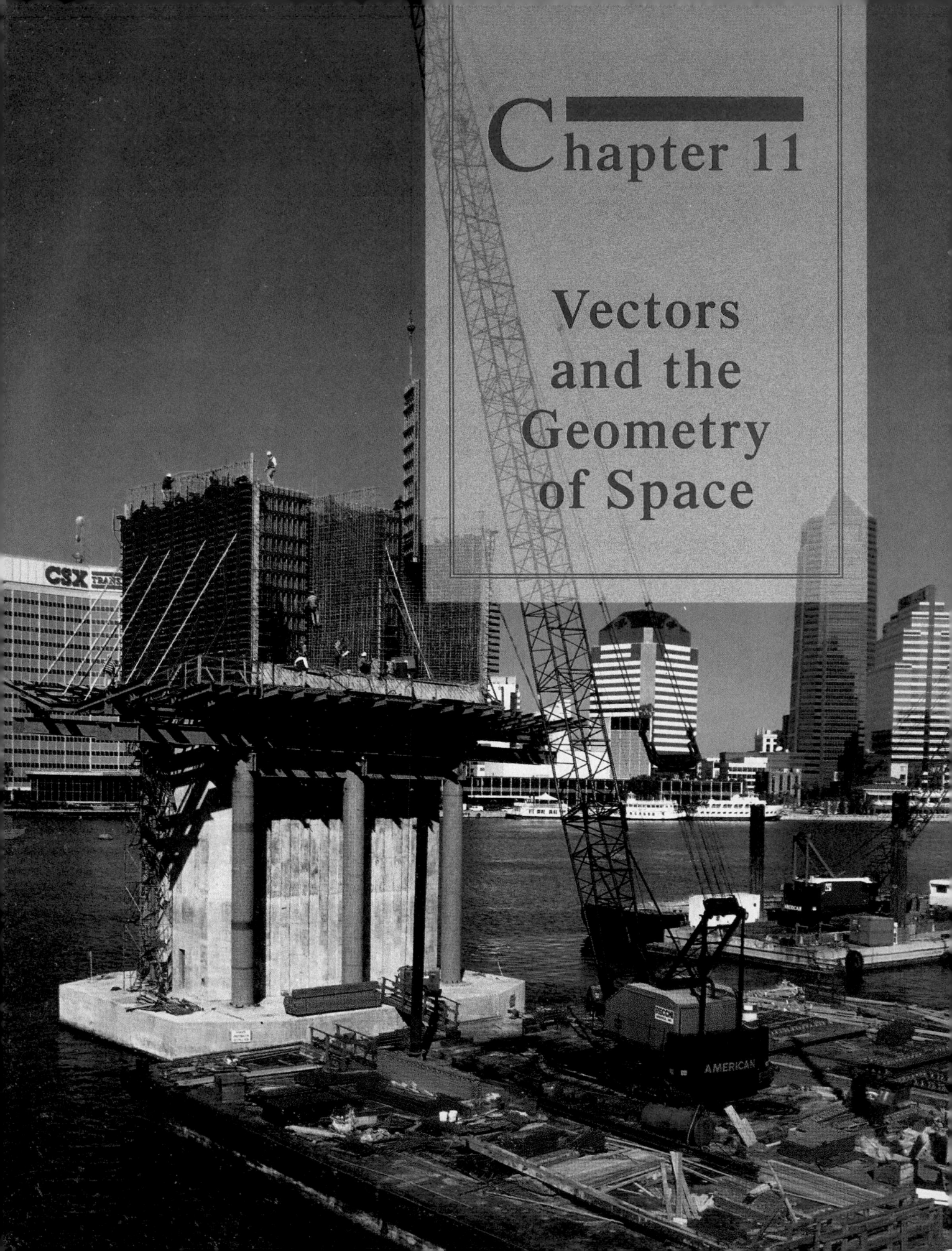

Chapter 11

Vectors and the Geometry of Space

11.1

Component Form of a Vector • Vector Operations • Standard Unit Vectors • Applications of Vectors

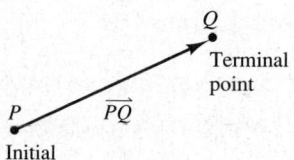

FIGURE 11.1
A directed line segment.

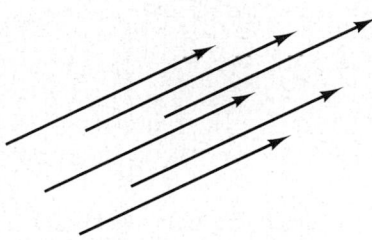

FIGURE 11.2
Equivalent directed line segments.

FIGURE 11.3
The vectors **u** and **v** are equal.

Component Form of a Vector

Many quantities in geometry and physics, such as area, volume, temperature, mass, and time, can be characterized by single real numbers scaled to appropriate units of measure. We call these **scalar quantities,** and the real number associated with each is called a **scalar.**

Other quantities, such as force and velocity, involve both magnitude and direction and cannot be characterized completely by single real numbers. A **directed line segment** is used to represent such a quantity, as shown in Figure 11.1. The directed line segment $\vec{PQ}$ has **initial point** P and **terminal point** Q, and its **length** is denoted by $\|\vec{PQ}\|$. Directed line segments that have the same length and direction are **equivalent,** as shown in Figure 11.2. The set of all directed line segments that are equivalent to a given directed line segment $\vec{PQ}$ is a **vector in the plane** and is denoted by $\mathbf{v} = \vec{PQ}$. In typesetting, vectors are usually denoted by lowercase, boldface letters such as $\mathbf{u}$, $\mathbf{v}$, and $\mathbf{w}$. When written by hand, however, vectors are often denoted by letters with arrows above them, such as $\vec{u}$, $\vec{v}$, and $\vec{w}$.

Be sure you see that a vector in the plane can be represented by many different directed line segments—all pointing in the same direction and all of the same length.

EXAMPLE 1   Vector Representation by Directed Line Segments

Let $\mathbf{v}$ be represented by the directed line segment from $(0, 0)$ to $(3, 2)$, and let $\mathbf{u}$ be represented by the directed line segment from $(1, 2)$ to $(4, 4)$. Show that $\mathbf{v} = \mathbf{u}$.

Solution   Let $P = (0, 0)$, $Q = (3, 2)$, $R = (1, 2)$, and $S = (4, 4)$, as shown in Figure 11.3. You can use the Distance Formula to show that $\vec{PQ}$ and $\vec{RS}$ have the *same length.*

$$\|\vec{PQ}\| = \sqrt{(3 - 0)^2 + (2 - 0)^2} = \sqrt{13}$$
$$\|\vec{RS}\| = \sqrt{(4 - 1)^2 + (4 - 2)^2} = \sqrt{13}$$

Both line segments have the *same direction,* because they both are directed toward the upper right on lines having the same slope.

$$\text{Slope of } \vec{PQ} = \frac{2 - 0}{3 - 0} = \frac{2}{3} \quad \text{and} \quad \text{Slope of } \vec{RS} = \frac{4 - 2}{4 - 1} = \frac{2}{3}$$

Because $\vec{PQ}$ and $\vec{RS}$ have the same length and direction, you can conclude that $\mathbf{v} = \mathbf{u}$.

◀ *This photo of a crane shows many intersecting lines. In Section 11.5, Exercises 15–18, you are asked to find the points of intersection of lines in space.*

FIGURE 11.4
The standard position of a vector.

The directed line segment whose initial point is the origin is often the most convenient representative of a set of equivalent directed line segments such as those shown in Figure 11.3. This representation of **v** is said to be in **standard position.** A directed line segment whose initial point is at the origin can be uniquely represented by the coordinates of its terminal point $Q = (v_1, v_2)$, as shown in Figure 11.4.

Definition of Component Form of a Vector in the Plane

If **v** is a vector in the plane whose initial point is the origin and whose terminal point is (v_1, v_2), then the **component form of v** is given by

$$\mathbf{v} = \langle v_1, v_2 \rangle.$$

The coordinates v_1 and v_2 are called the **components of v.** If both the initial point and the terminal point lie at the origin, then **v** is called the **zero vector** and is denoted by $\mathbf{0} = \langle 0, 0 \rangle$.

This definition implies that two vectors $\mathbf{u} = \langle u_1, u_2 \rangle$ and $\mathbf{v} = \langle v_1, v_2 \rangle$ are **equal** if and only if $u_1 = v_1$ and $u_2 = v_2$.

The following procedures can be used to convert directed line segments to component form or vice versa.

REMARK   It is important to understand that a vector represents a *set* of directed line segments (each having the same length and direction). In practice, however, it is common not to distinguish between a vector and one of its representatives.

1. If $P = (p_1, p_2)$ and $Q = (q_1, q_2)$, then the component form of the vector **v** represented by $\vec{PQ}$ is $\langle v_1, v_2 \rangle = \langle q_1 - p_1, q_2 - p_2 \rangle$. Moreover, the **length of v** is

$$\|\mathbf{v}\| = \sqrt{(q_1 - p_1)^2 + (q_2 - p_2)^2}$$
$$= \sqrt{v_1{}^2 + v_2{}^2}.$$

Length of a vector

2. If $\mathbf{v} = \langle v_1, v_2 \rangle$, then **v** can be represented by the directed line segment, in standard position, from $P = (0, 0)$ to $Q = (v_1, v_2)$.

The length of **v** is also called the **norm of v.** If $\|\mathbf{v}\| = 1$, then **v** is a **unit vector.** Moreover, $\|\mathbf{v}\| = 0$ if and only if **v** is the zero vector **0**.

EXAMPLE 2   Finding the Component Form and Length of a Vector

Find the component form and length of the vector **v** that has initial point $(3, -7)$ and terminal point $(-2, 5)$.

Solution   Let $P = (3, -7) = (p_1, p_2)$ and $Q = (-2, 5) = (q_1, q_2)$. Then the components of $\mathbf{v} = \langle v_1, v_2 \rangle$ are

$$v_1 = q_1 - p_1 = -2 - 3 = -5$$
$$v_2 = q_2 - p_2 = 5 - (-7) = 12.$$

Thus, as shown in Figure 11.5, $\mathbf{v} = \langle -5, 12 \rangle$, and the length of **v** is

$$\|\mathbf{v}\| = \sqrt{(-5)^2 + 12^2}$$
$$= \sqrt{169}$$
$$= 13.$$

FIGURE 11.5
Component form of **v**: $\mathbf{v} = \langle -5, 12 \rangle$.

Vector Operations

The two basic vector operations are **scalar multiplication** and **vector addition.**

FIGURE 11.6
The scalar multiplication of **v**.

Definitions of Vector Addition and Scalar Multiplication

Let $\mathbf{u} = \langle u_1, u_2 \rangle$ and $\mathbf{v} = \langle v_1, v_2 \rangle$ be vectors and let k be a scalar.

1. The **vector sum** of $\mathbf{u}$ and $\mathbf{v}$ is the vector $\mathbf{u} + \mathbf{v} = \langle u_1 + v_1, u_2 + v_2 \rangle$.
2. The **scalar multiple** of k and $\mathbf{u}$ is the vector $k\,\mathbf{u} = \langle ku_1, ku_2 \rangle$.
3. The **negative** of $\mathbf{v}$ is the vector $-\mathbf{v} = (-1)\mathbf{v} = \langle -v_1, -v_2 \rangle$.
4. The **difference** of $\mathbf{u}$ and $\mathbf{v}$ is $\mathbf{u} - \mathbf{v} = \mathbf{u} + (-\mathbf{v}) = \langle u_1 - v_1, u_2 - v_2 \rangle$.

Geometrically, the product of a vector $\mathbf{v}$ and a scalar k is the vector that is k times as long as $\mathbf{v}$, as shown in Figure 11.6. If k is positive, then $k\,\mathbf{v}$ has the same direction as $\mathbf{v}$. If k is negative, then $k\,\mathbf{v}$ has the opposite direction.

To add two vectors geometrically, position them (without changing their magnitudes or directions) so that the initial point of one coincides with the terminal point of the other, as shown in Figure 11.7. The vector $\mathbf{u} + \mathbf{v}$ (called the **resultant vector**) is the diagonal of a parallelogram having $\mathbf{u}$ and $\mathbf{v}$ as its adjacent sides.

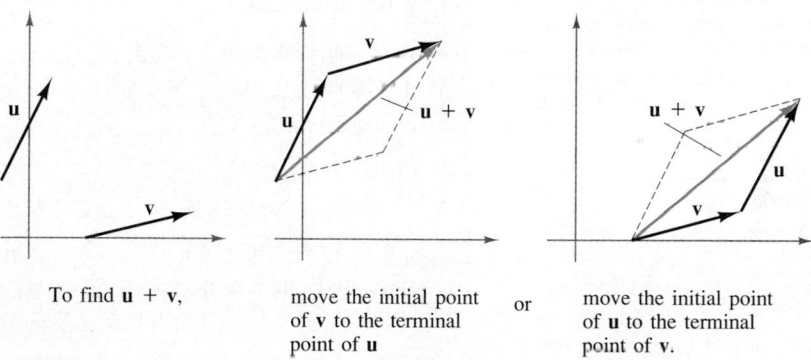

To find **u** + **v**,   move the initial point   or   move the initial point
                            of **v** to the terminal   of **u** to the terminal
                            point of **u**   point of **v**.

FIGURE 11.7

Figure 11.8 shows the equivalence of the geometric and algebraic definitions of vector addition and scalar multiplication. The right-most portion of the figure gives a geometric interpretation of $\mathbf{u} - \mathbf{v}$.

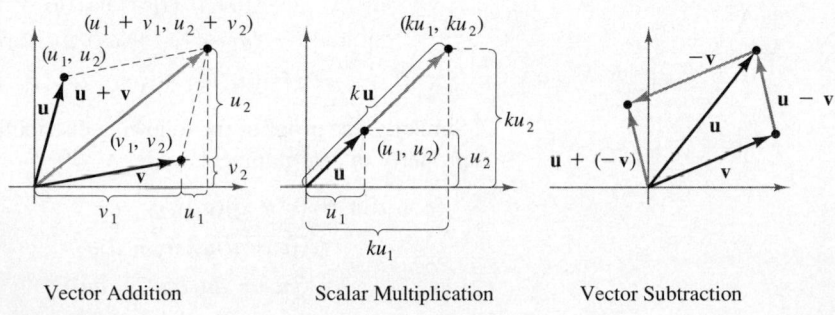

Vector Addition   Scalar Multiplication   Vector Subtraction

FIGURE 11.8

WILLIAM ROWAN HAMILTON

Some of the earliest work with vectors was done by the Irish mathematician William Rowan Hamilton (1805–1865). Hamilton spent many years developing a system of vector-like quantities called quaternions. Although Hamilton was convinced of the benefits of quaternions, the operations he defined did not produce good models for physical phenomena. It wasn't until the latter half of the nineteenth century that the Scottish physicist James Maxwell (1831–1879) restructured Hamilton's quaternions in a form useful for representing physical quantities such as force, velocity, and acceleration.

EXAMPLE 3   Vector Operations

Given $\mathbf{v} = \langle -2, 5 \rangle$ and $\mathbf{w} = \langle 3, 4 \rangle$, find the following vectors.

a. $\dfrac{1}{2}\mathbf{v}$ b. $\mathbf{w} - \mathbf{v}$ c. $\mathbf{v} + 2\mathbf{w}$

Solution

a. $\dfrac{1}{2}\mathbf{v} = \left\langle \dfrac{1}{2}(-2), \dfrac{1}{2}(5) \right\rangle = \left\langle -1, \dfrac{5}{2} \right\rangle$

b. $\mathbf{w} - \mathbf{v} = \langle w_1 - v_1, w_2 - v_2 \rangle = \langle 3 - (-2), 4 - 5 \rangle = \langle 5, -1 \rangle$

c. Using $2\mathbf{w} = \langle 6, 8 \rangle$, you have

$$\mathbf{v} + 2\mathbf{w} = \langle -2, 5 \rangle + \langle 6, 8 \rangle = \langle -2 + 6, 5 + 8 \rangle = \langle 4, 13 \rangle.$$

Vector addition and scalar multiplication share many properties of ordinary arithmetic, as shown in the following theorem.

THEOREM 11.1   Properties of Vector Operations

Let $\mathbf{u}$, $\mathbf{v}$, and $\mathbf{w}$ be vectors in the plane, and let c and d be scalars.

1. $\mathbf{u} + \mathbf{v} = \mathbf{v} + \mathbf{u}$ Commutative property
2. $(\mathbf{u} + \mathbf{v}) + \mathbf{w} = \mathbf{u} + (\mathbf{v} + \mathbf{w})$ Associative property
3. $\mathbf{u} + \mathbf{0} = \mathbf{u}$
4. $\mathbf{u} + (-\mathbf{u}) = \mathbf{0}$
5. $c(d\mathbf{u}) = (cd)\mathbf{u}$
6. $(c + d)\mathbf{u} = c\mathbf{u} + d\mathbf{u}$ Distributive property
7. $c(\mathbf{u} + \mathbf{v}) = c\mathbf{u} + c\mathbf{v}$ Distributive property
8. $1(\mathbf{u}) = \mathbf{u}, \; 0(\mathbf{u}) = \mathbf{0}$

Proof   The proof of the *associative property* of vector addition follows the associative property of addition of real numbers.

$$
\begin{aligned}
(\mathbf{u} + \mathbf{v}) + \mathbf{w} &= [\langle u_1, u_2 \rangle + \langle v_1, v_2 \rangle] + \langle w_1, w_2 \rangle \\
&= \langle u_1 + v_1, u_2 + v_2 \rangle + \langle w_1, w_2 \rangle \\
&= \langle (u_1 + v_1) + w_1, (u_2 + v_2) + w_2 \rangle \\
&= \langle u_1 + (v_1 + w_1), u_2 + (v_2 + w_2) \rangle \\
&= \langle u_1, u_2 \rangle + \langle v_1 + w_1, v_2 + w_2 \rangle = \mathbf{u} + (\mathbf{v} + \mathbf{w})
\end{aligned}
$$

Similarly, the proof of the following distributive property depends on the distributive property of real numbers.

$$
\begin{aligned}
(c + d)\mathbf{u} &= (c + d)\langle u_1, u_2 \rangle \\
&= \langle (c + d)u_1, (c + d)u_2 \rangle \\
&= \langle cu_1 + du_1, cu_2 + du_2 \rangle \\
&= \langle cu_1, cu_2 \rangle + \langle du_1, du_2 \rangle = c\mathbf{u} + d\mathbf{u}
\end{aligned}
$$

The other properties can be proved in a similar manner.

Any set of vectors (with an accompanying set of scalars) that satisfies the eight properties given in Theorem 11.1 is a **vector space.** The eight properties are the *vector space axioms.* Thus, this theorem states that the set of vectors in the plane (with the set of real numbers) forms a vector space. You will study vector spaces in detail if you take a course in *linear algebra.*

THEOREM 11.2   Length of a Scalar Multiple

Let **v** be a vector and c be a scalar. Then

$$\| c\,\mathbf{v} \| = | c | \, \| \mathbf{v} \|$$

where $| c |$ is the absolute value of c.

Proof   Because $c\,\mathbf{v} = \langle cv_1, cv_2 \rangle$, it follows that

$$
\begin{aligned}
\| c\,\mathbf{v} \| = \| \langle cv_1, cv_2 \rangle \| &= \sqrt{(cv_1)^2 + (cv_2)^2} \\
&= \sqrt{c^2 v_1{}^2 + c^2 v_2{}^2} \\
&= \sqrt{c^2(v_1{}^2 + v_2{}^2)} \\
&= | c | \sqrt{v_1{}^2 + v_2{}^2} \\
&= | c | \, \| \mathbf{v} \|.
\end{aligned}
$$

In many applications of vectors it is useful to find a unit vector that has the same direction as a given vector. The following theorem gives a procedure for doing this.

EMMY NOETHER

One person who contributed to our knowledge of axiomatic systems was the German mathematician Emmy Noether (1882–1935). Noether is generally recognized as the leading woman mathematician in recent history.

FOR FURTHER INFORMATION
For more information on Emmy Noether, see the article "Emmy Noether, Greatest Woman Mathematician" by Clark Kimberling in the March, 1982 issue of *Mathematics Teacher.*

THEOREM 11.3   Unit Vector in the Direction of v

If **v** is a nonzero vector in the plane, then the vector

$$\mathbf{u} = \frac{\mathbf{v}}{\| \mathbf{v} \|} = \frac{1}{\| \mathbf{v} \|}\,\mathbf{v}$$

and has length 1 and the same direction as **v.**

Proof   Because $1/\| \mathbf{v} \|$ is positive and $\mathbf{u} = (1/\| \mathbf{v} \|)\mathbf{v}$, you can conclude that **u** has the same direciton as **v.** To see that $\| \mathbf{u} \| = 1$, note that

$$
\begin{aligned}
\| \mathbf{u} \| &= \left\| \left(\frac{1}{\| \mathbf{v} \|} \right) \mathbf{v} \right\| \\
&= \left| \frac{1}{\| \mathbf{v} \|} \right| \, \| \mathbf{v} \| \\
&= \frac{1}{\| \mathbf{v} \|}\, \| \mathbf{v} \| = 1.
\end{aligned}
$$

Thus, **u** has length 1 and the same direction as **v.**

In Theorem 11.3, **u** is called a **unit vector in the direction of v.** The process of multiplying **v** by $1/\| \mathbf{v} \|$ to get a unit vector is called **normalization of v.**

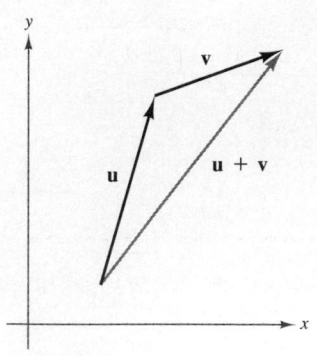

FIGURE 11.9
Triangle inequality

EXAMPLE 4   Finding a Unit Vector

Find a unit vector in the direction of $\mathbf{v} = \langle -2, 5 \rangle$ and verify that the result has length 1.

Solution From Theorem 11.3, the unit vector in the direction of $\mathbf{v}$ is

$$\frac{\mathbf{v}}{\|\mathbf{v}\|} = \frac{\langle -2, 5 \rangle}{\sqrt{(-2)^2 + (5)^2}} = \frac{1}{\sqrt{29}}\langle -2, 5 \rangle = \left\langle \frac{-2}{\sqrt{29}}, \frac{5}{\sqrt{29}} \right\rangle.$$

This vector has length 1, because

$$\sqrt{\left(\frac{-2}{\sqrt{29}}\right)^2 + \left(\frac{5}{\sqrt{29}}\right)^2} = \sqrt{\frac{4}{29} + \frac{25}{29}} = \sqrt{\frac{29}{29}} = 1.$$

Generally, the length of the sum of two vectors is not equal to the sum of their lengths. To see this, consider the vectors $\mathbf{u}$ and $\mathbf{v}$ as shown in Figure 11.9. By considering $\mathbf{u}$ and $\mathbf{v}$ as two sides of a triangle, you can see that the length of the third side is $\|\mathbf{u} + \mathbf{v}\|$, and you have

$$\|\mathbf{u} + \mathbf{v}\| \le \|\mathbf{u}\| + \|\mathbf{v}\|.$$

Equality occurs only if the vectors $\mathbf{u}$ and $\mathbf{v}$ have the *same* direction. This result is called the **Triangle Inequality** for vectors. (You are asked to prove this in Exercise 60, Section 11.3.)

Standard Unit Vectors

The unit vectors $\langle 1, 0 \rangle$ and $\langle 0, 1 \rangle$ are called the **standard unit vectors** in the plane and are denoted by

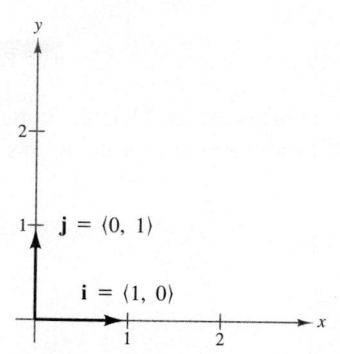

FIGURE 11.10
Standard unit vectors **i** and **j**.

$$\boxed{\mathbf{i} = \langle 1, 0 \rangle \quad \text{and} \quad \mathbf{j} = \langle 0, 1 \rangle} \qquad \text{Standard unit vectors}$$

as shown in Figure 11.10. These vectors can be used to represent any vector, as follows.

$$\mathbf{v} = \langle v_1, v_2 \rangle = \langle v_1, 0 \rangle + \langle 0, v_2 \rangle = v_1\langle 1, 0 \rangle + v_2\langle 0, 1 \rangle = v_1\mathbf{i} + v_2\mathbf{j}$$

The vector $\mathbf{v} = v_1\mathbf{i} + v_2\mathbf{j}$ is called a **linear combination** of $\mathbf{i}$ and $\mathbf{j}$. The scalars v_1 and v_2 are called the **horizontal** and **vertical components of v**, respectively.

EXAMPLE 5   Writing a Vector as a Linear Combination of Unit Vectors

Let $\mathbf{u}$ be the vector with initial point $(2, -5)$ and terminal point $(-1, 3)$, and let $\mathbf{v} = 2\mathbf{i} - \mathbf{j}$. Write each of the following vectors as a linear combination of the standard unit vectors $\mathbf{i}$ and $\mathbf{j}$.

a. u   **b. w = 2u − 3v**

Solution

a. $\mathbf{u} = \langle q_1 - p_1, q_2 - p_2 \rangle = \langle -1 - 2, 3 - (-5) \rangle$
$$= \langle -3, 8 \rangle = -3\mathbf{i} + 8\mathbf{j}$$

b. $\mathbf{w} = 2\mathbf{u} - 3\mathbf{v} = 2(-3\mathbf{i} + 8\mathbf{j}) - 3(2\mathbf{i} - \mathbf{j})$
$$= -6\mathbf{i} + 16\mathbf{j} - 6\mathbf{i} + 3\mathbf{j} = -12\mathbf{i} + 19\mathbf{j}$$

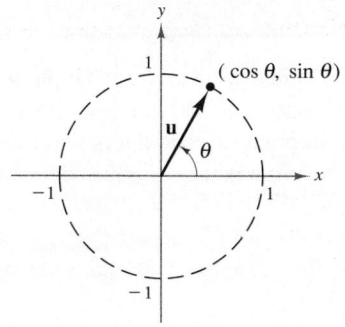

FIGURE 11.11
The angle θ from the positive x-axis to the vector $\mathbf{u}$.

If $\mathbf{u}$ is a unit vector such that θ is the angle (measured counterclockwise) from the positive x-axis to $\mathbf{u}$, then the terminal point of $\mathbf{u}$ lies on the unit circle, and you have

$$\mathbf{u} = \langle \cos\theta, \sin\theta \rangle = \cos\theta\,\mathbf{i} + \sin\theta\,\mathbf{j} \qquad \text{Unit vector}$$

as shown in Figure 11.11. Moreover, it follows that any other nonzero vector $\mathbf{v}$ making an angle θ with the positive x-axis has the same direction as $\mathbf{u}$, and you can write

$$\mathbf{v} = \|\mathbf{v}\|\langle \cos\theta, \sin\theta \rangle = \|\mathbf{v}\|\cos\theta\,\mathbf{i} + \|\mathbf{v}\|\sin\theta\,\mathbf{j}.$$

EXAMPLE 6   Writing a Vector of Given Length and Direction

The vector $\mathbf{v}$ has a length of 3 and makes an angle of $30° = \pi/6$ with the positive x-axis. Write $\mathbf{v}$ as a linear combination of the unit vectors $\mathbf{i}$ and $\mathbf{j}$.

Solution   Because the angle between $\mathbf{v}$ and the positive x-axis is $\theta = \pi/3$, you can write the following.

$$\mathbf{v} = \|\mathbf{v}\|\cos\theta\,\mathbf{i} + \|\mathbf{v}\|\sin\theta\,\mathbf{j} = 3\cos\frac{\pi}{6}\mathbf{i} + 3\sin\frac{\pi}{6}\mathbf{j}$$
$$= \frac{3\sqrt{3}}{2}\mathbf{i} + \frac{3}{2}\mathbf{j}$$

Applications of Vectors

There are many applications of vectors in physics and engineering. We conclude this section by considering two examples.

EXAMPLE 7   Finding the Resultant Force

Two tugboats are pushing an ocean liner, as shown in Figure 11.12. Each boat is exerting a force of 400 pounds. What is the resultant force on the ocean liner?

Solution   Using Figure 11.12, you can represent the forces exerted by the first and second tugboats as

$$\mathbf{F}_1 = 400\langle \cos 20°, \sin 20° \rangle$$
$$= 400\cos(20°)\mathbf{i} + 400\sin(20°)\mathbf{j}$$
$$\mathbf{F}_2 = 400\langle \cos(-20°), \sin(-20°) \rangle$$
$$= 400\cos(20°)\mathbf{i} - 400\sin(20°)\mathbf{j}.$$

To obtain the resultant force on the ocean liner, add these two forces.

$$\mathbf{F} = \mathbf{F}_1 + \mathbf{F}_2$$
$$= 400\cos(20°)\mathbf{i} + 400\sin(20°)\mathbf{j} + 400\cos(20°)\mathbf{i} - 400\sin(20°)\mathbf{j}$$
$$= 800\cos(20°)\mathbf{i}$$
$$\approx 752\mathbf{i}$$

Thus, the resultant force on the ocean liner is approximately 752 pounds in the direction of the positive x-axis.

FIGURE 11.12
The resultant force on the ocean liner is approximately 752 pounds in the positive x-direction.

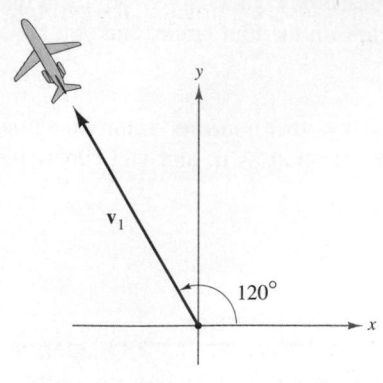

v₁

120°

Direction without wind

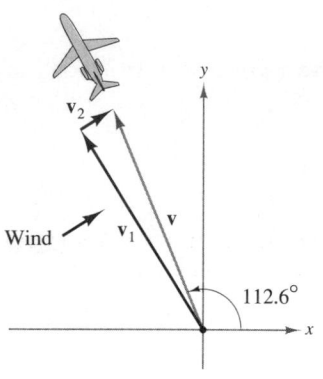

v₂

Wind **v₁** **v**

112.6°

Direction with wind

FIGURE 11.13

EXAMPLE 8 Finding a Velocity

An airplane is traveling at a fixed altitude with a negligible wind factor. The plane is headed N 30° W (which means 30° west of north) at a speed of 500 miles per hour, as shown in Figure 11.13. As the plane reaches a certain point, it encounters wind with a velocity of 70 miles per hour in the direction E 45° N. What is the resultant speed and direction of the plane?

Solution Using Figure 11.13, you can represent the velocity of the plane by the vector

$$\mathbf{v}_1 = 500\cos(120°)\mathbf{i} + 500\sin(120°)\mathbf{j}.$$

The velocity of the wind is represented by the vector

$$\mathbf{v}_2 = 70\cos(45°)\mathbf{i} + 70\sin(45°)\mathbf{j}.$$

The resultant velocity of the plane is

$$\mathbf{v} = \mathbf{v}_1 + \mathbf{v}_2$$
$$= 500\cos(120°)\mathbf{i} + 500\sin(120°)\mathbf{j} + 70\cos(45°)\mathbf{i} + 70\sin(45°)\mathbf{j}$$
$$\approx -200.5\mathbf{i} + 482.5\mathbf{j}.$$

To find the speed and direction, write $\mathbf{v} = \|\mathbf{v}\|(\cos\theta\,\mathbf{i} + \sin\theta\,\mathbf{j})$. Because

$$\|\mathbf{v}\| \approx \sqrt{(-200.5)^2 + (482.5)^2} \approx 522.5$$

you can write

$$\mathbf{v} \approx 522.5\left(\frac{-200.5}{522.5}\mathbf{i} + \frac{482.5}{522.5}\mathbf{j}\right) \approx 522.5[\cos(112.6°)\mathbf{i} + \sin(112.6°)\mathbf{j}].$$

Therefore, the speed of the plane, as altered by the wind, is approximately 522.5 miles per hour in a flight path that makes an angle of 112.6° with the positive x-axis.

EXERCISES for Section 11.1

In Exercises 1–4, (a) find the component form of the vector v and (b) sketch the vector with its initial point at the origin.

1.

2.

3.

4.

In Exercises 5–12, the initial and terminal points of a vector v are given. (a) Sketch the given directed line segment, (b) write the vector in component form, and (c) sketch the vector with its initial point at the origin.

	Initial point	Terminal point
5.	$(1, 2)$	$(5, 5)$
6.	$(3, -5)$	$(4, 7)$
7.	$(10, 2)$	$(6, -1)$
8.	$(0, -4)$	$(-5, -1)$
9.	$(6, 2)$	$(6, 6)$
10.	$(7, -1)$	$(-3, -1)$
11.	$(\frac{3}{2}, \frac{4}{3})$	$(\frac{1}{2}, 3)$
12.	$(0.12, 0.60)$	$(0.84, 1.25)$

In Exercises 13 and 14, sketch the scalar multiple of v.

13. $\mathbf{v} = \langle 2, 3 \rangle$

    a. $2\mathbf{v}$ **b.** $-3\mathbf{v}$ **c.** $\frac{7}{2}\mathbf{v}$ **d.** $\frac{2}{3}\mathbf{v}$

14. $v = \langle -1, 5 \rangle$
   a. $4v$ **b.** $-\frac{1}{2}v$ **c.** $0v$ **d.** $-6v$

In Exercises 15–20, find the vector v, where $u = \langle 2, -1 \rangle$ and $w = \langle 1, 2 \rangle$. Illustrate the indicated vector operations geometrically.

15. $v = \frac{3}{2}u$ **16.** $v = u + w$

17. $v = u + 2w$ **18.** $v = -u + w$

19. $v = \frac{1}{2}(3u + w)$ **20.** $v = u - 2w$

In Exercises 21–26, find a and b such that $v = au + bw$, where $u = \langle 1, 2 \rangle$ and $w = \langle 1, -1 \rangle$.

21. $v = \langle 2, 1 \rangle$ **22.** $v = \langle 0, 3 \rangle$

23. $v = \langle 3, 0 \rangle$ **24.** $v = \langle 3, 3 \rangle$

25. $v = \langle 1, 1 \rangle$ **26.** $v = \langle -1, 7 \rangle$

In Exercises 27 and 28, the vector v and its initial point are given. Find the terminal point.

27. $v = \langle -1, 3 \rangle$, initial point $(4, 2)$

28. $v = \langle 4, -9 \rangle$, initial point $(3, 2)$

In Exercises 29–34, find the magnitude of v.

29. $v = \langle 4, 3 \rangle$ **30.** $v = \langle 12, -5 \rangle$

31. $v = 6i - 5j$ **32.** $v = -10i + 3j$

33. $v = 4j$ **34.** $v = i - j$

In Exercises 35–38, find the following.
   a. $\| u \|$ **b.** $\| v \|$ **c.** $\| u + v \|$

   d. $\left\| \dfrac{u}{\| u \|} \right\|$ **e.** $\left\| \dfrac{v}{\| v \|} \right\|$ **f.** $\left\| \dfrac{u + v}{\| u + v \|} \right\|$

35. $u = \langle 1, -1 \rangle$ **36.** $u = \langle 0, 1 \rangle$
     $v = \langle -1, 2 \rangle$ $v = \langle 3, -3 \rangle$

37. $u = \langle 1, \frac{1}{2} \rangle$ **38.** $u = \langle 2, -4 \rangle$
     $v = \langle 2, 3 \rangle$ $v = \langle 5, 5 \rangle$

In Exercises 39 and 40, demonstrate the triangle inequality using the vectors u and v.

39. $u = \langle 2, 1 \rangle$ **40.** $u = \langle -3, 2 \rangle$
     $v = \langle 5, 4 \rangle$ $v = \langle 1, -2 \rangle$

In Exercises 41–44, find the vector v with the given magnitude and the same direction as u.

	Magnitude	Direction
41.	$\| v \| = 4$	$u = \langle 1, 1 \rangle$
42.	$\| v \| = 4$	$u = \langle -1, 1 \rangle$
43.	$\| v \| = 2$	$u = \langle \sqrt{3}, 3 \rangle$
44.	$\| v \| = 3$	$u = \langle 0, 3 \rangle$

In Exercises 45–48, find a unit vector (a) parallel to the graph of $f(x)$ at the indicated point, and (b) normal to the graph of $f(x)$ at the indicated point.

	Function	Point
45.	$f(x) = x^3$	$(1, 1)$
46.	$f(x) = x^3$	$(-2, -8)$
47.	$f(x) = \sqrt{25 - x^2}$	$(3, 4)$
48.	$f(x) = \tan x$	$\left(\dfrac{\pi}{4}, 1 \right)$

In Exercises 49–52, find the component form of v given its magnitude and the angle it makes with the positive x-axis.

	Magnitude	Angle
49.	$\| v \| = 3$	$\theta = 0°$
50.	$\| v \| = 1$	$\theta = 45°$
51.	$\| v \| = 2$	$\theta = 150°$
52.	$\| v \| = 1$	$\theta = 3.5°$

In Exercises 53–56, find the component form of u + v given the magnitudes of u and v and the angles that u and v make with the positive x-axis.

53. $\| u \| = 1, \theta_u = 0°$
    $\| v \| = 3, \theta_v = 45°$

54. $\| u \| = 4, \theta_u = 0°$
    $\| v \| = 2, \theta_v = 60°$

55. $\| u \| = 2, \theta_u = 4$
    $\| v \| = 1, \theta_v = 2$

56. $\| u \| = 5, \theta_u = -0.5$
    $\| v \| = 5, \theta_v = 0.5$

In Exercises 57 and 58, find the component form of v given the magnitudes of u and u + v and the angles that u and u + v make with the positive x-axis.

57. $\| u \| = 1, \theta = 45°$ **58.** $\| u \| = 4, \theta = 30°$
    $\| u + v \| = \sqrt{2}, \theta = 90°$ $\| u + v \| = 6, \theta = 120°$

*C **In Exercises 59 and 60, use a computer algebra system to find the magnitude and direction of the resultant of the vectors.**

59.   **60.**

*A C indicates an exercise in which you are instructed to use graphing technology or a symbolic computer algebra system. The solutions of other exercises may also be facilitated by appropriate technology.

61. *Resultant Force*   Forces with magnitudes of 35 and 50 pounds act on a hook (see figure). The angle between the two forces is 30°. Find the direction and magnitude of the resultant (vector sum) of these two forces.

62. *Resultant Force*   Forces with magnitudes of 500 and 200 pounds act on a machine part at angles of 30° and −45°, respectively, with the *x*-axis (see figure). Find the direction and magnitude of the resultant (vector sum) of these forces.

FIGURE FOR 61   **FIGURE FOR 62**

63. *Resultant Force*   Three forces with magnitudes of 75, 100, and 125 pounds act on an object at angles of 30°, 45°, and 120°, respectively, with the positive *x*-axis. Find the direction and magnitude of the resultant of these forces.

64. *Resultant Force*   Three forces with magnitudes of 70, 40, and 60 pounds act on an object at angles of −30°, 45°, and 135°, respectively, with the positive *x*-axis. Find the direction and magnitude of the resultant of these forces.

65. *Tether Ball*   A tether ball weighing 1 pound is pulled outward from the pole by a horizontal force **u** until the rope makes a 30° angle with the pole (see figure). Determine the resulting tension in the rope and the magnitude of **u**.

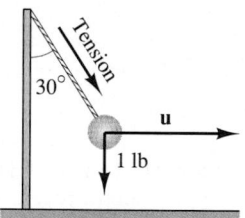

FIGURE FOR 65

66. *Barge Towing*   A loaded barge is being towed by two tugboats, and the magnitude of the resultant is 6000 pounds directed along the axis of the barge. Find the tension in the tow lines if they each make a 20° angle with the axis of the barge.

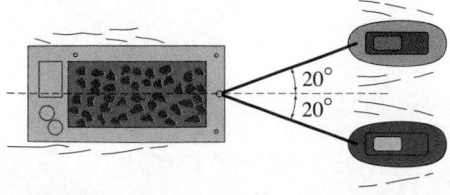

FIGURE FOR 66

67. *Shared Load*   To carry a 100-pound cylindrical weight, two workers lift on the ends of short ropes tied to an eyelet on the top center of the cylinder. One rope makes a 20° angle away from the vertical and the other a 30° angle (see figure).
   a. Find the tension in each rope if the resultant force is vertical.
   b. Find the vertical component of each worker's force.

68. *Navigation*   An airplane is headed 32° north of west. Its speed with respect to the air is 580 miles per hour. The wind at the plane's altitude is from the southwest at 60 miles per hour (see figure). What is the true direction of the plane, and what is its speed with respect to the ground?

FIGURE FOR 67   **FIGURE FOR 68**

69. *Navigation*   A plane flies at a constant ground speed of 450 miles per hour due east and encounters a 50 mile per hour wind from the northwest. Find the air speed and compass direction that will allow the plane to maintain its ground speed and eastward direction.

70. *Projectile Motion*   A gun with a muzzle velocity of 1200 feet per second is fired at an angle of 6° with the horizontal. Find the vertical and horizontal components of the velocity.

71. Three vertices of a parallelogram are (1, 2), (3, 1), and (8, 4). Find the three possible fourth vertices (see figure).

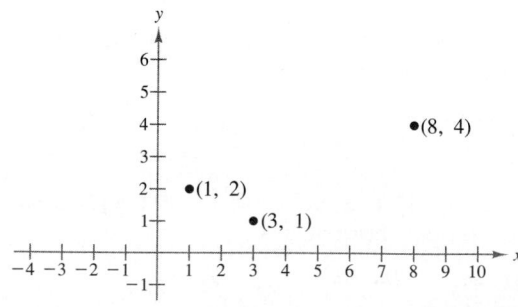

FIGURE FOR 71

72. Use vectors to find the points of trisection of the line segment with endpoints (1, 2) and (7, 5).

73. If $\mathbf{F}_1 + \mathbf{F}_2 + \mathbf{F}_3 = \mathbf{0}$, find T_2 and T_3.

$$\mathbf{F}_1 = -3600\mathbf{j}$$
$$\mathbf{F}_2 = T_2(\cos 35°\,\mathbf{i} - \sin 35°\,\mathbf{j})$$
$$\mathbf{F}_3 = T_3(\cos 92°\,\mathbf{i} + \sin 92°\,\mathbf{j})$$

74. Prove that the vectors

$$\mathbf{u} = (\cos\theta)\mathbf{i} - (\sin\theta)\mathbf{j}$$

and

$$\mathbf{v} = (\sin\theta)\mathbf{i} + (\cos\theta)\mathbf{j}$$

are unit vectors for any angle θ.

75. Using vectors, prove that the line segment joining the midpoints of two sides of a triangle is parallel to, and one-half the length of, the third side.

76. Using vectors, prove that the diagonals of a parallelogram bisect each other.

True or False In Exercises 77–82, determine whether the statement is true or false. If it is false, explain why or give an example that shows it is false.

77. If $\mathbf{u}$ and $\mathbf{v}$ have the same magnitude and direction, then $\mathbf{u} = \mathbf{v}$.

78. If $\mathbf{u}$ is a unit vector in the direction of $\mathbf{v}$, then $\mathbf{v} = \|\mathbf{v}\|\,\mathbf{u}$.

79. If $\mathbf{u} = a\mathbf{i} + b\mathbf{j}$ is a unit vector, then $a^2 + b^2 = 1$.

80. If $\mathbf{v} = a\mathbf{i} + b\mathbf{j} = \mathbf{0}$, then $a = -b$.

81. If $a = b$, then $\|a\mathbf{i} + b\mathbf{j}\| = \sqrt{2}a$.

82. If $\mathbf{u}$ and $\mathbf{v}$ have the same magnitude but opposite directions, then $\mathbf{u} + \mathbf{v} = \mathbf{0}$.

83. Prove that the vector $\mathbf{w} = \|\mathbf{u}\|\,\mathbf{v} + \|\mathbf{v}\|\,\mathbf{u}$ bisects the angle between $\mathbf{u}$ and $\mathbf{v}$.

SECTION

11.2 Space Coordinates and Vectors in Space

Coordinates in Space • Vectors in Space • Applications

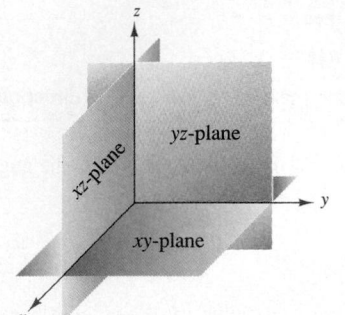

FIGURE 11.14
The three-dimensional
coordinate system.

Coordinates in Space

Up to this point in the text, we have been primarily concerned with the two-dimensional coordinate system. Much of the remaining part of your study of calculus will involve the three-dimensional coordinate system.

Before extending the concept of a vector to three dimensions, we introduce the **three-dimensional coordinate system.** You can construct this system by passing a z-axis perpendicular to both the x- and y-axes at the origin. Figure 11.14 shows the positive portion of each coordinate axis. Taken as pairs, the axes determine three **coordinate planes:** the **xy-plane,** the **xz-plane,** and the **yz-plane.** These three coordinate planes separate three-space into eight **octants.** The first octant is the one for which all three coordinates are positive. In this three-dimensional system, a point P in space is determined by an ordered triple (x, y, z), where x, y, and z are as follows.

$x =$ directed distance from yz-plane to P

$y =$ directed distance from xz-plane to P

$z =$ directed distance from xy-plane to P

Several points are shown in Figure 11.15.

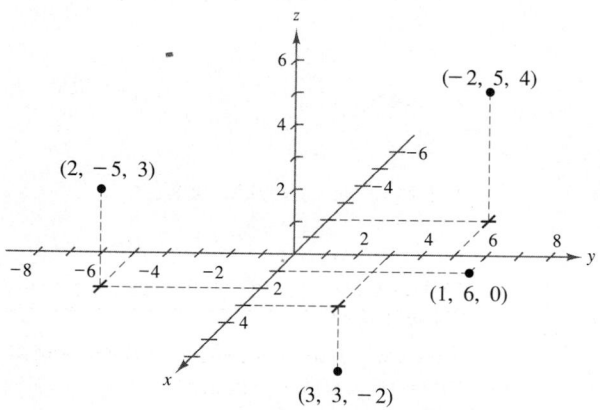

FIGURE 11.15
Points in the three-dimensional coordinate system are
represented by ordered triples.

Right-handed   Left-handed
System   System

FIGURE 11.16

A three-dimensional coordinate system can have either a **left-handed** or a **right-handed** orientation. To determine the orientation of a system, imagine that you are standing at the origin, with your arms pointing in the direction of the positive x- and y-axes, and with the z-axis pointing up, as shown in Figure 11.16. The system is right-handed or left-handed, depending on which hand points along the x-axis. In this text we work exclusively with the right-handed system.

Many of the formulas established for the two-dimensional coordinate system can be extended to three dimensions. For example, to find the distance between two points in space, you can use the Pythagorean Theorem twice, as shown in Figure 11.17. By doing this, you will obtain the formula for the distance between the points (x_1, y_1, z_1) and (x_2, y_2, z_2).

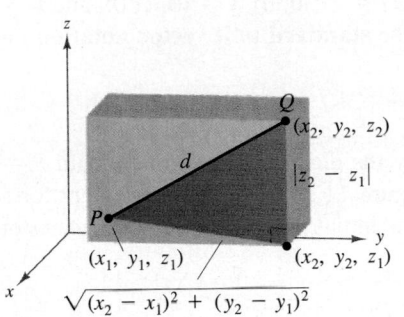

FIGURE 11.17
The distance between two points in space.

$$d = \sqrt{(x_2 - x_1)^2 + (y_2 - y_1)^2 + (z_2 - z_1)^2} \qquad \text{Distance Formula}$$

EXAMPLE 1 Finding the Distance Between Two Points in Space

The distance between the points $(2, -1, 3)$ and $(1, 0, -2)$ is

$$d = \sqrt{(1 - 2)^2 + (0 + 1)^2 + (-2 - 3)^2}$$
$$= \sqrt{1 + 1 + 25} = \sqrt{27} = 3\sqrt{3}.$$

A **sphere** with center at (x_0, y_0, z_0) and radius r is defined as the set of all points (x, y, z) such that the distance between (x, y, z) and (x_0, y_0, z_0) is r. You can use the Distance Formula to find the **standard equation of a sphere** of radius r, centered at (x_0, y_0, z_0). If (x, y, z) is an arbitrary point on the sphere, then the equation of the sphere is

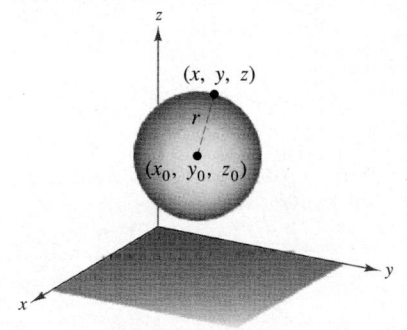

FIGURE 11.18

$$(x - x_0)^2 + (y - y_0)^2 + (z - z_0)^2 = r^2 \qquad \text{Equation of a sphere}$$

as shown in Figure 11.18. Moreover, the midpoint of the line segment joining the points (x_1, y_1, z_1) and (x_2, y_2, z_2) has coordinates

$$\left(\frac{x_1 + x_2}{2}, \frac{y_1 + y_2}{2}, \frac{z_1 + z_2}{2} \right). \qquad \text{Midpoint Rule}$$

The following example makes use of both of these formulas.

EXAMPLE 2 Finding the Equation of a Sphere

Find the standard equation for the sphere that has the points $(5, -2, 3)$ and $(0, 4, -3)$ as endpoints of a diameter.

Solution By the Midpoint Rule, the center of the sphere is

$$\left(\frac{5 + 0}{2}, \frac{-2 + 4}{2}, \frac{3 - 3}{2} \right) = \left(\frac{5}{2}, 1, 0 \right).$$

By the Distance Formula, the radius is

$$r = \sqrt{\left(0 - \frac{5}{2} \right)^2 + (4 - 1)^2 + (-3 - 0)^2} = \sqrt{\frac{97}{4}} = \frac{\sqrt{97}}{2}.$$

Therefore, the standard equation of the sphere is

$$\left(x - \frac{5}{2} \right)^2 + (y - 1)^2 + (z - 0)^2 = \frac{97}{4}.$$

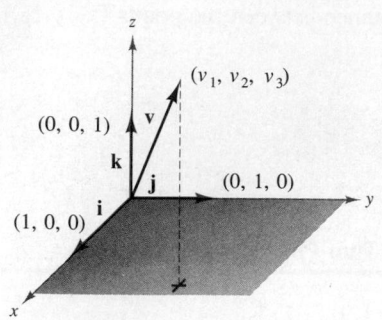

FIGURE 11.19
The standard unit vectors in space.

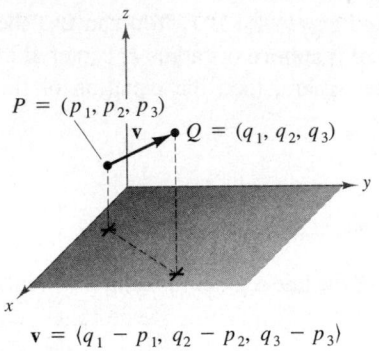

FIGURE 11.20

Vectors in Space

In space, vectors are denoted by ordered triples $\mathbf{v} = \langle v_1, v_2, v_3 \rangle$. The **zero vector** is denoted by $\mathbf{0} = \langle 0, 0, 0 \rangle$. Using the unit vectors $\mathbf{i} = \langle 1, 0, 0 \rangle$, $\mathbf{j} = \langle 0, 1, 0 \rangle$, and $\mathbf{k} = \langle 0, 0, 1 \rangle$ in the direction of the positive z-axis, the **standard unit vector notation** for $\mathbf{v}$ is

$$\mathbf{v} = v_1 \mathbf{i} + v_2 \mathbf{j} + v_3 \mathbf{k}$$

as shown in Figure 11.19. If $\mathbf{v}$ is represented by the directed line segment from $P = (p_1, p_2, p_3)$ to $Q = (q_1, q_2, q_3)$, as shown in Figure 11.20, then the component form of $\mathbf{v}$ is given by subtracting the coordinates of the initial point from the coordinates of the terminal point, as follows.

$$\mathbf{v} = \langle v_1, v_2, v_3 \rangle = \langle q_1 - p_1, q_2 - p_2, q_3 - p_3 \rangle$$

Vectors in Space

Let $\mathbf{u} = \langle u_1, u_2, u_3 \rangle$ and $\mathbf{v} = \langle v_1, v_2, v_3 \rangle$ be vectors in space and let c be a scalar.

1. *Equality of Vectors:* $\mathbf{u} = \mathbf{v}$ if and only if $u_1 = v_1$, $u_2 = v_2$, and $u_3 = v_3$.
2. *Component Form:* If $\mathbf{v}$ is represented by the directed line segment from $P = (p_1, p_2, p_3)$ to $Q = (q_1, q_2, q_3)$ then
$$\mathbf{v} = \langle v_1, v_2, v_3 \rangle = \langle q_1 - p_1, q_2 - p_2, q_3 - p_3 \rangle.$$
3. *Length:* $\|\mathbf{v}\| = \sqrt{v_1^2 + v_2^2 + v_3^2}$
4. *Unit Vector in the Direction of* $\mathbf{v}$: $\dfrac{\mathbf{v}}{\|\mathbf{v}\|} = \left(\dfrac{1}{\|\mathbf{v}\|}\right)\langle v_1, v_2, v_3 \rangle$, $\mathbf{v} \neq \mathbf{0}$
5. *Vector Addition:* $\mathbf{v} + \mathbf{u} = \langle v_1 + u_1, v_2 + u_2, v_3 + u_3 \rangle$
6. *Scalar Multiplication:* $c\mathbf{v} = \langle cv_1, cv_2, cv_3 \rangle$

REMARK   The properties of vector addition and scalar multiplication given in Theorem 11.1 are also valid for vectors in space.

EXAMPLE 3   Finding the Component Form of a Vector in Space

Find the component form and length of the vector $\mathbf{v}$ having initial point $(-2, 3, 1)$ and terminal point $(0, -4, 4)$. Then find a unit vector in the direction of $\mathbf{v}$.

Solution   The component form of $\mathbf{v}$ is

$$\mathbf{v} = \langle q_1 - p_1, q_2 - p_2, q_3 - p_3 \rangle = \langle 0 - (-2), -4 - 3, 4 - 1 \rangle$$
$$= \langle 2, -7, 3 \rangle$$

which implies that its length is

$$\|\mathbf{v}\| = \sqrt{(2)^2 + (-7)^2 + (3)^2} = \sqrt{62}.$$

The unit vector in the direction of $\mathbf{v}$ is

$$\mathbf{u} = \frac{\mathbf{v}}{\|\mathbf{v}\|} = \frac{1}{\sqrt{62}}\langle 2, -7, 3 \rangle.$$

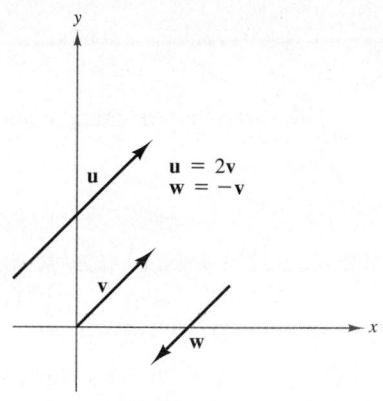

Parallel Vectors

FIGURE 11.21

Recall from the definition of scalar multiplication that positive scalar multiples of a nonzero vector **v** have the same direction as **v**, whereas negative multiples have the direction opposite that of **v**. In general, two nonzero vectors **u** and **v** are **parallel** if there is some scalar c such that **u** $= c\,$**v**.

> ### Definition of Parallel Vectors
>
> Two nonzero vectors **u** and **v** are **parallel** if there is some scalar c such that **u** $= c\,$**v**.

For example, in Figure 11.21, the vectors **u**, **v**, and **w** are parallel because **u** $= 2$**v** and **w** $= -$**v**.

EXAMPLE 4   Parallel Vectors

Vector **w** has initial point $(2, -1, 3)$ and terminal point $(-4, 7, 5)$. Which of the following vectors is parallel to **w**?

a. u $= \langle 3, -4, -1 \rangle$ **b. v** $= \langle 12, -16, 4 \rangle$

Solution   Begin by writing **w** in component form.

$$\mathbf{w} = \langle -4 - 2, 7 - (-1), 5 - 3 \rangle = \langle -6, 8, 2 \rangle$$

a. Because **u** $= \langle 3, -4, -1 \rangle = -\frac{1}{2}\langle -6, 8, 2 \rangle = -\frac{1}{2}$**w**, you can conclude that **u** *is* parallel to **w**.

b. In this case, you want to find a scalar c such that

$$\langle 12, -16, 4 \rangle = c\langle -6, 8, 2 \rangle.$$

However, when you equate corresponding components, you obtain $c = -2$ for the first two components and $c = 2$ for the third. Hence, the equation has no solution, and the vectors are *not* parallel.

EXAMPLE 5   Using Vectors to Determine Collinear Points

Determine whether the points $P = (1, -2, 3)$, $Q = (2, 1, 0)$, and $R = (4, 7, -6)$ lie on the same line.

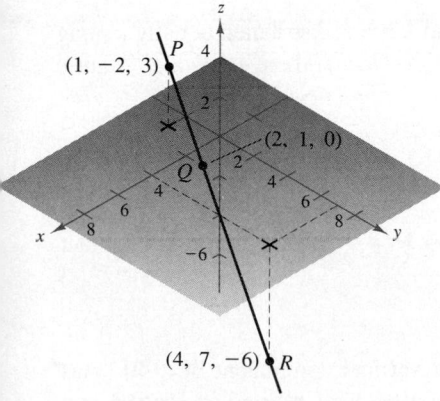

FIGURE 11.22
The points P, Q, and R lie on the same line.

Solution   The component forms of $\overrightarrow{PQ}$ and $\overrightarrow{PR}$ are

$$\overrightarrow{PQ} = \langle 2 - 1, 1 - (-2), 0 - 3 \rangle = \langle 1, 3, -3 \rangle$$

and

$$\overrightarrow{PR} = \langle 4 - 1, 7 - (-2), -6 - 3 \rangle = \langle 3, 9, -9 \rangle.$$

These two vectors have a common initial point. Hence, P, Q, and R lie on the same line if and only if $\overrightarrow{PQ}$ and $\overrightarrow{PR}$ are parallel—which they are because $\overrightarrow{PR} = 3\overrightarrow{PQ}$, as shown in Figure 11.22.

EXAMPLE 6 Standard Unit Vector Notation

a. Write the vector $\mathbf{v} = 4\mathbf{i} - 5\mathbf{k}$ in component form.
b. Find the terminal point of the vector $\mathbf{v} = 7\mathbf{i} - \mathbf{j} + 3\mathbf{k}$, given that the initial point is $P = (-2, 3, 5)$.

Solution

a. Because $\mathbf{j}$ is missing, its component is 0 and

$$\mathbf{v} = 4\mathbf{i} - 5\mathbf{k} = \langle 4, 0, -5 \rangle.$$

b. You need to find $Q = (q_1, q_2, q_3)$ such that $\mathbf{v} = \vec{PQ} = 7\mathbf{i} - \mathbf{j} + 3\mathbf{k}$. This implies that $q_1 - (-2) = 7$, $q_2 - 3 = -1$, and $q_3 - 5 = 3$. The solution of these three equations is $q_1 = 5$, $q_2 = 2$, and $q_3 = 8$. Therefore, $Q = (5, 2, 8)$.

Applications

EXAMPLE 7 Measuring Force

A television camera weighing 120 pounds is supported by a tripod, as shown in Figure 11.23. Represent the force exerted on each leg of the tripod as a vector. (Assume that the weight is distributed equally among the three legs.)

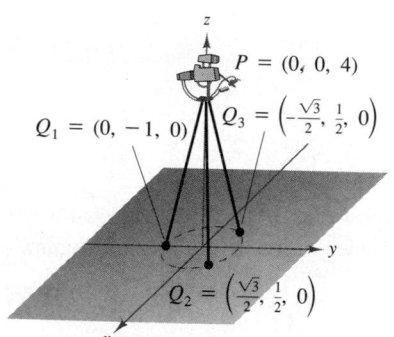

FIGURE 11.23

Solution Let the vectors $\mathbf{F}_1$, $\mathbf{F}_2$, and $\mathbf{F}_3$ represent the forces exerted on the three legs. From Figure 11.23, you can determine the directions of $\mathbf{F}_1$, $\mathbf{F}_2$, and $\mathbf{F}_3$ to be as follows.

$$\vec{PQ_1} = \langle 0 - 0, -1 - 0, 0 - 4 \rangle = \langle 0, -1, -4 \rangle$$

$$\vec{PQ_2} = \left\langle \frac{\sqrt{3}}{2} - 0, \frac{1}{2} - 0, 0 - 4 \right\rangle = \left\langle \frac{\sqrt{3}}{2}, \frac{1}{2}, -4 \right\rangle$$

$$\vec{PQ_3} = \left\langle -\frac{\sqrt{3}}{2} - 0, \frac{1}{2} - 0, 0 - 4 \right\rangle = \left\langle -\frac{\sqrt{3}}{2}, \frac{1}{2}, -4 \right\rangle$$

Because each leg has the same length, and the total force is distributed equally among the three legs, you know that $\| \mathbf{F}_1 \| = \| \mathbf{F}_2 \| = \| \mathbf{F}_3 \|$. Hence, there exists a constant c such that

$$\mathbf{F}_1 = c\langle 0, -1, -4 \rangle, \qquad \mathbf{F}_2 = c\left\langle \frac{\sqrt{3}}{2}, \frac{1}{2}, -4 \right\rangle, \qquad \mathbf{F}_3 = c\left\langle -\frac{\sqrt{3}}{2}, \frac{1}{2}, -4 \right\rangle.$$

Let the total force exerted by the object be given by $\mathbf{F} = -120\mathbf{k}$. Then, using the fact that

$$\mathbf{F} = \mathbf{F}_1 + \mathbf{F}_2 + \mathbf{F}_3$$

you can conclude that $\mathbf{F}_1$, $\mathbf{F}_2$, and $\mathbf{F}_3$ each has a vertical component of -40. This implies that $c(-4) = -40$ and $c = 10$. Therefore, the forces exerted on the legs can be represented by

$$\mathbf{F}_1 = \langle 0, -10, -40 \rangle, \qquad \mathbf{F}_2 = \langle 5\sqrt{3}, 5, -40 \rangle, \qquad \mathbf{F}_3 = \langle -5\sqrt{3}, 5, -40 \rangle.$$

EXERCISES for Section 11.2

In Exercises 1–4, plot the points on the same three-dimensional coordinate system.

1. a. $(2, 1, 3)$ **b.** $(-1, 2, 1)$

2. a. $(3, -2, 5)$ **b.** $\left(\frac{3}{2}, 4, -2\right)$

3. a. $(5, -2, 2)$ **b.** $(5, -2, -2)$

4. a. $(0, 4, -5)$ **b.** $(4, 0, 5)$

In Exercises 5–8, find the lengths of the triangle with the indicated vertices, and determine whether the triangle is a right triangle, an isosceles triangle, or neither.

5. $(0, 0, 0), (2, 2, 1), (2, -4, 4)$

6. $(5, 3, 4), (7, 1, 3), (3, 5, 3)$

7. $(1, -3, -2), (5, -1, 2), (-1, 1, 2)$

8. $(5, 0, 0), (0, 2, 0), (0, 0, -3)$

In Exercises 9 and 10, find the coordinates of the midpoint of the line segment joining the points.

9. $(5, -9, 7), (-2, 3, 3)$ **10.** $(4, 0, -6), (8, 8, 20)$

In Exercises 11–14, find the general form of the equation of the sphere.

11. Center: $(0, 2, 5)$
Radius: 2

12. Center: $(4, -1, 1)$
Radius: 5

13. Endpoints of a diameter: $(2, 0, 0), (0, 6, 0)$

14. Center: $(-2, 1, 1)$, tangent to the xy-coordinate plane

In Exercises 15–18, find the center and radius of the sphere.

15. $x^2 + y^2 + z^2 - 2x + 6y + 8z + 1 = 0$

16. $x^2 + y^2 + z^2 + 9x - 2y + 10z + 19 = 0$

17. $9x^2 + 9y^2 + 9z^2 - 6x + 18y + 1 = 0$

18. $4x^2 + 4y^2 + 4z^2 - 4x - 32y + 8z + 33 = 0$

In Exercises 19–22, (a) find the component form of the vector v and (b) sketch the vector with its initial point at the origin.

19. **20.**

21. **22.**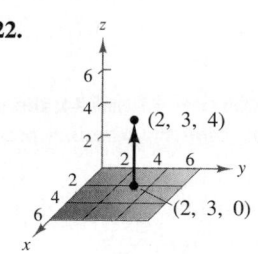

In Exercises 23 and 24, the initial and terminal points of a vector v are given. (a) Sketch the directed line segment, (b) find the component form of the vector, and (c) sketch the vector with its initial point at the origin.

23. Initial point: $(-1, 2, 3)$
Terminal point: $(3, 3, 4)$

24. Initial point: $(2, -1, -2)$
Terminal point: $(-4, 3, 7)$

In Exercises 25 and 26, sketch each scalar multiple of v.

25. $\mathbf{v} = \langle 1, 2, 2 \rangle$
a. $2\mathbf{v}$ **b.** $-\mathbf{v}$ **c.** $\frac{3}{2}\mathbf{v}$ **d.** $0\mathbf{v}$

26. $\mathbf{v} = \langle 2, -2, 1 \rangle$
a. $-\mathbf{v}$ **b.** $2\mathbf{v}$ **c.** $\frac{1}{2}\mathbf{v}$ **d.** $\frac{5}{2}\mathbf{v}$

In Exercises 27–32, find the vector z, given $\mathbf{u} = \langle 1, 2, 3 \rangle$, $\mathbf{v} = \langle 2, 2, -1 \rangle$, and $\mathbf{w} = \langle 4, 0, -4 \rangle$.

27. $\mathbf{z} = \mathbf{u} - \mathbf{v}$ **28.** $\mathbf{z} = \mathbf{u} - \mathbf{v} + 2\mathbf{w}$

29. $\mathbf{z} = 2\mathbf{u} + 4\mathbf{v} - \mathbf{w}$ **30.** $\mathbf{z} = 5\mathbf{u} - 3\mathbf{v} - \frac{1}{2}\mathbf{w}$

31. $2\mathbf{z} - 3\mathbf{u} = \mathbf{w}$ **32.** $2\mathbf{u} + \mathbf{v} - \mathbf{w} + 3\mathbf{z} = \mathbf{0}$

In Exercises 33–36, determine which of the vectors are parallel to z.

33. $\mathbf{z} = \langle 3, 2, -5 \rangle$
a. $\langle -6, -4, 10 \rangle$ **b.** $\left\langle 2, \frac{4}{3}, -\frac{10}{3} \right\rangle$
c. $\langle 6, 4, 10 \rangle$ **d.** $\langle 1, -4, 2 \rangle$

34. $\mathbf{z} = \frac{1}{2}\mathbf{i} - \frac{2}{3}\mathbf{j} + \frac{3}{4}\mathbf{k}$
a. $6\mathbf{i} - 4\mathbf{j} + 9\mathbf{k}$ **b.** $-\mathbf{i} + \frac{4}{3}\mathbf{j} - \frac{3}{2}\mathbf{k}$
c. $12\mathbf{i} + 9\mathbf{k}$ **d.** $\frac{3}{4}\mathbf{i} - \mathbf{j} + \frac{9}{8}\mathbf{k}$

35. z has initial point $(1, -1, 3)$ and terminal point $(-2, 3, 5)$.
a. $-6\mathbf{i} + 8\mathbf{j} + 4\mathbf{k}$ **b.** $4\mathbf{j} + 2\mathbf{k}$

36. z has initial point $(3, 2, -1)$ and terminal point $(-1, -3, 5)$.
a. $\langle 0, 5, -6 \rangle$ **b.** $\langle 8, 10, -12 \rangle$

In Exercises 37–40, use vectors to determine whether the given points lie in a straight line.

37. $(0, -2, -5), (3, 4, 4), (2, 2, 1)$

38. $(1, -1, 5), (0, -1, 6), (3, -1, 3)$

39. $(1, 2, 4), (2, 5, 0), (0, 1, 5)$

40. $(0, 0, 0), (1, 3, -2), (2, -6, 4)$

In Exercises 41 and 42, use vectors to show that the points form the vertices of a parallelogram.

41. (2, 9, 1), (3, 11, 4), (0, 10, 2), (1, 12, 5)

42. (1, 1, −3), (9, −1, 2), (11, 2, 1), (3, 4, −4)

In Exercises 43 and 44, the vector v and its initial point are given. Find the terminal point.

43. $\mathbf{v} = \langle 3, -5, 6 \rangle$
Initial point: (0, 6, 2)

44. $\mathbf{v} = \langle 0, \frac{1}{2}, -\frac{1}{3} \rangle$
Initial point: $\left(3, 0, -\frac{2}{3} \right)$

In Exercises 45–50, find the magnitude of v.

45. $\mathbf{v} = \langle 0, 0, 0 \rangle$ **46.** $\mathbf{v} = \langle 1, 0, 3 \rangle$

47. $\mathbf{v} = \mathbf{i} - 2\mathbf{j} - 3\mathbf{k}$ **48.** $\mathbf{v} = -4\mathbf{i} + 3\mathbf{j} + 7\mathbf{k}$

49. Initial point of **v**: (1, −3, 4)
Terminal point of **v**: (1, 0, −1)

50. Initial point of **v**: (0, −1, 0)
Terminal point of **v**: (1, 2, −2)

In Exercises 51–54, find a unit vector (a) in the direction of u and (b) in the direction opposite that of u.

51. $\mathbf{u} = \langle 2, -1, 2 \rangle$ **52.** $\mathbf{u} = \langle 6, 0, 8 \rangle$

53. $\mathbf{u} = \langle 3, 2, -5 \rangle$ **54.** $\mathbf{u} = \langle 8, 0, 0 \rangle$

C **In Exercises 55–58, use a computer algebra system to determine the specified quantity where $\mathbf{u} = \langle -1, 3, 4 \rangle$ and $\mathbf{v} = \langle 5, 4.5, -6 \rangle$.**

55. $6\mathbf{u} - 4\mathbf{v}$ **56.** $2\mathbf{u} + \frac{5}{2}\mathbf{v}$

57. $\| \mathbf{u} + \mathbf{v} \|$ **58.** $\dfrac{\mathbf{v}}{\| \mathbf{v} \|}$

In Exercises 59 and 60, determine the values of c that satisfy the equation. Let $\mathbf{u} = \mathbf{i} + 2\mathbf{j} + 3\mathbf{k}$ and $\mathbf{v} = 2\mathbf{i} + 2\mathbf{j} - \mathbf{k}$.

59. $\| c\mathbf{v} \| = 5$ **60.** $\| c\mathbf{u} \| = 3$

In Exercises 61 and 62, use vectors to find the point that lies two-thirds of the way from P to Q.

61. $P = (4, 3, 0)$, $Q = (1, -3, 3)$

62. $P = (1, 2, 5)$, $Q = (6, 8, 2)$

In Exercises 63 and 64, sketch the vector v and write its component form.

63. **v** lies in the yz-plane, has magnitude 2, and makes an angle of 30° with the positive y-axis.

64. **v** lies in the xz-plane, has magnitude 5, and makes an angle of 45° with the positive z-axis.

65. Let $\mathbf{u} = \mathbf{i} + \mathbf{j}$, $\mathbf{v} = \mathbf{j} + \mathbf{k}$, and $\mathbf{w} = a\mathbf{u} + b\mathbf{v}$.
  a. Sketch **u** and **v**.
  b. If $\mathbf{w} = \mathbf{0}$, show that a and b must both be 0.
  c. Find a and b such that $\mathbf{w} = \mathbf{i} + 2\mathbf{j} + \mathbf{k}$.
  d. Show that no choice of a and b yields $\mathbf{w} = \mathbf{i} + 2\mathbf{j} + 3\mathbf{k}$.

66. The initial and terminal points of the vector **v** are (x_1, y_1, z_1) and (x, y, z), respectively. Describe the set of all points (x, y, z) such that $\| \mathbf{v} \| = 4$.

67. *Diagonal of a Cube* Find the component form of the unit vector **v** in the direction of the diagonal of the cube shown in the accompanying figure.

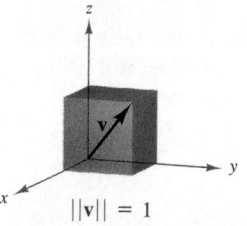

$\|\mathbf{v}\| = 1$ **FIGURE FOR 67**

68. *Tower Guy Wire* The guy wire to a 100-foot tower has a tension of 550 pounds. Using the distances shown in the figure, write the component form of the vector **F** respresenting the tension in the wire.

FIGURE FOR 68

69. *Light Installation* The lights in an auditorium are 25-pound discs of radius 18 inches. Each disc is supported by three equally spaced 48-inch wires to the ceiling (see figure). Find the tension in each wire.

48 in.

18 in.

FIGURE FOR 69

70. *Light Installation* Repeat Exercise 69 using r_0 inches as the radius of each light fixture. Determine the limit of the tension of each wire as $r_0 \rightarrow 48^-$.

71. *Load Supports*   Find the tension in each of the supporting cables in the accompanying figure if the weight of the crate is 500 newtons.

FIGURE FOR 71

72. *Building Construction*   A precast concrete wall is temporarily kept in its vertical position by ropes (see figure). Find the total force exerted on the pin at position A if the tensions in AB and AC are 420 and 650 pounds, respectively.

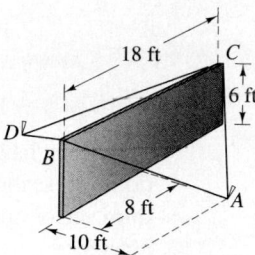

FIGURE FOR 72

73. Write an equation whose graph consists of the set of points $P = (x, y, z)$ that are twice as far from $A = (0, -1, 1)$ as from $B = (1, 2, 0)$.

The Dot Product of Two Vectors

The Dot Product • Angle Between Two Vectors • Direction Cosines • Projections and Vector Components • Work

The Dot Product

So far you have studied two operations with vectors—vector addition and multiplication by a scalar—each of which yields another vector. In this section you will study a third vector operation, called the **dot product.** This product yields a scalar, rather than a vector.

Definition of Dot Product

The **dot product** of $\mathbf{u} = \langle u_1, u_2 \rangle$ and $\mathbf{v} = \langle v_1, v_2 \rangle$ is

$\mathbf{u} \cdot \mathbf{v} = u_1 v_1 + u_2 v_2.$

The **dot product** of $\mathbf{u} = \langle u_1, u_2, u_3 \rangle$ and $\mathbf{v} = \langle v_1, v_2, v_3 \rangle$ is

$\mathbf{u} \cdot \mathbf{v} = u_1 v_1 + u_2 v_2 + u_3 v_3.$

REMARK The dot product of two vectors is also called the **inner product** (or scalar product) of two vectors.

THEOREM 11.4   Properties of the Dot Product

Let $\mathbf{u}$, $\mathbf{v}$, and $\mathbf{w}$ be vectors in the plane or in space and let c be a scalar.

1. $\mathbf{u} \cdot \mathbf{v} = \mathbf{v} \cdot \mathbf{u}$ Commutative property
2. $\mathbf{u} \cdot (\mathbf{v} + \mathbf{w}) = \mathbf{u} \cdot \mathbf{v} + \mathbf{u} \cdot \mathbf{w}$ Distributive property
3. $c(\mathbf{u} \cdot \mathbf{v}) = c\mathbf{u} \cdot \mathbf{v} = \mathbf{u} \cdot c\mathbf{v}$
4. $\mathbf{0} \cdot \mathbf{v} = 0$
5. $\mathbf{v} \cdot \mathbf{v} = \| \mathbf{v} \|^2$

Proof To prove the first property, let $\mathbf{u} = \langle u_1, u_2, u_3 \rangle$ and $\mathbf{v} = \langle v_1, v_2, v_3 \rangle$. Then

$$\mathbf{u} \cdot \mathbf{v} = u_1 v_1 + u_2 v_2 + u_3 v_3$$
$$= v_1 u_1 + v_2 u_2 + v_3 u_3$$
$$= \mathbf{v} \cdot \mathbf{u}.$$

For the fifth property, let $\mathbf{v} = \langle v_1, v_2, v_3 \rangle$. Then

$$\mathbf{v} \cdot \mathbf{v} = v_1^2 + v_2^2 + v_3^2$$
$$= \left(\sqrt{v_1^2 + v_2^2 + v_3^2} \right)^2 = \| \mathbf{v} \|^2.$$

Proofs of the other properties are left to you.

EXAMPLE 1   Finding Dot Products

Given $\mathbf{u} = \langle 2, -2 \rangle$, $\mathbf{v} = \langle 5, 8 \rangle$, and $\mathbf{w} = \langle -4, 3 \rangle$, find the following.

a. $\mathbf{u} \cdot \mathbf{v}$ **b.** $(\mathbf{u} \cdot \mathbf{v})\mathbf{w}$ **c.** $\mathbf{u} \cdot (2\mathbf{v})$ **d.** $\| \mathbf{w} \|^2$

Solution

a. $\mathbf{u} \cdot \mathbf{v} = \langle 2, -2 \rangle \cdot \langle 5, 8 \rangle = 2(5) + (-2)(8) = -6$

b. $(\mathbf{u} \cdot \mathbf{v})\mathbf{w} = -6\langle -4, 3 \rangle = \langle 24, -18 \rangle$

c. $\mathbf{u} \cdot (2\mathbf{v}) = 2(\mathbf{u} \cdot \mathbf{v}) = 2(-6) = -12$

d. $\| \mathbf{w} \|^2 = \mathbf{w} \cdot \mathbf{w} = \langle -4, 3 \rangle \cdot \langle -4, 3 \rangle = (-4)(-4) + (3)(3) = 25$

Notice that the result of part b is a *vector* quantity, whereas the results of the other three parts are *scalar* quantities.

Angle Between Two Vectors

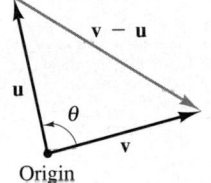

FIGURE 11.24
The angle between two vectors.

The **angle between two nonzero vectors** is the angle θ, $0 \le \theta \le \pi$, between their respective standard position vectors, as shown in Figure 11.24. The next theorem shows how to find this angle using the dot product. (Note that we do not define the angle between the zero vector and another vector.)

THEOREM 11.5   Angle Between Two Vectors

If θ is the angle between two nonzero vectors $\mathbf{u}$ and $\mathbf{v}$, then

$$\cos \theta = \frac{\mathbf{u} \cdot \mathbf{v}}{\| \mathbf{u} \| \, \| \mathbf{v} \|}.$$

Proof   Consider the triangle determined by vectors $\mathbf{u}$, $\mathbf{v}$, and $\mathbf{v} - \mathbf{u}$, as shown in Figure 11.24. By the Law of Cosines, you can write

$$\| \mathbf{v} - \mathbf{u} \|^2 = \| \mathbf{u} \|^2 + \| \mathbf{v} \|^2 - 2 \| \mathbf{u} \| \, \| \mathbf{v} \| \cos \theta.$$

Using the properties of the dot product, the left side can be rewritten as

$$\begin{aligned}
\| \mathbf{v} - \mathbf{u} \|^2 &= (\mathbf{v} - \mathbf{u}) \cdot (\mathbf{v} - \mathbf{u}) \\
&= (\mathbf{v} - \mathbf{u}) \cdot \mathbf{v} - (\mathbf{v} - \mathbf{u}) \cdot \mathbf{u} \\
&= \mathbf{v} \cdot \mathbf{v} - \mathbf{u} \cdot \mathbf{v} - \mathbf{v} \cdot \mathbf{u} + \mathbf{u} \cdot \mathbf{u} \\
&= \| \mathbf{v} \|^2 - 2\mathbf{u} \cdot \mathbf{v} + \| \mathbf{u} \|^2
\end{aligned}$$

and substitution back into the Law of Cosines yields

$$\| \mathbf{v} \|^2 - 2\mathbf{u} \cdot \mathbf{v} + \| \mathbf{u} \|^2 = \| \mathbf{u} \|^2 + \| \mathbf{v} \|^2 - 2 \| \mathbf{u} \| \, \| \mathbf{v} \| \cos \theta$$

$$-2\mathbf{u} \cdot \mathbf{v} = -2 \| \mathbf{u} \| \, \| \mathbf{v} \| \cos \theta$$

$$\cos \theta = \frac{\mathbf{u} \cdot \mathbf{v}}{\| \mathbf{u} \| \, \| \mathbf{v} \|}.$$

If the angle between two vectors is known, then rewriting Theorem 11.5 in the form

$$\mathbf{u} \cdot \mathbf{v} = \|\mathbf{u}\| \|\mathbf{v}\| \cos \theta$$ Alternative form of dot product

produces an alternative way to calculate the dot product. From this form, you can see that because $\|\mathbf{u}\|$ and $\|\mathbf{v}\|$ are always positive, $\mathbf{u} \cdot \mathbf{v}$ and $\cos \theta$ will always have the same sign. Figure 11.25 shows the five possible orientations of two vectors.

FIGURE 11.25

Definition of Orthogonal Vectors

The vectors $\mathbf{u}$ and $\mathbf{v}$ are **orthogonal** if $\mathbf{u} \cdot \mathbf{v} = 0$.

REMARK The terms "perpendicular," "orthogonal," and "normal" all mean essentially the same thing—meeting at right angles. However, we usually say that two vectors are *orthogonal*, two lines or planes are *perpendicular*, and a vector is *normal* to a given line or plane.

From this definition, it follows that the zero vector is orthogonal to every vector $\mathbf{u}$, because $\mathbf{0} \cdot \mathbf{u} = 0$. Moreover, for $0 \le \theta \le \pi$, you know that $\cos \theta = 0$ if and only if $\theta = \pi/2$. Thus, you can use Theorem 11.5 to conclude that two *nonzero* vectors are orthogonal if and only if the angle between them is $\pi/2$.

EXAMPLE 2   Finding the Angle Between Two Vectors

For $\mathbf{u} = \langle 3, -1, 2 \rangle$, $\mathbf{v} = \langle -4, 0, 2 \rangle$, $\mathbf{w} = \langle 1, -1, -2 \rangle$, and $\mathbf{z} = \langle 2, 0, -1 \rangle$, find the angles between the following pairs of vectors.

a. $\mathbf{u}$ and $\mathbf{v}$ **b.** $\mathbf{u}$ and $\mathbf{w}$ **c.** $\mathbf{v}$ and $\mathbf{z}$

Solution

a. $\cos \theta = \dfrac{\mathbf{u} \cdot \mathbf{v}}{\|\mathbf{u}\| \|\mathbf{v}\|} = \dfrac{-12 + 4}{\sqrt{14}\sqrt{20}} = \dfrac{-8}{2\sqrt{14}\sqrt{5}} = \dfrac{-4}{\sqrt{70}}$

Because $\mathbf{u} \cdot \mathbf{v} < 0$, $\theta = \arccos \dfrac{-4}{\sqrt{70}} \approx 2.069$ radians.

b. $\cos \theta = \dfrac{\mathbf{u} \cdot \mathbf{w}}{\|\mathbf{u}\| \|\mathbf{w}\|} = \dfrac{3 + 1 - 4}{\sqrt{14}\sqrt{6}} = \dfrac{0}{\sqrt{84}} = 0$

Because $\mathbf{u} \cdot \mathbf{w} = 0$, $\mathbf{u}$ and $\mathbf{w}$ are *orthogonal*. Furthermore, $\theta = \pi/2$.

c. $\cos \theta = \dfrac{\mathbf{v} \cdot \mathbf{z}}{\|\mathbf{v}\| \|\mathbf{z}\|} = \dfrac{-8 + 0 - 2}{\sqrt{20}\sqrt{5}} = \dfrac{-10}{\sqrt{100}} = -1$

Consequently, $\theta = \pi$.

Direction Cosines

For a vector in the plane, you have seen that it is convenient to measure direction in terms of the angle, measured counterclockwise, *from* the positive x-axis to the vector. In space it is more convenient to measure direction in terms of the angles *between* the nonzero vector **v** and the three unit vectors **i**, **j**, and **k**, as shown in Figure 11.26. The angles α, β, and γ are the **direction angles of v**, and $\cos \alpha$, $\cos \beta$, and $\cos \gamma$ are the **direction cosines of v**. Because

$$\mathbf{v} \cdot \mathbf{i} = \|\mathbf{v}\| \|\mathbf{i}\| \cos \alpha = \|\mathbf{v}\| \cos \alpha$$

and

$$\mathbf{v} \cdot \mathbf{i} = \langle v_1, v_2, v_3 \rangle \cdot \langle 1, 0, 0 \rangle = v_1$$

it follows that $\cos \alpha = v_1 / \|\mathbf{v}\|$. By similar reasoning with the unit vectors **j** and **k**, you have

$$\cos \alpha = \frac{v_1}{\|\mathbf{v}\|} \qquad \text{α is angle between v and i}$$

$$\cos \beta = \frac{v_2}{\|\mathbf{v}\|} \qquad \text{β is angle between v and j}$$

$$\cos \gamma = \frac{v_3}{\|\mathbf{v}\|}. \qquad \text{γ is angle between v and k}$$

Consequently, any nonzero vector **v** in space has the normalized form

$$\frac{\mathbf{v}}{\|\mathbf{v}\|} = \frac{v_1}{\|\mathbf{v}\|}\mathbf{i} + \frac{v_2}{\|\mathbf{v}\|}\mathbf{j} + \frac{v_3}{\|\mathbf{v}\|}\mathbf{k} = \cos \alpha \, \mathbf{i} + \cos \beta \, \mathbf{j} + \cos \gamma \, \mathbf{k}$$

and because $\mathbf{v}/\|\mathbf{v}\|$ is a unit vector, it follows that

$$\cos^2 \alpha + \cos^2 \beta + \cos^2 \gamma = 1.$$

EXAMPLE 3   Finding Direction Angles

Find the direction cosines and angles for the vector $\mathbf{v} = 2\mathbf{i} + 3\mathbf{j} + 4\mathbf{k}$, and show that $\cos^2 \alpha + \cos^2 \beta + \cos^2 \gamma = 1$.

Solution   Because $\|\mathbf{v}\| = \sqrt{2^2 + 3^2 + 4^2} = \sqrt{29}$, you can write the following.

$$\cos \alpha = \frac{v_1}{\|\mathbf{v}\|} = \frac{2}{\sqrt{29}} \qquad \rightarrow \qquad \alpha \approx 68.2°$$

$$\cos \beta = \frac{v_2}{\|\mathbf{v}\|} = \frac{3}{\sqrt{29}} \qquad \rightarrow \qquad \beta \approx 56.1°$$

$$\cos \gamma = \frac{v_3}{\|\mathbf{v}\|} = \frac{4}{\sqrt{29}} \qquad \rightarrow \qquad \gamma \approx 42.0°$$

Furthermore, the sum of the squares of the direction cosines is

$$\cos^2 \alpha + \cos^2 \beta + \cos^2 \gamma = \frac{4}{29} + \frac{9}{29} + \frac{16}{29}$$

$$= \frac{29}{29}$$

$$= 1.$$

(See Figure 11.27.)

FIGURE 11.26
Direction angles

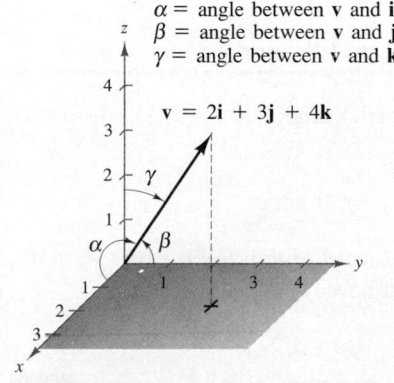

$\alpha = $ angle between **v** and **i**
$\beta = $ angle between **v** and **j**
$\gamma = $ angle between **v** and **k**

$\mathbf{v} = 2\mathbf{i} + 3\mathbf{j} + 4\mathbf{k}$

FIGURE 11.27
The direction angles of **v**.

FIGURE 11.28

The force due to gravity pulls the boat against the ramp and down the ramp.

Projections and Vector Components

You have already seen applications in which two vectors are added to produce a resultant vector. Many applications in physics and engineering pose the reverse problem—decomposing a given vector into the sum of two **vector components.** To see the usefulness of this procedure, we look at a physical example.

Consider a boat on an inclined ramp, as shown in Figure 11.28. The force **F** due to gravity pulls the boat *down* the ramp and *against* the ramp. These two forces, $\mathbf{w}_1$ and $\mathbf{w}_2$, are orthogonal—they are called the vector components of **F**.

$$\mathbf{F} = \mathbf{w}_1 + \mathbf{w}_2 \qquad \text{Vector components of } \mathbf{F}$$

The forces $\mathbf{w}_1$ and $\mathbf{w}_2$ help you analyze the effect of gravity on the boat. For example, $\mathbf{w}_1$ indicates the force necessary to keep the boat from rolling down the ramp, whereas $\mathbf{w}_2$ indicates the force that the tires must withstand.

Definition of Projection and Vector Components

Let **u** and **v** be nonzero vectors. Moreover, let $\mathbf{u} = \mathbf{w}_1 + \mathbf{w}_2$, where $\mathbf{w}_1$ is parallel to **v** and $\mathbf{w}_2$ is orthogonal to **v**, as shown in Figure 11.29.

1. $\mathbf{w}_1$ is called the **projection of u onto v** or the **vector component of u along v**, and is denoted by $\mathbf{w}_1 = \text{proj}_{\mathbf{v}}\,\mathbf{u}$.
2. $\mathbf{w}_2 = \mathbf{u} - \mathbf{w}_1$ is called the **vector component of u orthogonal to v.**

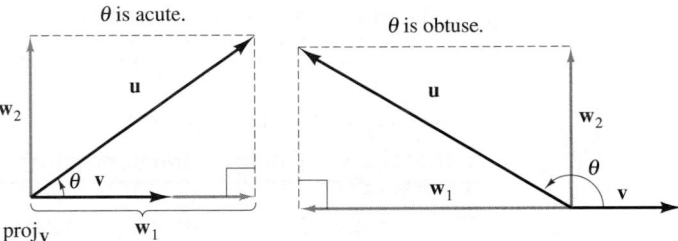

FIGURE 11.29

$\mathbf{w}_1 = \text{proj}_{\mathbf{v}}\,\mathbf{u} = $ projection of **u** onto **v** = vector component of **u** along **v**.

$\mathbf{w}_2 = $ vector component of **u** orthogonal to **v**.

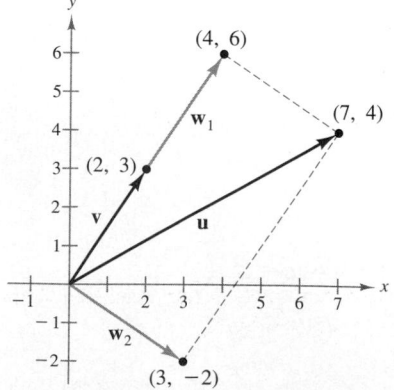

FIGURE 11.30

$\mathbf{u} = \mathbf{w}_1 + \mathbf{w}_2$

EXAMPLE 4   Finding a Vector Component of u Orthogonal to v

Find the vector component of $\mathbf{u} = \langle 7, 4 \rangle$ that is orthogonal to $\mathbf{v} = \langle 2, 3 \rangle$, given that $\mathbf{w}_1 = \langle 4, 6 \rangle$ and

$$\mathbf{u} = \langle 7, 4 \rangle = \mathbf{w}_1 + \mathbf{w}_2.$$

Solution   Because $\mathbf{u} = \mathbf{w}_1 + \mathbf{w}_2$, where $\mathbf{w}_1$ is parallel to **v**, it follows that $\mathbf{w}_2$ is the vector component of **u** orthogonal to **v**. Thus, you have

$$\mathbf{w}_2 = \mathbf{u} - \mathbf{w}_1 = \langle 7, 4 \rangle - \langle 4, 6 \rangle = \langle 3, -2 \rangle.$$

Check to see that $\mathbf{w}_2$ is orthogonal to **v**, as shown in Figure 11.30.

From Example 4, you can see that it is easy to find the vector component $\mathbf{w}_2$ once you have found the projection, $\mathbf{w}_1$, of $\mathbf{u}$ onto $\mathbf{v}$. To find this projection, use the dot product, as indicated in the next theorem.

THEOREM 11.6   Projection Using the Dot Product

If $\mathbf{u}$ and $\mathbf{v}$ are nonzero vectors, then the projection of $\mathbf{u}$ onto $\mathbf{v}$ is given by

$$\text{proj}_{\mathbf{v}}\,\mathbf{u} = \left(\frac{\mathbf{u}\cdot\mathbf{v}}{\|\mathbf{v}\|^2}\right)\mathbf{v}.$$

Proof   From Figure 11.29, let $\mathbf{w}_1 = \text{proj}_{\mathbf{v}}\,\mathbf{u}$. Because $\mathbf{w}_1$ is a scalar multiple of $\mathbf{v}$, you can write

$$\mathbf{u} = \mathbf{w}_1 + \mathbf{w}_2 = c\,\mathbf{v} + \mathbf{w}_2\,.$$

Taking the dot product of both sides with $\mathbf{v}$ produces

$$\begin{aligned}
\mathbf{u}\cdot\mathbf{v} &= (c\,\mathbf{v} + \mathbf{w}_2)\cdot\mathbf{v} \\
&= c\,\mathbf{v}\cdot\mathbf{v} + \mathbf{w}_2\cdot\mathbf{v} \\
&= c\,\|\mathbf{v}\|^2 + \mathbf{w}_2\cdot\mathbf{v}.
\end{aligned}$$

Because $\mathbf{w}_2$ and $\mathbf{v}$ are orthogonal, it follows that $\mathbf{w}_2\cdot\mathbf{v} = 0$ and you have

$$\mathbf{u}\cdot\mathbf{v} = c\,\|\mathbf{v}\|^2 \quad\rightarrow\quad c = \frac{\mathbf{u}\cdot\mathbf{v}}{\|\mathbf{v}\|^2}.$$

Therefore, you can write $\mathbf{w}_1$ as

$$\mathbf{w}_1 = \left(\frac{\mathbf{u}\cdot\mathbf{v}}{\|\mathbf{v}\|^2}\right)\mathbf{v}.$$

The projection of $\mathbf{u}$ onto $\mathbf{v}$ can be written as a scalar multiple of a unit vector in the direction of $\mathbf{v}$. That is,

$$\left(\frac{\mathbf{u}\cdot\mathbf{v}}{\|\mathbf{v}\|^2}\right)\mathbf{v} = \left(\frac{\mathbf{u}\cdot\mathbf{v}}{\|\mathbf{v}\|}\right)\frac{\mathbf{v}}{\|\mathbf{v}\|} = (k)\frac{\mathbf{v}}{\|\mathbf{v}\|}.$$

The scalar k is called the **component of $\mathbf{u}$ in the direction of $\mathbf{v}$**. Thus,

$$k = \frac{\mathbf{u}\cdot\mathbf{v}}{\|\mathbf{v}\|} = \|\mathbf{u}\|\cos\theta. \qquad \text{\small Component of } \mathbf{u} \text{ in the direction of } \mathbf{v}$$

REMARK   Note the distinction between the terms "component" and "vector component." For example, using the standard unit vectors with $\mathbf{u} = u_1\mathbf{i} + u_2\mathbf{j}$, you have

u_1 is the *component* of $\mathbf{u}$ in the direction of $\mathbf{i}$

and

$u_1\mathbf{i}$ is the *vector component* in the direction of $\mathbf{i}$.

EXAMPLE 5   Decomposing a Vector into Vector Components

Find the projection of **u** onto **v** and the vector component of **u** orthogonal to **v** for the vectors

$$\mathbf{u} = 3\mathbf{i} - 5\mathbf{j} + 2\mathbf{k} \quad \text{and} \quad \mathbf{v} = 7\mathbf{i} + \mathbf{j} - 2\mathbf{k}.$$

Solution   The projection of **u** onto **v** is

$$\mathbf{w}_1 = \left(\frac{\mathbf{u} \cdot \mathbf{v}}{\|\mathbf{v}\|^2}\right)\mathbf{v} = \left(\frac{12}{54}\right)(7\mathbf{i} + \mathbf{j} - 2\mathbf{k}) = \frac{14}{9}\mathbf{i} + \frac{2}{9}\mathbf{j} - \frac{4}{9}\mathbf{k}.$$

The vector component of **u** orthogonal to **v** is the vector

$$\begin{aligned}
\mathbf{w}_2 &= \mathbf{u} - \mathbf{w}_1 \\
&= (3\mathbf{i} - 5\mathbf{j} + 2\mathbf{k}) - \left(\frac{14}{9}\mathbf{i} + \frac{2}{9}\mathbf{j} - \frac{4}{9}\mathbf{k}\right) \\
&= \frac{13}{9}\mathbf{i} - \frac{47}{9}\mathbf{j} + \frac{22}{9}\mathbf{k}.
\end{aligned}$$

(See Figure 11.31.)

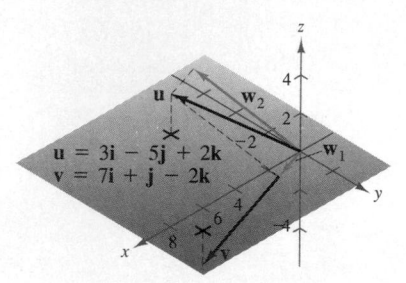

FIGURE 11.31
$\mathbf{u} = \mathbf{w}_1 + \mathbf{w}_2$

In the next example, we return to the problem involving the boat and the ramp to see how we can use vector components in a physical problem.

EXAMPLE 6   Finding a Force

A 600-pound boat sits on a ramp inclined at 30°, as shown in Figure 11.32. What force is required to keep the boat from rolling down the ramp?

Solution   Because the force due to gravity is vertical and downward, you can represent the gravitational force by the vector

$$\mathbf{F} = -600\mathbf{j}.$$

To find the force required to keep the boat from rolling down the ramp, we project **F** onto a unit vector **v** in the direction of the ramp, as follows.

$$\mathbf{v} = \cos 30°\mathbf{i} + \sin 30°\mathbf{j} = \frac{\sqrt{3}}{2}\mathbf{i} + \frac{1}{2}\mathbf{j} \qquad \text{Unit vector along ramp}$$

Therefore, the projection of **F** onto **v** is given by

$$\begin{aligned}
\mathbf{w}_1 = \text{proj}_\mathbf{v}\mathbf{F} &= \left(\frac{\mathbf{F} \cdot \mathbf{v}}{\|\mathbf{v}\|^2}\right)\mathbf{v} \\
&= (\mathbf{F} \cdot \mathbf{v})\mathbf{v} = (-600)\left(\frac{1}{2}\right)\mathbf{v} \\
&= -300\left(\frac{\sqrt{3}}{2}\mathbf{i} + \frac{1}{2}\mathbf{j}\right).
\end{aligned}$$

$\mathbf{w}_1 = \text{proj}_\mathbf{v}(\mathbf{F})$

FIGURE 11.32

The magnitude of this force is 300, and therefore a force of 300 pounds is required to keep the boat from rolling down the ramp.

Work

The work W done by a constant force $\mathbf{F}$ acting along the lines of motion of an object is given by

$$W = (\text{magnitude of force})(\text{distance})$$

$$= \|\mathbf{F}\| \, \|\overrightarrow{PQ}\|$$

as shown in Figure 11.33(a). If the constant force $\mathbf{F}$ is not directed along the line of motion, then you can see from Figure 11.33(b) that the work W done by the force is

$$W = \|\text{proj}_{\overrightarrow{PQ}} \mathbf{F}\| \, \|\overrightarrow{PQ}\|$$

$$= (\cos\theta) \, \|\mathbf{F}\| \, \|\overrightarrow{PQ}\|$$

$$= \mathbf{F} \cdot \overrightarrow{PQ}.$$

$$\text{Work} = \|\mathbf{F}\| \, \|\overrightarrow{PQ}\|$$

(a) Force acts along the line of motion.

$$\text{Work} = \|\text{proj}_{\overrightarrow{PQ}} \mathbf{F}\| \, \|\overrightarrow{PQ}\|$$

(b) Force acts at angle θ with the line of motion.

FIGURE 11.33

We summarize this notion of work in the following definition.

Definition of Work

The work W done by a constant force $\mathbf{F}$ as its point of application moves along the vector $\overrightarrow{PQ}$ is given by either of the following.

1. $W = \|\text{proj}_{\overrightarrow{PQ}} \mathbf{F}\| \, \|\overrightarrow{PQ}\|$ Projection form

2. $W = \mathbf{F} \cdot \overrightarrow{PQ}$ Dot product form

FIGURE 11.34

EXAMPLE 7   Finding Work

To close a sliding door, a person pulls on a rope with a constant force of 50 pounds at a constant angle of $60°$, as shown in Figure 11.34. Find the work done in moving the door 12 feet to its closed position.

Solution   Using a projection, you can calculate the work as follows.

$$W = \|\text{proj}_{\overrightarrow{PQ}} \mathbf{F}\| \, \|\overrightarrow{PQ}\|$$

$$= \cos(60°) \, \|\mathbf{F}\| \, \|\overrightarrow{PQ}\|$$

$$= \frac{1}{2}(50)(12)$$

$$= 300 \text{ ft} \cdot \text{lb}$$

1-55 odd except 35

In Exercises 1–6, find (a) $\mathbf{u} \cdot \mathbf{v}$, (b) $\mathbf{u} \cdot \mathbf{u}$, (c) $\|\mathbf{u}\|^2$, (d) $(\mathbf{u} \cdot \mathbf{v})\mathbf{v}$, and (e) $\mathbf{u} \cdot (2\mathbf{v})$.

1. $\mathbf{u} = \langle 3, 4 \rangle$
$\mathbf{v} = \langle 2, -3 \rangle$

2. $\mathbf{u} = \langle 5, 12 \rangle$
$\mathbf{v} = \langle -3, 2 \rangle$

3. $\mathbf{u} = \langle 2, -3, 4 \rangle$
$\mathbf{v} = \langle 0, 6, 5 \rangle$

4. $\mathbf{u} = \mathbf{i}$
$\mathbf{v} = \mathbf{i}$

5. $\mathbf{u} = 2\mathbf{i} - \mathbf{j} + \mathbf{k}$
$\mathbf{v} = \mathbf{i} - \mathbf{k}$

6. $\mathbf{u} = 2\mathbf{i} + \mathbf{j} - 2\mathbf{k}$
$\mathbf{v} = \mathbf{i} - 3\mathbf{j} + 2\mathbf{k}$

7. *Revenue* The vector $\mathbf{u} = \langle 3240, 1450, 2235 \rangle$ gives the number of units for products X, Y, and Z. The vector $\mathbf{v} = \langle 2.22, 1.85, 3.25 \rangle$ gives the price (in dollars) per unit for each product. Find the dot product $\mathbf{u} \cdot \mathbf{v}$, and explain what information it gives.

8. If $\mathbf{u} \cdot \mathbf{v} = \mathbf{u} \cdot \mathbf{w}$ and $\mathbf{u} \neq 0$, then is it true that $\mathbf{v} = \mathbf{w}$? If it is false, explain why or give an example that shows it is false.

In Exercises 9 and 10, find $\mathbf{u} \cdot \mathbf{v}$.

9. $\|\mathbf{u}\| = 8$, $\|\mathbf{v}\| = 5$, and the angle between $\mathbf{u}$ and $\mathbf{v}$ is $\pi/3$.

10. $\|\mathbf{u}\| = 40$, $\|\mathbf{v}\| = 25$, and the angle between $\mathbf{u}$ and $\mathbf{v}$ is $5\pi/6$.

In Exercises 11–18, find the angle θ between the given vectors.

11. $\mathbf{u} = \langle 1, 1 \rangle$
$\mathbf{v} = \langle 2, -2 \rangle$

12. $\mathbf{u} = \langle 3, 1 \rangle$
$\mathbf{v} = \langle 2, -1 \rangle$

13. $\mathbf{u} = 3\mathbf{i} + \mathbf{j}$
$\mathbf{v} = -2\mathbf{i} + 4\mathbf{j}$

14. $\mathbf{u} = \cos\left(\dfrac{\pi}{6}\right)\mathbf{i} + \sin\left(\dfrac{\pi}{6}\right)\mathbf{j}$
$\mathbf{v} = \cos\left(\dfrac{3\pi}{4}\right)\mathbf{i} + \sin\left(\dfrac{3\pi}{4}\right)\mathbf{j}$

15. $\mathbf{u} = \langle 1, 1, 1 \rangle$
$\mathbf{v} = \langle 2, 1, -1 \rangle$

16. $\mathbf{u} = 2\mathbf{i} + 3\mathbf{j} + \mathbf{k}$
$\mathbf{v} = -3\mathbf{i} + 2\mathbf{j}$

17. $\mathbf{u} = 3\mathbf{i} + 4\mathbf{j}$
$\mathbf{v} = -2\mathbf{j} + 3\mathbf{k}$

18. $\mathbf{u} = 2\mathbf{i} - 3\mathbf{j} + \mathbf{k}$
$\mathbf{v} = \mathbf{i} - 2\mathbf{j} + \mathbf{k}$

In Exercises 19–26, determine whether $\mathbf{u}$ and $\mathbf{v}$ are orthogonal or parallel, or neither.

19. $\mathbf{u} = \langle 4, 0 \rangle$
$\mathbf{v} = \langle 1, 1 \rangle$

20. $\mathbf{u} = \langle 2, 18 \rangle$
$\mathbf{v} = \langle \frac{3}{2}, -\frac{1}{6} \rangle$

21. $\mathbf{u} = \langle 4, 3 \rangle$
$\mathbf{v} = \langle \frac{1}{2}, -\frac{2}{3} \rangle$

22. $\mathbf{u} = -\frac{1}{3}(\mathbf{i} - 2\mathbf{j})$
$\mathbf{v} = 2\mathbf{i} - 4\mathbf{j}$

23. $\mathbf{u} = \mathbf{j} + 6\mathbf{k}$
$\mathbf{v} = \mathbf{i} - 2\mathbf{j} - \mathbf{k}$

24. $\mathbf{u} = -2\mathbf{i} + 3\mathbf{j} - \mathbf{k}$
$\mathbf{v} = 2\mathbf{i} + \mathbf{j} - \mathbf{k}$

25. $\mathbf{u} = \langle 2, -3, 1 \rangle$
$\mathbf{v} = \langle -1, -1, -1 \rangle$

26. $\mathbf{u} = \langle \cos\theta, \sin\theta, -1 \rangle$
$\mathbf{v} = \langle \sin\theta, -\cos\theta, 0 \rangle$

27. Use vectors to prove that the diagonals of a rhombus are perpendicular.

28. Consider a regular tetrahedron with vertices $(0, 0, 0)$, $(k, k, 0)$, $(k, 0, k)$, and $(0, k, k)$, where k is a positive real number.
a. Sketch the graph of the tetrahedron.
b. Find the length of each edge.
c. Use the dot product to find the angle between any two edges.
d. Find the angle between the line segments from the centroid $(k/2, k/2, k/2)$ to two vertices. This is the *bond angle* for a molecule such as CH_4 or $PbCl_4$, where the structure of the molecule is a tetrahedron.

29. What is known about θ, the angle between two nonzero vectors $\mathbf{u}$ and $\mathbf{v}$, if
a. $\mathbf{u} \cdot \mathbf{v} = 0$?   **b.** $\mathbf{u} \cdot \mathbf{v} > 0$?   **c.** $\mathbf{u} \cdot \mathbf{v} < 0$?

30. If $\mathbf{u}$ and $\mathbf{v}$ are orthogonal to $\mathbf{w}$, then is it true that $\mathbf{u} + \mathbf{v}$ is orthogonal to $\mathbf{w}$? If it is true, prove it. Otherwise explain why it is false or give an example that shows it is false.

In Exercises 31–34, find the direction cosines of u and demonstrate that the sum of the squares of the direction cosines is 1.

31. $\mathbf{u} = \mathbf{i} + 2\mathbf{j} + 2\mathbf{k}$ **32.** $\mathbf{u} = 3\mathbf{i} - \mathbf{j} + 5\mathbf{k}$

33. $\mathbf{u} = \langle 0, 6, -4 \rangle$ **34.** $\mathbf{u} = \langle a, b, c \rangle$

C In Exercises 35 and 36, use a computer algebra system or write a computer program to determine the magnitude and direction angles of the resultant of forces F_1 and F_2. The magnitude and terminal point of each vector is given. Assume that the initial point of the vectors is at the origin.

Vector	Magnitude	Terminal point
35. F_1	50 lb	$(10, 5, 3)$
F_2	80 lb	$(12, 7, -5)$
36. F_1	300 N	$(-20, -10, 5)$
F_2	100 N	$(5, 15, 0)$

37. Find the angle between the diagonal of a cube and one of its edges.

38. Find the angle between the diagonal of a cube and the diagonal of one of its sides.

39. *Load-Supporting Cables* A load is supported by three cables, as shown in the accompanying figure. Find the direction angles of the load supporting cable *OA*.

FIGURE FOR 39

40. Determine the weight of the load in Exercise 39 if the tension in the cable *OA* is 200 newtons.

In Exercises 41–48, (a) find the projection of u onto v, and (b) find the vector component of u orthogonal to v.

41. $\mathbf{u} = \langle 2, 3 \rangle$
   $\mathbf{v} = \langle 5, 1 \rangle$

42. $\mathbf{u} = \langle 2, -3 \rangle$
   $\mathbf{v} = \langle 3, 2 \rangle$

43. $\mathbf{u} = \langle 2, 1, 2 \rangle$
   $\mathbf{v} = \langle 0, 3, 4 \rangle$

44. $\mathbf{u} = \langle 0, 4, 1 \rangle$
   $\mathbf{v} = \langle 0, 2, 3 \rangle$

45. $\mathbf{u} = \langle 1, 1, 1 \rangle$
   $\mathbf{v} = \langle -2, -1, 1 \rangle$

46. $\mathbf{u} = \langle -2, -1, 1 \rangle$
   $\mathbf{v} = \langle 1, 1, 1 \rangle$

47. $\mathbf{u} = \langle 5, -4, 3 \rangle$
   $\mathbf{v} = \langle 1, 0, 0 \rangle$

48. $\mathbf{u} = \langle 5, -4, 3 \rangle$
   $\mathbf{v} = \langle 0, 1, 0 \rangle$

49. *Braking Load*   A truck with a gross weight of 32,000 pounds is parked on a 15° slope (see figure). Assuming that the only force to overcome is the force due to gravity,
   a. find the force required to keep the truck from rolling down the hill.
   b. find the force perpendicular to the hill.

50. *Load-Supporting Cables*   Find the magnitude of the projection of the load-supporting cable *OA* onto the positive *z*-axis as shown in the figure.

Weight = 32,000 lb

FIGURE FOR 49

FIGURE FOR 50

51. What can be said about the vectors **u** and **v** if
   a. the projection of **u** onto **v** equals **u**?
   b. the projection of **u** onto **v** equals **0**?

52. If the projection of **u** onto **v** has the same magnitude as the projection of **v** onto **u**, then is it true that $\| \mathbf{u} \| = \| \mathbf{v} \|$?

53. *Work*   An object is pulled 10 feet across a floor, using a force of 85 pounds. Find the work done if the direction of the force is 60° above the horizontal (see figure).

FIGURE FOR 53

FIGURE FOR 54

54. *Work*   A toy wagon is pulled by exerting a force of 15 pounds on a handle that makes a 30° angle with the horizontal. Find the work done in pulling the wagon 50 feet (see figure).

Work   In Exercises 55 and 56, find the work done in moving a particle from *P* to *Q* if the magnitude and direction of the force is given by v.

55. $P = (0, 0, 0)$, $Q = (4, 7, 5)$, $\mathbf{v} = \langle 1, 4, 8 \rangle$

56. $P = (1, 3, 0)$, $Q = (-3, 5, 10)$, $\mathbf{v} = -2\mathbf{i} + 3\mathbf{j} + 6\mathbf{k}$

57. Prove properties 2, 3, and 4 of Theorem 11.4.

58. Prove that
$$\| \mathbf{u} - \mathbf{v} \|^2 = \| \mathbf{u} \|^2 + \| \mathbf{v} \|^2 - 2\mathbf{u} \cdot \mathbf{v}.$$

59. Prove the **Cauchy-Schwarz Inequality**
$$\| \mathbf{u} \cdot \mathbf{v} \| \leq \| \mathbf{u} \| \| \mathbf{v} \|.$$

60. Prove the triangle inequality
$$\| \mathbf{u} + \mathbf{v} \| \leq \| \mathbf{u} \| + \| \mathbf{v} \|.$$

[*Hint:* Use the result of Exercise 59 and the fact that $\| \mathbf{u} + \mathbf{v} \|^2 = (\mathbf{u} + \mathbf{v}) \cdot (\mathbf{u} + \mathbf{v}).$]

61. Consider the vectors $\mathbf{u} = \langle \cos \alpha, \sin \alpha, 0 \rangle$ and $\mathbf{v} = \langle \cos \beta, \sin \beta, 0 \rangle$, where $\alpha > \beta$. Find the dot product of the vectors and use the result to prove the identity
$$\cos(\alpha - \beta) = \cos \alpha \cos \beta + \sin \alpha \sin \beta.$$

62. Consider the two curves $y_1 = x^2$ and $y_2 = x^{1/3}$. Find unit tangent vectors to each curve at their points of intersection. Find the angles between the curves at their points of intersection.

SECTION
11.4 The Cross Product of Two Vectors in Space

The Cross Product • The Triple Scalar Product

The Cross Product

Many applications in physics, engineering, and geometry involve finding a vector in space that is orthogonal to two given vectors. In this section you will study a product that will yield such a vector. It is called the **cross product,** and it is most conveniently defined and calculated using the standard unit vector form.

Definition of Cross Product of Two Vectors in Space

Let $\mathbf{u} = u_1\mathbf{i} + u_2\mathbf{j} + u_3\mathbf{k}$ and $\mathbf{v} = v_1\mathbf{i} + v_2\mathbf{j} + v_3\mathbf{k}$ be vectors in space. The **cross product** of $\mathbf{u}$ and $\mathbf{v}$ is the vector

$$\mathbf{u} \times \mathbf{v} = (u_2v_3 - u_3v_2)\mathbf{i} - (u_1v_3 - u_3v_1)\mathbf{j} + (u_1v_2 - u_2v_1)\mathbf{k}.$$

REMARK Be sure you see that this definition applies only to three-dimensional vectors. The cross product is not defined for two-dimensional vectors.

A convenient way to calculate $\mathbf{u} \times \mathbf{v}$ is to use the following *determinant form* with cofactor expansion. (This 3×3 determinant form is used simply to help remember the formula for the cross product—it is technically not a determinant, because the entries of the corresponding matrix are not all real numbers.)

$$\mathbf{u} \times \mathbf{v} = \begin{vmatrix} \mathbf{i} & \mathbf{j} & \mathbf{k} \\ u_1 & u_2 & u_3 \\ v_1 & v_2 & v_3 \end{vmatrix} \quad \begin{matrix} \leftarrow \text{Put ``}\mathbf{u}\text{'' in Row 2} \\ \leftarrow \text{Put ``}\mathbf{v}\text{'' in Row 3} \end{matrix}$$

$$= \begin{vmatrix} \mathbf{i} & \mathbf{j} & \mathbf{k} \\ u_1 & u_2 & u_3 \\ v_1 & v_2 & v_3 \end{vmatrix}\mathbf{i} - \begin{vmatrix} \mathbf{i} & \mathbf{j} & \mathbf{k} \\ u_1 & u_2 & u_3 \\ v_1 & v_2 & v_3 \end{vmatrix}\mathbf{j} + \begin{vmatrix} \mathbf{i} & \mathbf{j} & \mathbf{k} \\ u_1 & u_2 & u_3 \\ v_1 & v_2 & v_3 \end{vmatrix}\mathbf{k}$$

$$= \begin{vmatrix} u_2 & u_3 \\ v_2 & v_3 \end{vmatrix}\mathbf{i} - \begin{vmatrix} u_1 & u_3 \\ v_1 & v_3 \end{vmatrix}\mathbf{j} + \begin{vmatrix} u_1 & u_2 \\ v_1 & v_2 \end{vmatrix}\mathbf{k}$$

$$= (u_2v_3 - u_3v_2)\mathbf{i} - (u_1v_3 - u_3v_1)\mathbf{j} + (u_1v_2 - u_2v_1)\mathbf{k}$$

Note the minus sign in front of the **j**-component. Each of the three 2×2 determinants can be evaluated by using the following diagonal pattern.

$$\begin{vmatrix} a & b \\ c & d \end{vmatrix} = ad - bc$$

EXAMPLE 1   Finding the Cross Product

Given $\mathbf{u} = \mathbf{i} - 2\mathbf{j} + \mathbf{k}$ and $\mathbf{v} = 3\mathbf{i} + \mathbf{j} - 2\mathbf{k}$, find the following.

a. $\mathbf{u} \times \mathbf{v}$ **b.** $\mathbf{v} \times \mathbf{u}$ **c.** $\mathbf{v} \times \mathbf{v}$

Solution

a. $\mathbf{u} \times \mathbf{v} = \begin{vmatrix} \mathbf{i} & \mathbf{j} & \mathbf{k} \\ 1 & -2 & 1 \\ 3 & 1 & -2 \end{vmatrix} = \begin{vmatrix} -2 & 1 \\ 1 & -2 \end{vmatrix} \mathbf{i} - \begin{vmatrix} 1 & 1 \\ 3 & -2 \end{vmatrix} \mathbf{j} + \begin{vmatrix} 1 & -2 \\ 3 & 1 \end{vmatrix} \mathbf{k}$

$$= (4 - 1)\mathbf{i} - (-2 - 3)\mathbf{j} + (1 + 6)\mathbf{k}$$
$$= 3\mathbf{i} + 5\mathbf{j} + 7\mathbf{k}$$

b. $\mathbf{v} \times \mathbf{u} = \begin{vmatrix} \mathbf{i} & \mathbf{j} & \mathbf{k} \\ 3 & 1 & -2 \\ 1 & -2 & 1 \end{vmatrix} = \begin{vmatrix} 1 & -2 \\ -2 & 1 \end{vmatrix} \mathbf{i} - \begin{vmatrix} 3 & -2 \\ 1 & 1 \end{vmatrix} \mathbf{j} + \begin{vmatrix} 3 & 1 \\ 1 & -2 \end{vmatrix} \mathbf{k}$

$$= (1 - 4)\mathbf{i} - (3 + 2)\mathbf{j} + (-6 - 1)\mathbf{k}$$
$$= -3\mathbf{i} - 5\mathbf{j} - 7\mathbf{k}$$

Note that this result is the negative of that in part a.

c. $\mathbf{v} \times \mathbf{v} = \begin{vmatrix} \mathbf{i} & \mathbf{j} & \mathbf{k} \\ 3 & 1 & -2 \\ 3 & 1 & -2 \end{vmatrix} = \mathbf{0}$ ■

The results obtained in Example 1 suggest some interesting *algebraic* properties of the cross product. For instance, $\mathbf{u} \times \mathbf{v} = -(\mathbf{v} \times \mathbf{u})$ and $\mathbf{v} \times \mathbf{v} = \mathbf{0}$. These properties, and several others, are summarized in the following theorem.

THEOREM 11.7   Algebraic Properties of the Cross Product

Let $\mathbf{u}$, $\mathbf{v}$, and $\mathbf{w}$ be vectors in space and let c be a scalar.

1. $\mathbf{u} \times \mathbf{v} = -(\mathbf{v} \times \mathbf{u})$
2. $\mathbf{u} \times (\mathbf{v} + \mathbf{w}) = (\mathbf{u} \times \mathbf{v}) + (\mathbf{u} \times \mathbf{w})$
3. $c(\mathbf{u} \times \mathbf{v}) = (c\mathbf{u}) \times \mathbf{v} = \mathbf{u} \times (c\mathbf{v})$
4. $\mathbf{u} \times \mathbf{0} = \mathbf{0} \times \mathbf{u} = \mathbf{0}$
5. $\mathbf{u} \times \mathbf{u} = \mathbf{0}$
6. $\mathbf{u} \cdot (\mathbf{v} \times \mathbf{w}) = (\mathbf{u} \times \mathbf{v}) \cdot \mathbf{w}$

Proof   To prove the first property, let $\mathbf{u} = u_1\mathbf{i} + u_2\mathbf{j} + u_3\mathbf{k}$ and $v_1\mathbf{i} + v_2\mathbf{j} + v_3\mathbf{k}$. Then,

$$\mathbf{u} \times \mathbf{v} = (u_2 v_3 - u_3 v_2)\mathbf{i} - (u_1 v_3 - u_3 v_1)\mathbf{j} + (u_1 v_2 - u_2 v_1)\mathbf{k}$$

and

$$\mathbf{v} \times \mathbf{u} = (v_2 u_3 - v_3 u_2)\mathbf{i} - (v_1 u_3 - v_3 u_1)\mathbf{j} + (v_1 u_2 - v_2 u_1)\mathbf{k}$$

which implies that $\mathbf{u} \times \mathbf{v} = -(\mathbf{v} \times \mathbf{u})$. Proofs of properties 2, 3, 5, and 6 are left as exercises (see Exercises 39–42).   ■

Note that the first property listed in Theorem 11.7 indicates that the cross product is *not commutative*. In particular, this property indicates that the vectors $\mathbf{u} \times \mathbf{v}$ and $\mathbf{v} \times \mathbf{u}$ have equal lengths but opposite directions. The following theorem lists some other *geometric* properties of the cross product of two vectors.

FIGURE 11.35
The vectors **u** and **v** form adjacent sides of a parallelogram.

> **THEOREM 11.8   Geometric Properties of the Cross Product**
>
> Let $\mathbf{u}$ and $\mathbf{v}$ be nonzero vectors in space, and let θ be the angle between $\mathbf{u}$ and $\mathbf{v}$.
>
> 1. $\mathbf{u} \times \mathbf{v}$ is orthogonal to both $\mathbf{u}$ and $\mathbf{v}$.
> 2. $\|\mathbf{u} \times \mathbf{v}\| = \|\mathbf{u}\| \, \|\mathbf{v}\| \sin \theta$.
> 3. $\mathbf{u} \times \mathbf{v} = \mathbf{0}$ if and only if $\mathbf{u}$ and $\mathbf{v}$ are scalar multiples of each other.
> 4. $\|\mathbf{u} \times \mathbf{v}\|$ = area of parallelogram having $\mathbf{u}$ and $\mathbf{v}$ as adjacent sides.

Proof To prove the second property, note that because $\cos \theta = (\mathbf{u} \cdot \mathbf{v})/(\|\mathbf{u}\| \, \|\mathbf{v}\|)$, it follows that

$$
\|\mathbf{u}\| \, \|\mathbf{v}\| \sin \theta = \|\mathbf{u}\| \, \|\mathbf{v}\| \sqrt{1 - \cos^2 \theta}
$$

$$
= \|\mathbf{u}\| \, \|\mathbf{v}\| \sqrt{1 - \frac{(\mathbf{u} \cdot \mathbf{v})^2}{\|\mathbf{u}\|^2 \|\mathbf{v}\|^2}}
$$

$$
= \sqrt{\|\mathbf{u}\|^2 \|\mathbf{v}\|^2 - (\mathbf{u} \cdot \mathbf{v})^2}
$$

$$
= \sqrt{(u_1^2 + u_2^2 + u_3^2)(v_1^2 + v_2^2 + v_3^2) - (u_1 v_1 + u_2 v_2 + u_3 v_3)^2}
$$

$$
= \sqrt{(u_2 v_3 - u_3 v_2)^2 + (u_1 v_3 - u_3 v_1)^2 + (u_1 v_2 - u_2 v_1)^2}
$$

$$
= \|\mathbf{u} \times \mathbf{v}\|.
$$

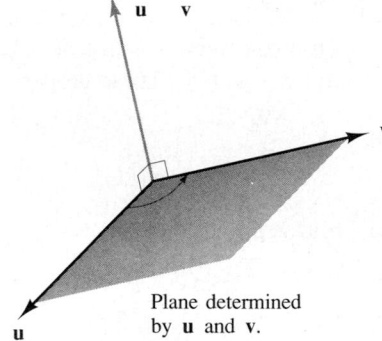

Plane determined by **u** and **v**.

To prove the fourth property, refer to Figure 11.35, which is a parallelogram having $\mathbf{v}$ and $\mathbf{u}$ as adjacent sides. Because the height of the parallelogram is $\|\mathbf{v}\| \sin \theta$, the area is

$$
\text{Area} = (\text{base})(\text{height}) = \|\mathbf{u}\| \, \|\mathbf{v}\| \sin \theta = \|\mathbf{u} \times \mathbf{v}\|.
$$

Proofs of the first and third properties are left as exercises (see Exercises 43 and 44.)

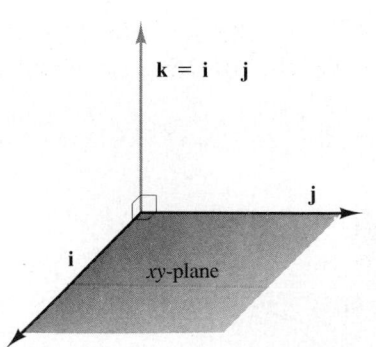

FIGURE 11.36
Right-handed systems

REMARK It follows from the first two properties listed in Theorem 11.8 that if $\mathbf{n}$ is a unit vector orthogonal to both $\mathbf{u}$ and $\mathbf{v}$, then

$$
\mathbf{u} \times \mathbf{v} = \pm\big(\|\mathbf{u}\| \, \|\mathbf{v}\| \sin \theta\big)\mathbf{n}.
$$

Both $\mathbf{u} \times \mathbf{v}$ and $\mathbf{v} \times \mathbf{u}$ are perpendicular to the plane determined by $\mathbf{u}$ and $\mathbf{v}$. One way to remember the orientation of the vectors $\mathbf{u}$, $\mathbf{v}$, and $\mathbf{u} \times \mathbf{v}$ is to compare them with the unit vectors $\mathbf{i}$, $\mathbf{j}$, and $\mathbf{k} = \mathbf{i} \times \mathbf{j}$, as shown in Figure 11.36. The three vectors $\mathbf{u}$, $\mathbf{v}$, and $\mathbf{u} \times \mathbf{v}$ form a *right-handed system,* whereas the three vectors $\mathbf{u}$, $\mathbf{v}$, and $\mathbf{v} \times \mathbf{u}$ form a *left-handed system.*

FIGURE 11.37
The vector $\mathbf{u} \times \mathbf{v}$ is orthogonal to both $\mathbf{u}$ and $\mathbf{v}$.

EXAMPLE 2   Using the Cross Product

Find a unit vector that is orthogonal to both

$$\mathbf{u} = \mathbf{i} - 4\mathbf{j} + \mathbf{k} \quad \text{and} \quad \mathbf{v} = 2\mathbf{i} + 3\mathbf{j}.$$

Solution   The cross product $\mathbf{u} \times \mathbf{v}$, as shown in Figure 11.37, is orthogonal to both $\mathbf{u}$ and $\mathbf{v}$.

$$\mathbf{u} \times \mathbf{v} = \begin{vmatrix} \mathbf{i} & \mathbf{j} & \mathbf{k} \\ 1 & -4 & 1 \\ 2 & 3 & 0 \end{vmatrix}$$

$$= -3\mathbf{i} + 2\mathbf{j} + 11\mathbf{k}$$

Because $\|\mathbf{u} \times \mathbf{v}\| = \sqrt{(-3)^2 + 2^2 + 11^2} = \sqrt{134}$, a unit vector orthogonal to both $\mathbf{u}$ and $\mathbf{v}$ is

$$\frac{\mathbf{u} \times \mathbf{v}}{\|\mathbf{u} \times \mathbf{v}\|} = -\frac{3}{\sqrt{134}}\mathbf{i} + \frac{2}{\sqrt{134}}\mathbf{j} + \frac{11}{\sqrt{134}}\mathbf{k}.$$

REMARK   In Example 2, note that you could have used the cross product $\mathbf{v} \times \mathbf{u}$ to form a unit vector that is orthogonal to both $\mathbf{u}$ and $\mathbf{v}$. With that choice, you would have obtained the negative of the unit vector found in the example.

EXAMPLE 3   Geometric Application of the Cross Product

Show that the quadrilateral with vertices at the following points is a parallelogram, and find its area.

$$A = (5, 2, 0) \qquad B = (2, 6, 1)$$
$$C = (2, 4, 7) \qquad D = (5, 0, 6)$$

Solution   From Figure 11.38 you can see that the sides of the quadrilateral correspond to the following four vectors.

$$\vec{AB} = -3\mathbf{i} + 4\mathbf{j} + \mathbf{k} \qquad \vec{CD} = 3\mathbf{i} - 4\mathbf{j} - \mathbf{k} = -\vec{AB}$$
$$\vec{AD} = 0\mathbf{i} - 2\mathbf{j} + 6\mathbf{k} \qquad \vec{CB} = 0\mathbf{i} + 2\mathbf{j} - 6\mathbf{k} = -\vec{AD}$$

Thus, $\vec{AB}$ is parallel to $\vec{CD}$ and $\vec{AD}$ is parallel to $\vec{CB}$, and you can conclude that the quadrilateral is a parallelogram with $\vec{AB}$ and $\vec{AD}$ as adjacent sides. Moveover, because

$$\vec{AB} \times \vec{AD} = \begin{vmatrix} \mathbf{i} & \mathbf{j} & \mathbf{k} \\ -3 & 4 & 1 \\ 0 & -2 & 6 \end{vmatrix}$$

$$= 26\mathbf{i} + 18\mathbf{j} + 6\mathbf{k}$$

the area of the parallelogram is

$$\|\vec{AB} \times \vec{AD}\| = \sqrt{1036} \approx 32.19.$$

Is the parallelogram a rectangle? You can tell whether it is by finding the angle between the vectors $\vec{AB}$ and $\vec{AD}$.

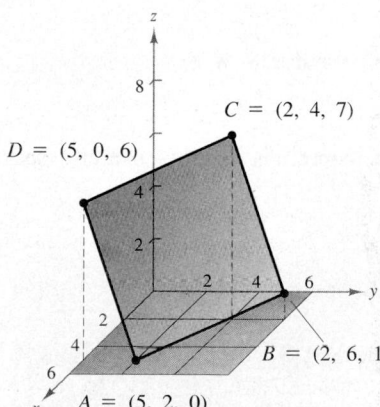

FIGURE 11.38
The area of the parallelogram is approximately 32.19.

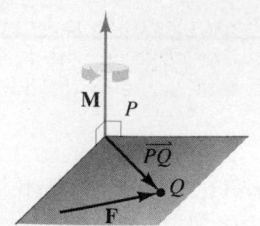

FIGURE 11.39
The moment of **F** about P.

In physics, the cross product can be used to measure **torque**—the **moment M of a force F about a point P,** as shown in Figure 11.39. If the point of application of the force is Q, then the moment of **F** about P is given by

$$\mathbf{M} = \vec{PQ} \times \mathbf{F}.$$

The magnitude of the moment **M** measures the tendency of the vector $\vec{PQ}$ to rotate counterclockwise (using the right-hand rule) about an axis directed along the vector **M**.

EXAMPLE 4 An Application of the Cross Product

A vertical force of 50 pounds is applied to the end of a 1-foot lever that is attached to an axle at point P, as shown in Figure 11.40. Find the moment of this force about the point P when $\theta = 60°$.

Solution If you represent the 50-pound force as $\mathbf{F} = -50\mathbf{k}$ and the lever as

$$\vec{PQ} = \cos(60°)\mathbf{j} + \sin(60°)\mathbf{k} = \frac{1}{2}\mathbf{j} + \frac{\sqrt{3}}{2}\mathbf{k}$$

then the moment of **F** about P is given by

$$\mathbf{M} = \vec{PQ} \times \mathbf{F} = \begin{vmatrix} \mathbf{i} & \mathbf{j} & \mathbf{k} \\ 0 & \dfrac{1}{2} & \dfrac{\sqrt{3}}{2} \\ 0 & 0 & -50 \end{vmatrix} = -25\mathbf{i}.$$

The magnitude of this moment is 25 foot-pounds. ∎

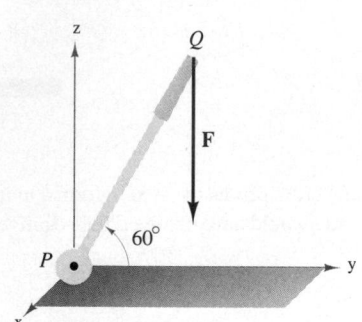

FIGURE 11.40
A vertical force of 50 pounds is applied at point Q.

REMARK In Example 4, note that the moment (the tendency of the lever to rotate about its axle) is dependent upon the angle θ. When $\theta = \pi/2$, the moment is **0**. The moment is greatest when $\theta = 0$.

The Triple Scalar Product

For vectors **u**, **v**, and **w** in space, the dot product of **u** and $\mathbf{v} \times \mathbf{w}$

$$\mathbf{u} \cdot (\mathbf{v} \times \mathbf{w})$$

is called the **triple scalar product.** The proof of this theorem is left as an exercise (see Exercise 47).

REMARK The value of a determinant is multiplied by -1 if two rows are interchanged. After two such interchanges, the value of the determinant will be unchanged. Thus, the following triple scalar products are equivalent.

$$\mathbf{u} \cdot (\mathbf{v} \times \mathbf{w})$$
$$= \mathbf{v} \cdot (\mathbf{w} \times \mathbf{u})$$
$$= \mathbf{w} \cdot (\mathbf{u} \times \mathbf{v})$$

THEOREM 11.9 The Triple Scalar Product

For $\mathbf{u} = u_1\mathbf{i} + u_2\mathbf{j} + u_3\mathbf{k}$, $\mathbf{v} = v_1\mathbf{i} + v_2\mathbf{j} + v_3\mathbf{k}$, and $\mathbf{w} = w_1\mathbf{i} + w_2\mathbf{j} + w_3\mathbf{k}$, the triple scalar product is given by

$$\mathbf{u} \cdot (\mathbf{v} \times \mathbf{w}) = \begin{vmatrix} u_1 & u_2 & u_3 \\ v_1 & v_2 & v_3 \\ w_1 & w_2 & w_3 \end{vmatrix}$$

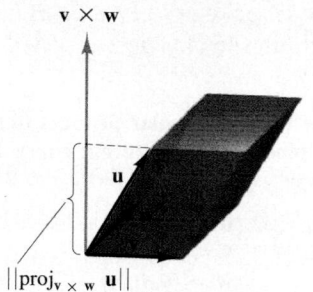

FIGURE 11.41
Area of base $= \| \mathbf{v} \times \mathbf{w} \|$
Volume of
parallelepiped $= | \mathbf{u} \cdot (\mathbf{v} \times \mathbf{w}) |$

If the vectors $\mathbf{u}$, $\mathbf{v}$, and $\mathbf{w}$ do not lie in the same plane, then the triple scalar product $\mathbf{u} \cdot (\mathbf{v} \times \mathbf{w})$ can be used to determine the volume of the parallelepiped with $\mathbf{u}$, $\mathbf{v}$, and $\mathbf{w}$ as adjacent sides, as shown in Figure 11.41. This is established in the following theorem.

THEOREM 11.10   Geometric Property of Triple Scalar Product

The volume of V of a parallelepiped with vectors $\mathbf{u}$, $\mathbf{v}$, and $\mathbf{w}$ as adjacent sides is given by
$$V = \left| \mathbf{u} \cdot (\mathbf{v} \times \mathbf{w}) \right|.$$

Proof   In Figure 11.41, note that
$$\| \mathbf{v} \times \mathbf{w} \| = \text{area of base}$$
and
$$\| \text{proj}_{\mathbf{v} \times \mathbf{w}} \mathbf{u} \| = \text{height of a parallelepiped}.$$
Therefore, the volume is
$$
\begin{aligned}
V &= (\text{height})(\text{area of base}) = \| \text{proj}_{\mathbf{v} \times \mathbf{w}} \mathbf{u} \| \, \| \mathbf{v} \times \mathbf{w} \| \\
&= \left| \frac{\mathbf{u} \cdot (\mathbf{v} \times \mathbf{w})}{\| \mathbf{v} \times \mathbf{w} \|} \right| \| \mathbf{v} \times \mathbf{w} \| \\
&= \left| \mathbf{u} \cdot (\mathbf{v} \times \mathbf{w}) \right|.
\end{aligned}
$$

FIGURE 11.42
The parallelepiped has a volume of 36.

EXAMPLE 5   Volume by the Triple Scalar Product

Find the volume of the parallelepiped having $\mathbf{u} = 3\mathbf{i} - 5\mathbf{j} + \mathbf{k}$, $\mathbf{v} = 2\mathbf{j} - 2\mathbf{k}$, and $\mathbf{w} = 3\mathbf{i} + \mathbf{j} + \mathbf{k}$ as adjacent edges (see Figure 11.42).

Solution   By Theorem 11.10, you have
$$
\begin{aligned}
V &= | \mathbf{u} \cdot (\mathbf{v} \times \mathbf{w}) | \\
&= \begin{vmatrix} 3 & -5 & 1 \\ 0 & 2 & -2 \\ 3 & 1 & 1 \end{vmatrix} \\
&= 3 \begin{vmatrix} 2 & -2 \\ 1 & 1 \end{vmatrix} - (-5) \begin{vmatrix} 0 & -2 \\ 3 & 1 \end{vmatrix} + (1) \begin{vmatrix} 0 & 2 \\ 3 & 1 \end{vmatrix} \\
&= 3(4) + 5(6) + 1(-6) \\
&= 36.
\end{aligned}
$$

A natural consequence of Theorem 11.10 is that the volume of the parallelepiped is 0 if and only if the three vectors are coplanar. That is, if the vectors $\mathbf{u} = \langle u_1, u_2, u_3 \rangle$, $\mathbf{v} = \langle v_1, v_2, v_3 \rangle$, and $\mathbf{w} = \langle w_1, w_2, w_3 \rangle$ have the same initial point, then they lie in the same plane if and only if
$$
\mathbf{u} \cdot (\mathbf{v} \times \mathbf{w}) = \begin{vmatrix} u_1 & u_2 & u_3 \\ v_1 & v_2 & v_3 \\ w_1 & w_2 & w_3 \end{vmatrix} = 0.
$$

EXERCISES for Section 11.4

In Exercises 1–6, find the cross product of unit vectors and sketch your result.

1. $\mathbf{j} \times \mathbf{i}$ **2.** $\mathbf{i} \times \mathbf{j}$

3. $\mathbf{j} \times \mathbf{k}$ **4.** $\mathbf{k} \times \mathbf{j}$

5. $\mathbf{i} \times \mathbf{k}$ **6.** $\mathbf{k} \times \mathbf{i}$

In Exercises 7–12, find $\mathbf{u} \times \mathbf{v}$ and show that it is orthogonal to both $\mathbf{u}$ and $\mathbf{v}$.

7. $\mathbf{u} = \langle 2, -3, 1 \rangle$
  $\mathbf{v} = \langle 1, -2, 1 \rangle$

8. $\mathbf{u} = \langle -1, 1, 2 \rangle$
  $\mathbf{v} = \langle 0, 1, 0 \rangle$

9. $\mathbf{u} = \langle 12, -3, 0 \rangle$
  $\mathbf{v} = \langle -2, 5, 0 \rangle$

10. $\mathbf{u} = \langle -10, 0, 6 \rangle$
  $\mathbf{v} = \langle 7, 0, 0 \rangle$

11. $\mathbf{u} = \mathbf{i} + \mathbf{j} + \mathbf{k}$
  $\mathbf{v} = 2\mathbf{i} + \mathbf{j} - \mathbf{k}$

12. $\mathbf{u} = \mathbf{j} + 6\mathbf{k}$
  $\mathbf{v} = \mathbf{i} - 2\mathbf{j} + \mathbf{k}$

C In Exercises 13–16, use a computer algebra system to find $\mathbf{u} \times \mathbf{v}$ and a unit vector orthogonal to $\mathbf{u}$ and $\mathbf{v}$.

13. $\mathbf{u} = \langle 4, -3.5, 7 \rangle$
  $\mathbf{v} = \langle -1, 8, 4 \rangle$

14. $\mathbf{u} = \langle -8, -6, 4 \rangle$
  $\mathbf{v} = \langle 10, -12, -2 \rangle$

15. $\mathbf{u} = -3\mathbf{i} + 2\mathbf{j} - 5\mathbf{k}$
  $\mathbf{v} = \frac{1}{2}\mathbf{i} - \frac{3}{4}\mathbf{j} + \frac{1}{10}\mathbf{k}$

16. $\mathbf{u} = \frac{2}{3}\mathbf{k}$
  $\mathbf{v} = \frac{1}{2}\mathbf{i} + 6\mathbf{k}$

In Exercises 17–20, find the area of the parallelogram that has the vectors as adjacent sides.

17. $\mathbf{u} = \mathbf{j}$
  $\mathbf{v} = \mathbf{j} + \mathbf{k}$

18. $\mathbf{u} = \mathbf{i} + \mathbf{j} + \mathbf{k}$
  $\mathbf{v} = \mathbf{j} + \mathbf{k}$

19. $\mathbf{u} = \langle 3, 2, -1 \rangle$
  $\mathbf{v} = \langle 1, 2, 3 \rangle$

20. $\mathbf{u} = \langle 2, -1, 0 \rangle$
  $\mathbf{v} = \langle -1, 2, 0 \rangle$

In Execises 21 and 22, verify that the points are the vertices of a parallelogram and find its area.

21. $(1, 1, 1)$, $(2, 3, 4)$, $(6, 5, 2)$, $(7, 7, 5)$

22. $(2, -1, 1)$, $(5, 1, 4)$, $(0, 1, 1)$, $(3, 3, 4)$

In Exercises 23–26, find the area of the triangle with the given vertices. ($\frac{1}{2} \| \mathbf{u} \times \mathbf{v} \|$ is the area of the triangle having $\mathbf{u}$ and $\mathbf{v}$ as adjacent sides.)

23. $(0, 0, 0)$, $(1, 2, 3)$, $(-3, 0, 0)$

24. $(2, -3, 4)$, $(0, 1, 2)$, $(-1, 2, 0)$

25. $(1, 3, 5)$, $(3, 3, 0)$, $(-2, 0, 5)$

26. $(1, 2, 0)$, $(-2, 1, 0)$, $(0, 0, 0)$

In Exercises 27–30, find $\mathbf{u} \cdot (\mathbf{v} \times \mathbf{w})$.

27. $\mathbf{u} = \mathbf{i}$
  $\mathbf{v} = \mathbf{j}$
  $\mathbf{w} = \mathbf{k}$

28. $\mathbf{u} = \langle 1, 1, 1 \rangle$
  $\mathbf{v} = \langle 2, 1, 0 \rangle$
  $\mathbf{w} = \langle 0, 0, 1 \rangle$

29. $\mathbf{u} = \langle 2, 0, 1 \rangle$
  $\mathbf{v} = \langle 0, 3, 0 \rangle$
  $\mathbf{w} = \langle 0, 0, 1 \rangle$

30. $\mathbf{u} = \langle 2, 0, 0 \rangle$
  $\mathbf{v} = \langle 1, 1, 1 \rangle$
  $\mathbf{w} = \langle 0, 2, 2 \rangle$

In Exercises 31 and 32, use the triple scalar product to find the volume of the parallelepiped having adjacent edges $\mathbf{u}$, $\mathbf{v}$, and $\mathbf{w}$ (see figures).

31. $\mathbf{u} = \mathbf{i} + \mathbf{j}$
  $\mathbf{v} = \mathbf{j} + \mathbf{k}$
  $\mathbf{w} = \mathbf{i} + \mathbf{k}$

32. $\mathbf{u} = \langle 1, 3, 1 \rangle$
  $\mathbf{v} = \langle 0, 5, 5 \rangle$
  $\mathbf{w} = \langle 4, 0, 4 \rangle$

FIGURE FOR 31

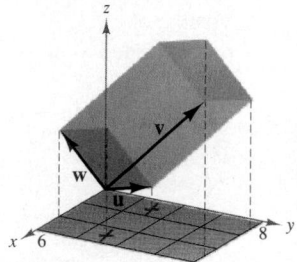

FIGURE FOR 32

In Exercises 33 and 34, find the volume of the parallelepiped with the given vertices (see figures).

33. $(0, 0, 0)$, $(3, 0, 0)$, $(0, 5, 1)$, $(3, 5, 1)$, $(2, 0, 5)$, $(5, 0, 5)$, $(2, 5, 6)$, $(5, 5, 6)$

34. $(0, 0, 0)$, $(1, 1, 0)$, $(1, 0, 2)$, $(0, 1, 1)$, $(2, 1, 2)$, $(1, 1, 3)$, $(1, 2, 1)$, $(2, 2, 3)$

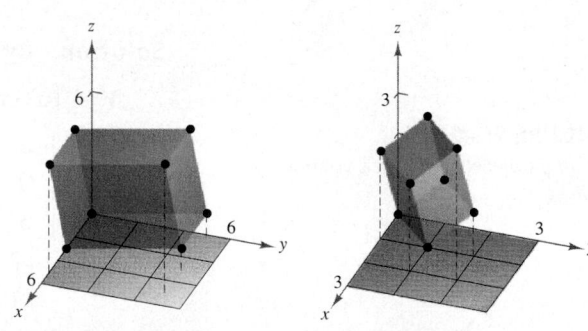

FIGURE FOR 33   **FIGURE FOR 34**

35. *Torque* A child applies the brakes on a bicycle by applying a downward force of 20 pounds on the pedal when the crank makes a 40° angle with the horizontal (see figure). Find the torque at P if the crank is 6 inches in length.

FIGURE FOR 35

36. *Torque*   Both the magnitude and direction of the force on a crankshaft change as the crankshaft rotates. Find the torque on the crankshaft using the position and data shown in the figure.

0.16 ft

2000 lb

60°

FIGURE FOR 36

37. *Torque*   A force of 200 pounds acts on the bracket shown in the figure.
   a. Determine the vector $\vec{AB}$ and the vector **F** representing the force. (**F** will be in terms of θ.)
   b. Find the magnitude of the moment about A by evaluating $\|\vec{AB} \times \mathbf{F}\|$.
   c. Use the result of part b to determine the magnitude of the moment when $\theta = 30°$.
   d. Use the result of part b to determine the angle θ when the magnitude of the moment is maximum.

200 lbs

F

θ

B

12 in.

A

15 in.

FIGURE FOR 37

FIGURE FOR 38

18 in.

A

θ

F

30°

O

38. *Torque*   A force of 60 pounds acts on the pipe wrench shown in the figure.
   a. Find the magnitude of the moment about O by evaluating $\|\vec{OA} \times \mathbf{F}\|$. (The result will be a function of θ.)
   b. Use the result of part a to determine the magnitude of the moment when $\theta = 45°$.
   c. Use the result of part a to determine the angle θ when the magnitude of the moment is maximum. Is the answer what you expected? Why or why not?

In Exercises 39–46, prove the property of the cross product.

39. $\mathbf{u} \times (\mathbf{v} + \mathbf{w}) = (\mathbf{u} \times \mathbf{v}) + (\mathbf{u} \times \mathbf{w})$

40. $(c\,\mathbf{u}) \times \mathbf{v} = c(\mathbf{u} \times \mathbf{v})$

41. $\mathbf{u} \times \mathbf{u} = \mathbf{0}$

42. $\mathbf{u} \cdot (\mathbf{v} \times \mathbf{w}) = (\mathbf{u} \times \mathbf{v}) \cdot \mathbf{w}$

43. $\mathbf{u} \times \mathbf{v}$ is orthogonal to both **u** and **v**.

44. $\mathbf{u} \times \mathbf{v} = \mathbf{0}$ if and only if **u** and **v** are scalar multiples of each other.

45. $\|\mathbf{u} \times \mathbf{v}\| = \|\mathbf{u}\|\,\|\mathbf{v}\|$ if **u** and **v** are orthogonal.

46. $\mathbf{u} \times (\mathbf{v} \times \mathbf{w}) = (\mathbf{u} \cdot \mathbf{w})\mathbf{v} - (\mathbf{u} \cdot \mathbf{v})\mathbf{w}$

47. Prove Theorem 11.9.

48. Consider the vectors $\mathbf{u} = \langle \cos\alpha, \sin\alpha, 0 \rangle$ and $\mathbf{v} = \langle \cos\beta, \sin\beta, 0 \rangle$, where $\alpha > \beta$. Find the cross product of the vectors and use the result to prove the identity

$$\sin(\alpha - \beta) = \sin\alpha\cos\beta - \cos\alpha\sin\beta.$$

49. *Essay*   Read the article "Tooth Tables: Solution of a Dental Problem by Vector Algebra" by Gary Hosler Meisters in the November, 1982 issue of *Mathematics Magazine*. Then write a paragraph explaining how vectors and vector algebra can be used in the construction of dental inlays.

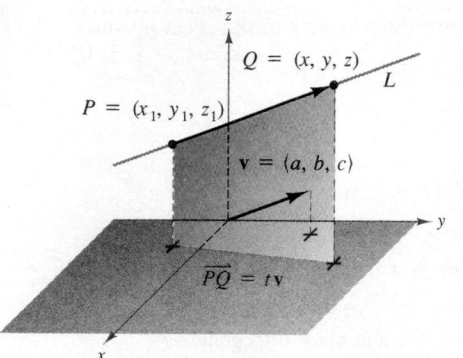

FIGURE 11.43
A line has direction vector **v**.

Lines in Space

In the plane, *slope* is used to determine an equation of a line. In space, it is more convenient to use *vectors* to determine the equation of a line.

In Figure 11.43, consider the line L through the point $P = (x_1, y_1, z_1)$ and parallel to the vector $\mathbf{v} = \langle a, b, c \rangle$. The vector **v** is the **direction vector** for the line L, and a, b, and c are the **direction numbers.** One way of describing the line L is to say that it consists of all points $Q = (x, y, z)$ for which the vector $\vec{PQ}$ is parallel to **v**. This means that $\vec{PQ}$ is a scalar multiple of **v**, and you can write $\vec{PQ} = t\mathbf{v}$, where t is a scalar.

$$\vec{PQ} = \langle x - x_1, y - y_1, z - z_1 \rangle = \langle at, bt, ct \rangle = t\mathbf{v}$$

By equating corresponding components, you can obtain the **parametric equations** of a line in space.

THEOREM 11.11   Parametric Equations of a Line in Space

A line L parallel to the vector $\mathbf{v} = \langle a, b, c \rangle$ and passing through the point $P = (x_1, y_1, z_1)$ is represented by the **parametric equations**

$$x = x_1 + at, \qquad y = y_1 + bt, \qquad z = z_1 + ct.$$

If the direction numbers a, b, and c are all nonzero, then you can eliminate the parameter t to obtain the **symmetric equations** of a line.

$$\frac{x - x_1}{a} = \frac{y - y_1}{b} = \frac{z - z_1}{c} \qquad \text{Symmetric equations}$$

EXAMPLE 1   Finding Parametric and Symmetric Equations

Find parametric and symmetric equations of the line L that passes through the point $(1, -2, 4)$ and is parallel to $\mathbf{v} = \langle 2, 4, -4 \rangle$.

Solution To find a set of parametric equations of the line, use the coordinates $x_1 = 1$, $y_1 = -2$, and $z_1 = 4$ and direction numbers $a = 2$, $b = 4$, and $c = -4$ (see Figure 11.44).

$$x = 1 + 2t, \qquad y = -2 + 4t, \qquad z = 4 - 4t \qquad \text{Parametric equations}$$

Because a, b, and c are all nonzero, a set of symmetric equations is

$$\frac{x - 1}{2} = \frac{y + 2}{4} = \frac{z - 4}{-4}. \qquad \text{Symmetric equations}$$

FIGURE 11.44
The vector **v** is parallel to the line.

Neither the parametric equations nor the symmetric equations of a given line are unique. For instance, in Example 1, by letting $t = 1$ in the parametric equations you would obtain the point $(3, 2, 0)$. Using this point with the direction numbers $a = 2$, $b = 4$, and $c = -4$ produces the parametric equations

$$x = 3 + 2t, \qquad y = 2 + 4t, \qquad z = -4t.$$

EXAMPLE 2   Parametric Equations of a Line Through Two Points

Find a set of parametric equations of the line that passes through the points $(-2, 1, 0)$ and $(1, 3, 5)$.

Solution   Begin by letting $P = (-2, 1, 0)$ and $Q = (1, 3, 5)$. Then a direction vector for the line passing through P and Q is given by

$$\mathbf{v} = \vec{PQ} = \langle 1 - (-2), 3 - 1, 5 - 0 \rangle = \langle 3, 2, 5 \rangle = \langle a, b, c \rangle.$$

Using the direction numbers $a = 3$, $b = 2$, and $c = 5$, with the point $P = (-2, 1, 0)$, you can obtain the parametric equations

$$x = -2 + 3t, \qquad y = 1 + 2t, \qquad z = 5t.$$

Planes in Space

You have seen how an equation of a line in space can be obtained from a point on the line and a vector *parallel* to it. You will now see that an equation of a plane in space can be obtained from a point in the plane and a vector *normal* (perpendicular) to it.

Consider the plane containing the point $P = (x_1, y_1, z_1)$ and having a nonzero normal vector $\mathbf{n} = \langle a, b, c \rangle$, as shown in Figure 11.45. This plane consists of all points $Q = (x, y, z)$ for which vector $\vec{PQ}$ is orthogonal to $\mathbf{n}$. Using the dot product, you can write the following.

$$\mathbf{n} \cdot \vec{PQ} = 0$$
$$\langle a, b, c \rangle \cdot \langle x - x_1, y - y_1, z - z_1 \rangle = 0$$
$$a(x - x_1) + b(y - y_1) + c(z - z_1) = 0$$

The third equation of the plane is said to be in **standard form.**

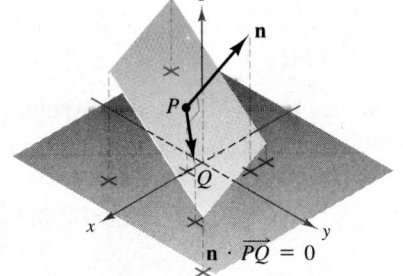

FIGURE 11.45
The normal vector $\mathbf{n}$ is orthogonal to each vector $\vec{PQ}$ in the plane.

THEOREM 11.12   Standard Equation of a Plane in Space

The plane containing the point (x_1, y_1, z_1) and having a normal vector $\mathbf{n} = \langle a, b, c \rangle$ can be represented, in **standard form,** by the equation
$$a(x - x_1) + b(y - y_1) + c(z - z_1) = 0.$$

By regrouping terms, you obtain the **general form** of the equation of a plane in space,

$$ax + by + cz + d = 0.$$

Given the general form of the equation of a plane, it is easy to find a normal vector to the plane. Simply use the coefficients of x, y, and z and write $\mathbf{n} = \langle a, b, c \rangle$.

EXAMPLE 3   Finding an Equation of a Plane in Three-Space

Find the general equation of the plane containing the points $(2, 1, 1)$, $(0, 4, 1)$, and $(-2, 1, 4)$.

Solution   To apply Theorem 11.12, you need a point in the plane and a vector that is normal to the plane. There are three choices for the point, but no normal vector is given. To obtain a normal vector, use the cross product of vectors $\mathbf{u}$ and $\mathbf{v}$ extending from the point $(2, 1, 1)$ to the points $(0, 4, 1)$ and $(-2, 1, 4)$, as shown in Figure 11.46. The component forms of $\mathbf{u}$ and $\mathbf{v}$ are

$$\mathbf{u} = \langle 0 - 2, 4 - 1, 1 - 1 \rangle = \langle -2, 3, 0 \rangle$$
$$\mathbf{v} = \langle -2 - 2, 1 - 1, 4 - 1 \rangle = \langle -4, 0, 3 \rangle$$

and it follows that

$$\mathbf{n} = \mathbf{u} \times \mathbf{v} = \begin{vmatrix} \mathbf{i} & \mathbf{j} & \mathbf{k} \\ -2 & 3 & 0 \\ -4 & 0 & 3 \end{vmatrix}$$
$$= 9\mathbf{i} + 6\mathbf{j} + 12\mathbf{k}$$
$$= \langle a, b, c \rangle$$

is normal to the given plane. Using the direction numbers for $\mathbf{n}$ and the point $(x_1, y_1, z_1) = (2, 1, 1)$, you can determine an equation of the plane to be

$$a(x - x_1) + b(y - y_1) + c(z - z_1) = 0$$
$$9(x - 2) + 6(y - 1) + 12(z - 1) = 0 \qquad \text{Standard form}$$
$$9x + 6y + 12z - 36 = 0$$
$$3x + 2y + 4z - 12 = 0. \qquad \text{General form}$$

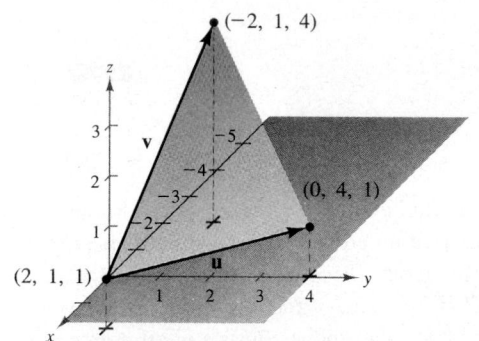

FIGURE 11.46
A plane determined by $\mathbf{u}$ and $\mathbf{v}$.

REMARK   In Example 3, check to see that each of the three points satisfies the equation $3x + 2y + 4z - 12 = 0$.

Two distinct planes in three-space either are parallel or intersect in a line. If they intersect, you can determine the angle between them from the angle between their normal vectors, as shown in Figure 11.47. Specifically, if vectors $\mathbf{n}_1$ and $\mathbf{n}_2$ are normal to two intersecting planes, then the angle θ between the normal vectors is equal to the angle between the two planes and is given by

$$\cos \theta = \frac{|\mathbf{n}_1 \cdot \mathbf{n}_2|}{\|\mathbf{n}_1\| \, \|\mathbf{n}_2\|}. \qquad \text{Angle between two planes}$$

Consequently, two planes with normal vectors $\mathbf{n}_1$ and $\mathbf{n}_2$ are

1. *perpendicular* if $\mathbf{n}_1 \cdot \mathbf{n}_2 = 0$.
2. *parallel* if $\mathbf{n}_1$ is a scalar multiple of $\mathbf{n}_2$.

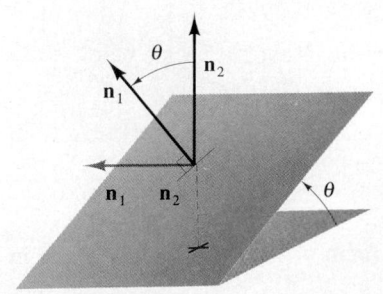

FIGURE 11.47
The angle between two planes.

EXAMPLE 4   Finding the Line of Intersection of Two Planes

Find the angle between the two planes given by

$$x - 2y + z = 0 \qquad \text{Equation for Plane 1}$$
$$2x + 3y - 2z = 0 \qquad \text{Equation for Plane 2}$$

and find parametric equations of their line of intersection (see Figure 11.48).

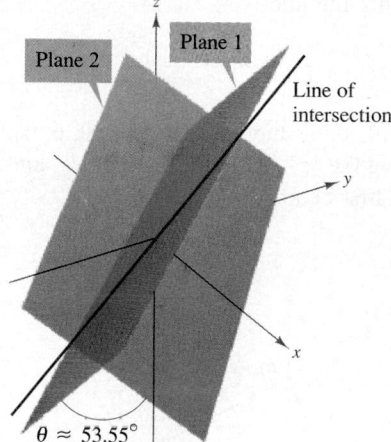

FIGURE 11.48
The angle between the planes is approximately 53.55°.

Solution The normal vectors for the planes are $\mathbf{n}_1 = \langle 1, -2, 1 \rangle$ and $\mathbf{n}_2 = \langle 2, 3, -2 \rangle$. Consequently, the angle between the two planes is determined as follows.

$$\cos \theta = \frac{|\mathbf{n}_1 \cdot \mathbf{n}_2|}{\|\mathbf{n}_1\| \|\mathbf{n}_2\|} \qquad \text{Cosine of angle between } \mathbf{n}_1 \text{ and } \mathbf{n}_2$$

$$= \frac{|-6|}{\sqrt{6}\sqrt{17}}$$

$$= \frac{6}{\sqrt{102}}$$

$$\approx 0.59409 \qquad \theta \approx \arccos 0.59409$$

This implies that the angle between the two planes is $\theta \approx 53.55°$. You can find the line of intersection of the two planes by simultaneously solving the two linear equations representing the planes. One way to do this is to multiply the first equation by -2 and add the result to the second equation.

$$
\begin{array}{lcl}
x - 2y + z = 0 & \rightarrow & -2x + 4y - 2z = 0 \\
2x + 3y - 2z = 0 & & \underline{2x + 3y - 2z = 0} \\
& & 7y - 4z = 0 \quad \rightarrow \quad y = \dfrac{4z}{7}
\end{array}
$$

Substituting $y = 4z/7$ back into one of the original equations, you can determine that $x = z/7$. Finally, by letting $t = z/7$, you obtain the parametric equations

$$x = t, \qquad y = 4t, \qquad z = 7t \qquad \text{Line of intersection}$$

which indicate that 1, 4, and 7 are direction numbers for the line of intersection.

Note that the direction numbers in Example 4 can be obtained from the cross product of the two normal vectors as follows.

$$\mathbf{n}_1 \times \mathbf{n}_2 = \begin{vmatrix} \mathbf{i} & \mathbf{j} & \mathbf{k} \\ 1 & -2 & 1 \\ 2 & 3 & -2 \end{vmatrix}$$

$$= \begin{vmatrix} -2 & 1 \\ 3 & -2 \end{vmatrix} \mathbf{i} - \begin{vmatrix} 1 & 1 \\ 2 & -2 \end{vmatrix} \mathbf{j} + \begin{vmatrix} 1 & -2 \\ 2 & 3 \end{vmatrix} \mathbf{k}$$

$$= \mathbf{i} + 4\mathbf{j} + 7\mathbf{k}$$

This means that the line of intersection of the two planes is parallel to the cross product of their normal vectors.

Sketching Planes in Space

If a plane in space intersects one of the coordinate planes, we call the line of intersection the **trace** of the given plane in the coordinate plane. To sketch a plane in space, it is helpful to find its points of intersection with the coordinate axes and its traces in the coordinate planes. For example, consider the plane given by

$$3x + 2y + 4z = 12. \quad \text{Equation of plane}$$

We find the xy-trace by letting $z = 0$ and sketching the line

$$3x + 2y = 12 \quad \text{xy-trace}$$

in the xy-plane. This line intersects the x-axis at $(4, 0, 0)$ and the y-axis at $(0, 6, 0)$. In Figure 11.49, we continue this process by finding the yz-trace and the xz-trace, and then shading in the triangular region lying in the first octant.

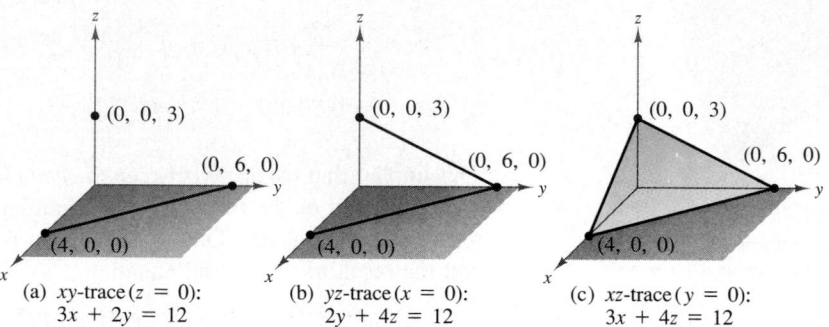

(a) xy-trace $(z = 0)$:
$3x + 2y = 12$

(b) yz-trace $(x = 0)$:
$2y + 4z = 12$

(c) xz-trace $(y = 0)$:
$3x + 4z = 12$

FIGURE 11.49
Traces of the plane:
$3x + 2y + 4z = 12$.

If the equation of a plane has a missing variable such as $2x + z = 1$, then the plane must be *parallel to the axis* represented by the missing variable, as shown in Figure 11.50. If two variables are missing from the equation of a plane, then it is *parallel to the coordinate plane* represented by the missing variables, as shown in Figure 11.51.

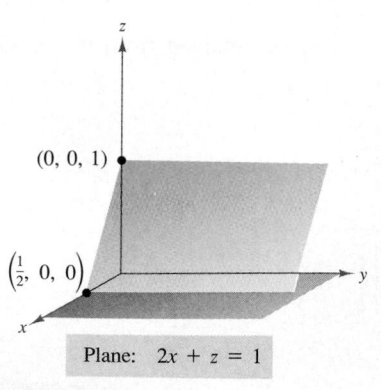

Plane: $2x + z = 1$

FIGURE 11.50

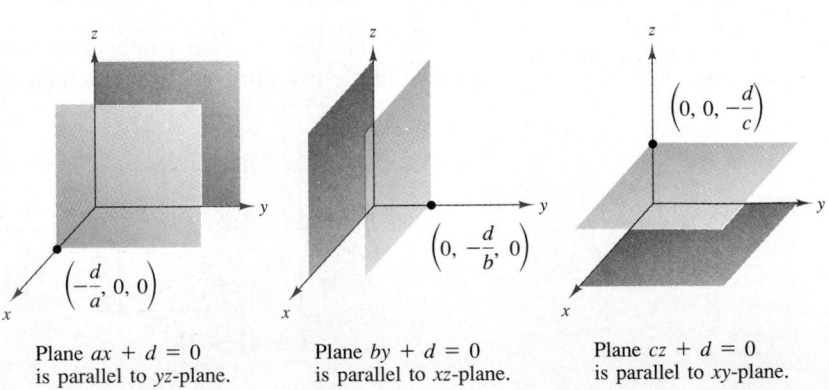

Plane $ax + d = 0$
is parallel to yz-plane.

Plane $by + d = 0$
is parallel to xz-plane.

Plane $cz + d = 0$
is parallel to xy-plane.

FIGURE 11.51

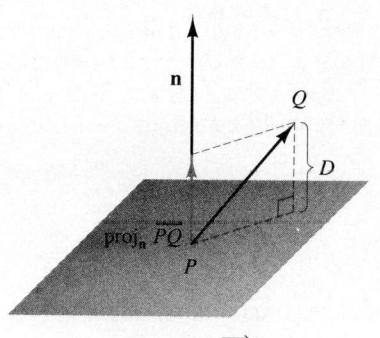

$$D = \|\text{proj}_{\mathbf{n}}\, \overrightarrow{PQ}\|$$

FIGURE 11.52
The distance between a point
and a plane.

Distances Between Points, Planes, and Lines

We conclude this section with a discussion of two basic types of distance problems in space.

1. Finding the distance between a point and a plane
2. Finding the distance between a point and a line

The solutions of these problems illustrate the versatility and usefulness of vectors in coordinate geometry—the first problem uses the *dot product* of two vectors, and the second problem uses the *cross product*.

The distance D between a point Q and a plane is the length of the shortest line segment connecting Q to the plane, as shown in Figure 11.52. If P is *any* point in the plane, you can find this distance by projecting the vector $\overrightarrow{PQ}$ onto the normal vector **n**. The length of this projection is the desired distance.

THEOREM 11.13   Distance Between a Point and a Plane

The distance between a plane and a point Q (not in the plane) is

$$D = \|\text{proj}_{\mathbf{n}}\, \overrightarrow{PQ}\| = \frac{|\overrightarrow{PQ} \cdot \mathbf{n}|}{\|\mathbf{n}\|}$$

where P is a point in the plane and **n** is normal to the plane.

To find a point in the plane given by $ax + by + cz + d = 0$ $(a \neq 0)$, let $y = 0$ and $z = 0$. Then, from the equation $ax + d = 0$, you can conclude that the point $(-d/a, 0, 0)$ lies in the plane.

EXAMPLE 5   Finding the Distance Between a Point and a Plane

Find the distance between the point $Q = (1, 5, -4)$ and the plane given by $3x - y + 2z = 6$.

REMARK The choice of the point P in Example 5 is arbitrary. Try choosing a different point to verify that you obtain the same distance.

Solution   You know that $\mathbf{n} = \langle 3, -1, 2 \rangle$ is normal to the given plane. To find a point in the plane, let $y = 0$ and $z = 0$, and obtain the point $P = (2, 0, 0)$. The vector from P to Q is given by

$$\overrightarrow{PQ} = \langle 1 - 2, 5 - 0, -4 - 0 \rangle$$
$$= \langle -1, 5, -4 \rangle.$$

Using the distance formula (given in Theorem 11.13) produces

$$D = \frac{|\overrightarrow{PQ} \cdot \mathbf{n}|}{\|\mathbf{n}\|} = \frac{|\langle -1, 5, -4 \rangle \cdot \langle 3, -1, 2 \rangle|}{\sqrt{9 + 1 + 4}}$$
$$= \frac{|-3 - 5 - 8|}{\sqrt{14}}$$
$$= \frac{16}{\sqrt{14}}.$$

From Theorem 11.13, you can determine that the distance between the point $Q = (x_0, y_0, z_0)$ and the plane given by $ax + by + cz + d = 0$ is

$$D = \frac{|a(x_0 - x_1) + b(y_0 - y_1) + c(z_0 - z_1)|}{\sqrt{a^2 + b^2 + c^2}}$$

$$D = \frac{|ax_0 + by_0 + cz_0 + d|}{\sqrt{a^2 + b^2 + c^2}} \qquad \text{Distance between point and plane}$$

where $P = (x_1, y_1, z_1)$ is a point on the plane and $d = -(ax_1 + by_1 + cz_1)$.

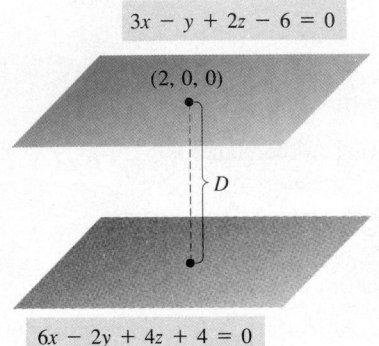

$3x - y + 2z - 6 = 0$

$(2, 0, 0)$

D

$6x - 2y + 4z + 4 = 0$

FIGURE 11.53
The distance between the parallel planes is approximately 2.14.

EXAMPLE 6   Finding the Distance Between Two Parallel Planes

Find the distance between the two parallel planes given by

$$3x - y + 2z - 6 = 0 \quad \text{and} \quad 6x - 2y + 4z + 4 = 0.$$

Solution   The two planes are shown in Figure 11.53. To find the distance between the planes, choose a point in the first plane, say $(x_0, y_0, z_0) = (2, 0, 0)$. Then, from the second plane, you can determine that $a = 6, b = -2, c = 4$, and $d = 4$, and conclude that the distance is

$$D = \frac{|ax_0 + by_0 + cz_0 + d|}{\sqrt{a^2 + b^2 + c^2}}$$

$$= \frac{|6(2) + (-2)(0) + (4)(0) + 4|}{\sqrt{6^2 + (-2)^2 + 4^2}} = \frac{16}{\sqrt{56}} = \frac{8}{\sqrt{14}} \approx 2.14.$$

The formula for the distance between a point and a line in space resembles that for the distance between a point and a plane—except that you replace the dot product by the cross product and replace the normal vector **n** by a direction vector for the given line.

THEOREM 11.14   Distance Between a Point and a Line in Space

The distance between a point Q and a line in space is given by

$$D = \frac{\|\overrightarrow{PQ} \times \mathbf{u}\|}{\|\mathbf{u}\|}$$

where **u** is the direction vector for the line and P is a point on the line.

Point Q

P

θ

u

$D = \|\overrightarrow{PQ}\| \sin\theta$

Line

FIGURE 11.54
The distance between a point and a line.

Proof   In Figure 11.54, let D be the distance between the point Q and the given line. Then $D = \|\overrightarrow{PQ}\| \sin\theta$, where θ is the angle between **u** and $\overrightarrow{PQ}$. By Theorem 11.8, you have

$$\|\mathbf{u}\| \|\overrightarrow{PQ}\| \sin\theta = \|\mathbf{u} \times \overrightarrow{PQ}\| = \|\overrightarrow{PQ} \times \mathbf{u}\|.$$

Consequently,

$$D = \|\overrightarrow{PQ}\| \sin\theta = \frac{\|\overrightarrow{PQ} \times \mathbf{u}\|}{\|\mathbf{u}\|}.$$

EXAMPLE 7   Finding the Distance Between a Point and a Line

Find the distance between the point $Q = (3, -1, 4)$ and the line given by

$$x = -2 + 3t, \quad y = -2t, \quad \text{and} \quad z = 1 + 4t.$$

Solution   Using the direction numbers 3, -2, and 4, you know that the direction vector for the line is

$$\mathbf{u} = \langle 3, -2, 4 \rangle. \qquad \text{Direction vector for line}$$

To find a point on the line, let $t = 0$ and obtain

$$P = (-2, 0, 1). \qquad \text{Point on the line}$$

Thus,

$$\overrightarrow{PQ} = \langle 3 - (-2), -1 - 0, 4 - 1 \rangle = \langle 5, -1, 3 \rangle$$

and you can form the cross product

$$\overrightarrow{PQ} \times \mathbf{u} = \begin{vmatrix} \mathbf{i} & \mathbf{j} & \mathbf{k} \\ 5 & -1 & 3 \\ 3 & -2 & 4 \end{vmatrix} = 2\mathbf{i} - 11\mathbf{j} - 7\mathbf{k} = \langle 2, -11, -7 \rangle.$$

Finally, using Theorem 11.14, you can find the distance to be

$$D = \frac{\|\overrightarrow{PQ} \times \mathbf{u}\|}{\|\mathbf{u}\|} = \frac{\sqrt{174}}{\sqrt{29}} = \sqrt{6} \approx 2.45.$$

(See Figure 11.55.)

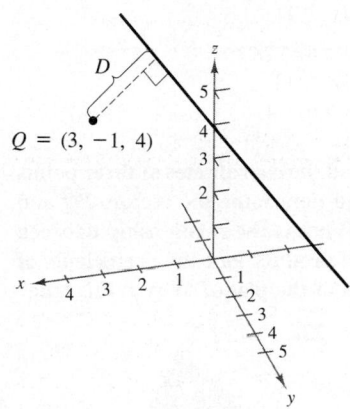

FIGURE 11.55
The distance between the point Q and the line is $\sqrt{6} \approx 2.45$.

EXERCISES for Section 11.5

> **TECHNOLOGY**
> Laboratory Guide
> Lab 11.5

1. The figure shows the line given by

$$x = 1 + 3t, \quad y = 2 - t, \quad \text{and} \quad z = 2 + 5t.$$

  a. Draw an arrow on the line to indicate its orientation.
  b. Find the coordinates of two points P and Q on the line. Determine the vector $\overrightarrow{PQ}$. What is the relationship of the components of the vector with the coefficients of t in the parametric equations? Why is this true?
  c. Determine the coordinates of the point where the line intersects the xz-plane.

2. The figure shows the line given by

$$x = 2 - 3t, \quad y = 2, \quad \text{and} \quad z = 1 - t.$$

  a. Draw an arrow on the line to indicate its orientation.
  b. Find the coordinates of two points P and Q on the line. Determine the vector $\overrightarrow{PQ}$. What is the relationship of the components of the vector with the coefficients of t in the parametric equations? Why is this true?

  c. Determine the coordinates of the point where the line intersects the xy-plane.
  d. Does the line intersect the xz-coordinate plane? Explain why or why not.

FIGURE FOR 1   **FIGURE FOR 2**

In Exercises 3–8, find a set of (a) parametric equations and (b) symmetric equations of the line through the point and parallel to the specified vector or line. (For each line, express the direction numbers as integers.)

Point	Parallel to
3. $(0, 0, 0)$	$\mathbf{v} = \langle 1, 2, 3 \rangle$
4. $(0, 0, 0)$	$\mathbf{v} = \langle -2, \frac{5}{2}, 1 \rangle$
5. $(-2, 0, 3)$	$\mathbf{v} = 2\mathbf{i} + 4\mathbf{j} - 2\mathbf{k}$
6. $(-2, 0, 3)$	$\mathbf{v} = 6\mathbf{i} + 3\mathbf{j}$
7. $(1, 0, 1)$	$\begin{aligned} x &= 3 + 3t \\ y &= 5 - 2t \\ z &= -7 + t \end{aligned}$
8. $(-3, 5, 4)$	$\dfrac{x - 1}{3} = \dfrac{y + 1}{-2} = z - 3$

In Exercises 9 and 10, find a set of (a) parametric equations and (b) symmetric equations of the line through the two points. (For each line, express the direction numbers as integers.)

9. $(5, -3, -2)$, $\left(-\frac{2}{3}, \frac{2}{3}, 1\right)$

10. $(1, 0, 1)$, $(1, 3, -2)$

In Exercises 11 and 12, find a set of parametric equations of the line.

11. The line passes through the point $(2, 3, 4)$ and is parallel to the xz-plane and the yz-plane.

12. The line passes through the point $(2, 3, 4)$ and is perpendicular to the plane given by $3x + 2y - z = 6$.

In Exercises 13 and 14, determine which of the points lie on the line L.

13. The line L passes through the point $(-2, 3, 1)$ and is parallel to the vector $\mathbf{v} = 4\mathbf{i} - \mathbf{k}$.
 a. $(2, 3, 0)$ **b.** $(-6, 3, 2)$
 c. $(2, 1, 0)$ **d.** $(6, 3, -2)$

14. The line L passes through the points $(2, 0, -3)$ and $(4, 2, -2)$.
 a. $(4, 1, -2)$ **b.** $\left(\frac{5}{2}, \frac{1}{2}, -\frac{11}{4}\right)$
 c. $(-1, -3, -4)$ **d.** $(0, -2, -4)$

In Exercises 15–18, determine whether the lines intersect, and if so, find the point of intersection and the cosine of the angle of intersection.

15. $\begin{aligned} x &= 4t + 2 \\ y &= 3 \\ z &= -t + 1 \end{aligned}$ $\qquad$ $\begin{aligned} x &= 2s + 2 \\ y &= 2s + 3 \\ z &= s + 1 \end{aligned}$

16. $\begin{aligned} x &= -3t + 1 \\ y &= 4t + 1 \\ z &= 2t + 4 \end{aligned}$ $\qquad$ $\begin{aligned} x &= 3s + 1 \\ y &= 2s + 4 \\ z &= -s + 1 \end{aligned}$

17. $\dfrac{x}{3} = \dfrac{y - 2}{-1} = z + 1$

$\dfrac{x - 1}{4} = y + 2 = \dfrac{z + 3}{-3}$

18. $\dfrac{x - 2}{-3} = \dfrac{y - 2}{6} = z - 3$

$\dfrac{x - 3}{2} = y + 5 = \dfrac{z + 2}{4}$

C In Exercises 19 and 20, use a computer algebra system to graph the pair of intersecting lines and find the point of intersection.

19. $\begin{aligned} x &= 2t + 3 \\ y &= 5t - 2 \\ z &= -t + 1 \end{aligned}$ $\qquad$ $\begin{aligned} x &= -2s + 7 \\ y &= s + 8 \\ z &= 2s - 1 \end{aligned}$

20. $\begin{aligned} x &= 2t - 1 \\ y &= -4t + 10 \\ z &= t \end{aligned}$ $\qquad$ $\begin{aligned} x &= -5s - 12 \\ y &= 3s + 11 \\ z &= -2s - 4 \end{aligned}$

In Exercises 21 and 22, (a) find the coordinates of three points P, Q, and R in the plane, and determine the vectors $\overrightarrow{PQ}$ and $\overrightarrow{PR}$ and (b) find $\overrightarrow{PQ} \times \overrightarrow{PR}$. What is the relationship between the components of the cross product and the coefficients of the variables in the equation of the plane? Why is this true?

21. **22.**

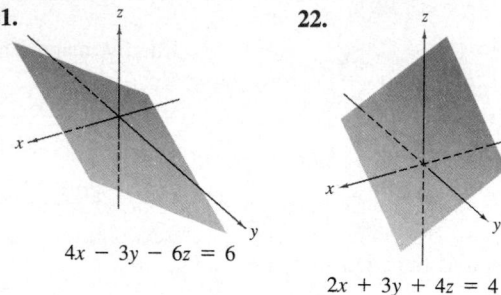

$4x - 3y - 6z = 6$

$2x + 3y + 4z = 4$

In Exercises 23–28, find an equation of the plane passing through the point and perpendicular to the specified vector or line.

Point	Perpendicular to
23. $(2, 1, 2)$	$\mathbf{n} = \mathbf{i}$
24. $(1, 0, -3)$	$\mathbf{n} = \mathbf{k}$
25. $(3, 2, 2)$	$\mathbf{n} = 2\mathbf{i} + 3\mathbf{j} - \mathbf{k}$
26. $(0, 0, 0)$	$\mathbf{n} = -3\mathbf{i} + 2\mathbf{k}$
27. $(0, 0, 6)$	$x = 1 - t, y = 2 + t, z = 4 - 2t$
28. $(3, 2, 2)$	$\dfrac{x - 1}{4} = y + 2 = \dfrac{z + 3}{-3}$

In Exercises 29–40, find an equation of the plane.

29. The plane passes through the points $(0, 0, 0)$, $(1, 2, 3)$, and $(-2, 3, 3)$.

30. The plane passes through the points $(1, 2, -3)$, $(2, 3, 1)$, and $(0, -2, -1)$.

31. The plane passes through the points $(1, 2, 3)$, $(3, 2, 1)$, and $(-1, -2, 2)$.

32. The plane passes through the point $(1, 2, 3)$ and is parallel to the yz-plane.

33. The plane passes through the point $(1, 2, 3)$ and is parallel to the xy-plane.

34. The plane contains the y-axis and makes an angle of $\pi/6$ with the positive x-axis.

35. The plane contains lines given by

$$\frac{x-1}{-2} = y - 4 = z$$

and

$$\frac{x-2}{-3} = \frac{y-1}{4} = \frac{z-2}{-1}.$$

36. The plane passes through the point $(2, 2, 1)$ and contains the line given by

$$\frac{x}{2} = \frac{y-4}{-1} = z.$$

37. The plane passes through the points $(2, 2, 1)$ and $(-1, 1, -1)$ and is perpendicular to the plane $2x - 3y + z = 3$.

38. The plane passes through the points $(3, 2, 1)$ and $(3, 1, -5)$ and is perpendicular to the plane $6x + 7y + 2z = 10$.

39. The plane passes through the points $(1, -2, -1)$ and $(2, 5, 6)$ and is parallel to the x-axis.

40. The plane passes through the points $(4, 2, 1)$ and $(-3, 5, 7)$ and is parallel to the z-axis.

In Exercises 41–46, determine whether the planes are parallel or orthogonal, or neither. If they are neither parallel nor orthogonal, find the angle of intersection.

41. $5x - 3y + z = 4$
    $x + 4y + 7z = 1$

42. $3x + y - 4z = 3$
    $-9x - 3y + 12z = 4$

43. $x - 3y + 6z = 4$
    $5x + y - z = 4$

44. $3x + 2y - z = 7$
    $x - 4y + 2z = 0$

45. $x - 5y - z = 1$
    $5x - 25y - 5z = -3$

46. $2x - z = 1$
    $4x + y + 8z = 10$

In Exercises 47–52, mark the intercepts and sketch a graph of the plane.

47. $4x + 2y + 6z = 12$

48. $3x + 6y + 2z = 6$

49. $2x - y + 3z = 4$

50. $2x - y + z = 4$

51. $y + z = 5$

52. $x + 2y = 4$

C **In Exercises 53–56, use a computer algebra system to obtain a graph of the plane.**

53. $2x + y - z = 6$

54. $x - 3z = 3$

55. $-5x + 4y - 6z + 8 = 0$

56. $2.1x - 4.7y - z + 3 = 0$

In Exercises 57 and 58, find a set of parametric equations for the line of intersection of the planes.

57. $3x + 2y - z = 7$
    $x - 4y + 2z = 0$

58. $x - 3y + 6z = 4$
    $5x + y - z = 4$

In Exercises 59–62, find the point of intersection (if any) of the plane and the line. Also determine whether the line lies in the plane.

59. $2x - 2y + z = 12$
    $$x - \frac{1}{2} = \frac{y + (3/2)}{-1} = \frac{z+1}{2}$$

60. $2x + 3y = -5$
    $$\frac{x-1}{4} = \frac{y}{2} = \frac{z-3}{6}$$

61. $2x + 3y = 10$
    $$\frac{x-1}{3} = \frac{y+1}{-2} = z - 3$$

62. $5x + 3y = 17$
    $$\frac{x-4}{2} = \frac{y+1}{-3} = \frac{z+2}{5}$$

In Exercises 63 and 64, find the distance between the point and the line.

63. $(10, 3, -2)$
    $x = 4t - 2$
    $y = 3$
    $z = -t + 1$

64. $(4, 1, -2)$
    $x = 2t + 2$
    $y = 2t$
    $z = t - 3$

In Exercises 65 and 66, find the distance between the point and the plane.

65. $(0, 0, 0)$
    $2x + 3y + z = 12$

66. $(1, 2, 3)$
    $2x - y + z = 4$

In Exercises 67 and 68, find the distance between the planes.

67. $x - 3y + 4z = 10$
    $x - 3y + 4z = 6$

68. $2x - 4z = 4$
    $2x - 4z = 10$

In Exercises 69 and 70, find the distance between the two skew lines (lines that are neither parallel nor intersecting).

69. $x = \dfrac{y}{2} = \dfrac{z}{3}$ $\dfrac{x-1}{-1} = y - 4 = z + 1$

70. $x = 3t$ $x = 4s + 1$
    $y = -t + 2$ $y = s - 2$
    $z = t - 1$ $z = -3s - 3$

71. a. Describe and find an equation for the surface generated by all points (x, y, z) that are four units from the point $(3, -2, 5)$.

   b. Describe and find an equation for the surface generated by all points (x, y, z) that are four units from the plane $4x - 3y + z = 10$.

72. Consider the two nonzero vectors **u** and **v**. Describe the geometric figure generated by the terminal points of the following vectors where s and t represent all real numbers.
   a. $t\mathbf{v}$ **b.** $\mathbf{u} + t\mathbf{v}$ **c.** $s\mathbf{u} + t\mathbf{v}$

73. *Mechanical Design*   A chute at the top of the grain elevator of a combine has the purpose of funneling the grain into a bin (see figure). Find the angle between two adjacent sides.

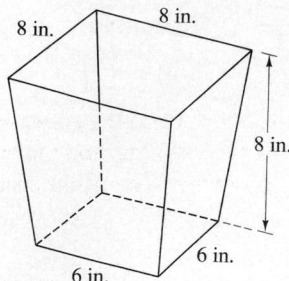

FIGURE FOR 73

74. If a_1, b_1, c_1 and a_2, b_2, c_2 are two sets of direction numbers for the same line, show that there exists a scalar d such that $a_1 = a_2 d$, $b_1 = b_2 d$, and $c_1 = c_2 d$.

True or False   **In Exercises 75 and 76, determine whether the statement is true or false. If it is false, explain why or give an example that shows it is false.**

75. If $\mathbf{v} = a_1 \mathbf{i} + b_1 \mathbf{j} + c_1 \mathbf{k}$ is any vector in the plane given by $a_2 x + b_2 y + c_2 z + d_2 = 0$, then $a_1 a_2 + b_1 b_2 + c_1 c_2 = 0$.

76. Every pair of lines in space are either intersecting or parallel.

77. Consider the plane that passes through the points P, R, and S. Show that the distance from a point Q to this plane is

$$\text{Distance} = \frac{|\mathbf{u} \cdot (\mathbf{v} \times \mathbf{w})|}{\|\mathbf{u} \times \mathbf{v}\|}$$

where $\mathbf{u} = \vec{PR}$, $\mathbf{v} = \vec{PS}$, and $\mathbf{w} = \vec{PQ}$.

78. Show that the distance between the parallel planes $ax + by + cz + d_1 = 0$ and $ax + by + cz + d_2 = 0$ is

$$\text{Distance} = \frac{|d_1 - d_2|}{\sqrt{a^2 + b^2 + c^2}}.$$

79. Prove that any two nonintersecting lines in space lie in parallel planes.

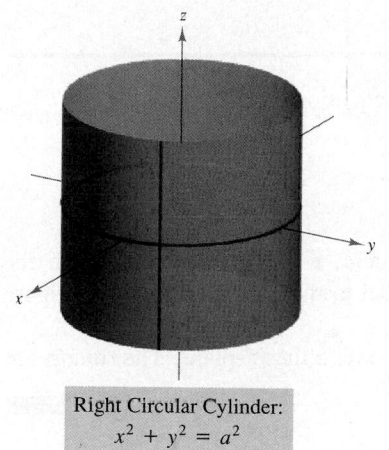

Right Circular Cylinder:
$$x^2 + y^2 = a^2$$

FIGURE 11.56
Rulings are parallel to z-axis.

Cylindrical Surfaces

The first five sections of this chapter contained the vector portion of the preliminary work necessary to study vector calculus and the calculus of space. In this and the next section, you will study surfaces in space and alternative coordinate systems for space. You have already studied two special types of surfaces.

1. **Spheres** $(x - x_0)^2 + (y - y_0)^2 + (z - z_0)^2 = r^2$ Section 11.2
2. **Planes** $ax + by + cz + d = 0$ Section 11.5

A third type of surface in space is called a **cylindrical surface,** or simply a **cylinder.** To define a cylinder, consider the familiar right circular cylinder shown in Figure 11.56. You can imagine that this cylinder is generated by a vertical line moving around the circle $x^2 + y^2 = a^2$ in the xy-plane. This circle is called a **generating curve** for the cylinder, as indicated in the following definition.

Definition of a Cylinder

Let C be a curve in a plane and let L be a line not in a parallel plane. The set of all lines parallel to L and intersecting C is called a **cylinder.** C is called the **generating curve** (or **directrix**) of the cylinder, and the parallel lines are called **rulings.**

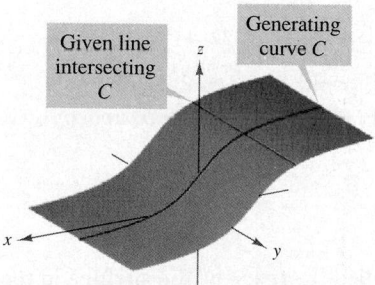

Given line intersecting C

Generating curve C

FIGURE 11.57
Cylinder: Rulings intersect C and are parallel to the given line.

REMARK   Without loss of generality, you can assume that C lies in one of the three coordinate planes. Moreover, in this text we restrict the discussion to *right* cylinders—cylinders whose rulings are perpendicular to the coordinate plane containing C, as shown in Figure 11.57.

For the right circular cylinder shown in Figure 11.56, the equation of the generating curve is

$$x^2 + y^2 = a^2.$$ Equation of generating curve in xy-plane

To find an equation for the cylinder, note that you can generate any one of the rulings by fixing the values of x and y and then allowing z to take on all real values. In this sense the value of z is arbitrary and is, therefore, not included in the equation. In other words, the equation of this cylinder is simply the equation of its generating curve.

$$x^2 + y^2 = a^2$$ Equation of cylinder in space

Equations of Cylinders

The equation of a cylinder whose rulings are parallel to one of the coordinate axes contains only the variables corresponding to the other two axes.

EXAMPLE 1   Sketching a Cylinder

Sketch the surfaces represented by the following equations.

a. $z = y^2$ **b.** $z = \sin x,$ $0 \leq x \leq 2\pi$

Solution

a. The graph is a cylinder whose generating curve, $z = y^2$, is a parabola in the yz-plane. The rulings of the cylinder are parallel to the x-axis, as shown in Figure 11.58.

b. The graph is a cylinder generated by the sine curve in the xz-plane. The rulings are parallel to the y-axis, as shown in Figure 11.59.

Generating curve C lies in yz-plane.

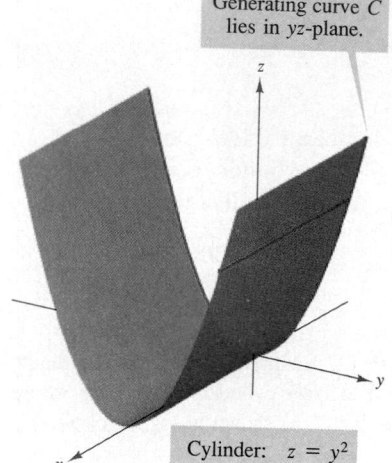

Cylinder: $z = y^2$

FIGURE 11.58
Rulings are parallel to x-axis.

Quadric Surfaces

The fourth basic type of surface in space is a **quadric surface.** Quadric surfaces are the three-dimensional analogs of conic sections.

Quadric Surface

The equation of a **quadric surface** in space is a second-degree equation of the form

$$Ax^2 + By^2 + Cz^2 + Dxy + Exz + Fyz + Gx + Hy + Iz + J = 0.$$

There are six basic types of quadric surfaces: **ellipsoid, hyperboloid of one sheet, hyperboloid of two sheets, elliptic cone, elliptic paraboloid,** and **hyperbolic paraboloid.**

Generating curve C lies in xz-plane.

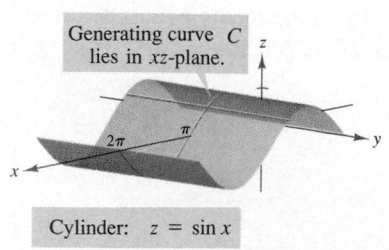

Cylinder: $z = \sin x$

FIGURE 11.59
Rulings are parallel to y-axis.

The intersection of a surface with a plane is called the **trace of the surface** in the plane. To visualize a surface in space, it is helpful to determine its traces in some well-chosen planes. The traces of quadric surfaces are conics. These traces, together with the **standard form** of the equation of each quadric surface, are shown in Table 11.1.

In Table 11.1, only one of several orientations of each quadric surface is shown. If the surface is oriented along a different axis, then its standard equation will change accordingly, as illustrated in Examples 2 and 3. The fact that the two types of paraboloids have one variable raised to the first power can be helpful in classifying quadric surfaces. The other four types of basic quadric surfaces have equations that are of *second degree* in all three variables.

TABLE 11.1   Quadric Surfaces

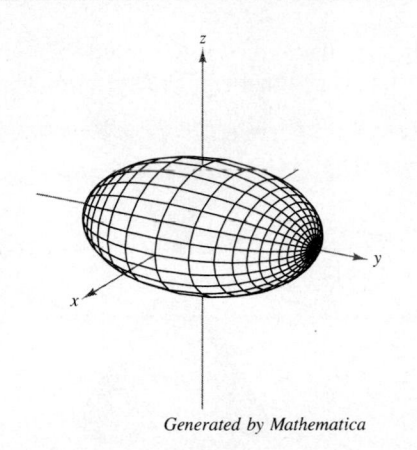

Generated by Mathematica

Ellipsoid

$$\frac{x^2}{a^2} + \frac{y^2}{b^2} + \frac{z^2}{c^2} = 1$$

Trace	Plane
Ellipse	Parallel to xy-plane
Ellipse	Parallel to xz-plane
Ellipse	Parallel to yz-plane

The surface is a sphere if
$a = b = c \neq 0$.

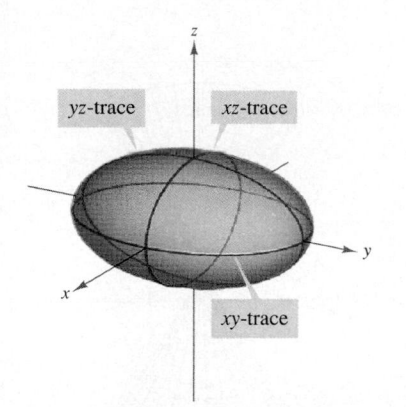

Hyperboloid of One Sheet

$$\frac{x^2}{a^2} + \frac{y^2}{b^2} - \frac{z^2}{c^2} = 1$$

Trace	Plane
Ellipse	Parallel to xy-plane
Hyperbola	Parallel to xz-plane
Hyperbola	Parallel to yz-plane

The axis of the hyperboloid
corresponds to the variable
whose coefficient is negative.

Generated by Mathematica

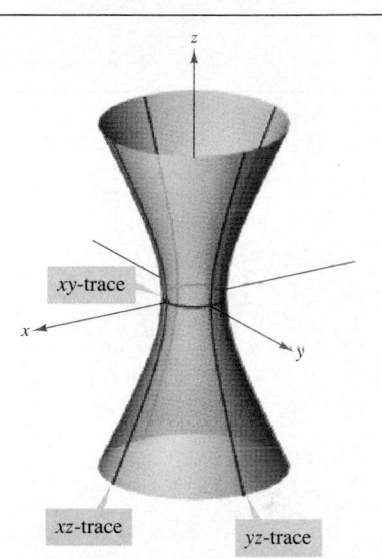

Hyperboloid of Two Sheets

$$\frac{z^2}{c^2} - \frac{x^2}{a^2} - \frac{y^2}{b^2} = 1$$

Trace	Plane
Ellipse	Parallel to xy-plane
Hyperbola	Parallel to xz-plane
Hyperbola	Parallel to yz-plane

The axis of the hyperboloid
corresponds to the variable
whose coefficient is positive.
There is no trace in the
coordinate plane perpendicular
to this axis.

Generated by Mathematica

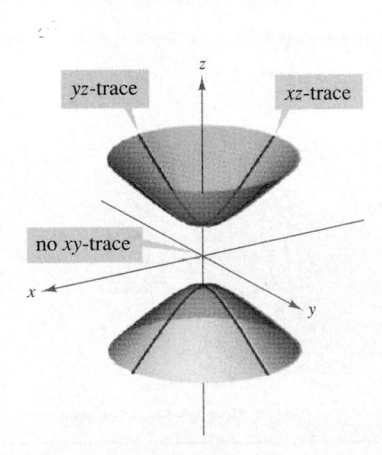

TABLE 11.1 (Continued)

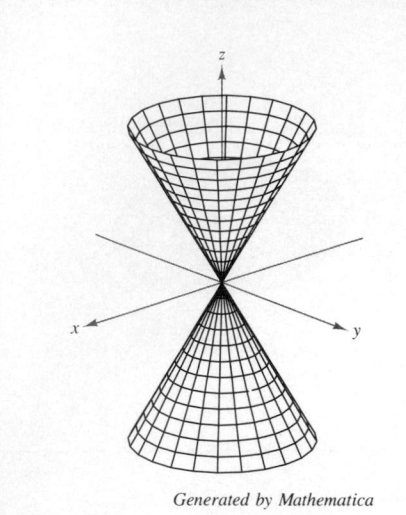 *Generated by Mathematica*	***Elliptic Cone*** $$\frac{x^2}{a^2} + \frac{y^2}{b^2} - \frac{z^2}{c^2} = 0$$ *Trace* — *Plane* Ellipse — Parallel to *xy*-plane Hyperbola — Parallel to *xz*-plane Hyperbola — Parallel to *yz*-plane The axis of the cone corresponds to the variable whose coefficient is negative. The traces in the coordinate planes parallel to this axis are intersecting lines.	
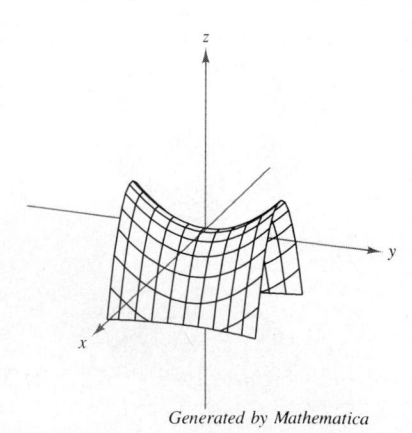 *Generated by Mathematica*	***Elliptic Paraboloid*** $$z = \frac{x^2}{a^2} + \frac{y^2}{b^2}$$ *Trace* — *Plane* Ellipse — Parallel to *xy*-plane Parabola — Parallel to *xz*-plane Parabola — Parallel to *yz*-plane The axis of the paraboloid corresponds to the variable raised to the first power.	
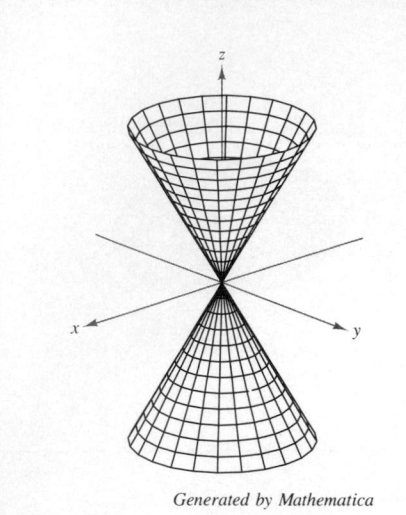 *Generated by Mathematica*	***Hyperbolic Paraboloid*** $$z = \frac{y^2}{b^2} - \frac{x^2}{a^2}$$ *Trace* — *Plane* Hyperbola — Parallel to *xy*-plane Parabola — Parallel to *xz*-plane Parabola — Parallel to *yz*-plane The axis of the paraboloid corresponds to the variable raised to the first power.	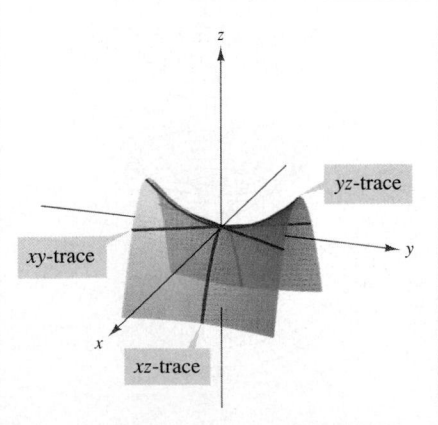

To classify a quadric surface, begin by writing the surface in standard form. Then, determine several traces taken in the coordinate planes *or* taken in planes that are parallel to the coordinate planes.

EXAMPLE 2 Sketching a Quadric Surface

Classify and sketch the surface given by $4x^2 - 3y^2 + 12z^2 + 12 = 0$.

Solution Begin by writing the given equation in standard form.

$$4x^2 - 3y^2 + 12z^2 + 12 = 0 \qquad \text{Given equation}$$

$$\frac{x^2}{-3} + \frac{y^2}{4} - z^2 - 1 = 0 \qquad \text{Divide by } -12$$

$$\frac{y^2}{4} - \frac{x^2}{3} - \frac{z^2}{1} = 1 \qquad \text{Standard form}$$

From Table 11.1, you can conclude that the surface is a hyperboloid of two sheets with the y-axis as its axis. To sketch the graph of this surface, it helps to find the traces in the coordinate planes.

xy-trace ($z = 0$)	$\dfrac{y^2}{4} - \dfrac{x^2}{3} = 1$	Hyperbola
xz-trace ($y = 0$)	$\dfrac{x^2}{3} + \dfrac{z^2}{1} = -1$	No trace
yz-trace ($x = 0$)	$\dfrac{y^2}{4} - \dfrac{z^2}{1} = 1$	Hyperbola

The graph is shown in Figure 11.60.

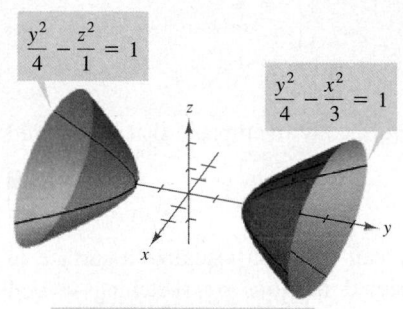

$\dfrac{y^2}{4} - \dfrac{z^2}{1} = 1$

$\dfrac{y^2}{4} - \dfrac{x^2}{3} = 1$

Hyperboloid of Two Sheets:

$$\frac{y^2}{4} - \frac{x^2}{3} - z^2 = 1$$

FIGURE 11.60

EXAMPLE 3 Sketching a Quadric Surface

Classify and sketch the surface given by $x - y^2 - 4z^2 = 0$.

Solution Because *x* is raised only to the first power, the surface is a paraboloid. The axis of the paraboloid is the x-axis. In the standard form, the equation is

$$x = y^2 + 4z^2. \qquad \text{Standard form}$$

Some convenient traces are as follows.

xy-trace ($z = 0$)	$x = y^2$	Parabola
xz-trace ($y = 0$)	$x = 4z^2$	Parabola
parallel to yz-plane ($x = 4$)	$\dfrac{y^2}{4} + \dfrac{z^2}{1} = 1$	Ellipse

Thus, the surface is an *elliptic* paraboloid, as shown in Figure 11.61.

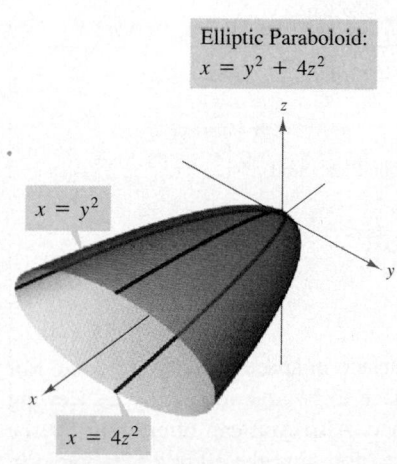

Elliptic Paraboloid:
$x = y^2 + 4z^2$

$x = y^2$

$x = 4z^2$

FIGURE 11.61

Some second-degree equations in *x*, *y*, and *z* do not represent quadric surfaces. For instance, the graph of

$$x^2 + y^2 + z^2 = 0$$

is a single point, and the graph of

$$x^2 + y^2 = 1$$

is a right circular cylinder.

For a quadric surface not centered at the origin, you can form the standard equation by completing the square, as demonstrated in Example 4.

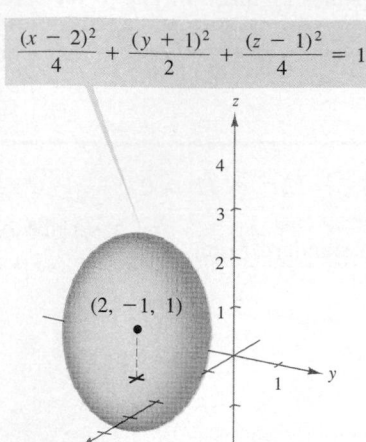

$$\frac{(x-2)^2}{4} + \frac{(y+1)^2}{2} + \frac{(z-1)^2}{4} = 1$$

$(2, -1, 1)$

FIGURE 11.62
An Ellipsoid centered at $(2, -1, 1)$.

EXAMPLE 4   A Quadric Surface Not Centered at the Origin

Classify and sketch the surface given by

$$x^2 + 2y^2 + z^2 - 4x + 4y - 2z + 3 = 0.$$

Solution   Completing the square for each variable produces the following.

$$(x^2 - 4x + \quad) + 2(y^2 + 2y + \quad) + (z^2 - 2z + \quad) = -3$$
$$(x^2 - 4x + 4) + 2(y^2 + 2y + 1) + (z^2 - 2z + 1) = -3 + 4 + 2 + 1$$
$$(x - 2)^2 + 2(y + 1)^2 + (z - 1)^2 = 4$$
$$\frac{(x - 2)^2}{4} + \frac{(y + 1)^2}{2} + \frac{(z - 1)^2}{4} = 1$$

From this equation, you can see that the quadric surface is an ellipsoid that is centered at $(2, -1, 1)$. Its graph is shown in Figure 11.62.

TECHNOLOGY   A computer graphing utility can help you visualize a surface in space.* Most such utilities create three-dimensional illusions by sketching several traces of the surface and then applying a "hidden-line" routine that blocks out portions of the surface that lie behind other portions of the surface. Figure 11.63 shows two different quadric surfaces that were generated by a computer.

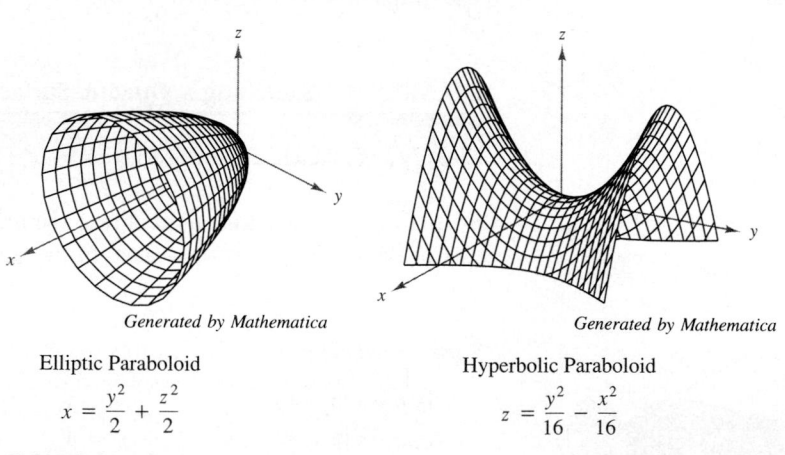

Generated by Mathematica   *Generated by Mathematica*

Elliptic Paraboloid   Hyperbolic Paraboloid

$$x = \frac{y^2}{2} + \frac{z^2}{2}$$ $$z = \frac{y^2}{16} - \frac{x^2}{16}$$

FIGURE 11.63

Using a computer to sketch the graph of a surface in space requires practice. For one thing, you must know enough about the surface to be able to specify a "viewing box" that gives a representative view of the surface. Also, you can often improve the view of a surface by rotating the axis. For instance, note that the elliptic paraboloid in Figure 11.63 is seen from a line of sight that is "higher" than the line of sight used to view the hyperbolic paraboloid.

*Some 3-D graphing utilities require surfaces to be entered with parametric equations. For a discussion of this technique, see Section 15.5.

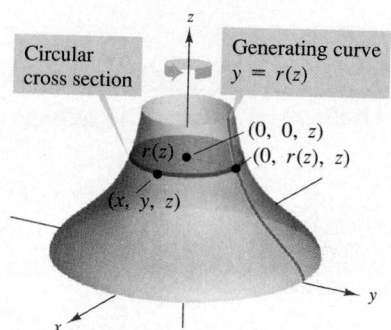

FIGURE 11.64
A surface of revolution.

Surfaces of Revolution

The fifth special type of surface we will study is called a **surface of revolution.** In Section 6.4 we looked at a method for finding the *area* of such a surface. We now look at a procedure for finding its *equation.*

Consider the graph of the **radius function**

$$y = r(z) \qquad \text{Generating curve}$$

in the yz-plane. If this graph is revolved about the z-axis, it forms a surface of revolution, as shown in Figure 11.64. The trace of the surface in the plane $z = z_0$ is a circle whose radius is $r(z_0)$ and whose equation is

$$x^2 + y^2 = [r(z_0)]^2. \qquad \text{Circular trace in plane } z = z_0$$

Replacing z_0 by z produces an equation that is valid for all values of z. In a similar manner we can obtain equations for surfaces of revolution for the other two axes, and we summarize the results as follows.

Surface of Revolution

If the graph of a radius function r is revolved about one of the coordinate axes, then the equation of the resulting surface of revolution has one of the following forms.

1. **Revolved about the x-axis** $y^2 + z^2 = [r(x)]^2$
2. **Revolved about the y-axis** $x^2 + z^2 = [r(y)]^2$
3. **Revolved about the z-axis** $x^2 + y^2 = [r(z)]^2$

EXAMPLE 5   Finding an Equation for a Surface of Revolution

a. An equation for the surface of revolution formed by revolving the graph of $y = 1/z$ about the z-axis is

$$x^2 + y^2 = [r(z)]^2 = \left(\frac{1}{z}\right)^2$$

b. To find an equation for the surface formed by revolving the graph of $9x^2 = y^3$ about the y-axis, solve for x in terms of y to obtain

$$x = \frac{1}{3}y^{3/2} = r(y). \qquad \text{Radius function}$$

Thus, the equation for this surface is

$$x^2 + z^2 = [r(y)]^2 \qquad \text{Revolved about the } y\text{-axis}$$

$$x^2 + z^2 = \left(\frac{1}{3}y^{3/2}\right)^2 \qquad \text{Substitute } \tfrac{1}{3}y^{3/2} \text{ for } r(y)$$

$$x^2 + z^2 = \frac{1}{9}y^3. \qquad \text{Equation of surface}$$

The graph is shown in Figure 11.65.

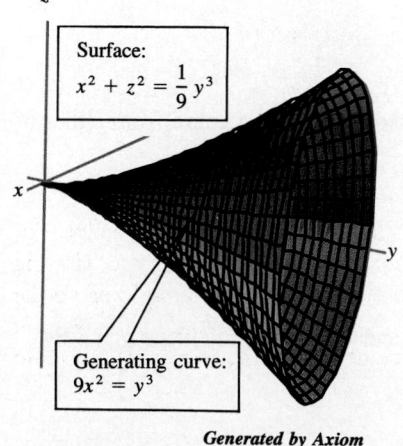

Surface:
$x^2 + z^2 = \dfrac{1}{9}y^3$

Generating curve:
$9x^2 = y^3$

Generated by Axiom

FIGURE 11.65

The generating curve for a surface of revolution is not unique. For instance, the surface

$$x^2 + z^2 = e^{-2y}$$

can be formed by revolving either the graph of $x = e^{-y}$ about the y-axis or the graph of $z = e^{-y}$ about the y-axis, as shown in Figure 11.66.

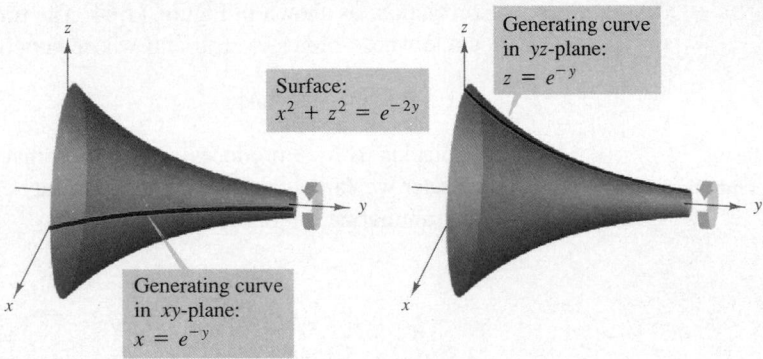

FIGURE 11.66

EXAMPLE 6   Finding a Generating Curve for a Surface of Revolution

Find a generating curve and the axis of revolution for the surface given by $x^2 + 3y^2 + z^2 = 9$.

Solution   You now know that the equation has one of the following forms.

$$x^2 + y^2 = [r(z)]^2 \qquad \text{Revolved about } z\text{-axis}$$
$$y^2 + z^2 = [r(x)]^2 \qquad \text{Revolved about } x\text{-axis}$$
$$x^2 + z^2 = [r(y)]^2 \qquad \text{Revolved about } y\text{-axis}$$

Because the coefficients of x^2 and z^2 are equal, you should choose the third form and write

$$x^2 + z^2 = 9 - 3y^2.$$

The y-axis is the axis of revolution. You can choose a generating curve from either of the following traces.

$$x^2 = 9 - 3y^2 \qquad \text{Trace in } xy\text{-plane}$$
$$z^2 = 9 - 3y^2 \qquad \text{Trace in } yz\text{-plane}$$

For example, using the first trace, the generating curve is the semiellipse given by

$$x = \sqrt{9 - 3y^2}. \qquad \text{Generating curve}$$

The graph of this surface is shown in Figure 11.67.

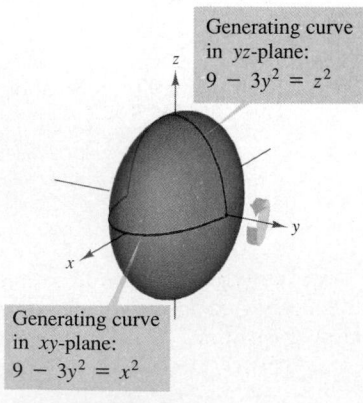

FIGURE 11.67

EXERCISES for Section 11.6

In Exercises 1–6, match the equation with its graph. [The graphs are labeled (a), (b), (c), (d), (e), and (f).]

1. $\dfrac{x^2}{9} + \dfrac{y^2}{16} + \dfrac{z^2}{9} = 1$

2. $15x^2 - 4y^2 + 15z^2 = -4$

3. $4x^2 - y^2 + 4z^2 = 4$ **4.** $y^2 = 4x^2 + 9z^2$

5. $4x^2 - 4y + z^2 = 0$ **6.** $4x^2 - y^2 + 4z = 0$

(a)

(b)

(c)

(d)

(e)

(f)

In Exercises 7–16, describe and sketch each surface.

7. $z = 3$ **8.** $x = 4$

9. $y^2 + z^2 = 9$ **10.** $x^2 + z^2 = 16$

11. $x^2 - y = 0$ **12.** $y^2 + z = 4$

13. $4x^2 + y^2 = 4$ **14.** $y^2 - z^2 = 4$

15. $z - \sin y = 0$ **16.** $z - e^y = 0$

17. The four figures are graphs of the quadric surface

$$z = x^2 + y^2.$$

Match each of the four graphs with the point in space from which the paraboloid is viewed. The four points are $(0, 0, 20)$, $(0, 20, 0)$, $(20, 0, 0)$, and $(10, 10, 20)$.

(a)

(b)

(c)

(d)
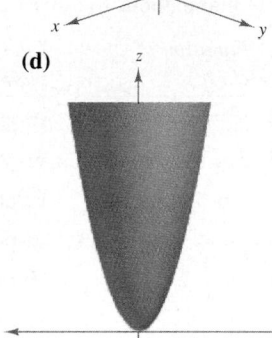

C **18** Use a graphing utility to sketch three views of the equation

$$y^2 + z^2 = 4$$

with views from the points
a. $(10, 0, 0)$ **b.** $(0, 10, 0)$ **c.** $(10, 10, 10)$.

In Exercises 19–30, identify and sketch the given quadric surface.

19. $x^2 + \dfrac{y^2}{4} + z^2 = 1$ **20.** $\dfrac{x^2}{9} + \dfrac{y^2}{16} + \dfrac{z^2}{16} = 1$

21. $16x^2 - y^2 + 16z^2 = 4$ **22.** $z^2 - x^2 - \dfrac{y^2}{4} = 1$

23. $x^2 - y + z^2 = 0$ **24.** $z = 4x^2 + y^2$

25. $x^2 - y^2 + z = 0$ **26.** $3z = -y^2 + x^2$

27. $z^2 = x^2 + \dfrac{y^2}{4}$ **28.** $x^2 = 2y^2 + 2z^2$

29. $16x^2 + 9y^2 + 16z^2 - 32x - 36y + 36 = 0$
30. $4x^2 + y^2 - 4z^2 - 16x - 6y - 16z + 9 = 0$

C In Exercises 31–38, use a computer algebra system to obtain a graph of the surface. Identify the surface as a cylinder, quadric surface, or surface of revolution. (*Hint:* It may be necessary to solve for z and acquire two equations to graph.)

31. $z = 2\sin x$ **32.** $z = x^2 + 0.5y^2$

33. $x^2 + z^2 = \left(\dfrac{2}{y}\right)^2$ **34.** $x^2 + y^2 = e^{-z}$

35. $z^2 = x^2 + 4y^2$ **36.** $4y = x^2 + z^2$

37. $4x^2 - y^2 + 4z^2 = -16$ **38.** $9x^2 + 4y^2 - 8z^2 = 72$

In Exercises 39–42, sketch the region bounded by the graphs of the equations.

39. $z = 2\sqrt{x^2 + y^2}, z = 2$

40. $z = \sqrt{4 - x^2}, y = \sqrt{4 - x^2}, x = 0, y = 0, z = 0$

41. $x^2 + y^2 = 1, x + z = 2, z = 0$

42. $z = \sqrt{4 - x^2 - y^2}, y = 2z, z = 0$

In Exercises 43–48, find an equation for the surface of revolution generated by revolving the curve in the specified coordinate plane about the given axis.

Equation of curve	Coordinate plane	Axis of revolution
43. $z^2 = 4y$	yz-plane	y-axis
44. $z = 2y$	yz-plane	y-axis
45. $z = 2y$	yz-plane	z-axis
46. $2z = \sqrt{4 - x^2}$	xz-plane	x-axis
47. $xy = 2$	xy-plane	x-axis
48. $z = \ln y$	yz-plane	z-axis

In Exercises 49 and 50, find an equation of a generating curve given the equation of its surface of revolution.

49. $x^2 + y^2 - 2z = 0$ **50.** $x^2 + z^2 = \sin^2 y$

In Exercises 51 and 52, use the Shell Method to find the volume of the solid below the surface of revolution and above the xy-coordinate plane.

51. The curve $z = 4x - x^2$ in the xz-plane is revolved about the z-axis.

52. The curve $z = \sin y$ $(0 \le y \le \pi)$ in the yz-plane is revolved about the z-axis.

In Exercises 53 and 54, analyze the trace when the surface given by

$$z = \frac{x^2}{2} + \frac{y^2}{4}$$

is intersected by the planes.

53. Find the length of the major and minor axes and the coordinates of the foci of the ellipse generated when the surface is intersected by the planes given by
a. $z = 2$.   **b.** $z = 8$.

54. Find the coordinates of the focus of the parabola formed when the surface is intersected by the planes given by
a. $y = 4$.   **b.** $x = 2$.

55. *Shape of the Earth* Because of the forces caused by its rotation, the earth is an oblate ellipsoid rather than a sphere. The equatorial radius is 3963 miles and the polar radius is 3942 miles. Find an equation of the ellipsoid. (Assume that the center of the earth is at the origin and the trace formed by the plane $z = 0$ corresponds to the equator.)

56. *Satellite Antenna* A satellite receiving dish is a paraboloid with the dimensions shown in the figure. Find the equation of the generating curve in the yz-plane and find the corresponding equation of the surface of revolution.

FIGURE FOR 56

57. *Inside or Outside* Three classic "topological" surfaces are shown below. The sphere and torus have both an "inside" and an "outside." Does the Klein bottle have both an inside and an outside? Explain.

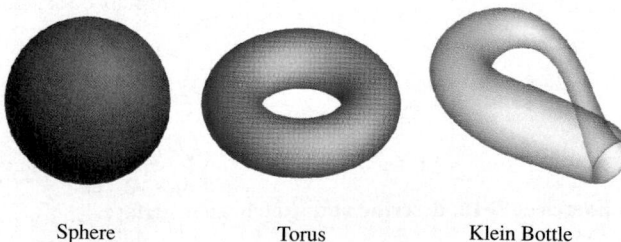

Sphere   Torus   Klein Bottle

FIGURE FOR 57

58. Determine the intersection of the hyperbolic paraboloid $z = y^2/b^2 - x^2/a^2$ with the plane $bx + ay - z = 0$. (Assume $a, b > 0$.)

Cylindrical and Spherical Coordinates

Cylindrical Coordinates • Spherical Coordinates

Cylindrical Coordinates

You have already seen that some two-dimensional graphs are easier to represent in polar coordinates than in rectangular coordinates. A similar situation exists for surfaces in space. In this section, you will study two alternative space-coordinate systems. The first, the **cylindrical coordinate system,** is an extension of polar coordinates to space.

Cylindrical coordinates:
$$r^2 = x^2 + y^2$$
$$\tan \theta = \frac{y}{x}$$
$$z = z$$

Rectangular coordinates:
$$x = r \cos \theta$$
$$y = r \sin \theta$$
$$z = z$$

$P \bullet (x, y, z)$
(r, θ, z)

FIGURE 11.68

The Cylindrical Coordinate System

In a **cylindrical coordinate system,** a point P in space is represented by an ordered triple (r, θ, z).

1. (r, θ) is a polar representation of the projection of P in the xy-plane.
2. z is the directed distance from (r, θ) to P.

To convert from rectangular to cylindrical coordinates (or vice versa), use the conversion guidelines for polar coordinates, as illustrated in Figure 11.68.

Cylindrical to rectangular

$$x = r \cos \theta, \; y = r \sin \theta, \; z = z$$

Rectangular to cylindrical

$$r^2 = x^2 + y^2, \; \tan \theta = \frac{y}{x}, \; z = z$$

The point $(0, 0, 0)$ is called the **pole.** Moreover, because the representation of a point in the polar coordinate system is not unique, it follows that the representation in the cylindrical coordinate system is also not unique.

EXAMPLE 1 Changing from Cylindrical to Rectangular Coordinates

Express the point $(r, \theta, z) = (4, 5\pi/6, 3)$ in rectangular coordinates.

Solution Using the *cylindrical-to-rectangular* conversion equations produces

$$x = 4 \cos \frac{5\pi}{6} = 4\left(-\frac{\sqrt{3}}{2}\right) = -2\sqrt{3}$$

$$y = 4 \sin \frac{5\pi}{6} = 4\left(\frac{1}{2}\right) = 2$$

$$z = 3.$$

Thus, in rectangular coordinates, the point is $(x, y, z) = \left(-2\sqrt{3}, 2, 3\right)$, as shown in Figure 11.69.

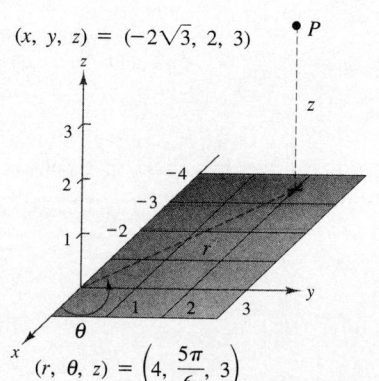

$(x, y, z) = (-2\sqrt{3}, 2, 3)$
P

$(r, \theta, z) = \left(4, \dfrac{5\pi}{6}, 3\right)$

FIGURE 11.69

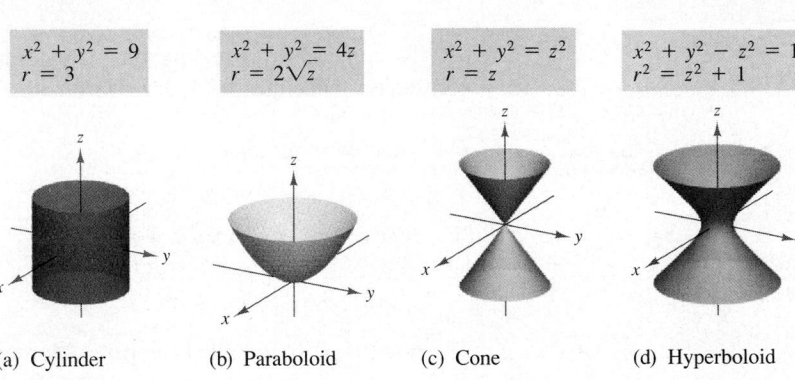

FIGURE 11.70

(x, y, z) = (1, √3, 2)

r = 2

z = 2

θ = π/3

(r, θ, z) = (2, π/3, 2)

EXAMPLE 2   Changing from Rectangular to Cylindrical Coordinates

Express the point $(x, y, z) = \left(1, \sqrt{3}, 2\right)$ in cylindrical coordinates.

Solution   Use the rectangular-to-cylindrical conversion equations.

$$r = \pm\sqrt{1 + 3} = \pm 2$$

$$\tan\theta = \sqrt{3} \quad\longrightarrow\quad \theta = \arctan\left(\sqrt{3}\right) + n\pi = \frac{\pi}{3} + n\pi$$

$$z = 2$$

You have two choices for r and infinitely many choices for θ. As shown in Figure 11.70, two convenient representations of the point are

$$\left(2, \frac{\pi}{3}, 2\right) \qquad r > 0 \text{ and } \theta \text{ in Quadrant I}$$

$$\left(-2, \frac{4\pi}{3}, 2\right). \qquad r < 0 \text{ and } \theta \text{ in Quadrant III}$$

Cylindrical coordinates are especially convenient for representing cylindrical surfaces and surfaces of revolution with the z-axis as the axis of symmetry, as shown in Figure 11.71.

$$x^2 + y^2 = 9$$
$$r = 3$$

$$x^2 + y^2 = 4z$$
$$r = 2\sqrt{z}$$

$$x^2 + y^2 = z^2$$
$$r = z$$

$$x^2 + y^2 - z^2 = 1$$
$$r^2 = z^2 + 1$$

(a) Cylinder   (b) Paraboloid   (c) Cone   (d) Hyperboloid

FIGURE 11.71

Vertical planes containing the z-axis and horizontal planes also have simple cylindrical coordinate equations, as shown in Figure 11.72.

Vertical plane: $\theta = c$

$\theta = c$

Horizontal plane: $z = c$

FIGURE 11.72

EXAMPLE 3 Rectangular-to-Cylindrical Conversion

Find equations in cylindrical coordinates for the surfaces whose rectangular equations are as follows.

a. $x^2 + y^2 = 4z^2$ **b.** $y^2 = x$

Solution

a. From the previous section, you know that the graph $x^2 + y^2 = 4z^2$ is a "double-napped" cone with its axis along the z-axis, as shown in Figure 11.73. If you replace $x^2 + y^2$ by r^2, the equation in cylindrical coordinates is

$$x^2 + y^2 + 4z^2. \qquad \text{Rectangular coordinate equation}$$
$$r^2 = 4z^2. \qquad \text{Cylindrical coordinate equation}$$

The equation $r = 2z$ represents the top half of the cone, and the equation $r = -2z$ represents the bottom half of the cone.

b. The graph of the surface $y^2 = x$ is a parabolic cylinder with rulings parallel to the z-axis, as shown in Figure 11.74. By replacing y^2 by $r^2 \sin^2 \theta$ and x by $r \cos \theta$, you obtain the following equation in cylindrical coordinates.

$$y^2 = x \qquad \text{Rectangular equation}$$
$$r^2 \sin^2 \theta = r \cos \theta \qquad \text{Substitute } r\sin\theta \text{ for } y \text{ and } r\cos\theta \text{ for } x$$
$$r(r \sin^2 \theta - \cos \theta) = 0 \qquad \text{Collect terms and factor}$$
$$r \sin^2 \theta - \cos \theta = 0 \qquad \text{Divide both sides by } r$$
$$r = \frac{\cos \theta}{\sin^2 \theta} \qquad \text{Solve for } r$$
$$r = \csc \theta \cot \theta \qquad \text{Cylindrical equation}$$

Note that this equation includes a point for which $r = 0$, so nothing was lost by dividing both sides by the factor r.

Rectangular coordinates: $x^2 + y^2 = 4z^2$

Cylindrical coordinates: $r = \pm 2z$

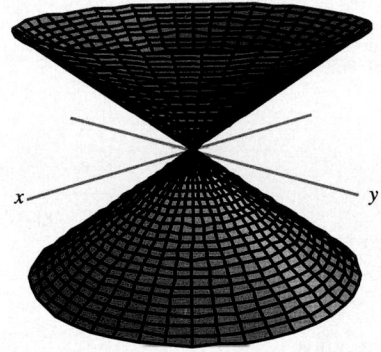

Generated by Axiom

FIGURE 11.73

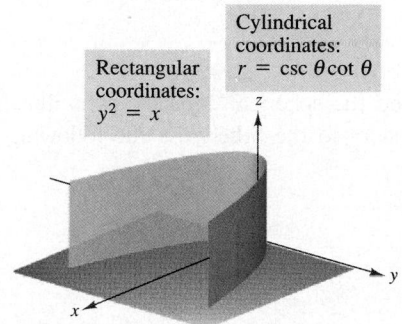

Cylindrical coordinates: $r = \csc \theta \cot \theta$

Rectangular coordinates: $y^2 = x$

FIGURE 11.74

EXAMPLE 4 Cylindrical-to-Rectangular Conversion

Find a rectangular equation for the graph represented by the cylindrical equation $r^2 \cos 2\theta + z^2 + 1 = 0$.

Solution

$$r^2 \cos 2\theta + z^2 + 1 = 0 \qquad \text{Cylindrical equation}$$
$$r^2(\cos^2 \theta - \sin^2 \theta) + z^2 + 1 = 0 \qquad \text{Trigonometric identity}$$
$$r^2 \cos^2 \theta - r^2 \sin^2 \theta + z^2 = -1$$
$$x^2 - y^2 + z^2 = -1 \qquad \text{Replace } r\cos\theta \text{ by } x \text{ and } r\sin\theta \text{ by } y$$
$$y^2 - x^2 - z^2 = 1 \qquad \text{Rectangular equation}$$

which is a hyperboloid of two sheets whose axis lies along the y-axis, as shown in Figure 11.75.

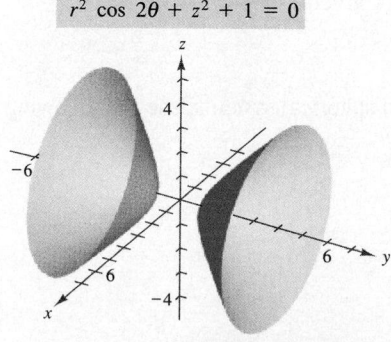

Cylindrical coordinates: $r^2 \cos 2\theta + z^2 + 1 = 0$

Rectangular coordinates: $y^2 - x^2 - z^2 = 1$

FIGURE 11.75

Spherical Coordinates

In the **spherical coordinate system,** each point is represented by an ordered triple— the first coordinate is a distance, and the second and third coordinates are angles. This system is similar to the latitude-longitude system used to identify points on the surface of the earth. For example, the point on the surface of the earth whose latitude is 40° N (north of the equator) and whose longitude is 80° W (west of the prime meridan) is shown in Figure 11.76. Assuming that the earth is spherical with a radius of 4000 miles, you would label this point as

$$(4000, \ -80°, \ 50°)$$

Radius ←   80° clockwise   →50° down from
            from prime   North Pole
            meridian

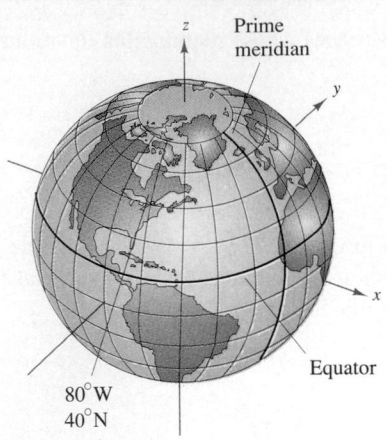

FIGURE 11.76

The Spherical Coordinate System

In a **spherical coordinate system,** a point P in space is represented by an ordered triple (ρ, θ, ϕ).

1. ρ is the distance between P and the origin, $\rho \geq 0$.
2. θ is the same angle used in cylindrical coordinates.
3. ϕ is the angle *between* the positive z-axis and the line segment $\overrightarrow{OP}$, $0 \leq \phi \leq \pi$.

Note that the first and third coordinates, ρ and ϕ, are nonnegative. ρ is the lowercase Greek letter rho, and ϕ is the lowercase Greek letter phi.

The relationship between the rectangular and the spherical coordinates is illustrated in Figure 11.77. To convert from one system to the other, use the following conversion equations.

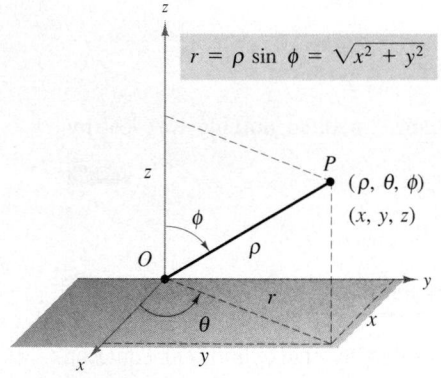

FIGURE 11.77
Spherical coordinates

Spherical to rectangular

$$x = \rho \sin \phi \cos \theta, \qquad y = \rho \sin \phi \sin \theta, \qquad z = \rho \cos \phi$$

Rectangular to spherical

$$\rho^2 = x^2 + y^2 + z^2, \qquad \tan \theta = \frac{y}{x}, \qquad \phi = \arccos\left(\frac{z}{\sqrt{x^2 + y^2 + z^2}}\right)$$

To change coordinates between the cylindrical and spherical systems, use the following conversion equations.

Spherical to cylindrical

$$r^2 = \rho^2 \sin^2 \phi, \qquad \theta = \theta, \qquad z = \rho \cos \phi$$

Cylindrical to spherical

$$\rho = \sqrt{r^2 + z^2}, \qquad \theta = \theta, \qquad \phi = \arccos\left(\frac{z}{\sqrt{r^2 + z^2}}\right)$$

The spherical coordinate system is useful primarily for surfaces in space that have a *point* or *center* of symmetry. For example, Figure 11.78 shows three surfaces with simple spherical equations.

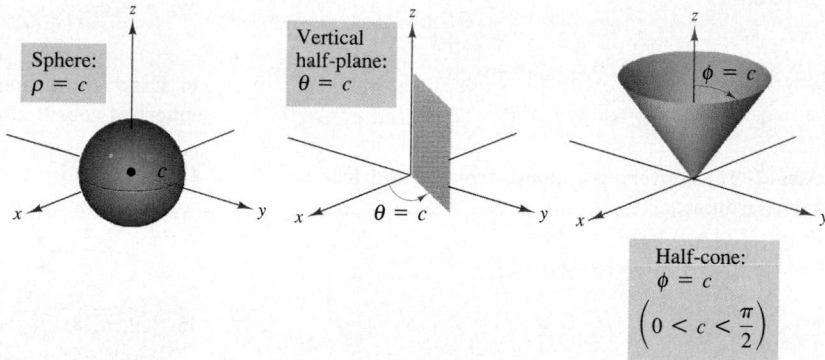

Sphere:
$\rho = c$

Vertical half-plane:
$\theta = c$

$\phi = c$

Half-cone:
$\phi = c$
$\left(0 < c < \dfrac{\pi}{2}\right)$

FIGURE 11.78

EXAMPLE 5   Rectangular-to-Spherical Conversion

Find an equation in spherical coordinates for the surfaces represented by the following rectangular coordinate equations.

a. Cone $x^2 + y^2 = z^2$ **b. Sphere** $x^2 + y^2 + z^2 - 4z = 0$

Solution

a. Making the appropriate replacements for x, y, and z in the given equation yields the following.

$$x^2 + y^2 = z^2$$
$$\rho^2 \sin^2 \phi \cos^2 \theta + \rho^2 \sin^2 \phi \sin^2 \theta = \rho^2 \cos^2 \phi$$
$$\rho^2 \sin^2 \phi(\cos^2 \theta + \sin^2 \theta) = \rho^2 \cos^2 \phi$$
$$\rho^2 \sin^2 \phi = \rho^2 \cos^2 \phi$$
$$\frac{\sin^2 \phi}{\cos^2 \phi} = 1 \qquad \rho \geq 0$$
$$\tan^2 \phi = 1 \qquad \phi = \pi/4 \text{ or } \phi = 3\pi/4$$

The equation $\phi = \pi/4$ represents the *upper* half-cone, and the equation $\phi = 3\pi/4$ represents the *lower* half-cone.

b. Because $\rho^2 = x^2 + y^2 + z^2$ and $z = \rho \cos \phi$, the given equation has the following spherical form.

$$\rho^2 - 4\rho \cos \phi = 0 \qquad \longrightarrow \qquad \rho(\rho - 4 \cos \phi) = 0$$

Temporarily discarding the possibility that $\rho = 0$, we have the spherical equation

$$\rho - 4 \cos \phi = 0 \quad \text{or} \quad \rho = 4 \cos \phi.$$

Note that the solution set for this equation includes a point for which $\rho = 0$, so nothing is lost by discarding the factor ρ. The sphere represented by the equation $\rho = 4 \cos \phi$ is shown in Figure 11.79.

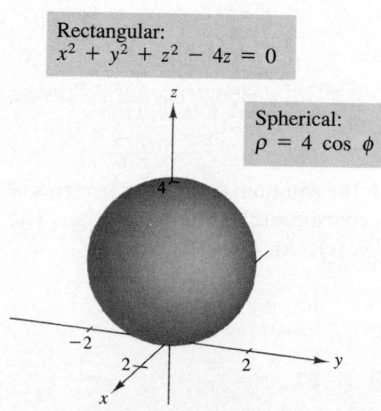

Rectangular:
$x^2 + y^2 + z^2 - 4z = 0$

Spherical:
$\rho = 4 \cos \phi$

FIGURE 11.79

EXERCISES for Section 11.7

In Exercises 1–6, convert the point from rectangular to cylindrical coordinates.

1. $(0, 5, 1)$ **2.** $\left(2\sqrt{2}, -2\sqrt{2}, 4\right)$

3. $\left(1, \sqrt{3}, 4\right)$ **4.** $\left(\sqrt{3}, -1, 2\right)$

5. $(2, -2, -4)$ **6.** $(-3, 2, -1)$

In Exercises 7–12, convert the point from cylindrical to rectangular coordinates.

7. $(5, 0, 2)$ **8.** $\left(4, \dfrac{\pi}{2}, -2\right)$

9. $\left(2, \dfrac{\pi}{3}, 2\right)$ **10.** $\left(3, -\dfrac{\pi}{4}, 1\right)$

11. $\left(4, \dfrac{7\pi}{6}, 3\right)$ **12.** $\left(1, \dfrac{3\pi}{2}, 1\right)$

In Exercises 13–20, find an equation in rectangular coordinates for the equation in cylindrical coordinates, and sketch its graph.

13. $r = 2$ **14.** $z = 2$

15. $\theta = \dfrac{\pi}{6}$ **16.** $r = \dfrac{z}{2}$

17. $r = 2\sin\theta$ **18.** $r = 2\cos\theta$

19. $r^2 + z^2 = 4$ **20.** $z = r^2\sin^2\theta$

In Exercises 21–26, convert the point from rectangular to spherical coordinates.

21. $(4, 0, 0)$ **22.** $(1, 1, 1)$

23. $\left(-2, 2\sqrt{3}, 4\right)$ **24.** $\left(2, 2, 4\sqrt{2}\right)$

25. $\left(\sqrt{3}, 1, 2\sqrt{3}\right)$ **26.** $(-4, 0, 0)$

In Exercises 27–32, convert the point from spherical to rectangular coordinates.

27. $\left(4, \dfrac{\pi}{6}, \dfrac{\pi}{4}\right)$ **28.** $\left(12, \dfrac{3\pi}{4}, \dfrac{\pi}{9}\right)$

29. $\left(12, -\dfrac{\pi}{4}, 0\right)$ **30.** $\left(9, \dfrac{\pi}{4}, \pi\right)$

31. $\left(5, \dfrac{\pi}{4}, \dfrac{3\pi}{4}\right)$ **32.** $\left(6, \pi, \dfrac{\pi}{2}\right)$

In Exercises 33–40, find an equation in rectangular coordinates for the equation in spherical coordinates, and sketch its graph.

33. $\rho = 2$ **34.** $\theta = \dfrac{3\pi}{4}$

35. $\phi = \dfrac{\pi}{6}$ **36.** $\phi = \dfrac{\pi}{2}$

37. $\rho = 4\cos\phi$ **38.** $\rho = 2\sec\phi$

39. $\rho = \csc\phi$ **40.** $\rho = 4\csc\phi\sec\theta$

In Exercises 41–46, convert the point from cylindrical to spherical coordinates.

41. $\left(4, \dfrac{\pi}{4}, 0\right)$ **42.** $\left(2, \dfrac{2\pi}{3}, -2\right)$

43. $\left(4, -\dfrac{\pi}{6}, 6\right)$ **44.** $\left(-4, \dfrac{\pi}{3}, 4\right)$

45. $(12, \pi, 5)$ **46.** $\left(4, \dfrac{\pi}{2}, 3\right)$

In Exercises 47–52, convert the point from spherical to cylindrical coordinates.

47. $\left(10, \dfrac{\pi}{6}, \dfrac{\pi}{2}\right)$ **48.** $\left(4, \dfrac{\pi}{18}, \dfrac{\pi}{2}\right)$

49. $\left(6, -\dfrac{\pi}{6}, \dfrac{\pi}{3}\right)$ **50.** $\left(5, -\dfrac{5\pi}{6}, \pi\right)$

51. $\left(8, \dfrac{7\pi}{6}, \dfrac{\pi}{6}\right)$ **52.** $\left(7, \dfrac{\pi}{4}, \dfrac{3\pi}{4}\right)$

C **In Exercises 53–58, use a computer algebra system or write a computer program to convert the coordinates of a point from one system to another among the rectangular, cylindrical, and spherical coordinate systems. Use the computer to complete the following table.**

	Rectangular	Cylindrical	Spherical
53.	$(4, 6, 3)$		
54.	$(6, -2, -3)$		
55.		$\left(5, \dfrac{\pi}{9}, 8\right)$	
56.		$(10, -0.75, 6)$	
57.			$\left(20, \dfrac{2\pi}{3}, \dfrac{\pi}{4}\right)$
58.			$(7.5, 0.25, 1)$

In Exercises 59–64, match the equation (expressed in terms of cylindrical or spherical coordinates) with its graph. [The graphs are labeled (a), (b), (c), (d), (e), and (f).]

59. $r = 5$ **60.** $\theta = \dfrac{\pi}{4}$

61. $\rho = 5$ **62.** $\phi = \dfrac{\pi}{4}$

63. $r^2 = z$

(a)

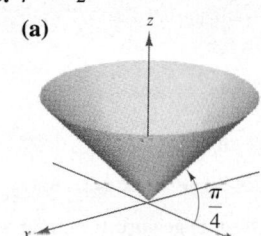

64. $\rho = 4 \sec \phi$

(b)

(c)

(d)

(e)

(f)

In Exercises 65–72, convert the rectangular equation to (a) cylindrical coordinates and (b) spherical coordinates.

65. $x^2 + y^2 + z^2 = 16$

66. $4(x^2 + y^2) = z^2$

67. $x^2 + y^2 + z^2 - 2z = 0$

68. $x^2 + y^2 = z$

69. $x^2 + y^2 = 4y$

70. $x^2 + y^2 = 16$

71. $x^2 - y^2 = 9$

72. $y = 4$

In Exercises 73–76, sketch the solid that has the given description in cylindrical coordinates.

73. $0 \leq \theta \leq \pi/2, 0 \leq r \leq 2, 0 \leq z \leq 4$

74. $-\pi/2 \leq \theta \leq \pi/2, 0 \leq r \leq 3, 0 \leq z \leq r \cos \theta$

75. $0 \leq \theta \leq 2\pi, 0 \leq r \leq a, r \leq z \leq a$

76. $0 \leq \theta \leq 2\pi, 2 \leq r \leq 4, z^2 \leq -r^2 + 6r - 8$

In Exercises 77 and 78, sketch the solid that has the given description in spherical coordinates.

77. $0 \leq \theta \leq 2\pi, 0 \leq \phi \leq \pi/6, 0 \leq \rho \leq a \sec \phi$

78. $0 \leq \theta \leq 2\pi, \pi/4 \leq \phi \leq \pi/2, 0 \leq \rho \leq 1$

In Exercises 79–82, find inequalities that describe the solid, and state the coordinate system used. Position the solid on the coordinate system you choose so that the inequalities are as simple as possible.

79. A cube with each edge 10 centimeters long

80. A cylindrical shell 8 feet long with an inside diameter of 0.75 inch and an outside diameter of 1.25 inches

81. A spherical shell with inside and outside radii of 4 and 6 inches, respectively

82. The solid that remains after a hole 1 inch in diameter is drilled through the center of a sphere 6 inches in diameter

83. *Latitude and Longitude* Los Angeles is located at 34.05° north latitude and 118.24° west longitude, and Rio de Janeiro, Brazil is located at 22.90° south latitude and 43.22° west longitude (see figure). Assume that the earth is spherical with a radius of 4000 miles.
   a. Find the spherical coordinates for the location of each city.
   b. Find the rectangular coordinates for the location of each city.
   c. Find the angle (in radians) between the vectors from the center of the earth to each city.
   d. Find the great-circle distance s between the cities. [*Hint:* $s = r\theta$]

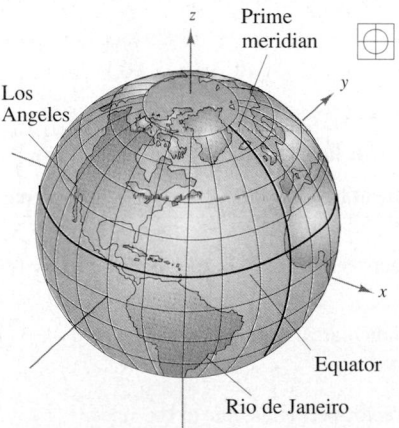

FIGURE FOR 83

84. Repeat Exercise 83 for the cities of Boston, located at 42.36° north latitude and 71.06° west longitude, and Honolulu, located at 21.31° north latitude and 157.86° west longitude.

85. Identify the curve of intersection of the surfaces (in cylindrical coordinates) $z = \sin \theta$ and $r = 1$.

REVIEW EXERCISES for Chapter 11

In Exercises 1 and 2, let u = $\overrightarrow{PQ}$ and v = $\overrightarrow{PR}$, and find (a) the component forms of u and v, (b) the magnitude of v, (c) u · v, (d) 2u + v, and (e) the vector component of u in the direction of v, and (f) the vector component of u orthogonal to v.

1. $P = (1, 2)$, $Q = (4, 1)$, $R = (5, 4)$

2. $P = (-2, -1)$, $Q = (5, -1)$, $R = (2, 4)$

In Exercises 3 and 4, let u = $\overrightarrow{PQ}$ and v = $\overrightarrow{PR}$, and find (a) the component forms of u and v, (b) u · v, (c) u × v, (d) an equation of the plane containing P, Q, and R, and (e) a set of parametric equations of the line through P and Q.

3. $P = (5, 0, 0)$, $Q = (4, 4, 0)$, $R = (2, 0, 6)$

4. $P = (2, -1, 3)$, $Q = (0, 5, 1)$, $R = (5, 5, 0)$

In Exercises 5 and 6, determine if the vectors are orthogonal or parallel, or neither.

5. $\langle -4, 3, -6 \rangle$ **6.** $\langle 7, -2, 3 \rangle$
   $\langle 16, -12, 24 \rangle$ $\langle -1, 4, 5 \rangle$

In Exercises 7–10, find the angle θ between the vectors u and v.

7. $u = 5 \left[\cos\left(\frac{3\pi}{4}\right) i + \sin\left(\frac{3\pi}{4}\right) j \right]$

   $v = 2 \left[\cos\left(\frac{2\pi}{3}\right) i + \sin\left(\frac{2\pi}{3}\right) j \right]$

8. $u = \langle 4, -1, 5 \rangle$
   $v = \langle 3, 2, -2 \rangle$

9. $u = \langle 10, -5, 15 \rangle$ **10.** $u = \langle 1, 0, -3 \rangle$
   $v = \langle -2, 1, -3 \rangle$ $v = \langle 2, -2, 1 \rangle$

In Exercises 11–14, find the component form of u.

11. The angle, measured counterclockwise, from the positive x-axis to u is 135°, and $\| u \| = 4$.

12. The angle between u and the positive x-axis is 180°, and $\| u \| = 8$.

13. u is perpendicular to the plane $x - 3y + 4z = 0$, and $\| u \| = 3$.

14. u is a unit vector perpendicular to the lines
   $\begin{array}{ll} x = 4 - t & x = -3 + 7s \\ y = 3 + 2t & y = -2 + s \\ z = 1 + 5t & z = 1 + 2s. \end{array}$

In Exercises 15–24, let u = $\langle 3, -2, 1 \rangle$, v = $\langle 2, -4, -3 \rangle$, and w = $\langle -1, 2, 2 \rangle$.

15. Find $\| u \|$.

16. Find the angle between u and v.

17. Show that $u \cdot u = \| u \|^2$.

18. Determine a unit vector perpendicular to the plane containing v and w.

19. Determine the projection of w onto u.

20. Show that $u \times v = -(v \times u)$.

21. Show that $u \cdot (v + w) = u \cdot v + u \cdot w$.

22. Show that $u \times (v + w) = (u \times v) + (u \times w)$.

23. Find the volume of the solid whose edges are u, v, and w.

24. Find the work done in moving an object along the vector u, if the applied force is w.

25. *Load Supports*   Find the tension in each of the supporting cables shown in the figure.

26. *Equilibrium*   A 100-pound collar slides on a frictionless vertical rod (see figure). Find the distance y for which the system is in equilibrium if the counterweight weighs 120 pounds.

FIGURE FOR 25   **FIGURE FOR 26**

27. *Minimum Length*   In a manufacturing process, an electric hoist lifts 500-pound ingots (see figure). Find the shortest cable connecting points P, O, and Q that can be used if the tension in the cable cannot exceed 750 pounds. (Assume that O is at the midpoint of the cable.)

28. *Torque*   The specifications for a tractor state that the torque on a bolt with head size $\frac{7}{8}$ inch cannot exceed 200 foot-pounds. Determine the maximum force $\| F \|$ that can be applied to the wrench in the figure.

FIGURE FOR 27   **FIGURE FOR 28**

In Exercises 29–32, find (a) a set of parametric equations for the given line, and (b) a set of symmetric equations for the given line.

29. The line passes through $(1, 2, 3)$ and is perpendicular to the xz-coordinate plane.

30. The line passes through $(1, 2, 3)$ and is parallel to the line given by $x = y = z$.

31. The line is the intersection of the planes given by

$$3x - 3y - 7z = -4$$

and

$$x - y + 2z = 3.$$

32. The line passes through the point $(0, 1, 4)$ and is perpendicular to $\mathbf{u} = \langle 2, -5, 1 \rangle$ and $\mathbf{v} = \langle -3, 1, 4 \rangle$.

In Exercises 33–36, find an equation of the plane.

33. The plane passes through the point $(1, 2, 3)$ and is orthogonal to the line given by $x = y = z$.

34. The plane passes through the point $(4, 2, 1)$ and is parallel to the yz-coordinate plane.

35. The plane contains the lines

$$\frac{x - 1}{-2} = y = z + 1$$

$$\frac{x + 1}{-2} = y - 1 = z - 2.$$

36. The plane passes through the points $(-3, -4, 2)$, $(-3, 4, 1)$, and $(1, 1, -2)$.

In Exercises 37 and 38, find the distance from the point to the plane.

37. $(1, 0, 2)$
$$2x - 3y + 6z = 6$$

38. $(0, 0, 0)$
$$4x - 7y + z = 2$$

In Exercises 39 and 40, find the distance between the parallel planes.

39. $5x - 3y + z = 2$
$5x - 3y + z = -3$

40. $3x - 2y - z = -1$
$6x + 4y - 2z = 1$

In Exercises 41 and 42, find the distance between the two skew lines.

41. $\dfrac{x}{1} = \dfrac{y}{2} + \dfrac{z}{3}$

$\dfrac{x + 1}{-1} = \dfrac{y}{3} = \dfrac{z + 2}{2}$

42. $x - 4 = \dfrac{y - 3}{-1} = \dfrac{z - 7}{3}$

$\dfrac{x + 3}{-5} = \dfrac{y - 7}{2} = \dfrac{z + 5}{-6}$

In Exercises 43–52, sketch a graph of the specified surface.

43. $x + 2y + 3z = 6$

44. $y = z^2$

45. $y = \frac{1}{2}z$

46. $y = \cos z$

47. $x^2 + z^2 = 4$

48. $x^2 + y^2 + z^2 - 2x + 4y - 6z + 5 = 0$

49. $16x^2 + 16y^2 - 9z^2 = 0$

50. $\dfrac{x^2}{16} + \dfrac{y^2}{9} + z^2 = 1$

51. $\dfrac{x^2}{16} - \dfrac{y^2}{9} + z^2 = -1$

52. $\dfrac{x^2}{25} + \dfrac{y^2}{4} - \dfrac{z^2}{100} = 1$

C **In Exercises 53–56, use a computer algebra system to graph the surface. (*Hint:* It may be necessary to solve for z and acquire two equations to graph.)**

53. $z = y^2 - 4x^2$

54. $z = 4y^2 - x^3$

55. $z = e^{-x/2} \sin y$

56. $\dfrac{x^2}{16} + \dfrac{y^2}{9} - z^2 = 1$

57. *Machine Design* The top of a rubber bushing designed to absorb vibrations in an automobile is the surface of revolution generated by revolving the curve $z = \frac{1}{2}y^2 + 1 (0 \le y \le 2)$ in the yz-plane about the z-axis (see figure).

a. Find an equation for the surface of revolution.

b. If all measurements are in centimeters and the bushing is set on the xy-plane, use the Shell Method to find its volume.

c. Suppose the bushing has a hole of diameter 1 cm through its center and parallel to the axis of revolution. Find the volume of rubber in the bushing.

58. *Machine Design* A tractor fuel tank has the shape and dimensions shown in the figure. In fabricating the tank, it is necessary to know the angle between two adjacent sides. Find the angle.

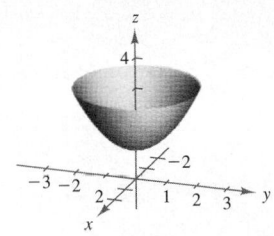

FIGURE FOR 57

FIGURE FOR 58

In Exercises 59 and 60, convert the given point from rectangular coordinates to (a) cylindrical coordinates and (b) spherical coordinates.

59. $\left(-2\sqrt{2},\ 2\sqrt{2},\ 2\right)$

60. $\left(\dfrac{\sqrt{3}}{4},\ \dfrac{3}{4},\ \dfrac{3\sqrt{3}}{2}\right)$

In Exercises 61 and 62, find an equation of the given surface in (a) cylindrical coordinates and (b) spherical coordinates.

61. $x^2 - y^2 = 2z$

62. $x^2 + y^2 + z^2 = 16$

In Exercises 63 and 64, find an equation in rectangular coordinates for the equation in cylindrical coordinates.

63. $r^2(\cos^2 \theta - \sin^2 \theta) + z^2 = 1$

64. $r^2 = 16z$

In Exercises 65 and 66, find an equation in rectangular coordinates for the equation in spherical coordinates.

65. $\rho = \csc \phi$

66. $\rho = 5$

Chapter 12

Vector-Valued Functions

12.1

Space Curves and Vector-Valued Functions • Limits and Continuity

Curve in plane

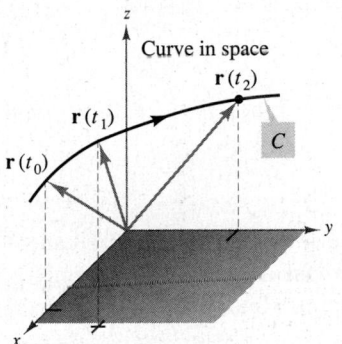

Curve in space

FIGURE 12.1
Curve C is traced out by the terminal point of position vector $\mathbf{r}(t)$.

Space Curves and Vector-Valued Functions

In Section 10.1, a *plane curve* was defined as the set of ordered pairs $(f(t), g(t))$ together with their defining parametric equations

$$x = f(t) \quad \text{and} \quad y = g(t)$$

where f and g are continuous functions of t on an interval I. This definition can be extended naturally to three-dimensional space as follows. A **space curve** C is the set of all ordered triples $(f(t), g(t), h(t))$ together with their defining parametric equations

$$x = f(t), \quad y = g(t), \quad \text{and} \quad z = h(t)$$

where f, g, and h are continuous functions of t on an interval I.

Before looking at examples of space curves, we introduce a new type of function, called a **vector-valued function,** that maps real numbers onto vectors.

Definition of a Vector-Valued Function

A function of the form

$$\mathbf{r}(t) = f(t)\mathbf{i} + g(t)\mathbf{j} \qquad \text{Plane}$$

or

$$\mathbf{r}(t) = f(t)\mathbf{i} + g(t)\mathbf{j} + h(t)\mathbf{k} \qquad \text{Space}$$

is a **vector-valued function,** where the **component functions** f, g, and h are real-valued functions of the parameter t. Vector-valued functions are sometimes denoted as $\mathbf{r}(t) = \langle f(t), g(t) \rangle$ or $\mathbf{r}(t) = \langle f(t), g(t), h(t) \rangle$.

REMARK   Note the distinction between the vector-valued function $\mathbf{r}$ and the real-valued functions f, g, and h. All are functions of the real variable t, but $\mathbf{r}(t)$ is a vector, whereas $f(t)$, $g(t)$, and $h(t)$ are real numbers (for each specific value of t).

Vector-valued functions serve dual roles in the representation of curves. By letting the parameter t represent time, you can use a vector-valued function to represent *motion* along a curve. Or, in the more general case, you can use a vector-valued function to *trace the graph* of a curve. In either case, the terminal point of the position vector $\mathbf{r}(t)$ coincides with the point (x, y) or (x, y, z) on the curve given by the parametric equations, as shown in Figure 12.1. The arrowhead on the curve indicates the curve's *orientation* by pointing in the direction of increasing values of t.

◄ *The angle at which a roadway should be banked so that no lateral frictional force is exerted on an automobile's tires is discussed in Section 12.3, Exercise 44.*

Unless stated otherwise, the **domain** of a vector-valued function **r** is considered to be the intersection of the domains of the component functions f, g, and h. For instance, the domain of

$$\mathbf{r}(t) = (\ln t)\mathbf{i} + \sqrt{1 - t}\,\mathbf{j} + t\mathbf{k}$$

is the interval $(0, 1]$.

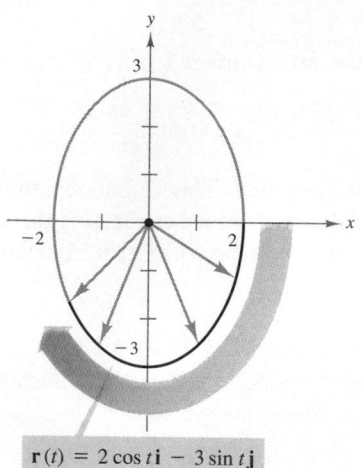

FIGURE 12.2
The ellipse is traced clockwise as the parameter t increases from 0 to 2π.

EXAMPLE 1   Sketching a Plane Curve

Sketch the plane curve represented by the vector-valued function

$$\mathbf{r}(t) = 2\cos t\,\mathbf{i} - 3\sin t\,\mathbf{j}, \qquad 0 \le t \le 2\pi.$$

Solution   From the position vector $\mathbf{r}(t)$, you can write the parametric equations $x = 2\cos t$ and $y = -3\sin t$. Solving for $\cos t$ and $\sin t$ and using the identity $\cos^2 t + \sin^2 t = 1$ produce the rectangular equation

$$\frac{x^2}{2^2} + \frac{y^2}{3^2} = 1.$$

The graph of this rectangular equation is the ellipse shown in Figure 12.2. Note that the curve has a *clockwise* orientation. That is, as t increases from 0 to 2π, the position vector $\mathbf{r}(t)$ moves clockwise, and its terminal point traces the ellipse.

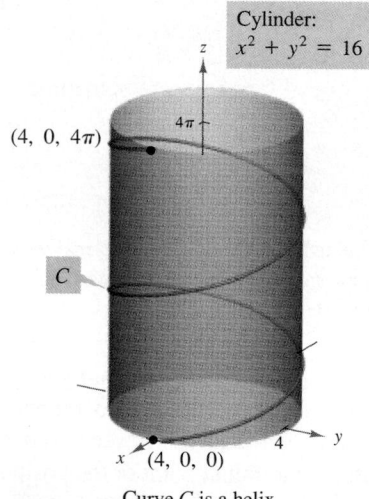

FIGURE 12.3
At t increases from 0 to 4π, two spirals on the helix are traced out.

EXAMPLE 2   Sketching a Space Curve

Sketch the space curve represented by the vector-valued function

$$\mathbf{r}(t) = 4\cos t\,\mathbf{i} + 4\sin t\,\mathbf{j} + t\mathbf{k}, \qquad 0 \le t \le 4\pi.$$

Solution   From the first two parametric equations $x = 4\cos t$ and $y = 4\sin t$, you can obtain

$$x^2 + y^2 = 16.$$

This means that the curve lies on a right circular cylinder of radius 4, centered about the z-axis. To locate the curve on this cylinder, you can use the third parametric equation $z = t$. In Figure 12.3, note that as t increases from 0 to 4π the point (x, y, z) spirals up the cylinder to produce a **helix.**

In Examples 1 and 2, you were given a vector-valued function and asked to sketch the corresponding curve. The next two examples concern the reverse problem—finding a vector-valued function to represent a given graph. Of course, if the graph is described parametrically, then representation by a vector-valued function is straightforward. For instance, to represent the line in space given by

$$x = 2 + t, \quad y = 3t, \quad \text{and} \quad z = 4 - t$$

you can simply use the vector-valued function given by

$$\mathbf{r}(t) = (2 + t)\mathbf{i} + 3t\mathbf{j} + (4 - t)\mathbf{k}.$$

If a set of parametric equations for the graph is not given, then the problem of representing the graph by a vector-valued function boils down to finding a set of parametric equations.

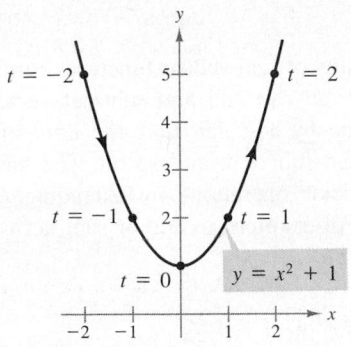

FIGURE 12.4
There are many ways to
parametrize this graph. One
way is to let $x = t$.

EXAMPLE 3   Representing a Graph by a Vector-Valued Function

Represent the parabola given by $y = x^2 + 1$ by a vector-valued function.

Solution   Although there are many ways to choose the parameter t, a natural choice is to let $x = t$. Then $y = t^2 + 1$ and you have $\mathbf{r}(t) = t\,\mathbf{i} + (t^2 + 1)\mathbf{j}$. Note in Figure 12.4 the orientation produced by this particular choice of parameter. Had you chosen $x = -t$ as the parameter, the curve would have been oriented in the opposite direction.

In the next example, you can see how to represent a graph in space that is defined as the intersection of two surfaces in space.

EXAMPLE 4   Representing a Graph by a Vector-Valued Function

Sketch the graph represented by the intersection of the semiellipsoid

$$\frac{x^2}{12} + \frac{y^2}{24} + \frac{z^2}{4} = 1, \quad z \geq 0$$

and the parabolic cylinder $y = x^2$. Then, find a vector-valued function to represent the graph.

Solution   The intersection of the two surfaces is shown in Figure 12.5. As in Example 3, a natural choice of parameter is $x = t$. For this choice, you can use the given equation $y = x^2$ to obtain $y = t^2$. Then, it follows that

$$\frac{z^2}{4} = 1 - \frac{x^2}{12} - \frac{y^2}{24} = 1 - \frac{t^2}{12} - \frac{t^4}{24} = \frac{24 - 2t^2 - t^4}{24}.$$

Because the curve lies above the xy-plane, you should choose the positive square root for z and obtain the following parametric equations.

$$x = t, \quad y = t^2, \quad \text{and} \quad z = \sqrt{\frac{24 - 2t^2 - t^4}{6}}$$

The resulting vector-valued function is

$$\mathbf{r}(t) = t\,\mathbf{i} + t^2\mathbf{j} + \sqrt{\frac{24 - 2t^2 - t^4}{6}}\,\mathbf{k}, \quad -2 \leq t \leq 2.$$

From the points $(-2, 4, 0)$ and $(2, 4, 0)$ shown in Figure 12.5, you can see that the curve is traced as t increases from -2 to 2.

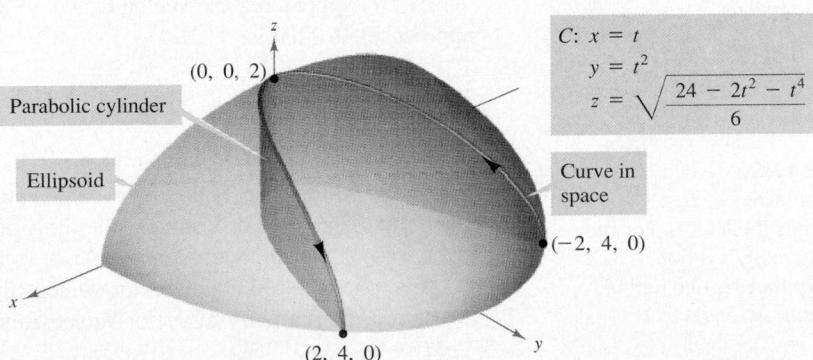

FIGURE 12.5
The curve C is the intersection of the
semiellipsoid and the parabolic cylinder.

Limits and Continuity

Many techniques and definitions used in the calculus of real-valued functions can be applied to vector-valued functions. For instance, you can add and subtract vector-valued functions, multiply a vector-valued function by a scalar, take the limit of a vector-valued function, differentiate a vector-valued function, and so on. The basic approach is to capitalize on the linearity of vector operations by extending the definitions on a component-by-component basis. For example, to add or subtract two vector-valued functions (in the plane), you can write

$$
\begin{aligned}
\mathbf{r}_1(t) + \mathbf{r}_2(t) &= [f_1(t)\mathbf{i} + g_1(t)\mathbf{j}] + [f_2(t)\mathbf{i} + g_2(t)\mathbf{j}] \quad &\text{Sum} \\
&= [f_1(t) + f_2(t)]\mathbf{i} + [g_1(t) + g_2(t)]\mathbf{j} \\
\mathbf{r}_1(t) - \mathbf{r}_2(t) &= [f_1(t)\mathbf{i} + g_1(t)\mathbf{j}] - [f_2(t)\mathbf{i} + g_2(t)\mathbf{j}] \quad &\text{Difference} \\
&= [f_1(t) - f_2(t)]\mathbf{i} + [g_1(t) - g_2(t)]\mathbf{j}.
\end{aligned}
$$

Similarly, to multiply a vector-valued function by a scalar, you can write

$$
\begin{aligned}
c\,\mathbf{r}(t) &= c[f_1(t)\mathbf{i} + g_1(t)\mathbf{j}] \quad &\text{Scalar multiple} \\
&= c f_1(t)\mathbf{i} + c g_1(t)\mathbf{j}.
\end{aligned}
$$

This component-by-component extension of operations with real-valued functions to vector-valued functions is further illustrated in the following definition of the limit of a vector-valued function.

Definition of the Limit of a Vector-Valued Function

1. If $\mathbf{r}$ is a vector-valued function such that $\mathbf{r}(t) = f(t)\mathbf{i} + g(t)\mathbf{j}$, then

$$
\lim_{t \to a} \mathbf{r}(t) = \left[\lim_{t \to a} f(t) \right]\mathbf{i} + \left[\lim_{t \to a} g(t) \right]\mathbf{j} \qquad \text{Plane}
$$

provided f and g have limits as $t \to a$.

2. If $\mathbf{r}$ is a vector-valued function such that $\mathbf{r}(t) = f(t)\mathbf{i} + g(t)\mathbf{j} + h(t)\mathbf{k}$, then

$$
\lim_{t \to a} \mathbf{r}(t) = \left[\lim_{t \to a} f(t) \right]\mathbf{i} + \left[\lim_{t \to a} g(t) \right]\mathbf{j} + \left[\lim_{t \to a} h(t) \right]\mathbf{j} \qquad \text{Space}
$$

provided f, g, and h have limits as $t \to a$.

If $\mathbf{r}(t)$ approaches the vector $\mathbf{L}$ as $t \to a$, then the length of the vector $\mathbf{r}(t) - \mathbf{L}$ approaches 0. That is,

$$
\| \mathbf{r}(t) - \mathbf{L} \| \to 0 \quad \text{as} \quad t \to a.
$$

This is illustrated graphically in Figure 12.6. With this definition of the limit of a vector-valued function, you can develop vector versions of most of the limit theorems given in Chapter 1. For example, the limit of the sum of two vector-valued functions is the sum of their individual limits. Also, you can use the orientation of the curve $\mathbf{r}(t)$ to define one-sided limits of vector-valued functions. The following definition extends the notion of continuity to vector-valued functions.

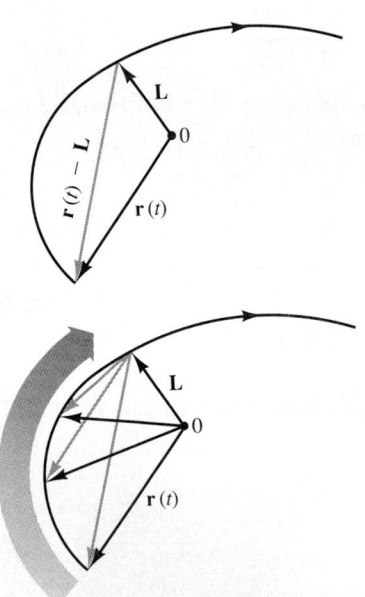

FIGURE 12.6

As t approaches a, $\mathbf{r}(t)$ approaches the limit $\mathbf{L}$. For the limit $\mathbf{L}$ to exist, it is not necessary that $\mathbf{r}(a)$ be defined or that $\mathbf{r}(a)$ be equal to $\mathbf{L}$.

> **Definition of Continuity of a Vector-Valued Function**
>
> A vector-valued function **r** is **continuous at the point** given by $t = a$ if the limit of $\mathbf{r}(t)$ exists as $t \to a$ and
>
> $$\lim_{t \to a} \mathbf{r}(t) = \mathbf{r}(a).$$
>
> A vector-valued function **r** is **continuous on an interval** I if it is continuous at every point in the interval.

From this definition, it follows that a vector-valued function is continuous at $t = a$ if and only if each of its component functions is continuous at $t = a$.

EXAMPLE 5   Continuity of Vector-Valued Functions

Discuss the continuity of the vector-valued function given by

$$\mathbf{r}(t) = t\mathbf{i} + a\mathbf{j} + (a^2 - t^2)\mathbf{k}, \qquad a \text{ is a constant}$$

at $t = 0$.

Solution   As t approaches 0, the limit is

$$\lim_{t \to 0} \mathbf{r}(t) = \left[\lim_{t \to 0} t \right]\mathbf{i} + \left[\lim_{t \to 0} a \right]\mathbf{j} + \left[\lim_{t \to 0}(a^2 - t^2) \right]\mathbf{k}$$
$$= 0\mathbf{i} + a\mathbf{j} + a^2\mathbf{k}$$
$$= a\mathbf{j} + a^2\mathbf{k}.$$

Because

$$\mathbf{r}(0) = (0)\mathbf{i} + (a)\mathbf{j} + (a^2)\mathbf{k} = a\mathbf{j} + a^2\mathbf{k}$$

you can conclude that **r** is continuous at $t = 0$. By similar reasoning, you can conclude that the vector-valued function **r** is continuous at all real-number values of t.

For each value of a, the curve represented by the vector-valued function in Example 5

$$\mathbf{r}(t) = t\mathbf{i} + a\mathbf{j} + (a^2 - t^2)\mathbf{k}, \qquad a \text{ is a constant}$$

is a parabola. You can think of each parabola as the intersection of the vertical plane $y = a$ and the hyperbolic paraboloid $y^2 - x^2 = z$, as shown in Figure 12.7.

TECHNOLOGY   Almost any type of three-dimensional sketch is difficult to do by hand, but sketching curves in space is especially difficult. The problem is in trying to create the illusion of three dimensions. Computer graphing utilities use a variety of techniques to add "three-dimensionality" to sketches of space curves—one way is to show the curve on a surface, as in Figure 12.7.

FIGURE 12.7
For each value of a, the curve represented by $\mathbf{r}(t) = t\mathbf{i} + a\mathbf{j} + (a^2 - t^2)\mathbf{k}$ is a parabola.

EXERCISES for Section 12.1

In Exercises 1–8, find the domain of the vector-valued function.

1. $\mathbf{r}(t) = 5t\,\mathbf{i} - 4t\,\mathbf{j} - \dfrac{1}{t}\mathbf{k}$

2. $\mathbf{r}(t) = \sqrt{4 - t^2}\,\mathbf{i} + t^2\,\mathbf{j} - 6t\,\mathbf{k}$

3. $\mathbf{r}(t) = \ln t\,\mathbf{i} - e^t\,\mathbf{j} - t\,\mathbf{k}$

4. $\mathbf{r}(t) = \sin t\,\mathbf{i} + 4\cos t\,\mathbf{j} + t\,\mathbf{k}$

5. $\mathbf{r}(t) = \mathbf{F}(t) + \mathbf{G}(t)$ where

$\mathbf{F}(t) = \cos t\,\mathbf{i} - \sin t\,\mathbf{j} + \sqrt{t}\,\mathbf{k}$

and

$\mathbf{G}(t) = \cos t\,\mathbf{i} + \sin t\,\mathbf{j}.$

6. $\mathbf{r}(t) = \mathbf{F}(t) - \mathbf{G}(t)$ where

$\mathbf{F}(t) = \ln t\,\mathbf{i} + 5t\,\mathbf{j} - 3t^2\,\mathbf{k}$

and

$\mathbf{G}(t) = \mathbf{i} + 4t\,\mathbf{j} - 3t^2\,\mathbf{k}.$

7. $\mathbf{r}(t) = \mathbf{F}(t) \times \mathbf{G}(t)$ where

$\mathbf{F}(t) = \sin t\,\mathbf{i} + \cos t\,\mathbf{j}$

and

$\mathbf{G}(t) = \sin t\,\mathbf{j} + \cos t\,\mathbf{k}.$

8. $\mathbf{r}(t) = \mathbf{F}(t) \times \mathbf{G}(t)$ where

$\mathbf{F}(t) = t^3\,\mathbf{i} - t\,\mathbf{j} + t\,\mathbf{k}$

and

$\mathbf{G}(t) = \sqrt[3]{t}\,\mathbf{i} + \dfrac{1}{t+1}\mathbf{j} + (t+2)\mathbf{k}.$

In Exercises 9 and 10, find $\|\mathbf{r}(t)\|$.

9. $\mathbf{r}(t) = \sin 3t\,\mathbf{i} + \cos 3t\,\mathbf{j} + t\,\mathbf{k}$

10. $\mathbf{r}(t) = \sqrt{t}\,\mathbf{i} + 3t\,\mathbf{j} - 4t\,\mathbf{k}$

In Exercises 11–14, match the equation with its graph. [The graphs are labeled (a), (b), (c), and (d).]

11. $\mathbf{r}(t) = t\,\mathbf{i} + 2t\,\mathbf{j} + t^2\,\mathbf{k}, \qquad -2 \le t \le 2$

12. $\mathbf{r}(t) = \cos(\pi t)\mathbf{i} + \sin(\pi t)\mathbf{j} + t^2\,\mathbf{k}, \qquad -1 \le t \le 1$

13. $\mathbf{r}(t) = t\,\mathbf{i} + t^2\,\mathbf{j} + e^{0.75t}\,\mathbf{k}, \qquad -2 \le t \le 2$

14. $\mathbf{r}(t) = t\,\mathbf{i} + \ln t\,\mathbf{j} + \dfrac{2t}{3}\mathbf{k}, \qquad 0.1 \le t \le 5$

(a)

(b)

(c)   (d)

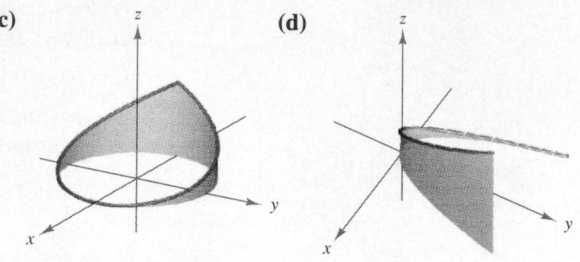

15. The four figures below are graphs of the vector-valued function

$$\mathbf{r}(t) = 4\cos t\,\mathbf{i} + 4\sin t\,\mathbf{j} + \dfrac{t}{4}\mathbf{k}.$$

Match the four graphs with the point in space from which the helix is viewed. The four points are $(0, 0, 20)$, $(20, 0, 0)$, $(-20, 0, 0)$, and $(10, 20, 10)$.

(a)   (b)

Generated by Mathematica   *Generated by Mathematica*

(c)   (d)

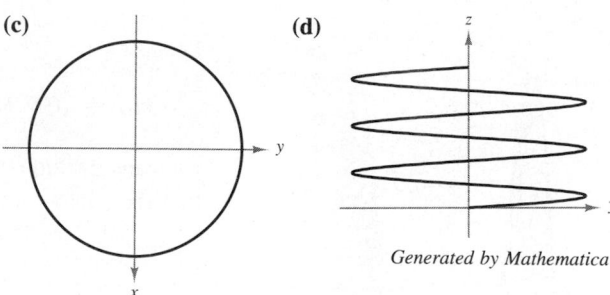

Generated by Mathematica   *Generated by Mathematica*

Generated by Mathematica

16. Sketch three graphs of the vector-valued function

$\mathbf{r}(t) = t\,\mathbf{i} + t\,\mathbf{j} + 2\mathbf{k}$

with views from the points
a. $(0, 0, 20)$ b. $(10, 0, 0)$ c. $(5, 5, 5)$.

In Exercises 17–26, sketch the curve represented by the vector-valued function and give the orientation of the curve.

17. $\mathbf{r}(t) = 3t\,\mathbf{i} + (t - 1)\mathbf{j}$

18. $\mathbf{r}(t) = 2\cos t\,\mathbf{i} + 2\sin t\,\mathbf{j}$

19. $\mathbf{r}(t) = (-t + 1)\mathbf{i} + (4t + 2)\mathbf{j} + (2t + 3)\mathbf{k}$

20. $\mathbf{r}(t) = t\,\mathbf{i} + (2t - 5)\mathbf{j} + 3t\,\mathbf{k}$

21. $\mathbf{r}(t) = 2\cos t\,\mathbf{i} + 2\sin t\,\mathbf{j} + t\,\mathbf{k}$

22. $\mathbf{r}(t) = 3\cos t\,\mathbf{i} + 4\sin t\,\mathbf{j} + \dfrac{t}{2}\mathbf{k}$

23. $\mathbf{r}(t) = 2\sin t\,\mathbf{i} + 2\cos t\,\mathbf{j} + e^{-t}\mathbf{k}$

24. $\mathbf{r}(t) = t\,\mathbf{i} + t^2\mathbf{j} + \dfrac{3t}{2}\mathbf{k}$

25. $\mathbf{r}(t) = \langle t, t^2, \frac{2}{3}t^3 \rangle$

26. $\mathbf{r}(t) = \langle \cos t + t\sin t, \sin t - t\cos t, t \rangle$

C In Exercises 27–30, use a computer algebra system to identify the common curve.

27. $\mathbf{r}(t) = -\dfrac{1}{2}t^2\mathbf{i} + t\,\mathbf{j} - \dfrac{\sqrt{3}}{2}t^2\mathbf{k}$

28. $\mathbf{r}(t) = t\,\mathbf{i} - \dfrac{\sqrt{3}}{2}t^2\mathbf{j} + \dfrac{1}{2}t^2\mathbf{k}$

29. $\mathbf{r}(t) = \sin t\,\mathbf{i} + \left(\dfrac{\sqrt{3}}{2}\cos t - \dfrac{1}{2}t\right)\mathbf{j} + \left(\dfrac{1}{2}\cos t + \dfrac{\sqrt{3}}{2}\right)\mathbf{k}$

30. $\mathbf{r}(t) = -\sqrt{2}\sin t\,\mathbf{i} + 2\cos t\,\mathbf{j} + \sqrt{2}\sin t\,\mathbf{k}$

In Exercises 31–34, represent the plane curve by a vector-valued function. (There are many correct answers.)

31. $y = 4 - x$

32. $y = 4 - x^2$

33. $x^2 + y^2 = 25$

34. $\dfrac{x^2}{25} + \dfrac{y^2}{16} = 1$

In Exercises 35–42, sketch the space curve represented by the intersection of the surfaces. Then, represent the curve by a vector-valued function using the given parameter.

Surfaces	Parameter
35. $z = x^2 + y^2,\ x + y = 0$	$x = t$
36. $z = x^2 + y^2,\ z = 4$	$x = 2\cos t$
37. $x^2 + y^2 = 4,\ z = x^2$	$x = 2\sin t$
38. $4x^2 + y^2 + 4z^2 = 16,\ x = y^2$	$y = t$
39. $x^2 + y^2 + z^2 = 4,\ x + z = 2$	$x = 1 + \sin t$
40. $x^2 + y^2 + z^2 = 10,\ x + y = 4$	$x = 2 + \sin t$
41. $x^2 + z^2 = 4,\ y^2 + z^2 = 4$ (first octant)	$x = t$
42. $x^2 + y^2 + z^2 = 16,\ xy = 4$ (first octant)	$x = t$

In Exercises 43–48, evaluate the limit.

43. $\displaystyle\lim_{t\to 2}\left(t\,\mathbf{i} + \dfrac{t^2 - 4}{t^2 - 2t}\mathbf{j} + \dfrac{1}{t}\mathbf{k} \right)$

44. $\displaystyle\lim_{t\to 0}\left(e^t\mathbf{i} + \dfrac{\sin t}{t}\mathbf{j} + e^{-t}\mathbf{k} \right)$

45. $\displaystyle\lim_{t\to 0}\left(t^2\mathbf{i} + 3t\,\mathbf{j} + \dfrac{1 - \cos t}{t}\mathbf{k} \right)$

46. $\displaystyle\lim_{t\to 1}\left(\sqrt{t}\,\mathbf{i} + \dfrac{\ln t}{t^2 - 1}\mathbf{j} + 2t^2\mathbf{k} \right)$

47. $\displaystyle\lim_{t\to 0}\left(\dfrac{1}{t}\mathbf{i} + \cos t\,\mathbf{j} + \sin t\,\mathbf{k} \right)$

48. $\displaystyle\lim_{t\to\infty}\left(e^{-t}\mathbf{i} + \dfrac{1}{t}\mathbf{j} + \dfrac{t}{t^2 + 1}\mathbf{k} \right)$

In Exercises 49–54, determine the interval(s) on which the vector-valued function is continuous.

49. $\mathbf{r}(t) = t\,\mathbf{i} + \dfrac{1}{t}\mathbf{j}$

50. $\mathbf{r}(t) = \sqrt{t}\,\mathbf{i} + \sqrt{t - 1}\,\mathbf{j}$

51. $\mathbf{r}(t) = t\,\mathbf{i} + \arcsin t\,\mathbf{j} + (t - 1)\mathbf{k}$

52. $\mathbf{r}(t) = \sin t\,\mathbf{i} + \cos t\,\mathbf{j} + \ln t\,\mathbf{k}$

53. $\mathbf{r}(t) = \langle e^{-t}, t^2, \tan t \rangle$

54. $\mathbf{r}(t) = \langle 8, \sqrt{t}, \sqrt[3]{t} \rangle$

55. Let $\mathbf{r}(t)$ and $\mathbf{u}(t)$ be vector-valued functions whose limits exist as $t \to c$. Prove that

$$\lim_{t\to c}[\mathbf{r}(t) \times \mathbf{u}(t)] = \lim_{t\to c}\mathbf{r}(t) \times \lim_{t\to c}\mathbf{u}(t).$$

56. Let $\mathbf{r}(t)$ and $\mathbf{u}(t)$ be vector-valued functions whose limits exist as $t \to c$. Prove that

$$\lim_{t\to c}[\mathbf{r}(t) \cdot \mathbf{u}(t)] = \lim_{t\to c}\mathbf{r}(t) \cdot \lim_{t\to c}\mathbf{u}(t).$$

57. Prove that if $\mathbf{r}$ is a vector-valued function that is continuous at c, then $\|\mathbf{r}\|$ is continuous at c.

58. Verify that the converse of Exercise 57 is not true by finding a vector-valued function $\mathbf{r}$ such that $\|\mathbf{r}\|$ is continuous at c but $\mathbf{r}$ is not continuous at c.

True or False In Exercises 59 and 60, determine whether the statement is true or false. If it is false, explain why or give an example that shows it is false.

59. If f, g, and h are first-degree polynomial functions, then the curve given by $x = f(t)$, $y = g(t)$, $z = h(t)$ is a line.

60. If the curve given by $x = f(t)$, $y = g(t)$, $z = h(t)$ is a line, then f, g, and h are first-degree polynomial functions of t.

---

*A **C** indicates an exercise in which you are instructed to use graphing technology or a symbolic computer algebra system. The solutions of other exercises may also be facilitated by appropriate technology.

SECTION 12.2 Differentiation and Integration of Vector-Valued Functions

Differentiation of Vector-Valued Functions • Integration of Vector-Valued Functions

Differentiation of Vector-Valued Functions

In Sections 12.3–12.5, you will study several important applications involving the calculus of vector-valued functions. In preparation for that study, this section is devoted to the mechanics of differentiation and integration of vector-valued functions.

The definition of the derivative of a vector-valued function parallels that given for real-valued functions.

Definition of the Derivative of a Vector-Valued Function

The **derivative of a vector-valued function r** is defined by

$$\mathbf{r}'(t) = \lim_{\Delta t \to 0} \frac{\mathbf{r}(t + \Delta t) - \mathbf{r}(t)}{\Delta t}$$

for all t for which the limit exists. If $\mathbf{r}'(c)$ exists, then $\mathbf{r}$ is **differentiable** at c. If $\mathbf{r}'(c)$ exists for all c in an open interval I, then $\mathbf{r}$ is **differentiable** on the interval I. Differentiability of vector-valued functions can be extended to closed intervals by considering one-sided limits.

REMARK   In addition to $\mathbf{r}'(t)$, other notations for the derivative of a vector-valued function are

$$D_t[\mathbf{r}(t)], \frac{d}{dt}[\mathbf{r}(t)], \quad \text{and} \quad \frac{d\mathbf{r}}{dt}.$$

Differentiation of vector-valued functions can be done on a *component-by-component basis*. To see why this is true, consider the function given by $\mathbf{r}(t) = f(t)\mathbf{i} + g(t)\mathbf{j}$. Applying the definition of the derivative produces the following.

$$\mathbf{r}'(t) = \lim_{\Delta t \to 0} \frac{\mathbf{r}(t + \Delta t) - \mathbf{r}(t)}{\Delta t}$$

$$= \lim_{\Delta t \to 0} \frac{f(t + \Delta t)\mathbf{i} + g(t + \Delta t)\mathbf{j} - f(t)\mathbf{i} - g(t)\mathbf{j}}{\Delta t}$$

$$= \lim_{\Delta t \to 0} \left\{ \left[\frac{f(t + \Delta t) - f(t)}{\Delta t} \right]\mathbf{i} + \left[\frac{g(t + \Delta t) - g(t)}{\Delta t} \right]\mathbf{j} \right\}$$

$$= \left\{ \lim_{\Delta t \to 0} \left[\frac{f(t + \Delta t) - f(t)}{\Delta t} \right] \right\}\mathbf{i} + \left\{ \lim_{\Delta t \to 0} \left[\frac{g(t + \Delta t) - g(t)}{\Delta t} \right] \right\}\mathbf{j}$$

$$= f'(t)\mathbf{i} + g'(t)\mathbf{j}$$

This important result is listed in the following theorem. Note that the derivative of the vector-valued function $\mathbf{r}$ is itself a vector-valued function. (A geometric interpretation of the derivative of a vector-valued function is discussed in the next section.)

THEOREM 12.1   Differentiation of Vector-Valued Functions

1. If $\mathbf{r}(t) = f(t)\mathbf{i} + g(t)\mathbf{j}$, where f and g are differentiable functions of t, then
$\mathbf{r}'(t) = f'(t)\mathbf{i} + g'(t)\mathbf{j}$.   Plane

2. If $\mathbf{r}(t) = f(t)\mathbf{i} + g(t)\mathbf{j} + h(t)\mathbf{k}$, where f, g, and h are differentiable functions of t, then
$\mathbf{r}'(t) = f'(t)\mathbf{i} + g'(t)\mathbf{j} + h'(t)\mathbf{k}$.   Space

EXAMPLE 1   Differentiation of Vector-Valued Functions

Find the derivative of each of the following vector-valued functions.

a. $\mathbf{r}(t) = t^2\mathbf{i} - 4\mathbf{j}$ **b.** $\mathbf{r}(t) = \dfrac{1}{t}\mathbf{i} + \ln t\,\mathbf{j} + e^{2t}\mathbf{k}$

Solution   Differentiating on a component-by-component basis produces the following.

a. $\mathbf{r}'(t) = 2t\,\mathbf{i} - 0\mathbf{j} = 2t\,\mathbf{i}$

b. $\mathbf{r}'(t) = -\dfrac{1}{t^2}\mathbf{i} + \dfrac{1}{t}\mathbf{j} + 2e^{2t}\mathbf{k}$

Higher-order derivatives of vector-valued functions are obtained by successive differentiation of each component function.

EXAMPLE 2   Higher-Order Differentiation

For the vector-valued function given by $\mathbf{r}(t) = \cos t\,\mathbf{i} + \sin t\,\mathbf{j} + 2t\,\mathbf{k}$, find the following.

a. $\mathbf{r}'(t)$ **b.** $\mathbf{r}''(t)$ **c.** $\mathbf{r}'(t) \cdot \mathbf{r}''(t)$ **d.** $\mathbf{r}'(t) \times \mathbf{r}''(t)$

Solution

a. $\mathbf{r}'(t) = -\sin t\,\mathbf{i} + \cos t\,\mathbf{j} + 2\mathbf{k}$

b. $\mathbf{r}''(t) = -\cos t\,\mathbf{i} - \sin t\,\mathbf{j} + 0\mathbf{k} = -\cos t\,\mathbf{i} - \sin t\,\mathbf{j}$

c. $\mathbf{r}'(t) \cdot \mathbf{r}''(t) = \sin t \cos t - \sin t \cos t = 0$

d. $\mathbf{r}'(t) \times \mathbf{r}''(t) = \begin{vmatrix} \mathbf{i} & \mathbf{j} & \mathbf{k} \\ -\sin t & \cos t & 2 \\ -\cos t & -\sin t & 0 \end{vmatrix}$

$= \begin{vmatrix} \cos t & 2 \\ -\sin t & 0 \end{vmatrix}\mathbf{i} - \begin{vmatrix} -\sin t & 2 \\ -\cos t & 0 \end{vmatrix}\mathbf{j} + \begin{vmatrix} -\sin t & \cos t \\ -\cos t & -\sin t \end{vmatrix}\mathbf{k}$

$= 2\sin t\,\mathbf{i} - 2\cos t\,\mathbf{j} + \mathbf{k}$

Note that the dot product in part c is a *real-valued* function, not a vector-valued function.

The curve represented by the vector-valued function

$$\mathbf{r}(t) = f(t)\mathbf{i} + g(t)\mathbf{j} + h(t)\mathbf{k}$$

is **smooth on an open interval** I if f', g', and h' are continuous on I and $\mathbf{r}'(t) \neq \mathbf{0}$ for any value of t in the interval I.

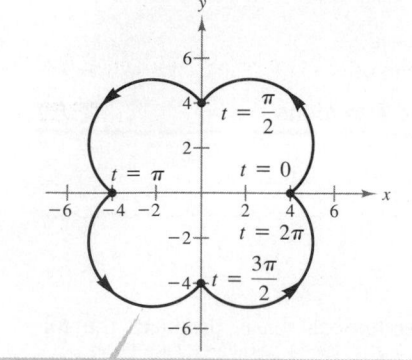

$$\mathbf{r}(t) = (5\cos t - \cos 5t)\mathbf{i} + (5\sin t - \sin 5t)\mathbf{j}$$

FIGURE 12.8
The epicycloid is not smooth at the points where it intersects the axes.

EXAMPLE 3   Finding Intervals on Which a Curve is Smooth

Find the intervals on which the epicycloid C given by

$$\mathbf{r}(t) = (5\cos t - \cos 5t)\mathbf{i} + (5\sin t - \sin 5t)\mathbf{j}, \qquad 0 \le t \le 2\pi$$

is smooth.

Solution   The derivative of $\mathbf{r}$ is

$$\mathbf{r}'(t) = (-5\sin t + 5\sin 5t)\mathbf{i} + (5\cos t - 5\cos 5t)\mathbf{j}.$$

In the interval $[0, 2\pi]$, the only values of t for which $\mathbf{r}'(t) = 0\mathbf{i} + 0\mathbf{j}$ are $t = 0$, $\pi/2$, π, $3\pi/2$, and 2π. Therefore, you can conclude that C is smooth in the intervals

$$\left(0, \frac{\pi}{2}\right), \quad \left(\frac{\pi}{2}, \pi\right), \quad \left(\pi, \frac{3\pi}{2}\right), \quad \text{and} \quad \left(\frac{3\pi}{2}, 2\pi\right)$$

as shown in Figure 12.8.

REMARK   In Figure 12.8, note that the curve is not smooth at points at which the curve makes an abrupt change in direction—such points are called **cusps** or **nodes.**

Most of the differentiation rules in Chapter 2 have counterparts for vector-valued functions, and several are listed in the following theorem. Note that the theorem contains three versions of "product rules." Property 3 gives the derivative of the product of a real-valued function f and a vector-valued function $\mathbf{r}$. Property 4 gives the derivative of the dot product of two vector-valued functions, and Property 5 gives the derivative of the cross product of two vector-valued functions, (in space). (Property 5 applies only to three-dimensional vector-valued functions, because the cross product is not defined for two-dimensional vectors.)

THEOREM 12.2   Properties of the Derivative

Let $\mathbf{r}$ and $\mathbf{u}$ be differentiable vector-valued functions of t, let f be a differentiable real-valued function of t, and let c be a scalar.

1. $D_t[c\,\mathbf{r}(t)] = c\,\mathbf{r}'(t)$
2. $D_t[\mathbf{r}(t) \pm \mathbf{u}(t)] = \mathbf{r}'(t) \pm \mathbf{u}'(t)$
3. $D_t[f(t)\mathbf{r}(t)] = f(t)\mathbf{r}'(t) + f'(t)\mathbf{r}(t)$
4. $D_t[\mathbf{r}(t) \cdot \mathbf{u}(t)] = \mathbf{r}(t) \cdot \mathbf{u}'(t) + \mathbf{r}'(t) \cdot \mathbf{u}(t)$
5. $D_t[\mathbf{r}(t) \times \mathbf{u}(t)] = \mathbf{r}(t) \times \mathbf{u}'(t) + \mathbf{r}'(t) \times \mathbf{u}(t)$
6. $D_t[\mathbf{r}(f(t))] = \mathbf{r}'(f(t))f'(t)$
7. If $\mathbf{r}(t) \cdot \mathbf{r}(t) = c$, then $\mathbf{r}(t) \cdot \mathbf{r}'(t) = 0$.

Proof To prove the fourth property, let

$$\mathbf{r}(t) = f_1(t)\mathbf{i} + g_1(t)\mathbf{j} \quad \text{and} \quad \mathbf{u}(t) = f_2(t)\mathbf{i} + g_2(t)\mathbf{j}$$

where f_i and g_i are differentiable functions of t. Then,

$$\mathbf{r}(t) \cdot \mathbf{u}(t) = f_1(t)f_2(t) + g_1(t)g_2(t)$$

and it follows that

$$
\begin{aligned}
D_t[\mathbf{r}(t) \cdot \mathbf{u}(t)] &= f_1(t)f_2'(t) + f_1'(t)f_2(t) + g_1(t)g_2'(t) + g_1'(t)g_2(t) \\
&= [f_1(t)f_2'(t) + g_1(t)g_2'(t)] + [f_1'(t)f_2(t) + g_1'(t)g_2(t)] \\
&= \mathbf{r}(t) \cdot \mathbf{u}'(t) + \mathbf{r}'(t) \cdot \mathbf{u}(t).
\end{aligned}
$$

Proofs of the other properties are left as exercises (see Exercises 45–49 and Exercise 52).

EXAMPLE 4 Using Properties of the Derivative

For the vector-valued functions given by

$$\mathbf{r}(t) = \frac{1}{t}\mathbf{i} - \mathbf{j} + (\ln t)\mathbf{k} \quad \text{and} \quad \mathbf{u}(t) = t^2\mathbf{i} - 2t\mathbf{j} + \mathbf{k}$$

find

a. $D_t[\mathbf{r}(t) \cdot \mathbf{u}(t)]$.   **b.** $D_t[\mathbf{u}(t) \times \mathbf{u}'(t)]$.

Solution

a. Because $\mathbf{r}'(t) = -\dfrac{1}{t^2}\mathbf{i} + \dfrac{1}{t}\mathbf{k}$ and $\mathbf{u}'(t) = 2t\mathbf{i} - 2\mathbf{j}$, you have

$$
\begin{aligned}
D_t[\mathbf{r}(t) \cdot \mathbf{u}(t)] &= \mathbf{r}(t) \cdot \mathbf{u}'(t) + \mathbf{r}'(t) \cdot \mathbf{u}(t) \\
&= \left(\frac{1}{t}\mathbf{i} - \mathbf{j} + \ln t\,\mathbf{k}\right) \cdot (2t\mathbf{i} - 2\mathbf{j}) \\
&\quad + \left(-\frac{1}{t^2}\mathbf{i} + \frac{1}{t}\mathbf{k}\right) \cdot (t^2\mathbf{i} - 2t\mathbf{j} + \mathbf{k}) \\
&= 2 + 2 + (-1) + \frac{1}{t} \\
&= 3 + \frac{1}{t}.
\end{aligned}
$$

b. Because $\mathbf{u}'(t) = 2t\mathbf{i} - 2\mathbf{j}$ and $\mathbf{u}''(t) = 2\mathbf{i}$, you have

$$
\begin{aligned}
D_t[\mathbf{u}(t) \times \mathbf{u}'(t)] &= [\mathbf{u}(t) \times \mathbf{u}''(t)] + [\mathbf{u}'(t) \times \mathbf{u}'(t)] \\
&= \begin{vmatrix} \mathbf{i} & \mathbf{j} & \mathbf{k} \\ t^2 & -2t & 1 \\ 2 & 0 & 0 \end{vmatrix} + \mathbf{0} \\
&= \begin{vmatrix} -2t & 1 \\ 0 & 0 \end{vmatrix}\mathbf{i} - \begin{vmatrix} t^2 & 1 \\ 2 & 0 \end{vmatrix}\mathbf{j} + \begin{vmatrix} t^2 & -2t \\ 2 & 0 \end{vmatrix}\mathbf{k} \\
&= 0\mathbf{i} - (-2)\mathbf{j} + 4t\,\mathbf{k} \\
&= 2\mathbf{j} + 4t\,\mathbf{k}.
\end{aligned}
$$

REMARK Try reworking parts a and b in Example 4 by first forming the dot and cross products and then differentiating to see that you obtain the same results.

Integration of Vector-Valued Functions

The following definition is a natural consequence of the definition of the derivative of a vector-valued function.

> **Definition of Integration of Vector-Valued Functions**
>
> 1. If $\mathbf{r}(t) = f(t)\mathbf{i} + g(t)\mathbf{j}$, where f and g are continuous on $[a, b]$, then the **indefinite integral (antiderivative)** of $\mathbf{r}$ is
>
> $$\int \mathbf{r}(t)\, dt = \left[\int f(t)\, dt\right]\mathbf{i} + \left[\int g(t)\, dt\right]\mathbf{j} \qquad \text{Plane}$$
>
> and its **definite integral** over the interval $a \le t \le b$ is
>
> $$\int_a^b \mathbf{r}(t)\, dt = \left[\int_a^b f(t)\, dt\right]\mathbf{i} + \left[\int_a^b g(t)\, dt\right]\mathbf{j}.$$
>
> 2. If $\mathbf{r}(t) = f(t)\mathbf{i} + g(t)\mathbf{j} + h(t)\mathbf{k}$, where f, g, and h are continuous on $[a, b]$, then the **indefinite integral (antiderivative)** of $\mathbf{r}$ is
>
> $$\int \mathbf{r}(t)\, dt = \left[\int f(t)\, dt\right]\mathbf{i} + \left[\int g(t)\, dt\right]\mathbf{j} + \left[\int h(t)\, dt\right]\mathbf{k} \qquad \text{Space}$$
>
> and its **definite integral** over the interval $a \le t \le b$ is
>
> $$\int_a^b \mathbf{r}(t)\, dt = \left[\int_a^b f(t)\, dt\right]\mathbf{i} + \left[\int_a^b g(t)\, dt\right]\mathbf{j} + \left[\int_a^b h(t)\, dt\right]\mathbf{k}.$$

The antiderivative of a vector-valued function is a family of vector-valued functions all differing by a constant vector $\mathbf{C}$. For instance, if $\mathbf{r}(t)$ is a three-dimensional vector-valued function, then for the indefinite integral $\int \mathbf{r}(t)\, dt$, you obtain three constants of integration

$$\int f(t)\, dt = F(t) + C_1, \quad \int g(t)\, dt = G(t) + C_2, \quad \int h(t)\, dt = H(t) + C_3$$

where $F'(t) = f(t)$, $G'(t) = g(t)$, and $H'(t) = h(t)$. These three *scalar* constants produce one *vector* constant of integration,

$$\begin{aligned}
\int \mathbf{r}(t)\, dt &= [F(t) + C_1]\mathbf{i} + [G(t) + C_2]\mathbf{j} + [H(t) + C_3]\mathbf{k} \\
&= [F(t)\mathbf{i} + G(t)\mathbf{j} + H(t)\mathbf{k}] + [C_1\mathbf{i} + C_2\mathbf{j} + C_3\mathbf{k}] \\
&= \mathbf{R}(t) + \mathbf{C}
\end{aligned}$$

where $\mathbf{R}'(t) = \mathbf{r}(t)$.

EXAMPLE 5 Integrating a Vector-Valued Function

Evaluate the indefinite integral

$$\int (t\mathbf{i} + 3\mathbf{j})\, dt.$$

Solution Integrating on a component-by-component basis produces

$$\int (t\mathbf{i} + 3\mathbf{j})\, dt = \frac{t^2}{2}\mathbf{i} + 3t\mathbf{j} + \mathbf{C}.$$

Example 6 shows how to evaluate the definite integral of a vector-valued function.

EXAMPLE 6   Definite Integral of a Vector-Valued Function

Evaluate the integral

$$\int_0^1 \mathbf{r}(t)\,dt = \int_0^1 \left(\sqrt[3]{t}\,\mathbf{i} + \frac{1}{t+1}\mathbf{j} + e^{-t}\mathbf{k} \right) dt.$$

Solution

$$\int_0^1 \mathbf{r}(t)\,dt = \left(\int_0^1 t^{1/3}\,dt \right)\mathbf{i} + \left(\int_0^1 \frac{1}{t+1}\,dt \right)\mathbf{j} + \left(\int_0^1 e^{-t}\,dt \right)\mathbf{k}$$

$$= \left[\left(\frac{3}{4} \right)t^{4/3} \right]_0^1 \mathbf{i} + \left[\ln|t+1| \right]_0^1 \mathbf{j} + \left[-e^{-t} \right]_0^1 \mathbf{k}$$

$$= \frac{3}{4}\mathbf{i} + (\ln 2)\mathbf{j} + \left(1 - \frac{1}{e} \right)\mathbf{k}$$

As with real-valued functions, you can narrow the family of antiderivatives of a vector-valued function $\mathbf{r}'$ down to a single antiderivative by imposing an initial condition on the vector-valued function $\mathbf{r}$. This is demonstrated in the next example.

EXAMPLE 7   The Antiderivative of a Vector-Valued Function

Find the antiderivative of

$$\mathbf{r}'(t) = \cos 2t\,\mathbf{i} - 2\sin t\,\mathbf{j} + \frac{1}{1+t^2}\mathbf{k}$$

that satisfies the initial condition $\mathbf{r}(0) = 3\mathbf{i} - 2\mathbf{j} + \mathbf{k}$.

Solution

$$\mathbf{r}(t) = \int \mathbf{r}'(t)\,dt$$

$$= \left(\int \cos 2t\,dt \right)\mathbf{i} + \left(\int -2\sin t\,dt \right)\mathbf{j} + \left(\int \frac{1}{1+t^2}\,dt \right)\mathbf{k}$$

$$= \left(\frac{1}{2}\sin 2t + C_1 \right)\mathbf{i} + (2\cos t + C_2)\mathbf{j} + (\arctan t + C_3)\mathbf{k}$$

Letting $t = 0$ and using the fact that $\mathbf{r}(0) = 3\mathbf{i} - 2\mathbf{j} + \mathbf{k}$, you have

$$\mathbf{r}(0) = (0 + C_1)\mathbf{i} + (2 + C_2)\mathbf{j} + (0 + C_3)\mathbf{k}$$
$$= 3\mathbf{i} + (-2)\mathbf{j} + \mathbf{k}.$$

Equating corresponding components produces

$$C_1 = 3, \quad 2 + C_2 = -2, \quad \text{and} \quad C_3 = 1.$$

Thus, the antiderivative that satisfies the given initial condition is

$$\mathbf{r}(t) = \left(\frac{1}{2}\sin 2t + 3 \right)\mathbf{i} + (2\cos t - 4)\mathbf{j} + (\arctan t + 1)\mathbf{k}.$$

EXERCISES for Section 12.2

TECHNOLOGY
Laboratory Guide
Lab 12.2

In Exercises 1–4, (a) sketch the plane curve represented by the vector-valued function and (b) sketch the vectors $\mathbf{r}(t_0)$ and $\mathbf{r}'(t_0)$ for the specified value of t_0. Position the vectors so that the initial point of $\mathbf{r}(t_0)$ is at the origin and the initial point of $\mathbf{r}'(t_0)$ is at the terminal point of $\mathbf{r}(t_0)$.

1. $\mathbf{r}(t) = t^2\mathbf{i} + t\mathbf{j}$ $\qquad t_0 = 2$

2. $\mathbf{r}(t) = t\mathbf{i} + t^3\mathbf{j}$ $\qquad t_0 = 1$

3. $\mathbf{r}(t) = \cos t\,\mathbf{i} + \sin t\,\mathbf{j}$ $\qquad t_0 = \dfrac{\pi}{2}$

4. $\mathbf{r}(t) = t^2\mathbf{i} + \dfrac{1}{t}\mathbf{j}$ $\qquad t_0 = 2$

5. Consider the vector-valued function $\mathbf{r}(t) = t\mathbf{i} + t^2\mathbf{j}$.
   a. Sketch the graph of $\mathbf{r}(t)$.
   b. Sketch the vectors $\mathbf{r}(\frac{1}{4})$, $\mathbf{r}(\frac{1}{2})$, and $\mathbf{r}(\frac{1}{2}) - \mathbf{r}(\frac{1}{4})$ on the graph in part a.
   c. Compare the vector $\mathbf{r}'(\frac{1}{4})$ with the vector
$$\frac{\mathbf{r}(\frac{1}{2}) - \mathbf{r}(\frac{1}{4})}{\frac{1}{2} - \frac{1}{4}}.$$

6. Consider the vector-valued function $\mathbf{r}(t) = t\mathbf{i} + (4 - t^2)\mathbf{j}$.
   a. Sketch the graph of $\mathbf{r}(t)$.
   b. Sketch the vectors $\mathbf{r}(1)$, $\mathbf{r}(1.25)$, and $\mathbf{r}(1.25) - \mathbf{r}(1)$ on the graph in part a.
   c. Compare the vector $\mathbf{r}'(1)$ with the vector
$$\frac{\mathbf{r}(1.25) - \mathbf{r}(1)}{1.25 - 1}.$$

In Exercises 7 and 8, (a) sketch the space curve represented by the vector-valued function. (b) sketch the vectors $\mathbf{r}(t_0)$ and $\mathbf{r}'(t_0)$ for the specified value of t_0.

7. $\mathbf{r}(t) = 2\cos t\,\mathbf{i} + 2\sin t\,\mathbf{j} + t\,\mathbf{k}$, $\quad t_0 = \dfrac{3\pi}{2}$

8. $\mathbf{r}(t) = t\mathbf{i} + t^2\mathbf{j} + \frac{3}{2}\mathbf{k}$, $\quad t_0 = 2$

In Exercises 9 and 10, a vector-valued function and its graph are given. The graph also shows the unit vectors $\mathbf{r}'(t_0)/\|\mathbf{r}'(t_0)\|$ and $\mathbf{r}''(t_0)/\|\mathbf{r}''(t_0)\|$. Find these two unit vectors and identify them on the graph.

9. $\mathbf{r}(t) = \cos(\pi t)\mathbf{i} + \sin(\pi t)\mathbf{j} + t^2\mathbf{k}$, $\quad t_0 = -\frac{1}{4}$

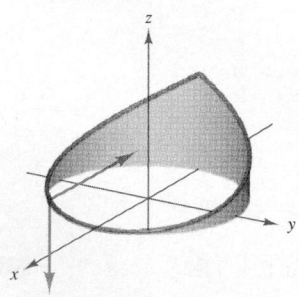

10. $\mathbf{r}(t) = t\mathbf{i} + t^2\mathbf{j} + e^{0.75t}\mathbf{k}$, $\quad t_0 = \frac{1}{4}$

In Exercises 11–18, find $\mathbf{r}'(t)$.

11. $\mathbf{r}(t) = 6t\mathbf{i} - 7t^2\mathbf{j} + t^3\mathbf{k}$

12. $\mathbf{r}(t) = \dfrac{1}{t}\mathbf{i} + 16t\mathbf{j} + \dfrac{t^2}{2}\mathbf{k}$

13. $\mathbf{r}(t) = a\cos^3 t\,\mathbf{i} + a\sin^3 t\,\mathbf{j} + \mathbf{k}$

14. $\mathbf{r}(t) = \sqrt{t}\,\mathbf{i} + t\sqrt{t}\,\mathbf{j} + \ln t\,\mathbf{k}$

15. $\mathbf{r}(t) = e^{-t}\mathbf{i} + 4\mathbf{j}$

16. $\mathbf{r}(t) = \langle \sin t - t\cos t, \cos t + t\sin t, t^2 \rangle$

17. $\mathbf{r}(t) = \langle t\sin t, t\cos t, t \rangle$

18. $\mathbf{r}(t) = \langle \arcsin t, \arccos t, 0 \rangle$

In Exercises 19 and 20, find the following.
   a. $\mathbf{r}'(t)$ $\qquad\qquad$ **b.** $\mathbf{r}''(t)$
   c. $D_t[\mathbf{r}(t) \cdot \mathbf{u}(t)]$ $\qquad$ **d.** $D_t[3\mathbf{r}(t) - \mathbf{u}(t)]$
   e. $D_t[\mathbf{r}(t) \times \mathbf{u}(t)]$ $\qquad$ **f.** $D_t[\|\mathbf{r}(t)\|]$, $\quad t > 0$

19. $\mathbf{r}(t) = t\mathbf{i} + 3t\mathbf{j} + t^2\mathbf{k}$
$\mathbf{u}(t) = 4t\mathbf{i} + t^2\mathbf{j} + t^3\mathbf{k}$

20. $\mathbf{r}(t) = t^2\mathbf{i} + \sin t\,\mathbf{j} + \cos t\,\mathbf{k}$
$\mathbf{u}(t) = \dfrac{1}{t^2}\mathbf{i} + \sin t\,\mathbf{j} + \cos t\,\mathbf{k}$

In Exercises 21–26, find the open interval(s) on which the curve given by the vector-valued function is smooth.

21. $\mathbf{r}(t) = t^2\mathbf{i} + t^3\mathbf{j}$

22. $\mathbf{r}(t) = \dfrac{1}{t - 1}\mathbf{i} + 3t\mathbf{j}$

23. $\mathbf{r}(\theta) = 2\cos^3\theta\,\mathbf{i} + 3\sin^3\theta\,\mathbf{j}$

24. $\mathbf{r}(\theta) = (\theta + \sin\theta)\mathbf{i} + (1 - \cos\theta)\mathbf{j}$

25. $\mathbf{r}(\theta) = (\theta - 2\sin\theta)\mathbf{i} + (1 - 2\cos\theta)\mathbf{j}$

26. $\mathbf{r}(t) = \dfrac{3t}{1 + t^3}\mathbf{i} + \dfrac{3t^2}{1 + t^3}\mathbf{j}$

In Exercises 27–34, evaluate the indefinite integral.

27. $\displaystyle\int (2t\mathbf{i} + \mathbf{j} + \mathbf{k})\,dt$

28. $\displaystyle\int (3t^2\mathbf{i} + 4t\mathbf{j} - 8t^3\mathbf{k})\,dt$

29. $\int \left(\frac{1}{t}\mathbf{i} + \mathbf{j} - t^{3/2}\mathbf{k} \right) dt$

30. $\int \left[(2t - 1)\mathbf{i} + 4t^3\mathbf{j} + 3\sqrt{t}\,\mathbf{k} \right] dt$

31. $\int (e^t\mathbf{i} + \sin t\,\mathbf{j} + \cos t\,\mathbf{k})\, dt$

32. $\int \left(\ln t\,\mathbf{i} + \frac{1}{t}\mathbf{j} + \mathbf{k} \right) dt$

33. $\int \left(\sec^2 t\,\mathbf{i} + \frac{1}{1 + t^2}\mathbf{j} \right) dt$

34. $\int (e^{-t}\sin t\,\mathbf{i} + e^{-t}\cos t\,\mathbf{j})\, dt$

In Exercises 35–40, find r(t) for the given conditions.

35. $\mathbf{r}'(t) = 4e^{2t}\mathbf{i} + 3e^t\mathbf{j}$
$\mathbf{r}(0) = 2\mathbf{i}$

36. $\mathbf{r}'(t) = 2t\mathbf{j} + \sqrt{t}\,\mathbf{k}$
$\mathbf{r}(0) = \mathbf{i} + \mathbf{j}$

37. $\mathbf{r}''(t) = -32\mathbf{j}$
$\mathbf{r}'(0) = 600\sqrt{3}\mathbf{i} + 600\mathbf{j}$
$\mathbf{r}(0) = \mathbf{0}$

38. $\mathbf{r}''(t) = -4\cos t\,\mathbf{j} - 3\sin 5\mathbf{k}$
$\mathbf{r}'(0) = 3\mathbf{k}$
$\mathbf{r}(0) = 4\mathbf{j}$

39. $\mathbf{r}'(t) = te^{-t^2}\mathbf{i} - e^{-t}\mathbf{j} + \mathbf{k}$
$\mathbf{r}(0) = \frac{1}{2}\mathbf{i} - \mathbf{j} + \mathbf{k}$

40. $\mathbf{r}'(t) = \frac{1}{1 + t^2}\mathbf{i} + \frac{1}{t^2}\mathbf{j} + \frac{1}{t}\mathbf{k}$
$\mathbf{r}(1) = 2\mathbf{i}$

In Exercises 41–44, evaluate the definite integral.

41. $\int_0^1 (8t\,\mathbf{i} + t\,\mathbf{j} - \mathbf{k})\, dt$

42. $\int_{-1}^1 \left(t\,\mathbf{i} + t^3\mathbf{j} + \sqrt[3]{t}\,\mathbf{k} \right) dt$

43. $\int_0^{\pi/2} [(a\cos t)\mathbf{i} + (a\sin t)\mathbf{j} + \mathbf{k}]\, dt$

44. $\int_0^3 (e^t\mathbf{i} + te^t\mathbf{k})\, dt$

In Exercises 45–52, prove the given property. In each case, assume that r, u, and v are differentiable vector-valued functions of t, f is a differentiable real-valued function of t, and c is a scalar.

45. $D_t[c\,\mathbf{r}(t)] = c\,\mathbf{r}'(t)$

46. $D_t[\mathbf{r}(t) \pm \mathbf{u}(t)] = \mathbf{r}'(t) \pm \mathbf{u}'(t)$

47. $D_t[f(t)\mathbf{r}(t)] = f(t)\mathbf{r}'(t) + f'(t)\mathbf{r}(t)$

48. $D_t[\mathbf{r}(t) \times \mathbf{u}(t)] = [\mathbf{r}(t) \times \mathbf{u}'(t)] + [\mathbf{r}'(t) \times \mathbf{u}(t)]$

49. $D_t[\mathbf{r}(f(t))] = \mathbf{r}'(f(t))f'(t)$

50. $D_t[\mathbf{r}(t) \times \mathbf{r}'(t)] = \mathbf{r}(t) \times \mathbf{r}''(t)$

51. $D_t[\mathbf{r}(t) \cdot (\mathbf{u}(t) \times \mathbf{v}(t))]$
$= \mathbf{r}'(t) \cdot [\mathbf{u}(t) \times \mathbf{v}(t)] + \mathbf{r}(t) \cdot [\mathbf{u}'(t) \times \mathbf{v}(t)] + \mathbf{r}(t) \cdot [\mathbf{u}(t) \times \mathbf{v}'(t)]$

52. If $\mathbf{r}(t) \cdot \mathbf{r}(t)$ is a constant, then $\mathbf{r}(t) \cdot \mathbf{r}'(t) = 0$.

True or False **In Exercises 53 and 54, determine whether the statement is true or false. If it is false, explain why or give an example that shows it is false.**

53. $\frac{d}{dt}\big[\|\mathbf{r}(t)\| \big] = \|\mathbf{r}'(t)\|$

54. If $\mathbf{r}$ and $\mathbf{u}$ are differentiable vector-valued functions of t, then $D_t[\mathbf{r}(t) \cdot \mathbf{u}(t)] = \mathbf{r}'(t) \cdot \mathbf{u}'(t)$.

SECTION

12.3 Velocity and Acceleration

Velocity and Acceleration • Projectile Motion

Velocity and Acceleration

You are now ready to combine your study of parametric equations, curves, vectors, and vector-valued functions to form a model for motion along a curve. We begin by looking at the motion of an object in the plane. (The motion of an object in space could be developed similarly.)

As an object moves along a curve in the plane, the coordinates x and y of its center of mass are both functions of time t. Rather than using f and g to represent these two functions, it is convenient to write

$$x = x(t) \quad \text{and} \quad y = y(t).$$

Thus, the position vector $\mathbf{r}(t)$ takes the form

$$\mathbf{r}(t) = x(t)\mathbf{i} + y(t)\mathbf{j}. \qquad \text{Position vector}$$

The beauty of this vector model for representing motion is that you can use the first and second derivatives of the vector-valued function $\mathbf{r}$ to find the object's velocity and acceleration. (Recall from the previous chapter that velocity and acceleration are both vector quantities having magnitude and direction.)

To find the velocity and acceleration vectors at a given time t, consider a point

$$Q = (x(t + \Delta t), y(t + \Delta t))$$

that is approaching the point

$$P = (x(t), y(t))$$

along the curve C given by

$$\mathbf{r}(t) = x(t)\mathbf{i} + y(t)\mathbf{j}$$

as shown in Figure 12.9. As $\Delta t \to 0$, the direction of the vector $\overrightarrow{PQ}$ (denoted by $\Delta \mathbf{r}$) approaches the *direction of motion* at the time t.

$$\Delta \mathbf{r} = \mathbf{r}(t + \Delta t) - \mathbf{r}(t)$$

$$\frac{\Delta \mathbf{r}}{\Delta t} = \frac{\mathbf{r}(t + \Delta t) - \mathbf{r}(t)}{\Delta t}$$

$$\lim_{\Delta t \to 0} \frac{\Delta \mathbf{r}}{\Delta t} = \lim_{\Delta t \to 0} \frac{\mathbf{r}(t + \Delta t) - \mathbf{r}(t)}{\Delta t}$$

If this limit exists (and is not equal to the zero vector), it is defined to be the **tangent vector** to the curve at the point P. Note that this is the same limit used to define $\mathbf{r}'(t)$. Thus, the direction of $\mathbf{r}'(t)$ gives the direction of motion at time t. Moreover, the magnitude of the vector $\mathbf{r}'(t)$

$$\| \mathbf{r}'(t) \| = \| x'(t)\mathbf{i} + y'(t)\mathbf{j} \|$$

$$= \sqrt{[x'(t)]^2 + [y'(t)]^2}$$

gives the **speed** of the object at time t. Similarly, you can use $\mathbf{r}''(t)$ to represent acceleration, as indicated in the following definition.

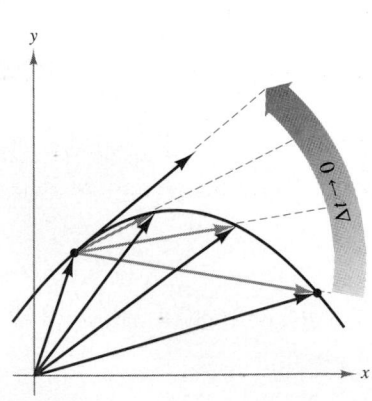

FIGURE 12.9

As $\Delta t \to 0$, $\dfrac{\Delta \mathbf{r}}{\Delta t}$ approaches tangent vector.

Definition of Velocity and Acceleration

If x and y are twice-differentiable functions of t, and $\mathbf{r}$ is a vector-valued function given by $\mathbf{r}(t) = x(t)\mathbf{i} + y(t)\mathbf{j}$, then the velocity vector, acceleration vector, and speed at time t are as follows.

$$\text{Velocity} = \mathbf{v}(t) \quad = \mathbf{r}'(t) \quad = x'(t)\mathbf{i} + y'(t)\mathbf{j}$$
$$\text{Acceleration} = \mathbf{a}(t) \quad = \mathbf{r}''(t) \quad = x''(t)\mathbf{i} + y''(t)\mathbf{j}$$
$$\text{Speed} = \|\mathbf{v}(t)\| = \|\mathbf{r}'(t)\| = \sqrt{[x'(t)]^2 + [y'(t)]^2}$$

For motion along a space curve, the definitions are similar. That is, if $\mathbf{r}(t) = x(t)\mathbf{i} + y(t)\mathbf{j} + z(t)\mathbf{k}$, then you have

$$\text{Velocity} = \mathbf{v}(t) \quad = \mathbf{r}'(t) \quad = x'(t)\mathbf{i} + y'(t)\mathbf{j} + z'(t)\mathbf{k}$$
$$\text{Acceleration} = \mathbf{a}(t) \quad = \mathbf{r}''(t) \quad = x''(t)\mathbf{i} + y''(t)\mathbf{j} + z''(t)\mathbf{k}$$
$$\text{Speed} = \|\mathbf{v}(t)\| = \|\mathbf{r}'(t)\| = \sqrt{[x'(t)]^2 + [y'(t)]^2 + [z'(t)]^2}.$$

EXAMPLE 1   Finding Velocity and Acceleration Along a Plane Curve

REMARK   In Example 1, note that the velocity and acceleration vectors are orthogonal at any point in time. This is characteristic of motion at a constant speed. (See Exercise 47.)

Find the velocity vector, speed, and acceleration vector of a particle that moves along the plane curve C described by

$$\mathbf{r}(t) = 2\sin\frac{t}{2}\mathbf{i} + 2\cos\frac{t}{2}\mathbf{j}. \qquad \text{Position vector}$$

Solution   The velocity vector is

$$\mathbf{v}(t) = \mathbf{r}'(t) = \cos\frac{t}{2}\mathbf{i} - \sin\frac{t}{2}\mathbf{j}. \qquad \text{Velocity vector}$$

The speed (at any time) is

$$\|\mathbf{r}'(t)\| = \sqrt{\cos^2\frac{t}{2} + \sin^2\frac{t}{2}} = 1. \qquad \text{Speed}$$

The acceleration vector is

$$\mathbf{a}(t) = \mathbf{r}''(t) = -\frac{1}{2}\sin\frac{t}{2}\mathbf{i} - \frac{1}{2}\cos\frac{t}{2}\mathbf{j}. \qquad \text{Acceleration vector}$$

The parametric equations for the curve in Example 1 are $x = 2\sin(t/2)$ and $y = 2\cos(t/2)$. By eliminating the parameter t, you can obtain the rectangular equation

$$x^2 + y^2 = 4. \qquad \text{Rectangular equation}$$

Thus, the curve is a circle of radius 2 centered at the origin, as shown in Figure 12.10. Because the velocity vector

$$\mathbf{v}(t) = \cos\frac{t}{2}\mathbf{i} - \sin\frac{t}{2}\mathbf{j}$$

has a constant magnitude but a changing direction as t increases, the particle moves around the circle at a constant speed.

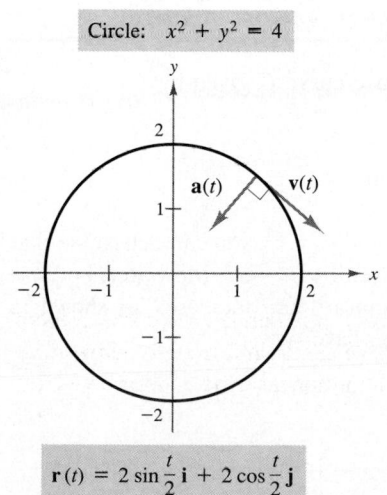

Circle: $x^2 + y^2 = 4$

$$\mathbf{r}(t) = 2\sin\frac{t}{2}\mathbf{i} + 2\cos\frac{t}{2}\mathbf{j}$$

FIGURE 12.10
The particle moves around the
circle at constant speed.

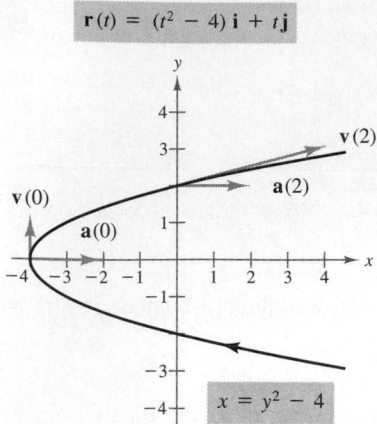

FIGURE 12.11

At each point on the curve, the acceleration vector points to the right.

EXAMPLE 2 Sketching Velocity and Acceleration Vectors in the Plane

Sketch the path of an object moving along the plane curve given by

$$\mathbf{r}(t) = (t^2 - 4)\mathbf{i} + t\,\mathbf{j} \qquad \text{Position vector}$$

and find the velocity and acceleration vectors when $t = 0$ and $t = 2$.

Solution Using the parametric equations $x = t^2 - 4$ and $y = t$, you can determine that the curve is a parabola given by $x = y^2 - 4$, as shown in Figure 12.11. The velocity vector (at any time) is

$$\mathbf{v}(t) = \mathbf{r}'(t) = 2t\,\mathbf{i} + \mathbf{j} \qquad \text{Velocity vector}$$

and the acceleration vector (at any time) is

$$\mathbf{a}(t) = \mathbf{r}''(t) = 2\,\mathbf{i}. \qquad \text{Acceleration vector}$$

When $t = 0$, the velocity and acceleration vectors are given by

$$\mathbf{v}(0) = 2(0)\mathbf{i} + \mathbf{j} = \mathbf{j} \quad \text{and} \quad \mathbf{a}(0) = 2\mathbf{i}.$$

When $t = 2$, the velocity and acceleration vectors are given by

$$\mathbf{v}(2) = 2(2)\mathbf{i} + \mathbf{j} = 4\mathbf{i} + \mathbf{j}^{\,\cdot} \quad \text{and} \quad \mathbf{a}(2) = 2\mathbf{i}.$$

For the object moving along the path shown in Figure 12.11, note that the acceleration vector is constant (it has a magnitude of 2 and points to the right). This implies that the speed of the object is decreasing as the object moves toward the vertex of the parabola, and the speed is increasing as the object moves away from the vertex of the parabola.

This type of motion is *not* characteristic of comets that travel on parabolic paths through our solar system. For such comets, the acceleration vector always points to the origin (the sun), which implies that the comet's speed increases as it approaches the vertex of the path and decreases as it moves away from the vertex. (See Figure 12.12.)

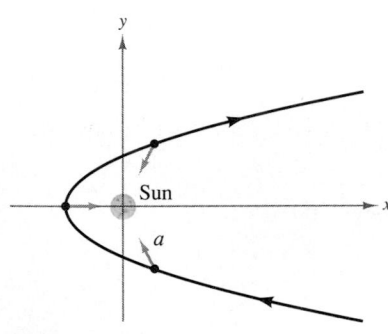

FIGURE 12.12

At each point in the comet's orbit, the acceleration vector points toward the sun.

EXAMPLE 3 Sketching Velocity and Acceleration Vectors in Space

Sketch the path of an object moving along the space curve C given by

$$\mathbf{r}(t) = t\,\mathbf{i} + t^3\mathbf{j} + 3t\,\mathbf{k}, \qquad t \geq 0 \qquad \text{Position vector}$$

and find the velocity and acceleration vectors when $t = 1$.

Solution Using the parametric equations $x = t$ and $y = t^3$, you can determine that the path of the object lies on the cubic cylinder given by $y = x^3$. Moreover, because $z = 3t$, the object starts at $(0, 0, 0)$ and moves upward as t increases, as shown in Figure 12.13. Because $\mathbf{r}(t) = t\,\mathbf{i} + t^3\mathbf{j} + 3t\,\mathbf{k}$, you have

$$\mathbf{v}(t) = \mathbf{r}'(t) = \mathbf{i} + 3t^2\mathbf{j} + 3\mathbf{k} \qquad \text{Velocity vector}$$

and

$$\mathbf{a}(t) = \mathbf{r}''(t) = 6t\,\mathbf{j}. \qquad \text{Acceleration vector}$$

When $t = 1$, the velocity and acceleration vectors are given by

$$\mathbf{v}(1) = \mathbf{r}'(1) = \mathbf{i} + 3\mathbf{j} + 3\mathbf{k} \quad \text{and} \quad \mathbf{a}(1) = \mathbf{r}''(1) = 6\mathbf{j}.$$

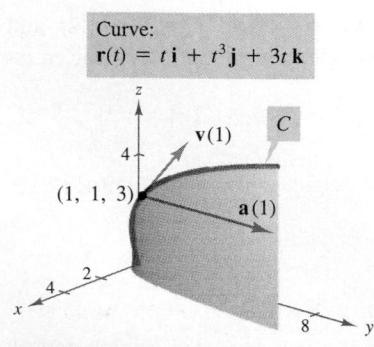

FIGURE 12.13

So far in this section, we have concentrated on finding the velocity and accelera-tion by differentiating the position function. Many practical applications involve the reverse problem—finding the position function for a given velocity or acceleration. This is demonstrated in the next example.

EXAMPLE 4   Finding a Position Function by Integration

An object starts from rest at the point $P = (1, 2, 0)$ and moves with an acceleration of

$$\mathbf{a}(t) = \mathbf{j} + 2\mathbf{k} \qquad \text{Acceleration vector}$$

where $\| \mathbf{a}(t) \|$ is measured in feet per second per second. Find the location of the object after $t = 2$ seconds.

Solution   From the description of the object's motion, you can deduce the following *initial conditions*. Because the object starts from rest, you have

$$\mathbf{v}(0) = \mathbf{0}.$$

Moreover, because the object starts at the point $(x, y, z) = (1, 2, 0)$, you have

$$\mathbf{r}(0) = x(0)\mathbf{i} + y(0)\mathbf{j} + z(0)\mathbf{k} = 1\mathbf{i} + 2\mathbf{j} + 0\mathbf{k} = \mathbf{i} + 2\mathbf{j}.$$

To find the position function, you should integrate twice, each time using one of the initial conditions to solve for the constant of integration. The velocity vector is

$$\mathbf{v}(t) = \int \mathbf{a}(t)\, dt = \int (\mathbf{j} + 2\mathbf{k})\, dt = t\mathbf{j} + 2t\mathbf{k} + \mathbf{C}$$

where $\mathbf{C} = C_1\mathbf{i} + C_2\mathbf{j} + C_3\mathbf{k}$. Letting $t = 0$ and applying the initial condition $\mathbf{v}(0) = \mathbf{0}$, you obtain

$$\mathbf{v}(0) = C_1\mathbf{i} + C_2\mathbf{j} + C_3\mathbf{k} = \mathbf{0} \quad \rightarrow \quad C_1 = C_2 = C_3 = 0.$$

Thus, the *velocity* at any time t is

$$\mathbf{v}(t) = t\mathbf{j} + 2t\mathbf{k}. \qquad \text{Velocity vector}$$

Integrating once more produces

$$\mathbf{r}(t) = \int \mathbf{v}(t)\, dt = \int (t\mathbf{j} + 2t\mathbf{k})\, dt = \frac{t^2}{2}\mathbf{j} + t^2\mathbf{k} + \mathbf{C}$$

where $\mathbf{C} = C_4\mathbf{i} + C_5\mathbf{j} + C_6\mathbf{k}$. Letting $t = 0$ and applying the initial condition $\mathbf{r}(0) = \mathbf{i} + 2\mathbf{j}$, you have

$$\mathbf{r}(0) = C_4\mathbf{i} + C_5\mathbf{j} + C_6\mathbf{k} = \mathbf{i} + 2\mathbf{j} \quad \rightarrow \quad C_4 = 1,\ C_5 = 2,\ C_6 = 0.$$

Thus, the *position* vector is

$$\mathbf{r}(t) = \mathbf{i} + \left(\frac{t^2}{2} + 2\right)\mathbf{j} + t^2\mathbf{k}. \qquad \text{Position vector}$$

The location of the object after 2 seconds is given by $\mathbf{r}(2) = \mathbf{i} + 4\mathbf{j} + 4\mathbf{k}$, as shown in Figure 12.14.

Curve:

$$\mathbf{r}(t) = \mathbf{i} + \left(\frac{t^2}{2} + 2\right)\mathbf{j} + t^2\mathbf{k}$$

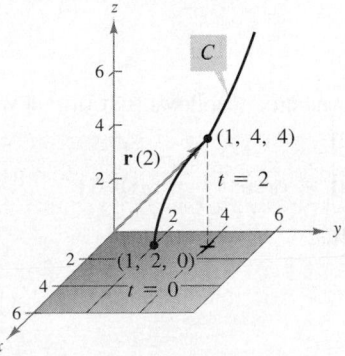

FIGURE 12.14

The object takes two seconds to move from the point $(1, 2, 0)$ to the point $(1, 4, 4)$.

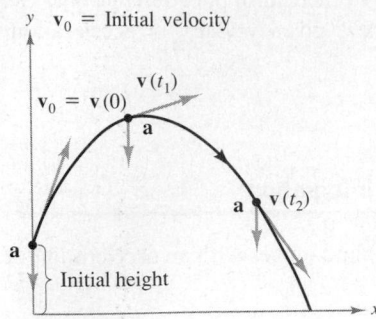

FIGURE 12.15
The parabolic path of a projectile.

Projectile Motion

We now have the machinery to derive the parametric equations for the path of a projectile. We assume that gravity is the only force acting on the projectile after it is launched. Hence, the motion occurs in a vertical plane, which we represent by the *xy*-coordinate system with the origin as a point on the earth's surface, as shown in Figure 12.15. For a projectile of mass *m*, the force due to gravity is

$$\mathbf{F} = -mg\mathbf{j} \qquad \text{Force due to gravity}$$

where the gravitational constant $g = 32$ feet per second per second, or 9.81 meters per second per second. By **Newton's Second Law of Motion,** this same force produces an acceleration $\mathbf{a} = \mathbf{a}(t)$, and satisfies the equation $\mathbf{F} = m\mathbf{a}$. Consequently, the acceleration of the projectile is given by $m\mathbf{a} = -mg\mathbf{j}$, which implies that

$$\mathbf{a} = -g\mathbf{j}. \qquad \text{Acceleration of projectile}$$

EXAMPLE 5   Derivation of the Position Function for a Projectile

A projectile of mass *m* is launched from an initial position $\mathbf{r}_0$ with an initial velocity $\mathbf{v}_0$. Find its position vector as a function of time.

Solution   Begin with the acceleration $\mathbf{a}(t) = -g\mathbf{j}$ and integrate twice.

$$\mathbf{v}(t) = \int \mathbf{a}(t)\,dt = \int -g\mathbf{j}\,dt = -gt\mathbf{j} + \mathbf{C}_1$$

$$\mathbf{r}(t) = \int \mathbf{v}(t)\,dt = \int (-gt\mathbf{j} + \mathbf{C}_1)\,dt = -\frac{1}{2}gt^2\mathbf{j} + \mathbf{C}_1 t + \mathbf{C}_2$$

You can use the facts that $\mathbf{v}(0) = \mathbf{v}_0$ and $\mathbf{r}(0) = \mathbf{r}_0$ to solve for the constant vectors $\mathbf{C}_1$ and $\mathbf{C}_2$. Doing this produces $\mathbf{C}_1 = \mathbf{v}_0$ and $\mathbf{C}_2 = \mathbf{r}_0$. Therefore, the position vector is

$$\mathbf{r}(t) = -\frac{1}{2}gt^2\mathbf{j} + t\mathbf{v}_0 + \mathbf{r}_0. \qquad \text{Position vector} \qquad \blacksquare$$

In many projectile problems, the constant vectors $\mathbf{r}_0$ and $\mathbf{v}_0$ are not given explicitly. Often you are given the initial height *h*, the initial speed v_0, and the angle θ at which the projectile is launched, as shown in Figure 12.16. From the given height, you can deduce that

$$\mathbf{r}_0 = h\mathbf{j}.$$

Because the speed gives the magnitude of the initial velocity, it follows that $v_0 = \|\mathbf{v}_0\|$ and you can write

$$\mathbf{v}_0 = x\mathbf{i} + y\mathbf{j} = (\|\mathbf{v}_0\|\cos\theta)\mathbf{i} + (\|\mathbf{v}_0\|\sin\theta)\mathbf{j} = v_0\cos\theta\mathbf{i} + v_0\sin\theta\mathbf{j}.$$

Thus, the position vector can be written in the form

$$\boxed{\mathbf{r}(t) = -\frac{1}{2}gt^2\mathbf{j} + t\mathbf{v}_0 + \mathbf{r}_0} \qquad \text{Position vector}$$

$$= -\frac{1}{2}gt^2\mathbf{j} + tv_0\cos\theta\mathbf{i} + tv_0\sin\theta\mathbf{j} + h\mathbf{j}$$

$$= (v_0\cos\theta)t\mathbf{i} + \left[h + (v_0\sin\theta)t - \frac{1}{2}gt^2\right]\mathbf{j}.$$

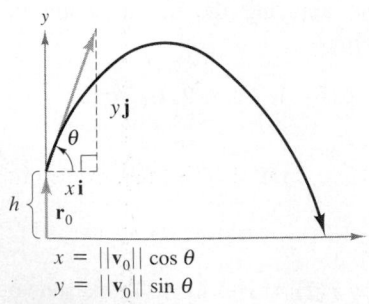

FIGURE 12.16
The initial conditions for a projectile.

> **THEOREM 12.3   Position Function for a Projectile**
>
> Neglecting air resistance, the path of a projectile launched from an initial height h with initial speed v_0 and angle of elevation θ is described by the vector function
>
> $$\mathbf{r}(t) = (v_0 \cos \theta)t\,\mathbf{i} + \left[h + (v_0 \sin \theta)t - \frac{1}{2}gt^2 \right]\mathbf{j}$$
>
> where g is the gravitational constant.

EXAMPLE 6   Describing the Path of a Baseball

A baseball is hit 3 feet above ground at 100 feet per second and at an angle of $\pi/4$ with respect to the ground, as shown in Figure 12.17. Find the maximum height reached by the baseball. Will it clear a 10-foot-high fence located 300 feet from home plate?

Solution   You are given $h = 3$, $v_0 = 100$, and $\theta = \pi/4$. Thus, using $g = 32$ feet per second per second produces

$$\mathbf{r}(t) = \left(100 \cos \frac{\pi}{4} \right)t\,\mathbf{i} + \left[3 + \left(100 \sin \frac{\pi}{4} \right)t - 16t^2 \right]\mathbf{j}$$

$$= \left(50\sqrt{2}\,t \right)\mathbf{i} + \left(3 + 50\sqrt{2}\,t - 16t^2 \right)\mathbf{j}$$

$$\mathbf{v}(t) = \mathbf{r}'(t) = 50\sqrt{2}\,\mathbf{i} + \left(50\sqrt{2} - 32t \right)\mathbf{j}.$$

The maximum height occurs when $y'(t) = 50\sqrt{2} - 32t = 0$, which implies that

$$t = \frac{25\sqrt{2}}{16} \approx 2.21 \text{ sec.}$$

Hence, the maximum height reached by the ball is

$$y = 3 + 50\sqrt{2}\left(\frac{25\sqrt{2}}{16} \right) - 16\left(\frac{25\sqrt{2}}{16} \right)^2$$

$$= \frac{649}{8}$$

$$\approx 81 \text{ ft.} \qquad \qquad \text{Maximum height when } t \approx 2.21$$

The ball is 300 feet from where it was hit when

$$300 = x(t) = 50\sqrt{2}\,t.$$

Solving this equation for t produces $t = 3\sqrt{2} \approx 4.24$ seconds. At this time, the height of the ball is

$$y = 3 + 50\sqrt{2}(3\sqrt{2}) - 16(3\sqrt{2})^2$$

$$= 303 - 288$$

$$= 15 \text{ ft.} \qquad \qquad \text{Height when } t \approx 4.24$$

Therefore, the ball clears the 10-foot fence for a home run.

FIGURE 12.17
The batter hits the ball with an initial speed of 100 feet per second.

EXERCISES for Section 12.3

In Exercises 1–8, the position function r describes the path of an object moving in the xy-plane. Sketch a graph of the path and sketch the velocity and acceleration vectors at the given point.

Position function	Point
1. $\mathbf{r}(t) = 3t\,\mathbf{i} + (t-1)\mathbf{j}$	$(3, 0)$
2. $\mathbf{r}(t) = (6-t)\mathbf{i} + t\,\mathbf{j}$	$(3, 3)$
3. $\mathbf{r}(t) = t^2\mathbf{i} + t\,\mathbf{j}$	$(4, 2)$
4. $\mathbf{r}(t) = t^3\mathbf{i} + t^2\mathbf{j}$	$(1, 1)$
5. $\mathbf{r}(t) = 2\cos t\,\mathbf{i} + 2\sin t\,\mathbf{j}$	$\left(\sqrt{2}, \sqrt{2}\right)$
6. $\mathbf{r}(t) = 2\cos t\,\mathbf{i} + 3\sin t\,\mathbf{j}$	$(2, 0)$
7. $\mathbf{r}(t) = \langle t - \sin t, 1 - \cos t \rangle$	$(\pi, 2)$
8. $\mathbf{r}(t) = \langle e^{-t}, e^{t} \rangle$	$(1, 1)$

In Exercises 9–16, the position function r describes the path of an object moving in space. Find the velocity, speed, and acceleration of the object.

9. $\mathbf{r}(t) = t\,\mathbf{i} + (2t - 5)\mathbf{j} + 3t\,\mathbf{k}$

10. $\mathbf{r}(t) = 4t\,\mathbf{i} + 4t\,\mathbf{j} + 2t\,\mathbf{k}$

11. $\mathbf{r}(t) = t\,\mathbf{i} + t^2\mathbf{j} + \dfrac{t^2}{2}\,\mathbf{k}$

12. $\mathbf{r}(t) = t\,\mathbf{i} + 3t\,\mathbf{j} + \dfrac{t^2}{2}\,\mathbf{k}$

13. $\mathbf{r}(t) = t\,\mathbf{i} + t\,\mathbf{j} + \sqrt{9 - t^2}\,\mathbf{k}$

14. $\mathbf{r}(t) = t^2\mathbf{i} + t\,\mathbf{j} + 2t^{3/2}\,\mathbf{k}$

15. $\mathbf{r}(t) = \langle 4t, 3\cos t, 3\sin t \rangle$

16. $\mathbf{r}(t) = \langle e^{t}\cos t, e^{t}\sin t, e^{t} \rangle$

Linear Approximation In Exercises 17 and 18, the graph of the vector-valued function r(t) and a tangent vector to the graph at $t = t_0$ are given. (a) Find a set of parametric equations for the tangent line to the graph at $t = t_0$ and (b) use the equations for the line to approximate $\mathbf{r}(t_0 + 0.1)$.

17. $\mathbf{r}(t) = \left\langle t, -t^2, \dfrac{t^3}{4} \right\rangle$, $t_0 = 1$

$\left(1, -1, \tfrac{1}{4}\right)$

18. $\mathbf{r}(t) = \left\langle t, \sqrt{25 - t^2}, \sqrt{25 - t^2} \right\rangle$, $t_0 = 3$

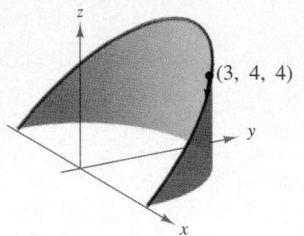

$(3, 4, 4)$

In Exercises 19–22, use the given acceleration function to find the velocity and position functions. Then find the position at time $t = 2$.

19. $\mathbf{a}(t) = \mathbf{i} + \mathbf{j} + \mathbf{k}$
$\mathbf{v}(0) = \mathbf{0}$
$\mathbf{r}(0) = \mathbf{0}$

20. $\mathbf{a}(t) = \mathbf{i} + \mathbf{k}$
$\mathbf{v}(0) = 5\mathbf{j}$
$\mathbf{r}(0) = \mathbf{0}$

21. $\mathbf{a}(t) = t\,\mathbf{j} + t\,\mathbf{k}$
$\mathbf{v}(1) = 5\mathbf{j}$
$\mathbf{r}(1) = \mathbf{0}$

22. $\mathbf{a}(t) = -\cos t\,\mathbf{i} - \sin t\,\mathbf{j}$
$\mathbf{v}(0) = \mathbf{j} + \mathbf{k}$
$\mathbf{r}(0) = \mathbf{i}$

Projectile Motion In Exercises 23–36, use the model for projectile motion and assume there is no air resistance.

23. Find the vector-valued function for the path of a projectile launched at a height of 10 feet above the ground with an initial velocity of 88 feet per second and at an angle of 30° above the horizontal. Use a graphing utility to sketch the path of the projectile.

24. Determine the maximum height and the range of a projectile fired at a height of 3 feet above the ground with an initial velocity of 900 feet per second and at an angle of 45° above the horizontal.

25. A baseball, hit 3 feet above the ground, leaves the bat at an angle of 45° and is caught by an outfielder 300 feet from home plate. What was the initial speed of the ball, and how high did it rise if it was caught 3 feet above the ground?

26. A baseball player at second base throws the ball 90 feet to the player at first base. The ball is thrown at 50 miles per hour at an angle of 15° above the horizontal. At what height does the first baseman catch the ball if the ball is thrown from a height of 5 feet?

27. The quarterback of a football team releases a pass at a height of 7 feet above the playing field, and the football is caught by a receiver 30 yards directly downfield at a height of 4 feet. The pass is released at an angle of 35° with the horizontal.
 a. Find the speed of the football when it is released.
 b. Find the maximum height of the ball.
 c. Find the time the receiver has to position himself after the quarterback releases the ball.

28. The center-field fence in a ballpark is 10 feet high and 400 feet from home plate. The ball is hit 3 feet above the ground and leaves the bat at a speed of 100 miles per hour. What is the minimum angle if the hit is a home run?

29. A bale ejector consists of two variable-speed belts at the end of a baler. Its purpose is to toss bales into a trailing wagon. In loading the back of the wagon, a bale must be thrown to a position 8 feet above and 16 feet behind the ejector. Find the minimum initial speed of the bale and the corresponding angle at which it must be ejected from the baler.

30. A bomber is flying at an altitude of 30,000 feet with a speed of 540 miles per hour (792 feet per second) (see figure). When should the bomb be released in order to hit the target? (Give your answer in terms of the angle of depression from the plane to the target.) What is the speed of the bomb at the time of impact?

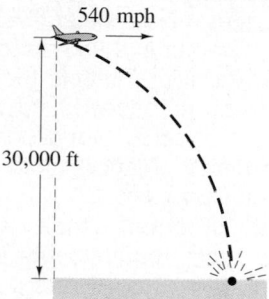

540 mph

30,000 ft

FIGURE FOR 30

31. Find the angle at which an object must be thrown to obtain
a. the maximum range.
b. the maximum height.

32. A shot fired from a gun with a muzzle velocity of 1200 feet per second is to hit a target 3000 feet away. Determine the minimum angle of elevation of the gun.

33. A projectile is fired from ground level at an angle of 10° with the horizontal. Find the minimum initial velocity necessary if the projectile is to have a range of 100 feet.

C **34.** Use a graphing utility to graph the paths of the projectile for the specified values of θ and v_0. For each case, use the graph to approximate the maximum height and the range of the projectile. (Assume the projectile is launched from ground level.)
a. $\theta = 10°$, $v_0 = 66$ ft/sec
b. $\theta = 10°$, $v_0 = 146$ ft/sec
c. $\theta = 45°$, $v_0 = 66$ ft/sec
d. $\theta = 45°$, $v_0 = 146$ ft/sec
e. $\theta = 60°$, $v_0 = 66$ ft/sec
f. $\theta = 60°$, $v_0 = 146$ ft/sec

35. Eliminate the parameter t from the position function for the motion of a projectile to show that the rectangular equation is

$$y = -\frac{16 \sec^2 \theta}{v_0^2} x^2 + (\tan \theta)x + h.$$

36. The path of a ball is given by the rectangular equation

$$y = x - 0.005x^2.$$

Use the result of Exercise 35 to find the position function. Then find the speed and direction of the ball at the point when it has traveled 60 feet horizontally.

Cycloidal Motion **In Exercises 37 and 38, consider the motion of a point (or particle) on the circumference of a rolling circle. As the circle rolls, it generates the cycloid**

$$\mathbf{r}(t) = b(\omega t - \sin \omega t)\mathbf{i} + b(1 - \cos \omega t)\mathbf{j}$$

where ω is the constant angular velocity of the circle.

37. Find the velocity and acceleration vectors of the particle. Use the results to determine the times that the speed of the particle will be
a. 0.   **b.** maximum.

38. Find the maximum speed of a point on the circumference of an automobile wheel of radius 1 foot when the automobile is traveling 55 miles per hour. Compare this speed with the speed of the automobile.

Circular Motion **In Exercises 39–42, consider a particle moving on a circular path of radius b described by**

$$\mathbf{r}(t) = b \cos \omega t\, \mathbf{i} + b \sin \omega t\, \mathbf{j}$$

where $\omega = d\theta/dt$ is the constant angular velocity.

39. Find the velocity vector and show that it is orthogonal to $\mathbf{r}(t)$.

40. Show that the speed of the particle is $b\omega$.

41. Find the acceleration vector and show that its direction is always toward the center of the circle.

42. Show that the magnitude of the acceleration vector is $\omega^2 b$.

Circular Motion **In Exercises 43 and 44, use the results of Exercises 39–42.**

43. A stone weighing 1 pound is attached to a 2-foot string and is whirled horizontally (see figure). The string will break under a force of 10 pounds. Find the maximum speed the stone can attain without breaking the string. (Use $\mathbf{F} = m\mathbf{a}$, where $m = \frac{1}{32}$.)

44. A 3000-pound automobile is negotiating a circular interchange of radius 300 feet at 30 miles per hour (see figure). Assuming the roadway to be level, find the force between the tires and the road so that the car stays on the circular path and does not skid. (Use $\mathbf{F} = m\mathbf{a}$, where $m = 3000/32$.) Find the angle at which the roadway should be banked so that no lateral frictional force is exerted on the tires of the automobile.

1 lb

2 ft

30 mph

300 ft

FIGURE FOR 43   **FIGURE FOR 44**

45. *Throwing a Shot-Put* The path of a shot-put thrown at an angle θ is given by

$$\mathbf{r}(t) = (v_0 \cos \theta)t\,\mathbf{i} + [h + (v_0 \sin \theta)t - \tfrac{1}{2}gt^2]\mathbf{j}$$

where v_0 is the initial speed, h is the initial height, t is time in seconds, and g is the acceleration due to gravity. Verify that the shot-put will remain in the air for a total of

$$t = \frac{v_0 \sin \theta + \sqrt{v_0^2 \sin^2 \theta + 2gh}}{g} \text{ sec}$$

and will travel a horizontal distance of

$$\frac{v_0^2 \cos \theta}{g}\left(\sin \theta + \sqrt{\sin^2 \theta + \frac{2gh}{v_0^2}} \right) \text{ ft.}$$

46. *Throwing a Shot-Put* A shot-up is thrown from a height of $h = 6$ feet with an initial speed of $v_0 = 45$ feet per second. Find the total time of travel and the total horizontal distance traveled if the shot-put was thrown at an angle of $\theta = 42.5°$ with the horizontal.

47. Prove that if an object is traveling at a constant speed, its velocity and acceleration vectors are orthogonal.

48. Prove that an object moving in a straight line at a constant speed has an acceleration of 0.

49. *Essay* Consider a particle moving on the path $\mathbf{r}_1(t) = x(t)\mathbf{i} + y(t)\mathbf{j} + z(t)\mathbf{k}$. Discuss any changes in the position, velocity, or acceleration of the particle if its position is given by the vector-valued function $\mathbf{r}_2(t) = \mathbf{r}_1(2t)$. Generalize the results for the position function $\mathbf{r}_3(t) = \mathbf{r}_1(\omega t)$.

CAREER INTERVIEW

Sarah L. Smith, Ph.D.
Sport Science and
Technology Division
U.S. Olympic Committee
Colorado Springs, CO 80909

As a sports biomechanist, I help elite athletes and their coaches improve technique through the use of video and the principles of physics and human anatomy. To achieve greater distance in the shot put, for example, an athlete's performances are captured on film, and then analyzed to yield shot velocity, angle of release, and release height values. These three factors determine the distance that the shot will travel. Data from multiple trials indicates where adjustments are needed and is compared with the statistics of other world-class performers as well as with mathematical models that suggest optimal technique values.

During my undergraduate program, I took calculus courses to fulfill my math requirements as a physical education major. However, it was not until I was enrolled in a doctoral program that I fully realized how fundamental calculus is to the understanding and application of biomechanics concepts.

Today, in all biomechanical analyses, it is necessary to deal with such measures as linear and angular displacement, velocity, acceleration, force, momentum, and impulse. Whether assisting a decathlete to put the shot farther, a figure skater to complete a triple axel, or a luge athlete to have an excellent start technique, math knowledge and application is a necessary tool.

Tangent Vectors and Normal Vectors

Tangent Vectors and Normal Vectors • Tangential and Normal Components of Acceleration

Tangent Vectors and Normal Vectors

In the previous section, you learned that the velocity vector points in the direction of motion. This observation leads to the following definition, which applies to any smooth curve—not just to those for which the parameter represents time.

Definition of Unit Tangent Vector

Let C be a smooth curve represented by $\mathbf{r}$ on an open interval I. The **unit tangent vector** $\mathbf{T}(t)$ at t is defined to be

$$\mathbf{T}(t) = \frac{\mathbf{r}'(t)}{\|\mathbf{r}'(t)\|}, \qquad \mathbf{r}'(t) \neq \mathbf{0}.$$

REMARK   Recall that a curve is *smooth* on an interval if $\mathbf{r}'$ is continuous and nonzero on the interval. Thus, "smoothness" is sufficient to guarantee that a curve has a unit tangent vector.

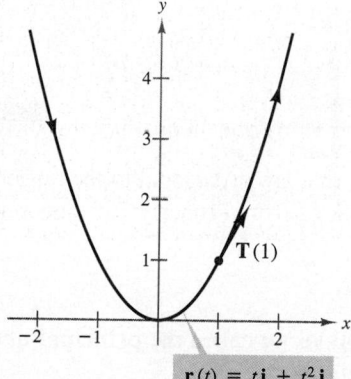

FIGURE 12.18
The direction of the unit tangent vector depends on the orientation of the curve.

EXAMPLE 1   Finding the Unit Tangent Vector

Find the unit tangent vector to the curve given by

$$\mathbf{r}(t) = t\mathbf{i} + t^2\mathbf{j}$$

when $t = 1$.

Solution   The derivative of $\mathbf{r}(t)$ is $\mathbf{r}'(t) = \mathbf{i} + 2t\mathbf{j}$. Thus, the unit tangent vector is

$$\mathbf{T}(t) = \frac{\mathbf{r}'(t)}{\|\mathbf{r}'(t)\|} = \frac{1}{\sqrt{1 + 4t^2}}(\mathbf{i} + 2t\mathbf{j}).$$

When $t = 1$, the unit tangent vector is

$$\mathbf{T}(1) = \frac{1}{\sqrt{5}}(\mathbf{i} + 2\mathbf{j})$$

as shown in Figure 12.18.

REMARK   In Example 1, note that the direction of the unit tangent vector depends on the orientation of the curve. For instance, if the parabola in Figure 12.18 were given by $\mathbf{r}(t) = -(t - 2)\mathbf{i} + (t - 2)^2\mathbf{j}$, then $\mathbf{T}(1)$ would still represent the unit tangent vector at the point $(1, 1)$, but it would point in the opposite direction. (Try verifying this.)

The **tangent line to a curve** at a point is the line passing through the point and parallel to the unit tangent vector. In Example 2, the unit tangent vector is used to find the tangent line at a point on a helix.

EXAMPLE 2   Finding the Tangent Line at a Point on a Curve

Find $\mathbf{T}(t)$ and then find a set of parametric equations for the tangent line to the helix given by

$$\mathbf{r}(t) = 2\cos t\,\mathbf{i} + 2\sin t\,\mathbf{j} + t\,\mathbf{k}$$

at the point corresponding to $t = \pi/4$.

Solution   The derivative of $\mathbf{r}(t)$ is $\mathbf{r}'(t) = -2\sin t\,\mathbf{i} + 2\cos t\,\mathbf{j} + \mathbf{k}$, which implies that $\|\mathbf{r}'(t)\| = \sqrt{4\sin^2 t + 4\cos^2 t + 1} = \sqrt{5}$. Therefore, the unit tangent vector is

$$\mathbf{T}(t) = \frac{\mathbf{r}'(t)}{\|\mathbf{r}'(t)\|} = \frac{1}{\sqrt{5}}(-2\sin t\,\mathbf{i} + 2\cos t\,\mathbf{j} + \mathbf{k}).$$

When $t = \pi/4$, the unit tangent vector is

$$\mathbf{T}\!\left(\frac{\pi}{4}\right) = \frac{1}{\sqrt{5}}\left(-2\frac{\sqrt{2}}{2}\mathbf{i} + 2\frac{\sqrt{2}}{2}\mathbf{j} + \mathbf{k}\right)$$

$$= \frac{1}{\sqrt{5}}(-\sqrt{2}\mathbf{i} + \sqrt{2}\mathbf{j} + \mathbf{k}).$$

Using the direction numbers $a = -\sqrt{2}$, $b = \sqrt{2}$, and $c = 1$, and the point $(x_1, y_1, z_1) = (\sqrt{2},\ \sqrt{2},\ \pi/4)$, you can obtain the following parametric equations (given with parameter s).

$$x = x_1 + as = \sqrt{2} - \sqrt{2}s$$
$$y = y_1 + bs = \sqrt{2} + \sqrt{2}s$$
$$z = z_1 + cs = \frac{\pi}{4} + s$$

This tangent line is shown in Figure 12.19.

Curve:
$\mathbf{r}(t) = 2\cos t\,\mathbf{i} + 2\sin t\,\mathbf{j} + t\,\mathbf{k}$

$\left(\sqrt{2},\ \sqrt{2},\ \dfrac{\pi}{4}\right)$

FIGURE 12.19
The tangent line at a point on a curve is determined by the unit tangent at the point.

In Example 2, there are infinitely many vectors that are orthogonal to the tangent vector $\mathbf{T}(t)$. One of these is the vector $\mathbf{T}'(t)$. This follows from Property 7 of Theorem 12.2. That is,

$$\mathbf{T}(t)\cdot\mathbf{T}(t) = \|\mathbf{T}(t)\|^2 = 1 \quad \longrightarrow \quad \mathbf{T}(t)\cdot\mathbf{T}'(t) = 0.$$

By normalizing the vector $\mathbf{T}'(t)$, you obtain a special vector called the **principal unit normal vector,** as indicated in the following definition.

Definition of Principal Unit Normal Vector

Let C be a smooth curve represented by $\mathbf{r}$ on an open interval I. If $\mathbf{T}'(t) \neq \mathbf{0}$, then the **principal unit normal vector** at t is defined to be

$$\mathbf{N}(t) = \frac{\mathbf{T}'(t)}{\|\mathbf{T}'(t)\|}.$$

EXAMPLE 3   Finding the Principal Unit Normal Vector

Find $\mathbf{N}(t)$ and $\mathbf{N}(1)$ for the curve represented by

$$\mathbf{r}(t) = 3t\,\mathbf{i} + 2t^2\,\mathbf{j}.$$

Solution   By differentiating, you obtain

$$\mathbf{r}'(t) = 3\mathbf{i} + 4t\,\mathbf{j} \quad \text{and} \quad \|\mathbf{r}'(t)\| = \sqrt{9 + 16t^2}$$

which implies that the unit tangent vector is

$$\mathbf{T}(t) = \frac{\mathbf{r}'(t)}{\|\mathbf{r}'(t)\|}$$

$$= \frac{1}{\sqrt{9 + 16t^2}}(3\mathbf{i} + 4t\,\mathbf{j}). \qquad \text{Unit tangent vector}$$

Using Theorem 12.2, differentiate $\mathbf{T}(t)$ with respect to t to obtain

$$\mathbf{T}'(t) = \frac{1}{\sqrt{9 + 16t^2}}(4\mathbf{j}) - \frac{16t}{(9 + 16t^2)^{3/2}}(3\mathbf{i} + 4t\,\mathbf{j})$$

$$= \frac{12}{(9 + 16t^2)^{3/2}}(-4t\,\mathbf{i} + 3\mathbf{j})$$

$$\|\mathbf{T}'(t)\| = 12\sqrt{\frac{9 + 16t^2}{(9 + 16t^2)^3}} = \frac{12}{9 + 16t^2}.$$

Therefore, the principal unit normal vector is

$$\mathbf{N}(t) = \frac{\mathbf{T}'(t)}{\|\mathbf{T}'(t)\|}$$

$$= \frac{1}{\sqrt{9 + 16t^2}}(-4t\,\mathbf{i} + 3\mathbf{j}). \qquad \text{Principal unit normal vector}$$

When $t = 1$, the principal unit normal vector is

$$\mathbf{N}(1) = \frac{1}{5}(-4\mathbf{i} + 3\mathbf{j})$$

as shown in Figure 12.20.

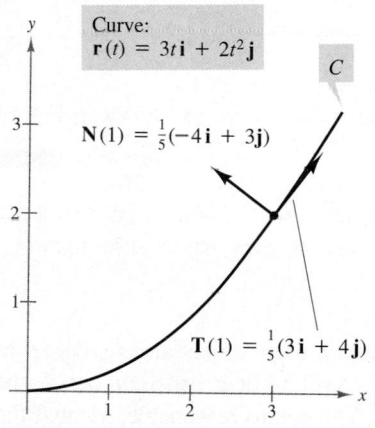

FIGURE 12.20

Principal unit normal vector points toward the concave side of the curve.

The principal unit normal vector can be difficult to evaluate algebraically. For plane curves, you can simplify the algebra by finding

$$\mathbf{T}(t) = x(t)\mathbf{i} + y(t)\mathbf{j} \qquad \text{Unit tangent vector}$$

and observing that $\mathbf{N}(t)$ must be either

$$\mathbf{N}_1(t) = y(t)\mathbf{i} - x(t)\mathbf{j} \qquad \text{or} \qquad \mathbf{N}_2(t) = -y(t)\mathbf{i} + x(t)\mathbf{j}.$$

FIGURE 12.21

At any point on a curve, a unit normal vector is orthogonal to the unit tangent vector. The *principal* unit normal vector points in the direction the curve is turning.

Because $\sqrt{[x(t)]^2 + [y(t)]^2} = 1$, it follows that both $\mathbf{N}_1(t)$ and $\mathbf{N}_2(t)$ are unit normal vectors. The *principal* unit normal vector $\mathbf{N}$ is the one that points toward the concave side of the curve, as indicated in Figure 12.20 (see Exercise 43). This also holds for curves in space. That is, for an object moving along a curve C in space, the vector $\mathbf{T}(t)$ points in the direction the object is moving, whereas the vector $\mathbf{N}(t)$ is orthogonal to $\mathbf{T}(t)$ and points in the direction the object is turning, as shown in Figure 12.21.

EXAMPLE 4   Finding the Principal Unit Normal Vector

Find the principal unit normal vector for the helix given by

$$\mathbf{r}(t) = 2\cos t\,\mathbf{i} + 2\sin t\,\mathbf{j} + t\,\mathbf{k}.$$

Solution   From Example 2, you know that the principal unit tangent vector is

$$\mathbf{T}(t) = \frac{1}{\sqrt{5}}(-2\sin t\,\mathbf{i} + 2\cos t\,\mathbf{j} + \mathbf{k}).$$

Thus, $\mathbf{T}'(t)$ is given by

$$\mathbf{T}'(t) = \frac{1}{\sqrt{5}}(-2\cos t\,\mathbf{i} - 2\sin t\,\mathbf{j}).$$

Because $\|\mathbf{T}'(t)\| = 2/\sqrt{5}$, it follows that the principal unit normal vector is

$$\mathbf{N}(t) = \frac{\mathbf{T}'(t)}{\|\mathbf{T}'(t)\|} = \frac{1}{2}(-2\cos t\,\mathbf{i} - 2\sin t\,\mathbf{j}) = -\cos t\,\mathbf{i} - \sin t\,\mathbf{j}.$$

Note that this vector is horizontal and points toward the z-axis, as shown in Figure 12.22.

Helix:
$\mathbf{r}(t) = 2\cos t\,\mathbf{i} + 2\sin t\,\mathbf{j} + t\,\mathbf{k}$

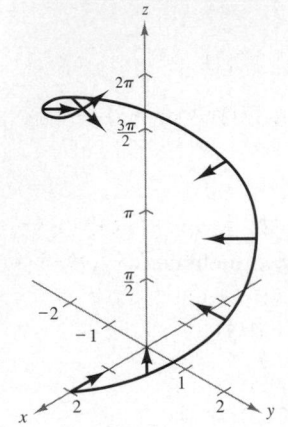

FIGURE 12.22
$\mathbf{N}(t)$ is normal and points toward the z-axis.

Tangential and Normal Components of Acceleration

We now return to the problem of describing the motion of an object along a curve. In the previous section, you saw that for an object traveling at a *constant speed,* the velocity and acceleration vectors are perpendicular. This seems reasonable, because the speed would not be constant if any acceleration were acting in the direction of motion. [You can verify this observation by noting that $\mathbf{r}''(t)\cdot\mathbf{r}'(t) = 0$ if $\|\mathbf{r}'(t)\|$ is a constant. See Property 7 of Theorem 12.2.]

However, for an object traveling at a *variable speed,* the velocity and acceleration vectors are not necessarily perpendicular. For instance, you saw that the acceleration vector for a projectile always points down, regardless of the direction of motion.

In general, part of the acceleration (the tangential component) acts in the line of motion, and part (the normal component) acts perpendicular to the line of motion. In order to determine these two components, you can use the unit vectors $\mathbf{T}(t)$ and $\mathbf{N}(t)$, which serve in much the same way as do $\mathbf{i}$ and $\mathbf{j}$ in representing vectors in the plane. The following theorem states that the acceleration vector lies in the plane determined by $\mathbf{T}(t)$ and $\mathbf{N}(t)$.

THEOREM 12.4   Acceleration Vector

If $\mathbf{r}(t)$ is the position vector for a smooth curve C and $\mathbf{N}(t)$ exists, then the acceleration vector $\mathbf{a}(t)$ lies in the plane determined by $\mathbf{T}(t)$ and $\mathbf{N}(t)$.

Proof To simplify the notation, we write $\mathbf{T}$ for $\mathbf{T}(t)$, $\mathbf{T}'$ for $\mathbf{T}'(t)$, and so on. Because $\mathbf{T} = \mathbf{r}'/\|\mathbf{r}'\| = \mathbf{v}/\|\mathbf{v}\|$, it follows that

$$\mathbf{v} = \|\mathbf{v}\|\, \mathbf{T}.$$

By differentiating, you obtain

$$\mathbf{a} = \mathbf{v}' = D_t[\|\mathbf{v}\|]\mathbf{T} + \|\mathbf{v}\|\,\mathbf{T}' \qquad \text{Product Rule}$$

$$= D_t[\|\mathbf{v}\|]\mathbf{T} + \|\mathbf{v}\|\,\mathbf{T}'\left(\frac{\|\mathbf{T}'\|}{\|\mathbf{T}'\|}\right)$$

$$= D_t[\|\mathbf{v}\|]\mathbf{T} + \|\mathbf{v}\|\,\|\mathbf{T}'\|\,\mathbf{N}. \qquad \mathbf{N} = \mathbf{T}'/\|\mathbf{T}'\|$$

Because $\mathbf{a}$ is written as a linear combination of $\mathbf{T}$ and $\mathbf{N}$, it must lie in the plane determined by $\mathbf{T}$ and $\mathbf{N}$.

The coefficients of $\mathbf{T}$ and $\mathbf{N}$ in the proof of this theorem are called the **tangential** and **normal components of acceleration** and are denoted by $a_\mathbf{T} = D_t[\|\mathbf{v}\|]$ and $a_\mathbf{N} = \|\mathbf{v}\|\,\|\mathbf{T}'\|$. Thus, you can write

$$\mathbf{a}(t) = a_\mathbf{T}\,\mathbf{T}(t) + a_\mathbf{N}\,\mathbf{N}(t).$$

The following theorem gives some convenient formulas for $a_\mathbf{N}$ and $a_\mathbf{T}$.

THEOREM 12.5   Tangential and Normal Components of Acceleration

If $\mathbf{r}(t)$ is the position vector for a smooth curve C [for which $\mathbf{N}(t)$ exists], then the tangential and normal components of acceleration are as follows.

$$a_\mathbf{T} = D_t[\|\mathbf{v}\|] = \mathbf{a} \cdot \mathbf{T} = \frac{\mathbf{v} \cdot \mathbf{a}}{\|\mathbf{v}\|}$$

$$a_\mathbf{N} = \|\mathbf{v}\|\,\|\mathbf{T}'\| = \mathbf{a} \cdot \mathbf{N} = \frac{\|\mathbf{v} \times \mathbf{a}\|}{\|\mathbf{v}\|} = \sqrt{\|\mathbf{a}\|^2 - a_\mathbf{T}^2}$$

Note that $a_\mathbf{N} \geq 0$. The normal component of acceleration is also called the **centripetal component of acceleration.**

Proof Note that $\mathbf{a}$ lies in the plane of $\mathbf{T}$ and $\mathbf{N}$. Thus, you can use Figure 12.23 to conclude that for any time t, the component of the projection of the acceleration vector onto $\mathbf{T}$ is given by $a_\mathbf{T} = \mathbf{a} \cdot \mathbf{T}$, and onto $\mathbf{N}$ is given by $a_\mathbf{N} = \mathbf{a} \cdot \mathbf{N}$. Moreover, because $\mathbf{a} = \mathbf{v}'$ and $\mathbf{T} = \mathbf{v}/\|\mathbf{v}\|$, you have

$$a_\mathbf{T} = \mathbf{a} \cdot \mathbf{T} = \mathbf{T} \cdot \mathbf{a} = \frac{\mathbf{v}}{\|\mathbf{v}\|} \cdot \mathbf{a} = \frac{\mathbf{v} \cdot \mathbf{a}}{\|\mathbf{v}\|}.$$

In Exercises 45 and 46, you are asked to prove the other parts of the theorem.

FIGURE 12.23
The tangential and normal components of acceleration are obtained by projecting $\mathbf{a}$ onto $\mathbf{T}$ and $\mathbf{N}$.

REMARK The formulas from Theorem 12.5, together with several other formulas from this chapter, are summarized on page 825.

EXAMPLE 5   Tangential and Normal Components of Acceleration

Find the tangential and normal components of acceleration for the position function given by $\mathbf{r}(t) = 3t\,\mathbf{i} - t\,\mathbf{j} + t^2\,\mathbf{k}.$

Solution   Begin by finding the velocity, speed, and acceleration.

$$\mathbf{v}(t) = \mathbf{r}'(t) = 3\mathbf{i} - \mathbf{j} + 2t\,\mathbf{k}$$
$$\|\mathbf{v}(t)\| = \sqrt{9 + 1 + 4t^2} = \sqrt{10 + 4t^2}$$
$$\mathbf{a}(t) = \mathbf{r}''(t) = 2\mathbf{k}$$

By Theorem 12.5, the tangential component of acceleration is

$$a_{\mathbf{T}} = \frac{\mathbf{v} \cdot \mathbf{a}}{\|\mathbf{v}\|} = \frac{4t}{\sqrt{10 + 4t^2}}$$

and because

$$\mathbf{v} \times \mathbf{a} = \begin{vmatrix} \mathbf{i} & \mathbf{j} & \mathbf{k} \\ 3 & -1 & 2t \\ 0 & 0 & 2 \end{vmatrix} = -2\mathbf{i} - 6\mathbf{j}$$

the normal component of acceleration is

$$a_{\mathbf{N}} = \frac{\|\mathbf{v} \times \mathbf{a}\|}{\|\mathbf{v}\|} = \frac{\sqrt{4 + 36}}{\sqrt{10 + 4t^2}} = \frac{2\sqrt{10}}{\sqrt{10 + 4t^2}}.$$

REMARK   In Example 5, you could have used the alternative formula for $a_{\mathbf{N}}$ as follows.

$$a_{\mathbf{N}} = \sqrt{\|\mathbf{a}\|^2 - a_{\mathbf{T}}^2} = \sqrt{(2)^2 - \frac{16t^2}{10 + 4t^2}} = \frac{2\sqrt{10}}{\sqrt{10 + 4t^2}}$$

EXAMPLE 6   Finding $a_{\mathbf{T}}$ and $a_{\mathbf{N}}$ for a Circular Helix

Find the tangential and normal components of acceleration for the helix given by $\mathbf{r}(t) = b\cos t\,\mathbf{i} + b\sin t\,\mathbf{j} + ct\,\mathbf{k}, \ b > 0.$

Solution

$$\mathbf{v}(t) = \mathbf{r}'(t) = -b\sin t\,\mathbf{i} + b\cos t\,\mathbf{j} + c\,\mathbf{k}$$
$$\|\mathbf{v}(t)\| = \sqrt{b^2\sin^2 t + b^2\cos^2 t + c^2} = \sqrt{b^2 + c^2}$$
$$\mathbf{a}(t) = \mathbf{r}''(t) = -b\cos t\,\mathbf{i} - b\sin t\,\mathbf{j}$$

By Theorem 12.5, the tangential component of acceleration is

$$a_{\mathbf{T}} = \frac{\mathbf{v} \cdot \mathbf{a}}{\|\mathbf{v}\|} = b^2\sin t\cos t - b^2\sin t\cos t + 0 = 0.$$

Moreover, because $\|\mathbf{a}(t)\| = \sqrt{b^2\cos^2 t + b^2\sin^2 t} = b$, you can use the alternative formula for the normal component of acceleration to obtain

$$a_{\mathbf{N}} = \sqrt{\|\mathbf{a}(t)\|^2 - a_{\mathbf{T}}^2} = \sqrt{b^2 - 0^2} = b.$$

Note that the normal component of acceleration is equal to the magnitude of the acceleration. In other words, because the speed is constant, the acceleration is perpendicular to the velocity (see Figure 12.24).

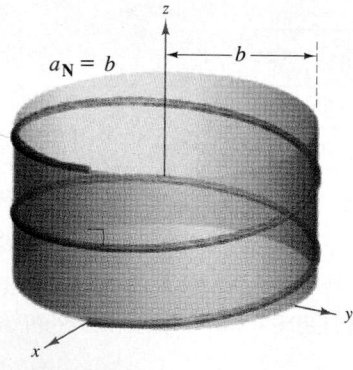

FIGURE 12.24
The normal component of acceleration is equal to the radius of the cylinder around which the helix is spiraling.

REMARK   The normal component of acceleration is equal to the radius of the cylinder around which the helix is spiraling.

EXAMPLE 7   Projectile Motion

The position function for the projectile shown in Figure 12.25 is given by

$$\mathbf{r}(t) = \left(50\sqrt{2}\,t\right)\mathbf{i} + \left(50\sqrt{2}\,t - 16t^2\right)\mathbf{j}. \qquad \text{Position vector}$$

Find the tangential component of acceleration when $t = 0$, 1, and $25\sqrt{2}/16$.

Solution

$$\mathbf{v}(t) = 50\sqrt{2}\,\mathbf{i} + (50\sqrt{2} - 32t)\mathbf{j} \qquad \text{Velocity vector}$$

$$\|\mathbf{v}(t)\| = 2\sqrt{50^2 - 16(50)\sqrt{2}\,t + 16^2 t^2} \qquad \text{Speed}$$

$$\mathbf{a}(t) = -32\mathbf{j} \qquad \text{Acceleration vector}$$

The tangential component of acceleration is

$$
a_{\mathrm{T}}(t) = \frac{\mathbf{v}(t) \cdot \mathbf{a}(t)}{\|\mathbf{v}(t)\|}
$$

$$
= \frac{-32\left(50\sqrt{2} - 32t\right)}{2\sqrt{50^2 - 16(50)\sqrt{2}\,t + 16^2 t^2}}.
$$

At the specified times, you have

$$
a_{\mathrm{T}}(0) = \frac{-32\left(50\sqrt{2}\right)}{100} = -16\sqrt{2} \approx -22.6
$$

$$
a_{\mathrm{T}}(1) = \frac{-32\left(50\sqrt{2} - 32\right)}{2\sqrt{50^2 - 16(50)\sqrt{2} + 16^2}} \approx -15.4
$$

$$
a_{\mathrm{T}}\!\left(\frac{25\sqrt{2}}{16}\right) = \frac{-32\left(50\sqrt{2} - 50\sqrt{2}\right)}{50\sqrt{2}} = 0.
$$

$\mathbf{r}(t) = (50\sqrt{2}\,t)\,\mathbf{i} + (50\sqrt{2}\,t - 16t^2)\,\mathbf{j}$

FIGURE 12.25
The path of a projectile.

You can see from Figure 12.25 that, at the maximum height, the tangential component is 0. This is reasonable because the direction of motion is horizontal at that point and the tangential component of the acceleration is equal to the horizontal component of the acceleration.

In Example 7, the alternative formula for a_{N} yields the following normal component of acceleration.

$$
a_{\mathrm{N}} = \sqrt{\|\mathbf{a}\|^2 - a_{\mathrm{T}}^2} = \sqrt{32^2 - \frac{32^2(50\sqrt{2} - 32t)^2}{4(50^2 - 16(50)\sqrt{2}\,t + 16^2 t^2)}}
$$

$$
= \frac{32(50)\sqrt{2}}{2\sqrt{50^2 - 16(50)\sqrt{2}\,t + 16^2 t^2}}
$$

When $t = 25\sqrt{2}/16$, you can determine that $a_{\mathrm{N}} = 32$, which makes sense because the tangential component is 0 at this point and therefore the entire acceleration is directed downward.

EXERCISES for Section 12.4

In Exercises 1–6, find the unit tangent vector $\mathbf{T}(t)$ and find a set of parametric equations for the line tangent to the space curve at the given point.

Function	Point
1. $\mathbf{r}(t) = t\mathbf{i} + t^2\mathbf{j} + t\mathbf{k}$	$(0,0,0)$
2. $\mathbf{r}(t) = t\mathbf{i} + t^2\mathbf{j} + \frac{2}{3}\mathbf{k}$	$(1,1,\frac{2}{3})$
3. $\mathbf{r}(t) = 2\cos t\,\mathbf{i} + 2\sin t\,\mathbf{j} + t\mathbf{k}$	$(2,0,0)$
4. $\mathbf{r}(t) = \langle t, t, \sqrt{4-t^2}\,\rangle$	$(1,1,\sqrt{3}\,)$
5. $\mathbf{r}(t) = \langle 2\cos t, 2\sin t, 4 \rangle$	$(\sqrt{2}, \sqrt{2}, 4)$
6. $\mathbf{r}(t) = \langle 2\sin t, 2\cos t, 4\sin^2 t \rangle$	$(1, \sqrt{3}, 1)$

[C] In Exercises 7 and 8, use a graphing utility to graph the space curve. Then find the tangent vector $\mathbf{T}(t)$ and find a set of parametric equations for the line tangent to the space curve at the given point. Sketch the graph of the line on the space curve.

Function	Point
7. $\mathbf{r}(t) = \langle t, t^2, \frac{2}{3}t^3 \rangle$	$(3, 9, 18)$
8. $\mathbf{r}(t) = 3\cos t\,\mathbf{i} + 4\sin t\,\mathbf{j} + \frac{t}{2}\mathbf{k}$	$\left(0, 4, \frac{\pi}{4}\right)$

In Exercises 9–12, find $\mathbf{v}(t)$, $\mathbf{a}(t)$, $\mathbf{T}(t)$, and $\mathbf{N}(t)$ (if it exists) for an object moving along the path given by the vector-valued function $\mathbf{r}(t)$. Use the results to determine the form of the path. Is the speed of the object constant or changing?

9. $\mathbf{r}(t) = 4t\,\mathbf{i}$ 10. $\mathbf{r}(t) = 4t\,\mathbf{i} - 2t\,\mathbf{j}$

11. $\mathbf{r}(t) = 4t^2\,\mathbf{i}$ 12. $\mathbf{r}(t) = t^2\mathbf{j} + \mathbf{k}$

In Exercises 13–18, find $\mathbf{T}(t)$, $\mathbf{N}(t)$, $a_{\mathbf{T}}$, and $a_{\mathbf{N}}$ at the given time t for the plane curve $\mathbf{r}(t)$.

Function	Time
13. $\mathbf{r}(t) = t\mathbf{i} + \frac{1}{t}\mathbf{j}$	$t = 1$
14. $\mathbf{r}(t) = t\mathbf{i} + t^2\mathbf{j}$	$t = 1$
15. $\mathbf{r}(t) = e^t\cos t\,\mathbf{i} + e^t\sin t\,\mathbf{j}$	$t = \frac{\pi}{2}$
16. $\mathbf{r}(t) = a\cos\omega t\,\mathbf{i} + b\sin\omega t\,\mathbf{j}$	$t = 0$
17. $\mathbf{r}(t) = \langle \cos\omega t + \omega t\sin\omega t, \sin\omega t - \omega t\cos\omega t \rangle$	$t = t_0$
18. $\mathbf{r}(t) = \langle \omega t - \sin\omega t, 1 - \cos\omega t \rangle$	$t = t_0$

Circular Motion In Exercises 19–22, consider an object moving according to the position function

$$r(t) = a\cos\omega t\,\mathbf{i} + a\sin\omega t\,\mathbf{j}.$$

19. Find $\mathbf{T}(t)$, $\mathbf{N}(t)$, $a_{\mathbf{T}}$, and $a_{\mathbf{N}}$.

20. Determine the directions of $\mathbf{T}$ and $\mathbf{N}$ relative to the position function $\mathbf{r}$.

21. Determine the speed of the object at any time t and explain its value relative to the value of $a_{\mathbf{T}}$.

22. If the angular velocity ω is halved, by what factor is $a_{\mathbf{N}}$ changed?

In Exercises 23 and 24, sketch the graph of the plane curve represented by the vector-valued function, and, at the point on the curve determined by $\mathbf{r}(t_0)$, sketch the vectors $\mathbf{T}$ and $\mathbf{N}$. Note that $\mathbf{N}$ points toward the concave side of the curve.

Function	Time
23. $\mathbf{r}(t) = t\mathbf{i} + \frac{1}{t}\mathbf{j}$	$t_0 = 2$
24. $\mathbf{r}(t) = 2\cos t\,\mathbf{i} + 2\sin t\,\mathbf{j}$	$t_0 = \frac{\pi}{4}$

25. *Motion of a Particle* The figure shows the path of a particle modeled by the vector-valued function

$$\mathbf{r}(t) = \langle \pi t - \sin\pi t, 1 - \cos\pi t \rangle.$$

The figure also shows the vectors $\mathbf{v}(t)/\|\mathbf{v}(t)\|$ and $\mathbf{a}(t)/\|\mathbf{a}(t)\|$ at the specified values of t.

a. Identify the graph and its orientation.

b. Find $a_{\mathbf{T}}$ and $a_{\mathbf{N}}$ at $t = \frac{1}{2}$, $t = 1$, and $t = \frac{3}{2}$.

c. Determine whether the speed of the particle is increasing, decreasing, or maximum at each of the specified values of t. Give reasons for your answers.

26. *Motion of a Particle* The figure shows a particle moving along the involute of a circle modeled by

$$\mathbf{r}(t) = \langle \cos\pi t + \pi t\sin\pi t, \sin\pi t - \pi t\cos\pi t \rangle.$$

It also shows the vectors $\mathbf{v}(t)$ and $\mathbf{a}(t)$ for $t = 1$ and $t = 2$.

a. Find $a_{\mathbf{T}}$ and $a_{\mathbf{N}}$ at $t = 1$ and $t = 2$.

b. Determine whether the speed of the particle is increasing or decreasing at each of the specified values of t. Give reasons for your answers.

FIGURE FOR 25

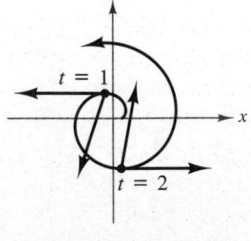

FIGURE FOR 26

In Exercises 27–30, find $T(t)$, $N(t)$, a_T, and a_N at the given time t for the space curve $r(t)$.

Function	Time
27. $r(t) = t\,i + 2t\,j - 3t\,k$	$t = 1$
28. $r(t) = 4t\,i - 4t\,j + 2t\,k$	$t = 2$
29. $r(t) = t\,i + t^2\,j + \dfrac{t^2}{2}\,k$	$t = 1$
30. $r(t) = e^t \cos t\,i + e^t \sin t\,j + e^t\,k$	$t = 0$

In Exercises 31 and 32, use a graphing utility to graph the space curve. Also find $T(t)$, $N(t)$, a_T, and a_N at the given time t. Sketch $T(t)$ and $N(t)$ on the space curve.

Function	Time
31. $r(t) = 4t\,i + 3\cos t\,j + 3\sin t\,k$	$t = \dfrac{\pi}{2}$
32. $r(t) = t\,i + 3t^2\,j + \dfrac{t^2}{2}\,k$	$t = 2$

Unit Binormal Vector In Exercises 33 and 34, find the vectors T, N, and $B = T \times N$ for the vector-valued function $r(t)$ at the specified value of t (see figures). The vector B is called the unit binormal vector to the curve.

33. $r(t) = 2\cos t\,i + 2\sin t\,j + \dfrac{t}{2}\,k$, $t_0 = \dfrac{\pi}{2}$

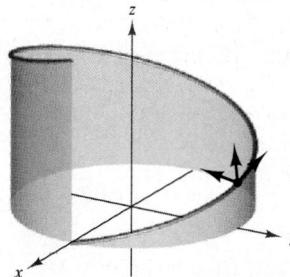

FIGURE FOR 33

34. $r(t) = t\,i + t^2\,j + \dfrac{t^3}{3}\,k$, $t_0 = 1$

FIGURE FOR 34

35. *Projectile Motion* Find the tangential and normal components of acceleration for a projectile fired at an angle θ with the horizontal at an initial speed of v_0. What are the components when the projectile is at its maximum height?

36. *Projectile Motion* A plane flying at an altitude of 30,000 feet and a speed of 540 miles per hour (792 feet per second) releases a bomb. Find the tangential and normal components of acceleration acting on the bomb.

37. *Centripetal Acceleration* An object is spinning at a constant speed on the end of a string, according to the position function given in Exercises 19–22.
   a. If the angular velocity ω is doubled, how is the centripetal component of acceleration changed?
   b. If the angular velocity is unchanged but the length of the string is halved, how is the centripetal component of acceleration changed?

38. *Centripetal Force* An object of mass m moves at a constant speed v in a circular path of radius r. The force required to produce the centripetal component of acceleration is called the centripetal force and is given by $F = mv^2/r$. Newton's Law of Universal Gravitation is given by $F = GMm/d^2$, where d is the distance between the centers of the two bodies of masses M and m, and G is a gravitational constant. Use this law to show that the speed required for circular motion is $v = \sqrt{GM/r}$.

Orbital Speed In Exercises 39–42, use the result of Exercise 38 to find the speed necessary for the given circular orbit around the earth. Let $GM = 9.56 \times 10^4$ cubic miles per second per second, and assume that the radius of the earth is 4000 miles.

39. The orbit of a space shuttle 100 miles above the surface of the earth

40. The orbit of a space shuttle 200 miles above the surface of the earth

41. The orbit of a heat capacity mapping satellite 385 miles above the surface of the earth (see figure)

FIGURE FOR 41

42. The orbit of a SYNCOM satellite r miles above the surface of the earth that is in geosynchronous orbit. [The satellite completes one orbit per sidereal day (23 hours, 56 minutes), and thus appears to remain stationary above a point on the earth.]

43. Prove that the principal unit normal vector N points toward the concave side of a plane curve.

44. Prove that the vector $T'(t)$ is 0 for an object moving in a straight line.

45. Prove that $a_N = \dfrac{\| v \times a \|}{\| v \|}$.

46. Prove that $a_N = \sqrt{\| a \|^2 - a_T^2}$.

12.5   Arc Length • Arc Length Parameter • Curvature • Applications

Arc Length

In Section 10.2, you saw that the arc length of a smooth *plane* curve C given by the parametric equations $x = x(t)$ and $y = y(t)$, $a \leq t \leq b$, is

$$s = \int_a^b \sqrt{[x'(t)]^2 + [y'(t)]^2} \, dt.$$

In vector form, where C is given by $\mathbf{r}(t) = x(t)\mathbf{i} + y(t)\mathbf{j}$, you can rewrite this equation for arc length as

$$s = \int_a^b \| \mathbf{r}'(t) \| \, dt.$$

EXAMPLE 1   Finding the Arc Length of a Plane Curve

Find the arc length of the curve given by $\mathbf{r}(t) = \cos t\, \mathbf{i} + \sin t\, \mathbf{j}$ from $t = 0$ to $t = 2\pi$.

Solution   Using $x(t) = \cos t$ and $y(t) = \sin t$, you obtain $x'(t) = -\sin t$ and $y'(t) = \cos t$. Thus, the arc length from $t = 0$ to $t = 2\pi$ is

$$s = \int_0^{2\pi} \sqrt{[x'(t)]^2 + [y'(t)]^2} \, dt = \int_0^{2\pi} \sqrt{\sin^2 t + \cos^2 t} \, dt$$

$$= \int_0^{2\pi} dt$$

$$= t \Big]_0^{2\pi} = 2\pi.$$

In Figure 12.26, notice that the curve is a circle of radius 1. Thus, the arc length formula confirms that the circumference of the circle is 2π.

FIGURE 12.26
The arc length of the circle is 2π.

$\mathbf{r}(t) = \cos t\, \mathbf{i} + \sin t\, \mathbf{j}$

The formula for the arc length of a plane curve has a natural extension to a smooth curve in *space*, as stated in the following theorem.

THEOREM 12.6   Arc Length of Space Curve

If C is a smooth curve given by $\mathbf{r}(t) = x(t)\mathbf{i} + y(t)\mathbf{j} + z(t)\mathbf{k}$, on an interval $[a, b]$, then the arc length of C on the interval is

$$s = \int_a^b \sqrt{[x'(t)]^2 + [y'(t)]^2 + [z'(t)]^2} \, dt = \int_a^b \| \mathbf{r}'(t) \| \, dt.$$

EXAMPLE 2   Finding the Arc Length of a Curve in Space

Find the arc length of the curve given by

$$\mathbf{r}(t) = t\,\mathbf{i} + \frac{4}{3}t^{3/2}\,\mathbf{j} + \frac{1}{2}t^2\,\mathbf{k}$$

from $t = 0$ to $t = 2$, as shown in Figure 12.27.

Solution   Using $x(t) = t$, $y(t) = \frac{4}{3}t^{3/2}$, and $z(t) = \frac{1}{2}t^2$, you obtain $x'(t) = 1$, $y'(t) = 2t^{1/2}$, and $z'(t) = t$. Thus, the arc length from $t = 0$ to $t = 2$ is given by

$$s = \int_0^2 \sqrt{[x'(t)]^2 + [y'(t)]^2 + [z'(t)]^2}\,dt$$

$$= \int_0^2 \sqrt{1 + 4t + t^2}\,dt$$

$$= \int_0^2 \sqrt{(t + 2)^2 - 3}\,dt \qquad \text{Integration Formula 26}$$

$$= \left[\frac{t + 2}{2}\sqrt{(t + 2)^2 - 3} - \frac{3}{2}\ln\left|(t + 2) + \sqrt{(t + 2)^2 - 3}\right|\right]_0^2$$

$$= 2\sqrt{13} - \frac{3}{2}\ln\left(4 + \sqrt{13}\right) - 1 + \frac{3}{2}\ln 3$$

$$\approx 4.816.$$

REMARK   The formula for arc length given in Theorem 12.6 is independent of the parameter used to represent C. To illustrate this independence, try using the vector-valued function

$$\mathbf{r}(t) = t^2\,\mathbf{i} + \frac{4}{3}t^3\,\mathbf{j} + \frac{1}{2}t^4\,\mathbf{k}$$

to represent the curve in Example 2. Then find the arc length from $t = 0$ to $t = \sqrt{2}$ and compare the result with that found in Example 2.

EXAMPLE 3   Finding the Arc Length of a Helix

Find the length of one turn of the helix given by

$$\mathbf{r}(t) = b\cos t\,\mathbf{i} + b\sin t\,\mathbf{j} + \sqrt{1 - b^2}\,t\,\mathbf{k}$$

as shown in Figure 12.28.

Solution   Because $\mathbf{r}'(t) = -b\sin t\,\mathbf{i} + b\cos t\,\mathbf{j} + \sqrt{1 - b^2}\,\mathbf{k}$, the arc length of one turn is

$$s = \int_0^{2\pi} \|\mathbf{r}'(t)\|\,dt$$

$$= \int_0^{2\pi} \sqrt{b^2(\sin^2 t + \cos^2 t) + (1 - b^2)}\,dt$$

$$= \int_0^{2\pi} dt$$

$$= 2\pi.$$

Curve: $x = t$
$y = \frac{4}{3}t^{3/2}$
$z = \frac{t^2}{2}$

FIGURE 12.27
As t increases from 0 to 2, the vector $\mathbf{r}(t)$ traces out a curve whose length is approximately 4.816.

Curve:
$\mathbf{r}(t) = b\cos t\,\mathbf{i} + b\sin t\,\mathbf{j} + \sqrt{1 - b^2}\,t\,\mathbf{k}$

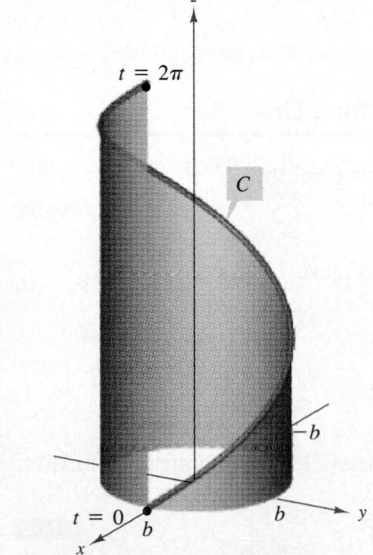

FIGURE 12.28
One turn of a helix.

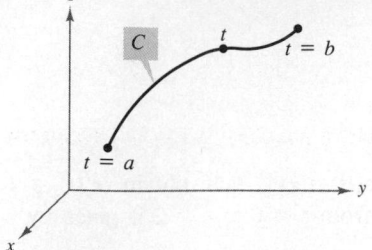

$$s(t) = \int_a^t \sqrt{[x'(u)]^2 + [y'(u)]^2 + [z'(u)]^2}\, du$$

FIGURE 12.29

REMARK The arc length function s is *nonnegative*. It measures the distance along C from the initial point $(x(a), y(a), z(a))$ to the point $(x(t), y(t), z(t))$.

Arc Length Parameter

You have seen that curves can be represented by vector-valued functions in different ways, depending on the choice of parameter. For *motion* along a curve, the convenient parameter is time, t. However, for studying the *geometric properties* of a curve, the convenient parameter is often arc length, s.

Definition of Arc Length Function

Let C be a smooth curve given by $\mathbf{r}(t)$ defined on the closed interval $[a, b]$. For $a \le t \le b$, the **arc length function** is given by

$$s(t) = \int_a^t \| \mathbf{r}'(u) \|\, du = \int_a^t \sqrt{[x'(u)]^2 + [y'(u)]^2 + [z'(u)]^2}\, du.$$

The arc length, s, is called the **arc length parameter** (see Figure 12.29).

Using the definition of the arc length function and the Second Fundamental Theorem of Calculus, you can conclude that

$$\frac{ds}{dt} = \| \mathbf{r}'(t) \|.$$

THEOREM 12.7 The Derivative of the Arc Length Function

Let C be a smooth curve given by $\mathbf{r}(t)$ on the closed interval $[a, b]$. The derivative of the arc length function s for this curve is given by

$$\frac{ds}{dt} = \| \mathbf{r}'(t) \|.$$

In differential form, you can write $ds = \| \mathbf{r}'(t) \|\, dt$.

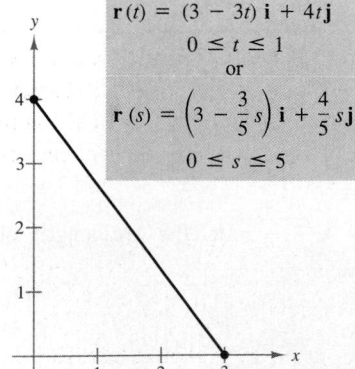

FIGURE 12.30
The line segment from (3, 0) to (0, 4) can be parametrized in many ways.

EXAMPLE 4   Finding the Arc Length Function for a Line

Find the arc length function $s(t)$ for the line segment given by

$$\mathbf{r}(t) = (3 - 3t)\mathbf{i} + 4t\,\mathbf{j}, \qquad 0 \le t \le 1$$

and express $\mathbf{r}$ as a function of the parameter s (see Figure 12.30).

Solution   Because $\mathbf{r}'(t) = -3\mathbf{i} + 4\mathbf{j}$ and $\| \mathbf{r}'(t) \| = \sqrt{3^2 + 4^2} = 5$, you have

$$s(t) = \int_0^t \| \mathbf{r}'(u) \|\, du = \int_0^t 5\, du = 5t.$$

Using $s = 5t$ (or $t = s/5$), you can rewrite $\mathbf{r}$ using the arc length parameter as follows.

$$\mathbf{r}(s) = \left(3 - \frac{3}{5}s\right)\mathbf{i} + \frac{4}{5}s\,\mathbf{j}, \qquad 0 \le s \le 5.$$

One of the advantages of writing a vector-valued function in terms of the arc length parameter is that $\| \mathbf{r}'(s) \| = 1$. For instance, in Example 4, you have

$$\| \mathbf{r}'(s) \| = \sqrt{\left(-\frac{3}{5} \right)^2 + \left(\frac{4}{5} \right)^2} = 1.$$

Thus, for a smooth curve C, represented by $\mathbf{r}(s)$, where s is the arc length parameter, the arc length between a and b is

$$\text{Length of arc} = \int_a^b \| \mathbf{r}'(s) \| \, ds$$

$$= \int_a^b ds = b - a$$

$$= \text{length of interval.}$$

Furthermore, if t is *any* parameter such that $\| \mathbf{r}'(t) \| = 1$, then t must be the arc length parameter. These results are summarized in the following theorem, which we state without proof.

THEOREM 12.8   Arc Length Parameter

If C is a smooth curve given by

$$\mathbf{r}(s) = x(s)\mathbf{i} + y(s)\mathbf{j} \quad \text{or} \quad \mathbf{r}(s) = x(s)\mathbf{i} + y(s)\mathbf{j} + z(s)\mathbf{k}$$

where s is the arc length parameter, then

$$\| \mathbf{r}'(s) \| = 1.$$

Moreover, if t is *any* parameter for the vector-valued function $\mathbf{r}$ such that $\| \mathbf{r}'(t) \| = 1$, then t must be the arc length parameter.

FIGURE 12.31
Curvature at P is greater than at Q.

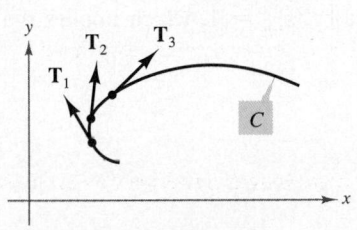

FIGURE 12.32
T changes as s changes.

Curvature

An important use of the arc length parameter is to find **curvature**—the measure of how sharply a curve bends. For instance, in Figure 12.31 the curve bends more sharply at P than at Q, and we say that the curvature is greater at P than at Q. You can calculate curvature by calculating the magnitude of the rate of change of the unit tangent vector **T** with respect to the arc length s, as indicated in Figure 12.32.

Definition of Curvature

Let C be a smooth curve (in the plane *or* in space) given by $\mathbf{r}(s)$, where s is the arc length parameter. The **curvature** at s is given by

$$K = \left\| \frac{d\mathbf{T}}{ds} \right\| = \| \mathbf{T}'(s) \|.$$

Because a straight line doesn't "curve," its curvature should be 0, as demonstrated in Example 5.

EXAMPLE 5   Curvature of a Straight Line

Find the curvature of the line given by

$$\mathbf{r}(s) = \left(3 - \frac{3}{5}s\right)\mathbf{i} + \frac{4}{5}s\mathbf{j}.$$

Solution   To begin, note that $\mathbf{r}$ is written in terms of the arc length parameter (see Example 4). Because $\mathbf{r}'(s) = -\frac{3}{5}\mathbf{i} + \frac{4}{5}\mathbf{j}$, and $\|\mathbf{r}'(s)\| = 1$, you have

$$\mathbf{T}(s) = \frac{\mathbf{r}'(s)}{\|\mathbf{r}'(s)\|} = -\frac{3}{5}\mathbf{i} + \frac{4}{5}\mathbf{j}.$$

Therefore, the curvature is

$$K = \|\mathbf{T}'(s)\| = \|0\mathbf{i} + 0\mathbf{j}\| = 0$$

at every point on the line.

A circle has the same curvature at any point. Moreover, the curvature and the radius of the circle are inversely related. That is, a circle with a large radius has a small curvature, and a circle with a small radius has a large curvature. This inverse relationship is made explicit in the following example.

EXAMPLE 6   Finding the Curvature of a Circle

Show that the curvature of a circle of radius r is $K = 1/r$.

Solution   Without loss of generality, you can consider the circle to be centered at the origin. Let (x, y) be any point on the circle and let s be the length of the arc from $(r, 0)$ to (x, y), as shown in Figure 12.33. By letting θ be the central angle of the circle, you can represent the circle by

$$\mathbf{r}(\theta) = r\cos\theta\,\mathbf{i} + r\sin\theta\,\mathbf{j}. \qquad \theta \text{ is the parameter}$$

Using the formula for the length of a circular arc $s = r\theta$, you can rewrite $\mathbf{r}(\theta)$ in terms of the arc length parameter as follows.

$$\mathbf{r}(s) = r\cos\frac{s}{r}\mathbf{i} + r\sin\frac{s}{r}\mathbf{j} \qquad \text{Arc length } s \text{ is the parameter}$$

Thus, $\mathbf{r}'(s) = -\sin\frac{s}{r}\mathbf{i} + \cos\frac{s}{r}\mathbf{j}$, and it follows that $\|\mathbf{r}'(s)\| = 1$, which implies that the unit tangent vector is

$$\mathbf{T}(s) = \frac{\mathbf{r}'(s)}{\|\mathbf{r}'(s)\|} = -\sin\frac{s}{r}\mathbf{i} + \cos\frac{s}{r}\mathbf{j}$$

and the curvature is given by

$$K = \|\mathbf{T}'(s)\| = \left\|-\frac{1}{r}\cos\frac{s}{r}\mathbf{i} - \frac{1}{r}\sin\frac{s}{r}\mathbf{j}\right\| = \frac{1}{r}$$

at every point on the circle.

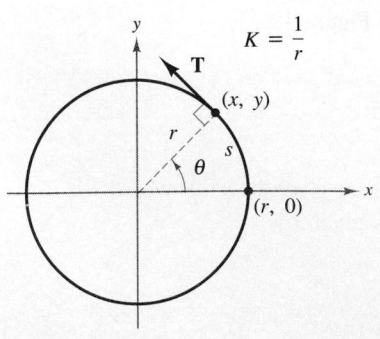

FIGURE 12.33
The curvature of a circle.

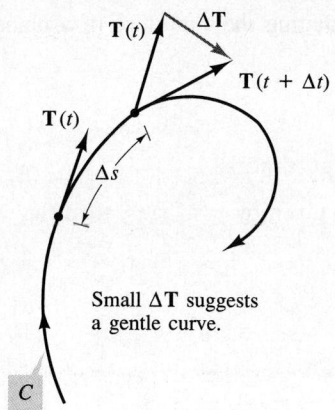

Small $\Delta \mathbf{T}$ suggests
a gentle curve.

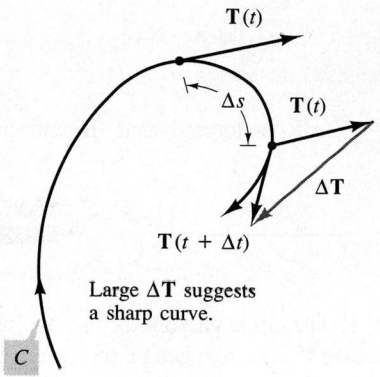

Large $\Delta \mathbf{T}$ suggests
a sharp curve.

FIGURE 12.34

In Examples 5 and 6, the curvature was found by applying the definition directly. This requires that the curve be written in terms of the arc length parameter s. The following theorem gives two other formulas for finding the curvature of a curve written in terms of an arbitrary parameter t. We leave the proof of this theorem as an exercise (see Exercise 70, parts a and b).

THEOREM 12.9   Formulas for Curvature

If C is a smooth curve given by $\mathbf{r}(t)$, then the curvature of C at t is given by

$$K = \frac{\| \mathbf{T}'(t) \|}{\| \mathbf{r}'(t) \|} = \frac{\| \mathbf{r}'(t) \times \mathbf{r}''(t) \|}{\| \mathbf{r}'(t) \|^3}.$$

Because $\| \mathbf{r}'(t) \| = ds/dt$, the first formula implies that curvature is the ratio of the rate of change in the tangent vector $\mathbf{T}$ to the rate of change in arc length. To see the reasonableness of this, let Δt be a "small number." Then,

$$\frac{\mathbf{T}'(t)}{ds/dt} \approx \frac{[\mathbf{T}(t + \Delta t) - \mathbf{T}(t)]/\Delta t}{[s(t + \Delta t) - s(t)]/\Delta t} = \frac{\mathbf{T}(t + \Delta t) - \mathbf{T}(t)}{s(t + \Delta t) - s(t)} = \frac{\Delta \mathbf{T}}{\Delta s}.$$

In other words, for a given Δs, the greater the value of $\Delta \mathbf{T}$, the more the curve bends at t, as shown in Figure 12.34.

EXAMPLE 7   Finding the Curvature of a Space Curve

Find the curvature of the curve given by $\mathbf{r}(t) = 2t\,\mathbf{i} + t^2\,\mathbf{j} - \frac{1}{3}t^3\,\mathbf{k}$.

Solution   It is not apparent whether this parameter is arc length, so you should use the formula $K = \| \mathbf{T}'(t) \| / \| \mathbf{r}'(t) \|$.

$$\mathbf{r}'(t) = 2\mathbf{i} + 2t\,\mathbf{j} - t^2\,\mathbf{k}$$

$$\| \mathbf{r}'(t) \| = \sqrt{4 + 4t^2 + t^4} = t^2 + 2 \qquad \text{Length of } \mathbf{r}'(t)$$

$$\mathbf{T}(t) = \frac{\mathbf{r}'(t)}{\| \mathbf{r}'(t) \|} = \frac{2\mathbf{i} + 2t\,\mathbf{j} - t^2\,\mathbf{k}}{t^2 + 2}$$

$$\mathbf{T}'(t) = \frac{(t^2 + 2)(2\mathbf{j} - 2t\,\mathbf{k}) - (2t)(2\mathbf{i} + 2t\,\mathbf{j} - t^2\,\mathbf{k})}{(t^2 + 2)^2}$$

$$= \frac{-4t\,\mathbf{i} + (4 - 2t^2)\mathbf{j} - 4t\,\mathbf{k}}{(t^2 + 2)^2}$$

$$\| \mathbf{T}'(t) \| = \frac{\sqrt{16t^2 + 16 - 16t^2 + 4t^4 + 16t^2}}{(t^2 + 2)^2}$$

$$= \frac{2(t^2 + 2)}{(t^2 + 2)^2} = \frac{2}{t^2 + 2}. \qquad \text{Length of } \mathbf{T}'(t)$$

Therefore,

$$K = \frac{\| \mathbf{T}'(t) \|}{\| \mathbf{r}'(t) \|} = \frac{2}{(t^2 + 2)^2}. \qquad \text{Curvature}$$

The following theorem gives a formula for calculating the curvature of a plane curve given by $y = f(x)$.

THEOREM 12.10   Curvature in Rectangular Coordinates

If C is the graph of a twice-differentiable function given by $y = f(x)$, then the curvature at the point (x, y) is given by

$$K = \frac{|y''|}{[1 + (y')^2]^{3/2}}.$$

Proof By representing the curve C by $\mathbf{r}(x) = x\mathbf{i} + f(x)\mathbf{j} + 0\mathbf{k}$ (where x is the parameter), you obtain $\mathbf{r}'(x) = \mathbf{i} + f'(x)\mathbf{j}$

$$\|\mathbf{r}'(x)\| = \sqrt{1 + [f'(x)]^2}$$

and $\mathbf{r}''(x) = f''(x)\mathbf{j}$. Because $\mathbf{r}'(x) \times \mathbf{r}''(x) = f''(x)\mathbf{k}$, it follows that the curvature is

$$K = \frac{\|\mathbf{r}'(x) \times \mathbf{r}''(x)\|}{\|\mathbf{r}'(x)\|^3} = \frac{|f''(x)|}{\{1 + [f'(x)]^2\}^{3/2}} = \frac{|y''|}{[1 + (y')^2]^{3/2}}.$$ ▬

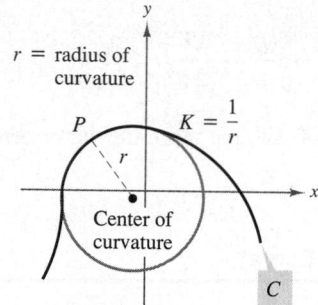

r = radius of curvature

P $K = \dfrac{1}{r}$

r

Center of curvature

C

FIGURE 12.35
The circle of curvature.

Let C be a curve with curvature K at the point P. The circle passing through point P with radius $r = 1/K$ is called the **circle of curvature** if the circle lies on the concave side of the curve and shares a common tangent line with the curve at point P. The radius is called the **radius of curvature** at P, and the center of the circle is called the **center of curvature.**

The circle of curvature gives us a nice way to graphically estimate the curvature K at a point P on a curve. Using a compass, you can sketch a circle that approximates the concave side of the curve at point P, as shown in Figure 12.35. If the circle has a radius of r, then you can estimate the curvature to be $K = 1/r$.

EXAMPLE 8   Finding Curvature in Rectangular Coordinates

Find the curvature of the parabola given by $y = x - \frac{1}{4}x^2$ at $x = 2$. Sketch the circle of curvature at $(2, 1)$.

Solution The curvature at $x = 2$ is as follows.

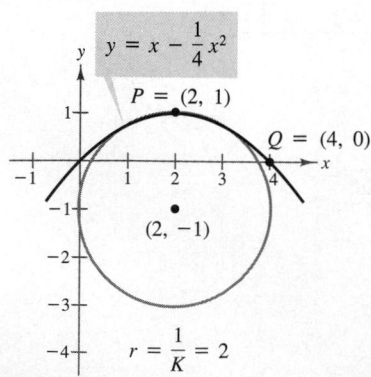

$y = x - \dfrac{1}{4}x^2$

$P = (2, 1)$

$Q = (4, 0)$

$(2, -1)$

$r = \dfrac{1}{K} = 2$

FIGURE 12.36
The circle of curvature.

$$y' = 1 - \frac{x}{2} \qquad\qquad y' = 0$$

$$y'' = -\frac{1}{2} \qquad\qquad y'' = -\frac{1}{2}$$

$$K = \frac{|y''|}{[1 + (y')^2]^{3/2}} \qquad K = \frac{1}{2}$$

Because the curvature at $(2, 1)$ is $\frac{1}{2}$, it follows that the radius of the circle of curvature at that point is 2. Thus, the center of curvature is $(2, -1)$, as shown in Figure 12.36. (In the figure, note that the curve has the greatest curvature at P. Try showing that the curvature at Q is $1/2^{5/2} \approx 0.177$.) ▬

FIGURE 12.37
The amount of thrust felt by passengers in a car that is turning depends on two things—the speed of the car and the sharpness of the turn.

Arc length and curvature are closely related to the tangential and normal components of acceleration. The tangential component of acceleration is the rate of change of the speed, which in turn is the rate of change of the arc length. This component is negative as a moving object slows down and positive as it speeds up—regardless of whether the object is turning or traveling in a straight line. Thus, the tangential component is solely a function of the arc length and is independent of the curvature. On the other hand, the normal component of acceleration is a function of **both** speed and curvature. This component measures the acceleration acting perpendicular to the direction of motion. To see why the normal component is affected by both speed and curvature, imagine that you are driving a car around a turn, as shown in Figure 12.37. If your speed is high and the turn is sharp, you feel yourself thrown against the car door. By lowering your speed *or* taking a more gentle turn, you are able to lessen this sideways thrust.

The next theorem explicitly states the relationship among speed, curvature, and the components of acceleration.

REMARK Note that Theorem 12.11 gives additional formulas for a_T and a_N.

THEOREM 12.11   Acceleration, Speed, and Curvature

If $\mathbf{r}(t)$ is the position vector for a smooth curve C, then the acceleration vector is given by

$$\mathbf{a}(t) = \frac{d^2 s}{dt^2}\,\mathbf{T} + K\left(\frac{ds}{dt}\right)^2\mathbf{N}$$

where K is the curvature of C and ds/dt is the speed.

Proof   For the position vector $\mathbf{r}(t)$, you have

$$\mathbf{a}(t) = a_T\,\mathbf{T} + a_N\,\mathbf{N} = D_t[\|\,\mathbf{v}\,\|]\mathbf{T} + \|\,\mathbf{v}\,\|\,\|\,\mathbf{T}'\,\|\,\mathbf{N}$$

$$= \frac{d^2 s}{dt^2}\,\mathbf{T} + \frac{ds}{dt}(\|\,\mathbf{v}\,\|K)\mathbf{N} = \frac{d^2 s}{dt^2}\,\mathbf{T} + K\left(\frac{ds}{dt}\right)^2\mathbf{N}.$$

EXAMPLE 9   Tangential and Normal Components of Acceleration

Find a_T and a_N for the curve given by $\mathbf{r}(t) = 2t\,\mathbf{i} + t^2\,\mathbf{j} - \frac{1}{3}t^3\,\mathbf{k}$.

Solution   From Example 7, you know that

$$\frac{ds}{dt} = \|\,\mathbf{r}'(t)\,\| = t^2 + 2 \quad \text{and} \quad K = \frac{2}{(t^2 + 2)^2}.$$

Therefore,

$$a_T = \frac{d^2 s}{dt^2} = 2t \qquad\qquad \text{Tangential component}$$

and

$$a_N = K\left(\frac{ds}{dt}\right)^2 = \frac{2}{(t^2 + 2)^2}(t^2 + 2)^2 = 2. \qquad \text{Normal component}$$

Applications

There are many applications in physics and engineering dynamics that involve the relationship among speed, arc length, curvature, and acceleration. One such application concerns frictional force.

Suppose a moving object with mass m is in contact with a stationary object. The total force required to produce an acceleration **a** along a given path is

$$\mathbf{F} = m\mathbf{a}$$

$$= m\left(\frac{d^2s}{dt^2}\right)\mathbf{T} + mK\left(\frac{ds}{dt}\right)^2\mathbf{N}$$

$$= ma_\mathbf{T}\mathbf{T} + ma_\mathbf{N}\mathbf{N}.$$

The portion of this total force that is supplied by the stationary object is called the **force of friction.** For example, if a car is rounding a turn, the roadway exerts a frictional force that keeps the car from sliding off the road. If the car is not sliding, then the frictional force is perpendicular to the direction of motion and has magnitude equal to the normal component of acceleration, as shown in Figure 12.38. The potential frictional force of a road around a turn can be increased by banking the roadway.

Force of friction

FIGURE 12.38
The force of friction is perpendicular to the direction of motion.

EXAMPLE 10   Frictional Force

A 360-kilogram go-cart is driven at a speed of 60 kilometers per hour around a circular racetrack of radius 12 meters. To keep the cart from skidding off course, what frictional force must the track surface exert on the tires?

Solution   The frictional force must equal the mass times the normal component of acceleration. For this circular path, we know that the curvature is $K = \frac{1}{12}$. Therefore, the frictional force is

$$ma_\mathbf{N} = mK\left(\frac{ds}{dt}\right)^2$$

$$= (360 \text{ kg})\left(\frac{1}{12 \text{ m}}\right)\left(\frac{60{,}000 \text{ m}}{3600 \text{ sec}}\right)^2$$

$$\approx 8333 \text{ (kg)(m)/sec}^2.$$

Several of the formulas dealing with vector-valued functions are summarized below.

Summary of Velocity, Acceleration, and Curvature

Let C be a curve given by the position function

$$\mathbf{r}(t) = x(t)\mathbf{i} + y(t)\mathbf{j} \qquad \text{Curve in the plane}$$

or

$$\mathbf{r}(t) = x(t)\mathbf{i} + y(t)\mathbf{j} + z(t)\mathbf{k}. \qquad \text{Curve in space}$$

Velocity vector, speed, and acceleration vector:

$$\mathbf{v}(t) = \mathbf{r}'(t)$$

$$\|\mathbf{v}(t)\| = \frac{ds}{dt} = \|\mathbf{r}'(t)\|$$

$$\mathbf{a}(t) = \mathbf{r}''(t) = a_\mathbf{T}\mathbf{T}(t) + a_\mathbf{N}\mathbf{N}(t)$$

Unit tangent vector and principal unit normal vector:

$$\mathbf{T}(t) = \frac{\mathbf{r}'(t)}{\|\mathbf{r}'(t)\|} \quad \text{and} \quad \mathbf{N}(t) = \frac{\mathbf{T}'(t)}{\|\mathbf{T}'(t)\|}$$

Components of acceleration:

$$a_\mathbf{T} = \mathbf{a} \cdot \mathbf{T} = \frac{\mathbf{v} \cdot \mathbf{a}}{\|\mathbf{v}\|} = \frac{d^2 s}{dt^2}$$

$$a_\mathbf{N} = \mathbf{a} \cdot \mathbf{N} = \sqrt{\|\mathbf{a}\|^2 - a_\mathbf{T}^2} = \frac{\|\mathbf{v} \times \mathbf{a}\|}{\|\mathbf{v}\|} = K\left(\frac{ds}{dt}\right)^2$$

Formulas for curvature in the plane:

$$K = \frac{|y''|}{[1 + (y')^2]^{3/2}} \qquad C \text{ given by } y = f(x)$$

$$K = \frac{|x'y'' - y'x''|}{[(x')^2 + (y')^2]^{3/2}} \qquad C \text{ given by } x = x(t), y = y(t)$$

Formulas for curvature in the plane or in space:

$$K = \|\mathbf{T}'(s)\| = \|\mathbf{r}''(s)\| \qquad s \text{ is arc length parameter}$$

$$K = \frac{\|\mathbf{T}'(t)\|}{\|\mathbf{r}'(t)\|} = \frac{\|\mathbf{r}'(t) \times \mathbf{r}''(t)\|}{\|\mathbf{r}'(t)\|^3} \qquad t \text{ is general parameter}$$

$$K = \frac{\mathbf{a}(t) \cdot \mathbf{N}(t)}{\|\mathbf{v}(t)\|^2}$$

Cross product formulas apply only to curves in space.

EXERCISES for Section 12.5

TECHNOLOGY
Laboratory Guide
Lab 12.5

In Exercises 1–4, sketch the plane curve and find its length over the indicated interval.

Function	Interval
1. $\mathbf{r}(t) = t\mathbf{i} + 3t\mathbf{j}$	$[0, 4]$
2. $\mathbf{r}(t) = t\mathbf{i} + t^2\mathbf{k}$	$[0, 4]$
3. $\mathbf{r}(t) = a\cos^3 t\,\mathbf{i} + a\sin^3 t\,\mathbf{j}$	$[0, 2\pi]$
4. $\mathbf{r}(t) = a\cos t\,\mathbf{i} + a\sin t\,\mathbf{j}$	$[0, 2\pi]$

In Exercises 5–8, sketch the space curve and find its length over the indicated interval.

Function	Interval
5. $\mathbf{r}(t) = 2t\mathbf{i} - 3t\mathbf{j} + t\mathbf{k}$	$[0, 2]$
6. $\mathbf{r}(t) = \langle 4t, 3\cos t, 3\sin t \rangle$	$\left[0, \dfrac{\pi}{2}\right]$
7. $\mathbf{r}(t) = a\cos t\,\mathbf{i} + a\sin t\,\mathbf{j} + bt\,\mathbf{k}$	$[0, 2\pi]$
8. $\mathbf{r}(t) = \langle \sin t - t\cos t, \cos t + t\sin t, t^2 \rangle$	$\left[0, \dfrac{\pi}{2}\right]$

[C] In Exercises 9 and 10, use a computer or calculator and Simpson's Rule with $n = 10$ to approximate the length of the space curve over the indicated interval.

Function	Interval
9. $\mathbf{r}(t) = t^2\mathbf{i} + t\mathbf{j} + \ln t\,\mathbf{k}$	$1 \le t \le 3$
10. $\mathbf{r}(t) = \sin \pi t\,\mathbf{i} + \cos \pi t\,\mathbf{j} + t^3\mathbf{k}$	$0 \le t \le 2$

11. Consider the helix represented by the vector-valued function

$\mathbf{r}(t) = \langle 2\cos t, 2\sin t, t \rangle$.

a. Express the length of the arc s on the helix as a function of t by evaluating the integral

$$s = \int_0^t \sqrt{[x'(u)]^2 + [y'(u)]^2 + [z'(u)]^2}\,du.$$

b. Solve for t in the relationship derived in part a, and substitute the result into the original set of parametric equations. This yields a parametrization of the curve in terms of the arc length parameter s.

c. Find the coordinates of the point on the helix when the length of the arc is $s = \sqrt{5}$ and $s = 4$.

d. Verify that $\| \mathbf{r}'(s) \| = 1$.

12. Repeat Exercise 11 for the curve represented by the vector-valued function

$\mathbf{r}(t) = \langle 4(\sin t - t\cos t), 4(\cos t + t\sin t), \frac{3}{2}t^2 \rangle$.

In Exercises 13–16, find the curvature K of the curve where s is the arc length parameter.

13. $\mathbf{r}(s) = [(2 - s)\mathbf{i} + (3 + 2s)] / \sqrt{5}$

14. $\mathbf{r}(s) = \dfrac{\sqrt{3}}{9}[(3 - 2s)^{3/2}\mathbf{i} + 2\sqrt{2}s^{3/2}\mathbf{j}]$

15. The arc length parametrization of the helix in Exercise 11

16. The arc length parametrization of the space curve in Exercise 12

In Exercises 17–20, find the curvature K of the plane curve at the specified value of the parameter.

17. $\mathbf{r}(t) = 4t\mathbf{i} - 2t\mathbf{j}$
$t = 1$

18. $\mathbf{r}(t) = t^2\mathbf{j} + \mathbf{k}$
$t = 0$

19. $\mathbf{r}(t) = t\mathbf{i} + \dfrac{1}{t}\mathbf{j}$
$t = 1$

20. $\mathbf{r}(t) = t\mathbf{i} + t^2\mathbf{j}$
$t = 1$

In Exercises 21–32, find the curvature K of the curve.

21. $\mathbf{r}(t) = 4\cos 2\pi t\,\mathbf{i} + 4\sin 2\pi t\,\mathbf{j}$

22. $\mathbf{r}(t) = 2\cos \pi t\,\mathbf{i} + \sin \pi t\,\mathbf{j}$

23. $\mathbf{r}(t) = a\cos \omega t\,\mathbf{i} + a\sin \omega t\,\mathbf{j}$

24. $\mathbf{r}(t) = a\cos \omega t\,\mathbf{i} + b\sin \omega t\,\mathbf{j}$

25. $\mathbf{r}(t) = e^t\cos t\,\mathbf{i} + e^t\sin t\,\mathbf{j}$

26. $\mathbf{r}(t) = \langle a(\omega t - \sin \omega t), a(1 - \cos \omega t) \rangle$

27. $\mathbf{r}(t) = \langle \cos \omega t + \omega t\sin \omega t, \sin \omega t - \omega t\cos \omega t \rangle$

28. $\mathbf{r}(t) = 4t\mathbf{i} - 4t\mathbf{j} + 2t\mathbf{k}$

29. $\mathbf{r}(t) = t\mathbf{i} + t^2\mathbf{j} + \dfrac{t^2}{2}\mathbf{k}$

30. $\mathbf{r}(t) = t\mathbf{i} + 3t^2\mathbf{j} + \dfrac{t^2}{2}\mathbf{k}$

31. $\mathbf{r}(t) = 4t\mathbf{i} + 3\cos t\,\mathbf{j} + 3\sin t\,\mathbf{k}$

32. $\mathbf{r}(t) = e^t\cos t\,\mathbf{i} + e^t\sin t\,\mathbf{j} + e^t\mathbf{k}$

In Exercises 33–38, find the curvature and radius of curvature of the plane curve at the specified value of x.

Function	Point
33. $y = 3x - 2$	$x = a$
34. $y = mx + b$	$x = a$
35. $y = 2x^2 + 3$	$x = -1$
36. $y = x + \dfrac{1}{x}$	$x = 1$
37. $y = \sqrt{a^2 - x^2}$	$x = 0$
38. $y = \frac{3}{4}\sqrt{16 - x^2}$	$x = 0$

Essay In Exercises 39 and 40, two circles of curvature to the graph of the function are given. (a) Find the equation of the smaller circle, and (b) write a short paragraph explaining why the circles have different radii.

39. $f(x) = \sin x$

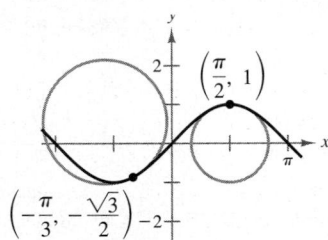

40. $f(x) = \dfrac{4x^2}{x^2 + 3}$

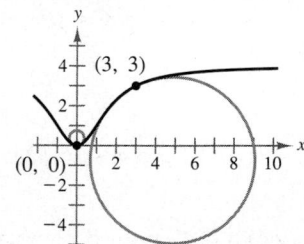

In Exercises 41–44, sketch the graph of the function and sketch the circle of curvature to the graph at the specified value of x.

Function	Point
41. $y = x + \dfrac{1}{x}$	$x = 1$
42. $y = \ln x$	$x = 1$
43. $y = e^x$	$x = 0$
44. $y = \frac{1}{3}x^3$	$x = 1$

In Exercises 45–48, (a) find the point on the curve at which curvature K is a maximum and (b) find the limit of K as $x \to \infty$.

45. $y = (x - 1)^2 + 3$ **46.** $y = x^3$

47. $y = x^{2/3}$ **48.** $y = \ln x$

49. Does the graph of $y = (x - 1)^3 + 3$ have points at which the curvature is 0? If so, find the points.

50. *Essay*   Use the result of Exercise 49 to write a paragraph describing the points on the graph of $y = f(x)$ at which the curvature is 0.

51. Show that the curvature is greatest at the endpoints of the major axis and least at the endpoints of the minor axis for the ellipse given by
$$x^2 + 4y^2 = 4.$$

52. Verify that the curvature at any point (x, y) on the graph of $y = \cosh x$ is y^2.

53. Find all a and b such that the two curves given by
$$y_1 = ax(b - x)$$
and
$$y_2 = \frac{x}{x + 2}$$
intersect at only one point and have a common tangent line and equal curvature at the point. Sketch a graph for each set of values for a and b.

54. Given a twice-differentiable function $y = f(x)$, determine its curvature at a relative extremum. Can the curvature ever be greater than it is at a relative extremum? Why or why not?

55. The smaller the curvature in a bend of a road, the faster a car can travel. Assume that the maximum speed around a turn is inversely proportional to the square root of the curvature. A car moving on the path $y = \frac{1}{3}x^3$ (x and y measured in miles) can safely go 30 miles per hour at $(1, \frac{1}{3})$. How fast can it go at $(\frac{3}{2}, \frac{9}{8})$?

56. The curve C is given by the polar equation $r = f(\theta)$. Show that the curvature K at the point (r, θ) is
$$K = \frac{|2(r')^2 - rr'' + r^2|}{[(r')^2 + r^2]^{3/2}}.$$
(*Hint:* Represent the curve by $\mathbf{r}(\theta) = r \cos\theta \, \mathbf{i} + r \sin\theta \, \mathbf{j}$.)

In Exercises 57–60, use the result of Exercise 56 to find the curvature of the polar curve.

57. $r = 1 + \sin\theta$ **58.** $r = \theta$

59. $r = a \sin\theta$ **60.** $r = e^\theta$

61. Given the polar curve $r = e^{a\theta}$, $a > 0$, find the curvature K and determine the limit of K as
  a. $\theta \to \infty$.   **b.** $a \to \infty$.

62. Show that the formula for the curvature of a polar curve $r = f(\theta)$ given in Exercise 56 reduces to
$$K = \frac{2}{|r'|}$$
for the curvature *at the pole*.

In Exercises 63 and 64, use the result of Exercise 62 to find the curvature of the rose curve at the pole.

63. $r = 4 \sin 2\theta$ **64.** $r = 6 \cos 3\theta$

65. For a smooth curve given by the parametric equations $x = f(t)$ and $y = g(t)$, prove that the curvature is
$$K = \frac{|f'(t)g''(t) - g'(t)f''(t)|}{\{[f'(t)]^2 + [g'(t)]^2\}^{3/2}}.$$

66. Use the result of Exercise 65 to find the curvature of the curve represented by the parametric equations
$$x(t) = t^3 \quad \text{and} \quad y(t) = \frac{1}{2}t^2.$$

In Exercises 67 and 68, use Theorem 12.11 to find a_T and a_N for the curve given by the vector-valued function.

67. $\mathbf{r}(t) = 3t^2\mathbf{i} + (3t - t^3)\mathbf{j}$

68. $\mathbf{r}(t) = t^2\mathbf{i} + 2t\mathbf{j} + t^2\mathbf{k}$

69. *Frictional Force*   A 4000-pound vehicle is driven at a speed of 30 miles per hour on a circular interchange of radius 100 feet. To keep the vehicle from skidding off course, what frictional force must the road surface exert on the tires?

70. Use the definition of curvature in space, $K = \|\mathbf{T}'(s)\| = \|\mathbf{r}''(s)\|$, to verify the following three formulas.
  a. $K = \dfrac{\|\mathbf{T}'(t)\|}{\|\mathbf{r}'(t)\|}$ **b.** $K = \dfrac{\|\mathbf{r}'(t) \times \mathbf{r}''(t)\|}{\|\mathbf{r}'(t)\|^3}$
  c. $K = \dfrac{\mathbf{a}(t) \cdot \mathbf{N}(t)}{\|\mathbf{v}(t)\|^2}$

71. *Cornu Spiral*   The cornu spiral is given by
$$x = \int_0^s \cos\frac{\pi u^2}{2}\, du, \qquad y = \int_0^s \sin\frac{\pi u^2}{2}\, du.$$

The spiral shown in the figure was plotted over the interval $-\pi \le x \le \pi$.
  a. Find the arc length of this curve from $s = 0$ to $s = a$.
  b. Find the curvature of the graph when $s = a$.
  c. The cornu spiral was discovered by James Bernoulli. He found that the spiral has an amazing relationship between curvature and arc length. What is this relationship?

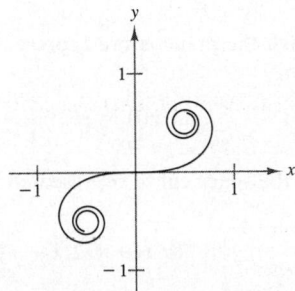

FIGURE FOR 71

72. A sphere of radius 4 is dropped into the paraboloid given by $z = x^2 + y^2$.
  a. How close will the sphere come to the vertex of the paraboloid?
  b. What is the radius of the largest sphere that will touch the vertex?

73. Find the curvature of the cycloid given by the parametric equations

$$x(\theta) = a(\theta - \sin\theta), \qquad y(\theta) = a(1 - \cos\theta).$$

What are the minimum and maximum values of K?

74. Let C be given by $y = f(x)$. Show that the center of curvature for (x, y) on C is $(x_0, y_0) = (x - y'z, y + z)$, where

$$z = \frac{1 + (y')^2}{y''}.$$

Find the center of curvature for $y = e^x$ at the point $(0, 1)$ on the curve.

In Exercises 75–82, you are asked to verify Kepler's Laws of Planetary Motion. For these exercises, assume that each planet moves in an orbit given by the vector-valued function r. Let $r = \|\mathbf{r}\|$, let G represent the universal gravitational constant, let M represent the mass of the sun, and let m represent the mass of the planet.

75. Prove that $\mathbf{r} \cdot \mathbf{r}' = r\dfrac{dr}{dt}$.

76. Using Newton's Second Law, $\mathbf{F} = m\mathbf{a}$, and Newton's Law of Gravitation,

$$\mathbf{F} = \frac{-GmM}{r^3}\mathbf{r},$$

show that $\mathbf{a}$ and $\mathbf{r}$ are parallel, and that

$$\mathbf{r}(t) \times \mathbf{r}'(t) = \mathbf{L}$$

is a constant vector. Hence, $\mathbf{r}(t)$ moves in a *fixed plane*, orthogonal to $\mathbf{L}$.

77. Prove that

$$\frac{d}{dt}\left[\frac{\mathbf{r}}{r}\right] = \frac{1}{r^3}\{[\mathbf{r} \times \mathbf{r}'(t)] \times \mathbf{r}\}.$$

78. Show that

$$\frac{\mathbf{r}'}{GM} \times \mathbf{L} - \frac{\mathbf{r}}{r} = \mathbf{e},$$

is a constant vector.

79. Prove Kepler's First Law: Each planet moves in an elliptical orbit with the sun as a focus.

80. Assume that the elliptical orbit

$$r = \frac{ed}{1 + e\cos\theta}$$

is in the xy-plane, with $\mathbf{L}$ along the z-axis. Prove that

$$\|\mathbf{L}\| = r^2\frac{d\theta}{dt}.$$

81. Prove Kepler's Second Law: Each ray from the sun to a planet sweeps out equal areas of the ellipse in equal times.

82. Prove Kepler's Third Law: The square of the period of a planet's orbit is proportional to the cube of the mean distance between the planet and the sun.

REVIEW EXERCISES for Chapter 12

In Exercises 1–4, (a) find the domain of r and (b) determine the values (if any) of t for which the function is discontinuous.

1. $\mathbf{r}(t) = t\,\mathbf{i} + \csc t\,\mathbf{k}$

2. $\mathbf{r}(t) = \sqrt{t}\,\mathbf{i} + \dfrac{1}{t-4}\mathbf{j} + \mathbf{k}$

3. $\mathbf{r}(t) = \ln t\,\mathbf{i} + t\,\mathbf{j} + t\,\mathbf{k}$

4. $\mathbf{r}(t) = (2t + 1)\mathbf{i} + t^2\mathbf{j} + t\,\mathbf{k}$

In Exercises 5 and 6, sketch the plane curve represented by the vector-valued function.

5. $\mathbf{r}(t) = \cos t\,\mathbf{i} + 2\sin^2 t\,\mathbf{j}$ **6.** $\mathbf{r}(t) = t\,\mathbf{i} + \dfrac{t}{t-1}\mathbf{j}$

In Exercises 7–10, sketch the space curve represented by the vector-valued function.

7. $\mathbf{r}(t) = \mathbf{i} + t\,\mathbf{j} + t^2\mathbf{k}$ **8.** $\mathbf{r}(t) = 2t\,\mathbf{i} + t\,\mathbf{j} + t^2\mathbf{k}$

9. $\mathbf{r}(t) = 2\cos t\,\mathbf{i} + t\,\mathbf{j} + 2\sin t\,\mathbf{k}$

10. $\mathbf{r}(t) = \mathbf{i} + \sin t\,\mathbf{j} + \mathbf{k}$

[C] In Exercises 11 and 12, use a graphing utility to graph the space curve represented by the vector-valued function.

11. $\mathbf{r}(t) = \frac{1}{2}t\,\mathbf{i} + \sqrt{t}\,\mathbf{j} + \frac{1}{4}t^3\mathbf{k}$

12. $\mathbf{r}(t) = t\,\mathbf{i} + \ln t\,\mathbf{j} + \frac{1}{2}t^2\mathbf{k}$

In Exercises 13 and 14, sketch the space curve represented by the intersection of the surfaces. Use the parameter $x = t$ to find a vector-valued function for the space curve.

13. $z = x^2 + y^2$ **14.** $x^2 + z^2 = 4$
     $x + y = 0$ $x - y = 0$

In Exercises 15 and 16, find the indicated limit.

15. $\lim\limits_{t \to 2}(t^2\mathbf{i} + \sqrt{4 - t^2}\,\mathbf{j} + \mathbf{k})$

16. $\lim\limits_{t \to 0}\left(\dfrac{\sin 2t}{t}\mathbf{i} + e^{-t}\mathbf{j} + e^t\mathbf{k}\right)$

In Exercises 17 and 18, find the following.
  a. $\mathbf{r}'(t)$ **b.** $\mathbf{r}''(t)$
  c. $D_t[\mathbf{r}(t) \cdot \mathbf{u}(t)]$ **d.** $D_t[\mathbf{u}(t) - 2\mathbf{r}(t)]$
  e. $D_t[\|\mathbf{r}(t)\|], \quad t > 0$ **f.** $D_t[\mathbf{r}(t) \times \mathbf{u}(t)]$

17. $\mathbf{r}(t) = 3t\,\mathbf{i} + (t - 1)\mathbf{j}$
    $\mathbf{u}(t) = t\,\mathbf{i} + t^2\mathbf{j} + \frac{2}{3}t^3\mathbf{k}$

18. $\mathbf{r}(t) = \sin t\,\mathbf{i} + \cos t\,\mathbf{j} + t\,\mathbf{k}$
    $\mathbf{u}(t) = \sin t\,\mathbf{i} + \cos t\,\mathbf{j} + \dfrac{1}{t}\mathbf{k}$

In Exercises 19 and 20, find a set of parametric equations for the line tangent to the space curve at the indicated point.

19. $\mathbf{r}(t) = 2\cos t\,\mathbf{i} + 2\sin t\,\mathbf{j} + t\,\mathbf{k}, \qquad t = \dfrac{3\pi}{4}$

20. $\mathbf{r}(t) = t\,\mathbf{i} + t^2\mathbf{j} + \frac{2}{3}t^3\mathbf{k}, \qquad t = 2$

In Exercises 21–24, find the indefinite integral.

21. $\displaystyle \int (\cos t\,\mathbf{i} + t\cos t\,\mathbf{j})\,dt$

22. $\displaystyle \int (\ln t\,\mathbf{i} + t\ln t\,\mathbf{j} + \mathbf{k})\,dt$

23. $\displaystyle \int \|\cos t\,\mathbf{i} + \sin t\,\mathbf{j} + t\,\mathbf{k}\|\,dt$

24. $\displaystyle \int (t\,\mathbf{j} + t^2\mathbf{k}) \times (\mathbf{i} + t\,\mathbf{j} + t\,\mathbf{k})\,dt$

In Exercises 25 and 26, find $\mathbf{r}(t)$ for the given condition.

25. $\mathbf{r}'(t) = 2t\,\mathbf{i} + e^t\mathbf{j} + e^{-t}\mathbf{k}$
$\mathbf{r}(0) = \mathbf{i} + 3\mathbf{j} - 5\mathbf{k}$

26. $\mathbf{r}'(t) = \sec t\,\mathbf{i} + \tan t\,\mathbf{j} + t^2\mathbf{k}$
$\mathbf{r}(0) = 3\mathbf{k}$

In Exercises 27 and 28, evaluate the definite integral.

27. $\displaystyle \int_{-2}^{2} (3t\,\mathbf{i} + 2t^2\mathbf{j} - t^3\mathbf{k})\,dt$

28. $\displaystyle \int_{0}^{1} \left(\sqrt{t}\,\mathbf{j} + t\sin t\,\mathbf{k}\right)\,dt$

Projectile Motion In Exercises 29–32, use the model for projectile motion and assume there is no air resistance.

29. *Projectile Motion* A projectile is fired from ground level at an angle of elevation of 30°. Find the range of the projectile if the initial velocity is 75 feet per second.

30. *Conveyor Belt* The center of a truckbed is 6 feet below and 4 feet horizontally from the end of a horizontal conveyor that is discharging gravel (see figure). Determine the speed ds/dt at which the conveyor belt should be moving so that the gravel falls onto the center of the truck bed.

FIGURE FOR 30

31. *Projectile Motion* A projectile is fired from ground level at an angle of 20° with the horizontal. Find the minimum initial velocity necessary if the projectile is to have a range of 250 feet.

C **32.** Use a graphing utility to graph the paths of a projectile if $v_0 = 50$ feet per second and
a. $\theta = 30°$.   **b.** $\theta = 45°$.   **c.** $\theta = 60°$.
Use the graphs to approximate the maximum height and range of the projectile for each case.

In Exercises 33–40, find the velocity, speed, and acceleration at time t. Then, find $\mathbf{a}\cdot\mathbf{T}$, $\mathbf{a}\cdot\mathbf{N}$, and the curvature at time t.

33. $\mathbf{r}(t) = (1 + 4t)\mathbf{i} + (2 - 3t)\mathbf{j}$

34. $\mathbf{r}(t) = 5t\,\mathbf{i}$

35. $\mathbf{r}(t) = 2(t + 1)\mathbf{i} + \dfrac{2}{t + 1}\mathbf{j}$

36. $\mathbf{r}(t) = t\,\mathbf{i} + \sqrt{t}\,\mathbf{j}$

37. $\mathbf{r}(t) = e^t\mathbf{i} + e^{-t}\mathbf{j}$

38. $\mathbf{r}(t) = t\cos t\,\mathbf{i} + t\sin t\,\mathbf{j}$

39. $\mathbf{r}(t) = t\,\mathbf{i} + t^2\mathbf{j} + \frac{1}{2}t^2\mathbf{k}$

40. $\mathbf{r}(t) = (t - 1)\mathbf{i} + t\,\mathbf{j} + \dfrac{1}{t}\mathbf{k}$

In Exercises 41 and 42, find the length of the space curve over the indicated interval.

41. $\mathbf{r}(t) = \frac{1}{2}t\,\mathbf{i} + \sin t\,\mathbf{j} + \cos t\,\mathbf{k}, \qquad 0 \le t \le \pi$

42. $\mathbf{r}(t) = e^t\sin t\,\mathbf{i} + e^t\cos t\,\mathbf{k}, \qquad 0 \le t \le \pi$

43. *Satellite Orbit* Find the speed necessary for a satellite to maintain a circular orbit 600 miles above the surface of the earth.

44. *Centripetal Force* An automobile is in a circular traffic exchange and its speed is twice that posted. By what factor is the centripetal force increased over that which would occur at the posted speed?

45. *Highway Construction* A civil engineer designs a highway as indicated in the figure. BC is an arc of a circle. AB and CD are straight lines tangent to the circular arc. Criticize the design.

46. *Exit Ramp* A highway has an exit ramp that begins at the origin of a coordinate system and follows the curve given by

$$y = \frac{1}{32}x^{5/2}$$

to the point $(4, 1)$ (see figure). Then it follows a circular path whose curvature is that given by the curve at $(4, 1)$. What is the radius of the circular arc?

FIGURE FOR 45

FIGURE FOR 46

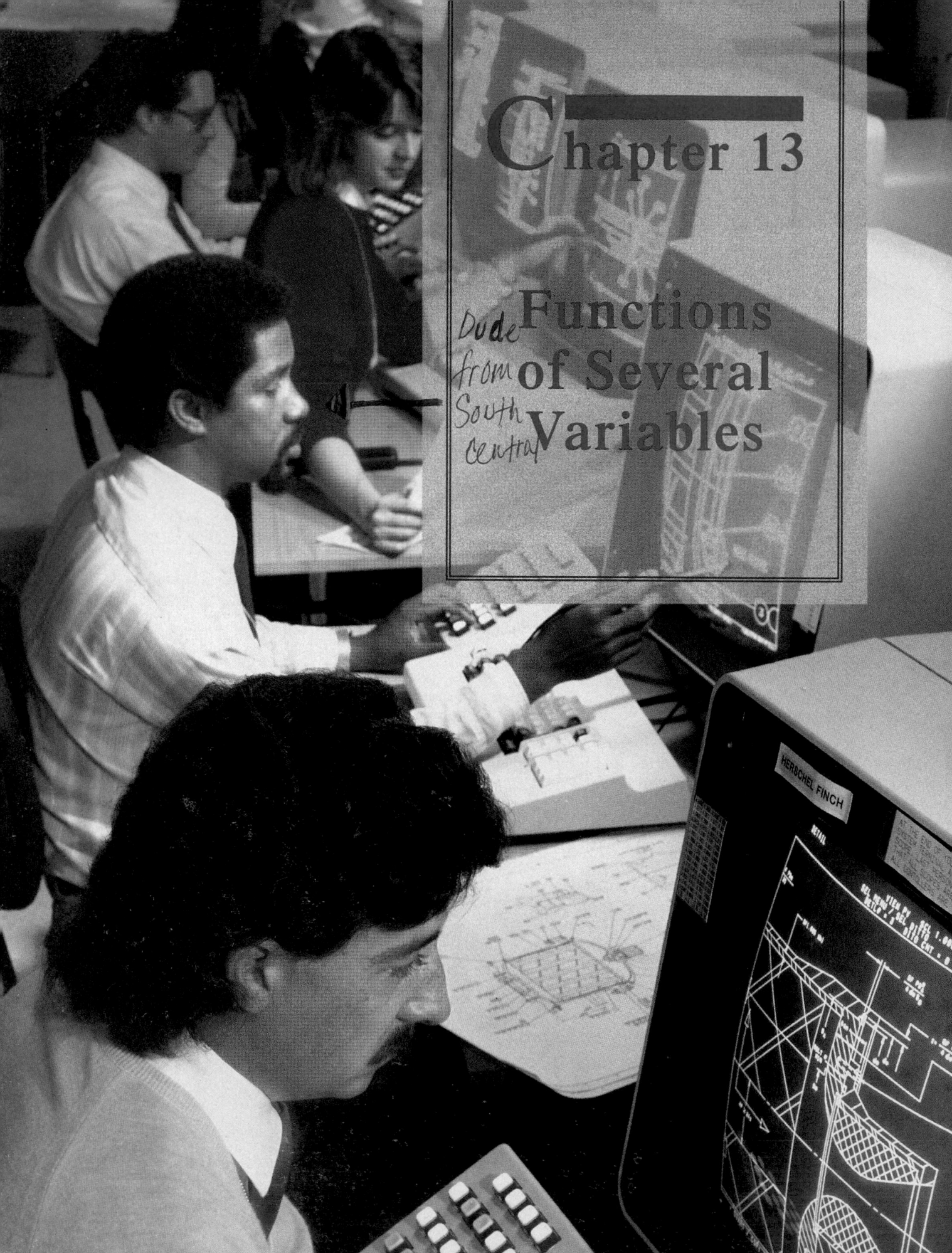

Chapter 13

Functions of Several Variables

13.1

Functions of Several Variables • The Graph of a Function of Two Variables • Level Curves • Level Surfaces • Computer Graphics

Functions of Several Variables

So far in the text, we have dealt only with functions of single (independent) variables. Many familiar quantities, however, are functions of two or more variables. For instance, the work done by a force ($W = FD$) and the volume of a right circular cylinder ($V = \pi r^2 h$) are both functions of two variables. The volume of a rectangular solid ($V = lwh$) is a function of three variables. The notation for a function of two or more variables is similar to that for a function of a single variable. Here are two examples.

$$z = \underbrace{f(x, y)}_{\text{2 variables}} = x^2 + xy$$

and

$$w = \underbrace{f(x, y, z)}_{\text{3 variables}} = x + 2y - 3z.$$

Definition of a Function of Two Variables

Let D be a set of ordered pairs of real numbers. If to each ordered pair (x, y) in D there corresponds a unique real number $f(x, y)$, then f is called a **function of x and y.** The set D is the **domain** of f, and the corresponding set of values for $f(x, y)$ is the **range** of f.

For the function given by $z = f(x, y)$, we call x and y the **independent variables** and z the **dependent variable.**

Similar definitions can be given to functions of three, four, or n variables, where the domains consist of ordered triples (x_1, x_2, x_3), quadruples (x_1, x_2, x_3, x_4), and n-tuples $(x_1, x_2, \ldots, x_n)$, respectively. In all cases, the range is a set of real numbers. In this chapter, we limit the discussion to functions of two or three variables.

As with functions of one variable, the most common way to describe a function of several variables is with an *equation,* and unless otherwise restricted, you can assume that the domain is the set of all points for which the equation is defined. For instance, the domain of the function given by

$$f(x, y) = x^2 + y^2$$

is assumed to be the entire xy-plane.

◀ *Computers can be used to sketch surfaces in space. In Section 13.1, Exercises 39–42, you are asked to use a computer graphing utility to sketch three-dimensional surfaces.*

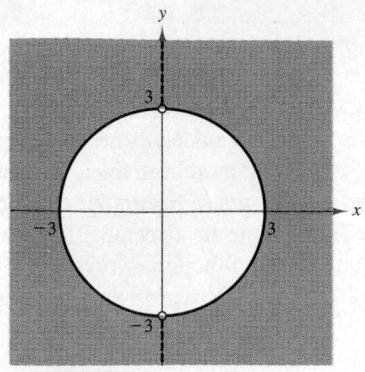

Domain of
$$f(x, y) = \frac{\sqrt{x^2 + y^2 - 9}}{x}$$

FIGURE 13.1

EXAMPLE 1   Finding the Domains of Functions of Several Variables

Find the domains of the following functions.

a. $f(x, y) = \dfrac{\sqrt{x^2 + y^2 - 9}}{x}$ **b.** $g(x, y, z) = \dfrac{x}{\sqrt{9 - x^2 - y^2 - z^2}}$

Solution

a. The function f is defined for all points (x, y) such that $x \neq 0$ and

$$x^2 + y^2 \geq 9.$$

Thus, the domain is the set of all points lying on or outside the circle $x^2 + y^2 = 9$, *except* those on the y-axis, as shown in Figure 13.1.

b. The function g is defined for all points (x, y, z) such that

$$x^2 + y^2 + z^2 < 9.$$

Consequently, the domain is the set of all points (x, y, z) lying inside a sphere of radius 3 that is centered at the origin.

Functions of several variables can be combined in the same ways as functions of single variables. For instance, you can form the sum, difference, product, and quotient of two functions of two variables as follows.

$$(f \pm g)(x, y) = f(x, y) \pm g(x, y)$$ Sum or difference
$$(fg)(x, y) = f(x, y)g(x, y)$$ Product
$$\frac{f}{g}(x, y) = \frac{f(x, y)}{g(x, y)}, \qquad g(x, y) \neq 0$$ Quotient

You cannot form the composite of two functions of several variables. However, if h is a function of several variables and g is a function of a single variable, then you can form the **composite** function $(g \circ h)(x, y)$ as follows.

$$(g \circ h)(x, y) = g(h(x, y))$$ Composition

The domain of this composite function consists of all (x, y) in the domain of h such that $h(x, y)$ is in the domain of g. For example, the function given by

$$f(x, y) = \sqrt{16 - 4x^2 - y^2}$$

can be viewed as the composite of the function of two variables given by $h(x, y) = 16 - 4x^2 - y^2$ and the function of a single variable given by $g(u) = \sqrt{u}$. The domain of this function is the set of all points lying on or inside the ellipse given by $4x^2 + y^2 = 16$.

A function that can be expressed as a sum of functions of the form $cx^m y^n$ (where c is a real number and m and n are nonnegative integers) is called a **polynomial function** of two variables. For instance, the functions given by

$$f(x, y) = x^2 + y^2 - 2xy + x + 2 \quad \text{and} \quad g(x, y) = 3xy^2 + x - 2$$

are polynomial functions of two variables. A **rational function** is the quotient of two polynomial functions. Similar terminology is used for functions of more than two variables.

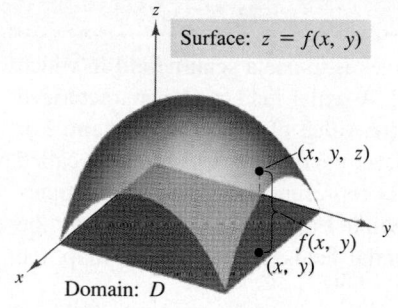

FIGURE 13.2

The Graph of a Function of Two Variables

As with functions of single variables, you can learn a lot about the behavior of a function of two variables by sketching its graph. The **graph** of a function f of two variables is the set of all points (x, y, z) for which $z = f(x, y)$ and (x, y) is in the domain of f. This graph can be interpreted geometrically as a *surface in space,* as discussed in Sections 11.5 and 11.6. In Figure 13.2, note that the graph of $z = f(x, y)$ is a surface whose projection onto the xy-plane is D, the domain of f. Consequently, to each point (x, y) in D there corresponds a point (x, y, z) on the surface, and, conversely, to each point (x, y, z) on the surface there corresponds a point (x, y) in D.

EXAMPLE 2   Describing the Graph of a Function of Two Variables

What is the range of $f(x, y) = \sqrt{16 - 4x^2 - y^2}$? Describe the graph of f.

Solution   The domain D implied by the equation for f is the set of all points (x, y) such that $16 - 4x^2 - y^2 \geq 0$. Thus, D is the set of all points lying on or inside the ellipse given by

$$\frac{x^2}{4} + \frac{y^2}{16} = 1.$$

The range of f is all values $z = f(x, y)$ such that $0 \leq z \leq \sqrt{16}$ or

$$0 \leq z \leq 4. \qquad \text{Range of } f$$

A point (x, y, z) is on the graph of f if and only if

$$z = \sqrt{16 - 4x^2 - y^2}$$
$$z^2 = 16 - 4x^2 - y^2$$
$$4x^2 + y^2 + z^2 = 16$$
$$\frac{x^2}{4} + \frac{y^2}{16} + \frac{z^2}{16} = 1, \qquad 0 \leq z \leq 4.$$

From Section 11.6, you know that the graph of f is the upper half of an ellipsoid, as shown in Figure 13.3.

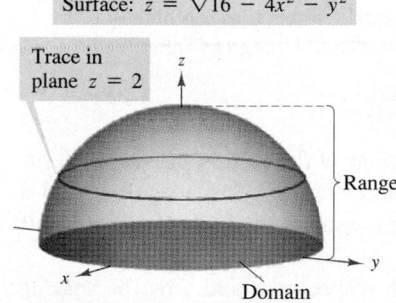

FIGURE 13.3
Graph of
$f(x, y) = \sqrt{16 - 4x^2 - y^2}$

To sketch a surface in space *by hand,* it helps to use traces in planes parallel to the coordinate planes, as shown in Figure 13.3. For example, to find the trace of the surface in the plane $z = 2$, substitute $z = 2$ in the equation $z = \sqrt{16 - 4x^2 - y^2}$ and obtain

$$2 = \sqrt{16 - 4x^2 - y^2} \quad \rightarrow \quad \frac{x^2}{3} + \frac{y^2}{12} = 1.$$

Thus, the trace is an ellipse centered at the point $(0, 0, 2)$ with major and minor axes of lengths $4\sqrt{3}$ and $2\sqrt{3}$, respectively.

Traces are also used with most three-dimensional graphing utilities. For instance, Figure 13.4 shows a computer-generated version of the surface given in Example 2. For this sketch, the computer took 20 traces parallel to the xy-plane and 10 traces in vertical planes.

If you have access to a three-dimensional graphing utility, try using it to sketch the graphs of several surfaces.

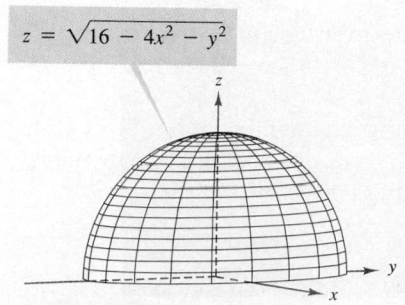

FIGURE 13.4
Computer-generated graph

Level Curves

A second way to visualize a function of two variables is to use a **scalar field** in which the scalar $z = f(x, y)$ is assigned to the point (x, y). A scalar field can be characterized by **level curves** (or **contour lines**) along which the value of $f(x, y)$ is constant. For instance, the weather map in Figure 13.5 shows level curves of equal pressure called **isobars.** In weather maps for which the level curves represent points of equal temperature, the level curves are called **isotherms,** as shown in Figure 13.6. Another common use of level curves is in representing electric potential fields. In this type of map, the level curves are called **equipotential lines.**

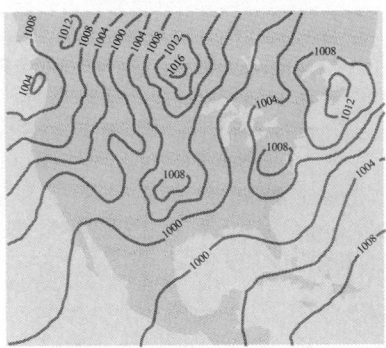

FIGURE 13.5
Level curves show the lines of equal pressure (isobars) measured in millibars.

FIGURE 13.6
Level curves show the lines of equal temperature (isotherms) measured in degrees Fahrenheit.

Contour maps are commonly used to show regions of the earth's surface, with the level curves representing the height above sea level. This type of map is called a **topographic map.** For example, the topographic map in Figure 13.7 represents Half Dome and vicinity in Yosemite Valley, California, as shown in Figure 13.8.

A contour map depicts the variation of z with respect to x and y by the spacing between level curves. Much space between level curves indicates that z is changing slowly, whereas little space indicates a rapid change in z. Furthermore, to give a good three-dimensional illusion in a contour map, it is important to choose c-values that are *evenly spaced.*

FIGURE 13.7

FIGURE 13.8

EXAMPLE 3   Sketching a Contour Map

The hemisphere given by $f(x, y) = \sqrt{64 - x^2 - y^2}$ is shown in Figure 13.9. Sketch a contour map for this surface using level curves corresponding to $c = 0, 1, 2, \ldots, 8$.

Solution   For each value of c, the equation given by $f(x, y) = c$ is a circle (or point) in the xy-plane. For example, when $c_1 = 0$ the level curve is

$$x^2 + y^2 = 64 \qquad \text{Circle of radius 8}$$

which is a circle of radius 8. Figure 13.10 shows the nine level curves for the hemisphere.

FIGURE 13.9
Level curves

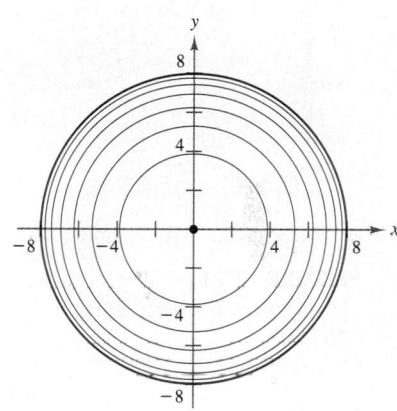

FIGURE 13.10
Contour map

EXAMPLE 4   Sketching a Contour Map

The hyperbolic paraboloid given by

$$z = y^2 - x^2$$

is shown in Figure 13.11. Sketch a contour map for this surface.

Solution   For each value of c, we let $f(x, y) = c$ and sketch the resulting level curve in the xy-plane. For this function, each of the level curves ($c \neq 0$) is a hyperbola whose asymptotes are the lines $y = \pm x$. If $c < 0$, the transverse axis is horizontal. For instance, the level curve for $c = -4$ is given by

$$\frac{x^2}{2^2} - \frac{y^2}{2^2} = 1. \qquad \text{Hyperbola with horizontal transverse axis}$$

If $c > 0$, the transverse axis is vertical. For instance, the level curve for $c = 4$ is given by

$$\frac{y^2}{2^2} - \frac{x^2}{2^2} = 1. \qquad \text{Hyperbola with vertical transverse axis}$$

If $c = 0$, the level curve is the degenerate conic representing the intersecting asymptotes, as shown in Figure 13.12.

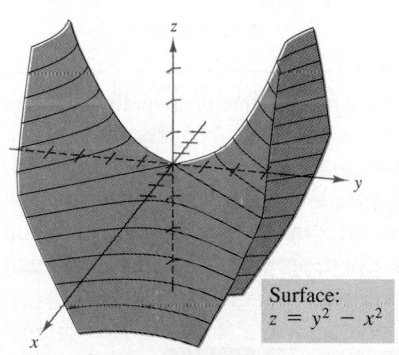

Surface:
$z = y^2 - x^2$

Hyperbolic Paraboloid

FIGURE 13.11

FIGURE 13.12
Hyperbolic level curves (at increments of 2)

One example of a function of two variables used in economics is the **Cobb-Douglas production function.** This function is used as a model to represent the number of units produced by varying amounts of labor and capital. If x measures the units of labor and y measures the units of capital, then the number of units produced is given by

$$f(x, y) = Cx^a y^{1-a}$$

where C is constant and $0 < a < 1$.

$$z = 100x^{0.6}y^{0.4}$$

$c = 80{,}000$ $c = 160{,}000$

(1000, 500)   (2000, 1000)

FIGURE 13.13
Level curves (at increments of 10,000)

EXAMPLE 5   The Cobb-Douglas Production Function

A manufacturer estimates a production function to be $f(x, y) = 100x^{0.6}y^{0.4}$, where x is the number of units of labor and y is the number of units of capital. Compare the production level when $x = 1000$ and $y = 500$ with the production level when $x = 2000$ and $y = 1000$.

Solution   When $x = 1000$ and $y = 500$, the production level is

$$f(1000,\ 500) = 100(1000^{0.6})(500^{0.4}) \approx 100(63.10)(12.01) \approx 75{,}786.$$

When $x = 2000$ and $y = 1000$, the production level is

$$f(2000,\ 1000) = 100(2000^{0.6})(1000^{0.4}) \approx 100(95.64)(15.85) \approx 151{,}572.$$

The level curves of the surface $z = f(x, y)$ are shown in Figure 13.13. Note that by doubling *both x and y*, you double the production level (see Exercise 66).

Level Surfaces

The concept of a level curve can be extended by one dimension to define a **level surface.** If f is a function of three variables and c is a constant, then the graph of the equation $f(x, y, z) = c$ is a **level surface** of the function f, as shown in Figure 13.14.

With computers, engineers and scientists have developed other ways to view functions of three variables. For instance, Figure 13.15 shows a computer simulation that uses color to represent the pressure waves of a high-speed train traveling through a tunnel.

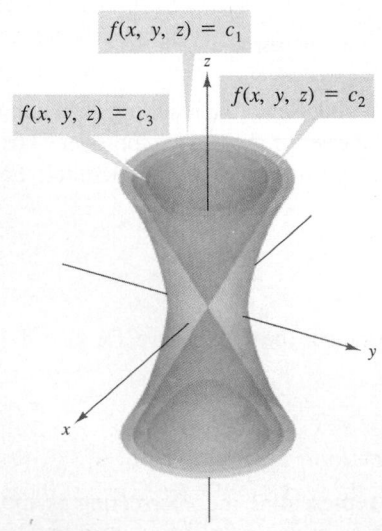

$f(x, y, z) = c_1$

$f(x, y, z) = c_3$ $f(x, y, z) = c_2$

FIGURE 13.14
Level surfaces of f

High–Speed Train (M=0.40)
Running in a Tunnel
Pressure Contour Plots
0.92   1.08

FIGURE 13.15

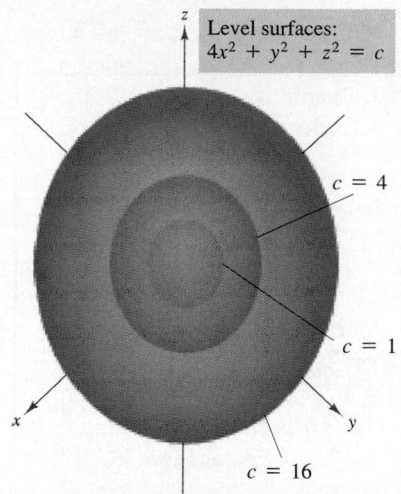

Level surfaces:
$4x^2 + y^2 + z^2 = c$

$c = 4$

$c = 1$

$c = 16$

FIGURE 13.16

MARY FAIRFAX SOMERVILLE

Somerville (1780–1872) was interested in the problem of creating geometric models for functions of several variables. Her most well-known book, *The Mechanics of the Heavens*, was published in 1831.

$f(x, y) = (x^2 + y^2)e^{1 - x^2 - y^2}$

FIGURE 13.17
Computer-generated graph

EXAMPLE 6   Level Surfaces

Describe the level surfaces of the function

$$f(x, y, z) = 4x^2 + y^2 + z^2.$$

Solution   Each level surface has an equation of the form

$$4x^2 + y^2 + z^2 = c. \qquad \text{Equation of level surface}$$

Therefore, the level surfaces are ellipsoids (whose cross sections parallel to the yz-plane are circles). As c increases, the radii of the circular cross sections increase according to the square root of c. For example, the level surfaces corresponding to the values $c = 1$, $c = 4$, and $c = 16$ are as follows.

$$\frac{x^2}{1/4} + \frac{y^2}{1} + \frac{z^2}{1} = 1 \qquad \text{Level surface for } c = 1 \text{ (Ellipse)}$$

$$\frac{x^2}{1} + \frac{y^2}{4} + \frac{z^2}{4} = 1 \qquad \text{Level surface for } c = 4 \text{ (Ellipse)}$$

$$\frac{x^2}{4} + \frac{y^2}{16} + \frac{z^2}{16} = 1 \qquad \text{Level surface for } c = 16 \text{ (Ellipse)}$$

These level surfaces are shown in Figure 13.16.   ■

REMARK   If the function in Example 6 represents the *temperature* at the point (x, y, z), then the level surfaces shown in Figure 13.16 would be called **isothermal surfaces.**

Computer Graphics

The problem of sketching the graph of a surface in space can be simplified by using a computer. Although there are several types of three-dimensional graphing utilities, most use some form of trace analysis to give the illusion of three dimensions. To use such a graphing utility, you usually need to enter the equation of the surface, the region in the xy-plane over which the surface is to be plotted, and the number of traces to be taken. For instance, to sketch the surface given by

$$f(x, y) = (x^2 + y^2)e^{1 - x^2 - y^2}$$

you might choose the following bounds for x, y, and z.

$$-3 \leq x \leq 3 \qquad \text{Bounds for } x$$
$$-3 \leq y \leq 3 \qquad \text{Bounds for } y$$
$$0 \leq z \leq 3 \qquad \text{Bounds for } z$$

Figure 13.17 shows a computer-generated sketch of this surface using 26 traces taken parallel to the yz-plane. To heighten the three-dimensional effect, the program uses a "hidden line" routine. That is, it begins by plotting the traces in the foreground (those corresponding to the largest x-values), and then, as each new trace is plotted, the program determines whether all or only part of the next trace should be shown.

Table 13.1 shows a variety of surfaces that were plotted by computer.

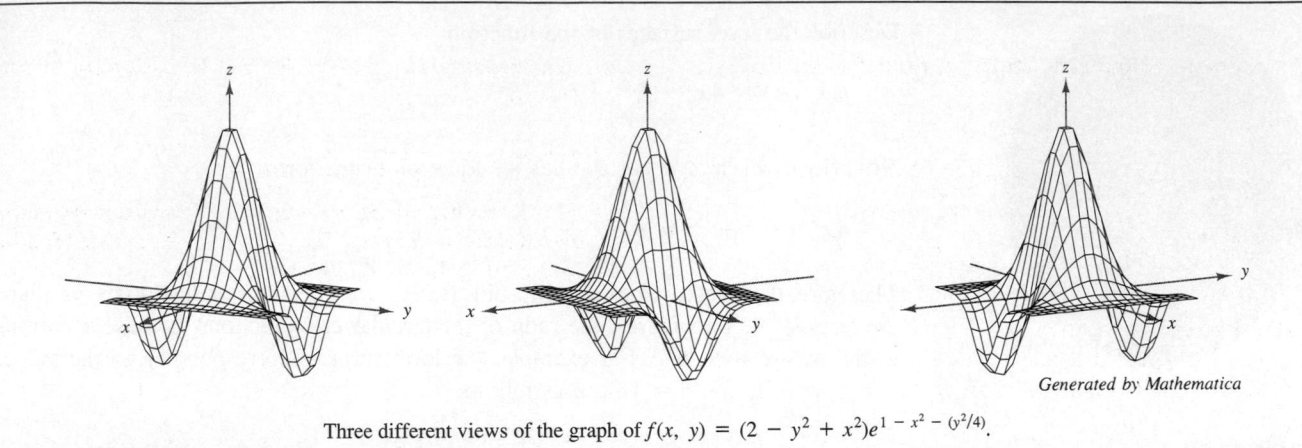

Three different views of the graph of $f(x, y) = (2 - y^2 + x^2)e^{1 - x^2 - (y^2/4)}$.

$$f(x, y) = \frac{-4x}{x^2 + y^2 + 1}$$

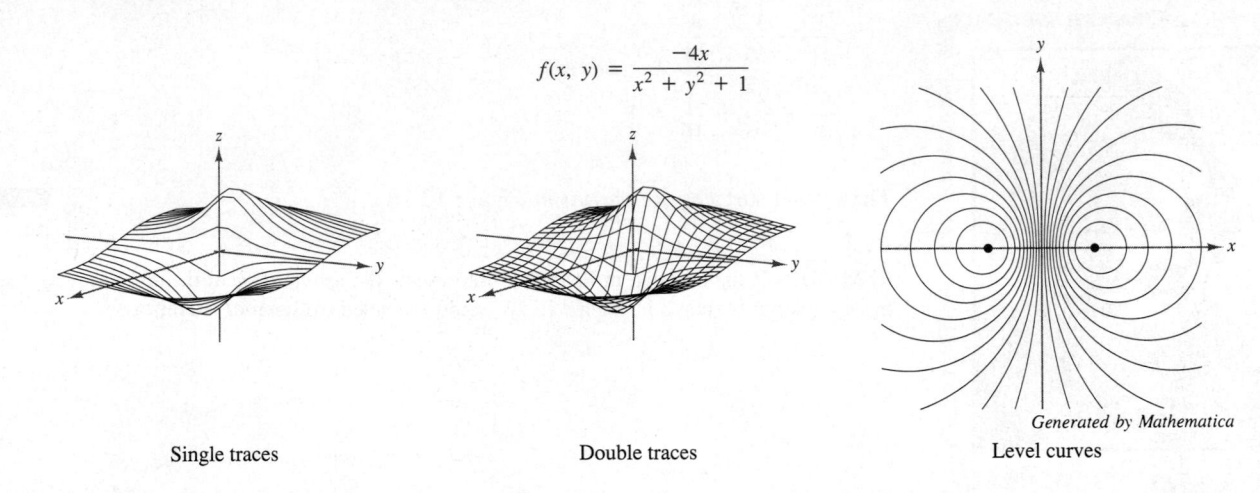

Single traces   Double traces   Level curves

$f(x, y) = \sin x \sin y$ $f(x, y) = -\dfrac{1}{\sqrt{x^2 + y^2}}$ $f(x, y) = \dfrac{1 - x^2 - y^2}{\sqrt{|1 - x^2 - y^2|}}$

EXERCISES for Section 13.1

TECHNOLOGY
Laboratory Guide
Lab 13.1

In Exercises 1–12, find and simplify the function values.

1. $f(x, y) = \dfrac{x}{y}$
   a. $(3, 2)$ b. $(-1, 4)$ c. $(30, 5)$
   d. $(5, y)$ e. $(x, 2)$ f. $(5, t)$

2. $f(x, y) = 4 - x^2 - 4y^2$
   a. $(0, 0)$ b. $(0, 1)$ c. $(2, 3)$
   d. $(1, y)$ e. $(x, 0)$ f. $(t, 1)$

3. $f(x, y) = xe^y$
   a. $(5, 0)$ b. $(3, 2)$ c. $(2, -1)$
   d. $(5, y)$ e. $(x, 2)$ f. (t, t)

4. $g(x, y) = \ln |x + y|$
   a. $(2, 3)$ b. $(5, 6)$ c. $(e, 0)$
   d. $(0, 1)$ e. $(2, -3)$ f. (e, e)

5. $h(x, y, z) = \dfrac{xy}{z}$
   a. $(2, 3, 9)$ b. $(1, 0, 1)$

6. $f(x, y, z) = \sqrt{x + y + z}$
   a. $(0, 5, 4)$ b. $(6, 8, -3)$

7. $f(x, y) = x \sin y$
   a. $\left(2, \dfrac{\pi}{4}\right)$ b. $(3, 1)$

8. $V(r, h) = \pi r^2 h$
   a. $(3, 10)$ b. $(5, 2)$

9. $g(x, y) = \displaystyle\int_x^y (2t - 3)\, dt$
   a. $(0, 4)$ b. $(1, 4)$

10. $g(x, y) = \displaystyle\int_x^y \dfrac{1}{t}\, dt$
    a. $(4, 1)$ b. $(6, 3)$

11. $f(x, y) = x^2 - 2y$
    a. $\dfrac{f(x + \Delta x, y) - f(x, y)}{\Delta x}$ b. $\dfrac{f(x, y + \Delta y) - f(x, y)}{\Delta y}$

12. $f(x, y) = 3xy + y^2$
    a. $\dfrac{f(x + \Delta x, y) - f(x, y)}{\Delta x}$ b. $\dfrac{f(x, y + \Delta y) - f(x, y)}{\Delta y}$

In Exercises 13–24, describe the domain and range of the function.

13. $f(x, y) = \sqrt{4 - x^2 - y^2}$

14. $f(x, y) = \sqrt{4 - x^2 - 4y^2}$

15. $f(x, y) = \arcsin(x + y)$ 16. $f(x, y) = \arccos(y/x)$

17. $f(x, y) = \ln(4 - x - y)$ 18. $f(x, y) = \ln(4 - xy)$

19. $z = \dfrac{x + y}{xy}$ 20. $z = \dfrac{xy}{x - y}$

21. $f(x, y) = e^{x/y}$ 22. $f(x, y) = x^2 + y^2$

23. $g(x, y) = \dfrac{1}{xy}$ 24. $g(x, y) = x\sqrt{y}$

25. The graphs labeled (a), (b), (c), and (d) are graphs of the function

$$f(x, y) = \frac{-4x}{x^2 + y^2 + 1}.$$

Match the four graphs with the points in space from which the surface is viewed. The four points are $(20, 15, 25)$, $(-15, 10, 20)$, $(20, 20, 0)$, and $(20, 0, 0)$.

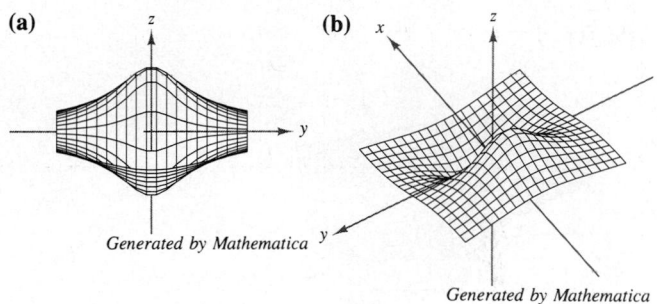

(a)   (b)

Generated by Mathematica

Generated by Mathematica

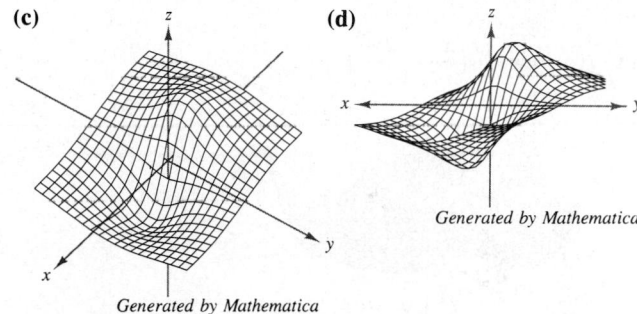

(c)   (d)

Generated by Mathematica

Generated by Mathematica

26. Use the function described in Exercise 25.
    a. Find the domain and range of the function.
    b. Identify the points in the xy-plane where the function value is 0.
    c. Does the surface pass through all the octants of the rectangular coordinate system? Give reasons for your answer.

In Exercises 27–30, match the graph of the given surface with one of the contour maps. [The contour maps are labeled (a), (b), (c), and (d).]

27. $f(x, y) = e^{1 - x^2 - y^2}$

28. $f(x, y) = e^{1-x^2 + y^2}$

29. $f(x, y) = \ln |y - x^2|$

30. $f(x, y) = \cos\left(\dfrac{x^2 + 2y^2}{4}\right)$

(a)

(b)

(c)

(d)

In Exercises 31–38, sketch the graph of the surface specified by the function.

31. $f(x, y) = 5$ **32.** $f(x, y) = 6 - 2x - 3y$

33. $f(x, y) = y^2$ **34.** $g(x, y) = \frac{1}{2}x$

35. $z = 4 - x^2 - y^2$ **36.** $z = \sqrt{x^2 + y^2}$

37. $f(x, y) = e^{-x}$

38. $f(x, y) = \begin{cases} xy, & x \geq 0, y \geq 0 \\ 0, & x < 0 \text{ or } y < 0 \end{cases}$

*C In Exercises 39–42, use a graphing utility to graph the function.

39. $z = y^2 - x^2 + 1$ **40.** $z = \frac{1}{12}\sqrt{144 - 16x^2 - 9y^2}$

41. $f(x, y) = x^2 e^{(-xy/2)}$ **42.** $f(x, y) = x \sin y$

43. a. Sketch the graph of the surface given by $f(x, y) = x^2 + y^2$.

   b. On the surface in part a, sketch the graphs of $z = f(1, y)$ and $z = f(x, 1)$.

44. a. Sketch the graph of the surface given by $f(x, y) = xy$ in the first octant.

   b. On the surface in part a, sketch the graph of $z = f(x, x)$.

In Exercises 45–52, describe the level curves of the function. Sketch the level curves for the given c-values.

45. $z = x + y$ $c = -1, 0, 2, 4$

46. $z = 6 - 2x - 3y$ $c = 0, 2, 4, 6, 8, 10$

47. $z = \sqrt{25 - x^2 - y^2}$ $c = 0, 1, 2, 3, 4, 5$

48. $f(x, y) = x^2 + y^2$ $c = 0, 2, 4, 6, 8$

49. $f(x, y) = xy$ $c = \pm 1, \pm 2, \ldots, \pm 6$

50. $f(x, y) = e^{xy}$ $c = 1, 2, 3, 4, \frac{1}{2}, \frac{1}{3}, \frac{1}{4}$

51. $f(x, y) = \dfrac{x}{x^2 + y}$ $c = \pm\dfrac{1}{2}, \pm 1, \pm\dfrac{3}{2}, \pm 2$

52. $f(x, y) = \ln(x - y)$ $c = 0, \pm\frac{1}{2}, \pm 1, \frac{3}{2}, \pm 2$

53. All of the level curves of the surface given by $z = f(x, y)$ are circles. Does this imply that the graph of f is a hemisphere? Illustrate your answer with an example.

54. Construct a function whose level curves are lines passing through the origin.

In Exercises 55–60, sketch the level surface $f(x, y, z) = c$ at the specified value of c.

55. $f(x, y, z) = x - 2y + 3z$ $c = 6$

56. $f(x, y, z) = 4x + y + 2z$ $c = 4$

57. $f(x, y, z) = x^2 + y^2 + z^2$ $c = 9$

58. $f(x, y, z) = x^2 + y^2 - z$ $c = 1$

59. $f(x, y, z) = 4x^2 + 4y^2 - z^2$ $c = 0$

60. $f(x, y, z) = \sin x - z$ $c = 0$

*A C indicates an exercise in which you are instructed to use graphing technology or a symbolic computer algebra system. The solutions of other exercises may also be facilitated by appropriate technology.

61. *Investment* In 1994, an investment of $1000 was made in a bond earning 10 percent compounded annually. Assume that the buyer pays tax at rate R and the annual rate of inflation is I. In the year 2004, the value V of the investment in constant 1994 dollars is

$$V(I, R) = 1000\left[\frac{1 + 0.10(1 - R)}{1 + I}\right]^{10}.$$

Use this function of two variables to complete the table.

	Inflation rate		
Tax rate	0	0.03	0.05
0			
0.28			
0.35			

62. *Investment* A principal of $1000 is deposited in a savings account that earns an interest rate of r (expressed as a decimal), compounded continuously. The amount $A(r, t)$ after t years is given by

$$A(r, t) = 1000e^{rt}.$$

Use this function of two variables to complete the table.

	Number of years			
Rate	5	10	15	20
0.08				
0.10				
0.12				
0.14				

63. *Forestry* The **Doyle Log Rule** is one of several methods used to determine the lumber yield of a log (in board-feet) in terms of its diameter d (in inches) and its length L (in feet). The number of board-feet is given by

$$N(d, L) = \left(\frac{d - 4}{4}\right)^2 L.$$

a. Find the number of board-feet of lumber in a log 22 inches in diameter and 12 feet in length.
b. Find $N(30, 12)$.

64. *Queuing Model* The average length of time that a customer waits in line for service is given by

$$W(x, y) = \frac{1}{x - y}, \qquad y < x$$

where y is the average arrival rate, expressed as the number of customers per unit of time, and x is the average service rate, expressed in the same units. Evaluate W at the following points.
a. $(15, 10)$ **b.** $(12, 9)$ **c.** $(12, 6)$ **d.** $(4, 2)$

65. *Temperature Distribution* The temperature T (in degrees Celsius) at any point (x, y) in a circular steel plate of radius 10 feet is

$$T = 600 - 0.75x^2 - 0.75y^2$$

where x and y are measured in feet. Sketch some of the isothermal curves.

66. *Cobb-Douglas Production Function* Use the Cobb-Douglas production function (see Example 5) to show that if the number of units of labor and the number of units of capital are doubled, then the production level is also doubled.

67. *Construction Cost* A rectangular box with an open top has a length of x feet, a width of y feet, and a height of z feet. Express the cost C of constructing the box as a function of x, y, and z if it costs $0.75 per square foot to build the base and $0.40 per square foot to build the sides.

68. *Volume* A propane tank is constructed by welding hemispheres to the ends of a right circular cylinder. Write the volume V of the tank as a function of r and l, where r is the radius of the cylinder and hemispheres and l is the length of the cylinder.

69. *Ideal Gas Law* According to the Ideal Gas Law, $PV = kT$, where P is pressure, V is volume, T is temperature (in Kelvins), and k is a constant of proportionality. A tank contains 2600 cubic inches of nitrogen at a pressure of 20 pounds per square inch and a temperature of $300°K$.
a. Determine k.
b. Express P as a function of V and T and describe the level curves.

70. *Equity* The table gives the total revenue x (in billions of dollars), the total assets in properties, plants, and equipment y (in billions of dollars), and the common stockholder's equity per share z for Phillips Petroleum Company for the years 1986 through 1991. (*Source:* 1992 Annual Report)

Year	1986	1987	1988
Revenue (x)	10.018	10.917	11.490
Assets (y)	9.186	8.772	8.417
Equity (z)	7.55	7.08	8.69

Year	1989	1990	1991
Revenue (x)	12.492	13.975	13.259
Assets (y)	7.832	8.301	8.298
Equity (z)	8.74	10.51	10.61

A model for these data is

$$z = f(x, y) = 1.085x + 0.779y - 10.778.$$

a. Use the model to approximate the equity per share if $x = 15$ and $y = 10$.
b. Which of the two variables in this model has the greater influence on the equity per share of common stock?

71. *Meteorology* Meteorologists measure the atmospheric pressure in units called millibars. From these observations they create weather maps on which the curves of equal atmospheric pressure (isobars) are drawn (see figure). If the closer the isobars the higher the wind speed, match points *A*, *B*, and *C* with

a. the highest pressure.
b. the lowest pressure.
c. the highest wind velocity.

FIGURE FOR 71

72. *Acid Rain* The acidity of rainwater is measured in units called pH. A pH of 7 is neutral, smaller values are increasingly acidic, and larger values are increasingly alkaline. The map shows the curves of equal pH and gives evidence that downwind of heavily industrialized areas the acidity has been increasing. Using the level curves on the map, determine the direction of the prevailing winds in the northeastern United States.

FIGURE FOR 72

13.2

Neighborhoods in the Plane • Limit of a Function of Two Variables • Continuity of a Function of Two Variables • Continuity of a Function of Three Variables

SONYA KOVALEVSKY

Much of the terminology used to define limits and continuity of a function of two or three variables was introduced by the German mathematician Karl Weierstrass (1815–1897). Weierstrass was a gifted teacher. One of his best known students was the Russian mathematician Sonya Kovalevsky (1850–1891), who applied many of Weierstrass's techniques to problems in mathematical physics and became one of the first women to gain acceptance as a research mathematician. (For more information on Sonya Kovalevsky, see the article "S. Kovalevsky: A Mathematical Lesson" by Karen D. Rappaport in the October, 1981 issue of the *American Mathematical Monthly*.)

Neighborhoods in the Plane

In this section, you will study limits and continuity involving functions of two or three variables. The section begins with functions of two variables. Then, at the end of the section, the concepts are extended to functions of three variables.

We begin our discussion of the limit of a function of two variables by defining a two-dimensional analog to an interval on the real line. Using the formula for the distance between two points (x, y) and (x_0, y_0) in the plane, you can define the **δ-neighborhood** about (x_0, y_0) to be the **disc** centered at (x_0, y_0) with radius $\delta > 0$

$$\left\{ (x, y) \colon \sqrt{(x - x_0)^2 + (y - y_0)^2} < \delta \right\} \qquad \text{Open disc}$$

as shown in Figure 13.18. When this formula contains the *less than* inequality, $<$, the disc is called **open,** and when it contains the *less than or equal to* inequality, $\leq$, the disc is called **closed.** This corresponds to the use of $<$ and $\leq$ to define open and closed intervals.

A point (x_0, y_0) in a plane region R is an **interior point** of R if there exists a δ-neighborhood about (x_0, y_0) that lies entirely in R, as shown in Figure 13.19. If every point in R is an interior point, then R is an **open region.** A point (x_0, y_0) is a **boundary point** of R if every open disc centered at (x_0, y_0) contains points inside R *and* points outside R. By definition, a region must contain its interior points, but it need not contain its boundary points. If a region contains all its boundary points, then the region is **closed.** A region that contains some but not all of its boundary points is neither open nor closed.

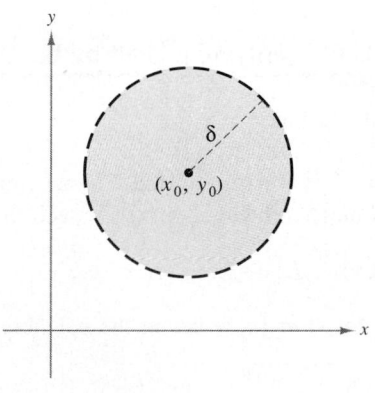

FIGURE 13.18
An open disc

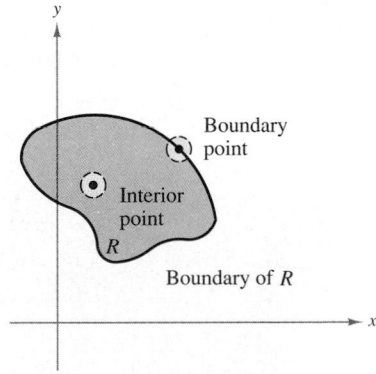

FIGURE 13.19
The boundary and interior points of a region R.

Limit of a Function of Two Variables

Definition of the Limit of a Function of Two Variables

Let f be a function of two variables defined on an open disc centered at (x_0, y_0), except possibly at (x_0, y_0), and let L be a real number. Then

$$\lim_{(x, y) \to (x_0, y_0)} f(x, y) = L$$

if for each $\varepsilon > 0$, there corresponds a $\delta > 0$ such that

$$|f(x, y) - L| < \varepsilon \quad \text{whenever} \quad 0 < \sqrt{(x - x_0)^2 + (y - y_0)^2} < \delta.$$

REMARK Graphically, this definition of a limit implies that for any point $(x, y) \neq (x_0, y_0)$ in the disc of radius δ, the value $f(x, y)$ lies between $L + \varepsilon$ and $L - \varepsilon$, as shown in Figure 13.20.

The definition of the limit of a function of two variables is similar to the definition of the limit of a function of a single variable, yet there is a critical difference. To determine whether a function of a single variable possesses a limit, you need only test the approach from two directions— from the left and from the right. If the function approaches the same limit from the right and from the left, you can conclude that the limit exists. However, for a function of two variables, the statement

$$(x, y) \to (x_0, y_0)$$

means that the point (x, y) is allowed to approach (x_0, y_0) from any "direction." If the value of

$$\lim_{(x, y) \to (x_0, y_0)} f(x, y)$$

is not the same for all possible approaches, or **paths,** to (x_0, y_0), then the limit does not exist.

FIGURE 13.20
For any (x, y) in the circle of radius δ, the value $f(x, y)$ lies between $L + \varepsilon$ and $L - \varepsilon$.

EXAMPLE 1 Verifying a Limit by the Definition

Show that $\displaystyle\lim_{(x, y) \to (a, b)} x = a$.

Solution   Let $f(x, y) = x$ and $L = a$. You need to show that for each $\varepsilon > 0$, there exists a δ-neighborhood about (a, b) such that

$$|f(x, y) - L| = |x - a| < \varepsilon$$

whenever $(x, y) \neq (a, b)$ lies in the neighborhood. You can first observe that from

$$0 < \sqrt{(x - a)^2 + (y - b)^2} < \delta$$

it follows that

$$|f(x, y) - a| = |x - a| = \sqrt{(x - a)^2} \leq \sqrt{(x - a)^2 + (y - b)^2} < \delta.$$

Thus, you can choose $\delta = \varepsilon$, and the limit is verified.

Limits of functions of several variables have the same properties regarding sums, differences, products, and quotients as do limits of functions of single variables. (See Theorem 1.2 in Chapter 1.) Some of these properties are used in the next example.

EXAMPLE 2   Verifying a Limit

Evaluate the following limits.

a. $\displaystyle\lim_{(x,y)\to(1,2)} \frac{5x^2y}{x^2+y^2}$ **b.** $\displaystyle\lim_{(x,y)\to(0,0)} \frac{5x^2y}{x^2+y^2}$

Solution

a. By using the properties of limits of products and sums, you obtain

$$\lim_{(x,y)\to(1,2)} 5x^2y = 5(1^2)(2) = 10$$

and

$$\lim_{(x,y)\to(1,2)} (x^2+y^2) = (1^2+2^2) = 5.$$

Because the limit of a quotient is equal to the quotient of the limits (and the denominator is not 0), you have

$$\lim_{(x,y)\to(1,2)} \frac{5x^2y}{x^2+y^2} = \frac{10}{5} = 2.$$

b. In this case, the limits of the numerator and of the denominator are both 0, and so you cannot determine the existence (or nonexistence) of a limit by taking the limits of the numerator and denominator separately and then dividing. However, from the graph of f in Figure 13.21, it seems reasonable that the limit might be 0. Thus, you try applying the definition to $L = 0$. First, note that

$$|y| \le \sqrt{x^2+y^2} \quad \text{and} \quad \frac{x^2}{x^2+y^2} \le 1.$$

Then, in a δ-neighborhood about $(0, 0)$, you have $0 < \sqrt{x^2+y^2} < \delta$, and it follows that for $(x, y) \ne (0, 0)$

$$\begin{aligned}
|f(x,y) - 0| &= \left| \frac{5x^2y}{x^2+y^2} \right| \\
&= 5|y| \left(\frac{x^2}{x^2+y^2} \right) \\
&\le 5|y| \\
&\le 5\sqrt{x^2+y^2} \\
&< 5\delta.
\end{aligned}$$

Thus, you can choose $\delta = \varepsilon/5$ and conclude that

$$\lim_{(x,y)\to(0,0)} \frac{5x^2y}{x^2+y^2} = 0.$$

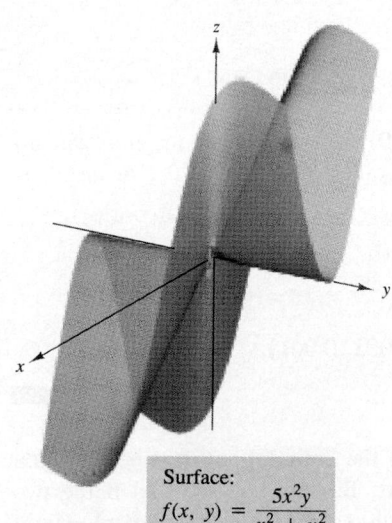

Surface:
$$f(x, y) = \frac{5x^2y}{x^2+y^2}$$

FIGURE 13.21

FIGURE 13.22

$\lim\limits_{(x,y)\to(0,0)} \dfrac{1}{x^2 + y^2}$ does not exist.

For some functions, it is easy to recognize that a limit does not exist. For instance, it is clear that the limit

$$\lim_{(x,y)\to(0,0)} \frac{1}{x^2 + y^2}$$

does not exist because the values of $f(x, y)$ increase without bound as (x, y) approaches 0 along *any* path (see Figure 13.22).

For other functions, it is not so easy to recognize that a limit does not exist. For instance, the next example describes a limit that does not exist because the function approaches different values along different paths.

EXAMPLE 3   A Limit that Does Not Exist

Show that the following limit does not exist.

$$\lim_{(x,y)\to(0,0)} \left(\frac{x^2 - y^2}{x^2 + y^2}\right)^2$$

Solution   The domain of the function given by

$$f(x, y) = \left(\frac{x^2 - y^2}{x^2 + y^2}\right)^2$$

consists of all points in the xy-plane except for the point $(0, 0)$. To show that the limit as (x, y) approaches $(0, 0)$ does not exist, consider approaching $(0, 0)$ along two different "paths," as shown in Figure 13.23. Along the x-axis, every point is of the form $(x, 0)$, and the limit along this approach is

$$\lim_{(x,0)\to(0,0)} \left(\frac{x^2 - 0^2}{x^2 + 0^2}\right)^2 = \lim_{(x,0)\to(0,0)} (1)^2$$

$$= 1. \qquad \text{Limit along line } y = x$$

However, if (x, y) approaches $(0, 0)$ along the line $y = x$, you obtain

$$\lim_{(x,x)\to(0,0)} \left(\frac{x^2 - x^2}{x^2 + x^2}\right)^2 = \lim_{(x,x)\to(0,0)} \left(\frac{0}{2x^2}\right)^2$$

$$= 0. \qquad \text{Limit along } x\text{-axis}$$

This means that in any open disc centered at $(0, 0)$ there are points (x, y) at which f takes on the value 1, and other points at which f takes on the value 0. For instance, $f(x, y) = 1$ at the points

$$(1, 0), \quad (0.1, 0), \quad (0.01, 0), \quad \text{and} \quad (0.001, 0)$$

and $f(x, y) = 0$ at the points

$$(1, 1), \quad (0.1, 0.1), \quad (0.01, 0.01), \quad \text{and} \quad (0.001, 0.001).$$

Hence, f does not have a limit as $(x, y) \to (0, 0)$. ∎

Along x-axis: $(x, 0) \to (0, 0)$
Limit is 1.

Along line $y = x$: $(x, x) \to (0, 0)$
Limit is 0.

FIGURE 13.23

$\lim\limits_{(x,y)\to(0,0)} \left(\dfrac{x^2 - y^2}{x^2 + y^2}\right)^2$ does not exist.

In Example 3, you were able to conclude that the limit did not exist because you found two approaches that yielded different limits. Be sure you see that if the two approaches had yielded the same limit, you still could not have concluded that the limit exists. To form such a conclusion, you must show that the limit is the same along *all* possible approaches.

Continuity of a Function of Two Variables

Notice in Example 2a that the limit of

$$f(x, y) = \frac{5x^2 y}{(x^2 + y^2)}$$

as $(x, y) \rightarrow (1, 2)$ can be evaluated by direct substitution. That is, the limit is $f(1, 2) = 2$. In such cases the function f is said to be **continuous** at the point $(1, 2)$.

Definition of Continuity of a Function of Two Variables

A function f of two variables is **continuous at a point** (x_0, y_0) in an open region R if $f(x_0, y_0)$ is equal to the limit of $f(x, y)$ as (x, y) approaches (x_0, y_0). That is,

$$\lim_{(x, y) \to (x_0, y_0)} f(x, y) = f(x_0, y_0).$$

The function f is **continuous in the open region R** if it is continuous at every point in R.

REMARK   This definition of continuity can be extended to *boundary points* of the open region R by considering a special type of limit in which (x, y) is allowed to approach (x_0, y_0) along paths lying in the region R. This notion is similar to that of one-sided limits, as discussed in Chapter 1.

In Example 2b, it is shown that the function $f(x, y) = 5x^2 y/(x^2 + y^2)$ is not continuous at $(0, 0)$. However, because the limit at this point exists, you can remove the discontinuity by defining f at $(0, 0)$ as being equal to its limit there. Such a discontinuity is called **removable**. As shown in Example 3, the function $f(x, y) = [(x^2 - y^2)/(x^2 + y^2)]^2$ is also not continuous at $(0, 0)$, but this discontinuity is **nonremovable**.

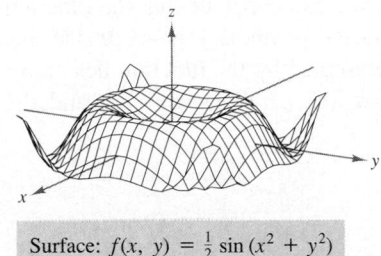

Surface: $f(x, y) = \frac{1}{2} \sin(x^2 + y^2)$

Generated by Mathematica

FIGURE 13.24
The function f is continuous at every point in the plane.

Surface:
$f(x, y) = \cos(y^2) e^{-\sqrt{x^2 + y^2}}$

FIGURE 13.25
The function f is continuous at every point in the plane.

THEOREM 13.1   Properties of Continuous Functions of Two Variables

If k is a real number and f and g are continuous at (x_0, y_0), then the following functions are continuous at (x_0, y_0).

1. Scalar multiple kf
2. Sum and difference $f \pm g$
3. Product fg
4. Quotient f/g,   if $g(x_0, y_0) \neq 0$

Theorem 13.1 establishes the continuity of *polynomial* and *rational* functions at every point in their domains. Furthermore, the continuity of other types of functions can be extended naturally from one to two variables. For instance, the functions whose graphs are shown in Figures 13.24 and 13.25 are continuous at every point in the plane.

The next theorem states conditions under which a composite function is continuous.

THEOREM 13.2 Continuity of a Composite Function

If h is continuous at (x_0, y_0) and g is continuous at $h(x_0, y_0)$, then the composite function given by $(g \circ h)(x, y) = g(h(x, y))$ is continuous at (x_0, y_0). That is,

$$\lim_{(x, y) \to (x_0, y_0)} g(h(x, y)) = g(h(x_0, y_0)).$$

REMARK Note in Theorem 13.2 that h is a function of two variables and g is a function of one variable.

EXAMPLE 4 Testing for Continuity

Discuss the continuity of the following functions.

a. $f(x, y) = \dfrac{x - 2y}{x^2 + y^2}$ **b.** $g(x, y) = \dfrac{2}{y - x^2}$

Solution

a. Because a rational function is continuous at every point in its domain, you can conclude that f is continuous at each point in the xy-plane except at $(0, 0)$, as shown in Figure 13.26.

b. The function given by $g(x, y) = 2/(y - x^2)$ is continuous except at the points at which the denominator is 0, $y - x^2 = 0$. Thus, you can conclude that the function is continuous at all points except those lying on the parabola $y = x^2$. Inside this parabola, you have $y > x^2$, and the surface represented by the function lies above the xy-plane, as shown in Figure 13.27. Outside the parabola, $y < x^2$, and the surface lies below the xy-plane.

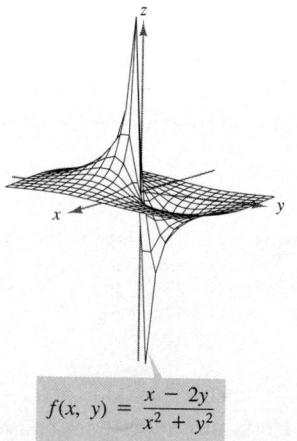

$f(x, y) = \dfrac{x - 2y}{x^2 + y^2}$

Generated by Mathematica

FIGURE 13.26
The function f is not continuous at $(0, 0)$.

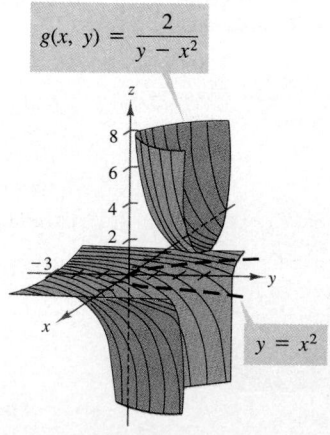

$g(x, y) = \dfrac{2}{y - x^2}$

FIGURE 13.27
The function g is not continuous. on the parabola $y = x^2$.

Continuity of a Function of Three Variables

The preceding definitions of limits and continuity can be extended to functions of three variables by considering points (x, y, z) within the *open sphere*

$$(x - x_0)^2 + (y - y_0)^2 + (z - z_0)^2 < \delta^2. \qquad \text{Open sphere}$$

The radius of this sphere is δ, and the sphere is centered at (x_0, y_0, z_0), as shown in Figure 13.28. A point (x_0, y_0, z_0) in a region R in space is an **interior point** of R if there exists a δ-sphere about (x_0, y_0, z_0), that lies entirely in R. If every point in R is an interior point, then R is called **open.**

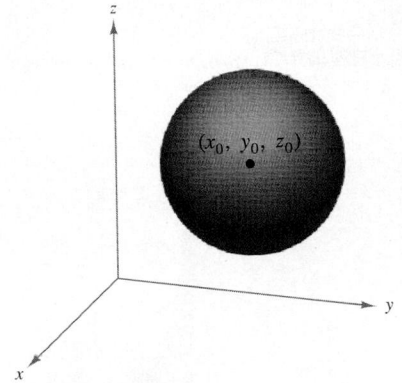

FIGURE 13.28
Open sphere in space

> ### Definition of Continuity of a Function of Three Variables
>
> A function f of three variables is **continuous at a point** (x_0, y_0, z_0) in an open region R if $f(x_0, y_0, z_0)$ is defined and equal to the limit of $f(x, y, z)$ as (x, y, z) approaches (x_0, y_0, z_0). That is,
>
> $$\lim_{(x, y, z) \to (x_0, y_0, z_0)} f(x, y, z) = f(x_0, y_0, z_0)$$
>
> The function f is **continuous in the open region R** if it is continuous at every point in R.

EXAMPLE 5   Testing Continuity of a Function of Three Variables

The function

$$f(x, y, z) = \frac{1}{x^2 + y^2 - z}$$

is continuous at each point in space except at the points on the paraboloid given by $z = x^2 + y^2$.

EXERCISES for Section 13.2

In Exercises 1–4, find the indicated limit by using the limits

$$\lim_{(x, y) \to (a, b)} f(x, y) = 5 \quad \text{and} \quad \lim_{(x, y) \to (a, b)} g(x, y) = 3.$$

1. $\displaystyle \lim_{(x, y) \to (a, b)} [f(x, y) - g(x, y)]$

2. $\displaystyle \lim_{(x, y) \to (a, b)} \left[\frac{4f(x, y)}{g(x, y)} \right]$

3. $\displaystyle \lim_{(x, y) \to (a, b)} [f(x, y) g(x, y)]$

4. $\displaystyle \lim_{(x, y) \to (a, b)} \left[\frac{f(x, y) - g(x, y)}{f(x, y)} \right]$

In Exercises 5–14, find the limit and discuss the continuity of the function.

5. $\displaystyle \lim_{(x, y) \to (2, 1)} (x + 3y^2)$

6. $\displaystyle \lim_{(x, y) \to (0, 0)} (5x + 3xy + y + 1)$

7. $\displaystyle \lim_{(x, y) \to (2, 4)} \frac{x + y}{x - y}$

8. $\displaystyle \lim_{(x, y) \to (1, 1)} \frac{x}{\sqrt{x + y}}$

9. $\displaystyle \lim_{(x, y) \to (0, 1)} \frac{\arcsin(x/y)}{1 + xy}$

10. $\displaystyle \lim_{(x, y) \to (\pi/4, 2)} y \sin xy$

11. $\lim\limits_{(x,y)\to(0,0)} e^{xy}$

12. $\lim\limits_{(x,y)\to(1,1)} \dfrac{xy}{x^2 + y^2}$

13. $\lim\limits_{(x,y,z)\to(1,2,5)} \sqrt{x + y + z}$

14. $\lim\limits_{(x,y,z)\to(2,0,1)} xe^{yz}$

In Exercises 15–18, discuss the continuity of the function and evaluate the limit of $f(x, y)$ (if it exists) as $(x, y) \to (0, 0)$.

15. $f(x, y) = e^{xy}$

16. $f(x, y) = \dfrac{x^2}{(x^2 + 1)(y^2 + 1)}$

17. $f(x, y) = \ln(x^2 + y^2)$

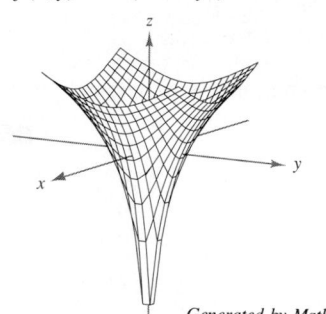

Generated by Mathematica

18. $f(x, y) = 1 - \dfrac{\cos(x^2 + y^2)}{x^2 + y^2}$

Generated by Mathematica

In Exercises 19–22, discuss the continuity of the function and evaluate the limit along the suggested paths. Does the limit exist? Why or why not?

19. $\lim\limits_{(x,y)\to(0,0)} \dfrac{xy}{x^2 + y^2}$

Paths: $y = 0$ and $y = x$

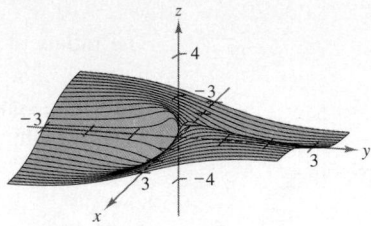

20. $\lim\limits_{(x,y)\to(0,0)} \dfrac{y}{x^2 + y^2}$

Paths: $y = 0$ and $y = x$

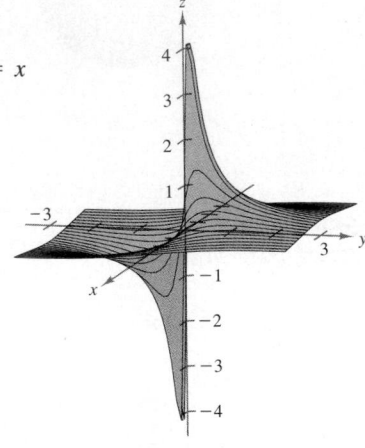

21. $\lim\limits_{(x,y)\to(0,0)} -\dfrac{xy^2}{x^2 + y^4}$

Paths: $x = y^2$ and $x = -y^2$

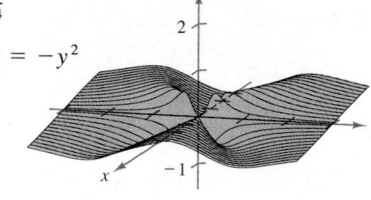

22. $\lim\limits_{(x,y)\to(0,0)} \dfrac{2x - y^2}{2x^2 + y}$

Paths: $y = 0$ and $y = x$

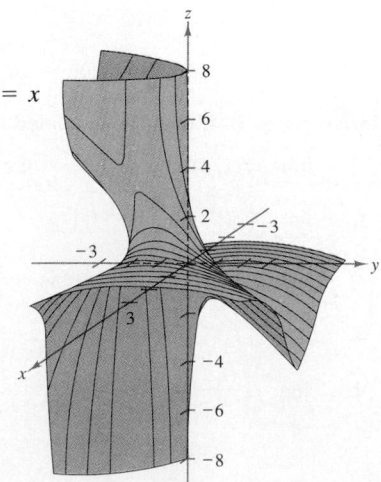

C In Exercises 23–26, use a graphing utility to graph the function and find (if it exists)

$$\lim_{(x,y)\to(0,0)} f(x,y).$$

23. $f(x,y)= \sin x + \cos y$

24. $f(x,y) = \sin \dfrac{1}{x} + \cos \dfrac{1}{x}$

25. $f(x,y) = \dfrac{x^2 y}{x^4 + 4y^2}$

26. $f(x,y) = \dfrac{x^2 + y^2}{x^2 y}$

In Exercises 27–30, use polar coordinates to find the limit. (*Hint*: Let $x = r \cos\theta$ and $y = r \sin\theta$, and note that $(x,y) \to (0, 0)$ is equivalent to $r \to 0$.)

27. $\displaystyle\lim_{(x,y)\to(0,\,0)} \dfrac{\sin(x^2 + y^2)}{x^2 + y^2}$

28. $\displaystyle\lim_{(x,y)\to(0,\,0)} \dfrac{xy^2}{x^2 + y^2}$

29. $\displaystyle\lim_{(x,y)\to(0,\,0)} \dfrac{x^3 + y^3}{x^2 + y^2}$

30. $\displaystyle\lim_{(x,y)\to(0,\,0)} \dfrac{x^2 y^2}{x^2 + y^2}$

In Exercises 31–34, discuss the continuity of the function.

31. $f(x,y,z) = \dfrac{1}{\sqrt{x^2 + y^2 + z^2}}$

32. $f(x,y,z) = \dfrac{z}{x^2 + y^2 - 4}$

33. $f(x,y,z) = \dfrac{\sin z}{e^x + e^y}$

34. $f(x,y,z) = xy \sin z$

In Exercises 35–38, discuss the continuity of the composite function $f \circ g$.

35. $f(t) = t^2$

$g(x,y) = 3x - 2y$

36. $f(t) = \dfrac{1}{t}$

$g(x,y) = x^2 + y^2$

37. $f(t) = \dfrac{1}{t}$

$g(x,y) = 3x - 2y$

38. $f(t) = \dfrac{1}{4 - t}$

$g(x,y) = x^2 + y^2$

In Exercises 39–42, find the following limits.

a. $\displaystyle\lim_{\Delta x\to 0} \dfrac{f(x + \Delta x, y) - f(x,y)}{\Delta x}$

b. $\displaystyle\lim_{\Delta y\to 0} \dfrac{f(x, y + \Delta y) - f(x,y)}{\Delta y}$

39. $f(x,y) = x^2 - 4y$

40. $f(x,y) = x^2 + y^2$

41. $f(x,y) = 2x + xy - 3y$

42. $f(x,y) = \sqrt{y}(y + 1)$

43. Prove that

$$\lim_{(x,y)\to(a,b)} [f(x,y) + g(x,y)] = L_1 + L_2$$

where $f(x,y)$ approaches L_1 and $g(x,y)$ approaches L_2 as $(x,y) \to (a,b)$.

44. Prove that if f is continuous and $f(a,b) < 0$, then there exists a δ-neighborhood about (a,b) such that $f(x,y) < 0$ for every point (x,y) in the neighborhood.

True or False In Exercises 45–48, determine whether the statement is true or false. If it is false, explain why or give an example that shows it is false.

45. If $\displaystyle\lim_{(x,y)\to(0,0)} f(x,y) = 0$, then $\displaystyle\lim_{x\to 0} f(x, 0) = 0$.

46. If $\displaystyle\lim_{(x,y)\to(0,0)} f(0, y) = 0$, then $\displaystyle\lim_{(x,y)\to(0,0)} f(x, y) = 0$.

47. If f is continuous for all nonzero x and y and $f(0, 0) = 0$, then $\displaystyle\lim_{(x,y)\to(0,0)} f(x,y) = 0$.

48. If g and h are continuous functions of x and y, respectively, and $f(x,y) = g(x) + h(y)$, then f is continuous.

49. Prove that

$$\lim_{(x,y)\to(0,0)} \dfrac{xy^3}{x^2 + 2y^6}$$

does not exist.

Partial Derivatives of a Function of Two Variables • Partial Derivatives of a Function of Three or More Variables • Higher-Order Partial Derivatives

JEAN LE ROND D'ALEMBERT

The introduction of partial derivatives followed Newton's and Leibniz's work in calculus by several years. Between 1730 and 1760, Leonhard Euler and Jean Le Rond d'Alembert (1717–1783) separately published several papers on dynamics, in which they established much of the theory of partial derivatives. These papers used functions of two or more variables to study problems involving equilibrium, fluid motion, and vibrating strings.

Partial Derivatives of a Function of Two Variables

In applications of functions of several variables, the question often arises, "How will a function be affected by a change in one of its independent variables?" You can answer this by considering the independent variables one at a time. For example, to determine the effect of a catalyst in an experiment, a chemist could conduct the experiment several times using varying amounts of the catalyst, while keeping constant other variables such as temperature and pressure. You can use a similar procedure to determine the rate of change of a function f with respect to one of its several independent variables. The process is called **partial differentiation,** and the result is referred to as the **partial derivative** of f with respect to the chosen independent variable.

Definition of Partial Derivatives of a Function of Two Variables

If $z = f(x, y)$, then the **first partial derivatives** of f with respect to x and y are the functions f_x and f_y defined by

$$f_x(x, y) = \lim_{\Delta x \to 0} \frac{f(x + \Delta x, y) - f(x, y)}{\Delta x}$$

$$f_y(x, y) = \lim_{\Delta y \to 0} \frac{f(x, y + \Delta y) - f(x, y)}{\Delta y}$$

provided the limits exist.

This definition indicates that if $z = f(x, y)$, then to find f_x you *consider y constant* and differentiate with respect to x. Similarly, to find f_y, you *consider x constant* and differentiate with respect to y.

EXAMPLE 1   Finding Partial Derivatives

Find f_x and f_y for $f(x, y) = 3x - x^2y^2 + 2x^3y$.

Solution   Considering y to be constant and differentiating with respect to x produces

$$f_x(x, y) = 3 - 2xy^2 + 6x^2y.$$

Considering x to be constant and differentiating with respect to y produces

$$f_y(x, y) = -2x^2y + 2x^3.$$

Notation for First Partial Derivatives

For $z = f(x, y)$, the partial derivatives f_x and f_y are denoted by

$$\frac{\partial}{\partial x} f(x, y) = f_x(x, y) = z_x = \frac{\partial z}{\partial x}$$

and

$$\frac{\partial}{\partial x} f(x, y) = f_x(x, y) = z_x = \frac{\partial z}{\partial x}$$

The first partials evaluated at the point (a, b) are denoted by

$$\frac{\partial z}{\partial x}\bigg|_{(a, b)} = f_x(a, b) \quad \text{and} \quad \frac{\partial z}{\partial y}\bigg|_{(a, b)} = f_y(a, b).$$

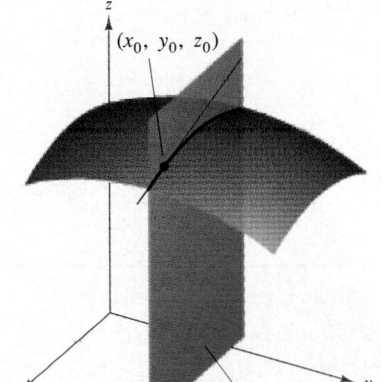

FIGURE 13.29

$\dfrac{\partial f}{\partial x} = $ (slope in x-direction)

FIGURE 13.30

$\dfrac{\partial f}{\partial y} = $ (slope in y-direction)

EXAMPLE 2   Finding and Evaluating Partial Derivatives

For $f(x, y) = xe^{x^2 y}$, find f_x and f_y, and evaluate each at the point $(1, \ln 2)$.

Solution   Because

$$f_x(x, y) = xe^{x^2 y}(2xy) + e^{x^2 y}$$

the partial derivative of f with respect to x at $(1, \ln 2)$ is

$$f_x(1, \ln 2) = e^{\ln 2}(2 \ln 2) + e^{\ln 2} = 4 \ln 2 + 2.$$

Because

$$f_y(x, y) = xe^{x^2 y}(x^2) = x^3 e^{x^2 y}$$

the partial derivative of f with respect to y at $(1, \ln 2)$ is

$$f_y(1, \ln 2) = e^{\ln 2} = 2.$$

The partial derivatives of a function of two variables, $z = f(x, y)$, have a useful geometric interpretation. If $y = y_0$, then $z = f(x, y_0)$ represents the curve formed by intersecting the surface $z = f(x, y)$ with the plane $y = y_0$, as shown in Figure 13.29. Therefore,

$$f_x(x_0, y_0) = \lim_{\Delta x \to 0} \frac{f(x_0 + \Delta x, y_0) - f(x_0, y_0)}{\Delta x}$$

represents the slope of this curve at the point $(x_0, y_0, f(x_0, y_0))$. Note that both the curve and the tangent line lie in the plane $y = y_0$. Similarly,

$$f_y(x_0, y_0) = \lim_{\Delta y \to 0} \frac{f(x_0, y_0 + \Delta y) - f(x_0, y_0)}{\Delta y}$$

represents the slope of the curve given by the intersection of $z = f(x, y)$ and the plane $x = x_0$ at $(x_0, y_0, f(x_0, y_0))$, as shown in Figure 13.30.

Informally, we say that the values of $\partial f / \partial x$ and $\partial f / \partial y$ at the point (x_0, y_0, z_0) denote the **slope of the surface in the x- and y-directions,** respectively.

EXAMPLE 3   Finding the Slope of a Surface in the x- and y-Directions

Find the slopes of the surface given by

$$f(x, y) = -\frac{x^2}{2} - y^2 + \frac{25}{8}$$

at the point $\left(\frac{1}{2}, 1, 2\right)$ in the x-direction and the y-direction.

Solution   The partial derivatives of f with respect to x and y are

$$f_x(x, y) = -x \quad \text{and} \quad f_y(x, y) = -2y.$$

Thus, in the x-direction, the slope is

$$f_x\left(\frac{1}{2}, 1\right) = -\frac{1}{2} \qquad \text{Figure 13.31(a)}$$

and in the y-direction, the slope is

$$f_y\left(\frac{1}{2}, 1\right) = -2. \qquad \text{Figure 13.31(b)}$$

Surface:
$$z = -\frac{x^2}{2} - y^2 + \frac{25}{8}$$

(a) Slope in x-direction:
$$f_x\left(\tfrac{1}{2}, 1\right) = -\tfrac{1}{2}$$

(b) Slope in y-direction:
$$f_y\left(\tfrac{1}{2}, 1\right) = -2$$

FIGURE 13.31

EXAMPLE 4   Finding the Slope of a Surface in the x- and y-Directions

Find the slope of the surface given by $f(x, y) = 1 - (x - 1)^2 - (y - 2)^2$ at the point $(1, 2, 1)$ in the x-direction and in the y-direction.

Solution   The partial derivatives of f with respect to x and y are

$$f_x(x, y) = -2(x - 1) \quad \text{and} \quad f_y(x, y) = -2(y - 2).$$

Thus, at the point $(1, 2, 1)$, the slopes in the x- and y-directions are

$$f_x(1, 2) = -2(1 - 1) = 0 \quad \text{and} \quad f_y(1, 2) = -2(2 - 2) = 0$$

as indicated in Figure 13.32.

Surface:
$$f(x, y) = 1 - (x - 1)^2 - (y - 2)^2$$

$f_x(1, 2)$

$(1, 2, 1)$

$f_y(1, 2)$

FIGURE 13.32

No matter how many variables are involved, partial derivatives of several variables can be interpreted as *rates of change.*

EXAMPLE 5   Using Partial Derivatives to Find Rates of Change

The area of a parallelogram with adjacent sides a and b and included angle θ is given by $A = ab \sin \theta$, as shown in Figure 13.33.

a. Find the rate of change of A with respect to a when $a = 10$, $b = 20$, and $\theta = \pi/6$.
b. Find the rate of change of A with respect to θ when $a = 10$, $b = 20$, and $\theta = \pi/6$.

Solution

a. To find the rate of change of the area with respect to a, hold b and θ constant and differentiate with respect to a to obtain

$$\frac{\partial A}{\partial a} = b \sin \theta.$$

When $a = 10$, $b = 20$, and $\theta = \pi/6$, you have

$$\partial A / \partial a = 20 \sin (\pi/6) = 10.$$

b. To find the rate of change of the area with respect to θ, hold a and b constant and differentiate with respect to θ to obtain

$$\frac{\partial A}{\partial \theta} = ab \cos \theta.$$

When $a = 10$, $b = 20$, and $\theta = \pi/6$, you have

$$\partial A / \partial \theta = 200 \cos(\pi/6) = 100\sqrt{3}.$$

$A = ab \sin \theta$

$a \sin \theta$

FIGURE 13.33
The area of the parallelogram is $ab \sin \theta$.

Partial Derivatives of a Function of Three or More Variables

The concept of a partial derivative can be extended naturally to functions of three or more variables. For instance, if $w = f(x, y, z)$, then there are three partial derivatives, each of which is formed by holding two of the variables constant. That is, to define the partial derivative of w with respect to x, consider y and z to be constant and differentiate with respect to x. A similar process is used to find the derivatives of w with respect to y and with respect to z.

$$\frac{\partial w}{\partial x} = f_x(x, y, z) = \lim_{\Delta x \to 0} \frac{f(x + \Delta x, y, z) - f(x, y, z)}{\Delta x}$$

$$\frac{\partial w}{\partial y} = f_y(x, y, z) = \lim_{\Delta y \to 0} \frac{f(x, y + \Delta y, z) - f(x, y, z)}{\Delta y}$$

$$\frac{\partial w}{\partial z} = f_z(x, y, z) = \lim_{\Delta z \to 0} \frac{f(x, y, z + \Delta z) - f(x, y, z)}{\Delta z}$$

In general, if $w = f(x_1, x_2, \ldots, x_n)$, then there are n partial derivatives denoted by

$$\frac{\partial w}{\partial x_k} = f_{x_k}(x_1, x_2, \ldots, x_n), \qquad k = 1, 2, \ldots, n.$$

To find the partial derivative with respect to one of the variables, hold the other variables constant and differentiate with respect to the given variable.

EXAMPLE 6   Finding Partial Derivatives

a. To find the partial derivative of $f(x, y, z) = xy + yz^2 + xz$ with respect to z, consider x and y to be constant and obtain

$$\frac{\partial}{\partial z}[xy + yz^2 + xz] = 2yz + x.$$

b. To find the partial derivative of $f(x, y, z) = z \sin(xy^2 + 2z)$ with respect to z, consider x and y to be constant. Then, using the Product Rule, you obtain

$$\frac{\partial}{\partial z}[z \sin(xy^2 + 2z)] = (z)\frac{\partial}{\partial z}[\sin(xy^2 + 2z)] + \sin(xy^2 + 2z)\frac{\partial}{\partial z}[z]$$

$$= (z)[\cos(xy^2 + 2z)](2) + \sin(xy^2 + 2z)$$

$$= 2z \cos(xy^2 + 2z) + \sin(xy^2 + 2z).$$

c. To find the partial derivative of $f(x, y, z, w) = (x + y + z)/w$ with respect to w, consider x, y, and z to be constant and obtain

$$\frac{\partial}{\partial w}\left[\frac{x + y + z}{w}\right] = -\frac{x + y + z}{w^2}.$$

Higher-Order Partial Derivatives

As is true for ordinary derivatives, it is possible to take second, third, and higher partial derivatives of a function of several variables, provided such derivatives exist. Higher-order derivatives are denoted by the order in which the differentiation occurs. For instance, the function $z = f(x, y)$ has the following second partial derivatives.

1. Differentiate twice with respect to x:

$$\frac{\partial}{\partial x}\left(\frac{\partial f}{\partial x}\right) = \frac{\partial^2 f}{\partial x^2} = f_{xx}.$$

2. Differentiate twice with respect to y:

$$\frac{\partial}{\partial y}\left(\frac{\partial f}{\partial y}\right) = \frac{\partial^2 f}{\partial y^2} = f_{yy}.$$

3. Differentiate first with respect to x and then with respect to y:

$$\frac{\partial}{\partial y}\left(\frac{\partial f}{\partial x}\right) = \frac{\partial^2 f}{\partial y \partial x} = f_{xy}.$$

4. Differentiate first with respect to y and then with respect to x:

$$\frac{\partial}{\partial x}\left(\frac{\partial f}{\partial y}\right) = \frac{\partial^2 f}{\partial x \partial y} = f_{yx}.$$

REMARK   Note that the two types of notation for mixed partials have different conventions for indicating the order of differentiation.

$$\frac{\partial}{\partial y}\left(\frac{\partial f}{\partial x}\right) = \frac{\partial^2 f}{\partial y \partial x} \qquad \text{Right-to-left order}$$

$$(f_x)_y = f_{xy} \qquad \text{Left-to-right order}$$

You can remember the order by observing that in both notations, you differentiate first with respect to the variable "nearest" f.

The third and fourth cases are called **mixed partial derivatives.**

EXAMPLE 7 **Finding Second Partial Derivatives**

Find the second partial derivatives of $f(x, y) = 3xy^2 - 2y + 5x^2y^2$, and determine the value of $f_{xy}(-1, 2)$.

Solution Begin by finding the first partial derivatives with respect to x and y.

$$f_x(x, y) = 3y^2 + 10xy^2 \quad \text{and} \quad f_y(x, y) = 6xy - 2 + 10x^2y$$

Then, differentiate each of these with respect to x and y.

$$f_{xx}(x, y) = 10y^2 \quad \text{and} \quad f_{yy}(x, y) = 6x + 10x^2$$

$$f_{xy}(x, y) = 6y + 20xy \quad \text{and} \quad f_{yx}(x, y) = 6y + 20xy$$

At $(-1, 2)$, the value of f_{xy} is $f_{xy}(-1, 2) = 12 - 40 = -28$. ▬

Notice in Example 7 that the two mixed partials are equal. Sufficient conditions for this occurrence are given in the next theorem.

THEOREM 13.3 **Equality of Mixed Partial Derivatives**

If f is a function of x and y such that f_{xy} and f_{yx} are continuous on an open disc R, then for every (x, y) in R,

$$f_{xy}(x, y) = f_{yx}(x, y).$$

Theorem 13.3 also applies to a function f of *three or more variables* so long as all second partial derivatives are continuous. For example, if $w = f(x, y, z)$ and all the second partial derivatives are continuous in an open region R, then at each point in R the order of differentiation in the mixed second partial derivatives is irrelevant. If the third partial derivatives of f are also continuous, then the order of differentiation of the mixed third partial derivatives is irrelevant.

EXAMPLE 8 **Finding Higher-Order Partial Derivatives**

Show that $f_{xz} = f_{zx}$ and $f_{xzz} = f_{zxz} = f_{zzx}$ for the function given by

$$f(x, y, z) = ye^x + x \ln z.$$

Solution
First partials:

$$f_x(x, y, z) = ye^x + \ln z, \qquad f_z(x, y, z) = \frac{x}{z}$$

Second partials: (Note that the first two are equal.)

$$f_{xz}(x, y, z) = \frac{1}{z}, \qquad f_{zx}(x, y, z) = \frac{1}{z}, \qquad f_{zz}(x, y, z) = -\frac{x}{z^2}$$

Third partials: (Note that all three are equal.)

$$f_{xzz}(x, y, z) = -\frac{1}{z^2}, \qquad f_{zxz}(x, y, z) = -\frac{1}{z^2}, \qquad f_{zzx}(x, y, z) = -\frac{1}{z^2}$$ ▬

TECHNOLOGY
Laboratory Guide
Lab 13.3

EXERCISES for Section 13.3

In Exercises 1–20, find the first partial derivatives with respect to x and with respect to y.

1. $f(x, y) = 2x - 3y + 5$

2. $f(x, y) = x^2 - 3y^2 + 7$

3. $z = x\sqrt{y}$

4. $z = x^2 - 3xy + y^2$

5. $z = x^2 e^{2y}$

6. $z = xe^{x/y}$

7. $z = \ln(x^2 + y^2)$

8. $z = \ln\sqrt{xy}$

9. $z = \ln\dfrac{x + y}{x - y}$

10. $z = \dfrac{x^2}{2y} + \dfrac{4y^2}{x}$

11. $h(x, y) = e^{-(x^2 + y^2)}$

12. $g(x, y) = \ln\sqrt{x^2 + y^2}$

13. $f(x, y) = \sqrt{x^2 + y^2}$

14. $f(x, y) = \dfrac{xy}{x^2 + y^2}$

15. $z = \tan(2x - y)$

16. $z = \sin 3x \cos 3y$

17. $z = e^y \sin xy$

18. $z = \cos(x^2 + y^2)$

19. $f(x, y) = \displaystyle\int_x^y (t^2 - 1)\, dt$

20. $f(x, y) = \displaystyle\int_x^y (2t + 1)\, dt + \int_y^x (2t - 1)\, dt$

In Exercises 21–24, use the limit definition of partial derivatives to find $f_x(x, y)$ and $f_y(x, y)$.

21. $f(x, y) = 2x + 3y$

22. $f(x, y) = \dfrac{1}{x + y}$

23. $f(x, y) = \sqrt{x + y}$

24. $f(x, y) = x^2 - 2xy + y^2$

In Exercises 25–28, find the slopes of the surface at the specified point in the x- and y-directions.

25. $g(x, y) = 4 - x^2 - y^2$
$(1, 1, 2)$

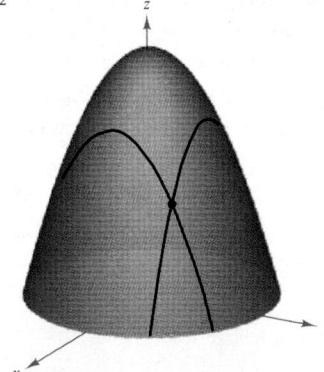

26. $h(x, y) = x^2 - y^2$
$(-2, 1, 3)$

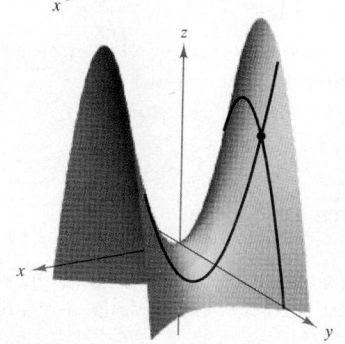

27. $z = e^{-x} \cos y$
$(0, 0, 1)$

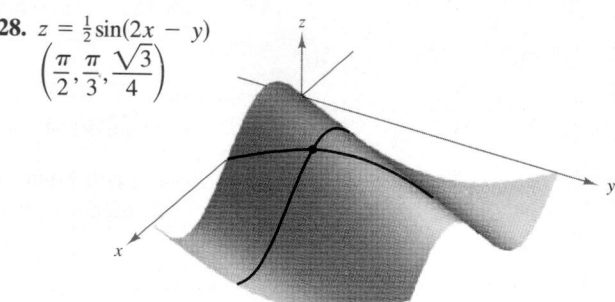

28. $z = \frac{1}{2}\sin(2x - y)$
$\left(\dfrac{\pi}{2}, \dfrac{\pi}{3}, \dfrac{\sqrt{3}}{4}\right)$

In Exercises 29–32, evaluate f_x and f_y at the specified point.

Function	Point
29. $f(x, y) = \arctan\dfrac{y}{x}$	$\left(2, -2, -\dfrac{\pi}{4}\right)$
30. $f(x, y) = \arcsin xy$	$\left(1, 0, \dfrac{\pi}{2}\right)$
31. $f(x, y) = \dfrac{xy}{x - y}$	$(2, -2, -1)$
32. $f(x) = \dfrac{4xy}{\sqrt{x^2 + y^2}}$	$(1, 0, 0)$

In Exercises 33–36, sketch the curve formed by the intersection of the given surface and plane. Find the slope of the curve at the given point.

Surface	Plane	Point
33. $z = \sqrt{49 - x^2 - y^2}$	$x = 2$	$(2, 3, 6)$
34. $z = x^2 + 4y^2$	$y = 1$	$(2, 1, 8)$
35. $z = 9x^2 - y^2$	$y = 3$	$(1, 3, 0)$
36. $z = 9x^2 - y^2$	$x = 1$	$(1, 3, 0)$

In Exercises 37–44, find the second partial derivatives

$$\frac{\partial^2 z}{\partial x^2}, \quad \frac{\partial^2 z}{\partial y^2}, \quad \frac{\partial^2 z}{\partial y\, \partial x}, \quad \text{and} \quad \frac{\partial^2 z}{\partial x\, \partial y}.$$

Observe that the second mixed partials are equal.

37. $z = x^2 - 2xy + 3y^2$

38. $z = x^4 - 3x^2 y^2 + y^4$

39. $z = \sqrt{x^2 + y^2}$

40. $z = \ln(x - y)$

41. $z = e^x \tan y$

42. $z = xe^y + ye^x$

43. $z = \arctan \dfrac{y}{x}$

44. $z = \sin(x - 2y)$

C In Exercises 45–48, use a symbolic differentiation utility to find the first and second partial derivatives of the function.

45. $z = x \sec y$

46. $z = \sqrt{9 - x^2 - y^2}$

47. $z = \ln \dfrac{x}{x^2 + y^2}$

48. $z = \dfrac{xy}{x - y}$

In Exercises 49–54, find the first partial derivatives with respect to x, y, and z.

49. $w = \sqrt{x^2 + y^2 + z^2}$

50. $w = \dfrac{xy}{x + y + z}$

51. $F(x, y, z) = \ln \sqrt{x^2 + y^2 + z^2}$

52. $G(x, y, z) = \dfrac{1}{\sqrt{1 - x^2 - y^2 - z^2}}$

53. $H(x, y, z) = \sin(x + 2y + 3z)$

54. $f(x, y, z) = 3x^2 y - 5xyz + 10yz^2$

In Exercises 55–58, show that the mixed partials f_{xyy}, f_{yxy}, and f_{yyx} are equal.

55. $w = xyz$

56. $w = x^2 - 3xy + 4yz + z^3$

57. $f(x, y, z) = e^{-x} \sin yz$

58. $f(x, y, z) = \dfrac{x}{y + z}$

Laplace's Equation   In Exercises 59–62, show that each function satisfies Laplace's equation

$$\frac{\partial^2 z}{\partial x^2} + \frac{\partial^2 z}{\partial y^2} = 0.$$

59. $z = 5xy$

60. $z = \frac{1}{2}(e^y - e^{-y}) \sin x$

61. $z = e^x \sin y$

62. $z = \arctan \dfrac{y}{x}$

Wave Equation   In Exercises 63 and 64, show that the function satisfies the wave equation

$$\frac{\partial^2 z}{\partial t^2} = c^2 \frac{\partial^2 z}{\partial x^2}.$$

63. $z = \sin(x - ct)$

64. $z = \sin \omega ct \sin \omega x$

Heat Equation   In Exercises 65 and 66, show that the function satisfies the heat equation

$$\frac{\partial z}{\partial t} = c^2 \frac{\partial^2 z}{\partial x^2}.$$

65. $z = e^{-t} \cos \dfrac{x}{c}$

66. $z = e^{-t} \sin \dfrac{x}{c}$

67. *Marginal Costs*   A company manufactures two types of wood-burning stoves: a freestanding model and a fireplace-insert model. The cost function for producing x freestanding and y fireplace-insert stoves is

$$C = 32\sqrt{xy} + 175x + 205y + 1050.$$

Find the marginal costs ($\partial C/\partial x$ and $\partial C/\partial y$) when $x = 80$ and $y = 20$.

68. *Marginal Productivity*   Consider the Cobb-Douglas production function

$$f(x, y) = 100x^{0.6} y^{0.4}.$$

When $x = 1000$ and $y = 500$, find
a. the marginal productivity of labor, $\partial f/\partial x$.
b. the marginal productivity of capital, $\partial f/\partial y$.

69. *University Admissions*   Let N be the number of applicants to a university, p the charge for food and housing at the university, and t the tuition. Suppose that N is a function of p and t such that $\partial N/\partial p < 0$ and $\partial N/\partial t < 0$. What information is gained by noticing that both partials are negative?

70. *Investment*   The value of an investment of \$1000 earning 10 percent compounded annually is given by

$$V(I, R) = 1000 \left[\frac{1 + 0.10(1 - R)}{1 + I} \right]^{10}$$

where I is the annual rate of inflation and R is the tax rate for the person making the investment. Calculate $V_I(0.03, 0.28)$ and $V_R(0.03, 0.28)$. Determine whether the tax rate or the rate of inflation has the greater "negative" influence on the growth of the investment.

71. *Temperature Distribution*   The temperature at any point (x, y) in a steel plate is

$$T = 500 - 0.6x^2 - 1.5y^2$$

where x and y are measured in feet. At the point $(2, 3)$, find the rate of change of the temperature with respect to the distance moved along the plate in the directions of the x- and y-axes, respectively.

72. *Ideal Gas Law*   According to the Ideal Gas Law, $PV = kT$, where P is pressure, V is volume, T is temperature, and k is a constant of proportionality. Find
a. $\partial P/\partial T$.   **b.** $\partial V/\partial P$.

73. *Projectile Motion*   The range of a projectile fired at an angle θ above the horizontal with veolcity v_0 is

$$R = \frac{v_0^2 \sin 2\theta}{32}.$$

Evaluate $\partial R/\partial v_0$ and $\partial R/\partial \theta$ when $v_0 = 2000$ feet per second and $\theta = 5°$.

74. Consider the function defined by

$$f(x, y) = \begin{cases} \dfrac{xy(x^2 - y^2)}{x^2 + y^2}, & (x, y) \neq (0, 0) \\ 0, & (x, y) = (0, 0). \end{cases}$$

a. Find $f_x(x, y)$ and $f_y(x, y)$ for $(x, y) \neq (0, 0)$.
b. Use the definition of partial derivatives to find $f_x(0, 0)$ and $f_y(0, 0)$.

$\left(Hint: f_x(0, 0) = \lim\limits_{\Delta x \to 0} \dfrac{f(\Delta x, 0) - f(0, 0)}{\Delta x}.\right)$

c. Use the definition of partial derivatives to find $f_{xy}(0, 0)$ and $f_{yx}(0, 0)$.

d. Using Theorem 13.3 and the result of part c, what can be said about f_{xy} or f_{yx}?

True or False **In Exercises 75–78, determine whether the statement is true or false. If it is false, explain why or give an example that shows it is false.**

75. If $z = f(x, y)$ and $\partial z/\partial x = \partial z/\partial y$, then $z = c(x + y)$.

76. If $z = f(x)g(y)$, then $(\partial z/\partial x) + (\partial z/\partial y) = f'(x)g(y) + f(x)g'(y)$.

77. If $z = e^{xy}$, then $\dfrac{\partial^2 z}{\partial y \, \partial x} = (xy + 1)e^{xy}$.

78. If a cylindrical surface, $z = f(x, y)$, has rulings parallel to the y-axis, then $\partial z/\partial y = 0$.

79. The ideal gas law is $PV = mRT$, where P is the pressure, V is the volume, m is the mass, R is a fixed constant ("the gas constant"), and T is the absolute temperature. Show that

$$\frac{\partial T}{\partial P} \frac{\partial P}{\partial V} \frac{\partial V}{\partial T} = -1.$$

80. Let $f(x, y) = \displaystyle\int_x^y \sqrt{1 + t^3} \, dt$. Find $\dfrac{\partial f}{\partial x}$ and $\dfrac{\partial f}{\partial y}$.

81. *Moiré Patterns* Read the article "Moiré Fringes and the Conic Sections" by Mike Cullen in the November, 1990 issue of the *College Mathematics Journal*. The article describes how two families of level curves given by

$$f(x, y) = a \quad \text{and} \quad g(x, y) = b$$

can form Moiré patterns. After reading the article, write a paper explaining how the expression

$$\frac{\partial f}{\partial x} \cdot \frac{\partial g}{\partial x} + \frac{\partial f}{\partial y} \cdot \frac{\partial g}{\partial y}$$

is related to the Moiré patterns formed by intersecting the two families of level curves. Use one of the following patterns as an example in your paper.

Differentials

13.4 Increments and Differentials • Differentiability • Approximation by Differentials

Increments and Differentials

In this section, we generalize the concepts of increments and differentials to functions of two or more variables. Recall from Section 3.9 that for $y = f(x)$, the differential of y was defined as $dy = f'(x)\,dx$. Similar terminology is used for a function of two variables, $z = f(x, y)$. That is, Δx and Δy are the **increments of x and y,** and the **increment of** z is given by

$$\Delta z = f(x + \Delta x, y + \Delta y) - f(x, y). \qquad \text{Increment of } z$$

Definition of Total Differential

If $z = f(x, y)$ and Δx and Δy are increments of x and y, then the **differentials** of the independent variables x and y are

$$dx = \Delta x \quad \text{and} \quad dy = \Delta y$$

and the **total differential** of the dependent variable z is

$$dz = \frac{\partial z}{\partial x}dx + \frac{\partial z}{\partial y}dy = f_x(x, y)\,dx + f_y(x, y)\,dy.$$

This definition can be extended to a function of three or more variables. For instance, if $w = f(x, y, z, u)$, then $dx = \Delta x$, $dy = \Delta y$, $dz = \Delta z$, $du = \Delta u$, and the total differential of w is

$$dw = \frac{\partial w}{\partial x}dx + \frac{\partial w}{\partial y}dy + \frac{\partial w}{\partial z}dz + \frac{\partial w}{\partial u}du.$$

EXAMPLE 1   Finding the Total Differential

a. The total differential dz for $z = 2x\sin y - 3x^2y^2$ is

$$dz = \frac{\partial z}{\partial x}dx + \frac{\partial z}{\partial y}dy$$

$$= (2\sin y - 6xy^2)\,dx + (2x\cos y - 6x^2y)\,dy.$$

b. The total differential dw for $w = x^2 + y^2 + z^2$ is

$$dw = \frac{\partial w}{\partial x}dx + \frac{\partial w}{\partial y}dy + \frac{\partial w}{\partial z}dz$$

$$= 2x\,dx + 2y\,dy + 2z\,dz.$$

Differentiability

In Section 3.9, you learned that for a *differentiable* function given by $y = f(x)$, you can use the differential $dy = f'(x)\,dx$ as an approximation (for small Δx) to the value $\Delta y = f(x + \Delta x) - f(x)$. When a similar approximation is possible for a function of two variables, the function is said to be **differentiable.** This is stated explicitly in the following definition.

Definition of Differentiability

A function f given by $z = f(x, y)$ is **differentiable** at (x_0, y_0) if Δz can be expressed in the form

$$\Delta z = f_x(x_0, y_0)\,\Delta x + f_y(x_0, y_0)\,\Delta y + \varepsilon_1\,\Delta x + \varepsilon_2\,\Delta y$$

where both ε_1 and $\varepsilon_2 \rightarrow 0$ as $(\Delta x, \Delta y) \rightarrow (0, 0)$. The function f is **differentiable in a region R** if it is differentiable at each point of R.

EXAMPLE 2   Showing that a Function is Differentiable

Show that the function given by $f(x, y) = x^2 + 3y$ is differentiable at every point in the plane.

Solution   Letting $z = f(x, y)$, the increment of z at an arbitrary point (x, y) in the plane is

$$\begin{aligned}
\Delta z &= f(x + \Delta x, y + \Delta y) - f(x, y) \\
&= (x^2 + 2x\,\Delta x + \Delta x^2) + 3(y + \Delta y) - (x^2 + 3y) \\
&= 2x\,\Delta x + \Delta x^2 + 3\,\Delta y \\
&= 2x(\Delta x) + 3(\Delta y) + \Delta x(\Delta x) + 0(\Delta y) \\
&= f_x(x, y)\,\Delta x + f_y(x, y)\,\Delta y + \varepsilon_1\,\Delta x + \varepsilon_2\,\Delta y
\end{aligned}$$

where $\varepsilon_1 = \Delta x$ and $\varepsilon_2 = 0$. Because $\varepsilon_1 \rightarrow 0$ and $\varepsilon_2 \rightarrow 0$ as $(\Delta x, \Delta y) \rightarrow (0, 0)$, it follows that f is differentiable at every point in the plane.   ■

Be sure you see that the term "differentiable" is used differently for functions of two variables than for one variable. A function of one variable is differentiable at a point if its derivative exists at the point. However, for a function of two variables, the existence of the partial derivatives f_x and f_y does not guarantee that the function is differentiable (see Example 5). The following theorem gives a *sufficient* condition for differentiability of a function of two variables.

THEOREM 13.4   Sufficient Condition for Differentiability

If f is a function of x and y, where f_x and f_y are continuous in an open region R, then f is differentiable on R.

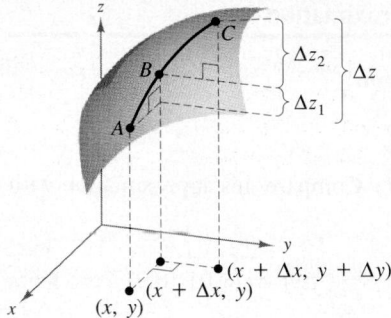

FIGURE 13.34
$\Delta z = f(x + \Delta x, y + \Delta y) - f(x, y)$

Proof　Let S be the surface defined by $z = f(x, y)$, where f, f_x, and f_y are continuous at (x, y). Let A, B, and C be points on surface S, as shown in Figure 13.34. From this figure, you can see that the change in f from point A to point C is given by

$$\begin{aligned} \Delta z &= f(x + \Delta x, y + \Delta y) - f(x, y) \\ &= [f(x + \Delta x, y) - f(x, y)] + [f(x + \Delta x, y + \Delta y) - f(x + \Delta x, y)] \\ &= \Delta z_1 + \Delta z_2. \end{aligned}$$

Between A and B, y is fixed and x changes. Hence, by the Mean Value Theorem, there is a value x_1 between x and $x + \Delta x$ such that

$$\Delta z_1 = f(x + \Delta x, y) - f(x, y) = f_x(x_1, y)\Delta x.$$

Similarly, between B and C, x is fixed and y changes, and there is a value y_1 between y and $y + \Delta y$ such that

$$\Delta z_2 = f(x + \Delta x, y + \Delta y) - f(x + \Delta x, y) = f_y(x + \Delta x, y_1)\, \Delta y.$$

By combining these two results, you can write

$$\Delta z = \Delta z_1 + \Delta z_2 = f_x(x_1, y)\, \Delta x + f_y(x + \Delta x, y_1)\, \Delta y.$$

If you define ε_1 and ε_2 by

$$\varepsilon_1 = f_x(x_1, y) - f_x(x, y) \quad \text{and} \quad \varepsilon_2 = f_y(x + \Delta x, y_1) - f_y(x, y)$$

it follows that

$$\begin{aligned} \Delta z = \Delta z_1 + \Delta z_2 &= [\varepsilon_1 + f_x(x, y)]\, \Delta x + [\varepsilon_2 + f_y(x, y)]\, \Delta y \\ &= [f_x(x, y)\, \Delta x + f_y(x, y)\, \Delta y] + \varepsilon_1\, \Delta x + \varepsilon_2\, \Delta y. \end{aligned}$$

By the continuity of f_x and f_y and the fact that $x \le x_1 \le x + \Delta x$ and $y \le y_1 \le y + \Delta y$, it follows that $\varepsilon_1 \to 0$ and $\varepsilon_2 \to 0$ as $\Delta x \to 0$ and $\Delta y \to 0$. Therefore, by definition, f is differentiable.

Approximation by Differentials

Theorem 13.4 tells you that you can choose $(x + \Delta x, y + \Delta y)$ close enough to (x, y) to make $\varepsilon_1\, \Delta x$ and $\varepsilon_2\, \Delta y$ insignificant. In other words, for small Δx and Δy, you can use the approximation

$$\Delta z \approx dz.$$

This approximation is illustrated graphically in Figure 13.35. Recall that the partial derivatives $\partial z/\partial x$ and $\partial z/\partial y$ can be interpreted as the slopes of the surface in the x and y directions, respectively. This means that

$$dz = \frac{\partial z}{\partial x} \Delta x + \frac{\partial z}{\partial y} \Delta y$$

$\Delta z = f(x + \Delta x, y + \Delta y) - f(x, y)$

FIGURE 13.35

The exact change in z is Δz. This change can be approximated by the differential dz.

represents the change in height of a plane that is tangent to the surface at the point $(x, y, f(x, y))$. Because a plane in space is represented by a linear equation in the variables x, y, and z, the approximation of Δz by dz is called a **linear approximation.** You will learn more about this geometric interpretation in Section 13.7.

EXAMPLE 3   Using a Differential as an Approximation

Use the differential dz to approximate the change in

$$z = \sqrt{4 - x^2 - y^2}$$

as (x, y) moves from the point $(1, 1)$ to $(1.01, 0.97)$. Compare this approximation with the exact change in z.

Solution   Letting $(x, y) = (1, 1)$ and $(x + \Delta x, y + \Delta y) = (1.01, 0.97)$ produces

$$dx = \Delta x = 0.01 \quad \text{and} \quad dy = \Delta y = -0.03.$$

Thus, the change in z can be approximated by

$$\Delta z \approx dz$$
$$= \frac{\partial z}{\partial x} dx + \frac{\partial z}{\partial y} dy$$
$$= \frac{-x}{\sqrt{4 - x^2 - y^2}} \Delta x + \frac{-y}{\sqrt{4 - x^2 - y^2}} \Delta y.$$

When $x = 1$ and $y = 1$, you have

$$\Delta z \approx -\frac{1}{\sqrt{2}}(0.01) - \frac{1}{\sqrt{2}}(-0.03)$$
$$= \frac{0.02}{\sqrt{2}}$$
$$= \sqrt{2}(0.01)$$
$$\approx 0.0141. \qquad \text{Approximation of } \Delta x \text{ by } dx$$

In Figure 13.36 you can see that the exact change corresponds to the difference in the heights of two points on the surface of a hemisphere. This difference is given by

$$\Delta z = f(1.01, 0.97) - f(1, 1)$$
$$= \sqrt{4 - (1.01)^2 - (0.97)^2} - \sqrt{4 - 1^2 - 1^2}$$
$$\approx 0.0137. \qquad \text{Actual value of } \Delta x \qquad \blacksquare$$

A function of three variables, $w = f(x, y, z)$, is called **differentiable** at (x, y, z) provided that

$$\Delta w = f(x + \Delta x, y + \Delta y, z + \Delta z) - f(x, y, z)$$

can be expressed in the form

$$\Delta w = f_x \Delta x + f_y \Delta y + f_z \Delta z + \varepsilon_1 \Delta x + \varepsilon_2 \Delta y + \varepsilon_3 \Delta z$$

where ε_1, ε_2, and $\varepsilon_3 \to 0$ as $(\Delta x, \Delta y, \Delta z) \to (0, 0, 0)$. With this definition of differentiability, Theorem 13.4 has the following extension for functions of three variables. If f is a function of x, y, and z, where f, f_x, f_y, and f_z are continuous in an open region R, then f is differentiable on R.

In Section 3.9, you used differentials to approximate the propagated error introduced by an error in measurement. This application of differentials is further illustrated in Example 4.

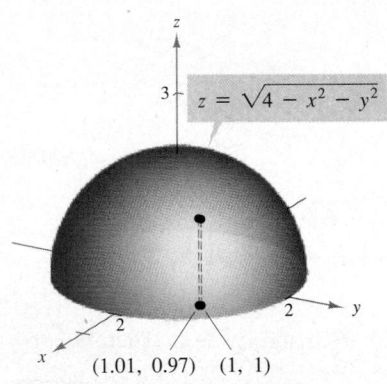

FIGURE 13.36
As (x, y) moves from $(1, 1)$ to the point $(1.01, 0.97)$, the value of $f(x, y)$ changes by about 0.0137.

EXAMPLE 4   Error Analysis

The possible error involved in measuring each dimension of a rectangular box is ± 0.1 millimeter. The dimensions of the box are $x = 50$ centimeters, $y = 20$ centimeters, and $z = 15$ centimeters, as shown in Figure 13.37. Use dV to estimate the propagated error and the relative error in the calculated volume of the box.

Solution   The volume of the box is given by $V = xyz$, and thus

$$dV = \frac{\partial V}{\partial x}dx + \frac{\partial V}{\partial y}dy + \frac{\partial V}{\partial z}dz = yz\,dx + xz\,dy + xy\,dz.$$

Using 0.1 millimeter = 0.01 centimeter, you have $dx = dy = dz = \pm 0.01$, and the propagated error is approximately

$$dV = (20)(15)(\pm 0.01) + (50)(15)(\pm 0.01) + (50)(20)(\pm 0.01)$$
$$= 300(\pm 0.01) + 750(\pm 0.01) + 1000(\pm 0.01)$$
$$= 2050(\pm 0.01) = \pm 20.5 \text{ cm}^3.$$

Because the measured volume is

$$V = (50)(20)(15) = 15,000 \text{ cm}^3$$

the relative error, $\Delta V/V$, is approximately

$$\frac{\Delta V}{V} \approx \frac{dV}{V} = \frac{20.5}{15,000} \approx 0.14\%.$$

FIGURE 13.37
Volume = xyz

As is true for a function of a single variable, if a function of two or more variables is differentiable at a point, then it is also continuous there.

THEOREM 13.5   Differentiability Implies Continuity

If a function of x and y is differentiable at (x_0, y_0), then it is continuous at (x_0, y_0).

Proof   Let f be differentiable at (x_0, y_0), where $z = f(x, y)$. Then

$$\Delta z = [f_x(x_0, y_0) + \varepsilon_1]\Delta x + [f_y(x_0, y_0) + \varepsilon_2]\Delta y$$

where both ε_1 and $\varepsilon_x \to 0$ as $(\Delta x, \Delta y) \to (0, 0)$. However, by definition, you know that Δz is given by

$$\Delta z = f(x_0 + \Delta x, y_0 + \Delta y) - f(x_0, y_0).$$

Letting $x = x_0 + \Delta x$ and $y = y_0 + \Delta y$ produces

$$f(x, y) - f(x_0, y_0) = [f_x(x_0, y_0) + \varepsilon_1]\Delta x + [f_y(x_0, y_0) + \varepsilon_2]\Delta y$$
$$= [f_x(x_0, y_0) + \varepsilon_1](x - x_0) + [f_y(x_0, y_0) + \varepsilon_2](y - y_0).$$

Taking the limit as $(x, y) \to (x_0, y_0)$, you have

$$\lim_{(x, y) \to (x_0, y_0)} f(x, y) = f(x_0, y_0)$$

which means that f is continuous at (x_0, y_0).

Remember that the existence of f_x and f_y is not sufficient to guarantee differentiability, as illustrated in the next example.

EXAMPLE 5   A Function That is Not Differentiable

Show that $f_x(0, 0)$ and $f_y(0, 0)$ both exist, but that f is not differentiable at $(0, 0)$ where f is defined as

$$f(x, y) = \begin{cases} \dfrac{-3xy}{x^2 + y^2}, & \text{if}(x, y) \neq (0, 0) \\ 0, & \text{if } (x, y) = (0, 0). \end{cases}$$

Solution   You can show that f is not differentiable at $(0, 0)$ by showing that it is not continuous at this point. To see that f is not continuous at $(0, 0)$, look at the values of $f(x, y)$ along two different approaches to $(0, 0)$, as shown in Figure 13.38. Along the line $y = x$, the limit is

$$\lim_{(x, x) \to (0, 0)} f(x, y) = \lim_{(x, x) \to (0, 0)} \frac{-3x^2}{2x^2} = -\frac{3}{2}$$

whereas along $y = -x$ you have

$$\lim_{(x, -x) \to (0, 0)} f(x, y) = \lim_{(x, -x) \to (0, 0)} \frac{3x^2}{2x^2} = \frac{3}{2}.$$

Thus, the limit of $f(x, y)$ as $(x, y) \to (0, 0)$ does not exist, and you can conclude that f is not continuous at $(0, 0)$. Therefore, by Theorem 13.5, you know that f is not differentiable at $(0, 0)$. On the other hand, by the definition of the partial derivatives f_x and f_y, you have

$$f_x(0, 0) = \lim_{\Delta x \to 0} \frac{f(\Delta x, 0) - f(0, 0)}{\Delta x} = \lim_{\Delta x \to 0} \frac{0 - 0}{\Delta x} = 0$$

and

$$f_y(0, 0) = \lim_{\Delta y \to 0} \frac{f(0, \Delta y) - f(0, 0)}{\Delta y} = \lim_{\Delta y \to 0} \frac{0 - 0}{\Delta y} = 0.$$

Thus, the partial derivatives at $(0, 0)$ exist.

(a)

(b)

Generated by Mathematica

(c)

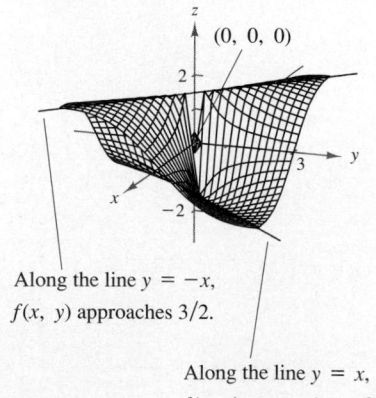

$$f(x, y) = \begin{cases} \dfrac{-3xy}{x^2 + y^2}, & (x, y) \neq (0, 0) \\ 0, & (x, y) = (0, 0) \end{cases}$$

Along the line $y = -x$, $f(x, y)$ approaches $3/2$.

Along the line $y = x$, $f(x, y)$ approaches $-3/2$.

FIGURE 13.38

(d)

EXERCISES for Section 13.4

TECHNOLOGY
Laboratory Guide
Labs 13.4–13.5

In Exercises 1–10, find the total differential.

1. $z = 3x^2y^3$

2. $z = \dfrac{x^2}{y}$

3. $z = \dfrac{-1}{x^2 + y^2}$

4. $z = e^x \sin y$

5. $z = x \cos y - y \cos x$

6. $z = \frac{1}{2}(e^{x^2+y^2} - e^{-x^2-y^2})$

7. $w = 2z^3y \sin x$

8. $w = e^x \cos y + z$

9. $w = \dfrac{x + y}{z - 2y}$

10. $w = x^2yz^2 + \sin yz$

In Exercises 11–16, (a) evaluate $f(1, 2)$ and $f(1.05, 2.1)$ and calculate Δz and (b) use the total differential dz to approximate Δz.

11. $f(x, y) = 9 - x^2 - y^2$

12. $f(x, y) = \sqrt{x^2 + y^2}$

13. $f(x, y) = x \sin y$

14. $f(x, y) = xy$

15. $f(x, y) = 3x - 4y$

16. $f(x, y) = \dfrac{x}{y}$

In Exercises 17–20, use a total differential to approximate the quantity.

17. $\sqrt{(5.05)^2 + (3.1)^2} - \sqrt{5^2 + 3^2}$

18. $(2.03)^2(1 + 8.9)^3 - 2^2(1 + 9)^3$

19. $\dfrac{1 - (3.05)^2}{(5.95)^2} - \dfrac{1 - 3^2}{6^2}$

20. $\sin[(1.05)^2 + (0.95)^2] - \sin(1^2 + 1^2)$

21. *Area* The area of the rectangle in the figure is $A = lh$. Find dA and identify the regions in the figure whose areas are given by its terms. What region represents the difference between ΔA and dA?

22. *Volume* The volume of the right circular cylinder in the figure is $V = \pi r^2 h$. Find dV and identify the regions in the figure whose volumes are given by its terms. What region represents the difference between ΔV and dV?

23. *Volume* The radius r and height h of a right circular cylinder are measured with possible errors of 4 percent and 2 percent, respectively. Approximate the maximum possible percentage error in measuring the volume.

24. *Volume* A right circular cone of height $h = 6$ and radius $r = 3$ is constructed, and in the process errors Δr and Δh are made in the radius and height, respectively. Complete the table to show the relationship between ΔV and dV for the given errors.

Δr	Δh	dV	ΔV	$\Delta v - dV$
0.1	0.1			
0.1	−0.1			
0.001	0.002			
−0.0001	0.0002			

25. *Volume* A trough is 16 feet long (see figure). Its cross sections are isosceles triangles—each of the two equal sides are 18 inches long. The angle between the two equal sides is θ.

   a. Express the volume of the trough as a function of θ and determine the value of θ such that the volume is maximum.

   b. Approximate the change from the maximum volume if the maximum error in the linear measurements is one-half inch and the maximum error in the angle measurement is 2°.

26. *Baseball* A baseball player in center field is playing approximately 330 feet from the television camera that is behind home plate. A batter hits a fly ball that goes to the wall that is 420 feet from the camera (see figure).

   a. Approximate the number of feet that the center fielder has to run to make the catch if the camera turns 9° in following the play.

   b. Approximate the maximum possible error in the result of part a if the position of the center fielder could be in error by as much as 6 feet and the maximum error in measuring the rotation of the camera is 1°.

FIGURE FOR 21   **FIGURE FOR 22**

FIGURE FOR 25   **FIGURE FOR 26**

27. *Area* A triangle is measured and two adjacent sides are found to be 3 and 4 inches long, with an included angle of $\pi/4$. The possible errors in measurement are $\frac{1}{16}$ inch in the sides and 0.02 radian in the angle. Approximate the maximum possible error in the computation of the area.

28. *Acceleration* The centripetal acceleration of a particle moving in a circle is $a = v^2/r$, where v is the velocity and r is the radius of the circle. Approximate the maximum percentage error in measuring the acceleration due to errors of 2 percent in v and 1 percent in r.

29. *Power* Electrical power P is given by $P = E^2/R$, where E is voltage and R is resistance. Approximate the maximum percentage error in calculating power if 200 volts is applied to a 4000-ohm resistor and the possible percentage errors in measuring E and R are 2 percent and 3 percent, respectively.

30. *Resistance* The total resistance R of two resistors connected in parallel is given by

$$\frac{1}{R} = \frac{1}{R_1} + \frac{1}{R_2}.$$

Approximate the change in R as R_1 is increased from 10 ohms to 10.5 ohms and R_2 is decreased from 15 ohms to 13 ohms.

31. *Inductance* The inductance L (in microhenrys) of a straight nonmagnetic wire in free space is given by

$$L = 0.00021\left(\ln\frac{2h}{r} - 0.75\right)$$

where h is the length of the wire in millimeters and r is the radius of a circular cross section. Approximate L when $r = 2 \pm \frac{1}{16}$ millimeters and $h = 100 \pm \frac{1}{100}$ millimeters.

32. *Pendulum* The period T of a pendulum of length L is given by $T = 2\pi\sqrt{L/g}$, where g is the acceleration due to gravity. A pendulum is moved from the Canal Zone, where $g = 32.09$ feet per second per second, to Greenland, where $g = 32.24$ feet per second per second. Because of the change in temperature, the length of the pendulum changes from 2.5 feet to 2.48 feet. Approximate the change in the period of the pendulum.

In Exercises 33–36, show that the given function is differentiable by finding values for ε_1 and ε_2 as designated in the definition of differentiability, and verify that both ε_1 and $\varepsilon_2 \to 0$ as $(\Delta x, \Delta y) \to (0, 0)$.

33. $f(x, y) = x^2 - 2x + y$ **34.** $f(x, y) = x^2 + y^2$

35. $f(x, y) = x^2 y$ **36.** $f(x, y) = 5x - 10y + y^3$

In Exercises 37 and 38, use the given function to prove that (a) $f_x(0, 0)$ and $f_y(0, 0)$ exist, and (b) f is not differentiable at $(0, 0)$.

37. $f(x, y) = \begin{cases} \dfrac{3x^2 y}{x^4 + y^2}, & (x, y) \neq (0, 0) \\ 0, & (x, y) = (0, 0) \end{cases}$

38. $f(x, y) = \begin{cases} \dfrac{2x^2 y^2}{x^4 + y^4}, & (x, y) \neq (0, 0) \\ 0, & (x, y) = (0, 0) \end{cases}$

CAREER INTERVIEW

S. S. Chen

Chemical Engineer

Director of Research and Development

Badger Company

Weymouth, MA 02189

At the chemical process laboratory, I lead a team of chemists and engineers in developing processes that produce chemicals used in products as diverse as soda bottles, automobile tires, and fertilizers. Working with raw materials derived from petroleum, we convert them into polymers, which you may know better as plastics. We create new processes by finding ways to employ less raw materials and energy for manufacturing products of higher purity, with smaller environmental impact.

Because calculus is the natural language for describing motion, it is utilized in our daily work to express the speed of chemical reactions. For example, you may have encountered in chemistry the first order rate equation that describes the rate of conversion of a substance into another substance; that is, if substance A is being converted irreversibly to substance B, then the rate at which A decreases and B increases is proportional to the concentration of A at any given moment:

$$A \to B$$

$$-\frac{dA}{dt} = \frac{dB}{dt} = kA$$

Chain Rules for Functions of Several Variables

13.5 Chain Rules for Functions of Several Variables • Implicit Partial Differentiation

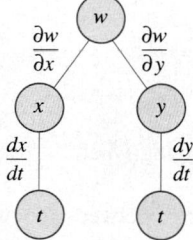

FIGURE 13.39
Chain Rule: one independent variable. w is a function of x and y, which are each functions of t. This diagram represents the derivative of w with respect to t.

Chain Rules for Functions of Several Variables

Your work with differentials in the previous section provides the basis for the extension of the Chain Rule to functions of two variables. There are two cases—the first case involves w as a function of x and y, where x and y are functions of a single independent variable t. (A proof of this theorem is given in Appendix A.)

THEOREM 13.6   Chain Rule: One Independent Variable

Let $w = f(x, y)$, where f is a differentiable function of x and y. If $x = g(t)$ and $y = h(t)$, where g and h are differentiable functions of t, then w is a differentiable function of t, and

$$\frac{dw}{dt} = \frac{\partial w}{\partial x}\frac{dx}{dt} + \frac{\partial w}{\partial y}\frac{dy}{dt}.$$

REMARK   The Chain Rule presented in this theorem can be represented schematically as shown in Figure 13.39.

EXAMPLE 1   Using the Chain Rule with One Independent Variable

Let $w = x^2 y - y^2$, where $x = \sin t$ and $y = e^t$. Find dw/dt when $t = 0$.

Solution   By the Chain Rule for one independent variable, you have

$$\frac{dw}{dt} = \frac{\partial w}{\partial x}\frac{dx}{dt} + \frac{\partial w}{\partial y}\frac{dy}{dt}$$

$$= 2xy(\cos t) + (x^2 - 2y)e^t.$$

When $t = 0$, $x = 0$, and $y = 1$, it follows that

$$\frac{dw}{dt} = 0 - 2 = -2.$$

The Chain Rules presented in this section provide alternative techniques for solving many problems in single-variable calculus. For instance, in Example 1, you could have used single-variable techniques to find dw/dt by first writing w as a function of t,

$$w = x^2 y - y^2 = (\sin t)^2(e^t) - (e^t)^2 = e^t \sin^2 t - e^{2t}$$

and then differentiating as usual. (Try doing this.)

The Chain Rule in Theorem 13.6 can be extended to any number of variables. For example, if each x_i is a differentiable function of a single variable t, then for

$$w = f(x_1, x_2, \ldots, x_n)$$

you have

$$\frac{dw}{dt} = \frac{\partial w}{\partial x_1}\frac{dx_1}{dt} + \frac{\partial w}{\partial x_2}\frac{dx_2}{dt} + \cdots + \frac{\partial w}{\partial x_n}\frac{dx_n}{dt}.$$

EXAMPLE 2   An Application of a Chain Rule to Related Rates

Two objects are traveling in elliptical paths given by the following parametric equations.

$$x_1 = 4\cos t \quad \text{and} \quad y_1 = 2\sin t \qquad \text{First object}$$
$$x_2 = 2\sin 2t \quad \text{and} \quad y_2 = 3\cos 2t \qquad \text{Second object}$$

At what rate is the distance between the two objects changing when $t = \pi$?

Solution   From Figure 13.40, you can see that the distance s between the two objects is given by

$$s = \sqrt{(x_2 - x_1)^2 + (y_2 - y_1)^2}$$

and when $t = \pi$, you have $x_1 = -4$, $y_1 = 0$, $x_2 = 0$, $y_2 = 3$, and

$$s = \sqrt{(0 + 4)^2 + (3 - 0)^2} = 5.$$

When $t = \pi$, the partial derivatives of s are as follows.

$$\frac{\partial s}{\partial x_1} = \frac{-(x_2 - x_1)}{\sqrt{(x_2 - x_1)^2 + (y_2 - y_1)^2}} = -\frac{1}{5}(0 + 4) = -\frac{4}{5}$$

$$\frac{\partial s}{\partial y_1} = \frac{-(y_2 - y_1)}{\sqrt{(x_2 - x_1)^2 + (y_2 - y_1)^2}} = -\frac{1}{5}(3 - 0) = -\frac{3}{5}$$

$$\frac{\partial s}{\partial x_2} = \frac{(x_2 - x_1)}{\sqrt{(x_2 - x_1)^2 + (y_2 - y_1)^2}} = \frac{1}{5}(0 + 4) = \frac{4}{5}$$

$$\frac{\partial s}{\partial y_2} = \frac{(y_2 - y_1)}{\sqrt{(x_2 - x_1)^2 + (y_2 - y_1)^2}} = \frac{1}{5}(3 - 0) = \frac{3}{5}$$

When $t = \pi$, the derivatives of x_1, y_1, x_2, and y_2 are

$$\frac{dx_1}{dt} = -4\sin t = 0 \qquad \frac{dy_1}{dt} = 2\cos t = -2$$

$$\frac{dx_2}{dt} = 4\cos 2t = 4 \qquad \frac{dy_2}{dt} = -6\sin 2t = 0.$$

Therefore, using the appropriate Chain Rule, you know that the distance is changing at the rate of

$$\frac{ds}{dt} = \frac{\partial s}{\partial x_1}\frac{dx_1}{dt} + \frac{\partial s}{\partial y_1}\frac{dy_1}{dt} + \frac{\partial s}{\partial x_2}\frac{dx_2}{dt} + \frac{\partial s}{\partial y_2}\frac{dy_2}{dt}$$

$$= \left(-\frac{4}{5}\right)(0) + \left(-\frac{3}{5}\right)(-2) + \left(\frac{4}{5}\right)(4) + \left(\frac{3}{5}\right)(0)$$

$$= \frac{22}{5}.$$

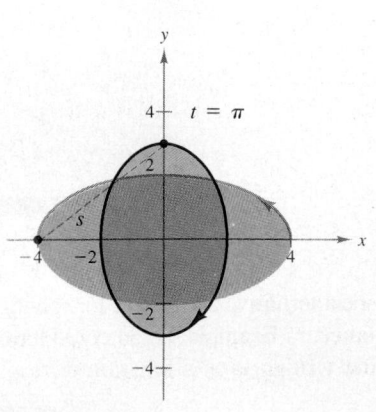

FIGURE 13.40
Paths of two objects traveling in elliptical orbits.

In Example 2, note that s is the function of four *intermediate* variables, $x_1, y_1, x_2,$ and y_2, which in turn are each functions of a single variable t. Another type of composite function is one in which the intermediate variables are themselves functions of more than one variable. For instance, if $w = f(x, y)$, where $x = g(s, t)$ and $y = h(s, t)$, then it follows that w is a function of s and t, and you can consider the partial derivatives of w with respect to s and t. One way to find these partial derivatives is to write w as a function of s and t explicitly by substituting the equations $x = g(s, t)$ and $y = h(s, t)$ into the equation $w = f(x, y)$. Then you can find the partial derivatives in the usual way, as demonstrated in the next example.

EXAMPLE 3 Finding Partial Derivatives by Substitution

Find $\partial w / \partial s$ and $\partial w / \partial t$ for $w = 2xy$, where $x = s^2 + t^2$ and $y = s/t$.

Solution Begin by substituting $x = s^2 + t^2$ and $y = s/t$ into the equation $w = 2xy$ to obtain

$$w = 2xy = 2(s^2 + t^2)\left(\frac{s}{t}\right) = 2\left(\frac{s^3}{t} + st\right).$$

Then, to find $\partial w / \partial s$, hold t constant and differentiate with respect to s.

$$\frac{\partial w}{\partial s} = 2\left(\frac{3s^2}{t} + t\right) = \frac{6s^2 + 2t^2}{t}$$

Similarly, to find $\partial w / \partial t$, hold s constant and differentiate with respect to t to obtain

$$\frac{\partial w}{\partial t} = 2\left(-\frac{s^3}{t^2} + s\right) = 2\left(\frac{-s^3 + st^2}{t^2}\right) = \frac{2st^2 - 2s^3}{t^2}.$$

Theorem 13.7 gives an alternative method for finding the partial derivatives in Example 3—without explicitly writing w as a function of s and t.

THEOREM 13.7 Chain Rule: Two Independent Variables

Let $w = f(x, y)$, where f is a differentiable function of x and y. If $x = g(s, t)$ and $y = h(s, t)$ such that the first partials $\partial x / \partial s$, $\partial x / \partial t$, $\partial y / \partial s$, and $\partial y / \partial t$ all exist, then $\partial w / \partial s$ and $\partial w / \partial t$ exist and are given by

$$\frac{\partial w}{\partial s} = \frac{\partial w}{\partial x}\frac{\partial x}{\partial s} + \frac{\partial w}{\partial y}\frac{\partial y}{\partial s} \quad \text{and} \quad \frac{\partial w}{\partial t} = \frac{\partial w}{\partial x}\frac{\partial x}{\partial t} + \frac{\partial w}{\partial y}\frac{\partial y}{\partial t}.$$

Proof To obtain $\partial w / \partial s$, hold t constant and apply Theorem 13.6 to obtain the desired result. Similarly, for $\partial w / \partial t$ hold s constant and apply Theorem 13.6.

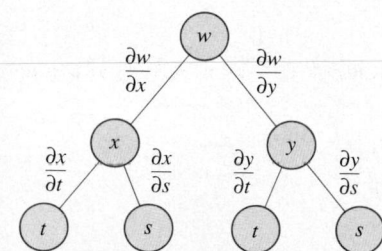

FIGURE 13.41
Chain Rule: two independent variables

REMARK The Chain Rule in this theorem is shown schematically in Figure 13.41.

EXAMPLE 4   The Chain Rule with Two Independent Variables

Use the Chain Rule to find $\partial w/\partial s$ and $\partial w/\partial t$ for

$$w = 2xy$$

where $x = s^2 + t^2$ and $y = s/t$.

Solution   (These same partials were found in Example 3.) Using Theorem 13.7, you can hold t constant and differentiate with respect to s to obtain

$$\frac{\partial w}{\partial s} = \frac{\partial w}{\partial x}\frac{\partial x}{\partial s} + \frac{\partial w}{\partial y}\frac{\partial y}{\partial s}$$

$$= 2y(2s) + 2x\left(\frac{1}{t}\right)$$

$$= 4\left(\frac{s^2}{t}\right) + \frac{2s^2 + 2t^2}{t} \qquad \text{Substitute } s/t \text{ for } y \text{ and } s^2 + t^2 \text{ for } x$$

$$= \frac{6s^2 + 2t^2}{t}.$$

Similarly, holding s fixed gives

$$\frac{\partial w}{\partial t} = \frac{\partial w}{\partial x}\frac{\partial x}{\partial t} + \frac{\partial w}{\partial y}\frac{\partial y}{\partial t}$$

$$= 2y(2t) + 2x\left(\frac{-s}{t^2}\right)$$

$$= 2\left(\frac{s}{t}\right)(2t) + 2(s^2 + t^2)\left(\frac{-s}{t^2}\right) \qquad \text{Substitute } s/t \text{ for } y \text{ and } s^2 + t^2 \text{ for } x$$

$$= 4s - \frac{2s^3 + 2st^2}{t^2}$$

$$= \frac{4st^2 - 2s^3 - 2st^2}{t^2}$$

$$= \frac{2st^2 - 2s^3}{t^2}.$$

The Chain Rule in Theorem 13.7 can also be extended to any number of variables. For example, if w is a differentiable function of the n variables $x_1, x_2, \ldots, x_n$ where each x_i is a differentiable function of the m variables $t_1, t_2, \ldots, t_m$, then for

$$w = f(x_1, x_2, \ldots, x_n)$$

you obtain the following.

$$\frac{\partial w}{\partial t_1} = \frac{\partial w}{\partial x_1}\frac{\partial x_1}{\partial t_1} + \frac{\partial w}{\partial x_2}\frac{\partial x_2}{\partial t_1} + \cdots + \frac{\partial w}{\partial x_n}\frac{\partial x_n}{\partial t_1}$$

$$\frac{\partial w}{\partial t_2} = \frac{\partial w}{\partial x_1}\frac{\partial x_1}{\partial t_2} + \frac{\partial w}{\partial x_2}\frac{\partial x_2}{\partial t_2} + \cdots + \frac{\partial w}{\partial x_n}\frac{\partial x_n}{\partial t_2}$$

$$\vdots$$

$$\frac{\partial w}{\partial t_m} = \frac{\partial w}{\partial x_1}\frac{\partial x_1}{\partial t_m} + \frac{\partial w}{\partial x_2}\frac{\partial x_2}{\partial t_m} + \cdots + \frac{\partial w}{\partial x_n}\frac{\partial x_n}{\partial t_m}$$

EXAMPLE 5   The Chain Rule for a Function of Three Variables

Find $\partial w/\partial s$ and $\partial w/\partial t$ when $s = 1$ and $t = 2\pi$ for the function given by

$$w = xy + yz + xz$$

where $x = s\cos t$, $y = s\sin t$, and $z = t$.

Solution   By extending the result of Theorem 13.7, you have

$$\frac{\partial w}{\partial s} = \frac{\partial w}{\partial x}\frac{\partial x}{\partial s} + \frac{\partial w}{\partial y}\frac{\partial y}{\partial s} + \frac{\partial w}{\partial z}\frac{\partial z}{\partial s}$$

$$= (y + z)(\cos t) + (x + z)(\sin t) + (y + x)(0)$$

$$= (y + z)(\cos t) + (x + z)(\sin t).$$

When $s = 1$ and $t = 2\pi$, you have $x = 1$, $y = 0$, and $z = 2\pi$. Therefore, $\partial w/\partial s = 2\pi(1) + (1 + 2\pi)(0) + 0 = 2\pi$. Furthermore,

$$\frac{\partial w}{\partial t} = \frac{\partial w}{\partial x}\frac{\partial x}{\partial t} + \frac{\partial w}{\partial y}\frac{\partial y}{\partial t} + \frac{\partial w}{\partial z}\frac{\partial z}{\partial t}$$

$$= (y + z)(-s\sin t) + (x + z)(s\cos t) + (y + x)(1)$$

and for $s = 1$ and $t = 2\pi$ it follows that

$$\frac{\partial w}{\partial t} = (0 + 2\pi)(0) + (1 + 2\pi)(1) + (0 + 1)(1)$$

$$= 2 + 2\pi.$$

Implicit Partial Differentiation

We conclude this section with an application of the Chain Rule to determine the derivative of a function defined *implicitly*. Suppose that x and y are related by the equation $F(x, y) = 0$, where it is assumed that $y = f(x)$ is a differentiable function of x. To find dy/dx, you could use the techniques discussed in Section 2.5. However, you will see that the Chain Rule provides a convenient alternative. If you consider the function given by

$$w = F(x, y) = F(x, f(x))$$

you can apply Theorem 13.6 to obtain

$$\frac{dw}{dx} = F_x(x, y)\frac{dx}{dx} + F_y(x, y)\frac{dy}{dx}.$$

Because $w = F(x, y) = 0$ for all x in the domain of f, you know that $dw/dx = 0$ and you have

$$F_x(x, y)\frac{dx}{dx} + F_y(x, y)\frac{dy}{dx} = 0.$$

Now, if $F_y(x, y) \neq 0$, you can use the fact that $dx/dx = 1$ to conclude that

$$\frac{dy}{dx} = -\frac{F_x(x, y)}{F_y(x, y)}.$$

A similar procedure can be used to find the partial derivatives of functions of several variables that are defined implicitly.

THEOREM 13.8   Chain Rule: Implicit Differentiation

If the equation $F(x, y) = 0$ defines y implicity as a differentiable function of x, then

$$\frac{dy}{dx} = -\frac{F_x(x, y)}{F_y(x, y)}, \qquad F_y(x, y) \neq 0.$$

If the equation $F(x, y, z) = 0$ defines z implicitly as a differentiable function of x and y, then

$$\frac{\partial z}{\partial x} = -\frac{F_x(x, y, z)}{F_z(x, y, z)} \quad \text{and} \quad \frac{\partial z}{\partial y} = -\frac{F_y(x, y, z)}{F_z(x, y, z)}, \qquad F_z(x, y, z) \neq 0.$$

This theorem can be extended to differentiable functions defined implicitly with any number of variables.

EXAMPLE 6   Finding a Derivative Implicitly

REMARK   Compare the solution of Example 6 with the solution of Example 2 in Section 2.5.

Find dy/dx, given $y^3 + y^2 - 5y - x^2 + 4 = 0$.

Solution   Begin by defining a function F as

$$F(x, y) = y^3 + y^2 - 5y - x^2 + 4.$$

Then, using Theorem 13.8, you have

$$F_x(x, y) = -2x \quad \text{and} \quad F_y(x, y) = 3y^2 + 2y - 5$$

and it follows that

$$\frac{dy}{dx} = -\frac{F_x(x, y)}{F_y(x, y)} = \frac{-(-2x)}{3y^2 + 2y - 5} = \frac{2x}{3y^2 + 2y - 5}.$$

EXAMPLE 7   Finding Partial Derivatives Implicitly

Find $\partial z/\partial x$ and $\partial z/\partial y$, given $3x^2z - x^2y^2 + 2z^3 + 3yz - 5 = 0$.

Solution   To apply Theorem 13.8, let

$$F(x, y, z) = 3x^2z - x^2y^2 + 2z^3 + 3yz - 5.$$

Then

$$F_x(x, y, z) = 6xz - 2xy^2$$
$$F_y(x, y, z) = -2x^2y + 3z$$
$$F_z(x, y, z) = 3x^2 + 6z^2 + 3y$$

and you obtain

$$\frac{\partial z}{\partial x} = -\frac{F_x}{F_z} = \frac{2xy^2 - 6xz}{3x^2 + 6z^2 + 3y}$$

$$\frac{\partial z}{\partial y} = -\frac{F_y}{F_z} = \frac{2x^2y - 3z}{3x^2 + 6z^2 + 3y}.$$

EXERCISES for Section 13.5

In Exercises 1–4, find dw/dt using the appropriate Chain Rule.

1. $w = x^2 + y^2$
$x = e^t, y = e^{-t}$

2. $w = \sqrt{x^2 + y^2}$
$x = \sin t, y = e^t$

3. $w = x \sec y$
$x = e^t, y = \pi - t$

4. $w = \ln\dfrac{y}{x}$
$x = \cos t, y = \sin t$

In Exercises 5–8, find $\partial w/\partial s$ and $\partial w/\partial t$ using the appropriate Chain Rule, and evaluate each partial derivative at the indicated values of s and t.

Function	Point
5. $w = x^2 + y^2$ $\quad x = s + t, y = s - t$	$s = 2, t = -1$
6. $w = y^3 - 3x^2 y$ $\quad x = e^s, y = e^t$	$s = 0, t = 1$
7. $w = x^2 - y^2$ $\quad x = s \cos t, y = s \sin t$	$s = 3, t = \dfrac{\pi}{4}$
8. $w = \sin(2x + 3y)$ $\quad x = s + t, y = s - t$	$s = 0, t = \dfrac{\pi}{2}$

In Exercises 9–14, find dw/dt (a) by the appropriate Chain Rule and (b) by converting w to a function of t before differentiating.

9. $w = xy$
$x = 2 \sin t, y = \cos t$

10. $w = \cos(x - y)$
$x = t^2, y = 1$

11. $w = x^2 + y^2 + z^2$
$x = e^t \cos t, y = e^t \sin t, z = e^t$

12. $w = xy \cos z$
$x = t, y = t^2, z = \arccos t$

13. $w = xy + xz + yz$
$x = t - 1, y = t^2 - 1, z = t$

14. $w = xyz$
$x = t^2, y = 2t, z = e^{-t}$

In Exercises 15–18, find $\partial w/\partial r$ and $\partial w/\partial\theta$ (a) by the appropriate Chain Rule and (b) by converting w to a function of r and θ before differentiating.

15. $w = x^2 - 2xy + y^2$
$x = r + \theta, y = r - \theta$

16. $w = \sqrt{4 - 2x^2 - 2y^2}$
$x = r \cos\theta, y = r \sin\theta$

17. $w = \arctan\dfrac{y}{x}$
$x = r \cos\theta, y = r \sin\theta$

18. $w = \dfrac{xy}{z}$
$x = r + \theta, y = r - \theta, z = \theta^2$

In Exercises 19–22, differentiate implicitly to find the first partial derivatives of z.

19. $x^2 + y^2 + z^2 = 25$

20. $xz + yz + xy = 0$

21. $\tan(x + y) + \tan(y + z) = 1$

22. $z = e^x \sin(y + z)$

In Exercises 23 and 24, differentiate implicitly to find all first and second partial derivatives of z.

23. $x^2 + 2yz + z^2 = 1$

24. $x + \sin(y + z) = 0$

In Exercises 25 and 26, differentiate implicitly to find the first partial derivatives of w.

25. $xyz + xzw - yzw + w^2 = 5$

26. $x^2 + y^2 + z^2 + 6xw - 8w^2 = 5$

27. *Volume and Surface Area* Let θ be the angle between the equal sides of an isosceles triangle, and x the length of these sides. If x is increasing at $\frac{1}{2}$ m/hr and θ is increasing at $\pi/90$ rad/hr, find the rate of increase of the area when $x = 6$ and $\theta = \pi/4$.

28. *Volume and Surface Area* The radius of a right circular cylinder is increasing at the rate of 6 inches per minute, and the height is decreasing at the rate of 4 inches per minute. What is the rate of change of the volume and surface area when the radius is 12 inches and the height is 36 inches?

29. *Volume and Surface Area* Repeat Exercise 28 for a right circular cone.

30. *Volume and Surface Area* The two radii of the frustum of a right circular cone are increasing at the rate of 4 centimeters per minute, and the height is increasing at the rate of 12 centimeters per minute (see figure). Find the rate at which the volume and surface area are changing when the two radii are 15 and 25 centimeters and the height is 10 centimeters.

31. *Moment of Inertia* An annular cylinder has an inside radius of r_1 and an outside radius of r_2 (see figure). The moment of inertia is given by

$$I = \frac{1}{2}m(r_1^2 + r_2^2)$$

where m is the mass. Find the rate at which I is changing at the instant the radii are 6 and 8 centimeters, respectively, if the two radii are increasing at the rate of 2 centimeters per second.

FIGURE FOR 30

FIGURE FOR 31

32. *Ideal Gas Law*   The Ideal Gas Law is $pV = RT$, where R is a constant. If p and V are functions of time, find dT/dt, the rate at which the temperature changes with respect to time.

[C] **33.** *Projectile Motion*   A projectile is launched at an angle of $45°$ with the horizontal and with an initial velocity of 64 feet per second. A television camera is located in the plane of the path of the projectile and 50 feet behind the launch site (see figure).

    a. Find parametric equations for the path of the projectile in terms of the parameter t representing time.

    b. Express the angle α that the camera makes with the horizontal in terms of x and y and in terms of the parameter t.

    c. Use the results of part b to find $d\alpha/dt$.

    d. Use a graphing utility to obtain a graph of α in terms of t. Is this graph symmetric with the axis of the parabolic arch of the projectile? Why or why not? At what time is the rate of change of α greatest?

    e. At what time is the angle α maximum? Does this occur when the projectile is at its greatest height?

FIGURE FOR 33

34. *Projectile Motion*   Consider the distance between the launch site and the projectile in Exercise 33 (see figure below).

    a. Express the distance d in terms of x and y and in terms of the parameter t.

    b. Use the results of part a to find the rate of change of d.

    c. Find the rate of change of the distance when $t = 2$.

    d. When is the rate of change of d minimum during the flight of the projectile? Does it occur at the time the projectile reaches its maximum height?

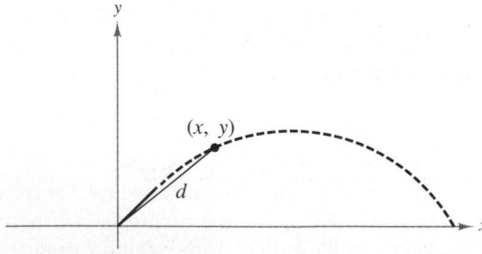

FIGURE FOR 34

Homogeneous Functions   **In Exercises 35–38, a function f in x and y is homogeneous of degree n, if**

$$f(tx, ty) = t^n f(x, y).$$

Determine the degree of the function, and show that

$$xf_x(x, y) + yf_y(x, y) = nf(x, y).$$

35. $f(x, y) = x^3 - 3xy^2 + y^3$

36. $f(x, y) = \dfrac{xy}{\sqrt{x^2 + y^2}}$

37. $f(x, y) = e^{x/y}$ **38.** $f(x, y) = \dfrac{x^2}{\sqrt{x^2 + y^2}}$

39. Show that if $f(x, y)$ is homogeneous of degree n, then

$$xf_x(x, y) + yf_y(x, y) = nf(x, y).$$

    (*Hint:* Let $g(t) = f(tx, ty) = t^n f(x, y)$. Find $g'(t)$ and then let $t = 1$.)

40. Show that

$$\frac{\partial w}{\partial u} + \frac{\partial w}{\partial v} = 0$$

    for $w = f(x, y)$, $x = u - v$, and $y = v - u$.

41. Demonstrate the result of Exercise 40 for the function given by

$$w = (x - y)\sin(y - x).$$

42. Consider the function $w = f(x, y)$ where $x = r\cos\theta$ and $y = r\sin\theta$. Prove the following.

    a. $\dfrac{\partial w}{\partial x} = \dfrac{\partial w}{\partial r}\cos\theta - \dfrac{\partial w}{\partial\theta}\dfrac{\sin\theta}{r}$

    $\dfrac{\partial w}{\partial y} = \dfrac{\partial w}{\partial r}\sin\theta + \dfrac{\partial w}{\partial\theta}\dfrac{\cos\theta}{r}$

    b. $\left(\dfrac{\partial w}{\partial x}\right)^2 + \left(\dfrac{\partial w}{\partial y}\right)^2 = \left(\dfrac{\partial w}{\partial r}\right)^2 + \left(\dfrac{1}{r^2}\right)\left(\dfrac{\partial w}{\partial\theta}\right)^2$

43. Demonstrate the result of Exercise 42b for the function given by

$$w = \arctan\frac{y}{x}.$$

44. *Cauchy-Riemann Equations*   Given the functions $u(x, y)$ and $v(x, y)$, show that the **Cauchy-Riemann differential equations**

$$\frac{\partial u}{\partial x} = \frac{\partial v}{\partial y} \quad\text{and}\quad \frac{\partial u}{\partial y} = -\frac{\partial v}{\partial x}$$

    can be written in polar coordinate form as

$$\frac{\partial u}{\partial r} = \frac{1}{r}\frac{\partial v}{\partial\theta} \quad\text{and}\quad \frac{\partial v}{\partial r} = -\frac{1}{r}\frac{\partial u}{\partial\theta}.$$

45. Demonstrate the result of Exercise 44 for the functions

$$u = \ln\sqrt{x^2 + y^2} \quad\text{and}\quad v = \arctan\frac{y}{x}.$$

46. *Wave Equation*   Show that

$$u(x, t) = \frac{1}{2}[f(x - ct) + f(x + ct)]$$

    is a solution to the one-dimensional wave equation

$$\frac{\partial^2 u}{\partial t^2} = c^2\frac{\partial^2 u}{\partial x^2}.$$

    (This equation describes the small transverse vibration of elastic strings such as those on certain musical instruments.)

13.6

Directional Derivative • The Gradient of a Function of Two Variables • Applications of the Gradient • Functions of Three Variables

Surface:
$z = f(x, y)$

FIGURE 13.42

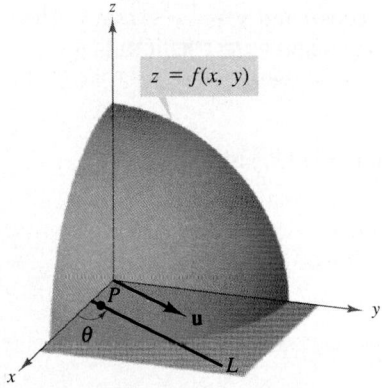

$z = f(x, y)$

FIGURE 13.43

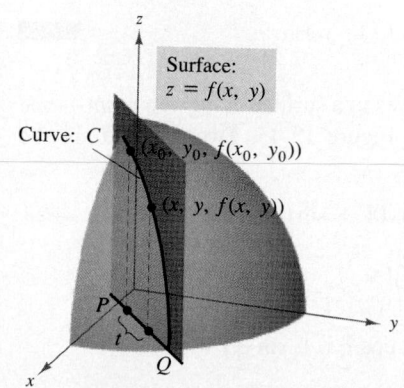

Surface:
$z = f(x, y)$

Curve: C

$(x_0, y_0, f(x_0, y_0))$

$(x, y, f(x, y))$

FIGURE 13.44

Directional Derivative

Suppose you are standing on the hillside pictured in Figure 13.42 and want to determine the hill's incline toward the z-axis. If the hill were represented by $z = f(x, y)$, then you would already know how to determine the slopes in two different directions—the slope in the y-direction would be given by the partial derivative $f_y(x, y)$, and the slope in the x-direction would be given by the partial derivative $f_x(x, y)$. In this section, you will see that these two partial derivatives can be used to find the slope in *any* direction.

To determine the slope at a point on a surface, we define a new type of derivative called a **directional derivative.** We begin by letting $z = f(x, y)$ be a *surface* and $P = (x_0, y_0)$ a *point* in the domain of f, as shown in Figure 13.43. The "direction" of the directional derivative is given by a unit vector

$$\mathbf{u} = \cos\theta\,\mathbf{i} + \sin\theta\,\mathbf{j}$$

where θ is the angle the vector makes with the positive x-axis. To find the desired slope, reduce the problem to two dimensions by intersecting the surface with a vertical plane passing through the point P and parallel to $\mathbf{u}$, as shown in Figure 13.44. This vertical plane intersects the surface to form a curve C. The slope of the surface at $(x_0, y_0, f(x_0, y_0))$ in the direction of $\mathbf{u}$ is defined as the slope of the curve C at that point.

Informally, you can write the slope of the curve C as a limit that looks much like those used in single-variable calculus. The vertical plane used to form C intersects the xy-plane in a line L, represented by the parametric equations

$$x = x_0 + t\cos\theta \quad \text{and} \quad y = y_0 + t\sin\theta$$

so that for any value of t, the point $Q = (x, y)$ lies on the line L. For each of the points P and Q, there is a corresponding point on the surface.

$(x_0, y_0, f(x_0, y_0))$	Point above P
$(x, y, f(x, y))$	Point above Q

Moreover, because the distance between P and Q is

$$\sqrt{(x - x_0)^2 + (y - y_0)^2} = \sqrt{(t\cos\theta)^2 + (t\sin\theta)^2} = |t|$$

you can write the slope of the secant line through $(x_0, y_0, f(x_0, y_0))$ and $(x, y, f(x, y))$ as

$$\frac{f(x, y) - f(x_0, y_0)}{t} = \frac{f(x_0 + t\cos\theta, y_0 + t\sin\theta) - f(x_0, y_0)}{t}.$$

Finally, by letting t approach 0, you arrive at the following definition.

Definition of Directional Derivative

Let f be a function of two variables x and y and let $\mathbf{u} = \cos\theta\,\mathbf{i} + \sin\theta\,\mathbf{j}$ be a unit vector. Then the **directional derivative of f in the direction of u**, denoted by $D_{\mathbf{u}}f$, is

$$D_{\mathbf{u}}f(x, y) = \lim_{t \to 0} \frac{f(x + t\cos\theta, y + t\sin\theta) - f(x, y)}{t}$$

provided this limit exists.

Calculating directional derivatives by this definition is similar to finding the derivative of a function of one variable by the limit process (given in Section 2.1). A simpler "working" formula for finding directional derivatives involves the partial derivatives f_x and f_y.

THEOREM 13.9   Directional Derivative

If f is a differentiable function of x and y, then the directional derivative of f in the direction of the unit vector $\mathbf{u} = \cos\theta\,\mathbf{i} + \sin\theta\,\mathbf{j}$ is

$$D_{\mathbf{u}}f(x, y) = f_x(x, y)\cos\theta + f_y(x, y)\sin\theta.$$

Proof   For a fixed point (x_0, y_0), let $x = x_0 + t\cos\theta$ and $y = y_0 + t\sin\theta$. Then, let $g(t) = f(x, y)$. Because f is differentiable, you can apply the Chain Rule given in Theorem 13.7 to obtain

$$g'(t) = f_x(x, y)\frac{\partial x}{\partial t} + f_y(x, y)\frac{\partial y}{\partial t} = f_x(x, y)\cos\theta + f_y(x, y)\sin\theta.$$

If $t = 0$, then $x = x_0$ and $y = y_0$, so

$$g'(0) = f_x(x_0, y_0)\cos\theta + f_y(x_0, y_0)\sin\theta.$$

By the definition of $g'(t)$, it is also true that

$$g'(0) = \lim_{t \to 0} \frac{g(t) - g(0)}{t} = \lim_{t \to 0} \frac{f(x_0 + t\cos\theta, y_0 + t\sin\theta) - f(x_0, y_0)}{t}.$$

Consequently, $D_{\mathbf{u}}f(x_0, y_0) = f_x(x_0, y_0)\cos\theta + f_y(x_0, y_0)\sin\theta.$ ■

There are infinitely many directional derivatives to a surface at a given point—one for each direction specified by $\mathbf{u}$, as indicated in Figure 13.45. Two of these are the partial derivatives f_x and f_y.

1. Direction of positive x-axis ($\theta = 0$): $\mathbf{u} = \cos 0\,\mathbf{i} + \sin 0\,\mathbf{j} = \mathbf{i}$

$$D_{\mathbf{i}}f(x, y) = f_x(x, y)\cos 0 + f_y(x, y)\sin 0 = f_x(x, y)$$

2. Direction of positive y-axis ($\theta = \pi/2$): $\mathbf{u} = \cos\dfrac{\pi}{2}\mathbf{i} + \sin\dfrac{\pi}{2}\mathbf{j} = \mathbf{j}$

$$D_{\mathbf{j}}f(x, y) = f_x(x, y)\cos\frac{\pi}{2} + f_y(x, y)\sin\frac{\pi}{2} = f_y(x, y)$$

Surface:
$z = f(x, y)$

(x, y)

The vector **u**

FIGURE 13.45

EXAMPLE 1   Finding a Directional Derivative

Find the directional derivative of

$$f(x, y) = 4 - x^2 - \frac{1}{4}y^2 \qquad \text{Surface}$$

at $(1, 2)$ in the direction of

$$\mathbf{u} = \left(\cos\frac{\pi}{3}\right)\mathbf{i} + \left(\sin\frac{\pi}{3}\right)\mathbf{j}. \qquad \text{Direction}$$

Solution   Because f_x and f_y are continuous, f is differentiable, and you can apply Theorem 13.9.

$$D_{\mathbf{u}}f(x, y) = f_x(x, y)\cos\theta + f_y(x, y)\sin\theta$$

$$= (-2x)\cos\theta + \left(-\frac{y}{2}\right)\sin\theta$$

Evaluating at $\theta = \pi/3$, $x = 1$, and $y = 2$ produces

$$D_{\mathbf{u}}f(1, 2) = (-2)\left(\frac{1}{2}\right) + (-1)\left(\frac{\sqrt{3}}{2}\right)$$

$$= -1 - \frac{\sqrt{3}}{2}$$

$$\approx -1.866.$$

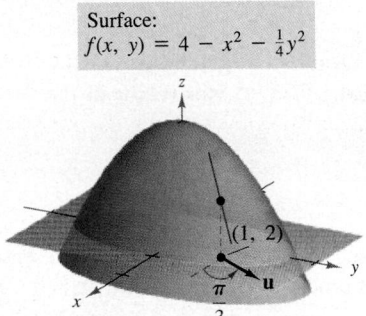

Surface:
$f(x, y) = 4 - x^2 - \frac{1}{4}y^2$

FIGURE 13.46

REMARK   Note in Figure 13.46 that you can interpret the directional derivative as giving the slope of the surface at the point $(1, 2, 2)$ in the direction of the unit vector $\mathbf{u}$.

We have been specifying direction by a unit vector $\mathbf{u}$. If the direction is given by a vector whose length is not 1, then we must normalize the vector before applying the formula in Theorem 13.9.

EXAMPLE 2   Finding a Directional Derivative

Find the directional derivative of

$$f(x, y) = x^2 \sin 2y \qquad \text{Surface}$$

at $(1, \pi/2)$ in the direction of

$$\mathbf{v} = 3\mathbf{i} - 4\mathbf{j}. \qquad \text{Direction}$$

Solution   Because f_x and f_y are continuous, f is differentiable, and you can apply Theorem 13.9. Begin by finding a unit vector in the direction of $\mathbf{v}$.

$$\mathbf{u} = \frac{\mathbf{v}}{\|\mathbf{v}\|} = \frac{3}{5}\mathbf{i} - \frac{4}{5}\mathbf{j} = \cos\theta\,\mathbf{i} + \sin\theta\,\mathbf{j}$$

Using this unit vector, you have

$$D_{\mathbf{u}}f(x, y) = (2x \sin 2y)(\cos\theta) + (2x^2 \cos 2y)(\sin\theta)$$

$$D_{\mathbf{u}}f\left(1, \frac{\pi}{2}\right) = (2 \sin \pi)\left(\frac{3}{5}\right) + (2 \cos \pi)\left(-\frac{4}{5}\right)$$

$$= (0)\left(\frac{3}{5}\right) + (-2)\left(-\frac{4}{5}\right)$$

$$= \frac{8}{5}.$$

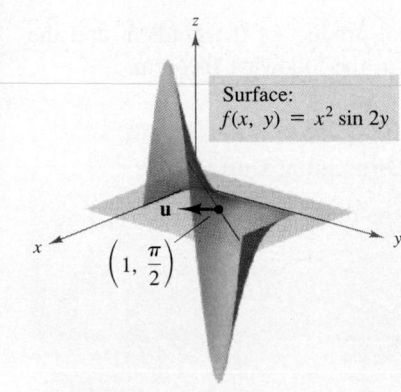

Surface:
$f(x, y) = x^2 \sin 2y$

FIGURE 13.47

(See Figure 13.47.)

The Gradient of a Function of Two Variables

The **gradient** of a function of two variables is a vector-valued function of two variables. This function has many important uses, some of which are described later in this section.

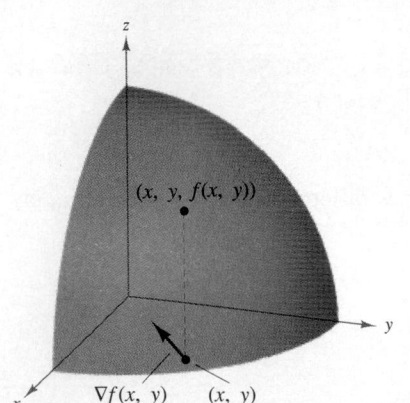

FIGURE 13.48
The gradient of f is a vector in the xy-plane.

> ### Definition of Gradient of a Function of Two Variables
>
> Let $z = f(x, y)$ be a function of x and y such that f_x and f_y exist. Then the **gradient of** f, denoted by $\nabla f(x, y)$, is the vector
>
> $$\nabla f(x, y) = f_x(x, y)\mathbf{i} + f_y(x, y)\mathbf{j}.$$
>
> We read ∇f as "del f." Another notation for the gradient is **grad** $f(x, y)$. In Figure 13.48, note that for each (x, y), the gradient $\nabla f(x, y)$ is a vector in the plane (not a vector in space).

REMARK   No value is assigned to the symbol ∇ by itself. It is an operator in the same sense that d/dx is an operator. When ∇ operates on $f(x, y)$, it produces the vector $\nabla f(x, y)$.

EXAMPLE 3   Finding the Gradient of a Function

Find the gradient of $f(x, y) = y \ln x + xy^2$ at the point $(1, 2)$.

Solution   Using $f_x(x, y) = \dfrac{y}{x} + y^2$ and $f_y(x, y) = \ln x + 2xy$, you have

$$\nabla f(x, y) = \left(\frac{y}{x} + y^2\right)\mathbf{i} + (\ln x + 2xy)\mathbf{j}.$$

At the point $(1, 2)$, the gradient is

$$\nabla f(1, 2) = \left(\frac{2}{1} + 2^2\right)\mathbf{i} + [\ln 1 + 2(1)(2)]\mathbf{j}$$
$$= 6\mathbf{i} + 4\mathbf{j}.$$

Because the gradient of f is a vector, you can write the directional derivative of f in the direction of $\mathbf{u}$ as

$$D_{\mathbf{u}} f(x, y) = [f_x(x, y)\mathbf{i} + f_y(x, y)\mathbf{j}] \cdot [\cos \theta \, \mathbf{i} + \sin \theta \, \mathbf{j}].$$

In other words, the directional derivative is the dot product of the gradient and the direction vector. This useful result is summarized in the following theorem.

> ### THEOREM 13.10   Alternative Form of the Directional Derivative
>
> If f is a differentiable function of x and y, then the directional derivative of f in the direction of the unit vector $\mathbf{u}$ is
>
> $$D_{\mathbf{u}} f(x, y) = \nabla f(x, y) \cdot \mathbf{u}.$$

EXAMPLE 4   Using $\nabla f(x, y)$ to Find a Directional Derivative

Find the directional derivative of

$$f(x, y) = 3x^2 - 2y^2$$

at $(-\frac{3}{4}, 0)$ in the direction from $P = (-\frac{3}{4}, 0)$ to $Q = (0, 1)$.

Solution   Because the partials of f are continuous, f is differentiable and you can apply Theorem 13.10. A vector in the specified direction is

$$\overrightarrow{PQ} = \mathbf{v} = \left(0 + \frac{3}{4}\right)\mathbf{i} + (1 - 0)\mathbf{j}$$

$$= \frac{3}{4}\mathbf{i} + \mathbf{j}$$

and a unit vector in this direction is

$$\mathbf{u} = \frac{\mathbf{v}}{\|\mathbf{v}\|} = \frac{3}{5}\mathbf{i} + \frac{4}{5}\mathbf{j}. \qquad \text{Unit vector in direction of } \overrightarrow{PQ}$$

Because $\nabla f(x, y) = f_x(x, y)\mathbf{i} + f_y(x, y)\mathbf{j} = 6x\,\mathbf{i} - 4y\,\mathbf{j}$, the gradient at $(-\frac{3}{4}, 0)$ is

$$\nabla f\left(-\frac{3}{4}, 0\right) = -\frac{9}{2}\mathbf{i} + 0\mathbf{j}. \qquad \text{Gradient at } (-\frac{3}{4}, 0)$$

Consequently, at $(-\frac{3}{4}, 0)$ the directional derivative is

$$D_{\mathbf{u}}f\left(-\frac{3}{4}, 0\right) = \nabla f\left(-\frac{3}{4}, 0\right) \cdot \mathbf{u}$$

$$= \left(-\frac{9}{2}\mathbf{i} + 0\mathbf{j}\right) \cdot \left(\frac{3}{5}\mathbf{i} + \frac{4}{5}\mathbf{j}\right)$$

$$= -\frac{27}{10}. \qquad \text{Directional derivative at } (-\frac{3}{4}, 0)$$

(See Figure 13.49.)

Surface:
$f(x, y) = 3x^2 - 2y^2$

FIGURE 13.49

Applications of the Gradient

You have already seen that there are many directional derivatives at the point (x, y) on a surface. In many applications we would like to know in which direction to move so that $f(x, y)$ increases most rapidly. This direction is called the direction of steepest ascent, and it is given by the gradient, as stated in the following theorem.

THEOREM 13.11   Properties of the Gradient

Let f be differentiable at the point (x, y).

1. If $\nabla f(x, y) = \mathbf{0}$, then $D_{\mathbf{u}}f(x, y) = 0$ for all $\mathbf{u}$.
2. The direction of *maximum* increase of f is given by $\nabla f(x, y)$. The maximum value of $D_{\mathbf{u}}f(x, y)$ is $\|\nabla f(x, y)\|$.
3. The direction of *minimum* increase of f is given by $-\nabla f(x, y)$. The minimum value of $D_{\mathbf{u}}f(x, y)$ is $-\|\nabla f(x, y)\|$.

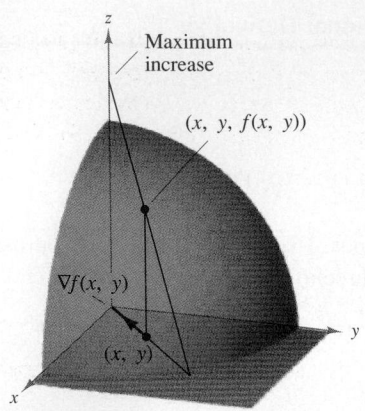

FIGURE 13.50
The gradient of f is a vector in the xy-plane that points in the direction of maximum increase on the surface given by $z = f(x, y)$.

Proof   If $\nabla f(x, y) = \mathbf{0}$, then for any direction (any $\mathbf{u}$), you have

$$D_{\mathbf{u}}f(x, y) = \nabla f(x, y) \cdot \mathbf{u} = (0\mathbf{i} + 0\mathbf{j}) \cdot (\cos\theta\,\mathbf{i} + \sin\theta\,\mathbf{j}) = 0.$$

If $\nabla f(x, y) \neq \mathbf{0}$, then let ϕ be the angle between $\nabla f(x, y)$ and a unit vector $\mathbf{u}$. Using the dot product, you can apply Theorem 11.6 to conclude that

$$\begin{aligned} D_{\mathbf{u}}f(x, y) &= \nabla f(x, y) \cdot \mathbf{u} \\ &= \| \nabla f(x, y) \| \, \| \mathbf{u} \| \cos\phi \\ &= \| \nabla f(x, y) \| \cos\phi \end{aligned}$$

and it follows that the maximum value of $D_{\mathbf{u}}f(x, y)$ will occur when $\cos\phi = 1$. Thus $\phi = 0$, and the maximum value for the directional derivative occurs when $\mathbf{u}$ has the same direction as $\nabla f(x, y)$. Moreover, this largest value for $D_{\mathbf{u}}f(x, y)$ is precisely

$$\| \nabla f(x, y) \| \cos\phi = \| \nabla f(x, y) \|.$$

Similarly, the minimum value of $D_{\mathbf{u}}f(x, y)$ can be obtained by letting $\phi = \pi$ so that $\mathbf{u}$ points in the direction opposite that of $\nabla f(x, y)$, as indicated in Figure 13.50.

To visualize one of the properties of the gradient, imagine a skier coming down a mountainside. If $f(x, y)$ denotes the altitude of the skier, then $-\nabla f(x, y)$ indicates the *compass direction* the skier should take to ski the path of steepest descent. (Remember that the gradient indicates direction in the xy-plane and does not itself point up or down the mountainside.)

As another illustration of the gradient, consider the temperature $T(x, y)$ at any point (x, y) on a flat metal plate. In this case, $\nabla T(x, y)$ gives the direction of greatest temperature increase at the point (x, y), as illustrated in the next example.

Level curves:
$T(x, y) = 20 - 4x^2 - y^2$

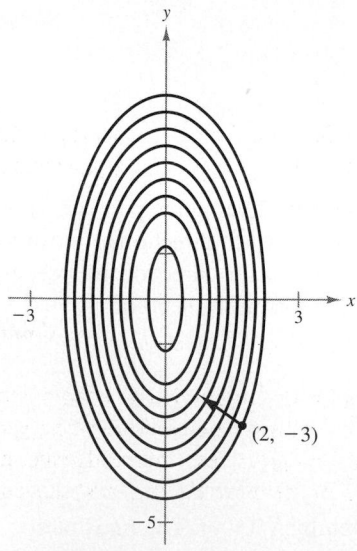

FIGURE 13.51
Direction of most rapid increase in temperature at $(2, -3)$.

EXAMPLE 5　Finding the Direction of Maximum Increase

The temperature in degrees Celsius on the surface of a metal plate is

$$T(x, y) = 20 - 4x^2 - y^2$$

where x and y are measured in inches. In what direction from $(2, -3)$ does the temperature increase most rapidly? What is this rate of increase?

Solution   The gradient is

$$\begin{aligned} \nabla T(x, y) &= T_x(x, y)\mathbf{i} + T_y(x, y)\mathbf{j} \\ &= -8x\,\mathbf{i} - 2y\,\mathbf{j}. \end{aligned}$$

It follows that the direction of maximum increase is given by

$$\nabla T(2, -3) = -16\mathbf{i} + 6\mathbf{j}$$

as shown in Figure 13.51, and the rate of increase is

$$\| \nabla T(2, -3) \| = \sqrt{256 + 36} = \sqrt{292} \approx 17.09° \text{ per inch.}$$

The solution presented in Example 5 can be misleading. Although the gradient points in the direction of maximum temperature increase, it does not necessarily point toward the hottest spot on the plate. In other words, the gradient provides a local solution to finding an increase relative to the temperature at the point $(2, -3)$. *Once you leave that position, the direction of maximum increase may change.*

EXAMPLE 6 Finding the Path of a Heat-Seeking Particle

A heat-seeking particle is located at the point $(2, -3)$ on a metal plate whose temperature at (x, y) is

$$T(x, y) = 20 - 4x^2 - y^2.$$

Find the path of the particle as it continuously moves in the direction of maximum temperature increase.

Solution Let the path be represented by the position function

$$\mathbf{r}(t) = x(t)\mathbf{i} + y(t)\mathbf{j}.$$

A tangent vector at each point $(x(t), y(t))$ is given by

$$\mathbf{r}'(t) = \frac{dx}{dt}\mathbf{i} + \frac{dy}{dt}\mathbf{j}.$$

Because the particle seeks maximum temperature increase, the directions of $\mathbf{r}'(t)$ and $\nabla T(x, y) = -8x\mathbf{i} - 2y\mathbf{j}$ are the same at each point of the path. Thus,

$$-8x = k\frac{dx}{dt} \quad \text{and} \quad -2y = k\frac{dy}{dt}$$

where k depends on t. By solving each equation for dt/k and equating the results, you obtain

$$\frac{dx}{-8x} = \frac{dy}{-2y}.$$

The solution of this differential equation is $x = Cy^4$. Because the particle starts at the point $(2, -3)$, you can determine that $C = \frac{2}{81}$. Thus, the path of the heat-seeking particle is

$$x = \frac{2}{81}y^4.$$

The path is shown in Figure 13.52. ▬▬

In Figure 13.52, the path of the particle (determined by the gradient at each point) appears to be orthogonal to each of the level curves. This becomes clear when you consider that the temperature $T(x, y)$ is constant along a given level curve. Hence, at any point (x, y) on the curve, the rate of change of T in the direction of a unit tangent vector $\mathbf{u}$ is 0, and you can write

$$\nabla f(x, y) \cdot \mathbf{u} = D_\mathbf{u}T(x, y) = 0. \qquad \text{$\mathbf{u}$ is a unit tangent vector}$$

Because the dot product of $\nabla f(x, y)$ and $\mathbf{u}$ is 0, you can conclude that they must be orthogonal. This result is stated in the following theorem.

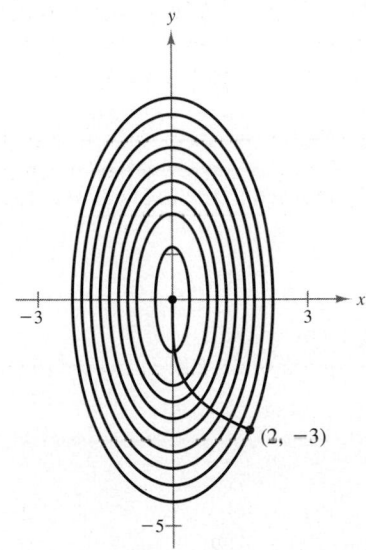

Level curves:
$T(x, y) = 20 - 4x^2 - y^2$

$(2, -3)$

FIGURE 13.52
Path followed by a heat-seeking particle.

> **THEOREM 13.12   Gradient is Normal to Level Curves**
>
> If f is differentiable at (x_0, y_0) and $\nabla f(x_0, y_0) \neq \mathbf{0}$, then $\nabla f(x_0, y_0)$ is normal to the level curve through (x_0, y_0).

EXAMPLE 7   Finding a Normal Vector to a Level Curve

Sketch the level curve corresponding to $c = 0$ for the function given by

$$f(x, y) = y - \sin x$$

and find a normal vector at several points on the curve.

Solution   The level curve for $c = 0$ is given by

$$0 = y - \sin x$$
$$y = \sin x$$

as shown in Figure 13.53. Because the gradient vector of f at (x, y) is

$$\nabla f(x, y) = f_x(x, y)\mathbf{i} + f_y(x, y)\mathbf{j}$$
$$= -\cos x\,\mathbf{i} + \mathbf{j}$$

you can use Theorem 13.12 to conclude that $\nabla f(x, y)$ is normal to the level curve at the point (x, y). Some gradient vectors are

$$\nabla f(-\pi, 0) = \mathbf{i} + \mathbf{j}$$
$$\nabla f\left(-\frac{2\pi}{3}, -\frac{\sqrt{3}}{2}\right) = \frac{1}{2}\mathbf{i} + \mathbf{j}$$
$$\nabla f\left(-\frac{\pi}{2}, -1\right) = \mathbf{j}.$$

Several others are shown in Figure 13.53.

(a) The surface is given by $f(x, y) = y - \sin x$.   (b) The level curve is given by $f(x, y) = 0$.

FIGURE 13.53

Functions of Three Variables

The definitions of the directional derivative and the gradient can be extended naturally to functions of three or more variables. As often happens, some of the geometric interpretation is lost in the generalization from functions of two variables to those of three variables. For example, you cannot interpret the directional derivative of a function of three variables to represent slope.

The definitions and properties of the directional derivative and the gradient of a function of three variables are given in the following summary.

Directional Derivative and Gradient for a Function of Three Variables

Let f be a function of x, y, and z, with continuous first partial derivatives. The **directional derivative of f** in the direction of a unit vector

$$\mathbf{u} = a\mathbf{i} + b\mathbf{j} + c\mathbf{k}$$

is given by

$$D_{\mathbf{u}}f(x,y,z) = af_x(x,y,z) + bf_y(x,y,z) + cf_z(x,y,z).$$

The **gradient of f** is defined to be

$$\nabla f(x,y,z) = f_x(x,y,z)\mathbf{i} + f_y(x,y,z)\mathbf{j} + f_z(x,y,z)\mathbf{k}.$$

Properties of the gradient are as follows.

1. $D_{\mathbf{u}}f(x,y,z) = \nabla f(x,y,z) \cdot \mathbf{u}$
2. If $\nabla f(x,y,z) = \mathbf{0}$, then $D_{\mathbf{u}}f(x,y,z) = 0$ for all $\mathbf{u}$.
3. The direction of *maximum* increase of f is given by $\nabla f(x,y,z)$. The maximum value of $D_{\mathbf{u}}f(x,y,z)$ is $\| \nabla f(x,y,z) \|$.
4. The direction of *minimum* increase of f is given by $-\nabla f(x,y,z)$. The minimum value of $D_{\mathbf{u}}f(x,y,z)$ is $-\| \nabla f(x,y,z) \|$.

EXAMPLE 8   Finding the Gradient for a Function of Three Variables

Find $\nabla f(x,y,z)$ for the function given by

$$f(x,y,z) = x^2 + y^2 - 4z$$

and find the direction of maximum increase of f at the point $(2, -1, 1)$.

Solution   The gradient vector is given by

$$\nabla f(x,y,z) = f_x(x,y,z)\mathbf{i} + f_y(x,y,z)\mathbf{j} + f_z(x,y,z)\mathbf{k}$$

$$= 2x\mathbf{i} + 2y\mathbf{j} - 4\mathbf{k}.$$

Hence, it follows that the direction of maximum increase at $(2, -1, 1)$ is

$$\nabla f(2, -1, 1) = 4\mathbf{i} - 2\mathbf{j} - 4\mathbf{k}.$$

EXERCISES for Section 13.6

In Exercises 1–12, find the directional derivative of the function at P in the direction of v.

1. $f(x, y) = 3x - 4xy + 5y$,
   $P = (1, 2)$, $\mathbf{v} = \frac{1}{2}(\mathbf{i} + \sqrt{3}\mathbf{j})$

2. $f(x, y) = x^2 - y^2$,
   $P = (4, 3)$, $\mathbf{v} = \frac{\sqrt{2}}{2}(\mathbf{i} + \mathbf{j})$

3. $f(x, y) = xy$,
   $P = (2, 3)$, $\mathbf{v} = \mathbf{i} + \mathbf{j}$

4. $f(x, y) = \dfrac{x}{y}$,
   $P = (1, 1)$, $\mathbf{v} = -\mathbf{j}$

5. $g(x, y) = \sqrt{x^2 + y^2}$,
   $P = (3, 4)$, $\mathbf{v} = 3\mathbf{i} - 4\mathbf{j}$

6. $g(x, y) = \arcsin xy$,
   $P = (1, 0)$, $\mathbf{v} = \mathbf{i} + 5\mathbf{j}$

7. $h(x, y) = e^x \sin y$,
   $P = \left(1, \dfrac{\pi}{2}\right)$, $\mathbf{v} = -\mathbf{i}$

8. $h(x, y) = e^{-(x^2+y^2)}$,
   $P = (0, 0)$, $\mathbf{v} = \mathbf{i} + \mathbf{j}$

9. $f(x, y, z) = xy + yz + xz$,
   $P = (1, 1, 1)$, $\mathbf{v} = 2\mathbf{i} + \mathbf{j} - \mathbf{k}$

10. $f(x, y, z) = x^2 + y^2 + z^2$,
    $P = (1, 2, -1)$, $\mathbf{v} = \mathbf{i} - 2\mathbf{j} + 3\mathbf{k}$

11. $h(x, y, z) = x \arctan yz$,
    $P = (4, 1, 1)$, $\mathbf{v} = \langle 1, 2, -1 \rangle$

12. $h(x, y, z) = xyz$,
    $P = (2, 1, 1)$, $\mathbf{v} = \langle 2, 1, 2 \rangle$

In Exercises 13–16, find the directional derivative of the function in the direction $\mathbf{u} = \cos\theta\,\mathbf{i} + \sin\theta\,\mathbf{j}$.

13. $f(x, y) = x^2 + y^2$

    $\theta = \dfrac{\pi}{4}$

14. $f(x, y) = \dfrac{y}{x + y}$

    $\theta = -\dfrac{\pi}{6}$

15. $f(x, y) = \sin(2x - y)$

    $\theta = -\dfrac{\pi}{3}$

16. $g(x, y) = xe^y$

    $\theta = \dfrac{2\pi}{3}$

In Exercises 17–20, find the directional derivative of the function at the point P in the direction of Q.

17. $f(x, y) = x^2 + 4y^2$
    $P = (3, 1)$, $Q = (1, -1)$

18. $f(x, y) = \cos(x + y)$
    $P = (0, \pi)$, $Q = \left(\dfrac{\pi}{2}, 0\right)$

19. $h(x, y, z) = \ln(x + y + z)$
    $P = (1, 0, 0)$, $Q = (4, 3, 1)$

20. $g(x, y, z) = xye^z$
    $P = (2, 4, 0)$, $Q = (0, 0, 0)$

In Exercises 21 and 22, use the graph to estimate the components of the vector in the direction of the maximum increase in the function at the specified point. Then find the gradient at the point and compare it with your estimate.

21. $f(x, y) = \frac{1}{10}(x^2 - 3xy + y^2)$
    $(1, 2)$

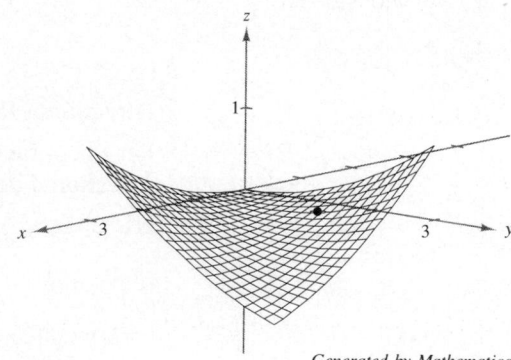

Generated by Mathematica

22. $f(x, y) = \frac{1}{2}y\sqrt{x}$
    $(1, 2)$

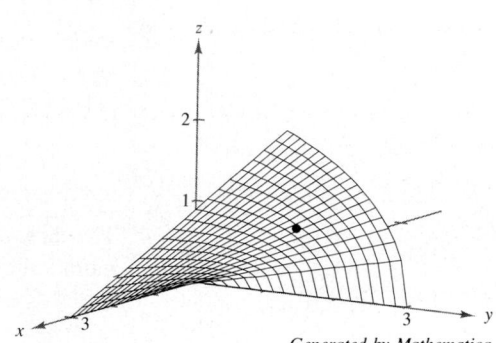

Generated by Mathematica

In Exercises 23–30, find the gradient of the function and the maximum value of the directional derivative at the indicated point.

Function	Point
23. $h(x, y) = x \tan y$	$\left(2, \dfrac{\pi}{4}\right)$
24. $h(x, y) = y \cos(x - y)$	$\left(0, \dfrac{\pi}{3}\right)$
25. $g(x, y) = \ln \sqrt[3]{x^2 + y^2}$	$(1, 2)$
26. $g(x, y) = ye^{-x^2}$	$(0, 5)$
27. $f(x, y, z) = \sqrt{x^2 + y^2 + z^2}$	$(1, 4, 2)$
28. $f(x, y, z) = xe^{yz}$	$(2, 0, -4)$
29. $w = \dfrac{1}{\sqrt{1 - x^2 - y^2 - z^2}}$	$(0, 0, 0)$
30. $w = xy^2z^2$	$(2, 1, 1)$

In Exercises 31–38, use the function given by

$$f(x,y) = 3 - \frac{x}{3} - \frac{y}{2}.$$

31. Sketch the graph of f in the first octant and plot the point $(3, 2, 1)$.

32. Find $D_{\mathbf{u}}f(3, 2)$ where $\mathbf{u} = \cos\theta\,\mathbf{i} + \sin\theta\,\mathbf{j}$.

   a. $\theta = \dfrac{\pi}{4}$ **b.** $\theta = \dfrac{2\pi}{3}$

33. Find $D_{\mathbf{u}}f(3, 2)$ where $\mathbf{u} = \cos\theta\,\mathbf{i} + \sin\theta\,\mathbf{j}$.

   a. $\theta = \dfrac{4\pi}{3}$ **b.** $\theta = -\dfrac{\pi}{6}$

34. Find $D_{\mathbf{u}}f(3, 2)$ where $\mathbf{u} = \mathbf{v}/\|\mathbf{v}\|$.
   a. $\mathbf{v} = \mathbf{i} + \mathbf{j}$ **b.** $\mathbf{v} = -3\mathbf{i} - 4\mathbf{j}$

35. Find $D_{\mathbf{u}}f(3, 2)$ where $\mathbf{u} = \mathbf{v}/\|\mathbf{v}\|$.
   a. $\mathbf{v}$ is the vector from $(1, 2)$ to $(-2, 6)$.
   b. $\mathbf{v}$ is the vector from $(3, 2)$ to $(4, 5)$.

36. Find $\nabla f(x, y)$.

37. Find the maximum value of the directional derivative at $(3, 2)$.

38. Find a unit vector $\mathbf{u}$ orthogonal to $\nabla f(3, 2)$ and calculate $D_{\mathbf{u}}f(3, 2)$. Discuss the geometric meaning of the result.

In Exercises 39–42, use the function given by

$$f(x, y) = 9 - x^2 - y^2.$$

39. Sketch the graph of f in the first octant and plot the point $(1, 2, 4)$ on the surface.

40. Find $D_{\mathbf{u}}f(1, 2)$ where $\mathbf{u} = \cos\theta\,\mathbf{i} + \sin\theta\,\mathbf{j}$.

   a. $\theta = -\dfrac{\pi}{4}$ **b.** $\theta = \dfrac{\pi}{3}$

41. Find $\nabla f(1, 2)$ and $\|\nabla f(1, 2)\|$.

42. Find a unit vector $\mathbf{u}$ orthogonal to $\nabla f(1, 2)$ and calculate $D_{\mathbf{u}}f(1, 2)$. Discuss the geometric meaning of the result.

In Exercises 43–46, find a normal vector to the level curve $f(x, y) = c$ at P.

43. $f(x, y) = x^2 + y^2$ 44. $f(x, y) = 6 - 2x - 3y$
   $c = 25,\ P = (3, 4)$ $c = 6,\ P = (0, 0)$

45. $f(x, y) = \dfrac{x}{x^2 + y^2}$ 46. $f(x, y) = xy$
   $c = \frac{1}{2},\ P = (1, 1)$ $c = -3,\ P = (-1, 3)$

In Exercises 47–50, use the gradient to find a unit normal vector to the graph of the equation at the indicated point. Sketch your results.

Equation	Point
47. $4x^2 - y = 6$	$(2, 10)$
48. $3x^2 - 2y^2 = 1$	$(1, 1)$
49. $9x^2 + 4y^2 = 40$	$(2, -1)$
50. $xe^y - y = 5$	$(5, 0)$

51. *Temperature Distribution* The temperature at the point (x, y) on a metal plate is given by

$$T = \frac{x}{x^2 + y^2}.$$

Find the direction of greatest increase in heat from the point $(3, 4)$.

52. *Topography* The surface of a mountain is described by the equation

$$h(x, y) = 4000 - 0.001x^2 - 0.004y^2.$$

Suppose a mountain climber is at the point $(500, 300, 3390)$. In what direction should the climber move in order to ascend at the greatest rate?

Heat-Seeking Path In Exercises 53 and 54, find the path followed by a heat-seeking particle placed at point P on a metal plate with a temperature field given by $T(x,y)$.

Temperature field	Point
53. $T(x, y) = 400 - 2x^2 - y^2$	$P = (10, 10)$
54. $T(x, y) = 50 - x^2 - 2y^2$	$P = (4, 3)$

55. *Meteorology* Meteorologists measure the atmospheric pressure in units called millibars. From these observations they create weather maps on which the curves of equal atmospheric pressure (isobars) are drawn (see figure). These are level curves of the function $P(x, y)$ yielding the pressure at any point. Sketch the gradients to the isobars at the points A, B, and C. Although the magnitudes of the gradients are unknown, their lengths relative to each other can be estimated. At which of the three points is the wind speed greatest if the speed increases as the pressure gradient increases?

FIGURE FOR 55

56. Let P be a point on the ellipse given by

$$\frac{x^2}{a^2} + \frac{y^2}{b^2} = 1$$

where $\mathbf{T}$ is a tangent vector to the ellipse at P and $f(x, y) = d_1 + d_2$ is the sum of the distances from the foci to P (see figure). Show that $\mathbf{T} \cdot \nabla f(x, y)$ is 0, and give a geometric interpretation of this result.

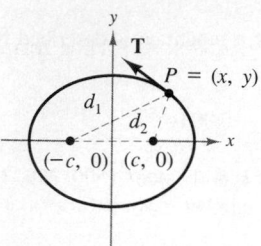

FIGURE FOR 56

True or False **In Exercises 57–60, determine whether the statement is true or false. If it is false, explain why or give an example that shows it is false.**

57. If $f(x, y) = \sqrt{1 - x^2 - y^2}$, then $D_{\mathbf{u}}f(0, 0) = 0$ for any unit vector $\mathbf{u}$.

58. If $f(x, y) = x + y$, then $-1 \le D_{\mathbf{u}}f(x, y) \le 1$.

59. If $D_{\mathbf{u}}f(x, y)$ exists, then $D_{\mathbf{u}}f(x, y) = -D_{-\mathbf{u}}f(x, y)$.

60. If $D_{\mathbf{u}}f(x_0, y_0) = c$ for any unit vector $\mathbf{u}$, then $c = 0$.

61. The computer-generated figure shows the graph of

$$f(x, y) = \frac{-4y}{1 + x^2 + y^2}$$

over the square domain in which $-4 \le x \le 4$ and $-4 \le y \le 4$. Use the graph to estimate the value of the gradient at the indicated point. Then, confirm your estimate analytically.

a. $(0, 1, -2)$ **b.** $(0, -1, 2)$
c. $(2, 0, 0)$ **d.** $(-2, 0, 0)$
e. $(0, 0, 0)$ **f.** $(1, 1, -\frac{4}{3})$

Generated by Maple

62. Let $f(x, y) = \begin{cases} \dfrac{4xy}{x^2 + y^2}, & \text{if } (x, y) \ne (0, 0) \\ 0, & \text{if } (x, y) = (0, 0) \end{cases}$

Let $\mathbf{u} = 1/\sqrt{2}\,(\mathbf{i} + \mathbf{j})$. Does the directional derivative of f at $P = (0, 0)$ in the direction of $\mathbf{u}$ exist? If $f(0, 0)$ were defined as 2 instead of 0, would $\mathbf{u}$ exist?

63. Find a function f such that

$$\nabla f = e^x \cos y\,\mathbf{i} - e^x \sin y\,\mathbf{j} + z\,\mathbf{k}.$$

Tangent Plane and Normal Line to a Surface • The Angle of Inclination of a Plane • A Comparison of the Gradients $\nabla f(x, y)$ and $\nabla F(x, y, z)$

Tangent Plane and Normal Line to a Surface

So far we have represented surfaces in space primarily by equations of the form

$$z = f(x, y). \qquad \text{Equation of a surface } S$$

In the development to follow, however, it is convenient to use the more general representation $F(x, y, z) = 0$. For a surface S given by $z = f(x, y)$, you can convert to the general form by defining F as

$$F(x, y, z) = f(x, y) - z.$$

Then, because $f(x, y) - z = 0$, you can consider S to be the level surface of F given by

$$F(x, y, z) = 0. \qquad \text{Alternative equation of surface } S$$

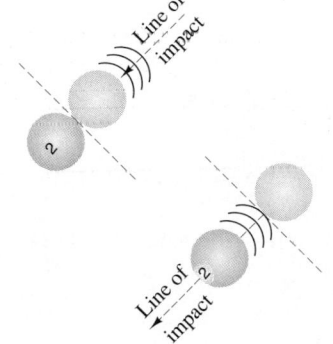

EXAMPLE 1   Writing an Equation of a Surface

For the function given by $F(x, y, z) = x^2 + y^2 + z^2 - 4$, describe the level surface given by $F(x, y, z) = 0$.

Solution   The level surface given by $F(x, y, z) = 0$ can be written as

$$x^2 + y^2 + z^2 = 4$$

which is a sphere of radius 2 whose center is at the origin.   ■

FIGURE 13.54

You have seen many examples of the usefulness of normal lines in applications involving curves. Normal lines are equally important in analyzing surfaces and solids. For example, consider the collision of two billiard balls. When a stationary ball is struck at a point P on its surface, it moves along the **line of impact** determined by P and the center of the ball. The impact can occur in *two* ways. If the cue ball is moving along the line of impact, it stops dead and imparts all of its momentum to the stationary ball, as shown in Figure 13.54. This kind of shot requires precision, because the line of impact must coincide exactly with the direction of the cue ball. More often, the cue ball is deflected to one side or the other and retains part of its momentum. That part of the momentum that is transferred to the stationary ball occurs along the line of impact, *regardless* of the direction of the cue ball, as shown in Figure 13.55. This line of impact is called the **normal line** to the surface of the ball at the point P.

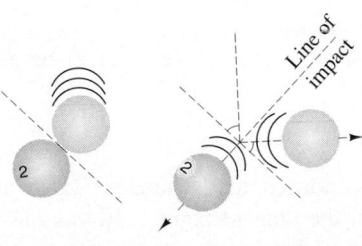

FIGURE 13.55

In the process of finding a normal line to a surface, you are also able to solve the problem of finding a **tangent plane** to the surface. Let S be a surface given by $F(x, y, z) = 0$, and let $P = (x_0, y_0, z_0)$ be a point on S. Let C be a curve on S through P that is defined by the vector-valued function $\mathbf{r}(t) = x(t)\mathbf{i} + y(t)\mathbf{j} + z(t)\mathbf{k}$. Then, for all t

$$F(x(t), y(t), z(t)) = 0.$$

If F is differentiable and $x'(t)$, $y'(t)$, and $z'(t)$ all exist, it follows from the Chain Rule that

$$0 = F'(t) = F_x(x, y, z)x'(t) + F_y(x, y, z)y'(t) + F_z(x, y, z)z'(t).$$

At (x_0, y_0, z_0), the equivalent vector form is

$$0 = \underbrace{\nabla F(x_0, y_0, z_0)}_{\text{Gradient}} \cdot \underbrace{\mathbf{r}'(t_0)}_{\substack{\text{Tangent} \\ \text{vector}}}.$$

This result means that the gradient at P is orthogonal to the tangent vector of every curve on S through P. Thus, all tangent lines at P lie in a plane that is normal to $\nabla F(x_0, y_0, z_0)$ and contains P, as shown in Figure 13.56.

Surface S: $F(x, y, z) = 0$

$P = (x_0, y_0, z_0)$ ∇F

FIGURE 13.56
Tangent plane to surface S at P

Definition of Tangent Plane and Normal Line

Let F be differentiable at the point $P = (x_0, y_0, z_0)$ on the surface S given by $F(x, y, z) = 0$ such that $\nabla F(x_0, y_0, z_0) \neq \mathbf{0}$.

1. The plane through P that is normal to $\nabla F(x_0, y_0, z_0)$ is called the **tangent plane to S at P.**
2. The line through P having the direction of $\nabla F(x_0, y_0, z_0)$ is called the **normal line to S at P.**

REMARK   In the remainder of this section, we assume $\nabla F(x_0, y_0, z_0)$ to be nonzero unless stated otherwise.

To find an equation for the tangent plane to S at (x_0, y_0, z_0), let (x, y, z) be an arbitrary point in the tangent plane. Then the vector

$$\mathbf{v} = (x - x_0)\mathbf{i} + (y - y_0)\mathbf{j} + (z - z_0)\mathbf{k}$$

lies in the tangent plane. Because $\nabla F(x_0, y_0, z_0)$ is normal to the tangent plane at (x_0, y_0, z_0), it must be orthogonal to every vector in the tangent plane, and you have $\nabla F(x_0, y_0, z_0) \cdot \mathbf{v} = 0$, which leads to the result in the following theorem.

THEOREM 13.13   Equation of Tangent Plane

If F is differentiable at (x_0, y_0, z_0), then an equation of the tangent plane to the surface given by $F(x, y, z) = 0$ at (x_0, y_0, z_0) is

$$F_x(x_0, y_0, z_0)(x - x_0) + F_y(x_0, y_0, z_0)(y - y_0) + F_z(x_0, y_0, z_0)(z - z_0) = 0.$$

EXAMPLE 2   Finding an Equation of a Tangent Plane

Find an equation of the tangent plane to the hyperboloid given by

$$z^2 - 2x^2 - 2y^2 = 12$$

at the point $(1, -1, 4)$.

Solution   Begin by writing the equation of the surface as

$$z^2 - 2x^2 - 2y^2 - 12 = 0.$$

Then, considering

$$F(x, y, z) = z^2 - 2x^2 - 2y^2 - 12$$

you have

$$F_x(x, y, z) = -4x, \qquad F_y(x, y, z) = -4y, \qquad F_z(x, y, z) = 2z.$$

At the point $(1, -1, 4)$, the partial derivatives are

$$F_x(1, -1, 4) = -4, \qquad F_y(1, -1, 4) = 4, \qquad F_z(1, -1, 4) = 8.$$

Therefore, an equation of the tangent plane at $(1, -1, 4)$ is

$$-4(x - 1) + 4(y + 1) + 8(z - 4) = 0$$
$$-4x + 4 + 4y + 4 + 8z - 32 = 0$$
$$-4x + 4y + 8z - 24 = 0$$
$$x - y - 2z + 6 = 0.$$

Figure 13.57 shows a portion of the hyperboloid and tangent plane.

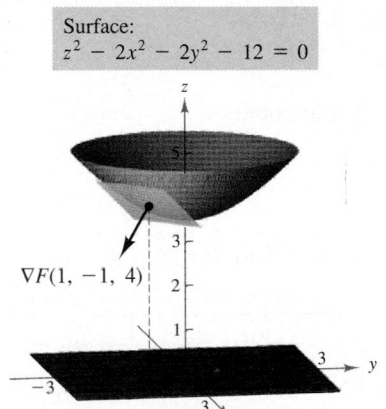

Surface:
$$z^2 - 2x^2 - 2y^2 - 12 = 0$$

$\nabla F(1, -1, 4)$

FIGURE 13.57
Tangent plane to surface.

TECHNOLOGY   Some three-dimensional graphing utilities are capable of sketching tangent planes to surfaces. Two examples are shown in Figure 13.58. The first of these computer drawings depicts a tangent plane to the sphere $x^2 + y^2 + z^2 = 1$, and the second depicts a tangent plane to the paraboloid $z = 2 - x^2 - y^2$.

Generated by Mathematica

FIGURE 13.58

To find the equation of the tangent plane at a point on a surface given by $z = f(x, y)$, you can define the function F by

$$F(x, y, z) = f(x, y) - z.$$

Then S is given by the level surface $F(x, y, z) = 0$, and by Theorem 13.13 an equation of the tangent plane to S at the point (x_0, y_0, z_0) is

$$f_x(x_0, y_0)(x - x_0) + f_y(x_0, y_0)(y - y_0) - (z - z_0) = 0.$$

EXAMPLE 3   Finding an Equation of the Tangent Plane

Find the equation of the tangent plane to the paraboloid

$$z = 1 - \frac{1}{10}(x^2 + 4y^2)$$

at the point $(1, 1, \frac{1}{2})$.

Solution   From $z = f(x, y) = 1 - \frac{1}{10}(x^2 + 4y^2)$, you obtain

$$f_x(x, y) = -\frac{x}{5} \quad \rightarrow \quad f_x(1, 1) = -\frac{1}{5}$$

and

$$f_y(x, y) = -\frac{4y}{5} \quad \rightarrow \quad f_y(1, 1) = -\frac{4}{5}.$$

Therefore, an equation of the tangent plane at $(1, 1, \frac{1}{2})$ is

$$f_x(1, 1)(x - 1) + f_y(1, 1)(y - 1) - \left(z - \frac{1}{2}\right) = 0$$

$$-\frac{1}{5}(x - 1) - \frac{4}{5}(y - 1) - \left(z - \frac{1}{2}\right) = 0$$

$$-\frac{1}{5}x - \frac{4}{5}y - z + \frac{3}{2} = 0.$$

This tangent plane is shown in Figure 13.59.   ■

In Section 13.4, you saw that the total differential

$$dz = f_x(x, y)dx + f_y(x, y)dy$$

can be used to approximate the increment Δz. (See Figure 13.35 in Section 13.4.) We now give a geometric interpretation of this approximation in terms of the tangent plane to a surface. The tangent plane to the surface $z = f(x, y)$ at (x_0, y_0, z_0) is given by

$$f_x(x_0, y_0)(x - x_0) + f_y(x_0, y_0)(y - y_0) - (z - z_0) = 0$$

or, equivalently,

$$z = z_0 + f_x(x_0, y_0)(x - x_0) + f_y(x_0, y_0)(y - y_0).$$

Thus, at (x_0, y_0), the *tangent plane* has height z_0 and at $(x_0 + dx, y_0 + dy)$ it has height

$$z = z_0 + f_x(x_0, y_0)(x_0 + dx - x_0) + f_y(x_0, y_0)(y_0 + dy - y_0)$$

$$= z_0 + f_x(x_0, y_0)\,dx + f_y(x_0, y_0)\,dy.$$

That is, $z = z_0 + dz$. You can see from Figure 13.60 that for points near (x_0, y_0, z_0) the increments $dx = \Delta x$ and $dy = \Delta y$ are small and $dz \approx \Delta z$.

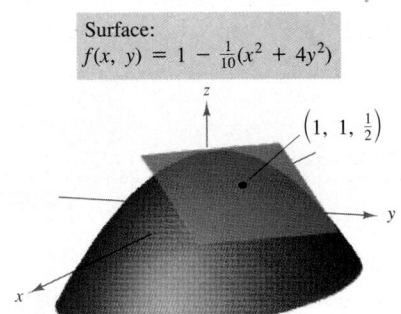

Surface:
$f(x, y) = 1 - \frac{1}{10}(x^2 + 4y^2)$

$(1, 1, \frac{1}{2})$

FIGURE 13.59
Tangent plane to surface

$\Delta z = f(x_0 + dx, y_0 + dy) - f(x_0, y_0)$

Surface:
$z = f(x, y)$

Δz dz

$(x_0 + dx, y_0 + dy)$

(x_0, y_0) dy dx

FIGURE 13.60
When dx and dy are small,
$dz \approx \Delta z$.

The gradient $\nabla F(x, y, z)$ gives a convenient way to find equations of normal lines, as shown in Example 4.

EXAMPLE 4   Finding an Equation of a Normal Line to a Surface

Find a set of symmetric equations for the normal line to the surface given by $xyz = 6$ at the point $(2, 3, 1)$.

Solution   Begin by letting $F(x, y, z) = xyz - 6$. Then, the gradient is given by

$$\nabla F(x, y, z) = F_x(x, y, z)\mathbf{i} + F_y(x, y, z)\mathbf{j} + F_z(x, y, z)\mathbf{k}$$
$$= yz\mathbf{i} + xz\mathbf{j} + xy\mathbf{k}.$$

and at the point $(2, 3, 1)$ you have

$$\nabla F(2, 3, 1) = (3)(1)\mathbf{i} + (2)(1)\mathbf{j} + (2)(3)\mathbf{k}$$
$$= 3\mathbf{i} + 2\mathbf{j} + 6\mathbf{k}.$$

The normal line at $(2, 3, 1)$ has direction numbers 3, 2, and 6, and the corresponding set of symmetric equations is

$$\frac{x - 2}{3} = \frac{y - 3}{2} = \frac{z - 1}{6}.$$

(See Figure 13.61.)

$\nabla F(2, 3, 1)$

Surface: $xyz = 6$

FIGURE 13.61

Knowing that the gradient $\nabla F(x, y, z)$ is normal to the surface given by $F(x, y, z) = 0$ allows you to solve a variety of problems dealing with surfaces and curves in space.

EXAMPLE 5   Finding the Equation of a Tangent Line to a Curve

Describe the tangent line to the curve of intersection of the surfaces

$$x^2 + 2y^2 + 2z^2 = 20 \qquad \text{Ellipsoid}$$

and

$$x^2 + y^2 + z = 4 \qquad \text{Paraboloid}$$

at the point $(0, 1, 3)$, as shown in Figure 13.62.

Solution   Begin by finding the gradients to both surfaces at the point $(0, 1, 3)$.

Ellipsoid	Paraboloid
$F(x, y, z) = x^2 + 2y^2 + 2z^2 - 20$	$G(x, y, z) = x^2 + y^2 + z - 4$
$\nabla F(x, y, z) = 2x\mathbf{i} + 4y\mathbf{j} + 4z\mathbf{k}$	$\nabla G(x, y, z) = 2x\mathbf{i} + 2y\mathbf{j} + \mathbf{k}$
$\nabla F(0, 1, 3) = 4\mathbf{j} + 12\mathbf{k}$	$\nabla G(0, 1, 3) = 2\mathbf{j} + \mathbf{k}$

The cross product of these two gradients is a vector that is tangent to both surfaces at the point $(0, 1, 3)$.

$$\nabla F(0, 1, 3) \times \nabla G(0, 1, 3) = \begin{vmatrix} \mathbf{i} & \mathbf{j} & \mathbf{k} \\ 0 & 4 & 12 \\ 0 & 2 & 1 \end{vmatrix} = -20\mathbf{i}.$$

Thus, the tangent line to the curve of intersection of the two surfaces at the point $(0, 1, 3)$ is a line that is parallel to the x-axis and passes through the point $(0, 1, 3)$.

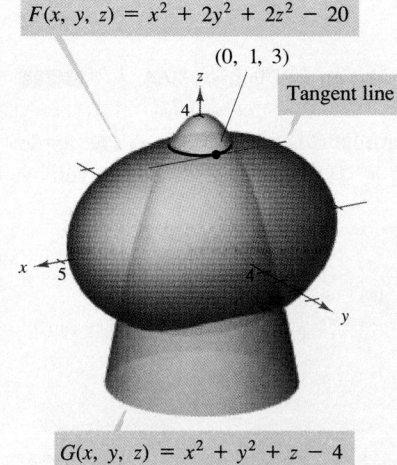

$F(x, y, z) = x^2 + 2y^2 + 2z^2 - 20$

$(0, 1, 3)$

Tangent line

$G(x, y, z) = x^2 + y^2 + z - 4$

FIGURE 13.62

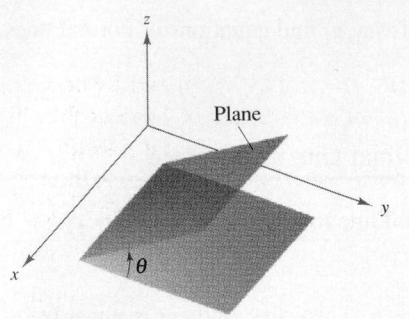

FIGURE 13.63
The angle of inclination.

The Angle of Inclination of a Plane

Another use of the gradient $\nabla F(x, y, z)$ is to determine the angle of inclination of the tangent plane to a surface. The **angle of inclination** of a plane is defined as the angle $\theta, 0 \le \theta \le \pi/2$, between the given plane and the xy-plane, as shown in Figure 13.63. (The angle of inclination of a horizontal plane is defined as 0.) Because the vector **k** is normal to the xy-plane, you can use the formula for the cosine of the angle between two planes (given in Section 11.5) to conclude that the angle of inclination of a plane with normal vector **n** is given by

$$\cos \theta = \frac{|\mathbf{n} \cdot \mathbf{k}|}{\|\mathbf{n}\| \|\mathbf{k}\|} = \frac{|\mathbf{n} \cdot \mathbf{k}|}{\|\mathbf{n}\|}.$$ Angle of inclination of a plane

EXAMPLE 6   Finding the Angle of Inclination of a Tangent Plane

Find the angle of inclination of the tangent plane to the ellipsoid given by

$$\frac{x^2}{12} + \frac{y^2}{12} + \frac{z^2}{3} = 1$$

at the point $(2, 2, 1)$.

Solution   If you let

$$F(x, y, z) = \frac{x^2}{12} + \frac{y^2}{12} + \frac{z^2}{3} - 1$$

then the gradient of F at the point $(2, 2, 1)$ is given by

$$\nabla F(x, y, z) = \frac{x}{6}\mathbf{i} + \frac{y}{6}\mathbf{j} + \frac{2z}{3}\mathbf{k}$$

$$\nabla F(2, 2, 1) = \frac{1}{3}\mathbf{i} + \frac{1}{3}\mathbf{j} + \frac{2}{3}\mathbf{k}.$$

Because $\nabla F(2, 2, 1)$ is normal to the tangent plane and **k** is normal to the xy-plane, it follows that the angle of inclination of the tangent plane is given by

$$\cos \theta = \frac{|\nabla F(2, 2, 1) \cdot \mathbf{k}|}{\|\nabla F(2, 2, 1)\|}$$

$$= \frac{2/3}{\sqrt{(1/3)^2 + (1/3)^2 + (2/3)^2}} = \sqrt{\frac{2}{3}}$$

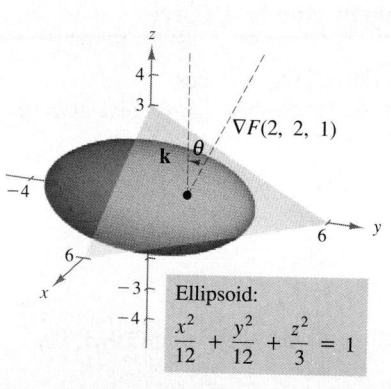

FIGURE 13.64

which implies that $\theta = \arccos \sqrt{2/3} \approx 35.3°$, as shown in Figure 13.64.   ◼

A special case of the procedure shown in Example 6 is worth noting. The angle of inclination θ of the tangent plane to the surface $z = f(x, y)$ at (x_0, y_0, z_0) is given by

$$\cos \theta = \frac{1}{\sqrt{[f_x(x_0, y_0)]^2 + [f_y(x_0, y_0)]^2 + 1}}.$$ Alternative formula for angle of inclination

(See Exercise 50.)

A Comparison of the Gradients $\nabla f(x, y)$ and $\nabla F(x, y, z)$

We conclude this section with a comparison of the gradients $\nabla f(x, y)$ and $\nabla F(x, y, z)$. In the previous section, you saw that the gradient of a function f of two variables is normal to the *level curves* of f. Specifically, Theorem 13.12 stated that if f is differentiable at (x_0, y_0) and $\nabla f(x_0, y_0) \neq \mathbf{0}$, then $\nabla f(x_0, y_0)$ is normal to the level curve through (x_0, y_0). Having developed normal lines to surfaces, you can now extend this result to a function of three variables.

THEOREM 13.14   Gradient is Normal to Level Surfaces

If F is differentiable at (x_0, y_0, z_0) and $\nabla F(x_0, y_0, z_0) \neq \mathbf{0}$, then $\nabla F(x_0, y_0, z_0)$ is normal to the level surface through (x_0, y_0, z_0).

When working with the gradients $\nabla f(x, y)$ and $\nabla F(x, y, z)$, be sure you remember that $\nabla f(x, y)$ is a vector in the xy-plane and $\nabla F(x, y, z)$ is a vector in space.

EXERCISES for Section 13.7

TECHNOLOGY
Laboratory Guide
Lab 13.7

In Exercises 1–10, find a unit normal vector to the surface at the indicated point. (*Hint:* Normalize the gradient vector $\nabla F(x, y, z)$.)

Surface	Point
1. $x + y + z = 4$	$(2, 0, 2)$
2. $x^2 + y^2 + z^2 = 11$	$(3, 1, 1)$
3. $z = \sqrt{x^2 + y^2}$	$(3, 4, 5)$
4. $z = x^3$	$(2, 1, 8)$
5. $x^2 y^4 - z = 0$	$(1, 2, 16)$
6. $x^2 + 3y + z^3 = 9$	$(2, -1, 2)$
7. $z - x \sin y = 4$	$\left(6, \dfrac{\pi}{6}, 7\right)$
8. $ze^{x^2 - y^2} - 3 = 0$	$(2, 2, 3)$
9. $\ln\left(\dfrac{x}{y - z}\right) = 0$	$(1, 4, 3)$
10. $\sin(x - y) - z = 2$	$\left(\dfrac{\pi}{3}, \dfrac{\pi}{6}, -\dfrac{3}{2}\right)$

In Exercises 11–14, find an equation of the tangent plane to the surface at the indicated point.

11. $z = 25 - x^2 - y^2$
$(3, 1, 15)$

12. $z = \sqrt{x^2 + y^2}$
$(3, 4, 5)$

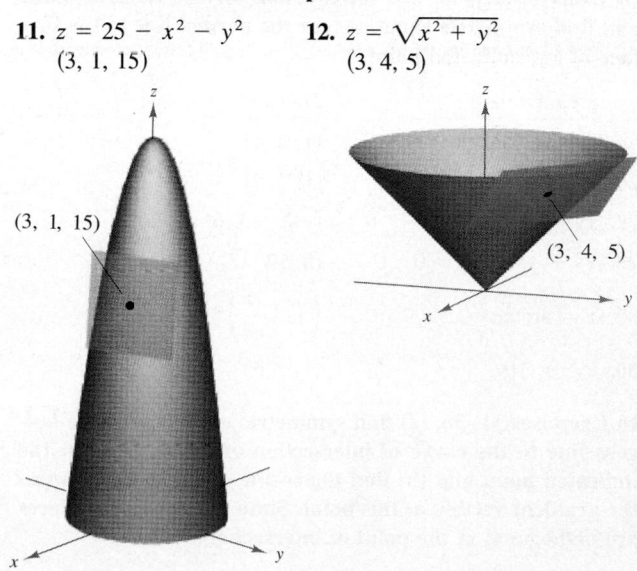

13. $f(x, y) = \dfrac{y}{x}$

$(1, 2, 2)$

14. $g(x, y) = \arctan\dfrac{y}{x}$

$(1, 0, 0)$

$(1, 2, 2)$

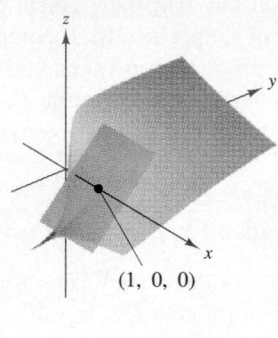

$(1, 0, 0)$

In Exercises 15–24, find an equation of the tangent plane to the surface at the indicated point.

Surface	Point
15. $g(x, y) = x^2 - y^2$	$(5, 4, 9)$
16. $f(x, y) = 2 - \frac{2}{3}x - y$	$(3, -1, 1)$
17. $z = e^x(\sin y + 1)$	$\left(0, \dfrac{\pi}{2}, 2\right)$
18. $z = x^3 - 3xy + y^3$	$(1, 2, 3)$
19. $h(x, y) = \ln \sqrt{x^2 + y^2}$	$(3, 4, \ln 5)$
20. $h(x, y) = \cos y$	$\left(5, \dfrac{\pi}{4}, \dfrac{\sqrt{2}}{2}\right)$
21. $x^2 + 4y^2 + z^2 = 36$	$(2, -2, 4)$
22. $x^2 + 2z^2 = y^2$	$(1, 3, -2)$
23. $xy^2 + 3x - z^2 = 4$	$(2, 1, -2)$
24. $y = x(2z - 1)$	$(4, 4, 1)$

In Exercises 25–30, find an equation for the tangent plane and find symmetric equations for the normal line to the surface at the indicated point.

Surface	Point
25. $x^2 + y^2 + z = 9$	$(1, 2, 4)$
26. $x^2 + y^2 + z^2 = 9$	$(1, 2, 2)$
27. $xy - z = 0$	$(-2, -3, 6)$
28. $x^2 + y^2 - z^2 = 0$	$(5, 12, 13)$
29. $z = \arctan\dfrac{y}{x}$	$\left(1, 1, \dfrac{\pi}{4}\right)$
30. $xyz = 10$	$(1, 2, 5)$

In Exercises 31–36, (a) find symmetric equations of the tangent line to the curve of intersection of the surfaces at the indicated point and (b) find the cosine of the angle between the gradient vectors at this point. State whether the surfaces are orthogonal at the point of intersection.

Surfaces	Point
31. $x^2 + y^2 = 5$	$(2, 1, 2)$
$z = x$	

Surfaces	Point
32. $z = x^2 + y^2$	$(2, -1, 5)$
$z = 4 - y$	
33. $x^2 + z^2 = 25$	$(3, 3, 4)$
$y^2 + z^2 = 25$	
34. $z = \sqrt{x^2 + y^2}$	$(3, 4, 5)$
$2x + y + 2z = 20$	
35. $x^2 + y^2 + z^2 = 6$	$(2, 1, 1)$
$x - y - z = 0$	
36. $z = x^2 + y^2$	$(1, 2, 5)$
$x + y + 6z = 33$	

In Exercises 37–40, find the angle of inclination θ of the tangent plane to the surface at the indicated point.

Surface	Point
37. $3x^2 + 2y^2 - z = 15$	$(2, 2, 5)$
38. $xy - z^2 = 0$	$(2, 2, 2)$
39. $x^2 - y^2 + z = 0$	$(1, 2, 3)$
40. $x^2 + y^2 = 5$	$(2, 1, 3)$

In Exercises 41 and 42, find the point on the surface where the tangent plane is horizontal.

41. $z = 3 - x^2 - y^2 + 6y$

42. $z = 3x^2 + 2y^2 - 3x + 4y - 5$

In Exercises 43 and 44, find the path of a heat-seeking particle in the temperature field T, starting at the specified point.

43. $T(x, y, z) = 400 - 2x^2 - y^2 - 4z^2,\ (4, 3, 10)$

44. $T(x, y, z) = 100 - 3x - y - z^2,\ (2, 2, 5)$

In Exercises 45 and 46, show that the tangent plane to the quadric surface at the point (x_0, y_0, z_0) can be written in the given form.

45. *Ellipsoid* $\dfrac{x^2}{a^2} + \dfrac{y^2}{b^2} + \dfrac{z^2}{c^2} = 1$

Plane $\dfrac{x_0 x}{a^2} + \dfrac{y_0 y}{b^2} + \dfrac{z_0 z}{c^2} = 1$

46. *Hyperboloid* $\dfrac{x^2}{a^2} + \dfrac{y^2}{b^2} - \dfrac{z^2}{c^2} = 1$

Plane $\dfrac{x_0 x}{a^2} + \dfrac{y_0 y}{b^2} - \dfrac{z_0 z}{c^2} = 1$

47. Show that any tangent plane to the cone $z^2 = a^2 x^2 + b^2 y^2$ passes through the origin.

48. Show that any line normal to a sphere passes through the center of the sphere.

49. Prove Theorem 13.14.

50. Prove that the angle of inclination θ of the tangent plane to the surface $z = f(x, y)$ at the point (x_0, y_0, z_0) is given by

$$\cos\theta = \frac{1}{\sqrt{[f_x(x_0, y_0)]^2 + [f_y(x_0, y_0)]^2 + 1}}.$$

C **51.** *Approximation*   Consider the following approximations centered at $(0, 0)$ for a function $f(x, y)$.

Linear Approximation

$$P_1(x, y) \approx f(0, 0) + f_x(0, 0)x + f_y(0, 0)y$$

Quadratic Approximation

$$P_2(x, y) \approx f(0, 0) + f_x(0, 0)x + f_y(0, 0)y$$
$$+ \frac{1}{2}f_{xx}(0, 0)x^2 + f_{xy}(0, 0)xy + \frac{1}{2}f_{yy}(0, 0)y^2$$

[Note that the linear approximation is the tangent plane to the surface at $(0, 0, f(0, 0))$.]

a. Find the linear approximation of $f(x, y) = e^{(x-y)}$ centered at $(0, 0)$.

b. Find the quadratic approximation of $f(x, y) = e^{(x-y)}$ centered at $(0, 0)$.

c. If $x = 0$ in the quadratic approximation, then you obtain the second-degree Taylor polynomial for what function? Answer the same question for $y = 0$.

d. Complete the table.

x	y	$f(x,y)$	$P_1(x,y)$	$P_2(x,y)$
0	0			
0	0.1			
0.2	0.1			
0.2	0.5			
1	0.5			

e. Use a graphing utility to graph the surfaces given by $z = f(x, y)$, $z = P_1(x, y)$, and $z = P_2(x, y)$.

52. *Approximation*   Repeat Exercise 51 for the function $f(x, y) = \cos(x + y)$.

53. Let f be a differentiable function, and consider the surface $z = xf(y/x)$. Show that the tangent plane at any point $P = (x_0, y_0, z_0)$ on the surface passes through the origin.

FOR FURTHER INFORMATION   Biologists use the concept of diversity to measure the proportions of different types of organisms within an environment. For more information on this, see the article "Information Theory and Biological Diversity" by Steven Kolmes and Kevin Mitchell in the 1990 *UMAP Modules*.

54. *Wildflower Diversity*   The diversity of wildflowers in a meadow can be measured by counting the number of daisies, buttercups, shooting stars, and so on. If there are n types of wildflowers, each with a proportion of p_i of the total population, then it follows that $p_1 + p_2 + \cdots + p_n = 1$. The measure of diversity of the population is defined as

$$H = -\sum_{i=1}^{n} p_i \log_2 p_i.$$

In this definition, it is understood that $p_i \log_2 p_i = 0$ when $p_i = 0$. The tables show the proportions of wildflowers in a meadow in May, June, August, and September.

a. Determine the wildflower diversity for each month. How would you interpret September's diversity? Which month had the greatest diversity?

b. If the meadow contains 10 types of wildflowers in roughly equal proportions, is the diversity of the population greater or less than the diversity of a similar distribution of four types of wildflowers? What type of distribution (of 10 types of wildflowers) would produce a maximum diversity?

c. Let H_n represent the maximum diversity of n types of wildflowers. Does H_n approach a limit as n approaches infinity?

May

Flower type	1	2	3	4
Proportion	$\frac{5}{16}$	$\frac{5}{16}$	$\frac{5}{16}$	$\frac{1}{16}$

June

Flower type	1	2	3	4
Proportion	$\frac{1}{4}$	$\frac{1}{4}$	$\frac{1}{4}$	$\frac{1}{4}$

August

Flower type	1	2	3	4
Proportion	$\frac{1}{4}$	0	$\frac{1}{4}$	$\frac{1}{2}$

September

Flower type	1	2	3	4
Proportion	0	0	0	1

KARL WEIERSTRASS

Although the Extreme Value Theorem had been used by earlier mathematicians, the first to provide a rigorous proof was the German mathematician Karl Weierstrass (1815–1897). Weierstrass also provided rigorous justifications for many other mathematical results already in common use. We are indebted to him for much of the logical foundation upon which modern calculus is built.

Absolute Extrema and Relative Extrema

In Chapter 3, you studied techniques for finding the extreme values of a function of a single variable. In this section, we extend these techniques to functions of two variables. For example, in Theorem 13.15 the Extreme Value Theorem for a function of a single variable is extended to a function of two variables.

The values $f(a, b)$ and $f(c, d)$ such that

$$f(a, b) \leq f(x, y) \leq f(c, d)$$

for all (x, y) in R are called the **minimum** and **maximum** of f in the region R, as shown in Figure 13.65.

Recall from Section 13.2 that a region in the plane is *closed* if it contains all of its boundary points. The Extreme Value Theorem deals with a region in the plane that is both closed and *bounded*. A region in the plane is called **bounded** if it is a subregion of a closed disc in the plane.

THEOREM 13.15   Extreme Value Theorem

Let f be a continuous function of two variables x and y defined on a closed bounded region R in the xy-plane.

1. There is at least one point in R where f takes on a minimum value.
2. There is at least one point in R where f takes on a maximum value.

A minimum is also called an **absolute minimum** and a maximum is also called an **absolute maximum.** As in single-variable calculus, we distinguish between absolute extrema and **relative extrema.**

Definition of Relative Extrema

Let f be a function defined on a region R containing (x_0, y_0).

1. The function f has a **relative minimum** at (x_0, y_0) if

$$f(x, y) \geq f(x_0, y_0)$$

for all (x, y) in an *open* disc containing (x_0, y_0).

2. The function f has a **relative maximum** at (x_0, y_0) if

$$f(x, y) \leq f(x_0, y_0)$$

for all (x, y) in an *open* disc containing (x_0, y_0).

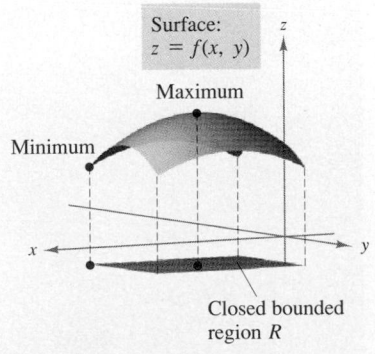

FIGURE 13.65
R contains point(s) at which $f(x, y)$ is minimum and point(s) at which $f(x, y)$ is maximum.

FIGURE 13.66
Relative extrema

To say that f has a relative maximum at (x_0, y_0) means that the point (x_0, y_0, z_0) is at least as high as all nearby points on the graph of $z = f(x, y)$. Similarly, f has a relative minimum at (x_0, y_0) if (x_0, y_0, z_0) is at least as low as all nearby points on the graph (see Figure 13.66).

To locate relative extreme of f, you can investigate the points at which the gradient of f is **0** or undefined. Such points are called **critical points** of f.

Definition of Critical Point

Let f be defined on an open region R containing (x_0, y_0). The point (x_0, y_0) is a **critical point** of f if one of the following is true.

1. $f_x(x_0, y_0) = 0$ and $f_y(x_0, y_0) = 0$.
2. $f_x(x_0, y_0)$ or $f_y(x_0, y_0)$ does not exist.

Recall from Theorem 13.11 that if f is differentiable and

$$\nabla f(x_0, y_0) = f_x(x_0, y_0)\mathbf{i} + f_y(x_0, y_0)\mathbf{j}$$
$$= 0\mathbf{i} + 0\mathbf{j}$$

then every directional derivative at (x_0, y_0) must be 0. In Exercise 46 you are asked to show that this implies that the function has a horizontal tangent plane at the point (x_0, y_0), as shown in Figure 13.67. It appears that such a point is a likely location of a relative extremum. This is confirmed by Theorem 13.16. (A proof of this theorem is given in Appendix A.)

FIGURE 13.67

THEOREM 13.16

If f has a relative extremum at (x_0, y_0) on an open region R, then (x_0, y_0) is a critical point of f.

EXAMPLE 1   Finding a Relative Extremum

Determine the relative extrema of

$$f(x, y) = 2x^2 + y^2 + 8x - 6y + 20.$$

Solution   Begin by finding the critical points of f. Because

$$f_x(x, y) = 4x + 8 \qquad \text{Partial with respect to } x$$

and

$$f_y(x, y) = 2y - 6 \qquad \text{Partial with respect to } y$$

are defined for all x and y, the only critical points are those for which both first partial derivatives are 0. To locate these points, let $f_x(x, y)$ and $f_y(x, y)$ be 0, and solve the system of equations

$$4x + 8 = 0 \quad \text{and} \quad 2y - 6 = 0$$

to obtain the critical point $(-2, 3)$. By completing the square, you can conclude that for all $(x, y) \neq (-2, 3)$,

$$f(x, y) = 2(x + 2)^2 + (y - 3)^2 + 3 > 3.$$

Therefore, a relative *minimum* of f occurs at $(-2, 3)$. The value of the relative minimum is $f(-2, 3) = 3$, as shown in Figure 13.68.   ■

Example 1 shows a relative minimum occurring at one type of critical point—the type for which both $f_x(x, y)$ and $f_y(x, y)$ are 0. The next example concerns a relative maximum that occurs at the other type of critical point—the type for which either $f_x(x, y)$ or $f_y(x, y)$ is undefined.

EXAMPLE 2   Finding a Relative Extremum

Determine the relative extrema of $f(x, y) = 1 - (x^2 + y^2)^{1/3}$.

Solution   Because

$$f_x(x, y) = -\frac{2x}{3(x^2 + y^2)^{2/3}} \qquad \text{Partial with respect to } x$$

and

$$f_y(x, y) = -\frac{2y}{3(x^2 + y^2)^{2/3}} \qquad \text{Partial with respect to } y$$

it follows that both partial derivatives are defined for all points in the xy-plane except for $(0, 0)$. Moreover, because the partial derivatives cannot both be 0 unless both x and y are 0, you can conclude that $(0, 0)$ is the only critical point. In Figure 13.69, note that $f(0, 0)$ is 1. For all other (x, y) it is clear that

$$f(x, y) = 1 - (x^2 + y^2)^{1/3} < 1.$$

Therefore, f has a relative *maximum* at $(0, 0)$.   ■

REMARK   In Example 2, $f_x(x, y) = 0$ for every point on the y-axis other than $(0, 0)$. However, because $f_y(x, y)$ is nonzero, these are not critical points. Remember that *one* of the partials must be undefined or *both* must be 0 in order to yield a critical point.

Surface:
$f(x, y) = 2x^2 + y^2 + 8x - 6y + 20$

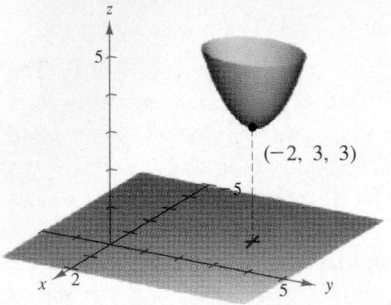

FIGURE 13.68
The function $z = f(x, y)$ has a relative minimum at $(-2, 3)$.

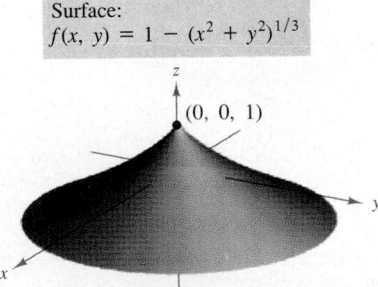

Surface:
$f(x, y) = 1 - (x^2 + y^2)^{1/3}$

FIGURE 13.69
$f_x(x, y)$ and $f_y(x, y)$ are undefined at $(0, 0)$.

The Second Partials Test

Theorem 13.16 tells you that to find relative extrema you need only examine values of $f(x, y)$ at critical points. However, as is true for a function of one variable, the critical points of a function of two variables do not always yield relative maxima or minima. Some critical points yield **saddle points,** which are neither relative maxima nor relative minima.

As an example of a critical point that does not yield a relative extremum, consider the surface given by

$$f(x, y) = y^2 - x^2 \qquad \text{Hyperbolic paraboloid}$$

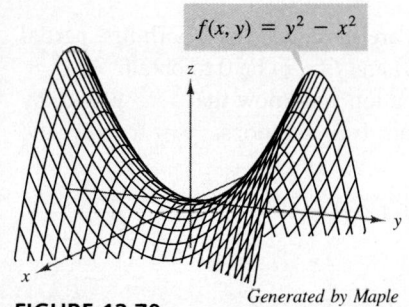

$f(x, y) = y^2 - x^2$

FIGURE 13.70
Saddle point at
$(0, 0, 0)$: $f_x(0, 0) = f_y(0, 0) = 0$

Generated by Maple

as shown in Figure 13.70. At the point $(0, 0)$, both partial derivatives are 0. The function f does not, however, have a relative extremum at this point because in any open disc centered at $(0, 0)$ the function takes on both negative values (along the x-axis) *and* positive values (along the y-axis). Thus, the point $(0, 0, 0)$ is a saddle point of the surface. (The name "saddle point" comes from the fact that the surface shown in Figure 13.70 resembles a saddle.)

For the functions in Examples 1 and 2, it is relatively easy to determine the relative extrema, because each function was either given, or able to be written, in completed square form. For more complicated functions, algebraic arguments are less convenient and it is better to rely on the analytic means presented in the following Second Partials Test. This is the two-variable counterpart of the Second Derivative Test for functions of one variable. The proof of this theorem is best left to a course in advanced calculus.

THEOREM 13.17   Second Partials Test

Let f have continuous second partial derivatives on an open region containing a point (a, b) for which

$$f_x(a, b) = 0 \quad \text{and} \quad f_y(a, b) = 0.$$

To test for relative extrema of f, we define the quantity

$$d = f_{xx}(a, b) f_{yy}(a, b) - [f_{xy}(a, b)]^2.$$

1. If $d > 0$ and $f_{xx}(a, b) > 0$, then f has a **relative minimum** at (a, b).
2. If $d > 0$ and $f_{xx}(a, b) < 0$, then f has a **relative maximum** at (a, b).
3. If $d < 0$, then $(a, b, f(a, b))$ is a **saddle point.**
4. The test is inconclusive if $d = 0$.

REMARK   If $d > 0$, then $f_{xx}(a, b)$ and $f_{yy}(a, b)$ must have the same sign. This means that $f_{xx}(a, b)$ can be replaced by $f_{yy}(a, b)$ in the first two parts of the test.

A convenient device for remembering the formula for d in the Second Partials Test is given by the 2×2 determinant

$$d = \begin{vmatrix} f_{xx}(a, b) & f_{xy}(a, b) \\ f_{yx}(a, b) & f_{yy}(a, b) \end{vmatrix}$$

where $f_{xy}(a, b) = f_{yx}(a, b)$ by Theorem 13.3.

EXAMPLE 3   Using the Second Partials Test

Find the relative extrema of $f(x, y) = -x^3 + 4xy - 2y^2 + 1$.

Solution   Begin by finding the critical points of f. Because

$$f_x(x, y) = -3x^2 + 4y \quad \text{and} \quad f_y(x, y) = 4x - 4y$$

are defined for all x and y, the only critical points are those for which both first partial derivatives are 0. To locate these points, let $f_x(x, y)$ and $f_y(x, y)$ be 0 to obtain $-3x^2 + 4y = 0$ and $4x - 4y = 0$. From the second equation you know that $x = y$, and, by substitution into the first equation, you obtain two solutions: $y = x = 0$ and $y = x = \frac{4}{3}$. Because

$$f_{xx}(x, y) = -6x, \quad f_{yy}(x, y) = -4, \quad \text{and} \quad f_{xy}(x, y) = 4$$

it follows that, for the critical point $(0, 0)$,

$$d = f_{xx}(0, 0)f_{yy}(0, 0) - [f_{xy}(0, 0)]^2 = 0 - 16 < 0$$

and, by the Second Partials Test, you can conclude that $(0, 0, 1)$ is a saddle point of f. Furthermore, for the critical point $\left(\frac{4}{3}, \frac{4}{3}\right)$,

$$d = f_{xx}\left(\tfrac{4}{3}, \tfrac{4}{3}\right) f_{yy}\left(\tfrac{4}{3}, \tfrac{4}{3}\right) - \left[f_{xy}\left(\tfrac{4}{3}, \tfrac{4}{3}\right) \right]^2$$
$$= -8(-4) - 16 = 16 > 0$$

and because $f_{xx}\left(\frac{4}{3}, \frac{4}{3}\right) = -8 < 0$ you can conclude that f has a relative maximum at $\left(\frac{4}{3}, \frac{4}{3}\right)$, as shown in Figure 13.71.

Relative maximum

Saddle point $(0, 0, 1)$

$\left(\frac{4}{3}, \frac{4}{3}\right)$

$f(x, y) = -x^3 + 4xy - 2y^2 + 1$

Generated by Maple

FIGURE 13.71

The Second Partials Test can fail to find relative extrema in two ways. If either of the first partial derivatives is undefined, then you cannot use the test. Also, if

$$d = f_{xx}(a, b)f_{yy}(a, b) - [f_{xy}(a, b)]^2 = 0$$

the test fails. In such cases, you can try a sketch or some other approach, as demonstrated in the next example.

EXAMPLE 4   Failure of the Second Partials Test

Find the relative extrema of $f(x, y) = x^2y^2$.

Solution   Because $f_x(x, y) = 2xy^2$ and $f_y(x, y) = 2x^2y$, you know that both partial derivatives are 0 if $x = 0$ or $y = 0$. That is, every point along the x- or y-axis is a critical point. Moreover, because

$$f_{xx}(x, y) = 2y^2, \quad f_{yy}(x, y) = 2x^2, \quad \text{and} \quad f_{xy}(x, y) = 4xy$$

you know that if either $x = 0$ or $y = 0$, then

$$d = f_{xx}(x, y)f_{yy}(x, y) - [f_{xy}(x, y)]^2$$
$$= 4x^2y^2 - 16x^2y^2 = -12x^2y^2 = 0.$$

$f(x, y) = x^2y^2$

If $y = 0$
then $f(x, y) = 0.$ If $x = 0$
then $f(x, y) = 0.$
Generated by Mathematica

FIGURE 13.72

Thus, the Second Partials Test fails. However, because $f(x, y) = 0$ for every point along the x- or y-axis and $f(x, y) = x^2y^2 > 0$ for all other points, you can conclude that each of these critical points yields an absolute minimum, as shown in Figure 13.72.

Absolute extrema of a function can occur in two ways. First, some relative extrema also happen to be absolute extrema. For instance, in Example 1, $f(-2, 3)$ is an absolute minimum of the function. (On the other hand, the relative maximum found in Example 3 is not an absolute maximum of the function.) Second, absolute extrema can occur at a boundary point of the domain. This is illustrated in Example 5.

EXAMPLE 5 Finding Absolute Extrema

Find the absolute extrema of the function

$$f(x, y) = \sin xy$$

on the closed region given by $0 \le x \le \pi$ and $0 \le y \le 1$.

Solution From the partial derivatives

$$f_x(x, y) = y \cos xy \quad \text{and} \quad f_y(x, y) = x \cos xy$$

you can see that each point lying on the hyperbola given by $xy = \pi/2$ is a critical point. These points each yield the value

$$f(x, y) = \sin\left(\frac{\pi}{2}\right) = 1$$

which you know is the absolute maximum, as shown in Figure 13.73. The only other critical point of f *lying in the given region* is $(0, 0)$. It yields an absolute minimum of 0, because

$$0 \le xy \le \pi$$

implies that

$$0 \le \sin xy \le 1.$$

To hunt for other absolute extrema, you should consider the four boundaries of the region formed by taking traces with the vertical planes $x = 0$, $x = \pi$, $y = 0$, and $y = 1$. In doing this, you will find that $\sin xy = 0$ at all points on the x-axis, the y-axis, and at the point $(\pi, 1)$. Each of these points yields an absolute minimum for the surface, as shown in Figure 13.73.

Surface:
$f(x, y) = \sin xy$

Absolute maximum

Absolute minimum

$xy = \dfrac{\pi}{2}$

$(\pi, 1)$

Absolute minimum

Domain:
$0 \le x \le \pi$
$0 \le y \le 1$

FIGURE 13.73

The concepts of relative extrema and critical points can be extended to functions of three or more variables. If all first partial derivatives of

$$w = f(x_1, x_2, x_3, \ldots, x_n)$$

exist, then it can be shown that a relative maximum or minimum can occur at $(x_1, x_2, x_3, \ldots, x_n)$ only if every first partial derivative is 0 at that point. This means that the critical points are obtained by solving the following system of equations.

$$f_{x_1}(x_1, x_2, x_3, \ldots, x_n) = 0$$
$$f_{x_2}(x_1, x_2, x_3, \ldots, x_n) = 0$$
$$\vdots$$
$$f_{x_n}(x_1, x_2, x_3, \ldots, x_n) = 0$$

The extension of Theorem 13.17 to three or more variables is also possible, although we will not consider such an extension in this text.

EXERCISES for Section 13.8

In Exercises 1–4, identify any extrema of the function by recognizing its given form or its form after completing the square. Verify your results by using the partial derivatives to locate any critical points and test for relative extrema.

1. $f(x, y) = x^2 + y^2 + 2x - 6y + 6$

2. $f(x, y) = -x^2 - y^2 + 4x + 8y - 11$

3. $f(x, y) = \sqrt{25 - (x - 2)^2 - y^2}$

4. $f(x, y) = \sqrt{x^2 + y^2 + 1}$

In Exercises 5–20, examine each function for relative extrema and saddle points.

5. $g(x, y) = (x - 1)^2 + (y - 3)^2$

6. $g(x, y) = 9 - (x - 3)^2 - (y + 2)^2$

7. $f(x, y) = 2x^2 + 2xy + y^2 + 2x - 3$

8. $f(x, y) = -x^2 - 5y^2 + 8x - 10y - 13$

9. $f(x, y) = -5x^2 + 4xy - y^2 + 16x + 10$

10. $f(x, y) = x^2 + 6xy + 10y^2 - 4y + 4$

11. $z = 2x^2 + 3y^2 - 4x - 12y + 13$

12. $z = -3x^2 - 2y^2 + 3x - 4y + 5$

13. $h(x, y) = x^2 - y^2 - 2x - 4y - 4$

14. $h(x, y) = x^2 - 3xy - y^2$

15. $g(x, y) = xy$

16. $g(x, y) = 120x + 120y - xy - x^2 - y^2$

17. $f(x, y) = x^3 - 3xy + y^3$

18. $f(x, y) = 4xy - x^4 - y^4$

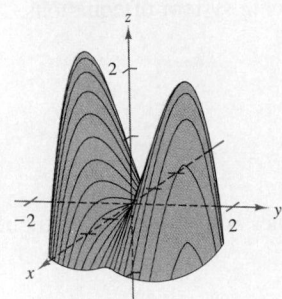

19. $z = e^{-x} \sin y$

20. $z = (\frac{1}{2} - x^2 + y^2)e^{1 - x^2 - y^2}$

C In Exercises 21–24, use a graphing utility to graph the surface and locate any relative extrema and saddle points.

21. $z = \dfrac{-4x}{x^2 + y^2 + 1}$

22. $f(x, y) = y^3 - 3yx^2 - 3y^2 - 3x^2 + 1$

23. $z = (x^2 + 4y^2)e^{1 - x^2 - y^2}$

24. $z = e^{xy}$

In Exercises 25–28, determine whether there is a relative maximum, a relative minimum, a saddle point, or insufficient information to determine the nature of the function $f(x, y)$ at the critical point (x_0, y_0).

25. $f_{xx}(x_0, y_0) = 9$
$f_{yy}(x_0, y_0) = 4$
$f_{xy}(x_0, y_0) = 6$

26. $f_{xx}(x_0, y_0) = -3$
$f_{yy}(x_0, y_0) = -8$
$f_{xy}(x_0, y_0) = 2$

27. $f_{xx}(x_0, y_0) = -9$
$f_{yy}(x_0, y_0) = 6$
$f_{xy}(x_0, y_0) = 10$

28. $f_{xx}(x_0, y_0) = 25$
$f_{yy}(x_0, y_0) = 8$
$f_{xy}(x_0, y_0) = 10$

In Exercises 29–36, find the absolute extrema of the function over the region R. (In each case, R contains the boundaries.)

29. $f(x, y) = 12 - 3x - 2y$
R: The triangular region in the xy-plane with vertices $(2, 0)$, $(0, 1)$, and $(1, 2)$.

30. $f(x, y) = (2x - y)^2$
R: The triangular region in the xy-plane with vertices $(2, 0)$, $(0, 1)$, and $(1, 2)$.

31. $f(x, y) = 3x^2 + 2y^2 - 4y$
R: The region in the xy-plane bounded by the graphs of $y = x^2$ and $y = 4$.

32. $f(x, y) = 2x - 2xy + y^2$
   R: The region in xy-plane bounded by the graphs of $y = x^2$
   and $y = 1$.

33. $f(x, y) = x^2 + xy$,
   $R = \{(x, y): |x| \leq 2, |y| \leq 1\}$

34. $f(x, y) = x^2 + 2xy + y^2$,
   $R = \{(x, y): |x| \leq 2, |y| \leq 1\}$

35. $f(x, y) = x^2 + 2xy + y^2$,
   $R = \{(x, y): x^2 + y^2 \leq 8\}$

36. $r(x, y) = x^2 - 4xy$,
   $R = \{(x, y): 0 \leq x \leq 4, 0 \leq y \leq \sqrt{x}\}$

In Exercises 37–42, find the critical points and test for relative extrema. List the critical points for which the Second Partials Test fails.

37. $f(x, y) = x^3 + y^3$

38. $f(x, y) = x^3 + y^3 - 3x^2 + 6y^2 + 3x + 12y + 7$

39. $f(x, y) = (x - 1)^2(y + 4)^2$

40. $f(x, y) = \sqrt{(x - 1)^2 + (y + 2)^2}$

41. $f(x, y) = x^{2/3} + y^{2/3}$

42. $f(x, y) = (x^2 + y^2)^{2/3}$

In Exercises 43 and 44, find the critical points of the function and from the form of the function determine whether each point is a relative maximum or a relative minimum.

43. $f(x, y, z) = x^2 + (y - 3)^2 + (x + 1)^2$

44. $f(x, y, z) = 4 - [x(y - 1)(z + 2)]^2$

45. The figure shows the level curves for an unknown function $f(x, y)$. What, if any, information can be given about the points A, B, C, and D?

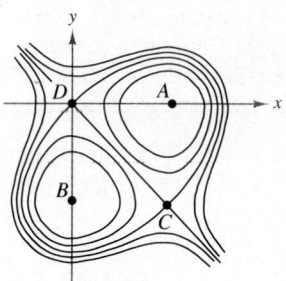

FIGURE FOR 45

46. Prove that if f is a differentiable function such that $\nabla f(x_0, y_0) = \mathbf{0}$, then the tangent plane at (x_0, y_0) is horizontal.

True or False **In Exercises 47–50, determine whether the statement is true or false. If it is false, explain why or give an example that shows it is false.**

47. If f has a relative maximum at (x_0, y_0, z_0), then $f_x(x_0, y_0) = f_y(x_0, y_0) = 0$.

48. The function given by $f(x, y) = \sqrt[3]{x^2 + y^2}$ has a relative minimum at the origin.

49. If f is continuous for all x and y and has two relative minima, then f must have at least one relative maximum.

50. If $f_x(x_0, y_0) = 0$ and $f_y(x_0, y_0) = 0$, then at the point (x_0, y_0, z_0), the tangent plane to the surface given by $z = f(x, y)$ is horizontal.

Applied Optimization Problems

In this section, we survey a few of the many applications of extrema of functions of two (or more) variables.

EXAMPLE 1 Finding Maximum Volume

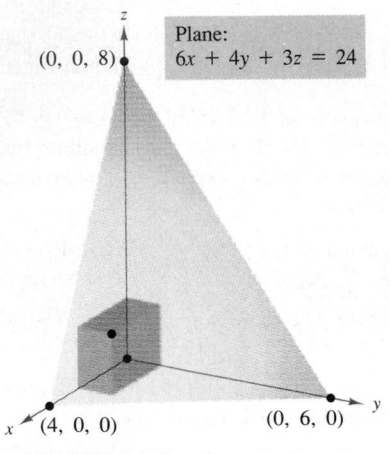

Plane:
$6x + 4y + 3z = 24$

$(0, 0, 8)$

x $(4, 0, 0)$ $(0, 6, 0)$ y

FIGURE 13.74
The maximum volume of the box is $\frac{64}{9}$ cubic units.

A rectangular box is resting on the xy-plane with one vertex at the origin. The opposite vertex lies in the plane $6x + 4y + 3z = 24$, as shown in Figure 13.74. Find the maximum volume of such a box.

Solution Let x, y, and z represent the length, width, and height of the box. Because one vertex of the box lies in the plane $6x + 4y + 3z = 24$, you know that $z = \frac{1}{3}(24 - 6x - 4y)$, and you can write the volume xyz of the box as a function of two variables.

$$V(x, y) = (x)(y)\left[\frac{1}{3}(24 - 6x - 4y)\right] = \frac{1}{3}(24xy - 6x^2y - 4xy^2)$$

By setting the first partial derivatives equal to 0

$$V_x(x, y) = \frac{1}{3}(24y - 12xy - 4y^2) = \frac{y}{3}(24 - 12x - 4y) = 0$$

$$V_y(x, y) = \frac{1}{3}(24x - 6x^2 - 8xy) = \frac{x}{3}(24 - 6x - 8y) = 0$$

you obtain the critical points $(0, 0)$ and $\left(\frac{4}{3}, 2\right)$. At $(0, 0)$ the volume is 0, so that point does not yield a maximum volume. At the point $\left(\frac{4}{3}, 2\right)$, you can apply the Second Partials Test.

$$V_{xx}(x, y) = -4y, \qquad V_{yy}(x, y) = \frac{-8x}{3}, \qquad V_{xy}(x, y) = \frac{1}{3}(24 - 12x - 8y)$$

Because

$$V_{xx}\left(\frac{4}{3}, 2\right)V_{yy}\left(\frac{4}{3}, 2\right) - \left[V_{xy}\left(\frac{4}{3}, 2\right)\right]^2 = (-8)\left(-\frac{32}{9}\right) - \left(-\frac{8}{3}\right)^2 = \frac{64}{3} > 0$$

and

$$V_{xx}\left(\frac{4}{3}, 2\right) = -8 < 0$$

you can conclude from the Second Partials Test that the maximum volume is

$$V\left(\frac{4}{3}, 2\right) = \frac{1}{3}\left[24\left(\frac{4}{3}\right)(2) - 6\left(\frac{4}{3}\right)^2(2) - 4\left(\frac{4}{3}\right)(2^2)\right]$$

$$= \frac{64}{9} \text{ cubic units.}$$

(Note that the volume is 0 at the boundary points of the triangular domain of V.)

REMARK In many applied problems, the domain of the function to be optimized is a closed bounded region. To find minimum or maximum points, you must not only test critical points, but also consider the values of the function at points on the boundary.

In Section 3.10, you studied several applications of extrema in economics and business. In practice, such applications often involve more than one independent variable. For instance, a company may produce several models of one type of product. The price per unit and profit per unit are usually different for each model. Moreover, the demand for each model is often a function of the prices of the other models (as well as its own price). The next example illustrates an application involving two products.

EXAMPLE 2   Finding the Maximum Profit

The profit obtained by producing x units of product A and y units of product B is approximated by the model

$$P(x, y) = 8x + 10y - (0.001)(x^2 + xy + y^2) - 10{,}000.$$

Find the production level that produces a maximum profit.

Solution   The partial derivatives of the profit function are

$$P_x(x, y) = 8 - (0.001)(2x + y)$$

and

$$P_y(x, y) = 10 - (0.001)(x + 2y).$$

By setting these partial derivatives equal to 0, you obtain the following system of equations.

$$8 - (0.001)(2x + y) = 0$$
$$10 - (0.001)(x + 2y) = 0$$

After simplifying, this system of linear equations can be written as

$$2x + y = 8000$$
$$x + 2y = 10{,}000.$$

Solving this system produces $x = 2000$ and $y = 4000$. The second partial derivatives of P are

$$P_{xx}(2000, 4000) = -0.002$$
$$P_{yy}(2000, 4000) = -0.002$$
$$P_{xy}(2000, 4000) = -0.001.$$

Moreover, because $P_{xx} < 0$ and

$$P_{xx}(2000, 4000)P_{yy}(2000, 4000) - [P_{xy}(2000, 4000)]^2$$
$$= (-0.002)^2 - (-0.001)^2 > 0$$

you can conclude that the production level of $x = 2000$ units and $y = 4000$ units yields a *maximum* profit.

FOR FURTHER INFORMATION
For more information on the use of mathematics in economics, see the article "Mathematical Methods of Economics" by Joel Franklin in the April, 1983 issue of the *American Mathematical Monthly*.

REMARK   In Example 2, we assumed that the manufacturing plant is able to produce the required number of units to yield a maximum profit. In actual practice, the production would be bounded by physical constraints. You will study such constrained optimization problems in the next section.

The Method of Least Squares

Many of the examples in this text have involved **mathematical models.** For instance, Example 2 involves a quadratic model for profit. There are several ways to develop such models—one is called the **method of least squares.**

In constructing a model to represent a particular phenomenon, the goals are simplicity and accuracy. Of course, these goals often conflict. For instance, a simple linear model for the points in Figure 13.75 is

$$y = 1.8566x - 5.0246.$$

However, Figure 13.76 shows that by choosing the slightly more complicated quadratic model*

$$y = 0.1996x^2 - 0.7281x + 1.3749$$

you can achieve greater accuracy.

FIGURE 13.75

FIGURE 13.76

As a measure of how well the model $y = f(x)$ fits the collection of points

$$\{(x_1, y_1), (x_2, y_2), (x_3, y_3), \ldots, (x_n, y_n)\}$$

you can add the squares of the differences between the actual y-values and the values given by the model to obtain the **sum of the squared errors**

$$S = \sum_{i=1}^{n} [f(x_i) - y_i]^2.$$

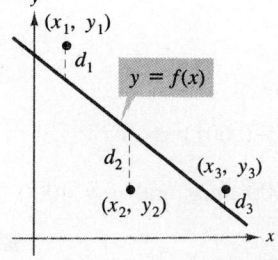

FIGURE 13.77
Sum of the squared errors:
$S = d_1^2 + d_2^2 + d_3^2$

Graphically, S can be interpreted as the sum of the squares of the vertical distances between the graph of f and the given points in the plane, as shown in Figure 13.77. If the model is perfect, then $S = 0$. However, when perfection is not feasible, we settle for a model that minimizes S. Statisticians call the *linear model* that minimizes S the **least squares regression line.** The proof that this line actually minimizes S involves the minimizing of a function of two variables.

*A method for finding the least squares quadratic model for a collection of data is described in Exercise 35.

> **THEOREM 13.18   Least Squares Regression Line**
>
> The **least squares regression line** for $\{(x_1, y_1), (x_2, y_2), \ldots, (x_n, y_n)\}$ is given by $f(x) = ax + b$, where
>
> $$a = \frac{n\sum_{i=1}^{n} x_i y_i - \sum_{i=1}^{n} x_i \sum_{i=1}^{n} y_i}{n\sum_{i=1}^{n} x_i^2 - \left(\sum_{i=1}^{n} x_i\right)^2} \quad \text{and} \quad b = \frac{1}{n}\left(\sum_{i=1}^{n} y_i - a\sum_{i=1}^{n} x_i\right).$$

ADRIEN-MARIE LEGENDRE

The method of least squares was introduced by the French mathematician Adrien-Marie Legendre (1752–1833). Legendre is best known for his work in geometry. In fact, his text *Elements of Geometry* was so popular in the United States that it continued to be used for 33 editions, spanning a period of more than 100 years.

Proof   Let $S(a, b)$ represent the sum of the squared errors for the model $f(x) = ax + b$ and the given set of points. That is,

$$S(a, b) = \sum_{i=1}^{n} [f(x_i) - y_i]^2$$

$$= \sum_{i=1}^{n} (ax_i + b - y_i)^2$$

where the points (x_i, y_i) represent constants. Because S is a function of a and b, you can use the methods discussed in the previous section to find the minimum value of S. Specifically, the first partial derivatives of S are

$$S_a(a, b) = \sum_{i=1}^{n} 2x_i(ax_i + b - y_i)$$

$$= 2a\sum_{i=1}^{n} x_i^2 + 2b\sum_{i=1}^{n} x_i - 2\sum_{i=1}^{n} x_i y_i$$

$$S_b(a, b) = \sum_{i=1}^{n} 2(ax_i + b - y_i)$$

$$- 2a\sum_{i=1}^{n} x_i + 2nb - 2\sum_{i=1}^{n} y_i.$$

By setting these two partial derivatives equal to 0, you obtain the values for a and b that are listed in the theorem. We leave it to you to apply the Second Partials Test (see Exercise 36) to verify that these values of a and b yield a minimum.   ▬

If the x-values are symmetrically spaced about the y-axis, then $\sum x_i = 0$, and the formulas for a and b simplify to

$$a = \frac{\sum_{i=1}^{n} x_i y_i}{\sum_{i=1}^{n} x_i^2} \quad \text{and} \quad b = \frac{1}{n}\sum_{i=1}^{n} y_i.$$

This simplification is often possible with a translation of the x-values. For instance, if the x-values in a data collection consist of the years 1980, 1981, 1982, 1983, and 1984, you could let 1982 be represented by 0.

EXAMPLE 3　Finding the Least Squares Regression Line

Find the least squares regression line for the following points.

$$(-3, 0), \qquad (-1, 1), \qquad (0, 2), \qquad (2, 3)$$

Solution　Table 13.2 shows the calculations involved in finding the least squares regression line using $n = 4$.

TABLE 13.2

x	y	xy	x^2
-3	0	0	9
-1	1	-1	1
0	2	0	0
2	3	6	4
$\displaystyle\sum_{i=1}^{n} x_i = -2$	$\displaystyle\sum_{i=1}^{n} y_i = 6$	$\displaystyle\sum_{i=1}^{n} x_i y_i = 5$	$\displaystyle\sum_{i=1}^{n} x_i^2 = 14$

$$f(x) = \frac{8}{13}x + \frac{47}{26}$$

FIGURE 13.78
Least squares regression line

Applying Theorem 13.18 produces

$$a = \frac{n\displaystyle\sum_{i=1}^{n} x_i y_i - \displaystyle\sum_{i=1}^{n} x_i \displaystyle\sum_{i=1}^{n} y_i}{n\displaystyle\sum_{i=1}^{n} x_i^2 - \left(\displaystyle\sum_{i=1}^{n} x_i\right)^2}$$

$$= \frac{4(5) - (-2)(6)}{4(14) - (-2)^2}$$

$$= \frac{8}{13}$$

and

$$b = \frac{1}{n}\left(\sum_{i=1}^{n} y_i - a\sum_{i=1}^{n} x_i\right)$$

$$= \frac{1}{4}\left[6 - \frac{8}{13}(-2)\right]$$

$$= \frac{47}{26}.$$

Thus, the least squares regression line is

$$y = \frac{8}{13}x + \frac{47}{26}$$

as shown in Figure 13.78.

TECHNOLOGY　Many calculators have "built-in" least squares regression programs. If your calculator has such a program, try using it to duplicate the results of Example 3.

EXERCISES for Section 13.9

In Exercises 1 and 2, find the minimum distance from the point to the plane given by $2x + 3y + z = 12$. (Hint: To simplify the computations, minimize the square of the distance.)

1. $(0, 0, 0)$ **2.** $(1, 2, 3)$

In Exercises 3 and 4, find the minimum distance from the point to the paraboloid given by $z = x^2 + y^2$.

3. $(5, 5, 0)$ **4.** $(5, 0, 0)$

In Exercises 5–8, find three positive numbers x, y, and z that satisfy the given conditions.

5. The sum is 30 and the product is maximum.

6. The sum is 32 and $P = xy^2z$ is maximum.

7. The sum is 30 and the sum of the squares is minimum.

8. The sum is 1 and the sum of the squares is minimum.

9. *Volume* The sum of the length and the girth (perimeter of a cross section) of packages carried by parcel post cannot exceed 108 inches. Find the dimensions of the rectangular package of largest volume that may be sent by parcel post.

10. *Volume* Repeat Exercise 9 under the condition that the sum of the two perimeters of the two cross sections shown in the figure cannot exceed 108 inches.

11. *Minimum Cost* A water line is to be built from point P to point S and must pass through regions where construction costs differ (see figure). Find x and y so that the total cost C will be minimum if the cost per mile in dollars is $3k$ from P to Q, $2k$ from Q to R, and k from R to S.

FIGURE FOR 10   **FIGURE FOR 11**

12. *Volume* The material for constructing the base of an open box costs 1.5 times as much per unit area as the material for constructing the sides. For a fixed amount of money C, find the dimensions of the box of largest volume that can be made.

13. *Volume* The volume of an ellipsoid

$$\frac{x^2}{a^2} + \frac{y^2}{b^2} + \frac{z^2}{c^2} = 1$$

is $4\pi abc/3$. For fixed $a + b + c$, show that the ellipsoid of maximum volume is a sphere.

14. *Volume* Show that the rectangular box of maximum volume inscribed in a sphere of radius r is a cube.

15. *Volume and Surface Area* Show that the rectangular box of given volume and minimum surface area is a cube.

16. *Area* A trough with trapezoidal cross sections is formed by turning up the edges of a 10-inch-wide sheet of aluminum. Find the cross section of maximum area.

17. *Area* Repeat Exercise 16 for a sheet of aluminum that is w inches wide.

18. *Revenue* A company manufactures two products. The total revenue from x_1 units of product 1 and x_2 units of product 2 is

$$R = -5x_1^2 - 8x_2^2 - 2x_1x_2 + 42x_1 + 102x_2.$$

Find x_1 and x_2 so as to maximize the revenue.

19. *Revenue* A retail outlet sells two competitive products, the prices of which are p_1 and p_2. Find p_1 and p_2 so as to maximize total revenue, where

$$R = 500p_1 + 800p_2 + 1.5p_1p_2 - 1.5p_1^2 - p_2^2.$$

20. *Profit* A corporation manufactures a product at two locations. The cost of producing x_1 units at location 1 is

$$C_1 = 0.02x_1^2 + 4x_1 + 500$$

and the cost of producing x_2 units at location 2 is

$$C_2 = 0.05x_2^2 + 4x_2 + 275.$$

If the product sells for \$15 per unit, find the quantity that should be produced at each location to maximize the profit, $P = 15(x_1 + x_2) - C_1 - C_2$.

21. *Distance* A company has retail outlets located at the points $(0, 0)$, $(2, 2)$, and $(-2, 2)$ (see figure). Management plans to build a distribution center so that the sum of the distances S from the center to the outlets is minimum. From the symmetry of the problem it is clear that the distribution center will be located on the y-axis, and therefore, S is a function of the single variable y. Using techniques of Chapter 3, find the required value of y.

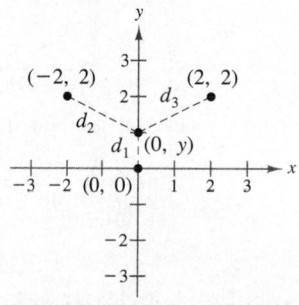

FIGURE FOR 21

C **22.** The retail outlets described in Exercise 21 are located at $(0, 0)$, $(4, 2)$, and $(-2, 2)$ (see figure). The location of the distribution center is (x, y), and therefore the sum of the distances S is a function of x and y.

  a. Write the expression giving the sum of the distances S. Use a computer algebra system to obtain a graph of the surface. Does the surface have a minimum?

  b. Use a computer algebra system to obtain S_x and S_y. Observe that solving the system $S_x = 0$ and $S_y = 0$ is very difficult. Therefore, we will approximate the location of the distribution center.

  c. An initial estimate of the critical point is $(x_1, y_1) = (1, 1)$. Calculate $-\nabla S(1, 1)$ with components $-S_x(1, 1)$ and $-S_y(1, 1)$. What direction is given by the vector $-\nabla S(1, 1)$?

  d. The second estimate of the critical point is

  $$(x_2, y_2) = (x_1 - S_x(x_1, y_1)t, \ y_1 - S_y(x_1, y_1)t).$$

  If these coordinates are substituted into $S(x, y)$, then S becomes a function of the single variable t. Find the value of t that minimizes S. Use this value of t to find the estimate (x_2, y_2).

  e. Complete two more iterations of the process in part d to obtain (x_4, y_4).

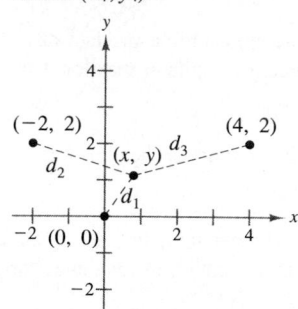

FIGURE FOR 22

In Exercises 23–26, (a) find the least squares regression line and (b) calculate S, the sum of the squared errors.

23.

24.

25.

26.

In Exercises 27–30, find the least squares regression line for the given points.

27. $(0, 0)$, $(1, 1)$, $(3, 4)$, $(4, 2)$, $(5, 5)$

28. $(1, 0)$, $(3, 3)$, $(5, 6)$

29. $(0, 6)$, $(4, 3)$, $(5, 0)$, $(8, -4)$, $(10, -5)$

30. $(5, 2)$, $(0, 0)$, $(2, 1)$, $(7, 4)$, $(10, 6)$, $(12, 6)$

31. *Demand Equation*   A store manager wants to know the demand for a certain product as a function of price. The daily sales for three different prices of the product are given in the following table.

Price (x)	\$1.00	\$1.25	\$1.50
Demand (y)	450	375	330

  a. Find the least squares regression line for these data.
  b. Estimate the demand when the price is \$1.40.

32. *Carcinogens*   After contamination with a carcinogen, people in different geographic regions were assigned an exposure index, which represented the degree of contamination. Using the following data, find a least squares regression line to estimate the mortality per 100,000 people for a given exposure.

Exposure (x)	2.67	1.35	3.93	5.14	7.43
Mortality (y)	135.2	118.5	167.3	197.6	204.7

33. *Agriculture*   An agronomist used four test plots to determine the relationship between the wheat yield (in bushels per acre) and the amount of fertilizer (in hundreds of pounds per acre). The results are given in the following table.

Fertilizer (x)	1.0	1.5	2.0	2.5
Yield (y)	32	41	48	53

Find the least squares regression line for these data, and estimate the yield for a fertilizer application of 160 pounds per acre.

34. *Women in the Work Force*   The following table gives the percentage and number (in millions) of women in the work force for selected years. (*Source:* U.S. Department of Labor)

Year	1940	1950	1960	1970
Percentage (x)	24.3	29.0	32.5	37.2
Number (y)	12.8	18.4	23.3	31.6

Year	1980	1988	1989	1990
Percentage (x)	42.0	44.5	44.8	45.3
Number (y)	45.6	54.9	56.2	56.6

  a. Find the least squares regression line for these data.
  b. According to this model, approximately how many women enter the labor force for each one point increase in the percentage of women in the labor force?

35. Find a system of equations whose solution yields the coefficients a, b, and c for the least squares regression quadratic $y = ax^2 + bx + c$ for the points

$(x_1, y_1), (x_2, y_2), \ldots, (x_n, y_n)$

by minimizing the sum

$$S(a, b, c) = \sum_{i=1}^{n} (y_i - ax_i^2 - bx_i - c)^2.$$

36. Use the Second Partials Test to verify that the formulas for a and b given in Theorem 13.18 yield a minimum.

$$\left(\textit{Hint: } \text{Use the fact that } n \sum_{i=1}^{n} x_i^2 \geq \left(\sum_{i=1}^{n} x_i \right)^2. \right)$$

In Exercises 37–40, use the result of Exercise 35 to find the least squares regression quadratic for the given points. Then plot the points and sketch the graph of the least squares quadratic.

37. $(-2, 0), (-1, 0), (0, 1), (1, 2), (2, 5)$

38. $(-4, 5), (-2, 6), (2, 6), (4, 2)$

39. $(0, 0), (2, 2), (3, 6), (4, 12)$

40. $(0, 10), (1, 9), (2, 6), (3, 0)$

41. *Engine Performance* After developing a new turbocharger for an automobile engine, the following experimental data were obtained for speed in miles per hour at 2-second intervals. Fit a least squares quadratic to the data.

Time (x)	0	2	4	6	8	10
Speed (y)	0	15	30	50	65	70

42. The following table gives the world population in billions for five different years. (*Source:* World Population Data Sheet)

Year	1970	1975	1980	1985	1990
Population (y)	3.7	4.1	4.5	4.8	5.3

Let $x = 0$ represent the year 1980.

a. Find the least squares regression quadratic for these data.

b. Use this quadratic to estimate the world population for the year 2010.

43. How would the procedure of Exercise 22 change if you are searching for a maximum value of a function $S = f(x, y)$?

13.10

Lagrange Multipliers • Constrained Optimization Problems • The Method of Lagrange Multipliers with Two Constraints

Lagrange Multipliers

The Method of Lagrange Multipliers is named after the French mathematician Joseph Louis Lagrange (1736–1813). Lagrange first introduced the method in his famous paper on mechanics, written when he was just 19 years old!

Many optimization problems have restrictions or **constraints** on the values that can be used to produce the optimal solution. Such constraints tend to complicate optimization problems because the optimal solution can occur at a boundary point of the domain. In this section, you will study an ingenious technique for solving such problems. It is called the **Method of Lagrange Multipliers.**

To see how this technique works, suppose you want to find the rectangle of maximum area that can be inscribed in the ellipse given by

$$\frac{x^2}{3^2} + \frac{y^2}{4^2} = 1.$$

Let (x, y) be the vertex of the rectangle in the first quadrant, as shown in Figure 13.79. Because the rectangle has sides of length $2x$ and $2y$, its area is given by

$$f(x, y) = 4xy. \qquad \text{Objective function}$$

You want to find x and y such that $f(x, y)$ is a maximum. Your choice of (x, y) is restricted to first-quadrant points that lie on the ellipse.

$$\frac{x^2}{3^2} + \frac{y^2}{4^2} = 1 \qquad \text{Constraint}$$

Now, consider the constraint equation to be a fixed level curve of

$$g(x, y) = \frac{x^2}{3^2} + \frac{y^2}{4^2}.$$

The level curves of f represent a family of hyperbolas

$$f(x, y) = 4xy = k.$$

In this family, the level curves that meet the given constraint correspond to the hyperbolas that intersect the ellipse. Moreover, to maximize $f(x, y)$, you want to find the hyperbola that just barely satisfies the constraint. The level curve that does this is the one that is *tangent* to the ellipse, as shown in Figure 13.80.

To find the appropriate hyperbola, use the fact that two curves are tangent at a point if and only if their gradient vectors are parallel. This means that $\nabla f(x, y)$ must be a scalar multiple of $\nabla g(x, y)$ at the point of tangency. In the context of constrained optimization problems, this scalar is denoted by λ.

$$\nabla f(x, y) = \lambda \nabla g(x, y)$$

The scalar λ is called a **Lagrange multiplier.** Theorem 13.19 gives the necessary conditions for the existence of such multipliers.

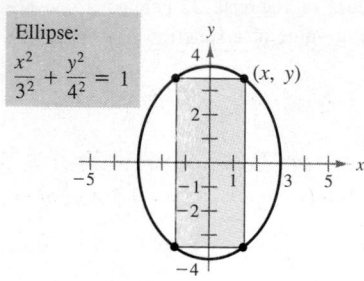

Ellipse:
$$\frac{x^2}{3^2} + \frac{y^2}{4^2} = 1$$

FIGURE 13.79
Objective function: $f(x, y) = 4xy$

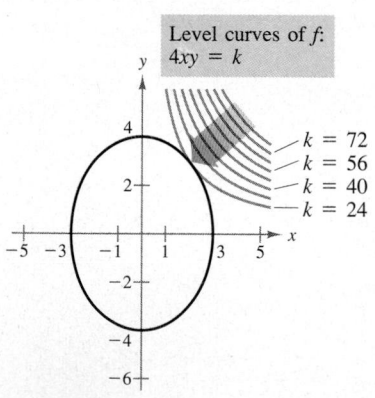

Level curves of f:
$4xy = k$

$k = 72$
$k = 56$
$k = 40$
$k = 24$

FIGURE 13.80
Constraint: $g(x, y) = 1$

THEOREM 13.19   Lagrange's Theorem

Let f and g have continuous first partial derivatives such that f has an extremum at a point (x_0, y_0) on the smooth constraint curve $g(x, y) = c$. If $\nabla g(x_0, y_0) \neq \mathbf{0}$, then there is a real number λ such that

$$\nabla f(x_0, y_0) = \lambda \nabla g(x_0, y_0).$$

Proof To begin, represent the smooth curve given by $g(x, y) = c$ by the vector-valued function

$$\mathbf{r}(t) = x(t)\mathbf{i} + y(t)\mathbf{j}, \qquad \mathbf{r}'(t) \neq 0$$

where x' and y' are continuous on an open interval I. Define the function h by $h(t) = f(x(t), y(t))$. Then, because $f(x_0, x_0)$ is an extreme value of f, you know that

$$h(t_0) = f(x(t_0), y(t_0)) = f(x_0, y_0)$$

is an extreme value of h. This implies that $h'(t_0) = 0$, and, by the Chain Rule,

$$h'(t_0) = f_x(x_0, y_0)x'(t_0) + f_y(x_0, y_0)y'(t_0) = \nabla f(x_0, y_0) \cdot \mathbf{r}'(t_0) = 0.$$

REMARK Lagrange's Theorem can be shown to be true for functions of three variables, using a similar argument with level surfaces and Theorem 13.14.

Therefore, $\nabla f(x_0, y_0)$ is orthogonal to $\mathbf{r}'(t_0)$. Moreover, by Theorem 13.12, $\nabla g(x_0, y_0)$ is also orthogonal to $\mathbf{r}'(t_0)$. Consequently, the gradients $\nabla f(x_0, y_0)$ and $\nabla g(x_0, y_0)$ are parallel, and there must exist a scalar λ such that

$$\nabla f(x_0, y_0) = \lambda \nabla g(x_0, y_0).$$

The Method of Lagrange Multipliers uses Theorem 13.19 to find the extreme values of a function f subject to a constraint.

Method of Lagrange Multipliers

Let f and g satisfy the hypothesis of Lagrange's Theorem, and let f have a minimum or maximum subject to the constraint $g(x, y) = c$. To find the minimum or maximum of f, use the following steps.

1. Simultaneously solve the equations

   $$\nabla f(x, y) = \lambda \nabla g(x, y) \quad \text{and} \quad g(x, y) = c$$

   by solving the following system of equations.

   $$f_x(x, y) = \lambda g_x(x, y)$$

   $$f_y(x, y) = \lambda g_y(x, y)$$

   $$g(x, y) = c$$

2. Evaluate f at each solution point obtained in the first step. The largest value yields the maximum of f subject to the constraint $g(x, y) = c$, and the smallest value yields the minimum of f subject to the constraint $g(x, y) = c$.

Constrained Optimization Problems

At the beginning of this section, we described a problem in which we wanted to maximize the area of a rectangle that is inscribed in an ellipse. Example 1 shows how to use Lagrange multipliers to solve this problem.

EXAMPLE 1 Using a Lagrange Multiplier with One Constraint

Find the maximum value of

$$f(x, y) = 4xy, \qquad x > 0, \, y > 0$$

subject to the constraint $(x^2/3^2) + (y^2/4^2) = 1$.

Solution   To begin, let

$$g(x, y) = \frac{x^2}{3^2} + \frac{y^2}{4^2} = 1.$$

By equating $\nabla f(x, y) = 4y\mathbf{i} + 4x\mathbf{j}$ and $\lambda \nabla g(x, y) = (2\lambda x/9)\mathbf{i} + (\lambda y/8)\mathbf{j}$, you can obtain the following system of equations.

$$4y = \frac{2}{9}\lambda x \qquad f_x(x, y) = \lambda g_x(x, y)$$

$$4x = \frac{1}{8}\lambda y \qquad f_y(x, y) = \lambda g_y(x, y)$$

$$\frac{x^2}{3^2} + \frac{y^2}{4^2} = 1 \qquad \text{Constraint}$$

REMARK   Writing the constraint as

$$g(x, y) = \frac{x^2}{3^2} + \frac{y^2}{4^2} = 1$$

or as

$$g(x, y) = \frac{x^2}{3^2} + \frac{y^2}{4^2} - 1 = 0$$

does not affect the solution—the constant is eliminated when you form ∇g.

From the first equation you obtain $\lambda = 18y/x$, and substitution into the second equation produces

$$4x = \frac{1}{8}\left(\frac{18y}{x}\right)y \quad \rightarrow \quad x^2 = \frac{9}{16}y^2.$$

Substituting this value for x^2 into the third equation produces

$$\frac{1}{9}\left(\frac{9}{16}y^2\right) + \frac{1}{16}y^2 = 1 \quad \rightarrow \quad y^2 = 8.$$

Thus, $y = \pm 2\sqrt{2}$. Because it is required that $y > 0$, choose the positive value and find that

$$x^2 = \frac{9}{16}y^2 = \frac{9}{16}(8) = \frac{9}{2} \quad \rightarrow \quad x^2 = \frac{9}{2} \quad \rightarrow \quad x = \frac{3}{\sqrt{2}}.$$

Thus, the maximum of f is

$$f\left(\frac{3}{\sqrt{2}}, 2\sqrt{2}\right) = 4\left(\frac{3}{\sqrt{2}}\right)(2\sqrt{2}) = 24.$$

Example 1 can also be solved using techniques you learned in Chapter 3. To see how, try to find the maximum value of $A = 4xy$ given that $(x^2/3^2) + (y^2/4^2) = 1$. To begin, solve the second equation for y to obtain $y = \frac{4}{3}\sqrt{9 - x^2}$. Then substitute into the first equation to obtain

$$A = 4x\left(\frac{4}{3}\sqrt{9 - x^2}\right).$$

Finally, use the techniques of Chapter 3 to maximize A.

EXAMPLE 2   A Business Application

The Cobb-Douglas production function (see Example 5, Section 13.1) for a particular manufacturer is given by

$$f(x, y) = 100x^{3/4}y^{1/4} \qquad \text{Objective function}$$

where x represents the units of labor (at \$150 per unit) and y represents the units of capital (at \$250 per unit). The total cost of labor and capital is limited to \$50,000. Find the maximum production level for this manufacturer.

Solution   From the given function, you have

$$\nabla f(x, y) = 75x^{-1/4}y^{1/4}\mathbf{i} + 25x^{3/4}y^{-3/4}\mathbf{j}.$$

FOR FURTHER INFORMATION
For more information on the use of Lagrange multipliers in economics, see the article "Lagrange Multiplier Problems in Economics" by John V. Baxley and John C. Moorhouse in the August-September, 1984 issue of the *American Mathematical Monthly*.

The limit on the cost of labor and capital produces the constraint

$$g(x, y) = 150x + 250y = 50,000. \qquad \text{Constraint}$$

Thus, $\lambda\nabla g(x, y) = 150\lambda\mathbf{i} + 250\lambda\mathbf{j}$. This gives rise to the following system of equations.

$$75x^{-1/4}y^{1/4} = 150\lambda \qquad f_x(x, y) = \lambda g_x(x, y)$$
$$25x^{3/4}y^{-3/4} = 250\lambda \qquad f_y(x, y) = \lambda g_y(x, y)$$
$$150x + 250y = 50,000 \qquad \text{Constraint}$$

By solving for λ in the first equation

$$\lambda = \frac{75x^{-1/4}y^{1/4}}{150} = \frac{x^{-1/4}y^{1/4}}{2}$$

and substituting into the second equation, you obtain

$$25x^{3/4}y^{-3/4} = 250\left(\frac{x^{-1/4}y^{1/4}}{2}\right) \qquad \text{Multiply by } x^{1/4}y^{3/4}$$
$$25x = 125y.$$

Thus, $x = 5y$. By substituting into the third equation, you have

$$150(5y) + 250y = 50,000$$
$$1000y = 50,000$$
$$y = 50 \text{ units of capital}$$
$$x = 250 \text{ units of labor.}$$

Thus, the maximum production is

$$f(250, 50) = 100(250)^{3/4}(50)^{1/4} \approx 16,719 \text{ product units.}$$

Economists call the Lagrange multiplier obtained in a production function the **marginal productivity of money.** For instance, in Example 2 the marginal productivity of money at $x = 250$ and $y = 50$ is

$$\lambda = \frac{x^{-1/4}y^{1/4}}{2} = \frac{(250)^{-1/4}(50)^{1/4}}{2} \approx 0.334$$

which means that for each additional dollar spent on production, 0.334 additional units of the product can be produced.

EXAMPLE 3   Lagrange Multipliers and Three Variables

Find the minimum value of

$$f(x, y, z) = 2x^2 + y^2 + 3z^2 \qquad \text{Objective function}$$

subject to the constraint $2x - 3y + 4z = 49$.

Solution   Let $g(x, y, z) = 2x - 3y + 4z = 49$. Then, because $\nabla f(x, y, z) = 4x\mathbf{i} + 2y\mathbf{j} + 6z\mathbf{k}$ and $\lambda\nabla g(x, y, z) = 2\lambda\mathbf{i} - 3\lambda\mathbf{j} + 4\lambda\mathbf{k}$, you obtain the following system of equations.

$$4x = 2\lambda \qquad\qquad f_x(x, y, z) = \lambda g_x(x, y, z)$$
$$2y = -3\lambda \qquad\qquad f_y(x, y, z) = \lambda g_y(x, y, z)$$
$$6z = 4\lambda \qquad\qquad f_z(x, y, z) = \lambda g_z(x, y, z)$$
$$2x - 3y + 4z = 49 \qquad\qquad \text{Constraint}$$

The solution of this system is $x = 3$, $y = -9$, and $z = 4$. Therefore, the optimum value of f is

$$f(3, -9, 4) = 2(3)^2 + (-9)^2 + 3(4)^2 = 147.$$

From the original function and constraint, it is clear that $f(x, y, z)$ has no maximum. Thus, the above optimum value of f is a minimum.

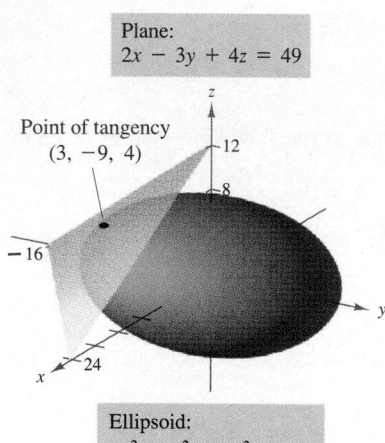

Plane:
$2x - 3y + 4z = 49$

Point of tangency
$(3, -9, 4)$

Ellipsoid:
$2x^2 + y^2 + 3z^2 = 147$

FIGURE 13.81

At the beginning of this section, we gave a graphical interpretation of constrained optimization problems in two variables. In three variables, the interpretation is similar, except that we use level surfaces instead of level curves. For instance, in Example 3, the level surfaces of f are ellipsoids centered at the origin, and the constraint $2x - 3y + 4z = 49$ is a plane. The minimum value of f is represented by the ellipsoid that is tangent to the constraint plane, as shown in Figure 13.81.

EXAMPLE 4   Optimization Inside a Region

Find the extreme values of

$$f(x, y) = x^2 + 2y^2 - 2x + 3 \qquad \text{Objective function}$$

subject to the constraint $x^2 + y^2 \leq 10$.

Solution   To solve this problem, you can break the constraint into two cases.

a. For points *on the circle,* $x^2 + y^2 = 10$, you can use Lagrange multipliers to find that the maximum value of $f(x, y)$ is 24—this value occurs at $(-1, 3)$ and at $(-1, -3)$. In a similar way, you can determine that the minimum value of $f(x, y)$ is 6.675—this value occurs at $\left(\sqrt{10}, 0\right)$.

b. For points *inside the circle,* you can use the techniques discussed in Section 13.8 to conclude that the function has a relative minimum of 2 at the point $(1, 0)$.

By combining these two results, you can conclude that f has a maximum of 24 at $(-1, \pm 3)$ and a minimum of 2 at $(1, 0)$.

The Method of Lagrange Multipliers with Two Constraints

For optimization problems involving *two* constraint functions g and h, we introduce a second Lagrange multiplier, μ (the lowercase Greek letter mu), and solve the equation

$$\nabla f = \lambda \nabla g + \mu \nabla h$$

as illustrated in Example 5.

EXAMPLE 5   Optimization with Two Constraints

Let $T(x, y, z) = 20 + 2x + 2y + z^2$ represent the temperature at each point on the sphere $x^2 + y^2 + z^2 = 11$. Find the extreme temperatures on the curve formed by the intersection of the plane $x + y + z = 3$ and the sphere.

Solution   The two constraints are $g(x, y, z) = x^2 + y^2 + z^2 = 11$ and $h(x, y, z) = x + y + z = 3$. Using

$$\nabla T(x, y, z) = 2\mathbf{i} + 2\mathbf{j} + 2z\,\mathbf{k}, \quad \lambda \nabla g(x, y, z) = 2\lambda x\,\mathbf{i} + 2\lambda y\,\mathbf{j} + 2\lambda z\,\mathbf{k}$$

and $\mu \nabla h(x, y, z) = \mu\mathbf{i} + \mu\mathbf{j} + \mu\mathbf{k}$, you can write the following system of equations.

$$\begin{aligned}
2 &= 2\lambda x + \mu & & T_x(x, y, z) = \lambda g_x(x, y, z) + \mu h_x(x, y, z) \\
2 &= 2\lambda y + \mu & & T_y(x, y, z) = \lambda g_y(x, y, z) + \mu h_y(x, y, z) \\
2z &= 2\lambda z + \mu & & T_z(x, y, z) = \lambda g_z(x, y, z) + \mu h_z(x, y, z) \\
x^2 + y^2 + z^2 &= 11 & & \text{Constraint 1} \\
x + y + z &= 3 & & \text{Constraint 2}
\end{aligned}$$

By subtracting the second equation from the first, you can obtain the following system.

$$\begin{aligned}
\lambda(x - y) &= 0 \\
2z(1 - \lambda) - \mu &= 0 \\
x^2 + y^2 + z^2 &= 11 \\
x + y + z &= 3
\end{aligned}$$

From the first equation, you can conclude that $\lambda = 0$ or $x = y$. If $\lambda = 0$, you can show that the critical points are $(3, -1, 1)$ and $(-1, 3, 1)$. (Try doing this—it takes a little work!) If $\lambda \neq 0$, then $x = y$ and you can show that the critical points occur when $x = y = (3 \pm 2\sqrt{3})/3$ and $z = (3 \mp 4\sqrt{3})/3$. Finally, to find the optimal solutions, compare the temperatures at the four critical points.

$$T(3, -1, 1) = T(-1, 3, 1) = 25$$

$$T\left(\frac{3 - 2\sqrt{3}}{3}, \frac{3 - 2\sqrt{3}}{3}, \frac{3 + 4\sqrt{3}}{3}\right) = \frac{91}{3} \approx 30.33$$

$$T\left(\frac{3 + 2\sqrt{3}}{3}, \frac{3 + 2\sqrt{3}}{3}, \frac{3 - 4\sqrt{3}}{3}\right) = \frac{91}{3} \approx 30.33$$

Thus, $T = 25$ is the minimum temperature and $T = \frac{91}{3}$ is the maximum temperature on the curve.

REMARK   The system of equations that arises in the method of Lagrange multipliers is not, in general, a linear system, and the solution often requires ingenuity.

In Exercises 1–12, use Lagrange multipliers to find the indicated extrema. In each case, assume that x and y are positive.

1. Maximize $f(x, y) = xy$
Constraint: $x + y = 10$

2. Maximize $f(x, y) = xy$
Constraint: $2x + y = 4$

3. Minimize $f(x, y) = x^2 + y^2$
Constraint: $x + y - 4 = 0$

4. Minimize $f(x, y) = x^2 + y^2$
Constraint: $-2x - 4y + 5 = 0$

5. Minimize $f(x, y) = x^2 - y^2$
Constraint: $x - 2y + 6 = 0$

6. Maximize $f(x, y) = x^2 - y^2$
Constraint: $y - x^2 = 0$

7. Maximize $f(x, y) = 2x + 2xy + y$
Constraint: $2x + y = 100$

8. Minimize $f(x, y) = 3x + y + 10$
Constraint: $x^2 y = 6$

9. Maximize $f(x, y) = \sqrt{6 - x^2 - y^2}$
Constraint: $x + y - 2 = 0$

10. Minimize $f(x, y) = \sqrt{x^2 + y^2}$
Constraint: $2x + 4y - 15 = 0$

11. Maximize $f(x, y) = e^{xy}$
Constraint: $x^2 + y^2 - 8 = 0$

12. Minimize $f(x, y) = 2x + y$
Constraint: $xy = 32$

In Exercises 13–16, use Lagrange multipliers to find the indicated extrema. In each case, assume that x, y, and z are positive.

13. Minimize $f(x, y, z) = x^2 + y^2 + z^2$
Constraint: $x + y + z - 6 = 0$

14. Maximize $f(x, y, z) = xyz$
Constraint: $x + y + z - 6 = 0$

15. Minimize $f(x, y, z) = x^2 + y^2 + z^2$
Constraint: $x + y + z = 1$

16. Minimize $f(x, y, z) = 2x + y + 2z$
Constraint: $x^2 + y^2 + z^2 = 4$

In Exercises 17–20, use Lagrange multipliers to find the indicated extrema of f subject to two constraints. In each case, assume x, y, and z are nonnegative.

17. Maximize $f(x, y, z) = xyz$
Constraints: $x + y + z = 32$
$x - y + z = 0$

18. Minimize $f(x, y, z) = x^2 + y^2 + z^2$
Constraints: $x + 2z = 4$
$x + y = 8$

19. Maximize $f(x, y, z) = xy + yz$
Constraints: $x + 2y = 6$
$x - 3z = 0$

20. Maximize $f(x, y, z) = xyz$
Constraints: $x^2 + z^2 = 5$
$x - 2y = 0$

In Exercises 21 and 22, use Lagrange multipliers to find any extrema of the function subject to the constraint $x^2 + y^2 \leq 1$.

21. $f(x, y) = x^2 + 3xy + y^2$

22. $f(x, y) = e^{-xy}$

In Exercises 23–26, use Lagrange multipliers to find the minimum distance from the curve or surface to the specified point. (*Hint:* In Exercise 23, minimize $f(x, y) = x^2 + y^2$ subject to the constraint $2x + 3y = -1$.)

Curve	*Point*
23. Line: $2x + 3y = -1$	$(0, 0)$
24. Circle: $(x - 4)^2 + y^2 = 4$	$(0, 10)$

Surface	*Point*
25. Plane: $x + y + z = 1$	$(2, 1, 1)$
26. Cone: $z = \sqrt{x^2 + y^2}$	$(4, 0, 0)$

In Exercises 27 and 28, find the highest point on the curve of intersection of the given surfaces.

27. Sphere: $x^2 + y^2 + z^2 = 36$
Plane: $2x + y - z = 2$

28. Cone: $x^2 + y^2 - z^2 = 0$
Plane: $x + 2z = 4$

29. *Volume* Find the dimensions of the rectangular package of largest volume subject to the constraint that the sum of the length and the girth cannot exceed 108 inches. (Maximize $V = xyz$ subject to the constraint $x + 2y + 2z = 108$.)

30. *Volume* The material for the base of an open box costs 1.5 times as much as the material for the sides. Find the dimensions of the box of largest volume that can be made for a fixed cost C. (Maximize $V = xyz$ subject to $1.5xy + 2xz + 2yz = C$.)

31. *Cost* A cargo container (in the shape of a rectangular solid) must have a volume of 480 cubic feet. Use Lagrange multipliers to find the dimensions of the container of this size that has minimum cost if the bottom will cost \$5 per square foot to construct and the sides and top will cost \$3 per square foot to construct.

32. *Surface Area* Use Lagrange multipliers to find the dimensions of the right circular cylinder with volume V_0 cubic units and minimum surface area.

33. *Volume* Use Lagrange multipliers to find the dimensions of the rectangular box of maximum volume that can be inscribed (with edges parallel to the coordinate axes) in the ellipsoid

$$\frac{x^2}{a^2} + \frac{y^2}{b^2} + \frac{z^2}{c^2} = 1.$$

34. *Geometric and Arithmetic Means*

a. Use Lagrange multipliers to prove that the product of three positive numbers, x, y, and z, whose sum has the constant value S, is maximum when the three numbers are equal. Use this result to prove that

$$\sqrt[3]{xyz} \le \frac{x + y + z}{3}.$$

b. Generalize the result of part a to prove that the product $x_1 x_2 x_3 \cdots x_n$ is maximum when $x_1 = x_2 = x_3 = \cdots = x_n$, $\sum_{i=1}^n x_i = S$, and all $x_i \ge 0$. Use this result to prove that

$$\sqrt[n]{x_1 x_2 x_3 \cdots x_n} \le \frac{x_1 + x_2 + x_3 + \cdots + x_n}{n}.$$

This shows that the geometric mean is never greater than the arithmetic mean.

35. *Refraction of Light* When light waves traveling in a transparent medium strike the surface of a second transparent medium, they tend to "bend" in order to follow the path of minimum time. This tendency is called refraction and is described by **Snell's Law of Refraction**,

$$\frac{\sin \theta_1}{v_1} = \frac{\sin \theta_2}{v_2}$$

where θ_1 and θ_2 are the magnitudes of the angles shown in the figure, and v_1 and v_2 are the velocities of light in the two media. Use Lagrange multipliers to derive this law using the constraint $x + y = a$.

36. *Area and Perimeter* A semicircle is situated on top of a rectangle (see figure). If the area is fixed and the perimeter is minimum, or if the perimeter is fixed and the area is maximum, use Lagrange multipliers to verify that the length of the rectangle is twice its height.

C **37.** Consider the function $g(\alpha, \beta, \gamma) = \cos \alpha \cos \beta \cos \gamma$ subject to the constraint that α, β and γ are the angles of a triangle.

a. Use Lagrange multipliers to maximize g.

b. Use the constraint to reduce the function g to a function of two independent variables. Use a computer algebra system to obtain a graph of the surface represented by g. Identify the maximum values on the graphs.

38. *Temperature Distribution* Let

$$T(x, y, z) = 100 + x^2 + y^2$$

represent the temperature at each point on the sphere $x^2 + y^2 + z^2 = 50$. Find the maximum temperature on the curve formed by the intersection of the sphere and the plane $x - z = 0$.

Production Level **In Exercises 39 and 40, find the maximum production level P if the total cost of labor (at \$48 per unit) and capital (at \$36 per unit) is limited to \$100,000.**

39. $P(x, y) = 100x^{0.25}y^{0.75}$

40. $P(x, y) = 100x^{0.6}y^{0.4}$

Cost **In Exercises 41 and 42, find the minimum cost of producing 20,000 units of a product, where x is the number of units of labor (at \$48 per unit) and y is the number of units of capital (at \$36 per unit).**

41. $P(x, y) = 100x^{0.25}y^{0.75}$

42. $P(x, y) = 100x^{0.6}y^{0.4}$

43. a. Maximize

$$\sum_{i=1}^n x_i y_i$$

subject to

$$\sum_{i=1}^n x_i^2 = 1 \quad \text{and} \quad \sum_{i=1}^n y_i^2 = 1.$$

b. Prove that

$$\sum_{i=1}^n a_i b_i \le \left(\sum_{i=1}^n a_i^2 \right)^{1/2} \left(\sum_{i=1}^n b_i^2 \right)^{1/2}$$

for any numbers $a_1, \ldots, a_n$, and $b_1, \ldots, b_n$.

FIGURE FOR 35

FIGURE FOR 36

REVIEW EXERCISES for Chapter 13

In Exercises 1–4, sketch several level curves for the given function.

1. $f(x, y) = e^{x^2 + y^2}$

2. $f(x, y) = \ln xy$

3. $f(x, y) = x^2 - y^2$

4. $f(x, y) = \dfrac{x}{x + y}$

C **In Exercises 5 and 6, use a graphing utility to graph the function.**

5. $f(x, y) = e^{-(x^2 + y^2)}$

6. $g(x, y) = |y|^{1 + |x|}$

In Exercises 7–10, discuss the continuity of the function and evaluate the limit, if it exists.

7. $\displaystyle \lim_{(x, y) \to (1,1)} \frac{xy}{x^2 + y^2}$

8. $\displaystyle \lim_{(x, y) \to (1,1)} \frac{xy}{x^2 - y^2}$

9. $\displaystyle \lim_{(x, y) \to (0,0)} \frac{-4x^2 y}{x^4 + y^2}$

10. $\displaystyle \lim_{(x, y) \to (0,0)} \frac{y + xe^{-y^2}}{1 + x^2}$

In Exercises 11–20, find all first partial derivatives.

11. $f(x, y) = e^x \cos y$

12. $f(x, y) = \dfrac{xy}{x + y}$

13. $z = xe^y + ye^x$

14. $z = \ln(x^2 + y^2 + 1)$

15. $g(x, y) = \dfrac{xy}{x^2 + y^2}$

16. $w = \sqrt{x^2 + y^2 + z^2}$

17. $f(x, y, z) = z \arctan \dfrac{y}{x}$

18. $f(x, y, z) = \dfrac{1}{\sqrt{1 - x^2 - y^2 - z^2}}$

19. $u(x, t) = ce^{-n^2 t} \sin nx$ **20.** $u(x, t) = c \sin(akx) \cos kt$

In Exercises 21 and 22, find $\partial z / \partial x$ and $\partial z / \partial y$.

21. $x^2 y - 2yz - xz - z^2 = 0$

22. $xz^2 - y \sin z = 0$

In Exercises 23–26, find all second partial derivatives and verify that the second mixed partials are equal.

23. $f(x, y) = 3x^2 - xy + 2y^3$

24. $h(x, y) = \dfrac{x}{x + y}$

25. $h(x, y) = x \sin y + y \cos x$

26. $g(x, y) = \cos(x - 2y)$

Laplace Equation **In Exercises 27–30, show that the function satisfies the Laplace equation**

$$\frac{\partial^2 z}{\partial x^2} + \frac{\partial^2 z}{\partial y^2} = 0.$$

27. $z = x^2 - y^2$ **28.** $z = x^3 - 3xy^2$

29. $z = \dfrac{y}{x^2 + y^2}$ **30.** $z = e^x \sin y$

In Exercises 31 and 32, find dz.

31. $z = x \sin \dfrac{y}{x}$ **32.** $z = \dfrac{xy}{\sqrt{x^2 + y^2}}$

33. *Error Analysis* The legs of a right triangle are measured at 5 inches and 12 inches, with a possible error of $\frac{1}{16}$ inch. Approximate the maximum possible error in computing the length of the hypotenuse. Approximate the maximum percentage error.

34. *Error Analysis* To determine the height of a tower, the angle of elevation to the top of the tower was measured from a point $100 \pm \frac{1}{2}$ foot from the base. The angle is measured at $33°$, with a possible error of $1°$. Assuming the ground to be horizontal, approximate the maximum error in determining the height of the tower.

35. *Error Analysis* The volume of a right circular cone is $V = \frac{1}{3}\pi r^2 h$. Find the approximate error in the volume due to possible errors of $\frac{1}{8}$ inch in the measured values of r and h, if these are found to be 2 and 5 inches, respectively.

36. *Error Analysis* Approximate the error in the lateral surface area of the cone of Exercise 35. (The lateral surface area is given by $A = \pi r \sqrt{r^2 + h^2}$.)

In Exercises 37 and 38, find the required derivatives (a) by the Chain Rule and (b) by substitution before differentiation.

37. $u = x^2 + y^2 + z^2$, $\dfrac{\partial u}{\partial r}, \dfrac{\partial u}{\partial t}$ **38.** $u = y^2 - x$, $\dfrac{du}{dt}$

$x = r \cos t, \; y = r \sin t, \; z = t$ $x = \cos t, \; y = \sin t$

In Exercises 39–42, find the directional derivative in the direction of v at the specified point.

Function	Direction	Point
39. $f(x, y) = x^2 y$	$\mathbf{v} = \mathbf{i} - \mathbf{j}$	$(2, 1)$
40. $f(x, y) = \frac{1}{4}y^2 - x^2$	$\mathbf{v} = 2\mathbf{i} + \mathbf{j}$	$(1, 4)$
41. $w = y^2 + xz$	$\mathbf{v} = 2\mathbf{i} - \mathbf{j} + 2\mathbf{k}$	$(1, 2, 2)$
42. $w = 6x^2 + 3xy - 4y^2 z$	$\mathbf{v} = \mathbf{i} + \mathbf{j} - \mathbf{k}$	$(1, 0, 1)$

In Exercises 43–46, find the gradient and the maximum value of the directional derivative of the function at the specified point.

Function	Point
43. $z = \dfrac{y}{x^2 + y^2}$	$(1, 1)$
44. $z = \dfrac{x^2}{x - y}$	$(2, 1)$
45. $z = e^{-x} \cos y$	$\left(0, \dfrac{\pi}{4}\right)$
46. $z = x^2 y$	$(2, 1)$

In Exercises 47–50, find an equation of the tangent plane and equations for the normal line to the given surface at the specified point.

Surface	Point
47. $f(x, y) = x^2 y$	(2, 1, 4)
48. $f(x, y) = \sqrt{25 - y^2}$	(2, 3, 4)
49. $z = -9 + 4x - 6y - x^2 - y^2$	(2, -3, 4)
50. $z = \sqrt{9 - x^2 - y^2}$	(1, 2, 2)

In Exercises 51 and 52, find symmetric equations of the tangent line to the curve of intersection of the given surfaces at the indicated point.

Surfaces	Point
51. $z = x^2 - y^2, z = 3$	(2, 1, 3)
52. $z = 25 - y^2, y = x$	(4, 4, 9)

In Exercises 53–56, locate and classify any extrema of the function.

53. $f(x, y) = x^3 - 3xy + y^2$

54. $f(x, y) = 2x^2 + 6xy + 9y^2 + 8x + 14$

55. $f(x, y) = xy + \dfrac{1}{x} + \dfrac{1}{y}$

56. $z = 50(x + y) - (0.1x^3 + 20x + 150) - (0.05y^3 + 20.6y + 125)$

Essay   In Exercises 57 and 58, write a short paragraph about the surface whose level curves are given. Comment on possible extrema, saddle points, the magnitude of the gradient, etc.

57.   **58.**

In Exercises 59 and 60, locate and classify any extrema of the function by using Lagrange multipliers.

59. $z = x^2 y$
Constraint: $x + 2y = 2$

60. $w = xy + yz + xz$
Constraint: $x + y + z = 1$

61. *Profit*   A corporation manufactures a product at two locations. The cost functions for producing x_1 units at location 1 and x_2 units at location 2 are given by

$$C_1 = 0.05x_1^2 + 15x_1 + 5400$$
$$C_2 = 0.03x_2^2 + 15x_2 + 6100$$

and the total revenue function is

$$R = [225 - 0.4(x_1 + x_2)](x_1 + x_2).$$

Find the production levels at the two locations that will maximize the profit $P(x_1, x_2) = R - C_1 - C_2$.

62. *Cost*   A manufacturer has an order for 1000 units that can be produced at two locations. Let x_1 and x_2 be the numbers of units produced at the two locations. Find the number that should be produced at each location to meet the order and minimize cost, if the cost function is

$$C = 0.25x_1^2 + 10x_1 + 0.15x_2^2 + 12x_2.$$

63. *Production Level*   The production function for a manufacturer is

$$f(x, y) = 4x + xy + 2y.$$

Assume that the total amount available for labor and capital is \$2000, and that units of labor and capital cost \$20 and \$4, respectively. Find the maximum production level for this manufacturer.

64. *Motor Vehicle Sales*   The following table gives passenger car retail sales (y in millions) and truck and bus retail sales (x in millions) in the United States for selected years. (*Source:* U.S. Department of Commerce)

Year	1975	1980	1983	1984
Trucks/Buses (x)	2.4	2.2	2.7	3.5
Cars (y)	8.6	9.0	9.2	10.4

Year	1985	1986	1987	1988
Trucks/Buses (x)	4.0	4.0	4.2	4.6
Cars (y)	11.0	11.5	10.3	10.6

a. Find the least squares regression line for these data.
b. What information does the slope of the regression line give about the relationship between the sales of passenger cars and trucks?

65. *Aerodynamics*   The following table gives the drag force y in kilograms for a certain motor vehicle at specified speeds x in kilometers per hour.

Speed (x)	25	50	75	100	125
Drag (y)	28	38	54	75	102

a. Find the least squares regression quadratic for these data.
b. Use the quadratic to estimate the total drag when the vehicle is moving at 80 kilometers per hour.

True or False   In Exercises 66–68, determine whether the statement is true or false. If it is false, explain why or give an example that shows it is false.

66. Of all parallelepipeds having a fixed surface area, the cube has the largest volume.

67. The gradient $\nabla f(x_0, y_0)$ is normal to the surface given by $z = f(x, y)$ at the point (x_0, y_0, z_0).

68. The plane $x_0 x + y_0 y + z_0 z = c^2$ is tangent to the sphere $x^2 + y^2 + z^2 = c^2$ at the point (x_0, y_0, z_0).

Chapter 14

Multiple Integration

Iterated Integrals

In Chapters 14 and 15, you will survey several applications of integration involving functions of several variables. Chapter 14 is much like Chapter 6 in that it surveys the use of integration to find plane areas, volumes, surface areas, moments, and centers of mass.

In Chapter 13, you saw that it is meaningful to differentiate functions of several variables with respect to one variable while holding the other variables constant. You can *integrate* functions of several variables by a similar procedure. For example, if you are given the partial derivative $f_x(x, y) = 2xy$, then, by considering y constant, you can integrate with respect to x to obtain

$$f(x, y) = \int f_x(x, y)\, dx \qquad \text{Integrate with respect to } x$$

$$= \int 2xy\, dx \qquad y \text{ is held constant}$$

$$= y \int 2x\, dx \qquad \text{Factor out constant } y$$

$$= y(x^2) + C(y) \qquad \text{Antiderivative of } 2x \text{ is } x^2$$

$$= x^2 y + C(y). \qquad C(y) \text{ is function of } y$$

Note that the "constant" of integration, $C(y)$, is a function of y. In other words, by integrating with respect to x, you are able to recover $f(x, y)$ only partially. The total recovery of a function of x and y from its partial derivatives is a topic you will study in Chapter 15. For now, we are more concerned with extending definite integrals to functions of several variables. For instance, by considering y constant, you can apply the Fundamental Theorem of Calculus to evaluate

$$\int_1^{2y} 2xy\, dx = x^2 y \Big]_1^{2y} = (2y)^2 y - (1)^2 y = 4y^3 - y.$$

x is the variable of integration and y is fixed.

Replace x by the limits of integration.

The result is a function of y.

Similarly, you can integrate with respect to y by holding x fixed. Both procedures are summarized as follows.

$$\int_{h_1(y)}^{h_2(y)} f_x(x, y)\, dx = f(x, y) \Big]_{h_1(y)}^{h_2(y)} = f(h_2(y), y) - f(h_1(y), y)$$

$$\int_{g_1(x)}^{g_2(x)} f_y(x, y)\, dy = f(x, y) \Big]_{g_1(x)}^{g_2(x)} = f(x, g_2(x)) - f(x, g_1(x))$$

Note that the variable of integration cannot appear in either limit of integration. For instance, it makes no sense to write $\int_0^x y\, dx$.

◄ *This photo shows the 1992 winner of the America's Cup. In Section 14.2, Exercise 62, you are asked to find the center of pressure of a sailboat's sail.*

EXAMPLE 1 Integrating with Respect to y

Evaluate $\displaystyle\int_1^x (2x^2y^{-2} + 2y)\,dy$.

Solution Considering x to be constant and integrating with respect to y produces

$$\int_1^x (2x^2y^{-2} + 2y)\,dy = \left[\frac{-2x^2}{y} + y^2\right]_1^x$$

$$= \left(\frac{-2x^2}{x} + x^2\right) - \left(\frac{-2x^2}{1} + 1\right)$$

$$= 3x^2 - 2x - 1.$$

Notice in Example 1 that the integral defines a function of x and can *itself* be integrated, as shown in the next example.

EXAMPLE 2 The Integral of an Integral

Evaluate $\displaystyle\int_1^2\left[\int_1^x (2x^2y^{-2} + 2y)\,dy\right]dx$.

Solution Using the result of Example 1, you have

$$\int_1^2\left[\int_1^x (2x^2y^{-2} + 2y)\,dy\right]dx = \int_1^2 (3x^2 - 2x - 1)\,dx$$

$$= \left[x^3 - x^2 - x\right]_1^2$$

$$= 2 - (-1)$$

$$= 3.$$

The integral in Example 2 is an **iterated integral.** The brackets used in Example 2 are normally not written. Instead, iterated integrals are usually written simply as

$$\int_a^b\int_{g_1(x)}^{g_2(x)} f(x, y)\,dy\,dx \quad \text{and} \quad \int_c^d\int_{h_1(y)}^{h_2(y)} f(x, y)\,dx\,dy.$$

The **inside limits of integration** can be variable with respect to the outer variable of integration. However, the **outside limits of integration** must be constant with respect to both variables of integration. After performing the inside integration, you obtain a definite integral, and the second integration produces a real number. The limits of integration for an iterated integral identify two sets of boundary intervals for the variables. For instance, in Example 2, the outside limits indicate that x lies in the interval $1 \le x \le 2$ and the inside limits indicate that y lies in the interval $1 \le y \le x$. Together, these two intervals determine the **region of integration** R of the iterated integral, as shown in Figure 14.1.

Because an iterated integral is just a special type of definite integral—one in which the integrand is also an integral—you can use the properties of definite integrals to evaluate iterated integrals.

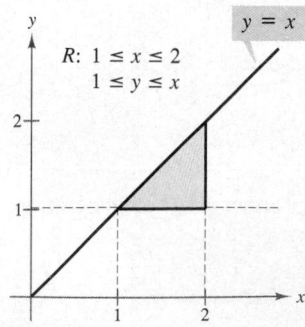

FIGURE 14.1

The region of integration for $\displaystyle\int_1^2\int_1^x f(x, y)\,dy\,dx$.

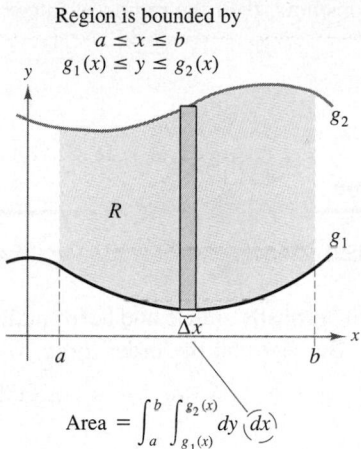

Region is bounded by
$$a \leq x \leq b$$
$$g_1(x) \leq y \leq g_2(x)$$

$$\text{Area} = \int_a^b \int_{g_1(x)}^{g_2(x)} dy\,dx$$

FIGURE 14.2
Vertically simple region

Region is bounded by
$$c \leq y \leq d$$
$$h_1(y) \leq x \leq h_2(y)$$

$$\text{Area} = \int_c^d \int_{h_1(y)}^{h_2(y)} dx\,dy$$

FIGURE 14.3
Horizontally simple region

Area of a Plane Region

In the remainder of this section, you will take a new look at an old problem—that of finding the area of a plane region. Consider the plane region R bounded by $a \leq x \leq b$ and $g_1(x) \leq y \leq g_2(x)$, as shown in Figure 14.2. The area of R is given by the definite integral

$$\int_a^b [g_2(x) - g_1(x)]\,dx. \qquad \text{Area of } R$$

Using the Fundamental Theorem of Calculus, you can rewrite the integrand $g_2(x) - g_1(x)$ as a definite integral. Specifically, if you consider x to be fixed and let y vary from $g_1(x)$ to $g_2(x)$, you can write

$$\int_{g_1(x)}^{g_2(x)} dy = y \Big]_{g_1(x)}^{g_2(x)} = g_2(x) - g_1(x).$$

Combining these two integrals, you can write the area of the region R as an iterated integral

$$\int_a^b \int_{g_1(x)}^{g_2(x)} dy\,dx = \int_a^b y \Big]_{g_1(x)}^{g_2(x)} dx \qquad \text{Area of } R$$

$$= \int_a^b [g_2(x) - g_1(x)]\,dx.$$

Placing a representative rectangle in the region R helps determine both the order and the limits of integration. A vertical rectangle implies the order $dy\,dx$, with the inside limits corresponding to the upper and lower bounds of the rectangle, as shown in Figure 14.2. This type of region is called **vertically simple,** because the outside limits of integration represent the vertical lines $x = a$ and $x = b$.

Similarly, a horizontal rectangle implies the order $dx\,dy$, with the inside limits determined by the left and right bounds of the rectangle, as shown in Figure 14.3. This type of region is called **horizontally simple,** because the outside limits represent the horizontal lines $y = c$ and $y = d$. The iterated integrals used for these two types of simple regions are summarized as follows.

Area of a Region in the Plane

1. If R is defined by $a \leq x \leq b$ and $g_1(x) \leq y \leq g_2(x)$, where g_1 and g_2 are continuous on $[a, b]$, then the area of R is given by

$$A = \int_a^b \int_{g_1(x)}^{g_2(x)} dy\,dx. \qquad \text{Figure 14.2 (vertically simple)}$$

2. If R is defined by $c \leq y \leq d$ and $h_1(y) \leq x \leq h_2(y)$, where h_1 and h_2 are continuous on $[c, d]$, then the area of R is given by

$$A = \int_c^d \int_{h_1(y)}^{h_2(y)} dx\,dy. \qquad \text{Figure 14.3 (horizontally simple)}$$

REMARK Be sure you see that the order of integration of these two integrals is different—the order $dy\,dx$ corresponds to a vertically simple region, and the order $dx\,dy$ corresponds to a horizontally simple region.

If all four limits of integration happen to be constants, then the region of integration is rectangular, as shown in Example 3.

FIGURE 14.4

$$\text{Area} = \int_a^b \int_c^d dy\,dx =$$
$$(d - c)(b - a).$$

EXAMPLE 3   The Area of a Rectangular Region

Use an iterated integral to represent the area of the rectangle shown in Figure 14.4.

Solution   The region shown in Figure 14.4 is both vertically simple and horizontally simple, so you can use either order of integration. By choosing the order $dy\,dx$, you obtain the following.

$$\int_a^b \int_c^d dy\,dx = \int_a^b y \Big]_c^d dx$$

$$= \int_a^b (d - c)\,dx$$

$$= (d - c)x \Big]_a^b$$

$$= (d - c)(b - a)$$

EXAMPLE 4   Finding Area by an Iterated Integral

Use an iterated integral to find the area of the region bounded by the graphs of $f(x) = \sin x$ and $g(x) = \cos x$ between $x = \pi/4$ and $x = 5\pi/4$.

Solution   Because f and g are given as functions of x, a vertical representative rectangle is convenient, and you can choose $dy\,dx$ as the order of integration, as shown in Figure 14.5. The outside limits of integration are $\pi/4 \le x \le 5\pi/4$. Moreover, because the rectangle is bounded above by $f(x) = \sin x$ and below by $g(x) = \cos x$, you have

$$\text{Area of } R = \int_{\pi/4}^{5\pi/4} \int_{\cos x}^{\sin x} dy\,dx$$

$$= \int_{\pi/4}^{5\pi/4} y \Big]_{\cos x}^{\sin x} dx$$

$$= \int_{\pi/4}^{5\pi/4} (\sin x - \cos x)\,dx$$

$$= \Big[-\cos x - \sin x \Big]_{\pi/4}^{5\pi/4}$$

$$= 2\sqrt{2}.$$

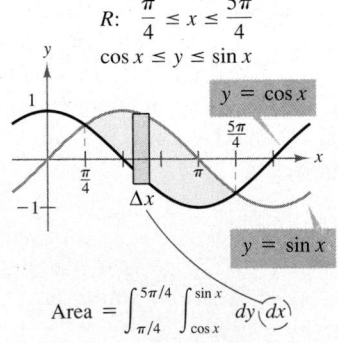

$$R: \quad \frac{\pi}{4} \le x \le \frac{5\pi}{4}$$
$$\cos x \le y \le \sin x$$

$$\text{Area} = \int_{\pi/4}^{5\pi/4} \int_{\cos x}^{\sin x} dy\,dx$$

FIGURE 14.5

REMARK   The region of integration of an iterated integral need not have any straight lines as boundaries. For instance, the region of integration shown in Figure 14.5 is *vertically simple* even though it has no vertical lines as left and right boundaries. The quality that makes the region vertically simple is that it is bounded above and below by the graphs of *functions of x*.

One order of integration will often produce a simpler integration problem than the other order. For instance, try reworking Example 4 with the order $dx\,dy$—you may be surprised to see that the task is formidable! However, if you succeed, you will see that the answer is the same. In other words, the order of integration affects the ease of integration, but not the value of the integral.

EXAMPLE 5   Comparing Different Orders of Integration

Sketch the region whose area is represented by the integral

$$\int_0^2 \int_{y^2}^4 dx\,dy.$$

Then find another iterated integral using the order $dy\,dx$ to represent the same area, and show that both integrals yield the same value.

Solution   From the given limits of integration, you know that

$$y^2 \le x \le 4 \qquad \text{Inner limits of integration}$$

which means that the region R is bounded on the left by the parabola $x = y^2$ and on the right by the line $x = 4$. Furthermore, because

$$0 \le y \le 2 \qquad \text{Outer limits of integration}$$

you know that R is bounded below by the x-axis, as shown in Figure 14.6(a). The value of this integral is

$$\int_0^2 \int_{y^2}^4 dx\,dy = \int_0^2 x\Big]_{y^2}^4 dy$$

$$= \int_0^2 (4 - y^2)\,dy$$

$$= \left[4y - \frac{y^3}{3}\right]_0^2 = \frac{16}{3}.$$

To change the order of integration to $dy\,dx$, place a vertical rectangle in the region, as shown in Figure 14.6(b). From this you can see that the constant bounds $0 \le x \le 4$ serve as the outer limits of integration. By solving for y in the equation $x = y^2$, you can conclude that the inner bounds are $0 \le y \le \sqrt{x}$. Therefore, the area of the region can also be represented by

$$\int_0^4 \int_0^{\sqrt{x}} dy\,dx.$$

By evaluating this integral, you can see that it has the same value as the original integral.

$$\int_0^4 \int_0^{\sqrt{x}} dy\,dx = \int_0^4 y\Big]_0^{\sqrt{x}} dx$$

$$= \int_0^4 \sqrt{x}\,dx$$

$$= \frac{2}{3}x^{3/2}\Big]_0^4 = \frac{16}{3}$$

(a)

(b)

FIGURE 14.6

TECHNOLOGY Some computer software can perform symbolic integration for integrals such as that in Example 6. If you have access to such software, try using it to evaluate the integrals in the exercises and examples given in this section.

Sometimes it is not possible to calculate the area of a region with a single iterated integral. In such cases you can divide the region into subregions such that the area of each subregion can be calculated by an iterated integral. The total area is then the sum of the iterated integrals.

EXAMPLE 6 An Area Represented by Two Iterated Integrals

Find the area of the region R that lies below the parabola

$$y = 4x - x^2 \qquad \text{Parabola forms upper boundary}$$

above the x-axis, and above the line

$$y = -3x + 6 \qquad \text{x-axis and line form lower boundary}$$

as shown in Figure 14.7.

Solution Begin by dividing R into the two subregions R_1 and R_2 shown in Figure 14.7. In both regions, it is convenient to use vertical rectangles, and you have

$$\text{Area} = \int_1^2 \int_{-3x+6}^{4x-x^2} dy\,dx + \int_2^4 \int_0^{4x-x^2} dy\,dx$$

$$= \int_1^2 (4x - x^2 + 3x - 6)\,dx + \int_2^4 (4x - x^2)\,dx$$

$$= \left[\frac{7x^2}{2} - \frac{x^3}{3} - 6x \right]_1^2 + \left[2x^2 - \frac{x^3}{3} \right]_2^4$$

$$= \left(14 - \frac{8}{3} - 12 - \frac{7}{2} + \frac{1}{3} + 6 \right) + \left(32 - \frac{64}{3} - 8 + \frac{8}{3} \right)$$

$$= \frac{15}{2}.$$

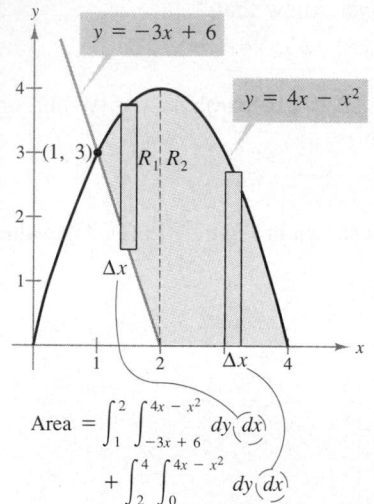

$y = -3x + 6$

$y = 4x - x^2$

$(1, 3)$

R_1 R_2

Δx

Δx

$$\text{Area} = \int_1^2 \int_{-3x+6}^{4x-x^2} dy\,dx$$
$$+ \int_2^4 \int_0^{4x-x^2} dy\,dx$$

FIGURE 14.7

REMARK In Examples 3–6, be sure you see the benefit of sketching the region of integration. We strongly recommend that you develop the habit of making sketches to help determine the limits of integration for all iterated integrals in this chapter.

At this point you may be wondering why you would need iterated integrals. After all, you already know how to use conventional integration to find the area of a region in the plane. (For instance, compare the solution of Example 4 in this section with that given in Example 3 in Section 6.1.) The need for iterated integrals will become clear in the next section. In this section, we have chosen to give primary attention to procedures for finding the limits of integration of the region of an iterated integral, and the following exercise set is designed to develop skill in this important procedure.

EXERCISES for Section 14.1

In Exercises 1–10, evaluate the integral.

1. $\displaystyle\int_0^x (2x - y)\, dy$

2. $\displaystyle\int_x^{x^2} \frac{y}{x}\, dy$

3. $\displaystyle\int_1^{2y} \frac{y}{x}\, dx$

4. $\displaystyle\int_0^{\cos y} y\, dx$

5. $\displaystyle\int_0^{\sqrt{4-x^2}} x^2 y\, dy$

6. $\displaystyle\int_{x^2}^{\sqrt{x}} (x^2 + y^2)\, dy$

7. $\displaystyle\int_{e^y}^{y} \frac{y \ln x}{x}\, dx$

8. $\displaystyle\int_{-\sqrt{1-y^2}}^{\sqrt{1-y^2}} (x^2 + y^2)\, dx$

$\qquad + y^2)\, dx$

9. $\displaystyle\int_0^{x^3} y e^{-y/x}\, dy$

10. $\displaystyle\int_y^{\pi/2} \sin^3 x \cos y\, dx$

In Exercises 11–20, evaluate the iterated integral.

11. $\displaystyle\int_0^1 \int_0^2 (x + y)\, dy\, dx$

12. $\displaystyle\int_0^1 \int_0^x \sqrt{1 - x^2}\, dy\, dx$

13. $\displaystyle\int_1^2 \int_0^4 (x^2 - 2y^2 + 1)\, dx\, dy$

14. $\displaystyle\int_0^1 \int_y^{2y} (1 + 2x^2 + 2y^2)\, dx\, dy$

15. $\displaystyle\int_0^1 \int_0^{\sqrt{1-y^2}} (x + y)\, dx\, dy$

16. $\displaystyle\int_0^2 \int_{3y^2-6y}^{2y-y^2} 3y\, dx\, dy$

17. $\displaystyle\int_0^2 \int_0^{\sqrt{4-y^2}} \frac{2}{\sqrt{4 - y^2}}\, dx\, dy$

18. $\displaystyle\int_0^{\pi/2} \int_0^{2\cos\theta} r\, dr\, d\theta$

19. $\displaystyle\int_0^{\pi/2} \int_0^{\sin\theta} \theta r\, dr\, d\theta$

20. $\displaystyle\int_0^{\pi/4} \int_0^{\cos\theta} 3r^2 \sin\theta\, dr\, d\theta$

In Exercises 21–24, evaluate the improper iterated integral.

21. $\displaystyle\int_1^{\infty} \int_0^{1/x} y\, dy\, dx$

22. $\displaystyle\int_0^3 \int_0^{\infty} \frac{x^2}{1 + y^2}\, dy\, dx$

23. $\displaystyle\int_0^{\infty} \int_0^{\infty} xy e^{-(x^2+y^2)}\, dx\, dy$

24. $\displaystyle\int_1^{\infty} \int_1^{\infty} \frac{1}{xy}\, dx\, dy$

In Exercises 25–28, sketch the region R of integration and switch the order of integration.

25. $\displaystyle\int_0^4 \int_0^y f(x, y)\, dx\, dy$

26. $\displaystyle\int_0^4 \int_{\sqrt{y}}^2 f(x, y)\, dx\, dy$

27. $\displaystyle\int_{-\pi/2}^{\pi/2} \int_0^{\cos x} f(x, y)\, dy\, dx$

28. $\displaystyle\int_{-1}^1 \int_{x^2}^1 f(x, y)\, dy\, dx$

In Exercises 29–36, sketch the region R whose area is given by the iterated integral. Then switch the order of integration and show that both orders yield the same area.

29. $\displaystyle\int_0^1 \int_0^2 dy\, dx$

30. $\displaystyle\int_1^2 \int_2^4 dx\, dy$

31. $\displaystyle\int_0^1 \int_{-\sqrt{1-y^2}}^{\sqrt{1-y^2}} dx\, dy$

32. $\displaystyle\int_0^2 \int_0^x dy\, dx + \int_2^4 \int_0^{4-x} dy\, dx$

33. $\displaystyle\int_0^2 \int_{x/2}^1 dy\, dx$

34. $\displaystyle\int_0^4 \int_{\sqrt{x}}^2 dy\, dx$

35. $\displaystyle\int_0^1 \int_{y^2}^{\sqrt[3]{y}} dx\, dy$

36. $\displaystyle\int_{-2}^2 \int_0^{4-y^2} dx\, dy$

In Exercises 37–42, use an iterated integral to find the area of the region.

37.

38.

39.

40.

41.

42.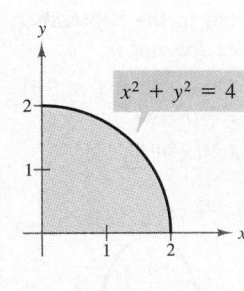

In Exercises 43–48, use an iterated integral to find the area of the region bounded by the graphs of the equations.

43. $\sqrt{x} + \sqrt{y} = 2,\ x = 0,\ y = 0$

44. $y = x^{3/2},\ y = x$

45. $2x - 3y = 0$, $x + y = 5$, $y = 0$

46. $xy = 9$, $y = x$, $y = 0$, $x = 9$

47. $\dfrac{x^2}{a^2} + \dfrac{y^2}{b^2} = 1$

48. $y = x$, $y = 2x$, $x = 2$

In Exercises 49–52, evaluate the iterated integral. (Note that it is necessary to switch the order of integration.)

49. $\displaystyle\int_0^2 \int_x^2 x\sqrt{1 + y^3}\,dy\,dx$ **50.** $\displaystyle\int_0^2 \int_x^2 e^{-y^2}\,dy\,dx$

51. $\displaystyle\int_0^1 \int_y^1 \sin x^2\,dx\,dy$ **52.** $\displaystyle\int_0^2 \int_{y^2}^4 \sqrt{x}\sin x\,dx\,dy$

*C In Exercises 53–56, use a computer to evaluate the iterated integral.

53. $\displaystyle\int_0^2 \int_{x^2}^{2x} (x^3 + 3y^2)\,dy\,dx$ **54.** $\displaystyle\int_0^1 \int_y^{2y} \sin(x + y)\,dx\,dy$

55. $\displaystyle\int_0^4 \int_0^y \dfrac{2}{(x + 1)(y + 1)}\,dx\,dy$ **56.** $\displaystyle\int_0^a \int_0^{a-x} (x^2 + y^2)\,dy\,dx$

C In Exercises 57 and 58, use a computer to approximate the iterated integral.

57. $\displaystyle\int_0^2 \int_0^{4-x^2} e^{xy}\,dy\,dx$

58. $\displaystyle\int_0^2 \int_x^2 \sqrt{16 - x^3 - y^3}\,dy\,dx$

True or False   **In Exercises 59 and 60, determine whether the statement is true or false. If it is false, explain why or give an example that shows it is false.**

59. $\displaystyle\int_a^b \int_c^d f(x, y)\,dy\,dx = \int_c^d \int_a^b f(x, y)\,dx\,dy$

60. $\displaystyle\int_0^1 \int_0^x f(x, y)\,dy\,dx = \int_0^1 \int_0^y f(x, y)\,dx\,dy$

Areas of Simple Closed Curves   **In Exercises 61–66, match the closed curve with its area. (These exercises were adapted from the article "The Surveyor's Area Formula" by Bart Braden in the September, 1986 issue of the *College Mathematics Journal*.)**

61. Ellipse: ($0 \le t \le 2\pi$)
$x = b\cos t$
$y = a\sin t$

62. Astroid: ($0 \le t \le 2\pi$)
$x = a\cos^3 t$
$y = a\sin^3 t$

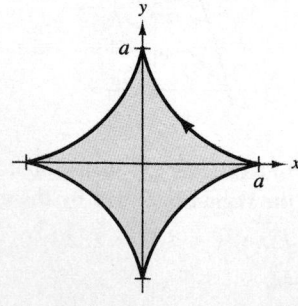

63. Cardioid: ($0 \le t \le 2\pi$)
$x = 2a\cos t - a\cos 2t$
$y = 2a\sin t - a\sin 2t$

64. Deltoid: ($0 \le t \le 2\pi$)
$x = 2a\cos t + a\cos 2t$
$y = 2a\sin t - a\sin 2t$

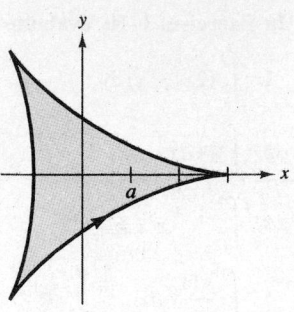

65. Hourglass: ($0 \le t \le 2\pi$)
$x = a\sin 2t$
$y = b\sin t$

66. Teardrop: ($0 \le t \le 2\pi$)
$x = 2a\cos t - a\sin 2t$
$y = b\sin t$

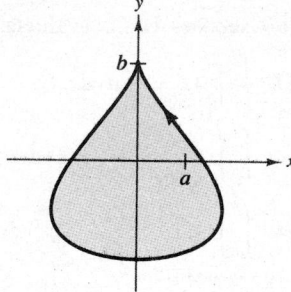

a. $\dfrac{8}{3}ab$ **b.** $\dfrac{3}{8}\pi a^2$ **c.** $2\pi a^2$

d. πab **e.** $2\pi ab$ **f.** $6\pi a^2$

*A C indicates an exercise in which you are instructed to use graphing technology or a symbolic computer algebra system. The solutions of other exercises may also be facilitated by appropriate technology.

Double Integrals and Volume

14.2 Double Integrals and Volume of a Solid Region • Properties of Double Integrals • Evaluation of Double Integrals

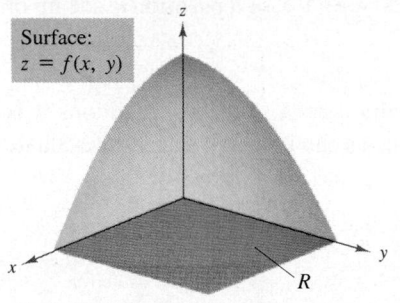

Surface:
$z = f(x, y)$

FIGURE 14.8

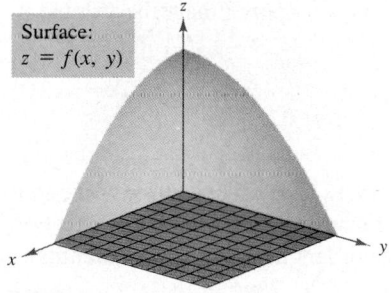

Surface:
$z = f(x, y)$

FIGURE 14.9
The inner partition of region R.

Double Integrals and Volume of a Solid Region

You already know that a definite integral over an *interval* uses a limit process to assign measure to quantities such as area, volume, arc length, and mass. In this section, we use a similar process to define the **double integral** of a function of two variables over a *region in the plane*.

Consider a continuous function f such that $f(x, y) \geq 0$ for all (x, y) in a region R in the xy-plane. The goal is to find the volume of the solid region lying between the surface given by $z = f(x, y)$ and the xy-plane, as shown in Figure 14.8. You can begin by superimposing a rectangular grid over the region, as shown in Figure 14.9. The rectangles lying entirely within R form an **inner partition**, Δ, whose **norm** $\| \Delta \|$ is defined as the length of the longest diagonal of the n rectangles. Next, choose a point (x_i, y_i) in each rectangle and form the rectangular prism whose height is $f(x_i, y_i)$, as shown in Figure 14.10. Because the area of the ith rectangle is $\Delta A_i = \Delta x_i \Delta y_i$, it follows that the volume of the ith prism is

$$f(x_i, y_i) \, \Delta A_i = f(x_i, y_i) \, \Delta x_i \, \Delta y_i \qquad \text{Volume of } i\text{th prism}$$

and you can approximate the volume of the solid region by the Riemann sum of the volumes of all n prisms,

$$\sum_{i=1}^{n} f(x_i, y_i) \, \Delta x_i \, \Delta y_i \qquad \text{Riemann sum}$$

as shown in Figure 14.11. This approximation can be improved by tightening the mesh of the grid to form smaller and smaller rectangles, as shown in Example 1.

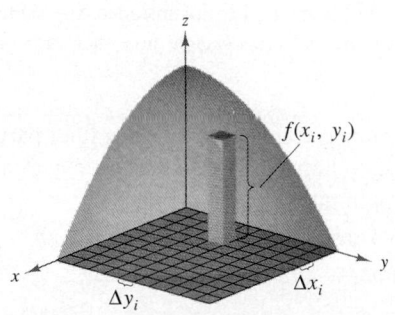

$f(x_i, y_i)$

Δy_i Δx_i

FIGURE 14.10

FIGURE 14.11
Volume approximated by rectangular prisms.

EXAMPLE 1   Approximating the Volume of a Solid

Approximate the volume of the solid lying between the paraboloid

$$f(x, y) = 1 - \frac{1}{2}x^2 - \frac{1}{2}y^2$$

and the square region R given by $0 \le x \le 1, 0 \le y \le 1$. Use a partition made up of squares whose edges have a length of $\frac{1}{4}$.

Solution   Begin by forming the specified partition of R. For this partition, it is convenient to choose the centers of the subregions as the points at which to evaluate $f(x, y)$.

$$(\tfrac{1}{8}, \tfrac{1}{8}) \quad (\tfrac{1}{8}, \tfrac{3}{8}) \quad (\tfrac{1}{8}, \tfrac{5}{8}) \quad (\tfrac{1}{8}, \tfrac{7}{8})$$
$$(\tfrac{3}{8}, \tfrac{1}{8}) \quad (\tfrac{3}{8}, \tfrac{3}{8}) \quad (\tfrac{3}{8}, \tfrac{5}{8}) \quad (\tfrac{3}{8}, \tfrac{7}{8})$$
$$(\tfrac{5}{8}, \tfrac{1}{8}) \quad (\tfrac{5}{8}, \tfrac{3}{8}) \quad (\tfrac{5}{8}, \tfrac{5}{8}) \quad (\tfrac{5}{8}, \tfrac{7}{8})$$
$$(\tfrac{7}{8}, \tfrac{1}{8}) \quad (\tfrac{7}{8}, \tfrac{3}{8}) \quad (\tfrac{7}{8}, \tfrac{5}{8}) \quad (\tfrac{7}{8}, \tfrac{7}{8})$$

Because the area of each square is $\Delta x_i \Delta y_i = \frac{1}{16}$, you can approximate the volume by the sum

$$\sum_{i=1}^{16} f(x_i, y_i) \, \Delta x_i \, \Delta y_i = \sum_{i=1}^{16} \left(1 - \frac{1}{2}x_i^2 - \frac{1}{2}y_i^2 \right) \left(\frac{1}{16} \right) \approx 0.672.$$

This approximation is shown graphically in Figure 14.12. The exact volume of the solid is $\frac{2}{3}$ (see Example 2). You can obtain a better approximation by using a finer partition. For example, with a partition of squares with sides of length $\frac{1}{10}$, the approximation is 0.668.   ▬

REMARK   Some three-dimensional graphing utilities are capable of sketching figures such as that shown in Figure 14.12. For instance, the sketch shown in Figure 14.13 was drawn with a computer program. In this sketch, note that each of the rectangular prisms lies within the solid region.

In Example 1, note that by using finer partitions, you can obtain better approximations of the volume. This observation suggests that you could obtain the exact volume by taking a limit. That is,

$$\text{Volume} = \lim_{\|\Delta\| \to 0} \sum_{i=1}^{n} f(x_i, y_i) \, \Delta x_i \, \Delta y_i.$$

The precise meaning of this limit is that the limit is equal to L if for every $\varepsilon > 0$ there exists a $\delta > 0$ such that

$$\left| L - \sum_{i=1}^{n} f(x_i, y_i) \, \Delta x_i \, \Delta y_i \right| < \varepsilon$$

for all partitions Δ of the plane region R (that satisfy $\|\Delta\| < \delta$) and for all possible choices of x_i and y_i in the ith region.

Using the limit of a Riemann sum to define volume is a special case of using the limit to define a **double integral.** The general case, however, does not require that the function be positive or continuous.

Surface:
$f(x, y) = 1 - \frac{1}{2}x^2 - \frac{1}{2}y^2$

FIGURE 14.12

FIGURE 14.13

REMARK   Having defined a double integral, we will occasionally refer to a definite integral as a **single integral.**

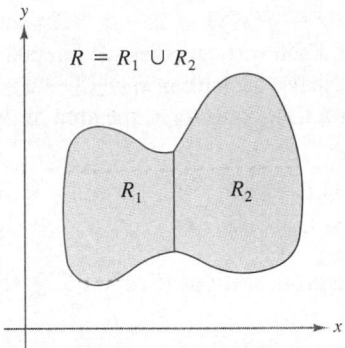

FIGURE 14.14
Two regions are nonoverlapping if their intersection is a set that has an area of 0. In this figure, the area of the line segment that is common to R_1 and R_2 is 0.

Definition of Double Integral

If f is defined on a closed, bounded region R in the xy-plane, then the **double integral of f over R** is given by

$$\int_R\!\!\int f(x, y)\, dA = \lim_{\|\Delta\| \to 0} \sum_{i=1}^{n} f(x_i, y_i)\, \Delta x_i \Delta y_i$$

provided the limit exists. If the limit exists, then f is **integrable** over R.

Sufficient conditions for the double integral of f on the region R to exist are that R can be written as the union of a finite number of nonoverlapping (see Figure 14.14) subregions that are vertically or horizontally simple *and* that f is continuous on the region R.

A double integral can be used to find the volume of a solid region that lies between the xy-plane and the surface given by $z = f(x, y)$.

Volume of a Solid Region

If f is integrable over a plane region R and $f(x, y) \geq 0$ for all (x, y) in R, then the volume of the solid region that lies above R and below the graph of f is defined as

$$V = \int_R\!\!\int f(x, y)\, dA.$$

Properties of Double Integrals

Double integrals share many properties of single integrals.

THEOREM 14.1   Properties of Double Integrals

Let f and g be continuous over a closed, bounded plane region R, and let c be a constant.

1. $\displaystyle\int_R\!\!\int c f(x, y)\, dA = c \int_R\!\!\int f(x, y)\, dA$

2. $\displaystyle\int_R\!\!\int [f(x, y) \pm g(x, y)]\, dA = \int_R\!\!\int f(x, y)\, dA \pm \int_R\!\!\int g(x, y)\, dA$

3. $\displaystyle\int_R\!\!\int f(x, y)\, dA \geq 0, \qquad \text{if } f(x, y) \geq 0$

4. $\displaystyle\int_R\!\!\int f(x, y)\, dA \geq \int_R\!\!\int g(x, y)\, dA, \qquad \text{if } f(x, y) \geq g(x, y)$

5. $\displaystyle\int_R\!\!\int f(x, y)\, dA = \int_{R_1}\!\!\int f(x, y)\, dA + \int_{R_2}\!\!\int f(x, y)\, dA$

where R is the union of two nonoverlapping subregions R_1 and R_2.

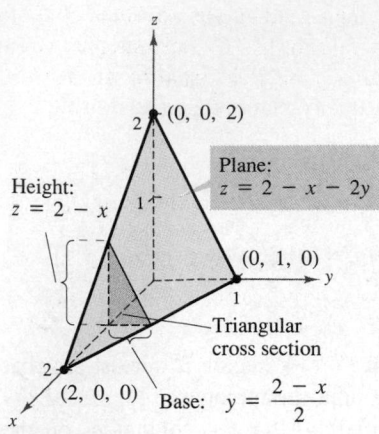

FIGURE 14.15

Volume: $\displaystyle\int_0^2 A(x)\,dx$

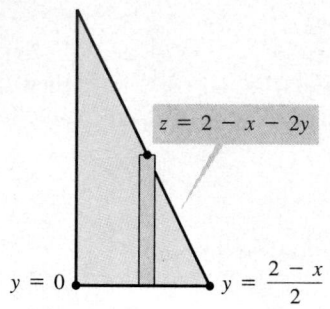

FIGURE 14.16

Triangular cross section

Evaluation of Double Integrals

Normally, the first step in evaluating a double integral is to rewrite it as an iterated integral. To show how this is done, we use a geometric model of a double integral as the volume of a solid.

Consider the solid region bounded by the plane $z = f(x, y) = 2 - x - 2y$ and the three coordinate planes, as shown in Figure 14.15. Each vertical cross section taken parallel to the yz-plane is a triangular region whose base has a length of $y = (2 - x)/2$ and whose height is $z = 2 - x$. This implies that for a fixed value of x, the area of the triangular cross section is

$$A(x) = \frac{1}{2}(\text{base})(\text{height}) = \frac{1}{2}\left(\frac{2 - x}{2}\right)(2 - x) = \frac{(2 - x)^2}{4}.$$

By the formula for the volume of a solid with known cross sections (Section 6.2), the volume of the solid is

$$\text{Volume} = \int_a^b A(x)\,dx = \int_0^2 \frac{(2 - x)^2}{4}\,dx = -\frac{(2 - x)^3}{12}\bigg]_0^2 = \frac{8}{12} = \frac{2}{3}.$$

This procedure works no matter how $A(x)$ is obtained. In particular, you can find $A(x)$ by integration, as indicated in Figure 14.16. That is, you consider x to be constant, and integrate $z = 2 - x - 2y$ from 0 to $(2 - x)/2$ to obtain

$$A(x) = \int_0^{(2-x)/2} (2 - x - 2y)\,dy$$

$$= (2 - x)y - y^2 \bigg]_0^{(2-x)/2}$$

$$= \frac{(2 - x)^2}{4}.$$

Combining these results, you have the *iterated integral*

$$\text{Volume} = \int_R\!\!\int f(x, y)\,dA = \int_0^2\int_0^{(2-x)/2} (2 - x - 2y)\,dy\,dx.$$

To better understand this procedure, it helps to imagine the integration as two sweeping motions. For the inner integration, a vertical line sweeps out the area of a cross section. For the outer integration, the triangular cross section sweeps out the volume, as shown in Figure 14.17.

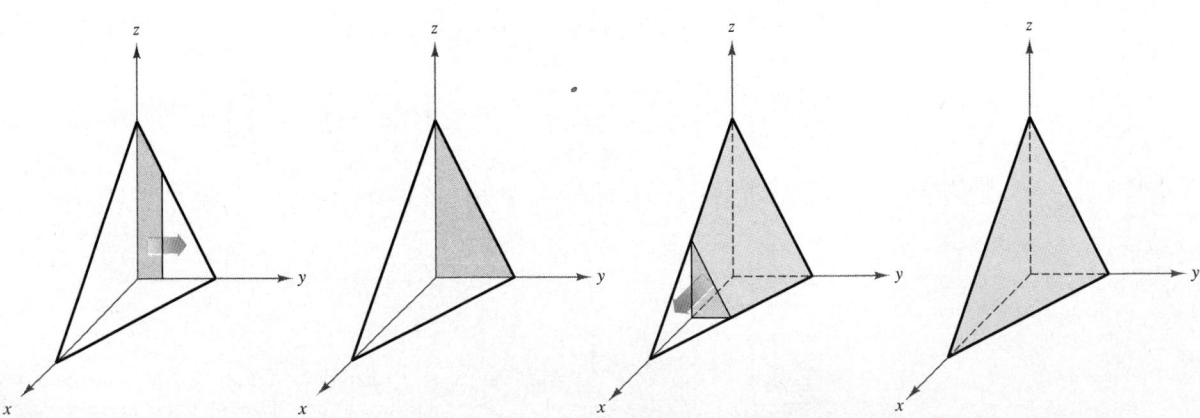

Integrate with respect to y to obtain the area of the cross section.

Integrate with respect to x to obtain the volume of the solid.

FIGURE 14.17

The following theorem was proved by the Italian mathematician Guido Fubini (1879–1943). The theorem states that if R is a vertically or horizontally simple region and f is continuous on R, then the double integral of f on R is equal to an iterated integral.

THEOREM 14.2   Fubini's Theorem

Let f be continuous on a plane region R.

1. If R is defined by $a \leq x \leq b$ and $g_1(x) \leq y \leq g_2(x)$, where g_1 and g_2 are continuous on $[a, b]$, then

$$\iint_R f(x, y)\, dA = \int_a^b \int_{g_1(x)}^{g_2(x)} f(x, y)\, dy\, dx.$$

2. If R is defined by $c \leq y \leq d$ and $h_1(y) \leq x \leq h_2(y)$, where h_1 and h_2 are continuous on $[c, d]$, then

$$\iint_R f(x, y)\, dA = \int_c^d \int_{h_1(y)}^{h_2(y)} f(x, y)\, dx\, dy.$$

EXAMPLE 2   Evaluating a Double Integral as an Iterated Integral

Evaluate

$$\iint_R \left(1 - \frac{1}{2}x^2 - \frac{1}{2}y^2\right) dA$$

where R is the region given by $0 \leq x \leq 1$, $0 \leq y \leq 1$.

Solution   Because the region R is a simple square, it is both vertically and horizontally simple, and you can use either order of integration. Suppose you choose $dy\, dx$ by placing a vertical representative rectangle in the region, as shown in Figure 14.18. This produces the following.

$$\iint_R \left(1 - \frac{1}{2}x^2 - \frac{1}{2}y^2\right) dA = \int_0^1 \int_0^1 \left(1 - \frac{1}{2}x^2 - \frac{1}{2}y^2\right) dy\, dx$$

$$= \int_0^1 \left[\left(1 - \frac{1}{2}x^2\right)y - \frac{y^3}{6}\right]_0^1 dx$$

$$= \int_0^1 \left(\frac{5}{6} - \frac{1}{2}x^2\right) dx = \left[\frac{5}{6}x - \frac{x^3}{6}\right]_0^1 = \frac{2}{3}$$

$0 \leq x \leq 1$
$0 \leq y \leq 1$

$$\iint_R f(x, y)\, dA = \int_0^1 \int_0^1 f(x, y)\, dy\, dx$$

FIGURE 14.18
The volume of the solid region is $\frac{2}{3}$.

REMARK   The double integral evaluated in Example 2 represents the volume of the solid region approximated in Example 1. Note that the approximation obtained in Example 1 is quite good (0.672 vs. $\frac{2}{3}$), even though we used a partition consisting of only 16 squares. The error resulted because we used the centers of the square subregions as the points in the approximation. This is comparable to the Midpoint Rule approximation of a single integral.

The difficulty of evaluating a single integral $\int_a^b f(x)\,dx$ usually depends on the function f, and not on the interval $[a, b]$. This is a major difference between single and double integrals. In the next example, we integrate the same function as that in Examples 1 and 2. Notice that a change in the region R produces a much more difficult integration problem.

EXAMPLE 3   Finding Volume by a Double Integral

Find the volume of the solid region bounded by the paraboloid $z = 4 - x^2 - 2y^2$ and the xy-plane.

Solution   By letting $z = 0$, you can see that the base of the region in the xy-plane is the ellipse $x^2 + 2y^2 = 4$, as shown in Figure 14.19. This plane region is both vertically and horizontally simple, so the order $dy\,dx$ is appropriate.

Variable bounds for y $-\sqrt{\dfrac{(4 - x^2)}{2}} \le y \le \sqrt{\dfrac{(4 - x^2)}{2}}$

Constant bounds for x $-2 \le x \le 2$

The volume is given by

$$
\begin{aligned}
V &= \int_{-2}^{2} \int_{-\sqrt{(4-x^2)/2}}^{\sqrt{(4-x^2)/2}} (4 - x^2 - 2y^2)\,dy\,dx \\[2mm]
&= \int_{-2}^{2} \left[(4 - x^2)y - \frac{2y^3}{3} \right]_{-\sqrt{(4-x^2)/2}}^{\sqrt{(4-x^2)/2}} dx \\[2mm]
&= \frac{4}{3\sqrt{2}} \int_{-2}^{2} (4 - x^2)^{3/2}\,dx \\[2mm]
&= \frac{4}{3\sqrt{2}} \int_{-\pi/2}^{\pi/2} 16\cos^4 \theta\,d\theta \qquad x = 2\sin\theta \\[2mm]
&= \frac{64}{3\sqrt{2}}(2) \int_{0}^{\pi/2} \cos^4 \theta\,d\theta \\[2mm]
&= \frac{128}{3\sqrt{2}} \left(\frac{3\pi}{16} \right) = 4\sqrt{2}\,\pi. \qquad \text{Wallis's Formula}
\end{aligned}
$$

REMARK   In Example 3, note the usefulness of Wallis's Formula. You may want to review this formula in Section 7.3.

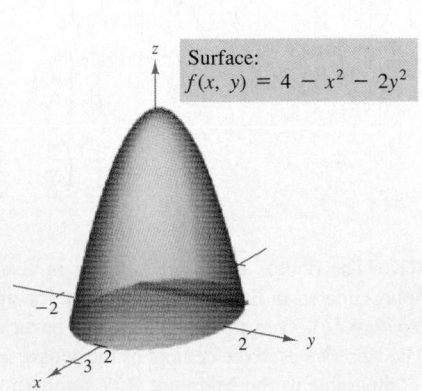

Surface:
$f(x, y) = 4 - x^2 - 2y^2$

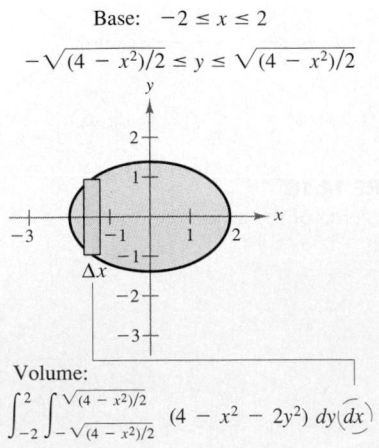

Base: $-2 \le x \le 2$
$-\sqrt{(4 - x^2)/2} \le y \le \sqrt{(4 - x^2)/2}$

Volume:
$$\int_{-2}^{2} \int_{-\sqrt{(4-x^2)/2}}^{\sqrt{(4-x^2)/2}} (4 - x^2 - 2y^2)\,dy\,dx$$

FIGURE 14.19

In Examples 2 and 3, the order of integration was optional, because the regions were both vertically and horizontally simple. Moreover, had you used the order $dx\,dy$, you would have obtained integrals of comparable difficulty. There are, however, some occasions in which one order of integration is much more convenient than the other. Example 4 shows such a case.

EXAMPLE 4   Comparing Different Orders of Integration

Find the volume of the solid region R bounded by the surface

$$f(x, y) = e^{-x^2} \qquad\qquad \text{Surface}$$

and the planes $y = 0$, $y = x$, and $x = 1$, as shown in Figure 14.20.

Solution   The base of R in the xy-plane is bounded by the lines $y = 0$, $x = 1$, and $y = x$. The two possible orders of integration are given in Figure 14.21.

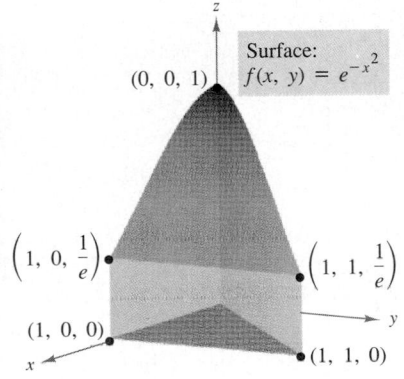

FIGURE 14.20
Base is bounded by
$y = 0$, $y = x$, $x = 1$.

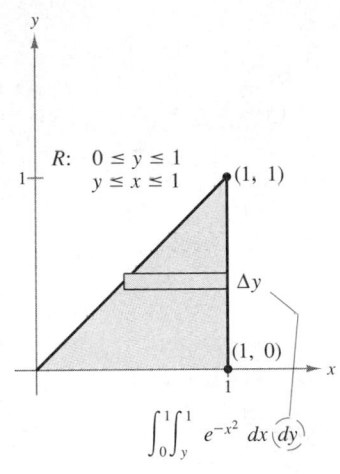

FIGURE 14.21

By setting up the corresponding iterated integrals, you can see that the order $dx\,dy$ requires the antiderivative $\int e^{-x^2}\,dx$, which is not an elementary function. On the other hand, the order $dy\,dx$ produces the integral

$$\int_0^1\int_0^x e^{-x^2}\,dy\,dx = \int_0^1 e^{-x^2}y\bigg]_0^x dx$$

$$= \int_0^1 xe^{-x^2}\,dx$$

$$= -\frac{1}{2}e^{-x^2}\bigg]_0^1$$

$$= -\frac{1}{2}\left(\frac{1}{e} - 1\right)$$

$$= \frac{e - 1}{2e}$$

$$\approx 0.316.$$

REMARK   Try using a symbolic integration utility to evaluate the integral in Example 4.

EXAMPLE 5   Volume of a Region Bounded by Two Surfaces

Find the volume of the solid region R bounded above by the paraboloid $z = 1 - x^2 - y^2$ and below by the plane $z = 1 - y$, as shown in Figure 14.22.

Solution Equating z-values, you can determine that the intersection of the two surfaces occurs on the right circular cylinder given by

$$1 - y = 1 - x^2 - y^2 \quad \rightarrow \quad x^2 = y - y^2.$$

Because the volume of R is the difference between the volume under the paraboloid and the volume under the plane, you have

$$\text{Volume} = \int_0^1 \int_{-\sqrt{y-y^2}}^{\sqrt{y-y^2}} (1 - x^2 - y^2)\, dx\, dy - \int_0^1 \int_{-\sqrt{y-y^2}}^{\sqrt{y-y^2}} (1 - y)\, dx\, dy$$

$$= \int_0^1 \int_{-\sqrt{y-y^2}}^{\sqrt{y-y^2}} (y - y^2 - x^2)\, dx\, dy$$

$$= \int_0^1 \left[(y - y^2)x - \frac{x^3}{3} \right]_{-\sqrt{y-y^2}}^{\sqrt{y-y^2}} dy$$

$$= \frac{4}{3} \int_0^1 (y - y^2)^{3/2}\, dy$$

$$= \left(\frac{4}{3} \right) \left(\frac{1}{8} \right) \int_0^1 [1 - (2y - 1)^2]^{3/2}\, dy$$

$$= \frac{1}{6} \int_{-\pi/2}^{\pi/2} \frac{\cos^4 \theta}{2}\, d\theta \qquad\qquad 2y - 1 = \sin\theta$$

$$= \frac{1}{6} \int_0^{\pi/2} \cos^4 \theta\, d\theta = \left(\frac{1}{6} \right) \left(\frac{3\pi}{16} \right)$$

$$= \frac{\pi}{32}. \qquad\qquad\qquad \text{Wallis's Formula}$$

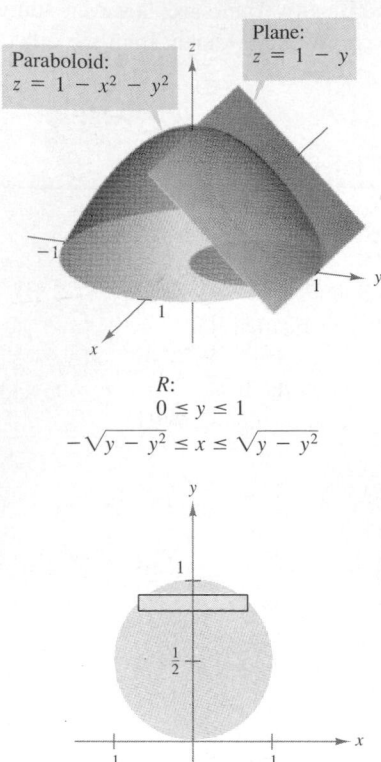

Paraboloid:
$z = 1 - x^2 - y^2$

Plane:
$z = 1 - y$

R:
$0 \leq y \leq 1$
$-\sqrt{y - y^2} \leq x \leq \sqrt{y - y^2}$

The volume of the solid region is $\pi/32$.

FIGURE 14.22

EXERCISES for Section 14.2

In Exercises 1–4, approximate the integral

$$\int_R \int f(x, y)\, dA$$

by dividing the rectangle R with vertices $(0, 0)$, $(4, 0)$, $(4, 2)$, and $(0, 2)$ into eight equal squares and finding the sum

$$\sum_{i=1}^{8} f(x_i, y_i)\, \Delta x_i\, \Delta y_i$$

where (x_i, y_i) is the center of the ith square. Evaluate the double integral and compare it with the approximation.

1. $\int_0^4 \int_0^2 (x + y)\, dy\, dx$

2. $\int_0^4 \int_0^2 xy\, dy\, dx$

3. $\int_0^4 \int_0^2 (x^2 + y^2)\, dy\, dx$

4. $\int_0^4 \int_0^2 \frac{1}{(x + 1)(y + 1)}\, dy\, dx$

5. Let R be a region in the xy-plane whose area is B. If $f(x, y) = k$ for every point (x, y) in R, then what is the value of

$$\int_R \int f(x, y)\, dA?$$

6. Let R represent a county in the northern part of the United States, and let $f(x, y)$ represent the total annual snowfall at the point (x, y) in R. Give an interpretation of each of the following.

a. $\int_R \int f(x, y)\, dA$

b. $\dfrac{\displaystyle\int_R \int f(x, y)\, dA}{\displaystyle\int_R \int dA}$

In Exercises 7–12, sketch the region R and evaluate the double integral

$$\int_R \int f(x,y)\, dA.$$

7. $\int_0^2 \int_0^1 (1 + 2x + 2y)\, dy\, dx$ **8.** $\int_0^\pi \int_0^{\pi/2} \sin^2 x \cos^2 y\, dy\, dx$

9. $\int_0^6 \int_{y/2}^3 (x + y)\, dx\, dy$ **10.** $\int_0^1 \int_y^{\sqrt{y}} x^2 y^2\, dx\, dy$

11. $\int_{-a}^a \int_{-\sqrt{a^2-x^2}}^{\sqrt{a^2-x^2}} (x + y)\, dy\, dx$

12. $\int_0^1 \int_{y-1}^0 e^{x+y}\, dx\, dy + \int_0^1 \int_0^{1-y} e^{x+y}\, dx\, dy$

In Exercises 13–18, set up the integral for both orders of integration, and use the more convenient order to evaluate the integral over the region R.

13. $\int_R \int xy\, dA$
R: rectangle with vertices $(0, 0)$, $(0, 5)$, $(3, 5)$, and $(3, 0)$

14. $\int_R \int \sin x \sin y\, dA$
R: rectangle with vertices $(-\pi, 0)$, $(\pi, 0)$, $(\pi, \pi/2)$, and $(-\pi, \pi/2)$

15. $\int_R \int \frac{y}{x^2 + y^2}\, dA$
R: triangle bounded by $y = x$, $y = 2x$, $x = 2$

16. $\int_R \int \frac{y}{1 + x^2}\, dA$
R: region bounded by $y = 0$, $y = \sqrt{x}$, $x = 4$

17. $\int_R \int x\, dA$
R: sector of a circle in the first quadrant bounded by $y = \sqrt{25 - x^2}$, $3x - 4y = 0$, $y = 0$

18. $\int_R \int (x^2 + y^2)\, dA$
R: semicircle bounded by $y = \sqrt{4 - x^2}$, $y = 0$

In Exercises 19–28, use a double integral to find the volume of the specified solid.

19.

$0 \le x \le 4$
$0 \le y \le 2$

20.
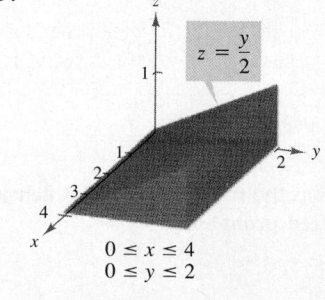
$0 \le x \le 4$
$0 \le y \le 2$

21.

22.

23.

24.

25.

26.

27. Improper integral
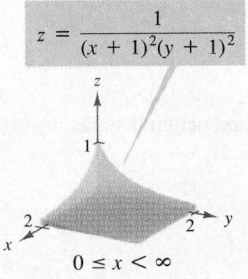
$0 \le x < \infty$
$0 \le y < \infty$

28. Improper integral
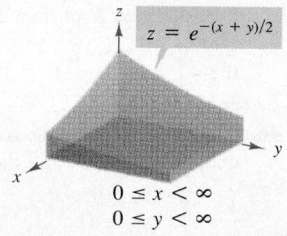
$0 \le x < \infty$
$0 \le y < \infty$

C In Exercises 29 and 30, use a symbolic integration utility to find the volume of the solid region.

29.

30.

In Exercises 31–38, use a double integral to find the volume of the solid bounded by the graphs of the equations.

31. $z = xy$, $z = 0$, $y = x$, $x = 1$ (first octant)

32. $y = 0$, $z = 0$, $y = x$, $z = x$, $x = 0$, $x = 5$

33. $z = 0$, $z = x^2$, $x = 0$, $x = 2$, $y = 0$, $y = 4$

34. $x^2 + y^2 + z^2 = r^2$

35. $x^2 + z^2 = 1$, $y^2 + z^2 = 1$ (first octant)

36. $y = 1 - x^2$, $z = 1 - x^2$ (first octant)

37. $z = x + y$, $x^2 + y^2 = 4$ (first octant)

38. $z = \dfrac{1}{1 + y^2}$, $x = 0$, $x = 2$, $y \geq 0$

Wallis's Formula   In Exercises 39 and 40, use Wallis's Formula as an aid in finding the volume of the solid bounded by the graph of the equations.

39. $z = x^2 + y^2$, $x^2 + y^2 = 4$, $z = 0$

40. $z = \sin^2 x$, $z = 0$, $0 \leq x \leq \pi$, $0 \leq y \leq 5$

C In Exercises 41–44, use a symbolic integration utility to find the volume of the solid bounded by the graphs of the equations.

41. $z = 4 - x^2 - y^2$, $z = 0$

42. $x^2 = 9 - y$, $z^2 = 9 - y$ (first octant)

43. $z = \dfrac{2}{1 + x^2 + y^2}$, $z = 0$, $y = 0$, $x = 0$, $y = -0.5x + 1$

44. $z = \ln(1 + x + y)$, $z = 0$, $y = 0$, $x = 0$, $x = 4 - \sqrt{y}$

45. If f is a continuous function such that $0 \leq f(x, y) \leq 1$ over a region R of area 1, prove that

$$0 \leq \int_R\!\!\int f(x, y)\, dA \leq 1.$$

46. Find the volume of the solid in the first octant bounded by the coordinate planes and the plane

$$\frac{x}{a} + \frac{y}{b} + \frac{z}{c} = 1$$

where $a > 0$, $b > 0$, and $c > 0$.

In Exercises 47–50, evaluate the iterated integral. (Note that it is necessary to switch the order of integration.)

47. $\displaystyle\int_0^1 \int_{y/2}^{1/2} e^{-x^2}\, dx\, dy$

48. $\displaystyle\int_0^1 \int_0^{\arccos y} \sin x \sqrt{1 + \sin^2 x}\, dx\, dy$

49. $\displaystyle\int_0^{\ln 10} \int_{e^x}^{10} \frac{1}{\ln y}\, dy\, dx$

50. $\displaystyle\int_0^2 \int_{x^2}^4 \sqrt{y} \cos y\, dy\, dx$

In Exercises 51–54, find the average value of $f(x, y)$ over the region R where

$$\text{Average} = \frac{1}{A} \int_R\!\!\int f(x, y)\, dA$$

and where A is the area of R.

51. $f(x, y) = x$
    R: rectangle with vertices $(0, 0)$, $(4, 0)$, $(4, 2)$, and $(0, 2)$

52. $f(x, y) = xy$
    R: rectangle with vertices $(0, 0)$, $(4, 0)$, $(4, 2)$, and $(0, 2)$

53. $f(x, y) = x^2 + y^2$
    R: square with vertices $(0, 0)$, $(2, 0)$, $(2, 2)$, and $(0, 2)$

54. $f(x, y) = e^{x+y}$
    R: triangle with vertices $(0, 0)$, $(0, 1)$, and $(1, 1)$

55. *Average Production*   The Cobb-Douglas production function for a company is

$$f(x, y) = 100x^{0.6}y^{0.4}.$$

Estimate the average production level if the number of units of labor varies between 200 and 250 and the number of units of capital varies between 300 and 325.

C 56. *Average Profit*   A firm's profit in marketing two products is

$$P = 192x + 576y - x^2 - 5y^2 - 2xy - 5000$$

where x and y represent the number of units of each product. Use a computer algebra system to evaluate the double integral yielding the average weekly profit if x varies between 40 and 50 units and y varies between 45 and 60 units.

Probability   A joint density function of the continuous random variables x and y is a function $f(x, y)$ satisfying the following properties.

a. $f(x, y) \geq 0$ for all (x, y)

b. $\displaystyle\int_{-\infty}^{\infty} \int_{-\infty}^{\infty} f(x, y)\, dA = 1$

c. $P[(x, y) \in R] = \displaystyle\int_R\!\!\int f(x, y)\, dA$

In Exercises 57–60, show that the function is a joint density function and find the required probability.

57. $f(x, y) = \begin{cases} \frac{1}{10}, & 0 \leq x \leq 5, 0 \leq y \leq 2 \\ 0, & \text{elsewhere} \end{cases}$
    $P(0 \leq x \leq 2, 1 \leq y \leq 2)$

58. $f(x, y) = \begin{cases} \frac{1}{4}xy, & 0 \leq x \leq 2, 0 \leq y \leq 2 \\ 0, & \text{elsewhere} \end{cases}$

$P(0 \leq x \leq 1, 1 \leq y \leq 2)$

59. $f(x, y) = \begin{cases} \frac{1}{27}(9 - x - y), & 0 \leq x \leq 3, 3 \leq y \leq 6 \\ 0, & \text{elsewhere} \end{cases}$

$P(0 \leq x \leq 1, 4 \leq y \leq 6)$

60. $f(x, y) = \begin{cases} e^{-(x-y)}, & x \geq 0, y \geq 0 \\ 0, & \text{elsewhere} \end{cases}$

$P(0 \leq x \leq 1, x \leq y \leq 1)$

61. *Center of Pressure on a Sail* Perform the required integrations to verify that the center of pressure on a triangular sail with vertices at $(0, 0)$, $(2, 1)$, and $(0, 5)$ is $(x_p, y_p) = (\frac{7}{12}, \frac{31}{12})$, where

$$x_p = \frac{\displaystyle\iint_R xy\, dA}{\displaystyle\iint_R y\, dA} \quad \text{and} \quad y_p = \frac{\displaystyle\iint_R y^2\, dA}{\displaystyle\iint_R y\, dA}.$$

[C] **62.** *Center of Pressure on a Sail* Repeat Exercise 61 for a sail with vertices at $(0, 0)$, $(3, 1)$, and $(0, 7)$. Use a computer algebra system to perform the required double integrations.

[C] **63.** Consider a continuous function $f(x, y)$ over the rectangular region R with vertices $(0, 0)$, $(a, 0)$, (a, b), and $(0, b)$. Partition the intervals $[0, a]$ and $[0, b]$ into m and n subintervals, respectively, so that the subintervals in a given direction are of equal length. Write a computer program to compute the sum

$$\sum_{i=1}^{m} \sum_{j=1}^{n} f(x_i, y_j)\, \Delta x_i\, \Delta y_j \approx \int_0^a \int_0^b f(x, y)\, dA$$

where (x_i, y_j) is the center of a representative rectangle in R.

[C] **64.** Consider the double integral

$$\int_0^1 \int_0^2 \sin\sqrt{x + y}\, dy\, dx.$$

a. Use a computer algebra system to approximate the double integral.
b. Use the program from Exercise 63 to approximate the integral when $m = 4$ and $n = 8$.

In Exercises 65 and 66, determine which value best approximates the volume of the solid between the xy-plane and the function over the region. (Make your selection on the basis of a sketch of the solid and *not* by performing any calculations.)

65. $f(x, y) = 4x$
R: square with vertices $(0, 0)$, $(4, 0)$, $(4, 4)$, and $(0, 4)$
 a. -200 **b.** 600 **c.** 50 **d.** 125 **e.** 1000

66. $f(x, y) = \sqrt{x^2 + y^2}$
R: circle bounded by $x^2 + y^2 = 9$
 a. 50 **b.** 500 **c.** -500 **d.** 5 **e.** 5000

True or False In Exercises 67 and 68, determine whether the statement is true or false. If it is false, explain why or give an example that shows it is false.

67. The volume of the sphere $x^2 + y^2 + z^2 = 1$ is given by the integral

$$V = 8 \int_0^1 \int_0^1 \sqrt{1 - x^2 - y^2}\, dx\, dy.$$

68. If $f(x, y) \leq g(x, y)$ for all (x, y) in R, and both f and g are continuous over R, then

$$\iint_R f(x, y)\, dA \leq \iint_R g(x, y)\, dA.$$

69. *Essay* From 1963 to 1986, the volume of the Great Salt Lake approximately tripled while its top surface area approximately doubled. Read the article "Relations Between Surface Area and Volume in Lakes" by Daniel Cass and Gerald Wildenberg in the November, 1990 issue of the *College Mathematics Journal*. Then give examples of solids that have "water levels" a and b such that $V(b) = 3V(a)$ and $A(b) = 2A(a)$, as indicated in the figure.

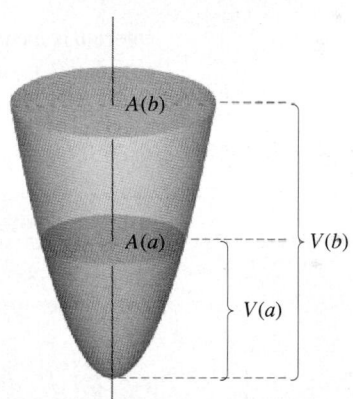

FIGURE FOR 69

70. Find $\displaystyle\int_0^\infty \frac{e^{-x} - e^{-2x}}{x}\, dx$.

$\left[\text{Hint: Evaluate } \displaystyle\int_1^2 e^{-xy}\, dy. \right]$

71. Let $f(x) = \displaystyle\int_1^x e^{t^2}\, dt$. Find the average value of f on the interval $[0, 1]$.

FIGURE 14.23
Area of R_i: $\Delta A_i = r_i \Delta r_i \Delta \theta_i$

Double Integrals in Polar Coordinates

Some double integrals are *much* easier to evaluate in polar form than in rectangular form. This is especially true for regions such as circles, cardioids, and petal curves, and for integrands that involve $x^2 + y^2$.

To define a double integral of a continuous function $z = f(r, \theta)$ in polar coordinates, consider a region R bounded by the graphs of $r = g_1(\theta)$ and $r = g_2(\theta)$ and by the lines $\theta = \alpha$ and $\theta = \beta$. On R, superimpose a polar grid made of rays and circular arcs, as shown in Figure 14.23. The polar sectors R_i lying entirely within R form an **inner polar partition,** Δ, whose **norm** $\| \Delta \|$ is the length of the longest diagonal of the n polar sectors. Recall from the development of area in polar coordinates that if you choose a point (r_i, θ_i) in R_i such that r_i is the average radius of R_i, then the area of R_i is

$$\Delta A_i = r_i \, \Delta r_i \, \Delta \theta_i \qquad \text{Area of } R_i$$

where θ is measured in radians (see Exercise 39). To define a double integral in the polar coordinate system, you can form the Riemann sum $\Sigma f(r_i, \theta_i) r_i \, \Delta r_i \, \Delta \theta_i$ and take the limit as $\| \Delta \| \to 0$.

Double Integral in Polar Coordinates

If f is a continuous function of r and θ on a closed bounded plane region R, then the **double integral of f over R** in polar coordinates is given by

$$\iint_R f(r, \theta) \, dA = \lim_{\| \Delta \| \to 0} \sum_{i=1}^{n} f(r_i, \theta_i) r_i \, \Delta r_i \, \Delta \theta_i = \iint_R f(r, \theta) r \, dr \, d\theta.$$

REMARK In polar coordinates, note that $dA = r \, dr \, d\theta$ or $dA = r \, d\theta \, dr$.

The region R is restricted to two basic types, **r-simple** regions and **θ-simple** regions, as shown in Figure 14.24.

FIGURE 14.24

You can evaluate a double integral in polar form with an iterated integral, as stated in the following polar form of Fubini's Theorem.

THEOREM 14.3   Polar Form of Fubini's Theorem

Let f be continuous on a plane region R.

1. If R is defined by $\theta_1 \leq \theta \leq \theta_2$ and $g_1(\theta) \leq r \leq g_2(\theta)$, where g_1 and g_2 are continuous on $[\theta_1, \theta_2]$, then

$$\iint_R f(r, \theta)\, dA = \int_{\theta_1}^{\theta_2} \int_{g_1(\theta)}^{g_2(\theta)} f(r, \theta) r\, dr\, d\theta.$$

2. If R is defined by $r_1 \leq r \leq r_2$ and $h_1(r) \leq \theta \leq h_2(r)$, where h_1 and h_2 are continuous on $[r_1, r_2]$, then

$$\iint_R f(r, \theta)\, dA = \int_{r_1}^{r_2} \int_{h_1(r)}^{h_2(r)} f(r, \theta) r\, d\theta\, dr.$$

REMARK Note that each integrand in Theorem 14.3 contains a factor r. This arises from the area of the ith polar sector $\Delta A_i = r_i\, \Delta r_i\, \Delta\theta_i$.

EXAMPLE 1   Evaluating a Double Polar Integral

Evaluate

$$\iint_R \sin\theta\, dA$$

where R is the first-quadrant region lying inside the circle given by $r = 4\cos\theta$ and outside the circle given by $r = 2$.

Solution In Figure 14.25, you can see that R is an r-simple region, and you can sketch a representative sector whose bounds yield the following limits of integration.

$$0 \leq \theta \leq \frac{\pi}{3} \qquad \text{Fixed bounds on } \theta$$

$$2 \leq r \leq 4\cos\theta \qquad \text{Variable bounds on } r$$

Thus, you obtain

$$\iint_R \sin\theta\, dA = \int_0^{\pi/3} \int_2^{4\cos\theta} (\sin\theta) r\, dr\, d\theta$$

$$= \int_0^{\pi/3} (\sin\theta)\frac{r^2}{2}\bigg]_2^{4\cos\theta} d\theta$$

$$= \frac{1}{2}\int_0^{\pi/3} (\sin\theta)(16\cos^2\theta - 4)\, d\theta$$

$$= 2\int_0^{\pi/3} \left[4\cos^2\theta(\sin\theta) - \sin\theta\right] d\theta$$

$$= 2\left[-\frac{4\cos^3\theta}{3} + \cos\theta\right]_0^{\pi/3}$$

$$= 2\left[\left(-\frac{4}{3}\left(\frac{1}{2}\right)^3 + \frac{1}{2}\right) - \left(-\frac{4}{3} + 1\right)\right]$$

$$= \frac{4}{3}.$$

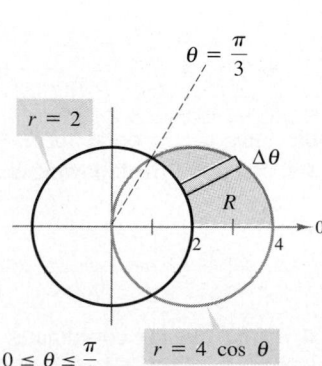

$\theta = \dfrac{\pi}{3}$

$r = 2$

$\Delta\theta$

R

$r = 4\cos\theta$

$R:\ 0 \leq \theta \leq \dfrac{\pi}{3}$
$\quad 2 \leq r \leq 4\cos\theta$

FIGURE 14.25
R is an r-simple region.

EXAMPLE 2   Finding Areas of Polar Regions

Use a double integral to find the area enclosed by the graph of $r = 3\cos 3\theta$.

Solution   Let R be one petal of the curve shown in Figure 14.26. This region is r-simple, and the boundaries are as follows.

$$-\frac{\pi}{6} \le \theta \le \frac{\pi}{6} \qquad \text{Fixed bounds on } \theta$$

$$0 \le r \le 3\cos 3\theta \qquad \text{Variable bounds on } r$$

Thus, the area of one petal is

$$\frac{1}{3}A = \iint_R dA = \int_{-\pi/6}^{\pi/6} \int_0^{3\cos 3\theta} r\,dr\,d\theta$$

$$= \int_{-\pi/6}^{\pi/6} \frac{r^2}{2} \Big]_0^{3\cos 3\theta} d\theta$$

$$= \frac{9}{2} \int_{-\pi/6}^{\pi/6} \cos^2 3\theta\,d\theta$$

$$= \frac{9}{4} \int_{-\pi/6}^{\pi/6} (1 + \cos 6\theta)\,d\theta$$

$$= \frac{9}{4} \left[\theta + \frac{1}{6}\sin 6\theta \right]_{-\pi/6}^{\pi/6}$$

$$= \frac{3\pi}{4}.$$

Therefore, the total area is $A = 9\pi/4$.

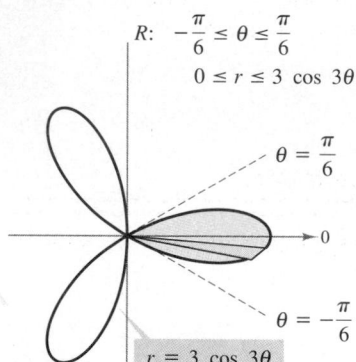

$$R: \quad -\frac{\pi}{6} \le \theta \le \frac{\pi}{6}$$
$$0 \le r \le 3\cos 3\theta$$

$$\theta = \frac{\pi}{6}$$

$$\theta = -\frac{\pi}{6}$$

$$r = 3\cos 3\theta$$

FIGURE 14.26
The area of R is $3\pi/4$, and the total area is $9\pi/4$.

TECHNOLOGY   Any symbolic integration utility that can handle double integrals in rectangular coordinates can also handle double integrals in polar coordinates. The reason this is true is that once you have formed the iterated integral, its value is not changed by using different variables. In other words, if you use a symbolic integrator to evaluate

$$\int_{-\pi/6}^{\pi/6} \int_0^{3\cos 3y} x\,dx\,dy$$

you should obtain the same value that was obtained in Example 2.

Change of Variables to Polar Form

You will now see the reason for introducing double integrals in polar form. With sufficient restrictions on a function f and a region R, you can make the following change of variables.

$$x = r\cos\theta, \qquad y = r\sin\theta$$
$$r^2 = x^2 + y^2, \qquad dA = r\,dr\,d\theta$$

This change of variables converts a double integral in rectangular coordinates to a double integral in polar coordinates, as stated in the following theorem. The proof of this theorem is discussed in Section 14.8.

THEOREM 14.4   Change of Variables to Polar Form

Let R be a plane region consisting of all points $(x, y) = (r \cos \theta, r \sin \theta)$ satisfying the condition

$$0 \le g_1(\theta) \le r \le g_2(\theta), \qquad \theta_1 \le \theta \le \theta_2$$

where $0 < (\theta_2 - \theta_1) \le 2\pi$. If g_1 and g_2 are continuous on $[\theta_1, \theta_2]$ and f is continuous on R, then

$$\iint_R f(x, y)\, dA = \int_{\theta_1}^{\theta_2} \int_{g_1(\theta)}^{g_2(\theta)} f(r \cos \theta, r \sin \theta) r\, dr\, d\theta.$$

If $z = f(x, y)$ is nonnegative on R, then the integral in Theorem 14.4 can be interpreted as the *volume* of the solid region between the graph of f and the region R. We use this interpretation in our next example.

EXAMPLE 3   Change of Variables to Polar Coordinates

Use polar coordinates to find the volume of the solid region bounded above by the hemisphere

$$z = \sqrt{16 - x^2 - y^2} \qquad \text{Hemisphere forms upper surface}$$

and below by the circular region R given by

$$x^2 + y^2 = 4 \qquad \text{Circular region forms lower surface}$$

as shown in Figure 14.27.

Solution   In Figure 14.27, you can see that R has the bounds

$$-\sqrt{4 - y^2} \le x \le \sqrt{4 - y^2}, \qquad -2 \le y \le 2$$

and that $0 \le z \le \sqrt{16 - x^2 - y^2}$. In polar coordinates the bounds are

$$0 \le r \le 2 \quad \text{and} \quad 0 \le \theta \le 2\pi$$

with height $z = \sqrt{16 - x^2 - y^2} = \sqrt{16 - r^2}$. Consequently, the volume V is given by

$$
\begin{aligned}
V = \iint_R f(x, y)\, dA &= \int_0^{2\pi} \int_0^2 \sqrt{16 - r^2}\, r\, dr\, d\theta \\
&= -\frac{1}{3} \int_0^{2\pi} (16 - r^2)^{3/2} \Big]_0^2 \, d\theta \\
&= -\frac{1}{3} \int_0^{2\pi} \left(24\sqrt{3} - 64 \right) d\theta \\
&= -\frac{8}{3} (3\sqrt{3} - 8)\theta \Big]_0^{2\pi} \\
&= \frac{16\pi}{3} \left(8 - 3\sqrt{3} \right) \\
&\approx 46.98.
\end{aligned}
$$

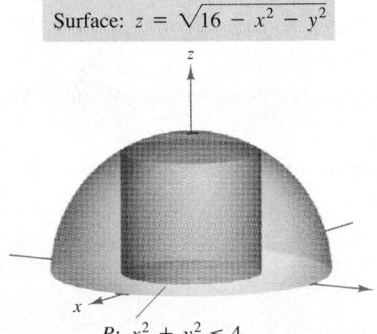

Surface: $z = \sqrt{16 - x^2 - y^2}$

$R: x^2 + y^2 \le 4$

FIGURE 14.27

REMARK   To see the benefit of polar coordinates in Example 3, you should try to evaluate the corresponding rectangular double integral

$$\int_{-2}^{2} \int_{-\sqrt{4-y^2}}^{\sqrt{4-y^2}} \sqrt{16 - x^2 - y^2}\, dx\, dy.$$

$R: \quad 1 \le r \le \sqrt{5}$
$\quad 0 \le \theta \le 2\pi$

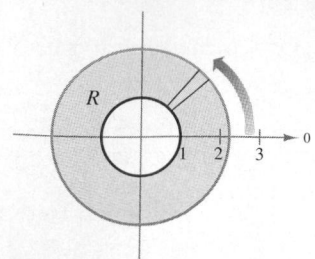

FIGURE 14.28
r-Simple region:
$dA = r\, dr\, d\theta$.

$R: \quad 1 \le r \le \sqrt{5}$
$\quad 0 \le \theta \le 2\pi$

FIGURE 14.29
θ-Simple region:
$dA = r\, d\theta\, dr$.

EXAMPLE 4   Change of Variables to Polar Coordinates

Let R be the annular region lying between the two circles $x^2 + y^2 = 1$ and $x^2 + y^2 = 5$, as shown in Figure 14.28. Evaluate the integral

$$\iint_R (x^2 + y)\, dA.$$

Solution   The polar boundaries are $1 \le r \le \sqrt{5}$ and $0 \le \theta \le 2\pi$. Thus, you have

$$\iint_R (x^2 + y)\, dA = \int_0^{2\pi} \int_1^{\sqrt{5}} (r^2 \cos^2 \theta + r \sin \theta) r\, dr\, d\theta.$$

For the sake of illustration, we change the order of integration, as indicated in Figure 14.29, and obtain

$$\iint_R (x^2 + y)\, dA = \int_1^{\sqrt{5}} \int_0^{2\pi} r^2 (r \cos^2 \theta + \sin \theta)\, d\theta\, dr$$

$$= \int_1^{\sqrt{5}} r^2 \left[\int_0^{2\pi} \left(\frac{r}{2} + \frac{r \cos 2\theta}{2} + \sin \theta \right) d\theta \right] dr$$

$$= \int_1^{\sqrt{5}} r^2 \left[\frac{r\theta}{2} + \frac{r \sin 2\theta}{4} - \cos \theta \right]_0^{2\pi} dr$$

$$= \int_1^{\sqrt{5}} r^2 (\pi r)\, dr = \pi \frac{r^4}{4} \bigg]_1^{\sqrt{5}}$$

$$= 6\pi.$$

Try integrating with $dA = r\, dr\, d\theta$ to see which order is more convenient. ▬

EXERCISES for Section 14.3

In Exercises 1–6, evaluate the double integral $\int_R \int f(r, \theta)\, dA$, and sketch the region R.

1. $\displaystyle \int_0^{2\pi} \int_0^6 3r^2 \sin \theta\, dr\, d\theta$

2. $\displaystyle \int_0^{\pi/4} \int_0^4 r^2 \sin \theta \cos \theta\, dr\, d\theta$

3. $\displaystyle \int_0^{\pi/2} \int_2^3 \sqrt{9 - r^2}\, r\, dr\, d\theta$

4. $\displaystyle \int_0^{\pi/2} \int_0^3 re^{-r^2}\, dr\, d\theta$

5. $\displaystyle \int_0^{\pi/2} \int_0^{1+\sin \theta} \theta\, dr\, d\theta$

6. $\displaystyle \int_0^{\pi/2} \int_0^{1-\cos \theta} \sin \theta\, dr\, d\theta$

In Exercises 7–12, use a double integral to find the area of the indicated region.

7.

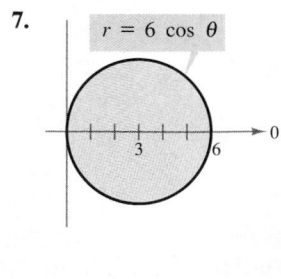

$r = 6 \cos \theta$

8.

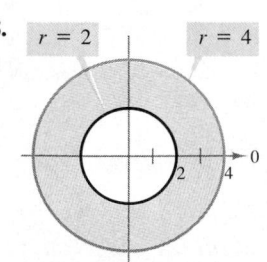

$r = 2$ $r = 4$

9.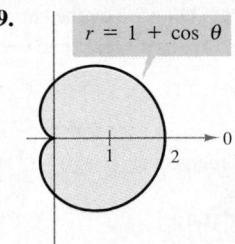

$r = 1 + \cos\theta$

10.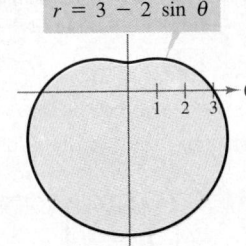

$r = 3 - 2\sin\theta$

11.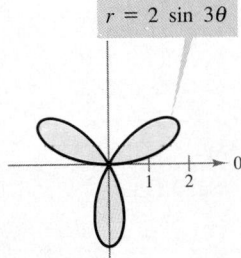

$r = 2\sin 3\theta$

12.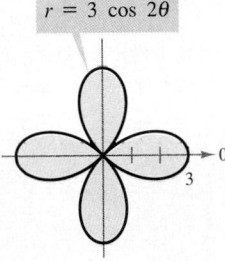

$r = 3\cos 2\theta$

In Exercises 13–18, evaluate the double integral by changing to polar coordinates.

13. $\displaystyle\int_0^a \int_0^{\sqrt{a^2-y^2}} y\,dx\,dy$

14. $\displaystyle\int_0^a \int_0^{\sqrt{a^2-x^2}} x\,dy\,dx$

15. $\displaystyle\int_0^3 \int_0^{\sqrt{9-x^2}} (x^2+y^2)^{3/2}\,dy\,dx$

16. $\displaystyle\int_0^2 \int_y^{\sqrt{8-y^2}} \sqrt{x^2+y^2}\,dx\,dy$

17. $\displaystyle\int_0^2 \int_0^{\sqrt{2x-x^2}} xy\,dy\,dx$

18. $\displaystyle\int_0^4 \int_0^{\sqrt{4y-y^2}} x^2\,dx\,dy$

In Exercises 19 and 20, combine the sum of the two double integrals into a single double integral by using polar coordinates. Evaluate the resulting double integral.

19. $\displaystyle\int_0^2 \int_0^x \sqrt{x^2+y^2}\,dy\,dx + \int_2^{2\sqrt{2}} \int_0^{\sqrt{8-x^2}} \sqrt{x^2+y^2}\,dy\,dx$

20. $\displaystyle\int_0^{5\sqrt{2}/2} \int_0^x xy\,dy\,dx + \int_{5\sqrt{2}/2}^5 \int_0^{\sqrt{25-x^2}} xy\,dy\,dx$

In Exercises 21–24, use polar coordinates to evaluate the double integral $\int_R \int f(x,y)\,dA$.

21. $f(x,y) = x + y$
$R: x^2 + y^2 \le 4,\ 0 \le x,\ 0 \le y$

22. $f(x,y) = e^{-(x^2+y^2)}$
$R: x^2 + y^2 \le 4,\ 0 \le x,\ 0 \le y$

23. $f(x,y) = \arctan\dfrac{y}{x}$
$R: x^2 + y^2 \ge 1,\ x^2 + y^2 \le 4,\ 0 \le y \le x$

24. $f(x,y) = 9 - x^2 - y^2$
$R: x^2 + y^2 \le 9,\ 0 \le x,\ 0 \le y$

Volume **In Exercises 25–30, use a double integral in polar coordinates to find the volume of the solid bounded by the graphs of the equations.**

25. $z = xy,\ x^2 + y^2 = 1$ (first octant)

26. $z = x^2 + y^2 + 1,\ z = 0,\ x^2 + y^2 = 4$

27. $z = \sqrt{x^2+y^2},\ z = 0,\ x^2 + y^2 = 25$

28. $z = \sqrt{x^2+y^2},\ z = 0,\ x^2 + y^2 \ge 4,\ x^2 + y^2 \le 16$

29. Inside the hemisphere $z = \sqrt{16 - x^2 - y^2}$ and inside the cylinder $x^2 + y^2 - 4x = 0$

30. Inside the hemisphere $z = \sqrt{16 - x^2 - y^2}$ and outside the cylinder $x^2 + y^2 = 1$

31. *Volume* Find a so that the volume inside the hemisphere $z = \sqrt{16 - x^2 - y^2}$ and outside the cylinder $x^2 + y^2 = a^2$ is one-half the volume of the hemisphere.

32. *Volume* Use a double integral in polar coordinates to find the volume of a sphere of radius a.

33. *Probability* The integral

$$I = \int_{-\infty}^{\infty} e^{-x^2/2}\,dx$$

is important in calculating probabilities when using the normal probability distribution.

a. Use polar coordinates to evaluate the double integral

$$I^2 = \left(\int_{-\infty}^{\infty} e^{-x^2/2}\,dx\right)\left(\int_{-\infty}^{\infty} e^{-y^2/2}\,dy\right)$$

$$= \int_{-\infty}^{\infty}\int_{-\infty}^{\infty} e^{-(x^2+y^2)/2}\,dA.$$

b. Use the result of part a to determine I.

FOR FURTHER INFORMATION For more information on this problem, see the article "Integrating e^{-x^2} Without Polar Coordinates" by William Dunham in the January, 1988 issue of *Mathematics Teacher*.

34. Use the result of Exercise 33 and change of variables to evaluate the following integrals. No integration is required.

a. $\displaystyle\int_{-\infty}^{\infty} e^{-x^2}\,dx$ **b.** $\displaystyle\int_{-\infty}^{\infty} e^{-4x^2}\,dx$

35. *Population* The population density of a city is approximated by the model

$$f(x,y) = 4000e^{-0.01(x^2+y^2)}, \qquad x^2 + y^2 \le 49$$

where x and y are measured in miles. Integrate the density function over the specified circular region to approximate the population of the city.

36. *Probability* Find k so that the function

$$f(x,y) = \begin{cases} ke^{-(x^2+y^2)}, & x \ge 0,\ y \ge 0 \\ 0, & \text{elsewhere} \end{cases}$$

is a probability density function.

37. Consider the region bounded by the graphs of $y = 2$, $y = 4$, $y = x$, and $y = \sqrt{3}x$ and the double integral $\int_R \int f\, dA$. Determine the limits of integration if the region R is divided into
  a. horizontal representative elements.
  b. vertical representative elements.
  c. polar sectors.

38. Repeat Exercise 37 for a region R bounded by the graphs of the equation $(x - 2)^2 + y^2 = 4$.

39. Show that the area of the polar sector R (see figure) is

$$\text{Area} = r\Delta r\Delta\theta$$

where $r = (r_1 + r_2)/2$ is the average radius of R.

FIGURE FOR 39

40. Let R be the region in the first quadrant bounded by the curves $x^2 + y^2 = 1$ and $x^2 + y^2 = 4$. Find

$$\int_R \int \ln(x^2 + y^2)\, dA.$$

41. Find the "volume" of the "four-dimensional sphere" $x^2 + y^2 + z^2 + w^2 = a^2$, by calculating

$$16 \int_0^a \int_0^{\sqrt{a^2-x^2}} \int_0^{\sqrt{a^2-x^2-y^2}} \int_0^{\sqrt{a^2-x^2-y^2-z^2}} dw\, dz\, dy\, dx.$$

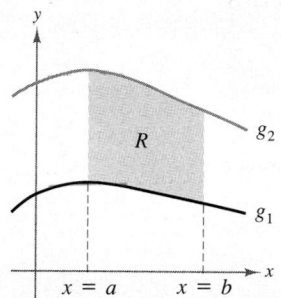

FIGURE 14.30
Lamina of constant density.

Mass

In Section 6.7, we discussed several applications of integration involving a lamina of *constant* density ρ. For example, if the lamina corresponding to the region R, as shown in Figure 14.30, has a constant density ρ, then the mass of the lamina is given by

$$\text{Mass} = \rho A = \rho \iint_R dA = \iint_R \rho \, dA. \qquad \text{Constant density}$$

The use of a double integral suggests a natural extension of the formula for finding the mass of a lamina of *variable* density, where the density at (x, y) is given by the **density function ρ.**

Definition of Mass of a Planar Lamina of Variable Density

If ρ is a continuous density function on the lamina corresponding to a plane region R, then the mass m of the lamina is given by

$$m = \iint_R \rho(x, y) \, dA. \qquad \text{Variable density}$$

REMARK Density is normally expressed as mass per unit volume. For a planar lamina, however, density is mass per unit surface area.

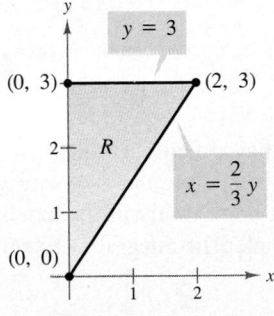

FIGURE 14.31
Lamina of variable density.

REMARK In Figure 14.31, note that the planar lamina is shaded so that the darkest shading corresponds to the densest part.

EXAMPLE 1   Finding the Mass of a Planar Lamina

Find the mass of the triangular lamina with vertices $(0, 0)$, $(0, 3)$, and $(2, 3)$, given that the density at (x, y) is $\rho(x, y) = 2x + y$.

Solution   As shown in Figure 14.31, region R has the boundaries $x = 0$, $y = 3$, and $y = 3x/2$ (or $x = 2y/3$). Therefore, the mass of the lamina is

$$
\begin{aligned}
m &= \iint_R (2x + y) \, dA = \int_0^3 \int_0^{2y/3} (2x + y) \, dx \, dy \\
&= \int_0^3 \left[x^2 + xy \right]_0^{2y/3} dy \\
&= \frac{10}{9} \int_0^3 y^2 \, dy = \frac{10}{9} \frac{y^3}{3} \Bigg]_0^3 = 10.
\end{aligned}
$$

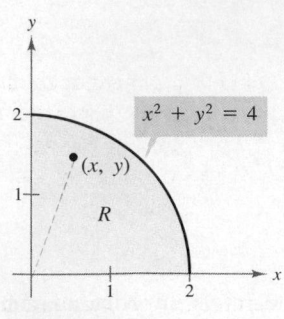

FIGURE 14.32
Density at (x, y):
$\rho(x, y) = k\sqrt{x^2 + y^2}$.

EXAMPLE 2   Finding Mass by Polar Coordinates

Find the mass of the lamina corresponding to the first-quadrant portion of the circle $x^2 + y^2 = 4$, where the density at the point (x, y) is proportional to the distance between the point and the origin, as shown in Figure 14.32.

Solution   At any point (x, y), the density of the lamina is

$$\rho(x, y) = k\sqrt{(x - 0)^2 + (y - 0)^2}$$
$$= k\sqrt{x^2 + y^2}.$$

Because $0 \le x \le 2$ and $0 \le y \le \sqrt{4 - x^2}$, the mass is given by

$$m = \int_R \int k\sqrt{x^2 + y^2}\, dA$$
$$= \int_0^2 \int_0^{\sqrt{4-x^2}} k\sqrt{x^2 + y^2}\, dy\, dx.$$

To simplify the integration, you can change to polar coordinates, using the bounds $0 \le \theta \le \pi/2$ and $0 \le r \le 2$. Thus, the mass is

$$m = \int_R \int k\sqrt{x^2 + y^2}\, dA = \int_0^{\pi/2} \int_0^2 k\sqrt{r^2}\, r\, dr\, d\theta$$
$$= \int_0^{\pi/2} \int_0^2 kr^2\, dr\, d\theta$$
$$= \int_0^{\pi/2} \frac{kr^3}{3}\Bigg]_0^2 d\theta$$
$$= \frac{8k}{3} \int_0^{\pi/2} d\theta$$
$$= \frac{8k}{3} \theta \Bigg]_0^{\pi/2}$$
$$= \frac{4\pi k}{3}.$$

TECHNOLOGY   On many occasions in this text, we have mentioned the benefits of computer programs that perform symbolic integration. Even if you use such a program regularly, you should remember that its greatest benefit comes only in the hands of a knowledgeable user. For instance, notice how much simpler the integral in Example 2 becomes when it is converted to polar form.

Rectangular form	*Polar form*
$\int_0^2 \int_0^{\sqrt{4-x^2}} k\sqrt{x^2 + y^2}\, dy\, dx$	$\int_0^{\pi/2} \int_0^2 kr^2\, dr\, d\theta$

If you have access to software that performs symbolic integration, try using it to evaluate both integrals. Some software programs cannot handle the first integral, but any program that can handle double integrals can evaluate the second integral.

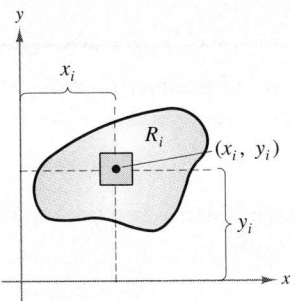

FIGURE 14.33
$M_x = (\text{mass})(y_i)$
$M_y = (\text{mass})(x_i)$

Moments and Center of Mass

For a lamina of variable density, moments of mass are defined in a manner similar to that used for the uniform density case. For a partition Δ of a lamina corresponding to a plane region R, consider the ith rectangle R_i of area ΔA_i, as shown in Figure 14.33. Assume that the mass of R_i is concentrated at one of its interior points (x_i, y_i). The moment of mass of R_i with respect to the x-axis can be approximated by

$$(\text{mass})(y_i) \approx [\rho(x_i, y_i)\,\Delta A_i](y_i).$$

Similarly, the moment of mass with respect to the y-axis can be approximated by

$$(\text{mass})(x_i) \approx [\rho(x_i, y_i)\,\Delta A_i](x_i).$$

By forming the Riemann sum of all such products and taking the limit as the norm of Δ approaches 0, you obtain the following definitions of moments of mass with respect to the x- and y-axes.

Moments and Center of Mass of a Variable Density Planar Lamina

Let ρ be a continuous density function on the planar lamina R. The **moments of mass** with respect to the x- and y-axes are

$$M_x = \iint_R y\rho(x, y)\,dA \quad \text{and} \quad M_y = \iint_R x\rho(x, y)\,dA.$$

If m is the mass of the lamina, then the **center of mass** is

$$(\bar{x}, \bar{y}) = \left(\frac{M_y}{m}, \frac{M_x}{m}\right).$$

If R represents a simple plane region rather than a lamina, the point $(\bar{x}, \bar{y})$ is called the **centroid** of the region.

For some planar laminas, you can determine the center of mass (or one of its coordinates) using symmetry rather than using integration. For instance, consider the laminas shown in Figure 14.34. Using symmetry, you can see that $\bar{x} = 0$ for the first lamina and $\bar{y} = 0$ for the second lamina.

$$R: 0 \leq x \leq 1$$
$$-\sqrt{1 - x^2} \leq y \leq \sqrt{1 - x^2}$$

$$R: -\sqrt{1 - y^2} \leq x \leq \sqrt{1 - y^2}$$
$$0 \leq y \leq 1$$

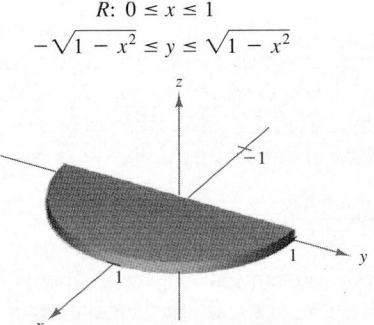

Symmetric with respect to the x-axis   Symmetric with respect to the y-axis

FIGURE 14.34

EXAMPLE 3   Finding the Center of Mass

Find the center of mass of the lamina corresponding to the parabolic region

$$0 \le y \le 4 - x^2 \qquad \text{Parabolic region}$$

where the density at the point (x, y) is proportional to the distance between (x, y) and the x-axis, as shown in Figure 14.35.

Solution   Because the lamina is symmetric with respect to the y-axis and $\rho(x, y) = ky$, the center of mass lies on the y-axis. Thus, $\bar{x} = 0$. To find $\bar{y}$, first find the mass of the lamina.

$$
\begin{aligned}
\text{Mass} &= \int_{-2}^{2} \int_{0}^{4-x^2} ky \, dy \, dx = \frac{k}{2} \int_{-2}^{2} y^2 \Big]_{0}^{4-x^2} dx \\
&= \frac{k}{2} \int_{-2}^{2} (16 - 8x^2 + x^4) \, dx \\
&= \frac{k}{2} \left[16x - \frac{8x^3}{3} + \frac{x^5}{5} \right]_{-2}^{2} \\
&= k\left(32 - \frac{64}{3} + \frac{32}{5} \right) \\
&= \frac{256k}{15}
\end{aligned}
$$

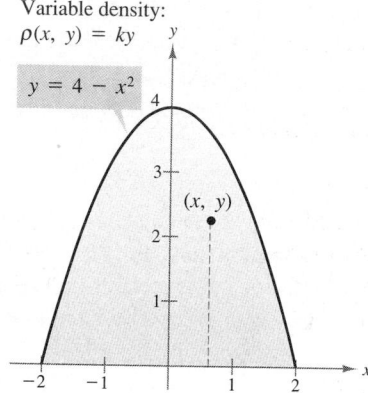

Variable density:
$\rho(x, y) = ky$

$y = 4 - x^2$

(x, y)

FIGURE 14.35
The parabolic region of variable density.

Next, find the moment about the x-axis.

$$
\begin{aligned}
M_x &= \int_{-2}^{2} \int_{0}^{4-x^2} (y)(ky) \, dy \, dx = \frac{k}{3} \int_{-2}^{2} y^3 \Big]_{0}^{4-x^2} dx \\
&= \frac{k}{3} \int_{-2}^{2} (64 - 48x^2 + 12x^4 - x^6) \, dx \\
&= \frac{k}{3} \left[64x - 16x^3 + \frac{12x^5}{5} - \frac{x^7}{7} \right]_{-2}^{2} \\
&= \frac{4096k}{105}
\end{aligned}
$$

Thus,

$$\bar{y} = \frac{M_x}{m} = \frac{4096k/105}{256k/15} = \frac{16}{7}$$

and the center of mass is $(0, \frac{16}{7})$.

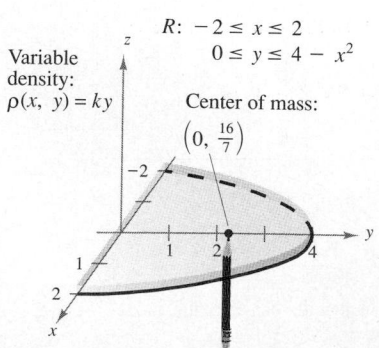

$R: -2 \le x \le 2$
$\quad\quad 0 \le y \le 4 - x^2$
Variable
density:
$\rho(x, y) = ky$

Center of mass:
$\left(0, \frac{16}{7}\right)$

FIGURE 14.36
The center of mass is the balancing point.

Although you can think of the moments M_x and M_y as measuring the tendency to rotate about the x- or y-axis, the calculation of moments is usually an intermediate step toward a more tangible goal. The use of the moments M_x and M_y in Example 3 is typical—to find the center of mass. Determination of the center of mass is useful in a variety of applications that allow you to treat a lamina as if its mass were concentrated at just one point. Intuitively, you can think of the center of mass as the balancing point of the lamina. For instance, the lamina in Example 3 should balance on the point of a pencil placed at $(0, \frac{16}{7})$, as shown in Figure 14.36.

Moments of Inertia

The moments M_x and M_y used in determining the center of mass of a lamina are sometimes called the **first moments** about the x- and y-axes. In each case, the moment is the product of a mass times a distance.

$$M_x = \int_R\!\!\int (y)\underbrace{\rho(x, y)\, dA}, \qquad M_y = \int_R\!\!\int (x)\underbrace{\rho(x, y)\, dA}$$

Distance to x-axis	Mass

Distance to y-axis	Mass

We now look at another type of moment—the **second moment,** or the **moment of inertia** of a lamina about a line. In the same way that mass is a measure of the tendency of matter to resist a change in straight-line motion, the moment of inertia about a line is a *measure of the tendency of matter to resist a change in rotational motion.* For example, if a particle of mass m is a distance d from a fixed line, then its moment of inertia about the line is defined as

$$I = md^2 = (\text{mass})(\text{distance})^2.$$

As with moments of mass, you can generalize this concept to obtain the moments of inertia about the x- and y-axes of a lamina of variable density. These second moments are denoted by I_x and I_y, and in each case the moment is the product of a mass times the square of a distance.

$$I_x = \int_R\!\!\int (y^2)\underbrace{\rho(x, y)\, dA}, \qquad I_y = \int_R\!\!\int (x^2)\underbrace{\rho(x, y)\, dA}$$

Square of distance to x-axis	Mass

Square of distance to y-axis	Mass

The sum of the moments I_x and I_y is called the **polar moment of inertia** and is denoted by I_0. For a lamina in the xy-plane, I_0 represents the moment of inertia of the lamina about the z-axis. The term "polar moment of inertia" stems from the fact that the square of the polar distance r is used in the calculation.

$$I_0 = \int_R\!\!\int (x^2 + y^2)\rho(x, y)\, dA = \int_R\!\!\int r^2\, \rho(x, y)\, dA$$

EXAMPLE 4   Finding the Moment of Inertia

Find the moment of inertia about the x-axis of the lamina in Example 3.

Solution   From the definition of moment of inertia, you have

$$\begin{aligned}
I_x &= \int_{-2}^{2} \int_{0}^{4-x^2} y^2(ky)\, dy\, dx \\
&= \frac{k}{4} \int_{-2}^{2} y^4 \Big]_0^{4-x^2} dx \\
&= \frac{k}{4} \int_{-2}^{2} (256 - 256x^2 + 96x^4 - 16x^6 + x^8)\, dx \\
&= \frac{k}{4} \left[256x - \frac{256x^3}{3} + \frac{96x^5}{5} - \frac{16x^7}{7} + \frac{x^9}{9} \right]_{-2}^{2} = \frac{32{,}768k}{315}.
\end{aligned}$$

The moment of inertia I of a revolving lamina can be used to measure its kinetic energy. For example, suppose a planar lamina is revolving about a line with an **angular speed** of ω radians per second, as shown in Figure 14.37. The kinetic energy of the revolving lamina is

$$E = \frac{1}{2}I\omega^2. \qquad \text{Kinetic energy for rotational motion}$$

On the other hand, the kinetic energy of a mass m moving in a straight line at a velocity v is

$$E = \frac{1}{2}mv^2. \qquad \text{Kinetic energy for linear motion}$$

Thus, the kinetic energy of a mass moving in a straight line is proportional to its mass, but the kinetic energy of a mass revolving about an axis is proportional to its moment of inertia.

The **radius of gyration** $\bar{\bar{r}}$ of a revolving mass m with moment of inertia I is defined to be

$$\bar{\bar{r}} = \sqrt{\frac{I}{m}}. \qquad \text{Radius of gyration}$$

If the entire mass were located at a distance $\bar{\bar{r}}$ from its axis of revolution, it would have the same moment of inertia and, consequently, the same kinetic energy. For instance, the radius of gyration of the lamina in Example 4 about the x-axis is given by

$$\bar{\bar{y}} = \sqrt{\frac{I_x}{m}} = \sqrt{\frac{32{,}768k/315}{256k/15}} = \sqrt{\frac{128}{21}} \approx 2.47.$$

FIGURE 14.37
Planar lamina revolving at ω radians per second.

EXAMPLE 5   Finding the Radius of Gyration

Find the radius of gyration about the y-axis for the lamina corresponding to the region $R: 0 \le y \le \sin x$, $0 \le x \le \pi$, where the density at (x, y) is given by $\rho(x, y) = x$.

Solution   The region R is shown in Figure 14.38. By integrating $\rho(x, y) = x$ over the region R, you can determine that the mass of the region is π. The moment of inertia about the y-axis is

$$
\begin{aligned}
I_y &= \int_0^\pi \int_0^{\sin x} x^3 \, dy \, dx \\
&= \int_0^\pi x^3 y \Big]_0^{\sin x} \, dx \\
&= \int_0^\pi x^3 \sin x \, dx \\
&= \left[(3x^2 - 6)(\sin x) - (x^3 - 6x)(\cos x) \right]_0^\pi \\
&= \pi^3 - 6\pi.
\end{aligned}
$$

Thus, the radius of gyration about the y-axis is

$$\bar{\bar{x}} = \sqrt{\frac{I_y}{m}} = \sqrt{\frac{\pi^3 - 6\pi}{\pi}} = \sqrt{\pi^2 - 6} \approx 1.97.$$

FIGURE 14.38
The radius of gyration about the y-axis is approximately 1.97.

TECHNOLOGY
Laboratory Guide
Labs 14.4–14.5

EXERCISES for Section 14.4

In Exercises 1–4, find the mass and center of mass of the lamina for the specified densities.

1. R: rectangle with vertices $(0, 0)$, $(a, 0)$, $(0, b)$, and (a, b)
   a. $\rho = k$ **b.** $\rho = ky$

2. R: rectangle with vertices $(0, 0)$, $(a, 0)$, $(0, b)$, and (a, b)
   a. $\rho = kxy$ **b.** $\rho = k(x^2 + y^2)$

3. R: triangle with vertices $(0, 0)$, $(b/2, h)$, and $(b, 0)$
   a. $\rho = k$ **b.** $\rho = ky$

4. R: triangle with vertices $(0, 0)$, $(0, a)$, and $(a, 0)$
   a. $\rho = k$ **b.** $\rho = x^2 + y^2$

In Exercises 5–16, find the mass and center of mass of the lamina bounded by the graphs of the given equations for the specified density or densities. (*Hint:* Some of the integrals are simpler in polar coordinates.)

5. $y = \sqrt{a^2 - x^2}$, $y = 0$
   a. $\rho = k$ **b.** $\rho = k(a - y)y$

6. $x^2 + y^2 = a^2$, $0 \le x$, $0 \le y$
   a. $\rho = k$ **b.** $\rho = k(x^2 + y^2)$

7. $y = \sqrt{x}$, $y = 0$, $x = 4$, $\rho = kxy$

8. $y = x^2$, $y = 0$, $x = 4$, $\rho = kx$

9. $y = \dfrac{1}{1 + x^2}$, $y = 0$, $x = -1$, $x = 1$, $\rho = k$

10. $xy = 4$, $x = 1$, $x = 4$, $\rho = kx^2$

11. $x = 16 - y^2$, $x = 0$, $\rho = kx$

12. $y = 9 - x^2$, $y = 0$, $\rho = ky^2$

13. $y = \sin\dfrac{\pi x}{L}$, $y = 0$, $x = 0$, $x = L$, $\rho = ky$

14. $y = \cos\dfrac{\pi x}{L}$, $y = 0$, $x = 0$, $x = \dfrac{L}{2}$, $\rho = k$

15. $y = \sqrt{a^2 - x^2}$, $0 \le y \le x$, $\rho = k$

16. $y = \sqrt{a^2 - x^2}$, $y = 0$, $y = x$, $\rho = k\sqrt{x^2 + y^2}$

C In Exercises 17–20, use a symbolic integration utility to find the mass and center of mass of the lamina bounded by the graphs of the given equations for the specified density.

17. $y = e^{-x}$, $y = 0$, $x = 0$, $x = 2$, $\rho = ky$

18. $y = \ln x$, $y = 0$, $z = 1$, $x = e$, $\rho = \dfrac{k}{x}$

19. $r = 2\cos 3\theta$, $-\dfrac{\pi}{6} \le \theta \le \dfrac{\pi}{6}$, $\rho = k$

20. $r = 1 + \cos\theta$, $\rho = k$

In Exercises 21–26, verify the given moment(s) of inertia and find $\bar{\bar{x}}$ and $\bar{\bar{y}}$. Assume each lamina has a density of $\rho = 1$. (These regions are common shapes used in engineering.)

21. Rectangle

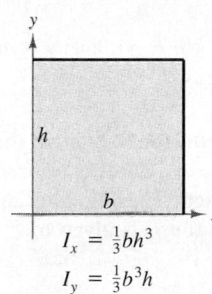

$$I_x = \tfrac{1}{3}bh^3$$
$$I_y = \tfrac{1}{3}b^3 h$$

22. Right triangle

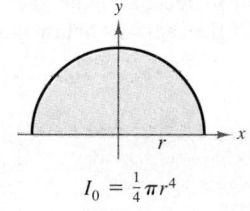

$$I_x = \tfrac{1}{12}bh^3$$
$$I_y = \tfrac{1}{12}b^3 h$$

23. Circle

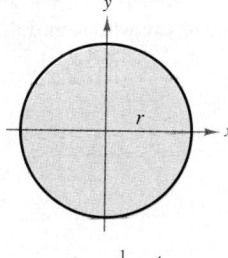

$$I_0 = \tfrac{1}{2}\pi r^4$$

24. Semicircle

$$I_0 = \tfrac{1}{4}\pi r^4$$

25. Quarter circle

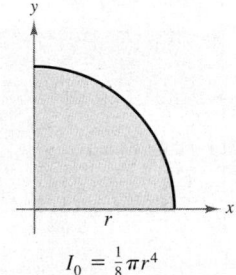

$$I_0 = \tfrac{1}{8}\pi r^4$$

26. Ellipse

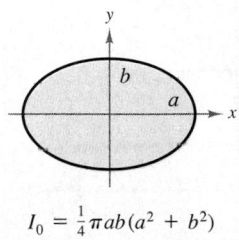

$$I_0 = \tfrac{1}{4}\pi ab(a^2 + b^2)$$

C In Exercises 27–34, find I_x, I_y, I_0, $\bar{x}$, and $\bar{y}$ for the lamina bounded by the graphs of the given equations. Use a symbolic integration utility to evaluate the double integrals.

27. $y = 0$, $y = b$, $x = 0$, $x = a$, $\rho = ky$

28. $y = \sqrt{a^2 - x^2}$, $y = 0$, $\rho = ky$

29. $y = 4 - x^2$, $y = 0$, $x > 0$, $\rho = kx$

30. $y = x$, $y = x^2$, $\rho = kxy$

31. $y = \sqrt{x}$, $y = 0$, $x = 4$, $\rho = kxy$

32. $y = x^2$, $y^2 = x$, $\rho = x^2 + y^2$

33. $y = x^2$, $y^2 = x$, $\rho = kx$

34. $y = x^3$, $y = 4x$, $\rho = ky$

C In Exercises 35–40, find the moment of inertia I, about the specified line, of the lamina bounded by the graphs of the equations. Use a symbolic integration utility to evaluate the double integrals.

35. $x^2 + y^2 = b^2$, $\rho = k$, line: $x = a(a > b)$

36. $y = 0$, $y = 2$, $x = 0$, $x = 4$, $\rho = k$, line: $x = 6$

37. $y = \sqrt{x}$, $y = 0$, $x = 4$, $\rho = kx$, line: $x = 6$

38. $y = \sqrt{a^2 - x^2}$, $y = 0$, $\rho = ky$, line: $y = a$

39. $y = \sqrt{a^2 - x^2}$, $y = 0$, $0 \leq x$, $\rho = k(a - y)$, line: $y = a$

40. $y = 4 - x^2$, $y = 0$, $\rho = k$, line: $y = 2$

Hydraulics   In Exercises 41–44, determine the location of the horizontal axis y_a at which a vertical gate in a dam is required to be hinged so that there is no moment causing rotation under the specified loading. The model for y_a is given by

$$y_a = \bar{y} - \frac{I_{\bar{y}}}{hA}$$

where $\bar{y}$ is the y-coordinate of the centroid of the gate, $I_{\bar{y}}$ is the moment of inertia of the gate about the line $y = \bar{y}$, h is the depth of the centroid below the surface of the water, and A is the area of the gate.

41.

42.

43.

44.

45. Prove the following Theorem of Pappus.

Let R be a region in a plane and let L be a line in the same plane such that L does not intersect the interior of R. If r is the distance between the centroid of R and the line, then the volume V of the solid of revolution formed by revolving R about the line is given by

$$V = 2\pi rA$$

where A is the area of R.

Surface Area

At this point you know a great deal about the solid region lying between a surface and a closed and bounded region R in the xy-plane, as shown in Figure 14.39. For example, you know how to find: the extrema of f on R (Section 13.8), the area of the base R of the solid (Section 14.1), the volume of the solid (Section 14.2), and the centroid of the base R (Section 14.4).

In this section, you will learn how to find the upper **surface area** of the solid. Later, you will learn how to find the centroid of the solid (Section 14.6) and the lateral surface area (Section 15.2).

To begin, consider a surface S given by $z = f(x, y)$ defined over a region R. Assume that R is closed and bounded and that f has continuous first partial derivatives. To find the surface area, construct an inner partition of R consisting of n rectangles, where the area of the ith rectangle R_i is $\Delta A_i = \Delta x_i \Delta y_i$, as shown in Figure 14.40. In each R_i, let (x_i, y_i) be the point that is closest to the origin. At the point $(x_i, y_i, z_i) = (x_i, y_i, f(x_i, y_i))$ on the surface S, construct a tangent plane T_i. The area of the portion of the tangent plane that lies directly above R_i is approximately equal to the area of the surface lying directly above R_i. That is, $\Delta T_i \approx \Delta S_i$. Hence, the surface area of S is given by $\sum \Delta S_i \approx \sum \Delta T_i$.

To find the area of the parallelogram ΔT_i, note that its sides are given by the vectors

$$\mathbf{u} = \Delta x_i \mathbf{i} + f_x(x_i, y_i) \Delta x_i \mathbf{k} \quad \text{and} \quad \mathbf{v} = \Delta y_i \mathbf{j} + f_y(x_i, y_i) \Delta y_i \mathbf{k}.$$

From Theorem 11.8, the area of ΔT_i is given by $\| \mathbf{u} \times \mathbf{v} \|$, where

$$\mathbf{u} \times \mathbf{v} = \begin{vmatrix} \mathbf{i} & \mathbf{j} & \mathbf{k} \\ \Delta x_i & 0 & f_x(x_i, y_i) \Delta x_i \\ 0 & \Delta y_i & f_y(x_i, y_i) \Delta y_i \end{vmatrix}$$

$$= -f_x(x_i, y_i) \Delta x_i \Delta y_i \mathbf{i} - f_y(x_i, y_i) \Delta x_i \Delta y_i \mathbf{j} + \Delta x_i \Delta y_i \mathbf{k}$$

$$= (-f_x(x_i, y_i)\mathbf{i} - f_y(x_i, y_i)\mathbf{j} + \mathbf{k}) \Delta A_i.$$

Thus, the area of ΔT_i is

$$\| \mathbf{u} \times \mathbf{v} \| = \sqrt{[f_x(x_i, y_i)]^2 + [f_y(x_i, y_i)]^2 + 1} \ \Delta A_i$$

and

$$\text{Surface area of } S \approx \sum_{i=1}^{n} \Delta S_i \approx \sum_{i=1}^{n} \sqrt{1 + [f_x(x_i, y_i)]^2 + [f_y(x_i, y_i)]^2} \ \Delta A_i.$$

This suggests the following definition of surface area.

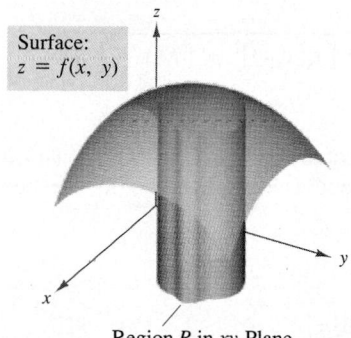

Surface: $z = f(x, y)$

Region R in xy-Plane

FIGURE 14.39

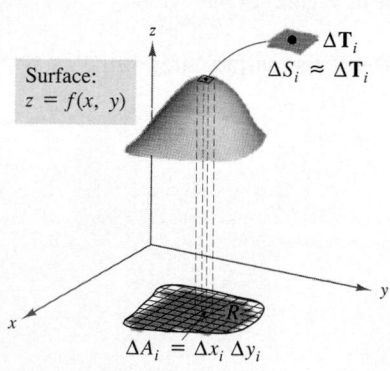

Surface: $z = f(x, y)$

ΔT_i

$\Delta S_i \approx \Delta T_i$

$\Delta A_i = \Delta x_i \Delta y_i$

FIGURE 14.40

Definition of Surface Area

If f and its first partial derivatives are continuous on the closed region R in the xy-plane, then the **area of the surface** $z = f(x, y)$ over R is given by

$$\text{Surface area} = \int_R\!\!\int dS = \int_R\!\!\int \sqrt{1 + [f_x(x, y)]^2 + [f_y(x, y)]^2}\, dA.$$

As an aid to remembering the double integral for surface area, it is helpful to note its similarity to the integral for arc length.

Length on x-axis $\qquad \displaystyle\int_a^b dx$

Arc length in xy-plane $\qquad \displaystyle\int_a^b ds = \int_a^b \sqrt{1 + [f'(x)]^2}\, dx$

Area in xy-plane $\qquad \displaystyle\int_R\!\!\int dA$

Surface area in space $\qquad \displaystyle\int_R\!\!\int dS = \int_R\!\!\int \sqrt{1 + [f_x(x, y)]^2 + [f_y(x, y)]^2}\, dA$

Like integrals for arc length, integrals for surface area are often very difficult to evaluate. However, one type that is easily evaluated is demonstrated in the next example.

EXAMPLE 1   The Surface Area of a Plane Region

Find the surface area of that portion of the plane $z = 2 - x - y$ that lies above the circle $x^2 + y^2 = 1$ in the first quadrant, as shown in Figure 14.41.

Solution   Because $f_x(x, y) = -1$ and $f_y(x, y) = -1$, the surface area is given by

$$S = \int_R\!\!\int \sqrt{1 + [f_x(x, y)]^2 + [f_y(x, y)]^2}\, dA$$

$$= \int_R\!\!\int \sqrt{3}\, dA$$

$$= \sqrt{3} \int_R\!\!\int dA.$$

Note that the integral on the right is simply $\sqrt{3}$ times the area of the region R. Thus, the area of S is

$$S = \sqrt{3}\,(\text{area of } R)$$

$$= \sqrt{3}\left(\frac{\pi}{4}\right)$$

$$= \frac{\sqrt{3}\,\pi}{4}.$$

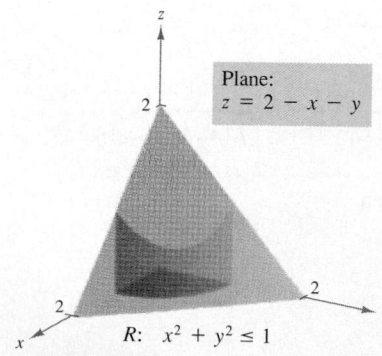

Plane: $z = 2 - x - y$

$R:\ x^2 + y^2 \le 1$

FIGURE 14.41
The surface area of the portion of the plane that lies above the quarter circle is $\sqrt{3}\pi/4$.

Surface:
$f(x, y) = 1 - x^2 + y$

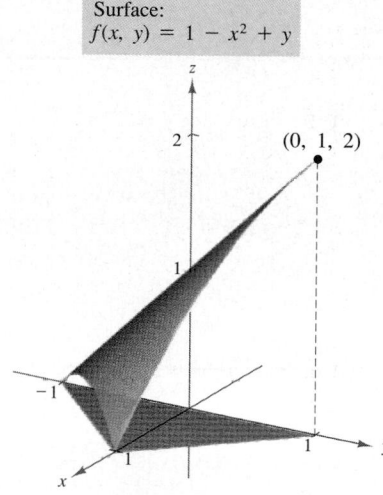

(0, 1, 2)

y R:

$0 \le x \le 1$
$x - 1 \le y \le 1 - x$

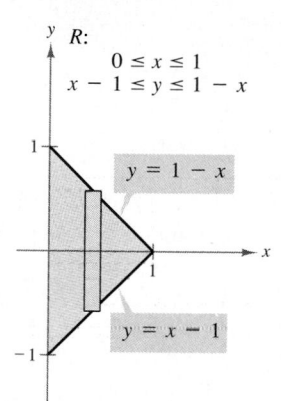

$y = 1 - x$

$y = x - 1$

FIGURE 14.42

EXAMPLE 2   Finding Surface Area

Find the area of that portion of the surface $f(x, y) = 1 - x^2 + y$ that lies above the triangular region with vertices $(1, 0, 0)$, $(0, -1, 0)$, and $(0, 1, 0)$, as shown in Figure 14.42.

Solution   Because $f_x(x, y) = -2x$ and $f_y(x, y) = 1$, you have

$$S = \iint_R \sqrt{1 + [f_x(x, y)]^2 + [f_y(x, y)]^2} \, dA$$

$$= \iint_R \sqrt{1 + 4x^2 + 1} \, dA.$$

In Figure 14.42, you can see that the bounds for R are $0 \le x \le 1$ and $x - 1 \le y \le 1 - x$. Thus, the integral becomes

$$S = \int_0^1 \int_{x-1}^{1-x} \sqrt{2 + 4x^2} \, dy \, dx$$

$$= \int_0^1 y\sqrt{2 + 4x^2} \Big]_{x-1}^{1-x} dx$$

$$= \int_0^1 \left(2\sqrt{2 + 4x^2} - 2x\sqrt{2 + 4x^2} \right) dx$$

$$= \left[x\sqrt{2 + 4x^2} + \ln\left(2x + \sqrt{2 + 4x^2}\right) - \frac{(2 + 4x^2)^{3/2}}{6} \right]_0^1$$

$$= \sqrt{6} + \ln\left(2 + \sqrt{6}\right) - \sqrt{6} - \ln\sqrt{2} + \frac{1}{3}\sqrt{2}$$

$$\approx 1.618.$$

EXAMPLE 3   Change of Variables to Polar Coordinates

Find the surface area of the paraboloid $z = 1 + x^2 + y^2$ that lies above the unit circle, as shown in Figure 14.43.

Solution   Because $f_x(x, y) = 2x$ and $f_y(x, y) = 2y$, you have

$$S = \iint_R \sqrt{1 + [f_x(x, y)]^2 + [f_y(x, y)]^2} \, dA$$

$$= \iint_R \sqrt{1 + 4x^2 + 4y^2} \, dA.$$

You can convert to polar coordinates by letting $x = r\cos\theta$ and $y = r\sin\theta$. Then, because the region R is bounded by $0 \le r \le 1$ and $0 \le \theta \le 2\pi$, you have

$$S = \int_0^{2\pi} \int_0^1 \sqrt{1 + 4r^2} \, r \, dr \, d\theta = \int_0^{2\pi} \frac{1}{12}(1 + 4r^2)^{3/2} \Big]_0^1 d\theta$$

$$= \int_0^{2\pi} \frac{5\sqrt{5} - 1}{12} \, d\theta$$

$$= \frac{\pi(5\sqrt{5} - 1)}{6}$$

$$\approx 5.33.$$

Paraboloid:
$z = 1 + x^2 + y^2$

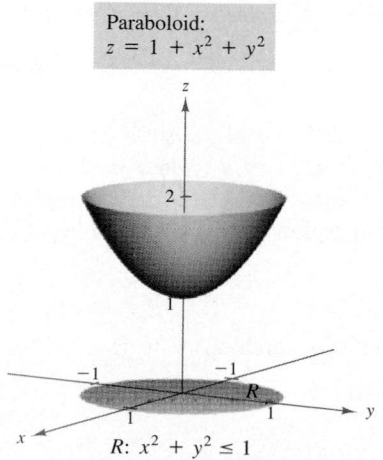

R

$R: x^2 + y^2 \le 1$

FIGURE 14.43
The surface area of the portion of the paraboloid that lies above the unit circle is approximately 5.33.

EXAMPLE 4   Finding Surface Area

Find the surface area S of that portion of the hemisphere

$$f(x, y) = \sqrt{25 - x^2 - y^2}$$

that lies above the region R bounded by the circle $x^2 + y^2 = 9$, as shown in Figure 14.44.

Solution   The first partial derivatives of f are

$$f_x(x, y) = \frac{-x}{\sqrt{25 - x^2 - y^2}} \quad \text{and} \quad f_y(x, y) = \frac{-y}{\sqrt{25 - x^2 - y^2}}$$

and from the formula for surface area, you have

$$dS = \sqrt{1 + [f_x(x, y)]^2 + [f_y(x, y)]^2} \, dA$$

$$= \frac{5}{\sqrt{25 - x^2 - y^2}} \, dA.$$

Therefore, the surface area is

$$S = \int\!\!\int_R \frac{5}{\sqrt{25 - x^2 - y^2}} \, dA.$$

You can convert to polar coordinates by letting $x = r\cos\theta$ and $y = r\sin\theta$. Then, because the region R is bounded by

$$0 \le r \le 3 \quad \text{and} \quad 0 \le \theta \le 2\pi$$

you obtain

$$S = \int_0^{2\pi} \int_0^3 \frac{5}{\sqrt{25 - r^2}} \, r \, dr \, d\theta$$

$$= 5 \int_0^{2\pi} -\sqrt{25 - r^2} \, \Big]_0^3 \, d\theta$$

$$= 5 \int_0^{2\pi} d\theta$$

$$= 10\pi.$$

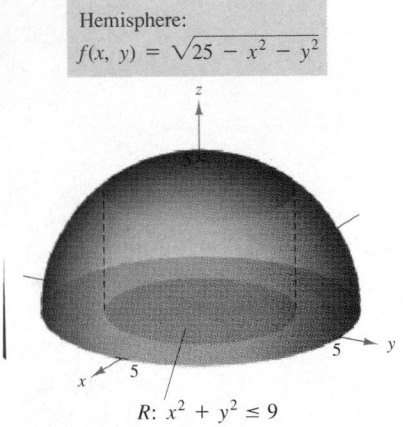

Hemisphere:
$$f(x, y) = \sqrt{25 - x^2 - y^2}$$

$R: x^2 + y^2 \le 9$

FIGURE 14.44
The surface area of the portion of the hemisphere that lies above the circle is 10π.

The procedure used in Example 4 can be extended to find the surface area of a sphere by using the region R bounded by the circle $x^2 + y^2 = a^2$, where $0 < a < 5$, as shown in Figure 14.45. The surface area of that portion of the hemisphere $f(x, y) = \sqrt{25 - x^2 - y^2}$ lying above the circular region can be shown to be

$$S = \int\!\!\int_R \frac{5}{\sqrt{25 - x^2 - y^2}} \, dA$$

$$= \int_0^{2\pi} \int_0^a \frac{5}{\sqrt{25 - r^2}} \, r \, dr \, d\theta$$

$$= 10\pi \left(5 - \sqrt{25 - a^2} \right).$$

By taking the limit as a approaches 5 and doubling the result, we obtain a total area of 100π. (The surface area of a sphere of radius r is $S = 4\pi r^2$.)

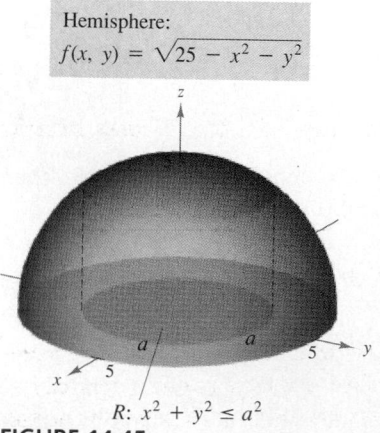

Hemisphere:
$$f(x, y) = \sqrt{25 - x^2 - y^2}$$

$R: x^2 + y^2 \le a^2$

FIGURE 14.45

You can use Simpson's Rule or the Trapezoidal Rule to approximate the value of a double integral, *provided* you can get through the first integration. This is demonstrated in the next example.

EXAMPLE 5   Approximating Surface Area by Simpson's Rule

Find the area of the surface of the paraboloid

$$f(x, y) = 2 - x^2 - y^2$$

that lies above the square region bounded by $-1 \le x \le 1$ and $-1 \le y \le 1$, as shown in Figure 14.46.

Solution Using the partial derivatives

$$f_x(x, y) = -2x \quad \text{and} \quad f_y(x, y) = -2y$$

you obtain a surface area of

$$S = \int\int_R \sqrt{1 + [f_x(x, y)]^2 + [f_y(x, y)]^2} \, dA$$

$$= \int\int_R \sqrt{1 + 4x^2 + 4y^2} \, dA.$$

In polar coordinates, the line $x = 1$ is given by $r \cos \theta = 1$ or $r = \sec \theta$, and you can determine from Figure 14.47 that one-fourth of the region R is bounded by

$$0 \le r \le \sec \theta \quad \text{and} \quad -\frac{\pi}{4} \le \theta \le \frac{\pi}{4}.$$

Letting $x = r \cos \theta$ and $y = r \sin \theta$ produces

$$\frac{1}{4} S = \frac{1}{4} \int\int_R \sqrt{1 + 4x^2 + 4y^2} \, dA$$

$$= \int_{-\pi/4}^{\pi/4} \int_0^{\sec \theta} \sqrt{1 + 4r^2} \, r \, dr \, d\theta$$

$$= \int_{-\pi/4}^{\pi/4} \frac{1}{12} (1 + 4r^2)^{3/2} \Big]_0^{\sec \theta} d\theta$$

$$= \frac{1}{12} \int_{-\pi/4}^{\pi/4} [(1 + 4 \sec^2 \theta)^{3/2} - 1] \, d\theta.$$

Finally, using Simpson's Rule with $n = 10$, you can approximate this *single* integral to be

$$S = \frac{1}{3} \int_{-\pi/4}^{\pi/4} [(1 + 4 \sec^2 \theta)^{3/2} - 1] \, d\theta$$

$$\approx 7.45.$$

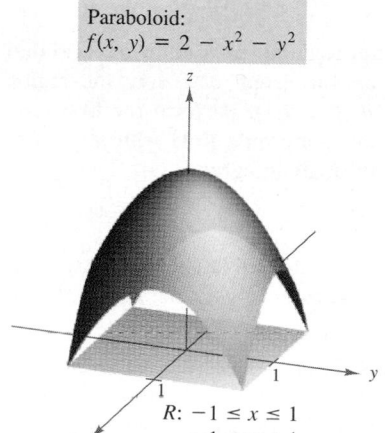

Paraboloid:
$f(x, y) = 2 - x^2 - y^2$

$R: -1 \le x \le 1$
$-1 \le y \le 1$

FIGURE 14.46

$r = \sec \theta$

$\theta = \frac{\pi}{4}$

$\theta = -\frac{\pi}{4}$

FIGURE 14.47
One-fourth of the region R is bounded by $0 \le r \le \sec \theta$ and $-\frac{\pi}{4} \le \theta \le \frac{\pi}{4}$.

TECHNOLOGY Most computer programs that are capable of performing symbolic integration for multiple integrals are also capable of performing numerical approximation techniques. If you have access to such software, try using it to approximate the value of the integral in Example 5.

EXERCISES for Section 14.5

TECHNOLOGY
Laboratory Guide
Labs 14.5–14.6

In Exercises 1–14, find the area of the surface given by $z = f(x,y)$ over the region R. (*Hint:* Some of the integrals are simpler in polar coordinates.)

1. $f(x, y) = 2x + 2y$
R: triangle with vertices $(0, 0)$, $(2, 0)$, and $(0, 2)$

2. $f(x, y) = 10 + 2x - 3y$
R: square with vertices $(0, 0)$, $(2, 0)$, $(0, 2)$ and $(2, 2)$

3. $f(x, y) = 8 + 2x + 2y$
$R = \{(x, y): x^2 + y^2 \le 4\}$

4. $f(x, y) = 10 + 2x - 3y$
$R = \{(x, y): x^2 + y^2 \le 9\}$

5. $f(x, y) = 9 - x^2$
R: square with vertices $(0, 0)$, $(3, 0)$, $(0, 3)$, and $(3, 3)$

6. $f(x, y) = y^2$
R: square with vertices $(0, 0)$, $(3, 0)$, $(0, 3)$, and $(3, 3)$

7. $f(x, y) = 2 + x^{3/2}$
R: rectangle with vertices $(0, 0)$, $(0, 4)$, $(3, 4)$, and $(3, 0)$

8. $f(x, y) = 2 + \frac{2}{3}x^{3/2}$
$R = \{(x, y): 0 \le x \le 1, 0 \le y \le 1 - x\}$

9. $f(x, y) = \ln |\sec x|$
$R = \left\{(x, y): 0 \le x \le \dfrac{\pi}{4}, 0 \le y \le \tan x\right\}$

10. $f(x, y) = 4 + x^2 - y^2$
$R = \{(x, y): x^2 + y^2 \le 1\}$

11. $f(x, y) = \sqrt{x^2 + y^2}$
$R = \{(x, y): 0 \le f(x, y) \le 1\}$

12. $f(x, y) = xy$
$R = \{(x, y): x^2 + y^2 \le 16\}$

13. $f(x, y) = \sqrt{a^2 - x^2 - y^2}$
$R = \{(x, y): x^2 + y^2 \le b^2, b < a\}$

14. $f(x, y) = \sqrt{a^2 - x^2 - y^2}$
$R = \{(x, y): x^2 + y^2 \le a^2\}$

In Exercises 15–18, find the area of the surface.

15. The portion of the plane $z = 24 - 3x - 2y$ in the first octant.

16. The portion of the paraboloid $z = 16 - x^2 - y^2$ in the first octant.

17. The portion of the sphere $x^2 + y^2 + z^2 = 25$ inside the cylinder $x^2 + y^2 = 9$.

18. The portion of the cone $z = \sqrt{x^2 + y^2}$ inside the cylinder $x^2 + y^2 = 1$.

C **In Exercises 19–24, write a double integral that represents the surface area of $z = f(x,y)$ over the region R. Use a symbolic integration utility to evaluate the double integral.**

19. $f(x, y) = 2y + x^2$
R: triangle with vertices $(0, 0)$, $(1, 0)$, and $(1, 1)$

20. $f(x, y) = 2x + y^2$
R: triangle with vertices $(0, 0)$, $(2, 0)$, and $(2, 2)$

21. $f(x, y) = 4 - x^2 - y^2$
$R = \{(x, y): 0 \le f(x, y)\}$

22. $f(x, y) = x^2 + y^2$
$R = \{(x, y): 0 \le f(x, y) \le 16\}$

23. $f(x, y) = 4 - x^2 - y^2$
$R = \{(x, y): 0 \le x \le 1, 0 \le y \le 1\}$

24. $f(x, y) = \frac{2}{3}x^{3/2} + \cos x$
$R = \{(x, y): 0 \le x \le 1, 0 \le y \le 1\}$

C **In Exercises 25 and 26, approximate the double integral that gives the surface area on the graph of f over the region $R = \{(x,y): 0 \le x \le 1, 0 \le y \le 1\}$. Perform the first integration by hand. Then use Simpson's Rule with $n = 10$ to approximate the resulting single integral.**

25. $f(x, y) = e^x$ **26.** $f(x, y) = \frac{2}{5}y^{5/2}$

In Exercises 27–32, set up the double integral that gives the area of the surface on the graph of f over the region R.

27. $f(x, y) = x^3 - 3xy + y^3$
R: square with vertices $(1, 1)$, $(-1, 1)$, $(-1, -1)$, and $(1, -1)$

28. $f(x, y) = e^{-x} \sin y$
$R = \{(x, y): 0 \le x \le 4, 0 \le y \le x\}$

29. $f(x, y) = e^{-x} \sin y$
$R = \{(x, y): x^2 + y^2 \le 4\}$

30. $f(x, y) = x^2 - 3xy - y^2$
$R = \{(x, y): 0 \le x \le 4, 0 \le y \le x\}$

31. $f(x, y) = e^{xy}$
$R = \{(x, y): 0 \le x \le 4, 0 \le y \le 10\}$

32. $f(x, y) = \cos(x^2 + y^2)$
$R = \left\{(x, y): x^2 + y^2 \le \dfrac{\pi}{2}\right\}$

33. Find the surface area of the solid of intersection of the cylinders $x^2 + z^2 = 1$ and $y^2 + z^2 = 1$ (see figure).

34. Show that the surface area of the cone
$$z = k\sqrt{x^2 + y^2}, \quad k > 0$$
over the circular region $x^2 + y^2 \le r^2$ in the xy-plane is $\pi r^2 \sqrt{k^2 + 1}$, as shown in the figure.

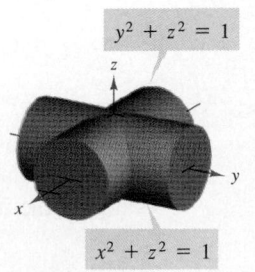

$y^2 + z^2 = 1$

$x^2 + z^2 = 1$

FIGURE FOR 33

$z = k\sqrt{x^2 + y^2}, k > 0$

FIGURE FOR 34

35. *Building Design*   A new auditorium is built with a foundation in the shape of $\frac{1}{4}$ of a circle of radius 50 feet. Therefore, it forms a region R bounded by the graph of $x^2 + y^2 = 50^2$ with $x \geq 0$ and $y \geq 0$. The equations for the floor and ceiling are given by the following.

$$z = \frac{x + y}{5} \qquad \text{Floor: incline plane}$$

$$z = 20 + \frac{xy}{100} \qquad \text{Ceiling}$$

a. Calculate the volume of the room. This is needed to determine the heating and cooling requirements.
b. Find the surface area of the ceiling.

36. *Product Design*   A company produces a spherical object whose radius is 25 centimeters. A hole, with radius 4 centimeters, is drilled through the center of the object. Find
a. the volume of the object.
b. the outer surface area of the object.

37. The angle between a plane P and the xy-plane is θ, where $0 \leq \theta < \pi/2$. The projection of a rectangular region in P onto the xy-plane is a rectangle whose sides have lengths of Δx and Δy, as shown in the figure. Prove that the area of the rectangular region in P is $\sec \theta \, \Delta x \, \Delta y$.

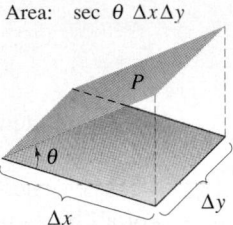

Area: $\sec \theta \, \Delta x \Delta y$

Area in xy-plane: $\Delta x \Delta y$

FIGURE FOR 37

38. *Capillary Action*   A well-known property of liquids is that they will rise in narrow vertical channels–this property is called "capillary action." The figure shows two plates, which form a narrow wedge, in a container of liquid. The upper surface of the liquid follows a hyperbolic shape given by

$$z = \frac{k}{\sqrt{x^2 + y^2}}$$

where x, y, and z are measured in inches. The constant k depends on the angle of the wedge, the type of liquid, and the material that comprises the flat plates.

$\theta = 2 \arctan (0.01)$

9 in.

13 in.

a. Find the volume of the liquid that has risen in the wedge. (Assume $k = 1$.)
b. Find the horizontal surface area of the liquid that has risen in the wedge.
(This problem was adapted from the article "Capillary Phenomena" by Thomas B. Greenslade, Jr. in the May, 1992 issue of the *Physics Teacher*.)

Solid Region Q

Volume of $Q \approx \sum\limits_{i=1}^{n} \Delta V_i$

FIGURE 14.48

Triple Integrals

The procedure used to define a **triple integral** follows that used for double integrals. Consider a function f of three variables that is continuous over a bounded solid region Q. Then, encompass Q with a network of boxes and form the **inner partition** consisting of all boxes lying entirely within Q, as shown in Figure 14.48. The volume of the ith box is

$$\Delta V_i = \Delta x_i \, \Delta y_i \, \Delta z_i. \qquad \text{Volume of } i\text{th box}$$

The **norm** $\|\Delta\|$ of the partition is the length of the longest diagonal of the n boxes in the partition. Choose a point (x_i, y_i, z_i) in each box and form the Riemann sum

$$\sum_{i=1}^{n} f(x_i, y_i, z_i) \, \Delta V_i.$$

Taking the limit as $\|\Delta\| \to 0$ leads to the following definition.

Definition of Triple Integral

If f is continuous over a bounded solid region Q, then the **triple integral of f over Q** is defined as

$$\iiint\limits_{Q} f(x, y, z) \, dV = \lim_{\|\Delta\| \to 0} \sum_{i=1}^{n} f(x_i, y_i, z_i) \, \Delta V_i$$

provided the limit exists. The **volume** of the solid region Q is given by

$$\text{Volume of } Q = \iiint\limits_{Q} dV.$$

Each of the properties of double integrals in Theorem 14.1 can be restated in terms of triple integrals

1. $\displaystyle \iiint\limits_{Q} c\,f(x, y, z) \, dV = c \iiint\limits_{Q} f(x, y, z) \, dV$

2. $\displaystyle \iiint\limits_{Q} [f(x, y, z) \pm g(x, y, z)] \, dV = \iiint\limits_{Q} f(x, y, z) \, dV \pm \iiint\limits_{Q} g(x, y, z) \, dV$

3. $\displaystyle \iiint\limits_{Q} f(x, y, z) \, dV = \iiint\limits_{Q_1} f(x, y, z) \, dV + \iiint\limits_{Q_2} f(x, y, z) \, dV$

where Q is the union of two nonoverlapping solid subregions Q_1 and Q_2.

If the solid region Q is simple, the triple integral $\iiint f(x, y, z) \, dV$ can be evaluated with an iterated integral using one of the six possible orders of integration:

$$dx \, dy \, dz \qquad dy \, dx \, dz \qquad dz \, dx \, dy$$
$$dx \, dz \, dy \qquad dy \, dz \, dx \qquad dz \, dy \, dx.$$

The following version of Fubini's Theorem describes a region that is considered simple with respect to the order $dz \, dy \, dx$. Similar descriptions can be given for the other five orders.

THEOREM 14.5   Evaluation by Iterated Integrals

Let f be continuous on a solid region Q defined by

$$a \leq x \leq b, \qquad h_1(x) \leq y \leq h_2(x), \qquad g_1(x, y) \leq z \leq g_2(x, y)$$

where h_1, h_2, g_1, and g_2 are continuous functions. Then,

$$\iiint\limits_{Q} f(x, y, z) \, dV = \int_a^b \int_{h_1(x)}^{h_2(x)} \int_{g_1(x,y)}^{g_2(x,y)} f(x, y, z) \, dz \, dy \, dx.$$

To evaluate a triple iterated integral in the order $dz \, dy \, dx$, hold *both* x and y constant for the innermost integration. Then, hold x constant for the second integration. This is demonstrated in Example 1.

EXAMPLE 1   Evaluating a Triple Iterated Integral

Evaluate the triple iterated integral

$$\int_0^2 \int_0^x \int_0^{x+y} e^x(y + 2z) \, dz \, dy \, dx.$$

Solution   For the first integration, hold x and y constant and integrate with respect to z.

$$\int_0^2 \int_0^x \int_0^{x+y} e^x(y + 2z) \, dz \, dy \, dx = \int_0^2 \int_0^x e^x(yz + z^2) \Big]_0^{x+y} \, dy \, dx$$

$$= \int_0^2 \int_0^x e^x(x^2 + 3xy + 2y^2) \, dy \, dx$$

REMARK Example 1 demonstrates the integration order $dz \, dy \, dx$. For other orders, you can follow a similar procedure. For instance, to evaluate a triple iterated integral in the order $dx \, dy \, dz$, hold both y and z constant for the innermost integration and integrate with respect to x. Then, for the second integration, hold z constant and integrate with respect to y. Finally, for the third integration, integrate with respect to z.

For the second integration, hold x constant and integrate with respect to y.

$$\int_0^2 \int_0^x e^x(x^2 + 3xy + 2y^2) \, dy \, dx = \int_0^2 \left[e^x \left(x^2 y + \frac{3xy^2}{2} + \frac{2y^3}{3} \right) \right]_0^x dx$$

$$= \frac{19}{6} \int_0^2 x^3 e^x \, dx$$

$$= \frac{19}{6} \left[e^x(x^3 - 3x^2 + 6x - 6) \right]_0^2$$

$$= 19 \left(\frac{e^2}{3} + 1 \right)$$

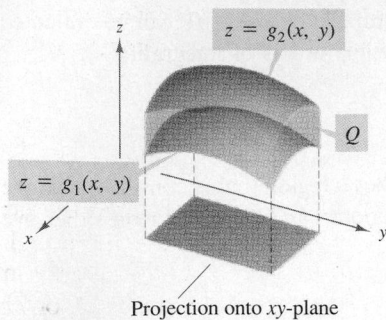

FIGURE 14.49
Solid region Q lies between two surfaces.

$0 \le z \le 2\sqrt{4 - x^2 - y^2}$

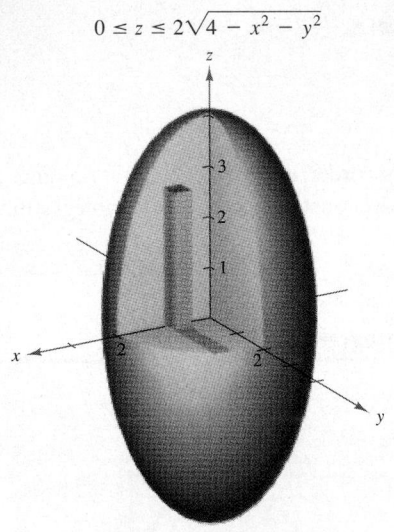

Ellipsoid: $4x^2 + 4y^2 + z^2 = 16$

FIGURE 14.50
The volume of the ellipsoid is $64\pi/3$.

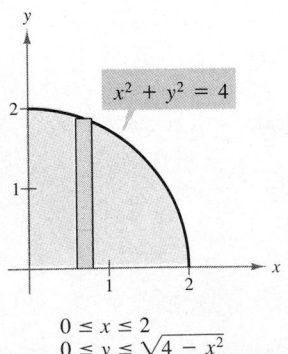

$0 \le x \le 2$
$0 \le y \le \sqrt{4 - x^2}$

FIGURE 14.51
The solid region Q has a triangular base and top with rectangular vertical sides.

To find the limits for a particular order of integration, it is generally advisable to first determine the innermost limits, which may be functions of the outer two variables. Then, by projecting the solid Q onto the coordinate plane of the outer two variables, you can determine their limits of integration by the methods used for double integrals. For instance, to evaluate

$$\iiint_Q f(x, y, z)\, dz\, dy\, dx$$

first determine the limits for z, and then the integral has the form

$$\iint \left[\int_{g_1(x,y)}^{g_2(x,y)} f(x, y, z)\, dz \right] dy\, dx.$$

By projecting the solid Q onto the xy-plane, you can determine the limits for x and y as we did for double integrals, as shown in Figure 14.49.

EXAMPLE 2   Using a Triple Integral to Find Volume

Find the volume of the ellipsoidal solid given by $4x^2 + 4y^2 + z^2 = 16$.

Solution   Because x, y, and z play similar roles in the equation, the order of integration is probably immaterial, and we arbitrarily choose $dz\, dy\, dx$. Moreover, you can simplify the calculation by considering only that portion of the ellipsoid lying in the first octant, as shown in Figure 14.50. From the order $dz\, dy\, dx$, you first determine the bounds for z:

$$0 \le z \le 2\sqrt{4 - x^2 - y^2}.$$

In Figure 14.51, you can see that the boundaries for x and y are $0 \le x \le 2$ and $0 \le y \le \sqrt{4 - x^2}$, so the volume of the ellipsoid is

$$V = \iiint_Q dV$$

$$= 8 \int_0^2 \int_0^{\sqrt{4-x^2}} \int_0^{2\sqrt{4-x^2-y^2}} dz\, dy\, dx$$

$$= 8 \int_0^2 \int_0^{\sqrt{4-x^2}} z \Big]_0^{2\sqrt{4-x^2-y^2}} dy\, dx$$

$$= 16 \int_0^2 \int_0^{\sqrt{4-x^2}} \sqrt{(4 - x^2) - y^2}\, dy\, dx$$

$$= 8 \int_0^2 \left[y\sqrt{4 - x^2 - y^2} + (4 - x^2)\arcsin\left(\frac{y}{\sqrt{4 - x^2}}\right) \right]_0^{\sqrt{4-x^2}} dx$$

$$= 8 \int_0^2 (4 - x^2)\left(\frac{\pi}{2}\right) dx$$

$$= 4\pi \left[4x - \frac{x^3}{3} \right]_0^2$$

$$= 4\pi \left[\left(8 - \frac{8}{3}\right) - (0 - 0) \right]$$

$$= \frac{64\pi}{3}.$$

Example 2 is unusual in that all six possible orders of integration produce integrals of comparable difficulty. Try setting up some other possible orders of integration to find the volume of the ellipsoid. For instance, the order $dx\,dy\,dz$ yields the integral

$$V = 8 \int_0^4 \int_0^{\sqrt{16-z^2}/2} \int_0^{\sqrt{16-4y^2-z^2}/2} dx\,dy\,dz.$$

If you solve this integral, you will obtain the same volume obtained in Example 2. This is always the case—the order of integration does not affect the value of the integral. However, the order of integration often does affect the complexity of the integral. In Example 3, the given order of integration is not convenient, and we change the order to simplify the problem.

EXAMPLE 3   Changing the Order of Integration

Evaluate $\displaystyle\int_0^{\sqrt{\pi/2}} \int_x^{\sqrt{\pi/2}} \int_1^3 \sin y^2\,dz\,dy\,dx.$

Solution   Note that after one integration in the given order, you would encounter the integral $2\int \sin(y^2)\,dy$, which is not an elementary function. To avoid this problem, change the order of integration to $dz\,dx\,dy$, so that y is the outer variable. The solid region Q is given by

$$0 \le x \le \sqrt{\frac{\pi}{2}}, \qquad x \le y \le \sqrt{\frac{\pi}{2}}, \qquad 1 \le z \le 3$$

as shown in Figure 14.52, and the projection of Q in the xy-plane yields the bounds

$$0 \le y \le \sqrt{\frac{\pi}{2}} \quad \text{and} \quad 0 \le x \le y.$$

Therefore, you have

$$\begin{aligned}
V &= \iiint_Q dV \\
&= \int_0^{\sqrt{\pi/2}} \int_0^y \int_1^3 \sin(y^2)\,dz\,dx\,dy \\
&= \int_0^{\sqrt{\pi/2}} \int_0^y z\sin(y^2)\Big]_1^3\,dx\,dy \\
&= 2\int_0^{\sqrt{\pi/2}} \int_0^y \sin(y^2)\,dx\,dy \\
&= 2\int_0^{\sqrt{\pi/2}} x\sin(y^2)\Big]_0^y\,dy \\
&= 2\int_0^{\sqrt{\pi/2}} y\sin(y^2)\,dy \\
&= -\cos(y^2)\Big]_0^{\sqrt{\pi/2}} \\
&= 1.
\end{aligned}$$

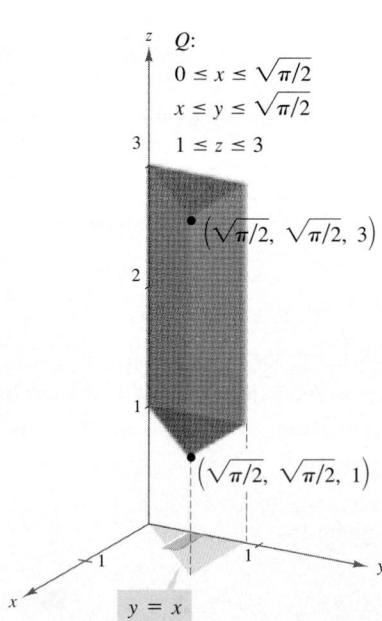

z Q:
$0 \le x \le \sqrt{\pi/2}$
$x \le y \le \sqrt{\pi/2}$
$1 \le z \le 3$

$\left(\sqrt{\pi/2},\ \sqrt{\pi/2},\ 3\right)$

$\left(\sqrt{\pi/2},\ \sqrt{\pi/2},\ 1\right)$

$y = x$

FIGURE 14.52
The volume of the solid region Q is 1.

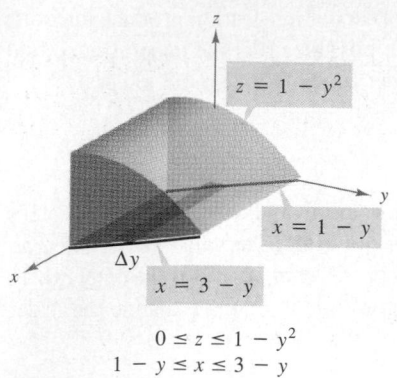

$$0 \le z \le 1 - y^2$$
$$1 - y \le x \le 3 - y$$
$$0 \le y \le 1$$

FIGURE 14.53

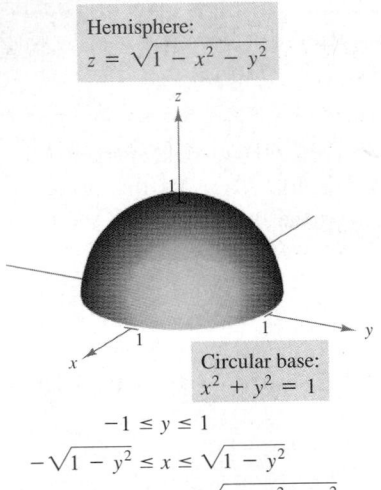

$$-1 \le y \le 1$$
$$-\sqrt{1 - y^2} \le x \le \sqrt{1 - y^2}$$
$$0 \le z \le \sqrt{1 - x^2 - y^2}$$

FIGURE 14.54

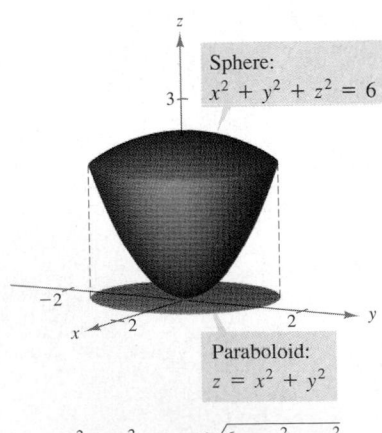

$$x^2 + y^2 \le z \le \sqrt{6 - x^2 - y^2}$$
$$-\sqrt{2 - x^2} \le y \le \sqrt{2 - x^2}$$
$$-\sqrt{2} \le x \le \sqrt{2}$$

FIGURE 14.55

EXAMPLE 4   Determining the Limits of Integration

Set up a triple integral for the volume of each solid region.

a. The region in the first octant bounded above by the cylinder $z = 1 - y^2$ and lying between the vertical planes $x + y = 1$ and $x + y = 3$.

b. The upper hemisphere given by $z = \sqrt{1 - x^2 - y^2}$.

c. The region bounded below by the paraboloid $z = x^2 + y^2$ and above by the sphere $x^2 + y^2 + z^2 = 6$.

Solution

a. In Figure 14.53, note that the solid is bounded below by the xy-plane ($z = 0$) and above by the cylinder $z = 1 - y^2$. Therefore,

$$0 \le z \le 1 - y^2. \qquad \text{Bounds for } z$$

Projecting the region onto the xy-plane produces a parallelogram. Because two sides of the parallelogram are parallel to the x-axis, you have the following bounds:

$$1 - y \le x \le 3 - y \quad \text{and} \quad 0 \le y \le 1.$$

Therefore, the volume of the region is given by

$$V = \iiint_Q dV = \int_0^1 \int_{1-y}^{3-y} \int_0^{1-y^2} dz\, dx\, dy.$$

b. For the upper hemisphere given by $z = \sqrt{1 - x^2 - y^2}$, you have

$$0 \le z \le \sqrt{1 - x^2 - y^2}. \qquad \text{Bounds for } z$$

In Figure 14.54, note that the projection of the hemisphere onto the xy-plane is the circle given by $x^2 + y^2 = 1$, and you can use either order $dx\,dy$ or $dy\,dx$. Choosing the first produces

$$-\sqrt{1 - y^2} \le x \le \sqrt{1 - y^2} \quad \text{and} \quad -1 \le y \le 1$$

which implies that the volume of the region is given by

$$V = \iiint_Q dV = \int_{-1}^1 \int_{-\sqrt{1-y^2}}^{\sqrt{1-y^2}} \int_0^{\sqrt{1-x^2-y^2}} dz\, dx\, dy.$$

c. For the region bounded below by the paraboloid $z = x^2 + y^2$ and above by the sphere $x^2 + y^2 + z^2 = 6$, you have

$$x^2 + y^2 \le z \le \sqrt{6 - x^2 - y^2}. \qquad \text{Bounds for } z$$

The sphere and the paraboloid intersect when $z = 2$. Moreover, you can see in Figure 14.55 that the projection of the solid region onto the xy-plane is the circle given by $x^2 + y^2 = 2$. Using the order $dy\,dx$ produces

$$-\sqrt{2 - x^2} \le y \le \sqrt{2 - x^2} \quad \text{and} \quad -\sqrt{2} \le x \le \sqrt{2}$$

which implies that the volume of the region is given by

$$V = \iiint_Q dV = \int_{-\sqrt{2}}^{\sqrt{2}} \int_{-\sqrt{2-x^2}}^{\sqrt{2-x^2}} \int_{x^2+y^2}^{\sqrt{6-x^2-y^2}} dz\, dy\, dx.$$

Center of Mass and Moments of Inertia

In the remainder of this section, we discuss two applications of triple integrals that are important in engineering. Consider a solid region Q whose density at (x, y, z) is given by the **density function ρ**. The **center of mass** of a solid region Q of mass m is given by $(\bar{x}, \bar{y}, \bar{z})$, where

$$m = \iiint_Q \rho(x, y, z)\,dV \qquad\qquad M_{xz} = \iiint_Q y\rho(x, y, z)\,dV$$

$$M_{yz} = \iiint_Q x\rho(x, y, z)\,dV \qquad\qquad M_{xy} = \iiint_Q z\rho(x, y, z)\,dV$$

and

$$\bar{x} = \frac{M_{yz}}{m}, \qquad \bar{y} = \frac{M_{xz}}{m}, \qquad \bar{z} = \frac{M_{xy}}{m}.$$

The quantities M_{yz}, M_{xz}, and M_{xy} are called the **first moments** of the region Q about the yz-, xz-, and xy-planes, respectively.

The first moments for solid regions are taken about a plane, whereas the second moments for solids are taken about a line. The **second moments** (or **moments of inertia**) about the x-, y- and z-axes are as follows.

$$I_x = \iiint_Q (y^2 + z^2)\rho(x, y, z)\,dV \qquad \text{Moment of inertia about } x\text{-axis}$$

$$I_y = \iiint_Q (x^2 + z^2)\rho(x, y, z)\,dV \qquad \text{Moment of inertia about } y\text{-axis}$$

$$I_z = \iiint_Q (x^2 + y^2)\rho(x, y, z)\,dV \qquad \text{Moment of inertia about } z\text{-axis}$$

For problems requiring the calculation of all three moments, considerable effort can be saved by applying the additive property of triple integrals and writing

$$I_x = I_{xz} + I_{xy}, \quad I_y = I_{yz} + I_{xy}, \quad \text{and} \quad I_z = I_{yz} + I_{xz}$$

where I_{xy}, I_{xz}, and I_{yz} are as follows.

$$I_{xy} = \iiint_Q z^2\rho(x, y, z)\,dV$$

$$I_{xz} = \iiint_Q y^2\rho(x, y, z)\,dV$$

$$I_{yz} = \iiint_Q x^2\rho(x, y, z)\,dV$$

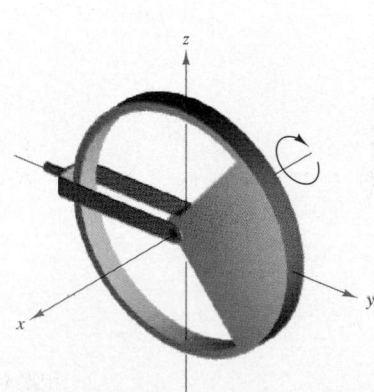

FIGURE 14.56

REMARK   In engineering and physics, the moment of inertia of a mass is used to find the time required for a mass to reach a given speed of rotation about an axis, as indicated in Figure 14.56. The greater the moment of inertia, the longer a force must be applied for the mass to reach the given speed.

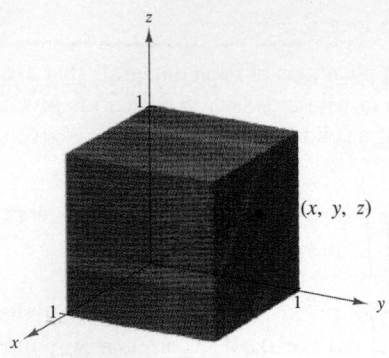

FIGURE 14.57
Variable density:
$\rho(x, y, z) = k(x^2 + y^2 + z^2)$.

EXAMPLE 5   Finding the Center of Mass of a Solid Region

Find the center of mass of the unit cube shown in Figure 14.57, given that the density at the point (x, y, z) is proportional to the square of its distance from the origin.

Solution   Because the density at (x, y, z) is proportional to the square of the distance between $(0, 0, 0)$ and (x, y, z), you have

$$\rho(x, y, z) = k(x^2 + y^2 + z^2).$$

You can use this density function to find the mass of the cube. Because of the symmetry of the region, any order of integration will produce an integral of comparable difficulty.

$$
\begin{aligned}
m &= \int_0^1 \int_0^1 \int_0^1 k(x^2 + y^2 + z^2)\, dz\, dy\, dx \\
&= k \int_0^1 \int_0^1 \left[(x^2 + y^2)z + \frac{z^3}{3} \right]_0^1 dy\, dx \\
&= k \int_0^1 \int_0^1 \left(x^2 + y^2 + \frac{1}{3} \right) dy\, dx \\
&= k \int_0^1 \left[\left(x^2 + \frac{1}{3} \right)y + \frac{y^3}{3} \right]_0^1 dx \\
&= k \int_0^1 \left(x^2 + \frac{2}{3} \right) dx \\
&= k \left[\frac{x^3}{3} + \frac{2x}{3} \right]_0^1 \\
&= k
\end{aligned}
$$

The first moment about the yz-plane is

$$
\begin{aligned}
M_{yz} &= k \int_0^1 \int_0^1 \int_0^1 x(x^2 + y^2 + z^2)\, dz\, dy\, dx \\
&= k \int_0^1 x \left[\int_0^1 \int_0^1 (x^2 + y^2 + z^2)\, dz\, dy \right] dx.
\end{aligned}
$$

Note that x can be factored out of the two inner integrals, because it is constant with respect to y and z. After factoring, the two inner integrals are the same as for the mass m. Hence, you have

$$
\begin{aligned}
M_{yz} &= k \int_0^1 x \left(x^2 + \frac{2}{3} \right) dx \\
&= k \left[\frac{x^4}{4} + \frac{x^2}{3} \right]_0^1 \\
&= \frac{7k}{12}.
\end{aligned}
$$

Therefore,

$$\overline{x} = \frac{M_{yz}}{m} = \frac{7k/12}{k} = \frac{7}{12}.$$

Finally, from the nature of ρ and the symmetry of x, y, and z in this solid region, you have $\overline{x} = \overline{y} = \overline{z}$, and the center of mass is $\left(\frac{7}{12}, \frac{7}{12}, \frac{7}{12} \right)$.

EXAMPLE 6   Moments of Inertia for a Solid Region

Find the moments of inertia about the x- and y-axes for the solid region lying between the hemisphere

$$z = \sqrt{4 - x^2 - y^2}$$

and the xy-plane, given that the density at (x, y, z) is proportional to the distance between (x, y, z) and the xy-plane.

Solution   The density of the region is given by $\rho(x, y, z) = kz$. Considering the symmetry of this problem, you know that $I_x = I_y$, and you need to compute only one moment, say I_x. From Figure 14.58, choose the order $dz\,dy\,dx$ and write

$$I_x = \iiint_Q (y^2 + z^2)\rho(x, y, z)\,dV$$

$$= \int_{-2}^{2} \int_{-\sqrt{4-x^2}}^{\sqrt{4-x^2}} \int_{0}^{\sqrt{4-x^2-y^2}} (y^2 + z^2)(kz)\,dz\,dy\,dx$$

$$= k \int_{-2}^{2} \int_{-\sqrt{4-x^2}}^{\sqrt{4-x^2}} \left[\frac{y^2 z^2}{2} + \frac{z^4}{4} \right]_{0}^{\sqrt{4-x^2-y^2}} dy\,dx$$

$$= k \int_{-2}^{2} \int_{-\sqrt{4-x^2}}^{\sqrt{4-x^2}} \left[\frac{y^2(4 - x^2 - y^2)}{2} + \frac{(4 - x^2 - y^2)^2}{4} \right] dy\,dx$$

$$= \frac{k}{4} \int_{-2}^{2} \int_{-\sqrt{4-x^2}}^{\sqrt{4-x^2}} [(4 - x^2)^2 - y^4]\,dy\,dx$$

$$= \frac{k}{4} \int_{-2}^{2} \left[(4 - x^2)^2 y - \frac{y^5}{5} \right]_{-\sqrt{4-x^2}}^{\sqrt{4-x^2}} dx$$

$$= \frac{k}{4} \int_{-2}^{2} \frac{8}{5}(4 - x^2)^{5/2}\,dx$$

$$= \frac{4k}{5} \int_{0}^{2} (4 - x^2)^{5/2}\,dx \qquad x = 2\sin\theta$$

$$= \frac{4k}{5} \int_{0}^{\pi/2} 64\cos^6\theta\,d\theta$$

$$= \left(\frac{256k}{5} \right)\left(\frac{5\pi}{32} \right) \qquad \text{Wallis's Formula}$$

$$= 8k\pi.$$

Thus, $I_x = 8k\pi$, and $I_y = 8k\pi$. ◼

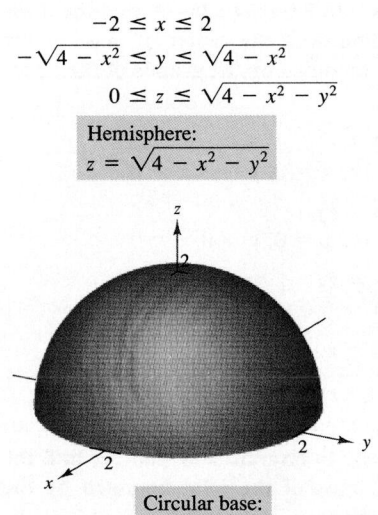

$-2 \le x \le 2$
$-\sqrt{4 - x^2} \le y \le \sqrt{4 - x^2}$
$0 \le z \le \sqrt{4 - x^2 - y^2}$

Hemisphere:
$z = \sqrt{4 - x^2 - y^2}$

Circular base:
$x^2 + y^2 = 4$

FIGURE 14.58
Variable density:
$\rho(x, y, z) = kz$

REMARK   In Example 6, notice that the moments of inertia about the x- and y-axes are equal to each other. The moment about the z-axis, however, is different. Does it seem that the moment of inertia about the z-axis should be less or greater than the moments calculated in Example 6? By performing the calculations, you can determine that $I_z = \frac{16}{3}\pi$. This tells you that the solid shown in Figure 14.58 has a greater resistance to rotation about the x- or y-axis than about the z-axis.

EXERCISES for Section 14.6

In Exercises 1–8, evaluate the triple integral.

1. $\int_0^3 \int_0^2 \int_0^1 (x + y + z)\, dx\, dy\, dz$

2. $\int_{-1}^1 \int_{-1}^1 \int_{-1}^1 x^2 y^2 z^2\, dx\, dy\, dz$

3. $\int_0^1 \int_0^x \int_0^{xy} x\, dz\, dy\, dx$

4. $\int_0^4 \int_0^\pi \int_0^{1-x} x\sin y\, dz\, dy\, dx$

5. $\int_1^4 \int_0^1 \int_0^x 2ze^{-x^2}\, dy\, dx\, dz$

6. $\int_1^4 \int_1^{e^2} \int_0^{1/xz} \ln z\, dy\, dz\, dx$

7. $\int_0^9 \int_0^{y/3} \int_0^{\sqrt{y^2 - 9x^2}} z\, dz\, dx\, dy$

8. $\int_0^{\pi/2} \int_0^{y/2} \int_0^{1/y} \sin y\, dz\, dx\, dy$

C **In Exercises 9 and 10, use a symbolic integration utility to evaluate the triple integral.**

9. $\int_0^2 \int_{-\sqrt{4-x^2}}^{\sqrt{4-x^2}} \int_0^{x^2} x\, dz\, dy\, dx$

10. $\int_0^{\sqrt 2} \int_0^{\sqrt{2-x^2}} \int_{2x^2+y^2}^{4-y^2} y\, dz\, dy\, dx$

C **In Exercises 11 and 12, use a symbolic integration utility to approximate the triple integral.**

11. $\int_0^2 \int_0^{\sqrt{4-x^2}} \int_1^4 \frac{x^2 \sin y}{z}\, dz\, dy\, dx$

12. $\int_0^3 \int_0^{2-(2y/3)} \int_0^{6-2y-3z} ze^{-x^2y^2}\, dx\, dz\, dy$

In Exercises 13–16, sketch the solid whose volume is given by the triple integral, and rewrite the integral with the specified order of integration.

13. $\int_0^4 \int_0^{(4-x)/2} \int_0^{(12-3x-6y)/4} dz\, dy\, dx$
Rewrite using the order $dy\, dx\, dz$.

14. $\int_0^4 \int_0^{\sqrt{16-x^2}} \int_0^{10-x-y} dz\, dy\, dx$
Rewrite using the order $dz\, dx\, dy$.

15. $\int_0^1 \int_y^1 \int_0^{\sqrt{1-y^2}} dz\, dx\, dy$
Rewrite using the order $dz\, dy\, dx$.

16. $\int_0^2 \int_{2x}^4 \int_0^{\sqrt{y^2-4x^2}} dz\, dy\, dx$
Rewrite using the order $dx\, dy\, dz$.

In Exercises 17 and 18, list the six possible orders of integration for the triple integral

$$\iiint_Q xyz\, dV$$

over the solid Q.

17. $Q = \{(x, y, z): 0 \le x \le 1, 0 \le y \le x, 0 \le z \le 3\}$

18. $Q = \{(x, y, z): 0 \le x \le 2, x^2 \le y \le 4, 0 \le z \le 2 - x\}$

Volume   In Exercises 19–24, use a triple integral to find the volume of the solid bounded by the graphs of the given equations.

19. $x = 4 - y^2,\ z = 0,\ z = x$

20. $z = xy,\ z = 0,\ x = 0,\ x = 1,\ y = 0,\ y = 1$

21. $x^2 + y^2 + z^2 = a^2$

22. $z = 9 - x^2 - y^2,\ z = 0$

23. $z = 4 - x^2,\ y = 4 - x^2$ (first octant)

24. $z = 9 - x^2,\ y = -x + 2,\ y = 0,\ z = 0,\ x \ge 0$

Mass and Center of Mass   In Exercises 25–28, find the mass and the indicated coordinates of the center of mass of the solid of specified density bounded by the graphs of the equations.

25. Find $\bar x$ using $\rho(x, y, z) = k$.
$Q: 2x + 3y + 6z = 12,\ x = 0,\ y = 0,\ z = 0$

26. Find $\bar y$ using $\rho(x, y, z) = ky$.
$Q: 2x + 3y + 6z = 12,\ x = 0,\ y = 0,\ z = 0$

27. Find $\bar z$ using $\rho(x, y, z) = kx$.
$Q: z = 4 - x,\ z = 0,\ y = 0,\ y = 4,\ x = 0$

28. Find $\bar y$ using $\rho(x, y, z) = k$.
$Q: \dfrac{x}{a} + \dfrac{y}{b} + \dfrac{z}{c} = 1\ (a, b, c > 0),\ x = 0,\ y = 0,\ z = 0$

Mass and Center of Mass   In Exercises 29 and 30, find the mass and the center of mass of the solid bounded by the graphs of the given equations.

29. $x = 0,\ x = b,\ y = 0,\ y = b,\ z = 0,\ z = b$,
$\rho(x, y, z) = kxy$

30. $x = 0,\ x = a,\ y = 0,\ y = b,\ z = 0,\ z = c,\ \rho(x, y, z) = kz$

C **Centroid**   In Exercises 31–36, find the centroid of the solid region bounded by the graphs of the given equations or described by the figure. Use a symbolic integration utility to evaluate the triple integrals. (Assume uniform density and find the center of mass.)

31. $z = \dfrac{h}{r}\sqrt{x^2 + y^2},\ z = h$

32. $y = \sqrt{4 - x^2},\ z = y,\ z = 0$

33. $z = \sqrt{4^2 - x^2 - y^2},\ z = 0$

34. $z = \dfrac{1}{y^2 + 1}$, $z = 0$, $x = -2$, $x = 2$, $y = 0$, $y = 1$

35.

36.

 Moments of Inertia   In Exercises 37–40, find I_x, I_y, and I_z for the indicated solid of specified density. Use a symbolic integration utility to evaluate the triple integrals.

37. **a.** $\rho = k$
    b. $\rho = kxyz$

38. **a.** $\rho(x, y, z) = k$
    b. $\rho(x, y, z) = k(x^2 + y^2)$

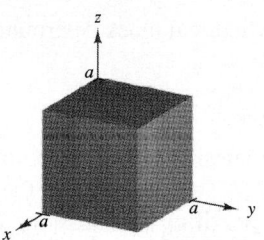

39. **a.** $\rho(x, y, z) = k$
    b. $\rho = ky$

40. **a.** $\rho = kz$
    b. $\rho = k(4 - z)$

 Moments of Inertia   In Exercises 41 and 42, verify the moments of inertia for the solids of uniform density. Use a symbolic integration utility to evaluate the triple integrals.

41. $I_x = I_z = \frac{1}{12}m(3a^2 + L^2)$
$I_y = \frac{1}{2}ma^2$

42. $I_x = \frac{1}{12}m(a^2 + b^2)$
$I_y = \frac{1}{12}m(b^2 + c^2)$
$I_z = \frac{1}{12}m(a^2 + c^2)$

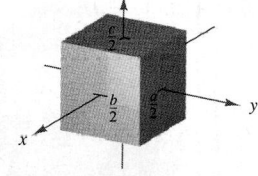

FIGURE FOR 41   **FIGURE FOR 42**
Right Circular Cylinder

Moment of Inertia   In Exercises 43 and 44, set up a triple integral that gives the moment of inertia about the z-axis of the solid Q of density ρ.

43. $Q = \{(x, y, z): -1 \le x \le 1, -1 \le y \le 1,$
$0 \le z \le 1 - x\}$
$\rho = \sqrt{x^2 + y^2 + z^2}$

44. $Q = \{(x, y, z): x^2 + y^2 \le 1, 0 \le z \le 4 - x^2 - y^2\}$
$\rho = kx^2$

SECTION
14.7   Triple Integrals in Cylindrical and Spherical Coordinates
Triple Integrals in Cylindrical Coordinates • Triple Integrals in Spherical Coordinates

PIERRE SIMON DE LAPLACE

One of the first to use a cylindrical coordinate system was the French mathematician Pierre Simon de Laplace (1749–1827). Laplace has been called the "Newton of France," and he published many important works in mechanics, differential equations, and probability.

Triple Integrals in Cylindrical Coordinates

Many common solid regions such as spheres, ellipsoids, cones, and paraboloids can yield difficult triple integrals in rectangular coordinates. In fact, it is precisely this difficulty that led to the introduction of nonrectangular coordinate systems. In this section, you will learn how to use *cylindrical* and *spherical* coordinates to evaluate triple integrals.

Recall from Section 11.7 that the rectangular conversion equations for cylindrical coordinates are

$$x = r\cos\theta, \quad y = r\sin\theta, \quad \text{and} \quad z = z.$$

In this coordinate system, the simplest solid region is a cylindrical block determined by

$$r_1 \le r \le r_2, \quad \theta_1 \le \theta \le \theta_2, \quad \text{and} \quad z_1 \le z \le z_2$$

as shown in Figure 14.59.

To obtain the cylindrical coordinate form of a triple integral, let f be a continuous function of r, θ, and z defined over a bounded solid region Q. Then, encompass the solid by a network of cylindrical blocks and form the **inner partition** Δ, consisting of all blocks lying entirely within Q. The **norm** $\|\Delta\|$ of the partition is the length of the longest diagonal of the n blocks in Δ. If you choose a point (r_i, θ_i, z_i) in the ith block such that r_i is the average radius of the block, then the block's volume is

$$\Delta V_i = (\text{area of base})(\text{height}) = (r_i \Delta r_i \Delta\theta_i) \Delta z_i.$$

Forming the sum

$$\sum_{i=1}^{n} f(r_i, \theta_i, z_i) r_i \Delta r_i \Delta\theta_i \Delta z_i$$

and taking the limit as $\|\Delta\| \to 0$ suggests the following definition.

Triple Integral in Cylindrical Coordinates

If f is a continuous function of r, θ, and z on a bounded solid region Q, then in cylindrical coordinates the **triple integral of f over Q** is

$$\iiint\limits_{Q} f(r, \theta, z) \, dV = \lim_{\|\Delta\| \to 0} \sum_{i=1}^{n} f(r_i, \theta_i, z_i) r_i \Delta r_i \Delta\theta_i \Delta z_i.$$

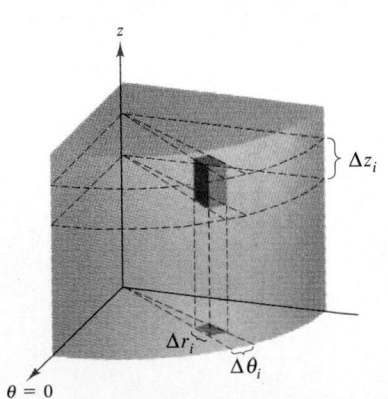

FIGURE 14.59
Volume of cylindrical block:
$\Delta V_i = r_i \Delta r_i \Delta\theta_i \Delta z_i$.

You can evaluate a triple integral in cylindrical coordinates, provided that the solid region Q is simple with respect to one of the six possible orders of integration. For example, if the region is bounded above and below by $h_1(r, \theta) \le z \le h_2(r, \theta)$, then you can write

$$\iiint_Q f(r, \theta, z)\, dV = \iint_R \left[\int_{h_1(r,\theta)}^{h_2(r,\theta)} f(r, \theta, z)\, dz \right] r\, dr\, d\theta$$

where the double integral over R is evaluated in polar coordinates. (That is, R is a plane region that is either r-simple or θ-simple, as discussed in Section 14.3.) If R is r-simple, then the iterated form is

$$\int_{\theta_1}^{\theta_2} \int_{g_1(\theta)}^{g_2(\theta)} \int_{h_1(r,\theta)}^{h_2(r,\theta)} f(r, \theta, z) r\, dz\, dr\, d\theta.$$

To visualize a particular order of integration, it helps to view the iterated integral in terms of three sweeping motions—each adding another dimension to the solid. For instance, in the order $dr\, d\theta\, dz$, the first integration occurs in the r-direction as a point sweeps out a ray. Then, as θ increases, the line sweeps out a sector. Finally, as z increases, the sector sweeps out a solid wedge, as shown in Figure 14.60.

Integrate with respect to r

Integrate with respect to θ

Integrate with respect to z

FIGURE 14.60

EXAMPLE 1   Finding Volume by Cylindrical Coordinates

Find the volume of the solid region Q cut from the sphere $x^2 + y^2 + z^2 = 4$ by the cylinder $r = 2\sin\theta$, as shown in Figure 14.61.

Solution   Because $x^2 + y^2 + z^2 = r^2 + z^2 = 4$, the bounds on z are

$$-\sqrt{4 - r^2} \le z \le \sqrt{4 - r^2}.$$

Let R be the circular projection of the solid onto the $r\theta$-plane. Then the bounds on R are $0 \le r \le 2\sin\theta$ and $0 \le \theta \le \pi$. Thus, the volume of Q is

$$
\begin{aligned}
V &= \int_0^\pi \int_0^{2\sin\theta} \int_{-\sqrt{4-r^2}}^{\sqrt{4-r^2}} r\, dz\, dr\, d\theta \\
&= \int_0^\pi \int_0^{2\sin\theta} rz \Big]_{-\sqrt{4-r^2}}^{\sqrt{4-r^2}} dr\, d\theta \\
&= 2 \int_0^{\pi/2} \int_0^{2\sin\theta} 2r\sqrt{4-r^2}\, dr\, d\theta \\
&= 2 \int_0^{\pi/2} -\frac{2}{3}(4-r^2)^{3/2} \Big]_0^{2\sin\theta} d\theta \\
&= \frac{4}{3} \int_0^{\pi/2} \left[8 - 8\cos^3\theta \right] d\theta \\
&= \frac{32}{3} \int_0^{\pi/2} \left[1 - (\cos\theta)(1 - \sin^2\theta) \right] d\theta \\
&= \frac{32}{3} \left[\theta - \sin\theta + \frac{\sin^3\theta}{3} \right]_0^{\pi/2} \\
&= \frac{16}{9}(3\pi - 4).
\end{aligned}
$$

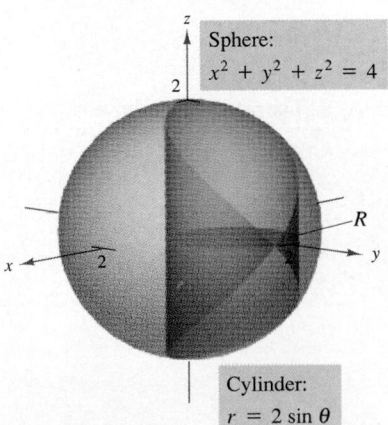

Sphere:
$x^2 + y^2 + z^2 = 4$

Cylinder:
$r = 2\sin\theta$

FIGURE 14.61
The volume of the solid region Q is $\frac{16}{9}(3\pi - 4)$.

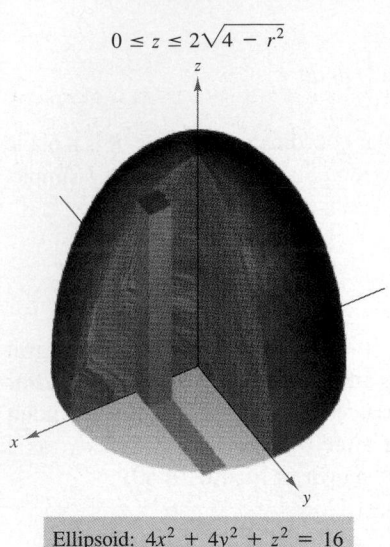

$$0 \le z \le 2\sqrt{4 - r^2}$$

Ellipsoid: $4x^2 + 4y^2 + z^2 = 16$

FIGURE 14.62
The mass of the ellipsoidal solid is $16\pi k$.

EXAMPLE 2   Finding Mass by Cylindrical Coordinates

Find the mass of the ellipsoidal solid Q given by $4x^2 + 4y^2 + z^2 = 16$, lying above the xy-plane. The density at a point in the solid is proportional to the distance between the point and the xy-plane.

Solution   The density function $\rho(r, \theta, z) = kz$. The bounds on z are

$$0 \le z \le \sqrt{16 - 4x^2 - 4y^2} = 2\sqrt{4 - r^2}$$

where $0 \le r \le 2$ and $0 \le \theta \le 2\pi$, as shown in Figure 14.62. The mass of the solid is

$$m = \int_0^{2\pi} \int_0^2 \int_0^{\sqrt{16-4r^2}} kzr\,dz\,dr\,d\theta$$

$$= \frac{k}{2} \int_0^{2\pi} \int_0^2 z^2 r \Big]_0^{\sqrt{16-4r^2}} dr\,d\theta$$

$$= \frac{k}{2} \int_0^{2\pi} \int_0^2 (16r - 4r^3)\,dr\,d\theta$$

$$= \frac{k}{2} \int_0^{2\pi} \left[8r^2 - r^4 \right]_0^2 d\theta$$

$$= 8k \int_0^{2\pi} d\theta = 16\pi k.$$

EXAMPLE 3   Finding a Moment of Inertia

Find the moment of inertia about the axis of symmetry of the solid bounded by the paraboloid $z = x^2 + y^2$ and the plane $z = 4$, as shown in Figure 14.63. The density at each point is proportional to the distance between the point and the z-axis.

Solution   Because the z-axis is the axis of symmetry, and $\rho(x, y, z) = k\sqrt{x^2 + y^2}$, it follows that

$$I_z = \iiint_Q k(x^2 + y^2)\sqrt{x^2 + y^2}\,dV.$$

In cylindrical coordinates, $0 \le r \le \sqrt{x^2 + y^2} = \sqrt{z}$. Therefore, you have

$$I_z = k \int_0^4 \int_0^{2\pi} \int_0^{\sqrt{z}} r^2(r)r\,dr\,d\theta\,dz$$

$$= k \int_0^4 \int_0^{2\pi} \frac{r^5}{5} \Big]_0^{\sqrt{z}} d\theta\,dz$$

$$= k \int_0^4 \int_0^{2\pi} \frac{z^{5/2}}{5}\,d\theta\,dz$$

$$= k \int_0^4 \frac{z^{5/2}}{5} (2\pi)\,dz$$

$$= k \left[\left(\frac{2\pi}{5}\right)\left(\frac{2}{7}\right) z^{7/2} \right]_0^4 = \frac{512k\pi}{35}.$$

Q: bounded by
$z = x^2 + y^2$
$z = 4$

FIGURE 14.63
The moment of inertia about the z-axis is $(512k\pi)/(35)$.

Triple Integrals in Spherical Coordinates

Recall from Section 11.7 that the rectangular conversion equations for spherical coordinates are

$$x = \rho \sin\phi \cos\theta, \quad y = \rho \sin\phi \sin\theta, \quad \text{and} \quad z = \rho \cos\phi.$$

For solids in spherical coordinates, the fundamental element of volume is a spherical block determined by $\rho_1 \le \rho \le \rho_2$, $\theta_1 \le \theta \le \theta_2$, and $\phi_1 \le \phi \le \phi_2$, as shown in Figure 14.64. If (ρ, θ, ϕ) is a point in the interior of such a block, then the volume of the block can be approximated as follows.

$$\Delta V \approx \rho^2 \sin\phi \, \Delta\rho \, \Delta\theta \, \Delta\phi$$

Using the usual inner partition–summation–limit process, you can develop the following version of a triple integral in spherical coordinates.

Triple Integral in Spherical Coordinates

If f is a continuous function of ρ, θ, and ϕ on a bounded solid region Q, then in spherical coordinates the **triple integral of f over Q** is

$$\iiint\limits_{Q} f(\rho, \theta, \phi) \, dV = \lim_{\|\Delta\| \to 0} \sum_{i=1}^{n} f(\rho_i, \theta_i, \phi_i) \rho_i^2 \sin\phi_i \, \Delta\rho_i \, \Delta\theta_i \, \Delta\phi_i.$$

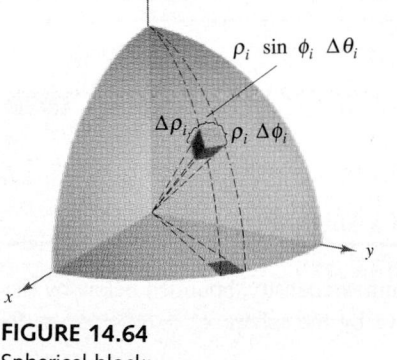

FIGURE 14.64
Spherical block:
$\Delta V_i \approx \rho_i^2 \sin\phi_i \, \Delta\rho_i \, \Delta\theta_i \, \Delta\phi_i.$

REMARK   The Greek letter ρ used in spherical coordinates is not related to density. Rather, it is the three-dimensional analog of the r used in polar coordinates. For problems involving spherical coordinates and a density function, we will use a different symbol to denote density.

Like triple integrals in cylindrical coordinates, triple integrals in spherical coordinates are evaluated with iterated integrals. As with cylindrical coordinates, you can visualize a particular order of integration by viewing the iterated integral in terms of three sweeping motions—each adding another dimension to the solid. For instance, the iterated integral

$$\int_0^{2\pi} \int_0^{\pi/4} \int_0^3 \rho^2 \sin\phi \, d\rho \, d\phi \, d\theta$$

(which is used Example 4) is illustrated in Figure 14.65.

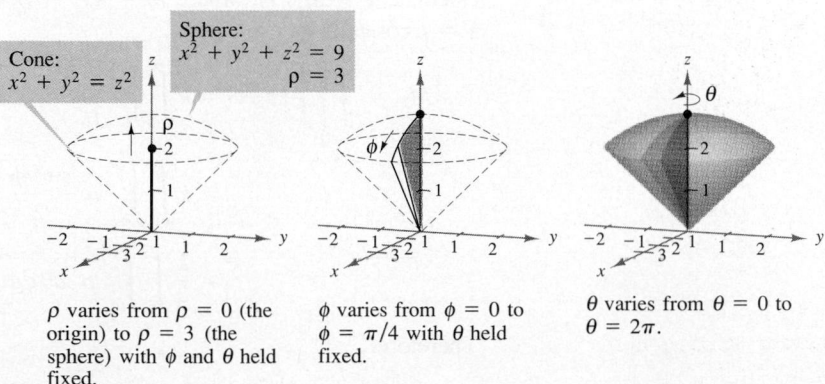

ρ varies from $\rho = 0$ (the origin) to $\rho = 3$ (the sphere) with ϕ and θ held fixed.

ϕ varies from $\phi = 0$ to $\phi = \pi/4$ with θ held fixed.

θ varies from $\theta = 0$ to $\theta = 2\pi$.

FIGURE 14.65

EXAMPLE 4   Finding Volume in Spherical Coordinates

Find the volume of the solid region Q bounded below by the upper nappe of the cone $z^2 = x^2 + y^2$ and above by the sphere $x^2 + y^2 + z^2 = 9$, as shown in Figure 14.66.

Solution   In spherical coordinates, the equation of the sphere is

$$\rho^2 = x^2 + y^2 + z^2 = 9 \quad \rightarrow \quad \rho = 3.$$

Furthermore, the sphere and cone intersect when

$$(x^2 + y^2) + z^2 = (z^2) + z^2 = 9 \quad \rightarrow \quad z = \frac{3}{\sqrt{2}}$$

and, because $z = \rho \cos \phi$, it follows that

$$\left(\frac{3}{\sqrt{2}}\right)\left(\frac{1}{3}\right) = \cos \phi \quad \rightarrow \quad \phi = \frac{\pi}{4}.$$

Consequently, you can use the integration order $d\rho\, d\phi\, d\theta$, where $0 \le \rho \le 3$, $0 \le \phi \le \pi/4$, and $0 \le \theta \le 2\pi$. The volume is

$$V = \iiint_Q dV = \int_0^{2\pi} \int_0^{\pi/4} \int_0^3 \rho^2 \sin\phi\, d\rho\, d\phi\, d\theta$$

$$= \int_0^{2\pi} \int_0^{\pi/4} 9 \sin\phi\, d\phi\, d\theta$$

$$= 9 \int_0^{2\pi} -\cos\phi \Big]_0^{\pi/4} d\theta$$

$$= 9 \int_0^{2\pi} \left(1 - \frac{\sqrt{2}}{2}\right) d\theta = 9\pi\left(2 - \sqrt{2}\right) \approx 16.56.$$

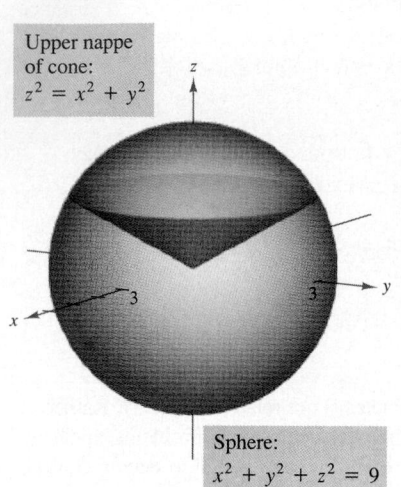

Upper nappe of cone: $z^2 = x^2 + y^2$

Sphere: $x^2 + y^2 + z^2 = 9$

FIGURE 14.66

EXAMPLE 5   Finding the Center of Mass of a Solid Region

Find the center of mass of the solid region Q of uniform density, bounded below by the upper nappe of the cone $z^2 = x^2 + y^2$ and above by the sphere $x^2 + y^2 + z^2 = 9$.

Solution   Because the density is uniform, you can consider the density at the point (x, y, z) to be k. By symmetry, the center of mass lies on the z-axis, and we need only calculate $\bar{z} = M_{xy}/m$, where $m = kV = 9k\pi\left(2 - \sqrt{2}\right)$ from Example 4. Because $z = \rho \cos \phi$, it follows that

$$M_{xy} = \iiint_Q kz\, dV = k \int_0^3 \int_0^{2\pi} \int_0^{\pi/4} (\rho \cos \phi)\rho^2 \sin \phi\, d\phi\, d\theta\, d\rho$$

$$= k \int_0^3 \int_0^{2\pi} \rho^3 \frac{\sin^2 \phi}{2} \Big]_0^{\pi/4} d\theta\, d\rho$$

$$= \frac{k}{4} \int_0^3 \int_0^{2\pi} \rho^3\, d\theta\, d\rho = \frac{k\pi}{2} \int_0^3 \rho^3\, d\rho = \frac{81k\pi}{8}.$$

Therefore,

$$\bar{z} = \frac{M_{xy}}{m} = \frac{81k\pi/8}{9k\pi(2 - \sqrt{2})} = \frac{9(2 + \sqrt{2})}{16} \approx 1.92$$

and the center of mass is approximately $(0, 0, 1.92)$.

EXERCISES for Section 14.7

TECHNOLOGY
Laboratory Guide
Labs 14.6–14.7

In Exercises 1–6, evaluate the triple integral.

1. $\displaystyle\int_0^4 \int_0^{\pi/2} \int_0^2 r\cos\theta\, dr\, d\theta\, dz$

2. $\displaystyle\int_0^{\pi/4} \int_0^2 \int_0^{2-r} rz\, dz\, dr\, d\theta$

3. $\displaystyle\int_0^{\pi/2} \int_0^{2\cos^2\theta} \int_0^{4-r^2} r\sin\theta\, dz\, dr\, d\theta$

4. $\displaystyle\int_0^{\pi/2} \int_0^{\pi} \int_0^2 e^{-\rho^3}\rho^2\, d\rho\, d\theta\, d\phi$

5. $\displaystyle\int_0^{2\pi} \int_0^{\pi/4} \int_0^{\cos\phi} \rho^2 \sin\phi\, d\rho\, d\phi\, d\theta$

6. $\displaystyle\int_0^{\pi/4} \int_0^{\pi/4} \int_0^{\cos\theta} \rho^2 \sin\phi\cos\phi\, d\rho\, d\theta\, d\phi$

C **In Exercises 7 and 8, use a symbolic integration utility to evaluate the triple integral.**

7. $\displaystyle\int_0^4 \int_0^z \int_0^{\pi/2} re^r\, d\theta\, dr\, dz$

8. $\displaystyle\int_0^{\pi/2} \int_0^{\pi} \int_0^{\sin\theta} (2\cos\phi)\rho^2\, d\rho\, d\theta\, d\phi$

In Exercises 9–12, sketch the solid region whose volume is given by the integral, and evaluate the integral.

9. $\displaystyle\int_0^{\pi/2} \int_0^3 \int_0^{e^{-r^2}} r\, dz\, dr\, d\theta$

10. $\displaystyle\int_0^{2\pi} \int_0^{\sqrt3} \int_0^{3-r^2} r\, dz\, dr\, d\theta$

11. $\displaystyle\int_0^{2\pi} \int_{\pi/6}^{\pi/2} \int_0^4 \rho^2 \sin\phi\, d\rho\, d\phi\, d\theta$

12. $\displaystyle\int_0^{2\pi} \int_0^{\pi} \int_2^5 \rho^2 \sin\phi\, d\rho\, d\phi\, d\theta$

In Exercises 13–16, convert the integral from rectangular coordinates to both cylindrical and spherical coordinates, and evaluate the simplest integral.

13. $\displaystyle\int_{-2}^2 \int_{-\sqrt{4-x^2}}^{\sqrt{4-x^2}} \int_{x^2+y^2}^4 x\, dz\, dy\, dx$

14. $\displaystyle\int_0^2 \int_0^{\sqrt{4-x^2}} \int_0^{\sqrt{16-x^2-y^2}} \sqrt{x^2+y^2}\, dz\, dy\, dx$

15. $\displaystyle\int_{-a}^a \int_{-\sqrt{a^2-x^2}}^{\sqrt{a^2-x^2}} \int_a^{a+\sqrt{a^2-x^2-y^2}} x\, dz\, dy\, dx$

16. $\displaystyle\int_0^1 \int_0^{\sqrt{1-x^2}} \int_0^{\sqrt{1-x^2-y^2}} \sqrt{x^2+y^2+z^2}\, dz\, dy\, dx$

In Exercises 17–22, use cylindrical coordinates to find the indicated characteristic of the cone (see figure).

17. *Volume*   Find the volume of the cone.

18. *Centroid*   Find the centroid of the cone.

C **19.** *Center of Mass*   Find the center of mass of the cone, assuming that its density at any point is proportional to the distance between the point and the axis of the cone. Use a symbolic integration utility to evaluate the triple integral.

C **20.** *Center of Mass*   Find the center of mass of the cone, assuming that its density at any point is proportional to the distance between the point and the base. Use a symbolic integration utility to evaluate the triple integral.

21. *Moment of Inertia*   Assume that the cone has uniform density, and show that the moment of inertia about the z-axis is

$$I_z = \frac{3}{10}mr_0^2.$$

22. *Moment of Inertia*   Assume that the density is given by

$$\rho(x,y,z) = k(x^2 + y^2)$$

and find the moment of inertia about the z-axis.

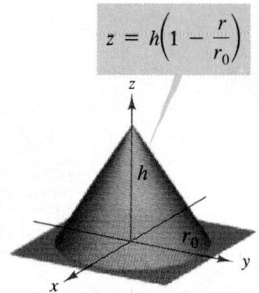

$$z = h\left(1 - \frac{r}{r_0}\right)$$

FIGURE FOR 17–22

Moment of Inertia   **In Exercises 23 and 24, use cylindrical coordinates to verify the given formula for the moment of inertia of the solid of uniform density.**

23. Cylindrical shell: $I_z = \frac12 m(a^2 + b^2)$

$0 < a \le r \le b, 0 \le z \le h$

C **24.** Right circular cylinder: $I_z = \frac32 ma^2$

$r = 2a\sin\theta, 0 \le z \le h$

Use a computer algebra system to evaluate the triple integral.

Volume   **In Exercises 25–28, use cylindrical coordinates to find the volume of the solid.**

25. Solid inside both $x^2 + y^2 + z^2 = a^2$

and $\left(x - \dfrac{a}{2}\right)^2 + y^2 = \left(\dfrac{a}{2}\right)^2$

26. Solid inside $x^2 + y^2 + z^2 = 16$ and outside $z = \sqrt{x^2 + y^2}$

27. *Volume*   Find the volume of the solid bounded by the graphs of the sphere $r^2 + z^2 = a^2$ and the cylinder $r = a\cos\theta$.

28. *Volume*   Find the volume of the solid inside the sphere $x^2 + y^2 + z^2 = 4$ and above the cone $z^2 = x^2 + y^2$.

C *Volume*   **In Exercises 29 and 30, use spherical coordinates to find the volume of the solid.**

29. The torus given by $\rho = 4\sin\phi$. Use a symbolic integration utility to evaluate the triple integral.

30. The solid between the spheres $x^2 + y^2 + z^2 = a^2$ and $x^2 + y^2 + z^2 = b^2$, $b > a$, and inside the cone $z^2 = x^2 + y^2$

Mass   **In Exercises 31 and 32, use spherical coordinates to find the mass of the sphere $x^2 + y^2 + z^2 = a^2$ with the specified density.**

31. The density at any point is proportional to the distance between the point and the origin.

32. The density at any point is proportional to the distance of the point from the z-axis.

Center of Mass   **In Exercises 33 and 34, use spherical coordinates to find the center of mass of the solid of uniform density.**

33. Hemispherical solid of radius r

34. Solid lying between two concentric hemispheres of radii r and R, where $r < R$

Moment of Inertia   **In Exercises 35 and 36, use spherical coordinates to find the moment of inertia about the z-axis of the solid of uniform density.**

35. Solid bounded by the hemisphere $\rho = \cos\phi$, $\pi/4 \le \phi \le \pi/2$, and the cone $\phi = \pi/4$

36. Solid lying between two concentric hemispheres of radii r and R, where $r < R$

Wrinkled and Bumpy Spheres   **In Exercises 37 and 38, find the volume of the wrinkled sphere or bumpy sphere. These solids are used as models for tumors. (For more information on this, see the article "Heat Therapy for Tumors" by Leah Edelstine-Keshet in the Summer, 1991 issue of the *UMAP Journal*.)**

37. Wrinkled sphere
$\rho = 1 + 0.2\sin 8\theta \sin\phi$
$(0 \le \theta \le 2\pi, 0 \le \phi \le \pi)$

38. Bumpy sphere
$\rho = 1 + 0.2\sin 8\theta \sin 4\phi$
$(0 \le \theta \le 2\pi, 0 \le \phi \le \pi)$

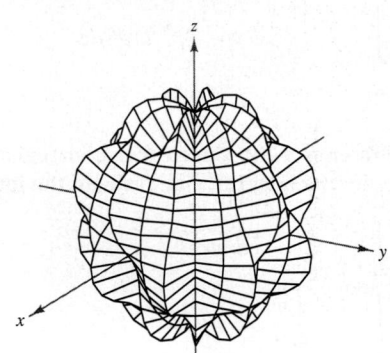

Change of Variables: Jacobians

Jacobians • Change of Variables for Double Integrals

Jacobians

For the single integral

$$\int_a^b f(x)\,dx$$

you can change variables by letting $x = g(u)$, so that $dx = g'(u)\,du$, and obtain

$$\int_a^b f(x)\,dx = \int_c^d f(g(u))g'(u)\,du$$

where $a = g(c)$ and $b = g(d)$. Note that the change of variables process introduces an additional factor $g'(u)$ into the integrand. This also occurs in the case of double integrals.

$$\iint_R f(x, y)\,dA = \iint_S f(g(u, v), h(u, v)) \underbrace{\left| \frac{\partial x}{\partial u}\frac{\partial y}{\partial v} - \frac{\partial y}{\partial u}\frac{\partial x}{\partial v} \right|}_{\text{Jacobian}} du\,dv$$

where the change of variables $x = g(u, v)$ and $y = h(u, v)$ introduces a factor called the **Jacobian** of x and y with respect to u and v. In defining the Jacobian, it is convenient to use the following determinant notation.

CARL GUSTAV JACOBI

The Jacobian is named after the German mathematician Carl Gustav Jacobi (1804–1851). Jacobi is known for his work in many areas of mathematics, but his interest in integration stemmed from the problem of finding the circumference of an ellipse.

Definition of the Jacobian

If $x = g(u, v)$ and $y = h(u, v)$, then the **Jacobian** of x and y with respect to u and v, denoted by $\partial(x, y)/\partial(u, v)$, is

$$\frac{\partial(x, y)}{\partial(u, v)} = \begin{vmatrix} \dfrac{\partial x}{\partial u} & \dfrac{\partial x}{\partial v} \\[2ex] \dfrac{\partial y}{\partial u} & \dfrac{\partial y}{\partial v} \end{vmatrix} = \frac{\partial x}{\partial u}\frac{\partial y}{\partial v} - \frac{\partial y}{\partial u}\frac{\partial x}{\partial v}.$$

EXAMPLE 1   The Jacobian for Rectangular-to-Polar Conversion

Find the Jacobian for the change of variables defined by

$$x = r\cos\theta \quad \text{and} \quad y = r\sin\theta.$$

Solution   From the definition of a Jacobian, you obtain

$$\frac{\partial(x, y)}{\partial(r, \theta)} = \begin{vmatrix} \dfrac{\partial x}{\partial r} & \dfrac{\partial x}{\partial \theta} \\[2ex] \dfrac{\partial y}{\partial r} & \dfrac{\partial y}{\partial \theta} \end{vmatrix} = \begin{vmatrix} \cos\theta & -r\sin\theta \\ \sin\theta & r\cos\theta \end{vmatrix} = r\cos^2\theta + r\sin^2\theta = r.$$

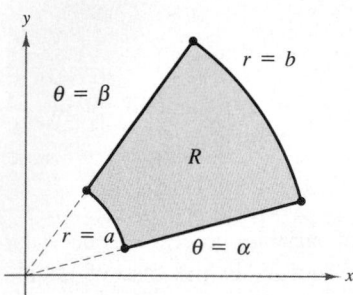

FIGURE 14.67

S in the $r\theta$-plane corresponds to R in the xy-plane.

Example 1 points out that the change of variables from rectangular to polar coordinates for a double integral can be written as

$$\iint_R f(x, y)\, dA = \iint_S f(r\cos\theta, r\sin\theta) r\, dr\, d\theta, \qquad r > 0$$

$$= \iint_S f(r\cos\theta, r\sin\theta) \left| \frac{\partial(x, y)}{\partial(r, \theta)} \right| \, dr\, d\theta$$

where S is the region in the $r\theta$-plane that corresponds to the region R in the xy-plane, as shown in Figure 14.67.

In general, a change of variables is given by a one-to-one **transformation** T from a region S in the uv-plane to a region R in the xy-plane, to be given by

$$T(u, v) = (x, y)$$
$$= (g(u, v), h(u, v))$$

where g and h have continuous first partial derivatives in the region S. Note that the point (u, v) lies in S and the point (x, y) lies in R. In most cases, you are hunting for a transformation for which the region S is simpler than the region R.

EXAMPLE 2 Finding a Change of Variables to Simplify a Region

Let R be the region bounded by the lines

$$x - 2y = 0, \quad x - 2y = -4, \quad x + y = 4, \quad \text{and} \quad x + y = 1$$

as shown in Figure 14.68. Find a transformation T from a region S to R such that S is a rectangular region (with sides parallel to the u- or v-axis).

Solution To begin, let $u = x + y$ and $v = x - 2y$. Solving this system of equations for x and y produces $T(u, v) = (x, y)$, where

$$x = \frac{1}{3}(2u + v) \quad \text{and} \quad y = \frac{1}{3}(u - v).$$

The four boundaries for R in the xy-plane give rise to the following bounds for S in the uv-plane.

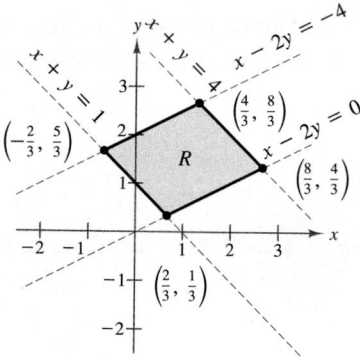

FIGURE 14.68

Region R in the xy-plane.

Bounds in xy-plane		Bounds in uv-plane
$x + y = 1$	$\rightarrow$	$u = 1$
$x + y = 4$	$\rightarrow$	$u = 4$
$x - 2y = 0$	$\rightarrow$	$v = 0$
$x - 2y = -4$	$\rightarrow$	$v = -4$

The region S is shown in Figure 14.69. Note that the transformation T maps the vertices of the region S onto the vertices of the region R.

$$T(1, 0) = \left(\frac{1}{3}[2(1) + 0], \frac{1}{3}[1 - 0] \right) = \left(\frac{2}{3}, \frac{1}{3} \right)$$

$$T(4, 0) = \left(\frac{1}{3}[2(4) + 0], \frac{1}{3}[4 - 0] \right) = \left(\frac{8}{3}, \frac{4}{3} \right)$$

$$T(1, -4) = \left(\frac{1}{3}[2(1) - 4], \frac{1}{3}[1 - (-4)] \right) = \left(-\frac{2}{3}, \frac{5}{3} \right)$$

$$T(4, -4) = \left(\frac{1}{3}[2(4) - 4], \frac{1}{3}[4 - (-4)] \right) = \left(\frac{4}{3}, \frac{8}{3} \right)$$

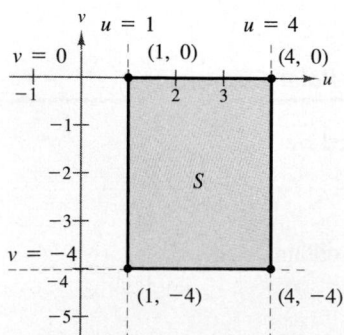

FIGURE 14.69

Region S in the uv-plane.

Change of Variables for Double Integrals

> **THEOREM 14.6   Change of Variables for Double Integrals**
>
> Let R and S be regions in the xy- and uv-planes that are related by the equations $x = g(u, v)$ and $y = h(u, v)$ such that each point in R is the image of a unique point in S. If f is continuous on R, g and h have continuous partial derivatives on S, and $\partial(x, y)/\partial(u, v)$ is nonzero on S, then
>
> $$\iint_R f(x, y)\, dA = \iint_S f(g(u, v), h(u, v)) \left| \frac{\partial(x, y)}{\partial(u, v)} \right| du\, dv.$$

Proof   Consider the case in which S is a rectangular region in the uv-plane with vertices (u, v), $(u + \Delta u, v)$, $(u + \Delta u, v + \Delta v)$, and $(u, v + \Delta v)$, as shown in Figure 14.70. The images of these vertices in the xy-plane are shown in Figure 14.71. If Δu and Δv are small, then the continuity of g and h implies that R is approximately a parallelogram determined by the vectors $\overrightarrow{MN}$ and $\overrightarrow{MQ}$. Thus, the area of R is $\Delta A \approx \| \overrightarrow{MN} \times \overrightarrow{MQ} \|$. Moreover, for small Δu and Δv, the partial derivatives of g and h with respect to u can be approximated by

$$g_u(u, v) \approx \frac{g(u + \Delta u, v) - g(u, v)}{\Delta u}$$

and

$$h_u(u, v) \approx \frac{h(u + \Delta u, v) - h(u, v)}{\Delta u}.$$

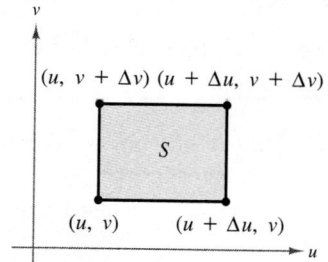

FIGURE 14.70
Area of $S = \Delta u \Delta v$
$\Delta u > 0$, $\Delta v > 0$.

Consequently,

$$\overrightarrow{MN} = [g(u + \Delta u, v) - g(u, v)]\mathbf{i} + [h(u + \Delta u, v) - h(u, v)]\mathbf{j}$$
$$\approx [g_u(u, v)\, \Delta u]\mathbf{i} + [h_u(u, v)\, \Delta u]\mathbf{j} = \frac{\partial x}{\partial u}\Delta u\, \mathbf{i} + \frac{\partial y}{\partial u}\Delta u\, \mathbf{j}.$$

Similarly, you can approximate $\overrightarrow{MQ}$ by $\dfrac{\partial x}{\partial v}\Delta v\, \mathbf{i} + \dfrac{\partial y}{\partial v}\Delta v\, \mathbf{j}$, which implies that

$$\overrightarrow{MN} \times \overrightarrow{MQ} \approx \begin{vmatrix} \mathbf{i} & \mathbf{j} & \mathbf{k} \\ \dfrac{\partial x}{\partial u}\Delta u & \dfrac{\partial y}{\partial u}\Delta u & 0 \\ \dfrac{\partial x}{\partial v}\Delta v & \dfrac{\partial y}{\partial v}\Delta v & 0 \end{vmatrix} = \begin{vmatrix} \dfrac{\partial x}{\partial u} & \dfrac{\partial y}{\partial u} \\ \dfrac{\partial x}{\partial v} & \dfrac{\partial y}{\partial v} \end{vmatrix} \Delta u\, \Delta v\, \mathbf{k}.$$

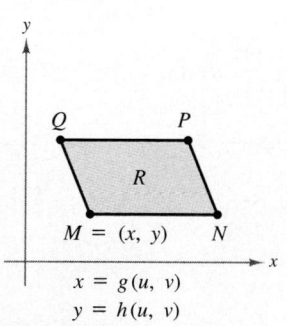

FIGURE 14.71
$M = (g(u, v), h(u, v))$
$N = (g(u + \Delta u, v), h(u + \Delta u, v))$
$P = (g(u + \Delta u, v + \Delta v),$
$\quad h(u + \Delta u, v + \Delta v))$
$Q = (g(u, v + \Delta v), h(u, v + \Delta v))$

It follows that, in Jacobian notation,

$$\Delta A \approx \| \overrightarrow{MN} \times \overrightarrow{MQ} \| \approx \left| \frac{\partial(x, y)}{\partial(u, v)} \right| \Delta u\, \Delta v.$$

Because this approximation improves as Δu and Δv approach 0, the limiting case can be written as

$$dA \approx \| \overrightarrow{MN} \times \overrightarrow{MQ} \| \approx \left| \frac{\partial(x, y)}{\partial(u, v)} \right| du\, dv.$$

The next two examples show how a change of variables can simplify the integration process. The simplification can occur in various ways. You can make a change of variables to simplify either the *region R* or the *integrand f(x, y)*, or both.

EXAMPLE 3   Using a Change of Variables to Simplify a Region

Let R be the region bounded by the lines

$$x - 2y = 0, \quad x - 2y = -4, \quad x + y = 4, \quad \text{and} \quad x + y = 1$$

as shown in Figure 14.72. Evaluate the double integral

$$\int_R\!\!\int 3xy\, dA.$$

Solution   From Example 2, you can use the following change of variables.

$$y = \frac{1}{3}(u - v) \quad \text{and} \quad x = \frac{1}{3}(2u + v)$$

The partial derivatives of x and y are

$$\frac{\partial x}{\partial u} = \frac{2}{3}, \quad \frac{\partial x}{\partial v} = \frac{1}{3}, \quad \frac{\partial y}{\partial u} = \frac{1}{3}, \quad \text{and} \quad \frac{\partial y}{\partial v} = -\frac{1}{3}$$

which implies that the Jacobian is

$$\frac{\partial(x, y)}{\partial(u, v)} = \begin{vmatrix} \dfrac{\partial x}{\partial u} & \dfrac{\partial x}{\partial v} \\[2mm] \dfrac{\partial y}{\partial u} & \dfrac{\partial y}{\partial v} \end{vmatrix}$$

$$= \begin{vmatrix} \dfrac{2}{3} & \dfrac{1}{3} \\[2mm] \dfrac{1}{3} & -\dfrac{1}{3} \end{vmatrix}$$

$$= -\frac{2}{9} - \frac{1}{9} = -\frac{1}{3}.$$

Therefore, by Theorem 14.6, you obtain

$$\int_R\!\!\int 3xy\, dA = \int_S\!\!\int 3\left[\frac{1}{3}(2u + v)\frac{1}{3}(u - v)\right]\left|\frac{\partial(x, y)}{\partial(u, v)}\right| dv\, du$$

$$= \int_1^4 \int_{-4}^0 \frac{1}{9}(2u^2 - uv - v^2)\, dv\, du$$

$$= \frac{1}{9}\int_1^4 \left[2u^2v - \frac{uv^2}{2} - \frac{v^3}{3}\right]_{-4}^0 du$$

$$= \frac{1}{9}\int_1^4 \left(8u^2 + 8u - \frac{64}{3}\right) du$$

$$= \frac{1}{9}\left[\frac{8u^3}{3} + 4u^2 - \frac{64}{3}u\right]_1^4$$

$$= \frac{164}{9}.$$

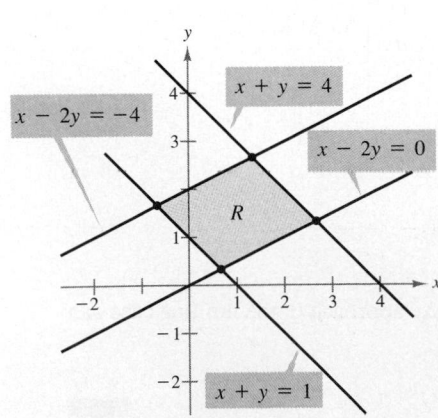

FIGURE 14.72

EXAMPLE 4   Using a Change of Variables to Simplify an Integrand

Let R be the region bounded by the square with vertices $(0, 1)$, $(1, 2)$, $(2, 1)$, and $(1, 0)$. Evaluate the integral

$$\int_R\int (x + y)^2 \sin^2(x - y)\, dA.$$

Solution   Note that the sides of R lie on the lines $x + y = 1$, $x - y = 1$, $x + y = 3$, and $x - y = -1$, as shown in Figure 14.73. Letting $u = x + y$ and $v = x - y$, you can determine the bounds for region S in the uv-plane to be

$$1 \le u \le 3 \quad \text{and} \quad -1 \le v \le 1$$

as shown in Figure 14.74. Solving for x and y in terms of u and v produces

$$x = \frac{1}{2}(u + v) \quad \text{and} \quad y = \frac{1}{2}(u - v).$$

The partial derivatives of x and y are

$$\frac{\partial x}{\partial u} = \frac{1}{2}, \quad \frac{\partial x}{\partial v} = \frac{1}{2}, \quad \frac{\partial y}{\partial u} = \frac{1}{2}, \quad \text{and} \quad \frac{\partial y}{\partial v} = -\frac{1}{2}$$

which implies that the Jacobian is

$$\frac{\partial(x, y)}{\partial(u, v)} = \begin{vmatrix} \dfrac{\partial x}{\partial u} & \dfrac{\partial x}{\partial v} \\[2mm] \dfrac{\partial y}{\partial u} & \dfrac{\partial y}{\partial v} \end{vmatrix} = \begin{vmatrix} \dfrac{1}{2} & \dfrac{1}{2} \\[2mm] \dfrac{1}{2} & -\dfrac{1}{2} \end{vmatrix}$$

$$= -\frac{1}{4} - \frac{1}{4} = -\frac{1}{2}.$$

By Theorem 14.6, it follows that

$$\int_R\int (x + y)^2 \sin^2(x - y)\, dA = \int_{-1}^{1}\int_{1}^{3} u^2 \sin^2 v \left(\frac{1}{2}\right) du\, dv$$

$$= \frac{1}{2}\int_{-1}^{1} (\sin^2 v)\frac{u^3}{3}\bigg]_{1}^{3} dv$$

$$= \frac{13}{3}\int_{-1}^{1} \sin^2 v\, dv$$

$$= \frac{13}{6}\int_{-1}^{1} (1 - \cos 2v)\, dv$$

$$= \frac{13}{6}\left[v - \frac{1}{2}\sin 2v\right]_{-1}^{1}$$

$$= \frac{13}{6}\left[2 - \frac{1}{2}\sin 2 + \frac{1}{2}\sin(-2)\right]$$

$$= \frac{13}{6}(2 - \sin 2)$$

$$\approx 2.363.$$

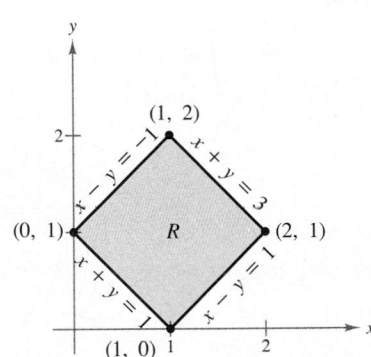

FIGURE 14.73
Region R in the xy-plane.

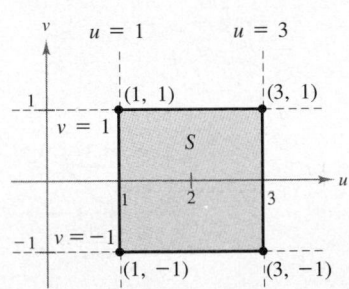

FIGURE 14.74
Region S in the uv-plane.

In each of the change-of-variable examples in this section, the region S has been a rectangle with sides parallel to the u- or v-axis. Occasionally, a change of variables can be used for other types of regions. For instance, Figure 14.75 shows a change of variables from a circular region S to an elliptical region R.

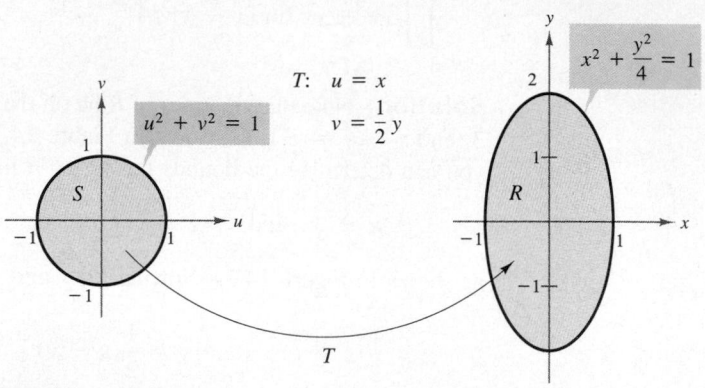

FIGURE 14.75
The change of variables from a circular region to an elliptical region.

EXERCISES for Section 14.8

In Exercises 1–8, find the Jacobian $\partial(x,y)/\partial(u,v)$ for the indicated change of variables.

1. $x = -\frac{1}{2}(u - v), \ y = \frac{1}{2}(u + v)$

2. $x = au + bv, \ y = cu + dv$

3. $x = u - v^2, \ y = u + v$

4. $x = u - uv, \ y = uv$

5. $x = u\cos\theta - v\sin\theta, \ y = u\sin\theta + v\cos\theta$

6. $x = u + a, \ y = v + a$

7. $x = e^u \sin v, \ y = e^u \cos v$

8. $x = \dfrac{u}{v}, \ y = u + v$

In Exercises 9 and 10, sketch the image S in the uv-plane of the region R in the xy-plane using the given transformations.

9. $x = 3u + 2v$
$y = 3v$

10. $x = 4u + v$
$y = u + 2v$

In Exercises 11–16, use the indicated change of variables to evaluate the double integral.

11. $\displaystyle\int\int_R 48xy\,dx\,dy$
$x = \frac{1}{2}(u + v)$
$y = \frac{1}{2}(u - v)$

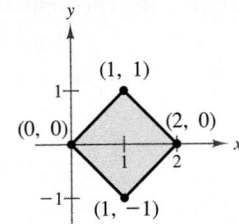

12. $\displaystyle\int\int_R 4(x^2 + y^2)\,dx\,dy$
$x = \frac{1}{2}(u + v)$
$y = \frac{1}{2}(u - v)$

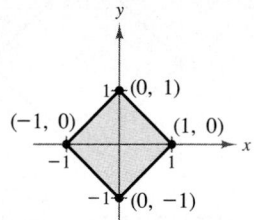

13. $\displaystyle\int\int_R 4(x + y)e^{x-y}\,dy\,dx$
$x = \frac{1}{2}(u + v)$
$y = \frac{1}{2}(u - v)$

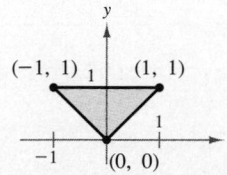

14. $\displaystyle\int\int_R y(x - y)\,dx\,dy$
$x = u + v$
$y = u$

15. $\displaystyle\iint_R \frac{\sqrt{x+y}}{x}\,dy\,dx$

$x = u, \; y = uv$

R: triangle with vertices $(1, 0)$, $(4, 0)$, and $(4, 4)$

16. $\displaystyle\iint_R y\sin xy\,dy\,dx$

$x = \dfrac{u}{v}, \; y = v$

R: region bounded by the graphs of $xy = 1$, $xy = 4$, $y = 1$, and $y = 4$

Volume **In Exercises 17–22, use a change of variables to find the volume of the solid region lying below the surface** $z = f(x, y)$ **and above the plane region** R.

17. $f(x, y) = (x + y)e^{x-y}$

R: region bounded by the square with vertices $(4, 0)$, $(6, 2)$, $(4, 4)$, and $(2, 2)$

18. $f(x, y) = (x + y)^2 \sin^2(x - y)$

R: region bounded by the square with vertices $(\pi, 0)$, $(3\pi/2, \pi/2)$, (π, π), and $(\pi/2, \pi/2)$

19. $f(x, y) = \sqrt{(x - y)(x + 4y)}$

R: region bounded by the parallelogram with vertices $(0, 0)$, $(1, 1)$, $(5, 0)$, and $(4, -1)$

20. $f(x, y) = (3x + 2y)^2\sqrt{2y - x}$

R: region bounded by the parallelogram with vertices $(0, 0)$, $(-2, 3)$, $(2, 5)$, and $(4, 2)$

21. $f(x, y) = \sqrt{x + y}$

R: region bounded by the triangle with vertices $(0, 0)$, $(a, 0)$, and $(0, a)$, where $0 < a$

22. $f(x, y) = \dfrac{xy}{1 + x^2 y^2}$

R: region bounded by the graphs of $xy = 1$, $xy = 4$, $x = 1$, $x = 4$. (*Hint:* Let $x = u$, $y = v/u$.)

23. Consider the region R in the xy-plane bounded by the ellipse

$$\frac{x^2}{a^2} + \frac{y^2}{b^2} = 1$$

and the transformation $x = au$ and $y = bv$.

a. Sketch the graph of the region R and its image S under the specified transformation.

b. Find $\partial(x, y)/\partial(u, v)$.

c. Find the area of the ellipse.

24. ***Volume*** Use the result of Exercise 23 to find the volume of each dome-shaped solid lying below the surface $z = f(x, y)$ and above the elliptical region R. [*Hint:* After making the change of variables specified by the results of Exercise 23, make a second change of variables to polar coordinates.]

a. $f(x, y) = 16 - x^2 - y^2$

$R: \dfrac{x^2}{16} + \dfrac{y^2}{9} \le 1$

b. $f(x, y) = A\cos\left(\dfrac{\pi}{2}\sqrt{\dfrac{x^2}{a^2} + \dfrac{y^2}{b^2}}\right)$

$R: \dfrac{x^2}{a^2} + \dfrac{y^2}{b^2} \le 1$

In Exercises 25–28, find the Jacobian $\partial(x, y, z)/\partial(u, v, w)$ **for the indicated change of variables. If** $x = f(u, v, w)$, $y = g(u, v, w)$, **and** $z = h(u, v, w)$, **then the Jacobian of** x, y, **and** z **with respect to** u, v, **and** w **is**

$$\frac{\partial(x, y, z)}{\partial(u, v, w)} = \begin{vmatrix} \dfrac{\partial x}{\partial u} & \dfrac{\partial x}{\partial v} & \dfrac{\partial x}{\partial w} \\[2mm] \dfrac{\partial y}{\partial u} & \dfrac{\partial y}{\partial v} & \dfrac{\partial y}{\partial w} \\[2mm] \dfrac{\partial z}{\partial u} & \dfrac{\partial z}{\partial v} & \dfrac{\partial z}{\partial w} \end{vmatrix}.$$

25. $x = u(1 - v), \; y = uv(1 - w), \; z = uvw$

26. $x = 4u - v, \; y = 4v - w, \; z = u + w$

27. *Spherical coordinates*
$x = \rho\sin\phi\cos\theta, \; y = \rho\sin\phi\sin\theta, \; z = \rho\cos\phi$

28. *Cylindrical coordinates*
$x = r\cos\theta, \; y = r\sin\theta, \; z = z$

REVIEW EXERCISES for Chapter 14

In Exercises 1 and 2, evaluate the integral.

1. $\displaystyle\int_1^{x^2} x\ln y\,dy$

2. $\displaystyle\int_y^{2y} (x^2 + y^2)\,dx$

In Exercises 3–12, evaluate the multiple integral. Change the coordinate system when convenient.

3. $\displaystyle\int_0^1\int_0^{1+x} (3x + 2y)\,dy\,dx$

4. $\displaystyle\int_0^2\int_{x^2}^{2x} (x^2 + 2y)\,dy\,dx$

5. $\displaystyle\int_0^3\int_0^{\sqrt{9-x^2}} 4x\,dy\,dx$

6. $\displaystyle\int_0^{\sqrt{3}}\int_{2-\sqrt{4-y^2}}^{2+\sqrt{4-y^2}} dx\,dy$

7. $\displaystyle\int_0^h\int_0^x \sqrt{x^2 + y^2}\,dy\,dx$

8. $\displaystyle\int_0^4\int_0^{\sqrt{16-y^2}} (x^2 + y^2)\,dx\,dy$

9. $\displaystyle\int_{-3}^3\int_{-\sqrt{9-x^2}}^{\sqrt{9-x^2}}\int_{x^2+y^2}^9 \sqrt{x^2 + y^2}\,dz\,dy\,dx$

10. $\displaystyle\int_{-2}^2\int_{-\sqrt{4-x^2}}^{\sqrt{4-x^2}}\int_0^{(x^2+y^2)/2} (x^2 + y^2)\,dz\,dy\,dx$

11. $\displaystyle\int_0^a\int_0^b\int_0^c (x^2 + y^2 + z^2)\,dx\,dy\,dz$

12. $\int_0^5 \int_0^{\sqrt{25-x^2}} \int_0^{\sqrt{25-x^2-y^2}} \dfrac{1}{1+x^2+y^2+z^2} \, dz \, dy \, dx$

C In Exercises 13–16, use a computer algebra system to evaluate the multiple integral.

13. $\int_{-2}^4 \int_{y^2/4}^{(4+y)/2} (x-y) \, dx \, dy$

14. $\int_{-2}^2 \int_0^{4-y^2} (8x - 2y^2) \, dx \, dy$

15. $\int_{-1}^1 \int_{-\sqrt{1-x^2}}^{\sqrt{1-x^2}} \int_{-\sqrt{1-x^2-y^2}}^{\sqrt{1-x^2-y^2}} (x^2+y^2) \, dz \, dy \, dx$

16. $\int_0^2 \int_0^{\sqrt{4-x^2}} \int_0^{\sqrt{4-x^2-y^2}} xyz \, dz \, dy \, dx$

In Exercises 17–24, write the limits for the double integral

$$\int_R \int f(x,y) \, dA$$

for both orders of integration. Compute the area of R by letting $f(x,y) = 1$ and integrating.

17. Triangle: vertices $(0,0)$, $(3,0)$, and $(0,1)$

18. Triangle: vertices $(0,0)$, $(3,0)$, and $(2,2)$

19. The larger area between the graphs of $x^2+y^2 = 25$ and $x = 3$

20. Region bounded by the graphs of $y = 6x - x^2$ and $y = x^2 - 2x$

21. Region enclosed by the graph of $y^2 = x^2 - x^4$

22. Region bounded by the graphs of $x = y^2 + 1$, $x = 0$, $y = 0$, and $y = 2$

23. Region bounded by the graphs of $x = y + 3$ and $x = y^2 + 1$

24. Region bounded by the graphs of $x = -y$ and $x = 2y - y^2$

Volume   In Exercises 25–30, use a multiple integral and a convenient coordinate system to find the volume of the solid.

25. Solid bounded by the graphs of $z = x^2 - y + 4$, $z = 0$, $x = 0$, and $x = 4$

26. Solid bounded by the graphs of $z = x + y$, $z = 0$, $x = 0$, $x = 3$, and $y = x$

27. Solid bounded by the graphs of $z = 0$ and $z = h$, outside the cylinder $x^2 + y^2 = 1$ and inside the hyperboloid $x^2 + y^2 - z^2 = 1$

28. Solid that remains after drilling a hole of radius b through the center of a sphere of radius R $(b < R)$

29. Solid inside the graphs of $r = 2 \cos \theta$ and $r^2 + z^2 = 4$

30. Solid inside the graphs of $r^2 + z = 16$ and $r = 2 \sin \theta$

Probability   In Exercises 31 and 32, find k so that the function is a joint density function and find the required probability, where $P(a \leq x \leq b, c \leq y \leq d) = \int_c^d \int_a^b f(x,y) \, dx \, dy$.

31. $f(x,y) = \begin{cases} kxye^{-(x+y)}, & x \geq 0, y \geq 0 \\ 0, & \text{elsewhere} \end{cases}$
    $P(0 \leq x \leq 1, 0 \leq y \leq 1)$

32. $f(x,y) = \begin{cases} kxy, & 0 \leq x \leq 1, 0 \leq y \leq x \\ 0, & \text{elsewhere} \end{cases}$
    $P(0 \leq x \leq 0.5, 0 \leq y \leq 0.25)$

C **Mass and Center of Mass**   In Exercises 33 and 34, find the mass and center of mass of the lamina bounded by the graphs of the equations of specified density. Use a symbolic integration utility to evaluate the multiple integrals.

33. $y = 2x$, $y = 2x^3$ (first quadrant)
    a. $\rho = kxy$ b. $\rho = k(x^2 + y^2)$

34. $y = \dfrac{h}{2}\left(2 - \dfrac{x}{L} - \dfrac{x^2}{L^2}\right)$ (first quadrant)
    $\rho = k$

C **Surface Area**   In Exercises 35 and 36, find the area of the surface on the function $f(x,y)$ over the region R.

35. $f(x,y) = 16 - x^2 - y^2$
    $R = \{(x,y): x^2 + y^2 \leq 16\}$

36. $f(x,y) = 16 - x - y^2$
    $R = \{(x,y): 0 \leq x \leq 2, 0 \leq y \leq x\}$
    Use a symbolic integration utility to evaluate the multiple integral.

Center of Mass   In Exercises 37–40, find the center of mass of the solid of uniform density bounded by the graphs of the given equations.

37. Solid inside the hemisphere $\rho = \cos \phi$, $\pi/4 \leq \phi \leq \pi/2$, and outside the cone $\phi = \pi/4$

38. Wedge: $x^2 + y^2 = a^2$, $z = cy(0 < c)$, $0 \leq y$, $0 \leq z$

39. $x^2 + y^2 + z^2 = a^2$ (first octant)

40. $x^2 + y^2 + z^2 = 25$, $z = 4$ (the larger solid)

Moment of Inertia   In Exercises 41 and 42, find the moment of inertia I_z of the solid of specified density.

41. The solid of uniform density inside the paraboloid $z = 16 - x^2 - y^2$ and outside the cylinder $x^2 + y^2 = 9$, $z \geq 0$.

42. $x^2 + y^2 + z^2 = a^2$, density is proportional to the distance from the center

In Exercises 43 and 44, give a geometric interpretation of the triple integral.

43. $\displaystyle\int_0^{2\pi}\int_{-\pi/2}^{\pi/2}\int_0^{6\sin\phi}\rho^2\sin\phi\,d\rho\,d\phi\,d\theta$

44. $\displaystyle\int_0^{\pi}\int_0^{2}\int_0^{1+r^2}r\,dz\,dr\,d\theta$

True or False **In Exercises 45–48, determine whether the statement is true or false. If it is false, explain why or give an example that shows it is false.**

45. $\displaystyle\int_a^b\int_c^d f(x)g(y)\,dy\,dx = \left[\int_a^b f(x)\,dx\right]\left[\int_c^d g(y)\,dy\right]$

46. If f is continuous over R_1 and R_2 and $\int_{R_1}\!\!\int dA = \int_{R_2}\!\!\int dA$, then $\int_{R_1}\!\!\int f(x,y)\,dA = \int_{R_1}\!\!\int f(x,y)\,dA$.

47. $\displaystyle\int_{-1}^{1}\int_{-1}^{1}\cos(x^2+y^2)\,dx\,dy = 4\int_0^1\int_0^1\cos(x^2+y^2)\,dx\,dy$

48. $\displaystyle\int_0^1\int_0^1\frac{1}{1+x^2+y^2}\,dx\,dy < \frac{\pi}{4}$

Chapter 15

Vector Analysis

Vector Fields

In Chapter 12, you studied vector-valued functions—functions that assign a vector to a *real number*. There you saw that vector-valued functions of a real number are useful in representing curves and motion along a curve. In this chapter, you will study two other types of vector-valued functions—functions that assign a vector to a *point in the plane* or to a *point in space*. Such functions are called **vector fields,** and they are useful in representing various types of **force fields** and **velocity fields.**

Definition of a Vector Field

Let M and N be functions of two variables x and y, defined on a plane region R. The function **F** defined by

$$\mathbf{F}(x, y) = M\mathbf{i} + N\mathbf{j} \qquad \text{Plane}$$

is called a **vector field over R.**

Let M, N, and P be functions of three variables x, y, and z, defined on a solid region Q in space. The function **F** defined by

$$\mathbf{F}(x, y, z) = M\mathbf{i} + N\mathbf{j} + P\mathbf{k} \qquad \text{Space}$$

is called a **vector field over Q.**

From this definition, you see that the *gradient* is one example of a vector field. For example, if

$$f(x, y) = x^2 + y^2$$

then the gradient of f

$$\nabla f(x, y) = f_x(x, y)\mathbf{i} + f_y(x, y)\mathbf{j}$$
$$= 2x\mathbf{i} + 2y\mathbf{j} \qquad \text{Vector field in the plane}$$

is a vector field in the plane. Similarly, if

$$f(x, y, z) = x^2 + y^2 + z^2$$

then the gradient of f

$$\nabla f(x, y, z) = f_x(x, y, z)\mathbf{i} + f_y(x, y, z)\mathbf{j} + f_z(x, y, z)\mathbf{k}$$
$$= 2x\mathbf{i} + 2y\mathbf{j} + 2z\mathbf{k} \qquad \text{Vector field in space}$$

is a vector field in space.

◄ *The velocity of swirling water can be modeled by a velocity field, as discussed in Section 15.8, Example 3.*

Velocity field

FIGURE 15.1
Rotating wheel

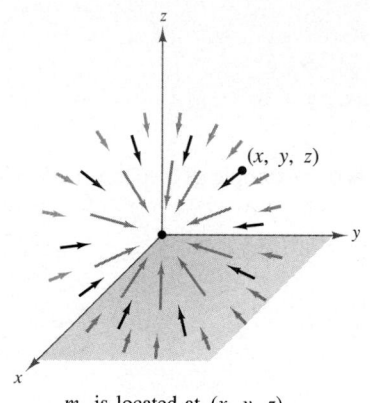

FIGURE 15.2
Air-flow vector field

z

(x, y, z)

y

x

m_1 is located at (x, y, z).
m_2 is located at $(0, 0, 0)$.

FIGURE 15.3
Gravitational force field

Some common *physical* examples of vector fields are **velocity fields, gravitational fields,** and **electric force fields.**

1. *Velocity fields* describe the motion of a system of particles in the plane or in space. For instance, Figure 15.1 shows the vector field determined by a wheel rotating on an axle. Notice that the velocity vectors are determined by the location of their initial points—the farther a point is from the axle, the greater its velocity. Velocity fields are also determined by the flow of liquids through a container or by the flow of air currents around a moving object, as shown in Figure 15.2.

2. *Gravitational fields* are defined by **Newton's Law of Gravitation,** which states that the force of attraction exerted on a particle of mass m_1 located at (x, y, z) by a particle of mass m_2 located at $(0, 0, 0)$ is given by

$$\mathbf{F}(x, y, z) = \frac{-Gm_1m_2}{x^2 + y^2 + z^2}\mathbf{u}$$

where G is the gravitational constant and $\mathbf{u}$ is the unit vector in the direction from the origin to (x, y, z). In Figure 15.3, you can see that the gravitational field $\mathbf{F}$ has the properties that $\mathbf{F}(x, y, z)$ always points toward the origin, and that the magnitude of $\mathbf{F}(x, y, z)$ is the same at all points equidistant from the origin. A vector field with these two properties is called a **central force field.** Using the position vector $\mathbf{r} = x\mathbf{i} + y\mathbf{j} + z\mathbf{k}$ for the point (x, y, z), we can express the gravitational field $\mathbf{F}$ as

$$\mathbf{F}(x, y, z) = \frac{-Gm_1m_2}{\|\mathbf{r}\|^2} \cdot \frac{\mathbf{r}}{\|\mathbf{r}\|} = \frac{-Gm_1m_2}{\|\mathbf{r}\|^2}\mathbf{u}.$$

3. *Electric force fields* are defined by **Coulomb's Law,** which states that the force exerted on a particle with electric charge q_1 located at (x, y, z) by a particle with electric charge q_2 located at $(0, 0, 0)$ is given by

$$\mathbf{F}(x, y, z) = \frac{cq_1q_2}{\|\mathbf{r}\|^2}\mathbf{u}$$

where $\mathbf{r} = x\mathbf{i} + y\mathbf{j} + z\mathbf{k}$, $\mathbf{u} = \mathbf{r}/\|\mathbf{r}\|$, and c is a constant that depends on the choice of units for $\|\mathbf{r}\|$, q_1, and q_2.

Note that an electric force field has the same form as a gravitational field. That is, $\mathbf{F}(x, y, z) = (k/\|\mathbf{r}\|^2)\mathbf{u}$. Such a force field is called an **inverse square field.**

Definition of Inverse Square Field

Let $\mathbf{r}(t) = x(t)\mathbf{i} + y(t)\mathbf{j} + z(t)\mathbf{k}$ be the position vector. The vector field $\mathbf{F}$ is an **inverse square field** if

$$\mathbf{F}(x, y, z) = \frac{k}{\|\mathbf{r}\|^2}\mathbf{u}$$

where k is a real number and $\mathbf{u} = \mathbf{r}/\|\mathbf{r}\|$ is a unit vector in the direction of $\mathbf{r}$.

Because vector fields consist of infinitely many vectors, it is not possible to actually create a sketch of the field. Instead, when you sketch a vector field, your goal is to sketch representative vectors that help you visualize the field.

EXAMPLE 1   Sketching a Vector Field

Sketch some vectors in the vector field given by

$$\mathbf{F}(x, y) = -y\mathbf{i} + x\mathbf{j}.$$

Solution   You could plot vectors at several random points in the plane. However, it is more enlightening to plot vectors of equal magnitude. This corresponds to finding level curves in scalar fields. In this case, vectors of equal magnitude lie on circles.

$$\|\mathbf{F}\| = c \qquad \text{Vectors of length } c$$
$$\sqrt{x^2 + y^2} = c$$
$$x^2 + y^2 = c^2 \qquad \text{Equation of circle}$$

To begin making the sketch, choose a value for c and plot several vectors on the resulting circle. For instance, the following vectors occur on the unit circle.

Point	Vector
$(1, 0)$	$\mathbf{F}(1, 0) = \mathbf{j}$
$(0, 1)$	$\mathbf{F}(0, 1) = -\mathbf{i}$
$(-1, 0)$	$\mathbf{F}(-1, 0) = -\mathbf{j}$
$(0, -1)$	$\mathbf{F}(0, -1) = \mathbf{i}$

These and several other vectors in the vector field are shown in Figure 15.4. Note in the figure that this vector field is similar to that illustrated by the rotating wheel shown in Figure 15.1.

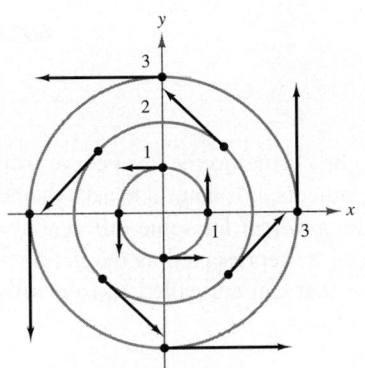

Vector field:
$\mathbf{F}(x, y) = -y\mathbf{i} + x\mathbf{j}$

FIGURE 15.4

EXAMPLE 2   Sketching a Vector Field

Sketch some vectors in the vector field given by

$$\mathbf{F}(x, y) = 2x\mathbf{i} + y\mathbf{j}.$$

Solution   For this vector field, vectors of equal length lie on ellipses given by

$$\|\mathbf{F}\| = \sqrt{(2x)^2 + (y)^2} = c \quad \longrightarrow \quad 4x^2 + y^2 = c^2.$$

For $c = 1$, sketch several vectors $2x\mathbf{i} + y\mathbf{j}$ of magnitude 1 at points on the ellipse given by

$$4x^2 + y^2 = 1.$$

For $c = 2$, sketch several vectors $2x\mathbf{i} + y\mathbf{j}$ of magnitude 2 at points on the ellipse given by

$$4x^2 + y^2 = 4.$$

These vectors are shown in Figure 15.5.

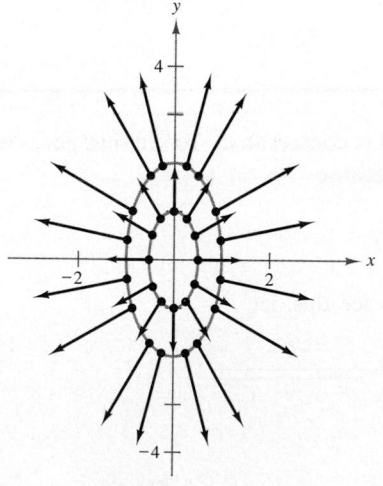

Vector field:
$\mathbf{F}(x, y) = 2x\mathbf{i} + y\mathbf{j}$

FIGURE 15.5

Velocity field:
$\mathbf{v}(x, y, z) = (16 - x^2 - y^2)\mathbf{k}$

FIGURE 15.6

EXAMPLE 3   Sketching a Velocity Field

Sketch some vectors in the velocity field given by

$$\mathbf{v}(x, y, z) = (16 - x^2 - y^2)\mathbf{k}$$

where $x^2 + y^2 \leq 16$.

Solution   You can imagine that $\mathbf{v}$ describes the velocity of a liquid flowing through a tube of radius 4. Vectors near the z-axis are longer than those near the edge of the tube. For instance, at the point $(0, 0, 0)$, the velocity vector is $\mathbf{v}(0, 0, 0) = 16\mathbf{k}$, whereas at the point $(0, 3, 0)$, the velocity vector is $\mathbf{v}(0, 3, 0) = 7\mathbf{k}$. Figure 15.6 shows these and several other vectors for the velocity field. From the figure, you can see that the speed of the liquid is greater near the center of the tube than near the edges of the tube.

Conservative Vector Fields

Notice in Figure 15.5 that all the vectors appear to be normal to the level curve from which they emanate. Because this is a property of gradients, it is natural to ask whether the vector field given by $\mathbf{F}(x, y) = 2x\,\mathbf{i} + y\,\mathbf{j}$ is the *gradient* for some differentiable function f. The answer is that some vector fields can be represented as the gradients of differentiable functions and some cannot—those that can are called **conservative** vector fields.

Definition of Conservative Vector Field

A vector field $\mathbf{F}$ is called **conservative** if there exists a differentiable function f such that $\mathbf{F} = \nabla f$. The function f is called the **potential function** for $\mathbf{F}$.

EXAMPLE 4   Conservative Vector Fields

a. The vector field given by $\mathbf{F}(x, y) = 2x\,\mathbf{i} + y\,\mathbf{j}$ is conservative. To see this, consider the potential function $f(x, y) = x^2 + \frac{1}{2}y^2$. Because

$$\nabla f = 2x\,\mathbf{i} + y\,\mathbf{j} = \mathbf{F}$$

it follows that $\mathbf{F}$ is conservative.

b. Every inverse square field is conservative. To see this, let

$$\mathbf{F}(x, y, z) = \frac{k}{\|\mathbf{r}\|^2}\mathbf{u} \quad \text{and} \quad f(x, y, z) = \frac{-k}{\sqrt{x^2 + y^2 + z^2}}$$

where $\mathbf{u} = \mathbf{r}/\|\mathbf{r}\|$. Because

$$
\begin{aligned}
\nabla f &= \frac{kx}{(x^2 + y^2 + z^2)^{3/2}}\mathbf{i} + \frac{ky}{(x^2 + y^2 + z^2)^{3/2}}\mathbf{j} + \frac{kz}{(x^2 + y^2 + z^2)^{3/2}}\mathbf{k} \\
&= \frac{k}{x^2 + y^2 + z^2}\left(\frac{x\,\mathbf{i} + y\,\mathbf{j} + z\,\mathbf{k}}{\sqrt{x^2 + y^2 + z^2}}\right) = \frac{k}{\|\mathbf{r}\|^2}\frac{\mathbf{r}}{\|\mathbf{r}\|} = \frac{k}{\|\mathbf{r}\|^2}\mathbf{u}
\end{aligned}
$$

it follows that $\mathbf{F}$ is conservative.

As can be seen in Example 4b, many important vector fields, including gravitational fields, magnetic fields, and electric force fields, are conservative. Most of the terminology in this chapter comes from physics. For example, the term "conservative" is derived from the classic physical law regarding the conservation of energy. This law states that the sum of the kinetic energy and the potential energy of a particle moving in a conservative force field is constant. (The kinetic energy of a particle is the energy due to its motion, and the potential energy is the energy due to its position in the force field.)

The following important theorem gives a necessary and sufficient condition for a vector field *in the plane* to be conservative.

THEOREM 15.1   Test for Conservative Vector Field in the Plane

Let M and N have continuous first partial derivatives on an open disc R. The vector field given by $\mathbf{F}(x, y) = M\mathbf{i} + N\mathbf{j}$ is conservative if and only if

$$\frac{\partial N}{\partial x} = \frac{\partial M}{\partial y}.$$

Proof   To prove that the given condition is necessary for $\mathbf{F}$ to be conservative, we suppose there exists a potential function f such that

$$\mathbf{F}(x, y) = \nabla f(x, y) = M\mathbf{i} + N\mathbf{j}.$$

Then, we have

$$f_x(x, y) = M \quad \rightarrow \quad f_{xy}(x, y) = \frac{\partial M}{\partial y}$$

$$f_y(x, y) = N \quad \rightarrow \quad f_{yx}(x, y) = \frac{\partial N}{\partial x}$$

and, by the equivalence of the mixed partials f_{xy} and f_{yx}, we conclude that $\partial N/\partial x = \partial M/\partial y$ for all (x, y) in R. The sufficiency of the condition is proved in Section 15.4.   ∎

REMARK   Theorem 15.1 requires that the domain of $\mathbf{F}$ be an open disc. If R is simply an open region, then the given condition is necessary but not sufficient to produce a conservative vector field.

EXAMPLE 5   Testing for Conservative Vector Fields in the Plane

a. The vector field given by $\mathbf{F}(x, y) = x^2 y\,\mathbf{i} + xy\,\mathbf{j}$ *is not* conservative because

$$\frac{\partial M}{\partial y} = \frac{\partial}{\partial y}[x^2 y] = x^2 \quad \text{and} \quad \frac{\partial N}{\partial x} = \frac{\partial}{\partial x}[xy] = y.$$

b. The vector field given by $\mathbf{F}(x, y) = 2x\,\mathbf{i} + y\,\mathbf{j}$ *is* conservative because

$$\frac{\partial M}{\partial y} = \frac{\partial}{\partial y}[2x] = 0 \quad \text{and} \quad \frac{\partial N}{\partial x} = \frac{\partial}{\partial x}[y] = 0.$$

Theorem 15.1 tells you whether a vector field is conservative. It does not tell you how to find a potential function of **F**. The problem is comparable to antidifferentiation. Sometimes you will be able to find a potential function by simple inspection. For instance, in Example 4 we observed that

$$f(x, y) = x^2 + \frac{1}{2}y^2$$

has the property that $\nabla f(x, y) = 2x\mathbf{i} + y\mathbf{j}$. We will discuss techniques for finding potential functions in detail in Chapter 16 (in the section on exact differential equations). For now, we outline the procedure by way of an example.

EXAMPLE 6   Finding a Potential Function for $\mathbf{F}(x, y)$

Find a potential function for

$$\mathbf{F}(x, y) = 2xy\,\mathbf{i} + (x^2 - y)\mathbf{j}.$$

Solution   From Theorem 15.1, it follows that **F** is conservative because

$$\frac{\partial}{\partial y}[2xy] = 2x \quad \text{and} \quad \frac{\partial}{\partial x}[x^2 - y] = 2x.$$

If f is a function whose gradient is equal to $\mathbf{F}(x, y)$, then $\nabla f(x, y) = 2xy\,\mathbf{i} + (x^2 - y)\mathbf{j}$ which implies that

$$f_x(x, y) = 2xy \quad \text{and} \quad f_y(x, y) = x^2 - y.$$

To reconstruct the function f from these two partial derivatives, integrate $f_x(x, y)$ with respect to x and $f_y(x, y)$ with respect to y, as follows.

$$f(x, y) = \int f_x(x, y)\, dx = \int 2xy\, dx = x^2 y + g(y)$$

$$f(x, y) = \int f_y(x, y)\, dy = \int (x^2 - y)\, dy = x^2 y - \frac{y^2}{2} + h(x)$$

Note that $g(y)$ is constant with respect to x and $h(x)$ is constant with respect to y. To find a single expression that represents $f(x, y)$, let $g(y) = -\frac{1}{2}y^2 + K$ and $h(x) = K$. Then, you can write

$$f(x, y) = x^2 y + g(y) + K$$

$$= x^2 y - \frac{y^2}{2} + K.$$

You can check this result by forming the gradient of f—it should be equal to the original function **F**.

REMARK   Note that the solution in Example 6 is comparable to that given by an indefinite integral. That is, the solution represents a family of potential functions, any two of which differ by a constant. To find a unique solution, you would have to be given an initial condition satisfied by the potential function.

Curl of a Vector Field

Theorem 15.1 has a counterpart for vector fields in space. Before stating that result, we define the **curl of a vector field** in space.

Definition of Curl of a Vector Field

The **curl** of $\mathbf{F}(x, y, z) = M\mathbf{i} + N\mathbf{j} + P\mathbf{k}$ is

$$\mathbf{curl}\ \mathbf{F}(x, y, z) = \nabla \times \mathbf{F}(x, y, z)$$

$$= \left(\frac{\partial P}{\partial y} - \frac{\partial N}{\partial z}\right)\mathbf{i} - \left(\frac{\partial P}{\partial x} - \frac{\partial M}{\partial z}\right)\mathbf{j} + \left(\frac{\partial N}{\partial x} - \frac{\partial M}{\partial y}\right)\mathbf{k}.$$

REMARK   If **curl F** $= \mathbf{0}$, then we say that **F** is **irrotational.**

The cross-product notation used for curl comes from viewing the gradient ∇f as the result of the **differential operator** ∇ acting on the function f. In this context, you can use the following determinant form as an aid in remembering the formula for curl.

$$\mathbf{curl}\ \mathbf{F}(x, y, z) = \nabla \times \mathbf{F}(x, y, z)$$

$$= \begin{vmatrix} \mathbf{i} & \mathbf{j} & \mathbf{k} \\ \dfrac{\partial}{\partial x} & \dfrac{\partial}{\partial y} & \dfrac{\partial}{\partial z} \\ M & N & P \end{vmatrix}$$

$$= \left(\frac{\partial P}{\partial y} - \frac{\partial N}{\partial z}\right)\mathbf{i} - \left(\frac{\partial P}{\partial x} - \frac{\partial M}{\partial z}\right)\mathbf{j} + \left(\frac{\partial N}{\partial x} - \frac{\partial M}{\partial y}\right)\mathbf{k}$$

EXAMPLE 7   Finding the Curl of a Vector Field

Find **curl F** for the vector field given by

$$\mathbf{F}(x, y, z) = 2xy\,\mathbf{i} + (x^2 + z^2)\mathbf{j} + 2zy\,\mathbf{k}.$$

Solution   The curl of **F** is given by

$$\mathbf{curl}\ \mathbf{F}(x, y, z) = \nabla \times \mathbf{F}(x, y, z)$$

$$= \begin{vmatrix} \mathbf{i} & \mathbf{j} & \mathbf{k} \\ \dfrac{\partial}{\partial x} & \dfrac{\partial}{\partial y} & \dfrac{\partial}{\partial z} \\ 2xy & x^2 + z^2 & 2zy \end{vmatrix}$$

$$= \begin{vmatrix} \dfrac{\partial}{\partial y} & \dfrac{\partial}{\partial z} \\ x^2 + z^2 & 2zy \end{vmatrix}\mathbf{i} - \begin{vmatrix} \dfrac{\partial}{\partial x} & \dfrac{\partial}{\partial z} \\ 2xy & 2zy \end{vmatrix}\mathbf{j} + \begin{vmatrix} \dfrac{\partial}{\partial x} & \dfrac{\partial}{\partial y} \\ 2xy & x^2 + z^2 \end{vmatrix}\mathbf{k}$$

$$= (2z - 2z)\mathbf{i} - (0 - 0)\mathbf{j} + (2x - 2x)\mathbf{k}$$

$$= \mathbf{0}.$$

Later in this chapter, we will assign a physical interpretation to the curl of a vector field. But for now, the primary use we make of curl is in the following test for conservative vector fields in space. The test states that for a vector field whose domain is all of three-dimensional space (or an open sphere), the curl is $\mathbf{0}$ at every point in the domain if and only if $\mathbf{F}$ is conservative. The proof is similar to that given in Theorem 15.1.

THEOREM 15.2   Test for Conservative Vector Field in Space

Suppose M, N, and P have continuous first partial derivatives in an open sphere Q in space. The vector field given by $\mathbf{F}(x, y, z) = M\mathbf{i} + N\mathbf{j} + P\mathbf{k}$ is conservative if and only if

$$\text{curl } \mathbf{F}(x, y, z) = \mathbf{0}.$$

That is, $\mathbf{F}$ is conservative if and only if

$$\frac{\partial P}{\partial y} = \frac{\partial N}{\partial z}, \quad \frac{\partial P}{\partial x} = \frac{\partial M}{\partial z}, \quad \text{and} \quad \frac{\partial N}{\partial x} = \frac{\partial M}{\partial y}.$$

From Theorem 15.2, you can see that the vector field given in Example 7 is conservative because $\text{curl } \mathbf{F}(x, y, z) = \mathbf{0}$. Try showing that the vector field $\mathbf{F}(x, y, z) = x^3y^2z\mathbf{i} + x^2z\mathbf{j} + x^2y\mathbf{k}$ is not conservative—you can do this by showing that its curl is $\text{curl } \mathbf{F}(x, y, z) = (x^3y^2 - 2xy)\mathbf{j} + (2xz - 2x^3yz)\mathbf{k} \neq \mathbf{0}$.

For vector fields in space that pass the test for being conservative, you can find a potential function by following the same pattern used in the plane (as demonstrated in Example 6).

EXAMPLE 8   Finding a Potential Function for F(x,y,z)

Find a potential function for $\mathbf{F}(x, y, z) = 2xy\mathbf{i} + (x^2 + z^2)\mathbf{j} + 2zy\mathbf{k}$.

Solution   From Example 7, you know that the vector field given by $\mathbf{F}$ is conservative. If f is a function such that $\mathbf{F}(x, y, z) = \nabla f(x, y, z)$, then

$$f_x(x, y, z) = 2xy, \quad f_y(x, y, z) = x^2 + z^2, \quad \text{and} \quad f_z(x, y, z) = 2zy$$

and by integrating with respect to x, y, and z separately, you obtain

$$f(x, y, z) = \int M\,dx = \int 2xy\,dx = x^2y + g(y, z)$$

$$f(x, y, z) = \int N\,dy = \int (x^2 + z^2)\,dy = x^2y + z^2y + h(x, z)$$

$$f(x, y, z) = \int P\,dz = \int 2zy\,dz = z^2y + k(x, y).$$

Comparing these three versions of $f(x, y, z)$, you can conclude that

$$g(y, z) = z^2y, \quad h(x, z) = 0, \quad \text{and} \quad k(x, y) = x^2y.$$

Therefore, $f(x, y, z)$ is given by

$$f(x, y, z) = x^2y + z^2y + K.$$

REMARK   Examples 6 and 8 are illustrations of a type of problem called *recovering a function from its gradient*. We will discuss other methods for solving this type of problem in Chapter 16. One popular method gives an interplay between successive "partial integrations" and partial differentiations.

Divergence of a Vector Field

You have seen that the curl of a vector field **F** is itself a vector field. Another important function defined on a vector field is **divergence**, which is a scalar function.

Definition of Divergence of a Vector Field

The **divergence** of $\mathbf{F}(x, y) = M\mathbf{i} + N\mathbf{j}$ is

$$\text{div } \mathbf{F}(x, y) = \nabla \cdot \mathbf{F}(x, y) = \frac{\partial M}{\partial x} + \frac{\partial N}{\partial y}. \qquad \text{Plane}$$

The **divergence** of $\mathbf{F}(x, y, z) = M\mathbf{i} + N\mathbf{j} + P\mathbf{k}$ is

$$\text{div } \mathbf{F}(x, y, z) = \nabla \cdot \mathbf{F}(x, y, z) = \frac{\partial M}{\partial x} + \frac{\partial N}{\partial y} + \frac{\partial P}{\partial z}. \qquad \text{Space}$$

If $\text{div } \mathbf{F} = 0$, then **F** is said to be **divergence free.**

REMARK Divergence can be viewed as a type of derivative of **F** in that, for vector fields representing velocities of moving particles, the divergence measures the rate of particle flow per unit volume at a point. In hydrodynamics (the study of fluid motion) a velocity field that is divergence free is called **incompressible.** In the study of electricity and magnetism, a vector field that is divergence free is called **solenoidal.**

The dot-product notation used for divergence comes from considering ∇ as a **differential operator,** as follows.

$$\nabla \cdot \mathbf{F}(x, y, z) = \left[\left(\frac{\partial}{\partial x}\right)\mathbf{i} + \left(\frac{\partial}{\partial y}\right)\mathbf{j} + \left(\frac{\partial}{\partial z}\right)\mathbf{k} \right] \cdot (M\mathbf{i} + N\mathbf{j} + P\mathbf{k})$$

$$= \frac{\partial M}{\partial x} + \frac{\partial N}{\partial y} + \frac{\partial P}{\partial z}$$

EXAMPLE 9   Finding the Divergence of a Vector Field

Find the divergence at $(2, 1, -1)$ for the vector field

$$\mathbf{F}(x, y, z) = x^3 y^2 z\,\mathbf{i} + x^2 z\,\mathbf{j} + x^2 y\,\mathbf{k}.$$

Solution   The divergence of **F** is

$$\text{div } \mathbf{F}(x, y, z) = \frac{\partial}{\partial x}[x^3 y^2 z] + \frac{\partial}{\partial y}[x^2 z] + \frac{\partial}{\partial z}[x^2 y] = 3x^2 y^2 z.$$

At the point $(2, 1, -1)$, the divergence is

$$\text{div } \mathbf{F}(2, 1, -1) = 3(2^2)(1^2)(-1) = -12.$$

There are many important properties of the divergence and curl of a vector field **F** (see Exercises 59–65). One that is used often is described in Theorem 15.3. You are asked to prove this theorem in Exercise 66.

THEOREM 15.3   Relationship Between Divergence and Curl

If $\mathbf{F}(x, y, z) = M\mathbf{i} + N\mathbf{i} + P\mathbf{k}$ is a vector field and M, N, and P have continuous second partial derivatives, then

$$\text{div } (\mathbf{curl\ F}) = 0.$$

EXERCISES for Section 15.1

In Exercises 1–14, sketch several representative vectors in the vector field.

1. $\mathbf{F}(x, y) = \mathbf{i} + \mathbf{j}$

2. $\mathbf{F}(x, y) = 2\mathbf{i}$

3. $\mathbf{F}(x, y) = x\mathbf{j}$

4. $\mathbf{F}(x, y) = y\mathbf{i}$

5. $\mathbf{F}(x, y) = x\mathbf{i} + y\mathbf{j}$

6. $\mathbf{F}(x, y) = -x\mathbf{i} + y\mathbf{j}$

7. $\mathbf{F}(x, y) = x\mathbf{i} + 3y\mathbf{j}$

8. $\mathbf{F}(x, y) = y\mathbf{i} - x\mathbf{j}$

9. $\mathbf{F}(x, y, z) = 3y\mathbf{j}$

10. $\mathbf{F}(x, y) = x\mathbf{i}$

11. $\mathbf{F}(x, y) = 4x\mathbf{i} + y\mathbf{j}$

12. $\mathbf{F}(x, y) = \mathbf{i} + (x^2 + y^2)\mathbf{j}$

13. $\mathbf{F}(x, y, z) = \mathbf{i} + \mathbf{j} + \mathbf{k}$

14. $\mathbf{F}(x, y, z) = x\mathbf{i} + y\mathbf{j} + z\mathbf{k}$

In Exercises 15–20, find the gradient vector field for the scalar function. (That is, find the conservative vector field for the given potential function.)

15. $f(x, y) = 5x^2 + 3xy + 10y^2$

16. $f(x, y) = \sin 3x \cos 4y$

17. $f(x, y, z) = z - ye^{x^2}$

18. $f(x, y, z) = \dfrac{y}{z} + \dfrac{z}{x} - \dfrac{xz}{y}$

19. $g(x, y, z) = xy \ln(x + y)$

20. $g(x, y, z) = x \arcsin yz$

In Exercises 21–28, determine whether the vector field is conservative. If it is, find a potential function for the vector field.

21. $\mathbf{F}(x, y) = 2xy\mathbf{i} + x^2\mathbf{j}$

22. $\mathbf{F}(x, y) = \dfrac{1}{y^2}(y\mathbf{i} - 2x\mathbf{j})$

23. $\mathbf{F}(x, y) = xe^{x^2y}(2y\mathbf{i} + x\mathbf{j})$

24. $\mathbf{F}(x, y) = 2xy^3\mathbf{i} + 3y^2x^2\mathbf{j}$

25. $\mathbf{F}(x, y) = \dfrac{x\mathbf{i} + y\mathbf{j}}{x^2 + y^2}$

26. $\mathbf{F}(x, y) = \dfrac{2y}{x}\mathbf{i} - \dfrac{x^2}{y^2}\mathbf{j}$

27. $\mathbf{F}(x, y) = e^x(\cos y\,\mathbf{i} + \sin y\,\mathbf{j})$

28. $\mathbf{F}(x, y) = \dfrac{2x\mathbf{i} + 2y\mathbf{j}}{(x^2 + y^2)^2}$

In Exercises 29–32, find the curl of the vector field F at the indicated point.

Vector field	Point
29. $\mathbf{F}(x, y, z) = xyz\mathbf{i} + y\mathbf{j} + z\mathbf{k}$	$(1, 2, 1)$
30. $\mathbf{F}(x, y, z) = x^2z\mathbf{i} - 2xz\mathbf{j} + yz\mathbf{k}$	$(2, -1, 3)$
31. $\mathbf{F}(x, y, z) = e^x \sin y\,\mathbf{i} - e^x \cos y\,\mathbf{j}$	$(0, 0, 3)$
32. $\mathbf{F}(x, y, z) = e^{-xyz}(\mathbf{i} + \mathbf{j} + \mathbf{k})$	$(3, 2, 0)$

In Exercises 33–36, find the curl of the vector field F.

33. $\mathbf{F}(x, y, z) = \arctan\dfrac{x}{y}\mathbf{i} + \ln\sqrt{x^2 + y^2}\mathbf{j} + \mathbf{k}$

34. $\mathbf{F}(x, y, z) = \dfrac{yz}{y - z}\mathbf{i} + \dfrac{xz}{x - z}\mathbf{j} + \dfrac{xy}{x - y}\mathbf{k}$

35. $\mathbf{F}(x, y, z) = \sin(x - y)\mathbf{i} + \sin(y - z)\mathbf{j} + \sin(z - x)\mathbf{k}$

36. $\mathbf{F}(x, y, z) = \sqrt{x^2 + y^2 + z^2}(\mathbf{i} + \mathbf{j} + \mathbf{k})$

In Exercises 37–42, determine whether the vector field F is conservative. If it is, find a potential function for the vector field.

37. $\mathbf{F}(x, y, z) = \sin y\,\mathbf{i} - x\cos y\,\mathbf{j} + \mathbf{k}$

38. $\mathbf{F}(x, y, z) = e^z(y\mathbf{i} + x\mathbf{j} + \mathbf{k})$

39. $\mathbf{F}(x, y, z) = e^z(y\mathbf{i} + x\mathbf{j} + xy\mathbf{k})$

40. $\mathbf{F}(x, y, z) = 3x^2y^2z\mathbf{i} + 2x^3yz\mathbf{j} + x^3y^2\mathbf{k}$

41. $\mathbf{F}(x, y, z) = \dfrac{1}{y}\mathbf{i} - \dfrac{x}{y^2}\mathbf{j} + (2z - 1)\mathbf{k}$

42. $\mathbf{F}(x, y, z) = \dfrac{x}{x^2 + y^2}\mathbf{i} + \dfrac{y}{x^2 + y^2}\mathbf{j} + \mathbf{k}$

In Exercises 43 and 44, find curl $(\mathbf{F} \times \mathbf{G})$.

43. $\mathbf{F}(x, y, z) = \mathbf{i} + 2x\mathbf{j} + 3y\mathbf{k}$
$\mathbf{G}(x, y, z) = x\mathbf{i} - y\mathbf{j} + z\mathbf{k}$

44. $\mathbf{F}(x, y, z) = x\mathbf{i} - z\mathbf{k}$
$\mathbf{G}(x, y, z) = x^2\mathbf{i} + y\mathbf{j} + z^2\mathbf{k}$

In Exercises 45 and 46, find curl (curl F) $= \nabla \times (\nabla \times \mathbf{F})$.

45. $\mathbf{F}(x, y, z) = xyz\mathbf{i} + y\mathbf{j} + z\mathbf{k}$

46. $\mathbf{F}(x, y, z) = x^2z\mathbf{i} - 2xz\mathbf{j} + yz\mathbf{k}$

In Exercises 47–50, find the divergence of the vector field F.

47. $\mathbf{F}(x, y) = 6x^2\mathbf{i} - xy^2\mathbf{j}$

48. $\mathbf{F}(x, y) = xe^x\mathbf{i} + ye^y\mathbf{j}$

49. $\mathbf{F}(x, y, z) = \sin x\,\mathbf{i} + \cos y\,\mathbf{j} + z^2\mathbf{k}$

50. $\mathbf{F}(x, y, z) = \ln(x^2 + y^2)\mathbf{i} + xy\mathbf{j} + \ln(y^2 + z^2)\mathbf{k}$

In Exercises 51–54, find the divergence of the vector field F at the indicated point.

Vector field	Point
51. $\mathbf{F}(x, y, z) = xyz\mathbf{i} + y\mathbf{j} + z\mathbf{k}$	$(1, 2, 1)$
52. $\mathbf{F}(x, y, z) = x^2z\mathbf{i} - 2xz\mathbf{j} + yz\mathbf{k}$	$(2, -1, 3)$
53. $\mathbf{F}(x, y, z) = e^x \sin y\,\mathbf{i} - e^x \cos y\,\mathbf{j}$	$(0, 0, 3)$
54. $\mathbf{F}(x, y, z) = e^{-xyz}(\mathbf{i} + \mathbf{j} + \mathbf{k})$	$(3, 2, 0)$

In Exercises 55 and 56, find div$(\mathbf{F} \times \mathbf{G})$.

55. $\mathbf{F}(x, y, z) = \mathbf{i} + 2x\mathbf{j} + 3y\mathbf{k}$
$\mathbf{G}(x, y, z) = x\mathbf{i} - y\mathbf{j} + z\mathbf{k}$

56. $F(x, y, z) = x\mathbf{i} - z\mathbf{k}$
  $G(x, y, z) = x^2\mathbf{i} + y\mathbf{j} + z^2\mathbf{k}$

In Exercises 57 and 58, find

  div (curl F) $= \nabla \cdot (\nabla \times F)$.

57. $F(x, y, z) = xyz\mathbf{i} + y\mathbf{j} + z\mathbf{k}$

58. $F(x, y, z) = x^2 z\mathbf{i} - 2xz\mathbf{j} + yz\mathbf{k}$

**In Exercises 59–65, prove the given property for vector fields
F and G and scalar function f. (Assume the required partial
derivatives are continuous.)**

59. curl $(F + G) =$ curl F $+$ curl G

60. curl $(\nabla f) = \nabla \times (\nabla f) = 0$

61. div $(F + G) =$ div F $+$ div G

62. div $(F \times G) = ($curl F$) \cdot G - F \cdot ($curl G$)$

63. $\nabla \times [\nabla f + (\nabla \times F)] = \nabla \times (\nabla \times F)$

64. $\nabla \times (fF) = f(\nabla \times F) + (\nabla f) \times F$

65. div $(fF) = f$ div F $+ \nabla f \cdot F$

66. Prove Theorem 15.3.

In Exercises 67–70, let $F(x, y, z) = x\mathbf{i} + y\mathbf{j} + z\mathbf{k}$, **and**
$f(x, y, z) = \| F(x, y, z) \|$.

67. Show that $\nabla(\ln f) = \dfrac{F}{f^2}$.

68. Show that $\nabla\left(\dfrac{1}{f}\right) = -\dfrac{F}{f^3}$.

69. Show that $\nabla f^n = nf^{n-2}F$.

70. The **Laplacian** is the differential operator

$$\nabla^2 = \nabla \cdot \nabla = \frac{\partial^2}{\partial x^2} + \frac{\partial^2}{\partial y^2} + \frac{\partial^2}{\partial z^2}$$

and Laplace's equation is

$$\nabla^2 w = \frac{\partial^2 w}{\partial x^2} + \frac{\partial^2 w}{\partial y^2} + \frac{\partial^2 w}{\partial z^2} = 0.$$

Any function that satisfies this equation is called **harmonic.**
Show that the function $1/f$ is harmonic.

71. *Magnetic Field* A cross section of the earth's magnetic field
can be represented as a vector field in which the center of the
earth is located at the origin and the positive y-axis points in
the direction of the magnetic north pole. The equation for this
field is

$$F(x, y) = M(x, y)\mathbf{i} + N(x, y)\mathbf{j}$$

$$= \frac{m}{(x^2 + y^2)^{5/2}}[3xy\,\mathbf{i} + (2y^2 - x^2)\mathbf{j}]$$

where m is the magnetic moment of the earth. Show that this
vector field is conservative.

15.2

Piecewise Smooth Curves • Line Integrals • Line Integrals of Vector Fields • Line Integrals in Differential Form

JOSIAH WILLARD GIBBS

Many physicists and mathematicians have contributed to the theory and applications described in this chapter—Newton, Gauss, Laplace, Hamilton, Maxwell, and many others. However, the use of vector analysis to describe the results is primarily due to the American mathematical physicist Josiah Willard Gibbs (1839–1903).

Piecewise Smooth Curves

A classical property of gravitational fields is that, subject to certain physical constraints, the work done by gravity on an object moving between two points in the field is independent of the path taken by the object. One of the constraints is that the **path** must be a piecewise smooth curve. Recall that a plane curve C given by

$$\mathbf{r}(t) = x(t)\mathbf{i} + y(t)\mathbf{j}, \qquad a \le t \le b$$

is **smooth** if dx/dt and dy/dt are continuous on $[a, b]$ and not simultaneously 0 on (a, b). Similarly, a space curve C given by

$$\mathbf{r}(t) = x(t)\mathbf{i} + y(t)\mathbf{j} + z(t)\mathbf{k}, \qquad a \le t \le b$$

is **smooth** if dx/dt, dy/dt, and dz/dt are continuous on $[a, b]$ and not simultaneously 0 on (a, b). A curve C is **piecewise smooth** if the interval $[a, b]$ can be partitioned into a finite number of subintervals, on each of which C is smooth.

EXAMPLE 1   Finding a Piecewise Smooth Parametrization

Find a piecewise smooth parametrization of the graph C shown in Figure 15.7.

Solution   Because C consists of three line segments C_1, C_2, and C_3, you can construct a smooth parametrization for each and piece them together by making the last t-value in C_i correspond to the first t-value in C_{i+1} as follows.

$$
\begin{aligned}
C_1 &: x(t) = 0, & y(t) &= 2t, & z(t) &= 0, & 0 \le t \le 1 \\
C_2 &: x(t) = t - 1, & y(t) &= 2, & z(t) &= 0, & 1 \le t \le 2 \\
C_3 &: x(t) = 1, & y(t) &= 2, & z(t) &= t - 2, & 2 \le t \le 3
\end{aligned}
$$

Therefore, C is given by

$$\mathbf{r}(t) = \begin{cases} 2t\,\mathbf{j}, & 0 \le t \le 1 \\ (t - 1)\mathbf{i} + 2\mathbf{j}, & 1 \le t \le 2 \\ \mathbf{i} + 2\mathbf{j} + (t - 2)\mathbf{k}, & 2 \le t \le 3. \end{cases}$$

Because C_1, C_2, and C_3 are smooth, it follows that C is piecewise smooth. ▬

Recall that the parametrization of a curve induces an **orientation** to the curve. For instance, in Example 1 the curve is oriented so that the positive direction is from $(0, 0, 0)$, following the curve to $(1, 2, 1)$. Try finding a parametrization that induces the opposite orientation.

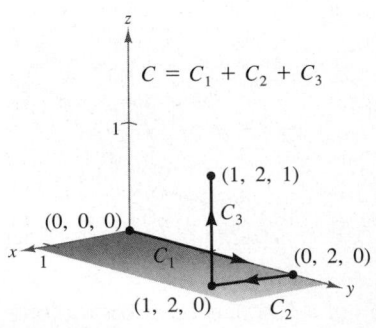

FIGURE 15.7
The graph C is piecewise smooth.

Line Integrals

Up to this point in the text, you have studied various types of integrals. For a single integral

$$\int_a^b f(x)\,dx \qquad \text{Integrate over interval } [a,b]$$

you integrated over the interval $[a,b]$. Similarly for a double integral

$$\iint_R f(x,y)\,dA \qquad \text{Integrate over region } R$$

you integrated over the region R in the plane. In this section, you will study a new type of integral called a **line integral**

$$\int_C f(x,y)\,ds \qquad \text{Integrate over curve } C$$

for which you integrate over a piecewise smooth curve C. (The terminology is somewhat unfortunate—this type of integral might be better described as a "curve integral.")

To introduce the concept of a line integral, consider the mass of a wire of finite length, given by a curve C in space. The density (mass per unit length) of the wire at the point (x,y,z) is given by $f(x,y,z)$. Partition the curve C by the points $P_0, P_1, \ldots, P_n$, producing n subarcs, as shown in Figure 15.8. The length of the ith subarc is given by Δs_i. Next, choose a point (x_i, y_i, z_i) in each subarc. If the length of each subarc is small, then the total mass of the wire can be approximated by the sum

$$\text{Mass of wire} \approx \sum_{i=1}^n f(x_i, y_i, z_i)\,\Delta s_i.$$

If you let $\|\Delta\|$ denote the length of the longest subarc and let $\|\Delta\|$ approach 0, it seems reasonable that the limit of this sum approaches the mass of the wire. This leads to the following definition.

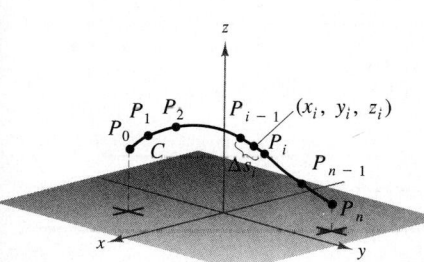

FIGURE 15.8
Partition of curve C

Definition of Line Integral

If f is defined in a region containing a smooth curve C of finite length, then the **line integral of f along C** is given by

$$\int_C f(x,y)\,ds = \lim_{\|\Delta\| \to 0} \sum_{i=1}^n f(x_i, y_i)\,\Delta s_i \qquad \text{Plane}$$

or

$$\int_C f(x,y,z)\,ds = \lim_{\|\Delta\| \to 0} \sum_{i=1}^n f(x_i, y_i, z_i)\,\Delta s_i \qquad \text{Space}$$

provided this limit exists.

As with the integrals in Chapter 14, evaluation of a line integral is best accomplished by converting to a definite integral. It can be shown that if f is *continuous*, then the above limit exists and is the same for all smooth parametrizations of C.

To evaluate a line integral over a plane curve C given by $\mathbf{r}(t) = x(t)\mathbf{i} + y(t)\mathbf{j}$, use the fact that

$$ds = \| \mathbf{r}'(t) \| \, dt = \sqrt{[x'(t)]^2 + [y'(t)]^2} \, dt.$$

A similar formula holds for a space curve, as indicated in the following theorem.

THEOREM 15.4 Evaluation of a Line Integral as a Definite Integral

Let f be continuous in a region containing a smooth curve C. If C is given by $\mathbf{r}(t) = x(t)\mathbf{i} + y(t)\mathbf{j}$, where $a \le t \le b$, then

$$\int_C f(x, y) \, ds = \int_a^b f(x(t), y(t))\sqrt{[x'(t)]^2 + [y'(t)]^2} \, dt.$$

If C is given by $\mathbf{r}(t) = x(t)\mathbf{i} + y(t)\mathbf{j} + z(t)\mathbf{k}$, where $a \le t \le b$, then

$$\int_C f(x, y, z) \, ds = \int_a^b f(x(t), y(t), z(t))\sqrt{[x'(t)]^2 + [y'(t)]^2 + [z'(t)]^2} \, dt.$$

Note that if $f(x, y, z) = 1$, then the line integral gives the arc length of the curve C, as defined in Section 12.5. That is,

$$\int_C 1 \, ds = \int_a^b \| \mathbf{r}'(t) \| \, dt = \text{length of curve } C.$$

REMARK The value of the line integral in Example 2 does not depend on the parametrization of the line segment C (any smooth parametrization will produce the same value). To convince yourself of this, try some other parametrizations, such as $x = 1 + 2t$, $y = 2 + 4t$, $z = 1 + 2t$, $-\frac{1}{2} \le t \le 0$, or $x = -t$, $y = -2t$, $z = -t$, $-1 \le t \le 0$.

EXAMPLE 2   Evaluating a Line Integral

Evaluate

$$\int_C (x^2 - y + 3z) \, ds$$

where C is the line segment shown in Figure 15.9.

Solution Begin by writing a parametric form of the equation of a line:

$$x = t, \quad y = 2t, \quad \text{and} \quad z = t, \qquad 0 \le t \le 1.$$

Hence, $x'(t) = 1$, $y'(t) = 2$, and $z'(t) = 1$, which implies that

$$\sqrt{[x'(t)]^2 + [y'(t)]^2 + [z'(t)]^2} = \sqrt{1^2 + 2^2 + 1^2} = \sqrt{6}.$$

Thus, the line integral takes the following form.

$$\int_C (x^2 - y + 3z) \, ds = \int_0^1 (t^2 - 2t + 3t)\sqrt{6} \, dt$$

$$= \sqrt{6} \int_0^1 (t^2 + t) \, dt$$

$$= \sqrt{6} \left[\frac{t^3}{3} + \frac{t^2}{2} \right]_0^1$$

$$= \frac{5\sqrt{6}}{6}$$

FIGURE 15.9

Suppose C is a path composed of smooth curves $C_1, C_2, \ldots, C_n$. If f is continuous on C, then it can be shown that

$$\int_C f(x, y)\, ds = \int_{C_1} f(x, y)\, ds + \int_{C_2} f(x, y)\, ds + \cdots + \int_{C_n} f(x, y)\, ds.$$

This property is used in Example 3.

EXAMPLE 3 Evaluating a Line Integral Over a Path

Evaluate the line integral

$$\int_C x\, ds$$

where C is the piecewise smooth curve shown in Figure 15.10.

Solution Begin by integrating up the line $y = x$, using the following parametrization.

$$C_1: x = t, y = t, \qquad 0 \le t \le 1.$$

For this curve, $\mathbf{r}(t) = t\mathbf{i} + t\mathbf{j}$, which implies that $x'(t) = 1$ and $y'(t) = 1$. Thus, $\sqrt{[x'(t)]^2 + [y'(t)]^2} = \sqrt{2}$, and you have

$$\int_{C_1} x\, ds = \int_0^1 t\sqrt{2}\, dt = \left. \frac{\sqrt{2}}{2} t^2 \right]_0^1 = \frac{\sqrt{2}}{2}.$$

Next integrate down the parabola $y = x^2$, using the parametrization

$$C_2: x = 1 - t, y = (1 - t)^2, \qquad 0 \le t \le 1.$$

For this curve, $\mathbf{r}(t) = (1 - t)\mathbf{i} + (1 - t)^2\mathbf{j}$, which implies that $x'(t) = -1$ and $y'(t) = -2(1 - t)$. Thus,

$$\sqrt{[x'(t)]^2 + [y'(t)]^2} = \sqrt{1 + 4(1 - t)^2},$$

and you have

$$\int_{C_2} x\, ds = \int_0^1 (1 - t)\sqrt{1 + 4(1 - t)^2}\, dt$$

$$= -\frac{1}{8} \left. \left[\frac{2}{3}[1 + 4(1 - t)^2]^{3/2} \right] \right]_0^1$$

$$= \frac{1}{12}(5^{3/2} - 1).$$

Consequently,

$$\int_C x\, ds = \int_{C_1} x\, ds + \int_{C_2} x\, ds = \frac{\sqrt{2}}{2} + \frac{1}{12}(5^{3/2} - 1) \approx 1.56.$$

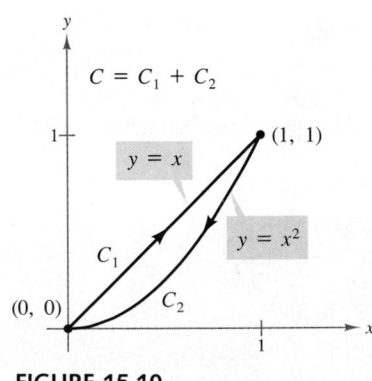

FIGURE 15.10

For parametrizations given by $\mathbf{r}(t) = x(t)\mathbf{i} + y(t)\mathbf{j} + z(t)\mathbf{k}$, it is helpful to remember the form of ds as

$$ds = \| \mathbf{r}'(t) \|\, dt = \sqrt{[x'(t)]^2 + [y'(t)]^2 + [z'(t)]^2}\, dt.$$

This is demonstrated in Example 4.

EXAMPLE 4   Evaluating a Line Integral

Evaluate $\displaystyle\int_C (x + 2)\, ds$, where C is the curve represented by

$$\mathbf{r}(t) = t\,\mathbf{i} + \frac{4}{3}t^{3/2}\mathbf{j} + \frac{1}{2}t^2\mathbf{k}, \qquad 0 \le t \le 2.$$

Solution   Because $\mathbf{r}'(t) = \mathbf{i} + 2t^{1/2}\mathbf{j} + t\,\mathbf{k}$, and

$$\|\mathbf{r}'(t)\| = \sqrt{[x'(t)]^2 + [y'(t)]^2 + [z'(t)]^2} = \sqrt{1 + 4t + t^2}$$

it follows that

$$
\begin{aligned}
\int_C (x + 2)\, ds &= \int_0^2 (t + 2)\sqrt{1 + 4t + t^2}\, dt \\
&= \frac{1}{2}\int_0^2 2(t + 2)(1 + 4t + t^2)^{1/2}\, dt \\
&= \frac{1}{3}\left[(1 + 4t + t^2)^{3/2} \right]_0^2 \\
&= \frac{1}{3}\left(13\sqrt{13} - 1\right) \\
&\approx 15.29.
\end{aligned}
$$

The next example shows how a line integral can be used to find the mass of a spring whose density varies. In Figure 15.11, note that the density of this spring increases as the spring spirals up the z-axis.

EXAMPLE 5   Finding the Mass of a Spring

Find the mass of a spring in the shape of the circular helix

$$\mathbf{r}(t) = \frac{1}{\sqrt{2}}(\cos t\,\mathbf{i} + \sin t\,\mathbf{j} + t\,\mathbf{k}), \qquad 0 \le t \le 6\pi$$

where the density of the wire is $\rho(x, y, z) = 1 + z$, as shown in Figure 15.11.

Solution   Because

$$\|\mathbf{r}'(t)\| = \frac{1}{\sqrt{2}}\sqrt{(-\sin t)^2 + (\cos t)^2 + (1)^2} = 1$$

it follows that the mass of the spring is

$$
\begin{aligned}
\text{Mass} &= \int_C (1 + z)\, ds = \int_0^{6\pi}\left(1 + \frac{t}{\sqrt{2}}\right) dt \\
&= \left[t + \frac{t^2}{2\sqrt{2}} \right]_0^{6\pi} \\
&= 6\pi\left(1 + \frac{3\pi}{\sqrt{2}}\right) \\
&\approx 144.47.
\end{aligned}
$$

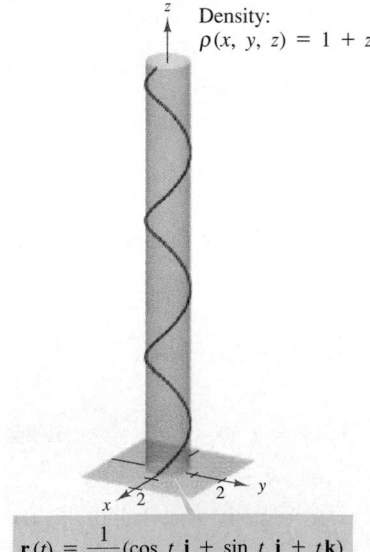

Density:
$\rho(x, y, z) = 1 + z$

$\mathbf{r}(t) = \dfrac{1}{\sqrt{2}}(\cos t\,\mathbf{i} + \sin t\,\mathbf{j} + t\,\mathbf{k})$

FIGURE 15.11
The mass of the spring is
approximately 144.47.

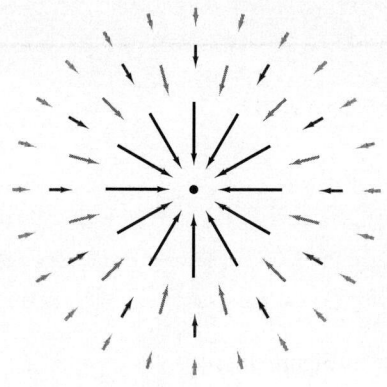

Inverse Square Force Field **F**

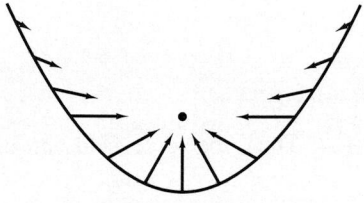

Vectors Along a Parabolic
Path in the Force Field **F**

FIGURE 15.12

Line Integrals of Vector Fields

One of the most important physical applications of line integrals is that of finding the **work** done on an object moving in a force field. For example, Figure 15.12 shows an inverse square field similar to the gravitational field of the sun. Note that the magnitude of the force along a circular path about the center is constant, whereas the magnitude of the force along a parabolic path varies from point to point.

To see how a line integral can be used to find work done in a force field **F**, consider an object moving along a path C in the field, as shown in Figure 15.13. To determine the work done by the force, you need consider only that part of the force that is acting in the same direction as that in which the object is moving (or the opposite direction). This means that at each point on C, you can consider the projection $\mathbf{F} \cdot \mathbf{T}$ of the force vector **F** onto the unit tangent vector **T**. On a small subarc of length Δs_i, the increment of work is

$$\Delta W_i = (\text{force})(\text{distance})$$
$$\approx \left[\mathbf{F}(x_i, y_i, z_i) \cdot \mathbf{T}(x_i, y_i, z_i) \right] \Delta s_i$$

where (x_i, y_i, z_i) is a point in the ith subarc. Consequently, the total work done is given by the following integral.

$$W = \int_C \mathbf{F}(x, y, z) \cdot \mathbf{T}(x, y, z) \, ds$$

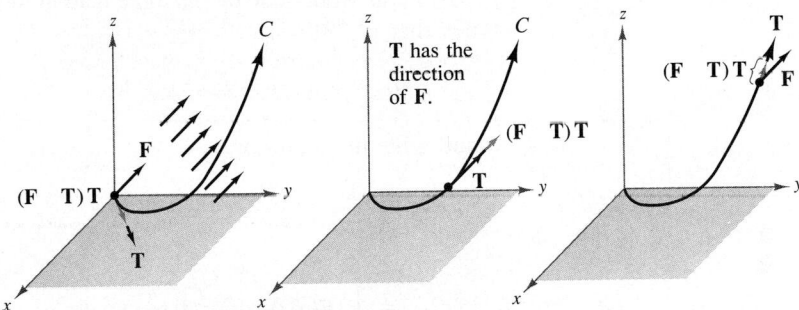

FIGURE 15.13
At each point of C, the force in the direction of motion is $(\mathbf{F} \cdot \mathbf{T})\mathbf{T}$.

This line integral appears in other contexts and is the basis of the following definition of the **line integral of a vector field.** Note in the definition that

$$\mathbf{F} \cdot \mathbf{T} \, ds = \mathbf{F} \cdot \frac{\mathbf{r}'(t)}{\|\mathbf{r}'(t)\|} \, \|\mathbf{r}'(t)\| \, dt = \mathbf{F} \cdot \mathbf{r}'(t) \, dt = \mathbf{F} \cdot d\mathbf{r}.$$

Definition of Line Integral of a Vector Field

Let **F** be a continuous vector field defined on a smooth curve C given by $\mathbf{r}(t)$. The **line integral** of **F** on C is given by

$$\int_C \mathbf{F} \cdot d\mathbf{r} = \int_C \mathbf{F} \cdot \mathbf{T} \, ds = \int_a^b \mathbf{F}(x(t), y(t), z(t)) \cdot \mathbf{r}'(t) \, dt.$$

EXAMPLE 6   Work Done by a Force

Find the work done by the force field

$$\mathbf{F}(x, y, z) = -\frac{1}{2}x\mathbf{i} - \frac{1}{2}y\mathbf{j} + \frac{1}{4}\mathbf{k} \qquad \text{Force field, } \mathbf{F}$$

on a particle as it moves along the helix given by

$$\mathbf{r}(t) = \cos t\,\mathbf{i} + \sin t\,\mathbf{j} + t\,\mathbf{k} \qquad \text{Space curve, } C$$

from the point $(1, 0, 0)$ to $(-1, 0, 3\pi)$, as shown in Figure 15.14.

Solution   Because

$$\mathbf{r}(t) = x(t)\mathbf{i} + y(t)\mathbf{j} + z(t)\mathbf{k}$$
$$= \cos t\,\mathbf{i} + \sin t\,\mathbf{j} + t\,\mathbf{k}$$

it follows that $x(t) = \cos t$, $y(t) = \sin t$, and $z(t) = t$. Thus, the force field can be written as

$$\mathbf{F}(x(t), y(t), z(t)) = -\frac{1}{2}\cos t\,\mathbf{i} - \frac{1}{2}\sin t\,\mathbf{j} + \frac{1}{4}\mathbf{k}.$$

To find the work done by the force field in moving a particle along the curve C, use the fact that

$$\mathbf{r}'(t) = -\sin t\,\mathbf{i} + \cos t\,\mathbf{j} + \mathbf{k}$$

and write the following.

$$W = \int_C \mathbf{F} \cdot d\mathbf{r}$$
$$= \int_a^b \mathbf{F}(x(t), y(t), z(t)) \cdot \mathbf{r}'(t)\, dt$$
$$= \int_0^{3\pi} \left(-\frac{1}{2}\cos t\,\mathbf{i} - \frac{1}{2}\sin t\,\mathbf{j} + \frac{1}{4}\mathbf{k}\right) \cdot (-\sin t\,\mathbf{i} + \cos t\,\mathbf{j} + \mathbf{k})\, dt$$
$$= \int_0^{3\pi} \left(\frac{1}{2}\sin t\cos t - \frac{1}{2}\sin t\cos t + \frac{1}{4}\right) dt$$
$$= \int_0^{3\pi} \frac{1}{4}\, dt$$
$$= \left[\frac{1}{4}t\right]_0^{3\pi}$$
$$= \frac{3\pi}{4}$$

FIGURE 15.14

Generated by Mathematica

FIGURE 15.15
This computer-generated view of the force field in Example 6 indicates that each vector in the force field points toward the z-axis.

REMARK   In Example 6, note that the x- and y-components of the force field end up contributing nothing to the total work. This occurs because *in this particular example* the z-component of the force field is the only portion of the force that is acting in the same (or opposite) direction in which the particle is moving (see Figure 15.15).

For line integrals of vector functions, the orientation of the curve C is important. If the orientation of the curve is reversed, then the unit tangent vector $\mathbf{T}(t)$ is changed to $-\mathbf{T}(t)$, and you obtain

$$\int_{-C} \mathbf{F} \cdot d\mathbf{r} = -\int_{C} \mathbf{F} \cdot d\mathbf{r}.$$

EXAMPLE 7   Orientation and Parametrization of a Curve

Let $\mathbf{F}(x, y) = y\mathbf{i} + x^2\mathbf{j}$ and evaluate the line integral $\int_C \mathbf{F} \cdot d\mathbf{r}$ for each of the following parabolic curves (see Figure 15.16).

a. C_1: $\mathbf{r}_1(t) = (4 - t)\mathbf{i} + (4t - t^2)\mathbf{j}$, $0 \le t \le 3$
b. C_2: $\mathbf{r}_2(t) = t\mathbf{i} + (4t - t^2)\mathbf{j}$, $1 \le t \le 4$

Solution

a. Because $\mathbf{r}_1'(t) = -\mathbf{i} + (4 - 2t)\mathbf{j}$ and

$$\mathbf{F}(x(t), y(t)) = (4t - t^2)\mathbf{i} + (4 - t)^2\mathbf{j}$$

the line integral is

$$\int_{C_1} \mathbf{F} \cdot d\mathbf{r} = \int_0^3 [(4t - t^2)\mathbf{i} + (4 - t)^2\mathbf{j}] \cdot [-\mathbf{i} + (4 - 2t)\mathbf{j}]\, dt$$

$$= \int_0^3 (-4t + t^2 + 64 - 64t + 20t^2 - 2t^3)\, dt$$

$$= \int_0^3 (-2t^3 + 21t^2 - 68t + 64)\, dt$$

$$= \left[-\frac{t^4}{2} + 7t^3 - 34t^2 + 64t \right]_0^3$$

$$= \frac{69}{2}.$$

b. Because $\mathbf{r}_2'(t) = \mathbf{i} + (4 - 2t)\mathbf{j}$ and

$$\mathbf{F}(x(t), y(t)) = (4t - t^2)\mathbf{i} + t^2\mathbf{j}$$

the line integral is

$$\int_{C_2} \mathbf{F} \cdot d\mathbf{r} = \int_1^4 [(4t - t^2)\mathbf{i} + t^2\mathbf{j}] \cdot [\mathbf{i} + (4 - 2t)\mathbf{j}]\, dt$$

$$= \int_1^4 (4t - t^2 + 4t^2 - 2t^3)\, dt$$

$$= \int_1^4 (-2t^3 + 3t^2 + 4t)\, dt$$

$$= \left[-\frac{t^4}{2} + t^3 + 2t^2 \right]_1^4$$

$$= -\frac{69}{2}.$$

The answer in part b is the negative of that in part a because C_1 and C_2 represent opposite orientations of the same parabolic segment. ▬▬

C_1: $\mathbf{r}_1(t) = (4 - t)\mathbf{i} + (4t - t^2)\mathbf{j}$
C_2: $\mathbf{r}_2(t) = t\mathbf{i} + (4t - t^2)\mathbf{j}$

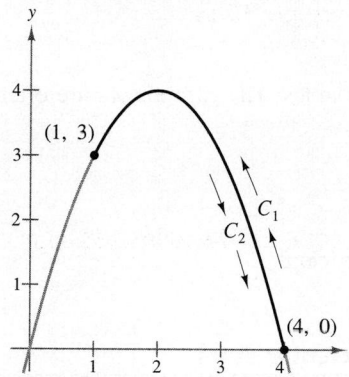

FIGURE 15.16

REMARK Although the value of the line integral in Example 7 does depend on the orientation of C, it does not depend on the parametrization of C. To see this, let C_3 be represented by

$$\mathbf{r}_3 = (t + 2)\mathbf{i} + (4 - t^2)\mathbf{j}$$

where $-1 \le t \le 2$. The graph of this curve is the same parabolic segment shown in Figure 15.16. Does the value of the line integral over C_3 agree with the value over C_1 or C_2? Why?

REMARK   The orientation of C affects the value of the differential form of a line integral. Specifically, if $-C$ has the opposite orientation as C, then

$$\int_{-C} M\,dx + N\,dy = -\int_{C} M\,dx + N\,dy.$$

Thus, of the three line integral forms presented in this section, the orientation of C does not affect the form $\int_C f(x, y)\,ds$, but it does affect the vector form and the differential form.

A second commonly used form of line integral is derived from the vector field notation used in the previous section. If $\mathbf{F}$ is a vector field of the form $\mathbf{F}(x, y) = M\mathbf{i} + N\mathbf{j}$, and C is given by $\mathbf{r}(t) = x(t)\mathbf{i} + y(t)\mathbf{j}$, then $\mathbf{F} \cdot d\mathbf{r}$ is often written as $M\,dx + N\,dy$.

$$\int_{C} \mathbf{F} \cdot d\mathbf{r} = \int_{C} \mathbf{F} \cdot \frac{d\mathbf{r}}{dt}\,dt = \int_{a}^{b} (M\mathbf{i} + N\mathbf{j}) \cdot (x'(t)\mathbf{i} + y'(t)\mathbf{j})\,dt$$

$$= \int_{a}^{b} \left(M\frac{dx}{dt} + N\frac{dy}{dt} \right) dt$$

$$= \int_{C} (M\,dx + N\,dy)$$

This **differential form** can be extended to three variables. The parentheses are often omitted, as follows.

$$\int_{C} M\,dx + N\,dy \quad \text{and} \quad \int_{C} M\,dx + N\,dy + P\,dz$$

Notice how this differential notation is used in Example 8.

EXAMPLE 8   Evaluating a Line Integral in Differential Form

Let C be the circle of radius 3 given by

$$\mathbf{r}(t) = 3\cos t\,\mathbf{i} + 3\sin t\,\mathbf{j}, \qquad 0 \le t \le 2\pi$$

(see Figure 15.17) and evaluate the line integral

$$\int_{C} y^3\,dx + (x^3 + 3xy^2)\,dy.$$

Solution   Because $x = 3\cos t$ and $y = 3\sin t$, you have $dx = -3\sin t\,dt$ and $dy = 3\cos t\,dt$. Thus, the line integral is

$$\int_{C} M\,dx + N\,dy$$

$$= \int_{C} y^3\,dx + (x^3 + 3xy^2)\,dy$$

$$= \int_{0}^{2\pi} \left[(27\sin^3 t)(-3\sin t) + (27\cos^3 t + 81\cos t\sin^2 t)(3\cos t) \right] dt$$

$$= 81 \int_{0}^{2\pi} (\cos^4 t - \sin^4 t + 3\cos^2 t\sin^2 t)\,dt$$

$$= 81 \int_{0}^{2\pi} \left(\cos^2 t - \sin^2 t + \frac{3}{4}\sin^2 2t \right) dt$$

$$= 81 \int_{0}^{2\pi} \left[\cos 2t + \frac{3}{4}\left(\frac{1 - \cos 4t}{2} \right) \right] dt$$

$$= 81 \left[\frac{\sin 2t}{2} + \frac{3}{8}t - \frac{3\sin 4t}{32} \right]_{0}^{2\pi}$$

$$= \frac{243\pi}{4}.$$

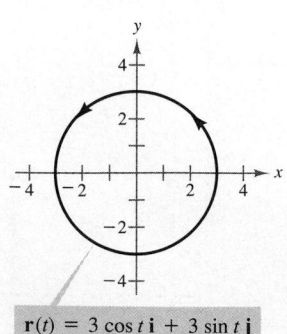

$\mathbf{r}(t) = 3\cos t\,\mathbf{i} + 3\sin t\,\mathbf{j}$

FIGURE 15.17

For curves represented by $y = g(x)$, $a \le x \le b$, you can let $x = t$ and obtain the parametric form

$$x = t \quad \text{and} \quad y = g(t), \qquad a \le t \le b.$$

Because $dx = dt$ for this form, you have the option of evaluating the line integral in the variable x or t. This is demonstrated in Example 9.

EXAMPLE 9   Evaluating a Line Integral in Differential Form

Evaluate

$$\int_C y\,dx + x^2\,dy$$

where C is the parabolic arc given by $y = 4x - x^2$ from $(4, 0)$ to $(1, 3)$, as shown in Figure 15.18.

Solution   Rather than converting to the parameter t, you can simply retain the variable x and write

$$y = 4x - x^2 \rightarrow dy = (4 - 2x)\,dx.$$

Then, in the direction from $(4, 0)$ to $(1, 3)$, the line integral is

$$\int_C y\,dx + x^2\,dy = \int_4^1 \left[(4x - x^2)\,dx + x^2(4 - 2x)\,dx \right]$$

$$= \int_4^1 (4x + 3x^2 - 2x^3)\,dx$$

$$= \left[2x^2 + x^3 - \frac{x^4}{2} \right]_4^1$$

$$= \frac{69}{2}.$$

(See Example 7.)

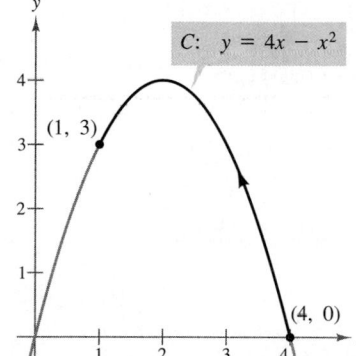

FIGURE 15.18
The line integral over C from $(4, 0)$ to $(1, 3)$ is $\dfrac{69}{2}$.

EXERCISES for Section 15.2

TECHNOLOGY
Laboratory Guide
Labs 15.1–15.2

In Exercises 1–6, find a piecewise smooth parametrization of the path C.

1.

2.

3.

4.

5.   **6.**

In Exercises 7–10, evaluate the line integral over the specified path.

7. $\int_C (x - y)\, ds$
  $C: \mathbf{r}(t) = 4t\,\mathbf{i} + 3t\,\mathbf{j}$
  $0 \leq t \leq 2$

8. $\int_C 4xy\, ds$
  $C: \mathbf{r}(t) = t\,\mathbf{i} + (1 - t)\mathbf{j}$
  $0 \leq t \leq 1$

9. $\int_C (x^2 + y^2 + z^2)\, ds$
  $C: \mathbf{r}(t) = \sin t\,\mathbf{i} + \cos t\,\mathbf{j} + 8t\,\mathbf{k}$
  $0 \leq t \leq \pi/2$

10. $\int_C 8xyz\, ds$
  $C: \mathbf{r}(t) = 3\,\mathbf{i} + 12t\,\mathbf{j} + 5t\,\mathbf{k}$
  $0 \leq t \leq 2$

In Exercises 11–14, evaluate

$$\int_C (x^2 + y^2)\, ds$$

along the path.

11. C: x-axis from $x = 0$ to $x = 3$

12. C: y-axis from $y = 1$ to $y = 10$

13. C: counterclockwise around the circle $x^2 + y^2 = 1$ from $(1, 0)$ to $(0, 1)$

14. C: counterclockwise around the circle $x^2 + y^2 = 4$ from $(2, 0)$ to $(0, 2)$

In Exercises 15–18, evaluate

$$\int_C (x + 4\sqrt{y})\, ds$$

along the path.

15. C: line from $(0, 0)$ to $(1, 1)$

16. C: line from $(0, 0)$ to $(3, 9)$

17. C: counterclockwise around the triangle with vertices $(0, 0)$, $(1, 0)$, and $(0, 1)$

18. C: counterclockwise around the square with vertices $(0, 0)$, $(1, 0)$, $(1, 1)$, and $(0, 1)$

Mass   **In Exercises 19 and 20, find the total mass of two turns of a spring with density ρ in the shape of the circular helix**

$$\mathbf{r}(t) = 3\cos t\,\mathbf{i} + 3\sin t\,\mathbf{j} + 2t\,\mathbf{k}.$$

19. $\rho(x, y, z) = \frac{1}{2}(x^2 + y^2 + z^2)$

20. $\rho(x, y, z) = 2$

In Exercises 21–26, evaluate

$$\int_C \mathbf{F} \cdot d\mathbf{r}$$

where C is represented by $\mathbf{r}(t)$.

21. $\mathbf{F}(x, y) = xy\,\mathbf{i} + y\,\mathbf{j}$
  $C: \mathbf{r}(t) = 4t\,\mathbf{i} + t\,\mathbf{j}$
  $0 \leq t \leq 1$

22. $\mathbf{F}(x, y) = xy\,\mathbf{i} + y\,\mathbf{j}$
  $C: \mathbf{r}(t) = 4\cos t\,\mathbf{i} + 4\sin t\,\mathbf{j}$
  $0 \leq t \leq \dfrac{\pi}{2}$

23. $\mathbf{F}(x, y) = 3x\,\mathbf{i} + 4y\,\mathbf{j}$
  $C: \mathbf{r}(t) = 2\cos t\,\mathbf{i} + 2\sin t\,\mathbf{j}$
  $0 \leq t \leq \dfrac{\pi}{2}$

24. $\mathbf{F}(x, y) = 3x\,\mathbf{i} + 4y\,\mathbf{j}$
  $C: \mathbf{r}(t) = t\,\mathbf{i} + \sqrt{4 - t^2}\,\mathbf{j}$
  $-2 \leq t \leq 2$

25. $\mathbf{F}(x, y, z) = x^2y\,\mathbf{i} + (x - z)\mathbf{j} + xyz\,\mathbf{k}$
  $C: \mathbf{r}(t) = t\,\mathbf{i} + t^2\,\mathbf{j} + 2\mathbf{k}$
  $0 \leq t \leq 1$

26. $\mathbf{F}(x, y, z) = x^2\,\mathbf{i} + y^2\,\mathbf{j} + z^2\,\mathbf{k}$
  $C: \mathbf{r}(t) = \sin t\,\mathbf{i} + \cos t\,\mathbf{j} + t^2\,\mathbf{k}$
  $0 \leq t \leq \dfrac{\pi}{2}$

* C In Exercises 27 and 28, use a symbolic integration utility to evaluate the integral

$$\int_C \mathbf{F} \cdot d\mathbf{r}$$

where C is represented by $\mathbf{r}(t)$.

27. $\mathbf{F}(x, y, z) = x^2z\,\mathbf{i} + 6y\,\mathbf{j} + yz^2\,\mathbf{k}$
  $C: \mathbf{r}(t) = t\,\mathbf{i} + t^2\,\mathbf{j} + \ln t\,\mathbf{k}$
  $1 \leq t \leq 3$

28. $\mathbf{F}(x, y, z) = \dfrac{x\,\mathbf{i} + y\,\mathbf{j} + z\,\mathbf{k}}{\sqrt{x^2 + y^2 + z^2}}$
  $C: \mathbf{r}(t) = t\,\mathbf{i} + t\,\mathbf{j} + e^t\,\mathbf{k}$
  $0 \leq t \leq 2$

Work   **In Exercises 29–34, find the work done by the force field $\mathbf{F}$ on an object moving along the specified path.**

29. $\mathbf{F}(x, y) = -x\,\mathbf{i} - 2y\,\mathbf{j}$
  $C: y = x^3$ from $(0, 0)$ to $(2, 8)$

30. $\mathbf{F}(x, y) = x^2\,\mathbf{i} - xy\,\mathbf{j}$
  $C: x = \cos^3 t,\ y = \sin^3 t$ from $(1, 0)$ to $(0, 1)$

*A C indicates an exercise in which you are instructed to use graphing technology or a symbolic computer algebra system. The solutions of other exercises may also be facilitated by appropriate technology.

31. $\mathbf{F}(x, y) = 2x\mathbf{i} + y\mathbf{j}$
   C: counterclockwise around the triangle whose vertices are
   $(0, 0)$, $(1, 0)$, and $(1, 1)$

32. $\mathbf{F}(x, y) = -y\mathbf{i} - x\mathbf{j}$
   C: counterclockwise along the semicircle $y = \sqrt{4 - x^2}$
   from $(2, 0)$ to $(-2, 0)$

33. $\mathbf{F}(x, y, z) = x\mathbf{i} + y\mathbf{j} - 5z\mathbf{k}$
   C: $\mathbf{r}(t) = 2\cos t\,\mathbf{i} + 2\sin t\,\mathbf{j} + t\,\mathbf{k}$
   $0 \le t \le 2\pi$

34. $\mathbf{F}(x, y, z) = yz\mathbf{i} + xz\mathbf{j} + xy\mathbf{k}$
   C: line from $(0, 0, 0)$ to $(5, 3, 2)$

35. *Work*   Find the work done by a person weighing 150 pounds walking exactly one revolution up a circular helical staircase of radius 3 feet if the person rises 10 feet.

36. *Work*   A particle moves along the path $y = x^2$ from the point $(0, 0)$ to the point $(1, 1)$. The force field $\mathbf{F}$ is measured at five points along the path, and the results are shown in the following table.

(x, y)	$(0, 0)$	$(\frac{1}{4}, \frac{1}{16})$	$(\frac{1}{2}, \frac{1}{4})$	$(\frac{3}{4}, \frac{9}{16})$	$(1, 1)$
$\mathbf{F}(x, y)$	$5\mathbf{i}$	$3.5\mathbf{i} + \mathbf{j}$	$2\mathbf{i} + 2\mathbf{j}$	$1.5\mathbf{i} + 3\mathbf{j}$	$\mathbf{i} + 5\mathbf{j}$

Use Simpson's Rule to approximate the work done by the force field.

In Exercises 37–40, demonstrate the property that

$$\int_C \mathbf{F} \cdot d\mathbf{r} = 0$$

regardless of the initial and terminal points of C, if the tangent vector $\mathbf{r}'(t)$ is orthogonal to the force vector $\mathbf{F}$.

37. $\mathbf{F}(x, y) = y\mathbf{i} - x\mathbf{j}$
   C: $\mathbf{r}(t) = t\mathbf{i} - 2t\mathbf{j}$

38. $\mathbf{F}(x, y) = -3y\mathbf{i} + x\mathbf{j}$
   C: $\mathbf{r}(t) = t\mathbf{i} - t^3\mathbf{j}$

39. $\mathbf{F}(x, y) = (x^3 - 2x^2)\mathbf{i} + \left(x - \dfrac{y}{2}\right)\mathbf{j}$
   C: $\mathbf{r}(t) = t\mathbf{i} + t^2\mathbf{j}$

40. $\mathbf{F}(x, y) = x\mathbf{i} + y\mathbf{j}$
   C: $\mathbf{r}(t) = 3\sin t\,\mathbf{i} + 3\cos t\,\mathbf{j}$

In Exercises 41–44, evaluate the line integral over the path C given by $x = 2t$, $y = 10t$, where $0 \le t \le 1$.

41. $\displaystyle\int_C (x + 3y^2)\,dy$

42. $\displaystyle\int_C (x + 3y^2)\,dx$

43. $\displaystyle\int_C xy\,dx + y\,dy$

44. $\displaystyle\int_C (y - 3x)\,dx + x^2\,dy$

In Exercises 45–50, evaluate the integral

$$\int_C (2x - y)\,dx + (x + 3y)\,dy$$

along the path.

45. C: x-axis from $x = 0$ to $x = 5$

46. C: y-axis from $y = 0$ to $y = 2$

47. C: line segments from $(0, 0)$ to $(3, 0)$ and from $(3, 0)$ to $(3, 3)$

48. C: line segments from $(0, 0)$ to $(0, -3)$ and from $(0, -3)$ to $(2, -3)$

49. C: parabolic path $x = t$, $y = 2t^2$, from $(0, 0)$ to $(2, 8)$

50. C: elliptic path $x = 4\sin t$, $y = 3\cos t$, from $(0, 3)$ to $(4, 0)$

Lateral Surface Area   **In Exercises 51–58, find the area of the lateral surface (see figure) over the curve C in the xy-plane and under the surface $z = f(x, y)$, where**

$$\text{Lateral surface area} = \int_C f(x, y)\,ds.$$

51. $f(x, y) = h$
   C: line from $(0, 0)$ to $(3, 4)$

52. $f(x, y) = y$
   C: line from $(0, 0)$ to $(4, 4)$

53. $f(x, y) = xy$
   C: $x^2 + y^2 = 1$ from $(1, 0)$ to $(0, 1)$

54. $f(x, y) = x + y$
   C: $x^2 + y^2 = 1$ from $(1, 0)$ to $(0, 1)$

55. $f(x, y) = h$
   C: $y = 1 - x^2$ from $(1, 0)$ to $(0, 1)$

56. $f(x, y) = y + 1$
   C: $y = 1 - x^2$ from $(1, 0)$ to $(0, 1)$

57. $f(x, y) = xy$
   C: $y = 1 - x^2$ from $(1, 0)$ to $(0, 1)$

58. $f(x, y) = x^2 - y^2 + 4$
   C: $x^2 + y^2 = 4$

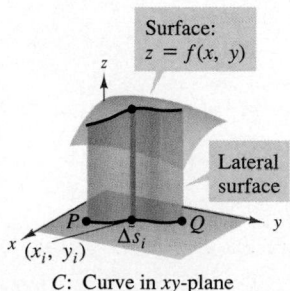

FIGURE FOR 51–58

59. *Engine Design*   A tractor engine has a steel component with a circular base modeled by the vector-valued function $\mathbf{r}(t) = 2\cos t\,\mathbf{i} + 2\sin t\,\mathbf{j}$. Its height is given by $z = 1 + y^2$. All measurements of the component are given in centimeters.
  a. Find the lateral surface area of the component.
  b. If the component is in the form of a shell of thickness 0.2 centimeter, use the result of part a to approximate the amount of steel used in its manufacture.
  c. Make a sketch of the component.

60. *Building Design*   The ceiling of a building has a height above the floor given by $z = 20 + \frac{1}{4}x$, and one of the walls follows a path given by $y = x^{3/2}$. Find the surface area of the wall if $0 \le x \le 40$. (All measurements are given in feet.)

Estimation   **In Exercises 61 and 62, determine which value best approximates the lateral surface area over the curve C in the xy-plane and under the surface $z = f(x,y)$. (Make your selection on the basis of a sketch of the surface and *not* by performing any calculations.)**

61. $f(x,y) = y$
  C: $y = x^2$ from $(0, 0)$ to $(2, 4)$
  a. 2   **b.** 4   **c.** 8   **d.** 16

62. $f(x,y) = e^{xy}$
  C: line from $(0, 0)$ to $(2, 2)$
  a. 54   **b.** 25   **c.** -250 **d.** 75   **e.** 100

True or False   **In Exercises 63–66, determine whether the statement is true or false. If it is false, explain why or give an example that shows it is false.**

63. If C is given by $x(t) = t$, $y(t) = t$, $0 \le t \le 1$, then

$$\int_C xy\,ds = \int_0^1 t^2\,dt.$$

64. If $C_2 = -C_1$, then $\displaystyle\int_{C_1} f(x, y)\,ds + \int_{C_2} f(x, y)\,ds = 0$.

65. The vector functions $\mathbf{r}_1 = t\,\mathbf{i} + t^2\,\mathbf{j}$, $0 \le t \le 1$, and $\mathbf{r}_2 = (1 - t)\,\mathbf{i} + (1 - t)^2\,\mathbf{j}$, $0 \le t \le 1$, define the same curve.

66. If $\displaystyle\int_C \mathbf{F} \cdot \mathbf{T}\,ds = 0$, then $\mathbf{F}$ and $\mathbf{T}$ are orthogonal.

Fundamental Theorem of Line Integrals

In the previous section, we pointed out that in a gravitational field the work done by gravity on an object moving between two points in the field is independent of the path taken by the object. In this section, you will study an important generalization of this result—it is called the **Fundamental Theorem of Line Integrals.**

We begin with an example in which the line integral of a *conservative vector field* is evaluated over three different paths.

EXAMPLE 1　Line Integral of a Conservative Vector Field

Find the work done by the force field

$$\mathbf{F}(x, y) = \frac{1}{2}xy\,\mathbf{i} + \frac{1}{4}x^2\mathbf{j}$$

on a particle that moves from $(0, 0)$ to $(1, 1)$ along the following paths.

a. $C_1: y = x$　　　**b.** $C_2: x = y^2$　　　**c.** $C_3: y = x^3$

Solution　(See Figure 15.19.)

a. Let $\mathbf{r}(t) = t\mathbf{i} + t\mathbf{j}$ for $0 \le t \le 1$, so that

$$d\mathbf{r} = (\mathbf{i} + \mathbf{j})\,dt \quad \text{and} \quad \mathbf{F}(x, y) = \frac{1}{2}t^2\mathbf{i} + \frac{1}{4}t^2\mathbf{j}.$$

Then, the work done is

$$W = \int_{C_1} \mathbf{F} \cdot d\mathbf{r} = \int_0^1 \frac{3}{4}t^2\,dt = \frac{1}{4}t^3\bigg]_0^1 = \frac{1}{4}.$$

b. Let $\mathbf{r}(t) = t\mathbf{i} + \sqrt{t}\,\mathbf{j}$ for $0 \le t \le 1$, so that

$$d\mathbf{r} = \left(\mathbf{i} + \frac{1}{2\sqrt{t}}\mathbf{j}\right)dt \quad \text{and} \quad \mathbf{F}(x, y) = \frac{1}{2}t^{3/2}\mathbf{i} + \frac{1}{4}t^2\mathbf{j}.$$

Then, the work done is

$$W = \int_{C_2} \mathbf{F} \cdot d\mathbf{r} = \int_0^1 \frac{5}{8}t^{3/2}\,dt = \frac{1}{4}t^{5/2}\bigg]_0^1 = \frac{1}{4}.$$

c. Let $\mathbf{r}(t) = \frac{1}{2}t\mathbf{i} + \frac{1}{8}t^3\mathbf{j}$ for $0 \le t \le 2$, so that

$$d\mathbf{r} = \left(\frac{1}{2}\mathbf{i} + \frac{3}{8}t^2\mathbf{j}\right)dt \quad \text{and} \quad \mathbf{F}(x, y) = \frac{1}{32}t^4\mathbf{i} + \frac{1}{16}t^2\mathbf{j}.$$

Then, the work done is

$$W = \int_{C_3} \mathbf{F} \cdot d\mathbf{r} = \int_0^2 \frac{5}{128}t^4\,dt = \frac{1}{128}t^5\bigg]_0^2 = \frac{1}{4}.$$

$C_1: y = x$

$C_2: x = y^2$

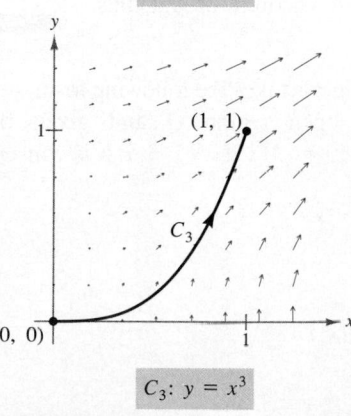

$C_3: y = x^3$

FIGURE 15.19
The work done by a conservative vector field is the same for all paths.

In Example 1, note that the vector field $\mathbf{F}(x, y) = \frac{1}{2}xy\,\mathbf{i} + \frac{1}{4}x^2\,\mathbf{j}$ is conservative because $\mathbf{F}(x, y) = \nabla f(x, y)$, where $f(x, y) = \frac{1}{4}x^2 y$. In such cases, the following theorem states that the value of $\int_C \mathbf{F} \cdot d\mathbf{r}$ is given by

$$\int_C \mathbf{F} \cdot d\mathbf{r} = f(x(1), y(1)) - f(x(0), y(0)) = \frac{1}{4} - 0 = \frac{1}{4}.$$

THEOREM 15.5   Fundamental Theorem of Line Integrals

Let C be a piecewise smooth curve lying in an open region R and given by

$$\mathbf{r}(t) = x(t)\mathbf{i} + y(t)\mathbf{j}, \qquad a \leq t \leq b.$$

If $\mathbf{F}(x, y) = M\mathbf{i} + N\mathbf{j}$ is conservative in R, and M and N are continuous in R, then

$$\int_C \mathbf{F} \cdot d\mathbf{r} = \int_C \nabla f \cdot d\mathbf{r} = f(x(b), y(b)) - f(x(a), y(a))$$

where f is a potential function of $\mathbf{F}$. That is, $\mathbf{F}(x, y) = \nabla f(x, y)$.

Proof   We provide a proof only for a smooth curve. For piecewise smooth curves, the procedure is carried out separately on each smooth portion. Because

$$\mathbf{F}(x, y) = \nabla f(x, y) = f_x(x, y)\mathbf{i} + f_y(x, y)\mathbf{j},$$

it follows that

$$\int_C \mathbf{F} \cdot d\mathbf{r} = \int_a^b \mathbf{F} \cdot \frac{d\mathbf{r}}{dt}\,dt = \int_a^b \left[f_x(x, y)\frac{dx}{dt} + f_y(x, y)\frac{dy}{dt} \right] dt$$

and, by the Chain Rule (Theorem 13.6), you have

$$\int_C \mathbf{F} \cdot d\mathbf{r} = \int_a^b \frac{d}{dt}[f(x(t), y(t))]\,dt = f(x(b), y(b)) - f(x(a), y(a)).$$

The last step is an application of the Fundamental Theorem of Calculus.

In space, the Fundamental Theorem of Line Integrals takes the following form. Let C be a piecewise smooth curve lying in an open region Q and given by $\mathbf{r}(t) = x(t)\mathbf{i} + y(t)\mathbf{j} + z(t)\mathbf{k}$, $a \leq t \leq b$. If $\mathbf{F}(x, y, z) = M\mathbf{i} + N\mathbf{j} + P\mathbf{k}$ is conservative and M, N, and P are continuous, then

$$\int_C \mathbf{F} \cdot d\mathbf{r} = \int_C \nabla f \cdot d\mathbf{r}$$
$$= f(x(b), y(b), z(b)) - f(x(a), y(a), z(a))$$

where $\mathbf{F}(x, y, z) = \nabla f(x, y, z)$.

The Fundamental Theorem of Line Integrals states that if the vector field $\mathbf{F}$ is conservative, then the line integral between any two points is simply the difference in the values of the *potential* function f at these points.

EXAMPLE 2   Using the Fundamental Theorem of Line Integrals

Evaluate $\int_C \mathbf{F} \cdot d\mathbf{r}$, where C is a piecewise smooth curve from $(-1, 4)$ to $(1, 2)$ and

$$\mathbf{F}(x, y) = 2xy\,\mathbf{i} + (x^2 - y)\mathbf{j}.$$

(See Figure 15.20.)

Solution   From Example 6 of Section 15.1, you know that $\mathbf{F}$ is the gradient of f where

$$f(x, y) = x^2 y - \frac{1}{2}y^2 + K.$$

Consequently, $\mathbf{F}$ is conservative, and by the Fundamental Theorem of Line Integrals, it follows that

$$\int_C \mathbf{F} \cdot d\mathbf{r} = f(1, 2) - f(-1, 4)$$

$$= \left[1^2(2) - \frac{1}{2}(2^2) \right] - \left[(-1)^2(4) - \frac{1}{2}(4^2) \right]$$

$$= 4.$$

Note that it is unnecessary to include a constant K as part of f, because it is canceled by subtraction.

FIGURE 15.20
Using the Fundamental Theorem of Line Integrals, $\int_C \mathbf{F} \cdot d\mathbf{r} = 4$.

EXAMPLE 3   Using the Fundamental Theorem of Line Integrals

Evaluate $\int_C \mathbf{F} \cdot d\mathbf{r}$, where C is a piecewise smooth curve from $(1, 1, 0)$ to $(0, 2, 3)$ and

$$\mathbf{F}(x, y, z) = 2xy\,\mathbf{i} + (x^2 + z^2)\mathbf{j} + 2zy\,\mathbf{k}.$$

(See Figure 15.21.)

Solution   From Example 8 of Section 15.1, you know that $\mathbf{F}$ is the gradient of f where

$$f(x, y, z) = x^2 y + z^2 y.$$

Consequently, $\mathbf{F}$ is conservative, and by the Fundamental Theorem of Line Integrals, it follows that

$$\int_C \mathbf{F} \cdot d\mathbf{r} = f(0, 2, 3) - f(1, 1, 0)$$

$$= [(0)^2(2) + (3^2)(2)] - [1^2(1) + (0^2)(1)]$$

$$= 17.$$

FIGURE 15.21
Using the Fundamental Theorem of Line Integrals, $\int_C \mathbf{F} \cdot d\mathbf{r} = 17$.

REMARK   In Examples 2 and 3, be sure you see that the value of the line integral is the same for any smooth curve C that has the given initial and terminal points. For instance, in Example 3, try evaluating the line integral for the curve given by

$$\mathbf{r}(t) = (1 - t)\mathbf{i} + (1 + t)\mathbf{j} + 3t\,\mathbf{k}.$$

You should obtain

$$\int_C \mathbf{F} \cdot d\mathbf{r} = \int_0^1 (30t^2 + 16t - 1)\,dt = 17.$$

R_1 is connected. R_2 is not connected.

FIGURE 15.22

Independence of Path

From the Fundamental Theorem of Line Integrals, it is clear that if **F** is continuous and conservative in an open region R, then the value of $\int_C \mathbf{F} \cdot d\mathbf{r}$ is the same for every piecewise smooth curve C from one fixed point in R to another fixed point in R. This result is described by saying that the line integral $\int_C \mathbf{F} \cdot d\mathbf{r}$ is **independent of path** in the region R.

A region in the plane (or in space) is **connected** if any two points in the region can be joined by a piecewise smooth curve lying entirely within the region, as shown in Figure 15.22. In open regions that are *connected,* the path independence of $\int_C \mathbf{F} \cdot d\mathbf{r}$ is equivalent to the condition that **F** is conservative.

THEOREM 15.6   Independence of Path and Conservative Vector Fields

If **F** is continuous on an open connected region, then the line integral

$$\int_C \mathbf{F} \cdot d\mathbf{r}$$

is independent of path if and only if **F** is conservative.

Proof   If **F** is conservative, then by the Fundamental Theorem of Line Integrals the line integral is independent of path. We establish the converse for a plane region R. Let $\mathbf{F}(x, y) = M\mathbf{i} + N\mathbf{j}$, and let (x_0, y_0) be a fixed point in R. If (x, y) is any point in R, then choose a piecewise smooth curve C running from (x_0, y_0) to (x, y), and define f by

$$f(x, y) = \int_C \mathbf{F} \cdot d\mathbf{r} = \int_C M\,dx + N\,dy.$$

The existence of C in R is guaranteed by the fact that R is connected. You can show that f is a potential function of **F** by considering two different paths between (x_0, y_0) and (x, y). For the *first* path, choose (x_1, y) in R such that $x \neq x_1$. This is possible because R is open. Then choose C_1 and C_2, as shown in Figure 15.23. Using the independence of path, it follows that

$$f(x, y) = \int_C M\,dx + N\,dy = \int_{C_1} M\,dx + N\,dy + \int_{C_2} M\,dx + N\,dy.$$

Because the first integral does not depend on x, and $dy = 0$ in the second integral, you have

$$f(x, y) = g(y) + \int_{C_2} M\,dx$$

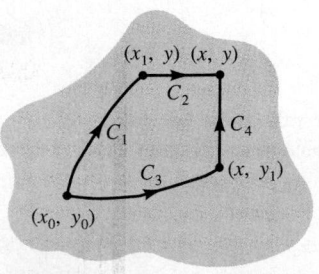

FIGURE 15.23

and it follows that the partial derivative of f with respect to x is $f_x(x, y) = M$. For the *second* path, choose a point (x, y_1). Using reasoning similar to that used for the first path, you can conclude that $f_y(x, y) = N$. Therefore,

$$\nabla f(x, y) = f_x(x, y)\mathbf{i} + f_y(x, y)\mathbf{j} = M\mathbf{i} + N\mathbf{j} = \mathbf{F}(x, y)$$

and it follows that **F** is conservative.

EXAMPLE 4   Finding Work in a Conservative Force Field

For the force field given by

$$\mathbf{F}(x, y, z) = e^x \cos y\, \mathbf{i} - e^x \sin y\, \mathbf{j} + 2\mathbf{k}$$

show that $\int_C \mathbf{F} \cdot d\mathbf{r}$ is independent of path, and calculate the work done by $\mathbf{F}$ on an object moving along a curve C from $(0,\ \pi/2,\ 1)$ to $(1,\ \pi,\ 3)$.

Solution   Writing the force field in the form $\mathbf{F}(x, y, z) = M\mathbf{i} + N\mathbf{j} + P\mathbf{k}$, you have $M = e^x \cos y$, $N = -e^x \sin y$, and $P = 2$, and it follows that

$$\frac{\partial P}{\partial y} = 0 = \frac{\partial N}{\partial z}$$

$$\frac{\partial P}{\partial x} = 0 = \frac{\partial M}{\partial z}$$

$$\frac{\partial N}{\partial x} = -e^x \sin y = \frac{\partial M}{\partial y}.$$

Hence, $\mathbf{F}$ is conservative. If f is a potential function of $\mathbf{F}$, then

$$f_x(x, y, z) = e^x \cos y$$
$$f_y(x, y, z) = -e^x \sin y$$
$$f_z(x, y, z) = 2.$$

By integrating with respect to x, y, and z separately, we obtain

$$f(x, y, z) = \int f_x(x, y, z)\, dx = \int e^x \cos y\, dx = e^x \cos y + g(y, z)$$

$$f(x, y, z) = \int f_y(x, y, z)\, dy = \int -e^x \sin y\, dy = e^x \cos y + h(x, z)$$

$$f(x, y, z) = \int f_z(x, y, z)\, dz = \int 2\, dz = 2z + k(x, y).$$

By comparing these three versions of $f(x, y, z)$, you conclude that

$$f(x, y, z) = e^x \cos y + 2z + K.$$

Therefore, the work done by $\mathbf{F}$ along *any* curve C from $(0,\ \pi/2,\ 1)$ to $(1,\ \pi,\ 3)$ is

$$W = \int_C \mathbf{F} \cdot d\mathbf{r} = \left[e^x \cos y + 2z \right]_{(0,\pi/2,1)}^{(1,\pi,3)}$$
$$= (-e + 6) - (0 + 2)$$
$$= 4 - e.$$

How much work would be done if the object in Example 4 moved from the point $(0,\ \pi/2,\ 1)$ to $(1,\ \pi,\ 3)$ and then back to the starting point $(0,\ \pi/2,\ 1)$? The Fundamental Theorem of Line Integrals states that there is zero work done. Remember that, by definition, work can be negative. Hence, by the time the object gets back to its starting point, the amount of work that registers positively is canceled out by the amount of work that registers negatively.

A curve C, given by $\mathbf{r}(t)$ for $a \leq t \leq b$, is **closed** if $\mathbf{r}(a) = \mathbf{r}(b)$. By the Fundamental Theorem of Line Integrals, you can conclude that if $\mathbf{F}$ is continuous and conservative on an open region R, then the line integral over every closed curve C is 0.

THEOREM 15.7   Equivalent Conditions

Let $\mathbf{F}(x, y, z) = M\mathbf{i} + N\mathbf{j} + P\mathbf{k}$ have continuous first partial derivatives in an open connected region R, and let C be a piecewise smooth curve in R. The following conditions are equivalent.

1. $\mathbf{F}$ is conservative. That is, $\mathbf{F} = \nabla f$ for some function f.

2. $\displaystyle\int_C \mathbf{F} \cdot d\mathbf{r}$ is independent of path.

3. $\displaystyle\int_C \mathbf{F} \cdot d\mathbf{r} = 0$ for every *closed* curve C in R.

REMARK Theorem 15.7 gives you options for evaluating a line integral involving a conservative vector field. You can use a potential function, or it might be more convenient to choose a particularly simple path, such as a straight line.

EXAMPLE 5   Evaluating a Line Integral

Evaluate $\int_{C_1} \mathbf{F} \cdot d\mathbf{r}$, where $\mathbf{F}(x, y) = (y^3 + 1)\mathbf{i} + (3xy^2 + 1)\mathbf{j}$ and C_1 is the semicircular path from $(0, 0)$ to $(2, 0)$, as shown in Figure 15.24.

Solution   You have the following three options.

a. You can use the method of the previous section to evaluate the line integral along the *given curve*. To do this, you can use the parametrization $\mathbf{r}(t) = (1 - \cos t)\mathbf{i} + \sin t\,\mathbf{j}$, where $0 \leq t \leq \pi$. For this parametrization, it follows that $d\mathbf{r} = \mathbf{r}'(t)\,dt = (\sin t\,\mathbf{i} + \cos t\,\mathbf{j})\,dt$, and

$$\int_{C_1} \mathbf{F} \cdot d\mathbf{r} = \int_0^\pi (\sin t + \sin^4 t + \cos t + 3\sin^2 t \cos t - 3\cos^2 t \sin^2 t)\,dt.$$

This integral ought to dampen your enthusiasm for this option.

b. You can try to find a *potential function* and evaluate the line integral by the Fundamental Theorem of Line Integrals. Using the technique demonstrated in Example 4, you can find the potential function to be $f(x, y) = xy^3 + x + y + K$, and, by the Fundamental Theorem,

$$W = \int_{C_1} \mathbf{F} \cdot d\mathbf{r} = f(2, 0) - f(0, 0) = 2.$$

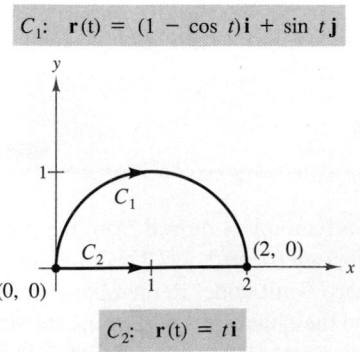

C_1: $\mathbf{r}(t) = (1 - \cos t)\mathbf{i} + \sin t\,\mathbf{j}$

C_2: $\mathbf{r}(t) = t\mathbf{i}$

FIGURE 15.24

c. Knowing that $\mathbf{F}$ is conservative, you have a third option. Because the value of the line integral is independent of path, you can replace the semicircular path with a *simpler path*. Suppose you choose the straight line path C_2 from $(0, 0)$ to $(2, 0)$. Then, $\mathbf{r}(t) = t\mathbf{i}$, where $0 \leq t \leq 2$. Thus, $d\mathbf{r} = \mathbf{i}\,dt$ and $\mathbf{F}(x, y) = (y^3 + 1)\mathbf{i} + (3xy^2 + 1)\mathbf{j} = \mathbf{i} + \mathbf{j}$, so that

$$\int_{C_1} \mathbf{F} \cdot d\mathbf{r} = \int_{C_2} \mathbf{F} \cdot d\mathbf{r} = \int_0^2 1\,dt = t\Big]_0^2 = 2.$$

Of the three options, obviously the third one is the easiest.

Conservation of Energy

In 1840, the English physicist Michael Faraday (1791–1867) wrote, "Nowhere is there a pure creation or production of power without a corresponding exhaustion of something to supply it." This statement represents the first formulation of one of the most important laws of physics—the **Law of Conservation of Energy.** In modern terminology, the law is stated as follows: *In a conservative force field, the sum of the potential and kinetic energies of an object remains constant from point to point.*

You can use the Fundamental Theorem of Line Integrals to derive this law. From physics, the **kinetic energy** of a particle of mass m and speed v is $k = \frac{1}{2}mv^2$. The **potential energy** p of a particle at point (x, y, z) in a conservative vector field $\mathbf{F}$ is defined as

$$p(x, y, z) = -f(x, y, z)$$

where f is the potential function for $\mathbf{F}$. Consequently, the work done by $\mathbf{F}$ along a smooth curve C from A to B is

$$W = \int_C \mathbf{F} \cdot d\mathbf{r} = f(x, y, z)\Big]_A^B = -p(x, y, z)\Big]_A^B = p(A) - p(B)$$

as indicated in Figure 15.25. In other words, work W is equal to the difference in the potential energies of A and B. Now, suppose that $\mathbf{r}(t)$ is the position vector for a particle moving along C from $A = \mathbf{r}(a)$ to $B = \mathbf{r}(b)$. At any time t, the particle's velocity, acceleration, and speed are $\mathbf{v}(t) = \mathbf{r}'(t)$, $\mathbf{a}(t) = \mathbf{r}''(t)$, and $v(t) = \|\mathbf{v}(t)\|$, respectively. Thus, by Newton's Second Law of Motion, $\mathbf{F} = m\,\mathbf{a}(t) = m(\mathbf{v}'(t))$, and the work done by $\mathbf{F}$ is

$$
\begin{aligned}
W &= \int_C \mathbf{F} \cdot d\mathbf{r} = \int_a^b \mathbf{F} \cdot \mathbf{r}'(t)\, dt \\
&= \int_a^b \mathbf{F} \cdot \mathbf{v}(t)\, dt = \int_a^b [m\,\mathbf{v}'(t)] \cdot \mathbf{v}(t)\, dt \\
&= \int_a^b m[\mathbf{v}'(t) \cdot \mathbf{v}(t)]\, dt \\
&= \frac{m}{2} \int_a^b \frac{d}{dt}[\mathbf{v}(t) \cdot \mathbf{v}(t)]\, dt \\
&= \frac{m}{2} \int_a^b \frac{d}{dt}[\|\mathbf{v}(t)\|^2]\, dt = \frac{m}{2} \|\mathbf{v}(t)\|^2 \Big]_a^b \\
&= \frac{m}{2}[v(t)]^2 \Big]_a^b = \frac{1}{2}m[v(b)]^2 - \frac{1}{2}m[v(a)]^2 \\
&= k(B) - k(A).
\end{aligned}
$$

Equating these two results for W produces

$$p(A) - p(B) = k(B) - k(A)$$
$$p(A) + k(A) = p(B) + k(B)$$

which implies that the sum of the potential and kinetic energies remains constant from point to point.

MICHAEL FARADAY

Several philosophers of science have considered Faraday's law of conservation of energy to be the greatest generalization ever conceived by humankind. Many physicists have contributed to our knowledge of this law. Two early and influential ones were James Prescott Joule (1818–1889) and Hermann Ludwig Helmholtz (1821–1894).

FIGURE 15.25

The work done by $\mathbf{F}$ along C is

$$W = \int_C \mathbf{F} \cdot d\mathbf{r} = p(A) - p(B).$$

EXERCISES for Section 15.3

In Exercises 1–4, show that the value of $\int_C \mathbf{F} \cdot d\mathbf{r}$ is the same for both parametric representations of C.

1. $\mathbf{F}(x, y) = x^2\mathbf{i} + xy\mathbf{j}$
  a. $\mathbf{r}_1(t) = t\mathbf{i} + t^2\mathbf{j}, \quad 0 \le t \le 1$

  b. $\mathbf{r}_2(\theta) = \sin\theta\,\mathbf{i} + \sin^2\theta\,\mathbf{j}, \quad 0 \le \theta \le \dfrac{\pi}{2}$

2. $\mathbf{F}(x, y) = (x^2 + y^2)\mathbf{i} - x\mathbf{j}$
  a. $\mathbf{r}_1(t) = t\mathbf{i} + \sqrt{t}\,\mathbf{j}, \quad 0 \le t \le 4$
  b. $\mathbf{r}_2(w) = w^2\mathbf{i} + w\mathbf{j}, \quad 0 \le w \le 2$

3. $\mathbf{F}(x, y) = y\mathbf{i} - x\mathbf{j}$

  a. $\mathbf{r}_1(\theta) = \sec\theta\,\mathbf{i} + \tan\theta\,\mathbf{j}, \quad 0 \le \theta \le \dfrac{\pi}{3}$

  b. $\mathbf{r}_2(t) = \sqrt{t+1}\,\mathbf{i} + \sqrt{t}\,\mathbf{j}, \quad 0 \le t \le 3$

4. $\mathbf{F}(x, y) = y\mathbf{i} + x^2\mathbf{j}$
  a. $\mathbf{r}_1(t) = (2 + t)\mathbf{i} + (3 - t)\mathbf{j}, \quad 0 \le t \le 3$
  b. $\mathbf{r}_2(w) = (2 + \ln w)\mathbf{i} + (3 - \ln w)\mathbf{j}, \quad 1 \le w \le e^3$

In Exercises 5–18, find the value of the line integral $\int_C \mathbf{F} \cdot d\mathbf{r}$. (*Hint:* If **F** is conservative, the integration may be easier on an alternative path.)

5. $\mathbf{F}(x, y) = 2xy\mathbf{i} + x^2\mathbf{j}$
  a. $\mathbf{r}_1(t) = t\mathbf{i} + t^2\mathbf{j}, \quad 0 \le t \le 1$
  b. $\mathbf{r}_2(t) = t\mathbf{i} + t^3\mathbf{j}, \quad 0 \le t \le 1$

6. $\mathbf{F}(x, y) = ye^{xy}\mathbf{i} + xe^{xy}\mathbf{j}$
  a. $\mathbf{r}_1(t) = t\mathbf{i} - \frac{3}{2}(t - 2)\mathbf{j}, \quad 0 \le t \le 2$
  b. Line segments from $(0, 3)$ to $(0, 0)$, and then from $(0, 0)$ to $(2, 0)$

7. $\mathbf{F}(x, y) = y\mathbf{i} - x\mathbf{j}$
  a. $\mathbf{r}_1(t) = t\mathbf{i} + t\mathbf{j}, \quad 0 \le t \le 1$
  b. $\mathbf{r}_2(t) = t\mathbf{i} + t^2\mathbf{j}, \quad 0 \le t \le 1$
  c. $\mathbf{r}_3(t) = t\mathbf{i} + t^3\mathbf{j}, \quad 0 \le t \le 1$

8. $\mathbf{F}(x, y) = xy^2\mathbf{i} + 2x^2y\mathbf{j}$

  a. $\mathbf{r}_1(t) = t\mathbf{i} + \dfrac{1}{t}\mathbf{j}, \quad 1 \le t \le 3$

  b. $\mathbf{r}_2(t) = (t + 1)\mathbf{i} - \frac{1}{3}(t - 3)\mathbf{j}, \quad 0 \le t \le 2$

9. $\displaystyle\int_C y^2\,dx + 2xy\,dy$

a.

b.

c.

d.
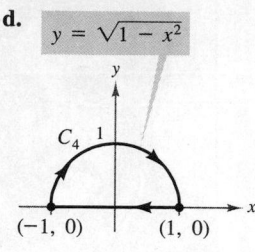

10. $\displaystyle\int_C (2x - 3y + 1)\,dx - (3x + y - 5)\,dy$

a.

b.

c.

d.
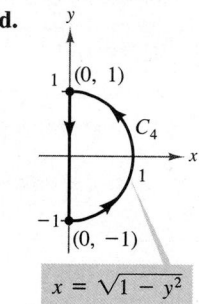

11. $\displaystyle\int_C 2xy\,dx + (x^2 + y^2)\,dy$

  a. C: ellipse $\dfrac{x^2}{25} + \dfrac{y^2}{16} = 1$ from $(5, 0)$ to $(0, 4)$

  b. C: parabola $y = 4 - x^2$ from $(2, 0)$ to $(0, 4)$

12. $\displaystyle\int_C (x^2 + y^2)\,dx + 2xy\,dy$

  a. $\mathbf{r}_1(t) = t^3\mathbf{i} + t^2\mathbf{j}, \quad 0 \le t \le 2$

  b. $\mathbf{r}_2(t) = 2\cos t\,\mathbf{i} + 2\sin t\,\mathbf{j}, \quad 0 \le t \le \dfrac{\pi}{2}$

13. $\mathbf{F}(x, y, z) = yz\mathbf{i} + xz\mathbf{j} + xy\mathbf{k}$
  a. $\mathbf{r}_1(t) = t\mathbf{i} + 2\mathbf{j} + t\mathbf{k}, \quad 0 \le t \le 4$
  b. $\mathbf{r}_2(t) = t^2\mathbf{i} + t\mathbf{j} + t^2\mathbf{k}, \quad 0 \le t \le 2$

14. $\mathbf{F}(x, y, z) = \mathbf{i} + z\mathbf{j} + y\mathbf{k}$
  a. $\mathbf{r}_1(t) = \cos t\,\mathbf{i} + \sin t\,\mathbf{j} + t^2\mathbf{k}, \quad 0 \le t \le \pi$
  b. $\mathbf{r}_2(t) = (1 - 2t)\mathbf{i} + \pi^2 t\,\mathbf{k}, \quad 0 \le t \le 1$

15. $F(x, y, z) = (2y + x)\mathbf{i} + (x^2 - z)\mathbf{j} + (2y - 4z)\mathbf{k}$
  a. $\mathbf{r}_1(t) = t\mathbf{i} + t^2\mathbf{j} + \mathbf{k},\quad 0 \le t \le 1$
  b. $\mathbf{r}_2(t) = t\mathbf{i} + t\mathbf{j} + (2t - 1)^2\mathbf{k},\quad 0 \le t \le 1$

16. $F(x, y, z) = -y\mathbf{i} + x\mathbf{j} + 3xz^2\mathbf{k}$
  a. $\mathbf{r}_1(t) = \cos t\,\mathbf{i} + \sin t\,\mathbf{j} + t\mathbf{k},\quad 0 \le t \le \pi$
  b. $\mathbf{r}_2(t) = (1 - 2t)\mathbf{i} + \pi t\,\mathbf{k},\quad 0 \le t \le 1$

17. $F(x, y, z) = e^z(y\mathbf{i} + x\mathbf{j} + xy\mathbf{k})$
  a. $\mathbf{r}_1(t) = 4\cos t\,\mathbf{i} + 4\sin t\,\mathbf{j} + 3\mathbf{k},\quad 0 \le t \le \pi$
  b. $\mathbf{r}_2(t) = (4 - 8t)\mathbf{i} + 3\mathbf{k},\quad 0 \le t \le 1$

18. $F(x, y, z) = y\sin z\,\mathbf{i} + x\sin z\,\mathbf{j} + xy\cos x\,\mathbf{k}$
  a. $\mathbf{r}_1(t) = t^2\mathbf{i} + t^2\mathbf{j},\quad 0 \le t \le 2$
  b. $\mathbf{r}_2(t) = 4t\,\mathbf{i} + 4t\,\mathbf{j},\quad 0 \le t \le 1$

In Exercises 19–28, evaluate the given line integral using the Fundamental Theorem of Line Integrals.

19. $\displaystyle\int_C (y\mathbf{i} + x\mathbf{j}) \cdot d\mathbf{r}$

  C: smooth curve from $(0, 0)$ to $(3, 8)$

20. $\displaystyle\int_C [2(x + y)\mathbf{i} + 2(x + y)\mathbf{j}] \cdot d\mathbf{r}$

  C: smooth curve from $(-1, 1)$ to $(3, 2)$

21. $\displaystyle\int_C \cos x \sin y\,dx + \sin x \cos y\,dy$

  C: smooth curve from $(0, -\pi)$ to $\left(\dfrac{3\pi}{2}, \dfrac{\pi}{2}\right)$

22. $\displaystyle\int_C \dfrac{y\,dx - x\,dy}{x^2 + y^2}$

  C: smooth curve from $(1, 1)$ to $\left(2\sqrt{3}, 2\right)$

23. $\displaystyle\int_C e^x \sin y\,dx + e^x \cos y\,dy$

  C: cycloid $x = \theta - \sin\theta,\ y = 1 - \cos\theta$ from $(0, 0)$ to $(2\pi, 0)$

24. $\displaystyle\int_C \dfrac{2x}{(x^2 + y^2)^2}\,dx + \dfrac{2y}{(x^2 + y^2)^2}\,dy$
  C: circle $(x - 4)^2 + (y - 5)^2 = 9$ clockwise from $(7, 5)$ to $(1, 5)$

25. $\displaystyle\int_C (z + 2y)\,dx + (2x - z)\,dy + (x - y)\,dz$
  a. C: line segment from $(0, 0, 0)$ to $(1, 1, 1)$
  b. C: line segments from $(0, 0, 0)$ to $(0, 0, 1)$ to $(1, 1, 1)$
  c. C: line segments from $(0, 0, 0)$ to $(1, 0, 0)$ to $(1, 1, 0)$ to $(1, 1, 1)$

26. Repeat Exercise 25 using the integral

  $\displaystyle\int_C zy\,dx + xz\,dy + xy\,dz.$

27. $\displaystyle\int_C -\sin x\,dx + z\,dy + y\,dz$

  C: smooth curve from $(0, 0, 0)$ to $\left(\dfrac{\pi}{2}, 3, 4\right)$

28. $\displaystyle\int_C 6x\,dx - 4z\,dy - (4y - 20z)\,dz$

  C: smooth curve from $(0, 0, 0)$ to $(3, 4, 0)$

Work In Exercises 29 and 30, find the work done by the force field F in moving an object from P to Q.

29. $F(x, y) = 9x^2y^2\mathbf{i} + (6x^3y - 1)\mathbf{j}$
  $P = (0, 0),\ Q = (5, 9)$

30. $F(x, y) = \dfrac{2x}{y}\mathbf{i} - \dfrac{x^2}{y^2}\mathbf{j}$
  $P = (-1, 1),\ Q = (3, 2)$

31. *Work* A stone weighing 1 pound is attached to the end of a 2-foot string and is whirled horizontally with one end held fixed. It makes one revolution per second. Find the work done by the force **F** that keeps the stone moving in a circular path. (*Hint:* Use Force = (Mass)(Centripetal Acceleration).)

32. *Work* If $\mathbf{F}(x, y, z) = a_1\mathbf{i} + a_2\mathbf{j} + a_3\mathbf{k}$ is a constant force vector field, show that the work done in moving a particle along any path from P to Q is

  $W = \mathbf{F} \cdot \overrightarrow{PQ}.$

33. *Kinetic and Potential Energy* The kinetic energy of an object moving through a conservative force field is decreasing at a rate of 10 units per minute. At what rate is the potential energy changing?

34. *Work* To allow a way of escape for workers in a hazardous job 50 meters above ground level, a slide wire has been installed. It runs from their position to a point on the ground 50 meters from the base of the installation where they are located. Show that the work done by the gravitational force field for a 150-pound man moving the length of the slide wire is the same for the following two paths.
  a. $\mathbf{r}(t) = t\mathbf{i} + (50 - t)\mathbf{j}$
  b. $\mathbf{r}(t) = t\mathbf{i} + \frac{1}{50}(50 - t)^2\mathbf{j}$

35. *Work* Can you find a path for the slide wire in Exercise 34 such that the work done by the gravitational force field would differ from the amounts of work done for the two paths given? Explain why or why not.

36. Let $\mathbf{F}(x, y) = \dfrac{y}{x^2 + y^2}\mathbf{i} - \dfrac{x}{x^2 + y^2}\mathbf{j}.$
  a. Show that

  $\dfrac{\partial N}{\partial x} = \dfrac{\partial M}{\partial y}$

  where

  $M = \dfrac{y}{x^2 + y^2}$

  and

  $N = \dfrac{-x}{x^2 + y^2}.$

  b. If $\mathbf{r}(t) = \cos t\,\mathbf{i} + \sin t\,\mathbf{j}$, for $0 \le t \le \pi$, find $\int_C \mathbf{F} \cdot d\mathbf{r}$.
  c. If $\mathbf{r}(t) = \cos t\,\mathbf{i} - \sin t\,\mathbf{j}$, for $0 \le t \le \pi$, find $\int_C \mathbf{F} \cdot d\mathbf{r}$.

d. If $\mathbf{r}(t) = \cos t\,\mathbf{i} + \sin t\,\mathbf{j}$, for $0 \leq t \leq 2\pi$, find $\int_C \mathbf{F} \cdot d\mathbf{r}$. Why doesn't this contradict Theorem 15.7?

e. Show that

$$\nabla\left(\arctan\frac{x}{y}\right) = \mathbf{F}.$$

True or False **In Exercises 37–40, determine whether the statement is true or false. If it is false, explain why or give an example that shows it is false.**

37. If C_1, C_2, and C_3 have the same initial and terminal points and $\int_{C_1} \mathbf{F} \cdot d\mathbf{r}_1 = \int_{C_2} \mathbf{F} \cdot d\mathbf{r}_2$, then $\int_{C_1} \mathbf{F} \cdot d\mathbf{r}_1 = \int_{C_3} \mathbf{F} \cdot d\mathbf{r}_3$.

38. If $\mathbf{F} = y\,\mathbf{i} + x\,\mathbf{j}$ and C is given by $\mathbf{r}(t) = (4\sin t)\mathbf{i} + (3\cos t)\mathbf{j}$, $0 \leq t \leq \pi$, then $\int_C \mathbf{F} \cdot d\mathbf{r} = 0$.

39. If $\mathbf{F}$ is conservative in a region R bounded by a simple closed path, and C lies within R, then $\int_C \mathbf{F} \cdot d\mathbf{r}$ is path-independent.

40. If $\mathbf{F} = M\mathbf{i} + N\mathbf{j}$ and $\partial M/\partial x = \partial N/\partial y$, then $\mathbf{F}$ is conservative.

41. f is called harmonic if

$$\frac{\partial^2 f}{\partial x^2} + \frac{\partial^2 f}{\partial y^2} = 0.$$

Prove that if f is harmonic, then

$$\int_C \left(\frac{\partial f}{\partial y}dx - \frac{\partial f}{\partial x}dy\right) = 0$$

where C is a smooth closed curve in the plane.

Simply Connected

Not Simply Connected

FIGURE 15.26

SECTION 15.4

Green's Theorem

Green's Theorem • Alternative Forms of Green's Theorem

Green's Theorem

In this section, you will study **Green's Theorem,** named after the English mathematician George Green (1793–1841). This theorem states that the value of a double integral over a *simply connected* plane region R is determined by the value of a line integral around the boundary of R.

A curve C given by $\mathbf{r}(t) = x(t)\mathbf{i} + y(t)\mathbf{j}$, where $a \le t \le b$, is **simple** if it does not cross itself—that is, $\mathbf{r}(c) \ne \mathbf{r}(d)$ for all c and d in the open interval (a, b). A plane region R is **simply connected** if its boundary consists of *one* simple closed curve, as shown in Figure 15.26.

THEOREM 15.8   Green's Theorem

Let R be a simply connected region with a piecewise smooth boundary C, oriented counterclockwise (that is, C is traversed *once* so that the region R always lies to the *left*). If M and N have continuous partial derivatives in an open region containing R, then

$$\int_C M\,dx + N\,dy = \iint_R \left(\frac{\partial N}{\partial x} - \frac{\partial M}{\partial y}\right) dA.$$

Proof   We give a proof only for a region that is both vertically simple and horizontally simple, as shown in Figure 15.27.

$$\int_C M\,dx = \int_{C_1} M\,dx + \int_{C_2} M\,dx$$
$$= \int_a^b M(x, f_1(x))\,dx + \int_b^a M(x, f_2(x))\,dx$$
$$= \int_a^b [M(x, f_1(x)) - M(x, f_2(x))]\,dx$$

On the other hand,

$$\iint_R \frac{\partial M}{\partial y}\,dA = \int_a^b \int_{f_1(x)}^{f_2(x)} \frac{\partial M}{\partial y}\,dy\,dx = \int_a^b M(x, y)\Big]_{f_1(x)}^{f_2(x)} dx$$
$$= \int_a^b [M(x, f_2(x)) - M(x, f_1(x))]\,dx.$$

Consequently, $\int_C M\,dx = -\int_R\int \partial M/\partial y\,dA$. Similarly, you can use $g_1(y)$ and $g_2(y)$ to show that $\int_C N\,dy = \int_R\int \partial N/\partial x\,dA$. By adding the integrals $\int_C M\,dx$ and $\int_C N\,dy$, you obtain the conclusion stated in the theorem.

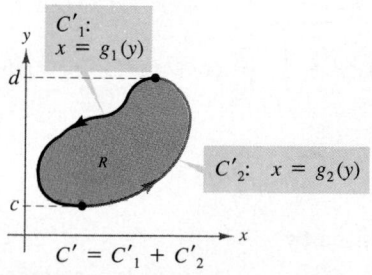

R is vertically simple.

R is horizontally simple.

FIGURE 15.27

EXAMPLE 1   Using Green's Theorem

Use Green's Theorem to evaluate the line integral

$$\int_C y^3 \, dx + (x^3 + 3xy^2) \, dy$$

where C is the path from $(0, 0)$ to $(1, 1)$ along the graph of $y = x^3$ and from $(1, 1)$ to $(0, 0)$ along the graph of $y = x$, as shown in Figure 15.28.

Solution   Because $M = y^3$ and $N = x^3 + 3xy^2$, it follows that

$$\frac{\partial N}{\partial x} = 3x^2 + 3y^2 \quad \text{and} \quad \frac{\partial M}{\partial y} = 3y^2.$$

Applying Green's Theorem, you then have

$$\int_C y^3 \, dx + (x^3 + 3xy^2) \, dy = \iint_R \left(\frac{\partial N}{\partial x} - \frac{\partial M}{\partial y} \right) dA$$

$$= \int_0^1 \int_{x^3}^x [(3x^2 + 3y^2) - 3y^2] \, dy \, dx$$

$$= \int_0^1 \int_{x^3}^x 3x^2 \, dy \, dx$$

$$= \int_0^1 3x^2 y \Big]_{x^3}^x \, dx$$

$$= \int_0^1 (3x^3 - 3x^5) \, dx$$

$$= \left[\frac{3x^4}{4} - \frac{x^6}{2} \right]_0^1$$

$$= \frac{1}{4}.$$

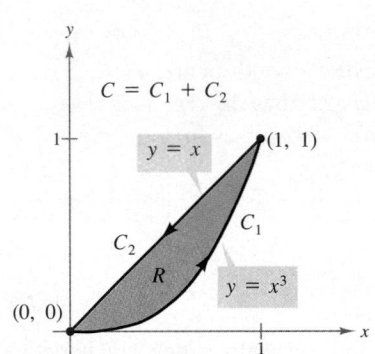

FIGURE 15.28
C is simple and closed, and the region R always lies to the left of C.

Green's Theorem cannot be applied to every line integral. Among other restrictions stated in Theorem 15.8, the curve C must be simple and closed. When Green's Theorem does apply, however, it can save time. To see this, try using the techniques described in Section 15.2 to evaluate the line integral in Example 1. To do this, you would need to write the line integral as

$$\int_C y^3 \, dx + (x^3 + 3xy^2) \, dy$$

$$= \int_{C_1} y^3 \, dx + (x^3 + 3xy^2) \, dy + \int_{C_2} y^3 \, dx + (x^3 + 3xy^2) \, dy$$

where C_1 is the cubic path given by

$$\mathbf{r}(t) = t\mathbf{i} + t^3\mathbf{j}$$

from $t = 0$ to $t = 1$, and C_2 is the line segment given by

$$\mathbf{r}(t) = (1 - t)\mathbf{i} + (1 - t)\mathbf{j}$$

from $t = 0$ to $t = 1$.

EXAMPLE 2   Using Green's Theorem to Calculate Work

While subject to the force $\mathbf{F}(x, y) = y^3\mathbf{i} + (x^3 + 3xy^2)\mathbf{j}$, a particle travels once around the circle of radius 3 shown in Figure 15.29. Use Green's Theorem to find the work done by $\mathbf{F}$.

Solution   From Example 1, you know by Green's Theorem that

$$\int_C y^3\,dx + (x^3 + 3xy^2)\,dy = \iint_R 3x^2\,dA.$$

In polar coordinates, using $x = r\cos\theta$ and $dA = r\,dr\,d\theta$, the work done is

$$
\begin{aligned}
W &= \iint_R 3x^2\,dA = \int_0^{2\pi}\int_0^3 3(r\cos\theta)^2 r\,dr\,d\theta \\
&= 3\int_0^{2\pi}\int_0^3 r^3\cos^2\theta\,dr\,d\theta \\
&= 3\int_0^{2\pi} \frac{r^4}{4}\cos^2\theta\Big]_0^3 d\theta \\
&= 3\int_0^{2\pi} \frac{81}{4}\cos^2\theta\,d\theta \\
&= \frac{243}{8}\int_0^{2\pi} (1 + \cos 2\theta)\,d\theta \\
&= \frac{243}{8}\left[\theta + \frac{\sin 2\theta}{2}\right]_0^{2\pi} \\
&= \frac{243\pi}{4}.
\end{aligned}
$$

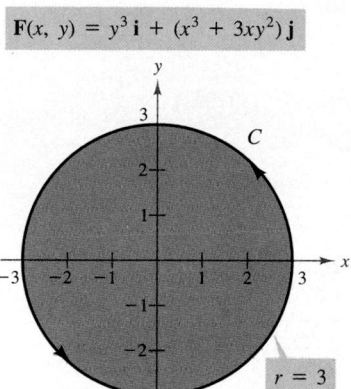

$$\mathbf{F}(x, y) = y^3\mathbf{i} + (x^3 + 3xy^2)\mathbf{j}$$

$$r = 3$$

FIGURE 15.29
The work done by $\mathbf{F}$ as a particle travels once around the circle is $\dfrac{243\pi}{4}$.

When evaluating line integrals over closed curves, remember that, for conservative vector fields (those for which $\partial N/\partial x = \partial M/\partial y$), the value of the line integral is 0. This is easily seen from the statement of Green's Theorem:

$$\int_C M\,dx + N\,dy = \iint_R \left(\frac{\partial N}{\partial x} - \frac{\partial M}{\partial y}\right) dA = 0.$$

EXAMPLE 3   Green's Theorem and Conservative Vector Fields

Evaluate the line integral

$$\int_C y^3\,dx + 3xy^2\,dy$$

where C is the path shown in Figure 15.30.

Solution   From this line integral, $M = y^3$ and $N = 3xy^2$. Thus, $\partial N/\partial x = 3y^2$ and $\partial M/\partial y = 3y^2$. This implies that the vector field $\mathbf{F} = M\mathbf{i} + N\mathbf{j}$ is conservative, and because C is closed, you can conclude that

$$\int_C y^3\,dx + 3xy^2\,dy = 0.$$

FIGURE 15.30
C is closed.

EXAMPLE 4   Using Green's Theorem for a Piecewise Smooth Curve

Evaluate

$$\int_C (\arctan x + y^2)\,dx + (e^y - x^2)\,dy$$

where C is the path enclosing the annular region shown in Figure 15.31.

Solution   In polar coordinates, R is given by $1 \le r \le 3$ for $0 \le \theta \le \pi$. Moreover,

$$\frac{\partial N}{\partial x} - \frac{\partial M}{\partial y} = -2x - 2y = -2(r\cos\theta + r\sin\theta).$$

Thus, by Green's Theorem,

$$\int_C (\arctan x + y^2)\,dx + (e^y - x^2)\,dy = \iint_R -2(x + y)\,dA$$

$$= \int_0^\pi \int_1^3 -2r(\cos\theta + \sin\theta)r\,dr\,d\theta$$

$$= \int_0^\pi -2(\cos\theta + \sin\theta)\frac{r^3}{3}\bigg]_1^3 \, d\theta$$

$$= \int_0^\pi \left(-\frac{52}{3}\right)(\cos\theta + \sin\theta)\,d\theta$$

$$= -\frac{52}{3}\bigg[\sin\theta - \cos\theta\bigg]_0^\pi$$

$$= -\frac{104}{3}.$$

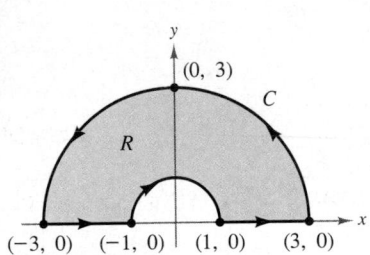

FIGURE 15.31
C is piecewise smooth.

In Examples 1, 2, and 4, Green's Theorem was used to evaluate line integrals as double integrals. You can also use the theorem to evaluate double integrals as line integrals. One useful application occurs when $\partial N/\partial x - \partial M/\partial y = 1$.

$$\int_C M\,dx + N\,dy = \int_C \left(\frac{\partial N}{\partial x} - \frac{\partial M}{\partial y}\right)dA$$

$$= \iint_R 1\,dA$$

$$= \text{area of region } R$$

Among the many choices for M and N satisfying the stated condition, the choice of $M = -y/2$ and $N = x/2$ produces the following line integral for the area of region R.

THEOREM 15.9   Line Integral for Area

If R is a plane region bounded by a piecewise smooth simple closed curve C, oriented counterclockwise, then the area of R is given by

$$A = \frac{1}{2}\int_C x\,dy - y\,dx.$$

EXAMPLE 5 **Finding Area by a Line Integral**

Use a line integral to find the area of the ellipse

$$\frac{x^2}{a^2} + \frac{y^2}{b^2} = 1.$$

Solution Using Figure 15.32, you can induce a counterclockwise orientation to the elliptical path by letting

$$x = a\cos t \quad \text{and} \quad y = b\sin t, \qquad 0 \le t \le 2\pi.$$

Therefore, the area is

$$
\begin{aligned}
A &= \frac{1}{2}\int_C x\,dy - y\,dx \\
&= \frac{1}{2}\int_0^{2\pi} \left[(a\cos t)(b\cos t)\,dt - (b\sin t)(-a\sin t)\,dt \right] \\
&= \frac{ab}{2}\int_0^{2\pi} (\cos^2 t + \sin^2 t)\,dt \\
&= \frac{ab}{2}t \Big]_0^{2\pi} \\
&= \pi ab.
\end{aligned}
$$

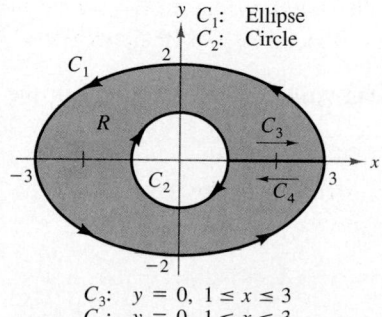

FIGURE 15.32
The area of the ellipse is πab.

Green's Theorem can be extended to cover some regions that are not simply connected. This is demonstrated in the next example.

EXAMPLE 6 **Green's Theorem Extended to a Region with a Hole**

Let R be the region inside the ellipse $(x^2/9) + (y^2/4) = 1$ and outside the circle $x^2 + y^2 = 1$. Evaluate the line integral

$$\int_C 2xy\,dx + (x^2 + 2x)\,dy$$

where $C = C_1 + C_2$ is the boundary of R, as shown in Figure 15.33.

Solution To begin, we introduce the line segments C_3 and C_4, as shown in Figure 15.33. Note that because the curves C_3 and C_4 have opposite orientations, the line integrals over them cancel. Furthermore, we can apply Green's Theorem to the region R using the boundary $C_1 + C_4 + C_2 + C_3$ to obtain

$$
\begin{aligned}
\int_C 2xy\,dx + (x^2 + 2x)\,dy &= \iint_R \left(\frac{\partial N}{\partial x} - \frac{\partial M}{\partial y} \right) dA \\
&= \iint_R (2x + 2 - 2x)\,dA \\
&= 2\iint_R dA = 2(\text{area of } R) \\
&= 2[\pi(3)(2) - \pi(1^2)] \\
&= 10\pi.
\end{aligned}
$$

C_1: Ellipse
C_2: Circle

C_3: $y = 0,\ 1 \le x \le 3$
C_4: $y = 0,\ 1 \le x \le 3$

FIGURE 15.33

In Section 15.1, we presented a theorem that provided a necessary and sufficient condition for conservative vector fields. There, we proved the theorem in only one direction. We now outline the other direction, using Green's Theorem. Let $\mathbf{F}(x, y) = M\mathbf{i} + N\mathbf{j}$ be defined on an open disc R. We want to show that if M and N have continuous first partial derivatives and

$$\frac{\partial M}{\partial y} = \frac{\partial N}{\partial x}$$

then $\mathbf{F}$ is conservative. Suppose that C is a closed path forming the boundary of a connected region lying in R. Then, using the fact that $\partial M / \partial y = \partial N / \partial x$, we can apply Green's Theorem to conclude that

$$\int_C \mathbf{F} \cdot d\mathbf{r} = \int_C M \, dx + N \, dy = \iint_R \left(\frac{\partial N}{\partial x} - \frac{\partial M}{\partial y} \right) dA = 0.$$

This, in turn, is equivalent to showing that $\mathbf{F}$ is conservative (see Theorem 15.7).

Alternative Forms of Green's Theorem

We conclude this section with the derivation of two vector forms of Green's Theorem for regions in the plane. The extension of these vector forms to three dimensions is the basis for the discussion in the remaining sections of this chapter. If $\mathbf{F}$ is a vector field in the plane, you can write

$$\mathbf{F}(x, y) = M\mathbf{i} + N\mathbf{j} + 0\mathbf{k}$$

so that the curl of $\mathbf{F}$, as described in Section 15.1, is given by

$$\mathbf{curl}\ \mathbf{F} = \nabla \times \mathbf{F} = \begin{vmatrix} \mathbf{i} & \mathbf{j} & \mathbf{k} \\ \dfrac{\partial}{\partial x} & \dfrac{\partial}{\partial y} & \dfrac{\partial}{\partial z} \\ M & N & 0 \end{vmatrix}$$

$$= -\frac{\partial N}{\partial z}\mathbf{i} + \frac{\partial M}{\partial z}\mathbf{j} + \left(\frac{\partial N}{\partial x} - \frac{\partial M}{\partial y} \right)\mathbf{k}.$$

Consequently,

$$(\mathbf{curl}\ \mathbf{F}) \cdot \mathbf{k} = \left[-\frac{\partial N}{\partial z}\mathbf{i} + \frac{\partial M}{\partial z}\mathbf{j} + \left(\frac{\partial N}{\partial x} - \frac{\partial M}{\partial y} \right)\mathbf{k} \right] \cdot \mathbf{k}$$

$$= \frac{\partial N}{\partial x} - \frac{\partial M}{\partial y}.$$

With appropriate conditions on $\mathbf{F}$, C, and R, we can write Green's Theorem in the vector form

$$\int_C \mathbf{F} \cdot d\mathbf{r} = \iint_R \left(\frac{\partial N}{\partial x} - \frac{\partial M}{\partial y} \right) dA$$

$$= \iint_R (\mathbf{curl}\ \mathbf{F}) \cdot \mathbf{k} \, dA. \qquad \text{First alternative form}$$

The extension of this vector form of Green's Theorem to surfaces in space produces **Stokes's Theorem,** which is discussed in Section 15.8.

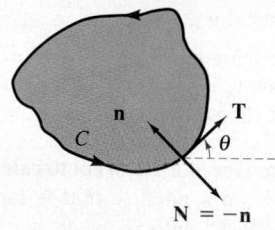

$T = \cos \theta \mathbf{i} + \sin \theta \mathbf{j}$

$\mathbf{n} = \cos\left(\theta + \dfrac{\pi}{2}\right)\mathbf{i} + \sin\left(\theta + \dfrac{\pi}{2}\right)\mathbf{j}$

$\quad = -\sin \theta \mathbf{i} + \cos \theta \mathbf{j}$

$\mathbf{N} = \sin \theta \mathbf{i} - \cos \theta \mathbf{j}$

FIGURE 15.34

For the second vector form of Green's Theorem, assume the same conditions for **F**, C, and R. Using the arc length parameter s for C, we have $\mathbf{r}(s) = x(s)\mathbf{i} + y(s)\mathbf{j}$. Thus, a unit tangent vector **T** to curve C is given by

$$\mathbf{r}'(s) = \mathbf{T} = x'(s)\mathbf{i} + y'(s)\mathbf{j}.$$

In Figure 15.34, we can see that the *outward* unit normal vector **N** can then be written as

$$\mathbf{N} = y'(s)\mathbf{i} - x'(s)\mathbf{j}.$$

Consequently, for $\mathbf{F}(x, y) = M\mathbf{i} + N\mathbf{j}$, we can apply Green's Theorem to obtain

$$\int_C \mathbf{F} \cdot \mathbf{N} \, ds = \int_a^b (M\mathbf{i} + N\mathbf{j}) \cdot (y'(s)\mathbf{i} - x'(s)\mathbf{j}) \, ds$$

$$= \int_a^b \left(M\frac{dy}{ds} - N\frac{dx}{ds} \right) ds$$

$$= \int_C M \, dy - N \, dx$$

$$= \int_C -N \, dx + M \, dy$$

$$= \int\!\!\int_R \left(\frac{\partial M}{\partial x} + \frac{\partial N}{\partial y} \right) dA \qquad \text{Green's Theorem}$$

$$= \int\!\!\int_R \operatorname{div} \mathbf{F} \, dA.$$

Therefore,

$$\int_C \mathbf{F} \cdot \mathbf{N} \, ds = \int\!\!\int_R \operatorname{div} \mathbf{F} \, dA. \qquad \text{Second alternative form}$$

The extension of this form to three dimensions is called the **Divergence Theorem,** which is discussed in Section 15.7. The physical interpretation of divergence and curl will be discussed in Sections 15.7 and 15.8.

EXERCISES for Section 15.4

TECHNOLOGY
Laboratory Guide
Labs 15.3–15.4

In Exercises 1–4, verify Green's Theorem by evaluating both integrals

$$\int_C y^2 \, dx + x^2 \, dy = \int\!\!\int_R \left(\frac{\partial N}{\partial x} - \frac{\partial M}{\partial y} \right) dA$$

for the specified path.

1. C: boundary of the square with vertices $(0, 0)$, $(4, 0)$, $(4, 4)$, and $(0, 4)$

2. C: boundary of the triangle with vertices $(0, 0)$, $(4, 0)$, and $(4, 4)$

3. C: boundary of the region lying between the graphs of $y = x$ and $y = x^2/4$

4. C: $x^2 + y^2 = 1$

C In Exercises 5 and 6, verify Green's Theorem by using a computer algebra system to evaluate both integrals

$$\int_C xe^y \, dx + e^x \, dy = \int\!\!\int_R \left(\frac{\partial N}{\partial x} - \frac{\partial M}{\partial y} \right) dA$$

for the specified path.

5. C: $x^2 + y^2 = 4$

6. C: boundary of the region bounded by the graphs of $y = x$, and $y = x^3$.

In Exercises 7–10, use Green's Theorem to evaluate the integral

$$\int_C (y - x)\,dx + (2x - y)\,dy$$

for the specified path.

7. C: boundary of the region lying between the graphs of $y = x$ and $y = x^2 - x$

8. C: $x = 2\cos\theta,\ y = \sin\theta$

9. C: boundary of the region lying inside the rectangle bounded by $x = -5$, $x = 5$, $y = -3$, and $y = 3$ and outside the square bounded by $x = -1$, $x = 1$, $y = -1$, and $y = 1$

10. C: boundary of the region lying inside the circle $x^2 + y^2 = 16$ and outside the circle $x^2 + y^2 = 1$

In Exercises 11–20, use Green's Theorem to evaluate the line integral.

11. $\displaystyle\int_C 2xy\,dx + (x + y)\,dy$

    C: boundary of the region lying between the graphs of $y = 0$ and $y = 4 - x^2$

12. $\displaystyle\int_C y^2\,dx + xy\,dy$

    C: boundary of the region bounded by the graphs of $y = 0$, $y = \sqrt{x}$, and $x = 4$

13. $\displaystyle\int_C (x^2 - y^2)\,dx + 2xy\,dy$

    C: $x^2 + y^2 = a^2$

14. $\displaystyle\int_C (x^2 - y^2)\,dx + 2xy\,dy$

    C: $r = 1 + \cos\theta$

15. $\displaystyle\int_C 2\arctan\frac{y}{x}\,dx + \ln(x^2 + y^2)\,dy$

    C: $x = 4 + 2\cos\theta,\ y = 4 + \sin\theta$

16. $\displaystyle\int_C e^x \sin 2y\,dx + 2e^x \cos 2y\,dy$

    C: $x^2 + y^2 = a^2$

17. $\displaystyle\int_C \sin x \cos y\,dx + (xy + \cos x \sin y)\,dy$

    C: boundary of the region lying between the graphs of $y = x$ and $y = \sqrt{x}$

18. $\displaystyle\int_C (e^{-x^2/2} - y)\,dx + (e^{-y^2/2} + x)\,dy$

    C: boundary of the region lying between the graphs of the circle $x = 5\cos\theta,\ y = 5\sin\theta$ and the ellipse $x = 2\cos\theta$, $y = \sin\theta$

19. $\displaystyle\int_C xy\,dx + (x + y)\,dy$

    C: boundary of the region lying between the graphs of $x^2 + y^2 = 1$ and $x^2 + y^2 = 9$

20. $\displaystyle\int_C 3x^2 e^y\,dx + e^y\,dy$

    C: boundary of the region lying between the squares with vertices $(1, 1)$, $(-1, 1)$, $(-1, -1)$, and $(1, -1)$, and $(2, 2)$, $(-2, 2)$, $(-2, -2)$, and $(2, -2)$

Work   **In Exercises 21–24, use Green's Theorem to calculate the work done by the force F on a particle that is moving counterclockwise around the closed path C.**

21. $\mathbf{F}(x, y) = xy\,\mathbf{i} + (x + y)\,\mathbf{j}$
    C: $x^2 + y^2 = 4$

22. $\mathbf{F}(x, y) = (e^x - 3y)\,\mathbf{i} + (e^y + 6x)\,\mathbf{j}$
    C: $r = 2\cos\theta$

23. $\mathbf{F}(x, y) = (x^{3/2} - 3y)\,\mathbf{i} + \left(6x + 5\sqrt{y}\right)\,\mathbf{j}$
    C: boundary of the triangle with vertices $(0, 0)$, $(5, 0)$, and $(0, 5)$

24. $\mathbf{F}(x, y) = (3x^2 + y)\,\mathbf{i} + 4xy^2\,\mathbf{j}$
    C boundary of the region bounded by the graphs of $y = \sqrt{x}$, $y = 0$, and $x = 4$

Area   **In Exercises 25–28, use a line integral to find the area of the region R.**

25. R: region bounded by the graph of $x^2 + y^2 = a^2$

26. R: triangle bounded by the graphs of $x = 0$, $2x - 3y = 0$, and $x + 3y = 9$

27. R: region bounded by the graphs of $y = 2x + 1$ and $y = 4 - x^2$

28. R: region inside the loop of the folium of Descartes bounded by the graph of

$$x = \frac{3t}{t^3 + 1}, \qquad y = \frac{3t^2}{t^3 + 1}.$$

In Exercises 29 and 30, use Green's Theorem to verify the line integral formulas.

29. *Centroid*   The centroid of the region having area A bounded by the simple closed path C is

$$\bar{x} = \frac{1}{2A}\int_C x^2\,dy, \qquad \bar{y} = -\frac{1}{2A}\int_C y^2\,dx.$$

30. *Area*   The area of a plane region bounded by the simple closed path C given in polar coordinates is

$$A = \frac{1}{2}\int_C r^2\,d\theta.$$

Centroid   **In Exercises 31–34, use the result of Exercise 29 to find the centroid of the region.**

31. R: region bounded by the graphs of $y = 0$ and $y = 4 - x^2$

32. R: region bounded by the graphs of $y = \sqrt{a^2 - x^2}$ and $y = 0$

33. R: region bounded by the graphs of $y = x^3$ and $y = x$, $0 \le x \le 1$

34. R: triangle with vertices $(-a, 0)$, $(a, 0)$, and (b, c), where $-a \le b \le a$

Area **In Exercises 35–38, use the result of Exercise 30 to find the area of the region bounded by the graph of the polar equation.**

35. $r = a(1 - \cos\theta)$ **36.** $r = a\cos 3\theta$

37. $r = 1 + 2\cos\theta$
(inner loop)

38. $r = \dfrac{3}{2 - \cos\theta}$

39. Let

$$I = \int_C \frac{y\,dx - x\,dy}{x^2 + y^2}$$

where C is a circle oriented counterclockwise. Show that $I = 0$ if C does not contain the origin. What is I if C contains the origin?

40. a. Let C be the line segment joining (x_1, y_1) and (x_2, y_2). Show that

$$\int_C -y\,dx + x\,dy = x_1 y_2 - x_2 y_1.$$

b. Let $(x_1, y_1), (x_2, y_2), \ldots, (x_n, y_n)$ be the vertices of a polygon. Prove that the area enclosed is

$$\frac{1}{2}[(x_1 y_2 - x_2 y_1) + (x_2 y_3 - x_3 y_2) + \cdots$$
$$+ (x_{n-1} y_n - x_n y_{n-1}) + (x_n y_1 - x_1 y_n)].$$

Area **In Exercises 41 and 42, find the area enclosed by the polygon with the given vertices.**

41. Pentagon: $(0, 0), (2, 0), (3, 2), (1, 4), (-1, 1)$

42. Hexagon: $(0, 0), (2, 0), (3, 2), (2, 4), (0, 3), (-1, 1)$

43. Use Green's Theorem to prove that

$$\int_C f(x)\,dx + g(y)\,dy = 0$$

if f and g are differentiable functions and C is a piecewise smooth, simple closed path.

44. Let $\mathbf{F} = M\mathbf{i} + N\mathbf{j}$, where M and N have continuous first partial derivatives in a simply connected region R. Prove that if C is simple, smooth, and closed, and $N_x = M_y$, then

$$\int_C \mathbf{F} \cdot d\mathbf{r} = 0.$$

In Exercises 45 and 46, prove the identity where R is a simply connected region with boundary C. Also assume that the required partial derivatives of the scalar functions f and g are continuous. The expressions $D_N f$ and $D_N g$ are the derivatives in the direction of the outward normal vector $\mathbf{N}$ of C and are defined by

$$D_N f = \nabla f \cdot \mathbf{N}, \qquad D_N g = \nabla g \cdot \mathbf{N}.$$

45. Green's first identity:

$$\iint_R (f\nabla^2 g + \nabla f \cdot \nabla g)\,dA = \int_C f D_N g\,ds$$

(*Hint:* Use the alternative form of Green's Theorem and the property

$$\text{div}(f\mathbf{G}) = f\,\text{div}\,\mathbf{G} + \nabla f \cdot \mathbf{G}.)$$

46. Green's second identity:

$$\iint_R (f\nabla^2 g - g\nabla^2 f)\,dA = \int_C (f D_N g - g D_N f)\,ds$$

(*Hint:* Use Exercise 45 twice.)

SECTION **Parametric Surfaces**

15.5 Parametric Surfaces • Finding Parametric Equations for Surfaces • Normal Vectors and Tangent Planes • Area of a Parametric Surface

Parametric Surfaces

You already know how to represent a curve in the plane or in space by a set of parametric equations—or, equivalently, by a vector-valued function.

$$\mathbf{r}(t) = x(t)\mathbf{i} + y(t)\mathbf{j} \qquad \text{Plane curve}$$

$$\mathbf{r}(t) = x(t)\mathbf{i} + y(t)\mathbf{j} + z(t)\mathbf{k} \qquad \text{Space curve}$$

In this section, you will learn how to represent a surface in space by a set of parametric equations—or by a vector-valued function. For curves, note that the vector-valued function $\mathbf{r}$ is a function of a *single* parameter t. For surfaces, the vector-valued function is a function of *two* parameters u and v.

TECHNOLOGY Some computer graphing utilities are capable of sketching surfaces that are represented parametrically. If you have access to such software, try using it to sketch some of the surfaces in the examples and exercises in this section.

Definition of Parametric Surface

Let x, y, and z be functions of u and v that are continuous on a domain D in the uv-plane. The set of points (x, y, z) given by

$$\mathbf{r}(u, v) = x(u, v)\mathbf{i} + y(u, v)\mathbf{j} + z(u, v)\mathbf{k} \qquad \text{Parametric surface}$$

is called a **parametric surface.** The equations

$$x = x(u, v), \quad y = y(u, v), \quad \text{and} \quad z = z(u, v)$$

are the **parametric equations** for the surface.

If S is a parametric surface given by the vector-valued function $\mathbf{r}$, then S is traced out by the position vector $\mathbf{r}(u, v)$ as the point (u, v) moves throughout the domain D, as shown in Figure 15.35.

FIGURE 15.35

FIGURE 15.36

EXAMPLE 1 Sketching a Parametric Surface

Identify and sketch the parametric surface S given by

$$\mathbf{r}(u, v) = 3\cos u\,\mathbf{i} + 3\sin u\,\mathbf{j} + v\,\mathbf{k}$$

where $0 \le u \le 2\pi$ and $0 \le v \le 4$.

Solution Because $x = 3\cos u$ and $y = 3\sin u$, you know that for each point (x, y, z) on the surface, x and y are related by the equation $x^2 + y^2 = 3^2$. In other words, each cross section of S taken parallel to the xy-plane is a circle of radius 3, centered on the z-axis. Because $z = v$, where $0 \le v \le 4$, you can see that the surface is a right circular cylinder of height 4. The radius of the cylinder is 3, and the z-axis forms the axis of the cylinder, as shown in Figure 15.36.

As with parametric representations of curves, parametric representations of surfaces are not unique. That is, there are many other sets of parametric equations that could be used to represent the surface shown in Figure 15.36.

EXAMPLE 2 Sketching a Parametric Surface

Identify and sketch the parametric surface S given by

$$\mathbf{r}(u, v) = \sin u \cos v\,\mathbf{i} + \sin u \sin v\,\mathbf{j} + \cos u\,\mathbf{k}$$

where $0 \le u \le \pi$ and $0 \le v \le 2\pi$.

Solution To identify the surface, you can try to use trigonometric identities to eliminate the parameters. After some experimentation, you can discover that

$$\begin{aligned}
x^2 + y^2 + z^2 &= (\sin u \cos v)^2 + (\sin u \sin v)^2 + (\cos u)^2 \\
&= \sin^2 u \cos^2 v + \sin^2 u \sin^2 v + \cos^2 u \\
&= \sin^2 u(\cos^2 v + \sin^2 v) + \cos^2 u \\
&= \sin^2 u + \cos^2 u \\
&= 1.
\end{aligned}$$

Thus, each point on S lies on the unit sphere, centered at the origin, as shown in Figure 15.37. For fixed $u = d_i$, $\mathbf{r}(u, v)$ traces out latitude (or meridian) circles

$$x^2 + y^2 = \sin^2 d_i, \qquad 0 \le d_i \le \pi$$

that are parallel to the xy-plane, and for fixed $v = c_i$, $\mathbf{r}(u, v)$ traces out longitude (or great) circles.

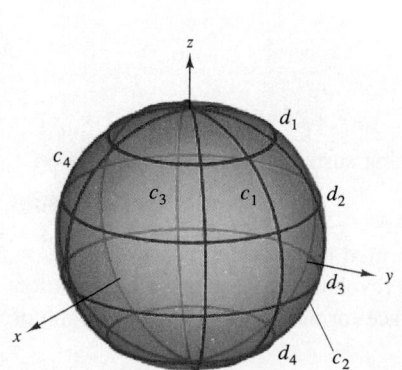

FIGURE 15.37

REMARK To further convince yourself that the vector-valued function in Example 2 traces out the entire unit sphere, recall that the parametric equations

$$x = \rho \sin \phi \cos \theta, \quad y = \rho \sin \phi \sin \theta, \quad \text{and} \quad z = \rho \cos \phi$$

where $0 \le \theta \le 2\pi$ and $0 \le \phi \le \pi$, describe the conversion from rectangular to spherical coordinates, as discussed in Section 11.7.

Finding Parametric Equations for Surfaces

In Examples 1 and 2, you were asked to identify the surface described by a given set of parametric equations. The reverse problem—that of writing a set of parametric equations for a given surface—is generally more difficult. One such problem that is straightforward, however, is finding the parametric equations for a surface that is given by $z = f(x, y)$. You can parametrize such a surface as

$$\mathbf{r}(x, y) = x\mathbf{i} + y\mathbf{j} + f(x, y)\mathbf{k}.$$

EXAMPLE 3   Representing a Surface Parametrically

Write a set of parametric equations for the cone given by

$$z = \sqrt{x^2 + y^2}$$

as shown in Figure 15.38.

Solution   Because this surface is given in the form $z = f(x, y)$, you can let x and y be the parameters. Then the cone is represented by the vector-valued function

$$\mathbf{r}(x, y) = x\mathbf{i} + y\mathbf{j} + \sqrt{x^2 + y^2}\,\mathbf{k}$$

where (x, y) varies over the entire xy-plane.

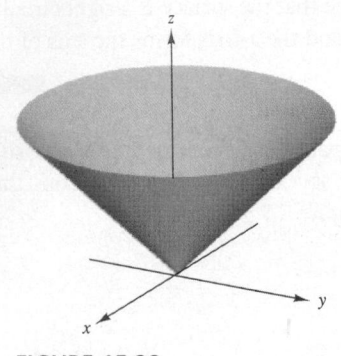

FIGURE 15.38

A second type of surface that is easily represented parametrically is a surface of revolution. For instance, to represent the surface formed by revolving the graph of $y = f(x)$, $a \leq x \leq b$, about the x-axis, use

$$x = u, \quad y = f(u)\cos v, \quad \text{and} \quad z = f(u)\sin v$$

where $a \leq u \leq b$ and $0 \leq v \leq 2\pi$.

EXAMPLE 4   Representing a Surface of Revolution Parametrically

Write a set of parametric equations for the surface of revolution obtained by revolving

$$f(x) = \frac{1}{x}, \qquad 1 \leq x \leq 10$$

about the x-axis.

Solution   Use the parameters u and v

$$x = u, \quad y = \frac{1}{u}\cos v, \quad \text{and} \quad z = \frac{1}{u}\sin v$$

where $1 \leq u \leq 10$ and $0 \leq v \leq 2\pi$. The resulting surface is a portion of *Gabriel's Horn,* as shown in Figure 15.39.

The surface of revolution in Example 4 is formed by revolving the graph of $y = f(x)$ about the x-axis. For other types of surfaces of revolution, a similar parametrization can be used. For instance, to parametrize the surface formed by revolving the graph of $x = f(z)$ about the z-axis, you can use

$$z = u, \quad x = f(u)\cos v, \quad \text{and} \quad y = f(u)\sin v.$$

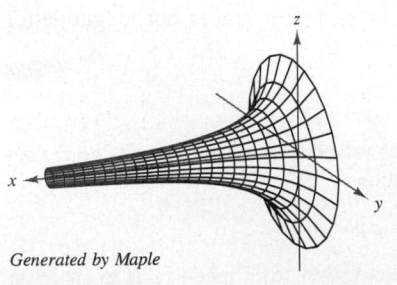

Generated by Maple

FIGURE 15.39

Normal Vectors and Tangent Planes

Let S be a parametric surface given by

$$\mathbf{r}(u, v) = x(u, v)\mathbf{i} + y(u, v)\mathbf{j} + z(u, v)\mathbf{k}$$

over an open region D such that x, y, and z have continuous partial derivatives on D. The **partial derivatives of r** with respect to u and v are defined as

$$\mathbf{r}_u = \frac{\partial x}{\partial u}(u, v)\mathbf{i} + \frac{\partial y}{\partial u}(u, v)\mathbf{j} + \frac{\partial z}{\partial u}(u, v)\mathbf{k}$$

and

$$\mathbf{r}_v = \frac{\partial x}{\partial v}(u, v)\mathbf{i} + \frac{\partial y}{\partial v}(u, v)\mathbf{j} + \frac{\partial z}{\partial v}(u, v)\mathbf{k}.$$

Each of these partial derivatives is a vector-valued function that can be interpreted geometrically in terms of tangent vectors. For instance, if $v = v_0$ is held constant, then $\mathbf{r}(u, v_0)$ is a vector-valued function of a single parameter and defines a curve C_1 that lies on the surface S. The tangent vector to C_1 at the point $(x(u_0, v_0), y(u_0, v_0), z(u_0, v_0))$ is given by

$$\mathbf{r}_u(u_0, v_0) = \frac{\partial x}{\partial u}(u_0, v_0)\mathbf{i} + \frac{\partial y}{\partial u}(u_0, v_0)\mathbf{j} + \frac{\partial z}{\partial u}(u_0, v_0)\mathbf{k}$$

as shown in Figure 15.40. In a similar way, if $u = u_0$ is held constant, then $\mathbf{r}(u_0, v)$ is a vector-valued function of a single parameter and defines a curve C_2 that lies on the surface S. The tangent vector to C_2 at the point $(x(u_0, v_0), y(u_0, v_0), z(u_0, v_0))$ is given by

$$\mathbf{r}_v(u_0, v_0) = \frac{\partial x}{\partial v}(u_0, v_0)\mathbf{i} + \frac{\partial y}{\partial v}(u_0, v_0)\mathbf{j} + \frac{\partial z}{\partial v}(u_0, v_0)\mathbf{k}.$$

If the normal vector $\mathbf{r}_u \times \mathbf{r}_v$ is not $\mathbf{0}$ for any (u, v) in D, then the surface S is called **smooth** and will have a tangent plane. Informally, a smooth surface is one that has no sharp points or cusps. For instance, spheres, ellipsoids, and paraboloids are smooth, whereas the cone given in Example 3 is not smooth.

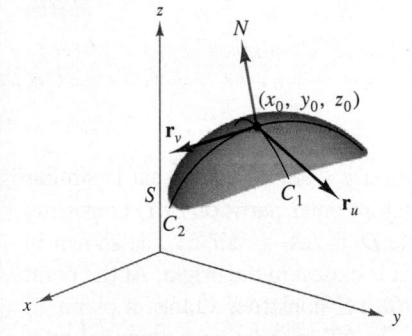

FIGURE 15.40

REMARK Figure 15.40 shows the normal vector $\mathbf{r}_u \times \mathbf{r}_v$. The vector $\mathbf{r}_v \times \mathbf{r}_u$ is also normal to S and points in the opposite direction.

Normal Vector to a Smooth Parametric Surface

Let S be a smooth parametric surface

$$\mathbf{r}(u, v) = x(u, v)\mathbf{i} + y(u, v)\mathbf{j} + z(u, v)\mathbf{k}$$

defined over an open region D in the uv-plane. Let (u_0, v_0) be a point in D. A normal vector at the point

$$(x_0, y_0, z_0) = (x(u_0, v_0), y(u_0, v_0), z(u_0, v_0))$$

is given by

$$\mathbf{N} = \mathbf{r}_u(u_0, v_0) \times \mathbf{r}_v(u_0, v_0) = \begin{vmatrix} \mathbf{i} & \mathbf{j} & \mathbf{k} \\ \dfrac{\partial x}{\partial u} & \dfrac{\partial y}{\partial u} & \dfrac{\partial z}{\partial u} \\ \dfrac{\partial x}{\partial v} & \dfrac{\partial y}{\partial v} & \dfrac{\partial z}{\partial v} \end{vmatrix}.$$

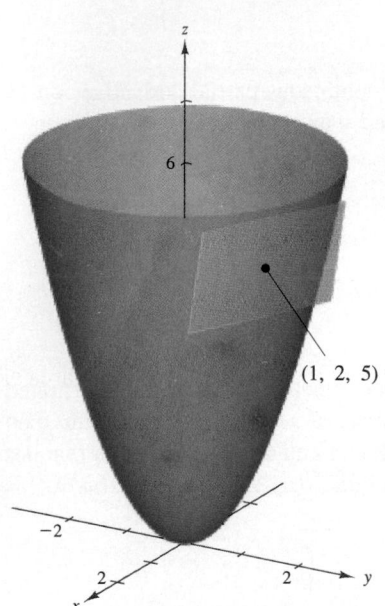

FIGURE 15.41

EXAMPLE 5   Finding a Tangent Plane to a Parametric Surface

Find an equation of the tangent plane to the paraboloid given by

$$\mathbf{r}(u, v) = u\mathbf{i} + v\mathbf{j} + (u^2 + v^2)\mathbf{k}$$

at the point (1, 2, 5).

Solution   The point in the uv-plane that is mapped to the point $(x, y, z) = (1, 2, 5)$ is $(u, v) = (1, 2)$. The partial derivatives of $\mathbf{r}$ are

$$\mathbf{r}_u = \mathbf{i} + 2u\mathbf{k} \quad \text{and} \quad \mathbf{r}_v = \mathbf{j} + 2v\mathbf{k}.$$

The normal vector is given by

$$\mathbf{r}_u \times \mathbf{r}_v = \begin{vmatrix} \mathbf{i} & \mathbf{j} & \mathbf{k} \\ 1 & 0 & 2u \\ 0 & 1 & 2v \end{vmatrix} = -2u\mathbf{i} - 2v\mathbf{j} + \mathbf{k}$$

which implies that the normal vector at (1, 2, 5) is $\mathbf{r}_u \times \mathbf{r}_v = -2\mathbf{i} - 4\mathbf{j} + \mathbf{k}$. Thus, an equation of the tangent plane at (1, 2, 5) is

$$-2(x - 1) - 4(y - 2) + (z - 5) = 0$$
$$-2x - 4y + z = -5.$$

The tangent plane is shown in Figure 15.41.

Area of a Parametric Surface

To define the area of a parametric surface, you can use a development that is similar to that given in Section 14.5. Begin by constructing an inner partition of D consisting of n rectangles, where the area of the ith rectangle D_i is $\Delta A_i = \Delta u_i \Delta v_i$, as shown in Figure 15.42. In each D_i, let (u_i, v_i) be the point that is closest to the origin. At the point $(x_i, y_i, z_i) = (x(u_i, v_i), y(u_i, v_i), z(u_i, v_i))$ on the surface S, construct a tangent plane T_i. The area of the portion of S that corresponds to D_i, ΔT_i, can be approximated by a parallelogram in the tangent plane. That is, $\Delta T_i \approx \Delta S_i$. Hence, the surface area of S is given by $\Sigma \Delta S_i \approx \Sigma \Delta T_i$. The area of the parallelogram in the tangent plane is

$$\| \Delta u_i \mathbf{r}_u \times \Delta v_i \mathbf{r}_v \| = \| \mathbf{r}_u \times \mathbf{r}_v \| \Delta u_i \Delta v_i$$

which leads to the following definition.

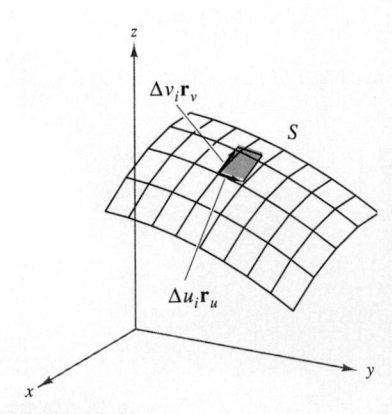

FIGURE 15.42

Area of a Parametric Surface

Let S be a smooth parametric surface

$$\mathbf{r}(u, v) = x(u, v)\mathbf{i} + y(u, v)\mathbf{j} + z(u, v)\mathbf{k}$$

defined over an open region D in the uv-plane. If each point on the surface S corresponds to exactly one point in the domain D, then the **surface area** of S is given by

$$\text{Surface area} = \iint_S dS = \iint_D \| \mathbf{r}_u \times \mathbf{r}_v \| \, dA$$

where $\mathbf{r}_u = \dfrac{\partial x}{\partial u}\mathbf{i} + \dfrac{\partial y}{\partial u}\mathbf{j} + \dfrac{\partial z}{\partial u}\mathbf{k}$ and $\mathbf{r}_v = \dfrac{\partial x}{\partial v}\mathbf{i} + \dfrac{\partial y}{\partial v}\mathbf{j} + \dfrac{\partial z}{\partial v}\mathbf{k}.$

For a surface S given by $z = f(x, y)$, this formula for surface corresponds to that given in Section 14.5. To see this, you can parametrize the surface using the vector-valued function

$$\mathbf{r}(x, y) = x\mathbf{i} + y\mathbf{j} + f(x, y)\mathbf{k}$$

defined over the region R in the xy-plane. Using

$$\mathbf{r}_x = \mathbf{i} + f_x(x, y)\mathbf{k} \quad \text{and} \quad \mathbf{r}_y = \mathbf{j} + f_y(x, y)\mathbf{k}$$

you have

$$\mathbf{r}_x \times \mathbf{r}_y = \begin{vmatrix} \mathbf{i} & \mathbf{j} & \mathbf{k} \\ 1 & 0 & f_x(x, y) \\ 0 & 1 & f_y(x, y) \end{vmatrix} = -f_x(x, y)\mathbf{i} - f_y(x, y)\mathbf{j} + \mathbf{k}$$

and $\|\mathbf{r}_x \times \mathbf{r}_y\| = \sqrt{[f_x(x, y)]^2 + [f_y(x, y)]^2 + 1}$. This implies that the surface area of S is

$$\text{Surface area} = \iint_R \|\mathbf{r}_x \times \mathbf{r}_y\| \, dA$$

$$= \iint_R \sqrt{1 + [f_x(x, y)]^2 + [f_y(x, y)]^2} \, dA.$$

EXAMPLE 6   Finding Surface Area

REMARK The surface in Example 6 does not quite fulfill the hypothesis that each point on the surface corresponds to exactly one point in D. For this surface, $\mathbf{r}(u, 0) = \mathbf{r}(u, 2\pi)$ for any fixed value of u. However, because the overlap consists of only a semicircle (which has no area), you can still apply the formula for the area of a parametric surface.

Find the surface area of the unit sphere given by

$$\mathbf{r}(u, v) = \sin u \cos v \, \mathbf{i} + \sin u \sin v \, \mathbf{j} + \cos u \, \mathbf{k}$$

where the domain D is given by $0 \le u \le \pi$ and $0 \le v \le 2\pi$.

Solution   Begin by calculating $\mathbf{r}_u$ and $\mathbf{r}_v$.

$$\mathbf{r}_u = \cos u \cos v \, \mathbf{i} + \cos u \sin v \, \mathbf{j} - \sin u \, \mathbf{k}$$

$$\mathbf{r}_v = -\sin u \sin v \, \mathbf{i} + \sin u \cos v \, \mathbf{j}$$

The cross product of these two vectors is

$$\mathbf{r}_u \times \mathbf{r}_v = \begin{vmatrix} \mathbf{i} & \mathbf{j} & \mathbf{k} \\ \cos u \cos v & \cos u \sin v & -\sin u \\ -\sin u \sin v & \sin u \cos v & 0 \end{vmatrix}$$

$$= \sin^2 u \cos v \, \mathbf{i} + \sin^2 u \sin v \, \mathbf{j} + \sin u \cos u \, \mathbf{k}$$

which implies that

$$\|\mathbf{r}_u \times \mathbf{r}_v\| = \sqrt{(\sin^2 u \cos v)^2 + (\sin^2 u \sin v)^2 + (\sin u \cos u)^2}$$

$$= \sqrt{\sin^4 u + \sin^2 u \cos^2 u}$$

$$= \sqrt{\sin^2 u}$$

$$= \sin u. \qquad \text{\small $\sin u \ge 0$ for $0 \le u \le \pi$}$$

Finally, the surface area of the sphere is

$$A = \iint_D \|\mathbf{r}_u \times \mathbf{r}_v\| \, dA = \int_0^{2\pi} \int_0^{\pi} \sin u \, du \, dv$$

$$= \int_0^{2\pi} 2 \, dv = 4\pi.$$

EXAMPLE 7   Finding Surface Area

Find the surface area of the torus given by

$$\mathbf{r}(u, v) = (2 + \cos u)\cos v\,\mathbf{i} + (2 + \cos u)\sin v\,\mathbf{j} + \sin u\,\mathbf{k}$$

where the domain D is given by $0 \le u \le 2\pi$ and $0 \le v \le 2\pi$. (The torus is shown in Figure 15.43.)

Solution   Begin by calculating $\mathbf{r}_u$ and $\mathbf{r}_v$.

$$\mathbf{r}_u = -\sin u\cos v\,\mathbf{i} - \sin u\sin v\,\mathbf{j} + \cos u\,\mathbf{k}$$
$$\mathbf{r}_v = -(2 + \cos u)\sin v\,\mathbf{i} + (2 + \cos u)\cos v\,\mathbf{j}$$

The cross product of these two vectors is

$$\mathbf{r}_u \times \mathbf{r}_v = \begin{vmatrix} \mathbf{i} & \mathbf{j} & \mathbf{k} \\ -\sin u\cos v & -\sin u\sin v & \cos u \\ -(2 + \cos u)\sin v & (2 + \cos u)\cos v & 0 \end{vmatrix}$$
$$= -(2 + \cos u)[\cos v\cos u\,\mathbf{i} + \sin v\cos u\,\mathbf{j} + \sin u\,\mathbf{k}]$$

which implies that

$$\begin{aligned} \|\mathbf{r}_u \times \mathbf{r}_v\| &= (2 + \cos u)\sqrt{(\cos v\cos u)^2 + (\sin v\cos u)^2 + \sin^2 u} \\ &= (2 + \cos u)\sqrt{\cos^2 u\,(\cos^2 v + \sin^2 v) + \sin^2 u} \\ &= (2 + \cos u)\sqrt{\cos^2 u + \sin^2 u} \\ &= 2 + \cos u. \end{aligned}$$

Finally, the surface area of the torus is

$$\begin{aligned} A &= \int\!\!\int_D \|\mathbf{r}_u \times \mathbf{r}_v\|\,dA = \int_0^{2\pi}\int_0^{2\pi} (2 + \cos u)\,du\,dv \\ &= \int_0^{2\pi} 4\pi\,dv = 8\pi^2. \end{aligned}$$

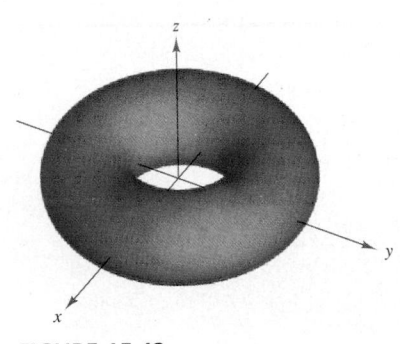

FIGURE 15.43

If the surface S is a surface of revolution, then you can show that the formula for surface area given in Section 6.4 is equivalent to the formula given in this section. For instance, suppose f is a nonnegative function such that f' is continuous over the interval $[a, b]$. Let S be the surface of revolution formed by revolving the graph of f, where $a \le x \le b$, about the x-axis. From Section 6.4, you know that the surface area is given by

$$\text{Surface area} = \int_a^b f(x)\sqrt{1 + [f'(x)]^2}\,dx.$$

To represent S parametrically, let $x = u$, $y = f(u)\cos v$, and $z = f(u)\sin v$, where $a \le u \le b$ and $0 \le v \le 2\pi$. Then,

$$\mathbf{r}(u, v) = u\,\mathbf{i} + f(u)\cos v\,\mathbf{j} + f(u)\sin v\,\mathbf{k}.$$

Try showing that the formula

$$\text{Surface area} = \int\!\!\int_D \|\mathbf{r}_u \times \mathbf{r}_v\|\,dA$$

is equivalent to the formula given above.

EXERCISES for Section 15.5

In Exercises 1–4, match the vector-valued function with its graph. [The graphs are labeled (a), (b), (c), and (d).]

1. $\mathbf{r}(u, v) = u\mathbf{i} + v\mathbf{j} + uv\mathbf{k}$

2. $\mathbf{r}(u, v) = u\cos v\mathbf{i} + u\sin v\mathbf{j} + u\mathbf{k}$

3. $\mathbf{r}(u, v) = 2\cos v\cos u\mathbf{i} + 2\cos v\sin u\mathbf{j} + 2\sin v\mathbf{k}$

4. $\mathbf{r}(u, v) = 4\cos u\mathbf{i} + 4\sin u\mathbf{j} + v\mathbf{k}$

a.

Generated by Maple

b.

Generated by Maple

c.

Generated by Maple

d.

Generated by Maple

In Exercises 5–8, find the rectangular equation for the surface by eliminating the parameters from the vector-valued function, identify the surface, and sketch its graph.

5. $\mathbf{r}(u, v) = u\mathbf{i} + v\mathbf{j} + \dfrac{v}{2}\mathbf{k}$

6. $\mathbf{r}(u, v) = u\cos v\mathbf{i} + u\sin v\mathbf{j} + u^2\mathbf{k}$

7. $\mathbf{r}(u, v) = 2\cos u\mathbf{i} + v\mathbf{j} + 2\sin u\mathbf{k}$

8. $\mathbf{r}(u, v) = 5\cos v\cos u\mathbf{i} + 5\cos v\sin u\mathbf{j} + 5\sin v\mathbf{k}$

In Exercises 9–12, determine how the graph of the surface $s(u, v)$ differs from the graph of $\mathbf{r}(u, v) = u\cos v\mathbf{i} + u\sin v\mathbf{j} + u^2\mathbf{k}$, where $0 \le u \le 2$ and $0 \le v \le 2\pi$. (It is not necessary to graph s.)

9. $s(u, v) = u\cos v\mathbf{i} + u\sin v\mathbf{j} - u^2\mathbf{k}$
$0 \le u \le 2, 0 \le v \le 2\pi$

10. $s(u, v) = u\cos v\mathbf{i} + u^2\mathbf{j} + u\sin v\mathbf{k}$
$0 \le u \le 2, 0 \le v \le 2\pi$

11. $s(u, v) = u\cos v\mathbf{i} + u\sin v\mathbf{j} + u^2\mathbf{k}$
$0 \le u \le 3, 0 \le v \le 2\pi$

12. $s(u, v) = 4u\cos v\mathbf{i} + 4u\sin v\mathbf{j} + u^2\mathbf{k}$
$0 \le u \le 2, 0 \le v \le 2\pi$

C In Exercises 13–18, use a graphing utility to graph the surface represented by the vector-valued function.

13. $\mathbf{r}(u, v) = 2u\cos v\mathbf{i} + 2u\sin v\mathbf{j} + u^4\mathbf{k}$
$0 \le u \le 1, 0 \le v \le 2\pi$

14. $\mathbf{r}(u, v) = 2\cos v\cos u\mathbf{i} + 4\cos v\sin u\mathbf{j} + \sin v\mathbf{k}$
$0 \le u \le 2\pi, 0 \le v \le 2\pi$

15. $\mathbf{r}(u, v) = 2\sinh u\cos v\mathbf{i} + \sinh u\sin v\mathbf{j} + \cosh u\mathbf{k}$
$0 \le u \le 2, 0 \le v \le 2\pi$

16. $\mathbf{r}(u, v) = 2u\cos v\mathbf{i} + 2u\sin v\mathbf{j} + v\mathbf{k}$
$0 \le u \le 1, 0 \le v \le 3\pi$

17. $\mathbf{r}(u, v) = (u - \sin u)\cos v\mathbf{i} + (1 - \cos u)\sin v\mathbf{j} + u\mathbf{k}$
$0 \le u \le \pi, 0 \le v \le 2\pi$

18. $\mathbf{r}(u, v) = \cos^3 u\cos v\mathbf{i} + \sin^3 u\sin v\mathbf{j} + u\mathbf{k}$
$0 \le u \le \dfrac{\pi}{2}, 0 \le v \le 2\pi$

C **19.** Use a graphing utility to graph the torus

$$\mathbf{r}(u, v) = (a + b\cos v)\cos u\mathbf{i}$$
$$+ (a + b\cos v)\sin u\mathbf{j} + b\sin v\mathbf{k}$$

for each set of values for a and b and where $0 \le u \le 2\pi$ and $0 \le v \le 2\pi$. Use the results to describe the effect of a and b on the shape of the torus.

a. $a = 4, b = 1$ **b.** $a = 4, b = 2$
c. $a = 8, b = 1$ **d.** $a = 8, b = 3$

C **20.** Consider the vector-valued function in Exercise 16.
a. Sketch a graph of the function where u is held constant at $u = 1$. Identify the graph.
b. Sketch a graph of the function where v is held constant at $v = 2\pi/3$. Identify the graph.
c. If a surface is represented by the vector-valued function $\mathbf{r} = \mathbf{r}(u, v)$, what generalization can you make about the graph of the function if one of the parameters is held constant?

In Exercises 21–28, find a vector-valued function whose graph is the specified surface.

21. The plane $z = y$

22. The plane $x + y + z = 6$

23. The cylinder $x^2 + y^2 = 16$

24. The cylinder $x^2 + 4y^2 = 16$

25. The cylinder $z = x^2$

26. The ellipsoid $\dfrac{x^2}{9} + \dfrac{y^2}{4} + \dfrac{z^2}{1} = 1$

27. The part of the plane $z = 4$ that lies inside the cylinder $x^2 + y^2 = 9$

28. The part of the paraboloid $z = x^2 + y^2$ that lies inside the cylinder $x^2 + y^2 = 9$

Surface of Revolution　In Exercises 29–32, write a set of parametric equations for the surface of revolution obtained by revolving the graph of the function about the specified axis.

Function	Axis of revolution
29. $y = \dfrac{x}{2}, \quad 0 \le x \le 6$	x-axis
30. $y = \sqrt{x}, \quad 0 \le x \le 4$	x-axis
31. $x = \sin z, \quad 0 \le z \le \pi$	z-axis
32. $z = 4 - y^2, \quad 0 \le y \le 2$	y-axis

Tangent Plane　In Exercises 33–36, find an equation of the tangent plane to the surface given by the vector-valued function at the specified point.

Function	Point
33. $\mathbf{r}(u,v) = (u+v)\mathbf{i} + (u-v)\mathbf{j} + v\mathbf{k}$	$(1, -1, 1)$

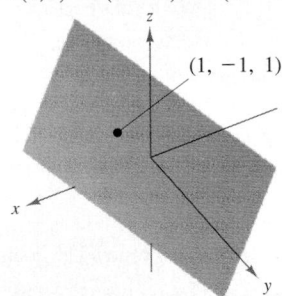

(1, −1, 1)

FIGURE FOR 33

34. $\mathbf{r}(u,v) = u\mathbf{i} + v\mathbf{j} + \sqrt{uv}\,\mathbf{k}$　　$(1, 1, 1)$

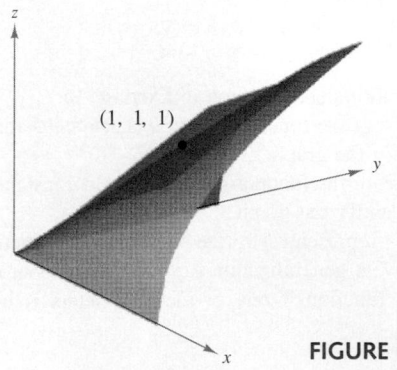

(1, 1, 1)

FIGURE FOR 34

Function	Point
35. $\mathbf{r}(u,v) = 2u\cos v\,\mathbf{i} + 3u\sin v\,\mathbf{j} + u^2\mathbf{k}$	$(0, 6, 4)$
36. $\mathbf{r}(u,v) = u\cosh v\,\mathbf{i} + u\sinh v\,\mathbf{j} + u^2\mathbf{k}$	$(-2, 0, 4)$

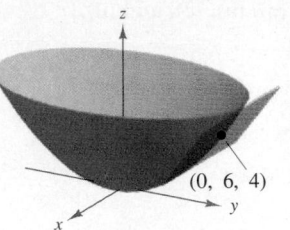

(0, 6, 4)

FIGURE FOR 35　　　　**FIGURE FOR 36**

(−2, 0, 4)

Area　In Exercises 37–44, find the area of the surface over the specified region.

37. The part of the plane
$$\mathbf{r}(u,v) = 2u\mathbf{i} - \frac{v}{2}\mathbf{j} + \frac{v}{2}\mathbf{k}$$
where $0 \le u \le 2$ and $0 \le v \le 1$

38. The part of the paraboloid
$$\mathbf{r}(u,v) = 2u\cos v\,\mathbf{i} + 2u\sin v\,\mathbf{j} + u^2\mathbf{k}$$
where $0 \le u \le 2$ and $0 \le v \le 2\pi$

39. The part of the cylinder
$$\mathbf{r}(u,v) = a\cos u\,\mathbf{i} + a\sin u\,\mathbf{j} + v\mathbf{k}$$
where $0 \le u \le 2\pi$ and $0 \le v \le b$

40. The sphere
$$\mathbf{r}(u,v) = a\sin u\cos v\,\mathbf{i} + a\sin u\sin v\,\mathbf{j} + a\cos u\,\mathbf{k}$$
where $0 \le u \le \pi$ and $0 \le v \le 2\pi$

41. The part of the cone
$$\mathbf{r}(u,v) = au\cos v\,\mathbf{i} + au\sin v\,\mathbf{j} + u\mathbf{k}$$
where $0 \le u \le b$ and $0 \le v \le 2\pi$

42. The torus
$$\mathbf{r}(u,v) = (a + b\cos v)\cos u\,\mathbf{i} + (a + b\cos v)\sin u\,\mathbf{j} + b\sin v\,\mathbf{k}$$
where $a > b$, $0 \le u \le 2\pi$, and $0 \le v \le 2\pi$

43. The surface of revolution
$$\mathbf{r}(u,v) = \sqrt{u}\cos v\,\mathbf{i} + \sqrt{u}\sin v\,\mathbf{j} + u\mathbf{k}$$
where $0 \le u \le 4$ and $0 \le v \le 2\pi$

44. The surface of revolution
$$\mathbf{r}(u,v) = u\mathbf{i} + \sin u\cos v\,\mathbf{j} + \sin u\sin v\,\mathbf{k}$$
where $0 \le u \le \pi$ and $0 \le v \le 2\pi$

45. Graph and find the area of one turn of the spiral ramp
$$\mathbf{r}(u,v) = u\cos v\,\mathbf{i} + u\sin v\,\mathbf{j} + 2v\mathbf{k}, \quad 0 \le u \le 3, \quad 0 \le v \le 2\pi.$$

46. Find a set of parametric equations for the hyperboloid $x^2 + y^2 - z^2 = 1$, and determine the tangent plane at $(1, 0, 0)$.

47. Let f be a nonnegative function such that f' is continuous over the interval $[a, b]$. Let S be the surface of revolution formed by revolving the graph of f, where $a \le x \le b$, about the x-axis. Let $x = u$, $y = f(u) \cos v$, and $z = f(u) \sin v$, where $a \le u \le b$ and $0 \le v \le 2\pi$. Then, S is represented parametrically by

$$\mathbf{r}(u, v) = u\,\mathbf{i} + f(u)\cos v\,\mathbf{j} + f(u)\sin v\,\mathbf{k}.$$

Show that the following formulas are equivalent.

Surface area $= \displaystyle\int_a^b f(x)\sqrt{1 + [f'(x)]^2}\,dx$

Surface area $= \displaystyle\int_D\int \|\mathbf{r}_u \times \mathbf{r}_v\|\,dA$

48. Show that the cone in Example 3 can be represented parametrically by

$$\mathbf{s}(u, v) = u\cos v\,\mathbf{i} + u\sin v\,\mathbf{j} + u\,\mathbf{k}$$

where $0 \le v \le 2\pi$ and $0 \le u$.

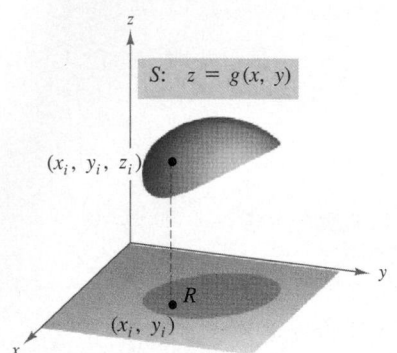

FIGURE 15.44
Scalar function f assigns a number to
each point of S.

Surface Integrals

The remainder of this chapter deals primarily with **surface integrals.** We begin by
considering surfaces given by $z = g(x, y)$. Later in the section we will consider more
general surfaces given in parametric form.

Let S be a surface given by $z = g(x, y)$ and R its projection on the xy-plane, as
shown in Figure 15.44. Suppose g, g_x, and g_y are continuous at all points in R, and f
is defined on S. Employing the procedure used to find surface area in Section 14.5,
evaluate f at (x_i, y_i, z_i) and form the sum

$$\sum_{i=1}^{n} f(x_i, y_i, z_i)\, \Delta S_i$$

where $\Delta S_i \approx \sqrt{1 + [g_x(x_i, y_i)]^2 + [g_y(x_i, y_i)]^2}\, \Delta A_i$. Provided the limit as $\| \Delta \|$ approaches 0 exists, the **surface integral of f over S** is defined as

$$\iint_S f(x, y, z)\, dS = \lim_{\|\Delta\| \to 0} \sum_{i=1}^{n} f(x_i, y_i, z_i)\, \Delta S_i.$$

This integral can be evaluated by a double integral.

THEOREM 15.10   Evaluating a Surface Integral

Let S be a surface with equation $z = g(x, y)$ and R its projection on the xy-plane. If g, g_x, and g_y are continuous on R and f is continuous on S, then the surface integral of f over S is

$$\iint_S f(x, y, z)\, dS = \iint_R f(x, y, g(x, y))\sqrt{1 + [g_x(x, y)]^2 + [g_y(x, y)]^2}\, dA.$$

For surfaces described by functions of x and z (or y and z), you can make the
following adjustments to Theorem 15.10. If S is the graph of $y = g(x, z)$ and R is its
projection on the xz-plane, then

$$\iint_S f(x, y, z)\, dS = \iint_R f(x, g(x, z), z)\sqrt{1 + [g_x(x, z)]^2 + [g_z(x, z)]^2}\, dA.$$

If S is the graph of $x = g(y, z)$ and R is its projection on the yz-plane, then

$$\iint_S f(x, y, z)\, dS = \iint_R f(g(y, z), y, z)\sqrt{1 + [g_y(y, z)]^2 + [g_z(y, z)]^2}\, dA.$$

EXAMPLE 1 **Evaluating a Surface Integral**

Evaluate the surface integral

$$\iint_S (y^2 + 2yz)\, dS$$

where S is the first-octant portion of the plane $2x + y + 2z = 6$.

Solution Begin by writing S as

$$z = \frac{1}{2}(6 - 2x - y)$$

$$g(x, y) = \frac{1}{2}(6 - 2x - y).$$

Using the partial derivatives $g_x(x, y) = -1$ and $g_y(x, y) = -\frac{1}{2}$, you can write

$$\sqrt{1 + [g_x(x, y)]^2 + [g_y(x, y)]^2} = \sqrt{1 + 1 + \frac{1}{4}} = \frac{3}{2}.$$

Using Figure 15.45 and Theorem 15.10, we obtain

$$\iint_S (y^2 + 2yz)\, dS = \iint_R f(x, y, g(x, y))\sqrt{1 + [g_x(x, y)]^2 + [g_y(x, y)]^2}\, dA$$

$$= \iint_R \left[y^2 + 2y\left(\frac{1}{2}\right)(6 - 2x - y) \right]\left(\frac{3}{2}\right) dA$$

$$= 3\int_0^3 \int_0^{2(3-x)} y(3 - x)\, dy\, dx$$

$$= 6\int_0^3 (3 - x)^3\, dx$$

$$= -\frac{3}{2}(3 - x)^4 \bigg]_0^3$$

$$= \frac{243}{2}.$$

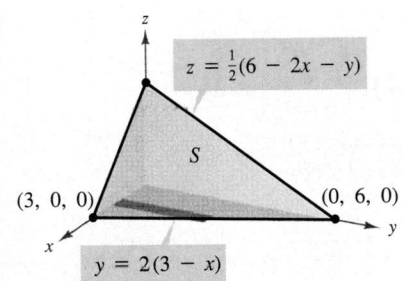

$z = \frac{1}{2}(6 - 2x - y)$

S

$(3, 0, 0)$ $(0, 6, 0)$

$y = 2(3 - x)$

FIGURE 15.45

An alternative solution to Example 1 would be to project S onto the yz-plane, as shown in Figure 15.46. Then, $x = \frac{1}{2}(6 - y - 2z)$, and

$$\sqrt{1 + [g_y(y, z)]^2 + [g_z(y, z)]^2} = \sqrt{1 + \frac{1}{4} + 1} = \frac{3}{2}.$$

Thus, the surface integral is

$$\iint_S (y^2 + 2yz)\, dS = \iint_{R'} f(g(y, z), y, z)\sqrt{1 + [g_y(y, z)]^2 + [g_z(y, z)]^2}\, dA$$

$$= \int_0^6 \int_0^{(6-y)/2} (y^2 + 2yz)\left(\frac{3}{2}\right) dz\, dy$$

$$= \frac{3}{8}\int_0^6 (36y - y^3)\, dy$$

$$= \frac{243}{2}.$$

$(0, 0, 3)$

$z = \dfrac{6 - y}{2}$

$(0, 6, 0)$

$x = \frac{1}{2}(6 - y - 2z)$

FIGURE 15.46

Try reworking Example 2 by projecting S onto the xz-plane.

In Example 1, you could have projected the surface S onto any one of the three coordinate planes. In the next example, the surface is a portion of a cylinder centered about the z-axis, and you have the option of projecting S onto either the xz-plane or the xy-plane.

EXAMPLE 2   Evaluating a Surface Integral

Evaluate the surface integral

$$\iint_S (x + z)\, dS$$

where S is the first-octant portion of the cylinder $y^2 + z^2 = 9$ between $x = 0$ and $x = 4$, as shown in Figure 15.47.

Solution   Begin by projecting onto the xy-plane, so that $z = g(x, y) = \sqrt{9 - y^2}$, and obtain

$$\sqrt{1 + [g_x(x, y)]^2 + [g_y(x, y)]^2} = \sqrt{1 + \left(\frac{-y}{\sqrt{9 - y^2}}\right)^2}$$

$$= \frac{3}{\sqrt{9 - y^2}}.$$

Theorem 15.10 does not apply directly, because g_y is not continuous when $y = 3$. However, you can apply the theorem for $0 \le b < 3$ and then take the limit as b approaches 3, as follows.

$$\iint_S (x + z)\, dS = \lim_{b \to 3^-} \int_0^b \int_0^4 \left(x + \sqrt{9 - y^2}\right) \frac{3}{\sqrt{9 - y^2}}\, dx\, dy$$

$$= \lim_{b \to 3^-} 3 \int_0^b \int_0^4 \left(\frac{x}{\sqrt{9 - y^2}} + 1\right) dx\, dy$$

$$= \lim_{b \to 3^-} 3 \int_0^b \left[\frac{x^2}{2\sqrt{9 - y^2}} + x\right]_0^4 dy$$

$$= \lim_{b \to 3^-} 3 \int_0^b \left(\frac{8}{\sqrt{9 - y^2}} + 4\right) dy$$

$$= \lim_{b \to 3^-} 3 \left[4y + 8\arcsin\frac{y}{3}\right]_0^b$$

$$= \lim_{b \to 3^-} 3\left(4b + 8\arcsin\frac{b}{3}\right)$$

$$= 36 + 24\left(\frac{\pi}{2}\right)$$

$$= 36 + 12\pi$$

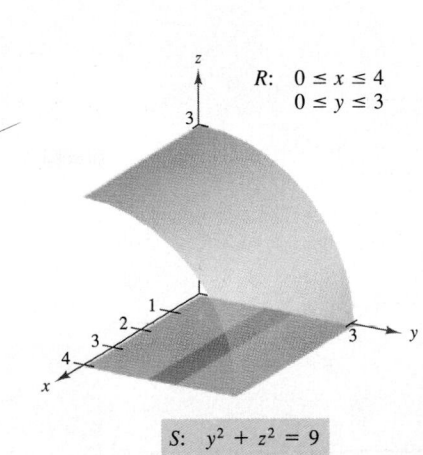

$R:\ \ 0 \le x \le 4$
$\quad\ \ 0 \le y \le 3$

$S:\ y^2 + z^2 = 9$

FIGURE 15.47

TECHNOLOGY   Some symbolic integration utilities are capable of evaluating improper integrals. If you have access to such computer software, try using it to evaluate the improper integral

$$\int_0^3 \int_0^4 \left(x + \sqrt{9 - y^2}\right) \frac{3}{\sqrt{9 - y^2}}\, dx\, dy.$$

Do you obtain the same result that was obtained in Example 2?

If the function f defined on the surface S is simply $f(x, y, z) = 1$, then the surface integral yields the *surface area* of S.

$$\text{Area of surface} = \iint_S 1 \, dS$$

On the other hand, if S is a lamina of variable density and $\rho(x, y, z)$ is the density at the point (x, y, z), then the *mass* of the lamina is given by

$$\text{Mass of lamina} = \iint_S \rho(x, y, z) \, dS.$$

EXAMPLE 3   Finding the Mass of a Surface Lamina

A cone-shaped surface lamina S is given by

$$z = 4 - 2\sqrt{x^2 + y^2}, \qquad 0 \le z \le 4$$

as shown in Figure 15.48. At each point on S, the density is proportional to the distance between the point and the z-axis. Find the mass m of the lamina.

Solution   Projecting S onto the xy-plane produces the following.

$$S: z = 4 - 2\sqrt{x^2 + y^2} = g(x, y), \qquad 0 \le z \le 4$$
$$R: x^2 + y^2 \le 4$$

Density: $\rho(x, y, z) = k\sqrt{x^2 + y^2}$

Using a surface integral, you can find the mass to be

$$
\begin{aligned}
m &= \iint_S \rho(x, y, z) \, dS \\
&= \iint_R k\sqrt{x^2 + y^2}\sqrt{1 + [g_x(x, y)]^2 + [g_y(x, y)]^2} \, dA \\
&= k\iint_R \sqrt{x^2 + y^2}\sqrt{1 + \frac{4x^2}{x^2 + y^2} + \frac{4y^2}{x^2 + y^2}} \, dA \\
&= k\iint_R \sqrt{5}\sqrt{x^2 + y^2} \, dA \\
&= k\int_0^{2\pi}\int_0^2 (\sqrt{5}\,r) r \, dr \, d\theta \\
&= \frac{\sqrt{5}\,k}{3}\int_0^{2\pi} r^3 \Big]_0^2 \, d\theta \\
&= \frac{8\sqrt{5}\,k}{3}\int_0^{2\pi} d\theta \\
&= \frac{8\sqrt{5}\,k}{3}\Big[\theta\Big]_0^{2\pi} \\
&= \frac{16\sqrt{5}\,k\pi}{3}.
\end{aligned}
$$

(a)   *Generated by Axiom*

Cone:
$z = 4 - 2\sqrt{x^2 + y^2}$

(b) x $R: \; x^2 + y^2 \le 4$

FIGURE 15.48
Density: $\rho(x, y, z) = k\sqrt{x^2 + y^2}$

Parametric Surfaces and Surface Integrals

REMARK   Notice that ds and dS can be written as
$$ds = \| \mathbf{r}'(t) \| \, dt$$
and
$$dS = \| \mathbf{r}_u(u, v) \times \mathbf{r}_v(u, v) \| \, dA.$$

For a surface S given by the vector-valued function

$$\mathbf{r}(u, v) = x(u, v)\mathbf{i} + y(u, v)\mathbf{j} + z(u, v)\mathbf{k} \qquad \text{Parametric surface}$$

defined over a region D in the uv-plane, you can show that the surface integral of $f(x, y, z)$ over S is given by

$$\iint_S f(x, y, z) \, dS = \iint_D f(x(u, v), y(u, v), z(u, v)) \, \| \mathbf{r}_u(u, v) \times \mathbf{r}_v(u, v) \| \, dA.$$

Note the similarity to a line integral over a space curve C.

$$\int_C f(x, y, z) \, ds = \int_a^b f(x(t), y(t), z(t)) \, \| \mathbf{r}'(t) \| \, dt \qquad \text{Line integral}$$

EXAMPLE 4   Evaluating a Surface Integral

Example 2 demonstrated an evaluation of the surface integral

$$\iint_S (x + z) \, dS$$

where S is the first-octant portion of the cylinder $z^2 + y^2 = 9$ between $x = 0$ and $x = 4$, as shown in Figure 15.49. Reevaluate this integral in parametric form.

Solution   In parametric form, the surface is given by

$$\mathbf{r}(x, \theta) = x\mathbf{i} + 3\cos\theta\,\mathbf{j} + 3\sin\theta\,\mathbf{k}$$

where $0 \le x \le 4$ and $0 \le \theta \le \pi/2$. To evaluate the surface integral in parametric form, begin by calculating the following.

$$\mathbf{r}_x = \mathbf{i}$$
$$\mathbf{r}_\theta = -3\sin\theta\,\mathbf{j} + 3\cos\theta\,\mathbf{k}$$

$$\mathbf{r}_x \times \mathbf{r}_\theta = \begin{vmatrix} \mathbf{i} & \mathbf{j} & \mathbf{k} \\ 1 & 0 & 0 \\ 0 & -3\sin\theta & 3\cos\theta \end{vmatrix} = -3\cos\theta\,\mathbf{j} - 3\sin\theta\,\mathbf{k}$$

$$\| \mathbf{r}_x \times \mathbf{r}_\theta \| = \sqrt{9\cos^2\theta + 9\sin^2\theta} = 3$$

Therefore, the surface integral can be evaluated as follows.

$$\iint_S (x + 3\sin\theta)3 \, dA = \int_0^4 \int_0^{\pi/2} (3x + 9\sin\theta) \, d\theta \, dx$$

$$= \int_0^4 \left[3x\theta - 9\cos\theta \right]_0^{\pi/2} dx$$

$$= \int_0^4 \left(\frac{3\pi}{2}x + 9 \right) dx$$

$$= \left[\frac{3\pi}{4}x^2 + 9x \right]_0^4$$

$$= 12\pi + 36$$

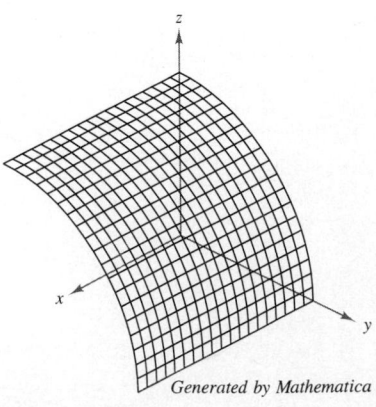

Generated by Mathematica

FIGURE 15.49

Orientation of a Surface

Unit normal vectors are used to induce an orientation to a surface S in space. A surface is called **orientable** if a unit normal vector $\mathbf{N}$ can be defined at every nonboundary point of S in such a way that the normal vectors vary continuously over the surface S. If this is possible, then S is called an **oriented surface.**

An orientable surface S has two distinct sides. Thus, when you orient a surface, you are selecting one of the two possible unit normal vectors. If S is a closed surface such as a sphere, it is customary to choose the unit normal vector $\mathbf{N}$ to be the one that points outward from the sphere.

Most common surfaces, such as spheres, paraboloids, ellipses, and planes, are orientable. (See Exercise 37 for an example of a surface that is *not* orientable.) Moreover, for an orientable surface, the gradient vector provides a convenient way to find a unit normal vector. That is, for an orientable surface S given by

$$z = g(x, y) \qquad \text{Orientable surface}$$

let

$$G(x, y, z) = z - g(x, y).$$

Then, S can be oriented by either of the following unit normal vectors

$$\mathbf{N} = \frac{\nabla G(x, y, z)}{\|\nabla G(x, y, z)\|}$$

$$= \frac{-g_x(x, y)\mathbf{i} - g_y(x, y)\mathbf{j} + \mathbf{k}}{\sqrt{1 + [g_x(x, y)]^2 + [g_y(x, y)]^2}} \qquad \text{Upward unit normal}$$

$$\mathbf{N} = \frac{-\nabla G(x, y, z)}{\|\nabla G(x, y, z)\|}$$

$$= \frac{g_x(x, y)\mathbf{i} + g_y(x, y)\mathbf{j} - \mathbf{k}}{\sqrt{1 + [g_x(x, y)]^2 + [g_y(x, y)]^2}} \qquad \text{Downward unit normal}$$

as shown in Figure 15.50. Similarly, you can use the gradient vectors

$$\nabla G(x, y, z), \qquad G(x, y, z) = y - g(x, z)$$

and

$$\nabla G(x, y, z), \qquad G(x, y, z) = x - g(y, z)$$

to orient surfaces given by $y = g(x, z)$ or $x = g(y, z)$.

If the smooth orientable surface S is given in parametric form by

$$\mathbf{r}(u, v) = x(u, v)\mathbf{i} + y(u, v)\mathbf{j} + z(u, v)\mathbf{k} \qquad \text{Parametric surface}$$

then the unit normal vectors are given by

$$\mathbf{N} = \frac{\mathbf{r}_u \times \mathbf{r}_v}{\|\mathbf{r}_u \times \mathbf{r}_v\|}$$

and

$$\mathbf{N} = \frac{\mathbf{r}_v \times \mathbf{r}_u}{\|\mathbf{r}_v \times \mathbf{r}_u\|}.$$

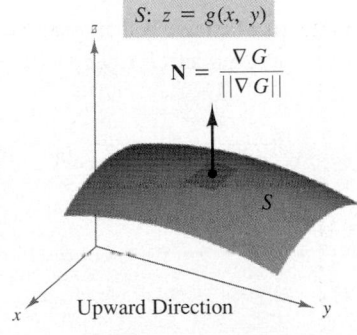

$S: z = g(x, y)$

$\mathbf{N} = \dfrac{\nabla G}{\|\nabla G\|}$

Upward Direction

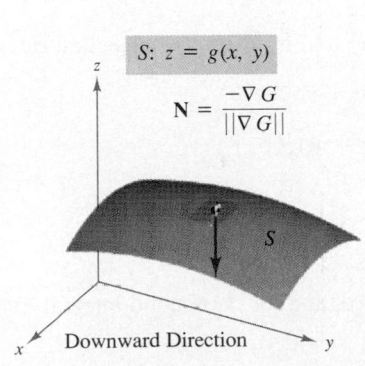

$S: z = g(x, y)$

$\mathbf{N} = \dfrac{-\nabla G}{\|\nabla G\|}$

Downward Direction

FIGURE 15.50

FIGURE 15.51
Velocity field **F** has direction of fluid flow.

Flux Integrals

One of the principal applications involving the vector form of a surface integral relates to the flow of a fluid through a surface S. Suppose an oriented surface S is submerged in a fluid having a continuous velocity field **F**. Let ΔS be the area of a small patch of the surface S over which **F** is nearly constant. Then the amount of fluid crossing this region per unit of time is approximated by the volume of the column of height $\mathbf{F} \cdot \mathbf{N}$, as shown in Figure 15.51. That is,

$$\Delta V = (\text{height})(\text{area of base}) = (\mathbf{F} \cdot \mathbf{N}) \, \Delta S.$$

Consequently, the volume of fluid crossing the surface S per unit of time (called the **flux of F across S**) is given by the surface integral in the following definition.

Definition of Flux Integral

Let $\mathbf{F}(x, y, z) = M \mathbf{i} + N \mathbf{j} + P \mathbf{k}$, where M, N, and P have continuous first partial derivatives on the surface S oriented by a unit normal vector **N**. The **flux integral of F across S** is given by

$$\iint_S \mathbf{F} \cdot \mathbf{N} \, dS.$$

Geometrically, a flux integral is the surface integral over S of the *normal component* of **F**. If $\rho(x, y, z)$ is the density of the fluid at (x, y, z), then the flux integral

$$\iint_S \rho \, \mathbf{F} \cdot \mathbf{N} \, dS$$

represents the *mass* of the fluid flowing across S per unit of time.

To evaluate a flux integral for a surface given by $z = g(x, y)$, let $G(x, y, z) = z - g(x, y)$. Then, $\mathbf{N} \, dS$ can be written as follows.

$$\mathbf{N} \, dS = \frac{\nabla G(x, y, z)}{\|\nabla G(x, y, z)\|} \, dS = \frac{\nabla G(x, y, z)}{\sqrt{[g_x]^2 + [g_y]^2 + 1}} \sqrt{(g_x)^2 + (g_y)^2 + 1} \, dA$$
$$= \nabla G(x, y, z) \, dA$$

THEOREM 15.11 Evaluating a Flux Integral

Let S be an oriented surface given by $z = g(x, y)$ and let R be its projection on the xy-plane.

$$\iint_S \mathbf{F} \cdot \mathbf{N} \, dS = \iint_R \mathbf{F} \cdot (-g_x(x, y)\mathbf{i} - g_y(x, y)\mathbf{j} + \mathbf{k}) \, dA$$

$$\iint_S \mathbf{F} \cdot \mathbf{N} \, dS = \iint_R \mathbf{F} \cdot (g_x(x, y)\mathbf{i} + g_y(x, y)\mathbf{j} - \mathbf{k}) \, dA$$

For the first integral, the surface is oriented upward, and for the second integral, the surface is oriented downward.

EXAMPLE 5   Using a Flux Integral to Find the Rate of Mass Flow

Let S be that portion of the paraboloid

$$z = g(x, y) = 4 - x^2 - y^2$$

lying above the xy-plane, oriented by an upward unit normal vector, as shown in Figure 15.52. A fluid of constant density ρ is flowing through the surface S according to the velocity field

$$\mathbf{F}(x, y, z) = x\mathbf{i} + y\mathbf{j} + z\mathbf{k}.$$

Find the rate of mass flow through S.

Solution   Begin by computing the partial derivatives of g.

$$g_x(x, y) = -2x \quad \text{and} \quad g_y(x, y) = -2y$$

The rate of mass flow through the surface S is

$$\iint_S \rho \mathbf{F} \cdot \mathbf{N} \, dS = \rho \iint_R \mathbf{F} \cdot (-g_x(x, y)\mathbf{i} - g_y(x, y)\mathbf{j} + \mathbf{k}) \, dS$$

$$= \rho \iint_R (x\mathbf{i} + y\mathbf{j} + (4 - x^2 - y^2)\mathbf{k}) \cdot (2x\mathbf{i} + 2y\mathbf{j} + \mathbf{k}) \, dS$$

$$= \rho \iint_R [2x^2 + 2y^2 + (4 - x^2 - y^2)] \, dA$$

$$= \rho \iint_R [4 + x^2 + y^2] \, dA$$

$$= \rho \int_0^{2\pi} \int_0^2 [4 + r^2] r \, dr \, d\theta \qquad \text{Polar coordinates}$$

$$= \rho \int_0^{2\pi} 12 \, d\theta$$

$$= 24\pi\rho.$$

FIGURE 15.52

$$z = 4 - x^2 - y^2$$

For an oriented surface S given by the vector-valued function

$$\mathbf{r}(u, v) = x(u, v)\mathbf{i} + y(u, v)\mathbf{j} + z(u, v)\mathbf{k} \qquad \text{Parametric surface}$$

defined over a region D in the uv-plane, you can define the flux integral of $\mathbf{F}$ across S as

$$\iint_S \mathbf{F} \cdot \mathbf{N} \, dS = \iint_D \mathbf{F} \cdot \left(\frac{\mathbf{r}_u \times \mathbf{r}_v}{\| \mathbf{r}_u \times \mathbf{r}_v \|} \right) \| \mathbf{r}_u \times \mathbf{r}_v \| \, dA$$

$$= \iint_D \mathbf{F} \cdot (\mathbf{r}_u \times \mathbf{r}_v) \, dA.$$

Note the similarity of this integral to the line integral

$$\int_C \mathbf{F} \cdot d\mathbf{r} = \int_C \mathbf{F} \cdot \mathbf{T} \, ds.$$

A summary of formulas for line integrals and surface integrals is listed at the end of this section.

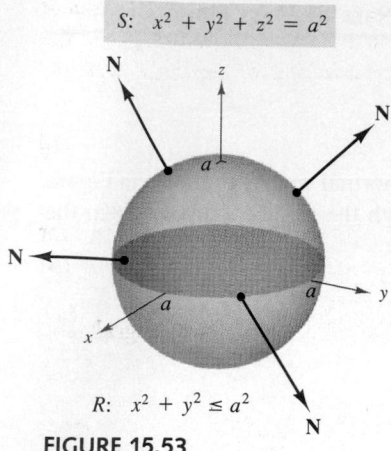

$S:\ x^2 + y^2 + z^2 = a^2$

$R:\ x^2 + y^2 \le a^2$

FIGURE 15.53

EXAMPLE 6   Finding the Flux of an Inverse Square Field

Find the flux over the sphere S given by $x^2 + y^2 + z^2 = a^2$, where $\mathbf{F}$ is an inverse square field given by

$$\mathbf{F}(x, y, z) = \frac{q}{\|\mathbf{r}\|^2}\frac{\mathbf{r}}{\|\mathbf{r}\|} = \frac{q\,\mathbf{r}}{\|\mathbf{r}\|^3}$$

where $\mathbf{r} = x\mathbf{i} + y\mathbf{j} + z\mathbf{k}$. Assume S is oriented outward, as shown in Figure 15.53.

Solution   The sphere is given by

$$\mathbf{r}(u, v) = x(u, v)\mathbf{i} + y(u, v)\mathbf{j} + z(u, v)\mathbf{k}$$
$$= a\sin u\cos v\,\mathbf{i} + a\sin u\sin v\,\mathbf{j} + a\cos u\,\mathbf{k}$$

where $0 \le u \le \pi$ and $0 \le v \le 2\pi$. The partial derivatives of $\mathbf{r}$ are

$$\mathbf{r}_u(u, v) = a\cos u\cos v\,\mathbf{i} + a\cos u\sin v\,\mathbf{j} - a\sin u\,\mathbf{k}$$

and

$$\mathbf{r}_v(u, v) = -a\sin u\sin v\,\mathbf{i} + a\sin u\cos v\,\mathbf{j}$$

which implies that the normal vector $\mathbf{r}_u \times \mathbf{r}_v$ is

$$\mathbf{r}_u \times \mathbf{r}_v = \begin{vmatrix} \mathbf{i} & \mathbf{j} & \mathbf{k} \\ a\cos u\cos v & a\cos u\sin v & -a\sin u \\ -a\sin u\sin v & a\sin u\cos v & 0 \end{vmatrix}$$
$$= a^2(\sin^2 u\cos v\,\mathbf{i} + \sin^2 u\sin v\,\mathbf{j} + \sin u\cos u\,\mathbf{k}).$$

Now, using

$$\mathbf{F}(x, y, z) = \frac{q\,\mathbf{r}}{\|\mathbf{r}\|^3}$$
$$= q\frac{x\mathbf{i} + y\mathbf{j} + z\mathbf{k}}{\|x\mathbf{i} + y\mathbf{j} + z\mathbf{k}\|^3}$$
$$= \frac{q}{a^3}(a\sin u\cos v\,\mathbf{i} + a\sin u\sin v\,\mathbf{j} + a\cos u\,\mathbf{k})$$

it follows that

$$\mathbf{F} \cdot (\mathbf{r}_u \times \mathbf{r}_v) = \frac{q}{a^3}[(a\sin u\cos v\,\mathbf{i} + a\sin u\sin v\,\mathbf{j} + a\cos u\,\mathbf{k}) \cdot$$
$$a^2(\sin^2 u\cos v\,\mathbf{i} + \sin^2 u\sin v\,\mathbf{j} + \sin u\cos u\,\mathbf{k})$$
$$= q(\sin^3\cos^2 v + \sin^3 u\sin^2 v + \sin u\cos^2 u)$$
$$= q\sin u.$$

Finally, the flux over the sphere S is given by

$$\iint_S \mathbf{F} \cdot \mathbf{N}\, dS = \iint_D (q\sin u)\, dA$$
$$= \int_0^{2\pi}\int_0^{\pi} q\sin u\, du\, dv$$
$$= 4\pi q.$$

The result in Example 6 shows that the flux across a sphere S in an inverse square field is independent of the radius of S. In particular, if $\mathbf{E}$ is an electric field, then the result in Example 6, along with Coulomb's Law, yields one of the basic laws of electrostatics, known as **Gauss's Law:**

$$\iint_S \mathbf{E} \cdot \mathbf{N} \, dS = 4\pi q \qquad \text{Gauss's Law}$$

where q is a point charge located at the center of the sphere. Gauss's Law is valid for more general closed surfaces that enclose the origin, and relates the flux out of the surface to the total charge q inside the surface.

We conclude this section with a summary of different forms of line integrals and surface integrals.

Summary of Line and Surface Integrals

Line Integrals

$$ds = \| \mathbf{r}'(t) \| \, dt = \sqrt{[x'(t)]^2 + [y'(t)]^2 + [z'(t)]^2} \, dt$$

$$\int_C f(x, y, z) \, ds = \int_a^b f(x(t), y(t), z(t)) \, ds \qquad \text{Scalar form}$$

$$\int_C \mathbf{F} \cdot d\mathbf{r} = \int_C \mathbf{F} \cdot \mathbf{T} \, ds = \int_a^b \mathbf{F}(x(t), y(t), z(t)) \cdot \mathbf{r}'(t) \, dt \qquad \text{Vector form}$$

Surface Integrals $(z = g(x, y))$

$$dS = \sqrt{1 + [g_x(x, y)]^2 + [g_y(x, y)]^2} \, dA$$

$$\iint_S f(x, y, z) \, dS = \iint_R f(x, y, g(x, y)) \sqrt{1 + [g_x(x, y)]^2 + [g_y(x, y)]^2} \, dA \qquad \text{Scalar form}$$

$$\iint_S \mathbf{F} \cdot \mathbf{N} \, dS = \iint_R \mathbf{F} \cdot (-g_x(x, y)\mathbf{i} - g_y(x, y)\mathbf{j} + \mathbf{k}) \, dA \qquad \text{Vector form (upward normal)}$$

Surface Integrals   *(parametric form)*

$$dS = \| \mathbf{r}_u(u, v) \times \mathbf{r}_v(u, v) \| \, dA$$

$$\iint_S f(x, y, z) \, dS = \iint_D f(x(u, v), y(u, v), z(u, v)) \, dS \qquad \text{Scalar form}$$

$$\iint_S \mathbf{F} \cdot \mathbf{N} \, dS = \iint_D \mathbf{F} \cdot (\mathbf{r}_u \times \mathbf{r}_v) \, dA \qquad \text{Vector form}$$

EXERCISES for Section 15.6

TECHNOLOGY
Laboratory Guide
Lab 15.6

In Exercises 1–4, evaluate

$$\iint_S (x - 2y + z) \, dS.$$

1. $S\colon z = 4 - x,\, 0 \le x \le 4,\, 0 \le y \le 4$

2. $S\colon z = 10 - 2x + 2y,\, 0 \le x \le 2,\, 0 \le y \le 4$

3. $S\colon z = 10,\, x^2 + y^2 \le 1$

4. $S\colon z = \frac{2}{3}x^{3/2},\, 0 \le x \le 1,\, 0 \le y \le x$

In Exercises 5 and 6, evaluate

$$\iint_S xy \, dS.$$

5. $S\colon z = 6 - x - 2y$ (first octant)

6. $S\colon z = h,\, 0 \le x \le 2,\, 0 \le y \le \sqrt{4 - x^2}$

C In Exercises 7 and 8, use a symbolic integration utility to evaluate

$$\iint_S xy \, dS.$$

7. $S: z = 9 - x^2, 0 \le x \le 2, 0 \le y \le x$

8. $S: z = xy, 0 \le x \le 2, 0 \le y \le 2$

C In Exercises 9 and 10, use a symbolic integration utility to evaluate

$$\iint_S (x^2 - 2xy) \, dS$$

over the specified surface S.

9. $S: z = 10 - x^2 - y^2, 0 \le x \le 2, 0 \le y \le 2$

10. $S: z = \cos x, 0 \le x \le \dfrac{\pi}{2}, 0 \le y \le \dfrac{1}{2}x$

In Exercises 11–14, evaluate

$$\iint_S f(x, y) \, dS$$

over the specified surface S.

11. $f(x, y) = y + 5$

$S: \mathbf{r}(u, v) = u\mathbf{i} + v\mathbf{j} + \dfrac{v}{2}\mathbf{k}$

$0 \le u \le 1, 0 \le v \le 2$

12. $f(x, y) = x + y$

$S: \mathbf{r}(u, v) = 2\cos u\,\mathbf{i} + 2\sin u\,\mathbf{j} + v\mathbf{k}$

$0 \le u \le \dfrac{\pi}{2}, 0 \le v \le 2$

13. $f(x, y) = xy$

$S: \mathbf{r}(u, v) = 2\cos u\,\mathbf{i} + 2\sin u\,\mathbf{j} + v\mathbf{k}$

$0 \le u \le \dfrac{\pi}{2}, 0 \le v \le 2$

14. $f(x, y) = x + y$

$S: \mathbf{r}(u, v) = 2u\cos v\,\mathbf{i} + 2u\sin v\,\mathbf{j} + u\mathbf{k}$

$0 \le u \le 4, 0 \le v \le \pi$

In Exercises 15–20, evaluate

$$\iint_S f(x, y, z) \, dS.$$

15. $f(x, y, z) = x^2 + y^2 + z^2$

$S: z = x + 2, x^2 + y^2 \le 1$

16. $f(x, y, z) = \dfrac{xy}{z}$

$S: z = x^2 + y^2, 4 \le x^2 + y^2 \le 16$

17. $f(x, y, z) = \sqrt{x^2 + y^2 + z^2}$

$S: z = \sqrt{x^2 + y^2}, x^2 + y^2 \le 4$

18. $f(x, y, z) = \sqrt{x^2 + y^2 + z^2}$

$S: z = \sqrt{x^2 + y^2}, (x - 1)^2 + y^2 \le 1$

19. $f(x, y, z) = x^2 + y^2 + z^2$

$S: x^2 + y^2 = 9, 0 \le x \le 3, 0 \le y \le 3, 0 \le z \le 9$

20. $f(x, y, z) = x^2 + y^2 + z^2$

$S: x^2 + y^2 = 9, 0 \le x \le 3, 0 \le z \le x$

In Exercises 21–26, find the flux of **F** through S,

$$\iint_S \mathbf{F} \cdot \mathbf{N} \, dS$$

where **N** is the upper unit normal vector to S.

21. $\mathbf{F}(x, y, z) = 3z\mathbf{i} - 4\mathbf{j} + y\mathbf{k}$

$S: x + y + z = 1$ (first octant)

22. $\mathbf{F}(x, y, z) = x\mathbf{i} + y\mathbf{j}$

$S: 2x + 3y + z = 6$ (first octant)

23. $\mathbf{F}(x, y, z) = x\mathbf{i} + y\mathbf{j} + z\mathbf{k}$

$S: z = 9 - x^2 - y^2, 0 \le z$

24. $\mathbf{F}(x, y, z) = x\mathbf{i} + y\mathbf{j} + z\mathbf{k}$

$S: x^2 + y^2 + z^2 = 16$ (first octant)

25. $\mathbf{F}(x, y, z) = 4\mathbf{i} - 3\mathbf{j} + 5\mathbf{k}$

$S: z = x^2 + y^2, x^2 + y^2 \le 4$

26. $\mathbf{F}(x, y, z) = x\mathbf{i} + y\mathbf{j} - 2z\mathbf{k}$

$S: z = \sqrt{a^2 - x^2 - y^2}$

In Exercises 27 and 28, find the flux of **F** over the closed surface. (Let **N** be the outward unit normal vector of each surface.)

27. $\mathbf{F}(x, y, z) = 4xy\mathbf{i} + z^2\mathbf{j} + yz\mathbf{k}$

$S:$ unit cube bounded by $x = 0, x = 1, y = 0, y = 1,$ $z = 0, z = 1$

28. $\mathbf{F}(x, y, z) = (x + y)\mathbf{i} + y\mathbf{j} + z\mathbf{k}$

$S: z = 1 - x^2 - y^2, z \ge 0$

Mass In Exercises 29 and 30, find the mass of the surface lamina S of density ρ.

29. $S: 2x + 3y + 6z = 12$ (first octant), $\rho(x, y, z) = x^2 + y^2$

30. $S: z = \sqrt{a^2 - x^2 - y^2}, \rho(x, y, z) = kz$

Moment of Inertia In Exercises 31 and 32, use the following formulas for the moments of inertia about the coordinate axes of a surface lamina of density ρ.

$$I_x = \iint_S (y^2 + z^2)\rho(x, y, z) \, dS$$

$$I_y = \iint_S (x^2 + z^2)\rho(x, y, z) \, dS$$

$$I_z = \iint_S (x^2 + y^2)\rho(x, y, z) \, dS$$

31. Show that the moment of inertia of a conical shell about its axis is $\frac{1}{2}ma^2$, where m is the mass and a is the radius.

32. Show that the moment of inertia of a spherical shell of uniform density about its diameter is $\frac{2}{3}ma^2$, where m is the mass and a is the radius.

Moment of Inertia In Exercises 33 and 34, find I_z for the given lamina with uniform density of 1.

33. $x^2 + y^2 = a^2, 0 \le z \le h$ **34.** $z = x^2 + y^2, 0 \le z \le h$

Flow Rate **In Exercises 35 and 36, find the rate of mass flow of a fluid of density ρ through the surface S oriented upwards if the velocity field is given by $\mathbf{F}(x, y, z) = 0.5z\,\mathbf{k}$.**

35. $S: z = 16 - x^2 - y^2, z \geq 0$

36. $S: z = \sqrt{16 - x^2 - y^2}$

37. The surface shown in the figure is called a Möbius strip. Explain why this surface is not orientable. (To create a Möbius strip, cut a strip of paper, make a single twist, and paste the ends together.)

38. Is the surface shown in the figure orientable?

Single twist

Double twist

FIGURE FOR 37   **FIGURE FOR 38**

39. *Electrical Charge* Let $\mathbf{E} = yz\,\mathbf{i} + xz\,\mathbf{j} + xy\,\mathbf{k}$ be an electrostatic field. Use Gauss's Law to find the total charge enclosed by the closed surface consisting of the hemisphere $z = \sqrt{1 - x^2 - y^2}$ and its circular base in the xy-plane.

CARL FRIEDRICH GAUSS

The Divergence Theorem is also called Gauss's Theorem, after the famous German mathematician Carl Friedrich Gauss (1777–1855). Gauss is recognized, with Newton and Archimedes, as one of the three greatest mathematicians in history. One of his many contributions to mathematics was made at the age of 22, when, as part of his doctoral dissertation, he proved the *Fundamental Theorem of Algebra*.

Divergence Theorem

Recall from Section 15.4 that an alternative form of Green's Theorem is

$$\int_C \mathbf{F} \cdot \mathbf{N}\,ds = \int\int_R \left(\frac{\partial M}{\partial x} + \frac{\partial N}{\partial y}\right) dA$$

$$= \int\int_R \operatorname{div}\mathbf{F}\,dA.$$

In an analogous way, the **Divergence Theorem** gives the relationship between a triple integral over a solid region Q and a surface integral over the surface of Q. In the statement of the theorem, the surface S is **closed** in the sense that it forms the complete boundary of the solid Q. Regions bounded by spheres, ellipsoids, cubes, tetrahedrons, or some combination of these surfaces are typical examples of closed surfaces. Assume that Q is a solid region on which a triple integral can be evaluated, and that the closed surface S is oriented by *outward* unit normal vectors, as shown in Figure 15.54. With these restrictions on S and Q, we state the following theorem.

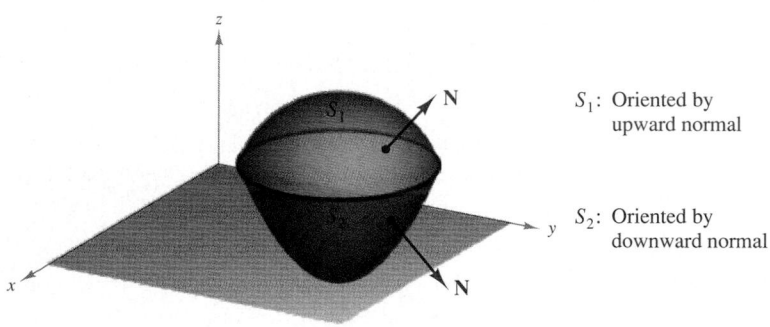

S_1: Oriented by upward normal

S_2: Oriented by downward normal

FIGURE 15.54

THEOREM 15.12   The Divergence Theorem

Let Q be a solid region bounded by a closed surface S oriented by a unit normal vector directed outward from Q. If $\mathbf{F}$ is a vector field whose component functions have continuous partial derivatives in Q, then

$$\int\int_S \mathbf{F} \cdot \mathbf{N}\,dS = \int\int\int_Q \operatorname{div}\mathbf{F}\,dV.$$

Proof By letting $\mathbf{F}(x, y, z) = M\mathbf{i} + N\mathbf{j} + P\mathbf{k}$, the theorem takes the form

$$\iint_S \mathbf{F} \cdot \mathbf{N}\, dS = \iint_S (M\mathbf{i} \cdot \mathbf{N} + N\mathbf{j} \cdot \mathbf{N} + P\mathbf{k} \cdot \mathbf{N})\, dS$$

$$= \iiint_Q \left(\frac{\partial M}{\partial x} + \frac{\partial N}{\partial y} + \frac{\partial P}{\partial z} \right) dV.$$

You can prove this by verifying the following three equations.

$$\iint_S M\mathbf{i} \cdot \mathbf{N}\, dS = \iiint_Q \frac{\partial M}{\partial x}\, dV$$

$$\iint_S N\mathbf{j} \cdot \mathbf{N}\, dS = \iiint_Q \frac{\partial N}{\partial y}\, dV$$

$$\iint_S P\mathbf{k} \cdot \mathbf{N}\, dS = \iiint_Q \frac{\partial P}{\partial z}\, dV$$

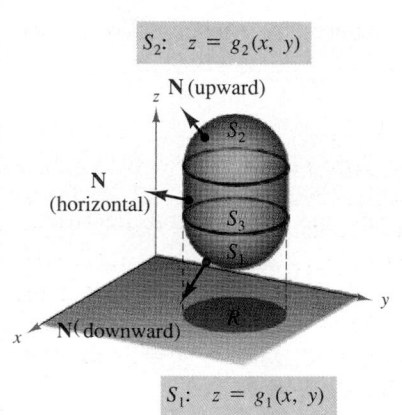

$S_2: \quad z = g_2(x, y)$

$\mathbf{N}$ (upward)

$\mathbf{N}$ (horizontal)

$\mathbf{N}$ (downward)

$S_1: \quad z = g_1(x, y)$

FIGURE 15.55

Because the verifications of the three equations are similar, we will discuss only the third. We restrict the proof to a **simple solid** region with upper surface $z = g_2(x, y)$ and lower surface $z = g_1(x, y)$, whose projections on the xy-plane coincide and form region R. If Q has a lateral surface such as S_3 in Figure 15.55, then a normal vector is horizontal, which implies that $P\mathbf{k} \cdot \mathbf{N} = 0$. Consequently, you have

$$\iint_S P\mathbf{k} \cdot \mathbf{N}\, dS = \iint_{S_1} P\mathbf{k} \cdot \mathbf{N}\, dS + \iint_{S_2} P\mathbf{k} \cdot \mathbf{N}\, dS + 0.$$

On the upper surface S_2, the outward normal vector is upward, whereas on the lower surface S_1, the outward normal vector is downward. Therefore, by Theorem 15.11, you have the following.

$$\iint_{S_1} P\mathbf{k} \cdot \mathbf{N}\, dS = \iint_R P(x, y, g_1(x, y))\mathbf{k} \cdot \left(\frac{\partial g_1}{\partial x}\mathbf{i} + \frac{\partial g_1}{\partial y}\mathbf{j} - \mathbf{k} \right) dA$$

$$= -\iint_R P(x, y, g_1(x, y))\, dA$$

$$\iint_{S_2} P\mathbf{k} \cdot \mathbf{N}\, dS = \iint_R P(x, y, g_2(x, y))\mathbf{k} \cdot \left(-\frac{\partial g_2}{\partial x}\mathbf{i} - \frac{\partial g_2}{\partial y}\mathbf{j} + \mathbf{k} \right) dA$$

$$= \iint_R P(x, y, g_2(x, y))\, dA$$

Adding these results, you obtain

$$\iint_S P\mathbf{k} \cdot \mathbf{N}\, dS = \iint_R [P(x, y, g_2(x, y)) - P(x, y, g_1(x, y))]\, dA$$

$$= \iint_R \left[\int_{g_1(x,y)}^{g_2(x,y)} \frac{\partial P}{\partial z}\, dz \right] dA$$

$$= \iiint_Q \frac{\partial P}{\partial z}\, dV.$$

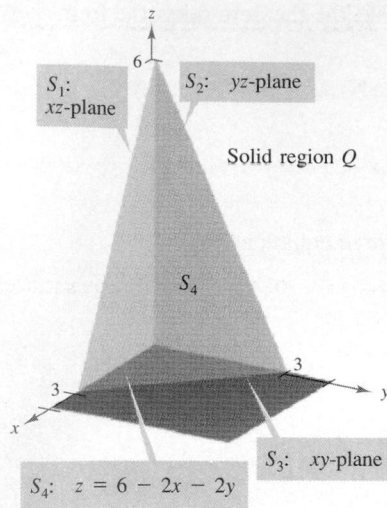

S_1: xz-plane

S_2: yz-plane

Solid region Q

S_4

3

3

S_3: xy-plane

S_4: $z = 6 - 2x - 2y$

FIGURE 15.56

EXAMPLE 1   Using the Divergence Theorem

Let Q be the solid region bounded by the coordinate planes and the plane $2x + 2y + z = 6$, and let $\mathbf{F} = x\mathbf{i} + y^2\mathbf{j} + z\mathbf{k}$. Find

$$\iint_S \mathbf{F} \cdot \mathbf{N}\, dS$$

where S is the surface of Q.

Solution   In Figure 15.56, you can see that Q is bounded by four subsurfaces. Thus, you would need four *surface integrals* to evaluate

$$\iint_S \mathbf{F} \cdot \mathbf{N}\, dS.$$

However, by the Divergence Theorem, you need only one triple integral. Because

$$\operatorname{div} \mathbf{F} = \frac{\partial M}{\partial x} + \frac{\partial N}{\partial y} + \frac{\partial P}{\partial z} = 1 + 2y + 1 = 2 + 2y$$

you have

$$\iint_S \mathbf{F} \cdot \mathbf{N}\, dS = \iiint_Q \operatorname{div} \mathbf{F}\, dV$$

$$= \int_0^3 \int_0^{3-y} \int_0^{6-2x-2y} (2 + 2y)\, dz\, dx\, dy$$

$$= \int_0^3 \int_0^{3-y} (2z + 2yz) \Big]_0^{6-2x-2y} dx\, dy$$

$$= \int_0^3 \int_0^{3-y} (12 - 4x + 8y - 4xy - 4y^2)\, dx\, dy$$

$$= \int_0^3 \left[12x - 2x^2 + 8xy - 2x^2y - 4xy^2 \right]_0^{3-y} dy$$

$$= \int_0^3 (18 + 6y - 10y^2 + 2y^3)\, dy$$

$$= \left[18y + 3y^2 - \frac{10y^3}{3} + \frac{y^4}{2} \right]_0^3$$

$$= \frac{63}{2}.$$

TECHNOLOGY   In you have access to a symbolic integration utility that can evaluate triple-iterated integrals, try using it to verify the result of Example 1. When you are using such a utility, note that the first step is to convert the triple integral to an iterated integral—this step must be done by hand. To give yourself some practice with this important step, try finding the limits of integration for the following iterated integrals. Then use a computer to verify that the value is the same as that obtained in Example 1.

$$\int_?^? \int_?^? \int_?^? (2 + 2y)\, dy\, dz\, dx, \qquad \int_?^? \int_?^? \int_?^? (2 + 2y)\, dx\, dy\, dz$$

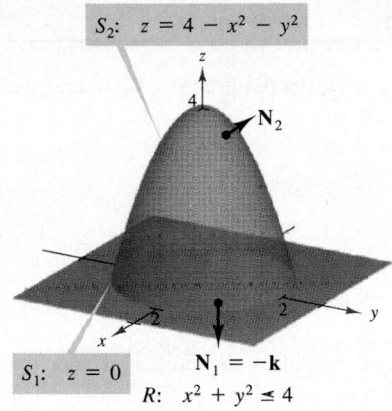

$S_2: \ z = 4 - x^2 - y^2$

$\mathbf{N}_2$

$S_1: \ z = 0$

$\mathbf{N}_1 = -\mathbf{k}$

$R: \ x^2 + y^2 \le 4$

FIGURE 15.57

EXAMPLE 2   Verifying the Divergence Theorem

Let Q be the solid region between the paraboloid

$$z = 4 - x^2 - y^2$$

and the xy-plane. Verify the Divergence Theorem for

$$\mathbf{F}(x, y, z) = 2z\mathbf{i} + x\mathbf{j} + y^2\mathbf{k}.$$

Solution   In Figure 15.57, you can see that the outward normal vector for the surface S_1 is $\mathbf{N}_1 = -\mathbf{k}$, whereas the outward normal vector for the surface S_2 is

$$\mathbf{N}_2 = \frac{2x\mathbf{i} + 2y\mathbf{j} + \mathbf{k}}{\sqrt{4x^2 + 4y^2 + 1}}.$$

Thus, by Theorem 15.11, you have

$$\iint_S \mathbf{F} \cdot \mathbf{N}\, dS$$

$$= \iint_{S_1} \mathbf{F} \cdot \mathbf{N}_1\, dS + \iint_{S_2} \mathbf{F} \cdot \mathbf{N}_2\, dS$$

$$= \iint_{S_1} \mathbf{F} \cdot (-\mathbf{k})\, dS + \iint_{S_2} \mathbf{F} \cdot (2x\mathbf{i} + 2y\mathbf{j} + \mathbf{k})\, dA$$

$$= \iint_R -y^2\, dA + \iint_R (4xz + 2xy + y^2)\, dA$$

$$= -\int_{-2}^{2} \int_{-\sqrt{4-y^2}}^{\sqrt{4-y^2}} y^2\, dx\, dy + \int_{-2}^{2} \int_{-\sqrt{4-y^2}}^{\sqrt{4-y^2}} (4xz + 2xy + y^2)\, dx\, dy$$

$$= \int_{-2}^{2} \int_{-\sqrt{4-y^2}}^{\sqrt{4-y^2}} (4xz + 2xy)\, dx\, dy$$

$$= \int_{-2}^{2} \int_{-\sqrt{4-y^2}}^{\sqrt{4-y^2}} [4x(4 - x^2 - y^2) + 2xy]\, dx\, dy$$

$$= \int_{-2}^{2} \int_{-\sqrt{4-y^2}}^{\sqrt{4-y^2}} (16x - 4x^3 - 4xy^2 + 2xy)\, dx\, dy$$

$$= \int_{-2}^{2} \left[8x^2 - x^4 - 2x^2y^2 + x^2y \right]_{-\sqrt{4-y^2}}^{\sqrt{4-y^2}} dy$$

$$= \int_{-2}^{2} 0\, dy$$

$$= 0.$$

On the other hand, because

$$\text{div } \mathbf{F} = \frac{\partial}{\partial x}[2z] + \frac{\partial}{\partial y}[x] + \frac{\partial}{\partial z}[y^2] = 0 + 0 + 0 = 0$$

you can apply the Divergence Theorem to obtain the equivalent result

$$\iint_S \mathbf{F} \cdot \mathbf{N}\, dS = \iiint_Q \text{div } \mathbf{F}\, dV = \iiint_Q 0\, dV = 0.$$

EXAMPLE 3   Using the Divergence Theorem

Let Q be the solid bounded by the cylinder $x^2 + y^2 = 4$, the plane $x + z = 6$, and the xy-plane, as shown in Figure 15.58. Find

$$\iint_S \mathbf{F} \cdot \mathbf{N} \, dS$$

where S is the surface of Q and

$$\mathbf{F}(x, y, z) = (x^2 + \sin z)\mathbf{i} + (xy + \cos z)\mathbf{j} + e^y \mathbf{k}.$$

Solution   Direct evaluation of this surface integral would be difficult. However, by the Divergence Theorem, you can evaluate the integral as follows.

$$
\begin{aligned}
\iint_S \mathbf{F} \cdot \mathbf{N} \, dS &= \iiint_Q \operatorname{div} \mathbf{F} \, dV \\
&= \iiint_Q (2x + x + 0) \, dV \\
&= \iiint_Q 3x \, dV \\
&= \int_0^{2\pi} \int_0^2 \int_0^{6-r\cos\theta} (3r\cos\theta) r \, dz \, dr \, d\theta \\
&= \int_0^{2\pi} \int_0^2 (18r^2 \cos\theta - 3r^3 \cos^2\theta) \, dr \, d\theta \\
&= \int_0^{2\pi} (48\cos\theta - 12\cos^2\theta) \, d\theta \\
&= \left[48\sin\theta - 6\left(\theta + \frac{1}{2}\sin 2\theta\right) \right]_0^{2\pi} \\
&= -12\pi
\end{aligned}
$$

Notice that cylindrical coordinates with $x = r\cos\theta$ and $dV = r \, dz \, dr \, d\theta$ were used to evaluate the triple integral.

Plane:
$x + z = 6$

Cylinder:
$x^2 + y^2 = 4$

FIGURE 15.58

Even though we stated the Divergence Theorem for a simple solid region Q bounded by a closed surface, the theorem is also valid for regions that are the finite unions of simple solid regions. For example, let Q be the solid bounded by the closed surfaces S_1 and S_2, as shown in Figure 15.59. To apply the Divergence Theorem to this solid, let $S = S_1 \cup S_2$. The normal vector $\mathbf{N}$ to S is given by $-\mathbf{N}_1$ on S_1 and by $\mathbf{N}_2$ on S_2. Thus, you can write

$$
\begin{aligned}
\iiint_Q \operatorname{div} \mathbf{F} \, dV &= \iint_S \mathbf{F} \cdot \mathbf{N} \, dS \\
&= \iint_S \mathbf{F} \cdot (-\mathbf{N}_1) \, dS + \iint_S \mathbf{F} \cdot \mathbf{N}_2 \, dS \\
&= -\iint_S \mathbf{F} \cdot \mathbf{N}_1 \, dS + \iint_S \mathbf{F} \cdot \mathbf{N}_2 \, dS.
\end{aligned}
$$

FIGURE 15.59

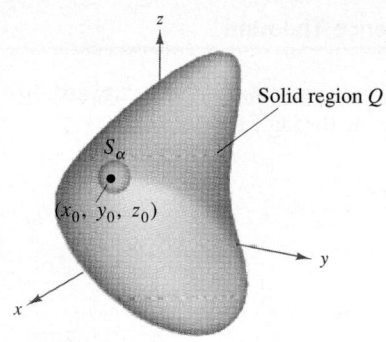

FIGURE 15.60

Flux and the Divergence Theorem

To help understand the Divergence Theorem, consider the two sides of the equation

$$\iint_S \mathbf{F} \cdot \mathbf{N}\, dS = \iiint_Q \operatorname{div} \mathbf{F}\, dV.$$

You know from Section 15.6 that the flux integral on the left determines the total fluid flow across the surface S per unit of time. This can be approximated by summing the fluid flow across small patches of the surface. The triple integral on the right measures this same fluid flow across S, but from a very different perspective—namely, by calculating the flow of fluid into (or out of) small *cubes* of volume ΔV_i. The flux of the ith cube is approximately

$$\text{Flux of } i\text{th cube} \approx \operatorname{div} \mathbf{F}(x_i, y_i, z_i)\, \Delta V_i$$

for some point (x_i, y_i, z_i) in the ith cube. Note that for a cube in the interior of Q, the gain (or loss) of fluid through any one of its six sides is offset by a corresponding loss (or gain) through one of the sides of an adjacent cube. After summing over all the cubes in Q, the only fluid flow that is not canceled by adjoining cubes is that on the outside edges of the cubes on the boundary. Thus, the sum

$$\sum_{i=1}^{n} \operatorname{div} \mathbf{F}(x_i, y_i, z_i)\, \Delta V_i$$

approximates the total flux into (or out of) Q, and therefore through the surface S.

To see what is meant by the divergence of $\mathbf{F}$ at a point, we consider ΔV_α to be the volume of a small sphere S_α of radius α and center (x_0, y_0, z_0), contained in region Q, as shown in Figure 15.60. Applying the Divergence Theorem to S_α produces

$$\text{Flux of } \mathbf{F} \text{ across } S_\alpha = \iiint_{Q_\alpha} \operatorname{div} \mathbf{F}\, dV \approx \operatorname{div} \mathbf{F}(x_0, y_0, z_0)\, \Delta V_\alpha$$

where Q_α is the interior of S_α. Consequently, you have

$$\operatorname{div} \mathbf{F}(x_0, y_0, z_0) \approx \frac{\text{flux of } \mathbf{F} \text{ across } S_\alpha}{\Delta V_\alpha}$$

and, by taking the limit as $\alpha \to 0$, you obtain the divergence of $\mathbf{F}$ at the point (x_0, y_0, z_0).

$$\operatorname{div} \mathbf{F}(x_0, y_0, z_0) = \lim_{\alpha \to 0} \frac{\text{flux of } \mathbf{F} \text{ across } S_\alpha}{\Delta V_\alpha}$$
$$= \text{flux per unit volume at } (x_0, y_0, z_0).$$

(a) Source: div $\mathbf{F} > 0$

(b) Sink: div $\mathbf{F} < 0$

(c) Incompressible: div $\mathbf{F} = 0$

FIGURE 15.61

The point (x_0, y_0, z_0) in a vector field is classified as a source, a sink, or incompressible, as follows.

1. **Source,** if div $\mathbf{F} > 0$ (see Figure 15.61a).
2. **Sink,** if div $\mathbf{F} < 0$ (see Figure 15.61b).
3. **Incompressible,** if div $\mathbf{F} = 0$ (see Figure 15.61c).

EXAMPLE 4　Calculating Flux by the Divergence Theorem

Let Q be the region bounded by the sphere $x^2 + y^2 + z^2 = 4$. Find the outward flux of the vector field $\mathbf{F}(x, y, z) = 2x^3\mathbf{i} + 2y^3\mathbf{j} + 2z^3\mathbf{k}$ through the sphere.

Solution　By the Divergence Theorem, you have

$$
\begin{aligned}
\text{Flux across } S &= \iint_S \mathbf{F} \cdot \mathbf{N}\,dS \\
&= \iiint_Q \text{div } \mathbf{F}\,dV \\
&= \iiint_Q 6(x^2 + y^2 + z^2)\,dV \\
&= 6\int_0^2\int_0^\pi\int_0^{2\pi} \rho^4 \sin\phi\,d\theta\,d\phi\,d\rho \qquad \text{Spherical coordinates}\\
&= 6\int_0^2\int_0^\pi 2\pi\rho^4 \sin\phi\,d\phi\,d\rho \\
&= 12\pi\int_0^2 2\rho^4\,d\rho \\
&= 24\pi\left(\frac{32}{5}\right) \\
&= \frac{768\pi}{5}.
\end{aligned}
$$

EXERCISES for Section 15.7

In Exercises 1–4, verify the Divergence Theorem by evaluating

$$\iint_S \mathbf{F} \cdot \mathbf{N}\,dS$$

as a surface integral and as a triple integral.

1. $\mathbf{F}(x, y, z) = 2x\mathbf{i} - 2y\mathbf{j} + z^2\mathbf{k}$
   S: cube bounded by the planes $x = 0$, $x = a$, $y = 0$, $y = a$, $z = 0$, and $z = a$

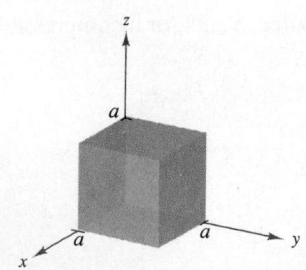

2. $\mathbf{F}(x, y, z) = 2x\mathbf{i} - 2y\mathbf{j} + z^2\mathbf{k}$
   S: cylinder $x^2 + y^2 = 1$, $0 \le z \le h$

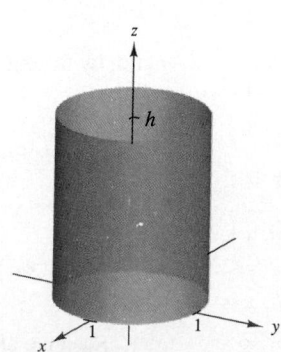

3. $\mathbf{F}(x, y, z) = (2x - y)\mathbf{i} - (2y - z)\mathbf{j} + z\mathbf{k}$
S: surface bounded by the plane $2x + 4y + 2z = 12$ and the coordinate planes

4. $\mathbf{F}(x, y, z) = xy\mathbf{i} + z\mathbf{j} + (x + y)\mathbf{k}$
S: surface bounded by the planes $y = 4$ and $z = 4 - x$, and the coordinate planes

FIGURE FOR 3   **FIGURE FOR 4**

In Exercises 5–14, use the Divergence Theorem to evaluate

$$\iint_S \mathbf{F} \cdot \mathbf{N} \, dS$$

and find the outward flux of $\mathbf{F}$ through the surface of the solid bounded by the graphs of the equations.

5. $\mathbf{F}(x, y, z) = x^2\mathbf{i} + y^2\mathbf{j} + z^2\mathbf{k}$
S: $x = 0$, $x = a$, $y = 0$, $y = a$, $z = 0$, $z = a$

6. $\mathbf{F}(x, y, z) = x^2z\mathbf{i} - y\mathbf{j} + xyz\mathbf{k}$
S: $x = 0$, $x = a$, $y = 0$, $y = a$, $z = 0$, $z = a$

7. $\mathbf{F}(x, y, z) = x^2\mathbf{i} - 2xy\mathbf{j} + xyz^2\mathbf{k}$
S: $z = \sqrt{a^2 - x^2 - y^2}$, $z = 0$

8. $\mathbf{F}(x, y, z) = xy\mathbf{i} + yz\mathbf{j} - yz\mathbf{k}$
S: $z = \sqrt{a^2 - x^2 - y^2}$, $z = 0$

9. $\mathbf{F}(x, y, z) = x\mathbf{i} + y\mathbf{j} + z\mathbf{k}$
S: $x^2 + y^2 + z^2 = 4$

10. $\mathbf{F}(x, y, z) = xyz\mathbf{j}$
S: $x^2 + y^2 = 9$, $z = 0$, $z = 4$

11. $\mathbf{F}(x, y, z) = x\mathbf{i} + y^2\mathbf{j} - z\mathbf{k}$
S: $x^2 + y^2 = 9$, $z = 0$, $z = 4$

12. $\mathbf{F}(x, y, z) = (xy^2 + \cos z)\mathbf{i} + (x^2y + \sin z)\mathbf{j} + e^z\mathbf{k}$
S: $z = \sqrt{x^2 + y^2}$, $z = 4$

13. $\mathbf{F}(x, y, z) = x^3\mathbf{i} + x^2y\mathbf{j} + x^2e^y\mathbf{k}$
S: $z = 4 - y$, $z = 0$, $x = 0$, $x = 6$, $y = 0$

14. $\mathbf{F}(x, y, z) = xe^z\mathbf{i} + ye^z\mathbf{j} + e^z\mathbf{k}$
S: $z = 4 - y$, $z = 0$, $x = 0$, $x = 6$, $y = 0$

15. Use the Divergence Theorem to show that the volume of the solid bounded by a surface S is

$$\iint_S x \, dy \, dz = \iint_S y \, dz \, dx = \iint_S z \, dx \, dy.$$

16. Verify the result of Exercise 15 for the cube bounded by $x = 0$, $x = a$, $y = 0$, $y = a$, $z = 0$, and $z = a$.

In Exercises 17 and 18, evaluate

$$\iint_S \text{curl } \mathbf{F} \cdot \mathbf{N} \, dS$$

where S is the closed surface of the solid bounded by the graphs of $x = 4$ and $z = 9 - y^2$, and the coordinate planes.

17. $\mathbf{F}(x, y, z) = (4xy + z^2)\mathbf{i} + (2x^2 + 6yz)\mathbf{j} + 2xz\mathbf{k}$

18. $\mathbf{F}(x, y, z) = xy\cos z\mathbf{i} + yz\sin x\mathbf{j} + xyz\mathbf{k}$

19. Show that

$$\iint_S \text{curl } \mathbf{F} \cdot \mathbf{N} \, dS = 0$$

for any closed surface S.

20. For the constant vector field given by

$$\mathbf{F}(x, y, z) = a_1\mathbf{i} + a_2\mathbf{j} + a_3\mathbf{k}$$

show that

$$\iint_S \mathbf{F} \cdot \mathbf{N} \, dS = 0$$

where V is the volume of the solid bounded by the closed surface S.

21. Given the vector field

$$\mathbf{F}(x, y, z) = x\mathbf{i} + y\mathbf{j} + z\mathbf{k}$$

show that

$$\iint_S \mathbf{F} \cdot \mathbf{N} \, dS = 3V$$

where V is the volume of the solid bounded by the closed surface S.

22. Given the vector field

$$\mathbf{F}(x, y, z) = x\mathbf{i} + y\mathbf{j} + z\mathbf{k}$$

show that

$$\frac{1}{\|\mathbf{F}\|} \iint_S \mathbf{F} \cdot \mathbf{N} \, dS = \frac{3}{\|\mathbf{F}\|} \iiint_Q dV.$$

In Exercises 23 and 24, prove the identity, assuming that Q, S, and $\mathbf{N}$ meet the conditions of the Divergence Theorem and that the required partial derivatives of the scalar functions f and g are continuous. The expressions $D_\mathbf{N}f$ and $D_\mathbf{N}g$ are the derivatives in the direction of the vector $\mathbf{N}$ and are defined by

$$D_\mathbf{N}f = \nabla f \cdot \mathbf{N}, \qquad D_\mathbf{N}g = \nabla g \cdot \mathbf{N}.$$

23. $\iiint_Q (f\nabla^2 g + \nabla f \cdot \nabla g) \, dV = \iint_S f D_\mathbf{N}g \, dS$

(*Hint*: Use div $(f\mathbf{G}) = f \text{div } \mathbf{G} + \nabla f \cdot \mathbf{G}$.)

24. $\iiint_Q (f\nabla^2 g - g\nabla^2 f) \, dV = \iint_S (f D_\mathbf{N}g - g D_\mathbf{N}f) \, dS$

(*Hint*: Use Exercise 23 twice.)

Stokes's Theorem

A second higher-dimension analog of Green's Theorem is called **Stokes's Theorem,** after the English mathematical physicist George Gabriel Stokes (1819–1903). Stokes was part of a group of English mathematical physicists referred to as the Cambridge School, which included William Thompson (Lord Kelvin) and James Clerk Maxwell. In addition to making contributions to physics, Stokes worked with infinite series and differential equations, as well as with the integration results presented in this section.

Stokes's Theorem gives the relationship between a surface integral over an oriented surface S and a line integral along a closed space curve C forming the boundary of S, as shown in Figure 15.62. The positive direction along C is counterclockwise relative to the normal vector **N**. That is, if you imagine grasping the normal vector **N** with your right hand, with your thumb pointing in the direction of **N**, then your fingers will point toward the positive direction C, as shown in Figure 15.63.

GEORGE GABRIEL STOKES

Direction along C is counterclockwise relative to **N**.

FIGURE 15.62   **FIGURE 15.63**

THEOREM 15.13   Stokes's Theorem

Let S be an oriented surface with unit normal vector **N**, bounded by a piece-wise smooth, simple closed curve C. If **F** is a vector field whose component functions have continuous partial derivatives on an open region containing S and C, then

$$\int_C \mathbf{F} \cdot d\mathbf{r} = \iint_S (\operatorname{curl} \mathbf{F}) \cdot \mathbf{N}\, dS.$$

REMARK   The line integral may be expressed in the differential form $\int_C M\,dx + N\,dy + P\,dz$ or in the vector form $\int_C \mathbf{F} \cdot \mathbf{T}\, ds$.

EXAMPLE 1   Using Stokes's Theorem

Let C be the oriented triangle lying in the plane $2x + 2y + z = 6$, as shown in Figure 15.64. Evaluate

$$\int_C \mathbf{F} \cdot d\mathbf{r}$$

where $\mathbf{F}(x, y, z) = -y^2\mathbf{i} + z\mathbf{j} + x\mathbf{k}$.

Solution   Using Stokes's Theorem, you can begin by finding the curl of $\mathbf{F}$.

$$\text{curl } \mathbf{F} = \begin{vmatrix} \mathbf{i} & \mathbf{j} & \mathbf{k} \\ \dfrac{\partial}{\partial x} & \dfrac{\partial}{\partial y} & \dfrac{\partial}{\partial z} \\ -y^2 & z & x \end{vmatrix} = -\mathbf{i} - \mathbf{j} + 2y\mathbf{k}$$

Considering $z = 6 - 2x - 2y = g(x, y)$, you can use Theorem 15.11 for an upward normal vector to obtain

$$\int_C \mathbf{F} \cdot d\mathbf{r} = \iint_S (\text{curl } \mathbf{F}) \cdot \mathbf{N} \, dS$$

$$= \iint_R (-\mathbf{i} - \mathbf{j} + 2y\mathbf{k}) \cdot [-g_x(x, y)\mathbf{i} - g_y(x, y)\mathbf{j} + \mathbf{k}] \, dA$$

$$= \iint_R (-\mathbf{i} - \mathbf{j} + 2y\mathbf{k}) \cdot (2\mathbf{i} + 2\mathbf{j} + \mathbf{k}) \, dA$$

$$= \int_0^3 \int_0^{3-y} (2y - 4) \, dx \, dy$$

$$= \int_0^3 (-2y^2 + 10y - 12) \, dy$$

$$= \left[-\frac{2y^3}{3} + 5y^2 - 12y \right]_0^3$$

$$= -9.$$

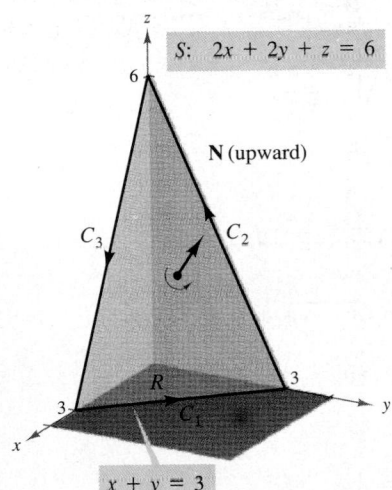

$S: \ 2x + 2y + z = 6$

$\mathbf{N}$ (upward)

C_3 C_2

R C_1

$x + y = 3$

FIGURE 15.64

Try evaluating the line integral in Example 1 directly, *without* using Stokes's Theorem. One way to do this would be to consider C as the union of C_1, C_2, and C_3, as follows.

$$C_1: \mathbf{r}_1(t) = (3 - t)\mathbf{i} + t\mathbf{j}, \qquad 0 \le t \le 3$$
$$C_2: \mathbf{r}_2(t) = (6 - t)\mathbf{j} + (2t - 6)\mathbf{k}, \qquad 3 \le t \le 6$$
$$C_3: \mathbf{r}_3(t) = (t - 6)\mathbf{i} + (18 - 2t)\mathbf{k}, \qquad 6 \le t \le 9$$

The value of the line integral is

$$\int_C \mathbf{F} \cdot d\mathbf{r} = \int_{C_1} \mathbf{F} \cdot \mathbf{r}_1{}'(t) \, dt + \int_{C_2} \mathbf{F} \cdot \mathbf{r}_2{}'(t) \, dt + \int_{C_3} \mathbf{F} \cdot \mathbf{r}_3{}'(t) \, dt$$

$$= \int_0^3 t^2 \, dt + \int_3^6 (-2t + 6) \, dt + \int_6^9 (-2t + 12) \, dt$$

$$= 9 - 9 - 9$$

$$= -9.$$

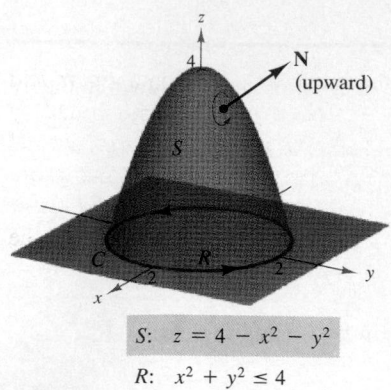

$S: \; z = 4 - x^2 - y^2$

$R: \; x^2 + y^2 \leq 4$

FIGURE 15.65

EXAMPLE 2   Verifying Stokes's Theorem

Verify Stokes's Theorem for $\mathbf{F}(x, y, z) = 2z\mathbf{i} + x\mathbf{j} + y^2\mathbf{k}$, where S is the surface of the paraboloid $z = 4 - x^2 - y^2$ and C is the trace of S in the xy-plane, as shown in Figure 15.65.

Solution   As a *surface integral,* you have $z = g(x, y) = 4 - x^2 - y^2$ and

$$\mathbf{curl\ F} = \begin{vmatrix} \mathbf{i} & \mathbf{j} & \mathbf{k} \\ \dfrac{\partial}{\partial x} & \dfrac{\partial}{\partial y} & \dfrac{\partial}{\partial z} \\ 2z & x & y^2 \end{vmatrix} = 2y\mathbf{i} + 2\mathbf{j} + \mathbf{k}.$$

By Theorem 15.11 for an upward normal vector $\mathbf{N}$, you obtain

$$
\begin{aligned}
\iint_S (\mathbf{curl\ F}) \cdot \mathbf{N}\, dS &= \iint_R (2y\mathbf{i} + 2\mathbf{j} + \mathbf{k}) \cdot (2x\mathbf{i} + 2y\mathbf{j} + \mathbf{k})\, dA \\
&= \int_{-2}^{2} \int_{-\sqrt{4-y^2}}^{\sqrt{4-y^2}} (4xy + 4y + 1)\, dx\, dy \\
&= \int_{-2}^{2} \left[2x^2 y + (4y + 1)x \right]_{-\sqrt{4-y^2}}^{\sqrt{4-y^2}} dy \\
&= \int_{-2}^{2} 2(4y + 1)\sqrt{4 - y^2}\, dy \\
&= \int_{-2}^{2} (8y\sqrt{4 - y^2} + 2\sqrt{4 - y^2})\, dy \\
&= \left[-\frac{8}{3}(4 - y^2)^{3/2} + y\sqrt{4 - y^2} + 4\arcsin\frac{y}{2} \right]_{-2}^{2} \\
&= 4\pi.
\end{aligned}
$$

As a *line integral,* you can parametrize C by

$$\mathbf{r}(t) = 2\cos t\,\mathbf{i} + 2\sin t\,\mathbf{j} + 0\mathbf{k}, \qquad 0 \leq t \leq 2\pi.$$

For $\mathbf{F}(x, y, z) = 2z\mathbf{i} + x\mathbf{j} + y^2\mathbf{k}$, you obtain

$$
\begin{aligned}
\int_C \mathbf{F} \cdot d\mathbf{r} &= \int_C M\, dx + N\, dy + P\, dz \\
&= \int_C 2z\, dx + x\, dy + y^2\, dz \\
&= \int_0^{2\pi} [0 + 2\cos t(2\cos t) + 0]\, dt \\
&= \int_0^{2\pi} 4\cos^2 t\, dt \\
&= 2\int_0^{2\pi} (1 + \cos 2t)\, dt \\
&= 2\left[t + \frac{1}{2}\sin 2t \right]_0^{2\pi} \\
&= 4\pi.
\end{aligned}
$$

FIGURE 15.66

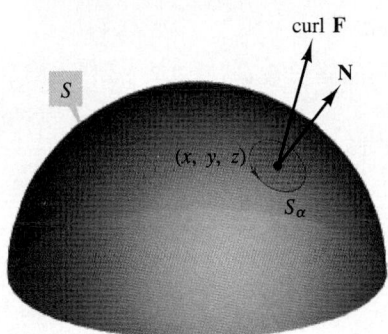

FIGURE 15.67

Physical Interpretation of Curl

Stokes's Theorem provides insight into a physical interpretation of curl. In a vector field **F**, let S_α be a *small* circular disc of radius α, centered at (x, y, z) and with boundary C_α, as shown in Figure 15.66. At each point on the circle C_α, **F** has a normal component **F** $\cdot$ **N** and a tangential component **F** $\cdot$ **T**. The more closely **F** and **T** are aligned, the greater the value of **F** $\cdot$ **T**. Thus, a fluid tends to move along the circle rather than across it. Consequently, we say that the line integral around C_α measures the **circulation of F around C_α.** That is,

$$\int_{C_\alpha} \mathbf{F} \cdot \mathbf{T} \, ds = \text{circulation of } \mathbf{F} \text{ around } C_\alpha.$$

Now consider a small disc S_α to be centered at some point (x, y, z) on the surface S, as shown in Figure 15.67. On such a small disc, **curl F** is nearly constant, because it varies little from its value at (x, y, z). Moreover, **curl F** $\cdot$ **N** is also nearly constant on S_α, because all unit normals to S_α are about the same. Consequently, Stokes's Theorem yields

$$\int_{C_\alpha} \mathbf{F} \cdot \mathbf{T} \, ds = \int\int_{S_\alpha} (\mathbf{curl\ F}) \cdot \mathbf{N} \, ds$$

$$\approx (\mathbf{curl\ F}) \cdot \mathbf{N} \int\int_{S_\alpha} dS$$

$$\approx (\mathbf{curl\ F}) \cdot \mathbf{N}(\pi\alpha^2).$$

Therefore,

$$(\mathbf{curl\ F}) \cdot \mathbf{N} \approx \frac{\displaystyle\int_{C_\alpha} \mathbf{F} \cdot \mathbf{T} \, ds}{\pi\alpha^2}$$

$$= \frac{\text{circulation of } \mathbf{F} \text{ around } C_\alpha}{\text{area of disc } S_\alpha}$$

$$= \text{rate of circulation.}$$

Assuming conditions are such that the approximation improves for smaller and smaller discs ($\alpha \rightarrow 0$), it follows that

$$(\mathbf{curl\ F}) \cdot \mathbf{N} = \lim_{\alpha \to 0} \frac{1}{\pi\alpha^2} \int_{C_\alpha} \mathbf{F} \cdot \mathbf{T} \, ds$$

which is referred to as the **rotation of F about N.** That is,

$$\mathbf{curl\ F}(x, y, z) \cdot \mathbf{N} = \text{rotation of } \mathbf{F} \text{ about } \mathbf{N} \text{ at } (x, y, z).$$

In this case, the rotation of **F** is maximum when **curl F** and **N** have the same direction. Normally, this tendency to rotate will vary from point to point on the surface S, and Stokes's Theorem

$$\underbrace{\int\int_S (\mathbf{curl\ F}) \cdot \mathbf{N} \, dS}_{\text{Surface integral}} = \underbrace{\int_C \mathbf{F} \cdot d\mathbf{r}}_{\text{Line integral}}$$

says that the collective measure of this *rotational* tendency taken over the entire surface S (surface integral) is equal to the tendency of a fluid to *circulate* around the boundary C (line integral).

FIGURE 15.68

REMARK   If **curl F** = **0** throughout region Q, then the rotation of **F** about each unit normal **N** is zero. That is, **F** is irrotational. From earlier work, you know that this is a characteristic of conservative vector fields.

EXAMPLE 3   Motion of a Swirling Liquid

A liquid is swirling around in a cylindrical container of radius 2 such that its motion is described by the velocity field

$$\mathbf{F}(x, y, z) = -y\sqrt{x^2 + y^2}\,\mathbf{i} + x\sqrt{x^2 + y^2}\,\mathbf{j}$$

as shown in Figure 15.68. Find

$$\iint_S (\mathbf{curl\ F}) \cdot \mathbf{N}\,dS$$

where S is the upper surface of the cylindrical tank.

Solution   The curl of **F** is given by

$$\mathbf{curl\ F} = \begin{vmatrix} \mathbf{i} & \mathbf{j} & \mathbf{k} \\ \dfrac{\partial}{\partial x} & \dfrac{\partial}{\partial y} & \dfrac{\partial}{\partial z} \\ -y\sqrt{x^2 + y^2} & x\sqrt{x^2 + y^2} & 0 \end{vmatrix}$$

$$= 3\sqrt{x^2 + y^2}\,\mathbf{k}.$$

Letting **N** = **k**, you have

$$\iint_S (\mathbf{curl\ F}) \cdot \mathbf{N}\,dS = \iint_R 3\sqrt{x^2 + y^2}\,dA$$

$$= \int_0^{2\pi} \int_0^2 (3r)r\,dr\,d\theta$$

$$= \int_0^{2\pi} r^3 \Big]_0^2\,d\theta$$

$$= \int_0^{2\pi} 8\,d\theta$$

$$= 16\pi.$$

Summary of Integration Formulas

Fundamental Theorem of Calculus

$$\int_a^b F'(x)\,dx = F(b) - F(a)$$

Fundamental Theorem of Line Integrals

$$\int_C \mathbf{F} \cdot d\mathbf{r} = \int_C \nabla f \cdot d\mathbf{r} = f(x(b), y(b)) - f(x(a), y(a))$$

Green's Theorem

$$\int_C M\,dx + N\,dy = \iint_R \left(\frac{\partial N}{\partial x} - \frac{\partial M}{\partial y}\right) dA = \int_C \mathbf{F} \cdot \mathbf{T}\,ds = \int_C \mathbf{F} \cdot d\mathbf{r} = \iint_R (\mathbf{curl\ F}) \cdot \mathbf{k}\,dA$$

$$\int_C \mathbf{F} \cdot \mathbf{N}\,ds = \iint_R \text{div}\,\mathbf{F}\,dA$$

Divergence Theorem

$$\iint_S \mathbf{F} \cdot \mathbf{N}\,dS = \iiint_Q \text{div}\,\mathbf{F}\,dV$$

Stokes's Theorem

$$\int_C \mathbf{F} \cdot d\mathbf{r} = \iint_S (\mathbf{curl\ F}) \cdot \mathbf{N}\,dS$$

EXERCISES for Section 15.8

In Exercises 1–6, find the curl of the vector field F.

1. $\mathbf{F}(x, y, z) = (2y - z)\mathbf{i} + xyz\mathbf{j} + e^z\mathbf{k}$

2. $\mathbf{F}(x, y, z) = x^2\mathbf{i} + y^2\mathbf{j} + x^2\mathbf{k}$

3. $\mathbf{F}(x, y, z) = 2z\mathbf{i} - 4x^2\mathbf{j} + \arctan x\,\mathbf{k}$

4. $\mathbf{F}(x, y, z) = x\sin y\,\mathbf{i} - y\cos x\mathbf{j} + yz^2\mathbf{k}$

5. $\mathbf{F}(x, y, z) = e^{x^2+y^2}\mathbf{i} + e^{y^2+z^2}\mathbf{j} + xyz\mathbf{k}$

6. $\mathbf{F}(x, y, z) = \arcsin y\,\mathbf{i} + \sqrt{1 - x^2}\,\mathbf{j} + y^2\mathbf{k}$

In Exercises 7–10, verify Stokes's Theorem by evaluating

$$\int_C \mathbf{F} \cdot \mathbf{T}\, ds$$

as a line integral and as a double integral.

7. $\mathbf{F}(x, y, z) = (-y + z)\mathbf{i} + (x - z)\mathbf{j} + (x - y)\mathbf{k}$
$S: z = \sqrt{1 - x^2 - y^2}$

8. $\mathbf{F}(x, y, z) = (-y + z)\mathbf{i} + (x - z)\mathbf{j} + (x - y)\mathbf{k}$
$S: z = 4 - x^2 - y^2,\ 0 \le z$

9. $\mathbf{F}(x, y, z) = xyz\mathbf{i} + y\mathbf{j} + z\mathbf{k}$
$S: 3x + 4y + 2z = 12$ (first octant)

10. $\mathbf{F}(x, y, z) = z^2\mathbf{i} + x^2\mathbf{j} + y^2\mathbf{k}$
$S: z = x^2,\ 0 \le x \le a,\ 0 \le y \le a$

In Exercises 11–20, use Stokes's Theorem to evaluate

$$\int_C \mathbf{F} \cdot d\mathbf{r}.$$

11. $\mathbf{F}(x, y, z) = 2y\mathbf{i} + 3z\mathbf{j} + x\mathbf{k}$
C: triangle with vertices $(0, 0, 0)$, $(0, 2, 0)$, and $(1, 1, 1)$

12. $\mathbf{F}(x, y, z) = \arctan \dfrac{x}{y}\mathbf{i} + \ln \sqrt{x^2 + y^2}\mathbf{j} + \mathbf{k}$
C: triangle with vertices $(0, 0, 0)$, $(1, 1, 1)$, and $(0, 0, 2)$

13. $\mathbf{F}(x, y, z) = z^2\mathbf{i} + x^2\mathbf{j} + y^2\mathbf{k}$
$S: z = 4 - x^2 - y^2,\ 0 \le z$

14. $\mathbf{F}(x, y, z) = 4xz\mathbf{i} + y\mathbf{j} + 4xy\mathbf{k}$
$S: z = 4 - x^2 - y^2,\ 0 \le z$

15. $\mathbf{F}(x, y, z) = z^2\mathbf{i} + y\mathbf{j} + xz\mathbf{k}$
$S: z = \sqrt{4 - x^2 - y^2}$

16. $\mathbf{F}(x, y, z) = x^2\mathbf{i} + z^2\mathbf{j} - xyz\mathbf{k}$
$S: z = \sqrt{4 - x^2 - y^2}$

17. $\mathbf{F}(x, y, z) = -\ln \sqrt{x^2 + y^2}\mathbf{i} + \arctan \dfrac{x}{y}\mathbf{j} + \mathbf{k}$
$S: z = 9 - 2x - 3y$ over one petal of $r = 2\sin 2\theta$ in the first octant

18. $\mathbf{F}(x, y, z) = yz\mathbf{i} + (2 - 3y)\mathbf{j} + (x^2 + y^2)\mathbf{k}$
S: the first-octant portion of $x^2 + z^2 = 16$ over $x^2 + y^2 = 16$

19. $\mathbf{F}(x, y, z) = xyz\mathbf{i} + y\mathbf{j} + z\mathbf{k}$
$S: z = x^2,\ 0 \le x \le a,\ 0 \le y \le a$

20. $\mathbf{F}(x, y, z) = xyz\mathbf{i} + y\mathbf{j} + z\mathbf{k}$
S: the first-octant portion of $z = x^2$ over $x^2 + y^2 = a^2$

Motion of a Liquid In Exercises 21 and 22, the motion of a liquid in a cylindrical container of radius 1 is described by the velocity field $\mathbf{F}(x, y, z)$. Find

$$\int_S \int (\text{curl } \mathbf{F}) \cdot \mathbf{N}\, dS$$

where S is the upper surface of the cylindrical tank.

21. $\mathbf{F}(x, y, z) = \mathbf{i} + \mathbf{j} - 2\mathbf{k}$

22. $\mathbf{F}(x, y, z) = -y\mathbf{i} + x\mathbf{j}$

23. Let f and g be scalar functions with continuous partial derivatives, and let C and S satisfy the conditions of Stokes's Theorem. Verify each of the following identities.

a. $\displaystyle\int_C (f\nabla g) \cdot d\mathbf{r} = \int_S \int (\nabla f \times \nabla g) \cdot \mathbf{N}\, dS$

b. $\displaystyle\int_C (f\nabla f) \cdot d\mathbf{r} = 0$

c. $\displaystyle\int_C (f\nabla g + g\nabla f) \cdot d\mathbf{r} = 0$

24. Demonstrate the result of Exercise 23 for the functions $f(x, y, z) = xyz$ and $g(x, y, z) = z$. Let S be the hemisphere $z = \sqrt{4 - x^2 - y^2}$.

25. Let $\mathbf{C}$ be a constant vector. Let S be an oriented surface with a unit normal vector $\mathbf{N}$, bounded by a smooth curve C. Prove that

$$\int_S \int \mathbf{C} \cdot \mathbf{N}\, dS = \frac{1}{2}\int_C (\mathbf{C} \times \mathbf{r})\, d\mathbf{r}.$$

REVIEW EXERCISES for Chapter 15

In Exercises 1 and 2, sketch several representative vectors in the given vector field.

1. $\mathbf{F}(x, y, z) = x\mathbf{i} + \mathbf{j} + 2\mathbf{k}$

2. $\mathbf{F}(x, y) = \mathbf{i} - 2y\mathbf{j}$

In Exercises 3 and 4, find a three-dimensional vector field that has the potential function f.

3. $f(x, y, z) = 8x^2 + xy + z^2$

4. $f(x, y, z) = x^2 e^{yz}$

In Exercises 5–12, determine whether F is conservative. If it is, find the potential function f.

5. $\mathbf{F}(x, y) = \dfrac{1}{y}\mathbf{i} - \dfrac{y}{x^2}\mathbf{j}$

6. $\mathbf{F}(x, y) = -\dfrac{y}{x^2}\mathbf{i} + \dfrac{1}{x}\mathbf{j}$

7. $\mathbf{F}(x, y) = (6xy^2 - 3x^2)\mathbf{i} + (6x^2y + 3y^2 - 7)\mathbf{j}$

8. $\mathbf{F}(x, y) = (-2y^3 \sin 2x)\mathbf{i} + 3y^2(1 + \cos 2x)\mathbf{j}$

9. $\mathbf{F}(x, y, z) = (4xy + z)\mathbf{i} + (2x^2 + 6y)\mathbf{j} + 2z\mathbf{k}$

10. $\mathbf{F}(x, y, z) = (4xy + z^2)\mathbf{i} + (2x^2 + 6yz)\mathbf{j} + 2xz\mathbf{k}$

11. $\mathbf{F}(x, y, z) = \dfrac{yz\mathbf{i} - xz\mathbf{j} - xy\mathbf{k}}{y^2z^2}$

12. $\mathbf{F}(x, y, z) = \sin z(y\mathbf{i} + x\mathbf{j} + \mathbf{k})$

In Exercises 13–20, find (a) the divergence of the vector field F, and (b) the curl of the vector field F.

13. $\mathbf{F}(x, y, z) = x^2\mathbf{i} + y^2\mathbf{j} + z^2\mathbf{k}$

14. $\mathbf{F}(x, y, z) = xy^2\mathbf{j} - zx^2\mathbf{k}$

15. $\mathbf{F}(x, y, z) = (\cos y + y\cos x)\mathbf{i} + (\sin x - x\sin y)\mathbf{j} + xyz\mathbf{k}$

16. $\mathbf{F}(x, y, z) = (3x - y)\mathbf{i} + (y - 2z)\mathbf{j} + (z - 3x)\mathbf{k}$

17. $\mathbf{F}(x, y, z) = \arcsin x\,\mathbf{i} + xy^2\mathbf{j} + yz^2\mathbf{k}$

18. $\mathbf{F}(x, y, z) = (x^2 - y)\mathbf{i} - (x + \sin^2 y)\mathbf{j}$

19. $\mathbf{F}(x, y, z) = \ln(x^2 + y^2)\mathbf{i} + \ln(x^2 + y^2)\mathbf{j} + z\mathbf{k}$

20. $\mathbf{F}(x, y, z) = \dfrac{z}{x}\mathbf{i} + \dfrac{z}{y}\mathbf{j} + z^2\mathbf{k}$

In Exercises 21–26, evaluate the line over the specified path(s).

21. $\displaystyle\int_C (x^2 + y^2)\,ds$

   a. C: line segment from $(-1, -1)$ to $(2, 2)$
   b. C: $x^2 + y^2 = 16$, one revolution counterclockwise, starting at $(4, 0)$

22. $\displaystyle\int_C xy\,ds$

   a. C: line segment from $(0, 0)$ to $(5, 4)$
   b. C: counterclockwise around the triangle with vertices $(0, 0)$, $(4, 0)$, and $(0, 2)$

23. $\displaystyle\int_C (x^2 + y^2)\,ds$

   C: $\mathbf{r}(t) = (\cos t + t\sin t)\mathbf{i} + (\sin t - t\cos t)\mathbf{j}$,
   $0 \le t \le 2\pi$

24. $\displaystyle\int_C x\,ds$

   C: $\mathbf{r}(t) = (t - \sin t)\mathbf{i} + (1 - \cos t)\mathbf{j}, \ 0 \le t \le 2\pi$

25. $\displaystyle\int_C (2x - y)\,dx + (x + 3y)\,dy$

   a. C: line segment from $(0, 0)$ to $(2, -3)$
   b. C: counterclockwise around the circle $x = 3\cos t$, $y = 3\sin t$

26. $\displaystyle\int_C (2x - y)\,dx + (x + 3y)\,dy$

   C: $\mathbf{r}(t) = (\cos t + t\sin t)\mathbf{i} + (\sin t - t\sin t)\mathbf{j}, \ 0 \le t \le \pi/2$

Lateral Surface Area   In Exercises 27 and 28, find the lateral surface area over the curve C in the xy-plane and under the surface $z = f(x, y)$.

27. $f(x, y) = 5 + \sin(x + y)$
   C: $y = 3x$ from $(0, 0)$ to $(2, 6)$

28. $f(x, y) = 12 - x - y$
   C: $y = x^2$ from $(0, 0)$ to $(2, 4)$

In Exercises 29–34, evaluate

$$\int_C \mathbf{F} \cdot d\mathbf{r}.$$

29. $\mathbf{F}(x, y) = xy\mathbf{i} + x^2\mathbf{j}$
   C: $\mathbf{r}(t) = t^2\mathbf{i} + t^3\mathbf{j}, \ 0 \le t \le 1$

30. $\mathbf{F}(x, y) = (x - y)\mathbf{i} + (x + y)\mathbf{j}$
   C: $\mathbf{r}(t) = 4\cos t\,\mathbf{i} + 3\sin t\,\mathbf{j}, \ 0 \le t \le 2\pi$

31. $\mathbf{F}(x, y, z) = x\mathbf{i} + y\mathbf{j} + z\mathbf{k}$
   C: $\mathbf{r}(t) = 2\cos t\,\mathbf{i} + 2\sin t\,\mathbf{j} + t\mathbf{k}, \ 0 \le t \le 2\pi$

32. $\mathbf{F}(x, y, z) = (2y - z)\mathbf{i} + (z - x)\mathbf{j} + (x - y)\mathbf{k}$
   C: curve of intersection of $x^2 + z^2 = 4$ and $y^2 + z^2 = 4$ from $(2, 2, 0)$ to $(0, 0, 2)$

33. $\mathbf{F}(x, y, z) = (y - z)\mathbf{i} + (z - x)\mathbf{j} + (x - y)\mathbf{k}$
   C: curve of intersection of $z = x^2 + y^2$ and $x + y = 0$ from $(-2, 2, 8)$ to $(2, -2, 8)$

34. $\mathbf{F}(x, y, z) = (x^2 - z)\mathbf{i} + (y^2 + z)\mathbf{j} + x\mathbf{k}$
   C: curve of intersection of $z = x^2$ and $x^2 + y^2 = 4$ from $(0, -2, 0)$ to $(0, 2, 0)$

35. *Work*   Find the work done by the force field $\mathbf{F} = x\mathbf{i} - \sqrt{y}\,\mathbf{j}$ along the path $y = x^{3/2}$ from $(0, 0)$ to $(4, 8)$.

36. *Work*   Find the work done by the engines on a 20-ton aircraft if it climbs 2000 feet while making a 90° turn in a circular arc of radius 10 miles.

In Exercises 37 and 38, use the Fundamental Theorem of Line Integrals to evaluate the given integral.

37. $\displaystyle\int_C 2xyz\,dx + x^2z\,dy + x^2y\,dz$
   C: smooth curve from $(0, 0, 0)$ to $(1, 4, 3)$

38. $\displaystyle\int_C y\,dx + x\,dy + \dfrac{1}{z}\,dz$
   C: smooth curve from $(0, 0, 1)$ to $(4, 4, 4)$

In Exercises 39–44, use Green's Theorem to evaluate the line integral.

39. $\displaystyle\int_C y\,dx + 2x\,dy$
   C: boundary of the square with vertices $(0, 0)$, $(0, 2)$, $(2, 0)$, and $(2, 2)$

40. $\displaystyle\int_C xy\,dx + (x^2 + y^2)\,dy$
 C: boundary of the square with vertices $(0, 0)$, $(0, 2)$, $(2, 0)$, and $(2, 2)$

41. $\displaystyle\int_C xy^2\,dx + x^2y\,dy$
 C: $x = 4\cos t$, $y = 2\sin t$

42. $\displaystyle\int_C (x^2 - y^2)\,dx + 2xy\,dy$
 C: $x^2 + y^2 = a^2$

43. $\displaystyle\int_C xy\,dx + x^2\,dy$
 C: boundary of the region between the graphs of $y = x^2$ and $y = x$

44. $\displaystyle\int_C y^2\,dx + x^{2/3}\,dy$
 C: $x^{2/3} + y^{2/3} = 1$

C In Exercises 45 and 46, use a graphing utility to graph the surface represented by the vector-valued function.

45. $\mathbf{r}(u, v) = \sec u\cos v\,\mathbf{i} + (1 + 2\tan u)\sin v\,\mathbf{j} + 2u\,\mathbf{k}$
 $0 \le u \le \dfrac{\pi}{3},\ 0 \le v \le 2\pi$

46. $\mathbf{r}(u, v) = e^{-u/4}\cos v\,\mathbf{i} + e^{-u/4}\sin v\,\mathbf{j} + \dfrac{u}{6}\,\mathbf{k}$
 $0 \le u \le 4,\ 0 \le v \le 2\pi$

47. Evaluate the surface integral

$$\iint_S z\,dS$$

over the surface

S: $\mathbf{r}(u, v) = (u + v)\mathbf{i} + (u - v)\mathbf{j} + \sin v\,\mathbf{k}$

where $0 \le u \le 2$ and $0 \le v \le \pi$.

C **48.** Use a computer to graph the surface S and to approximate the surface integral

$$\iint_S (x + y)\,dS$$

where S is the surface
S: $\mathbf{r}(u, v) = u\cos v\,\mathbf{i} + u\sin v\,\mathbf{j} + (u - 1)(2 - u)\mathbf{k}$ over
 $0 \le u \le 2$ and $0 \le v \le 2\pi$.

In Exercises 49 and 50, verify the Divergence Theorem by evaluating

$$\iint_S \mathbf{F}\cdot\mathbf{N}\,dS$$

as a surface integral and as a triple integral.

49. $\mathbf{F}(x, y, z) = x^2\mathbf{i} + xy\mathbf{j} + z\mathbf{k}$
 Q: solid region bounded by the coordinate planes and the plane $2x + 3y + 4z = 12$

50. $\mathbf{F}(x, y, z) = x\mathbf{i} + y\mathbf{j} + z\mathbf{k}$
 Q: solid region bounded by the coordinate planes and the plane $2x + 3y + 4z = 12$

In Exercises 51 and 52, verify Stokes's Theorem by evaluating

$$\int_C \mathbf{F}\cdot d\mathbf{r}$$

as a line integral and as a double integral.

51. $\mathbf{F}(x, y, z) = (\cos y + y\cos x)\mathbf{i} + (\sin x - x\sin y)\mathbf{j} + xyz\,\mathbf{k}$
 S: portion of $z = y^2$ over the square in the xy-plane with vertices $(0, 0)$, $(a, 0)$, (a, a), and $(0, a)$

52. $\mathbf{F}(x, y, z) = (x - z)\mathbf{i} + (y - z)\mathbf{j} + x^2\mathbf{k}$
 S: first-octant portion of the plane $3x + y + 2z = 12$

53. Consider the integral

$$\int_C \mathbf{F}\cdot d\mathbf{r}$$

where

$$\mathbf{F}(x, y) = \frac{-y}{x^2 + y^2}\mathbf{i} + \frac{x}{x^2 + y^2}\mathbf{j}$$

and

$$\mathbf{r}(t) = \cos t\,\mathbf{i} + \sin t\,\mathbf{j}.$$

Because $N_x = M_y$ and C is a circle, it is expected that the line integral will have a value of 0. However, upon direct integration we obtain

$$\int_C \mathbf{F}\cdot d\mathbf{r} = 2\pi.$$

Which is correct, and why?

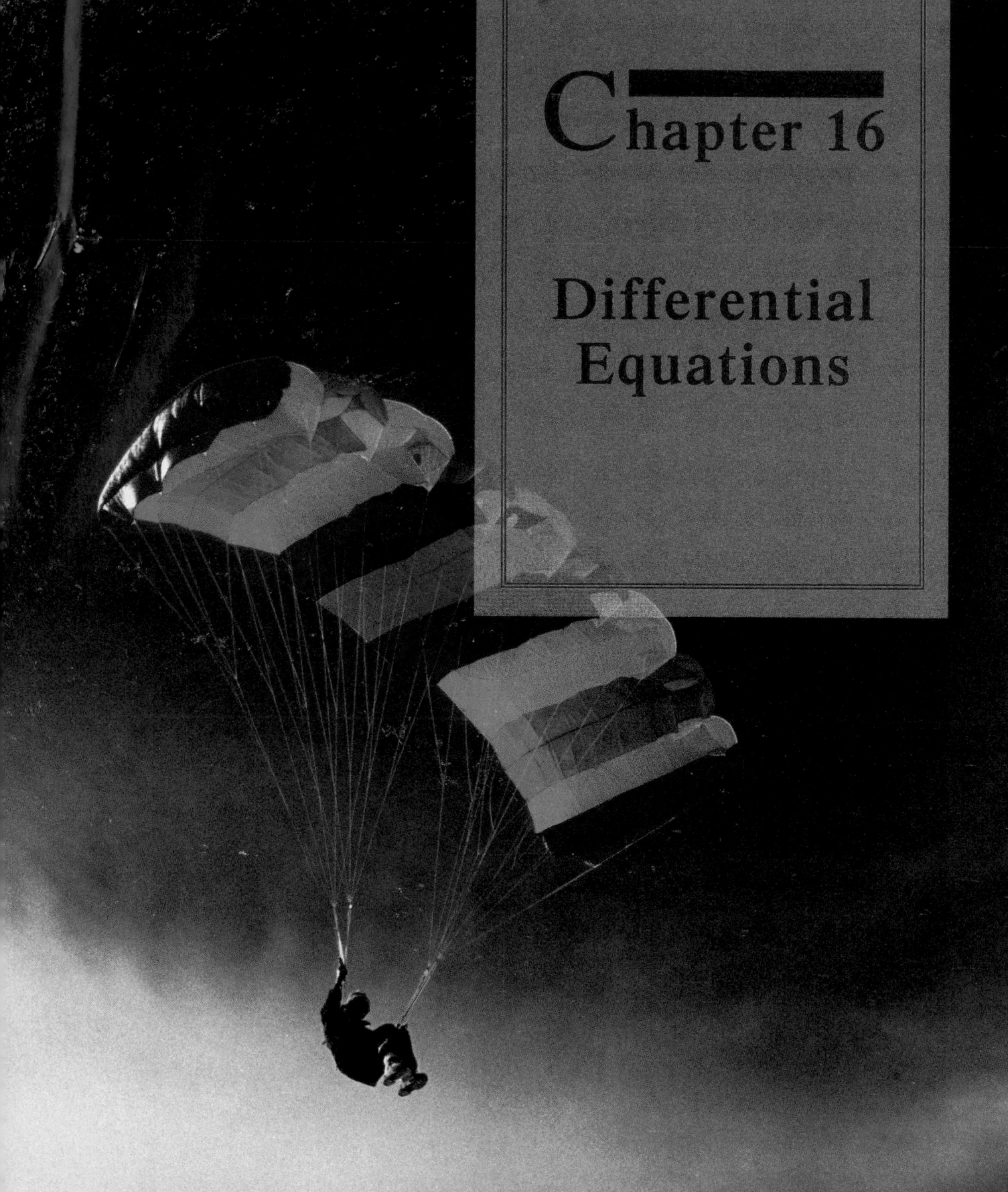

Chapter 16

Differential Equations

Type and Order

Several times in the text, we have identified physical phenomena that can be described by differential equations. For example, we saw that problems involving radioactive decay, population growth, chemical reactions, Newton's Law of Cooling, and gravitational force can be formulated in terms of differential equations. In this chapter, we review the terminology and techniques of differential equations discussed earlier in the text and introduce new terminology and techniques.

A **differential equation** is an equation involving a function and one or more of its derivatives. If the function has only one independent variable, the equation is called an **ordinary differential equation.** For instance,

$$\frac{d^2y}{dx^2} + 3\frac{dy}{dx} - 2y = 0$$

is an ordinary differential equation in which the dependent variable $y = f(x)$ is a twice-differentiable function of x. A differential equation involving a function of several variables and its partial derivatives is called a **partial differential equation.** In this chapter, the discussion is restricted to ordinary differential equations.

In addition to **type** (ordinary or partial), differential equations are classified by order. The **order** of a differential equation is determined by the highest-order derivative in the equation. For instance,

$$y' = y \qquad \text{First-order differential equation}$$

is a first-order differential equation and

$$\frac{d^2y}{dx^2} = 4y \qquad \text{Second-order differential equation}$$

is a second-order differential equation.

EXAMPLE 1   Classifying Differential Equations

Equation	Type	Order
a. $y''' + 4y = 2$	Ordinary	3
b. $\dfrac{d^2s}{dt^2} = -32$	Ordinary	2
c. $(y')^2 - 3y = e^x$	Ordinary	1
d. $\dfrac{\partial^2 u}{\partial x^2} + \dfrac{\partial^2 u}{\partial y^2} = 0$	Partial	2
e. $y - \sin y' = 0$	Ordinary	1

◀ *The fall of a parachutist can be modeled by a differential equation, as discussed in Section 16.6, Exercise 35.*

General and Particular Solutions

A function $y = f(x)$ is called a **solution** of a differential equation if the equation is satisfied when y and its derivatives are replaced by $f(x)$ and its derivatives, respectively. For example, differentiation and substitution would show that $y = e^{-2x}$ is a solution of the differential equation

$$y' + 2y = 0.$$

It can be shown that every solution of this differential equation is of the form $y = Ce^{-2x}$, where C is any real number, and we call $y = Ce^{-2x}$ the **general solution.** (Some differential equations have **singular solutions** that cannot be written as special cases of the general solution. However, we will not consider such solutions in this text.)

In Section 4.1, Example 8, you saw that the second-order differential equation $s''(t) = -32$ has the general solution

$$s(t) = -16t^2 + C_1 t + C_2$$

which contains two arbitrary constants. It can be shown that a differential equation of order n has a general solution with n arbitrary constants.

EXAMPLE 2   Verifying Solutions

Determine whether or not the given functions are solutions of the differential equation $y'' - y = 0$.

a. $y = \sin x$ **b.** $y = e^{2x}$ **c.** $y = 4e^{-x}$ **d.** $y = Ce^x$

Solution

a. Because $y = \sin x$, $y' = \cos x$, and $y'' = -\sin x$, it follows that

$$y'' - y = -\sin x - \sin x = -2\sin x \neq 0.$$

Hence, $y = \sin x$ *is not* a solution.

b. Because $y = e^{2x}$, $y' = 2e^{2x}$, and $y'' = 4e^{2x}$, it follows that

$$y'' - y = 4e^{2x} - e^{2x} = 3e^{2x} \neq 0.$$

Hence, $y = e^{2x}$ *is not* a solution.

c. Because $y = 4e^{-x}$, $y' = -4e^{-x}$, and $y'' = 4e^{-x}$, it follows that

$$y'' - y = 4e^{-x} - 4e^{-x} = 0.$$

Hence, $y = 4e^{-x}$ *is* a solution.

d. Because $y = Ce^x$, $y' = Ce^x$, and $y'' = Ce^x$, it follows that

$$y'' - y = Ce^x - Ce^x = 0.$$

Hence, $y = Ce^x$ *is* a solution for any value of C.

Later in this chapter, you will see that the general solution of the differential equation in Example 2 is $y = C_1 e^x + C_2 e^{-x}$. A **particular solution** of a differential equation is any solution that is obtained by assigning specific values to the constants in the general solution.

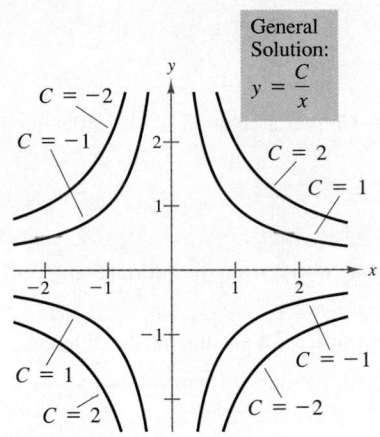

FIGURE 16.1
Solution curves for $xy' + y = 0$.

Geometrically, the general solution of a first-order differential equation represents a family of curves known as **solution curves,** one for each value assigned to the arbitrary constant. For instance, you can easily verify that every function of the form $y = C/x$ is a solution of the differential equation $xy' + y = 0$. Figure 16.1 shows some of the solution curves corresponding to different values of C.

Particular solutions of a differential equation are obtained from **initial conditions** that give the value of the dependent variable or one of its derivatives for a particular value of the independent variable. The term "initial condition" stems from the fact that, often in problems involving time, the value of the dependent variable or one of its derivatives is known at the *initial* time $t = 0$. For instance, the second-order differential equation $s''(t) = -32$, which has the general solution

$$s(t) = -16t^2 + C_1 t + C_2 \qquad \text{General solution}$$

might have the following initial conditions.

$$s(0) = 80, \qquad s'(0) = 64 \qquad \text{Initial conditions}$$

In this case, the initial conditions yield the particular solution

$$s(t) = -16t^2 + 64t + 80. \qquad \text{Particular solution}$$

EXAMPLE 3   Finding a Particular Solution

For the differential equation

$$xy' - 3y = 0$$

verify that $y = Cx^3$ is a solution and find the particular solution determined by the initial condition $y = 2$ when $x = -3$.

Solution   You know that $y = Cx^3$ is a solution because $y' = 3Cx^2$ and

$$xy' - 3y = x(3Cx^2) - 3(Cx^3) = 0.$$

Furthermore, the initial condition $y = 2$ when $x = -3$ yields

$$y = Cx^3 \quad \rightarrow \quad 2 = C(-3)^3 \quad \rightarrow \quad C = -\frac{2}{27}$$

and you can conclude that the particular solution is $y = -2x^3/27$.

REMARK   To determine a particular solution, the number of initial conditions must match the number of constants in the general solution.

In the remainder of this chapter, you will study procedures for finding the general solutions of several classes of differential equations. Each section will include practical applications of that particular type of differential equation. The discussion focuses mainly on first- and second-order equations and linear equations with constant coefficients. By a **linear differential equation with constant coefficients** $a_0, a_1, a_2, \ldots, a_n$, we mean an equation of the form

$$a_n y^{(n)} + a_{n-1} y^{(n-1)} + \cdots + a_2 y'' + a_1 y' + a_0 y = F(x)$$

which is linear in y and its derivatives.

EXERCISES for Section 16.1

In Exercises 1–12, classify the differential equation according to type and order.

1. $\dfrac{dy}{dx} + 3xy = x^2$

2. $y'' + 2y' + y = 1$

3. $\dfrac{d^2x}{dt^2} + 2\dfrac{dx}{dt} - 4x = e^t$

4. $\dfrac{d^2u}{dt^2} + \dfrac{du}{dt} = \sec t$

5. $y^{(4)} + 3(y')^2 - 4y = 0$

6. $x^2y'' + 3xy' = 0$

7. $(y'')^2 + 3y' - 4y = 0$

8. $\dfrac{\partial u}{\partial t} = C^2 \dfrac{\partial^2 u}{\partial x^2}$

9. $\dfrac{\partial u}{\partial t} + \dfrac{\partial u}{\partial y} = 2u$

10. $\dfrac{\partial^2 u}{\partial x \, \partial y} = \dfrac{\partial u}{\partial y}$

11. $\dfrac{d^2y}{dx^2} = \sqrt{1 + \left(\dfrac{dy}{dx}\right)^2}$

12. $\sqrt{\dfrac{d^2y}{dx^2}} = \dfrac{dy}{dx}$

In Exercises 13–18, verify the given solution of the differential equation.

Solution	Differential equation
13. $y = Ce^{4x}$	$\dfrac{dy}{dx} = 4y$
14. $x^2 + y^2 = Cy$	$y' = \dfrac{2xy}{x^2 - y^2}$
15. $y = C_1 \cos x + C_2 \sin x$	$y'' + y = 0$
16. $y = C_1 e^{-x} \cos x + C_2 e^{-x} \sin x$	$y'' + 2y' + 2y = 0$
17. $u = e^{-t} \sin bx$	$b^2 \dfrac{\partial u}{\partial t} = \dfrac{\partial^2 u}{\partial x^2}$
18. $u = \dfrac{y}{x^2 + y^2}$	$\dfrac{\partial^2 u}{\partial x^2} + \dfrac{\partial^2 u}{\partial y^2} = 0$

In Exercises 19–24, determine whether the function is a solution of the differential equation

$$y^{(4)} - 16y = 0.$$

19. $y = 3\cos x$ **20.** $y = 3\cos 2x$

21. $y = e^{-2x}$ **22.** $y = 5\ln x$

23. $y = C_1 e^{2x} + C_2 e^{-2x} + C_3 \sin 2x + C_4 \cos 2x$

24. $y = 5e^{-2x} + 3\cos 2x$

In Exercises 25–30, determine whether the function is a solution of the differential equation

$$x\dfrac{\partial u}{\partial x} - y\dfrac{\partial u}{\partial y} = 0.$$

25. $u = e^{x+y}$ **26.** $u = 5$

27. $u = x^2 y^2$ **28.** $u = \sin xy$

29. $u = (xy)^n$ **30.** $u = x^2 + y^2$

31. It is known that $y = Ce^{kx}$ is a solution of the differential equation

$$\dfrac{dy}{dx} = 0.07y.$$

Is it possible to determine C or k from the information given? If so, find its value.

32. It is known that $y = A \sin \omega t$ is a solution of the differential equation

$$\dfrac{d^2y}{dt^2} + 16y = 0.$$

Find the value of ω.

In Exercises 33 and 34, some of the curves corresponding to different values of C in the general solution of the differential equation are given. Find the particular solution that passes through the point indicated on each graph.

	Solution	Differential equation
33.	$y^2 = Cx^3$	$2xy' - 3y = 0$
34.	$2x^2 - y^2 = C$	$yy' - 2x = 0$

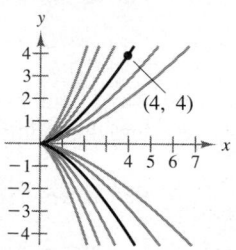

FIGURE FOR 33   **FIGURE FOR 34**

In Exercises 35 and 36, the general solution of the differential equation is given. Sketch the graph of the particular solutions for the given values of C.

35. $4yy' - x = 0$
$2y^2 - x^2 = C$
$C = 0,\ C = \pm 1,\ C = \pm 4$

36. $yy' + x = 0$
$x^2 + y^2 = C$
$C = 0,\ C = 1,\ C = 4$

In Exercises 37–42, verify that the general solutions satisfy the differential equation. Then find the particular solution that satisfies the given initial condition.

37. $y = Ce^{-2x}$
$y' + 2y = 0$
$y = 3$ when $x = 0$

38. $2x^2 + 3y^2 = C$
$2x + 3yy' = 0$
$y = 2$ when $x = 1$

39. $y = C_1 \sin 3x + C_2 \cos 3x$
$y'' + 9y = 0$
$y = 2$ and $y' = 1$ when $x = \dfrac{\pi}{6}$

40. $y = C_1 + C_2 \ln x$
$xy'' + y' = 0$
$y = 0$ and $y' = \frac{1}{2}$ when $x = 2$

41. $y = C_1 x + C_2 x^3$
$x^2 y'' - 3xy' + 3y = 0$
$y = 0$ and $y' = 4$ when $x = 2$

42. $y = e^{2x/3}(C_1 + C_2 x)$
$9y'' - 12y' + 4y = 0$
$y = 4$ when $x = 0$, and $y = 0$ when $x = 3$

In Exercises 43–50, use integration to find a general solution of the differential equation.

43. $\dfrac{dy}{dx} = 3x^2$ **44.** $\dfrac{dy}{dx} = \dfrac{1}{1 + x^2}$

45. $\dfrac{dy}{dx} = \dfrac{x - 2}{x}$ **46.** $\dfrac{dy}{dx} = x \cos x$

47. $\dfrac{dy}{dx} = e^x \sin 2x$ **48.** $\dfrac{dy}{dx} = \tan^2 x$

49. $\dfrac{dy}{dx} = x\sqrt{x - 3}$ **50.** $\dfrac{dy}{dx} = xe^x$

51. *Population Growth*   The limiting capacity of the habitat for a particular wildlife herd is L. The growth rate dN/dt of the herd is proportional to the unused opportunities for growth, as described by the differential equation

$$\frac{dN}{dt} = k(L - N).$$

The general solution of this differential equation is

$N = L - Ce^{-kt}.$

Suppose 100 animals are released into a tract of land that can support 750 of these animals. After 2 years, the herd has grown to 160 animals.
a. Find the population function in terms of the time t in years.
b. Verify that the function from part a is a solution of the differential equation.
c. Sketch the graph of this population function.

52. *Investment*   The rate of growth of an investment is proportional to the amount of the investment at any time t. That is,

$$\frac{dA}{dt} = kA.$$

a. Show that $A = Ce^{kt}$ is a solution of this differential equation.
b. Find the particular solution of this differential equation if the initial investment is \$1000, and 10 years later the balance is \$3320.12.

Direction Fields   **Consider the first-order differential equation**

$$\frac{dy}{dx} = F(x, y).$$

The function F assigns to each point (x_0, y_0) in its domain a direction (slope) of the solution to the differential equation. To visualize this, draw a short mark at each of several points to indicate the direction associated with each point and obtain a direction field. This field can be used to approximate graphically the solutions of the differential equation. In Exercises 53–56, use the given differential equation and direction field to answer questions about the solutions of the differential equation.

53. $\dfrac{dy}{dx} = x$

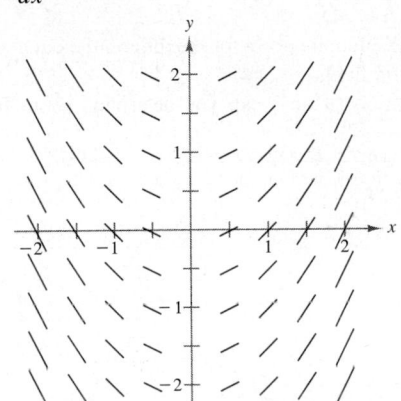

a. Sketch several solution curves for the differential equation on the direction field.
b. Use integration to find the solution of the differential equation. Compare the result with the sketches from part a.

54. $\dfrac{dy}{dx} = -\dfrac{x}{y}$

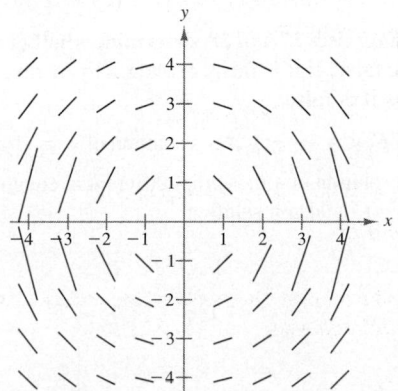

a. Sketch several solution curves for the differential equation on the direction field.
b. If possible, identify the solutions. Then verify your guesses by differentiation.

55. $\dfrac{dy}{dx} = 4 - y$

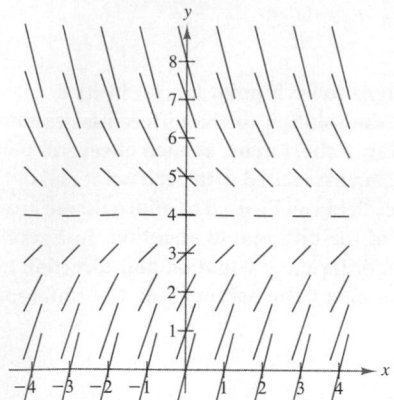

    a. Sketch several solution curves for the differential equation on the direction field.

    b. What information (if any) can you determine about the solution as $x \to \infty$?

56. $\dfrac{dy}{dx} = 0.25y(4 - y)$

    a. Sketch several solution curves for the differential equation on the direction field.

    b. What information (if any) can you determine about the solution as $x \to \infty$?

    c. What information (if any) can you determine about the solution as $x \to -\infty$?

True or False   **In Exercises 57 and 58, determine whether the statement is true or false. If it is false, explain why or give an example that shows it is false.**

57. For any value of C, $y = (x - C)^3$ is a solution of $y' = 3y^{2/3}$.

58. If $y = f(x)$ is a solution of a first-order differential equation, then $y = f(x) + C$ is also a solution.

CAREER INTERVIEW

George M. Drakeley, III
Electrical Systems Branch Head
Submarine Program Office
Naval Sea Systems Command
Arlington, VA 05250

My organization helps the navy buy, maintain, and improve its ships. Among our responsibilities is the problem of noise generation; we devise ways of minimizing the possibility of detection by other vessels underwater.

It is central to the analysis of acoustics to understand the vibration patterns of different machinery and the amount of time a machine can operate before its vibrations begin to increase. (A new car may run smoothly, but over time the parts will wear, resulting in larger vibrations and more noise.)

Using second order differential equations to model equipment behavior, we analyze the generation of acoustic energy and its transmission into the water outside. At present, we are evaluating an improved, quieter system of magnetic bearings for various rotating machinery.

Separation of Variables in First-Order Equations

Separation of Variables • Homogeneous Differential Equations • Applications

Separation of Variables

In this section, you will study techniques for solving differential equations that can be written in the form

$$M(x) + N(y)\frac{dy}{dx} = 0$$

where M is a continuous function of x alone and N is a continuous function of y alone. For this type of equation, all x terms can be collected with dx and all y terms with dy, and a solution can be obtained by integration. Such equations are said to be **separable,** and the solution procedure is called **separation of variables.**

EXAMPLE 1   Separation of Variables

Find the general solution of

$$(x^2 + 4)\frac{dy}{dx} = xy.$$

Solution   To begin, note that $y = 0$ is a solution. To find other solutions, assume $y \neq 0$ and separate variables as follows.

$$(x^2 + 4)\,dy = xy\,dx \qquad \text{Differential form}$$

$$\frac{dy}{y} = \frac{x}{x^2 + 4}\,dx \qquad \text{Separate variables}$$

Now, integrate to obtain

$$\int \frac{dy}{y} = \int \frac{x}{x^2 + 4}\,dx \qquad \text{Integrate}$$

$$\ln |y| = \frac{1}{2}\ln(x^2 + 4) + C_1$$

$$= \ln \sqrt{x^2 + 4} + C_1$$

$$|y| = e^{C_1}\sqrt{x^2 + 4}$$

$$y = \pm e^{C_1}\sqrt{x^2 + 4}.$$

Because $y = 0$ is also a solution, you can write the general solution as

$$y = C\sqrt{x^2 + 4}. \qquad \text{General solution}$$

REMARK   We encourage you to *check your solutions* throughout this chapter. For instance, in Example 1 you can check the solution $y = C\sqrt{x^2 + 4}$ by differentiating and substituting into the original equation.

In some cases, it is not feasible to write the general solution in the explicit form $y = f(x)$. The next example is a case in point; implicit differentiation can be used to verify this solution.

EXAMPLE 2   Finding a Particular Solution

Given the initial condition $y(0) = 1$, find the particular solution of the equation

$$xy \, dx + e^{-x^2}(y^2 - 1) \, dy = 0.$$

Solution   Note that $y = 0$ is a solution of the differential equation—but this solution does not satisfy the initial condition. Hence, you can assume $y \neq 0$. To separate variables, you must rid the first term of y and the second of e^{-x^2}. Thus, you should multiply by e^{x^2}/y and obtain the following.

$$\left(\frac{e^{x^2}}{y}\right) xy \, dx + \left(\frac{e^{x^2}}{y}\right) e^{-x^2}(y^2 - 1) \, dy = 0, \qquad y \neq 0$$

$$xe^{x^2} \, dx + \left(y - \frac{1}{y}\right) dy = 0$$

$$\int xe^{x^2} \, dx + \int \left(y - \frac{1}{y}\right) dy = 0$$

$$\frac{1}{2}e^{x^2} + \frac{y^2}{2} - \ln|y| = C_1$$

$$e^{x^2} + y^2 - \ln y^2 = 2C_1 = C$$

From the given initial condition, use the fact that $y = 1$ when $x = 0$ to obtain $1 + 1 + 0 = 2 = C$. Thus, the particular solution has the implicit form $e^{x^2} + y^2 - \ln y^2 = 2$.

EXAMPLE 3   Finding a Particular Solution Curve

Find the equation of the curve that passes through the point $(1, 3)$ and has a slope of y/x^2 at the point (x, y), as shown in Figure 16.2.

Solution   Because the slope of the curve is given by y/x^2, you have

$$\frac{dy}{dx} = \frac{y}{x^2}$$

with the initial condition $y(1) = 3$. Separating variables and integrating produces

$$\int \frac{dy}{y} = \int \frac{dx}{x^2}, \qquad y \neq 0$$

$$\ln|y| = -\frac{1}{x} + C_1$$

$$y = e^{-(1/x)+C_1} = Ce^{-1/x}.$$

Because $y = 3$ when $x = 1$, it follows that $3 = Ce^{-1}$ and $C = 3e$. Therefore, the equation of the specified curve is

$$y = (3e)e^{-1/x} = 3e^{(x-1)/x}, \qquad x > 0.$$

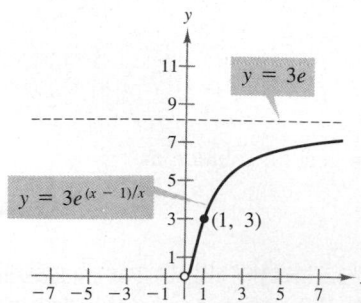

$y = 3e$

$y = 3e^{(x-1)/x}$

$(1, 3)$

FIGURE 16.2

Homogeneous Differential Equations

Some differential equations that are not separable in x and y can be made separable by a change of variables. This is true for differential equations of the form $y' = f(x, y)$, where f is a **homogeneous function.**

The function given by $z = f(x, y)$ is said to be **homogeneous of degree n** if

$$f(tx, ty) = t^n f(x, y)$$

where n is a real number.

EXAMPLE 4   Verifying Homogeneous Functions

a. $f(x, y) = x^2 y - 4x^3 + 3xy^2$ is a homogeneous function of degree 3 because

$$
\begin{aligned}
f(tx, ty) &= (tx)^2(ty) - 4(tx)^3 + 3(tx)(ty)^2 \\
&= t^3(x^2 y) - t^3(4x^3) + t^3(3xy^2) \\
&= t^3(x^2 y - 4x^3 + 3xy^2) \\
&= t^3 f(x, y).
\end{aligned}
$$

b. $f(x, y) = xe^{x/y} + y\sin(y/x)$ is a homogeneous function of degree 1 because

$$f(tx, ty) = txe^{tx/ty} + ty\sin\frac{ty}{tx} = t\left(xe^{x/y} + y\sin\frac{y}{x}\right) = tf(x, y).$$

c. $f(x, y) = x + y^2$ is not a homogeneous function because

$$f(tx, ty) = tx + t^2 y^2 = t(x + ty^2) \neq t^n(x + y^2).$$

d. $f(x, y) = x/y$ is a homogeneous function of degree 0 because

$$f(tx, ty) = \frac{tx}{ty} = t^0\frac{x}{y}.$$

Definition of Homogeneous Differential Equation

A **homogeneous differential equation** is an equation of the form

$$M(x, y)\, dx + N(x, y)\, dy = 0$$

where M and N are homogeneous functions of the same degree.

EXAMPLE 5   Homogeneous Differential Equations

a. $(x^2 + xy)\, dx + y^2\, dy = 0$ is homogeneous of degree 2.

b. $(x^2 + 1)\, dx + y^2\, dy = 0$ is *not* a homogeneous differential equation.

To solve a homogeneous differential equation by the method of separation of variables, we use the following change of variables theorem.

THEOREM 16.1   Change of Variables for Homogeneous Equations

If $M(x, y)\,dx + N(x, y)\,dy = 0$ is homogeneous, then it can be transformed into a differential equation whose variables are separable by the substitution

$$y = vx$$

where v is a differentiable function of x.

EXAMPLE 6   Solving a Homogeneous Differential Equation

Find the general solution of $(x^2 - y^2)\,dx + 3xy\,dy = 0$.

Solution   Because $(x^2 - y^2)$ and $3xy$ are both homogeneous of degree 2, let $y = vx$ to obtain $dy = x\,dv + v\,dx$. Then, by substitution, you have

$$(x^2 - v^2x^2)\,dx + 3x(vx)\overbrace{(x\,dv + v\,dx)}^{dy} = 0$$
$$(x^2 + 2v^2x^2)\,dx + 3x^3v\,dv = 0$$
$$x^2(1 + 2v^2)\,dx + x^2(3vx)\,dv = 0.$$

Dividing by x^2 and separating variables produces

$$(1 + 2v^2)\,dx = -3vx\,dv$$
$$\int \frac{dx}{x} = \int \frac{-3v}{1 + 2v^2}\,dv$$
$$\ln|x| = -\frac{3}{4}\ln(1 + 2v^2) + C_1$$
$$4\ln|x| = -3\ln(1 + 2v^2) + \ln|C|$$
$$\ln x^4 = \ln\left|C(1 + 2v^2)^{-3}\right|$$
$$x^4 = C(1 + 2v^2)^{-3}.$$

Substituting for v produces the following general solution.

$$x^4 = C\left[1 + 2\left(\frac{y}{x}\right)^2\right]^{-3}$$
$$\left(1 + 2\frac{y^2}{x^2}\right)^3 x^4 = C$$
$$(x^2 + 2y^2)^3 = Cx^2 \qquad \text{General solution}$$

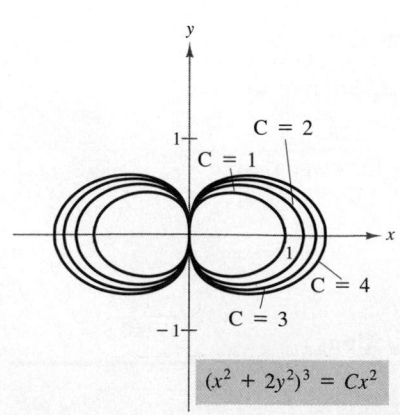

$(x^2 + 2y^2)^3 = Cx^2$

FIGURE 16.3

TECHNOLOGY   If you have access to a graphing utility, try using it to graph several of the solutions in Example 6. For instance, Figure 16.3 shows the graphs of $(x^2 + 2y^2)^3 = Cx^2$ for $C = 1, 2, 3,$ and 4.

Applications

EXAMPLE 7   Application to Food Preservation

In the preservation of food, cane sugar is broken down (inverted) into two simpler sugars: glucose and fructose. In dilute solutions, the inversion rate is proportional to the concentration $y(t)$ of unaltered sugar. If the concentration is $\frac{1}{50}$ when $t = 0$ and $\frac{1}{200}$ after 3 hours, find the concentrations of unaltered sugar after 6 hours and after 12 hours.

Solution   Because the rate of inversion is proportional to $y(t)$, you can write the following differential equation.

$$\frac{dy}{dt} = ky$$

Separating the variables and integrating produces

$$\int \frac{1}{y} dy = \int k \, dt$$

$$\ln|y| = kt + C_1$$

$$y = Ce^{kt}.$$

From the given conditions, you can obtain the following.

$$y(0) = \frac{1}{50} \quad \rightarrow \quad C = \frac{1}{50}$$

$$y(3) = \frac{1}{200} \quad \rightarrow \quad \frac{1}{200} = \left(\frac{1}{50}\right)e^{3k} \quad \rightarrow \quad k = -\frac{\ln 4}{3}$$

Therefore, the concentration of unaltered sugar is given by

$$y(t) = \frac{1}{50} e^{-(\ln 4)t/3} = \frac{1}{50}(4^{-t/3}).$$

When $t = 6$ and $t = 12$, you have the following concentrations of unaltered sugar.

$$y(6) = \frac{1}{50}(4^{-2}) = \frac{1}{800} \qquad \text{After 6 hours}$$

$$y(12) = \frac{1}{50}(4^{-4}) = \frac{1}{12,800} \qquad \text{After 12 hours}$$

(See Figure 16.4)

FIGURE 16.4

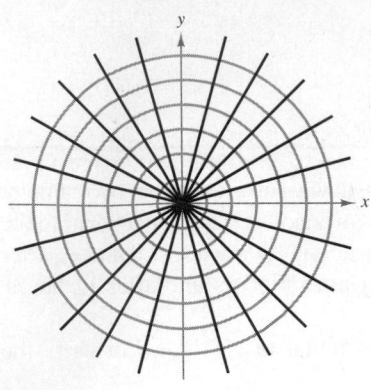

FIGURE 16.5

Each line $y = Kx$ is an orthogonal trajectory to the family of circles.

A common problem in electrostatics, thermodynamics, and hydrodynamics involves finding a family of curves, each of which is orthogonal to all members of a given family of curves. For example, Figure 16.5 shows a family of circles

$$x^2 + y^2 = C \qquad \text{Family of circles}$$

each of which intersects the lines in the family

$$y = Kx \qquad \text{Family of lines}$$

at right angles. Two such families of curves are said to be **mutually orthogonal,** and each curve in one of the families is called an **orthogonal trajectory** of the other family. In electrostatics, lines of force are orthogonal to the *equipotential curves.* In thermodynamics, the flow of heat across a plane surface is orthogonal to the *isothermal curves.* In hydrodynamics, the flow (stream) lines are orthogonal trajectories of the *velocity potential curves.*

EXAMPLE 8   Finding Orthogonal Trajectories

Describe the orthogonal trajectories for the family of curves given by $y = C/x$ for $C \neq 0$. Sketch several members of each family.

Solution   First, solve the given equation for C and write $xy = C$. Then, by differentiating implicitly with respect to x, you obtain the differential equation

$$xy' + y = 0 \qquad \text{Differential equation}$$

$$x\frac{dy}{dx} = -y$$

$$\frac{dy}{dx} = -\frac{y}{x}. \qquad \text{Slope of given family}$$

Because y' represents the slope of the given family of curves at (x, y), it follows that the orthogonal family has the negative reciprocal slope x/y, and we write

$$\frac{dy}{dx} = \frac{x}{y}. \qquad \text{Slope of orthogonal family}$$

Now you can find the orthogonal family by separating variables and integrating.

$$\int y\,dy = \int x\,dx$$

$$\frac{y^2}{2} = \frac{x^2}{2} + C_1$$

Therefore, each orthogonal trajectory is a hyperbola given by

$$\frac{y^2}{K} - \frac{x^2}{K} = 1$$

$$2C_1 = K \neq 0.$$

The centers are at the origin, and the transverse axes are vertical for $K > 0$ and horizontal for $K < 0$. Several trajectories are shown in Figure 16.6.

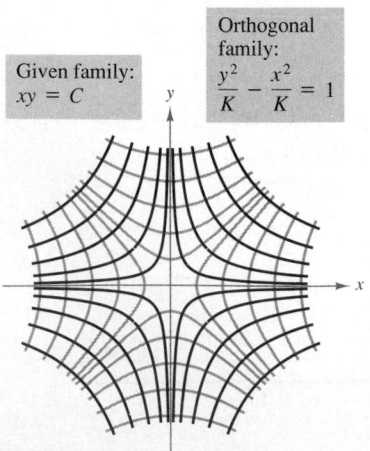

Given family: $xy = C$

Orthogonal family: $\dfrac{y^2}{K} - \dfrac{x^2}{K} = 1$

FIGURE 16.6

EXERCISES for Section 16.2

TECHNOLOGY
Laboratory Guide
Labs 16.2–16.3

In Exercises 1–10, find the general solution of the differential equation.

1. $\dfrac{dy}{dx} = \dfrac{x}{y}$

2. $\dfrac{dy}{dx} = \dfrac{x^2 + 2}{3y^2}$

3. $\dfrac{dr}{ds} = 0.05r$

4. $\dfrac{dr}{ds} = 0.05s$

5. $(2 + x)y' = 3y$

6. $xy' = y$

7. $yy' = \sin x$

8. $\sqrt{1 - 4x^2}\,y' = 1$

9. $y \ln x - xy' = 0$

10. $yy' - 2xe^x = 0$

In Exercises 11–20, find the particular solution of the differential equation that satisfies the initial condition.

Differential equation	Initial condition
11. $yy' - e^x = 0$	$y(0) = 4$
12. $\sqrt{x} + \sqrt{y}\,y' = 0$	$y(1) = 4$
13. $y(x + 1) + y' = 0$	$y(-2) = 1$
14. $xyy' - \ln x = 0$	$y(1) = 0$
15. $(1 + x^2)y' - (1 + y^2) = 0$	$y(0) = \sqrt{3}$
16. $\sqrt{1 - x^2}\,y' - \sqrt{1 - y^2} = 0$	$y(0) = 1$
17. $\dfrac{du}{dv} = uv \sin v^2$	$u(0) = 1$
18. $\dfrac{dr}{ds} = e^{r+s}$	$r(1) = 0$
19. $dP - kP\,dt = 0$	$P(0) = P_0$
20. $dT + k(T - 70)\,dt = 0$	$T(0) = 140$

In Exercises 21 and 22, find an equation for the curve that passes through the given point and has the specified slope.

Point	Slope
21. $(1, 1)$	$y' = -\dfrac{9x}{16y}$
22. $(8, 2)$	$y' = \dfrac{2y}{3x}$

In Exercises 23 and 24, find all functions f having the specified property.

23. The tangent to the graph of f at the point (x, y) intersects the x-axis at $(x + 2, 0)$.

24. All tangents to the graph of f pass through the origin.

In Exercises 25 and 26, sketch a few solutions of the differential equation on the given direction field and then find the general solution analytically.

25. $\dfrac{dy}{dx} = 0.5y$

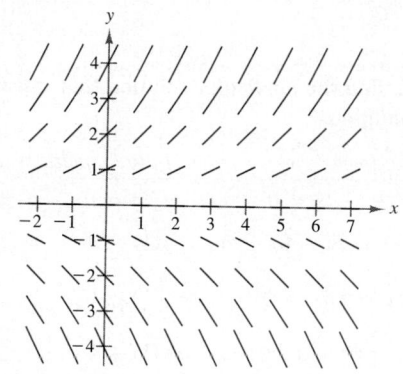

26. $\dfrac{dy}{dx} = 1 + y^2$

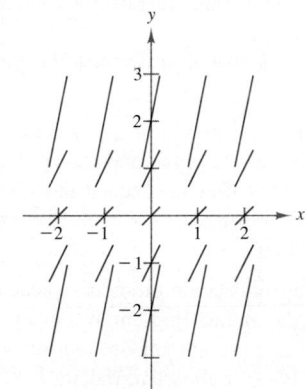

In Exercises 27–32, determine whether the function is homogeneous, and if so, determine its degree.

27. $f(x, y) = x^3 - 4xy^2 + y^3$

28. $f(x, y) = \dfrac{xy}{\sqrt{x^2 + y^2}}$

29. $f(x, y) = 2 \ln xy$

30. $f(x, y) = \tan(x + y)$

31. $f(x, y) = 2 \ln \dfrac{x}{y}$

32. $f(x, y) = \tan \dfrac{y}{x}$

In Exercises 33–38, solve the homogeneous differential equation.

33. $y' = \dfrac{x + y}{2x}$

34. $y' = \dfrac{2x + y}{y}$

35. $y' = \dfrac{x - y}{x + y}$

36. $y' = \dfrac{x^2 + y^2}{2xy}$

37. $y' = \dfrac{xy}{x^2 - y^2}$

38. $y' = \dfrac{3x + 2y}{x}$

In Exercises 39–42, find the particular solution that satisfies the given initial condition.

Differential equation	Initial condition
39. $x\,dy - (2xe^{-y/x} + y)\,dx = 0$	$y(1) = 0$
40. $-y^2\,dx + x(x + y)\,dy = 0$	$y(1) = 1$
41. $\left(x\sec\dfrac{y}{x} + y\right)dx - x\,dy = 0$	$y(1) = 0$
42. $\left(y - \sqrt{x^2 - y^2}\right)dx - x\,dy = 0$	$y(1) = 0$

43. *Investment Growth* The amount A of an investment P increases at a rate proportional to A at any instant of time t.
  a. Find an equation for A as a function of t.
  b. The initial investment is $1000 and the rate is 11 percent. Find the balance after 10 years.
  c. Find the time necessary to double the investment if the rate is 11 percent.

44. *Population Growth* The rate of growth of a population of fruit flies is proportional to the size of the population at any instant. If there were 180 flies after the second day of the experiment and 300 after the fourth day, how many flies were there in the original population?

45. *Radioactive Decay* The rate of decomposition of radioactive radium is proportional to the amount present at a given instant. Find the percentage of a present amount that remains after 25 years, if the half-life of radioactive radium is 1600 years.

46. *Chemical Reaction* In a chemical reaction, a certain compound changes into another compound at a rate proportional to the unchanged amount. If initially there is 20 grams of the original compound, and there is 16 grams after 1 hour, when will 75 percent of the compound be changed?

47. *Newton's Law of Cooling* Newton's Law of Cooling states that the rate of change in the temperature of an object is proportional to the difference between its temperature and the temperature of the surrounding air. Suppose a room is kept at a constant temperature of 70°, and an object cools from 350° to 150° in 45 minutes. How long will it take for the object to cool to a temperature of 80°?

  REMARK   To see how Newton's Law of Cooling is applied to the question "How long should I leave a thermometer in my mouth to take my body temperature accurately?", read the article "Fever" by Elmo Moore and Charles M. Biles in the *UMAP Journal*, Volume 4, Number 3, 1983.

C **48.** *Radio Reception* In hilly areas, radio reception may be poor. Consider a situation where an FM transmitter is located at the point $(-1, 1)$ behind a hill modeled by the graph of $y = x - x^2$, and a radio receiver is on the opposite side of the hill. (Assume the x-axis is ground level at the base of the hill.)
  a. What is the closest the radio can be to the hill so that reception is unobstructed?
  b. Write the closest position x of the radio as a function of h if the transmitter is located at $(-1, h)$.
  c. Use a graphing utility to graph the function from part b. Determine the vertical asymptote of the function and interpret the result.

C **49.** *Sales Increase* Let S (in thousands of units) represent the sales of a new product, let L (in thousands of units) represent the maximum level of sales, and let t represent the time (in months). The rate of change of S with respect to t varies jointly as the product of S and $L - S$.
  a. Write and solve the differential equation for this sales model if $L = 100$, $S = 10$ when $t = 0$, and $S = 20$ when $t = 1$.
  b. At what time is the growth in sales increasing most rapidly?
  c. Use a graphing utility to graph the sales function.
  d. The figure shows the direction field for the differential equation. Sketch the solution of part a on the direction field.
  e. If the estimated maximum level of sales is correct, use the direction field to describe the shape of the solution curves for sales if at some period of time sales exceed L.

FIGURE FOR 49

C **50.** *Learning Theory* The rate of change in the proportion P of correct responses after n trials varies jointly as P and $L - P$, where L is the limiting proportion of correct responses.
  a. Write and solve the differential equation for this learning theory model.
  b. Write P as a function of n if $L = 1.00$, $P = 0.25$ when $n = 0$, and $P = 0.60$ when $n = 10$. Use a graphing utility to graph the solution.
  c. Are there any inflection points on the graph from part b? What (if any exist) do they mean in the context of this model?

*A **C** indicates an exercise in which you are instructed to use graphing technology or a symbolic computer algebra system. The solutions of other exercises may also be facilitated by appropriate technology.

51. *Sailing*   Ignoring resistance, a sailboat starting from rest accelerates (dv/dt) at a rate proportional to the difference between the velocities of the wind and the boat.

a. Write the velocity as a function of time if the wind is blowing at 20 knots, and after 1 minute the boat is moving at 5 knots.

b. Use the result of part a to write the distance traveled by the boat as a function of time.

52. *Boating*   If s gives the position of an object moving in a straight line, the velocity and acceleration are given by

$$v = \frac{ds}{dt}, \qquad a = \frac{dv}{dt} = \frac{ds}{dt}\frac{dv}{ds} = v\frac{dv}{ds}.$$

Suppose that a boat weighing 640 pounds has a motor that exerts a force of 60 pounds. If the resistance (in pounds) to the motion of the boat is 3 times the velocity (in feet per second) and if the boat starts from rest, find its maximum speed. Use the equation

$$\text{Force} = F = ma = mv\frac{dv}{ds} = \frac{w}{32}v\frac{dv}{ds}$$

where w is the weight of the boat.

In Exercises 53–58, find the orthogonal trajectories of the given family and sketch several members of each family.

53. $x^2 + y^2 = C$

54. $2x^2 - y^2 = C$

55. $x^2 = Cy$

56. $y^2 = 2Cx$

57. $y^2 = Cx^3$

58. $y = Ce^x$

True or False   **In Exercises 59–62, determine whether the statement is true or false. If it is false, explain why or give an example that shows it is false.**

59. The differential equation $y' = xy - 2y + x - 2$ can be written in separated variables form.

60. The differential equation $y' = \sin(x + y) + \sin(x - y)$ can be written in separated variables form.

61. The function $f(x, y) = x^2 + xy + 2$ is homogeneous.

62. The families $x^2 + y^2 = 2Cy$ and $x^2 + y^2 = 2Kx$ are mutually orthogonal.

63. *Weight Loss*   A person's weight depends on both the amount of calories consumed and the energy used. Moreover, the amount of energy used depends on a person's weight—the average amount of energy used by a person is 17.5 calories per pound per day. Thus, the more weight a person loses, the less energy the person uses (assuming that the person maintains a constant level of activity). An equation that can be used to model weight loss is

$$3500\left(\frac{dw}{dt}\right) = C - 17.5w$$

where w is the person's weight (in pounds), t is the time in days, and C is the constant daily calorie consumption.

a. Find the general solution of the differential equation.

b. Consider a person who weighs 180 pounds and begins a diet of 2500 calories per day. How long will it take the person to lose 10 pounds? How long will it take the person to lose 35 pounds?

c. The graph shows the person's weight as a function of time. What is the "limiting" weight of the person?

This exercise is based on the article "A Linear Diet Model" by Arthur C. Segal in the January, 1987 issue of the *College Mathematics Journal*.

Exact Differential Equations

In this section, you will study a method for solving the first-order differential equation

$$M(x, y)\, dx + N(x, y)\, dy = 0$$

for the special case in which this equation represents the exact differential of a function $z = f(x, y)$.

Definition of an Exact Differential Equation

The equation $M(x, y)\, dx + N(x, y)\, dy = 0$ is an **exact differential equation** if there exists a function f of two variables x and y having continuous partial derivatives such that

$$f_x(x, y) = M(x, y) \quad \text{and} \quad f_y(x, y) = N(x, y).$$

The general solution of the equation is $f(x, y) = C$.

From Section 14.3, you know that if f has continuous second partials, then

$$\frac{\partial M}{\partial y} = \frac{\partial^2 f}{\partial y\, \partial x} = \frac{\partial^2 f}{\partial x\, \partial y} = \frac{\partial N}{\partial x}.$$

This suggests the following test for exactness.

THEOREM 16.2   Test for Exactness

Let M and N have continuous partial derivatives on an open disc R. The differential equation $M(x, y)\, dx + N(x, y)\, dy = 0$ is exact if and only if

$$\frac{\partial M}{\partial y} = \frac{\partial N}{\partial x}.$$

Exactness is a fragile condition in the sense that seemingly minor alterations in an exact equation can destroy its exactness. This is demonstrated in the following example.

EXAMPLE 1   Testing for Exactness

a. The differential equation $(xy^2 + x)\,dx + yx^2\,dy = 0$ is exact because

$$\frac{\partial M}{\partial y} = \frac{\partial}{\partial y}[xy^2 + x] = 2xy \quad \text{and} \quad \frac{\partial N}{\partial x} = \frac{\partial}{\partial x}[yx^2] = 2xy.$$

But the equation $(y^2 + 1)\,dx + xy\,dy = 0$ is not exact, even though it is obtained by dividing both sides of the first equation by x.

REMARK   Every differential equation of the form

$$M(x)\,dx + N(y)\,dy = 0$$

is exact. In other words, a separable variables equation is actually a special type of an exact equation.

b. The differential equation $\cos y\,dx + (y^2 - x\sin y)\,dy = 0$ is exact because

$$\frac{\partial M}{\partial y} = \frac{\partial}{\partial y}[\cos y] = -\sin y \quad \text{and} \quad \frac{\partial N}{\partial x} = \frac{\partial}{\partial x}[y^2 - x\sin y] = -\sin y.$$

But the equation $\cos y\,dx + (y^2 + x\sin y)\,dy = 0$ is not exact, even though it differs from the first equation only by a single sign.

Note that the test for exactness of $M(x, y)\,dx + N(x, y)\,dy = 0$ is the same as the test for determining whether $\mathbf{F}(x, y) = M(x, y)\mathbf{i} + N(x, y)\mathbf{j}$ is the gradient of a potential function (Theorem 15.1). This means that a general solution $f(x, y) = C$ to an exact differential equation can be found by the method used to find a potential function for a conservative vector field.

EXAMPLE 2   Solving an Exact Differential Equation

Solve the differential equation $(2xy - 3x^2)\,dx + (x^2 - 2y)\,dy = 0$.

Solution   The given differential equation is exact because

$$\frac{\partial M}{\partial y} = \frac{\partial}{\partial y}[2xy - 3x^2] = 2x = \frac{\partial N}{\partial x} = \frac{\partial}{\partial x}[x^2 - 2y].$$

The general solution, $f(x, y) = C$, is given by

$$f(x, y) = \int M(x, y)\,dx$$

$$= \int (2xy - 3x^2)\,dx = x^2 y - x^3 + g(y).$$

In Section 15.1, you determined $g(y)$ by integrating $N(x, y)$ with respect to y and reconciling the two expressions for $f(x, y)$. An alternative method is to partially differentiate this version of $f(x, y)$ with respect to y and compare the result with $N(x, y)$. In other words,

$$f_y(x, y) = \frac{\partial}{\partial y}[x^2 y - x^3 + g(y)] = x^2 + g'(y) = \overbrace{x^2 - 2y}^{N(x,\,y)}.$$

$$\boxed{g'(y) = -2y}$$

Thus, $g'(y) = -2y$, and it follows that $g(y) = -y^2 + C_1$. Therefore,

$$f(x, y) = x^2 y - x^3 - y^2 + C_1$$

and the general solution is $x^2 y - x^3 - y^2 = C$. Figure 16.7 shows the solution curves that correspond to $C = 1, 10, 100,$ and 1000.

FIGURE 16.7

EXAMPLE 3   Solving an Exact Differential Equation

Find the particular solution of

$$(\cos x - x \sin x + y^2)\,dx + 2xy\,dy = 0$$

that satisfies the initial condition $y = 1$ when $x = \pi$.

Solution   The differential equation is exact because

$$\overbrace{\frac{\partial}{\partial y}[\cos x - x\sin x + y^2]}^{\frac{\partial M}{\partial y}} = 2y = \overbrace{\frac{\partial}{\partial x}[2xy]}^{\frac{\partial N}{\partial x}}.$$

Because $N(x, y)$ is simpler than $M(x, y)$, it is better to begin by integrating $N(x, y)$.

$$f(x, y) = \int N(x, y)\,dy = \int 2xy\,dy = xy^2 + g(x)$$

$$f_x(x, y) = \frac{\partial}{\partial x}[xy^2 + g(x)] = y^2 + g'(x) = \overbrace{\cos x - x\sin x + y^2}^{M(x, y)}$$

$$\boxed{g'(x) = \cos x - x\sin x}$$

Thus, $g'(x) = \cos x - x\sin x$ and

$$g(x) = \int (\cos x - x\sin x)\,dx$$

$$= x\cos x + C_1$$

which implies $f(x, y) = xy^2 + x\cos x + C_1$, and the general solution is

$$xy^2 + x\cos x = C. \qquad \text{General solution}$$

Applying the given initial condition produces

$$\pi(1)^2 + \pi\cos\pi = C$$

which implies that $C = 0$. Hence, the particular solution is

$$xy^2 + x\cos x = 0. \qquad \text{Particular solution}$$

The graph of the particular solution is shown in Figure 16.8. Notice that the graph consists of two parts: the ovals are given by $y^2 + \cos x = 0$, and the y-axis is given by $x = 0$.

FIGURE 16.8

In Example 3, note that if $z = f(x, y) = xy^2 + x\cos x$, then the total differential of z is given by

$$dz = f_x(x, y)\,dx + f_y(x, y)\,dy$$

$$= (\cos x - x\sin x + y^2)\,dx + 2xy\,dy$$

$$= M(x, y)\,dx + N(x, y)\,dy.$$

In other words, $M\,dx + N\,dy = 0$ is called an *exact* differential equation because $M\,dx + N\,dy$ is exactly the differential of $f(x, y)$.

Integrating Factors

If the differential equation $M(x, y)\,dx + N(x, y)\,dy = 0$ is not exact, it may be possible to make it exact by multiplying by an appropriate factor $u(x, y)$, which is called an **integrating factor** for the differential equation.

EXAMPLE 4   Multiplying by an Integrating Factor

a. If the differential equation

$$2y\,dx + x\,dy = 0 \qquad \text{Not an exact equation}$$

is multiplied by the integrating factor $u(x, y) = x$, the resulting equation

$$2xy\,dx + x^2\,dy = 0 \qquad \text{Exact equation}$$

is exact—the left side is the total differential of $x^2 y$.

b. If the equation

$$y\,dx - x\,dy = 0 \qquad \text{Not an exact equation}$$

is multiplied by the integrating factor $u(x, y) = 1/y^2$, the resulting equation

$$\frac{1}{y}\,dx - \frac{x}{y^2}\,dy = 0 \qquad \text{Exact equation}$$

is exact—the left side is the total differential of x/y.

Finding an integrating factor can be difficult. However, there are two classes of differential equations whose integrating factors can be found routinely—namely, those that possess integrating factors that are functions of either x alone or y alone. The following theorem, which we present without proof, outlines a procedure for finding these two special categories of integrating factors.

THEOREM 16.3   Integrating Factors

Consider the differential equation $M(x, y)\,dx + N(x, y)\,dy = 0$.
1. If

$$\frac{1}{N(x, y)}[M_y(x, y) - N_x(x, y)] = h(x)$$

is a function of x alone, then $e^{\int h(x)\,dx}$ is an integrating factor.
2. If

$$\frac{1}{M(x, y)}[N_x(x, y) - M_y(x, y)] = k(y)$$

is a function of y alone, then $e^{\int k(y)\,dy}$ is an integrating factor.

REMARK   If either $h(x)$ or $k(y)$ is constant, Theorem 16.3 still applies. As an aid to remembering these formulas, note that the subtracted partial derivative identifies both the denominator and the variable for the integrating factor.

EXAMPLE 5   Finding an Integrating Factor

Solve the differential equation $(y^2 - x)\,dx + 2y\,dy = 0$.

Solution   The given equation is not exact because $M_y(x, y) = 2y$ and $N_x(x, y) = 0$. However, because

$$\frac{M_y(x, y) - N_x(x, y)}{N(x, y)} = \frac{2y - 0}{2y} = 1 = h(x)$$

it follows that $e^{\int h(x)\,dx} = e^{\int dx} = e^x$ is an integrating factor. Multiplying the given differential equation by e^x produces the exact differential equation

$$(y^2 e^x - xe^x)\,dx + 2ye^x\,dy = 0$$

whose solution is obtained as follows.

$$f(x, y) = \int N(x, y)\,dy = \int 2ye^x\,dy = y^2 e^x + g(x)$$

$$f_x(x, y) = y^2 e^x + g'(x) = \overbrace{y^2 e^x - xe^x}^{M(x, y)}$$

$$\boxed{g'(x) = -xe^x}$$

Therefore, $g'(x) = -xe^x$ and $g(x) = -xe^x + e^x + C_1$, which implies that

$$f(x, y) = y^2 e^x - xe^x + e^x + C_1.$$

The general solution is $y^2 e^x - xe^x + e^x = C$, or $y^2 - x + 1 = Ce^{-x}$.   ▪

In the next example, we show how a differential equation can help in sketching a force field given by $\mathbf{F}(x, y) = M(x, y)\mathbf{i} + N(x, y)\mathbf{j}$.

EXAMPLE 6   An Application to Force Fields

Sketch the force field given by

$$\mathbf{F}(x, y) = \frac{2y}{\sqrt{x^2 + y^2}}\mathbf{i} - \frac{y^2 - x}{\sqrt{x^2 - y^2}}\mathbf{j}$$

by finding and sketching the family of curves tangent to $\mathbf{F}$.

Solution   At the point (x, y) in the plane, the vector $\mathbf{F}(x, y)$ has a slope of

$$\frac{dy}{dx} = \frac{-(y^2 - x)/\sqrt{x^2 - y^2}}{2y/\sqrt{x^2 + y^2}} = \frac{-(y^2 - x)}{2y}$$

which, in differential form, is

$$2y\,dy = -(y^2 - x)\,dx$$

$$(y^2 - x)\,dx + 2y\,dy = 0.$$

From Example 5, we know that the general solution of this differential equation is $y^2 - x + 1 = Ce^{-x}$, or $y^2 = x - 1 + Ce^{-x}$. Figure 16.9 shows several representative curves from this family. Note that the force vector at (x, y) is tangent to the curve passing through (x, y).   ▪

Force field:
$$\mathbf{F}(x, y) = \frac{2y}{\sqrt{x^2 + y^2}}\mathbf{i} - \frac{y^2 - x}{\sqrt{x^2 + y^2}}\mathbf{j}$$

Family of tangent curves to $\mathbf{F}$:
$$y^2 = x - 1 + Ce^{-x}$$

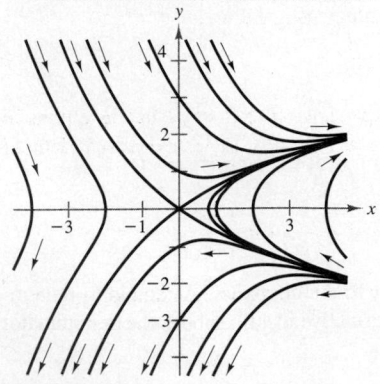

FIGURE 16.9

EXERCISES for Section 16.3

TECHNOLOGY
Laboratory Guide
Labs 16.2–16.3

In Exercises 1–10, determine whether the differential equation is exact. If it is, find the general solution.

1. $(2x - 3y)\,dx + (2y - 3x)\,dy = 0$

2. $ye^x\,dx + e^x\,dy = 0$

3. $(3y^2 + 10xy^2)\,dx + (6xy - 2 + 10x^2y)\,dy = 0$

4. $2\cos(2x - y)\,dx - \cos(2x - y)\,dy = 0$

5. $(4x^3 - 6xy^2)\,dx + (4y^3 - 6xy)\,dy = 0$

6. $2y^2e^{xy^2}\,dx + 2xye^{xy^2}\,dy = 0$

7. $\dfrac{1}{x^2 + y^2}(x\,dy - y\,dx) = 0$

8. $e^{-(x^2+y^2)}(x\,dx + y\,dy) = 0$

9. $\dfrac{1}{(x - y)^2}(y^2\,dx + x^2\,dy) = 0$

10. $e^y\cos xy\,[y\,dx + (x + \tan xy)\,dy] = 0$

In Exercises 11–16, find the particular solution that satisfies the initial condition.

Differential equation	Initial condition
11. $\dfrac{y}{x - 1}\,dx + [\ln(x - 1) + 2y]\,dy = 0$	$y(2) = 4$
12. $\dfrac{1}{\sqrt{x^2 + y^2}}(x\,dx + y\,dy) = 0$	$y(4) = 3$
13. $\dfrac{1}{x^2 + y^2}(x\,dx + y\,dy) = 0$	$y(0) = 4$
14. $e^{3x}(\sin 3y\,dx + \cos 3y\,dy) = 0$	$y(0) = \pi$
15. $(2x\tan y + 5)\,dx + (x^2\sec^2 y)\,dy = 0$	$y(0) = 0$
16. $(x^2 + y^2)\,dx + 2xy\,dy = 0$	$y(3) = 1$

In Exercises 17–26, find the integrating factor that is a function of x or y alone and use it to find the general solution of the differential equation.

17. $y\,dx - (x + 6y^2)\,dy = 0$ **18.** $(2x^3 + y)\,dx - x\,dy = 0$

19. $(5x^2 - y)\,dx + x\,dy = 0$

20. $(5x^2 - y^2)\,dx + 2y\,dy = 0$

21. $(x + y)\,dx + \tan x\,dy = 0$

22. $(2x^2y - 1)\,dx + x^3\,dy = 0$

23. $y^2\,dx + (xy - 1)\,dy = 0$

24. $(x^2 + 2x + y)\,dx + 2\,dy = 0$

25. $2y\,dx + \left(x - \sin\sqrt{y}\right)dy = 0$

26. $(-2y^3 + 1)\,dx + (3xy^2 + x^3)\,dy = 0$

In Exercises 27–30, use the integrating factor to find the general solution of the differential equation.

27. $(4x^2y + 2y^2)\,dx + (3x^3 + 4xy)\,dy = 0$
$u(x, y) = xy^2$

28. $(3y^2 + 5x^2y)\,dx + (3xy + 2x^3)\,dy = 0$
$u(x, y) = x^2y$

29. $(-y^5 + x^2y)\,dx + (2xy^4 - 2x^3)\,dy = 0$
$u(x, y) = x^{-2}y^{-3}$

30. $-y^3\,dx + (xy^2 - x^2)\,dy = 0$
$u(x, y) = x^{-2}y^{-2}$

31. Show that each of the following are integrating factors for the differential equation

$$y\,dx - x\,dy = 0.$$

a. $\dfrac{1}{x^2}$ **b.** $\dfrac{1}{y^2}$ **c.** $\dfrac{1}{xy}$ **d.** $\dfrac{1}{x^2 + y^2}$

32. Show that the differential equation

$$(axy^2 + by)\,dx + (bx^2y + ax)\,dy = 0$$

is exact only if $a = b$. If $a \neq b$, show that x^my^n is an integrating factor, where

$$m = -\frac{2b + a}{a + b}, \qquad n = -\frac{2a + b}{a + b}.$$

C In Exercises 33–36, use a graphing utility to graph the family of tangent curves to the given force field.

33. $\mathbf{F}(x, y) = \dfrac{y}{\sqrt{x^2 + y^2}}\mathbf{i} - \dfrac{x}{\sqrt{x^2 + y^2}}\mathbf{j}$

34. $\mathbf{F}(x, y) = \dfrac{x}{\sqrt{x^2 + y^2}}\mathbf{i} - \dfrac{y}{\sqrt{x^2 + y^2}}\mathbf{j}$

35. $\mathbf{F}(x, y) = 4x^2y\mathbf{i} - \left(2xy^2 + \dfrac{x}{y^2}\right)\mathbf{j}$

36. $\mathbf{F}(x, y) = (1 + x^2)\mathbf{i} - 2xy\mathbf{j}$

In Exercises 37 and 38, find an equation for the curve passing through the point with the specified slope.

Point	Slope
37. $(2, 1)$	$\dfrac{dy}{dx} = \dfrac{y - x}{3y - x}$
38. $(0, 2)$	$\dfrac{dy}{dx} = \dfrac{-2xy}{x^2 + y^2}$

39. *Cost*   If $y = C(x)$ represents the cost of producing x units in a manufacturing process, then the **elasticity of cost** is defined as

$$E(x) = \frac{\text{marginal cost}}{\text{average cost}} = \frac{C'(x)}{C(x)/x} = \frac{x}{y}\frac{dy}{dx}.$$

Find the cost function if the elasticity function is

$$E(x) = \frac{20x - y}{2y - 10x}$$

where $C(100) = 500$ and $100 \le x$.

40. *Euler Method* Consider the differential equation $y' = F(x, y)$ with the initial condition $y(x_0) = y_0$. At any point (x_k, y_k) in the domain of F, $F(x_k, y_k)$ yields the slope of the solution at that point. The Euler Method gives a discrete set of estimates of the y values of a solution of the differential equation using the iterative formula

$$y_{k+1} = y_k + F(x_k, y_k) \Delta x$$

where $\Delta x = x_{k+1} - x_k$.

a. Write a short paragraph describing the general idea of how the Euler Method works.

b. How will decreasing the magnitude of Δx affect the accuracy of the Euler Method?

41. *Euler Method* Use the Euler Method to approximate $y(1)$ for the values of Δx given in the table if $y' = x + \sqrt{y}$ and $y(0) = 2$. (Note that the number of iterations increases as Δx decreases.) Sketch a graph of the approximate solution on the direction field in the figure.

Δx	0.50	0.25	0.10
Estimate of $y(1)$			

The value of $y(1)$, accurate to three decimal places, is 4.213.

FIGURE FOR 41

42. a. *Euler Method* Write a computer program or use a spreadsheet to perform the calculations of the Euler Method. The output should be two columns giving x_k and y_k, respectively. Use the program to approximate the solution of the differential equation $y' = x\sqrt[3]{y}$ over the interval $[1, 2]$ if $\Delta x = 0.01$ and $y(1) = 1$.

b. Solve the differential equation analytically, and compare the exact value of $y(2)$ with the estimate found in part a.

True or False **In Exercises 43–46, determine whether the statement is true or false. If it is false, explain why or give an example that shows it is false.**

43. The differential equation $2xy\,dx + (y^2 - x^2)\,dy = 0$ is exact.

44. If $M\,dx + N\,dy = 0$ is exact, then $xM\,dx + xN\,dy = 0$ is also exact.

45. If $M\,dx + N\,dy = 0$ is exact, then $[f(x) + M]\,dx + [g(y) + N]\,dy = 0$ is also exact.

46. The differential equation $f(x)\,dx + g(y)\,dy = 0$ is exact.

16.4
First-Order Linear Differential Equations • Bernoulli Equations • Applications

First-Order Linear Differential Equations

In this section, you will see how integrating factors help to solve a very important class of first-order differential equations—first-order *linear* differential equations.

> #### Definition of First-Order Linear Differential Equation
>
> A first-order linear differential equation is an equation of the form
>
> $$\frac{dy}{dx} + P(x)y = Q(x)$$
>
> where P and Q are continuous functions of x. This first-order linear differential equation is said to be in **standard form.**

To solve a first-order linear differential equation, you can use an integrating factor $u(x)$, which converts the left side into the derivative of the product $u(x)y$. That is, you need a factor $u(x)$ such that

$$u(x)\frac{dy}{dx} + u(x)P(x)y = \frac{d[u(x)y]}{dx}$$

$$u(x)y' + u(x)P(x)y = u(x)y' + yu'(x)$$

$$u(x)P(x)y = yu'(x)$$

$$P(x) = \frac{u'(x)}{u(x)}$$

$$\ln|u(x)| = \int P(x)\,dx + C_1$$

$$u(x) = Ce^{\int P(x)\,dx}.$$

Because you don't need the most general integrating factor, let $C = 1$. Multiplying the original equation $y' + P(x)y = Q(x)$ by $u(x) = e^{\int P(x)\,dx}$ produces

$$y'e^{\int P(x)\,dx} + yP(x)e^{\int P(x)\,dx} = Q(x)e^{\int P(x)\,dx}$$

$$\frac{d}{dx}\left[ye^{\int P(x)\,dx}\right] = Q(x)e^{\int P(x)\,dx}.$$

The general solution is given by

$$ye^{\int P(x)\,dx} = \int Q(x)e^{\int P(x)\,dx}\,dx + C.$$

THEOREM 16.4   Solution of a First-Order Linear Differential Equation

An integrating factor for the first-order differential equation

$$y' + P(x)y = Q(x)$$

is $u(x) = e^{\int P(x)\,dx}$. The solution of the differential equation is

$$ye^{\int P(x)\,dx} = \int Q(x)e^{\int P(x)\,dx}\,dx + C.$$

REMARK   Rather than memorizing this formula, just remember that multiplication by the integrating factor $e^{\int P(x)\,dx}$ converts the left side of the differential equation into the derivative of the product $ye^{\int P(x)\,dx}$.

EXAMPLE 1   Solving a First-Order Linear Differential Equation

Find the general solution of

$$xy' - 2y = x^2.$$

Solution   The *standard form* of the given equation is

$$y' + P(x)y = Q(x)$$

$$y' - \left(\frac{2}{x}\right)y = x. \qquad \text{Standard form}$$

Thus, $P(x) = -2/x$, and you have

$$\int P(x)\,dx = -\int \frac{2}{x}\,dx = -\ln x^2$$

$$e^{\int P(x)\,dx} = e^{-\ln x^2} = \frac{1}{x^2}. \qquad \text{Integrating factor}$$

Therefore, multiplying both sides of the standard form by $1/x^2$ yields

$$\frac{y'}{x^2} - \frac{2y}{x^3} = \frac{1}{x}$$

$$\frac{d}{dx}\left[\frac{y}{x^2}\right] = \frac{1}{x}$$

$$\frac{y}{x^2} = \int \frac{1}{x}\,dx$$

$$\frac{y}{x^2} = \ln |x| + C$$

$$y = x^2(\ln |x| + C). \qquad \text{General solution}$$

Several solution curves (for $C = -2, -1, 0, 1, 2, 3,$ and 4) are shown in Figure 16.10.

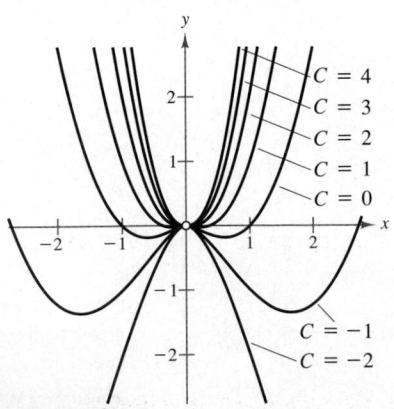

FIGURE 16.10

EXAMPLE 2   Solving a First-Order Linear Differential Equation

Find the general solution

$$y' - y\tan t = 1, \qquad -\frac{\pi}{2} < t < \frac{\pi}{2}.$$

Solution   The equation is already in the standard form $y' + P(t)y = Q(t)$. Thus, $P(t) = -\tan t$, and

$$\int P(t)\,dt = -\int \tan t\,dt = \ln|\cos t|$$

which implies that the integrating factor is

$$e^{\int P(t)\,dt} = e^{\ln|\cos t|}$$
$$= |\cos t|. \qquad \text{Integrating factor}$$

A quick check shows that $\cos t$ is also an integrating factor. Thus, multiplying $y' - y\tan t = 1$ by $\cos t$ produces

$$\frac{d}{dt}[y\cos t] = \cos t$$

$$y\cos t = \int \cos t\,dt$$

$$y\cos t = \sin t + C$$

$$y = \tan t + C\sec t. \qquad \text{General solution}$$

Several solution curves are shown in Figure 16.11.

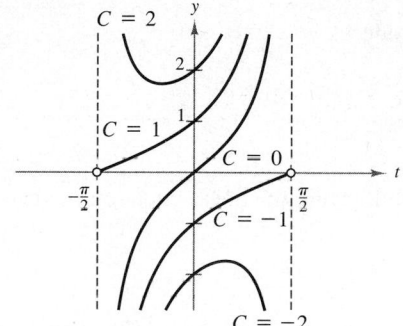

$C = 2$

$C = 1$

$C = 0$

$C = -1$

$C = -2$

FIGURE 16.11

Bernoulli Equations

A well-known *nonlinear* equation that reduces to a linear one with an appropriate substitution is the **Bernoulli equation,** named after James Bernoulli (1654–1705).

$$y' + P(x)y = Q(x)y^n \qquad \text{Bernoulli equation}$$

This equation is linear if $n = 0$, and has separable variables if $n = 1$. Thus, in the following development, assume that $n \neq 0$ and $n \neq 1$. Begin by multiplying by y^{-n} and $(1 - n)$ to obtain

$$y^{-n}y' + P(x)y^{1-n} = Q(x)$$
$$(1 - n)y^{-n}y' + (1 - n)P(x)y^{1-n} = (1 - n)Q(x)$$
$$\frac{d}{dx}[y^{1-n}] + (1 - n)P(x)y^{1-n} = (1 - n)Q(x)$$

which is a linear equation in the variable y^{1-n}. Letting $z = y^{1-n}$ produces the linear equation

$$\frac{dz}{dx} + (1 - n)P(x)z = (1 - n)Q(x).$$

Finally, by Theorem 16.4, the *general solution of the Bernoulli equation* is

$$y^{1-n}e^{\int (1-n)P(x)\,dx} = \int (1 - n)Q(x)e^{\int (1-n)P(x)\,dx}\,dx + C.$$

EXAMPLE 3   Solving a Bernoulli Equation

Find the general solution of $y' + xy = xe^{-x^2}y^{-3}$.

Solution   For this Bernoulli equation, let $n = -3$, and use the substitution

$$z = y^4 \qquad\qquad \text{Let } z = y^{1-n} = y^{1-(-3)}$$
$$z' = 4y^3 y'. \qquad\qquad \text{Differentiate}$$

Multiplying the original equation by $4y^3$ produces

$$y' + xy = xe^{-x^2}y^{-3} \qquad\qquad \text{Original equation}$$
$$4y^3 y' + 4xy^4 = 4xe^{-x^2} \qquad\qquad \text{Multiply both sides by } 4y^3$$
$$z' + 4xz = 4xe^{-x^2}. \qquad\qquad \text{Linear equation: } z' + P(x)z = Q(x)$$

This equation is linear in z. Using $P(x) = 4x$ produces

$$\int P(x)\,dx = \int 4x\,dx = 2x^2$$

which implies that e^{2x^2} is an integrating factor. Multiplying the linear equation by this factor produces

$$z' + 4xz = 4xe^{-x^2} \qquad\qquad \text{Linear equation}$$
$$z'e^{2x^2} + 4xze^{2x^2} = 4xe^{x^2} \qquad\qquad \text{Exact equation}$$
$$\frac{d}{dx}\left[ze^{2x^2}\right] = 4xe^{x^2} \qquad\qquad \text{Write left side as total differential}$$
$$ze^{2x^2} = \int 4xe^{x^2}\,dx \qquad\qquad \text{Integrate both sides}$$
$$ze^{2x^2} = 2e^{x^2} + C$$
$$z = 2e^{-x^2} + Ce^{-2x^2}. \qquad\qquad \text{Divide both sides by } e^{2x^2}$$

Finally, substituting $z = y^4$, the general solution is

$$y^4 = 2e^{-x^2} + Ce^{-2x^2}. \qquad\qquad \text{General solution}$$

So far you have studied several types of first-order differential equations. Of these, the separable variables case is usually the simplest, and solution by an integrating factor is usually a last resort.

Summary of First-Order Differential Equations

Method	*Form of equation*
1. Separable variables:	$M(x)\,dx + N(y)\,dy = 0$
2. Homogeneous:	$M(x, y)\,dx + N(x, y)\,dy = 0$, where M and N are nth-degree homogeneous.
3. Exact:	$M(x, y)\,dx + N(x, y)\,dy = 0$ where $\partial M/\partial y = \partial N/\partial x$.
4. Integrating factor:	$u(x, y)\,M(x, y)\,dx + u(x, y)N(x, y)\,dy = 0$ is exact.
5. Linear:	$y' + P(x)y = Q(x)$
6. Bernoulli equation:	$y' + P(x)y = Q(x)y^n$

Applications

One type of problem that can be described in terms of a differential equation involves chemical mixtures, as illustrated in the next example.

4 gal/min

5 gal/min

FIGURE 16.12

EXAMPLE 4   A Mixture Problem

A tank contains 50 gallons of a solution composed of 90 percent water and 10 percent alcohol. A second solution containing 50 percent water and 50 percent alcohol is added to the tank at the rate of 4 gallons per minute. As the second solution is being added, the tank is being drained at the rate of 5 gallons per minute, as shown in Figure 16.12. Assuming the solution in the tank is stirred constantly, how much alcohol is in the tank after 10 minutes?

Solution   Let y be the number of gallons of alcohol in the tank at any time t. You know that $y = 5$ when $t = 0$. Because the number of gallons of solution in the tank at any time is $50 - t$, and the tank loses 5 gallons of solution per minute, it must lose

$$\left(\frac{5}{50 - t}\right)y$$

gallons of alcohol per minute. Furthermore, because the tank is gaining 2 gallons of alcohol per minute, the rate of change of alcohol in the tank is given by

$$\frac{dy}{dt} = 2 - \left(\frac{5}{50 - t}\right)y \quad \rightarrow \quad \frac{dy}{dt} + \left(\frac{5}{50 - t}\right)y = 2.$$

To solve this linear equation, let $P(t) = 5/(50 - t)$ and obtain

$$\int P(t)\, dt = \int \frac{5}{50 - t}\, dt = -5 \ln |50 - t|.$$

Because $t < 50$, you can drop the absolute value signs and conclude that

$$e^{\int P(t)\, dt} = e^{-5\ln(50-t)} = \frac{1}{(50 - t)^5}.$$

Thus, the general solution is

$$\frac{y}{(50 - t)^5} = \int \frac{2}{(50 - t)^5}\, dt = \frac{1}{2(50 - t)^4} + C$$

$$y = \frac{50 - t}{2} + C(50 - t)^5.$$

Because $y = 5$ when $t = 0$, you have

$$5 = \frac{50}{2} + C(50)^5 \quad \rightarrow \quad -\frac{20}{50^5} = C$$

which means that the particular solution is

$$y = \frac{50 - t}{2} - 20\left(\frac{50 - t}{50}\right)^5.$$

Finally, when $t = 10$, the amount of alcohol in the tank is

$$y = \frac{50 - 10}{2} - 20\left(\frac{50 - 10}{50}\right)^5 = 13.45 \text{ gal}$$

which represents a solution containing 33.6 percent alcohol.   ■

In most falling-body problems discussed so far in the text, we have neglected air resistance. The next example includes this factor. In the example, the air resistance on the falling object is assumed to be proportional to its velocity v. If g is the gravitational constant, then the downward force F on a falling object of mass m is given by the difference $mg - kv$. But by Newton's Second Law of Motion, you know that $F = ma = m(dv/dt)$, which yields the following differential equation.

$$m\frac{dv}{dt} = mg - kv \quad \rightarrow \quad \frac{dv}{dt} + \frac{k}{m}v = g$$

EXAMPLE 5   A Falling Object with Air Resistance

An object of mass m is dropped from a hovering helicopter. Find its velocity as a function of time t, assuming the air resistance is proportional to the velocity of the object.

Solution   The velocity v satisfies the equation

$$\frac{dv}{dt} + \frac{kv}{m} = g$$

where g is the gravitational constant and k is the constant of proportionality. Letting $b = k/m$, you can *separate variables* to obtain

$$dv = (g - bv)\,dt$$
$$\int \frac{dv}{g - bv} = \int dt$$
$$-\frac{1}{b}\ln|g - bv| = t + C_1$$
$$\ln|g - bv| = -bt - bC_1$$
$$g - bv = Ce^{-bt}.$$

Because the object was dropped, $v = 0$ when $t = 0$; thus $g = C$, and it follows that

$$-bv = -g + ge^{-bt} \quad \rightarrow \quad v = \frac{g - ge^{-bt}}{b} = \frac{mg}{k}(1 - e^{-kt/m}). \qquad \blacksquare$$

REMARK   Note in Example 5 that the velocity approaches a limit of mg/k. This limit is due to the air resistance. For falling-body problems in which air resistance is neglected, the velocity increases without bound.

A simple electric circuit consists of electric current I (in amperes), a resistance R (in ohms), an inductance L (in henrys), and a constant electromotive force E (in volts), as shown in Figure 16.13. According to Kirchhoff's Second Law, if the switch S is closed when $t = 0$, then the applied electromotive force (voltage) is equal to the sum of the voltage drops in the rest of the circuit. This in turn means that the current I satisfies the differential equation

$$L\frac{dI}{dt} + RI = E.$$

FIGURE 16.13

EXAMPLE 6   An Electric Circuit Problem

Find the current I as a function of time t (in seconds), given that I satisfies the differential equation $L(dI/dt) + RI = \sin 2t$, where R and L are nonzero constants.

Solution   In standard form, the given linear equation is

$$\frac{dI}{dt} + \frac{R}{L}I = \frac{1}{L}\sin 2t.$$

Let $P(t) = R/L$, so that $e^{\int P(t)\,dt} = e^{(R/L)t}$, and, by Theorem 16.4,

$$Ie^{(R/L)t} = \frac{1}{L}\int e^{(R/L)t}\sin 2t\,dt$$

$$= \frac{1}{4L^2 + R^2}e^{(R/L)t}(R\sin 2t - 2L\cos 2t) + C.$$

Thus, the general solution is

$$I = e^{-(R/L)t}\left[\frac{1}{4L^2 + R^2}e^{(R/L)t}(R\sin 2t - 2L\cos 2t) + C\right]$$

$$I = \frac{1}{4L^2 + R^2}(R\sin 2t - 2L\cos 2t) + Ce^{-(R/L)t}.$$

EXERCISES for Section 16.4

TECHNOLOGY
Laboratory Guide
Lab 16.4

True or False   In Exercises 1 and 2, determine whether the statement is true or false. If it is false, explain why or give an example that shows it is false.

1. $y' + x\sqrt{y} = x^2$ is a first-order linear differential equation.

2. $y' + xy = e^x y$ is a first-order linear differential equation.

In Exercises 3–12, solve the first-order linear differential equation.

3. $\dfrac{dy}{dx} + \left(\dfrac{1}{x}\right)y = 3x + 4$ **4.** $\dfrac{dy}{dx} + \left(\dfrac{2}{x}\right)y = 3x + 1$

5. $\dfrac{dy}{dx} = e^x - y$ **6.** $y' + 2y = \sin x$

7. $y' - y = \cos x$ **8.** $y' + 2xy = 2x$

9. $(3y + \sin 2x)\,dx - dy = 0$

10. $(y - 1)\sin x\,dx - dy = 0$

11. $(x - 1)y' + y = x^2 - 1$

12. $y' + 5y = e^{5x}$

In Exercises 13–18, find the particular solution of the differential equation that satisfies the boundary condition.

Differential equation	Boundary condition
13. $y'\cos^2 x + y - 1 = 0$	$y(0) = 5$
14. $x^3y' + 2y = e^{1/x^2}$	$y(1) = e$
15. $y' + y\tan x = \sec x + \cos x$	$y(0) = 1$

Differential equation	Boundary condition
16. $y' + y\sec x = \sec x$	$y(0) = 4$
17. $y' + \left(\dfrac{1}{x}\right)y = 0$	$y(2) = 2$
18. $y' + (2x - 1)y = 0$	$y(1) = 2$

In Exercises 19–24, solve the Bernoulli differential equation.

19. $y' + 3x^2 y = x^2 y^3$ **20.** $y' + 2xy = xy^2$

21. $y' + \left(\dfrac{1}{x}\right)y = xy^2$ **22.** $y' + \left(\dfrac{1}{x}\right)y = x\sqrt{y}$

23. $y' - y = x^3\sqrt[3]{y}$ **24.** $yy' - 2y^2 = e^x$

Electrical Circuits   In Exercises 25–28, use the differential equation for electrical circuits given by

$$L\frac{dI}{dt} + RI = E.$$

In this equation, I is the current, R is the resistance, L is the inductance, and E is the electromotive force (voltage).

25. Solve the differential equation given a constant voltage E_0.

26. Use the result of Exercise 25 to find the equation for the current if $I(0) = 0$, $E_0 = 110$ volts, $R = 550$ ohms, and $L = 4$ henrys. When does the current reach 90 percent of its limiting value?

27. Solve the differential equation given a periodic electromotive force $E_0\sin\omega t$.

28. Verify that the solution of Exercise 27 can be written in the form

$$I = ce^{-(R/L)t} + \frac{E_0}{\sqrt{R^2 + \omega^2 L^2}} \sin(\omega t + \phi)$$

where ϕ, the phase angle, is given by $\arctan(\omega L/R)$. (Note that the exponential term approaches 0 as $t \to \infty$. This implies that the current approaches a periodic function.

29. *Population Growth*   When predicting population growth, demographers must consider birth and death rates as well as the net change caused by the difference between the rates of immigration and emigration. Let P be the population at time t and N be the net increase per unit time due to the difference between immigration and emigration. Thus, the rate of growth of the population is given by

$$\frac{dP}{dt} = kP + N, \qquad N \text{ is constant.}$$

Solve this differential equation to find P as a function of time if at time $t = 0$ the size of the population is P_0.

30. *Investment Growth*   A large corportion starts at time $t = 0$ to continuously invest part of its receipts at a rate of P dollars per year in a fund for future corporate expansion. Assume that the fund earns r percent interest per year compounded continuously. Thus, the rate of growth of the amount A in the fund is given by

$$\frac{dA}{dt} = rA + P$$

where $A = 0$ when $t = 0$. Solve this differential equation for A as a function of t.

***Investment Growth*   In Exercises 31 and 32, use the result of Exercise 30.**

31. Find A for the following.
   a. $P = \$100,000$, $r = 12\%$, and $t = 5$ years
   b. $P = \$250,000$, $r = 15\%$, and $t = 10$ years

32. Find t if the corporation needs $\$800,000$ and it can invest $\$75,000$ per year in a fund earning 13% interest compounded continuously.

33. *Intravenous Feeding*   Glucose is added intravenously to the bloodstream at the rate of q units per minute, and the body removes glucose from the bloodstream at a rate proportional to the amount present. Assume $Q(t)$ is the amount of glucose in the bloodstream at time t.
   a. Determine the differential equation describing the rate of change with respect to time of glucose in the bloodstream.
   b. Solve the differential equation from part a, letting $Q = Q_0$ when $t = 0$.
   c. Find the limit of $Q(t)$ as $t \to \infty$.

34. *Learning Curve*   The management at a certain factory has found that the maximum number of units a worker can produce in a day is 30. The rate of increase in the number of units N produced with respect to time t in days by a new employee is proportional to $30 - N$.

a. Determine the differential equation describing the rate of change of performance with respect to time.
b. Solve the differential equation from part a.
c. Find the particular solution for a new employee who produced 10 units on the first day at the factory and 19 units on the twentieth day.

***Mixture*   In Exercises 35–38, consider a tank that at time $t = 0$ contains v_0 gallons of a solution of which, by weight, q_0 pounds is soluble concentrate. Another solution containing q_1 pounds of the concentrate per gallon is running into the tank at the rate of r_1 gallons per minute. The solution in the tank is kept well stirred and is withdrawn at the rate of r_2 gallons per minute.**

35. If Q is the amount of concentrate in the solution at any time t, show that

$$\frac{dQ}{dt} + \frac{r_2 Q}{v_0 + (r_1 - r_2)t} = q_1 r_1.$$

36. A 200-gallon tank is full of a solution containing 25 pounds of concentrate. Starting at time $t = 0$, distilled water is admitted to the tank at the rate of 10 gallons per minute, and the well-stirred solution is withdrawn at the same rate.
   a. Find the amount of the concentrate in the solution as a function of t.
   b. Find the time at which the amount of concentrate in the tank reaches 15 pounds.
   c. Find the quantity of the concentrate in the solution as $t \to \infty$.

37. Repeat Exercise 36, assuming that the solution entering the tank contains 0.05 pounds of concentrate per gallon.

38. A 200-gallon tank is half full of distilled water. At time $t = 0$, a solution containing 0.5 pound of concentrate per gallon enters the tank at the rate of 5 gallons per minute, and the well-stirred mixture is withdrawn at the rate of 3 gallons per minute.
   a. At what time will the tank be full?
   b. At the time the tank is full, how many pounds of concentrate will it contain?

In Exercises 39–42, match the differential equation with its solution.

Differential equation	Solution
39. $y' - 2x = 0$	**a.** $y = Ce^{x^2}$
40. $y' - 2y = 0$	**b.** $y = -\frac{1}{2} + Ce^{x^2}$
41. $y' - 2xy = 0$	**c.** $y = x^2 + C$
42. $y' - 2xy = x$	**d.** $y = Ce^{2x}$

In Exercises 43–58, solve the first-order differential equation by any appropriate method.

43. $\dfrac{dy}{dx} = \dfrac{e^{2x+y}}{e^{x-y}}$ **44.** $\dfrac{dy}{dx} = \dfrac{x+1}{y(y+2)}$

45. $(1 + y^2) \, dx + (2xy + y + 2) \, dy = 0$

46. $(1 + 2e^{2x+y}) \, dx + e^{2x+y} \, dy = 0$

47. $y \cos x - \cos x + \dfrac{dy}{dx} = 0$

48. $(x + 1) \dfrac{dy}{dx} = e^x - y$

49. $(x^2 + \cos y) \dfrac{dy}{dx} = -2xy$

50. $y' = 2x \sqrt{1 - y^2}$

51. $(3y^2 + 4xy) \, dx + (2xy + x^2) \, dy = 0$

52. $(x + y) \, dx - x \, dy = 0$

53. $(2y - e^x) \, dx + x \, dy = 0$

54. $(y^2 + xy) \, dx - x^2 \, dy = 0$

55. $(x^2 y^4 - 1) \, dx + x^3 y^3 \, dy = 0$

56. $y \, dx + (3x + 4y) \, dy = 0$

57. $3y \, dx - (x^2 + 3x + y^2) \, dy = 0$

58. $x \, dx + (y + e^y)(x^2 + 1) \, dy = 0$

Second-Order Linear Differential Equations

In this section and the following section, we discuss methods for solving higher-order linear differential equations.

Definition of Linear Differential Equation of Order n

Let $g_1, g_2, \ldots, g_n$ and f be functions of x with a common (interval) domain. An equation of the form

$$y^{(n)} + g_1(x)y^{(n-1)} + g_2(x)y^{(n-2)} + \cdots + g_{n-1}(x)y' + g_n(x)y = f(x)$$

is called a **linear differential equation of order n.** If $f(x) = 0$, the equation is **homogeneous;** otherwise, it is **nonhomogeneous.**

REMARK   Note that this use of the term *homogeneous* differs from that in Section 16.2.

We discuss homogeneous equations in this section, and leave the nonhomogeneous case for the next section.

The functions $y_1, y_2, \ldots, y_n$ are **linearly independent** if the *only* solution of the equation

$$C_1 y_1 + C_2 y_2 + \cdots + C_n y_n = 0$$

is the trivial one, $C_1 = C_2 = \cdots = C_n = 0$. Otherwise, this set of functions is **linearly dependent.**

EXAMPLE 1   Linearly Independent and Dependent Functions

a. The functions $y_1(x) = \sin x$ and $y_2 = x$ are linearly independent because the only values of C_1 and C_2 for which

$$C_1 \sin x + C_2 x = 0$$

for all x are $C_1 = 0$ and $C_2 = 0$.

b. It can be shown that two functions form a linearly dependent set if and only if one is a constant multiple of the other. For example, $y_1(x) = x$ and $y_2(x) = 3x$ are linearly dependent because

$$C_1 x + C_2(3x) = 0$$

has the nonzero solutions $C_1 = -3$ and $C_2 = 1$.

The following theorem points out the importance of linear independence in constructing the general solution of a second-order linear homogeneous differential equation with constant coefficients.

THEOREM 16.5 Linear Combinations of Solutions

If y_1 and y_2 are linearly independent solutions of the differential equation $y'' + ay' + by = 0$, then the general solution is

$$y = C_1 y_1 + C_2 y_2$$

where C_1 and C_2 are constants.

Proof   We prove the theorem in only one direction. If y_1 and y_2 are solutions, then you can obtain the following system of equations.

$$y_1''(x) + ay_1'(x) + by_1(x) = 0$$
$$y_2''(x) + ay_2'(x) + by_2(x) = 0$$

Multiplying the first equation by C_1, multiplying the second by C_2, and adding produces

$$[C_1 y_1''(x) + C_2 y_2''(x)] + a[C_1 y_1'(x) + C_2 y_2'(x)] + b[C_1 y_1(x) + C_2 y_2(x)] = 0$$

which means that

$$y = C_1 y_1 + C_2 y_2$$

is a solution, as desired. The proof that all solutions are of this form is best left to a full course on differential equations.

This theorem states that if you can find two linearly independent solutions, then you can obtain the general solution by forming a **linear combination** of the two solutions.

To find two linearly independent solutions, note that the nature of the equation $y'' + ay' + by = 0$ suggests that it may have solutions of the form $y = e^{mx}$. If so, then $y' = me^{mx}$ and $y'' = m^2 e^{mx}$. Thus, by substitution, $y = e^{mx}$ is a solution if and only if

$$y'' + ay' + by = 0$$
$$m^2 e^{mx} + ame^{mx} + be^{mx} = 0$$
$$e^{mx}(m^2 + am + b) = 0.$$

Because e^{mx} is never 0, $y = e^{mx}$ is a solution if and only if

$$m^2 + am + b = 0. \qquad \text{Characteristic equation}$$

This is the **characteristic equation** of the differential equation

$$y'' + ay' + by = 0.$$

Note that the characteristic equation can be determined from its differential equation simply by replacing y'' by m^2, y' by m, and y by 1.

EXAMPLE 2   Characteristic Equation with Distinct Real Roots

Solve the differential equation

$$y'' - 4y = 0.$$

Solution   In this case, the characteristic equation is

$$m^2 - 4 = 0 \qquad \text{Characteristic equation}$$

so $m = \pm 2$. Thus, $y_1 = e^{m_1 x} = e^{2x}$ and $y_2 = e^{m_2 x} = e^{-2x}$ are particular solutions of the given differential equation. Furthermore, because these two solutions are linearly independent, you can apply Theorem 16.5 to conclude that the general solution is

$$y = C_1 e^{2x} + C_2 e^{-2x}. \qquad \text{General solution}$$

The characteristic equation in Example 2 has two distinct real roots. From algebra, you know that this is only one of *three* possibilities for quadratic equations. In general, the quadratic equation $m^2 + am + b = 0$ has roots

$$m_1 = \frac{-a + \sqrt{a^2 - 4b}}{2} \quad \text{and} \quad m_2 = \frac{-a - \sqrt{a^2 - 4b}}{2}$$

which fall into one of three cases.

1. Two distinct real roots, $m_1 \neq m_2$
2. Two equal real roots, $m_1 = m_2$
3. Two complex conjugate roots, $m_1 = \alpha + \beta i$ and $m_2 = \alpha - \beta i$

In terms of the differential equation $y'' + ay' + by = 0$, these three cases correspȯ to three different types of general solutions.

THEOREM 16.6   Solutions of $y'' + ay' + by = 0$

The solutions of
$$y'' + ay' + by = 0$$
fall into one of the following three cases, depending on the solutions of the characteristic equation, $m^2 + am + b = 0$.

1. *Distinct Real Roots*   If $m_1 \neq m_2$ are distinct real roots of the characteristic equation, then the general solution is
$$y = C_1 e^{m_1 x} + C_2 e^{m_2 x}.$$

2. *Equal Real Roots*   If $m_1 = m_2$ are equal real roots of the characteristic equation, then the general solution is
$$y = C_1 e^{m_1 x} + C_2 x e^{m_1 x} = (C_1 + C_2 x) e^{m_1 x}.$$

3. *Complex Roots*   If $m_1 = \alpha + \beta i$ and $m_2 = \alpha - \beta i$ are complex roots of the characteristic equation, then the general solution is
$$y = C_1 e^{\alpha x} \cos \beta x + C_2 e^{\alpha x} \sin \beta x.$$

FOR FURTHER INFORMATION
For more information on Theorem 16.6, see "A Note on a Differential Equation" by Russell Euler in the 1989 winter issue of the *Missouri Journal of Mathematical Sciences*.

EXAMPLE 3 Characteristic Equation with Complex Roots

Find the general solution of the differential equation

$$y'' + 6y' + 12y = 0.$$

Solution The characteristic equation

$$m^2 + 6m + 12 = 0$$

has two complex roots, as follows.

$$m = \frac{-6 \pm \sqrt{36 - 48}}{2}$$

$$= \frac{-6 \pm \sqrt{-12}}{2}$$

$$= -3 \pm \sqrt{-3}$$

$$= -3 \pm \sqrt{3}i$$

Thus, $\alpha = -3$ and $\beta = \sqrt{3}$, and the general solution is

$$y = C_1 e^{-3x} \cos \sqrt{3}x + C_2 e^{-3x} \sin \sqrt{3}x.$$

REMARK In Example 3, note that although the characteristic equation has two *complex* roots, the solution of the differential equation is *real*.

EXAMPLE 4 Characteristic Equation with Repeated Roots

Solve the differential equation

$$y'' + 4y' + 4y = 0$$

subject to the initial conditions $y(0) = 2$ and $y'(0) = 1$.

Solution The characteristic equation

$$m^2 + 4m + 4 = (m + 2)^2 = 0$$

has two equal real roots given by $m = -2$. Thus, the general solution is

$$y = C_1 e^{-2x} + C_2 x e^{-2x}. \qquad \text{General solution}$$

Now, because $y = 2$ when $x = 0$, we have

$$2 = C_1(1) + C_2(0)(1) = C_1.$$

Furthermore, because $y' = 1$ when $x = 0$, we have

$$y' = -2C_1 e^{-2x} + C_2(-2xe^{-2x} + e^{-2x})$$

$$1 = -2(2)(1) + C_2[-2(0)(1) + 1]$$

$$5 = C_2.$$

Therefore, the solution is

$$y = 2e^{-2x} + 5xe^{-2x}. \qquad \text{Particular solution}$$

Try checking this solution in the original differential equation.

Higher-Order Linear Differential Equations

For higher-order homogeneous linear differential equations, you can find the general solution in much the same way as you do for second-order equations. That is, you begin by determining the n roots of the characteristic equation. Then, based on these n roots, you form a linearly independent collection of n solutions. The major difference is that with equations of third or higher order, roots of the characteristic equation may occur more than twice. When this happens, the linearly independent solutions are formed by multiplying by increasing powers of x, as demonstrated in Examples 5–7.

EXAMPLE 5   Solving a Third-Order Equation

Find the general solution of $y''' - y' = 0$.

Solution   The characteristic equation is

$$m^3 - m = 0$$
$$m(m - 1)(m + 1) = 0$$
$$m = 0, 1, -1.$$

Because the characteristic equation has three distinct roots, the general solution is

$$y = C_1 + C_2 e^{-x} + C_3 e^x. \qquad \text{General solution}$$

EXAMPLE 6   Solving a Third-Order Equation

Find the general solution of $y''' + 3y'' + 3y' + y = 0$.

Solution   The characteristic equation is

$$m^3 + 3m^2 + 3m + 1 = 0$$
$$(m + 1)^3 = 0$$
$$m = -1.$$

Because the root $m = -1$ occurs three times, the general solution is

$$y = C_1 e^{-x} + C_2 x e^{-x} + C_3 x^2 e^{-x}. \qquad \text{General solution}$$

EXAMPLE 7   Solving a Fourth-Order Equation

Find the general solution of $y^{(4)} + 2y'' + y = 0$.

Solution   The characteristic equation is as follows.

$$m^4 + 2m^2 + 1 = 0$$
$$(m^2 + 1)^2 = 0$$
$$m = \pm i$$

Because each of the roots $m_1 = \alpha + \beta i = 0 + i$ and $m_2 = \alpha - \beta i = 0 - i$ occurs twice, the general solution is

$$y = C_1 \cos x + C_2 \sin x + C_3 x \cos x + C_4 x \sin x. \qquad \text{General solution}$$

FIGURE 16.14
A rigid object of mass m attached to the end of the spring causes a displacement of y.

l = natural length

y = displacement

Applications

One of the many applications of linear differential equations is describing the motion of an oscillating spring. According to Hooke's Law, a spring that is stretched (or compressed) y units from its natural length l tends to *restore* itself to its natural length by a force F that is proportional to y. That is, $F(y) = -ky$, where k is the **spring constant** and indicates the stiffness of a given spring.

Suppose a rigid object of mass m is attached to the end of the spring and causes a displacement, as shown in Figure 16.14. Assume that the mass of the spring is negligible compared to m. If the object is pulled down and released, then the resulting oscillations are a product of two opposing forces—the spring force $F(y) = -ky$ and the weight mg of the object. Under such conditions, you can use a differential equation to find the position y of the object as a function of time t. According to Newton's Second Law of Motion, the force acting on the weight is $F = ma$, where $a = d^2y/dt^2$ is the acceleration. Assuming the motion is **undamped**—that is, there are no other external forces acting on the object—it follows that $m(d^2y/dt^2) = -ky$, and you have

$$\frac{d^2y}{dt^2} + \left(\frac{k}{m}\right)y = 0. \qquad \text{Undamped motion of a spring}$$

EXAMPLE 8   Undamped Motion of a Spring

Suppose a 4-pound weight stretches a spring 8 inches from its natural length. The weight is pulled down an additional 6 inches and released with an initial upward velocity of 8 feet per second. Find a formula for the position of the weight as a function of time t.

Solution By Hooke's Law, $4 = k(\frac{2}{3})$, so $k = 6$. Moreover, because the weight w is given by mg, it follows that $m = w/g = \frac{4}{32} = \frac{1}{8}$. Hence, the resulting differential equation for this undamped motion is

$$\frac{d^2y}{dt^2} + 48y = 0.$$

Because the characteristic equation $m^2 + 48 = 0$ has complex roots $m = 0 \pm 4\sqrt{3}\,i$, the general solution is

$$y = C_1 e^0 \cos 4\sqrt{3}\,t + C_2 e^0 \sin 4\sqrt{3}\,t = C_1 \cos 4\sqrt{3}\,t + C_2 \sin 4\sqrt{3}\,t.$$

Using the initial conditions, you have

$$\frac{1}{2} = C_1(1) + C_2(0) \quad \rightarrow \quad C_1 = \frac{1}{2} \qquad y(0) = \tfrac{1}{2}$$

$$y'(t) = -4\sqrt{3}\,C_1 \sin 4\sqrt{3}\,t + 4\sqrt{3}\,C_2 \cos 4\sqrt{3}\,t$$

$$8 = -4\sqrt{3}\left(\frac{1}{2}\right)(0) + 4\sqrt{3}\,C_2(1) \quad \rightarrow \quad C_2 = \frac{2\sqrt{3}}{3}. \qquad y'(0) = 8$$

Consequently, the position at time t is given by

$$y = \frac{1}{2}\cos 4\sqrt{3}\,t + \frac{2\sqrt{3}}{3}\sin 4\sqrt{3}\,t.$$

FIGURE 16.15
A damped vibration could be caused by friction and movement through a liquid.

Suppose the object in Figure 16.15 undergoes an additional damping or frictional force that is proportional to its velocity. A case in point would be the damping force due to friction and movement through a fluid. Considering this damping force, $-p(dy/dt)$, the differential equation for the oscillation is

$$m\frac{d^2y}{dt^2} = -ky - p\frac{dy}{dt}$$

or, in standard linear form,

$$\frac{d^2y}{dt^2} + \frac{p}{m}\left(\frac{dy}{dt}\right) + \frac{k}{m}y = 0. \qquad \text{Damped motion of a spring}$$

EXERCISES for Section 16.5

TECHNOLOGY
Laboratory Guide
Labs 16.5–16.6

In Exercises 1–4, verify that the equation is a solution of the differential equation.

Solution	Differential equation
1. $y = (C_1 + C_2x)e^{-3x}$	$y'' + 6y' + 9y = 0$
2. $y = C_1e^{2x} + C_2e^{-2x}$	$y'' - 4y = 0$
3. $y = C_1\cos 2x + C_2\sin 2x$	$y'' + 4y = 0$
4. $y = e^{-x}\sin 3x$	$y'' + 2y' + 10y = 0$

In Exercises 5–30, find the general solution of the linear differential equation.

5. $y'' - y' = 0$

6. $y'' + 2y' = 0$

7. $y'' - y' - 6y = 0$

8. $y'' + 6y' + 5y = 0$

9. $2y'' + 3y' - 2y = 0$

10. $16y'' - 16y' + 3y = 0$

11. $y'' + 6y' + 9y = 0$

12. $y'' - 10y' + 25y = 0$

13. $16y'' - 8y' + y = 0$

14. $9y'' - 12y' + 4y = 0$

15. $y'' + y = 0$

16. $y'' + 4y = 0$

17. $y'' - 9y = 0$

18. $y'' - 2y = 0$

19. $y'' - 2y' + 4y = 0$

20. $y'' - 4y' + 21y = 0$

21. $y'' - 3y' + y = 0$

22. $3y'' + 4y' - y = 0$

23. $9y'' - 12y' + 11y = 0$

24. $2y'' - 6y' + 7y = 0$

25. $y^{(4)} - y = 0$

26. $y^{(4)} - y'' = 0$

27. $y''' - 6y'' + 11y' - 6y = 0$

28. $y''' - y'' - y' + y = 0$

29. $y''' - 3y'' + 7y' - 5y = 0$

30. $y''' - 3y'' + 3y' - y = 0$

31. Consider the differential equation $y'' + 100y = 0$ and the solution $y = C_1\cos 10x + C_2\sin 10x$. Find the particular solution satisfying each of the following initial conditions.
    a. $y(0) = 2, \; y'(0) = 0$
    b. $y(0) = 0, \; y'(0) = 2$
    c. $y(0) = -1, \; y'(0) = 3$

32. Determine C and ω such that $y = C\sin\sqrt{3}\,t$ is a particular solution of the differential equation $y'' + \omega y = 0$, where $y'(0) = -5$.

In Exercises 33 and 34, find the particular solution of the linear differential equation.

33. $y'' - y' - 30y = 0$
    $y(0) = 1, \; y'(0) = -4$

34. $y'' + 2y' + 3y = 0$
    $y(0) = 2, \; y'(0) = 1$

Vibrating Spring **In Exercises 35–40, describe the motion of a 32-pound weight suspended on a spring. Assume that the weight stretches the spring $\frac{2}{3}$ foot from its natural position.**

35. The weight is pulled $\frac{1}{2}$ foot below the equilibrium position and released.

36. The weight is raised $\frac{2}{3}$ foot above the equilibrium position and released.

37. The weight is raised $\frac{2}{3}$ foot above the equilibrium position and started off with a downward velocity of $\frac{1}{2}$ foot per second.

38. The weight is pulled $\frac{1}{2}$ foot below the equilibrium position and started off with an upward velocity of $\frac{1}{2}$ foot per second.

39. The weight is pulled $\frac{1}{2}$ foot below the equilibrium position and released. The motion takes place in a medium that furnishes a damping force of magnitude $\frac{1}{8}$ speed at all times.

40. The weight is pulled $\frac{1}{2}$ foot below the equilibrium position and released. The motion takes place in a medium that furnishes a damping force of magnitude $\frac{1}{4}|v|$ at all times.

Vibrating Spring **In Exercises 41–44, match the differential equation with the graph of a particular solution. [The graphs are labeled (a), (b), (c), and (d).] The correct match can be made by comparing the frequency of the oscillations or the rate at which the oscillations are being damped with the appropriate coefficient in the differential equation.**

41. $y'' + 9y = 0$

42. $y'' + 25y = 0$

43. $y'' + 2y' + 10y = 0$ **44.** $y'' + y' + \frac{37}{4}y = 0$

(a)

(b)

(c)

(d)

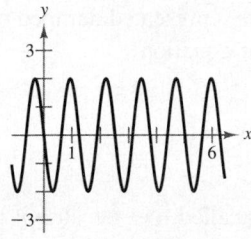

45. If the characteristic equation of the differential equation

$$y'' + ay' + by = 0$$

has two equal real roots given by $m = r$, show that

$$y = C_1 e^{rx} + C_2 x e^{rx}$$

is a solution.

46. If the characteristic equation of the differential equation

$$y'' + ay' + by = 0$$

has complex roots given by $m_1 = \alpha + \beta i$ and $m_2 = \alpha - \beta i$, show that

$$y = C_1 e^{\alpha x} \cos \beta x + C_2 e^{\alpha x} \sin \beta x$$

is a solution.

True or False   In Exercises 47–50, determine whether the statement is true or false. If it is false, explain why or give an example that shows it is false.

47. $y = C_1 e^{3x} + C_2 e^{-3x}$ is the general solution of $y'' - 6y' + 9 = 0$.

48. $y = (C_1 + C_2 x)\sin x + (C_3 + C_4 x)\cos x$ is the general solution of $y^{(4)} + 2y' + y = 0$.

49. $y = x$ is a solution of $a_n y^{(n)} + a_{n-1} y^{(n-1)} + \cdots + a_1 y' + a_0 y = 0$ if and only if $a_1 = a_0 = 0$.

50. It is possible to choose a and b so that $y = x^2 e^x$ is a solution of $y'' + ay' + by = 0$.

The *Wronskian* of two differentiable functions f and g, denoted by $W(f, g)$, is defined as the function given by the determinant

$$W(f, g) = \begin{vmatrix} f & g \\ f' & g' \end{vmatrix}.$$

The functions f and g are linearly independent if there exists at least one value of x for which $W(f, g) \neq 0$. In Exercises 51–54, use the Wronskian to verify the linear independence of the two functions.

51. $y_1 = e^{ax}$
$y_2 = e^{bx}$ $(a \neq b)$

52. $y_1 = e^{ax}$
$y_2 = x e^{ax}$

53. $y_1 = e^{ax} \sin bx$
$y_2 = e^{ax} \cos bx$, $b \neq 0$

54. $y_1 = x$
$y_2 = x^2$

55. Euler's differential equation is of the form

$$x^2 y'' + axy' + by = 0, \qquad x > 0$$

where a and b are constants.
a. Show that this equation can be transformed into a second-order linear equation with constant coefficients by using the substitution $x = e^t$.
b. Solve $x^2 y'' + 6xy' + 6y = 0$.

56. Solve

$$y'' + Ay = 0$$

where A is constant, subject to the conditions $y(0) = 0$ and $y(\pi) = 0$.

Second-Order Nonhomogeneous Linear Equations

Nonhomogeneous Equations • Method of Undetermined Coefficients • Variation of Parameters

Nonhomogeneous Equations

In the previous section, we represented damped oscillations of a spring by the *homogeneous* second-order linear equation

$$\frac{d^2y}{dt^2} + \frac{p}{m}\left(\frac{dy}{dt}\right) + \frac{k}{m}y = 0. \qquad \text{Free motion}$$

This type of oscillation is called **free** because it is determined solely by the spring and gravity and is free of the action of other external forces. If such a system is also subject to an external periodic force such as $a \sin bt$, caused by vibrations at the opposite end of the spring, then the motion is called **forced,** and it is characterized by the *nonhomogeneous* equation

$$\frac{d^2y}{dt^2} + \frac{p}{m}\left(\frac{dy}{dt}\right) + \frac{k}{m}y = a \sin bt. \qquad \text{Forced motion}$$

In this section, you will study two methods for finding the general solution of a nonhomogeneous linear differential equation. In both methods, the first step is to find the general solution of the corresponding homogeneous equation.

$$y = y_h \qquad \text{General solution of homogeneous equation}$$

Having done this, you try to find a particular solution of the nonhomogeneous equation.

$$y = y_p \qquad \text{Particular solution of nonhomogeneous equation}$$

By combining these two results, you can conclude that the general solution of the nonhomogeneous equation is $y = y_h + y_p$, as stated in the following theorem.

SOPHIE GERMAIN

Many of the early contributors to calculus were interested in forming mathematical models for vibrating strings and membranes, oscillating springs, and elasticity. One of these was the French mathematician Sophie Germain (1776–1831), who in 1816 was awarded a prize by the French Academy for a paper entitled "Memoir on the Vibrations of Elastic Plates."

THEOREM 16.7 Solution of Nonhomogeneous Linear Equation

Let

$$y'' + ay' + by = F(x)$$

be a second-order nonhomogeneous linear differential equation. If y_p is a particular solution of this equation and y_h is the general solution of the corresponding homogeneous equation, then

$$y = y_h + y_p$$

is the general solution of the nonhomogeneous equation.

Method of Undetermined Coefficients

You already know how to find the solution y_h of a linear *homogeneous* differential equation. The remainder of this section looks at ways to find the particular solution y_p. If $F(x)$ in

$$y'' + ay' + by = F(x)$$

consists of sums or products of x^n, e^{mx}, $\cos \beta x$, or $\sin \beta x$, then you can find a particular solution y_p by the method of **undetermined coefficients.** The gist of this method is to guess that the solution y_p is a generalized form of $F(x)$. Here are some examples.

a. If $F(x) = 3x^2$, choose $y_p = Ax^2 + Bx + C$.
b. If $F(x) = 4xe^x$, choose $y_p = Axe^x + Be^x$.
c. If $F(x) = x + \sin 2x$, choose $y_p = (Ax + B) + C\sin 2x + D\cos 2x$.

Then, by substitution, determine the coefficients for the generalized solution.

EXAMPLE 1   Method of Undetermined Coefficients

Find the general solution of the equation

$$y'' - 2y' - 3y = 2\sin x.$$

Solution   To find y_h, solve the characteristic equation.

$$m^2 - 2m - 3 = 0$$
$$(m + 1)(m - 3) = 0$$
$$m = -1 \quad \text{or} \quad m = 3$$

Thus, $y_h = C_1 e^{-x} + C_2 e^{3x}$. Next, let y_p be a generalized form of $2\sin x$.

$$y_p = A\cos x + B\sin x$$
$$y_p' = -A\sin x + B\cos x$$
$$y_p'' = -A\cos x - B\sin x$$

Substitution into the original differential equation yields

$$y'' - 2y' - 3y = 2\sin x$$
$$-A\cos x - B\sin x + 2A\sin x - 2B\cos x - 3A\cos x - 3B\sin x = 2\sin x$$
$$(-4A - 2B)\cos x + (2A - 4B)\sin x = 2\sin x.$$

By equating coefficients of like terms, you obtain

$$-4A - 2B = 0 \quad \text{and} \quad 2A - 4B = 2$$

with solutions $A = \frac{1}{5}$ and $B = -\frac{2}{5}$. Therefore,

$$y_p = \frac{1}{5}\cos x - \frac{2}{5}\sin x$$

and the general solution is

$$y = y_h + y_p$$
$$= C_1 e^{-x} + C_2 e^{3x} + \frac{1}{5}\cos x - \frac{2}{5}\sin x.$$

In Example 1, the form of the homogeneous solution

$$y_h = C_1 e^{-x} + C_2 e^{3x}$$

has no overlap with the function $F(x)$ in the equation

$$y'' + ay' + by = F(x).$$

However, suppose the given differential equation in Example 1 were of the form

$$y'' - 2y' - 3y = e^{-x}.$$

Now, it would make no sense to guess that the particular solution were $y = Ae^{-x}$, because you know that this solution would yield 0. In such cases, you should alter your guess by multiplying by the lowest power of x that removes the duplication. For this particular problem, you would guess

$$y_p = Axe^{-x}.$$

EXAMPLE 2   Method of Undetermined Coefficients

Find the general solution of

$$y'' - 2y' = x + 2e^x.$$

Solution   The characteristic equation $m^2 - 2m = 0$ has solutions $m = 0$ and $m = 2$. Thus,

$$y_h = C_1 + C_2 e^{2x}.$$

Because $F(x) = x + 2e^x$, your first choice for y_p would be $(A + Bx) + Ce^x$. However, because y_h *already* contains a constant term C_1, you should multiply the *polynomial part* by x and use

$$y_p = Ax + Bx^2 + Ce^x$$
$$y_p' = A + 2Bx + Ce^x$$
$$y_p'' = 2B + Ce^x.$$

Substitution into the differential equation produces

$$y'' - 2y' = x + 2e^x$$
$$(2B + Ce^x) - 2(A + 2Bx + Ce^x) = x + 2e^x$$
$$(2B - 2A) - 4Bx - Ce^x = x + 2e^x.$$

Equating coefficients of like terms yields the system

$$2B - 2A = 0, \qquad -4B = 1, \qquad -C = 2$$

with solutions $A = B = -\frac{1}{4}$ and $C = -2$. Therefore,

$$y_p = -\frac{1}{4}x - \frac{1}{4}x^2 - 2e^x$$

and the general solution is

$$y = y_h + y_p$$
$$= C_1 + C_2 e^{2x} - \frac{1}{4}x - \frac{1}{4}x^2 - 2e^x.$$

In Example 2, the polynomial part of the initial guess

$$(A + Bx) + Ce^x$$

for y_p overlapped by a constant term with $y_h = C_1 + C_2e^{2x}$, and it was necessary to multiply the polynomial part by a power of x that removed the overlap. The next example further illustrates some choices for y_p that eliminate overlap with y_h. Remember that in all cases the first guess for y_p should match the types of functions occurring in $F(x)$.

EXAMPLE 3   Choosing the Form of the Particular Solution

Determine a suitable choice for y_p for the following.

$y'' + ay' + by = F(x)$	y_h
a. $y'' = x^2$	$C_1 + C_2x$
b. $y'' + 2y' + 10y = 4\sin 3x$	$C_1e^{-x}\cos 3x + C_2e^{-x}\sin 3x$
c. $y'' - 4y' + 4 = e^{2x}$	$C_1e^{2x} + C_2xe^{2x}$

Solution

a. Because $F(x) = x^2$, the normal choice for y_p would be $A + Bx + Cx^2$. However, because $y_h = C_1 + C_2x$ already contains a linear term, you should multiply by x^2 to obtain

$$y_p = Ax^2 + Bx^3 + Cx^4.$$

b. Because $F(x) = 4\sin 3x$ and each term in y_h contains a factor of e^{-x}, you can simply let

$$y_p = A\cos 3x + B\sin 3x.$$

c. Because $F(x) = e^{2x}$, the normal choice for y_p would be Ae^{2x}. However, because $y_h = C_1e^{2x} + C_2xe^{2x}$ already contains an xe^{2x} term, you should multiply by x^2 to get

$$y_p = Ax^2e^{2x}.$$ ■

EXAMPLE 4   Solving a Third-Order Equation

Find the general solution of

$$y''' + 3y'' + 3y' + y = x.$$

Solution   From Example 6 in the previous section, you know that the homogeneous solution is

$$y_h = C_1e^{-x} + C_2xe^{-x} + C_3x^2e^{-x}.$$

Because $F(x) = x$, let $y_p = A + Bx$ and obtain $y_p' = B$ and $y_p'' = 0$. Thus, by substitution, you have

$$(0) + 3(0) + 3(B) + (A + Bx) = (3B + A) + Bx = x.$$

Thus, $B = 1$ and $A = -3$, which implies that $y_p = -3 + x$. Therefore, the general solution is

$$y = y_h + y_p$$
$$= C_1e^{-x} + C_2xe^{-x} + C_3x^2e^{-x} - 3 + x.$$

Variation of Parameters

The method of undetermined coefficients works well if $F(x)$ is made up of polynomials or functions whose successive derivatives have a cyclic pattern. For functions such as $1/x$ and $\tan x$, which do not possess such characteristics, it is better to use a more general method called **variation of parameters.** In this method, you assume that y_p has the same *form* as y_h, except that the constants in y_h are replaced by variables.

Variation of Parameters

To find the general solution to the equation $y'' + ay' + by = F(x)$, use the following steps.

1. Find $y_h = C_1 y_1 + C_2 y_2$.
2. Replace the constants by variables to form $y_p = u_1 y_1 + u_2 y_2$.
3. Solve the following system for u_1' and u_2'.

$$u_1' y_1 + u_2' y_2 = 0$$
$$u_1' y_1' + u_2' y_2' = F(x)$$

4. Integrate to find u_1 and u_2. The general solution is $y = y_h + y_p$.

EXAMPLE 5   Variation of Parameters

Solve the differential equation

$$y'' - 2y' + y = \frac{e^x}{2x}, \qquad x > 0.$$

Solution   The characteristic equation $m^2 - 2m + 1 = (m - 1)^2 = 0$ has one solution, $m = 1$. Thus, the homogeneous solution is

$$y_h = C_1 y_1 + C_2 y_2 = C_1 e^x + C_2 x e^x.$$

Replacing C_1 and C_2 by u_1 and u_2 produces

$$y_p = u_1 y_1 + u_2 y_2 = u_1 e^x + u_2 x e^x.$$

The resulting system of equations is

$$u_1' e^x + u_2' x e^x = 0$$
$$u_1' e^x + u_2'(x e^x + e^x) = \frac{e^x}{2x}.$$

Subtracting the second equation from the first produces $u_2' = 1/2x$. Then, by substi- in the first equation, you have $u_1' = -\frac{1}{2}$. Finally, integration yields

$$u_1 = -\int \frac{1}{2} dx = -\frac{x}{2} \quad \text{and} \quad u_2 = \frac{1}{2} \int \frac{1}{x} dx = \frac{1}{2} \ln x = \ln \sqrt{x}.$$

From this result it follows that a particular solution is

$$y_p = -\frac{1}{2} x e^x + \left(\ln \sqrt{x} \right) x e^x$$

and the general solution is

$$y = C_1 e^x + C_2 x e^x - \frac{1}{2} x e^x + x e^x \ln \sqrt{x}.$$

EXAMPLE 6 Variation of Parameters

Solve the differential equation

$$y'' + y = \tan x.$$

Solution Because the characteristic equation $m^2 + 1 = 0$ has solutions $m = \pm i$, the homogeneous solution is

$$y_h = C_1 \cos x + C_2 \sin x.$$

Replacing C_1 and C_2 by u_1 and u_2 produces

$$y_p = u_1 \cos x + u_2 \sin x.$$

The resulting system of equations is

$$u_1{}' \cos x + u_2{}' \sin x = 0$$
$$-u_1{}' \sin x + u_2{}' \cos x = \tan x.$$

Multiplying the first equation by $\sin x$ and the second by $\cos x$ produces

$$u_1{}' \sin x \cos x + u_2{}' \sin^2 x = 0$$
$$-u_1{}' \sin x \cos x + u_2{}' \cos^2 x = \sin x.$$

Adding these two equations produces $u_2{}' = \sin x$, which implies that

$$u_1{}' = -\frac{\sin^2 x}{\cos x}$$
$$= \frac{\cos^2 x - 1}{\cos x}$$
$$= \cos x - \sec x.$$

Integration yields

$$u_1 = \int (\cos x - \sec x)\,dx$$
$$= \sin x - \ln |\sec x + \tan x|$$

and

$$u_2 = \int \sin x\,dx$$
$$= -\cos x$$

so that

$$y_p = \sin x \cos x - \cos x \ln |\sec x + \tan x| - \sin x \cos x$$
$$= -\cos x \ln |\sec x + \tan x|$$

and the general solution is

$$y = y_h + y_p$$
$$= C_1 \cos x + C_2 \sin x - \cos x \ln |\sec x + \tan x|.$$

EXERCISES for Section 16.6

In Exercises 1–4, verify that the equation is a solution of the differential equation.

Solution	Differential equation
1. $y = 2(e^{2x} - \cos x)$	$y'' + y = 10e^{2x}$
2. $y = (2 + \frac{1}{2}x)\sin x$	$y'' + y = \cos x$
3. $y = 3\sin x - \cos x \ln \lvert \sec x + \tan x \rvert$	$y'' + y = \tan x$
4. $y = (5 - \ln\lvert \sin x \rvert)\cos x - x\sin x$	$y'' + y = \csc x \cot x$

In Exercises 5–20, solve the differential equation by the method of undetermined coefficients.

5. $y'' - 3y' + 2y = 2x$

6. $y'' - 2y' - 3y = x^2 - 1$

7. $y'' + y = x^3$
$y(0) = 1,\ y'(0) = 0$

8. $y'' + 4y = 4$
$y(0) = 1,\ y'(0) = 6$

9. $y'' + 2y' = 2e^x$

10. $y'' - 9y = 5e^{3x}$

11. $y'' - 10y' + 25y = 5 + 6e^x$

12. $16y'' - 8y' + y = 4(x + e^x)$

13. $y'' + y' = 2\sin x$
$y(0) = 0,\ y'(0) = -3$

14. $y'' + y' - 2y = 3\cos 2x$
$y(0) = -1,\ y'(0) = 2$

15. $y'' + 9y = \sin 3x$

16. $y'' + 4y' + 5y = \sin x + \cos x$

17. $y''' - 3y' + 2y = 2e^{-2x}$

18. $y''' - y'' = 4x^2$
$y(0) = 1,\ y'(0) = 1,\ y''(0) = 1$

19. $y' - 4y = xe^x - xe^{4x}$
$y(0) = \dfrac{1}{3}$

20. $y' + 2y = \sin x$
$y\left(\dfrac{\pi}{2}\right) = \dfrac{2}{5}$

In Exercises 21–26, solve the differential equation by the method of variation of parameters.

21. $y'' + y = \sec x$

22. $y'' + y = \sec x \tan x$

23. $y'' + 4y = \csc 2x$

24. $y'' - 4y' + 4y = x^2 e^{2x}$

25. $y'' - 2y' + y = e^x \ln x$

26. $y'' - 4y' + 4y = \dfrac{e^{2x}}{x}$

Electrical Circuits   **In Exercises 27 and 28, use the electrical circuit differential equation**

$$\frac{d^2q}{dt^2} + \left(\frac{R}{L}\right)\frac{dq}{dt} + \left(\frac{1}{LC}\right)q = \left(\frac{1}{L}\right)E(t)$$

where R is the resistance (in ohms), C is the capacitance (in farads), L is the inductance (in henrys), $E(t)$ is the electromotive force (in volts), and q is the charge on the capacitor (in coulombs). Find the charge q as a function of time for the electrical circuit described. Assume that $q(0) = 0$ and $q'(0) = 0$.

27. $R = 20,\ C = 0.02,\ L = 2$
$E(t) = 12\sin 5t$

28. $R = 20,\ C = 0.02,\ L = 1$
$E(t) = 10\sin 5t$

C *Vibrating Spring*   **In Exercises 29–32, find the particular solution of the differential equation**

$$\frac{w}{g}y''(t) + by'(t) + ky(t) = \frac{w}{g}F(t)$$

for the oscillating motion of an object on the end of a spring. Use a graphing utility to graph the solution. In the equation, y is the displacement from the equilibrium (positive direction is downward) measured in feet, and t is time in seconds (see figure). The constant w is the weight of the object, g is the acceleration due to gravity, b is the magnitude of the resistance to the motion, k is the spring constant from Hooke's Law, and $F(t)$ is the acceleration imposed on the system.

29. $\frac{24}{32}y'' + 48y = \frac{24}{32}(48\sin 4t)$
$y(0) = \frac{1}{4},\ y'(0) = 0$

30. $\frac{2}{32}y'' + 4y = \frac{2}{32}(4\sin 8t)$
$y(0) = \frac{1}{4},\ y'(0) = 0$

31. $\frac{2}{32}y'' + y' + 4y = \frac{2}{32}(4\sin 8t)$
$y(0) = \frac{1}{4},\ y'(0) = -3$

32. $\frac{4}{32}y'' + \frac{1}{2}y' + \frac{25}{2}y = 0$
$y(0) = \frac{1}{2},\ y'(0) = -4$

$l = $ natural length

$y = $ displacement

m

Spring displacement

FIGURE FOR 29–32

33. *Vibrating Spring*   Rewrite y_h in the solution for Exercise 29 by using the identity

$$a\cos \omega t + b\sin \omega t = \sqrt{a^2 + b^2}\,\sin(\omega t + \phi)$$

where $\phi = \arctan a/b$.

34. *Vibrating Spring*   The figure shows the particular solution of the differential equation

$$\frac{4}{32}y'' + by' + \frac{25}{2}y = 0$$

$$y(0) = \frac{1}{2},\ y'(0) = -4$$

for values of the resistance component b in the interval $[0, 1]$. (Note that when $b = \frac{1}{2}$, the problem is identical to that of Exercise 32.)

a. If there is no resistance to the motion ($b = 0$), describe the motion.

b. If $b > 0$, what is the ultimate effect of the retarding force?

c. Is there a real number M such that there will be no oscillations of the spring if $b > M$? Explain your answer.

FIGURE FOR 34   *Generated by Mathematica*

35. *Parachute Jump*   The fall of a parachutist is described by the second-order linear differential equation

$$\frac{w}{g}\frac{d^2y}{dt^2} - k\frac{dy}{dt} = w$$

where w is the weight of the parachutist, y is the height at time t, g is the acceleration due to gravity, and k is the drag factor of the parachute. If the parachute is opened at 2000 feet,

$y(0) = 2000$, and at that time the velocity is $y'(0) = -100$ feet per second, then for a 160-pound parachutist, using $k = 8$, the differential equation is

$$-5y'' - 8y' = 160.$$

Using the given initial conditions, verify that the solution of the differential equation is

$$y = 1950 + 50e^{-1.6t} - 20t.$$

36. *Parachute Jump*   Repeat Exercise 35 for a parachutist who weighs 192 pounds and has a parachute with a drag factor of $k = 9$.

37. Solve the differential equation

$$x^2y'' - xy' + y = 4x\ln x$$

given that $y_1 = x$ and $y_2 = x\ln x$ are solutions of the corresponding homogeneous equation.

Power Series Solution of a Differential Equation

We conclude this chapter by showing how power series can be used to solve certain types of differential equations. We begin with the general **power series solution** method.

Recall from Chapter 8 that a power series represents a function f on an interval of convergence, and that you can successively differentiate the power series to obtain a series for f', f'', and so on. These properties are used in the power series solution method demonstrated in the first two examples.

EXAMPLE 1   Power Series Solution

Use a power series to solve the differential equation $y' - 2y = 0$.

Solution   Assume that $y = \sum a_n x^n$ is a solution. Then, $y' = \sum n a_n x^{n-1}$. Substituting for y' and $-2y$, you obtain the following series form of the differential equation. (Note that from the third step to the fourth, the index of summation is changed to ensure that x^n occurs in both sums.)

$$y' - 2y = 0$$

$$\sum_{n=1}^{\infty} n a_n x^{n-1} - 2\sum_{n=0}^{\infty} a_n x^n = 0$$

$$\sum_{n=1}^{\infty} n a_n x^{n-1} = \sum_{n=0}^{\infty} 2 a_n x^n$$

$$\sum_{n=0}^{\infty} (n + 1)a_{n+1} x^n = \sum_{n=0}^{\infty} 2 a_n x^n$$

Now, by equating coefficients of like terms, you obtain the **recursion formula** $(n + 1)a_{n+1} = 2a_n$, which implies that

$$a_{n+1} = \frac{2a_n}{n + 1}, \qquad n \geq 0.$$

This formula generates the following results.

a_0	a_1	a_2	a_3	a_4	a_5	$\ldots$
a_0	$2a_0$	$\dfrac{2^2 a_0}{2}$	$\dfrac{2^3 a_0}{3!}$	$\dfrac{2^4 a_0}{4!}$	$\dfrac{2^5 a_0}{5!}$	$\ldots$

Using these values as the coefficients for the *solution* series, you have

$$y = \sum_{n=0}^{\infty} \frac{2^n a_0}{n!} x^n = a_0 \sum_{n=0}^{\infty} \frac{(2x)^n}{n!} = a_0 e^{2x}.$$

In Example 1, the differential equation could be solved easily without using a series. The differential equation in Example 2 cannot be solved by any of the methods discussed in previous sections.

EXAMPLE 2   Power Series Solution

Use a power series to solve the differential equation $y'' + xy' + y = 0$.

Solution   Assume that $\sum_{n=0}^{\infty} a_n x^n$ is a solution. Then you have

$$y' = \sum_{n=1}^{\infty} na_n x^{n-1}, \qquad xy' = \sum_{n=1}^{\infty} na_n x^n, \qquad y'' = \sum_{n=2}^{\infty} n(n-1)a_n x^{n-2}.$$

Substituting for y'', xy', and y in the given differential equation, you obtain the following series.

$$\sum_{n=2}^{\infty} n(n-1)a_n x^{n-2} + \sum_{n=0}^{\infty} na_n x^n + \sum_{n=0}^{\infty} a_n x^n = 0$$

$$\sum_{n=2}^{\infty} n(n-1)a_n x^{n-2} = -\sum_{n=0}^{\infty} (n+1)a_n x^n$$

To obtain equal powers of x, adjust the summation indices by replacing n by $n + 2$ in the left-hand sum, to obtain

$$\sum_{n=0}^{\infty} (n+2)(n+1)a_{n+2} x^n = -\sum_{n=0}^{\infty} (n+1)a_n x^n.$$

By equating coefficients, you have $(n + 2)(n + 1)a_{n+2} = -(n + 1)a_n$, from which you obtain the recursion formula

$$a_{n+2} = -\frac{(n+1)}{(n+2)(n+1)} a_n = -\frac{a_n}{n+2}, \qquad n \geq 0,$$

and the coefficients of the solution series are as follows.

$$a_2 = -\frac{a_0}{2} \qquad\qquad\qquad a_3 = -\frac{a_1}{3}$$

$$a_4 = -\frac{a_2}{4} = \frac{a_0}{2 \cdot 4} \qquad\qquad a_5 = -\frac{a_3}{5} = \frac{a_1}{3 \cdot 5}$$

$$a_6 = -\frac{a_4}{6} = -\frac{a_0}{2 \cdot 4 \cdot 6} \qquad\qquad a_7 = -\frac{a_5}{7} = -\frac{a_1}{3 \cdot 5 \cdot 7}$$

$$\vdots \qquad\qquad\qquad\qquad \vdots$$

$$a_{2k} = \frac{(-1)^k a_0}{2 \cdot 4 \cdot 6 \cdots (2k)} = \frac{(-1)^k a_0}{2^k(k!)} \qquad a_{2k+1} = \frac{(-1)^k a_1}{3 \cdot 5 \cdot 7 \cdots (2k+1)}$$

Thus, you can represent the general solution as the sum of two series—one for the even-powered terms with coefficients in terms of a_0 and one for the odd-powered terms with coefficients in terms of a_1.

$$y = a_0\left(1 - \frac{x^2}{2} + \frac{x^4}{2 \cdot 4} - \cdots\right) + a_1\left(x - \frac{x^3}{3} + \frac{x^5}{3 \cdot 5} - \cdots\right)$$

$$= a_0 \sum_{k=0}^{\infty} \frac{(-1)^k x^{2k}}{2^k(k!)} + a_1 \sum_{k=0}^{\infty} \frac{(-1)^k x^{2k+1}}{3 \cdot 5 \cdot 7 \cdots (2k+1)}$$

The solution has two arbitrary constants, a_0 and a_1, as you would expect in the general solution of a second-order differential equation.

Approximation by Taylor Series

A second type of series solution method involves a differential equation *with initial conditions* and makes use of Taylor series, as given in Section 8.10.

EXAMPLE 3   Approximation by Taylor Series

Use a Taylor series to find the series solution of

$$y' = y^2 - x$$

given the initial condition $y = 1$ when $x = 0$. Then, use the first six terms of this series solution to approximate values of y for $0 \leq x \leq 1$.

Solution   Recall from Section 8.10 that, for $c = 0$,

$$y = y(0) + y'(0)x + \frac{y''(0)}{2!}x^2 + \frac{y'''(0)}{3!}x^3 + \cdots .$$

Because $y(0) = 1$ and $y' = y^2 - x$, you obtain the following.

$$y(0) = 1$$
$$y' = y^2 - x \qquad\qquad y'(0) = 1$$
$$y'' = 2yy' - 1 \qquad\qquad y''(0) = 2 - 1 = 1$$
$$y''' = 2yy'' + 2(y')^2 \qquad\qquad y'''(0) = 2 + 2 = 4$$
$$y^{(4)} = 2yy''' + 6y'y'' \qquad\qquad y^{(4)}(0) = 8 + 6 = 14$$
$$y^{(5)} = 2yy^{(4)} + 8y'y''' + 6(y'')^2 \qquad\qquad y^{(5)}(0) = 28 + 32 + 6 = 66$$

Therefore, you can approximate the values of the solution from the series

$$y = y(0) + y'(0)x + \frac{y''(0)}{2!}x^2 + \frac{y'''(0)}{3!}x^3 + \frac{y^{(4)}(0)}{4!}x^4 + \frac{y^{(5)}(0)}{5!}x^5 + \cdots$$

$$= 1 + x + \frac{1}{2}x^2 + \frac{4}{3!}x^3 + \frac{14}{4!}x^4 + \frac{66}{5!}x^5 + \cdots .$$

Using the first six terms of this series, you can compute values for y in the interval $0 \leq x \leq 1$, as shown in Table 16.1.

TABLE 16.1

x	y
0.0	1.0000
0.1	1.1057
0.2	1.2264
0.3	1.3691
0.4	1.5432
0.5	1.7620
0.6	2.0424
0.7	2.4062
0.8	2.8805
0.9	3.4985
1.0	4.3000

EXERCISES for Section 16.7

In Exercises 1–6, verify that the power series solution of the differential equation is equivalent to the solution found using the techniques in Sections 16.2–16.6.

1. $y' - y = 0$ **2.** $y' - ky = 0$

3. $y'' - 9y = 0$ **4.** $y'' - k^2y = 0$

5. $y'' + 4y = 0$ **6.** $y'' + k^2y = 0$

In Exercises 7–10, use power series to solve the differential equation and find the interval of convergence of the series.

7. $y' + 3xy = 0$ **8.** $y' - 2xy = 0$

9. $y'' - xy' = 0$ **10.** $y'' - xy' - y = 0$

In Exercises 11 and 12, find the first three terms of each of the power series representing independent solutions of the differential equation.

11. $(x^2 + 4)y'' + y = 0$ **12.** $y'' + x^2y = 0$

In Exercises 13 and 14, use Taylor's Theorem to find the series solution of the differential equation under the specified initial conditions. Use n terms of the series to approximate y for the given value of x and compare the result with the approximation given by the Euler Method for $\Delta x = 0.1$.

13. $y' + (2x - 1)y = 0$, $y(0) = 2$, $n = 5$, $x = \frac{1}{2}$

14. $y' - 2xy = 0$, $y(0) = 1$, $n = 4$, $x = 1$

In Exercises 15 and 16, use Taylor's Theorem to find the series solution of the differential equation under the specified initial conditions. Use n terms of the series to approximate y for the given value of x.

15. $y'' - 2xy = 0$, $y(0) = 1$, $y'(0) = -3$, $n = 6$, $x = \frac{1}{4}$

16. $y'' - 2xy' + y = 0$, $y(0) = 1$, $y'(0) = 2$, $n = 8$, $x = \frac{1}{2}$

In Exercises 17–20, verify that the series converges to the given function on the indicated interval. (*Hint:* Use the given differential equation.)

17. $\displaystyle\sum_{n=0}^{\infty} \frac{x^n}{n!} = e^x$, $(-\infty, \infty)$

   Differential equation: $y' - y = 0$

18. $\displaystyle\sum_{n=0}^{\infty} \frac{(-1)^n x^{2n}}{(2n)!} = \cos x$, $(-\infty, \infty)$

   Differential Equation: $y'' + y = 0$

19. $\displaystyle\sum_{n=0}^{\infty} \frac{(-1)^n x^{2n+1}}{2n + 1} = \arctan x$, $(-1, 1)$

   Differential Equation: $(x^2 + 1)y'' + 2xy' = 0$

20. $\displaystyle\sum_{n=0}^{\infty} \frac{(2n)! x^{2n+1}}{(2^n n!)^2(2n + 1)} = \arcsin x$, $(-1, 1)$

   Differential equation: $(1 - x^2)y'' - xy' = 0$

21. Find the first six terms in the series solution of Airy's equation $y'' - xy = 0$.

REVIEW EXERCISES for Chapter 16

In Exercises 1–4, classify the differential equation according to type and order.

1. $\dfrac{\partial^2 u}{\partial t^2} = c^2 \dfrac{\partial^2 u}{\partial x^2}$

2. $yy'' = x + 1$

3. $y'' + 3y' - 10 = 0$

4. $(y'')^2 + 4y' = 0$

In Exercises 5 and 6, use the given differential equation and its direction field to answer questions about the solutions of the differential equation.

5. $\dfrac{dy}{dx} = \dfrac{y}{x}$

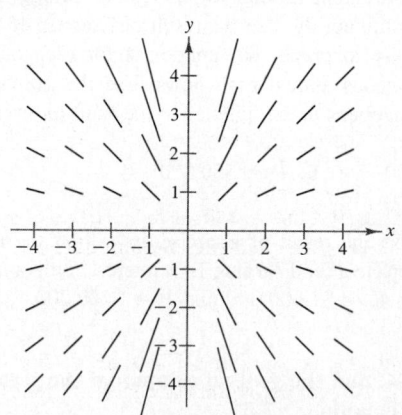

a. Sketch several solution curves for the differential equation on the direction field.

b. Find the general solution of the differential equation. Compare the result with the sketches from part a.

6. $\dfrac{dy}{dx} = \sqrt{1 - y^2}$

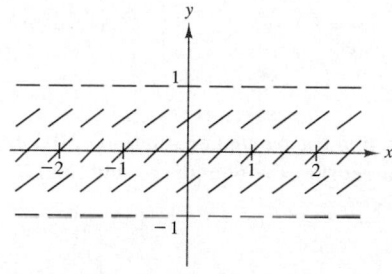

a. Sketch several solution curves for the differential equation on the direction field.

b. When is the rate of change of the solution greatest? When is it least?

c. Find the general solution of the differential equation. Compare the result with the sketches from part a.

In Exercises 7–10, match the differential equation with its solution.

Differential equation	Solution
7. $y' - 4 = 0$	a. $y = C_1 e^{2x} + C_2 e^{-2x}$
8. $y' - 4y = 0$	b. $y = 4x + C$
9. $y'' - 4y = 0$	c. $y = C_1 \cos 2x + C_2 \sin 2x$
10. $y'' + 4y = 0$	d. $y = Ce^{4x}$

In Exercises 11–32, find the general solution of the first-order differential equation.

11. $\dfrac{dy}{dx} - \dfrac{y}{x} = 2 + \sqrt{x}$ **12.** $\dfrac{dy}{dx} + xy = 2y$

13. $y' - \dfrac{2y}{x} = \dfrac{y'}{x}$ **14.** $\dfrac{dy}{dx} - 3x^2 y = e^{x^3}$

15. $\dfrac{dy}{dx} - \dfrac{y}{x} = \dfrac{x}{y}$ **16.** $\dfrac{dy}{dx} - \dfrac{3y}{x^2} = \dfrac{1}{x^2}$

17. $(10x + 8y + 2)\,dx + (8x + 5y + 2)\,dy = 0$

18. $(y + x^3 + xy^2)\,dx - x\,dy = 0$

19. $(2x - 2y^3 + y)\,dx + (x - 6xy^2)\,dy = 0$

20. $3x^2y^2\,dx + (2x^3y + x^3y^4)\,dy = 0$

21. $dy = (y\tan x + 2e^x)\,dx$

22. $y\,dx - \left(x + \sqrt{xy}\right)dy = 0$

23. $(x - y - 5)\,dx - (x + 3y - 2)\,dy = 0$

24. $y' = 2x\sqrt{1 - y^2}$

25. $x + yy' = \sqrt{x^2 + y^2}$

26. $xy' + y = \sin x$

27. $yy' + y^2 = 1 + x^2$

28. $2x\,dx + 2y\,dy = (x^2 + y^2)\,dx$

29. $(1 + x^2)\,dy = (1 + y^2)\,dx$

30. $x^3yy' = x^4 + 3x^2y^2 + y^4$

31. $xy' - ay = bx^4$

32. $y' = y + 2x(y - e^x)$

In Exercises 33–40, find the particular solution of the differential equation that satisfies the boundary condition.

33. $y' - 2y = e^x$
 $y(0) = 4$

34. $y' + \dfrac{2y}{x} = -x^9y^5$
 $y(1) = 2$

35. $x\,dy = (x + y + 2)\,dx$
 $y(1) = 10$

36. $ye^{xy}\,dx + xe^{xy}\,dy = 0$
 $y(-2) = -5$

37. $(1 + y)\ln(1 + y)\,dx + dy = 0$
 $y(0) = 2$

38. $(2x + y - 3)\,dx + (x - 3y + 1)\,dy = 0$
 $y(2) = 0$

39. $y' = x^2y^2 - 9x^2$
 $y(0) = \dfrac{3(1 + e)}{1 - e}$

40. $2xy' - y = x^3 - x$
 $y(4) = 2$

In Exercises 41 and 42, find the orthogonal trajectories of the given family and sketch several members of each family.

41. $(x - C)^2 + y^2 = C^2$

42. $y - 2x = C$

43. *Snow Removal* Assume that the rate of change in the number of miles s of road cleared per hour by a snowplow is inversely proportional to the height h of snow.
 a. Write and solve the differential equation to find s as a function of h.
 b. Find the particular solution if $s = 25$ miles when $h = 2$ inches and $s = 12$ miles when $h = 10$ inches ($2 \le h \le 15$).

44. *Growth Rate* Let x and y be the sizes of two internal organs of a particular mammal at time t. Empirical data indicate that the relative growth rates of these two organs are equal, and hence we have
$$\frac{1}{x}\frac{dx}{dt} = \frac{1}{y}\frac{dy}{dt}.$$
Solve this differential equation, writing y as a function of x.

45. *Population Growth* The rate of growth in the number N of deer in a state park varies jointly over time t as N and $L - N$, where L is the estimated limiting size of the herd. Write N as a function of t if $N = 100$ when $t = 0$ and $N = 200$ when $t = 4$.

46. *Investment* Let $A(t)$ be the amount in a fund earning interest at an annual rate r compounded continuously. If a continuous cash flow of P dollars per year is withdrawn from the fund, then the rate of change of A is given by the differential equation
$$\frac{dA}{dt} = rA - P$$
where $A = A_0$ when $t = 0$. Solve this differential equation for A as a function of t.

C **47.** *Investment* A retired couple plans to withdraw P dollars per year from a retirement account of $500,000 earning 10% compounded continuously. Use the result of Exercise 46 and a graphing utility to graph the function A for each of the following continuous annual cash flows. Use the graphs to describe what happens to the balance in the fund for each of the cases.
 a. $P = 40,000$ **b.** $P = 50,000$
 c. $P = 60,000$

48. *Investment* Use the result of Exercise 46 to find the time necessary to deplete a fund earning 14% interest compounded continuously if $A_0 = 1,000,000$ and $P = 200,000$.

In Exercises 49–54, find the general solution of the second-order differential equation.

49. $y'' + y = x^3 + x$

50. $y'' + 2y = e^{2x} + x$

51. $y'' + y = 2\cos x$

52. $y'' + 5y' + 4y = x^2 + \sin 2x$

53. $y'' - 2y' + y = 2xe^x$

54. $y'' + 2y' + y = \dfrac{1}{x^2 e^x}$

In Exercises 55–58, find the particular solution of the differential equation that satisfies the initial conditions.

Differential equation	Initial conditions
55. $y'' - y' - 2y = 0$	$y(0) = 0,\ y'(0) = 3$
56. $y'' + 4y' + 5y = 0$	$y(0) = 2,\ y'(0) = -7$
57. $y'' + 4y = \cos x$	$y(0) = 6,\ y'(0) = -6$
58. $y'' + 3y' = 6x$	$y(0) = 2,\ y'(0) = \frac{10}{3}$

Vibrating Spring **In Exercises 59 and 60, describe the motion of a 64-pound weight suspended on a spring. Assume that the weight stretches the spring $\frac{4}{3}$ feet from its natural position.**

59. The weight is pulled $\frac{1}{2}$ foot below the equilibrium position and released.

60. The weight is pulled $\frac{1}{2}$ foot below the equilibrium position and released. The motion takes place in a medium that furnishes a damping force of magnitude $\frac{1}{8}$ speed at all times.

In Exercises 61 and 62, find the series solution of the differential equation.

61. $(x - 4)y' + y = 0$

62. $y'' + 3xy' - 3y = 0$

APPENDIX A

Proofs of Selected Theorems

**THEOREM 1.2
PROPERTIES OF LIMITS
(Properties 3, 4, and 5)
(page 71)**

Let b and c be real numbers, let n be a positive integer, and let f and g be functions with the following limits.

$\lim\limits_{x \to c} f(x) = L$ and $\lim\limits_{x \to c} g(x) = K$

3. $\lim\limits_{x \to c} f(x)g(x) = LK$

4. $\lim\limits_{x \to c} \dfrac{f(x)}{g(x)} = \dfrac{L}{K}$, provided $K \neq 0$

5. $\lim\limits_{x \to c} [f(x)]^n = L^n$

Proof   Given that

$$\lim_{x \to c} f(x) = L \quad \text{and} \quad \lim_{x \to c} g(x) = K,$$

to prove Property 3, we write

$$f(x)g(x) = [f(x) - L][g(x) - K] + [Lg(x) + Kf(x)] - LK.$$

Because the limit of $f(x)$ is L, and the limit of $g(x)$ is K, we have

$$\lim_{x \to c} [f(x) - L] = 0 \quad \text{and} \quad \lim_{x \to c} [g(x) - K] = 0.$$

Let $0 < \varepsilon < 1$. Then there exists $\delta > 0$ such that if $0 < |x - c| < \delta$, then

$$|f(x) - L - 0| < \varepsilon \quad \text{and} \quad |g(x) - K - 0| < \varepsilon,$$

which implies that

$$|[f(x) - L][g(x) - K] - 0| = |f(x) - L| \, |g(x) - K| < \varepsilon\varepsilon < \varepsilon.$$

Hence,

$$\lim_{x \to c} [f(x) - L][g(x) - K] = 0.$$

Furthermore, by Property 1 of Theorem 1.2, we have

$$\lim_{x \to c} Lg(x) = LK \quad \text{and} \quad \lim_{x \to c} Kf(x) = KL.$$

Finally, by Property 2 of Theorem 1.2, we obtain

$$\lim_{x \to c} f(x)g(x) = \lim_{x \to c} [f(x) - L][g(x) - K] + \lim_{x \to c} Lg(x) + \lim_{x \to c} Kf(x) - \lim_{x \to c} LK$$
$$= 0 + LK + KL - LK$$
$$= LK.$$

To prove Property 4, we note that it is sufficient to prove that

$$\lim_{x \to c} \frac{1}{g(x)} = \frac{1}{K}.$$

Then we can use Property 3 to write

$$\lim_{x \to c} \frac{f(x)}{g(x)} = \lim_{x \to c} f(x) \frac{1}{g(x)} = \lim_{x \to c} f(x) \cdot \lim_{x \to c} \frac{1}{g(x)} = \frac{L}{K}.$$

Let $\varepsilon > 0$. Because $\lim_{x \to c} g(x) = K$, there exists $\delta_1 > 0$ such that if

$$0 < |x - c| < \delta_1, \text{ then } |g(x) - K| < \frac{|K|}{2},$$

which implies that

$$|K| = |g(x) + (|K| - g(x))| \le |g(x)| + ||K| - g(x)| < |g(x)| + \frac{|K|}{2}.$$

That is, for $0 < |x - c| < \delta_1$,

$$\frac{|K|}{2} < |g(x)| \qquad \text{or} \qquad \frac{1}{|g(x)|} < \frac{2}{|K|}.$$

Similarly, there exists a $\delta_2 > 0$ such that if $0 < |x - c| < \delta_2$, then

$$|g(x) - K| < \frac{|K|^2}{2} \varepsilon.$$

Let δ be the smaller of δ_1 and δ_2. For $0 < |x - c| < \delta$, we have

$$\left| \frac{1}{g(x)} - \frac{1}{K} \right| = \left| \frac{K - g(x)}{g(x)K} \right| = \frac{1}{|K|} \cdot \frac{1}{|g(x)|} |K - g(x)|$$

$$< \frac{1}{|K|} \cdot \frac{2}{|K|} \frac{|K|^2}{2} \varepsilon = \varepsilon.$$

Thus,

$$\lim_{x \to c} \frac{1}{g(x)} = \frac{1}{K}.$$

Finally, the proof of Property 5 can be obtained by a straightforward application of mathematical induction coupled with Property 3.   ▬

**THEOREM 1.4
THE LIMIT OF A FUNCTION
INVOLVING A RADICAL
(page 73)**

Let n be a positive integer. The following limit is valid for all c if n is odd, and is valid for $c > 0$ if n is even.

$$\lim_{x \to c} \sqrt[n]{x} = \sqrt[n]{c}.$$

Proof   Consider the case for which $c > 0$ and n is any positive integer. For a given $\varepsilon > 0$, we need to find $\delta > 0$ such that

$$|\sqrt[n]{x} - \sqrt[n]{c}| < \varepsilon \qquad \text{whenever} \qquad 0 < |x - c| < \delta,$$

which is the same as saying

$$-\varepsilon < \sqrt[n]{x} - \sqrt[n]{c} < \varepsilon \qquad \text{whenever} \qquad -\delta < x - c < \delta.$$

We assume $\varepsilon < \sqrt[n]{c}$, which implies that $0 < \sqrt[n]{c} - \varepsilon < \sqrt[n]{c}$. Now, we let δ be the smaller of the two numbers.

$$c - (\sqrt[n]{c} - \varepsilon)^n \qquad \text{and} \qquad (\sqrt[n]{c} + \varepsilon)^n - c$$

Then we have

$$-\delta < x - c \qquad\qquad < \delta$$

$$-[c - (\sqrt[n]{c} - \varepsilon)^n] < x - c \qquad < (\sqrt[n]{c} + \varepsilon)^n - c$$

$$(\sqrt[n]{c} - \varepsilon)^n - c < x - c \qquad < (\sqrt[n]{c} + \varepsilon)^n - c$$

$$(\sqrt[n]{c} - \varepsilon)^n < x \qquad\qquad < (\sqrt[n]{c} + \varepsilon)^n$$

$$\sqrt[n]{c} - \varepsilon < \sqrt[n]{x} \qquad\qquad < \sqrt[n]{c} + \varepsilon$$

$$-\varepsilon < \sqrt[n]{x} - \sqrt[n]{c} < \varepsilon$$

THEOREM 1.5
THE LIMIT OF A COMPOSITE
FUNCTION (page 73)

If f and g are functions such that

$$\lim_{x \to c} g(x) = L \qquad \text{and} \qquad \lim_{x \to L} f(x) = f(L)$$

then

$$\lim_{x \to c} f(g(x)) = f(L).$$

Proof   For a given $\varepsilon > 0$, we must find $\delta > 0$ such that

$$|f(g(x)) - f(L)| < \varepsilon \qquad \text{whenever} \qquad 0 < |x - c| < \delta.$$

Because the limit of $f(x)$ as $x \to L$ is $f(L)$, we know there exists $\delta_1 > 0$ such that

$$|f(u) - f(L)| < \varepsilon \qquad \text{whenever} \qquad |u - L| < \delta_1.$$

Moreover, because the limit of $g(x)$ as $x \to c$ is L, we know there exists $\delta > 0$ such that

$$|g(x) - L| < \delta_1 \qquad \text{whenever} \qquad 0 < |x - c| < \delta.$$

Finally, letting $u = g(x)$, we have

$$|f(g(x)) - f(L)| < \varepsilon \qquad \text{whenever} \qquad 0 < |x - c| < \delta.$$

THEOREM 1.8
THE SQUEEZE THEOREM
(page 80)

If $h(x) \le f(x) \le g(x)$ for all x in an open interval containing c, except possibly at c itself, and if

$$\lim_{x \to c} h(x) = L = \lim_{x \to c} g(x)$$

then

$$\lim_{x \to c} f(x) \text{ exists and is equal to } L.$$

Proof   For $\varepsilon > 0$ there exist δ_1 and δ_2 such that

$$|h(x) - L| < \varepsilon \qquad \text{whenever} \qquad 0 < |x - c| < \delta_1$$

and

$$|g(x) - L| < \varepsilon \qquad \text{whenever} \qquad 0 < |x - c| < \delta_2.$$

Because $h(x) \le f(x) \le g(x)$ for all x in an open interval containing c, except possibly at c itself, there exists $\delta_3 > 0$ such that $h(x) \le f(x) \le g(x)$ for $0 < |x - c| < \delta_3$. Let δ be the smallest of δ_1, δ_2, and δ_3. Then, if $0 < |x - c| < \delta$, it follows that $|h(x) - L| < \varepsilon$ and $|g(x) - L| < \varepsilon$, which implies that

$$-\varepsilon < h(x) - L < \varepsilon \qquad \text{and} \qquad -\varepsilon < g(x) - L < \varepsilon$$
$$L - \varepsilon < h(x) \qquad \text{and} \qquad g(x) < L + \varepsilon.$$

Now, because $h(x) \le f(x) \le g(x)$, it follows that $L - \varepsilon < f(x) < L + \varepsilon$, which implies that $|f(x) - L| < \varepsilon$. Therefore,

$$\lim_{x \to c} f(x) = L.$$

THEOREM 1.14
VERTICAL ASYMPTOTES
(page 98)

Let f and g be continuous on an open interval containing c. If $f(c) \ne 0$, $g(c) = 0$, and there exists an open interval containing c such that $g(x) \ne 0$ for all $x \ne c$ in the interval, then the graph of the function given by

$$h(x) = \frac{f(x)}{g(x)}$$

has a vertical asymptote at $x = c$.

Proof   We consider the case for which $f(c) > 0$, and there exists $b > c$ such that $c < x < b$ implies $g(x) > 0$. Then for $M > 0$, we choose δ_1 such that

$$0 < x - c < \delta_1 \qquad \text{implies that} \qquad \frac{f(c)}{2} < f(x) < \frac{3f(c)}{2}$$

and δ_2 such that

$$0 < x - c < \delta_2 \qquad \text{implies that} \qquad 0 < g(x) < \frac{f(c)}{2M}.$$

Now we let δ be the smaller of δ_1 and δ_2. Then, it follows that

$$0 < x - c < \delta \qquad \text{implies that} \qquad \frac{f(x)}{g(x)} > \frac{f(c)}{2}\left(\frac{2M}{f(c)}\right) = M.$$

Therefore, it follows that

$$\lim_{x \to c^+} \frac{f(x)}{g(x)} = \infty$$

and the line $x = c$ is a vertical asymptote of the graph of h.

ALTERNATE FORM OF THE
DERIVATIVE (page 110)

The derivative of f at c is given by

$$f'(c) = \lim_{x \to c} \frac{f(x) - f(c)}{x - c}$$

provided this limit exists.

Proof   The derivative of f at c is given by

$$f'(c) = \lim_{\Delta x \to 0} \frac{f(c + \Delta x) - f(c)}{\Delta x}.$$

Let $x = c + \Delta x$. Then $x \to c$ as $\Delta x \to 0$. Thus, replacing $c + \Delta x$ by x, we have

$$f'(c) = \lim_{\Delta x \to 0} \frac{f(c + \Delta x) - f(c)}{\Delta x} = \lim_{x \to c} \frac{f(x) - f(c)}{x - c}.$$

**THEOREM 2.10
THE CHAIN RULE
(page 138)**

If $y = f(u)$ is a differentiable function of u, and $u = g(x)$ is a differentiable function of x, then $y = f(g(x))$ is a differentiable function of x and

$$\frac{dy}{dx} = \frac{dy}{du} \cdot \frac{du}{dx} \quad \text{or equivalently,} \quad \frac{d}{dx}[f(g(x))] = f'(g(x))g'(x).$$

Proof   In Section 2.4, we let $h(x) = f(g(x))$ and used the alternate form of the derivative to show that $h'(c) = f'(g(c))g'(c)$, provided $g(x) \neq g(c)$ for values of x other than c. We now consider a more general proof. We begin by considering the derivative of f.

$$f'(x) = \lim_{\Delta x \to 0} \frac{f(x + \Delta x) - f(x)}{\Delta x} = \lim_{\Delta x \to 0} \frac{\Delta y}{\Delta x}$$

For a fixed value of x, we define a function η such that

$$\eta(\Delta x) = \begin{cases} 0, & \Delta x = 0 \\ \dfrac{\Delta y}{\Delta x} - f'(x), & \Delta x \neq 0. \end{cases}$$

Because the limit of $\eta(\Delta x)$ as $\Delta x \to 0$ doesn't depend upon the value of $\eta(0)$, we have

$$\lim_{\Delta x \to 0} \eta(\Delta x) = \lim_{\Delta x \to 0} \left[\frac{\Delta y}{\Delta x} - f'(x) \right] = 0$$

and we can conclude that η is continuous at 0. Moreover, because $\Delta y = 0$ when $\Delta x = 0$, the equation

$$\Delta y = \Delta x \eta(\Delta x) + \Delta x f'(x)$$

is valid whether Δx is zero or not. Now, by letting $\Delta u = g(x + \Delta x) - g(x)$, we use the continuity of g to conclude that

$$\lim_{\Delta x \to 0} \Delta u = \lim_{\Delta x \to 0} [g(x + \Delta x) - g(x)] = 0,$$

which implies that

$$\lim_{\Delta x \to 0} \eta(\Delta u) = 0.$$

Finally,

$$\Delta y = \Delta u \eta(\Delta u) + \Delta u f'(u) \rightarrow \frac{\Delta y}{\Delta x} = \frac{\Delta u}{\Delta x} \eta(\Delta u) + \frac{\Delta u}{\Delta x} f'(u), \quad \Delta x \neq 0$$

and taking the limit as $\Delta x \to 0$, we have

$$\frac{dy}{dx} = \frac{du}{dx} \left[\lim_{\Delta x \to 0} \eta(\Delta u) \right] + \frac{du}{dx} f'(u) = \frac{dy}{dx}(0) + \frac{du}{dx} f'(u)$$

$$= \frac{du}{dx} f'(u) = \frac{du}{dx} \cdot \frac{dy}{du}.$$

**CONCAVITY
INTERPRETATION
(page 189)**

1. Let f be differentiable at c. If the graph of f is concave upward at $(c, f(c))$, then the graph of f lies *above* the tangent line at $(c, f(c))$ on some open interval containing c.
2. Let f be differentiable at c. If the graph of f is concave downward at $(c, f(c))$, then the graph of f lies *below* the tangent line at $(c, f(c))$ on some open interval containing c.

Proof Assume f is concave upward at c. Then by definition, there exists an interval (a, b) containing c such that f' is increasing on (a, b). The equation of the tangent line to the graph of f at c is given by

$$g(x) = f(c) + f'(c)(x - c).$$

If x is in the open interval (c, b), then the directed distance from the point $(x, f(x))$ (on the graph of f) to the point $(x, g(x))$ (on the tangent line) is given by

$$d = f(x) - [f(c) + f'(c)(x - c)]$$
$$= f(x) - f(c) - f'(c)(x - c).$$

Moreover, by the Mean Value Theorem there exists a number z in (c, x) such that

$$f'(z) = \frac{f(x) - f(c)}{x - c}.$$

Thus, we have

$$d = f(x) - f(c) - f'(c)(x - c)$$
$$= f'(z)(x - c) - f'(c)(x - c)$$
$$= [f'(z) - f'(c)](x - c).$$

The second factor $(x - c)$ is positive because $c < x$. Moreover, because f' is increasing, it follows that the first factor $[f'(z) - f'(c)]$ is also positive. Therefore, $d > 0$ and we conclude that the graph of f lies above the tangent line. If x is in the open interval (a, c), a similar argument can be given. $\blacksquare$

THEOREM 3.10
LIMITS AT INFINITY
(page 198)

If r is a positive rational number, and c is any real number, then

$$\lim_{x \to \infty} \frac{c}{x^r} = 0.$$

Furthermore, if x^r is defined when $x < 0$, then

$$\lim_{x \to -\infty} \frac{c}{x^r} = 0.$$

Proof We begin by proving that

$$\lim_{x \to \infty} \frac{1}{x} = 0.$$

For $\varepsilon > 0$, let $M = 1/\varepsilon$. Then, for $x > M$, we have

$$x > M = \frac{1}{\varepsilon} \quad \to \quad \frac{1}{x} < \varepsilon \quad \to \quad \left| \frac{1}{x} - 0 \right| < \varepsilon.$$

Therefore, by the definition of a limit at infinity, we conclude that the limit of $1/x$ as $x \to \infty$ is 0. Now, using this result, and letting $r = m/n$, we can write the following.

$$\lim_{x \to \infty} \left(\frac{c}{x^r} \right) = \lim_{x \to \infty} \left(\frac{c}{x^{m/n}} \right)$$
$$= c \left[\lim_{x \to \infty} \left(\frac{1}{\sqrt[n]{x}} \right)^m \right]$$
$$= c \left[\lim_{x \to \infty} \sqrt[n]{\frac{1}{x}} \right]^m$$

$$= c \left[\sqrt[n]{\lim_{x \to \infty} \frac{1}{x}} \right]^m$$

$$= c \left[\sqrt[n]{0} \right]^m$$

$$= 0$$

The proof of the second part of the theorem is similar.

THEOREM 4.2
SUMMATION FORMULAS
(page 260)

1. $\displaystyle\sum_{i=1}^{n} c = cn$ **2.** $\displaystyle\sum_{i=1}^{n} i = \frac{n(n+1)}{2}$

3. $\displaystyle\sum_{i=1}^{n} i^2 = \frac{n(n+1)(2n+1)}{6}$ **4.** $\displaystyle\sum_{i=1}^{n} i^3 = \frac{n^2(n+1)^2}{4}$

Proof The proof of part 1 is straightforward. By adding c to itself n times, we obtain a sum of nc.

To prove part 2, we write the sum in increasing and decreasing order and add corresponding terms as follows.

$$\sum_{i=1}^{n} i = \quad 1 \quad + \quad 2 \quad + \quad 3 \quad + \cdots + (n-1) + \quad n$$

$$\downarrow \qquad \downarrow \qquad \downarrow \qquad\qquad \downarrow \qquad \downarrow$$

$$\sum_{i=1}^{n} i = \quad n \quad + (n-1) + (n-2) + \cdots + \quad 2 \quad + \quad 1$$

$$\downarrow \qquad \downarrow \qquad \downarrow \qquad\qquad \downarrow \qquad \downarrow$$

$$2\sum_{i=1}^{n} i = \underbrace{(n+1) + (n+1) + (n+1) + \cdots + (n+1) + (n+1)}_{n \text{ terms}}$$

Therefore,

$$\sum_{i=1}^{n} i = \frac{n(n+1)}{2}.$$

To prove part 3, we use mathematical induction. First, if $n = 1$, the result is true because

$$\sum_{i=1}^{1} i^2 = 1^2 = 1 = \frac{1(1+1)(2+1)}{6}.$$

Now, assuming the result is true for $n = k$, we show that it is true for $n = k + 1$, as follows.

$$\sum_{i=1}^{k+1} i^2 = \sum_{i=1}^{k} i^2 + (k+1)^2$$

$$= \frac{k(k+1)(2k+1)}{6} + (k+1)^2$$

$$= \frac{k+1}{6}(2k^2 + k + 6k + 6)$$

$$= \frac{k+1}{6}[(2k+3)(k+2)]$$

$$= \frac{(k+1)(k+2)[2(k+1)+1]}{6}$$

Part 4 can be proved using a similar argument with mathematical induction.

◼

**THEOREM 4.8
PRESERVATION OF
INEQUALITY
(page 277)**

1. If f is integrable and nonnegative on the closed interval $[a, b]$, then

$$0 \leq \int_a^b f(x)dx.$$

2. If f and g are integrable on the closed interval $[a, b]$, and $f(x) \leq g(x)$ for every x in $[a, b]$, then

$$\int_a^b f(x)dx \leq \int_a^b g(x)dx.$$

Proof   To prove part 1, we suppose, on the contrary, that

$$\int_a^b f(x)dx = I < 0.$$

Then, let $a = x_0 < x_1 < x_2 < \cdots < x_n = b$ be a partition of $[a, b]$, and let

$$R = \sum_{i=1}^n f(c_i)\Delta x_i$$

be a Riemann sum. Because $f(x) \geq 0$, it follows that $R \geq 0$. Now, for $\|\Delta\|$ sufficiently small, we have $|R - I| < -I/2$, which implies that

$$\sum_{i=1}^n f(c_i)\Delta x_i = R < I - \frac{I}{2} < 0,$$

which is not possible. From this contradiction, we can conclude that

$$0 \leq \int_a^b f(x)dx.$$

To prove the second part of the theorem, we note that $f(x) \leq g(x)$ implies that $g(x) - f(x) \geq 0$. Hence, we can apply the result of part 1 to conclude that

$$0 \leq \int_a^b (g(x) - f(x))dx$$

$$0 \leq \int_a^b g(x)dx - \int_a^b f(x)dx$$

$$\int_a^b f(x)dx \leq \int_a^b g(x)dx.$$

◼

**PROPERTIES OF THE
NATURAL LOGARITHMIC
FUNCTION (page 318)**

$$\lim_{x \to 0^+} \ln x = -\infty \qquad \text{and} \qquad \lim_{x \to \infty} \ln x = \infty$$

Proof   To begin, we show that $\ln 2 \geq \frac{1}{2}$. From the Mean Value Theorem of Integrals, we can write

$$\ln 2 = \int_1^2 \frac{1}{x} dx = (2 - 1)\frac{1}{c} = \frac{1}{c}$$

where c is in $[1, 2]$. This implies that

$$1 \leq c \leq 2$$

$$1 \geq \frac{1}{c} \geq \frac{1}{2}$$

$$1 \geq \ln 2 \geq \frac{1}{2}.$$

Now, let N be any positive (large) number. Because $\ln x$ is increasing, it follows that if $x > 2^{2N}$, then

$$\ln x > \ln 2^{2N} = 2N \ln 2.$$

However, because $\ln 2 \geq \frac{1}{2}$, it follows that

$$\ln x > 2N \ln 2 \geq 2N\left(\frac{1}{2}\right) = N.$$

This verifies the second limit. To verify the first limit, we let $z = 1/x$. Then, $z \to \infty$ as $x \to 0^+$, and we can write

$$\lim_{x \to 0^+} \ln x = \lim_{x \to 0^+} \left(-\ln \frac{1}{x}\right) = \lim_{z \to \infty} (-\ln z) = -\lim_{z \to \infty} \ln z = -\infty. \qquad \blacksquare$$

THEOREM 5.8
CONTINUITY AND
DIFFERENTIABILITY OF
INVERSE FUNCTIONS
(page 339)

Let f be a function whose domain is an interval I. If f possesses an inverse, then the following statements are true.

1. If f is continuous on its domain, then f^{-1} is continuous on its domain.
2. If f is increasing on its domain, then f^{-1} is increasing on its domain.
3. If f is decreasing on its domain, then f^{-1} is decreasing on its domain.
4. If f is differentiable at c and $f'(c) \neq 0$, then f^{-1} is differentiable at $f(c)$.

Proof To prove part 1, we first show that if f is continuous on I, and has an inverse, then f is strictly monotonic on I. Suppose that f were not strictly monotonic. Then there would exist numbers x_1, x_2, x_3 in I such that $x_1 < x_2 < x_3$ but $f(x_2)$ is not between $f(x_1)$ and $f(x_3)$. Without loss of generality, assume $f(x_1) < f(x_3) < f(x_2)$. By the Intermediate Value Theorem, there exists a number x_0 between x_1 and x_2 such that $f(x_0) = x_3$. Thus, f is not one-to-one, and cannot have an inverse. Hence f must be strictly monotonic.

Because f is continuous, the Intermediate Value Theorem implies that the set of values of f,

$$\{f(x): x \in I\}, \text{ forms an interval } J.$$

Assume that a is an interior point of J. From the previous argument, $f^{-1}(a)$ is an interior point of I. Let $\varepsilon > 0$. There exists $0 < \varepsilon_1 < \varepsilon$ such that

$$I_1 = (f^{-1}(a) - \varepsilon_1, f^{-1}(a) + \varepsilon_1) \subseteq I.$$

Because f is strictly monotonic on I_1, the set of values $\{f(x): x \in I_1\}$ forms an interval $J_1 \subseteq J$. Let $\delta > 0$ such that $(a - \delta, a + \delta) \subseteq J_1$. Finally, if

$$|y - a| < \delta, \text{ then } |f^{-1}(y) - f^{-1}(a)| < \varepsilon_1 < \varepsilon.$$

Hence, f^{-1} is continuous at a. A similar proof can be given if a is an endpoint.

To prove part 2, let y_1 and y_2 be in the domain of f^{-1}, with $y_1 < y_2$. Then, there exist x_1 and x_2 in the domain of f such that

$$f(x_1) = y_1 < y_2 = f(x_2).$$

Since f is increasing, $f(x_1) < f(x_2)$ holds precisely when $x_1 < x_2$. Therefore,

$$f^{-1}(y_1) = x_1 < x_2 = f^{-1}(y_2),$$

which implies that f^{-1} is increasing. (Part 3 can be proved in a similar way.)
    Finally, to prove part 4, we consider the limit

$$(f^{-1})'(a) = \lim_{y \to a} \frac{f^{-1}(y) - f^{-1}(a)}{y - a}$$

where a is in the domain of f^{-1} and $f^{-1}(a) = c$. Because f is differentiable at c, f is continuous at c, and so is f^{-1} at a. Thus, $y \to a$ implies that $x \to c$, and we have

$$(f^{-1})'(a) = \lim_{x \to c} \frac{x - c}{f(x) - f(c)}$$

$$= \lim_{x \to c} \frac{1}{\dfrac{f(x) - f(c)}{x - c}}$$

$$= \frac{1}{\displaystyle\lim_{x \to c} \frac{f(x) - f(c)}{x - c}}$$

$$= \frac{1}{f'(c)}.$$

Hence, $(f^{-1})'(a)$ exists, and f^{-1} is differentiable at $f(c)$.   ▬

**THEOREM 5.9
THE DERIVATIVE OF AN
INVERSE FUNCTION
(page 339)**

Let f be a function that is differentiable on an interval I. If f possesses an inverse function g, then g is differentiable at any x for which $f'(g(x)) \neq 0$. Moreover,

$$g'(x) = \frac{1}{f'(g(x))}, \quad f'(g(x)) \neq 0.$$

Proof   From the proof of Theorem 5.8, letting $a = x$, we know that g is differentiable. Using the Chain Rule we differentiate both sides of the equation $x = f(g(x))$ to obtain

$$1 = f'(g(x)) \frac{d}{dx}[g(x)].$$

Because $f'(g(x)) \neq 0$, we can divide by this quantity to obtain

$$\frac{d}{dx}[g(x)] = \frac{1}{f'(g(x))}.$$ ▬

**THEOREM 5.15
A LIMIT INVOLVING e
(page 357)**

$$\lim_{x \to \infty} \left(1 + \frac{1}{x}\right)^x = \lim_{x \to \infty} \left(\frac{x + 1}{x}\right)^x = e$$

Proof   Let $y = \lim_{x \to \infty} \left(1 + \frac{1}{x}\right)^x$. Taking the natural log of both sides we have

$$\ln y = \ln \left[\lim_{x \to \infty} \left(1 + \frac{1}{x}\right)^x \right].$$

Because the natural logarithmic function is continuous, we can write

$$\ln y = \lim_{x \to \infty} \left[x \ln \left(1 + \frac{1}{x} \right) \right]$$

$$= \lim_{x \to \infty} \left[\frac{\ln \left[1 + (1/x) \right]}{1/x} \right].$$

Letting $x = \dfrac{1}{t}$ we have

$$\lim_{t \to 0^+} \frac{\ln (1 + t)}{t}$$

$$= \lim_{t \to 0^+} \frac{\ln (1 + t) - \ln 1}{t}$$

$$= \frac{d}{dx} \ln x \text{ at } x = 1$$

$$= \frac{1}{x} \text{ at } x = 1$$

$$= 1.$$

Finally, because $\ln y = 1$, we know that $y = e$ and we conclude that

$$\lim_{x \to \infty} \left(1 + \frac{1}{x} \right)^x = e.$$

**THEOREM 7.3
THE EXTENDED MEAN VALUE
THEOREM
(page 526)**

If f and g are differentiable on an open interval (a, b) and continuous on $[a, b]$ such that $g'(x) \neq 0$ for any x in (a, b), then there exists a point c in (a, b) such that

$$\frac{f'(c)}{g'(c)} = \frac{f(b) - f(a)}{g(b) - g(a)}.$$

Proof We can assume that $g(a) \neq g(b)$, since otherwise by Rolle's Theorem it would follow that $g'(x) = 0$ for some x in (a, b). Now, we define $h(x)$ to be

$$h(x) = f(x) - \left(\frac{f(b) - f(a)}{g(b) - g(a)} \right) g(x).$$

Then

$$h(a) = f(a) - \left(\frac{f(b) - f(a)}{g(b) - g(a)} \right) g(a) = \frac{f(a)g(b) - f(b)g(a)}{g(b) - g(a)}$$

and

$$h(b) = f(b) - \left(\frac{f(b) - f(a)}{g(b) - g(a)} \right) g(b) = \frac{f(a)g(b) - f(b)g(a)}{g(b) - g(a)}$$

and by Rolle's Theorem there exists a point c in (a, b) such that

$$h'(c) = f'(c) - \frac{f(b) - f(a)}{g(b) - g(a)} g'(c) = 0,$$

which implies that

$$\frac{f'(c)}{g'(c)} = \frac{f(b) - f(a)}{g(b) - g(a)}.$$

We can use the Extended Mean Value Theorem to prove L'Hôpital's Rule. Of the several different cases of this rule, we illustrate the proof of only one case and leave the remaining cases for you to prove.

THEOREM 7.4
L'HÔPITAL'S RULE
(page 526)

Let f and g be functions that are differentiable on an open interval (a, b) containing c, except possibly at c itself. Assume that $g'(x) \neq 0$ for all x in (a, b), except possibly at c itself. If the limit of $f(x)/g(x)$ as x approaches c produces the indeterminate form $0/0$, then

$$\lim_{x \to c} \frac{f(x)}{g(x)} = \lim_{x \to c} \frac{f'(x)}{g'(x)}$$

provided the limit on the right exists (or is infinite). This result also applies if the limit of $f(x)/g(x)$ as x approaches c produces the indeterminate forms ∞/∞, $(-\infty)/\infty$, $\infty/(-\infty)$, or $(-\infty)/(-\infty)$.

Proof   We consider the case for which

$$\lim_{x \to c^+} f(x) = 0 \qquad \text{and} \qquad \lim_{x \to c^+} g(x) = 0.$$

Define the following new functions:

$$F(x) = \begin{cases} f(x), & x \neq c \\ 0, & x = c \end{cases} \qquad \text{and} \qquad G(x) = \begin{cases} g(x), & x \neq c \\ 0, & x = c \end{cases}.$$

For any $x, c < x < b$, F and G are differentiable on $(c, x]$ and continuous on $[c, x]$. We can apply the Extended Mean Value Theorem to conclude that there exists a number z in (c, x) such that

$$\frac{F'(z)}{G'(z)} = \frac{F(x) - F(c)}{G(x) - G(c)} = \frac{F(x)}{G(x)} = \frac{f'(z)}{g'(z)} = \frac{f(x)}{g(x)}.$$

Finally, by letting x approach c from the right, $x \to c^+$, we have $z \to c^+$ because $c < z < x$, and

$$\lim_{x \to c^+} \frac{f(x)}{g(x)} = \lim_{x \to c^+} \frac{f'(z)}{g'(z)} = \lim_{z \to c^+} \frac{f'(z)}{g'(z)} = \lim_{x \to c^+} \frac{f'(x)}{g'(x)}.$$

The proof for the case where $x \to c^-$ and $x \to c$ are left to the reader.   ▪

THEOREM 8.19
TAYLOR'S THEOREM
(page 602)

If a function f is differentiable through order $n + 1$ in an interval I containing c, then for each x in I, there exists z between x and c such that

$$f(x) = f(c) + f'(c)(x - c) + \frac{f''(c)}{2!}(x - c)^2 + \cdots + \frac{f^{(n)}(c)}{n!}(x - c)^n + R_n(x)$$

where

$$R_n(x) = \frac{f^{(n+1)}(z)}{(n + 1)!}(x - c)^{n+1}.$$

Proof   To find $R_n(x)$ we fix x in I ($x \neq c$) and write

$$R_n(x) = f(x) - P_n(x)$$

where $P_n(x)$ is the nth Taylor polynomial for $f(x)$. Then we let g be a function of t defined by

$$g(t) = f(x) - f(t) - f'(t)(x - t) - \cdots - \frac{f^{(n)}(t)}{n!}(x - t)^n - R_n(x)\frac{(x - t)^{n+1}}{(x - c)^{n+1}}.$$

The reason for defining g this way is that differentiation with respect to t has a telescoping effect. For example, we have

$$\frac{d}{dt}[-f(t) - f'(t)(x - t)] = -f'(t) + f'(t) - f''(t)(x - t)$$

$$= -f''(t)(x - t).$$

The result is that the derivative $g'(t)$ simplifies to

$$g'(t) = -\frac{f^{(n+1)}(t)}{n!}(x - t)^n + (n + 1)R_n(x)\frac{(x - t)^n}{(x - c)^{n+1}}$$

for all t between c and x. Moreover, for a fixed x

$$g(c) = f(x) - [P_n(x) + R_n(x)] = f(x) - f(x) = 0$$

and

$$g(x) = f(x) - f(x) - 0 - \cdots - 0 = f(x) - f(x) = 0.$$

Therefore, g satisfies the conditions of Rolle's Theorem, and it follows that there is a number z between c and x such that $g'(z) = 0$. Substituting z for t in the equation for $g'(t)$ and then solving for $R_n(x)$, we obtain

$$g'(z) = -\frac{f^{(n+1)}(z)}{n!}(x - z)^n + (n + 1)R_n(x)\frac{(x - z)^n}{(x - c)^{n+1}} = 0$$

$$R_n(x) = \frac{f^{(n+1)}(z)}{(n + 1)!}(x - c)^{n+1}.$$

Finally, because $g(c) = 0$, we have

$$0 = f(x) - f(c) - f'(c)(x - c) - \cdots - \frac{f^{(n)}(c)}{n!}(x - c)^n - R_n(x)$$

$$f(x) = f(c) + f'(c)(x - c) + \cdots + \frac{f^{(n)}(c)}{n!}(x - c)^n + R_n(x).$$

**THEOREM 9.7
ROTATION OF AXES
(page 660)**

$Ax^2 + Bxy + Cy^2 + Dx + Ey + F = 0$ where $B \neq 0$, can be rewritten as
$A'(x')^2 + C'(y')^2 + D'x' + E'y' + F' = 0$

by rotating the coordinate axes through an angle θ, where

$$\cot 2\theta = \frac{A - C}{B}.$$

The coefficients of the new equation are obtained by making the substitutions

$x = x' \cos \theta - y' \sin \theta$
$y = x' \sin \theta + y' \cos \theta.$

Proof   To discover how the coordinates in the xy-system are related to the coordinates in the $x'y'$-system, we choose a point $P = (x, y)$ in the original system and attempt to find its coordinates (x', y') in the rotated system. In either system the distance r between the point P and the origin is the same, thus the equations for x, y, x', and y' are those given in Figure A.1. Using the formulas for the sine and cosine of the difference of two angles, we obtain

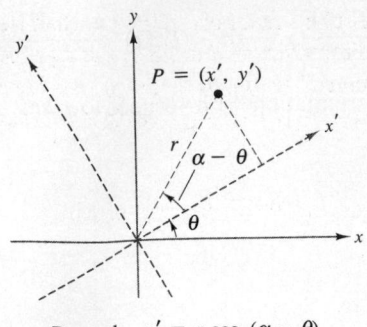

Rotated: $x' = r \cos (\alpha - \theta)$
$y' = r \sin (\alpha - \theta)$

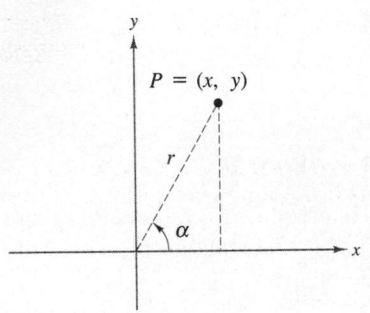

Original: $x = r \cos \alpha$
$y = r \sin \alpha$

$$x' = r \cos (\alpha - \theta) = r(\cos \alpha \cos \theta + \sin \alpha \sin \theta)$$
$$= r \cos \alpha \cos \theta + r \sin \alpha \sin \theta = x \cos \theta + y \sin \theta$$
$$y' = r \sin (\alpha - \theta) = r(\sin \alpha \cos \theta - \cos \alpha \sin \theta)$$
$$= r \sin \alpha \cos \theta - r \cos \alpha \sin \theta = y \cos \theta - x \sin \theta.$$

Solving this system for x and y yields

$$x = x' \cos \theta - y' \sin \theta \quad \text{and} \quad y = x' \sin \theta + y' \cos \theta.$$

Finally, by substituting these values for x and y into the original equation and collecting terms, we obtain the following.

$$A' = A \cos^2 \theta + B \cos \theta \sin \theta + C \sin^2 \theta$$

$$C' = A \sin^2 \theta - B \cos \theta \sin \theta + C \cos^2 \theta$$

$$D' = D \cos \theta + E \sin \theta$$

$$E' = -D \sin \theta + E \cos \theta$$

$$F' = F$$

Now, in order to eliminate the $x'y'$ term, we must select θ so that $B' = 0$ as follows.

$$B' = 2(C - A) \sin \theta \cos \theta + B(\cos^2 \theta - \sin^2 \theta)$$

$$= (C - A) \sin 2\theta + B \cos 2\theta$$

$$= B(\sin 2\theta) \left(\frac{C - A}{B} + \cot 2\theta \right) = 0, \quad \sin 2\theta \neq 0$$

If $B = 0$, no rotation is necessary since the xy term is not present in the original equation. If $B \neq 0$, then the only way to make $B' = 0$ is to let

$$\cot 2\theta = \frac{A - C}{B}, \quad B \neq 0.$$

Thus, we have established the desired results.

<hr />

| THEOREM 10.10 CLASSIFICATION OF CONICS BY ECCENTRICITY (page 705) | The locus of a point in the plane whose distance from a fixed point (*focus*) has a constant ratio to its distance from a fixed line (*directrix*) is a conic. The constant ratio e is the *eccentricity* of the conic.

1. The conic is an ellipse if $0 < e < 1$.
2. The conic is a parabola if $e = 1$.
3. The conic is a hyperbola if $e > 1$. |

Proof If $e = 1$, then by definition, the conic must be a parabola. If $e \neq 1$, then we consider the focus F to lie at the origin and the directrix $x = d$ to lie to the right of the origin, as shown in Figure A.2. For the point $P = (r, \theta) = (x, y)$, we have $|PF| = r$ and $|PQ| = d - r \cos\theta$. Given that $e = |PF|/|PQ|$, it follows that

$$|PF| = |PQ|e \rightarrow r = e(d - r \cos \theta).$$

By converting to rectangular coordinates and squaring both sides, we obtain

$$x^2 + y^2 = e^2(d - x)^2 = e^2(d^2 - 2dx + x^2).$$

Completing the square produces

$$\left(x + \frac{e^2 d}{1 - e^2}\right)^2 + \frac{y^2}{1 - e^2} = \frac{e^2 d^2}{(1 - e^2)^2}.$$

If $e < 1$, then this equation represents an ellipse. If $e > 1$, then $1 - e^2 < 0$, and the equation represents a hyperbola. ∎

**THEOREM 13.6
CHAIN RULE: ONE
INDEPENDENT VARIABLE
(page 869)**

Let $w = f(x, y)$, where f is a differentiable function of x and y. If $x = g(t)$ and $y = h(t)$, where g and h are differentiable functions of t, then w is a differentiable function of t, and

$$\frac{dw}{dt} = \frac{\partial w}{\partial x}\frac{dx}{dt} + \frac{\partial w}{\partial y}\frac{dy}{dt}.$$

Proof Because g and h are differentiable functions of t, we know that both Δx and Δy approach zero as Δt approaches zero. Moreover, because f is a differentiable function of x and y, we know that

$$\Delta w = \frac{\partial w}{\partial x}\Delta x + \frac{\partial w}{\partial y}\Delta y + \varepsilon_1 \Delta x + \varepsilon_2 \Delta y$$

where both ε_1 and $\varepsilon_2 \rightarrow 0$ as $(\Delta x, \Delta y) \rightarrow (0, 0)$. Thus, for $\Delta t \neq 0$, we have

$$\frac{\Delta w}{\Delta t} = \frac{\partial w}{\partial x}\frac{\Delta x}{\Delta t} + \frac{\partial w}{\partial y}\frac{\Delta y}{\Delta t} + \varepsilon_1 \frac{\Delta x}{\Delta t} + \varepsilon_2 \frac{\Delta y}{\Delta t}$$

from which it follows that

$$\frac{dw}{dt} = \lim_{\Delta t \to 0} \frac{\Delta w}{\Delta t} = \frac{\partial w}{\partial x}\frac{dx}{dt} + \frac{\partial w}{\partial y}\frac{dy}{dt} + 0\left(\frac{dx}{dt}\right) + 0\left(\frac{dy}{dt}\right)$$

$$= \frac{\partial w}{\partial x}\frac{dx}{dt} + \frac{\partial w}{\partial y}\frac{dy}{dt}. \qquad ∎$$

APPENDIX B

Basic Differentiation Rules for Elementary Functions

1. $\dfrac{d}{dx}[cu] = cu'$

2. $\dfrac{d}{dx}[u \pm v] = u' \pm v'$

3. $\dfrac{d}{dx}[uv] = uv' + vu'$

4. $\dfrac{d}{dx}\left[\dfrac{u}{v}\right] = \dfrac{vu' - uv'}{v^2}$

5. $\dfrac{d}{dx}[c] = 0$

6. $\dfrac{d}{dx}[u^n] = nu^{n-1}u'$

7. $\dfrac{d}{dx}[x] = 1$

8. $\dfrac{d}{dx}[|u|] = \dfrac{u}{|u|}(u'),\ u \neq 0$

9. $\dfrac{d}{dx}[\ln u] = \dfrac{u'}{u}$

10. $\dfrac{d}{dx}[e^u] = e^u u'$

11. $\dfrac{d}{dx}[\sin u] = (\cos u)u'$

12. $\dfrac{d}{dx}[\cos u] = -(\sin u)u'$

13. $\dfrac{d}{dx}[\tan u] = (\sec^2 u)u'$

14. $\dfrac{d}{dx}[\cot u] = -(\csc^2 u)u'$

15. $\dfrac{d}{dx}[\sec u] = (\sec u \tan u)u'$

16. $\dfrac{d}{dx}[\csc u] = -(\csc u \cot u)u'$

17. $\dfrac{d}{dx}[\arcsin u] = \dfrac{u'}{\sqrt{1 - u^2}}$

18. $\dfrac{d}{dx}[\arccos u] = \dfrac{-u'}{\sqrt{1 - u^2}}$

19. $\dfrac{d}{dx}[\arctan u] = \dfrac{u'}{1 + u^2}$

20. $\dfrac{d}{dx}[\text{arccot } u] = \dfrac{-u'}{1 + u^2}$

21. $\dfrac{d}{dx}[\text{arcsec } u] = \dfrac{u'}{|u|\sqrt{u^2 - 1}}$

22. $\dfrac{d}{dx}[\text{arccsc } u] = \dfrac{-u'}{|u|\sqrt{u^2 - 1}}$

APPENDIX C

Integration Tables

Forms Involving u^n

1. $\displaystyle\int u^n \, du = \frac{u^{n+1}}{n+1} + C, \, n \neq -1$

2. $\displaystyle\int \frac{1}{u} \, du = \ln|u| + C$

Forms Involving $a + bu$

3. $\displaystyle\int \frac{u}{a + bu} \, du = \frac{1}{b^2}\left(bu - a \ln|a + bu|\right) + C$

4. $\displaystyle\int \frac{u}{(a + bu)^2} \, du = \frac{1}{b^2}\left(\frac{a}{a + bu} + \ln|a + bu|\right) + C$

5. $\displaystyle\int \frac{u}{(a + bu)^n} \, du = \frac{1}{b^2}\left[\frac{-1}{(n-2)(a+bu)^{n-2}} + \frac{a}{(n-1)(a+bu)^{n-1}}\right] + C, \, n \neq 1, 2$

6. $\displaystyle\int \frac{u^2}{a + bu} \, du = \frac{1}{b^3}\left(-\frac{bu}{2}(2a - bu) + a^2 \ln|a + bu|\right) + C$

7. $\displaystyle\int \frac{u^2}{(a + bu)^2} \, du = \frac{1}{b^3}\left(bu - \frac{a^2}{a + bu} - 2a \ln|a + bu|\right) + C$

8. $\displaystyle\int \frac{u^2}{(a + bu)^3} \, du = \frac{1}{b^3}\left[\frac{2a}{a + bu} - \frac{a^2}{2(a + bu)^2} + \ln|a + bu|\right] + C$

9. $\displaystyle\int \frac{u^2}{(a + bu)^n} \, du = \frac{1}{b^3}\left[\frac{-1}{(n-3)(a+bu)^{n-3}} + \frac{2a}{(n-2)(a+bu)^{n-2}} - \frac{a^2}{(n-1)(a+bu)^{n-1}}\right] + C,$
$$n \neq 1, 2, 3$$

10. $\displaystyle\int \frac{1}{u(a + bu)} \, du = \frac{1}{a} \ln\left|\frac{u}{a + bu}\right| + C$

11. $\displaystyle\int \frac{1}{u(a + bu)^2} \, du = \frac{1}{a}\left(\frac{1}{a + bu} + \frac{1}{a} \ln\left|\frac{u}{a + bu}\right|\right) + C$

12. $\displaystyle\int \frac{1}{u^2(a + bu)} \, du = -\frac{1}{a}\left(\frac{1}{u} + \frac{b}{a} \ln\left|\frac{u}{a + bu}\right|\right) + C$

13. $\int \dfrac{1}{u^2(a + bu)^2}\, du = -\dfrac{1}{a^2}\left[\dfrac{a + 2bu}{u(a + bu)} + \dfrac{2b}{a} \ln \left| \dfrac{u}{a + bu} \right| \right] + C$

Forms Involving $a + bu + cu^2$, $b^2 \neq 4ac$

14. $\int \dfrac{1}{a + bu + cu^2}\, du = \begin{cases} \dfrac{2}{\sqrt{4ac - b^2}} \arctan \dfrac{2cu + b}{\sqrt{4ac - b^2}} + C, & b^2 < 4ac \\[4mm] \dfrac{1}{\sqrt{b^2 - 4ac}} \ln \left| \dfrac{2cu + b - \sqrt{b^2 - 4ac}}{2cu + b + \sqrt{b^2 - 4ac}} \right| + C, & b^2 > 4ac \end{cases}$

15. $\int \dfrac{u}{a + bu + cu^2}\, du = \dfrac{1}{2c} \left(\ln |a + bu + cu^2| - b \int \dfrac{1}{a + bu + cu^2}\, du \right)$

Forms Involving $\sqrt{a + bu}$

16. $\int u^n \sqrt{a + bu}\, du = \dfrac{2}{b(2n + 3)} \left[u^n(a + bu)^{3/2} - na \int u^{n-1} \sqrt{a + bu}\, du \right]$

17. $\int \dfrac{1}{u\sqrt{a + bu}}\, du = \begin{cases} \dfrac{1}{\sqrt{a}} \ln \left| \dfrac{\sqrt{a + bu} - \sqrt{a}}{\sqrt{a + bu} + \sqrt{a}} \right| + C, & 0 < a \\[4mm] \dfrac{2}{\sqrt{-a}} \arctan \sqrt{\dfrac{a + bu}{-a}} + C, & a < 0 \end{cases}$

18. $\int \dfrac{1}{u^n \sqrt{a + bu}}\, du = \dfrac{-1}{a(n - 1)} \left[\dfrac{\sqrt{a + bu}}{u^{n-1}} + \dfrac{(2n - 3)b}{2} \int \dfrac{1}{u^{n-1}\sqrt{a + bu}}\, du \right]$, $n \neq 1$

19. $\int \dfrac{\sqrt{a + bu}}{u}\, du = 2\sqrt{a + bu} + a \int \dfrac{1}{u\sqrt{a + bu}}\, du$

20. $\int \dfrac{\sqrt{a + bu}}{u^n}\, du = \dfrac{-1}{a(n - 1)} \left[\dfrac{(a + bu)^{3/2}}{u^{n-1}} + \dfrac{(2n - 5)b}{2} \int \dfrac{\sqrt{a + bu}}{u^{n-1}}\, du \right]$, $n \neq 1$

21. $\int \dfrac{u}{\sqrt{a + bu}}\, du = \dfrac{-2(2a - bu)}{3b^2} \sqrt{a + bu} + C$

22. $\int \dfrac{u^n}{\sqrt{a + bu}}\, du = \dfrac{2}{(2n + 1)b} \left(u^n \sqrt{a + bu} - na \int \dfrac{u^{n-1}}{\sqrt{a + bu}}\, du \right)$

Forms Involving $a^2 \pm u^2$, $0 < a$

23. $\int \dfrac{1}{a^2 + u^2}\, du = \dfrac{1}{a} \arctan \dfrac{u}{a} + C$

24. $\int \dfrac{1}{u^2 - a^2}\, du = -\int \dfrac{1}{a^2 - u^2}\, du = \dfrac{1}{2a} \ln \left| \dfrac{u - a}{u + a} \right| + C$

25. $\int \dfrac{1}{(a^2 \pm u^2)^n}\, du = \dfrac{1}{2a^2(n - 1)} \left[\dfrac{u}{(a^2 \pm u^2)^{n-1}} + (2n - 3) \int \dfrac{1}{(a^2 \pm u^2)^{n-1}}\, du \right]$, $n \neq 1$

Forms Involving $\sqrt{u^2 \pm a^2}$, $0 < a$

26. $\int \sqrt{u^2 \pm a^2}\, du = \dfrac{1}{2} \left(u\sqrt{u^2 \pm a^2} \pm a^2 \ln |u + \sqrt{u^2 \pm a^2}| \right) + C$

27. $\displaystyle\int u^2\sqrt{u^2 \pm a^2}\, du = \frac{1}{8}\Big[u(2u^2 \pm a^2)\sqrt{u^2 \pm a^2} - a^4 \ln\big|u + \sqrt{u^2 \pm a^2}\big|\Big] + C$

28. $\displaystyle\int \frac{\sqrt{u^2 + a^2}}{u}\, du = \sqrt{u^2 + a^2} - a\ln\left|\frac{a + \sqrt{u^2 + a^2}}{u}\right| + C$

29. $\displaystyle\int \frac{\sqrt{u^2 - a^2}}{u}\, du = \sqrt{u^2 - a^2} - a\,\mathrm{arcsec}\,\frac{|u|}{a} + C$

30. $\displaystyle\int \frac{\sqrt{u^2 \pm a^2}}{u^2}\, du = \frac{-\sqrt{u^2 \pm a^2}}{u} + \ln\big|u + \sqrt{u^2 \pm a^2}\big| + C$

31. $\displaystyle\int \frac{1}{\sqrt{u^2 \pm a^2}}\, du = \ln\big|u + \sqrt{u^2 \pm a^2}\big| + C$

32. $\displaystyle\int \frac{1}{u\sqrt{u^2 + a^2}}\, du = \frac{-1}{a}\ln\left|\frac{a + \sqrt{u^2 + a^2}}{u}\right| + C$

33. $\displaystyle\int \frac{1}{u\sqrt{u^2 - a^2}}\, du = \frac{1}{a}\,\mathrm{arcsec}\,\frac{|u|}{a} + C$

34. $\displaystyle\int \frac{u^2}{\sqrt{u^2 \pm a^2}}\, du = \frac{1}{2}\Big(u\sqrt{u^2 \pm a^2} \mp a^2\ln\big|u + \sqrt{u^2 \pm a^2}\big|\Big) + C$

35. $\displaystyle\int \frac{1}{u^2\sqrt{u^2 \pm a^2}}\, du = \mp\,\frac{\sqrt{u^2 \pm a^2}}{a^2 u} + C$

36. $\displaystyle\int \frac{1}{(u^2 \pm a^2)^{3/2}}\, du = \frac{\pm u}{a^2\sqrt{u^2 \pm a^2}} + C$

Forms Involving $\sqrt{a^2 - u^2},\ 0 < a$

37. $\displaystyle\int \sqrt{a^2 - u^2}\, du = \frac{1}{2}\Big(u\sqrt{a^2 - u^2} + a^2\arcsin\frac{u}{a}\Big) + C$

38. $\displaystyle\int u^2\sqrt{a^2 - u^2}\, du = \frac{1}{8}\Big[u(2u^2 - a^2)\sqrt{a^2 - u^2} + a^4\arcsin\frac{u}{a}\Big] + C$

39. $\displaystyle\int \frac{\sqrt{a^2 - u^2}}{u}\, du = \sqrt{a^2 - u^2} - a\ln\left|\frac{a + \sqrt{a^2 - u^2}}{u}\right| + C$

40. $\displaystyle\int \frac{\sqrt{a^2 - u^2}}{u^2}\, du = \frac{-\sqrt{a^2 - u^2}}{u} - \arcsin\frac{u}{a} + C$

41. $\displaystyle\int \frac{1}{\sqrt{a^2 - u^2}}\, du = \arcsin\frac{u}{a} + C$

42. $\displaystyle\int \frac{1}{u\sqrt{a^2 - u^2}}\, du = \frac{-1}{a}\ln\left|\frac{a + \sqrt{a^2 - u^2}}{u}\right| + C$

43. $\displaystyle\int \frac{u^2}{\sqrt{a^2 - u^2}}\, du = \frac{1}{2}\Big(-u\sqrt{a^2 - u^2} + a^2\arcsin\frac{u}{a}\Big) + C$

44. $\displaystyle\int \frac{1}{u^2\sqrt{a^2 - u^2}}\, du = \frac{-\sqrt{a^2 - u^2}}{a^2 u} + C$

45. $\displaystyle\int \frac{1}{(a^2 - u^2)^{3/2}}\, du = \frac{u}{a^2\sqrt{a^2 - u^2}} + C$

Forms Involving $\sin u$ or $\cos u$

46. $\displaystyle\int \sin u \, du = -\cos u + C$

47. $\displaystyle\int \cos u \, du = \sin u + C$

48. $\displaystyle\int \sin^2 u \, du = \frac{1}{2}(u - \sin u \cos u) + C$

49. $\displaystyle\int \cos^2 u \, du = \frac{1}{2}(u + \sin u \cos u) + C$

50. $\displaystyle\int \sin^n u \, du = -\frac{\sin^{n-1} u \cos u}{n} + \frac{n-1}{n}\int \sin^{n-2} u \, du$

51. $\displaystyle\int \cos^n u \, du = \frac{\cos^{n-1} u \sin u}{n} + \frac{n-1}{n}\int \cos^{n-2} u \, du$

52. $\displaystyle\int u \sin u \, du = \sin u - u \cos u + C$

53. $\displaystyle\int u \cos u \, du = \cos u + u \sin u + C$

54. $\displaystyle\int u^n \sin u \, du = -u^n \cos u + n\int u^{n-1} \cos u \, du$

55. $\displaystyle\int u^n \cos u \, du = u^n \sin u - n\int u^{n-1} \sin u \, du$

56. $\displaystyle\int \frac{1}{1 \pm \sin u} \, du = \tan u \mp \sec u + C$

57. $\displaystyle\int \frac{1}{1 \pm \cos u} \, du = -\cot u \pm \csc u + C$

58. $\displaystyle\int \frac{1}{\sin u \cos u} \, du = \ln |\tan u| + C$

Forms Involving $\tan u$, $\cot u$, $\sec u$, $\csc u$

59. $\displaystyle\int \tan u \, du = -\ln |\cos u| + C$

60. $\displaystyle\int \cot u \, du = \ln |\sin u| + C$

61. $\displaystyle\int \sec u \, du = \ln |\sec u + \tan u| + C$

62. $\displaystyle\int \csc u \, du = \ln |\csc u - \cot u| + C$

63. $\displaystyle\int \tan^2 u \, du = -u + \tan u + C$

64. $\displaystyle\int \cot^2 u \, du = -u - \cot u + C$

65. $\displaystyle\int \sec^2 u \, du = \tan u + C$

66. $\displaystyle\int \csc^2 u \, du = -\cot u + C$

67. $\displaystyle\int \tan^n u \, du = \frac{\tan^{n-1} u}{n-1} - \int \tan^{n-2} u \, du, \; n \neq 1$

68. $\displaystyle\int \cot^n u \, du = -\frac{\cot^{n-1} u}{n-1} - \int (\cot^{n-2} u) \, du, \; n \neq 1$

69. $\displaystyle\int \sec^n u \, du = \frac{\sec^{n-2} u \tan u}{n-1} + \frac{n-2}{n-1} \int \sec^{n-2} u \, du, \; n \neq 1$

70. $\displaystyle\int \csc^n u \, du = -\frac{\csc^{n-2} u \cot u}{n-1} + \frac{n-2}{n-1} \int \csc^{n-2} u \, du, \; n \neq 1$

71. $\displaystyle\int \frac{1}{1 \pm \tan u} \, du = \frac{1}{2}\left(u \pm \ln|\cos u \pm \sin u|\right) + C$

72. $\displaystyle\int \frac{1}{1 \pm \cot u} \, du = \frac{1}{2}\left(u \mp \ln|\sin u \pm \cos u|\right) + C$

73. $\displaystyle\int \frac{1}{1 \pm \sec u} \, du = u + \cot u \mp \csc u + C$

74. $\displaystyle\int \frac{1}{1 \pm \csc u} \, du = u - \tan u \pm \sec u + C$

Forms Involving Inverse Trigonometric Functions

75. $\displaystyle\int \arcsin u \, du = u \arcsin u + \sqrt{1 - u^2} + C$

76. $\displaystyle\int \arccos u \, du = u \arccos u - \sqrt{1 - u^2} + C$

77. $\displaystyle\int \arctan u \, du = u \arctan u - \ln \sqrt{1 + u^2} + C$

78. $\displaystyle\int \text{arccot } u \, du = u \text{ arccot } u + \ln \sqrt{1 + u^2} + C$

79. $\displaystyle\int \text{arcsec } u \, du = u \text{ arcsec } u - \ln|u + \sqrt{u^2 - 1}| + C$

80. $\displaystyle\int \text{arccsc } u \, du = u \text{ arccsc } u + \ln|u + \sqrt{u^2 - 1}| + C$

Forms Involving e^u

81. $\displaystyle\int e^u \, du = e^u + C$

82. $\displaystyle\int u e^u \, du = (u - 1)e^u + C$

83. $\displaystyle\int u^n e^u \, du = u^n e^u - n \int u^{n-1} e^u \, du$

84. $\displaystyle\int \frac{1}{1 + e^u} \, du = u - \ln(1 + e^u) + C$

85. $\displaystyle\int e^{au} \sin bu \, du = \frac{e^{au}}{a^2 + b^2}(a \sin bu - b \cos bu) + C$

86. $\displaystyle\int e^{au} \cos bu \, du = \frac{e^{au}}{a^2 + b^2}(a \cos bu + b \sin bu) + C$

Forms Involving ln u

87. $\displaystyle\int \ln u \, du = u(-1 + \ln u) + C$

88. $\displaystyle\int u \ln u \, du = \frac{u^2}{4}(-1 + 2 \ln u) + C$

89. $\displaystyle\int u^n \ln u \, du = \frac{u^{n+1}}{(n+1)^2}[-1 + (n+1) \ln u] + C, \; n \neq -1$

90. $\displaystyle\int (\ln u)^2 \, du = u[2 - 2 \ln u + (\ln u)^2] + C$

91. $\displaystyle\int (\ln u)^n \, du = u(\ln u)^n - n \int (\ln u)^{n-1} \, du$

Answers to Odd-Numbered Exercises

PREREQUISITES CHAPTER

Section P.1 (*page 8*)

1. Rational **3.** Irrational **5.** Rational

7. Rational **9.** Rational **11.** $\frac{4}{11}$ **13.** $\frac{11}{37}$

15. a. True **b.** False **c.** True **d.** False
e. False **f.** False

17. $x < -\frac{1}{2}$

19. $x \geq \frac{1}{2}$

21. $-\frac{1}{2} < x < \frac{7}{2}$

23. $x < -4$

25. $x > 6$

27. $-1 < x < 1$

29. $x \geq 13,\ x \leq -7$

31. $a - b < x < a + b$

33. $-3 < x < 2$

35. $0 < x < 3$

37. $-3 \leq x \leq 1$

39. $-3 \leq x \leq 2$

41. 4, −4, 4

43. a. −51, 51, 51 **b.** 51, −51, 51 **45.** 1

47. a. 14 **b.** 10 **49.** $|x| \leq 2$ **51.** $|x - 2| > 2$

53. a. $|x - 12| \leq 10$ **b.** $|x - 12| \geq 10$

55. $r > 12.5\%$ **57.** $m < 24{,}062.5$ miles

59. $x \leq 41$ or $x \geq 59$

61. a. $\frac{355}{112} > \pi$ **b.** $\frac{22}{7} > \pi$ **63.** b

73. False. The reciprocal of 2 is $\frac{1}{2}$ which is not an integer.

74. True **75.** True **76.** False, $|0| = 0$.

77. True **78.** True

79. $|-3 - 1| > |-3| - |1|$
$|3 - 1| = |3| - |1|$

Section P.2 (*page 15*)

1. $d = 2\sqrt{5}$

3. $d = 2\sqrt{10}$

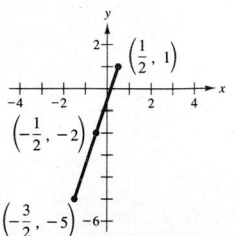

5. $d = \sqrt{8 - 2\sqrt{3}}$

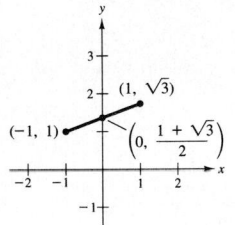

7. Right triangle
$d_1 = \sqrt{45},\ d_2 = \sqrt{5}$
$d_3 = \sqrt{50}$
$(d_1)^2 + (d_2)^2 = (d_3)^2$

9. Rhombus, the length of each side is $\sqrt{5}$.

11.

13. $d_1 = 2\sqrt{5}$, $d_2 = \sqrt{5}$, $d_3 = 3\sqrt{5}$
Collinear, since $d_1 + d_2 = d_3$

15. $d_1 = \sqrt{2}$, $d_2 = \sqrt{13}$, $d_3 = 5$
Not collinear, since $d_1 + d_2 > d_3$

17. $x = \pm 3$ **19.** $y = \pm\sqrt{55}$ **21.** $3x - 2y - 1 = 0$

23. $\left(\dfrac{3x_1 + x_2}{4}, \dfrac{3y_1 + y_2}{4}\right)$ $\left(\dfrac{x_1 + x_2}{2}, \dfrac{y_1 + y_2}{2}\right)$
$\left(\dfrac{x_1 + 3x_2}{4}, \dfrac{y_1 + 3y_2}{4}\right)$

25. a. $\left(x + \frac{5}{2}\right)^2 - \frac{25}{4}$ **b.** $(x + 4)^2 - 9$ **27.** c   **28.** b

29. a   **30.** d   **31.** $x^2 + y^2 - 9 = 0$

33. $x^2 + y^2 - 4x + 2y - 11 = 0$

35. $x^2 + y^2 + 2x - 4y = 0$

37. $x^2 + y^2 - 6x - 4y + 3 = 0$

39. $x^2 + y^2 - 6x - 8y = 0$

41. $x^2 + y^2 = 26,000^2$

43. $(x - 1)^2 + (y + 3)^2 = 4$

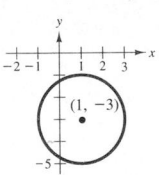

45. $(x - 1)^2 + (y + 3)^2 = 0$

47. $\left(x - \frac{1}{2}\right)^2 + \left(y - \frac{1}{2}\right)^2 = 2$

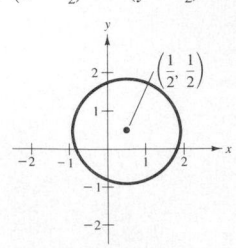

49. $\left(x + \frac{1}{2}\right)^2 + \left(y + \frac{5}{4}\right)^2 = \frac{9}{4}$

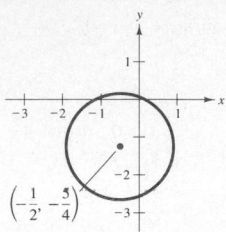

51. $y = \dfrac{-6 + \sqrt{99 + 4x - 4x^2}}{2}$

$y = \dfrac{-6 - \sqrt{99 + 4x - 4x^2}}{2}$

53.

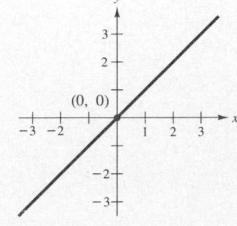

61. True   **62.** False, the distance is $|2b|$.   **63.** True

64. True

Section P.3   (*page 23*)

1. c   **2.** d   **3.** b   **4.** f   **5.** a   **6.** e

7. $(0, -3)$, $\left(\frac{3}{2}, 0\right)$ **9.** $(0, -2)$, $(-2, 0)$, $(1, 0)$

11. $(0, 0)$, $(-3, 0)$, $(3, 0)$ **13.** $\left(0, \frac{1}{2}\right)$, $(1, 0)$ **15.** $(0, 0)$

17. Symmetric with respect to the y-axis

19. Symmetric with respect to the y-axis

21. Symmetric with respect to the x-axis

23. Symmetric with respect to the origin

25. Symmetric with respect to the origin

27. $(1, 2)$ is not on the graph.
$(1, -1)$ is on the graph.
$(4, 5)$ is on the graph.

29. $\left(1, \frac{1}{5}\right)$ is on the graph.
$\left(2, \frac{1}{2}\right)$ is on the graph.
$(-1, -2)$ is not on the graph.

31. $y = x$ **33.** $y = x + 3$
Symmetry: origin   Symmetry: none

35. $y = -3x + 2$
Symmetry: none

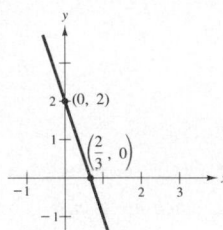

37. $y = \frac{1}{2}x - 4$
Symmetry: none

39. $y = 1 - x^2$
Symmetry: y-axis

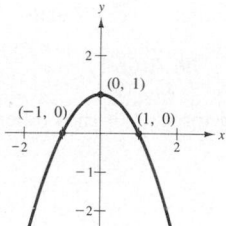

41. $y = x^3 + 2$
Symmetry: none

43. $y = (x + 2)^2$
Symmetry: none

45. $y = \dfrac{1}{x}$
Symmetry: origin

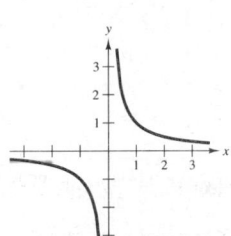

47. $y = -2x^2 + x + 1$
Symmetry: none

49. $y = x\sqrt{25 - x^2}$
Symmetry: origin

51. $x^2 + 4y^2 = 4$
Symmetry: origin and both axes

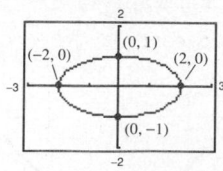

53. $y = (x + 2)(x - 4)(x - 6)$ **55.** $y = x$
57. $(1, 1)$ **59.** $(5, 2)$ **61.** $(-1, -2), (2, 1)$
63. $(-1, -1), (0, 0), (1, 1)$
65. $(-1, -5), (0, -1), (2, 1)$

67. $x \approx 192$ units
69. a. $k = 4$ **b.** $k = -1/8$
 c. $k = $ any real number   **d.** $k = 1$

71. a.

73. a.

 b. 3.37   **c.** 191.6   **b.** 211.6°   **c.** 24.7 lb/in.²
77. False, $(1, 2)$ is also a point on the graph.   **78.** True
79. True   **80.** True
81. $(1 - K^2)x^2 + (1 - K^2)y^2 + 4K^2x - 4K^2 = 0$

Section P.4 (*page 31*)

1. $m = 1$ **3.** $m = 0$ **5.** $m = -12$
7.

9. $m = 3$

11. $m = -\frac{2}{3}$

13. $(0, 1), (1, 1), (3, 1)$ **15.** $(0, 10), (2, 4), (3, 1)$

17. a.

b. 0.185, $1.82
19. $m = -\frac{1}{5}$, $(0, 4)$ **21.** m is undefined, no y-intercept
23. $2x - y - 3 = 0$ **25.** $3x + y = 0$

27. $y + 2 = 0$ **29.** $3x - 4y + 12 = 0$

31. $2x - 3y = 0$ **33.** $4x - y + 2 = 0$

35. $x - 3 = 0$ **37.** $3x + 2y - 6 = 0$
39. $x + y - 3 = 0$
41. a. $2x - y - 3 = 0$ **b.** $x + 2y - 4 = 0$
43. a. $40x + 24y - 53 = 0$ **b.** $24x - 40y + 9 = 0$
45. a. $x - 2 = 0$ **b.** $y - 5 = 0$
47.   **49.**

51.

53. $2x - y = 0$ **55.** Not collinear, since $m_1 \neq m_2$
57. $\left(0, \dfrac{-a^2 + b^2 + c^2}{2c}\right)$ **59.** $\left(b, \dfrac{a^2 - b^2}{c}\right)$
61. $5F - 9C - 160 = 0$ **63.** $C = 0.27x + 125$
65. a. $S = 2200t + 28,500$ **b.** $39,500
67. a. $x = \frac{1}{15}(1130 - p)$ **b.** 45 units   **c.** 49 units
69. 2   **71.** $\dfrac{5\sqrt{2}}{2}$ **73.** $2\sqrt{2}$ **79.** True

80. False, their slopes are negative reciprocals of each other.

81. $y = -\dfrac{\sqrt{2}}{4}x + \sqrt{2}$

$y = \dfrac{\sqrt{2}}{4}x - \sqrt{2}$

Section P.5 (*page 42*)

1. a. -3 **b.** -9 **c.** $2b - 3$ **d.** $2x - 5$
3. a. 1   **b.** 3   **c.** $\sqrt{c + 3}$ **d.** $\sqrt{x + \Delta x + 3}$
5. a. -1 **b.** 2   **c.** 6   **d.** $2t^2 + 4$
7. $3 + \Delta x$, $\Delta x \neq 0$ **9.** $3x^2 + 3x\,\Delta x + (\Delta x)^2$, $\Delta x \neq 0$
11. $\dfrac{-1}{\sqrt{x - 1}\left(1 + \sqrt{x - 1}\right)}$
13. $f(x) = 4 - x$ **15.** $f(x) = \sqrt{x - 1}$
   Domain: $(-\infty, \infty)$ Domain: $[1, \infty)$
   Range: $(-\infty, \infty)$ Range: $[0, \infty)$

     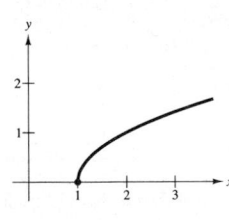

17. $f(x) = \sqrt{9 - x^2}$ **19.** $f(x) = |x - 2|$
   Domain: $[-3, 3]$ Domain: $(-\infty, \infty)$
   Range: $[0, 3]$ Range: $[0, \infty)$

21. Domain: $[-2, 2]$
Range: $\left[-2, 2\sqrt{2}\,\right]$
Intercepts: $\left(-\sqrt{2}, 0\right), (0, 2)$

23.

p_1 has one zero and p_2 has three zeros. As x moves away from the origin, the right and left-hand behavior of the graph of a cubic will differ. In the one direction it will rise and in the other direction it will fall. Therefore, a cubic must always have at least one zero.

25. y is not a function of x.   **27.** y is a function of x.
29. y is not a function of x.   **31.** y is a function of x.
33. y is a function of x.   **35.** y is not a function of x.

37. a.

b. L is a function of d with domain and range $(0, \infty)$.
39. a. A vertical translation   **b.** Reflection (about the x-axis)

c. A horizontal translation

d. Reflection (about the y-axis)

41. a. A reflection in the x-axis followed by a vertical translation.
b. A horizontal translation followed by a vertical translation.
43. a. 0   **b.** 0   **c.** -1 **d.** $\sqrt{15}$
e. $\sqrt{x^2 - 1}$ **f.** $x - 1$

45. $(f \circ g)(x) = x$ **47.** $(f \circ g)(x) = \dfrac{x + 1}{x}$
Domain: $[0, \infty)$
$(g \circ f)(x) = |x|$ Domain: $(-\infty, 0), (0, \infty)$
Domain: $(-\infty, \infty)$ $(g \circ f)(x) = \dfrac{1}{x + 1}$
Domain: $(-\infty, -1), (-1, \infty)$

49. $(A \circ r)(t) = 0.36\pi t^2$
$A \circ r$ represents the area of the circle at time t.
51. $\frac{10}{7}$ **53.** Even   **55.** Odd
61. a. $f(x) = x^2(4 - x^2)$ **b.** $f(x) = x(4 - x^2)$

63. $R(x) = 4 - x^2/2, \ r(x) = 2$ **65.** $h(x) = x^2, \ p(x) = x$

67. $A = xy = x\left(\dfrac{100 - 2x}{2}\right) = x(50 - x)$

69. $V = x(12 - 2x)^2 = 4x(6 - x)^2$
71. $V = 108x^2 - 4x^3$

73. $T = \dfrac{\sqrt{x^2 + 4}}{2} + \dfrac{\sqrt{x^2 - 6x + 10}}{4}$

75. False, if $f(x) = x^2$, then $f(-1) = f(1)$.   **76.** True
77. True
78. False, if $f(x) = x^2$, then $f(2x) = 4x^2 \neq 2f(x)$.
79. $f(x) = \frac{1}{2}[f(x) + f(-x)] + \frac{1}{2}[f(x) - f(-x)]$

81. $f(x) = \begin{cases} 1 - x, & x < 3 \\ x - 5, & x \geq 3 \end{cases}$

$g(x) = \begin{cases} 2(1 - x), & x < 0 \\ 2, & 0 \leq x < 2 \\ 2(x - 1), & x \geq 2 \end{cases}$

Section P.6   (*page 53*)

1. a. $396°, -324°$ **b.** $240°, -480°$
3. a. $\dfrac{19\pi}{9}, -\dfrac{17\pi}{9}$ **b.** $\dfrac{10\pi}{3}, -\dfrac{2\pi}{3}$

5. a. $\dfrac{\pi}{6}$, 0.524 **b.** $\dfrac{5\pi}{6}$, 2.618

  c. $\dfrac{7\pi}{4}$, 5.498 **d.** $\dfrac{2\pi}{3}$, 2.094

7. a. 270° **b.** 210° **c.** −105° **d.** −135.6°

9.

r	8 ft	15 in.	85 cm	24 in.	$\dfrac{12{,}963}{\pi}$ mi
s	12 ft	24 in.	63.75π cm	96 in.	8642 mi
θ	1.5	1.6	$\dfrac{3\pi}{4}$	4	$\dfrac{2\pi}{3}$

11. 171.89°

13. a. $\sin\theta = \frac{4}{5}$ $\csc\theta = \frac{5}{4}$ **b.** $\sin\theta = -\frac{15}{17}$ $\csc\theta = -\frac{17}{15}$

  $\cos\theta = \frac{3}{5}$ $\sec\theta = \frac{5}{3}$ $\cos\theta = \frac{8}{17}$ $\sec\theta = \frac{17}{8}$

  $\tan\theta = \frac{4}{3}$ $\cot\theta = \frac{3}{4}$ $\tan\theta = -\frac{15}{8}$ $\cot\theta = -\frac{8}{15}$

15. a. Quadrant III **b.** Quadrant IV

17. $\dfrac{\sqrt{3}}{2}$ **19.** $\dfrac{4}{3}$ **21.** $\dfrac{17}{15}$

23. a. $\sin 60° = \dfrac{\sqrt{3}}{2}$ **b.** $\sin 120° = \dfrac{\sqrt{3}}{2}$

  $\cos 60° = \dfrac{1}{2}$ $\cos 120° = -\dfrac{1}{2}$

  $\tan 60° = \sqrt{3}$ $\tan 120° = -\sqrt{3}$

  c. $\sin\dfrac{\pi}{4} = \dfrac{\sqrt{2}}{2}$ **d.** $\sin\dfrac{5\pi}{4} = -\dfrac{\sqrt{2}}{2}$

  $\cos\dfrac{\pi}{4} = \dfrac{\sqrt{2}}{2}$ $\cos\dfrac{5\pi}{4} = -\dfrac{\sqrt{2}}{2}$

  $\tan\dfrac{\pi}{4} = 1$ $\tan\dfrac{5\pi}{4} = 1$

25. a. $\sin 225° = -\dfrac{\sqrt{2}}{2}$ **b.** $\sin(-225°) = \dfrac{\sqrt{2}}{2}$

  $\cos 225° = -\dfrac{\sqrt{2}}{2}$ $\cos(-225°) = -\dfrac{\sqrt{2}}{2}$

  $\tan 225° = 1$ $\tan(-225°) = -1$

  c. $\sin\dfrac{5\pi}{3} = -\dfrac{\sqrt{3}}{2}$ **d.** $\sin\dfrac{11\pi}{6} = -\dfrac{1}{2}$

  $\cos\dfrac{5\pi}{3} = \dfrac{1}{2}$ $\cos\dfrac{11\pi}{6} = \dfrac{\sqrt{3}}{2}$

  $\tan\dfrac{5\pi}{3} = -\sqrt{3}$ $\tan\dfrac{11\pi}{6} = -\dfrac{\sqrt{3}}{3}$

27. a. 0.1736 **b.** 5.759 **29. a.** 0.3640 **b.** 0.3640

31. a. $\theta = \dfrac{\pi}{4}, \dfrac{7\pi}{4}$ **b.** $\theta = \dfrac{3\pi}{4}, \dfrac{5\pi}{4}$

33. a. $\theta = \dfrac{\pi}{4}, \dfrac{5\pi}{4}$ **b.** $\theta = \dfrac{5\pi}{6}, \dfrac{11\pi}{6}$

35. $\theta = \dfrac{\pi}{4}, \dfrac{3\pi}{4}, \dfrac{5\pi}{4}, \dfrac{7\pi}{4}$ **37.** $\theta = 0, \dfrac{\pi}{4}, \pi, \dfrac{5\pi}{4}$

39. $\theta = \dfrac{\pi}{3}, \dfrac{5\pi}{3}$ **41.** $\theta = 0, \dfrac{\pi}{2}, \pi$ **43.** $y = \dfrac{100\sqrt{3}}{3}$

45. $x = \dfrac{25\sqrt{3}}{3}$ **47.** 5099 ft **49.** 2.63 in.

51. a. Period: π **b.** Period: 2 **53.** Period: $\frac{1}{2}$
  Amplitude: 2 Amplitude: $\frac{1}{2}$ Amplitude: 3

55. Period: $\dfrac{\pi}{2}$ **57.** Period: $\dfrac{2\pi}{5}$

59. a. Change in amplitude **b.** Change in period

  c. Horizontal translation

61.   **63.**

65.   **67.**

69.   **71.**

73. a. Period of f: 2π
  Period of g: π

  b. $h(x) = A \cos \alpha x + B \sin \beta x$ is periodic. The period of $a \cos \alpha x$ is $2\pi/\alpha$. The period of $B \sin \beta x$ is $2\pi/\beta$. The period of h is the least common multiple of these two periods.

75. $y = 3 \cos\left(\dfrac{x}{2} - \dfrac{\pi}{2}\right)$

77. a.

b. $T(t) = 70 + 15 \sin\left(\dfrac{\pi t}{12} - \dfrac{2\pi}{3}\right)$

c.

79. a. $\frac{1}{440}$

b. 440   **c.**

81. $f(x) = \dfrac{4}{\pi}\left(\sin \pi x + \dfrac{1}{3}\sin 3\pi x + \dfrac{1}{5}\sin 5\pi x + \cdots\right)$

83.

a. Greatest: Summer
Smallest: Winter

b. 15 days

85. If α is the angle the line makes with the positive x-axis, then the slope of the line is $m = \tan \alpha$.

87.

The graph of $|f(x)|$ will reflect any parts of the graph below the x-axis about the x-axis. The graph of $f(|x|)$ will reflect the part of the graph left of the y-axis about the y-axis.

89. a. 0.8845

c. $\frac{1}{262}$

d. 262

e. The sound whose wave is given by p sounds like high C because $30\,p_2(t)$ has a frequency of 524 and it is the loudest component of the sound.

Review Exercises for the Prerequisites Chapter

1. $-1 \le x \le 5$ **3.** $x < -5,\ x > -1$

5. $\frac{27}{16}$ **7.** $(-1, 3),\ (3, 2),\ (1, 1)$

9. Center: $(-3, 1)$ **11.** Point: $(-3, 1)$
Radius: 3

13. $c = -21$

15. $x^2 + y^2 - 2x - 4y - 4 = 0$

a. on the circle

b. inside the circle

c. outside the circle

d. inside the circle

17. $t = \frac{7}{3}$

19.   **21.**

23.   **25.**

27. a. $7x - 16y + 78 = 0$

b. $5x - 3y + 22 = 0$

c. $y + 2x = 0$

d. $x + 2 = 0$

29. $(-4, 5)$ **31.** $(4, 1)$

33. The speed of the plane is 560 mph.

35. $A = x(12 - x)$
Domain: $\{x: 0 < x < 12\}$

37. $d = 45t$

Domain: $\{t: t \geq 0\}$

39. Not a function

41. Function

43. a.

b.

c.

d.

45. a.

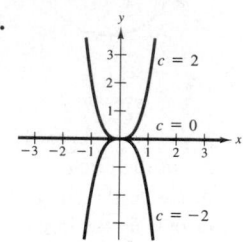

b.

47. a. $f(x) - g(x) = -x^2 - 2x$

b. $f(x)g(x) = (1 - x^2)(2x + 1)$

c. $g(f(x)) = 3 - 2x^2$

49. a. 3, Minus   **b.** 2, Plus

51. 47,285 ft $\approx$ 9 mi

53.

55.

57.

59.

61.

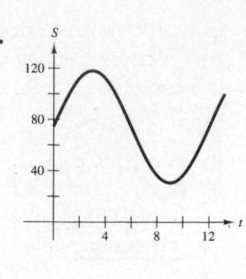

CHAPTER 1

Section 1.1   (*page 68*)

1.

x	1.9	1.99	1.999	2.001	2.01	2.1
$f(x)$	0.3448	0.3344	0.3334	0.3332	0.3322	0.3226

$$\lim_{x \to 2} \frac{x - 2}{x^2 - x - 2} \approx .0.3333 \quad \left(\text{Actual limit is } \frac{1}{3}. \right)$$

3.

x	-0.1	-0.01	-0.001	0.001	0.01	0.1
$f(x)$	0.2911	0.2889	0.2887	0.2887	0.2884	0.2683

$$\lim_{x \to 0} \frac{\sqrt{x + 3} - \sqrt{3}}{x} \approx 0.2887 \quad \left(\text{Actual limit is } \frac{1}{2\sqrt{3}}. \right)$$

5.

x	2.9	2.99	2.999	3.001	3.01	3.1
$f(x)$	-0.0641	-0.0627	-0.0625	-0.0625	-0.0623	-0.06

$$\lim_{x \to 3} \frac{[1/(x + 1)] - (1/4)}{x - 3} \approx -0.0625 \quad \left(\text{Actual limit is } -\frac{1}{16}. \right)$$

7.

x	-0.1	-0.01	-0.001	0.001	0.01	0.1
$f(x)$	0.9983	0.99998	1.0000	1.0000	0.99998	0.9983

$$\lim_{x \to 0} \frac{\sin x}{x} \approx 1.0000 \quad (\text{Actual limit is 1.})$$

9. 1   **11.** 2   **13.** Limit does not exist.

15. Limit does not exist.   **17.** Limit does not exist.

19. $L = 8$. Let $\delta = \dfrac{0.01}{3} \approx 0.0033$

21. $L = 1$.

Assume $1 < x < 3$ and let $\delta = \dfrac{0.01}{5} = 0.002$.

23. Limit is 5. Let $\delta = \varepsilon$.   **25.** Limit is 3. Any δ will work.

27. Limit is 0. Let $\delta = \varepsilon^3$.

29. Limit is 2. Let $\delta = \dfrac{\varepsilon}{3}$ for $0 < x < 2$.

31. False, the existence or nonexistence of $f(x)$ at $x = c$ has no bearing on the existence of the limit of $f(x)$ as $x \to c$.

32. True   **33.** False, see Exercise 31.

34. False, the existence or nonexistence of the limit of $f(x)$ as $x \to c$ has no bearing on the existence of $f(c)$.

35. It is not obvious from the graph that the function does not exist at $x = 4$.

$$\lim_{x \to 4} f(x) = \frac{1}{6}$$

37.

n	$4 \pm [0.1]^n$	$f(4 \pm [0.1]^n)$
1	3.9000	6.9000
2	3.9900	6.9900
3	3.9990	6.9990
4	3.9999	6.9999
4	4.0001	7.0001
3	4.0010	7.0010
2	4.0100	7.0100
1	4.1000	7.1000

41. $\lim_{x \to 0}(1 + x)^{1/x} \approx 2.71828$

(You will study this limit in Chapter 5.)

Section 1.2 (*page 74*)

1. a. 1 **b.** 3 **3. a.** $\dfrac{\sqrt{3}}{2}$ **b.** 1 **5.** 16

7. -1 **9.** -4 **11.** 2 **13.** 1 **15.** $\frac{1}{2}$

17. -2 **19.** 1 **21.** -1 **23.** 1 **25.** $\frac{1}{2}$ **27.** -1

29. a. 15 **b.** 5 **c.** 6 **d.** $\frac{2}{3}$

31. a. 64 **b.** 2 **c.** 12 **d.** 8

33.

Domain: $\{x : 0 \le x < 9, x > 9\}$

$\lim_{x \to 9} f(x) = 6$

It is not obvious from the graph that the function does not exist at $x = 9$.

35. $f(x) = -\dfrac{1}{x}$, $g(x) = \dfrac{1}{x}$ (Other solutions possible.)

Section 1.3 (*page 82*)

1. a. 1 **b.** 3

$$g(x) = \frac{-2x^2 + x}{x} = -2x + 1, x \ne 0$$

3. a. 2 **b.** 0

$$g(x) = \frac{x^3 - 1}{x - 1} = x^2 + x, x \ne 1$$

5. -2

$$f(x) = \frac{x^2 - 1}{x + 1} = x - 1 = g(x), x \ne -1$$

7. 12

$$f(x) = \frac{x^3 + 8}{x + 2} = x^2 - 2x + 4, x \ne -2$$

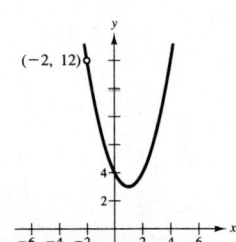

9. $\dfrac{1}{10}$ **11.** $\dfrac{3}{2}$ **13.** $\dfrac{\sqrt{3}}{6}$ **15.** $-\dfrac{1}{4}$

17. 2 **19.** $2x - 2$

21.

x	-0.1	-0.01	-0.001	0.001	0.01	0.1
$f(x)$	0.358	0.354	0.354	0.354	0.353	0.349

$$\lim_{x \to 0} \frac{\sqrt{x + 2} - \sqrt{2}}{x} \approx 0.354 \quad \left(\text{Actual limit is } \frac{1}{2\sqrt{2}}.\right)$$

23.

x	1.9	1.99	1.999	1.9999	2.0001	2.001	2.01	2.1
$f(x)$	72.39	79.20	79.92	79.99	80.01	80.08	80.80	88.41

$$\lim_{x \to 2} \frac{x^5 - 32}{x - 2} \approx 80 \quad \text{(Actual limit is 80.)}$$

25. $\frac{1}{5}$ **27.** 0   **29.** 0   **31.** 0   **33.** 1   **35.** 1
37. 3   **39.** 0   **41.** 4
43. 0   **45.** 0

47. 0

49.

The magnitudes of $f(x)$ and $g(x)$ are approximately equal when x is "close to" 0. Therefore, their ratio is approximately 1.

51. -160 ft/sec   **57.** $f(x) = \begin{cases} 4, & x \geq 0 \\ -4, & x < 0 \end{cases}$

59. $\lim_{x \to 0} f(x)$ does not exist.
$\lim_{x \to 0} g(x) = 0$ **61.** $\frac{1}{2}$; $\cos(0.1) \approx .995$

Section 1.4   (*page 92*)

1. a. 1   **b.** 1   **c.** 1
3. a. 0   **b.** 0   **c.** 0
5. a. 3   **b.** -3 **c.** Limit does not exist.
7. $\frac{1}{10}$ **9.** Limit does not exist.
11. Limit does not exist.
13. $-\dfrac{1}{x^2}$ **15.** Limit does not exist.
17. 2   **19.** Limit does not exist.
21. 3   **23.** Discontinuous at $x = -2$ and $x = 2$

25. Discontinuous at every integer
27. Continuous for all real x
29. Continuous for all real x
31. Nonremovable discontinuity at $x = 1$
33. Continuous for all real x
35. Removable discontinuity at $x = 1$;
     Nonremovable discontinuity at $x = -2$
37. Nonremovable discontinuity at $x = -2$
39. Continuous for all real x
41. Nonremovable discontinuity at $x = 2$
43. Continuous for all real x
45. Nonremovable discontinuities at integer multiples of $\dfrac{\pi}{2}$
47. Nonremovable discontinuities at each integer
49. $a = 2$ **51.** $a = 4$
53. Continuous for all real x
55. Nonremovable discontinuities at $x = 1$ and $x = -1$
57. Continuous on $(-\infty, \infty)$
59. Continuous on $\ldots, (-2\pi, 0), (0, 2\pi), (2\pi, 4\pi), \ldots$
61. Nonremovable discontinuity at each integer.

63. Discontinuous at $x = 3$

65.

It is not obvious from the graph that the function is discontinuous at $x = 0$.

67. $f(x)$ is continuous on $[2, 4]$.
     $f(2) = -1$ and $f(4) = 3$
     By the Intermediate Value Theorem, $f(c) = 0$ for at least one value c between 2 and 4.
69. 0.68   **71.** $f(3) = 11$ **73.** $f(2) = 4$

75. Discontinuous at every positive integer

77. Discontinuous at every even positive integer

81. True   **82.** True
83. False, the rational function $f(x) = p(x)/q(x)$ has at most n discontinuities where n is the degree of $q(x)$.
84. False, it is discontinuous at $x = 1$.
89. Discontinuous at $x = \pm 1, \pm 2, \pm 3, \ldots$

Section 1.5 (page 101)

1. $\displaystyle\lim_{x \to 2^+} \frac{1}{(x+2)^2} = \infty$ **3.** $\displaystyle\lim_{x \to -2^+} \tan \frac{\pi x}{4} = -\infty$

$\displaystyle\lim_{x \to -2^-} \frac{1}{(x+2)^2} = \infty$ $\displaystyle\lim_{x \to -2^-} \tan \frac{\pi x}{4} = \infty$

5. $\displaystyle\lim_{x \to -3^+} \frac{1}{x^2 - 9} = -\infty$ **7.** $\displaystyle\lim_{x \to -3^+} \frac{x^2}{x^2 - 9} = -\infty$

$\displaystyle\lim_{x \to -3^-} \frac{1}{x^2 - 9} = \infty$ $\displaystyle\lim_{x \to -3^-} \frac{x^2}{x^2 - 9} = \infty$

9. $x = 0$ **11.** $x = 2, x = -1$ **13.** $x = \pm 1$
15. $x = 0$ **17.** $x = -2, x = 1$

19. $x = \dfrac{\pi}{4} + \dfrac{n\pi}{2}$, n an integer

21. No vertical asymptote
23. $x = n\pi$, n a nonzero integer
25. Removable discontinuity at $x = -1$
27. Vertical asymptote at $x = -1$
29. $-\infty$ **31.** ∞
33. $\frac{4}{5}$ **35.** $-\infty$ **37.** ∞ **39.** $\frac{1}{2}$ **41.** ∞
43. $-\infty$

45. ∞ if the constant of proportionality k is positive and $-\infty$ if k is negative.
47. a. $\frac{7}{12}$ ft/sec   **b.** $\frac{3}{2}$ ft/sec   **c.** ∞
49. a. $176 million   **b.** $528 million
      c. $1584 million   **d.** ∞

51.

53. False, let $p(x) = x^2 - 1$.
54. False, let $f(x) = \dfrac{1}{x^2}$ and $g(x) = \dfrac{1}{x^4}$.

55. False, let $f(x) = \dfrac{1}{x^2 + 1}$.

56. True

57. False, let $f(x) = \begin{cases} \dfrac{1}{x}, & x \neq 0 \\ 0, & x = 0 \end{cases}$.

58. True

Review Exercises for Chapter 1

1. 7   **3.** 77   **5.** $\frac{10}{3}$ **7.** $-\frac{1}{4}$ **9.** -1 **11.** 75

13. $-\infty$ **15.** $\dfrac{\sqrt{3}}{2}$ **17.** $-\infty$ **19.** $\dfrac{1}{3}$ **21.** $-\infty$

23. $\frac{4}{5}$ **25.** ∞

27.

x	1.1	1.01	1.001	1.0001
$f(x)$	0.5680	0.5764	0.5772	0.5773

$\displaystyle\lim_{x \to 1^+} \frac{\sqrt{2x+1} - \sqrt{3}}{x - 1} \approx 0.577$ $\left(\text{Actual limit is } \dfrac{\sqrt{3}}{3}.\right)$

29. Nonremovable discontinuity at each integer
Continuous on $(k, k+1)$ for all integers k
31. Removable discontinuity at $x = 1$
Continuous on $(-\infty, 1) \cup (1, \infty)$
33. Nonremovable discontinuity at $x = 2$
Continuous on $(-\infty, 2) \cup (2, \infty)$
35. Nonremovable discontinuity at $x = -1$
Continuous on $(-\infty, -1) \cup (-1, \infty)$
37. Nonremovable discontinuity at each even integer
Continuous on $(2k, 2k+2)$ for all integers k
39. $c = -\frac{1}{2}$
41. Nonremovable discontinuity every 6 months

43. $x = 0$ **45.** $x = 10$
47. a. $14,117.65   **b.** $80,000.00
      c. $720,000.00   **d.** ∞

49. False $\lim\limits_{x \to 0^-} \dfrac{|x|}{x} = -1$ **50.** True   **51.** True

52. False, the existence of the limit of $f(x)$ as $x \to c$ has no bearing on the existence of $f(c)$

53. True   **54.** False, the limit does not exist.   **55.** True

57. Domain: $(-\infty, 0] \cup [1, \infty)$

$$\lim\limits_{x \to 0^-} f(x) = 0$$
$$\lim\limits_{x \to 1^+} f(x) = 0$$

CHAPTER 2

Section 2.1   (*page 112*)

1. a. $m = 0$ **b.** $m = -3$

3.
$$y = \frac{f(4) - f(1)}{4 - 1}(x - 1) + f(1) = x + 1$$

5. $f'(x) = 0$ **7.** $f'(x) = -5$

9. $f'(x) = 4x + 1$ **11.** $f'(x) = 3x^2 - 12$

13. $f'(x) = \dfrac{-1}{(x - 1)^2}$ **15.** $f'(x) = \dfrac{1}{2\sqrt{x - 4}}$

17. $f'(x) = 2x$
Tangent line: $y = 4x - 3$

19. $f'(x) = 3x^2$
Tangent line: $y = 12x - 16$

21. $f'(x) = 1 - \dfrac{1}{x^2}$
Tangent line: $y = 2$

23. $y = 3x - 2$ **25.** $y = 2x + 1$
$y = 3x + 2$ $y = -2x + 9$

27. b   **28.** e   **29.** c   **30.** a   **31.** f   **32.** d

33.

x	-2	-1.5	-1	-0.5	0	0.5	1	1.5	2
$f(x)$	-2	$-\frac{27}{32}$	$-\frac{1}{4}$	$-\frac{1}{32}$	0	$\frac{1}{32}$	$\frac{1}{4}$	$\frac{27}{32}$	2
$f'(x)$	3	$\frac{27}{16}$	$\frac{3}{4}$	$\frac{3}{16}$	0	$\frac{3}{16}$	$\frac{3}{4}$	$\frac{27}{16}$	3

35. a.

b. The graphs of S for decreasing values of Δx are secant lines approaching the tangent line to the graph of f at the point $(2, f(2))$.

37. $f(x) = x^2 - 1$
$$f'(2) = \lim_{x \to 2} \frac{f(x) - f(2)}{x - 2}$$
$$= \lim_{x \to 2} \frac{(x^2 - 1) - 3}{x - 2}$$
$$= \lim_{x \to 2} (x + 2) = 4$$

39. $f(x) = x^3 + 2x^2 + 1$
$$f'(-2) = \lim_{x \to -2} \frac{f(x) - f(-2)}{x + 2}$$
$$= \lim_{x \to -2} \frac{(x^3 + 2x^2 + 1) - 1}{x + 2}$$
$$= \lim_{x \to -2} x^2 = 4$$

41. $f(x) = (x - 1)^{2/3}$
$$f'(1) = \lim_{x \to 1} \frac{f(x) - f(1)}{x - 1}$$
$$= \lim_{x \to 1} \frac{(x - 1)^{2/3} - 0}{x - 1}$$
$$= \lim_{x \to 1} \frac{1}{(x - 1)^{1/3}}$$
Limit does not exist.
f is not differentiable at $x = 1$.

43. $(-\infty, -3), (-3, \infty)$

45. $(-\infty, -1), (-1, \infty)$ **47.** $(-\infty, 3), (3, \infty)$

49. $(1, \infty)$ **51.** $(-\infty, 0), (0, \infty)$

53. f is not differentiable at $x = 1$.

55. $f'(1) = 0$ **57.** f is differentiable at $x = 2$.

59. a. 3   **b.** -3 **61.** True

62. False, let $f(x) = |x|$ at $x = 0$.   **63.** True

64. False, let $f(x) = \begin{cases} x, & x \le 0 \\ 2x, & x > 0 \end{cases}$ at $x = 0$

Section 2.2   (*page 124*)

1. a. $\frac{1}{2}$ **b.** $\frac{3}{2}$ **c.** 2   **d.** 3

3. 0   **5.** 1   **7.** $2x$ **9.** $-4t + 3$ **11.** $3t^2 - 2$

13. $2x + \dfrac{1}{2}\sin x$ **15.** $-\dfrac{1}{x^2} - 3\cos x$

Function	Rewrite	Derivative	Simplify
17. $y = \dfrac{1}{3x^3}$	$y = \dfrac{1}{3}x^{-3}$	$y' = -x^{-4}$	$y' = -\dfrac{1}{x^4}$
19. $y = \dfrac{1}{(3x)^3}$	$y = \dfrac{1}{27}x^{-3}$	$y' = -\dfrac{1}{9}x^{-4}$	$y' = -\dfrac{1}{9x^4}$
21. $y = \dfrac{\sqrt{x}}{x}$	$y = x^{-1/2}$	$y' = -\dfrac{1}{2}x^{-3/2}$	$y' = -\dfrac{1}{2x^{3/2}}$

23. -1 **25.** 0 **27.** 4 **29.** 3

31. $3x^2 - 3 + \dfrac{8}{x^5}$ **33.** $2t + \dfrac{4}{t^2}$

35. $\dfrac{x^3 - 8}{x^3}$ **37.** $3x^2 + 1$ **39.** $\dfrac{4}{5s^{1/5}}$

41. $\dfrac{2}{\sqrt{x}} - 3\sin x$ **43.** $2x + y - 2 = 0$

45. $x + 48y - 20 = 0$

47. $(0, 2), \left(\sqrt{\tfrac{3}{2}}, -\tfrac{1}{4}\right), \left(-\sqrt{\tfrac{3}{2}}, -\tfrac{1}{4}\right)$

49. No horizontal tangents   **51.** (π, π)

53.

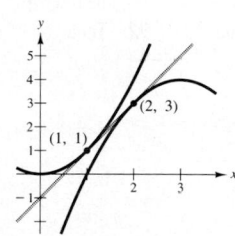

The rate of change of f is constant and therefore f' is a constant function.

55. $y = 2x - 1$ $y = 4x - 4$

57. $x - 4y + 4 = 0$

59. a.

   c. It becomes less accurate.

b. $T(x) = 3(x - 4) + 8 = 3x - 4$

d.

Δx	-3	-2	-1	-0.5	-0.1	0	0.1	0.5	1	2	3
$f(4 + \Delta x)$	1	2.828	5.196	6.458	7.702	8	8.302	9.456	11.180	14.697	18.520
$T(4 + \Delta x)$	-1	2	5	6.5	7.7	8	8.3	9.5	11	14	17

61. False, let $f(x) = x$ and $g(x) = x + 1$.   **62.** True

63. False, $dy/dx = 0$ **64.** True

65. a. A and B **b.** Greater

c.

d. B and C, D and E

67. Average rate: 2 **69.** Average rate: $\tfrac{1}{2}$

Instantaneous rates: Instantaneous rates:

$f'(1) = 2$ $f'(1) = 1$

$f'(2) = 2$ $f'(2) = \tfrac{1}{4}$

71. a. $s(t) = -16t^2 + 1362$

      $v(t) = -32t$

b. -48 ft/sec

c. $s'(1) = -32$ ft/sec

     $s'(2) = -64$ ft/sec

d. $t = \dfrac{\sqrt{1362}}{4} \approx 9.226$ sec

e. -295.242 ft/sec

73. $v(5) = 224$ ft/sec

$v(10) = 64$ ft/sec

75.   **77.**

79. a. $T(x) = 0.14x^2 - 3.33x + 58.40$

b.

c. $T'(30) = 5.07$ ft/mi/hr

     $T'(40) = 7.87$ ft/mi/hr

     $T'(55) = 12.07$ ft/mi/hr

81. 8m^2 **83. a.** $\tfrac{4}{9}$ **b.** $\tfrac{1}{3}$ **c.** 0 **d.** $-\tfrac{5}{9}$

85. $-\$1.91, -\$1.93$

87. $\dfrac{dF}{dr} = -2K\dfrac{m_1 m_2}{r^3}$

Force decreases as the distance between the particles increases.

89. $y = 2x^2 - 3x + 1$ **91.** $y = -9x, \ y = -\tfrac{9}{4}x - \tfrac{27}{4}$

93. $a = \tfrac{1}{3}, \ b = -\tfrac{4}{3}$

Section 2.3 (*page 135*)

1. $f'(x) = 2x^2$
$f'(0) = 0$

3. $f'(x) = (x^3 - 3x)(4x + 3) + (2x^2 + 3x + 5)(3x^2 - 3)$
$= 10x^4 + 12x^3 - 3x^2 - 18x - 15$
$f'(0) = -15$

5. $f'(x) = \cos x - x \sin x$
$f'\left(\dfrac{\pi}{4}\right) = \dfrac{\sqrt{2}}{8}(4 - \pi)$

	Function	Rewrite	Derivative	Simplify
7.	$y = \dfrac{x^2 + 2x}{x}$	$y = x + 2$	$y' = 1$	$y' = 1$
9.	$y = \dfrac{7}{3x^3}$	$y = \dfrac{7}{3}x^{-3}$	$y' = -7x^{-4}$	$y' = -\dfrac{7}{x^4}$
11.	$y = \dfrac{3x^2 - 5}{7}$	$y = \dfrac{1}{7}(3x^2 - 5)$	$y' = \dfrac{1}{7}(6x)$	$y' = \dfrac{6x}{7}$

13. $\dfrac{(2x - 3)(3) - (3x - 2)(2)}{(2x - 3)^2} = -\dfrac{5}{(2x - 3)^2}$

15. $\dfrac{(x^2 - 1)(-2 - 2x) - (3 - 2x - x^2)(2x)}{(x^2 - 1)^2} = \dfrac{2}{(x + 1)^2}, \ x \neq 1$

17. $\dfrac{\sqrt{x}(1) - (x + 1)[1/(2\sqrt{x})]}{x} = \dfrac{x - 1}{2x^{3/2}}$

19. $6s^2(s^3 - 2)$

21. $\dfrac{(t^2 + 2t + 2)(1) - (t + 1)(2t + 2)}{(t^2 + 2t + 2)^2} = \dfrac{-t^2 - 2t}{(t^2 + 2t + 2)^2}$

23. $(3x^3 + 4x)[(x - 5)\cdot 1 + (x + 1)\cdot 1]$
$+ [(x - 5)(x + 1)](9x^2 + 4)$
$= 15x^4 - 48x^3 - 33x^2 - 32x - 20$

25. $\left[\dfrac{x + 1}{x + 2}\right](2) + (2x - 5)\left[\dfrac{(x + 2)(1) - (x + 1)(1)}{(x + 2)^2}\right]$
$= \dfrac{2x^2 + 8x - 1}{(x + 2)^2}$

27. $\dfrac{(x^2 - c^2)(2x) - (x^2 + c^2)(2x)}{(x^2 - c^2)^2} = -\dfrac{4xc^2}{(x^2 - c^2)^2}$

29. $t(t \cos t + 2 \sin t)$ **31.** $-\dfrac{t \sin t + \cos t}{t^2}$

33. $-1 + \sec^2 x = \tan^2 x$ **35.** $\dfrac{1}{2\sqrt{t}} + 4 \sec t \tan t$

37. $-5x \csc x \cot x + 5 \csc x = 5 \csc x(1 - x \cot x)$

39. $\csc x \cot x - \cos x = \cos x \cot^2 x$

41. $x^2 \cos x + 2x \sin x - 2x \sin x + 2 \cos x$
$= x^2 \cos x + 2 \cos x$

43. $x(x \sec^2 x + 2 \tan x)$ **45.** $\dfrac{1 - \sin\theta + \theta\cos\theta}{(1 - \sin\theta)^2}$

47. $y' = \dfrac{-2 \csc x \cot x}{(1 - \csc x)^2}, \ -4\sqrt{3}$

49. $h'(t) = \dfrac{\sec t(t \tan t - 1)}{t^2}, \ \dfrac{1}{\pi^2}$

51. $y = -x + 4$ **53.** $y = -x - 2$

55. $4x - 2y - \pi + 2 = 0$ **57.** $(0, 0), (2, 4)$

59. 0 **61.** -10

63. $n = 1, f'(x) = x \cos x + \sin x$
$n = 2, f'(x) = x^2 \cos x + 2x \sin x$
$n = 3, f'(x) = x^3 \cos x + 3x^2 \sin x$
$n = 4, f'(x) = x^4 \cos x + 4x^3 \sin x$
$f'(x) = x^n \cos x + nx^{n-1} \sin x$

65. a. $-\$38.13$ **b.** $-\$10.37$ **c.** $-\$3.80$
The costs decrease with increased order size.

67. 31.55 **71.** $\dfrac{3}{\sqrt{x}}$ **73.** $\dfrac{2}{(x - 1)^3}$

75. $-3 \sin x$ **77.** $2x$ **79.** $\dfrac{1}{\sqrt{x}}$

81. a. $f''(x) = g(x)h''(x) + 2g'(x)h'(x) + g''(x)h(x)$
$f'''(x) = g(x)h'''(x) + 3g'(x)h''(x)$
$\qquad\qquad + 3g''(x)h'(x) + g'''(x)h(x)$
$f^{(4)}(x) = g(x)h^{(4)}(x) + 4g'(x)h'''(x) + 6g''(x)h''(x)$
$\qquad\qquad + 4g'''(x)h'(x) + g^{(4)}(x)h(x)$

b. $f^{(n)}(x) = g(x)h^{(n)}(x) + \dfrac{n!}{1!(n - 1)!}g'(x)h^{(n-1)}(x)$
$+ \dfrac{n!}{2!(n - 2)!}g''(x)h^{(n-2)}(x)$
$+ \cdots + \dfrac{n!}{(n - 1)!1!}g^{(n-1)}(x)h'(x) + g^{(n)}(x)h(x)$

83. $v(3) = 27$ m/sec
$a(3) = -6$ m/sec^2
The speed of the object is decreasing, but the rate of that decrease is increasing.

85.

t	0	1	2	3	4
$s(t)$	0	57.75	99	123.75	132
$v(t)$	66	49.5	33	16.5	0
$a(t)$	-16.5	-16.5	-16.5	-16.5	-16.5

Average velocity on:
[0, 1] is 57.75
[1, 2] is 41.25
[2, 3] is 24.75
[3, 4] is 8.25

87. False, $dy/dx = f(x)g'(x) + g(x)f'(x)$ **88.** True
89. True   **90.** True   **91.** True   **92.** True
93. $f'(x) = 2|x|$
$f''(x)$ does not exist.
95.

Section 2.4 (*page 144*)

$y = f(g(x))$	$u = g(x)$	$y = f(u)$
1. $y = (6x - 5)^4$	$u = 6x - 5$	$y = u^4$
3. $y = \sqrt{x^2 - 1}$	$u = x^2 - 1$	$y = \sqrt{u}$
5. $y = \csc^3 x$	$u = \csc x$	$y = u^3$

7. $6(2x - 7)^2$ **9.** $-108(4 - 9x)^3$

11. $\dfrac{2}{3}(9 - x^2)^{-1/3}(-2x) = -\dfrac{4x}{3(9 - x^2)^{1/3}}$

13. $\dfrac{1}{2}(1 - t)^{-1/2}(-1) = -\dfrac{1}{2\sqrt{1 - t}}$

15. $\frac{1}{3}(9x^2 + 4)^{-2/3}(18x) = \frac{6x}{(9x^2 + 4)^{2/3}}$

17. $(4 - x^2)^{-1/2}(-2x) = -\frac{2x}{\sqrt{4 - x^2}}$

19. $-\frac{1}{(x - 2)^2}$ **21.** $-2(t - 3)^{-3}(1) = -\frac{2}{(t - 3)^3}$

23. $-\frac{1}{2(x + 2)^{3/2}}$

25. $x^2[4(x - 2)^3(1)] + (x - 2)^4(2x) = 2x(x - 2)^3(3x - 2)$

27. $x\left(\frac{1}{2}\right)(1 - x^2)^{-1/2}(-2x) + (1 - x^2)^{1/2}(1) = \frac{1 - 2x^2}{\sqrt{1 - x^2}}$

29. $\frac{(x^2 + 1)^{1/2}(1) - x(1/2)(x^2 + 1)^{-1/2}(2x)}{x^2 + 1} = \frac{1}{(x^2 + 1)^{3/2}}$

31. $\frac{1 - 3x^2 - 4x^{3/2}}{2\sqrt{x}(x^2 + 1)^2}$

The zero of $g'(x)$ corresponds to the point on the graph of the function where the tangent line is horizontal.

33. $\frac{3t(t^2 + 3t - 2)}{(t^2 + 2t - 1)^{3/2}}$

The zero of $g'(x)$ corresponds to the point on the graph of the function where the tangent line is horizontal.

35. $-\frac{\sqrt{\dfrac{x + 1}{x}}}{2x(x + 1)}$

y' has no zeros.

37. $\frac{t}{\sqrt{1 + t}}$

The zero of $s'(t)$ corresponds to the point on the graph of the function where the tangent line is horizontal.

39. a. 1   **b.** 2   **41.** $-3\sin 3x$

43. $12\sec^2 4x$ **45.** $\sin 2\theta \cos 2\theta = \frac{1}{2}\sin 4\theta$

47. $\frac{1}{2\sqrt{x}} + 2x\cos(2x)^2$ **49.** $-\sin x \cos(\cos x)$

51. $y' = -\dfrac{\pi x \sin \pi x + \cos \pi x + 1}{x^2}$

The zeros of y' correspond to the points on the graph of the function where the tangent lines are horizontal.

53. $s'(t) = \dfrac{t + 1}{\sqrt{t^2 + 2t + 8}}, \dfrac{3}{4}$

55. $f'(x) = \dfrac{-9x^2}{(x^3 - 4)^2}, -\dfrac{9}{25}$

57. $f'(t) = \dfrac{-5}{(t - 1)^2}, -5$

59. $y' = -6\sec^3(2x)\tan(2x), 0$

61. $9x - 5y - 2 = 0$ **63.** $2x - y - 2\pi = 0$

65.

The zeros of f' correspond to the points where the graph of f has horizontal tangents.

67.

The zeros of f' correspond to the points where the graph of f has horizontal tangents.

69. $12(5x^2 - 1)(x^2 - 1)$ **71.** $2[\cos x^2 - 2x^2 \sin x^2]$

73. a. 24   **b.** Not possible since $g'h(5)$ is not known.
     c. $\frac{4}{3}$ **d.** 162

75. $\dfrac{2(2x - 3)}{|2x - 3|}$ **77.** $-|x|\sin x + \dfrac{x}{|x|}\cos x$

79. a. 1.461   **b.** -1.016

81. 0.2 rad, 1.45 rad/sec   **83.** 0.04224

85. a. $f'(x) = \beta\cos\beta x$
        $f''(x) = -\beta^2\sin\beta x$
        $f'''(x) = -\beta^3\cos\beta x$
        $f^{(4)}(x) = \beta^4\sin\beta x$
     c. $f^{(2k)}(x) = (-1)^k\beta^{2k}\sin\beta x$
        $f^{(2k-1)}(x) = (-1)^{k+1}\beta^{2k-1}\cos\beta x$

87. False, $y' = \frac{1}{2}(1 - x)^{-1/2}(-1)$.

88. False, $f'(x) = 2(\sin 2x)(\cos 2x)(2)$
        $= 4\sin 2x\cos 2x$.

89. True

Section 2.5 (page 152)

1. $-\dfrac{x}{y}$ **3.** $-\sqrt{\dfrac{y}{x}}$ **5.** $\dfrac{y - 3x^2}{2y - x}$ **7.** $\dfrac{1 - 3x^2y^3}{3x^3y^2 - 1}$

9. $\dfrac{4xy - 3x^2 - 3y^2}{2x(3y - x)}$ **11.** $\dfrac{\cos x}{4\sin 2y}$

13. $\dfrac{\cos x - \tan y - 1}{x\sec^2 y}$ **15.** $\dfrac{y\cos(xy)}{1 - x\cos(xy)}$

17. $-\dfrac{y}{x}, -\dfrac{1}{4}$ **19.** $\dfrac{18x}{(x^2 + 9)^2y}$, undefined

21. $-\sqrt[3]{\dfrac{y}{x}}$, $-\dfrac{1}{2}$ **23.** $-\dfrac{x^2}{x^2+1}$, 0 **25.** $-\dfrac{1}{2}$ **27.** 0

29. $x + 2y - 6 = 0$

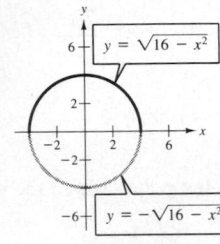

31. a. $y_1 = \sqrt{16 - x^2}$
$y_2 = -\sqrt{16 - x^2}$

b.

$y = \sqrt{16 - x^2}$

$y = -\sqrt{16 - x^2}$

c. $y' = \mp \dfrac{x}{\sqrt{16 - x^2}} = -\dfrac{x}{y}$ **d.** $y' = -\dfrac{x}{y}$

33. a. $y_1 = \dfrac{3}{4}\sqrt{16 - x^2}$
$y_2 = -\dfrac{3}{4}\sqrt{16 - x^2}$

b.

$y = \dfrac{3}{4}\sqrt{16 - x^2}$

$y = -\dfrac{3}{4}\sqrt{16 - x^2}$

c. $y' = \mp \dfrac{3x}{4\sqrt{16 - x^2}} = -\dfrac{9x}{16y}$ **d.** $y' = -\dfrac{9x}{16y}$

35. $\dfrac{10}{x^3}$ **37.** $-\dfrac{16}{y^3}$ **39.** $\dfrac{3y}{4x^2} = \dfrac{3x}{4y}$

41. a. At $(4, 3)$:
Tangent line: $4x + 3y - 25 = 0$
Normal line: $3x - 4y = 0$
b. At $(-3, 4)$:
Tangent line: $3x - 4y + 25 = 0$
Normal line: $4x + 3y = 0$

45. Horizontal tangents: $(-4, 0)$, $(-4, 10)$
Vertical tangents: $(0, 5)$, $(-8, 5)$

47.

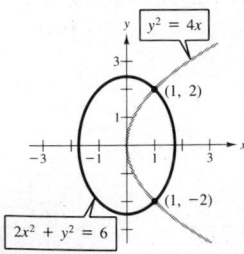

$y^2 = 4x$

$(1, 2)$

$(1, -2)$

$2x^2 + y^2 = 6$

At $(1, 2)$:
Slope of ellipse: -1
Slope of parabola: 1
At $(1, -2)$:
Slope of parabola: -1

49.

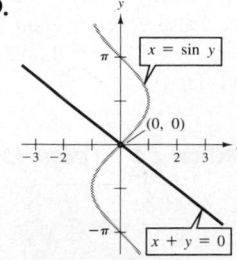

$x = \sin y$

$(0, 0)$

$x + y = 0$

At $(0, 0)$:
Slope of line: -1
Slope of sine curve: 1

51.

$C = 1$

$K = -1$

$C = 4$

$K = 2$

Derivatives: $\dfrac{dy}{dx} = -\dfrac{y}{x}$, $\dfrac{dy}{dx} = \dfrac{x}{y}$

53. $-\dfrac{\pi}{2} < y < \dfrac{\pi}{2}$, $\dfrac{dy}{dx} = \dfrac{1}{1 + x^2}$

57. $-\dfrac{3}{4}$ **59.** $\sqrt{3}$

Section 2.6 (*page 159*)

1. a. $\dfrac{3}{4}$ **b.** 20   **3. a.** $-\dfrac{5}{8}$ **b.** $\dfrac{3}{2}$

5. a. -4 cm/sec   **b.** 0 cm/sec   **c.** 4 cm/sec
d. 12 cm/sec

7. a. 8 cm/sec   **b.** 4 cm/sec   **c.** 2 cm/sec
d. 6.851 cm/sec

9. a. $-2\sqrt{5} \approx -4.472$ cm/sec   **b.** 0 cm/sec
c. $2\sqrt{5} \approx 4.472$ cm/sec
d. $\dfrac{126}{\sqrt{109}} \approx 12.069$ cm/sec

11. a. 24π in.^{2}/min   **b.** 96π in.^{2}/min

13. b. When $\theta = \dfrac{\pi}{6}$, $\dfrac{dA}{dt} = \dfrac{\sqrt{3}}{8}s^2$
When $\theta = \dfrac{\pi}{3}$, $\dfrac{dA}{dt} = \dfrac{1}{8}s^2$
c. If s and $d\theta/dt$ are constant, dA/dt is proportional to $\cos\theta$.

15. a. $\dfrac{5}{\pi}$ ft/min   **b.** $\dfrac{5}{4\pi}$ ft/min

17. a. 36 cm^2/sec   **b.** 360 cm^2/sec

19. $\dfrac{8}{405\pi}$ ft/min   **21. a.** 24.6% **b.** $\dfrac{1}{64}$ ft/min

23. a. $-\dfrac{7}{12}$ ft/sec
$-\dfrac{3}{2}$ ft/sec
$-\dfrac{48}{7}$ ft/sec
b. $\dfrac{527}{24}$ ft^2/sec
c. $\dfrac{1}{12}$ rad/sec

25. $\dfrac{dy}{dt} = \dfrac{1}{2}$ ft/sec
$\dfrac{dx}{dt} = -\dfrac{\sqrt{3}}{6}$ ft/sec

27. a. -750 mph   **b.** 20 min

29. $-\dfrac{28}{\sqrt{10}} \approx -8.85$ ft/sec

31. a. $\dfrac{25}{3}$ ft/sec   **b.** $\dfrac{10}{3}$ ft/sec

33. a. 12 sec   **b.** $(0, \sqrt{5})$ **c.** $\dfrac{\sqrt{6}\pi}{24} \approx 0.32$ ft/sec

37. $v^{0.3}\left(1.3p\dfrac{dv}{dt} + v\dfrac{dp}{dt}\right) = 0$

39. $\dfrac{1}{20}$ rad/sec

41. a. $\frac{1}{2}$ rad/min **b.** $\frac{3}{2}$ rad/min **c.** 1.87 rad/min
43. a. 0 ft/sec **b.** 10π ft/sec **c.** $10\sqrt{3}\pi$ ft/sec
45. -0.1808 ft/sec^2

Review Exercises for Chapter 2

1. $f'(x) = 2x - 2$
3. a.

  b. $y = 2x$

c.

Δx	$f(x + \Delta x)$	$f(x)$	$\dfrac{f(x + \Delta x) - f(x)}{\Delta x}$
2	2.8284	0	1.4142
1	1.7321	0	1.7321
0.5	0.9354	0	1.8708
0.1	0.1975	0	1.9748

d. The last column gives the slopes of secant lines. As Δx approaches 0, these slopes approach the slope of the tangent line.

5. $3x(x - 2)$ **7.** $\dfrac{2(x^3 + 1)}{x^3}$ **9.** $-\dfrac{4}{3t^3}$

11. $\dfrac{-3x^2}{2\sqrt{1 - x^3}}$ **13.** $2(6x^3 - 9x^2 + 16x - 7)$

15. $s(s^2 - 1)^{3/2}(8s^3 - 3s + 25)$

17. $-\dfrac{x^2 + 1}{(x^2 - 1)^2}$ **19.** $\dfrac{6x}{(4 - 3x^2)^2}$

21. $-9\sin(3x + 1)$ **23.** $-\csc 2x \cot 2x$
25. $\frac{1}{2}(1 - \cos 2x) = \sin^2 x$
27. $\sin^{1/2} x \cos x - \sin^{5/2} x \cos x = \cos^3 x \sqrt{\sin x}$

29. $-x\sec^2 x - \tan x$ **31.** $\dfrac{x\cos x - 2\sin x}{x^3}$

33. $t(t - 1)^4(7t - 2)$

The zeros of f' correspond to the points on the graph of the function where the tangent line is horizontal.

35. $\dfrac{x + 2}{(x + 1)^{3/2}}$

g' is not equal to zero for any x.

37. $\dfrac{5}{6(t + 1)^{1/6}}$ **39.** $-\dfrac{\sec^2\sqrt{1 - x}}{2\sqrt{1 - x}}$

f' has no zeros. y' has no zeros.

41. $4 - 4\sin 2x$ **43.** $2\csc^2 x \cot x$ **45.** $\dfrac{2(t + 2)}{(1 - t)^4}$

47. $2\sec^2 x(x\tan x + 1)$ **49.** $-\dfrac{2x + 3y}{3(x + y^2)}$

51. $\dfrac{2y\sqrt{x} - y\sqrt{y}}{2x\sqrt{y} - x\sqrt{x}}$ **53.** $\dfrac{y\sin x + \sin y}{\cos x - x\cos y}$

55. Tangent line: $3x - y + 7 = 0$
    Normal line: $x + 3y - 1 = 0$
57. Tangent line: $x + 2y - 10 = 0$
    Normal line: $2x - y = 0$
59. Tangent line: $2x - 3y - 3 = 0$
    Normal line: $3x + 2y - 11 = 0$

61. a. $(0, -1), \left(-2, \dfrac{7}{3}\right)$ **b.** $(-3, 2), \left(1, -\dfrac{2}{3}\right)$

  c. $\left(-1 + \sqrt{2}, \dfrac{2[1 - 2\sqrt{2}]}{3}\right)$

   $\left(-1 - \sqrt{2}, \dfrac{2[1 + 2\sqrt{2}]}{3}\right)$

63. a. Yes

  b. No. The derivative from the right and left are not equal.
65. $y'' + y = -(2\sin x + 3\cos x) + (2\sin x + 3\cos x) = 0$
67. a. -18.667 deg/hr **c.** -3.240 deg/hr
  b. -7.284 deg/hr **d.** -0.747 deg/hr
69. a. 50 vib/sec/lb **b.** 33.33 vib/sec/lb
71. 56 ft/sec
73. a.

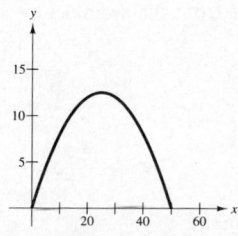

  b. 50 **c.** $x = 25$
  d. $y' = 1 - 0.04x$

x	0	10	25	30	50
y'	.1	0.6	0	-0.2	-1

  e. $y'(25) = 0$

77. a. $2\sqrt{2}$ units/sec   **b.** 4 units/sec   **c.** 8 units/sec
79. $\frac{2}{25}$ ft/min   **81.** -120 ft/sec

CHAPTER 3
Section 3.1   *(page 170)*

1. $f'(0) = 0$ **3.** $f'(4) = 0$
5. $f'(-2)$ is undefined   **7.** $x = 0, x = 2$
9. $t = \dfrac{8}{3}, t = 4$ **11.** $x = 0, x = \dfrac{\pi}{3}, \pi, \dfrac{5\pi}{3}$
13. Minimum: $(2, 2)$
Maximum: $(-1, 8)$
15. Minimum: $(0, 0)$ and $(3, 0)$
Maximum: $(\frac{3}{2}, \frac{9}{4})$
17. Minimum: $(-1, -4)$ and $(2, -4)$
Maximum: $(0, 0)$ and $(3, 0)$
19. Minimum: $(0, 0)$
Maximum: $(-1, 5)$
21. Minimum: $(1, 1)$
Maximum: $(4, 4)$
23. Minimum: $(1, -1)$
Maximum: $(0, -\frac{1}{2})$
25. Minimum: $\left(\dfrac{1}{6}, \dfrac{\sqrt{3}}{2}\right)$
Maximum: $(0, 1)$
27. Continuous on $\left[0, \dfrac{\pi}{4}\right]$
Not continuous on $[0, \pi]$
29. a. Yes   **b.** No
31. a. No   **b.** Yes
33. a. Minimum: $(0, -3)$ Maximum: $(2, 1)$
b. Minimum: $(0, -3)$
c. Maximum: $(2, 1)$
d. No extrema
35. Minimum: $(0, 2)$ Maximum: $(3, 36)$
37. Minimum: $(4, 1)$
39. Maximum: $\left| f''\left(\sqrt[3]{-10 + \sqrt{108}}\right) \right| \approx 1.47$
41. Maximum: $\left| f^{(4)}(0) \right| = \frac{56}{81}$
43. Maximum: $P(12) = 72$
45. The part of the lawn farthest from the sprinkler.
47. a.

$(0.4398, -1.0613)$

b. $(0.4398, -1.0613)$ Minimum

49. True   **50.** True   **51.** True
52. False, let $f(x) = x^2$ and $g(x) = (x - 1)^2$
53. All real numbers
(Slope is zero at nonintegers and undefined at integers.)

Section 3.2   *(page 177)*

1. $f(0) = f(2) = 0$
f is not differentiable on $(0, 2)$.
3. $f'(1) = 0$
5. $f'\left(\dfrac{6 - \sqrt{3}}{3}\right) = 0$ **7.** Not differentiable at $x = 0$
$f'\left(\dfrac{6 + \sqrt{3}}{3}\right) = 0$
9. Not differentiable at $x = 0$ **11.** $f'\left(-2 + \sqrt{5}\right) = 0$
13. $f'\left(\dfrac{\pi}{2}\right) = 0$ **15.** $f'\left(\dfrac{\pi}{4}\right) = 0$
$f'\left(\dfrac{3\pi}{2}\right) = 0$
17. $f'(-0.5756) = 0$ **19.** Not continuous on $[0, \pi]$
21. a. $f(1) = f(2) = 64$
b. Velocity $= 0$ for some t in $[1, 2]$
23. No, let $f(x) = x^2$ on $[-1, 2]$.
25. $f'(-\frac{1}{2}) = -1$ **27.** $f'(\frac{8}{27}) = 1$
29. $f'\left(\dfrac{-2 + \sqrt{6}}{2}\right) = \dfrac{2}{3}$ **31.** $f'\left(\dfrac{\sqrt{3}}{3}\right) = 1$
33. $f'\left(-\dfrac{\pi}{2}\right) = f'\left(\dfrac{\pi}{2}\right) = 1$
35. The function is discontinuous on $[0, 6]$.
37.

(−5, 5) $f(x) = |x|$ (5, 5)

39. a. -48 ft/sec   **b.** $t = \frac{3}{2}$ sec
41. Secant line: $x - 4y + 3 = 0$
Tangent line: $c = 4, x - 4y + 4 = 0$

tangent
f
secant

43. False, f is not continuous on $[-1, 1]$.
44. False, let $f(x) = \dfrac{x^3 - 4x}{x^2 - 1}$.
45. True   **46.** True   **47.** True   **48.** True

57. $f(x) = \cos x$, $g(x) = \sin x$

Section 3.3 (*page 186*)

1. Increasing on $(3, \infty)$
Decreasing on $(-\infty, 3)$
3. Increasing on $(-\infty, -2)$ and $(2, \infty)$
Decreasing on $(-2, 2)$
5. Increasing on $(-\infty, 0)$
Decreasing on $(0, \infty)$
7. Critical number: $x = 1$
Increasing on $(-\infty, 1)$
Decreasing on $(1, \infty)$
Relative maximum: $(1, 5)$
9. Critical number: $x = 3$
Increasing on $(3, \infty)$
Decreasing on $(-\infty, 3)$
Relative minimum: $(3, -9)$
11. Critical numbers: $x = -2, 1$
Increasing on $(-\infty, -2)$ and $(1, \infty)$
Decreasing on $(-2, 1)$
Relative maximum: $(-2, 20)$
Relative minimum: $(1, -7)$
13. Critical numbers: $x = -1, 1$
Increasing on $(-\infty, -1)$ and $(1, \infty)$
Decreasing on $(-1, 1)$
Relative maximum: $(-1, \frac{4}{5})$
Relative minimum: $(1, -\frac{4}{5})$
15. Critical number: $x = 0$
Increasing on $(-\infty, \infty)$
No relative extrema
17. Critical number: $x = 5$
Increasing on $(-\infty, 5)$
Decreasing on $(5, \infty)$
Relative maximum: $(5, 5)$
19. Critical number: $x = 0$
Discontinuities: $x = -3, 3$
Increasing on $(-\infty, -3)$ and $(-3, 0)$
Decreasing on $(0, 3)$ and $(3, \infty)$
Relative maximum: $(0, 0)$
21. Critical numbers: $x = 0, 4$
Increasing on $(-\infty, 0)$ and $(4, \infty)$
Decreasing on $(0, 4)$
Relative maximum: $(0, 15)$
Relative minimum: $(4, -17)$

23. Critical number: $x = 1$
Increasing on $(1, \infty)$
Decreasing on $(-\infty, 1)$
Relative minimum: $(1, 0)$

25. Critical numbers: $x = -1, 1$
Discontinuity: $x = 0$
Increasing on $(-\infty, -1)$ and $(1, \infty)$
Decreasing on $(-1, 0)$ and $(0, 1)$
Relative maximum: $(-1, -2)$
Relative minimum: $(1, 2)$

27. Critical numbers: $x = -3, 1$
Discontinuity: $x = -1$
Increasing on $(-\infty, -3)$ and $(1, \infty)$
Decreasing on $(-3, -1)$ and $(-1, 1)$
Relative maximum: $(-3, -8)$
Relative minimum: $(1, 0)$

29. Critical numbers: $x = \dfrac{\pi}{6}, \dfrac{5\pi}{6}$

Increasing on $\left(0, \dfrac{\pi}{6}\right)$, $\left(\dfrac{5\pi}{6}, 2\pi\right)$

Decreasing on $\left(\dfrac{\pi}{6}, \dfrac{5\pi}{6}\right)$

Relative maximum: $\left(\dfrac{\pi}{6}, \dfrac{[\pi + 6\sqrt{3}]}{12}\right)$

Relative minimum: $\left(\dfrac{5\pi}{6}, \dfrac{[5\pi - 6\sqrt{3}]}{12}\right)$

31. Critical numbers: $x = \dfrac{\pi}{2}, \dfrac{7\pi}{6}, \dfrac{3\pi}{2}, \dfrac{11\pi}{6}$

Increasing on $\left(0, \dfrac{\pi}{2}\right)$, $\left(\dfrac{7\pi}{6}, \dfrac{3\pi}{2}\right)$, $\left(\dfrac{11\pi}{6}, 2\pi\right)$

Decreasing on $\left(\dfrac{\pi}{2}, \dfrac{7\pi}{6}\right)$, $\left(\dfrac{3\pi}{2}, \dfrac{11\pi}{6}\right)$

Relative maxima: $\left(\dfrac{\pi}{2}, 2\right)$, $\left(\dfrac{3\pi}{2}, 0\right)$

Relative minima: $\left(\dfrac{7\pi}{6}, -\dfrac{1}{4}\right)$, $\left(\dfrac{11\pi}{6}, -\dfrac{1}{4}\right)$

33. a. $f'(x) = \dfrac{2(9 - 2x^2)}{\sqrt{9 - x^2}}$

b.

c. Critical numbers: $x = \pm \dfrac{3\sqrt{2}}{2}$

d. $f' > 0$ on $\left(-\dfrac{3\sqrt{2}}{2}, \dfrac{3\sqrt{2}}{2}\right)$

$f' < 0$ on $\left(-3, -\dfrac{3\sqrt{2}}{2}\right), \left(\dfrac{3\sqrt{2}}{2}, 3\right)$

35. a. $f'(t) = t(t\cos t + 2\sin t)$

b.

c. Critical numbers: $x = 2.2889, 5.0870$

d. $f' > 0$ on $(0, 2.2889), (5.0870, 2\pi)$

$f' < 0$ on $(2.2889, 5.0870)$

37. $g'(0) < 0$ **39.** $g'(-6) < 0$ **41.** $g'(0) > 0$

43.

45.

Minimum at the approximate critical number $x = -0.40$.
Maximum at the approximate critical number $x = 0.48$.

47. Yes; $\theta = \dfrac{\pi}{2}$

49. $r = \dfrac{2R}{3}$

51. Decreasing when $0 < t < 6.02$ days
Increasing when $6.02 < t < 14$ days

53. a. $T = 10°, R \approx 8.3666 \ \Omega$

b.

55. $f(x) = -\dfrac{1}{2}x^3 + \dfrac{3}{2}x^2$ **57.** True

58. False, let $f(x) = x$, and $g(x) = x$ on the interval $(-\infty, 0)$.

59. True **60.** False, let $f(x) = x^3$. **61.** True

62. False, let $f(x) = x^3$ and $c = 0$.

Section 3.4 (*page 194*)

1. Concave upward: $(-\infty, \infty)$

3. Concave upward: $(-\infty, 1)$
Concave downward: $(1, \infty)$

5. Concave upward: $(-\infty, -1), (1, \infty)$
Concave downward: $(-1, 1)$

7. Relative maximum: $(3, 9)$

9. Relative minimum: $(5, 0)$

11. Relative maximum: $(0, 3)$
Relative minimum: $(2, -1)$

13. Relative minimum: $(3, -25)$

15. Relative minimum: $(0, -3)$

17. Relative maximum: $(-2, -4)$
Relative minimum: $(2, 4)$

19. No relative extrema since f is non-increasing.

21. Relative maximum: $(-2, 16)$
Relative minimum: $(2, -16)$
Point of inflection: $(0, 0)$

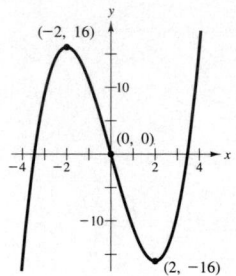

23. Point of inflection: $(2, 0)$

25. Relative minima: $(\pm 2, -4)$
Relative maximum: $(0, 0)$

Points of inflection: $\left(\pm\dfrac{2}{\sqrt{3}}, -\dfrac{20}{9}\right)$

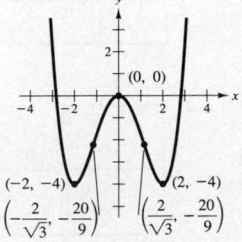

27. Relative minimum: $(1, -27)$
Points of inflection: $(2, -16)$, $(4, 0)$

29. Relative minimum: $(-2, -2)$

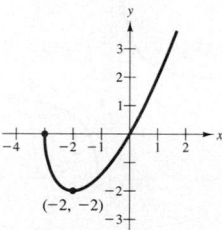

31. Relative minimum: $(3\pi, -1)$
Relative maximum: $(\pi, 1)$
Point of inflection: $(2\pi, 0)$

33. Relative minima: $\left(\dfrac{\pi}{2}, 1\right)$, $\left(\dfrac{5\pi}{2}, 1\right)$
Relative maxima: $\left(\dfrac{3\pi}{2}, -1\right)$, $\left(\dfrac{7\pi}{2}, -1\right)$

35. Relative minimum: $\left(\dfrac{5\pi}{3}, -2.598\right)$
Relative maximum: $\left(\dfrac{\pi}{3}, 2.598\right)$
Points of inflection: $(\pi, 0)$,
$(1.823, 1.452)$, $(4.46, -1.452)$

37. a. $f'(x) = 0.2x(x - 3)^2(5x - 6)$
$f''(x) = 0.4(x - 3)(10x^2 - 24x + 9)$
b. Relative maximum: $(0, 0)$
Relative minimum: $(1.2, -1.6796)$
Points of inflection: $(0.4652, -0.7049)$,
$(1.9348, -0.9049)$, $(3, 0)$
c.

39. a. $f'(x) = \cos x - \cos 3x + \cos 5x$
$f''(x) = -\sin x + 3 \sin 3x - 5 \sin 5x$
b. Relative maximum: $(\pi/2, 1.53333)$
Points of inflection:
$(0.5236, 0.2667)$, $(1.1731, 0.9638)$,
$(1.9685, 0.9637)$, $(2.6180, 0.2667)$
c.

41.

43.

45.

47.

49.

51.

53. $f(x) = (x - c)^n$ has a point of inflection at $(c, 0)$ if n is odd and $n \geq 3$.

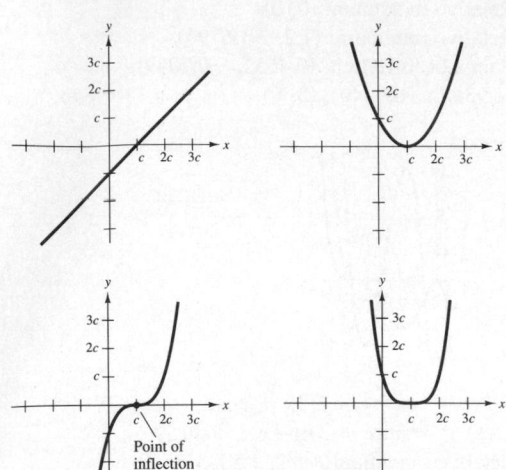

55. $f(x) = \frac{1}{2}x^3 - 6x^2 + \frac{45}{2}x - 24$

57. a. $f(x) = \frac{1}{32}x^3 + \frac{3}{16}x^2$

  b. Two miles from touchdown

61. $x = \left(\dfrac{15 - \sqrt{33}}{16}\right) L \approx 0.578 \, L$ **63.** $x = 100$ units

65. $\theta = \dfrac{\pi}{2} + 2n\pi,\ \theta = \dfrac{3\pi}{2} + 2n\pi$

67.

The values of f, P_1, and P_2 and their first derivatives are equal to $x = 0$.

69.

The values of f, P_1, and P_2 and their first derivatives are equal to $x = \pi/4$.

71. a. $S'' > 0$ **b.** $S'' < 0$ **c.** $S' = C,\ S'' = 0$

  d. $S' = 0,\ S'' = 0$ **e.** $S' < 0,\ S'' > 0$ **f.** $S' > 0$

73. True   **74.** False, 0 is not in the domain of f.

75. False, the maximum value is $\sqrt{13} \approx 3.60555$.

76. True   **79.** $(1, 0)$

Section 3.5 (*page 203*)

1. h   **2.** c   **3.** e   **4.** a   **5.** d

6. g   **7.** b   **8.** f   **9.** $\frac{2}{3}$ **11.** 0

13. Limit does not exist.   **15.** Limit does not exist.

17. 5   **19.** -1 **21.** 2   **23.** 1

25. 0   **27.** 0   **29.** 1   **31.** 0

33. $-\frac{1}{2}$

35.

37.

39.

41.

43.

45.

47.

49.

51.

53.

55. a.

  c.

The slant asymptote $y = x$.

57.

x	10^0	10^1	10^2	10^3	10^4	10^5	10^6
$f(x)$	1.000	0.513	0.501	0.500	0.500	0.500	0.500

$$\lim_{x \to \infty} x - \sqrt{x(x - 1)} = \tfrac{1}{2}$$

59.

x	10^0	10^1	10^2	10^3	10^4	10^5	10^6
$f(x)$	0.479	0.500	0.500	0.500	0.500	0.500	0.500

$$\lim_{x \to \infty} x \sin \frac{1}{2x} = \frac{1}{2}$$

61. 0.5   **63.** 100%

65. False, let $f(x) = \dfrac{2x}{\sqrt{x^2 + 2}}$

$f'(x) > 0$ for all real numbers.

66. False, let $f(x) = \begin{cases} -\frac{1}{8}x^2 + \frac{1}{2}x + 1, & x < 0 \\ \sqrt{x + 1}, & x \geq 0 \end{cases}$

$f''(x) < 0, f(x)$ increases without bound.

69.

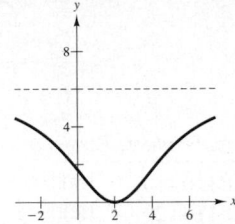

Section 3.6 (*page 211*)

1.

3.

5.

7.

9.

11.

13.

15.

17.

19.

21.

23.

25.

27.

29.

31.

33.

35.

37.

39.

The graph crosses the horizontal asymptote $y = 4$. The graph of f does not cross its vertical asymptote $x = c$ since $f(c)$ does not exist.

41.

The rational function is not reduced to lowest terms.

43.

The graph appears to approach the line $y = -x + 1$ which is the slant asymptote.

45. $y = \dfrac{1}{x - 5}$ **47.** $y = \dfrac{3x^2 - 13x - 9}{x - 5}$

49. a. The horizontal asymptote is $y = a$.
   b. The vertical asymptote is $x = b$.
51. $a < 0$ and $b^2 < 3ac$
53. $a < 0$ and $b^2 = 3ac$
55. $a < 0$ and $b^2 > 3ac$

57.

59.

61.

63.

Section 3.7 (*page 218*)

1. a.

First number, x	Second number	Product, P
10	110−10	$10(110-10) = 1000$
20	110−20	$20(110-20) = 1800$
30	110−30	$30(110-30) = 2400$
40	110−40	$40(110-40) = 2800$
50	110−50	$50(110-50) = 3000$
60	110−60	$60(110-60) = 3000$

b. $P = x(110 - x)$ **c.** 55 and 55
d.

First number, x	Second number	Product, P
10	110−10	$10(110-10) = 1000$
20	110−20	$20(110-20) = 1800$
30	110−30	$30(110-30) = 2400$
40	110−40	$40(110-40) = 2800$
50	110−50	$50(110-50) = 3000$
60	110−60	$60(110-60) = 3000$
70	110−70	$70(110-70) = 2800$
80	110−80	$80(110-80) = 2400$
90	110−90	$90(110-90) = 1800$
100	110−100	$100(110-100) = 1000$

e.

3. $\sqrt{192}$ and $\sqrt{192}$ **5.** 1 and 1   **7.** $l = w = 25$ ft
9. $l = w = 8$ ft   **11.** $\left(\dfrac{7}{2}, \sqrt{\dfrac{7}{2}}\right)$ **13.** $x = \dfrac{Q_0}{2}$
15. 600 m × 300 m
17. a. $A = 150$ in.^{2}
      $V = 99$ in.^{3}
   b. $A = 150$ in.^{2}
      $V = 125$ in.^{3}
   c. $A = 150$ in.^{2}
      $V = 117$ in.^{3}

19. a.

Height, x	Length and width	Volume, V
1	$24 - 2(1)$	$1[24 - 2(1)]^2 = 484$
2	$24 - 2(2)$	$2[24 - 2(2)]^2 = 800$
3	$24 - 2(3)$	$3[24 - 2(3)]^2 = 972$
4	$24 - 2(4)$	$4[24 - 2(4)]^2 = 1024$
5	$24 - 2(5)$	$5[24 - 2(5)]^2 = 980$
6	$24 - 2(6)$	$6[24 - 2(6)]^2 = 864$

b. $V = x(24 - 2x)^2$ **c.** $x = 4$, $V = 1024$

d.

21. $\dfrac{5 - \sqrt{7}}{6}$ ft $\times \dfrac{1 + \sqrt{7}}{3}$ ft $\times \dfrac{4 + \sqrt{7}}{3}$ ft

23. Rectangular portion: $\dfrac{16}{\pi + 4}$ ft $\times \dfrac{32}{\pi + 4}$ ft

25. $x = 3$, $y = \frac{3}{2}$ **27.** $(0, 0)$, $(4, 0)$, $(0, 6)$

29. Width: $\dfrac{5\sqrt{2}}{2}$; Length: $5\sqrt{2}$

31. Bases: r and $2r$; altitude: $\dfrac{\sqrt{3}r}{2}$

33. a.

Radius, r	Height	Surface area, S
0.2	$\dfrac{22}{\pi(0.2)^2}$	$2\pi(0.2)\left[0.2 + \dfrac{22}{\pi(0.2)^2}\right] \approx 220.3$
0.4	$\dfrac{22}{\pi(0.4)^2}$	$2\pi(0.4)\left[0.4 + \dfrac{22}{\pi(0.4)^2}\right] \approx 111.0$
0.6	$\dfrac{22}{\pi(0.6)^2}$	$2\pi(0.6)\left[0.6 + \dfrac{22}{\pi(0.6)^2}\right] \approx 75.6$
0.8	$\dfrac{22}{\pi(0.8)^2}$	$2\pi(0.8)\left[0.8 + \dfrac{22}{\pi(0.8)^2}\right] \approx 59.0$

b. $S = 2\pi r\left(r + \dfrac{22}{nr^2}\right)$ **c.** $r = \sqrt[3]{\dfrac{11}{\pi}}$, $h = 2r$

d.

Radius, r	Height	Surface area, S
0.2	$\dfrac{22}{\pi(0.2)^2}$	$2\pi(0.2)\left[0.2 + \dfrac{22}{\pi(0.2)^2}\right] \approx 220.3$
0.4	$\dfrac{22}{\pi(0.4)^2}$	$2\pi(0.4)\left[0.4 + \dfrac{22}{\pi(0.4)^2}\right] \approx 111.0$
0.6	$\dfrac{22}{\pi(0.6)^2}$	$2\pi(0.6)\left[0.6 + \dfrac{22}{\pi(0.6)^2}\right] \approx 75.6$
0.8	$\dfrac{22}{\pi(0.8)^2}$	$2\pi(0.8)\left[0.8 + \dfrac{22}{\pi(0.8)^2}\right] \approx 59.0$
1.0	$\dfrac{22}{\pi(1.0)^2}$	$2\pi(1.0)\left[1.0 + \dfrac{22}{\pi(1.0)^2}\right] \approx 50.3$
1.2	$\dfrac{22}{\pi(1.2)^2}$	$2\pi(1.2)\left[1.2 + \dfrac{22}{\pi(1.2)^2}\right] \approx 45.7$
1.4	$\dfrac{22}{\pi(1.4)^2}$	$2\pi(1.4)\left[1.4 + \dfrac{22}{\pi(1.4)^2}\right] \approx 43.7$

Radius, r	Height	Surface area, S
1.6	$\dfrac{22}{\pi(1.6)^2}$	$2\pi(1.6)\left[1.6 + \dfrac{22}{\pi(1.6)^2}\right] \approx 43.6$
1.8	$\dfrac{22}{\pi(1.8)^2}$	$2\pi(1.8)\left[1.8 + \dfrac{22}{\pi(1.8)^2}\right] \approx 44.8$
2.0	$\dfrac{22}{\pi(2.0)^2}$	$2\pi(2.0)\left[2.0 + \dfrac{22}{\pi(2.0)^2}\right] \approx 47.1$

e.

35. 18 in. $\times$ 18 in. $\times$ 36 in.

37. $\dfrac{32\pi r^3}{81}$ **39.** $r = \sqrt[3]{\dfrac{9}{\pi}} \approx 1.42$ in.

41. Side of square: $\dfrac{10\sqrt{3}}{9 + 4\sqrt{3}}$

Side of triangle: $\dfrac{30}{9 + 4\sqrt{3}}$

43. a. $\dfrac{10\pi}{\pi + 3 + 2\sqrt{2}} \approx 3.5$ ft   **b.** 10 ft

45. $w = 8\sqrt{3}$, $h = 8\sqrt{6}$ **47.** $h = \sqrt{2}$ ft

49. 1 mile from the nearest point on the coast

53. $\theta = \arctan k$, $F = \dfrac{kW}{\sqrt{k^2 + 1}}$

55. $\theta = \dfrac{2\pi}{3}\left(3 - \sqrt{6}\right) \approx 66°$

57. 14.1421×7.071

59. 42.1°

Section 3.8 (*page 228*)

1.

n	x_n	$f(x_n)$	$f'(x_n)$	$\dfrac{f(x_n)}{f'(x_n)}$	$x_n - \dfrac{f(x_n)}{f'(x_n)}$
1	1.7000	−0.1100	3.4000	−0.0324	1.7324
2	1.7324	0.0012	3.4648	0.0003	1.7321

3.

n	x_n	$f(x_n)$	$f'(x_n)$	$\dfrac{f(x_n)}{f'(x_n)}$	$x_n - \dfrac{f(x_n)}{f'(x_n)}$
1	3	0.1411	−0.9900	−0.1425	3.1425
2	3.1425	−0.0009	−1.0000	0.0009	3.1416

5. 0.682   **7.** 1.146   **9.** 3.317   **11.** −0.489

13. 0.569   **15.** 4.493   **17.** 0.74

19. a.

b. 1.333   **c.** 2.405

d. $y = -3x + 4$

$y = -1.313x + 3.156$

e. If the initial estimate $x = x_1$ is not sufficiently close to the desired zero of a function, the x-intercept of the corresponding tangent line to the function may approximate a second zero of the function.

21. $f'(x_1) = 0$ **23.** $1 = x_1 = x_3 = \dots$
$0 = x_2 = x_4 = \dots$

25. $x_{i+1} = \dfrac{x_i^2 + a}{2x_i}$ **27.** 2.646   **29.** 1.565   **33.** 3.141

35. 0.860

37. (1.939, 0.240)   **39.** $x \approx 1.563$ mi   **41.** \$384,356

43. $x \approx 11.803$ **45.** False, let $f(x) = \dfrac{x^2 - 1}{x - 1}$

46. True   **47.** True   **48.** True   **49.** 7.7253

Section 3.9 (*page 236*)

1.

x	1.9	1.99	2	2.01	2.1
$f(x)$	3.610	3.960	4	4.040	4.410
$T(x)$	3.600	3.960	4	4.040	4.400

3.

x	1.9	1.99	2	2.01	2.1
$f(x)$	24.761	31.208	32	32.808	40.841
$T(x)$	24.000	31.200	32	32.800	40.000

5.

x	1.9	1.99	2	2.01	2.1
$f(x)$	0.946	0.913	0.909	0.905	0.863
$T(x)$	0.951	0.913	0.909	0.905	0.868

7. $\Delta y = 0.331,\ dy = 0.300$

9. $\Delta y = -0.039,\ dy = -0.040$

11. $6x\,dx$ **13.** $-\dfrac{3}{(2x - 1)^2}\,dx$

15. $\dfrac{1 - 2x^2}{\sqrt{1 - x^2}}\,dx$ **17.** $\dfrac{2\sec^2 x(x^2\tan x + \tan x - x)}{(x^2 + 1)^2}\,dx$

19. $-\pi\sin\!\left(\dfrac{6\pi x - 1}{2}\right)dx$

21. $\pm\frac{3}{8}$ in.^{2} **23.** $\pm 7\pi$ in.^{2} **25. a.** $\frac{2}{3}\%$ **b.** 1.25%

27. a. $\pm 2.88\pi$ in.^{3} **b.** $\pm 0.96\pi$ in.^{2} **c.** 1%, $\frac{2}{3}\%$

29. 80π cm^3 **31. a.** $\frac{1}{4}\%$ **b.** 216 sec = 3.6 min

35. a. 0.87%   **b.** 2.16%   **37.** 4961 ft

39. $f(x) = \sqrt{x},\ dy = \dfrac{1}{2\sqrt{x}}\,dx$

$f(99.4) \approx \sqrt{100} + \dfrac{1}{2\sqrt{100}}(-0.6) \approx 9.97$

Calculator: 9.97

41. $f(x) = \sqrt[4]{x},\ dy = \dfrac{1}{4x^{3/4}}\,dx$

$f(624) \approx \sqrt[4]{625} + \dfrac{1}{4(625)^{3/4}}(-1) \approx 4.998$

Calculator: 4.998

43. $f(x) = x^4,\ dy = 4x^3\,dx$
$f(0.99) \approx 1^4 + 4(1)(-0.01) = 1 - 4(0.01)$

45. $f(x) = \sec x,\ dy = \sec x\tan x\,dx$
$f(0.03) \approx \sec 0 + \sec 0\tan 0(0.03) = 1 + 0(0.03)$

47. True   **48.** True   **49.** True

50. False, if $f(x) = \sqrt{x},\ x = 1$, and $\Delta x = 3$, then $dy = \frac{3}{2}$ and $\Delta y = 1$.

51. a. $dA = 2x\Delta x,\ \Delta A = 2x\Delta x + (\Delta x)^2$

b.

c. $\Delta A - dA = (\Delta x)^2$

53.

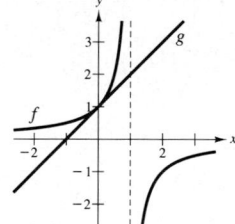

Section 3.10 (*page 242*)

1. a. Fixed cost

b.

c. Yes, it occurs when production costs are increasing at their slowest rate.

3. 4500   **5.** 300   **7.** 200   **9.** 200

11. $x = 30$ **13.** $x = 1500$ **15.** $x = 3$
$p = 60$ $p = 35$

19. a. *Order*

size, x	Price	Profit, P
102	$90 - 2(0.15)$	$102[90 - 2(0.15)] - 102(60) = 3029.40$
104	$90 - 4(0.15)$	$104[90 - 4(0.15)] - 104(60) = 3057.60$
106	$90 - 6(0.15)$	$106[90 - 6(0.15)] - 106(60) = 3084.60$
108	$90 - 8(0.15)$	$108[90 - 8(0.15)] - 108(60) = 3110.40$

b. $P = x[90 - (x - 100)(0.15)] - x(60) = x(45 - 0.15x), x \geq 100$

c. 150 units

d. *Order*

size, x	Price	Profit, P
⋮	⋮	⋮
146	$90 - 46(0.15)$	$146[90 - 46(0.15)] - 146(60) = 3372.60$
148	$90 - 48(0.15)$	$148[90 - 48(0.15)] - 148(60) = 3374.40$
150	$90 - 50(0.15)$	$150[90 - 50(0.15)] - 150(60) = 3375.00$
152	$90 - 52(0.15)$	$152[90 - 52(0.15)] - 152(60) = 3374.40$
154	$90 - 54(0.15)$	$154[90 - 54(0.15)] - 154(60) = 3372.60$
⋮	⋮	⋮

e.

21. $10\sqrt{30} \approx 54.8$ mph

23. Line should run from the power station to a point across the river $3/(2\sqrt{7})$ mile downstream.

25. $y = \dfrac{64}{141}x$, 6.1 mi   **27.** $y \approx 0.3x$, 4.5 mi

29. 8%   **31.** $x \approx 40$ units   **33.** \$30,000

35. a. $t = 151.25$, May 31
   b. $t = 333.75$, Nov. 30

37. $\eta = -\frac{17}{3}$, elastic   **39.** $\eta = -\frac{1}{2}$, inelastic

41. Maximum: $s = \$10,000$
   Point of Diminishing Returns: $s = \$5833$

Review Exercises for Chapter 3

1. Critical numbers: $x = 1, \frac{7}{3}$
   Increasing on $(-\infty, 1)$, $(\frac{7}{3}, \infty)$
   Decreasing on $(1, \frac{7}{3})$

3. Critical numbers: $x = 1, x = 0$
   Increasing on $(1, \infty)$
   Decreasing on $(0, 1)$

5. Maximum: $(2\pi, 17.57)$
   Minimum: $(2.73, 0.88)$

7. Minimum: $(2, -12)$

9. $\left(\dfrac{\pi}{2}, \dfrac{\pi}{2}\right)$, $\left(\dfrac{3\pi}{2}, \dfrac{3\pi}{2}\right)$

11. Vertical asymptote: $x = 4$
   Horizontal asymptote: $y = 2$

13. Vertical asymptote: $x = 0$
   Horizontal asymptote: $y = -2$

15.

17.

19.

21.

23.

25.

27.

29.

31.

33.

Relative minimum: (3, 108)
Relative maximum: (−3, −108)

35.

Relative minimum: (−0.155, −1.077)
Relative maximum: (2.155, 0.077)

37. $f'\left(\dfrac{2744}{729}\right) = \dfrac{3}{7}$

39. $f'(0) = 1$ **41.** No, f is discontinuous at $x = 0$

43. $c = \dfrac{x_1 + x_2}{2}$ **45.** $t \approx 4.92 \approx 4\!:\!55$ P.M.
   $d \approx 64$ mi

47. (0, 0), (5, 0), (0, 10)   **51.** 14.05 ft

53. $3(3^{2/3} + 2^{2/3})^{3/2} \approx 21.07$ ft

55. a. $y = \dfrac{1}{4}$ in., $v = 4$ in./sec

   c. Period: $\dfrac{\pi}{6}$ Frequency: $\dfrac{6}{\pi}$

57. −0.347, −1.532, 1.879   **59.** −1.164, 1.453

61. $dS = \pm1.8\pi$ in.^{2}, $\dfrac{dS}{S} \times 100 \approx \pm0.56\%$

   $dV = \pm 8.1\pi$ in.^{3}, $\dfrac{dV}{V} \times 100 \approx \pm0.83\%$

63. \$48   **65.** 120   **67.** $x = \sqrt{\dfrac{2Qs}{r}}$

69. Maximum: (1, 3)
   Minimum: (1, 1)

71. False, let $f(x) = x^3$ and $c = 0$.

72. False, the horizontal asymptotes of
   $f(x) = 2x/\sqrt{x^2 + 2}$ are $y = 2$ and $y = -2$.

73. The first and second derivatives of the deficit
   function are both positive.

CHAPTER 4 *(page 256)*

Section 4.1

1. $y = t^3 + C$ **3.** $y = \frac{2}{5}x^{5/2} + C$

Given	Rewrite	Integrate	Simplify
5. $\displaystyle\int \sqrt[3]{x}\,dx$	$\displaystyle\int x^{1/3}\,dx$	$\dfrac{x^{4/3}}{4/3} + C$	$\dfrac{3}{4}x^{4/3} + C$
7. $\displaystyle\int \dfrac{1}{x\sqrt{x}}\,dx$	$\displaystyle\int x^{-3/2}\,dx$	$\dfrac{x^{-1/2}}{-1/2} + C$	$-\dfrac{2}{\sqrt{x}} + C$
9. $\displaystyle\int \dfrac{1}{2x^3}\,dx$	$\dfrac{1}{2}\displaystyle\int x^{-3}\,dx$	$\dfrac{1}{2}\left(\dfrac{x^{-2}}{-2}\right) + C$	$-\dfrac{1}{4x^2} + C$

11. $\frac{1}{4}x^4 + 2x + C$ **13.** $\frac{2}{5}x^{5/2} + x^2 + x + C$

15. $\dfrac{3}{5}x^{5/3} + C$ **17.** $-\dfrac{1}{2x^2} + C$

19. $\frac{2}{15}x^{1/2}(3x^2 + 5x + 15) + C$ **21.** $x^3 + \frac{1}{2}x^2 - 2x + C$

23. $\frac{2}{7}y^{7/2} + C$ **25.** $x + C$

27. $-2\cos x + 3\sin x + C$ **29.** $t + \csc t + C$

31. $\tan\theta + \cos\theta + C$ **33.** $\tan y + C$

35.   **37.**

39.

41. $y = x^2 - x + 1$ **43.** $y = \sin x + 4$

45. $f(x) = x^2 + x + 4$

47. $f(x) = -4x^{1/2} + 3x = -4\sqrt{x} + 3x$

49. a. $h(t) = \dfrac{t^2}{4} + 2t + 5$ **b.** 26 in.

53. $v_0 \approx 187.617$ ft/sec

55. a. $\frac{154}{39} \approx 3.95$ ft/sec^2 **b.** $\frac{1859}{3} \approx 619.67$ ft

57. a. 300 ft   **b.** 60 ft/sec ≈ 41 mph

61. $C(x) = x^2 - 12x + 50$ **63.** $R = x(100 - \frac{5}{2}x)$

   $\overline{C}(x) = x - 12 + \dfrac{50}{x}$ $p = 100 - \frac{5}{2}x$

65. True   **66.** True   **67.** True   **68.** True

69. 320 m, −32 m/sec

71.

Section 4.2 *(page 267)*

1. 35   **3.** $\frac{158}{85}$ **5.** $4c$ **7.** $\displaystyle\sum_{i=1}^{9} \dfrac{1}{3i}$

9. $\displaystyle\sum_{j=1}^{8}\left[2\left(\dfrac{j}{8}\right) + 3\right]$ **11.** $\dfrac{2}{n}\displaystyle\sum_{i=1}^{n}\left[\left(\dfrac{2i}{n}\right)^3 - \left(\dfrac{2i}{n}\right)\right]$

13. $\dfrac{3}{n}\displaystyle\sum_{i=1}^{n}\left[2\left(1 + \dfrac{3i}{n}\right)^2\right]$ **15.** 420   **17.** 2470

19. $\dfrac{1015}{n^3}$ **21.** $\dfrac{8}{3}$ **23.** $\dfrac{81}{4}$ **25.** 9

27. $\lim\limits_{n \to \infty}\left[8\left(\dfrac{n^2 + n}{n^2}\right)\right] = 8$

29. $\lim\limits_{n \to \infty}\dfrac{1}{6}\left[\dfrac{2n^3 - 3n^2 + n}{n^3}\right] = \dfrac{1}{3}$

31. $\lim\limits_{n \to \infty}2\left[\dfrac{10n^4 + 13n^3 + 4n^2}{n^4}\right] = 20$

33. $S \approx 0.768$ **35.** $S \approx 0.746$ **37.** $S \approx 0.859$
  $s \approx 0.518$ $s \approx 0.646$ $s \approx 0.659$

39. a.

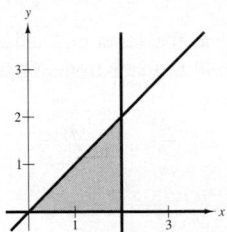

e.

n	5	10	50	100
$s(n)$	1.6	1.8	1.96	1.98
$S(n)$	2.4	2.2	2.04	2.02

41. $A = 2$

43. $A = \dfrac{7}{3}$

45. $A = 34$

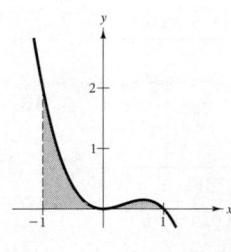

47. $A = \dfrac{2}{3}$

49. $A = 6$

51. $\dfrac{69}{8}$ **53.** 0.345

55.

n	4	8	12	16	20
Approximate area	5.3838	5.3523	5.3439	5.3403	5.3384

57.

n	4	8	12	16	20
Approximate area	2.2223	2.2387	2.2418	2.2430	2.2435

59. b   **61.** True   **62.** True

Section 4.3 (*page 277*)

1. $\displaystyle\int_0^5 3\,dx$ **3.** $\displaystyle\int_{-4}^4 (4 - |x|)\,dx$ **5.** $\displaystyle\int_{-2}^2 (4 - x^2)\,dx$

7. $\displaystyle\int_0^\pi \sin x\,dx$ **9.** $\displaystyle\int_0^2 y^3\,dy$

11. $A = 12$

13. $A = 8$

15. $A = 14$

17. $A = 1$

19. $A = \dfrac{9\pi}{2}$

21. a. 13   **b.** -10 **c.** 0   **d.** 30
23. a. 8   **b.** -12 **c.** -4 **d.** 30
25. 36   **27.** 0   **29.** $\dfrac{10}{3}$

31. $\displaystyle\int_{-1}^5 (3x + 10)\,dx$ **33.** $\displaystyle\int_0^3 \sqrt{x^2 + 4}\,dx$

35.

n	4	8	12	16	20
$L(n)$	3.6830	3.9956	4.0707	4.1016	4.1177
$M(n)$	4.3082	4.2076	4.1838	4.1740	4.1690
$R(n)$	3.6830	3.9956	4.0707	4.1016	4.1177

37.

n	4	8	12	16	20
$L(n)$	0.5890	0.6872	0.7199	0.7363	0.7461
$M(n)$	0.7854	0.7854	0.7854	0.7854	0.7854
$R(n)$	0.9817	0.8836	0.8508	0.8345	0.8247

39. a   **41.** 272   **43.** True

44. False, $\displaystyle\int_0^1 x\sqrt{x}\,dx \neq \left(\int_0^1 x\,dx\right)\left(\int_0^1 \sqrt{x}\,dx\right)$

45. True   **46.** True

47. False, $\displaystyle\int_0^2 (-x)\,dx = -2$ **48.** False, $\displaystyle\int_{-2}^4 x\,dx = 6.$

49. $\dfrac{4\sqrt{2}}{3}$

53. $f(x)$ is not integrable on the interval $[0, 1]$. As $\|\Delta\| \to 0$, $f(c_i) = 1$ or $f(c_i) = 0$ in each subinterval since there are an infinite number of both rational and irrational numbers in any interval, no matter how small.

55. 1   **57.** $\frac{1}{3}$

Section 4.4   (*page 289*)

1. 1   **3.** $-\frac{5}{2}$ **5.** $-\frac{10}{3}$ **7.** $\frac{1}{3}$ **9.** $\frac{1}{2}$

11. $\frac{2}{3}$ **13.** -4 **15.** $-\frac{1}{18}$ **17.** $-\frac{27}{20}$

19. $\dfrac{9}{2}$ **21.** $\pi + 2$ **23.** $\dfrac{2\sqrt{3}}{3}$ **25.** 0

27. $\displaystyle\int_0^3 10{,}000(t - 6)\,dt = -\$135{,}000$ **29.** $\dfrac{1}{6}$

31. $\dfrac{12\sqrt{3}}{5}$ **33.** 1   **35.** 10   **37.** 6

39. 0.4380, 1.7908   **41.** $\pm\arccos\dfrac{\sqrt{\pi}}{2} \approx \pm 0.4817$

43. Average value $= \dfrac{8}{3}$ **45.** Average value $= \dfrac{2}{\pi}$

     $x = \pm\dfrac{2\sqrt{3}}{3} \approx \pm 1.155$ $x \approx 0.690, \; x \approx 2.451$

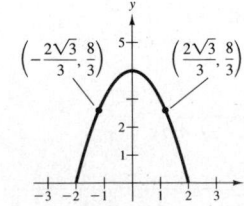

47. a. 8   **b.** $\frac{4}{3}$ **c.** 20, $\frac{10}{3}$

49. a.

b. $x = 40$: 0.16 $x = 50$: 0.52

c. $30 - 40$: 1.61 $50 - 60$: 8.65

51. a. $F(x) = 100\sec^2 x$ **b.** 165.4 lb

53. $\frac{1}{2}x^2 + 2x$ **55.** $\frac{3}{4}x^{4/3} - 12$ **57.** $\tan x - 1$

59. $x^2 - 2x + 5$ **61.** $\sqrt{x^4 + 1}$ **63.** $x\cos x$ **65.** 8

67. $\cos x\sqrt{\sin x}$ **69.** $3x^2\sin x^6$

71.

The extrema of F corresponds to the zeros of f and the inflection points of F correspond to the extrema of f.

73. a. $C(x) = 1000(12x^{5/4} + 125)$

   b. $C(1) = \$137{,}000$

     $C(5) = \$214{,}721$

     $C(10) = \$338{,}394$

75. True   **76.** True

77. False, $f(x) = x^{-2}$ has a nonremovable discontinuity at $x = 0$.

Section 4.5   (*page 301*)

$\displaystyle\int f(g(x))g'(x)\,dx$	$u = g(x)$	$du = g'(x)$
1. $\displaystyle\int (5x^2 + 1)^2(10x)\,dx$	$5x^2 + 1$	$10x\,dx$
3. $\displaystyle\int \dfrac{x}{\sqrt{x^2 + 1}}\,dx$	$x^2 + 1$	$2x\,dx$
5. $\displaystyle\int \tan^2 x\sec^2 x\,dx$	$\tan x$	$\sec^2 x\,dx$

7. $\dfrac{(1 + 2x)^5}{5} + C$ **9.** $\dfrac{2}{3}(9 - x^2)^{3/2} + C$

11. $\dfrac{(x^3 - 1)^5}{15} + C$ **13.** $-\dfrac{15}{8}(1 - x^2)^{4/3} + C$

15. $-\dfrac{1}{3(1 + x^3)} + C$ **17.** $-\dfrac{1}{4}\left(1 + \dfrac{1}{t}\right)^4 + C$

19. $\sqrt{2x} + C$

21. $\frac{2}{5}x^{5/2} + 2x^{3/2} + 14x^{1/2} + C = \frac{2}{5}\sqrt{x}(x^2 + 5x + 35) + C$

23. $\frac{1}{4}t^4 - t^2 + C$

25. $6y^{3/2} - \frac{2}{5}y^{5/2} + C = \frac{2}{5}y^{3/2}(15 - y) + C$

27. $2x^2 - 4\sqrt{16 - x^2} + C$

29. $-\dfrac{1}{2(x^2 + 2x - 3)} + C$ **31.** $-\dfrac{1}{2}\cos 2x + C$

33. $-\sin\dfrac{1}{\theta} + C$ **35.** $\dfrac{1}{4}\sin^2 2x + C$ or $-\dfrac{1}{4}\cos^2 2x + C_1$

37. $\frac{1}{5}\tan^5 x + C$ **39.** $\frac{1}{2}\tan^2 x + C$ or $\frac{1}{2}\sec^2 x + C_1$

41. $-\cot x - x + C$

43. $\frac{2}{15}(x + 2)^{3/2}(3x - 4) + C$

45. $-\frac{2}{105}(1 - x)^{3/2}(15x^2 + 12x + 8) + C$

47. $\dfrac{\sqrt{2x - 1}}{15}(3x^2 + 2x - 13) + C$

49. $-x - 1 - 2\sqrt{x + 1} + C$ or $-(x + 2\sqrt{x + 1}) + C_1$

51. 0   **53.** 2   **55.** $\frac{1}{2}$ **57.** $\frac{4}{15}$ **59.** $\frac{1209}{28}$

61. $\frac{3\sqrt{3}}{4}$ **63.** $2(\sqrt{3} - 1)$

65. $\frac{1}{3}\sqrt{2x + 1}(x - 1) + C$ **67.** $\frac{144}{5}$ **69.** $\frac{7088}{105}$

71. $\frac{1}{6}(2x - 1)^3 + C_1 = \frac{4}{3}x^3 - 2x^2 + x - \frac{1}{6} + C_1$

or

$\frac{4}{3}x^3 - 2x^2 + x + C_2$

Answers differ by a constant: $C_2 = C_1 - \frac{1}{6}$

73. 4   **75. a.** $\frac{8}{3}$ **b.** $\frac{16}{3}$ **c.** $-\frac{8}{3}$ **d.** 8

77. $2\int_0^4 (6x^2 - 3)\,dx = 232$

79. $f(x) = -\frac{1}{3}[(1 - x^2)^{3/2} - 8]$

81. $V(t) = \dfrac{200,000}{t + 1} + 300,000$

$340,000

83. a. $C(x) = \frac{3}{2}(12x + 1)^{2/3} + 56.35$

b.

85. a. 102.352 thousand units
   b. 102.352 thousand units
   c. 74.5 thousand units

87. False, $\displaystyle\int (2x + 1)^2\,dx = \frac{1}{6}(2x + 1)^3 + C.$

88. False, $\displaystyle\int x(x^2 + 1)\,dx = \frac{1}{4}(x^2 + 1)^2 + C.$

89. True   **90.** True   **91.** True

92. False, $\displaystyle\int \sin^2 2x \cos 2x\,dx = \frac{1}{6}\sin^3 2x + C.$

Section 4.6 (page 311)

	Exact	Trapezoidal	Simpson's
1.	2.6667	2.7500	2.6667
3.	4.0000	4.2500	4.0000
5.	4.0000	4.0625	4.0000
7.	12.6667	12.6640	12.6667
9.	0.1667	0.1676	0.1667

	Trapezoidal	Simpson's
11.	3.283	3.240
13.	0.342	0.372
15.	0.957	0.978
17.	0.089	0.089
19.	0.194	0.186
21.	0.500	0.000
23.	$n = 366$	$n = 26$
25.	$n = 130$	$n = 12$
27.	$n = 643$	$n = 48$

29. Proof

31.

n	$L(n)$	$M(n)$	$R(n)$	$T(n)$	$S(n)$
4	12.7771	15.3965	18.4340	15.6055	15.4845
8	14.0868	15.4480	16.9152	15.5010	15.4662
10	14.3569	15.4544	16.6197	15.4883	15.4658
12	14.5386	15.4578	16.4242	15.4814	15.4657
16	14.7674	15.4613	16.1816	15.4745	15.4657
20	14.9056	15.4628	16.0370	15.4713	15.4657

33.

n	$L(n)$	$M(n)$	$R(n)$	$T(n)$	$S(n)$
4	2.8163	3.5456	3.7256	3.2709	3.3996
8	3.1809	3.5053	3.6356	3.4083	3.4541
10	3.2478	3.4990	3.6115	3.4296	3.4624
12	3.2909	3.4952	3.5940	3.4425	3.4674
16	3.3431	3.4910	3.5704	3.4568	3.4730
20	3.3734	3.4888	3.5552	3.4643	3.4759

35. 0.701   **37.** 3.14159   **39.** 89,250 ft^2

41. a. 9920 ft^2 **b.** $10,413\frac{1}{3}$ ft^2 **43.** 2.478

45. a. 0   **b.** $L'(x) = \dfrac{1}{x}$, $L'(1) = 1$ **c.** 2.718

Review Exercises for Chapter 4

1.

3. $\frac{2}{3}x^3 + \frac{1}{2}x^2 - x + C$ **5.** $\frac{1}{7}x^7 + \frac{3}{5}x^5 + x^3 + x + C$

7. $\frac{1}{2}x^2 - \frac{1}{x} + C$ **9.** $\frac{2}{3}\sqrt{x^3 + 3} + C$

11. $\dfrac{(3x^2 - 1)^5}{30} + C$ **13.** $\frac{1}{4}\sin^4 x + C$

15. $2\sqrt{1 - \cos\theta} + C$ **17.** $\dfrac{\tan^{n+1} x}{n + 1} + C$ $n \neq -1$

19. $\dfrac{1}{3\pi}(1 + \sec \pi x)^3 + C$ **21.** $y = 2 - x^2$

23. 240 ft/sec

25. a. 3 sec   **b.** 144 ft   **c.** $\frac{3}{2}$ sec   **d.** 108 ft

27. a. $\displaystyle\sum_{i=1}^{10} (2i - 1)$ **b.** $\displaystyle\sum_{i=1}^{n} i^3$ **c.** $\displaystyle\sum_{i=1}^{10} (4i + 2)$

29. a. $S = \dfrac{5mb^2}{8}$, $s = \dfrac{3mb^2}{8}$

   b. $S(n) = \dfrac{mb^2(n + 1)}{2n}$, $s(n) = \dfrac{mb^2(n - 1)}{2n}$

   c. $\dfrac{1}{2}mb^2$ **d.** $\dfrac{1}{2}mb^2$

31. a. 13   **b.** 7   **c.** 11   **d.** 50

33. 16   **35.** 0   **37.** 2   **39.** $\dfrac{422}{5}$ **41.** $\dfrac{28\pi}{15}$ **43.** 2

45. 6   **47.** $\frac{10}{3}$

49. $\frac{1}{4}$ **51.** 16

53. $\sqrt{3}$

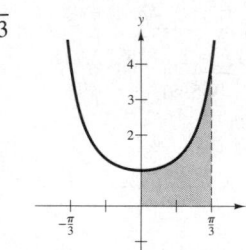

55. Average value $= \frac{2}{5}$, $x = \frac{29}{4}$ **57.** Average value $= 2$, $x = 2$

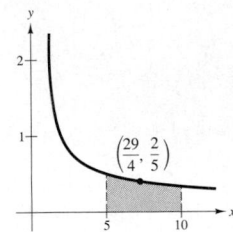

59. 0.254   **61.** 1.6234 liters

63. a. $C \approx \$9.17$ **b.** $C \approx \$3.14$ Savings $\approx \$6.03$

65. a. $\dfrac{19{,}800}{M}$ **b.** $\dfrac{22{,}800}{M}$

67. a. $0.025 = 2.5\%$ **b.** $0.736 = 73.6\%$

68. False, only constants can be taken through the integral sign.

69. False, $\dfrac{d}{dx}\left[-\dfrac{1}{x^2}\right] \neq \dfrac{1}{x}$.   **70.** True   **71.** True

72. False, $\displaystyle\int \tan x\,dx = -\ln|\cos x| + C$.

CHAPTER 5

Section 5.1   *(page 324)*

1.

x	0.5	1.5	2	2.5	3	3.5	4
$\int_1^x 1/t\,dt$	-0.6932	0.4055	0.6932	0.9163	1.0987	1.2529	1.3865

3. b   **4.** d   **5.** a   **6.** c

7. $f(x) = 3\ln x$ **9.** $f(x) = \ln 2x$
     Domain: $x > 0$ Domain: $x > 0$

11. $f(x) = \ln(x - 1)$
     Domain: $x > 1$

13. a. 1.7917 **b.** -0.4055 **c.** 4.3944 **d.** 0.5493

15. $\ln 2 - \ln 3$ **17.** $\ln x + \ln y - \ln z$ **19.** $\frac{3}{2}\ln 2$

21. $3[\ln(x+1) + \ln(x-1) - 3\ln x]$

23. $\ln z + 2\ln(z-1)$ **25.** $\ln\dfrac{x-2}{x+2}$

27. $\ln\sqrt[3]{\dfrac{x(x+3)^2}{x^2-1}}$ **29.** $\ln\dfrac{9}{\sqrt{x^2+1}}$ **31.** 3   **33.** 2

35. $\dfrac{2}{x}$ **37.** $\dfrac{4(\ln x)^3}{x}$ **39.** $\dfrac{2x^2-1}{x(x^2-1)}$ **41.** $\dfrac{1-x^2}{x(x^2+1)}$

43. $\dfrac{1-2\ln t}{t^3}$ **45.** $\dfrac{2}{x\ln x^2} = \dfrac{1}{x\ln x}$ **47.** $\dfrac{1}{1-x^2}$

49. $\dfrac{-4}{x(x^2+4)}$ **51.** $\dfrac{\sqrt{x^2+1}}{x^2}$ **53.** $\cot x$

55. $-\tan x + \dfrac{\sin x}{\cos x - 1}$ **57.** $\dfrac{3\cos x}{(\sin x - 1)(\sin x + 2)}$

59. $\dfrac{2}{x}(\sin 2x + x\cos 2x\ln x^2)$ **61.** $\dfrac{2xy}{3-2y^2}$

63. $5x - y - 2 = 0$ **65.** $xy'' + y' = x\left(\dfrac{-2}{x^2}\right) + \dfrac{2}{x} = 0$

67. Relative minimum: $(1, \frac{1}{2})$

69. Relative minimum: $(e^{-1}, -e^{-1})$

71. Relative minimum: (e, e)

Point of inflection: $\left(e^2, \dfrac{e^2}{2}\right)$

73.

The values of f, P_1, and P_2, and their first derivatives agree at $x = 1$. The values of the second derivatives of f and P_2 agree at $x = 1$.

75. 0.567   **77.** $\dfrac{2x^2 - 1}{\sqrt{x^2 - 1}}$ **79.** $\dfrac{3x^3 - 15x^2 + 8x}{2(x - 1)^3\sqrt{3x - 2}}$

81. $\dfrac{(2x^2 + 2x - 1)\sqrt{x - 1}}{(x + 1)^{3/2}}$

83. $\beta = \dfrac{10}{\ln 10}(\ln I + 16\ln 10)$

60 db

85. a.

b. $p = 10$: 4.75 deg/lb/in.^{2}
$p = 70$: 0.97 deg/lb/in.^{2}

87. $g'(x) = \dfrac{f'(x)}{f(x)}$, $f(x) > 0$ **a.** yes   **b.** yes

Therefore the signs of $g'(x)$ and f' are the same.

89. For large values of x, g increases at a higher rate than f in both cases. The natural logarithm function increases very slowly for large values of x.

a.

b.

91.

93. False, $\ln x + \ln 25 = \ln 25x$.   **94.** True   **95.** True

96. False, $y' = 0$.

Section 5.2   (*page 333*)

1. $\ln|x + 1| + C$ **3.** $-\frac{1}{2}\ln|3 - 2x| + C$

5. $\ln\sqrt{x^2 + 1} + C$ **7.** $\dfrac{x^2}{2} - 4\ln|x| + C$

9. $\frac{1}{3}\ln|x^3 + 3x^2 + 9x| + C$ **11.** $\frac{1}{3}(\ln x)^3 + C$

13. $2\sqrt{x + 1} + C$

15. $x + 6\sqrt{x} + 18\ln|\sqrt{x} - 3| + C$

17. $2\ln|x - 1| - \dfrac{2}{x - 1} + C$

19. $\ln|\sin\theta| + C$ **21.** $-\frac{1}{2}\ln|\csc 2x + \cot 2x| + C$

23. $\ln|1 + \sin t| + C$

25. $\ln|\sec x - 1| + C$ **27.** $-3\ln|2 - x| + C$

29. $-\ln\sqrt{|\cos 2\theta|} + C$ **31.** $\frac{5}{3}\ln 13 \approx 4.275$ **33.** $\frac{7}{3}$

35. $-\ln 3$ **37.** $\ln\left|\dfrac{2 - \sin 2}{1 - \sin 1}\right| \approx 1.929$

39. $2[\sqrt{x} - \ln(1 + \sqrt{x})] + C$ **41.** $-\sin(1 - x) + C$

43. $\ln(\sqrt{2} + 1) - \dfrac{\sqrt{2}}{2} \approx 0.174$ **49.** $\dfrac{1}{x}$ **51.** 0

53. d   **55.** $\frac{15}{2} + 8\ln 2 \approx 13.045$ square units

57. $P(t) = 1000[12\ln|1 + 0.25t| + 1]$
$P(3) \approx 7715$

59. \$168.27   **61.** False, $\dfrac{1}{2}(\ln x) = \ln x^{1/2}$.

62. False, $\dfrac{d}{dx}[\ln x] = \dfrac{1}{x}$.   **63.** True

64. False, the integrand, $1/x$, has a nonremovable discontinuity in the interval $[-1, e]$.

65. $\lim_{k \to 0} f_k(x) = \ln x$

Section 5.3   (*page 341*)

1.

3.

5.

7.

9. $f^{-1}(x) = \dfrac{x + 3}{2}$ **11.** $f^{-1}(x) = x^{1/5}$

13. $f^{-1}(x) = x^2, \quad x \geq 0$

15. $f^{-1}(x) = \sqrt{4 - x^2}, \quad 0 \leq x \leq 2$

17. $f^{-1}(x) = x^3 + 1$ **19.** $f^{-1}(x) = x^{3/2}, \quad x \geq 0$

21. $f^{-1}(x) = \dfrac{\sqrt{7}x}{\sqrt{1 - x^2}}, \quad -1 < x < 1$

23. $f^{-1}(x) = \begin{cases} \dfrac{1 - \sqrt{1 + 16x^2}}{2x} & \text{if } x \neq 0 \\ 0 & \text{if } x = 0 \end{cases}$

25.

x	1	2	3	4
$f^{-1}(x)$	0	1	2	4

27. Inverse exists   **29.** Inverse does not exist
31. One-to-one   **33.** One-to-one

35. Not one-to-one

37. Inverse exists   **39.** Inverse does not exist
41. Inverse exists   **43.** $f'(x) = 2(x - 4) > 0$ on $(4, \infty)$

45. $f'(x) = -\dfrac{8}{x^3} < 0$ on $(0, \infty)$

47. $f'(x) = -\sin x < 0$ on $(0, \pi)$

49. Not continuous at $\dfrac{(2n - 1)\pi}{2}$

51. One-to-One   **53.** One-to-one
$f^{-1}(x) = x^2 + 2, x \geq 0$ $f^{-1}(x) = 2 - x, x \geq 0$
55. $f^{-1}(x) = \sqrt{x} + 3, x \geq 0$

57. $f^{-1}(x) = x - 3, x \geq 0$ **59.** $\dfrac{1}{5}$ **61.** $\dfrac{2\sqrt{3}}{3}$

63. $\frac{1}{13}$ **65.** $f'(\frac{1}{2}) = \frac{3}{4}, (f^{-1})'(\frac{1}{8}) = \frac{4}{3}$
67. $f'(5) = \frac{1}{2}, (f^{-1})'(1) = 2$ **69.** 32   **71.** 600

73. $(g^{-1} \circ f^{-1})(x) = \dfrac{x + 1}{2}$ **75.** $(f \circ g)^{-1}(x) = \dfrac{x + 1}{2}$

81. False, let $f(x) = x^2$.   **82.** True   **83.** True

84. False, let $f(x) = \dfrac{1}{x}$.

85. No, let $f(x) = \begin{cases} x, & 0 \leq x \leq 1 \\ 1 - x, & 1 < x \leq 2 \end{cases}$.   **87.** $\sqrt{17}$

Section 5.4 (*page 350*)

1. a. $\ln 1 = 0$ **b.** $\ln 7.389 \ldots = 2$
3. a. $e^{0.6931\cdots} = 2$ **b.** $e^{2.128\cdots} = 8.4$
5. a. $x = 4$ **b.** $x = \frac{3}{2}$ **7. a.** $x = e^2$ **b.** $x = \ln 4$
9.   **11.**

13. a.

Translation 2 units to the right

b.

Reflection in the x-axis and a vertical shrink

c.

Reflection in the y-axis and a translation 3 units upward

15.   **17.**

19. c   **20.** d   **21.** a   **22.** b   **23.** $2.7182805 < e$

25. a. 3   **b.** -3 **27.** $2e^{2x}$ **29.** $2(x-1)e^{-2x+x^2}$

31. $\dfrac{e^{\sqrt{x}}}{2\sqrt{x}}$ **33.** $3(e^{-t}+e^t)^2(e^t-e^{-t})$ **35.** $2x$

37. $\dfrac{2e^{2x}}{1+e^{2x}}$ **39.** $\dfrac{-2(e^x-e^{-x})}{(e^x+e^{-x})^2}$ **41.** x^2e^x

43. $e^{-x}\left(\dfrac{1}{x}-\ln x\right)$ **45.** $2e^x\cos x$ **47.** $\dfrac{10-e^y}{xe^y+3}$

49. $3(6x+5)e^{-3x}$

53. Relative maximum: $\left(0,\,1/\sqrt{2\pi}\right)$

Points of inflection: $\left(\pm1,\,1/\sqrt{2\pi e}\right)$

55. Relative minimum: $(0,1)$

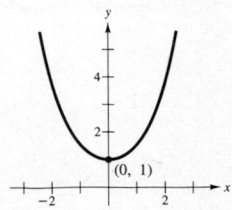

57. Relative minimum: $(0,0)$

Relative maximum: $(2, 4e^{-2})$

Points of inflection: $\left(2\pm\sqrt{2},\,\left(6\pm4\sqrt{2}\right)e^{-(2\pm\sqrt{2})}\right)$

59.

The values of f, P_1, and P_2, and their first derivatives agree at $x = 0$. The values of the second derivatives of f and P_2 agree at $x = 0$.

61. $A = \sqrt{2}e^{-1/2}$ **63.** 0.567

65. a.

b. When x increases without bound, $1/x$ approaches zero, and $e^{1/x}$ approaches 1. Therefore, $f(x)$ approaches $\frac{2}{1+1} = 1$. Thus, $f(x)$ has a horizontal asymptote at $y = 1$. As x approaches zero from the right, $1/x$ approaches ∞, $e^{1/x}$ approaches ∞ and $f(x)$ approaches 0. As x approaches zero from the left, $1/x$ approaches $-\infty$, $e^{1/x}$ approaches 0, and $f(x)$ approaches 2. The limit does not exist since the left limit does not equal the right limit. Therefore, $x = 0$ is a nonremovable discontinuity.

67. $e^{5x} + C$ **69.** $\dfrac{e^2-1}{2e^2}$

71. $x - \ln(e^x + 1) + C_1$ or $-\ln(1 + e^{-x}) + C_2$

73. $\dfrac{e}{3}(e^2 - 1)$ **75.** $-\dfrac{2}{3}(1 - e^x)^{3/2} + C$

77. $\ln|e^x - e^{-x}| + C$ **79.** $-\frac{5}{2}e^{-2x} + e^{-x} + C$

81. $\dfrac{1}{\pi}e^{\sin\pi x} + C$ **83.** $\ln|\cos e^{-x}| + C$

85. $\dfrac{1}{2a}e^{ax^2} + C$ **87.** $f(x) = \dfrac{1}{2}(e^x + e^{-x})$

89. $e^5 - 1 \approx 147.413$ **91.** $1 - e^{-1} \approx 0.632$

93. a. Midpoint Rule: 92.1898

Trapezoidal Rule: 93.8371

Simpson's Rule: 92.7385

b. Midpoint Rule: 1.1906

Trapezoidal Rule: 1.1827

Simpson's Rule: 1.1880

Section 5.5 (*page 359*)

1. -3 **3.** 0   **5. a.** $\log_2 8 = 3$ **b.** $\log_3(1/3) = -1$

7. a. $10^{-2} = 0.01$ **b.** $(1/2)^{-3} = 8$

9. a. $x = 3$ **b.** $x = -1$ **11. a.** $x = \frac{1}{3}$ **b.** $x = \frac{1}{16}$
13. a. $x = -1, 2$ **b.** $x = \frac{1}{3}$

15.   **17.**

19.   **21.**

23. a. False   **b.** True   **c.** True   **d.** False
25. $(\ln 4)4^x$ **27.** $(\ln 5)5^{x-2}$ **29.** $t2^t(t \ln 2 + 2)$
31. $\dfrac{1}{x(\ln 3)}$ **33.** $\dfrac{x - 2}{(\ln 2)x(x - 1)}$ **35.** $\dfrac{x}{(\ln 5)(x^2 - 1)}$
37. $2(1 - \ln x)x^{(2/x)-2}$
39. $(x - 2)^{x+1}\left[\dfrac{x + 1}{x - 2} + \ln(x - 2)\right]$
41. $g(x) = x^x, k(x) = 2^x, h(x) = x^2, f(x) = \log_2 x$
43. a. \$40.64
    b. $C'(1) \approx 0.051P, C'(8) \approx 0.072P$
    c. $\ln 1.05$

45.

n	1	2	4	12	365	Continuous compounding
A	\$2061.03	\$2088.15	\$2102.35	\$2112.06	\$2116.84	\$2117.00

47.

n	1	2	4	12	365	Continuous compounding
A	\$17,449.40	\$18,679.19	\$19,358.15	\$19,837.40	\$20,077.29	\$20,085.54

49.

t	1	10	20	30	40	50
P	\$91,393.12	\$40,656.97	\$16,529.89	\$6720.55	\$2732.37	\$1110.90

51.

t	1	10	20	30	40	50
P	\$90,521.24	\$36,940.70	\$13,646.15	\$5040.98	\$1862.17	\$687.90

53. c

55. a. \$332.01   **b.** \$604.96   **c.** \$1102.32

57. $p'(1) \approx 113.5$ fish/month
    $p'(10) \approx 403.2$ fish/month
    $t = 5 \ln 19 \approx 14.72$

59. a.

b. 80.3%   **c.** $x \approx 1.546$ or 1546 egg masses
d. $x \approx 1.105$ or 1105 egg masses
61. $-\dfrac{4^{-x}}{\ln 4} + C$ **63.** 4   **65.** $-\dfrac{7^{(3 - x^2)}}{2 \ln 7} + C$
67. \$2825.51

69.

x	1	10^{-1}	10^{-2}	10^{-4}	10^{-6}
$(1 + x)^{1/x}$	2	2.594	2.705	2.718	2.718

71. $y = 600(1.05)^t$ **72.** False, e is an irrational number.
73. True   **74.** True   **75.** True   **76.** True   **77.** True

Section 5.6 (*page 367*)

1. $y^2 - 5x^2 = C$ **3.** $y = Ce^{(2x^{3/2})/3}$

5. $y = C(1 + x^2)$

7. $\dfrac{dQ}{dt} = \dfrac{k}{t^2}$ **9.** $\dfrac{dN}{ds} = k(250 - s)$

$Q = -\dfrac{k}{t} + C$ $N = -\dfrac{k}{2}(250 - s)^2 + C$

11. $y = \frac{1}{4}t^2 + 10$ **13.** $y = 10e^{-t/2}$

15. $y = \frac{1}{2}e^{0.4605t}$ **17.** $y = 0.6687e^{0.4024t}$

19. Amount after 1000 years: 6.52 grams
Amount after 10,000 years: 0.14 gram

21. Initial quantity: 6.70 grams
Amount after 1000 years: 5.94 grams

23. Initial quantity: 2.16 grams   **25.** 95.81%
Amount after 10,000 years: 1.63 grams

27. Time to double: 5.78 years
Amount after 10 years: $3320.12

29. Annual rate: 8.94%
Amount after 10 years: $1833.67

31. Annual rate: 9.50%   **33.** $112,087.09
Time to double: 7.30 years

35. a. 6.64 years   **b.** 6.33 years
c. 6.30 years   **d.** 6.30 years

37. $y \approx 4.22e^{0.0430t}$
9.97 million

39. k. The larger the value of k, the greater the rate of growth
of the population.

41. 527.06 mm Hg

43. a. $N \approx 30(1 - e^{-0.0502t})$ **b.** 36 days

45. a. $S \approx 30e^{-1.7918/t}$ **b.** 20,965 units

c.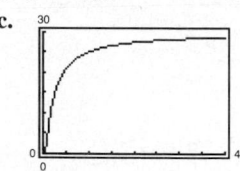

47. 2010 ($t = 16$)

49. a. 20 db   **b.** 70 db   **c.** 95 db   **d.** 120 db

51. a. $10^{8.3} \approx 199,526,231.5$ **b.** 10^R **c.** $\dfrac{1}{I \ln 10}$

53. 22.35°

Section 5.7 (*page 377*)

1. $\dfrac{\pi}{6}$ **3.** $\dfrac{\pi}{3}$ **5.** $\dfrac{\pi}{6}$ **7.** $-\dfrac{\pi}{4}$

9. 2.50   **11.** $\arccos(\frac{1}{1.269}) \approx 0.66$

13. The range of $y = \arctan x$ is $-\dfrac{\pi}{2} < y < \dfrac{\pi}{2}$.

15.

17. -0.1

19. a. $\dfrac{1}{2}$ **b.** $\dfrac{\sqrt{3}}{2}$

21. a. $\frac{3}{5}$ **b.** $\frac{5}{3}$

23. a. $-\sqrt{3}$

b. $-\frac{13}{5}$

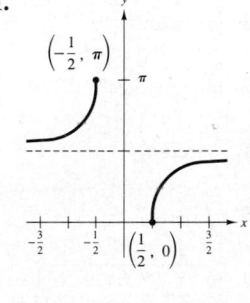

25. x **27.** $\sqrt{1 - 4x^2}$ **29.** $\dfrac{\sqrt{x^2 - 1}}{x}$

31. $\dfrac{\sqrt{x^2 - 9}}{3}$ **33.** $\dfrac{\sqrt{x^2 + 2}}{x}$ **35.** $\arcsin\left(\dfrac{9}{\sqrt{x^2 + 81}}\right)$

39.   **41.**

43. $x = \frac{1}{3}\left[\sin(\frac{1}{2}) + \pi\right] \approx 1.207$ **45.** $x = \frac{1}{3}$

47. $\dfrac{2}{\sqrt{2x - x^2}}$ **49.** $-\dfrac{3}{\sqrt{4 - x^2}}$

51. $\dfrac{a}{a^2 + x^2}$ **53.** $\dfrac{3x - \sqrt{1 - 9x^2}\arcsin 3x}{x^2\sqrt{1 - 9x^2}}$

55. $\dfrac{-6}{1 + 36x^2}$ **57.** $-\dfrac{t}{\sqrt{1 - t^2}}$ **59.** 0   **61.** $\dfrac{1}{1 - x^4}$

63. $\arcsin x$

65. $y = -2x + \left(\dfrac{5\pi}{6} - \sqrt{3}\right)$ **67.**

$y = -2x + \left(\dfrac{\pi}{6} + \sqrt{3}\right)$

69. Relative maximum: $(1.272, -0.606)$
Relative minimum: $(-1.272, 3.747)$
71. $(0.7862, 0.6662)$
73. a. $h(t) = -16t^2 + 256$
     $t = 4$ sec
   b. $t = 1$: -0.0520 rad/sec
     $t = 2$: -0.1116 rad/sec
77. a. The range of $y = \arcsin x$ is $-\dfrac{\pi}{2} \le y \le \dfrac{\pi}{2}$.

   b. Maximum: $\dfrac{\pi}{2}$
       Minimum: $-\dfrac{\pi}{2}$

79. True   **80.** False, the range is $\left[\dfrac{-\pi}{2}, \dfrac{\pi}{2}\right]$.   **81.** True

82. False, $\arcsin^2 0 + \arccos^2 0 = \left(\dfrac{\pi}{2}\right)^2 \ne 1$.

Section 5.8   (*page 386*)

1. $\dfrac{\pi}{18}$ **3.** $\dfrac{\pi}{6}$
5. $2 \operatorname{arcsec} |2x| + C$ **7.** $\frac{1}{2}x^2 - \frac{1}{2}\ln(x^2 + 1) + C$
9. $\arcsin(x + 1) + C$ **11.** $\frac{1}{2}\arcsin t^2 + C$
13. $\frac{1}{2}(\arctan x)^2 + C$ **15.** $\dfrac{\pi^2}{32} \approx 0.308$
17. $\dfrac{\sqrt{3} - 2}{2} \approx -0.134$ **19.** $\arcsin e^x + C$
21. $\dfrac{1}{3}\arctan\left(\dfrac{x - 3}{3}\right) + C$ **23.** $\dfrac{1}{4}\arctan\dfrac{e^{2x}}{2} + C$
25. $\dfrac{\pi}{4}$ **27.** $\dfrac{\pi}{2}$
29. $\ln|x^2 + 6x + 13| - 3\arctan\left(\dfrac{x + 3}{2}\right) + C$
31. $\arcsin\left(\dfrac{x + 2}{2}\right) + C$ **33.** $-\sqrt{-x^2 - 4x} + C$
35. $4 - 2\sqrt{3} + \dfrac{\pi}{6} \approx 1.059$
37. $\frac{1}{2}\arctan(x^2 + 1) + C$ **39.** $\frac{1}{4}\arcsin(4x - 2) + C$
41. $2\left(\sqrt{x - 1} - \arctan\sqrt{x - 1}\right) + C$
43. $2\sqrt{e^t - 3} - 2\sqrt{3}\arctan\left(\dfrac{\sqrt{e^t - 3}}{\sqrt{3}}\right) + C$
45. $y = \arctan x^3 + C$ **47.** $y = C\sqrt{1 + x^2}$
49. $\dfrac{\pi}{4}$ **51.** $\dfrac{\pi}{8}$
53. a. Average value of $f(x)$ over the interval $[x, x + 2]$
      Maximum at $x = -1$
   b. $F(x) = \arctan(x + 2) - \arctan x$
      $F'(x) = \dfrac{-4(x + 1)}{(x^2 + 1)(x^2 + 4x + 5)}$, $F'(-1) = 0$

55. a. $v(t) = -32t + 500$

   b. $s(t) = -16t^2 + 500t$, 3906.25 ft
   c. $v(t) = \sqrt{\dfrac{32}{k}} \tan\left[\arctan\left(500\sqrt{\dfrac{k}{32}}\right) - \sqrt{32k}\, t\right]$
   d.

     6.86 sec
   e. 1088 ft
   f. When air resistance is taken into account, the maximum height of the object is not as great.
57. b. 3.14159
59. a. $\arcsin x + C$
   b. $-\sqrt{1 - x^2} + C$
   c. Does not fit Basic Formulas
61. a. $\frac{2}{3}(x - 1)^{3/2} + C$
   b. $\frac{2}{15}(x - 1)^{3/2}(3x + 2) + C$
   c. $\frac{2}{3}\sqrt{x - 1}(x + 2) + C$
63.

     0.57

Section 5.9   (*page 396*)

1. a. 10.018   **b.** -0.964
3. a. $\frac{4}{3}$ **b.** $\frac{13}{12}$
5. a. 1.317   **b.** 0.962
7. $\left(\dfrac{e^x - e^{-x}}{e^x + e^{-x}}\right)^2 + \left(\dfrac{2}{e^x + e^{-x}}\right)^2 = 1$
9. $\left(\dfrac{e^x - e^{-x}}{2}\right)\left(\dfrac{e^y + e^{-y}}{2}\right) + \left(\dfrac{e^x + e^{-x}}{2}\right)\left(\dfrac{e^y - e^{-y}}{2}\right)$
   $= \dfrac{e^{(x+y)} - e^{-(x+y)}}{2}$
   $= \sinh(x + y)$

11. $\left(\dfrac{e^x - e^{-x}}{2}\right)\left[3 + 4\left(\dfrac{e^x - e^{-x}}{2}\right)^2\right]$

$= \dfrac{e^{3x} - e^{-3x}}{2}$

$= \sinh(3x)$

13. $\sinh x = \dfrac{3}{2}$

$\cosh x = \dfrac{\sqrt{13}}{2}$

$\tanh x = \dfrac{3\sqrt{13}}{13}$

$\operatorname{csch} x = \dfrac{2}{3}$

$\operatorname{sech} x = \dfrac{2\sqrt{13}}{13}$

$\coth x = \dfrac{\sqrt{13}}{3}$

15. $-2x\cosh(1 - x^2)$ **17.** $\coth x$

19. $\operatorname{csch} x$ **21.** $\sinh^2 x$ **23.** $\operatorname{sech} t$

25. $\dfrac{y}{x}[\cosh x + x(\sinh x)\ln x] = \dfrac{x^{\cosh x}}{x}[\cosh x + x(\sinh x)\ln x]$

27. $-2(\cosh x - \sinh x)^2 = -2e^{-2x}$

29. Relative maxima: $(\pm\pi, \cosh\pi)$
Relative minimum: $(0, -1)$

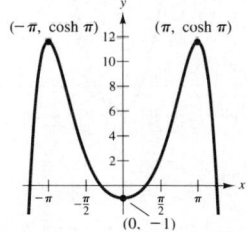

31. Relative maximum: $(1.20, 0.66)$
Relative minimum: $(-1.20, -0.66)$

35. $-\frac{1}{2}\cosh(1 - 2x) + C$

37. $\frac{1}{3}\cosh^3(x - 1) + C$

39. $\ln|\sinh x| + C$ **41.** $-\coth\dfrac{x^2}{2} + C$

43. $\operatorname{csch}\dfrac{1}{x} + C$ **45.** $\dfrac{1}{5}\ln 3$ **47.** $\dfrac{\pi}{4}$

49. $\frac{1}{2}\arctan x^2 + C$

51. $\dfrac{3}{\sqrt{9x^2 - 1}}$ **53.** $|\sec x|$

55. $2\sec 2x$ **57.** $2\sinh^{-1}(2x)$ **59.** $-\dfrac{\sqrt{a^2 - x^2}}{x}$

61. $-\operatorname{csch}^{-1}(e^x) + C = -\ln\left(\dfrac{1 + \sqrt{1 + e^{2x}}}{e^x}\right) + C$

63. $2\sinh^{-1}\sqrt{x} + C = 2\ln\left(\sqrt{x} + \sqrt{1 + x}\right) + C$

65. $-\ln\left[\dfrac{1 + \sqrt{(x - 1)^2 + 1}}{|x - 1|}\right] + C$

67. $\dfrac{1}{2\sqrt{6}}\ln\left|\dfrac{\sqrt{2}(x + 1) + \sqrt{3}}{\sqrt{2}(x + 1) - \sqrt{3}}\right| + C$

69. $\dfrac{1}{4}\arcsin\left(\dfrac{4x - 1}{9}\right) + C$

71. $-\dfrac{x^2}{2} - 4x - \dfrac{10}{3}\ln\left|\dfrac{x - 5}{x + 1}\right| + C$

73. $8\arctan(e^2) - 2\pi \approx 5.207$

75. $\frac{5}{2}\ln(\sqrt{17} + 4) \approx 5.237$ **77.** $\frac{52}{31}$ kg

79. If k is increased the time of descent will increase.

Review Exercises for Chapter 5

1. Vertical asymptote: $x = 0$

3. $\frac{1}{5}[\ln(2x + 1) + \ln(2x - 1) - \ln(4x^2 + 1)]$

5. $\ln\left(\dfrac{3\sqrt[3]{4 - x^2}}{x}\right)$

7. False, the domain of $f(x) = \ln x$ is the set of all positive real numbers.

8. False, $\ln x + \ln y = \ln xy$.   **9.** $e^4 - 1 \approx 53.598$

11. $\dfrac{1}{2x}$ **13.** $\dfrac{1 + 2\ln x}{2\sqrt{\ln x}}$ **15.** $\dfrac{x}{(a + bx)^2}$

17. $\dfrac{1}{x(a + bx)}$ **19.** $\dfrac{1}{7}\ln|7x - 2| + C$

21. $-\ln|1 + \cos x| + C$ **23.** $3 + \ln 4$

25. $\ln(2 + \sqrt{3})$ **27.** $\dfrac{x^2}{2} + 3\ln|x| + C$

29. $f^{-1}(x) = 2x + 6$ **31.** $f^{-1}(x) = x^2 - 1, \quad x \geq 0$

33. $f^{-1}(x) = \sqrt{x + 5}$ **35.** $f^{-1}(x) = e^{2x}$

37. $f^{-1}(x) = \sqrt{\dfrac{x}{2} + 4}$ **39.** $-2x$

41. $te^t(t + 2)$

43. $\dfrac{e^{2x} - e^{-2x}}{\sqrt{e^{2x} + e^{-2x}}}$ **45.** $3^{x-1} \ln 3$

47. $\dfrac{x(2 - x)}{e^x}$ **49.** $(1 - x^2)^{-3/2}$

51. $\dfrac{x}{|x|\sqrt{x^2 - 1}} + \operatorname{arcsec} x$ **53.** $(\arcsin x)^2$

55. $2 - \dfrac{\sinh \sqrt{x}}{2\sqrt{x}}$ **57.** $\dfrac{-y}{x(2y + \ln x)}$

59. $-\dfrac{2x \sin x^2 + e^y}{xe^y}$

61. a. ax^{a-1} **b.** $(\ln a)a^x$
  c. $x^x(1 + \ln x)$ **d.** 0

63. $-\dfrac{1}{6}e^{-3x^2} + C$

65. $\dfrac{e^{4x} - 3e^{2x} - 3}{3e^x} + C$

67. $\ln|e^x - 1| + C$ **69.** $\dfrac{1}{2}\arctan(e^{2x}) + C$
71. $\dfrac{1}{2}\arcsin x^2 + C$ **73.** $\dfrac{1}{2}\ln(16 + x^2) + C$
75. $\dfrac{1}{4}\left(\arctan \dfrac{x}{2}\right)^2 + C$ **77.** $\dfrac{1}{2}\ln\left(\sqrt{x^4 - 1} + x^2\right) + C$
79. $y = -e^{-t^2/2} + C$ **81.** $y = -\dfrac{1}{2}\ln(1 + e^{-2x}) + C$
83. $A = \ln 2$

85. $-\dfrac{1}{2}(e^{-16} - 1) \approx 0.500$ **87.** 1.346
89. a. $\$525.64$ **b.** $\$824.36$ **c.** $\$74,206.58$
91. $\$3499.38$ **93.** $A = 500e^{-0.0128t}$
95. a. -24.26% **b.** -14.72%
97. a. 0.3935 **b.** 0.7769
  c. 0.2492 **d.** 0.9502
99. a. $v(t) = \dfrac{1}{k}[32 + (kv_0 - 32)e^{kt}],\ k < 0$

  b. $\dfrac{32}{k}$

  c. $s(t) = \dfrac{32}{k}t - \dfrac{1}{k^2}(kv_0 - 32)(1 - e^{kt}) + s_0$

101. a. 157 words/min
  b. $t = 5$: 3.55 words/min/week
    $t = 14$: 4.71 words/min/week

CHAPTER 6
Section 6.1 (*page 409*)

1. $A = 36$ **3.** $A = 9$ **5.** $A = \dfrac{3}{2}$

11. d
13. $A = \dfrac{32}{3}$ **15.** $A = \dfrac{9}{2}$

     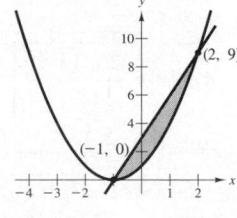

17. $A = 1$ **19.** $A = \dfrac{37}{12}$

21. $A = \dfrac{64}{3}$ **23.** $A = \dfrac{3}{2}$

     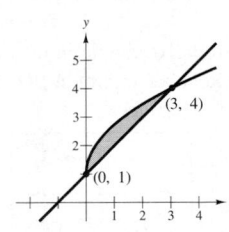

25. $A = \dfrac{\pi}{2} - \dfrac{1}{3} \approx 1.237$ **27.** $A = \dfrac{9}{2}$

29. $A = 6$

31. $A = 8\ln 2 \approx 5.545$

33. $A \approx 10.612$

35. $A = 2(1 - \ln 2) \approx 0.614$

37. $A = 4$

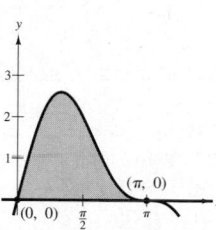

39. $A = \dfrac{1}{2}\left(1 - \dfrac{1}{e}\right) \approx 0.316$

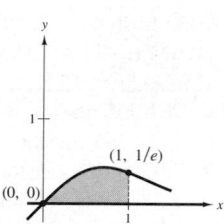

41. $A = 2(\ln 5)^2 \approx 5.181$

43. $A = \frac{1}{2}ac$

45. $b = 9\left(1 - \dfrac{1}{\sqrt[3]{4}}\right) \approx 3.330$

47. $x^4 - 2x^2 + 1 \le 1 - x^2$ on $[-1, 1]$
$$A = \int_{-1}^{1} \left[(1 - x^2) - (x^4 - 2x^2 + 1)\right] dx = \frac{4}{15}$$

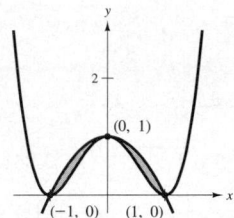

49. $A = \displaystyle\int_{-2}^{1} \left[x^3 - (3x - 2)\right] dx = \dfrac{27}{4}$

51. 1.7587   **53.** 4.7731

55. $\frac{1}{6}$ **57.** \$1.625 billion

59. a.

b. 8.9583 billion

61. a. 6.031 m²   **b.** 12.062 m³   **c.** 60,310 lb

63. Consumer surplus = 1600
Producer surplus = 400

65. Consumer surplus = 50,000
Producer surplus = 25,497

67. True   **68.** True   **69.** $|c - d|$

Section 6.2 (*page 420*)

1. $\dfrac{\pi}{3}$ **3.** $\dfrac{16\pi}{3}$ **5.** $\dfrac{15\pi}{2}$ **7.** $\dfrac{2\pi}{35}$ **9.** 8π

11. $\dfrac{\pi}{4}$ **13. a.** 8π **b.** $\dfrac{128\pi}{5}$ **c.** $\dfrac{256\pi}{15}$ **d.** $\dfrac{192\pi}{5}$

15. a. $\dfrac{32\pi}{3}$ **b.** $\dfrac{64\pi}{3}$ **17.** 18π

19. $\pi(8\ln 4 - \frac{3}{4}) \approx 32.49$

21. $\dfrac{208\pi}{3}$ **23.** $\dfrac{384\pi}{5}$ **25.** $\pi\ln 4$ **27.** $\dfrac{3\pi}{4}$

29. $\dfrac{\pi}{2}\left(1 - \dfrac{1}{e^2}\right) \approx 1.358$ **31.** π **33.** a

35. a. $\dfrac{512\pi}{15}$

b. No, the solid has only been translated horizontally.

37. 18π **41.** $\pi r^2 h\left(1 - \dfrac{h}{H} + \dfrac{h^2}{3H^2}\right)$ **43.** $\dfrac{\pi}{30}$

45. One-fourth: 32.64 ft
Three-fourths: 82.64 ft

47. 1.969   **49.** 49.023

51. a. $\dfrac{128}{3}$ **b.** $\dfrac{32\sqrt{3}}{3}$ **c.** $\dfrac{16\pi}{3}$ **d.** $\dfrac{32}{3}$

53. a. $\dfrac{1}{10}$ **b.** $\dfrac{\pi}{80}$ **c.** $\dfrac{\sqrt{3}}{40}$ **d.** $\dfrac{3}{80}$ **e.** $\dfrac{\pi}{20}$

55. a. $\dfrac{2r^3}{3}$ **b.** $\dfrac{2r^3 \tan\theta_0}{3}$ **59.** $5\sqrt{1 - 2^{-2/3}} \approx 3.0415$

Section 6.3   (*page 429*)

1. $\dfrac{16\pi}{3}$ **3.** $\dfrac{128\pi}{5}$ **5.** 8π **7.** $\dfrac{16\pi}{3}$

9. $\dfrac{8\pi}{3}$ **11.** $\sqrt{2\pi}\left(1 - \dfrac{1}{\sqrt{e}}\right) \approx 0.986$

13. $\dfrac{8\pi}{3}$ **15.** $\dfrac{\pi}{2}$ **17.** 16π **19.** 64π

21. a. $\dfrac{128\pi}{7}$ **b.** $\dfrac{64\pi}{5}$ **c.** $\dfrac{96\pi}{5}$ **d.** $\dfrac{320\pi}{7}$

23. a. $\dfrac{\pi a^3}{15}$ **b.** $\dfrac{\pi a^3}{15}$ **c.** $\dfrac{4\pi a^3}{15}$ **d.** $\dfrac{4\pi a^3}{15}$ **25.** d

27. Diameter $= 2\sqrt{4 - 2\sqrt{3}} \approx 1.464$ **29.** $\dfrac{\pi h^3}{6}$

31. $2\pi^2 r^2 R$

33. Both integrals yield the volume of the solid generated by revolving the region bounded by the graphs of $y = \sqrt{x - 1}$, $y = 0$, and $x = 5$ about the x-axis.

35. 1.5016   **37.** 186.0552   **39.** 122,313 ft^3

41. a. $2\pi \displaystyle\int_0^r hx\left(1 - \dfrac{x}{r}\right)dx$

   b. $2\pi \displaystyle\int_{-r}^r (R - x)\left(2\sqrt{r^2 - x^2}\right)dx$

   c. $2\pi \displaystyle\int_0^r 2x\sqrt{r^2 - x^2}\,dx$ **d.** $2\pi \displaystyle\int_0^r hx\,dx$

   e. $2\pi \displaystyle\int_0^b 2ax\sqrt{1 - \dfrac{x^2}{b^2}}\,dx$

Section 6.4   (*page 439*)

1. 13   **3.** $\dfrac{2}{3}\left(\sqrt{8} - 1\right) \approx 1.219$

5. $5\sqrt{5} - 2\sqrt{2} \approx 8.352$ **7.** $\dfrac{33}{16}$

9. $s = \displaystyle\int_{-2}^{1} \sqrt{2 + 4x + 4x^2}\,dx$

11. $s = \displaystyle\int_0^2 \sqrt{1 + 4x^2}\,dx$

13. $s = \displaystyle\int_0^2 \dfrac{\sqrt{e^{2y} + 1}}{e^y}\,dy$ **15.** b

17. a. 64.125   **b.** 64.525   **c.** 64.666

19. 2.1468   **21.** 3.8202   **23.** $\dfrac{2}{3}$

25. $60[\sinh 1 - \sinh(-1)] \approx 141$ ft

27. $3 \arcsin\dfrac{2}{3} \approx 2.1892$

29. $\dfrac{\pi}{9}\left(82\sqrt{82} - 1\right) \approx 258.85$ **31.** $\dfrac{47\pi}{16}$

33. $\dfrac{\pi}{27}\left(145\sqrt{145} - 10\sqrt{10}\right) \approx 199.48$ **35.** 14.4260

39. $6\pi(3 - \sqrt{5}) \approx 14.40$

41. Surface area $= \dfrac{\pi}{27}$ ft$^2 \approx 16.8$ in.^{2}

   Amount of glass $= \dfrac{\pi}{27}\left(\dfrac{0.015}{12}\right)$
               ≈ 0.00015 ft^3
               ≈ 0.25 in.^{3}

49. a. $\dfrac{ds}{dx} = \sqrt{1 + [f'(x)]^2}$

   b. $ds = \sqrt{1 + [f'(x)]^2}\,dx$
     $(ds)^2 = \left[1 + \left(\dfrac{dy}{dx}\right)^2\right]dx = (dx)^2 + (dy)^2$

   c. $s(x) = \displaystyle\int_1^x \sqrt{1 + \dfrac{9}{4}t}\,dt$
     $s(2) = \dfrac{22}{27}\sqrt{22} - \dfrac{13}{27}\sqrt{13}$

Section 6.5   (*page 448*)

1. 1000 ft·lb   **3.** 300 ft·lb

5. 30.625 in·lb ≈ 2.55 ft·lb   **7.** 360 in·lb $= 30$ ft·lb

9. 180 in·lb $= 15$ ft·lb

11. a. 761.905 mi·ton $\approx 8.05 \times 10^9$ ft·lb
   b. 1454.545 mi·ton $\approx 1.54 \times 10^{10}$ ft·lb

13. a. 2.93×10^4 mi·ton $\approx 3.10 \times 10^{11}$ ft·lb
   b. 3.38×10^4 mi·ton $\approx 3.57 \times 10^{11}$ ft·lb

15. a. 2496 ft·lb   **b.** 9984 ft·lb

17. 431,308.8π ft·lb   **19.** 20,217.6π ft·lb

21. 2995.2π ft·lb   **23.** 2457π ft·lb   **25.** 337.5 ft·lb

27. 300 ft·lb   **29.** 168.75 ft·lb   **31.** 7987.5 ft·lb

33. $2000 \ln\dfrac{3}{2} \approx 810.93$ ft·lb   **35.** $\dfrac{3k}{4}$ **37.** 3250 ft·lb

39. 10,203 ft·lb

Section 6.6   (*page 456*)

1. 936 lb   **3.** 748.8 lb   **5.** 1123.2 lb

7. 748.8 lb   **9.** 1064.96 lb   **11.** 748.8 lb

13. 15,163.2 lb   **15.** 2814 lb   **17.** 3376.8 lb

19. 94.5 lb   **23.** 960 lb

25. 9984 lb (on the wall at the shallow end)
   39,936 lb (on the wall at the deep end)
   46,592 lb (on the side walls)

27. 6483 lb   **29.** 536.4 lb

Section 6.7   (*page 466*)

1. $\bar{x} = -\dfrac{6}{7}$ **3.** $\bar{x} = 12$ **5.** $\bar{x} = 17$

7. $(\bar{x}, \bar{y}) = \left(\dfrac{10}{9}, -\dfrac{1}{9}\right)$ **9.** $(\bar{x}, \bar{y}) = \left(-\dfrac{7}{8}, -\dfrac{7}{16}\right)$

11. $M_x = 4\rho$, $M_y = \dfrac{64\rho}{5}$

$(\bar{x}, \bar{y}) = \left(\dfrac{12}{5}, \dfrac{3}{4}\right)$

13. $M_x = \dfrac{\rho}{12}$, $M_y = \dfrac{\rho}{15}$ 15. $M_x = \dfrac{99\rho}{5}$, $M_y = \dfrac{27\rho}{4}$

$(\bar{x}, \bar{y}) = \left(\dfrac{2}{5}, \dfrac{1}{2}\right)$ $(\bar{x}, \bar{y}) = \left(\dfrac{3}{2}, \dfrac{22}{5}\right)$

17. $M_x = \dfrac{192\rho}{7}$, $M_y = 96\rho$ 19. $M_x = 0$, $M_y = \dfrac{256\rho}{15}$

$(\bar{x}, \bar{y}) = \left(5, \dfrac{10}{7}\right)$ $(\bar{x}, \bar{y}) = \left(\dfrac{8}{5}, 0\right)$

21. $M_x = \dfrac{27\rho}{4}$, $M_y = -\dfrac{27\rho}{10}$ 23. $(\bar{x}, \bar{y}) = \left(\dfrac{1}{2}, \dfrac{2}{5}\right)$

$(\bar{x}, \bar{y}) = \left(-\dfrac{3}{5}, \dfrac{3}{2}\right)$

25. $(\bar{x}, \bar{y}) = \left(\dfrac{12}{7}, \dfrac{26}{7}\right)$ 27. $(\bar{x}, \bar{y}) = \left(\dfrac{b}{3}, \dfrac{c}{3}\right)$

29. $(\bar{x}, \bar{y}) = \left(\dfrac{(a + 2b)c}{3(a + b)}, \dfrac{a^2 + ab + b^2}{3(a + b)}\right)$

31. $(\bar{x}, \bar{y}) = \left(0, \dfrac{4b}{3\pi}\right)$ 33. $(\bar{x}, \bar{y}) = (3.0, 127.6)$

35. $(\bar{x}, \bar{y}) = (0, 16.5)$ 37. $(\bar{x}, \bar{y}) = (0, 12.98)$

39. $(\bar{x}, \bar{y}) = \left(\dfrac{4 + 3\pi}{4 + \pi}, 0\right)$ 41. $(\bar{x}, \bar{y}) = \left(0, \dfrac{135}{34}\right)$

43. $(\bar{x}, \bar{y}) = \left(\dfrac{2 + 3\pi}{2 + \pi}, 0\right)$ 45. $V = 160\pi^2 \approx 1579.14$

47. $V = \dfrac{128\pi}{3} \approx 134.04$ 49. $(\bar{x}, \bar{y}) = \left(0, \dfrac{2r}{\pi}\right)$

51. $(\bar{x}, \bar{y}) = \left(\dfrac{n + 1}{n + 2}, \dfrac{n + 1}{4n + 2}\right)$

As $n \to \infty$, $(\bar{x}, \bar{y}) \to \left(1, \dfrac{1}{4}\right)$

Review Exercises for Chapter 6

1. $A = \dfrac{4}{5}$ 3. $A = \dfrac{\pi}{2}$

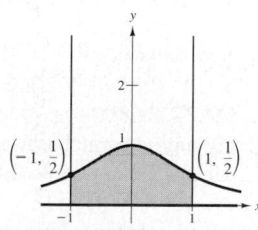

5. $A = \dfrac{1}{2}$ 7. $A = \dfrac{512}{3}$

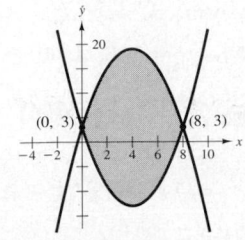

9. $A = \dfrac{1}{6}$ 11. $A = e^2 + 1$

13. $A = 2\sqrt{2}$

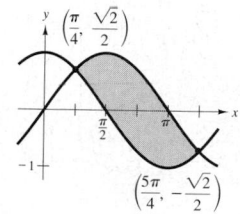

15. $A = \displaystyle\int_0^2 [0 - (y^2 - 2y)]\,dy$

$= \displaystyle\int_{-1}^0 2\sqrt{x + 1}\,dx = \dfrac{4}{3}$

17. $A = \displaystyle\int_0^2 (y^2 + 1)\,dy$

$= \displaystyle\int_0^1 2\,dx + \int_1^5 \left(2 - \sqrt{x - 1}\right) dx = \dfrac{14}{3}$

19. a. $\dfrac{64\pi}{3}$ b. $\dfrac{128\pi}{3}$ c. $\dfrac{64\pi}{3}$ d. $\dfrac{160\pi}{3}$

21. a. 64π b. 48π

23. $\dfrac{\pi^2}{4}$ 25. $\dfrac{4\pi}{3}(20 - 9\ln 3) \approx 42.359$ 27. $\dfrac{4}{15}$

29. 1.958 ft   31. $\dfrac{8}{15}\left(1 + 6\sqrt{3}\right) \approx 6.076$ 33. 15π

35. 50 in · lb $\approx$ 4.167 ft · lb

37. $104{,}000\pi$ ft · lb $\approx$ 163.4 ft · ton   39. 250 ft · lb

41. 72,800 lb (on side walls)
    62,400 lb (on wall at deep end)
    15,500 lb (on wall at shallow end)

43. 4992π lb   45. $(\bar{x}, \bar{y}) = \left(\dfrac{a}{5}, \dfrac{a}{5}\right)$

47. $(\bar{x}, \bar{y}) = \left(0, \dfrac{2a^2}{5}\right)$ 49. $(\bar{x}, \bar{y}) = \left(\dfrac{2(9\pi + 49)}{3(\pi + 9)}, 0\right)$

51. 4018.2 ft

CHAPTER 7

Section 7.1 (page 476)

1. b   3. c

5. $\displaystyle\int u^n\,du$ 7. $\displaystyle\int \dfrac{du}{u}$

$u = 3x - 2,\ n = 4$ $u = 1 - 2\sqrt{x}$

9. $\displaystyle\int \frac{du}{\sqrt{a^2 - u^2}}$ **11.** $\displaystyle\int \sin u \, du$

$u = t, \ a = 1$ $\qquad u = t^2$

13. $\displaystyle\int e^u \, du$ **15.** $-\dfrac{1}{5}(-2x + 5)^{5/2} + C$

$u = \sin x$

17. $\dfrac{1}{2}v^2 - \dfrac{1}{6(3v - 1)^2} + C$

19. $-\dfrac{1}{3}\ln|-t^3 + 9t + 1| + C$

21. $\dfrac{1}{2}x^2 + x + \ln|x - 1| + C$ **23.** $\dfrac{1}{3}\ln\left|\dfrac{3x - 1}{3x + 1}\right| + C$

25. $\dfrac{x}{15}(12x^4 + 20x^2 + 15) + C$ **27.** $\dfrac{1}{4\pi}\sin 2\pi x^2 + C$

29. $-\dfrac{1}{\pi}\csc \pi x + C$ **31.** $\dfrac{1}{5}e^{5x} + C$

33. $2\ln(1 + e^x) + C$ **35.** $\ln|\sec x(\sec x + \tan x)| + C$

37. $\ln(t^2 + 4) - \dfrac{1}{2}\arctan\dfrac{t}{2} + C$

39. $-\dfrac{1}{2}\arcsin(2t - 1) + C$ **41.** $\dfrac{1}{2}\ln\left|\cos\dfrac{2}{t}\right| + C$

43. $3\arcsin\dfrac{x - 3}{3} + C$ **45.** $\dfrac{1}{4}\arctan\dfrac{2x + 1}{8} + C$

47. $y = \dfrac{1}{2}e^{2x} + 2e^x + x + C$ **49.** $s = \dfrac{1}{2}\arcsin t^2 + C$

51. $y = \dfrac{1}{2}\arctan\dfrac{\tan x}{2} + C$ **53.** $\dfrac{1}{2}$

55. $\dfrac{1}{2}(1 - e^{-1}) \approx 0.316$ **57.** 4   **59.** $\dfrac{\pi}{18}$

61. $\dfrac{1}{3}\arctan\left(\dfrac{x + 2}{3}\right) + C$ **63.** $2 - \sqrt{2}$ **65.** a   **67.** $\dfrac{4}{3}$

69. $a = \dfrac{1}{2}$

71. a. $\pi(1 - e^{-1}) \approx 1.986$

b. $b = \sqrt{\ln\left(\dfrac{3\pi}{3\pi - 4}\right)} \approx 0.743$

73. $\dfrac{2}{\arcsin(4/5)} \approx 2.157$ **75.** 1.0320   **77.** True

78. False, if $u = \sin x$, then $du = \cos x \, dx$.

79. False, $\displaystyle\int u^n \, du$, $n = -\dfrac{1}{2}$.   **80.** True

81. $a = \sqrt{2}, \ b = \dfrac{\pi}{4}$

$-\dfrac{1}{\sqrt{2}}\ln\left|\csc\left(x + \dfrac{\pi}{4}\right) + \cot\left(x + \dfrac{\pi}{4}\right)\right| + C$

Section 7.2   (page 486)

1. a. $x \sin x$ **b.** $x^2 \cos x$ **c.** $x^2 e^x$ **d.** $\ln x$

3. $u = x, \ dv = e^{2x} \, dx$ **5.** $u = (\ln x)^2, \ dv = dx$

7. $u = x, \ dv = \sec^2 x \, dx$ **9.** $-\dfrac{1}{4e^{2x}}(2x + 1) + C$

11. $e^x(x^3 - 3x^2 + 6x - 6) + C$ **13.** $\dfrac{1}{3}e^{x^3} + C$

15. $\dfrac{1}{4}[2(t^2 - 1)\ln|t + 1| - t^2 + 2t] + C$

17. $\dfrac{(\ln x)^3}{3} + C$ **19.** $\dfrac{e^{2x}}{4(2x + 1)} + C$

21. $(x - 1)^2 e^x + C$ **23.** $\dfrac{2(x - 1)^{3/2}}{15}(3x + 2) + C$

25. $x \sin x + \cos x + C$

27. $x \arctan x - \dfrac{1}{2}\ln(1 + x^2) + C$

29. $\dfrac{1}{5}e^{2x}(2\sin x - \cos x) + C$ **31.** $y = \dfrac{1}{2}e^{x^2} + C$

33. $y = \dfrac{2}{405}(27t^2 - 24t + 32)\sqrt{2 + 3t} + C$

35. $\sin y = x^2 + C$ **37.** $-\dfrac{\pi}{2}$

39. $\dfrac{e[\sin(1) - \cos(1)] + 1}{2} \approx 0.909$

41. $\dfrac{\pi}{2} - 1$ **43.** $\dfrac{e^{2x}}{4}(2x^2 - 2x + 1) + C$

45. $(3x^2 - 6)\sin x - (x^3 - 6x)\cos x + C$

47. $x \tan x + \ln|\cos x| + C$

49. $-\dfrac{e^{-4t}}{128}(32t^3 + 24t^2 + 12t + 3) + C$

51. $\dfrac{1}{13}(2e^{-\pi} + 3) \approx 0.2374$ **53.** $\dfrac{2}{5}(2x - 3)^{3/2}(x + 1) + C$

55. $\dfrac{1}{3}\sqrt{4 + x^2}(x^2 - 8) + C$

57. $n = 0$: $x(\ln x - 1) + C$ **65.** $\dfrac{x^4}{16}(4\ln x - 1) + C$

$n = 1$: $\dfrac{x^2}{4}(2\ln x - 1) + C$

$n = 2$: $\dfrac{x^3}{9}(3\ln x - 1) + C$

$n = 3$: $\dfrac{x^4}{16}(4\ln x - 1) + C$

$n = 4$: $\dfrac{x^5}{25}(5\ln x - 1) + C$

$\displaystyle\int x^n \ln x \, dx = \dfrac{x^{n+1}}{(n + 1)^2}[(n + 1)\ln x - 1] + C$

67. $\dfrac{e^{2x}}{13}(2\cos 3x + 3\sin 3x) + C$ **69.** $1 - \dfrac{5}{e^4} \approx 0.908$

71. $\dfrac{\pi}{1 + \pi^2}\left(\dfrac{1}{e} + 1\right) \approx 0.395$

73. a. 1

b. $\pi(e - 2) \approx 2.257$

c. $\dfrac{(e^2 + 1)\pi}{2} \approx 13.177$

d. $\left(\dfrac{e^2 + 1}{4}, \dfrac{e - 2}{2}\right) \approx (2.097, 0.359)$

75. a. $3.2(\ln 2) - 0.2 \approx 2.018$

b. $12.8(\ln 4) - 7.2(\ln 3) - 1.8 \approx 8.035$

77. \$131,528.68

81. For any integratable function, $\int f(x) \, dx = C + \int f(x) \, dx$, but this cannot be used to imply that $C = 0$.

83. $s(t) = -16t^2 + 12{,}000t\left[1 + \ln\dfrac{50{,}000}{50{,}000 - 400t}\right]$

$+ 1{,}500{,}000 \ln\dfrac{50{,}000 - 400t}{50{,}000}$

When $t = 100$, $s(100) \approx 557{,}168.626$ feet.

Section 7.3   (page 496)

1. $-\dfrac{1}{4}\cos^4 x + C$ **3.** $\dfrac{1}{12}\sin^6 2x + C$

5. $-\dfrac{1}{3}\cos^3 x + \dfrac{2}{5}\cos^5 x - \dfrac{1}{7}\cos^7 x + C$

7. $\dfrac{1}{12}(6x + \sin 6x) + C$

9. $\dfrac{1}{8}(2x^2 - 2x \sin 2x - \cos 2x) + C$

15. $\frac{1}{3}\ln|\sec 3x + \tan 3x| + C$

17. $\frac{1}{15}\tan 5x(3 + \tan^2 5x) + C$

19. $\frac{1}{2\pi}(\sec \pi x \tan \pi x + \ln|\sec \pi x + \tan \pi x|) + C$

21. $\tan^4\left(\frac{x}{4}\right) - 2\tan^2\left(\frac{x}{4}\right) - 4\ln\left|\cos\frac{x}{4}\right| + C$

23. $\frac{1}{2}\tan^2 x + C$ **25.** $\frac{\tan^3 x}{3} + C$ **27.** $\frac{\sec^6 4x}{24} + C$

29. $\frac{1}{3}\sec^3 x + C$

31. $r = \frac{1}{32\pi}(12\pi\theta - 8\sin 2\pi\theta + \sin 4\pi\theta) + C$

33. $y = \frac{1}{9}\sec^3 3x - \frac{1}{3}\sec 3x + C$

35. $-\frac{1}{10}(\cos 5x + 5\cos x) + C$

37. $\frac{1}{8}(2\sin 2\theta - \sin 4\theta) + C$

39. $\frac{1}{4}(\ln|\csc^2 2x| - \cot^2 2x) + C$

41. $-\cot\theta - \frac{1}{3}\cot^3\theta + C$

43. $\ln|\csc t - \cot t| + \cos t + C$

45. $\ln|\csc x - \cot x| + \cos x + C$ **47.** $t - 2\tan t + C$

49. π **51.** $\frac{1}{2}(1 - \ln 2)$ **53.** $\ln 2$ **55.** $\frac{4}{3}$

57. $\frac{1}{16}(6x + 8\sin x + \sin 2x) + C$

59. $\frac{1}{4\pi}[\sec^3 \pi x \tan \pi x$

$+ \frac{3}{2}(\sec \pi x \tan \pi x + \ln|\sec \pi x + \tan \pi x|)] + C$

61. $\frac{1}{5\pi}\sec^5 \pi x + C$ **63.** $\frac{3\sqrt{2}}{10}$ **65.** $\frac{3\pi}{16}$

67. $\frac{\tan^6 3x}{18} + \frac{\tan^4 3x}{12} + C_1$

$= \frac{\sec^6 3x}{18} - \frac{\sec^4 3x}{12} + C_2$

69. $\frac{1}{2}$ **71. a.** $\frac{\pi^2}{2}$ **b.** $(\bar{x}, \bar{y}) = \left(\frac{\pi}{2}, \frac{\pi}{8}\right)$

77. $-\frac{1}{15}\cos x(3\sin^4 x + 4\sin^2 x + 8) + C$

79. $\frac{5}{6\pi}\tan\frac{2\pi x}{5}\left(\sec^2\frac{2\pi x}{5} + 2\right) + C$

83. a. $H(t) = 55.46 - 23.88\cos\frac{\pi t}{6} - 3.34\sin\frac{\pi t}{6}$

b. $L(t) = 39.34 - 20.78\cos\frac{\pi t}{6} - 4.33\sin\frac{\pi t}{6}$

c. Summer

85. a. $\frac{2T}{ug}\sinh\frac{ugL}{2T}$ feet

b. $\frac{4}{g}v\sinh\frac{ugL}{2T}$

Section 7.4 (page 506)

1. $\displaystyle\int\frac{\sqrt{x^2 + 16}}{x}dx$ **3.** $\displaystyle\int \arcsin\frac{x}{4}dx$

---

5. $\frac{x}{25\sqrt{25 - x^2}} + C$

7. $5\ln\left|\frac{5 - \sqrt{25 - x^2}}{x}\right| + \sqrt{25 - x^2} + C$

9. $\ln\left|x + \sqrt{x^2 - 4}\right| + C$

11. $\frac{1}{15}(x^2 - 4)^{3/2}(3x^2 + 8) + C$ **13.** $\frac{1}{3}(1 + x^2)^{3/2} + C$

15. $\frac{1}{2}\left(\arctan x + \frac{x}{1 + x^2}\right) + C$

17. $\frac{1}{2}x\sqrt{4 + 9x^2} + \frac{2}{3}\ln\left|3x + \sqrt{4 + 9x^2}\right| + C$

19. $\sqrt{x^2 + 9} + C$ **21.** 2π

23. $\ln\left|x + \sqrt{x^2 - 9}\right| + C$ **25.** $-\frac{(1 - x^2)^{3/2}}{3x^3} + C$

27. $-\frac{1}{3}\ln\left|\frac{\sqrt{4x^2 + 9} + 3}{2x}\right| + C$ **29.** $-\frac{1}{\sqrt{x^2 + 3}} + C$

31. $\frac{1}{3}(1 + e^{2x})^{3/2} + C$

33. $\frac{1}{2}\left(\arcsin e^x + e^x\sqrt{1 - e^{2x}}\right) + C$

35. $\frac{1}{4}\left(\frac{x}{x^2 + 2} + \frac{1}{\sqrt{2}}\arctan\frac{x}{\sqrt{2}}\right) + C$

37. $x\,\text{arcsec}\,2x - \frac{1}{2}\ln\left|2x + \sqrt{4x^2 - 1}\right| + C$

39. $\arcsin\left(\frac{x - 2}{2}\right) + C$

41. $\sqrt{x^2 + 4x + 8} - 2\ln\left|\sqrt{x^2 + 4x + 8} + (x + 2)\right| + C$

43. $\sqrt{3} - \frac{\pi}{3} \approx 0.685$ **45.** $9(2 - \sqrt{2}) \approx 5.272$

47. $\frac{1}{2}(x - 15)\sqrt{x^2 + 10x + 9}$

$+ 33\ln\left|\sqrt{x^2 + 10x + 9} + (x + 5)\right| + C$

49. $\frac{1}{2}(x\sqrt{x^2 - 1} + \ln|x + \sqrt{x^2 - 1}|) + C$

51. πr^2 **53.** $6\pi^2$

55. $\ln\left[\frac{5(\sqrt{2} + 1)}{\sqrt{26} + 1}\right] + \sqrt{26} - \sqrt{2} \approx 4.367$

57. $100\sqrt{2} + 50\ln\left(\frac{\sqrt{2} + 1}{\sqrt{2} - 1}\right) \approx 229.559$

59. 187.2π lb   **61. a.** 121.3 lb   **b.** 92.98 lb

63. $\frac{\pi}{32}\left[102\sqrt{2} - \ln(3 + 2\sqrt{2})\right] \approx 13.989$

65. b. $y = -12\ln\left(\frac{12 - \sqrt{144 - x^2}}{x}\right) - \sqrt{144 - x^2}$

c. $x = 0$ **d.** 5.2 ft

67. True   **68.** False, $\displaystyle\int\frac{\sqrt{x^2 - 1}}{x}dx = \int\tan^2\theta\,d\theta$

69. False, $\displaystyle\int_0^{\sqrt{3}}\frac{dx}{(1 + x^2)^{3/2}} = \int_0^{\pi/3}\cos\theta\,d\theta$.   **70.** True

Section 7.5 (page 517)

1. $\frac{A}{x} + \frac{B}{x - 10}$ **3.** $\frac{A}{x} + \frac{Bx + C}{x^2 + 10}$

5. $\dfrac{A}{x} + \dfrac{B}{x^2} + \dfrac{C}{x-10}$ **7.** $\dfrac{1}{2}\ln\left|\dfrac{x-1}{x+1}\right| + C$

9. $\ln\left|\dfrac{x-1}{x+2}\right| + C$

11. $\dfrac{3}{2}\ln|2x-1| - 2\ln|x+1| + C$

13. $5\ln|x-2| - \ln|x+2| - 3\ln|x| + C$

15. $x^2 + \dfrac{3}{2}\ln|x-4| - \dfrac{1}{2}\ln|x+2| + C$

17. $\dfrac{1}{x} + \ln|x^4 + x^3| + C$ **19.** $\ln\left|\dfrac{x^2+1}{x}\right| + C$

21. $\dfrac{1}{6}\left[\ln\left|\dfrac{x-2}{x+2}\right| + \sqrt{2}\arctan\left(\dfrac{x}{\sqrt{2}}\right)\right] + C$

23. $\dfrac{1}{16}\ln\left|\dfrac{4x^2-1}{4x^2+1}\right| + C$

25. $\ln|x+1| + \sqrt{2}\arctan\left(\dfrac{x-1}{\sqrt{2}}\right) + C$

27. $\dfrac{1}{2}x^2 + 3x + 6\ln|x-1| - \dfrac{4}{x-1} - \dfrac{1}{2(x-1)^2} + C$

29. $4\ln|x-2| + \ln(x^2+4) + \dfrac{1}{2}\arctan\dfrac{x}{2} + C$ **31.** $\ln 2$

33. $\dfrac{1}{2}\ln\left(\dfrac{8}{5}\right) - \dfrac{\pi}{4} + \arctan 2 \approx 0.557$

35. $3\ln|x-3| - \dfrac{9}{x-3} + C$

37. $\dfrac{\sqrt{2}}{2}\arctan\dfrac{x}{\sqrt{2}} - \dfrac{1}{2(x^2+2)} + C$

39. $\ln|x-2| + \dfrac{1}{2}\ln|x^2+x+1|$
$$- \sqrt{3}\arctan\left(\dfrac{2x+1}{\sqrt{3}}\right) + C$$

41. $\ln 2 + \arctan 2 - \arctan 3 \approx 0.551$

43. $\ln\left|\dfrac{\cos x}{\cos x - 1}\right| + C$ **45.** $\ln\left|\dfrac{-1+\sin x}{2+\sin x}\right| + C$

47. $\dfrac{1}{5}\ln\left|\dfrac{e^x-1}{e^x+4}\right| + C$ **53.** c

55. a. $2\pi\left[\arctan 3 - \dfrac{3}{10}\right] \approx 5.963$

b. $(\bar{x}, \bar{y}) \approx (1.521, 0.412)$

57. $x = \dfrac{n[e^{(n+1)kt} - 1]}{n + e^{(n+1)kt}}$ **59.** $\dfrac{\pi}{8}$

61. $\dfrac{1/12}{x} + \dfrac{1/10}{x-1} + \dfrac{111/140}{x+4} + \dfrac{1/42}{x-3}$

Section 7.6 (page 523)

1. $-\dfrac{1}{2}x(2-x) + \ln|1+x| + C$

3. $\dfrac{1}{2}\left[e^x\sqrt{e^{2x}+1} + \ln\left(e^x + \sqrt{e^{2x}+1}\right)\right] + C$

5. $-\dfrac{\sqrt{1-x^2}}{x} + C$

7. $\dfrac{1}{16}(6x - 3\sin 2x\cos 2x - 2\sin^3 2x\cos 2x) + C$

9. $-2\left(\cot\sqrt{x} + \csc\sqrt{x}\right) + C$

11. $x - \dfrac{1}{2}\ln(1+e^{2x}) + C$ **13.** $\dfrac{1}{16}x^4(4\ln x - 1) + C$

15. $\dfrac{1}{2}e^{x^2} + C$

17. $\dfrac{1}{2}\big[(x^2+1)\text{arcsec}(x^2+1)$
$$- \ln\big((x^2+1) + \sqrt{x^4+2x^2}\big)\big] + C$$

19. $\dfrac{1}{9}x^3(-1+3\ln x) + C$ **21.** $\dfrac{\sqrt{x^2-4}}{4x} + C$

23. $\dfrac{2}{9}\left(\ln|1-3x| + \dfrac{1}{1-3x}\right) + C$

25. $e^x\arccos(e^x) - \sqrt{1-e^{2x}} + C$

27. $\dfrac{1}{2}(x^2 + \cot^2 x + \csc^2 x) + C$

29. $\arctan(\sin x) + C$ **31.** $\dfrac{\sqrt{2}}{2}\arctan\left(\dfrac{1+\sin\theta}{\sqrt{2}}\right) + C$

33. $-\dfrac{\sqrt{2+9x^2}}{2x} + C$

35. $(t^4 - 12t^2 + 24)\sin t + (4t^3 - 24t)\cos t + C$

37. $\dfrac{1}{4}(2\ln|x| - 3\ln|3 + 2\ln|x||) + C$

39. $\dfrac{3x-10}{2(x^2-6x+10)} + \dfrac{3}{2}\arctan(x-3) + C$

41. $\dfrac{1}{2}\ln|x^2 - 3 + \sqrt{x^4 - 6x^2 + 5}| + C$

43. $-\dfrac{1}{3}\sqrt{4-x^2}(x^2+8) + C$

45. $\dfrac{2}{1+e^x} - \dfrac{1}{2(1+e^x)^2} + \ln(1+e^x) + C$

53. $\dfrac{1}{\sqrt{5}}\ln\left|\dfrac{2\tan(\theta/2) - 3 - \sqrt{5}}{2\tan(\theta/2) - 3 + \sqrt{5}}\right| + C$

55. $\ln 2$ **57.** $\dfrac{1}{2}\ln(3 - 2\cos\theta) + C$

59. $2\sin\sqrt{\theta} + C$ **61.** $-\dfrac{2\sqrt{1-x}}{\sqrt{x}} + C$

63. $\sqrt{2-2x-x^2} - \sqrt{3}\ln\left|\dfrac{\sqrt{3} + \sqrt{2-2x-x^2}}{x+1}\right| + C$

65. $-\csc\theta + C$ **67.** $\dfrac{40}{3}$ **69.** $1919.145 \text{ ft}\cdot\text{lb}$

71. a. $V = 80\ln\left(\sqrt{10} + 3\right) \approx 145.5 \text{ ft}^3$
$W = 11{,}840\ln\left(\sqrt{10} + 3\right) \approx 21{,}530.4 \text{ lb}$

b. $(0, 1.19)$

Section 7.7 (page 532)

1. $\dfrac{1}{3}$ **3.** $\dfrac{1}{4}$ **5.** $\dfrac{5}{3}$ **7.** 3   **9.** 0   **11.** 2

13. $n = 1: 0$ **15.** $\dfrac{2}{3}$ **17.** 1   **19.** $\dfrac{3}{2}$ **21.** ∞ **23.** 1
$n = 2: \dfrac{1}{2}$
$n \geq 3: \infty$

25. 0   **27.** $0\cdot\infty$, 0   **29.** $0\cdot\infty$, 1   **31.** Not indeterminate, 0

33. ∞^0, 1   **35.** 1^∞, e   **37.** $\infty - \infty$, $-\dfrac{3}{2}$ **39.** $\infty - \infty$, ∞

41. a. $f(x) = x^2 - 25$, $g(x) = x - 5$
b. $f(x) = (x-5)^2$, $g(x) = x^2 - 25$
c. $f(x) = x^2 - 25$, $g(x) = (x-5)^3$

43. 0   **45.** 0   **47.** 0

49.

x	10	10^2	10^4	10^6	10^8	10^{10}
$\dfrac{(\ln x)^4}{x}$	2.811	4.498	0.720	0.036	0.001	0.000

53. Horizontal asymptote: $y = 1$
Relative maximum: $(e, e^{1/e})$

55. Horizontal asymptote: $y = 0$

Relative minimum: $\left(1, \dfrac{2}{e}\right)$

57. Limit is not of the form $0/0$ or ∞/∞.

59. Limit is not of the form $0/0$ or ∞/∞.

61. 1   **63.** $v = 32t + v_0$ **65.** $\frac{3}{4}$

67. a.

b. $\frac{1}{2}$

69. a.

b. $\frac{5}{2}$

71. $c = \dfrac{2}{3}$ **73.** $c = \dfrac{\pi}{4}$

75. False, L'Hôpital's Rule does not apply since $\lim\limits_{x \to 0} (x^2 + x + 1) \neq 0$.

76. False, $y' = \dfrac{e^x(x - 2)}{x^3}$.   **77.** True

78. False, let $f(x) = x$ and $g(x) = x + 1$.

79. a. $\frac{1}{2}\sin\theta - \frac{1}{2}\sin\theta\cos\theta$ **c.** $\dfrac{\sin\theta - \sin\theta\cos\theta}{\theta - \sin\theta\cos\theta}$

  b. $\frac{1}{2}\theta - \frac{1}{2}\sin\theta\cos\theta$ **d.** $\frac{3}{4}$

Section 7.8 (*page 543*)

1. Infinite discontinuity at $x = 0$

  4

3. Infinite discontinuity at $x = 1$

  Diverges

5. Infinite limit of integration   **7.** Diverges   **9.** 2

11. 1   **13.** $\frac{1}{2}$ **15.** π **17.** $\dfrac{\pi}{4}$ **19.** Diverges

21. Diverges   **23.** 6   **25.** $-\frac{1}{4}$ **27.** Diverges

29. $\ln(2 + \sqrt{3})$ **31.** 0   **33.** $p > 1$ **37.** Diverges

39. Converges   **41.** Converges   **43.** Diverges

45. Converges   **47.** $\dfrac{1}{s}$ **49.** $\dfrac{2}{s^3}$ **51.** $\dfrac{s}{s^2 + a^2}$

53. $\dfrac{s}{s^2 - a^2}$ **55. a.** 1   **b.** $\dfrac{\pi}{3}$ **c.** Diverges

57. 6

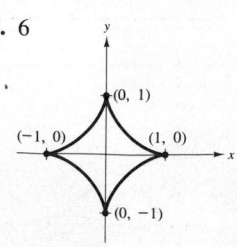

59. a. $\Gamma(1) = 1$, $\Gamma(2) = 1$, $\Gamma(3) = 2$ **c.** $\Gamma(n) = (n - 1)!$

61. b. 43.53%   **c.** 7

63. a. \$743,997.58   **b.** \$795,584.54   **c.** \$858,333.33

65. $\dfrac{k(\sqrt{a^2 + 1} - 1)}{a^2\sqrt{a^2 + 1}}$ **67.** False, let $f(x) = \dfrac{1}{x + 1}$.

68. True   **69.** True   **70.** True

71. a. 2

  b. ∞

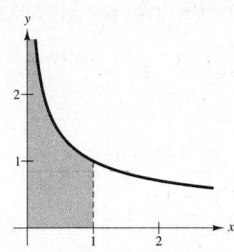

Review Exercises for Chapter 7

1. $\dfrac{e^{2x}}{13}(2\sin 3x - 3\cos 3x) + C$

3. $\dfrac{1}{3\pi}\sin(\pi x - 1)[\cos^2(\pi x - 1) + 2] + C$

5. $\dfrac{2}{3}\left[\tan^3\left(\dfrac{x}{2}\right) + 3\tan\left(\dfrac{x}{2}\right)\right] + C$ **7.** $\dfrac{3\sqrt{4 - x^2}}{x} + C$

9. $\frac{1}{4}[6\ln|x - 1| - \ln(x^2 + 1) + 6\arctan x] + C$

11. $x + \frac{9}{8}\ln|x - 3| - \frac{25}{8}\ln|x + 5| + C$

13. $\tan\theta + \sec\theta + C$ **15.** $-\dfrac{1}{x}(1 + \ln 2x) + C$

17. $\dfrac{1}{2}\left(4\arcsin\dfrac{x}{2} + x\sqrt{4 - x^2}\right) + C$

19. $\dfrac{3}{2}\ln(x^2 + 1) - \dfrac{1}{2(x^2 + 1)} + C$ **21.** $16\arcsin\dfrac{x}{4} + C$

23. $\dfrac{1}{2}\ln|x^2 + 4x + 8| - \arctan\left(\dfrac{x + 2}{2}\right) + C$

25. $\frac{1}{8}(\sin 2\theta - 2\theta\cos 2\theta) + C$ **27.** $\frac{1}{2}(2\theta - \cos 2\theta) + C$

29. $2\sqrt{1 - \cos x} + C$ **31.** $\sin x \ln(\sin x) - \sin x + C$

33. $\frac{1}{16}[(8x^2 - 1)\arcsin 2x + 2x\sqrt{1 - 4x^2}] + C$

35. $\frac{4}{3}[x^{3/4} - 3x^{1/4} + 3\arctan(x^{1/4})] + C$

37. $y = \dfrac{3}{2}\ln\left|\dfrac{x - 3}{x + 3}\right| + C$

39. $y = x\ln|x^2 + x| - 2x + \ln|x + 1| + C$

41. $-\dfrac{\sqrt{4 + x^2}}{4x} + C$ **43.** $\dfrac{1}{3}\sqrt{4 + x^2}(x^2 - 8) + C$

45. $\frac{1}{5}$ **47.** $\dfrac{1}{2}(\ln 4)^2 \approx 0.961$ **49.** π **51.** $\frac{128}{15}$

53. $(\bar{x}, \bar{y}) = \left(0, \dfrac{4}{3\pi}\right)$

55. a. $e - 1 \approx 1.72$

  b. 1

  c. $\frac{1}{2}(e - 1) \approx 0.86$

  d. 1.46 (Simpson's Rule)

57. 3.82   **59.** 0   **61.** ∞ **63.** 1

65. $1000e^{0.09} \approx 1094.17$ **67.** $\frac{32}{3}$ **69.** Diverges

71. a. \$4,637,228.40   **b.** \$5,555,555.56

73. $0.015846 < \int_2^\infty \frac{1}{x^5 - 1}\,dx < 0.015851$

77. False, $\int \frac{\ln x^2}{x}\,dx = \frac{1}{2}\int (\ln x^2)\frac{2}{x}\,dx = \frac{1}{2}\int u\,du.$ **78.** True

79. False, $\int_{-1}^1 \sqrt{x^2 - x^3}\,dx = \int_{-1}^1 |x|\sqrt{1 - x}\,dx.$ **80.** True

CHAPTER 8

Section 8.1 (*page 557*)

1. $2, 4, 8, 16, 32$ **3.** $-\frac{1}{2}, \frac{1}{4}, -\frac{1}{8}, \frac{1}{16}, -\frac{1}{32}$

5. $-1, -\frac{1}{4}, \frac{1}{9}, \frac{1}{16}, -\frac{1}{25}$ **7.** $3, \frac{9}{2}, \frac{27}{6}, \frac{81}{24}, \frac{243}{120}$

9. $3, 4, 6, 10, 18$ **11.** $14, 17$ **13.** $\frac{3}{16}, -\frac{3}{32}$

15. $10 \cdot 9$ **17.** $n + 1$ **19.** $\frac{1}{(2n + 1)(2n)}$ **21.** $3n - 2$

23. $n^2 - 2$ **25.** $\frac{n + 1}{n + 2}$ **27.** $\frac{(-1)^{n-1}}{2^{n-2}}$ **29.** $\frac{n + 1}{n}$

31. $\frac{n}{(n + 1)(n + 2)}$ **33.** $\frac{(-1)^{n-1}}{1 \cdot 3 \cdot 5 \cdots (2n - 1)} = \frac{(-1)^{n-1}2^n n!}{(2n)!}$

35. Converges to 1 **37.** Diverges **39.** Converges to $\frac{3}{2}$

41. Diverges **43.** Converges to 0 **45.** Converges to 0

47. Diverges **49.** Converges to 0 **51.** Converges to 0

53. Converges to e^k **55.** Monotonic, bounded

57. Not monotonic, bounded **59.** Not monotonic, bounded

61. Monotonic, bounded **63.** Not monotonic, bounded

65. 5 **67.** $\frac{1}{3}$

69. a. $a_n = 10 - \frac{1}{n}$

b. A monotonic, bounded sequence must converge (Theorem 8.5)

c. $a_n = \frac{3n}{4n + 1}$

d. The unbounded sequence does not converge.

71. a. No

b.

n	1	2	3	4	5
A_n	\$9086.25	\$9173.33	\$9261.24	\$9349.99	\$9439.60

n	6	7	8	9	10
A_n	\$9530.06	\$9621.39	\$9713.59	\$9806.68	\$9900.66

73. a. $\$2,500,000,000(0.8)^n$

b.

Year	1	2	3	4
Budget	\$2,000,000,000	\$1,600,000,000	\$1,280,000,000	\$1,024,000,000

c. Converges to 0

75.

n	0	1	2	3	4	5
a_n	\$127.42	\$144.79	\$162.16	\$179.53	\$196.90	\$214.27

n	6	7	8	9	10
a_n	\$231.64	\$249.01	\$266.38	\$283.75	\$301.12

1995 Cost: \$387.97

77. $S_6 = 240, S_7 = 440, S_8 = 810, S_9 = 1490, S_{10} = 2740$

79. a. $a_9 = a_{10} = \frac{1,562,500}{567}$

b. Decreasing

c. Factorials increase more rapidly than exponentials.

81. $1, 1.4142, 1.4422, 1.4142, 1.3797, 1.3480$
Converges to 1

83. a. $1, 1, 2, 3, 5, 8, 13, 21, 34, 55, 89, 144$

b. $1, 2, 1.5, 1.6667, 1.6, 1.6250, 1.6154, 1.6190, 1.6176, 1.6182$

c. $\rho = \frac{1 + \sqrt{5}}{2} \approx 1.6180$

85. True **86.** True **87.** True **88.** True

89. $1.4142, 1.8478, 1.9616, 1.9904, 1.9976$

$\lim_{x \to \infty} a_n = 2$

Section 8.2 (*page 566*)

1. $1, 1.25, 1.361, 1.424, 1.464$

3. $3, -1.5, 5.25, -4.875, 10.3125$

5. $3, 4.5, 5.25, 5.625, 5.8125$ **7.** $\lim_{n \to \infty} a_n = 1 \neq 0$

9. $\lim_{n \to \infty} a_n = 1 \neq 0$ **11.** Geometric series: $r = \frac{3}{2} > 1$

13. Geometric series: $r = 1.055 > 1$ **15.** $\lim_{n \to \infty} a_n = \frac{1}{2} \neq 0$

17. Geometric series: $r = \frac{3}{4} < 1$

19. Geometric series: $r = 0.9 < 1$

21. Telescoping series: $a_n = \frac{1}{n} - \frac{1}{n + 1}$ **23.** 2

25. $\frac{2}{3}$ **27.** $\frac{10}{9}$ **29.** $\frac{9}{4}$ **31.** $\frac{3}{4}$ **33.** 3 **35.** $\frac{1}{2}$

37. $\frac{4}{9}$ **39.** $\sum_{n=0}^\infty \frac{3}{40}(0.01)^n = \frac{5}{66}$ **41.** Diverges

43. Converges **45.** Diverges **47.** Converges

49. Diverges   **51.** Diverges   **53.** $\dfrac{1}{1-x}$, $|x| < 1$

55.

Horizontal asymptote: $y = 6$
The horizontal asymptote is the sum of the series.

57. The required term for each series is $n = 100$ and $n = 5$, respectively. The second series converges at a faster rate.

59. 20

n	5	10	20	50	100
S_n	8.1902	13.0264	17.5685	19.8969	19.9995

The terms of the series decrease in magnitude relatively slowly and the sequence of partial sums approaches the sum of the series relatively slowly.

61. $\dfrac{40}{3}$

n	5	10	20	50	100
S_n	13.3203	13.3333	13.3333	13.3333	13.3333

The terms of the series decrease in magnitude relatively rapidly and the sequence of partial sums approaches the sum of the series relatively rapidly.

63. $80{,}000(1 - 0.9^n)$ **65.** 152.42 ft   **67.** $\frac{1}{8}$
69. a. 126 in.^{2} **b.** 128 in.^{2} **71.** \$3,623,993.23
73. \$3048.1 million   **75. a.** \$26,046.33   **b.** \$26,111.12
77. a. \$632,407.96   **b.** \$640,501.61

83. $\displaystyle\sum_{n=0}^{\infty} 1$, $\displaystyle\sum_{n=0}^{\infty}(-1)$ (Answer is not unique.)

85. True   **86.** True
87. False, the series must begin with $n = 0$ in order for the limit to be $a/(1-r)$.
88. True

91. $L - \dfrac{1}{4}L - 2\left(\dfrac{1}{16}\right)L - 4\left(\dfrac{1}{64}\right)L - 8\left(\dfrac{1}{256}\right)L$

$= L - L\left(\dfrac{1}{4} + \dfrac{1}{8} + \dfrac{1}{16} + \dfrac{1}{32} + \cdots\right)$

$= L - L\displaystyle\sum_{n=0}^{\infty}\dfrac{1}{4}\left(\dfrac{1}{2}\right)^n = L\left(1 - \dfrac{1/4}{1 - 1/2}\right) = \dfrac{1}{2}L$

The remaining pieces are getting smaller and smaller, thus making the table appear to disappear.

Section 8.3   (page 573)

1. Diverges   **3.** Converges   **5.** Converges
7. Diverges   **9.** Diverges   **11.** Converges
13. Diverges   **15.** Diverges   **17.** Converges
19. Converges   **21.** Diverges   **23.** Converges
25. Converges   **27.** Diverges   **29.** Diverges
31. Converges   **33.** $p > 1$

35. a.

M	2	4	6	8
N	4	31	226	1673

b. No, since the magnitude of the terms of the series are approaching zero, it requires more and more terms to increase the partial sum by 2.

39. $R_6 \approx 0.0015$
$S_6 \approx 1.0811$

41. $R_{10} \approx 0.0997$ **43.** $R_4 \approx 5.6 \times 10^{-8}$
$S_{10} \approx 0.9818$ $S_6 \approx 0.4049$

45. $N \geq 7$ **47.** $N \geq 2$

49. a.

n	5	10	20	50	100
S_n	3.7488	3.7500	3.7500	3.7500	3.7500

b.

n	5	10	20	50	100
S_n	1.4636	1.5498	1.5962	1.6251	1.6350

In part a, the magnitude of the terms of the series approaches 0 more rapidly and the sequence of partial sums approaches the sum of the series more quickly.

51. b. $\displaystyle\sum_{n=2}^{\infty}\dfrac{1}{n^{1.1}} = 0.4665 + 0.2987 + 0.2176$
$+ 0.1703 + 0.1393 + \cdots$

$\displaystyle\sum_{n=2}^{\infty}\dfrac{1}{n \ln n} = 0.7213 + 0.3034 + 0.1803$
$+ 0.1243 + 0.0930 + \cdots$

c. $n > e^{40}$
53. $-\ln 2$

Section 8.4   (page 579)

1. Converges   **3.** Diverges   **5.** Converges
7. Diverges   **9.** Converges   **11.** Converges
13. Diverges   **15.** Diverges   **17.** Converges
19. Diverges   **21.** Converges   **23.** Diverges
25. Diverges   **27.** Diverges; p-Series Test
29. Converges; Comparison Test with
$\displaystyle\sum_{n=1}^{\infty}\left(\dfrac{1}{3}\right)^n$

31. Diverges; nth Term Test
33. Converges; Integral Test   **37.** Diverges
39. Converges

43. b.

n	5	10	20	50	100
S_n	1.1839	1.2087	1.2212	1.2287	1.2312

c. 0.1226
d. 0.0277
e. No, only a finite sum has been subtracted from the sum of the series.

47. False, let $a_n = \dfrac{1}{n^3}$ and $b_n = \dfrac{1}{n^2}$.   **48.** True   **49.** True

50. False, let $a_n = \dfrac{1}{n}$, $b_n = \dfrac{1}{n^2}$, and $c_n = \dfrac{10}{n}$.

Section 8.5   (page 587)

1. Converges   **3.** Converges   **5.** Diverges
7. Converges   **9.** Diverges   **11.** Diverges
13. Converges   **15.** Converges   **17.** Converges

19. Converges   **21.** Converges absolutely
23. Converges conditionally   **25.** Diverges
27. Converges conditionally   **29.** Converges absolutely
31. Converges absolutely   **33.** Converges conditionally
35. Converges absolutely   **37.** 0.368   **39.** 0.842
41. 0.693   **43.** 7

45. a.

(graph of S_n vs n)

b.

n	1	2	3	4	5	6	7	8
S_n	1.0000	0.6667	0.8667	0.7238	0.8349	0.7440	0.8209	0.7543

d. The points alternate sides of the horizontal line that represents the sum of the series. The distance between the successive points and the line decreases. That distance is always less than the magnitude of the next term of the series.

49. a. $-1 < x < 1$
   b. $-1 \le x < 1$

51. True   **52.** True   **53.** False, let $a_n = \dfrac{(-1)^n}{n}$.

54. True

Section 8.6 (*page 594*)

5. Diverges   **7.** Converges   **9.** Converges
11. Diverges   **13.** Converges   **15.** Diverges
17. Converges   **19.** Converges   **21.** Diverges
23. Converges   **27.** Converges   **29.** Converges
31. Diverges   **33.** Converges
35. Converges; Alternating Series Test
37. Converges; *p*-Series Test   **39.** Diverges; *n*th-Term Test
41. Diverges; Ratio Test
43. Converges; Limit Comparsion Test with $b_n = \dfrac{1}{2^n}$
45. Converges; Comparision Test with $b_n = \dfrac{1}{2^n}$
47. Converges; Ratio Test   **49.** Converges; Ratio Test
51. Converges; Ratio Test   **53.** a. and c.
55. $\displaystyle\sum_{n=0}^{\infty} \dfrac{n+1}{4^{n+1}}$ **57.** -0.7769
59. No, the series $\displaystyle\sum_{n=1}^{\infty} \dfrac{1}{n+10{,}000}$ diverges.

Section 8.7 (*page 604*)

1. d   **2.** c   **3.** a   **4.** b
5. $1 - x + \frac{1}{2}x^2 - \frac{1}{6}x^3$ **7.** $1 + 2x + 2x^2 + \frac{4}{3}x^3 + \frac{2}{3}x^4$
9. $x - \frac{1}{6}x^3 + \frac{1}{120}x^5$ **11.** $x + x^2 + \frac{1}{2}x^3 + \frac{1}{6}x^4$
13. $1 - x + x^2 - x^3 + x^4$ **15.** $2 - 3x^3 + x^4$
17. $1 - (x-1) + (x-1)^2 - (x-1)^3 + (x-1)^4$
19. $(x-1) - \frac{1}{2}(x-1)^2 + \frac{1}{3}(x-1)^3 - \frac{1}{4}(x-1)^4$

21. a. $P_3(x) = x + \dfrac{1}{3}x^3$

   b. $P_5(x) = x + \dfrac{1}{3}x^3 + \dfrac{2}{15}x^5$

   c. $Q_3(x) = 1 + 2\left(x - \dfrac{\pi}{4}\right) + 2\left(x - \dfrac{\pi}{4}\right)^2 + \dfrac{8}{3}\left(x - \dfrac{\pi}{4}\right)^3$

23. a.

x	0	0.25	0.50	0.75	1.00
sin x	0	0.2474	0.4794	0.6816	0.8415
$P_1(x)$	0	0.25	0.50	0.75	1.00
$P_3(x)$	0	0.2474	0.4792	0.6797	0.8333
$P_5(x)$	0	0.2474	0.4794	0.6817	0.8417
$P_7(x)$	0	0.2474	0.4794	0.6816	0.8415

b.

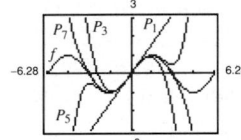

25. a. $P_3(x) = x + \frac{1}{6}x^3$

b.

x	−1	−0.75	−0.50	−0.25
$f(x)$	−1.571	−0.848	−0.524	−0.253
$P_3(x)$	−1.167	−0.820	−0.521	−0.253

x	0	0.25	0.50	0.75	1
$f(x)$	0	0.253	0.524	0.848	1.571
$P_3(x)$	0	0.253	0.521	0.820	1.167

c.

27. 0.6042   **29.** 0.1823   **31.** $R_4 \le 2.03 \times 10^{-5}$

33. $R_3 \le 7.82 \times 10^{-3}$ **35.** 3   **37.** 9

39. $-0.3936 < x < 0$

41. $e^x : P_4(x) = 1 + x + \frac{1}{2}x^2 + \frac{1}{6}x^3 + \frac{1}{24}x^4$

$xe^x : Q_4(x) = x + x^2 + \frac{1}{2}x^3 + \frac{1}{6}x^4$

$= xP_4(x) - \frac{1}{24}x^4$

47.

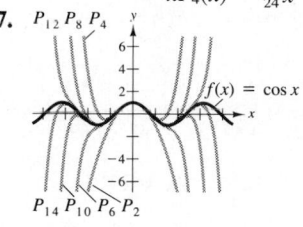

Section 8.8   (*page 613*)

1. $R = 1$ **3.** $R = \frac{1}{2}$ **5.** $R = \infty$ **7.** $(-2, 2)$

9. $(-1, 1]$ **11.** $(-\infty, \infty)$ **13.** $x = 0$ **15.** $(-4, 4)$

17. $(0, 10]$ **19.** $(0, 2]$ **21.** $(0, 2c)$ **23.** $(-\frac{1}{2}, \frac{1}{2})$

25. $(-\infty, \infty)$ **27.** $(-1, 1)$ **29.** $x = 3$

31. a. $(-2, 2)$ **b.** $(-2, 2)$ **c.** $(-2, 2)$ **d.** $[-2, 2)$

33. a. $(0, 2]$ **b.** $(0, 2)$ **c.** $(0, 2)$ **d.** $[0, 2]$

35. a. For $f(x)$: $(-\infty, \infty)$

For $g(x)$: $(-\infty, \infty)$

d. $f(x) = \sin x$

$g(x) = \cos x$

39. c.

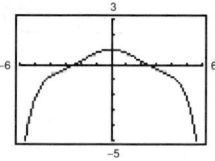

d. 0.92

41. a. $\frac{8}{5}$ **b.** $\frac{8}{11}$

c. The alternating series converges more rapidly. The partial sums of the series of positive terms approach the sum from below. The partial sums of the alternating series alternate sides of the horizontal line representing the sum.

d.

M	10	100	1000	10,000
N	4	9	15	21

43. $f(x) = \dfrac{1}{1 + x}$ **45.** $f(x) = \cos x$

47. False, let $a_n = \dfrac{(-1)^n}{n2^n}$.   **48.** True   **49.** True

50. True

Section 8.9   (*page 620*)

1. $\displaystyle\sum_{n=0}^{\infty} \frac{x^n}{2^{n+1}}$ **3.** $\displaystyle\sum_{n=0}^{\infty} \frac{(-1)^n x^n}{2^{n+1}}$

$(-2, 2)$ $(-2, 2)$

5. $\displaystyle\sum_{n=0}^{\infty} \frac{(x - 5)^n}{(-3)^{n+1}}$ **7.** $-3 \displaystyle\sum_{n=0}^{\infty} (2x)^n$

$(2, 8)$ $(-\frac{1}{2}, \frac{1}{2})$

9. $-\dfrac{1}{11} \displaystyle\sum_{n=0}^{\infty} \left[\dfrac{2}{11}(x + 3)\right]^n$ **11.** $\dfrac{3}{2} \displaystyle\sum_{n=0}^{\infty} \left(\dfrac{x}{-2}\right)^n$

$\left(-\dfrac{17}{2}, \dfrac{5}{2}\right)$ $(-2, 2)$

13. $\displaystyle\sum_{n=0}^{\infty} \left[\dfrac{1}{(-2)^n} - 1\right] x^n$ **15.** $2 \displaystyle\sum_{n=0}^{\infty} x^{2n}$

$(-1, 1)$ $(-1, 1)$

17. $2 \displaystyle\sum_{n=0}^{\infty} x^{2n}$ **19.** $\displaystyle\sum_{n=1}^{\infty} n(-1)^n x^{n-1}$

$(-1, 1)$ $(-1, 1)$

21. $\displaystyle\sum_{n=0}^{\infty} \dfrac{(-1)^n x^{n+1}}{n + 1}$ **23.** $\displaystyle\sum_{n=0}^{\infty} (-1)^n x^{2n}$ **25.** $\displaystyle\sum_{n=0}^{\infty} (-1)^n (2x)^{2n}$

$(-1, 1]$ $(-1, 1)$ $(-\frac{1}{2}, \frac{1}{2})$

27.

x	0.0	0.2	0.4	0.6	0.8	1.0
S_2	0.000	0.180	0.320	0.420	0.480	0.500
$\ln(x + 1)$	0.000	0.182	0.336	0.470	0.588	0.693
S_3	0.000	0.183	0.341	0.492	0.651	0.833

29. 0.245   **31.** 0.125

33. $\displaystyle\sum_{n=0}^{\infty} nx^{n-1}$

$(-1, 1)$

35. 2, since the probability of obtaining a head on a single toss is $\frac{1}{2}$, it is expected that, on average, a head will be obtained in 2 tosses.

39. 3.14   **41.** $\ln \frac{3}{2} \approx 0.4055$ **43.** $\ln \frac{7}{5} \approx 0.3365$

45. $\arctan \frac{1}{2} \approx 0.4636$

47. The series of Exercise 44 converges to its sum at a slower rate because its terms approach 0 to a much slower rate.

Section 8.10   (*page 630*)

1. $\displaystyle\sum_{n=0}^{\infty} \frac{(2x)^n}{n!}$ **3.** $\dfrac{\sqrt{2}}{2} \displaystyle\sum_{n=0}^{\infty} \frac{(-1)^{n(n+1)/2}}{n!} \left(x - \frac{\pi}{4}\right)^n$

5. $\displaystyle\sum_{n=0}^{\infty} \frac{(-1)^n (x-1)^{n+1}}{n+1}$ **7.** $\displaystyle\sum_{n=0}^{\infty} \frac{(-1)^n (2x)^{2n+1}}{(2n+1)!}$

9. $1 + \dfrac{x^2}{2!} + \dfrac{5x^4}{4!} + \cdots$

11. $\displaystyle\sum_{n=0}^{\infty} (-1)^n (n+1) x^n$

13. $\dfrac{1}{2}\left[1 + \displaystyle\sum_{n=1}^{\infty} \dfrac{(-1)^n 1\cdot 3\cdot 5 \cdots (2n-1) x^{2n}}{2^{3n} n!} \right]$

15. $1 + \dfrac{x^2}{2} + \displaystyle\sum_{n=2}^{\infty} \dfrac{(-1)^{n+1} 1\cdot 3\cdot 5 \cdots (2n-3) x^{2n}}{2^n n!}$

17. $1 + \dfrac{x^2}{2} + \dfrac{x^4}{2^2 2!} + \dfrac{x^6}{2^3 3!} + \cdots$ **19.** $\displaystyle\sum_{n=0}^{\infty} \dfrac{(-1)^n (2x)^{2n+1}}{(2n+1)!}$

21. $\displaystyle\sum_{n=0}^{\infty} \dfrac{(-1)^n x^n}{(2n)!}$ **23.** $\displaystyle\sum_{n=0}^{\infty} \dfrac{(-1)^n x^{2n}}{(2n+1)!}$ **25.** $\displaystyle\sum_{n=0}^{\infty} \dfrac{x^{2n+1}}{(2n+1)!}$

27. $\dfrac{1}{2}\left[1 + \displaystyle\sum_{n=0}^{\infty} \dfrac{(-1)^n (2x)^{2n}}{(2n)!} \right]$ **29.** $\displaystyle\sum_{n=0}^{\infty} \dfrac{(-1)^n x^{2n+1}}{(2n+1)!}$

31. $P_5(x) = x + x^2 + \dfrac{1}{3}x^3 - \dfrac{1}{30}x^5 + \cdots$

33. $P_5(x) = x - \dfrac{1}{2}x^2 - \dfrac{1}{6}x^3 + \dfrac{3}{40}x^5 + \cdots$

35. $P_4(x) = x - x^2 + \dfrac{5}{6}x^3 - \dfrac{5}{6}x^4 + \cdots$

37. $\displaystyle\sum_{n=0}^{\infty} \dfrac{(-1)^{(n+1)} x^{2n+3}}{(2n+3)(n+1)!}$ **39.** 0.8415   **41.** 0.6931

43. 0   **45.** 0.9461   **47.** 0.5312   **49.** 0.2010

51. 0.3413

53. $P_5(x) = x - 2x^3 + \dfrac{2}{3}x^5$
$\left[-\dfrac{3}{4}, \dfrac{3}{4} \right]$

55. $P_5(x) = (x-1) - \dfrac{1}{24}(x-1)^3 + \dfrac{1}{24}(x-1)^4$
$\qquad\qquad\qquad\qquad\qquad - \dfrac{71}{1920}(x-1)^5$
$\left[\dfrac{1}{4}, 2 \right]$

59. a.

c. $\displaystyle\sum_{n=0}^{\infty} 0 x^n = 0 \neq f(x)$

Review Exercises for Chapter 8

1. $a_n = \dfrac{1}{n!}$ **3.** Converges to 0   **5.** Diverges

7. Converges to 0   **9.** Converges to 0

11. a.

n	1	2	3	4	5	6	7	8
A_n	\$5100.00	\$5202.00	\$5306.04	\$5412.16	\$5520.40	\$5630.81	\$5743.43	\$5858.30

b. \$11,040.20

13. 1, 2.5, 4.75, 8.125, 13.3875

15. 0.5, 0.45833, 0.45972, 0.45970, 0.45970

17. 3   **19.** $\frac{1}{2}$ **21.** $\frac{1}{11}$ **23.** $45\frac{1}{3}$ ft

25. \$5087.14   **27.** Diverges   **29.** Converges

31. Converges   **33.** Diverges   **35.** Diverges

37. Converges

39. a.

N	5	10	20	30	40
$\sum_{n=1}^{N} \frac{1}{n^p}$	1.4636	1.5498	1.5962	1.6122	1.6202
$\int_{N}^{\infty} \frac{1}{x^p} dx$	0.2000	0.1000	0.0500	0.0333	0.0250

b.

N	5	10	20	30	40
$\sum_{n=1}^{N} \frac{1}{n^p}$	1.0367	1.0369	1.0369	1.0369	1.0369
$\int_{N}^{\infty} \frac{1}{x^p} dx$	0.0004	0.0000	0.0000	0.0000	0.0000

The series of part b converges more rapidly. This is evident from the integrals which give the remainders of the partial sums.

41. $(-10, 10)$ **43.** $[1, 3]$ **45.** Converges only at $x = 2$

47. $\frac{\sqrt{2}}{2} \sum_{n=0}^{\infty} \frac{(-1)^{n(n+1)/2}}{n!} \left(x - \frac{3\pi}{4}\right)^n$ **49.** $\sum_{n=0}^{\infty} \frac{(x \ln 3)^n}{n!}$

51. $-\sum_{n=0}^{\infty} (x + 1)^n$ **53.** $\sum_{n=0}^{\infty} \frac{2}{3}\left(\frac{x}{3}\right)^n$

55. $1 + 2x + 2x^2 + \frac{4}{3}x^3$ **57.** $f(x) = \dfrac{3}{3 - 2x}$

$\left(-\frac{3}{2}, \frac{3}{2}\right)$

59. $\sum_{n=0}^{\infty} \frac{(-1)^n x^{2n+1}}{(2n + 1)(2n + 1)!}$ **61.** $\sum_{n=0}^{\infty} \frac{(-1)^n x^{n+1}}{(n + 1)^2}$

63. 0   **65.** 0.996   **67.** 0.560

69. a. 4   **b.** 6   **c.** 5   **d.** 10

CHAPTER 9

Section 9.1   (*page 640*)

1. e   **3.** a   **5.** d

7. Vertex: $(0, 0)$
   Focus: $(0, \frac{1}{16})$
   Directrix: $y = -\frac{1}{16}$

9. Vertex: $(0, 0)$
   Focus: $(-\frac{3}{2}, 0)$
   Directrix: $x = \frac{3}{2}$

11. Vertex: $(0, 0)$
   Focus: $(0, -2)$
   Directrix: $y = 2$

13. Vertex: $(1, -2)$
   Focus: $(1, -4)$
   Directrix: $y = 0$

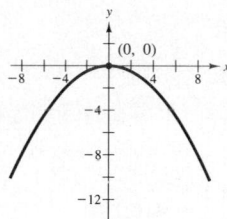

15. Vertex: $(5, -\frac{1}{2})$
   Focus: $(\frac{11}{2}, -\frac{1}{2})$
   Directrix: $x = \frac{9}{2}$

17. Vertex: $(1, 1)$
   Focus: $(1, 2)$
   Directrix: $y = 0$

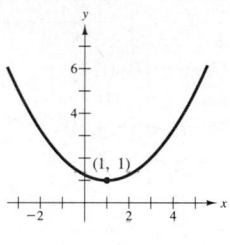

19. Vertex: $(-1, 2)$
   Focus: $(0, 2)$
   Directrix: $x = -2$

21. Vertex: $(2, -2)$
   Focus: $(0, -2)$
   Directrix: $x = 4$

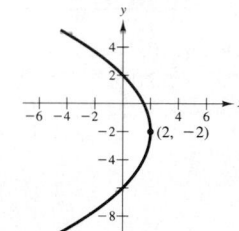

23. Vertex: $(-2, 1)$
   Focus: $(-2, -\frac{1}{2})$
   Directrix: $y = \frac{5}{2}$

25. Vertex: $(\frac{1}{4}, -\frac{1}{2})$
   Focus: $(0, -\frac{1}{2})$
   Directrix: $x = \frac{1}{2}$

27. $x^2 + 6y = 0$ **29.** $y^2 - 4y + 8x - 20 = 0$

31. $x^2 + 24y + 96 = 0$ **33.** $y^2 - 8x - 4y + 4 = 0$

35. $x^2 + y - 4 = 0$ **37.** $5x^2 - 14x - 3y + 9 = 0$

39. $(x - h)^2 = -8(y + 1)$ **41.** $3x - 2y^2 = 0$

43. $\frac{9}{4}$ ft   **45.** $4x - y - 8 = 0$ **47.** $y = 2ax_0 x - ax_0^2$

49. Tangent lines: $2x + y - 1 = 0$
$2x - 4y - 1 = 0$
Point of intersection: $(\frac{1}{2}, 0)$
(on the directrix)

51. $y = \dfrac{1}{4}x + 2$ **53.** $x_0 = \dfrac{2\sqrt{3}}{3}$

$y = -\dfrac{1}{2}x - 1$ Distance from hill: $\dfrac{2\sqrt{3}}{3} - 1$

55. $\dfrac{16(4 + 3\sqrt{3} - 2\pi)}{3} \approx 15.536 \text{ ft}^2$

57. $2[\sqrt{2} + \ln(\sqrt{2} + 1)] \approx 4.591$

59. $2\sqrt{5} + \ln(\sqrt{5} + 2) \approx 5.916$

61. a. $y = \dfrac{1}{800}x^2$

    b. $100\left[\sqrt{5} + 4\ln\left(\dfrac{1 + \sqrt{5}}{2}\right)\right] \approx 416.1 \text{ ft}$

66. False, see the definition of a parabola.

67. True

Section 9.2 (*page 649*)

1. e   **3.** c   **5.** f

7. Center: $(0, 0)$
Foci: $(\pm 3, 0)$
Vertices: $(\pm 5, 0)$
$e = \dfrac{3}{5}$

9. Center: $(0, 0)$
Foci: $(0, \pm 3)$
Vertices: $(0, \pm 5)$
$e = \dfrac{3}{5}$

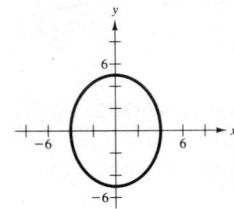

11. Center: $(0, 0)$
Foci: $(\pm 2, 0)$
Vertices: $(\pm 3, 0)$
$e = \dfrac{2}{3}$

13. Center: $(0, 0)$
Foci: $(\pm\sqrt{3}, 0)$
Vertices: $(\pm 2, 0)$
$e = \dfrac{\sqrt{3}}{2}$

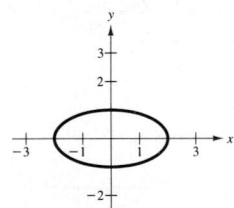

15. Center: $(0, 0)$
Foci: $(0, \pm 1)$
Vertices: $(0, \pm\sqrt{3})$
$e = \dfrac{\sqrt{3}}{3}$

17. Center: $(1, 5)$
Foci: $(1, 9), (1, 1)$
Vertices: $(1, 10), (1, 0)$
$e = \dfrac{4}{5}$

19. Center: $(-2, 3)$
Foci: $(-2, 3 \pm \sqrt{5})$
Vertices: $(-2, 6), (-2, 0)$
$e = \dfrac{\sqrt{5}}{3}$

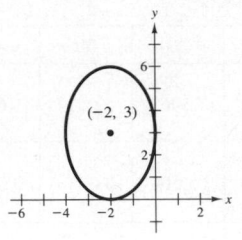

21. Center: $(1, -1)$
Foci: $\left(1 \pm \dfrac{3\sqrt{10}}{20}, -1\right)$
Vertices: $\left(1 \pm \dfrac{\sqrt{10}}{4}, -1\right)$
$e = \dfrac{3}{5}$

23. Center: $(0, 0)$
Foci: $\left(0, \pm\dfrac{\sqrt{3}}{2}\right)$
Vertices: $(0, \pm 1)$

25. Center: $\left(\dfrac{1}{2}, -1\right)$
Foci: $\left(\dfrac{1}{2} \pm \sqrt{2}, -1\right)$
Vertices: $\left(\dfrac{1}{2} \pm \sqrt{5}, -1\right)$

27. Center: $\left(\dfrac{3}{2}, -1\right)$
Foci: $\left(\dfrac{3}{2} - \sqrt{2}, -1\right), \left(\dfrac{3}{2} + \sqrt{2}, -1\right)$
Vertices: $\left(-\dfrac{1}{2}, -1\right), \left(\dfrac{7}{2}, -1\right)$

29. $\dfrac{x^2}{9} + \dfrac{y^2}{5} = 1$ **31.** $\dfrac{x^2}{25} + \dfrac{y^2}{16} = 1$

33. $\dfrac{(x - 3)^2}{9} + \dfrac{(y - 5)^2}{16} = 1$

35. $\dfrac{x^2}{24} + \dfrac{y^2}{49} = 1$ **37.** $\dfrac{x^2}{16} + \dfrac{7y^2}{16} = 1$

39.

41.

43.

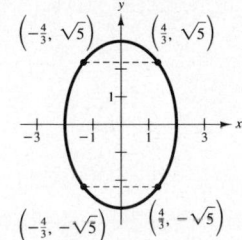

$\left(-\frac{4}{3}, \sqrt{5}\right)$ $\left(\frac{4}{3}, \sqrt{5}\right)$

$\left(-\frac{4}{3}, -\sqrt{5}\right)$ $\left(\frac{4}{3}, -\sqrt{5}\right)$

45. a. $2a$

    b. The two thumbtacks are located at the foci and the length of the string is the constant sum of the distances from the foci.

47. $3\sqrt{19} \approx 13$ ft

49. Perihelion: 2,756,832,000 mi
    Aphelion: 4,575,168,000 mi

53. 0.9367

57. Minor axis endpoints: $(-6, -2)$, $(0, -2)$
    Major axis endpoints: $(-3, -6)$, $(-3, 2)$

59. $\left(0, \frac{25}{3}\right)$ **61.** $\sqrt{2}a \times \sqrt{2}b$ **63.** 60π

65. a. 2π

    b. Volume $= \dfrac{8\pi}{3}$

       Surface area $= \dfrac{2\pi\left(9 + 4\sqrt{3}\pi\right)}{9} \approx 21.48$

    c. Volume $= \dfrac{16\pi}{3}$

       Surface area $= \dfrac{4\pi\left[6 + \sqrt{3}\ln\left(2 + \sqrt{3}\right)\right]}{3} \approx 34.69$

67. 22.10   **71.** $\dfrac{x^2}{36} + \dfrac{y^2}{9} = 1$ **72.** True

73. False, the equation of an ellipse is second degree in x and y.

Section 9.3   (*page 657*)

1. e   **3.** f   **5.** d

7. Center: $(0, 0)$
    Vertices: $(\pm 1, 0)$
    Foci: $\left(\pm\sqrt{2}, 0\right)$

9. Center: $(0, 0)$
    Vertices: $(0, \pm 1)$
    Foci: $\left(0, \pm\sqrt{5}\right)$

11. Center: $(0, 0)$
    Vertices: $(0, \pm 5)$
    Foci: $(0, \pm 13)$

13. Center: $(0, 0)$
    Vertices: $(0, \pm 2)$
    Foci: $(0, \pm 3)$

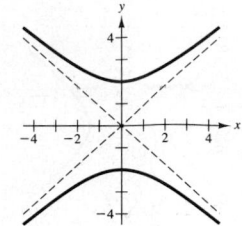

15. Center: $(1, -2)$
    Vertices: $(-1, -2)$, $(3, -2)$
    Foci: $\left(1 \pm \sqrt{5}, -2\right)$

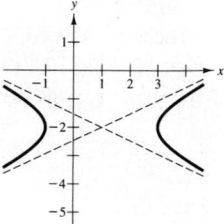

17. Center: $(2, -6)$
    Vertices: $(2, -5)$, $(2, -7)$
    Foci: $\left(2, -6 \pm \sqrt{2}\right)$

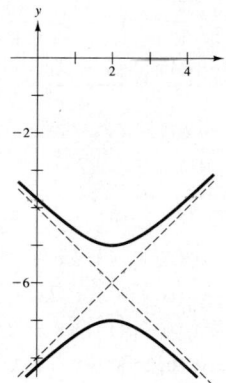

19. Center: $(2, -3)$
    Vertices: $(1, -3)$, $(3, -3)$
    Foci: $\left(2 \pm \sqrt{10}, -3\right)$

21. Degenerate hyperbola
    Graph is two lines intersecting at $(-1, -3)$.

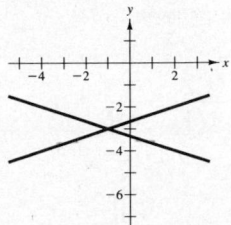

23. Center: $(0, 0)$
Vertices: $(\pm\sqrt{3}, 0)$
Foci: $(\pm\sqrt{5}, 0)$

25. Center: $(1, -3)$
Vertices: $(1, -3 \pm \sqrt{2})$
Foci: $(1, -3 \pm 2\sqrt{5})$

27. Center: $(1, -3)$
Vertices: $(-1, -3), (3, -3)$
Foci: $(1 \pm \sqrt{10}, -3)$

29. $\dfrac{y^2}{4} - \dfrac{x^2}{12} = 1$ **31.** $\dfrac{x^2}{1} - \dfrac{y^2}{9} = 1$

33. $\dfrac{(x-3)^2}{9} - \dfrac{(y-2)^2}{4} = 1$ **35.** $\dfrac{y^2}{9} - \dfrac{(x-2)^2}{9/4} = 1$

37. $\dfrac{(x-6)^2}{9} - \dfrac{(y-2)^2}{7} = 1$

41. $x \approx 110.3$ mi   **43. a.** $\dfrac{64x^2}{225} - \dfrac{64y^2}{31} = 1$ **b.** No

45. a. At $(6, \sqrt{3})$: $2x - 3\sqrt{3}y - 3 = 0$
At $(6, -\sqrt{3})$: $2x + 3\sqrt{3}y - 3 = 0$
b. At $(6, \sqrt{3})$: $9x + 2\sqrt{3}y - 60 = 0$
At $(6, -3)$: $9x - 2\sqrt{3}y - 60 = 0$

47. Volume $= \dfrac{4\pi}{3}$

Surface area $= \pi\left[2\sqrt{7} - 1 + \dfrac{\sqrt{2}}{2}\ln\left(\dfrac{\sqrt{2}+1}{2\sqrt{2}+\sqrt{7}}\right)\right]$
≈ 11.66

51. False, $y^2 - x^2 + 2x + 2y = 0$ yields two intersecting lines.
52. True
53. False, the equation of a hyperbola is second degree in both variables.
54. True   **57.** Parabola   **59.** Circle
61. Circle   **63.** Hyperbola

Section 9.4 (*page 664*)

1. $\dfrac{(y')^2}{2} - \dfrac{(x')^2}{2} = 1$ **3.** $\dfrac{(x')^2}{1/4} - \dfrac{(y')^2}{1/6} = 1$

5. $\dfrac{(x' - 3\sqrt{2})^2}{16} - \dfrac{(y' - \sqrt{2})^2}{16} = 1$

7. $\dfrac{(x')^2}{3} + \dfrac{(y')^2}{2} = 1$ **9.** $x' = -(y')^2$

11. $y' = \dfrac{(x')^2}{6} - \dfrac{x'}{3}$ **13.** $\theta = 45°$

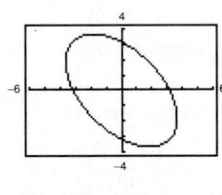

15. $\theta \approx 26.57°$ **17.** $\theta \approx 31.72°$

19. $\left(0, \pm\dfrac{2\sqrt{15}}{5}\right), \dfrac{dy}{dx} = \dfrac{1}{5}$ **21.** Parabola   **23.** Ellipse

25. Hyperbola   **27.** Parabola
29. Two lines   **31.** Two parallel lines

33. Two lines

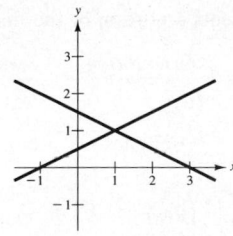

39. $x^2 + y^2 - 2xy - 2x - 2y + 1 = 0$
$0 \leq x \leq 1, 0 \leq y \leq 1$

Review Exercises for Chapter 9

1. h   **2.** a   **3.** e   **4.** i   **5.** f
6. b   **7.** c   **8.** j   **9.** g   **10.** d
11. Circle   **13.** Hyperbola
Center: $(\frac{1}{2}, -\frac{3}{4})$ Center: $(-4, 3)$
Radius: 1   Vertices: $\left(-4 \pm \sqrt{2}, 3\right)$

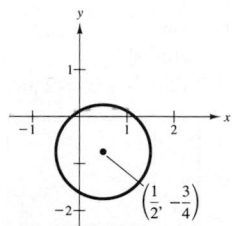

15. Ellipse   **17.** Parabola
Center: $(2, -3)$ Vertex: $(3, 0)$
Vertices: $\left(2, -3 \pm \dfrac{\sqrt{2}}{2}\right)$

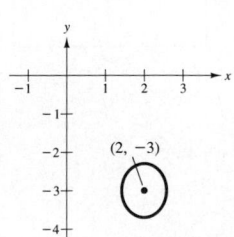

19. Ellipse
Center: $(1, -\frac{1}{2})$
Vertices: $(1 \pm \frac{3}{4}, -\frac{1}{2})$

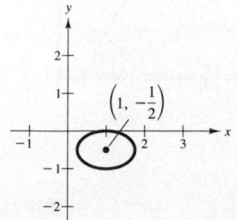

21. Parabola
Vertex: $\left(-\dfrac{\sqrt{2}}{4}, \dfrac{\sqrt{2}}{4}\right)$

23. $x^2 - 8x + 8y = 0$ **25.** $y^2 - 4y - 12x + 4 = 0$
27. $4x + 4y - 7 = 0$ **29.** $\dfrac{(x - 2)^2}{25} + \dfrac{y^2}{21} = 1$
31. $\dfrac{2x^2}{9} + \dfrac{y^2}{36} = 1$ **33.** $\dfrac{x^2}{4} + \dfrac{3y^2}{16} = 1$
35. 3 feet on either side of center   **41.** 192π ft^2
43. 4.212 ft   **45.** 15.87
47. $\dfrac{y^2}{1} - \dfrac{x^2}{8} = 1$ **49.** $\dfrac{x^2}{4} - \dfrac{y^2}{12} = 1$
51. $\dfrac{5(x - 4)^2}{16} - \dfrac{5y^2}{64} = 1$ **53.** $a = \sqrt{5}$

CHAPTER 10
Section 10.1 (*page 674*)

1. a.

t	0	1	2	3	4
x	0	1	$\sqrt{2}$	$\sqrt{3}$	2
y	1	0	-1	-2	-3

b.

c. $y = 1 - x^2, x \geq 0$
3. $2x - 3y + 5 = 0$ **5.** $y = (x - 1)^2$

7. $y = \dfrac{1}{2}x^{2/3}$ **9.** $y = \dfrac{x + 1}{x}$

11. $y = \dfrac{|x - 4|}{2}$

13. $y = \dfrac{1}{x}$, $|x| \geq 1$

15. $x^2 + y^2 = 9$

17. $\dfrac{x^2}{16} + \dfrac{y^2}{4} = 1$

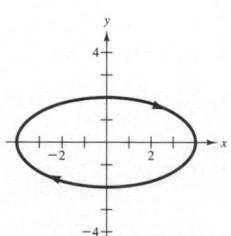

19. $\dfrac{(x - 4)^2}{4} + \dfrac{(y + 1)^2}{1} = 1$

21. $\dfrac{(x - 4)^2}{4} + \dfrac{(y + 1)^2}{16} = 1$

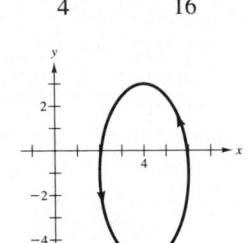

23. $\dfrac{x^2}{16} - \dfrac{y^2}{9} = 1$

25. $y = \ln x$

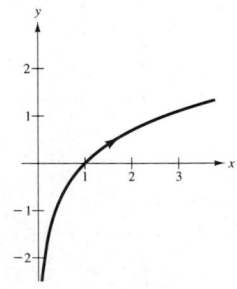

27. $y = \dfrac{1}{x^3}$, $x > 0$

29. Each curve represents a portion of the line $y = 2x + 1$.

	Domain	Orientation	Smooth
a.	$-\infty < x < \infty$	Up	Yes
b.	$-1 \leq x \leq 1$	Oscillates	No, $\dfrac{dx}{d\theta} = \dfrac{dy}{d\theta} = 0$ when $\theta = 0$, π
c.	$0 < x < \infty$	Down	Yes
d.	$0 < x < \infty$	Up	Yes

31. a., b., and d. represent the circle $x^2 + y^2 = 1$.

	Orientation	Smooth
a.	Counterclockwise	Yes
b.	Clockwise	Yes
c.	Clockwise	Yes
d.	Clockwise	Yes

33. $y - y_1 = \dfrac{y_2 - y_1}{x_2 - x_1}(x - x_1)$

35. $(x - h)^2 + (y - k)^2 = r^2$

37. $\dfrac{(x - h)^2}{a^2} - \dfrac{(y - k)^2}{b^2} = 1$

39. $x = 5t$
$y = -2t$
(Solution is not unique.)

41. $x = 2 + 4\cos\theta$
$y = 1 + 4\sin\theta$
(Solution is not unique.)

43. $x = 5\cos\theta$
$y = 3\sin\theta$
(Solution is not unique.)

45. $x = 4\sec\theta$
$y = 3\tan\theta$
(Solution is not unique.)

47. $x = t$
$y = 3t - 2$
$x = t - 3$
$y = 3t - 11$
(Solution is not unique.)

49. $x = t$
$y = t^3$
$x = \tan t$
$y = \tan^3 t$
(Solution is not unique.)

51.

53.

55.

57.

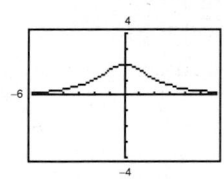

59. b   **60.** c   **61.** d   **62.** a

63. a. Maximum height: 14.2 ft
Range: 155.6 ft

b. Maximum height: 31.8 ft
   Range: 350.0 ft

c. Maximum height: 60.5 ft
   Range: 242.0 ft

d. Maximum height: 136.1 ft
   Range: 544.5 ft

e. Maximum height: 90.8 ft
   Range: 209.6 ft

f. Maximum height: 204.2 ft
   Range: 471.6 ft

65. $x = a\theta - b\sin\theta$
   $y = a - b\cos\theta$

67. True

68. False, since neither x nor y can be negative, the graph consists of only the portion of the line $y = x$ in the first quadrant.

69. False, let $x = t^2$ and $y = t$.

70. False, let $x = \sin t$ and $y = \cos t$

71. $t = 2 : (-3)^2 + 4^2 = 5^2$
   $t = 3 : (-8)^2 + 6^2 = 10^2$

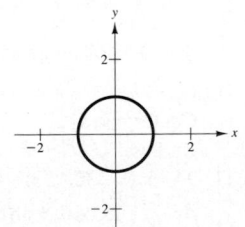

72. d   **73.** c   **74.** a   **75.** e   **76.** f   **77.** b

Section 10.2 (*page 684*)

1. $\dfrac{dy}{dx} = \dfrac{3}{2}$, $\dfrac{d^2y}{dx^2} = 0$

3. $\dfrac{dy}{dx} = 2t + 3$, $\dfrac{d^2y}{dx^2} = 2$

   At $t = -1, \dfrac{dy}{dx} = 1$, $\dfrac{d^2y}{dx^2} = 2$

5. $\dfrac{dy}{dx} = -\cot\theta$, $\dfrac{d^2y}{dx^2} = -\dfrac{\csc^3\theta}{2}$

   At $\theta = \dfrac{\pi}{4}, \dfrac{dy}{dx} = -1$, $\dfrac{d^2y}{dx^2} = -\sqrt{2}$

7. $\dfrac{dy}{dx} = 2\csc\theta$, $\dfrac{d^2y}{dx^2} = -2\cot^3\theta$

   At $\theta = \dfrac{\pi}{6}, \dfrac{dy}{dx} = 4$, $\dfrac{d^2y}{dx^2} = -6\sqrt{3}$

9. $\dfrac{dy}{dx} = -\tan\theta$, $\dfrac{d^2y}{dx^2} = \dfrac{\sec^4\theta\csc\theta}{3}$

   At $\theta = \dfrac{\pi}{4}, \dfrac{dy}{dx} = -1$, $\dfrac{d^2y}{dx^2} = \dfrac{4\sqrt{2}}{3}$

11. $\left(-\dfrac{2}{\sqrt{3}}, \dfrac{3}{2}\right), 3\sqrt{3}x - 8y + 18 = 0$

   $(0, 2), y - 2 = 0$

   $\left(2\sqrt{3}, \dfrac{1}{2}\right), \sqrt{3}x + 8y - 10 = 0$

13. $2x - y - 5 = 0$ **15.** $y - 2 = 0$

17. Horizontal: $(1, 0), (-1, \pi), (1, -2\pi)$

   Vertical: $\left(\dfrac{\pi}{2}, 1\right), \left(-\dfrac{3\pi}{2}, -1\right), \left(\dfrac{5\pi}{2}, 1\right)$

19. Horizontal: $(1, 0)$
   Vertical: None

21. Horizontal: $(0, -2), (2, 2)$
   Vertical: None

23. Horizontal: $(0, 3), (0, -3)$
   Vertical $(3, 0), (-3, 0)$

25. Horizontal: $(4, 0), (4, -2)$
   Vertical: $(2, -1), (6, -1)$

27. Horizontal: None
   Vertical: $(1, 0), (-1, 0)$

29. $\sqrt{2}(1 - e^{-\pi/2}) \approx 1.12$

31. $2\sqrt{5} + \ln(2 + \sqrt{5}) \approx 5.916$

33. $\dfrac{1}{12}[\ln(\sqrt{37} + 6) + 6\sqrt{37}] \approx 3.249$ **35.** $6a$

37. $8a$

39. a.

b. 219.2 ft   **c.** 230.8 ft

41. a.

b. $(0, 0), (1.6799, 2.1165)$ **c.** 6.557

43. a. $32\pi\sqrt{5}$ **b.** $16\pi\sqrt{5}$

45. 32π **47.** $\dfrac{12\pi a^2}{5}$

49. $2\pi r^2(1 - \cos\theta)$

51. $\dfrac{3\pi}{8}$ **53.** $\dfrac{3\pi}{2}$ **55.** $\left(\dfrac{3}{4}, \dfrac{8}{5}\right)$ **57.** 36π

59. a.

b. The parameter can be considered as time and the orientation of the graph gives the direction of movement of the weight.

61. False,

$$\dfrac{d^2y}{dx^2} = \dfrac{\dfrac{d}{dt}\left[\dfrac{g'(t)}{f'(t)}\right]}{f'(t)} = \dfrac{f'(t)g''(t) - g'(t)f''(t)}{[f'(t)]^3}.$$

62. False, the graph of $y = x^{2/3}$ does not have a horizontal asymptote at the origin.

63. True   **64.** True

Section 10.3   (*page 694*)

1.

3.

5.

7. $\left(\sqrt{2}, \dfrac{\pi}{4}\right), \left(-\sqrt{2}, \dfrac{5\pi}{4}\right)$

9. $(5, 2.214), (-5, 5.356)$

11. $r = a$ **13.** $r = 4\csc\theta$

15. $r = -\dfrac{2}{3\cos\theta - \sin\theta}$ **17.** $r = 9\csc^2\theta\cos\theta$

19. $x^2 + y^2 = 9$ **21.** $x^2 + y^2 - y = 0$

   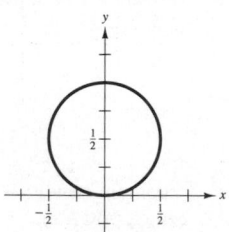

23. $\sqrt{x^2 + y^2} = \arctan\dfrac{y}{x}$ **25.** $x - 3 = 0$

27.

$-\dfrac{\pi}{2} \le \theta \le \dfrac{\pi}{2}$

Horizontal tangents: 2

29.

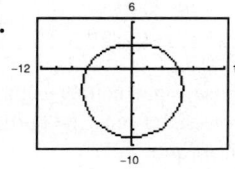

$0 \le \theta \le 2\pi$

Horizontal tangents: 2

31. $(x - h)^2 + (y - k)^2 = h^2 + k^2$

Center: (h, k)

Radius: $\sqrt{h^2 + k^2}$

33. $2\sqrt{5}$ **35.** 5.6

37. $\dfrac{dy}{dx} = \dfrac{2\cos\theta(3\sin\theta + 1)}{6\cos^2\theta - 2\sin\theta - 3}$

$\left(5, \dfrac{\pi}{2}\right), \dfrac{dy}{dx} = 0$

$(2, \pi), \dfrac{dy}{dx} = -\dfrac{2}{3}$

$\left(-1, \dfrac{3\pi}{2}\right), \dfrac{dy}{dx} = 0$

39. $\dfrac{(1 + 2\cos\theta)(1 - \cos\theta)}{\sin\theta(2\cos\theta - 1)}$, -1

41. $\dfrac{2\sin\theta\cos\theta}{1 - 2\sin^2\theta}$, $-\sqrt{3}$

43. Horizontal: $\left(2, \dfrac{\pi}{2}\right)$, $\left(\dfrac{1}{2}, \dfrac{7\pi}{6}\right)$, $\left(\dfrac{1}{2}, \dfrac{11\pi}{6}\right)$

Vertical: $\left(\dfrac{3}{2}, \dfrac{\pi}{6}\right)$, $\left(\dfrac{3}{2}, \dfrac{5\pi}{6}\right)$

45. $\left(5, \dfrac{\pi}{2}\right)$, $\left(1, \dfrac{3\pi}{2}\right)$

47. $(0, 0)$, $(1.4142, 0.7854)$, $(1.4142, 2.3562)$

49. $(7, 1.5708)$, $(3, 4.7124)$ **51.** $\theta = 0$

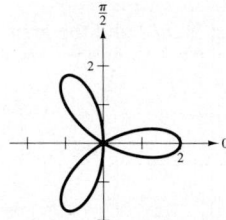

53. $\theta = \dfrac{\pi}{6}, \dfrac{\pi}{2}, \dfrac{5\pi}{6}$ **55.** $\theta = 0, \dfrac{\pi}{2}$

57.   **59.**

61.   **63.**

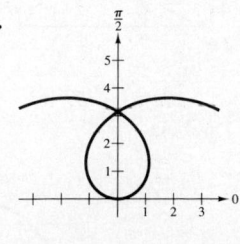

65. $0 \le \theta \le 2\pi$ **67.** $0 \le \theta < 2\pi$

69. $0 \le \theta < 4\pi$ **71.** $0 \le \theta < \pi$

73.   **75.**

77. a.   **b.**

c.   **d.**

81. a. $r = 2 - \sin\left(\theta - \dfrac{\pi}{4}\right)$ **b.** $r = 2 + \cos\theta$

$= 2 - \dfrac{\sqrt{2}(\sin\theta - \cos\theta)}{2}$

c. $r = 2 + \sin\theta$ **d.** $r = 2 - \cos\theta$

83. a.

b.

85. True   **86.** True   **87.** True   **88.** True

89. False, the graph of $r = \cos\theta$ has a vertical tangent at $(1, 0)$.

90. False, the diameter is 1.

93.

95.

Section 10.4 (*page 703*)

1. 16π **3.** $\dfrac{\pi}{3}$ **5.** $\dfrac{\pi}{8}$ **7.** $\dfrac{3\pi}{2}$

9. $\dfrac{2\pi - 3\sqrt{3}}{2}$ **11.** $\pi + 3\sqrt{3}$

13. $\left(1, \dfrac{\pi}{2}\right), \left(1, \dfrac{3\pi}{2}\right), (0, 0)$

15. $\left(\dfrac{2 - \sqrt{2}}{2}, \dfrac{3\pi}{4}\right), \left(\dfrac{2 + \sqrt{2}}{2}, \dfrac{7\pi}{4}\right), (0, 0)$

17. $\left(\dfrac{3}{2}, \dfrac{\pi}{6}\right), \left(\dfrac{3}{2}, \dfrac{5\pi}{6}\right), (0, 0)$ **19.** $(2, 4), (-2, -4)$

21. $\left(2, \dfrac{\pi}{12}\right), \left(2, \dfrac{5\pi}{12}\right), \left(2, \dfrac{7\pi}{12}\right), \left(2, \dfrac{11\pi}{12}\right)$

$\left(2, \dfrac{13\pi}{12}\right), \left(2, \dfrac{17\pi}{12}\right), \left(2, \dfrac{19\pi}{12}\right), \left(2, \dfrac{23\pi}{12}\right)$

23. $(0.581, \pm 0.535), (2.581, \pm 1.376)$

25. $(0, 0), (0.935, 0.363), (0.535, -1.006)$

The graphs reach the pole at different times (θ-values).

27. $\dfrac{4}{3}\left(4\pi - 3\sqrt{3}\right)$ **29.** $11\pi - 24$

31. $\dfrac{2}{3}\left(4\pi - 3\sqrt{3}\right)$

33. $\dfrac{5\pi a^2}{4}$ **35.** $\dfrac{a^2}{2}(\pi - 2)$

37. a. $(x^2 + y^2)^{3/2} = ax^2$

b.

c. $\dfrac{15\pi}{2}$

39. $2\pi a$ **41.** 8

43. $\dfrac{\pi}{4}\sqrt{\pi^2 + 4} + \ln\left(\dfrac{\pi + \sqrt{\pi^2 + 4}}{2}\right)$

45. $\ln\left(\dfrac{\sqrt{4\pi^2 + 1} + 2\pi}{\sqrt{\pi^2 + 1} + \pi}\right) + \dfrac{2\sqrt{\pi^2 + 1} - \sqrt{4\pi^2 + 1}}{2\pi}$

47. $A \approx 0.667, s \approx 4.439$

49. 4π **51.** $\dfrac{2\pi\sqrt{1 + a^2}}{1 + 4a^2}(e^{\pi a} - 2a)$ **53.** 21.87

55. $4\pi^2 ab$

57. False, the area is given by $\displaystyle\int_0^{\pi/2} \sin^2\theta\, d\theta$.

58. False, the graphs of $f(\theta) = 1$ and $g(\theta) = -1$ coincide.

59. False, if $f(\theta) = 0$ and $g(\theta) = \sin 2\theta$, then there is only one point of intersection.

60. True, the area enclosed by the first is $\pi/2$ and the area enclosed by the second is $\pi/4$.

63. $y^2 = \dfrac{x^2(1 + x)}{1 - x}$

$A = 2 - \dfrac{\pi}{2}$

Section 10.5 (*page 710*)

1.

3.

5. c   **7.** a   **9.** b

11. Parabola

13. Ellipse

15. Ellipse

17. Hyperbola

19. Hyperbola

21. $\pi/4$ radians counterclockwise

23. $\pi/6$ radians clockwise

25. $r = \dfrac{1}{1 - \cos\theta}$

27. $r = \dfrac{1}{2 + \sin\theta}$ **29.** $r = \dfrac{2}{1 + 2\cos\theta}$

31. $r = \dfrac{2}{1 - \sin\theta}$ **33.** $r = \dfrac{16}{5 + 3\cos\theta}$

35. $r = \dfrac{9}{4 - 5\sin\theta}$ **39.** $r^2 = \dfrac{9}{1 - (16/25)\cos^2\theta}$

41. $r^2 = \dfrac{-16}{1 - (25/9)\cos^2\theta}$ **43.** 10.88

47. $\dfrac{\pi}{2}$ **49.** 0   **51.** $\dfrac{\pi}{3}$ **53.** $r = \dfrac{8200}{1 + \sin\theta}$, 1466.67 mi

57. $r = \dfrac{92,931,075.2223}{1 - 0.0167\cos\theta}$
Perihelion: 91,404,618 mi
Aphelion: 94,509,382 mi

59. $r = \dfrac{5.537 \times 10^9}{1 - 0.2481\cos\theta}$
Perihelion: 4.436×10^9 k
Aphelion: 7.364×10^9 k

61. a. 9.341×10^{18} k^2
    21.867 yr
b. 0.899 rad, larger
c. a. 2.559×10^9 k, 1.17×10^8 k/yr
    b. 4.118×10^9 k, 1.88×10^8 k/yr

Review Exercises for Chapter 10

1. a. $\dfrac{dy}{dx} = -\dfrac{3}{4}$
Horizontal tangents: None
b. $y = \dfrac{-3x + 11}{4}$
c.

3. a. $\dfrac{dy}{dx} = -2t^2$
Horizontal tangents: None
b. $y = 3 + \dfrac{2}{x}$
c.

5. a. $\dfrac{dy}{dx} = \dfrac{(t - 1)(2t + 1)^2}{t^2(t - 2)^2}$
Horizontal tangents: $\left(\dfrac{1}{3}, -1\right)$
b. $y = \dfrac{4x^2}{(5x - 1)(x - 1)}$
c.

7. a. $\dfrac{dy}{dx} = -\dfrac{5}{2}\cot\theta$
Horizontal tangents: (3, 7), (3, −3)
b. $\dfrac{(x - 3)^2}{4} + \dfrac{(y - 2)^2}{25} = 1$
c.

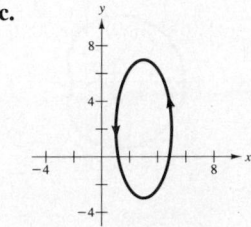

9. a. $\dfrac{dy}{dx} = -4\tan\theta$

Horizontal tangents: $(\pm 1, 0)$

b. $x^{2/3} + (y/4)^{2/3} = 1$

c.

11. a. $\dfrac{dy}{dx} = \dfrac{2\cos 2\theta}{-\csc^2\theta}$

Horizontal tangents: $(1, 1), (-1, -1)$

b.

13. $x = -3 + 4\cos\theta$
$y = 4 + 3\sin\theta$

15.

19. $\dfrac{\pi^2 r}{2}$

21. Circle   **23.** Line

25. Cardioid   **27.** Limaçon

   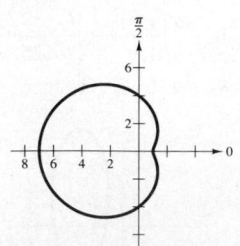

29. Rose curve   **31.** Rose curve

33. Parabola   **35.**

37.

39. $x^2 + y^2 - 3x = 0$

41. $(x^2 + y^2 + 2x)^2 = 4(x^2 + y^2)$

43. $(x^2 + y^2)^2 = x^2 - y^2$

45. $y^2 = x^2\left(\dfrac{4-x}{4+x}\right)$ **47.** $r = a\cos^2\theta\sin\theta$

49. $r^2 = a^2\theta^2$ **51.** $r = 10\sin\theta$

53. $r = \dfrac{4}{1 - \cos\theta}$ **55.** $r = \dfrac{5}{3 - 2\cos\theta}$

57. a. $\pm\dfrac{\pi}{3}$

b. Vertical: $(-1, 0), (3, \pi), (\frac{1}{2}, \pm 1.318)$
Horizontal: $(-0.686, \pm 0.568), (2.186, \pm 2.206)$

c.

61. $\dfrac{9\pi}{2}$ **63.** $\dfrac{\pi}{32}$ **65.** a^2 **67.** $\dfrac{8\pi - 6\sqrt{3}}{3}$

69. $8a$ **71.** $\arctan\left(\dfrac{2\sqrt{3}}{3}\right) \approx 49.1°$

CHAPTER 11
Section 11.1   (*page 722*)

1. $\langle 4, 2 \rangle$

3. $\langle -7, 0 \rangle$

5. $\langle 4, 3 \rangle$

7. $\langle -4, -3 \rangle$

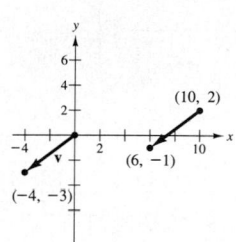

9. $\langle 0, 4 \rangle$

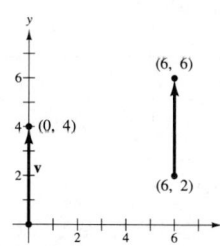

11. $\langle -1, \frac{5}{3} \rangle$

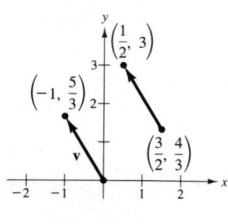

13. a. $\langle 4, 6 \rangle$

b. $\langle -6, -9 \rangle$

c. $\langle 7, \frac{21}{2} \rangle$

d. $\langle \frac{4}{3}, 2 \rangle$

15. $\langle 3, -\frac{3}{2} \rangle$

17. $\langle 4, 3 \rangle$

19. $\langle \frac{7}{2}, -\frac{1}{2} \rangle$

21. $a = 1, b = 1$

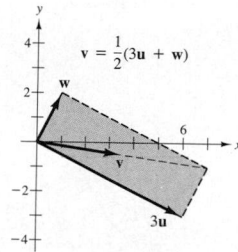

23. $a = 1, b = 2$ **25.** $a = \frac{2}{3}, b = \frac{1}{3}$

27. $(3, 5)$ **29.** 5   **31.** $\sqrt{61}$ **33.** 4

35. a. $\sqrt{2}$ **b.** $\sqrt{5}$ **c.** 1   **d.** 1   **e.** 1   **f.** 1

37. a. $\sqrt{5}/2$ **b.** $\sqrt{13}$ **c.** $\sqrt{85}/2$
   d. 1   **e.** 1   **f.** 1

41. $\langle 2\sqrt{2}, 2\sqrt{2} \rangle$ **43.** $\langle 1, \sqrt{3} \rangle$

45. a. $\pm \dfrac{1}{\sqrt{10}} \langle 1, 3 \rangle$ **b.** $\pm \dfrac{1}{\sqrt{10}} \langle 3, -1 \rangle$

47. a. $\pm \dfrac{1}{5} \langle -4, 3 \rangle$ **b.** $\pm \dfrac{1}{5} \langle 3, 4 \rangle$

49. $\langle 3, 0 \rangle$ **51.** $\langle -\sqrt{3}, 1 \rangle$ **53.** $\left(\dfrac{3 + \sqrt{2}}{\sqrt{2}} \right) \mathbf{i} + \left(\dfrac{3}{\sqrt{2}} \right) \mathbf{j}$

55. $(2 \cos 4 + \cos 2) \mathbf{i} + (2 \sin 4 + \sin 2) \mathbf{j}$

57. $-\dfrac{\sqrt{2}}{2} \mathbf{i} + \dfrac{\sqrt{2}}{2} \mathbf{j}$ **59.** 1.33, 132.5°

61. 12.3°, 82.2 lb   **63.** 71.3°, 228.5 lb

65. Tension in the rope: 1.1547 lb
   Magnitude of $\mathbf{u}$: 0.5774 lb

67. Tensions: 44.65 lb and 65.27 lb
   Vertical components: 38.67 lb and 61.33 lb

69. N 85.13° E, 416.15 mph   **71.** $(-4, -1), (6, 5), (10, 3)$

73. $T_2 = 157.316$
   $T_3 = 3692.482$

77. True   **78.** True   **79.** True   **80.** False, $a = b = 0$

81. False, $\| a\mathbf{i} + b\mathbf{j} \| = \sqrt{2}|a|$ **82.** True

Section 11.2 (*page 731*)

1.

3.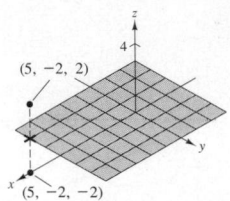

5. $3, 3\sqrt{5}, 6$ **7.** $6, 6, 2\sqrt{10}$
Right triangle   Isosceles triangle
9. $(\frac{3}{2}, -3, 5)$ **11.** $x^2 + y^2 + z^2 - 4y - 10z + 25 = 0$
13. $x^2 + y^2 + z^2 - 2x - 6y = 0$
15. Center: $(1, -3, -4)$ **17.** Center: $(\frac{1}{3}, -1, 0)$
Radius: 5   Radius: 1
19. a. $\langle -2, 2, 2 \rangle$ **21. a.** $\langle -3, 0, 3 \rangle$
  b.   **b.**

23. $\langle 4, 1, 1 \rangle$

25. a.   **b.**

  c.   **d.**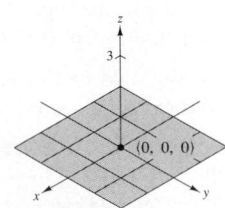

27. $\langle -1, 0, 4 \rangle$ **29.** $\langle 6, 12, 6 \rangle$ **31.** $\langle \frac{7}{2}, 3, \frac{5}{2} \rangle$
33. a and b   **35.** a   **37.** Collinear
39. Not collinear
43. $(3, 1, 8)$ **45.** 0   **47.** $\sqrt{14}$ **49.** $\sqrt{34}$
51. a. $\frac{1}{3}\langle 2, -1, 2 \rangle$ **b.** $-\frac{1}{3}\langle 2, -1, 2 \rangle$
53. a. $\frac{1}{\sqrt{38}}\langle 3, 2, -5 \rangle$ **b.** $-\frac{1}{\sqrt{38}}\langle 3, 2, -5 \rangle$
55. $\langle -26, 0, 48 \rangle$ **57.** $\frac{\sqrt{305}}{2}$

59. $\pm\frac{5}{3}$ **61.** $(2, -1, 2)$
63. $\langle 0, \sqrt{3}, \pm 1 \rangle$

65. a. 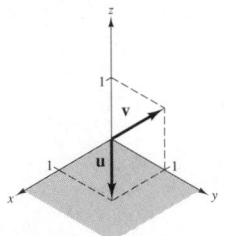   **c.** $a = b = 1$

67. $\frac{\sqrt{3}}{3}\langle 1, 1, 1 \rangle$ **69.** 8.99 lb
71. $\overrightarrow{AB}$: 202.920 N
$\overrightarrow{AC}$: 157.909 N
$\overrightarrow{AD}$: 226.520 N
73. $\left(x - \frac{4}{3}\right)^2 + (y - 3)^2 + \left(z + \frac{1}{3}\right)^2 = \frac{44}{9}$

Section 11.3 (*page 742*)

1. a. -6 **b.** 25   **c.** 25   **d.** $\langle -12, 18 \rangle$ **e.** -12
3. a. 2   **b.** 29   **c.** 29   **d.** $\langle 0, 12, 10 \rangle$ **e.** 4
5. a. 1   **b.** 6   **c.** 6   **d.** $\mathbf{i} - \mathbf{k}$ **e.** 2
7. $17,139.05$
Total revenue

9. 20   **11.** $\frac{\pi}{2}$ **13.** $\arccos\left(-\frac{1}{5\sqrt{2}}\right) \approx 98.1°$
15. $\arccos\left(\frac{\sqrt{2}}{3}\right) \approx 61.9°$ **17.** $\arccos\left(-\frac{8\sqrt{13}}{65}\right) \approx 116.3°$
19. Neither   **21.** Orthogonal
23. Neither   **25.** Orthogonal
29. a. $\theta = \frac{\pi}{2}$ **b.** $0 < \theta < \frac{\pi}{2}$ **c.** $\frac{\pi}{2} < \theta < \pi$
31. $\cos\alpha = \frac{1}{3}$ **33.** $\cos\alpha = 0$
$\cos\beta = \frac{2}{3}$ $\cos\beta = \frac{3}{\sqrt{13}}$
$\cos\gamma = \frac{2}{3}$ $\cos\gamma = -\frac{2}{\sqrt{13}}$
35. Magnitude: 124.309 lb   **37.** $\arccos\left(\frac{1}{\sqrt{3}}\right) \approx 54.7°$
$\alpha = 29.48°$
$\beta = 61.39°$
$\gamma = 96.53°$

39. $\alpha = 90°$
$\beta = 45°$
$\gamma = 45°$

41. a. $\langle \frac{5}{2}, \frac{1}{2} \rangle$ **b.** $\langle -\frac{1}{2}, \frac{5}{2} \rangle$

43. a. $\langle 0, \frac{33}{25}, \frac{44}{25} \rangle$ **b.** $\langle 2, -\frac{8}{25}, \frac{6}{25} \rangle$

45. a. $\langle \frac{2}{3}, \frac{1}{3}, -\frac{1}{3} \rangle$ **b.** $\langle \frac{1}{3}, \frac{2}{3}, \frac{4}{3} \rangle$

47. a. $\langle 5, 0, 0 \rangle$ **b.** $\langle 0, -4, 3 \rangle$

49. a. 8282.2 lb   **b.** 30,909.6 lb

51. a. **u** and **v** are parallel   **b.** **u** and **v** are orthogonal

53. 425 ft·lb   **55.** 72

Section 11.4   (*page 750*)

1. $-\mathbf{k}$ **3.** $\mathbf{i}$

5. $-\mathbf{j}$

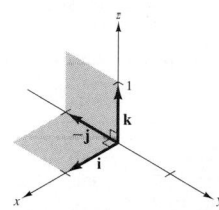

7. $\langle -1, -1, -1 \rangle$

9. $\langle 0, 0, 54 \rangle$ **11.** $\langle -2, 3, -1 \rangle$

13. $\left\langle -70, -23, \frac{57}{2} \right\rangle$

$\left\langle \dfrac{-140}{\sqrt{24{,}965}}, \dfrac{-46}{\sqrt{24{,}965}}, \dfrac{57}{\sqrt{24{,}965}} \right\rangle$

15. $\left\langle -\dfrac{71}{20}, -\dfrac{11}{5}, \dfrac{5}{4} \right\rangle$

$\left\langle \dfrac{-71}{\sqrt{7602}}, \dfrac{44}{\sqrt{7602}}, \dfrac{25}{\sqrt{7602}} \right\rangle$

17. 1   **19.** $6\sqrt{5}$ **21.** $2\sqrt{83}$ **23.** $\dfrac{3\sqrt{13}}{2}$

25. $\dfrac{9\sqrt{6}}{2}$ **27.** 1   **29.** 6   **31.** 2

33. 75   **35.** $10 \cos 40° \approx 7.66$ ft·lb

37. a. $\overrightarrow{AB} = -\dfrac{5}{4}\mathbf{j} + \mathbf{k}$

$\mathbf{F} = -200(\cos\theta\,\mathbf{j} + \sin\theta\,\mathbf{k})$

   b. $25(8\cos\theta + 10\sin\theta)$

   c. $25(4\sqrt{3} + 5)$ **d.** $\theta = 51.34°$

Section 11.5   (*page 759*)

1. a.

   b. $P = (1, 2, 2)$, $Q = (10, -1, 17)$, $\overrightarrow{PQ} = \langle 9, -3, 15 \rangle$
(There are many correct answers.) The components of
the vector and the coefficients of t are proportional
since the line is parallel to $\overrightarrow{PQ}$.

   c. $\left(-\frac{1}{5}, \frac{12}{5}, 0\right)$

	Parametric Equations	Symmetric Equations	Direction Numbers
3.	$x = t$ $y = 2t$ $z = 3t$	$x = \dfrac{y}{2} = \dfrac{z}{3}$	$1, 2, 3$
5.	$x = -2 + 2t$ $y = 4t$ $z = 3 - 2t$	$\dfrac{x+2}{2} = \dfrac{y}{4} = \dfrac{z-3}{-2}$	$2, 4, -2$
7.	$x = 1 + 3t$ $y = -2t$ $z = 1 + t$	$\dfrac{x-1}{3} = \dfrac{y}{-2} = \dfrac{z-1}{1}$	$3, -2, 1$
9.	$x = 5 + 17t$ $y = -3 - 11t$ $z = -2 - 9t$	$\dfrac{x-5}{17} = \dfrac{y+3}{-11} = \dfrac{z+2}{-9}$	$17, -11, -9$

11. $x = 2$ **13.** a, b
$y = 3$
$z = 4 + t$

15. $(2, 3, 1)$, $\cos\theta = \dfrac{7\sqrt{17}}{51}$ **17.** Nonintersecting

19. $(7, 8, -1)$

21. a. $P = (0, 0, -1)$, $Q = (0, -2, 0)$, $R = (3, 4, -1)$
$\overrightarrow{PQ} = \langle 0, -2, 1 \rangle$, $\overrightarrow{PR} = \langle 3, 4, 0 \rangle$
(There are many correct answers.)

   b. $\overrightarrow{PQ} \times \overrightarrow{PR} = \langle -4, 3, 6 \rangle$
The components of the cross product are proportional
to the coefficients of the variables in the equation. The
cross product is parallel to the normal vector.

23. $x - 2 = 0$ **25.** $2x + 3y - z = 10$

27. $3x - 2z = 0$ **29.** $3x + 9y - 7z = 0$

31. $4x - 3y + 4z = 10$ **33.** $z = 3$
35. $x + y + z = 5$ **37.** $7x + y - 11z = 5$
39. $y - z = -1$
41. Orthogonal   **43.** $83.5°$ **45.** Parallel
47.

49.

51.

53.

55.

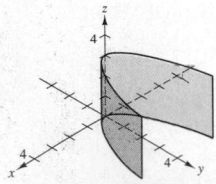

57. $x = 2$
     $y = 1 + t$
     $z = 1 + 2t$

59. $(2, -3, 2)$ **61.** Nonintersecting   **63.** 0
65. $\dfrac{6\sqrt{14}}{7}$ **67.** $\dfrac{2\sqrt{26}}{13}$ **69.** $\dfrac{10\sqrt{26}}{13}$
71. a. Sphere
     $x^2 + y^2 + z^2 - 6x + 4y - 10z + 22 = 0$
    b. Planes
     $4x - 3y + z = 10 \pm 4\sqrt{26}$
73. $\arccos \frac{1}{65} \approx 89.1°$ **75.** True
76. False, they may be nonintersecting.
    (See Exercise 61.)

Section 11.6   (*page 771*)

1. c   **2.** e   **3.** f   **4.** b   **5.** d   **6.** a
7. Plane   **9.** Right circular cylinder

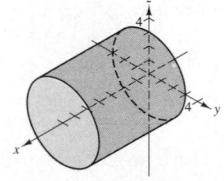

11. Parabolic cylinder   **13.** Elliptic cylinder

15. Cylinder

17. a. $(20, 0, 0)$
    b. $(10, 10, 20)$
    c. $(0, 0, 20)$
    d. $(0, 20, 0)$

19. Ellipsoid   **21.** Hyperboloid of one sheet

23. Elliptic paraboloid   **25.** Hyperbolic paraboloid

27. Cone   **29.** Ellipsoid

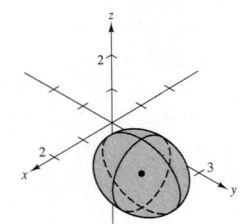

31. Cylinder   **33.** Surface of Revolution

35. Elliptic cone

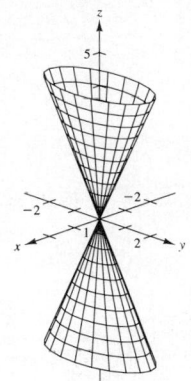

37. Hyperboloid of two sheets

17. $x^2 + y^2 - 2y = 0$ **19.** $x^2 + y^2 + z^2 = 4$

   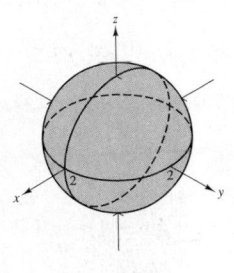

21. $\left(4, 0, \dfrac{\pi}{2}\right)$ **23.** $\left(4\sqrt{2}, \dfrac{2\pi}{3}, \dfrac{\pi}{4}\right)$ **25.** $\left(4, \dfrac{\pi}{6}, \dfrac{\pi}{6}\right)$

27. $\left(\sqrt{6}, \sqrt{2}, 2\sqrt{2}\right)$ **29.** $(0, 0, 12)$

31. $\left(\dfrac{5}{2}, \dfrac{5}{2}, -\dfrac{5\sqrt{2}}{2}\right)$

33. $x^2 + y^2 + z^2 = 4$ **35.** $3x^2 + 3y^2 - z^2 = 0$

39.

41.

43. $x^2 + z^2 = 4y$ **45.** $4x^2 + 4y^2 = z^2$

47. $y^2 + z^2 = \dfrac{4}{x^2}$ **49.** $y = \sqrt{2z}$ **51.** $\dfrac{128\pi}{3}$

53. a. Major axis: $4\sqrt{2}$
   Minor axis: 4
   Foci: $(0, \pm 2, 2)$
   b. Major axis: $8\sqrt{2}$
   Minor axis: 8
   Foci: $(0, \pm 4, 8)$

55. $\dfrac{x^2}{3963^2} + \dfrac{y^2}{3963^2} + \dfrac{z^2}{3942^2} = 1$ **57.** No

37. $x^2 + y^2 + (z - 2)^2 = 4$ **39.** $x^2 + y^2 = 1$

   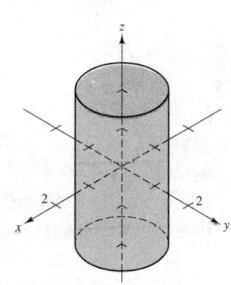

41. $\left(4, \dfrac{\pi}{4}, \dfrac{\pi}{2}\right)$ **43.** $\left(2\sqrt{13}, -\dfrac{\pi}{6}, \arccos\left[\dfrac{3}{\sqrt{13}}\right]\right)$

45. $\left(13, \pi, \arccos\left[\dfrac{5}{13}\right]\right)$ **47.** $\left(10, \dfrac{\pi}{6}, 0\right)$

49. $\left(3\sqrt{3}, -\dfrac{\pi}{6}, 3\right)$ **51.** $\left(4, \dfrac{7\pi}{6}, 4\sqrt{3}\right)$

Section 11.7   (*page 778*)

1. $\left(5, \dfrac{\pi}{2}, 1\right)$ **3.** $\left(2, \dfrac{\pi}{3}, 4\right)$ **5.** $\left(2\sqrt{2}, -\dfrac{\pi}{4}, -4\right)$

7. $(5, 0, 2)$ **9.** $\left(1, \sqrt{3}, 2\right)$ **11.** $\left(-2\sqrt{3}, -2, 3\right)$

13. $x^2 + y^2 = 4$ **15.** $x - \sqrt{3}y = 0$

   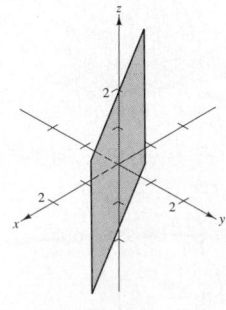

Rectangular	*Cylindrical*	*Spherical*
53. (4, 6, 3)	(7.211, 0.983, 3)	(7.810, 0.983, 1.177)
55. (4.698, 1.710, 8)	$\left(5, \dfrac{\pi}{9}, 8\right)$	(9.434, 0.349, 0.304)
57. (−7.071, 12.247, 14.142)	(14.142, 2.094, 14.142)	$\left(20, \dfrac{2\pi}{3}, \dfrac{\pi}{4}\right)$

59. d   **60.** e   **61.** c   **62.** a   **63.** f   **64.** b

65. a. $r^2 + z^2 = 16$ **b.** $\rho = 4$

67. a. $r^2 + (z - 1)^2 = 1$ **b.** $\rho = 2 \cos \phi$

69. a. $r = 4 \sin \theta$ **b.** $\rho = \dfrac{4 \sin \theta}{\sin \phi} = 4 \sin \theta \csc \phi$

71. a. $r^2 = \dfrac{9}{\cos^2 \theta - \sin^2 \theta}$ **b.** $\rho^2 = \dfrac{9 \csc^2 \phi}{\cos^2 \theta - \sin^2 \theta}$

73.

75.

77.

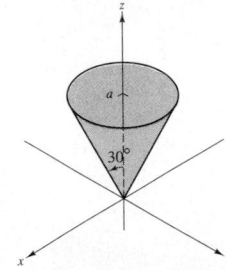

79. Rectangular: $0 \le x \le 10$
$0 \le y \le 10$
$0 \le z \le 10$

81. Spherical: $4 \le \rho \le 6$

83. a. Los Angeles: (4000, −118.24°, 55.95°)
Rio de Janeiro: (4000, −43.22°, 112.90°)

b. Los Angeles: (−1568.2, −2919.7, 2239.7)
Rio de Janeiro: (2685.2, −2523.3, −1556.5)

c. 91.18° ≈ 1.59 radians

d. 6366 mi

85. Ellipse

Review Exercises for Chapter 11

1. a. $\mathbf{u} = 3\mathbf{i} - \mathbf{j}$
$\mathbf{v} = 4\mathbf{i} + 2\mathbf{j}$
b. $2\sqrt{5}$ **c.** 10   **d.** 10**i**   **e.** 2**i** + **j**   **f.** **i** − 2**j**

3. a. $\mathbf{u} = -\mathbf{i} + 4\mathbf{j}$ **b.** 3
$\mathbf{v} = -3\mathbf{i} + 6\mathbf{j}$
c. 24**i** + 6**j** + 12**k**   **d.** $4x + y + 2z = 20$
e. $x = 4 - t, y = 4 + 4t, z = 0$

5. Parallel   **7.** $\theta = \arccos\left(\dfrac{\sqrt{2} + \sqrt{6}}{4}\right) = 15°$

9. π **11.** $-2\sqrt{2}\mathbf{i} + 2\sqrt{2}\mathbf{j}$

13. $\dfrac{3}{\sqrt{26}}\mathbf{i} - \dfrac{9}{\sqrt{26}}\mathbf{j} + \dfrac{12}{\sqrt{26}}\mathbf{k}$ **15.** $\sqrt{14}$

19. $\left\langle -\dfrac{15}{14}, \dfrac{10}{14}, -\dfrac{1}{14} \right\rangle$ **23.** 4

25. $\overline{AC}$: 21.7 lb   **27.** $\dfrac{27\sqrt{2}}{2} \approx 19.1$ in.
$\overline{BC}$: 17.7 lb

29. a. $x = 1, y = 2 + t, z = 3$ **b.** None

31. a. $x = t, y = -1 + t, z = 1$
b. $x = y + 1, z = 1$

33. $x + y + z = 6$

35. $x + 2y = 1$ **37.** $\dfrac{8}{7}$ **39.** $\dfrac{\sqrt{35}}{7}$ **41.** $\dfrac{\sqrt{3}}{3}$

43.

45.

47.

49.

51.

53.

55.

57. a. $x^2 + y^2 - 2z + 2 = 0$
b. $8\pi \approx 25.1$ cm³
c. $\dfrac{495\pi}{64} \approx 24.3$ cm³

59. a. $\left(4, \dfrac{3\pi}{4}, 2\right)$
b. $\left(2\sqrt{5}, \dfrac{3\pi}{4}, \arccos\left[\dfrac{\sqrt{5}}{5}\right]\right)$

61. a. $r^2 \cos 2\theta = 2z$
 b. $\rho = 2 \sec 2\theta \cos \phi \csc^2 \phi$
63. $x^2 - y^2 + z^2 = 1$ **65.** $x^2 + y^2 = 1$

CHAPTER 12

Section 12.1   (*page 788*)

1. $(-\infty, 0), (0, \infty)$ **3.** $(0, \infty)$
5. $[0, \infty)$ **7.** $(-\infty, \infty)$
9. $\sqrt{1 + t^2}$ **11.** b   **13.** d
15. a. $(-20, 0, 0)$ **b.** $(10, 20, 10)$
  c. $(0, 0, 20)$ **d.** $(20, 0, 0)$

17.

19.

21.

23.

25.
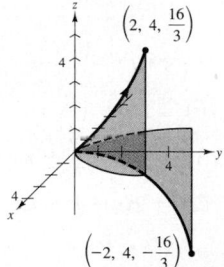

27. Parabola   **29.** Helix   **31.** $\mathbf{r}(t) = t\mathbf{i} + (4 - t)\mathbf{j}$
33. $\mathbf{r}(t) = 5 \cos t\,\mathbf{i} + \sin t\,\mathbf{j}$
35. $\mathbf{r}(t) = t\mathbf{i} - t\mathbf{j} + 2t^2\mathbf{k}$ **37.** $\mathbf{r}(t) = 2 \sin t\,\mathbf{i} + 2 \cos t\,\mathbf{j}$
                              $+ 4 \sin^2 t\,\mathbf{k}$

39. $\mathbf{r}(t) = (1 + \sin t)\mathbf{i} + \sqrt{2} \cos t\,\mathbf{j} + (1 - \sin t)\mathbf{k}$
and
$\mathbf{r}(t) = (1 + \sin t)\mathbf{i} - \sqrt{2} \cos t\,\mathbf{j} + (1 - \sin t)\mathbf{k}$

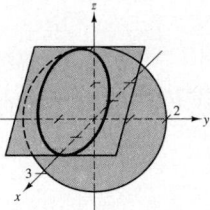

41. $\mathbf{r}(t) = t\mathbf{i} + t\mathbf{j} + \sqrt{4 - t^2}\,\mathbf{k}$

43. $2\mathbf{i} + 2\mathbf{j} + \frac{1}{2}\mathbf{k}$ **45.** 0
47. Limit does not exist   **49.** $(-\infty, 0), (0, \infty)$
51. $[-1, 1]$ **53.** $\left(-\dfrac{\pi}{2} + n\pi, \dfrac{\pi}{2} + n\pi\right)$ **59.** True

Section 12.2   (*page 796*)

1. $\mathbf{r}(2) = 4\mathbf{i} + 2\mathbf{j}$
  $\mathbf{r}'(2) = 4\mathbf{i} + \mathbf{j}$

3. $\mathbf{r}\left(\dfrac{\pi}{2}\right) = \mathbf{j}$
  $\mathbf{r}'\left(\dfrac{\pi}{2}\right) = -\mathbf{i}$

5. a.,   b.

c. The vector
$$\dfrac{\mathbf{r}\left(\frac{1}{2}\right) - \mathbf{r}\left(\frac{1}{4}\right)}{\frac{1}{2} - \frac{1}{4}}$$
approximates the tangent vector $\mathbf{r}'\left(\frac{1}{4}\right)$.

7. $\mathbf{r}\left(\dfrac{3\pi}{2}\right) = -2\mathbf{j} + \left(\dfrac{3\pi}{2}\right)\mathbf{k}$

$\mathbf{r}'\left(\dfrac{3\pi}{2}\right) = 2\mathbf{j} + \mathbf{k}$

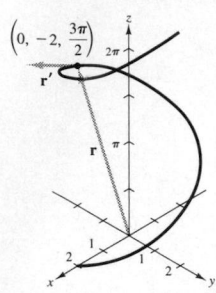

9. $\dfrac{\mathbf{r}'(-\frac{1}{4})}{\|\mathbf{r}'(-\frac{1}{4})\|} = \dfrac{1}{\sqrt{4\pi^2+1}}(\sqrt{2}\pi\mathbf{i} + \sqrt{2}\pi\mathbf{j} - \mathbf{k})$

$\dfrac{\mathbf{r}''(-\frac{1}{4})}{\|\mathbf{r}''(-\frac{1}{4})\|} = \dfrac{1}{2\sqrt{\pi^2+4}}(-\sqrt{2}\pi^2\mathbf{i} + \sqrt{2}\pi^2\mathbf{j} + 4\mathbf{k})$

11. $6\mathbf{i} - 14t\mathbf{j} + 3t^2\mathbf{k}$

13. $-3a\cos^2 t\,\mathbf{i} + 3a\sin^2 t\cos t\,\mathbf{j}$ **15.** $-e^{-t}\mathbf{i}$

17. $\langle \sin t + t\cos t,\ \cos t - t\sin t,\ 1\rangle$

19. a. $\mathbf{i} + 3\mathbf{j} + 2t\mathbf{k}$ **b.** $2\mathbf{k}$ **c.** $8t + 9t^2 + 5t^4$

d. $-\mathbf{i} + (9 - 2t)\mathbf{j} + (6t - 3t^2)\mathbf{k}$

e. $8t^3\mathbf{i} + (12t^2 - 4t^3)\mathbf{j} + (3t^2 - 24t)\mathbf{k}$

f. $\dfrac{10 + 2t^2}{\sqrt{10 + t^2}}$

21. $(-\infty, 0),\ (0, \infty)$ **23.** $\left(\dfrac{n\pi}{2},\ \dfrac{(n+1)\pi}{2}\right)$

25. $(-\infty, \infty)$ **27.** $t^2\mathbf{i} + t\mathbf{j} + t\mathbf{k} + \mathbf{C}$

29. $\ln t\,\mathbf{i} + t\mathbf{j} - \frac{2}{5}t^{5/2}\mathbf{k} + \mathbf{C}$

31. $e^t\mathbf{i} - \cos t\,\mathbf{j} + \sin t\,\mathbf{k} + \mathbf{C}$

33. $\tan t\,\mathbf{i} + \arctan t\,\mathbf{j} + \mathbf{C}$ **35.** $2e^{2t}\mathbf{i} + 3(e^t - 1)\mathbf{j}$

37. $600\sqrt{3}t\,\mathbf{i} + (-16t^2 + 600t)\mathbf{j}$

39. $\left(\dfrac{2 - e^{-t^2}}{2}\right)\mathbf{i} + (e^{-t} - 2)\mathbf{j} + (t + 1)\mathbf{k}$

41. $4\mathbf{i} + \dfrac{1}{2}\mathbf{j} - \mathbf{k}$ **43.** $a\mathbf{i} + a\mathbf{j} + \dfrac{\pi}{2}\mathbf{k}$

53. False, let $\mathbf{r}(t) = \cos t\,\mathbf{i} + \sin t\,\mathbf{j} + \mathbf{k}$, then $\dfrac{d}{dt}[\|\mathbf{r}(t)\|] = 0$, but $\|\mathbf{r}'(t)\| = 1$.

54. False, $D_t[\mathbf{r}(t)\cdot\mathbf{u}(t)] = \mathbf{r}(t)\cdot\mathbf{u}'(t) + \mathbf{r}'(t)\cdot\mathbf{u}(t)$

Section 12.3   (*page 804*)

1. $\mathbf{v}(t) = 3\mathbf{i} + \mathbf{j}$
$\mathbf{a}(t) = \mathbf{0}$

3. $\mathbf{v}(t) = 4\mathbf{i} + \mathbf{j}$
$\mathbf{a}(t) = 2\mathbf{i}$

5. $\mathbf{v}(t) = -\sqrt{2}\mathbf{i} + \sqrt{2}\mathbf{j}$
$\mathbf{a}(t) = -\sqrt{2}\mathbf{i} - \sqrt{2}\mathbf{j}$

7. $\mathbf{v}(t) = 2\mathbf{i}$
$\mathbf{a}(t) = -\mathbf{j}$

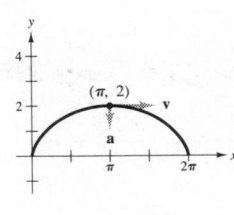

9. $\mathbf{v}(t) = \mathbf{i} + 2\mathbf{j} + 3\mathbf{k}$
$s(t) = \sqrt{14}$
$\mathbf{a}(t) = \mathbf{0}$

11. $\mathbf{v}(t) = \mathbf{i} + 2t\mathbf{j} + t\mathbf{k}$
$s(t) = \sqrt{1 + 5t^2}$
$\mathbf{a}(t) = 2\mathbf{j} + \mathbf{k}$

13. $\mathbf{v}(t) = \mathbf{i} + \mathbf{j} - \dfrac{t}{\sqrt{9 - t^2}}\mathbf{k}$

$s(t) = \sqrt{\dfrac{18 - t^2}{9 - t^2}}$

$\mathbf{a}(t) = \dfrac{-9}{(9 - t^2)^{3/2}}\mathbf{k}$

15. $\mathbf{v}(t) = 4\mathbf{i} - 3\sin t\,\mathbf{j} + 3\cos t\,\mathbf{k}$
$s(t) = 5$
$\mathbf{a}(t) = -3\cos t\,\mathbf{j} - 3\sin t\,\mathbf{k}$

17. $x = 1 + t$
$y = -1 - 2t$
$z = \dfrac{1}{4} + \dfrac{3}{4}t$
$(1.100, -1.200, 0.325)$

19. $\mathbf{v}(t) = t(\mathbf{i} + \mathbf{j} + \mathbf{k})$
$\mathbf{r}(t) = \dfrac{t^2}{2}(\mathbf{i} + \mathbf{j} + \mathbf{k})$
$\mathbf{r}(2) = 2(\mathbf{i} + \mathbf{j} + \mathbf{k})$

21. $\mathbf{v}(t) = \left(\dfrac{t^2}{2} + \dfrac{9}{2}\right)\mathbf{j} + \left(\dfrac{t^2}{2} - \dfrac{1}{2}\right)\mathbf{k}$

$\mathbf{r}(t) = \left(\dfrac{t^3}{6} + \dfrac{9}{2}t - \dfrac{14}{3}\right)\mathbf{j} + \left(\dfrac{t^3}{6} - \dfrac{1}{2}t + \dfrac{1}{3}\right)\mathbf{k}$

$\mathbf{r}(2) = \dfrac{17}{3}\mathbf{j} + \dfrac{2}{3}\mathbf{k}$

23. $\mathbf{r}(t) = 44\sqrt{3}t\,\mathbf{i} + (10 + 44t - 16t^2)\mathbf{j}$

25. $v_0 = 40\sqrt{6}$ ft/sec, 78 ft

27. a. 54.1 ft/sec   **b.** 22.0 ft   **c.** 2.0 sec

29. $v_0 = 32$ ft/sec, $\theta = 45°$

31. a. $\theta = \dfrac{\pi}{4}$ **b.** $\theta = \dfrac{\pi}{2}$ **33.** 96.7 ft/sec

37. $\mathbf{v}(t) = b\omega[(1 - \cos\omega t)\mathbf{i} + \sin\omega t\,\mathbf{j}]$

$\mathbf{a}(t) = b\omega^2[\sin\omega t\,\mathbf{i} + \cos\omega t\,\mathbf{j}]$

   a. $\|\mathbf{v}(t)\| = 0$ when $\omega t = 0, 2\pi, 4\pi, \ldots$

   b. $\|\mathbf{v}(t)\|$ is maximum when $\omega t = \pi, 3\pi, \ldots$

39. $\mathbf{v}(t) = -b\omega\sin\omega t\,\mathbf{i} + b\omega\cos\omega t\,\mathbf{j}$

$\mathbf{v}(t)\cdot\mathbf{r}(t) = 0$

41. $\mathbf{a}(t) = -b\omega^2[\cos\omega t\,\mathbf{i} + \sin\omega t\,\mathbf{j}] = -\omega^2\mathbf{r}(t)$

43. $8\sqrt{10}$ ft/sec   **49.** $\mathbf{r}_3{}'(t) = \omega\mathbf{r}_1{}'(t)$

$\mathbf{r}_3{}''(t) = \omega^2\mathbf{r}_1{}''(t)$

Section 12.4   (*page 814*)

1. $\mathbf{T}(0) = \dfrac{\sqrt{2}}{2}(\mathbf{i} + \mathbf{k})$ **3.** $\mathbf{T}(0) = \dfrac{\sqrt{5}}{5}(2\mathbf{j} + \mathbf{k})$

$\begin{array}{ll} x = t & x = 2 \\ y = 0 & y = 2t \\ z = t & z = t \end{array}$

5. $\mathbf{T}\left(\dfrac{\pi}{4}\right) = \dfrac{1}{2}\langle -\sqrt{2}, \sqrt{2}, 0\rangle$

$x = \sqrt{2} - \sqrt{2}t$

$y = \sqrt{2} + \sqrt{2}t$

$z = 4$

7. $\mathbf{T}(3) = \frac{1}{19}\langle 1, 6, 18\rangle$

$x = 3 + t$

$y = 9 + 6 + 6t$

$z = 18 + 18t$

9. $\mathbf{v}(t) = 4\mathbf{i}$

$\mathbf{a}(t) = \mathbf{0}$

$\mathbf{T}(t) = \mathbf{i}$

$\mathbf{N}(t) =$ undefined

The path is a line and the speed is constant.

11. $\mathbf{v}(t) = 8t\,\mathbf{i}$

$\mathbf{a}(t) = 8\mathbf{i}$

$\mathbf{T}(t) = \mathbf{i}$

$\mathbf{N}(t) = \mathbf{0}$

The path is a line and the speed is variable.

13. $\mathbf{T} = \dfrac{\sqrt{2}}{2}(\mathbf{i} - \mathbf{j})$ **15.** $\mathbf{T} = \dfrac{\sqrt{2}}{2}(-\mathbf{i} + \mathbf{j})$

$\mathbf{N} = \dfrac{\sqrt{2}}{2}(\mathbf{i} + \mathbf{j})$ $\mathbf{N} = -\dfrac{\sqrt{2}}{2}(\mathbf{i} + \mathbf{j})$

$a_{\mathbf{T}} = -\sqrt{2}$ $a_{\mathbf{T}} = \sqrt{2}e^{\pi/2}$

$a_{\mathbf{N}} = \sqrt{2}$ $a_{\mathbf{N}} = \sqrt{2}e^{\pi/2}$

17. $\mathbf{T} = (\cos\omega t_0)\mathbf{i} + (\sin\omega t_0)\mathbf{j}$

$\mathbf{N} = (-\sin\omega t_0)\mathbf{i} + (\cos\omega t_0)\mathbf{j}$

$a_{\mathbf{T}} = \omega^2$

$a_{\mathbf{N}} = \omega^3 t_0$

19. $\mathbf{T}(t) = -\sin(\omega t)\mathbf{i} + \cos(\omega t)\mathbf{j}$

$\mathbf{N}(t) = -\cos(\omega t)\mathbf{i} - \sin(\omega t)\mathbf{j}$

$a_{\mathbf{T}} = 0$

$a_{\mathbf{N}} = a\omega^2$

21. $\|\mathbf{v}(t)\| = a\omega$

The speed is constant since $a_{\mathbf{T}} = 0$.

23. $\mathbf{r}(2) = 2\mathbf{i} + \dfrac{1}{2}\mathbf{j}$

$\mathbf{T}(2) = \dfrac{\sqrt{17}}{17}(4\mathbf{i} - \mathbf{j})$

$\mathbf{N}(2) = \dfrac{\sqrt{17}}{17}(\mathbf{i} + 4\mathbf{j})$

25. a. Cycloid

b. $t = \dfrac{1}{2}: a_{\mathbf{T}} = \dfrac{\sqrt{2}\pi^2}{2}, a_{\mathbf{N}} = \dfrac{\sqrt{2}\pi^2}{2}$

$t = 1: a_{\mathbf{T}} = 0, a_{\mathbf{N}} = \pi^2$

$t = \dfrac{3}{2}: a_{\mathbf{T}} = \dfrac{\sqrt{2}\pi^2}{2}, a_{\mathbf{N}} = \dfrac{\sqrt{2}\pi^2}{2}$

c. $t = \frac{1}{2}$: Increasing

$t = 1$: Maximum

$t = \frac{1}{2}$: Decreasing

27. $\mathbf{T}(t) = \dfrac{\sqrt{14}}{14}(\mathbf{i} + 2\mathbf{j} - 3\mathbf{k})$

$\mathbf{N}(t) = \mathbf{0}$

$a_{\mathbf{T}} = 0$

$a_{\mathbf{N}} = 0$

29. $T = \dfrac{\sqrt{6}}{6}(i + 2j + k)$

$N = \dfrac{\sqrt{30}}{30}(-5i + 2j + k)$

$a \cdot T = \dfrac{5\sqrt{6}}{6}$

$a \cdot N = \dfrac{\sqrt{30}}{6}$

31. $T = \dfrac{1}{5}(4i - 3j)$

$N = -k$

$a \cdot T = 0$

$a \cdot N = 3$

33. $T\left(\dfrac{\pi}{2}\right) = \dfrac{\sqrt{17}}{17}(-4i + k)$

$N\left(\dfrac{\pi}{2}\right) = -j$

$B\left(\dfrac{\pi}{2}\right) = \dfrac{\sqrt{17}}{17}(i + 4k)$

35. $a_T = \dfrac{-32(v_0 \sin\theta - 32t)}{\sqrt{v_0^2 \cos^2\theta + (v_0 \sin\theta - 32t)^2}}$

$a_N = \dfrac{32 v_0 \cos\theta}{\sqrt{v_0^2 \cos^2\theta + (v_0 \sin\theta - 32t)^2}}$

37. **a.** Centripetal component is quadrupled
  b. Centripetal component is halved
39. 4.83 mi/sec   **41.** 4.67 mi/sec

Section 12.5 (*page 825*)

1. $4\sqrt{10}$ **3.** $6a$

    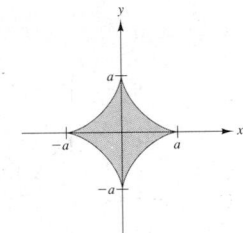

5. $2\sqrt{15}$ **7.** $2\pi\sqrt{a^2 + b^2}$

    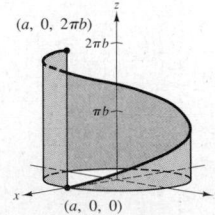

9. 8.37
11. **a.** $s = \sqrt{5}\,t$

  b. $r(t) = 2\cos\dfrac{s}{\sqrt{5}}i + 2\sin\dfrac{s}{\sqrt{5}}j + \dfrac{s}{\sqrt{5}}k$

  c. $s = \sqrt{5}$: (1.081, 1.683, 1.000)
    $s = 4$: (−0.433, 1.953, 1.789)

13. $K = 0$ **15.** $K = \frac{2}{5}$ **17.** $K = 0$

19. $\dfrac{\sqrt{2}}{2}$ **21.** $\dfrac{1}{4}$ **23.** $\dfrac{1}{a}$ **25.** $\dfrac{\sqrt{2}}{2}e^{-t}$ **27.** $\dfrac{1}{\omega t}$

29. $\dfrac{\sqrt{5}}{(1 + 5t^2)^{3/2}}$ **31.** $\dfrac{3}{25}$ **33.** $K = 0$, undefined

35. $K = \dfrac{4}{(17)^{3/2}}$ **37.** $K = \dfrac{1}{a}, \dfrac{1}{K} = a$

39. **a.** $\left(x - \dfrac{\pi}{2}\right)^2 + y^2 = 1$

  b. Since the curvature is not as great, the radius of curvature is greater.

41. $(x - 1)^2 + \left(y - \dfrac{5}{2}\right)^2 = \left(\dfrac{1}{2}\right)^2$

43. $(x + 2)^2 + (y - 3)^2 = 8$

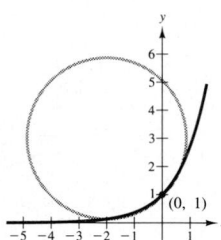

45. **a.** $(1, 3)$ **b.** 0 **47.** **a.** $K \to \infty$ as $x \to 0$ **b.** 0
49. Yes $(-1, 3)$
53. $a = \frac{1}{4}, b = 2$

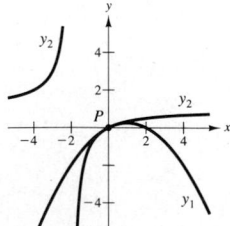

55. 56.27 mph   **57.** $\dfrac{3}{2\sqrt{2}(1 + \sin\theta)}$

59. $\dfrac{2}{|a|}$ **61.** **a.** 0 **b.** 0 **63.** $\dfrac{1}{4}$

67. $a_T = 6t$ **69.** 55 lb
  $a_N = 6$

71. **a.** a **73.** $K = \dfrac{1}{4a}\sec\dfrac{\theta}{2}$
  b. πa
  c. $K = \pi a$

  Minimum: $K = \dfrac{1}{4a}$

  There is no maximum.

Review Exercises for Chapter 12

1. a. All reals except $n\pi$, n is an integer
   b. $t = n\pi$, n is an integer
3. a. $(0, \infty)$
   b. Continuous

5.   **7.**

9.   **11.**

13. $x = t$, $y = -t$, $z = 2t^2$ **15.** $4\mathbf{i} + \mathbf{k}$

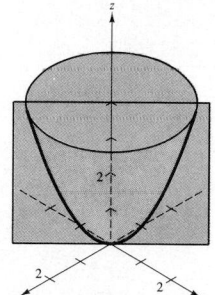

17. a. $3\mathbf{i} + \mathbf{j}$ **b.** $\mathbf{0}$ **c.** $4t + 3t^2$
   d. $-5\mathbf{i} + (2t - 2)\mathbf{j} + 2t^2\mathbf{k}$ **e.** $\dfrac{10t - 1}{\sqrt{10t^2 - 2t + 1}}$
   f. $\left(\dfrac{8}{3}t^3 - 2t^2\right)\mathbf{i} - 8t^3\mathbf{j} + (9t^2 - 2t + 1)\mathbf{k}$
19. $x = -\sqrt{2} - \sqrt{2}\,t$
   $y = \sqrt{2} - \sqrt{2}\,t$
   $z = \dfrac{3\pi}{4} + t$
21. $\sin t\,\mathbf{i} + (t\sin t - \sin t)\mathbf{j} + \mathbf{C}$
23. $\frac{1}{2}\left(t\sqrt{1 + t^2} + \ln\left|t + \sqrt{1 + t^2}\right|\right) + \mathbf{C}$
25. $\mathbf{r}(t) = (t^2 + 1)\mathbf{i} + (e^t + 2)\mathbf{j} - (e^{-t} + 4)\mathbf{k}$
27. $\frac{32}{3}\mathbf{j}$ **29.** 152 ft   **31.** 111.56 ft/sec
33. $\mathbf{v} = 4\mathbf{i} - 3\mathbf{j}$
   $\|\mathbf{v}\| = 5$
   $\mathbf{a} = \mathbf{0}$
   $\mathbf{a} \cdot \mathbf{T} = 0$
   $\mathbf{a} \cdot \mathbf{N} = 0$
   $K = 0$

35. $\mathbf{v} = 2\mathbf{i} - \dfrac{2}{(t + 1)^2}\mathbf{j}$
   $\|\mathbf{v}\| = \dfrac{2\sqrt{(t + 1)^4 + 1}}{(t + 1)^2}$
   $\mathbf{a} = \dfrac{4}{(t + 1)^3}\mathbf{j}$
   $\mathbf{a} \cdot \mathbf{T} = \dfrac{-4}{(t + 1)^3\sqrt{(t + 1)^4 + 1}}$
   $\mathbf{a} \cdot \mathbf{N} = \dfrac{4}{(t + 1)\sqrt{(t + 1)^4 + 1}}$
   $K = \dfrac{(t + 1)^3}{[(t + 1)^4 + 1]^{3/2}}$

37. $\mathbf{v} = e^t\mathbf{i} - e^{-t}\mathbf{j}$
   $\|\mathbf{v}\| = \sqrt{e^{2t} + e^{-2t}}$
   $\mathbf{a} = e^t\mathbf{i} + e^{-t}\mathbf{j}$
   $\mathbf{a} \cdot \mathbf{T} = \dfrac{e^{2t} - e^{-2t}}{\sqrt{e^{2t} + e^{-2t}}}$
   $\mathbf{a} \cdot \mathbf{N} = \dfrac{2}{\sqrt{e^{2t} + e^{-2t}}}$
   $K = \dfrac{2}{(e^{2t} + e^{-2t})^{3/2}}$

39. $\mathbf{v} = \mathbf{i} + 2t\mathbf{j} + t\mathbf{k}$
   $\|\mathbf{v}\| = \sqrt{1 + 5t^2}$
   $\mathbf{a} = 2\mathbf{j} + \mathbf{k}$
   $\mathbf{a} \cdot \mathbf{T} = \dfrac{5t}{\sqrt{1 + 5t^2}}$
   $\mathbf{a} \cdot \mathbf{N} = \dfrac{-5\sqrt{5}}{\sqrt{1 + 5t^2}}$
   $K = \dfrac{\sqrt{5}}{(1 + 5t^2)^{3/2}}$

41. $\dfrac{\sqrt{5}\pi}{2}$ **43.** 4.56 mi/sec

45. The curvature changes abruptly from zero to a nonzero constant.

CHAPTER 13

Section 13.1 (*page 839*)

1. a. $\dfrac{3}{2}$ **b.** $-\dfrac{1}{4}$ **c.** 6   **d.** $\dfrac{5}{y}$ **e.** $\dfrac{x}{2}$ **f.** $\dfrac{5}{t}$
3. a. 5   **b.** $3e^2$ **c.** $\dfrac{2}{e}$ **d.** $5e^y$ **e.** xe^2 **f.** te^t
5. a. $\frac{2}{3}$ **b.** 0   **7. a.** $\sqrt{2}$ **b.** $3\sin 1$
9. a. 4   **b.** 6   **11. a.** $2x + \Delta x$ **b.** -2
13. Domain: $\{(x, y): x^2 + y^2 \le 4\}$
   Range: $0 \le z \le 2$
15. Domain: $\{(x, y): -1 \le x + y \le 1\}$
   Range: $-\dfrac{\pi}{2} \le z \le \dfrac{\pi}{2}$
17. Domain: $\{(x, y): y < -x + 4\}$
   Range: All real numbers
19. Domain: $\{(x, y): x \ne 0, y \ne 0\}$
   Range: All real numbers
21. Domain: $\{(x, y): y \ne 0\}$
   Range: $0 < z$
23. Domain: $\{(x, y): x \ne 0, y \ne 0\}$
   Range: $0 < |z|$
25. a. $(20, 0, 0)$ **b.** $(-15, 10, 20)$
   c. $(20, 15, 25)$ **d.** $(20, 20, 0)$
27. c   **29.** b

31.

33.

35.

37.

39.

41.

43.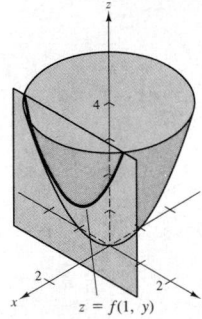

45. Lines: $x + y = c$

47. Circles centered at $(0, 0)$
Radius ≤ 5

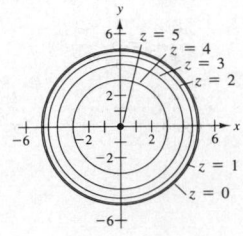

49. Hyperbolas: $xy = c$

51. Circles passing through $(0, 0)$
Centered at $\left(\dfrac{1}{2c}, 0\right)$

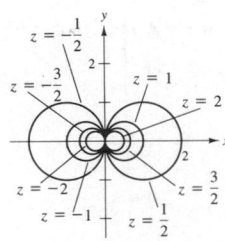

53. No. $z = e^{-(x^2 + y^2)}$

55.

57.   **59.**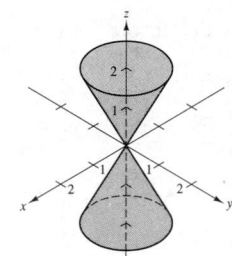

61.

	Inflation Rate		
Tax Rate	**0**	**0.03**	**0.05**
0	$2593.74	$1929.99	$1592.33
0.28	$2004.23	$1491.34	$1230.42
0.35	$1877.14	$1396.77	$1152.40

63. a. 243 board-feet   **b.** 507 board-feet

65.

67. $C = 0.75xz + 0.80(xz + yz)$

69. a. $k = 1300$ **b.** $P = \dfrac{1300T}{V}$

The level curves are lines.

71. a. C   **b.** A   **c.** B

Section 13.2 (*page 849*)

1. 2   **3.** 15   **5.** 5, Continuous

7. -3, Continuous for $x \neq y$

9. 0, Continuous for $xy \neq -1$, $y \neq 0$, $\left|\dfrac{x}{y}\right| \leq 1$

11. 1, Continuous

13. $2\sqrt{2}$, Continuous for $x + y + z \geq 0$

15. 1, Continuous   **17.** Continuous except at $(0, 0)$; $-\infty$

19. Continuous except at $(0, 0)$

$y = 0 : 0$

$y = x : \frac{1}{2}$

Limit does not exist.

21. Continuous except at $(0, 0)$

$x = y^2 : -\frac{1}{2}$

$x = -y^2 : \frac{1}{2}$

Limit does not exist.

23. 1   **25.** Limit does not exist.

27. 1   **29.** 0   **31.** Continuous except at $(0, 0, 0)$

33. Continuous   **35.** Continuous

37. Continuous for $y \neq \dfrac{3x}{2}$ **39. a.** $2x$ **b.** -4

41. a. $2 + y$ **b.** $x - 3$ **45.** True

46. False, let $f(x, y) = \dfrac{xy}{x^2 + y^2}$.

47. False, let $f(x, y) = \begin{cases} \ln(x^2 + y^2), & x \neq 0, y \neq 0 \\ 0, & x = 0, y = 0. \end{cases}$

48. True

Section 13.3 (*page 858*)

1. $f_x(x, y) = 2$ **3.** $f_x(x, y) = \sqrt{y}$

$f_y(x, y) = -3$ $f_y(x, y) = \dfrac{x}{2\sqrt{y}}$

5. $f_x(x, y) = 2xe^{2y}$

$f_y(x, y) = 2x^2 e^{2y}$

7. $f_x(x, y) = \dfrac{2x}{x^2 + y^2}$ **9.** $f_x(x, y) = \dfrac{-2y}{x^2 - y^2}$

$f_y(x, y) = \dfrac{2y}{x^2 + y^2}$ $f_y(x, y) = \dfrac{2x}{x^2 - y^2}$

11. $f_x(x, y) = -2xe^{-(x^2+y^2)}$

$f_y(x, y) = -2ye^{-(x^2+y^2)}$

13. $f_x(x, y) = \dfrac{x}{\sqrt{x^2 + y^2}}$

$f_y(x, y) = \dfrac{y}{\sqrt{x^2 + y^2}}$

15. $f_x(x, y) = 2\sec^2(2x - y)$

$f_y(x, y) = -\sec^2(2x - y)$

17. $f_x(x, y) = ye^y \cos xy$

$f_y(x, y) = e^y(x \cos xy + \sin xy)$

19. $f_x(x, y) = 1 - x^2$ **21.** $f_x(x, y) = 2$

$f_y(x, y) = y^2 - 1$ $f_y(x, y) = 3$

23. $f_x(x, y) = \dfrac{1}{2\sqrt{x + y}}$ **25.** $g_x(1, 1) = -2$

$g_y(1, 1) = -2$

$f_y(x, y) = \dfrac{1}{2\sqrt{x + y}}$

27. $\dfrac{\partial z}{\partial x} = -1$ **29.** $\dfrac{\partial z}{\partial x} = \dfrac{1}{4}$ **31.** $\dfrac{\partial z}{\partial x} = -\dfrac{1}{4}$

$\dfrac{\partial z}{\partial y} = 0$ $\dfrac{\partial z}{\partial y} = \dfrac{1}{4}$ $\dfrac{\partial z}{\partial y} = \dfrac{1}{4}$

33. $-\frac{1}{2}$ **35.** 18

37. $\dfrac{\partial^2 z}{\partial x^2} = 2$

$\dfrac{\partial^2 z}{\partial y^2} = 6$

$\dfrac{\partial^2 z}{\partial y \partial x} = \dfrac{\partial^2 z}{\partial x \partial y} = -2$

39. $\dfrac{\partial^2 z}{\partial x^2} = \dfrac{y^2}{(x^2 + y^2)^{3/2}}$

$\dfrac{\partial^2 z}{\partial y^2} = \dfrac{x^2}{(x^2 + y^2)^{3/2}}$

$\dfrac{\partial^2 z}{\partial y \partial x} = \dfrac{\partial^2 z}{\partial x \partial y} = \dfrac{-xy}{(x^2 + y^2)^{3/2}}$

41. $\dfrac{\partial^2 z}{\partial x^2} = e^x \tan y$

$\dfrac{\partial^2 z}{\partial y^2} = 2e^x \sec^2 y \tan y$

$\dfrac{\partial^2 z}{\partial y \partial x} = \dfrac{\partial^2 z}{\partial x \partial y} = e^x \sec^2 y$

43. $\dfrac{\partial^2 z}{\partial x^2} = \dfrac{2xy}{(x^2 + y^2)^2}$

$\dfrac{\partial^2 z}{\partial y^2} = \dfrac{-2xy}{(x^2 + y^2)^2}$

$\dfrac{\partial^2 z}{\partial y \partial x} = \dfrac{\partial^2 z}{\partial x \partial y} = \dfrac{y^2 - x^2}{(x^2 + y^2)^2}$

45.
$$\frac{\partial z}{\partial x} = \sec y$$
$$\frac{\partial z}{\partial y} = x \sec y \tan y$$
$$\frac{\partial^2 z}{\partial x^2} = 0$$
$$\frac{\partial^2 z}{\partial y^2} = x \sec y (\sec^2 y + \tan^2 y)$$
$$\frac{\partial^2 z}{\partial y \, \partial x} = \frac{\partial^2 z}{\partial x \, \partial y} = \sec y \tan y$$

47.
$$\frac{\partial z}{\partial x} = \frac{y^2 - x^2}{x(x^2 + y^2)}$$
$$\frac{\partial z}{\partial y} = \frac{-2y}{x^2 + y^2}$$
$$\frac{\partial^2 z}{\partial x^2} = \frac{x^4 - 4x^2 y^2 - y^4}{x^2(x^2 + y^2)^2}$$
$$\frac{\partial^2 z}{\partial y^2} = \frac{2(y^2 - x^2)}{(x^2 + y^2)^2}$$
$$\frac{\partial^2 z}{\partial y \, \partial x} = \frac{\partial^2 z}{\partial x \, \partial y} = \frac{4xy}{(x^2 + y^2)^2}$$

49.
$$\frac{\partial w}{\partial x} = \frac{x}{\sqrt{x^2 + y^2 + z^2}}$$
$$\frac{\partial w}{\partial y} = \frac{y}{\sqrt{x^2 + y^2 + z^2}}$$
$$\frac{\partial w}{\partial z} = \frac{z}{\sqrt{x^2 + y^2 + z^2}}$$

51.
$$F_x(x, y, z) = \frac{x}{x^2 + y^2 + z^2}$$
$$F_y(x, y, z) = \frac{y}{x^2 + y^2 + z^2}$$
$$F_z(x, y, z) = \frac{z}{x^2 + y^2 + z^2}$$

53. $H_x(x, y, z) = \cos(x + 2y + 3z)$
$H_y(x, y, z) = 2\cos(x + 2y + 3z)$
$H_z(x, y, z) = 3\cos(x + 2y + 3z)$

55. $f_{xyy}(x, y, z) = f_{yxy}(x, y, z)$
$\qquad = f_{yyx}(x, y, z) = 0$

57. $f_{xyy}(x, y, z) = f_{yxy}(x, y, z)$
$\qquad = f_{yyz}(x, y, z) = z^2 e^{-x} \sin yz$

59. $\dfrac{\partial^2 z}{\partial x^2} + \dfrac{\partial^2 z}{\partial y^2} = 0 + 0 = 0$

61. $\dfrac{\partial^2 z}{\partial x^2} + \dfrac{\partial^2 z}{\partial y^2} = e^x \sin y - e^x \sin y = 0$

63. $\dfrac{\partial^2 z}{\partial t^2} = -c^2 \sin(x - ct) = \dfrac{\partial^2 z}{\partial x^2}$

65. $\dfrac{\partial z}{\partial t} = -e^{-t} \cos \dfrac{x}{c} = c^2 \dfrac{\partial^2 z}{\partial x^2}$

67. $\dfrac{\partial C}{\partial x} = 183$
$\dfrac{\partial C}{\partial y} = 237$

69. An increase in either price will cause a decrease in demand.

71. $\dfrac{\partial T}{\partial x} = -2.4°/\text{ft}$ **73.** $\dfrac{\partial R}{\partial v_0} = 21.7$
$\dfrac{\partial T}{\partial y} = -9°/\text{ft}$ $\dfrac{\partial R}{\partial \theta} = 246{,}201.9$

75. False, let $z = x + y + 1$.
76. True
77. True
78. True

Section 13.4 (*page 867*)

1. $dz = 6xy^3 dx + 9x^2 y^2 dy$

3. $dz = \dfrac{2}{(x^2 + y^2)^2}(x \, dx + y \, dy)$

5. $dz = (\cos y + y \sin x)dy - (x \sin y + \cos x)dy$
7. $dw = 2z^3 y \cos x \, dx + 2z^3 \sin x \, dy + 6z^2 y \sin x \, dz$

9. $dw = \dfrac{1}{z - 2y} dx + \dfrac{z + 2x}{(z - 2y)^2} dy - \dfrac{x + y}{(z - 2y)^2} dz$

11. a. -0.5125 **b.** -0.5
13. a. -0.00293 **b.** 0.00385
15. a. -0.25 **b.** -0.25
17. 0.094 **19.** -0.012
21. $dA = h \, dl + l \, dh$ **23.** 10%

25. a. $V = 2592 \sin \theta$, $\theta = \dfrac{\pi}{2}$ **b.** 17.3 ft^3

27. $\pm 0.24 \text{ in.}^2$ **29.** 7%
31. $L \approx 8.096 \times 10^{-4} \pm 6.6 \times 10^{-6}$ micro-henrys
33. $\varepsilon_1 = \Delta x$ **35.** $\varepsilon_1 = y\Delta x$
$\quad \varepsilon_2 = 0$ $\quad \varepsilon_2 = 2x\Delta x + (\Delta x)^2$
(Answer is not unique.)   (Answer is not unique.)

Section 13.5 (*page 875*)

1. $2(e^{2t} - e^{-2t})$ **3.** $e^t \sec(\pi - t)[1 - \tan(\pi - t)]$

5. $\dfrac{\partial w}{\partial s} = 4s$, 8 **7.** $\dfrac{\partial w}{\partial s} = 2s \cos 2t$, 0
$\quad \dfrac{\partial w}{\partial t} = 4t$, -4 $\quad \dfrac{\partial w}{\partial t} = -2s^2 \sin 2t$, -18

9. $2 \cos 2t$ **11.** $4e^{2t}$ **13.** $3(2t^2 - 1)$

15. $\dfrac{\partial w}{\partial r} = 0$ **17.** $\dfrac{\partial w}{\partial r} = 0$ **19.** $\dfrac{\partial z}{\partial x} = \dfrac{-x}{z}$
$\quad \dfrac{\partial w}{\partial \theta} = 4(x - y)$ $\quad \dfrac{\partial w}{\partial \theta} = 1$ $\quad \dfrac{\partial z}{\partial y} = \dfrac{-y}{z}$

21. $\dfrac{\partial z}{\partial x} = \dfrac{-\sec^2(x + y)}{\sec^2(y + z)}$
$\quad \dfrac{\partial z}{\partial y} = -1 - \dfrac{\sec^2(x + y)}{\sec^2(y + z)}$

23. $\dfrac{\partial z}{\partial x} = \dfrac{-x}{y + z}$

$\dfrac{\partial z}{\partial y} = \dfrac{-z}{y + z}$

$\dfrac{\partial^2 z}{\partial x^2} = -\dfrac{(y + z)^2 + x^2}{(y + z)^3}$

$\dfrac{\partial^2 z}{\partial y^2} = \dfrac{z(2y + z)}{(y + z)^3}$

$\dfrac{\partial^2 z}{\partial x\,\partial y} = \dfrac{\partial^2 z}{\partial y\,\partial x} = \dfrac{xy}{(x + y)^3}$

25. $\dfrac{\partial w}{\partial z} = \dfrac{yw - xy + xw}{xz - yz + 2w}$　**27.** $\dfrac{3}{2}\sqrt{2} + \dfrac{\sqrt{2}}{10}\pi$

$\dfrac{\partial w}{\partial y} = \dfrac{-xz + zw}{xz - yz + 2w}$

$\dfrac{\partial w}{\partial x} = \dfrac{-yz - zw}{xz - yz + 2w}$

29. $\dfrac{dV}{dt} = 1536\pi \text{ in.}^3/\text{min}$　**31.** $28m \text{ cm}^2/\text{sec}$

$\dfrac{dS}{dt} = \dfrac{36\pi}{5}(20 + 9\sqrt{10}) \text{ in.}^2/\text{min}$

(Surface area includes base.)

33. a. $x = 32\sqrt{2}t$

$y = 32\sqrt{2}t - 16t^2$

b. $\alpha = \arctan\left(\dfrac{y}{x + 50}\right)$

$= \arctan\left(\dfrac{32\sqrt{2}t - 16t^2}{32\sqrt{2}t + 50}\right)$

c. $\dfrac{d\alpha}{dt} = \dfrac{-16(8\sqrt{2}t + 25t - 25\sqrt{2})}{64t^4 - 256\sqrt{2}t^3 + 1024t^2 + 800\sqrt{2}t + 625}$

d.

No. The rate of change of α is greatest when the projectile is closest to the camera.

e. α is maximum when $t = 0.98$.

The projectile is at its maximum height when $t = \sqrt{2} \approx 1.41$ sec.

35. 0　**37.** 0

Section 13.6　(*page 886*)

1. $\dfrac{\sqrt{3} - 5}{2}$　**3.** $\dfrac{5\sqrt{2}}{2}$　**5.** $-\dfrac{7}{25}$　**7.** $-e$

9. $\dfrac{2\sqrt{6}}{3}$　**11.** $\dfrac{(8 + \pi)\sqrt{6}}{24}$　**13.** $\sqrt{2}(x + y)$

15. $\left(\dfrac{2 + \sqrt{3}}{2}\right)\cos(2x - y)$　**17.** $-7\sqrt{2}$　**19.** $\dfrac{7\sqrt{19}}{19}$

21. $-\dfrac{2}{5}\mathbf{i} + \dfrac{1}{10}\mathbf{j}$　**23.** $\tan y\,\mathbf{i} + x\sec^2 y\,\mathbf{j}$, $\sqrt{17}$

25. $\dfrac{2}{3(x^2 + y^2)}(x\mathbf{i} + y\mathbf{j})$, $\dfrac{2\sqrt{5}}{15}$　**27.** $\dfrac{x\mathbf{i} + y\mathbf{j} + z\mathbf{k}}{\sqrt{x^2 + y^2 + z^2}}$, 1

29. $\dfrac{x\mathbf{i} + y\mathbf{j} + z\mathbf{k}}{(1 - x^2 - y^2 - z^2)^{3/2}}$, 0

31.

33. a. $\dfrac{2 + 3\sqrt{3}}{12}$　**b.** $\dfrac{3 - 2\sqrt{3}}{12}$

35. a. $-\dfrac{1}{5}$　**b.** $-\dfrac{11\sqrt{10}}{60}$

37. $\dfrac{\sqrt{13}}{6}$

39.

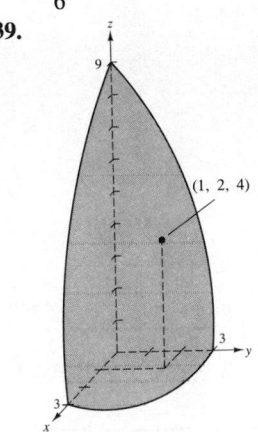

41. $-2\mathbf{i} - 4\mathbf{j}$, $2\sqrt{5}$　**43.** $6\mathbf{i} + 8\mathbf{j}$　**45.** $-\dfrac{1}{2}\mathbf{j}$

47. $\dfrac{\sqrt{257}}{257}(16\mathbf{i} - \mathbf{j})$　**49.** $\dfrac{\sqrt{85}}{85}(9\mathbf{i} - 2\mathbf{j})$

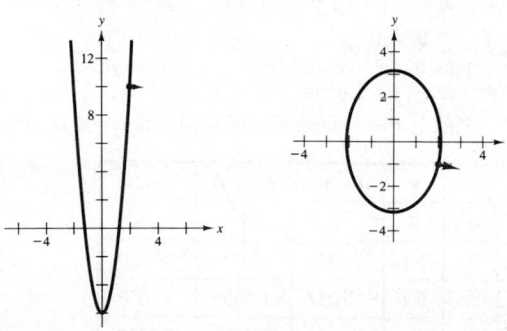

51. $\dfrac{1}{625}(7\mathbf{i} - 24\mathbf{j})$　**53.** $y^2 = 10x$　**55.** A　**57.** True

58. False, $D_{\mathbf{u}}f(x, y) = \sqrt{2} > 1$ where $\mathbf{u} = \cos\dfrac{\pi}{4}\mathbf{i} + \sin\dfrac{\pi}{4}\mathbf{j}$.

59. True　**60.** True

61. a. 0　**b.** 0　**c.** $-\dfrac{4}{5}\mathbf{j}$

d. $-\dfrac{4}{5}\mathbf{j}$　**e.** $-4\mathbf{j}$　**f.** $\dfrac{8}{9}\mathbf{i} - \dfrac{4}{9}\mathbf{j}$

63. $f(x, y, z) = e^x \cos y + 2z + C$

Section 13.7 (*page 895*)

1. $\dfrac{\sqrt{3}}{3}(\mathbf{i} + \mathbf{j} + \mathbf{k})$ 3. $\dfrac{\sqrt{2}}{10}(3\mathbf{i} + 4\mathbf{j} - 5\mathbf{k})$

5. $\dfrac{\sqrt{2049}}{2049}(32\mathbf{i} + 32\mathbf{j} - \mathbf{k})$

7. $\dfrac{\sqrt{113}}{113}(-\mathbf{i} - 6\sqrt{3}\mathbf{j} + 2\mathbf{k})$

9. $\dfrac{\sqrt{3}}{3}(\mathbf{i} - \mathbf{j} + \mathbf{k})$ 11. $6x + 2y + z = 35$

13. $2x - y + z = 2$ 15. $10x - 8y - z = 9$

17. $2x - z = -2$ 19. $3x + 4y - 25z = 25(1 - \ln 5)$

21. $x - 4y + 2z = 18$ 23. $x + y + z = 1$

25. $2x + 4y + z = 14$

$\dfrac{x - 1}{2} = \dfrac{y - 2}{4} = \dfrac{z - 4}{1}$

27. $3x + 2y + z = -6$

$\dfrac{x + 2}{3} = \dfrac{y + 3}{2} = \dfrac{z - 6}{1}$

29. $x - y + 2z = \dfrac{\pi}{2}$

$\dfrac{x - 1}{1} = \dfrac{y - 1}{-1} = \dfrac{z - (\pi/4)}{2}$

31. **a.** $\dfrac{x - 2}{1} = \dfrac{y - 1}{-2} = \dfrac{z - 2}{1}$ **b.** $\dfrac{\sqrt{10}}{5}$, Not orthogonal

33. **a.** $\dfrac{x - 3}{4} = \dfrac{y - 3}{4} = \dfrac{z - 4}{-3}$ **b.** $\dfrac{16}{25}$, Not orthogonal

35. **a.** $\dfrac{y - 1}{1} = \dfrac{z - 1}{-1}, x = 2$ **b.** 0, Orthogonal

37. $86.0°$ 39. $77.4°$ 41. $(0, 3, 12)$

43. $x = 4e^{-4t}, y = 3e^{-2t}, z = 10e^{-8t}$

51. **a.** $P_1(x, y) = 1 + x - y$

b. $P_2(x, y) = 1 + x - y + \frac{1}{2}x^2 - xy + \frac{1}{2}y^2$

c. If $x = 0$, $P_2(0, y) = 1 - y + \frac{1}{2}y^2$

This is the second-degree Taylor polynomial for e^{-y}
If $y = 0$, $P_2(0, y) = 1 + x + \frac{1}{2}x^2$

This is the second-degree Taylor polynomial for e^x

d.

x	y	$f(x,y)$	$P_1(x,y)$	$P_2(x,y)$
0	0	1	1	1
0	0.1	0.9048	0.9000	0.9050
0.2	0.1	1.1052	1.1000	1.1050
0.2	0.5	0.7408	0.7000	0.7450
1	0.5	1.6487	1.5000	1.6250

e.

Section 13.8 (*page 904*)

1. Relative minimum: $(-1, 3, -4)$
3. Relative maximum: $(2, 0, 5)$
5. Relative minimum: $(1, 3, 0)$
7. Relative minimum: $(-1, 1, -4)$
9. Relative maximum: $(8, 16, 74)$
11. Relative minimum: $(1, 2, -1)$
13. Saddle point: $(1, -2, -1)$ 15. Saddle point: $(0, 0, 0)$
17. Saddle point: $(0, 0, 0)$
Relative minimum: $(1, 1, -1)$
19. There are no critical numbers.
21. Relative maximum: $(-1, 0, 2)$
Relative minimum: $(1, 0, -2)$

23. Relative maxima: $(0, \pm 1, 4)$
Relative minimum: $(0, 0, 0)$
Saddle points: $(\pm 1, 0, 1)$

25. Insufficient information   27. Saddle point
29. Absolute maximum: $(0, 1, 10)$
Absolute minimum: $(2, 0, 6)$
31. Absolute maxima: $(\pm 2, 4, 28)$
Absolute minimum: $(0, 1, -2)$
33. Absolute maxima: $(2, 1, 6), (-2, -1, 6)$
Absolute minima: $(-\frac{1}{2}, 1, -\frac{1}{4}), (\frac{1}{2}, -1, -\frac{1}{4})$
35. Absolute maxima: $(2, 2, 16), (-2, -2, 16)$
Absolute minima: $(x, -x, 0), |x| \le 2$
37. Saddle point: $(0, 0, 0)$
Test fails
39. Absolute minima: $(1, a, 0), (b, -4, 0)$
Test fails
41. Absolute minimum: $(0, 0, 0)$
Test fails
43. Relative minimum: $(0, 3, -1)$
45. Relative extrema at A and B
Saddle points at C and D
47. False, let $f(x, y) = |1 - x - y|$ at the point $(0, 0, 1)$.
48. True
49. False, let $f(x, y) = x^4 - 2x^2 + y^2$.   50. True

Section 13.9 (*page 911*)

1. $\dfrac{6\sqrt{14}}{7}$ 3. 6   5. 10, 10, 10   7. 10, 10, 10

9. $36 \times 18 \times 18$ in.

11. $x = \dfrac{\sqrt{2}}{2} \approx 0.707$ mi

$y = \dfrac{3\sqrt{2} + 2\sqrt{3}}{6} \approx 1.284$ mi

17. Each edge of $w/3$ inches is turned up $60°$ from the horizontal.

19. $p_1 = \$586.67, p_2 = \$840.00$

21. $y = \dfrac{6 - 2\sqrt{3}}{3} \approx 0.845$

23. a. $y = \dfrac{3}{4}x + \dfrac{4}{3}$ **b.** $\dfrac{1}{6}$

25. a. $y = -2x + 4$ **b.** 2

27. $y = \dfrac{37}{43}x + \dfrac{7}{43}$ **29.** $y = -\dfrac{175}{148}x + \dfrac{945}{148}$

31. a. $y = -240x + 685$ **b.** $y = 349$

33. $y = 14x + 19$

    41.4 bushels/acre

35. $a\sum_{i=1}^{n} x_i^4 + b\sum_{i=1}^{n} x_i^3 + c\sum_{i=1}^{n} x_i^2 = \sum_{i=1}^{n} x_i^2 y_i$

    $a\sum_{i=1}^{n} x_i^3 + b\sum_{i=1}^{n} x_i^2 + c\sum_{i=1}^{n} x_i = \sum_{i=1}^{n} x_i y_i$

    $a\sum_{i=1}^{n} x_i^2 + b\sum_{i=1}^{n} x_i + \quad cn = \sum_{i=1}^{n} y_i$

37. $y = \dfrac{3}{7}x^2 + \dfrac{6}{5}x + \dfrac{26}{35}$ **39.** $y = x^2 - x$

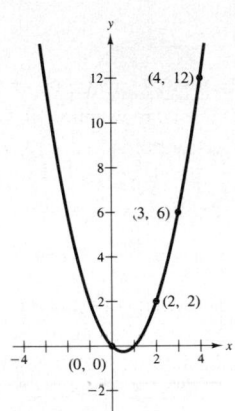

41. $y = -\dfrac{25}{112}x^2 + \dfrac{541}{56}x - \dfrac{25}{14}$

43. Use ∇S rather than $-\nabla S$.

Section 13.10   (*page 920*)

1. $f(5, 5) = 25$ **3.** $f(2, 2) = 8$ **5.** $f(2, 4) = -12$

7. $f(25, 50) = 2600$ **9.** $f(1, 1) = 2$

11. $f(2, 2) = e^4$ **13.** $f(2, 2, 2) = 12$

15. $f(\frac{1}{3}, \frac{1}{3}, \frac{1}{3}) = \frac{1}{3}$ **17.** $f(8, 16, 8) = 1024$

19. $f(3, \frac{3}{2}, 1) = 6$

21. Maxima: $f\left(\dfrac{\sqrt{2}}{2}, \dfrac{\sqrt{2}}{2}\right) = \dfrac{5}{2}$

      $f\left(-\dfrac{\sqrt{2}}{2}, -\dfrac{\sqrt{2}}{2}\right) = \dfrac{5}{2}$

   Minima: $f\left(-\dfrac{\sqrt{2}}{2}, \dfrac{\sqrt{2}}{2}\right) = -\dfrac{1}{2}$

      $f\left(\dfrac{\sqrt{2}}{2}, -\dfrac{\sqrt{2}}{2}\right) = -\dfrac{1}{2}$

23. $\dfrac{\sqrt{13}}{13}$ **25.** $\sqrt{3}$

27. $x = \dfrac{10 + 2\sqrt{265}}{15}$

    $y = \dfrac{5 + \sqrt{265}}{15}$

    $z = \dfrac{-1 + \sqrt{265}}{3}$

29. $36 \times 18 \times 18$ in.   **31.** $\sqrt[3]{360} \times \sqrt[3]{360} \times \frac{4}{3}\sqrt[3]{360}$ ft

33. $\dfrac{2\sqrt{3}a}{3} \times \dfrac{2\sqrt{3}b}{3} \times \dfrac{2\sqrt{3}c}{3}$

37. a. $g\left(\dfrac{\pi}{3}, \dfrac{\pi}{3}, \dfrac{\pi}{3}\right) = \dfrac{1}{8}$ **b.**

39. $P\left(\dfrac{3125}{6}, \dfrac{6250}{3}\right) = 147{,}314$

41. $x = 50\sqrt{2}$

     $y = 200\sqrt{2}$

     cost = \$13,576.45

43. a. $x_1 = y_1 = 1$, $x_i = y_i = 0$ for $i > 1$

Review Exercises for Chapter 13

1.

3.

5.

7. Continuous except at $(0, 0)$

   Limit: $\frac{1}{2}$

9. Continuous except at $(0, 0)$

   Limit does not exist.

11. $f_x(x, y) = e^x \cos y$ **13.** $f_x(x, y) = e^y + ye^x$

    $f_y(x, y) = -e^x \sin y$ $f_y(x, y) = xe^y + e^x$

15. $f_x(x, y) = \dfrac{y(y^2 - x^2)}{(x^2 + y^2)^2}$ **17.** $f_x(x, y, z) = \dfrac{-yz}{x^2 + y^2}$

    $f_y(x, y) = \dfrac{x(x^2 - y^2)}{(x^2 + y^2)^2}$ $f_y(x, y, z) = \dfrac{xz}{x^2 + y^2}$

                      $f_z(x, y, z) = \arctan\dfrac{y}{x}$

19. $u_x(x, t) = cne^{-n^2 t} \cos nx$

    $u_t(x, t) = -cn^2 e^{-n^2 t} \sin nx$

21. $-\dfrac{2yz + z - 2xy}{2xy + x + 2z}$ **23.** $f_{xx}(x, y) = 6$

                     $f_{yy}(x, y) = 12y$

                     $f_{xy}(x, y) = f_{yx}(x, y) = -1$

25. $f_{xx}(x, y) = -y \cos x$

    $f_{yy}(x, y) = -x \sin y$

    $f_{xy}(x, y) = f_{yx}(x, y) = \cos y - \sin x$

31. $\left(\sin\dfrac{y}{x} - \dfrac{y}{x}\cos\dfrac{y}{x}\right) dx + \left(\cos\dfrac{y}{x}\right) dy$

33. 0.082 in., 0.63% **35.** π in.^{3}

37. $\dfrac{\partial u}{\partial r} = 2r$

$\dfrac{\partial u}{\partial t} = 2t$

39. 0 **41.** $\frac{2}{3}$ **43.** $\left\langle -\frac{1}{2}, 0 \right\rangle, \frac{1}{2}$ **45.** $\left\langle -\frac{\sqrt{2}}{2}, -\frac{\sqrt{2}}{2} \right\rangle, 1$

47. $4x + 4y - z = 8$

$\dfrac{x-2}{4} = \dfrac{y-1}{4} = \dfrac{z-4}{-1}$

49. Tangent plane: $z = 4$

Normal line: $x = 2, y = -3, z = 4 + t$

51. $\dfrac{x-2}{1} = \dfrac{y-1}{2}, z = 3$

53. Relative minimum: $\left(\frac{3}{2}, \frac{9}{4}, -\frac{27}{16} \right)$

Saddle point: $(0, 0, 0)$

55. Relative minimum: $(1, 1, 3)$

57. The level curves are hyperbolas. There is a saddle point at the critical point $(0, 0)$, but there are no extrema.

59. Relative maximum: $\left(\frac{4}{3}, \frac{1}{3}, \frac{16}{27} \right)$

Relative minimum: $(0, 1, 0)$

61. $x_1 = 94$ **63.** $f(49.4, 253) = 13{,}189.8$

$x_2 = 157$

65. a. $y = 0.0045x^2 + 0.0717x + 23.3914$ **b.** 57.6 k

67. False, $\nabla F(x_0, y_0, z_0)$ is normal.

CHAPTER 14

Section 14.1 (*page 931*)

1. $\dfrac{3x^2}{2}$ **3.** $y \ln(2y)$ **5.** $\dfrac{4x^2 - x^4}{2}$

7. $\dfrac{y}{2}[(\ln y)^2 - y^2]$ **9.** $x^2(1 - e^{-x^2} - x^2 e^{-x^2})$ **11.** 3

13. $\frac{20}{3}$ **15.** $\frac{2}{3}$ **17.** 4

19. $\dfrac{\pi^2}{32} + \dfrac{1}{8}$ **21.** $\dfrac{1}{2}$ **23.** $\dfrac{1}{4}$

25. $\displaystyle\int_0^4 \int_0^y f(x, y)\,dx\,dy = \int_0^4 \int_x^4 f(x, y)\,dy\,dx$

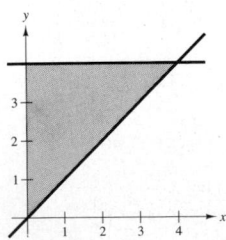

27. $\displaystyle\int_{-\pi/2}^{\pi/2} \int_0^{\cos x} f(x, y)\,dy\,dx$

$= \displaystyle\int_0^1 \int_{-\arccos x}^{\arccos x} f(x, y)\,dx\,dy$

29. $\displaystyle\int_0^1 \int_0^2 dy\,dx = \int_0^2 \int_0^1 dx\,dy = 2$

31. $\displaystyle\int_0^1 \int_{-\sqrt{1-y^2}}^{\sqrt{1-y^2}} dy\,dx = \int_{-1}^1 \int_0^{\sqrt{1-x^2}} dy\,dx = \dfrac{\pi}{2}$

33. $\displaystyle\int_0^2 \int_{x/2}^1 dy\,dx = \int_0^1 \int_0^{2y} dx\,dy = 1$

35. $\displaystyle\int_0^1 \int_{y^2}^{\sqrt[3]{x}} dx\,dy = \int_0^1 \int_{x^3}^{\sqrt{x}} dy\,dx = \dfrac{5}{12}$

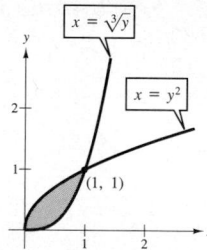

37. 24 **39.** $\frac{16}{3}$ **41.** $\frac{9}{2}$ **43.** $\frac{8}{3}$ **45.** 5 **47.** πab

49. $\frac{26}{9}$ **51.** $\frac{1}{2}(1 - \cos 1) \approx 0.230$ **53.** $\frac{1664}{105}$

55. $(\ln 5)^2$ **57.** 20.5648 **59.** True

60. False, let $f(x, y) = x$.

61. d **62.** b **63.** f **64.** c **65.** a **66.** e

Section 14.2 (*page 940*)

1. 24 (Approximation is exact)

3. Approximation: 52 **5.** kB

Exact: $\frac{160}{3}$

7. 8

9. 36

11. 0

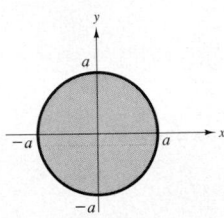

13. $\displaystyle\int_0^3\int_0^5 xy\,dy\,dx = \frac{225}{4}$

$\displaystyle\int_0^5\int_0^3 xy\,dx\,dy = \frac{225}{4}$

15. $\displaystyle\int_0^2\int_x^{2x} \frac{y}{x^2+y^2}\,dy\,dx = \ln\frac{5}{2}$

$\displaystyle\int_0^2\int_{y/2}^y \frac{y}{x^2+y^2}\,dx\,dy + \int_2^4\int_{y/2}^2 \frac{y}{x^2+y^2}\,dx\,dy = \ln\frac{5}{2}$

17. $\displaystyle\int_0^3\int_{4y/3}^{\sqrt{25-y^2}} x\,dx\,dy = 25$

19. 4

$\displaystyle\int_0^4\int_0^{3x/4} x\,dy\,dx + \int_4^5\int_0^{\sqrt{25-x^2}} x\,dy\,dx = 25$

21. 8   **23.** 12   **25.** $\frac{3}{8}$ **27.** 1   **29.** 8π **31.** $\frac{1}{8}$

33. $\frac{32}{3}$ **35.** $\frac{2}{3}$ **37.** $\frac{16}{3}$ **39.** 8π **41.** 8π

43. 1.2315   **47.** $1 - e^{-1/4} \approx 0.221$ **49.** 9   **51.** 2

53. $\frac{8}{3}$ **55.** 25,645.24   **57.** $\frac{1}{5}$ **59.** $\frac{7}{27}$ **65.** d

67. False, $V = 8\displaystyle\int_0^1\int_0^{\sqrt{1-y^2}} \sqrt{1-x^2-y^2}\,dx\,dy.$

69. $A(z) = \pi(z+1)^2,\ V = \left(\dfrac{\pi}{3}\right)[(z+1)^3 - 1]$

$a \approx 1.2674,\ b \approx 2.2066$

71. $\frac{1}{2}(1-e)$

Section 14.3 (*page 948*)

1. 0

3. $\dfrac{5\sqrt{5}\pi}{6}$

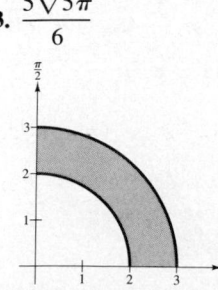

5. $\dfrac{\pi^2}{8} + 1$

7. 9π **9.** $\dfrac{3\pi}{2}$

11. π **13.** $\dfrac{a^3}{3}$ **15.** $\dfrac{243\pi}{10}$ **17.** $\dfrac{2}{3}$

19. $\displaystyle\int_0^{\pi/4}\int_0^{2\sqrt{2}} r^2\,dr\,d\theta = \dfrac{4\sqrt{2}\pi}{3}$ **21.** $\dfrac{16}{3}$ **23.** $\dfrac{\pi^2}{16}$

25. $\dfrac{1}{8}$ **27.** $\dfrac{250\pi}{3}$ **29.** $\dfrac{64}{9}(3\pi - 4)$

31. $2\sqrt{4 - 2\sqrt[3]{2}}$

33. a. 2π **b.** $\sqrt{2\pi}$ **35.** 486,789

37. a. $\displaystyle\int_2^4\int_{\sqrt{3}x}^x f\,dx\,dy$ **b.** $\displaystyle\int_{2/\sqrt{3}}^2\int_2^{\sqrt{3}x} f\,dy\,dx$

c. $\displaystyle\int_{\pi/4}^{\pi/3}\int_{2\csc x}^{4\csc x} f r\,dr\,d\theta$

41. $\frac{1}{2}\pi^2 a^4$

Section 14.4 (*page 957*)

1. a. $m = kab,\ \left(\dfrac{a}{2},\dfrac{b}{2}\right)$ **b.** $m = \dfrac{kab^2}{2},\ \left(\dfrac{a}{2},\dfrac{2b}{3}\right)$

3. a. $m = \dfrac{kbh}{2},\ \left(\dfrac{b}{2},\dfrac{h}{3}\right)$ **b.** $m = \dfrac{kh^2b}{6},\ \left(\dfrac{b}{2},\dfrac{h}{2}\right)$

5. a. $m = \dfrac{k\pi a^2}{2},\ \left(0,\dfrac{4a}{3\pi}\right)$ **b.** $m = \dfrac{ka^4}{24}(16 - 3\pi)$

$\left(0,\dfrac{a}{5}\left[\dfrac{15\pi - 32}{16 - 3\pi}\right]\right)$

7. $m = \dfrac{32k}{3},\ \left(3,\dfrac{8}{7}\right)$ **9.** $m = \dfrac{k\pi}{2},\ \left(0,\dfrac{\pi+2}{4\pi}\right)$

11. $m = \dfrac{8192k}{15},\ \left(\dfrac{64}{7},0\right)$ **13.** $m = \dfrac{kL}{4},\ \left(\dfrac{L}{2},\dfrac{16}{9\pi}\right)$

15. $m = \dfrac{k\pi a^2}{8},\ \left(\dfrac{4\sqrt{2}a}{3\pi},\dfrac{4a(2-\sqrt{2})}{3\pi}\right)$

17. $m = \dfrac{k}{4}(1 - e^{-4}),\ \left(\dfrac{e^4-5}{2(e^4-1)},\dfrac{4}{9}\left[\dfrac{e^6-1}{e^6-e^2}\right]\right)$

19. $m = \dfrac{k\pi}{3},\ (1.12, 0)$ **21.** $\bar{\bar{x}} = \dfrac{\sqrt{3}b}{3},\ \bar{\bar{y}} = \dfrac{\sqrt{3}h}{3}$

23. $\bar{\bar{x}} = \dfrac{r}{2},\ \bar{\bar{y}} = \dfrac{r}{2}$ **25.** $\bar{\bar{x}} = \dfrac{r}{2},\ \bar{\bar{y}} = \dfrac{r}{2}$

27. $I_x = \dfrac{kab^4}{4}$

$I_y = \dfrac{kb^2a^3}{6}$

$I_0 = \dfrac{3kab^4 + 2ka^3b^2}{12}$

$\bar{\bar{x}} = \dfrac{\sqrt{3}a}{3}$

$\bar{\bar{y}} = \dfrac{\sqrt{2}b}{2}$

29. $I_x = \dfrac{32k}{3}$

$I_y = \dfrac{16k}{3}$

$I_0 = 16k$

$\bar{\bar{x}} = \dfrac{2\sqrt{3}}{3}$

$\bar{\bar{y}} = \dfrac{2\sqrt{6}}{3}$

31. $I_x = 16k$

$I_y = \dfrac{512k}{5}$

$I_0 = \dfrac{592k}{5}$

$\bar{\bar{x}} = \dfrac{4\sqrt{15}}{5}$

$\bar{\bar{y}} = \dfrac{\sqrt{6}}{2}$

33. $I_x = \dfrac{3k}{56}$

$I_y = \dfrac{k}{18}$

$I_0 = \dfrac{55k}{504}$

$\bar{\bar{x}} = \dfrac{\sqrt{30}}{9}$

$\bar{\bar{y}} = \dfrac{\sqrt{70}}{14}$

35. $\dfrac{k\pi b^2}{4}(b^2 + 4a^2)$ **37.** $\dfrac{42{,}752k}{315}$

39. $a^5\left(\dfrac{7\pi}{16} - \dfrac{17}{15}\right)k$ **41.** $\dfrac{L}{3}$ **43.** $\dfrac{L}{2}$

Section 14.5 (*page 964*)

1. 6   **3.** 12π **5.** $\frac{3}{4}\left[6\sqrt{37} + \ln\left(\sqrt{37} + 6\right)\right]$

7. $\frac{4}{27}\left(31\sqrt{31} - 8\right)$ **9.** $\sqrt{2} - 1$ **11.** $\sqrt{2}\pi$

13. $2\pi a\left(a - \sqrt{a^2 - b^2}\right)$ **15.** $48\sqrt{14}$ **17.** 10π

19. $\dfrac{27 - 5\sqrt{5}}{12}$ **21.** $\dfrac{\pi}{6}\left(17\sqrt{17} - 1\right)$ **23.** 1.8616

25. 2.0035

27. $\displaystyle\int_{-1}^{1}\int_{-1}^{1}\sqrt{1 + 9(x^2 - y^2)^2 + 9(y^2 - x^2)^2}\,dy\,dx$

29. $\displaystyle\int_{-2}^{2}\int_{-\sqrt{4 - x^2}}^{\sqrt{4 - x^2}}\sqrt{1 + e^{-2x}}\,dy\,dx$

31. $\displaystyle\int_{0}^{4}\int_{0}^{10}\sqrt{1 + e^{2xy}(x^2 + y^2)}\,dy\,dx$

33. 16   **35. a.** 30,415.74 ft^3 **b.** 2081.53 ft^2

Section 14.6 (*page 974*)

1. 18   **3.** $\dfrac{1}{10}$ **5.** $\dfrac{15}{2}\left(1 - \dfrac{1}{e}\right)$ **7.** $\dfrac{729}{4}$

9. $\frac{128}{15}$ **11.** 2.44166

13. $\displaystyle\int_{0}^{3}\int_{0}^{(12-4z)/3}\int_{0}^{(12-4z-3x)/6} dy\,dx\,dz$

15. $\displaystyle\int_{0}^{1}\int_{0}^{x}\int_{0}^{\sqrt{1-y^2}} dz\,dy\,dx$

17. $\displaystyle\int_{0}^{1}\int_{0}^{x}\int_{0}^{3}xyz\,dz\,dy\,dx \quad \int_{0}^{1}\int_{y}^{1}\int_{0}^{3}xyz\,dz\,dx\,dy$

$\displaystyle\int_{0}^{1}\int_{0}^{3}\int_{0}^{x}xyz\,dy\,dz\,dx \quad \int_{0}^{3}\int_{0}^{1}\int_{0}^{x}xyz\,dy\,dx\,dz$

$\displaystyle\int_{0}^{3}\int_{0}^{1}\int_{y}^{1}xyz\,dx\,dy\,dz \quad \int_{0}^{1}\int_{0}^{3}\int_{y}^{1}xyz\,dx\,dz\,dy$

19. $\dfrac{256}{15}$ **21.** $\dfrac{4\pi r^3}{3}$ **23.** $\dfrac{256}{15}$

25. $m = 8k$ **27.** $m = \dfrac{128k}{3}$

$\bar{x} = \frac{3}{2}$ $\bar{z} = 1$

29. $m = \dfrac{kb^5}{4}$

$\left(\dfrac{2b}{3}, \dfrac{2b}{3}, \dfrac{b}{2}\right)$

31. $\left(0, 0, \dfrac{3h}{4}\right)$ **33.** $\left(0, 0, \dfrac{3}{2}\right)$ **35.** $\left(5, 6, \dfrac{5}{4}\right)$

37. a. $I_x = \dfrac{3ka^5}{3}$

$I_y = \dfrac{2ka^5}{3}$

$I_z = \dfrac{2ka^5}{3}$

b. $I_x = \dfrac{ka^8}{8}$

$I_y = \dfrac{ka^8}{8}$

$I_z = \dfrac{ka^8}{8}$

39. a. $I_x = 256k$

$I_y = \dfrac{512k}{3}$

$I_z = 256k$

b. $I_x = \dfrac{2048k}{3}$

$I_y = \dfrac{1024k}{3}$

$I_z = \dfrac{1048k}{3}$

43. $\displaystyle\int_{-1}^{1}\int_{-1}^{1}\int_{0}^{1-x}(x^2 + y^2)\sqrt{x^2 + y^2 + z^2}\,dz\,dy\,dx$

Section 14.7 (*page 981*)

1. 8   **3.** $\dfrac{52}{45}$ **5.** $\dfrac{\pi}{8}$ **7.** $\pi(e^4 + 3)$

9. $\dfrac{\pi}{4}(1 - e^{-9})$ **11.** $\dfrac{64\sqrt{3}\pi}{3}$

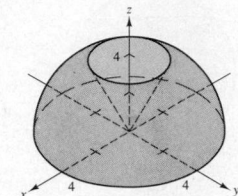

13. Cylindrical: $\int_0^{2\pi} \int_0^2 \int_{r^2}^4 r^2 \cos\theta \, dz \, dr \, d\theta = 0$

Spherical: $\int_0^{2\pi} \int_0^{\arctan(1/2)} \int_0^{4\sec\phi} \rho^3 \sin^2\phi \cos\theta \, d\rho \, d\phi \, d\theta$

$+ \int_0^{2\pi} \int_{\arctan(1/2)}^{\pi/2} \int_0^{\cot\phi\csc\phi} \rho^3 \sin^2\phi \cos\phi \, d\rho \, d\phi \, d\theta = 0$

15. Cylindrical: $\int_0^{2\pi} \int_0^a \int_a^{a+\sqrt{a^2-r^2}} r^2 \cos\theta \, dz \, dr \, d\theta = 0$

Spherical: $\int_0^{\pi/4} \int_0^{2\pi} \int_{a\sec\phi}^{2a\cos\phi} \rho^3 \sin^2\phi \cos\theta \, d\rho \, d\theta \, d\phi = 0$

17. $\dfrac{\pi r_0^2 h}{3}$ **19.** $\left(0, 0, \dfrac{h}{5}\right)$ **25.** $\dfrac{4}{3} a^3 \left(\dfrac{\pi}{2} - \dfrac{2}{3}\right)$

27. $\dfrac{2a^3}{9}(3\pi - 4)$ **29.** $16\pi^2$ **31.** $k\pi a^4$

33. $\left(0, 0, \dfrac{3r}{8}\right)$ **35.** $\dfrac{k\pi}{192}$ **37.** $\dfrac{104\pi}{75}$

Section 14.8 (page 988)

1. $-\dfrac{1}{2}$ **3.** $1 + 2v$ **5.** 1 **7.** $-e^{2u}$

9. **11.** 0

13. $2(1 - e^{-2})$ **15.** $\dfrac{32}{9}(2\sqrt{2} - 1)$ **17.** $12(e^4 - 1)$

19. $\dfrac{100}{9}$ **21.** $\dfrac{2}{5} a^{5/2}$

23. a.

b. ab **c.** πab

25. $u^2 v$ **27.** $-\rho^2 \sin\phi$

Review Exercises for Chapter 14

1. $(x^3 - x)\ln x$ **3.** $\dfrac{29}{6}$ **5.** 36

7. $\dfrac{h^3}{6}\left[\ln(\sqrt{2} + 1) + \sqrt{2}\right]$ **9.** $\dfrac{324\pi}{5}$

11. $\dfrac{abc}{3}(a^2 + b^2 + c^2)$ **13.** $\dfrac{27}{5}$ **15.** $\dfrac{8\pi}{15}$

17. $\int_0^3 \int_0^{(3-x)/3} dy \, dx = \int_0^1 \int_0^{3-3y} dx \, dy = \dfrac{3}{2}$

19. $\int_{-5}^3 \int_{-\sqrt{25-y^2}}^{\sqrt{25-y^2}} dy \, dx = \int_{-5}^{-4} \int_{-\sqrt{25-y^2}}^{\sqrt{25-y^2}} dx \, dy$

$+ \int_{-4}^4 \int_{-\sqrt{25-y^2}}^3 dx \, dy + \int_4^5 \int_{-\sqrt{25-y^2}}^{\sqrt{25-y^2}} dx \, dy$

$= \dfrac{25\pi}{2} + 12 + 25 \arcsin\dfrac{3}{5} \approx 66.36$

21. $4 \int_0^1 \int_0^{x\sqrt{1-x^2}} dy \, dx = 4 \int_0^{1/2} \int_{\sqrt{(1-\sqrt{1-4y^2})/2}}^{\sqrt{(1+\sqrt{1-4y^2})/2}} dx \, dy = \dfrac{4}{3}$

23. $\int_2^5 \int_{x-3}^{\sqrt{x-1}} dy \, dx + 2 \int_1^2 \int_0^{\sqrt{x-1}} dy \, dx = \int_{-1}^2 \int_{y^2+1}^{y+3} dx \, dy = \dfrac{9}{2}$

25. $\dfrac{3296}{15}$ **27.** $\dfrac{\pi h^3}{3}$ **29.** $\dfrac{16}{3}\left(\dfrac{\pi}{2} - \dfrac{2}{3}\right)$

31. $k = 1, 0.070$

33. a. $m = \dfrac{k}{4}, \left(\dfrac{32}{45}, \dfrac{64}{55}\right)$

b. $m = \dfrac{17k}{30}, \left(\dfrac{936}{1309}, \dfrac{784}{663}\right)$

35. $\dfrac{\pi}{6}(65\sqrt{65} - 1)$ **37.** $\left(0, 0, \dfrac{1}{4}\right)$

39. $\left(\dfrac{3a}{8}, \dfrac{3a}{8}, \dfrac{3a}{8}\right)$ **41.** $\dfrac{833k\pi}{3}$

43. Volume of a torus:
Formed by a circle of radius 3
Centered at $(0, 3, 0)$
Revolved about the z-axis

45. True

46. False, $\int_0^1 \int_0^1 x \, dx \, dy \ne \int_1^2 \int_1^2 x \, dx \, dy$

47. True

48. True $\int_0^1 \int_0^1 \dfrac{1}{1 + x^2 + y^2} dx \, dy < \int_1^2 \int_1^2 \dfrac{1}{1 + x^2} dx \, dy = \dfrac{\pi}{4}$.

CHAPTER 15

Section 15.1 (page 1002)

1. **3.**

5. **7.**

9. **11.**

13.

15. $(10x + 3y)\mathbf{i} + (3x + 20y)\mathbf{j}$

17. $-2xye^{x^2}\mathbf{i} - e^{x^2}\mathbf{j} + \mathbf{k}$

19. $\left[\dfrac{xy}{x + y} + y\ln(x + y)\right]\mathbf{i} + \left[\dfrac{xy}{x + y} + y\ln(x + y)\right]\mathbf{j}$

21. Conservative: $f(x, y) = x^2 y + K$

23. Conservative: $f(x, y) = e^{x^2 y} + K$

25. Conservative: $f(x, y) = \frac{1}{2}\ln(x^2 + y^2) + K$

27. Not conservative   **29.** $2\mathbf{j} - \mathbf{k}$ **31.** $-2\mathbf{k}$

33. $\dfrac{2x\mathbf{k}}{x^2 + y^2}$

35. $\cos(y - z)\mathbf{i} + \cos(z - x)\mathbf{j} + \cos(x - y)\mathbf{k}$

37. Not conservative

39. Conservative: $f(x, y, z) = xye^z + K$

41. Conservative: $f(x, y, z) = \dfrac{x}{y} + z^2 - z + K$

43. $6x\mathbf{j} - 3y\mathbf{k}$ **45.** $z\mathbf{j} + y\mathbf{k}$ **47.** $12x - 2xy$

49. $\cos x - \sin y + 2z$ **51.** 4   **53.** 0

55. $2z + 3x$ **57.** 0

Section 15.2   (*page 1013*)

1. $3\cos t\,\mathbf{i} + 3\sin t\,\mathbf{j}, \ 0 \le t \le 2\pi$

3. $\mathbf{r}(t) = \begin{cases} t\,\mathbf{i}, & 0 \le t \le 3 \\ 3\,\mathbf{i} + (t - 3)\mathbf{j}, & 3 \le t \le 6 \\ (9 - t)\mathbf{i} + 3\mathbf{j}, & 6 \le t \le 9 \\ (12 - t)\mathbf{j}, & 9 \le t \le 12 \end{cases}$

5. $\mathbf{r}(t) = \begin{cases} t\,\mathbf{i} + \sqrt{t}\,\mathbf{j}, & 0 \le t \le 1 \\ (2 - t)\mathbf{i} + (2 - t)\mathbf{j}, & 1 \le t \le 2 \end{cases}$

7. 10   **9.** $\dfrac{\sqrt{65}\pi}{6}(3 + 16\pi^2)$ **11.** 9   **13.** $\dfrac{\pi}{2}$

15. $\dfrac{19\sqrt{2}}{6}$ **17.** $\dfrac{19}{6}(1 + \sqrt{2})$

19. $\dfrac{2\sqrt{13}}{3}(27 + 64\pi^2) \approx 4973.8$

21. $\frac{35}{6}$ **23.** 2   **25.** $-\frac{17}{15}$ **27.** 249.49   **29.** -66

31. 0   **33.** $-10\pi^2$ **35.** 1500 ft·lb   **41.** 1010

43. $\frac{190}{3}$ **45.** 25   **47.** $\frac{63}{2}$ **49.** $\frac{316}{3}$ **51.** $5h$

53. $\dfrac{1}{2}$ **55.** $\dfrac{h}{4}\left[2\sqrt{5} + \ln(2 + \sqrt{5})\right]$

57. $\dfrac{1}{120}(25\sqrt{5} - 11)$

59. a. $12\pi \approx 37.70$ cm²   **b.** $\dfrac{12\pi}{5} \approx 7.54$ cm³

c.

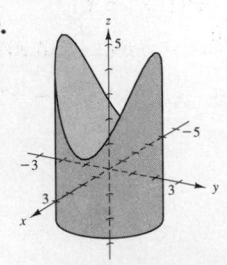

61. c   **63.** False, $\displaystyle\int_C xy\,dx = \sqrt{2}\int_0^1 t^2\,dt$.

64. False, the orientation of C does not affect the form $\int_C f(x, y)\,ds$.

65. False, the orientations are different.

66. False, see Exercise 32.

Section 15.3   (*page 1024*)

1. $\frac{11}{15}$ **3.** $-\ln(2 + \sqrt{3}) = -\frac{1}{2}\ln(7 + 4\sqrt{3})$ **5.** 1

7. a. 0   **b.** $-\frac{1}{3}$ **c.** $-\frac{1}{2}$

9. a. 64   **b.** 0   **c.** 0   **d.** 0   **11. a.** $\frac{64}{3}$ **b.** $\frac{64}{3}$

13. a. 32   **b.** 32   **15. a.** $\frac{2}{3}$ **b.** $\frac{17}{6}$

17. a. 0   **b.** 0   **19.** 24   **21.** -1 **23.** 0

25. a. 2   **b.** 2   **c.** 2   **27.** 11   **29.** 30,366   **31.** 0

33. Increasing: 10 units/min

35. No, the force field is conservative.

37. False, it would be true if $\mathbf{F}$ were conservative.

38. True   **39.** True

40. False, the requirement is $\dfrac{\partial M}{\partial y} = \dfrac{\partial N}{\partial x}$.

Section 15.4   (*page 1033*)

1. 0   **3.** $\frac{32}{15}$ **5.** 19.99   **7.** $\frac{4}{3}$ **9.** 56   **11.** $\frac{32}{3}$

13. 0   **15.** 0   **17.** $\frac{1}{12}$ **19.** 8π **21.** 4π **23.** $\frac{225}{2}$

25. πa^2 **27.** $\frac{32}{3}$ **31.** $(0, \frac{8}{5})$

33. $\left(\dfrac{8}{15}, \dfrac{8}{21}\right)$ **35.** $\dfrac{3\pi a^2}{2}$ **37.** $\pi - \dfrac{3\sqrt{3}}{2}$

39. $I = -2\pi$ when C is a circle that contains the origin.

41. $\frac{19}{2}$

Section 15.5   (*page 1043*)

1. c   **2.** d   **3.** b   **4.** a

5. $y - 2z = 0$ **7.** $x^2 + z^2 = 4$

    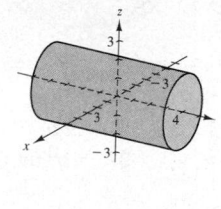

9. The cone is inverted through the xy-plane.

11. The height of the cone is increased from 2 to 3 units.

13.   **15.**

17.

19. a.

b.

c.

d.

The radius of the generating circle that is revolved about the z-axis is b, and its center is a units from the axis of revolution.

21. $\mathbf{r}(u, v) = u\mathbf{i} + v\mathbf{j} + v\mathbf{k}$
23. $\mathbf{r}(u, v) = 4\cos u\,\mathbf{i} + 4\sin u\,\mathbf{j} + v\mathbf{k}$
25. $\mathbf{r}(u, v) = u\mathbf{i} + v\mathbf{j} + u^2\mathbf{k}$
27. $\mathbf{r}(u, v) = v\cos u\,\mathbf{i} + v\sin u\,\mathbf{j} + 4\mathbf{k}$
  $0 \le v \le 3$
29. $x = u,\ y = \dfrac{u}{2}\cos v,\ z = \dfrac{u}{2}\sin v$
  $0 \le u \le 6,\ 0 \le v \le 2\pi$
31. $x = \sin u\cos v,\ y = \sin u\sin v,\ z = u$
  $0 \le u \le \pi,\ 0 \le v \le 2\pi$
33. $x - y - 2z = 0$ **35.** $4y - 3z = 12$ **37.** $2\sqrt{2}$
39. $2\pi ab$ **41.** $\pi ab^2\sqrt{a^2 + 1}$
43. $\dfrac{\pi}{6}\left(17\sqrt{17} - 1\right) \approx 36.177$
45. $2\pi\left[\frac{3}{2}\sqrt{13} + 2\ln\left(3 + \sqrt{13}\right) - 2\ln 2\right]$

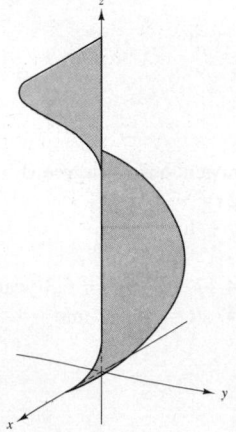

Section 15.6 (*page 1055*)

1. 0 **3.** 10π **5.** $\dfrac{27\sqrt{6}}{2}$ **7.** $\dfrac{391\sqrt{12} + 1}{240}$

9. -11.47 **11.** $6\sqrt{5}$ **13.** 8 **15.** $\dfrac{19\sqrt{2}\pi}{4}$

17. $\dfrac{32\pi}{3}$ **19.** 486π **21.** $-\dfrac{4}{3}$ **23.** $\dfrac{243\pi}{2}$ **25.** 20π

27. $\frac{5}{2}$ **29.** $\frac{364}{3}$ **33.** $2\pi a^3 h$ **35.** $64\pi\rho$
37. If a normal vector at a point P on the surface is moved around the Möbius strip once, it will point in the opposite direction.
39. 0

Section 15.7 (*page 1064*)

1. a^4 **3.** 18 **5.** $3a^4$ **7.** 0
9. 32π **11.** 0 **13.** 2304 **17.** 0

Section 15.8 (*page 1071*)

1. $-xy\mathbf{i} - \mathbf{j} + (yz - 2)\mathbf{k}$ **3.** $\left(2 - \dfrac{1}{1 + x^2}\right)\mathbf{j} - 8x\mathbf{k}$
5. $z(x - 2e^{y^2+z^2})\mathbf{i} - yz\mathbf{j} - 2ye^{x^2+y^2}\mathbf{k}$ **7.** 2π
9. 0 **11.** 1 **13.** 0 **15.** 0
17. $\dfrac{8}{3}$ **19.** $\dfrac{a^5}{4}$ **21.** 0

Review Exercises for Chapter 15

1.

3. $(16x + y)\mathbf{i} + x\mathbf{j} + 2z\mathbf{k}$

5. Not conservative
7. Conservative: $f(x, y) = 3x^2y^2 - x^3 + y^3 - 7y + K$
9. Not conservative
11. Conservative: $f(x, y, z) = \dfrac{x}{yz} + K$
13. a. div $\mathbf{F} = 2x + 2y + 2z$ **b.** curl $\mathbf{F} = 0$
15. a. div $\mathbf{F} = -y\sin x - x\cos y + xy$
  b. curl $\mathbf{F} = xz\mathbf{i} - yz\mathbf{j}$
17. a. div $\mathbf{F} = \dfrac{1}{\sqrt{1 - x^2}} + 2xy + 2yz$
  b. curl $\mathbf{F} = z^2\mathbf{i} + y^2\mathbf{k}$
19. a. div $\mathbf{F} = \dfrac{2x + 2y}{x^2 + y^2} + 1$ **b.** curl $\mathbf{F} = \dfrac{2x - 2y}{x^2 + y^2}\mathbf{k}$
21. a. $6\sqrt{2}$ **b.** 128π **23.** $2\pi^2(1 + 2\pi^2)$
25. a. $\dfrac{35}{2}$ **b.** 18π **27.** $\dfrac{\sqrt{10}}{4}(1 - \cos 8) \approx 10.906$
29. $\frac{5}{7}$ **31.** $2\pi^2$ **33.** $\frac{64}{3}$ **35.** $\frac{8}{3}(3 - 4\sqrt{2}) \approx -7.085$
37. 12 **39.** 4 **41.** 0 **43.** $\frac{1}{12}$

45.

47. 0   **49.** 66

51. $-\dfrac{2a^6}{5}$ **53.** $\displaystyle\int_C \mathbf{F}\cdot d\mathbf{r} = 2\pi$
$\mathbf{F}$ is undefined at the point $(0, 0)$.

CHAPTER 16
Section 16.1   (*page 1078*)

Type	Order
1. Ordinary	1
3. Ordinary	2
5. Ordinary	4
7. Ordinary	2
9. Partial	1
11. Ordinary	2

19. Not a solution   **21.** Solution
23. Solution   **25.** Not a solution   **27.** Solution
29. Solution   **31.** $k = 0.07$ **33.** $4y^2 = x^3$
35.

$C = 0$

$C = 1$

$C = -1$

$C = 4$

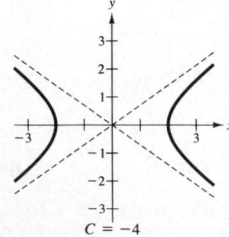

$C = -4$

37. $y = 3e^{-2x}$ **39.** $y = 2\sin 3x - \frac{1}{3}\cos 3x$
41. $y = -2x + \frac{1}{2}x^3$ **43.** $y = x^3 + C$
45. $y = x - \ln x^2 + C$
47. $y = \dfrac{e^x}{5}(\sin 2x - 2\cos 2x) + C$

49. $y = \frac{2}{5}(x - 3)^{5/2} + 2(x - 3)^{3/2} + C$
51. a. $N = 750 - 650e^{-(t/2)\ln(65/59)}$
$\quad\quad = 750 - 650e^{-0.0484t}$

c.

53. a.

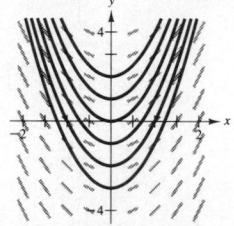

b. $y = \frac{1}{2}x^2 + C$

55. a.

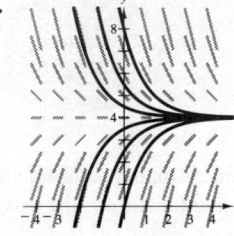

b. $y \rightarrow 4$

57. True   **58.** True

Section 16.2   (*page 1087*)

1. $y^2 - x^2 = C$ **3.** $r = Ce^{0.05s}$ **5.** $y = C(x + 2)^3$
7. $y^2 = C - 2\cos x$ **9.** $y = Ce^{(\ln x)^2/2}$
11. $y^2 = 2e^x + 14$ **13.** $y = e^{-(x^2+2x)/2}$
15. $\dfrac{y - x}{1 + xy} = \sqrt{3}$
17. $u = e^{(1-\cos v^2)/2}$ **19.** $P = P_0e^{kt}$
21. $9x^2 + 16y^2 = 25$ **23.** $f(x) = Ce^{-x/2}$
25. $y = Ce^{0.05x}$ **27.** Homogeneous of degree 3

29. Not homogeneous   **31.** Homogeneous of degree 0
33. $x = C(x - y)^2$ **35.** $y^2 + 2xy - x^2 = C$
37. $y = Ce^{-x^2/2y^2}$ **39.** $e^{y/x} = 1 + \ln x^2$
41. $x = e^{\sin(y/x)}$
43. a. $A = Pe^{kt}$ **b.** $A = \$3004.17$ **c.** $t = 6.3$ years
45. 98.9% of original amount   **47.** $t = 119.7$ min

49. a. $\dfrac{dS}{dt} = kS(L - S)$

$$S = \dfrac{100}{1 + 9e^{-0.8109t}}$$

b. $t = 2.7$ months

c.

d.

e. Sales will decrease toward the line $S = L$.

51. a. $\dfrac{dv}{dt} = k(W - v)$

$v = 20(1 - e^{-0.2877t})$

b. $s = 20t + 65.9(e^{-0.2877t} - 1)$

53. Circles: $x^2 + y^2 = C$
Lines: $y = Kx$

55. Parabolas: $x^2 = Cy$
Ellipses: $x^2 + 2y^2 = K$

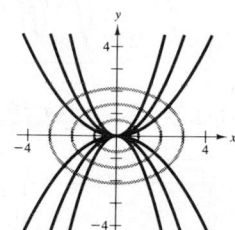

57. Curves: $y^2 = Cx^3$
Ellipses: $2x^2 + 3y^2 = K$

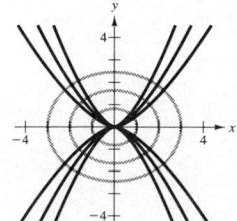

59. True
$y' = xy - 2y + x - 2 = (x - 2)(y + 1)$

60. True,
$y' = \sin(x + y) + \sin(x - y) = 2\sin x \cos y$

61. False, $f(tx, ty) \neq t^n f(x, y)$. **62.** True

63. a. $W(t) = \frac{1}{35}(2C + ke^{-t/200})$

b. $W(t) = \frac{20}{7}(50 + 13e^{-t/200})$
Lose 10 lb: 63 days
Lose 35 lb: 570 days

c. 142.9 lb

Section 16.3   (page 1095)

1. $x^2 - 3xy + y^2 = C$ **3.** $3xy^2 + 5x^2y^2 - 2y = C$

5. Not exact   **7.** $\arctan\dfrac{x}{y} = C$ **9.** Not exact

11. $y\ln(x - 1) + y^2 = 16$ **13.** $x^2 + y^2 = 16$

15. $x^2\tan y + 5x = 0$

17. Integrating factor: $\dfrac{1}{y^2}$ **19.** Integrating factor: $\dfrac{1}{x^2}$

$\dfrac{x}{y} - 6y = C$ $\dfrac{y}{x} + 5x = C$

21. Integrating factor: $\cos x$ **23.** Integrating factor: $\dfrac{1}{y}$
$y\sin x + x\sin x + \cos x = C$
$xy - \ln y = C$

25. Integrating factor: $\dfrac{1}{\sqrt{y}}$ **27.** Integrating factor: xy^2
$x\sqrt{y} + \cos\sqrt{y} = C$ $x^4y^3 + x^2y^4 = C$

29. Integrating factor: $\dfrac{1}{x^2y^3}$

$\dfrac{y^2}{x} + \dfrac{x}{y^2} = C$

33. $x^2 + y^2 = C$ **35.** $2x^2y^4 + x^2 = C$

37. $x^2 - 2xy + 3y^2 = 3$

39. $C = \dfrac{5\left(x^2 + \sqrt{x^4 - 1{,}000{,}000x}\right)}{x}$

41.

Δx	0.50	0.25	0.10
Estimate	3.7798	3.9875	4.1207

43. False, $\dfrac{\partial M}{\partial y} = 2x$, $\dfrac{\partial N}{\partial x} = -2x$.

44. False, $y\,dx + x\,dy = 0$ is exact, but $xy\,dx + x^2\,dy = 0$ is not exact.

45. True   **46.** True

Section 16.4   (page 1103)

1. False, $y' + xy = x^2$ is linear.   **2.** True

3. $y = x^2 + 2x + \dfrac{C}{x}$ **5.** $y = \dfrac{1}{2}e^x + Ce^{-x}$

7. $y = \frac{1}{2}(\sin x - \cos x) + Ce^x$

9. $y = -\frac{1}{13}(3\sin 2x + 2\cos 2x) + Ce^{3x}$

11. $y = \dfrac{x^3 - 3x + C}{3(x - 1)}$ **13.** $y = 1 + 4e^{-\tan x}$

15. $y = \sin x + (x + 1)\cos x$ **17.** $xy = 4$

19. $\dfrac{1}{y^2} = Ce^{2x^3} + \dfrac{1}{3}$ **21.** $y = \dfrac{1}{Cx - x^2}$

23. $y^{2/3} = Ce^{2x/3} - \frac{1}{4}(4x^3 + 18x^2 + 54x + 81)$

25. $I = \dfrac{E_0}{R} + Ce^{-Rt/L}$

27. $I = Ce^{-(R/L)t} + \dfrac{E_0}{\sqrt{R^2 + \omega^2 L^2}}(R \sin \omega t - \omega L \cos \omega t)$

29. $P = -\dfrac{N}{k} + \left(\dfrac{N}{k} + P_0\right)e^{kt}$

31. a. \$685,099.00   **b.** \$5,802,815.12

33. a. $\dfrac{dQ}{dt} = q - kQ$

   b. $Q = \dfrac{q}{k} + \left(Q_0 - \dfrac{q}{k}\right)e^{-kt}$

   c. $\dfrac{q}{k}$

37. a. $Q = 10 + 15e^{-0.05t}$

   b. $t = -20 \ln\frac{1}{3} \approx 21.97$ min

   c. 10 pounds

39. c   **40.** d   **41.** a   **42.** b   **43.** $2e^x + e^{-2y} = C$

45. $x + xy^2 + \frac{1}{2}y^2 + 2y = C$

47. $y = Ce^{-\sin x} + 1$ **49.** $x^2y + \sin y = C$

51. $x^3y^2 + x^4y = C$ **53.** $y = \dfrac{e^x(x - 1) + C}{x^2}$

55. $x^4y^4 - 2x^2 = C$ **57.** $3 \arctan\dfrac{x}{y} - y = C$

Section 16.5   *(page 1112)*

5. $y = C_1 + C_2 e^x$ **7.** $y = C_1 e^{3x} + C_2 e^{-2x}$

9. $y = C_1 e^{x/2} + C_2 e^{-2x}$ **11.** $y = C_1 e^{-3x} + C_2 x e^{-3x}$

13. $y = C_1 e^{x/4} + C_2 x e^{x/4}$ **15.** $y = C_1 \sin x + C_2 \cos x$

17. $y = C_1 e^{3x} + C_2 e^{-3x}$

19. $y = e^x\left(C_1 \sin \sqrt{3}x + C_2 \cos \sqrt{3}x\right)$

21. $y = C_1 e^{(3+\sqrt{5})x/2} + C_2 e^{(3-\sqrt{5})x/2}$

23. $y = e^{2x/3}\left(C_1 \sin \dfrac{\sqrt{7}x}{3} + C_2 \cos \dfrac{\sqrt{7}x}{3}\right)$

25. $y = C_1 e^x + C_2 e^{-x} + C_3 \sin x + C_4 \cos x$

27. $y = C_1 e^x + C_2 e^{2x} + C_3 e^{3x}$

29. $y = C_1 e^x + e^x(C_2 \sin 2x + C_3 \cos 2x)$

31. a. $y = 2 \cos 10x$ **b.** $y = \frac{1}{5}\sin 10x$

   c. $y = -\cos 10x + \frac{3}{10}\sin 10x$

33. $y = \frac{1}{11}(e^{6x} + 10e^{-5x})$ **35.** $y = \frac{1}{2}\cos 4\sqrt{3}t$

37. $y = \dfrac{2}{3}\cos 4\sqrt{3}t - \dfrac{\sqrt{3}}{24}\sin 4\sqrt{3}t$

39. $y = \dfrac{e^{-t/16}}{2}\left(\cos\dfrac{\sqrt{12{,}287}t}{16} + \dfrac{\sqrt{12{,}287}}{12{,}287}\sin\dfrac{\sqrt{12{,}287}t}{16}\right)$

41. b   **42.** d   **43.** c   **44.** a

47. False, the general solution is $y = C_1 e^{3x} + C_2 x e^{3x}$.

48. True   **49.** True

50. False, the solution $y = x^2 e^x$ requires that $m = 1$ is a triple root of the characteristic equation. Since the characteristic equation is quadratic, $m = 1$ can be at most a double root.

55. b. $y = \dfrac{C_1}{x^3} + \dfrac{C_2}{x^2}$

Section 16.6   *(page 1120)*

5. $y = C_1 e^x + C_2 e^{2x} + x + \frac{3}{2}$

7. $y = \cos x + 6 \sin x + x^3 - 6x$

9. $y = C_1 + C_2 e^{-2x} + \frac{2}{3}e^x$

11. $y = (C_1 + C_2 x)e^{5x} + \frac{3}{8}e^x + \frac{1}{5}$

13. $y = -1 + 2e^{-x} - \cos x - \sin x$

15. $y = \left(C_1 - \dfrac{x}{6}\right)\cos 3x + C_2 \sin 3x$

17. $y = C_1 e^x + C_2 x e^x + \left(C_3 + \dfrac{2x}{9}\right)e^{-2x}$

19. $y = \left(\dfrac{4}{9} - \dfrac{1}{2}x^2\right)e^{4x} - \dfrac{1}{9}(1 + 3x)e^x$

21. $y = (C_1 + \ln|\cos x|)\cos x + (C_2 + x)\sin x$

23. $y = \left(C_1 - \dfrac{x}{2}\right)\cos 2x + \left(C_2 + \dfrac{1}{4}\ln|\sin 2x|\right)\sin 2x$

25. $y = (C_1 + C_2 x)e^x + \dfrac{x^2 e^x}{4}(\ln x^2 - 3)$

27. $q = \frac{3}{25}(e^{-5t} + 5te^{-5t} - \cos 5t)$

29. $y = \frac{1}{4}\cos 8t - \frac{1}{2}\sin 8t + \sin 4t$

31. $y = (\frac{9}{32} - \frac{3}{4}t)e^{-8t} - \frac{1}{32}\cos 8t$

33. $y = \dfrac{\sqrt{5}}{4}\sin\left(8t - \arctan\dfrac{1}{2}\right)$

     $= \dfrac{\sqrt{5}}{4}\sin(8t - 0.4636)$

37. $y = C_1 x + C_2 x \ln x + \frac{2}{3}x(\ln x)^3$

Section 16.7   *(page 1124)*

1. $y = a_0 \sum_{n=0}^{\infty} \dfrac{x^n}{n!} = a_0 e^x$

3. $y = a_0 \sum_{k=0}^{\infty} \dfrac{(3x)^{2k}}{(2k)!} + \dfrac{a_1}{3}\sum_{k=0}^{\infty} \dfrac{(3x)^{2k+1}}{(2k + 1)!}$

     $= C_0 \sum_{k=0}^{\infty} \dfrac{(3x)^k}{k!} + C_1 \sum_{k=0}^{\infty} \dfrac{(-3x)^k}{k!}$

     $= C_0 e^{3x} + C_1 e^{-3x}$

where $C_0 + C_1 = a_0$ and $C_0 - C_1 = \dfrac{1}{3}a_1$.

5. $y = a_0 \sum_{k=0}^{\infty} \frac{(-1)^k (2x)^{2k}}{(2k)!} + \frac{a_1}{4} \sum_{k=0}^{\infty} \frac{(-1)^k (2x)^{2k+1}}{(2k+1)!}$

$\quad\quad = C_0 \cos 2x + C_1 \sin 2x$

7. $y = a_0 \sum_{k=0}^{\infty} \frac{(-3)^k}{2^k k!} x^{2k} + a_1 \sum_{k=0}^{\infty} \frac{(-3)^k}{1 \cdot 3 \cdot 5 \cdots (2k+1)} x^{2k+1}$

Interval of Convergence: $(-\infty, \infty)$

9. $y = a_1 \sum_{k=0}^{\infty} \frac{x^{2x+1}}{2^k (k!)(2k+1)}$

Interval of Convergence: $(-\infty, \infty)$

11. $y = a_0 \left(1 - \frac{x^2}{8} + \frac{x^4}{128} - \cdots \right)$

$\quad\quad + a_1 \left(x - \frac{x^3}{24} + \frac{7x^5}{1920} - \cdots \right)$

13. Taylor's Theorem: $y = 2 + \frac{2x}{1!} - \frac{2x^2}{2!} - \frac{10x^3}{3!} + \frac{2x^4}{4!}$

$\quad\quad\quad\quad + \cdots$

$$y\left(\frac{1}{2}\right) \approx 2.552$$

Euler Method: $y\left(\frac{1}{2}\right) \approx 2.672$

15. $y = 1 - \frac{3x}{1!} + \frac{2x^3}{3!} - \frac{12x^4}{4!} + \frac{16x^6}{6!} - \frac{120x^7}{7!} + \cdots$

$y\left(\frac{1}{4}\right) \approx 0.253$

21. $y = a_0 + a_1 x + \frac{a_0}{6} x^3 + \frac{a_1}{12} x^4 + \frac{a_0}{180} x^6 + \frac{a_1}{504} x^7$

Review Exercises for Chapter 16

1. Type: Partial   **3.** Type: Ordinary
Order: 2   Order: 2

5. a.   **b.** $y = Cx$

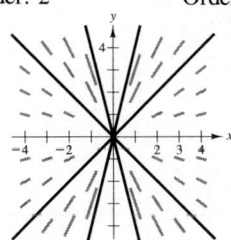

7. b   **8.** d   **9.** a   **10.** c

11. $y = x \ln x^2 + 2x^{3/2} + C$ **13.** $y = C(1-x)^2$

15. $y^2 = x^2 \ln x^2 + Cx^2$

17. $5x^2 + 8xy + 2x + \frac{5}{2}y^2 + 2y = C$

19. $xy - 2xy^3 + x^2 = C$

21. $y = e^x(1 + \tan x) + C \sec x$

23. $x^2 - 2xy - 10x - 3y^2 + 4y = C$

25. $y^2 = 2Cx + C^2$

27. $y^2 = x^2 - x + \frac{3}{2} + Ce^{-2x}$

29. $\frac{y - x}{1 + xy} = C$ **31.** $y = \frac{bx^4}{4 - a} + Cx^a$

33. $y = 5e^{2x} - e^x$ **35.** $y = x \ln|x| - 2 + 12x$

37. $\ln|1 + y| = \ln 3e^{-x}$ **39.** $y = \frac{3(1 + e^{2x^2+1})}{1 - e^{2x^2+1}}$

41. Family of circles: $x^2 + (y - K)^2 = K^2$

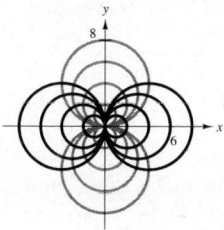

43. a. $\frac{ds}{dh} = \frac{k}{h}$

$\quad\quad s = k \ln h + C$

b. $s = 25 - \frac{13 \ln(h/2)}{\ln 5}, \ 2 \leq h \leq 15$

45. $N = \frac{500}{1 + 4e^{-0.2452t}}$

47. a. Balance increases

b. Balance remains $500,000

c. Balance depleted in 17.9 years

49. $y = C_1 \sin x + C_2 \cos x - 5x + x^3$

51. $y = (C_1 + x)\sin x + C_2 \cos x$

53. $y = \left(C_1 + C_2 x + \frac{x^3}{3} \right) e^x$ **55.** $y = e^{2x} - e^{-x}$

57. $y = \frac{17}{3} \cos 2x - 3 \sin 2x + \frac{1}{3} \cos x$

59. $y = \frac{1}{2} \cos(2\sqrt{6}t)$ **61.** $y = a_0 \sum_{n=0}^{\infty} \frac{x^n}{4^n}$

Photograph Credits

Index

ALGEBRA

Factors and Zeros of Polynomials

Let $p(x) = a_n x^n + a_{n-1} x^{n-1} + \cdots + a_1 x + a_0$ be a polynomial. If $p(a) = 0$, then a is a *zero* of the polynomial and a solution of the equation $p(x) = 0$. Furthermore, $(x - a)$ is a *factor* of the polynomial.

Fundamental Theorem of Algebra

An nth degree polynomial has n (not necessarily distinct) zeros. Although all of these zeros may be imaginary, a real polynomial of odd degree must have at least one real zero.

Quadratic Formula

If $p(x) = ax^2 + bx + c$, and $0 \le b^2 - 4ac$, then the real zeros of p are $x = (-b \pm \sqrt{b^2 - 4ac})/2a$.

Special Factors

$$x^2 - a^2 = (x - a)(x + a) \qquad\qquad x^3 - a^3 = (x - a)(x^2 + ax + a^2)$$
$$x^3 + a^3 = (x + a)(x^2 - ax + a^2) \qquad\qquad x^4 - a^4 = (x^2 - a^2)(x^2 + a^2)$$

Binomial Theorem

$$(x + y)^2 = x^2 + 2xy + y^2 \qquad\qquad (x - y)^2 = x^2 - 2xy + y^2$$
$$(x + y)^3 = x^3 + 3x^2 y + 3xy^2 + y^3 \qquad\qquad (x - y)^3 = x^3 - 3x^2 y + 3xy^2 - y^3$$
$$(x + y)^4 = x^4 + 4x^3 y + 6x^2 y^2 + 4xy^3 + y^4 \qquad (x - y)^4 = x^4 - 4x^3 y + 6x^2 y^2 - 4xy^3 + y^4$$
$$(x + y)^n = x^n + nx^{n-1} y + \frac{n(n-1)}{2!} x^{n-2} y^2 + \cdots + nxy^{n-1} + y^n$$
$$(x - y)^n = x^n - nx^{n-1} y + \frac{n(n-1)}{2!} x^{n-2} y^2 - \cdots \pm nxy^{n-1} \mp y^n$$

Rational Zero Theorem

If $p(x) = a_n x^n + a_{n-1} x^{n-1} + \cdots + a_1 x + a_0$ has integer coefficients, then every *rational zero* of p is of the form $x = r/s$, where r is a factor of a_0 and s is a factor of a_n.

Factoring by Grouping

$$acx^3 + adx^2 + bcx + bd = ax^2(cx + d) + b(cs + d) = (ax^2 + b)(cx + d)$$

Arithmetic Operations

$$ab + ac = a(b + c) \qquad \frac{a}{b} + \frac{c}{d} = \frac{ad + bc}{bd} \qquad \frac{a + b}{c} = \frac{a}{c} + \frac{b}{c}$$

$$\frac{\left(\dfrac{a}{b}\right)}{\left(\dfrac{c}{d}\right)} = \left(\frac{a}{b}\right)\left(\frac{d}{c}\right) = \frac{ad}{bc} \qquad \frac{\left(\dfrac{a}{b}\right)}{c} = \frac{a}{bc} \qquad \frac{a}{\left(\dfrac{b}{c}\right)} = \frac{ac}{b}$$

$$a\left(\frac{b}{c}\right) = \frac{ab}{c} \qquad \frac{a - b}{c - d} = \frac{b - a}{d - c} \qquad \frac{ab + ac}{a} = b + c$$

Exponents and Radicals

$$a^0 = 1, \quad a \ne 0 \qquad (ab)^x = a^x b^x \qquad a^x a^y = a^{x+y} \qquad \sqrt{a} = a^{1/2} \qquad \frac{a^x}{a^y} = a^{x-y} \qquad \sqrt[n]{a} = a^{1/n}$$
$$\left(\frac{a}{b}\right)^x = \frac{a^x}{b^x} \qquad \sqrt[n]{a^m} = a^{m/n} \qquad a^{-x} = \frac{1}{a^x} \qquad \sqrt[n]{ab} = \sqrt[n]{a}\sqrt[n]{b} \qquad (a^x)^y = a^{xy} \qquad \sqrt[n]{\frac{a}{b}} = \frac{\sqrt[n]{a}}{\sqrt[n]{b}}$$